WHO'S WHO IN ATOMS

WHO'S WHO IN ATOMS

AN INTERNATIONAL REFERENCE BOOK

5th Edition

VOLUME I—A-K

HARRAP RESEARCH PUBLICATIONS
182 HIGH HOLBORN, LONDON, W.C.1

Published in Great Britain by
GEORGE G. HARRAP & CO. LTD.,
182 High Holborn, London, W.C.1

FIRST EDITION 1959
SECOND EDITION 1960
THIRD EDITION 1962
FOURTH EDITION 1965
FIFTH EDITION 1969

© *Copyright 1959, 1960, Vallancey Press Ltd.*
© *Copyright 1962, 1965, 1969, George G. Harrap & Co. Ltd.*
SBN 245 58682 2

*Printed in Great Britain by the Redwood Press Ltd.
London and Trowbridge
Made in Great Britain*

Publishers' Introduction

Since the publication of the first edition of *Who's Who in Atoms* in April 1959 there has been with each new edition a considerable increase in the number of entries included. Now in this the fifth edition we publish over 22,000 entries as compared with about 8,500 in the first edition. Although obsolete entries have been removed there is still an overall increase of 1,280 entries in this edition over the previous one.

As new organisations are formed in this field they are invited to supply biographical details of the most important people they employ. We offer our thanks to all organisations, of which there are 3,262 drawn from 78 countries, for their assistance in providing information.

This edition has been entirely revised and we thank each individual who has either checked or corrected his or her existing entry or has provided biographical details for the first time for this edition. We should also like to thank those who have helped us to trace the addresses of their former colleagues who had moved to new positions.

We should like in particular to thank Mr. R. W. B. Truscott, Chief Press Officer, United Kingdom Atomic Energy Authority, London; Dr. Arnost Honig, Head, Central Defectoscopic Centre for Building Industry and Materials, Technical University, Brno; and Mr. Bruce Adkins, European Nuclear Energy Agency of the Organisation for Economic Cooperation and Development, Paris. We are grateful also for the help we continue to receive from the United States Atomic Energy Commission, the American Nuclear Society, the Atomic Industrial Forum and the International Atomic Energy Agency.

It has been our aim to publish a book as free from error as is humanly possible, but it is almost inevitable that some errors will have occurred. Names which might have been included may have been omitted. However, we are always pleased to receive notification of errors and suggestions for new names. These will be considered for the sixth edition, which is now in preparation.

LONDON, *July* 1969.

Alphabetical Arrangement

Considerable thought has been given to the alphabetical arrangement of the names in this book. With entries from over seventy countries, it would only cause confusion to follow the rules of each individual country. Although the *I.S.O. Draft Recommendations on Bibliographical References* have been studied, they have not been adopted in full, as it has been necessary to bear in mind that those who use this book will not always be familiar with the structure of the names or the practice followed in the country of origin. To make it as simple as possible to find a required name the following procedures have been adopted:

Compound surnames or family names, whether hyphenated or not, are entered under the first part of the surname, but cross-references are given under the other part or parts of the name. Surnames beginning with prefixes such as d', de, le, van, von, are entered under the prefix—*e.g.*, de Gaulle would be entered under the letter D and von Brentano under the letter V. Cross-references, however, would be given—under the letters G and B in these examples.

In all cases where the surname is normally preceded by a first name the names are transposed so that the entry appears under the surname. In certain countries—for example, China—it is the practice for the surname to be written first, and therefore transposition is unnecessary.

Abbreviations

A.A.	Associate in Arts
A.A.A.S.	American Association for the Advancement of Science
A.A.E.C.	Australian Atomic Energy Commission
A.A.P.G.	American Association of Petroleum Geologists
A.A.P.T.	American Association of Physics Teachers
A.A.U.P.	American Association of University Professors
A.B.	Bachelor of Arts
A.C.C. and C.E.	Association of Consulting Chemists and Chemical Engineers
A.C.E.C.	Ateliers de Constructions Electriques de Charleroi
A.C.F.A.S.	Association Canadienne Française pour l'Advancement des Sciences
A.C.G.I.	Associate of the City and Guilds Institute
A.C.I.	American Concrete Institute
A.C.I.S.	Associate of the Chartered Institute of Secretaries
A.C.R.	American College of Radiology
A.C.S.	American Chemical Society
A.C.S.M.	Associate of the Camborne School of Mines
A.C.W.A.	Associate of the Institute of Cost and Works Accountants
A.E.C.	Atomic Energy Commission
A.E.C.L.	Atomic Energy of Canada Ltd.
A.E.E.T.	Atomic Energy Establishment, Trombay
A.E.I.	Associated Electrical Industries Ltd.
A.E.R.E.	Atomic Energy Research Establishment
A.F.B.	Air Force Base
A.F.R.Ae.S.	Associate Fellow of the Royal Aeronautical Society
A.G.	Aktiengesellschaft
A.G.R.	Advanced Gas Cooled Reactor
A.G.U.	American Geophysical Union
A.I.A.A.	American Institute of Aeronautics and Astronautics
A.I.B.S.	American Institute of Biological Sciences
A.I.Br.	Association des Ingenieurs sortis de l'Universite de Bruxelles
A.I.Ch.E.	American Institute of Chemical Engineers
A.I.H.A.	American Industrial Hygiene Association
A.I.Lg.	Association des Ingenieurs, Liege
A.I.M.	Associate of the Institute of Metals
A.I.M.E.	American Institute of Mining Engineers; Associate of the Institution of Mining Engineers
A.I.M.M.E.	American Institute of Mining, Metallurgical and Petroleum Engineers
A.I.N.A.M.E.	American Institute of Naval Architects and Marine Engineers
A.I.P.	American Institute of Physics
A.I.P.G.	American Institute of Petroleum Geologists
A.I.R.I.	Associate of the Institute of the Rubber Industry
A.I.S.E.	Association of Iron and Steel Engineers
A.I.S.M.	Associate of the Indian School of Mining
A.Inst.P.	Associate of the Institute of Physics
A.K.C.	Associate of King's College
A.L.A.	Associate of the Library Association
A.L.P.R.	Argonne Low-Power Reactor
A.M.	Associate Member; Master of Arts
A.M.A.	American Medical Association
A.M.C.T.	Associate of Manchester College of Technology
A.M.I.C.E.	Associate Member of the Institution of Civil Engineers
A.M.I.Chem.E.	Associate Member of the Institution of Chemical Engineers
A.M.I.E.E.	Associate Member of the Institution of Electrical Engineers
A.M.I.Mar.E.	Associate Member of the Institute of Marine Engineers
A.M.I.Mech.E.	Associate Member of the Institution of Mechanical Engineers
A.M.Inst.F.	Associate Member of the Institute of Fuel
A.M.Inst.W.	Associate Member of the Institute of Welding

A.N.I.D.E.L.	Associazione Nazionale Imprese Produttrici Distributrici di Energia Elletrica
A.N.P.D.	Aircraft Nuclear Propulsion Department (of G.E.C.)
A.N.S.	American Nuclear Society
A.P.E.O.	Association of Professional Engineers of the Province of Ontario
A.P.H.A.	American Public Health Association
A.P.S.	American Physical Society
A.Ph.A.	American Pharmaceutical Association
A.R.C.	Agricultural Research Council
A.R.C.S.	Associate of the Royal College of Science
A.R.C.S.T.	Associate of the Royal College of Science and Technology
A.R.D.E.	Armament Research and Development Establishment (Woolwich)
A.R.I.C.	Associate of the Royal Institute of Chemistry
A.R.S.	American Rocket Society
A.R.S.M.	Associate of the Royal School of Mines
A.R.T.C.	Associate of the Royal Technical College
A.S.A.	Acoustical Society of America; American Standards Association; American Statistical Association
A.S.B.C.	American Society of Biophysics and Cosmology
A.S.C.E.	American Society of Civil Engineers
A.S.E.A.	Allmanna Svenska Electriska A.B.
A.S.E.E.	American Society for Engineering Education
A.S.M.	American Society for Metals
A.S.M.E.	American Society of Mechanical Engineers
A.S.Q.C.	American Society for Quality Control
A.S.T.C.	Associate of the Sydney Technical College
A.S.T.M.	American Society for Testing Materials
A.T.E.N.	Association Technique pour la Production et l'Utilisation de l'Energie Nucléaire
A.W.R.E.	Atomic Weapons Research Establishment
A.W.S.	American Welding Society
Acad.	Academy
Appl.	Applied
Assoc.	Associate, Associated, Association
Asst.	Assistant
B.A.	Bachelor of Arts
B.A.I.	Bachelor of Engineering (Baccalarius In Arts Ingeniara)
B.A.Sc.	Bachelor of Applied Science
B.B.A.	Bachelor of Business Administration
B.C.E.	Bachelor of Civil Engineering
B.C.L.	Bachelor of Civil Law; Bachelor of Canon Law
B.C.S.	Bachelor of Commercial Science
B.Ch.E.	Bachelor of Chemical Engineering
B.Chir.	Bachelor of Surgery
B.E.	Bachelor of Engineering; Bachelor of Education
B.E.E.	Bachelor of Electrical Engineering
B.E.P.O.	British Experimental Pile O
B.Ed.	Bachelor of Education
B.J.	Bachelor of Journalism
B.M.E.	Bachelor of Mining Engineering
B.Mech.Eng.	Bachelor of Mechanical Engineering
B.Met.E.	Bachelor of Metallurgical Engineering
B.N.E.	Bachelor of Nuclear Engineering
B.N.E.S.	British Nuclear Energy Society
B.P.	Boîte Postale
B.S.	Bachelor of Science
B.S.A.	Bachelor of Scientific Agriculture
B.S.C.	Bachelor of Science in Commerce
B.S.C.E.	Bachelor of Science in Civil Engineering
B.S.E.	Bachelor of Science in Engineering
B.S.E.E.	Bachelor of Science in Electrical Engineering
B.S.I.	British Standards Institution
B.S. (I.E.)	Bachelor of Science (Industrial Engineering)
B.S.M.	Bachelor of Science in Medicine
B.S.M.E.	Bachelor of Science in Mechanical Engineering
B.Sc.	Bachelor of Science
B.Sc.Tech.	Bachelor of Technical Science
B. Technol.	Bachelor of Technology

Biochem.	Biochemical, Biochemist, Biochemistry
Biol.	Biological, Biology, Biologist
Biophys.	Biophysical, Biophysics, Biophysicist
Bt.	Baronet
Bul.	Bulletin
C.C.	Companion of the Order of Canada
C.C.N.Y.	College of the City of New York
C.C.R.	Common Centre for Research
C.C.T.A.	Commission pour la Coopération Technique en Afrique
C.E.	Civil Engineer
C.E.A.	Commissariat à l'Energie Atomique
C.E.C.A.	Communauté Européenne du Charbon et de l'Acier
C.E.E.A.	Communauté Européenne de l'Energie Atomique
C.E.G.B.	Central Electricity Generating Board
C.E.I.	Commission Electrotechnique Internationale
C.E.N.	Centre d'Etudes de l'Energie Nucléaire
C.E.R.N.	Conseil Européen pour la Recherche Nucléaire
C.G.I.A.	City and Guilds Institute Award
C.I.G.R.E.	International Conference on Large Electric Systems
C.I.M.	Canadian Institute of Metals
C.I.O.M.S.	Council for International Organisations of Medical Science
C.I.S.E.	Centro Informazioni Studi Esperienze
C.M.	Master of Surgery
C.M.I.	Comité Maritime International
C.N.A.M.	Conservatoire National des Arts et Métiers
C.N.E.N.	Comitato Nazionale per l'Energia Nucleare
C.N.R.N.	Comitato Nazionale per le Richerche Nucleari
C.N.R.S.	Centre National de la Recherche Scientifique
C.O.	Commanding Officer
C.P.A.	Chartered Patent Agent
C.S.I.R.O.	Commonwealth Scientific and Industrial Research Organisation
C.Sc.	Candidate of Science
C.T.R.	Controlled Thermonuclear Research
C.U.P.M.	Committee on Undergraduate Programme in Mathematics
Ch.	Chapter
Ch.B.	Bachelor of Surgery
Ch.E.	Chemical Engineer
Chem.	Chemical, Chemistry, Chemist
Co.	Company
Coll.	College
Com.	Committee
Compl.	Compliance
Corp.	Corporation
D.A.W.	Deutsche Akademie der Wissenschaften
D. and E.	Development and Engineering
D.At.En.	Department of Atomic Energy
D.C.H.	Diploma in Child Health
D.C.L.	Doctor of Civil Law
D.C.S.O.	Deputy Chief Scientific Officer
D.D.R.	Deutsches Democratisches Republik
D.E.R.E.	Dounreay Experimental Research Establishment
D.E.S.	Doctor of Engineering Science
D. Eng.	Doctor of Engineering
D. ès Sc.	Docteur ès Sciences
D.F.H.	Diploma of Faraday House
D.G.M.	Deutsche Gesellschaft für Metallkunde
D.G.O.	Diploma in Gynaecology and Obstetrics
D.I.C.	Diploma of Imperial College
D.I.H.	Diploma of Industrial Health
D.L.C.	Diploma of Loughborough College
D.M.A.	Dental Manufacturers of America
D.M.D.	Doctor of Dental Medicine
D.M.R.	Diploma in Medical Radiology
D.M.R.D.	Diploma in Medical Radiological Diagnosis
D.M.R.E.	Diploma in Medical Radiology and Electrology
D.M.R.T.	Diploma in Radiotherapy
D.Met.	Doctor of Metallurgy
D.N.A.	desoxyribonucleic acid

D.P.A.	Diploma in Public Administration
D.P.H.	Diploma in Public Health
D.P.M.	Diploma in Psychological Medicine
D.Ph.	Doctor of Philosophy
D.R.	Diploma in Radiology
D.R.B.	Defence Research Board (Canada)
D.R.T.C.	Diploma of the Royal Technical College
D.S.I.R.	Department of Scientific and Industrial Research
D.Sc.	Doctor of Science
D.T.H.	Diploma in Tropical Hygiene
D.T.M.	Diploma in Tropical Medicine
D.T.M.H.	Diploma in Tropical Medicine and Hygiene
D.T.R.	Diploma in Therapeutic Radiology
D.T.R.E.	Diploma in Therapeutic Radiology and (Medical) Electricity
D.Tech.Sc.	Doctor in Technical Science
D.Techn.b.c.	Doctor of Technology honoris causa
D.V.M.	Doctor of Veterinary Medicine
Deleg.	Delegate, Delegation
Dept.	Department
Dip. Ed.	Diploma in Education
Dipl.Chem.	Diploma in Chemistry
Dipl.Ing.	Diploma in Engineering
Dipl. Phys.	Diploma in Physics
Div.	Division
Doz.	Dozent
Dr.	Doctor
Dr.agr.	Doctor of Agriculture
Dr. Eng.	Doctor of Engineering
Dr. h.c.	Doctor honoris causa
Dr.-Ing.	Doctor of Engineering
Dr.Ing. e.h.	Doctor of Engineering (honorary)
Dr.-Ir.	Doctor of Engineering
Dr.iur.	Doctor of Laws
Dr. jur.	Doctor of Laws
Dr.med.	Doctor of Medicine
Dr.med.dent.	Doctor of Dentistry
Dr. med. h.c.	Honorary Doctor of Medicine
Dr. med. vet.	Doctor of Veterinary Medicine
Dr. oec.	Doctor of Commerce
Dr. Phil.	Doctor of Philosophy
Dr.rer.agr.	Doctor of Agricultural Science
Dr.rer.hort.	Doctor of Horticulture
Dr. rer. nat.	Doctor of Natural Science
Dr. rer. pol.	Doctor of Political Science
Dr. sc. tech.	Doctor of Technical Science
Drs.	Doctorandus
E.A.E.S.	European Atomic Energy Society
E.B.W.R.	Experimental Boiling Water Reactor
E.C.S.C.	European Coal and Steel Community
E.D.	Doctor of Engineering
E.E.	Electrical Engineering
E.E.C.	European Economic Community
E.M.S.A.	Electron Microscope Society of America
E. Met.	Engineer of Metallurgy
E.N.E.A.	European Nuclear Energy Agency
E.N.S.	Ecole Normale Supérieure
E.N.S.E.E.G.	Ecole Nationale Supérieure d'Electrochimie et d'Electrométallurgie, Grenoble
E.N.S.I.	Energia Nucleare Sud Italia
E.N.S.I.A.M.	Ecole Nationale d'Ingenieurs, Arts et Métiers
E.P.A.	European Productivity Agency
E.P.R.	electron paramagnetic resonance
E.P.U.L.	Ecole Polytechnique de l'Université de Lausanne
E.S.E.	Ecole Superieure d'Electricité
E.S.P.C.I.	Ecole Supérieure de Physique et de Chimie Industrielles de la Ville de Paris
E.T.H.	Eidgenössische Technische Hochschule
Educ.	Educated

Elec.	Electric, Electrical, Electricity
Electrochem.	Electrochemical
Eng.	Engineer, Engineers, Engineering
Exec.	Executive
Exptl.	Experimental
F.A.C.C.	Fellow of the American College of Cardiology
F.A.C.O.I.	Fellow of the American College of Osteopathic Internists
F.A.C.R.	Fellow of the American College of Radiology
F.A.I.P.	Fellow of the Australian Institute of Physics
F.A.O.	Food and Agriculture Organisation
F.A.S.	Federation of American Scientists
F.C.A.	Fellow of the Institute of Chartered Accountants
F.C.G.I.	Fellow of the City and Guilds Institute
F.C.I.I.	Fellow of the Chartered Insurance Institute
F.C.P.S.	Fellow of the Cambridge Philosophical Society
F.C.Path.	Fellow of the College of Pathologists
F.C.R.A.	Fellow of the College of Radiologists of Australia
F.C.T.	Federal Capital Territory
F.D.	Doctor of Philosophy
F.F.A.R.C.S.	Fellow of the Faculty of Anaesthetics, Royal College of Surgeons
F.F.R.	Fellow of the Faculty of Radiologists
F.F.R.(T)	Fellow of the Faculty of Radiologists (London) in Radiotherapy
F.F.S.	Fellow of the Faculty of Secretaries
F.G.M.S.	Fellow of the Geological, Mining and Metallurgical Society, India
F.G.S.	Fellow of the Geological Society
F.I. Arb.	Fellow of the Institute of Arbitrators
F.I.C.I.	Fellow of the Institute of Chemistry of Ireland
F.I.E.N.	Forum Italiano dell'Energia Nuclear
F.I.M.	Fellow of the Institute of Metals
F.I.P.A.C.E.	Fédération Internationale des Producteurs Autoconsommateurs Industriele d'Electricité
F.I.R.	Fellow of the Institute of Radiology
F.Inst.F.	Fellow of the Institute of Fuel
F.Inst.P.	Fellow of the Institute of Physics
F.N.I.	Fellow of the National Institute of Sciences, India
F.R.A.C.I.	Fellow of the Royal Australian Chemical Institute
F.R.A.C.P.	Fellow of the Royal Australian College of Physicians
F.R.A.S.	Fellow of the Royal Astronomical Society
F.R.C.P. (C)	Fellow of the Royal College of Physicians (Canada)
F.R.C.S.	Fellow of the Royal College of Surgeons
F.R.I.C.	Fellow of the Royal Institute of Chemistry
F.R.M.S.	Fellow of the Royal Microscopical Society
F.R.S.	Fellow of the Royal Society
F.R.S.A.	Fellow of the Royal Society of Arts
F.R.S.C.	Fellow of the Royal Society of Canada
F.R.S.E.	Fellow of the Royal Society of Edinburgh
F.R.S.H.	Fellow of the Royal Society of Health
F.S.A.	Fellow of the Society of Arts
F.S.G.T.	Fellow of the Society of Glass Technology
F.S.S.	Fellow of the Royal Statistical Society
F.T.I.	Fellow of the Textile Institute
Fil. Hed. Dr.	Filosofie Hedersdoktor (Honorary Ph. D.)
Fil. lic.	Licentiate in Philosophy
G.A.M.M.	Gesellschaft für angewandte Mathematik und Mechanik
G.A.T.T.	General Agreement on Tariffs and Trade
G.E.C.	General Electric Co.
Gen.	General
Govt.	Government
H.E.	His Excellency
H.N.C.	Higher National Certificate
H.P.S.	Health Physics Society
Hon.	Honorary
Hons.	Honours
I.A.E.A.	International Atomic Energy Agency
I.A.R.I.	Indian Agricultural Research Institute
I.A.U.	International Astronomical Union

i/c	in charge
I.C.A.	Institute of Chartered Accountants
I.C.A.I.	Instituto Catolico de Artes e Industries, Madrid
I.C.A.O.	International Civil Aviation Organisation
I.C.E.	Institute of Civil Engineers
I.C.P.U.A.E.	International Conference on the Peaceful Uses of Atomic Energy
I.C.R.P.	International Commission on Radiological Protection
I.C.R.U.	International Commission on Radiological Units and Measurements
I.C.S.U.	International Council of Scientific Unions
I.E.E.	Institute of Electrical Engineers
I.E.E.E.	Institute of Electrical and Electronic Engineers
I.E.R.E.	Institution of Electronic and Radio Engineers
I.E.S.S.	Institution of Engineers and Shipbuilders in Scotland
I.I.T.	Illinois Institute of Technology
I.L.O.	International Labour Organisation
I.M.M.	Institute of Mining and Metallurgy
I.Mar.E.	Institute of Marine Engineers
I.N.A.	Institution of Naval Architects; Institut National Agronomique
I.N.E.	Institution of Nuclear Engineers
I.N.F.N.	Instituto Nazionale di Fisica Nucleare
I.N.M.M.	Institute of Nuclear Materials Management
I.N.S.T.N.	Institut National des Sciences et Techniques Nucléaires
I.Q.S.Y.	International Quiet Sun Year
I.R.S.I.A.	Institut pour l'Encouragement de la Recherche Scientifique dans l'Industrie et l'Agriculture
I.S.A.	Instrument Society of America
I.S.N.S.E.	International School of Nuclear Science and Engineering
I.S.O.	International Organisation for Standardisation
I.U.P.A.C.	International Union of Pure and Applied Chemistry
I.U.P.A.P.	International Union of Pure and Applied Physics
I.V.I.C.	Instituto Venezolano de Investigaciones Cientificas
Inc.	Incorporated
Ind.	Industrial
Ing.	Ingénieur
Ing.Chim.E.N.S.C.P.	Ingénieur Chimie, Ecole Nationale Supérieure de Chimie, Paris (Univ.)
Ing.E.E.M.I.	Ingénieur Ecole d'Electricite et de Mécanique Industrielles
Ing.E.N.S.E.E.G.	Ingénieur, Ecole Nationale Supérieure d'Electrochimie et d'Electrometallurgie, Grenoble (Univ.)
Ing.E.N.S.G. (Nancy)	Ingénieur, Ecole Nationale Supérieure de Geologie, Nancy (Univ.)
Ing.E.S.C.I.L. (Lyon)	Ingénieur Ecole Supérieure de Chimie Industrielle de Lyon
Ing.E.S.E.	Ingénieur, Ecole Supérieure d'Electricité
Ing.E.S.M.E.	Ingénieur de Ecole Spéciale de Mécanique et d'Electricité A.M. Ampere
Ing.I.F.F.F.	Ingénieur Institut Français du Froid Industrial
Ing. Mag.	Magister in Engineering
Inorg.	Inorganic
Inst.	Institute, Institution, Institut
Instr.	Instruments
Internat.	International
Ir.	Engineer
J.	Journal
J.A.I.F.	Japan Atomic Industrial Forum
J.D.	Doctor of Jurisprudence
J.S.D.	Doctor of Juristic Science
J.U.D.	Juris utruisque Doctor (Doctor of both civil and canon law)
Jun.	Junior
K.A.P.L.	Knolls Atomic Power Laboratory
L.D.	Libero Docente o Professore
L.E.T.	linear energy transfer
L. ès Sc.	Licencié ès Sciences
L.H.D.	Doctor of Humane Letters
LL.B.	Bachelor of Laws
L.M.S.	Licentiate in Medicine and Surgery
L.R.C.P.S.	Licentiate of the Royal College of Physicians in Scotland
L.R.S.R.	Laboratori Riuniti Studie Ricerche
Lab.	Laboratory

Lic. tech.	Licentiate in Technology
Ltd.	Limited
M.A.	Master of Arts
M.A.O.Coll.	Master of Arts (Oriental College)
M.A.Sc.	Master of Applied Science
M.B.	Bachelor of Medicine
M.B.A.	Master of Business Administration
M.B.B.S.	Conjoint degree Bachelor of Medicine, Bachelor of Surgery
M.B.I.M.	Member of the British Institute of Management
M.Biorad.	Master of Bioradiation
M.C.E.	Master of Civil Engineering
M.C.P.A.	Member of the College of Pathologists of Australia
M.C.R.A.	Member of the College of Radiologists of Australasia
M.C.S.	Master of Commercial Science
M.Ch.E.	Master of Chemical Engineering
M.D.	Doctor of Medicine
M.E.	Mechanical Engineer
M.E.E.	Master of Electrical Engineering
M.Eng.	Master of Engineering
M. Eng. Sc.	Master of Engineering Science
M.F.	Master of Forestry
MHD	magnetohydrodynamics
M.I.A.A. (S.A.)	Member of the Institute of Assayers and Analysists (South Africa)
M.I.E. Aust.	Member of the Institution of Engineers, Australia
M.I.E.S.	Member of the Institution of Engineers and Shipbuilders in Scotland
M.I.E.R.E.	Member of the Institution of Electronic and Radio Engineers
M.I.H.V.E.	Member of the Institution of Heating and Ventilating Engineers
M.I. Loco. E.	Member of the Institute of Locomotive Engineers
M.I.M.	Master of Industrial Management
M.I.M.M.	Member of the Institution of Mining and Metallurgy
M.I.Mar.E.	Member of the Institute of Marine Engineers
M.I.Mech.E.	Member of the Institution of Mechanical Engineers
M.I.N.A.	Member of the Institution of Naval Architecture
M.I.Nuc.E.	Member of the Institute of Nuclear Engineers
M.I.P.S.	Member of the Institute of Purchasing and Supply
M.I.Prod.E.	Member of the Institution of Production Engineers
M.I.T.	Massachusetts Institute of Technology
M.Inst.B.E.	Member of the Institute of British Engineers
M. Inst. F.	Member of the Institute of Fuel
M.Inst.P.	Member of the Institute of Physics
M.M.E.	Master of Mechanical Engineering
M.Med.Rad.T.	Master of Medicine in Radiotherapy
M.N.E.	Master of Naval Engineering
M.O.P.	Ministry of Power
M.O.S.	Ministry of Supply
M.O.T.	Ministry of Transport
M.P.O.A.	Member of the Purchasing Officers Association
M.P.H.	Master of Public Health
M.R.A.C.P.	Member of the Royal Australian College of Physicians
M.R.C.	Medical Research Council
M.R.C.P.	Member of the Royal College of Physicians
M.R.C.V.S.	Member of the Royal College of Veterinary Surgeons
M.R.I.A.	Member of the Royal Irish Academy
M.R.I.N.A.	Member of the Royal Institute of Naval Architects
M.R.S.H.	Member of the Royal Society of Health
M.S.	Master of Science
M.S.A.Chem.I.	Member of the South African Chemical Institute
M.S.C.E.	Master of Science in Civil Engineering
M.S.E.	Master of Science in Chemical Engineering
M.S.E.E.	Master of Science in Electrical Engineering
M.S.J.	Master of Science in Journalism
M.S.L.S.	Master of Science in Library Science
M.S.N.E.	Master of Science in Nuclear Engineering
M.S.P.H.	Master of Science in Public Health
M.Sc.	Master of Science
MTR	materials testing reactor

M.U.Dr.	Medicinae Universae Doctor
M.V.Dr.	Medicus Veterinarius Doctor (Doctor of Veterinary Medicine)
Mag.	Magazine
mag.scient.	magister scientiarum (Copenhagen)
Mass. Inst. Technol.	Massachusetts Institute of Technology
Maths.	Mathematics
Mech.	Mechanical
Metal.	Metallurgical, Metallurgy, Metallurgist
N.A.B.M.	National Association of British Manufacturers
N.A.C.A.	National Advisory Committee for Aeronautics
N.A.C.E.	National Association of Corrosion Engineers
N.A.E.C.	National Association of Electric Companies
N.A.M.	National Association of Manufacturers
N.A.S.	National Academy of Sciences
N.A.S.A.	National Aeronautics and Space Administration
N.A.T.O.	North Atlantic Treaty Organisation
N.B.R.I.	National Building Research Institute
N.B.S.	National Bureau of Standards
N.C.B.	National Coal Board
N.C.R.P.	National Committee on Radiation Protection
N.C.S.C.	North Carolina State College
N.D.A.	National Diploma in Agriculture; Nuclear Development Corporation of America
N.D.H.	National Diploma in Horticulture
N.E.R.V.A.	Nuclear Engine for Rocket Vehicle Application
N.I.H.	National Institutes of Health
N.I.R.N.S.	National Institute for Research in Nuclear Science
NMR	nuclear magnetic resonance
N.P.P.C.	Nuclear Power Plant Co. Ltd.
N.P.R.L.	National Physical Research Laboratory
N.Q.R.	nuclear quadrupole resonance
N.R.C.	National Research Council of Canada
N.R.L.	Naval Research Laboratory
N.R.T.S.	National Reactor Testing Station
N.S.F.	National Science Foundation
N.S.P.E.	National Society of Professional Engineers
N.Y.C.	New York City
N.Y.O.O.	New York Operations Office
N.Y.U.	New York University
Nat.	National
Nucl.	Nuclear
O.C.D.R.	Office of Civilian Defence Research
O.E.C.D.	Organisation for Economic Co-operation and Development
O.E.E.C.	Organisation for European Economic Co-operation
O.M.R.E.	Organic Moderated Reactor Experiment
O.N.R.	Office of Naval Research
O.R.I.N.S.	Oak Ridge Institute of Nuclear Studies
O.R.N.L.	Oak Ridge National Laboratory
O.R.S.O.R.T.	Oak Ridge School of Reactor Technology
P.E.	Professional Engineer
P.H.S.	Public Health Service
P.R.S.	President of the Royal Society
P.S.A.	Pacific Science Association
P.S.O.	Principal Scientific Officer
P.T.T.	Postes, Télégraphes et Téléphones
Ph.A.Mr.	Magister of Applied Physics
Ph.B.	Bachelor of Philosophy
Ph.D.	Doctor of Philosophy
Ph. Lic.	Licentiate of Philosophy
Phys.	Physical, Physics, Physicist
Polytech.	Polytechnic
Pres.	President
Proc.	Proceedings
Prof.	Professor
Pty.	Proprietary
R.A.E.	Royal Aircraft Establishment
R.A.S.C.	Royal Astronomical Society of Canada
R. and D.	Research and Development

R.C.A.	Royal Canadian Academy
R.E.A.	Rural Electrification Administration
R.E.M.	Röntgen equivalent man
RESA	Scientific Research Society of America
R.E.T.M.A.	Radio Electronics Television Manufacturers Association
r.f.	radio frequency
R.M.C.S.	Royal Military College of Science
R.N.Dr.	Doctor of Natural Science
Rad.	Radiation
Radiobiol.	Radiobiological
Radiochem.	Radiochemical
Radiol.	Radiology, Radiological, Radiologist
Rep.	Representative
Res.	Research
Rev.	Review
Roy.	Royal
S.A.B.V.	Société Anglo-Belge Vulcain
S.B.	Bachelor of Science
S.E.L.N.I.	Societa Elettronucleare Italiana
S.E.N.T.A.	Société d'Etudes Nucléaires et de Techniques Advancées
S.E.R.A.I.	Société d'Etudes de Recherches et d'Applications pour l'Industrie
S.E.S.A.	Society for Experimental Stress Analysis
S.F.I.T.V.	Société Française des Ingenieurs Technique du Vide
S.G.H.W.	steam graphite heavy water reactor
S.G.H.W.R.	steam graphite heavy water reactor
S.I.M.E.A.	Societa Italiana Meriodionale Energia Atomica
S.J.	Society of Jesus
S.M.	Master of Science
S.N.T.L.	Státni Nakladatelstvi Technické Literatury
S.P.R.L.	Société de Personnes à Responsabilité Limitée
S.P.S.O.	Senior Principal Scientific Officer
S.R.C.	Science Research Council
S.S.N.D.	School Sisters of Notre Dame
S.S.O.	Senior Scientific Officer
S.T.L.	Reader (or Professor) of Sacred Theology
S.T.R.	Submarine Thermal Reactor
Sc.B.	Bachelor of Science
Sci.	Science, Sciences, Scientific, Scientist
Sec.	Secretary
Sect.	Section
Sen.	Senior
Soc.	Society
Sta.	Societa
Sté.	Société
Subcom.	Subcommittee
Supt.	Superintendent
Supv.	Supervisor
T.A.P.P.I.	Technical Association of the Pulp and Paper Industry
T.H.	Technische Hochschule
T.I.S.	Technical Information Service
T.N.O.	Central National Council for Applied Scientific Research in the Netherlands
T.T.D.	Transvaal Teachers Diploma
Teach.	Teaching
Tech.	Technical
Technol.	Technological, Technology
Trans.	Transactions
Treas.	Treasurer
U.C.L.A.	University of California at Los Angeles
U.I.C.C.	Union Internationale Contre le Cancer
U.K.A.E.A.	United Kingdom Atomic Energy Authority
U.N.	United Nations
U.N.A.M.	United Nations Association of Mexico
U.N.E.S.C.O.	United Nations Educational Scientific and Cultural Organisation
U.N.I.C.E.F.	United Nations Children's Fund
U.N.I.P.E.D.E.	International Union of Producers and Distributors of Electric Power

U.R.S.I.	Union Radio-Scientifique Internationale
U.S.A.E.C.	United States Atomic Energy Commission
U.S.A.F.	United States Air Force
U.S.A.S.I.	United States of America Standards Institute
U.S.C.	University of South Carolina; University of Southern California
U.S.M.A.	United States Military Academy
U.S.N.	United States Navy
U.S.P.H.S.	United States Public Health Service
U.S.S.I.	Société de Construction d'Une Usine de Séparation Isotopique
Univ.	University
V.D.E.	Verband Deutscher Elektrotechniker
V.D.I.	Verein Deutscher Ingenieure
V.M.D.	Doctor of Veterinary Medicine
W.G.L. or W.G.L.R.	Wissenschaftlicher Gesellschaft für Luft- und Raumfahrt e.V.
W.H.O.	World Health Organisation

AABAKKEN, Bjarne. Born 1931. Educ.: Graz Tech. Univ. Chem. (1952-55), Head, Tech. Information Service (1955-58), Tech. Sec. (1958), Exec. Sec. (1958-9), Head, Tech. Secretariat (1961-67), Inst. for Atomic Energy, Norway; Exec. Sec., Norwegian Atomic Energy Council 1958-59, 1961-67. Manager, Indo-Norwegian Project, 1967-. Tech. Sec., O.E.C.D. High Temperature Reactor Project (Dragon), 1959-61. Chairman, Norwegian Atomic Energy Soc. Societies: Norwegian Polytech. Soc.; Norwegian Atomic Energy Soc.
Nuclear interests: Management; Technical administration; Public relations; Technical information.
Address: Indo-Norwegian Project, P.O. Box 137, Ernakulam, Kerala, India.

AALTO, Erkki Juhani, Dipl.ing. (Civil Eng.), M.S. (Nucl. Eng.), Teknologielicentiat (Ph.D.) (Reactor Phys.). Born 1934. Educ.: Finnish Inst. Technol., Wisconsin Univ. and Chalmers Inst. of Technol., Sweden. Res. Phys. (Shielding), Swedish Atomic Energy Co., Studsvik Res. Centre, 1960-62; Supervisor, Shielding and Rad. Analysis, 1963-. Societies: A.N.S.; Eng. Soc. in Finland.
Nuclear interests: Shielding, radiation damage and materials technology, reactor and station design and lay-out, economy.
Address: AB Atomenergi, Box 43041, Stockholm 43, Sweden.

AAMODT, Nils Godtfred, Dipl. Ing. Born 1926. Educ.: Federal Tech. Coll., Zürich. Formerly Director of Administration. now Asst. Managing Director, Inst. for Atomenergi.
Address: Institutt for Atomenergi, Postboks 175, Lillestrom, Norway.

AARKROG, Asker, M.S. (Chem. Eng.). Born 1932. Educ.: Denmark Tech. Univ. Sect. Leader, Environmental Control Sect., Health Phys. Dept., Danish A.E.C. Societies: Inst. Danish Civil Eng.; Nordic Soc. of Rad. Protection; H.P.S.
Nuclear interests: Fallout; Radioecology, especially ^{90}Sr in the human foodchain; Low level radiochemistry; Environmental control.
Address: Health Physics Department, Danish Atomic Energy Commission, Riso, Roskilde, Denmark.

AARSET, B. Sen. Eng., Development Div., O.E.C.D. Halden Reactor Project.
Address: O.E.C.D. Halden Reactor Project, P.O. Box 173, Halden, Norway.

AAS, Steinar, Berging. N.T.H., M.Sc. Born 1925. Educ.: Norway Tech. Univ., Carnegie Inst. Technol., Pittsburg, Pa. Project Manager, Halden Reactor Project, O.E.C.D. Society: Inst. of Metals, London.
Nuclear interests: Fuel and reactor materials.
Address: P.O. Box 173, Halden, Norway.

ABAIGAR, H. E. Joaquin Buxe-Dulce de
See de ABAIGAR, H. E. Joaquin Buxe-Dulce.

ABBADESSA, John Peter, B.S. (Business Administration, cum laude), M.B.A. Born: 1920. Educ.: American and Pennsylvania Univs. and Harvard Graduate School of Business Administration. Director, Transportation Div. (1959-60), Deputy Director, Civil Accounting and Auditing Div. (1960-62), U.S. Gen. Accounting Office; Controller, U.S. A.E.C., 1962-. Society: American Inst. of Certified Public Accountants.
Nuclear interests: As chief financial officer of the U.S. Atomic Energy Commission respon sible to the General Manager and also direct responsibilities to the five Commissioners for advice and counsel on financial matters.
Address: 6900 Old Gate Lane, Rockville, Maryland, U.S.A.

ABBATT, John Dilworth, M.D., Ch.B., D.M.R. Born 1923. Educ.: Edinburgh Univ. Roy. Infirmary, Edinburgh, 1948-50; Member, M.R.C., Ext. Sci. Staff, British Post Graduate Medical School, Hammersmith Hospital, London, 1950-57; Medical Radiobiol., Atomic Energy Div., G.E.C., and Chief Medical Officer, G.E.C. Heavy Eng. Div., England, 1957-62; Asst. Chief, Rad. Protection Div., Dept. of Nat. Health and Welfare, Ottawa. Societies: British Inst. of Radiol.; Assoc. Rad. Res.; British Medical Assoc.
Nuclear interests: Radiobiology and radiation safety, with particular reference to both acute and long term radiation effects in man and their mechanism of production. Areas of work: radiation epidemiology, cytogenetics, thyroid function, leukaemia, and acute radiation syndrome (radiation sickness).
Address: Radiation Protection Division, Department of National Health and Welfare, Brookfield Road, Ottawa 8, Ontario, Canada.

ABBOSOV, O. Paper: Co-author, Determination of Indium by (γ, γ)- Reaction with 5 MeV Linear Electron Accelerator (letter to the Editor, Atomnaya Energiya,vol. 24, No. 3, 1968).
Address: Academy of Sciences of the U.S.S.R., 14 Leninsky Prospekt, Moscow V-71, U.S.S.R.

ABBOTT, William Edward, B.S. (Phys., M.I.T.). Born 1915. Educ.:Williams Coll. and M.I.T. Standard Oil Co. of California, 1948-50; Fay, Spofford and Thorndike, Eng., 1950-51; Eng., K.A.P.L., 1951-52; Group Leader, Reactor Eng. Analysis, Atomics Internat., a Div. of North American Aviation, 1953-56; Asst. to Manager (Tech.),Reactor Eng., Westinghouse Atomic Power Dept., 1956-57; Manager, Reactor Development, Atomic Power Div., Westinghouse Elec. Corp., 1957-. Societies: A.P.S.; A.N.S.; I.E.E.E.
Nuclear interests: Reactor design, management of engineering and research and development, nuclear physics and reactor codes.

Address: Westinghouse Atomic Power Department, Westinghouse Electric Corporation. P.O. Box 355, Pittsburgh 30, Pa., U.S.A.

ABDEL-RASSOUL, A.A., B.Sc., M.Sc., Dr. rer. nat. Born 1929 Educ.: Alexandria and Johannes Gutenberg (Mainz) Univs. Demonstrator (-1959), Lecturer (-1964), now Asst. Prof., Nucl. Chem. Dept., United Arab Atomic Energy Establishment. Sci. Member, Middle Eastern Radioisotope Centre for the Arab Countries; Member, Radiochem. Lab. Board. Society: United Arab Republic Chem. Soc. Nuclear interests: Production of radioisotopes, Nuclear activation; Nuclear chemical reactions.
Address: Nuclear Chemistry Department, Atomic Energy Establishment. P.O.B., Cairo, United Arab Republic.

ABDEL RAZIK, A. H. At Atomic Energy Establishment, Cairo. Deleg., Third I.C.P.U.A.-E., Geneva, Sept. 1964.
Address: Atomic Energy Establishment, Inshas, Nr. Cairo, Egypt, United Arab Republic.

ABDEL-WAHAB, Mohamed Fathy, B.Sc., D.B.A., M.Sc., Dr. Phil.,D.Sc. (Radiobiochem.). Born 1925. Educ.: Cairo and Vienna Univs. Prof. Radiobiol., A.E.E., U.A.R.; Deleg. to Khartoum Univ. as Prof. of Biochem. Member, Syndicate of Scientists. Book: Uses of Co 60 (In Arabic). Societies: Assoc. of Radioisotopes, Cairo; German Academic Soc., Leipzig. Nuclear interests: Radiobiochemistry; Use of tracers in metabolism studies of proteins and nucleoproteins.
Address: Faculty of Veterinary Science, Khartoum University, P.O.B.32, Khartoum, Sudan.

ABE, Kyuzi. Director, Rad. Lab., Biol. Res. Lab., Tanabe Seiyaku Co. Ltd.
Address: Tanabe Seiyaku Co. Ltd., Biological Research Laboratory, Shimotoda, Todamachi, Kitaadachi-gun, Saitama-ken, Japan.

ABE, Shigetada. Director, Power Reactor Development, Japan Atomic Energy Res. Inst., 1966-.
Address: Japan Atomic Energy Research Institute, 1-1-13 Shinbashi, Minato-ku, Tokyo, Japan.

ABE, Shizuo. Chief, Planning Sect., Architectural Eng. Div., Nishimatsu Construction Co. Ltd.
Address: Nishimatsu Construction Co. Ltd., 13 Nishikubo-Sakuragawa-cho, Shiba, Minato-ku, Tokyo, Japan.

ABE, Tadashi. Sen. Eng. Officer, Nucl. Ship Res. Office, Kawasaki Kisen Kaisha Ltd.
Address: Kawasaki Kisen Kaisha Ltd., Nuclear Ship Research Office, Tokyo Head Branch Office, 6-1, 1-chome Marunouchi, Chiyoda-ku, Tokyo, Japan.

ABECASIS, Fernando Maria MANZANARES. See MANZANARES ABECASIS, Fernando Maria.

ABEGG, K. Director, Nucl. Dept., Oerlikon Eng. Co.
Address: Oerlikon Engineering Company, Nuclear Department, Zürich 11/50, Switzerland.

ABEGG, Roland, Dr. Prof. and Head, Dept. of Zoology, Formerly Member, Nucl. Sci. and Eng. Development Com., now Member, Nucl. Advisory Com., Louisiana Polytech. Inst.
Address: Louisiana Polytechnic Institute, Tech. Station, Ruston, Louisiana, U.S.A.

ABEL, Derek Edmund, C.Eng., A.M.I.Mech.E. Born 1920. Sen. Design Eng., Risley (1950-60), Sen. Design Eng., Windscale (1960-), U.K.A.-E.A. Society: Windscale and Calder Branch, Nucl. Eng. Soc.
Nuclear interest: Fuel element reprocessing plant design, including decanning, chemical separation purification and recovery, active waste treatment, storage and disposal, shielded cells, glove boxes, remote handling equipment and containers for radioactive and fissile materials.
Address: 'Blengfell', Wasdale Road, Gosforth, Cumberland, England.

ABEL, Emil, Dipl. Eng. (Chem.). Born 1933. Educ.: Bratislava Tech. High School. Lab. of the Hygiene of Rad., Inst. of Hygiene, 1958-. Nuclear interests: Radiochemical analysis of foodstuffs; Radiochemical methods; Dosimetry.
Address: Research Institute of Hygiene, Bratislava, Czechoslovakia.

ABELE, Manlio, Dr. E.E. Born 1920. Educ.: Roy. Polytech. Inst., Turin. Prof. Electromagnetic Theory, Cordoba Nat. Univ. and Advanced School of Aerotechnique, Cordoba, Argentina, 1948-55; Prof. Exptl. Phys., Inst. Phys., San Carlos de Bariloche, Argentina, 1956-58; Tech. Adviser, Aerotech. Inst. of Cordoba, Argentina, 1948-55; Head, Res. Dept., Bariloche Atomic Centre of Argentine Nat. A.E.C., 1956-58; Prof. Aeronautical Eng., Brooklyn Polytech. Inst., 1959-66; Gen. Appl. Sci. Labs., Inc., 1966-. Societies: A.P.S.; Argentine Phys. Assoc.; A.I.A.A.
Nuclear interests: Theoretical and experimental research in plasma physics; Interaction between a plasma and high power electromagnetic fields.
Address: 100 S. Ocean Avenue, Freeport, New York, U.S.A.

ABELLARD, J. Managing Director, Sté de Exploitation de Produits Industriels.
Address: Société d' Exploitation de Produits Industriels, 47 rue du Faubourg, St-Honoré, Paris 8, France.

ABELSON, Philip H., Dr. Formerly Member Gen. Advisory Com. to U.S.A.E.C.; Formerly Member, Plowshare Advisory Com., U.S.A.E.C.

Director, Geophys. Lab., Carnegie Inst., Washington.
Address: Carnegie Institute of Washington, 2801 Upton Street, NW, Washington D.C. 20008, U.S.A.

ABERCROMBIE, Stanley Douglas, H.N.C. (Elec. Eng.) and other Diplomas. Born 1920. Educ.: Tech. College. Elec. Res. Assoc.; M. of A.P., T.R.E. later M.O.S., 1941; M. of S., C.I.A. Woolwich, London, 1950; M. of S., H.E.R., Fort Halstead, Kent, 1952; Senior Supt., A.W.R.E., U.K.A.E.A., 1967. Societies: Sen. Member, Inst. Radio Eng. (S.M.I.R.E.) (U.S.A.); F.I.E.E. (U.K.); M. Brit. R.E.
Nuclear interests: Detection techniques and effects of nuclear weapons fired underground, in the atmosphere, ionosphere, and in outer space. Study of the electromagnetic pulse (EMP) radiated by the weapon and its effects. Weapon ionospheric effects including radio blackout. Geophysical and ionospheric measurements. Radio propagation study and measurement.
Address: 11 St. Peter's Hill, Caversham, Reading, Berks., England.

ABEYRATNA, E., Dr. Member, Atomic Energy Com., Nat. Planning Com. of Ceylon.
Address: Atomic Energy Committee, National Planning Council of Ceylon, 5 Galle Buck Road, Colombo 1, Ceylon.

ABIL'DAEV, A. Kh. Paper: Co-author, Determination of Uranium and Plutonium Content in Samples of Mountain Rocks and Ores by means of Account of Fission Fragments (letter to the Editor, Atomnaya Energiya, vol. 23, No. 6, 1967).
Address: Academy of Sciences of the U.S.S.R., 14 Leninsky Prospekt, Moscow V-71, U.S.S.R.

ABKOWITZ, Stanley, B.S. (Chem. Eng.). Born 1927. Educ.: M.I.T. Manager, New Development Dept., Nucl. Metals, Inc., West Concord. Book: Titanium in Industry (Van Nostrand, 1955). Societies: A.S.M.; A.I.M.M.E.
Nuclear interests: Metallurgy of superalloys, reactive and refractory metals.
Address: Nuclear Metals, Inc., West Concord, Massachusetts, U.S.A.

ABLITT, John Frederick, B.Sc. Tech. (Hons.). Born 1919. Educ.: Manchester Univ. Eng. Control Officer (B.E.P.O.) (1951-53), Senior Eng. (Design) (1953-58). A.E.R.E., Harwell; Asst. Chief Eng., Health and Safety Branch, Power Reactors Group, U.K.A.E.A., 1958-. Societies: Assoc. Member, Inst. Mech. Eng.; British Nucl. Eng. Soc.
Nuclear interests: Reactor design and operation; Experimental equipment design; Safety features of reactor design and operation and of experimental equipment design.
Address: 15 Castleway, Hale Barns, Altrincham, Cheshire, England.

ABOU SINNA, Ibrahim A., M.B., Ch.B. (Cairo), D.M.R. (London). Born 1910. Educ.: Cairo and London Univs. Prof. Radiol., Faculty of Medicine, Ain Shams Univ., Abbassia, Cairo; Director, Dept. of Radiol. and Radioisotope Lab., Demerdash Univ. Hospital, Cairo. Chairman, Egyptian Soc. of Radioisotopes. Societies: Egyptian Radiol. Soc.; Egyptian Medical Assoc.; Soc. Nucl. Medicine, U.S.A.
Nuclear interests: Radioactive isotopes in the fields of diagnosis, therapy and research.
Address: 2 Mazloum Street, Cairo, U.A.R.

ABOV, Yury Georgiyevich. At Inst. of Theoretical and Exptl. Phys. Papers: Co-author, Critical Test conducted on a Heavy-water Exptl. Reactor (2nd I.C.P.U.A.E., Geneva, Sept. 1958); co-author, Criticality Stand Tests of a Heavy-Water Reactor with Fuel Elements of Rod Type (Letter to the Editor, Atomnaya Energiya, vol. 12, No. 2, 1962).
Address: Institute of Theoretical and Experimental Physics, Academy of Sciences of the U.S.S.R., Post Box 3315,89 Ulitsa Cheremushkinskaya, Moscow, U.S.S.R.

ABOWD, Richard George, Jr., B.S. (Mech. Eng.), M.S. (Mech. Eng.). Born 1924. Educ.: Chicago, Notre Dame and Michigan Univs. Project Eng. and Health Phys. Officer, Petroleum Chemicals Res., R. and D. Dept., Ethyl Corp. Society: Soc. of Automotive Eng.
Nuclear interests: Radiotracers; Radiation effects on materials.
Address: 1600 West Eight Mile Road, Ferndale, Michigan 48220, U.S.A.

ABRAGAM, Anatole, D.Phil. (Oxon). Born 1915. Educ.: Paris and Oxford Univs. Phys., Centre for Nucl. Studies, Saclay, 1947-; Res. Fellow, Harvard Univ., 1952-53; Prof., Nucl. Magnetism, Coll. de France; Director, Phys., French A.E.C. Book: Principles of Nucl. Magnetism. Papers: Papers on resonance, solid state phys., nucl. phys. Society: Member of Council, French Phys. Soc.
Address: 33 rue Croulebarbe, Paris, France.

ABRAHAMS, Albert Philip, B.S. Born 1919. Educ.: Columbia and New York Univs., and C.C.N.Y. Radiophys. (1954-58), Sen. Radiophys. (1958-60), Assoc. Radiophys. (1960-62), Principal Radiophys. (1962-), New York State Labor Dept. Member, Nucl. Standards Board, American Standards Assoc. Societies: A.N.S.; American Conference of Governmental Industrial Hygienists; H.P.S.
Nuclear interests: Radiation protection; Health physics; Atomic and nuclear physics; Legislation and regulation of atomic energy and radiation dosimetry.
Address: 1592 Jesup Avenue, Bronx 52, New York, N.Y., U.S.A.

ABRAHAMSEN, Egil, B.Sc. Born 1923. Educ.: Trondheim Tech. Univ. Surveyor, Classification Soc. (1952), Chief of Res.

ABR

Dept. (1955), Managing Director (1967), Det norske Veritas. Chairman, Norwegian Ship Model Testing Basin; Chairman, Branch Council, Norwegian Shipbuilding Industry; Pres., Internat. Ship Structures Congress, 1967; Board Member, Norwegian Ship Res. Inst.; Board Member, Norwegian Inst. of Nucl. Energy. Societies: Inst. of Naval Architects, London; Soc. of Naval Architects and Marine Eng., New York; Soc. for Exptl. Stress Analysis, Conn., U.S.A.
Nuclear interest: Design, construction and classification of nuclear power ships.
Address: Det norske Veritas, P.O. Box 6060, Oslo 6, Norway.

ABRAMOV, V. A. Paper: Co-author, Optimum Parameters of Arc as Source of Ionising Rad. (letter to the Editor, Atomnaya Energiya, vol. 23, No. 6, 1967).
Address: Academy of Sciences of the U.S.S.R., 14 Leninsky Prospekt, Moscow V-71, U.S.S.R.

ABRAMOVITCH, M. D. Papers: Co-author, On the Corrosion Resistance of Some Materials in Sodium and Lithium (2nd I.C.P.U.A.E., Geneva, Sept. 1958); co-author, Thermal Diffusivity for Sodium and Lithium (Atomnaya Energiya, vol. 11, No. 3, 1961); co-author, Evaluation of Brittle Fracture Behaviour of Thick Sheet Materials (letter to the Editor, ibid, vol. 23, No. 6, 1967).
Address: Nuclear Energy Institute, Academy of Sciences of the U.S.S.R., 14 Leninsky Prospekt, Moscow V-71, U.S.S.R.

ABRAMS, I. A. Paper: Co-author, Measurement of Great Gamma - Rays Doses and Fluxes by Using Photoactivation of Nucl. Isomeric States (letter to the Editor, Atomnaya Energiya, vol. 20, No. 5, 1966).
Address: Academy of Sciences of the U.S.S.R., 14 Leninsky Prospekt, Moscow V-71, U.S.S.R.

ABRAMSON, H. Norman, B.S. (Mech. Eng.), M.S. (Eng. Mech.), Ph.D. (Eng. Mech.). Born 1926. Educ.: Stanford and Texas Univs. Project Analytical Eng., Chance - Vought Aircraft Corp., 1951-52; Assoc. Prof., Aeronautical Eng., Texas A and M Coll., 1952-56; Director, Mech. Sci., Southwest Res. Inst., 1956-. Exec. Com.,Appl. Mech. Div., A.S.M.E. Societies: A.I.A.A.; Soc. Naval Architects and Marine Eng.
Nuclear interest: Reactor design problems involving dynamics and fluid mechanics.
Address: Southwest Research Institute, Department of Mechanical Sciences, 8500 Culebra Road, San Antonio, Texas 78206, U.S.A.

ABRAMYAN, E. A. Papers: Co-author, Vacuum Tubes for High-Current Accelerators (Atomnaya Energiya, vol. 22, No. 1, 1967); co-author, Intensity Proton Beam Acceleration on Transformer-Type Set (ibid, vol. 22, No. 5, 1967); co-author, High-Current Impulse Electron Accelerator (ibid, vol. 23, No. 1, 1967); co-author, High Intensity Electron Beams in Strong Focusing Acceleration Tubes (ibid, vol. 25, No. 2, 1968).
Address: Academy of Sciences of the U.S.S.R., 14 Leninsky Prospekt, Moscow V-71, U.S.S.R.

ABREU, Maria Helena de. See de ABREU, Maria Helena.

ABREU FARO, Manuel Jose de. See de ABREU FARO, Manuel Jose.

ABS, Hermann J., Dr. rer. pol. h.c. stellv. Vorsitzender, Fachkommission V Wirtschaftliche, finanzielle und soziale Probleme, and Mitglied, Deutsche Atomkommission.
Address: Deutsche Atomkommission, 5300 Bonn 9, Postfach, Germany.

ABSON, Wilfred, B.Sc. Born 1916. Educ.: Hull Univ. Coll. Head of Rad. Detectors Group, Electronics Div. A.E.R.E., Harwell, 1952-. Society: A.M.I.E.E.
Nuclear interests: Nuclear radiation measurements and instrumentation.
Address: Brecon, Orchard Road, East Hendred, Berks., England.

ACCARY, André, Ing. E.N.S.C.P., Docteur d'Etat. Born 1926. Educ.: Paris Univ. Lab. de Phys. de Rayons X, Paris, 1950-51; Centre d'Étude de Chimie Métallurgique, C.N.R.S., 1951-55; Metal Res. Lab., Carnegie Inst. Technol., 1955-58; Centre d'Etudes Nucl. de Saclay, C.E.A. (France), 1958-; Chargé de cours à l'Ecole Nationale Supérieure de Céramique de Sèvres, 1959-; Chargé de conférences spécialisées, (Saclay), (1959-), Maître de conférences (1962), I.N.S.T.N. Chef de Section, C.E.A., 1961. Book: Editor, Génie Atomique, tome IV (Les Matériaux Nucléaires) (I.N.S.T.N. et Presses Universitaires de France, 1961). Societies: Sté. Française de Métallurgie; Sté Chimique de France; Inst. Metals.
Nuclear interest: Refractory materials (basically nuclear fuels).
Address: 15 quai de la Gironde, Paris 19, France.

ACCINNI, Filippo, Nucl. Eng. Born 1936. Educ.: Milan Politecnico. C.I.S.E., Milan. Nuclear interest: Experimental nuclear reactor physics.
Address: Centro Informazioni Studi Esperienze, C.P. 3986, Milan, Italy.

ACCOYER, L. Ing., Nucl. energy, Bureau Veritas, S.A.
Address: Bureau Veritas, S.A., 31 rue Henri Rochefort, Paris 17, France.

ACEVEDO, Alejandro MARQUEZ. See MARQUEZ ACEVEDO, Alejandro.

ACHE, Hans-Joachim, Dr. rer. nat., Diplom-Chemiker. Born 1931. Educ.: Cologne Univ. Assoc. Prof., Virginia Polytech. Inst. Societies: Gesellschaft Deutscher Chemiker, Fachgruppe Kernchemie; A.C.S.; Rad. Res. Soc.

Nuclear interests: Radio- and radiation-chemistry; Hot atom chemistry; Labelling of organic substances with radioactive isotopes, especially with tritium and carbon-14; Radiation chemistry of organic compounds.
Address: Department of Chemistry, Virginia Polytechnic Institute, Blacksburg, Virginia 24061, U.S.A.

ACHECAR, Abelardo ELIAS DIAZ. See ELIAS DIAZ ACHECAR, Abelardo.

ACHMATOWICZ, Osman, Prof., D.phil. Council Member, State Council for the Peaceful Use of Nucl. Energy, Poland.
Address: App. 23, 22 Nowowiejska str., Warsaw, Poland.

ACIOLI, J. L., Asst. Prof. Theoretical Phys. Dept., Centro Brasileiro de Pesquisas Fisicas.
Address: Centro Brasileiro de Pesquisas Fisicas, 71 Avenida Wenceslau Braz, Rio de Janeiro, Brazil.

ACKER, Ernest R. Vice-Pres., Power Reactor Development Co.; Pres., Empire State Atomic Development Associates Inc.; Chairman of the Board, Central Hudson Gas and Electric Corp.
Address: Power Reactor Development Co., 1911 First Street, Detroit 26, Michigan, U.S.A.

ACKERET, Jakob, Prof. Dr. Prof. of Aerodynamics, Swiss Federal Inst. of Technol. Adviser, Third I.C.P.U.A.E., Geneva, Sept. 1964.
Address: Swiss Federal Institute of Technology, 33 Leonhardstrasse, Zürich 6, Switzerland.

ACKERMANN, J. Member, Conseil d'Administration, Energie Nucléaire S.A.
Address: Energie Nucléaire S.A., 10 avenue de la Gare, Lausanne, Switzerland.

ACKERMANN, P. Member, Federal Commission for Radioactivity Control, Physikalische Anstalt, Basel Univ.; Member, Alarmausschuss, Eidg. Kommission des Innern.
Address: Basel University, 82 Klingelbergstrasse, Basel, Switzerland.

ACKROYD, Ronald Tunstall, Ph.D., D.Eng. Born 1921. Educ.: Liverpool Univ. Entered Dept. of Atomic Energy M.O.S. in 1949 a S.S.O. Various appointments since in reactor phys. and design. Tech. Manager (Gen. Neutronics), Central Tech. Service, Reactor Group, U.K.A.E.A., Risley. Society: F.I.E.E.
Nuclear interests: Fast reactor design; Reactor physics and applied mathematics.
Address: Bienn Cianta, Heath Road, Upton-by-Chester, Cheshire, England.

ACONE, Gerardo. Vice Pres., Tecnider-Tecnica Idrocarburi e Derivati S.p.A.
Address: Tecnider-Tecnica, Idrocarburi e Derivati, S.p.A., 4 Piazza Velasca, Milan, Italy.

ACOSTA, Virgilio, Dr. in Phys. and Maths. Born 1921. Educ.: Havana Univ. Prof. of Phys. and Prof. in charge, Dept. Nucl. Eng., Villanueva Univ., 1951-; Prof. Phys., Vibora Inst. 1941-. Books: Experiments in Phys., Editorial Selecta (1956); Gen. Phys. (1954); Physic Introduction (2 vol. 1950); Ampliación de Fisica (1952).
Address: 9 Parque, Nuevo Vedado, Havana, Cuba.

ACTON, Eric William Vickers, B.Sc. (Gen.), B.Sc. (Special) (Phys.), M.Sc.,C.Eng. Born 1920. Educ.: London Univ. Sci., G.E.C., Res. Lab., Wembley, Mddx., 1950-56; Sect. Head, Heat Transfer, G.E.C., Atomic Energy Div., Erith, 1956-59; Sect. Head, Heat Transfer, C.E.G.B., Berkeley Nuclear Labs., 1960-65; Sen. Sci., Heat Transfer, E.S.R.O., Noordwijk, Netherlands, 1965-. Societies: I.E.E.; A.Inst.P.
Nuclear interests: Reactor core and fuel element heat transfer; Reactor safety and control instrumentation.
Address: 11 Bijdorpstraat, Katwijk-Binnen, Netherlands.

ADACHI, Takeshi, Eng. Dr. Born 1913. Educ.: Tokyo Univ. of Sci. and Literature (now Tokyo Univ. of Educ.). Res. Fellow (1942-66), Asst. Manager, Development Staff, Gen. Eng. Group (1966-), Central Res. Lab., Tokyo Shibaura Elec. Co. Ltd. Com., Japan Ind. Standard Com., 1958-60. Societies: Phys.Soc. of Japan; Japan Soc. of Appl. Phys.; Atomic Energy Soc. of Japan; A.N.S.
Nuclear interests: Neutron generator, development and its application, especially pulsed neutron technique and activation analysis.
Address: 271 Kizuki Gioncho, Kawasaki city, Kanagawa-ken, Japan.

ADAIR, Robert Kemp, Ph.D. Born 1924. Educ.: Wisconsin Univ. Ford Fellow, C.E.R.N., 1963; Chairman, Phys. Dept., Yale Univ., 1967-. Exec. Com., Div. of Particles and Fields, and Fellow, A.P.S., 1967. Book: Coauthor, Strange Particles (New York, Wiley, 1965).
Nuclear interest: Elementary particles and fields.
Address: 88 Kildeer Road, Hamden, Connecticut, U.S.A.

ADAM, Hans, Dr. phil., physiker, Ing. V.D.I. Born 1907. Educ.: Kiel Univ. Co-Editor, Kerntechnik. Book: Einführung in der Kerntechnik (1967).
Nuclear interests: The use of isotopes; Reactor techniques; Education of the nuclear engineer.
Address: 16 Wilhelminenstr., Kiel, Germany.

ADAM, Pierre Henri. At Batignolles-Chatillon (Mécanique Générale) S.A.
Address: Batignolles-Chatillon (Mécanique Générale) S.A., 5 rue de Monttessuy, Paris 7, France.

ADA

ADAMCHUK, Yu. V. Papers: Co-author, The Total Neutron Cross-section of Np^{237} between 2 and 10,000 eV (letter to the Editor, Atomnaya Energiya, vol. 6, No. 5, 1959); co-author, Radiative Capture Cross-Sections for Disprosium isotopes in the Energy Region of 0.023 - lev (letter to the Editor, ibid, vol. 16, No. 1, 1964). Address: Academy of Sciences of the U.S.S.R., 14 Leninsky Prospekt, Moscow V-71, U.S.S.R.

ADAMCZEWSKI, Ignacy, mgr. phil., Dr. of Phys. Born 1907. Educ.: Warsaw Univ. Full Prof.Phys., Head of Phys. Dept. (1945-), Dean of Chem. Faculty, Tech. Univ. and Medical Acad., Gdańsk. Vice-Pres., Polish Phys. Soc. Group, Gdańsk. Books: Health Protection against Ionising Rad. (P.Z.W.L., 1959); Course of Modern Phys. (1967); Physics for Medicians P.Z.W.L., 1962); Biophysics (P.Z.W.L., 1967); Ionisation and Conductivity in Liquid Dielectrics (P.W.N., 1965) (in French: Paris, Masson 1968) (in English: London, Taylor and Francis, 1968). Societies: Polish Phys. Soc., National Com. Radiol. Protection.
Nuclear interests: Physics of dielectrics; Health physics; Radiation dosimetry; Biophysics; Applications of radioisotopes, accelerators of low energy.
Address: 5 Fahrenheita, Gdańsk, Poland.

ADAMEC, Vladimir, Ing., C.Sc. Born 1928. Educ.: Bratislava Tech. Univ. Cables and Insulating Materials Res. Inst., Bratislava, 1951-.
Nuclear interests: Radiation effects on physical properties of electrical insulating materials, especially polymers.
Address: 12 VÚKI, Továrenská, Bratislava, Czechoslovakia.

ADAMOV, I. Yu. Paper: Co-author, Microwave Rad. of Electrodeless Induction Discharge (Atomnaya Energiya, vol. 16, No. 2, 1964).
Address: Academy of Sciences of the U.S.S.R., 14 Leninsky Prospekt, Moscow V-71, U.S.S.R.

ADAMOVICH, V. K. Papers: Evaluation of Neutron Irradiation Dose Exciting Changes in Mechanical Properties of Pure Metals (letter to the Editor, Atomnaya Energiya, vol. 15, No. 5, 1963); On Neutron Rad. Dose Characterising Metal Diffusion Change (letter to the Editor ibid., vol. 24, No. 3, 1968). On Neutron Dose Characterising Rad. Embrittlement Ferritic Steels (letter to the Editor, ibid, vol. 25, No. 1, 1968).
Address: Academy of Sciences of the U.S.S.R., 14 Leninsky Prospekt, Moscow V-71, U.S.S.R.

ADAMOVSKII, L. A. Paper: Co-author, Exptl. Res. of Boiling Water Reactor Stability (Atomnaya Energiya, vol. 24, No. 4, 1968).
Address: Academy of Sciences of the U.S.S.R., 14 Leninsky Prospekt, Moscow V-71, U.S.S.R.

ADAMS, Charles Albert, B.Sc. (Hons. Mathematics), B.Sc. (Hons. Phys.). Born 1907. Educ.: London Univ. Chief Nucl. Health and Safety Officer, C.E.G.B., 1958-; previously Chief of Trials and Safety, Atomic Weapons Res. Establishment. Society. Fellow, Inst. Physics.
Nuclear interests: Health physics and reactor physics.
Address: 30 Clevedon Road, Tilehurst, nr. Reading, Berks., England.

ADAMS, Gail Dayton, B.S., M.S., Ph.D. Born 1918. Educ.: Case Inst. Technol. and Illinois Univ. Res. Phys. and Formerly Assoc. Director, Radiol. Lab., Clin. Prof. Phys. (Radiol.), School of Medicine, Univ. California Medical Centre, San Francisco, 1953-. Director (1958-63), Pres. (1958-60), American Assoc. of Physicists in Medicine; Sec.-Treas., Rad. Res. Soc., 1963-. Member, Phys. Com., Radiol. Soc. of North America. Societies: Fellow, A.P.S.; Rad. Res. Soc.; American Assoc. of Physicists in Medicine; Radiol. Soc. of North America; H.P.S.; American Coll. of Radiol. Certifications: Radiolog. phys. by American Board of Radiol.; Health phys. by American Board of Health Phys.
Nuclear interests: Radiological physics, especially dosimetry of X- and gamma-rays, instrument standardisation, and isotope measurements; Health physics (design and protection considerations for radiation enclosures); Nuclear physics (absorption of high energy photons; design, construction, and operation of betatrons and synchrotrons).
Address: Radiological Laboratory, University of California Medical Centre, San Francisco, California 94122, U.S.A.

ADAMS, Gerald Edward, B.Sc., Ph.D. Born 1930. Educ.: London and Manchester Univs. Argonne Nat. Lab., 1958-60; C.E.N., Saclay, 1960-61; Res. Unit in Radiobiol., British Empire Cancer Campaign for Res. Societies: Faraday Soc.; Assoc. for Rad. Res.
Nuclear interest: Application of pulse methods (pulse radiolysis, etc.) to radiation chemistry, radiation biology and the general field of free radical chemistry.
Address: British Empire Cancer Campaign for Research, Research Unit in Radiobiology, Mount Vernon Hospital, Northwood, Middlesex, England.

ADAMS, John Bertram, F.R.S., D.Sc. (h.c.) Geneva Univ., 1960; Birmingham Univ., 1961; Surrey Univ., 1966. M.A., Oxford Univ., 1967. Born 1920. Educ.: Eltham Coll. A.E.R.E. 1945-53; Director, Proton Synchroton Div., (1954-60), Director-Gen. (1960-61), C.E.R.N.; Director, Culham Lab., A.E.A., 1960-67; Controller, Ministry of Technol., 1965-66; Member for Res., U.K.A.E.A., 1966-. Society: I.E.E. Nuclear interests: Nuclear particles acceleration design and construction; High energy nuclear

physics; Plasma physics; Thermonuclear reactions.
Address: The Grey House, Lincombe Lane, Boar's Hill, Oxford, England.

ADAMS, Richard James, Ph.D. (British Columbia), M.Sc. (Lond.), B.Sc. (Lond.). Born 1928.
Educ.: London and British Columbia Univs. Chemist, Packaging and Allied Trades Res. Assoc., Leatherhead, Surrey, 1953-58; Sen. Sci. Officer, Water Pollution Labs. D.S.I.R. Stevenage, 1958-61; Postdoctoral Res. Fellow, British Columbia Univ., 1963-64; Assoc. Res. Officer, Appl. Phys. Div., Nat. Res. Council, Ottawa, Canada. Society: F.R.I.C.
Nuclear interest: Standardisation of radionuclides.
Address: Box 18, Wren Road, R.R.1., Ottawa, Ontario, Canada.

ADAMS, Richard M., Ph.D. Born 1916. Educ.: Chicago Univ. and I.I.T. Asst. Lab. Director, Argonne Nat. Lab. Societies: A.N.S.; A.C.S.; American Inst. Chems.; A.A.A.S.
Address: Argonne National Laboratory, 9700 South Cass Avenue, Argonne, Illinois 60439, U.S.A.

ADAMS, Robert B. Rad. Phys., Res. Assoc. Wearn Lab. for Medical Res., Western Reserve Univ.
Address: Western Reserve University, 2064 Abington Road, Cleveland 6, Ohio, U.S.A.

ADAMS, Robert Edward, B.S., M.S. Born 1929. Educ.: Memphis State and Mississippi Univs. O.R.N.L. 1956-. Societies: A.N.S.; A.C.S.
Nuclear interests: Methods for the containment and disposal of both gaseous and solid fission product aerosols released from the fuel of nuclear reactors under accident conditions.
Address: Reactor Chemistry Division, Oak Ridge National Laboratory, Oak Ridge, Tennessee, U.S.A.

ADAMS, Robert W. Pres., Western Nucl., Inc. Member, Board of Directors, Atomic Ind. Forum, 1966-67.
Address: Western Nuclear, Inc., 1700 Broadway, Denver, Colorado 80202, U.S.A.

ADAMS, Ruth. Managing Editor, Bulletin of the Atomic Scientists.
Address: 935 E. 60th Street, Chicago 37, Illinois, U.S.A.

ADAMSKA, Mrs. Bozena, M.Sc. (Phys.). Born: 1936. Educ.: Warsaw Univ. Res. worker, Warsaw Tech. Univ., 1959-61; Res. worker, Inst. Nucl. Res., Warsaw, 1961-.
Nuclear interests: Health physics: whole body counting; Neutron dosimetry; Radiation detection.
Address: 17 M.154 Stoleczna, Warsaw 86, Poland.

ADAMSKI, Leslaw, M.Sc. Phys. Born 1933. Educ.: Warsaw Univ. Sci. worker, Dept. Exptl. Phys., Warsaw Univ., 1955-58; Sci. worker, Dept. of Reactor Phys., Inst. Nucl. Res., 1958-.
Nuclear interests: Reactor physics and engineering; Radiation detectors; Activation analysis.
Address: 17 M.154 Stoleczna, Warsaw 86, Poland.

ADAMSKI, Tadeusz, Docent, Dipl. Chem. Ing. Born 1907. Educ.: Warsaw Tech. Inst. Inst. Inorg. Chem., Gliwice, 1949-54; Inst. Organic Synthesis, Oswiecim, 1955-; Inst. Nucl. Res., Warsaw. 1956-67; Head, Dept. of Chem. Technol. Polish Acad. of Sci. Inst. of Phys. Chem., 1967-. Society: Polish Chem. Soc.
Nuclear interests: Development of advanced technological methods; Improvement of unit processes; Process economics.
Address: 14 M.81 Nowolipki-Street, Warsaw, Poland.

ADAMSON, George M., Jr., B.S. (Metal. Eng.). Born 1919. Educ.: Utah Univ. Sen. Metal., O.R.N.L. 1946-. Member, A.S.T.M. Coms. on Zirconium and Titanium. Societies: A.S.M.; American Inst. Metal. Eng.
Nuclear interests: Metallurgy - fuel element development; Alloy research.
Address: 382 East Drive, Oak Ridge, Tennessee, U.S.A.

ADAMSON, J. E. Group Sec., Weapons Group, U.K.A.E.A.
Address: Atomic Weapons Research Establishment, U.K.A.E.A., Aldermaston, Berks., England.

ADANY, Leon. Head, Centre de Recherches et d'Irradiations L. Adany. Deleg., I.A.E.A. Conference on the Use of Radioisotopes in the Physical Sciences and Industry, Copenhagen, 6-17 September 1960.
Address: 6 Chemin de la Croix, Champigny sur Marne, (Seine), France.

ADELINE, A. R. Manager, Finance and Contract Administration, Hanfield Environmental Health Foundation, Hanford Facilities, U.S.A.E.C.
Address: U.S.A.E.C., Hanfield Environmental Health Foundation, Richland, Washington. U.S.A.

ADELMAN, Frank L. Formerly with Inst. for Defence Analysis, Arlington; Sci. Advisor, Geophys. Div. (1966-67), Asst. Director (1967-), I.I.T. Res. Inst.; Consultant, Nucl. Effects, Office of D.O.D. Res. and Eng.; Member, several Joint Coms., U.S.A.E.C. and D.O.D.
Address: IIT Research Institute, Geophysics Division, Technology Centre, 10 West 35th Street, Chicago, Illinois 60616, U.S.A.

ADERHOLD, Howard C., Assoc. Degree in Radio Communication. Born 1929. Educ.: Williamsport Community College. Systems Eng., Curtiss-Wright Corp., 1956-61; Reactor Supervisor, Cornell Univ., 1962-. Vice-Chairman, Niagara-Finger Lakes Section, A.N.S. Nuclear interests: Reactor operations and management.
Address: 23 Muriel Street, Ithaca, New York 14850, U.S.A.

ADERHOLD, O. C., Dr. Education, State of Georgia Nucl. Advisory Commission.
Address: State of Georgia Nuclear Advisory Commission, Office of Publications, Georgia Institute of Technology, Atlanta 13, Georgia, U.S.A.

ADIWIKARTA, Sukeni. Lecturer, Nucl. Chem. Dept., Bandung Inst. of Technol.
Address: Bandung Institute of Technology, 10 Djl. Ganeca, Bandung, Indonesia.

ADKINS, Bruce M. Born 1921. Editor, R.N.S.S. Journal, Royal Naval Scientific Service, 1947-53; Publicity Manager, Elliott Brothers (London) Ltd. later Elliott-Automation Ltd., 1953-58; Head, Public Relations and Information, European Nuclear Energy Agency (O.E.C.D.), 1958-. Society: Founder Member, Nuclear Public Relations Contact Group. Nuclear interests: Nuclear public relations.
Address: European Nuclear Energy Agency, O.E.C.D., Paris, France.

ADKINS, Rutherford Hadley, B.S. (Phys.), M.S. (Phys.), Ph.D. (Phys.). Born 1924. Educ.: Virginia State Coll., Howard (Washington) and Catholic (Washington) Univs. Instructor, Phys. (1949-51), Res. Assoc. Phys. (1953-55), Assoc. Prof., Phys. (1955-58), Prof.,Phys., and Chairman,Dept. (1958-62), Prof. Phys. and Mathematics (1962-), Acting Chairman,Dept. of Phys. (1966-), Fisk Univ., Director, Academic Year Inst., N.S.F., 1967-. Societies: A.P.S.; A.A.P.T.
Nuclear interest: Nuclear physics.
Address: 2017 Clintondale Drive, Nashville, Tennessee, U.S.A.

ADLER, Felix T., Ph.D. (Theoretical Phys.). Born 1915. Educ.: Zurich Univ. and E.T.H. Assoc. Prof., Phys., Carnegie Inst. of Technol., 1950-56; O.R.N.L., 1955-56; Staff Member, Gen. Atomic/Gen. Dynamics Corp., 1956-58; Prof., Phys. and Nucl. Eng., Illinois Univ., Urbana, 1958-; Consultant, Los Alamos Sci. Lab., 1961-; O.R.N.L., 1955-56, 1966-. Member, and Past Chairman, Reactor Eng. Advisory Com. (1960-65), Member, and Past Chairman, Reactor Phys. Advisory Com. (1965-), Idaho Div. (1960-), Argonne Nat. Lab. Societies: A.N.S.; Fellow, A.P.S.
Nuclear interests: Neutron physics; Nuclear cross sections; Reactor kinetics.
Address: Physics Department, Illinois University, Urbana, Illinois 61801, U.S.A.

ADLER, Richard. Architect, Brodsky, Hopf, and Adler.
Address: Brodsky, Hopf, and Adler, 235 E. 42nd Street, New York 17, New York. U.S.A.

ADLER, Ronald. Res. Faculty, Dept. of Phys., Washington Univ., Seattle.
Nuclear interest: Theoretical physics.
Address: Washington University, Department of Physics, Seattle, Washington 98105, U.S.A.

ADLER-RACZ, Joseph H. d'. See d'ADLER-RACZ, Joseph H.

ADLEY, Frank Edgar, B.S. (Mech. Eng.). Born 1912. Educ.: Northeastern and Harvard Univs. and M.I.T. Manager, Environmental Health Sciences, Hanford Environmental Health Foundation, 1965-.
Nuclear interest: Environmental health sciences.
Address: Hanford Environmental Health Foundation, Richland, Washington 99352, U.S.A.

ADLOFF, Jean-Pierre, D. ès Sc. Educ.: Strasbourg Univ. Prof., Nucl. Chem., Strasbourg Univ.; Assoc. Director, Nucl. Chem. Dept., Nucl. Res. Centre, Strasbourg. Societies: Sté. Chimique de France; Sté. Française de Physique; Sté. de Chimie Physique; A.A.A.S.
Nuclear interests: Hot atom chemistry; Chemical effects of nuclear transformations; Mössbauer spectroscopy; Beam techniques; Radiation chemistry; Radiochemistry.
Address: Centre de Recherches Nucléaires, Département de Chimie Nucléaire, rue du Loess, Strasbourg-Cronenbourg, Bas-Rhin, France.

ADO, Yu. M. Papers: Co-author, Coherent Rad. from the Electrons in a Synchrotron (Atomnaya Energiya, vol. 9, No. 6, 1960); Particle Storage in the Synchrotron (Letter to the Editor, ibid, vol. 12, No. 1, 1962); co-author, Bremsstrahlung Spectrum for Electrons of Energy 260 MeV (ibid., No. 3); co-author, Positron Stacking and Obtaining of Electron-Positron Beams in Synchrotron (ibid, vol. 23, No. 1, 1967).
Address: Academy of Sciences of the U.S.S.R., 14 Leninsky Prospekt, Moscow V-71, U.S.S.R.

ADOLPH, Eivind, M.S. Born 1932. Educ.: Danish Tech. Univ. Struers Sci. Instruments, 1958-59; Danish A.E.C., 1959-.
Nuclear interest: Metallurgy; Mechanical testing; Metallography.
Address: Metallurgy Department, Research Establishment Riso, Atomic Energy Commission, Roskilde, Denmark.

ADORNI, Napoleone, Ind. Expert Elec. Technol. Born 1923. Educ.: Tech. Inst. Feltrinelli, Milan. Asst. and Lab. Instructor, Elec. Technol., Inst. Feltrinelli, Milan, 1946-53; Chief, Exptl. Plants Construction and Instruments Service, C.I.S.E. Labs.

Nuclear interests: Design and construction of plants and connected instrumentation for experiments in the field of reactor engineering.
Address: C.I.S.E., Casella Postale 3986, 20100 Milan, Italy.

ADRIEN, Gaston. Born 1914. Educ.: Ecole Navale. Officier de Marine, 1933-59; Public Relations à Marcoule.
Nuclear interests: Research in and development of public relations in the atomic field.
Address: Centre de Marcoule, B.P. 106, Bagnols-sur-Cèze, (Gard), France.

AEBI, H., Prof. Dr. Member, Swiss Nat. Sci. Foundation Director, Medizinisch-chemisches Inst., Bern Univ.
Address: Universität Bern, 28 Bühlstrasse, Bern, Switzerland.

AERNI, Paul, Dr. iur. Born 1918. Educ.: Berne and Zurich Univs. Gen. Manager, Accident and Casualty Insurance Company of Winterthur. Pres., Unfalldirektoren - Konferenz, Swiss Federal Commission for Atomic Energy. Society: Schweizerische Vereinigung für Atomenergie.
Nuclear interest: Insurance.
Address: 40 General Guisanstrasse, Winterthur, Switzerland.

AESCHIMANN, Charles, Elec. Eng. Born 1908. Educ.: Federal Inst. Technol., Zürich. Member, Exec. Com. Aare-Tessin A.G. for Elec., 1959-; Reactor A.G., 1955-; Suisatom A.G., 1957-; Federal Commission for Atomic Energy; Swiss Federation for Atomic Energy. Societies: Atomic Ind. Forum, Inc.; Swiss Soc. of Specialists of Nucl. Technics.
Nuclear interests: Nuclear management and economic criteria.
Address: 12 Bahnhofquai, Olten, Switzerland.

AFABLE, Pedro G., B.S.(C.E.). Born 1912. Educ.: Philippines Univ. Tech. Asst., Nat. Economic Council, 1953-55; Part-time Prof., Polytech. Colleges of the Philippines, 1953-; Chief of Branch, Nat. Economic Council, 1956-58; Deputy Commissioner (1958-64), Acting Commissioner (1964-), Philippine A.E.C. Chairman, Nucl. Power Study Com.; Chairman, Nat. Com. on Irrigation Res.; Vice-Chairman, Nat. Com. on World Power Conference; Sec., Marikina Project Co-ordinating Com.; Vice-Pres., Assoc. of Govt. Civil Eng. of the Philippines. Societies: Fellow, Philippine Assoc. for the Advancement of Sci.; Philippine Soc. of Civil Eng.; Supporting Member, U.S. Highway Res. Board; A.N.S.; H.P.S.; Assoc. Member, N.R.C. of the Philippines.
Nuclear interests: Management; Industrial applications of radioisotopes; Planning nuclear development; Economics of nuclear power.
Address: c/o Philippine Atomic Energy Commission, 727 Herran Street, Manila, Philippines.

AFANAS'EV, V. A. Paper: Co-author, Exptl. Res. of Boiling Water Reactor Stability (Atomnaya Energiya, vol. 24, No. 4, 1968).
Address: Academy of Sciences of the U.S.S.R., 14 Leninsky Prospekt, Moscow V-71, U.S.S.R.

AFANAS'EV, V. N. Paper: Co-author, Surface Relief Determination by Means of Back-Scattered Gamma-Rad. (Abstract, Atomnaya Energiya, vol. 19, No. 6, 1965).
Address: Academy of Sciences of the U.S.S.R., 14 Leninsky Prospekt, Moscow V-71, U.S.S.R.

AFANAS'YEV, V.P. Paper: Co-author, A Technique for Dubna Synchrocyclotron Neutron Beam Study of Material Shielding Properties (Atomnaya Energiya, vol. 16, No. 5, 1964).
Address: Academy of Sciences of the U.S.S.R., 14 Leninsky Prospekt, Moscow V-71, U.S.S.R.

AFONIN, G.S. Deputy Head, Dept. Internat. Relations and Sci. Tech. Information U.S.S.R. State Com. for the Utilisation of Atomic Energy; Gen. Sec. of Deleg. to 2nd I.C.P.U.A.E., Geneva, Sept. 1958.
Address: U.S.S.R. State Committee for the Utilisation of Atomic Energy, 26 Staromonetnii Pereulok, Moscow, U.S.S.R.

AFRIKANTOV, Igor Ivanovich. Born 1916. Dr. Tech. Sci. Sci. Consultant, U.S.S.R. State Com. on Atomic Energy; Design of Reactors, Principal, Dept. for the Exploitation of Atomic Energy, U.S.S.R. Member Fast Reactor Team, visiting U.K., Jan. 1960.
Nuclear interest: Power reactors.
Address: U.S.S.R. State Committee on Atomic Energy, Moscow, U.S.S.R.

AGAEV, N. A. Paper: Co-author, Viscosity of Heavy Water at High Pressure for Interval Temperature 4-100°C (letter to the Editor, Atomnaya Energiya, vol. 23, No. 2, 1967).
Address: Academy of Sciences of the U.S.S.R., 14 Leninsky Prospekt, Moscow V-71, U.S.S.R.

AGAR, Alan W., B.Sc. (Special Phys.). Born 1920. Educ.: London Univ. Phys., Res. Dept., Metropolitan-Vickers Elec. Co., Ltd., 1950-54; Phys. A.E.I. Res. Lab., Aldermaston Court, 1954-56; Director and Head of Lab., Aeon Labs., 1957-61; Consultant Eng., A.E.I. Instrumentation Div., 1961-62; Eng.-in-Charge, Electron Microscopy, A.E.I. Sci. Apparatus Dept., 1963-64; Manager, Electron Microscopes, 1964-. Societies: F.Inst. P.; Fellow, I.E.E.
Nuclear interests: Design of electron microscopes for metallurgical purposes; Provision of a consultant service with electron microscopes for metallurgical and physical problems.
Address: A.E.I. Scientific Apparatus Department, P.O. Box 1, Harlow, Essex, England.

AGEEV, E. A. KRAMER-. See KRAMER-AGEEV, E. A.

AGELAO, G., Dr. Problemi di sicurezza e Dosimetria, Facolta d'Ingegneria, Istituto di Applicazioni e Impianti Nucleari, Palermo Univ.
Address: Palermo University, Viale delle Scienze, Palermo, Sicily, Italy.

AGENO, Mario, Degree in Phys., Ph.D. Born 1915. Educ.: Genoa and Rome Univs. Head, Phys. Labs., Istituto Superiore di Sanità, Rome; Prof., Phys., Faculty of Medicine, Rome Univ. Pres., Italian Soc. of Biophys. and Molecular Biol., 1968-70. Societies: Italian Phys. Soc.; European Soc. for Rad. Biol.
Nuclear interest: Nuclear physics.
Address: Laboratori di Fisica, Istituto Superiore di Sanità, 299 Viale Regina Elena, 00161 Rome, Italy.

AGER, Eric Sidney, A.M.I.M.E. Born 1923. Sen. Project Eng., Bristol Aerojet Ltd., Banwell.
Nuclear interests: Reactor design; In-pile test rigs; Instrumentation.
Address: 'Galena', 17 Glover's Field, Shipham, Somerset, England.

AGER-HANSSEN, Henrik Julius, B.S. (Phys.), Nucl. Eng.). Born 1930. Educ.: Oslo and New York Univs. Instructor, Nucl. Eng., N.Y.U., 1955-57; Res. Phys., Kjeller, 1957-58; Chief Phys. (1958-60), Chief of R. and D. (1960-62), Deputy Project Manager (1962-). O.E.C.D. Halden Reactor Project; Project Manager (1965-66), Assistant Director (1967-), Power Reactor Project, Kjeller. Consultant, Nucl. Propelled Rockets, Reaction Motors, New Jersey, 1956-57. Member, E.N.E.A. Group of Experts on Nucl. Data; Member, I.A.E.A. NORA Com. Society: Norsk Fysisk Selskap.
Nuclear interests: Project management; Nuclear power economy; Reactor strategies.
Address: Institutt for Atomenergi, Kjeller, Norway.

AGETA, Yuzi. Asst. Chief, Rad. Safety Sect., Atomic Energy Bureau, Atomic Energy Commission of Japan.
Address: Atomic Energy Commission of Japan, 3-4 Kasumigaseki, Chiyoda-ku, Tokyo, Japan.

AGINSKI, Paul. Tech. Director, Lab. Industriel d'Electronique Belin.
Address: Laboratoire Industriel d'Electronique Belin, 296 avenue Napoléon Bonaparte, Rueil-Malmaison, (Seine et Oise), France.

AGNEW, Harold Melvin, M.S., Ph.D. (Phys.). E.O.Lawrence Award, 1966. Born 1921. Educ.: Denver and Chicago Univs. Sci. Advisory Board, U.S.A.F., 1957; Member (1964), Chairman (1966), Sci. Advisory Panel U.S. Army; President's Sci. Advisory Com., 1965; Defence Sci. Board, 1966; Advisory Com. for Civil Defence, O.R.N.L., 1967-. Society: Fellow, A.P.S.
Nuclear interest: Design of nuclear weapons and nuclear energy sources.
Address: Weapons Division Leader, Los Alamos Scientific Laboratory, P.O. Box 1663, Los Alamos, New Mexico 87544, U.S.A.

AGNEW, Robert J. Supervisor and Director, Nucl. Lab., Texaco Inc.
Address: Research and Technical Department, Texaco Inc., P.O. Box 509, Beacon, New York, U.S.A.

AGO, Roberto. Member, Com. Scientifico, Diritto ed Economia Nucleare.
Address: Diritto ed Economia Nucleare, 68 Viale Bruno Buozzi, Rome, Italy.

AGODI, A., Gruppo Teorico, Centro Siciliano de Fisica Nucleare.
Address: Centro Siciliano de Fisica Nucleare, Istituto di Fisica, Universita de Catania, 57 Corso Italia, Catania, Italy.

AGOSHKOV, Mikhail I. Deputy Chief Learned Sec., Acad. Sci. of the U.S.S.R.
Address: Academy of Sciences of the U.S.S.R., 14 Leninsky Prospekt, Moscow V-71, U.S.S.R.

AGOSTINO, Vincent d' See d'AGOSTINO, Vincent.

AGRANOFF, Bernard William, B.S. (Chem.), M.D. Born 1926. Educ.: Michigan Univ., Wayne State Medical School and M.I.T. Medical Officer, U.S.P.H.S., Nat. Inst. of Neurological Diseases and Blindness, Bethesda, Maryland. Prof. Biol. Chem.,Dept. of Biol. Chem., and Res. Biochem., Mental Health Res. Inst., Michigan Univ. Book: Low Level Tritium Counting Techniques (chapter in: Liquid Scintillation Counting, by C. G. Bell and E. N. Hayes, Pergamon Press). Societies: American Soc. of Biol. Chemists; A.C.S.; Michigan Nucleonic Soc.; Internat. Neurochemical Soc.; Assoc. for Res. in Nervous and Mental Diseases.
Nuclear interest: Use of isotopic tracers in biochemical research, particularly in the nervous system.
Address: Mental Health Research Institute, Michigan University, Ann Arbor, Michigan 48104, U.S.A.

AGRAWAL, Krishna Das, M.Sc. Born 1925. Educ.: Banaras Hindu Univ. Asst. Geologist (1950-52), Geologist (1952-58), Sen. Geologist (1958-), Dept. of Atomic Energy, Govt. of India.
Address: 9-B/6, N.E.A. Poorvi Marg, Rajinder Nagar, New Delhi-5, India.

AGRESTA, Joseph, B.E.E., M.S., Ph.D. Born 1929. Educ.: Cooper Union, New York Univ. and O.R.S.O.R.T. Graduate Asst., New York Univ., 1950-53; Phys., Curtiss-Wright Res. Div., 1953-56; Res. Assoc., New York Univ., 1956-58; Lecturer, C.C. N.Y., 1958-62; Phys., United Nucl. Corp., 1958-62; Adj. Asst. Prof. Phys., New York Univ., 1959-62. Phys. Union Carbide Res. Inst., 1962-. Societies: A.N.S.; A.A.P.T.; New York Acad. Sci.

Nuclear interests: Reactor theory; Neutron physics; Numerical methods; Direct energy conversion.
Address: 78 Joyce Road, Hartsdale, N.Y., U.S.A.

AGT, A. J. W. van. See van AGT, A. J. W.

AGUAYO, Rafael. Tech. Adviser, Comision Nacional de Investigaciones Atomicas.
Address: Comision Nacional de Investigaciones Atomicas, Ciudad Trujillo, Republica Dominicana.

AGUIAR, Nelson VISTA-. See VISTA-AGUIAR, Nelson.

AGUILA GOICOECHEA, Jose Maria. Vocale, Comision Interministerial de Conservacion de Alimentos por Irradiacion, Junta de Energia Nucl.
Address: Junta de Energia Nuclear, Ciudad Universitaria, Madrid 3, Spain.

AGUILAR, Mario PIZARRO. See PIZARRO AGUILAR, Mario.

AGUILAR PAZ, Jesus, Dr. Pres., Comision Hondurena de Energia Atomica; Participant, 3rd Inter-American Symposium on the Peaceful Application of Nucl. Energy, Rio de Janeiro, July, 1960.
Address: Comision Hondurena de Energia Atomica, Tegucigalpa, D.C., Honduras.

AGUILERA, Don Cibar Caceras, Navy Lieutenant Eng. Member, Nat. A.E.C., Paraguay.
Address: National Atomic Energy Commission, Department of Organisations, Treaties and International Agreements, Ministry of Foreign Affairs, Asuncion, Paraguay.

AGUIRRE-BATRES, Francisco, Prof. San Carlos Univ., Guatemala. Born 1924. Educ.: San Carlos (Guatemala), Texas, Vanderbilt and Havana Univs. and O.R.I.N.S. Chief of Labs., Central American Inst. of Res. and Technol., 1956-. Vice-Pres., Nat. Commission of Nucl. Energy, Guatemala, 1956-. Societies: Coll. of Chems., Guatemala; A.A.A.S.
Nuclear interests: Radioisotopes in the biological field.
Address: 1a. Avenida 9-23, Zona 10, Guatemala City, Guatemala, Central America.

AGUIRRE GONZALO, Jose Maria. Formerly Vocal, Comision Asesora de Equipo Ind., now Vocale, Comision Asesora de Centrales Nucleares, Junta de Energia Nucl., Ministerio de Industria.
Address: Junta de Energia Nuclear, Ciudad Universitaria, Madrid 3, Spain.

AH, M. V. RAMANI-. See RAMANI-AH, M. V.

AHLSTRÖM, Per-Eric. Born 1932. Educ.: Roy. Inst. of Technol. Head, Reactor Phys. Group, Swedish State Power Board, 1963-.
Nuclear interests: Reactor physics; Fuel management.
Address: Swedish State Power Board, Fack, Vällingby 1, Sweden.

AHLUWALIA, Harjit Singh, B.Sc. (Hons.), M.Sc., Ph.D. Born 1934. Educ.: Panjab Univ. Tech. Assistance Expert of U.N.E.S.C.O. to Laboratorio de Fisica Cosmica, Universidad Mayor de San Andres, La Paz, 1962; Res. Assoc., Southwest Centre for Advanced Studies, Dallas, Texas, 1963-64; Visiting Prof. of I.A.-E.A. to Comision Boliviana de Energia Nucl., La Paz, 1965-67; Sci. Director, Laboratorio de Fisica Cosmica, Universidad Mayor de San Andres, La Paz, 1965-67; Assoc. Prof., Phys. and Astronomy Dept., New Mexico Univ., 1968-; Corresponding Member, Internat. Cosmic Ray Commission, I.U.P.A.P., 1966-. Societies: A.G.U.; I.E.E.E.; A.M.S.; A.A.A.S.; Italian Phys. Soc.; Phys. Soc. of Japan.; I.U.P.A.P.
Nuclear interests: Teaching of nuclear physics. Active research interest in geo-, helio- and astrophysical aspects of cosmic radiation; Nuclear electronics and general instrumentation.
Address: Department of Physics and Astronomy, New Mexico University, 800 Yale Boulevard, N.E., Albuquerque, New Mexico 87106, U.S.A.

AHLZEN, Gunnar., M.Sc. Eng. Member, Board of Directors, AKK Atomic Power Group.
Address: AKK Atomic Power Group, 19 Stureplan, Stockholm C, Sweden.

AHMAD, Ishfaq, M.Sc. (Lahore), D.Sc. (Montreal). Born 1930. Educ.: Panjab and Montreal Univs. Lecturer, West Pakistan Educ. Dept., 1952-59; S.S.O. (1961-66), P.S.O. (1966-), Sec. (1967-), Pakistan A.E.C.; Res. Fellow, Univ. Inst. Theoretical Phys., Copenhagen, Denmark, 1961-62; Attaché des Recherches, Montreal and Ottawa Univs., 1963-65.
Nuclear interest: Experimental nuclear physics.
Address: Pakistan Atomic Energy Commission, P.O. Box No. 3112, Karachi, Pakistan.

AHMED, Riad Aly. Member, Lebanese A.E.C.
Address: Lebanese Atomic Energy Commission, Ministry of Public Works, Beirut, Lebanon.

AHMED, Shaukat, B.Sc. (Agriculture), M.Sc. Agri. (Agricultural Chem.), Ph.D. (Plant Physiology). Born 1925. Educ.: Bombay, Sind, and North Carolina Univs. Director, Atomic Energy Agricultural Res. Centre, Tandojam, 1962-. Societies: Pakistan Assoc. for the Advancement of Sci.; American Soc. of Plant Physiologists.
Nuclear interests: Use of radioisotopes and atomic radiations in soil plant relationships, uptake and distribution studies etc.
Address: Atomic Energy Agricultural Research Centre, Tandojam, West Pakistan.

AHN, Chi Yul, Ph.D. (Medical Sci.). Born 1922. Educ.: Chunnam Univ. Chief, Radiol. Dept., First Army Hospital, 1952; Chief, Radiol. Dept., Seoul Red Cross Hospital, 1956; Chief, Radiol. Dept., St. Mary's Hospital, 1957; Director, Radiol. Res. Inst., Office of Atomic Energy, Ministry of Sci. and Technol., 1963. Society: Korean Radiol. Soc.
Address: Kwanghwamun, P.O. Box 142, Seoul, Korea.

AHN, Dong Hyuck. Director, Dept. of Nucl. Eng., Hanyang Inst. Technol., Hanyang Univ., Korea.
Address: Department of Nuclear Engineering, Hanyang Institute of Technology, Hanyang University, Seoul, Korea.

AHN, Se Hee, B.S., M.S., Ph.D. Born 1928. Educ.: Chosun Christian (Seoul) and Northwestern (Illinois) Univs. Chairman, Phys. Dept. (1961-62), Prof., Phys. (1962-), Dean, School of Sci. (1962-), Yonsei Univ. Counsellor of Sci., Korean Office of Atomic Energy; Director, Korean Phys. Soc.; Director, Sci. Educ. Study Com. of Korea. Societies: A.P.S.; Phys. Soc. of Japan.
Nuclear interest: Nuclear physics (nuclear reaction and nuclear spectroscopy).
Address: Yonsei University, Seoul, Korea.

AHRLAND, Sten H., Ph.D. Born 1921. Educ.: Lund Univ. Docent, Lund Univ., 1952; Res. Sci., Eurochemic Co, Mol, Belgium, 1958; Res. Docent, Lund Univ., 1960; Res. Assoc., Swedish Natural Sci. Res. Council, 1965 (afilliated with the Dept. of Inorganic and Phys. Chem., Lund Univ.).
Nuclear interests: Nuclear chemistry, especially the chemistry of the actinoides, and separation processes for fission products.
Address: Department of Inorganic and Physical Chemistry, Lund University, Lund, Sweden.

AHRONSON, A. G. Stores, Finance Div., A.E.C.L.
Address: Chalk River Nuclear Research Laboratories, Chalk River, Ontario, Canada.

AICARDI, Maurice. Born 1919. Sec. Gen. du Plan de Modernisation et d'Equipement; Member, Commission Consultative pour la Production d'Electricité d'Origine Nucleaire and Member, Comité de l'Equipement Industriel, C.E.A., France; Member, Economic and Social Com., Euratom, 1958-66.
Address: Commissariat à l'Energie Atomique, 29-33 rue de la Federation, Paris 15, France.

AIELLO, Paolo, Dr. Electronic Eng. Born 1934. Educ.: Bologna Univ. Asst., Reactor Control and Instrumentation Dept., Scuola di Specializzazione in Ingegneria Nucleare di Bologna, 1958; Vice-Director, Exptl. Control Lab., Computing Centre, C.N.E.N., Bologna.
Nuclear interests: Reactor control, instrumentation and simulation.
Address: Centro di Calcolo del Comitato, Nazionale per l'Energia Nucleare, 2 Via Mazzini, Bologna, Italy.

AIGINGER, Hannes, Dipl. Ing., Dr. tech. Born 1937. Educ.: Vienna T.H. and Vienna Tech. Univ: Wissenschaftliche Hilfskraft, Inst. für Experimentalphysik, Vienna T.H., 1961; Hochschulassistent, Atominst. der Österreichische Hochschulen, 1962-. Austrian Delegate, Com. for Res. Co-operation, O.E.C.D., 1965-. Societies: Osterreichische Physikalische Gesellschaft; Österreichischer Fachverband für Strahlenschutz.
Nuclear interests: Detection of nuclear radiation; Bremsstrahlung and Cerenkov radiation; Radiation induced colour centres; Management; Elastic and inelastic electron scattering.
Address: Atominstitut der Österreichischen Hochschulen, 115 Schüttelstr., Vienna 2, Austria.

AIGRAIN, Pierre. Prof., Faculté des Sci. de Paris; Directeur Sci. des Recherches et Moyens d'Essais, Ministère des Armées. Member, Conseil Sci., and Member, Conseil d'Enseignement de l'I.N.S.T.N. (1965-), C.E.A.
Address: Commissariat à l'Energie Atomique, 29-33 rue de la Federation, Paris 15, France.

AIKEN, Senator George D. Member, Joint Com. on Atomic Energy, U.S.A.E.C.
Address: Senate Office Building, Washington D.C., U.S.A.

AIKIN, Archibald McKinlay, B.Sc., Ph.D. Born 1918. Educ.: McGill Univ. Head, Chem. Eng. Branch (1956-63), Head, Fuel and Fuel Channel Development Branch (1963-65), Director, Development Eng. Div. (1965-67), Director, Advanced Projects and Reactor Phys. Div. (1967-68), Gen. Manager, Nucl. Power Marketing (1968-), A.E.C.L. Societies: Fellow, Chem. Inst. of Canada; A.P.E.O.
Nuclear interests: Development of D_2O moderated reactors, including optimization, economics, physics, heat transfer, engineering, fuel and fuel cycles; Operation of test facilities, including reactors and loops.
Address: Atomic Energy of Canada Limited, 150 Kent Street, Ottawa 4, Ontario, Canada.

AILLERET, Pierre-Marie, Prof. of Energetics (Ecole Nationale des Ponts et Chaussees). Born 1900. Educ.: Ecole Polytechnique. Membre du Comité de l'Energie Atomique (1951-67), now Member, Conseil Scientifique, C.E.A.; Conseiller Scientifique et Technique, Electricité de France. Vice-Pres. Hispano-Francesa d'Energia Nuclear (Hifrensa); Membre, Comite Scientifique et Technique, Euratom; Pres., Internat. Electrotech. Commission (I.E.C.); Member, Directing Committee and Pres. Study Com. on Atomic Energy, Unipede.
Nuclear interests: Production of electrical energy.
Address: Electricite de France, 12 Place des Etats Unis, Paris 16, France.

AINSWORTH, Robert Lea, B.Sc., A.R.S.M. Born 1917. Educ.: Roy. School of Mines and London Univ. Chem. Group Manager (-1963), Asst. to the Managing Director (1963-66), Director (1966-), Murex Ltd. Society: F.I.M. Nuclear interests: Metallurgy: Production of tantalum, niobium, tungsten, molybdenum, zirconium and beryllium.
Address: Murex Ltd., Rainham, Essex, England.

AISENBERG, Sol, Dr. Phys., Phys. Dept., Space Sci., Inc.
Address: Space Sciences, Incorporated, Physics Department, 301 Bear Hill Road, Waltham, Massachusetts, U.S.A.

AITCHISON, Leslie, D. Met., M.Sc., B.Sc. Born 1891. Educ.: Sheffield, Birmingham and London Univs. Consultant Metal., various metal and eng. Companies.; Formerly Prof., Ind. Metal., now Hon. Res. Fellow, Birmingham Univ. Societies: F.R.I.C.; Fellow, Roy. Aeronautical Soc.; Inst. Mech. Eng.; Fellow, Inst. of Metal.; Hon. Member, Inst. of Metals; C.G.I.A.
Nuclear interest: Metallurgy.
Address: 138 Banbury Road, Stratford-upon-Avon, Warwickshire, England.

AITCHISON, Ronald Ernest, M.Sc. Born 1921. Educ.: Sydney Univ. Assoc. Prof. Communication Eng., Sydney Univ. Societies: F. Inst.P.; Member, Inst. of Eng., Australia; Sen. Member, Inst. of Radio Eng., Australia; Sen. Member, I.E.E.E. (U.S.A.).
Nuclear interests: Transistor circuits; Nuclear magnetic resonance spectroscopy.
Address: Electrical Engineering School, Sydney University, Sydney, New South Wales, Australia.

AITKEN, J. H. Formerly Assoc. Res. Officer, X-rays and Nucl. Radiations Sect., N.R.C., Canada; Low Energy Nucl. Phys., Dept. of Phys., Toronto Univ.
Address: Toronto University, Department of Physics, Toronto 5, Ontario, Canada.

AITKEN, Martin Jim, M.A., D.Phil. Born 1922. Educ.: Oxford Univ. Sen. Res. Officer, Oxford Univ. Book: Phys. and Archaeology (Intersci., 1961). Society: Phys. Soc.
Nuclear interests: Archaeological applications of nuclear physics, particularly thermoluminescence dating of ancient pottery.
Address: 6 Keble Road, Oxford, England.

AITZETMÜLLER, Kurt, Sen. Officer, Atominst. der Osterreichischen Hochschulen.
Address: Atominstitut der Osterreichischen Hochschulen, 115 Schuttelstrasse, Vienna 2, Austria.

AJDACIC, Vladimir, Diploma of Nat. Sci., Ph.D. (Phys.). Born 1933. Educ.: Belgrade Univ. Sen. Res. Assoc., Boris Kidrich Inst. Nucl. Sci., Belgrade-Vincha. Society: Yugoslav Soc. Physiciens and Mathematiciens.
Nuclear interests: Methods of detection and nuclear reactions.
Address: Boris Kidrich Institute of Nuclear Sciences, Vincha, P.O.B. 522, Belgrade, Yugoslavia.

AJTAI, Miklós, Dr. chem. Born 1914. Educ.: Eötvos Lorand Univ., Budapest. Deputy Prime Minister; Pres. Nat. A.E.C., Hungary.
Address: Parlament, Kossuth Lajos tér, Budapest 5, Hungary.

AJZENBERG-SELOVE, Fay, B.S.E., M.S., Ph.D. Born 1926. Educ.: Michigan and Wisconsin Univs. Res. Fellow, Caltech, 1952, 1954; Lecturer, Smith Coll., 1952-53; Visiting Fellow, Consultant, Staff member, M.I.T., 1952-53; Asst. Prof., then Assoc. Prof., Boston Univ., 1953-57; Visiting Asst. Prof., Columbia Univ., 1955; Visiting Prof. and Smith-Mundt Fellow, U.S. Dept. of State, Nat. Univ., Mexico, 1955; Visiting Assoc. Phys., Brookhaven, 1956; Lecturer, Pennsylvania Univ., 1957; Assoc. Prof., (1957-62), Prof. (1962-), Haverford Coll. Guggenheim Fellow, 1965-66. Book: Editor, Nucl. Spectroscopy, Vols. A and B (Academic Press, 1960). Society: Fellow, A.P.S.
Nuclear interest: Nuclear physics (nuclear structure, detection of neutrons).
Address: Haverford College, Haverford, Pa., U.S.A.

AKA, Esref Zeki, B.S., Ph.D. (Metal Eng.). Born 1924. Educ.: Mining Acad., Freiberg and Missouri Univ. Metal Eng. (1952-54), Smelter Supt. (1954-55), Ergani Copper Works, Maden, Turkey; Chief Metal Eng., Etibank, Ankara, 1956-61; Asst. Manager for Mine Res., 1961. Pres., Internat. Club, Rolla, Missouri, 1950. Societies: Assoc. Turkish Eng. and Architects
Nuclear interests: Raw materials for atomic energy; Basic research in nuclear metallurgy; Nuclear fuels and radiation damage.
Address: 54 cu Sokak No. 8, Bahcelievler, Ankara, Turkey.

AKABASHI, Mitsuya. Director, Div. of Pharmaceutical Sci., Nat. Inst. of Radiol. Sci., Atomic Energy Commission of Japan.
Address: National Institute of Radiological Sciences, 9-1, 4-chome, Anagawa, Chiba-shi, Chiba-ken, Japan.

AKALAYEV, G. G. Paper: Co-author, Low-Energy Gamma-Ray Transition to ^{238}Pu and ^{240}Pu (letter to the Editor, Atomnaya Energiya, vol. 16, No. 5, 1964).
Address: Academy of Sciences of the U.S.S.R., 14 Leninsky Prospekt, Moscow V-71, U.S.S.R.

AKAMATSU, Yasuyuki, M.D., Ph.D. Born 1928. Educ.: Nara Medical Coll. Intern, Osaka Nissei Hospital, 1950-51; Resident Osaka Univ. Medical School, 1951-55; Physician, Pathology, Atomic Bomb Casualty Commission, Hiroshima, 1955-57; Asst. Instructor, Osaka Univ. Medical School, 1957-61; Instructor, Inst. Cancer Res., Osaka Univ., 1961-. Societies: Japanese Foun-

dation for Cancer Res.; Japanese Pathological Soc.; Internat. Acad. of Pathology; American Board of Pathologists.
Nuclear interests: Medical research especially cancer research, metabolism of nucleoprotein, and trace of viruses.
Address: D40-202, Tsukumodai 5-11, Suita, Japan.

AKAP'EV, G. N. Papers: Co-author, Synthesis and Determination of Radioactive Properties of Some Isotopes of Fermium (Atomnaya Energiya, vol. 21, No. 4, 1966); co-author, On Nucl. Properties of Isotopes of Element 102 with Mass Numbers 255 and 256 (letter to the Editor, ibid, vol. 22, No. 2, 1967).
Address: Academy of Sciences of the U.S.S.R., 14 Leninsky Prospekt, Moscow V-71, U.S.S.R.

AKAR, Philippe, Eng. Born 1919. Educ.: Ecole Nationale Supérieure des Mines de Paris, Ancien élève de l'Ecole des Sciences Politiques. Ets. Hutchinson, 1945-54; Sté. l'Air Liquide, 1955-; Administrateur Directeur Gén. Adjoint, Sté. des Très Basses Températures. Vice-Pres., Commission 10, Internat. Inst. Refrigeration; Sec., Deep Low Temperatures Com., European Federation of Chem Eng. Societies: Sté. de Chimie Industrielle; Assoc. Française du Caoutchouc; Internat. Inst. Refrigeration; European Federation of Chem. Eng.
Nuclear interests: Management; Isotope Separation; Cryogenics.
Address: 10 Quai de la Megisserie, Paris 6, France.

AKCASU, Ziya A., B.S., M.S. (Electronics), Ph.D. (Nucl. Eng.). Born 1925. Educ.: Istanbul Tech. and Michigan Univs. Asst., Istanbul Tech. Univ., 1950-53; Turkish Army, 1953-55; Docent, Istanbul Tech. Univ., 1955-59; Res. Assoc., Argonne Nat. Lab., 1959-61; Asst. Prof. (1963-65), Assoc. Prof. (1965-68), Prof. (1968-), Michigan Univ., Ann Arbor. Societies: A.N.S.; A.P.S.
Nuclear interests: Reactor dynamics; Fluctuation analysis with application to reactor noise analysis; Neutron wave analysis; Inelastic neutron scattering; Interaction of radiation with plasmas.
Address: 901 Westwood, Ann Arbor, Michigan 48103, U.S.A.

AKERHIELM, Fredrik Samuel, M.S. (Phys. Eng.). Born 1929. Educ.: Roy. Inst. Technol., Stockholm. In AB Atomenergi, Stockholm, 1956-; Swedish rep. (1961-63), and Head, Reactor Dynamics Group (1962-63), O.E.C.D. Halden Reactor Project. At present group leader for dynamic experiments, AB Atomenergi. Society: A.N.S.
Nuclear interest: Reactor dynamics.
Address: AB Atomenergi, Studsvik, Nyköping, Sweden.

AKERLOW, R. W. Born 1907. Manager, Special Projects, Stearns-Roger Corp.
Nuclear interest: Management.
Address: Stearns-Roger, Corporation, P.O. Box 5888, Denver 80217, Colorado, U.S.A.

AKERMAN, Karol, Dr. of Chem. Born 1913. Educ.: Jagieloński Univ., Cracow. Chief, Dept. XVI, Inst. of Nucl. Res., Warsaw, 1960-; Prof. of Technol., Chair of Technol., Marie Curie-Sklodowska Univ., Lublin, Poland, 1958-.
Nuclear interests: Applied radiochemistry.
Address: 11,m. 103 ul.Mazowiecka, Warsaw, Poland.

AKERREN, Bengt Olof, L.L.B. Born 1922. Educ.: Uppsala Univ. Swedish Foreign Service, serving in Stockholm, Moscow, Berlin, Frankfurt, Rome, Torquay (G.A.T.T.), Brussels, Luxemburg, London, Antwerp. Leopoldville. Brazzaville, 1944-. Counsellor, Swedish Embassy in Vienna and Deputy Resident Rep. for Sweden to the I.A.E.A.; now Permanent Swedish Delegate to the Internat. Organisations at Geneva.
Address: 91-93 rue de la Servette, Geneva 7, Switzerland.

AKERS, Lawrence Keith, B.S., M.S., Ph.D. Born 1919. Educ.: Florida, Georgia, and Vanderbilt Univs. Sen. Sci. (1956-58), Res. Sci. (1958-61), Acting Chairman, Univ. Relations Div. (1958-59), O.R.I.N.S.; Head, Training Unit, Div. of Exchange and Training, I.A.E.A., Vienna, 1959-60; Principal Sci. (1961-64), Asst. Chairman, Special Training Div. (1964-65), Chairman Special Training Div. (1965-), O.R.I.N.S. (now Oak Ridge Assoc. Univs.). Societies: A.P.S.; RESA; Tennessee Acad. Sci.; A.A.P.T.; A.N.S.
Nuclear interests: Nuclear physics; Radioactive geochronology; Management of nuclear-related activities.
Address: Oak Ridge Associated Universities, Oak Ridge, Tennessee, U.S.A.

AKERS, William Walter, B.S., M.S., Ph.D. Born 1922. Educ.: Texas Technol. Coll. and Texas and Michigan Univs. Chairman, Chem. Eng. Dept., Rice Univ., 1956-. Formerly Vice Chairman, Council, Formerly Rep. for Rice Inst., and Formerly Member, Board Directors, O.R.I.N.S.; Member, Board of Directors, Oak Ridge Assoc. Universities, U.S.A.E.C.
Address: Chemical Engineering Department, Rice University, P.O. Box 1892, Houston, Texas, U.S.A.

AKHROMENKO, A. I. Prof. Director, Lab. of All-Union Inst. for Sci. Res. in Forestry and in the Mechanisation of Forest Management.
Nuclear interests: Plant physiology, forest science.
Address: Laboratory of the All-Union Institute for Scientific Research in Forestry and in the Mechanisation of Forest Management, 12 ul. Pisarevskaya Pushkino, Moscow province, U.S.S.R.

AKIBA, Ryukichi. Dean, Eng. Dept., and Vice-Pres., Atomic Energy Res. Assoc., Kanto Gakuin Univ.
Address: Kanto Gakuin University, Mutsuura, Kanazawa-ku, Yokohama, Japan.

AKIMOV, I. S. Paper: Co-author, The Analytical Method of Calculation: Irregular Burn out Fuel in Reactors (Atomnaya Energiya, vol. 16, No. 6, 1964).
Address: Academy of Sciences of the U.S.S.R., 14 Leninsky Prospekt, Moscow V-71, U.S.S.R.

AKIN, Erol, B.S., M.S., Ph.D., Assoc. Prof. Born 1934. Educ.: Ankara and Maryland Univs. Nucl. Reactor Director, Coll. of Eng., and member, Reactor Safety Com., Maryland Univ. Books: Sub-Critical Reactor Lab. Manual (1965); Nucl. Reactor Operator's Training Manual (1968); Nucl. Reactor Lab. Experiments (1968). Society: A.N.S.
Nuclear interests: Nuclear reactor operations, training, teaching and research on neutron diffraction.
Address: Department of Chemical Engineering, Maryland University, College Park, Maryland 20742, U.S.A.

AKPINAR, Sait, Dr. rer. nat. (Göttingen). Born 1913. Educ.: Frankfurt, Göttingen and Istanbul Univs. Resident Res. Assoc., Argonne Nat. Lab., 1958-60; Ind. Reactor Labs., Princeton, N.J.; Prof. Exptl. Phys., Istanbul Univ.; Director, Cekmece Nucl. Res. and Training Centre. Society: Turkish Phys. Soc.
Nuclear interests: Nuclear instrumentation; Nuclear physics; Reactor physics; Management.
Address: Cekmece Nükleer Merkezi, P.K. 1, Hava Alani, Istanbul, Turkey.

AKROYD, Bayly Gilbey. Born 1915. Publicity Officer, National Coal Board, 1947-59; Deputy Director of Public Relations, U.K.A.E.A., 1959-.
Nuclear interest: Public relations.
Address: Touchen End Pottery, Maidenhead, Berks., England.

AKSHANOV, B. S. Papers: Co-author, Exptl. Investigation of Hot Plasma Excited by Electron Beams (Atomnaya Energiya, vol. 23, No. 3, 1967); co-author, Investigation of Ion's Heating in Magnetic Mirror Trap (ibid, vol. 25, No. 2, 1968).
Address: Academy of Sciences of the U.S.S.R., 14 Leninsky Prospekt, Moscow V-71, U.S.S.R.

AKWEI, H. E. Richard M. Ambassador for Ghana to Switzerland; Permanent Rep. for Ghana to U.N. Office, Geneva; Resident Rep. for Ghana to I.A.E.A.
Address: 11 Belpstrasse, Berne, Switzerland.

AKYEL, Nuri, Dr. Member, Radioisotope Com., Turkish Assoc. for Cancer Res. and Control; Gynaecologist, Turkish Cancer Hospital.
Address: Turkish Cancer Hospital, No. 2. Imrahor Caddesi, Ankara, Turkey.

AL-FALAHI, S., B.Sc. Chem., Radioisotope Dept., Ministry of Health Republic Hospital.
Address: Ministry of Health Republic Hospital, Radioisotope Department, Baghdad, Iraq.

AL-HASHIMY, S. T., M.D. Phys., Radioisotope Dept., Ministry of Health Republic Hospital.
Address: Ministry of Health Republic Hospital, Radioisotope Department, Baghdad, Iraq.

AL-HINDAWI, Ali Yahya, M.B., Ch.B. (Baghdad), Ph.D. (Sheffield). Born 1931. Educ.: Baghdad and Sheffield Univs. Physician (1958), Head (1958-), Radioisotopes Dept., Republic Hospital, Baghdad; Member, Iraqi A.E.C. Societies: Iraq Medical Soc.; Iraq. Cancer Soc.
Nuclear interests: Thyroid function tests; Red cell survival studies and in vivo counting techniques in haemolytic anaemias; Liver scanning.
Address: Radioisotopes Department, Republic Hospital, Baghdad, Iraq.

AL-HITI, Thabit NAMAN. See NAMAN AL-HITI, Thabit.

AL-HUSSEINI, Jassem, Dr. Member, Iraqi A.E.C.
Address: Iraqi Atomic Energy Commission, Baghdad, Iraq.

AL-RUBAYI, Nouriddin, B.Sc., M.Sc., Ph.D. Born 1929. Educ.: Wales and Wisconsin Univs. Eng., Ever Ready, 1950-51; Res. Asst., Wisconsin Univ., 1952-56; Chief Elec. Eng., Ministry of Communication, 1956-; Lecturer, Baghdad Univ., 1957-62, 1965-. Member, Iraq. A.E.C., 1963-; Member, Board, Iraq Nat. Elec. Administration; Member, Board, Iraq Railways; Member, Iraq Specification and Standards Organisation; Consultant Member, Central Purchasing Com. of Iraq. Societies. I.E.E.E.; I.E.E.
Nuclear interest: Management and nuclear physics.
Address: Iraqi Railways, Schachyah, Baghdad,

AL-SAIGH, Miss N., M.Sc. Chem., Radioisotope Dept., Ministry of Health Republic Hospital.
Address: Ministry of Health Republic Hospital, Radioisotope Department, Baghdad, Iraq.

AL-TA'I, Fadhil. Vice Chairman, Iraqi A.E.C.
Address: Iraqi Atomic Energy Commission, Baghdad, Iraq.

AL-WAHBI, Sabih, M.D. Born 1905. Educ.: American Univ., Beirut, Post Graduate School, London, and Edinburgh Univ. Director, Karkh Hospital, 1941-54; Acting Director Gen. of Health on four occasions, 1941-54; Minister of Health, 1954; Physician Specialist and Director, Hospital, 1954-59; Acting Director Gen. of Health for a period, 1954-59; Chief, Education in Medicine and Allied Subjects,

ALA

W.H.O., Geneva, 1959-60; Specialist, Ministry of Health, 1960-; Vice Chairman, Iraqi A.E.C., 1960-; Pres., Iraqi Red Crescent Soc.; Vice-Pres., Anti-Tuberculosis Soc.; Vice-Pres., Child Welfare Soc.; Deleg. and Chief Deleg. of Iraq, 2nd-11th, 14th and 16th Assemblies, Vice Chairman, Regional Com. (1954), Member, U.N.I.C.E.F./W.H.O. Health Policy Com., Chairman, Exec. Board (1955), Vice-Pres., Assembly (1955), Pres., 10th Assembly (1957), Pres., Special Commemorative Session, Assembly, U.S.A. (1958), Chairman, Regional Com. (1958), Member, Exec. Board (1963), Expert Advisory Panel, W.H.O.; Chief Deleg., 7th Internat. Atomic Energy Conference, Vienna, 1963; Member, Internat. Quarantine Com. Societies: Iraq Medical Soc.; Fellow, American Coll. of Chest Physicians.
Address: Iraq Atomic Energy Commission, Ministry of Industry, Baghdad, Iraq.

ALABASTER, A. J., Instructor Commander R.N., M.A. Principal Lecturer, Dept. of Nucl. Sci. and Technol., Roy. Naval Coll.
Address: Royal Naval College, Department of Nuclear Science and Technology, Greenwich, London S.E.10, England.

ALADYEV, Ivan T., D.Sc. Born 1915. Educ.: Mech. Eng. Dept. of Acad., Moscow. Head, Atomic Energy Dept., Acad. of Sci. of the U.S.S.R.; Manager of a Lab., Inst. of Energetics. Books: Primenenie atomnoi energii v miznych celiach (Akademia Nauk S.S.S.R., 1956; Russian); Atom for peace (Moscow, Oriental Literature Publishing House, 1958; English). Nuclear interests: Nuclear power plants; Power reactors; Control fusion; Reactor design; Reactor technology; Heat transfer.
Address: Academy of Sciences, 14 Leninsky Prospekt, Moscow, U.S.S.R.

ALAGA, Gaja, Dr. of Phys. Born 1924. Educ.: Zagreb Univ. Prof., Theoretical Phys., Zagreb Univ. and Ruder Boskovic Inst., Zagreb. Book: Chapter XI, Nuclear Physics, I. Supek, Teorijska fizika (Struktura materije) (Zagreb, Skolska Knjiga, 1964).
Nuclear interest: Nuclear physics.
Address: Institute Ruder Boskovic, 54 Bijenicka cesta, Zagreb, Yugoslavia.

ALAJOKI, P. Director, Elec. Dept., Imatran Voima Osakeyhtio.
Address: Imatran Voima Osakeyhtio, 16 Malminkatu, Helsinki, Finland.

ALAM, Muhammad Nurul, Ph.D. (Phys., London), M.Sc. and B.Sc. (Hons., Phys., Dacca). Born 1920. Educ.: London and Dacca Univs. Head, Dept. of Phys., Rajshahi Coll., Pakistan, 1951-55; Head, Phys. Div., and Leader, Res. Reactor Group, Pakistan A.E.C., Karachi, 1955-; Phys., Cento Inst. Nucl. Sci. Society: A.N.S.
Nuclear interests: Research experiments with a research reactor and operation and administration of a research reactor.
Address: Cento Institute of Nuclear Science, P.O. Box 1828, Tehran, Iran.

ALANQUAND, Camille. Fabrication, Usine Siersatom.
Address: Siersatom, Usine Siersatom, Dourdan, (Seine et Oise), France.

ALARCON, Abelardo. Lab. de Fisica Cosmica de Chacaltaya, Univ. Mayor de S. Andres.
Address: Universidad Mayor de San Andres, Laboratorio de Fisica Cosmica de Chacaltaya, La Paz, Bolivia.

ALARCON, Adolfo OCAMPO. See OCAMPO ALARCON, Adolfo.

ALBA-ANDRADE, Fernando, M.Sc., Sc.D. Born 1919. Educ.: Mexico Nat. Univ. Director, Inst. of Phys., Mexico Nat. Univ., 1956-. Member, Consulting Board, Mexican Nucl. Energy Commission, 1957-. Pres., Acad. of Sci. Res.; Member, Mexican Deleg., Sci. Com. for the Study of Atomic Rad. of the U.N. Societies: A.P.S.; Soc. Mexicana de Física.
Nuclear interest: Experimental nuclear physics.
Address: Instituto de Física Torre de Ciencias 8o. piso, Ciudad Universitaria, México 20, D.F., Mexico.

ALBANO, Alejandro Reynaldo, A.A. (Ind. Chem.), B.S. (Chem. Eng.). Born 1925. Educ.: Adamson Univ., Manila. Armed Forces Officer (2nd Lieutenant to Captain), Chem. Warfare Service (Reserve Force), 1951-64, Nucl. Materials Custodian (1958), Acting Asst. Head, Health Phys. Dept., Chief, Rad. Exposure Control Group, and Ind. Safety Officer, Philippine Atomic Res. Centre (1961-), Philippine A.E.C.
Nuclear interests: Radiological health and safety.
Address: Philippine Atomic Enery Commission, 727 Herran Street, Manila, Philippines.

ALBAREDA HERRERA, Jose Maria. Vocale, De Biologia Vegetal y Aplicaciones Industriales, Junta de Energia Nucl. Madrid.
Address: Junta de Energia Nuclear, Ciudad Universitaria, Madrid, Spain.

ALBAUGH, Frederick William, A.B. (Chem.), M S. (Chem.), Ph.D. (Phys. Chem.). Born 1913. Educ.: U.C.L.A. and Michigan Univ. Head, Analytical Chem., Eng. Dept. (1950-52), Manager, Appl. Res., Eng. Dept. (1952-55), Manager, Advance Eng., Eng. Dept. (1955-56), Manager, Reactor and Fuels Lab. (1956-64), Hanford Labs., G.E.C.; Manager, Reactor and Fuels Dept. (1965-66), Assoc. Director (1966-67), Director (1967-), Pacific Northwest Lab., Battelle Memorial Inst. Societies: A.N.S.; A.S.M.
Nuclear interests: Management of chemical, metallurgical and ceramics research and development on nuclear fuels and materials; Plutonium recycle technology is of particular interest; Secondary interest in technology of

heavy-water reactors.
Address: Pacific Northwest Laboratory, Battelle Memorial Institute, P.O. Box 999, Richland, Washington, U.S.A.

ALBERIGI-QUARANTA, Alessandro, Dr. in Phys. Born 1927. Educ.: Rome Univ. Director and Electronics Prof., Centro di Elettronica, Inst. Fisico, Modena Univ. Books: Co-author, Lezioni di elettronica dei transienti (1956), Elettronica (1960). Societies: A.P.S.; Soc. Italiana di Fisica; I.E.E.E.
Nuclear interests: Nuclear detectors; Nuclear electronic instrumentation.
Address: Istituto Fisico dell'Università, 4 Via Universita, 41100 Modena, Italy.

ALBERS-SCHOENBERG, Heinz, Dr. sc. nat. Born 1926. Educ.: Swiss Federal Inst. of Technol. Head of operation of Swiss D_2O Res. Reactor DIORIT (20 MW), 1958-63.
Nuclear interest: Operation of nuclear power plants.
Address: Nordostschweizerische Kraftwerke A.G. Baden, Switzerland.

ALBERT, Dietmar, Dipl. Phys., Dr.rer.nat. Born 1935. Educ.: Karl-Marx-Univ., Leipzig. Zentralinst. für Kernforschung, Rossendorf near Dresden. Society: Physikalische Gesellschaft in der D.D.R.
Nuclear interests: Reactor physics: neutron spectrum in epithermal and thermal energy range, thermalisation, foil activations, lattice parameters, pulsed neutron source.
Address: 135 Holbeinstr., 8019 Dresden, German Democratic Republic.

ALBERT, Donald. Instructor in Phys., Adelphi Coll.
Address: Physics Department, Adelphi College, Garden City, New York, U.S.A.

ALBERT, Henri. Director Gen., Sté. de Recherches Techniques et Industrielles (S.R.T.I.).
Address: Societe de Recherches Techniques et Industrielles (S.R.T.I.), 17 route de la Reine, Boulogne-Billancourt, France.

ALBERT, Philippe Edouard, D. es Sc., Born 1922. Educ.: Paris Univ. Chercheur (1946-58), Maître de Recherches (1958-62), Directeur de Recherche (1962-), C.N.R.S.; Lecturer, I.N.S.- T.N., Saclay and E.N.S.C.P., Paris. Editorial Advisory Board, J. Nucl. Materials and J. Radioanalytical Chem. Book: L'analyse par radioactivité (Paris, Gauthier-Villars et Cie, 1964). Societies: Sté. chimique de France; Sté. Française de Métallurgique; Inst. Metals; A.N.S. Nuclear interests: Research on high purity metals. Activation analysis by thermal neutrons activation, charged particles activation and gamma photon activation. Systematic analysis by these different methods of very large number of impurities ($\geqslant 60$) in very pure Al-Fe-Zr Cu-Ni-Mg-Cr-Mo.
Address: Laboratoire d'Analyse par activation du Centre d'Etudes de Chimie Metallurgique, 15 rue Georges Urbain, 94 Vitry-sur-Seine, France.

ALBERTINOLI, P. Sen. Sci., Centre Sci. de Monaco.
Address: Centre Scientifique de Monaco, 16 boulevard de Suisse, Monte Carlo, Monaco.

ALBISU, Francisco, Dr.-Ing. Ind., M.S. (Nucl. Eng.). Born 1929. Educ.: Escuela de Ingenieros de Bilbao, and M.I.T. Now in Nucl. Technol. Dept., Laboratorios de Ensayos e Investigacion Industrial; Prof. Nucl. Technol., Bilbao School of Eng.
Nuclear interests: Reactor physics; Reactor design; Training and university reactors.
Address: Apartado 1234, Bilbao, Spain.

ALBONETTI, Achille, Dr. of Laws. Born 1927. Educ.: Rome Univ., London School of Economics, Columbia (N.Y.) and St. Francis Xavier (Canada) Univs. Editorial Sec., Produttività (Rome), 1950; Economic Expert to Italian Technical Deleg., Washington, 1951; Economic Expert to Italian Deleg. to O.E.E.C., Paris, 1953; Member Drafting Coms., Euratom and Common Market Treaties, Brussels, 1955-57; Consultant, Italian Nat. Com. for Nucl. Research, 1956; Economic Counsellor to Italian Deleg. to O.E.E.C., 1958; Director, Cabinet of the Vice-Pres. of the E.E.C. Commission, 1959- 60; Director, Div. for Internat. Affairs and Economic Studies, Nat. Com. for Nucl. Energy, Rome, 1961; Director, Eurochemic. Books: Euratom e Sviluppo Nucleare (1958); Preistoria degli Stati Uniti d'Europa (1961); La Collaborazione Internazionale Nucleare (1963); L'Europa e la Questione Nucleare (1964); Divario Tecnologico, Ricerca Scientifica e Produttività in Europa e negli Stati Uniti (1967). Nuclear interests: Management; Economics, International affairs.
Address: National Committee for Nuclear Energy, 15 Via Belisario, Rome, Italy.

ALBRECHT, Robert W., B.S.E.E., M.S.N.E., Ph.D. Born 1935. Educ.: Purdue and Michigan Univs. Asst. Prof., (Reactor Kinetics), Washington Univ., Seattle. Society: A.N.S.
Nuclear interests: Nuclear reactor dynamics including reactor noise and pulsed neutron. Cold neutron physics and general reactor theory.
Address: 210 Electrical Engineering Building, Washington University, Seattle, Washington 98105, U.S.A.

ALBRECHTSEN, Svend Erik, M.Sc. (Elec. Eng.). Born 1915. Educ.: Copenhagen Tech. Univ. Managing Director, South East Zealand Elec. Co. Ltd. Pres., Danish Assoc. of Elec. Supply Undertakings; Vice Chairman, Danatom, Copenhagen.
Nuclear interest: Nuclear power plants, design and operation, management.
Address: 7 Tingvej, Haslev, Denmark.

ALBRIGHT, Lyle Frederick, B.S. (Ch.E.), M.S. (Ch.E.), Ph.D. (Ch.E.). Born 1921. Educ.: Michigan Univ. Assoc. Prof., Oklahoma Univ., 1951-55; Prof., Purdue Univ., 1955-. Res. Consultant, M.W. Kellogg Co., Abbott Labs., and Morton Chem. Co. Societies: A.I.Ch.E.; A.C.S.; Combustion Inst.; American Oil Chemists' Soc.
Nuclear interests: Effect of radiation on chemical reactions such as hydrogenation, oxidation, and vapor phase nitration.
Address; School of Chemical Engineering, Purdue University, Lafayette, Indiana, U.S.A.

ALCALA, Proceso E. Sen. Sci. (Plant Pathology), Philippine Atomic Res. Centre.
Address: Philippine Atomic Research Centre, Diliman, Quezon City, Philippines.

ALDER, Fritz, Dr. Phil. Born 1916. Educ.: Federal Inst. Technol., Switzerland and Basle Univ. Brown, Boveri et Cie, Baden, 1946-55; Stanford Univ., U.S.A., 1950-51; Head, Swimming Pool Reactor SAPHIR, Reactor Ltd., 1955-59; Head, Health and Safety Dept., Federal Inst. Reactor Res., 1959-. Chairman, Federal Commission for Nucl. Safety. Societies: Swiss Phys. Soc.; Fachverband für Strahlenschutz; Gesellschaft für Strahlenbiol.; Schweiz. Gesellschaft von Fachleuten der Kerntechnik.
Nuclear interests: Nuclear safety; Health physics.
Address: Im Sand, Gebenstorf/AG, Switzerland.

ALDER, Keith Frederick, M.Sc. Born 1921. Educ.: Melbourne Univ. At Armament Res. Establishment, Woolwich, 1949-50; Lecturer in Metal., Univ. Technol. (N.S.W.), Newcastle, Univ. Coll., 1951; Sen. Lecturer in Phys. Metal., Melbourne Univ., 1952-53; Head, Metal. Sect., (1954-60), Acting Director (1961-62), Director (1962-), Res. Establishment (Attached A.E.R.E. Harwell, 1954-57), Commissioner (1968-), A.A.E.C. Societies: Fellow, Inst. of Metallurgists; Assoc. Member, Australasian Inst. of Mining and Metal.; Assoc. Member, Inst. of Radio and Electronics Eng., Australia.
Nuclear interests: Metallurgy, Reactor design.
Address: Australian Atomic Energy Commission Research Establishment, Private Mail Bag, Sutherland, New South Wales, Australia.

ALDERHOUT, Jack J. H., B.S. (Chem.) Doctoral examination in Radiochem. and Microbiol. Born 1932. Educ.: Amsterdam Univ. Safety Officer, European Co. for Chem. Processing of Irradiated Fuels (Eurochemic), Mol., Belgium.
Nuclear interests: Health and safety; Bio-assay; Meteorology and air pollution; Ecology; Civil defence and fire-fighting.
Address: 5 Luxemburglaan, Geel, Belgium.

ALDERMAN, John Edward. Born 1926. Educ.: Acton Tech. Coll. Designer/Draughtsman (1955-59), Project Eng., Windscale (1959-65), Patents Licensing Dept., London office (1965-68), U.K.A.E.A. Society: A.M.I.Mech.E.
Nuclear interests: Commercial exploitation of inventions, designs in nuclear field generally, apart from reactor applications and fuel element manufacture.
Address: 26 Five Oaks Close, St. John's, Woking, Surrey, England.

ALDERS, Jacobus A. G. Tweede Voorzitter, Nederlands Katholiek Vakverbond. Netherlands member, Specialised Nucl. Section for Economic Problems and Specialised Nucl. Section for Social Health and Development Problems, Economic and Social Com., Euratom.
Address: 15 Celsiusstraat, Amersfoort, Netherlands.

ALDERTON, Douglas Alan, H.N.C. (Appl. Phys.). Born 1934. Development Manager, Isotope Developments Ltd., 1960-64; Sen. Nucl. Sales Eng. (1964-66), Manager, Nucleonics Div. (1966-), E.M.I. Electronics Ltd. Nucleonics Steering Com., Sci. Instrument Manufacturers' Assoc. Society: Assoc. Member, Inst. of Nucl. Eng.
Nuclear interests: Instrumentation required for nuclear physics experiments, health physics protection and nuclear medicine.
Address: E.M.I. Electronics Limited, Nucleonics Division, Hayes, Middlesex, England.

ALDINGTON, The Rt. Hon. Lord, Toby Austin Richard William, K.C.M.G., M.A. Born 1914. Educ.: New College, Oxford Univ. Chairman, United Power Co. Ltd., 1962-; Chairman, Gen. Elec. Co. Ltd., 1964-.
Address: 19 Cadogan Gardens, London, S.W.3, England.

ALDRICH, Lyman Thomas, M.A., Ph.D. Born 1917. Educ.: Minnesota Univ. Staff Member, Dept. of Terrestrial Magnetism, Carnegie Inst., Washington. Societies: A.P.S.; A.G.U.; A.A.A.S.; Geochemical Soc.; Seismological Soc. of America.
Nuclear interests: Mass spectrometry; Nuclear geology.
Address: 5241 Broad Branch Road, N.W., Washington, D.C. 20015, U.S.A.

ALDWORTH, James P., B.S. Born 1925. Educ.: Northwestern Univ. Advertising Director then Director of Market Planning, Nucl.-Chicago Corp., -1963; Pres., Aldworth Instruments. Inc.
Nuclear interests: Management and design of products for schools, particularly in nuclear field. Current products include radiation cloud chamber and radioactivity experiment system.
Address: Aldworth Instruments, 17 W. Harrison Street, Oak Park, Illinois, U.S.A.

ALECK, Benjamin J., C.E., M.C.E. Born 1917. Educ.: Cornell Univ. Vice-Pres. Res., Arde, Inc., 1961-. Societies: A.S.M.E.; A.I.A.A.
Nuclear interests: Reactor design and associated hardware design.
Address: Arde, Inc., 100 West Century Road, Paramus, New Jersey, U.S.A.

ALEGRIA, Jose Luiz. Jefe del Departmento de Planificacion y Control de Programas, Jefe de Planes de Presidencia, Comision Nacional de la Energia Atomica.
Address: 8250, Avda. del Libertador Gral San Martin, Buenos Aires, Argentina.

ALEJANDRO, Alejandro. Technical adviser, Comision Nacional de Investigaciones Atomicas.
Address: Comision Nacional de Investigaciones Atomicas, Ciudad Trujillo, Republica Dominicana.

ALEKSANDROV, Anatoly P., Dr. Physico-Mathematical Sci., Academician of the Acad. of Sci. of the U.S.S.R.; Director, Kurchatov Inst. of Atomic Energy; Member of the Praesidium, U.S.S.R. Acad. Sci. Deleg. to 2nd I.C.P.U.A.E., Geneva, Sept. 1958; Deleg to 3rd I.C.P.U.A.E., Geneva, 1964; Member, Deleg. of Soviet Atomic Experts Visiting U.S.A., Nov. 1959. Papers: Co-author, A Nucl. Ice-breaker (Atomnaya Energiya, vol. 5, No. 3, 1958); The Problems of Nucl. Power (ibid, vol. 13, No. 2, 1962).
Address: Kurchatov Institute of Atomic Energy, 46 Ulitsa Kurchatova, Post Box 3402, Moscow, U.S.S.R.

ALEKSANDROVA, G. I. Paper: Co-author, Determination of Oxygen in Germanium and Silicon Using Activation by Helium-3 (Atomnaya Energiya, vol. 23, No. 1, 1967).
Address: Academy of Sciences of the U.S.S.R., 14 Leninsky Prospekt, Moscow V-71, U.S.S.R.

ALEKSANDROVA, L. S. KOZYREVA - See KOZYREVA-ALEKSANDROVA, L. S.

ALEKSANDROVA, V. N. Papers: Co-author, The Effects of Rad. on the Electrochem. Behaviour of IXI8H9T Steel (letter to the Editor, Atomnaya Energiya, vol. 10, No. 2, 1961); co-author, Investigation of Long-Lived Radioisotopes in Coolant at Kurchatov Nucl. Power Plant (ibid, vol. 24, No. 3, 1968).
Address: Academy of Sciences of the U.S.S.R., 14 Leninsky Prospekt, Moscow V-71, U.S.S.R.

ALEKSANDROWICZ, Jerzy, M.Sc. Formerly Chief, Reactor Exploitation Sect., Nucl. Res. Inst., now Head, Reactor Operation Dept., Soltan Nucl. Res. Centre, Warsaw; Deleg. to 2nd I.C.P.U.A.E., Geneva, Sept. 1958; Member, Editorial Board, Nukleonika.
Address: Instytut Baden Jadrowych, Soltan Nuclear Research Centre, Swierk near Warsaw, Poland.

ALEKSANKIN, M. M. Pisarzhevskij Inst. for Phys. Chem., Acad. of Sci. of the Ukrainian S.S.R. Paper: Co-author, Investigation on the Oxidation Mechanism of Pyruvic Acid, Phenole and Salicyle Adelhyde by Hydrogen Peroxide by using Isotopes (Kernenergie, vol. 5, No.4/5, 1962).
Address: Pisarzhevskij Institute for Physical Chemistry, 15 Ulitsa Lenina, Kiev, Ukrainian S.S.R., U.S.S.R.

ALEKSEENKO, R. N. At Inst. of Atomic Energy. Adviser, Third I.C.P.U.A.E., Geneva, Sept. 1964.
Address: Kurchatov Institute of Atomic Energy, 46 Ulitsa Kurchatova, Post Box 3402, Moscow, U.S.S.R.

ALEKSEENKO, V. A. Paper: Co-author, Ion-Exchange Separation of Uranium and Rare-Earth Elements (Atomnaya Energiya, vol. 20, No. 1, 1966).
Address: Academy of Sciences of the U.S.S.R., 14 Leninsky Prospekt, Moscow V-71, U.S.S.R.

ALEKSEEV, N. V. Papers: Co-author, A Unit for fast Neutron Polarisation Study (letter to the Editor, Atomnaya Energiya, vol. 15, No. 1, 1963); co-author, Calculating Spectra of Neutrino Produced by 70 GeV Protons (letter to the Editor, ibid, vol. 23, No. 1, 1967).
Address: Academy of Sciences of the U.S.S.R., 14 Leninsky Prospekt, Moscow V-71, U.S.S.R.

ALEKSEEV, O. A. Paper: Co-author, Preparation and Studies of some Plutonium Monocarbide Properties (Atomnaya Energiya, vol. 22, No. 6, 1967).
Address: Academy of Sciences of the U.S.S.R., 14 Leninsky Prospekt, Moscow V-71, U.S.S.R.

ALEKSEEV, V. I. Papers: Radioisotope Studies of Scrap Fusion Kinetics and Slag Formation in the Scrap-ore Process (letter to the Editor, Atomnaya Energiya, vol. 3, No. 10, 1957); co-author, Pulse Miniature Fission Chambers (letter to the Editor, ibid, vol. 20, No. 3, 1966).
Address: Academy of Sciences of the U.S.S.R., 14 Leninsky Prospekt, Moscow V-71, U.S.S.R.

ALEKSENKO, Yu. N. Papers: Co-author, Some Remarks on Rad. Stability Low Melting Coolants in Liquid and Solid State (Atomnaya Energiya, vol. 22, No. 1, 1967); co-author, Circulating Rate Effect on Rad. - Induced Conversion of Organic Coolants at Elevated Temperatures (ibid, vol. 22, No. 5, 1967); co-author, Thermal Neutrons Fluxes Distributions in Various Reflectors with Canals (letter to the Editor, ibid, vol. 24, No. 5, 1968).
Address: Academy of Sciences of the U.S.S.R., 14 Leninsky Prospekt, Moscow V-71, U.S.S.R.

ALEKSEYEVSKI, N. E. Member, Russian deleg. visiting Britain July 18-28, 1961, to discuss solid state phys. res.
Address: Vavilov Institute of Physical Problems, U.S.S.R. Academy of Sciences, 2 Vorobevskoye Chaussee, Moscow, U.S.S.R.

ALEKSIC, Radomir. Adviser, Federal Nucl. Energy Commission.

Address: Federal Nuclear Energy Commission, 29 Kosancicev Venac, P.O. Box 353, Belgrade, Yugoslavia.

ALEMANY, Joaquin CATALA de. See CATALA de ALEMANY, Joaquin.

ALEN, Léon Joseph, Ing. civil, Général-Major. Born 1913. Educ.: Ecole Royale Militaire. Societies: Inst. interuniversitaire des sci. nucléaires; Centre d'études nucléaires (Mol). Address: Commandant de l'Ecole Royale Militaire, 30, avenue de la Renaissance, Brussels 4, Belgium.

ALENA, Franco CANCELLARIO d'. See CANCELLARIO d'ALENA, Franco.

ALER, Bo Aron Albert, Ph.D. Born 1926. Educ.: Stockholm Univ. Head, Nucl. Phys. Dept., Res. Inst. of Nat. Defence, Sweden, 1952-57; Head, Planning Office, (1957-), Director of Administration, A.B. Atomenergi, Stockholm. Member, Steering Com. for Nucl. Energy, and Member, Group on Co-operation in the Field of Nucl. Reactors, E.N.E.A.; Swedish Alternate, Board of Directors, Eurochemic; Expert, Swedish Atomic Energy Board.
Address: A.B. Atomenergi, P.O. Box 9042, Stockholm 9, Sweden.

ALESHCHENKOV, P. I. Eng., Acad. of Sci. of the U.S.S.R.; Deleg. to 2nd I.C.P.U.A.E., Geneva, Sept. 1958. Papers: Co-author, A Uranium-graphite Reactor giving Superheated High-pressure Steam (Atomnaya Energiya, vol. 5, No. 3, 1958); co-author, I.V. Kurchatov Beloyarsk Nucl. Power Station (ibid, vol. 16, No. 6, 1964).
Address: Kurchatov Beloyarsk Nuclear Power Station, Zarechny Settlement, Sverdlovsk Region, U.S.S.R.

ALESSIO, Juan Tomas D'. See D'ALESSIO, Juan Tomas.

ALEXANDER, Edward Lawson, B.S., M.S., Ph.D. Born 1925. Educ.: Maine and Vanderbilt Univs. Res. Assoc., Nucl. and Rad. Chem. and Project Sci., K.A.P.L., G.E.C., Schenectady, New York, 1955-57; Assoc. Prof., Chem., and Asst. Director, Res.,Reactor Project, Georgia Inst. Technol., Atlanta, 1957-58; Manager, Radiol. Sci., Ind. Reactor Labs., Inc., Plainsboro, New Jersey, 1958-62. Prof., Rad. Sci. and Director, Rad. Sci. Centre, Rutgers Univ., New Brunswick, N.J., 1962-67; Coordinator of Graduate Studies, and Head, Dept. of Radiol. Sci., Lowell Technol. Inst., Massachusetts, 1967-. Societies: A.C.S.; A.N.S.; H.P.S.; A.A.A.S.; A.A.U.P.
Nuclear interests: Radiation chemistry; Radiochemical separations; Nuclear spectroscopy; Radiation physics; Health physics; Public health physics.

Address: 11 Oriole Drive, Andover, Massachusetts, U.S.A.

ALEXANDER, Gideon, M.Sc., Ph.D. Born 1932. Educ.: Hebrew Univ., Jerusalem, Dublin Univ., and Dublin Inst. for Advanced Studies. Sen. Sci., Israel A.E.C., 1958-60; Res. Assoc., Lawrence Rad. Lab., California Univ., Berkeley, 1960-62; Sen. Sci. (1963), now Assoc. Prof., Nucl. Phys. Dept., Weizmann Inst., Rehovoth, Israel. Council member, Israel Phys. Soc. Societies: Israel Phys. Soc.; A.P.S.
Nuclear interests: High energy physics; Elementary particles; Cosmic rays.
Address: Nuclear Physics Department, Weizmann Institute of Science, Rehovoth, Israel.

ALEXANDER, John Malcolm, D.Sc. (Eng.), Ph.D., D.I.C., F.C.G.I. Born 1921. Educ.: London Univ. Head, Metal Deformation Sect., Aluminium Labs., Ltd., Banbury, 1952-55; Head, Reactor Theoretical Mech. Design Sect., Atomic Power Dept., English Elec. Co. Ltd., Whetstone, near Leicester, 1955-57; Reader in Plasticity (1957-63), Prof. of Eng. Plasticity (1963-), Dept. Mech. Eng., City and Guilds Coll., London Univ. Vice-Pres. and Member of Council, Met. Eng. Com., Publications Com., Chairman, Met. Reviews Com., Inst. Metals. Societies: M.I.Mech.E.; M.I.Prod.E.; Inst. Phys., British Soc. Rheology; F.I.M.
Nuclear interests: Properties and uses of reactor materials, especially for fuel elements and graphite moderator.
Address: Mansard Cottage, Shere Road, West Horsley, Surrey, England.

ALEXANDER, K. F., Dr.rer.nat.habil. Head, Reactor Div., Zentralinst. für Kernforschung. Papers: Co-author, Messung des Aufslussspektrums thermischer Neutronen aus Graphit unter verschiedenen geometrischen Bedingungen (Kernenergie, vol. 3, No. 12, 1960); co-author, Ein mechanisches Filter für Strahlexperimente mit thermischen Neutronen am Reaktor (ibid, vol. 4, No. 9, 1961); Problems of Low-Energy Nucl. Phys. (ibid, vol. 8, No. 7, 1965).
Address: Zentralinstitut für Kernforschung, Bereich Reaktortechnik und Neutronenphysik, Rossendorf bei Dresden, German Democratic Republic.

ALEXANDER, Karl, Dr. jur. Gäste, Arbeitskreis I/1, Haftung und Verischerung, Federal Ministry for Sci. Res., Germany.
Address: Federal Ministry for Scientific Research, 46 Luisenstrasse, Bad Godesberg, Germany.

ALEXANDER, Lloyd George, B.S., Ph.D. (Chem. Eng.). Born 1919. Educ.: Purdue Univ. Development Eng., O.R.N.L. Society: A.N.S.
Nuclear interests: Analysis and evaluation of nuclear, thermal, and chemical systems in proposed nuclear power plants and estimation of nuclear power costs in advanced systems.
Address: Oak Ridge National Laboratory, P.O. Box Y, Oak Ridge, Tennessee, U.S.A.

ALEXANDER, Peter, D.Sc., Ph.D., D.I.C. Born 1922. Educ.: London Univ. Manager, Res. Dept., Wolsey, Ltd., 1944-50; Head, Radiobiol. Dept., Chester Beatty Res. Inst., Inst. of Cancer Res., London; Prof. Radiobiol., London Univ. Books: Fundamentals of Radiobiol. (2nd edition, Pergamon, 1961); Atomic Rad. and Life (Penguin, 1958). Societies: Faraday Soc.; Biochem. Soc.; Rad. Res. Soc.
Nuclear interests: Radiobiology; Radiation chemistry of macromolecules.
Address: Chester Beatty Research Institute, Clifton Avenue, Belmont, Sutton, Surrey, England.

ALEXANDER, R. R. Res. Assoc., Rad. Biol. Lab., New York State Veterinary Coll., Cornell Univ.
Address: New York State Veterinary College, Cornell University, Ithaca, New York, U.S.A.

ALEXANDER, Robert Charles, M.I.Mech.E. Asst. Chief Eng. (1951), Deputy Chief Eng. (1955), Chief Eng. (1957-), Chem. Plants Design, U.K.A.E.A., Risley. Societies: B.N.E.S.; I.Mech.E.
Nuclear interests: Nuclear chemical processing plant; High activity remote handling and examination facilities.
Address: Woodside, Hall Drive, Appleton, Nr. Warrington, Lancs, England.

ALEXANDER, Samuel Nathan, A.B. (Phys.), B.S. (Elec. Eng.), M.S. (Elec. Eng.). Born 1910. Educ.: Oklahoma Univ. and M.I.T. Sen. Res. Fellow, Eng. Sci., N.B.S., 1966-. Member, Mathematics and Computer Sci. Res. Advisory Com., U.S.A.E.C., -1968. Book: Co-author, Systems Engineering Handbook. Societies: Fellow, I.E.E.E.; Assoc. for Computing Machinery; A.P.S.; American Documentation Inst.
Nuclear interest: Application of data processing and computation to problems of nuclear research and data gathering.
Address: 4120 Stanford Street, Chevy Chase, Maryland, U.S.A.

ALEXANDERSON, Eldon L., M.S. (Phys.). Born 1919. Educ.: Hillsdale Coll. and Michigan Univ. Head, Reactor and System Analysis Sect., Atomic Power Development Assocs., 1958-; Reactor Eng., Power Reactor Development Co., 1959-. Society: A.N.S.
Nuclear interests: Reactor physics.
Address: Power Reactor Development Co., 1911 First, Detroit, Michigan, U.S.A.

ALEXANDRE, J. R. Director, Nucl. Dept., Degremont Sobelco S.A.
Address: Degremont Sobelco S.A., 37 place Xavier Neujean, Liège, Belgium.

ALEXANDROV, Anatoly P. See ALEKSANDROV, Anatoly P.

ALEXANDROV, B. M. Paper: Co-author, Determination of Probabilities of Spontaneous Fission of U^{233}, U^{235} and Am^{243} (Atomnaya Energiya, vol. 20, No. 4, 1966).
Address: Academy of Sciences of the U.S.S.R., 14 Leninsky Prospekt, Moscow V-71, U.S.S.R.

ALEXANDROVA, V. N. Paper: Co-author, Effect of Gamma - Rad. on Deposition (letter to the Editor, Atomnaya Energiya, vol. 20, No. 5, 1966).
Address: Academy of Sciences of the U.S.S.R., 14 Leninsky Prospekt, Moscow V-71, U.S.S.R.

ALEXANKIN, M. M. See ALEKSANKIN, M. M.

ALEXIOU, Athanasios. Head, Sect. of New Energy Sources, Public Power Corp.
Address: Public Power Corp., 30 Chalcocondyli Street, Athens, Greece.

ALEXOPOULOS, Kessar D., Dr. nat. sci. Born 1909. Educ.: Federal Inst. Technol., Zurich. Prof. Phys., Athens Univ. Member, Roy. Res. Foundation. Book: General Phys. (5 Vol.).
Nuclear interests: Measurement of soft β particles; Radiation damage.
Address: Department of Physics, Athens University, 104 Sononos Str., Athens, Greece.

ALFIMOV, N. N. Paper: Co-author, The Results of Investigation of Water and Air β-Activity in Some Areas of the Pacific Ocean (letter to the Editor, Atomnaya Energiya, vol. 16, No. 3, 1964).
Address: Academy of Sciences of the U.S.S.R., 14 Leninsky Prospekt, Moscow V-71, U.S.S.R.

ALFVEN, Hannes, Prof. Member, Statens Rad. for Atomforskning. Papers: Magneto-Hydrodynamics and Thermonuclear Problems (2nd I.C.P.U.A.E., Geneva, Sept. 1958); Magneto-Hydrodynamics and Thermonuclear Problems (Progress in Nucl. Energy, Ser.XI, Vol. 1, 1958.); co-author, Gas insulation of a hot plasma (letter to the Editor, Nature, vol. 188, No. 4753, 1960 and Nucl. Fusion, Suppl., Part 1, 1962).
Address: Royal Institute of Technology, Stockholm 70, Sweden.

ALI, Mohammad Innas, B.Sc. (Hons.), M.Sc., M.E.E., Ph.D. Born 1916. Educ.: Dacca New York and London Univs. Formerly Prof. and Formerly Head, Dept. of Phys., Dacca Univ. Member, Nat. Com. for Geodesy and Geophys. in Pakistan. Socicties: Fellow, Pakistan Acad. of Sci.; Pakistan Assoc. for the Advancement of Sci.
Nuclear interests: Nuclear physics: high energy accelerators, study of elementary particles and their interactions with emulsion technique, nuclear scattering, nuclear models, plasma physics, and nuclear magnetic resonance.
Address: Department of Physics, Dacca University, Dacca-2, East Pakistan.

ALI, H. E. Osman. Head, Pakistan Mission to Euratom, 1966-.
Address: c/o Euratom, 51-53 rue Belliard, Brussels, Belgium.

ALI, S. Marghoob, Dr., M.Sc., D.Phil. (Oxon). Reader, Radiochem. Sect., Peshawar Univ.
Address: Peshawar University, Department of Chemistry, Peshawar, Pakistan.

ALIA, Manuel, B.S.C. (Geology), Dr. S. (Geology). Born 1917. Geological Adviser, Junta de Energia Nucl. de Espana; Director, Geological Service, Spanish West Africa, 1945-57; Prof. Geology, Valladolid Univ., 1942-53; Prof. Geology, Madrid Univ., 1958-. Books: Caracteristicas morfológicas y geológicas de la zona septentrional del Sahara espanol (1945); co-author, El Sahara espanol (1949); Contribución al conocimiento geomorfológico de las zonas medias del Sahara espanol (1949).
Nuclear interests: Everything related to geology.
Address: 44 Avenida Miraflores, Madrid, Spain.

ALIAGA-KELLY, Denis Thomas William, B.Sc. Born 1916. Educ.: Nat. Univ. Ireland. Sci. Asst. to Director, Inst. for Ind. Res. and Standards, Dublin, 1950-51; Tech. Manager, C. Warren and Co., Ltd., Dublin, 1951-56; Special Projects and Health Phys., Plessey Nucleonics Ltd., 1956-63; Chief Phys., Nucl. Enterprises (G.B.) Ltd., Edinburgh, 1963-. Societies: A.I.P.; A.P.S.; I.E.R.E.; Inst. Phys. and Phys. Soc.; Soc. for Radiol. Protection.
Nuclear interests: Research in nuclear radiation detectors and systems; Neutron activation analysis; Gamma ray spectrometry, health physics instruments, and reactor faulty fuel element detection.
Address: 13 West Castle Road, Edinburgh 10, Scotland.

ALIAN, Atef Mahmoud, B.Sc. (1st class Hons., Special Chem.), Ph.D. Born 1936. Educ.: Cairo Univ. Head, Analytical Chem. Div., and Lecturer United Arab Atomic Energy Establishment, Cairo, 1965-; Researcher, Nat. Centre for Criminological and Social Res., Cairo. Society: Sci. Assoc., United Arab Republic.
Nuclear interests: Neutron activation analysis; Radiochemical separation procedures; Quality control analysis of nuclear materials.
Address: Atomic Energy Establishment, Atomic Energy Post Office, Cairo, United Arab Republic.

ALIKHANI, H. E. A. N., Dr. Member, Nat. Iranian A.E.C.
Address: National Iranian Atomic Energy Commission, Ministry of Economy, Tehran, Iran.

ALIKHANOV, Abram Isaakovich. Born 1904. Educ.: Polytech. Inst., Leningrad. Director, Power Eng. Inst., Acad. of Sci. of the U.S.S.R. Member, Editorial Board, Atomnaya Energiya; Director, Inst. Theoretical and Exptl. Phys., Academy of Sci. of the U.S.S.R. Deleg., 2nd I.C.P.U.A.E., Geneva, Sept. 1958. Books: Slabye vzaimodeistviya. Noveishie issledovaniya betaraspada (Sovremennye problemy fiziki). (Weak Reactions, New Investigations into Beta-decay. Contemporary problems of Phys.) (Fizmatgiz, M, 1960); Recent Res. on Beta-Disintegration (London, Pergamon). Societies: Acad. of Sci. of the U.S.S.R.; Acad. of Sci. of the Armenian S.S.R.
Nuclear interests: Reactor design, nuclear physics.
Address: Institute of Theoretical and Experimental Physics, 89 Ulitsa Cheremushkinskaya, Post Box 3315, Moscow, U.S.S.R.

ALIKHANOV, S. G. Papers: Co-author, Plasma Diffusion in a Magnetic Field due to Coulomb Collision (Atomnaya Energiya, vol. 14, No. 2, 1963); Production of Megagauss Fields by Magnetodynamics Cumulation (ibid, vol. 23, No. 6, 1967); co-author, Pulsed High Density Thermonucl. System (ibid, vol. 25, No. 1, 1968).
Address: Academy of Sciences of the U.S.S.R., 14 Leninsky Prospekt, Moscow V-71, U.S.S.R.

ALIKHAN'YAN, Artemii Isaakovich. Born 1908. Educ.: Leningrad Univ. Director, Inst. Phys., Acad. of Sci. of the Armenian S.S.R. Societies: Corr. Member, Acad. of Sci. of the U.S.S.R.; Member, Acad. of Sci. of the Armenian S.S.R.
Nuclear interests: Cosmic rays, nuclear physics.
Address: Academy of Sciences of the Armenian S.S.R., Institute of Physics, 61 Ulitsa Abovyana, Yerevan, Armenian S.S.R.

ALIKHANYAN, Sergei. Geneticist. Staff Member, Inst. of Atomic Energy, Moscow.
Address: Kurchatov Institute of Atomic Energy, 46 Ulitsa Kurchatova, Post Box 3402, Moscow, U.S.S.R.

ALINE, Peter G., B.A., M.A., Ph.D. Born 1926. Educ.: Reed Coll. and Oregon Univ. Phys., Atomic Power Equipment Dept., Gen. Elec. Co., 1956-. Societies: A.N.S.; A.P.S.
Nuclear interests: Reactor physics; Development of computer-oriented systems for the design of light water power reactors; Verification of these design methods utilising operating power reactor data.
Address: Atomic Power Equipment Department, General Electric Company, P.O. Box 254, San Jose, Calif., U.S.A.

ALISERIS, Carlos W. Minister Counsellor, Embassy of Uruguay, Austria; Uruguay. Alternate Resident Rep., I.A.E.A.
Address: 4/III/10 Opernring, Vienna 1, Austria.

ALIVERTI, Guiseppina Angela Giuditta, Dr. Phys. Born 1994. Educ.: Turin Univ. Prof. Meteorology and Oceanography, Istituto Universitario Navale di Napoli, 1950-. Dean, Naval Sci. Faculty, Istituto Universitario Navale di Napoli, 1960-. Book: Esercizi di Fisica Pratica, 8th edition (Manuale Hoepli, 1967). Society: Soc. Italiana di Fisica.

Nuclear interests: Radioactivity of air and water; Oceanography.
Address: Istituto di Meteorologia ed Oceanografia dell'Istituto Universitario Navale, Naples, Italy.

ALIX, Pierre. Inspecteur Gén. des Travaux, C.E.A. Member, Commission Consultative des Marchés, C.E.A.
Address: Commissariat à l'Energie Atomique, 29-33 rue de la Federation, Paris 15, France.

ALKHAZASHVILI, G. M. Papers: Co-author, The Role of Redox Processes when Oxides of Uranium Dissolve in Acids (Atomnaya Energiya, vol. 8, No. 4, 1960); co-author, Some Features of Uranite Dissolution in H_2SO_4 Solutions with Oxidizing Agent (ibid., vol. 13, No. 2, 1962); co-author, The Influence of Iron Minerals in Ores on Uranium Oxidation in Acid Media (ibid., vol. 15, No. 4, 1963); co-author, A Study of Uranium Sorption by Minerals of Wall Rocks (ibid., vol. 16, No. 1, 1964).
Address: Academy of Sciences of the U.S.S.R., 14 Leninsky Prospekt, Moscow V-71, U.S.S.R.

ALKHAZOV, I. D. Papers: Co-author, Spontaneous Fission Periods for ^{240}Pu and ^{242}Pu (letter to the Editor, Atomnaya Energiya, vol. 15, No. 2, 1963); co-author, Energy Distribution of Spontaneous Fission Fragments of ^{244}Cm (ibid., No. 3); co-author, Cm^{241} Spontaneous Fission Accompanied by Long Range α- Particle Emission (ibid., vol. 16, No. 2, 1964).
Address: Academy of Sciences of the U.S.S.R., 14 Leninsky Prospekt, Moscow V-71, U.S.S.R.

ALLAN, Charles Lewis Cuthbert, M.A. Born 1911. Educ.: Cambridge Univ. Chief Elec. and Mech. Eng., North of Scotland Hydro-Elec. Board, 1954-63; Deputy Chairman, (1964-67), Chairman (1968-), South of Scotland Elec. Board. Societies: I.E.E.; Inst. of Civil Eng. Nuclear interest: Construction and operation of nuclear power stations for electricity supply.
Address: "Woodcliffe", Sutherland Crescent, Helensburgh, Dunbartonshire, England.

ALLAN, Douglas Lacey, B.Sc. Born 1920. Educ.: London Univ. P.S.O., U.K.A.E.A., A.E.R.E., Harwell. Society: A.Inst.P. Nuclear interest: Nuclear physics.
Address: The Barn, Church Street, Sutton Courtenay, Abingdon, Berks., England.

ALLARDICE, Corbin, A.B. (Maths). Born 1921. Educ.: Seton Hall Univ. Director of Information, N.Y.O.O. (1946-52), Asst. to Manager, N.Y.O.O. (1952-53), U.S.A.E.C.; Exec. Director, Joint Com. on Atomic Energy (U.S. Congress), 1953-55; Adviser on Atomic Energy, International Bank for Reconstruction and Development, 1955-. Vice-Pres., Nucl. Energy Writers Assoc.; World Bank Member, Steering Com., Project E.N.S.I., 1957-58. Book: Atomic Power - an Appraisal. Societies: A.N.S.; Nat. Assoc. Sci. Writers; Nucl. Energy Writers Assoc.; Cosmos Club. Nuclear interests: Economic aspects of atomic energy.
Address: 300 Mansion Drive, Alexandria, Virginia, U.S.A.

ALLARDICE, E. C., I.C.S. Controller, Atomic Energy Establishment, Dept. of Atomic Energy, Govt. of India.
Address: Atomic Energy Establishment, Trombay, Apollo Pier Road, Bombay, India.

ALLDAY, Coningsby, B.Sc. (Hons., Chem.). Commercial Director, Production Group, U.K.A.E.A.
Address: 1 Hadrian Way, Sandiway, Northwich, Cheshire, England.

ALLEGRETTI, Niksa, Prof. Formerly Head, Biol. and Bio-chem. Dept., Rudjer Boskovic Nucl. Inst. Member, Sci. Advisory Com., Federal Nucl. Energy Commission.
Address: Rudjer Boskovic Nuclear Institute, 54 Bijenicka Cesta, P.O. Box 171, Zagreb, Yugoslavia.

ALLEN, A. Jack. Managing Director, Athole G. Allen (Stockton) Ltd.
Address: Athole G. Allen (Stockton) Ltd., Stockton-on-Tees Chemical Works, Bowesfield Industrial Estate, Bowesfield Lane, Stockton-on-Tees, Co. Durham, England.

ALLEN, Alexander John, M.A. (New York), Ph.D. (New York). Born 1900. Educ.: Colorado and New York Univs. Asst. Director, Biochem. Res. Foundation, Delaware, 1944; Prof. Phys. and Director Rad. Lab., Pittsburgh Univ., and Westinghouse Graduate Prof. Eng., 1944-. Societies: Fellow, A.P.S.; A.A.U.P.
Address: Radiation Laboratory, Pittsburgh University, Pittsburgh 13, Pennsylvania, U.S.A.

ALLEN, Augustine Oliver, B.S., Ph.D. Born 1910. Educ.: California and Harvard Univs. Sen. Chem., Brookhaven Nat. Lab., 1948-. Book: Rad. Chemistry of Water and Aqueous Solutions (D. Van Nostrand Co., 1961). Societies: A.C.S.; Radiation Res. Soc. Nuclear interest: Radiation chemistry.
Address: Department of Chemistry, Brookhaven National Laboratory, Upton, Long Island, New York, U.S.A.

ALLEN, Donald G., A.B., Ll.B. Born 1913. Educ.: Dartmouth Coll. and Harvard Univ. Pres., Yankee Atomic Elec. Co.
Address: Yankee Atomic Electric Co., 441 Stuart Street, Boston, Mass. 02116, U.S.A.

ALLEN, Douglas Frederick, Dipl., Ryerson Inst. Technol. Born 1929. Educ.: Ryerson Inst. Technol., Toronto. Res. Technician, Reactor R. and D., A.E.C.L., 1952-58; Reactor Supervisor, Sub-critical Assembly, Toronto Univ., 1958-62; Canadian Service Manager, Nuclear-Chicago Corp., 1962-.

Nuclear interests: Nuclear reactor physics and technology; Nuclear instrumentation, particularly as used in the application of radioisotopes in life and physical sciences, clinical medicine and industrial process control; Health radiation monitoring.
Address: 430 Hounslow Avenue, Willowdale, Ontario, Canada.

ALLEN, Edmund Peter, M.B., F.R.C.P., Edinburgh, F.F.R. Born 1911. Educ.: Otago Univ. Visiting Radiol., New Plymouth Hospital; Member, Radiol. Advisory Council, New Zealand. Pres., and Fellow, Coll. of Radiol. of Australasia. Societies: British Inst. of Radiol.; Faculty of Radiologists.
Nuclear interests: Radiation therapy; Radiation health legislation.
Address: 109 Powderham Street, New Plymouth, New Zealand.

ALLEN, Herbert C., Jr., M.D. Born 1917. Educ.: Richmond Univ. and Medical Coll. of Virginia. Director, Nucl. Medicine Labs. of Texas; Director, Atomic Energy Industrial Labs. of the Southwest; Director, Dept. Nucl. Medicine, Hermann Hospital; Director, Dept. of Nucl. Medicine, Memorial Baptist Hospital; Pres.-Treas., Atomic Food Processing Corp. of America, Houston. Chairman, Rad. Medicine Advisory Com. of Harris County, Texas Medical Soc., 1959-; Sec.-Treas., Houston Thyroid Club, 1960-; Chairman, Com. for Beneficial Uses of Nucl. Medicine, Houston, 1959-.
Nuclear interests: Nuclear medicine, diagnosis and therapy; Thyroid physiology and pathology, instrumentation designation for localisation of organs, especially brain tumors; Radiotherapy for prostatic diseases; Industrial applications of atomic energy; Atomic food processing.
Address: 4010 Martinshire Drive, Houston 25, Texas, U.S.A.

ALLEN, James Alfred Van. See Van ALLEN, James Alfred.

ALLEN, James E. Dr. Member, Coordinating Council, New York State Office of Atomic Development.
Address: New York State Office of Atomic Development, The Alfred E. Smith State Office Building, Albany, New York 12225, U.S.A.

ALLEN, John Edward, Ph.D., D.Eng. Born 1928. Educ.: Liverpool Univ. Formerly at A.E.R.E. Harwell; now at Laboratorio Gas Ionizzati (Euratom-C.N.E.N.), also Scuola di Perfezionamento in Fisica, Università di Roma. Societies: A.M.I.E.E.; F.Inst.P.
Nuclear interests: Plasma physics; Thermonuclear reactions.
Address: Laboratorio Gas Ionizzati, Euratom-C.N.E.N., Casella Postale 65, Frascati, Rome, Italy.

ALLEN, John W. Manager, Nucl. Transportation, Stanray Products Div. Co.
Address: Stanray Products Division, Nuclear Department, 4527 Columbia Avenue, Hammond, Indiana, U.S.A.

ALLEN, K. W., Dr. Formerly Head, Exptl. Div. C, Culham Lab., U.K.A.E.A.; Prof., of Nucl. Structure, Nucl. Phys. Lab., Dept. of Nucl. Phys., Oxford Univ. Member, Board of Editors, Nucl. Fusion.
Address: Oxford University, Department of Nuclear Physics, 21 Banbury Road, Oxford, England.

ALLEN, Kenneth D. A., M.D. Member, Consulting Editorial Board, J. Nucl. Medicine.
Address: 333 N. Michigan Avenue, Chicago 1, Illinois, U.S.A.

ALLEN, Myron S., B.Sc. (Tufts), M.A. (Southern California), Ph.D. (Southern California). Born 1901. Educ.: Tufts and Southern California Univs. Member, Board of Trustees, H. A. Shiffer Associates Sci. Foundation, 1957-; Prof., Dept. of Phys., Long Beach City Coll., 1929-; Pres., Tech. Services Res. (Solar Dynamics); Chairman, Solar Dynamics Applications Com. for Nucl. Space Stations and Vehicles, Pac. Inst. of Earth Sci. Res. Staff (Visiting), Psychological Lab., Southern California Univ. Dean, Dept. of Phys., Pac. Inst. Earth Sciences. Exec. Director, Creative Growth Centre. Director, Inst. of Sci. Creativity. Book: Morphological Creativity (Prentice-Hall, 1962). Societies: A.A.P.T.; National Educational Assoc. (U.S.A.).
Address: 15850 West Road, Los Gatos, California, U.S.A.

ALLEN, Philip Wymer, B.Sc. Born 1918. Educ.: Parsons Coll., Fairfield, Iowa, Iowa State and Chicago Univs. Res. Meteorologist, U.S. Weather Bureau, 1949-53; Leading Analyst, U.S. Nat. Meteorological Centre, 1953-56; Chief, Air Resources, Las Vegas Lab., Las Vegas, Nevada. Societies: American Meteorological Soc.; A.G.U.
Nuclear interests: Meteorology and nuclear safety: mechanics of production of radioactive pollution; characteristics of radioactive clouds; diffusion and dispersion; particle deposition; wind and weather prediction; air trajectories.
Address: 817 Lacy Lane, Las Vegas, Nevada, 89107, U.S.A.

ALLEN, Robert C., Dr. Leader, Thermionics Group (-1962), Head, Thermionics and Thermoelectrics Dept. (1962-), Atomics International.
Address: Atomics International, P.O. Box 309, Canoga Park, California, U.S.A.

ALLEN, Wells P., Jr. Vice-Pres., New York State Elec. and Gas Corp.
Address: New York State Electric and Gas Corporation, Vestal Parkway East, Binghamton, New York 13902, U.S.A.

ALLEN, William Douglas, B.Sc., D.Phil. Born 1914. Educ.: Adelaide and Oxford Univs. Head, Proton Linear Accelerator Div., Rutherford High Energy Lab.; Joint Appointment with Dept. of Appl. Phys. Sci., Reading Univ. as Visiting Prof. Book: Neutron Detection (Newnes, 1960). Societies: Fellow, Inst. Phys. and Phys. Soc.; A.I.P.
Nuclear interests: Nuclear physics (neutrons); Accelerating machines: electrostatic generators, linacs.
Address: Science Research Council, Rutherford High Energy Laboratory, Chilton, Didcot, Berks., England.

ALLEN-WILLIAMS, David John Francis, M.A., Ph.D. Born 1918. Educ.: Cambridge Univ. Junior British Empire Cancer Campaign Fellow, 1950-54; Sen. (then Chief) Res. Eng., Davey, Paxman and Co., 1954-58; Prof. Mech. Eng., Western Australia Univ., 1958-. Western Australian Univ. Councillor, Australian Inst. Nucl. Sci. and Eng. Societies: M.I.E.Aust.; M.I.Mech. E.; I.E.E.; A.M.I.C.E.
Nuclear interests: Basic scientific and engineering research in all fields covered by the University.
Address: Department of Mechanical Engineering, University of Western Australia, Nedlands, Western Australia 6009, Australia.

ALLEST, Jean d'. See d'ALLEST, Jean.

ALLGEIER, Joseph P. Res. Sci.-Eng., Appl. Nucl. Phys. Div., Battelle-Columbus Labs. Address: Battelle-Columbus Laboratories, 505 King Avenue, Columbus, Ohio 43201, U.S.A.

ALLIBONE, Thomas Edward, D.Sc., Ph.D. (Sheffield), Ph.D. (Cantab.), F.R.S. Born 1903. Educ.: Sheffield and Cambridge Univs. Chief Sci., C.E.G.B.; formerly Director of Res., A.E.I. Res. Lab., Aldermaston; Director, Res. and Education, A.E.I. (Woolwich) Ltd. Book: Release and Use of Atomic Energy (Chapman and Hall, 1961). Societies: Roy. Soc.; Fellow, I.E.E.; Fellow, I.E.E.E.; F.Inst.P.; Inst. of Phys. and Phys. Soc.
Nuclear interests: Reactors; Research and power; Nuclear physics; Fusion.
Address: Round Hill, Enborne, Newbury, Berks., England.

ALLIO, Robert J., B.Met.E. (Rensselaer), M.Sc. (Ohio State), Ph.D. (Rensselaer). Born 1931. Educ.: Rensselaer Polytech. Inst. and Ohio State Univ. Manager, Materials and Processes Development, Atomic Power Div. (1962-66), Eng. Manager, Nucl. Fuel Div., (1966-.) Westinghouse. Societies: A.S.M.; A.N.S.; A.I.M.E.; A.A.A.S.; N.Y. Acad. of Sci.
Nuclear interests: Metallurgy and ceramics; Nuclear fuel cycle; Reactor technology.
Address: Westinghouse Electric Corp., Nuclear Fuel Division, P.O. Box 355, Pittsburgh, Pennsylvania 15230, U.S.A.

ALLIS, William Phelps, B.S., M.S. (M.I.T.), D.Sc. (Nancy). Born 1901. Educ.: M.I.T. and Nancy Univ. Emeritus Prof., M.I.T.; Visiting Fellow, St. Catherine's Coll., Oxford, 1968; Consultant, Los Alamos Sci. Lab.; Consultant, Argonne Nat. Lab.; Consultant, Nat. Bureau of Standards; Asst. Sec. Gen. for Sci. Affairs, N.A.T.O., 1962-64. Chairman, Gaseous Electronics Conference; Council, American Acad. Arts and Sci. Books: Nuclear Fusion (D. Van Nostrand Co., 1960); Motions of Ions and Electrons (Handbuch der Physik, vol. XXI, 1956); Co-author, Waves in Anisotropic Plasmas (M.I.T. Press Research Monograph Series, 1963). Societies: Fellow, A.P.S.; Fellow, Phys. Soc. London; A.A.P.T.; A.A.A.S.
Nuclear interest: Plasma physics with possible application to nuclear fusion.
Address: 33 Reservoir Street, Cambridge 02138, Mass., U.S.A.

ALLISY, André, L. ès Sc. Physiques. Born 1924. Educ.: Paris Univ. Member, I.C.R.U.; Member, Nat. Com. on Rad. Protection; Chief, Ionising Rad. Sect., Bureau Internat. des Poids et Mesures; Member, Joint Commission on Radioactivity, I.C.S.U. Societies: British Inst. of Radiol.; Sté. Française de Physique; H.P.S.
Nuclear interests: X and gamma ray standardisation; Standard measurements of radionuclides and neutron fields.
Address: 41 rue de Boulainvilliers, Paris 16, France.

ALLKOFER, Otto Claus, Priv.-Doz., Dr., Dipl. Phys. Born 1929. Educ.: Hamburg, Kiel and Regensburg Univs. Inst. für Reine und Angewandte Kernphysik, Christian-Albrechts Univ., Kiel.
Nuclear interests: Cosmic rays; Particle detectors; Radioactivity.
Address: 12 Königsbergerstrasse, Kiel, Germany.

ALLRED, John Caldwell, B.A., M.A., Ph.D. Born 1926. Educ.: Texas Christian Univ. and Texas Univ. Assoc. Dean of Arts and Sci. (1959-61), Asst. to Pres. (1961-62), Vice-Pres. (1962-), Assoc. Prof. Phys. (1956-61), Prof. (1961-), Houston Univ.
Nuclear interests: Reactor design, nuclear physics, solid state.
Address: University of Houston, Houston, Texas, U.S.A.

ALLRED, William B., B.S. (Chem. Eng.). Born 1920. Educ.: North Carolina State Univ. and Polytech. Inst. Brooklyn. Supervising Reactor Eng., U.S.A.E.C., Oak Ridge, 1948-54; Project Manager, Nucl. Div. Combustion Eng. Inc., 1954-62; Process Eng., Nucl. Div., Stone and Webster Eng. Corp., 1962-. Societies: A.I.Ch.E., Past Sec., Subcom. on Nucl. Power to A.S.M.E. Boiler and Pressure Vessel Com., A.N.S.; Atomic Ind. Forum.
Nuclear interests: Management; Reactor design.
Address: 12 Whittier Drive, Acton, Mass., U.S.A.

ALMEIDA, Yvone G. de See de ALMEIDA, Yvone G.

ALMEIDA SANTOS, Joao Rodrigues, Ph.D. Born 1906. Educ.: Coimbra and Manchester Univs. Prof. Phys., Coimbra Univ., 1941-; Member, Comissao de Estudos da Energia Nuclear, Portugal.
Nuclear interest: Nuclear physics.
Address: Laboratório de Fisica da Universidade, Coimbra, Portugal.

ALMELA SAMPER, Antonio, Mining Eng. Born 1903. Educ.: Escuela Especial de Ingenieros de Minas de Madrid. Consejero de la Junta de Energia Nuclear, 1959.
Nuclear interests: Mining and geological prospection.
Address: Consejo de Mineria, 34 Cristobal Bordiu, Madrid 3, Spain.

ALMOND, Peter Richard, B.Sc., M.A., Ph.D. Born 1937. Educ.: Nottingham, Bristol and Rice Univs. Fellow in Medical Phys. (1959-60), Rice Univ. Fellow in Phys. (1960-64), Postdoctoral Fellow in Biophys. (1964-65), Asst. Prof. of Biophys. and Asst. Phys. (1965-), Texas Univ. M.D. Anderson Hospital and Tumour Inst.; Assoc., Graduate Faculty, Texas Univ. Graduate School of Biomedical Sci., Houston, 1966-. Member, Subcom. on Rad. Dosimetry, American Assoc. Phys. in Medicine, 1965-.; Sci. Com. No. 26, High Energy X-ray Dosimetry, N.C.R.P. Societies: A.P.S.; American Assoc. of Phys. in Medicine.
Nuclear interest: Dosimetry.
Address: Department of Physics, University of Texas M.D. Anderson Hospital and Tumour Institute at Houston, Houston, Texas 77025, U.S.A.

ALONSO, Marcelo, Dr. in Phys. (Havana). Born 1921. Educ.: Havana and Yale Univs. Prof. Theoretical Phys., Havana Univ. Tech. Director, Nucl. Energy Commission of Cuba; Adviser on Nucl. Energy, Nat. Bank for Social and Economic Development. Member, Board of Directors, Cuban Phys. and Mathematical Soc.; Deputy Director, Dept. of Sci. Affairs, Pan American Union, 1963-. Book: Atomic Phys. (1958, Cultural S.A., printed by the University of Havana). Societies: Cuban Phys. and Math. Soc.; A.P.S.; A.N.S.
Nuclear interests: Theoretical nuclear physics, particularly at low energies, nuclear reactor theory.
Address: Departamento de Fisica, Universidad de la Havana, Havana, Cuba.

ALONSO SANTOS, A. Member, Editorial Board, Energia Nuclear.
Address: Energia Nuclear, 22 Avda. Complutense, Ciudad Universitaria, Madrid 3, Spain.

ALPAN, Hasan Sadrettin, B.Sc. (Hons.), Ph.D. (Mining). Born 1924. Educ.: Birmingham Univ. Turkish mining eng.; Member, A.E.C. of Turkey; Member, Board of Trustees, Middle East Tech. Univ.; Asst. Prof. Mining Exploration, Istanbul Tech. Univ., Mining Faculty, 1954; Chief of Exploration Project of Üköprü chromite deposit (1953), Chief of Exploration Project of Uludag scheelite deposit, (1954), Chief Eng., Mining Dept., (1954), Director, Atomic Energy Raw Materials Dept. (1956), Director, Mining Dept. (1958), Asst. Gen. Director (1959), now Gen. Director, M.T.A. Inst. Society: Chamber of Eng. of Turkey.
Nuclear interest: Atomic energy raw materials.
Address: 30 Sokak, No. 8 Bahçelievler, Ankara, Turkey.

ALPER, Tikvah, M.A., M.S. (Ed.), Washington. Born 1909. Educ.: Cape Town, Berlin and Washington (St. Louis) Univs. Director, Medical Res. Council, Exptl. Radiopathology Res. Unit, Hammersmith Hospital, London. Societies: F. Inst.P.; Fellow, Phys. Soc.; Inst. of Radiol.; Hospital Physicists' Assoc.; Assoc. for Rad. Res.; Rad. Res. Soc.; Photobiol. Group.
Nuclear interests: Radiation biology, chemistry and physics; Use of radioisotopes in biological and radiobiological investigations.
Address: Medical Research Council, Experimental Radiopathology Research Unit, Hammersmith Hospital, Ducane Road, London, W.12, England.

ALPHAND, Hervé. Ambassador and Sec.-Gen., Ministry of Foreign Affairs; Member, Atomic Energy Com., France.
Address: Comité de L'Energie Atomique, Commissariat à L'Energie Atomique, 29-33 rue de la Federation, Paris 15, France.

ALS-NIELSEN, J. Danish Member, Study Group on the High Flux Reactor, O.E.C.D., E.N.E.A.
Address: O.E.C.D. European Nuclear Energy Agency, 38 boulevard Suchet, Paris 16, France.

AL'SHEVSKII, L. E. Paper: Co-author, Ultrasonic Effect on Plasticity of High Boron Stainless Steel (letter to the Editor, Atomnaya Energiya, vol. 20, No. 5, 1966).
Address: Academy of Sciences of the U.S.S.R., 14 Leninsky Prospekt, Moscow V-71, U.S.S.R.

ALSINA, F. A. Member, Editorial Board, Nucl. Instruments and Methods,.
Address: c/o Nuclear Instruments and Methods, North-Holland Publishing Co., P.O. Box 103, Amsterdam, Netherlands.

ALSTAD, Joralf, Cand. Real. Sen. Officer, Div. Nucl. Chem., Oslo Univ.
Address: Oslo University, Forsknings v., Blindern, Norway.

ALSTER, Jonas, Ir., Ph.D. Born 1933. Educ.: Delft Tech. Univ. Assoc. Prof., Northeastern University. Societies: A.P.S.; Phys. Soc. of the Netherlands.
Nuclear interests: Low energy nuclear physics; Reactions and spectroscopy.

Address: Northeastern University, Boston, Massachusetts, U.S.A.

ALTER, H. Ward, A.B., Ph.D. Born 1923. Educ.: California Univ., Berkeley. Manager, Separations Chem., K.A.P.L., 1948-57; Manager, Nucl. and Inorg. Chem. (1957-66), Manager, Advanced Nucl. Applications (1966-), Vallecitos Atomic Lab. G.E.C. Councillor, California Sect., A.C.S. Society: A.N.S.
Nuclear interests: Reactor fuel cycle; Analytical and nuclear chemistry; Heavy element chemistry; Track etch technology.
Address: 7 Kirkcrest Lane, Danville, California, U.S.A.

ALTER, Harry, M.S., A.B. Born 1927. Educ.: New York, Pittsburgh, and Pennsylvania State Univs., and O.R.S.O.R.T. Phys. Bettis Atomic Power Lab.,-Westinghouse Elec. Corp., 1954-56; Phys., Curtiss Wright Corp., 1956-59; Phys., (1959-66), Supervisor (1966-), Atomics Internat. Neutron Cross Section Advisory Com., U.S.A.E.C.; Exec. Com., Reactor Phys. Div., A.N.S.; Cross Section Evaluation Working Group, U.S.A.E.C. Societies: A.N.S.; A.P.S.; A.A.A.S.
Nuclear interests: Reactor physics; Neutron physics; Nuclear physics, Mathematics; Computers.
Address: Atomics International, 8900 DeSoto Avenue, Canoga Park, Calif. 91304, U.S.A.

ALTET, Vicente Rogla. Vocale, Comisiones Asesoras de Biologia Vegetal y Aplicaciones Industriales, Junta de Energia Nucl.
Address: Junta de Energia Nuclear, Ciudad Universitaria, Madrid 3, Spain.

ALTHAMMER, Walter, Dr. jur. Born 1928. Educ.: Munich Univ. Society: Deutsches Atom forum.
Address: 18 Mössmannstrasse, 8901 Bergheim, Germany.

ALTHER, Peter, Dr. Member, Board of Directors, Atomkraft A.G.
Address: Atomkraft A.G., 29 Bärengasse, Zürich, Switzerland.

ALTHOUSE, Ernest E. Member, Operating Com., Power Reactor Development Co.; Sen. Vice Pres. (-1967), Pres. (1967-), Central Hudson Gas and Elec. Co.
Address: Central Hudson Gas and Electric Company, South Road, Poughkeepsie, New York, U.S.A.

ALTPETER, R. J. Formerly Prof., and Member, Nucl. Eng. Programme Com., now Prof. of Chem. Eng., Wisconsin Univ.
Address: Wisconsin University, Madison 4, Wisconsin 53706, U.S.A.

ALVA SALDANA, Luis, Ing. Delegado, Junta de Control de Energia Atomica del Peru.

Address: Junta de Control de Energia Atomica del Peru, 3420 Avenida Arequipa, San Isidro, Apartado 914, Lima, Peru.

ALVARDO, Sergio, Civil Eng., M.Sc., Born 1934. Educ.: Chile and Rochester Univs. Empresa Nacional de Electricidad, 1958-61; Chile Univ. 1963-; Servicio Nacional de Salud, 1963-65; Comisión Chilena de Energia Nucl., 1964. Society: Soc. Chilena de Medicina Nucl. Nuclear interests: Radiation physics. Utilisation of research reactors.
Address: Comision Chilena de Energia Nuclear, Miraflores 138, Fono 32135, Santiago, Chile.

ALVAREZ, Luis Walter, S.B., S.M., Ph.D. Born 1911. Educ.: Chicago Univ. Prof. Phys. (1946-), Assoc. Director, Rad. Lab. (1954-), California Univ. Societies: N.A.S.; A.P.S.; American Philosophical Soc.; American Acad. of Arts and Sci.
Address: 14 Northampton Road, Berkeley, California, U.S.A.

ALVAREZ, Silvia STANTIC. See STANTIC ALVAREZ, Silvia.

ALVAREZ, Vicente, Ing. Member, Nat. Energy Commission.
Address: Comisión Nacional de Energia, Ministerio de Fomento y Obras Públicas, Managua, D.N., Nicaragua.

ALVAREZ GARCILLAN, Mario. Vocale, Comisiones Asesoras de Biologia Vegetal y Aplicaciones Industriales, Junta de Energia Nucl.
Address: Junta de Energia Nuclear, Ciudad Universitaria, Madrid 3, Spain.

ALVES, Maria Alice F., M.Sc. Centro de Estudos de Fisica Nucl., Coimbra Univ.
Address: Centro de Estudos de Fisica Nuclear, Laboratório de Fisica da Universidade, Coimbra, Portugal.

ALVIAL, C. Gabriel. Born 1922. Educ.: Chile Univ. Prof. Phys., Chile Univ., 1951-; Director, Cosmic Ray Centre, Maths. and Phys. Faculty, Chile Univ., 1957-. Has worked in Prof. Giuseppe Occhialini's group of Milan Univ. and in Cosmic Ray Lab. directed by Prof. Marcel Schein of Ryerson Phys. Lab., Chicago Univ. as Fellow of John Guggenheim Memorial Foundation. Societies: Soc. Italiana di Fisica; Fellow, John Simon Memorial Guggenheim Foundation.
Nuclear interests: Cosmic ray physics; Nuclear emulsion methods and techniques; Elemetary particles.
Address: Bravo 943, Dep. 303, Santiago, Chile.

ALZMANN, Günther, Dipl. Ing., Dr.rer.nat. Born 1930. Educ.: Munich T.H. Reactor supv., Reactor station Phys.-Dept., Munich T.H. Nuclear interests: Health physics; Nuclear physics, especially nuclear fission; Reactor physics.
Address: Reaktorstation Garching, Garching bei Munich, Germany.

AMADO, Ralph David, Ph.D. Born 1932. Educ.: Stanford and Oxford Univs. Prof. Phys., Pennsylvania Univ. Society: Fellow, A.P.S. Nuclear interest: Nuclear and elementary particle theory, mostly three-body problem. Address: Pennsylvania University, Physics Department, Philadelphia, Pa. 19104, U.S.A.

AMAEV, A. D. At. Inst. of Atomic Energy. Adviser, Third I.C.P.U.A.E., Geneva, Sept. 1964. Address: Kurchatov Institute of Atomic Energy, 46 Ulitsa Kurchatova, Post Box 3402, Moscow, U.S.S.R.

AMALDI, Edoardo, Dr. phys. Born 1908. Educ.: Rome Univ. Prof. Rome Univ. Formerly Pres. now Vice Pres., I.N.F.N. Formerly Member, C.N.E.N. (Commissione Direttiva); Former Vice-Pres. now Member, C.E.R.N. Council. Member, Editorial Board, Annals of Phys. Book: The production and slowing down of neutrons in Vol. 38/2 Handbuch der Physik, (Heidelberg, Springer, 1959). Societies: Academia dei Lincei; Nat. Acad. Sci. U.S.A.; Reale Stà. Sci. Upsala; Accademia Sci. U.S.S.R.; Roy. Inst. Great Britain; F.R.S. Nuclear interest: High energy nuclear physics. Address: Istituto di Fisica dell'Università di Roma, Rome, Italy.

AMALDI, U., Jr., Dr. Head, Nucl. Phys. Sect., Laboratori di Fisica, Istituto Superiore di Sanita'. Address: Istituto Superiore di Sanita', 299 Viale Regina Elena, Rome, Italy.

AMANO, Makio. Born 1926. Educ.: Tokyo Univ. Chief, No. 3 Sect., Atomic Energy Dept., Ishikawajima Harima Heavy Industries Co., Ltd. Societies: Japan Soc. of Mech. Eng.; Japan High Pressure Inst. Nuclear interests: Design and fabrication of components for nuclear power plant such as nuclear pressure vessels, heat exchangers and pipings; Establishment of pressure vessel engineering. Address: 1-44-9, Nishikubo, Musashino-city, Tokyo, Japan.

AMARAL, C. M. do. See do AMARAL, C. M.

AMARI, Shohichi. Gen. Manager, Japan Nucl. Ship Development Agency. Address: Japan Nuclear Ship Development Agency, 35 Shiba-kotohira-cho, Minato-ku, Tokyo, Japan.

AMATO, Francesco D'. See D'AMATO, Francesco.

AMAVIS, Rene Jean, Chem. Eng. Born 1927. Educ.: Marseille Univ. Private firms, 1950-55; C.E.A. Français, 1955-60; In charge of waste problems, Euratom, C.C.R.-Ispra (Italia), 1960-. Society: Sté. de Radioprotection, Paris. Nuclear interests: All problems of treatment and disposal of radioactive wastes. Address: Chemical Group, Euratom-Ispra, Varese, Italy.

AMAYA LECLAIR, Manuel, Ing. Pres., Nat. Energy Commission. Address: Comisión Nacional de Energia, Ministerio de Fomento y Obras Públicas, Managua, D.N., Nicaragua.

AMBROS, Ignacio RENERO-. See RENERO-AMBROS, Ignacio.

AMBROSEN, Johan Peter, cand. mag. Born 1905. Educ.: Copenhagen Univ. Formerly at Niels Bohr Inst.; Principal Phys., Radiophys. Lab., Radium Centre, Copenhagen, 1954-. Member, Danish Defence Res. Board; Consultant, Danish Directorate of Civil Defence. Societies: British Hospital Physicist's Assoc.; Danish Phys. Soc. Nuclear interests: Radiation physics. Health physics; Hospital physics. Address: Radium Centre, 49 Strandboulevard, Copenhagen, Ø, Denmark.

AMBROSINO, Georges, Agr. es Sci. Phys. Born 1912. Educ.: Paris Univ. Charge Recherches, C.N.R.S.; Directeur Recherches, I.N.S.T.N. and C.E.A.; Directeur, Lab. Maurice de Broglie. Book: Co-author, Elements de Physique Nucléaire (Paris, Masson, 1960). Societies: Sté. Française Phys.; A.P.S. Nuclear interests: Nuclear reactions at low and middle energies; Beta radioactivity. Address: 5 rue St. Claude, Paris 3, France.

AMELINCKX, Severin, Dr. in Phys. Born 1922. Educ.: Ghent Univ. Deputy Director Gen., S.C.K.-C.E.N., Mol; Prof., Antwerp and Brussels Univs. Books: Les dislocations et la croissance des cristaux (Editions Masson, 1956); The Direct Observation of Dislocations (Academic Press, 1964). Nuclear interests: Radiation damage; Defects in solids; Crystal growth. Address: 34 Gouverneur Holvoetlaan, Deurne, Antwerpen, Belgium.

AMELL, Alexander Renton, B.S., Ph.D. Born 1923. Educ.: Massachusetts and Wisconsin Univs. Instructor, Hunter Coll., 1950-52; Asst. Prof., Lebanon Valley Coll., 1952-55; Res. Collaborator, Brookhaven Nat. Lab., 1955; Assoc. Prof. (1955-62), Prof. and Chairman (1962-), New Hampshire Univ. Society: A.C.S. Nuclear interests: Exchange reactions; Radiation chemistry. Address: Department of Chemistry, University of New Hampshire, Durham, N.H., U.S.A.

AMELOT, Pierre O.A., Electro-Mech. Civil Eng. Born 1921. Educ.: Brussels and Liege Univs. Pres., Carnoy-Vandensteen S.A. Societies: Inst. Belge de la Soudure; Comité Européen de la Chaudronnerie et de la Tôlerie. Nuclear interest: Nuclear piping, primary and

auxiliary piping, especially in stainless steel for P.W.R. and B.W.R. reactors.
Address: 56 Koopvaardijlaan, Ghent, Belgium.

AMELOT, Robert, Ing. des Arts et Manufactures. Born 1901. Educ.: Ghent Univ. Chairman, Carnoy-Vandensteen S.A. Society: Inst. Belge de Normalisation.
Nuclear interest: Nuclear piping, primary and auxiliary piping, especially in stainless steel for P.W.R. and B.W.R. reactors.
Address: 56 Koopvaardijlaan, Ghent, Belgium.

AMEMIYA, Ayao, Ph.D. (Phys.). Born 1907. Educ.: Tokyo Imperial Univ. Prof. Appl. Phys., Tokyo Univ., 1961-. Special Com. Rad. Chem., A.E.C., Japan; Com. Rad. Chem., Japan Atomic Ind. Forum, I.N.C.; Board of directors, Japanese Assoc. Rad. Res. Polymers. Books: Introduction to Rad. Chem. (in Japanese) (Mazuren Publ. Co., 1962); Tables of Molecular Integrals (in English) (Maruzen Publ. Co., 1955).
Societies: Phys. Soc., Japan; Soc. Appl. Phys., Japan; Atomic Energy Soc., Japan; Soc. Polymer Sci., Japan.
Nuclear interests: Radiation chemistry, specially radiation effects on polymers, simple organic molecules, etc., and its industrial application; Radiation sources for radiation-chemical research as well as for its industrial use such as accelerators, chemical piles. Theoretical problems on the interaction between radiation and molecular compounds.
Address: Shin-Ogawamachi Shinjuku, Tokyo, Japan.

AMES, Adelbert III, M.D. Member, Isotope Com., Massachusetts Gen. Hospital.
Address: Massachusetts General Hospital, Boston 14, Massachusetts, U.S.A.

AMEYE, Louis C. Pres., Federation de l'Industrie du Verre. Pres. and Member for Belgium, Specialised Nucl. Sect. for Economic Problems, and Member, Specialised Nucl. Sect. for Social Health and Development Problems, Economic and Social Com., Euratom.
Address: c/o Fédération de l'Industrie du Verre, 5 boulevard de l'Empereur, Brussels 1, Belgium.

AMICK, Arthur D. Member, Standing Com. on Nucl. Energy in the Elec. Industry, Nat. Assoc. of Railroad and Utilities Commissioners; Commissioner, South Carolina Public Service Commission.
Address: South Carolina Public Service Commission, 325 Wade Hampton Building, Columbia, S. C. 29201, U.S.A.

AMIEL, Saadia, M.Sc., Ph.D. (Jerusalem). Born 1929. Educ.: Hebrew Univ., Jerusalem. Res. Chem. (1952-60), Head, Nucl. Chem. Dept. (1960-), Israel A.E.C.; Visiting Lecturer, Graduate School, Hebrew Univ., 1962-; Visiting Sci., Rad. Lab., California Univ., Berkeley, 1956-57; Res. Assoc., Brookhaven Nat. Lab., 1957-58. Societies: Israel Chem. Assoc.; A.P.S.;

Israel Assoc. for the Advancement of Sci.
Nuclear interests: Nuclear chemistry; nuclear reactions and properties, special isotope production, analysis by radioactivation and nuclear reactions.
Address: Nuclear Chemistry Department, Israel Atomic Energy Commission, Soreq Research Establishment, Yavne, Israel.

AMIET, Jean-Pierre, Dr. in Phys. Born 1932. Educ.: Geneva Univ. Inst. für theoretische Physik, Heidelberg Univ. Prof. Theoretical Phys., Neuchâtel Univ., 1968-. Society: Phys. Soc. of Switzerland.
Nuclear interests: Nuclear theoretical physics and N-body problems.
Address: Neuchâtel University, Institut de Physique, 1 rue Breguel, CH200 Neuchâtel, Switzerland.

AMIOT, Henri, Ing. Gen.du Genie Maritime (C.R.). Directeur de l'Ecole, P.R.A.I.E.N.
Address: 19 avenue Niel, Paris 17, France.

AMIRHANOVA, I. B. Papers: Co-author, Relative Difference of $B^{11}F_3 - B^{10}F_3$ Steam Pressure (Atomnaya Energiya, vol. 19, No. 1, 1965); co-author, Investigation of Pressure Influence on Isotope Separation B^{11} and B^{10} Process by Distillation of BF_3 (ibid, vol. 23, No. 4, 1967).
Address: Academy of Sciences of the U.S.S.R., 14 Leninsky Prospekt, Moscow V-71, U.S.S.R.

AMIRUDDIN, Achmad, Dr. of Philosophy, Prof. of Inorg. and Nucl. Chem. Born 1932. Educ.: Indonesia and Kentucky Univs. Director, Pasar Djumat Res. Centre, Indonesian Atomic Energy Agency, Kebajoran, Djakarta, 1967-. Past. Pres., Indonesian Chem. Soc. Book: Introduction to Radiochem. and the Application of Radioisotopes (in Indonesian) (Dept. of Chem., Bandung Inst. of Technol., 1965).
Nuclear interests: Activation analysis; Chemical application of radioisotopes especially in geology and agriculture and management.
Address: Pasar Djumat Research Centre, Indonesian Atomic Energy Agency, Box Kbj/2, Kebajoran Lama, Djakarta, Indonesia.

AMIS, Edward Stephen, B.S., M.S. (Kentucky), Ph.D. (Columbia). Born 1905. Educ.: Kentucky, Chicago and Columbia Univs. Res. Assoc., Argonne Nat. Lab., 1955; Special Asst. to Director Res., Ethyl Corp., Baton Rouge, Louisiana, 1956; Director, Chem. Div., Baylor Univ., Summer Inst. for High School Sci. Teachers, 1958; Prof. Chem., Arkansas Univ., 1950-. Councillor, A.C.S. Books: Kinetics of Chemical Change in Solution (Macmillan Co., 1949); Solvent Effect on Reaction Rates and Mechanisms (Academic Press, 1966); NMR Studies of Ions in Pure and Mixed Solvents (Holden-Day, 1968). Society: Fellow, N.Y. Acad. Sci.
Nuclear interests: Tracer techniques using alpha and beta emitters in chemical processes.
Address: Department of Chemistry, University of Arkansas, Fayetteville, Ark. 72701, U.S.A.

AMLAUER, Karl. Senior officer, U.S. Nucl. Corp.
Address: U.S. Nuclear Corporation, P.O. Box 208, 801 N. Lake Street, Burbank, California, U.S.A.

AMMANN, Fred. Editor, New Techniques.
Address: New Techniques Publishing Ltd., 11 Tiefenhöfe, 8001 Zürich, Switzerland.

AMMERLAAN, Cornelis A. J., Drs. Born 1934. Educ.: Amsterdam Univ. Inst. for Nucl. Phys. Res., Amsterdam, 1960-64; Phys. Lab., Amsterdam Univ., 1964-. Society: Netherlands Phys. Soc.
Nuclear interests: Radiation detection; Radiation effects in solids.
Address: Physical Laboratory, Amsterdam University, 65 Valckenierstraat, Amsterdam, Netherlands.

AMMERS, Hugo van. See van AMMERS, Hugo.

AMMONS, James Earl, B.A., M.A. Born 1927. Educ.: Virginia and Ceylon Univs. Office of the Sec. (1955-59), Div. of Internat. Affairs (1959-60), A.E.C. Liaison Office, Paris (1960-63), Div. of Internat. Affairs (1963-), U.S.A.E.C.
Nuclear interests: Management and administration, particularly in international atomic energy cooperation in the peaceful uses of atomic energy.
Address: U.S. Atomic Energy Commission, Washington, D.C. 20545, U.S.A.

AMMUNDSEN, Miss E., M.D. Director Gen., Nat. Health Service of Denmark. Member, Departmental Co-ordination Com., Danish Atomic Energy Commission.
Address: National Health Service of Denmark, 1 Store Kongensgade, 1264 Copenhagen K, Denmark.

AMOIGNON, Jacques, Dr. Born 1928. Educ.: Paris Univ. With Sté. l'Air Liquide. Sec. Gén., Sté. Française des Ing. et Techniciens du Vide; Délégué de la France, Comité de Normalisation Européen (Pneurop) et à l'International Organisation for Standardisation (I.S.O.).
Nuclear interests: Toutes les applications du vide à l'énergie nucléaire.
Address: 25 rue de la Libération, 92, Rueil Malmaison, France.

AMONENKO, V. M. Papers: Co-author, Refining Beryllium and other Metals by Condensation on Heated Surfaces (2nd I.C.P.U.-A.E., Geneva, Sept. 1958); co-author, Deformation and Fracture of Rolled Beryllium of Various Purity (Atomnaya Energiya, vol. 16, No. 5, 1964).
Address: Nuclear Energy Institute, Academy of Sciences of the U.S.S.R. 14 Leninsky Prospekt, Moscow V-71, U.S.S.R.

AMOROSI, Alfred, M.E., B.S. (Chem.). Born 1913. Educ.: Cornell Univ. and Carnegie Inst. Technol. Materials Sect. Head (1948), Assoc. Director, Reactor Eng. Div. (1952), Argonne Nat. Lab.; Tech. Director, Atomic Power Development Assoc., Inc., 1954; Director, Liquid Metal Fast Breeder Programme Office, Argonne Nat. Lab., 1966. Board of Directors, A.N.S. Books: Gas Turbine Gas Charts; chapter on Selection of Reactors for Nucl. Eng. Handbook, by Harold Etherington; Assoc. Editor, Fast Reactor Technol. Plant Design. Society: A.S.M.E.
Nuclear interest: Reactor design.
Address: 5742 Dearborn Parkway, Downers Grove, Illinois 60515, U.S.A.

AMOUR, R. Bren D'. See D'AMOUR, R. Bren.

AMOUYAL, Albert, Ing. E.S.E. Born 1920. Chef, Département de Calcul Electronique, C.E.A. Society: A.N.S.
Address: Centre d'Etudes Nucléaires de Saclay, Département de Calcul Electronique, B.P. No. 2, 91 Gif-sur-Yvette, France.

AMPHLETT, Colin Bernard, B.Sc., Ph.D., D.Sc. Born 1920. Educ.: Birmingham Univ. Assist. to Div. Head, Chem. Div. (1958-61); Head, Rad. Chem. Group (1961-), A.E.R.E., Harwell. Editor, J. Inorg. Nucl. Chem. Books: Treatment and Disposal of Radioactive Wastes (Pergamon, 1961); Inorganic Ion Exchanges (Elsevier, 1964). Societies: Chem. Soc.; Faraday Soc.
Nuclear interests: Radiation chemistry; Ion exchange.
Address: Farm Field, Wootton Village, Boars Hill, Oxford, England.

AMSTEIN, Edmund Hollis, B.Sc., Ph.D., A.R.C.S. Born 1917. Educ.: London Univ. Res. Investigator, British Aluminium Co. Ltd., 1950-58; Chief Chem., Chester Works (1958-63), Res. Manager, Perivale (1963-), Assoc. Lead Manufacturers Ltd.
Nuclear interests: Metallurgical, particularly the applications of lead in radiation shielding.
Address: Associated Lead Manufacturers Limited, Research Laboratories, 7 Wadsworth Road, Perivale, Greenford, Middlesex, England.

AMSTER, Harvey Jerome, B.S., Ph.D. Born 1928. Educ.: California Inst. Technol. and M.I.T. Assoc. Prof. Nucl. Eng., California Univ. at Berkeley. Member, Tech. Group for Reactor Phys., A.N.S. Book: Chapter in Fast Neutron Physics (ed. Marion and Fowler, Interscience Publishers, 1959). Societies: A.P.S.; A.N.S.
Nuclear interests: Nuclear physics; Reactor theory; Radiation effects.
Address: 17250 Parkland Drive, Cleveland 20, Ohio, U.S.A.

AMSTUTZ, Edward D., Prof. Head, Chem. Dept., Lehigh Univ.
Address: Lehigh University, Bethlehem, Pennsylvania, U.S.A.

AN, Seung Bong, M.D. Born 1925. Educ.: Yon-Sei Univ. Medical Coll. Prof. Radiol., Yon-Sei Univ. Medical Coll., Seoul, and Director, Dept. of Radiol. and Nucl. Medicine, Severance Hospital, Seoul, 1958-. Societies: American Coll. of Radiol.; Korean Coll. of Radiol.
Nuclear interest: Radioactive isotopes in general medical fields.
Address: c/o Department of Radiology and Nuclear Medicine, Yon-Sei University Medical College, International P.O. Box 1010, Seoul, Korea.

AN, Shigehiro. Asst. Prof., Dept. of Nucl. Eng., Faculty of Eng., Tokyo Univ. Adviser, Third I.C.P.U.A.E., Geneva, Sept. 1964.
Address: Tokyo University, Tanashi-machi, Kitatama-gun, Tokyo, Japan.

AN, Thio Poo, Dr. Lecturer, Nucl. Chem. Dept., Bandung Inst. of Technol.
Address: Bandung Institute of Technology, 10 Djl. Ganeca, Bandung, Indonesia.

ANAND, Balmokand M., M.Sc., Ph.D. Born 1905. Educ.: Punjab and Bristol Univs. Reader in Phys. (1947-55), Prof., Phys. (1955-67), Head Phys. Dept., Punjab Univ. Pres., Punjab Univ. Phys. Assoc.; Chairman, Board of Control, Phys. Hons. School; Convener, Board of Studies in Phys.; Member, Academic Council, Sci. Faculty, Library Com., Board of Control Chem. Hons. School, Board of Control Maths. Hons. School, Combined Res. Board; Visiting Prof., Phys., California State Coll., Los Angeles, 1963-64; Member, Board of Governors, Regional, Eng. Coll., Kurukshitra and Indian Inst. of Technol., Kanpur. Papers: Author and co-author, many contributions to sci. journals, including Observations on the Life-time of the Neutral π-Meson (Proc. Roy. Soc., A Vol. 220, pp. 183-202, 1953); The Collision Mean Free Path of the Heavy Nuclei with $Z \geqslant 3$ in G-5 Nucl. Emulsions (Indian Sci. Congress, 1956). Society: Fellow, Indian Phys. Soc.
Address: D-6, Maharani Bagh, Ringh Road, New Delhi-14, India.

ANAN'EV, V. D. Paper: Co-author, 30 MeV Microtron-Injector for Pulsed Fast Reactor (Atomnaya Energiya, vol. 20, No. 2, 1966).
Address: Academy of Sciences of the U.S.S.R., 14 Leninsky Prospekt, Moscow V-71, U.S.S.R.

ANANYEV, V. D. Paper: Co-author, The Pulsed Reactor of the Lab. of Neutron Phys. in the Joint Inst. of Nucl. Res. (Kernenergie, vol. 9, No. 12, 1966).
Address: Vereinigten Institut fur Kernforschung, Laboratorium fur Neutronenphysik, Dubna, U.S.S.R.

ANASTASIJEVIC, Predrag, degree in Elec. and Nucl. Eng. Born 1923. Educ.: Belgrade Univ. Asst. Prof. Thermodynamics, Belgrade Univ., 1954-63; Head, Reactor and Neutron Phys. Dept. (1958-61), Head, Reactor Heat Transfer Dept. (1964-), Boris Kidric Inst. of Nucl. Sci.; Director, Training and Foreign Relations Dept., Sec., Sci. Advisory Com., Yugoslav Federal Nucl. Energy Commission, 1961-64; Pres., Yugoslav Project of Nucl. Fuel Development, 1964-; Sec. Gen., Yugoslav Assoc. of Heat Eng., 1964-.
Nuclear interests: Research and power reactor design; Heat and mass transfer of water and gas cooled reactors; Fuel elements research, development and fabrication.
Address: Boris Kidric Institute of Nuclear Sciences, Belgrade, P.O.B. 522, Yugoslavia.

ANATSKII, A. I. Paper: Co-author, Linear Induction Accelerator (Atomnaya Energiya, vol. 21, No. 6, 1966).
Address: Academy of Sciences of the U.S.S.R., 14 Leninsky Prospekt, Moscow V-71, U.S.S.R.

ANBAR, Michael, M.Sc., Ph.D. Born 1927. Educ.: Hebrew Univ., Jerusalem. Assoc. Prof., Weizmann Inst. of Sci., Rehovoth, Israel; Head, Dept. of Rad. Chem. Soreq Res. Establishment, Israel A.E.C., Yavne, Israel. Formerly Director, Radioisotope Training Centre, Rehovoth, Israel. Societies: A.C.S.; Chem. Soc., London; Israel Chem. Soc.
Nuclear interests: Radioisotope applications in chemical, biological and medical research; Radiation chemistry; Radiobiology; Nuclear education; Applications and analysis of stable isotopes.
Address: The Weizmann Institute of Science, Rehovoth, Israel.

ANCONA, Alessandro, Dott. Ing. Member, Steering Com., "Pool" Italiano per l'Assicurazione dei Rischi Atomici.
Address: "Pool" Italiano per l'Assicurazione dei Rischi Atomici, 10 Piazza San Bernardo, Rome, Italy.

ANDA, Ricardo, Medium Energies Dept., Lab. de Fisica Cosmica.
Address: Medium Energies Department, Laboratorio de Fisica Cosmica, La Paz, Bolivia.

ANDERSEN, Carl, M.Sc. (Chem. Eng.). Born 1915. Educ.: Denmark Tech. Univ.Pres., North Zealand Elec. and Tramway Co., 1959-; Pres., Isefjord Power Co., 1959-. Member, Danish A.E.C.; Official Rep. for Isefjord Power Co., Atomic Ind. Forum.
Nuclear interest: Power for production of electricity.
Address: Isefjord Power Company, 102 Strandvejen, Hellerup, Denmark.

ANDERSEN, Eigil, Cand. real. Born 1927. Educ.: Oslo Univ. Physicist, Inst. for Atomenergi, Kjeller, Norway.
Nuclear interests: Nuclear physics, reactor physics, and reactor design (core design).
Address: Institutt for Atomenergi, P.O. Box 40, Kjeller, Norway.

ANDERSEN, Pauli. Member, Board of Directors, Dansk Atomforsikrings Pool.
Address: Dansk Atomforsikrings Pool, 23 Gronningen, Copenhagen K, Denmark.

ANDERSEN, Richard Kendall, B.S. (E.E.), M.B.A. Born 1924. Educ.: Utah and Stanford Univs. Nucl. Reactor Eng. (1950-63), Manager, Product Planning and Marketing Res. (1963-67), Manager, Planning and Development (1967-), Nucl. Energy Div. G.E.C.
Nuclear interests: Nuclear power plant design and economics.
Address: General Electric Company, Nuclear Energy Division, 175 Curtner Avenue, San Jose, California 95125, U.S.A.

ANDERSEN, Svenn Lilledal, Dr. phil. Born 1926. Educ.: Oslo Univ. At present working on low-energy nuclear spectroscopy (experimental); Lecturer in phys. and crystal phys., Oslo Univ.
Nuclear interest: Nuclear physics.
Address: Institute of Physics, Oslo University, Blindern, Norway.

ANDERSON, Alfred Ronald. B.Sc., Ph.D. Born 1929. Educ.: Durham Univ. Principal Sci. Officer, A.E.R.E., Harwell, U.K.A.E.A., Exchange post at Argonne Nat. Lab., 1958-60. Societies: Faraday Soc.; Royal Inst. of Chem.; Chem. Soc.
Nuclear interests: Radiation chemistry; Chemical problems in reactors.
Address: Chemistry Division, A.E.R.E., Harwell, Didcot, Berks., England.

ANDERSON, Arthur N. Vice Pres. and Head, Mech., Nucl. and Design Eng. Div., Consolidated Edison of New York, 1968-.
Address: Consolidated Edison of New York, Inc., Mechanical, Nuclear and Design Engineering Division, 4 Irving Place, New York 3, New York, U.S.A.

ANDERSON, Clinton P. Born 1895. Educ.: Dakota Wesley and Michigan Univs.
Former Chairman now Member, Joint Com. on Atomic Energy, and Chairman, Com. on Aeronautical and Space Sci., U.S. Congress. Rep. to 77th U.S. Congress, State of New Mexico; re-elected to 78th and 79th Congresses; U.S. Senate, 1948; re-elected 1954 and 1960.
Address: Albuquerque, New Mexico, U.S.A.

ANDERSON, Denton Cosman, A.B. Born 1930. Educ.: Columbia Univ. Phys., Johns-Manville Res. Centre, 1951-55; Sci. (1957-60), Sen. Sci. (1960-), Bettis Atomic Power Lab. Society: A.N.S.
Nuclear interest: Transport theory.
Address: Bettis Atomic Power Laboratory, Westinghouse Electric Corporation, P.O. Box 1468, Pittsburgh 30, Pa., U.S.A.

ANDERSON, Donald Rex, B.A., M.A., Ph.D. Born 1916. Educ.: Utah Univ. Head, Biochem. Group, Radiobiol. Lab., Austin, Texas, 1956-60; Chief, Exptl. Sect. Radiobiol. Branch, U.S.A.F. School of Aerospace Medicine, 1960-. Societies: Rad. Res. Soc.; A.A.A.S.; N.Y. Acad. Sci.; Transplantation Soc.
Nuclear interests: Radiobiology. Chemical protection; Bone marrow therapy; Molecular and enzymatic inactivation.
Address: 3715 Invicta, San Antonio, Texas, U.S.A.

ANDERSON, Edward Leonard, Jr., M.S., Ph.D. Born 1928. Educ.: Chicago and Washington State Univs. Res. Asst., Washington State, 1949-50; Predoctoral Fellow, U.S.A.E.C., 1950-52; Res. Chem., California R. and D. Corp., 1952-53; Group Leader in Chem., Atomic Energy Div., Phillips Petroleum Co., 1953-56; Chem. Eng. Branch, U.S.A.E.C., 1956-59; Chief, Chem. Separations and Development Branch, U.S.A.E.C., 1959-. Societies: A.C.S.; A.N.S.; Nucl. Div., A.I.Ch.E.
Nuclear interests: Chemical processing and irradiated fuels, including aqueous, volatility and pyrometallurgical or closed cycle technology; High level waste disposal; Chemical plant safety and economics. Chemonuclear reactions and reactors. Fission product recovery. General chemistry and engineering.
Address: Chemical Separations and Development Branch, Division of Reactor Development, U.S.A.E.C., Washington, D.C. 20545, U.S.A.

ANDERSON, Elizabeth Blagg (Mrs. Edward H. Anderson), B.S., M.S. Born 1905. Educ.: Iowa Wesleyan and Iowa State Colls. Res. Assoc. (with emphasis on handling publications - editing, distribution), Medical Div., O.R.I.N.S. (now Oak Ridge Assoc. Univs.), 1953-. Societies: Soc. of Nucl. Medicine; American Medical Writers' Assoc.; A.A.A.S.; Conference of Biol. Editors; A.I.B.S.
Nuclear interests: The new language of nuclear studies; New meanings of terms; New abbreviations.
Address: Oak Ridge Associated Universities, Medical Division, P.O. Box 117, Oak Ridge, Tennessee 37830, U.S.A.

ANDERSON, Eric, B.Sc. Born 1912. Educ.: Birmingham Univ. Deputy Chief Eng., Parolle Elec. Plant Co., 1948-56; Chief Contracts Eng. (1956-62), Chief Operations Eng. (1962-), Nucl. Power Plant Co., Ltd. Societies: M.I.Mech-E.; A.M.I.E.E.
Nuclear interest: Project management.
Address: Slade Lane, Mobberley, Cheshire, England.

ANDERSON, Ernest C., A.B. (Augustana College), Ph.D. (Chicago). E.O. Lawrence Memorial Award, 1966. Born 1920. Educ.: Augustana Coll., Illinois and Chicago Univ. Staff Member, Los Alamos Sci. Lab., California Univ., Los Alamos, New Mexico, 1949-. Rask-Orsted Fellow, Copenhagen Univ., 1951. Societies: A.A.A.S.; Geochemical Soc.
Nuclear interest: Low-level counting techniques.

Address: Biomedical Research Group, Los Alamos Scientific Laboratory, Los Alamos, New Mexico, U.S.A.

ANDERSON, Francis H. Supt., El Segundo Works, Chem. Div., Allied Chem. Corp.; Vice-Pres. designate, Deputy Manager, and Acting Manager, Operations Div., Idaho Nucl. Corp. Address: Idaho Nuclear Corporation, Idaho Falls, Idaho, U.S.A.

ANDERSON, Frank, Ph.D., M.Sc. Born 1930. Educ.: Nat. Univ., Ireland and Univ. Coll., Dublin. Asst., Univ. Coll., Dublin, 1954-57; Res. Assoc., Johns Hopkins Univ., 1957-59; Visiting Asst. Phys., Brookhaven Nat. Lab., 1959-60; Coll. Lecturer (1960-), Director, Computation Lab. (1963-), Univ. Coll., Dublin. Societies: A.P.S.; Soc. Italiana di Fisica.
Nuclear interest: Elementary particle physics.
Address: Physics Dept., University College, Dublin, Ireland.

ANDERSON, Frank Abel, B.S., M.S., Ph.D. Born 1914. Educ.: Southern California, Maine, and Louisiana State Univs. Prof. Chem. Eng. and Chairman, Dept. Chem. Eng. (1948-63), Prof. Chem. Eng. and Assoc. Dean of Eng. (1963-), Mississippi Univ.; Sen. Project Eng., O.R.N.L., summers of 1954, 1955, 1956, 1959, and 1961; Theoretical Chem. Eng., Chemstrand Corp., summer of 1957. Societies: A.I.Ch.E.; A.C.S.; A.S.E.E.
Nuclear interests: Fluid mechanics, heat transfer.
Address: Department of Chemical Engineering, Mississippi University, Mississippi, U.S.A.

ANDERSON, Frank Struan, B.Mech.E. Born 1909. Educ.: Melbourne Univ. Director, Conzinc Riotinto of Australia Ltd. Member, Advisory Com. on Uranium Mining to Australian A.E.C. Societies: M.I.M.M.; M.Aus.I.M.M.; M.I.E.Aust.
Nuclear interests: Uranium mining and uranium oxide production.
Address: c/o Conzinc Riotinto of Australia Ltd., P.O. Box 384D, G.P.O., Melbourne, Victoria, Australia.

ANDERSON, George A. Manager, Nucl. Sci. and Eng. Depts., Curtiss-Wright Corp.
Address: Curtiss-Wright Corporation, Research Division, Quehanna, Pennsylvania, U.S.A.

ANDERSON, George Arthur, B.M.E. Born 1918. Educ.: Montana State Coll. and Minnesota Univ. Argonne Nat. Lab., 1948-55; Nucl. Energy Products, E.R.C.D. Div., A.C.F. Industries, Inc., 1955-. Books: Section of Nucl. Eng. Handbook; Pilot Channel Test Programme. Society: A.N.S.
Address: 10140 Riggs Road, Adelphi, Maryland, U.S.A.

ANDERSON, Guveren Montgomery, B.S. (Mech. Eng.), M.S. (Nucl. Sci.). Born 1918. Educ.: Illinois Inst. Technol. and Iowa State Coll. Sen. Tech. Asst., Div. of Reactor Development and Technol., U.S.A.E.C. Chairman, Advanced Converter and Alternate Coolant Task Forces; Steering Group Member, Interagency Advanced Power Group; Assoc. Fellow, Space Power Com. American Rocket Soc. Society: American Rocket Soc.
Nuclear interests: Evaluation of atomic energy programmes to develop economic advanced converters and breeders for civilian power; small reactors for terrestrial, oceanographic, maritime, and space uses: Development of power conversion systems including turboelectric, thermoelectric, and thermionic systems
Address: United States Atomic Energy Commission, Washington, D.C. 20545, U.S.A.

ANDERSON, Hamilton Holland, A.B., M.S., M.D. Born 1902. Educ.: California Univ. (San Francisco and Berkeley), Prof. Pharmacology and Exptl. Therapeutics, School of Medicine, California Univ., San Francisco, 1944- (Emeritus since 1960); Consultant to Graduate School (Hawaii Marine Lab.), Hawaii Univ., 1961-. Editor, American Lecture Series in Tropical Medicine; Editorial Board, Archives Internat. de Pharmacodynamie et de Therapie, Gand, Belge; Medicina Experimentalis, Bern, Switzerland; Exptl. Parasitology, N.Y.C. Societies: American Soc. Pharmacology and Exptl. Therapeutics; A.A.A.S.; American Soc. Tropical Medicine and Hygiene.
Nuclear interests: Tracer studies on metabolism of ^{14}C labelled morphine and its relatives in mammalian tissues; and of ^{14}C labelled nutrients and of rare earth metals utilised in the metabolism and for growth studies of Entamoebae (Notably Entamoeba histolytica).
Address: Lahaina, Box 205, Maui, Hawaii 96761, U.S.A.

ANDERSON, Harold Edward, B.Sc., M.A., M.Sc. Born 1916. Educ.: Indiana Univ., G.E. Graduate School of Nucl. Eng., South California Univ. Sen. Supervisor, Nucleonics Dept., Hanford Eng. Works, G.E.C., 1948-50; Lecturer, Eng. Dept., California Univ., 1951; Sen. Chem. Process Eng., Asst. Project Eng., Bechtel Corp., 1951-54; Consultant and Manager, Nucl. Eng., Paul Hardeman, Inc., 1956-58; Vice-Pres., Director and Consultant, Tech. Operations, Aerospace-Nuclear, 1960-. Societies: A.N.S.; A.I.Ch.E.; A.C.S.; Soc. for the Advancement of Management.
Nuclear interests: Radioisotopes: application in medicine, research and industry; Biomedical engineering.
Address: Aerospace-Nuclear, P.O. Box 112, Encino, Calif., U.S.A.

ANDERSON, Herbert L., A.B. (Columbia College), B.S. (Elec. Eng.) (Columbia), Ph.D. (Phys.) (Columbia). Born 1914. Prof. Phys., Enrico Fermi Inst. for Nucl. Studies and Dept. of Phys., Chicago Univ., 1950; Director, Enrico Fermi Inst. for Nucl. Studies, 1958-63;

Guggenheim Fellow, 1956-57; Fulbright Lecturer in Italy, 1956-57. Societies: A.P.S.; Italian Phys. Soc.; N.A.S.
Nuclear interests: High-energy physics and elementary particle physics.
Address: 4923 Kimbark Avenue, Chicago, Illinois, U.S.A.

ANDERSON, J. S. Manager, Uranium Operations, Vice-Pres. (1965-), Utah Construction and Mining Co.
Address: Utah Construction and Mining Co., 550 California Street, San Francisco 4, California, U.S.A.

ANDERSON, John B., A.B., J.D., Master of Laws. Born 1922. Educ.: Illinois and Harvard Univs. Rep. for Illinois, House of Representatives. Member, Joint Com. on Atomic Energy and House Com. on Armed Services, 1963-.
Address: 1225 Longworth House Office Building, Washington, D.C. 20515, U.S.A.

ANDERSON, John Bernard, B.Aero.Eng. Born 1917. Educ.: Minnesota Univ. Argonne Nat. Lab., 1949-54; Combustion Eng., Inc., 1954-. Societies: A.S.M.E.; A.N.S.
Nuclear interests: Direction of design, development, construction and test operation of complete nuclear plants - in particular, pressurised and boiling water type reactors for naval, army remote sites, maritime and utility applications.
Address: 58 Woodridge Circle, West Hartford, Connecticut, U.S.A.

ANDERSON, John Stuart, Ph.D. Born 1908. Educ.: London and Heidelberg Univs. Deputy Chief Sci. Officer, A.E.R.E., Harwell, 1947-53; Prof. Chem. and Head of Chem. Dept., Melbourne Univ., 1954-59; Director, Nat. Chem. Lab., London, 1959-63; Prof. of Inorganic Chem., Oxford Univ., 1963-. Treas., Australian Acad. Sci., 1956-59. Societies: F.R.S.; Fellow, Australian Acad. Sci.; Faraday Soc.; Chem. Soc. of London.
Nuclear interests: Chemistry of the heavy elements; Materials; Problems of uranium extraction.
Address: Inorganic Chemistry Laboratory, South Parks Road, Oxford, England.

ANDERSON, Jon D., B.S. (Civil Eng.). Born 1915. Educ.: Purdue Univ. Manager, Schenectady Naval Reactors Operation Office, U.S.A.-E.C., 1949-59; Deputy Director, New York State Office of Atomic and Space Development, Albany, New York, 1959-; Gen. Manager, New York State Atomic and Space Development Authority, 1962-. Societies: A.S.C.E.; American Soc. Military Eng.; A.N.S.; Atomic Ind. Forum.
Nuclear interests: Management; Promotion of atomic energy; Research, test and power reactors; Reprocessing of nuclear fuels; Regulation of atomic energy.
Address: Riverview Road, R.D. 1, Rexford, New York, U.S.A.

ANDERSON, Lewis. Botany Dept., Duke Univ.; Duke Univ. Rep., Council, Oak Ridge Assoc. Univs., 1965-.
Address: Botany Department, Duke University, Durham, North Carolina, U.S.A.

ANDERSON, R. L. Assoc. Prof., Nucl. Eng. Dept., Purdue Univ.
Address: Purdue University, Lafayette, Indiana, U.S.A.

ANDERSON, Robert Christian, A.B. (cum laude), M.A., Ph.D. Born 1918. Educ.: Middlebury Coll., M.I.T. and Princeton Univ. Assoc. Chem. (1948-56), Asst. Director (1956-), Brookhaven National Lab. Member, Sci. Advisory Council, New York State Legislature. Societies: A.C.S.; A.A.A.S.; A.S.E.E.; A.N.S.
Nuclear interest: Chemical effects of nuclear transformations in organic systems.
Address: Brookhaven National Laboratory, Upton, Long Island, New York, U.S.A.

ANDERSON, Roger H., B.Sc., Ph.D. Born 1930. Educ.: Washington Univ., Seattle. Asst. Prof. Math. Phys.
Nuclear interest: Nuclear many-body theory.
Address: Institute for Research, Seattle Pacific College, Seattle 99, Washington, U.S.A.

ANDERSON, Roy V., B.S. Ch.E. Born 1919. Educ.: Delaware Univ. Manager, Portsmouth Area Office, Piketon, and Deputy Manager, Fernald Area Office, Fernald, U.S.A.E.C.
Nuclear interests: Management; Uranium production.
Address: United States Atomic Energy Commission, Portsmouth Area Office, Piketon, Ohio, U.S.A.

ANDERSON, W., Ph.D. Member, Managing Subcom. on Monitoring, M.R.C.
Address: Medical Research Council, 20 Park Crescent, London W.1, England.

ANDERSON, William Kermit, B.S., M.S., Ph.D. Born 1909. Educ.: Agricultural and Mech Coll., Texas. Head, Inorganic Chem. NEPA Project, Oak Ridge, 1947-51; Assoc. Sci. and Project Eng., Submarine Reactor Control Mechanisms Development, and Head, Chem. Technol. Group (1951-54), Argonne Nat. Lab.; Manager, SAR Reactor Materials Development, and, later, Consulting Eng., Nucl. Materials, K.A.P.L., Schenectady, New York, 1954-. Member, Exec. Board, and Chairman, Programme Com., Materials Technol. Div., A.N.S.; Member, Administrative Com. on Papers and Publications, and Member, Tech. Coms., A.S.T.M. Books: Co-author, Neutron Absorber Materials for Reactor Control (U.S. Govt. Printing Office, 1962); co-author, Reactor Structural Materials (A.S.T.M.-S.T.P. 314). Societies: A.C.S.; A.S.M.; New York Acad. of Sci.
Nuclear interests: Development of nuclear core materials, particularly high temperature ceramic fuels and materials applicable to reactor control; Also organic and liquid metal

coolants for reactor application; Corrosion in nuclear environments.
Address: 3407 Rosendale Road, Schenectady, New York, U.S.A.

ANDERSSON, Gunnar Ivar Curt, Fil. mag. Born 1930. Educ.: Lund Univ. With A.B. Atomenergi, Stockholm, 1956-.
Nuclear interests: Reactor physics and reactor design.
Address: Atomenergi, Box 9042, Stockholm 9, Sweden.

ANDERSSON, Karl Göran, Dr. phil. Born 1920. Educ.: Uppsala Univ. Asst. Prof., Uppsala Univ., 1957-62. Consultant (1959-60), Res. Assoc. (1961-62), C.E.R.N., Geneva; Assoc. Prof., Chalmers Univ. of Technol. Gothenburg, 1962-.
Nuclear interests: Nuclear physics; Nuclear chemistry.
Address: Department of Physics, Chalmers University of Technology, Gothenburg, Sweden.

ANDERSSON, Olle, Dr. Society: Vice-Pres., Svenska Radiologförbundet.
Address: The Hospital, Norrtälje, Sweden.

ANDJELKOVIC, M. Head, Analytical Dept., Dept. for Processing of Nucl. Raw Materials, Inst. Nucl. Raw Materials.
Address: 86 Franse Deperea, Belgrade, Yugoslavia.

ANDLAU, Charles d'. See d'ANDLAU, Charles.

ANDO, Toyoroku, B. Eng. Born 1897. Educ.: Tokyo Imperial Univ. Pres., Onoda Cement Co., Ltd., 1948-; Director, Nippon Atomic Industry Group; Exec. Director, First Atomic Power Industry Group.
Address: c/o Onoda Cement Co., 1-1 Marunouchi Chiyodaku, Tokyo, Japan.

ANDO, Yoshio, Dr. Eng. Born 1922. Educ.: Tokyo Univ. Assoc. Prof., Dept. of Naval Architecture, Tokyo Univ., 1948; Assoc. Prof., Inst. of Ind. Sci., 1953; Prof., Dept. of Nucl. Eng., Tokyo Univ., 1962. Member, Advisory Com. of Reactor Safety, and Member, Safety Standard of Nucl. Ship, A.E.C.; Member, Tech. Com. of Nucl. Ship; Res. Adviser, Japan Atomic Energy Res. Inst.; Chairman, Reactor Fabrication and Inspection Com., Japan Welding Eng. Soc.; Chairman, Marine Reactor Com., Atomic Energy Soc. of Japan; Reactor Pressure Vessel Com., Japan Soc. Mech. Eng. Societies: Japan Welding Soc.; A.W.S.; Soc. of Naval Architects of Japan.
Nuclear interests: Nuclear reactor structure; Welding of reactor material and structure; Nuclear propulsion.
Address: Department of Nuclear Engineering, Tokyo University, 7-3-1, Hongo, Bunkyoku, Tokyo, Japan.

ANDRADE, Ayres Cunha DE. See DE ANDRADE, Ayres Cunha.

ANDRADE, Fernando ALBA-. See ALBA-ANDRADE, Fernando.

ANDRE, A. B. Rep. of Ribet-Desjardins, Commission Intersyndicale de l'Instrumentation et de la Mesure Nucléaire Française.
Address: Ribet-Desjardins, 13-17 rue Perier, Montrouge, (Seine), France.

ANDRE, Gaston E. Directeur adjoint, Union Minière du Haut-Katanga; Conseiller, Sté. Belgonucléaire; Chairman, Management Com., Bureau Belge des Radioisotopes; Administrateur, Groupement Professionnel de l'Industrie Nucléaire. Deleg. to 2nd I.C.P.U.A.E., Geneva, Sept. 1958.
Address: 6 rue Montagne du Parc, Brussels, Belgium.

ANDREEV, O. L. Papers: Co-author, Absolute Measurement of Neutron Source Yield by Gold Foil Method (letter to the Editor, Atomnaya Energiya, vol. 16, No. 3, 1964); co-author, Internat. Comparison of Neutron Sources (letter to the Editor, ibid, vol. 19, No. 1, 1965); co-author, Internat. Collation of Neutron Sources (letter to the Editor, ibid, No. 2); co-author, Exptl. Determination of Optimal Geometry in Thermal Neutron Flux Density Standard (letter to the Editor, ibid, vol. 23, No. 4, 1967).
Address: Academy of Sciences of the U.S.S.R., 14 Leninsky Prospekt, Moscow V-71, U.S.S.R.

ANDREEV, S. V. Candidate of Biol. Sci.; biophys. Director, Biophys. Lab., All-Union Inst. for Sci. Res. on Plant Protection, Leningrad. Deleg., I.A.E.A. Symposium on Radioisotopes and Rad. in Entomology, Bombay, Dec. 5-9, 1960.
Address: Biophysical Laboratory, All-Union Institute for Scientific Research on Plant Protection, 42 u. Gertsena, Leningrad, U.S.S.R.

ANDREEVA, M. M. Paper: Co-author, Determination of Uranium and Plutonium in Organic Solutions (letter to the Editor, Atomnaya Energiya, vol. 24, No. 1, 1968).
Address: Academy of Sciences of the U.S.S.R., 14 Leninsky Prospekt, Moscow V-71, U.S.S.R.

ANDREEVA, N. Z. Paper: Co-author, Investigation of Scintillation Characteristics of Lithium Glasses by Detection of Neutron, Beta and Gamma Rad. (Atomnaya Energiya, vol. 24, No. 2, 1968).
Address: Academy of Sciences of the U.S.S.R., 14 Leninsky Prospekt, Moscow V-71, U.S.S.R.

ANDREEVSKII, A. A. Papers: Heat Transfer to Molten Sodium Flowing Transversely across a Single Cylinder (letter to the Editor, Atomnaya Energiya, vol. 7, No. 3, 1959); co-author, Heat Transfer from a Bundle of Tubes Arranged in Staggered Order in the Cross-Flow of Molten

Sodium (letter to the Editor, ibid, vol. 13, No. 3, 1962); co-author, Alkali Metals Boiling Heat Transfer (letter to the Editor, ibid, vol. 19, No. 2, 1965); co-author, Potassium Boiling Heat Transfer in Tube at Moderate Steam Content Values (letter to the Editor, ibid, vol. 21, No. 1, 1966).
Address: Academy of Sciences of the U.S.S.R., 14 Leninsky Prospekt, Moscow V-71, U.S.S.R.

ANDREOTTI, Giulio, On. Minister of Industry and Trade; Chairman, C.N.E.N.
Address: Ministero dell'Industria e del Commercio, 33 Via Veneto, Rome, Italy.

ANDRES, Louis Emile, Chem. Eng., D.Sc. Born 1907. Educ.: Strasbourg Univ. Directeur Gén., Potasse et Engrais Chimiques (P.E.C.); Pres. Sté. d'Etudes et de Travaux pour l'Uranium (S.E.T.U.). Societies: Sté. Chimique de France; A.C.S.
Nuclear interests: Uranium and thorium ore treatment; Uranium metallurgy.
Address: 10 avenue George V, Paris 8, France.

ANDRESEN, K. Specialist on Nucl. Instr., Norsk A/S Philips.
Address: Norsk A/S Philips, P.B. 5040, Oslo 3, Norway.

ANDREW, Edward Raymond, M.A., Ph.D., Sc.D. (Cantab). Born 1921. Educ.: Cambridge Univ. Lecturer, St. Andrews Univ., 1949-54; Prof. Phys., Univ. Coll. of North Wales, Bangor, 1954-64; Prof. Phys., Nottingham Univ., 1964-.
Book: Nucl. Magnetic Resonance (Cambridge University Press, 1955). Societies: F.R.S.E.; Fellow, Inst. of Physics and Phys. Soc.; A.P.S.
Nuclear interest: Nuclear magnetic resonance.
Address: Physics Department, Nottingham University, Nottingham, England.

ANDREW, M. B. Staff Hydro-Mech. Eng., Resources Agency of California, Dept. of Water Resources.
Address: The Resources Agency of California, Department of Water Resources, 1416-9th Street, Sacramento, California, U.S.A.

ANDREWS, Douglas Guy, M.A. Born 1917. Educ.: Cambridge Univ. Eng. II and I, U.K.A.-E.A., 1949-57; Prof. Nucl. Eng., Toronto Univ., 1957-. Societies: Inst. Mech. Eng., London; Professional Eng. Assoc., Ontario; Canadian Nucl. Assoc.; A.N.S.; B.N.E.S.
Nuclear interests: Nuclear education and research; Heat transfer and fluid flow; Subcritical, heavy water assemblies; Radiation.
Address: Toronto University, Toronto 5, Ontario, Canada.

ANDREWS, Gould Arthur, A.B., M.D. Born 1918. Educ.: Michigan Univ., Certified by American Board of Internal Medicine. Chairman, Medical Div., Oak Ridge Associated Universities, 1962-. Books: Co-editor, Radioisotopes in Medicine (U.S.A.E.C., 1955), Progress in Medical Radioisotope Scanning (U.S.A.E.C., 1963), Radioactive Pharmaceuticals (U.S.A.E.C., 1966). Societies: American Coll. of Physicians; Soc. for Exptl. Biol. and Medicine; American Thyroid Assoc.; American Assoc. for Cancer Res.
Nuclear interests: Whole body irradiation and management of radiation accidents; Radioisotopes in haematology; Radioactive colloids; Carcinoma of the thyroid.
Address: Oak Ridge Associated Universities, Oak Ridge, Tennessee 37830, U.S.A.

ANDREWS, Howard Lucius, Ph.D. Born 1906. Educ.: Brown Univ. Chief, Rad. Safety Dept., Nat. Insts. of Health. Books: Nucl. Rad. Phys. (3rd ed. 1963, Prentice-Hall); Rad. Biophys. (Prentice-Hall, 1961). Societies: A.P.S.; Rad. Res. Soc.; H.P.S.
Nuclear interests: Radiation health protection; Radiobiology.
Address: National Institutes of Health, Bethesda, Maryland, U.S.A.

ANDREWS, J. N., B.Sc., D.I.C., A.R.I.C. Sen. Lecturer, Nucl. Technol. Sect., Bristol Coll. of Sci. and Technol.
Address: Bristol College of Science and Technology, Ashley Down, Bristol 7, England.

ANDREWS, James Crandall, B.S.(Phys.), M.S., L.S. Born 1921. Educ.: North Carolina and Illinois Univs. Tech. Information Service Supv., E.I. du Pont de Nemours and Co., Savannah River Lab., Aiken, S. Carolina, 1952-58; Library Supv. (1959-62), Library Director (1962-), Argonne Nat. Lab.; Nucl. Director, Sci. Sect., Sci.-Tech. Div., Special Libraries Assoc. Societies: Special Libraries Assoc.; American Documentation Inst.
Nuclear interests: Processing and dissemination of technical information. Library administration and operation.
Address: Argonne National Laboratory, 9700 South Cass Avenue, Argonne, Illinois 60439, U.S.A.

ANDREWS, John Robert, Ph.B., M.D., D.Sc. (Medicine). Born 1906. Educ.: Brown, Western Reserve and Pennsylvania Univs. Prof. and Director, Radiol., Bowman Gray School of Medicine of Wake Forest Coll. and North Carolina Baptist Hospitals, Inc., Winston-Salem, North Carolina, 1950-55; Chief, Rad. Branch, Nat. Cancer Inst., Nat. Insts. of Health, Bethesda, 1955-64; Prof., Radiol., and Director, Radiotherapy, Georgetown Univ. Medical Centre, Washington, D.C.
Societies: A.M.A.; American Roentgen Ray Soc.; Soc. of Nucl. Medicine; Rad. Res. Soc.; Radiol. Soc. of North America; American Assoc. for Cancer Res.
Nuclear interests: Human cancer radiotherapy research; Radioisotopes applications and methodology; Human physiological and metabolic effects of radiation; in vitro and in vivo quantitive mammalian cell radiobiology; Fast neutron beam radiobiology and radiotherapy.
Address: 3800 Reservoir Road, N.W., Washington, D.C. 20007, U.S.A.

ANDREWS, John Stewart Vincent, B.Eng., M.I.Mech.E. Born 1914. Educ.: Liverpool Univ. Works Director, Mersey Cable Works Ltd.; Eng. Controller, James A. Jobling and Co., Sunderland; Deputy Chief Eng., U.K.A.E.A. Reactor Group, Risley. Societies: B.N.E.S.; Plastics Inst.
Nuclear interests: General reactor engineering; Design of gas cooled, graphite moderated, reactors, and associated gas circuits and heat exchangers.
Address: 2 Wood Lane, Appleton, Cheshire, England.

ANDREWS, John Thomas, M.B., B.S. (London), L.R.C.P., M.R.C.S. (England). M.R.A.C.P., M.C.R.A., D.Obst. R.C.O.G. Born 1927. Educ.: London Univ. (The London Hospital). Radioisotope Specialist in charge of the Dept. of Nucl. Medicine; The Roy. Melbourne Hospital, Melbourne, Australia. Consultant in nucl. medicine, Victorian Hospitals and Charities Commission. Society: Melbourne Radioisotope Study Group.
Nuclear interests: Chemical applications of radioisotopes. Mainly organ scanning, circulation and kinetic studies and organ function tests.
Address: Department of Nuclear Medicine, Royal Melbourne Hospital, Melbourne, Australia.

ANDREWS, Trevor, M.Sc.Tech. Born 1921. Educ.: Manchester Univ. Asst. Eng., Liverpool Group, N.W. Gas Board, 1950-55; Chief Project Eng., Nucl. Power Plant Co., 1955-60; Partner, C. Andrews and Sons, Consulting Eng., 1960-. Council, Manchester Federation Sci. Soc.; Hon. Sec., Manchester Technol. Assoc. Societies: A.M.I.C.E.; A.M.I.Mech.E.
Nuclear interests: Project design and management; Siting, economics and layout; Services design.
Address: 23 Maitland Avenue, Chorlton-cum-Hardy, Manchester 21, England.

ANDREWS, Walter C. Member, Council, Inst. of Nucl. Eng.
Address: Institution of Nuclear Engineers, 147 Victoria Street, Westminster, London S.W.1, England.

ANDREWS, Warren McCormick, B.S. (Eng. Phys.), M.S. (Phys.), M.S. (Nucl. Eng.), Ph.D. (Nucl. Eng.). Educ.: Auburn, Vanderbilt and California Univs. O.R.N.L., 1954; Nucl. Eng., Aerojet Gen. Nucleonics, San Ramon, 1957-60; Lawrence Rad. Lab., Livermore, 1958-60; Special Asst. to the Pres. for Nucl. Sci. Centre (1961-65), Director, Nucl. Sci. Centre (1965-), Auburn Univ. Society: A.N.S.
Nuclear interests: Reactor physics; Pulsed neutron techniques for measuring neutron properties; Reactor, accelerator and radiation source facilities design and operation for university research programmes; Applied radiation techniques.
Address: Nuclear Science Centre, Auburn University, Auburn, Alabama, U.S.A.

ANDREWS, William Thomas. Born 1907. Sen. Eng. Asst. Central Electricity Authority, 1948-56; Sen. Mech. Elec. Eng. (1956-57), Head, Mech. Group, Station Eng. Dept., Atomic Power Projects (1957-), English Elec. Co. Societies: A.M.I. Mech.E.; A.M.I.E.E.; Whitworth Prizeman.
Nuclear interests: Design of systems and plant layouts associated with nuclear power stations.
Address: 28 St. Philips Road, Leicester, England.

ANDREYANOV, V. S. Paper: Co-author, Calculation of Neutron Fluxes in Cell of Intermediate Reactor with Control Rod (letter to the Editor, Atomnaya Energiya, vol. 24, No. 3, 1968).
Address: Academy of Sciences of the U.S.S.R., 14 Leninsky Prospekt, Moscow V-71, U.S.S.R.

ANDRIANOV, M.A. Paper: Co-author, Phase Diagrams of Plutonium with IIIA, IVA, VIII and IB Group-Metals (Atomnaya Energiya, vol. 23, No. 6, 1967).
Address: Academy of Sciences of the U.S.S.R., 14 Leninsky Prospekt, Moscow V-71, U.S.S.R.

ANDRIESSE, Dirk Cornelis, electrotech. eng. Born 1908. Educ.: Delft Univ. Staff Eng. (-1957), Sub-director (-1967), director (1967-), Provincial Elec. Board of Friesland. Member of Assoc. of Managing Directors of Electricity Supply Undertaking in the Netherlands.
Address: 59 Emmakade, Leeuwarden, Netherlands.

ANDRIESSEN, C. J., Ir. Managing Director, Bronswerk-Fijenoord N.V.
Address: Bronswerk-Fijenoord N.V., P.O. Box 28, Amersfoort, Netherlands.

ANDRIEU, Jean-Louis. Adjoint au Chef du Département de propulsion nucléaire, C.E.A. Adviser, Third I.C.P.U.A.E., Geneva, Sept. 1964.
Address: Commissariat a l'Energie Atomique, 29-33 rue de la Federation, Paris 15, France.

ANDRIOLA, A. D. Formerly Division Eng., Special Products Division, Farrel-Birmingham Co. Inc.; Manager, Value Analysis, Farrel Corp.
Address: Farrel Corporation, 25 Main Street, Ansonia, Connecticut 06401, U.S.A.

ANDRIOT, Jean, Ing. en chef du Corps des Mines. Born 1927. C.E.A.- Secretariat général de l'Energie - Electricité de France.
Nuclear interests: Etudes économiques; Programmes atomiques.
Address: 82 rue Pierre Brossolette, Chatillon, Seine, France.

ANDRLE, Zdenek, Ing. Born 1929. Educ.: Prague Tech. Univ. Head of Res. Group, Nucl. Power Div., Skoda, Plzen. Society: Czechoslovak Sci. and Tech. Soc.

Nuclear interests: Reactor design; Nuclear power station design; Fuel cycles.
Address: 87 Koterovská, Plzen, Czechoslovakia.

ANDRO, J. Sen. Officer, Sté. Stein et Roubaix.
Address: Société Stein et Roubaix, Nuclear Department, 24 rue Erlanger, Paris 16, France.

ANDRONIKASHVILI, Elevter Luarsabovich. Director, Inst. Phys., Acad. Sci. of the Georgian S.S.R. Paper: Co-author, The Model of a Radiative Indium-Gallium Loop of the Tbilisi IRT-2000 Reactor (Atomnaya Energiya, vol. 13, No. 4, 1962).
Address: Academy of Sciences of the Georgian S.S.R., 8 Ulitsa Dzerzhinskaya, Tbilisi, Georgian S.S.R.

ANDROSOV, A. V. Paper: Co-author, Gas Supply System for Ion Source of Electrostatic Generator (letter to the Editor, Atomnaya Energiya, vol. 20, No. 2, 1966).
Address: Academy of Sciences of the U.S.S.R., 14 Leninsky Prospekt, Moscow V-71, U.S.S.R.

ANDRY, Fernand. Ing., Nucl. Div., Electro-Entreprise, S.A.
Address: Electro-Entreprise, S.A., Division Nucleaire, 32 rue de Mogador, Paris 9, France.

ANDRYSEK, Oskar, M.D., C.Sc. Born 1931. Educ.: Charles Univ., Prague. Asst. Prof., Head of Clinical Isotope Dept., Inst. Biophys. Nucl. medicine, Charles Univ. Books: Textbook of Radiological Phys. for X-ray Technicians SZdN, Prague); Textbook of Nucl. Medicine for Medical Students (SZdN, Prague). Society: Czechoslovak Medical Soc. J. E. Purkyne (in sections: nucl. medicine, physiological, gastroenterology, nephrology and oncology).
Nuclear interests: Medical use of radioisotopes in pathophysiological research; Diagnostic methods in gastroenterology, nephrology and oncology.
Address: 15 Jugoslavská, Prague 2, Czechoslovakia.

ANDRYUSHIN, I. A. Paper: Co-author, On Expediency of Radiometric Uranium Ore Dressing and on Optimum Value of Separation during Dressing (letter to the Editor, Atomnaya Energiya, vol. 19, No. 1, 1965).
Address: Academy of Sciences of the U.S.S.R., 14 Leninsky Prospekt, Moscow V-71, U.S.S.R.

ANDRYUSHIN, N. F. Papers: Co-author, Scattering of Collimated Gamma Rad. into Rooms (letter to the Editor, Atomnaya Energiya, vol. 22, No. 5, 1967); co-author, Gamma-Field of Dipped in Water Isotropic Co^{60} Source near Interface of Water and Air (letter to the Editor, ibid vol. 23, No. 1, 1967); co-author, Scattering of Collimated Co^{60}, Cs^{137} and Au^{198} Gamma Rad. near Interface between Two Media (ibid, vol. 23, No. 2, 1967); co-author, Back-Scattering of Gamma-Rays Isotropic Sources from Cylindrical Barriers (ibid, vol. 23, No. 3, 1967).
Address: Academy of Sciences of the U.S.S.R., 14 Leninsky Prospekt, Moscow V-71, U.S.S.R.

ANDRZEJEWSKI, Stanislaw, Prof. Born 1908. Educ.: Warsaw Tech. Univ. Director, Inst. of Heat Eng., Warsaw Tech. Univ. Chairman, State Council for Power and Elec.; Member, State Council for Peaceful Utilisation of Atomic Energy.
Nuclear interests: Nuclear power economics; Nuclear power stations design.
Address: Instytut Techniki Cieplnej, 25 ul. Nowowiejska, Warsaw, Poland.

ANESINI, Andor. Member, Editorial Board, Energia es Atomtechnika.
Address: Muszaki Konyvkiado Bajcsy-Zsilinszky ut 22, Budapest 5, Hungary.

ANGELA, Gaetano, Dr. Div. Director, Inspectorate of Private Insurance, Ministry of Industry and Commerce, Rome.
Address: Ministry of Industry and Commerce, 33 Via Veneto, Rome, Italy.

ANGELBAUER, Ewald. Member, Board, Dinglerwerke A.G.
Address: Dinglerwerke A.G., Zweibrücken, Germany.

ANGELELLI, Victorio, Mining Eng. Born 1908. Educ.: Bergakademie,Freiberg, Germany. Military Factories 1944-58; Argentine A.E.C. 1958-; Prof., La Plata Univ., 1957-. Book: Los minerales de uranio, sus yacimientos y prospeccion (Buenos Aires, C.N.E.A., 1958).
Nuclear interests: Radioactive minerals geology.
Address: 8250 Avda. del Libertador, Buenos Aires, Argentina.

ANGELINI, Arnaldo Maria, Prof., Univ. Rome. Born 1909. Member (1952-), Vice-Pres. (1956-), Italian Com. for Nucl. Researches; Director, Nucl. Eng. Course Rome Univ. 1957-. Formerly Managing Director, Terni Co.; Now Gen. Manager, E.N.E.L. and Pres., C.I.S.E. Book: Funzionamento e Costruzione di Macchine Elettriche. Societies: I.E.E.E.; A.N.S.; Assoc. Elettrotecnica Italiana; Conférence Internationale des Grands Réseaux Electriques; Accademia Nazionale Lincei.
Nuclear interests: Metallurgy; Reactor design; Management.
Address: 160 Via Cassia Vecchia, Rome, Italy.

ANGELINO, Gian Carlo, Perito Industriale elettrotecnico. Born 1933. Educ.: Feltrinelli (diplomato, 1954). At C.I.S.E., 1956-; Instructor, Nucl. reactor, Feltrinelli Inst., 1957-.
Nuclear interest: Isotope separation.
Address: 5 Via Martiri Triestini, Milan, Italy.

ANGELL, Craig W., B.S. (Chem. Eng.). Born 1913. Educ.: Rensselaer Polytec. Inst. and Clarkson Memorial Coll. of Technol. Vice-Pres. and Sales Eng., Artisan Industries, Inc., Waltham, Mass. Member Admissions Com., A.I.Ch.E.; Gen. Chairman, A.N.S., Annual Meeting, Boston,

Mass. Societies: A.I.Ch.E.; A.N.S.; Inst of Food Technol.
Nuclear interests: Design and fabrication of process equipment such as test loops.
Address: Huckleberry Hill, Lincoln R.F.D.1., Massachusetts, U.S.A.

ANGELOPOULOS, Michael, Dipl.-Eng., Dr.-Ing., D.I.C. Born 1926. Educ.: Athens Tech. Univ., Braunschweig T.H., Germany, and London Univ. Tippetts-Abbett-McCarthy-Stratton Engs., (Athens); Power Plants Sect. (1956-59), Nucl. Power Office (1960-63), Tech. Adviser (1963-67), Public Power Corp. of Greece; Lecturer (1955-56, 1960-62), Assoc. Prof. Nucl. Eng., Athens Tech. Univ.; Member of the Board, Greek A.E.C., 1964-65. Society: Chambre Technique de Grèce.
Nuclear interests: Reactor physics; Reactor technology; Nuclear power plants.
Address: National Technical University, 42 28th October Street, Athens-147, Greece.

ANGELOVIC, Joseph William, B.S., M.S., Ph.D. Born 1930. Educ.: Utah State Univ. Chief, Rad. Effects Programme, Radiobiol. Lab., Bureau of Commercial Fisheries, Beaufort, North Carolina. Society: Rad. Res. Soc.
Nuclear interest: Radioecology and effects of the interactions of radiation with environmental factors on estuarine organisms.
Address: Bureau of Commercial Fisheries, Radiobiological Laboratory, Beaufort, North Carolina 28516, U.S.A.

ANGELUCCI, Giuseppe, Mech. Eng. Born 1895. Educ.: Milan Univ. Manager, Azienda Comunale Elettricità et Acque, 1943-51; Managing Director, Soc. Trentina di Elettricità, 1952-57; Managing Director, Soc. Elettronucleare Nazionale, 1958-; Formerly Director, C.I.S.E.; Formerly Member, Comitato Nazionale Associazione Nazionale di Ingegneria Nucleare. Societies: Assoc. Elettrotecnica Italiana; Rome Water Supply Com.; Euratom Commission of the Internat. Confederation of Business Executives.
Address: 37 Via Eleonora Duse, Rome, Italy.

ANGIER, Derek. Director of Res., Rad. Applications Inc., 1963-.
Address: Radiation Applications Inc., 36-40 37th Street, Long Island City 1, New York, U.S.A.

ANGONI, H.E.M.K. Albanian Ambassador to Austria; Albanian Resident Rep., I.A.E.A., 1967-.
Address: 39 Jacquingasse, Vienna 3, Austria.

ANGSTADT, C., Dr. Asst., Biophys. Sect., Biol. Chem. Dept., Hahnemann Medical Coll. and Hospital.
Address: Department of Biological Chemistry, Hahnemann Medical College and Hospital, 235 North 15th Street, Philadelphia 2, Pennsylvania, U.S.A.

ANGULO, Ignacio de PINEDO y. See de PINEDO y ANGULO, Ignacio.

ANIANSSON, Erik Axel Gunnar, Tekn. dr. Born 1924. Educ.: Roy. Inst. Technol., Stockholm. Prof., Dept. of Phys. Chem., Gothenburg Univ.
Nuclear interests: Radiation chemistry, isotope applications.
Address: 33^2 Aschebergsgatan, Göteborg C, Sweden.

ANNO, James Nelson, Jr., B.Sc. (Phys.), M.S. (Phys.), Ph.D. (Phys.). Born 1934. Educ.: Ohio State Univ. Asst. Div. Chief, Appl. Nucl. Phys. Div., Battelle-Columbus Labs. Book: Co-author, U.S. Res. Reactor Operation and Use (Addison-Wesley, 1958). Societies: A.N.S.; A.P.S.
Nuclear interest: Atomic physics.
Address: 4242 Clairmont Road, Columbus 21, Ohio, U.S.A.

ANNO, Yasuro, Prof. Head, Dept. of Radiol., Tottori Univ. School of Medicine.
Address: Tottori University School of Medicine, 86 Nishimachi, Yonago, Japan.

ANOHIN, V. P. Paper: Co-author, 300 MeV Electron Synchrotron at the Polytech. Inst. of Tomsk (Atomnaya Energiya, vol. 21, No. 6, 1966).
Address: Academy of Sciences of the U.S.S.R., 14 Leninsky Prospekt, Moscow V-71, U.S.S.R.

ANONSEN, S. H. Vice-Pres. and Gen. Manager, Operations Div., Weldon Spring Feed Materials Plant, operated for U.S.A.E.C. by Mallinckrodt Chem. Works.
Address: Weldon Spring Feed Materials Plant, Mallinckrodt Chemical Works, Uranium Division, P.O. Box 472, St. Charles, Missouri 63302, U.S.A.

ANOUTI, Jamil, Dr. Member, Lebanese A.E.C.
Address: Lebanese Atomic Energy Commission, Ministry of Public Works, Beirut, Lebanon.

ANSELL, Herbert Victor, M.A. Born 1908. Educ.: Oxford Univ. Sen. Lecturer, Brighton Coll. of Technol.
Nuclear interests: Radiochemistry; Isotopic exchange reactions.
Address: 15 Florence Road, Brighton BN1 6DL, Sussex, England.

ANSELMETTI, Gian Carlo, Dr. Eng. Born 1904. Educ.: Turin Polytech. Electrotech. eng. Administrator-Delegate and Director-Gen., Cogne S.p.A. Pres. Stà. Nazionale Officine di Savigliano; Counsellor Centro Italiano Studi Esperienze (C.I.S.E.); Teacher, technol. of elec. machines, Turin Polytechnic. Book: Trattato di misure elettriche.
Nuclear interest: Metallurgy.
Address: 38 Via Sacchi, Turin, Italy.

ANSTEE, John Walter, B.A. (Maths. and Phys.), M.S. (Electronics). Born 1925. Educ.: Toronto Univ. Measurement res., followed by supervision of radium operations, Eldorado Mining and Refining (1944), Ltd., Port Hope, Ontario, 1948-52; Commercial Products Div., Ottawa, A.E.C.L., 1952-; Branch Head, Chem. Production Operations, 1952-. Society: Canadian Assoc. of Physicists.
Nuclear interests: Radioisotope production and utilisation.
Address: 1889 Garfield Avenue, Ottawa 5, Ontario, Canada.

ANTHEUNISSEN, Willem, LL.B. Born 1927. Educ.: Leiden Univ. Deputy Chief, Nucl. Energy and Rad. Protection Dept., Ministry of Social Affairs and Public Health, -1962; Sec., Sci. Council for Nucl. Affairs, Netherlands, 1962-.
Nuclear interests: Dutch and international legal, organisational and institutional aspects of the peaceful uses of nuclear energy, especially scientific research activities, both pure and applied.
Address: 220ʰ Laan van Meerdervoort, The Hague, Netherlands.

ANTHOINE, Roger Marc. Elec., Mech. and Journalism degrees. Born 1925. Educ.: Univ. du Travail, Charleroi, and Inst. pour Journalistes, Brussels. Press Office, A.C.E.C., Belgium; Sen. Information Officer, C.E.R.N., Geneva, Switzerland. Society: Sté. Belge de Physique.
Address: c/o C.E.R.N., 1211 Geneva 23, Switzerland.

ANTHONY, C. W. Vice-Pres., Oklahoma Gas and Elec.-Co.
Address: Oklahoma Gas and Elec Co., 321 North Harvey, Oklahoma City 1, Oklahoma, U.S.A.

ANTHONY, Donald Joseph, B.S. (Chem.), Ph.D. (Phys.). Born 1922. Educ.: Siena Coll., Loudonville, New York and Notre Dame Univ. Manager, Reactor Phys., K.A.P.L., G.E.C., 1957-. Society: A.N.S.
Nuclear interests: Reactor physics; Reactor design; Management; Reactor operation.
Address: Knolls Atomic Power Laboratory, General Electric Company, Schenectady, New York, U.S.A.

ANTHONY, Michael, B.S. Born 1910. Educ.: M.I.T. Manager of Operations (-1962), Group Manager (1962-), Nucl. Products Group, Metals and Controls Inc. (Corp. Div.), Texas Instr. Inc. Societies: Atomic Ind. Forum; A.N.S.; A.S.M.
Nuclear interest: Management.
Address: 173 County Street, Attleboro, Massachusetts, U.S.A.

ANTIN, Gilberto LOPEZ-. See LOPEZ-ANTIN, Gilberto.

ANTOGNETTI, Lorenzo, M.D. Born 1898. Educ.: Genoa Univ. Director, Clinica Medica, Genoa Univ., 1952-. Vice-Pres., Italian Soc. of Nucl. Medicine and of Soc. of Endocrinology; Counsellor, Italian Soc. of Metabolic Diseases; co-Editor, Minerva Nucleare. Book: Co-author, Diagnostica e terapia con i radioisotopi (Minerva Medica). Societies: Italian Soc. of Internal Medicine; Italian Soc. of Rheumatology.
Nuclear interest: Nuclear medicine.
Address: 8 Via Fieschi, Genoa, Italy.

ANTOINE, Jean. Directeur Gen., Sté. Gen. d'Entreprises.
Address: Société Génerale d'Entreprises, 56 rue du Faubourg Saint-Honore, Paris 8, France.

ANTOKU, S. Lecturer, Res. Inst. for Nucl. Medicine and Biol., Hiroshima Univ.
Address: Hiroshima University, Higashisendamachi, Hiroshima, Japan.

ANTON, Nicholas Guy. Born 1906. Educ.: Tech. Inst. Leonardo da Vinci, Trieste and Columbia Univ., New York. Pres. and Director R. and D., Anton Electronic Labs., U.S.A., 1948-61; Pres. and Director of Res., E.O.N. Corp., U.S.A., 1961-. Chairman, Nucl. Instrumentation and Controls Com., Atomic Ind. Forum; Chairman, G-1 Com. on Nucl. Instrumentation, Electronic Industries Assoc.; Nucl. Standards Board, American Standards Assoc.; U.S. tech. deleg., I.C.P.U.A.E., Geneva, 1955. Observer, Federal Civil Defence Administration's Programme on Evaluation of Civil Defence Radiol. Defence Instruments; Co-ordinator for lab. equipment in 1955 Federal Civil Defence Bomb Test Programme; Chairman, Industrial Tech. Advisory Board, Federal Civil Defence Administration. Societies: F.I.E.E.E.; Fellow, A.P.S.; Fellow, New York Acad. of Sci.; Fellow, A.A.A.S.; American Mathematical Soc.; Atomic Ind. Forum; A.S.M.E.; Soc. for Nondestructive Testing; Electronic Industries Assoc.
Nuclear interests: Pioneered in design and development of nuclear detectors such as geiger and proportional counters, ionisation chambers, neutron boron trifluoride and fission detectors, ultraviolet and photon counters and related instrumentation for health monitoring reactor control and nuclear and medical research.
Address: 3172 Bedford Avenue, Brooklyn 10, New York, U.S.A.

ANTONANZAS, José L., M.Sc. (Chem.). Born 1934. Educ.: Madrid Univ. Tech. Sec., Centrales Nucleares del Norte, S.A., 1957-.
Nuclear interest: Nuclear production of electricity.
Address: Nuclenor, 12 Medio, Santander, Spain.

ANTONIO, Miss Pacita. Medical Technician, Radioisotope Lab., Philippine Gen. Hospital.
Address: Philippine General Hospital, Radioisotope Laboratory, Manila, Philippines.

ANTONOV, Anatoly V. Deleg., I.A.E.A. Symposium on Inelastic Scattering of Neutrons in Solids and Liquids, Vienna, Oct. 11-14, 1960. Deleg., I.A.E.A. Symposium on Pile Neutron Res. in Phys., Vienna, Oct. 17-21, 1960.

Papers: Co-author, The Investigation of Neutron Diffusion for Water and Ice at Temperatures near 0°C and -80°C (letter to the Editor, Atomnaya Energiya, vol. 13, No. 4, 1962); co-author, Temperature Dependence of Neutron Diffusion Parameters in Water and Ice (letter to the Editor, ibid, vol. 20, No. 2, 1966).
Address: Institute of Physics, Academy of Sciences of the U.S.S.R., Obninsk, near Maloyaroslavets, Moscow, U.S.S.R.

ANTONOV, I. N. Paper: Co-author, Dependence of Heat Conductibility Sodium from Concentration of Oxides (letter to the Editor, Atomnaya Energiya, vol. 19, No. 4, 1965).
Address: Academy of Sciences of the U.S.S.R., 14 Leninsky Prospekt, Moscow V-71, U.S.S.R.

ANTSIPOV, P. S. Paper: Co-author, 30 MeV Microtron-Injector for Pulsed Fast Reactor (Atomnaya Energiya, vol. 20, No. 2, 1966).
Address: Academy of Sciences of the U.S.S.R., 14 Leninsky Prospekt, Moscow V-71, U.S.S.R.

ANTTILA, J., M.Sc. Sen. Officer, Lab. of Phys. Metal., Dept. of Mining and Metal. Industry, Inst. of Technol.
Address: Institute of Technology, Otaniemi Finland.

ANTUNES, Manuel TELLES. See TELLES ANTUNES, Manuel.

ANTUNEZ, Antonio RODRIGUEZ-. See RODRIGUEZ-ANTUNEZ, Antonio.

ANTUNEZ, Francisco, Director of Exploration, Comision Nacional de Energia Nucl., Mexico.
Address: Av. Insurgentes Sur 1079 Tercer piso, Mexico 18, D.F., Mexico.

ANXIONNAZ, Rene. Pres., Sté. de Recherches Techniques et Industrielles
Address: 111 rue la Boëtie, Paris, France.

ANZAI, Masao, Bachelor of Economics, Bachelor of Laws. Born 1904. Educ.: Tokyo Univ. Pres., Showa Denko K. K., 1959-; Pres., Japan Electrodes Export Co., 1958-; Pres., Showa Neoprene K. K., 1960-; Pres., Tokuyama Petrochem. Co., 1963-; Pres., Japan Olefin Chem. Co., 1965-. Director, Japan Atomic Ind. Forum, 1959-.; Director, Tokyo Atomic Ind. Consortium, 1959-; Chairman, Japan Chem. Industry Assoc., 1966-.
Nuclear interests: Management; Reactor materials; Metallurgy; Utilisation of radioisotopes and radiation chemistry.
Address: 34 Shiba Miyamoto-cho, Minatoku, Tokyo, Japan.

ANZANI, Alberto, Elec. Eng. Doc. Born 1933. Educ.: Politecnico di Milan. Ercole Marelli S.p.A., Milan, 1958; C.E.S.I., Milan, 1958; C.N.R.N., Atomic Energy Commission, Ispra, 1959-61; Euratom, C.C.R. Ispra, 1961-.
Nuclear interests: Nuclear instrumentation; Electronics.
Address: Euratom, C.C.R., Ispra Varese, Italy.

AOKI, Junji, Dr. Senior Officer, Dept. of Sci., Toho Univ.
Address: Toho University, 4-77 Omori, Otaku, Tokyo, Japan.

AOKI, Kiyokaza. Formerly Director, Nucl. Ship Res. Sect., Kawasaki Kissen Kaisha Ltd.; Pres., Nihon Kisen Ltd.
Address: Nihon Kisen Ltd., 6-10-1 Sakae Machi Dori, Ikuta-ku, Kobe, Japan.

AOKI, S., Asst. Prof. Reactor Eng. Div. II, Res. Lab. of Nucl. Reactors, Tokyo Inst. Technol.
Address: Research Laboratory of Nuclear Reactors, Tokyo Institute of Technology, 1 Ohokayama, Keguro-ku, Tokyo, Japan.

AOKI, Tohru, M.S. Asst., Lab. of Radiochem., Inst. for Chem. Res., Kyoto Univ.
Address: Kyoto University, Institute of Chemical Research, Yoshida-Honamachi, Sakyo-ku, Kyoto, Japan.

AOUST, J. D. D'. See D'AOUST, J. D.

APARICI, Pedro CARDA. See CARDA APARICI, Pedro.

APOLLONOV, G. I. Member, Dept. of Internat. Relations and Sci.-Tech. Information, U.S.S.R. State Com. for the Utilisation of Atomic Energy.
Address: U.S.S.R. State Committee for the Utilisation of Atomic Energy, 26 Staromonetnii Pereulok, Moscow, U.S.S.R.

APPAPILLAI, Veluppillai, B.Sc., Ph.D., F.Inst.P. (London). Born 1913. Educ.: Univ. Coll., Colombo, and Manchester Univ. Reader in Phys., Ceylon Univ. Member of Council, Ceylon Assoc. for the Advancement of Sci.; Standards Advisory Council, Ceylon; Member, A.E.C., Ceylon. Societies: Ceylon Assoc. for the Advancement of Science; Inst. of Phys. and Phys. Soc., London.
Nuclear interests: Nuclear physics; Instrumentation.
Address: Department of Physics. Ceylon University, Peradeniya, Ceylon.

APPEL, Helmut, Dipl.-Ing., Dr. rer. nat. Born 1929. Educ.: Munich T.H. and Erlangen Univ. Wissenschaftlicher Asst. (1955-60), Kustos (1960-64), Wissenschaftlicher Rat (1964-), Karlsruhe T.H.
Nuclear interest: Nuclear physics.
Address: Institut für Experimentelle Kernphysik, Technische Hochschule Karlsruhe, Postfach Kernreaktor, Germany.

APPELT, Dietmar, Dr. Res. Fellow, Nucl. Phys. and Electronic Res. Dept., Oporto Univ.
Address: Oporto University, Nuclear Physics and Electronic Research Department, Oporto, Portugal.

APPLEBY, E. Chairman, Nucl. R. and D. Dept., Appleby and Ireland Ltd.
Address: Appleby and Ireland Ltd., Basingstoke, Hampshire, England.

APPLEBY, E. G. Formerly Works Director, now Managing Director, Nucl. R. and D. Dept., Appleby and Ireland Ltd.
Address: Appleby and Ireland Ltd., Basingstoke, Hampshire, England.

APPLEBY, E. M. H. Managing Director. Nucl. R. and D. Dept., Appleby and Ireland Ltd.
Address: Appleby and Ireland Ltd., Basingstoke, Hampshire, England.

APPLEYARD, Raymond Kenelm, M.A., Ph.D. Born 1922. Educ.: Cambridge Univ. Asst. Res. Officer, A.E.C.L., 1953-56; Sec. Sci. Com. on the Effects of Atomic Rad., U.N., New York, 1956-61; Director, Biol. Div., European Atomic Energy Community (Euratom), Brussels, 1961-. Societies: Soc. Gen. Microbiol.; American Genetical Soc.
Nuclear interest: Biological effects of ionising radiation, especially at the cellular and genetic levels.
Address: European Atomic Energy Community, 51-53 rue Belliard, Brussels, Belgium.

APRAIZ BARREIRO, Ramon. Director of Res. and Nucl. Eng., Sociedad Espanole de Construction Naval. Vôcale, Comision Asesoras de Centrales Nucl., Junta de Energia Nucl.
Address: 9-2 Alameda Mararredo, Bilbao, Spain.

APRIETO, Pacifico N., B.A. Born 1928. Educ,: Columbia and Philippines Univs. Sci. (Chief, Classification and Information Branch), Philippine A.E.C.; On leave of absence as Public Information Officer, I.A.E.A., Vienna. Societies: Radioisotope Soc. of the Philippines; Philippine Assoc. for the Advancement of Sci.
Nuclear interest: Popularisation of nuclear science.
Address: 14/6/4 Weinberggasse, 1190 Vienna, Austria.

APRO, Antal, Pres., Nat. A.E.C., Hungary. First Deputy of the Prime Minister.
Address: National Atomic Energy Commission, 1-3 Kossuth Lajos-ter, Budapest 5, Hungary.

APTEL, Jean, Ing. E.S.E. Born 1925. Educ.: Paris Univ. Chief Eng., Société Parisienne pour l'Industrie Electrique, -1962; Asst. Gen. Manager, Société d'Etudes pour le Developpement de l'Automatisme and Société Contrôle Industriel Regulation Matériels Automatiques, 1962-66. Director of Res. and Development, S. A. Heurtey, 1964-. Societies: Soc. Francaise des Electriciens; Assoc. Francaise de Regulation et Automatisme.
Nuclear interests: Magnetohydrodynamics; Heat exchangers; Control.
Address: 154 rue de Vaugirard, Paris 15, France.

AQUIJE, Carlos HERNANDEZ. See HERNANDEZ AQUIJE, Carlos.

ARA, Antonino Oscar PEREZ. See PEREZ ARA, Antonino Oscar.

ARAGONA, Massimo CASILLI D'. See CASILLI D'ARAGONA, Massimo.

ARAGONES PUIG, Damian. Director, Forum Atomico Espanol.
Address: Forum Atomico Espanol, 38 General Goded, Madrid 4, Spain.

ARAGOU, Maurice. Ing. en Chef Génie Maritime. Born 1906. Educ.: Ecole Polytechnique and Ecole Génie Maritime. Director, Sté. des Forges et Ateliers du Creusot; Pres., Sté. Framatome; Pres., Cie. pour l'Etude et la Realisation de Combustibles Atomiques.; Gerant, S.E.N.-T.A. Society: A.T.E.N.
Nuclear interests: Industrial architecture; Reactor design; Management; Fuel.
Address: Societé des Forges et Ateliers du Creusot, 15 rue Pasquier, Paris 8, France.

ARAKAWA, Masashi. Chief Eng., Film Badge Service Dept., Japan Safety Appliances Assoc.
Address: Japan Safety Appliances Association, c/o Kyoiku Building, 3-1 Sanken-cho, Bunkyo-ku, Tokyo, Japan.

ARAKAWA, Noboru. Manager, Radioactive Div., Chiyoda Safety Appliances Co.
Address: Chiyoda Safety Appliances Co., Radioactive Division, 5-2 Chome Yaesu, Chuo-ku, Tokyo, Japan.

ARAKI, Gentaro, D.Sc. Born 1902. Educ.: Tokyo Univ. In Dept. of Nucl. Eng., Emeritus Prof. (1966-), Kyoto Univ. Books: Meson Theory (Akitaya, 1948); Quantum Mechanics (Baihukan, 1961); Atomic Physics (Baihukan, 1963). Society: Phys. Soc. of Japan.
Nuclear interest: Theoretical nuclear physics.
Address: Department of Nuclear Engineering, Kyoto University, Yosida, Kyoto, Japan.

ARAKI, Huzihiro, Ph.D. (Princeton), D.Sc. (Kyoto). Born 1932. Educ.: Kyoto, Princeton Univs. Asst. (1957-61), Lecturer (1961-64), Asst. Prof. (1964-65), Prof. (1966-), Kyoto Univ. Res. Assoc. (1960-62), Asst. Prof. (1962-64), Illinois Univ.; Lecturer, Eidgenossische T.H. Zürich, 1961-62.
Nuclear interest: Theoretical physics, especially axxiomatic approach in quantum field theory.
Address: c/o Research Institute for Mathematical Sciences, Kyoto University, Kyoto, Japan.

ARAMBURU, Enrique SENDAGORTA. See SENDAGORTA ARAMBURU, Enrique.

ARANA, Miss Julita, Chem., Radioisotope Lab., Philippine Gen. Hospital.
Address: Philippine General Hospital, Manila, Philippines.

ARANGIO-RUIZ, Gaetano, Dr. in Law. Born 1919. Educ.: Naples Univ. Prof. Internat. Law, Faculty of Law, Padua Univ. Consultant and Director of Internat. Legal Affairs and Legislative Studies, C.N.E.N. Societies: Soc. Italiana per l'organizzazione internazionale, Rome; American Soc. of Internat. Law; Internat. Law Assoc.
Nuclear interests: International and national legal problems connected with nuclear energy.
Address: 51 Corso Trieste, Rome, Italy.

ARAUJO, José M. R. M., B.Sc. (Porto), Ph.D. (Theoretical Phys., Manchester). Born 1928. Educ.: Porto Univ. Lecturer, Phys. Dept. (1958-60), Prof. Phys. (1960-), Director, Nucl. Phys. and Electronics Res. Dept., Porto Univ. Co-editor, Nucl. Phys. (journal).
Nuclear interest: Nuclear physics.
Address: Laboratorio de Fisica, Universidade do Porto, Portugal.

ARAUJO PENNA, Monica de. See de ARAUJO PENNA, Monica.

ARAUNABENA, Pedro de ARIETA. See de ARIETA ARAUNABENA, Pedro.

ARBEL, Amram, B.Sc. (Mech. Eng.). Born 1935. Educ.: Technion, Israel Inst. Technol. Tech. Director, Res. Reactor, Soreq Res. Establishment; At present on loan to the Joint Sea Water Desalting Project. Societies: Inst. Nucl. Eng.; A.N.S.
Nuclear interest: Reactor operations.
Address: The Joint Sea Water Desalting Project, P.O. Box 20147, Tel Aviv, Israel.

ARBOLEDA, Carlos MONSALVE-. See MONSALVE-ARBOLEDA, Carlos.

ARBOLEDA de URIBE, Mrs. Esmeralda. Colombian Ambassador to Austria; Colombian Resident Rep., I.A.E.A.
Address: 25 Gregor Mendelstrasse, Vienna 18, Austria.

ARBON, Eric Raymond. Born 1914. Head, Eng. Development Lab., Bristol Aircraft Ltd., -1959; Eng. and Lab. Manager, Reactor Group Lab., U.K.A.E.A., Windscale, 1959-. Societies: M.I.Mech.E.; A.F.R.Ae.S.
Nuclear interests: Mechanical engineering services; Engineering and laboratory management; Remote handling techniques for irradiated materials and components.
Address: 5 Wastwater Rise, Seascale, Cumberland, England.

ARBUCKLE, W. A. Director, Rio Algom Mines Ltd.
Address: Rio Algom Mines Ltd., 335 Bay Street, Toronto, Ontario, Canada.

ARCANGELO, Tomas A. D'. See D'ARCANGELO, Tomas A.

ARCHANGELSKI, U. V., Dr. Dept. Chief, State Com. for Utilisation of Atomic Energy, U.S.S.R. Adviser, Third I.C.P.U.A.E., Geneva, Sept. 1964.
Address: U.S.S.R. State Committee for the Utilisation of Atomic Energy, 26 Staromonetnii Pereulok, Moscow, U.S.S.R.

ARCHER, Bryan J., Jr. Chief Geologist, Uranium Instruments Co.
Address: Uranium Instruments Co., P.O. Box 1191, Grand Junction, Colorado, U.S.A.

ARCHER, Leonard George. Born 1916. Chief Inspector, Atomic Weapons Res. Establishment, Aldermaston (-1960), Chief Inspector (Fuel Elements), Eng. Group, Springfields Works and Dounreay Exptl. Reactor Establishment, (1960-), U.K.A.E.A. Societies: A.M.I.Mech.E.; European Organisation for Quality Control.
Nuclear interests: Metrology of particle accelerators; Automatic and programmed machines.
Address: 12 Ribby Avenue, Wrea Green, Preston, Lancs., England.

ARCHER, Maurice Clifford, A.B., M.D., D.A.-B.R., M.A.C.R. Born 1905. Educ.: Ohio Wesleyan Univ. and Ohio State Medical School. Past Pres., Rad. Centre. Society: Soc. of Nucl. Medicine.
Nuclear interests: Radioactive phosphorus; Cobalt therapy.
Address: 800 Fifth Avenue, Fort Worth, Texas, U.S.A.

ARCY, J. D'. See D'ARCY, J.

ARDANUY, Joachim M. Editor, Atomic Energy Clearing House.
Nuclear interests: All phases; interest is in news.
Address: 401 Colorado Building, 1341 G St., N.W., Washington, D.C. 20005, U.S.A.

ARDELEANU, T., Ing. Sci. Director, Icechim, Inst. for Chem. Res.
Address: Icechim, Institute for Chemical Research, 202-204 Splaiul Independentei, Bucharest, Roumania.

ARDEN, Thomas Victor, D.Sc., Ph.D. (London). Born 1918. Educ.: London Univ. Head, Concentration Sect. (P.S.O.), Chem. Res. Lab. (now N.C.L.), 1947-52; Chief Development Chem., The Permutit Co., Ltd., 1952-60. Director, 1961-. Member, Advisory Com., Chem. Dept., Battersea Coll. of Technol.; Member, Chem. Panel, Nat. Council for Technol. Awards. Societies: F.R.I.C.; M.I.M.M.; Fellow, Chem. Soc. and S:C.I.
Nuclear interests: Recovery of uranium from ores, water and effluent treatment for atomic reactors.
Address: Hatherley, Icklingham Road, Cobham, Surrey, England.

ARDENNE, Manfred VON. See VON ARDENNE, Manfred.

ARDITTI, R., Prof. Dr. Director, Nucl. Dept., Soc. d'Exploitation des Matériels Hispano-Suiza. Address: Société d'Exploitation des Matériels Hispano-Suiza, rue du Capitaine Guynemer, Bois Colombes, Seine, France.

ARDITTI, Sol J., B.S., M.B.A. (Corp. Finance and Management). Born 1928. Educ.: New York Univ. Pres., R.A.I. Research Corp. Nuclear interests: Management in the chemical and nuclear research and development field. Address: R.A.I. Research Corporation., 36-40 37th Street, Long Island City 1, New York, U.S.A.

ARDLEY, George William, B.Sc. (Phys.), Ph.D. Born 1928. Educ.: Bristol and Birmingham Univs. G.E.C., U.S.A., 1952-54; English Electric, 1954-66; Nucl. Design and Construction, Ltd., 1966-. Nuclear interests: Fuel element design and development; Materials development for reactors; Heat transfer and fluid-flow problems in reactors. Address: 83 Carisbrooke Road, Leicester, England.

ARELLI, A., M.A., Assoc. Prof., Turku Univ. Address: Wihuri Physical Laboratory, University of Turku, Vesilinnantie 5, Turku, Finland.

ARELLANO SUAREZ, Carlos, Dr. Sec. for Sci. Procedures, Soc. Peruana de Radioligia. Address: Sociedad Peruana de Radiologia, Casilla No. 2306, Lima, Peru.

ARENA, Nicola, Dr. of Phys. Born 1939. Educ.: Catania Univ. Society: Soc. Italiana di Fisica. Nuclear interests: Nuclear physics in the field of low energy particles with particular interest in radiation detection and electronic equipment. Address: Catania University, Istituto di Fisica, 57 Corso Italia, I 95129 Catania, Italy.

ARENA, Romolo, Degree in Philosophy. Born 1920. Educ.: Milan Univ. Director, Istituto per la Ricostruzione Industriale. Member, Sect. of Economic questions in the nucl. field and Social and health affairs and information in the nucl. field, Economic and Social Com., European Communities Commission; Member, Board of Directors, Società Italiana Impianti. Address: c/o Istituto per la Ricostruzione Industriale, 89 Via Veneto, Rome, Italy.

ARENBERG, David Lewis. Born 1915. Educ.: Massachusetts and Clark Univs., and M.I.T. Phys. GS-11, N.R.L., 1948-50; Pres., Arenberg Ultrasonic Lab., Inc., 1950-. Member, Awards Com., Professional Group on Ultrasonic Eng., I.R.E. Societies: A.P.S.; I.R.E.; Acoustical Soc. America; Assoc. Member, Optical Soc. America. Nuclear interests: Storage devices for pulse height analysis (ultrasonic delay lines); Measurement of nuclear magnetic moments. Address: 94 Green Street, Jamaica Plain, Massachusetts 02130, U.S.A.

ARENDS, Tulio, M.D. Born 1918. Educ.: Univ. Central de Venezuela, Caracas. Chief, Haematology Sect. and Blood Bank, Hospital Universitario, Caracas, 1956-; Head, Exptl. Haematology Dept., I.V.I.C., Caracas, 1958-. Member, Editorial Commission, journal Acta Cientifica Venezolana, 1960-; Coordinator, Library, I.V.I.C., 1960-. Societies: Associación Venezolana para el Avance de la Ciencia (Venezuelan Assoc. for the Advance of Sci.); N.Y. Acad. Sci.; American Assoc. for Cancer Res.; Internat. Soc. of Blood Transfusion; Internat. Soc. of Haematology. Nuclear interests: Use of radioactive isotopes (especially ^{59}Fe) in biochemical genetic traits. The effect of irradiation on blood cells. Address: Departamento Hematologia Experimental, I.V.I.C., Apartado 1827, Caracas, Venezuela.

ARENDT, Paul Richard, Ph.D. (Phys.) (Berlin). Born 1900. Educ.: Jena and Berlin Univs. A.E.G., 1935-57; U.S. Army Electronics Command, Inst. for Exploratory Res., Fort Monmouth, New Jersey, 1957-; Director, Satellite and Space Communications Systems, Inc. Book: Reactor Technology (Physik Verlag, 1957). Societies: Deutsche Physikalische Gesellschaft; A.N.S.; A.G.U.; Nat. Com. 3, Internat. Sci. Radio Union. Nuclear interests: Research on radiation damage; Outer space radiation effects on space vehicles. Address: 60 Reynolds Drive, Eatontown, New Jersey 07724, U.S.A.

AREZZO, Bartira. Chem., Dept. of Res., Radioisotope Lab., Ministerio da Saude D.N.S.-Servico Nacional de Cancer. Address: Ministerio da Saude D.N.S.-Servico Nacional de Cancer, 23 Praca de Cruz Vermelha, Rio de Janeiro, Brazil.

ARGON, Rauf. Manager of Eng., Nuclide Corp. Address: Nuclide Corporation, Alloyd-General Vacuum Division, 81 Hicks Avenue, Medford, Massachusetts 02155, U.S.A.

ARGYLE, D. P. Manager, Radiochem., United States Testing Co., Inc. Address: United States Testing Co., Inc., 2800 George Washington Way, Richland, Washington 99352, U.S.A.

ARGYLE, H. J. Director, Security Div., Idaho Operations Office, U.S.A.E.C. Address: United States Atomic Energy Commission, Idaho Operations Office, P.O. Box 2108, Idaho Falls, Idaho 83401, U.S.A.

ARHANGEL'SKII, A. A. Paper: Co-author, Sensibility of the Scintillation Method in Gamma-Ray Defectoscopy (letter to the Editor, Atomnaya Energiya, vol. 19, No. 3, 1965). Address: Academy of Sciences of the U.S.S.R., 14 Leninsky Prospekt, Moscow V-71, U.S.S.R.

ARIAS, Ramon SIMON. See SIMON ARIAS, Ramon.

ARIAS, Rogelio Ernesto, M.D. Born 1920. Educ.: George Washington Univ. Asst. Chief, Obstetrics and Gynecology Dept., Gorgas Hospital, Ancon, Panama Canal Zone, 1953-; Clinical Prof. Obstetrics and Gynecology, Panama Univ., 1957-. Societies: Fellow, American Coll. of Surgeons; Fellow, American Coll. of Obstetricians and Gynecologists; Soc. Panamena de Obstetricia and Ginecologia; Associacion Medica Nacional de Panama. Nuclear interests: Use of radioactive isotopes in medicine, especially in the treatment of pelvic malignancies; Irradiation therapy for cancer.
Address: Box "O", Balboa Heights, Panama Canal Zone.

ARICAN, Kemal, Dr. Vice-Pres., Turkish Cancer and Radiobiol. Soc.
Address: Turkish Cancer and Radiobiological Society, 6 Tura Apt., 1 Bilge Sokak, Ankara-Bakanliklar, Turkey.

ARIETA ARAUNABENA, Pedro de. See de ARIETA ARAUNABENA, Pedro.

ARIFOV, O. Deleg., I.A.E.A. Conference on the Use of Radioisotopes in the Phys. Sci. and Industry, Copenhagen, Sept. 6-17, 1960.
Address: Academy of Sciences of the U.S.S.R., 14 Leninsky Prospekt, Moscow V-71, U.S.S.R.

ARIFOV, Ubai Arifovich, D. of Phys. and Maths., Prof., Academician. Born 1909. Educ.: Uzbek State Univ. in Samarkand. Director, Physico-Tech. Inst., Director, Nucl. Phys., Inst., Head, Electronic Lab., Chairman, joint Learned Council, Dept. of Phys. and Maths., Chairman, Learned Council, Nucl. Phys. Inst., Member, Learned Council of the Physico-Tech. Inst. and Member of the Presidium, Uzbek Acad. of Sci.; Member, Learned Council, Dept. of Phys. and Maths., Central Asian State Univ.; Pres. Uzbek Acad. of Sci., 1962. Papers: 40 papers on electronics and the utilisation of isotopes and radioactivity in science and engineering.
Address: Academy of Sciences of the Uzbek Republic, Kibrari, Near Tashkent, U.S.S.R.

ARIMA, Toshihiko. Dept. Director, Industrial Instrument R. and D. Dept., Yokogawa Electric Works Ltd.
Address: Yokogawa Electric Works Ltd., 3000 Kichijoji, Musashino-shi, Tokyo, Japan.

ARISAKOV. Eng., Nucl. Power Station. Inst. of Phys. and Power, Acad. Sci. of the U.S.S.R.
Address: Institute of Physics and Power, Obninsk, near Maloyaroslavets, Moscow, U.S.S.R.

ARISAWA, Hiromi, D. of Economics. Born 1896. Educ.: Tokyo Univ. Hon. Prof., Tokyo Univ. Commissioner, (1956-), and Acting Chairman, A.E.C. of Japan. Publications: Cartel-Trust and Combination; Control of the Industry in Japan; Elements of Statistics; Lectures on Atomic Energy "Economics".
Address: 2-109 Mabashi, Suginami-ku, Tokyo, Japan.

ARISPE, Enrique. Fellow, Peruvian Soc. of Radiol.
Address: Sociedad Peruana de Radiologia, Casilla No. 2306, Lima, Peru.

ARISTARHOV, N. N. Paper: Co-author, BR-5 Reactor Operating Experience in 1959-66 (Atomnaya Energiya, vol. 23, No. 6, 1967).
Address: Academy of Sciences of the U.S.S.R., 14 Leninsky Prospekt, Moscow V-71, U.S.S.R.

ARKADYEV, Georgi. U.S.S.R. Resident Rep., I.A.E.A.
Address: 4 Wohllebengasse, Vienna 4, Austria.

ARKHAROVA, I. I. Paper: Co-author, A Study of Uranium Sorption by Minerals of Wall Rocks (Atomnaya Energiya, vol. 16, No. 1, 1964).
Address: U.S.S.R. Academy of Sciences, 14 Leninsky Prospekt, Moscow V-71, U.S.S.R.

ARLEY, Niels H., Dr. phil. Born 1911. Educ.: Copenhagen Univ. Formerly Res. Asst. Phys., Bristol Univ., Asst. Prof. Phys., Princeton Univ.; Assoc. Prof. Physics, Niels Bohr Inst., Copenhagen Univ., 1954-; Sci., Dept. of Biophys., Norsk Hydro's Inst. for Cancer Res., 1957-. Sec., First Scandinavian Symposium on Carcinogenesis. Books: Theory of Stochastic Processes and their Application to the Theory of Cosmic Rad. (Copenhagen, Gad, 1943; New York, Wiley, 1949); Lectures on Radiobiol. at the Univs. of Copenhagen, Oslo and Uppsala (Norw. Cancer Soc. Pub., 1954); co-author, Atomkraft, Eine Einführung in die Probleme der Atomzeit (Scand. Univ. Books, 1959; Springer-Verlag, 1960).
Nuclear interest: Radiobiology, especially problems of carcinogenesis.
Address: Department of Biophysics, Norsk Hydro's Institute for Cancer Research, Montebello, Norway.

ARLMAN, Jan J., Ir. Born 1919. Educ.: Delft Tech. Univ. Radioisotopes Sales Manager, N.V. Philips-Duphar.
Nuclear interests: Radioisotopes production and application.
Address: N.V. Philips-Duphar, Cyclotron and Isotope Laboratories, Petten, Netherlands.

ARMBRUSTER, Hubert F., D. in Law. Born 1911. Educ.: Berlin, Heidelberg, Paris and Freiburg i. Br. Univs. Prof. Public Law, Mainz Univ., 1946-. Member 2 expert committees, German Ministry for Sci. Res., Director, Inst. of Atomic Energy Law, Mainz Univ. Managing Director, Kernkraftwerk Betriebsgellschaft m.b.H., 1966-. Member, Steering Com., Deutsches Atomforum e.V. Societies: Deutsche Gesellschaft für Auswärtige Politik; International Law Assoc.
Address: 69 An der Allee, 6500 Mainz, Germany.

ARMBRUSTER, Raymond, D. ès Sc. Born 1926. Educ.: Strasbourg Univ. Prof., Faculté des Sci., Strasbourg Univ.
Nuclear interest: Nuclear physics.
Address: Centre de Recherches Nucléaires, Département de Physique Nucléaire, Strasbourg-Cronenbourg, France.

ARMENGAUD, D. Diego GALVEZ. See GALVEZ ARMENGAUD, D. Diego.

ARMENGAUD, Francisco GALVEZ. See GALVEZ ARMENGAUD, Francisco.

ARMENTEROS, Rafael, B.Sc. (Hons.) (London), Ph.D. (Paris). Born 1922. Educ.: London Univ. Maitre de Recherches, C.N.R.S. (France); Sen. Phys., C.E.R.N. (Geneva). Society: A.P.S.
Nuclear interests: Nuclear physics; Elementary particles and high-energy nuclear interactions.
Address: T. C. Division, C.E.R.N., Geneva, Switzerland.

ARMINIO-MONFORTE, Giovanni d'. See d'ARMINIO-MONFORTE, Giovanni.

ARMISTEAD, William Houston, Ph.D. (Chem.). Born 1916. Educ.: Vanderbilt Univ. Vice Pres. and Director, Tech. Staffs, Corning Glass Works. Societies: Fellow, American Ceramic Soc.; A.C.S.
Nuclear interest: Materials research.
Address: Corning Glass Works, Sullivan Park, Corning, New York 14830, U.S.A.

ARMS, Henry Shull, B.A., B.S., Ph.D. Born 1912. Educ.: Idaho and Oxford (Rhodes Scholar) Univs. With English Elec. Co., Ltd., Rugby, 1950-53; Formerly Chief Eng., Atomic Power Projects of English Elec., Babcock and Wilcox, Ltd.; Chief Eng., Taylor Woodrow Construction Ltd., Whetstone, 1956. Societies: M.I.M.E.; F.Inst.P.
Address: 7 Newbold Road, Rugby, England.

ARMSTRONG, Arthur Henry, M.A. Born 1921. Educ.: Cambridge Univ. At M.O.S., 1949-55; U.K.A.E.A., 1955-. Societies: British Computer Soc.; Fellow, Inst. of Maths. and its Applications.
Nuclear interests: Mathematical theory of neutronics with particular interest in numerical methods; Computers.
Address: "Everglades", Brimpton Common, Reading, Berks., England.

ARMSTRONG, Colin, Ph.D., B.Sc. (Eng.), A.C.G.I. Born 1933. Educ.: London and Durham Univs. In R. and D. Dept., Clarke, Chapman and Co., Ltd., 1958-63; Sen. Lect., Rutherford Coll., Newcastle, 1963-. Societies: A.M.I. Chem. E.; Assoc. Member, A.I.Ch.E.
Nuclear interests: All aspects of heat removal systems for both commercial and research reactors.
Address: 12 Greystoke Gardens, Whickham, Newcastle-on-Tyne, England.

ARMSTRONG, Frank W. Member, Nucl. Components Com., Soc. for Nondestructive Testing, Inc.
Address: Society for Nondestructive Testing, Inc., Nuclear Components Committee, 914 Chicago Avenue, Evanston, Illinois 60202, U.S.A.

ARMSTRONG, Frederick E. Born 1920. Educ.: Oklahoma State and California Univs., O.R.I.N.S. Project Leader, Radiotracer and Nucl. Res., U.S. Dept. of the Interior, Bureau of Mines, Bartlesville, Oklahoma. Societies: A.I.M.M.E.; A.N.S.
Nuclear interests: Radiotracer application and instrumentation in petroleum engineering.
Address: P.O. Box 1398, Bartlesville, Okla., U.S.A.

ARMSTRONG, George Robert. Born 1920. Ministry of Works, 1950-56; H.M. Treasury, 1956-57; U.K.A.E.A., 1957-.
Address: United Kingdom Atomic Energy Authority, Risley, Nr. Warrington, Lancs., England.

ARMSTRONG, Gerald Rogers, Master's Degree (Nucl. Eng.), B.S. (Chem. Eng.). Born 1931. Educ.: Texas A. and M. and Southern Methodist Univs. Member, Advisory Com. to Nucl. Eng. Dept., Nevada Univ., 1966-67. Chairman, Southern Nevada Sect., A.N.S.
Nuclear interests: Nuclear rocket reactor development and testing; Assembly and remote disassembly; Technical and administrative management.
Address: 4120 West Bonanza, Las Vegas, Nevada 89107, U.S.A.

ARMSTRONG, Councillor, Hon. J. I. Member, Council, Nucl. Res. Foundation, Sydney Univ.
Address: Nuclear Research Foundation, Sydney University, Sydney, New South Wales, Australia.

ARMSTRONG, James Calvin, B.S., Ph.D. Born 1931. Educ.: Duke and Pittsburgh Univs. Res. Assoc. (1960-61), Asst. Prof. (1961-64), Visiting Lecturer (1964-), Maryland Univ. Societies: A.P.S.; A.A.A.S.; A.A.P.T.
Nuclear interests: Nuclear physics; Nuclear structure; Nuclear reaction mechanisms; Experimental techniques; Accelerators.
Address: Office of the Postmaster General, U.S. Post Office Department, Washington, D.C., U.S.A.

ARMSTRONG, Robert D, B.Comm. (Queen's), C.A. Educ.: Queen's Univ. Formerly Pres., Canadian Foundation Co. Ltd.; Formerly Vice-Pres., Finance, Canadian Nat. Railways; Formerly Pres., Chrysler Leasing Corp.; Pres., Chief Exec. Officer, and Director, Rio Algom Mines Ltd., 1966-.
Address: Rio Algom Mines Ltd., 335 Bay Street, Toronto 1, Canada.

ARMSTRONG, Wallace D., B.A., M.S., Ph.D., M.D. Born 1905. Educ.: Minnesota Univ., Minneapolis. Prof. and Head, Dept. of Biochem., Coll. of Medical Sci., Minnesota Univ., 1946-. Member, Advisory Com. on Medical Uses of Isotopes, U.S.A.E.C. Societies: American Soc. of Biol. Chems.; American Physiological Soc.; Biochem. Soc.
Nuclear interests: Radiofluoride; Studies on metabolism of calcified tissues through use of radioisotopes.
Address: Department of Biochemistry, 227 Millard Hall, University of Minnesota, Minneapolis, Minnesota 55455, U.S.A.

ARNDT, E. D. Pres., B.D. Bohna and Co. Inc., Nucl. Design.
Address: B.D. Bohna and Co. Inc., 515 Market Street, San Francisco 5, California, U.S.A.

ARNELL, Tage, M.Sc.Eng. Member, Board of Directors, AKK Atomic Power Group; Member, Board of Directors, Oskarshamnsverkets Kraftgruppe A.B.
Address: AKK Atomic Power Group, 19 Stureplan, Stockholm C, Sweden.

ARNESEN, Thomas, B.A. Born 1920. Educ.: Oslo Univ. Director, Nucl. Dept., Sarnia Inspection Co. Ltd. Societies: American Petroleum Inst.; N.A.C.E.
Nuclear interests: Isotopes; Metallurgy; Management.
Address: Sarnia Inspection Company Limited, 190 Hymus Road, Scarborough, Ontario, Canada.

ARNETT, William S. Formerly Asst. Gen. Manager, Plasti-Line Inc.; Pres., American Nucl. Corp., 1965-.
Address: American Nuclear Corp., Box 246, Oak Ridge, Tennessee, U.S.A.

ARNOLD, Donald S., B.Ch.E., M.Sc., Ph.D. (Chem. Eng.). Born 1920. Educ.: Ohio State Univ. Asst. Prof. Chem. Eng., North Carolina State Coll., 1948-53; Supervisor of Process Development, Chem. Dept. (1953-55), Head, Chem. Dept. (1955-59), Nat. Lead Co. of Ohio; Head, High Energy Fuels Sect. (1959), Manager, Res. (1959-), American Potash and Chem. Corp. Societies: A.I.Ch.E.; A.S.E.E.; Nucl. Eng. Div., A.I.Ch.E.; A.C.S.; A.I.M.M.E.; A.A.-A.S.
Nuclear interest: Production of nuclear fuels and materials.
Address: American Potash and Chemical Corporation, Trona, California, U.S.A.

ARNOLD, Eldon Drewes, B.S. (Ch.E.), M.S. (Ch.E.), Regius Prof. Eng. (Tennessee Univ.). Born 1927. Educ.: West Virginia and Tennessee Univs. and O.R.S.O.R.T., Design Chem. Eng., Long Range Planning Group, Chem. Technol. Div., O.R.N.L. Book: Co-author, Chapters on Neutron Sources, Prompt Fission Sources, Shipping Containers for Irradiated Fuel Elements, and Shipping Containers for Fission Product Waste in I.A.E.A. Eng. Compendium on Rad. Shielding (to be published by Springer-Verlag Publishing Co.). Societies: A.N.S.; American Soc. of Chem. Eng.; A.C.S.; N.S.P.E.
Nuclear interests: Engineering evaluations and mathematical determinations of the effects of reactor irradiation on production of transmutation isotopes; Effects of these isotopes on fuel handling processes, facility design, and operations; Mathematical determination and engineering evaluation of shielding requirements for facilities and isotopic power generators; Applied computer programming in fields of nuclear chemical engineering.
Address: 110 Everest Circle, Oak Ridge, Tennessee, U.S.A.

ARNOLD, H. Phys., Inst. for Nucl. Phys. Res., Amsterdam.
Address: Institute for Nuclear Physics Research, 18 Oosterringdijk, Amsterdam 0, Netherlands.

ARNOLD, Captain J. D. Asst. Chief for Production and Quality Control, Bureau of Naval Weapons, Dept. of the Navy.
Address: Bureau of Naval Weapons, Department of the Navy, Washington 25, D.C., U.S.A.

ARNOLD, James R., Ph.D. E.O. Lawrence Memorial Award, 1968. Born 1923. Educ.: Princeton Univ. Chem. Dept., Princeton Univ., 1955-58; Prof. Chem., California Univ., San Diego, 1958-; Consultant, N.A.S.A.; Consultant, Los Alamos Sci. Lab. Societies: N.A.S.; A.C.S.; A.A.A.S.
Nuclear interests: Cosmochemistry; radiation effects in meteorites.
Address: Department of Chemistry, California University, San Diego, La Jolla, California, U.S.A.

ARNOLD, P. M. Vice-Pres., R. and D., Phillips Petroleum Co.; Member, Board of Directors, Atomic Ind. Forum, 1966-69.
Address: Phillips Petroleum Co., Bartlesville, Oklahoma, U.S.A.

ARNOLD, Robert O. Education, State of Georgia Nucl. Advisory Commission.
Address: State of Georgia Nuclear Advisory Commission, Office of Publications, Georgia Institute of Technology, Atlanta 13, Georgia, U.S.A.

ARNOLD, Walter Paige. Born 1909. Asst. Gen. Manager, Randfontein Estates, 1948-53; Gen. Manager, Freddies Consolidated (Orange Free State), 1953-57; Managing Director, Algom Uranium Mines Ltd. and Northspan Uranium Mines Ltd., 1957-60; Gen. Manager (1961), Gen. Manager and Vice-Pres. (Mining Div.) (1964-), Rio Algom Mines Ltd., Member of exec., Chairman, Rock Burst Com., and Chairman, Uranium Com., Ontario Mining Assoc.; Formerly Director, Canadian Nucl. Assoc.
Address: 59 Russell Hill Road, Toronto, Ontario, Canada.

ARNOLD, William Howard, A.B., A.M., Ph.D. Born 1931. Educ.: Cornell and Princeton Univs. Instructor, Princeton Univ., 1955; Supv., Reactor Phys. Group, Westinghouse Atomic Power Dept., 1955-59; Manager, Reactor Phys. Design Sect., Westinghouse, 1959-61; Director, Nucl. Fuel Management, Nucl. Utilities Services, Inc., 1961-. Deputy Manager, Eng. and Development, Manager, Test Systems and Operations, now Programme Manager, Westinghouse Astronuclear Lab., 1962-. Societies: Fellow, A.N.S.; A.P.S.; Assoc. Fellow, A.I.A.A.
Nuclear interests: Elementary particle physics; Reactor physics; Nuclear core design; Management of power reactor projects.
Address: 433 Maple Avenue, Pittsburgh, Pennsylvania 15218, U.S.A.

ARNOL'DOV, M. N. Papers: Co-author, Indication of Oxide and Hydride Content in Melted Sodium by Measuring Elec. Resistivity (letter to the Editor, Atomnaya Energiya, vol. 21, No. 6, 1966); co-author, On Behaviour of Impurities in Na-K Heat-Transfer Medium (letter to the Editor, ibid, vol. 23, No. 1, 1967); co-author, Quantitative Determination of Hydrogen Impurity in Liquid Sodium by Method of Hydride Thermal Dissociation (ibid, vol. 24, No. 3, 1968); co-author, Conductivity Indicator of Admixtures in Liquid Metal Coolants (letter to the Editor, ibid, vol. 24, No. 4, 1968).
Address: Academy of Sciences of the U.S.S.R., 14 Leninsky Prospekt, Moscow V-71, U.S.S.R.

ARNOUX, H. Ing., Bureau Veritas.
Address: Bureau Veritas, 31 rue Henri Rochefort, Paris 17, France.

ARNS, Robert G., B.S. (Canisius Coll.), M.S. (Michigan), Ph.D. (Michigan). Born 1933. Educ.: Canisius Coll. and Michigan Univ. Res. Assoc. (1959-60), Instructor (1960), Michigan Univ.; Asst. Prof. Phys. (1960-63), Assoc. Prof. Phys. (1963-64), Buffalo Univ., Assoc. Prof. Phys., Ohio State Univ., 1964-. Society: A.P.S.
Nuclear interests: Nuclear physics, especially nuclear structure and reactions; Nuclear resonance fluorescence; Mössbauer effect; Nuclear reactor physics.
Address: Department of Physics, Ohio State University, Columbus 10, Ohio, U.S.A.

ARNTH-JENSEN, Niels. Member, Danish A.E.C.
Address: Danish Atomic Enery Commission, 29 Strandgade, Copenhagen K, Denmark.

ARON, P. M. Papers: Separation of Tracer Amounts of Lanthanoides from Plutonium using CaF_2 (letter to the Editor, Atomnaya Energiya, vol. 5, No. 2, 1958); co-author, Delayed Neutron Emission Probability from Halogen (letter to the Editor, ibid, vol. 16, No. 4, 1964); co-author, Radiochem. Determination of $^{27}Al\ (n,a)\ ^{24}Na$ Reaction Cross Section with 14.6 Mev Neutrons (letter to the Editor, ibid).
Address: Academy of Sciences of the U.S.S.R., 14 Leninsky Prospekt, Moscow V-71, U.S.S.R.

ARONIN, Lewis R., B.S. (Phys.). Born 1919. Educ.: M.I.T. Staff Member, M.I.T. Metal. Project and its successor Nucl. Metals, Inc., 1949-57; Group Leader, Phys. Metal. (1957-58), Manager, Metal. R. and D. Dept. (1958-66), Nucl. Metals, Inc., Concord, Mass.; Consultant, Tech. and Economic Evaluations, Ledgemont Lab., Kennecott Copper Corp., Lexington, Mass., 1966-67; Materials Eng., U.S. Army Materials and Mechanics Res. Centre, Watertown, Mass., 1967-. Book: Contributor, Nucl. Reactor Fuel Elements, Metal. and Fabrication (A.R. Kaufmann, Editor, Intersci. Publishers, 1961; U.S.-A.E.C. sponsorship). Societies: A.I.M.M.E.; A.S.M.
Nuclear interests: Metallurgy of nuclear reactor materials, including studies of uranium-base alloys, aluminium-uranium alloys, aluminium-lithium alloys, fabrication of clad nuclear fuel elements, physical metallurgy and fabrication of beryllium and high-temperature materials.
Address: 20 Ingleside Road, Lexington, Massachusetts 02173, U.S.A.

ARONSON, Raphael F. B.Phys. (Minnesota), M.A. (Harvard), Ph.D. (Phys., Harvard). Born 1928. Educ.: Minnesota and Harvard Univs. A.E.C. Predoctoral Fellow, 1949-51. With Nucl. Development Corp. of America, 1951-55; Head, Nucl. Dept. and Sen. Sci., TRG, Inc., 1955-63.; Sen. Res. Sci., (1963-), Assoc. Prof., Nucl. Eng. (1966-), School of Eng. and Sci., New York Univ. Societies: A.P.S.; A.N.S.; American Rocket Soc.
Nuclear interests: Nuclear physics; Reactor shielding; Reactor physics; Transport theory.
Address: 4 Hemsley Lane, Great Neck, New York, U.S.A.

ARONSON, Seymour, B.S., Ph.D. (Chem.), M.S. (Maths.). Educ.: Yeshiva Coll., Polytech. Inst. Brooklyn and Pittsburgh Univ. Res. Chem., Bettis Atomic Power Lab., Pittsburgh, 1955-61; Assoc. Phys. Chem., Brookhaven Nat. Lab., Upton, N.Y., 1961-. Societies: A.C.S.; American Ceramic Soc.; A.A.A.S.
Nuclear interests: Physical chemistry of nuclear materials; Thermodynamic and electrical properties of actinide compounds; Electrical and thermal properties of graphite; Diffusion in solids; Gas-solid kinetics.
Address: Brookhaven National Laboratory, Upton, Long Island, New York, U.S.A.

AROSEMENA, Jorge GARCIA. See GARCIA AROSEMENA, Jorge.

ARREGHINI, Antonio, Dr. Ing. Managing Director and Gen. Manager, Franco Tosi S.p.A.
Address: Franco Tosi S.p.A., Corso Italia 27, Legnano, Milan, Italy.

ARRHENIUS, Gustaf Olof, D.Sc. Born 1922. Educ.: Stockholm Univ. Prof., Marine Geology (1960-), and Asst. Director, Inst. for Pure and Appl. Phys. Studies, California Univ., La Jolla.

Nuclear interests: Metallurgy; Solid-state chemistry; Radiation physics.
Address: Scripps Institution of Oceanography, California University, La Jolla, California 92037, U.S.A.

ARRIAGA, Kaulza de. See de ARRIAGA, Kaulza.

ARRIETA SANCHEZ, Luis, Dr. Sec. Editor, Soc. Radiologica Panamena.
Address: Sociedad Radiologica Panamena, Box No. 6323, Panama, Republic of Panama.

ARRISON, Robert Alexander, Jr., S.B. and S.M. (Elec. Eng.). Born 1925. Educ.: M.I.T. Project Eng., Medical Products, and Project Eng., Ind. X-Ray Products, X-ray Dept., Gen. Elec. Co., Milwaukee, 1949-56; Vice-Pres., Eng., Keleket X-ray Corp., Mass., 1956-62; Vice-Pres., Eng., Veeco Inc., 1962-64; Vice-Pres. and Gen. Manager, Picker Inst. operation, Picker Corp., 1965-. Societies: I.E.E.E.; American Vacuum Soc.
Nuclear interests: Gamma imaging devices; X-ray diffraction; Nuclear instruments; Diagnostic and therapeutic x-ray equipment, including use of Cobalt-60 and other isotopes; Mass spectrometers for leak detection and residual gas analysis; Instrumentation and systems for high and ultra high vacuum.
Address: 32660 Creekside Drive, Pepper Pike, Ohio 44124, U.S.A.

ARROE, Hack, Dr. Phil. Born 1920. Educ.: Copenhagen Univ. At Wisconsin Univ., 1947-50, 1952-55; Consultant and Principal Spectroscopist, Project Matterhorn, Princeton Univ., 1956-58; Prof. Phys., Montana State Univ., 1955-58; Head, Phys. and Maths. Div., Denver Res. Inst., Denver Univ., 1958-61; Prof. Phys., Montana State Univ., 1961-63; Chairman, Prof. Phys., New York State Univ., Fredonia, 1963-.
Book: Studier over spektralliniers struktur (Copenhagen, Nordisk Bogtrykkeri, 1951). Societies: A.I.P.; A.P.S.; Math. Assoc. of America.
Nuclear interests: Uranium fission; Thermonuclear energy; Nuclear structure.
Address: State University of New York, College at Fedonia, Fredonia, New York, U.S.A.

ARROL, William John, B.Sc., Ph.D. Born 1915. Educ.: London Univ. Director of Group Res., Joseph Lucas, Ltd., Birmingham. Societies: Chem. Soc.; Soc. of Chem. Industry.
Nuclear interest: Nuclear chemistry.
Address: 240 Station Road, Knowle, Warwickshire, England.

ARROWSMITH, William Rankin, B.S., M.D. Born 1910. Educ.: Muskingum Coll., New Concord, and Ohio State Univ. Head, Dept. of Hematology (1947-), Head, Dept. of Internal Medicine (1951-), Ochsner Clinic, New Orleans; Prof., Clinical Medicine, School of Medicine, Tulane Univ.; Consultant (1950-62), Member, Board of Directors (1962-65), O.R.I.N.S.; Diplomate, American Board of Internal Medicine; Member, Staff, Ochsner Foundation Hospital; Consultant, Dept. of Medicine, Touro Infirmary; Sen. Visiting Physician, Charity Hospital, Louisiana; Polycythemia Vera Study Group, N.I.H. Societies: Fellow, Internat. Soc. of Hematology; Fellow, American Coll. of Physicians; A.M.A.; Southern Medical Assoc.; Southeastern Clinical Club; A.A.A.S.; American Federation for Clinical Res.; American Soc. of Hematology; Emeritus Member, Central Soc. for Clinical Res.; Orleans Parish Medical Soc.; Louisiana State Medical Soc.; New Orleans Graduate Medical Assembly; New Orleans Acad. of Internat. Medicine.
Nuclear interest: Medical therapeutic uses.
Address: Ochsner Clinic, 1514 Jefferson Highway, New Orleans, Louisiana 70121, U.S.A.

ARROYAVE-BORJES, Guillermo, Chemist-Pharmacist (S. Carlos, Guatemala); Ph.D. (Biochem., Rochester). Born 1922. Educ.: S. Carlos Univ., M.I.T., and Rochester Univ. Lecturer in Bacteriology (1949-50), Lecturer in Nutritional Biochem. (1953-55), S. Carlos Univ; Chief, Physiological Chem. Div., Inst. of Nutrition of Central America and Panama, 1955-. Member, Nucl. Energy Commission. Societies: Colegio de Farmacéuticos y Quimicos de Guatemala; Academia de Ciencias Médicas, Fisicas y Naturales de Guatemala; N.Y. Acad. Sci.; American Inst. of Nutrition.
Nuclear interest: Application of radioisotope techniques to biochemical research.
Address: Institute of Nutrition of Central America and Panama, Guatemala, Central America.

ARRUDA, Paulo Ribeiro de. See de ARRUDA, Paulo Ribeiro.

ARRUDA TRINDADE, Mrs. Helena. Res. Asst., Radioisotope Lab., Inst. de Biofisica, Brazil Univ.
Address: Brazil University, 458 Avenida Pasteur, Rio de Janeiro, Gb., Brazil.

ARSENIN, V. V. Papers: Co-author, On Possibility of Suppression of Inhomogeneous Plasma Drift Instability by Feedback (Atomnaya Energiya, vol. 24, No. 4, 1968); co-author, On Possibility of Plasma Current Column Stabilisation by Feedbacks (letter to the Editor, ibid, vol. 25, No. 2, 1968).
Address: Academy of Sciences of the U.S.S.R., 14 Leninsky Prospekt, Moscow V-71, U.S.S.R.

ARSENYEVA, Militsa. Rad. Genetics Lab., U.S.S.R. Acad. of Sci., Biological Phys. Inst. Adviser, Third I.C.P.U.A.E., Geneva, Sept. 1964.
Address: Radiation Genetics Laboratory, Biological Physical Institute, U.S.S.R. Academy of Sciences, Moscow, U.S.S.R.

ARTAMKIN, V. N. Papers: Co-author, Burnup in Plane Layer (letter to the Editor, Atomnaya Energiya, vol. 20, No. 1, 1966); co-author, Laminar Burnable Poison Usage (ibid, vol. 22,

No. 3, 1967); co-author, Optimal Reactor Shutdown for Short-Term Work (letter to the Editor, ibid, vol. 23, No. 2, 1967).
Address: Academy of Sciences of the U.S.S.R., 14 Leninsky Prospekt, Moscow V-71, U.S.S.R.

ARTAMONOV, V. V. Paper: Co-author, Phase Diagrams of UC_{14}-UO_2, UC_{14}-$PbCl_2$, UC_{14}-$MgCl_2$ Systems (Atomnaya Energiya, vol. 22, No. 6, 1967).
Address: Academy of Sciences of the U.S.S.R., 14 Leninsky Prospekt, Moscow V-71, U.S.S.R.

ARTAMONOV, Y. Y. Paper: Co-author, Preparation and Studies of some Plutonium Monocarbide Properties (Atomnaya Energiya, vol. 22, No. 6, 1967).
Address: Academy of Sciences of the U.S.S.R., 14 Leninsky Prospekt, Moscow V-71, U.S.S.R.

ARTANDI, Charles, Ph.D. Director of Res. Ethicon Inc.
Nuclear interests: Electron beam and cobalt sterilisation.
Address: Ethicon Inc., Somerville, New Jersey, U.S.A.

ARTAUD, Jean Georges, Ing. E.N.S.I.C., L. ès Sci. Born 1921. Educ.: Nancy Univ. Ing. Chargé de Mission, Département de Chimie, C.E.A., 1946-. Book: Analyse de l'Uranium par Spectroscopie d'Emission in Traité de Chimie Minérale (Prof. Paul Pascal, 1959). Societes: Sté. Chimique de France; Groupement pour l'Advancement des Méthodes Spectrographiques; Soc. for Appl. Spectroscopy.
Nuclear interests: Analytical chemistry, mainly instrumental and emission spectroscopy.
Address: Centre d'Etudes Nucléaires de Grenoble, B.P. No. 269, Grenoble, Isère, France.

ARTEAGA, Ricardo. Director, Geological Div., Junta de Energia Nucl., Madrid.
Address: Junta de Energia Nuclear, Ciudad Universitaria, Madrid 3, Spain.

ARTEM'EVA, N. A. Paper: Co-author, On Using of Different Approximations of Spheric Harmonic Method for Calculation of Neutron-Penetration through Shielding (Abstract, Atomnaya Energiya, vol. 19, No. 6, 1965).
Address: Academy of Sciences of the U.S.S.R., 14 Leninsky Prospekt, Moscow V-71, U.S.S.R.

ARTEMOV, A. J. Responsible Sec., Atomnaya Energiya.
Address: Atomnaya Energiya, 18 Kirov Ulitsa, Moscow, U.S.S.R.

ARTERBURN, Jesse O., B.Sc. (Phys.). Born 1919. Educ.: Georgia Inst. of Technol. Manager, Southwest Exptl. Fast Oxide Reactor Facility, 1966. Society: A.N.S.
Nuclear interest: Management and technical direction in research and development of advanced nuclear reactor projects, equipment and components.
Address: General Electric Company, Fayetteville, Arkansas, U.S.A.

ARTHOLD, W., Dir.Dkfm.Dr.jur. Managing Director, Dampfkraftwerk Korneuburg G.m.b.H.
Address: Dampfkraftwerk Korneuburg G.m.b.H., 6A Am Hof, Vienna 1, Austria.

ARTHUR, B. Wayne, B.S., M.S., Ph.D. Born 1925. Educ.: Auburn and Wisconsin Univs. Asst. Entomologist (1957-60), Assoc. Entomologist (1960-63), Auburn Univ.; Director, CIBA Corp. Agricultural Chems. Testing Labs. (1963-66), Director R. and D., CIBA Agrochemical Co. (1966-67), Vero Beach, Fla.; Pres., CIBA Agrochemical Co., Summit, N.J., 1967-. Toxicology Study Sect., U.S. P.H.S., N.I.H. Societies: Entomological Soc. America; A.C.S.
Nuclear interests: Absorption, distribution and metabolism of radiolabelled organophosphate, carbamate, and chlorinated hydrocarbon insecticides in insects and mammals.
Address: 556 Morris Avenue, Summit, N.J. 07901, U.S.A.

ARTHUR, George, B.Sc., Ph.D. Born 1928. Educ.: Roy. Tech. Coll., Glasgow. A.E.R.E., Harwell, 1952-55; J.G. Stein, Stirlingshire, 1955-58; C.A. Parsons, Newcastle, 1958-.
Nuclear interests: Ceramic fuels, UO_2, UC, UO_2 cermets; Preparation and properties; Fuel element development; Ceramic control materials.
Address: International Research and Development Co. Ltd., Fossway, Newcastle upon Tyne, 6, England.

ARTHUR, Herbert, B.Sc., M.Sc. Born 1914. Educ.: Sheffield Univ. Deputy Borough Education Officer, Tottenham, 1950-52; Sec. and Clerk to the Governors, Polytech., Regent Street, London, 1952-56; Div. Staff Training Officer, Northumberland and Cumberland Div., N.C.B., 1956-59; Group Training Officer, Production and Eng. Groups, U.K.A.E.A., 1959-. Societies: A.Inst. Phys.; Assoc. Member, I.E.R.E.
Nuclear interests: Broad field of training in nuclear industry, including reactor technology and management.
Address: White Gables, 32 Alexandra Road, Stockton Heath, Warrington, Lancs., England.

ARTHUR, Wallace, B. Elec. Eng., B. Eng. Sci., Ph.D. Born 1932. Educ.: New York Univ. Res. Asst., New York Univ., 1957-61; Lecturer in Phys., Rutgers Univ., 1961-62; Assoc. Prof. Phys., Chairman Phys. Dept., Fairleigh Dickinson Univ., 1962-. Pres., Fairleigh Dickinson Univ Chapter A.A.U.P. Societies: A.A.U.P.; A.A.A.S.; A.P.S.; A.A.P.T.; A.G.U.
Nuclear interests: Nuclear physics; Neutron physics; Cosmic radiation.
Address: 15-48 B Plaza Road, Fair Lawn, J.N., U.S.A.

ARTIGAS SANZ, D. Jose Antonio de. See DE ARTIGAS SANZ, D. Jose Antonio.

ARTSDALEN, Ervin R. VAN. See VAN ARTSDALEN, Ervin R.

ARTSIMOVICH, Lev Andreevich, Prof. Lenin Prize, 1958. Born 1909. Educ.: Minsk Univ. Worker, Inst. of Tech. Phys., Acad. of Sci. of the U.S.S.R., 1938-48. Head (1959), Academician Sec. (1963), Dept. of Phys. and Maths., U.S.S.R. Acad. of Sci., Director, Thermonucl. Lab., Kurchatov Inst. Atomic Energy. Deleg. to Conference sur la Physique des Plasmas et la Recherche concernant la Fusion Nucléaire Controlée, Salzbourg, Sept. 1961. Society: Member, Acad. of Sci. of the U.S.S.R.
Nuclear interests: Nuclear physics; Separation of isotopes.
Address: Kurchatov Institute of Atomic Energy, 46 Ulitsa Kurchatova, Post Box 3402, Moscow, U.S.S.R.

ARTUNKAL, Suphi, M.D. Born 1907. Educ.: Istanbul Univ. Lecturer (1944), Prof. (1958), Therapeutic Clinic and Pharmacology Dept. Member, Turkish A.E.C. Societies: Turkish Acad. Medicine; Turkish Soc. Medicine; New York Acad. Sci.; American Thyroid Assoc.; Panel Member, Ciba Foundation, London. Book: Use of Radioisotopes in Clinical Medicine (1962).
Nuclear interest: Application in clinical medicine mainly in thyroid disorders.
Address: Tedavi Klinigi, Haseki Hastanesi, Istanbul, Turkey.

ARYKOVA, Maya. Kazakh Pathology Inst., Dept. of Biol. and Medical Sci., Acad. Sci. of the Kazakh S.S.R.
Address: Kazakh Pathology Institute, 8 Marta 27, Alma-Ata, Kazakh S.S.R., U.S.S.R.

ASADA, Chohei, Univ. graduate. Born 1887. Educ.: Kyoto Imperial Univ. Sen. Official Adviser, Kobe Steel, Ltd.; Chairman, Shinko Trading Co., Ltd.; Director, Federation of Economic Organisations; Director, Kansai Economic Organisation; Chairman, Kobe Chamber of Commerce and Industry. Society: Japan Inst. of Metals (Councillor).
Nuclear interest: Material supply to atomic field.
Address: 5-1, Aza Nishimatsumoto, Nishihirano Mikage-cho, Higashinada-ku, Kobe, Japan.

ASADA, Tsunesaburo, Dr. Sci., Prof. Emeritus, Osaka Univ. Born 1900. Educ.: Tokyo Univ. Prof., Phys., Osaka Univ., 1930-64; Central Res. Lab., Kobe Steel Ltd., 1964. Societies: Soc. Atomic Energy of Japan (Vice-Pres.); American Soc. of Phys.; A.N.S.
Nuclear interest: Research in metals for nuclear reactor.
Address: Fundamental Research Laboratory, Kobe Steel Limited, Maruyama Gomo Nada-ku, Kobe, Japan.

ASAI, Takuo, M.D. Born 1918. Educ.: Osaka Univ. Chairman, Dept. of Medicine and Hygiene, Rad. Centre of Osaka Prefecture. Societies: H.P.S.; Atomic Energy Soc. of Japan; Nippon Societas Radiologica.
Nuclear interests: Medical radiology and radiation hygiene.
Address: Radiation Centre of Osaka Prefecture, Shinke-Cho, Sakai, Osaka, Japan.

ASANBAEV, Yu. A. Paper: Co-author, Start-Stop Simulation Problems of Nucl. Reactors (letter to the Editor, Atomnaya Energiya, vol. 22, No. 6, 1967).
Address: Academy of Sciences of the U.S.S.R., 14 Leninsky Prospekt, Moscow V-71, U.S.S.R.

ASANO, Junichi. Ship Surveyor, Ship Classification Soc. of Japan.
Address: Ship Classification Soc. of Japan, 1 Akasaka-Fukuyoshi-cho, Minato-ku, Tokyo, Japan.

ASANTI, Paavo, Prof. Born 1916. Educ.: Inst. of Technol. Director, Lab. of Metal., State Inst. for Tech. Res.
Address: State Institute for Technical Research, Laboratory of Metallurgy, Otaniemi, Finland.

ASANUMA, Fukumatsu. Safeguard Sect. Chief, Ship Res. Inst., Ministry of Transportation, Japan.
Address: Ministry of Transportation, Ship Research Institute, 1-1057 Mejiro-cho, Toshima-ku, Tokyo, Japan.

ASATIANI, P. Ya. Paper: Co-author, Investigation of Pressure Influence on Isotope Separation B^{11} and B^{10} Process by Distillation of BF_3 (Atomnaya Energiya, vol. 23, No. 4, 1967).
Address: Academy of Sciences of the U.S.S.R., 14 Leninsky Prospekt, Moscow V-71, U.S.S.R.

ASBURY, Mrs. Carole. Sec., Nucl. Reactor Project, Washington State Univ.
Address: Washington State University, Pullman, Washington, U.S.A.

ASCARELLI, Tullio. Member, Com. Scientifico, Diritto ed Economia Nucleare.
Address: Diritto ed Economia Nucleare, 68 Viale Bruno Buozzi, Rome, Italy..

ASCARI, Aldo, Dr. in Mathematics. Born 1924. Educ.: Milan Univ. Researcher, C.I.S.E., Milan, 1953-56; Chief, Appl. Mathematics Group, SORIN Nucl. Res. Centre, Saluggia, 1957-.
Nuclear interests: Reactor physics, mainly dynamics. Mathematical treatment of reactor physics and technology problems, with emphasis on numerical analysis.
Address: Sorin, Saluggia, Vercelli, Italy.

ASCHENBRENNER, Frank A., B.S. (Hons., Phys.), Ph.D. (Phys.). Born 1924. Educ.: Kansas State Coll. and M.I.T. Teaching Fellow (1950-52), Res. Asst. (1952-54), M.I.T. Sen. Eng., Nucl. Analysis (1954-56), Supervisor, Shielding (1956-59), Manager, Phys. and Maths. (1959-60), Manager, Shield Design (1960-61), A.N.P.D., G.E.C.; Director, Power and

Environments Systems (1961-63), Asst. Div. Director, Aerospace Sci. (1963-65), Asst. Div. Director, Res. Eng. and Test (1965-), Space Div., North American Rockwell Corp.; Captain, U.S. Naval Reserve, Commander Naval Air Reserve Staff 77(L). Vice Chairman, Orange County Sect., A.I.A.A. Societies: A.P.S.; A.N.S.; A.I.A.A.; Naval Reserve Assoc.
Nuclear interests: Nuclear physics and all applications of nuclear energy, including directing systems management activities for design of nuclear systems in space vehicles - e.g., nuclear rockets, space nuclear power systems, nuclear shielding.
Address: 716 Rancho Circle, Fullerton, California, U.S.A.

ASCHNER, Fritz Simon, Dipl. -Ing. (Mech.), Dipl. -Ing. (Elec.), M.Sc., Dr. -Ing. Born 1903. Educ.: Munich, Zürich, and Breslau T.H. Assoc. Prof., Technion, Israël Inst. of Technol., Haifa; Head, Mech. Planning and Design Dept., Israel Elec. Corp., Ltd., Head Office, Haifa, -1966. Societies: M.I.Mech.E. (London); Fellow, I.E.E. (London); Member Assoc. of Eng. and Architects in Israel.
Nuclear interests: Power reactor design; Design and economics of nuclear power stations; Dual-purpose plants for power generation and sea water desalination.
Address: 26 Shoshanat Hacarmel Street, Haifa, Israel.

ASCHOFF, Albrecht, Dr. Rechtsanwalt; Vorsitzender, Wirtschaftsausschusses, Deutschen Bundestages, 1963-65. Member, Specialised Nucl. Sect. for Economic Problems., Economic and Social Com., Euratom, 1966-.
Address: 58-64 Huyssenallee, 43 Essen, Germany.

ASCOLI, Prof. Prof. of Theoretical Phys., Palermo Univ.; Member, Sicilian Nucl. Res. Com.
Address: Palermo University, Via Maqueda, Palermo, Italy.

ASCOLI, Aurelio, Dr. of Elec. Eng., Libero Docente in Struttura della Materia. Born 1929. Educ.: Pavia Univ. and Milan Polytech. Lecturer, Exptl. Phys. (1963-65), Lecturer, Phys. of Condensed States (1965-67), Pavia Univ.; Lecturer, Phys., Milan Univ., 1965-. Book: Effetti Fisici delle Radiazioni sui Materiali (Istituto di Fisica del Politecnico di Milan, 1959). Society: Soc. Lombarda di Fisica.
Nuclear interests: Solid state physics, mainly: metal physics, lattice defects, diffusion, radiation damage, crystal growth and related problems.
Address: Centro Informazioni Studi Esperienze, Casella Postale 3986, Milan, Italy.

ASCOLI - BARTOLI, Ugo, Laurea in Fysica, Perfezionamento in Fysica Nucleare, Libera Docenza. Born 1923. Educ.: Rome Univ. Consultant, C.N.E.N., -1965; Chief, Thermonucl. Reaction Div., Euratom, Frascati, 1965-; Lecturer, Rome and Aquila Univs.
Nuclear interests: Plasma physics. Lasers in connection with controlled thermonuclear reactions.
Address: Laboratori Gas Ionizzati, Associazione Euratom-C.N.E.N., C.P. 65, Frascati Rome, Italy.

ASDENTE, Maria Pierpaoli, Prof., Dr. Member Editorial Board, Energia Nucleare,
Address: Energia Nucleare, C.I.S.E., Casella Postale 3986, Milan, Italy.

ASENJO, Jose Angel CERROLAZA. See CERROLAZA ASENJO, Jose Angel.

ASFIA, Safi, Eng. Born 1916. Educ.: Paris Polytech. and Paris Univ. Managing Director, Plan Organization; Also responsible for Atomic energy matters in the country.
Nuclear interests: General; Management; Nuclear materials; Economics.
Address: Plan Organization, Teheran, Iran.

ASH, Clifford L., M.D. First Vice Pres., American Radium Soc., 1968-69.
Address: Ontario Cancer Institute, 500 Sherbourne Street, Toronto 5, Ontario, Canada.

ASH, Milton Samuel, B.S.(E.E.), M.S., Ph.D. (Phys.). Born 1921. Educ.: California Univ., Berkeley and I.I.T. Sen. Sci., E.H. Plesset Associates; Israeli A.E.C. Fellowship 1964-65; Lecturer, U.C.L.A., Eng., Phys. Dept., 1958-. Exec. Consul, A.N.S., Los Angeles. Books: Nucl. Reactor Kinetics (McGraw-Hill, 1965); Optimal Shutdown Control in Nucl. Reactors (Academic Pres, 1966). Societies: Los Angeles Chapter, A.N.S.; American Assoc. for Atomic Sci.; A.P.S.
Nuclear interests: Reactor kinetics, control, especially fast reactors, fuel cycle optimal programming and economics.
Address: E.H. Plesset Associates, 2444 Wilshire Boulevard, Santa Monica, California 90403, U.S.A.

ASHAHARA, Yoshishige. Director, Japan Atomic Power Co. Ltd. Councillor, Fund for Peaceful Atomic Development of Japan.
Address: Japan Atomic Power Co. Ltd., Otemachi 1-chome, Chiyoda-ku, Tokyo, Japan.

ASHBY, George E., B.S. (Phys.). Born 1925. Educ.: Maryland Univ. Solid State Phys., U.S. Geological Survey, 1954-57, Res. Phys. (1957-64), Manager, Nucl. Res. (1964-67), Director, Inorganic and Nucl. Res. (1967-), W.R.Grace and Co. Society: A.N.S.
Nuclear interest: Nuclear fuels and reprocessing.
Address: Washington Research Centre, W.R. Grace and Company, Clarksville, Maryland 21029, U.S.A.

ASHFORD, Oliver Martin, B.Sc. (1st class hons. Natural Philosophy). Born 1915. Educ.: Glasgow Univ. Chief, Res. Section (1952-66), Chief, Planning Unit, (1966-), World Meteorological

Organisation. Societies: Fellow, Roy. Meteorological Soc.; A. Inst. P.
Nuclear interests: Uses of radioisotopes in meteorology.
Address: World Meteorological Organisation, Geneva, Switzerland.

ASHIHMIN, V. P. Paper: Co-author, Impurity State Effect on Deformation Texture of Low-Alloyed Alpha-Uranium (letter to the Editor, Atomnaya Energiya, vol. 24, No. 1, 1968).
Address: Academy of Sciences of the U.S.S.R., 14 Leninsky Prospekt, Moscow V-71, U.S.S.R.

ASHKINADZE, G. Sh. Paper: Co-author, Isotopic Composition of Xenon, Krypton and Argon Extracted from Oxide of Cm^{244}_{96} (Atomnaya Energiya, vol. 22, No. 6, 1967).
Address: Academy of Sciences of the U.S.S.R., 14 Leninsky Prospekt, Moscow V-71, U.S.S.R.

ASHLEY, Edward Lionel, B.Sc., A.M.I.C.E. Born 1911. Educ.: Tech. College with London External Degree. Chief Construction Eng., U.K.A.E.A., Risley.
Nuclear interests: Responsible for the supervision of construction work throughout the U.K. Atomic Energy Authority. This work embodies building experimental reactors, chemical plants and fuel element manufacturing facilities.
Address: Barnacre, Marlborough Crescent, Grappenhall, Nr. Warrington, Lancs., England.

ASHLEY, Ramon Lloyd, B.S. (Phys.). Born 1928. Educ.: Clarkson Coll. and Tufts Coll. Member, Exec. Com., Chairman of Programme Com., Div. Chairman, Shielding Div., and Member, Programme and Publications Com., A.N.S.
Nuclear interests: Shielding (reactors, isotope sources, and space), reactor safety and safeguards.
Address: Atomics International Division, P.O. Box 309, Canoga Park, California, U.S.A.

ASINGER, Friedrich, Prof., Dr.tech., Dipl. Ing., Dr.phil.habil. Mitglied, Arbeitskreis 11/2, Chemie, Federal Ministry for Sci. Res.
Address: Federal Ministry for Scientific Research, 46 Luisenstrasse, 532 Bad Godesberg, Germany.

ASPELUND, Olav, Master of Phys. Eng. Born 1930. Educ.: Norway Tech. Univ. Res. Asst., Theoretical Electrotechnique, Elec. Eng. Dept. (1955-56), Res. Asst., Phys. Eng. Dept. (1956-57), Tech. Univ. Norway; Res. Officer, Phys. Dept., Joint Establishment for Nucl. Energy Res., 1957-59; Res. Officer, Dept. of Reactor Development and Construction, A.B. Atomenergi, Studsvik, Sweden, 1959-. Societies: Norwegian Phys. Soc.; Swedish Phys. Soc.; Norwegian Soc. of Professional Eng.; Polytech. Soc. of Norway.
Nuclear interests: Reactor physics; Nuclear physics (especially fast-neutron physics).
Address: A.B. Atomenergi, Studsvik, Nyköping, Sweden.

ASPINALL, Wayne Norviel, A.B., LL.B. Born 1896. Educ.: Denver Univ. and Denver Law School. Member, Joint Com. on Atomic Energy. Member, Colorado State House of Representatives; Member, Colorado State Senate; elected to 81st and succeeding Congresses of the United States;
Address: Palisade, Colorado, U.S.A.

ASPREY, Larned B., B.S., Ph.D. Born 1919. Educ.: California (Berkeley) Univ. Staff member, Los Alamos Sci. Lab. Sec.-Treas., Div. of Inorganic Chem., A.C.S. Society: A.A.A.S.
Nuclear interests: Chemistry and metallurgy of actinide elements, particularly transplutonium elements.
Address: Box 1663, California University, Los Alamos Scientific Laboratory, Los Alamos, New Mexico, U.S.A.

ASTAF'EV, Z. P. Paper: Cop-author, Inversion Solution of Thermal Equations for Two-Component Nucl. Reactor (Atomnaya Energiya, vol. 24, No. 4, 1968).
Address: Academy of Sciences of the U.S.S.R., 14 Leninsky Prospekt, Moscow V-71, U.S.S.R.

ASTAHOV, V. V. Papers: Co-author, Gamma-Field of Dipped in Water Isotropic Co^{60} Source near Interface of Water and Air (letter to the Editor, Atomnaya Energiya, vol. 23, No. 1, 1967); co-author, Scattering of Collimated Co^{60} Cs^{137} and Au^{198} Gamma Rad. near Interface between Two Media (ibid, vol. 23, No. 2, 1967).
Address: Academy of Sciences of the U.S.S.R., 14 Leninsky Prospekt, Moscow V-71, U.S.S.R.

ASTIER, André Lucien, L. ès Sc., D. ès Sc. Born 1922. Educ.: Ecole Polytechnique, Paris Univ. Ingénieur des Ponts et Chaussées; Maître de Conférence à l'Ecole Polytechnique; Sous-Directeur, Lab. de Phys. Nucléaire, Collège de France. Societies: Membre Sté. Française de Physique; Soc. Italiana di Fisica.
Nuclear interest: High energy nuclear physics.
Address: Laboratoire de Physique Nucléaire, Collège de France, Paris 5, France.

ASTRAHAN, Dora Bencionovna, Candidate of Medical Sci. Born 1900. Educ.: 2nd Moscow Medical Inst. Medical practitioner, Radium Dept., Hertzen's Oncological Inst., 1925-27; Physician-radiologist, 1927-31; Sci. worker (1931-43), Sen. sci. worker (1943-58; more than 10 years work with radioactive isotopes), Hertzen's Oncological Inst. Member editorial staff of journal Voprosi Oncologii. Papers: 22 papers concerning radiotherapy of tumours. The principal papers deal with radiotherapy, X-ray therapy and the combined method of irradiation in treatment of skin tumours, the uterine cervix and the oral cavity. Society: Member, All-union Oncological Soc. (member Inspective Com.).
Address: 3 2nd Botkinski Proezd, Moscow D-284, U.S.S.R.

ASTRÖM, Björn, Ph.D. Born 1917. Educ.: Roy. Inst. Technol., Stockholm and Uppsala Univ. At Forsknings Inst. for Atomfysik, Stockholm. Book: Nucl. Spectroscopic Investigations by Means of Energy Selecting Electronic Devices (thesis, Uppsala, 1957).
Nuclear interests: Nuclear physics; Electronic instrumentation.
Address: 6 Bergavägen, Danderyd, Sweden.

ASTUNI, E., Prof. Member, Comitato Nazionale, Assoc. Nazionale di Ingegneria Nucl.
Address: Associazione Nazionale di Ingegneria Nucleare, 24 Piazza Sallustio, Rome, Italy.

ASTY, Jacques Robert Charles, Ing. diplômé de l'Ecole Polytechnique. Born 1917. Educ.: Ecole Polytechnique, Paris. Directeur (1952-), now Directeur Charge des Centres d'Etudes Nucléaires de Fontenay-aux-Roses et de Cadarache, C.E.A.
Address: 1 rue Cardinal Mercier, Paris 9, France.

ASUNDI, Rango Krishna, B.A. (Hons.), M.Sc., Ph.D. Born 1895. Educ.: Bombay and London Univs. Prof. Phys. and later Prof. Spectroscopy and Head, Dept. Spectroscopy, Banaras Hindu Univ., 1939-56; Phys. Bhabha Atomic Res. Centre, Bombay, India, 1957-. Chairman, Convention of Spectroscopists, India. Societies: Fellow, Nat. Inst. Sci., India; Fellow, Indian Acad. Sci.
Nuclear interests: Spectroscopy of atoms and molecules, of materials of nuclear interest, and of radiation damage.
Address: Pragati Flat 11, 60 Green Street, Santa Cruz West, Bombay 54, India.

ASWATHANARAYANA, Uppugunduri, D.Sc. Born 1928.. Educ.: Andhra Univ., India. Lecturer (1954-59), Reader (1959-66), Andhra Univ., India; Visiting Prof., Western Ontario Univ., 1966-67; Prof., Centre of Advanced Study in Geology, Saugar Univ., India, 1967-. Member, Exec. Council, Geological Soc. of India; Member, Exec. Council, Indian Geo-sci. Assoc.; Member, Editorial Board, Geochem. Soc. of India. Societies: A.G.U.; Indian Geophys. Union; Geological Soc. of America; Geochem. Soc., U.S.A.
Nuclear interests: Radiometric dating of rocks and minerals by Rb-Sr,K-Ar,Pb-U-Th fission-track techniques; Natural radiation environment; Radioactive elements in possible Upper Mantle materials, determined by neutron activation analysis techniques.
Address: Centre of Advanced Study in Geology, Saugar University, Sagar, M.P., India.

ATABAKI, H. E. Rahmat. Ambassador for Iran to Austria; Iran Resident Rep., I.A.E.A.
Address: 2 Wenzgasse, Vienna 13, Austria.

ATAMANOV, V. M. Paper: Co-author, Toroidal Chambers for Investigation of Plasma-RF Fields Interaction (letter to the Editor, Atomnaya Energiya, vol. 23, No. 1, 1967).
Address: Academy of Sciences of the U.S.S.R., 14 Leninsky Prospekt, Moscow V-71, U.S.S.R.

ATEN, Adriaan Hendrik Willem, Jr., D.Sc. (Utrecht). Born 1908. Educ.: Utrecht, Johns Hopkins and Columbia Univs. Head, Chem. Dept., Inst. for Nucl. Phys. Res., Amsterdam, 1947-; Extraordinary Prof. Radiochem., Municipal Univ., Amsterdam, 1949-. Societies: Koninklyke Nederlandse Chemische Vereniging; Nederlandse Natuurkundige Vereniging; Nederlandse Vereniging Voor Radiobiologie; A.C.S.; A.P.S.
Nuclear interests: New isotopes; Szilard Chalmers processes; Uses of tracers in the study of exchange reactions and reaction mechanisms; Chemistry of astatine; Calibration of radioisotopes; Radioisotope tolerances; Radiochemical analysis; Neutron cross-sections; Neutron dosimetry.
Address: Instituut voor Kernphysisch Onderzoek, 18 Oosterringdijk, Amsterdam, Netherlands.

ATHAVALE, Vishnu Traymbak, M.Sc., Ph.D. Born 1909. Educ.: Bombay Univ. Chem., Rare Minerals Sect., Geological Survey of India, 1948-51; i/c Analytical Sect., Chem. Div. (1951-57), Head, Analytical Div., (1957-), Bhabha Atomic Res. Centre. Societies: A.R.I.C., Great Britain and Ireland; Fellow, Indian Chem. Soc.
Nuclear interests: Analytical aspects of atomic energy materials such as ores, fuel elements, clads, moderators, coolants, fission products; Analytical control on processing plants and reactors using conventional and modern instrumental methods.
Address: Bhabha Atomic Research Centre, Modular Laboratories, Trombay, Bombay 74, India.

ATHERTON, Dennis, B.Sc. Born 1932. Educ.: Nottingham Univ. Lecturer, Kodak School of Ind. and Eng. Radiography, 1957-59; Tech. Rep., Inspection Equipment Ltd., 1959-61; Self-employed Consultant and Manufacturer of Radiographic Accessories, 1962-65; Managing Director, Intex (Radiographic) Ltd., 1965-. Society: Non-Destructive Testing Soc.
Nuclear interest: Non-destructive testing, particularly radiography.
Address: 81 Chaplin Road, Wembley, Middlesex, England.

ATKINS, Kenneth Robert, B.A., M.A., Ph.D. Born 1920. Educ.: Cambridge Univ. Assoc. Prof. Phys., Toronto Univ., 1951-54; Prof. Phys., Pennsylvania Univ. 1954-. Book: Liquid Helium (Cambridge Univ. Press, 1959). Societies: Phys. Soc. of London; A.P.S.
Nuclear interests: Isotopes of helium, ions in liquid helium.
Address: Department of Physics, Pennsylvania University, Philadelphia, Pennsylvania 19104, U.S.A.

ATKINS, Marvin Cleveland, B.S., M.S., Ph.D. Born 1931. Educ.: Texas A. and M., Illinois and Michigan Univs. Project Eng. and Group Leader (1st Lieut., U.S.A.F.), Materials Lab., Wright Air Development Centre, 1953-57; Group Leader (Captain and Major U.S.A.F.), Air Force Special Weapons Centre and Air Force Weapons Lab., 1960-64; Sen. Project Sci., Sect. Chief and Manager of Eng. Technol., Avco Corp., Space Systems Div., 1964-. Society: A.N.S.
Nuclear interests: Radiation sources and effects; Nuclear weapon effects; Research management; Re-entry systems; Composite materials.
Address: Avco Corporation, Lowell Industrial Park, Lowell, Massachusetts, U.S.A.

ATKINSON, H. R. Formerly Sen. Phys., now Principal Rad. Officer, Nat. Rad. Lab.
Address: National Radiation Laboratory, 108 Victoria Street, P. O. Box 1456, Christchurch, New Zealand.

ATMORE, Milton Graeme Montgomery, B.Sc. (Eng., Chem., with distinction), Ph.D. Born 1924. Educ.: Witwatersrand. Sen. Officer, Leaching Sect. Rand Refinery, seconded to Govt. Metal. Lab., 1951-52; Asst. Works Manager, Calcined Products (Pty.) Ltd., 1952-56; Project Metal., Rhodesian Anglo American, 1956-57; Cobalt Plant Supt., Cobalt Project Manager, Concentrator Supt., Rhokana Corp., Ltd., 1957-59; Res. Metal., Anglo American Corp., seconded to Atomic Energy Board, 1959-. Societies: F.R.I.C.; M.I.M.M.; M.S.A.-I.M.M.
Nuclear interests: Extraction and processing of uranium and other nuclear materials.
Address: 25 Wexford Avenue, Westcliff, Johannesburg, South Africa.

ATSUMI, T. Pres., Board of Directors, Kajima Construction Co.; Director, Tokyo Atomic Ind. Consortium.
Address: Kajima Construction Co., Ltd., 3 5-chome Yaesu, Chuo ku, Tokyo, Japan.

ATTA, Chester Murray VAN. See VAN ATTA, Chester Murray.

ATTIX, Frank Herbert, A.B., M.S. Born 1925. Educ.: California and Maryland Univs. N.B.S., Washington, D.C., 1950-57; Head, Dosimetry Branch, U.S.N.R.L., Washington, D.C., 1958-. Societies: A.P.S.; H.P.S.
Nuclear interest: Dosimetry of ionizing radiations.
Address: U.S. Naval Research Laboratory, Code 7620, Washington, D.C. 20390, U.S.A.

ATTWENGER, Wolfgang Johannes, Dipl. Ing. Born 1933. Educ.: Vienna T.H. Eng. for Nucl. Electronics, Österreichische Studiengesellschaft für Atomenergie, Vienna.
Nuclear interest: Nuclear electronics.
Address: Österreichische Studiengesellschaft für Atomenergie, 10 Lenaugasse, Vienna 8, Austria.

ATWOOD, J. L. Member, Board of Directors, Dynatom.
Address: Dynatom, 100 avenue Edouard Herriot, 92 le Plessis Robinson, France.

AUBEAU, Raymond, L. ès Sc. Born 1929. Educ.: Clermont-Ferrand Univ. Ing. de recherches, Service de Chimie Physique (1957-59), Service de Chimie des Solides, Sect. d'Etude de la Corrosion par Gaz et Métaux Liquides (1960-), CEN-Saclay, C.E.A.
Nuclear interests: Metallurgy; Physical chemistry; Thermodynamics.
Address: Commissariat à l'Energie Atomique, CEN-Saclay, Département de Métallurgie, B.P. No 2, Gif-sur-Yvette, S. et O., France.

AUBONE-QUIROGA, Marcelo, Aeronautic Eng. Born 1901. Educ.: Turin, Politecnic and Rome Univ. Board of Directors, C.N.E.A.
Nuclear interests: Power reactors; Management.
Address: 8250 Libertador Ave., Buenos Aires, Argentina.

AUBRETON, Marcel Yves LECLERCQ-. See LECLERCQ-AUBRETON, Marcel Yves.

AUCLAIR, Jean Michel. Born 1931. Educ.: Ecole Polytechnique, Paris. Eng., C.E.A., 1953-57; Chief Eng., Caratom, France, 1957-. Society: Sté. des Ing. Civils de France.
Nuclear interests: Nuclear reactor design; Nuclear instrumentation; Radioisotopes.
Address: 25 avenue de la Gare, Limours, Seine et Oise, France.

AUDERBRAND, H. Concerned with Nucl. Work, Sté. des Condenseurs Delas.
Address: Société des Condenseurs Delas, 38 avenue Kléber, Paris 16, France.

AUDERSET, P. At Stephan S.A.
Address: Stephan S.A., Fribourg, Switzerland.

AUDOUARD, Robert. Administrateur Directeur Gén., Les Ateliers de Provence, Sté. Filiale.
Address: Société des Chantiers et Ateliers de Provence, S.A., Les Ateliers de Provence, Société Filiale, 130 Chemin de la Madrague, Marseille 15, (B.D.R.), France.

AUDREN, Victor, Ing. Diplômé Ecole Polytechnique de Paris, M.S.(M.I.T.). Born 1918. Educ.: Ecole Polytechnique and M.I.T. Asst. Gen. Manager, Constructions Navales et Industrielles de la Mediterranée (C.N.I.M.) Formerly Asst. to Tech. Manager, Cie. des Messageries Maritimes, Paris. Assoc. Member, S.N.A.M.E.
Nuclear interest: Management of C.N.I.M.'s nuclear business in the technical and commercial fields.
Address: 50 Champs Elysees, Paris 8, France.

AUDSLEY, Arnold, B.Sc., Ph.D., F.R.I.C. Born 1917. Educ.: Leeds Univ. Principal Res. Chem., M.O.S., Dept. of Atomic Energy, R. and D. Branch, Springfields Works, Salwick, Preston, Lancs., 1952-56; P.S.O., Radiochem. Sect., Nat. Chem. Lab. (now Nat. Phys. Lab.), Teddington, Middlesex, 1956-67; P.S.O., Div. of Inorganic and Metal Structures, Nat. Phys. Lab., 1967-. Societies: F.R.I.C.; Fellow, Chem. Soc.; Fellow, Soc. Chem. Industry.
Nuclear interest: Extraction of metals from ores.
Address: 42 The Street, Little Bookham, Leatherhead, Surrey, England.

AUDY, J. Directeur, Région d'Equipement Nucléaire No. 1, Electricité de France. Expert, Third I.C.P.U.A.E., Geneva, Sept. 1964.
Address: Electricité de France, 2-4 avenue de la Libération, Boite Postale No. 47, Clamart, (Seine), France.

AUERBACH, Carl Richard Rudolf, Dr. phil. Born 1899. Educ.: Leipzig Univ. Sci. collaborator, Pintsch Bamag AG. Butzbach, Germany. Societies: Gesellschaft Deutscher Chemiker; Kolloid-Gesellschaft.
Nuclear interest: Nuclear chemistry.
Address: 12 Hauptstrasse, Kirch-Göns bei Butzbach (Hessen) Germany.

AUERBACH, Miss Charlotte, D.Sc., Ph.D., F.R.S. Book: Genetics in the Atomic Age (Oliver and Boyd Ltd.).
Address: Institute of Animal Genetics, West Mains Road, Edinburgh 9, Scotland.

AUERBACH, Jerzy Emilian, M.Sc. Born 1920. Educ.: Warsaw Tech. Inst. Tech. Director, Electronic Industry Trust, Warsaw; Deputy Commissioner of Government for Nucl. Energy.
Nuclear interests: Electronics; Apparatus of application of isotopes.
Address: Palac Kultury i Nauki, Warsaw, Poland.

AUERBACH, Stanley Irving, B.S., M.S., Ph.D. Born 1921. Educ.: Illinois and Northwestern Univs. At Roosevelt Univ., Chicago, 1950-54; at O.R.N.L., 1954-. Societies: H.P.S.; Ecological Soc. America; British Ecological Soc.; American Soc. Zoologists.
Nuclear interests: Health physics, problems of environmental contamination by radioactive wastes, radiation effects on populations, biological cycling of fission products.
Address: Health Physics Division, P.O. Box "X," Oak Ridge National Laboratory, Oak Ridge, Tennessee, U.S.A.

AUERBACH, Th. Member for Switzerland, E.N.E.A. Computer Programme Library Com.
Address: E.N.E.A. Computer Programme Library, Euratom Joint Research Centre, Ispra, (Varese), Italy.

AUFHAMMER, Gustav, Dr. techn. Wiss., o. Professor. Born 1899. Educ.: Munich T.H. Prof. Agronomy and Plant Breeding; Director, Inst. für Acker und Pflanzenbau Munich T.H. in Freising-Weihenstephan. Member Barley Com., European Brewery Convention, 1952; Dekan der Fakultät für Landwirtschaft in Weihenstephan, 1954-56; Rektor, Munich T.H., 1960-62.
Nuclear interests: Use of ionising radiation for the improvement of crop plants. Special interest in technics and genetic background of methods and results in mutation breeding work and stimulation effects after radiation.
Address: Institut für Acker- und Pflanzenbau, Freising-Weihenstephan, Germany.

AUGER, Pierre Victor, D. ès Sci. Born 1899. Educ.: Paris Univ. Prof., Paris Univ., 1937-; Director, Dept. of Natural Sci., Unesco, 1948-58; Special Consultant to U.N., 1959. Chairman, French Com. on Space Res.; Chairman, French Com. on Corpuscular Phys.; Director Gen., European Space Res. Organisation 1964-67. Societies: Sté. Francaise de Physique; Fellow, A.P.S.
Nuclear interests: Nuclear physics; Cosmic rays.
Address: 12 rue Emile Faguet, Paris, France.

AUGIER de CREMIERS, Roger Jacques, Ing., Ecole Centrale des Arts et Manufactures. Born 1921. Commissaire de la Marine. Directeur Adjoint, Centre de Marcoule, C.E.A., France.
Address: Villa "La Tonnelle", avenue Pasteur, Villeneuve-les Avignon, Gard, France.

AUGUSTINE, Robert J. Tech. Director, Appl. Health Phys., Inc., 1965-.
Address: Applied Health Physics Inc., 2986 Industrial Boulevard, P.O. Box 197, Bethel Park, Pa. 15102, U.S.A.

AUJALEU, Eugene. Member, Conseil d'Enseignement de l' I.N.S.T.N., C.E.A.
Address: Commissariat a l'Energie Atomique, 29-33 rue de la Federation, Paris 15, France.

AULT, John Douglas, Ph.D., D.I.C., A.R.I.C., B.Sc., A.R.C.S. Born 1932. Educ.: London Univ. A.W.R.E. Aldermaston, 1955-59; Deputy Gen. Manager, formerly Chief Chem., Saunders-Roe and Nucl. Enterprises Ltd.
Nuclear interests: Radioluminescence; Radiochemistry.
Address: Saunders-Roe and Nuclear Enterprises Ltd., North Hyde Road, Hayes, Middx., England.

AUNG, San Tha. Phys. Lecturer, Rangoon Univ.
Address: Rangoon University P.O., Rangoon, Burma.

AUNG, Tin, M.B., B.S., D.M.R.T. Born 1925. Educ.: Rangoon Univ. and Postgraduate Medical School, London. Radiotherapist, Head, Dept., Rangoon Gen. Hospital; Honorary Radiotherapist, Ramakrishna Mission Hospital, Rangoon, Burma.

Address: Radiotherapy Department, Rangoon General Hospital, Rangoon, Burma.

AUPETIT, M. Permanent Consultant, Sté. pour les Transports de l'Industrie Nucleaire.
Address: Societe pour les Transports de l'Industrie Nucleaire, 89 rue du Faubourg Saint-Honore, Paris 8, France.

AURELA, Asko Mikael, M.A., Phil. Lic. (Turku), Ph.D. (Durham). Born 1935. Educ.: Turku Univ. (Finland), and Durham Univs. Sen. Res. Phys., Wihuri Phys. Lab., 1966-; Docent of Phys., Turku Univ., 1967-. Society: Phys. Soc. Finland.
Nuclear interest: Cosmic rays.
Address: Wihuri Physical Laboratory, Turku University, 5 Vesilinnantie, Turku 2, Finland.

AURICOSTE. Jean. Gen. Manager. Cie. Européenne d'Automatisme Electronique.
Address: Cie. Européenne d'Automatisme Electronique, 27 rue de Marignan, Paris 8, France.

AUSLÄNDER, Joseph S., Dr. phil., D.Sc. Born 1911. Educ.: Berlin-Charlottenburg T.H., Berlin and Zurich Univs. Prof. Phys., Polytech. Inst., Bucharest.
Nuclear interests: Nuclear physics; Elementary particles.
Address: Laboratorul de Fizica, Inst. Politechnic Bucuresti, 123 Calea Grivitei, Bucharest, Roumania.

AUSLENDER, V. L. Papers: Co-author, Study of Self-Excitation and Rapid Damping of Coherent Transverse Oscillations in VEPP-2 Storage Ring (Atomnaya Energiya, vol. 22, No. 3, 1967); co-author, Lifetime and Intensity Measurements for Electron Beams in Storage Rings (ibid, vol. 22, No. 3, 1967); co-author, Lifetime and Beam Dimensions in Storage Rings as Function of Number of Particles (ibid, vol. 22, No. 3, 1967); co-author, Injection and Accumulation of Positron (ibid, vol. 22, No. 3, 1967).
Address: Novosibirsk Science Centre, 20 Sovietskaya Ulitsa, Novosibirsk, Siberia, U.S.S.R.

AUSSET, Lieutenant Commander Robert. Born 1923. Educ.: Ecole Navale. Marine Nationale, 1942-63; C.E.A. 1963-. Society: Assoc. des Ing. en Génie Atomique.
Nuclear interests: Safety in nuclear plants. Waste disposal in sea.
Address: 1 rue de l'Alma, Cherbourg, (Manche), France.

AUSTEL, Willi B.A., Dipl.-Ing., Dr.-Ing. Born 1932. Educ.: Aachen T.H. Societies: Verein Deutscher Eisenhüttenleute; Deutsche Gesellschaft für Metallkunde.
Nuclear interests: Properties of material; Testing of material; Metallurgy.

Address: Klöckner-Werke AG., Georgsmarienwerke, 1 Bessemer Strasse., 45 Osnabrück, Germany.

AUSTEN, Douglas Charles, B.A. (Hons. physiology and biochem.). Born 1909. Educ.: Toronto Univ. Asst. Supt., NRX Reactor (1948-50), Supt., Chem. Control Branch (1950-), Group Head of Universal Cells and Chem. Control Labs. (-1961), Group Head, Chem. Control Labs. and Heavy Water Reconcentration Plant (1961-), Group Head, Operations Labs., Chalk River Nucl. Labs., Ontario (1962-), A.E.C.L.
Nuclear interests: Reactor chemistry, particularly heavy water analyses and the automation of instrumental analyses.
Address: 41 Alder Crescent, Deep River, Ontario, Canada.

AUSTIN, Arthur Francis Gardiner, M.A. Born 1920. Educ.: Cambridge Univ. Director, Head Wrightson Teesdale Ltd.
Address: Head Wrightson Teesdale Limited, G.P.O. Box 10, Teesdale Iron Works, Thornaby-on-Tees, England.

AUSTIN, James B., C.h.E., Ph.D., D.Sc. (Hons., Lehigh). Born 1904. Educ.: Lehigh and Yale Univs. Director of Res., U.S. Steel Corp., 1946-54; Asst. Vice Pres., 1954-56; Vice Pres., Fundamental Res., 1956-57; Administrative Vice-Pres., Res. and Tech., 1957-. Director, Atomic Ind. Forum, 1957. Pres., A.S.M., 1954. Director, Metal Soc., A.I.M.M. Eng. Book: The Flow of Heat in Metals (A.S.M.). Societies: A.C.S.; American Ceramic Soc.; Iron and Steel Inst. (Hon. Vice Pres.); American Iron and Steel Inst.
Nuclear interest: In the field of the application of metals.
Address: 525 Wm. Penn Place, Pittsburgh 30, Pa., U.S.A.

AUSTIN, Sam M., B.S., M.S., Ph.D. Born 1933. Educ.: Wisconsin Univ. Res. Assoc., Wisconsin Univ., 1960; N.S.F. Postdoctoral Fellow, Oxford Univ., 1960-61; Asst. Prof., Stanford Univ., 1961-65; Assoc. Prof., Michigan State Univ., 1965-. Society: A.P.S.
Nuclear interests: Nuclear physics, especially the study of reaction mechanisms and the energy levels of light nuclei.
Address: Cyclotron Laboratory, Michigan State University, East Lansing, Michigan 48823, U.S.A.

AUSTIN, Walter Dawson, Fellow Inst. Chartered Accountants in England and Wales. Born 1906. Works Accountant, U.K.A.E.A., Production Group, Springfields Works, Salwick, near Preston, Lancs.
Address: 13 Cyprus Avenue, Fairhaven, Lytham St. Annes, Lancs., England.

AUSTONI, M. Member, Editorial Board, Atompraxis.
Address: c/o Atompraxis, Verlag G. Braun, 14-18 Karl-Friedrich Strasse, Karlsruhe, Germany.

AUZAC DE LA MARTINIE, Gerard d'. See d'AUZAC DE LA MARTINIE, Gerard.

AVAEV, V. N. Papers: Co-author, Reactor Neutron Flux Attenuation in Polyethylene (Atomnaya Energiya, vol. 15, No. 1, 1963); co-author, Spectra for Reactor Fast Neutrons Passed through Polyethylene (ibid); co-author, ⟨1.5 Mev Neutron Attenuation in Fe (letter to the Editor, ibid, vol. 16, No. 4, 1964).
Address: Academy of Sciences of the U.S.S.R., 14 Leninsky Prospekt, Moscow V-71, U.S.S.R.

AVAKUMOVIC, Vojislav Gregor, Prof. Dr. phil. Born 1910. Educ.: Berlin-Charlottenburg T.H. and Belgrade Univ. Prof., Belgrade Univ., 1949-52; Ordinarius, Sarajevo Univ., 1952-57; now Leiter der Arbeitsgruppe, Inst. für angewandte Mathematik, Kernforschungsanlage Jülich des Landes Nordrhein-Westfalen e.V.; Mitglied, Akademie der Wissenschaften in Belgrade.
Address: 116/118 Adalbertstr., Aachen, Germany.

AVAN, Louis, Dr. Sci. Prof. Lab. de Physique Nucléaire et Corpusculaire, Clermont Univ. Paper: Co-author, Course and Problems of Nucl. Phys. (Chapter II) (Industries Atomiques, vol. 5, No. 3-4, 1961).
Address: Laboratoire de Physique Nucléaire et Corpusculaire, Université de Clermont, 34 avenue Carnot, Clermont-Ferrand, France.

AVAN, Madeleine, Dr. Sci. Co-director, Lab. de Physique Nucléaire et Corpusculaire, Clermont Univ. Papers: Development of Varied Targets incorporated with Nucl. Emulsions (Industries Atomiques, vol. 4, No. 11-12, 1960); co-author, Course and Problems of Nucl. Physics (Chapter II) (ibid, vol. 5, No. 3-4, 1961).
Address: Laboratoire de Physique Nucléaire et Corpusculaire, Université de Clermont, 34 avenue Carnot, Clermont-Ferrand, France.

AVANZI, Silvana, Dr. Biol. Born 1922. Educ.: Pisa Univ. Cytologist, C.N.E.N., Rome, 1959-. Nucl. interests: Radiation and chemically induced chromosome changes in plants; Varietal differences in radiation resistance in plants.
Address: Laboratorio di applicazioni nucleari all'agricoltura, Centro Studi Nucleari della Casaccia, C.P. I-S. Maria di Galeria, Rome, Italy.

AVERBAH, B. A. Paper: Co-author, On Doppler Temperature Coefficient Evaluation for Homogeneous Reactors (letter to the Editor, Atomnaya Energiya, vol. 21, No. 5, 1966).
Address: Academy of Sciences of the U.S.S.R., 14 Leninsky Prospekt, Moscow V-71, U.S.S.R.

AVERDUNG, Dr. Inst. fur Kernphysik, Johannes Gutenberg Univ.
Address: Johannes Gutenberg University, Institut fur Kernphysik, 21 Saarstrasse, 65 Mainz, Germany.

AVERY, Burton Albert, B.A.Sc. (Mech. Eng.). Born 1922. Educ.: Toronto Univ. Sect. Leader, Design Eng., (1948), Chief Eng., (1959), Vice Pres. and Gen. Manager (1960), Orenda Engines Ltd.; Gen. Manager, Ind. Gas Turbine Div., Hawker Siddeley Canada Ltd, 1965; Director of Eng., Orenda Ltd., 1967. Member, Board of Directors, Canadian Nucl. Assoc. Societies: A.S.M.E.; Assoc. of Professional Eng. of Toronto; Soc. Automotive Eng.
Nuclear interests: Mechanical design of nuclear power plants and components; Economic power programmes; Research and development in nuclear power reactors; Compilation and dissemination of design and manufacturing standards for nuclear power plants.
Address: 11 Mountbatten Road, Weston, Ontario, Canada.

AVERY, Robert, B.S., M.S., Ph.D. Born 1921. Educ.: Illinois and Wisconsin Univs. Assoc. Phys. (1950-56), Sect. Head, Theoretical Phys. Sect. (1956-63), Director, Reactor Phys. Div. (1963-), Argonne Nat. Lab. Tech. Advisor, U.S. Deleg. 2nd I.C.P.U.A.E., Geneva, 1958, and 3rd I.C.P.U.A.E., Geneva, 1964; Member, Advisory Com. on Reactor Phys., U.S.A.E.C. Societies: Fellow, A.N.S.; Fellow, A.P.S.
Address: 1255 N. Sandburg Terrace, Chicago, Illinois, U.S.A.

AVERYANOV, P. G. Paper: Co-author, Experience in Operating the SM-2 Exptl. Reactor (Kernenergie, vol. 9, No. 10, 1966).
Address: Academy of Sciences of the U.S.S.R., 14 Leninsky Prospekt, Moscow V-71, U.S.S.R.

AVILA, Aureo Fernandez DE. See DE AVILA, Aureo Fernandez.

AVILA, Charles Francis, B.S. (Elec. Eng. and Business Administration), LL.D. Born 1906. Educ.: Harvard Univ. Chairman, Board, Boston Edison Co., Director, John Hancock Mutual Life Insurance Co.; Director, Liberty Mutual Insurance Co.; Director, Nat. Shawmut Bank of Boston; Director, Raytheon Co.; Director, Yankee Atomic Elec. Co.; Director, Connecticut Yankee Atomic Power Co.; Pres., Edison Elec. Inst. Society: Atomic Ind. Forum.
Nuclear interest: Nuclear power generation, design and management.
Address: 800 Boylston Street, Boston, Massachusetts 02199, U.S.A.

AVIVI, Pinhas, Ph.D., M.Sc. Born 1921. Educ.: London and Hebrew (Jerusalem) Univs. Sen. Lecturer, Hebrew Univ., Jerusalem, 1954-; Princeton Univ., 1960-61; U.K.A.E.A., 1965-67. Societies: I.P.S.; A.P.S.
Nuclear interest: Laser produced plasmas.
Address: Hebrew University, Physics Department, Jerusalem, Israel.

AVRAHAMI, Menashe, M.Sc., D.Sc. Born 1924. Educ.: Hebrew Univ., Jerusalem. Res. Council of Israel, 1949-53; Inst. for Fibres and Forest Products Res., 1953-54; Hebrew Univ.,

Jerusalem, 1954-55; Israel Inst. of Technol., Haifa, 1956-60; Alberta Univ., Canada, 1960-62; Nigeria Univ., 1962-64.
Nuclear interests: Metallurgy; Radiochemistry.
Address: Chemistry Department, Nigeria University, Nsukka, Nigeria.

AVRAMI, Louis, B.S. (Phys.), M.S. (Appl. Mechanics). Born 1922. Educ.: Rutgers Univ. and Stevens Inst. Technol. Supervisory Phys. (1950-), now Chief, Rad. Effects Res. and Facilities Branch, Explosives Lab., Picatinny Arsenal, Dover, New Jersey; Participant, Internat. Inst. Nucl. Sci. and Eng., Argonne Nat. Labs., 1960-61. Societies: A.N.S.; A.P.S.
Nuclear interests: Radiation effects in explosives using reactors, gamma-ray sources and accelerators; Design of large explosive containment irradiation capsules and shielding for accelerator facility; Development of radiation effects research programmes; Site surveys; Nuclear reactor hazard analyses.
Address: 4 Paula Court, Morristown, New Jersey 07960, U.S.A.

AVRAMITO, Maurice, Eng. Born 1931. Educ.: Ecole Nat. Supérieure Arts and Metiers, Paris and I.N.S.T.N. Chef, Département Nucléaire and Ingénierie, Sud Aviation S.A. Societies: Assoc. des Ingénieurs en Génie Atomique; Sté. Française d'Astronautique.
Nuclear interests: Management; Reactor design.
Address: 12 rue Pasteur, 92 Suresnes, France.

AVRIL, Robert André, Metal. and Mining Eng., Geologist. Born 1919. Educ.: Nancy Univ. Chief Eng., 1965; Chef, Div. minière de La Crouzille, C.E.A.
Address: Commissariat à l'Energie Atomique, BP No. 1, Razes, France.

AVZYANOV, V. S. Papers: Co-author, The Spatial and Energy Distributions of the Neutrons in a Stratum Containing a Borehole (Atomnaya Energiya, vol. 10, No. 5, 1961); co-author, Metal-Water Shielding for Point Neutron Source (ibid, vol. 16, No. 1, 1964); co-author, Two-Layer Shielding Optimisation Against Isotopic Fast Neutron Sources (ibid, vol. 23, No. 6, 1967).
Address: Academy of Sciences of the U.S.S.R., 14 Leninsky Prospekt, Moscow V-71, U.S.S.R.

AWNI, Adnan Hussain, B.A., M.Sc., Ph.D. (Chem.). Born 1924. Educ.: American Univ. of Beirut, Georgia Inst. Technol. and Virginia Polytech. Inst. Asst. Chem. Analyst, Ind. Res. Inst., Baghdad, 1950-52; Teaching and Res. Asst., Virginia Polytech. Inst., 1954-56; Res. Chem., Nat. Paint, Varnish and Laquer Assoc., Washington D.C., 1956-57; Ind. Chem., Ministry of Industry, Baghdad, 1957-63; Acting Sec. Gen., A.E.C., Iraq, 1964-.
Nuclear interest: Management.
Address: Iraq Atomic Energy Commission, Ministry of Industry, Baghdad, Iraq.

AWSHALOM, Miguel, Dr. Head, Health Phys. Group, Princeton Univ. Member, Advisory Panel on Accelerator Rad. Safety, U.S. A.E.C.
Address: Princeton University, Princeton, New Jersey, U.S.A.

AXEL, Peter, B.A., M.S., Ph.D. Born 1923. Educ.: Brooklyn Coll. and Illinois Univ. Asst. Prof. Phys. (1949-55), Assoc. Prof. Phys. (1955-59), Prof. Phys. (1959-), Illinois Univ. Sec. and Member of Board of Directors, Educational Foundation for Nucl. Sci. Society: Fellow, A.P.S.
Nuclear interest: Nuclear physics.
Address: Physics Department, Illinois University, Urbana, Illinois, U.S.A.

AXELRAD, Maurice, B.S. (Chem. Eng.), LL.B. Born 1934. Educ.: Purdue and Pennsylvania Univs. Attorney, Office of Gen. Counsel, U.S.A.E.C., 1958-61; Counsel, Office of Atomic and Space Development, New York State, 1961-66; Sec. and Assoc. Counsel, Atomic and Space Development Authority, New York State, 1962-. Coms. on Atomic Energy, Assoc. of the Bar of the City of New York. Society: New York State Bar Assoc.
Nuclear interest: Atomic energy law.
Address: New York State Atomic and Space Development Authority, 230 Park Avenue, New York, N.Y. 10017, U.S.A.

AXELSEN, Rolf, Siv.ing. Chairman, Norwegian Atomic Energy Soc.
Address: Norwegian Atomic Energy Society, c/o Den Polytekniske Forening, 7 Rosenkrantzgt, Oslo, Norway.

AXFORD, Roy A., Dr. Assoc. Prof. of Nucl. Eng., Technol. Inst., Northwestern Univ.
Address: Northwestern University, Technological Institute, Evanston, Illinois, U.S.A.

AXTON, Edward James, B.Sc. (Special Phys.), M.Sc. (Nucl. Phys.). Born 1923. Educ.: London Univ. Phys., Nat. Phys. Lab., Teddington, Middlesex. Societies: Assoc. Member, Inst. Phys.; Fellow, British Interplanetary Soc.
Nuclear interest: Neutron physics.
Address: 36 Hurst Road, East Molesey, Surrey, England.

AYBERS, Mehmet Nejat. Born 1917. Educ.: Liège, Istanbul Tech. and California (Berkeley) Univs. Prof. Mech. and Nucl. Eng. and Director, Inst. for Nucl. Energy, Istanbul Tech. Univ.; Member, Turkish A.E.C. Society: A.N.S.
Nuclear interests: Heat transfer in nuclear reactors; Design and operation of nuclear reactors; Nuclear power analysis and reactor safety.
Address: Kadiköy-Kiziltoprak, Fener Cad. Yelken Sak. No. 1, Nur Apt. D-9, Istanbul, Turkey.

AYENGAR, A. R. GOPAL-. See GOPAL-AYENGAR, A. R.

AYER, James E., B.S., M.S. Born 1924. Educ.: Maine Univ. Assoc. Chem. Eng., and Leader, Plutonium-Materials Fabrication Group, Argonne Nat. Lab.
Nuclear interests: Design and development of equipment for remote refabrication of reactor fuels; Radiation effects; Fuel process development; and related fields.
Address: 2619 Glasgow Street, Joliet, Ill., U.S.A.

AYERS, Arnold Leslie, B.S. (Chem. Eng.). Born 1917. Educ.: Iowa State Coll. Shift Supervisor, Materials Testing Reactor (1951-53), Asst. to Asst. Manager for Operations (1953-56), Asst. Supt. (1956-59), Supt., Idaho Chem. Processing Plant Operations (1959-66), Atomic Energy Div., Phillips Petroleum Co.; Manager, Idaho Chem. Processing Plant Operations, Idaho Nucl. Corp., 1966-67; Director, Nucl. Technol., Nucl. Fuels Dept., Allied Chem. Corp. Societies: A.I.Ch.E.; A.C.S.; A.A.A.S.
Nuclear interests: Management and operation of fuel reprocessing plants and nuclear testing reactors.
Address: 40 Bradwane Drive, Convent Station, New Jersey 07961, U.S.A.

AYERS, W.R. Vice Pres. and Sec., Bechtel Nucl. Corp.
Address: Bechtel Nuclear Corp., 220 Bush Street, San Francisco 4, California, U.S.A.

AYLEN, Rear Admiral I. G., C.B., D.Sc., M.I.Mech.E. Member, Rep. of the Inst. of Mech. Eng., British Nucl. Energy Soc.
Address: British Nuclear Energy Society, 1-7 Great George Street, London S.W.1, England.

AYOUB, A. K. O. At Atomic Energy Establishment, Cairo. Deleg., Third I.C.P.U.A.E., Geneva, Sept. 1964.
Address: Atomic Energy Establishment, Inshas, Nr. Cairo, Egypt, United Arab Republic.

AYRES, P. J., A.R.I.C. Formerly at Inst. Nucl. Medicine, Barnato Joel Labs., Middlesex Hospital Medical School.
Address: S.A. Courtauld Institute of Biochemistry, Middlesex Hospital Medical School, London W.1, England.

AYZYANOV, V. S. Paper: Co-author, Study of Spectra and Doses Generated by Monoenergetic Neutron Source in Iron-Water Shielding (Atomnaya Energiya, vol. 21, No. 1, 1966).
Address: Academy of Sciences of the U.S.S.R., 14 Leninsky Prospekt, Moscow V-71, U.S.S.R.

AZAD, Ali Asghar, B.Sc. (Phys.), M.Sc. (Phys. Chem.), Ph.D. (Phys.), Nucl. Sci. Diploma (I.S.N.S.E., Argonne Nat. Lab.). Born 1917. Educ.: Tehran, Lausanne and Pennsylvania State Univs. Prof. Phys. (1945), Prof. Atomic Phys. (1957), Member, Council of Univ. (1957), Tehran Univ.; Reactor Project Director, Tehran Univ., and i/c establishment of a one MW Res. Reactor in the Univ. Member, National Iranian A.E.C.; Member, Board of Governors, L./E.A., 1962. Book: Sur les Lois des effet Transversaux Mecanogalavaniques et Mecanothermoelectriques. Societies: Iranian Physical Soc.; A.N.S.
Address: Tehran University, Faculty of Science, P.O. Box 2989, Tehran, Iran.

AZARENKO, A. V. Paper: Co-author, Dependence of Hardened Uranium Texture from Heating Nature and Other Parameters of Heat Treatment (Atomnaya Energiya, vol. 16, No. 4, 1964).
Address: Academy of Sciences of the U.S.S.R., 14 Leninsky Prospekt, Moscow V-71, U.S.S.R.

AZEVEDO COUTINHO BRAGA, Carlos, D.Sc. (Elec. Eng.) (Oporto), Ph.D. (Phys.) (Oporto). Born 1899. Educ.: Oporto Univ. Prof. Phys. (Faculty of Sci.); Member, Nucl. Energy Commission. Book: Noçoes fundamentais de Fisica Atómica e Nuclear (Porto, Portucalense Editora, 1944). Societies: Ordem dos Engenheiros; Soc. Portuguesa de Quimica e Fisica.
Nuclear interest: Nuclear physics.
Address: 1271 R. Pedro Hispano, Porto, Portugal.

AZHAZHA, V. M. Papers: Co-author, Deformation and Fracture of Rolled Beryllium of Various Purity (Atomnaya Energiya, vol. 16, No. 5, 1964); co-author, Change of Beryllium Properties after Overaging Treatment (ibid, vol. 19, No. 3, 1965).
Address: Academy of Sciences of the U.S.S.R., 14 Leninsky Prospekt, Moscow V-71, U.S.S.R.

AZIMOV, C. A., D.Sc. Asst. Director (Res. Dept.), Inst. of Nucl. Phys. (Uzbek).
Address: Institute of Nuclear Physics (Uzbek), Uzbek Academy of Sciences, Kibrari, near Tashkent, U.S.S.R.

AZIMOV, Sadik, Dr. of Phys. and Maths. Prof. Born 1914. Educ.: Central Asian State Univ. Tashkent. Head, Lab. of Cosmic Rad., Physico-Tech. Inst., Acad. of Sci. of Uzbekistan. Asst. Director, Res. Dept., Nucl. Phys. Inst., Uzbek Acad. of Sci. Head, Chair of Nucl. Phys., Central Asian State Univ. Member, Learned Council, Dept. of Phys. and Maths., Central Asian State Univ.; Member, Learned Council, Nucl. Phys. Inst.; Member, Learned Council, Physico-Tech. Inst. Papers: 26 papers on cosmic radiation.
Address: Academy of Sciences of the Uzbek Republic, Kibrari, near Tashkent, U.S.S.R.

AZIZ, Farhat, M.Sc. (Hon. Panjab), Ph.D. (Cantab.). Born 1932. Educ.: Cambridge and Panjab Univs. Lecturer, Pharmaceutical Chem., Panjab Univ., Lahore, Pakistan, 1954-56; Overseas Res. Scholar of Roy. Commission for Exhibition of 1851, Phys. Chem. Labs., Cambridge Univ., 1956-59; Lecturer, Phys. Chem., Panjab Univ., 1959-60. Sen. Sci. Officer, Pakistan A.E.C., 1960-. Res. Fellow, A.E.R.E., Harwell, U.K., 1961.

Nuclear interest: Nuclear chemistry.
Address: Pakistan Atomic Energy Commission, Karachi, Pakistan.

AZIZ, Mohammed Ezzat Abdel. Asst. Prof. Dr. Head, Plasma Phys. and Accelerators Div., United Arab Atomic Energy Establishment.
Address: United Arab Atomic Energy Establishment, Inshas, Nr. Cairo, Egypt, United Arab Republic.

AZULA, Coronel (r) Gerardo GUERRERO.
See GUERRERO AZULA, Coronel (r) Gerardo.

AZUMA, R. E. Low Energy Nucl. Phys., Dept. of Phys., Toronto Univ.
Address: Toronto University, Department of Physics, Toronto 5, Ontario, Canada.

AZUMA, Taneaki, Asst. Prof. Inst. for Cancer Res., Faculty of Medicine, Osaka Univ.
Address: Osaka University, Institute for Cancer Research, Dozimahama-Dori, Fukusima-ku, Osaka, Japan.

AZUMA, Toshio, Ph.D. Born 1915. Educ.: Kyoto Univ. Chairman, Dept. of Phys., Rad. Centre of Osaka Prefecture. Societies: A.P.S.; A.N.S.; Japan Phys. Soc.
Nuclear interest: Nuclear physics.
Address: 44-7, Hantaji-Cho, Sakai, Osaka, Japan.

AZZAM, Reda Aly Ibrahim, B.Sc., Dipl. Nucl. Eng., Candidate of Chem. Sci. Born 1935. Educ.: Ein Shams Univ., Cairo. Nucl. Chem. (1959), Responsible Chem., Radioisotopes Production Labs. (1961), Sen. Res., R. and D. Div., (1963), Aqueous Reprocessing of Nucl. Fuels Unit and Ind. Applications of Rad. and Radioisotopes Unit, Lecturer, Rad. and Radioisotopes Technol. (1965), Nucl. Eng. Dept., Faculty of Eng., Alexandria Univ. Societies: Egyptian Soc. for Radioisotopes; Organisation of Sci. Professions.
Nuclear interests: Aqueous reprocessing of nuclear fuels, especially the chemistry of ruthenium; Solving industrial problems by radiation and radioisotopes techniques.
Address: United Arab Atomic Energy Establishment, Cairo, Egypt, United Arab Republic.

BA, Truc LE-. See LE-BA, Truc.

BA HLI, Freddy, Sc.D., M.S. (E.E.), B.Sc. Born 1922. Educ.: M.I.T., Lehigh and Rangoon Univs. Asst. Director of Telecommunications, Govt. of the Union of Burma, 1950-51; Res. Asst., Res. Lab. of Electronics, M.I.T. 1951-53; Sen. Res. Officer, (1953-55), Acting Director, Union of Burma Atomic Energy Centre, (1955-57), Director of Res. (1957-60), Director Gen. (1960-), Union of Burma Appl. Res. Inst. Chairman, Editorial Com., Union of Burma J. of Sci. and Technol. Society: Fellow, British Interplanetary Soc.
Nuclear interests: General.
Address: 327 U Wisara Road, Rangoon, Burma.

BAAEM, R. Administrateur, Fondation Nucleaire.
Address: Fondation Nucleaire, 4 rue de la Chancellerie, Brussels 1, Belgium.

BAARLI, Johan Vidar, Dr. phil. Born 1921. Educ.: Oslo Univ. Visiting Res. Fellow, Chicago Univ. 1952-53; Head, Dept. Biophys., Norsk Hydros Inst. for Cancer Res., Oslo, 1953; Fellow, Norwegian Cancer Soc., 1953-61; Sec. Com. on Nucl. Phys. of Norway, 1955-63. Member, Editorial Advisory Board, J. of Nucl. Energy (Pergamon Press), 1958-61; Adviser of Res. in Phys., Roy. Norwegian Council for Sci. and Ind. Res., 1959-60; Head, Health Phys., C.E.-R.N., 1961-; Sec., Com. of Nucl. Energy, Polytech. Assoc. of Norway. Societies: Norwegian Phys. Soc.; Hospital Physicists' Assoc.; H.P.S.; New York Acad. Sci.
Nuclear interests: High-energy radiation dosimetry, radiobiology and radiation protection.
Address: C.E.R.N., European Organisation for Nuclear Research, Geneva 23, Switzerland.

BABA, Arimasa, Dr. Director, Resources Res. Inst., Agency of Ind. Sci. and Technol., Ministry of Internat. Trade and Industry.
Address: Resources Research Institute, Ministry of International Trade and Industry, Kawaguchi-Saitama, Japan.

BABA, Ko, Prof. Agriculture Dept. Member, Radioisotope Res. Com., Niigata Univ. School of Medicine.
Address: Department of Agriculture, Niigata University School of Medicine, 757, 1-Bancho Asahimachi-Street, Niigata, Japan.

BABA, Toshio, B.E. Born 1930. Educ.: Tokyo Univ. Deputy Manager, Atomic Power Progressive Dept., Chugoku Elec. Power Co., Inc. Society: Atomic Energy Soc. of Japan.
Nuclear interest: Design and construction of nuclear power station.
Address: Atomic Power Progressive Department, Chugoku Electric Power Company Incorporated, 4-33 Konachi, Hiroshima, Japan.

BABALA, Dusan, Nucl. Eng. Born 1937. Educ.: Czech Tech. Univ. Nucl. Res. Inst., Prague, 1960-; I.A.E.A. Fellow in Inst. for Atomenergi, Kjeller, Norway, 1962-64.
Nuclear interests: Reactor kinetics; Reactor noise theory; Nuclear physics.
Address: Nuclear Research Institute, Rez, Czechoslovakia.

BABAYAN, C. P. Member, Inst. Nucl. Phys., Armenian Acad. Sci. Papers: Co-author, Integral Spectrum of Ionisation Pulses caused by Nucl. Active Particles of Cosmic Rad. at Mountain Altitudes (Nukleonika vol. 7, No. 2, 1962); co-author, Investigations of High Energy Particles Interactions with Atomic Nuclei at the Mountain Altitudes (ibid., No. 12).

BABB, Albert Leslie, B. Appl. Sci., M.S., Ph.D. Born 1925. Educ.: British Columbia and Illinois Univs. Asst. Prof. (1952-56), Assoc. Prof. (1956-60), Prof. (1960-), Chairman, Nucl. Eng.(1957-), Director, Nucl. Reactor Labs. (1960-), Head, Dept. of Nucl. Eng. (1965-), Washington Univ., Seattle. Vice Chairman, Nucl. Eng. Div., A.S.-E.E.; Member, Com. on A.E.C. Relations, A.S.E.E.; Member, Programme Com., A.N.S. Societes: A.I.C.E.; A.C.S.
Nuclear interests: Nuclear fuel processing; Nuclear fuel cycles and economics; Interactions of radiation with matter; Reactor operations.
Address: Nuclear Reactor Building, Washington University, Seattle, Washington 98105, U.S.A.

BABB, H. T. Eng. Supervisor, Carolinas Virginia Nucl. Power Assocs., Inc.
Address: Carolinas Virginia Nuclear Power Associates, Inc., Parr, South Carolina, U.S.A.

BABCOCK, Dale F., Ph.D., B.S. Born 1906. Educ.: Illinois Univ. Present Position: Director, Nucl. Eng. Sect. Tech. Div. (Atomic Energy), Du Pont Co., Wilmington, Delaware. Societies: A.C.S.; A.I.Ch.E.; A.N.S.
Nuclear interests: Reactor development and operation; Isotope separation processes.
Address: Explosives Department, Atomic Energy Division, E. I. du Pont de Nemours and Co., Wilmington, Delaware, U.S.A.

BABCOCK, Richard Vincent, B.S. (Ch.E.). Born 1933. Educ.: Drexel Inst. of Technol. (Chem. Eng.) and Pittsburgh Univ. (Phys.). Sen. Sci., Res. Labs., Westinghouse Elec. Co., 1956-. Societies: A.N.S.; Phys. Soc. of Pittsburgh; F.A.S.
Nuclear interests: Radiation damage, radiation detection, and biological effects.
Address: 4842 Ellsworth Avenue, Pittsburgh 13, Pa., U.S.A.

BABECKI, Jan, Dr. of Phys. and Mathematical Sci. Born 1932. Educ.: Jagiellonian Univ., Cracow, Acad. of Mining and Metal., Cracow, 1952-61; Inst. of Nucl. Res., Cracow, 1955-. Book: Co-author, Narzedzia nowej fizyki (Warsaw, Państwowe Zaklady Wydawnictw Szkolnych, 1960). Society: Polish Phys. Soc.
Nuclear interests: High energy physics; Elementary particles; Cosmic rays.
Address: Instytut Badań Jadrowych, Zaklad Fizyki Wysokich Energii, 30 al. Mickiewicza, Cracow, Poland.

BABICKY, Arnost, RNDr. Born 1923. Educ.: Charles Univ., Prague. Isotope Lab., Institutes for Biol. Res., Czechoslovak Acad. Sci., 1952-. Book: Radioisotopes in Biology and Medicine (in Czech) (Prague, S.Z.N., 1960). Society: Sect., Nucl. Medicine, Czechoslovak Medical Soc.
Nuclear interests: Use of radioisotopes in biology and medicine. Measurement of radioisotopes. Mineral metabolism and trace elements in physiology.
Address: Zbraslav n/Vlt. I, Smetanova 230, Czechoslovakia.

BABIKOV, Yu. M. Paper: Co-author, Thermal Properties of Hexafluorocarbon (letter to the Editor, Atomnaya Energiya, vol. 24, No. 3, 1968).
Address: Academy of Sciences of the U.S.S.R., 14 Leninsky Prospekt, Moscow V-71, U.S.S.R.

BABIKOVA, L. P. Paper: Co-author, Optimal Reactor Shutdown for Short-Term Work (letter to the Editor, Atomnaya Energiya, vol. 23, No. 2, 1967).
Address: Academy of Sciences of the U.S.S.R., 14 Leninsky Prospekt, Moscow V-71, U.S.S.R.

BABIKOVA, Yu. F. Paper: Co-author, Investigation of Scintillation Characteristics of Lithium Glasses by Detection of Neutron, Beta and Gamma Rad. (Atomnaya Energiya, vol. 24, No. 2, 1968).
Address: Academy of Sciences of the U.S.S.R., 14 Leninsky Prospekt, Moscow V-71, U.S.S.R.

BABIO, Felipe LAFITA. See LAFITA BABIO, Felipe.

BABKIN, G. I. Paper: Calculation of Fission Fragments Excitation Energy (Atomnaya Energiya, vol. 16, No. 4, 1964).
Address: Academy of Sciences of the U.S.S.R., 14 Leninsky Prospekt, Moscow V-71, U.S.S.R.

BACA, Carlos LABARTHE. See LABARTHE BACA, Carlos.

BACASTOW, Robert, B.S., M.S., Ph.D. Born 1930. Educ.: M.I.T. and California Univ., Berkeley. Res. Assoc., Lawrence Rad. Lab., 1963-64; Lecturer (1964-65), Asst. Prof. (1965-), California Univ., Riverside. Society: A.P.S.
Nuclear interest: Elementary particle physics.
Address: Physics Department, California University, Riverside, California, U.S.A.

BACCAREDDA, Mario, Dr. Chem. Eng. Born 1907. Educ.: Milan Polytech. Full Prof. of Ind. Chem. and Director, Inst. of Ind. and Appl. Chem., Pisa Univ., 1953-. Book: Materiali per Reattori Nucleari (Pisa, Nistri-Lischi, 1960). Society: Società Chimica Italiana, Rome.
Nuclear interests: Nuclear materials and reprocessing.
Address: Facoltà di Ingegneria, Università di Pisa, 2 Via Diotisalvi, 56100 Pisa, Italy.

BACH, David R., B.A., M.Sc., D.Ph. (Phys.). Born 1924. Educ.: Michigan Univ. Res. Assoc. (Phys.), K.A.P.L., Gen. Elec. Co. Assoc. Prof., Coll. of Eng., Michigan Univ. Society: A.P.S.
Nuclear interests: Primary experience and interest with experimental reactor physics; in particular, pulsed neutron, time of flight spectrum, and critical experiments.

Address: College of Engineering, Michigan University, Ann Arbor, Michigan.

BACH, Joseph H., B.S. (Met. Eng.), M.S. (Met. Eng.). Born 1921. Educ.: Purdue and Idaho Univs. Metal., Hanford Works, G.E.C. 1948-52; Sect. Head, Fuel Div., Sylvania Elec., 1952-54; Advisory Eng., Bettis Plant (1954-59), Advisory Eng., Atomic Power Dept. (1959-61), Advisory Eng., Astronucl. Lab. (1961-), Westinghouse Elec. Corp. Societies: A.I.M.E.; A.N.S.; A.S.M. Nuclear interests: Nuclear fuels; Metallurgy; Reactor design; Nuclear space technology.
Address: Westinghouse Electric Corp., Astronuclear Laboratory, P.O.B. 10864, Pittsburgh 36, Pa., U.S.A.

BACH, L. L. Plant Supt., Pathfinder Atomic Power Plant.
Address: Pathfinder Atomic Power Plant, Box 115, RFD No. 2, Sioux Falls, South Dakota 57100, U.S.A.

BACHARACH, Stephen S. NERVA Project, Aerojet (Azusa), 1963-. Chairman, Programme Com. (1962), Treas. (1963), A.N.S.
Address: Aerojet-NERVA, Azusa, California, U.S.A.

BACHER, Robert Fox, B.S., Ph.D., Sc.D. (Hon.). Born 1905. Educ.: Michigan Univ. Chairman, Div. of Phys., Maths. and Astronomy, and Director, Norman Bridge Lab. (1949-62), Prof., Phys. (1949-), Provost (1962-), California Inst. Technol. Trustee: Atoms for Peace Awards, Carnegie Corp., Inst. for Defence Analyses, Univs. Res. Assoc., Inc.; Chairman, U.S. Nat. Com., I.U.P.A.P.; Member, Board of Directors: Bell and Howell Co., Detroit Edison Co., Power Reactor Development Corp., TRW Inc.; Member, Editorial Council, Annals of Phys. Societies: A.P.S.; N.A.S.
Address: 345 S. Michigan, Pasadena, California 91106, U.S.A.

BACHERIKOV, I. P. Paper: Xenon Oscillations in Nucl. Reactor (letter to the Editor, Atomnaya Energiya, vol. 19, No. 4, 1965).
Address: Academy of Sciences of the U.S.S.R., 14 Leninsky Prospekt, Moscow V-71, U.S.S.R.

BACHMAN, Stefania, Dr. Phil. At Inst. of Appl. Rad. Chem., Tech. Univ., Lodz.
Address: Technical University, Institute of Applied Radiation Chemistry, 15 Wroblewskiego, Lodz, Poland.

BACKLUND, H. E. Sven E. Head, Swedish Mission to the E.E.C. and Euratom, 1968-.
Address: Swedish Mission, c/o Euratom, 51-53 rue Belliard, Brussels, Belgium.

BACKMAN, Gustaf Malkolm, Civilingenjör. Born 1935. Educ.: Chalmers Tech. Univ. Naval Architect,; Director Res., Swedish Ship Res. Foundation, 1964-.
Nuclear interest: Nuclear ship propulsion.
Address: Fack, Göteborg 8, Sweden.

BACKSTRÖM, Johan Willem VON. See VON BACKSTRÖM, Johan Willem.

BACLESSE Francois, M.D. Born 1896. Educ.: Paris Univ. Chef de service de Radiothérapie et de radio-diagnostic, Fondation Curie, 1926-61; Chef de service de Radiothérapie, American Hospital of Paris, 1947-64; Actuellement Consulting à cet hopital. Books: Les tumeurs malignes du pharynx et du larynx - Etude radiographique (Masson et Cie, 1960); co-author, Atlas de mammographie (1965).
Nuclear interest: Nuclear physics.
Address: 61 rue Claude Bernard, Paris, France.

BACON, George Edward, M.A., Sc.D. (Cantab.), Ph.D. (London). Born 1917. Educ.: Cambridge Univ. At A.E.R.E. Harwell, 1946-63; Prof. Phys., Sheffield Univ., 1963-. Books: Neutron Diffraction, 2nd edition (Oxford Univ. Press, 1962); Applications of Neutron Diffraction in Chem. (Pergamon Press, 1963); X-ray and Neutron Diffraction (Pergamon Press, 1966). Societies: Inst. Phys.; American Crystallographic Assoc. Nuclear interest: Neutron diffraction.
Address: 28 Whirlow Park Road, Sheffield 11, England.

BACQ, Zénon-Marcel, M.D. Born 1903. Prof. Radiobiol., Liège Univ. Books: Fundamentals of Radiobiol. (Pergamon); Chem. Protection against Ionising Rad. (Thomas). Societies: European Assoc. Radiobiol.; Assoc. Belge Radioprotection.
Address: Laboratoire de Physiopathologie, 32 boulevard de la Constitution, Liège, Belgium.

BADAWI, Ahmed, Dr. Rector, Ein Shams Univ., Cairo.
Address: Ein Shams University, Abbassia, Cairo, Egypt, United Arab Republic.

BADKIN, R. L. Paper: Co-author, Determination of Sum of Corrosion Products in Nucl. Reactor Waters (letter to the Editor, Atomnaya Energiya, vol. 20, No. 4, 1966).
Address: Academy of Sciences of the U.S.S.R., 14 Leninsky Prospekt, Moscow V-71, U.S.S.R.

BADLAND, Percy Alfred, B.Sc. (Eng.), C. Eng. Born 1908. Educ.: London Univ. Structural Eng., Ministry of Works, 1950-54; Sen. Structural Eng. (1954-56), Asst. Chief Structural Eng. (1956-), U.K.A.E.A., Risley. Societies: M.I.-Struct.E.; A.M.I.C.E.
Nuclear interest: Structural engineering.
Address: 5 Wood Lane, Appleton, Warrington, Lancashire, England.

BADMAYEV, Kirill Nikolayevich. Head, Radiol. Dept., Leningrad Prof. A. L. Polenov Neurosurgical Res. Inst. Paper: Co-author, Application of Isotope Encephalography and Electroencephaloscopy for Localization of Brain Tumours (2nd I.C.P.U.A.E., Geneva, Sept. 1958).
Address: Nuclear Energy Institute, Academy of Sciences of the U.S.S.R., 14 Leninsky Prospekt, Moscow V-71, U.S.S.R.

BADUKOVA, V. D. Paper: Co-author, Ceramic Blocks of Slurry Produced by Coagulation of Radioactive Sewage Wastes (Atomnaya Energiya, vol. 23, No. 6, 1967).
Address: Academy of Sciences of the U.S.S.R., 14 Leninsky Prospekt, Moscow V-71, U.S.S.R.

BAECKELAND, Edgard Gerard, Dr. en Médicine. Born 1929. Educ.: Liège Univ. Elève-Asst. d'Histologie, 1950-52; Asst. d'Histologie, 1956-60; Conservateur d'Histologie et d'Embryologie, 1960-. Societies: Stè. Belge de Biochimie; Assoc. des Anatomistes (Paris); European Tissue Culture Club.
Nuclear interest: Autoradiography.
Address: 20 rue de Pitteurs, Liège, Belgium.

BAENNINGER, Wilhelm, Dipl. Ing. Born 1902. Educ.: Swiss Inst. Technol., Zurich. Vice-Pres., Electro-Watt Elec. and Ind. Management Co., Ltd., Zurich; Director, Atomelectra, Ltd., Zurich. Swiss Federal Commission for Atomic Energy; Com. member, Assoc. Suisse pour l'energie atomique. Society: Assoc. Suisse pour l'energie atomique.
Nuclear interests: Plant design; Economics; Management.
Address: Electro-Watt Electrical and Industrial Management Company, Limited, 16 Talacker, Zurich, Switzerland.

BAER, Robert Lloyd, B.Mech.Eng., M.Sc. (Mech. Eng.). Born 1931. Educ.: Clarkson Coll. and Rensselaer Polytech. Inst. Eng. (1954-57), Sen. Eng. (1957-59), Nucl. Div., Pratt and Whitney Aircraft; Eng. Specialist (1959-61), Project Eng. (1961-63), Nucl. Div., Martin Co.; Chief, Reactor Systems, Hittman Assocs., 1963-; Lecturer, Catholic Univ. of America, 1963-. Chairman, Hons. and Awards Com., Baltimore Sect., A.S.M.E. Society: A.N.S.
Nuclear interests: Overall reactor systems design; Core thermal and fluid flow analysis; Fuel cycle management and economic studies of nuclear power plants used for commercial production of electricity and water; Remote military bases; Space power systems.
Address: 5800 Greenspring Avenue, Baltimore, Maryland 21209, U.S.A.

BAER, William, A.B., A.M., M.S. Born 1924. Educ.: Amherst Coll., Dartmouth Coll., Yale Univ. Fellow Sci., Westinghouse Elec. Corp., Bettis Atomic Power Lab., 1953-. Societies: A.P.S.; A.N.S.
Nuclear interests: Critical assembly measurements and cross sections; Data storage and retrieval; On-Line data analysis.
Address: P.O. Box 79, West Mifflin, Pennsylvania 15122, U.S.A.

BAERDEMAEKER, Adolphe J. L. de. See DE BAERDEMAEKER, Adolphe J. L.

BAERG, A. P., Dr. Formerly Assoc. Res. Officer, now Sen. Res. Officer, X-rays and Nucl. Rad. Sect., N.R.C., Canada. Member, Subcom. on use of Radioactivity Standards, U.S. Acad. Sci.-N.R.C.
Address: X-rays and Nuclear Radiations Section, National Research Council, Montreal Road Laboratories, Building M-35, Ottawa 7, Ontario, Canada.

BAERTSCHI, Peter, Dr. sc. nat. Born 1919. Educ.: E.T.H. Zürich, Res. Assoc., Inst. for Phys. Chem., Basel Univ., 1945-56; Head, Dept. of Chem., Federal Inst. for Reactor Res., Wuerenlingen, 1956-.
Nuclear interests: Nuclear and reactor chemistry and chemical engineering; Isotope separation, mass-spectrometry.
Address: 1 Blauenstrasse, Birsfelden, Switzerland.

BAES, Charles F., Jr., B.Sc. (Rutgers), M.Sc. (So. California), Ph.D. (So. California). Born 1924. Educ.: Rutgers and Southern California Univs. Group Leader, Reactor Chem. Div., O.R.N.L. 1959-. Society: A.C.S.
Nuclear interests: High temperature aqueous chemistry; Solvent extraction processes; Chemistry of pressurised water reactors; Chemistry of molten salts.
Address: Oak Ridge National Laboratory, Reactor Chemistry Division, P.O. Box X, Oak Ridge, Tennessee, U.S.A.

BAETJER, Anna, Dr. Member, Rad. Control Advisory Board, Maryland State.
Address: Radiation Control Advisory Board, Governor's Office, State House, Annapolis, Maryland 21404, U.S.A.

BAEZ, Juan N. Lab. de Investigaciones Nucleares, Inst. de Investigaciones, Facultad de Ingenieria Quimica, Univ. Nacional del Litoral, Santa Fe.
Address: Instituto de Investigaciones, Facultad de Ingenieria Quimica, 2829 Santiago del Estero, Santa Fe, Argentina.

BAEZ, Manuel MARTINEZ-. See MARTINEZ-BAEZ, Manuel.

BAGDASAROV, Yu. E. Adviser, Third I.C.P.-U.A.E., Geneva, Sept. 1964. Papers: Co-author, Calculation of the Nonstationary Temperature Field in a Reactor Channel and Thermoelastic Stresses in a Fuel Element Can (Atomnaya Energiya, vol. 13, No. 3, 1962); co-author, Studies of Transient Regimes of Natural Circulation in Multicircuit Nucl. Reactor Systems (ibid. vol. 16, No. 5, 1964); co-author, BN-350 and BOR Fast Reactors (ibid, vol. 21, No. 6, 1966); co-author, Sodium Technol. and Equipment of BN-350 Reactor (ibid, vol. 22, No. 1, 1967).
Address: Academy of Sciences of the U.S.S.R., 14 Leninsky Prospekt, Moscow V-71, U.S.S.R.

BAGGE, Erich, Prof., Dr.rer.nat. Born 1912. Educ.: Munich, Berlin, and Leipzig Univs. a.o. Prof., Hamburg Univ. 1948; Sci. Director, Gesellschaft für Kernenergieverwertung in Schiffbau und Schiffahrt Hamburg, 1956; o. Prof. and

Director, Inst. of Pure and Appl. Nucl. Physics, Kiel Univ., 1957. On West German Commission for Nucl. Energy, Com. III and Arbeitskreis III/1; Member for Germany, Nucl. Ship Propulsion Com., Euratom. Editor, Atomkernenergie. Books: Von der Uranspaltung bis Calderhall (1957); Ursprung und Eigenschaften der komischen Strahlung (1948). Jahrbuch der Studiengesellschaft für Kernenergieverwertung in Schiffbau und Schiffahrt, Hamburg (Munich Verlag Thiemig, 1958); similarly Jahrbuch (as above, 1961).
Nuclear interests: Cosmic ray physics, nuclear physics and nuclear energy.
Address: Institute for Pure and Applied Nuclear Physics, 40-60 Olshausenstrasse, Kiel, Germany.

BAGGERLY, Leo Lon, B.S., M.S., Ph.D. Born 1928. Educ.: California Inst. Technol. Sen. Res. Eng., Jet Propulsion Lab., Pasadena, 1955-56; Fulbright Lecturer in Phys., Ceylon Univ., Colombo, 1956-59; Assoc. Prof. Phys. (1959-66), Prof. Phys. (1966-), Texas Christian Univ., Fort Worth, Texas; on leave at Ling-Temco-Vought Res. Centre, Dallas, Texas, 1963-64. Societies: American Geophys. Union; A.P.S.; A.A.P.T.
Nuclear interests: Gamma spectroscopy; Bremsstrahlung production; Electron interactions below 10 MeV.
Address: Physics Department, Texas Christian University, Fort Worth, Texas 76129, U.S.A.

BAGLEY, Kenneth Quinton, B.Sc., Ph.D. Born 1927. Educ.: Durham Univ. Res. Manager, Radioactive Metal., U.K.A.E.A. (D.E.R.E.).
Nuclear interests: Metallurgy; Irradiation damage effects and mechanisms in cladding and structural materials; Investigation of fuel and fuel pin behaviour under irradiation in a fast neutron flux; Development of fuel elements for liquid metal cooled fast reactors.
Address: 13 Brims Road, Thurso, Caithness, Scotland.

BAGLIO, Antonio, Dr. Ing. Born 1908. Educ.: Rome Univ. and Polytechnicum, Milan. Inspector-Gen., Ministry of Industry and Commerce, Rome; Director, Nucl. Energy Div., 1955-. Deleg. to C.E.C.A. Energy Com. Luxembourg; Member, Nucl. Energy Steering Com., O.E.C.-E's Agency, Paris. Society: Assoc. Nazionale Ingegneria Nucleare, Rome.
Nuclear interest: Management.
Address: 2 Via Molise, Rome, Italy.

BAGNALLI, R. M. Managing Director, T. I. Stainless Tubes Ltd.
Address: T. I. Stainless Tubes Ltd., Oldbury, Birmingham, England.

BAGRETSOV, V. F. Papers: Co-author, Use of Flotation in Treating Radioactive Effluents (Atomnaya Energiya, vol. 9, No. 3, 1960); Phys. and Chem. Conditions of Underground Burying of Radioactive Wastes (ibid, vol. 24, No. 2, 1968).
Address: Academy of Sciences of the U.S.S.R., 14 Leninsky Prospekt, Moscow V-71, U.S.S.R.

BAGROVA, V. I. Papers: Co-author, Preparation and Studies of Some Plutonium Monocarbide Properties (Atomnaya Energiya, vol. 22, No. 6, 1967); co-author, Phase Diagrams of Plutonium with IIIA, IVA, VIII and IB Group-Metals (ibid, vol. 23, No. 6, 1967).
Address: Academy of Sciences of the U.S.S.R., 14 Leninsky Prospekt, Moscow V-71, U.S.S.R.

BAHADUR, Kanwar, Dr. Sen. Officer, Inst. of Nucl. Medicine and Allied Sci.
Address: Institute of Nuclear Medicine and Allied Sciences, Metcalfe House, Delhi-6, India.

BAHR, Henrik Eiler Storen. Born 1902. Educ.: Oslo Univ. Justice of Supreme Court, 1945-. Member, Norwegian Atomic Energy Council; Member, Permanent Internat. Court of Arbitration at The Hague; Member, European Nucl. Energy Tribunal in Paris; Norwegian Deleg., Hague Conference on Private Internat. Law. Society: Internat. Law Soc.
Nuclear interest: Legal problems.
Address: Hoyesterett, Oslo-Dep., Norway.

BAHUROV, V. G. Paper: Co-author, Underground Leaching of Poor Uranium Ores (Atomnaya Energiya, vol. 24, No. 2, 1968).
Address: Academy of Sciences of the U.S.S.R., 14 Leninsky Prospekt, Moscow V-71, U.S.S.R.

BAIBORODOV, Yury. Thermonucl. Lab., Kurchatov Inst. of Atomic Energy.
Address: Kurchatov Institute of Atomic Energy, 46 Ulitsa Kurchatova, Post Box 3402, Moscow, U.S.S.R.

BAIKIE, A. G., Dr. At Dept. of Medicine, St. Vincent's Hospital, Melbourne. Member, Radio-isotopes Standing Com., Nat. Health and Medical Res. Council.
Address: St. Vincent's Hospital, Department of Medicine, Melbourne, Australia.

BAILEY, Brian Edward, Ph.D., B.Sc. Born 1934. Educ.: Southampton Univ. Res. Chem., Courtaulds Fundamental Res. Lab., Maidenhead, 1957-62; R. and D. Lab., Internat. Synthetic Rubber Co. Ltd., 1962-67; Res. Manager, British Oxygen Chem., 1967-.
Nuclear interest: Application of radiotracer techniques to chemical research problems, particularly polymer kinetics.
Address: British Oxygen Chemicals, Chester-le-Street, County Durham, England.

BAILEY, Colin Alfred, M.A., D.Phil. Born 1931. Educ.: Oxford Univ. Post-Doctoral Res., Clarendon Lab. (1959), Development Officer, Dept. of Nucl. Phys. (1960-65), Lecturer, Eng. Sci., Dept. of Eng. Sci., Cryogenics Lab. (1965-), Lecturer in Eng., Keble Coll. (1965-), Oxford Univ.; S.S.O., Dept. of Sci. Adviser, Air Ministry, Whitehall, 1959-60.

Nuclear interests: Design, construction and operation of cryogenic bubble chambers.
Address: Oxford University, Department of Engineering Science, Parks Road, Oxford, England.

BAILEY, Kenneth Desmond, C. Eng., Fellow, Inst. of Elec. Eng. Born 1922. Elec. Eng., C.E.G.B., Southern Div., Portsmouth Generating Station, Portsmouth, 1949-55; Elec. Eng., Capenhurst Works (1955-59), Works Eng., Services, Windscale and Calder Works (1959-), U.K.A.E.A. (P.G.).
Nuclear interest: Engineering application to all branches of nuclear work.
Address: 39 Coniston Avenue, Seascale, Cumberland, England.

BAILEY, Lowell F. Arkansas Univ. representative on Council, O.R.I.N.S.
Address: Arkansas University, Department of Chemistry, Fayetteville, Arkansas, U.S.A.

BAILEY, N. T. J., D.Sc. Member, Managing Subcom: on Monitoring, M.R.C.
Address: Medical Research Council, 20 Park Crescent, London W.1, England.

BAILEY, R. F. Works Manager, Capenhurst Works, U.K.A.E.A. (Production Group).
Address: U.K. Atomic Energy Authority, Capenhurst Works, Chester, England.

BAILEY, Robert Eugene, B.S. (Chem. Eng.), M.S. (Chem. Eng.), Ph.D. (Nucl. Eng.). Born 1931. Educ.: Illinois and Purdue Univs. At Argonne Nat. Lab., 1956-59; U.S.A.E.C. Fellow, 1959-60; Purdue Univ.; Societies: A.N.S.; American Soc. of Eng. Sci.; A.S.E.E.
Nuclear interests: Reactor kinetics and control, and time dependent neutronics.
Address: 112 Tecumseh Park Court, W. Lafayette, Ind., U.S.A.

BAILEY, Sydney William, A.R.C.S., B.Sc. Born 1918. Educ.: London Univ. Principal Res. Officer, Div. of Entomology, Commonwealth Sci. and Ind. Res. Organisation, 1950-.
Nuclear interests: Interested in the use of radiation for killing and for causing sterility in insect pests, particularly those attacking stored foodstuffs.
Address: P. O. Box 109, City, Canberra, A.C.T., Australia.

BAILEY, William Howard, A.Met., F.I.M. Born 1923. Educ.: Sheffield Univ. Tech. Director, Jessop-Saville Ltd., 1959-67; Gen. Manager, Special Materials Div., English Steel Forge and Eng. Corp., 1968-.
Nuclear interests: Metallurgical, with particular reference to alloy steels, stainless and heat resisting materials produced by air melting and vacuum melting techniques.
Address: English Steel Forge and Engineering Corporation, River Don Works, Sheffield, England.

BAILIE, R. C. Instructor, Chem. Eng., Dept. of Nucl. Eng., Kansas State Coll.
Address: Department of Nuclear Engineering, Kansas State College, Manhattan, Kansas, U.S.A.

BAILLY DU BOIS, Bernard, Deleg. to 2nd I.C.P.U.A.E., Geneva, Sept. 1958.; Expert, Third I.C.P.U.A.E., Geneva, Sept. 1964; Member, Study Group on Long Term Role of Nucl. Energy in Western Europe, O.E.C.D., E.N.E.A. Paper: Evolution des etudes neutroniques en relation avec les projets de reactuer (Energia Nucleare, vol. 8, No. 1, 1961).
Address: Commissariat à l'Energie Atomique, C.E.N. Saclay, B.P.2, Gif-sur-Yvette, (S.et O.), France.

BAILY, Norman Arthur, B.S., M.A., Ph.D., Born 1915. Educ.: St. John's, New York and Columbia Univs. Assoc. Clinical Prof. Radiol., U.C.L.A. Manager, Space Sci. Dept., Hughes Res. Labs., Malibu, California, 1959-. Societies: A.P.S.; A.A.A.S.; Biophys. Soc.; American Coll. Radiol.; American Assoc. Phys. in Medicine; Res. Soc.; New York Acad. Sci.; I.R.E. (P.G.-N.S.).
Nuclear interests: Interaction of radiation with matter; Uses of isotopes in biology and medicine; Dosimetry; Health physics.
Address: Hughes Research Laboratories, 3011 Malibu Canyon Road, Malibu, California, U.S.A.

BAIN, James A., B.S., Ph.D. Born 1918. Educ.: Wisconsin Univ. Prof. Pharmacology (1954-), Director, Div. Basic Health Sci. (1960-), Emory Univ. Society: American Soc. Pharmacology and Exptl. Therapeutics.
Nuclear interest: Use of radioisotopes in biological research.
Address: Division of Basic Health Sciences, Emory University, Atlanta, Georgia 30322, U.S.A.

BAINBRIDGE, G. M. Mine Geologist, Denison Mines Ltd.
Address: Denison Mines Ltd., 4 King Street West, Toronto 1, Ontario, Canada.

BAINBRIDGE, Gilbert Ronald, C. Eng., B.Sc., Ph.D. Born 1925. Educ.: Durham Univ. Lecturer, Paisley Tech. Coll., 1952-54; Fellow, McMaster Univ., Canada, 1954-56; U.K.A.E.A., 1956-68. Societies: F.Inst.P.; I.E.E.
Nuclear interest: Reactor design.
Address: Reactor Group, United Kingdom Atomic Energy Authority, Risley, Warrington, Lancs., England.

BAINBRIDGE, K. T., Prof. Phys. Dept., Harvard Univ. Member, Editorial Board. Nucl. Instruments and Methods, Netherlands.
Address: Department of Physics, Harvard University, Cambridge 38, Massachusetts, U.S.A.

BAINES, Bernard Deryck. Born 1922. Educ.: London Univ. Project Eng., Cyanamid Products Ltd., 1952-56; Chem. Eng. and Tech. Sales Manager, Hawker Siddeley Nucl. Power Co. Ltd., 1956-61; Sales Manager and Irradiations Plant Manager, Nucl. Chem. Plant Ltd., 1961-67; Director, CON-RAD Eng., Ltd., 1967-. Societies: B.N.E.S.; Assoc. for Rad. Res.
Nuclear interests: Nuclear engineering with special interest in irradiation technology; Economic evaluation of medical, chemical and food irradiation processes; Nuclear heat in process industries.
Address: 31 The Fairway, Burnham, Bucks., England.

BAINES, Gordon Owen, M.A. (Maths. and Phys.), Ph.D. (Phys.). Born 1909. Educ.: St. Andrews Univ.; with U.K. Dept. of Atomic Energy (now U.K.A.E.A.) in Tech. Policy Div., London, 1946-50; with A.E.C.L., Chalk River, Ont. (formerly N.R.C. of Canada, Atomic Energy Div.), initially as Asst. to Director of Phys. Research Div., Asst. to Vice-Pres. (R. and D.), Manager, Administration, 1950-. Societies: Fellow, Phys. Soc. London; Fellow, Inst. Phys. (U.K.).
Nuclear interest: Management.
Address: 51 Hillcrest Avenue, Deep River, Ontario, Canada.

BAINES, N. D., B.Sc. Radiochem. Inspector, Ministry of Housing and Local Govt.
Address: Ministry of Housing and Local Government, Whitehall, London, S.W.1, England.

BAIR, Joe Keagy, B.A. Born 1918. Educ.: Rice Inst. and Columbia Univ. Fairchild Eng. and Air. Co., 1947-51; O.R.N.L. 1951-. Societies: Tennessee Acad. of Sci.; A.P.S.
Nuclear interest: Energy levels in light nuclei.
Address: 200 W. Fairview Road, Oak Ridge, Tennessee, U.S.A.

BAIRD, Walter S. Former Chairman, now Pres. (1964-), Baird-Atomic Inc.
Address: Baird Atomic Inc., 33 University Road, Cambridge 38, Massachusetts, U.S.A.

BAISIER, Pierre, Ing. civil mécanicien et électricien. Born 1923. Educ.: Liège and Brussels Univs. Manager, Mercantile Marine Eng. and Graving Docks Co. S.A. Member, Comité de Direction, Assoc. Industrielle Belge pour Centrales de Puissance à Réacteurs Rapides.
Nuclear interests: Realisations in the field of plate and boilerwork (containment buildings, vessels, heat exchangers, shielding) for nuclear installations. Study fabrication and erection.
Address: Mercantile Marine Engineering and Graving Docks Company, 403 Hansadok, Antwerp, Belgium.

BAISSAS, Henri, Agrégé de Physique. Born 1899. Educ.: Toulouse and Rennes Univs. Inspecteur Général de l'Instruction Publique; Directeur de Centre d'Etudes Nucléaires de Fontenay-aux-Roses, 1957; Directeur du Cabinet du Haut Commissaire à l'Energie Atomique, 1959; Directeur de la Physique au Commissariat á l'Energie Atomique. 1962. Conseiller scientifique auprès du Haut Commissaire à l'Energie Atomique. Book: Génie Atomique 1960 (Radioactivité) et divers. Societies: Sté. de Physique et du Radium; Union des Physiciens.
Nuclear interest: Nuclear physics.
Address: 11 bis rue Villebois Mareuil, Paris 17, France.

BAJOREK, Antoni, Ph.D. Born 1934. Educ.: Jagiellonian Univ., Cracow. Chair of Structure Res. Jagiellonian Univ.
Nuclear interests: Scattering of neutrons by solids and molecules.
Address: 101 ul. Czarnowiejska, Cracow, Poland.

BAK, A. Kristian, Cand. polyt., M.Sc.M.E., Dr. Tech. (Hon.). Born 1892. Educ.: Denmark Tech. Univ., Copenhagen. Chairman, Danish Board for Testing and Approval of Elec. Equipment; Pres., Danish Nat. Com. for World Power Conference; Hon. Member, Union Internationale des Distributeurs de Chaleur, Unichal, Paris; Member, Danish A.E.C. Societies: Inst. Danish Civil Eng.; Danish Acad. Tech. Sci.; A.S.M.E.
Nuclear interest: Reactor design.
Address: 23 C.F. Richsvej, Copenhagen F, Denmark.

BAK, M. A. Papers: Co-author, Yields of Delayed Neutrons from Fission in ^{239}Pu and ^{232}Th as Induced by Neutrons of Energy 14.5 MeV (letter to the Editor, Atomnaya Energiya, vol. 11, No. 6, 1961); co-author, Cross-Section and Resonance Integrals of Capture and Fission of Long-Lived Americium (ibid, vol. 23, No. 4, 1967); co-author, Production of Pu^{238} by U^{235} and Np^{237} Neutron Irradiation (ibid, vol. 23, No. 6, 1967); co-author, Formation of Cm^{242} and Cm^{244} by Reactors Neutron Irradiation of Cm^{241} (ibid, vol. 24, No. 3, 1968).
Address: Academy of Sciences of the U.S.S.R., 14 Leninsky Prospekt, Moscow V-71, U.S.S.R.

BAKAI, A. S. Papers: Particles Leakage from Storage Ring Due to Random Variations in Amplitude and Frequency of Compensating Field (letter to the Editor, Atomnaya Energiya, vol. 19, No. 4, 1965); On Capture Coefficient of Particles in Accelerator (letter to the Editor, ibid, vol. 21, No. 4, 1966).
Address: Academy of Sciences of the U.S.S.R., 14 Leninsky Prospekt, Moscow V-71, U.S.S.R.

BAKER, Adolph, B.A., M.S., Ph.D. Prof., Dept. of Nucl. Sci. and Eng., Lowell Technol. Inst.
Address: Lowell Technological Institute, Department of Nuclear Science and Engineering, Lowell, Massachusetts 01854, U.S.A.

BAKER, Arthur Lemprière Lancey, D.Sc.(Eng.). Born 1905. Educ.: Manchester Univ. Prof., Concrete Structures and Technol., Imperial Coll. of Sci. and Technol. Member, Nucl. Safety

Advisory Com., Ministry of Power, 1963-. Nuclear interest: Prestressed concrete pressure vessels.
Address: Imperial College of Science and Technology, Department of Civil Engineering, London, S.W.7, England.

BAKER, Broughton Leonard, B.S. (Chem Eng.), Ph.D. (Chem. Eng.). Born 1912. Educ.: South Carolina Univ. and North Carolina State Coll. Assoc. Prof. Chem. Eng., 1946-56; Prof. Chem. Eng., and Head, Dept. of Chem. Eng., South Carolina Univ., 1956-. Societies: Nucl. Eng. Div., A.I.Ch.E.; Atomic Ind. Forum; A.S.E.E.; A.C.S.
Nuclear interests: Radioactive waste disposal; Nuclear engineering education.
Address: 819 Burwell Lane, Columbia, South Carolina, U.S.A.

BAKER, David Kenneth, Ph.D. Born 1923. Educ.: Pennsylvania Univ. Consultant, U.S. State Dept., -1963; Prof. Phys., Union Coll., Schenectady, N. Y., 1953-65; Gen. Elec. R. and D. Centre, 1965-67; Academic Dean, St. Lawrence Univ., 1967-. Book: Elements of Modern Phys. Societies: A.A.U.P.; A.I.P.; A.A.P.T.
Nuclear interest: Education.
Address: St. Lawrence University Canton, New York, U.S.A.

BAKER, George, F.C.A. Born 1910. Administration and Sec., and Director, Rio Algom Mines Ltd.
Address: Rio Algom Mines Ltd., 335 Bay Street, Toronto 1, Ontario, Canada.

BAKER, J. R. S. Tech. Director, Nucl. R. and D. Dept., Appleby and Ireland Ltd.
Address: Appleby and Ireland, Ltd., Basingstoke, Hampshire, England.

BAKER, John Robert WILCOX-. See WILCOX-BAKER, John Robert.

BAKER, Lieutenant-Colonel John W., B.S. (Geological Eng.), B.S. (Elec. Eng.), M.S. (Nucl. Eng.). Born 1925. Educ.: Montana School of Mines and Air Force Inst. Technol. Regular Officer, U.S. Air Force 1951-; Chief, Operations and Maintenance Div., Air Force Nucl. Eng. Test Facility, Ohio, 1966-. Society: A.N.S.
Nuclear interests: Applied research and engineering development of nuclear power reactors.
Address: Air Force Nuclear Engineering Test Facility, Wright-Patterson Air Force Base, Ohio, U.S.A.

BAKER, Marshall. Assoc. Prof., Phys. Dept., Assoc. Prof., Nucl. Phys. Lab and Cosmic Ray and High Energy Lab., Washington Univ.
Nuclear interest: Theoretical physics.
Address: Washington University, Department of Physics, Seattle, Washington 98105, U.S.A.

BAKER, Matthew William, M.Eng., F.I.E.E. Born 1916. Educ.: Liverpool Univ. Gen. Manager, Stonehouse Operations. Sperry Gyroscope Div. Sperry Rand, Ltd., Stonehouse, Glos., 1966-.
Nuclear interests: Reactor control and instrumentation.
Address: Sperry Gyroscope Division, Sperry Rand Limited, Bonds Mill, Stonehouse, Gloucestershire, England.

BAKER, Milton Robert, B.A., M.A., Ph.D. Born 1932. Educ.: Harvard Univ.Res. Fellow, Harvard Univ., 1960-61; Asst. Prof., Brandeis Univ., 1961-63; Asst. Prof., The Johns Hopkins Univ., 1963-66; Res. Phys., Hewlett-Packard Co., 1966-. Societies: A.P.S.; A.A.A.S.
Nuclear interests: Nuclear magnetic interactions in molecules; Hyperfine interactions; Masers; Frequency standards; Atomic and molecular beams.
Address: Hewlett-Packard Company, Frequency and Time Division, East, Beverly, Massachusetts 01915, U.S.A.

BAKER, Philip S., A.B., M.A., Ph.D. Born 1916. Educ.: DePauw, Arkansas and Illinois Univs. Assoc. Prof. Chem., Bradley Univ., 1948-52; Supt., Isotopes Separations Dept. (1956-58), Supt. Isotopes Sales Dept. (1958-61), Supt. Advanced Isotopes Development Dept. (1961-62), Asst. Supt., Isotopes Development Centre and Director, Isotopes Information Centre., (1962-), O.R.N.L. Editor, Isotopes and Rad. Technol. Societies: A.C.S.; A.N.S.; A.A.A.S.; Fellow, American Inst. of Chemists.
Nuclear interests: Isotope separation; Radioisotopes from enriched stable isotopes; Isotope applications.
Address: Building 3047, Oak Ridge National Laboratory, Oak Ridge, Tennessee, U.S.A.

BAKER, R. A. Formerly Chief Eng., Elec Eng. Dept., now Vice Pres. in Charge of Elec. Operation, Public Service Elec. and Gas Co., New Jersey.
Address: Public Service Electric and Gas Co., 80 Park Place, Room 7200, Newark, New Jersey 07101, U.S.A.

BAKER, Richard Dean, B.S. (Chem.Eng.), Ph.D. (Phys. Chem.). Born 1913. Educ.: South Dakota School of Mines and Technol. and Iowa State Coll., Group Leader (1943-56), Div. Leader, Chem. and Metal. (1956-), Los Alamos Sci. Lab. Member, N.A.S.A. Materials Res. Advisory Com. Society: A.C.S.
Nuclear interests: Chemistry and metallurgy.
Address: Los Alamos Scientific Laboratory, Box 1663, Los Alamos, New Mexico, U.S.A.

BAKER, Rowland. Born 1908. Educ.: Roy. Naval Coll., Greenwich. Naval Constructor-in-Chief, Roy. Canadian Navy, Ottawa, 1948-56; Tech. Chief Executive, Dreadnought Project Team, Admiralty, 1957-. Societies: Inst. of Naval Architects; Soc. of Naval Architects and Marine Eng.
Nuclear interest: Production of nuclear submarines.
Address: Admiralty, Foxhill, Bath, England.

BAKER, Selwyn John, M.D. Born 1925. Educ.: Melbourne Univ. Lecturer in Medicine (1955-57), Assoc. Prof., Medicine (1957-62), Prof., Medicine (1962-), Head, Wellcome Res. Unit (1956-), Christian Medical Coll. Councillor, Indian Soc. of Haematology; Com., Indian Soc. of Gastroenterology. Societies: Internat. Soc. Haematology; New York Acad. Sci.
Nuclear interests: Application of radioisotopes and radioisotope techniques in experimental medicine; Biochemistry and physiology with special reference to tropical anaemias and the malabsorption syndrome.
Address: Wellcome Research Unit, Christian Medical College, Vellore, South India.

BAKER, W. A. Chief Development Eng., Bristol Aeroplane Plastics Ltd.
Address: Bristol Aeroplane Plastics Ltd., Filton House, Bristol, England.

BAKER, W. M., Prof., Phys. Dept., Detroit Univ.
Address: Detroit University, 4001 W. McNichols Road, Detroit 21, Michigan, U.S.A.

BAKER, Wilfred Edmund, B.Eng., M.S. (Eng.), D.Eng. Born 1924. Educ.: Johns Hopkins Univ. Supervisory Phys., Ballistic Res. Labs., Aberdeen Proving Ground, Maryland, 1949-61; Principal Development Eng., Aircraft Armaments, Inc., Cockeysville, Maryland, 1961-. Societies: A.S.M.E.; A.N.S.; American Ordnance Assoc.
Nuclear interests: Effects of reactor runaway on containment structures; Safety of nuclear plants; Dynamic modeling applied to response of containment structures.
Address: 401 Rosebank Avenue, Baltimore 12, Maryland, U.S.A.

BAKER, William Oliver, B.S., Ph.D. (Phys. Chem.), D.Sc. (Hon.), Washington Coll., Georgetown, Pittsburgh and Seton Hall Univs.), D.Eng. (Hon., Stevens Inst. Technol.), D.Laws. (Hon., Glasgow Univ.). Born 1915. Educ.: Washington Coll. and Princeton Univ. Vice Pres., Res., Bell Telephone Labs., 1955-; Director, Babcock and Wilcox Corp., 1962-; Director, Mead Johnson and Co., 1966-; Director,Summit and Elizabeth Trust Co., 1958-. Trustee, Aerospace Corp., 1961-; Trustee, Carnegie-Mellon Univ., 1958-; Trustee, Old Dominion Foundation, 1966-; Trustee, Princeton Univ., 1964-; Trustee, Rockefeller Univ., 1960-. Societies: N.A.S.; A.C.S.; Fellow, A.P.S.
Nuclear interests: Administration of low-energy nuclear physics programme, including use of special on-line data processing equipment.
Address: Bell Tellephone Laboratories, Murray Hill, New Jersey 07974, U.S.A.

BAKHUIZEN van den BRINK, J. N., Prof. Dr. Member, Sci. Council for Nucl.Affairs.
Address: Scientific Council for Nuclear Affairs, 24 Duinweg, The Hague, Netherlands.

BAKISH, Robert, B.S. (Met. Eng.), M.S. (Met.), D. Eng. (Met.). Born 1926. Educ.: Columbia and Yale Univs. Sen. Metal. i/c Metal., Sprague Elec. Co., 1955-57; Consultant i/c Phys. Metal., C.I.B.A., 1957-59; Director of Res., Alloyd Corp. and Vice-Pres., Alloyd Electronics Corp., 1959-62; Exec. Vice-Pres., Director R. and D., Electronics and Alloys Inc., 1962-. Books: Proc. 2nd Electron Beam Symposium, 1960 (Alloyd Corp.); Proc. 3rd Electron Beam Symposium, 1961 (Alloyd Electronics Corp.); Proc. 4th Electron Beam Symposium, 1962 (Alloyd Electronics Corp.); Introduction to Electron Beams Technol. (Wiley, 1962); Handbook on Electron Beam Welding (Wiley, 1964).
Nuclear interests: Metallurgy of corrosion; High temperature metallurgy; Electron beam applications.
Address: Electronics and Alloys Inc., 435 Victoria Terrace, Ridgefield, N.J., U.S.A.

BAKKER, E. K. den. See den BAKKER, E. K.

BAKKER, Johannes H., Electrotech. Eng. Born 1917. Educ.: Delft Tech. Univ. Managing Director, N.V. Gemeenschappelijke Kernenergie centrale Nederland, building the 50 MW Nucl. Power Station. Adviser, Ind. Council for Nucl. Energy.
Nuclear interest: Energy production.
Address: 310 Utrechtsew. Arnhem, Netherlands.

BAKLUSHIN, R. P. Papers: Co-author, BN-350 and BOR Fast Reactors (Atomnaya Energiya, vol. 21, No. 6, 1966); co-author, Sodium Technol. and Equipment of BN-350 Reactor (ibid, vol. 22, No. 1, 1967); co-author, The Nucl. Power Plant with BN-350 Reactor (ibid, vol. 23, No. 5, 1967).
Address: Academy of Sciences of the U.S.S.R., 14 Leninsky Prospekt, Moscow V-71, U.S.S.R.

BAKULIN, N. V. Paper: Co-author, Investigation of Heat Transfer in Alkali Metals Condensation (letter to the Editor, Atomnaya Energiya, vol. 22, No. 5, 1967).
Address: Academy of Sciences of the U.S.S.R., 14 Leninsky Prospekt, Moscow V-71, U.S.S.R.

BALACHANDRAN, Aiyalam P. B.Sc.(Hons.), M.Sc., Ph.D. Born 1938. Educ.: Madras Univ. Assoc. Prof. of Phys., Syracuse Univ. Societies: A.P.S.; American Assoc. of Univ. Profs.
Nuclear interest: Elementary particle physics.
Address: Physics Department, Syracuse University, Syracuse, New York 13210, U.S.A.

BALAKRISHAN, I. S., M.Sc. Phys., Cancer Inst., India.
Address: Cancer Institute, Adyar, Madras 20, India.

BALANIK, I. S. Paper: Co-author, Effect of Titanium on Phase Constitution and Ductivity of Chrome-Nickel-Steel with High Content to Bore (Atomnaya Energiya, vol. 22, No. 5, 1967). Address: Academy of Sciences of the U.S.S.R., 14 Leninsky Prospekt, Moscow V-71, U.S.S.R.

BALARIN, Manfred, Dr.rer.nat. Born 1934.
Educ.: Halle and Leningrad Univs. Society:
Physikalische Gesellschaft der Deutschen Demokratischen Republik.
Nuclear interests: Solid state physics; Radiation damage.
Address: Deutsche Akademie der Wissenschaften zu Berlin,Forschungsgemeinschaft, Zentralinstitut für Kernforschung,Rossendorf, 8051-Dresden, PF 19, German Democratic Republic.

BALASHOV, V. D. Paper: Co-author, Investigation of PuO_2 Fuel Element Assembly of BR-5 Reactor (Atomnaya Energiya, vol. 24, No. 2, 1968).
Address: Academy of Sciences of the U.S.S.R., 14 Leninsky Prospekt, Moscow V-71, U.S.S.R.

BALBEKOV, V. I. Papers: Co-author, On Coherent Instability of Betatron Oscillations in Accelerators and Accumulators (Atomnaya Energiya, vol. 19, No. 2, 1965); co-author, Influence of Resistive Walls on Dimensions and Shape Oscillations of Beam in Particle Accelerators and Storage Rings (letter to the Editor, ibid, vol. 20, No. 3, 1966); Frequency speed Influence on Stability of Coherent Oscillations in Accelerators (letter to the Editor,ibid, vol. 23, No. 6, 1967).
Address: Academy of Sciences of the U.S.S.R., 14 Leninsky Prospekt, Moscow V-71, U.S.S.R.

BALCAR, Frederick Rhinehart, B.S., Ph.D. Born 1899. Educ.: Iowa and Pittsburgh Univs. Special Consultant, U.S. Air Force Office of Atomic Energy, 1950-52; Asst. Director Tech. Activities, Air Reduction Co., Inc., 1962-. Society: A.R.S.
Nuclear interests: Fission product recovery and purification; Industrial applications of radioisotopes.
Address: 1670 Valley Road, Millington, New Jersey, U.S.A.

BALCAR, Josef, M.Sc. (Agriculture), Dr. Phil. (Agriculture and Forestry), Dr. sc.tech. Born 1920. Educ.: Prague Agricultural Univ. Sci. worker and Member, Sci. Council, Res. Inst. of Tobacco Industry, Báb. Sci. worker, Czechoslovak Acad. Sci.
Nuclear interests: Application of radioisotopes in tobacco growing and in crop production; Fission products in the soil.
Address: 29 Libocká,Prague-Liboc, Czechoslovakia.

BALCH, William R. Business Manager, Cambridge Electron Accelerator, operated for U.S.A.E.C. by M.I.T. and Harvard Univ.
Address: U.S.A.E.C., Cambridge Electron Accelerator, Cambridge, Massachusetts, U.S.A.

BALDI, Francesco, Graduate in Phys. and U Univ. degree in Electrotechnics. Born 1917. Educ.: Pisa Univ. and Turin Eng. School. Chief, Appl. Phys. Group, Centro Sperimentale Metallurgico, Rome. Tech. Com., Nondestructive Testing Centre, Italian Metall. Assoc., Milan. Society: Assoc. Italiana di Metallurgia, Milan.
Nuclear interests: Metallurgy; Nuclear physics.
Address: 26 Via G. Trevis, 00147 Rome, Italy.

BALDO, M., Sen. Officer, Nucl. Emulsion Group, Inst. di Fisica, Padua Univ.
Nuclear interest: Study of strange particle interactions.
Address: Istituto di Fisica, Universita di Padua, Via F. Marzolo 8, Padua, Italy.

BALDOCK, Charlie Russell, B.A. (Lynchburg Coll.), M.S. (Virginia Polytech. Inst.), M.A. (Cornell), Ph.D. (Cornell). Born 1909. Physicist, 1943-. Societies: A.P.S.; A.P.S., Southeastern Sect.; A.C.S.; RESA; Tennessee Acad. Sci.; Virginia Acad. Sci.
Nuclear interests: Mass spectrometry and neutron cross sections.
Address: Oak Ridge National Laboratory, Oak Ridge, Tennessee, U.S.A.

BALDUS, Wolfgang, Doctor's degree in phys. Born 1921. Educ.: Munich T.H. Head, Development Dept. Gesellschaft für Linde's Eismaschinen A.G. Society: Deutsche Physikalische Gesellschaft.
Nuclear interests: Application of cryogenics for nucleonics - e.g., for reactor gas purification, cold neutron sources, bubble chambers.
Address: Linde A.G., 8021 Hoellriegelskreuth, near Munich, Germany.

BALDWIN, Elmer E., B.S. (Metal.), M.S. (Metal. Eng.). Born 1919. Educ.: Columbia UNiv. and Rensselaer Polytech. Inst. Manager, Mech. Metal., K.A.P.L., Schenectady, New York, 1956-. Societies: A.S.M.; A.S.M.E.
Nuclear interests: Metallurgy, mechanics of materials, and effects of environment (temperature, radiation, coolant) on the structural and dimensional stability of reactor core and structural components.
Address: 2220 Barcelona Road, Schenectady, New York, U.S.A.

BALE, William Freer, A.B., Ph.D. Born 1911. Educ.: Cornell and Rochester Univs. Assoc. Prof. Radiol. (1946-), Chief, Div. of Radiol. and Biophys., Atomic Energy Project (1947-), Prof. Rad. Biol. (1948-), Rochester Univ. Member, Advisory Com. for Biol. and Medicine, U.S.A.E.C.; Member, Board of Sci. Consultants, Sloan-Kettering Inst. for Cancer Res.; Member, Advisory Panel on Radiol. Instr., U.S. Dept. of Defence. Societies: A.P.S.; Rad. Res. Soc.; American Physiological Soc.; Soc. Exptl. Biol. and Medicine.
Nuclear interests: Studies directed toward the use of antibodies as carriers of radioactivity for therapy of cancer; Biological effects of radiation; Radioactive tracers in medical research
Address: University of Rochester, Atomic Energy Project, Box 187, Station 3, Rochester, New York, U.S.A.

BALEBANOV, V. M. Papers: Co-author, A Study of Single Particle Motion in Corragurated Magnetic Fields (letter to the Editor, Atomnaya Energiya, vol. 15, No. 4, 1963); co-author, Motion of Single Charged Particles in Magnetic Field with Helical Symmetry (letter to the Editor, ibid, vol. 15, No. 5, 1963); co-author, Azimuthal Drift of Charged Particles in Axially-Symmetric magnetic Field with Mirrors (letter to the Editor, ibid, vol. 21, No. 6, 1966).
Address: Academy of Sciences of the U.S.S.R., 14 Leninsky Prospekt, Moscow V-71, U.S.S.R.

BALENT, Ralph, B.S. (M.E.), M.S.(M.E.), Born 1923. Educ.: Denver and California Univs. Instructor (M.E.), Denver Univ., 1949-50; Res. Eng. (1950-54), Group Leader (1954-58), Deputy Director, Organic Reactors Dept. (1959), Director, Compact Power Systems Dept. (1960), Assoc. Manager, Compact Systems Div. (1961), Exec. Director, Compact Systems Div. (1963), Director, Eng. Management Div. (1965-66), Director, Power Systems (1966-), Atomics Internat. Div. of North American Aviation. Member, Board of Directors, A.N.S. Societies: A.S.M.E.; A.I.A.A.; A.N.S.; N.S.P.E.
Nuclear interests: Reactor engineering, nuclear power systems economics and management.
Address: 4754 Viviana, Tarzana, California, U.S.A.

BALESCU, R. At Univ. Libre, Brussels.
Nuclear interest: Plasma physics.
Address: Universite Libre de Bruxelles, 50 avenue F. D. Roosevelt, Brussels, Belgium.

BALFOUR, John Gordon, B.Sc., Ph.D. Born 1929. Educ.: St. Andrews and Glasgow Univs. Phys., E.M.I. Res. Labs., Ltd., 1953-55; Phys., then Head Phys. and Performance Group, Atomic Power Div., English Elec. Co. Ltd., 1955-66; Chief Phys., Nucl. Design and Construction Ltd. 1966-.
Nuclear interest: Nuclear power.
Address: 154 Leicester Road, Oadby, Leicestershire, England.

BALIGA, B., Dr. Reader, Saha Inst. of Nucl. Phys., Calcutta Univ.
Address: Calcutta University, Saha Institute of Nuclear Physics, 92 Acharya Prafulla Chandra Road, Calcutta-9, India.

BALIGA, S. P. Member, Phys. Dept., Ceylon Univ.
Address: Ceylon University, Thurstan Road, Colombo 5, Ceylon.

BALKE, Siegfried, Dipl. Ing., Dr.-Ing. Born 1902. Educ.: Munich T. H. Ind. Chem., 1925-53; Federal Minister for Posts and Telecommunications, 1953-56; Federal Minister for Nucl. Energy, 1956-62; Managing Director, Sigri-Kohlefabrikate-G.m.b.H./Meitingen, 1962-. Hon. Prof. Chem. Economy, Munich Univ. Coeditor, Ullmanns Encyclopaedia of Tech. Chem., 3rd ed.; Co-editor, Chemische Industrie and Die Atomwirtschaft; Chairman, Verein der Bayerischen Chemischen Industrie e.V., Munich; Vice-Pres., Verband der Chemische Industrie Frankfurt am Main; Member, Board of Directors, Berufsgenossenschaft der Chemischen Industrie, Heidelberg; Member, Praesidium of the Landersverband der Bayerischen Industrie; Senator, Max-Planck-Soc.; Pres., Board of Directors, Inst. für Plasmaphysik G.m.b.H.; Member, Steering Com., Deutsches Atomforum e.V.
Nuclear interests: Nuclear chemistry and management.
Address: 9 Lerchenfeldstrasse, Munich 22, Germany.

BALL, John Geoffrey, B.Sc., F.I.M. Born 1916. Educ.: Birmingham Univ. P.S.O. i/c Plutonium Metal. (1949-53); Head, Reactor Metal.Group (1953-56), A.E.R.E., Harwell; Prof. Phys. Metal. (1956-57), Head, Dept. of Metal. (1957-), Imperial Coll., London Univ. Council, Inst. Metal., 1958-; Advisory Com., Harwell Ceramics Centre; Council, British Nucl. Forum. Society: Fellow, Inst. Metal.;Inst. of Welding.
Nuclear interests: Physical metallurgy of nuclear fuels and construction materials.
Address: 3 Sylvan Close, Limpsfield, Surrey, England.

BALL, John Sigler, B.S. (Chem. Eng.), M.S. Born 1914. Educ.: Texas Technol. Coll. and Colorado Univ. At Laramie (1938-63), Res. Director, Bartlesville Petroleum Res. Centre (1963-), Bureau of Mines. Society: A.C.S.
Nuclear interests: Management; Use of nuclear devices for stimulation of gas production; Use of radioisotopes for tracing flow of fluids and for studying reactions of petroleum products.
Address: Box 1398, Bartlesville, Oklahoma 74003, U.S.A.

BALL, Ray, M. B.S.E.E. Born 1904. Educ.: Montana State Univ. Chief Eng., Montana Power Co. Society: I.E.E.E.
Nuclear interest: General.
Address: Montana Power Company, 40 East Broadway, Butte, Montana, U.S.A.

BALL, Raymond, B.Sc.(Eng.). Born 1924. Educ.: Battersea Polytech., London. Sen. Elec. Eng., Gen. Eng. Projects Dept. (1953-57), Head, Elec. Eng. Office of Power Station Eng. Dept., Atomic Power Div. (1957-62), English Elec. Co., Ltd.; Asst. Project Eng., Midlands Project Group, C.E.G.B., 1963-. Society: A.M.I.E.E.
Nuclear interest: Design and construction of power stations.
Address: 74 Buryfield Road, Solihull, Warwickshire, England.

BALL, Russell Martin, B.S., M.S. Born 1927. Educ.: Illinois and Chicago Univs. Sect. Chief, The Babcock and Wilcox Co.; Chief Eng., Nucl.-Chicago. Tech. Programme Chairman, Reactor Operations Div. and Chairman, North Carolina-Virginia Chapter, A.N.S. Book: Section contributor to Noise Analysis in Nucl. Systems

(A.E.C. Symposium Series No. 4, 1964).
Societies: I.E.E.E.; A.P.S.
Nuclear interests: Nuclear instrumentation; Detection and analysis of nuclear particles; Critical experiments on power reactors.
Address: Nuclear Development Centre, Babcock and Wilcox Co., P.O. Box 1260, 1201 Kemper Street, Lynchburg, Virginia, U.S.A.

BALLADARES, Noel Jerez, Dr. Member, Nat. Energy Commission.
Address: Comisión Nacional de Energia, Ministerio de Fomento y Obras Públicas, Managua, D. N., Nicaragua.

BALLADORE PALLIERI, G., Prof., Internat. and Constitutional Law, Sacro Cuore Univ., Milan, Italian Member, Nucl. Supply Agency, Euratom.
Address: Universita Sacro Cuore, 4 Stefano Jacini, Milan, Italy.

BALLAM, Joseph, B.S., Ph.D. (Phys.). Born 1917. Educ.: Michigan and California (Berkeley) Univs. Formerly Res. Assoc., Princeton Univ. and Prof. Phys., Michigan State Univ.; now Prof. and Assoc. Director (Res.) Stanford Linear Accelerator Centre, Stanford Univ. Res. collaborator, Brookhaven Nat. Lab. Book: High Energy Nucl. Phys. (Inter Sci. Publishers, 1956). Society: A.P.S.
Nuclear interests: High energy physics; Construction and design of hydrogen bubble chambers; Data analysis.
Address: Stanford Linear Accelerator Centre, Stanford University, Stanford, California, U.S.A.

BALLARD, Bristow Guy, B.Sc., D.Sc. (Hon.), D.Eng. (Hon.), LL.D. (Hon.). Born 1902. Educ.: Queen's Univ. Director, Radio and Elec. Eng. Div., (1948-63), Vice-Pres. (Sci.) (1954-63), Pres. (1963-67), N.R.C.; Pres., Canadian Patents and Development Ltd., 1967-. Member, Atomic Energy Control Board, 1963-67.
Address: 390 Cloverdale Road, Ottawa 2, Canada.

BALLARD, Geoffrey George, H.N.C. Born 1931. Educ.: Enfield Tech. Coll. Manager, Reactor Control Div., Elliott Process Automation.
Nuclear interests: Reactor instrumentation and control equipment - both electrical and mechanical (e.g., control rod mechanisms and allied control circuits).
Address: Appledore, Orchard Close, New Barn, Longfield, Kent, England.

BALLARD, Ray L. Sen. Officer, United States Nucl. Corp.
Address: United States Nuclear Corporation, P.O. Box 208, 801 N. Lake Street, Burbank, California, U.S.A.

BALLARD, S. S., Dr. Head Prof. Phys., Phys. Dept., Florida Univ.
Address: Florida University, Gainesville, Florida, U.S.A.

BALLARD, William Edward. Born 1896. Managing Director, Metallisation Ltd., 1922-61. Pres., Inst. Metal., 1958-59; At present W. E. Ballard, Consultants Ltd. Book: Metal Spraying and the Flame Deposition of Ceramics and Plastics. Societies: F.R.I.C.; Fellow, Inst. Metallurgists.
Nuclear interests: General metallurgy and corrosion; Welding.
Address: 38 Grassmoor Road, King's Norton, Birmingham 30, England.

BALLESIO, Pier Luigi, Dr. Director, Italelettronica.
Address: Italelettronica, 72 Via Ignazio Pettinengo, Rome, Italy.

BALLESTEROS, Alfonso MONDRAGON.
See MONDRAGON BALLESTEROS, Alfonso.

BALLESTEROS, Romulo. Asst. Investigator, Nucl. Phys: Sect., Dept. of Phys., Faculty of Mathematical Phys., Univ. Nacional de La Plata.
Address: Universidad Nacional de La Plata, La Plata, Argentina.

BALLHAUS, William F., A.B., M.E., Ph.D. Born 1918. Educ.: Stanford Univ. and California Inst. of Technol. Chief Eng., Norair Div. (1953), Vice Pres. and Chem. Eng., Nortronics Div. (1957), Corp. Vice Pres. and Gen. Manager, Nortronics Div. (1957), Exec. Vice Pres. and Director (1961), Northrop Corp.; Pres. and Director, Beckman Instr., Inc., 1965. Council of Trustees, Assoc. of U.S. Army; Advisory Council, School of Eng., Stanford Univ.; Trustee, Northrop Inst. of Technol.; Chairman, Analytical Instrument Sect., Sci. Apparatus Makers Assoc.
Address: 2500 Harbor Boulevard, Fullerton, California, U.S.A.

BALLIGAND, Pierre Louis, Ecole Polytechnique Diplom. Born 1916. Educ.: Lille Univ. Head of Service des Grandes Piles de Saclay, 1957-59; Head of Service des Grandes Piles expérimentales C.E.A., Saclay, 1959-61; Former Deputy Director Gen., I.A.E.A.; now Deputy Director, Nucl. Res. Centre, C.E.A., Grenoble.
Nuclear interests: Reactor design and operations; Health and safety.
Address: Centre d'Etudes Nucléaires, B.P. 269, (38) Grenoble, France.

BALLINGER, Dennis Arthur James, B.Eng. Born 1921. Educ.: Liverpool Univ. Agent, Oxford Contract (pre-stressed concrete bridge) (1949-50), Chief Eng., London Airport Contract (main access tunnel and ancillary work) (1950-52), Agent, London Airport Contract (central buildings, foundations, external services and fire main) (1952-53), Agent, Calder Hall Atomic Power Station Contract (1953-55), Contracts Manager, Atomic Power Dept.

(1955-56), Director (1957-65), Deputy Managing Director (1965-67), Managing Director (1967-), Taylor Woodrow Construction, Ltd.; Director, Nucl. Design and Construction, Ltd., 1966-. Society: Inst. of Civil Engineers.
Address: Taylor Woodrow Construction, Ltd., Ruislip Road, Southall, Middlesex, England.

BALLINGER, Lieutenant Colonel Edwin Ray, B.A., M.D. Born 1920. Educ.: George Washington Univ. School of Medicine, Maryville Coll. and Reed Coll. Chief, Radiobiol. Branch, U.S.A.F. School of Aerospace Medicine. Societies: A.M.A.; Aerospace Medical Assoc. Nuclear interests: Acute effects of protons, neutrons and gamma rays.
Address: U.S.A.F. School of Aerospace Medicine, Brooks Air Force Base, Texas, U.S.A.

BALLOU, Nathan E., B.S., M.S., Ph.D. Born 1919. Educ.: State Teachers Coll. (Duluth, Minnesota), Illinois and Chicago Univs. Head, Nucl. Chem. Branch, U.S.N. Radiol. Defence Lab., San Francisco, California, 1948-; Chief, Chem. Dept., C.E.N., Mol, Belgium, 1959-61. Chairman, Subcom. on Radiochem., Nat. Res. Council - N.A.S. Society: A.C.S.
Nuclear interests: Fission process; Nuclear reactions; Radiochemical techniques; High temperature chemistry.
Address: U.S. Naval Radiological Defence Laboratory, San Francisco, California 94135, U.S.A.

BALLREICH, Hans, Dr. jur. Mitglieder, Fachkommission I, Recht und Verwaltung, Federal Ministry for Sci. Res., Germany.
Address: Federal Ministry for Scientific Research, Fachkommission I, Kernenergierecht, 46 Luisenstrasse, 532 Bad Godesberg, Germany.

BALLY, Dorel, Dr. (Physics). Born 1923. Educ.: Bucharest and Moscow State Univs. Bucharest Univ., 1949-53; Inst. of Atomic Phys., Bucharest, 1953-. Book: Scattering of Thermal Neutrons (Rumanian Acad. of Sci., 1966). Nuclear interests: Neutron physics. Scattering of neutrons in solids and liquids. Neutron and X-ray diffraction.
Address: Institute of Atomic Physics, P.O. Box 35, Bucharest, Roumania.

BALNER, Hans, M.D. Born 1925. Educ.: Amsterdam Univ. Staff Member, Biol. Div., Euratom, 1961-. Treas., Transplantation Soc. Society: Netherlands Radiobiol. Soc.
Nuclear interest: Exclusively radio and biological subjects.
Address: Euratom, Biology Division, 51-53 rue Belliard, Brussels, Belgium.

BALTASAR, Benjamin S., Dr. of Medicine. Born 1934. Educ.: Philippines Univ. Instructor, Medicine, Philippines Univ., 1963-; Consultant, Medicine (1963-), Physician, Radioisotope Lab., Philippine Gen. Hospital. Societies: Radioisotope Soc. of the Philippines; Philippine Soc. of Nucl. Medicine.
Nuclear interest: Nuclear medicine;
Address: 105 University Avenue, Caloocan City, Philippines.

BALTMUGUR, K. K. Paper: Co-author, Fast Neutron Flux Measurement for IRT-200 Reactor (letter to the Editor, Atomnaya Energiya, vol. 20, No. 2, 1966).
Address: Academy of Sciences of the U.S.S.R., 14 Leninsky Prospekt, Moscow V-71, U.S.S.R.

BALUKOVA, V. D. Paper: Co-author, Phys. and Chem. Conditions of Underground Burying of Radioactive Wastes (Atomnaya Energiya, vol. 24, No. 2, 1968).
Address: Academy of Sciences of the U.S.S.R., 14 Leninsky Prospekt, Moscow V-71, U.S.S.R.

BALYASNYI, N. D. Paper: Co-author, Experience of Spectral Gamma-Log with Multichannel Spectrometers (letter to the Editor, Atomnaya Energiya, vol. 23, No. 1, 1967).
Address: Academy of Sciences of the U.S.S.R., 14 Leninsky Prospekt, Moscow V-71,U.S.S.R.

BALZHISER, Richard E., B.S. (Chem. Eng.), M.S. (Nucl. Eng.), Ph.D. (Chem. Eng.). Born 1932. Educ.: Michigan Univ. Instructor (1958-60), Asst. Prof. (1960-63), Assoc. Prof. (1963-), Chem. Eng. Dept. Michigan Univ. Societies: A.I.Ch.E.; A.I.M.E.; A.A.A.S.
Nuclear interests: Fuels and fuel processing; Liquid metal thermodynamics and heat transfer.
Address: Chemical Engineering Department, Michigan University, Ann Arbor, Michigan, U.S.A.

BAMFORD, Frank Wallace, B.Sc. (Econ.). Born 1922. Educ.: London Univ. Administration including Asst. Private Sec., Minister of Works (1951-52), Ministry of Works, 1949-56; Civil Service Pay Res. Unit, 1956-58; Deputy Establishments Officer (1958-63), Chief Commercial Officer, Production Group (1963-), U.K.A.E.A. Nuclear interest: Commercial with reference to all aspects of nuclear fuel services.
Address: Girdwood, Greenside Drive, Hale, Altrincham, Cheshire, England.

BAMFORD, Ronald, B.S., M.S., Ph.D. Born 1901. Educ.: Connecticut, Vermont, and Columbia Univs. Prof. Botany, Maryland Univ. Societies: Botanical Soc. of America; American Genetics Assoc.
Nuclear interest: Administration.
Address: Department of Botany, Maryland University, College Park, Maryland, U.S.A.

BAMFORD, William, B.Sc., C. Eng. Fellow, I.E.E. Born 1905. Educ.: Manchester Univ. Director of Sales, Everett, Edgcumbe, 1950-55; Gen. Home Sales Manager, Hilger and Watts, Ltd., 1955-; Director, Hilger and Watts International, Ltd.; Director, Milcrowave Instruments, Ltd.; Director, Hilger-I.R.D., Ltd. Society: Soc. of Instrument Technol.
Nuclear interest: Nuclear physics.

BAM

Address: Hilger and Watts, Ltd., 98 St. Pancras Way, Camden Road, London, N.W.1, England.

BAMMERT, Karl, Prof. Dr.-Ing. Born 1908. Educ.: Karlsruhe Tech. Univ., Leiter der Entwicklungsabteilung für Strömungsmaschinen der Gutehoffnungshütte Sterkrade AG. Oberhausen-Sterkrade, 1948-55; ord. Prof. für Strömungsmaschinen und Direktor des Inst. für Strömungsmaschinen Hanover T.H., 1965-. Societies: V.D.I.; F.I.P.A.C.E.; W.G.L.; G.A.M.M.; Schiffbautechnische Gesellschaft (STG).
Nuclear interests: Reactor design; Nuclear power plants.
Address: Institut für Strömungsmaschinen der Technischen Hochschule Hanover, 25 Appelstrasse, 3 Hanover, Germany.

BAN, Nguyen Tu, Ph.D. Deputy Director, and in Dept. of Nucl. Phys., Dalat Nucl. Res. Centre, Atomic Energy Office of the Republic of Vietnam.
Address: Atomic Energy Office of the Republic of Vietnam, Dalat Nuclear Research Centre, Dalat, Vietnam.

BANCORA, Mario Eduardo, B.Sc., Dipl. Ing. Born 1918. Educ.: Litoral Univ., Argentina, and California Univ. Phys. Prof., Litoral Univ., Instituto Tecnologico Buenos Aires, 1946-; Director, Fundamental Res., Argentine A.E.C.; 1953; Sen. Sci. Adviser, Preparatory Com. (1957), Director, Tech. Supplies (1958-60), Alternate to Governor, I.A.E.A.; Director, Argentine A.E.C.; Participant, 3rd Inter-American Symposium on the Peaceful Application of Nucl. Energy, Rio de Janeiro, 1960. Phys. Sci. Com., Nat. Res. Council; Commission for Planification of Sci., Education and Culture, Alliance for Progress; Expert Com., Inter American A.E.C.; I.A.E.A. Expert to Guatemala 1967 for overall planning of atomic energy programme. Societies: Assoc. Fisica Argentina; I.E.R.E.; Union Argentina de Ing.; Asociacion Universitaria Argentino - Norteamericana.
Nuclear interests: Management; Nuclear engineering; Direct energy conversion; Nuclear physics.
Address: 1540 Chacabuco, Rosario, Argentina.

BANDO, Masako. Asst., Theory of Elementary Particles, Dept. of Phys. II, Faculty of Sci., Kyoto Univ.
Address: Kyoto University, Department of Physics II, Yoshida-Honmachi, Sakyo-ku, Kyoto, Japan.

BANERIAN, Gordon, M.S. (Eng. Mechanics), all graduate work beyond that in Phys. Ph.D. Born 1922. Educ.: Wayne State Univ. Vice-Pres., Eng., Aerojet-General Nucleonics, San Ramon, California, 1959-; formerly Manager, Turbo-Machinery Div., Azusa, now Sen. Sci., Liquid Rocket Operations, Sacramento, Aerojet-Gen. Corp. Chairman, Advisory Com. on Nucl. Energy Systems, Member, Power Plant Advisory Com., N.A.S.A.; Member, Fluid Mechanics Subcom., A.S.M.E.
Nuclear interests: Nuclear engineering as well as management.
Address: 1122 Stewart Road, Sacramento, California, U.S.A.

BANERJEE, B. M., Head of Instrumentation Saha Inst. of Nucl. Phys.
Address: Saha Institute of Nuclear Physics, 92 Acharya Prafulla Chandra Road, Calcutta 9, India.

BANERJEE, Manoj K., B.Sc., M.Sc., Ph.D. Born 1931. Educ.: Patna and Calcutta Univs. Prof., Maryland Univ., 1967-. Editorial Advisory Board, Phys. Letters, North Holland Publishing Co. Society: A.P.S.
Address: Department of Physics and Astronomy, Maryland University, College Park, Maryland, U.S.A.

BANERJEE, Major R. N. Sen. Officer, Inst. of Nucl. Medicine and Allied Sci.
Address: Institute of Nuclear Medicine and Allied Sciences, Metcalfe House, Delhi-6, India.

BANERJEE, S. P., Dr. Asst. Prof., Sect. of Nucl. Chem., Dept. of Chem., Saugar Univ.
Address: P.O: Saugar University, Department of Chemistry, Saugar, M.P., India.

BANKS, Charles Vandiver, B.Ed. (Hons.), M.S., Ph.D. Born 1919. Educ.: Western Illinois and Iowa State Univs. Assoc. Prof. Chem., Chem. Dept., and Assoc. Chem. (1949-54), Prof, Chem. (1954-), and Sect. Chief, Inst. for Atomic Res. and Ames Lab. of A.E.C., Iowa State Univ. Books: Analytical Chem. of the Rare Earths, in The Rare Earths, edited F.H. Spedding and A.H. Daane (New York, John Wiley, 1961); Cobalt in Treatise on Analytical Chem., edited I.M. Kolthoff, P.J. Elving, and E.B. Sandell (New York, Interscience Publishers, 1962); Nickel (ibid). Societies: A.C.S.; A.A.A.S.; Iowa Acad. Sci.
Nuclear interests: Analytical chemistry of titanium, vanadium, chromium, zirconium, niobium, molybdenum, hafnium, tantalum, tungsten, rare earths, thorium, and uranium; Determination of gases in metals and metal salts; Organophosphorus compounds as extractants for heavy metals.
Address: 2019 Ashmore Drive, Ames, Iowa, U.S.A.

BANKS, William F., B.S., M.S. (M.E.), M.S. (E.E.). Born 1924. Educ.: Texas A. and M., California (Los Angeles) and George Washington Univs. Programme Manager, Snap-8 Programme, Aerojet-Gen. Corp., Azusa, California.
Nuclear interest: Design and development of gas-cooled power reactors; Development of liquid metal power conversion systems.
Address: Aerojet-General Corporation, Azusa, California 91702, U.S.A.

BANNIER, Jan Hendrik, Dr. Phys. Born 1909. Educ.: Utrecht Univ., Director, Netherlands Organisation for the Advancement of Pure Res. (Z.W.O.), 1948-. Council, C.E.R.N. Societies: Netherlands Phys. Assoc.; Roy. Netherlands Chem. Assoc.; Roy. Inst. Eng.
Nuclear interests: Nuclear physics; Management.
Address: 6 van Bommellaan, Wassenaar, The Netherlands.

BÄNNINGER, Wilhelm. See BAENNINGER, Wilhelm.

BANOS, Rafael, Member, Grupo de Trabajo del Riesgo Atomico, Spain.
Address: Sindicato Nacional del Seguro, Comision del Riesgo Atomico, 4 Avenida Calvo Sotelo, Madrid 1, Spain.

BANTEGUI, Mrs. Celia Gallego, B.S., B.S.Ch.E., M.A. (Statistics). Born 1926. Educ.: Chicago, Harvard, Nat. and Philippines Univs. Meteorologist, Philippine Weather Bureau, 1949-56; Asst. Prof., Maths., Univ. of the East, 1949-; Lecturer, Statistics, Graduate School, Philippines Univ., 1958-62; Asst. Sci. (1959-62), Sen. Sci. (1962-67), Supervising Sci. (1967-), Philippine A.E.C. Member, Tech. Com. on Safe Transport of Radioactive Materials in Philippines, 1963-; Member, Board of Examiners for Chem. Eng., 1966-. Societies: H.P.S.; Internat. Rad. Protection Assoc.; Philippine Assoc. for Rad. Protection.
Nuclear interest: Health physics activities with emphasis on radiological health and safety procedures.
Address: 85 Sta. Catalina, Mesa Heights, Quezon City, Philippines.

BANTYSH, A. N. Papers: Co-author, Some Laws in the Thermodynamics of Isotope Exchange (Kernenergie, vol. 10, No. 4, 1967); co-author, The Partition Function Ratios of isotopically Substituted Molecules calculated by using the Urey-Bradley-Simanouti Potential Function (ibid, vol. 10, No. 5, 1967).
Address: Academy of Sciences of the U.S.S.R., 14 Leninsky Prospekt, Moscow V-71, U.S.S.R.

BAONZA DEL PRADO, Enrique, Civil Eng. Born 1927. Educ.: School of Civil Eng., Madrid. Chief, Lab. for Applications of Radioisotopes to Civil Eng., Ministry of Public Works, Madrid.
Nuclear interests: The use of radioisotopes in civil engineering; Development of routine method for the application of radiotracers in streamflow measurement; Dating of ground water with tritium and ^{14}C; Mixing of inflowing stream with the water in reservoirs; Prototype for determination of asphalt of bituminous concrete (in situ and/or in plant).
Address: 15 Avenida Bonn, Madrid 2, Spain.

BAPTIST, John Paul, A.B., M.S. Born 1919. Educ.: Maryville Coll. and Bucknell Univ. Fishery Biol. (Res.), Bureau of Commercial Fisheries, U.S. Dept. of the Interior, 1949-. Societies: American Fisheries Soc.; Assoc. of Southeastern Biol.
Nuclear interests: Marine radiobiology. Uptake of radionuclides by fishes. Application of radioisotope techniques to fishery biology.
Address: Bureau of Commercial Fisheries, Radiobiological Laboratory, Beaufort, N.C. 28516,U.S.A.

BAPTISTA, Antonio Manuel, Lic.Fis.Quim. Born 1924. Educ.: Lisbon Univ. Director, Isotope Lab., Inst. Portugues de Oncologia, 1953; Prof. Phys., Military Acad., 1961; Director, Centro de Estudos de Energia Nucl. Comissao de Estudos de Energia Nucl., 1966. Societies: Foreign Member, Real Academia de Ciencias, Spain; Sociedade Portuguesa de Radiologia e Medicina Nucl.; Sociedade Portuguesa de Quimica e Fisica; Internat. Soc. of Haematology.
Nuclear interests: Absolute measurements of radionuclides; Nuclear radiation detectors; Medical and biological applications of radionuclides; Nuclear reactor physics.
Address: Laboratorio de Isótopes, Instituto Portugues de Oncologia, Lisbon 4, Portugal.

BAR-OR, Abraham, M.Sc., Ph.D. Born 1924. Educ.: Hebrew (Jerusalem) and Birmingham Univs. Director, Metal. Labs., A.E.C., Israel. Society: Inst. of Metals., England.
Nuclear interest: Research and development in the field of nuclear materials.
Address: 10 Rehov Dvora, Beer-Sheva, Israel.

BARABANOV, I. R. Paper: Co-author, The Reliability Analysis of Methods for Studying Fast Neutron and γ-Ray Continuous Spectra (Atomnaya Energiya, vol. 16, No. 3, 1964).
Address: Academy of Sciences of the U.S.S.R., 14 Leninsky Prospekt, Moscow V-71, U.S.S.R.

BARABASCHI, S., Univ. degree in Physics with specialisation in Electronics; Ph.D. (Appl. Electronics). Born 1930. Educ.: Milan State Univ., At Labs., C.I.S.E., 1953-57; Head, Reactor Control and Servomechanism Group, Centro di Ispra, 1957-; Head, Control and Reactor Eng. Lab. (1961), Head, Technol. Div. (1963-), Director, Reactor Sector (1964-), C.N.E.N., Centro della Casaccia, Rome, Prof. i/c Reactor Control, Milan Politech. School, 1957-; Prof. i/c Appl. Electronics, Rome Univ., 1961-.
Nuclear interests: Reactor engineering; Reactor control and servo-mechanisms.
Address: 240 Via Cassia Antica, Rome, Italy.

BARAJAS, D. Alberto. Pres., Advisory Board, Comision Nacional de Energie Nuclear, Mexico.
Address: Av. Insurgentes Sur 1079 Tercer piso, Mexico 18, D.F., Mexico.

BARAN, James A., B.Sc. (Phys.), M.Sc. (Nucl. Eng.), Ph.D. (Nucl. Eng.). Born 1938. Educ.: Case Inst. Technol. and Kansas State Univ. Staff Member, Los Alamos Sci. Lab. Societies: A.N.S.; H.P.S.; A.A.A.S.; A.S.E.E.
Nuclear interests: Nuclear weapons effects;

Radiation shielding; Nuclear space applications and gamma ray spectroscopy; Presently involved in fast pulse and collimator techniques associated with nuclear weapon diagnostics.
Address: Los Alamos Scientific Laboratory, Group J-14, P.PO. Box 1663, Los Alamos, New Mexico 87544, U.S.A.

BARAN, Václav. Born 1932. Educ.: Military Coll. of Technol. Postgraduate Fellowship, (Prof. A.N. Nes meyanov), Moscow State Univ., 1955-56; Inst. of Inorganic Chem., Usti n.L., 1957-59; Inst. Nucl. Res., Rez, 1959-. Nuclear interest: Chemistry of nuclear materials.
Address: Nuclear Research Institute, Academy of Sciences, Rez, Prague, Czechoslovakia.

BARANOV, E. N. Papers: About Genesis of Redden Host Rocks of Hydrothermal Uranium Deposits (letter to the Editor, Atomnaya Energiya, vol. 5, No. 6, 1958); co-author, Content of Uranium in Sulphides as Indicator of Uranium Mineralisation (letter to the Editor, ibid, vol. 20, No. 2, 1966); co-author, To Problems of Moving of Hydrothermal Solutions. (ibid, vol. 20, No. 6, 1966).
Address: Academy of Sciences of the U.S.S.R., 14 Leninsky Prospekt, Moscow V-71, U.S.S.R.

BARANOV, I. A. Papers: Co-author, Comparison of the Kinetic Energies of the Fission Fragments produced from ^{238}U by Neutrons of Energies 3 and 15 Mev (letter to the Editor, Atomnaya Energiya, vol. 12, No. 2, 1962); co-author, Surface Barrier Silicon Detectors for Measurements in Neutron and Fission Fragment Fluxes (ibid, vol. 15, No. 1, 1964).
Address: Academy of Sciences of the U.S.S.R., 14 Leninsky Prospekt, Moscow V-71, U.S.S.R.

BARANOV, V. F. Paper: Dependence of Energy Loss Averaged over Electron Spectrum from End-Point Energy of Beta Spectrum, Atomic Number of Beta Emitter and Mode of Junction (letter to the Editor, Atomnaya Energiya,, vol. 19, No. 5, 1965).
Address: Academy of Sciences of the U.S.S.R., 14 Leninsky Prospekt, Moscow V-71, U.S.S.R.

BARANOV, V. I. Papers: The Use of Nuclear Emulsions to determine the Radiochemical Purity of α-emitters (letter to the Editor, Atomnaya Energiya, vol. 4, No. 2, 1958); co-author, Exact Measurements of the Concentrations of Radioisotopes of Lead and Bismuth in the Air of Underground Workings (letter to the Editor, ibid, vol. 9, No. 1, 1960); Development of Radiogeology in U.S.S.R. (ibid, vol. 24, No. 5, 1968).
Address: Academy of Sciences of the U.S.S.R., 14 Leninsky Prospekt, Moscow V-71, U.S.S.R.

BARANOV, Vladimir, Dr. ès Sc., Ing. des Arts and Manufactures. Born 1897. Educ.: Paris and Poitiers Univs. Ing. Conseil, Compagnie Générale de Géophysiques.
Address: 156 rue d'Aulnay, 92 Châtenay-Malabey, France.

BARANOWSKI, Frank Paul, B.Chem. Eng., M. Chem. Eng. Born 1921. Educ.: New York and Tennessee Univs. U.S. Army Lieutenant assigned to Manhattan Eng. District project, Oak Ridge, 1945-46; Chem. Eng. (1947-48) Physical Staff (1950-51) U.S.A.E.C., Oak Ridge; Union Carbide Corp. at A.E.C., O.R.N.L., 1948-50; Chem. Process Eng., Industrial Eng., Chief of Isotopes Separation Branch, Chief Chem. Processing Branch Deputy Director (1951-61), Director (1961-), Division of Production, U.S.A.E.C.
Nuclear interests: Management of production and related development programmes.
Address: 9206 Laurel Oak Drive, Bethesda, Maryland, U.S.A,

BARANSKI, Leslaw Antoni, B.M. Born 1942. Educ.: Acad. of Mining and Metal., Cracow. Res. Worker, Inst. of Nucl. Techniques, Lab. of Nucl. Instr.
Nuclear interests: Nuclear electronics; Application of isotopes in hydrogeology and mining.
Address: 54/9 Konarskiego str., Cracow, Poland.

BARASHENKOV, V. S.
Nuclear interest: High-energy nuclear physics.
Address: Academy of Sciences of the U.S.S.R., 14 Leninsky Prospekt, Moscow V-71, U.S.S.R.

BARATTA, Edmond John, B.S. Born 1928. Educ.: Washington and Jefferson Coll. and Northeastern Univ. Societies: A.C.S.; H.P.S. Nuclear interests: Radiochemical assays of environmental and biological media: Polonium-210, Caesium-137 and Strontium-90 in selected human tissues; Strontium-90 in human bones in infants and young adults; Polonium-210 in cigarette smoke and tobacco products.
Address: Northeastern Radiological Health Laboratory, United States Public Health Service, 109 Holton Street, Winchester, Massachusetts, U.S.A.

BARBACCIA Franco, M.D., Prof. Radiol. Born 1915. Educ.: Bologna and Pavia Univs. Primario Radiologo Inc., Ospedale Maggiore, Sesto S. Giovanni, Milan. Society: Soc. Italiana Radiol. Medica.
Nuclear interest: Radiodiagnostics.
Address: Ospedale Maggiore, Sesto S. Giovanni, Milan, Italy.

BARBANY, Gregorio MILLAN. See MILLAN BARBANY, Gregorio.

BARBARIC, Ljubomir, Dipl. eng. Born 1908. Educ.: Ljubliana Univ. Director, Sci. Res. Dept., Isotope Application and Ionising Rad. Protection, Yugoslav Nucl. Energy Commission; Prof. Phys., Faculty of Agriculture, Zemun near Belgrade. Director, Yugoslav Summer School of Nucl. Phys. Book: Nuklearna energija i njena primjena (Nucl. Energy and its

Applications) (Belgrade, 1961).
Nuclear interest: Nuclear physics.
Address: 7/11, Proleterski brigada, Belgrade, Yugoslavia.

BARBARO, Domenico, Prof. Prof. of Tech. Phys., Eng. Dept., Palermo Univ.; Member, Sicilian Nucl. Res. Com.
Address: Engineering Department, Palermo University, Villa delle Scienze, Palermo, Sicily, Italy.

BARBER, Robert Charles, B.Sc., Ph.D. Born 1936. Educ.: McMaster Univ. Post Doctorate Fellow, McMaster Univ., 1962-65; Asst. Prof., Manitoba Univ., 1965-. Societies: A.P.S.; Canadian Assoc. of Physicists.
Nuclear interest: Precise atomic mass determinations by high resolution mass spectrometer.
Address: Manitoba University, Winnipeg 19, Manitoba, Canada.

BARBIERI, Renato. Reader in Chem., Centro Chimica Nucl., Padua Univ.
Address: Padua University, Centro Chimica Nucleare, 6 Via Loredan, Padua, Italy.

BARBOUR, R. A. Head, Chem. Operations Div., Atomic Energy Board, South Africa. Adviser, Third I.C.P.U.A.E., Geneva, Sept. 1964.
Address: Atomic Energy Board, Private Bag 256, Pretoria, South Africa.

BARBOUR, William Ernest, Jr., B.S. (Elec. Eng., M.I.T.). Born 1909. Educ.: Michigan State Coll. and M.I.T., Founder, Pres. and Treas., Controls for Rad., Inc., 1957-58; Pres., Magnion, Inc., 1960-65; Director, Gen. Vacuum Corp., 1959-65; Executive Manager, Assoc. Nucl. Instrument Manufacturers, 1966-; Nucl. Consultant, 1965-; Consultant to N.A.S.A. 1965-. Nucl. Standards Board, U.S.A.S.I. Societies: A.I.E.E.; A.N.S.
Nuclear interests: Plasma, M.H.D. and fusion apparatus, especially in connection with high magnetic fields; Superconductivity and superconducting magnets; Nuclear instruments; Nuclear industrial controls.
Address: Box 460, Concord, Massachusetts 07142, U.S.A.

BARCELO, Gabriel, Civil Eng. Born 1909. Educ.: Escuela de Ingenieros de Camonos; Grenoble and Oxford Univs. Managing Director, Hidroeléctrica Moncabril, S.A. World Power Conference, Vienna, 1957; Course of Paris, 1957. Book: La Vibration du Beton. Societies: Assoc. para el Progreso de la dirección; Comisión Nacional de Productividad.
Nuclear interests: Management; Design, construction and operation of nuclear power plant.
Address: 157 Velazquez, Madrid 2, Spain.

BARCELO, Héctor. Head, Reactor Div., Puerto Rico Nucl. Centre.
Address: Puerto Rico Nuclear Centre, College Station, Mayaguez, Puerto Rico.

BARCHUGOV, V. V. Paper: Co-author, Neutron Chamber on Basis of "Long" Counter (letter to the Editor, Atomnaya Energiya, vol. 16, No. 4, 1964).
Address: Academy of Sciences of the U.S.S.R., 14 Leninsky Prospekt, Moscow V-71, U.S.S.R.

BARCHUK, I. F. Papers: Co-author, Fast Neutron Scattering and Capture by Atomic Nuclei (2nd I.C.P.U.A.E., Geneva, Sept. 1958); The Gamma-ray Spectra produced by Inelastic Fast Neutron Scattering in Mg, Al, Fe, Cu, Sn and Sb (Atomnaya Energiya, vol. 4, No. 2, 1958); co-author, Spectrum of γ-Rays Emerging from the Horizontal Channel in the VVR-M Reactor (letter to the Editor, ibid, vol. 12, No. 3, 1962); co-author, Fuel Elements Leak Control Method at WWR-M Reactor of the Physics Institute of the Ukrainian Academy of Sciences (letter to the Editor, ibid, vol. 21, No. 6, 1966).
Address: Nuclear Energy Institute, Academy of Sciences of the U.S.S.R., 14 Leninsky Prospekt, Moscow V-71, U.S.S.R.

BARCLAY, Frederick John, B.Sc., A.R.C.S.T. Born 1924. Educ.: Roy. Coll. Sci. Technol., Glasgow. Group Leader (Reactor Eng.), A.W.R.E., Aldermaston; Eng., Reactor Div., A.E.R.E., Harwell; Operations Manager (Dragon), A.E.E. Winfrith; Manager, Eng. Development, D.E.R.E. Societies: F.I.E.E.; Inst. of Mech. Eng.
Nuclear interest: Reactor design, operation and development.
Address: 25 Miller Place, Scrabster, Caithness, Scotland.

BARCLAY, John A. Manager, Huntsville, Northrop Space Labs., Northrop Corp.
Address: Northrop Space Laboratories, Hawthorne, California, U.S.A.

BARDEN, Charles R. Director, Div. Occupational Health and Rad. Control, Texas State Health Dept.
Address: Division of Occupational Health and Radiation Control, Texas State Department of Health, 1100 West 49th Street, Austin, Texas 78756, U.S.A.

BARDON, Georges. Directeur, Direction Gen., Elec. de France. Rep. of Elec. de France, Conseil d'Administration, A.T.E.N.
Address: Electricité de France, 2 rue Louis Murat, Paris 8, France.

BARDONE, Carlo. Managing Director, Tecnider-Tecnica Idrocarburi e Derivati S.p.A.
Address: Tecnider-Tecnica Idrocarburi e Derivati S.p.A., 4 Piazza Velasca, Milan, Italy.

BAREFORD, Christopher Frederic, M.Sc., Ph.D. Managing Director, Vickers Res. Establishment.
Address: Vickers Research Establishment, Sunninghill, nr. Ascot, Berks., England.

BAREIS, David W. Manager, Power Systems, Aero Space Technol. Res. Office, Marquardt Corp.
Address: Marquardt Corp., 16555 Saticoy Street, Van Nuys, California, U.S.A.

BARELKO, E. P. Paper: Co-author, Radiation-Chem. Stability of TBP in Hydrocarbons Solutions (Atomnaya Energiya, vol. 21, No. 4, 1966).
Address: Academy of Sciences of the U.S.S.R., 14 Leninsky Prospekt, Moscow V-71, U.S.S.R.

BARENBORG, Gilbert J. Manager, D2G Test and Field Eng., Knolls Atomic Power Lab., 1963-.
Address: Knolls Atomic Power Laboratory, Schenectady, New York, U.S.A.

BARENDREGT, Teun Johan, Dr. (Maths. and Phys.). Born 1920. Educ.: Utrecht State Univ. Managing Director, Comprimo N.V. Delegate to 1st, 2nd and 3rd Geneva Conferences; Conference of the Soviet Acad. Sci. on the Peaceful Use of Atomic Energy, Moscow 1955. Societies: Roy. Dutch Chem. Soc.; A.N.S.; E.A.E.S.; Netherlands Atomic Forum.
Nuclear interests: Design, construction and operation of reprocessing plants; Fabrication of low-cost fuel elements for water-cooled reactors; Economics of fuel cycles.
Address: Comprimo N. V., P.O. Box 4129, Amsterdam, Netherlands.

BARENDSEN, Gerrit Willem, Ph.D. (Phys.). Born 1927. Educ.: Groningen Univ. Res. Asst., Yale Univ., 1955-56; Radiobiol. Inst., Nat. Health Res. Council T.N.O., Netherlands, 1956-; Leiden Univ., 1958-.
Nuclear interests: Radiocarbon dating; Health physics; Radiobiology.
Address: Radiobiological Institute, National Health Research Council T.N.O., 151 Lange Kleiweg, Rijswijk, Netherlands.

BARFFORT, P. Euratom Member, Study Group on Digital Techniques, O.E.C.D., E.N.E.A.
Address: O.E.C.D., European Nuclear Energy Agency, 38 boulevard Suchet, Paris 16, France.

BARJON, Robert André Marie Joseph, M.Sc. (Pittsburgh), D. ès. Sc. (Paris). Born 1923. Educ.: Paris and Pittsburgh Univs. Prof. Nucl. Phys., Algiers Univ.; Vice-Director, Univ. and Cultural French Office for Algiers Inst. d'Etudes Nucléaires; Sen. Officer, Aldocatom. Book: Translation in French of Introductory Nucl. Phys. (Halliday).
Nuclear interests: Nuclear reactions at low energy, fundamental particles.
Address: Section de Génie Atomique, Institut Polytechnique, Grenoble, France.

BARKAN-MERIC, Süreyya, B.Sc. (Phys. - Mathematics), B.Sc. (Astronomy), Ph.D. (Nucl. Phys.), Prof. (Nucl. Phys). Born 1927. Educ.: Istanbul and Glasgow Univs. Istanbul Univ; Glasgow Univ., 1953; Argonne Nat. Lab., 1959; Saclay Centre d'Etudes Nucleaires, 1961; Strathclyde Univ. 1967; Harwell Atomic Energy Establishment 1968. Panel Member, Fellowships Com., Internat. Federation of Univ. Women. Book: Measurement of Radioisotopes (translated into Turkish) (Turkey, 1965). Societies: Turkish Phys. Soc.; Turkish-American Univ. Assoc.
Nuclear interests: Nuclear physics: mainly annihilation of single quantum positrons; Pair production; branching ratios stopping power and dE/dx for α- particles in solids; polarization of thermal neutrons; neutron capture. Ternary fission associated with long range particles such as α- tritium emission. Construction of a few special types of counters, and cloud chambers (multiparallel - plate γ- counter and diffusion cloud chamber suited to operate at very low pressure).
Address: Nuclear Physics Department, Istanbul University, Turkey.

BARKAS, Walter. Prof. of Phys., Dept. of Phys., Riverside, California Univ. Member, Subcom. on Penetration of Charged Particles in Matter, Com. on Nucl. Sci., Nat. Acad. Sci.-N.R.C.
Address: National Academy of Sciences - National Research Council, Committee on Nuclear Sciences, 2101 Constitution Avenue, N.W., Washington 25, D.C., U.S.A.

BARKER, J. A., B.Sc. Eng. (London), A.M.I.C.E. Member, Council, Inst. of Nucl. Eng.
Address: Institution of Nuclear Engineers, 147 Victoria Street, Westminster, London S.W.1, England.

BARKER, James J., B.Ch.E., M.Ch.E., D.Eng. Sci., Consulting Eng. Born 1922. Educ.: New York Univ. Prof., C.W. Post Coll. Sec., A.N.S. Societies: A.N.S.; A.I.Ch.E.; A.C.S.; N.Y. Acad. Sci.; A.A.A.S.; A.S.T.M.
Nuclear interests: Reactor and other nuclear plant and equipment design; Stable isotope separation; Adsorption of radioactive gases; Heat transfer.
Address: 10 Walden Avenue, Jericho, L.I., N.Y. 11753, U.S.A.

BARKER, N. T., M.Sc. (N.S.W.). Dept. of Nucl. and Rad. Chem., New South Wales Univ.
Address: New South Wales University, Box 1, Post Office, Kensington, New South Wales, Australia.

BARKER, Robert Francis, B.S. (E.E.), M.S. (Radiol. Phys.). Born 1922. Educ.: Ohio and Rochester univs. At Los Alamos Sci. Lab., California Univ., 1950-55; Isotopes Div., U.S.A.E.C., Oak Ridge, 1955-56; Div. of Licensing and Regulation, U.S.A.E.C., Washington, D.C., 1956-59; Sen. Officer, Div. of Health Safety and Waste Disposal, I.A.E.A., Vienna, 1959-61; Div. of Licensing and Regulation, U.S.A.E.C., 1961-67; Div. of Rad. Protection Standards, U.S.A.E.C., 1967-.

Book: Chapter on Rad. Protection in the Handbook on Non-destructive Testing (A.S.T.M. publication). Societies: H.P.S.; Soc. for Nondestructive Testing.
Nuclear interest: Health physics.
Address: United States Atomic Energy Commission, Washington, D.C. 20545, U.S.A.

BARKER, Robert Sidney, B.Sc. Born 1930. Educ.: Nottingham Univ. Group Leader, Rad. and Radiochem. Group, Res. Lab., Pilkington Brothers Ltd., Lathom, Ormskirk.
Nuclear interests: Radiation effects in solids, particularly glass; Use of radioisotopes to study ionic adsorption and diffusion; Industrial applications of radioisotopes; Health physics and safety.
Address: 15 Ormskirk Road, Rainford, near St. Helens, Lancs., England.

BARKER, Sidney Alan, D.Sc., Ph.D., B.Sc. (Hons.), F.R.I.C. Born 1926. Educ.: Birmingham Univ. Reader in Chem., Birmingham Univ. Societies: Chem. Soc.; F.R.I.C.
Nuclear interest: γ-Irradiation of carbohydrates.
Address: 1 Abdon Avenue, Selly Oak, Birmingham 29, England.

BARKOV, S. N. Papers: Turning Regulators Efficiency with Absorption Layer in Reactor Radial Reflector (letter to the Editor, Atomnaya Energiya, vol. 23, No. 4, 1967); Multigroup Analytical Method of Calculation of Heterogeneous Nucl. Reactor (ibid, vol. 24, No. 4, 1968).
Address: Academy of Sciences of the U.S.S.R., 14 Leninsky Prospekt, Moscow V-71, U.S.S.R.

BARKOVSKII, V. N. Paper: Co-author, 68.5 cm Sector - Focused Cyclotron (letter to the Editor, Atomnaya Energiya, vol. 20, No. 5, 1966).
Address: Academy of Sciences of the U.S.S.R., 14 Leninsky Prospekt, Moscow V-71, U.S.S.R.

BARLOCCI, A., Br. Sen. Officer, Lab. de Isotopes, Inst. de Radiologia y Centro de Lucha contra el Cancer, Hospital Pereyra Rossell, Montevideo.
Address: Laboratorio de Isotopes del Instituto de Radiologia y Centro de Lucha contra el Cancer, Hospital Pereyra Rossell, 1550 Blv. Artigas, Montevideo, Uruguay.

BARLOW, E. A. Civil and Mech. Design, Chalk River Nucl. Labs., A.E.C.L.
Address: Atomic Energy of Canada Ltd., Chalk River Nuclear Laboratories, Chalk River, Ontario, Canada.

BARLOW, Erasmus Darwin, M.A., M.B., B.Chir., D.P.M. Born 1915. Educ.: Cambridge Univ. Consultant, Psychological Medicine, St. Thomas's Hospital, 1950-66; Director, Group Investors Ltd., 1956-; Director (1958-), Chairman (1963-), Cambridge Instrument Co. Ltd.; Chairman, Electronic Instr. Ltd., 1962-. Member, Council, Sci. Instrument Manufacturers' Assoc., 1967.
Nuclear interest: Manufacture of instruments with wide applications in the nuclear field.
Address: Cambridge Instrument Company Limited, 13 Grosvenor Place, London, S.W.1, England.

BARLOW, Henry Septimus, B.Sc. (Hons., Phys.), M.Sc. (Res. Phys.). Born 1903. Educ.: Manchester Univ. Head, Phys. and Mathematics Dept., S.E. Essex Regional Coll. of Technol., -1964; Head, Elec. Eng. and Appl. Phys. Dept., Barking Regional Coll. of Technol., 1964-. Member, Joint Com. for Higher Nat. Diplomas or H.N.C.'s in Appl. Phys.; Member, Subject Panels for Degrees in Phys., Gen. Sci. and Photography, Council for Nat. Academic Awards. Society: F. Inst. P.
Address: 52 Goldstone Crescent, Hove, Sussex, BN3 6BE, England.

BARLOW, James Robert, A.B., M.P.A. Born 1926. Educ.: Princeton Univ. Special Asst. to Director, Sci. and Technol.Office, Exec. Office of the Pres., Washington. Societies: Nat. Assoc. Sci. Writers, New York; A.N.S.; A.A.A.S.
Nuclear interests: All aspects of nuclear applications having interest or concern for management.
Address: Science and Technology Office, Executive Office of the President, Washington, D.C. 20506, U.S.A.

BARMAN, H. Lowis, M.A. Educ.: Cambridge Univ. Formerly Director, Rolls-Royce and Assocs., Ltd. and Director, Nuclear Developments, Ltd.; Consultant, U.K.A.E.A.; Director, The Amersham Searle Corp., 1968-.
Address: c/o U.K.A.E.A., 11 Charles II Street, London S.W.1, England.

BARMAWI, M. Lecturer, Nucl. Phys. Dept., Bandung Inst. of Technol.
Address: Bandung Institute of Technology, 10 Djl. Ganeca, Bandung, Indonesia.

BARNABY, Charles Frank, M.Sc., Ph.D. Born 1927. Educ.: London Univ. Phys., Dept. of Clinical Res., Univ. Coll. Hospital Medical School, 1957-.
Nuclear interests: Scintillation counters, radiation physics, low-level counters.
Address: 7a George Street, London, W.1, England.

BARNARD, Anthony Charles Langrish, B.Sc., Ph.D. Born 1932. Educ.: Birmingham Univ. Res. Phys., Assoc. Elec. Industries, Ltd., 1956-58; Res. Assoc., State Univ. Iowa, 1958-60; Asst. Prof., Rice Univ., 1960-. Societies: A. Inst. P.; Fellow, Phys. Soc.; A.P.S.
Nuclear interests: Nuclear structure and reactions.
Address: Physics Department, Rice University, Houston, Texas, U.S.A.

BARNELL, Herbert Rex, M.A., Ph.D. (Cantab.), B.Sc. (London). Born 1907. Educ.: Cambridge and London Univs. Various scientific appointments, Ministry of Food and Ministry of Agriculture, Fisheries and Food; Chief Sci. Adviser (Food), Ministry of Agriculture, Fisheries and Food, 1959-. Societies: Soc. Chem. Ind.; Nutrition Soc.; Inst. Biol.; Inst. Food Sci. and Technol.
Nuclear interests: Fall-out contamination of foods; Treatment of foods with ionising radiations.
Address: Ministry of Agriculture, Fisheries and Food, Great Westminster House, Horseferry Road, London, S.W.1, England.

BARNES, Benjamin Ayer, S.B., M.D. Born 1919. Educ.: Harvard Coll. and Columbia Univ. Medical School. Asst. Surgical Resident, Memorial Hospital, New York, 1952; Surgical Res. Unit, U.S. Army Medical Corps, San Antonio, Texas, 1953; Clinical Assoc. in Surgery, Harvard Medical School, 1959-; Asst. Surgeon, Massachusetts Gen. Hospital, 1960-.
Nuclear interests: Biological applications of nuclear physics to biology and medicine.
Address: Massachusetts General Hospital, Boston 14, Massachusetts, U.S.A.

BARNES, Charles A., B.A., M.A., Ph.D. Born 1921. Educ.: McMaster, Toronto and Cambridge Univs. Prof. Phys., California Inst. Technol. Society: A.P.S.
Nuclear interest: Nuclear physics.
Address: Department of Physics, California Institute of Technology, Pasadena, California, U.S.A.

BARNES, Charles M., D.V.M., Ph.D. Born 1922. Educ.: Texas A. and M. Coll. and California Univ. Lt.-Col., U.S.A.F., Veterinary Corps, 1950-; Life Sci. and Safety Officer, Div. of Reactor Development, U.S.A.E.C., Washington 25, D.C., 1956-62; Armed Forces Inst. of Pathology, 1963-66; Manned Spacecraft Centre, N.A.S.A., 1967-. Societies: American Veterinary Medical Assoc.; American Veterinary Radiol. Soc.; H.P.S.
Nuclear interests: Toxicology of fission products; Aerospace application of nuclear energy.
Address: Rt 1 Box 92, Gladewater, Texas, U.S.A.

BARNES, David Walter Hugh, B.M., B.Ch. (Oxford), M.A. (Oxford). Born 1923. Educ.: Oxford Univ., and Univ. Coll. Hospital Medical School, London. Sci. Staff, M.R.C., Radiobiol. Res. Unit, Harwell, 1949-. Societies: Fellow, Roy. Soc. of Medicine, London; Fellow, Coll. of Pathologists.
Nuclear interests: Radiobiology, particularly the effects and treatment of whole body irradiation in the lethal range, and the use of radiation to achieve tissue grafting, especially of haematopoietic tissues; Investigation of the cellular kinetics and immunological problems associated with myeloid grafts.
Address: White Lodge, Boars Hill, Oxford, England.

BARNES, Edgar Charles, B.S. Born 1909. Educ.: Pennsylvania State Univ. Now at Westinghouse Elec. Corp., Bettis Atomic Power Div., Pittsburgh, Pa. Exec. Com., Nat. Com. on Rad. Protection; Nucl. Standards Board; American Standards Assoc.; Chairman, Gen. and Administrative Standards Com., Atomic Ind. Forum, Inc. Societies: A.I.H.A.; American Soc. of Safety Eng.; A.A.A.S.; H.P.S.
Nuclear interests: Health protection and safety in nuclear energy facilities.
Address: 124 Maple Avenue, Pittsburgh 18, Pennsylvania, U.S.A.

BARNES, Robert Sandford, B.Sc. (Hons. Phys.), M.Sc., D.Sc. Born 1924. Educ.: Manchester Univ. With U.K.A.E.A., 1948-, now Head, Metal. Div. Societies: Fellow, Inst. Phys.; Fellow, Inst. Metal.; Inst. of Metals.
Nuclear interests: Radiation effects and defects in crystals; Physical metallurgy and solid state physics.
Address: United Kingdom Atomic Energy Authority, Metallurgy Division, Atomic Energy Research Establishment, Harwell, Berkshire, England.

BARNES, Roderick Kirkwood, B.Sc., A.R.C.S. Educ.: Roy. Coll. of Sci. S.P.S.O., Nat. Phys. Lab. Society: Fellow, Inst. of Phys. and Phys. Soc.
Nuclear interests: Radiation measurements.
Address: National Physical Laboratory, Teddington, Middlesex, England.

BARNETT, Charles S., B.S. (Eng. Phys.), M.S. (Eng. Sci.). Born 1929. Educ.: California Univ., Berkeley. Phys., California Univ. Rad. Lab., 1953-55; Nucl. Eng., Mare Island Naval Shipyard, 1956-58; now Phys., Lawrence Rad. Lab., Livermore, California. Society: A.N.S.
Nuclear interests: Underground testing of nuclear devices.
Address: Lawrence Radiation Laboratory, P.O. Box 808, Livermore, California, U.S.A.

BARNETT, Clarence F. Director, Atomic and Molecular Processes Information Centre, Oak Ridge, 1965-. Papers: Co-author, The Oak Ridge Thermonucl. Experiment (2nd I.C.P.U.A.E., Geneva, Sept. 1958); co-author, Energy Distributions of Protons in DCX (Nucl. Fusion, vol. 1, No. 4, 1961).
Address: Oak Ridge National Laboratory, Post Office Box Y, Oak Ridge, Tennessee, U.S.A.

BARNETT, Claude C., B.S. (Walla Walla Coll.), M.S. (State Coll. of Washington), Ph.D. (Washington State Univ.). All degrees Phys. Major, Mathematics Minor. Born 1928. Educ.: Walla Walla Coll., State Coll. of Washington, Washington State Univ. Instructor in Phys. (1957), Asst. Prof. Phys. (1958), Assoc. Prof. Phys., Head of Phys. Dept. (1961-63), Prof. Phys. (1964-), Walla Walla Coll., Washington.

Society: A.P.S.
Nuclear interests: Theoretical nuclear physics; Applied nuclear physics.
Address: Physics Department, Walla Walla College, College Place, Washington, U.S.A.

BARNETT, Air Chief Marshal Sir Denis Hensley Fulton, G.C.B. Member, Weapons R. and D., U.K.A.E.A., 1965-.
Address: United Kingdom Atomic Energy Authority, 11 Charles II Street, London, S.W.1, England.

BARNETT, James Nutcombe, C. Eng., Fellow, I.E.E. Born 1918. Educ.: London Univ. Head, Eng. Services, A.E.I. Res. Lab., Aldermaston, 1947-63; J. J. Thomson Phys. Lab. (1963-65), Dept. of Appl. Phys. Sci. (1965-), Reading Univ. Society: Fellow, I.E.E.
Nuclear interests: Design of apparatus for plasma physics experiments, especially control systems and power supplies; Reactor management and safety.
Address: Boundary Hall, Tadley, near Basingstoke, Hampshire, England.

BARNETT, R. L. Deputy Chairman, Management Com., British Insurance (Atomic Energy) Com.
Address: British Insurance (Atomic Energy) Committee, Aldermary House, Queen Street, London E.C.4, England.

BARNETT, Steele, B.Sc. (Forestry). Born 1922. Educ.: Idaho Univ. Tech. Service Manager, Building Products, Boise Cascade Corp. Chairman, State of Idaho Nucl. Energy Commission.
Nuclear interests: Industrial promotion: Use of nuclear energy within Idaho; Location in Idaho of nuclear oriented industries.
Address: 6525 Robertson Drive, Boise, Idaho 83705, U.S.A.

BARNICK, Max, Dipl. Chem., Dr. phil., Prof. Born 1909. Educ.: Berlin Univ. Director, Battelle-Institut e.V., 1952-.
Nuclear interests: Management; Nuclear physics; Radiochemistry.
Address: Battelle-Institut e.V., Wiesbadener Strasse, Frankfurt am Main-W13, Germany.

BARNOTHY, Jeno M., Ph.D. Born 1904. Educ.: Pazmany Peter Univ., Budapest. Prof. Phys., Budapest Univ. Owner and Tech. Director, Forro Scientific Co. Society: A.P.S.
Nuclear interest: Theoretical nuclear physics.
Address: 833 Lincoln Street, Evanston, Illinois, U.S.A.

BARNOTHY, Madeline F., Ph.D. Born 1904. Educ.: Pazmany Peter Univ., Budapest and Gottingen Univ., Germany. Prof. Mathematics, Barat Coll., Lake Forest, Illinois, 1948-53; Res. Assoc., Northwestern Univ., 1953-59; Prof. Phys., Coll. of Pharmacy, Illinois Univ., 1955-. Societies: A.P.S.; A.A.A.S.
Nuclear interests: Nuclear physics. Methods to analyse samples of many radioactive isotopes.
Address: 833 Lincoln Street, Evanston, Illinois 60201, U.S.A.

BARNOWSKI, Frank P. Chief, Isotopes Separation Branch, then Chief, Chem. Processing Branch, Deputy Director (1960-61), Acting Director (1961-62), Director (1962-), Production Div., U.S.A.E.C.
Address: Production Division, U.S.A.E.C., Washington 25, D.C., U.S.A.

BARO, Gregorio B., Dr. in Chem. Born 1928. Educ.: Buenos Aires Univ. Member, Nat. A.E.C.; N.R.C. Fellow, 1958. Societies: Argentine Phys. Assoc.; Nucl. Soc.
Nuclear interests: Nuclear chemistry; Szilard-Chalmers effects; Applications of radioisotopes.
Address: 1272 Calle A. Alvarez, Vicente Lopez, Argentina.

BARON, H. W. Gen. Manager. Nucl. and Cryogenic Eng. Dept., Werkspoor Amsterdam.
Address: Werkspoor Amsterdam, Nuclear and Cryogenic Engineering Department, Oostenburgermiddenstraat, 62 Amsterdam, Netherlands.

BARON, Jean-Jacques, Ing. l'Ecole Centrale des Arts et Manufactures, Paris. Born 1909. Directeur, Ecole Centrale Arts et Manufactures Directeur, Div. Energie et Applications Atomiques, Cie. Pechiney; Vice-Prés., Cie. d'Etudes et de Réalisation de Combustibles Atomiques. Formerly Chairman, Sci. and Tech. Com., Euratom. Societies: Indatom; S.O.C.I.A.; C.F.M.U.; Groupe Intersyndical de l'Industrie Nucléaire.
Nuclear interests: Production of nuclear energy; Management: Fuel elements, new metals, reactor core and structural elements.
Address: 31 avenue Georges-Mandel, Paris 16, France.

BARON, Hon. Robert LE. See LE BARON, Hon. Robert.

BARON, Seymour, B.S. and M.S. (Chem. Eng., Johns Hopkins), Ph.D. (Chem. Eng., Columbia). Born 1923. Educ.: Johns Hopkins and Columbia Univs. Instructor and Adjunct Prof. in Nucl. Eng., Polytech. Inst. of Brooklyn and Columbia Univ. U.S. Rep. to 6th Nucl. Congress, Rome. I.S.A. Rep. 1959 Nucl. Congress; Chairman, Nucl. Energy Div., A.S.M.E. Societies: A.I.Ch.E.; A.S.M.E.; Fellow, A.N.S.; A.A.A.S. Nuclear interests: Engineering and design of nuclear power plants, reactor systems, and nuclear facilities.
Address: Vice Pres., Burns and Roe, Inc., 700 Kinderkamack Road, Oradell, New Jersey, U.S.A.

BARON, V. V. Paper: Co-author, Niobium-Base Alloys and Their Properties (Atomnaya Energiya, vol. 23, No. 1, 1967).
Address: Academy of Sciences of the U.S.S.R., 14 Leninsky Prospekt, Moscow V-71, U.S.S.R.

BARONI, Annetta. See LUCIANO, Annetta.

BARONI, Giustina, Dr. in Chem., Dr. in Phys., Libera Docenza in Fisica Sperimentale. Born 1923. Educ.: Rome Univ. Researcher, Istituto Nazionale di Fisica Nucleare; Prof., Rome Univ. Society: Società Italiana di Fisica.
Nuclear interest: Elementary particle physics.
Address: 32 Via Ovidio, Rome, Italy.

BARR, Frank Thomas, Ph.D., M.S., B.S. Born 1910. Educ.: Washington and Illinois Univs. Sen. Eng. Assoc., Esso Res. and Eng. Co., 1952-; Special assignment, Standard Oil Co. (N.J.), 1963. Chairman, Economics Area Programme Com., A.I.C.E. Societies: A.N.S.; Nucl. Eng. Div., A.I.C.E.
Nuclear interests: Nuclear technology and economics generally; Radiation processing, radiation supply; Competitive energy sources; Nuclear fuel cycle, reactor design, heavy water production.
Address: Petroleum Staff, Esso Research and Engineering Co., P.O. Box 215, Linden, N.J., U.S.A.

BARR, John H., Dr. Member, Nat. Advisory Com. on Rad.; Prof. of Radiol., Tufts Univ.
Address: Tufts University, 136 Harrison Avenue, Boston, Massachusetts 02111, U.S.A.

BARR, Nathaniel F., B.S., A.M., Ph.D. Born 1927. Educ.: Columbia Univ. Res. Assoc., Brookhaven Nat. Lab., 1954-56; Asst., Div. of Biophys., Sloan-Kettering Inst., 1956-61; Asst. Prof., Cornell Univ. Medical School, 1958-61; Lecturer, Gen. Studies, Columbia Univ., 1957-61; Chief, Radiol. Phys. and Instrumentation Branch, Biol. and Medicine Div. (1961-67), Tech. Adviser to Asst. Gen. Manager for R. and D., Office of the Gen. Manager (1967-), U.S.A.E.C. Assoc. Editor, Rad. Res. Societies: Rad. Res. Soc.; A.C.S.; H.P.S.
Nuclear interests: Radiation chemistry of aqueous solution; Dosimetry of ionising radiation.
Address: Office of the General Manager, B-312, United States Atomic Energy Commission, Washington D.C. 20545, U.S.A.

BARRAGAN M., Luis Fernando, Dr. in Medicine. Born 1932. Educ.: Univ. Mayor de San Andrés, La Paz, Bolivia. Chief, Radioisotopes Application Dept., Bolivian Commission of Nuclear Energy, 1963-. Teaching Asst. in Radioisotopes of the Propedentic Clinic Professorship (Univ. Mayor de San Andrés); Sci.-Director, Nucl. Medicine Centre; Member, Nat. Council of Sci. and Tech. Investigations, Nat. Sci. Acad., Bolivia; Member, Sci. Council, Bolivian High Altitude Inst. Societies: Founding Member, Bolivian Biol. and Nucl. Medicine Soc.; Representing Member, Latinoamerican Soc. of Biol. and Nucl. Medicine Assoc.; Founding Member, Bolivian Soc. of Endocrinology; Bolivian Medicine Athenaeum.
Nuclear interests: Study of the iron metabolism in altitude (4000 mts. height); Study of cardiopulmonary haemodynamics with radioisotopes in altitude; Blood volume determination and red cell survival in altitude; Study of endemic goiter and iodine metabolism at various altitudes.
Address: Centro de Medidina Nuclear, No. 4093 Casilla, La Paz, Bolivia.

BARRAGUE, Jorge E., Dr. Centro de Radiooncologia, Argentina.
Address: Centro de Radiooncologia, 888 Calle Azcuenaga, Buenos Aires, Argentina.

BARRALL, Raymond C., B.S., M.S. Born 1930. Western New Mexico and Rochester Univs. Supervisory Health Phys., Mare Island Naval Shipyard, 1957-59; Chief Health Phys., I.I.T. Res. Inst., 1959-65. Director, Health Phys., Stanford Univ., 1965-. Chairman, Student Activities Com., A.N.S. Societies: H.P.S.; A.N.S.; A.S.T.M.; American Assoc. of Physicists in Medicine.
Nuclear interests: All phases of health physics; Determination of neutron spectra in and near reactors and accelerators; Measurement of radioactive materials dispersed in air; Induced activity and absolute measurement of radioactivity; Levels and implications for patient dosimetry of radioactive impurities in radiopharmaceuticals; High energy neutron cross-sections.
Address: 978 Blair Court, Palo Alto, Calif., U.S.A.

BARRE, Raymond, Prof. Born 1924. Vice-Pres., Commission of the European Communities, 1967-.
Address: Commission of the European Communities, 23 avenue de la Joyeuse Entree, Brussels 4, Belgium.

BARREIRO, Ramon APRAIZ. See APRAIZ BARREIRO, Ramon.

BARRELL, Frank. Born 1904. Educ.: Manchester Coll. of Technol., and Woolwich Polytech. Generation Construction Eng., Yorkshire Div. (1948-57), Generation Construction Eng. H.Q. (1957-58), C.E.A.; Asst. Chief Project Eng. (1958-64), Deputy Project Group Director (1964-66), Project Group Director (1966-), Midlands Project Group, C.E.G.B. Societies: Inst. Mech. Eng.; I.E.E.; British Inst. of Management.
Nuclear interests: Responsible for nuclear projects of Midlands Project Group of C.E.G.B. i.e., Sizewell and Hinkley Point 'B'.
Address: Highlands, School Lane, Alvechurch, Worcestershire, England.

BARRENENGOA, Emilio de USAOLA. See de USAOLA BARRENENGOA, Emilio.

BARRER, Lester A., B.S. (Rutgers), M.Sc. (Yale), M.S. (New York). Born 1923. Educ.: Rutgers, Yale and New York Univs. Instructor, Hunter Coll., 1952-55; Res. Asst., Post Graduate Medical School, New York Univ., 1955-56; Sen.

Public Health Eng. (1956-58), Rad. Phys. (1958-60), Director of Res., Radium Res. Project (1960-63), New Jersey State Dept. Health; Tech. Adviser, New Jersey Commission on Rad. Protection, 1958-63; Private Consultant, 1959; Lecturer, Industrial Medicine, Post Graduate Medical School, New York Univ., 1959-; Formerly with Nucl. Safeguards Group, Nucl. Utility Services, Inc.; Manager, Bio-Medical Sci. Dept., Documentation Inc., 1965-. Fellow and member Res. Policy Com., Occupational Health Sect., A.P.H.A.; Rep., A.P.H.A., to Nucl. Standards Board, Subcom. on Nucl. Fuel Eng. (N5); Liaison Rep., Reactor Safety (N6) to Rad. Safety (Personnel) (N7) A.S.M.E.; Com. on Waste Disposal, Recording Units and Current Practice of Disposal of Radioactive Wastes by Isotope Users (N5.2); Com. on Design of Radioisotope Lab. (N5.4). Societies: H.P.S.; A.P.H.A.; American Conference Governmental Ind. Hygienists; American Ind. Hygiene Assoc.; New York Acad. Sci.
Nuclear interests: Epidemiology of radiation effects; Standardisation of nuclear activities; Design of experiments; Radiobiology; Hazards analysis for space nuclear application; Health physics; Environmental safety aspects of nuclear facilities.
Address: EBS Management Consultants Inc., 1625 Eye Street, N.W. Washington D.C. 20006, U.S.A.

BARRETEAU, M. Council Member representing Carbonisation et Charbons Actifs (C.E.C.A.), Groupement d'Industries Atomiques (GIAT).
Address: Carbonisation et Charbons Actifs (C.E.C.A.), 24 rue Murillo, Paris 8, France.

BARRETT, John Henry. Born 1930. Exptl. Officer, Isotope Applications Unit, Nat. Chem. Lab., D.S.I.R., 1957-65; Rad. Sci. Div., Nat. Phys. Lab., 1965-.
Nuclear interests: Radiation chemistry and chemical dosimetry.
Address: Radiation Science Division, National Physical Laboratory, Teddington, Middlesex, England.

BARRETT, Joseph Leo, B.A. (Geology - Chem.) Born 1932. Educ.: Western Reserve Univ. Operation Manager, Semi-Elements, Inc.
Nuclear interests: Spin resonance studies; Research of radioactive crystals; Neutron monochromatic crystals.
Address: 229 Haymont Drive, Gibsonia, Pennsylvania, U.S.A.

BARRETT, Lawrence G., B.S. (Phys. and Mathematics). Born 1923. Educ.: Trinity Coll., Hartford. Reactor Phys., K.A.P.L., G.E.C., 1953-57; Supt., Exptl. Eng. Phys. (1957-62), Project Manager and Tech. Manager, Nucl. Development Centre (1962-64), Babcock and Wilcox Co.; Pres., American Novawood Corp., 1964-. Society: A.N.S.
Nuclear interests: Process radiation; Reactor physics.
Address: 2432 Lakeside Drive, Lynchburg, Virginia 24501, U.S.A.

BARRETT, Lionel Richard, M.A., B.Sc., M.S. Born 1911. Educ.: Oxford Univ. Lecturer in Ceramic and silicate technol., Imperial Coll., London. Member of Council, British Ceramic Soc. Societies: British Ceramic Soc.; Soc. of Glass Technol.; Inst. of Fuel.
Nuclear interest: Ceramic materials for reactors.
Address: Department of Chemical Engineering and Chemical Technology, Imperial College, London, S.W.7, England.

BARRETT, Matthew J. Director, Rad. Shielding Group, Tech. Operations Inc., 1964-.
Nuclear interest: Radiation shielding.
Address: Radiation Shielding Group, Technical Operations Inc., Burlington, Massachusetts, U.S.A.

BARRETT, Norman Thomas, B.Sc.(Eng., First Class Hons.), C. Eng., A.M.I. Struct. E. Born 1923. Educ.: London Univ. Asst. Chief Structural Eng., U.K.A.E.A. Societies: Nucl. Eng. Soc.; A.M.I. Struct. E.
Nuclear interests: Structural engineering in atomic energy, particularly reactor containment structures and pressure vessels in both steel and concrete.
Address: 40 Culcheth Hall Drive, Culcheth, Warrington, Lancashire, England.

BARRETT, Paul H., Ph.D.,B.S. Born 1922. Educ.: Montana State and California (Berkeley) Univs. Prof., Phys., California Univ., Santa Barbara.
Nuclear interest: Nuclear structure.
Address: California University, Santa Barbara, Goleta, California, U.S.A.

BARRETT, Peter John, B.Sc. (Oxon.), M.A. (Oxon.), B.Sc. (Eng., Cape Town). Born 1934. Educ.: Cape Town and Oxford Univs. Lectureship in Phys., Cape Town Univ. Society: Fellow, Phys. Soc. (England).
Nuclear interest: Plasma physics.
Address: Department of Physics, Imperial College, London, S.W.7, England.

BARRETT, Richard Ethelred, B.Sc. (Mining). Born 1905. Educ.: McGill Univ. Head, Dept. Mining Eng., Toronto Univ., 1948-53; Manager, Beaverlodge Operation (1953-55), Director of Ore Procurement (1955-63), Eldorado Mining and Refining, Ltd., Ottawa; Exec. Director, Canadian Inst. of Mining and Metal., Montreal, 1963-. Societies: Canadian Inst. Mining and Metal.; A.I.M.M.E.
Nuclear interest: Raw materials.
Address: Apt. 201, West Wing, 4300 Western Avenue, Montreal 6, P.Q., Canada.

BARRETT, Terence, B.Sc. (Eng.). Born 1924. Educ.: South Wales and Monmouthshire Univ. Eng. Lab. (1948-53), Design Sect., Reactor Services Group, Eng. Div. (1953-57), Chief Pile

Operator, Pluto Reactor (1957-59), Group Leader, Design Group, Res. Reactors Div. (1959-61), Sen. Eng., Res. Reactors Div. (1961-65), Adviser on irrad. equipment to K.F.A. Jülich, West Germany (1965-67), Head Design Dept., Res. Reactors Div. (1967-), A.E.R.E. Society: Inst. Mech. Eng.
Nuclear interests: Reactor design, operations and maintenance; Experimental irradiation; Management.
Address: Research Reactors Division, United Kingdom Atomic Energy Authority, Atomic Energy Research Establishment, Harwell, Berkshire, England.

BARRIGA VILLALBA, Antonio Ma., Dr.
Pres., Board of Directors, Instituto de Asuntos Nucleares.
Address: Instituto de Asuntos Nucleares, Apartado Aereo 8595, Bogota D.E., Colombia.

BARRIOL, Jean, Prof. Prof., 3e cycle de Physique Theorique (Physique des Plasmas), Ecole Nat. Superieure d'Electricite et de Mecanique, Faculte des Sci., Nancy Univ.
Address: Nancy University, 2 rue de la Citadelle, Nancy, (Meurthe-et-Moselle), France.

BARROS, F. SOUZA. See SOUZA BARROS, F.

BARROS, Jose Antonio QUEIROZ de. See QUEIROZ de BARROS, Jose Antonio.

BARROS, L. A. DE. See DE BARROS, L. A.

BARROS, Manuel Gaspar DE. See DE BARROS, Manuel Gaspar.

BARROS, Solange M. C. de. See de BARROS, Solange M. C.

BARRUEL, Georges Michel. Dragon Project, 1960-63; Bureau Veritas, 1963-.
Nuclear interests: Nuclear engineering; Inspection of materials and components for reactors; Erection survey.
Address: Bureau Veritas, 17 Battery Place, New York, U.S.A.

BARRY, Jacques. Docteur en Droit. Diplômé d'Etudes Supérieures de Droit Public. Licencié ès-Lettres. Born 1929. Educ.: Faculté de Droit et Faculté de Lettres de Nancy. Avocat Stagiaire à la Cour de Nancy 1952. Responsable des questions Juridiques Commerciales et Financières du Centre de Marcoule. Chargé d'Enseignement à l'Institut d'Etudes Politiques de la Faculté de Droit d'Aix-en-Provence, 1957-
Nuclear interests: Questions d'ordre Juridique; Responsabilté en matière atomique; Protection des travailleurs; Fonctionnement des organismes nationaux et internationaux; Economie; Aspect industriel et commercial de l'energie atomique; Finance.
Address: Cour de Turroye, Villeneuve lès Avignon, Gard, France.

BARS, B., M.Sc. Sen. Officer, Reactor Lab., Dept. of Tech. Phys., Inst. of Technol.
Address: Institute of Technology, Helsinki, Finland.

BARSCHALL, Henry H., Ph.D. Born 1915. Educ.: Princeton Univ. Asst. Div. Leader, Los Alamos Sci. Lab., 1951-52; Prof. Phys., Wisconsin Univ., 1950-. Assoc. Editor, Nucl. Phys., 1959-; Chairman, Nucl. Phys. Div., A.P.S., 1968-69. Books: Detection of Neutrons in Encyclopedia of Phys., vol. 45, (Springer-Verlag, 1958); Fast Neutron Resonances in Nuclear Spectroscopy, vol. A, (Academic Press, 1960). Society: Fellow, A.P.S.
Nuclear interest: Neutron physics.
Address: Sterling Hall, Madison, Wisconsin 53706, U.S.A.

BARSHAY, Saul, A.B. (Cornell), M.S., Ph.D. (California, Berkeley). Born 1933. Educ.: Cornell and California (Berkeley) Univs. Prof. Phys., Rutgers, State Univ.
Nuclear interest: Nuclear theory.
Address: Physics Department, Rutgers State University, New Brunswick, New Jersey, U.S.A.

BARSUKOV, O. A. Papers: Co-author, The Spatial and Energy Distributions of the Neutrons in a Stratum containing a Borehole (Atomnaya Energiya, vol. 10, No. 5, 1961); co-author, Metal-Water Shielding for Point Neutron Source (ibid, vol. 15, No. 1, 1964); co-author. Study of Spectra and Doses Generated by Monoenergetic Neutron Source in Iron-Water Shielding (ibid, vol. 21, No. 1, 1966); co-author, Two-Layer Shielding Optimisation Against Isotopic Fast Neutron Sources (ibid, vol. 23, No. 6, 1967).
Address: Academy of Sciences of the U.S.S.R., 14 Leninsky Prospekt, Moscow V-71, U.S.S.R.

BARTECKI, Adam, Doz. Dr. Born 1920. Educ.: Wroclaw Tech Univ.; Inst. Chem. Technol. of Coal (1950-54), Inst. Inorg. Chem. II (1954-), Wroclaw Tech. Univ.; Polish Acad. Sci., Inst. Phys. Chem., Wroclaw, 1956-62. Editorial Board of Absorption Spectra in the Ultraviolet and Visible Region, Hungarian Acad. Sci., Budapest; Secretary, Spectroscopy Com., Polish Acad. Sci. Societies: Polish Chem. Soc.; Hungarian Acad. Sci.; Polish Acad. Sci.
Nuclear interests: Chemistry of uranium and physico-chemical researches in this field, especially molecular spectroscopy and absorption spectra theory of inorganic compounds as applied to oxycations and uranyl compounds, and organic solvo-chemistry of uranium compounds.
Address: Politechnika Wroclawska, Katedra Chemii Pierwiastków Rzadkich, 27 Wybrzeze Wyspiańskiego, Wroclaw, Poland.

BARTELS, William Charles, B.S., M.Sc. Born 1923. Educ.: Fordham Univ. and Polytech. Inst. of Brooklyn. Guest Assoc. Eng., Brookhaven Nat. Lab., 1951-52; Development Eng. at G.E.C., at K.A.P.L. and at H.K. Ferguson Co.,

Sen. Sci. and a manager, Assoc. Nucleonics, Inc.; Reactor Phys., Div. of Reactor Development, U.S.A.E.C., 1958-61; Tech. Asst. to various individual commissioners, U.S.A.E.C. Office of the Commissioners, 1961-67; Branch Chief, Terrestrial Low Power Reactors, U.S.A.E.C., 1967-. Society: A.N.S.
Nuclear interests: Research, development and uses of nucleonics.
Address: U.S. Atomic Energy Commission, Washington, D.C. 20545, U.S.A.

BARTH, Earl E., B.A., M.D. Born 1901. Educ.: Northwestern Univ. Chief Consultant, Veterans Administration Res. Hospital, 1954-; Chairman, Dept. of Radiol., and Prof., Northwestern Univ. Medical School, 1957-. Member, American Board of Radiol., 1959-65; Pres., A.C.R., 1960-61; Pres., American Roentgen Ray Soc., 1962.
Nuclear interest: Nuclear medicine.
Address: 670 N. Michigan Avenue, Chicago, Illinois 60611, U.S.A.

BARTH, J. Ancien Elève de l'Ecole Polytechnique. Born 1918. Directeur, Services d'Etudes Nucléaires, Sté. Générale de Constructions Electriques et Mécaniques, Alsthom. Societies: A.T.E.N.; Brevatom; Bureau Veritas; Groupe Intersyndical de l'Industrie Nucléaire.
Nuclear interests: Génie Atomique; Installations complète de production d'énergie nucléaire; Calcul de réacteurs; Mécanique et thermique nucléaire; Transfert de chaleur; Laboratoire d'études et de technique.
Address: Société Alsthom, 20 rue d'Athènes, 75-Paris, 9, France.

BARTHELS, M. Member, Management Com., Bureau Belge des Radioisotopes.
Address: Bureau Belge des Radioisotopes, 14 rue de la Chancellerie, Brussels 1, Belgium.

BARTHOLOME, Ernst, Prof. Dr. Mitglied, Arbeitskreis II/2, Chemie, Deutsche Atomkommission, Federal Ministry for Sci. Res.
Address: Federal Ministry for Scientific Research, 46 Luisenstrasse, 532 Bad Godesberg, Germany.

BARTHOLOMEW, Gilbert Alfred, B.A., Ph.D. Born 1922. Educ.: British Columbia and McGill Univs. P.R.O., Head Neutron Phys. Branch, A.E.C.L., 1948-. Societies: Canadian Assoc. Phys.; A.P.S.
Nuclear interests: Nuclear physics; Neutron physics.
Address: Neutron Physics Branch, Atomic Energy of Canada Limited, Chalk River, Ontario, Canada.

BARTHOLOMEW, N. C. Vice-Pres., Carborundum Co.
Address: Carborundum Company, P.O. Box 337, Niagara Falls, New York, U.S.A.

BARTKE, Jerzy, Ph.D. (Phys.,) Born 1936. Educ.: Jagellonian Univ. Asst., Jagellonian Univ., 1956-57; Asst. (1958-62), Sen. Asst. (1962-64), Res. Assoc. (1964-68), Inst. Nucl. Res.; At CERN, Geneva, 1959-61; At Inst. de Physique Nucléaire, Paris, France, 1966-68. Society: Polish Phys. Soc.
Nuclear interests: High energy reactions of strongly interacting particles, especially multiple production of mesons. Studies by means of the bubble chamber and nuclear emulsion techniques.
Address: Institute of Nuclear Research, 30 Al.Mickiewicza, Cracow, Poland.

BARTL, Daniel Otakar. Born 1921. Educ.: Prague Tech. Chem. High School. Specialist for corrosion, surface finishing and decontamination, Nucl. Dept., Skoda-Works, Pilsen, Czechoslovakia; Lecturer, Czechoslovak Res. and Tech. Soc., Nucl. Sect., Sect. for Corrosion and Metal Protection. Society: Czechoslovak Res. and Tech. Soc.
Nuclear interests: Corrosion and methods of protection of nuclear power equipments, decontamination of item, decontamination of radioactive liquid wastes.
Address: 8 nám. Hrdinu, Prague 4, Czechoslovakia.

BARTLETT, C. P., B.Sc. (E.E.). Deputy Chief Elec. Eng., Eng. Group, U.K.A.E.A. Society: I.E.E.
Nuclear interest: Electrical engineering.
Address: 34 Meadway, Sale, Cheshire, England.

BARTLETT, Sidney C. L. Born 1910. Educ.: Northampton Polytech. Inst. Head, Patent Dept., A.E.R.E., Harwell, 1950-. Society: Fellow, Chartered Inst. Patent Agents.
Nuclear interests: Overall interest in all nuclear developments with a view to establishing industrial property rights.
Address: Fieldside, Landon Road, Blewbury, Berks., England.

BARTO, Robert M. Member, Nucl. Components Com., Soc. for Nondestructive Testing Inc.
Address: Society for Nondestructive Testing, Inc., Nuclear Components Committee, 914 Chicago Avenue, Evanston, Illinois 60202, U.S.A.

BARTOCCI Aldo, Prof. Metal. Ing. Born 1909. Educ.: Rome Univ. Prof. Metal., Direttore Generale Centro Sperimentale Metallurgico, Rome, 1967. Book: Co-author, I Metalli e l'Acciaio, 3rd edition (Alterocca-Terni, 1963). Society: F.I.E.N.
Nuclear interests: Argomenti metallurgici; Costruzione di nuclear vessels.
Address: 6 Via O. Tommasini, 00162 Rome, Italy.

BARTOLI, Ugo ASCOLI-. See ASCOLI-BARTOLI, Ugo.

BARTOLOME, Zoilo M. Sen. Sci. (Reactor Phys.), Philippine Atomic Res. Centre.
Address: Philippine Atomic Research Centre, Diliman, Quezon City, Philippines.

BARTOLOMEI, Giancarlo, Medical Doctor. Born 1931. Educ.: Pisa Univ. Res. Asst., Medical Clinic, Pisa Univ., 1956-. Papers: 15 papers on experimental and clinical uses of radioisotopes in cardiology and endocrinology. Society: Italian Soc. for Nucl. Biol. and Medicine.
Address: Centro di Medicina Nucleare, Università di Pisa, Italy.

BARTOSEK, Jiri, R.N.Dr. Born 1936. Educ.: J. Ev. Purkyne Univ., Brno.
Nuclear interests: Research on the natural radioactivity of rocks; Radiometric analysis of Th, U and K content in rocks; Activation analysis by means of a neutron generator and its application in geology and geophysics.
Address: Institute of Applied Geophysics, 102 Podebradova, Brno, Czechoslovakia.

BARTOSEK, Václav, Candidate of Phys. and Mathematical Sci. Born 1928. Educ.: Charles Univ., Prague.
Nuclear interest: Reactor theory and analysis.
Address: Institute of Nuclear Research, Czechoslovakian Academy of Sciences, Prague-Rez, Czechoslovakia.

BARTOSZEK, Bronislaw, M.Sc. Member, State Council for the Peaceful Use of Nucl. Energy.
Address: State Council for the Peaceful Use of Nuclear Energy, Room 1819, Palace of Culture and Science, Warsaw, Poland.

BARTZ, M. H., Supt., Operations Eng., Phillips Petroleum Co., Atomic Energy Div., Idaho Falls.
Address: Phillips Petroleum Co., Atomic Energy Division, Idaho Falls, Idaho, U.S.A.

BARUCH, Pierre, Agrégé de Physique, D. és Sc. Born 1927. Educ.: Ecole Normale Supérieure, Paris Univ. At Ecole Normal Supérieure, Paris, 1952-59; at Bell Telephone Lab., Murray Hill, New Jersey, 1959-60; Prof., Faculty of Sci., Paris Univ. Societies: Sté. Française de Physique; A.P.S.
Nuclear interests: Radiation damage; Solid-state electronics applied to particle detection; Use of nuclear techniques in solid-state physics.
Address: Laboratoire de Physique, Ecole Normale Supérieure, 24 rue Lhomond, Paris 5, France.

BARUT, Asim O., Diploma and Dr. Sc. Born 1926. Educ.: Swiss Federal Inst. Technol., Zürich. Prof. Theoretical Phys., Colorado Univ. Books: Elektronenoptisches und statistisches Verhalten der Gittervervielfacher (Druekerei Lehmann A.G., Zürich, 1951); Electrodynamics and Classical Theory of fields and particles (Macmillan, 1964); The Theory of the scattering Matrix (Macmillan, 1967). Societies: A.P.S.; Swiss Phys. Soc.; A.A.A.S.
Nuclear interests: Nuclear physics - elementary particles.
Address: Physics Department, Colorado University, Boulder, Colorado 80302, U.S.A.

BARZ, Hans Ulrich, Dr. rer. nat.,Dipl.-Phys. Born 1933. Educ.: Greifswald Univ.
Nuclear interests: Reactor theory; Time dependent problems; Fast reactor.
Address: Deutsche Akademie der Wissenschaften zu Berlin Forschungsgemeinschaft, Zentralinstitut für Kernforschung, 8051 Dresden, PF 19, German Democratic Republic.

BARZELATTO, Jose, M.D. Born 1926. Educ.: Chile Catholic and Chile Univs. Asst. Prof., Medicine (1952-), and Member, Council, Chile Univ.; i/c of training course for Latin American physicians in the clinical use of radioisotopes, sponsor American Health Organisation; Member, Board, Chilean Nucl. Energy Com., 1957-; Exec. Sec., Comisión Nacional de Investigación Cientifica y Tecnológica. Societies: Sociedad Chilena de Medicina Nucl.; Endocrine Soc.; Corresponding Member, Argentine Soc. of Nucl. Medicine.
Address: 70-D Casilla, Santiago, Chile.

BASARGIN, Yu. G. Paper: Co-author, 68.5 cm Sector - Focused Cyclotron (letter to the Editor, Atomnaya Energiya, vol. 20, No. 5, 1966).
Address: Academy of Sciences of the U.S.S.R., 14 Leninsky Prospekt, Moscow V-71, U.S.S.R.

BASCHIERI Ivo, M.D., specialist in radiology. Born 1926. Educ.: Pisa Univ. Asst. at Clinica Medica, Pisa Univ.,-1957; Asst. at II Clinica Medica, Rome Univ. 1957-. Societies: Soc. Italiana di Endocrinologia; Soc. Italiana di Biologia e Medicina Nucleare; European Thyroid Assoc.; A.N.S.
Nuclear interest: Nuclear medicine (scintigraphy rénography, thyroid researches).
Address: 26 via della Mendola, 00135 Rome, Italy.

BASCOM, Willard. Born 1916. Educ.: Springfield Coll., Colorado School of Mines and California Univ. Res. Eng., California Univ.; 1950-54; Director of various jobs including Mohole Project, Nat. Acad. Sci., 1954-62; Pres., Ocean Sci. and Eng. Inc., 1962-64; Member, Plowshare Advisory Com., U.S.A.E.C. Societies: A.G.U.; Seismological Soc.
Nuclear interests: Peaceful use of nuclear energy in excavation; Creation of isotopes.
Address: 5001 Earlston Drive, Washington, D.C. 20016, U.S.A.

BASERGA, Renato L., M.D. Born 1925. Educ.: Milan Univ. Res. Assoc., Chicago Medical School, 1952-54; Resident in Pathology St. Luke's Hospital, Chicago, 1955-58; Instructor in Pathology (1958-1960), Asst. Prof. in Pathology (1960-), Northwestern

Univ. Medical School. Societies: American Assoc. for Cancer Res.; American Soc. Exptl. Pathology.
Nuclear interests: Biologic effects of ionising radiations; Isotopic tracer studies in biology and medicine.
Address: Department of Pathology, Northwestern University Medical School, 303 E. Chicago Avenue, Chicago 11, Ill., U.S.A.

BASHKIN, Stanley, B.A., Ph.D. Born 1923. Educ.: Brooklyn Coll., and Wisconsin Univ. Asst. Prof. Louisiana State Univ., 1950-53; Res. Assoc. (1953-56), Asst. Prof. (1956-59), Assoc. Prof. (1959-62), Iowa State Univ; Res. Fellow, Cal. Inst. Tech., summer 1959; Prof., Arizona Univ., 1962-; Director, Van de Graaff Laboratory. Societies: A.P.S.; Roy. Soc. of Arts and Sci.
Nuclear interest: Nuclear astrophysics.
Address: Physics Department, Arizona University, Tucson, Arizona, U.S.A.

BASILE, Robert Joseph Marcel, D. ès. Sc. Born 1919. Educ.: Paris Univ. Prof., Faculté des Sciences, Orléans. Societies: A.I.P.; Sté. Française de Physique.
Nuclear interests: Nuclear physics on heavy ions.
Address: Faculté des Sciences, 45 Orléans la Source, France.

BASIN, Ya. N. Papers: Co-author, New Models of Porons Stratum for Neutron Carrotage (letter to the Editor, Atomnaya Energiya, vol. 15, No. 4, 1963); co-author, On the Use of Ac-Be Neutron Sources in Trade Geophysics (letter to the Editor, ibid, vol. 16, No. 3, 1964).
Address: Academy of Sciences of the U.S.S.R., 14 Leninsky Prospekt, Moscow V-71, U.S.S.R.

BASMANOV, P. I. Papers: Co-author, Trapping of Shortlived Radon Decay Daughter Products with Fibre Filters (Atomnaya Energiya, vol. 15, No. 3, 1963); co-author, Miniature Instrument for Measurement of Mean Total Radon Concentration (letter to the Editor, ibid, vol. 21, No. 3, 1966).
Address: Academy of Sciences of the U.S.S.R., 14 Leninsky Prospekt, Moscow V-71, U.S.S.R.

BASOV, Nikolai Gennadievich. Nobel Prize for Phys., 1964; Lenin Prize, 1959. Born 1922. Educ.: Moscow Phys. Eng. Inst. Lab. Head, Deputy Director, Lebedev Inst. of Phys. Corres. Member, Acad. of Sci. of U.S.S.R.
Address: Lebedev Institute of Physics, 53 Leninsky Prospekt, Moscow, U.S.S.R.

BASS, Nathan W., B.Sc. Born 1917. Educ.: Cornell Univ. Vice-Pres., Brush Beryllium Co., Cleveland, Ohio. Book: Chapter II The Role of Beryllium in Industry in American Soc. for Metals Book. Societies: A.S.M., Electrochem. Soc.; A.A.A.S.; American Ceramic Soc.
Nuclear interests: Use of beryllium metal, beryllium oxide and beryllium intermetallics as a moderator, reflector and structural component in all forms of atomic devices and reactors.
Address: 2644 West Park Boulevard, Cleveland 20, Ohio, U.S.A.

BASS, Reiner Alfred, Dr. rer. nat., Ph.D., D.I.C. Born 1930. Educ.: Stuttgart T.H. and London Univ. Asst., Frankfurt a. Main Univ., 1958 and 1960-; Res. Assoc., Rice Univ., Houston, Texas, 1958-60. Society: Verband Deutscher Physikalischer Gesellschaften.
Nuclear interests: Low energy nuclear physics (nuclear reactions and spectroscopy, mainly fast neutron work).
Address: 82 Melibocusstrasse, Frankfurt a. Main, Germany.

BASSAN, Mayer ZAHARIA-. See ZAHARIA-BASSAN, Mayer.

BASSANI, G. F., Prof. Prof. of Theoretical Phys., Messina Univ. Member, Sicilian Nucl. Res. Com.
Address: Messina University, Via Tommaso Cannizzaro, Messina, Sicily, Italy.

BASSER, Sir Adolph, D.Sc. (Hon.). Born 1889. Member of Council, Nucl. Res. Foundation, Sydney Univ.
Address: 343 Edgecliff Road, Edgecliff, New South Wales, Australia.

BASSETT, Thomas Geer, B.S. (Mech. Eng.). M.S. (Nucl. Eng.). Born 1929. Educ.: Michigan Univ. Res. Asst., Eng. Res. Inst., Michigan Univ., 1955-57; Staff Member, Ford Nucl. Reactor, Michigan Univ., 1955-57; Project Eng., Power Reactor Development Co., 1957-60; Consultant to Missouri Univ., 1960; Reactor Eng., Big Rock Point Nucl. Power Plant, Consumers Power Co., 1960-65; Sen. Nucl. Eng., Niagara Mohawk Power Corp., 1965-. Society: A.N.S.
Nuclear interests: Power reactor plant design, engineering, and physics; Hazards and accident analyses.
Address: 4111 Wildwood Drive, Williamsville, New York 14221, U.S.A.

BASSI, Pietro, degree in Physics. Born 1922. Educ.: Padua Univ. Prof. Phys. Faculty of Eng. Bologna Univ.
Nuclear interest: Nuclear physics.
Address: Istituto di Fisica A. Righi, Università degli Studi di Bologna, Italy.

BASSLEER, Roger José Benoit, Doctor in Medicine. Born 1931. Educ.: Liège Univ. Asst. in Histology, Liège Univ.
Nuclear interest: Histo-autoradiography.
Address: 20 rue de Pitteurs, Liège, Belgium.

BASSO, G. Member for Italy, Halden Programme Group, O.E.C.D., E.N.E.A.
Address: O.E.C.D. European Nuclear Energy Agency, 38 boulevard Suchet, Paris 16, France.

BASSON, Johan Kristof, M.Sc., D.Sc. Born 1928. Educ.: Stellenbosch and Pretoria Univs. N.P.R.L., Council for Sci. and Ind. Res. Pretoria, 1947-58 (Head, Radioactivity Sect., from 1954); Sen. Lecturer, Stellenbosch Univ., 1958-59; Director, Isotopes and Rad., Atomic Energy Board, Pelindaba, South Africa, 1959-. Attached to: Brookhaven Nat. Lab., 1955; Royal Marsden Hospital, London, 1956 (5 months); A.E.R.E., Harwell, 1960-61. Societies: South African Inst. of Phys.; H.P.S. Nuclear interests: Applications of isotopes and radiation; Health physics; Standardisation of radioisotopes; Dosimetry.
Address: Isotopes and Radiation Division, Atomic Energy Board, Pretoria, South Africa.

BAST, Sister Eileen Marie, S.S.N.D., Ph.D. Prof. of Biol. and Director of Res. in Radiobiol., Nucl. Sci. Dept., Notre Dame Coll.
Address: Notre Dame College, Nuclear Science Department, 320 East Ripa Avenue, St. Louis, Missouri 63125, U.S.A.

BASTAGLI, Luciano. Economic and Commercial Sci. degree. Born 1905. Educ.: Florence Univ. Gen. Manager and Director, Compagnia Anonima Assicurazione Torino, 1959-; Gen. Manager, Vittoria Riassicurazione, Gen. Agent for Italy, Ancienne Mutuelle Accidents.
Address: Compagnia Anonima d'Assicurazione di Torino s.p.a., 16 Via Arcivescovado, 10121 Turin, Italy.

BASTAI, Pio, Prof. Co-Editor, Minerva Nucleare. Society: Pres., Soc. Italiana di Biologia e, Medicina Nucleare.
Address: 3 Via Assarotti, Turin, Italy.

BASTARD, C. Maitre Asst., Inst. de Phys. Nucl., Lyon Univ.
Address: Institut de Physique Nucléaire, Lyon University, 43 boulevard de l'Hippodrome, Villeurbanne, (Rhone), France.

BASTIANI, M. J. de. See de BASTIANI, M. J.

BASTO, E. LIMA. See LIMA BASTO, E.

BASTOS MARTINS, E. C. de. See de BASTOS MARTINS, E. C.

BASU, B., Formerly Lecturer, now Reader, Saha Inst. of Nucl. Physics.
Address: Saha Institute of Nuclear Physics, 92 Acharya Prafulla Chandra Road, Calcutta 9, India.

BASU, Jayanta, B.Sc. (Hons.), M.Sc. (Tech.), Ph.D. Born 1933. Educ.: Calcutta and Manchester Univs. Res. Asst., Indian Assoc. for the Cultivation of Sci., Calcutta, 1958; Sen. Res. Asst., D.S.I.R., Manchester Univ., 1960-61; Pool Officer, Council of Sci. and Ind. Res., Indian Inst. Technol. Bombay and Calcutta Univ., 1961-64; Reader, Saha Inst. of Nucl. Phys., Calcutta, 1964-. Sec., Bangiya Bijnan Parishad (Sci. Assoc. of Bengal); Member, Exec. Com., Sci. for Children, Calcutta. Societies: Sen. Member, I.E.E.E.; Fellow, Indian Phys. Soc.
Nuclear interests: Study of plasma in the context of thermonuclear fusion experiments. The study involves, in particular, plasma diagnostics and the behaviour of plasma in a magnetic field.
Address: Saha Institute of Nuclear Physics, 92 Acharya Prafulla Chandra Road, Calcutta-9, India.

BASYN, Jacques, Docteur en droit. Born 1901. Educ.: Louvain Univ. Prof., Internat. law of insurance, Louvain Univ.; Chairman, Insurers of the six Euratom countries; Chairman, Working group studying insurance problems in relation with nucl. risks, European Insurance Com., C.E.A., Paris.
Address: 189 avenue Brugmann, Brussels, Belgium.

BATAILLE, René P.J., Civil constructions Eng. Nucl. Phys. Born 1912. Educ.: Ghent and Brussels Univs. Director, Nucl. Applications Sect., Direction Gén. Energie, Brussels 1956-. Nuclear interests: Technique et économie de toutes les applications nucléaires.
Address: Administration de l'Energie, Ministére des Affaires économiques, 24-26, rue De Mot, Brussels 4, Belgium.

BATALIN, V. A. Paper: Co-author, Medium Energy Scattering by Atomic Nuclei (Atomnaya Energiya, vol. 16, No. 3, 1964).
Address: Academy of Sciences of the U.S.S.R., 14 Leninsky Prospekt, Moscow V-71, U.S.S.R.

BATALOV, A. A. Papers: Co-author, New Chem. Method for Determination of Rad. Dose in Reactor (Atomnaya Energiya, vol. 20, No. 6, 1966); co-author, Circulating Rate Effect on Rad. - Induced Conversion of Organic Coolants at Elevated Temperatures (ibid, vol. 22, No. 5, 1967); co-author, Thermal Neutrons Fluxes Distributions in Various Reflectors with Canals (letter to the Editor, ibid, vol. 24, No. 5, 1968).
Address: Academy of Sciences of the U.S.S.R., 14 Leninsky Prospekt, Moscow V-71, U.S.S.R.

BATCHELOR, Robert, M.A., Ph.D. Born 1924. Educ.: Cambridge Univ. Supt., Neutron Phys., U.K.A.E.A., Aldermaston. Book: Advisory Editor, Concise Encyclopaedia of Nucl. Energy (George Newnes Ltd., 1962).
Nuclear interest: Basic cross-section data for nuclear reactors.
Address: 112 Chapel Hill, Tilehurst, Reading, Berks., England.

BATENIN, I. V. Papers: X-ray Examination of Irradiated Uranium to study with Irradiation Growth (letter to the Editor, Atomnaya Energiya, vol. 3, No. 9, 1957); co-author, Dilatometric Studies of Rolled Uranium Rods (letter to the Editor, ibid, vol. 6, No. 5, 1959); co-author, The Growth of Uranium Rods in Corrosive Gas (ibid, vol. 7, No. 4, 1959);

co-author, Texture of Hardening of Uranium Rods (letter to the Editor, ibid, vol. 16, No. 4, 1964).
Address: Academy of Sciences of the U.S.S.R., 14 Leninsky Prospekt, Moscow V-71, U.S.S.R.

BATES, Charles Carpenter, B.A., M.A., Ph.D. Born 1918. Educ.: DePauw Univ., U.C.L.A. and A. and M. Coll.,Texas. Branch Head and Deputy Director, Div. of Oceanography, U.S.N. Hydrographic Office, 1950-57; Environmental Systems Co-ordinator, Office of Development Co-ordinator, Navy, 1957-60; Chief, Seismic Programmes, Nucl. Test Detection Office, Advanced Res. Projects Agency, 1960-64; Tech. Director U.S.N. Oceanographic Office, 1964-. Exec. Com., A.G.U., 1964-. Societies: Soc. of Exploration Geophysicists; Seismological Soc. of America; A.G.U.
Nuclear interests: Detection and identification of large-scale underwater and underground explosions; Distribution of low-level radioisotopes in the sea.
Address: 5807 Massachusetts Avenue, NW, Washington D.C. 20016, U.S.A.

BATES, J. L., Pres., Central Power and Light Co. Vice-Pres., Texas Atomic Energy Research Foundation.
Address: Texas Atomic Energy Research Foundation, P.O. Box 970, Fort Worth, Texas, U.S.A.

BATES, J. Lambert, B.S., Ph.D. Born 1928. Educ.: Brigham Young and Utah Univs. Sen. Sci., Hanford Labs., G.E.C., 1957-64; Sen. Sci., Pacific Northwest Lab. (formerly Hanford Labs.), 1965-. Societies: A.N.S.; American Ceramic Soc.
Nuclear interests: Nuclear ceramics with emphasis on thermal, electrical and optical properties at high temperatures; Irradiation damage.
Address: Pacific Northwest Laboratory, Battelle Memorial Institute, P.O. Box 999, Richland, Washington 99352, U.S.A.

BATES, William H. Member, Joint Com. on Atomic Energy, U.S.A.E.C.
Address: United States House of Representatives, Washington 25, D.C., U.S.A.

BATH, Gordon D. Geophys., Geological Survey, U.S. Dept. of the Interior.
Address: U.S. Department of the Interior, Geological Survey, Building 25, Federal Centre, Denver, Colorado, U.S.A.

BATLER, Emanuel, B.Sc. (Phys.). Born 1925. Educ.: City Univ. of New York. Vice Pres., Philips Electronics Industries Ltd., 1966-.
Nuclear interest: Application and marketing of nuclear detection instruments.
Address: Philips Electronics Industries Limited, 116 Vanderhoof Avenue, Toronto 17, Ontario, Canada.

BATOR, Bela. Member, Editorial Board, Energia es Atomtechnika.
Address: Muszaki Konyvkiado, Bajcsy-Zsilinszky ut 22, Budapest 5, Hungary.

BATOROV, B. B. At State Com. for Utilisation of Atomic Energy. Adviser, Third I.C.P.U.A.E., Geneva, Sept. 1964.
Address: State Committee for the Utilisation of Atomic Energy, 26 Staromonetnii Pereulok, Moscow, U.S.S.R.

BATOV. V. V. Papers: Co-author, Some Aspects of Nucl. Power Economics Incentive (Atomnaya Energiya, vol. 20, No. 5, 1966); co-author, Nucl. Fuel Efficiency Criterion (ibid, vol. 21, No. 3, 1966).
Address: Academy of Sciences of the U.S.S.R., 14 Leninsky Prospekt, Moscow V-71, U.S.S.R.

BATRA, Mrs. B. K., M.A. (Mount Holyoke), Ph.D. Res. Officer, Indian Cancer Res. Centre.
Address: Indian Cancer Research Centre, Parel, Bombay 12, India.

BATRES, Francisco AGUIRRE-. See AGUIRRE-BATRES, Francisco.

BATTEN, Archibald George Mount, F.C.I.I., F.I. Arb. Born 1902. Director, Alliance Assurance Co., Ltd.; Director, Sun Insurance Office Ltd.; Director, The London Assurance. Chairman, Advisory Com., British Insurance (Atomic Energy), 1957-59.
Nuclear interest: Nuclear insurance.
Address: Alliance Assurance Co., Ltd., Bartholomew Lane, London, E.C.2, England.

BATTIST, Lewis, B.A., Ph.D. (Phys. Chem.). Born 1929. Educ.: New York and Texas Univs. A.E.C. Contract Res. Fellow, Notre Dame Univ., 1951-52; Res. Sci., Texas Univ., 1951-52, Chem. and Radiol. Div., U.S. Army Chem. Corps, 1952-54; Res. Sci., Texas Univ., 1954-58; Sen. Sci. and Tech. Subcontract Administrator, Bettis Atomic Power Lab., Westinghouse Elec. Corp., 1958-61; Director, Analytical Div., Phys. Sci. Dept., Nucl. Sci. and Eng. Corp. 1961-62; Sen. Tech. Associate Nucl. Utility Services Inc., 1962-65; Assoc. Prof. Nucl. Sci. and Eng. and Radiol. Safety Official, Catholic Univ. of America, 1965-. Treas., Local Chapter, A.N.S. Societies: A.C.S.; A.N.S.; Fellow, A.A.A.S.; A.A.U.P.; New York Acad. Sci.; H.P.S.
Nuclear interests: Radiation dosimetry; Radiation detection and interactions; Biological effects of radiation; Reactor chemistry and radiochemistry.
Address: Room B-19, Pangborn Building, Catholic University of America, Washington D.C. 20017, U.S.A.

BATTISTINI, Francesco Domenico, Prof. Sen. Officer, Centro di Medicina Nucleare, Istituto di Clinica Pediatrica dell' Universita, Parma.

BAT

Address: Centro di Medicina Nucleare, Istituto di Clinica Pediatrica dell' Universita, Parma, Italy.

BATTISTINI, Giulio, Prof. Director, Elec. Technol. Inst., Eng. Dept., Pisa Univ. Member, Steering Commission, C.N.E.N. Ing. Pres., Assoc. Termotecnica Italiana, Comitato No. 13, Impianti Nucleari. Formerly Member, Assoc. Nazionale di Ingegneria Nucleare.
Address: Electrical Technology Institute, Engineering Department, Pisa University, Pisa, Italy.

BATURIN, G. N. Paper: On Uranium Content in Caspian Sea Sediments (letter to the Editor, Atomnaya Energiya, vol. 21, No. 6, 1966).
Address: Academy of Sciences of the U.S.S.R., 14 Leninsky Prospekt, Moscow V-71, U.S.S.R.

BATUROV, Boris Borisovich. Born 1928. Deputy Head, Directorate of Power Installation. Member, Soviet Fast Reactor Team visiting Britain, Jan. 1960. Member, Russian deleg. visiting Britain July 18-28, 1961, to discuss solid state physics research.
Address: State Committee for Utilisation of Atomic Energy, 26 Staromonetny Pereulok, Moscow, U.S.S.R.

BATZEL, Roger E., Ph.D. Born 1921. Educ.: California Univ., Berkeley. Assoc. Director, Chem. and Space Reactors, Lawrence Rad. Lab., California Univ., Livermore. Society: A.P.S.
Nuclear interests: Management responsibility for work in radiochemistry and general areas of chemistry and metallurgical work related to nuclear weapons. Responsibility for high temperature reactor design and materials research.
Address: Lawrence Radiation Laboratory, P.O. Box 808, Livermore, California, U.S.A.

BAUDEUF, A. Dépt. "Entreprises Générale, Groupement Atomique Alsacienne Atlantique.
Address: Groupement Atomique Alsacienne Atlantique, 100 avenue Edouard Herriot, Le Plessis Robinson, (Seine), France.

BAUDINET ROBINET, Yvette, Dr. en Sc. Physiques. Born 1933. Educ.: Liège Univ. At Inst. Interuniv. des Sciences Nucléaires.
Nuclear interest: Nuclear photographic plates.
Address: Institut de Physique, la Quai Roosevelt, Liège, Belgium.

BAUDOUX, Pierre, Prof., Director, Service de Physique Nucléaire and Director, Lab. de Physique de Plasmas, Univ. Libre de Bruxelles. Près. Sci. Commission, Vice-Pres., Conseil d'Administration, Inst. Interuniversitaire des Sciences Nucléaires.
Address: 50 avenue F. D. Roosevelt, Brussels, Belgium.

BAUER, Bruno, Prof. Dr. Swiss Federal Inst. Technol.; Tech. Director, Nat. Co. for Promotion of Ind. Uses of Atomic Energy. Editorial Board, Atompraxis. Societies: Swiss Socs. of Eng. and Architects; Swiss Assoc. for Atomic Energy.
Nuclear interest: Nuclear pilot plant design.
Address: Suisatom AG, 3 Bahnhofplatz, Zürich, Switzerland.

BAUER, Etienne Robert, licenié. Born 1918. Educ.: Paris Univ. Asst. Director, Inst. Nat. des Sci. et Techniques Nucléaires Saclay.
Nuclear interest: Education.
Address: 61 rue de Varenne, Paris 7, France.

BAUER, G. Pres., Federation Suisse des Associations de Fabricants d'Horlogerie.
Address: Federation Suisse des Associations de Fabricants d'Horlogerie, 6 rue d'Argent, Bienne, Switzerland.

BAUER, Helmut Wilhelm Armin, Dipl. Ing. (Tech. Chem.). Born 1934. Educ.: Vienna T.H.
Nuclear interests: Chemical dosimetry; Production of radioactive isotopes.
Address: 16/29 Nordbahnstrasse, Vienna 2, Austria.

BAUER, Robert, Prof. Dr. med. Member, Fachkommission für Strahlenschutz, Wirtschaftsministerium Baden-Württemberg; Director, Medizinisches Strahleninst., Tübingen Univ.
Address: Medizinischen Strahleninstitut, Universität Tübingen, Röntgenweg, Tübingen, Germany.

BAUGHAN, Edward Christopher, M.A., B.Sc. Born 1913. Educ.: Oxford Univ. Prof. Chem., Roy. Military Coll. Sci. Societies: Chem. Soc.; Faraday Soc.; Phys. Soc.
Nuclear interest: Teaching, particularly to the Government Services.
Address: Royal Military College of Science, Shrivenham, Swindon, Wilts., England.

BAULANGER, Werner, Dr. Director, Legal Div., I.A.E.A.
Address: International Atomic Energy Agency, 11 Kaerntnerring, Vienna 1, Austria.

BAULIER, Henri. Sen. Officer, Installation Dept., Ets. Garczynski et Traploir.
Address: Ets Garczynski et Traploir, 34 rue du Pavé, Le Mans (Sarthe), France.

BAULIES, Oscar Leopoldo, Dr. in Nat. Sci. Born 1923. Educ.: Cordoba Univ., Argentina. Head, Exploration Dept. of the Managership of Raw Materials (-1962), Adviser of the Managership of Raw Materials (1962-), Comision Nacional de Energia Atomica, Argentina.
Nuclear interest: Everything related to the geology of uranium.
Address: 9th floor, ap. "B", 891 Rioja, Rosario, Argentina.

BAULNY, L. Director of Nucl. Activities, Auxeltra Électrification et Travaux Speciaux S.A.
Address: Auxeltra Electrification et Travaux Speciaux S.A., 12 avenue de l'Astronomie, Brussels 3, Belgium.

BAUM, Siegmund J., B.A., M.A., Ph.D. Born 1920. Educ.: California Univ. (Los Angeles and Berkeley). Sen. Project Leader, U.S. Naval Radiol. Defence Lab.,1950-60; Group Leader, Radiobiol. Group., Douglas Missiles and Space Div., 1960-62; Head, Cellular Radiobiol. Div. (1962-64), Chairman, Exp. Pathology Dept. (1964-), Armed Forces Radiobiol. Res. Inst. Societies: American Physiological Soc.; Rad. Res. Soc.; Transplantation Soc.
Nuclear interests: Effects of ionizing radiation on mammals; Utilisation of isotopic tracer materials in biomedical research.
Address: Armed Forces Radiobiology Research Institute, National Naval Medical Centre, Bethesda, Maryland, U.S.A.

BAUMAN, Frederick Adams, Jr., B.S. (Chem., Northeastern), M.S. (Chem., Northeastern). Born 1923. Educ.: Northeastern Univ. and M.I.T. Team Leader, Analytical Chem., Flight Propulsion Div., Gen. Elec., Lynn, Massachusetts. Society: A.C.S.
Nuclear interest: Research in analytical methods for reactor materials.
Address: 61 Kingsbury Street, Needham 02192, Massachusetts, U.S.A.

BAUMANN, Carl D., B.S. M.S. Born 1915. Educ.: Albright Coll., Lehigh Univ. and O.R.S.O.R.T. Assoc. Phys., Union Carbide Nucl. Corp., 1947-50; Phys., Oak Ridge Nat. Lab., 1950-. Societies: A.P.S.; A.N.S.
Nuclear interest: Radiation damage to reactor fuels.
Address: Reactor Chemistry Division, Oak Ridge National Laboratory, P.O. Box X, Oak Ridge, Tennessee, U.S.A.

BAUMANN, Ernst, Dipl. Ing., Dr.s.c. tech. h.c. Born 1909. Educ.: E.T.H. and Lausanne Univ. o.Prof., Appl. Phys., Head, Inst. of Appl. Phys., Director Dept. of Ind. Res., Swiss Federal Inst. of Technol. Forschungsrat, Swiss Nat. Foundation.
Address: Institut für Technische Physik, E.T.H.-Aussenstation Hönggerberg, Postfach, 8049 Zürich, Switzerland.

BAUMANN, Germain, Dr. of Sci. Born 1931. Educ.: Strasbourg Univ.
Nuclear interests: Elementary particles and nuclear physics; Hyperfragments; Interactions of protons π and K mesons and of hyperons in nuclear emulsions.
Address: Départment de Physique Corpusculaire, Centre de Recherches Nucléaires, Strasbourg-Cronenbourg, Bas-Rhin, France.

BAUMEISTER, Herbert, Dr. Res. Div., Cobalt 60 Sterilisation Plant, Ethicon G.m.b.H.
Address: Ethicon G.m.b.H., Cobalt 60 Sterilisation Plant, 1 Robert-Koch-Str., 2 Glashutte/-Holst., Germany.

BAUMEISTER, Theodore, B.S., M.E. Born 1897. Educ.: Columbia Univ. Stevens Prof. Mech. Eng., Columbia Univ.; Consulting Eng., Gen. Public Utilities Corp., South Carolina Elec. and Gas Co., Babcock and Wilcox Co.; Director, Atomic Power Development Associates. Books: Power Sect. Bonilla's Nucl. Eng.; Editor-in-Chief, Mech. Eng. Handbook (Marks; McGraw-Hill). Societies: A.S.M.E.; I.Mech.E.; A.A.A.S.; New York Acad. Sci.
Nuclear interests: Generation of electric power from nuclear sources; competitive with and complementary to generation from conventional sources and by conventional means - i.e., hydro, steam and internal combustion.
Address: 4711 Iselin Avenue, Bronx, N.Y., 10471, U.S.A.

BAUMGARTNER, Eugen, Ph.D. Born 1926. Educ.: Basel Univ. Res. Assoc., Basel, until 1955; at Rochester Univ., N.Y., 1955-56; Priv. Doz., Basel, 1956. a.O. Prof., Basel, o.Prof. für Experimentalphysik 1960. Societies: A.P.S.; A.A.A.S.; Schweizerische Physikalische Gesellschaft; Schweizerische Naturforschende Gesellschaft.
Nuclear interest: Nuclear physics; Polarization phenomena; Nuclear structure.
Address: Physikalische Anstalt, der Universität, Basle, Switzerland.

BAUMGARTNER, Franz, Dr. rer. nat. Born 1929. Educ.: Munich T.H. Priv.-Dozent. Mitglied, Arbeitskreis II/2, Chemie, and Mitglied, Arbeitskreis III/5, Aufarbeitung Bestrahlter Brennstoffe, Deutsche Atomkommission, Federal Ministry for Sci. Res. Book: Co-author, Chemie radioaktiver Substanzen (in: Kerntechnik; ed. Riezler, Walcher). Society: Gesellschaft Deutscher Chemiker.
Nuclear interest: Radiochemistry.
Address: Institut für Radiochemie, Technische Hochschule Munich, 21 Arcisstrasse, Munich, Germany.

BAUMGARTNER, Richard. Born 1903. Educ.: Ecole Polytech., Paris.Ancien Gérant de U.S.S.I.; Past Pres., A.T.E.N.;
Nuclear interests: Construction d'une usine de séparation isotopique (engineering).
Address: Société Alsacienne de Constructions Mecaniques, 157 avenue de Neuilly, Neuilly-sur-Seine, (Haute-de-Seine), France.

BAUMGARTNER, William Vincent, B.S. (Chem.). Born 1929. Educ.: Seattle Univ. Rad. Monitoring Specialist and Sen. Eng. Radiol. Application (1952-67), Manager, External Dosimetry (1967-), U.S.T.C., Inc. Societies: A.N.S.; H.P.S.
Nuclear interests: Management of radiological application engineering groups for the

BAU

development and operation of radiation protection systems for people working in and around radiation zones such as reactors, separation facilities and laboratories.
Address: 1635 Alder, Richland, Washington, U.S.A.

BAUMINGER, E. R., Dr. Lecturer, Dept. of Phys., Exptl. Nucl. Phys. Sect., Hebrew Univ.
Address: Hebrew University, Department of Physics, Jerusalem, Israel.

BAUTIN, A. V. Papers: Co-author, Calculation of Yield and Meansquare Deflection of Positrons in Case of Passing of Electrons through Thick Foil (Atomnaya Energiya, vol. 20, No. 5, 1966); co-author, Production of Photons and Positrons from Bombardment of Thick Foil by Fast Electrons (ibid, vol. 21, No. 2, 1966).
Address: Academy of Sciences of the U.S.S.R., 14 Leninsky Prospekt, Moscow V-71, U.S.S.R.

BAUTISTA, Miss Fe. Medical Technician, Radioisotope Lab., Philippine Gen. Hospital.
Address: Philippine General Hospital, Manila, Philippines.

BAXTER, Alexander Duncan, M. Eng., C. Eng. Educ.: Liverpool Univ. Prof. Aircraft Propulsion (1950-57), Deputy Principal (1954-57), Coll. of Aeronautics, Cranfield; Exec. Director (Eng.), De Havilland Engine Co., Ltd., 1957-62; Tech. Exec. Bristol Siddeley Engines, Ltd., 1963-. Member of Govt. advisory committees; Member of R.A.F. Educ. Advisory Com. Societies: M.I.Mech.E.; Fellow, Roy. Aeronautical Soc.; F. Inst. Petroleum.
Nuclear interest: Application of reactors to marine and other mobile power plants.
Address: Bristol Siddeley Engines, Ltd., P.O. Box 3, Filton, Bristol, England.

BAXTER, Allin P. Sec., Atomics, Phys. and Sci. Fund, Inc.
Address: Atomics, Physics and Science Fund, Inc., 1033 30th Street, N.W., Washington 7, D.C., U.S.A.

BAXTER, Sir John Philip, K.B.E., B.Sc., Ph.D. Born 1905. Educ.: Birmingham Univ. I.C.I., Great Britain, latterly Res. Director, Gen. Chemicals Division, 1928-50; Prof. Chem. Eng. (1950-53), Vice-Chancellor (1953-), N.S.W. Univ. Member (1950-56), Chairman (1956-), A.A.E.C. Societies: Fellow, Australian Acad. of Sci.; F.R.A.C.I.; Inst. Eng., Australia; F.R.S.A.; M. I. Chem. E.; A.I.Ch.E.
Address: 1 Kelso Street, Enfield, N.S.W., Australia.

BAXTER, William J., M.B.A. Born 1899. Educ.: Clark Coll. and Harvard Univ. Pres., Nucl. Energy Res. Bureau.
Nuclear interest: Financial research of companies specialising in nuclear energy.
Address: Nuclear Energy Research Bureau, 68 William Street, New York 5, New York, U.S.A.

BAXTER, William J., Jr., M.S. Born 1934. Educ.: Colgate, New York and Columbia Univs. Vice-Pres., Nucl. Energy Res. Bureau. Society: Nucl. Energy Writers Assoc.
Nuclear interests: Research in new developments of commercial value; All phases of nuclear energy.
Address: Nuclear Energy Research Bureau, 68 William Street, New York 5, New York, U.S.A.

BAYARD, Robert Thomas, B.S. (Mech. Eng.), M.S. (Phys.), Ph.D. (Phys.). Born 1920. Educ.: Southern California and Pittsburgh Univs. Res. Phys., Res. Lab. (1947-50), Sect. Manager, Neutron Detector Development, Bettis Atomic Power Lab. (1950-56), Manager, P.W.R. Phys. Subdiv., Bettis Atomic Power Lab. (1956-62), Westinghouse Electric Corp.; I.A.E.A. Tech. Assistance Expert, Thailand, 1962-63; Manager, Central Phys. (1963-65), Manager, Reactor Development and Analysis Dept. (1965-), Bettis Atomic Power Lab., Westinghouse Electric Corp. Society: A.N.S.
Nuclear interests: Nuclear design; Reactor physics.
Address: Westinghouse Electric Corporation, Bettis Atomic Power Laboratory, P.O. Box 79, West Mifflin, Pennsylvania 15122, U.S.A.

BAYBARZ, Russell Dale, B.A. Born 1937. Educ.: Walla Walla Coll. Problem leader in transplutonium element chemistry, 1959. Society: A.C.S.
Nuclear interests: Separation chemistry of Am, Cm, Bk, Cf, Es, and Fm from each other and from contaminant lanthanides by the use of liquid-liquid solvent extraction methods.
Address: Oak Ridge National Laboratory, Oak Ridge, Tenn., U.S.A.

BAYER, Ryszard, Dr. nauk technicznych, inzynier elektronik. Born 1925. Educ.: Politechnika Warszawska. Head, Analog-digital Circuits Lab., Instytut Badań Jadrowych, Swierk, 1956-.
Nuclear interest: Nuclear electronic instruments.
Address: 20 m 140 Wiejska, Warsaw, Poland.

BAYLIN, Teodomiro GONZALEZ. See GONZALES BAYLIN, Teodomiro.

BAYLIS, Peter Roy. Born 1932. Chief Project Eng., Spembly Tech. Products Ltd.
Nuclear interests: Reactor design; Thermocouple instrumentation; Protective clothing and waste disposal.
Address: Spembly Technical Products Limited, Trinity Trading Estate, Sittingbourne, Kent, England.

BAYOL, Rodolfo, Civil Eng. Born 1912. Educ.: Buenos Aires Nat. Univ. Chief of Building Div., Civil Eng. Dept. (1957-), Member, Board of Directors, (1959-), Comision Nacional de Energia Atomica.
Nuclear interests: Building of reactors and nuclear laboratories (civil works).

Address: 8250 Avenida Libertador General San Martin, Buenos Aires, Argentina.

BAZ, A. I. Soviet Deleg., Convention on Thermonucl. Processes, Inst. of Elec. Eng., London, April 29-30, 1959. Papers: Co-author, The Angular Distribution of Fission Fragments in the Photofission of U^{238} (2nd I.C.P.U.A.E., Geneva, Sept. 1958); On the Existence of Light Nuclei with Large Proton or Neutron Excesses (letter to the Editor, Atomnaya Energiya, vol. 6, No. 5, 1959).
Address: Nuclear Energy Institute, Academy of Sciences of the U.S.S.R., 14 Leninsky Prospekt, Moscow V-71, U.S.S.R.

BAZ, Edgard EL-. See EL-BAZ, Edgard.

BAZARGAN, G. A. Dr. Pres., Tabriz Univ.; Prof., Tech. Coll., Tehran Univ.; Member, Nat. Iranian Atomic Energy Commission; Deleg. Third. I.C.P.U.A.E., Geneva, Sept. 1964.
Address: National Iranian Atomic Energy Commission, Ministry of Industry and Mines, Tehran, Iran.

BAZET, Juan E., Lawyer. Born 1921. Educ.: Buenos Aires Univ. Legal adviser and director of various ind. and commercial companies, 1948-; Co-drafter Argentine legislation on atomic energy and nucl. minerals, 1956; Legal adviser (1956-58), Head of Legal Dept. (1958-61), Sec., Board of Directors (1961-67), Head, Public Relations Div. (1967-), Argentine Nat. A.E.C. Argentine Deleg. to First Interamerican Symposium on Legal and Administrative Problems of Peaceful Uses of Atomic Energy, San Juan, Puerto Rico, 1959. Member, Argentine Subcom. on Atomic Energy, Interamerican Bar Assoc.
Nuclear interests: All legal questions arising from nuclear energy and its applications.
Address: Florida 440, Buenos Aires, Argentina.

BAZHANOVA, A. E. Paper: Co-author, Effect of Finite Conductivity of Metal on Equilibrium of Plasma Column in "Tokamak" (letter to the Editor, Atomnaya Energiya, vol. 20, No. 2, 1965).
Address: Academy of Sciences of the U.S.S.R., 14 Leninsky Prospekt, Moscow V-71, U.S.S.R.

BAZHENOV, V. A. Paper: Co-author, Measurements of Activity of Radioactive Gases by Means of Spherical Ionisation Chamber (letter to the Editor, Atomnaya Energiya, vol. 21, No. 2, 1966).
Address: Academy of Sciences of the U.S.S.R., 14 Leninsky Prospekt, Moscow V-71, U.S.S.R.

BAZZANO, Elsa, Biol. Dr. Born 1933. Educ.: Milan Univ. Voluntary Asst., Inst. of General Pathology, Milan Univ., 1959-61; With C.I.S.E., Segrate-Milan, Italy, 1960-.
Nuclear interests: Medical and haematological supervision in radiation work. Radiotossicology. Radiation protection.
Address: 23 via Francesco Soave, 20135, Milan, Italy.

BEACH, Louis Andrew, B.S., M.S., Ph.D. Born 1925. Educ.: Indiana Univ. Head, Reactor Shielding Sect. (1953-55), Head, Nucl. Reactions Branch (1955-66), Head, Phys. I Sect. of Cyclotron Branch (1966-), U.S. Naval Res. Lab. Societies: A.P.S.; A.A.A.S.; RESA; Washington Acad. Sci.; Philosophical Soc. of Washington.
Nuclear interests: Research with positive ion beam from a sector-focused cyclotron; Besides basic nuclear physics measurements of short lifetimes by Doppler Shift Attenuation, studies are made of neutron production by proton bombardment of materials for applications in accelerator shielding and for space shielding.
Address: 1200 Waynewood Boulevard, Alexandria, Virginia 22308, U.S.A.

BEACH, Norman F. Vice Pres., Eastman Kodak Co. of Rochester. Member, Gen. Advisory Com., State Office of Atomic Development, New York.
Address: Eastman Kodak Co., Rochester, New York 14650, U.S.A.

BEACOM, Seward E., B.S., M.S., Ph.D. Born 1912. Educ.: Mount Union Coll., Michigan and Connecticut Univs. Assoc. Prof. Chem., Central Connecticut State Coll., New Britain, 1950-57; Sen. Res. Chem. (1957-61), Head (1962-), Electrochem. Dept., Res. Lab. Gen. Motors Corp., Warren, Mich. Societies: Electrochem. Soc.; A.C.S.; American Electroplaters' Soc.; Soc. of Automotive Eng.; Inst. of Metal Finishing.
Nuclear interests: The use of radioactively tagged organic compounds to study the influence of organic addition agents on the leveling phenomenon; interaction of additives and the cause of bright deposits which accompany the electro-deposition of certain metals.
Address: 38621 Jonathan Drive, Mount Clemens, Mich., U.S.A.

BEADLE, George W. Member, Board of Directors, Educ. Foundation for Nucl. Sci.
Address: Educational Foundation for Nuclear Science, 935 E. 60th Street, Chicago 37, Illinois, U.S.A.

BEAL, Wing Commander D. G., A.R.Ae.S. Member, Council, Inst. of Nucl. Eng.
Address: Institution of Nuclear Engineers, 147 Victoria Street, Westminster, London S.W.1, England.

BEALE, Reginald John, Ph.D., B.Sc.(Eng., Hons.). Born 1922. Educ.: Battersea Polytech. and London Univ. Clayton Fellow, Queen Mary Coll., London, 1950-54; Res. Fellow, Nat. Coll. of Refrigerating Eng, 1954-55; Head, Heat Transfer Sect., and subsequently a Deputy Chief Eng., G.E.C. - Simon Carves Atomic Energy Group, 1955-58; Head, Reactor Design Dept., Atomic Power Constructions, 1958-60;

Head, Mechanical Eng. Dept., Hendon Tech. Coll., 1960-66; Vice-Principal, Borough Polytechn., 1966-. Society: M.I.Mech.E.
Nuclear interests: Reactor design, heat transfer, and economics of nuclear plant; Technical education in the nuclear field, and heat transfer research of both applied and academic nature into the problems met in nuclear reactors.
Address: 13 Cornwall Gardens Walk, Kensington, London S.W.7, England.

BEALEM, Jean-Marcel. Prés. Directeur Gén., Constructions Radio-Electriques et Electroniques du Centre.
Address: Constructions Radio-Electriques et Electroniques du Centre, 5 rue Daguerre, St. Etienne (Loire), France.

BEALL, Samuel E., Jr., B.S. Born 1919. Educ.: Tennessee Univ. Project Eng., Homogeneous Reactor Experiments (1950-60), Head, Reactor Operations Dept., Reactor Div. (1960-63), Director, Reactor Div., (1963-), O.R.N.L.
Address: Post Office Box Y, Oak Ridge National Laboratory, Oak Ridge, Tennessee, U.S.A.

BEAMS, Jesse Wakefield, A.B., M.A., Ph.D., Sc.D. Born 1898. Educ.: Fairmount Coll., Wisconsin and Virginia Univs. Francis Henry Smith Prof. Phys. and Chairman, Phys. Dept., Virginia Univ., 1945-61. Member, Gen. Advisory Com., U.S.A.E.C., 1954-60; Pres., A.P.S., 1958-59; Member, Com. of Sen. Reviewers, U.S.A.E.C. Societies: A.P.S.; American Philosophical Soc.; Nat. Acad. Sci.; American Acad. Arts and Sci.; Optical Soc. America.
Nuclear interest: Isotope separation.
Address: Department of Physics, Virginia University, McCormick Road, Charlottesville, Virginia, U.S.A.

BEARD, Raymond R. Formerly Acting Sci. Rep. of U.S.A.E.C., Buenos Aires; Chief, Procurement Branch, Div. of Raw Materials, U.S.A.E.C., 1964-.
Address: Division of Raw Materials, U.S.A.E.C., Washington D.C. 20545, U.S.A.

BEARE, John Wallace, B.A.Sc. (Chem. Eng.). Born 1936. Educ.: Roy. Military Coll., Canada and British Columbia Univ. Roy. Canadian Navy, 1953-61; Operations Supv., A.E.C.L., 1961-65; Assoc. Sci. Adviser, Atomic Energy Control Board, 1965-.
Nuclear interest: Nuclear safety in design, construction and operation of reactors.
Address: c/o Atomic Energy Control Board, P.O. Box 1046, Ottawa 4, Ontario, Canada.

BEATON, Charles R., B.S.S., LL.B. Born 1911. Educ.: C.C.N.Y. and Fordham Univ. Formerly Pres. and Director, American Tradair Corp. U.S. Subsidiary, E. K. Cole Ltd.; Pres., Amtradair Inc.; Treas., Pye Corp. America.
Nuclear interests: Management and technical sales and service direction; Instrumentation and equipment.
Address: Amtradair Inc., 30-95 32nd Street, at 31st Avenue, Long Island City, New York 11102, U.S.A.

BEATON, James R. Policy Exec., Nucl. Energy Com., Nat. Assoc. of Manufacturers.
Address: National Association of Manufacturers, 277 Park Avenue, New York, New York 10017, U.S.A.

BEATON, Roy Howard, B.S. (Ch.E.), D.Eng. (Ch.E.), D.Sc. (Hon.). Born 1916. Educ.: Northeastern and Yale Univs. Manager, Separations Tech. (1946-52), Manager, Design Hanford Works (1952-56), Manager, Construction Eng., Hanford Works (1956-57), G.E.C., Richland, Washington; Gen. Manager, Defence Products, X-Ray Dept., G.E.C., Milwaukee, Wisconsin, 1957-63; Gen. Manager, Spacecraft Dept., G.E.C., Philadelphia, 1963-64; Gen. Manager, Apollo Systems Dept., G.E.C., 1964-. Societies: A.I.Ch.E.; Atomic Ind. Forum; A.I.A.A.; American Astronautical Soc.
Nuclear interests: Separations plant design and operation; Reactor design and operation; Nuclear electronics; Atomic energy management; Nuclear propulsion of spacecraft.
Address: General Electric Company, Apollo Systems Department, Box 2500, Daytona Beach, Florida, U.S.A.

BEATTIE, William Cecil, Mech. Eng. Born 1904. Educ.: Stevens Inst. of Technol. Vice Pres., Consolidated Edison Co. of New York, Inc. Society: A.N.S.
Address: 4 Irving Place, New York, N.Y. 10003, U.S.A.

BEATTY, A. V., Dr. Nucl. Research, Biol. Dept., Coll. of Arts and Sci., Emory Univ.
Address: College of Arts and Sciences, Emory University, Atlanta 22, Georgia, U.S.A.

BEAUCHAMP, Edward Eli, B.S. (Chem. Eng.). Born 1918. Educ.: Rhode Island Univ. Supt., Isotopes Sales Dept., Isotopes Development Centre, O.R.N.L., 1952-. Chairman, American Standards Assoc. Com. N5.4. Societies: A.C.S.; A.N.S.
Nuclear interests: Radioisotopes, production, development, application, shipping and management in these fields.
Address: 116 Orchard Circle, Oak Ridge, Tennessee, U.S.A.

BEAUDOIN, Harold A., LL.B. (cum laude). Educ.: Suffolk Univ., Boston. Employed (1935-), Atomic Power Equipment Dept. (1955-) now Manager of Public Information, Atomic Power Equipment Dept., Gen. Elec. Co., San Jose. Society: Atomic Ind. Forum.
Address: 13205 Pierce Road, Saratoga, California, U.S.A.

BEAUDOUIN, Paul. Director, Ets. Beaudouin S.A.
Address: Ets. Beaudouin S.A., 1 et 3 rue Rataud, Paris 5, France.

BEAUFAIT, Loren J., Jr., B.S. (Chem.). Born 1921. Educ.: California Univ. Dept. Head, Radiochem. Operations, Tracerlab, Inc. 1948-60; Chief, Safety and Tech. Services, San Francisco Operations Office, U.S.A.E.C., 1960-. Books: Handbook of Radiochem. Techniques, vol. i, and Handbook of Radiochem. Procedures, vol. ii (1957). Societies: A.C.S.; H.P.S.; A.N.S.
Nuclear interests: Technical liaison for atomic energy programmes in physical research; Biology, medicine and isotope development programmes; Co-ordination of radiation and nuclear safety programmes.
Address: U.S. Atomic Energy Commission, San Francisco Operations Office, 2111 Bancroft Way, Berkeley, California, U.S.A.

BEAUFAYS, O., Dr. Sen. Officer, Lab. de Physique des Plasmas, Inst. de Physique, Brussels Univ.
Address: Institut de Physique, Brussels University, 50 avenue F. D. Roosevelt, Brussels 5, Belgium.

BEAUGE, Roger, L. ès Sc., Ing. des Arts et Manufactures. Educ.: Paris Univ. Ing., C.E.A., Centre d'Etudes Nucléaires de Saclay.
Nuclear interest: Research reactors.
Address: Centre d'Etudes Nucléaires de Saclay, B.P. 2, 91Gif-sur-Yvette, France.

BEAULIEU, Henri LEROY-. See LEROY-BEAULIEU, Henri.

BEAUMONT, Donald. Born 1924. Asst. Design Eng., U.K.A.E.A., Risley, 1956. Society: Risley Branch, Nucl. Eng. Soc.
Address: 29 Latham Avenue, Runcorn, Cheshire, England.

BEAUMONT, René. Sen. Officer, Antwerpse Buizen Maatschappij "Anbuma" S.A.
Address: Antwerpse Buizen Maatschappij "Anbuma" S.A., 147 rue de Breda, Antwerp, Belgium.

BEAVER, Wallace Widmer, B. Metal. Eng., M.Sc. (Chem. Eng.). Born 1919. Educ.: Rensselaer Polytech. Inst. and Ohio State Univ. Asst. to Director of R. and D., Director of Process Development, Director of R. and D. (1949-), Vice-Pres. (1960-), Brush Beryllium Co., Cleveland. Books: Chapter author, The Metal Beryllium (A.S.M., 1955); Powder Metal. in Nucl. Eng. (A.S.M., 1958); Progress in Nucl. Energy, Series V; Metal. and Fuels, vol. i (1956). Societies: A.S.M.; A.I.M.M.E.; A.N.S.; A.S.T.M.; Electrochemical Soc.; American Ceramic Soc.; British Inst. of Metals; A.I.A.A.
Address: 22349 E. Byron Road, Shaker Heights, Ohio 44122, U.S.A.

BEBBINGTON, William Pearson, B. Chem., Ph.D. Born 1915. Educ.: Cornell Univ. Gen. Supt., Works Tech. Dept., E.I. duPont de Nemours and Co., 1962-. Societies: A.I.Ch.E.; A.C.S.; A.A.A.S.; South Carolina Acad. Sci.
Nuclear interests: Heavy water production; Chemical separations.
Address: E.I. duPont de Nemours and Co., Savannah River Plant, Aiken, South Carolina 29801, U.S.A.

BEBBS, Elbert Howlett, B.S., M.S. Born 1917. Educ.: Virginia Union Univ. and Catholic Univ. America, Washington, D.C. Instructor Phys., Howard Univ., Washington, D.C., 1947-50; Phys., Naval. Res. Lab., Washington, D.C., 1951-. Societies: A.N.S.; RESA; A.P.S.
Nuclear interests: Nuclear physics; In particular, the measurement of transition probabilities for fast isomeric transitions ($\tau \sim 10^{-9}$ sec. or shorter) as a means of investigating nuclear wave functions and nuclear models. Reactor physics; Measurement of reactor lattice parameters; Neutron flux measurements; Critical experiments.
Address: Apartment 204, 2104 Shipley Terrace S.E., Washington 20, D.C., U.S.A.

BEBIN, Jean Paul Marie Victor, Ing. Born 1918. Educ.: Ecole Navale, Centre de Saclay. Ministère de la Défense Nationale, -1958; Sté. l'Air Liquide, 1958-. Society: A.N.S.
Nuclear interests: Low-temperature equipment in nuclear research; Gas purification plants for gas-cooled reactors; Gas control systems in metallurgy; Gas analysing; Gas liquefaction and separation plants.
Address: L'Air Liquide, 75 Quai d'Orsay, Paris 7, France.

BECAREVIC, Aleksandar D. Editor and Member, Editorial Board, Bulletin of the Boris Kidrich Inst. of Nucl. Sci. Papers: Co-author, The fate of the liver highly polymerised labelled deoxyribonucleic acid inject into the X-irradiated rats (Bulletin of the Institute of Nuclear Sciences Boris Kidrich, Vol. 10, March 1960); Co-author, Effects of Actinomycin-D on the Incorporation of 6-[14] C-orotic Acid into the Ribonucleic Acid Fractions of Resting Rat Liver (ibid, Vol. 17, No. 1, 1966).
Address: Laboratory of Radiobiology, Institute of Nuclear Sciences Boris Kidrich, P.O. Box 522, Belgrade, Yugoslavia.

BECCARI, E. Member, Editorial Com., Minerva Nucleare.
Address: 83-85 Corso Bramante, Turin, Italy.

BECCARIA, Giorgio, Chem. Dott. Eng. Born 1934. Educ.: Milan Univ. Officer concerned with nucl. affairs, Union Carbide Italia S.p.A.
Address: Union Carbide Italia S.p.A., 28 Via Durini, Milan, Italy.

BECCHINI, Franco, M.D. Born 1930. Educ.: Pisa Univ. Res. Asst., Medical Clinic, Pisa Univ., 1955-. Society: Italian Soc. for Nucl. Biol. and Medicine.
Nuclear interests: Experimental and clinical uses of radioisotopes in endocrinology.
Address: Centro di Medicina Nucleare, Universita di Pisa, Pisa, Italy.

BECHERT, K., Prof., Dr. Vorsitzender, Bundesrat Ausschuss für Atomkernenergie und Wasserwirtschaft; Formerly Member, Steering Com., German Atomic Forum.
Address: Bundersrat Ausschuss für Atomkernenergie und Wasserwirtschaft, Bundeshaus Zi. S 141, 53 Bonn, Germany.

BECHTEL., Stephen Davison, Hon. LL.D. (California and Loyola), Hon. D. Eng. (U. of the Pacific). Born 1900. Educ.: California Univ. Sen. Director, Bechtel Corp.; Pres. and Director, Lakeside Corp.; Director: Bechtel Nucl. Corp., Industrial Indemnity Co., Southern Pacific Co. and Canadian Bechtel Ltd.; Graduate Member, The Business Council; Sen. Member, Nat. Ind. Conference Board; Member, Directors Advisory Council and Internat. Council, Morgan Guaranty Trust Co. of N.Y.; Trustee, Ford Foundation and San Francisco Bay Area Council; Director, Stanford Res. Inst. Societies: A.S.C.E.; Soc. of American Military Eng.; American Petroleum Inst.; Consulting Constructors' Council of America; Soc. of Naval Architects and Marine Eng.
Nuclear interests: Engineer, constructor, manager of nuclear power generating plants, nuclear fuel plants and nuclear desalting plants in more than 25 states and throughout the world. Nuclear power generating plants include over 40 facilities with a total capacity in excess of 15,000 Mw(e).
Address: 155 Sansome Street, San Francisco, California 94104, U.S.A.

BECHTEL., Stephen Davison, Jr., B.S. (Civil Eng.), M.B.A. Born 1925. Educ.: Colorado, Purdue, and Stanford Univs. Director (1951), Chairman of Exec. Com. (1959), Pres. (1961), Bechtel Corp., San Francisco; Chairman, Canadian Bechtel, Ltd., Toronto, 1957; Chairman, Bechtel Internat. Corp., San Francisco, 1958. Director, Industrial Indemnity Co., San Francisco 1957; Director, Crocker Citizens National Bank, San Francisco 1958; Director, Tenneco, Inc., Houston, 1958; Director, Hanna Mining Co; Vice-Chairman and Trustee, Nat. Ind. Conference Board; Board of Trustees, California Acad. Sci.; Trustee California Inst. Technol.; Director, Southern Pacific Co.; Business Council. Society: A.C.S.E.
Nuclear interests: Management; Engineering and construction services for nuclear power stations, chemical processing plants, special test reactors, and laboratories, and all types of atomic energy facilities for industry and governments.
Address: 220 Bush Street, San Francisco 4, California, U.S.A.

BECK, C. U.S. Member, Com. on Reactor Safety Technol., O.E.C.D., E.N.E.A.
Address: O.E.C.D. European Nuclear Energy Agency, 38 boulevard Suchet, Paris 16, France.

BECK, Clarence J., B.S. (Metal. Eng.). Born 1920. Educ.: Kansas Univ. Process Metal. (1949-56), Manager, Process Metal. and Ceramics (1956-59), Manager, Metal. Process Development (1959-66), Manager, Irradiation Sample Processing (1966-), K.A.P.L. Society: A.S.M.
Nuclear interests: Has interest in the management of technical leadership of groups developing processes for the manufacture and evaluation of nuclear core components and materials. In addition, has specialised technical interest in the application of the hot extrusion process to the fabrication of such materials.
Address: Knolls Atomic Power Laboratory, General Electric Company, P.O. Box 1072, Schenectady, N.Y., U.S.A.

BECK, Clifford Keith, B.A. (Sci.) (Catawba Coll.), M.S. (Science) (Vanderbilt Univ.), Ph.D. (Phys.) (North Carolina Univ.); D.Sc. (hon.; Catawba Coll.). Born 1913. Educ.: Catawba Coll., Vanderbilt and North Carolina Univs. Head, Phys. Dept., and Director, Nucl. Projects, North Carolina State Coll., 1949-56; Chief (1956-60), Asst. Director (1960-62); Reactor Hazards Evaluation Branch, Div. of Licensing and Regulation, U.S.A.E.C.; Deputy Director of Regulation, U.S.A.E.C., 1962-; Member, U.S. Deleg. to Geneva Conference on Peaceful Uses of Atomic Energy (1955, '58, '64). Book: Res. Reactors (van Nostrand, 1956). Societies: Fellow, A.P.S.; A.N.S.; A.A.A.S.
Nuclear interests: Education and research in nuclear technology; Safety in nuclear and aerospace facilities.
Address: Route 1, Shiloh Church Road, Boyds, Maryland, U.S.A.

BECK, Friedrich, Dr. rer. nat. Born 1927. Educ.: Göttingen Univ. Res. Assoc., Fritz Haber-Inst. der Max-Planck-Gessellschaft, Berlin-Dahlem, 1952-54; Visiting Fellow, M.I.T., 1954-56; Res. Assoc., Munich Univ., 1956-58; Privat-Dozent, Heidelberg Univ., 1958-60; Prof. extraord., Frankfurt am Main Univ., 1960-63; o. Prof., Darmstadt T.H., 1963-. Society: Deutsche Physikalische Gesellschaft.
Nuclear interests: Theoretical nuclear physics; Correlation structure of nucleons in nuclei; Reaction mechanisms.
Address: Institut für theoretische Kernphysik, Technische Hochschule Darmstadt, 9 Schlossgartenstr., Darmstadt, Germany.

BECK, Guido, Ph.D. Born 1903. Educ.: Vienna Univ. Prof. Theoretical Phys. (Prof. Titular), Centro Brasileiro de Pesquisas Fisicas, Rio de Janeiro, 1951-. Book: Kernbau und Quantenmechanik, Handbuch der Radiologie, Bd. VI/1 (Leipsiz, Akademische Verlagsgessellschaft 1931). Societies: Assoc. Fisica Argentina; Acad. Brasileira de Ciências.

Nuclear interests: Nuclear physics (theory): Nuclear resonances, Collision problems. Theory of the beta decay, Nuclear shell model, n-Body problem.
Address: Centro Brasileiro de Pesquisas Fisicas, Av. Wenceslau Braz 71 (fundos), Rio de Janeiro, Brazil.

BECK, W. S., M.D. Member, Isotope Com., Massachusetts Gen. Hospital.
Address: Massachusetts General Hospital, Boston 14, Massachusetts, U.S.A.

BECKENBAUER, Franz, Member, Consultative Com., Euratom Supply Agency. Stv. Vorsitzender, Arbeitskreis III/4, Versorgung mit Brennstoffen, Deutsche Atomkommission Federal Ministry for Sci. Res.
Address: 4 Höhenweg, (13a) Sulzbach-Rosenberg, Germany.

BECKER, Erwin Willy, o. Prof., Dr. rer. nat., Dipl. Chem. Born 1920. Educ.: Göttingen, Jena and Munich Univs. Inst. Phys., Marburg/Lahn Univ., 1947-58; Director, Inst. Nucl. Eng., Karlsruhe Tech. Univ. Society: Deutsche Physikalische Gesellschaft.
Nuclear interests: Isotope separation; Nuclear fusion; Low temperature physics.
Address: 18 Strählerweg, Karlsruhe-Durlach, Germany.

BECKER, Harry C., B.S., M.S., Ph.D. Born 1913. Educ.: Illinois Univ. Res. Supv. (1953-62), Res. Assoc. (1962-), Texaco Res. Centre, Texaco, Inc. Societies: A.N.S.; A.C.S.
Nuclear interests: Radiation chemistry of organic systems; Neutron and charted particle activation studies.
Address: Texaco, Incorporated, Texaco Research Centre, P.O. Box 509, Beacon, New York 12508, U.S.A.

BECKER, Josef, Dr. h.c. Med. Born 1905. Educ.: Budapest, Vienna, Berlin and Heidelberg Univs. Prof. Radiol., Heidelberg Univ.; Direktor der Universitäts-Strahlenklinik, Heidelberg Univ. Editor of: Strahlentherapie; Nuclear-Medizin; Zentralblatt für die gesamte Radiologie. Books: Lokalisierte Applikation künstlich radioaktiver Isotope (Springer-Verlag, 1961); Radioisotope in Geburtshilfe und Gynäkologie (Basel, New York, S. Karger-Verlag, 1956); Betatron und Telekobalttherapie (Berlin, Göttingen, Heidelberg, Springer-Verlag, 1958); Die Supervolttherapie (Goerg Thieme Verlag, 1961). Society: Deutsche Atomkommission und Strahlenschutzkommission.
Nuclear interests: Local therapeutic applications of radioisotopes and supervolt therapy.
Address: Universitäts-Strahlenklinik (Czerny-Krankenhaus), 3 Voss Strasse, Heidelberg, Germany.

BECKER, Klaus, Dr. rer. nat., Dipl. Chem. Born 1933. Educ.: Berlin-Dahlem Free Univ. and Munich T.H. Book: Photographic Film Dosimetry (Springer 1962, Focal Press 1967). Societies: H.P.S.; Fachverband f. Strahlenschutz; Ges. Deutsch. Naturforsch. u. Artze; Deutsche Ges. Photogr.
Nuclear interests: Health physics; Personnel radiation dosimetry; Radiation effects on photographic emulsions; Radiophotoluminescence in glasses; Nuclear track registration in solids.
Address: Applied Dosimetry Research Group, Health Physics Division, Oak Ridge National Laboratory, Oak Ridge, Tennessee, U.S.A.

BECKER, Kurt, Dipl.-Ing. Born 1913. Educ.: Munich T.H. Geschäftsführer, Firma Ingenieurbüro Dipl.-Ing. Kurt Becker G.m.b.H.
Nuclear interests: Projektierung von Anlagen der Anwendung radio-aktiver Erkenntnisse und Materie.
Address: 15 Pienzenauerstrasse, Munich 27, Germany.

BECKER, Manfred, Dr. rer. nat., Dipl. Phys. Born 1921. Educ.: Stuttgart T.H. With Nukem, Nuklear- Chemie und-Metallurgie GmbH, Wolfgang bei Hanau, Main (former Degussa Nucl. Group). Societies: Gesellschaft für Kernenergieverwertung in Schiffbau und Schiffahrt mbH, Hamburg; Deutsche Gesellschaft für Metallkunde; Inst. Metals, London.
Nuclear interest: Fuel elements.
Address: 27 Am Eichenhain, Niederrodenbach bei Hanau, Germany.

BECKER, Rudolf, Dipl. Ing., Dr. Ing. E.h. Born 1905. Educ.: Munich T.H. Generalbevollmächtigter, Linde Aktiengesellschaft. Societies: V.D.I.; Deutscher Kälteverein.
Nuclear interests: Wärmeübertragung; Reaktor gas-reinigung.
Address: Linde A.G., 8021 Höllriegelskreuth bei Munich, Germany.

BECKERLEY, James Gwavas, A.B., Ph.D. Born 1915. Educ.: Stanford Univ. Director of Classification, U.S.A.E.C. Washington, 1949-54; George Washington Univ. 1950-54; Head, Eng. Phys. Dept., Schlumberger Well Surveying Corp., 1954-59; Phys., I.A.E.A. 1959-62; Siftor Project, M.I.T., 1962-. Editor, Annual Rev. of Nucl. Sci., 1952-58; Supervisory Sci., Radioptics, Inc. Editor, A.N.S. 1955-59. Member, Publications Com., A.N.S. Books: Editor and editorial adviser for P. Van Nostrand Co., Nucl. Energy Series. Society: A.N.S.
Nuclear interests: Neutron physics; Nuclear well logging.
Address: 9 Jerusalem Lane, Cohasset, Massachusetts 02025, U.S.A.

BECKJORD, Eric Stephen, A.B. (Cum Laude Phys.), M.S.E.E. Born 1929. Educ.: Harvard Coll. and M.I.T. Lt. (jg), U.S. Naval Reserve, 1951-54; G.E.C., 1956-63, including assignment at Argonne Nat. Lab. on EBWR 1956-57; Westinghouse Elec. Corp., 1963-. Lecturer: Kjeller Inst. and A.B. Atomenergi, BWR Stability (1962); M.I.T. Nucl. Reactor Safety Course, Reactor Dynamics (summers 1966 and 1967);

Member, A.N.S. Standards Com., 1964-; Chairman, A.N.S. Systems Eng. Subcom., 1967-. Societies: A.N.S.; I.E.E.E.
Nuclear interests: Nuclear power reactor technology, including reactor systems design, control and electrical systems design, reactor core thermal and hydraulic design, and reactor safety and licensing; Power plant engineering; Director energy conversion.
Address: Westinghouse Atomic Power Division, PWR Plant Division, Box 355, Pittsburgh, Pennsylvania 15230, U.S.A.

BECKMAN, L., Dr. Sen. Sci., Nucl. Dept., Res. Inst. of Nat. Defence, Sweden.
Address: Research Institute of National Defence, 10 Gyllenstiernsgatan, Stockholm 80, Sweden.

BECKMANN, Prof., Inst. für Kernphysik, Johannes Gutenberg Univ.
Address: Johannes Gutenberg University, Institut für Kernphysik, 21 Saarstrasse, 65 Mainz, Germany.

BECKMANN, Robert B., Prof. Chairman, Chem. Eng. Dept., and Dean, Coll. of Eng., Maryland Univ.; Maryland Univ. rep. on Council, Oak Ridge Assoc. Univs.
Address: Maryland University, Department of Chemical Engineering, College Park, Maryland, U.S.A.

BECKURTS, Kark-Heinz Fritz Ferdinand, Dr. rer. nat. (Göttingen), apl. Prof. (Karksruhe Univ.). Born 1930. Educ.: Göttingen Univ. Max Planck Institut für Physik, Göttingen, 1952-58; Kernforschungszentrum Karlsruhe, 1958-. European-American Nucl. Data Com., E.N.E.A. Book: Co-author, Elementare Neutronenphysik (Springer-Verlag, 1958); Neutron Physics (Springer-Verlag, 1964).
Nuclear interests: Neutron physics; Nuclear data; High flux reactors and their use in nuclear and solid state physics.
Address: 5 Osteroder Strasse, Karlsruhe, Germany.

BECQUEMONT, Jacques-Michel, Ing. Born 1906. Educ.: E.S.P.C.I. Head, Electronic Tube Div., Lab. Central Télécommunications. Societies: Sté. Française Electroniciens et Radioélectriciens; Sté. Ing. Techniciens du Vide.
Nuclear interest: Manufacture of Geiger Muller and neutron counter tubes.
Address: Laboratoire Central de Telecommunications, 46 avenue de Breteuil, Paris 7, France.

BEDA, A.G. Paper: Co-author, Thermal Neutron Activation Cross-Sections for Cd^{108} (letter to the Editor, ibid, vol. 16, No. 2, 1964)
Address: Academy of Sciences of the U.S.S.R., 14 Leninsky Prospekt, Moscow V-71, U.S.S.R.

BEDARIDA, Alberto. Pres., Betron-Industria e Ricerche Elettroniche S.p.A.
Address: Betron-Industria e Ricerche Elettroniche S.p.A., 21 and 23 Via Montebello, Livorno, Italy.

BEDARIDA, Roberto, Dr. Eng. Tech. Director, Betron-Industria e Ricerche Elettroniche S.p.A.
Address: Betron-Industria e Ricerche Elettroniche S.p.A., 21 and 23 Via Montebello, Livorno, Italy.

BEDEWI, F. EL-. See EL-BEDEWI, F.

BEDFORD, A. W. Manager, NR1 Project, Knolls Atomic Power Lab.
Address: Knolls Atomic Power Laboratory, Schenectady, New York, U.S.A.

BEDFORD, C. F. Drawing Office, Design Eng. Div., Chalk River Nucl. Labs., A.E.C.L.
Address: Atomic Energy of Canada Ltd., Chalk River Nuclear Laboratories, Chalk River, Ontario, Canada.

BEDINAR, Antonio PLATA. See PLATA BEDINAR, Antonio.

BEDNAR, Jaroslav, Ing. C.Sc. Born 1929. Educ.: Brno Tech. Univ. Inst. of Nucl. Res., Dept. of Rad. Chem., 1960-. Member, Nat. Com. on Rad. Res.
Nuclear interests: Radiation chemistry, radiation dosimetry.
Address: Institute of Nuclear Research, Rez, Czechoslovakia.

BEDNARZ, Roman, M.Sc. Born 1936. Educ.: Warsaw Univ. Asst., Warsaw Univ., 1958-59; Asst., Nucl. Res. Inst., 1959-62.
Nuclear interests: Transport theory; Reactor theory; Plasma physics.
Address: Reactor Theory Group, Institute of Nuclear Research, Swierk by Warsaw, Poland.

BEDOYA, P., Julio A., M.D. Facultad de Medicina de San Marcos de Lima, Dr. en Medicina, Curso de Especializacion en Metodologia de Radioisótopos, Inst. de Biofisica de Brazil, Rio de Janeiro. Born 1913. Educ.: Facultad de Medicina de Lima; Escuela de Medicina de Santiago de Chile. Member, Peruvian deleg., Inter-American Congress of Radiology and Nucl. Medicine; O.R.I.N.S. special course for non-U.S. citizens, 1955. Sec.-Gen. de la Liga Peruana de la Lucha Contra el Cancer. Jefe del Servicio de Medicina Nucl., Facultad de Medicina, Hospital Obrero de Lima, Perú; Prof. de Medicina Nucl. Pres., Associacion Latino-americana de Sociedades de Biologia y Medicina Nucl., 1967. Societies: Soc. Peruana de Radiologia; Colegio Inter-americano de Radiologia; Soc. Cubana de Cancerologia; Soc. Colombiana de Radiologia; International Club of Radiotherapists; Soc. Peruana de Biologia y Medicina Nucl.
Address: Hospital Obrero de Lima, Peru.

BEDROSIAN, Paul Harry, B.S., M.S., Ph.D. Born 1928. Educ.: Northeastern and Florida Univs. Commissioned Officer (1961-), Project Officer, Tech. Operations Branch, Div. of Radiol. Health, Rockville, Md. (1961-62), Dep. Director, Eng. Programme, Northeastern Radiol. Health Lab., Winchester, Mass. (1962-65), Chief, Radium Technol. Unit, Southeastern Radiol. Health Lab., Montgomery, Ala. (1965-), U.S.P.H.S. Societies: A.S.C.E.; H.P.S.; American Water Works Assoc.; A.P.H.A.; Radiol. Health Conference; Conference of Federal Sanitary Engs.
Nuclear interests: Studies on the fate and environmental distribution of radioactive materials resulting from nuclear operations and the health implications of associated radiations; Development of safeguards; handling of radionuclides, particularly radium, in medical and non-medical applications, together with dosimetric evaluations.
Address: Southeastern Radiological Health Laboratory, U.S. Public Health Service, P.O. Box 61, Montgomery, Alabama, U.S.A.

BEEGHLY, Hugh F., B.S. (Chem. Eng.). Born 1912. Educ.: West Virginia Univ. Head, Appl. Sci. (1949-56), Sen. Res. Assoc., Nucl. Technol. (1956-65), Jones and Laughlin Steel Corp.; Asst. Sec., A.C.S., 1965-66; N.B.S., U.S. Dept. of Commerce, 1966-. Chairman, Standardization Relations Com.; Member, Subcom. on Rad. Effects, A.S.T.M. Books: Contributor to: Nucl. Eng., Part IV. (A.I.Ch.E., 1956); Problems in Nucl. Eng., Vol I. (Pergamon Press, 1957); Corrosion and Wear Handbook for Water-cooled Reactors (McGraw Hill, 1957); Gases in Metals (Iron and Steel Inst., 1960). Treatise on Analytical Chem. (Wiley-Interscience, 1961); Standard Methods of Chem. Analysis (Van Nostrand, 1963); Handbook of Analytical Chem. (cGraw Hill, 1963). Societies: A.N.S.; A.C.S.; A.I.Ch.E.; A.I.M.M.E.
Nuclear interests: Consultation on research programmes; Reactor coolant technology; Materials for reactors and shielding; Chemistry of metals; Corrosion; Metallurgy; Direction of research.
Address: 9331 E. Parkhill Drive, Bethesda, Md. 20014, U.S.A.

BEEK, Josienus van. See van BEEK, Josienus.

BEEKMAN, Myron C., B.S.M.E. Born 1921. Educ.: Michigan Univ. Asst. to Exec. Vice Pres. for Production, Detroit Edison Co. Member, (1962-67), Chairman (1966-67), Exec. Com., Nucl. Eng. Div., Member, Policy Board, Codes and Standards, Member, Nucl. Eng. Div., Chairman, Hons. Com., A.S.M.E.; Member, Professional Div. Com., A.N.S.; Member, Accident Prevention Com., Rad. Task Force, Member, Prime Movers Com., Nucl. Task Force, Edison Elec. Inst. Book: Atomic Energy Chapter in Energy and the Michigan Economy - a Forecast (Michigan Univ., 1967). Society: Atomic Ind. Forum.
Nuclear interest: The application of nuclear energy as a means of electrical power generation and a means of more fully utilising our energy resources.
Address: The Detroit Edison Company, 2000 Second Avenue, Detroit, Michigan 48226, U.S.A.

BEEKMAN, Richard W. Formerly Res. Assoc., Nucl. Eng. Group Faculty, Eng. College, Washington Univ.
Nuclear interest: Reactor physics.
Address: Department of Nuclear Engineering, College of Engineering, Washington University, Seattle, Washington 98105, U.S.A.

BEELER, Joe R., Jun. B.S., M.S., Ph.D. Born 1924. Educ.: Kansas Univ. Sandia Corp., 1952-55; Gen. Elec. Co., 1955-. Societies: A.N.S.; A.P.S.
Nuclear interests: Reactor theory; Radiation effects in solids.
Address: 10909 Carnegie Drive, Cincinnati 40, Ohio, U.S.A.

BEELER, R., Dr. Collaborateur, Lab. de Physique Nucléaire Expérimentale, Geneva Univ.
Address: Geneva University, 32 boulevard d'Yvoy, 1211 Geneva 4, Switzerland.

BEEN, Ulf, Sivilingenior (Chem. Eng.). Born 1926. Educ.: Norwegian Tech. Univ., Trondheim. Chem. (1950), Manager, Isotope Labs. (1958-), Inst. for Atomenergi, Kjeller, Norway. Societies: Den Norske Ingeniorforening, Oslo; Norsk Kjemisk Selskap, Oslo.
Nuclear interests: Radioisotopes; Laboratory techniques; Power production.
Address: Institutt for Atomenergi, P.O. Box 40, Kjeller, Norway.

BEENKEN, Carl-Dietrich, Dipl. Ing. Born 1893. Educ.: T.H. Aachen and T.H. Munich. Direktor (Generaldirektor, 1949-), der Schleswig-Holsteinischen Landesbrandkasse in Kiel, 1938-58. Mitglied der Technischen Kommission des Verbands der Sachversicherer e.V. Societies: Verein Deutscher Ingeniuere, Düsseldorf; V.D.E., Frankfurt-Main. Vereinigung zur Förderung des Deutschen Brandschutzes, Cologne; Nat. Fire Protection Assoc., Boston; Deutscher Verein für Verischerungs Wissenschaft, Berlin.
Nuclear interests: Sachverisherung von Versuchs- und Leistungs-Reaktoren, Versicherung von Isotopengefahren in der Industrie; Probleme der Entseuchung und Schadenverhütung.
Address: 16 Trainsjochstrasse, Munich 82, Germany.

BEER, J. J. de. See de BEER, J. J.

BEER, Janusz Zygmunt, Dr. Born 1930. Educ.: Warsaw Univ. Dept. of Organic Chem., Warsaw Univ., 1951-59; Dept. of Radiobiol. and Health Protection, Inst. Nucl. Res., 1959-. Visiting worker, Chester Beatty Res. Inst., London

1962-63. Book: Co-author, Plant Metabolism studied by Isotopic Methods (Poland, Państwowe Wydawnictwo Rolnicze i Leśne 1960). Societies: Polish Chem. Soc.; Polish Biochem. Soc.; Assoc. for Rad. Res.; European Soc. for Rad. Biol.
Nuclear interests: Cellular radiobiology and radiation genetics, in particular radiation-induced heritable damage on cellular and subcellular level; Radiometric analysis in biochemical research, especially radiochromatography.
Address: Dept. of Radiobiology and Health Protection, Institute of Nuclear Research, Warsaw 91, Poland.

BEER, Johannes Frederik de. See de BEER, Johannes Frederik.

BEERE, T. Corresponding Member for Ireland, Com. on Reactor Safety Technol., O.E.C.D., E.N.E.A.
Address: O.E.C.D. European Nuclear Energy Agency, 38 boulevard Suchet, Paris 16, France.

BEERMANN, Hermann, Mitglied, Bundersvorstand, Deutsche Gewerkschaftsbund. German member, Specialised Nucl. Section for Social Health and Development Problems, Economic and Social Com., Euratom. Mitglied, Präsidium, and Mitglied, Fachkommission IV, Deutsche Atomkommission, Federal Ministry for Sci. Res.
Address: 8 Stromstrasse, Düsseldorf, Germany.

BEESTON, Joseph Mack, B.S. (Chem. Eng.), Ph.D. (Metal. Eng.). Born 1918. Educ.: Utah Univ. Asst. Prof., Washington State Univ. 1953-58; Sen. Metal.,Phillips Petroleum Co., Atomic Energy Div., 1958-66, Materials Res. Sect. Chief, Idaho Nucl. Corp., 1966-.
Societies: A.S.M.; A.I.M.E.; A.S.T.M.; Idaho Acad. Sci.
Nuclear interests: Physical metallurgy of solid state with special interest in radiation effects in solids.
Address: 1604 Charlene Street, Idaho Falls, Idaho, U.S.A.

BEETEM, W. Arthur. Chem., Geological Survey, U.S. Dept. of the Interior.
Address: U.S. Department of the Interior, Geological Survey, Building 25, Federal Centre, Denver, Colorado, U.S.A.

BEETLESTONE, Andrew, C. Eng., Fellow, I.E.E., Assoc. Manchester Coll. Technol. Born 1906. Educ.: Manchester Coll. Technol. Head, X-Ray Div., Mullard Ltd. Society: Fellow, I.E.E.
Nuclear interests: Measuring instruments; Radiation hazards and protection.
Address: Mullard Ltd., X-Ray Division, New Road, Mitcham Junction, Surrey, England.

BEETS, Ch. Chef du Département Spectrometrie Nucléaire des Reacteurs, Labs., C.E.N.
Address: C.E.N., Mol-Donk, Belgium.

BEGEMANN, Friedrich, Ph.D. Born 1927. Educ.:Göttingen and Berne Univs. Res. Assoc., Enrico Fermi Inst. for Nucl. Studies, 1954-57; Asst., Max Planck Inst. für Chemie, Mainz, 1957-.
Nuclear interests: Nuclear physics; Meteorites.
Address: Max Planck Institut für Chemie, Mainz, Germany.

BEGER, H., Obering. Vice-Pres., Fachnormenausschuss Radiologie, Germany.
Address: i. Fa. VEM Transformatoren- und Röntgenwerk, 48 Overbeckstr., Dresden N 30, German Democratic Republic.

BEGHIAN, Leon E., B.A., Ph.D. Prof., Dept. of Nucl. Sci. and Eng., Lowell Technol. Inst.
Address: Lowell Technological Institute, Department of Nuclear Science and Engineering, Lowell, Massachusetts 01854, U.S.A.

BEGUE, L., Ing. Gérant, Sté. Internat. d'Etudes et de Participations Electroniques et Nucléaires.
Address: Société Internationale d'Etudes et de Participations Electroniques et Nucléaires, 4 rue de Téhéran, Paris 8, France.

BEGUM, Mrs. Qumrunnesa, Dr. Lecturer, Nucl. Phys., Dept. of Phys., Dacca Univ.
Address: Dacca University, Dacca 2, Pakistan.

BEGUN, George Murray, A.B., M.A., Ph.D. Born 1921. Educ.: Colorado Coll., Columbia and Ohio State Univs. Chem., O.R.N.L., 1951-.
Societies: A.C.S.; RESA.
Nuclear interests: Isotope separation; Chemical isotope effects; Infrared and Raman spectra of isotopic molecules; Xenon compounds.
Address: Oak Ridge National Laboratory, Oak Ridge, Tennessee, U.S.A.

BEGZHANOV, R. B. Paper: Co-author, Interaction Cross Section of Neutrons with ^{149}Sm and ^{115}In Nuclei (letter to the Editor, Atomnaya Energiya, vol. 16, No. 6, 1964).
Address: Academy of Sciences of the U.S.S.R., 14 Leninsky Prospekt, Moscow V-71, U.S.S.R.

BEHLMER, Sidney G., B.S. (Chem. Eng.). Born 1904. Educ.: Purdue Univ., With Hartford Fire Insurance Co. Chairman, Governing Com., Nucl. Energy Property Insurance Assoc.; Exec. Com., Nucl. Insurance Rating Bureau.
Address: 36 Belcrest Road, West Hartford, Connecticut 06107, U.S.A.

BEHOUNEK, Francis, Dr. of Sci. Born 1898. Educ.: Charles Univ., Prague, Sorbonne, Paris Univ. Res. Dept., Radiotherapeutical Inst., Praha-Bulovka, 1945-54; Prof., Faculty of Tech. and Nucl. Phys., Charles Univ. and Chief, Dept. of Dosimetry, Inst. of Nucl. Phys., Czechoslovak Acad. of Sci., 1954-. Books: Textbooks of Radiology, etc. Societies: Czechoslovak Acad. Sci.; Corresponding Member, Academy Nuovi Lyncei, Roma; Hon. Member, Soc. Romana Balneo-Climatica, Bucuresti; Member, Com. on the Effects of

Atomic Rad. of U.N.
Nuclear interests: Protection dosimetry;
Management of radioactive wastes.
Address: 24 Sinkulova, Prague 4, Czechoslovakia.

BEHRENS, Harold, Chem. Eng., Ph.D. Born
1925. Educ.: Uruguay Univ. Lecturer, Phys.,
Chem. Dept., Uruguay Univ., 1946-55; Prof.
Chem., Concepcion Univ.; 1955-58; Res.
Assoc. (1958-60), Head, Radiochem. Lab.
(1960-), Prof. Gen. Chem., School of Eng.
(1963-), Faculty of Phys. and Mathematical
Sci., Chile Univ. Member, Board of Directors,
Santiago Sect., Chilean Chem. Soc. Book:
Guia de estudios para el trabajo con isotopes
radioactivos (Concepcion, 1956).
Nuclear interests: Hot atom chemistry;
Radiation chemistry; Reaction kinetics.
Address: Laboratorio de Radioquimica, Centro
de Quimica, Facultad de Ciencias Fisicas y
Matematicas, Universidad de Chile, Casilla
2777, Santiago, Chile.

BEHRENS, Heinz, Dipl.-Phys., Dr. rer. nat.
Born 1927. Educ.: Karl-Marx-Univ., Leipzig.
Society: Phys. Soc. of the D.D.R.
Nuclear interests: Radiation-induced polymerization, radiation chemistry of high polymers,
energy-transfer.
Address: Institut für organische Hochpolymere
der Deutschen Akademie der Wissenschaften,
15 Permoserstr., 705 Leipzig, German Democratic Republic.

BEHRENS, Johan D. Member, Board of
Directors, A/S Noratom.
Address: A/S Noratom, 20 Holmenveien, Oslo,
Norway.

BEHRMAN, Irvin E. In charge, Nucl. Dept.,
Varlacoid Chem. Co.
Address: Varlacoid Chemical Co., 116 Broad
Street, New York 4, U.S.A.

BEI, Shi-zhang. Vice-Chairman, Editorial Com.,
Scientia Sinica.
Address: Scientia Sinica, Academia Sinica,
Peking, China.

BEIDUK, Felix M. Sen. Faculty Member, Phys.
Dept., Villanova Univ.
Address: Villanova University, Villanova, Pennsylvania, U.S.A.

BEIERLE, Frederick P. Born 1931. Educ.:
San Jose State and Minnesota Univs. Vice-Pres.,
Nucl. Operators; Vice-Pres., California Nucl.,
Inc.
Nuclear interests: Reactor operations, supervisory and managerial. Radioactive waste
disposal operations.
Address: Box 638, Richland, Washington,
U.S.A.

BEIERWALTES, William Henry, A.B., M.D.
Born 1916. Educ.: Michigan Univ. Director,
Nucl. Medicine (1952), Prof. Medicine (1959),
Medical School, Michigan Univ. Pres., Soc.
Nucl. Medicine, 1965-66; Vice-Pres., American
Thyroid Assoc., 1966-67; Councellor, Central
Soc. Clinical Res., 1964-67. Book: Co-author,
Clinical Use of Radioisotopes (Philadelphia,
W.B. Saunders Co., 1957).
Nuclear interest: Nuclear medicine.
Address: University Hospital, University
Medical Centre, Ann Arbor, Michigan, U.S.A.

BEINER, Marcel, Dr. Inst. für Theoretische
Kernphysik, Bonn Univ.
Address: Bonn University, Institut für Theoretische Kernphysik, 16 Nussallee, Bonn,
Germany.

BEK-UZAROV, Diordie, M. of Sc. Graduated
from the Faculty of Sci. (Phys.). Born 1929.
Educ.: Belgrade Univ. Head, Absolute Measurement Group, Dept. of Phys., Inst. Nucl. Sci.
"Boris Kidrich," Belgrade; Hon. Asst. Electronics Dept., Faculty of Sci., Belgrade Univ. 1958-.
Society: Mathematical and Phys. Soc. of
Yugoslavia.
Nuclear interests: Nuclear physics; Problem of
measurement of the radioactivity of isotopes;
Dosimetric measurements; Nuclear electronics
and equipment; Methods of measurements.
Address: Institute of Nuclear Sciences, P.O.
Box 522, Belgrade, Yugoslavia.

BEKEFI, George, B.Sc., M.Sc., Ph.D. Born
1925. Educ.: London and McGill Univs. Asst.
Prof., McGill Univ., 1956-57; Res. Assoc., Res.
Lab. of Electronics (1957-60), Asst. Prof.
(1960-64), Assoc. Prof. (1964-), M.I.T. Member,
Exec. Com., Div. of Plasma Phys. A.P.S.
Societies: A.A.P.T.; Fellow, A.P.S.
Nuclear interests: Plasma physics and thermonuclear fusion.
Address: Massachusetts Institute of Technology,
77 Massachusetts Avenue, Cambridge, Massachusetts 02139, U.S.A.

BEKKUM, D.W. VAN. See VAN BEKKUM,
D.W.

BEL, Gerard LE. See LE BEL, Gerard.

BELAN, V. G. Paper: Co-author, Production of
Megagauss Fields by Magnetodynamics Cumulation (Atomnaya Energiya, vol. 23, No. 6, 1967).
Address: Academy of Sciences of the U.S.S.R.,
14 Leninsky Prospekt, Moscow V-71, U.S.S.R.

BELANE, Mrs. DEBRECZENI. See DEBRECZENI BELANE, Mrs.

BELANOVA, T. S. Papers: Fast-Neutron
Absorption Cross-sections (letter to the Editor,
Atomnaya Energiya, vol. 8, No. 6, 1960); co-author, On the Effect of Nuclon Number Parity
on the Magnitude of Rad. Capture (ibid, vol. 14,
No. 2, 1963); co-author, A Photoneutron Lab.
for Work with High Activity Sources (ibid, vol.
15, No. 6, 1963); co-author, Absolute Measurement 24 Kev Neutron Absorption Cross
Section (ibid, vol. 19, No. 1, 1965).

Address: Academy of Sciences of the U.S.S.R., 14 Leninsky Prospekt, Moscow V-71, U.S.S.R.

BELBEOCH, Roger. Sect. Accelerateur Linéaire et Techniques Annexes, Lab. de l'Accelerateur Linéaire, Ecole Normale Superieure, Paris Univ.
Address: Laboratoire de l'Accelerateur Lineaire, Ecole Normale Superieure, Batiment 200, Faculte des Sciences, Orsay (Seine et Oise), France.

BELCHER, Ernest Hugh, B.A. (Hons.) (Cantab.), M.A. (Hons.) (Cantab.), Ph.D. (London), B.Sc. (Hons.) (London). Born 1920. Educ.: Cambridge and London Univs. Lecturer in Biophys., Inst. Cancer Res., 1948-58; Sen. Lecturer in Medical Phys., Director, Radioisotope Unit, Postgraduate Medical School, London, 1958-63; Head, Sect. of Nucl. Medicine, I.A.E.A., Vienna, 1963-. Societies: British Inst. Radiol.; Hospital Phys. Assoc.; Physiological Soc.; Soc. for Rad. Res. Nuclear interests: Applications of radioisotopes in medical diagnosis and research; Radiation disimetry.
Address: International Atomic Energy Agency, Vienna I, Austria.

BELCHER, Philip F., A.B., LL.B. Educ.: Yale Univ. Alternate Documentary Div. Leader (1952-57), Asst. Director for Classification and Security and Documentary Div. Leader (1957-), Los Alamos Sci. Lab.
Address: 80 Obsidian Loop, Los Alamos, New Mexico, U.S.A.

BELDECOS, Nicholas Andrew, B.S. (Mech. Eng.), M.S. (Mech. Eng.). Born 1922. Educ.: Swarthmore Coll. and Pennsylvania Univ. Gen. Manager, Bettis Atomic Power Lab. Societies: A.S.M.E.; American Soc. of Naval Architects and Marine Eng.; A.N.S.; Atomic Ind. Forum. Nuclear interests: Performance of reactors and plants, with specific interest in reactor design.
Address: Westinghouse Electric Corporation, Bettis Atomic Power Laboratory, P.O. Box 79, West Mifflin, Pennsylvania, U.S.A.

BELELLI, Umberto. Chef du Secretariat technique, S.E.N.N. Adviser, Third I.C.P.U.A.E., Geneva, Sept. 1964; Member, Study Group on Long Term Role of Nucl. Energy in Western Europe, O.E.C.D., E.N.E.A.
Address: Societa Elettronucleare Nazionale, 6 Via Torino, Rome, Italy.

BELIK, Yu. A. Paper: Co-author, Exptl. Investigation of Variation of Electron Beam Cross Dimensions in the 1.5 GeV Synchrotron (Atomnaya Energiya, vol. 24, No. 1, 1968).
Address: Academy of Sciences of the U.S.S.R., 14 Leninsky Prospekt, Moscow V-71, U.S.S.R.

BELIN, Roger. Member, Comité des Mines, C.E.A., France.

Address: Commissariat a l'Energie Atomique, 29-33 rue de la Federation, Paris 15, France.

BELKIN, V. F. Papers: Co-author, Criticality Stand Tests of a Heavy-Water Reactor with Fuel Elements of Rod Type (letter to the Editor, Atomnaya Energiya, vol. 12, No. 2, 1962); co-author, The Determination of External Blocking-Effect for Multiplying Assemblies with D_2O Moderator (ibid., vol. 14, No. 3, 1963); co-author, Measurements of Thermal Neutron Density Distribution along Radius of Plug Fuel Elements (ibid., vol. 15, No. 5, 1963); co-author, The Determination of Migration Length and Multiplication Coefficient in a Heavy-Water Natural Uranium Lattice (Jaderna Energie, vol. 13, No. 11, 1967).
Address: Institute of Theoretical and Experimental Physics, Academy of Sciences of the U.S.S.R., 14 Leninsky Prospekt, Moscow V-71, U.S.S.R.

BELKNAP, Paul A., B.S. (Chem., Ohio), M.S. (Chem., Iowa State). Born 1916. Educ.: Ohio and Iowa State Univs. Pres. and Treas., Charleston Rubber Co. Member, A.C.S. Nuclear interests: Engineering and design of protective clothing, including Dry Box gloves, lead-filled gloves and sheeting, gloves for protection against radiation, shoe covers, and other articles from plastics and rubber such as covers for mechanical manipulations.
Address: Charleston Rubber Company, Stark Industrial Park, Charleston, South Carolina, 29405, U.S.A.

BELL, Carlos Goodwin, Jr., B.S., S.M., S.D. Born 1922. Educ.: A. and M. Coll. of Texas and Harvard Univ. Assoc. Prof., C.E., N.T.I. Northwestern Univ., Illinois. Book: Co-editor, Liquid Scintillation Counting (Pergamon Press, 1958). Societies: A.N.S.; A.S.C.E. Nuclear interest: Education.
Address: Oak Ridge School of Reactor Technology, Oak Ridge National Laboratory, Oak Ridge, Tennessee, U.S.A.

BELL, D. D. Staff Eng., Eldorado Mining and Refining Ltd. Deleg., Third I.C.P.U.A.E., Geneva, Sept. 1964; Chairman, Uranium Com., Canadian Nucl. Assoc.
Address: Eldorado Mining and Refining Ltd., Port Hope, Ontario, Canada.

BELL, Eric Brayton, B.Sc.,F. Inst. P. Born 1924. Educ.: Durham Univ. Sci. Officer, British Iron and Steel Res. Assoc., 1950-52; Phys. (1952-56), Head, Res. Dept. (1956-58) Isotope Developments Ltd.; Asst. Manager (Health Phys.), Springfield Works, U.K.A.E.A., 1958-59; Chief Phys. (1959-61), Gen. Manager (1961-), Saunders-Roe and Nucl. Enterprises. Nuclear interests: Industrial nucleonics; Isotope-activated light sources.
Address: Saunders-Roe and Nuclear Enterprises Limited, North Hyde Road, Hayes, Middlesex, England.

BELL, G. F., M.A., Ph.D. Born 1920. Educ.: Marquess Univ. Asst. Prof. (1950-55), Prof. (1955-), Nucl. Phys., Cherry Coll. Papers: 24 in field of nuclear physics. Societies: Rad. Phys. Soc.; Nu Omicron Tau.
Address: BM/ZWMX, London, W.C.1, England.

BELL, George I., A.B., Ph.D. Born 1926. Educ.: Harvard and Cornell Univs. U.S. A.E.C. Pre-Doctoral Fellow, 1950-51; Staff member, Los Alamos Sci. Lab., California Univ., 1951-; Prof. New Mexico Univ. (part time), 1961-; Visiting Lecturer, Harvard Univ., 1962-63. Societies: Fellow, A.P.S.; Fellow, A.N.S.; A.A.A.S.
Nuclear interests: Reactor physics; Neutronics; Theoretical physics in general; Synthesis and properties of heavy nuclei.
Address: Box 1663, Los Alamos, N.M., U.S.A.

BELL, Graydon D., B.S., M.S., Ph.D. Born 1923. Educ.: Kentucky Univ. and California Technol. Inst. Instructor, Kentucky Univ., Summers 1949, 1950; Asst. Prof. Phys., Robert Coll., Istanbul, 1951-54; Graduate Res. Asst. (1954-56), Res. Fellow (1956-57), California Technol. Inst.; Asst. Prof. Phys. (1957-60), Assoc. Prof. Phys. (1960-), Harvey Mudd Coll.; Phys. (Atomic and Nucl.)., Nat. Bureau of Standards, 1963-64(Sabbatical Year). Societies: American Astronomical Soc.; A.A.P.T.; A.A.A.S.
Nuclear interests: Teaching of nuclear physics; Astrophysical application of nuclear theory.
Address: Department of Physics, Harvey Mudd College, Claremont, California, U.S.A.

BELL, J., B.E.(Hons.). Dept. of Nucl. and Rad. Chem., New South Wales Univ.
Address: New South Wales University, Box 1, Post Office, Kensington, New South Wales, Australia.

BELL, Kenneth John, B.S.Ch.E., M.Ch.E., Ph.D. Born 1930. Educ.: Case Inst. Technol. and Delaware Univ. Eng., Hanford Operations, G.E.C., 1955-56; Asst. Prof., Case Inst. Technol., 1956-61; Visiting Faculty O.R.S.O.R.T., 1958; Assoc. Prof. (1961-67), Prof. (1967-), Oklahoma State Univ. Chairman, Energy Conversion and Transport Div., A.I.Ch.E.
Nuclear interests: Heat transfer (especially boiling) and fluid dynamics, with particular interest in thermal design of reactors and associated power recovery equipment.
Address: School of Chemical Engineering, Oklahoma State University, Stillwater, Oklahoma, U.S.A.

BELL, Lawrence Gerald, M.A.Sc. Born 1931. Educ.: British Columbia Univ. With A.E.C.L., 1955-62; Supv., Development Metal., Orenda Ltd., 1963-. Society: A.P.E.O.
Nuclear interests: Reactor metallurgy; Mechanical properties of zirconium alloys; Development of reactor components; Fuel alloy research; In-pile creep measurement and interpretation.
Address: Engineering Laboratories, Orenda Limited, P.O. Box 6001, International Airport, Toronto, Ontario, Canada.

BELL, M. A. Operating Manager, Carolinas Virginia Nucl. Power Assocs., Inc.
Address: Carolinas Virginia Nuclear Power Associates, Inc., Parr, South Carolina, U.S.A.

BELL, Persa Raymond, B.S. (Phys.), D.S. (hon.). Born 1913. Educ.: Howard Coll. Head, Phys. Electronics Group (1946-57), Asst. Director, Thermonucl. Exptl. Div. (1957-67), O.R.N.L.; Chief, Lunar and Earth Sci. Div. and Manager of the Lunar Receiving Lab., N.A.S.A.-Manned Spacecraft Centre, Houston, Texas, 1967-. Member, Editorial Board, Nucl. Instruments and Methods, Netherlands. Societies: Fellow, A.P.S.; Fellow, N.Y. Acad. Sci.; Soc. Nucl. Medicine; A.G.U.
Nuclear interests: Lunar and planetary studies; Thermonuclear research (plasma physics); Physics instrumentation; Nuclear medical instrumentation; Sonar and microwave radar.
Address: 202 Sleepy Hollow Ct., (Taylor Lake Village), Seabrook, Texas 77586, U.S.A.

BELL, Robert Edward, B.A., M.A., Ph.D. Born 1918. Educ.: British Columbia and McGill Univs. Sen. Res. Officer, Chalk River Labs. Canadian Atomic Energy Project, 1946-56; Prof. Phys. (1956-60), Rutherford Prof. Phys. and Director Rad. Lab. (1960-), McGill Univ., Montreal. Books: Contributions to: Beta and Gamma-ray Spectroscopy (North-Holland Pub. Co., 1955, pp. 494-520, 680-688, 696-706). Annual Reviews of Nuclear Science (Stanford, 1954, pp. 93-110). Societies: Fellow, Roy. Soc. of Canada; Fellow, A.P.S.; Vice Pres., Canadian Assoc. of Phys.
Nuclear interests: Nuclear physics and electronics; Laboratory administration.
Address: Radiation Laboratory, McGill University, Montreal 2, Canada.

BELL, Robert Younger, B.A. Born 1921. Educ.: Cambridge Univ. Design Eng., Gas Turbine Dept., W. H. Allen, Sons and Co., Ltd., Bedford, 1950-55; Manager, Nucl. Projects, Yarrow-Admiralty Res. Dept., Yarrow and Co., Ltd., Glasgow, 1955-. Societies: A.M.I.Mech.E.; M.I.Mar.E.
Nuclear interest: Marine propulsion installations.
Address: 2 Ardenconnel Way, Rhu, Dunbartonshire, Scotland.

BELL, Wayne Elliot. Asst. Sect. Leader, Gulf Gen. Atomic. Sec., San Diego Sect., A.N.S., 1966-.
Address: 408 Canyon Druve, Solana Beach, California 92075, U.S.A.

BELLE, Jack, B.S., Ph.D. Born 1921. Educ.: Pittsburgh Univ. and Brooklyn Polytech. Inst. Manager, Fuel Materials, Bettis Atomic Power Lab., 1955-. Book: Editor, Uranium Dioxide: Properties and Nuclear Applications. Societies: A.C.S.; American Ceramic Soc.; A.A.A.S.

BEL

Nuclear interest: Metallurgy.
Address: 1330 Varner Road, Pittsburgh 27, Pennsylvania, U.S.A.

BELLET, H. A., Capitaine de Vaisseau. Commandant, Ecole d'Application Maritime de l'Energie Atomique.
Address: Ecole d'Application Maritime de l'Energie Atomique, Cherbourg (Manche), France.

BELLIER, Jean. Born 1905. Educ.: Ecole Nationale des Ponts et Chaussées, Paris. Directeur Général du Bureau d'Etudes A. Coyne et J. Bellier.
Nuclear interest: Prestressed concrete caissons for atomic reactors.
Address: 19 rue Alphonse de Neuville, Paris 17, France.

BELLINGER, R., A.M.I.Mech.E. Director, Strachan and Henshaw Ltd.
Address: Strachan and Henshaw Ltd., Ashton Works, P.O. Box No. 103, Bristol 3, England.

BELLINO, Carmine S. Certified Public Accountant, Wright, Long and Co., Washington. Member, Board of Contract Appeals, U.S.A.E.C.
Address: U. S. Atomic Energy Commission, Board of Contract Appeals, Washington D.C. 20545, U.S.A.

BELLION, B. Asst. Editor, Minerva Nucleare.
Address: 83-85, Corso Bramante, Turin, Italy.

BELLUCO, Umberto. Reader in Chem., Centro Chimica Nucleare, Padua Univ.
Address: Universita di Padua, 6 Via Loredan, Padua, Italy.

BELMORE, Frederick Martin, B.S. (Chem.), B.S.E. (Chem. Eng.). Born 1915. Educ.: Virginia Univ. Vice-Pres., Gen. Manager and Director, Mallinckrodt Nucl. Corp., 1959-63; Special Asst. to Pres., Mallinckrodt Chem. Works, 1955-; Pres. and Chief Exec. Officer, The Matheson Co., Inc., 1963-. Societies: A.N.S.; A.C.S.; A.I.Ch.E.
Nuclear interest: Management in connection with production of nuclear fuels and fuel elements.
Address: 57 Joy Avenue, Webster Groves 19, Missouri, U.S.A.

BELOT, Félicien René, Civil Electrician and Mech. Eng. (Belgium), D.Sc. (Paris), Ph.D. Born 1922. Educ.: Polytechnical (Mons) and Paris Univs. Production Asst. Manager, Union des Centrales Electriques, 1946-; Belgian Deleg., Shippingport Atomic Plant, 1956-58. Societies: Sté. Belge des Electriciens; A.N.S.
Nuclear interests: Nuclear reactor economics; Reactor design and operation.
Address: 43 rue des Colonies, Brussels, Belgium.

BELOTTI, Agostino, Dr. Ing. Director, Sci. Instruments for Nucl. Applications Dept., Ingg. S. and Agostino Belotti.
Address: Ingg. S. and Agostino Belotti, 8 Piazza Trento, Milan, Italy.

BELOUS, V. N. Paper: Co-author, Corrosion Resistance of Structural Materials in Bor Contained Solutions (letter to the editor, Atomnaya Energiya, vol. 19, No. 6, 1965).
Address: Academy of Sciences of the U.S.S.R., 14 Leninsky Prospekt, Moscow V-71, U.S.S.R.

BELOUSEV, E. M. Inst. for Sci. Res. in Cotton-growing.
Nuclear interest: Plant physiology.
Address: Institute for Scientific Research in Cotton-growing, 41 Pushkinskaya ul., Tashkent, U.S.S.R.

BELOV, L. M. Papers: Co-author, Energy Distribution of Spontaneous Fission Fragments of ^{244}Cm (letter to the Editor, Atomnaya Energiya, vol. 15, No. 3, 1963); co-author, Cm^{241} Spontaneous Fission Accompanied by Long Range a Particle Emission (letter to the Editor, ibid, vol. 16, No. 2, 1964).
Address: Academy of Sciences of the U.S.S.R., 14 Leninsky Prospekt, Moscow V-71, U.S.S.R.

BELOV, S. P. Papers: Co-author, Angular and Energy Distributions of Gamma-rays scattered by Fe and Pb (letter to the Editor, Atomnaya Energiya, vol. 5, No. 4, 1958); co-author, The Space and Energy Distributions of Neutrons in Boron Carbide (letter to the Editor, ibid, vol. 6, No. 6, 1959); co-author, The Yield of Rad.-Capture γ-Rays from Iron (ibid, vol. 13, No. 1, 1962); co-author, Secondary Gamma-Rays Co-efficients for Aluminium, Copper and Tungsten (letter to the Editor, ibid, vol. 19, No. 5, 1965).
Address: Academy of Sciences of the U.S.S.R., 14 Leninsky Prospekt, Moscow V-71, U.S.S.R.

BELOV, S. V. Paper: Co-author, Determination of Average Energy β-Rad. of Small Activity Samples (letter to the Editor, Atomnaya Energiya, vol. 16, No. 6, 1964).
Address: Academy of Sciences of the U.S.S.R., 14 Leninsky Prospekt, Moscow V-71, U.S.S.R.

BELOVINSTEV, K. A. Papers: Co-author, A 6.5 MeV Microtron for Electron Injection into a Synchrotron (Atomnaya Energiya, vol. 14, No. 4, 1963); co-author, On the New Possibilities for Increasing the Efficiency of a Microtron (letter to the Editor, ibid., vol. 15, No. 1); co-author, About Possibility of Positron Production and Acceleration in Microtron (letter to the Editor, vol. 16, No. 4, 1964); co-author, Positron Stacking and Obtaining of Electron-Positron Beams in Synchrotron (ibid, vol.23, No. 1, 1967).
Address: Academy of Sciences of the U.S.S.R., 14 Leninsky Prospekt, Moscow V-71, U.S.S.R.

BELPOMME, Maurice, Eng. (Arts et Métiers), Eng. (Ecole Supérieure d'Electricité). Born 1918. Educ.: Arts et Métiers, and Ecole Supérieure d'Electricité. Manager, Derveaux Labs., 1952-58; From 1958: Manager, Compagnie Française Thomson Houston; Gen. Manager, Sodeteg (Sté. d'Etudes Techniques et d'Entreprises Générales); Managing Director, Sodetra (Sté. d'Etudes Techniques et de Travaux); Gen. Manager, Somes (Sté. d'Exploitation pour la Mise en oeuvre d'Engins Spatiaux) Director, Sté. Aurore. Societies: A.T.E.N.; SYNTEC (Chambre Syndicale des Bureaux d'Etudes Techniques de France).
Nuclear interests: Industrial architecture and technical studies on the following: all nuclear installations, nuclear physics, electronics, thermodynamics, liquid metals.
Address: le Mesnil Fleuri, 31 avenue Georges Clémenceau, Villennes sur Seine, (S. et O.), France.

BEL'SKAYA, Zh. N. Paper: Co-author, Application of Eigenfunction Expansions Method to Multidimensional Reactor Computations (Atomnaya Energiya, vol. 23, No. 2, 1967).
Address: Academy of Sciences of the U.S.S.R., 14 Leninsky Prospekt, Moscow V-71, U.S.S.R.

BELTER, Walter G., B.Sc. (Civil Eng.), M.S. Born 1923. Educ.: Wisconsin Univ. Sanitary Eng., Indiana State Board of Health, 1950-55; Nucl. Waste Decontamination Eng., U.S. Air Force 1955-57; Sanitary Eng., U.S.A.E.C., 1957- Vice Pres., Federal Water Quality Assoc., U.S.A. Book: Chapter on Atomic Energy in Industrial Waste-Water Control. Society: A.N.S.
Nuclear interest: Radioactive waste management research and development.
Address: Environmental and Sanitary Engineering Branch, Division of Reactor Development and Technology, United States Atomic Energy Commission, Washington, D.C. 20545, U.S.A.

BELTRAM, C. Manager, Production, Apollo Facility, Nucl. Materials and Equipment Corp.
Address: Nuclear Materials and Equipment Corp., Apollo, Pennsylvania 15613, U.S.A.

BELTRANENA-SINIBALDI, Julio, Chem. Eng. Born 1903. Prof., Univ. S. Carlos, Guatemala, 1934-; Owner of Chem. Ind. Lab., 1928-. Tech. Adviser, Dirrección General de Mineria e Hidrocarburos. Books: Clasificación Mineralógica; Apuntes sobre Quimica Aplicada para Ingenieros Civiles; Geologia para Ingenieros Civiles. Societies: Member (and ex-Pres.) Comisión Nacional de Energia Nucl., Guatemala; Member (and ex-Pres.) Chem. Eng. Assoc. of Guatemala; Phys. and Medical Acad. of Sci. of Guatemala; Chem. Eng. Assoc. of Mexico; Colegio de Farmaceuticos y Quimicos, Guatemala. Univ. Nacional de México, Univ. de S. Carlos, Guatemala. Founder, 1st Museum of Dirección General de Mineria e Hidrocarburos; Deleg. of Guatemala to Inter-American Symposium on the Peaceful Applications of Nucl. Energy, 1957; Deleg. of Guatemala to 1st Ordinary Meeting of Board of Governors, I.A.E.A., 1957.
Nuclear interests: Radioactive minerals and metallurgy.
Address: Laboratorio Quimico Industrial Comercial y Agricola, 10a. Calle 3-39, Zona 1, Guatemala City, Guatemala, Central America.

BELYAEVA, I. A. Paper: Co-author, Rad. near Reactor VVR-M (letter to the Editor, Atomnaya Energiya, vol. 19, No. 1, 1965).
Address: Academy of Sciences of the U.S.S.R., 14 Leninsky Prospekt, Moscow V-71, U.S.S.R.

BELYAKOV, A. N. Paper: Co-author, On Uniform Irradiation of Objects across Their Surface with Pulse Electron Beam (letter to the Editor, Atomnaya Energiya, vol. 19, No. 6, 1965).
Address: Academy of Sciences of the U.S.S.R., 14 Leninsky Prospekt, Moscow V-71, U.S.S.R.

BEN, W. R. van der. See van der BEN, W. R.

BEN-DAVID (DAVIS), Gerald, B.Sc., A.R.C.S., Ph.D., D.I.C. Born 1928. Educ.: Imperial Coll., London. Res. Fellowship, 1950- 53; Phys. Lecturer (1954-56), Acting Head Nucl. Eng. (1956-58), Technion Israel Inst. Technol., Haifa; Head, Nucl. Phys. Dept. (1959-), Head, Phys. Div. (1966-), Soreq Nucl. Res. Centre, Yavne; Bar-Ilan Univ., Ramat-Gan, 1963-.
Nuclear interests: Cosmic ray, reactor and neutron physics (including capture gamma-rays); Neutron flux determination fission.
Address: Nuclear Physics Department, Israel Atomic Energy Commission, Soreq Nuclear Research Centre, Yavne, Israel.

BENASSI, Enrico. Born 1901. Educ.: Parma Univ. Prof. Medical radiol. and Director Istituto di radiologia, Turin Univ. Book: Cenni sull'impiego degli isotopi radioattivi nella semeiotica clinica (Rome, Istituto Nazionale per l'Assicurazione contre Malattie, 1959). Societies: Stà. Italiana di Radiologia Medica e Medicina Nucleare; Stà. Italiana di Biologia e Medicina Nucleare.
Nuclear interest: Nuclear medicine.
Address: 3 Via Genova, Turin, Italy.

BENAVENT, Ferdinand Leopold F. J. DE. See DE BENAVENT, Ferdinand Leopold F. J.

BENDEL, David, B.Sc., DipEd. Born 1928. Educ.: Melbourne Univ. Phys., Commonwealth X-ray and Radium Labs., Melbourne, 1949-51; Sen. Lecturer, Rad. Phys., Roy. Melbourne Inst. of Technol., 1951-. Educ. Advisor, and Hon. Member, Australasian Inst. of Radiography; Rep. of Roy. Melbourne Inst. of Technol. on Victorian Conjoint Com. of Radiol. and Radiographers. Society: Assoc., Australian Inst. of Phys.
Nuclear interests: Lecturing in nuclear radiation physics; In charge of technician level course in

radioisotope work for medical isotope units; Consultant in industrial applications of radioisotopes.
Address: Royal Melbourne Institute of Technology, 124 Latrobe Street, Melbourne, Australia.

BENDER, Merrill Arthur, A.B., M.D. Born 1923. Educ.: Middlebury Coll., Vermont, and Harvard Medical School. Chief, Dept. of Nucl. Medicine,Head, Div. of Radiol., Roswell Park Memorial Inst. Pres., Soc. of Nucl. Medicine (1967-68). Societies: Soc. of Nucl. Medicine; American Assoc. for Cancer Res.
Nuclear interest: Nuclear medicine.
Address: Department of Nuclear Medicine, Roswell Park Memorial Institute, Buffalo, New York, U.S.A.

BENDER, Welcome. Director of Res., Space Exploration Group, Martin Co.
Address: Martin Company, Mail No. 7, Baltimore, Maryland 21203, U.S.A.

BENE, Georges J., D. ès Sc. Born 1919. Educ.: Faculty of Sci., Lyon and Inst. of Exptl. Phys., Geneva. Prof. extraordinaire, Geneva Univ.; Prof. titulaire, Grenoble Univ. Books: Co-author, Résonance paramagnétique nucléaire (1955). Societies: Sté. Suisse de Physique; Sté. Française de Physique; A.P.S.; Sec. Groupement Ampere.
Nuclear interests: Nuclear magnetic resonance in the low frequency range (terrestrial magnetic field) and on metallic samples.
Address: Laboratoire de Spectroscopie Hertzienne, Institut de Physique Expérimentale, Bd. d'Yvoy, Geneva, Switzerland.

BENEDEK, Laszlo. Member, Editorial Board, Energia es Atomtechnika.
Address: Muszaki Konyvkiado, Bajcsy-Zsilinszky ut 22, Budapest 5, Hungary.

BENEDICT, Karl Georg, Dipl. Phys. Born 1933. Educ.: Mainz Univ. Phys. i/c of Triga Mainz Reactor, Johannes Gutenberg Univ., Mainz, 1962-. Society: Deutsche Phys. Gesellschaft.
Nuclear interest: Reactor and neutron physics.
Address: Institut für Anorganische Chemie und Kernchemie, Johannes Gutenberg Universität, Mainz, Germany.

BENEDICT, Manson, B.Chem. (Cornell), S.M., Ph.D. (M.I.T.). Born 1907. Sci. Director (1951-58), Director (1963-), Nat. Res. Corp.; Prof. Nucl. Eng. (1951-58), Head, Nucl. Eng. Dept. (1958-), M.I.T.; Head, Office of Operations Analysis, U.S.A.E.C., 1951-52; Advisory Com. on Reactor Safeguards, U.S.A.E.C., 1947-58; Member (1958-), Chairman (1962-64), Gen. Advisory Com., U.S.A.E.C.; Director, Atomic Ind. Forum 1966-. Books: Co-author, Nucl. Chem. Eng. (McGraw-Hill, 1957). Societies: A.I.Ch.E.; A.N.S.; N.A.S.; Nat. Acad. Eng.; Fellow, American Acad. Arts and Sci.; A.C.S. Nuclear interests: Fuel cycles; Reactor engineering; Isotope separation.
Address: 25 Byron Road, Weston, Massachusetts, U.S.A.

BENENATI, Robert F., B.Ch.E., M.Ch.E., Dr. Ch.E. Born 1921. Educ.: Brooklyn Polytech. Inst. Consultant to Atomic Energy Div., H.K. Ferguson Co., 1950-52; Consultant to Water Kidde Nucl. Lab., 1952-54; Prof. Chem. Eng. and Director of Computer Centre; Nucl. Consultant to firm of Sanderson and Porter; Consultant to Union Carbide at Oak Ridge. Consultant to Nucl. Eng. Dept. of Brookhaven Nat. Lab. First Sec., Programme Chairman, and one of founders, N.Y. Sect., A.N.S. Societies: A.I.Ch.E.; A.C.S.; A.N.S.
Nuclear interests: Reactor analysis and design; Fluid flow and heat transfer; Fuel cycles; Fuel reprocessing.
Address: Polytechnic Institute of Brooklyn, 333 Jay Street, Brooklyn, New York, U.S.A.

BENENSON, W. Exptl. Nucl. Phys. Cyclotron, Dept. of Phys. and Astronomy, Michigan State Univ.
Address: Department of Physics and Astronomy, Michigan State University, East Lansing, Michigan, U.S.A.

BENES, Jaroslav, Dr. C.Sc. Born 1929. Educ.: Charles Univ., Prague. Nucl. Res. Inst., Czechoslovak Acad. Sci., 1956-. Book: Fundamentals of Autoradiography, (Prague, S.N.T.L., 1965).
Nuclear interests: Radiochemistry; Dosimetry.
Address: Nuclear Research Institute, 100 Na Truhlarce, Prague 8, Czechoslovakia.

BENES, Josef, RNDr. Born 1917. Educ.: Charles Univ., Prague. Asst. Phys. Dept. (1945-), Doc. of Exptl. Phys., Faculty of Maths. and Phys. (1951-), Asst. Prof. of Exptl. Phys., Faculty of Tech. and Nucl. Phys. (1959-), Prague.
Nuclear interests: Nuclear physics, especially beta and gamma spectroscopy.
Address: 7 Brehova, Prague 1, Czechoslovakia.

BENESOVSKY, Friedrich, Dr. -tech., Dipl.-Ing. Born 1914. Educ.: Brünn and Graz Univs. Director of Res., Metallwerk Plansee, Reutte, Tyrol. Books: Hartstoffe (Vienna, Springer Verlag, 1963); Hartmetalle (Vienna, Springer Verlag, 1965). Society: D.G.M.
Nuclear interests: Metallurgy of nuclear fuels especially carbides, nitrides, oxides; Compatibility with transition metals.
Address: Metallwerk Plansee, A6600 Reutte, Tyrol, Austria.

BENGSTON, Joel., B.S. (Trinity Coll., Conn.), M.S. and Ph.D. (Yale). Born 1926. Educ.: Trinity Coll., Connecticut,and Yale Univ. Phys., O.R.N.L., 1952-55; Phys., Theoretical Div., California Univ. Rad. Lab., 1955-58; Phys., Gen. Manager's Staff, Aerojet-General Nucleonics, San Ramon, California, 1958-60; Lecturer in Nucl. Eng., California Univ.

Berkeley, 1957-60; Phys. and Manager, Phys. Theory and Math., Nucl. Div., Combustion Eng. Inc., Windsor, Conn., 1960-62; Staff Member, Sci. and Technol. Div., Inst. for Defence Analyses, Arlington, 1962-; Lecturer Phys., Maryland Univ. 1966-67. Societies: A.P.S.; A.N.S.
Nuclear interests: Reactor analysis, weapons effects and nuclear physics.
Address: c/o Science and Technology Division, Institute for Defence Analyses, 400 Army-Navy Drive, Arlington, Virginia 22202, U.S.A.

BENGTSON, Kermit B., B.S. (Chem. Eng.), M.S., Ph.D. (Chem. Eng.). Born 1922. Educ.: Washington Univ. Group Leader in Chlorine Metal. Res., Kaiser Aluminum and Chem. Corp., Permanente, California, 1957-58; now Director, Washington Univ. Centre for Graduate Study at Hanford, Richland, Washington. Societies: A.I.Ch.E.; A.C.S.; A.S.E.E.; Glaciological Soc.
Nuclear interests: Development of graduate curriculum for students working with nuclear technology; Chemical processing technology.
Address: Washington University, Centre for Graduate Study, Richland, Washington, U.S.A.

BENINSON, Dan J., Dr.; U.N. Sci. Com. on Effects of Atomic Rad.; Nat. Commission of Atomic Energy, Argentina.
Address: Comision Nacional de la Energia Atomica, 8250 Avda. del Libertador Gral. San Martin, Buenos Aires, Argentina.

BENISZ, Jerzy, Dr. Born 1931. Educ.: Jagiellonian Univ., Cracow. Higher Pedagogical Coll., (Govt.), Katowice. Society: Polish Phys. Soc.
Nuclear interests: Mechanism of ternary fission of heavy nuclei (with the emission of a third particle with $Z > 2$); Investigation of the nuclear structure by means of direct reactions; Optical model for deuterons; Statistical model of the nucleus; The technique of the application of nuclear emulsions.
Address: Higher Pedagogical College, 9 street Szkolna, Katowice, Poland.

BENITEZ, Rodrigo FIERRO-. See FIERRO-BENITEZ, Rodrigo.

BENNETOT, Michel de. See de BENNETOT, Michel.

BENNETT, Carl Allen, Ph.D. (Maths.). Born 1921. Educ.: Bucknell and Michigan Univs. Manager, Appl. Maths, and Manager, Maths, Hanford Atomic Products Operation, G.E.C., Richland, 1950-65; On leave, Res. Assoc., Maths, Princeton Univ., 1950-51; On leave, Visiting Prof., Statistics, Stanford Univ., 1964; Manager, Maths Dept., Pacific Northwest Labs., Battelle Memorial Inst., 1965-. Societies: A.A.A.S.; A.C.S.; Fellow, A.S.A.; American Soc. for Quality Control; Biometric Soc.; Inst. for Mathematical Statistics; Internat. Assoc. for Statistics in the Phys. Sci.; Operations Res. Soc. of America; Soc. for Ind. and Appl. Maths.
Nuclear interests: Application of operations research and statistical methodology to nuclear research and development and production processes; Inspection and control procedures for domestic and international nuclear materials safeguards.
Address: 3115 West Canal Drive, Kennewick, Washington 99336, U.S.A.

BENNETT, Clarence Edwin, Ph.B., Sc.M., Ph.D. Born 1902. Educ.: Brown Univ. Prof. and Head, Phys. Dept., Chairman, Atomic Energy Com., Maine Univ.; Chairman, New England Sect., A.P.S.; Editor, Phys. Div., A.S.E.E.; Counsellor for Maine, A.I.P. Societies: Fellow, A.A.A.S.; American Optical Soc.
Nuclear interest: Administration of a Physics Department involving undergraduate and graduate work in nuclear physics.
Address: 65 Forest Avenue, Orono, Maine, U.S.A.

BENNETT, Edward F. P., B.Sc. (Eng.), A.C.G.I. Born 1914. Educ.: London Univ. Managing Director, Internat. Combustion Ltd.
Address: International Combustion Limited, Derby, England.

BENNETT, Edward L., B.A., Ph.D. Born 1921. Educ.: Reed Coll., Oregon and California Inst. Technol. Res. Biochem. and Assoc. Director, Lab. Chem. Biodynamics, Lawrence Rad. Lab., California Univ., Berkeley. Book: D_2O in Biol. Systems (Pergamon). Societies: American Soc. of Biol. Chem.; A.C.S.; New York Acad. of Sci.
Nuclear interest: Use of tracers in biological studies particularly biochemistry of learning.
Address: Laboratory of Chemical Biodynamics, Lawrence Radiation Laboratory, California University, Berkeley, California 94720, U.S.A.

BENNETT, G. P. Director of Eng., Worthington-Simpson Ltd.
Address: Worthington-Simpson Ltd., Lowfield Works, P.O. Box 17, Newark, Nottinghamshire, England.

BENNETT, George Alan, B.Ch.E., M.S. (Ch.E.), S.M. Born 1918. Educ.: Syracuse Univ., Case Inst. Technol., and M.I.T., Sen. Eng., Atomic Energy Dept., North American Aviation, Downey, Calif., 1951-52; Res. Fellow, Chem. Eng. Dept., Delaware Univ., Newark, 1952-54; Assoc. Chem. Eng., Argonne Nat. Laboratory, 1954-. Society: A.N.S.
Nuclear interests: Sodium technology as related to the programme planning necessary to the development of the liquid metal fast breeder reactor; Pyrochemistry, high temperature materials development, radioactive laboratory design and nuclear engineering.
Address: 1219 E. Willow Avenue, Wheaton, Ill., U.S.A.

BENNETT, J. Managing Director, A. Reyrolle and Co. Ltd.
Address: A. Reyrolle and Co. Ltd., Hebburn, Co. Durham, England.

BENNETT, John M., Prof. Head, Adolph Basser Computing Dept., School of Phys., Sydney Univ.
Address: Basser Computing Department, School of Physics, Sydney University, Sydney, New South Wales, Australia.

BENNETT, Ralph Decker, B.S. (E.E.), M.S. (E.E.), Ph.D. (Phys.). Born 1900. Educ.: Union Coll., Chicago and Princeton Univs. and California Inst. of Technol. Manager, G.E.C. Vallecitos Atomic Lab., 1956-61; Vice-Pres. (1961-67), Consultant (1967-), The Martin Co. Societies: A.N.S.; I.E.E.E.; A.P.S.; Atomic Ind. Forum.
Nuclear interest: Research and development.
Address: The Martin Co., Friendship International Airport, Maryland 21240, U.S.A.

BENNETT, Robert W. Formerly Director, R. and D. Eng., Alco Products Inc.; Asst. to Gen. Manager, Atomic Fuel Div., Westinghouse Elec. Corp., 1963-.
Address: Atomic Fuel Division, Westinghouse Electric Corp., P.O. Box 217, Cheswick, Pennsylvania, U.S.A.

BENNETT, Wallace Foster, A.B. Born 1898. Educ.: Utah Univ. Finance Com., Banking - Currency Com., Joint Com. on Atomic Energy, Select Com. Standards-Conduct, Joint Com. on Defence Production, Republican Policy Com., Joint Com. on Coinage, James Madison Memorial Commission; Member, Joint Com. on Atomic Energy, U.S.A.E.C., 1959-.
Address: 4201 Massachusetts Avenue, N.W., Washington, D.C., U.S.A.

BENNETT, William E. Ph.D. Born 1907. Educ.: Queen's Univ. Cambridge. Prof. Phys., Illinois Inst. Technol., 1947-60; Prof. Phys. Buffalo Univ., 1960-. Societies: A.A.A.S.; A.P.S.
Nuclear interest: Electrostatic generators.
Address: Physics Dept., Buffalo University, Buffalo 14, N.Y., U.S.A.

BENNETT, William M. Member, Standing Com. on Nucl. Energy in the Elec. Ind., Nat. Assoc. of Railroad and Utilities Commissioners.
Address: National Association of Railroad and Utilities Commissioners, Interstate Commerce Commission Building, P.O. Box 684, Washington, D.C. 20044, U.S.A.

BENNETT, William R., Jr., A.B., Ph.D., M.A. (Hon., Yale Univ.). Born 1930. Educ.: Princeton and Columbia Univs. Member, Tech. Staff Bell Telephone Labs., 1959-62; Assoc. Prof. (1962-65), Prof. (1965-), Phys. and Appl. Sci., Yale Univ.; Consultant: Inst. for Defence Analyses, 1962; Tech. Res. Group, 1962-; Army Res. Office, 1965-; CBS Labs., 1966-. Member, Editorial Advisory Board, J. Appl. Phys. and Appl. Phys. Letters, 1965-68; Member, Editorial Advisory Board, J. Quantum Electronics, 1965-68. Societies: Fellow, Optical Soc. of America; Sen. Member, I.E.E.E.; A.P.S.; Connecticut Acad. of Arts and Sci.; Fellow, Berkeley Coll., Yale Univ.
Nuclear interest: Applications of quantum electronics to atomic physics.
Address: Physics Department, Yale University, 217 Prospect Street, New Haven 11, Connecticut, U.S.A.

BENO, Joseph Henry, B.E.E. (Rensselaer). Born 1919. Educ.: Rensselaer Polytech. Inst., Atomic Energy for Management (Nat. Ind. Conference Board), and California Univ. Asst. Manager, Nucleonics Lab. (1957-), Acting Manager, Remote Handling Systems (1960-), Hughes Aircraft Co., Culver City, California, and Fullerton, California. Societies: A.N.S.; American Inst. of Elec. Eng.; A.A.A.S.; RESA.
Nuclear interests: Radiation effects on electronics; Development and manufacture of linear accelerators and cyclotrons; Development and manufacture of remote handling equipment for nuclear facilities and propulsion units; Management of nuclear oriented engineering activities.
Address: 12182 Southwest Country Lane, Santa Ana, California, U.S.A.

BENOIST-GUEUTAL, Mme. Pierrette. Born 1923. Educ.: Paris Univ. Prof., Faculté Sci., Paris. Societies: A.P.S.; Sté. Française Phys.
Nuclear interest: Nuclear theoretical physics.
Address: Institut de Physique Nucléaire, 91-Orsay, France.

BENOIT, André. Tech. Director, Nucl. Dept. Intertechniques S.A. Rep. of Intertechnique, Conseil d'Administration, A.T.E.N.
Address: Intertechnique S. A., B.P.1., Plaisir, (Seine-et-Oise), France.

BENOIT, J. Pres., Internat. Soc. for Cell Biol.
Address: International Society for Cell Biology, c/o Professor M. Chèvremont, Institut d'Histologie, 20 rue de Pitteurs, Liège, Belgium.

BENOIT, Jean Claude. Information Direction des Relations Exterieures et des Programmes, C.E.A. Member, Nucl. Public Relations Contact Group.
Address: Commissariat à L'Energie Atomique, 29-33 rue de la Federation, Paris 15, France.

BENOIT, Marc. Director of Projects, Asselin, Benoit, Boucher, Ducharme, Lapointe.
Address: Asselin, Benoit, Boucher, Ducharme, Lapointe, 4200 Dorchester Boulevard West, Montreal 6, Canada.

BENOIT-CATTIN, Pierre, L. es Sc. Born 1938. Educ.: Toulouse Univ. Gouvernement: Eng. in Nucl. Phys.; Eng., Nucl. Phys. Centre, Faculty of Sci., Toulouse Univ.

Nuclear interests: Nuclear physics: particle detectors and neutronic.
Address: Centre de physique nucléaire, Faculté des sciences, Toulouse University, 118 route de Narbonne, 31-Toulouse, (Haute-Garonne), France.

BENSADOUN, Maurice, Ancien élève de l'Ecole Polytechnique, Ing. de l'Ecole Nationale Supérieure des Télécommunications. Born 1923. Educ.: Paris Univ. Ing. militaire des télécommunications; Ing., C.E.A.; Directeur Gen., Sodern. Societies: Sté. Française Electroniciens et des Radioélectriciens; Sté. Française d'Astronautique.
Nuclear interest: Nuclear and spatial electronics.
Address: 2 avenue Fayolle, 94 Vincennes, France.

BENSCH, Friedrich, Ph.D. Born 1931. Educ.: Vienna Univ. First Phys. Inst., Vienna Univ.; Res. fellow, Atominst. der Oesterreichischen Hochschulen. Society: Austrian Phys. Soc.
Nuclear interest: Neutron physics.
Address: Atominstitut der Oesterreichischen Hochschulen, 115, Schüttelstrasse, Vienna 2, Austria.

BENSCHOP, J. P., Ing. Managing Director, Equipment and Boiler Manufacturing Co. den Haag Ltd.
Address: Equipment and Boiler Manufacturing Co. den Haag Ltd., Industrieterrein Noord, P.O. Box 69, Goes, Netherlands.

BENSIMON, Roland Louis, Ing. Educ.: Ecole Polytechnique et Ecole Supérieure d'Electricité. Born 1903. Res. Eng. Societies: Sté. Française de Métallurgie; Sté. de Chimie Industrielle; A.I.M.E.
Nuclear interests: Nuclear and ancillary metallurgy.
Address: 2 rue Victorien Sardou, Paris 16, France.

BENSON, Andrew A., Ph.D. (Org. Chem.), (Cal. Tech.). E.O. Lawrence Memorial Award, 1962. Born 1916. Educ.: California Univ. (Berkeley). Res. Assoc. BioOrganic Group, Rad. Lab., Berkeley, California, 1946-54; Assoc. Prof. Agriculture and Biol. Chem. (1955-59), Prof. (1959-61), Pennsylvania State Univ.; Prof. in Res., Biophys. and Nucl. Medicine and of Physiological Chem., U.C.L.A., 1961-62; Prof. Biol., Calif. Univ., San Diego, 1962-.
Nuclear interests: Tracer methodology; Neutron activation chromatography; Chemistry of recoil atoms.
Address: Department of Marine Biology, Scripps Institution of Oceanography, California University, San Diego, La Jolla, California, U.S.A.

BENSON, Dean Fred J. Director, Texas Eng. Experiment Station.
Address: Texas Engineering Experiment Station, College Station, Texas, U.S.A.

BENSON, Richard E., D.V.M., M.S. Born 1925. Educ.: Auburn and Rochester Univs. Lieutenant-Colonel, U.S.A.F. Veterinary Corps, 1950-; Command Veterinarian, Headquarters, 17th Air Force, 1967-. Societies: American Veterinary Medical Assoc.; American Veterinary Public Health Assoc.; American Veterinary Radiol. Soc.; H.P.S.
Nuclear interests: Biological effects of radiations; Metabolism of radionuclides; Diagnostic and therapeutic radiology.
Address: Headquarters, 17th Air Force, Box 9205, A.P.O., New York 09012, U.S.A.

BENSON, Willard R., M.E., M.S. (Appl. Mechanics), M.S. (Statistics). Born 1926. Educ.: Stevens Inst. of Technol., Virginia and Stanford Univs. Chief, Eng. Sci. Lab., Feltman Res. Labs., Picatinny Arsenal, U.S. Army Munitions Command. Societies: New York Acad. of Sci.; RESA.
Nuclear interests: Management; Warhead effects; Vulnerability of structures.
Address: United States Army Munitions Command, Picatinny Arsenal, Feltman Research Laboratories, Dover, New Jersey 07801, U.S.A.

BENSON GYLES, Thomas, M.Sc., M.A. Born 1903. Educ.: Melbourne Univ. Military Security Board, Germany, including Nucl. Controls, -1952; Tech. Adviser to Govt. of India for U.N., 1953-55; Patents Exploitation Officer, U.K.A.E.A., 1956-58; Tech. Adviser to Govt. of India for Colombo Plan, 1958-1960; Chief Eng. Nucl. and Chem. Eng. Dept., and Gen. Adviser and Consultant, W. S. Atkins and Partners, 1960-64. Societies: M.I.Chem.E.; Founder Fellow, British Inst. of Management.
Nuclear interests: General consulting including W. S. Atkins' interest in nuclear projects and laboratories.
Address: Fair Lawn, Ormond Avenue, Richmond, Surrey, England.

BENSTED, John Patrick Macrae, M.A., M.B., B.Chir. Born 1920. Educ.: Cambridge and London Univs. Fellow in Pathology, Minnesota Univ.; Hon. Lecturer in Pathology, Guy's Hospital, London; Res. Pathologist, Phys. Dept., Inst. Cancer Res., London. Societies: British Soc. Immunology; Assoc. Rad. Res.; Exptl. Pathology Club.
Nuclear interests: Mainly Pathological effects of radiation: Carcinogenic effects of bone-seeking isotopes, and effects of continuous γ-radiation, particularly on immune mechanisms.
Address: Department of Physics, Institute of Cancer Research, Clifton Avenue, Belmont, near Sutton, Surrey, England.

BENT, Ronald Albert. Born 1922. Managing Director, Lancashire Dynamo Electronic Products, Ltd. Societies: Fellow, I.E.E.; M.I.E.R.E.
Nuclear interests: Instrumentation and automatic control.
Address: Summercourt, 64 Brocton Road, Milford, Staffordshire, England.

BENTINCK, V. F. W. CAVENDISH-. See CAVENDISH-BENTINCK, V.F.W.

BENTLE, Gordon George, Ph.D., M.S., B.S. Born 1922. Educ.: Vanderbilt Univ. Member, Tech. Staff. Rocketdyne, 1954-. Society: American Ceramic Soc.
Nuclear interest: Metallurgy; Metal fuel swelling; Deformation of ceramic fuels; Diffusion in ceramic fuels.
Address: Rocketdyne, 6633 Canoga Avenue, Canoga Park, Calif., U.S.A.

BENTLEY, Roy Edward, B.Sc., Ph.D. Born 1930. Educ.: Birmingham Univ. Res. Student, Pharmacology Dept., Medical School, Birmingham Univ. 1951-55; Royal Air Force 1955-57; Phys., Inst. Cancer Res., Roy. Marsden Hospital, London, 1957-. Society: Assoc., Inst. of Physics.
Nuclear interests: Measurements of small amounts of radioactivity by β- and γ-counting; Naturally occurring radioactivity in the human environment; γ-spectrometry.
Address: Physics Department, Institute of Cancer Research, Clifton Avenue, Belmont, near Sutton, Surrey, England.

BENTO de CARMAGO, Pedro, Dr. Head, Nucl. Eng. Div., Inst. de Energia Atomica.
Address: Instituto de Energia Atomica, Cidade Universitaria "Armando de Salles Oliveira", 9, S.P., Caixa Postal 11049 - Pinheiros, Sao Paulo, Brazil.

BENTZEN, Francis L., B.S., M.S. (Phys.). Born 1920. Educ.: New Mexico Univ. With Sandia Corp., Albuquerque, New Mexico, 1949-54; Asst. Manager, Eng. and Test Branch, Atomic Energy Div., Phillips Petroleum Co., Idaho Falls, 1954-. Society: A.N.S.
Nuclear interests: Reactor safety; Kinetics studies; Instrumentation.
Address: 393 Hemlock Circle, Idaho Falls, Idaho, U.S.A.

BENVENISTE, Jacob, B.A., Ph.D. Born 1921. Educ.: Reed Coll., and California (Berkeley) Univ. Phys., The Aerospace Corp. Societies: A.P.S.; A.A.A.S.
Nuclear interests: Nuclear physics; Radiation effects on the solid state.
Address: The Aerospace Corporation, 1111 East Mill Street, San Bernardino, California, U.S.A.

BENVENUTI, Evandro, Dr. Sec. Gen., Internat. Exhibition of Electronics, Nucl. Energy, Wireless, T.V. and Cine.
Address: International Exhibition of Electronics, Nuclear Energy, Wireless, T.V. and Cine, 9 Via Crescenzio, Rome, Italy.

BENZI, Valerio, Dr. in Phys., Libero Docente in Fisica del Reattore Nucleare. Born 1928. Educ.: Rome Univ. Asst. Prof. Theoretical Phys. and Spectroscopy, Cagliari Univ., Sardinia; Asst. Prof. Nucl. Phys., Bologna Univ.; Deputy Director, Centro di Calcolo, C.N.E.N., Bologna. Member, Group of Specialists on Neutron Data Compilation, E.N.E.A., O.E.C.D. Book: Le Reazioni Nucleari (Cappelli Ed., 1961).
Nuclear interests: Fuel cycles and neutron cross-sections calculations.
Address: Centro di Calcolo del C.N.E.N., 2 Via Mazzini, Bologna, Italy.

BENZIGER, Theodore Michell, B.S. (Chem. Eng.). Born 1922. Educ.: Iowa State Univ. Res. Eng., Linde Air Products Co., 1946-52; Staff Member, Los Alamos Sci. Lab. 1952-.
Nuclear interests: Development of extruded, graphite-matrix fuel elements.
Address: Los Alamos Scientific Laboratory, P.O. Box 1663, Los Alamos, New Mex., U.S.A.

BERAK, Lubomir, Dr. rer. nat. (Chem.), candidate of tech. sci. Born 1923. Educ.: Caroline Univ., Prague and Czechoslovak Acad. Sci. Various appointments Czechoslovak Chem. industries, -1951; Res. Inst., CKD, Prague, -1957; Nucl. Res. Inst., Czech. Acad. Sci. (1957-), now Res. Group Leader. Society: Czechoslovak Chem. Soc., Radiochem. Sect.
Nuclear interests: Radioactive waste disposal, fission products, processes of sorption, solid state chemistry.
Address: Nuclear Research Institute, Czechoslovak Academy of Sciences, Rez, Czechoslovakia.

BERAN, Denis Carl, A.B. Born 1935. Educ.: Michigan Univ. District Manager, Nucleonics Magazine. Society: A.N.S.
Address: McGraw-Hill, Inc., 645 North Michigan Avenue, Chicago, Illinois 60611, U.S.A.

BERAN, Milos, RNDr. Born 1938. Educ.: Moscow State Univ. At Nucl. Res. Inst., Czechoslovak Acad. Sci. 1961-. Society: Czechoslovak Chem. Soc.
Nuclear interest: Nuclear chemistry.
Address: 121 Sidliste, Rez near Prague, Czechoslovakia.

BERANEK, Jiri. Born 1929. Educ.: Technical Univ., Prague, and Charles Univ., Prague (postgraduate study). Ministry of Chem. Industry, 1955; Chemoprojekt, 1955-59; Czechoslovak A.E.C., 1959-61; I.A.E.A., Vienna, 1961-67; Czechoslovak A.E.C., 1967-. Member, Editorial Board, Jaderna Energie. Book: Radioisotope Lab. Handbook (S.N.T.L. 1961). Societies: Czechoslovak Scientific Associations - Commission for Nucl. Energy; Czechoslovak Inst. of Chem. Eng.; A.N.S.
Nuclear interests: Siting and safety analyses of nuclear reactors and power plants; Safety evaluation of nuclear fuel processing and reprocessing; Radioactive waste treatment and disposal; Environmental surveillance.
Address: Czechoslovak Atomic Energy Commission, 9 Slezska, Prague 2, Czechoslovakia.

BERARDINELLI, Luciano. D. Chem. Eng. Born 1931. Educ.: Bologna Univ. Harwell Reactor School, 1958; Servizio P.R.O., C.N.E.N., Vice Tech. Correspondent of C.N.E.N. for Dragon Project, 1960-; Responsible for Co-ordination Office, Organic Reactor Programme (P.R.O.), 1963; Head, Co-ordination Office of Nuclear Desalination Programme (R.O.V.I.) 1965.
Nuclear interests: Reactor design and development.
Address: 4 Largo Forano, Rome, Italy.

BERBASOV, V. P. At Inst. of Atomic Energy, Phys.-Chem. Inst. Adviser, Third I.C.P.U.A.E., Geneva, Sept. 1964.
Address: Kurchatov Institute of Atomic Energy, Physical-Chemical Institute, 46 Ulitsa Kurchatova, Post Box 3402, Moscow, U.S.S.R.

BERDOT, J. Technician, Toulouse Univ.
Address: Toulouse University, Centre de Physique Nucléaire de la Faculte des Sciences, 118 Route de Narbonne, 31-Toulouse (Haut-Garonne), France.

BERENBLUM, Isaac, M.D., M.Sc. Born 1903. Educ.: Leeds Univ. Prof. and Head Dept. Exptl. Biol., Weizmann Inst. Sci., Rehovoth, 1950-. Book: Man against Cancer (Baltimore, Johns Hopkins Press, 1952); Cancer Res. Today (Oxford, Pergamon Press, 1967). Societies: Pathological Soc. of Great Britain and Ireland; American Assoc. for Cancer Res.
Nuclear interests: Experimental study of leukaemia induction by radiation, and other problems of radiation effects in animals, as part of the more general problem of experimental carcinogenesis.
Address: The Weizmann Institute of Science, Rehovoth, Israel.

BERENS, Tadeusz, Head, Dept. for Design and Construction of the Second Reactor, Soltan Nucl. Res. Centre, Inst. of Nucl. Res. Paper: Co-author, Polish High-Flux Res. Reactor R-II. Design Criteria and Functional Description (Nukleonika, vol. 13, No. 3, 1968).
Address: Institute of Nuclear Research, Swierk near Warsaw, Poland.

BERESOVSKI, Theodore, B.S. (Mech. Eng.), M.S. (Eng. Phys.). Educ.: Rensselaer Polytech. Inst. and Cornell UNiv. Res. Asst., Cornell Univ., 1950-52; Sen. Eng., Nucl. Aircraft Propulsion Project, Pratt and Whitney Aircraft, 1952-57; Group Leader, Nucl. Div., Allis Chalmers Manufacturing Co., 1957-59; Sci. Adviser, Brussels, Belgium, Office of the U.S.A.E.C., 1959-63; Sen. Evaluation Eng., U.S.A.E.C., 1963-.
Nuclear interests: Development of various nuclear reactor types including thermal, mechanical and physics design aspects and improved operation characteristics.
Address: U.S. Atomic Energy Commission, Washington, D.C. 20545, U.S.A.

BERETTA, A. Member, Editorial Com., Minerva Nucleare.
Address: 83-85 Corso Bramante, Turin, Italy.

BERETTA, R., Dr. Tech. Director, Ambrosini Acciai di Ambrosini e Beretta.
Address: Ambrosini Accaia di Ambrosini e Beretta, 28 Via Reina, Milan, Italy.

BEREZIN, A. K. Papers: Co-author, Interaction of Strong Electron Beams with Plasma (Atomnaya Energiya, vol. 11, No. 6, 1961); co-author, On the Interaction of High-Current Beams with a Plasma in a Magnetic Field (Atomnaya Energiya, vol. 14, No. 3, 1963); co-author, Exptl. Res. of Fast Ion Fluxes, being Generated in Plasma-Beam Discharges (letter to the Editor, ibid, vol. 24, No. 5, 1967).
Address: Academy of Sciences of the U.S.S.R., 14 Leninsky Prospekt, Moscow V-71, U.S.S.R.

BEREZINA, G. P. Papers: Co-author, Interaction of Strong Electron Beams with Plasma (Atomnaya Energiya, vol. 11, No. 6, 1961); co-author, On the Interaction of High-Current Beams with a Plasma in a Magnetic Field (Atomnaya Energiya, vol. 14, No. 3, 1963); co-author, Exptl. Res. of Fast Ion Fluxes, being Generated in Plasma-Beam Discharges (letter to the Editor, ibid, vol. 24, No. 5, 1968).
Address: Academy of Sciences of the U.S.S.R., 14 Leninsky Prospekt, Moscow V-71, U.S.S.R.

BERG, L. C. H. VAN DEN. See VAN DEN BERG, L. C. H.

BERG, Ole, M.Sc. (Elec. Eng.). Born 1928. Educ.: Tech. Univ. Denmark, Copenhagen. Isotopcentralen, Div. of the Danish Acad. of the Tech. Sci., 1958. Formerly Member, Selskabet for Anvendt Kernefysik. Society: Inst. of Danish Civil Eng.
Nuclear interest: Application of radioisotopes in technology.
Address: Isotopcentralen, 2 Skelbaekgade, Copenhagen V, Denmark.

BERG, Olof, M.Sc. (Elec. Eng.). Born 1901. Educ.: Stockholm Roy. Tech. Univ. Chairman, Atomkraftkonsortiet; Director, A.B. Atomenergi. Society: Swedish Assoc. Elec. Eng.
Address: 7 Ostermalmsgat., Stockholm, Sweden.

BERG, Richard M., A.B. Born 1924. Educ.: Rice Univ. Special Studies Office, Nat. Sci. Foundation, 1953-55; Various positions with Tech. Information Div., U.S.A.E.C., 1956-63; Dept. of State, 1963-. Society: A.N.S.
Address: 12918 Valleywood Drive, Silver Spring, U.S.A.

BERG, Selwyn Solomon, B.S., M.S., Ph.D. Born 1931. Educ.: Reed Coll., San Diego State Coll., Washington and Cornell Univs. Teaching Asst., Washington Univ., 1954-56; Sci., Westinghouse Atomic Power Dept., 1956-58; Res. Asst., Cornell Univ. 1958-61; Acting Reactor Supv.,

Cornell Nucl. Reactor Lab., 1961-62; Chief Responsible Person, Cornell Zero Power Reactor, 1962-63; Sen. Eng.,Von Karmen Centre, Aerojet-General, 1963-65; Principal Phys., Aerospace Dept., Perkin-Elmer, 1965-68; Sen Specialist, Pomona Div., Conductron, 1968-; Biophys., City of Hope Nat. Medical Res. Centre, 1968-.
Nuclear interest: Experimental applied physics.
Address: 1048 Lake Forest Drive, Claremont, California 91711, U.S.A.

BERGE, Grim, Cand. rea. Born 1922. Educ.: Oslo Univ. Sen. Sci., Inst. of Marine Res., Directorate of Fisheries, 1953-. Society: Rad. Advisory Council.
Nuclear interests: Marine radioactivity; Waste Disposal; Contamination problems; Biological uptake mechanisms; Effects of radiation on marine organisms.
Address: Institute of Marine Research, Directorate of Fisheries, Bergen, Norway.

BERGEL'SON, Boris Rafailovic. Papers: Calculation of the Passage of Fast Neutrons through Cylindrical Channels in a Biological Shield (letter to the Editor, Atomnaya Energiya, vol. 10, No. 4, 1961); Reactor Vessel Rad. Damage Shielding (Jaderná Energie, vol. 10, No. 5, 1964); Reactor Neutron Shield without Hydrogen (ibid., vol. 10, No. 8, 1964); co-author, Multigroup System of Integral Equation for Neutron Transport Description (Atomnaya Energiya, vol. 24, No. 1, 1968).
Address: Institute of Theoretical and Experimental Physics, Academy of Sciences of the U.S.S.R., 89 Ulitsa Cheremushkinskaya, Post Box 3315, Moscow, U.S.S.R.

BERGER, C. Rainer A., Ph.D. (Illinois). Born 1930. Educ.: Cambridge, Kiel and Illinois Univs. Aero Res. Ltd., Duxford, Cambs., 1951-53; Illinois Univ., 1955-60; Staff Sci. Convair Sci. Res. Dept., 1960-61. Societies: A.A.A.S.; A.C.S.
Nuclear interests: C-14 and H-3 applications in organic and biochemistry; Radiation chemistry (exobiochemistry and biology, the evolution of organic compounds).
Address: Office of the Chief Scientist, Lockheed California Company, Burbank, Calif., U.S.A.

BERGER, E. Sen. Officer, Camille Bauer A.G. Address: Camille Bauer A.G., 18 Dornacherstrasse, Basel, Switzerland.

BERGER, Frantisek Petr, Mech. Eng., C.Sc. Born 1925. Educ.: Prague Tech. Univ. Sci. worker, Inst. Nucl. Res., Czechoslovak Acad. Sci., 1955-; Visiting Lecturer Nucl. Eng., Queen Mary Coll., London Univ., 1967-68. Book: Vypocet a stavba jadernych reaktoru (Design of Nuclear Reactors) (Prague, S.N.T.L., 1963).
Nuclear interests: Reactor design, especially research and design of fuel elements; Heat transfer.
Address: Ustav jaderneho vyzkumu CSAV, Rez, Czechoslovakia.

BERGER, Hans, Dipl. Phys. Born 1920. Educ.: Berlin Tech. and Göttingen Univs. Siemens-Reiniger-Werke A.G., Erlangen (Sci. information), 1948-. Vice-chairman, Fachnormenausschuss Radiologie; Member, Kommission für Dosimetrie, radiologische Einheiten und Strahlenschutz in Deutsche Röntgengesellschaft. Societies: Deutsche Röntgengesellschaft; Deutsche Physikalische Gesellschaft; Gesellschaft Deutscher Naturforscher und Arzte.
Nuclear interests: Particle accelerators; Radiation physics; Dosimetry; Medical application of radioisotopes; Radiation protection.
Address: 27 Komotauer Strasse, Erlangen, Germany.

BERGER, J. Mechanics, Nucl. Dept., Soc. d'Exploitation des Materiels Hispano-Suiza.
Address: Société d'Exploitation des Materiels Hispano-Suiza, rue du Capitaine Guynemer, Bois Colombes, Seine, France.

BERGER, Klaus Hans, Ing. Ind., Dipl. Ing. Born 1927. Educ.: Buenos Aires Univ. Studiengesellschaft zur Förderung Kernenergieverwertung in Schiffbau und Schiffahrt e.V., Hamburg.
Nuclear interest: Utilisation in commercial ships.
Address: 19-21 Liesborner Weg, 1 Berlin 27, Germany.

BERGER, M., Dr. Book: Contributor to Eng. Compendium on Rad. Shielding (I.A.E.A.).
Address: National Bureau of Standards, Washington D.C., 20234, U.S.A.

BERGER, Martin J. Formerly Member, now Chairman, Subcom. on Penetration of Charged Particles in Matter, Com. on Nucl. Sci., Nat. Acad. Sci.-N.R.C.
Address: National Academy of Sciences - National Research Council, Committee on Nuclear Sciences, 2101 Constitution Avenue, N.W., Washington 25, D.C. U.S.A.

BERGER, Rene Louis, Ing. Chimiste., L. ès Sc. Born 1925. Educ.: Clermont-Ferrand Univ. and Ecole. Nat. Superieure de Chimie. Ing., C.E.A., 1956-. Member, Steering Com. for Nucl. Energy, E.N.E.A., 1965-.
Nuclear interest: Transuranium elements chemistry, production and applications.
Address: Centre d'Etudes Nucleaires de Fontenay Aux Roses, Departement de Chimie, B.P. No. 6, 92 Fontenay Aux Roses, France.

BERGER, Robert L., Ph.D., B.S., M.S. Born 1925. Educ.: Colorado A. and M. Coll., Pennsylvania State Univ., and exchange fellow at Cambridge Univ. Instructor in Phys., Park Coll., 1950-51; Asst. Prof. Phys. (1957-60), Assoc. Prof. (1960-62), Utah State Univ.; Sen. Phys., Lab. Tech. Development, Nat. Heart Inst., Bethesda, Maryland, 1962-. Societies: Biophys. Soc.; A.P.S.; A.A.P.T.; A.A.A.S.; Faraday Soc.;

Soc. Gen. Physiologists.
Nuclear interests: Radiobiology of enzymes and cell membranes; Digital computer finite element simulation of transport problems in fluids.
Address: Building 10, Room 5D06, National Heart Institute, Bethesda, Maryland, U.S.A.

BERGERARD, Joseph. Member, Conseil d'Enseignement I.N.S.T.N., C.E.A.
Address: Commissariat a l'Energie Atomique, 29-33 rue de la Federation, Paris 15, France.

BERGES, G. A. French Member, Health and Safety Sub-Com., O.E.C.D., E.N.E.A.
Address: O.E.C.D. European Nuclear Energy Agency, 38 boulevard Suchet, Paris 16, France.

BERGESEN, Sigval d. y. Born 1893. Sen. Partner, Sig. Bergesen d.y. and Co. (Oslo and Stavanger) shipowners, shipbuilders, insurance brokers and Lloyd's agents; Chairman of Board, Skibsaktieselskapet Snefonn; Skipsaksjeselskapet Bergehus; Financial and Commercial Chairman of Board, A/S Rosenberg Mekaniske Verksted, Stavanger; Chairman of Board, Teknisk Bureau A/S; Stavanger Preserving Co. A/S; Member of Board, Norske Fina A/S; Member of Board of Trustees, Norwegian Shipowners' Assoc.; Norwegian Shipowners Employers' Assoc.; Norges Skibshypothek A/S; Den norske Creditbank; Storebrand; Member, Norwegian A.E.C.; Pres., Norwegian Shipowner Atomic Com.; Formerly Member of the Board, Noratom A/S; Formerly Chairman, now Vice Chairman, Rederiatom (Nucl. Res. Group of Norwegian Shipowners).
Address: 15 Huk Avenue, Bygdoy, Oslo, Norway.

BERGGREN, Reynold G., B.S. Born 1923. Educ.: Illinois Inst. of Technol. and Tennessee Univ. Eng. Development, O.R.N.L. Societies: A.S.M.; A.S.T.M.
Nuclear interests: Metallurgy; Radiation effects in structural metals.
Address: Solid State Division, Oak Ridge National Laboratory, P.O. Box X, Oak Ridge, Tennessee, U.S.A.

BERGH, Anton Dirk Johan DE. See DE BERGH, Anton Dirk Johan.

BERGH, Helge, Chem. Eng., Agricultural Chem. Born 1911. Educ.: Trondheim Technol. and California Univs. Agricultural chem. Norwegian State Agricultural Lab., Bergen and Trondheim 1933-52 (Head, Soil and Plant Div., 1945-52). Sen. Res. Chem., Head Isotope Lab., Norwegian Defence Res. Establishment, 1952-60; Head Isotope Lab., Faculty of Agriculture, Ankara Univ., Ankara, working for I.A.E.A., Vienna. Book: Determination of Macro- and Micronutriments in Soil and Plant Materials (Oslo, 1952). Societies: Fellow, Norwegian Chem. Soc.; Fellow, Scandinavian Soil Res. Assoc.
Nuclear interest: Agricultural application of atomic energy, especially agricultural research applying radioactive isotopes.
Address: Isotope Laboratory, Faculty of Agriculture, University of Ankara, Ankara, Turkey.

BERGIER, P. Sen. Tech. Officer, Giovanola Frères S. A.
Address: Giovanola Frères S. A., Monthey, (Valais), Switzerland.

BERGLIN, Carl Laurentius William, B. Eng. (Hons., Chem. Eng., Queensland). Born 1906. Educ.: Queensland and Leeds Univs. Sen. Lecturer in Chem. Eng., Queensland Univ., 1947-54; Head, Chem. Eng. Sect. (1954-61), Operations Manager, Res. Establishment (1961-64), Consultant (1964-), A.A.E.C.; Principal, Australian School of Nucl. Technol., 1967-. Societies: Inst. Chem. Eng.; Australasian Inst. Mining and Metallurgy; Assoc. Member, Inst. Eng. Australia.
Nuclear interest: Chemical engineering.
Address: 11/35 The Esplanade, Cronulla, N.S.W. 2230, Australia.

BERGLUND, K. G. Manager, Svenska Atomforsakringspoolen, 1967-.
Address: Svenska Atomforsakringspoolen, 2 Birger Jarlsgatan, Stockholm 5, Sweden.

BERGMANN, Ernst David, Ph.D. Born 1903. Educ.: Berlin Univ. Director, Sci. Dept., Israel Ministry of Defence, 1948-66; Chairman, Israel A.E.C., 1952-66; Prof. Organic Chem., Hebrew Univ., 1952-66. Deleg. to 2nd I. C.P.U.A.E., Geneva, Sept. 1958. Book: Co-author, Textbook of Organic Chem. Societies: London Chem. Soc.; A.C.S.; Israel Chem. Soc.; N.Y. Acad. Sci.; Hon. Member, Argentine Chem. Soc.
Nuclear interests: Synthesis of labelled organic compounds; Effect of ionising radiation on insects; Protection against radiation by chemical compounds.
Address: 8 Keren Kayemeth Street, Jerusalem, Israel.

BERGMANN, Otto, Ph.D. Born 1925. Educ.: Vienna Univ. Scholar, Dublin Inst. for Advanced Studies, 1951-52; Sen. Res. Fellow, Adelaide Univ., 1952-55; Sen. Res. Fellow, New England Univ., N.S.W., 1956-58; Res. Sci., R.I.A.S., Baltimore, 1958-60; Assoc. Prof., Alabama Univ., 1960-62; Assoc. Prof. (1962-67), Prof. (1967-), George Washington Univ. Society: A.P.S.
Nuclear interest: Theoretical nuclear physics.
Address: Physics Department, George Washington University, Washington D.C. 20006, U.S.A.

BERGMANN, Peter Gabriel, Dr. rer. nat. Born 1915. Educ.: Dresden, Freiburg and Prague Univs. Adjunct Prof., Brooklyn Polytech. Inst. 1947-57; Assoc. Prof. (1947-50), Prof. (1950-), Syracuse Univ. Chairman, Phys. Dept., Yeshiva Univ. 1963-64. Internat. Com. on Relativity and Gravitation, 1959-. Books: Introduction

to the Theory of Relativity (Prentice-Hall, 1942); Basic Theories of Phys. (2 Vols.) (Dover Publications, 1962); The Riddle of Gravitation (Scribner, 1968). Societies: A.P.S.; American Maths. Soc.; A.A.A.S.; German Phys. Soc.; F.A.S.
Nuclear interests: General relativity; Quantum field theory; Statistical mechanics; Foundations of physical theories.
Address: Physics Department, Syracuse University, Syracuse, New York 13210, U.S.A.

BERGOGNON, Pierre. Born 1918. Educ.: Prytanée Militaire de la Flèche and Ecole Polytech., Paris. Ing. en Chef, Service Graphite, Cie. Pechiney. Society: B.N.E.S.
Nuclear interests: Manufacture of nuclear graphite.
Address: Cie. Pechiney, 23 rue Balzac, Paris 8, France.

BERGSJÖ, Per Bjarne, Cand. med (equivalent to M.D.). Born 1932. Educ.: Oslo Univ. Ordinary hospital jobs (clinical medicine). Society: Fellow, Norwegian Cancer Soc. (Res.).
Nuclear interests: Gynaecological radiotherapy. Conducting a clinical trial to test a possible oxygen effect in radiation treatment of carcinoma of the cervix uteri.
Address: Norwegian Radium Hospital, Oslo, Norway.

BERGSMA, Jitze, Drs. Born 1932. Educ.: State Univ. Groningen. Res. Sci. Reactor Centrum Nederland, Petten, 1957-. Society: Dutch Phys. Soc.
Nuclear interests: Inelastic scattering of neutrons in solids and liquids.
Address: Reactor Centrum Nederland, Petten, Netherlands.

BERGSTRALH, Thor A. Head, Weapons Effects Dept., San Bernardino Operations, Aerospace Corp. 1963-.
Address: Aerospace Corporation, San Bernardino Operations, San Bernardino, California, U.S.A.

BERGSTRÖM, Ingmar Lars, Prof. Born 1921. Educ.: Chalmers Inst. of Technol., Gothenburg and Roy. Inst. of Technol., Stockholm. Assoc. Prof. (1958-66), Head, Res. Inst. for Phys. (1966-), Roy. Inst. of Technol., Stockholm.
Address: Research Institute for Physics, Stockholm 50, Sweden.

BERGSTROM, Richard N., B.S., M.S. Born 1921. Educ.: I.I.T. Partner, Sargent and Lundy Eng., 1966-. Chairman, Subcom. N45, Sectional Com., A.S.A.; Member, Com. 349, Concrete Containment Standards, A.C.I. Societies: A.N.S.; A.S.C.E.; A.S.M.E.
Nuclear interest: Design of nuclear power stations, containment structures and reactor structures.
Address: Sargent and Lundy Engineers, 140 South Dearborn, Chicago, Illinois 60603, U.S.A.

BERGSTRÖM, Stig Olof Wilhelm, M. Eng. (naval architecture). Born 1925.Educ.: Roy. Inst. Technol., Stockholm. Sen. Eng., Torpedo Dept., Roy. Navy Board, 1951-55; Res. Fellow sponsored by American-Scandinavian Foundation, Hydrodynamics Lab., Calif. Inst. Technol., 1952-53; Assoc. Prof., res. on hydrodynamics, Res. Inst. Nat. Defence, 1955-57; Tech. Sec., Reactor Safeguards Com. (1957-59), Head of Sect. for Health and Safety (1960-), A.B. Atomenergi. Societies: H.P.S.; Svenska Teknologföreningen; Nordic Soc. for Rad. Protection.
Nuclear interests: Health physics; Environmental hygiene; Hazards evaluation.
Address: A.B. Atomenergi, Studsvik, Nyköping, Sweden.

BERINGER, E. Robert, Prof. Director, Heavy-Ion Accelerator Lab., Yale Univ.
Address: Physics Department, Yale University, Sloane Laboratory, 217 Prospect Street, New Haven, Connecticut 06520, U.S.A.

BERK, Sigmund, B.S., M.A. Born 1915. Educ.: Temple and Pennsylvania Univs. Head, Radioisotope and Rad. Chem. Labs., Pitman-Dunn Res. Labs., Frankford Arsenal, Philadelphia. Societies: A.N.S.; A.C.S.; RESA; H.P.S.; Rad. Res. Soc.
Nuclear interests: Radiation chemistry of organic systems; Effect of radiation on optical glass; Radiation backscattering; Effects of ionising radiation on fungi; Application of radioisotopes to nondestructive testing and industrial uses.
Address: 807 Larkspur street, Philadelphia 19116, Pennsylvania, U.S.A.

BERKE, Claus L., Dr. jur. Born 1928. Educ.: Mainz and Cologne Univs. Asst. Attorney; Sales Manager, Interatom, 1962-.
Nuclear interests: Management; Nuclear law and economics.
Address: 10 Lärchenweg, Bensberg/Köln, Germany.

BERKEFELD, Herbert SOSTMANN-. See SOSTMANN-BERKEFELD, Herbert.

BERKELMAN, Karl, B.S., Ph.D. Born 1933. Educ.: Rochester and Cornell Univs. Instructor and Res. Assoc., Phys. Dept., Cornell Univ., 1959-60; Visiting Res. Fellow (N.S.F.), Istituto Superiore di Sanita, Roma, 1960-61; Asst. Prof., Phys. Dept. (1961-63), Assoc. Prof. Phys. Dept. (1963-67), Prof. Phys. Dept. (1967-), Cornell Univ.; Visiting Sci. (N.S.F.), C.E.R.N., Geneva, 1967-68. Societies: A.P.S.; Stà Italiana di Fisica.
Nuclear interests: Experimental high-energy physics (with accelerators); Fundamental particles.
Address: 971 East State Street, Ithaca, N.Y., U.S.A.

BERKHOUT, Ulrich MEYER-. See MEYER-BERKHOUT, Ulrich.

BERKMAN, A. Tevfik. Born 1900. Educ.: Istanbul and Berlin Univs. Director, Inst. and Clinic of Radiotherapy, Faculty of Medicine, Istanbul Univ. Member, Turkish A.E.C., 1956,60 and 1963. Books: Problems in Radiotherapy and New Methods (1950); Radiotherapy of Lung Cancers (1958); Problems in Nucl. Medicine (1960); Radioactive Isotopes and their Medical Uses (1960). Lectures delivered at Cambridge University (1960). Societies: New York Acad. Sci.; Membre du Comite Internat. de Radiologie; British Inst. of Radiol.; Roy. Soc. of Radiol.; A.N.S. Nuclear interests: Radiotherapy; Nuclear medicine.
Address: 7 Cumhuriyet Caddesi, Taksim, Istanbul, Turkey.

BERKO, Stephan. Prof., Phys. Dept., Brandeis Univ. Nuclear interests: Low energy neutron scattering (nuclear spectroscopy of intermediate A nuclei). Interaction of positrons with electrons in metals (as a tool in solid state physics). Mössbauer effect experiments (measurement of nuclear sizes in excited states, etc.).
Address: Department of Physics, Brandeis University, Waltham 54, Massachusetts, U.S.A.

BERLIN, E. M. Paper: Co-author, Start-Stop Simulation Problems of Nucl. Reactors (letter to the Editor, Atomnaya Energiya, vol. 22, No. 6, 1967).
Address: Academy of Sciences of the U.S.S.R., 14 Leninsky Prospekt, Moscow V-71, U.S.S.R.

BERLIN, Nathaniel I., B.S., M.D., Ph.D. Born 1920. Educ.: Western Reserve Univ. and Long Island Coll. Medicine; California Univ. Donner Lab., California Univ., 1947-54; Medical Officer, Analysis Branch, Effects Div., Armed Forces Special Weapons Project, 1954-56; Head, Metabolism Service, (1956-), Chief, Gen. Medicine Branch, (1959-61), Clinical Director, (1961-), Nat. Cancer Inst. Societies: American Physiological Soc.; Rad. Res. Soc.; Biochemical Soc.; American Soc. Clinical Investigation.
Nuclear interest: Application of isotopic materials and methods to biological and medical research.
Address: National Cancer Institute, Bethesda, Maryland, U.S.A.

BERLINE, S. Gen. Manager, Sté. d'Exploitation et de Recherches Electroniques S.A.
Address: Société d'Exploitation et de Recherches Electroniques S.A., boulevard de Mantes, Aubergenville, (S. et O.), France.

BERLOVSKII, A. Ya. Paper: Co-author, Scintillation Detectors Energy Diapason Expansion in Gamma-Ray Dosimetry (letter to the Editor, Atomnaya Energiya, vol. 24, No. 3, 1968).
Address: Academy of Sciences of the U.S.S.R., 14 Leninsky Prospekt, Moscow V-71, U.S.S.R.

BERLYANT, S. M. Paper: Co-author, Rad. Cross-Linking of Polyethylene Insulation of Cable Articles on Large Scale (letter to the Editor, Atomnaya Energiya, vol. 21, No. 1, 1966).
Address: Academy of Sciences of the U.S.S.R., 14 Leninsky Prospekt, Moscow V-71, U.S.S.R.

BERMAN, A. U.S. Member, Group of Experts on Production of Energy from Radioisotopes, O.E.C.D., E.N.E.A.
Address: O.E.C.D. European Nuclear Energy Agency, 38 boulevard Suchet, Paris 16, France.

BERMAN, Abraham S., B.Ch.E., Ph.D. Born 1921. Educ.: C.C.N.Y. and Ohio State Univ. Dept. Head, Flow Res. Dept., Tech. Div., Oak Ridge Gaseous Diffusion Plant, Union Carbide Corp. Nucl. Div., Oak Ridge, Tenn., 1950-66; Prof., Dept. of Aeronautics and Eng. Mech., Minnesota Univ. Inst. of Technol., 1966-. Societies: A.C.S.; A.P.S.; A.A.A.S. Nuclear interest: Gaseous diffusion.
Address: Department of Aeronautics and Engineering Mechanics, Institute of Technology, Minnesota University, Minneapolis, Minnesota 55455, U.S.A.

BERMAN, Arthur Irwin, A.B., M.S., Ph.D. Born 1925. Educ.: Stanford Univ. Prof.Phys. Rensselaer Polytech. Inst. Book: The Physical Principles of Astronautics(New York, John Wiley). Societies: A.P.S.; A.A.P.T.; A.A.A.S.; A.I.A.A.; American Astronautical Soc. Nuclear interests: Photonuclear reactions; X-ray and gamma-ray absorption; Microradiography; Electron radiography.
Address: Rensselaer Polytechnic Institute, Hartford Graduate Division, East Windsor Hill, Connecticut, U.S.A.

BERMAN, I. B. Papers: Co-author, Determination of Uranium Clarke Concentration in Ionic Crystals (letter to the Editor, Atomnaya Energiya, vol. 22, No. 6, 1967); co-author, Determination of Uranium Concentration and its Spatial Distribution in Minerals and Rocks (ibid, vol. 23, No. 6, 1967).
Address: Academy of Sciences of the U.S.S.R., 14 Leninsky Prospekt, Moscow V-71, U.S.S.R.

BERMAN, William Howard, A.B., LL.B. Born 1924. Educ.: Harvard Univ. Special Asst. to Gen. Manager of A.E.C. for Res. and Ind. Development, 1957-58; Director, Atomic Energy Res. Project, Michigan Univ. Law School, 1958-61. Member, Law firm of Sharlitt, Hydeman and Berman, 1961. Deputy Gen. Counsel, U.S. Arms Control and Disarmament Agency, 1962-64. Books: Federal and State Responsibilities for Rad. Protection (Michigan Univ., 1958); Internat. Control of Nucl. Maritime Activities (Michigan Univ., 1959); The Atomic Energy Commission and Regulating Nucl. Facilities (Michigan UNiv., 1960). Society: Atomic Ind. Forum.

BER

Nuclear interests: Principal interests are in connection with the legal and policy problems posed by the development of atomic energy, both domestically and internationally. Have a special interest in regulatory problems and in the general area of developing and imposing controls to protect public health and safety.
Address: 3322 Dent Place, N.W., Washington 7, D.C., U.S.A.

BERMINGHAM, Anne, B.Sc. Born 1925. Educ.: Melbourne Univ. In charge, Carbon-14 Dating Lab., Inst. of Appl. Sci. of Victoria. Society: Assoc. Roy. Australian Chem. Inst.
Nuclear interests: Low level natural radioactivity.
Address: Carbon-14 Dating Laboratory, Institute of Applied Science of Victoria, 304-328 Swanston Street, Melbourne, Victoria, Australia.

BERMOND, Jacques, D. ès Sc. Born 1921. Educ.: Caen Univ. Lab. Phys. Corpusculaire, Caen Univ.
Nuclear interests: Physique corpusculaire; Hautes energies technique des emulsions nucléaires.
Address: 19 rue des Carmélites, Caen, France.

BERNARD, Michel Yves, Agrégé de Physique, D. es Sc. Physiques. Born 1927. Educ.: Paris Univ. Maitre de Conférences, Faculté des Sci. de Caen; Prof., I.N.S.T.N., Saclay; Prof. au Conservatoire National des Arts et Metiers. Books: Initiation à la mécanique quantique (Hachette, 1960), Précis d'optique electronique (Bordas, 1958); Masers et lasers (Presses Universitaires de France, 1964).
Nuclear interests: Optique corpusculaire; Accélérateurs de particules; Physique des plasma.
Address: 292 rue St. Martin, Paris 3, France.

BERNARD, Pierre Denis, Dr. Pres., Physiotechnie.
Address: Physiotechnie, 34 avenue Aristide Briand, Arcueil, (Seine), France.

BERNARDINI Gilberto, Prof. Born 1906. Educ.: Pisa and Florence Univs. Visiting Prof., Columbia Univ., 1949-50; Res. Prof., Illinois Univ., 1951-56; Director of Res., C.E.R.N., Geneva, 1953-60; Prof., Rome Univ. Vice-President Italian Inst. for Nucl. Res.; Pres., Società Italiana di Fisica. Societies: Soc. Nazionale Acc. Lincei; Fellow, A.P.S.; Soc. Acc. Scienze, Bologna.
Nuclear interest: Research in elementary particles physics with particular reference to weak interactions and high energy electrical magnetic processes.
Address: Istituto Fisica, Università Rome, Rome, Italy.

BERNARDO, Benito C., B.S. (Civil Eng.), M.S. (Sanitary Eng.). Born 1922. Educ.: Mapua Inst. Technol., Manila, North Carolina and Rochester Univs. Field Supervisor, U.S. Public Health Service, 1946-50; Sen. Sanitary Eng., Bureau of Health, Philippine Govt., 1951-59; Sci. and Chief of Classification and Information Branch (1959-60), Sen. Sci. (1960-), Philippine A.E.C. Societies: Philippine Assoc. for the Advancement of Science; Philippine Public Health Assoc.; Philippine Soc. of Sanitary Eng. and Public Health.
Nuclear interest: Health physics.
Address: Philippine Atomic Energy Commission, Atomic Research Centre, Diliman, Quezon City, Philippines.

BERNAS, René. D. ès Sc. Born 1920. Educ.: Sorbonne, Paris. Res. Fellow, C.N.R.S.; Directeur de Recherches, Lab. de Physique Nucl., Faculté des Sciences Orsay, 1950-. Societies: Sté. Française de Physique; A.P.S.
Nuclear interests: Separation of isotopes by electromagnetic means - with emphasis on achievement of very high purity; Nuclear reaction cross-section measurements by mass-spectrometry; Isotope dilution techniques.
Address: Laboratoire de Physique Nucléaire, Faculté des Sciences, B.P. No. 1, Orsay, (S. et O.), France.

BERNAS, Robert. Manager, Appareils Gamma-Siar; Sté. Qualitest.
Address: Appareils Gamma-Siar, 52 rue de Dunkerque, Paris 9, France.

BERNAUD, C. Chef des Services de productions de Pierrelatte. Paper: Quelques Problemes industriels posés par le traitement des effluents liquides radioactifs solutions particulières retenues à Marcoule (2nd I.C.P.U.A.E., Geneva, Sept. 1958). Expert, Third I.C.P.U.A.E., Geneva, Sept. 1964.
Address: Commissariat à l'Energie Atomique, Pierrelatte, France.

BERNER, Alojs, Dipl. Ing., Dr. techn. Sen. Officer, Atominst. der Osterreichischen Hochschulen.
Address: Atominstitut der Osterreichischen Hochschulen, 115 Schuttelstrasse, Vienna 2, Austria.

BERNER, Frau, G., Dr. Konservator, Dept. of Clinical-Exptl. Radiol., Univ. Strahlenklinik (Czerny Krankenhaus).
Address: Universitats Strahlenklinik (Czerny Krankenhaus), 3 Voss Strasse, Heidelberg, Germany.

BERNERT-CLESS, Traude, Ph.D. Born 1915. Educ.: Vienna Univ. Res. Assoc., Inst. of Radium Res. Nucl. Phys., Austrian Acad. Sci., Vienna, 1939-; Radioisotope in Biologie and Medizin (1949) Zählrohmethode; Head of isotope distribution centre of Austria (1949); Consultant to the Austrian Studiengesellschaft für Atomenergie for production and application of radioisotopes. Societies: Austrian Phys. Soc.; Phys. Chem. Soc. of Vienna.

Nuclear interests: Nuclear physics; Production and application of radioisotopes; Insurance problems concerning the risk involved in the use of radioactive substances.
Address: Institut für Radiumforschung und Kernphysik, 3 Boltzmanngasse, Vienna 9, Austria.

BERNEY, A. Phys., Lab. de Recherches sur la Physique des Plasmas, Swiss Nat. Foundation for Sci. Res.
Address: Swiss National Foundation for Scientific Research, Laboratoire de Recherches sur la Physique des Plasmas, 2 avenue Ruchonnet, Lausanne, Switzerland.

BERNHARD, Fritz G. H., Prof., Dr. rer. nat. habil. Dipl. Ing. Born 1913. Educ.: Leipzig Univ. and Berlin-Charlottenburg T.H. Vice-Director, Inst. Phys. Chem. German Acad. of Sci., Berlin; Director, I. Physikalisches Inst., Humboldt Univ. Book: Co-author, Multipliers and their Application in Nucl. Phys. (Berlin, Akademie Verlag. 1957). Societies: Physikalische Gesellschaft, Eastern and Western Germany.
Nuclear interests: Nuclear and solid state electronics and instrumentation; Mass spectroscopy and electromagnetic isotope separation; Scintillation techniques; Cathode physics.
Address: 19 Ohm-Krueger-strasse, Berlin-Karlshorst, German Democratic Republic.

BERNHARD, Karl, Dr. phil. Born 1904. Educ.: Zurich and Geneva Univs. Ordentlicher Prof., Physiologisch-Chem. Inst., Basle Univ. Gast. Arbeitskreis II/3, Biologie und Medizin, Deutsche Atomkommission, Federal Ministry for Sci. Res.
Address: Physiologisch-Chemisches Institut der Universität, 1 Vesalgasse, Basle, Switzerland.

BERNHARD, Michael, Dr. of biol. sci. Born 1925. Educ.: Humboldt (Berlin), Philipps (Marburg/L.) and Naples Univs. Aquarium Fisheries Biol., Food and Agriculture Organisation, U.N., Rome, 1955; Postdoctorate fellow, Scripps Inst. Oceanography, La Jolla, California, 1955-58; Director, Laboratorio per lo studio della contaminazione radioattiva del mare (Lab. of the study of the radioactive contamination of the sea), Divisione Biologica e di Protezione Sanitaria, C.N.E.N., Italy.
Nuclear interests: Oceanography; Radiobiology; Radioactive contamination of the sea, with special consideration of marine ecology.
Address: Laboratorio per lo Studio della Contaminazione Radioattiva del Mare, I 19030 Fiascherino, La Spezia, Italia.

BERNIER, G., Dr. Sci. concerned with Radio-isotope Work, Centre de Recherches des Hormones Vegetales de l'I.R.S.I.A.
Address: Centre de Recherches des Hormones Vegetales de l'I.R.S.I.A., Institut de Botanique, 3 rue Fusch, Liege, Belgium.

BERNOT, Jacques Jean Pierre. Ing. de l'Ecole Polytechnique, Paris. Born 1929. Eng. group leader, Res. Reactor Div., C.E.N., Saclay. Society: A.N.S.
Nuclear interests: Reactor physics and research reactor design and operation.
Address: Centre d'Etudes Nucléaires de Saclay, B.P. Bo. 2, Gif-sur-Yvette, Seine et Oise, France.

BERNS, Mathias. Chairman, Economic and Social Com., Euratom, 1968-70.
Address: 38 rue Joseph Junck, Luxembourg.

BERNSTEIN, David Maurice, B.Sc. (Hons. Natural Philosophy). Born 1938. Educ.: Glasgow Univ. Res. Phys., Plessey U.K. Ltd., 1959-61; Tech. Director, Miles Hivolt Ltd., 1961-67; Managing Director, Intertechnique Ltd., 1967-. Society: A. Inst. Phys.
Nuclear interests: Activation analysis, nucleonic instrumentation with particular reference to digital data processing and fast counting techniques in nuclear physics and nuclear medicine. Application of digital measurement techniques to scanning and clinical applications of radiation.
Address: Coppertop, 11 Kearsley Drive, The Heights, Findon Valley, Worthing, Sussex, England.

BERNSTEIN, E. M. Nucl. Reactions, Phys. Dept., Texas Univ.
Address: Physics Department, Texas University, Austin 12, Texas, U.S.A.

BERNSTEIN, Herbert I., B.A., M.Sc., Ph.D. Born 1914. Educ.: Swarthmore Coll. and Pennsylvania State Univ. Director, Nucl. Dept., Chemicals and Phosphates Ltd. Societies: A.C.S.; Israel Chem. Soc.
Nuclear interest: Management.
Address: Chemicals and Phosphates Ltd., P.O.B. 1428, Haifa, Israel.

BERNSTEIN, Rosa. Lecturer, Dept. of Molecular Phys. and Solid State, Centro Brasileiro de Pesquisas Fisicas.
Address: Centro Brasileiro de Pesquisas Fisicas, 71 Avenida Wenceslau Braz, Rio de Janeiro, Brazil.

BERRIDGE, Donald Roy. Born 1922. Chief Generation Design Eng., C.E.G.B. Society: M.I.Mech.E.
Nuclear interest: Power reactors.
Address: Lower Broadoak, Seale, Farnham, Surrey, England.

BERRIER, Raymond S. Res. and Effects Test Branch, Directorate of Ranges and Test Support DCS/Systems, Nucl. Test Div., U.S. Air Force Systems Command.
Address: U.S. Air Force Systems Command, Andrews Air Force Base, Washington 25, D.C., U.S.A.

BERRY, Eugene L., B.S. (Civil Eng.). Born 1913. Educ.: South Dakota State School Mines and Technol. Chief, Contracts Branch, U.S.A.E.C.
Nuclear interests: Management; Nuclear reactor projects and chemical plants, for technical contracts.
Address: 314 Redwood Drive, Idaho Falls, Idaho, U.S.A.

BERRY, Melville Douglas, B.Sc. Born 1908. Educ.: Manitoba Univ. Manager, Development Div., Power Projects, A.E.C.L.
Nuclear interest: Development.
Address: 126 Sabrina Drive, Weston, Ontario, Canada.

BERRY, Robert Malcolm, B.Sc. (Chem. Eng.). Born 1917. Educ.: Tufts Univ. Refinery Manager, Eldorado Nucl. Ltd., 1962-.
Nuclear interests: Management; Chemical and metallurgical processing of nuclear fuel materials.
Address: Eldorado Nuclear Limited, Port Hope, Ontario, Canada.

BERRY, William H., B.S. Born 1908. Educ.: Illinois Inst. Technol. and Missouri Inst. Law and Accountancy. Vice-Pres. and Manager, Pacific Coast Dept., Continental Insurance Companies, property damage and casualty insurance companies.
Nuclear interests: Property damage insurance for risks using nuclear materials ranging from the large nuclear power reactors down to the use of small quantities of radioactive isotopes.
Address: 160 Pine Street, San Francisco 11, California, U.S.A.

BERSON, Solomon A., B.S. (C.C.N.Y.), M.Sc. (New York), M.D. (New York), D.Sc. (Hon., Long Island). Born 1918. Educ.: C.C.N.Y. and N.Y.U. Chairman, Dept. Medicine, Mount Sinai School of Medicine, New York. Editorial Board, J. of Clinical Investigation. Societies: American Soc. Clin. Invest.; American Physiological Soc.; Harvey Soc. of N.Y.; Endocrine Soc.; Soc. for Exptl. Biol. and Medicine; Fellow, N.Y. Acad. Medicine; N.Y. Acad. Sci.; A.A.A.S.; A.H.A.; Assoc. of American Physicians; American Diabetes Soc.; New York Diabetes Assoc.; Diplomate of Pan American Medical Assoc. (Hon.); Peruvian Soc. of Endocrinology (Hon.); Chilean Soc. of Diabetes and Metabolic Diseases (Hon.); Chilean Soc. of Endocrinology (Hon.); Argentine Soc. of Endocrinology and Metabolism (Hon.).; American Medical Assoc.
Nuclear interest: Medical uses of radioisotopes.
Address: Department of Medicine, Mount Sinai School of Medicine, New York, U.S.A.

BERNSTEIN, Irving Aaron, Sc.B., Ph.D. Born 1926. Educ.: Cornell and Brown Univs. Vice-Pres. and Tech. Director (1957-58), Pres. and Tech. Director (1958-67), Controls for Radiation Inc.; Vice-Pres., Isotopes Inc., 1967-. Societies: A.N.S.; Rad. Res. Soc.; A.C.S.; H.P.S.
Nuclear interest: Radiation dosimetry; Radiation physics.
Address: Con-Rad Laboratories, Isotopes Incorporated, 130 Alewife Brook Pkwy., Cambridge, Massachusetts, U.S.A.

BERTAUT, Erwin Félix, Dr. in Sci. Born 1913. Educ.: Bordeaux and Grenoble Univs. Res. Director, C.N.R.S., 1955-; Sci. Advisor for the Nucl. Centre of Grenoble, 1957-. Society: Sté. française de Minéralogie et de Cristallographie.
Nuclear interest: Neutron diffraction applied to crystallographic and magnetic structure determination.
Address: Centre d'Etudes Nucléaires, B.P. 269, Grenoble, France.

BERTEIG, Leiv Audun, M.Sc. Born 1923. Educ.: Oslo Univ. Health Phys. (1956-), Div. Chief, (Rad. Hygiene in industry, res. and atomic energy) (1967-), State Inst. of Rad. Hygiene; Health Phys. Adviser, I.A.E.A., Iraq, 1963-65. Society: Nordiska Sällskapet för Strålskydd.
Nuclear interest: Radiation safety. Systems of control and their application.
Address: State Institute of Radiation Hygiene, Montebello, Oslo 3, Norway.

BERTELS, H. Netherlands Member, Group of Experts on Third Party Liability, O.E.C.D., E.N.E.A.
Address: O.E.C.D. European Nuclear Energy Agency, 38 boulevard Suchet, Paris 16, France.

BERTELSON, Peter Clark, Registered Professional Eng., B.S. (Mech. Eng.), M.S. (Eng. Mechanics), D.Sc. (Eng.). Born 1928. Educ.: Carnegie Inst. Technol., and O.R.S.O.R.T., Michigan and Washington Univs. With Ford Motor Co., 1948-51; Gen. Dynamics Corp., 1955-57; Internuclear Co., 1957-61; Ford Motor Co. 1962-. Societies: A.N.S.; A.S.M.E.; Ind. Maths. Soc.; S.A.E.
Nuclear interests: Reactor design and nuclear engineering; Reactor heat transfer, fluid flow.
Address: 30325 Ponds View Drive, Franklin, Michigan, U.S.A.

BERTERO, Mario, Dr. Inst. für Theoretische Kernphysik, Bonn Univ.
Address: Bonn University, Institut für Theoretische Kernphysik, 16 Nussallee, Bonn, Germany.

BERTHELOT, André Jean Louis, D. ès Sc. Born 1912. Educ.: Paris Univ. Head, High Energy Phys. Dept., Nucl. Centre of Saclay; Chef du Departement de Physique des Particules Elementaires, C.E.A.; Professor, Paris Univ. Societies: Sté. française de Physique; British Phys. Soc.; A.P.S.; Soc. Italiana di Fisica.
Nuclear interest: Elementary particle physics.
Address: C.E.N. de Saclay, B.P. No. 2, Gif-sur-Yvette, Seine et Oise, France.

BERTHELOT, Miss. Administrative Asst., Internat. Lab. of Marine Radioactivity.
Address: International Laboratory of Marine Radioactivity, Musée Océanographique, Monaco.

BERTHOD, L. Chef du Service Energie Atomique, Sté. Grenobloise d'Etudes et d'Applications Hydrauliques (Sogreah); Dept. Genie Atomique, Establissements Neyrpic, Grenoble.
Address: Société Grenobloise d'Etudes et d'Applications Hydrauliques (Sogreah), B.P. 145, Grenoble (Isere), France.

BERTHOLD, Fritz, Dr. rer. nat., Dipl. Phys. Born 1929. Educ.: Freiburg, Duke and Stanford Univs. Director, Laboratorium Prof. Dr. Berthold. Society: Deutsche Phys. Gesellschaft. Nuclear interests: Nuclear physics; Design of nuclear instruments; Management.
Address: 22 Calmbacher Strasse, 7547 Wildbad im Schwarzwald, Germany.

BERTHON, Rene. Directeur Gen., Dept. Entreprises, Sté. Parisienne pour l'Industrie Electrique.
Address: Societe Parisienne pour l'Industrie Electrique, Departement Entreprises, 85 boulevard Haussman, Paris, France.

BERTI, Mario, Ing. Power Remote Control Dept., Officine Galileo.
Address: Officine Galileo, 44 Via Carlo Bini, Florence, Italy.

BERTIN, Jean Henri, Ing. Ecole Polytechnique, Ing. Ecole Nationale Supérieure d'Aéronautique, Licencié en droit. Born 1917. Educ.: Ecole Polytechnique and Ecole Nationale Supérieure d'Aéronautique. Directeur Technique Adjoint, S.N.E.C.M.A. (Sté. Nationale d'Etudes et de Construction de Moteurs d'Avions), -1956; Prés. Directeur Gén., Sté. Berlin et Cie, 1956-. Member, Sci. and Tech. Com., Euratom, 1968-. Nuclear interests: Etude des réacteurs; Problèmes aérodynamiques et thermiques; Compresseurs; Isolation thermique; Protection biologique.
Address: Société Bertin et Cie, B.P. No. 3, Secteur Industriel de Plaisir, (S. et O.), France.

BERTINCHAMPS, Albert Jacques, B.S., M.D. Born 1924. Educ.: Brussels Univ. Inst. Interuniversitaire des Sci. Nucl., Belgium, 1952-57; Brookhaven Nat. Lab., Upton, 1957-59; Centre d'Etude de l'Energie Nucl., Mol, Belgium, 1959-60; Communauté Européene de l'Energie Atomique, (Euratom), Brussels, 1961-.
Nuclear interests: Radiobiology; Radiation biophysics.
Address: Euratom, 51 rue Belliard, Bruxelles, Belgium.

BERTINELLI, Enrico F., Dr. Civil Eng. Born 1924. Educ.: Padua Univ. Director and Manager, Dott. Ing. Giuseppe Torno and C. S.p.A. Society: A.S.C.E.
Address: Dott. Ing. Giuseppe Torno and C. S.p.A., 7 Via Albricci, Milan, Italy.

BERTINI, Adolfo, degree in Industrial Eng. Born 1925. Educ.: Turin Polytech. Univ. Eng., Steam Power Dept., Soc. Idroelettrica Piemonte, Turin, 1949-57; Chief Elec. Eng., Agip Nucleare, Milan, 1957-61; Station Supt., Latina Nucl. Power Station, Simea-Enel; Member, Comitate Nazionale, Assoc. Nazionale di Ingegneria Nucleare. Society: Assoc. Elettrotecnica Italiana.
Nuclear interest: Nuclear plant design and operation.
Address: 2 Via G. Cena, Latina, Italy.

BERTINI, S. Gruppo van Graaff, Istituto di Fisica, Catania Univ.
Address: Catania University, Istituto di Fisica, 57 Corso Italia, Catania, Sicily, Italy.

BERTINO, Romolo B. Formerly Tech. and Maintenance Eng., now Asst. Supt., Enrico Fermi Atomic Power Plant.
Address: Enrico Fermi Atomic Power Plant, Post Office Box 725, Monroe, Michigan, U.S.A.

BERTOLUS, Marcel. Prés. Establissements Charles Bertolus.
Nuclear interests: Revêtement de nickel chimique non poreux, anti-corrosion, dureté maximum, soudabilité électrique et électronique.
Address: Etablissements Charles Bertolus, 25 rue de Courcelles, Paris 8, France.

BERTONA, Luigi. Sen. Officer, Carlo Gavazzi S.p.A.
Address: Carlo Gavazzi S.p.A., 9 Via G. Ciardi, Milan, Italy.

BERTONAZZI, G. Chef du Département Génie Atomique, Services d'Etudes Nucléaires, Soc. Alsthom.
Address: Services d'Etudes Nucléaires, Société Alsthom, 20 rue d'Athènes, Paris 9, France.

BERTONI, Carlo Cesare, Dr. in Jurisprudence. Born 1935. Educ.: Naples Univ. Head, Intergovernmental Organisation Office, Internat. Affairs and Economic Studies Div., C.N.E.N., 1962-65; Alternate Governor to the I.A.E.A., 1962-64.; Asst. to the Vice-Chairman, C.N.E.N. 1965-; Sec., Consorzio Italiano ROVI, 1968-. Italian Member, Finance Com. C.E.R.N. 1961-65; Member, Working Group, European Atomic Energy Soc. Societies: F.I.E.N.; Italian Tech. Internat. Cooperation Experts Assoc.; Italian Internat. Organisations Soc. Nuclear interests: Nuclear industry development and international collaboration.
Address: c/o C.N.E.N., 15 Via Belisario, 00187 Rome, Italy.

BERTOTTI, Bruno, degrees in Phys. and Mathematics. Born 1930. Educ.: Pavia Univ. and Dublin Inst. for Advanced Studies. Asst. Prof., Pavia Univ., 1956-58; Member, Inst. for

Advanced Study, Princeton, N.J., 1958-59; Sen. Sci., Plasma Phys. Lab., Princeton Univ., 1959-61; Sen. Sci., Lab. Gas Ionizzati Frascati, Rome,1961-67; Prof. Theoretical Phys., Messina Univ.
Nuclear interest: Plasma physics.
Address: Istituto di Fisica, Messina University, Italy.

BERTRAM, Dieter, Dr. Dipl. Chem. Born 1934. Educ.: Humboldt Univ., Berlin. Zentralinst. für Kernforschung, 1958-.
Nuclear interests: Radiation chemistry; Radiation dosimetry; Radiation catalysis.
Address: Zentralinstitut für Kernforschung Rossendorf, Rossendorf bei Dresden, Dresden-Bad Weisser Hirsch, Postfach 19, German Democratic Republic.

BERZIN, A. K. Paper: Co-author, Method for Determination of Oil-Water Interface Based on Delayed Neutrons Detection (Atomnaya Energiya, vol. 21, No. 1, 1966).
Address: Academy of Sciences of the U.S.S.R., 14 Leninsky Prospekt, Moscow V-71, U.S.S.R.

BERZINA, I. G. Papers: Co-author, Uranium Nucleus Fission Fragment Tracks Determination of Age of Muscovites (letter to the Editor, Atomnaya Energiya, vol. 21, No. 4, 1966); co-author,Determination of Uranium Concentration and Its Spatial Distribution in Minerals and Rocks (ibid, vol. 23, No. 6, 1967).
Address: Academy of Sciences of the U.S.S.R., 14 Leninsky Prospekt, Moscow V-71, U.S.S.R.

BERZOLARI, Alberto GIGLI-. See GIGLI-

BESSA MENEZES e SOUSA, J., Eng. Res. Fellow, Nucl. Phys. and Electronics Res. Dept., Oporto Univ.
Address: Oporto University, Nuclear Physics and Electronics Research Department, Oporto, Portugal.

BESSE, Georges, Ing. du Corps des Mines. Born 1927. Educ.: École Polytechnique and Ecole des Mines. Gérant, U.S.S.I.; Directeur Gen., Groupement Atomique Alsacienne Atlantique.
Nuclear interests: Separation des isotopes de l'uranium.
Address: 22 avenue E. Herriot, Le Plessis Robinson, 92, France.

BESSEDE, R. Directeur des Services Commerciaux, Sté. Alsacienne de Constructions Atomiques de Telecommunications et d'Electronique.
Address: Societe Alsacienne de Constructions Atomiques de Telecommunications et d'Electronique, 32 rue de Lisbonne, Paris 8, France.

BESSLER, C. Italian Member, Study Group on Food Irradiation, O.E.C.D., E.N.E.A.
Address: O.E.C.D. European Nuclear Energy Agency, 38 boulevard Suchet, Paris 16, France.

BESSON, Jean. Directeur, Inst. d'Electrométallurgie et d'Electrochimie de Grenoble. Member, Commission Scientifique, Centre d'Etudes Nucléaires de Grenoble, C.E.A.
Address: Institut d'Electrometallurgie et d'Electrochimie, Grenoble University, Grenoble, France.

BESSON, M. Manager, Sté. d'Etudes et d'Equipements d'Entreprises.
Address: Société d'Etudes et d'Equipements d'Entreprises, 25 rue de Courcelles, Paris 8, France.

BESSONOV, E. G. Paper: Co-author, Positron Stacking and Obtaining of Electron-Positron Beams in Synchrotron (Atomnaya Energiya, vol. 23, No. 1, 1967).
Address: Academy of Sciences of the U.S.S.R., 14 Leninsky Prospekt, Moscow V-71, U.S.S.R.

BEST, G. C. Deputy Chief Eng., Eng. (Toronto) Div., A. V. Roe Canada Ltd.
Address: Engineering (Toronto) Division, A. V. Roe Canada Ltd., Box 4004, Terminal "A", Toronto, Ontario, Canada.

BESTAGNO, Maurizio Francesco, M.D. Born 1930. Educ.: Genoa Univ. Assistant of Clinica Medica Generale, Genoa Univ., 1957; now Chief, Nucl. Medicine Dept., City Hospital, Brescia. Society: Società Italiani di Biologia e Medicina Nucleare.
Nuclear interests: Isotope applications in medicine, particularly diagnosis and treatment of thyroid diseases, and gastroenterological studies with colloidal radiogold, labelled rose bengal, radioactive fats, etc.
Address: 3 Via Molinari, I-25100 Brescia, Italy.

BETEL, Isaäc, Ph.D. Born 1934. Educ.: Amsterdam Univ. Res. Assoc., Radiobiol. Inst., Organisation for Health Res. T.N.O. Society: European Soc. for Rad. Biol.
Nuclear interest: Biochemical aspects of radiobiology.
Address: 151 Lange Kleiweg, Rijswijk Z.H., Netherlands.

BETHE, Hans Albrecht, Ph.D., Hon. Dr. Sci.: Brooklyn Polytech. Inst. (1951); Denver Univ. (1952); Chicago Univ. (1953); Birmingham Univ. (1956); Harvard Univ. (1958). Enrico Fermi Prize (1961). Educ.: Munich Univ. Instructor in Theoretical Phys., Frankfurt, Stuttgart, Munich and Tubingen Univs., 1928-33; Lecturer, Manchester and Bristol Univs., 1933-35; Asst. Prof. of Phys., Cornell Univ., 1935-37; Professor of Phys., Cornell Univ., 1937-. Member, Editorial Council, Annals of Phys.; Member, Advisory Panel, F.A.S.; Member, U.S. Delegation to the Conference on the Discontinuation of Nucl. Tests, 1958-59. Books: Elementary Nucl. Theory (2nd ed.

1956); Mesons and Fields (with F. de Hoffmann and S. S. Schweber, 1955); Encyclopedia of Phys. (Handbuch der Physik, Berlin, 1932-33; 2nd ed., 1956). Societies: American Philosophical Soc.; N.A.S.; A.P.S. (pres., 1954); American Astronomical Soc.; Foreign Member, Roy Soc. (London).
Nuclear interests: Theoretical physics of the atom and the atomic nucleus, also energy production in stars.
Address: Laboratory of Nuclear Studies, Cornell University, Ithaca, New York, U.S.A.

BETHEL, Albert Lambert, B.S., M.A. (Phys.). Born 1921. Educ.: U.S. Military Acad., West Point, and Johns Hopkins Univ. post-graduate training 1948-51; Special Asst. to Chief, Armed Forces Special Weapons Project, Washington, 1951-55; Asst. to Manager, P.W.R. Project (1955-57), Manager P.W.R. Test and Operations Group and Manager, P.W.R. Shippingport Activity, Pennsylvania (1957-58), Manager, P.W.R. Power Plant Eng., Bettis Plant, (1958-60), Westinghouse Elec. Corp.; Director of Nucl. Activities, Consumers Power Co., 1960; Sen. Dept. Manager, Westinghouse Astronucl. Lab., Pittsburgh, Pennsylvania, 1960-63; Vice-Pres., Westinghouse Defence and Space Centre, Baltimore, Maryland, 1963-.
Nuclear interests: Management of integrated engineering design and development organisations engaged in the application of nuclear energy to meet aerospace and space propulsion and electrical requirements.
Address: P.O. Box 1693, Baltimore, Maryland 21203, U.S.A.

BETHGE, Klaus Heinrich Wilhelm, Dr. rer. nat., Dipl. -Phys. Born 1931. Educ.: Berlin Tech. and Heidelberg Univs. Res. Asst., II. Phys. Inst., Heidelberg Univ., 1961-67; Res. Assoc., Pennsylvania Univ., 1967-. Society: German Phys. Soc.
Nuclear interests: Nuclear physics (heavy ion nuclear reactions); Ion sources; Accelerator design.
Address: 55 Keplerstr., 69 Heidelberg, Germany.

BETIN, Yu. P. Paper: Co-author, The Efficiency of H^3/Zr Sources in Nondispersive X-Ray Spectrometric Analysis (letter to the Editor, Atomnaya Energiya, vol. 19, No. 3, 1965).
Address: Academy of Sciences of the U.S.S.R., 14 Leninsky Prospekt, Moscow V-71, U.S.S.R.

BETT, Frank Lincoln, B. Met. E. (Final Hons.), M. Eng. Sc. (1st Hons.), Management Certificate. Born 1926. Educ.: Melbourne Univ. Res. Officer, Melbourne Univ., 1955-56; Sen. Res. Officer, Australian A.E.C. Now seconded to Prime Minister's Dept. Societies: Australian Inst. Metals; Australian Inst. Welding; A.N.S.
Nuclear interests: Welding, non-destructive testing, materials engineering, liquid metals, ceramics, administration.
Address: c/o Prime Minister's Department, Canberra, A.C.T. 2600, Australia.

BETTERTON, Jesse Oatman, Jr., B.Sc. (Met. Eng.), D.Phil. Born 1920. Educ.: Lehigh and Oxford Univs. Metal., O.R.N.L., 1951-64. Societies: RESA; A.S.M.; A.I.M.E.; A.P.S.
Nuclear interests: Early transition metals Groups IV-VIA and alloys, especially zirconium; Studies of phase diagrams, lattice constants, low temperature properties such as specific heats, electrical and magnetoresistivity and Fermi surfaces, hard superconductors for high field solenoids; Atom vacancies in 21/13 electron compounds.
Address: 216 Outer Drive, Oak Ridge, Tennessee, U.S.A.

BETTOLO, G. B. MARINI-. See MARINI-BETTOLO, G. B.

BETTS, Austin W., Lieutenant Gen., B.S., M.S. Born 1912. Educ.: U.S. Military Acad., and M.I.T. In Office of Special Asst. for Guided Missiles (1956-59), Office of Director of R. and D. (1959), Director, Advanced Res. Projects Agency (1960-61), Office, Sec. of Defence; Director of Military Application, U.S.A.E.C., Washington, D.C., 1961-64; Deputy Chief (1964-66), Chief (1966-), R. and D., Dept. of the Army. Societies: Soc. of American Eng.; A.I.A.A.; A.N.S.
Nuclear interest: Management of development of weapons systems using nuclear weapons, as well as management of development of portable nuclear power systems.
Address: Qrs 12A, Ft Myer, Virginia 22211, U.S.A.

BETTS, R. H., Formerly Res. Chem., now Asst. Director, Chem. and Metal. Div., A.E.C.L.
Address: Atomic Energy of Canada Ltd., Chalk River Project, Chalk River, Ontario, Canada.

BETZ, Emile-Hippolyte, Dr. in medicine. Born 1919. Educ.: Liège Univ. Associé du Fonds Nat. de la Recherche Sci., 1951; Agrégé de Faculté (1957), Chargé de Cours à la Faculté de Médecine, (1959), Prof. Ordinaire, (1960), and Directeur, Centre Anticancéreux, Liège Univ. Sec.-Treas., Sté. Europeenne de Radiobiologie. Commission Sci., Inst. Interuniversitaire des Sci. Nucl. Book: Contribution à l'étude du syndrome endocrinien provoqué par l'irradiation totale de l'organisme (Paris, Masson, 1957).
Nuclear interests: Radiobiology and tumour research; Effects of local irradiation and whole-body irradiation on the organism, effects of chemical protectors on the cells.
Address: Institut de Pathologie, 1 rue des Bonnes Villes, Liège, Belgium.

BETZ, P. L., Dr. Director, Res. Dept., Baltimore Gas and Elec. Co.
Address: Baltimore Gas and Electric Co., Research Department, Gas and Electric Building, Baltimore 3, Maryland, U.S.A.

BETZLER, Karl-Erwin, Ing. des Maschinenbaus-Flugzeugbaues. Born 1915. Educ.: Akademie für Technik Chemnitz. Independent consultant Eng. 1957-.
Nuclear interests: Erection of nuclear installations, specialising in hot laboratories; Applying the results of nuclear research to industry, in an effort to solve technical problems in industry with nuclear techniques.
Address: Ingenieurbüro, 15 Viktor Scheffelstrasse, Munich 23, Germany.

BEUKEN, C. L., Dr. Ing. Ir. Formerly Member, Study Com. Regarding the Building of the First Dutch Nucl. Power Station, now Member, Advisory Com., N.V. Gemeenschappelijke Kernenergiecentrale Nederland.
Address: N.V. Gemeenschappelijke, Kernenergiecentrale Nederland, 310 Utrechtseweg, Arnhem, Netherlands.

BEULAYGUE, Marc. Chef Div. Minière de Vendèe, Direction des Productions, C.E.A.
Address: Commissariat à l'Energie Atomique, Direction des Productions, B.P. No. 5, Mortagne-sur-Sevre, (Vendèe), France.

BEUMER, Ths. J. J. A., Drs. Director, Internat. Social and Public Health Affairs, Ministry of Social Affairs and Public Health.
Address: Ministry of Social Affairs and Public Health, 73 Zeestraat, The Hague, Netherlands.

BEUTLER, Ernest, Ph.B., B.S., M.S. Born 1928. Educ.: Chicago Univ. Chairman, Div. Medicine and Director, Hospital for Blood Diseases, City of Hope Medical Centre, Duarte, California; Clinical Prof. of Medicine, Southern California Univ. Societies: American Soc. for Clinical Investigation; Central Soc. for Clinical Research; American Coll. Physicians.
Nuclear interest: Medical applications of tracer techniques.
Address: 1500 E. Duarte Road, Duarte, California, U.S.A.

BEVAN, D. J. M., Dr. Formerly Sen. Lecturer in Inorganic Chem., now Sen. Lecturer in Phys. Chem., Western Australia Univ. Paper: The Solution of Lime in Liquid Calcium and its Effects on the Reducing Properties of the Metal (Symposium on the Peaceful Uses of Atomic Energy in Australia, 1958).
Address: Western Australia University, Nedlands, Western Australia, Australia.

BEVERIDGE, Clifford. Born 1915. Educ.: Leeds Coll. of Technol. Development Eng., A. Reyrolle and Co., Ltd., Hebburn, 1946-51; Eng. and Sen. Eng., Power Services, Industrial Group, Risley (1951-57), Asst. Chief Eng., Reactor Group, Risley (1957-68), Asst. Chief Eng., Elec. Design Office, Eng. Group (1968-), U.K.A.E.A. Society: I.E.E.
Nuclear interests: Specialisation: design and construction.
Address: 104 Wellfield Road, Culcheth, Lancs., England.

BEVERLY, Robert Gene, B.S. (Chem. Eng.). Born 1923. Educ.: Denver Univ. Director, Rad. and Pollution Control, Mining and Metals Div., Union Carbide Corp., Grand Junction, Colorado, 1959-. Chairman, Rad. Subcom., Atomic Ind. Forum. Societies: A.I.M.M.E.; Air Pollution Control Assoc.; Water Pollution Control Federation.
Nuclear interests: Management; Raw materials metallurgy; Radiation control in mines and mills.
Address: 106 Columbine Court, Grand Junction, Colorado, U.S.A.

BEYER, Gerhard H., B.S., M.S., Ph.D. Born 1923. Educ.: Wisconsin Univ. Asst. Prof. Chem. Eng. (1949-52), Assoc. Prof. (1952-55), Iowa State Univ.; Assoc. Eng., Ames Lab., A.E.C. 1950-55; Prof. and Chairman, Dept. of Chem. Eng., Missouri Univ., 1956-64. Prof. and Chairman, Dept. of Chem. Eng., Virginia Polytechnic Inst., 1964-. Consultant to various companies in nucl. field. Societies: A.I.Ch.E.; A.C.S.; A.S.E.E.
Nuclear interest: Extractive metallurgy.
Address: 1415 Highland Avenue, Blacksburg, Va., U.S.A.

BEYER, Peder C., M.Sc. Born 1918. Educ.: Copenhagen Tech. Univ. Vice-Pres., Disa Elektronik A.S. Pres., Assoc. of Danish Electronics Manufactures; Vice Pres., Danish Res. Centre for Appl. Electronics. Societies: Inst. of Danish Eng.; Danish Acad. of Tech. Sci.
Nuclear interest: Electronic instrumentation.
Address: Disa Elektronik, A.S. 17 Herlev Hovedgade, Herlev, Denmark.

BEYERLE, Konrad, Dr.-Ing. (Aachen T.H.). Born 1900. Educ.: Munich and Aachen. Director, Inst. für Instrumentenkunde in der Max Planck Gesellschaft, 1946-57; Director, Inst. für Wissenschaftliche Apparate, Mess- und Regeltechnik der Kernforschungsanlage Jülich des Landes Nordrhein-Westfalen e.V., 1957-. Society: Physikalische Gesellschaft Deutschlands.
Nuclear interests: Forschung und Entwicklung, physikalischer Gerätebau.
Address: 14 Charlottenstrasse, Aachen, Germany.

BEYSTER, John Robert, Ph.D., M.S., B.S.E. Born 1924. Educ.: Michigan Univ. Phys. in Charge, Linear Accelerator, Gen. Atomic. Societies: A.N.S.; A.P.S.
Nuclear interests: Nuclear physics; Reactor physics; Neutron thermalisation; High-power accelerators; Pulsed reactors.
Address: P.O. Box 608, San Diego, California, U.S.A.

BEZDEK, Miroslav, RNDr., Deleg. to 2nd I.C.P.U.A.E., Geneva, Sept. 1958. Papers: Radiochem. Problems of Nucl. Fuels (Jaderna Energie, vol. 7, No. 10, 1961); co-author, Radiochem. Methods of Nucl. Fuel Burn-up (ibid, vol. 13, No. 5, 1967).

Address: Institute of Nuclear Physics, Czechoslovak Academy of Sciences, Rez u Prahy, Czechoslovakia.

BEZEL', V. S. Paper: Co-author, On Application of DK-0.2 Dosimeters for Fast Neutrons (Atomnaya Energiya, vol. 24, No. 3, 1968). Address: Academy of Sciences of the U.S.S.R., 14 Leninsky Prospekt, Moscow V-71, U.S.S.R.

BEZMATERNYH, A. S. Papers: Co-author, Al^{26} Production during Mg Irradiation with 20 Mev Deuterons (letter to the Editor, Atomnaya Energiya, vol. 19, No. 1, 1965); co-author, Na^{22} Separation from Magnesium Target, Irradiated with Deuterons (letter to the Editor, ibid). Address: Academy of Sciences of the U.S.S.R., 14 Leninsky Prospekt, Moscow V-71, U.S.S.R.

BEZNOSIKIVA, A. V. Papers: Co-author, Study of System UF_4 - CaF_2 (Atomnaya Energiya, vol. 22, No. 4, 1967); co-author, Phase Diagrams of Plutonium with IIIA, IVA, VIII and IB Group-Metals (ibid, vol. 23, No. 6, 1967). Address: Academy of Sciences of the U.S.S.R., 14 Leninsky Prospekt, Moscow V-71, U.S.S.R.

BEZUCHA, Jaromir. Born 1935. Nucl. Res. Inst., Czechoslovak Acad. of Sci., Rez, 1960-. Nuclear interest: Ceramic nuclear fuels. Address: 11 Malinová, Prague 10, Czechoslovakia.

BHAKTAVATSALU, R. Joint Sec., Dept. of Atomic Energy, A.E.C. Address: Atomic Energy Commission, Apollo Pier Road, Bombay 1, India.

BHANOT, Vidya Bhushan, B.Sc. (Hons. School), M.Sc., Ph.D. Born 1927. Educ.: Panjab (Lahore), Panjab (India) and Minnesota Univs. Lecturer, Phys., Aligarh Muslim Univ., 1951-54; Asst., Minnesota Univ., 1954-59; Reader (1960-66), Prof. (1966-), Panjab Univ., India. Vice-Pres., Panjab Univ. Phys. Assoc.; Asst. Editor, Everyday Sci. Societies: Fellow, Indian Phys. Soc.; N;A.S., India; A.P.S.; Northern India Sci. Assoc. Nuclear interests: Systematics of nuclear binding energies; Calculation of nuclear energy levels; Nuclear geophysics; Mass-spectrometry. Address: Physics Department, Panjab University, Chandigarh-14, India.

BHATNAGAR, Dharma Veer, M.Sc. Born 1922. Has worked at Kamani Metal Industries Jaipur, Dungar Coll., Bikaner, in Radiochem. Div. of Chem. Res. Lab., Teddington, London, and Bhabha Atomic Res. Centre, 1949-65. (At present i/c of Control and Res.). Nuclear interests: Problems connected with (i) uranium recovery from low grade ores, (ii) natural and percolation leaching of uranium in mines and waste dumps, (iii) uranium tetrafluoride reduction slag treatment for calcium fluoride and uranium recovery, (iv) recovery of Ni, Mo, and Cu as by-products from uranium ore processing and (v) separation and identification of minerals. Address: Uranium Corporation of India, P.O. Jadugoda, Dist. Singhbhum, Bihar, India.

BHATTACHARJEE, Sukhendu Bikash, M.Sc., D.Phil. (Sci.). Born 1933. Educ.: Gauhati and Calcutta Univs. Res. Biophys., Lawrence Rad. Lab., California Univ., Berkeley; Lecturer, Saha Inst. of Nucl. Phys., Calcutta, 1964-. Hon. Lecturer, Phys., Calcutta Univ. Societies: Indian Biophys. Soc.; Biophys. Soc., U.S.A. Nuclear interest: Interested in the influence of nuclear radiation on living cells. Address: Biophysics Division, Saha Institute of Nuclear Physics, 37 Belgachia Road, Calcutta 37, India.

BHATTACHARYYA, Rangalal, M.Sc. (Phys.), D.Phil. (Nucl. Phys.). Born 1931. Educ.: Calcutta Univ. Lecturer, Nucl. Phys. (1957-64), Reader, Nucl. Phys. (1964-), Saha Inst. of Nucl. Phys. Society: Indian Phys. Soc. Nuclear interests: Beta and gamma ray spectroscopy; Nuclear instrumentation; Experimental nuclear physics. Address: Saha Institute of Nuclear Physics, 92 Acharya P.C. Road, Calcutta-9, India.

BHATTACHARYYA, S. D., Dr. Lecturer, Saha Inst. of Nucl. Phys., Calcutta Univ. Address: Calcutta University, Saha Institute of Nuclear Physics, 92 Acharya Prafulla Chandra Road, Calcutta-9, India.

BHATTACHARYYA, Sudhindra Nath, M.Sc., D.Phil. Born 1932. Educ.: Calcutta and Moscow State Univs. Reader, Saha Inst. of Nucl. Phys., Calcutta Univ. Nuclear interest: Nuclear and radiation chemistry. Address: Saha Institute of Nuclear Physics, 92 Acharya Prafulla Chandra Road, Calcutta-9, India.

BHATTACHARYYA, Tarun Kumar, M.Sc. (Calcutta), Dr. es Sc. (Nancy). Born 1923. Educ.: Calcutta and Nancy Univs. Lecturer in Geology, Jadavpur Univ., Calcutta. Sen. Petrologist (Sci. Officer-SD2) (Petrology), In-Charge, Petrology Lab., Atomic Minerals Div., Dept. of Atomic Energy, Govt. of India. Society: Fellow, Geological, Mining and Metal. Soc. of India. Nuclear interests: Radioactive rocks and minerals. Address: 43 Pusa Road, New Delhi-5, India.

BHATTACHERJEE, Satyendra Kumar, M.Sc., Ph.D. Born 1926. Educ.: Calcutta and Notre Dame (Indiana) Univs. Fellow, Tata Inst. Fundamental Res., 1954-. Societies: A.P.S.; Fellow, Indian Phys. Soc. Nuclear interest: Nuclear spectroscopy (β,γ spectroscopy). Address: Tata Institute of Fundamental Research, Colaba, Bombay 5, India.

BHAVILAI, Rawi, Physicist, Chulalongkorn Univ.
Address: Chulalongkorn University, Bangkok, Thailand.

BHAVSAR, P. D., Dr. Assoc. Prof., Phys. Res. Lab.
Address: Physical Research Laboratory, Navrangpura, Ahmedabad 9, India.

BHIDE, Manohar Gopal, B.Sc. (principal, Maths., 1st class Hons.), M.Sc. (Phys.).Born 1935. Educ.: Bombay Univ. Fellow, Phys. Dept. of Ramnarain Ruia Coll., Bombay, 1954-56; Res. Asst. in Theoretical Phys. Div. (1956-61), Sci. Officer in Reactor Eng. Div. (1961-67), A.E.E.T.; now at Tata Inst. of Fundamental Res., Bombay; Deputed to work at A.E.R.E. Harwell on fast reactors, 1958-60; Deputed to Argonne Nat. Lab., 1960-62. Deleg. to 2nd I.C.P.U.A.E., 1958-; Indian deleg. to Seminar on Phys. of Fast and Intermediate Reactors at Vienna, Aug, 1961, organised by I.A.E.A.; Disarmament Study Group, Dept. of Atomic Energy, Govt. of India, 1962-; Sci. Sec., Twelfth Pugwash Conference, Udaipur, India, 1964.
Nuclear interests: Theoretical physics; Reactor design; Solid state physics.
Address: 76 Kokan Nagar, Bhandar Galli, Mahim, Bombay 16, India.

BIAGINI, Carissimo, M.D., L.D. (Gen. Pathology), L.D. (Radiol.), L.D. (Radiobiol.). Born 1923. Educ.: Florence Univ. At Centre of Radioisotopes and High Energies, Rome Univ., 1957; Chief Investigator, Res. Unit of C.N.E.N., Rome, 1958; Asst. to Prof., Inst. Radiol. (1959), Prof. of Nucl. Phys. Appl. to Medicine (1962), Rome Univ.; Prof. of Radiol., Sassari Univ., 1964. Book: Introduzione allo studio della radiobiologia (Rome, Universio, 1955); co-author, Le radiazioni di alta energia (Rome, Universo, 1959).Societies: Ital. Soc. Radiol. and Nucl. Medicine; Assòc. Internat. Radiobiol. Pays C.E.E.A.
Nuclear interests: Radioisotopes and accelerating machines in radiation therapy; Biological effects of high energy radiations; Health physics problems in medical uses of sources of radiation.
Address: Istituto di Radiologia dell'Università, 10 Viale San Pietro, Sassari, Italy.

BIALAS, Andrzej Stefan, Docent Dr. Born 1936. Educ.: Jagellonian Univ., Cracow. Jagellonian Univ. Cracow, 1958-; C.E.R.N., Geneva, 1964-65; Inst. of Nucl. Phys., Cracow, 1966-. Society: Polish Phys. Soc.
Nuclear interests: High-energy physics.
Address: Institute of Physics, Jagellonian University, 4 Reymonta, Cracow, Poland.

BIANCHERI, R. Member, Board, Centre Scientifique de Monaco.
Address: Centre Scientifique de Monaco, 16 boulevard de Suisse, Monte Carlo, Monaco.

BIANCHI, Angelo, Dr. Natural Sci., Dr. Biol. Sci., Libero docente in Genetics. Born 1926. Educ.: Pavia and Milan Univs. and Catholic Univ. in Piacenza. Council Member, Italian Soc. for Agricultural Genetics; Director, Agricultural Inst. Societies: Radiobiol. Soc. of Euratom countries; Sec., Italian Genetic Assoc.
Nuclear interest: Radiation genetics.
Address: Istituto di Allevamento Vegetale per la Cerealicoltura, 33 Via di Corticella , Bologna, Italy.

BIANCHI, Bruno, Dr. Eng. Born 1901. Educ.: Bologna Univ. Manager, Soc. Romana di Elettricità, 1939-52; Pres. and Gen. Manager,Soc. Finanziaria Elettrica Nazionale, 1952-. Member, Nat. Council of Economy and Work; Vice-Pres., Nat. Federation of Business Executives; Vice-Pres., Soc. Elettro-Nucleare Nazionale; Vice-Pres., Soc. Italiana Meridionale Energia Atomica.
Nuclear interests: Economy of reactors, and any problem relating to top-level management of nuclear enterprises.
Address: 14 Via Aniene, Rome, Italy.

BIANCHI, Claudio, M.D. Born 1930. Educ.: Pisa Univ. Res. Asst., Medical Clinic, Pisa Univ., 1956-. Societies: Italian Soc. Nucl. Biol. and Medicine; Italian Soc. Exptl. Biol.; Italian Soc. Nephrology; American Federation for Clinical Res.
Nuclear interest: Nuclear medicine and nephrology.
Address: Centro di Medicina Nucleare, Università di Pisa, Italy.

BIANCHI, Giuseppe, Eng. Born 1933. Educ.: Milan Polytech. At (1958-), now Director, Reactor Technol. Lab., C.N.E.N.
Address: Casaccia Nucleare Studies Centre, S.P. Anguillarese km1 + 300, Casaccia, Rome, Italy.

BIANCHI, Romano, M.D. Born 1929. Educ.: Pisa Univ. Res. Asst., Medical Clinic, Pisa Univ., 1956-. Papers: 14 papers on experimental and clinical uses of radioisotopes in cardiology. Society: Italian Soc. for Nucl. Biology and Medicine.
Address: Centro di Medicina Nucleare, Università di Pisa, Italy.

BIANCHINI, Armando. Born 1923. Educ.: Padua Univ. At Montevecchio S.I.P.Z., Milan, 1947-59; Centro Ricerche Metallurgiche S.p.A., Turin, 1959-. Societies: Assoc. Italiana di Metallurgia; Assoc. Mineraria Sarda; Geochem. Soc.
Nuclear interests: Metallurgy, isotopes.
Address: 77 Lungopo Antonelli, Turin, Italy.

BIASSONI, Paolo, Dr. in Medicine and Semiology Prof. Born 1929. Educ.: Genoa Univ. Asst., Clinica Medica Generale, Genoa Univ. Books: On thyroid physiology, radiobiology, hematology and gastroenterology. Society: Società Italiana di medicina nucleare.

Nuclear interests: Applications of nuclear power and isotopes in medicine.
Address: Clinica Medica dell'Università, 15 viale Benedetto, Genoa, Italy.

BIBERGAL, A. V.
Nuclear interest: Design of high-intensity irradiation sources.
Address: Academy of Sciences of the U.S.S.R., 14 Leninsky Prospekt, Moscow V-71, U.S.S.R.

BIBIKOVA, E. V. Paper: Co-author, On Metamorphism of Uranium Deposits (Atomnaya Energiya, vol. 16, No. 4, 1964).
Address: Academy of Sciences of the U.S.S.R., 14 Leninsky Prospekt, Moscow V-71, U.S.S.R.

BICE, Richard Avery, B.S. (M.E.). Born 1917. Educ.: Colorado State and Pittsburgh Univs. Manager, Eng. Dept. (1949-54), Director, Field Testing (1954-59), Vice-Pres., (1959-), Sandia Labs., Albuquerque, New Mexico. Societies: A.S.M.E.; American Ordnance Assoc.; American Management Assoc.
Nuclear interests: Management in the fields of design, testing and production of nuclear ordnance, including engineering activities in the fields of mechanical, electronic, computer technology and test instrumentation.
Address: Sandia Corporation, Albuquerque, New Mexico, U.S.A.

BICHEL, Jorgen, M.D. Born 1909. Educ.: Copenhagen Univ. Prof. Gen. Pathology, Århus Univ.; Head of a lab. under Danish Civil Defence, 1954-; Head, Anti-cancer Chemotherapy Sect., Radiumstation for Jutland Aarhus.
Nuclear interest: Chemical radioprotection.
Address: Kraeftforskningsinstitutet, Radiumstationen, Aarhus C., Denmark.

BICHKOV, I. F. At U.S.S.R. State Com. for Utilisation of Atomic Energy. Adviser, Third I.C.P.U.A.E., Geneva, Sept. 1964.
Address: U.S.S.R. State Committee for the Utilisation of Atomic Energy, 26 Staromonetnii Pereulok, Moscow, U.S.S.R.

BICHSEL, Hans. Member, Subcom. on Penetration of Charged Particles in Matter, Com. on Nucl. Sci., Nat. Acad. of Sci. – N.R.C.
Address: National Academy of Sciences – National Research Council, Committee on Nuclear Sciences, 2101 Constitution Avenue, N.W., Washington 25, D.C., U.S.A.

BICKENDORF, Otto. Mitglied, Fachkommission 1 Recht und Verwaltung, Mitglied, Arbeitskreis 1/1, Haftung und Versicherung, Federal Ministry for Sci. Res.
Address: Federal Ministry for Scientific Research, 46 Luisenstrasse, 532 Bad Godesberg, Germany.

BICKERTON, R. J., Member, Board of Editors, Nucl. Fusion.
Address: Nuclear Fusion, International Atomic Energy Agency, 11 Karntner Ring, Vienna 1, Austria.

BICKOV, V. N. At Phys.-Energetical Inst., U.S.S.R. State Atomic Energy Com. Adviser, Third I.C.P.U.A.E., Geneva, Sept. 1964.
Address: U.S.S.R. State Committee for the Utilisation of Atomic Energy, 26 Staromonetnii Pereulok, Moscow, U.S.S.R.

BIEHL, Arthur Trew, B.S., M.S., Ph.D. (Phys.). Born 1924. Educ.: Idaho Univ. Res. Fellow, California Inst. of Technol., 1949-50; North American Aviation, Downey, California, 1950-52; Univ. California Rad. Lab., 1952-56; Vice-Pres. and Tech. Director, Aerojet-Gen. Nucleonics, San Ramon, California, 1956-60; Pres., MB Associates, Walnut Creek, Calif., 1960-; Chairman of the Board, ORDTECH Corp., Walnut Creek, Calif., 1960-; Formerly Special Asst. to Director, now Assoc. Director, Advanced Studies, (1966-), Lawrence Rad. Lab., California Univ. Book: Co-author, Modern Nucl. Technol. Societies: A.N.S.; A.P.S.; A.S.E.E., I.R.E.
Address: Country Club Grounds, Diablo, California, U.S.A.

BIEMILLER, Andrew J., A.B. Born 1906. Educ.: Cornell and Pennsylvania Univs. Legislative Director, and Chairman, Staff Com. on Atomic Energy and Natural Resources, AFL-CI0. Labour Member, Labour-Management Advisory Com., U.S.A.E.C.
Address: AFL-CI0, 815 Sixteenth Street, N.W., Washington 6, D.C., U.S.A.

BIENLEIN, H. Johann K. Dr. rer. nat., Dipl. Phys. Born 1930. Educ.: Erlangen Univ. Wissenschaftlicher Asst., Physikalischen Inst., Erlangen Univ.
Nuclear interests: Nuclear physics, especially beta-decay (electron polarisation, internal bremsstrahlung) and nuclear spectroscopy.
Address: Physikalisches Institut der Universität, 6 Glükstrasse, Erlangen, Germany.

BIENVENU, Claude, Ancien élève de l'Ecole Polytechnique, Ing. Civil de l'Aéronautique. Born 1927. Educ.: Alger Univ. Ing. aux Etudes et Recherches, Service Thermique (1951-55), Chef des Services Etudes à la Région d'Equipment Thermiques Nucléaires No. 1 (1955-60), Directeur-Adjoint à la Région d'Equipement Thermique Nucléaire, (1960-62), Directeur-Adjoint à la Région d'Equipement Nucléaire No. 2, (1962-64), now Directeur, Region d'Equipement Nucl. No. 2, Electricite de France. Society: Ingénieurs Civils de France.
Nuclear interests: Conception et construction des centrales nucléaires.
Address: Parc de Grandmont, B.P. 26, Tours, Indre-et-Loire, France.

BIER, Konrad, Dr. phil. Born 1926. Educ.: Marburg-Lahn Univ. Director, Thermodynamics Inst. Karlsruhe Univ. (TH.).
Nuclear interest: Isotope separation.

Address: Universität Karlsruhe (T.H.), 75 Karlsruhe, Germany.

BIERLEIN, Theo Karl, Ph.D. (Phys. Chem.). Born 1924. Educ.: Portland Univ. and Catholic Univ. America. Manager, Phys. Metal., Washington Univ. 1963-. Societies: A.C.S.; A.S.M.; E.M.S.A.
Nuclear interests: Metallurgy; Radiation damage; Optical and electron metallography.
Address: 1005 Warren Ct., Richland, Washington, U.S.A.

BIERLY, James N., Jr., B.S., M.S., Ph.D. Born 1922. Educ.: Kutztown State Teachers Coll., Bucknell and Temple Univs. Assoc. Prof., Phys., Philadelphia Coll. of Pharmacy and Sci., 1961-; Consultant, Radiol. Phys., Misericordia Hospital, Philadelphia. Societies: A.P.S.; H.P.S.; American Assoc. of Phys. in Medicine; American Radiol. Soc.
Nuclear interest: Teaching radioisotope and radiological physics.
Address: 4627 Hazel Avenue, Philadelphia, Pennsylvania 19143, U.S.A.

BIERMAN, Charles Oliver, B.S. (E.E.). Born 1914. Educ.: U.S. Naval Acad. U.S. Marine Corps (retired 1955 as Brigadier Gen.). Asst. Manager, Product Development Dept., The Budd Co., 1956-.
Nuclear interests: Design and fabrication of reactor parts and nuclear engines, especially in development and prototype stages.
Address: 4037 Kottler Drive, Lafayette Hills, Pa., U.S.A.

BIERMANN, Ludwig Franz Benedikt, Dr. phil. Born 1907. Educ.: Munich, Freiburg and Göttingen Univs. Head, Astrophys. Sect., Max-Planck-Inst. f. Physik, Göttingen, 1947-58; Director, Inst. f. Astrophysik at Max-Planck Inst. f. Physik u. Astrophysik, Munich, 1958-. Prof. Munich Univ.; Member, Sci. Direction, Inst. für Plasmaphysik Garching. Societies: Max-Planck-Ges.; Astronomische Ges.; Physikalische Ges.; Internat. Astronomical Union; Internat. Acad. of Astronautics; Bayer. Akademie der Wissenschaften; Groupe de Liaison-Fusion, C.E.E.A.; Assoc., Roy. Astronomical Soc., London; Corresponding Member, Sté. Roy. des Sci. de Liège.
Nuclear interests: Astrophysics; Physics of high temperature and applications to astrophysics and to theory of controlled thermonuclear fusion.
Address: 12 Rohmederstr., Munich 23, Germany.

BIGELEISEN, Jacob, A.B. (New York Univ.), M.S. (Washington State Univ.), Ph.D. (California Univ., Berkeley). E.O. Lawrence Memorial Award, U.S.A.E.C. 1964. Born 1919. Chem. (1949-52), Sen. Chem. (1952-68), Brookhaven Nat. Lab.; Prof. Chem., Rochester Univ., New York, 1968-. Visiting Prof., Cornell Univ. 1953-54; Nat. Sci. Foundation Fellow and Hon. Visiting Prof. E.T.H., Zurich, 1962-63; Gilbert N. Lewis Lecturer, 1963; Visiting Distinguished Prof., State Univ. of New York, Buffalo, New York, 1966. Editorial Board, Annual Revs. of Phys. Chem., 1959-64; Editorial Board, J. Phys. Chem., 1961-65; Assoc. Editor, J. Chem. Phys., 1955-58; N.R.C.-N.A.S. Com. on Chem. Kinetics, 1958-64.
Book: Co-editor, Proc. of the Internat. Symposium on Isotope Separation (North-Holland Publishing Co., 1958). Societies: N.A.S.; Fellow, A.P.S.; Fellow, A.A.A.S.; A.C.S.
Nuclear interests: Physical chemistry; Isotope separation.
Address: 835 Allens Creek Road, Rochester, New York 14618, U.S.A.

BIGGERS, Robert Edward, A.B. (Emory), A.M. and Ph.D. (Princeton). Born 1930. Educ.: Emory and Princeton Univs. Chem., Chem. Div., Ga., State Dept. of Agriculture, Atlanta, 1948-54; Asst. in Instruction, Dept. of Chem., Princeton Univ., 1952-56; Res. Chem., Analytical Chem. Div., Oak Ridge Nat. Lab., 1956-. Societies: American Inst. of Chemists; Polarographic Soc., Int. (London): F.A.S.; A.C.S.; A.N.S.; A.A.A.S.
Nuclear interests: Physical, inorganic, and analytical chemistry of the heavy elements and actinides, particularly Pu; Absorption spectroscopy of aqueous systems at high temperature and high pressures, to the water critical point; Electrochemistry, spectrophotometry, polarography; Analysis and resolution of complex spectra and fine structure; Application of high-speed computing to chemical and particularly spectroscopic problems.
Address: 29 Jolly Drive, Rt. #17, Knoxville 21, Tennessee, U.S.A.

BIGGS, Burnard S., Ph.D. Born 1907. Educ.: Texas Univ. Asst. Chem. Director, Bell Labs., 1954-58; Director, Materials and Standards Eng. (1958-61), Vice Pres., Livermore Lab. (1961-), Sandia Corp. Societies: A.C.S.; American Ordnance Assoc.
Address: Sandia Corporation, Livermore Laboratory, P.O. Box 969, Livermore, California, U.S.A.

BIGGS, William Derrick, A. Met. (Sheffield), B.Sc. (Lond.), M.A. (Cantab.), Ph.D. (Birmingham). Born 1923. Educ.: London and Birmingham Univs. Birmingham Univ., 1952-55; Head of Fundamental Res., Murex Ltd., Waltham Cross, 1955-58; Lecturer, Eng. Dept., Cambridge Univ., 1958-.
Nuclear interest: Metallurgy.
Address: Pennymeads, London Road, Harston, Cambridge, England.

BIGHAM, Clifford Bruce, B.Sc., M.Sc. (Queen's, Kingston), Ph.D. (Liverpool). Born 1928. Educ.: Queen's (Kingston) and Liverpool Univ. Sen. Res. Officer, A.E.C.L.
Nuclear interest: Measurement of nuclear constants for reactor designs.
Address: Box 523, Deep River, Ontario, Canada.

BIGNON, Michel Jean Louis Marie, Ancien Elève de l'Ecole Polytechnique (Paris). Born 1915. Educ.: Ecole Polytechnique. Chief Eng., Sté. Ame, Heurtey (Paris); appointed by same to U.S.S.I. (Plessis Robinson) as Chief Eng., Process Dept. Societies: Sté. de Chimie Industrielle; Assoc. Française des Techniciens du Pétrole.
Nuclear interests: Physical and chemical processing of nuclear materials.
Address: 23 Hameau des Chardonnerets, La Celle St. Cloud, Seine et Oise, France.

BIGOTTE, Georges Gabriel, D ès Sc., Ing.-géologue. Born 1925. Educ.: Ecole Nationale Supérieure de Géologie Appliquée et de Prospection Minière. Chief geologist for overseas prospections by the French A.E.C., 1967.
Nuclear interests: Prospection, resources, dressing and uses of nuclear raw materials; Management in prospection.
Address: 41 rue H. Cloppet, Le Vesinet (78), France.

BIGOURDAN, Gérard. Gen. Sec., Entrepôts Frigorifiques Lyonnaise S.A.
Address: Entrepôts Frigorifiques Lyonnaise S.A., 18 rue Séguin, Lyon,(Rhône), France.

BIHET, O. L., Dr. Directeur, Centre National de Recherches Métallurgiques.
Address: Centre National de Recherches Métallurgiques, Abbaye du Val-Benoît, Liège, Belgium.

BIKMATOV, R. G. Paper: Co-author, Indication of Electron Beam by Residual Gas light (Atomnaya Energiya, vol. 24, No. 1, 1968).
Address: Academy of Sciences of the U.S.S.R., 14 Leninsky Prospekt, Moscow V-71, U.S.S.R.

BILDSTEIN, Hubert, Dipl. Ing., Dr. tech. sci., Ing. en Genie Atomique. Born 1929. Educ.: Vienna Tech. Univ. Asst., Technol. Inst., Vienna Tech. Univ., 1955-57; Project Manager, Reactor Installation (1957-60), Head, Chem. Dept., Reaktorzentrum Seibersdorf (1960-.), Osterreichische Studiengesellschaft für Atomenergie G.m.b.H. Societies: Verein Österreichischer Chemiker; Österreichische Chem.-Phys. Gesellschaft.
Nuclear interests: Radiation and radiochemistry; Reactor materials.
Address: Österreichische Studiengesellschaft für Atomenergie G.m.b.H., 10 Lenaugasse, Vienna 8, Austria.

BILES, Martin B., B.S., M.S., Ph.D. Born 1919. Educ.: California Univ. and N.C.S.C. At California Univ., 1950-51; U.S. Air Force officer, 1951-58; Reactor Phys. (1958-61), Chief, Test and Power Reactor Safety Branch, Licencing and Regulations Div. (1962-63), Asst. Director for Reactor Safety, Operational Safety Div. (1963-65), A.E.C. Sci. Rep., Brussels (1965-66), A.E.C. Sci. Rep., Paris (1966-), now Director, Operational Safety Div., U.S.A.E.C. Society: A.N.S.
Nuclear interests: Reactor design; Reactor operation; Reactor safety.
Address: 9600 E. Bexhill Drive, Kensington, Maryland, U.S.A.

BILGE, S. Turkish Member, European Nucl. Energy Tribunal, O.E.C.D., E.N.E.A.
Address: O.E.C.D. European Nuclear Energy Agency, 38 boulevard Suchet, Paris 16, France.

BILLIG, Ernst, B.Sc. (Mech. Eng.), B.Sc. (Elec. Eng.), Dipl. Ing., Dr. rer-tech., C. Eng. Educ.: Vienna Univ. Sect. Leader, Solid State Phys. (1946-58), Consultant (1958-), A.E.I. Res. Lab., Aldermaston; Elec. Res. Assoc., Leatherhead; Assoc. Director, Phys. Res. Wheatstone Lab., Kings Coll. London; Principal Lecturer, Borough Polytech. Semi Conductor Panel (No. 7) of C.I.T.C.E. Book: Advances in Electronics and Electro Physics, vol. 10 (1958). Societies: Fellow, I.E.E..; F. Inst. P.; Verb. Deutscher Phys. Ges.; Aslib; Roy. Institution; Fellow, A.I.P.
Nuclear interests: Solid state physics; Semiconductors and devices; Crystallography; Metallurgy; Electrical engineering.
Address: 25 Woodside Avenue, Esher, Surrey, England.

BILLIG, Kurt, Dr. Tech.Sc., Dipl. Ing. Born 1907. Educ.: Vienna Inst. of Technol. Prof. Civil Eng., Hongkong Univ. 1950-52; Director of Building Res. of India, 1952-55; Consultant to U.N., 1956-67; at present Chief Eng., R. and D., Taylor Woodrow Construction, Ltd. Societies: Inst. Civil Eng., London; Inst. Structural Eng., London; Inst. Civil Eng., Ireland; A.S.C.E.
Nuclear interests: Civil engineering; Pressure vessels; Shield design.
Address: 52 The Glen, Green Lane, Northwood, Middlesex, England.

BILLING, Heinz Eduard, Dr. rer. nat. Born 1914. Educ.: Göttingen and Munich Univs. Inst. für Instrumentenkunde, Göttingen; Inst. for Advanced Studies, Princeton, N.J.; Max-Planck-Inst. für Physik und Astrophysik, Munich. Wissenschaftliches Mitglied, Max-Planck-Inst. für Physik; Honorar-Prof., Erlangen-Nürnberg Univ. Society: Verband Deutscher Physikalischer Gesellschaften.
Nuclear interests: Development of electronic computers; Fundamental research on elements that could become useful for computers.
Address: Max-Planck-Institut für Physik und Astrophysik, Institut für Astrophysik, Abteilung Numerische Rechenmaschinen, 6 Föhringer Ring, Munich 23, Germany.

BILLINGTON, Douglas Sheldon, B.A., M.S., Ph.D. Born 1912. Educ.: Yankton Coll. and Iowa State Univ. Director, Solid State Div., O.R.N.L., 1952-. Member, Editorial Advisory Boards of: Internat. J. of Phys. and Chem. of Solids; J. of Nucl. Materials; Member, Solid State Sci. Advisory Panel, N.A.S.- Nat. Res. Council. Books: Co-author, Rad. Damage in

Solids (Princeton University Press, 1960); Editor, Phys. of Solids (Rad. Damage) (Italian Phys. Soc., Academic Press, 1962). Societies: Fellow, A.P.S.; Fellow, A.N.S.; Fellow, A.A.A.S.; A.C.S.; A.S.M.
Nuclear interests: Radiation damage; solid state physics; magnetic alloys; Be and Be alloys.
Address: 35 Outer Drive, Oak Ridge, Tennessee 37830, U.S.A.

BILLS, C. Wayne, B.S., M.S., Ph.D. Born 1924. Educ.: Colorado State and Colorado Univs. Res. Chem., Los Alamos Sci. Labs., 1947-50; Geophys. R. and D., Grand Junction Operations Office (1956-57), Chem. Separation of Irradiated Fuel, Richland Operations Office (1957-58), U.S.A.E.C.; Deputy Director, Health and Safety Div. (1959-61), SL-1 Reactor Accident Recovery Co-ordination (1961-62), Director, Nucl. Technol. Div. (1962-), Idaho Operations Office, U.S.A.E.C.; Princeton Fellow in Public Affairs, 1962-63. Societies: A.N.S.; A.C.S.
Nuclear interests: Radiochemistry; Reactor physics; Nuclear accident investigation; Chemical processing of nuclear fuels; Uranium exploration methods.
Address: U.S. Atomic Energy Commission, P.O. Box 2108, Idaho Falls, Idaho 83401, U.S.A.

BILOUS, Olegh, Ph.D., Chem. E. Born 1927. Educ.: Ecole Polytech., Paris,and Minnesota Univ. At Lab. Central des Poudres, Paris, 1951-55; C.E.A., Paris, 1955-. Dept. Chem. Eng., Monash Univ., Melbourne.
Nuclear interest: Nuclear engineering.
Address: 15 Milton Street, Elwood, Victoria, Australia.

BILPUCH, Edward George, B.S., M.S., Ph.D. Born 1927. Educ.: North Carolina Univ. Res. Assoc. (1956-58) Visiting Asst. Prof. (1958-61), Asst. Director (1961-), Nucl. Structure Lab., Duke Univ. Society: A.P.S.
Nuclear interests: Total and capture cross-sections for neutrons in the kev region; Nuclear level spacings at and above the neutron binding energy; (p,γ) reactions for nuclei with $A < 60$.
Address: Box 674, Chapel Hill, N.C., U.S.A.

BILS, O. O., Dipl. Ing. School of Nucl. Eng., New South Wales Univ.
Address: New South Wales University, Box 1, Post Office, Kensington, New South Wales, Australia.

BIMBHAT, K. S. In Charge of Heavy Water Plant, Fertilizer and Heavy Water Factory, Fertilizer Corp. of India.
Address: Fertilizer and Heavy Water Factory, Naya Nangal, Punjab, India.

BINAY, Charles. Chairman, Cie. Française des Produits Liebig S.A.
Address: Cie. Française des Produits Liebig S.A., 15 rue de Genève, La Courneuve, (Seine), France.

BINET, Francis Emeric, Dr. Med., Cand. Phil. Born 1913. Educ.: Budapest and Szeged (Hungary) Univs. Sen. Tutor in Statistics, Melbourne Univ., 1950-56; Res. Officer (now Sen. Res. Sci.), Div. Animal Genetics (Poultry Res. Centre), C.S.I.R.O., 1955-. Societies: F.S.S., London; Member of the "Pugwash Seminar" (New South Wales); Statistical Soc. of Australia, New South Wales Branch; Foundation Member, Australian Mathematical Soc.; Genetics Soc. Australia; Australasian Div. Biometric Soc.
Nuclear interests: Influence of radiation on human and on plant-gene-pools; Disarmament and international co-operation for welfare and peace.
Address: Commonwealth Scientific and Industrial Research Organisation, Division of Animal Genetics, P.O. Box 90, Epping 2121, New South Wales, Australia.

BINET, Léon. Member, Conseil d'Enseignement, I.N.S.T.N., C.E.A.
Address: Commissariat à l'Energie Atomique, Conseil d'Enseignement de l'Institut National des Sciences et Techniques Nucléaires, B.P. No. 6, Gif-sur-Yvette (S. et O.), France.

BINFORD, Franklin T., B.S. Born 1917. Educ.: Pennsylvania State and Tennessee Univs. Consultant, Reactor Centrum Nederland; Supt., Development Dept., Operations Div., Oak Ridge Nat. Lab. Member, Exec. Board, Reactor Operations Div., and Member, Standards Com., A.N.S. Books: Contributed to Nucl. Eng. Handbook (McGraw Hill, 1958); Programming and Utilisation of Res. Reactors (I.E.A.E.) (Academic Press, 1962).
Nuclear interests: Reactor design, operation, and safety; Isotope production.
Address: 105 Enfield Lane, Oak Ridge, Tennessee, U.S.A.

BINGGELI, Edmond Marcel, Eng.-chem., D. Sci. Born 1924. Educ.: Inst. of Technol., Lausanne Univ. Asst., Lab. of Nucl. Res., E.P.U.L., 1949-53; Scholarship Fellow, Swiss Nat. Fund of Sci. Res., 1953-54; Attaché of Res., French Nat. Centre of Sci. Res., Paris, 1956-57; Eng., Bonnard et Gardel, Consulting Engineers, Lausanne. Societies: Swiss Soc. Phys.; Soc. Natural Sci.
Nuclear interests: Cosmic rays; Health physics; Site location; Nuclear power plant safety; Underground containment; Shielding.
Address: 29 Ch. Montolivet, Lausanne, Switzerland.

BINGHAM, Gordon. Res. Faculty, Dept. of Phys., Washington Univ., Seattle.
Nuclear interest: High energy physics.
Address: Washington University, Department of Physics, Seattle, Washington 98105, U.S.A.

BINKERT, Eduard, Dipl. Elec. Eng. Born 1899. Educ.: E.T.H., Zürich. Director, Elec. Supply, city of Berne. Member, Swiss Federal Atomic Commission; Administrative Councillor,

Suisatom S.A.; Pres., Schweizerischer Elektrotechnischer Verein; Member, Swiss Assoc. for Atomic Energy. Societies: S.E.V.; S.I.A.
Nuclear interests: Reactor design and management.
Address: 1 Meisenweg, 3303 Jegenstorf, Switzerland.

BINKLEY, Francis. Dr. Nucl. Res. Biochem. Dept., School of Medicine, Emory Univ.
Address: School of Medicine, Emory University, Atlanta 22, Georgia, U.S.A.

BINKS, Walter, M.Sc. Born 1904. Educ.: Manchester Univ. Director, Radiol. Protection Service (Ministry of Health and M.R.C.), Sutton, Surrey, 1953-; Sec., Protection against Ionizing Rad., M.R.C.; Sec., Advisory Com. (Radioactive Substances Act), Educ. and Sci. Dept. Societies: Fellow, Inst. Phys. and Phys. Soc.; British Inst. of Radiology; Hon. Member, Faculty of Radiologists.
Nuclear interests: Protection of workers and general public against somatic and genetic effects of ionising radiations.
Address: 69 Downs Wood, Epsom Downs, Surrey, England.

BINNER, W. Dipl. Ing. Eng., Direktion Technische-Wissenschaftliches, Referat "DTW", Elin-Union A.G.; Deleg. I.A.E.A. Seminar on Codes for Reactor Computations, Vienna, 25-29 April 1960; Deleg., I.A.E.A. Conference on Small and Medium Power Reactors, Vienna, 5-9 September, 1960.
Address: Elin-Union, 1-5 Volksgartenstrasse, Vienna 1, Austria.

BINNEY, Harry Augustus Roy, B.Sc.(Eng.). Born 1907. Educ.: London Univ. Director, B.S.I.
Nuclear interests: Preparation of national and international standards for atomic energy.
Address: 2 Park Street, London, W.1, England.

BINNING, Kenneth George Henry, M.A. Born 1928. Educ.: Oxford Univ. Administrative class, Home Civil Service 1950-52; H.M. Treasury, 1952-56; Private Sec. to Financial Sec. to Treasury, 1956-57; Administration Manager, Authority Health and Safety Branch, (1959-62), Deputy Sec., Culham Lab. (1962-65), U.K.A.E.A.; Special Asst. to Controller, Ministry of Technology, 1965-67; Sec. Mintech/U.K.A.E.A. Programmes Analysis Unit, 1967-.
Address: 72 Five Mile Drive, Oxford, England.

BINOPOULOS, D., Ph.D. Director, Phys. Dept., Alexandra Hospital, Isotope Lab., Athens.
Address: Alexandra Hospital, Isotope Laboratory, Vas. Sophias-K. Lourou Str., Athens, Greece.

BINSON, Boonrod, B.Eng., S.M. (M.I.T.), D.Sc. (Harvard). Born 1915. Educ.: Chulalongkorn, Wisconsin and Harvard Univs. and M.I.T. Head, Elec. Eng. Dept., Coll. of Eng., Chulalongkorn Univ., 1953-; Sec.-Gen., Nat. Energy Authority, 1956-. Society: A.I.E.E.
Nuclear interest: Nuclear power reactors.
Address: National Energy Authority, Pibultham Villa, Kasatsuk Bridge, Bangkok, Thailand.

BINSTOCK, Martin H., B.S. (Chem.), Master of Metal. Eng. Born 1922. Educ.: Rensselaer Polytech. Inst. Past Pres., local chapter, A.S.M. Societies: A.N.S.; RESA.
Nuclear interests: Metallurgy of nuclear fuels and other materials for reactor and thermoelectric use.
Address: P.O. Box 309, Canoga Park, California, U.S.A.

BIONDI, Leonardi. Dr. in Eng. Born 1923. Educ.: Politecnico of Milan. Montecatini officer in charge for Nucl. Sect. Tech. Director, Italatom, 1960-. Society: Italian Nucl. Technicians Assoc.
Nuclear interests: Fuel elements manufacturing; Core design.
Address: 35 Viale Filippetti, Milan, Italy.

BIRCHER, Louis J., A.B., B.S. (Ed.), M.A., Ph.D. Born 1892. Educ.: Missouri and Chicago Univs. Prof. of Phys. Chem. and Dept. Chairman (1921-67), now Prof. Emeritus, Vanderbilt Univ. Chairman Nashville Sect. and Counsellor, A.A.A.S. Book: Physical Chem. - A Brief Course. Societies: A.C.S.; A.A.A.S.; Fellow, Tennessee Acad. Sci.
Nuclear interest: Teaching.
Address: Chemistry Department, Vanderbilt University, Nashville 5, Tennessee, U.S.A.

BIRD, Peter M., M.Sc. (Queen's), Ph.D. (Leeds). Born 1927. Educ.: Queen's (Kingston, Ontario) and Leeds Univs. Chief, Rad. Protection Div., Dept. Nat. Health and Welfare, 1950-. Societies: A.P.E.O.; Canadian Assoc. Radiologists; H.P.S.; Hospital Physicists Assoc.; Canadian Assoc. Physicists.
Nuclear interests: Radiation protection activities on a national basis designed to ensure the health and safety of Canadians working with radioisotopes and X-rays, and to assess and make recommendations concerning the radiation exposure of members of the public.
Address: Radiation Protection Division, Department of National Health and Welfare, Brookfield Road, Confederation Heights, Ottawa 8, Ontario, Canada.

BIRD, Raymond H., B.Sc. Tech. (Hons.). Born 1926. Educ.: Manchester Univ. Tech. Asst. (1947), Asst. Development Eng. (1950), Simon-Carves, Ltd.; Deputy Chief Eng., Steam Raising Plant (1955), Chief Performance Eng. (1958), G.E.C.-Simon-Carves Atomic Energy Group; Manager, Rep. Dept. (1961), Tokai-Mura Project Design Eng. (1963-), United Power Co. Ltd. Society: A.M.I.Mech.E.
Nuclear interests: Reactor and steam plant performances.
Address: 2 Sylvester Avenue, Chislehurst, Kent, England.

BIRK, Meir, Ing. (E.E.), Ph.D. Born 1924. Educ.: Technion (Haifa), Weizmann Inst. of Sci. (Rehovoth), and Hebrew Univ. (Jerusalem). Res. Officer, Ballistic measurements, Sci. Dept., Israeli Ministry of Defence, 1949-54; Member (1954), now Assoc. Prof. Phys. Dept., Weizmann Inst. of Sci., Rehovoth, Israel.
Nuclear interests: Nuclear electronic instrumentation; Nuclear physics.
Address: Physics Department, Weizmann Institute of Science, Rehovoth, Israel.

BIRKEL, Ralph Anton, B.Sc., M.S. Born 1928. Educ.: Northwestern and Pittsburgh Univs. Reactor Operations Supv., A.E.C., Hanford Works, G.E.C., 1950-57; Sen. Eng., Large Plant Eng. (1957-58), Tech. Asst. to Manager, Plant Development Sect. (1958-60), Sen. Supervisory Field Eng. and A.E.C. Licensed Power Reactor Operator, Nucl. Power Service Sect. (1960-61), Westinghouse Atomic Power Dept.; Reactor Project Manager, Allis-Chalmers Atomic Power Dept., 1961-62; U.S.A.E.C., 1962-, Now Deputy Chief, Reactor Operations Sect., Reactor Licensing Div. Societies: Atomic Industrial Forum; A.N.S.
Nuclear interests: Research and development management; Reactor operation.
Address: 9729 Connecticut Avenue, Kensington, Maryland 20795, U.S.A.

BIRKELUND, Oscar R. Born 1925. Educ.: Bergen Tekniske Skole, N.K.I., Stockholm. Chemist, Inst. for Atomenergi, Kjeller, Norway. Books: Investigations of the Chem. States of Carrierfree Phosphorus-32 as extracted into water from Pile-Irradiated sulphur (RICC/262). Preparation of Chromium-51 by Szilard-Chalmers Reaction and Separation by Ion Exchange Techniques (KR-40); Tagging of Human Red Cells with Chromium-51 (KR-63). Society: Norwegian Chem. Soc.
Nuclear interest: Development and production of isotopes.
Address: 3 Forskerhagen, Skedsmokorset, Norway.

BIRKHOFER, Adolf, Dipl. Ing. Electronics, theoretical phys. Born 1934. Educ.: Munich T.H. and Innsbruck Univ. Siemens-Halske A.G., Munich, 1957-58; Technische Uberwachungsverein, Bayern, 1959-; Mitglied, Reaktor-Sicherkeitskommission, Federal Ministry for Sci. Res. Societies: V.D.I.; V.D.E.
Nuclear interests: Reactor safety; Reactor physics; Reactor design (engineering).
Address: 65 Prinzenstrasse, Munich 19, Germany.

BIRKHOFF, Garrett, A.B. (Harvard), Hon. Ph.D. (Mexico, Lille). Born 1911. Educ.: Harvard and Cambridge Univs. Prof. Pure and Appl. Maths., Harvard Univ., 1947-. Consultant: Los Alamos Sci. Labs., Westinghouse Atomic Power Div., Argonne Nat. Lab., Brookhaven Nat. Labs., Gen. Motors Res. Labs. Pres., Soc. Ind. Appl. Maths., 1967-68. Books: 5 books; co-editor, Symposium Nucl. Reactor Theory (American Maths. Soc., 1961), etc. Societies: American Mathematical Soc.; A.P.S.; A.N.S.
Nuclear interests: Reactor calculations; Boiling reactors.
Address: 45 Fayerweather Street, Cambridge 38, Massachusetts, U.S.A.

BIRKHOFF, Robert D., B.S., Ph.D. Born 1925. Educ.: Northwestern Univ. and M.I.T. Assoc. Prof. Phys., Tennessee Univ., 1949-55; Sen. Phys., O.R.N.L., 1955-. Book: The Passage of Fast Electrons through Matter, vol. 34, Handbuch der Physik (1958). Societies: A.P.S.; H.P.S.
Nuclear interests: Interaction of radiation in matter; Health physics; Nuclear instrumentation.
Address: Health Physics Division, Oak Ridge National Laboratory, P.O. Box X, Oak Ridge, Tennessee, U.S.A.

BIRKNER, Rudolf, Prof. Dr. med. Born 1911. Educ.: Medizinstudium, Berlin Univ. Dirigierender Arzt. Society: Deutsche Röntgengesellschaft.
Nuclear interests: Radioisotopes for diagnostic and therapy; Telecobalt therapy; Development of a complete automatic method to the establishment and drawing by body-isodoses of stationary and movable source of radiation.
Address: Röntgeninstitut und Strahlentherapeutische Abteilung am Städt, Krankenhaus Moabit, 21 Turmstrasse, Berlin NW 21, Germany.

BIRKS, John Betteley, M.A., D.Sc. (Oxon.), Ph.D., (Glas.). Born 1920. Educ.: Oxford Univ. Prof. Phys., Rhodes Univ., S. Africa, 1951-54; Res. Manager, British Dielectric Res., London, 1954-57; Reader in Phys., Manchester Univ., 1957-; Visiting Prof., Louisiana State Univ., 1965-66. Books: Scintillation Counters (Pergamon Press, 1953); Theory and Practice of Scintillation Counting (Pergamon Press, 1964) Societies: F. Inst. P.; A.M.I.E.E.
Nuclear interests: Nuclear instrumentation, particularly scintillation counters; Research on the scintillation process in organic systems.
Address: 39 Spath Road, Didsbury, Manchester 20, England.

BIRNBAUM, Milton, A.B., M.S., Ph.D. Born 1920. Educ.: Brooklyn Coll. and Maryland Univ. Formerly with N.R.L., Washington, D.C.; Bulova R. and D. Lab., Woodside, N.Y.; Polytech. Inst. Brooklyn; At present with Aerospace Corp., Los Angeles. Societies: A.P.S.; A.A.P.T.; Fellow, Phys. Soc., London.
Nuclear interests: Mössbauer effects; Nuclear magnetic resonance; Nuclear hyperfine structure in electron paramagnetic resonance; Nuclear spectra; Neutron binding energy; Cosmic ray studies using nuclear emulsions; Electron synchrotron design; Lasers.
Address: 4904 Elkridge Drive, Palos Verdes Peninsula, California, U.S.A.

BIRO, George G., M.S.C.E., M.S.N.E., E.Sc.D. (Columbia) Born 1911. Educ.: Deutsche T.H., Prague, Brno and Columbia Univs. Concerned with shielding and special nucl. and structural problems in nucl. power generating stations, Consulting Eng. in Nucl. and Structural Projects (1968-), Gibbs and Hill, Inc. Consulting Engs., New York. Treasurer (1962-65), Chairman (1965-66) and presently director, N.Y. Metropolitan Sect., Treasurer Shielding Div. (1966-67), Exec. Com. of A.N.S. Intersect.Com. (1966-69), A.N.S.; Meritorious Service Award, N.Y. Metropolitan Sect., A.N.S., 1966. Books: Modern Methods of Calculation of Continuous Structures (Istanbul Univ. Edition, 1949); Simulation of Rad. Transport, A Semi-Analytical Model (Columbia Univ., 1967); Co-author, Eng. Compendium on Rad. Shielding (Sponsored by I.A.E.A., Springer Verlag, 1968). Societies: A.N.S.; A.P.S.; A.S.C.E.; N.Y. Acad. of Sci.
Nuclear interests: Nuclear shielding, and structural problems, design, operation and accident analysis in nuclear power plants. Development of scientific-engineering programmes.
Address: 41 Fairview Avenue, Great Neck, New York 11023, U.S.A.

BIROT, André, L. ès Sc. Born 1936. Educ.: Toulouse Univ. Asst. Nucl. Eng., Faculty of Science, Toulouse Univ.
Nuclear interest: Radioactivity in the atmosphere;
Address: Centre de physique nucléaire, Faculté des Sciences, Toulouse, France.

BIRTS, Leslie, B.Sc. (Eng.). Born 1915. Educ.: London Univ. Development Eng., Marine R. and D., Shell Petroleum Co.; Asst. Chief Eng., Reactor Technol. Div., Reactor Group, U.K.A.E.A., Risley. Societies: A.M.I.Mech.E.; M.I.Mar.E.
Nuclear interest: Marine reactor plants.
Address: Alpbach, Stocks Lane, Over Peover, Knutsford, Cheshire, England.

BIRYUKOV, O. V. Papers: Co-author, Closed Magnetic Trap with "Sirius" Type Helical Field (Atomnaya Energiya, vol. 23, No. 2, 1967); co-author, Longitudinal Magnetic Field of the Sirius Stellarator Reecetrack (ibid, vol. 23, No. 2, 1967); co-author, Divertor of the Sirius Stellarator (ibid, vol. 23, No. 2, 1967); co-author, On Possibility of Plasma Injection into Closed Magnetic Trap through Divertor (ibid, vol. 24, No. 1, 1968).
Address: Academy of Sciences of the U.S.S.R., 14 Leninsky Prospekt, Moscow V-71, U.S.S.R.

BIS, Jiri, M.M.E. Born 1941. Educ.: Moscow Power Inst. Res. Group, Nucl. Power Plant Div. Skoda Concern, Pilzen. Society: Czechoslovak Tech. and Sci. Assoc.
Nuclear interests: Heat transfer calculations; Transient processes of nuclear power plants; Reactor safety.
Address: 16 Zahradni, Pilzen, Czechoslovakia.

BISBY, Harry, B.Sc. (Special Phys. Hons.), A.K.C., C. Eng. Born 1915. Educ.: London Univ. Senior Principal Sci., Electronics Div., U.K.A.E.A., A.E.R.E., Harwell. Authority Rep. with British Scientific Instrument Res. Assoc. Society: I.E.E.
Nuclear interests: Analytical techniques and instrumentation and modular unit systems.
Address: Barcroft House, 59 High Street, Drayton, Berks., England.

BISCAIA, Aires Antonio Argel MELO. See MELO BISCAIA, Aires Antonio Argel.

BISCHOF, Werner, Assessor. Born 1928. Educ.: Munster and Göttingen Univs. Asst., Inst. of Public Internat. Law, Göttingen Univ., Department: Atomic Energy Law, 1958-. Books: Co-author, Göttingen Atomrechtskatalog; Responsible Editor, Series Kernenergierecht; Co-author, Internationale Atomhaftungskonventionen.
Nuclear interests: National and international atomic energy law; Comparative law; Legal problems of international co-operation in the field of atomic energy.
Address: 215 Holbornstrasse, Sudheim bei Göttingen, Germany.

BISHOP, Amasa Stone, B.S., Ph.D. Born 1920. Educ.: California Inst. Technol., California (Berkeley), Univ. Director, Controlled Thermonucl. Branch, U.S.A.E.C., 1953-56; European Sci. Rep., U.S.A.E.C., 1956-58; Head of U.S. Deleg. to Euratom, 1958-60; Staff of Plasma Phys. Lab., Princeton Univ., 1961-. Chairman, Standing Com. for Controlled Thermonucl. Res., U.S.A.E.C. Book: Project Sherwood — The U.S. Programme in Controlled Fusion (Addison Wesley, 1958). Societies: A.P.S.; A.A.A.S.
Nuclear interests: Meson physics; Controlled fusion; Nuclear structure; Reactor technology.
Address: Plasma Physics Laboratory, Princeton University, Box 451, Princeton, N.J., U.S.A.

BISHOP, George Robert, M.A., D.Phil. Born 1927. Educ.: Oxford Univ. I.C.I. Fellowship, Clarendon Lab., Oxford, 1951-53; Res. Fellow, St. Antony's Coll., Oxford, 1953-56; Boursier, C.E.A.,Ecole Normale Supérieure, Paris, 1955-59; Adjoint Scientifique du Directeur du Laboratoire de l'Accélèrateur Linéaire, Orsay, France, 1959-62. Professeur Associé, Faculté des Sci. Centre d'Orsay, 1962-64; Kelvin Prof. of Natural Philosophy, Glasgow Univ., 1964-; Director, Kelvin Lab., N.E.L., East Kilbride, Scotland, 1964-. Books: Contribution to Beta and Gamma Ray Spectroscopy (North Holland Publishing Co., 1955); Handbook of Phys. (Springer-Verlag, 1957).
Nuclear interests: Nuclear structure and electromagnetic interactions; Nuclear reactions; Oriented nuclei and polarisation in reactions; Photonuclear reactions; High energy physics; Radioactivity; Electron accelerators.

BIS

Address: Department of Natural Philosophy, Glasgow University, Glasgow, W. 2, Scotland.

BISKUPSKI, Olgierd, M.S. Formerly Head, Sci. Res. Dept., now Sci. Sec. of the State Council, Office of Govt. High Commissioner for Atomic Energy, Poland.
Address: Office of Government High Commissioner for Atomic Energy, Palace of Culture and Science, 18th floor, Warsaw, Poland.

BISMARCK, Fürst Otto VON. See VON BISMARCK, Fürst Otto.

BISPLINGHOFF, Raymond L. Former Assoc. Administrator for Advanced Res. and Technol., N.A.S.A.; Pres., Case Inst. of Technol., 1965-.
Address: Case Institute of Technology, University Circle, Cleveland 6, Ohio, U.S.A.

BISSET, George. Formerly Trustee, Power Reactor Development Co.; Director, Atomic Power Development Associates Inc.
Address: Atomic Power Development Associates Inc., 1911 First Street, Detroit 26, Michigan, U.S.A.

BISSONNETTE, Paul E., B.M.E. Born 1918. Educ.: Rensselaer Polytech. Inst. Society: A.S.M.E.
Nuclear interests: Responsible for all activities of a new concept of nuclear power plant located at the National Reactor Testing Station. Activities include construction of facilities, test operation of the propulsion plant, and the personnel training programme; and responsible for the nuclear safety and safeguards of the reactor plant, as reflected in the design, maintenance, test and utilisation of all equipment and systems.
Address: General Electric Company, Knolls Atomic Power Laboratory, P.O. Box 1846, Idaho Falls, Idaho, U.S.A.

BISWAS, Sukumar, M.Sc., D. Phil. (Calcutta), Ph.D. (Melbourne). Born 1924. Educ.: Calcutta and Melbourne Univs. Unesco Fellowship, Govt. of Australia, 1950-52; Res. Fellow, (1952-57) Fellow (1957-63), Reader (1963-65), Assoc. Prof. (1965-), Tata Inst. of Fundamental Res., Bombay; Res. Assoc. Minnesota Univ., 1959-61; N.A.S.A.- N.A.S. Sen. Postdoctoral Res. Assoc., Goddard Space Flight Centre, U.S.A., 1961-63. Societies: A.P.S.; A.G.U.
Nuclear interests: High energy nuclear interactions; Elementary particle physics; Cosmic ray physics and energetic solar particle physics.
Address: Tata Institute of Fundamental Research, Homi Bhabha Road, Colaba, Bombay-5, India.

BITTENCOURT, Epifanio F.S. Administration Dept., Nat. Nucl. Energy Commission of Brazil.
Address: National Nuclear Energy Commission of Brazil, 90 (ZC-82) Rua General Severiano, Rio de Janeiro - GB, Brazil.

BITTENCOURT, Helio F.S., LL.B. Born 1923. Educ.: Brazil Univ. Career diplomat (Washington 1950-53 and 1955-58; Costa Rica 1953-54; Vienna 1960-61; Prague 1962-64); Asst. to the Chairman, Nat. Nucl. Energy Commission, 1958-60; Formerly Chef-de-Cabinet to the Chairman, Nat. Nucl. Energy Commission, Brazil; Alternate Governor for Brazil, Board of Governors, I.A.E.A., 1959-61 and 1964-; Resident Rep. to I.A.E.A., 1960-61 and 1966-.
Nuclear interests: International relations, nuclear energy legislation, organisation and administrative matters.
Address: Permanent Mission of Brazil to the International Atomic Energy Agency, 1 Am Lugeck, Vienna I, Austria.

BIXEL, David. Formerly Phys., Battelle Memorial Inst., now Res. Sci.-Eng., Battelle-Columbus Labs.
Address: Battelle-Columbus Laboratories, Applied Nucl. Physics Division, 505 King Avenue, Columbus, Ohio 43201, U.S.A.

BIZOT, Jean Claude. Maitre de Conferences, Inst. du Radium, Faculte des Sci., Paris Univ.
Address: Institut du Radium, Laboratoire Curie, 11 rue Pierre Curie, Paris 5, France.

BIZZELL, Oscar M. Formerly Chief, Isotope Technol. Development Branch, Office of Isotope Development, now Chief, Tech. Utilisation Branch, Civilian Application Div., U.S.A.E.C. Delegate, I.A.E.A. Conference on the use of Radioisotopes in the Phys. Sci. and Ind., Copenhagen, 6-17 September, 1960. Society: A.N.S.
Address: 8801 Walnut Hill Road, Gaithersburg, Maryland 20760, U.S.A.

BJARNGARD, Bengt E., Dr. Director of Rad. Phys., Controls for Rad., Inc.
Address: Controls for Radiation, Inc., 130 Alewife Brook Parkway, Cambridge 40, Massachusetts, U.S.A.

BJERGE, Torkild, Chem. Eng., M.Sc. (Physics), Ph.D. Born 1902. Educ.: Tech. Univ. Denmark and Copenhagen Univ. Prof. Phys., Tech. Univ. Denmark, 1939-56; Administrative Head, Atomic Energy Res. Station, Riscoe, Denmark, 1956-. Books: Several textbooks on physics (electricity, light, atomic physics). Society: Acad. Tech. Sciences, Copenhagen.
Address: Atomic Energy Research Station, Risoe, Roskilde, Denmark.

BJERKEBO, Harry W. Direktor, Nat. Atomic Energy Council.
Address: National Atomic Energy Council, P.O. Box 40, Kjeller, Norway.

BJORCK, Olle, Sec., Reactor group, Swedish Assoc. of Metal Working Industries. Sec., Swedish Nucl. Ind. Group.
Address: Sveriges Mekanförbund, Fack, Stockholm 16, Sweden.

BJORGERD, Anders L., E.E. Born 1920. Educ.: Roy. Inst. of Technol. Elec. Eng. (1948-58), Chief Eng. (1958-63), Director, Operations (1963-), and Vice Pres., South Swedish Power Co., Ltd.
Address: South Swedish Power Company Limited, 1 Carl Gustafs väg, Malmö 5, Sweden.

BJORKLAND, John Alexander, B.S. (Phys.), M.S. (Phys.), Ph.D. (Eng. Phys.). Born 1923. Educ.: Chicago Univ. and N.C.S.C. Societies: A.P.S.; A.N.S.
Nuclear interest: Reactor instrumentation and control.
Address: 1005 Harvard Terrace, Evanston, Illinois 60202, U.S.A.

BJORKLUND, Frank E., B.S., M.S. Born 1922. Educ.: Utah Univ. Phys., California R. and D., 1952-54; Phys., Lawrence Rad. Lab., 1954-61. Society: A.P.S.
Nuclear interest: Nuclear physics.
Address: 788 Estates, Livermore, California, U.S.A.

BJÖRNERSTEDT, Rolf Gustaf, Ph.D. Born 1926. Educ.: Uppsala and Stockholm Univs. Head, Nucl. Phys. Dept., Res. Inst. of Nat. Defence, Sweden, 1957-. Book: Bone and Radiostrontium (Stockholm, Almqvist and Wiksell, 1958). Papers: Papers on health hazards from fission products and fallout. Society: H.P.S.
Address: Research Institute of National Defence, Stockholm 80, Sweden.

BLACK, Alexander A. Vice Pres. Acting Gen. Manager of Military Products Div., Fruehauf Corp.
Address: Fruehauf Corp., Detroit 32, Michigan 48232, U.S.A.

BLACK, Archibald Niel, M.A. Born 1912. Educ.: Cambridge Univ. Prof. Mech. Eng., Southampton Univ., 1950-. Society: M.I.Mech.E.
Nuclear interest: Nuclear engineering as part of under-graduate and post-graduate studies.
Address: 11 Pine Road, Eastleigh, Hampshire SO5 1LQ, England.

BLACK, Arthur L., B.Sc., Ph.D. Born 1922. Educ.: California Univ. Instructor and Jun. Biochem. (1951-52), Asst. Prof. and Asst. Biochem. (1952-57), Assoc. Prof. and Assoc. Biochem. (1957-62), Prof., Physiol. Chem. (1962-), School of Veterinary Medicine, California Univ., Davis. Societies: American Soc. Biol. Chem.; A.C.S.; American Inst. Nutrition.
Nuclear interests: Using isotopes as tracers to study intermediary metabolism in intact animals, especially ruminants; Utilization of fatty acids, carbohydrates and amino acids for energy and for biosynthesis.
Address: 891 Linden Lane, Davis, California, U.S.A.

BLACK, Gordon, B.Sc., Ph.D., D.I.C. Born 1923. Educ.: Durham and London Univs. Res. Phys., British Sci. Instrument Res. Assoc., 1950-56; Tech. Manager, Central Tech. Services, Reactor Group, U.K.A.E.A., Risley, 1956-; Part-time Prof. Automatic Data Processing, Faculty of Technol., Manchester Univ., 1964-. Society: F. Inst. P.
Nuclear interests: Electronic computers; Use of large-scale computing machinery in system optimisation.
Address: Overdale, Greenhill Road, Timperley, Cheshire, England.

BLACK, James Francis, Ph.D. (Princeton), M.A. (Princeton), B.S. (California). Born 1919. Educ.: Princeton and California Univs. Group Leader (1955-57), Res. Assoc. (1957-60), Sen. Res. Assoc. (1960-67), Sci. Adviser (1967-), Esso Res. and Eng. Co. Vice-Chairman, U.S.A.E.C. Advisory Com. on Isotopes and Rad. Development; Vice-Chairman, New Jersey State Commission on Rad. Protection; Vice-Chairman, Isotopes and Rad. Div., A.N.S. Societies: A.C.S.; A.N.S.; Nucl. Eng. Div., A.I.Ch.E.; American Meteorological Soc.
Nuclear interests: Radiation chemistry and radioisotope applications; Organic moderator-coolants.
Address: Products Research Division, Esso Research and Engineering Company, P.O. Box 51, Linden, New Jersey 07036, U.S.A.

BLACK, Paul Joseph, B.Sc., Ph.D. Born 1930. Educ.: Manchester and Cambridge Univs. Mr. and Mrs. John Jaffe Donation Student, Roy. Soc., 1953-56; Lecturer (1956-64), Sen. Lecturer (1964-66), Reader (1966-), Birmingham Univ. Society: F. Inst. P.
Nuclear interest: Gamma and X-ray resonance scattering.
Address: Physics Department, The University, Birmingham 15, England.

BLACK, Robert Earl, B.A., M.S. Born 1931. Educ.: Manchester Coll. and Vanderbilt Univ. Res. Phys., Gen. Motors Res. Lab., 1955-. Now Sen. Res. Assoc. Phys., Bendix Res. Lab.
Nuclear interest: Industrial applications of radioisotopes.
Address: 32366 Tareyton Avenue, Farmington, Michigan 48024, U.S.A.

BLACK, Rudolph A., B.S. (Geological Eng.), M.S. (Geophys.). Born 1924. Educ.: Washington Univ., St. Louis. Geophys., U.S. Geological Survey, 1950-57; Assoc. Prof., Mining Eng. Missouri School of Mines, Rolla, 1957-60; Geophys., U.S. Geological Survey, 1960-63; Programme Manager, Vela Uniform, Advanced Res. Projects Agency, U.S. Dept. of Defence, 1963-. Alternate Member, Federal Council Sci. and Technol. Com. on Solid Earth Sci.; Member, Intersoc. Com. for Rock Mechanics. Societies: A.G.U.; Soc. of Exploration Geophys.
Nuclear interest: Phenomenology of nuclear explosions, their effects on geologic materials and seismic signals produced by such

explosions.
Address: Advanced Research Projects Agency, Nuclear Test Detection Office, Pentagon, Washington, D.C., U.S.A.

BLACKBURN, Charles Ruthven Bickerton, M.D. (Sydney), F.R.C.P. (London), F.R.A.C.P. Educ.: Sydney Univ. Prof. Medicine, Sydney Univ., 1956-.
Nuclear interests: Research; Diagnosis; Treatment; Teaching in medicine.
Address: Department of Medicine, Sydney University, Sydney, N.S.W., Australia.

BLACKBURN, David Alexander, B.Sc., Ph.D. Born 1933. Educ.: Edinburgh Univ. Res. Assoc., Illinois Univ.; now Lecturer, Phys. Dept., York Univ. Society: Assoc. Inst. Phys. and Phys. Soc.
Nuclear interests: Application of tracer techniques to studies of diffusion in metals; Particular interest is in diffusion in grain-boundaries, and in diffusion in electric fields and thermal gradients.
Address: Department of Physics, York University, Heslington, York, England.

BLACKBURN, R. W. Asst. Sci. Adviser, Atomic Energy Control Board, Canada.
Address: Atomic Energy Control Board, P.O. Box 1046, Ottawa, Ontario, Canada.

BLACKETT, Patrick Maynard Stuart, C.H., M.A., Hon. D.Sc. (New Delhi, Reading, Strasbourg, Belfast, Leeds, Durham, Manchester, Oxford), Hon. Sc.D. (Cantab.), Hon. LL.D. (Glasgow, Dalhousie, St. Andrews), Hon. F. Inst. P. Born 1897. Educ.: Cambridge Univ. Langworthy Prof. Phys., Manchester Univ., 1937-53; Prof. Phys., Imperial Coll., London Univ., 1953-; Pro-Rector, 1961-. Board Member, Nat. Res. Development Corp.; Councillor, Overseas Development Inst.; Pres., British Assoc. for the Advancement of Sci., 1957-58.; Member, Advisory Council on R. and D., Ministry of Power. Books: Rayons Cosmiques (1934); Lectures on Rock Magnetism (1956). Societies: F.R.S.; Hon. Fellow, Indian Acad. of Sci.; Member, Berlin Acad. of Sci.; Hon. Fellow, Weizmann Inst. of Sci., Israel.
Nuclear interests: Plasma physics, etc.
Address: Department of Physics, Imperial College of Science and Technology, London S.W.7, England.

BLACKLEDGE, James P. Assoc. Director, Denver Res. Inst., Denver Univ.
Address: Denver University, University Park, Denver 10, Colorad, U.S.A.

BLACKMAN, Alfred C., B. Mech. Eng. Born 1908. Educ.: Cornell Univ. Chief, Div. of Ind. Safety, State of California, 1947-59; Consultant, Joint Congressional Com. on Atomic Energy, U.S. Govt., 1959; now Managing Director, American Soc. of Safety Eng., Chicago. Trustee, Veterans of Safety. Society: A.I.H.A.

Nuclear interests: Applications in industry; Nuclear protection design; Management; Education and training programmes for safety.
Address: 920 Sheridan Road, Wilmette, Illinois 60091, U.S.A.

BLACKWELL, H. Jack. Formerly Manager, Medina (San Antonio, Texas) Area, now Manager, Amarillo (Tex.) Area Office, U.S.A.E.C.
Address: c/o Albuquerque Operations Office, U.S.A.E.C., Post Office Box 5400, Albuquerque, New Mexico, U.S.A.

BLADY, John V., M.D. Sec., American Radium Soc., 1968-69.
Address: The Parkway House, 2201 Benjamin Franklin Parkway, Philadelphia, Pennsylvania 19130, U.S.A.

BLAGOVOLIN, S. M. Papers: Co-author, BN-350 and BOR Fast Reactors (Atomnaya Energiya, vol. 21, No. 6, 1966); co-author, Sodium Technol. and Equipment of BN-350 Reactor (ibid, vol. 22, No. 1, 1967).
Address: Academy of Sciences of the U.S.S.R., 14 Leninsky Prospekt, Moscow V-71, U.S.S.R.

BLAHD, William H., M.D. Born 1921. Educ.: Western Reserve, Arizona and Tulane Univs. Chief, Radioisotope Service, Veterans Administration, Los Angeles, 1956-.; Prof., Medicine, School of Medicine, California Univ. Board of Trustees, Soc. of Nucl. Medicine. Books: Co-author, The Practice of Nuclear Medicine (C.C. Thomas, 1958); Nuclear Medicine (McGraw-Hill, 1965). Societies: Rad. Res. Soc.; H.P.S.
Nuclear interests: Development of techniques for the application of radioisotope tracers to clinical medicine; Basic research studies on the cause of primary muscle diseases; Study of body composition.
Address: Veterans Administration Centre, Los Angeles, California 90073,U.S.A.

BLAINEY, Alan, B. Eng., Ph.D. Born 1915. Educ.: Liverpool Univ. A.E.R.E.,Harwell, 1946-55; De Beers Consolidated Mines Ltd., South Africa, 1955-61, Boart and Hard Metal Products (S.A.) Ltd., 1962-. Society: Fellow, Inst. Metallurgists, London.
Nuclear interests: Metallurgy, in particular the extraction and fabrication of metals and materials for nuclear fuel elements and reactor construction.
Address: High Pressure Materials Laboratory, Box 104, Crown Mines, Transvaal, S. Africa.

BLAINVILLAIN, Maurice. Directeur Technique, Sté. de Raffinage d'Uranium (S.R.U.).
Address: Société de Raffinage d'Uranium (S.R.U.), 23 boulevard Georges Clemenceau, Courbevoie, (Seine), France.

BLAIR, Henry Alexander, B.Sc. (Manitoba), M.Sc. (Manitoba), Ph.D. (Princeton). Born 1900. Educ.: Manitoba and Princeton Univs. Prof., Rochester Univ. School of Medicine,

1948-; Director, Dept. of Rad. Biol. and Atomic Energy Project, 1948-65. Editorial Board, Proc. Soc. Exptl. Biol. and Medicine, 1965-. Book: Editor, Biol. Effects of External Rad. (1954). Societies: A.P.S.; American Physiological Soc.; Soc. for Exptl. Biol. and Medicine; Rad. Res. Soc.; H.P.S.
Nuclear interest: Biological effects of radiation.
Address: 1392 Clover Road, Rochester 10, New York, U.S.A.

BLAIR, John S., B.S. (Yale), M.S. (Illinois), Ph.D. (Illinois). Born 1923. Educ.: Yale and Illinois Univs. Prof., (1952-),Formerly in Nucl. Phys. Lab. and Cosmic Ray and High Energy Lab., Washington Univ., Seattle. Society: A.P.S.
Nuclear interests: Theoretical nuclear physics; Nuclear reactions.
Address: Department of Physics, University of Washington, Seattle, Washington 98105, U.S.A.

BLAIR, Robert Collyer. Formerly Manager, Savannah River Operations Office, U.S.A.E.C. Now Consultant.
Address: P.O. Box 280, Aiken, South Carolina 29803, U.S.A.

BLAIR, Sidney Martin, B.Sc. M.Sc. Born 1898. Educ.: Birmingham and Alberta Univs. Joined (1949), now Vice Chairman and Director, Canadian Bechtel Ltd. Director: Bechtel Internat. Ltd.; Eastern and Chartered Trust Co.; Alberta and Southern Gas Co. Ltd.; Alberta Natural Gas Co.; Pres. (-1968) Director, Canadian Nucl. Assoc. Societies: A.I.Ch.E.; A.S.M.E.; Inst. Chem. Eng., London.
Address: Canadian Bechtel Ltd., 25 King Street West, Toronto 1, Ontario, Canada.

BLAIZOT, Jean. Member, Conseil d'Enseignement, I.N.S.T.N., C.E.A.
Address: Commissariat a l'Energie Atomique, 29-33 rue de la Federation, Paris 15, France.

BLAKE, Charles Andrew, Jr., B.S., Ph.D. Born 1922. Educ.: Yale UNiv. With Union Carbide Nucl. Co. 1950-. Society: A.C.S.
Nuclear interests: Chemistry of reactor systems, solvent extraction technology.
Address: Union Carbide Nuclear Company, Oak Ridge National Laboratory, Oak Ridge, Tennessee, U.S.A.

BLAKEBOROUGH, Christopher David, B.Sc. Born 1933. Educ.: Durham Univ. Chief Eng. (1961-67), Gen. Manager Control Valve Div. (1967-), J. Blakeborough and Sons Ltd. Society: A.M.I.Mech.E.
Nuclear interests: Manufacture of valves for nuclear power stations.
Address: Sherwood House, Brighouse, Yorkshire, England.

BLAKELY, John James Andrew, M.B., B.S., M.D., D.P.H., D.I.H. Born 1919. Educ.: London Univ. (Middlesex Hospital). Medical Officer Windscale Works (1949-58), Medical Officer in Charge, Chapelcross Works (1958-), U.K.A.E.A. Society: Soc. Occupational Medicine.
Nuclear interest: Radiation medicine.
Address: Chapelcross Works, Annan, Dumfriesshire, Scotland.

BLAKESLEE, T. M. Chief Elec. Eng. and Asst. Manager, Los Angeles Dept. of Water and Power.
Address: Los Angeles Department of Water and Power, P.O. Box 111, Los Angeles, California 90054, U.S.A.

BLANC, Daniel Louis Antoine, Agrégé de l'Université, Dr. es Sci. Born 1927. Educ.: Caen and Lille Univs., Collège de France. Prof. Nucl. Phys., Faculté des Sciences Toulouse, 1960-. Director, Centre of Atomic and Nucl. Phys. of the Faculty, 1958-. Chairman, Sté. française de Radioprotection, 1968. Books: Detecteurs de particules, compteurs et scintillateurs (Paris,Masson, 1959); co-auteur, Eléments de Physique Nucléaire (Paris, Masson, 1960); Propriétés physiques des cristaux (traduction de l'ouvrage de Nye) (Paris, Dunod, 1961); Les Radioéléments (Paris, Masson, 1966) Societies: Sté. française de Physique; Sté. italienne de Physique; Sté. française de Radioprotection.
Nuclear interests: Detection and dosimetry of nuclear particles; Neutron physics; Mass spectroscopy, ion sources; Atomic physics.
Address: Faculté des Sciences, 118 route de Narbonne, Toulouse, France.

BLANC, Gene A., B.S. (Phys. and Mathematics). Born 1927. Educ.: San Jose State Coll. Phys., Health Phys., and Asst. Safety Eng., Livermore Res. Lab., 1951-54; Staff, Isotopes Div. (1954-57), Asst. Director, Region IV, Div. of Compliance (1957-61), Director, Region II, Div. of Compliance (1961-63), U.S.A.E.C.; Coordinator, Atomic Energy Development and Rad. Protection, State of California, 1963-. Societies: Atomic Ind. Forum; A.N.S.; H.P.S.; A.A.A.S.
Nuclear interests: Promoting the use of atomic technology in all fields of endeavour; Nuclear safety; Management.
Address: Office of Atomic Energy Development and Radiation Protection, Room 1311, 1416 Ninth Street, Sacramento, California 95814, U.S.A.

BLANC, P. Directeur, Bureau Veritas.
Address: Bureau Veritas, 31 rue Henri, Rochefort, Paris 17, France.

BLANC LAPIERRE, André Joseph, Agrégé des Sci. Phys., D. ès Sc. Phys., D. ès Sc. Mathématiques. Born 1915. Educ.: E.N.S. and Paris Univ. Prof., Algiers Univ., 1949-61; Director, Inst. of Nucl. Phys. of Algiers, 1956-61; Prof. and Director, Laboratoire de l'Accélérateur Linéaire, E.N.S. and Paris Univ., 1961-. Comité Nat. Français de Radioelec. Sci. and of Internat. Sci. Radio Union. Books: Theorie

des fonctions aléatoires; Electronique générale; Méthodes mathématiques de la Mécanique statistique; Propriétés statistiques du Bruit de fond; Modèles statistiques pour l'étude de phénomènes de fluctuations; Mécanique Statistique. Societies: Sté. Française de Physique; Sté. Mathématique de France; Sté. Française des Electriciens; S.F.I.T.V.; Sté. des Radio-électriciens; A.P.S.
Nuclear interest: Nuclear physics and high energy physics.
Address: Laboratoire de l'Accélérateur Linéaire, Faculté des Sciences, Bâtiment 200, 91 Orsay, France.

BLANCARD, Jean. Member, Comité Financier, Member, Commission Consultative pour la Production d'Electricité d'Origine Nucléaire, and Pres., Comité de l'Equipement Industriel, C.E.A.
Address: Commissariat à l'Energie Atomique, 29-33 rue de la Federation, Paris 15, France.

BLANCHARD, P. A. Vice-Pres., Utah Power and Light Co.
Address: Utah Power and Light Company, 1407 West North Temple Street, P.O. Box 899, Salt Lake City 10, Utah, U.S.A.

BLANCHARD, Richard Lee, B.S., M.S., Ph.D. Born 1933. Educ.: Fort Hays Kansas State Coll., Vanderbilt and Washington Univs. At O.R.I.N.S., 1955-56; U.S. Public Health Service, 1956-. Societies: H.P.S.; New York Acad. of Sci.; RESA.
Nuclear interests: Radiochemistry; Natural radioactivity in the environment and in biological materials; Nuclear geochronology.
Address: Department of Health, Education and Welfare, United States Public Health Service, National Centre for Radiological Health, 4676 Columbia Parkway, Cincinnati, Ohio 45226, U.S.A.

BLANCHART, Camille. Member, Conseil d'Administration, Centre et Sud.
Address: Centre et Sud, 20 avenue de la Toison, Brussels 5, Belgium.

BLANCHET, Alain. Gen. Manager, Cie. pour l'Etude d'Equipements Industriels S.A.
Address: Cie. pour l'Etude d'Equipements Industriels S.A., 12 rue de la Pierre Levée, Paris, France.

BLANCO, Ralph Bernardo, B.S. (E.E.). Born 1923. Educ.: Mapua Inst. Technol., Inst. of Nucl. Sci. and Eng., Argonne Nat. Lab. and O.R.S.O.R.T. Instructor, School of Eng., De La Salle Coll., 1951-53; Sci. and Chief of Phys. Sect., Nat. Inst. of Sci. and Technol., 1954-58; Supervising Sci. and Head of Exptl. Services Dept., (1958-), now Chief Sci. (Deputy Director), Atomic Res. Centre, Philippine A.E.C. Society: Philippine Assoc. for the Advancement of Science.
Nuclear interests: Design, development and fabrication of special scientific instruments and equipment needed by research staff of Philippine Atomic Research Centre; Instrumentation and control of reactors; Hazards evaluation of reactors.
Address: 40 South Maya, Phil-Am. Life, Quezon City, Philippines.

BLANCO, Raymond Eugene, B.S. (Chem.). Born 1918. Educ.: North Dakota State Univ. Graduate work at Brooklyn Polytech. Inst. Sect. Chief, O.R.N.L., operated by Union Carbide Nucl. Co., 1946-. Societies: A.N.S.; A.C.S.
Nuclear interests: Reprocessing of nuclear fuels; Radioactive waste processing and disposal; Isotope recovery; Solvent extraction and ion exchange as separation methods.
Address: Chemical Technology Division, Oak Ridge National Laboratory, P.O. Box X, Oak Ridge, Tennessee, U.S.A.

BLANCO LOIZELIER, Enrique. Vocale, Comision Interministerial de Conservacion de Alimentos por Irradiacion, Junta de Energia Nucl.
Address: Junta de Energia Nucleare, Ciudad Universitaria, Madrid 3, Spain.

BLAND, John, Dr. Medium Energies Dept., Lab. de Fisica Cosmica, Univ. Mayor de San Andres.
Address: Laboratorio de Fisica Cosmica, Universidad Mayor de San Andres, La Paz, Bolivia.

BLANDIN, Pierre André. Ing. Civil des Mines. Born 1889. Educ.: Ecole Nationale Supérieure des Mines (Paris). Chairman, tech. com., French pool for insurances against atomic risks; Member, insurance com. against atomic risks, C.E.A. (Paris); Member, permanent com. against atomic risks; Comité Européen des Assurances.
Nuclear interests: Reactor design; Safety of reactors.
Address: 11 rue Pillet Will, Paris 9, France.

BLANKART, J., Dipl. Ing. Direktor, Centralschweizerische Kraftwerke A.G.
Address: Centralschweizerische Kraftwerke A.G., 33 Hirschengraben, Lucerne, Switzerland.

BLANKENSHIP, Forrest F., M.A., Ph.D. Born 1913. Educ.: Texas Univ. Prof. Chem. Oklahoma Univ., 1943-51; Sen. Sci. Adviser, Reactor Chem. Div., O.R.N.L., 1951-. Societies: A.C.S.; A.A.A.S.; A.N.S.
Nuclear interest: High temperature reactor materials.
Address: 627 Lake Shore Drive, Kingston, Tennessee, U.S.A.

BLANKENSHIP, James Lynn, B.S. (Eng. Phys.), M.S. (Phys.). Born 1931. Educ.: Tennesee Univ. O.R.N.L., 1955-67. Society: A.P.S. Nuclear interests: Radiation detection devices and electronics for nuclear instrumentation; Semiconductor nuclear radiation detectors; Solid state and semiconductor physics.

Address: Instrumentation and Controls Division, Oak Ridge National Laboratory, Post Office Box X, Oak Ridge, Tenn., U.S.A.

BLANKS, Bruce L., M.S. (Phys.). Born 1930. Educ.: Emory Univ., Atlanta. Res. Asst., Georgia Inst. Technol., 1957-58; Phys., Emory Univ., 1958-60; Staff Member, Los Alamos Sci. Lab., 1960-. Societies: Inst. of Radio Eng. (U.S.A.); A.P.S.; Soc. for Non-destructive Testing.
Nuclear interests: Radiation detection; Instrumentation for reactor materials evaluation.
Address: Los Alamos Scientific Laboratory, P.O. Box 1663, Los Alamos, New Mexico, U.S.A.

BLAREAU, Rene. Administrateur, Bureau d'Etudes Nucléaires (B.E.N.) S.A.; Pres., and Member, Conseil d'Administration, Centre et Sud; Manager, Cies. Reunis d'Electricité et de Transports S.A. Electrorail
Address: Centre et Sud, 20 avenue de la Toison, Brussels 5, Belgium.

BLASCH, Earl B., Metal. Eng. Born 1923. Educ.: Cincinnati Univ. Chief Development Eng., Ind. Dept., Nat. Lead Co. Inc. Societies: A.S.M.; A.N.S.
Nuclear interests: Concerned with metallurgical process of uranium for fuel and shielding application; Responsible for the fabrication of H.N.P.F. first core and for casting large shielding sections to 8,000 pounds.
Address: National Lead Company, Nuclear Division, 1130 Central Avenue, Albany, New York 12205, U.S.A.

BLASEK, Gerhard, Dipl. Phys. Born 1937. Educ.: Dresden Tech. High School. Sci. Asst. (Fellow) Dept. of ray measuring technic, Faculty of nucl. technics (1961-62), Sci. Asst., Inst. of exptl. phys. (1962), Dresden Tech. Univ.
Nuclear interests: Measuring technic.
Address: 22 Weisseritzstrasse, 801 Dresden, German Democratic Republic.

BLASER, Jean-Pierre, Prof. Dr. Head, Lab. High Energy Phys., Swiss Federal Inst. of Technol.
Address: Swiss Federal Institute of Technology, Building of Physics, 35 Gloriastrasse, 8006 Zürich, Switzerland.

BLASER, Robert Ulrich, B.Sc. (Phys., cum laude). Born 1916. Educ.: Akron Univ. Chief, Nucl. Eng. Sect., R. and D. Div. (1949-51), Asst. Supt., Materials Dept., R. and D. Div. (1951-61), Staff Asst. (1961-67), Chief, Plans (1967-), Babcock and Wilcox Co. Member of several A.S.T.M. coms. Book: Chapter 5, Corrosion and Wear Handbook for Water-cooled Reactors. Societies: A.S.M.E.; N.A.C.E.; A.P.S.; A.N.S.
Nuclear interests: Properties of materials such as strength, corrosion, thermal conductivity; Engineering systems with reference to heat transfer, mass transfer and deposition; Steam generators, separation and two-phase flow, Economics and long-range management planning —reference facilities, costs, personnel, organization.
Address: 115 Vincent Boulevard, Alliance, Ohio, U.S.A.

BLASQUEZ L., Teodor GARCIA. See GARCIA BLASQUEZ L., Teodoro.

BLASS, Gerhard, Dr. rer. nat. Born 1916. Educ.: Leipzig Univ. Asst. Prof. (1951-53), Assoc. Prof. (1953-61), Prof. (1961), Chairman, Phys. Dept. (1962-), Detroit Univ. Pres., Detroit Astronomical Soc. Book: Theoretical Phys. (Appleton-Century-Crofts, 1962). Society: A.P.S.
Nuclear interests: Nuclear theory.
Address: 16636 Wildemere, Detroit, Michigan, 48221, U.S.A.

BLATZ, Hanson, E.E., P.E. Born 1907. Educ.: Brooklyn Polytech. Inst., Columbia and Rochester Univs. Director, N.Y.C. Rad Control Office; Assoc. Prof. Environmental Medicine, N.Y.U. Medical Centre. Books: Rad. Hygiene Handbook (McGraw-Hill, 1959); Introduction to Radiol. Health (McGraw-Hill, 1964). Societies: Radiol. Soc. of North America; A.A.A.S.; Radiation Res. Soc.; A.P.H.A.; H.P.S.; Assoc. Fellow, American Coll. Radiol.
Nuclear interests: Radiation protection and control.
Address: 325 Broadway, New York 10007, U.S.A.

BLAU, Monte, Ph.D. (Chem.) Born 1926. Educ.: Wisconsin Univ. Principal Cancer Res. Sci., Roswell Park Memorial Inst., 1954-. Societies: American Assoc. for Cancer Research; Soc. of Nucl. Medicine; A.C.S.
Nuclear interest: Nuclear medicine.
Address: Department of Nuclear Medicine, Roswell Park Memorial Institute, Buffalo, New York, U.S.A.

BLAXTER, K. L., D.Sc., N.D.A. Member, Managing Subcom. on Monitoring, M.R.C.
Address: Medical Research Council, 20 Park Crescent, London W.1, England.

BLAY, J. A. Chairman, Public Relations and Publications Com., Canadian Nucl. Assoc.
Address: Canadian Nuclear Association, 19 Richmond Street West, Toronto 1, Ontario, Canada.

BLEAKNEY, Walker, B.S., Ph.D. (Phys.). Born 1901. Educ.: Whitman Coll. and Minnesota Univ. Chairman, Phys. Dept. (1960-67), Class of 1909 Prof. Phys. (1963-), Princeton Univ. Societies: N.A.S.; A.P.S.; American Acad. Arts and Sci.
Nuclear interest: Investigation of the nuclear masses by means of a mass spectrometer of high precision.

Address: Palmer Physical Laboratory, Princeton University, Princeton, New Jersey, U.S.A.

BLEARS, Jack, B.Sc. (Eng.). Born 1912.
Educ.: External Graduate, London Univ. Div. Chief Eng., A.E.I. Instrumentation Div.; Member, Editorial Advisory Com., Vacuum. I.E.E. rep. on Nat. Com. for Vacuum Sci. and Eng. Society: Inst. of Elec. Eng.
Nuclear interests: Management – Instruments and apparatus for studying the nucleus, the atom and the molecule.
Address: 484 London Road, Davenham, Cheshire, England.

BLEEK, Friedrich Karlheinz SCHMIDT-. See SCHMIDT-BLEEK, Friedrich Karlheinz.

BLEHA, Otakar, MUDr., C.Sc., Docent (Internal Diseases). Born 1916. Educ.: Charles Univ., Prague. Lecturer in Internal Diseases (1946), Docent (Asst. Prof.) (1961), also in charge of Clinical Radioisotope Univ, III. Medical Dept., Faculty of Gen. Medicine, Charles Univ., Prague.
Nuclear interests: Medical application of radioisotopes, endocrinology. Diagnostic and therapy of thyroid diseases.
Address: 1 U nemocnice, Prague 1, Czechoslovakia.

BLEIBERG, Melvin L., B.S., M.S. Born 1927. Educ.: Pittsburgh Univ. Manager, N.F.D. Fuel Element Technol., Westinghouse Elec. Corp. Societies: A.S.M.; A.I.M.E.; A.N.S.
Nuclear interests: Metallurgy; Irradiation effects.
Address: Nuclear Fuel Division, Westinghouse Electric Corporation, P.O. Box 355, Pittsburgh, Pennsylvania 15230, U.S.A.

BLEULER, Ernst, Dipl. Phys., Dr. sc. nat. Born 1916. Educ.: Swiss Federal Inst. Technol. Prof. Phys., Purdue Univ., 1950-. Books: Co-author, Exptl. Nucleonics (New York, Rinehart, 1952); Co-editor, Methods of Exptl. Phys., vol. II: Electronic Methods (New York, Academic Press, 1964). Societies: A.P.S.; A.A.P.T.; Schweizerische Phys. Gesellschaft.
Nuclear interests: Nuclear reactions; Nuclear physics.
Address: Department of Physics, Purdue University, Lafayette, Indiana, U.S.A.

BLEULER, Konrad, Prof. Inst. für Theoretische Kernphysik, Bonn Univ.
Address: Institut für Theoretische Kernphysik, Bonn University, 16 Nussallee, Bonn, Germany.

BLEWETT, John P., B.A., M.A., Ph.D. Born 1910. Educ.: Toronto and Princeton Univs. Sen. Phys., Brookhaven Nat. Lab., 1947-.
Book: Co-author, Particle Accelerators (McGraw-Hill, 1962). Societies: A.P.S.; I.E.E.E.; New York Acad. of Sci.
Nuclear interests: Particle accelerators; High energy physics.
Address: Brookhaven National Laboratory, Upton, New York, U.S.A.

BLEWITT, Thomas Hugh, D.Sc. Born 1921. Educ.: Case Inst. Technol. and Carnegie Inst. Technol. A.E.C. Fellow, 1948-50; Lecturer, Tennessee Univ., 1950-61; Prof., Enrico Fermi Internat. School of Phys., Italy, 1960; Consultant, Japanese Atomic Energy Res. Inst., 1961; Prof., Pan American School of Nucl. Metal., Buenos Aires, 1962; Prof., Internat. Summer School on Solid State Phys., Mol, Belgium, 1963; Expert for I.A.E.A. at the Argentina Nat. A.E.C., Buenos Aires, 1964, 1965, 1966; Prof., Metal. in Materials Eng., Illinois Univ., 1965-; Exchange Sci., U.K.A.E.R.E., Harwell, England, 1957-58. Societies: A.P.S.; Internat. Inst. of Refrigeration.
Nuclear interests: Metallurgy and radiation damage; Low-temperature irradiation studies and irradiation facilities, and experimental techniques in nuclear reactors.
Address: Metallurgy Division, Argonne National Laboratory, Argonne, Illinois 60439, U.S.A.

BLIEK, A. P. Ing. Principal, Cie. Maritime Belge S.A. (Lloyd Royal).
Address: Cie. Maritime Belge S.A. (Lloyd Royal), 61 Rempart Ste. Catherine, Antwerp, Belgium.

BLIKSTAD, Finn. Member, Board of Rep., A/S Noratom.
Address: A/S Noratom, 20 Holmenveien, Oslo, Norway.

BLIN, Jean C., Prof. Dr. Director, European Transuranium Inst. (Euratom), Karlsruhe Nucl. Res. Centre.
Address: Karlsruhe Nuclear Research Centre, Kernreaktor Bau- und Betriebs- Gesellschaft m.b.H., 5 Weberstrasse, Karlsruhe, Germany.

BLIN-STOYLE, Roger John, M.A., D. Phil. Born 1924. Educ.: Oxford Univ. Fellow, Wadham Coll., Oxford, 1955-62; Visiting Assoc. Prof. of Phys., M.I.T., 1959-60; Prof. of Theoretical Phys. and Dean of the School of Maths. and Phys. Sci., Sussex Univ., 1962-. Books: Theories of Nucl. Moments (Oxford Univ. Press, 1957). Societies: Inst. of Phys. and Phys. Soc.; A.P.S.
Nuclear interests: Theoretical nuclear physics and elementary particle physics.
Address: School of Mathematical and Physical Sciences, Sussex University, Brighton, Sussex, England.

BLINC, Robert, Ph.D. (Phys.). Born 1933. Educ.: Ljubljana Univ. and M.I.T. Formerly Head, Solid State Phys. Dept,, Nucl. Inst. "J. Stefan", Ljubljana; Lecturer, Ljubljana Univ. Member, Advisory Board on Phys. Sci., Yugoslav Nucl. Energy Commission. Visiting Faculty, (Solid State Phys.), Dept. of Phys., Washington Univ. Book: Editor, Lectures on Solid State Phys. (Hercegnovi, Federal Nucl. Energy Commission, Yugoslavia, 1963).

Society: A.C.S.
Nuclear interests: Electronic structure of radiation induced damage centres in solids; crystal structure of reactor materials, in particular fluorides of actinide elements.
Address: Washington University, Department of Physics, Seattle, Washington 98105, U.S.A.

BLINCOE, Clifton (Robert), B.S. (Chem.), M.A., Ph.D. Born 1926. Educ.: Missouri Univ. Asst. Prof. and Instructor, Missouri Univ., 1949-56; Assoc. Prof., Nevada Univ., 1956-. Societies: A.C.S.; Biophys. Soc.
Nuclear interests: Fallout accumulation in the biosphere; Radiotracer techniques.
Address: Nevada University, Reno, Nevada, U.S.A.

BLINOV, G. A. Paper: Co-author, Status Report of Positron-Electron Storage Ring VEPP-2 (Atomnaya Energiya, vol.19, No. 6, 1965).
Address: Novosibirsk Science Centre, 20 Sovietskaya Ulitsa, Novosibirsk, Siberia, U.S.S.R.

BLINOV, P. I. Papers: Co-author, "Run Away" Electron Collective Interactions with Plasma in the Stellarator S-1 (Atomnaya Energiya, vol. 19, No. 3, 1965); co-author, Effect of Helical Magnetic Field on Plasma Ohmic Heating in S-1 Machine (ibid, vol. 20, No. 4, 1966).
Address: Academy of Sciences of the U.S.S.R., 14 Leninsky Prospekt, Moscow V-71, U.S.S.R.

BLINOV, V. A. Papers: Co-author, Radioactive Fallout in the Areas around Leningrad (letter to the Editor, Atomnaya Energiya, vol. 5, No. 5, 1958); co-author, A Polygonal Magnetic Analyzer-Multispectrograph (letter to the Editor, ibid, vol. 13, No. 1, 1962); co-author, On Some Peculiarities of Measuring the Energy Spectra of α-Particles and Fission Products with Semiconductor Detectors (ibid., No. 5); co-author, Use of Gamma-Gamma Coincidence Spectrometer with Summation of Pulse Amplitudes for Analysis of Mixture Radioactive Isotopes (ibid, vol. 19, No. 4, 1965).
Address: Academy of Sciences of the U.S.S.R., 14 Leninsky Prospekt, Moscow V-71, U.S.S.R.

BLINOWSKI, Konrad, B.Sc. Born 1929. Educ.: Warsaw Univ. Asst. Warsaw Univ., 1951-55; Sen. Asst., Head of Neutron Phys. Lab., Inst. Nucl. Res., Swierk, 1955-. Society: Polish Phys. Soc.
Nuclear interest: Neutron diffraction.
Address: Nuclear Physics Department I-B, Institute of Nuclear Research, Swierk k/Otwocka, Poland.

BLIZARD, John, M.Sc. Born 1882. Educ.: Durham Univ. (Durham Coll. of Sci.). Director, Res. and Eng. (1945-59), Res. Consultant (1959-), Power Specialty Co. and Successor Foster Wheeler Corp., 1923-. Societies: A.S.M.E.; A.I.N.A.M.E.; I.N.E.; Atomic Ind. Forum.
Nuclear interests: Design of reactor pressure vessels, boilers and heat exchangers for nuclear power plants.
Address: Foster Wheeler Corporation, 666 Fifth Avenue, New York 19, New York, U.S.A.

BLIZNAKOV, Georgi, Prof. Member, Com. for the Peaceful Uses of Atomic Energy, Council of Ministers of the Bulgarian People's Republic.
Address: Council of Ministers, Sofia, Bulgaria.

BLOCH, B., Cand. Pharm. Co-worker, Isotope Lab., Roy. Veterinary and Agricultural Coll.
Address: Royal Veterinary and Agricultural College, 13 Bülowsvej, Copenhagen, Denmark.

BLOCH, E.. Dr. Member, Conseil d'Administration, Energie Nucléaire S.A.; Stellvertreter, Federal Commission for Atomic Energy.
Address: Energie Nucléaire S.A., 10 Avenue de la Gare, Lausanne, Switzerland.

BLOCH, Edward J. Born 1912. Educ.: Washington Univ. Deputy Director, Production Div. (1948-51), Director, Div. Construction and Supply (1951-54), Director, Div. Production (1954-59), Asst. Gen. Manager, Operations (1959-64), Deputy Gen. Manager (1965-), U.S.A.E.C.
Address: U.S. Atomic Energy Commission, Washington, D.C. 20545, U.S.A.

BLOCH, Felix, Ph.D. Nobel Prize in Phys., 1952. Born 1905. Educ.: Swiss Federal Inst. Technol. and Leipzig Univ. Director Gen., C.E.R.N., Geneva, 1954-55; Prof. Phys., Stanford Univ. Vice-Pres., A.P.S.; Member, Editorial Council, Annals of Phys. Societies: Fellow, A.P.S.; N.A.S.
Nuclear interest: Nuclear magnetism.
Address: Department of Physics, Stanford University, California, U.S.A.

BLOCH, Michel A., Ing. (Ecole Polytechnique, Paris), Ph.D. (Phys.). D. ès Sc. (Phys.) (Paris Univ.). Born 1930. Educ.: Ecole Polytechnique, Paris, and California Inst. Technol. Ing., C.E.A., 1958-.
Nuclear interests: Basic research in the field of high-energy physics. Experimental work using bubble-chambers.
Address: Laboratoire de Physique Nucléaire, Collège de France, Paris 5, France.

BLOCK, Pierre Marie Hubert Michel, Tech. Eng., Elec. Eng. Born 1919. Educ.: Ecole Centrale des Arts et Métiers, Brussels and Nancy Univ. Manager, Nucl. Dept. and Director, Electro-Navale et Ind. S. A. (member, Belgonucléaire); Fondation Nucl., Brussels; Groupement Professional Ind. Nucl.; Groupe Ind. Belge Réacteurs Rapides. Society: Sté. Roy. Belge Ing. et Industriels.
Nuclear interests: All electrical services and instrumentation, associated with the design, construction, operation and maintenance of

nuclear facilities; pneumatic sample irradiation transfer system, remote control systems and componenets, glove boxes, etc.
Address: E.N.I.-L'Electro-Navale et Industrielle S.A., 17 Kontichsesteenweg, Aartselaar, Antwerp, Belgium.

BLOCK, Robert Charles, B.S. (Elec. Eng.), M.A. (Phys.), Ph.D. (Phys.). Born 1929. Educ.: Newark Coll. of Eng., Columbia and Duke Univs. Electronic eng., Nat. Union Radio Corp., 1950-51; Electronic eng., Bendix Aviation Corp., 1951-; Phys., O.R.N.L., 1955-66; Exchange sci., A.E.R.E., Harwell, 1962-63; Prof., Nucl. Eng. and Sci., Rensselaer Polytech. Inst., 1966-. Societies: A.P.S.; A.N.S.; N.Y. Acad. Sci.
Nuclear interests: Nuclear structure physics; Low energy neutron physics; Electron linear accelerators; Fast chopper time-of-flight spectrometeters; Van de Graaff accelerators.
Address: 1163 Fernwood Drive, Schenectady, New York, U.S.A.

BLOCK, Wolfram, Dr. rer. nat., Dipl. -Chem. Born 1918. Educ.: Munich Univ. Max-Planck Inst. für Hirnforschung, Marburg/Lahn, 1950-54; Chemische Fabrik Promonta, Hamburg, 1954-60; Kernreaktor, Karlsruhe, 1960-61; Bundesanstalt fuer Gewasserkunde, Koblenz, 1961-. Society: Gesellschaft Deutscher Chemiker, Fachgruppe Kerm-Radio- und Strahlenchemie.
Nuclear interest: Isotopes in hydrology.
Address: Bundesanstalt fuer Gewaesserkunde, Postfach, 54 Koblenz, Germany.

BLOEDORN, Fernando G., M.D. Born 1913. Educ.: Univ. del Litoral, Rosario, Argentina. Curie Inst., Paris, Radiostationen, Copenhagen, Radiumhemmet, Stockholm, Christie Hospital, Manchester and Royal Marsden Hospital, London, 1949-51; Sen. Resident, Radiotherapy, Francis Delafield Hospital, Columbia-Presbyterian Medical Centre, New York, 1951; Assoc. Radiotherapist, M.D. Anderson Hospital and Tumour Inst., Houston, Texas, 1951-55; Assoc. Prof. Radiol., Postgraduate School of Medicine, Texas Univ., 1951-55; Head, Div. of Radiotherapy, Radiol. Dept., Maryland Univ. Hospital, 1955-; Assoc.Prof. (1955-60), Prof. (1960-), Radiol., Maryland Univ. School of Medicine; Asst. Prof. Radiol., Johns Hopkins Univ. School of Medicine, 1958-; Part-time Radiol., Johns Hopkins Hospital, 1958-. Consultant to 13 Hospitals. Medical Advisory Com., American Cancer Soc.; Pres.-Elect, American Radium Soc. Books: Contribution: Therapy of Cancer and Allied Diseases by Interstitial Intracavitary and Surface Gamma-Ray Therapy (in Science of Ionizing Radiation: Modes of Application. Charles C. Thomas) (Springfield, Illinois, 1965); Contributor to: Textbook of Radiotherapy (Philadelphia, Lea and Febiger, 1966). Societies: American Board of Radiol.; Rad. Study Sect., Nat. Inst. of Health; Inter-American Coll. Radiol.; A.M.A.; American Soc. Therapeutic Radiologists; American Coll. Radiol.; Radiol. Soc. North America; Hon Member, Soc. Argentina de Radiol.; Assoc. des Medicins Radiol. et Radiobiol., Fondation Curie; Radiol. Soc., N.Y. Medical Coll.
Address: Department of Radiology, Division of Radiotherapy, University of Maryland Hospital, Baltimore, Maryland 21201, U.S.A.

BLOEMERS, Hugo Willebrord, LL.D. Born 1908. Educ.: Gymnasium, LL.D. Rijksuniversiteit Utrecht. Pres., Board of Directors, K.E.M.A., 1959-; Pres., Board of Directors, G.K.N., 1965-. Member, Board of Trustees, Inst. for the Development of Nucl. Sci. for Peaceful Purposes (Reactor Centrum Nederland), 1959-; Vice-Pres., Ind. Council for Nucl. Energy; Commissioner of the Queen for the Province of Gelderland.
Address: 1 Braamweg, Arnhem, Netherlands.

BLOKHIN, Nikolai N., Prof. Director, Inst. of Exptl. and Clinical Oncology; Pres., U.S.S.R. Acad. of Medical Sci., 1960-68; Vice-Pres., Internat. Union against Cancer.
Address: Institute of Experimental and Clinical Oncology, Moscow, U.S.S.R.

BLOKHINTSEV, Dmitrii Ivanovich. Born 1908. Educ.: Moscow Univ. Prof., Moscow Univ., 1936; Director (1956-65), Director, Lab. of Theoretical Phys., Joint Inst. for Nucl. Research. Member, Commission de Travail, I.U.P.A.P.; Member, U.N. Sci. Advisory Com. Societies: Corr. Member, Acad. of Sci. of the Ukrainian S.S.R.; Corr. Member, U.S.S.R. Acad. of Sci.
Nuclear interests: Nuclear power stations and high-energy nuclear physics.
Address: Joint Institute for Nuclear Research, Dubna, Nr. Moscow, U.S.S.R.

BLOMBERG, Pehr Elvington, M.S. (Phys. Eng.). Born 1924. Educ.: Roy. Inst. Technol., Stockholm. In Res. Inst. Nat. Defence, Sweden, 1950-53; AB Atomenergi, Stockholm, 1954-; Swedish rep. and head of phys. group at O.E.C.D. Halden Reactor Project, 1958-60. At present Head, Sect. for Appl. Reactor Phys., A.B. Atomenergi.
Nuclear interest: Experimental power reactor physics and operation.
Address: A.B. Atomenergi, 32 Liljeholmsvägen, Box 43041, Stockholm 43, Sweden.

BLOOD, Charles Martin, B.S. (Chem. Eng.). Born 1917. Educ.: Michigan State and Tennessee Univs. Chem. Eng. (1950-52), Chem. (1952-), Oak Ridge Nat. Lab. Societies: A.S.M.; A.N.S.; A.C.S.
Nuclear interests: High temperature thermodynamics of structural metal fluorides in molten flourides; Solid-state diffusion of uranium and non-volatile fission products in graphites.
Address: 105 Plymouth Circle, Oak Ridge, Tennessee, U.S.A.

BLOOM, Justin L., B.S. Born 1924. Educ.: Calif. Inst. Technol. Chem. Eng., Shell Chem. Corp., 1950-52; Chem. Eng., Calif. Res. and Development Co., 1952-53; Chem. Eng., Lawrence Rad. Lab., 1953-56; Chief, Materials Management Branch, U.S. A.E.C., San Francisco Operations Office, 1956-60; Manager, Process Development, The Martin Co., 1960-63; Chief, Rad. and Thermal Applications Branch, (1963-66), Tech. Asst. to Commissioner, (1966-67), Asst. Director, Div. of Nucl. Materials Safeguards (1967-), U.S.A.E.C. Societies: A.N.S.; Nucl. Eng. Div., A.I.Ch.E.; H.P.S.; I.N.M.M.
Nuclear interests: Utilisation of radioisotopes as sources of heat and radiation; Development of separation, purification and compound formation processes for radioisotopes; Pilot plant and hot cell operations; Chemistry and metallurgy of radioactive materials; Safeguarding of fissionable materials.
Address: U.S. Atomic Energy Commission, Washington, D.C. 20545, U.S.A.

BLOOM, Martin H., Ph.D. Born 1921. Educ.: Brooklyn Polytech. Inst. Consultant, Inst. for Defence Analyses and Army Res. Office, etc.; Dean of Eng., Director of Gas Dynamics Res., Brooklyn Polytech. Inst., 1964-. Fluid Dynamics Com., A.I.A.A. Societies: Assoc. Fellow, A.I.A.A.; A.P.S.
Nuclear interest: Fluid mechanical aspects of reactor technology and diffusive phenomena.
Address: Brooklyn Polytechnic Institute, 333 Jay Street, Brooklyn 11201, New York, U.S.A.

BLOOM, Stewart Dave, B.Sc., M.Sc., Ph.D. Born 1923. Educ.: Chicago Univ. Phys., Lawrence Rad. Lab., Livermore, Calif., 1957-. Society: A.P.S.
Nuclear interest: Nuclear physics research in direct reactions and weak-interactions.
Address: California University, Lawrence Radiation Laboratory, P.O. Box 808, Livermore, California, U.S.A.

BLOOMFIELD, Philip E. Asst. Prof., Solid State Theory and Transport Phenomena, Phys. Dept., Pennsylvania Univ.
Address: Pennsylvania University, Physics Department, Philadelphia, Pennsylvania 19104, U.S.A.

BLOSS, Karl H., B.S., M.S., Ph.D. Born 1926. Educ.: Johann-Wolfgang-Goethe Univ., Frankfurt, and Auburn Univ. Asst. Prof., Chem., Tuskegee Inst., Alabama; Res. Assoc., Carver Res. Foundation, Tuskegee Inst., Alabama. Head, Dept. of Isotopes and Nucl. Technol., Battelle-Inst. e.V., Germany. Societies: Gesellschaft Deutscher Chemiker, Fachgruppe Kern-, Radio- und Strahlenchemie; A.C.S.
Nuclear interests: Labelled compounds; Tracer chemistry; Radiation chemistry.
Address: c/o Battelle-Institut e.V., Abt. Isotopen- und Kerntechnik, Wiesbadener Strasse, 6 Frankfurt am Main W 13, Germany.

BLOSSER, Henry Gabriel, B.A., M.S., Ph.D. Born 1928. Educ.: Virginia Univ. N.S.F. Fellow, 1953-54; Phys., Oak Ridge Nat. Lab., 1954-58; Assoc. Prof. (1958-61), Prof. (1961-), Michigan State Univ. Society: A.P.S.
Nuclear interests: Accelerator design, cyclotron resonant beam extraction, nuclear reaction studies.
Address: Cyclotron Laboratory, Michigan State University, East Lansing, Michigan, U.S.A.

BLOTCKY, Alan J., B.S. (Phys.). Born 1930. Educ.: Carnegie Inst. Technol. Graduate Res. Asst., Carnegie Tech., 1952-53; Head, Mass Spectrometer Maintenance Sect., Goodyear Atomic Corp., 1953-55; Nucl. Phys., Veterans Hospital, Omaha,Nebraska, 1956-. Instructor, Omaha Univ., 1960-61; Lecturer in Phys., Creighton Univ., 1962-; Instructor, Nebraska Univ. Coll. of Medicine, 1963-. Nat. Deleg. and Sect. Pres., Instrument Soc. of America, 1956. Societies: Instrument Soc. of America; Soc. of Nucl. Medicine; A.P.S.; H.P.S.; A.N.S.
Nuclear interests: Supervision and operation of Triga type nuclear reactor and nuclear physics programme aimed toward the use of a reactor in the activation analysis of trace elements in blood and tissue.
Address: 306 N. 92nd Street, Omaha 14, Nebraska, U.S.A.

BLOUNT, Bertie Kennedy, C.B., M.A., B.Sc. (Oxon.), Dr. phil. nat. (Frankfurt). Born 1907. Educ.: Oxford and Frankfurt Univs. Deputy Sec. D.S.I.R., 1952-66. Pres., Exec. Com., Internat. Inst. Refrigeration, 1963-. Societies: F.R.I.C.; Internat. Inst. Refrigeration.
Nuclear interest: General interest in the support, development and utilisation of science, including nuclear science.
Address: Athenaeum, London, S.W.1, England.

BLUDMAN, Sidney Arnold, A.B., M.S., Ph.D. Born 1927. Educ.: Cornell and Yale Univs. Lawrence Rad. Lab., Berkeley, 1952-61; Prof. of Phys., Pennsylvania Univ., 1961-. Society: A.P.S.
Nuclear interests: High energy nuclear physics.
Address: Department of Physics, Pennsylvania University, Philadelphia, Pa. 19104, U.S.A.

BLUE, Allan G., M.A. (Syracuse). Born 1928. Educ.: Union Coll., Cornell and Syracuse Univs. With U.S.A.E.C., 1953-. Society: A.N.S.
Nuclear interest: Management.
Address: 6515 Marywood Road, Bethesda, Maryland, U.S.A.

BLUE, James W., Rad. Phys. Branch., N.A.S.A., Lewis Res. Centre.
Address: Radiation Physics Branch, National Aeronautics and Space Administration, Lewis Research Centre, 21000 Brookpark Road, Cleveland 35, Ohio, U.S.A.

BLUE, Larry Ray, B.S., M.S. Born 1930. Educ.: Illinois and Colorado Univs. With Gen. Elec. Co., Richland, Washington, 1954-55; Atomics Internat., Canoga Park, California, 1955-64. Societies: A.N.S.; Assoc. for Computing Machinery.
Nuclear interest: Application of computers to reactor problems.
Address: 22632 Hatteras, Woodland Hills, California, U.S.A.

BLUHM, Max Michael, B.Sc., Ph.D. Born 1923. Educ.: University and Imperial Colls., London Univ. Sci., M.R.C. Unit, Cavendish Lab., Cambridge, 1952-55; Principal Phys., Regional Phys. Dept., Western Regional Hospital Board, Scotland, 1955-63; Principal Phys., North Middlesex Hospital, London, 1963-. Societies: Fellow, Inst. of Phys. and Phys. Soc.; British Inst. of Radiol.; Roy. Soc. Medicine.
Nuclear interests: Applications of radioactive isotopes to medicine; Use of nuclear radiations in the treatment of neoplasms.
Address: Medical Physics Department, North Middlesex Hospital, Silver Street, London N.18, England.

BLUM, A. Supt., Metals Plant, Nucl. Materials and Equipment Corp.
Address: Nuclear Materials and Equipment Corp., Apollo, Pennsylvania 15613, U.S.A.

BLUM, Alexandre. Born 1914. Educ.: Dipl. H.E.C. (Ecole des Hautes Etudes Commerciales). Society: Alsthom, Nucl. Studies Services.
Address: 20 rue d'Athènes, 75 Paris 9, France.

BLUM, Jacques M., L. ès Sc., D. ès Sc. Born 1924. Educ.: Paris Univ. Chargé de Recherches, Centre Nat. de la Recherche Scientifique, Paris, 1948-56; Assoc. Res., Rochester Univ., 1955-56; Sci. Adviser and Director, Dept. of Ind. Applications of Nucl. Phys., Intertechnique, Paris, 1956-58; Consulting Eng., Sté. Ugine-Kuhlmann,Paris; Chairman, Legislative Group, A.T.E.N. Editorial Adviser, Energie Nucleaire. Book: Co-author, Atome et Industrie. Society: Royal Photographic Soc. of Great Britain.
Nuclear interests: Management, research and engineering in the fields of: uranium cycle (milling, purification, fuel-element reprocessing); waste treatments and disposal; plants and reactors construction; metallurgy of less common metals; applications of modern and nuclear physics.
Address: 5 rue Voltaire, Levallois-Perret, 92, France.

BLUM, Pierre Lazare, Dr. Eng., D.Sc. Born 1922. Educ.: Grenoble Univ. Lab. Director for Solid-state Chemistry., 1956-; Director for Metal., 1956. Societies: Assoc. française de Cristallographie; Sté. Scientifique du Dauphiné; A.C.S.; Sté. française de Métallurgie.
Nuclear interests: Metallurgy; Solid-state chemistry; Radiation effects in refractory nuclear fuel materials.
Address: Centre d'Etudes Nucléaires de Grenoble, B.P. 169, Chemin des Martyrs, Grenoble, Isère, France.

BLUM, Robert, Pres. Director, Hispano-Suiza. Vice Prés. (1962), Prés. (1963), now Prés. d'Honneur, A.T.E.N.
Address: Societe d'Exploitations Materiels Hispano-Suiza, rue du Capitaine Guynemer, Bois Colombes, (Seine), France.

BLUM, Walter J., Legal Counsel, Educational Foundation for Nucl. Sci.
Address: Educational Foundation for Nuclear Science, 935 East 60th Street, Chicago 37, Illinois, U.S.A.

BLUMBERG, Herbert, Dipl. Phys. Inst. für Strahlen- und Kernphysik, Bonn Univ.
Address: Institut für Strahlen- und Kernphysik, Bonn University, 16 Nussallee, Bonn, Germany.

BLUMENTHAL, Bernhard, Dipl. -Ing., Dr. -Ing. Born 1901. Educ.: Berlin Tech. Univ. Argonne Nat. Lab., 1948-66.; Consultant, C.E.N., Mol. Belgium, 1959-61; U.S.A.E.C., Washington, D.C., 19660. Societies: A.I.M.E.; A.S.M.; A.N.S.; RESA.
Nuclear interests: Nuclear physical metallurgy; Preparation and properties of high purity metals.
Address: U.S. Atomic Energy Commission, Division of Reactor Standards, Washington, D.C. 20545, U.S.A.

BLUMENTRITT, Gerd, Dipl. -Ing. Born 1934. Educ.: Leipzig Univ. and Dresden T.H. Zentralinstitut für Kernphysik Dresden, 1959-.
Nuclear interests: Reactor physics; Accelerator design: Pulsed neutron research; Noise analysis.
Address: Deutsche Akademie der Wissenschaften zu Berlin, Forschungsgemeinschaft, Zentralinstitut für Kernforschung, Dresden 8051, P.F.19, German Democratic Republic.

BLUMER, Hans, Dipl. Ing. ETH, Dr. Sc. Tech. Asst. to Gen. Manager, Condensateurs Fribourg S.A.
Address: Condensateurs Fribourg S.A., Fribourg, Switzerland.

BLUMFIELD, C. W. Asst. Director, Eng. and Operation, D.E.R.E., U.K.A.E.A.
Address: United Kingdom Atomic Energy Authority, Dounreay Experimental Research Establishment, Dounreay, Caithness, Scotland.

BLUMKINA, Yu. A. Papers: Co-author, Prompt Neutron Number and the Kinetic Energy of Fragments in ^{235}U Low Energy Fission (letter to the Editor, Atomnaya Energiya, vol. 15, No. 1, 1963); Electronic Instrumentation for Control and Shielding of Fast Phys. Res. Reactor BR-I (letter to the Editor, ibid, vol. 16, No. 3, 1964); co-author, Angular Distributions of Fragments from Fission of ^{235}U, ^{239}Pu by 0.08-1.25 Mev Neutrons (letter to the Editor, ibid, No. 6).

Address: Academy of Sciences of the U.S.S.R., 14 Leninsky Prospekt, Moscow V-71, U.S.S.R.

BLYTHE, Harold John, B.Sc. Born 1906. Educ.: Manchester Univ. Joined A.E.R.E. July 1948, now Group Leader, Ind. Chem. Branch, Chem. Eng. and Process Technol. Div. Nuclear interests: Radioactive waste disposal; Irradiation effects on non-metallic materials; Decontamination; Control aspects of pile radiation chemistry.
Address: 34 Park Road, Abingdon, Berkshire, England.

BLYUMKINA, Yu.A. Papers: Co-author, A Fast Neutron Burst Reactor (Atomnaya Energiya, vol. 10, No. 5, 1961); Electronic Instrumentation for Control and Shielding of Fast Phys. Res. Reactor BR-1 (letter to the Editor, ibid, vol. 16, No. 3 1964); co-author, Angular Distributions of Fragments from Fission of U^{235}, Pu^{239} by 0.08-1.25 Mev Neutrons (ibid, vol. 16, No. 6, 1964).
Address: Academy of Sciences of the U.S.S.R., 14 Leninsky Prospekt, Moscow V-71, U.S.S.R.

BO, Bernard, Ing. E.C.P., M.S. Born 1930. Educ.: Ecole Centrale des Arts et Manufactures, Paris and M.I.T. Chief Eng., COCEI Eng. Nuclear interests: Automation and control of nuclear power plants.
Address: c/o COCEI, 22 rue de Clichy, Paris 9, France.

BOAG, John Wilson, D.Sc. Born 1911. Educ.: Glasgow and Cambridge Univs. T.H. Brunswick, Phys., Radiotherapeutic Res. Unit, M.R.C., 1942-53; Phys., N.B.S., Washington, D.C., 1953; Phys., Res. Unit in Radiobiol., B.E.C.C., 1954-65; Prof. Phys. appl. to Medicine, Inst. of Cancer Res., London Univ., 1965-. Book: Chapter on Ionization Chambers in Rad. Dosimetry (Academic Press, 1967). Societies: I.E.E.; Inst. Phys.; British Inst. of Radiol.; Roy. Statistical Soc.; Faraday Soc. Nuclear interests: Radiobiology; Radiation chemistry; Radiation physics.
Address: Department of Physics, Institute of Cancer Research, Sutton, Surrey, England.

BOALO, Bruno. Tech. Director, Veam S.r.L.
Address: Veam S.r.L., 16 Via Ambrogio Figino, Milan, Italy.

BOARDMAN, Brewer Francis, B.A. Born 1904. Educ.: Colorado State Coll., de Montpellier and Stanford Univs. Chief, Tech. Information Div., U.S.A.E.C. (Oak Ridge), 1947-52. Formerly Director, Tech. Information Branch, Phillips Petroleum Co., Atomic Energy Div., 1952-. Director, Tech. Information, Idaho Nucl. Corp., -1968. Member, Tech. Information Panel, U.S.A.E.C. Societies: A.N.S.; A.A.P.T.; Idaho Acad. of Sciences.
Nuclear interest: Information handling systems.
Address: 3011 29th Street, San Diego, California 92104, U.S.A.

BOBBITT, John Thomas. Born 1906. Educ.: Georgia Inst. Technol. Director, Special Materials Div. (1949), Asst. Lab. Director (1952), Manager, Tech. Services (1957), Asst. Lab. Director for Personnel Management (1964), Argonne Nat. Lab. Societies: RESA; A.N.S.; A.A.A.S.; American Management Assoc.
Nuclear interest: Management.
Address: 213 West 9th Street, Hinsdale, Illinois, U.S.A.

BOBKOV, Yu. V. Papers: Co-author, Changes of Internal Friction with Temperature for Polycrystalline Uranium (Atomnaya Energiya, vol. 8, No. 4, 1960); co-author, The Increase in Internal Friction Associated with Temperature Changes in Uranium (ibid, vol. 9, No. 5, 1960); co-author, Effect of Reactor Irradiation on Constitution of some Diluted Uranium Alloys (ibid, vol. 22, No. 6, 1967).
Address: Academy of Sciences of the U.S.S.R., 14 Leninsky Prospekt, Moscow V-71, U.S.S.R.

BOBLETER, Ortwin, Ph.D., Dozent. Born 1924. Educ.: Innsbruck Univ. At Innsbruck Univ., 1951-58; now Atominstitut der Österreichischen Hochschulen, Vienna. Society: Chem. Soc. of Austria; Deutsche Bunsengesellschaft für Physik. Chem.
Nuclear interests: Nuclear and radiochemistry; Reactor design; Fuel reprocessing; Education.
Address: Atominstitut der Österreichischen Hochschulen, 115 Schüttelstr., A- 1020 Vienna, Austria.

BOBROV-EGOROV, N. N. Paper: Co-author, On Some Rapidity Criteria of Neutron-Activation Analysis (letter to the Editor, Atomnaya Energiya, vol. 22, No. 1, 1967).
Address: Academy of Sciences of the U.S.S.R., 14 Leninsky Prospekt, Moscow V-71, U.S.S.R.

BOBROVSKII, G. A. Paper: Co-author, Insulation of Plasma in Tokamak' Installation (Atomnaya Energiya, vol. 22, No. 4, 1967).
Address: Academy of Sciences of the U.S.S.R., 14 Leninsky Prospekt, Moscow V-71, U.S.S.R.

BOCCIARELLI, D. Steve, Prof. Head, Electron Miscroscopy Sect., Laboratori di Fisica, Istituto Superiore di Sanita'.
Address: Istituto Superiore di Sanita', 299 Viale Regina Elena, Rome, Italy.

BOCH, Alfred L., B.S. (Elec.Eng.). Born 1914. Educ.: Northeastern and Tennessee Univs. Project Eng., 22″ and 86″ fixed frequency cyclotrons (1950-53), Asst. Director, Electronucl. Res. Div. (1954-57), Assoc. Director, Reactor Div. (1957-60), Project Eng., Molten Salt Reactor Experiment (1960-62), Project Director, High Flux Isotope Reactor (1962-), O.R.N.L., Nucl. Div., Union Carbide Corp. Chairman, Oak Ridge Sect., A.N.S.
Nuclear interests: Nuclear reactor design, development, construction, and operation;

Project management of cyclotrons and reactor projects.
Address: Oak Ridge National Laboratory, Union Carbide Corporation — Nuclear Division, P.O. Box X, Oak Ridge, Tennessee, U.S.A.

BOCHENEK, Bronisław, Prof. Head, Dept. of Nucl. Fuels and Structural Materials, Sołtan Nucl. Res. Centre, Inst. of Nucl. Res.
Address: Institute of Nuclear Research, Sołtan Nuclear Research Centre, Swierk near Warsaw, Poland.

BOCHER, P. Member, Study Group on Digital Techniques, O.E.C.D., E.N.E.A.
Address: O.E.C.D. European Nuclear Energy Agency, 38 boulevard Suchet, Paris 16, France.

BOCHIN, V. P. Paper: Co-author, Measurement of Neutron Temperature with Isotopes of Lutecium (letter to the Editor, Atomnaya Energiya, vol. 24, No. 4, 1968).
Address: Academy of Sciences of the U.S.S.R., 14 Leninsky Prospekt, Moscow V-71, U.S.S.R.

BOCHKAREV, V. V., B.Sc., Acad. of Sci. of the U.S.S.R.; Deleg. to 2nd I.C.P.U.A.E., Geneva, Sept. 1958. Papers: Co-author, Some Eng. and Technological Aspects of Radioisotope and Labelled Compound Production in the U.S.S.R. (2nd I.C.P.U.A.E., Geneva, Sept. 1958); co-author, Metrology Applied to Radioactivity in the U.S.S.R. (Atomnaya Energiya, vol. 8, No. 4, 1960); co-author, Attenuation of the γ-Rays from ^{60}Co, ^{137}Cs, and ^{198}Au by a Lead Shield of Cylindrical Form (letter to the Editor, ibid, vol. 11, No. 2, 1961); co-author, Measurements of Activity of Radioactive Gases by Means of Spherical Ionisation Chamber (letter to the Editor, ibid, vol. 21, No. 2, 1966).
Address: Nuclear Energy Institute, Academy of Sciences of the U.S.S.R., 14 Leninsky Prospekt, Moscow V-71, U.S.S.R.

BOCHKOV, A. L. Paper: Co-author, On Input and Output Concentration Ratio of Radioactive Gases Flowing through Dosimetric Camera DZ-70 (letter to the Editor, Atomnaya Energiya, vol. 21, No. 2, 1966).
Address: Academy of Sciences of the U.S.S.R., 14 Leninsky Prospekt, Moscow V-71, U.S.S.R.

BOCHVAR, Andrei A., Metallurgist, Dr. of Chem. Sci. Academician of the Acad. of Sci. of the U.S.S.R.; Deleg. to 2nd I.C.P.U.A.E., Geneva, Sept. 1958; Member, Editorial Board, Atomnaya Energiya.
Nuclear interest: Metallurgy of uranium and plutonium.
Address: 18 Kirov Ulitsa, Moscow, U.S.S.R.

BOCHVAR, I. A. Papers: Co-author, Artificial Radioisotopes for Use in Medical Radiography (letter to the Editor, Atomnaya Energiya, vol. 8, No. 4, 1960); co-author, Ionising Rad. Dosimeters Based on Alumophosphate Glass Thermoluminescence Measurements (ibid, vol. 15, No. 1, 1963); co-author, The External Background Exposure Measurements of U.S.S.R. Townspeople (letter to the Editor, ibid, vol. 19, No. 3, 1965); co-author, External Background Exposure Measurements of U.S.S.R. Townspeople in 1964-65 (letter to the Editor, ibid, vol. 22, No. 1, 1967).
Address: Academy of Sciences of the U.S.S.R., 14 Leninsky Prospekt, Moscow V-71, U.S.S.R.

BOCK, Ernst G.W., Dr. rer. nat. Born 1928. Educ.: Clausthal T.H., Munich Univ. and Roy. Inst. Technol. Stockholm. Uranium Prospection, Germany, 1955-56; Abem, Stockholm, 1956-58; A.B. Atomenergi, Stockholm, 1958-62; European Atomic Energy Community, Brussels, 1962-. Book: Chapter in Kärnteknik, Technical handbook FAKTA (Stockholm, 1960).
Nuclear interests: Nuclear physics, research planning and nuclear scientific and technical information.
Address: 57 avenue du Derby, Brussels, Belgium.

BOCK, Immo Eurich, B.Sc. (Mech. Eng.), M.Sc. (Mech. Eng.), D. Phil. Born 1934. Educ.: Witwatersrand and Oxford Univs. Head, Reactor Thermodynamics Subdiv., South African Atomic Energy Board, 1964-67; Manager, Factory Products, Thomson Electronics S.A. (Pty.). Ltd., 1967-.
Nuclear interests: Heat transfer; Power cycles. Instrumentation; Components; Management.
Address: Thomson Electronics S.A. (Pty.) Limited, Box 123, Newville, Transvaal, South Africa.

BOCKELMAN, Charles Kincaid, Ph.B., Ph.D. Born 1922. Educ.: Wisconsin Univ. Res. Assoc., M.I.T., 1951-55; Asst. Prof. (1955-58), Assoc. Prof. (1958-65), Prof. (1965-), Phys. Dept., Director, Div. of Phys. Sci. (1966-), Yale Univ. Visiting Fellow, Inst. for Theoretical Phys., Copenhagen, 1958-59; Visiting Lecturer, Inst. of Phys., Nat. Univ. of Mexico, Summer, 1964; Member, Sub-Com. on Nucl. Structure, N.A.S., Nat. Res. Council. Society: Fellow, A.P.S.
Nuclear interests: Nuclear structure; Nuclear reaction mechanisms.
Address: Physics Department, Yale University, New Haven, Connecticut, U.S.A.

BÖCKHOFF, Karl Heinz, Dipl. Phys. Born 1927. Educ.: Göttingen Univ. Head, Neutron Dept. (3 MeV Van de Graaff and 60 MeV electron linear accelerator), Central Bureau for Nucl. Measurement, Euratom.
Nuclear interests: Neutron physics; Accelerator design.
Address: Central Bureau for Nuclear Measurements, Euratom, Steenweg naar Retie, Geel, Belgium.

BODANSKY, David, Ph.D., M.A., B.S. Born 1927. Educ.: Harvard Univ. Columbia Univ., 1950; Asst. Prof. (1954), Prof. (1963-),

Washington Univ. Society: Fellow, A.P.S.
Nuclear interest: Experimental nuclear physics.
Address: Department of Physics, Washington University, Seattle, Washington 98105, U.S.A.

BODDY, Keith, B.Sc. (Chem. and Phys.), M.Sc. (Rad. Phys.), Ph.D. Born 1937. Educ.: Liverpool, London and Glasgow Univs. Radiol. Protection Officer, Res. Lab. A.E.I. Ltd., Aldermaston, 1959-63; Sen. Lecturer in Health Phys. and Nucl. Medicine, Scottish (Univs.) Res. Reactor Centre, East Kilbride, 1963-. Societies: Hospital Phys. Assoc.; Soc. for Radiol. Protection; B.R.P.A.; I.R.P.A.
Nuclear interests: Research in nuclear medicine: Development of shadow-shield and mobile whole-body monitors and their application in medical research — studies include measurement of natural body potassium, iron metabolism and vitamin B_{12} metabolism; In vivo activation analysis of intrathyroidal iodine; Activation analysis - trace elements in medicine; Health physics; Teaching.
Address: Scottish Research Reactor Centre, East Kilbride, Glasgow, Scotland.

BODE, H., Prof. Dr. Manager, R. and D. Lab., Varta A.G.
Address: Varta A.G., 72 Gundelhardstr. Kelkheim/Ts., Germany.

BODEN, Hans C., Dr. jur. et. rer. pol. Vorsitzender, Kernreaktor Finanzierungs-Gesellschaft mbH; Mitglieder, Deutsche Atomkommission, Bundesministerium für Wissenschaftliche Forschung.
Address: Kernreaktor-Finanzierungs-Gesellschaft mbH, 45 Brüningstrasse, Frankfurt (M)-Hoechst, Germany.

BODEN, J. Director, Johnson's Ethical Plastics Ltd.
Address: Johnson's Ethical Plastics Ltd., 32 Ajax Avenue, Trading Estate, Slough, Buckinghamshire, England.

BODENSTEDT, Erwin, Dr. habil. Born 1926. Educ.: Bonn Univ. Assistentenstelle, Bonn Univ., 1952-54; Cornell Univ. (post-doctoral fellowship) 1954-55; Forschungslabor der SSW Erlangen, 1955-56; wiss. Rat, Hamburg Univ., 1956-63; ord., Prof. Bonn Univ., 1963-. Society: Deutsche Physikalische Gesellschaft.
Nuclear interest: Low energy nuclear physics (nuclear spectroscopy, gamma-gamma angular correlations, nuclear magnetic moments).
Address: Institut für Strahlen- und Kernphysik, 14-16 Nussallee, Bonn, Germany.

BODHIRATNE, K. A. Sec., Atomic Energy Com., Nat. Planning Council of Ceylon.
Address: Atomic Energy Committee, National Planning Council of Ceylon, 5 Galle Buck Road, Colombo 1, Ceylon.

BODIN, Hugh Arthur Boyd, B.Sc., Ph.D. Born 1930. Educ.: Glasgow Univ. A.W.R.E. Aldermaston (1955-62); Culham Lab. (1963-), U.K.A.E.A.
Nuclear interests: Plasma physics, in particular studies of properties of dense plasmas produced by rapid magnetic compression with various magnetic field configurations; Experimental studies of the stability of such systems.
Address: 79 Enborne Road, Newbury, Berkshire, England.

BODMER, Jean-Jacques, Mech. Eng. Born 1930. Educ.: Inst. of Technol., Lausanne Univ. Asst., Lab. of thermic machines, E.P.U.L., Prof. C. Colombi, 1953-55; Eng., Brown, Boveri and Co., Baden, 1955-57; Eng., Bonnard and Gardel, Consulting Engineers, Lausanne; i/c equipment and construction of nuclear power plant, Lucens, Switzerland.
Nuclear interests: Site location; Nuclear power plant design; Architect-engineering; Economics evaluation.
Address: 27 Clos d'Aubonne, La Tour-de-Peilz, Vaud, Switzerland.

BODSON, G. Spécialiste dans les questions nucléaires, Ets. J. Bodson et Fils, S. A.
Address: Ets. J. Bodson et Fils, S.A., 14 bis rue Denis-Papin, Puteaux (Seine), France.

BODSON, Hubert Joseph Victor, Dr. jur. Born 1902. Ministre des Transports et de l'Energie. Pres., Conseil National de l'Energie Nucléaire; Member, Commission of the European Communities, 1967-.
Address: Commission of the European Communities, 23 avenue de la Joyeuse Entree, Brussels 4, Belgium.

BODSON, J. Prés. Directeur Gén., Ets. J. Bodson et Fils, S.A.
Address: Ets. J. Bodson et Fils, S.A., 14 bis rue Denis-Papin, Puteaux (Seine), France.

BODU, Robert Louis, Ing. Géologue-L. en Sci. Born 1925. Educ.: Nancy Univ. Metal., then Asst. to Mill Supt., Sté. des Mines de Zellidja (Morocco), 1948-55; Chef du Service des Traitements des Minerais, Direction des Productions, C.E.A., Paris, 1955-. Societies: A.I.M.M.E.; South Africa Inst. Mining and Metal.
Nuclear interests: Metallurgy. Milling of uranium ores to get yellow cake. Refining of yellow cake to obtain pure uranium metal and some other pure nuclear products such as UF_4, UO_2, etc.
Address: Commissariat a l'Energie Atomique, Direction des Productions, B.P. No. 4, (92) Chatillon-sous-Bagneux, France.

BOECK, Pedro de. See DE BOECK, Pedro.

BOEGNER, Jean-Marc, Diplômé Ecole Sci. Politiques, L. ès lettres. Born 1913. Ambassadeur Représentant Permanent de la France auprès des Communautés Européennes.
Address: 382 avenue de Tervueren, Brussels, Belgium.

BOEKHOLT, Dipl. Ing. Member, Fachausschuss für Strahlenschutz (Joint Kernenergieverwertung Gesellschaft and Deutsches Atomforum group).
Address: Fa. P. Hammers A.G. - Strahlbetonbau, 17 Burchhardstrasse, 2000 Hamburg 1, Germany.

BOER, Abraham Adolf DE. See DE BOER, Abraham Adolf.

BOER, Jan Hendrik de. See de BOER, Jan Hendrik.

BOERBOOM, A. J. H., Dr.-Ir. Born 1922.
Educ.: Delft and Amsterdam Univs. Leiden Univ., 1952: Lab. voor Massaspectrografie, Amsterdam, 1952-. Book: Ion optics of the Mass Spectrometer (Dutch) (Amsterdam, De Poortpers, 1957). Society: Netherlands Phys. Soc.
Nuclear interests: Mass spectrometry; Plasma physics and related topics such as ion optics and vacuum technique.
Address: 25^{II} Albrecht Dürerstrasse, Amsterdam Z., Netherlands.

BOERI, Christian. Chief Electronics, Mecaserto S.A.
Address: Mecaserto S.A., 126 boulevard d'Alsace-Lorraine, Le Perreux (Seine), France.

BOERMA, Marius. Born 1921. Dept. for Gen. Affairs. and Sen. Managing Officer, Netherlands Atomforum.
Nuclear interest: Information and public relations.
Address: Reactor Centrum Nederland, Atoomforum, 112 Scheveningseweg, The Hague, Netherlands.

BOERO, M. T. de. See de BOERO, M. T.

BOERSTED, Johannes den. See den BOERSTED, Johannes.

BOETTCHER, Alfred Richard, Dr. rer. techn. habil. Born 1913. Educ.: Freiburg Univ. and T.H. Danzig. Wissenschaftl. Vorstandsmitglied der Kernforschungsanlage Jülich; a. pl. Prof. T.H. Aachen. Mitglied vershiedener Gremien der Deutschen Atomkommission. Societies: Deutsche Gesellschaft für Metallkunde; A.N.S.
Nuclear interests: Metallurgy; Fuel elements; Isotopes separation.
Address: Kernforschungsanlage Jülich des Landes Nordrhein-Westfalen-e.V., Jülich/Rhld., Germany.

BOEUF, Guy LE. See LE BOEUF, Guy.

BOEVER, Alexandre, Dr.-Ing. Born 1892.
Educ.: Polytechnical School, Aachen. Society: Assoc. Luxembourgeoise des Ing. et Industriels, Luxembourg.
Nuclear interests: Nuclear physics and chemistry.
Address: 7 Rue Jules Wilhelm, Luxembourg.

BOFFI, Vinicio Claudio, Dr. Elec. Eng., Dr. Maths. Born 1927. Educ.: Rome Univ. Azienda Comunale Elettricita ed Acqua (Rome), 1953-57; Nucl. Energy Nat. Com., C.N.E.N. (Rome), 1957-.
Nuclear interest: Reactor and neutron physics.
Address: 43 Via La Spezia, Rome, Italy.

BOGAARDT, Maarten, Dr. Born 1922. Educ.: Groningen Univ., Coll. de France and Utrecht Univ. Foundation for fundamental res. for mat. (F.O.M.) at Utrecht, 1950-55; Bataafsche Petroleum Mij at The Hague, 1955-59; Head, Design and Evaluation Group, Reactor Centrum Nederland, The Hague, 1959-. Prof. Heat Technol. and Reactor Eng., Technol. Univ. Eindhoven, 1959-. Books: Kernenergie, een inleiding tot de reactorkunde; Zó zijn onze atomen (Wolters, 1958). Societies: K.I.V.I.; Ned. Natuurkundige Vereniging: Bataafs Genootschap Proefond. Wijsbegeerte.
Nuclear interests: Heat technology; Reactor physics; Reactor engineering.
Address: 43 van Hoornbeekstraat, The Hague, Netherlands.

BOGAIEVSKY, Michel, L. ès Sc., Ing. Civil de l'Aéronautique. Born 1912. Educ.: Paris Univ. and Ecole Nationale Supérieure de l'Aéronautique. With Bureau Veritas.
Nuclear interests: Strength of materials.
Address: 142 rue de Rennes, Paris 6, France.

BOGAN, Richard H., B.S.C.E., S.M., Sc.D. Born 1926. Educ.: Washington Univ., Seattle and M.I.T. Res. Asst., M.I.T., 1951-54; Asst. Prof. (1954-57), Assoc. Prof. (1957-65), Prof., Civil Eng. (1965-), Washington Univ., Seattle. Societies: A.S.C.E.; A.S.E.E.; A.A.A.S.
Nuclear interests: Control, management, treatment and disposal of radioactive wastes; Environmental engineering systems analysis and design.
Address: Washington University, College of Engineering, Seattle 5, Washington, U.S.A.

BOGARD, Benjamin Taylor, B.S. (Mech. Eng.), M.S. (Mech. Eng.). Born 1914. Educ.: Louisiana Polytech. Inst., Louisiana State and Texas Univs. Dean Eng. and Prof. Mech. Eng., (1952-), Formerly Member, Nucl. Sci. and Eng. Development Com., Louisiana Polytech. Inst. Societies: A.S.E.E.; A.S.M.E.
Nuclear interest: Management.
Address: 36 University Drive, Ruston, La., U.S.A.

BOGART, Donald. Reactor Phys. Branch, Nat. Aeronautics and Space Administration, Lewis Res. Centre.
Address: Reactor Physics Branch, National Aeronautics and Space Administration, Lewis Research Centre, 21000 Brookpark Road, Cleveland 35, Ohio, U.S.A.

BOGDANOV, A. P. Paper: Co-author, Gamma Rad. from AN BSSR Reactor IRT-2000 (letter to the Editor, Atomnaya Energiya, vol. 16, No. 4, 1964).
Address: Academy of Sciences of the U.S.S.R., 14 Leninsky Prospekt, Moscow V-71, U.S.S.R.

BOGDANOV, F. F. Papers: Co-author, Intensification of Heat Transfer in Channel (Atomnaya Energiya, vol. 22, No. 6, 1967); Investigation of Hydraulic Resistance in Bundle of Smooth Tubes (ibid, vol. 23, No. 1, 1967); co-author, Effect of Surface Boiling on Deposition Velocities of Gasoil Decomposition Products (ibid, vol. 25, No. 1, 1968).
Address: Academy of Sciences of the U.S.S.R., 14 Leninsky Prospekt, Moscow V-71, U.S.S.R.

BOGDANOV, G. F. Deleg. to Conference sur la Physique des Plasmas et la Recherche concernant la Fusion Nucléaire Contrôlée, Salzbourg, Sept. 1961. Papers: Co-author, Use of Silicon Surface Barrier Detectors for Measurement of Spectra of Fast Particles (letter to the Editor, Atomnaya Energiya, vol. 19, No. 5, 1965); co-author, Stopping Power of Ni for Proton and He_4^+ - Ions of Energies 20-90 KeV (letter to the Editor, ibid, vol. 22, No. 2, 1967); Vacuum Conditions and Properties of Hat Ions Plasma (ibid, vol. 23, No. 3, 1967); co-author, Scattering of Protons with Energy of 15-90 KeV in Thin Nickel Foils (letter to the Editor, ibid, vol. 23, No. 4, 1967).
Address: Academy of Sciences of the U.S.S.R., 14 Leninsky Prospekt, Moscow V-71, U.S.S.R.

BOGDANOV, O. S. Paper: Co-author, Linear Induction Accelerator (Atomnaya Energiya, vol. 21, No. 6, 1966).
Address: Academy of Sciences of the U.S.S.R., 14 Leninsky Prospekt, Moscow V-71, U.S.S.R.

BOGERD, Thomas, Eng. Born 1923. Educ.: Delft Univ. Director, N.V. Provinciale Zeeuwse Elekriciteits Maatschappij. Society: N.V. Gemeenschappelijke Kernenergiecentrale, Nederland.
Nuclear interests: Management; Production of energy by nuclear means.
Address: N.V. Provinciale Zeeuwse Elektriciteits Mij., Postbus 48, Middelburg, Netherlands.

BOGGIANO, Juan José Touya (h). Born 1933. Educ.: School of Medicine, Montevideo, Uruguay. Asst. Prof., School of Medicine, Tech. Director, Centro de Medicina Nucl., Montevideo, Uruguay; Chief Sci., Res. Contract of the I.A.E.A. with the School of Medicine, Montevideo (No. 332/OB on Radioisotopic Diagnosis of Hydatid Disease). Symposium on Atomic Energy and Latin-American Development, U.S.A.E.C. San Juan, Puerto Rico, 1967; Deputy Director, Advanced Course on Radioisotopic Application in Medicine for Latinamerica, I.A.E.A., Montevideo, 1967. Asociación Latim-americana de Sociedades de Biologia y Medicina Nucl. Societies: Soc. Uruguaya de Biologia y Medicina Nucl., Montevideo; Soc. Chilena de Biologia y Medicina Nucl.
Nuclear interests: Nuclear medicine.
Address: 3305 Echevarriarza, Montevideo, Uruguay.

BØGGILD, Jorgen K., Ph.D. Director, High Energy Phys. Dept., Copenhagen Univ. Member, Danish A.E.C.; Prof., Deleg. for Denmark to the Council C.E.R.N.
Address: Niels Bohr Institutet, Copenhagen University, 15-17 Blegdamsvej, Copenhagen Ø, Denmark.

BOGNER, Robert Leonard, B.M., M.S., Ph.D. Born 1927. Educ.: St. Louis Coll. of Pharmacy and Allied Sci., and Purdue Univ. Assoc. Prof. Pharmacology, St. John's Univ., 1952-54; Res. Assoc., Brookhaven Nat. Lab., 1952-54; Res. Pharmacologist, Walter Reed Army Inst. of Res., 1954-57; Sen. Pharmacologist (1957-), Formerly Manager Biomedical Isotope Applications Dept., now Assoc. Tech. Director, Nucl. Sci. and Eng. Corp. Societies: A.A.A.S.; A.C.S.; A.Ph.A.; New York Acad. Sci.
Corporation, P.O. Box 10901, Pittsburgh, Pennsylvania 15236, U.S.A.

BOGOLIUBOV, Nikolai Nikolaievich, Prof. Lenin Prize. Academician of the Acad. of Sci. of the U.S.S.R.; Deleg. to 2nd I.C.P.U.A.E., Geneva, Sept. 1958. Formerly Director, Lab. of Theoretical Phys., now Director, Joint Inst. for Nucl. Res.
Nuclear interests: High energy physics and mathematics.
Address: Joint Institute for Nuclear Research, Head Post Office, P.O. Box 79, Moscow, U.S.S.R.

BOGORODOTSKI, Prof. Director, Electro-Tech. Inst.
Address: Electro-Technical Institute, Leningrad, U.S.S.R.

BOGRAN FORTIN, Luis, Ing.; Voter, Comision Hondurena de Energia Atomica.
Address: Comision Hondurena de Energia Atomica, Tegucigalpa, D.C., Honduras.

BOHAL, Ladislav, Ph.D. (Eng.). Born 1930. Educ.: Prague Tech. Univ. Prague Tech. Univ. 1953-55; "Energoprojekt" Project Inst. of Power Plants 1956-65; Director, Power Res. Inst., 1965-. Society: Member, Central Com. for the Nucl. Sci., Czechoslovak Sci. and Tech. Soc. Book: The Economy of the Thermic Cycles in Nucl. Power Plants (Prague, UTEIN, 1964).
Nuclear interests: The economy of nuclear power plants; Reactor design.
Address: 7a Partyzánská, Prague 7, Czechoslovakia.

BOHANNON, James Raymond, Jr., Lt. Col., U.S. Air Force; B.S., M.S. Born 1921. Educ.: North Carolina State Coll. Staff Asst., Phys.

Dept., North Carolina State Coll. (Reactor Staff); Special Project Officer, Oak Ridge Nat. Lab.; Chief of Eng., Air Force Nucl. Eng. Test Facility; Assoc. Prof. Phys., Air Force Inst. Technol.; Project Eng., HQ. U.S.A.F.; Director, A.F.I.T.-Air Force Nucl. Eng. Test Facility; Registered Professional Eng.
Nuclear interests: Nuclear power plant design for military application; Facility and system designs for nuclear environment (physical and nuclear); Investigative engineering for nuclear systems improvements.
Address: 519 N. Lombardy Street, Arlington 3, Virginia, U.S.A.

BOHL, Robert Walter, B.S., M.S., Ph.D. (Metal. Eng.). Born 1925. Educ.: Illinois Univ. Prof. Metal. Eng., Illinois Univ. Chairman, Sangamon Valley Chapter, A.S.M. Societies: A.I.M.M.E.; A.S.E.E.
Nuclear interests: Metallurgy-alloy theory; Fuel element design.
Address: 214 Nuclear Engineering Laboratory, Illinois University, Urbana, Illinois, U.S.A.

BÖHM, Edgar Gerhard, Dr. rer. nat. Dipl. Phys. Born 1928. Educ.: Hannover Univ. Director, Nucl. Div., Gutehoffnungshütte Sterkrade A.G. (GHH), Oberhausen (Rhld.). Societies: Deutsche Atomkommission, Arbeitskreis Kernreaktoren; Physikalische Gesellschaft; Deutsches Atomforum; Kernenergie-Studiengesellschaft Hamburg; V.D.I.; B.N.E.S., London.
Nuclear interests: Nuclear power plants; Reactors for ship propulsion; Research reactors; Plants for accelerators; Nuclear equipment (pressure vessels, blowers, steam generators, closed-cycle gas turbines, etc.); Reactor design and research work.
Address: Gutehoffnungshütte, 42 Oberhausen-Sterkrade, Germany.

BÖHM, Federico (Friedrich), Eng. Born 1896. Educ.: Vienna Univ. Consulting Eng., Duperial S.A., Buenos Aires. Consulting Eng., Comision Nacional de Energia Atomica, Argentina.
Nuclear interests: Nuclear physics; Detectors; Counters.
Address: 1664 Sgo. Del Estero, Buenos Aires, Rep. Argentina.

BOHM, Henry Victor, Ph.D., M.S., B.A. Born 1929. Educ.: Brown, Illinois and Harvard Univs. Assoc. Prof. (1959-64), Prof., Dept. of Phys. (1964-), Wayne State Univ. Visiting Prof.: Phys. Dept., Cornell Univ., 1966-67; Phys. Dept., Lancaster Univ. (U.K.), Summer 1967. Societies: A.P.S.; A.A.A.S.
Nuclear interests: Cryogenics; Physics of metals.
Address: Department of Physics, Wayne State University, Detroit, Michigan, U.S.A.

BÖHM, Peter Paul, Doctor juris, Diplôme d'etudes supérieures européennes. Born 1934. Educ.: Bonn, Saarbrücken and Berlin Free Univs. Res. Fellow, Inst. for European Res., Saarbrücken, 1957-58; Res. Asst., Inst.

Comparative Law, Saarbrücken Univ., 1959-60; Consultant, legal div., I.A.E.A., Vienna, 1959; Res. Assoc., Atomic Energy Res. Project, Michigan Univ. Law School, 1960-61; Asst. Prof. (i.e., Wiss. Ass.), 1962-. Book: Die internationale Regelung der Eigentumsverhältnisse im Bereich der friedlichen Verwendung der Atomenergie (Saarbrücken, 1959).
Nuclear interests: Law concerning international organisations in the field of nuclear energy; International and domestic patent law and property law (with respect to ores, special nuclear materials and reactors); Safety control and liability for damages caused by nuclear vessels.
Address: 16 Breitestrasse, Saarbrücken, Germany.

BÖHNECKE, Günther, Dr. phil. Born 1896. Educ.: Berlin Univ. Pres., Deutsches Hydrographisches Inst., Hamburg (rtd.); Exec. Sec., Studiengesellschaft zur Förderung der Kernenergieverwertung in Schiffbau und Schiffahrt e.V., Hamburg. Society: Deutsche Geophysikalische Gesellschaft, Hamburg.
Nuclear interest: Management.
Address: 35 Ohlendorff's Tannen, Hamburg-Volksdorf, Germany.

BOHR, Aage Niels, Dr. phil (Copenhagen). Born 1922. Educ.: Copenhagen Univ. Prof. Phys., Copenhagen Univ., 1956-; Director, Niels Bohr Inst., 1963-. Book: Rotational States of Atomic Nuclei (Copenhagen, Ejnar Munksgaards Forlag, 1954). Societies: Det Kongelige Danske Videnskabernes Selskab, Copenhagen; Kungliga Fysiografiska Sällskapet, Lund, Sweden; Det Kongelige Norske Videnskabers Selskab, Trondheim; American Acad. of Arts and Sciences; American Philisophical Soc.
Nuclear interest: Nuclear physics.
Address: The Niels Bohr Institute, 17 Blegdamsvej, Copenhagen Ø, Denmark.

BOHR, E. Danish Director, Eurochemic.
Address: Eurochemic, Mol, Belgium.

BOHUMIR, Chutny, Dr. Ing., C.Sc. Born 1922. Educ.: Tech. and Charles Univs., Prague. Leader, Sci. Group, Inst. of Nucl. Res., Czechoslovak Acad. of Sci. Book: Chapter on Synthesis of Labelled Organic Compounds in Foundations of Radiochem. (J. Majer) (S.N.T.L., 1961). Society: Assoc. for Rad. Res., England.
Nuclear interests: Radiation chemistry; Syntheses of labelled organic compounds.
Address: Institute of Nuclear Research, Czechoslovak Academy of Sciences, Rez, near Prague, Czechoslovakia.

BOIS, Bernard BAILLY DU. See BAILLY DU BOIS, Bernard.

BOIS, Marcel DU. See DU BOIS, Marcel.

BOISARD, Pierre. R. and D. Manager, Cie. Française des Produits Liebig S.A.
Address: Cie. Française des Produits Liebig S.A., 15 rue de Geneve, La Courneuve, (Seine), France.

BOISOT, Marcel Henri, L. ès Sc. Born 1917. Educ.: Sorbonne. Deputy Director, Res. Lab. (1952), Maitre de Conférences (1956), Ecole Polytechnique; Prés. Directeur Gen., Cie. Technique pour l'Industrie Pétrolière, C.T.I.P.; Prés., Assoc. pour le Developpement des Applications Industrielles des Radiations, A.D.A.I.R.; Prés. Fondateur du Mouvement Jeunes-Science, 1955; Member, Study Group on Food Irradiation, O.E.C.D. Societies: Sté. française de Physique; Sté. française des Radio-Electriciens; Sté. française des Ingénieurs Techniciens du Vide.
Nuclear interests: Nuclear engineering; Nuclear irradiations in western countries and in underdeveloped areas; Nuclear detection by semiconductors.
Address: 7 rue de Verneuil, Paris 7, France.

BOISSEAU, René. Sen. Officer, Pompes Delasco S.A.R.L.
Address: Pompes Delasco S.A.R.L., 27 boulevard des Italiens, Paris 2, France.

BOISSONNET, Maurice. Directeur des Fabrications, Ets. G.B.G.
Address: Ets. G.B.G., 30 rue Eugène-Caron, Courbevoie, (Seine), France.

BOITEUX, Marcel Paul, Agrégé de l'Univ. (Mathématiques); Diplômé de l'Institut d'Etudes Politiques (Economie). Born 1922. Educ.: Ecole Normale Supérieure (Sci.) et Inst. d'Etudes Politiques. Directeur Général, Electricite de France. Prés. Comité Consultatif de la Recherche scientifique et technique, 1966-67; Member, Com. on Atomic Energy, (1968-), and Member, Commission Consultative pour la Production d'Electricite d'Origine Nucleaire, C.E.A.; Past-Pres., Internat. Federation of Operational Res. Socs.; Membre du Conseil, Econometric Soc. Societies: Inst. of Management Sci. (U.S.A.); Internat. Statistical Inst.; Assoc. Francaise de Sci. Economique; Assoc. Francaise d'Informatique et de Recherche Operationnelle.
Address: 2 rue Louis Murat, Paris 8, France.

BOITSOV, V. E. Paper: Association Pitchblende and Selenides in Ore of Hydrothermal Uranium Deposits (Atomnaya Energiya, vol. 20, No. 1, 1966).
Address: Academy of Sciences of the U.S.S.R., 14 Leninsky Prospekt, Moscow V-71, U.S.S.R.

BOIVIN, Sherman Burns, A.B., M.A. Born 1916. Educ.: Stanford Univ. Planning and Evaluation Officer, Federal Airways Administration, Los Angeles, Calif., 1949-51; Management Eng., U.S. Naval Missile Test Centre, Pt. Mugu, Calif., 1951-56; Budget Officer, Div. of Reactor Development (1956-57), Management Specialist, Div. of Reactor Development (1957-58), Management Asst., Office of the Gen. Manager (1958-62), U.S.A.E.C., Washington, D.C.; Asst. Manager, Administration, Idaho Operations Office, U.S.A.E.C., Idaho Falls, Idaho, 1962-67; Deputy Asst. for Economic and Community Affairs, Office of the Gen. Manager, U.S.A.E.C., Washington, D.C., 1967-.
Nuclear interests; Management, with particular interest in programme review and appraisal of performance; Technical manpower utilisation and requirements; Facility planning.
Address: 9233 Singleton Drive, Bethesda, Maryland 20034, U.S.A.

BOKHARI, Mohammed Sibtain, M.Sc., Ph.D. Born 1928. Educ.: Panjab and Liverpool Univs. Lecturer, Panjab Univ., 1952-56; Res. Asst., Western Reserve Univ., 1956-57; Sen. Sci. Officer, Pakistan A.E.C., 1960-. Sec., Electronic Com.; Sec. CENTO Sci. Symposium; Advisory Board, J. of Pakistan's Mineral Fuel, Water and Power Resources. Books: The Dependence of Slow Neutron Flux in Cosmic Rays (Panjab Univ., 1952); The Polarisation of Elastically Scattered Protons in the Medium Range of Energies (Liverpool Univ., 1959).
Nuclear interest: Double scattering experiments in the medium range of energies.
Address: Pakistan Atomic Energy Commission, Karachi, Pakistan.

BOLD, Hermanus Johannes VAN DEN. See VAN DEN BOLD, Hermanus Johannes.

BOLENDER, D. F. Resident Manager, Astronucl. Lab., Nucl. Rocket Development Station, Westinghouse Elec. Corp. Delegate, I.A.E.A. Symposium on Nucl. Ship Propulsion with Special Reference to Nucl. Safety, Taormina, 14-18 November, 1960.
Address: Westinghouse Electric Corporation, Astronuclear Laboratory, Nuclear Rocket Development Station, P.O. Box 2028 Jackass Flats, Nevada 89023, U.S.A.

BOLEY, Forrest Irving, B.S., Ph.D. Born 1925. Educ.: Iowa State Univ. At Wesleyan Univ., Middletown Conn., 1951-61; California Univ. Rad. Lab., Berkeley, 1961-64; Prof. Phys., Dartmouth Coll., 1964-. Editor, American J. of Phys. Societies: A.A.A.S.; A.P.S.; A.G.U.
Nuclear interests: Plasma physics; Controlled thermonuclear reactions; Extensive cosmic ray showers.
Address: Department of Physics, Dartmouth College, Hanover, New Hampshire, U.S.A.

BOLHAR-NORDENKAMPF, Ferdinand G., Dip.-Ing., Dr. techn. Born 1930. Educ.: Tech. Univ., Vienna and I.S.N.S.E., Argonne Nat. Lab. Demonstrator for Electrotechnics (1954-57), Lecturer in Nucl. Eng. and Electrotechnics (1958-60), Tech. Univ. Vienna; Asst. Counsellor, Austrian Federal Ministry of Trade Sect. III. Div. 18a (Nucl. Energy and Rad.), 1960-62, Sect. VI, Div. 36 (Special Tools, Machines and Equipment for Industries), 1962-66; Sect IV.

Div. 27 (Metals and Alloys, Special Tools, Elec. Equipment for Industries; Rad. Safety), 1966-. Preparatory Planning Group for first Austrian Power Reactor, 1959-62.
Nuclear interests: Management, safety of nuclear facilities, legal problems; Technical problems of particle accelerators; Electrical power production by nuclear reactors, control problems; Economics of nuclear power production.
Address: 12 Daponteg., Vienna 3, Austria.

BOLLA, Giuseppe, Dr. of Phys. Founder (1946), Director (-1955), Labs. C.I.S.E. (Centro Informazioni Studi Esperienze), Milan; Founder (1952), Director, Energia Nucleare; Director, Improvement Course in Appl. Phys. (1950-60) and Courses in Radioisotope Techniques, Polytech. School of Milan; Prof. of Exptl. Phys., Palermo Univ.; Prof. Superior Phys., Milan Univ.; Prof. Exptl. Phys., and Prof. Atomic Phys., Polytech. School of Milan; Founder and Director, and Director, Eng. School for Nucl. Technol. (1968-), Centro Studi Nucleari Enrico Fermi (CESNEF), Polytech. School of Milan. Papers: Several on spectroscopy and physical optics from 1932-45; Others on research problems and nuclear energy from 1952-62. Societies: Inst. Lombardo di Sci. e Lettere; Accademia della Sci. di Torino.
Address: Milan Polytechnic, 32 Piazza L. da Vinci, Milan, Italy.

BOLLAY, William, B.S., M.S., Ph.D., D.Sc. (Hon.). Born 1911. Educ.: Northwestern Univ. and California Inst. Technol. Tech. Director, Aerophys. Lab., North American Aviation, 1945-51; Pres. and Tech. Director, Aerophys. Development Corp., 1951-58; Consulting Eng., 1958-; Director, Western Nucl. Corp., 1961-62; Advisory Panel on Aeronautics, Defence Dept. 1957-60. Societies: Fellow, A.I.A.A.; Hon. Life Member, Air Force Assoc.
Nuclear interest: Nuclear propulsion systems.
Address: 4592 Via Vistosa, Santa Barbara, California, U.S.A.

BOLLENRATH, Franz Josef Michael. Dipl. Ing., Dr. Ing. Born 1898. Educ.: Aachen T.H. Director, Inst. für Werkstoffkunde, Rheinische-Westfälische T.H. Aachen, 1946-. Nat. Delegate, Member, Structures and Materials Panel, AGARD, N.A.T.O. Societies: Deutsche Gesellschaft für Metallkunde; Verein Deutscher Eisenhüttenleute; Inst. Metals; Wissenschaftliche Gesellschaft für Luftfahrt.
Nuclear interests: Technology. Radiation damage of reactor materials.
Address: 79 Pruessweg, Aachen, Germany.

BOLLER, U. MEYER-. See MEYER-BOLLER, U.

BOLLINGER, Lowell Moyer, B.A. (Oberlin), Ph.D. (Cornell). Born 1923. Educ.: Oberlin Coll. and Cornell Univ. Phys., Argonne Nat. Lab. Society: A.P.S.

Nuclear interests: Nuclear physics, especially neutron cross sections and the technology of neutron spectroscopy.
Address: Argonne National Laboratory, Argonne, Illinois, U.S.A.

BOLOTIN, L. I. Papers: Co-author, A High-current Electron Accelerator (Atomnaya Energiya, vol. 11, No. 1, 1961); co-author, Interaction of Strong Electron Beams with Plasma (ibid., No. 6); co-author, On the Interaction of High-Current Beams with a Plasma in a Magnetic Field (ibid., vol. 14, No. 3, 1963); co-author, High Intensity Electromagnetic Waves Preparation in Plasma Waveguides (ibid, vol. 25, No. 1, 1968).
Address: Academy of Sciences of the U.S.S.R., 14 Leninsky Prospekt, Moscow V-71, U.S.S.R.

BOLOTIN, V. F. Paper: Co-author, Spectrometric Method for Measurement of Long-Lived Alpha Activity Aerosols (Atomnaya Energiya, vol. 24, No. 3, 1968).
Address: Academy of Sciences of the U.S.S.R., 14 Leninsky Prospekt, Moscow V-71, U.S.S.R.

BOLT, Robert O., B.S. (James Millikin), M.S. and Ph.D. (Purdue). Born 1917. Res. Chem. (1946-56), Sen. Res. Chem. (1956-62), Supv. Res. Chem. (1962-65), Sen. Res. Assoc. (1965-), Chevron Res. Co., Richmond, California. Book: Co-editor and co-author, Rad. Effects on Organic Materials (U.S.A.E.C. sponsored, 1963). Societies: A.C.S.; American Soc. of Lubrication Eng.
Nuclear interests: Development of radiation resistant organic materials; Reactor coolant-moderators, oils, greases, hydrocarbon fuels, shield materials; Research on organic cooled and/or moderated reactors; Radiation chemistry of organics, particularly the polyphenyls.
Address: 55 Culloden Park Road, San Rafael, California 94901, U.S.A.

BOLTAX, Alvin, S.B., Sc.D. Born 1930. Educ.: M.I.T. Group Leader and Project Manager, Nucl. Metals Inc., 1954-60; Manager, Westinghouse Elec. Corp., 1960-. Societies: A.S.M.; A.I.M.E.; A.N.S.
Nuclear interests: Fuel element development and fabrication; Irradiation behaviour of fuel and structural materials; Basic mechanisms of fuel behaviour.
Address: Westinghouse Electric Corporation, Advanced Reactor Division, Walts Mill Site, P.O. Box 158, Madison, Pennsylvania 15698, U.S.A.

BOLTE, John R., B.A., M.A., M.S., Ph.D. Born 1929. Educ.: Iowa Univ. Prof., Phys., San Diego State Coll.
Nuclear interest: Teaching of courses in nuclear and modern physics.
Address: Department of Physics, San Diego State College, San Diego, California 92115, U.S.A.

BOLTON, Patrick J. Born 1915. Educ.: Cambridge Univ. U.N. Secretariat, -1957; Secretariat, I.A.E.A., 1957-.
Address: 3a/36/IV Annagasse, A-1010 Vienna, Austria.

BOLYATKO, A. V. Papers: Co-author, Penetration of Neutrons in Air (Atomnaya Energiya, vol. 21, No. 4, 1966); co-author, Distribution of Neutrons in Air Close to the Earth Surface (ibid, vol. 25, No. 2, 1968).
Address: Academy of Sciences of the U.S.S.R., 14 Leninsky Prospekt, Moscow V-71, U.S.S.R.

BOLZ, W., Dipl. Ing. Director, Technischer Uberwachungs-Verein Pfalz e.V.
Address: Technischer Uberwachungs-Verein Pfalz e.V., 62 Pirmasenser Strasse, 675 Kaiserslautern, Germany.

BOMAN, Kjell Gustaf, Civil Eng. Born 1917. Educ.: Stockholm Roy. Inst. Technol. Vice-Pres., SENTAB, 1962-, consultants to Swedish Atomic Power Group regarding preparations for construction of future Swedish nucl. power plants.
Address: Svenska Entreprenad A.B., SENTAB, 47 Brahegatan, Stockholm 0, Sweden.

BOMBASSEI FRASCANI de VETTOR, Giorgio, Dr. in Law, Dr. in Political, Economic and Social Sci. Born 1910. Educ.: Florence Univ. and "Cesare Alfieri" Inst., Florence. Permanent Rep. of Italy to the Council of Europe, 1956-60; Ambassador of Italy to Luxembourg and Permanent Member, Italian Deleg., E.C.S.C. 1961-64; Ambassador of Italy in Holland, 1965-67; Permanent Rep. of Italy to the European Communities.
Address: 62 rue Belliard, Brussels, Belgium.

BOMFORD, Kenneth David. Deputy Chief, Appl. Phys. (-1967), Chief, Appl. Phys. (1967-), Member, Weapons Group Board of Management (1967-), A.W.R.E., U.K.A.E.A.
Address: United Kingdom Atomic Energy Authority, Atomic Weapons Research Establishment, Aldermaston, Berks., England.

BOMPIANI, C., Dr. Sen. Officer, Radio-isotopes and High Energies Dept., Inst. Radiologia Medico, Rome Univ.
Address: Istituto di Radiologia Medica, Università di Roma, Ospedale Policlinico Umberto 1, Rome, Italy.

BONAFFOS, J. de. See de BONAFFOS, J.

BONALUMI, Riccardo Angelo, Nucl. eng., Ph.D. (Reactor Phys.). Born 1935. Educ.: Milan Politecnico,and Rome Univ. Reactor Phys. Researcher, C.I.S.E., 1959-; Asst. in Nucl. Reactor Phys. (1960-65), Asst. Prof. Reactor Phys., (1966-), Milan Politecnico; Reactor Phys. Supv., C.I.R.E.N.E. Project. Society: A.N.S.
Nuclear interests: Reactor design, with a particular concern in reactor physics (theory and experiments); Reactor analysis.
Address: Centro Informazioni Studi ed Esperienze, Casella Postale 3986, Milan, Italy.

BONANNI, Mario, Dr., Eng. Born 1927. Educ.: Florence, Bologna and Milan Polytech. Univs. and North Carolina State Coll., Raleigh. Rep. in U.S.A. of Sta. "La Centrale", Milan, -1959; Regional Rep. (1959-63), Pres. (1963-), Westinghouse Internat. Atomic Power Co. Ltd., Geneva. Chairman, Board, Combustibil per Reattori Nucleari S.p.A., Saluggia. Societies: Swiss Assoc. for Nucl. Energy; F.I.E.N.
Nuclear interests: Nuclear power stations; Nuclear fuel elements fabrication and sales.
Address: Westinghouse International Atomic Power Company Limited, 31 rue du Rhone, Geneva, Switzerland.

BONAZZI, Darius Simon, Ing. Civil des Ponts et Chaussées. Born 1932. Coyne et Bellier, Bureau d'Ingénieurs-Conseil, Paris.
Nuclear interest: Reactor design.
Address: 6 rue de Rueil Sèvres, (Seine et Oise), France.

BONCH-OSMOLOVSKII, A. G. Paper: Co-author, Collective Linear Acceleration of Ions (Atomnaya Energiya, vol. 24, No. 4, 1968).
Address: Academy of Sciences of the U.S.S.R., 14 Leninsky Prospekt, Moscow V-71, U.S.S.R.

BOND, Victor Potter, A.B., M.D., Ph.D. Born 1919. Educ.: California Univ. (Berkeley and San Francisco). Head, Exptl. Pathology Branch, U.S. Naval Radiol. Defence Lab., 1948-54; Chairman, Medical Dept., Medical Res. Centre, (1962-67), Assoc. Director for Life Sciences (1967-), Brookhaven Nat. Lab. Panel on Radiol. Instruments; Consultant, U.N. Sci. Com. on the Effects of Atomic Rad.; Director, Comm. on Rad. and Infection, Armed Forces Epidemiological Board; Lecturer, American Inst. of Biol. Sci. Books: Chapter in Atomic Medicine (Williams and Wilkins Co., 1959); Chapter in Symposium Radioisotopes in the Biosphere (Minnesota Univ., Centre for Continuation Study, 1959); Chapter in the Kinetics of Cellular Proliferation (New York, Grune and Stratton, 1959); co-author, Rad. Injury in Man (Springfield, Ill., Charles C. Thomas Publishers, 1960); Mammalian Radiation Lethality: A Disturbance in Cellular Kinetics, (Academic Press, 1965). Societies: American Physiological Soc.; Soc. for Exptl. Biol. and Medicine; Rad. Res. Soc.; N.Y. Acad. Sci.; Intern. Soc. Hematology; Nucl. Medicine Soc.; A.A.A.S.
Nuclear interests: Medicine; Radiobiology; Radiology; Dosimetry.
Address: Director's Office, Brookhaven National Laboratory, Upton, New York 11973, U.S.A.

BONDARENKO, V. M. Paper: Co-author, On Use of Cosmic Rays for Estimation of Effectiveness of Biol. Defences (Atomnaya Energiya, vol. 24, No. 4, 1968). Address: Academy of Sciences of the U.S.S.R., 14 Leninsky Prospekt, Moscow V-71,U.S.R.R.

BONDARENKO, V. V. Paper: Co-author, Comparison of Measurement Methods of Fission Rates (letter to the Editor, Atomnaya Energiya, vol. 24, No. 1, 1968). Address: Academy of Sciences of the U.S.S.R., 14 Leninsky Prospekt, Moscow V-71, U.S.S.R.

BONDAREV, B. I. Papers: Co-author, On Self-Consistent Particle Distribution and Limit Current in Linear Accelerator (Atomnaya Energiya, vol. 19, No. 5, 1965); co-author, R.F. Quadrupole Focusing in High Current Linear Accelerators (ibid, vol. 22, No. 6, 1967). Address: Academy of Sciences of the U.S.S.R., 14 Leninsky Prospekt, Moscow V-71, U.S.S.R.

BONDAREV, V. D. Paper: Co-author, Investigation of Fuel Elements Thermal Bowing (Atomnaya Energiya, vol. 21, No. 1, 1966). Address: Academy of Sciences of the U.S.S.R., 14 Leninsky Prospekt, Moscow V-71, U.S.S.R.

BONDELID, Rollon Oscar, B.S., M.S., Ph.D. Born 1923. Educ.: North Dakota and Washington (St. Louis) Univs. U.S. N.R.L., Washington, D.C. 20390, 1952-. Societies: A.P.S.; Washington Acad. Sci.; RESA. Nuclear interests: Nuclear scattering; Electrostatic generator development; Absolute energy measurement of nuclear reactions; Development of sector-focusing cyclotron research facility. Address: 6431 Cedar Lane, Temple Hills, Maryland 20031, U.S.A.

BONET-MAURY, Paul, D. ès. Sc. Maître de Recherches-Pharmacien. Born 1910. Educ.: Paris Univ., Faculté de Pharmacie, Faculté des Sci., and Faculté de Medicine. Directeur, Service de Radioprotection, Inst. du Radium. Prof. l'Inst. Nat. des Sci. et Techniques Nucl. Secretaire-Gen., Assoc. Internationale de Radioprotection. Societies: Sté. Française de Physique; Sté. de Biologie; Sté. de Radioprotection; H.P.S.
Nuclear interests: Radioprotection; Radiobiologie.
Address: Institut du Radium, 11 rue Pierre et Marie Curie, Paris 5, France.

BONETTI, Alberto Mario, Dottore in Fisica. Born 1920. Educ.: Genoa Univ. Asst., Genoa Univ., 1948-52; Asst.,Milan Univ., 1952-62; Res. Assoc., Milan Sect., I.N.F.N. (of the Italian Nat. Com. for Nucl. Energy), 1952-61; Chair Prof. of Exptl. Phys., Phys. Inst., Bari Univ., 1962-66; Chair Prof. of Space Phys., Florence Univ., 1967-. Head, Bari-Florence Sect., Nat. Cosmic Phys. Group, Italian Nat. Res. Council, 1963-. Societies: Italian Phys. Soc.; A.G.U.; A.A.A.S.
Nuclear interests: Nuclear physics (elementary particle and high-energy interaction physics); Cosmic ray physics; Space physics.
Address: Cattedra di Fisica dello Spazio, Università di Firenze, c/o Officine Galileo, 50134 Florence, Italy.

BONGARD, W. Eng., Dept. Head, Sté. Belge de Condensation et de Mecanique, S.A. Address: Societe Belge de Condensation et de Mecanique S.A., 58 rue Capouillet, Brussels 6, Belgium.

BONHAM, Lawrence D. Geologist, Geological Survey, U.S. Dept. of the Interior. Address: U.S. Department of the Interior, Geological Survey, Washington D.C. 20242, U.S.A.

BONI, Piero. With Sindicalista. Italian Member, Specialised Nucl. Sect. for Economic Problems, Economic and Social Com., Euratom, 1966-.
Address: 100 Circonvallazione Gianicolense, 152 Rome, Italy.

BONIFACIO, Renato, Ing. Agip Nucleare. Managing Director and Gen. Manager, S.G.S. Fairchild - Sta. Gen. Semi conduttori S.p.A. Address: S.G.S. Fairchild – Societa Generale Semiconduttori S.p.A., 1 Via Olivetti, Agrate B, Milan, Italy.

BONILLA, Charles F., A.B., B.S. (E.E.), Ph.D. (Ch.E.). Born 1909. Educ.: Columbia Univ. Prof. Chem. Eng. and Nucl. Sci. and Eng. Programme, Columbia Univ., 1948-; Director, Puerto Rico Nucl. Centre (A.E.C.) 1957-59; Consultant: Brookhaven Nat. Lab., 1948-56; K.A.P.L., 1952-55; E.I. duPont de Nemours and Co., Inc., 1955-62; Westinghouse Atomic Power Div., 1955-62; A.M.F. Atomics, 1956-58; Thompson Ramo Wooldridge, 1959-61; Space Power and Propulsion Sect. (G.E.), 1960-; Gen. Atomic, (Gen. Dynamics) 1960-; Atomic Power Development Assoc., 1960-; Nat. Lead Co., 1961-. Nucleonics Heat Transfer Com., and Thermophys. Properties Com. A.S.M.E.; Chairman, Reactor Safety Com., I.R.L. Reactor Advisory Com., Nucl. Eng. Com., Columbia Univ. Books: Editor and co-author, Nucl. Eng. (McGraw-Hill, 1957); co-author, Nucl. Eng. Handbook (McGraw-Hill, 1958); co-editor, Reactor Handbook (A.E.C. 1964); Fast Reactor Technol. (M.I.T., 1966); co-editor, Nucl. Eng. and Design (1967-). Societies: A.N.S.; A.S.E.E.; A.I.Ch.E.; A.C.S.
Nuclear interests: Thermal and hydrodynamic design of nuclear reactor cores and systems; Overall evaluations; High temperature reactor coolants (liquid metals and gases).
Address: Department of Chemical Engineering, Columbia University, New York 10027, New York, U.S.A.

BONNARD, Daniel, Civil Eng. Born 1907. Educ.: Polytech. School, Lausanne Univ. (E.P.U.L.). Prof., Polytech. School, Lausanne Univ.; Sen. Officer, Bonnard et Gardel; Gen. Sec., Communauté d'Interêts pour l'Etude de la Production et de l'Utilisation Industrielles de l'Energie Nucléaire; Member, Conseil d'Administration, Energie Nucléaire S.A. Lausanne. Societies: Swiss Soc. of Eng. and Architects; Swiss Assoc. for Atomic Energy.
Address: 10 avenue de la Gare, Lausanne, Switzerland.

BONNARD, Yvon. Contrôleur gén. Direction des Etudes et Recherches, Elec. de France.
Address: Electricité de France, 3 rue de Messine, Paris 8, France.

BONNARDOT, Georges. Directeur Tech., Syneravia, Teleflex-Syneravia.
Address: Teleflex-Syneravia, 32-34 rue Robert Witchitz, Ivry sur Seine, France.

BONNAURE Pierre, Ing. E.S.E., Ing. en Génie Atomique. Born 1929. Educ.: Faculté des Sci. de Paris, Ecole Supérieure d'Electricité de Paris, and I.N.S.T.N., Saclay. D.C. amplifiers and reactor controls (1955-57), Asst. to Head of Phys. Electronics Div. (1957-59), C.E.A. (France); Head, Regulation and Automation Div. (1959-64), Head, Critical Experiments Div. (1961-64), Head, E.S.S.O.R. Operations Reactor and Hot Cells (1964-67), Euratom. Nuclear interests: Reactor control and operation; Irradiation experiments.
Address: Euratom - C.C.R., Ispra, Casella Postale 1, Ispra, Varese, Italy.

BONNELL, John Aubrey Luther, M.B., B.S. (London), M.R.C.S. (Eng.), L.R.C.P. (London). Born 1924. Educ.: King's Coll. Hospital, London. Asst. Physician, Dept. for Res. in Ind. Medicine, M.R.C., 1950-57; Sen. Medical Officer (1957-58), Deputy Chief Nucl. Health and Safety Officer (Medical) (1957-), C.E.G.B. Societies: Fellow, Roy. Soc. of Medicine (London); H.P.S.; British Inst. of Radiol. Nuclear interests: Biological effects of ionising radiations; Radiation protection — study of methods for the protection of persons exposed to ionising radiations or radioactive materials and the medical supervision of occupationally exposed persons.
Address: Nuclear Health and Safety Department, Central Electricity Generating Board, Laud House, 20 Newgate Street, London E.C.1, England.

BONNELLY, Juan Ulises GARCIA. See GARCIA BONNELLY, Juan Ulises.

BONNEMORT, Philippe Michel, Ing., Génie Atomique. Born 1930. Educ.: E.S.E., and I.N.S.T.N. Chief Eng., C.O.C.E.I. Society: A.T.E.N. Nuclear interests: Nuclear power plant; Reactor design; Safety.
Address: C.O.C.E.I., 22 rue de Clichy, Paris 9, France.

BONNER, John F. Vice-Pres., Eng. (-1963), Vice-Pres. and Asst. Gen. Manager (1963-), Pacific Gas and Elec. Co.
Address: Pacific Gas and Electric Co., 245 Market Street, San Francisco 6, California, U.S.A.

BONNET, William A. Formerly Deputy Manager, Nevada Operations Office, U.S.A.E.C.
Address: Honolulu Area Office, U.S.A.E.C., P.O. Box 580, Honolulu, Hawaii 96809, U.S.A.

BONNEVIE-SVENDSEN, M. Sen. Chem., Institutt for Atomenergi.
Address: Institutt for Atomenergi (I.F.A.), Chemistry Division, Kjeller, Near Lillestrom, Norway.

BONNIAUD, Roger. Deleg., I.A.E.A. Sci. Conference on the Disposal of Radioactive Wastes, Monaco, 16-21 November, 1959.
Address: Commissariat à l'Energie Atomique, Centre d'Etudes Nucléaires, Fontenay-aux-Roses B.P. No. 6, Fontenay-aux-Roses, (Seine), France.

BONNYNS, Marc Andre, Medical Dr. Born 1937. Educ.: Univ. Libre de Brussels. Nuclear interest: Thyroid research.
Address: Laboratoire de médicine experimentale, 115 boulevard de Waterloo, Brussels, Belgium.

BONSDORFF, Magnus von. See von BONSDORFF, Magnus.

BONSIGNORE, A. Member, Editorial Com., Minerva Nucleare.
Address: Minerva Nucleare, Corso Bramante 83-85, Turin, Italy.

BONSIGNORE, P. Centro Calcolo Numerico, Centro Siciliano di Fisica Nucleare, Catania Univ.
Address: Catania University, Centro Siciliano di Fisica Nucleare, 57 Corso Italia, Catania, Sicily, Italy.

BONTA, János, Dr., Physician, Hygienist. Born 1903. Educ.: Pécs Univ. Vice-Head Inspector for Public Health, Ministry of Health, Hungary; Lecturer of labour health, High School of Trade Union, Hungary; Country Com. of Peaceful Uses of Atomic Energy; Country Com. of Labour Protection. Book: The Technical and Hygienic Protection against Ionising Radiations (Müszaki Könyvkiadó, 1962). Societies: Hygienic Soc.; Radiolog. Soc. Nuclear interests: The public health aspects of nuclear energy; Law, organisation, protection of workers against ionising radiations, etc.
Address: 27 Katona J. utca, Budapest 13, Hungary.

BONTE, Frederick J., B.S., M.D. Born 1922. Educ.: Western Reserve Univ. Prof. and Chairman, Dept. Radiol., Texas Univ. Southwestern Medical School, 1956-; Director, Dept. of Radiol., Parkland Memorial Hospital, Dallas, Texas, 1956-. Societies: American Coll. of Radiol.; Radiol. Soc. of North America; American Roentgen Ray Soc.; Nucl. Medicine Soc.
Nuclear interests: Nuclear medicine, with special reference to internal organ scanning.
Address: University of Texas Southwestern Medical School, 5323 Harry Hines Boulevard, Dallas, Texas, U.S.A.

BÖÖK, Jan, M.D., Ph.D. Born 1915. Educ.: Lund and Uppsala Univs. Prof. of Medical Genetics, Uppsala Univ., 1957-.
Nuclear interest: Radiation genetics in man.
Address: 24 V. Ågatan, Uppsala, Sweden.

BOOMAN, Glenn Lawrence, B.A., Ph.D. (Chem.). Born 1929. Educ.: Western Washington Coll. and Washington Univ. Asst. Group Leader, Analytical Development, Phillips Petroleum Co., Atomic Energy Div., Idaho, 1954-. Book: Co-author, Chapter on Analytical Chem. of Uranium for Treatise of Analytical Chem., edit. I. M. Kolthoff and P. J. Elving. Society: Analytical Div., A.C.S.
Nuclear interests: Analytical chemistry; Instrumental and electrochemical analysis.
Address: Phillips Petroleum Co., Atomic Energy Div., P.O. Box 2067, Idaho Falls, Idaho, U.S.A.

BOONSAITH, Tongrakon, Reactor Supervisor and Reactor Building and Equipment Designer, Chulalongkorn Univ.
Address: Chulalongkorn University, Bangkok, Thailand.

BOORMAN, Colin, B.Sc. (Phys.). Born 1924. Educ.: Bristol Univ. Res. Manager, Reactor Technol. Branch, U.K.A.E.A., 1959-. Society: Fellow, Inst. Phys. and Phys. Soc.
Nuclear interest: Classical physics problems in reactor design, and in associated plants.
Address: 159 Oldfield Road, Altrincham, Cheshire, England.

BOOTH, Adrian Hadfield, B.Sc., M.Sc. (Manitoba), Ph.D. (Manchester) Born 1917. Educ.: Manitoba and Manchester Univs. Assoc. Res. Officer, A.E.C.L., Chalk River, 1944-58; Asst. Chief, Rad. Protection Div., Nat. Health and Welfare Dept., Ottawa, 1958-.
Nuclear interests: Reactor control; Safety in industrial uses of isotopes; Radioisotopes in the environment.
Address: Radiation Protection Division, National Health and Welfare Department, Ottawa, Canada.

BOOTH, Rear Admiral Charles Thomas, U.S. Navy. B.S. (U.S. Naval Acad.), S.M. (M.I.T.). Born 1910. Educ.: U.S. Naval Acad., U.S. Naval Postgraduate School, M.I.T. and Nat. War Coll. C.O., VC-4 (All-Weather Fighters), 1950-51; Director, Electronics Test, Naval Air Test Centre, 1951-53; C.O., U.S.S. Badoeng Strait, 1954-55; Asst. Chief of Staff (Operations), Staff CinCPacFlt., 1955-57; C.O., U.S.S. Ranger (CVA-61), 1957-58; Chief of Staff, ComNavAirLant, 1958-59; Asst. Chief, Bureau of Naval Weapons, 1959-61; Commander, Carrier Div. FIVE, 1961-62. Director, Development Programmes Div., Opnav, 1962-63; Deputy CNO (Development), 1963-65. ComNavAirLant, 1965-.
Nuclear interests: Planning, management, and operational interests only as relate to military assignments within headquarters, bureaux of the Navy, or operational commands or staffs at sea, respectively.
Address: Commander Naval Air Force, U.S. Atlantic Fleet, Naval Air Station, Norfolk, Virginia 23511, U.S.A.

BOOTH, David Lawry, M.A., D.Phil. Born 1931. Educ.: Oxford Univ. Atomic Power Div., English Elec. Co. Ltd., 1957-66; Principal Eng., Nucl. Design and Construction Ltd., 1966-. Society: F.Inst.P.
Nuclear interest: The safety, engineering and economics of advanced reactors.
Address: Nuclear Design and Construction Ltd., Whetstone, Leicester, England.

BOOTH, Edward Allan, M.B., B.S., D.D.R., F.C.R.A. Born 1915. Educ.: Sydney Univ. Clinical Lecturer on Obstetric Radiol., Sydney Univ.; Consultant Radiol. (Lieutenant/Colonel), Eastern Command; Hon. Fellow, Inst. Radiography; Hon. Radiol., R.P.A. Hospital, Camperdown, N.S.W.; Hon. Consultant Radiol., Mater Hospital, Crows Nest. Council Member, and Past Pres., A.M.A. (N.S.W. Branch). Societies: Fellow, Coll. of Radiol. of Australasia; Faculty of Radiol. of Gt. Britain.
Nuclear interests: Nuclear physics and medical radiology.
Address: 748 Pacific Highway, Gordon 2072, N.S.W., Australia.

BOOTH, Edward C., Dr., Formerly Asst. Prof., now Assoc. Prof., Dept. Physics, Boston Univ.
Address: Department of Physics, Boston University, 675 Commonwealth Avenue, Boston 15, Massachusetts, U.S.A.

BOOTH, Eric Stuart, M. Eng. Born 1914. Educ.: Liverpool Univ. Generation Construction Eng., H.Q. (1951), Chief Eng., H.Q. (1957), Member for Eng. (1959), C.E.G.B.; Part-time Member, U.K.A.E.A. Societies: F.R.S.; F.I.E.E.; M. Inst. Mech. Eng.
Nuclear interests: Responsible for the planning, design and construction of the C.E.G.B. nuclear power stations in England and Wales.
Address: Sudbury House, 15 Newgate Street, London E.C.1, England.

BOOTH, Ray Sturgis, B.E.E., M.S.N.E., Ph.D. (Nucl. Eng.). Born 1938. Educ.: Florida Univ. 1st Lieutenant, U.S.A.F., with academic rank of Asst. Prof., Air Force Inst. of Technol.

Societies: A.N.S.; I.E.E.E.
Nuclear interests: Neutron wave propagation in nuclear systems; Dynamic behaviour and control of nuclear systems; Neutron and photon transport.
Address: Air Force Institute of Technology, Wright-Patterson Air Force Base, Ohio 45431, U.S.A.

BOOZ, Jochen, Dr. rer. nat., Dipl.-Phys. Born 1934. Educ.: Georgia Augusta Univ., Göttingen. Inst. für Medizinische Physik. und Biophys., Göttingen Univ., 1960-64; Cyclotron Unit, M.R.C., London, 1965; Leader, Dosimetry Sect., Biol. Service, Euratom Biol. Div., C.C.R. Ispra, 1965-. Societies: Fachverband für Strahlenschutz; Internat. Rad. Protection Assoc.
Nuclear interests: Nuclear physics; Dosimetry; Radiobiology; Radiation protection. Main research field: investigation of energy transfer and energy absorption in the sensitive sites of biological tissue (microdosimetry).
Address: Euratom Biology Service, C.C.R. Ispra, Varese, Italy.

BOPP, Charles Dan, B.S. (Chem. Eng.). Born 1923. Educ.: Purdue Univ. Development Eng., O.R.N.L. Societies: A.C.S.; American Ceramic Soc.; A.N.S.; A.I.Ch.E.
Nuclear interest: Radiation effects on ceramic and organic materials.
Address: 306 Virginia Road, Oak Ridge, Tennessee 37830, U.S.A.

BOPP, Friedrich Arnold, Dr. Born 1909. Educ.: Frankfurt (Main) and Göttingen Univs. o. Prof., Munich Univ. Nucl. Phys. Com., group 2, German A.E.C. Books: Co-editor, Lectures on Theoretical Phys., vols. 3, 4 and 5. Papers: On non-local electrodynamics, fundamentals of quantum mechanics and lattice space theory of elementary particles. Society: Bavarian Acad. Sci., Munich.
Address: 3 Sulzbacher, 8 Munich 23, Germany.

BORA, Kumud Chandra, M.Sc., Ph.D. (Cantab.). Born 1921. Educ.: Calcutta and Cambridge Univs. Post-Graduate Lecturer, Gauhati Univ.; Chief, Radiobiol. and Cytogenetics Group, Atomic Energy Establishment, Trombay. Societies: A.I.B.S.; Genetical Societies of U.K., U.S.A. and India.
Nuclear interests: Dosimetry, R.B.E., radiobiology and radiation cytogenetics.
Address: Biology Division, Atomic Energy Establishment, Trombay, Bombay, India.

BORDE, Anthony Hartley de. See de BORDE, Anthony Hartley.

BORDEN, Henry. Director, Rio Algom Mines Ltd.
Address: Rio Algom Mines Ltd., 120 Adelaide Street West, Toronto 1, Ontario, Canada.

BORELI, Fedor, Ph.D. Born 1923. Educ.: Belgrade Univ. Head, Phys. Lab., Inst. Nucl. Sci. "Boris Kidrich," Vincha–Belgrade, 1960-; Asst. Prof., Belgrade Univ. Society: Mathematical and Phys. Soc. of Yugoslavia.
Nuclear interests: Nuclear physics; Neutrons in nuclear reactions and neutron detection technique.
Address: Institute of Nuclear Sciences "Boris Kidrich," P.O. Box 522, Belgrade, Yugoslavia.

BORER, Anton. Born 1934. Director, Borer and Co. Electronics. Society: Swiss Federation of Nucl. Energy.
Nuclear interest: Equipment for experimental nuclear physics.
Address: Borer and Co. Electronics, 24, Heidenhubelstrasse, 4500 Solothurn, Switzerland.

BORESI, Arthur Peter, Ph.D. (Theoretical Mechanics). Born 1924. Educ.: Illinois Univ. Prof. Theoretical Mechanics, Illinois Univ., 1959-. Societies: A.N.S.; A.S.E.E.
Nuclear interests: Reactor design; Reactor structural considerations; Thermomechanics problems, such as creep thermoelasticity, etc.
Address: 107 Talbot Lab., Illinois University, Urbana, Illinois, U.S.A.

BORGHESE, Elio, M.D., Dr. Biol. Sci. Born 1909. Educ.: Pavia Univ. Prof., Human Anatomy, Naples Univ., 1962. Societies: Socs. of Anatomy, Embryology and Tissue Culture; European Soc. for Rad. Biol.
Nuclear interest: Research on effect of ionising radiation on embryonic development.
Address: Istituto di Anatomia topografica, 5 via L. Armanni, 80138 Naples, Italy.

BORGHGRAEF, R., Prof. i/c of Radiobiol. Lab. and Related Matters, Faculty of Medicine, Trico Centre, Lovanium Univ.
Address: Lovanium University, B.P. 247, Leopoldville 11, Congolese Republic.

BORGHINI, A. Board Member, Centre Scientifique de Monaco.
Address: Centre Scientifique de Monaco, 16 boulevard de Suisse, Monte Carlo, Monaco.

BORGNON, G. Lab. Chief, Eng., Ateliers de Montages Electriques S.A.
Address: Ateliers de Montages Electriques S.A., 77 rue Saint-Charles, Paris 15, France.

BORICK, Paul M., B.S., M.S., Ph.D. Born 1924. Educ.: Scranton and Syracuse Univs. Instructor, Le Moyne Coll. Head, Microbiol., Wallace and Tiernan; Sci. i/c, Bristol Myers; Manager Res., Ethicon, Inc. Pres. and N.J. Ch. Vice-Pres., American Soc. Microbiol. Societies: A.C.S.; A.A.A.S.; A.S.M.; S.I.M.
Nuclear interests: Biological irradiation, sterilisation and sterility of medical products by irradiation.
Address: 105 Forest Avenue, Cranford, N.J. 07016, U.S.A.

BORISEVICH, Nikolai Aleksandrovich, Candidate in Mathematical Phys. Deputy Director, Inst. of Phys., Acad. of Sci. of Byelorussian S.S.R. Byelorussian S.S.R. alternate deleg. I.A.E.A. Gen. Conference, Vienna, Sept. 1960.
Address: Institute of Physics, Academy of Sciences of Byelorussian S.S.R., Minsk, Byelorussian S.S.R.

BORISHANSKII, V. M. Papers: Co-author, Potassium Boiling Heat Transfer in Tube at Moderate Steam Content Values (letter to the Editor, Atomnaya Energiya, vol. 21, No. 1, 1966); co-author, High Conductivity Fluid Flow Turbulent Heat Transfer (letter to the Editor, ibid, vol. 22, No. 2, 1967); co-author, Oxide Effect on Heat Transfer to Sodium Flow across Staggered Tube Bank (letter to the Editor, ibid, vol. 22, No. 4, 1967); co-author, Heat Transfer to Longitudinal-Flowing Liquid Metal in Triangular Tube Banks (letter to the Editor, ibid, vol. 22, No. 4, 1967).
Address: Nuclear Energy Institute, Academy of Sciences of the U.S.S.R., 14 Leninsky Prospekt, Moscow V-71, U.S.S.R.

BORISKIN, A. G. Paper: Co-author, Indication of Electron Beam by Residual Gas light (Atomnaya Energiya, vol. 24, No. 1, 1968).
Address: Academy of Sciences of the U.S.S.R., 14 Leninsky Prospekt, Moscow V-71, U.S.S.R.

BORISOV, A. V. Papers: Co-author, Relative Difference of $B^{11}F_3$ - $B^{10}F_3$ Steam Pressure (Atomnaya Energiya, vol. 19, No. 1, 1965); co-author, Investigation of Pressure Influence on Isotope Separation B^{11} and B^{10} Process by Distillation of BF_3 (ibid, vol. 23, No. 4, 1967).
Address: Academy of Sciences of the U.S.S.R., 14 Leninsky Prospekt, Moscow V-71, U.S.S.R.

BORISOV, B. N. Paper: Co-author, Miniature Instrument for Measurement of Mean Total Radon Concentration (letter to the Editor, Atomnaya Energiya, vol. 21, No. 3, 1966).
Address: Academy of Sciences of the U.S.S.R., 14 Leninsky Prospekt, Moscow V-71, U.S.S.R.

BORISOV, V. A. Paper: Co-author, Rad. Chem. Stand Plant for Thermo-rad. Vulcanization of Shims with Emitter from Spent Fuel Elements (letter to the Editor, Atomnaya Energiya, vol. 23, No. 2, 1967).
Address: Academy of Sciences of the U.S.S.R., 14 Leninsky Prospekt, Moscow V-71, U.S.S.R.

BORJES, Guillermo ARROYAVE-. See ARROYAVE-BORJES, Guillermo.

BORKE, Mitchell L., B.S., M.S., Ph.D. Born 1919. Educ.: Warsaw and Illinois Univs. Asst. in Chem. (1951-55), Instructor in Chem. (1955-57), Illinois Univ.; Asst. Prof. (1957-61), Assoc. Prof. (1961-64), Prof. (1964-), Director, Bionucleonics Lab., Duquesne Univ. Societies: A.C.S.; A.Ph.A.; Acad. of Pharmaceutical Sci.
Nuclear interests: Education and training; Radiochemistry.
Address: Duquesne University, School of Pharmacy, Pittsburgh, Pennsylvania 15219, U.S.A.

BORKOWSKI, A., Dr. Service des Medicine, Inst. Jules Bordet (Centre des Tumeurs).
Address: Service des Medicine, Institut Jules Bordet (Centre des Tumeurs), 1 rue Heger-Bordet, Brussels, Belgium.

BORKOWSKI, C. J., Dr. Member, Com. on Nucl. Sci., Chairman, Subcom. on Instruments and Techniques Com. on Nucl. Sci., N.A.S.
Address: Oak Ridge National Laboratory, P.O. Box X, Oak Ridge, Tennessee, U.S.A.

BORMAN, V. D. Papers: Co-author, Measurement of Pressure Distribution behind Strong Shock Wave (letter to the Editor, Atomnaya Energiya, vol. 19, No. 5, 1965); Electrical Field of Plasma behind Strong Shock Waves (Abstract, ibid, vol. 20, No. 4, 1966).
Address: Academy of Sciences of the U.S.S.R., 14 Leninsky Prospekt, Moscow V-71, U.S.S.R.

BORN, Hans Joachim, Dr. phil. Born 1909. Prof. Radiochem., Tech. Univ. Munich. Member, committees and sub-committees, German Atomic Commission; Member, Sci. and Tech. Com., Euratom, 1964-68. Book: Beitrag Radiochemie in G. Hertz, Grundlagen und Arbeitsmethoden der Kernphysik (1957). Societies: Münchener Chemische Gesellschaft; Gesellschaft Deutscher Chemiker.
Nuclear interests: Radiochemistry; Application of radioactive isotopes.
Address: Institut für Radiochemie, Munich Technische Hochschule, 21 Arcisstrasse, Munich, Germany.

BORN, Max, Dr. Phil. (Göttingen), D.Sc.h.c. (Bristol), D. ès. Sc.h.c. (Bordeaux), Sc.D.h.c. (Oxford), Dr. rer. nat.h.c. (Freiburg i.B.), LL.D.h.c. (Edinburgh), Dr. ing.h.c. (Stuttgart), Dr. Phil.h.c. (Oslo), Dr. Sc.h.c. (Brussels), Dr. rer. nat.h.c. (Humboldt Univ., Berlin), M.A. (Cambridge). Nobel Prize for Physics, 1954. Born 1882. Educ.: Breslau, Heidelberg, Zürich and Göttingen Univs. Tait Prof. Appl. Mathematics, Edinburgh, 1936-53. Prof. Emeritus, Göttingen, Edinburgh Univs. Books: Atomic Physics (1956: one chapter on nuclear physics) and 20 other books. Societies: Roy. Soc., London; Roy. Soc., Edinburgh; Academies of Berlin, Göttingen, U.S.S.R., Copenhagen, Stockholm, Dublin, Washington, Boston, Lima, Bucharest and Bangalore.
Address: 4 Marcardstrasse, Bad Pyrmont, Germany.

BORNARD, Jean. Vice-Pres., Conseil d'Administration, Charbonnages de France. Member, Specialised Nucl. Sect. for Economic Problems, and Member, Specialised Nucl. Sect. for Social Health and Development Problems, Economic and Social Com., Euratom, 1966-.

Address: Charbonnages de France, 26, rue de Montholon, 75 Paris 9, France.

BORNEMISZA, Elemer, Lic. Quim., M.A., Ph.D. Born 1930. Educ.: Costa Rica and Florida Univs. Instructor up to Assoc. Prof., Costa Rica Univ., 1953-60; Asst. up to Sen. Soil Chem., Interamerican Inst. for Agricultural Sci. of the Organisation of American States, Nucl. Energy Programme, 1959-. Societies: New York Acad. of Sci.; Internat. Soil Sci. Soc.; Soil Sci. Soc. of America; Costa Rican Chem. Soc. Nuclear interests: Application of radioisotopes to agricultural and biological research, particularly in the tropics. Tracer work with plants and soils.
Address: Interamerican Institute for Agricultural Sciences, Turrialba, Costa Rica.

BORNSCHEUER, Eberhard, Dipl. Ing. Managing Director, Aug. Klonne.
Address: Aug. Klonne, 1 Kornebach Strasse, Dortmund, Germany.

BORODINA, Yu. I. Temporary Director, Inst. of Exptl. Biol. and Medicine, Novosibirsk Sci. Centre, Acad. Sci. of the U.S.S.R.
Address: Novosibirsk Science Centre, 20 Sovietskaya Ulitsa, Novosibirsk, Siberia, U.S.S.R.

BOROK, V. I. KEILIS-. See KEILIS-BOROK, V.I.

BOROWITZ, Sidney, B.S., M.S., Ph.D. Born 1918. Educ.: City College of New York and New York Univ. Asst. Prof. of Phys. (1950-55), Assoc. Prof. Phys. (1955-59), Prof. of Phys. (1959), Chairman, Phys. Dept. (1961-), New York Univ. Society: Fellow, A.P.S. Nuclear interests: Theoretical analysis of nuclear and atomic structure. Many body problems in inhomogeneous systems.
Address: New York University, Department of Physics, Gould Hall of Technology, University Heights, New York, New York 10453, U.S.A.

BORRACHERO, José. Born 1907. Member of the Board, Reunion, Central, and Plus Ultra Insurance Cos.; Gen. Manager, Plus Ultra. Chairman, Atom Risk Commission, Sindicato Nacional del Seguro; Member, Spanish Parliament, Voter of the Consorcio de Compensación de Seguros, in connection with atom risk insurance (Law 29.4.64).
Nuclear interest: Atom risk insurance.
Address: 19 Antonio Acuna, Madrid, Spain.

BORRELLI, Frank James, B.S., M.D. Born 1905. Educ.: Pittsburgh Univ. and New York Medical Coll. Prof. and Chairman in Radiol., Flower-Fifth Avenue Hospitals, 1937-; Director of Radiol., Metropolitan Hospital; Director of Radiol., Bird S. Coler Hospital; Consulting Director in Radiol., Columbus Hospital, New York, 1958-; Prof., New York Medical Coll., 1942-; Chairman, Isotope Com., New York Medical Coll., Flower-Fifth Avenue Hospitals. Deleg., Sect. for Radiol., New York State Medical Soc.; Chairman, Special Com. on Radiol., New York County Medical Soc., 1955-; Councillor, A.C.R. Societies: Fellow, A.M.A.; New York Roentgen Soc.; Radiol. Soc. of North America; Inter-American Coll. of Radiol.; Diplomate, American Board of Radiol.
Address: Flower-Fifth Avenue Hospitals, New York 29, U.S.A.

BORSCHETTE, Albert, Dr. Dr. ès Lettres. Born 1920. Educ.: Aix-en-Provence, Innsbruck, Munich and Paris Univs. Sec. à la Légation du Luxembourg à Bonn, 1950-53; Sec. puis Conseiller à l'Ambassade du Luxembourg à Bruxelles, 1953-58; Représentant du Luxembourg auprès de la Commission C.E.E.A., 1958-. Ambassadeur, Représentant Permanent du Luxembourg auprès des Communautés Européennes, 1958.
Nuclear interest: Nuclear physics.
Address: 26 Avenue Louis Vercauteren, Brussels, Belgium.

BORSETTO, P., Ing. Working in Nucl. Field, Dott. Ing. Giuseppe Torno and C. S.p.A.
Address: Dott. Ing. Giuseppe Torno and C. S.p.A., 7 Via Albricci, Milan, Italy.

BORST, Lyle Benjamin, A.B., A.M., Ph.D. Born 1912. Educ.: Illinois and Chicago Univs. Chairman, Dept. of Reactor Sci. and Eng., Brookhaven Nat. Lab., 1946-51; Prof. Phys., Utah Univ., 1951-54; Chairman, Dept. of Phys., N.Y.U., 1954-61; Prof. Phys., New York State Univ., Buffalo, 1962-. Books: Manhattan Project Record; Editor,Plutonium Production Reactors,; Editor, Feasibility of Inspection as a Method of Internat. Control of Atomic Energy (1946). Societies: A.P.S.; A.A.A.S. Nuclear interests: Neutron physics; Neutron interactions with solids; Coupled reactor systems; Nuclear astrophysics.
Address: 17 Twin Bridge Lane, Williamsville, New York 14221, U.S.A.

BORT, W. F. Vice-Pres. Construction, Kaiser Engineers.
Address: Kaiser Engineers, Division of Henry J. Kaiser Co., Kaiser Centre, 300 Lakeland Drive, Oakland 12, California, U.S.A.

BORTLIK, Jiri, Ing., C.Sc. Born 1937. Educ.: Prague Tech. Univ. Energoprojekt, Prague, 1960; Power Res. Inst., (Vyzkumny Ustav Energeticky-E.G.U.), Prague, 1967. Nuclear interests: Problems of nuclear source utilisation in heating — factors of safety, hygiene and economy; Project and working conditions of nuclear heat and power plants, district heating plants; Reactor design.
Address: 139/2536 Hlavni, Prague 4-Sporilov II, Czechoslovakia.

BORUCHOWSKI, Sabina. Eng., Chem. Born 1925. Educ.: Politechnika Lódska, Poland. Asst., Dept. of Radiotherapy and Isotopes,

BOR

Tel-Aviv Univ. Medical School.
Nuclear interests: Nuclear medicine and biochemistry.
Address: Ramat-Gen, San Martin 7, Israel.

BORY, George. Born 1888. Educ.: Ecole Polytechnique, Paris and Inst. de Saclay. Eng. of the French Navy (retired). Founder and Pres. of the partnership Bureau ECI. Vice-Pres. Internat. Federation of Consulting Eng.; Vice-Pres. Assoc. of French Eng. for Atomic Eng.
Nuclear interests: Nuclear energy; Reactor design; Electricity production, combined with treatment of sea water.
Address: 34 rue de Ponthieu, Paris 8, France.

BORZINA, I. G. Paper: Co-author, Determination of Uranium Clarke Concentration in Ionic Crystals (letter to the Editor, Atomnaya Energiya, vol. 22, No. 6, 1967).
Address: Academy of Sciences of the U.S.S.R., 14 Leninsky Prospekt, Moscow V-71, U.S.S.R.

BOSCH, Miss C. Asst., Internat. Lab. of Marine Radioactivity.
Address: International Laboratory of Marine Radioactivity, Musée Océanographique, Monaco.

BOSCH, Horacio Ernesto, Prof. Maths., Dr. in Phys., Diplomé Etudes Supérieures Sciences Physiques. Born 1927. Educ.: Buenos Aires and Paris Univs. Researcher at Argentine A.E.C. 1951-; i/c Nucl. Physics Div., 1957-. Prof. Nucl. Physics, La Plata Univ., 1957-; Formerly Director, Nucl. Dept., Coasin S.R.L. Society: Asociación Fisica Argentina.
Address: 8250 Av. Libertador Gral San Martin, Buenos Aires, Argentina.

BOSCH, J. VAN DEN. See VAN DEN BOSCH, J.

BOSCH, Julius, Dipl. Ing. Born 1917. Educ.: Munich T.H., In Rad. Measuring Instruments Dept. (1949-), Chief of Dept. (1955-), Frieseke and Hoepfner G.m.b.H. Erlangen, Germany.
Nuclear interests: Instrumentation for application of radionuclides in industry, science and research; Health physics instrumentation.
Address: 7 Zum Aussichtsturm, Rathsberg über Erlangen, Germany.

BOSCH CHAFER, Francisco. Head, Nucl. Dept., Hidroelectrica Espanola, S.A.; Tech. Advisor, Forum Atomico Espanol.
Address: Hidroelectrica Espanola, S.A., Nuclear Department, 1 Hermosilla, Madrid 1, Spain.

BOSE, Asok, M.Sc., Dr. rer. nat. Born 1922. Educ.: Calcutta and Göttingen Univs. Lecturer, Calcutta Univ., 1951-53; Alexander von Humboldt Fellow, Göttingen Univ., 1955-57; N.R.C. of Canada Post-doctoral Fellow, Montreal Univ., 1958-; Assoc. Prof., Dept. of Phys., Montreal Univ. Society: A.P.S.

Nuclear interest: Theoretical nuclear physics.
Address: Université de Montréal, Département de Physique, C. P. 6128, Montréal 3, P.Q., Canada.

BOSE, Mrs. M., Dr. Formerly Reader, now Assoc. Prof., Saha Inst. of Nucl. Phys., Calcutta Univ.
Address: Calcutta University, 92 Acharya Prafulla Chandra Road, Calcutta-9, India.

BOSKOVIC, N. Head, Rare Earth Dept., Dept. for Processing of Nucl. Raw Materials, Inst. Nucl. Raw Materials.
Address: 86 Franse Deperea, Belgrade, Yugoslavia.

BOSLEY, Wiliam, M.Sc., Ph.D. Born 1925. Educ.: Sheffield and Glasgow Univs. Principal, Mander Coll., Bedford. Societies: Fellow, Inst. Phys. and Phys. Soc.
Nuclear interests: Particle accelerators; High energy scattering; Cosmic rays.
Address: Mander College, Cauldwell Street, Bedford, England.

BOSSUAT, Theodule. Pres., Chambre Honoraire à la Cour des Comptes. Member, Comité Financier, and Pres., Commission Consultative des Marchés, C.E.A.
Address: c/o Comissariat a l'Energie Atomique, 29-33 rue de la Federation, Paris 15, France.

BOST, David E. Formerly Tech. Editor, now Assoc. Editor, Nucl. Sci. Abstracts.
Address: Nuclear Science Abstracts, Division of Technical Information Extension, U.S. Atomic Energy Commission, P.O. Box 62, Oak Ridge, Tennessee, U.S.A.

BOSTICK, Winston Harper, B.S., Ph.D. Born 1916. Educ.: Chicago Univ. Assoc. Prof., Phys., Tufts Univ., 1948-54; Staff Member, Lawrence Rad. Lab., Livermore, 1954-56; Prof., Head Phys. Dept., Stevens Inst. of Technol., 1956-. N.S.F. Sen. Post-Doctoral Fellowship, 1961-62. Books: Contribution to Rad. Lab. M.I.T. Series, Volume Pulse Generators (1946); Contribution on Acceleration of Plasma, to Plasma Phys. and Thermonuclear Res. Volume 2. (Pergamon Press). Society: Fellow, A.P.S.
Nuclear interests: Plasma physics; Controlled thermonuclear power.
Address: P.O. Box 167, Chester, New Jersey, U.S.A.

BOSTOCK, Donald E. Born 1909. Asst. Director, Programmes and Standards (1959-66); Director, Div. of Personnel (1966-), U.S.A.E.C.
Address: U.S. Atomic Energy Commission, Division of Personnel, Washington D.C. 20545, U.S.A.

BOSTROM, Norman Alvin, M.A., Ph.D. Born 1927. Educ.: Suffolk (Boston, Mass.,) and Texas Univs. Res. Sci., Texas Univ., 1950-56; Contract Administrator and Pres., Texas Nucl.

Corp., 1956-61; Pres., Texas Bio-Nucl. Corp., 1961-62; Gen. Manager, Kaman Instruments, A. Div. of Kaman Aircraft Corp., 1962-. Society: A.P.S.
Nuclear interests: Management; Particle accelerators; Activation analysis; Carbon-14 dating; Particle detectors; Neutron physics; Dosimetry; Nuclear physics; Bio-physics.
Address: P.O. Box 9431, Austin, Texas 78756, U.S.A.

BOSWELL, G. G. J., B.Sc., Ph.D., A.R.I.C. Radiol. Health and Safety, Roy. Coll. of Advanced Technol., Salford.
Address: Royal College of Advanced Technology, Salford 5, Lancashire, England.

BOSWELL, R. W. Former Director, Weapons Res. Establishment, Salisbury; Sec., Dept. of Nat. Development; Member, A.A.E.C.
Address: c/o Australian Atomic Energy Commission, 45 Beach Street, Coogee, New South Wales, Australia.

BOSWORTH, G. H. Senior Nucl. Eng., Bechtel Nucl. Corp.
Address: Bechtel Nuclear Corporation, 220 Bush Street, San Francisco 4, California, U.S.A.

BOTHE, Hans-Karl, Dr. rer. nat. Born 1929. Educ.: Martin-Luther-Univ., Halle (Saale). Karl-Marx-Univ., Leipzig, 1954; Inst. für angewandte Radioaktivität der Forschungsgemeinschaft der Deutschen Akademie der Wissenschaften zu Berlin, Leipzig, 1955-.
Nuclear interests: Technische Anwendungen radioaktiver Nuklide, insbesondere Ionisationsgasanalyse und Ionisationsdetektoren; Anwendung radioaktiver Tracer in der Industrie.
Address: 18 Manetstrasse, 7022 Leipzig N 22, German Democratic Republic.

BOTHE, Werner, Dipl. Ing. Born 1921. Educ.: Dresden and Stuttgart T.H. Siemens-Schuckert-Werke A.G., A.E.G., Neunkirchener Eisenwerk (Saar), 1949-57; Energieversorgung, Deutsches Electronen-Synchrotron (DESY). Society: V.D.E.
Nuclear interests: Electric power supply and cooling systems for particle accelerators.
Address: 17 Riesserstr., Hamburg-26, Germany.

BOTTA, Guido, M.A. (Naples). Born 1918. Educ.: Naples, Debrecen and N. Carolina Univs. Press Chief, U.S.I.S. Naples, 1950-51, 1954-56; Press and P.R. Chief, C.N.E.N. Pres. S.I.P.R.I. (Sta. per l'Incremento delle Publiche Relazioni in Italia); Vice-Pres. Italian Internat. Com. for Public Relations in Public Management. Society: F.I.E.N.
Nuclear interest: Techniques and use of mass media for the dissemination of information concerning nuclear activities.
Address: 15 Via Belisario, Rome, Italy.

BOTTANI, Nello. Tech. Manager, Ed. Zublin et Cie. A.G.
Address: Ed. Zublin et Cie. A. G., 6 Okenstrasse, 8031 Zürich, Switzerland.

BÖTTCHER, Hermann, Dipl. -Ing. Member, Fachkommission für Strahlenschutz, Wirtschaftsministerium Baden-Württemberg; Ministerialrat, Arbeitsministerium Baden-Württemberg.
Address: 30 Rotebühlstr., Stuttgart W., Germany.

BOTTLE, David Wallace, B.Sc., A.C.G.I. Born 1910. Educ.: London Univ. Chief Aerodynamicist, Short Bros. and Harland, Ltd., Belfast, 1955-58; Atomic Energy and Power Plant Divs., G.E.C. Erith, 1958-66; Turbine Div., C.A. Parsons, 1966-. Societies: A.M.I. Mech E.; A.F.R.Ae.S.
Nuclear interest: Reactor and turbine design and development.
Address: 1 Halewood Avenue, Newcastle-upon-Tyne 3, England.

BOTTOM, Virgil E. Dr. Phys., Nucl. Dept., Shiflet Bros.
Address: Shiflet Bros., P.O. Box 399, Abilene, Texas, U.S.A.

BOTTONI, Giorgio, Degree in Elec. Eng., Specialisation in Nucl. Eng. Born 1929. Educ.: Bologna Univ., Milan Politecnico and I.S.N.S.E. Argonne, Illinois. Thermal Power Plants Div., Edisonvolta S.p.A., 1954-; Tech. Staff, S.E.L.N.I., 1957-; Responsible for co-ordination of design and construction of 270 MWe Enrico Fermi Nucl. Power Plant; Thermal Power Plant Div., E.N.E.L., 1966-. Society: F.I.E.N.
Nuclear interest: Design and construction of nuclear power plants.
Address: E.N.E.L. - C.P.C.T., 10 Via Cardano, 20124 Milan, Italy.

BOTTRELL, Charles. Born 1913. Power Station Design and Layout Eng., Central Elec. Authority, later British Elec. Authority, 1950-56; Chief Power Station Eng., English Elec. Co., 1957; Chief Project Eng., Wylfa, 1963-66; Project Manager (Contracts), Nucl. Design and Construction Ltd., 1966-. Society: Corporate Member and Fellow, Inst. Elec. Eng.
Nuclear interests: Power generation; design and construction of nuclear power stations.
Address: 9 Highland Avenue, Kirby Muxloe, Leics., England.

BOUCHAYER, Robert, Ing. (Inst. Electrotechnique de Grenoble). Born 1919. Educ.: Grenoble Univ. C.E.A., Electricite de France and U.S.S.I., 1961-64. President and Directeur Gen., Sté. d'Entreprise de Montages.
Nuclear interest: Erection of nuclear equipments.
Address: 22 rue Ampère, Grenoble 38, France.

BOUCHER, Thomas, W., B.S. Born 1927. Educ.: Murray State Coll. Union Carbide Nucl. Co., Paducah, 1951-. Member: Kentucky

Advisory Com. on Nucl. Energy, Kentucky Development Council, Exec. Board of Kentucky State AFL-CIO; Sec.-Treas., OCAW, AFL-CIO Atomic Energy Workers Council; Officer, Paducah Atomic Local OCAW 3-550.
Address: Route 2, Harris Road, Paducah, Kentucky, U.S.A.

BOUCHET, Jacques, Eng. (French Polytech. School), Eng. (Atomic Eng.). Born 1922. Educ.: Ecole Polytechnique, Paris and Inst. Nat. des Sci. et Techniques Nucléaires, Saclay. Deputy Director, Region d'Equipement Thermique II (-1966), Head, Groupe Régional de Production Thermique OISE, (1966-), E.D.F. Society: French Assoc. of Nucl. Eng. (former Pres.).
Address: 9 Square Gabriel Fauré, 75 Paris 17, France.

BOUCHEZ, Robert Elie, D. ès Sc. Born 1921. Educ.: Paris Univ. Societies: Sté. Française de Phys.; A.P.S.; The Phys. Soc.; Sté. Italienne de Phys.
Nuclear interest: Nuclear physics theoretical and experimental nuclear reactions.
Address: Institut des Sciences Nucléaires, B.P. 21, rue des Martyrs, 38 Grenoble, France.

BOUCHINET-SERREULLES, C. Chairman Sté. d'Exploitation de Produits Industriels.
Address: 47 rue du Faubourg St-Honoré, Paris 8, France.

BOUCKE, Gerhard, Dipl. Ing. Born 1930. Educ.: Aachen T.H. Developing Eng., AEG - Telefunken Forschunginst., Ulm, 1956-.
Nuclear interests: Nuclear electronics, especially for space application.
Address: 43 Schoener-Berg-Weg, 79 Ulm, Donau, Germany.

BOUDINOV, Ivan. Member, Com. for the Peaceful uses of Atomic Energy, Council of Ministers of the Bulgarian People's Republic.
Address: Council of Ministers, Sofia, Bulgaria.

BOUDOT-WAHL, Jeanne, Agrégée de mathématiques. Born 1930. Educ.: La Sorbonne, Ecole Normale Superieure de Jeunes Filles. Prof. aux lycées de Dijon, Chartres et Sèvres, 1952-58; Ing., C.E.A. (C.E.N., Saclay), 1958-. Society: A.F.C.A.L.T.I.
Nuclear interests: Analyse numérique appliquée aux problemes nucléaires; Equation de transport; Equation de Thomas-Fermi-Dirac; Calculs de sections efficaces.
Address: Chateau de Merzé, Merzé-Cortambert par Cluny 71, France.

BOUDRANT, Robert Jean Charles. Born 1902. Educ.: Ecole Polytech., Paris. Chef du Service, Production Thermique (1946-55), Directeur, Production et Transport (1955-66), Elec. France; Prés. Directeur Gén., Sté. Nucl. France Belge Ardennes.
Nuclear interest: Management.
Address: Electricité de France, 3 rue de Mesine, Paris 8, France.

BOUDREAU, William Francis, M.Sc. (Phys. Metal.), B.Sc. (Metal. Eng.). Born 1914. Educ.: M.I.T. and Case Inst. Technol. Supt., Compressor Dept., Oak Ridge Gaseous Diffusion Plant, Union Carbide Nucl. Co., 1949-55; Development Eng., Reactor Div., Oak Ridge Nat. Lab., Union Carbide Nucl. Co., 1955-63; Manager, Compressor Development, York Div., Borg-Warner Corp., 1963-. Societies: A.N.S.; A.S.M.E.
Nuclear interests: Design and development of mechanical equipment for nuclear reactors, particularly rotating machinery, such as blowers for gas-cooled reactors and in-pile loops, or pumps and other devices for molten salts and liquid metals.
Address: 1391 Stratford Road, York, Pennsylvania, U.S.A.

BOUE, Paul. Treas. (-1967), Pres. (1967-), Assoc. des Ing. en Genie Atomique.
Address: Association des Ingenieurs en Genie Atomique, 6 rue de Castellane, Paris 8, France.

BOUILLENNE-COMHAIRE, Mrs. F. Sci. concerned with Radioisotope Work, Centre de Recherches des Hormones Vetales de l'I.R.S.I.A.
Address: Centre de Recherches des Hormones Vegetales de l'I.R.S.I.A., Institut de Botanique, 3 rue Fusch, Liège, Belgium.

BOUISSIERES, Georges, Dr. ès Sci. Born 1917. Educ.: Paris Univ. Maître de Conférence (1957), now Prof., Faculté des Sci., Paris Univ. Book: Contribution to Traité de chimie minérale (by P. Pascal): Radium, Tome IV, p. 929-955 (1958); Actinium, Tome VII, p. 1413-1446 (1959); co-author, Radon, Tome I, p. 1071-1083 (1955); co-author, Protactinium Tome XII, p. 617-680 (1958). Societies: A.C.S.; Sté. Chimique de France; Sté. de Chimie-Physique; Sté. française de Physique.
Nuclear interest: Chimie des radioéléments lourds et réactions nucléaires.
Address: Institut de Physique Nucléaire, Laboratoire Curie, 11 rue Pierre et Marie Curie, Paris 5, France.

BOUKIS, Sotos D., B.Sc., M.B.A. Born 1933. Educ.: Graduate School Ind. Studies, Athens, and Kansas Univ. Chief Administration and Classification Sect., 7206 Support Group, U.S. Air Force, 1959-60; Part-time Lecturer of Business Administration, Overseas Branch, Maryland Univ., 1959-67; Public Relations Officer, Greek A.E.C., 1960-67; Prof., Business Administration, Pierce Coll., Athens, 1962-67. Member, Nucl. Public Relations Contact Group. Books: Exploring the Atom (Greek) (translation from the English) (Greek A.E.C., 1960); A Study of Atomic Energy in Greece (Greek) (Greek A.E.C., 1961); The New Power (Greek) (translation from the English).
Address: 8 Vrioulon Street, Athens 505, Greece.

BOULADOUX, Maurice. Born 1907. Pres., French Federation of Christian Workers; Member, Branch of Internat. and Economic Institutions, French Economic and Social Council; Formerly Member, Economic and Social Com. to the European Community; Workers' Rep., and Pres., Euratom Economic and Social Com.
Address: 26 rue de Montholon, Paris 9, France.

BOULANGER, Robert. Production Manager, Mecaserto S.A.
Address: Mecaserto S.A., 126 boulevard d'Alsace-Lorraine, Le Perreux, (Seine), France.

BOULANGER, Warner, Dr. iur. Born 1920. Educ.: Heidelberg Univ. Regierungsrat, Ministry of Economics, Baden-Württemberg, 1956-58; Ministerialrat, Federal Ministry for Sci. Res. (former Ministry of Nucl. Energy), 1958-; Director, Legal Division, I.A.E.A., 1965-.
Society: International Law Assoc.
Nuclear interests: International co-operation in nuclear field; Atomic energy law.
Address: 12 Grüner Weg, 532 Pech über Bad Godesberg, Germany.

BOULAY, G. H. du. See du BOULAY, G. H.

BOULDIN, Walter. Financial Com., Power Reactor Development Co.; Pres., Alabama Power Co.; Pres., Edison Elec. Inst., 1963-.
Address: Alabama Power Co., 600 N 18th Street, Birmingham 2, Alabama, U.S.A.

BOULENGER, René Raphael, L. en Sc. physiques. Born 1919. Educ.: Brussels Univ., Health phys., Union Minière du Haut Katanga, 1947-53; Head, Rad. Measurement and Control Dept., Centre d'Etudes de l'Energie Nucléaire, 1955. Member, Advisory Board, journal Health Physics. Societies: H.P.S.; A.P.S.; A.N.S.; Sté Française de Physique; Sté. Belge de Physique.
Address: Villa 15, C.E.N., Mol-Donk, Belgium.

BOULY, Georges. Pres., Conseil d'Administration, Soc. Industrielle de Combustible Nucléaire.
Address: Societe Industrielle de Combustible Nucléaire, 32 rue de Lisbonne, Paris 8, France.

BOUQUIAUX, Jacques Joseph. Chem. Sci. Health Physicist. Born 1921. Educ.: Brussels Univ. Lab. de Chimie et de Physique, Inst. d'Hygiène et d'Epidémiologie, Ministry of Health, 1946-.
Nuclear interests: Protection against ionising radiation; Health physics.
Address: 56 Square Marie-Louise, Brussels 4, Belgium.

BOURAOUI, Habib. Chief, Radioisotope Sect., Tunisian A.E.C.
Address: Commissariat à l'Energie Atomique, Secrétariat d'Etat au Plan et aux Finances, Tunis, Tunisia.

BOURAT, G. Head, Nucl. Dept., Sté. des Usines Chimiques Rhône-Poulenc.
Address: Société des Usines Chimiques Rhône-Poulenc, Service des Recherches Nucléaires, 9 quai Jules Guesde, Vitry, (Seine), France.

BOURCEAU, Gérard Adrien, Ing. Dipl. Born 1914. Educ.: Ecole Nat. Supérieure Arts et Métiers, Ecole Navale and Ecole Supérieure d'Electricité. Director, Marine Dept. Bureau Veritas. Book: Gen. Tech. Conditions for Nucl. Ships (Bureau Veritas, 1962). Societies: Roy. Inst. Naval Architects; I.Mar.E.; Schiffbau Technische Gesellschaft; Assoc. Tech. Maritime et Aéronautique; Sté. Française Electriciens.
Nuclear interests: Metallurgy; Reactor design; General arrangements for nuclear ships and land-based power plants.
Address: Bureau Veritas, 31 rue Henri Rochefort, Paris 17, France.

BOUREK, Rudolf. Head. Ing., Lab. Measuring Instruments, Siemens and Halske G.m.b.H.
Address: Siemens und Halske G.m.b.H., Wiener Schwachstromwerke, 12 Apostelgasse, Vienna 3, Austria.

BOURGEOIS, Benjamin. Sen. Officer, Pignons S.A.
Address: Pignons S.A., Ballaigues, Switzerland.

BOURGEOIS, David. Sen. Officer, Pignons S.A.
Address: Pignons S.A., Ballaigues, Switzerland.

BOURGEOIS, Jean André Léon, Ing. Born 1914. Educ.: Ecole Polytechnique and E.S.E. Chef du Service de la Pile de Fontenay-Roses, 1954-58; Adjoint au Directeur de la Physiques et des Piles Atomiques, 1959-62; Chef du Département des Etudes de Piles, 1963.
Nuclear interests: Reactor shielding; Reactor studies; Reactor safety.
Address: Commissariat à l'Energie Atomique, B.P. 2, 91 Gif-sur-Yvette, France.

BOURGEOIS, Michel Robert, Eng., L. ès Sc. Born 1930. Educ.: Nancy Univ. Eng. (1957), Chief, Sect. (1967), C.E.A.
Nuclear interest: Research and development for reprocessing of spent nuclear fuels by unaqueous methods.
Address: Commissariat a l'Energie Atomique, Centre d'Etudes Nucléaires de Fontenay-aux-Roses, B.P. 6, 92 Fontenay-aux-Roses, France.

BOURGEOIS, Samuel. Sen. Officer, Pignons S.A.
Address: Pignons S.A., Ballaigues, Switzerland.

BOURGES, Yvon. Minister of State responsible for Sci. Res. and Atomic and Space Questions.
Address: 2 rue Royale, Paris 8, France.

BOURNIA, Anthony, B.S. (Chem. Eng.), M.S. (Chem. Eng.), Ph.D. Born 1925. Educ.: Idaho and Pittsburgh Univs. Societies: A.N.S.;

A.I.Ch.E.; A.R.S.; A.I.A.A.
Nuclear interests: Reactor design; Heat transfer; Fluid flow; Nuclear rockets.
Address: Astronuclear Laboratory, P.O. Box 10864, Westinghouse Electric Corporation, Pittsburgh, Pennsylvania 15236, U.S.A.

BOURNS, Arthur Newcombe, B.Sc. (Hons.) (Chem.), Ph.D. (Chem.). Born 1919. Educ.: Acadia and McGill Univs. Prof. Chem., and Vice-Pres., Sci., McMaster Univ. Societies: Fellow, Chem. Inst. of Canada; F.R.S.C.; A.C.S.
Nuclear interests: Kinetic isotope effects and tracer studies in the organic reaction mechanism field.
Address: McMaster University, Hamilton, Ontario, Canada.

BOURSNELL, John Colin, Ph.D. (London and Cambridge), A.R.C.S. Educ.: London Univ. S.P.S.O., Agricultural Res. Council, Unit of Reproductive Physiology and Biochem., Animal Res. Station, Cambridge. Book: Safety Techniques for Radioactive Tracers (Cambridge Univ. Pres, 1958). Societies: F.R.I.C.; Biochem. Soc.; Soc. for the Study of Fertility.
Nuclear interest: The use of isotopes in the solution of biochemical and physiological problems.
Address: Agricultural Research Council. Unit of Reproductive Physiology and Biochemistry, Animal Research Station, Huntingdon Road, Cambridge, England.

BOUSSARD, Roger, Ing. Formerly Chef, Département de Construction des Piles, Direction des Piles Atomiques, now Chef du Centre de la Hague, C.E.A. Delegate, I.A.E.A. Symposium on Fuel Element Fabrication, with Special Emphasis on Cladding Material, Vienna, 10-13 May, 1960. Expert, Third I.C.P.U.A.E., Geneva, Sept. 1964.
Address: Commissariat à l'Energie Atomique, Centre de la Hague, La Hague, Cherbourg, France.

BOUTOUYRIE, Bernard, Ing. (E.S.E., Paris). Born 1933. Educ.: Paris Univ. Dept. Manager, Lab. de l'Accelerateur Linéaire, Orsay, France. Society: Sté. Française des Electriciens.
Nuclear interests: Magnets (conventional and superconducting) vacuum tanks, power supplies, magnetic measurements (NMR, Hall effect).
Address: 12 avenue de la Promenade, Burs sur Yvette, (Seine et Oise), France.

BOUTRY, Georges, Ph.D., Agrégé de l'Université. Born 1904. Educ.: Paris Univ. Prof., Electronics, Conservatoire National des Arts et Métiers; Prés. Laboratoires d'Electroniques et de Physique Appliquée.
Address: Laboratoires d'Electroniques et de Physique Appliquées, 3 avenue Descartes, 94 Limeil Brevannes, France.

BOUVIER, Georg, Dipl. -Ing., Dr. tech. Born 1923. Educ.: Vienna T.H. and Graz T.H. Member, Managing Com., Veitscher Magnesitwerke A.G.
Nuclear interest: Investigation of tracer methods, radiation methods, Emanier methods.
Address: Veitscher Magnesitwerke A.G., Forschungsinstitut, 2 Magnesitstrasse, A-8707 Leoben, Austria.

BOUVILLE, André Charles, Dr. de Specialite (Phys. Nucl.). Born 1939. Educ.: Toulouse Univ. Asst., Faculté des Sci., Toulouse Univ. Society: Soc. Française de Phys.
Nuclear interest: Natural and artificial radioactivity of the atmospheric air.
Address: Centre de Physique Atomique et Nucléaire, 118 Route de Narbonne, 31 Toulouse, France.

BOUVY, Charles Franciscus, Masterdegree Netherlands laws. Born 1916. Educ.: Leiden State Univ. In private enterprise in Indonesia (management rubber and tea estates) 1950-58; working at (1958-63), Deputy Director (1964-), Directorate of Nucl. Energy, Minister of Economic Affairs, Netherlands. Sec., Ind. Nucl. Energy Council; Formerly Sec., Interdepartmental Nucl.Energy Com. (Neth.).
Address: 188 Leyweg, The Hague, Netherlands.

BOUW, Poey Seng, Dr. Lecturer, Nucl. Chem. Dept., Bandung Inst. of Technol.
Address: Bandung Institute of Technology, 10 Djl. Ganeca, Bandung, Indonesia.

BOVAY, H. E., Jr. Partner, Bovay Engs. Inc.
Address: Bovay Engineers Inc., 5009 Caroline Street, Houston 4, Texas, U.S.A.

BOVERI, Walter, Dr.; Brown, Boveri et Cie; Pres., Force Atomique S.A. Deleg. to 2nd I.C.P.U.A.E., Geneva, Sept. 1958.
Address: Brown, Boveri et Cie, Baden, Switzerland.

BOVEY, L. Dr., F. Inst. P. Hon. Sec., Spectroscopy Group, Inst. Phys. and the Phys. Soc.
Address: Spectroscopy Group, Institute of Physics and the Physical Society, 47 Belgrave Square, London S.W.1, England.

BOVIER, Ralph Frederick, Ind. Mech. Eng. Born 1908. Educ.: Pratt Inst. Pres., Pennsylvania Elec. Co. Vice Pres., Pennsylvania Elec. Assoc.
Address: Pennsylvania Electric Company, 1001 Broad Street, Johnstown, Pennsylvania, U.S.A.

BOWDEN, Andrew Thomson, Ph.D., B.Sc. Born 1900. Educ.: Edinburgh Univ. Chief Res. Eng. (1945), Director (1955), C.A. Parsons and Co. Ltd.; Tech. Director, Nucl. Power Plant Co., Ltd., 1955; Director, Internat. R. and D. Co. Ltd., 1963. Societies: Inst. Mech. Eng.; Inst. Marine Eng.; Iron and Steel Inst.; N.E. Coast Inst. of Eng. and Shipbuilders.
Nuclear interests: Design, manufacture and testing of nuclear power station equipment

at home and abroad.
Address: C.A. Parsons and Co., Ltd., Heaton Works, Newcastle upon Tyne, 6, England.

BOWDEN, Athelstan Claude Muir CORNISH-. See CORNISH-BOWDEN, Athelstan Claude Muir.

BOWEN, Boone M. Member, Board of Directors, Dynatomics Inc., 1963-.
Address: Dynatomics Inc., 180 Mills Street, N.W., Atlanta, Georgia 30313, U.S.A.

BOWEN, Carroll G. Director, M.I.T. Press, M.I.T. Member, Advisory Com. on Tech. Information, U.S.A.E.C.
Address: M.I.T. Press, Massachusetts Institute of Technology, Cambridge 39, Massachusetts, U.S.A.

BOWEN, John Henry, B.Sc. (Hons.). Born 1918. Educ.: Wales Univ. Res. into control characteristics of reactors, A.E.R.E. (1950), Tech. design of Calder Hall (1951), Asst. Chief Eng., Calder Project, Risley. Design of control and instrumentation of Calder Hall (1953), Asst. Chief Eng., Ind. Power Branch (1955), Chief Tech. Officer, Safeguards Div., Authority Health and Safety Branch, Risley (1957-), U.K.A.E.A. Book: Instrumentation and Control of Nucl. Reactors (Temple Press, 1959). Societies: A.M.I.Mech.E.; A.M.I.E.E. Nuclear interests: Safety of design and operation of nuclear reactors.
Address: 5 Leys Road, Timperley, Cheshire, England.

BOWEN, R. L. Vice-Pres., Texas Atomic Energy Research Foundation.
Address: Texas Atomic Energy Research Foundation, P.O. Box 970, Fort Worth, Texas, U.S.A.

BOWEN, Vaughan Tabor, B.A., Ph.D. Born 1915. Educ.: Yale Univ. Biol., Brookhaven Nat. Lab., 1948-54; Geochem., (1954-63), Sen. Sci. (1963-), Woods Hole Oceanographic Inst.; Lecturer in Zoology (1952-63), Lecturer in Biol. (1963-), Yale Univ. Societies: American Soc. of Naturalists; American Soc. of Zoologists; Corporation Marine Biol. Lab., Woods Hole; Corp. Bermuda Biol. Station.
Nuclear interests: Radionuclides as tracers in geochemistry, metabolism, ecology; Radiochemical analyses; Fallout and disposal in the oceans; Isotope determination techniques.
Address: Woods Hole Oceanographic Institution, Woods Hole, Massachusetts, U.S.A.

BOWES, D. R. Director, Johnson's Ethical Plastics Ltd.
Address: Johnson's Ethical Plastics Ltd., 32 Ajax Avenue, Trading Estate, Slough, Buckinghamshire, England.

BOWES, John, B.Sc. (hons.). Born 1926. Educ.: Roy. Coll. of Sci. and Technol., Glasgow. Air Conditioning Engineer, Thermotank, Ltd., 1949-53; Naval Machinery Design Eng., Yarrow-Admiralty Res. Dept., Yarrow and Co., Ltd., Glasgow, 1953-55; Chem. Plant Design Eng., I.C.I., Ltd., Glasgow 1955-57; Naval Machinery Design Eng. (1957-61), Chief Steam Eng. (1961-64), Project Chief Eng. (1964-67), Project Manager (1967-). Yarrow-Admiralty Res. Dept., Yarrow and Co., Ltd., Glasgow. Society: A.M.I.Mech.E.
Address: 11 Antonine Road, Bearsden, Glasgow, Scotland.

BOWIE, Stanley Hay Umphray, B.Sc. (1st Class Hons.). Born 1917. Educ.: Aberdeen Univ. Geologist, Sen. Geologist, Principal Geologist, Chief Geologist, Atomic Energy Div., Geological Survey of Great Britain, 1946-67; Chief Geological Consultant to U.K.A.E.A. 1955-; Head, Geochemical Div., Inst. of Geological Sci. 1967-; Member, Appl. Earth Sci. Editorial Board, Inst. of Mining and Metal.; Chairman, Com. on Geochemistry, Mineralogical Soc.; Sec., Com. on Ore Mineralogy, Mineralogical Soc.; Sec., Internat. Mineralogical Assoc., Commission on Ore Microscopy. Books: Contribution to Nucl. Geology (ed. J. Faul, 1954); Physical Methods in Determinative Mineralogy (ed. J. Zussman, 1967). Societies: Inst. Mining and Metal.; Soc. Economic Geologists: Mineralogical Soc. Great Britain; Geologists' Assoc. Great Britain; Fellow, Geological Soc. London; Mineralogical Soc. America.
Nuclear interests: Supply of raw materials of atomic energy; Research in fields of radiogeology, isotope geology, geochemistry and ore microscopy.
Address: Institute of Geological Sciences, Geochemical Division, 64-78 Gray's Inn Road, London W.C.1, England.

BOWIE, Thomas B., B.S., M.S. Born 1922. Educ.: Teacher's Coll. of Connecticut and N.Y.U. Nucl. Materials, Combustion Eng., Inc., Nucl. Div., Windsor, Connecticut, 1956-. Society: Inst. of Nucl. Materials Management. Nuclear interests: Advancement of nuclear materials management in industry and by government agencies and academic institutions through the establishment and encouragement of prescribed standards in the management of nuclear materials.
Address: Combustion Engineering, Incorporated, Nuclear Power Department P.O. Box 500, Windsor, Connecticut 06095, U.S.A.

BOWKER, Harry Whalley, M.A. (Cantab.), C. Eng., M.I.Mech.E. Born 1918. Educ.: Cambridge Univ. Asst. Director (Design), Risley, U.K.A.E.A., 1954; Feasibility Studies Group Leader (1954-58), Deputy Head (1958-59), Reactor Div., A.E.R.E., Harwell; Head of Eng. Services Div., Atomic Energy Establishment, Winfrith Heath, 1959-.
Nuclear interests: Management of a division responsible for the design and procurement of zero energy reactors and experimental equipment; Operation, maintenance and extension of site services; Training of apprentices.

BOW

Address: Tornell House, Leigh, Nr. Sherborne, Dorset, England.

BOWLER, M. G. Res. Officer, Nucl. Phys. Lab., Dept. of Nucl. Phys., Oxford Univ.
Address: Oxford University, Department of Nuclear Physics, 21 Banbury Road, Oxford, England.

BOWLES, B. J., B.Sc., Ph.D. Chem., CENTO Inst. of Nucl. Sci.
Address: CENTO Institute of Nuclear Science, P.O. Box 1828, Tehran, Iran.

BOWLES, Percy, B.Sc. (Eng., 1st class Hons.), M.Sc., M.I.Mech.E., Fellow, I.E.E. Born 1914. Educ.: Victoria Univ., Manchester. Head, Eng. Services Div. (1954-58), Deputy Chief Eng. (1958-61), A.E.R.E.; Chief Eng. Rutherford Lab., N.I.R.N.S., 1961-. Southern Branch Com., Inst. Mech. Eng.; Southern Elec. Consultative Council; I.E.E. Power Board. Societies: M.I.Mech.E.; Fellow, I.E.E.
Nuclear interests: Management; Engineering design, development and construction.
Address: 11 South Drive, Atomic Energy Research Establishment, Harwell, Berkshire, England.

BOWLES, Ronald Edward, B.S. (Mech. Eng.), M.S., Ph.D. Born 1924. Educ.: Maryland Univ. Mech. Eng., Naval Ordnance Lab., White Oak, Maryland, 1948-50; Sen. Eng., Johns Hopkins Univ., Appl. Phys. Lab., Silver Spring, Maryland, 1950-54; Aeronautical Power Plant Res. Sci., Res. Div., O.M.L., Redstone Arsenal, Alabama, 1954; Branch Chief, Non-Radio Systems Branch, Diamond Ordnance Fuze Labs., Washington, D.C., 1954-61; Pres., Bowles Eng. Co., Silver Spring, Md., 1961-. Societies: Inst. of the Aerospace Sciences; A.N.S.; American Ordnance Assoc.; I.S.A.; Washington Acad. Sci.; A.S.H.E.
Nuclear interests: Pure fluid systems and nuclear vulnerability.
Address: 12712 Meadowood Drive, Silver Spring, Maryland, U.S.A.

BOWLUS, Sanford S. Copes-Vulcan Div., Blaw-Knox Co.
Address: Blaw-Knox Co., Copes-Vulcan Division, 939 West 26th Street, Erie, Pennsylvania, U.S.A.

BOWMAN, D. E., M.D. Prof. Biochemistry and Chairman, Res. Unit in Radiobiol. and Radiochem., Indiana Univ.
Address: Indiana University, Medical Centre, 1100 West Michigan Street, Indianapolis Indiana 46207, U.S.A.

BOWMAN, Robert A. Vice Pres. and Chief Eng., Power and Ind. Div., and Director (1967-), Bechtel Corp.
Address: Bechtel Corporation, 220 Bush Street, San Francisco 4, California, U.S.A.

BOWMAN, Roy N., B.S. (Elec. Eng.). Born 1909. Educ.: California Univ., Berkeley. Manager, Bingham Pump Co. Societies: A.N.S.; Professional Eng. of California.
Nuclear interests: Reactor circulating pumps and associated pumping equipment.
Address: Bingham Pump Company, 2800 N.W. Front Avenue, Portland, Oregon 97210, U.S.A.

BOWN, John Edwin, B.Sc. (Hons.) (Eng.). Born 1914. Educ.: Bristol Univ. A.E.R.E., Harwell, since 1948; Head, Eng. Support Div., Harwell, 1966-. Societies: A.M.I.E.E.; A.M.I. Mech.E.
Nuclear interests: General organisation and management of design and engineering support services to nuclear research organisations; Special technical interest in design and development of shielded caves and cells and remote handling equipment.
Address: 3 South Drive, Atomic Energy Research Establishment, Harwell, Berks., England.

BOXER, Leslie Walter. Born 1925. Educ.: Borough Polytechnic, London, and Harvard Business School. Elec. Eng., Plant Div., Crompton Parkinson, -1957; Commercial Dept., Atomic Power Div., English Elec., 1957-60; Counsellor, E.N.E.A., O.E.C.D., 1960-. Societies: B.N.E.S.; Fellow, I.E.E.
Nuclear interests: International scientific and technical collaboration, especially in reactor technology and nuclear ship propulsion; General development of reactor engineering and new applications.
Address: European Nuclear Energy Agency, 38 boulevard Suchet, Paris 16, France.

BOYA, Luis Joaquin. Licenciado en Ciencias Fisicas (B.Sc.), Doctor en Ciencias Fisicas (Ph.D.). Born 1936. Educ.: Zaragoza Univ. Prof. encargado de Fisica Nucl. Zaragoza Univ., 1958-60; Prof. Adjunto de Fisica Matemática (1963-65), Prof. Agregado de Fisica Atómica y Nucl. (1966-), Barcelona Univ. Society: Instituto de Fisica Teórica, Barcelona.
Nuclear interests: Fisica nuclear teórica; Particulas elementales por métados grupateóricas.
Address: 2 José Antonio, Zaragoza, Spain.

BOYADJIYEV, A. V. Paper: Co-author, Cascade Interactions of Particles with Nuclei in High Energy Region (letter to the Editor, Atomnaya Energiya, vol. 16, No. 6, 1964).
Address: Academy of Sciences of the U.S.S.R., 14 Leninsky Prospekt, Moscow V-71, U.S.S.R.

BOYARINOV, V. S. Paper: Co-author, On Approximate Reactor Kinetics Description Applied in Stability Investigation (Atomnaya Energiya, vol. 21, No. 4, 1966).
Address: Academy of Sciences of the U.S.S.R., 14 Leninsky Prospekt, Moscow V-71, U.S.S.R.

BOYARSHINOV, L. M. Papers: Reflection of 250-1200 keV Energy Electrons (Atomnaya Energiya, vol. 21, No. 1, 1966); Angular Distribution of Secondary Reflected Beta-Rad. (ibid, vol. 21, No. 6, 1966); Back-Scattering of Positrons from Targets of Various Compositions (ibid, vol. 22, No. 5, 1967).
Address: Academy of Sciences of the U.S.S.R., 14 Leninsky Prospekt, Moscow V-71, U.S.S.R.

BOYCE, Joseph C., B.A. (Princeton), Ph.D. (Princeton). Born 1903. Educ.: Princeton, London, and Cambridge Univs. Prof. and Chairman, Dept. of Phys., Coll. of Eng., New York Univ., 1945-50; Assoc. Lab. Director, Argonne Nat. Lab., 1950-55; Academic Vice-Pres. (1955-61), Vice-Pres. for Graduate Studies and Res. (1961-62), Illinois Inst. Technol.; Asst. Director, Office of Sci. Personnel, N.A.S.-N.R.C. Washington, 1963-. Book: New Weapons for Air Warfare (Editor, 1947).
Societies: A.P.S.; Royal Astronomical Soc.
Nuclear interests: Nuclear engineering education.
Address: 906 New Hampshire Avenue N.W., Washington 37, D.C., U.S.A.

BOYD, A. D. McN. Director, Atomic Power Constructions Ltd.; Director, United Power Co. Ltd.; Vice-Chairman, Richardsons, Westgarth and Co. Ltd.
Address: Atomic Power Constructions Ltd., Vigilant House, 6-14 Sutton Court Road, Sutton, Surrey, England.

BOYD, Frederick Charles, B.A.Sc. Born 1926. Educ.: Toronto Univ. Phys., Commercial Products Div., A.E.C.L., 1951-55; Reactor phys., Civilian Atomic Power Dept., Canadian G.E.C., 1955-58; Sci. Adviser, Reactors, Atomic Energy Control Board, 1958-. Sec., Reactor Safety Advisory Com. Societies: A.N.S.; A.P.E.O.
Nuclear interest: Nuclear safety.
Address: 9 Sandwell Crescent, P.O. Box 145, Kanata, Ontario, Canada.

BOYD, George Edward, A.Chem.E., B.S., Ph.D. Born 1911. Educ.: Chicago Univ. Asst. Lab. Director, O.R.N.L., 1955-; Chairman, Div. Nucl. Chem. and Technol., A.C.S., 1966-. A.N.S. Rep., N.A.S. -Nat. Res. Council, 1966-69. Hon. Editorial Board, Radiokemiia (Radiochemistry, U.S.S.R.) translation, 1961-; Editorial Board, Radiochimica Acta, 1963-. Societies: A.C.S.; A.P.S.; Fellow, A.N.S.; A.A.A.S.
Nuclear interests: Pure and applied nuclear chemistry; Isotopes; Reactor chemistry; Nuclear fuel reprocessing; Nuclear and inorganic chemistry of technetium; Physical chemistry of ion exchangers; Production of high specific activity isotopes by Szilard-Chalmers reactions in nuclear reactors; Radiation chemistry of inorganic solids.
Address: Oak Ridge National Laboratory, P.O. Box X, Oak Ridge, Tennessee 37831, U.S.A.

BOYD, J. Cookman, Jr., B.A., LL.B. Born 1906. Educ.: Johns Hopkins and Maryland Univs. Chairman, Board of Trustees, Maryland Acad. of Sci.; Rad. Control Advisory Board, State of Maryland.
Address: 900 Aurora Federal Building, Baltimore, Maryland 21201, U.S.A.

BOYD, James Emory, A.B., M.A., Ph.D. Born 1906. Educ.: Georgia, Duke, and Yale Univs. Prof. Phys., Georgia Inst. Technol. 1946-; Res. Assoc., Director of Radar Projects (1946-50), Chief of Phys. Sci. Div. (1950-57), Asst. Director, Res. (1954-55), Assoc. Director (1955-57), Director (1957-61), Eng. Experiment Station, Pres. (1961-), West Georgia Coll. Member, U.R.S.I. Nat. Com.; Radar Subpanel of R.D.B. Commanding Officer, Naval Reserve Res. Co. 6-1, 1949-57; Chairman, Georgia Tech.'s Nucl. Sci. Com., 1955-; Chairman, Res. Com., Georgia Nucl. Advisory Commission, 1957-.
Societies: Inst. of Radio Eng.; A.P.S.; A.A.A.S.
Nuclear interests: Nuclear physics; Radioactivity; Neutron diffraction; Management.
Address: West Georgia College, Carrollton, Georgia, U.S.A.

BOYER, Mrs. M. Asst., Internat. Lab. of Marine Radioactivity.
Address: International Laboratory of Marine Radioactivity, Musée Océanographique, Monaco.

BOYER, Robert Lee, B.Sc. (Mech. Eng.). Born 1931. Educ.: Lehigh Univ. Analytical Eng. (1957-66), Group Supv. (1966-), Babcock and Wilcox Co. Society: American Management Soc.
Nuclear interests: Mechanical, systems, heat transfer, fluid flow, instrumentation, electrical, and control design for fast breeder reactor plants.
Address: Babcock and Wilcox Company, Atomic Energy Division 5061 Fort Avenue, P.O. Box 1260, Lynchburg, Virginia 24505, U.S.A.

BOYESEN, Jens. Member, Norwegian Atomic Energy Council.
Address: Foreign Office, 7 Juni plassen, Oslo-Dep., Norway.

BOYLE, Archibald Raymond, B.Sc., Ph.D. Born 1920. Educ.: Birmingham Univ. Tech. and Managing Director, Dobbie McInnes (Electronics) Ltd., 1958-65; Prof., Elec.Eng., Saskatchewan Univ., 1965-.
Address: Department of Electrical Engineering, Saskatchewan University, Saskatoon, Saskatchewan, Canada.

BOYLE, John Gerard, B.Sc. (Appl. Chem.). Born 1916. Educ.: Glasgow Univ. With Ministry of Supply (Dept. Atomic Energy) and U.K.A.E.A., 1947-; on loan to I.A.E.A., Vienna, 1963-64. Societies: A.R.I.C.; A.M.I.Chem.E.
Nuclear interest: Chemical plant management.

BOY

Address: Windscale Works, Seascale, Cumberland, England.

BOYLE, P. Chief Eng., Baldwin Instrument Co. Ltd.
Address: Baldwin Instrument Co. Ltd., Lowfield Street, Dartford, Kent, England.

BOYRIE DE MOYA, Emil, Member, Comision Nacional de Investigaciones Atomicas, Dominican Republic.
Address: Comision Nacional de Investigaciones Atomicas, Ciudad Trujillo, Dominican Republic.

BOZIN, G. M. Papers: Co-author, Penetration of Fast Neutrons through Thick Lithium Hydride Slabs (letter to the Editor, Atomnaya Energiya, vol. 21, No. 5, 1966); co-author, Transmission of Fast Neutrons through Cylindrical Duct Partially Penetrating Slab Shield (letter to the Editor, ibid, vol.24, No. 1, 1968).
Address: Academy of Sciences of the U.S.S.R., 14 Leninsky Prospekt, Moscow V-71, U.S.S.R.

BOZOKI, Eva Susan, Ph.D. Born 1938. Educ.: Roland Eötvös Univ., Budapest. Jun. Res. Assoc. (1959-), Sen. Res. Assoc. (1965-), Central Res. Inst. of Phys., Budapest, Sen. Res. Assoc., Joint Inst. of Nucl. Res., Dubna, 1964-65. Society: Roland Eötvös Phys. Soc.
Nuclear interest: High energy physics.
Address: Central Research Institute of Physics, P.O.B. 49, Budapest 114, Hungary.

BOZOKI, George, Ph.D., Cand. of Phys. Sci. Born 1930. Educ.: Roland Eötvös Univ., Budapest. Res. Assoc. (1953), Sen. Res. Assoc. (1962), Central Res. Inst. of Phys., Budapest; Deputy Leader, Lab. for Cosmic Rays, 1962; Group Leader, Lab. for High Energy Phys., Joint Inst. of Nucl. Res., Dubna, 1964-65; Asst. Prof., Faculty of Atomic Phys., Roland Eötvös Univ., 1959-66. Member, Subcom. for High Energy and Elementary Particle Phys., Hungarian Acad. of Sci. Society: Roland Eötvös Phys. Soc.
Nuclear interest: Cosmic ray and high energy physics.
Address: Central Research Institute for Physics, P.O.B. 49, Budapest 114, Hungary.

BOZOKY, László, Ph.D., Doctor of phys. sci. Born 1911. Educ.: Pázmány Péter Univ., Budapest. Head, Phys. Dept., Central Inst. for Oncology 1937-; Head, Radiol. Dept., Central Inst. for Phys. Res., 1952-59; Univ. lecturer, 1951-; Univ. prof., 1965-. Pres., Hungarian Electrotech. Soc.; Pres., E.L. Phys. Soc.; Exec. Council, Internat. Rad. Protection Assoc. Book: Protection against Nucl. Rad. (Budapest, Müszaki Könyvkiadó, 1960). Societies: Hungarian Electrotech. Soc.; Hungarian Biophys. Soc.; E.L.Phys. Soc.; Internat. Rad. Protection Soc.
Nuclear interests: Radiation measurements; Dosimetry; Teletherapy; Radiation sources;

Health physics; Whole body counting.
Address: Budapest XI, Szabolcska M. u. 1., Hungary.

BRAAM HOUCKGEEST, Floris A. van. See van BRAAM HOUCKGEEST, Floris A.

BRAAMS, Cornelis M., Dr. Born 1925. Educ.: Utrecht Univ. Prof. Plasma Phys., Utrecht Univ.; Director, F.O.M. Inst. Plasma Phys., 1958-.
Nuclear interest: Thermo-nuclear research.
Address: Instituut voor Plasma Fysica, Rijnhuizen, Jutphaas, Netherlands.

BRABEC, Vlastislav, R.N.Dr., C.Sc. Born 1932. Educ.: Charles Univ., Prague. Society: Assoc. of Czechoslovak Mathematicians and Phys.
Nuclear interest: Nuclear physics (beta-and gamma-ray spectroscopy).
Address: Ustav jaderného výzkumu C.S.A.V., Rez, Czechoslovakia.

BRABEN, Donald Walter, B.Sc., Ph.D. Born 1935. Educ.: Liverpool Univ. Demonstrator, Liverpool Univ., 1960-61; Res. Assoc. (1961-62), Asst. Prof.(1962-63), Alberta Univ.; S.S.O., Daresbury Nucl. Phys. Lab., Sci. Res. Council, 1963-. Society: Inst. of Phys. and the Phys. Soc.
Nuclear interest: High energy photon and electron interactions.
Address: Science Research Council, Daresbury Nuclear Physics Laboratory, Keckwick Lane, Daresbury, Nr. Warrington, Lancs., England.

BRABERS, Martin Joannes, Dr. Ir., Phys. Metal., Chem. Eng. Born 1926. Educ.: Delft Tech. Univ., Foundation for Fundamental Res. of Matter, Delft Univ., 1951-54; Asst. Metal. Dept., Princeton Univ., 1954-56; Netherlands-Norwegian Joint Establishment for Nucl. Energy Res., Lillestrøm, Norway, 1956-59; Reactor Centre, Petten, Netherlands, 1959; Studiecentrum voor Kernenergie (Centre d'Etude de l'Energie Nucléaire) Mol. Belgium, 1959-; Prof., Delft Tech. Univ., 1963-.
Nuclear interest: Physical metallurgy; Physical chemistry.
Address: Studiecentrum voor Kernenergie, Centre d'Etude de l'Energie Nucléaire, Mol, Belgium.

BRACCI, Alberto, Dr. in Phys. Born 1928. Educ.: Milan Univ. Director, Lab. for Reactor Phys. and Computation; Director, Appl. Nucl. Res. Branch, C.N.E.N. Society: A.N.S.
Nuclear interests: Reactor design; Management.
Address: C.N.E.N. 15 Via Belisario, Rome, Italy.

BRACE, Kirkland Clifford, B.A., M.D. Born 1921. Educ.: Iowa and Illinois Univs.; Medical Director, U.S. Public Health Service. Societies: Rad. Res. Soc.; American Medical Assoc.; American Coll. of Radiol.

Nuclear interests: Radiation biology; Radiation therapy.
Address: Radiation Branch, National Cancer Institute, Bethesda 14, Maryland, U.S.A.

BRACEWELL, George Marshall, A.M.I.E.E. Born 1923. Educ.: Burnley Tech. Coll. Res. Eng., C.A. Parsons Ltd., Newcastle-on-Tyne, 1949-51; Sen. Elec. Eng., Tweedales and Smalley (Platts) Ltd., 1951-54; Third Asst. Eng., H.Q. Res. Sect. (London), Central Elec. Authority, 1954-56; Head of Systems Analysis and Controls Group and later Head, C.T.R. Lab., Hawker Siddeley Nucl. Power Co. Ltd., 1956-59; Head, Safety Dept., Atomic Power Constructions Ltd.
Nuclear interests: Transient analysis of reactor fault conditions and control problems; Safety aspects of reactor operation and commissioning.
Address: 139 Mulgrave Road, Cheam, Surrey, England.

BRACHET, P. Head, Bureau d'Etudes Techniques P. Brachet.
Address: Bureau d'Etudes Techniques P. Brachet, 12 rue du Havre, Paris 9, France.

BRADBURN, Richard Neville, B.Sc. (Eng.). Born 1923. Asst. Chief Eng., Fuel Element Plants Design. Society: A.M.I.Mech.E.
Nuclear interest: Nuclear fuel processing and fabrication plants.
Address: 10 Lime Grove, Lowton, Warrington, Lancs., England.

BRADBURY, Norris E., B.A. (Chem.), Ph.D. (Phys.), D.Sc. (Hon.), LL.D. (Hon.). Born 1909. Educ.: Pomona Coll., California, and California Univ. Prof., Stanford Univ. 1942-50; Director, Los Alamos Sci. Lab. (1945-), Prof. (1951-), California Univ. Societies: A.P.S.; N.A.S.
Nuclear interests: Nuclear rocket propulsion; Experimental power reactors; Direct energy conversion; Atomic weapons; Direction and management nuclear research.
Address: Los Alamos Scientific Laboratory, P.O. Box 1663, Los Alamos, New Mexico, U.S.A.

BRADBURY-WILLIAMS, John Clifford, C. Eng., A.M.I.Mech.E., Assoc. Member, Inst. Elec. Eng., M.R.S.H. Born 1925. Educ.: Wandsworth Tech. Coll. Eng. Gen. Eng. Dept., Metropolitan-Vickers Elec. Co. Ltd., Manchester, 1950-53; Eng. II, A.E.R.E. Harwell, Design and Construction Group, 1953-57; Sen. Project Eng., Atomic Energy Dept., Babcock and Wilcox Ltd., 1957-61; Inspector, Nucl. Health and Safety Dept. (1961-64), Sen. Asst. Eng., Quality Control Unit (1964-67), C.E.G.B.; Independent Consultant and Lecturer in Safety.
Nuclear interests: Quality surveillance; Safety; Reliability.
Address: 44 Birdwood Road, Maidenhead, Berks., England.

BRADDYLL, John Richmond Gale, B.Sc., C. Eng. Born 1919. Educ.: Durham Univ. Manager, Nucl. Power Div. (1960-63), Deputy Manager, Marine Div. (1963-64), Manager, Marine Div. (1964-67), Manager, Marine and Diesel Div. (1967-) Vickers-Armstrongs (Eng.) Ltd. Societies: A.M.I.Mech.E.; M.I.Mar.E.
Nuclear interest: Manufacture of components for reactor systems including those for current nuclear submarine programme.
Address: Vickers Limited, Engineering Group, Barrow-in-Furness, Lancs., England.

BRADISH, John Patrick, B.S.M.E., M.S.M.E., Ph.D. Born 1923. Educ.: Lowell Inst. School (M.I.T.), Marquette and Wisconsin Univs. Instructor (1946), Asst. Prof. (1953), Director, Mech. Eng. Dept. (1958-61), Assoc. Prof. (1964), Marquette Univ. Internat. Editor, Nucl. Energy Eng.; Consultant, Argonne Nat. Lab., 1964-. Societies: A.S.M.E.; A.S.E.E.
Address: Marquette University, 1515 W. Wisconsin Avenue, Milwaukee, Wisconsin 53233, U.S.A.

BRADLEY, Arthur, Ph.D. Born 1926. Educ.: Columbia Univ. Director, Res., Rad. Res. Corp., 1958-.
Nuclear interests: Radiation effects on insulating materials; Activation of phosphors by β-emitters; Tritiated titanium foil.
Address: Radiation Research Corporation, 1150 Shames Drive, Westbury, Long Island, New York, U.S.A.

BRADLEY, George Edgar, Ph.D. Born 1924. Educ.: Miami and Michigan Univs. Prof. Phys., Western Michigan Univ. Pres., Michigan Phys. Teachers. Societies: A.P.S.; Nucleonics Soc.
Nuclear interests: Education; Beta spectroscopy.
Address: Physics Department, Western Michigan University, Kalamazoo, Mich., U.S.A.

BRADLEY, J. E. Director, Explosives Operations, Mound Lab.
Address: Mound Laboratory, Mound Road, Miamisburg, Ohio, U.S.A.

BRADLEY, John Ernest Stobart, B.Sc., Ph.D. Born 1927. Educ.: London Univ. In Biophys. Div., Nat. Inst. for Medical Res., 1952-53; Asst. Lecturer in Biophys., Radioisotopes Dept., Postgraduate Medical School of London, 1953-56; Lecturer, Dept. of Phys. Applied to Medicine, Middlesex Hospital Medical School, 1956-.
Books: Chinese-English Glossary of Mineral Names (N.Y., Plenum Press, 1963); Translator, Phys. of Nucl. Fission (Pergamon Press, 1957); translator, Principles of Radiochem. (Pergamon Press, 1962). Societies: Chem. Soc. (for isotope chem., nucl. chem., etc.).
Nuclear interests: Radiochemistry (especially cyclotron-made isotopes), Mossbauer effect (principally with chemical slant), low-level counting (unconventional techniques), radiation damage in organic solids.
Address: 83 Muswell Hill Road, London, N.10, England.

BRADLEY, John Martin, M.D., M.R.A.C.P., D.T.R., M.C.R.A., F.F.R. Born 1925. Educ.: Melbourne Univ. Consultant Radiotherapist, Peter MacCallum Clinic, 1965-; Visiting Radiotherapist, Repatriation Gen. Hospital, Heidelberg, 1962-67; Honorary Radiotherapist, Royal Melbourne Hospital, 1966-. Society: Coll. of Radiologists of Australasia.
Nuclear interests: Radiotherapy and radiation biology.
Address: 278 William Street, Melbourne 3000, Australia.

BRADLEY, Norman, B.Sc. Born 1925. Educ.: Manchester Univ. Project Designer (1950-53), Asst. Chief Designer, Land Boilers (1953-57), Foster Wheeler Ltd.; Sen. Design Eng. (1957-59), Asst. Chief Eng. (1959-63), Deputy Chief Eng. (1963-), U.K.A.E.A., Risley. Society: Inst. of Mech. Eng.
Nuclear interest: Reactor design directed to low-cost nuclear power.
Address: 14 Wellfield Road, Culcheth, Warrington, Lancs., England.

BRADNER, Hugh, A.B., Ph.D., D. Born 1915. Educ.: Miami Univ. and California Inst. Technol. Prof., Eng. Phys. and Geophysics, California Univ., San Diego. Society: Fellow, A.P.S.
Nuclear interests: Fundamental particle physics; Particle accelerators.
Address: California University, San Diego, La Jolla, California, U.S.A.

BRADSHAW, H. Sec., Nucl. Reactor Sect., British Elec. and Allied Manufacturers' Assoc. Inc.
Address: British Electrical and Allied Manufacturers' Association Inc., Nuclear Reactor Section, 8 Leicester Street, Leicester Square, London W.C.2, England.

BRADY, Edward Lewis, B.A. (California), M.A. (California), Ph.D. (M.I.T.). Born 1919. Educ.: California Univ. and M.I.T. Res. Assoc., Gen. Elec. Res. Lab., 1948-55; Manager, Coolant Chem., K.A.P.L., Gen. Elec. Co., 1955-56; U.S.A.E.C. Sci. Representative, American Embassy, London, 1956-58; Manager, Exptl. Equipment Development, K.A.P.L., Gen. Elec. Co., 1958-59; Sen. Sci. Adviser, United States Mission to I.A.E.A., Vienna, 1959-61; Chem. Dept., Gen. Atomic, San Diego, California, 1961-; Member, Subcom. on Techniques for the Distribution of Sci. Information, N.A.S. - N.R.C. Societies: A.C.S.; A.N.S.; A.P.S.
Nuclear interests: Materials; Radiochemistry.
Address: Chemistry Department, General Atomic, P.O. Box 608, San Diego 12, Calif., U.S.A.

BRADY, F. P., Prof. Sen. Officer, Crocker Nucl. Lab., Phys. Dept., California Univ., Davis.
Address: Crocker Nuclear Laboratory, California University, Davis, California, U.S.A.

BRADY, James J., A.B., M.A., Ph.D. Born 1904. Educ.: Indiana and California Univs. Prof. Phys., Oregon State Univ., 1943-; Phys., Rad. Lab., Calif. Univ., Summer 1958; Phys., Navy Electronics Lab., San Diego, Summer 1961. Societies: Fellow, A.P.S.; Fellow, A.A.A.S.
Nuclear interest: Worked on first disintegration by means of cyclotron (11″).
Address: Department of Physics, Oregon State University, Corvallis, Oregon, U.S.A.

BRAEKKEN, Tore. Director, Reactor A/S.
Address: Reactor A/S, 4 Tollbugata, Oslo, Norway.

BRAEM, Lucien, A. G. A., Ing. Civil des Mines, Ing. Electricien. Born 1902. Educ.: Brussels Univ. and Ecole Superieure d'Electricite, Paris. Hon. Manager and Consultant, Bell Telephone Mfg. Co.; Pres. Gen., Ste. Royale Belge des Electriciens; Administrateur, Groupement Professionnel de l'Industrie Nucléaire; Administrateur, Fondation Nucleaire.
Nuclear interests: Reactors instrumentation; Radioisotopes application.
Address: 156 Chaussée de Boitsitsfort, Brussels 17, Belgium.

BRAESTRUP, Carl Bjorn, B.S., P.E. Assoc. Prof. Born 1897. Educ.: M.I.T. Assoc. in Radiol., Columbia Univ., 1947-; Director, Phys. Lab., Delafield Hospital, 1950-. Director,and Chairman, Sci. Com. No. 9, N.C.R.P.; Expert Advisory Panel on Rad., W.H.O. Book: Co-author, Rad. Protection (Illinois, Charles C. Thomas, 1958). Societies: Rad. Res. Soc.; American Radium Soc.; Radiol. Soc. of North America.
Nuclear interests: Radiation protection and measurements.
Address: P.O. Box 230, Guildford, Connecticut 06437, U.S.A.

BRAGA, Carlos AZEVEDO COUTINHO. See AZEVEDO COUTINHO BRAGA, Carlos.

BRAGA, Elde Pires, Electronics and Elec. Eng. Born 1925. Educ.: Brazil Univ. Associated Eng., Syncrocyclotron Res., Conselho Nacional de Pesquisas, 1952-55; Chief Eng., Electronics and Res. Centre Brasileiro de Pesquisas Fisicas, 1956-58; Chief Eng., Electronics and Elec. Measurements, Inst. Nacional de Tecnologia, 1961-; Lectures on Electromagnetism, Escola Nacional de Engenharia. Books: Co-author, Singularities of Linear System functions. (Editora: Elservier); co-author, Euler'continuous fraction expansion and its applications in electrical circuit theory. (Air Force contract No. AF49(638)-648).
Nuclear interest: Nuclear physics.
Address: Instituto Nacional de Tecnologia, 82 Avenida Venezuela, Rio de Janeiro, Brazil.

BRAGANCA, Beatriz M., M.Sc., Ph.D., Educ.: McGill Univ. Head, Biochem. Div., and Dean, Cancer Res. Inst.; Lecturer, Bombay Univ. Societies: Biochem. Soc. (U.K.).; Soc. of Biol.

Chemists (India).
Nuclear interest: Application of radioisotopes to the study of biological problems, especially folic acid metabolism.
Address: Cancer Research Institute, Tata Memorial Centre, Bombay, India.

BRAJOVIĆ, Veljko, Dr. Head, Dept. of Solid State Phys., Boris Kidric Inst. of Nucl. Sci.
Address: Boris Kidric Institute of Nuclear Sciences, P.O. Box 522, Belgrade, Yugoslavia.

BRAKEL, Henry L. Prof. Emeritus, Phys. Dept. Faculty, Washington Univ.
Address: Washington University, Physics Department, Seattle, Washington 98105, U.S.A.

BRAMHAM, Stanley Frank. Born 1920. Educ.: Birmingham Central Tech. Coll. Nucl. Applications Eng., Leeds and Northrup Ltd. Society: Inst. Measurement and Control.
Nuclear interests: Rig control systems and special instrumentation.
Address: Leeds and Northrup Ltd., Wharfdale Road, Tyseley, Birmingham 11, England.

BRANCA, Guido. Graduated in Maths., Phys., and Civil Eng. (specialisation in Hydraulics). Born 1926. Educ.: Naples Univ. Formerly Asst. Prof., Inst. of Phys., Naples; now Chairman, Sanitary Eng. Lab., Nat. Com. for Nucl. Energy (C.N.E.N.). Society: Assoc. Nazionale di Ingegneria Sanitaria.
Nuclear interests: Water supply; Air and water pollution; treatment of domestic and industrial waste (liquid, solid and gaseous); Treatment, storage and disposal of radioactive waste; Decontamination of areas and equipment.
Address: 15 Via Belisario, Rome, Italy.

BRANCH, Harllee, Jr., A.B. (Davidson Coll., North Carolina), LL.B. (Emory Univ., Atlanta). Born 1906. Educ.: Emory Univ. (Law School). Member (1956-), Vice-Chairman (1957-63), Georgia Nucl. Energy Advisory Commission. U.S. Observer, Atoms-For-Peace Conference, Geneva, Switzerland, 1958. Member, Atomic Energy Com., Chamber of Commerce of the United States 1958-62.
Address: 3106 Nancy Creek Road, N.W., Atlanta, Georgia, U.S.A.

BRAND, Friedrich. German Member, Economic and Social Com. Specialised Nucl. Sect. for Economic Problems, Euratom.
Address: 7 Jahnstrasse, Oldenburg i.O., Germany.

BRAND, Glenn Eldon, B.S.(Ch.E.), M.S., (Ch.E.), Ph.D. Born 1917. Educ.: Missouri School of Mines, Washington State Coll. Asst. Prof. Chem. Eng., Missouri School of Mines, 1952-56; Group Leader, Reactor Chem., Atomics Internat., North American Rockwell, 1956-.
Nuclear interest: Fuel reprocessing.
Address: 196 Colt Lane, Thousand Oaks, California 91360, U.S.A.

BRAND, Max von. See von BRAND, Max.

BRANDAU, Carlos. Born 1908. Educ.: Madrid and Berlin Univs. Member, Comisión del Riesgo Atómico, Sindicato Espanol de Seguros.
Address: Sindicato Nacional del Seguro, Comisión del Riesgo Atómico, 4 Avenida Calvo Sotelo, Madrid 1, Spain

BRANDBERG, Sven, M. of Technol. Born 1923. Educ.: Stockholm Roy. Univ. of Technol. Alby New Chlorate, Avesta, 1955-59; A.S.E.A., Vasteras, 1959-. Societies: Swedish Chem. Soc.; Swedish Assoc. of Eng. and Architects.
Nuclear interests: Manufacture of reactor fuel including the raw materials concerned; Sales and marketing questions regarding reactor fuel.
Address: Nuclear Power Department, Reactor Fuel Factory, Allmanna Svenska Elektriska A.B., Vasteras, Sweden.

BRANDE, J. van den. See van den BRANDE, J.

BRANDES, David, M.D. Born 1918. Educ.: Buenos Aires Univ. Asst. Prof. Pathology, Johns Hopkins School of Medicine.
Address: Johns Hopkins School of Medicine, Baltimore, Maryland, U.S.A.

BRANDES, Eric A., B.Sc., A.R.C.S. Born 1911. Educ.: London Univ. Process Metal., Fulmer Res. Inst. Ltd., 1946-. Societies: Inst. of Metals; Iron and Steel Inst.; Fellow, Inst. of Metal.
Nuclear interests: Metallurgy; Ceramics.
Address: 46 Fieldway, Chalfont St. Peter, Buckinghamshire, England.

BRANDI, Hermann Theodor, Dr. mont., Dr. Ing. e.h. Born 1908. Educ.: Aachen, Berlin and Leoben T.H. Hüttendirektor, Vorstandsmitglied der August Thyssen-Hütte A.G., Duisburg; Vorstand Wirtschaftsvereinigung; Vorstand Verein Deutscher Eisenhüttenleute; Fachkommission III, Tech. Wirtschaftliche Fragen bei Reaktoren, Ministerium für wissenschaftliche Forschung.
Nuclear interest: Werkstoff-Fragen.
Address: 4 Wichernstrasse, Mülheim-Ruhr-Speldorf 433, Germany.

BRANDT, Horst, Dr. rer. pol. Geschaftsfuhrer, Kernkraftwerk Lingen G.m.b.H.
Address: Kernkraftwerk Lingen G.m.b.H., A.E.G. Building, Rheinlanddamm, 46 Dortmund, Germany.

BRANDT, Juergen, Dr. -Ing., Dr. rer. pol. Born 1885. Educ.: Munich, Berlin and Hamburg Univs. Member, Hamburg Res. Council. Societies: Study Assoc. for the Use of Nucl. Energy in Navigation and Industry, Hamburg.

Nuclear interests: Design and construction of reactors.
Address: 113 Rothenbaumchaussee, Hamburg 13, Germany.

BRANDT, Wolfgang Leo, Prof. Dr. med. h.c., Dr. -Ing.E.h., Dipl. Ing., Ehrensenator der Tech Univ. Berlin. Born 1908. Educ.: Lessing-Oberrealschule, Düsseldorf; Tech. Hochschule, Aachen; Tech. Univ. Berlin. Staatssekretär im Ministerium für Wirtschaft und Verkehr des Landes Nordrhein-Westfalen, 1953; Staatssekretär und Leiter des Landesamtes für Forschung bei dem Ministerpräsidenten des Landes Nordrhein-Westfalen, 1953; Honorar-prof. Aachen T.H.; Stellv. Vorsitzender der Deutschen Atomkommission und der Deutschen Kommission für Weltraumforschung; Ehrenpräsident der Deutschen Gesellschaft für Ortung und Navigation; Geschäftsführendes Präsidialmitglied der Arbeitsgemeinschaft für Forschung des Landes Nordrhein-Westfalen; Mitglied des Verwaltungsrates des Westdeutschen Rundfunks; Mitglied des Aufsichtsrates der Hoesch-Werkes A.G., Dortmund, des Gussstahlwerkes Witten A.G., der Rheinmetall Berlin A.G. Books: Staat und friedliche Atomforschung (Köln/Opladen,Westdeutscher Verlag); Die zweite industrielle Revolution (Munich, Paul List Verlag,); Forschen und Gestalten (Köln/Opladen,Westdeutscher Verlag). Societies: Arbeitsgemeinschaft für Forschung des Landes Nordrhein-Westfalen Akademie für Stadt und Landesplanung; Deutsche Atomkommission; Deutsche Kommission für Weltraumforschung.
Address: Landesamt für Forschung bei dem Ministerpräsidenten des Landes Nordrhein-Westfalen, 41 Cecilienallee, 4 Düsseldorf, Germany.

BRANDT REHBERG, Poul Kristian, Ph.D., M.D. Born 1895. Educ.: Copenhagen Univ., Prof. Zoophysiology, Copenhagen Univ. Chairman, Danish A.E.C. Societies: Roy. Danish Acad.; Acad. Tech Services.
Nuclear interests: Biological effects of radiation; Fallout problems; Radiation protection.
Address: Zoophysiological Lab. A, 34 Juliane Maries Vej, Copenhagen, Danmark.

BRANELL, Karl Erik. Born 1906. Educ.: Chalmers Tech. Inst., Göteborg. Manager, Dept. for Gen. Services, Studsvik Res. Station, A.B. Atomenergi, 1957-.
Address: A.B. Atomenergi, Studsvik, Nyköping, Sweden.

BRANNEN, E., Prof. Dr. Engaged on nucl. res. work, Dept. of Phys., Western Ontario Univ.
Address: Physics Department, Western Ontario University, London, Ontario, Canada.

BRANSCOMB, Lewis M., Dr. U.S.N.B.S., Washington. Member, Standing Com. for Controlled Thermunucl. Res., U.S.A.E.C.
Address: U.S. National Bureau of Standards, Washington D.C. 20234, U.S.A.

BRANSON, Byron Monroe, B.S. (Phys.). Born 1929. Educ.: Guildford Coll. Phys., (1957-60), Supervisory Phys., Chief, Nuclide Analysis (1961-63), Supervisory Phys., In Charge, Radiometrics, (1963-66), Radiol. Health Res. Activities, U.S.P.H.S.; Phys., Nucl. Medicine Unit, Nat. Centre for Radiol. Health, U.S.P.H.S. Societies: H.P.S.; Conference on Radiol. Health.
Nuclear interests: Whole body counting; Dose reduction in nuclear medicine; Whole body scanning.
Address: National Centre for Radiological Health, Nuclear Medicine Unit, Building T-2, 1090 Tusculum Avenue, Cincinnati, Ohio 45226, U.S.A.

BRANSON, Herman, Head, Dept. Phys., Howard Univ.
Address: Department of Physics, Howard University, Washington 1, D.C., U.S.A.

BRANSON, Norman G. Pres., Branson Instruments Inc.
Address: Branson Instruments Inc., 76 Progress Drive, Stamford, Connecticut, U.S.A.

BRANTLEY, John Calvin, B.S., Ph.D. Born 1921. Educ.: Northeast Missouri State Teachers Coll. and Illinois Univ. Res. Sci. (1949-56), Asst. Manager, Res. (1956-59), Director, Res. (1959-66), Union Carbide Corp., Phys., Nucl. Sci. and Eng. Corp., 1966-67; Vice-Pres., Internat. Chem. and Nucl. Corp., 1967-; Manager, NEN - Pharmaceutical Div., New England Nucl. Corp., 1968-. Member, Advisory Com. on Isotopes and Rad., U.S.A.E.C.; Chairman, Ad Hoc Com. on Isotopes, Atomic Ind. Forum. Societies: A.C.S.; A.N.S.
Nuclear interests: Reactors; Isotopes; Isotope applications; Management.
Address: International Chemical and Nuclear Corporation, 801 N. Lake Street, Burbank, California, U.S.A.

BRASHKIN, M. A. Papers: Co-author, High Power Gamma Irradiation Univ UK-30000 (letter to the Editor, Atomnaya Energiya, vol. 19, No. 1, 1965); co-author, Gamma-Irradiation Facility with Complex Source (letter to the Editor, ibid, vol. 24, No. 2, 1968).
Address: Academy of Sciences of the U.S.S.R., 14 Leninsky Prospekt, Moscow V-71, U.S.S.R.

BRASWELL, R. W. Vice President in charge of Eng., Mississippi Power and Light Co.
Address: Mississippi Power and Light Company, Jackson 5, Mississippi, U.S.A.

BRAUDO, Charles Joseph, M.A., Ph.D. Born 1918. Educ.: Cambridge Univ. Sen. Sci.,Weizmann Inst. Sci., 1952-59; Director, Radioisotope Training Centre, Rehovot, 1959-. Society: Phys. Soc. and Inst. Phys.

Nuclear interests: Education and training in use of radioisotopes; Industrial and applied research applications of radioisotopes.
Address: Weizmann Institute of Science, P.O. Box 26, Rehovot, Israel.

BRAUER, Heinz, Prof. Dr.-Ing. Born 1923. Educ.: Hanover T.H. Max-Planck-Inst. für Strömungsforschung, 1953-59; Mannesmann-Forschungsinst., 1960-63. Tech. Univ. Berlin, Lehrstuhl für Thermodynamik und Verfahrenstechnik. Society: Verfahrenstechnische Gesellschaft, V.D.I.
Nuclear interests: Heat transfer; Mass transfer; Fluid flow; Rectification.
Address: 19 Boltzmannstrasse, Berlin 33, Germany.

BRAUER, Wolfram, Prof. Dr. habil. Born 1925. Educ.: Berlin Univ. Kommis, Director, Inst. für Theoretische Physik, Humboldt Univ., 1962; Leader, Theoretical Solid State Group, German Acad. Sci.
Address: Physikalische-Technisches Institut der D.A.W., 40-41 Mohrenstrasse, Berlin W 8, German Democratic Republic.

BRAUKMANN, Karl. Born 1907. Bundesvorstand, Deutscher Gewerkschaftsbund. Arbeitskreis II/2 Nachwuchs, Deutsche Atomkommission.
Nuclear interest: Management.
Address: 4 Düsseldorf, Germany.

BRAUN, C. B. Exec. Vice Pres., Nucl. Fuel Services, Inc.
Address: Nuclear Fuel Services, Inc., Central Sales Office, Suite 906, Wheaton Plaza Building, Wheaton, Maryland 20902, U.S.A.

BRAUN, Heinz W. A., Dipl. Ing., D. Eng. Born 1922. Educ.: Berlin Tech. Univ. Development eng. for control circuits and reactor components, A.E.G., Germany, 1949-59; Head of Elec. Eng., Halden Reactor Project (H.B.W.R.), Norway, 1959-62; Head, Res. Div., E.N.E.A., Paris, 1962-63; Vice-Pres., Gerling-Konzern, Cologne, 1963-65; Vice-Pres., Allianz Versicherungs-A.G., Munich, 1966-.
Nuclear interests: Insurance of nuclear power plants.
Address: Allianz Versicherungs-A.G., Munich, Germany.

BRAUN, Henri, D. ès Sc. Phys. Born 1926. Educ.: Strasbourg Univ. Maître de Recherches, C.N.R.S., 1963.
Nuclear interests: Nuclear photographic plates; High-energy physics; Heavy fragments; Hyperfragments.
Address: Département de Physique Corpusculaire du Centre de Recherches Nucleaires, rue du Loess, Strasbourg-Cronenbourg, France.

BRAUN, Josef, M.Sc., Ph.D. Born 1919. Educ.: Roy. Tech. Inst., Stockholm, and Upsala Univ. Head, Radio Sect., Roy. Board Civil Air Administration, 1952-55; Head, Sec. for Gen. Phys., A.B. Atomenergi, Stockholm. Society: A.N.S.
Nuclear interests: Radiation shielding; Radiation detection and measurement; Magnetohydrodynamics; Direct conversion.
Address: A.B. Atomenergi, Studsvik, Nykoping, Sweden.

BRAUN, Ludwig, Jr., B.E.E., M.E.E., D.E.E. Born 1926. Educ.: Brooklyn Polytech. Inst. Gen. Elec.; Grumman Aircraft; Arma Corp.; Prof. Elec. Eng., Brooklyn Polytech. Inst. Professional tech. group on control, I.E.E.E. Book: On Adaptive Control Systems (Brooklyn Polytech. Inst., 1959); co-author, On the Approximate Identification of Process Dynamics in Adaptive Control (1960). Societies: Inst. Radio Eng.; Assoc. Member, I.E.E.E.
Nuclear interest: Control systems.
Address: 333 Jay Street, Brooklyn 1, New York, U.S.A.

BRAUN, Otto Peter, Obering. Born 1903. Educ.: Lage Technikum.
Nuclear interests: Electric heating cables for pipes; Electro heated apparatus for liquids air and gases; Temperature measuring; Project installation.
Address: 86 Ludwigsburgerstrasse, Postfach 65, 7141 Beihingen/N, Germany.

BRAUN, Wolfgang, Dr. rer. nat., Dipl.-Phys. Born 1927. Educ.: Stuttgart Tech. Univ. Atomic Power Dept., Siemens-Schuckertwerke A.G., Erlangen; Member for Germany, Halden Programme Group, O.E.C.D., E.N.E.A. Books: Co-author, Kerntechnik (Teubner, 1958).
Nuclear interests: Thermo- and hydrodynamic layout of reactor core; Reactor physics; Design of propulsion reactors.
Address: Siemens-Schuckertwerke A.G., Abt. Reaktor-Entwicklung, Erlangen/Bay, Germany.

BRAUNBEHRENS, H. VON. See VON BRAUNBEHRENS, H.

BRAUNER, Alvin Robert, B.S. Born 1924. Educ.: American Inst. of Electronics and Eng. At (1950-), now Reactor Supv., I.I.T. Res. Inst.
Nuclear interests: Reactor operations, instrumentation and experimental research.
Address: 711 W. 129th Place, Chicago, Illinois 60628, U.S.A.

BRAVO F. de CASTRO Y. See de CASTRO y BRAVO, F.

BRAY, J. Chef de Sect., Lab. de Recherches et Applications, Glaverbel S.A.
Address: Glaverbel S.A., rue de la Discipline, Gilly, Belgium.

BRAY, P. J., Barrister-at-Law, Commercial Director, Fairey Eng. Ltd.
Address: Fairey Engineering Ltd., Heston, Middlesex, England.

BRAZHNIKOV, E. M. Paper: Co-author, Lab. Installation for Investigation of Cheminucl.

Synthesis (letter to the Editor, Atomnaya Energiya, vol. 20, No. 3, 1966).
Address: Academy of Sciences of the U.S.S.R., 14 Leninsky Prospekt, Moscow V-71, U.S.S.R.

BRAZZINI, Ezzio. Fellow, Peruvian Soc. of Radiol.
Address: Sociedad Peruana de Radiologia, Casilla No. 2306, Lima, Peru.

BREARLY, A. Joint Managing Director, Adamson and Hatchett Ltd.
Address: Nuclear Engineering Department, Adamson and Hatchett Ltd., Engineering Works, Dukinfield, Cheshire, England.

BREATHITT, Edward T. Chairman, Kentucky Atomic Energy and Space Authority.
Address: Kentucky Atomic Energy and Space Authority, Room 28 State Capital Euilding, Frankfort, Kentucky, U.S.A.

BREAZEALE, William McSwain, B.S. (E.E.), Ph.D. (Phys.). Born 1908. Educ.: Rutgers and Virginia Univs. Principal Phys., O.R.N.L. 1949-53; Prof. Nucl. Eng., Pennsylvania State Univ., 1953-56; Asst. Manager, Eng. Dept., Atomic Energy Div., Babcock and Wilcox Co., 1956-63; Director, Nucl. Development Centre, 1963-. Member, U.S. Deleg. to 1960 Internat. Conference for Safety of Life at Sea, and Member, Nucl. Com.; Honors and Awards Com. (1964-), Elected Fellow (1963), A.N.S. Book: Co-author, Transmission Lines, Network and Filters (1949). Societies: A.N.S.; Inst. of Radio Eng.; Soc. of Naval Architects and Marine Eng.
Nuclear-interests: Design and operation of power plants for nuclear merchant vessels; Design and operation of research reactors.
Address: 1112 Villa Road, Lynchburg, Virginia, U.S.A.

BRECH, Frederick. Vice-Pres., Director of Res., Jarrell-Ash Co.
Address: Jarrell-Ash Co., 590 Lincoln Street, Waltham, Massachusetts 02154, U.S.A.

BRECHER, George, M.D. Born 1913. Educ.: Prague, Zurich and London Univs. Chief, Hematology Sect., Dept. Clinical Pathology, Clinical Centre, N.I.H., 1953-66; Prof. and Chairman, Div. Clinical Pathology and Lab. Medicine, Medical Centre, California Univ., 1966-. Society: Rad. Res. Soc.
Nuclear interest: Radiobiology (recovery of hemopoietic tissues from radiation, experimental therapy of X-radiation injury).
Address: California University, Medical Centre, San Francisco, California 94122, U.S.A.

BRECKENRIDGE, Robert George, B.A., M.A., Ph.D. Born 1915. Educ.: Cornell Univ. and M.I.T. Chief, Solid State Phys. Sect., U.S. N.B.S., 1949-55; Director, Res. Lab. (1955-63), Director, Res. Inst. (1963-64), Union Carbide Corp.; Director, Phys. Technol. Dept., Atomics Internat., 1964-. Societies: Fellow, A.P.S.; Washington Acad. of Sci.; A.C.S.; New York Acad. of Sci.
Nuclear interests: Radiation effects in solids, reactor materials, energy conversion, semiconductors, superconductivity.
Address: Atomics International, 8900 DeSoto, Canoga Park, California, U.S.A.

BRECKINRIDGE, John Bayne, A.B., LL.B. Born 1913. Educ.: Kentucky Univ. Formerly Attorney Gen. of Kentucky, 1960-; Formerly Member, Advisory Com. of State Officials, U.S.A.E.C.; Vice-Chairman, Nat. Assoc. of Attorneys Gen. Com. on Atomic Energy; Formerly Vice-Chairman, Southern Interstate Nucl. Board; Chairman, Kentucky Advisory Com. on Nucl. Energy; Member, Kentucky Atomic Energy and Space Authority; Special Com. on Atomic Energy Law, American Bar Assoc.
Address: 1100 Fincastle Road, Lexington, Kentucky, U.S.A.

BRECKON, Douglas Giles, B.A.Sc. Born 1924. Educ.: Toronto Univ. Operations Supv. (1946-54), Project Eng. N.R.X. Reactor (1954-55), Asst. Supt. N.R.X. Reactor (1955-59), Supt. N.R.X. Reactor (1959-62), Supt. N.R.U. Reactor (1962-67), Manager General Services Div. (1967-), Chalk River Nucl. Lab.
Nuclear interest: Radiation and industrial safety.
Address: Chalk River Nuclear Laboratories, Chalk River, Ontario, Canada.

BREDBERG, Carl D., M.Sc.Eng. Member, Board of Directors, AKK Atomic Power Group.
Address: AKK Atomic Power Group, 19 Stureplan, Stockholm C, Sweden.

BREE, Rudolf. Born 1907. Tech. Manager, Microcopy G.m.b.H.; Tech. Manager, Vermittlungsstelle für Vertragsforschung; Consultant for promoting Res. for small and medium Ind., Bundesverband der deutschen Industrie; Director, Centre for Information and Documentation, Euratom. Societies: V.D.I.; Wissenschaftliche Gesellschaft für Luft - und Raumfahrt; Deutsche Gesellschaft für Dokumentation.
Nuclear interest: Nuclear information.
Address: 75 rue Joseph II, Brussels 4, Belgium.

BREEN, Robert James, B.S., M.S. Born 1929. Educ.: St. John's (Minnesota) and St. Louis (Missouri) Univs. Res. Phys., Mound Lab., Monsanto Chem. Co., Miamisburg, Ohio, 1952-54; Instructor, Coll. of St. Thomas, St. Paul, Minnesota, 1954-55; Fellow Sci. (1955-) now Advisory Sci., Bettis Atomic Power Lab., Westinghouse Elec. Corp., Pittsburgh. Societies: A.P.S.; A.N.S.
Nuclear interests: Reactor physics; Reactor design methods; Neutron and gamma spectroscopy.
Address: 254 Orchard Drive, Pittsburgh, Pennsylvania 15228, U.S.A.

BREGADZE, Yu. I. Papers: Co-author, Ionization Methods of Measuring the Energy Absorbed from a Rad. Flux Consisting of Neutrons Mixed with γ-rays (Atomnaya Energiya, vol. 9, No. 2, 1960) ; co-author, Production of Pure Fluxes of Fast Neutrons for Radiobiologic Res. on IRT-1000 Reactor (letter to the Editor, ibid, vol. 12, No. 6, 1962); The Influence of Conducting Layer on Ion Chamber Homogeneity (letter to the Editor, ibid, vol. 19, No. 3, 1965).
Address: Academy of Sciences of the U.S.S.R., 14 Leninsky Prospekt, Moscow V-71, U.S.S.R.

BREGER, A. Kh.
Nuclear interests: High-intensity γ-ray sources and radiation-source reactors.
Address: Academy of Sciences of the U.S.S.R., 14 Leninsky Prospekt, Moscow V-71, U.S.S.R.

BREGNI CHACON, Eduardo, Licenciado en Biologiá. Member, Inst. Nacional de Energia Nuclear.
Address: Instituto Nacional de Energia Nuclear, Apartado Postal 1421, Guatemala, C.A.

BREH, K., Dipl. Phys. Co-Editor, Atompraxis.
Address: Verlag G. Braun, 14-18 Karl Friedrich Strasse, Karlsruhe, Germany.

BREHM, E., Dipl.-Phys. Member, Fachausschuss für Strahlenschutz, (joint group of Deutsches Atomforum and Kernenergieverwertung Gesellschaft).
Address: Brown-Boveri-Krupp Reaktorbau G.m.b.H., 40 Augusta Anlage, 6800 Mannheim, Germany.

BREHM, Richard L., B.S.E., M.S.E., Ph.D. Born 1933. Educ.: Michigan and California (Los Angeles) Univs. Assoc. Prof., Dept. of Nucl. Eng., Arizona Univ., 1965-; Consultant to U.S.A.E.C., 1967; Consultant to Jet Propulsion Lab., 1966. Society: A.N.S.
Nuclear interests: Fast reactor nuclear analysis; Reactor kinetics and nonlinear reactor stability; Thermionic space reactor dynamic analysis; Spatially-dependent neutron dynamic analysis; Neutron transport theory.
Address: 3325 E. Edgemont, Tucson, Arizona 85716, U.S.A.

BREIDENBACH, Ernest C. G. STUEKELBERG de. See STUECKELBERG de BREIDENBACH, Ernest C. G.

BREIT, Gregory, A.B. (Johns Hopkins), A.M. (Johns Hopkins), Ph.D. (Johns Hopkins). D.Sc. (hon., Wisconsin). Born 1899. Educ.: Johns Hopkins Univ. Prof. Phys., Yale Univ. 1947- (Donner professorship since 1958). Books: Author and co-author articles as follows in Vol. XVI/1. Ency. Physics (Handbuch der Physik): Theory of Resonance Reactions and Allied Topics; Coulomb Wave Functions; Polarisation of Nucleons Scattered by Nuclei; Coulomb Excitation (Springer Verlag, 1959); Physical Review, Reviews of Modern Physics. Societies: Nat. Acad. of Sci.; American Acad. of Arts and Sci.; A.P.S.; Inst. of Radio Eng.
Nuclear interests: Nuclear physics; Reactor design.
Address: Sloane Physics Laboratory, Yale University, 2014 Yale Station, New Haven, Connecticut, U.S.A.

BREITENBERGER, Ernst, Dr. phil. (Vienna), Ph.D. (Cambridge). Born 1924. Educ.: Graz and Vienna Univs. Prof. Phys. Ohio Univ., 1963-; Consultant, U.S. Naval Proving Ground, Dahlgren, Va., summer 1959. Societies: A.P.S.; Math. Assoc. (London).
Nuclear interests: Nuclear physics; Stochastic processes.
Address: Department of Physics, Ohio University, Athens, Ohio 45701, U.S.A.

BREITENHUBER, Ludwig, Ph.D. Born 1926. Educ.: Graz Univ. Prof., Inst. for Theoretical Phys. and Reactor Phys., Graz Tech. Univ.; Lecturer for Theoretical Phys., Graz Univ. Societies: Österreichische Physikalische Gesellschaft; A.N.S.
Nuclear interests: Theoretical nuclear physics; Neutron transport theory; Experimental neutron physics.
Address: 27 Merangasse, Graz, Austria.

BREITLING, G., Dozent Dr. rer. nat. Strahlenphysiker, Medizinisches Strahlen-Inst., Tübingen Univ.
Address: Medizinisches Strahlen-Institut, Universität Tübingen, Tübingen, Germany.

BREIVIK, Franz O., Cand. real. Born 1925. Educ.: Oslo Univ. Res. Fellowship, Nucl. Phys. Lab., Oslo Univ. Society: Norwegian Phys. Soc.
Nuclear interest: Nuclear physics (elementary particle physics).
Address: Oslo University, Institute of Physics, Blindern, Norway.

BREMMER, H., Dr. Group Leader, Theory, Inst. voor Plasma Fysica, Foundation for Fundamental Res. on Matter.
Address: Instituut voor Plasma Fysica, Rijnhuizen, Jutphaas, Netherlands.

BREMMERS, L. H. A., Dr. Director of Health Protection, Ministry of Social Affairs and Public Health.
Address: Ministry of Social Affairs and Public Health, 73 Zeestraat, The Hague, Netherlands.

BRENDAKOV, V. F. Paper: Co-author, Radioactive Fallouts on the U.S.S.R. Area in 1963 Year (Atomnaya Energiya, vol. 19, No. 1, 1965).
Address: Academy of Sciences of the U.S.S.R., 14 Leninsky Prospekt, Moscow V-71, U.S.S.R.

BRENIG, Wilhelm, o. Prof. Dr., Inst. für theoretische Phys., Technische Hochschule, and Max-Planck-Inst. für Phys. und Astrophys.
Address: Max-Planck-Institut für Physik und Astrophysik, 6 Föhringer Ring, Munich 23, Germany.

BRENK, Hendrik Athos Sydney van den. See van den BRENK, Hendrik Athos Sydney.

BRENNAN, Donald George, Ph.D. (Mathematics), B.S. (Mathematics). Born 1926. Educ.: M.I.T. Group Leader, Lincoln Lab. (1953-62), Res. Assoc., Mathematics Dept. (1959-61), M.I.T.; Sen. Scholar (1962-), Pres. (1962-64), Hudson Inst. Consultant at various times to U.S. Depts. of State and Defence, Arms Control and Disarmament Agency, and Exec. Office of the Pres. Book: Editor, Arms Control, Disarmament and National Security (New York, George Braziller, 1961). Societies: I.E.E.E.; American Math. Soc.
Nuclear interests: Nuclear weapons and nuclear technology as related to problems of war and peace via military policy, foreign policy, and arms control and disarmament.
Address: Hudson Institute, Quaker Ridge Road, Croton-on-Hudson, New York 10520, U.S.A.

BRENNAN, J. Manager, Production, Plutonium Plant, Nucl. Materials and Equipment Corp.
Address: Nuclear Materials and Equipment Corp., Apollo, Pennsylvania 15613, U.S.A.

BRENNAN, James T., M.C. U.S. Army, M.D., B.A. Born 1916. Educ.: Minnesota and Illinois Univs. Chief, Rad. Therapy, Walter Reed Gen. Hospital, 1960-61; Director, Armed Forces Radiobiol. Res. Inst., Nat. Naval Medical Centre, 1961-66; Prof., Radiol., Pennsylvania Univ. Societies: Radiol. Soc. of North America; H.P.S.; A.M.S.
Nuclear interest: Radiation therapy and neutron radiobiology.
Address: Donner Centre, Department of Radiology, Pennsylvania University, 3400 Spruce Street, Philadelphia, Pennsylvania 19104, U.S.A.

BRENNAN, John Edward Peter, B.E.E., M.B.A. Born 1928. Educ.: Manhattan Coll. and N.Y.U. Nucl. Cost Analyst (1955-56), Nucl. Economist (1956-57), Chief, Market and Economic Analysis (1957-58), Sales Liaison Eng., New York (1958-59), Staff Asst. to Vice-Pres. and Manager, Atomic Energy Div. (1959-60), Babcock and Wilcox; Asst. to Pres. (1960-64), Vice-Pres. (1966-), Garlock Inc. Nucleonics Com., I.E.E.E.; Membership Com., A.N.S. Societies: I.E.E.E.; A.N.S.
Nuclear interests: Nuclear economics; Sales; Management.
Address: Garlock Incorporated, Palmyra, New York, U.S.A.

BRENNER, Alfred Ephraim, Ph.D. Born 1931. Educ.: M.I.T. Ford Foundation Fellow, C.E.R.N., 1958; Instructor (1959), Asst. Prof. (1962), Sen. Res. Assoc. (1966), Harvard Univ. Society: A.P.S.
Nuclear interest: High energy experimental physics.
Address: Harvard University, Lyman Laboratory of Physics, Cambridge 38, Massachusetts, U.S.A.

BRENNER, Marten Withmar, Fil. dr. Born 1926. Educ.: Helsinki Univ. and Argonne Nat. Lab. Assoc. Prof., Inst. of Technol., Helsinki, 1962; Chief Phys., Rad. Therapy Clinic, Univ. Central Hospital, Helsinki 1962-66; Prof. Phys., Abo Akademi, Abo, Finland, 1966-. Chairman, Physicists Club of Turku. Societies: Finnish Sci. Soc.; Nordic Soc. for Rad. Protection; Nordic Soc. for Clinical Phys.
Nuclear interests: Nuclear spectroscopy; Nuclear reactions; Reactor physics; Application of radiation and isotopes in medicine.
Address: 2 B 7 Svartmunkegränd, Abo, Finland.

BRENNER, Otto. Born 1907. Pres., German Metal Workers Union. German member, and Vice Chairman (1968-70), Euratom Social and Economic Council, Nucl. Energy Sect.
Nuclear interests: Economic and social prospects and problems of peaceful use of nuclear energy. The Union is running a special department for research and activities in automation and nuclear energy.
Address: 70-76 Untermainkai, Frankfurt am Main, Germany.

BRESCANCIN, P., Ing. Tech. Manager, Dott. Ing. Giuseppe Torno and C.S.p.A.
Address: Dott. Ing. Giuseppe Torno and C.S.p.A., 7 Via Albricci, Milan, Italy.

BRESEE, James Collins, B.S., M.S., Sc.D. Born 1925. Educ.: Illinois Univ. and M.I.T. Asst. Prof. Chem. Eng., M.I.T., 1951-54; Development Eng., Sect. Chief and Asst. Div. Director, Chem. Technol. Div., (1954-65), Director, Civil Defence Study Group (1965-), O.R.N.L. Societies: A.C.S.; A.I.Ch.E.
Nuclear interests: Supervision of engineering design and development and pilot plant programs for radiochemical processing. Hazards evaluation for chemical operations.
Address: 122 Wendover Circle, Oak Ridge, Tennessee 37130, U.S.A.

BRESLAUER, Stephen K., A.B. (Phys.). Born 1934. Educ.: Cornell and Cincinnati Univs. Nucl. Eng., A.N.P.D., G.E.C., 1958-61; Reactor Phys., Reactor Licensing, U.S.A.E.C., 1961-65; Director, Nucl. Technol., Atomic and Space Development Authority, New York State, 1965-.
Nuclear interests: Nuclear engineering; Nuclear safety; Fast reactors.
Address: New York State Atomic and Space Development Authority, 230 Park Avenue, New York, N.Y. 10017, U.S.A.

BRESLAVALES, K. G. Paper: Co-author, Investigation of Separation of Neon Isotopes in Rectification Film Column (Atomnaya Energiya, vol. 20, No. 6, 1966).
Address: Academy of Sciences of the U.S.S.R., 14 Leninsky Prospekt, Moscow V-71, U.S.S.R.

BRESSON, Gilbert. Adjoint au Chef du Départment de Protection Sanitaire, C.E.A. Expert, Third I.C.P.U.A.E., Geneva, Sept. 1964; Member, Health and Safety Sub-Com., O.E.C.D., E.N.E.A.

Address: Commissariat a l'Energie Atomique, 29-33 rue de la Federation, Paris 15, France.

BRESSON, M. Chef du Service Tech. Metal., Cie. Gen. du Duralumin et du Cuivre.
Address: Cie. Generale du Duralumin et du Cuivre, 66 avenue Marceau, Paris 8, France.

BRETON, Denis, Dr. Phys., Elec. Eng. Born 1923. Educ.: Paris Univ. Chef de Service, Direction des Piles Atomiques, C.E.A., 1957-. Member, European Nucl. Data Com. Society: A.N.S.
Nuclear interests: Reactor physics; Nuclear physics; Plasma physics; Criticality.
Address: 86 avenue de la République, 92 Montrouge, France.

BRETON, Mlle. E. le. See le BRETON, Mlle. E.

BRETSCHER, Manuel Martin, Ph.D. (Phys.). Born 1928. Educ.: Washington Univ., St. Louis. Prof. Phys. and Co-Chairman, Dept. of Phys., Valparaiso Univ., 1956-67; Assoc. Phys., Argonne Nat. Lab., 1967-. Also summers of 1958, 1962, 1963, 1964, 1965. Res. Assoc., O.R.N.L. summers of 1955, 1956, 1957, 1959, 1960. Societies: A.P.S.; A.N.S.; A.a.P.T.
Nuclear interests: The physics of fast nuclear reactors, neutron interactions, and gamma ray spectroscopy.
Address: Argonne National Laboratory, 9700 South Cass Avenue, Argonne, Illinois 60440, U.S.A.

BREVNOV, N. N. Papers: Co-author, The Effects of Magnetic Field Local Disturbances on Plasma Confinement in a Magnetic Adiabatic Trap (Atomnaya Energiya, vol. 13, No. 5, 1962); co-author, The Electrostatic Channel to Inject Ion Beam in the Magnetic Trap (letter to the Editor, ibid., No. 6); co-author, The Passage of a Plasmoid through an Adiabatic Trap with Magnetic Mirrors (ibid, vol.14, No. 4, 1963); co-author, Registration of Hydrogen Ion Streams by Semiconductor Rad. Detector (letter to the Editor, ibid, vol. 20, No. 2, 1966).
Address: Academy of Sciences of the U.S.S.R., 14 Leninsky Prospekt, Moscow V-71, U.S.S.R.

BREWER, Leo, B.S., Ph.D. Born 1919. Educ.: California Inst. Technol., and California Univ., (Berkeley). Asst. Prof. (1945-50), Assoc. Prof. (1950-55), Prof. (1955-), Head, Inorganic Materials Div. (1961-), Dept. Chem. and Lawrence Radiation Lab., Assoc. Director, Lawrence Rad. Lab. (1967-), California Univ. Editorial Advisory Board, J. Phys. and Chem. of Solids; Commission on High Temperatures, I.U.P.A.C., 1957-61. Books: Author, or co-author, Chapters 3-8, The Chemistry and Metal. of Miscellaneous Materials; Thermodynamics (Vol. 19B, Nat. Nucl. Energy Series) (McGraw-Hill, 1950); co-author, Thermodynamics, 2nd edition (McGraw-Hill, 1961).
Nuclear interests: Chemical reactions in high-temperature reactors; Fabrication and chemical processing of fuel elements and other materials of high-temperature reactors.
Address: Chemistry Department, California University, Berkeley 4, California, U.S.A.

BREWER, R. F. Vice Pres., Iowa Southern Utilities Co.
Address: Iowa Southern Utilities Co., Centerville, Iowa, U.S.A.

BREY, Robert Newton, B.S. (Chem. Eng.). Born 1920. Educ.: Pennsylvania and Pennsylvania State Univs. Administrator, Project Operations, Radio Corp. America. Societies: A.I.Ch.E.; A.N.S.; Franklin Inst. American Rocket Society.
Nuclear interests: Management; Nuclear space power systems; Nuclear reactor instrumentation and control; Nuclear weapons.
Address: Radio Corporation of America, Major Systems Div., P.O. Box 147 Moorestown, N.J., U.S.A.

BREZHEVA, N. E., D.Sc. Member, Inst. Phys. Chem., Acad. of Sci. of the U.S.S.R., Deleg. to 2nd I.C.P.U.A.E., Geneva, Sept. 1958. Papers: Co-author, The Dependence of Glass Melt Chem. Resistance and Crystallization Power upon its Composition and Preparation Conditions (Atomnaya Energiya, vol. 15, No. 2, 1963); co-author, An Investigation of Volatilization Process at High Temperatures for Ruthenium Coprecipitated with Different Deposits (ibid., No. 3); co-author, An Investigation of Volatilization Process of Cesium Coprecipitated with Nickel and Calcium Double Ferrocyanide at High Temperatures (letter to the Editor, ibid.); co-author, Res. of Electrodeposition Kinetics of Hydroxide Deposits of Rare-Earths Elements (ibid, vol. 24, No. 5, 1968).
Address: Institute of Physical Chemistry, Academy of Sciences of the U.S.S.R., Moscow, U.S.S.R.

BREZINA, J. J. Chief Mech. Eng., B.D. Bohna and Co. Inc.
Address: B.D. Bohna and Co. Inc., 515 Market Street, San Francisco 5, California, U.S.A.

BRIAND, Jean F. Tech. Manager, Nucl. Dept., Centre d'Etudes et de Réalisations Electroniques S.A.R.L.
Address: Centre d'Etudes et de Réalisations Electroniques S.A.R.L., 7 rue Moret, Paris 11, France.

BRIAND, Jean Pierre, Maître Asst., Sect. Physique, Faculté des Sciences, Inst. du Radium, Paris Univ.
Address: Institut du Radium Laboratoire Curie, 11 rue Pierre Curie, Paris 5, France.

BRIANTI, G. Leader, Synchrotron Injector Div., Proton Synchrotron Dept., C.E.R.N.
Address: European Organisation for Nuclear Research, Meyrin, 1211 Geneva 23, Switzerland.

BRICKER, John W., A.B., LL.B. Born 1893. Educ.: Ohio State Univ. Member, Joint Com.

on Atomic Energy; Member, Joint Com. on Defence Production. Attorney-Gen., State of Ohio, for 2 terms; Governor, State of Ohio, for 3 terms; elected to U.S. Senate in 1946; re-elected 1952; Member of the Board, Ohio State Univ., 1961-62.
Address: Columbus, Ohio, U.S.A.

BRIDGES, The Right Hon. The Lord Edward, K.G., P.C., M.A., Hon. D.C.L. Oxford. Born 1892. Educ.: Oxford Univ. Permanent Sec., H.M. Treasury, 1945-56. Chairman, Governing Body, N.I.R.N.S., 1957-65. Chairman, British Council; Chairman, Fine Art Commission. Society: F.R.S.
Address: Goodmans Furze, Headley, Nr. Epsom, Surrey, England.

BRIDGES, Robert Lysle, B.A., LL.B. Born 1909. Educ.: California Univ. Director, Bechtel Nucl. Corp.
Address: 111 Sutter Street, San Francisco 4, California, U.S.A.

BRIDGMAN, Charles James, B.S., M.S., Ph.D. Born 1930. Educ.: U.S. Naval Academy, Annapolis, New Mexico, Colorado and North Carolina State Univs. Assoc. Prof. of Nucl. Eng. Dept. of Phys. Air Force Inst. of Technol. 1959-., and Chairman, Faculty Continuing Education Com. Faculty Sponsor, Air Force Inst. Technol. Student Branch, A.N.S. Societies: A.N.S.; A.I.A.A.; A.S.E.E.; A.A.U.P.
Nuclear interests: Neutral particle transport in both nuclear reactor and nuclear weapon environment, especially digital computer calculations; Space applications of nuclear energy.
Address: Physics Department, Air Force Institute of Technology, Wright-Patterson Air Force Base, Dayton, Ohio 45433, U.S.A.

BRIEFS, H., Dr. Ing. Quality Dept. Deutsche Edelstahlwerke, A.G.
Address: Deutsche Edelstahlwerke, A.G., Krefeld, Germany.

BRIEGLEB, Günther, Prof. Dr. phil. Born 1905. Vorstand, Inst. für Physikalische Chemie, Würzburg Univ.
Address: Würzburg University, Institut für Physikalische Chemie, 9-11 Marcusstrasse, 87 Würzburg, Germany.

BRIERLEY, Geoffrey, M.I.Mech.E. Born 1924. Educ.: Oldham Tech. Coll. Design Eng. (1954-60), Asst. Chief Eng., Ind. Power, Reactor Group (1960-63), Asst. Chief Eng., Gas-Cooled Reactors Directorate (1963-), U.K.A.E.A., Risley.
Address: 10 Bowers Avenue, Davyhulme, near Manchester, England.

BRIERWALTES, William., M.D. Pres., Soc. of Nucl. Medicine.
Address: Society of Nuclear Medicine, 333 North Michigan Avenue, Chicago 1, Illinois, U.S.A.

BRIESKORN, H. H. German Member, Health and Safety Sub-com., O.E.C.D., E.N.E.A.
Address: O.E.C.D. European Nuclear Energy Agency, 38 boulevard Suchet, Paris 16 France.

BRIEZ, G. Responsable du Dept., Nucl., Materiel Medical, Massiot-Philips S.A.
Address: Material Medical, Massiot-Philips S.A., 40 avenue Hoche, Paris 8, France.

BRIGGS, George Edward, M.A. Born 1893. Educ.: Cambridge Univ. Prof. Botany (1948-60), Prof. Emeritus of Botany (1960-), Cambridge Univ. Book: Co-author, Electrolytes and Plant Cells (Blackwell, 1961); Movement of Water in Plants (Blackwell, 1967). Society: F.R.S.
Nuclear interest: Use of radioactive isotopes in salt relation of plants.
Address: Botany School, Downing Street, Cambridge, England.

BRIGGS, Robert Beecher, B.S. (Chem. Eng.). Born 1918. Educ.: Wayne Univ. Assoc. Director, Molten-Salt Reactor Programme, O.R.N.L. Member, Atomic Safety and Licensing Panel, U.S.A.E.C. Book: Contributor to Nuclear Engineering Handbook (Editor, Harold Etherington) (McGraw-Hill Book Co., Inc., 1958). Societies: A.I.Ch.E.; A.C.S.
Nuclear interest: Design of nuclear reactor plants and management of reactor projects.
Address: Oak Ridge National Laboratory, Oak Ridge, Tennessee, U.S.A.

BRIGHT, Glenn O., B.S. (Eng. Phys.), M. Eng. Phys. Born 1922. Educ.: Oklahoma Univ. Phys. (1950-61), Programmes Consultant (1962-), Phillips Petroleum Co.; Special Instructor, Nucl. Eng., Oklahoma Univ., 1961-62. Society: A.N.S.
Nuclear interest: Reactor safety.
Address: Phillips Petroleum Co., P.O. Box 2067, Idaho Falls, Idaho, U.S.A.

BRIGHT, Herbert Samuel, B.S. (Phys.), B.S. (Maths.), M.S. (E.E.). Born 1919. Educ.: Michigan and California Univs. Supervising Sci., Computation Planning Sect., Bettis Atomic Power Div., Westinghouse Elec., Pittsburgh, Pa.; Pres., Transac. Computer Users' Group; Assoc. Editor (Standards), A.C.M. Communications; Chairman, A.C.M. Standards Com. Societies: A.P.S.; A.A.A.S.; Assoc. Computing Machinery; Inst. Radio Eng.
Nuclear interests: Nuclear physics measurements; Digital computation.
Address: P.O. Box 1468, Pittsburgh 30, Philadelphia, U.S.A.

BRIGHT, Norman Francis Henry, B.Sc. (Hons.), Ph.D. Born 1913. Educ.: Bristol Univ. Head, Phys. Chem. Sect., Mines Branch, Dept. of Energy, Mines and Resources, Govt. of Canada, 1953-. Councillor, Inorg. Chem. Div., Treas. and Director, Chem. Inst. of Canada. Societies: F.R.I.C.; Fellow, Chem. Inst. Canada; Fellow,

American Ceramic Soc.; American Crystallographic Assoc.
Nuclear interests: Technology of uranium dioxide preparation, fabrication, and sintering; Chemistry of uranium oxy-compounds.
Address: Mines Branch, 555 Booth Street, Ottawa 4, Ontario, Canada.

BRIGHTSEN, Ronald Armund, B.S., M.S. Born 1925. Educ.: Michigan Univ. and M.I.T. Senior Sci., Westinghouse Atomic Power Div., 1950-54; Pres., Nucl. Sci. and Eng. Corp., 1954-66; Pres., Panatomics, Inc., 1966-; Advisory Com., Axe Sci. Corp., 1954-; Director, Hazleton Nucl. Sci. Corp., 1960-65; Director, Radioactive Materials Corp., 1965-66. Book: Editorial Asst., Radiochem. Studies: The Fission Products. Societies: A.N.S.; N.Y. Acad. Sci.
Nuclear interests: Production and applications of radioactive materials; Nuclear investment analysis.
Address: 3333 Ivanhoe Road, Pittsburgh, Pennsylvania 15241, U.S.A.

BRIGOLI, Bruno, Dr. Chem. Born 1924. Educ.: Pisa Univ. Farmitalia (Montecatini), 1951-53; C.I.S.E., 1953-; Now Lecturer at Politecnico Milan. Book: Separazione degli isotopi: deuterio e acqua pesante (Politecnico Milan, 1959).
Nuclear interest: Physical chemistry: isotope separation.
Address: 29 Via Pinturicchio, Milan, Italy.

BRIKKER, I. N. Papers: Co-author, On the Frequency Analysis of Schemes with Goingup Reactor (letter to the Editor, Atomnaya Energiya, vol. 15, No. 1, 1963); Inversion Solution of Equations of Reactor Kinetics (ibid, vol. 21, No. 1, 1966); co-author, Inversion Solution of Thermal Equations for Two-Component Nucl. Reactor (ibid, vol. 24, No. 4, 1968).
Address: Academy of Sciences of the U.S.S.R., 14 Leninsky Prospekt, Moscow V-71, U.S.S.R.

BRIL, Kazimierz Jozef, Licencié en Sci. Chimiques. Born 1924. Educ.: Univ. Libre de Brussels. Head, Res. Lab. Orquima S.A., 1951-62; Head, Chem. Eng. Div., Inst. de Energia Atomica, 1962-. Book: Contribution to Progress in the Chem. and Technol. of the Rare Earths (ed. Pergamon Press, 1964). Society: A.C.S.
Nuclear interests: Technology of uranium, thorium and rare earth minerals and salts.
Address: Caixa Postal 21.031, Sao Paulo 17, (Brooklyn Paulista), Brazil.

BRILL, O. D. Papers: Co-author, Excitation Curves for the Reaction $B^{11}(d,2n)C^{11}$, $Be^9(a,2n)C^{11}$, $B^{10}(d,n)C^{11}$ and $C^{12}(d,n)N^{13}$ (letter to the Editor, Atomnaya Energiya, vol. 7, No. 4, 1959); Cross-Sections of $T(d,n)He^4$ and $D(n,d)He^3$ in Deuton Energy Range of 3-19 Mev (letter to the Editor, ibid, vol. 16, No. 2, 1964).
Address: Academy of Sciences of the U.S.S.R., 14 Leninsky Prospekt, Moscow V-71, U.S.S.R.

BRIMBERG, Stig A. S., Dr. Techn. Born 1924. Educ.: Roy. Inst. of Technol., Stockholm. Dept. of Theoretical Phys., Royal Inst. Technol, 1949-55; Reactor Development, Asea, Vasteras, 1955-62; Reactor Eng. Dept., AB Atomenergi, Stockholm. Book: On the Scattering of Slow Neutrons by Hydrogen Molecules (1956).
Address: AB Atomenergi, Box 27163, Stockholm 27, Sweden.

BRINCH, Chr., Member, Statens Atomenergirad, Norway.
Address: Statens Atomenergirad, P.O. Box 40, Kjeller, Norway.

BRINCK, J. Member for Euratom, Study Group on Long Term Role of Nucl. Energy in Western Europe, O.E.C.D., E.N.E.A.
Address: O.E.C.D. European Nuclear Energy Agency, 38 boulevard Suchet, Paris 16, France.

BRINK, Andries Jacob, B. Comm. Born 1908. Educ.: Pietermaritzburg Coll. and Pretoria Univ. Asst. Sec. (1955-60), Sec. (1960-), Atomic Energy Board; Sec., Commission of Enquiry into the Application of Nucl. Power in South Africa, 1956-61.
Nuclear interests: Technical administration, organisation and marketing of nuclear materials.
Address: 197 Marais Street, Brooklyn, Pretoria, South Africa.

BRINK, J. N. BAKHUIZEN van den. See BAKHUIZEN van den BRINK, J. N.

BRINKERHOFF, Charles M. Formerly Vice-Chairman and Chief Exec. Officer, now Chairman (1965-), Anaconda Co.
Address: Anaconda Co., 25 Broadway, New York, New York 10004, U.S.A.

BRINKMAN, Gerardus Antonius, Chem. Dr. Born 1933. Educ.: Amsterdam Univ. Radiochem., Inst. Nucl. Phys. Res. (I.K.O.). Societies: Roy. Dutch Chem. Assoc.; Dutch Phys. Assoc.
Nuclear interests: Radiochemistry; Absolute standardisations; Hot atom chemistry.
Address: 40 v. Hilligaertstraat, Amsterdam, Netherlands.

BRINKMAN, Harold L., Mech. Eng., M.S. (Ind. Eng.). Born 1924. Educ.: Stevens Inst. Technol. American Machine and Foundry Co., 1951-64; Manager, Eng. and Lab. Services, Gen. Telephone and Electronics Labs. Society: A.N.S.
Nuclear interests: Management of nuclear equipment development, including design and manufacture of automatic remote handling and control devices.
Address: 171-19 Crocheron Avenue, Flushing, New York 11358, U.S.A.

BRINKMAN, Hendrik, Ph.D. (Phys.). Born 1909. Educ.: Utrecht Univ. Prof., Exptl. Phys.,

and Director, Cyclotron Inst. for Nucl. Res., Groningen Univ., 1950-. Sec., Foundation for Fundamental Res. of Matter; Curator, Reactor Centrum Nederland; Editorial Board, Nucl. Instr. and Methods; Editorial Board, Nucl. Phys.; Editorial Board, Atoomenergie en haar Toepassingen. Societies: Roy. Netherlands Acad. of Sci.; Netherlands Phys. Soc.; A.P.S. Nuclear interests: Nuclear physics; Nuclear accelerators; Neutron physics; Plasma physics; Nuclear fusion; Reactor physics.
Address: 11 Botanicuslaan, Haren (Gr), Netherlands.

BRINKMANN, Dr. Inst. für Kernphysik, Johannnes Gutenberg Univ.
Address: Johannes Gutenberg University, Institut für Kernphysik, 21 Saarstrasse, 65 Mainz, Germany.

BRISKMAN, B. A. Paper: Co-author, Investigation of Fuel Elements Thermal Bowing (Atomnaya Energiya, vol. 21, No. 1, 1966).
Address: Academy of Sciences of the U.S.S.R., 14 Leninsky Prospekt, Moscow V-71, U.S.S.R.

BRITO, Sergio de SALVO. See de SALVO BRITO, Sergio.

BRITO MARIANO, Maria Helena de. See de BRITO MARIANO, Maria Helena.

BRITTAN, Raymond O., A.B. (Eng.) (Stanford), M.E. (Aero.) (Stanford). Born 1917. Educ.: Wisconsin, Minnesota, and Stanford Univs. Head, Critical Experiments Facility (1949-52), Head, Appl. Mechanics Sect. (1959-63), Head, Reactor Dynamics Sect. (1963-66), and Sen. Eng., Argonne Nat.Lab. Member, A.S.A. N6 Steering Com., also member Subcom. No. 1; Member, A.N.S. Standards Com.; Member, Hon. Advisory Board, Chicago Sect., Inst. of the Aerospace Sciences. Societies: A.I.A.A.; A.N.S.
Nuclear interests: Reactor technology — physical behaviour, materials compatibility, structures, fluid mechanics, safety, kinetics; Applications in space technology.
Address: Box 323, Route 3, Plainfield, Illinois, U.S.A.

BRITTEN, Henry, B.Sc.(Eng.). Born 1907. Educ.: Witwatersrand Univ., Johannesburg. Asst. Consulting Metal., Union Corp., Ltd., 1948-54; Consulting Metal., Anglo-Transvaal Cons. Inv. Co., Ltd., 1955-67. Member, Council, (as past Pres.), South African Inst. of Mining and Metal.; Member, Metal. Com., South African Atomic Energy Board. Book: Chairman, Publications Com. and contributor, Uranium in South Africa (The Assoc. Sci. and Tech. Socs. of South Africa, 1957). Societies: Life Member, South African Inst. of Mining and Metal.; Inst. of Mining and Metal.
Nuclear interests: Extraction metallurgy of uranium; Purification and processing of uranium for fuels.
Address: Anglo-Transvaal Cons. Inv. Company Limited, P.O. Box 7727, Johannesburg, Transvaal, South Africa.

BRITTIN, Wesley E., B.S., M.S., M.A., Ph.D. Born 1917. Educ.: Colorado, Princeton and Alaska Univs. Prof. and Chairman, Nucl. Phys. Lab., Phys. and Astrophys. Dept., Colorado Univ. Book: Editor, Lectures in Theoretical Phys., Vols. 1-5 (Wylie Interscience, 1958-63), Vols. 6-8 (Colorado Press), Vols. 9-10 (Gordon and Breach). Societies: A.P.S.; A.A.P.T.
Nuclear interests: Statistical mechanics - many body problems.
Address: Department of Physics and Astrophysics, Colorado University, Boulder, Colorado, U.S.A.

BRIX, Peter, Dr. Born 1918. Prof. Phys. and Director, Inst. für Tech. Kernphysik, Darmstadt T.H.
Nuclear interests: Nuclear physics; Radiation (60 MeV -electron linear accelerator).
Address: Institut für Technische Kernphysik der T.H., 9 Schlossgartenstrasse, Darmstadt, Germany.

BROAD, Richard. Chief of Nucl. Eng. Operations, Atomic Power Div., Newport News Shipbuilding and Dry Dock Co.
Address: Newport News Shipbuilding and Dry Dock Co., Atomic Power Division, Newport News, Virginia, U.S.A.

BROADBANK, Robert William Clayson, B.Sc., Ph.D., A.R.I.C. Born 1923. Educ.: Rugby Coll. of Technol. and Art and London Univ. Principal Lecturer, Chem., Leicester Regional Coll. of Technol., 1960-66; Head, Sci. Dept., Bolton Inst. of Technol., 1966-. Deleg., I.A.E.A. Conference on the Use of Radioisotopes in the Phys. Sci. and Ind., Copenhagen, 6-17 Sept. 1960.
Nuclear interest: Radiochemistry.
Address: Bolton Institute of Technology, Bolton, Lancs., England.

BROADBENT, D. T. Exec. Director and Chief Eng., Elliott (Treforest) Ltd.
Address: Elliott (Treforest) Ltd., Treforest, Pontypridd, Glamorgan, Wales.

BROADHURST, James Stanley, B.A., B.Sc. Born 1911. Educ.: Oxford Univ. Metal., John Taylor and Sons, Kolar Gold Field, S. India, -1955: Works Manager, Cotton's Colours, Ltd., Stoke on Trent, 1955-56; Programme Manager (Works) Production Group, U.K.A.E.A. Nuclear interests: Technical management; Production planning.
Address: 14 Wood Lane, Timperley, Cheshire, England.

BROADLEY, Jack Stewart, B.Sc., Ph.D. Born 1923. Educ.: Glasgow Univ. P.S.O., R. and D. Branch, U.K.A.E.A., Culcheth, 1950-55; Res. Manager, Chem. Development Group (1956-58), Head of Lab. (1959-61), Head of Chem. Div.

(1961-), D.E.R.E. Societies: Roy. Inst. of Chem.; Inst. of Metallurgists.
Nuclear interests: Processing of active fuel materials, preparation of new reactor fuel materials, general plutonium chemistry, and the development of alpha, beta, gamma handling techniques and the operation of active facilities; Control and management of active laboratories and plant.
Address: Dounreay Experimental Reactor Establishment, Dounreay, Thurso, Caithness, Scotland.

BROBST, William A., B.S. (Chem.). Born 1930. Educ.: Northwestern and Chicago Univs. Radiol. Phys., U.S.A.E.C. U.S. Deleg. to I.A.E.A. for transportation matters. Books: COO-212, Health and Safety Considerations for Uranium Fuel Fabrication Facilities (U.S.A.E.C., 1958); COO-228, External Rad. Problems from Plutonium (U.S.A.E.C., 1959); COO-272, Nucl. Fallout Prediction Plotting (U.S.A.E.C., 1963). Societies: H.P.S.; A.N.S.; Nat. Pilots Assoc.
Nuclear interests: Development of national regulations for transportation of radioactive materials, and standards for container testing and design.
Address: Deputy Director, Office of Hazardous Materials, U.S. Department of Transportation, Washington D.C. 20590, U.S.A.

BROCARD, Joseph Jean Marie Rodolphe, Born 1903. Educ.: Ecole Polytechnique. Conseiller Technique des Chantiers Navals de la Ciotat.
Address: 10 boulevard Flandrin, Paris 16, France.

BROCCARDO, Umberto, Dr. Elec. Eng. Born 1926. Educ.: Padova Univ. Nucl. Dept., Kennedy and Donkin, London, 1960-63; South African Council for Scientific and Industrial Research, Pretoria, 1963-. I.A.E.A. Fellow, 1959-60.
Nuclear interests: Reactor design and nuclear power stations; Nuclear electronics.
Address: 19 11th Street, Parkhurst, Johannesburg, South Africa.

BROCK, Peter, B. Met., Ph.D., F.I.M. Born 1926. Educ.: Sheffield Univ. British Non-Ferrous Metals Res. Assoc., 1952-58; Fuel Element Group, D.E.R.E., U.K.A.E.A., 1958-.
Nuclear interests: Fuels and fuel elements (particularly for fast reactors) including design, irradiation testing, out-of-pile testing, and fabrication methods.
Address: Dounreay Experimental Reactor Establishment, Thurso, Caithness, Scotland.

BROCK, Richard Donald, M. Agr. Sci., Ph.D. Born 1921. Educ.: Melbourne and London Univs. Principal Res. Officer, Genetics Sect., Div. Plant Industry, Commonwealth Sci. and Ind. Res. Organisation, Canberra.
Nuclear interests: Use of high-energy radiation for the induction of mutation in plants and general studies in radiation biology.
Address: C.S.I.R.O., Box 109 City, Canberra, Australia.

BROCKDORFF, Cay Baron. Director, Graetz Raytronik G.m.b.H.
Address: Graetz Raytronik G.m.b.H., P.O.B. 57, 50 Graetzstrasse, Altena/Westfalen, Germany.

BROCKETT, Clarence T. Head, Tech. Information, Lawrence Rad. Lab. Member, Tech. Information Panel, U.S.A.E.C.
Address: Lawrence Radiation Laboratory, Technical Information, Berkeley 4, California, U.S.A.

BROCKHOUSE, Bertram Neville, M.A., Ph.D. Born 1918. Educ.: British Columbia and Toronto Univs. Res. Officer (1950-62), Head, Neutron Phys. Branch (1960-62), A.E.C.L.; Prof. Phys. (1962-), Chairman, Phys. Dept. (1967-), McMaster Univ. Societies: A.P.S.; Canadian Assoc. of Phys.; Roy. Soc. of Canada; Roy. Soc. of London.
Nuclear interest: Neutron physics.
Address: McMaster University, Hamilton, Ontario, Canada.

BROCKLESBY, Richard Shearwood, A.R.I.B.A. Born 1911. Educ.: Bartlett School of Architecture and London Univ. With Ministry of Works, 1938-54; U.K.A.E.A., 1954-.
Nuclear interests: Design and construction of all buildings and structures connected with nuclear engineering.
Address: Millbank, Tattenhall, Cheshire, England.

BRODA, Engelbert, Dr. phil. (Vienna). Born 1910. Educ.: Vienna and Berlin Univs. Prof., Head of Radiochem. Dept., Inst. of Phys. Chem., Vienna Univ. Books: Advances in Radiochem. (Cambridge Univ. Press, 1950); co-author, Application of Radiochem. in Microchem. (German) in Handbuch der mikrochemischen Methoden, Vol ii (Vienna, Springer, 1955); Ludwig Boltzmann (German) (Vienna, Deuticke, 1955); Radioactive Isotopes in Biochem. (German: Vienna, Deuticke, 1958; English: Amsterdam, Elsevier, 1960); co-author, Tech. Applications of Radioactivity (German: Leipzig, Deutscher Verlag für Grundstoffindustrie, 1962; English: Oxford, Pergamon Press, 1966). Societies: Physikalisch-chemische Gesellschaft, Vienna; Verein oesterreichischer Chemiker, Vienna; Oesterreichische biochemische Gesellschaft, Vienna.
Nuclear interests: Applications of isotopes and radiations to chemistry, technology and biology; Radiation protection; Social consequences of nuclear development.
Address: Institut für Physikalische Chemie, University, Waehringer Strasse 42, Vienna 9, Austria.

BRODER, D. L.
Nuclear interests: Neutron absorption and

secondary γ-ray emission.
Address: Academy of Sciences of the U.S.S.R., 14 Leninsky Prospekt, Moscow V-71, U.S.S.R.

BRODIN, B. V. Papers: Co-author, The Reddening of Minerals in Uranium-bearing Veins (Atomnaya Energiya, vol. 10, No. 1, 1961); co-author, Hard Bitumen in Uraniumbearing Veins (ibid, vol. 16, No. 5, 1964).
Address: Academy of Sciences of the U.S.S.R., 14 Leninsky Prospekt, Moscow V-71, U.S.S.R.

BRODSKII, A. M. Papers: Co-author, Circulating Rate Effect on Rad-Induced Conversion of Organic Coolants at Elevated Temperature (Atomnaya Energiya, vol. 22, No. 5, 1967).
Address: Academy of Sciences of the U.S.S.R., 14 Leninsky Prospekt, Moscow V-71, U.S.S.R.

BRODSKY, A. E. Director, Pisarzhevski Inst. Phys. and Chem. Papers: Co-author, Investigation on the Oxidation Mechanism of Pyruvic Acid, Phenole and Salicyle Aldehyde by Hydrogen Peroxide by using Isotopes (Kernenergie, vol. 5, No. 4/5, 1962); co-author, The Mechanism of Anodic Ozone Formation in Aqueous Sulphate Solutions (ibid.).
Address: Pisarzhevski Institute of Physics and Chemistry, Academy of Sciences of the Ukrainian S.S.R., Kiev, Ukrainian S.S.R.

BRODSKY, F. J. Manufacturing Div. A Manager, Burlington A.E.C. Plant, operated for U.S.A.E.C. by Mason and Hanger-Silas Mason Co. Inc.
Address: U.S.A.E.C., Burlington A.E.C. Plant, Burlington, Iowa, U.S.A.

BRODSKY, Michel, Eng. E.N.S.C.P., Dr. of Laws. Born 1917. Educ.: Paris Univ. Directeur des Etudes et Constructions, Potasse et Engrais Chimiques. Directeur techniques, Sté. d'Etudes et de Travaux pour l'Uranium; Administrateur, Sté. de Raffinage d'Uranium.
Nuclear interests: Uranium industry and radioactive waste treatment and disposal.
Address: 16 Parc de Noailles, Saint-Germain en Laye, Seine et Oise, France.

BRODSKY, Robert, B.Ch.E., B.E.E. Born 1926. Educ.: Pratt Inst., Brooklyn, N.Y. and Maine and Columbia Univs. Partner, Brodsky, Hopf and Adler, New York.
Nuclear interests: Reactor design; Test loop design.
Address: 235 East 42nd Street, New York 17, New York, U.S.A.

BRODTKORB, Thor, Cand. jur. Born 1907. Educ.: Oslo Univ. Consul Gen., New York, 1952-58; Director Gen., Ministry of Foreign Affairs, Oslo, 1958-65; Norwegian Ambassador Prague, 1965-66; Ambassador, Vienna and Budapest, 1965-; Permanent Norwegian Rep., I.A.E.A., Vienna, 1965-.
Address: Royal Norwegian Embassy, Vienna, Austria.

BRODY, Howard M., Formerly Asst. Prof., now Assoc. Prof., High Energy Phys., Phys. Dept., Pennsylvania Univ.
Address: Pennsylvania University, Physics Department, Philadelphia, Pennsylvania 19104, U.S.A.

BRODY, Thomas, Chem. Eng. Born 1922. Educ.: Cambridge and Lausanne Univs. At Inst. Phys., Nat. Univ. Mexico, 1954-; Tech. Adviser, Nat. Nucl. Energy Commission, 1959-; Societies: Mexican Phys. Soc.; A.C.S.; Mexican Acad. for Sci. Res.
Nuclear interests: Radioisotope applications and low-energy nuclear physics; Computer techniques in nuclear physics.
Address: Instituto de Fisica de la Universidad Nacional Autónoma de México, Torre de Ciencias 8. piso, Ciudad Universitaria, Mexico 20, D.F., Mexico.

BROECKER, Wallace S. Prof. Geochem. and Director, Geochem. Lab., Lamont Geological Observatory.
Address: Lamont Geological Observatory, Columbia University, Palisades, New York, U.S.A.

BROEKMAN, H. van MOURIK. See van MOURIK BROEKMAN, H.

BROERSE, J. J. Staff Member, Radiobiol. Inst., Organisation for Health Res. T.N.O.
Address: Organisation for Health Research T.N.O., 151 Lange Kleiweg, Rijswijk (ZH), Netherlands.

BROERTJES, Cornelis, Ing. (degree of the Agricultural Univ. at Wageningen). Born 1923. Educ.: Wageningen Agricultural Univ. Staff member, Experiment Station for Arboriculture, Boskoop; Sen. staff member, Inst. for Atomic Sci. in Agriculture, 1959-.
Nuclear interest: Application of atomic energy in agriculture, especially for mutation-breeding.
Address: Postbus 48, Wageningen, Netherlands.

BROGLIE, Louis de. See de BROGLIE, Louis.

BRÖHL, Friedrich Wilhelm Philip, Dipl.-Ing., Dr.-Ing. Born 1912. Educ.: Aachen Tech. Univ. Mannesmann A.G., Düsseldorf; Klöckner-Werke A.G., Georgsmarienwerke, Osnabrück, 1955-. Societies: Verein Deutscher Eisenhüttenleute; A.S.T.M.
Nuclear interests: Metallurgy; Properties of reactor materials.
Address: Klöckner-Werke A.G., Georgsmarienwerke, 1 Bessemerstrasse, 45 Osnabrück, Germany.

BROIDO, Jeffry H. R. and D. Staff, General Atomic. Treas., San Diego Sect., A.N.S., 1966-.
Address: 8811 Robinhood Lane, La Jolla, California 92037, U.S.A.

BROLLEY, John Edward, Jr., S.B., M.S., Ph.D. Born 1919. Educ.: Chicago, Indiana and Princeton Univs., and M.I.T. At Los Alamos Sci. Lab., 1949-. Books: Author, co-author several

chapters in monographs. Societies: Fellow, A.P.S.; Fellow, A.A.A.S.; Archeological Inst. of America.
Nuclear interest: Primarily concerned with experimental and theoretical studies of elementary particle interactions.
Address: Los Alamos Scientific Laboratory, Los Alamos, New Mexico, U.S.A.

BROMLEY, D. (David) Allan, B.Sc. (Hon.), Eng. Phys.), M.Sc. (Phys.), Ph.D. (Phys.), M.A. Born 1926. Educ.: Queen's (Kingston, Ontario) and Rochester Univs. Instructor, Asst. Prof., Rochester Univ., 1950-55; Sen. Res. Officer, Sect. Head-Accelerators, A.E.C.L., Chalk River, 1955-60; Prof. Phys., Director, A.W. Wright Nucl. Structure Lab., Yale Univ. 1960-. Consultant: Brookhaven Nat. Lab., 1961-; O.R.N.L., 1961; Los Alamos Nat. Lab., 1966-; McGraw Hill Univ. 1965-; Director, United Nucl. Corp., 1967-. Editor, Academic Paperbacks in Phys.; Assoc. Editor, Phys. Review; Chairman, N.A.S. Com. Nucl. Sci., 1965-; Councillor, Div. Nucl. Phys., A.P.S., 1967-. Societies: Canadian Assoc. Phys.; Fellow, A.P.S.; A.A.A.S.; A.P.E.O.
Nuclear interests: Nuclear physics: accelerator research, nuclear spectroscopy, and nuclear reaction mechanisms.
Address: 35 Tokeneke Drive, Millbrook, North Haven, Connecticut 06473, U.S.A.

BRONDEL, Georges, Ing. Civil des Mines. Born 1920. Educ.: Paris and Oxford Univs. Chief of Energy Div., Communauté Economique Européenne, Bruxelles. Societies: Econometric Soc.; Sté. de Recherche operationelle; Sté. des Ingénieurs civils de France.
Nuclear interests: Economic aspects of nuclear energy.
Address: Communauté Economique Européenne, 80 rue d'Arlon, Brussels 4, Belgium.

BRONDI, Marcos, Ing. agrónomo. Born 1911. Educ.: Montevideo Univ. Formerly Consul Gen., London; Formerly Ambassador, Prague; Alternate Deleg., I.A.E.A., 1964-; Uruguayan Ambassador to Germany.
Nuclear interest: Problemas de organización.
Address: Ing. Marcos Brondi, Ambassador del Uruguay, 5 Zittelmanstrasse, Bonn, Germany.

BRONK, Detlev W., Pres., Rockefeller Inst. Trustee, Atoms for Peace Awards.
Address: Rockefeller Institute, New York 21, New York, U.S.A.

BROODMAN, J. J., Ir. Member, Staff for Nucl. Work, Equipment and Boiler Manufacturing Co. den Haag Ltd.
Address: Equipment and Boiler Manufacturing Co. den Haag Ltd., Industrieterrein Noord, P.O. Box 69, Goes, Netherlands.

BROOK, Greville Bertram, B. Met. (Hons.). Born 1926. Educ.: Sheffield Univ. Head, Phys. Metal. Sect., Fulmer Res. Inst. Ltd., 1957-. Societies: Inst. of Metals: Inst. of Metal.; A.S.M.
Nuclear interests: Physical metallurgy of magnesium, uranium and beryllium; Precipitation in stainless steels.
Address: 129 Heath Road, Beaconsfield, Buckinghamshire, England.

BROOK, Leopold, B.Sc. (Eng.). Born 1912. Educ.: London Univ. Deputy Manager, Power Plant Dept. (1947-56), Tech. Director, (1956). Managing Director (1962-63), Deputy Chairman (1963-65), Chairman (1965-67), Simon-Carves, Ltd.; Deputy Chairman and Chief Exec., Simon Engineering, Ltd., 1967-. Societies: Inst. Civil Eng.; M.I.Mech.E.; Sté. des Ing. Civils de France.
Address: Simon Engineering Limited, Cheadle Heath, Stockport, Cheshire, England.

BROOKE, Claude, Dipl. Phys. Born 1924. Educ.: Swiss Federal Polytech. Inst. Zurich. Chargé de Conférences, Univ. Libre de Bruxelles; Head, Study Group in Nucl. Phys., M.B.L.E. (Manufacture Belge de Lampes et de Matériel Electronique, S.A.), Brussels.
Nuclear interests: Nuclear instrumentation and detectors; Industrial applications of isotopes; Health physics; Reactor control.
Address: 184 avenue Blücher, Brussels, Belgium.

BROOKES, J. D. Member, Council, Nucl. Res. Foundation, Sydney Univ.
Address: Council, Nuclear Research Foundation, Sydney University, Sydney, New South Wales, Australia.

BROOKES, Leonard George, Dipl. Economics. Born 1919. Educ.: London Univ. Various appointments with Ministry of Agriculture and Fisheries; Formerly Personnel Manager, Atomic Energy Establishment, now Economics and Planning Branch, U.K.A.E.A.
Address: 25 Nelson Road, Westbourne, Bournemouth, Hants., England.

BROOKS, Frederick Phillip, Jr., B.A., M.S., Ph.D. Born 1931. Educ.: Duke and Harvard Univs. Eng. and Manager, OS/360 (1956-65), Consultant (1965), I.B.M. Chairman and Prof., Dept. of Information Sci., North Carolina Univ., 1964-. Member, Nat. Council, Assoc. for Computing Machinery; Member, Mathematics and Computer Sci. Res. Advisory Com., A.E.C. Book: Co-author, Automatic Data Processing (John Wiley and Sons, 1963).
Nuclear interests: Computer systems design, including very highspeed systems.
Address: Department of Information Science, Phillips Annex, North Carolina University, Chapel Hill, North Carolina 17514, U.S.A.

BROOKS, Harvey, Ph.D. Born 1915. Educ.: Yale and Harvard Univs. Consultant to Reactor Div., U.S.A.E.C.; Prof. Appl. Phys. and Dean of Eng. and Appl. Phys., Harvard Univ.; Member, Nat. Sci. Board, Naval Res. Advisory Co.; Trustee, Woods Hole Oceanographic Inst., Case Western Reserve Univ.; Chairman, Com.

on Sci. and Public Policy, N.A.S. Societies: Fellow, A.P.S.; A.N.S.; American Acad. of Arts and Sciences; American Philosophical Soc.; N.A.S.
Nuclear interests: Reactor theory; Reactor design and safety; Nuclear metallurgy; Radiation damage.
Address: 46 Brewster Street, Cambridge, Massachusetts 02138, U.S.A.

BROOKS, Kennedy Carr. Born 1903. Educ.: Cincinnati and Tennessee Univs. Eng., U.S.A.E.C., 1947-55; Manager, Paducah Area Office, Oak Ridge Operations, (1955-62), Asst. Director, Construction and Eng. Headquarters (1962-), U.S.A.E.C.
Address: U.S. Atomic Energy Commission, Washington, D.C., 20545, U.S.A.

BROOKSBANK, William A., Jr., B.S., M.S. Born 1923. Educ.: Tennessee Univ. Chem., O.R.N.L., 1949-58; N.A.S.A., 1958-.
Nuclear interest: Project management of the design, development, fabrication and flight testing of a nuclear propelled space vehicle, including all necessary nuclear research programmes.
Address: George C. Marshall Space Flight Centre, Propulsion and Vehicle Engineering Laboratory, Projects Office, Huntsville, Alabama, U.S.A.

BROPHY, James John, B.S.(E.E.), M.S. (Phys.), Ph.D.(Phys.). Born 1926. Educ.: I.I.T. Vice Pres., I.I.T. Res. Inst., 1961-67; Academic Vice Pres., I.I.T., 1967-. Societies: Fellow, A.P.S.; A.A.A.S.
Address: Illinois Institute of Technology, Chicago, Illinois 60616, U.S.A.

BROSI, Albert Ralph, Ph.D. Born 1907. Educ.: Chicago Univ. At O.R.N.L. 1943-. Societies: A.C.S.; A.P.S.; A.A.A.S.
Nuclear interests: Nuclear chemistry; Nuclear physics.
Address: Oak Ridge National Laboratory, P.O. Box X, Oak Ridge, Tennessee, U.S.A.

BROSZKIEWICZ, Roman, Ph.D., D.Sc. Born 1931. Educ.: Warsaw Univ. Res. Sci., Military Tech. Acad., 1951-56; Head, Lab. of Radioisotopes Distribution, Inst. of Nucl. Res., 1956-59; Sen. Res. Sci., Dept. of Rad. Chem., Inst. of Nucl. Res., 1959-. Member, Nat. Com. on Radiol. Protection. Societies: Miller Trust for Rad. Chem.; Polish Assoc. for Rad. Res.
Nuclear interests: Radiation chemistry; Radiation sources; Radiological protection.
Address: Instytut Badan Jadrowych, 16 Dorodna, Warsaw, Poland.

BROTHERS, LeRoy A., B.S. (Civil Eng.), C.E., D.Sc. (Hon.). Born 1904. Educ.: North Carolina State Univ. Dean, Eng. Coll., Drexel Inst. Technol., 1958-. Societies: A.S.C.E.; A.S.E.E.
Nuclear interest: Development of courses in nuclear engineering and nuclear science at Drexel Institute of Technology.

Address: 1040 Great Springs Road, Rosemont, Pennsylvania, U.S.A.

BROUARD, Pierre Edouard. Born 1937. Export Manager, Mecaserto.
Nuclear interest: Nuclear medicine.
Address: 31 avenue des Chalets, F-94 Choisy-le-Roi, France.

BROUET, Georges. Member, Conseil d'Enseignement de l' I.N.S.T.N., C.E.A.
Address: Commissariat a l'Energie Atomique, 29-33 rue de la Federation, Paris 15, France.

BROUN, G. O., Jr., M.D. Assoc. Editor, J. Nucl. Medicine.
Address: 333 N. Michigan Avenue, Chicago 1, Illinois, U.S.A.

BROUS, Chris James, B.S. (Mech. Eng.). Born 1919. Educ.: Johns Hopkins and New York Univs. Manager, Nucl. Eng. Lab., American Machine and Foundry Co., 1951-58; Asst. to Chief of Eng. Dept. (1958), Director, Product Eng. Dept. (1959), Director, Product Eng. and Manufacturing Dept. (1960-63), Atomics Internat.; Manager, Product Operations, Gen. Atomic, 1963-67. Societies: A.S.M.E.; A.N.S.; Atomic Industrial Forum; American Ordnance Assoc.; American Management Assoc.
Nuclear interest: Management of nuclear power reactor and major nuclear components programmes.
Address: P.O. Box 608, San Diego, California, U.S.A.

BROUSAIDES, Frederick J., A.B., M.S. Born 1926. Educ.: Boston and Northeastern Univs. American Cyanamid Co., 1951-54; Nat. Lead Co., 1954-56; American Cyanamid Co., 1956-59; Tracerlab, Inc., 1959-63; Parametrics, Inc., 1963-.
Nuclear interests: Applications of radioactive isotopes to industrial problems; Particular emphasis on low-level concentration detection and estimation of air pollutants and toxic gases.
Address: 20 Daniels Road, Framingham, Mass., U.S.A.

BROUSIL, Jindrich, M.D., Candidate of Medical Sci. Born 1931. Educ.: Charles Univ., Prague. Asst. Prof. Nucl. Medicine, Inst. of Biophys., 1953-. Book: Co-author, Radioisotopic Diagnostic Methods (Prague, The State Health Publishing House, 1965). Society: Sect. of Nucl. Medicine, Czechoslovak Medical Soc. of J. Ev. Purkyne.
Nuclear interests: Diagnostics by means of radioisotopes. Special interest in isotopic methods of investigation of blood formation disturbances.
Address: Institute of Biophysics, Faculty of General Medicine, Charles University, 3 Salmovska, Prague 2, Czechoslovakia.

BROUWER, Louis de. See de BROUWER, Louis.

BROWER, Pompilio, Dr. Supervisor de Minas y Petroleo de la Secretaria de Estado de Fomento; Technical Adviser, Comision Nacional de Investigaciones Atomicas, Dominican Republic.
Address: Comision Nacional de Investigaciones Atomicas, Ciudad Trujillo, Dominican Republic.

BROWN, Arthur F., M.A., Ph.D. Born 1920. Educ.: Edinburgh and Birmingham Univs. I.C.I. Fellow, Cavendish Lab., Cambridge. -1952: Lecturer in Natural Philosophy (1952-59), Sen. Lecturer (1959-63), Reader (1963-67), Edinburgh Univ. Prof. Phys., City Univ., London, 1967-. Societies: F.R.S.E.; Fellow, Inst. of Phys. and Phys. Soc.; Fellow, Inst. of Metal.
Nuclear interest: Application of tracer techniques to physical metallurgy.
Address: 36, Lings Coppice, London S.E.21, England.

BROWN, Arthur William, B.S. and M.S. (Chem. Eng.). Born 1928. Educ.: Colorado Univ. Reactor Eng. Phys. (1955-61), Group Leader, Reactor Phys. Group (1961-), Idaho Nucl. Corp. Society: A.N.S.
Nuclear interests: Reactor design; Development of reactor codes.
Address: 2940 East 17th Street, Idaho Falls, Idaho 83401, U.S.A.

BROWN, Basil, B.Sc., Ph.D. Sen. Lecturer Phys., Salford Univ. Society: A.Inst.P.
Nuclear interest: Nuclear science education.
Address: Beach House, 82 Cambridge Road, Southport, Lancashire, England.

BROWN, Ben L. Controller, Atomics, Phys. and Sci. Fund, Inc.
Address: Atomics, Physics and Science Fund, Inc., 1033, 30th Street, Northwest, Washington 7, D.C., U.S.A.

BROWN, Charles S. Formerly Vice-Pres., now Exec. Vice-Pres. (1965-), Abbott Labs.
Address: Abbott Laboratories, North Chicago, Illinois, U.S.A.

BROWN, Colin Leslie, B.Sc. Born 1929. Educ.: Birmingham Univ. Sci. Officer, A.E.R.E., Harwell, 1953; Group Leader, Exptl. Operations, O.E.E.C., Halden Reactor, Norway, 1958; Operations Eng. (1961), Sen. Technologist (1965-67), Dragon Operations Group, Atomic Energy Establishment, Winfrith. Society: A.M.I.Mech.E.
Nuclear interest: Reactor development and operation.
Address: United Kingdom Atomic Energy Authority, Atomic Energy Establishment, Winfrith, Nr. Dorchester, Dorset, England.

BROWN, Donald W. Educ.: Colorado Univ. Head, Div. of Nucl. Medicine, Colorado Univ., Denver.
Address: Colorado University, Division of Nuclear Medicine, Denver, Colorado, U.S.A.

BROWN, Sir Frederick Herbert Stanley, B.Sc. Born 1910. Educ.: Birmingham Univ. Chief Generation Eng. (Construction), Merseyside and N. Wales Div., (1949-51), Deputy Generation Design Eng. (1951-54), Generation Design Eng. (1954-56), Chief Eng. (1956-57), Member for Eng. (1957-59), Deputy Chairman (1959-64), Chairman (1965-), C.E.G.B. Pres., I.E.E., 1967-68. Societies: Inst. Mech. Eng.; I.E.E.
Nuclear interest: Operation of nuclear reactors for electricity supply.
Address: Central Electricity Generating Board, Sudbury House, 15 Newgate Street, London E.C.1, England.

BROWN, G. G. Sales Manager, Nucl. Process, Honeywell Controls Ltd.
Address: Honeywell Controls Ltd., Ruislip Road East, Greenford, Middlesex, England.

BROWN, George Edwin, Jr., A.B. Born 1923. Educ.: Harvard Univ. Professional Staff Member, R. and D., Joint Com. on Atomic Energy, U.S. Congress, 1956-59; Administrative Projects Manager, Atomic Ind. Forum, Inc., 1959-. Society: A.I.A.A.
Nuclear interests: Problems of concern to management in the peaceful applications of atomic energy. Special interest: space and marine applications of nuclear energy; Commercial use of nuclear explosives for oil and gas stimulation.
Address: Atomic Industrial Forum, Incorporated, 850 Third Avenue, New York, New York 10022, U.S.A.

BROWN, Gerald Edward, Ph.B., M.S., Ph.D., D.Sc. Born 1926. Educ.: South Dakota State, Iowa, Wisconsin, and Yale Univs. Book: Unified Theory of Nucl. Models (North Holland Publishing Co., 1964).
Nuclear interests: Theoretical nuclear physics.
Address: Nordic Institute of Theoretical Atomic Physics, 17 Blegdamsvej, Copenhagen Ø, Denmark.

BROWN, Glenn Halstead, B.S., Ph.D. Born 1915. Educ.: Ohio and Iowa State Univs. Vermont Univ., 1950-52; Cincinnati Univ.,1952-60; Kent State Univ., 1960-. Chairman, Akron Section, A.C.S. Book: Quantitative Chemistry (Prentice-Hall, Inc., 1963). Societies: A.C.S.; A.A.A.S.; American Inst. Chem.; Ohio Acad. Sci.
Nuclear interests: Neutron diffraction in organic liquid crystalline compounds. Also, radiotracer studies in quantitative chemistry.
Address: Department of Chemistry, Kent State University, Kent, Ohio 44240, Ohio, U.S.A.

BROWN, Gordon, Ph.D., B.Sc. (Eng.), D.I.C. Born 1921. Educ.: London Univ. Dep. Chief Eng., Calder Hall (1956), Chief Eng., Advanced

Gas-cooled Reactor (1957-61), Deputy Director, Industrial Power (1961-63), Deputy Director, Reactor Design (1963-64), Director, Gas-cooled Reactors (1964-65), Reactor Group, U.K.A.E.A., Tech. Director (1965-), Managing Director (1967-), Atomic Power Construction Ltd. Life Member, Nucl. Eng. Soc., Risley Branch; Member of Board, British Nucl. Energy Soc. Society: M.I.Mech.E.
Nuclear interests: Reactor design; Power generation.
Address: 28 Camborne Road, Sutton, Surrey, England.

BROWN, Gordon Clair, B.Sc. Born 1920. Educ.: Queen's Univ., Canada. Refinery Manager, Eldorado Mining and Refining Ltd., -1962; Gen. Manager, A.M.F. Atomics Canada Co. Ltd., 1962-; Manager, Atomic Fuel Dept., Canadian Westinghouse Co. Ltd., 1964-. Society: A.P.E.O.
Nuclear interests: Nuclear fuel development; Nuclear fuel manufacture.
Address: Port Hope, Ontario, Canada.

BROWN, H. Programme Management, Northrop Nortronics, Northrop Corp.
Address: Northrop Corporation, Northrop Nortronics, Applied Research Department, 2323 Teller Road, Newbury Park, California 91320, U.S.A.

BROWN, Harold, A.B., A.M. (Phys.), Ph.D. (Phys.). Born 1927. Educ.: Columbia Univ. Lecturer in Phys., Stevens Inst. Technol., 1949-50, California Univ. (Berkeley), 1951-52; Res. Sci., California Univ., E.O. Lawrence Rad. Lab., Berkeley 1950-52, Livermore 1952-61 (Group Leader 1953-56, Div. Leader 1956-58, Assoc. Director 1958-59, Deputy Director, 1959-60, Director 1960-61). Director of Defence Res. and Eng. 1961-65; Sec. of the Air Force, 1965-. Member, Polaris Steering Com., 1956-58; Consultant (1957), Member (1958-61) Air Force Sci. Advisory Board; Member, Sci. Advisory Com. on Ballistic Missiles to Sec. of Defence, 1958-61; Panel Consultant (1958-60), Member (1961), President's Sci. Advisory Com. Adviser to U.S. Deleg., Conference of Experts on Detection of Nucl. Tests, 1958; Sen. Sci. Adviser to Conference on Discontinuance of Nucl. Tests, 1958-59.
Nuclear interests: Nuclear physics; Neutron physics; Nuclear reactors and weapons.
Address: 416 Argyle Drive, Alexandria, Virginia, U.S.A.

BROWN, Harold Dean, B.S., M.S., Ph.D. Born 1927. Educ.: South Dakota State Coll., and Kansas Univ. Phys., E.I. Du Pont de Nemours, Savannah River Lab., 1952-; Sen. Officer, I.A.E.A., 1959.
Nuclear interests: Neutron thermalisation; Reactor dynamics; Reactor safety.
Address: Computer Usage Co. Inc., 1825 Connecticut Avenue, N.W., Washington 9, D.C., U.S.A.

BROWN, Howard C., B.S., LL.B. Born 1919. Educ.: Rockhurst Coll., Kansas City and Georgetown Univ. Asst. to Manager, Oak Ridge Operations Office (1948-50); Organisation and Methods Examiner, Div. Organisation and Personnel, Washington (1950-51), Exec. Officer later Asst. Director of Administration, Div. Biol. and Medicine (1951-57), Director of Administration, Office for Internat. Conference (1957-58) (2nd I.C.P.U.A.E., Geneva, Sept. 1958), Special Asst. Gen. Manager for Res. and Ind. Development (1958-59), Special Asst. to Chairman (1959-64), Asst. Gen. Manager for Administration (1964-66), Asst. Gen. Manager, (1966-), U.S.A.E.C.
Address: 9618 Carriage Road, Kensington, Maryland, U.S.A.

BROWN, J. A. Local Director and Tech. Manager, Shipbuilding Group, Vickers Ltd.
Address: Vickers-Armstrongs Naval Yard, Walker, Newcastle-upon-Tyne 6, England.

BROWN, Jack Harold Upton, B.S., Ph.D. (Rutgers). Born 1918. Educ.: Texas and Rutgers Univs. Director, Physiology Labs., Mellon Inst., Pittsburgh, 1948-50; Lecturer in Endocrinology, Pittsburgh Univ., 1948-50; Asst. Prof. Physiology, North Carolina Univ., 1950-52; Res. Sci., O.R.I.N.S., 1952; Prof. Physiology (1952-59), Chairman of Dept. (1956-57), Emory Univ.; Sci. Administrator, Div. of Gen. Medical Sci. (1961-62), Asst. Chief, Div. Res. Facilities and Resources (1962-64), Assoc. Director, Nat. Inst. Gen. Medical Sci., (1964-), Nat. Inst. Health, Bethesda, Maryland. Consultant to O.R.I.N.S., W.H.O., Veterans Administration, Lockheed Aircraft Co. Societies: American Physiological Soc.; Endocrine Soc.; Soc. for Exptl. Biol. and Medicine; A.C.S.
Nuclear interests: Use of various isotopes in tracer work in endocrinology; Intermediary metabolism.
Address: National Institute of General Medical Science, National Institutes of Health, Bethesda, Maryland, U.S.A.

BROWN, James Robert, B.Sc. Born 1927. Educ.: Queens Univ., Belfast. Eng. Manager, Equipment, E.M.I. Electronics, Ltd. Society: Associate Member, I.E.E.
Nuclear interest: All types of nuclear instruments, with special emphasis on instruments for the protection of personnel.
Address: 52 Ash Lane, Wells, Somerset, England.

BROWN, Keith B., B.S. (Chem., South Dakota). Born 1920. Educ.: South Dakota and Princeton Univs. Asst. Div. Director, O.R.N.L. Book: Ch. 1, Process Chem., Vol. 2, Progress in Nucl. Energy (London, Pergamon Press, 1958). Societies: A.C.S.; RESA.
Nuclear interests: Ore processing; Uranium and thorium chemistry and processing; Radiochemical separations and processing.
Address: 103 Villanova Road, Oak Ridge, Tennessee, U.S.A.

BROWN, Captain L. T. Medical Corps., U.S. Navy. Chief of Radiol. and Head of Training and Res., Nat. Naval Medical Centre.
Address: National Naval Medical Centre, Bethesda 14, Maryland, U.S.A.

BROWN, Rev. Leo C., S.J., A.B., A.M. (Phil.), S.T.L., A.M. (Economics), Ph.D. (Economics). Born 1900. Educ.: St. Louis and Harvard Univs. and St. Mary's Coll., St. Mary's, Kansas. Director, Inst. Social Order, 1947-62; Prof. Economics, St. Louis Univ. Member (1953-68), Chairman (1968-), Atomic Energy Labour Management Relations Panel, U.S.A.E.C.
Nuclear interest: Labour-management field.
Address: St. Louis University, St. Louis, Missouri 63103, U.S.A.

BROWN, N. J. C., B.A. At Inst. of Nucl. Medicine, Middlesex Hospital Medical School.
Address: Middlesex Hospital Medical School, Institute of Nuclear Medicine, Barnato Joel Laboratories, The Middlesex Hospital, London W.1, England.

BROWN, R. HANBURY. See HANBURY BROWN, R.

BROWN, Randall Emory, A.B., M.S. Born 1917. Educ.: Oregon, Stanford and Yale Univs. Geologist Supv., Unit Head (1947-55), Sen. Geologist (1956-65), G.E.C., Hanford Works; Sen. Res. Sci., Battelle Memorial Inst., Pacific Northwest Laboratory, Richland, Washington, 1965-. Societies: A.A.A.S.; Nat. Water Well Assoc.; Geochemical Soc.; Fellow, Geological Soc. America; A.I.M.M.E.; Trustee, Northwest Sci. Assoc.
Nuclear interests: Radioactive waste management; Disposal of radioactive liquid wastes to the ground.
Address: 504 Road 49 North, Pasco, Washington, U.S.A.

BROWN, Reynold F. Born 1916. Educ.: California Univ., Berkeley and California Univ. School of Medicine, San Francisco. Assoc. Clinical Prof. of Radiol.; Asst. to Chancellor, Environmental Health and Safety, and Director, Radiol. Sci. Educ. Project, Dept. of Radiol., California Univ. Medical Centre. Member, Board of Directors, Radiol. Soc. of North America; Councillor, A.C.R.; Member, Board of Directors, Nat. Council on Rad. Protection; Chairman, Task Force on Protection of the Patient, Internat. Commission on Rad. Protection; Member, Board of Directors, California Radiol. Soc.; Councillor, San Francisco Radiol. Soc.; Isotope Com., State Dept. of Public Health.
Nuclear interest: Radiation protection.
Address: Department of Radiology, California University Medical School, San Francisco, California, U.S.A.

BROWN, Robert H., Dr. Prof., Phys. Dept., Walla Walla Coll.
Address: Walla Walla College, Box 1047, College Place, Washington, U.S.A.

BROWN, Robert L., A.B., M.D. Born 1908. Educ.: Michigan Univ. and Harvard Medical School. Prof. Surgery, Neoplastic Disease, Emory Univ. School of Medicine, Atlanta, Georgia; Director, Emory Univ. Clinic. Cancer Com., American Coll. Surgeons. Societies: American Radium Soc.; James Ewing Soc.; American Coll. Surgeons.
Nuclear interest: Medical applications of atomic energy.
Address: Emory University Clinic, Emory University, Atlanta 22, Georgia, U.S.A.

BROWN, Roger C. Health Phys., Nucl. Reactor Project, Washington State Univ.
Address: Washington State University, Pullman, Washington, U.S.A.

BROWN, Ronald E., B.Sc., Ph.D. Born 1934. Educ.: Washington (Seattle) Univ. and California Inst. Technol. Asst. Prof., Phys., Minnesota Univ., 1964-. Society: Div. of Nucl. Phys., A.P.S.
Nuclear interests: Nuclear reactions and scattering; Nuclear structure and nuclear spectroscopy.
Address: School of Physics, Minnesota University, Minneapolis, Minnesota 55455, U.S.A.

BROWN, S. U., Dr. Head, Rad. Biol. Lab., Texas Eng. Experiment Station.
Address: Texas Engineering Experiment Station, College Station, Texas, U.S.A.

BROWN, Rear Admiral Samuel R., Jr. Member, Military Liaison Com., U.S.A.E.C.
Address: United States Atomic Energy Commission, Military Liaison Committee, Washington, D.C. 20545, U.S.A.

BROWN, Sanborn Conner, A.B., M.A., Ph.D. (Phys.). Born 1913. Educ.: Dartmouth Coll. and M.I.T. Assoc. Prof. Phys. (1949-62), Prof. Phys. (1962-), Assoc. Dean, Graduate School (1963-), M.I.T. Tech. Adviser, U.S. Deleg. to 2nd U.N. I.C.P.U.A.E., Geneva, 1958; U.S. Deleg. to I.A.E.A. Conference on Plasma Phys. and Controlled Thermonucl. Fusion, Salzburg, 1961. Societies: Fellow, Div. Plasma Phys., A.P.S.; A.A.P.T.; Fellow, American Acad. Arts and Sci.; Fellow, A.A.A.S.; Nat. Sci. Teachers Assoc.; History of Sci. Soc.; Internat. Union Pure and Appl. Phys.; Internat. Council Sci. Unions; Roy. Inst. of Great Britain.
Nuclear interest: Thermonuclear plasma research.
Address: Room 20A-125, Research Laboratory of Electronics, Massachusetts Institute of Technology, Cambridge 39, Massachusetts, U.S.A.

BROWN, Sherwood Fiske, S.B., S.M., Sc.D. Born 1902. Educ.: M.I.T. Chairman, Phys. Dept., Colby Coll., 1942-59; Dept. Nucl. Sci. and Eng. (1959-61), Head, Mech. and Textile

Eng. Dept. (1961-65), Head, Dept. Phys. (1965-), Lowell Technol. Inst.; U.S. Army Quartermaster Command, 1959-61; U.S. Air Force Cambridge Res. Lab., 1961-64. Societies: A.A.P.T.; American Ceramic Soc.; A.S.E.E.
Nuclear interest: Radiation damage.
Address: 3 Lancaster Avenue, Chelmsford, Massachusetts, U.S.A.

BROWN, Steven H., B.Sc. Born 1922. Educ.: M.I.T. and Yale Univ. Manager, New Projects Development. Societies: A.N.S.; A.I.Ch.E.
Nuclear interest: Reprocessing and shipping of radioactive material.
Address: National Lead Company, 111 Broadway, Room 1400, New York, N.Y. 10006, U.S.A.

BROWN, T. L. Special Metals Development Metal., Murex Ltd.
Address: Murex Ltd., Rainham, Essex, England.

BROWN, Thomas Walter Falconer, B.Sc. (Special Distinction Nat. Phil., Glasgow), D.Sc. (Glasgow), S.M. (Harvard). Born 1901. Educ.: Glasgow and Harvard Univs. Director, Pametrada Res. Station, Wallsend, 1944-62; Director of Marine Eng. Res., British Ship Res. Assoc., Wallsend Res. Station, 1962-; Formerly Member, Joint Panel on Nucl. Marine Propulsion; Chairman, Board of Governors, Rutherford Coll. of Sci. and Technol.; Member of Government Committees in Admiralty, Ministry of Transport. Societies: Inst. Mech. Engineers; Roy; Inst. Naval Architects (Member of Council); Fellow, North-east Coast Inst. of Eng. and Shipbuilders (Member of Council); F. Inst. P.; Fellow, R. Aero Soc.; A.S.M.E.; A.N.S.; American Soc. of Naval Architects and Marine Engineers.
Address: Dumbreck, Wylam, Northumberland, England.

BROWN, W. M. COURT. See COURT BROWN, W. M.

BROWN, Wade Lynn, B.A., Ph.D. Born 1906. Educ.: Texas Univ. Prof. Psychology, Texas Univ.; Head, Exptl. Psychology, Radiobiol. Lab., Texas Univ. Book: Chapter 48, Responses of the Nervous System to Ionizing Rad. (Academic Press, 1961). Society: Rad. Res. Soc.
Nuclear interest: Effects of ionizing radiations on behaviour.
Address: Department of Psychology, Texas University, Austin, Tex., U.S.A.

BROWN, William Marriott, B.A.Sc. (Hons., Mech. Eng.). Born 1921. Educ.: British Columbia Univ. Design Eng., A.E.C.L., 1950-55; Reactor Equipment Eng. (1956-60), Manager, Equipment Design (1960-62), Manager, Equipment Eng. (1962-67), Manager, Nucl. Projects Eng. (1967-), Canadian Gen. Elec. Co. Member, Tech. Council, Canadian Standards Assoc.; Chairman, Standing Papers Com., Canadian Nucl. Assoc.; Sec.-Treas., Peterborough Chapter, A.P.E.O.
Nuclear interest: Engineering management and design of nuclear power plants.
Address: Canadian General Electric Company Limited, Atomic Power Department, 107 Park Street, Peterborough, Ontario, Canada.

BROWNE, Cornelius Payne, B.A., Ph.D. Born 1923. Educ.: Wisconsin Univ. Res. Assoc., Phys. Dept., M.I.T., 1951-56; Asst. Prof. (1956-58), Assoc. Prof. (1958-64), Prof. (1964-), Notre Dame Univ. Society: Fellow, A.P.S.
Nuclear interests: Nuclear physics precision measurements on reaction energies and reaction mechanisms.
Address: Department of Physics, University of Notre Dame, Notre Dame, Indiana, U.S.A.

BROWNE, Harry L. Born 1919. Educ.: Manhattan Coll. and New York Univ. Control Officer, Manhattan Project, 1943-47; various managerial posts, U.S.A.E.C., 1947-55; Manager, Nucl. Products Dept., Thompson Products, Inc., 1955-57; Asst. to Gen. Manager, Gen. Atomic, 1957-60; Pres., Hazleton Nucl. Science Corp. 1960-64; Pres., U.S. Nucl. Corp., 1966-; Pres., Volk Radiochem. Co., 1960-. Societies: A.N.S.; Atomic Ind. Forum.
Nuclear interests: Management of nuclear R. and D. activities, especially in nuclear power plants and in the application of nuclear technique to industrial, medical and agricultural field.
Address: 4062 Fabian Way, Palo Alto, California 94303, U.S.A.

BROWNELL, George McLeod, B.Sc. (Manitoba), M.Sc. (Manitoba), Ph.D. (Minnesota). Born 1899. Educ.: Manitoba and Minnesota Univs. Prof., Chairman (1945-65), Emeritus Prof. (1965-), Dept. Geology, Manitoba Univ. Societies: F.R.S.C.; Fellow, Geological Soc. America; Soc. Economic Geologists; Geological Assoc. Canada; Canadian Inst. Mining and Metal.; Assoc. Professional Eng. Manitoba.
Nuclear interest: Nuclear physics applied to geology.
Address: 61 Cordova Street, Winnipeg 9, Manitoba, Canada.

BROWNELL, Gordon Lee, B.S., Ph.D. (Phys.). Born 1922. Educ.: Bucknell Univ. and M.I.T. Head, Phys. Res. Lab., Mass. Gen. Hospital; Res. Assoc. in Medicine, Harvard Medical School; Assoc. Prof., M.I.T.; Consultant-Lecturer in Rad. Phys., U.S. Chelsea Naval Hospital. Director, American Assoc. Phys. in Medicine. Member, Exec. Com., Isotopes and Rad. Div., A.N.S., 1966-. Book: Co-editor, Rad. Dosimetry (Academic Press, 1956). Societies: A.P.S.; Rad. Res. Soc.; H.P.S.; A.A.A.S.
Nuclear interests: Biological science; Health physics; Physics.
Address: Physics Research Laboratory, Massachusetts General Hospital, Boston, Massachusetts, U.S.A.

BROWNELL, Lloyd Earl, B.Ch.E. (Clarkson Coll. of Technol.), B.M.E. (Clarkson Coll. of Technol.), M.S. (Chem. Eng., Michigan), Ph.D. (Chem. Eng., Michigan). Born 1915. Educ.: Clarkson Coll. of Technol., New York, and Michigan Univ. Asst. Prof., Dept. of Chem. and Met. Eng. (1947-51), Assoc. Prof., Dept. of Chem. and Met. Eng. (1951-54), Supervisor, Fission Products Lab. (1951-60), Prof., Dept. of Chem. and Nucl. Eng. (1954-), Michigan Univ.; Consultant in Nucl. Sci., 1952-; Senior Res. Assoc., Battelle Northwest Lab., Richland, Washington, 1965-. Books: Co-author, Unit Operations (1950); co-author, Design of Process Equipment (1959); Rad. Uses in Ind. and Sci. (1961); co-editor, Chem. and Food Applications of Rad. (A.I.Ch.E. Symposium Series 83, Vol. 64, 1968); contributor to I.A.E.A. Compendium on Rad. Shielding, 1968. Societies: A.I.Ch.E.; A.C.S.; American Soc. of Refrigerating Eng.; New York Acad. Sci.; Inst. Food Technologists; A.A.A.S.; A.N.S.
Nuclear interests: Uses of nuclear radiation, utilisation of gross fission products, processing of foods with gamma radiation and flow through porous media and internal ballistics.
Address: 2219 E. Engineering Building, Michigan University, Ann Arbor, Michigan, U.S.A.

BROWNING, Colonel Louis E. Head, Administrative Depts., Armed Forces Radiobiol. Res. Inst., 1966-.
Address: Armed Forces Radiobiology Research Institute, National Naval Medical Centre, Bethesda 14, Maryland, U.S.A.

BROWNING, William Elgar, Jr., B.A. Born 1923. Educ.: Ohio State Univ. Sen. Chem., Fairchild Engine and Airplane Corp., 1947-51; Lead Eng., California R. and D. Corp., 1951-53; Chem., O.R.N.L., 1953-. Societies: A.C.S.; A.A.A.S.; A.N.S.; RESA.
Nuclear interests: Computer processing of reactor safety design data; Radioactive materials in gases, behaviour and methods of removal; Release of fission products in nuclear accidents; Effects of radiation on materials; Measurement of temperatures in nuclear reactors.
Address: Reactor Chemistry Division, Oak Ridge National Laboratory, P.O. Box Y, Oak Ridge, Tennessee, U.S.A.

BROWNLEE, William R. Exec. Vice Pres. and Director, (1966-), Chief Exec. Officer (1967-), Southern Services, Inc. Member, Tech. Development Com., H.T.D.A. Fuel Development Programme; Member, Advisory Com. on Toll Enrichment, Atomic Ind. Forum; Member, Nucl. Fuels Com., Edison Elec. Inst.
Address: Southern Services, Inc., 600 North 18th Street, Birmingham 2, Alabama, U.S.A.

BRUCE, Alan K., B.S., M.S., Ph.D. Born 1927. Educ.: New Hampshire and Rochester Univs. Res. Assoc., Biol. Div., O.R.N.L. 1956-57; Radiol. Safety Officer and Assoc. Prof. Biol., Buffalo Univ., 1957-. Societies: Rad. Res. Soc.; American Soc. for Microbiology; H.P.S.
Nuclear interests: Radiation microbiology; Isotopic tracer techniques; Health physics.
Address: Biology Department, State University of New York at Buffalo, Buffalo, New York 14214, U.S.A.

BRUCE, Francis Robert, B.Sc. (Chem.). Born 1919. Educ.: Tufts Coll. Assoc. Director, specialising in waste disposal and chem. processing of irradiated fuels, Chem. Technol. Div. (1944-59), Director, Rad. Safety and Control, specialising in rad. protection, Director's Div. (1960-), O.R.N.L. Sci. Sec. for 2nd I.C.P.U.A.E. Book: Co-editor, Process Chem., vol. 1 (1956), vol. 2 (1958), and vol. 3 (1961). Papers: Solvent Extraction Chem. (p. 100, vol. 7, I.C.P.U.A.E., Geneva, 1955); The Behaviour of Fission Products in Solvent Extraction Processes (Process Chem., vol. 1, 1956); Chem. Processing of Aqueous Blanket and Fuel from Thermal Breeder Reactors (Chem. Eng. Progr., 52, 347, 1956); Look for Nucl. Fuel Reprocessing by 1965 (Chem. Eng., July 1957, p. 202); co-author, Operating Experience with Two Direct-maintenance Radiochem. Pilot Plants (2nd I.C.P.U.A.E., Geneva, Sept. 1958). Societies: A.C.S.; A.I.Ch.E.; A.N.S.
Address: Euclid Circle, Oak Ridge, Tennessee, U.S.A.

BRUCE-KERR, Adrian George, H.N.Diploma (Mech. Eng.), A.M.I.Mech.E., M.I.Nucl. E. Born 1923. Educ.: Kingston Coll. of Technol. Kingston-on-Thames, Sen. Design Eng., Nucl. Power Div., Mitchell Eng. Ltd., 1957-62; Sen. Eng., Nucl. and Chem. Eng. Dept., W.S. Atkins and Partners (Consulting Eng.), 1962-64; U.N. Technical Assistance Expert, attached to Central Government of India, 1964-. Society: B.N.E.S.
Nuclear interests: Management, engineering design, inspection.
Address: c/o National and Grindlays Bank Ltd., 13 St. James's Square, London, S.W.1, England.

BRUCER, Marshall, S.B., M.D. Born 1913. Educ.: Chicago Univ. Visiting Prof. Physiology, Texas Univ., 1948-; Visiting Prof. Physiology, Duke Univ. School of Medicine, 1953-; Chairman, Medical Div., O.R.I.N.S., 1949-62. Member, Editorial Board, Internat. Journal of Appl. Rad. and Isotopes. Books: Co-author, Radioisotopes in Medicine (U.S.A.E.C., Sept. 1953); co-author, Roentgens, Rads, and Riddles (U.S.A.E.C., 1958); author, Thyroid Radioiodine Uptake (U.S.A.E.C., 1959). Societies: A.M.A.; Soc. for Exptl. Biol. and Medicine; Soc. of Nucl. Medicine; Radiol. Soc. of North America; American Statistical Assoc.
Nuclear interests: Nuclear medicine; Radiation protection; Radiation dosimetry; Radiation physiology; Radiation physics.
Address: Route 4, Box 203, Tuscon, Arizona, U.S.A.

BRÜCHNER, Hans Joachim, Dr. rer. nat., Diplom-Physiker. Born 1930. Educ.: Munich T.H. and California Inst. Technol. Chief phys. (1956-59), in charge of liaison to internat. atomic organisations and marketing (1959-63), Manager of reactor sales, (1963-65), Marketing Manager, (1965-), A.E.G. Nucl. Energy Div. Book: Chapter on Reactor Theory and Nucl. Fuel Cycles in Riezler Walcher, Einführung in die Kerntechnik (Stuttgart, B.G. Teubner, 1958).
Nuclear interests: Reactor design; Economics of nuclear power; international aspects of nuclear energy.
Address: 13 Heinrich-von-Kleiststrasse, Bad Homburg VdH, Germany.

BRUCK, Henri, D. ès Sci. Born 1909. Educ.: Göttingen and Paris Univs. Prof., Inst. Nat. Sci. et Tech. Nucléaires, Head, Sect. d'Optique Corpusculaire, Centre d'Etudes Nucléaires de Saclay. Book: Circular particle accelerators (Paris, Presses Universitaires de France, 1966).
Nuclear interests: Accelerators; Particle optics.
Address: 11 bis Avenue Carnot, Sceaux, Seine, France.

BRUECKNER, Keith Allan, B.A. (Minnesota), M.A. (Minnesota), Ph.D. (California). Born 1924. Educ.: Minnesota and California Univs. Prof. Phys., Pennsylvania Univ. 1956-59; Prof. Phys. and Chairman Phys. Dept., California Univ., San Diego, 1959-61; Vice-Pres. and Director Res., Inst. Defence Analyses, Washington D.C., 1961-62; Dean Letters and Science (1963-64), Dean Graduate Studies (1965), Prof. Phys. and Director, Inst. Rad. Phys. and Aerodynamics (1965-67), Prof. Phys. and Director, Inst. Pure and Appl. Phys. Sci. (1967-), California Univ., San Diego; Consultant: Los Alamos Sci. Lab. A.R.P.A., L.R.L., Gen. Motors and S.R.I. Books: Editor, Advances in Theoretical Phys. and Pure and Applied Phys. (Academic Press). Societies: A.P.S.; A.R.S.; A.I.A.A.
Nuclear interest: Theoretical nuclear physics.
Address: Pure and Applied Physical Sciences Institute, California University, San Diego, P.O. Box 109, La Jolla, California 92037, U.S.A.

BRUES, Austin M., A.B., M.D. (Harvard Medical School). Born 1906. Educ.: Harvard Univ. and Medical School. Director, Biol. and Medical Res. Div. (1946-62), Sen. Biol. (1962-), Argonne Nat. Lab. Consultant to Office of Surgeon-Gen., 1958-. Nat. Res. Council Com. on Pathological Effects of Atomic Rad.; U.S. Deleg. to U.N. Sci. Com. on Effects of Atomic Rad.; Advisory Com., Atomic Bomb Casualty Commission; W.H.O., Mental Health Com.; Com. on Rad. Effects, Internat. Commission on Radiol. Protection; Chairman, Advisory panel on environmental biol. to N.A.S.A. Societies: N.Y. Acad. Sci.; American Acad. Arts and Sci.; American Assoc. for Cancer Res.; Rad. Res. Soc.
Nuclear interests: Radiation biology and nuclear medicine; Mechanisms of carcinogenesis; Metabolism and action of internal emitters; Cellular effects of radiation.
Address: Argonne National Laboratory, Argonne, Illinois 60439, U.S.A.

BRUGGEMAN, W. H. Formerly Manager, DIG Project, now Manager, West Milton Site Operation, K.A.P.L.
Address: Knolls Atomic Power Laboratory, Schenectady, New York, U.S.A.

BRUGGER, Robert M., M.A., Ph.D. Born 1929. Educ.: Colorado Coll. and Rice Univ. Joint Honor Scholarship, Colorado Coll., 1947-51; Electronics Technician, White Sands Proving Ground, New Mexico, 1951; Graduate Asst., Rice Univ., 1951-54; Summer Res. Phys., E.I. du Pont, Waynesboro, Virginia, 1953; Humble Oil Co. Fellow, Rice Univ., 1954-55; Res. Phys. (1955-57), Head, Solid State Phys. Sect. (1957-), Phillips Petroleum Co., Atomic Energy Div., assigned to Nucl. Phys. Div., United Kingdom A.E.R.E., Harwell (1962-63). Society: A.P.S.
Nuclear interests: Nuclear physics; Slow neutron inelastic scattering.
Address: Phillips Petroleum Company, Atomic Energy Division, P.O. Box 2067, Idaho Falls, Idaho, U.S.A.

BRUIL, William Austin, Drs. Born 1932. Educ.: Amsterdam Univ. Wetenschappelijk Medewerker, Phys., Inst. voor Kernfysisch Onderzoek (I.K.O.), Amsterdam, 1955-60; Wetenschapplijk Hoofdmedewerker, Physicus, Tech. Hogeschool (T.H.E.), Eindhoven, 1960-.
Nuclear interest: Nuclear physics.
Address: Technische Hogeschool, 2, Insulindelaan, Eindhoven, Netherlands.

BRUIN, Margaret, Dr. High Energy Particles, Dept. of Physics, American Univ. of Beirut.
Address: American University of Beirut, Department of Physics, Beirut, Lebanon.

BRUINSMA, Pieter J. T., M.Sc. Born 1938. Educ.: Delft Tech. Univ. Res. Eng.,Michigan Univ., 1960-64; Chief Linac Eng., I.K.O. (Inst. voor Kernphysisch Onderzoek). Societies: I.E.E.E.; Koninklijk Inst. van Ingenieurs.
Nuclear interest: Accelerator design.
Address: I.K.O., 18 Ooster Ringdijk, Amsterdam, Netherlands.

BRUKL, Alfred, Prof. Head, Inst. of Inorg. Chem., Vienna Univ.
Address: Institute of Inorganic Chemistry, 42 Währingerstrasse, Vienna 9, Austria.

BRUN, Ernst, Prof. Dr. Sen. Officer, Physik. Inst., Zürich Univ.
Address: Physikalisches Institut, Universität Zürich, 9 Schonberggasse, 8001 Zürich, Switzerland.

BRUNDRETT, Sir Frederick, K.C.B., K.B.E., B.A. (Hons. Mathematics Wrangler), M.A., D.Sc. (Manchester). Born 1894. Educ.: Cambridge Univ. Sci. Adviser, Ministry of Defence, and Chairman, Defence Res. Policy Com., 1954-59; Chairman, Air Traffic Control Board, 1959-; Civil Service Commissioner and Sci. Adviser to Commission, 1960-67; Chairman, Naval Aircraft Res. Com., 1960-66; White Fish Authority; Chairman, Rural Ind. Bureau Trustees, 1961-65.
Nuclear interest: Nuclear weapons.
Address: Prinsted, Emsworth, Hampshire, England.

BRUNEAU, Jean-Luc. Tech. Counsellor to Minister for Sci., Nucl. and Space Affairs.
Address: Ministry of Scientific, Nuclear and Space Affairs, 2 rue Royale, Paris 8, France.

BRUNEAU, Marcel Louis Adolphe, Ing. Agronome. Born 1903. Educ.: Faculté Agronomique, Gembloux. Inst. Nat. pour l'Amélioration des Conserves de Légumes.
Nuclear interest: Recherches agronomiques.
Address: 78 rue du Long Chêne, Wezembeek-Oppem, Belgium.

BRUNEDER, Heinrich, Dr. phil. (Theoretical Phys. and Maths.). Born 1936. Educ.: Vienna Univ. Seconded from Reaktorzentrum Seibersdorf, Austria, to O.E.C.D. Dragon Project, Winfrith; now Leader, Maths. Div., Reaktorzentrum Seibersdorf.
Nuclear interests: Reactory theory; Programming reactor problems.
Address: Reaktorzentrum Seibersdorf, 10 Lenaugasse, 1082 Vienna, Austria.

BRUNELLI, Bruno, Prof. Dr. Director, Lab. Gas Ionizzati, C.N.E.N.
Address: Casella Postale 65, 00044 Frascati, Rome, Italy.

BRUNENKANT, Edward James, Jr., LL.B. Born 1921. Educ.: George Washington Univ. Director,Tech. Information Div., and Chairman, Advisory Com. on Tech. Information, U.S.A.E.C., Washington, D.C. Com. on Sci. and Tech. Information; Chairman, Panel on Legal Aspects of Information Systems, Com. on Sci. and Tech. Information.
Address: 8904 Honeybee Lane, Bethesda, Maryland 20034, U.S.A.

BRUNER, Harry Davis, B.S., M.S., M.D., Ph.D. Born 1911. Educ.: Louisville and Chicago Univs. Chief Sci., O.R.I.N.S. 1949-52; Prof. Physiology, Emory Univ., 1952-56; Chief, Medical Res. Branch, U.S.A.E.C. 1956-60; Asst. Director for Medical and Health Res., Div. Biol. and Medicine, U.S.A.E.C. Vice-Pres., Mid-Atlantic States Ch., American Soc. Nucl. Medicine. Book: Revised Tables to Correct for Physical Decay of Some Frequently used Radioisotopes (Oak Ridge, U.S.A.E.C., 1956). Societies: American Physiological Soc.; American Soc. for Pharmacology and Exptl. Therapeutics; Soc. for Exptl. Biol. and Medicine; American Soc. for Nucl. Medicine.
Nuclear interests: Biomedical research on radiation sickness and its treatment; Administration of biomedical research in general.
Address: U.S. Atomic Energy Commission, Washington D.C., 20545, U.S.A.

BRUNET, Georges. Directeur, Sté. d'Etudes Industrielles de Villefuif, Brunet Pramaggiore et Cie.
Address: Société d'Etudes Industrielles de Villefuif, Brunet Pramaggiore et Cie., 21 rue Georges le Bigot, Villejuif, (Seine), France.

BRUNET, R. Pres., Sté. Francaise d'Electro-Radiologie Medicale.
Address: Societe Francaise d'Electro-Radiologie Medicale, 9 rue Daru, Paris 8, France.

BRUNINX, Edward, Dr. Sc. Born 1929. Educ.: Ghent Univ. Res. Assoc., M.I.T., 1956-57; Res. Assoc., Michigan Univ.; Staff member, C.E.R.N., 1958-65; Exchange Fellow, Weizmann Inst., 1962-63; Staff Member, Philips Res. Labs., Eindhoven, 1965-.
Nuclear interests: Nuclear- and Radiochemistry.
Address: Philips Research Laboratories, Eindhoven, Netherlands.

BRUNNER, Alfred, Dr. Sc. tech., Dipl.masch. ing. Born 1918. Educ.: E.T.H., Zürich. Patent Coordinator, Sulzer Brothers Ltd., Winterthur.
Nuclear interest: Nuclear patents.
Address: 21 Wylandstrasse, Winterthur, Switzerland.

BRUNNER, Johannes Gerhard, Dr. rer. nat. habil. Born 1932. Educ.: Leipzig, Karl-Marx-Univ. Inst. fuer Kernphysik, Zeuthen (1955-56), Inst. fuer angewandte Radioaktivität, Leipzig (1956-), Deutsche Akademie der Wissenschaften zu Berlin. Docent for Radioactivity, Karl-Marx-Univ., Leipzig, 1964-. Books: Coauthor, Messverfahren unter Anwendung ionisierender Strahlung volume V of Handbuch der Technischen Betriebskontrolle (Leipzig, Akademische Verlagsgesellschaft, 1968); co-author, Analyse von Mineral- und Syntheseölen mit radiometrischen Methoden (Berlin, Akademieverlag, 1968). Society: Physikalische Gesellschaft in der D.D.R.
Nuclear interests: Non-destructive analysis with nuclear methods; Interaction between nuclear radiation and matter: Scintillation spectroscopy.
Address: 28 Gudrunstrasse, 703 Leipzig, German Democratic Republic.

BRUNNER, Johannes Heinrich, dipl. Phys. E.T.H. Born 1934. Educ.: E.T.H. (Swiss Federal Inst. Technol.), Zürich, and Argonne Nat. Lab. Health Phys. Asst. Head, Health Phys. Div., Eidg. Institut für Reaktorforschung (former Reactor Ltd.) at Würenlingen, 1958-.
Nuclear interests: Health physics in general; Dosimetry; Radiobiology; Health physics training.

Address: 26 Möhrlistrasse, Ch-8006 Zürich, Switzerland.

BRUNO, André, Ing. Born 1912. Educ.: Ecole Centrale des Arts et Manufactures. Directeur de l'Exploitation, Sté. des Acieries de Pompey. Address: Société Aciéries de Pompey, 61 rue de Monceau, Paris 8, France.

BRUNS, Robert Stewart III, B.S. (Phys.). Born 1930. Educ.: Middlebury Coll., Vermont. Eng. (1956-57), Group Supv., Eng. Dept. (1957-58), Group Supv., Sales Dept. (1958-59), Asst. Manager, Sales Dept. (1959-62), Manager, Administration (1962-64), A.M.F. Atomics; Assoc.,Morey Assocs., 1965-66; Data Management, Perkin-Elmer Corp., Wilton, Connecticut, 1967-. Address: 302 Cannon Road, Wilton, Connecticut, U.S.A.

BRUNSCHWIG, Gerard. Ing. des Ponts et Chaussees, Groupe d'Applications des Radioisotopes, Bridges and Roads Central Lab. Address: Bridges and Roads Central Laboratory, B.P. 155, 75 Orly-Aerogare, France.

BRUNSON, Glenn Samuel, B.S., M.S. Born 1922. Educ.: Princeton Univ. Assoc. Phys., Argonne National Lab. Society: A.N.S. Nuclear interests: Fast reactor physics; Specifically statistical population methods; Fast neutron spectroscopy kinetics and hazards analysis; Gamma spectroscopy, detector design and standardization, Utilisation of research reactors in developing countries. Address: International Atomic Energy Agency, Kaerntnerring 11, Vienna 1010, Austria.

BRUNTON, Donald, B.Sc., Ph.D. Born 1916. Educ.: Queen's Univ., Kingston, Ontario, California Inst. Technol., and McGill Univ. Manager, Nucl. Production Dept., Curtiss-Wright Corp., Princeton Div., Princeton, N.J., 1957-60; Manager, Control Res. and Eng., Ind. Nucleonics Corp., Columbus, Ohio, 1960-63; Pres., BRUN Corp., Columbus, Ohio, 1964-. Book: Chapter on Nucl. Rad. Detectors in Handbook of Applied Instrumentation (McGraw-Hill, 1961). Societies: A.P.S.; A.N.S.; Instrument Soc. of America; A.P.E.O. Nuclear interests: Industrial measurement and control with radioisotopes. Address: 1150 W. 3rd Avenue, Columbus, Ohio 43212, U.S.A.

BRUSCHETTI, Sergio, Degree in elec. eng. Born 1933. Educ.: Genoa Univ. Pisa Univ., 1957-58; C.I.S.E., Milano, 1959-60; Ansaldo S.P.A., Genoa. Nuclear interests: Engineering of light water reactors; Thermal cycles of power plant. Address: 5-14 Via Fratelli Dagnino, Genoa, Pegli, Italy.

BRUSCHI, Colombo, Nucl. eng. Born 1935. Educ.: Politecnico di Milano. C.I.S.E., 1960-. Nuclear interests: Reactor design from the viewpoint of nuclear reactor physics and analysis; Study and optimisation of fuel cycles. Address: Centro Informazioni Studi Esperienze, C.P. 3986, Milan, Italy.

BRUSH, Harvey F., B.S.(Chem. Eng.). Born 1920. Educ.: Pennsylvania State Univ. Employed (1946-), Chief Mech. Eng. (1960-61), Manager, Nucl. Eng. (1962-64), Manager Eng. (1964-), Bechtel Corp. Societies: A.S.M.E.; A.I.Ch.E.; A.N.S. Nuclear interests: Engineering and construction of nuclear and conventional power plants and special facilities. Address: Bechtel Corporation, P.O. Box 3965, 50 Beale Street, San Francisco, California 94119, U.S.A.

BRUSTAD, Tor, Cand. Real. Member, Sci. Advisory Council, Biophys., Norsk Hydro's Inst. for Cancer Res. Member, Rad. Planning Board, I.C.R.U. Society: Fellow, Norwegian Cancer Soc. Address: Norsk Hydro's Institute for Cancer Research, Norwegian Radium Hospital, Oslo, Norway.

BRUUSGAARD, Arne, M.D. (Oslo). Born 1906. Sen. Med. Inspector of Labour, 1946-; Member,Govt. Council on Rad. Protection; Member of Com. Norwegian Assoc. Ind. Medical Officers; Member, Rad. Advisory Council. Societies: Norwegian Medical Soc.; Hon. Member, Finnish Ind. Medical Soc.; Hon. Member, Assoc. Ind. Medical Officers (England). Nuclear interests: Radiation protection and health control. Address: P.B. 8103, Oslo, Dep., Norway.

BRY, Paul. Directeur Gén., Sté. Générale du Vide S.A. Address: Société Générale du Vide S.A., 186 rue du Faubourg Saint Honôre, Paris 8, France.

BRYAN, Frederick Allen, Jr., B.S., M.S. (Phys.), Ph.D. (Phys. and Maths.). Born 1933. Educ.: Michigan and North Carolina State Univs. Res. Asst., Eng. Res. Inst., Michigan Univ., 1954-57; Project Phys., Atomic Energy Div., Babcock and Wilcox Co., 1957-58; Instructor, Virginia Univ. Extension Div., 1957-58; Instructor, North Carolina State Univ., 1958-60; Sen. Phys., ASTRA, Inc., 1960-65; Group Leader, Res. Triangle Inst., 1965-. Nuclear interests: Theoretical and experimental aspects of reactor physics; Reactor design; Radiation transport and shielding; Nuclear physics and cryogenic neutron physics. Address: Research Triangle Institute, P.O. Box 12194, Research Triangle Park, North Carolina, U.S.A.

BRYAN, Geoffrey William, B.Sc., Ph.D. Born 1934. Educ.: Bristol Univ. Zoologist, Plymouth Lab., Marine Biological Assoc. of United

Kingdom, 1958-.
Nuclear interests: Biological aspects of radioactive waste disposal in the sea and in fresh water with particular reference to ^{137}Cs and neutron induced nuclides.
Address: The Laboratory, Citadel Hill, Plymouth, Devon, England.

BRYAN, Morris M., Jr., Res. Com., State of Georgia Nucl. Advisory Commission.
Address: State of Georgia Nuclear Advisory Commission, Office of Publications, Georgia Institute of Technology, Atlanta 13, Georgia, U.S.A.

BRYANT, Emmons. Director, Tracerlab-Keleket.
Address: Tracerlab-Keleket, 1601 Trapelo Road, Waltham 54, Massachusetts, U.S.A.

BRYANT, Frederick James, Ph.D. (Electrochem.), B.Sc. (Chem.), D.I.C., A.R.C.S. Born 1914. Educ.: London Univ. P.S.O., M.O.S., 1950-55; Principal Sci. (1955-58), Sen. Principal Sci. (1958-67), U.K.A.E.A. Group Leader, Woolwich and Chatham Outstations, Chem. Div., A.E.R.E.; Sen. Principal Sci., U.K.A.E.A. Group Leader, Materials Analysis Group, Analytical Sci. Div., A.E.R.E., 1967-. Society: Soc. for Analytical Chem.
Nuclear interests: Analytical and radiochemistry, with particular reference to reactor materials and systems and worldwide fallout.
Address: Hurstmere, Upper Basildon, Reading, Berkshire, England.

BRYANT, Gerrett H., Eng. of Mines. Born 1928. Educ.: Puget Sound Univ. and Colorado School of Mines, Mineral Ind. Consultant, G.H. Bryant and Assocs., Denver, Colorado. Society: A.I.M.E.
Nuclear interest: Land acquisition, exploration and project development consulting services for companies engaged in the search for source material ore reserves.
Address: 560 E. Amherst Place, Englewood, Colorado, U.S.A.

BRYANT, Howard Carnes, Ph.D., M.S., B.A. Born 1933. Educ.: California (Berkeley) and Michigan Univs. Asst. Prof. (1960-65), Assoc. Prof. (1965-67), New Mexico Univ.; Visiting Staff Member, Los Alamos Sci. Lab., 1961-67. Society: A.P.S.
Nuclear interest: Nuclear physics.
Address: 800 Yale Boulevard N.E., Department of Physics and Astronomy, New Mexico University, Albuquerque, New Mexico 87106, U.S.A.

BRYANT, Philip Stephen, B.Sc. Born 1905. Educ.: Bristol Univ.Manager, Chem. Group (1946-55), Sen. Metal (1955-60) Director (1960-), Murex Ltd. Societies: Fellow, Inst. Metal.; Inst. Metals; Iron and Steel Inst.; Inst. Chem. Eng.; Stainless Steel Development Assoc.
Nuclear interests: Was responsible for beryllium factory erected and operated at Milford Haven by Murex Ltd. on behalf of U.K.A.E.A.; Controlled zirconium pilot plant and tantalum/-niobium plants at Murex Rainham Works.
Address: Murex Limited, Rainham, Essex, England.

BRYANT, T. H. E., B.Sc. Dept. of Phys. applied to Medicine, Middlesex Hospital Medical School.
Address: The Middlesex, Hospital Medical School, Cleveland Street, London, W.1, England.

BRYER, R. G. R. and D. Eng., R.A. Stephen and Co. Ltd.
Address: R. A. Stephen and Co. Ltd., Miles Road, Mitcham, Surrey, England.

BRYNIELSSON, Harry Anders Bertil. Born 1914. Educ.: Roy. Inst. of Technol., Stockholm. Managing Director, L.K.B.-Produkter Fabriks A.B., Stockholm, 1943-51; Managing Director, A.B. Atomenergi, Stockholm, 1951-. Member, Roy. Swedish Acad. Eng. Sci.
Nuclear interest: Management.
Address: A.B. Atomenergi, P.O. Box 43041, Stockholm 43, Sweden.

BRYNJOLFSSON, Ari, M.Sc. Born 1926. Niels Bohr Inst., Copenhagen, 1948-54; Iceland Univ., 1954-55; Göttingen Univ., 1955-57; Res. Establishment Risö, Danish Atomic Commission, 1957-62; U.S. Army Natick Labs., Massachusetts, 1962-63; Res. Establishment Risö, Danish Atomic Commission, 1963-; Now Chief, Rad. Sources Div., U.S. Army Natick Labs. Books: On rock magnetism; On technochronological studies in Iceland; On absolute calibration of radiation; Irradiation facilities.
Nuclear interests: Gamma-electron irradiation technique; Irradiation sources and instruments; Irradiation monitors and standardisation.
Address: Radiation Sources Division, United States Army Natick Laboratories, Natick, Massachusetts 01760, U.S.A.

BRYSON, Theordore C., B.S. (Chem.). Born 1912. Educ.: Carnegie Inst. Technol. Lab. Supervisor (1943-52), Manager (1952-), Analytical Chem., Bettis Atomic Power Div., Westinghouse Elec. Corp. Societies: A.C.S.; A.S.T.M.; Soc. for Analytical Chemists of Pittsburgh.
Nuclear interests: The analytical chemistry of materials used for nuclear fuel elements, reactor control rods, fuel element cladding materials and the less familiar chemical elements; the chemical processing and reactions of these materials.
Address: Westinghouse-Bettis Atomic Power Laboratory, P.O. Box 1468, Pittsburgh 30, Pennsylvania, U.S.A.

BRZHECHKO, L. V. Paper: Co-author, Influence of Metal Chamber on Measurements of Plasma Parameters by Diamagnetic Probe

(Abstract, Atomnaya Energiya, vol. 20, No. 1, 1966).
Address: Academy of Sciences of the U.S.S.R., 14 Leninsky Prospekt, Moscow V-71, U.S.S.R.

BRZOSTOWSKI, Witold Konrad, M. Chem., D.Sci., D.habil. Born 1934. Educ.: Warsaw and Reading (England) Univs. Asst., Sen. Asst., Adiunkt., Warsaw Univ., 1955-61; Asst., Sen. Asst., Adiunkt (1956-65), Docent, Inst. of Phys. Chem. (1965-), Polish Acad. of Sci. Societies: The Faraday Soc.; Polish Chem. Soc. Nuclear interests: Application of labelled substances in the study of thermodynamics of liquids; Detection of weak beta radiation.
Address: 12 Brzozowa, Warsaw 40, Poland.

BRZOZOWSKI, Wojciech, Prof., Dr. Vice Director for Nucl. Eng., Inst. of Nucl. Res., Warsaw. Member, State Council for the Peaceful Use of Nucl. Energy.
Address: Institute of Nuclear Research, Swierk near Warsaw, Poland.

BUBEN, N. Ya. Paper: Co-author. Some Remarks on Rad. Stability Low Melting Coolants in Liquid and Solid State (Atomnaya Energiya, vol. 22, No. 1, 1967).
Address:Academy of Sciences of the U.S.S.R., 14 Leninsky Prospekt, Moscow V-71, U.S.S.R.

BUBNER, Martha Marianne, Dr. rer. nat., Dipl. Chem. Born 1935. Educ.: Dresden T.U. Book: Co-author, Die Synthese Kohlenstoff-14-markierter Organischer Verbindungen.(Leipzig, VEB G. Thieme, 1966).
Nuclear interests: Herstellung und anwendung radioaktiv markierter Verbindungen.
Address: Zentralinstitut für Kernforschung, 8051 Dresden, Postfach 19, German Democratic Republic.

BUCCINO, Salvatore George, B.S. (Phys.), Ph.D. (Phys.). Born 1933. Educ.: Yale and Duke Univs. Asst. Phys., Argonne Nat. Lab., 1963-65; Asst. Prof., (1965-67), Assoc. Prof., (1967-), Tulane Univ. Societies: A.P.S.; A.A.A.S. Nuclear interests: Low energy nuclear physics; Neutron scattering, stripping reactions, nuclear lifetimes and angular correlations.
Address: Physics Department, Tulane University, New Orleans, Louisiana 70118, U.S.A.

BUCHALET, Albert Louis Georges, General de brigade. B. Mathématiques et Philosophie. Born 1911. Educ.: Paris Univ.; Saint Cyr; Ecole Supérieure de Guerre; Diplomé de Centre des hautes Etudes Militaires. Directeur des applications militaires au C.E.A., 1955-60; Directeur chez M. M. Schneider et Cie., 1960-. Nuclear interests: Responsable de la fabrication de la première bombe atomique française.
Address: Schneider S.A., 42 rue d'Anjou, Paris 8, France.

BUCHANAN, John D., B.S. (Chem.). Born 1927. Educ.: Arizona Univ. Tracerlab., Inc., Richmond, California, 1950-59; Gen. Atomic Div., Gen. Dynamics Corp., San Diego., California, 1959-62; Isotopes Inc., Palo Alto Lab., (formerly Hazleton-Nucl. Sci. Corp.), 1962-. Societies: A.C.S.; A.N.S.; A.A.A.S.; H.P.S. Nuclear interests: Radiochemistry; Radiochemical analysis; Radioactivation analysis; Utilisation of radioisotopes.
Address: Isotopes Incorporated, Palo Alto Laboratory, 4062 Fabian St., Palo Alto, California 94303, U.S.A.

BUCHANAN, Wallace Davis, B.Sc., Dr. Med. Born 1907. Educ.: Indiana, Pennsylvania, and Northwestern Univs. Past. Pres., American Coll. of Radiol.; Past Pres., Indiana Roentgen Soc.; Diplomate, American Board of Radiol., 1947-. Societies: Chicago Roentgen Soc.; Radiol. Soc. of North America; American Roentgen Ray and Radium Soc.; Detroit Roentgen Ray and Radium Soc. Nuclear interests: Those related to the diagnostic and therapeutic clinical applications in medical practice.
Address: 1326 East Wayne Street, North, South Bend, Indiana 46615, U.S.A.

BUCHECKER, William A. Vice-Pres., Mining and Milling Div., United Nucl. Corp.
Address: United Nuclear Corp., 1730 K. Street, N.W., Washington 6, D.C., U.S.A.

BUCHETMANN, Franz, Dr. iur. Born 1908. Educ.: Munich Univ. Member of Management, Munich Reinsurance Co., Munich, 1950-; Member of Management, Deutsch, Kernreaktor - Versicherungsgemeinschaft 1959-.
Nuclear interests: Insurance and reinsurance of nuclear risks.
Address: 107 Königinstr., Munich 23, Germany.

BÜCHLER, Carlos L. A., Communications Eng. Born 1927. Educ.: Buenos Aires Univ. Researcher, Comisión Nacional de Energiá Atómica, Buenos Aires, 1953-56, Elec. Eng., Argonne Nat. Lab., 1956-59; Dept. of Safeguards and Inspection, I.A.E.A., 1959-. Nuclear interests: Reactor operations; Nuclear materials management and control; Inspection systems.
Address: International Atomic Energy Agency, 11 Kärntnerring, Vienna I, Austria.

BUCHLER, Walter Hartwig, Dr. oec. Born 1926. Educ.: Wirtschafts- und Sozialwissen schaft I. Fäkultat, Friedrich Alexander Univ. Nürnberg. Chief of Div., Buchler and Co., Braunschweig. Society: Deutsche Roentgengesellschaft.
Nuclear interests: Management of radioactive isotopes and radiation protection.
Address: 294 Frankfurter Str., Braunschweig, Germany.

BUCHLER, Walther, Dr. -Ing. Born 1900. Educ.: Munich Univ. und Braunschweig T.H. Managing partner, Buchler and Co., Braunschweig. Beirat: Staatl. Hochschule für Bild. Künste Braunschweig. Societies: Gesellschaft Deutscher Chemiker, Fachgruppe Kern- und

Strahlenchemie; Deutsche Roentgengesellschaft.
Nuclear interests: General application of radioisotopes and radiation protection.
Address: 19 Löwenwall, Braunschweig, Germany.

BUCHSBAUM, Solomon Jan, B.S., M.Sc., Ph.D. Born 1929. Educ.: McGill Univ. and M.I.T. Member, Tech. Staff (1958-61), Head, Solid State and Plasma Phys. Res. Dept. (1961-65), Director, Electronics Res. Labs. (1965-), Bell Telephone Labs. Chairman, Div. of Plasma Phys., A.P.S.; Consultant, Inst. of Defence Analyses; Consultant, U.S.A.E.C. Book: Waves in Anisotropic Plasmas (M.I.T. Press, 1963).
Nuclear interest: Thermonuclear fusion.
Address: Bell Telephone Laboratories, Holmdel, New Jersey 07733, U.S.A.

BUCHTELA, Karl, Dr. Phil. Born 1932. Educ.: Vienna Univ. Vienna Univ., 1958-59; Hochschulasst., Atominst. der Österreichischen Hochschulen. Sec., Arbeitsgruppe für Radiochemie und Strahlenchemie, Verein Österreichischen Chemiker. Society:Österreichischen Fachverband für Strahlenschutz.
Nuclear interest: Radiochemistry.
Address: 3-4 Schwedenplatz, A 1010 Vienna, Austria.

BUCINA, Ivan, Ing. Born 1929. Educ.: Prague Tech. Univ. and Charles' Univ., Prague. Chief, Dept. Dosimetry, Ustav pro vyzkum, vyrobu a vyuziti radioisotopu, Praha (Inst. for Res., Production and Application of Radioisotopes, Prague).
Nuclear interests: Metrology of radionuclides; Standardisation; Absolute and relative measurements of radioisotopes; Labelled compounds and sources; Dosimetry; Spectrometry.
Address: 24 Pristavni, Prague 7, Czechoslovakia.

BUCK, B. Res. Assoc., Nucl. Phys. Lab., Dept. of Nucl. Phys., Oxford Univ.
Address:Oxford University, Department of Nuclear Physics, 21 Banbury Road, Oxford, England.

BUCK, Burton R. Pres., Mining and Metals Div., Union Carbide Corp., 1965-.
Address: Mining and Metals Division, Union Carbide Corp., 270 Park Avenue, New York, New York 10017, U.S.A.

BUCK, Cyril. Born 1916. Educ.: Birmingham Tech. Coll. Director, Chem. Plant Design, Eng. Group, U.K.A.E.A. Eng. Practice Com., Inst. Chem. Eng. Societies: M.I.Mech.E.; M.I.Chem.E.
Nuclear interests: All plant associated with the nuclear fuel cycle; Dual purpose desalinisation plants.
Address: 63 Rainford Road, St. Helens, Lancashire, England.

BUCK, John Henry, B.Sc., M.Sc., Ph.D. (Phys.). Born 1912. Educ.: Saskatchewan and Rochester Univs. Phys. and Assoc. Director, Union Carbide and Carbon Chem. Co., O.R.N.L., 1950-53; Vice-Pres. and Tech. Director, Wells Surveys, Inc., Tulsa, Oklahoma, 1953-58; Vice-Pres., Eng., B. J. Electronics, Santa Ana, California, 1958-59; Vice-Pres. and Gen. Manager, Instr. Div., The Budd Co., Phoenixville, Pennsylvania, 1959-. Member, Atomic Safety and Licensing Board Panel, U.S.A.E.C. Societies: A.P.S.; A.A.A.S.; Soc. for Nondestructive Testing; A.N.S.
Nuclear interest: Management.
Address: Box 245, Phoenixville, Pennsylvania 19460, U.S.A.

BUCK, Walter Frank, B.S. Born 1933. Educ.: M.I.T. Development Eng. (1956-62), Manager, Radioactive Products, (1962-), U.S. Radium Corp.
Nuclear interests: Research and development of radiation and self-luminous sources, equipment and facilities. Management of source production facility.
Address: U.S. Radium Corp., Bloomsburg, Pennsylvania, U.S.A.

BUCKA, Hans, Dr. rer. nat. Born 1925. Educ.: Göttingen and Heidelberg Univs. Res. Asst. (1954-61), Privatdozent in Phys. (1961-63), Heidelberg Univ.; ordentlicher Prof.Phys., Berlin Tech. Univ.
Nuclear interests: Quadrupole moments; High frequency spectroscopy of excited atomic levels; Optical pumping methods; Transition probabilities between atomic levels.
Address: Technical University Berlin, 17-20 Marchstrasse, Berlin 10, Germany.

BUCKHAM, James Andrew, B.S., M.S. (Ch.E.), Ph.D. Born 1925. Educ.: Washington Unuv., Seattle. Instructor in Chem. Eng., Washington Univ., 1951-53; Res. Eng., California R. and D. Co., Livermore, 1950, 1953-54; Group Leader, Chem. Eng. Development (1954-60), Chief Tech. Projects Sect. (1960-61), Sen. Tech. Staff, CPP Tech. Branch (1961-62), Chief, Development Eng. Sect. (1962-63), Manager, ICPP Tech. Branch (1963-66), Atomic Energy Div., Phillips Petroleum Co., Idaho Falls; Manager, Chem. Technol. Branch, Idaho Nucl. Corp., Idaho Falls, 1966-; Prof., Idaho Univ. N.R.T.S. Education Programme, 1954-. Director, Nucl. Eng. Div., A.I.Ch.E. Societies: A.I.Ch.E.; A.C.S.; A.N.S.
Nuclear interests: Chemical reprocessing of reactor fuels; Treatment and disposal of radioactive wastes.
Address: 225 South Lloyd Circle, Idaho Falls, Idaho, U.S.A.

BUCKLEY, George Eric, B.Sc. Born 1916. Educ.: Manchester Univ. Manager, Health Phys. and Safety, Windscale Works, (1950), Works Manager, Capenhurst Works, (1952), Sen. Production Controller, Reactors, I.G.H.Q., (1956), Chief Operations Phys. Diffusion

Plants, (1958), Chief Tech. Manager, Windscale and Calder Works (1959), Superintendent (Reactors) Windscale (A.G.R.) and Calder Works (1964), U.K.A.E.A. Society: Inst. of Phys.
Nuclear interest: Increasing the efficiency of operation of all types of nuclear plants.
Address: 12 Wastwater Rise, Seascale, Cumberland, England.

BUCOVE, Bernard, M.D., D.P.H. Educ.: Toronto Univ. District Health Officer, Washington State, 1949-54; Chief, Div. Local Health Services, Washington Health Dept., 1954-55; Director, State Health Dept. and Chairman, State Health Board, Washington, 1955-. Member, Advisory Com. of State Officials to U.S.A.E.C.; Chairman, Tech. Advisory Board, Rad. Control; Member, Governor's Advisory Council, Nucl. Energy and Rad.; Responsibility for control of rad. hazards in Washington State. Societies: Fellow, A.P.H.A.; Charter Member, American Assoc. Public Health Physicians; A.M.A.; Roy. Soc. Health.
Address: 4715 Boston Harbor Road, Olympia, Washington, U.S.A.

BUCZKO, M. Paper: Investigations on Delayed Neutrons Emitted in Fission of U^{238} Induced by 14.7 MeV Neutrons (letter to the Editor, Atomnaya Energiya, vol. 20, No. 2, 1966).
Address: Academy of Sciences of the U.S.S.R., 14 Leninsky Prospekt, Moscow V-71, U.S.S.R.

BUDARAS, Adnan, Formerly Asst. Prof. Dr. Nucl. Dept., now Prof. Dr., Inst. and Clinic of Radiotherapy, Faculty of Medicine, Istanbul Univ.
Address: Faculty of Medicine, Institute and Clinic of Radiotherapy, Istanbul University, Istanbul, Turkey.

BUDDE, Reinhard, D.Ph. Born 1926. Educ.: Basel Univ. C.E.R.N., Geneva 1955-; Columbia Univ., New York, 1955-56. Societies: A.P.S.; Schweizerische Physikalische Gesellschaft.
Nuclear interests: High energy; Bubble chambers; Data handling; Vacuum physics; Low temperatures.
Address: Begnins, Vaud, Switzerland.

BUDDERY, John Harold, A.R.C.S., B.Sc., D.I.C., Ph.D. Born 1928. Educ.: London Univ. Head, Extraction Metal. Sect., Metal. Div. A.E.R.E., Harwell, 1951-59; Head, Chem. Sect., Materials Div., Berkeley Nucl. Labs., C.E.G.B., 1959-. Book: Co-author, Beryllium (Butterworths, 1959).
Nuclear interest: Chemical metallurgy.
Address: Central Electricity Generating Board, Berkeley Laboratories, Gloucestershire, England.

BÜDELER, Werner Wilhelm. Born 1928. Editor-in-chief, rocket-magazine "Weltraumfahrt". Council member, Deutsche Gesellschaft für Raketentechnik und Raumfahrt; Permanent deleg. to annual conferences of Internat. Astronautical Federation; Studiengesellschaft Verwendung Kernenergie in Schiffbau und Schiffahrt; Lecturer, Deutsches Atomforum, e.V. Book: Das Atom, Energie quelle der Zukunft (Müller and Kiepenheuer-Verlag, 1954). Society: F.R.A.S.
Nuclear interests: Dissemination of information on nuclear research, nuclear energy, etc.
Address: 8011 Vaterstetten bei Munich, Germany.

BUDESINSKY, Bretislav, Dipl. Ing., C.Sc. Born 1928. Educ.: Prague Tech. Univ. and Chem. Univ., Pardubice. Vice Head, Analytical Chem. Dept., Inst. Pharmacy and Biochem., 1956; Head, Analytical Chem. Dept., Nucl. Res. Inst., Czechoslovak Acad. Sci., 1963; Member, Analytical Div., I.U.P.A.C.; Member, Analytical Div., I.A.E.A., Vienna.
Nuclear interest: Analytical chemistry of nuclear materials – i.e., nuclear fuels etc.
Address: Nuclear Research Institute, Czechoslovak Academy of Sciences, Rez, Czechoslovakia.

BUDGE, William Lennox, B.S.(M.E.). Born 1915. Educ.: Colorado Univ. Manager, STR, Westinghouse Elec. Corp., Reactor Testing Station, Idaho Falls, 1951-53; Management, Westinghouse Elec. Corp., South Philadelphia, Pa., 1953-55; Sales Manager, Maryland Shipbuilding and Drydock Co., New York, 1955-56; Manager of Marketing, Westinghouse Elec. Corp., Atomic Power Div., Pittsburgh, 1956-. Societies: Soc.of Naval Architects and Marine Eng.; A.N.S.
Nuclear interest: Advance of atomic power for utilities, industry, and propulsion.
Address: Westinghouse Astronuclear Laboratory, P.O. Box 10864, Pittsburgh, Pennsylvania 15236, U.S.A.

BUDINI, Paolo, Prof. Born 1916. Educ.: Pisa Univ. Director, I.N.F.N. 1953-63; Director, Istituto Fisica (1956-64), Director, Istituto Fisica Teorica (1964-), Trieste Univ.; Deputy Director, I.A.E.A. Theoretical Phys. Centre, Trieste, 1964-. Book: Contributed to Kosmische Strahlung (Ed. W. Heisenberg, Springer, 1953). Societies: Stà Italiana Fisica; A.P.S.
Nuclear interests: Cosmic rays and field theory.
Address: Istituto di Fisica Teorica, Università degli Studi, Trieste, Italy.

BUDKER, Gersh Itskovich. Lenin Prize, 1967. Born 1918. Educ.: Moscow Univ. Inst. of Atomic Energy, Acad. of Sciences of the U.S.S.R., 1946; Prof., Inst. of Tech. Phys., Moscow, 1956; Director, Inst. of the Siberian Branch of the Acad. of Sciences of the U.S.S.R., 1957. Society: Corr. Member, Acad. of Sci. of the U.S.S.R.
Nuclear interest: Nuclear physics.
Address: S.I. Vavilov Inst. of Physical Problems, Novosibirsk Science Centre, 20 Sovietskaya Ulitsa, Novosibirsk, Siberia, U.S.S.R.

BUELL, Elton H., B.S. (Mech. Eng.). Born 1928. Educ.: Arizona and Idaho Univs. Manager, Sci. Res., Arizona Public Service Co.; Director and Vice Pres., Resources Co. (Arizona Public Service Co. Subsidiary). Societies: A.N.S.; A.S.M.E.
Nuclear interests: Core physics; Fuel cycle management; Power reactor systems.
Address: 17413 North 21st Drive, Phoenix, Arizona 85023, U.S.A.

BUENO LANTERO, Alejandro, Dr. Eng. Born 1934. Educ.: Bilbao Ind. Eng. School. Manager of Auxitrol Iberico S.A., Madrid; Prof. Servomechanisms, Ind. Eng. School, Madrid.
Nuclear interest: Reactor control.
Address: 22 Agustin de Foxá, Madrid, Spain.

BUEREN, H. G. van. See van BUEREN, H. G.

BUFFO, William J. Vice Pres., Nucl. Data Inc., -1963; Pres. and Director, Rad. Counter Labs. Inc., 1963-64; Member of Board and Pres., Northern Scientific, Inc., 1964-.
Address: Northern Scientific Inc., 303 Price Place, Madison, Wisconsin 53711, U.S.A.

BUFORD, William H., Jr., B.S. (Phys.). Born 1934. Educ.: Louisiana State Univ. Manager, Nucleonics and Phys., Aero Space Technol. Res. Office, Marquardt Corp. Member, Tech. Com. on Nucl. Propulsion, A.I.A.A. Society: A.N.S.
Nuclear interests: Nuclear propulsion; Radiation shielding; Radioisotope applications.
Address: Marquardt Corporation, 16555 Saticoy Street, Van Nuys, California, U.S.A.

BUGG, David V., B.A., M.A., Ph.D. Born 1935. Educ.: Cambridge Univ. Fellow, Emmanuel Coll. (1959-63), Demonstrator, Phys. Dept. (1960-63), Cambridge Univ.; P.S.O., Rutherford High Energy Lab., Chilton.
Nuclear interest: Experimental and theoretical high energy nuclear physics.
Address: The Old Bakehouse, Aston Tirrold, Berks., England.

BUGHER, John Clifford, B.S., B.A., M.D., M.S., Sc.D. (Hon.). Born 1901. Educ.: Taylor and Michigan Univs. Rockefeller Foundation, 1937-66; Labs. in N.Y. City, virus and phys. sects., 1949-51; leave of absence from Foundation to serve with U.S.A.E.C., 1951-55; Director for Medical Education and Public Health, U.S.A.E.C. (on leave of absence from Rockefeller Foundation), 1955-; Deputy Director, Div. of Biology and Medicine, 1951; Director, Div. of Biology and Medicine, 1952-55; Director, Puerto Rico Nucl. Centre, 1960-66; Chairman, Advisory Com. for Biol. and Medicine, U.S.A.E.C.; Com. on Atomic Casualties, U.S.N.A.S. and Nat. Res. Council; Nat. Com. on Rad. Protection. Papers: On biological effects of nuclear radiation and medical and biological aspects of atomic warfare. Societies: Federation of American Societies for Exptl. Pathology; American Assoc. of Pathologists and Bacteriologists; American Cancer Soc.; American Assoc. of Cancer Research; A.M.A.; American Soc. of Tropical Medicine and Hygiene; Royal Soc. of Tropical Medicine; Royal Geographical Soc.; A.P.H.A.; A.N.S.; Rad. Res. Soc.; N.Y. Acad. Sci.
Address: 946 Eve Street, Delray Beach, Florida 33444, U.S.A.

BUGNARD, Louis Camille, Ancien Elève de l'Ecole Polytechnique, M.D., Prof. de Biophysique. Born 1901. Educ.: Toulouse and Paris Univs. Directeur, Inst. Nat. d'Hygiène, 1946; Directeur, Service Central de Protection contre les Radiations Ionisantes, 1956; Director, Centre des Faibles Radioactivités, Gif-sur-Yvette, formerly Directeur, Direction de la Biologie et de la Santé, now Member, Comite de Biologie, C.E.A.; Member, Sci. and Tech. Com., Euratom, -1968; Formerly Vice-Chairman, now Member, I.C.R.P.; Member, U.N. Sci. Com. on the Effects of Atomic Rad. Societies: Membre correspondant, Nat. de l'Académie de Médecine; Membre, Académie de Chirurgie; Membre affilié, Royal Soc. of Medicine; Membre, Comité pour l'étude de l'action biologique des Radiations Ionisantes, O.N.U.; Membre, Commission Internationale de Protection contre les Radiations.
Address: 3 rue Léon Bonnat, Paris 16, France.

BUGORKOV, S. S. Papers: Co-author, The Yields of Ru^{103} and Ru^{106} in the Fast Neutron Fission of U^{235} and Pu^{239} (letter to the Editor, Atomnaya Energiya, vol. 6, No. 5, 1959); co-author, Radiochemical Determination of $^{27}Al(n,a)^{24}$ Na Reaction Cross-Section with 14.6 Mev Neutrons (letter to the Editor, ibid, vol. 16, No. 4, 1964); co-author, About Measurement of Thermal Neutron Flux and Cadmium Ratio by Gold Activation (letter to the Editor, ibid, vol. 21, No. 6, 1966).
Address: Academy of Sciences of the U.S.S.R., 14 Leninsky Prospekt, Moscow V-71, U.S.S.R.

BUHL, Siegfried, Dr. rer. nat. Born 1930. Educ.: Heidelberg Univ. Wissenschaftlich Asst. II. Physikalisches Inst., Heidelberg Univ., 1962. Society: Deutsche Physikalische Gesellschaft.
Nuclear interest: Nuclear physics especially spectroscopy and conversion of electrons.
Address: II Physikalisches Institute, Heidelberg University, 12 Philosophenweg, D-69 Heidelberg, Germany.

BÜHLER, Hans, Dr. Ing., Dr. rer. nat. habil. Born 1907. o. Prof., Hanover T.H. Papers: Co-author, Eigenspannungen in metallischem Uran durch unsachgemässe spanende Bearbeitung (Z. f. Metallk., vol. 52, No. 4, 1961); co-author, Dauerschwingfestigkeit von Uran mit 0.35 Gew.-Nb (ibid, vol. 52, No. 4, 1961); co-author, Die Staucheigenschaften von natürlichem Uran (ibid., vol. 57, No. 11, 1966); co-author, Die Staucheigenschaften von Zirkonium und Zircaloy 2 (ibid., vol. 58, No. 2, 1967).

Address: Technische Hochschule 1 Welfengarten, 3 Hanover, Germany.

BÜHLER, Mauricio Fridolin, Dr. in Chem. Born 1912. Educ.: Univ. Nacional de Buenos Aires, Prof., Organic Chem., Univ. Nacional de Tucumán, 1948-50; Res. Chem., Inst. Nacional de Tecnologia, 1950-53; Employed (1953-), Head, Organic Chem. Div. (1956-60), Head, Dept. of Chem. (1960-). Comision Nacional de Energia Atómica. Society: Asociación Quimica Argentina.
Nuclear interests: Use and application of radioactive tracers to problems of organic chemistry.
Address: 8250 Avda. del Libertador, Buenos Aires, Argentina.

BUHLER, R. D. Dr. Vice Pres. in charge of res., Plasmadyne Corp.
Address: Plasmadyne Corporation, 3839 South Main Street, Santa Ana, California, U.S.A.

BÜHLER, Rolf Ernst, Dr. sc. tech., Dipl. Ing. Chem. (E.T.H.). Born 1930. Educ.: Eidgenössische T.H., Zürich. Phys. Chem., Eidgenössische T.H., 1958-59; Nucl. Phys. Div., A.E.R.E A.E.R.E., Harwell (Swiss National Funds), 1959-60; Chem. Div., Argonne Nat. Lab., Argonne, Ill., 1960-61; Phys. Chem., Eidgenössische T.H., 1962-. Societies: A.P.S.; Faraday Soc., England.
Nuclear interests: Radiation chemistry; Direct observation of transients.
Address: Eidgenössische Technische Hochschule, Physikalisch Chemisches Laboratorium, 22 Universitätsstrasse, Zürich 6, Switzerland.

BÜHRING, Wolfgang, Dipl.-Phys., Dr. rer. nat. Born 1932. Educ.: Hanover T.H. and Heidelberg Univ. Res. Asst. (Wissenschaftlicher Assistent), Heidelberg Univ., 1959-.
Nuclear interest: Nuclear physics.
Address: Zweites Physikalisches Institut der Universität Heidelberg, 12 Philosophenweg, 69 Heidelberg, Germany.

BUIE, Bennett Frank, B.S., M.S., M.A., Ph.D. Born 1910. Educ.: South Carolina, Lehigh and Harvard Univs. Formerly geologist, U.S. Geological Survey; Prof. Geology and formerly Chairman, Geology Dept., Florida State Univ.; Consulting geologist for J. M. Huber Corp., and others. Societies: Geological Soc. of America; Mineralogical Soc. of America.; A.I.M.E.; Soc. of Economic Geologists; A.A.P.G.; Société Géologique de France.
Nuclear interest: Nuclear geology, including source materials, shielding, and effects of nuclear radiation on optical properties of minerals.
Address: Department of Geology, Florida State University, Tallahassee, Florida, U.S.A.

BUITRAGO, Cesar. Director, Comision de Energia Atomica de El Salvador.

Address: Comision de Energia Atomica de El Salvador, c/o Ministerio del Economia, San Salvador, El Salvador, C.A.

BUJALSKI, Cezary K. Born 1921. Educ.: Lwów and Rome Univs. Partner and managing director, Casa Italiana Commercio Estero, Rome, Special Rep., Internat. Assignments, Nucl.-Chicago Corp. Societies: F.I.E.N.; A.N.S.
Nuclear interests: Instrumentation; Applications of isotopes in all fields.
Address: 7 Via Umbria, Rome, Italy.

BUJDOSO, Erno, Dr. rer. nat. Born 1932. Educ.: Kossuth Lajos Sci. Univ., Debrecen. Nucl. Res. Inst., Debrecen; Res. Inst. for Nonferrous Metals, Budapest. Societies: Internat. Rad. Protection Assoc.; Lorand Eötvös Phys. Soc.
Nuclear interests: Activation analysis; Applications of isotopes in industry.
Address: Mártirok utja 41. V. 15, Budapest 2, Hungary.

BUKAEV, P. V. Paper: Co-author, Linear Induction Accelerator (Atomnaya Energiya, vol. 21, No. 6, 1966).
Address: Academy of Sciences of the U.S.S.R., 14 Leninsky Prospekt, Moscow V-71, U.S.S.R.

BUKAREV, V. A. Paper: Co-author, Measurement of dT-Neutron Yield with Auxiliary Annihilation Rad. Source (letter to the Editor, Atomnaya Energiya, vol. 22, No. 5, 1967).
Address: Academy of Sciences of the U.S.S.R., 14 Leninsky Prospekt, Moscow V-71, U.S.S.R.

BUKREEV, Yu. F. Paper: Co-author, Determination of Uranium Solubility in Bismuth by E.M.F. Method (Abstract, Atomnaya Energiya, vol. 20, No. 4, 1966).
Address: Academy of Sciences of the U.S.S.R., 14 Leninsky Prospekt, Moscow V-71, U.S.S.R.

BULAT, Thomas J., B.S., M.S., Ph.D. Born 1926. Educ.: Iowa Univ. Manager, Sonic Energy Eng., Pioneer-Central Div., Bendix Corp., Davenport, Iowa, 1950-. Chairman, Com. on Radiotracer Techniques in Electronic Tubes and Devices, A.S.T.M. Society: A.S.T.M. Nuclear interests: Basic radiotracer studies in metallurgy and chemistry; Radioactive decontamination techniques.
Address: Instruments and Life Support Division, Bendix Corporation, Hickory Grove Road, Davenport, Iowa, U.S.A.

BULATOV, B. P.
Nuclear interests: γ-ray measurements and albedo effects.
Address: Academy of Sciences of the U.S.S.R., 14 Leninsky Prospekt, Moscow V-71, U.S.S.R.

BULAVIN, P. E. Papers: Co-author, Calculation of Doppler Temperature Coefficient for Individual Resonances in Homogeneous Medium (letter to the Editor, Atomnaya Energiya, vol. 21, No. 1, 1966); co-author, On Calculation

of Relative Importance of Delayed Neutrons (letter to the Editor, ibid, vol. 23, No. 2, 1967).
Address: Academy of Sciences of the U.S.S.R., 14 Leninsky Prospekt, Moscow V-71, U.S.S.R.

BULEEV, N. I. Adviser, Third I.C.P.U.A.E., Geneva, Sept. 1964. Paper: A Numerical Method for Solving of Two-dimensional Diffusion Equations (letter to the Editor, Atomnaya Energiya, vol. 6, No. 3, 1959).
Address: Academy of Sciences of the U.S.S.R., 14 Leninsky Prospekt, Moscow V-71, U.S.S.R.

BULETTE, W. C., Jr. Controller, Bettis Atomic Power Lab., U.S.A.E.C.
Address: U.S.A.E.C., Bettis Atomic Power Laboratory, P.O. Box 79, West Mifflin, Pennsylvania 15122, U.S.A.

BULKIN, Yu. M. Papers: Co-author, The 50 MW Res. Reactor SM (Atomnaya Energiya, vol. 8, 1960); co-author, Res. and Training Reactor IR-100 (ibid, vol. 21, No. 5, 1966).
Address: Academy of Sciences of the U.S.S.R., 14 Leninsky Prospekt, Moscow V-71, U.S.S.R.

BULL, Arne, M.Sc. Head, Health Phys. Service, Oslo Univ. Sec., Rad. Advisory Council, Norway.
Address: Oslo University, Blindern, Norway.

BULL, James William Douglas, M.A., M.D., F.R.C.P., F.F.R. Born 1911. Educ.: Cambridge Univ. Consultant Radiologist, St. George's Hospital, London, and Nat. Hospital for Nervous Diseases, London. Teacher at London Univ. Pres., British Inst. of Radiology.
Nuclear interest: Isotopes used for the localisation of intracranial tumours.
Address: 20 Devonshire Place, London, W.1, England.

BULLA, Ricardo, Prof. Dr. Centro de Radiooncologia, Argentina.
Address: Centro de Radiooncologia, 888 Calle Azcuenaga, Buenos Aires, Argentina.

BULLIO, Pietro, LL.D. Born 1921. Educ.: Rome Univ. Internat. Counsellor-at-Law, N.Y.C., 1948-54; Editor (1957-58), Editor and Publisher (1961-), Atomo e Industria, Rome. Special consultant to Exec. Secretariat of Euratom Commission, Brussels, 1958; Sec.-Gen., F.I.E.N., Rome, 1958-; Member, Exec. Com. (1960-), Forum Atomique Européen (FORATOM), Paris; Working Group Member, Nucl. Public Relations Contact Group, Rome, 1961-; Member, Steering Com., Assoc. des Journalistes Européens, Paris, 1963-.
Nuclear interests: Law; Economics; International affairs; Management; Information; Public relations.
Address: 26-28 Via Paisiello, 00198 Rome, Italy.

BULLOCK, D. H. Sales Asst. to Vice Pres., Eldorado Mining and Refining Ltd. Chairman, Internat. Affairs Com., Canadian Nucl. Assoc.
Address: Eldorado Mining and Refining Ltd., Port Hope, Ontario, Canada.

BULLOCK, Frederick William, B.Sc., Ph.D. Born 1932. Educ.: London Univ. Lecturer, Univ. Coll., London, 1962-. Societies: Inst. of Phys. and Phys. Soc.
Nuclear interest: Elementary particle physics.
Address: 37 Cranbrook Drive, Saint Albans, Herts., England.

BULLOCK, Geoffrey Francis, M.A. Born 1909. Educ.: Cambridge Univ. Gen. Manager, Nat. and Vulcan Eng. Insurance Group. Nucl. Safety Advisory Com., M.O.P.; Advisory Com., British Insurance (Atomic Energy) Com.
Address: 14 Saint Mary's Parsonage, Manchester 3, England.

BULLOCK, J. B., B.S. (Aeronautical Eng.), M.S. (Nucl. Eng.). Born 1935. Educ.: Michigan Univ. Reactor Supervisor and Director, Ford Nucl. Reactor Facility, Phoenix Memorial Lab. 1961-65; Reactor Controls Development Group Eng., 1965-.
Nuclear interest: Nuclear reactor digital computer control development.
Address: Oak Ridge National Laboratory, Oak Ridge, Tennessee 37830, U.S.A.

BULMER, Joseph J., B.Ch.E., M.S.E. Ph.D. (Nucl. Sci. and Eng.). Born 1929. Educ.: Rensselaer Polytech. Inst., Michigan Univ. and O.R.S.O.R.T. Nucl. Eng., K.A.P.L.; NPE Project, Manager, NPE Nucl. Analysis; Adjunct Prof., Union Coll., Schenectady; Adjunct Asst. Prof., Rensselaer Polytech. Inst. Societies: A.N.S.; N.Y. Acad. Sci.
Nuclear interests: Reactor nuclear analysis; Reactor design.
Address: Knolls Atomic Power Laboratory, Schenectady, New York, U.S.A.

BÜLOW, Hans Reimar VON. See VON BÜLOW, Hans Reimar.

BUMM, H., Dr. Head, Metal. Lab., Kernforschungszentrum Karlsruhe.
Address: Gesellschaft für Kernforschung m.b.H., 4 Friedrichsplatz, Karlsruhe, Germany.

BÜNAU, Günther von. See von BÜNAU, Günther.

BUNCE, Stanley Chalmers, B.S., M.A., Ph.D. Born 1917. Educ.: Lehigh Univ. and Rensselaer Polytech. Inst. Prof., Organic Chem., Rensselaer Polytech. Inst. Society: A.C.S.
Nuclear interest: Tracer studies of organic reaction mechanisms.
Address: Department of Chemistry, Rensselaer Polytechnic Institute, Troy, New York, U.S.A.

BUNCH, Wilbur Lyle, B.S., M.S. Born 1925. Educ.: Wyoming Univ. Hanford Atomic Products Operation, Gen. Elec. Co., 1951-65; Battelle Memorial Inst., Pacific Northwest

Lab., 1965-. Books: Contributor to Reactor Handbook (2nd Edition, Volume IV; New York, Interscience Publishers (John Wiley and Sons), 1964); Contributor, Eng. Compendium on Rad. Shielding (Berlin, Springer-Verlag, 1968). Society: A.N.S.
Nuclear interests: Fast reactor shielding; Nuclear heating; Isotope production; Nuclear reactor design and analyses; Nuclear instrumentation; In-reactor neutron flux monitoring.
Address: 2403 Pullen Street, Richland, Washington, U.S.A.

BUNCHE, Ralph J., A.B., M.A. and Ph.D. degrees (major in Government and Internat. Relations). Born 1904. Educ.: California (Los Angeles) Harvard and Cape Town Univs. and London School of Economics. Under-Sec., United Nations. Member Trustee of the Rockefeller Foundation.
Nuclear interests: Responsibility for United Nations programme on peaceful uses of atomic energy, including United Nations Scientific Committee on the Effects of Atomic Radiation, the Advisory Committee on the Peaceful Uses of Atomic Energy and the United Nations relations with the International Atomic Energy Agency; Responsibility for organisation and conduct of the two UN sponsored International Conferences on the Peaceful Uses of Atomic Energy, held in Geneva in August 1955 and September 1958.
Address: 115-24 Grosvenor Road, Kew Gardens, New York, U.S.A.

BUND, Karlheinz, Dr.-Ing., Dr. h.nat. Vorstandsmitglied der STEAG. Member, Steering Com., Deutsches Atomforum e.V.
Address: 53 Bismarckstrasse, 4300 Essen, Germany.

BUNDE, Erich, Dr. phil. nat. Phys., Isotopenabteilung, Inst. and Poliklinik für physikalische Therapie und Röntgenologie, Munich Univ.
Address: Institut und Poliklinik für physikalische Therapie und Röntgenologie Universität München, 1 Ziemenstrasse, Munich 15, Germany.

BUNDY, Edwin S. Director, Atomic Power Development Associates, Inc.; Member, Exec. Com., High Temperature Reactor Development Associates, Inc.
Address: Atomic Power Development Associates, Inc., 1911 First Street, Detroit 26, Michigan, U.S.A.

BUNDY, P. A. Finance Director, Production Group, U.K.A.E.A.
Address: Production Group, U.K.A.E.A., Risley, Nr.Warrington, Lancs., England.

BÜNEMANN, Dietrich, Dipl. Phys., Dr. rer. nat. Born 1922. Educ.: Hamburg Univ. Theoretical Phys., Inst. Reaktorphysik, 2057 Geesthacht-Tesperhude, Gesellschaft Kernenergie in Schiffbau und Schiffahrt m.b.H. Society: Deutsche Phys. Gesellschaft.
Nuclear interests: Reactor physics; Transport theory; Plasma physics.
Address: 2051 Escheburg, 11 Götensberg, Germany.

BUNGARDT, Karl Georg, Prof., Dr.-Ing. Born 1911. Educ.: Aachen T.H. Director, Forschungsinst., Deutsche Edelstahlwerke Aktiengesellschaft, Krefeld. Mitglied, Arbeitskreis Brenn- und Baustoffe, Deutschen Atomkommission.
Nuclear interests: Metallurgie und Metallkunde der Stähle für den Reaktorbau.
Address: Deutsche Edelstahlwerke Aktiengesellschaft, 16 Oberschlesienstr., Krefeld, Germany.

BUNJI, Bela, Prof., Eng. Technol. Born 1910. Educ.: Belgrade Univ. Director, Inst. for Technol. of Nucl. and other Mineral Raw Materials. Councillor, Federal Commission for Nucl. Energy of Yugoslavia, 1955-. Societies: Assoc. of Mining and Metal. Eng. and Technicians; A.I.M.E.
Nuclear interests: Chemistry and metallurgy of nuclear raw materials.
Address: Institute for Technology of Nuclear and other Mineral Raw Materials, 86 Franchet Desperey Street, Belgrade, Yugoslavia.

BUNKE, Edward William Diedrich, Mech. Eng. Born 1915. Educ.: Stevens Inst. Technol. Manager, Aerospace and Ground Support Equipment Eng., Apollo Support Dept., G.E.C., 1963-. Society: A.S.M.E.
Nuclear interests: Reactor design; Management.
Address: General Electric Co., ASD, P.O. Box 294, Huntsville, Alabama, U.S.A.

BUNKER, Carl M., B.Sc. Born 1925. Educ.: Dayton Univ., Ohio. Geophys., U.S. Geological Survey. Society: Soc. of Exploration Geophys. Nuclear interest: Gamma-ray spectrometry.
Address: Isotope Geology Branch, United States Geological Survey, Federal Centre, Building 15, Denver, Colorado 80225, U.S.A.

BUNNAG, Sombhandu, Rear-Admiral, Member, Thai Atomic Energy Commission for Peace (Mech. Eng.); Technical Adviser, Royal Thai Navy.
Address: Department of Science, Rama V1 Road, Bangkok, Thailand.

BUNO, Washington, Dr. Member, Comision Nacional de Energia Atomica, Uruguay.
Address: Comision Nacional de Energia Atomica, 565 J. Herrera y Reissig, P.2., Montevideo, Uruguay.

BUONOMINI, G. Member, Editorial Board, Minerva Nucleare.
Address: 83-85 Corso Bramante, Turin, Italy.

BUPP, Lamar Paul, Ph.D., B.S. Born 1921. Educ.: Oregon State Univ. and California Univ. Berkeley Manager, Reactor Eng. Development (1956-57), Manager, Chem. R. and D. (1957-61), Hanford Labs., G.E.C., Richland, Washington; Manager, Gen. Elec. Nucleonics

Lab., Vallecitos Nucl. Centre, Pleasanton, California, 1961-. Societies: A.N.S.; A.C.S.; A.S.E.E.; Atomic Ind. Forum.
Nuclear interests: Management of research and development activities associated with physical sciences and engineering testing of reactor materials and processes.
Address: 4571 Grover Drive, Fremont, California 94536, U.S.A.

BURAN, T., Cand. real. Member, Res. on Elementary Particles and Cosmic Rad., Nucl. Phys. Lab., Oslo Univ.
Address: Oslo University, Nuclear Physics Laboratory, Institute of Physics, Blindern, Norway.

BURAS, Bronislaw, Prof. (Warsaw Univ.). Born 1915. Educ.: Warsaw Univ. Instructor in Phys. (1945-51), Lecturer (1953-54), Asst. Prof. (1954-59), Warsaw Univ.; Lecturer, Warsaw Agricultural Univ., 1951-53; Deputy Director Gen. for Res. (1955-57), Head, Lab. for Nucl. Phys. (1957-59), Inst. for Nucl. Res., Warsaw; Head, Phys. Sect., Div. of Res. and Labs., I.A.E.A., 1959-61; Head, Lab. for Nucl. Phys., Inst. for Nucl. Res., Swierk k. Otwocka, and Prof. (1961-), Head, Chair for Nucl. Methods in Solid State Phys. (1967-), Warsaw Univ. Society: Polish Phys. Soc.
Nuclear interests: Nuclear methods in solid state physics, in particular: elastic and inelastic scattering of neutrons in solids, structure and lattice dynamic studies by means of the time-of-flight method, Mössbauer effect, radiation damage in semiconductors, semiconductor detectors for nuclear radiation, growing of large metallic single crystals.
Address: Institute for Nuclear Research, Swierk k. Otwocka, Poland.

BURAVAS, Saman, B.Sc. (Hons., London), A.R.S.M. Born 1918. Educ.: Roy. School Mines, London. Deputy Director-Gen., Dept. of Mineral Resources, Bangkok. Pacific Sci. Congress, Member Standing Com. on Geology; Member, Sub-com. on Nucl. Power.
Nuclear interests: Geological application of radioactivity; Nuclear raw materials survey; Management.
Address: Department of Mineral Resources, Ministry of National Development, Bangkok, Thailand.

BURCH, Philip R. J., M.A., Ph.D. Born 1920. Educ.: Cambridge Univ. M.R.C. Fellow (1954), Deputy Director, M.R.C. Environmental Rad. Res. Unit (1959-), Hon. Reader, Dept. of Medical Phys., Leeds Univ. M.R.C. Radiobiol. Forum Subcom. Societies: British Inst. of Radiol.; Hospital Phys. Assoc.
Nuclear interests: Radiobiology; Radiation carcinogenesis and ageing; Radiation dosimetry.
Address: Medical Research Council, Environmental Radiation Research Unit, Department of Medical Physics, Leeds University, The General Informary, Leeds 1, England.

BURCH, William D., B.S. (Chem. Eng.), M.S. (Chem. Eng.). Born 1929. Educ.: Missouri Univ. and School of Mines and Metal. O.R.S.O.R.T. Development Eng., O.R.N.L., 1952-67; Group Leader i/c pilot plant development of Homogeneous Reactor Test Chem. Processing Plant, 1956-60; Group Leader i/c design of chem. processing equipment for the Transuranium Processing Plant, 1961-64; Group Leader i/c operation of TPP, 1965-. Society: A.N.S.
Nuclear interest: Chemical processing of reactor fuels; Production of the heavy actinide elements.
Address: Oak Ridge National Laboratory, Oak Ridge, Tennessee, U.S.A.

BURCHAM, William Ernest, M.A., Ph.D., F.R.S. Born 1913. Educ.: Cambridge Univ. Oliver Lodge Prof. Phys., Birmingham Univ., 1951-. Societies: Roy. Soc. of London; Phys. Soc. of London.
Nuclear interests: Cyclotrons, proton synchrotrons, nuclear reactions; Nuclear accelerators; Nuclear reactions at energies up to 1000 MeV; Teaching of nuclear physics.
Address: Department of Physics, The University, Edgbaston, Birmingham 15, England.

BURCHENKO, P. Ya. Papers: Co-author, Study on Electromagnetic Rad. from a Plasma of a Linear Pinch Discharge (Atomnaya Energiya, vol. 14, No. 4, 1963); co-author, Closed Magnetic Trap with "Sirius" Type Helical Field (ibid, vol. 23, No. 2, 1967); co-author, Magnetic System Adjustment of the Sirius Stellarator (ibid, vol. 23, No. 2, 1967); co-author, Shielding of Thermonucl. Device Magnetic Systems from Accidental Overloadings (ibid, vol. 23, No. 2, 1967).
Address: Academy of Sciences of the U.S.S.R., 14 Leninsky Prospekt, Moscow V-71, U.S.S.R.

BURCKHARDT, Christof Walter, dipl. phys. E.T.H., Dr. phil. nat. Born 1927. Educ.: Federal Tech. High School, Zürich, and Bern Univ. Society: Schweizerische Physikalische Gesellschaft.
Nuclear interest: Application of radioisotopes in research and industry.
Address: 39, Chemin Saussac CH 1256 Troinex - Geneva, Switzerland.

BURCKHARDT, Helmut, Dr.-Ing. E.h. With Unternehmensverband Ruhrbergbau. Member, Steering Com., Deutsches Atomforum e.V.
Address: 1 Friedrichstrasse, 4300 Essen, Germany.

BURCKHARDT, J., Dr. jur. Minister, Präsident Schweiz. Schulrat, E.T.H. Member, Swiss Federal Commission for Atomic Energy.
Address: 33 Leonhardstrasse, 8006 Zurich, Switzerland.

BURD, A. N., B.Sc. Joint Hon. Sec., Acoustics Group, Inst. of Phys. and The Phys. Soc.

BURDE, Jacob, M.Sc., Ph.D. Born 1923.
Educ.: Hebrew Univ., Jerusalem. U.N.E.S.C.O.
Fellow, 1957-58; Phys., Lawrence Rad. Lab.,
California Univ., Berkeley, 1963-64; Sen.
Lecturer, Nucl. Phys. Sect., Phys. Dept.,
Hebrew Univ., Jerusalem.
Nuclear interest: Nuclear structure; Investigation through lifetime measurements of excited nuclear levels, decay scheme investigation.
Address: Department of Physics, Hebrew University, Jerusalem, Israel.

BURDET, A. Maitre Asst. Inst. de Phys. Nucl.,
Lyon Univ.
Address: Institut de Physique Nucleaire, Lyon University, 43 boulevard de l'Hippodrome, Villeurbanne, (Rhone), France.

BURDETT, Robert Henry, B.Sc. (Eng.),
A.C.G.I. Born 1924. Educ.: London Univ.
Nucl. Systems Eng., Chief Eng.'s Dept., (1955),
Deputy Director (Mech.), Applications Branch,
R. and D. Dept. (1959), Special Developments
Eng., Design and Construction Dept. (1962),
Plant Development Eng., Generation Design
Dept., (1966-), C.E.G.B. Society: Inst. of
Mech. Eng.
Nuclear interest: Future reactor systems.
Address: Walden House, 24 Cathedral Place,
London, E.C.4, England.

BURDICK, Earl Edward, B.S. (Phys. and Math.,
Idaho State Coll.), M.S. (Phys., Idaho Univ.).
Born 1928. Educ.: Idaho State Coll. and Idaho
Univ. Society: A.N.S.
Nuclear interests: Reactor design and experiments; Management.
Address: 260 Evergreen Drive, Idaho Falls,
Idaho, U.S.A.

BUREAU du COLOMBIER, Louis. Chef du
Dept. Tech., Chantiers de l'Atlantique Penhoet-Loire.
Address: Chantiers de l'Atlantique Penhoet-Loire, 7 rue Auber, Paris 9, France.

BURG, Constant. Member, Comite de Biol.,
C.E.A.
Address: Commissariat a l'Energie Atomique,
29-33 rue de la Federation, Paris 15, France.

BURGA, Jorge DAVILA. See DAVILA BURGA, Jorge.

BURGE, Edward James, B.Sc., Ph.D. (Bristol),
M.A. (Oxford). Born 1925. Educ.: Bristol,
Uppsala and Oxford Univs. Sen. Demonstrator
(1955-56), Lecturer (1956-62), Reader (1962-),
King's Coll., London; Visiting Assoc. Prof.,
Manitoba Univ., 1962-63. Society: F.Inst.P.
Nuclear interests: Nuclear structure experiments at medium and low energies; Optimum design of experiments; Optical model analyses of elastic and inelastic scattering of 10-50 MeV protons; Teaching of nuclear physics.
Address: Physics Department, King's College, London University, Strand, London, W.C.2, England.

BURGER, Alwyn Johannes, M.Sc., Ph.D.
Born 1926. Educ.: Stellenbosch and Cape Town
Univs. Chief Res. Officer, N.P.R.L., Council
for Sci. and Ind. Res. Society: South African
Inst. Phys.
Nuclear interest: Application of advanced
mineral separation and purification, ion
exchange, isotope dilution and surface ion
emission techniques in dating zircon, sphere
and apatite from extrusive and intrusive rock
samples.
Address: P.O. Box 395, Pretoria, South Africa.

BURGER, John Clarence, B.Sc. (Chem. Eng.).
Born 1916. Educ.: Alberta Univ. Vice-Pres.
i/c of Refining and Sales, Eldorado Nucl. Ltd.,
Port Hope, Ontario. Societies: Canadian Nucl.
Assoc.; Atomic Ind. Forum.
Nuclear interests: Uranium refining and fuel
element fabrication.
Address: Eldorado Nuclear Limited, Port Hope,
Ontario, Canada.

BURGERS, William Gerard, Prof. Dr. Born
1897. Educ.: Groningen Univ. Prof. Phys.
Chem., Delft Technol. Univ., 1940-. Book:
Handbuch der Metallphysik (Ed. Prof. Dr. G.
Masing), vol. iii, 2nd part; Rekristallisation verformter Zustand und Erholung (Leipzig, Akademische Verlagsgesellschaft Becker und Erler
Kom.-Ges., 1941). Society: Acad. of Sci.,
Amsterdam.
Nuclear interest: Physical metallurgy.
Address: Gebouw voor Metaalkunde, 137
Rotterdamseweg, Delft, Netherlands.

BURGESS, Cyril Duncan, B.Sc., A.R.I.C. Born
1929. Educ.: London Univ. H.M. Chemical
Inspector of Factories. Societies: Soc. of
Radiological Protection; British Occupational
Hygiene Soc.
Nuclear interest: Radiation protection in
industries.
Address: H.M. Factory Inspectorate, 1 Chepstow Place, London, W.2, England.

BURGESS, H. L. J., F.I.E.S., A.M.I.Prod.E.,
M.I.Inf.Sc. Born 1905. Editor, Vacuum, 1948-58; Temporary Editorial, Nucl. Eng., 1959.
Editor, Nucl. Eng. Abstracts.
Nuclear interests: Editorial (and information)
covering nuclear physics and all applications of
atomic energy.
Address: 37 Arundel Gardens, London, W.11,
England.

BURGESS, K. Managing Director, Borer
Electronics Co. Ltd.; Managing Director, Nutec
Electronics Ltd.
Address: Borer Electronics Co. Ltd., 36 East
Street, Shoreham-by-Sea, Sussex, England.

BURGHOFF, H., Dr. Asst., Inst. für Kernverfahrenstechnik, Karlsruhe T.H.
Address: Institut für Kernverfahrenstechnik, Technische Hochschule Karlsruhe, Reaktorstation Leopoldshafen, 18, Strählerweg, Karlsruhe/Durlach, Germany.

BURGOV, N. A. Member, Soviet deleg. Visiting U.S. low energy phys. labs., January 1966. Papers: Resonance Scattering of γ-rays in Mg^{24} (Atomnaya Energiya, vol. 2, No. 6, 1957); co-author, The Gamma-Ray Spectrum of the TVR Reactor (letter to the Editor, ibid, vol.9, No. 3, 1960).
Address: Institute of Theoretical and Experimental Physics, Leningrad, U.S.S.R.

BURGUS, Warren Harold, B.S. (Iowa State Univ.), Ph.D. (Washington Univ., St. Louis). Born 1919. Educ.: Iowa State, Colorado and Washington (St. Louis) Univs. Staff member, Los Alamos Sci. Lab., 1948-54; Res. Chem., Atomic Energy Div. (1954-57), Head of MTR-ETR Chemistry Section (1957-66), Sen. Tech. Consultant, Water Reactor Safety Programme Office (1966-), Phillips Petroleum Co. Book: Contributor to Radioactivity appl. to Chem. (John Wiley, 1951). Societies: A.C.S.; A.A.A.S.; A.N.S.
Nuclear interests: Radioactive decay schemes; Integral and differential neutron cross section measurements; Fission products; Mechanism of the fission process; Power reactor safety.
Address: Atomic Energy Division, Phillips Petroleum Company, P.O. Box 2067, Idaho Falls, Idaho, U.S.A.

BURHOP, Eric Henry Stoneley, B.A., M.Sc., Ph.D., F.R.S. Born 1911. Educ.: Melbourne and Cambridge Univs. Prof. Phys., University Coll., London. Books: The Auger Effect (Cambridge University Press); Co-author, Electronic and Ionic Impact Phenomena (Oxford University Press). Societies: Phys. Soc., London; F.R.S.
Nuclear interests: High energy physics; Strange particle physics; Nuclear collisions.
Address: 39 Templemere, Oatlands Drive, Weybridge, Surrey.

BURK, Werner, Dr. rer. nat., Dipl.-Chem. Born 1932. Educ.: Leipzig and Dresden Univs.
Nuclear interests: Radiochemistry; Reprocessing.
Address: Zentralinstitut für Kernforschung, Rossendorf, Dresden, German Democratic Republic.

BURK, William Emmett, Jr., B.A. Born 1909. Educ.: Cornell and Southern California Univs. Consultant, Rand Corp., 1956-; Consultant, Mitre Corp., 1960. Pres., Assoc. Res. Design, Inc.; Member, American Inst. of Architects Com. on Sci. and Architecture. Societies: American Inst. of Architects; American Soc. of Military Eng.
Nuclear interest: Protective design and construction.
Address: 512 Yale Boulevard, S.E., Albuquerque, New Mexico, U.S.A.

BURKE, Coleman, Director, Tracerlab-Keleket.
Address: Tracerlab-Keleket, 1601 Trapelo Road, Waltham 54, Massachusetts, U.S.A.

BURKE, E. A., Dr. Chairman, Phys. Dept., St. John's Univ.
Address: St. John's University, Grand Central and Utopia Parkways, Jamaica 23, New York, U.S.A.

BURKE, John J., B.S. (Phys.). Born 1921. Educ.: Boston Coll. Chief, Organisation and Personnel (1953-55), Deputy Area Manager (1955-60), Manager (1960-62), Los Alamos Area Office, U.S.A.E.C.; Asst. to Asst. Gen. Manager (1962-64), Asst. Director for Economic Impact, Office of Economic Impact and Conversion (1964-65), Director of Congressional Relations, (1965-67), Special Asst. to Asst. Gen. Manager for Operations (1967-), U.S.A.E.C.
Address: United States Atomic Energy Commission, Washington D.C. 20545, U.S.A.

BURKE, Joseph Eldrid, B.S., Ph.D. (Chem.). Born 1914. Educ.: McMaster and Cornell Univs. Manager Metal., K.A.P.L. (1951-54), Res. Assoc., Res. Lab. (1954-55), Manager, Ceramics Branch, R. and D. Centre (1955-), G.E.C.; Adjunct Prof. Metal., Rensselaer Polytech. Inst., 1957-64. Editorial Advisory Board, J. Nucl. Materials; Gen. Editor, Progress in Ceramic Sci. Series. Books: Co-author, Procedures in Exptl. Metal.; Co-editor, The Metal Beryllium; Societies: A.S.M.; American Ceramic Soc.; A.I.M.E.; British Ceramic Soc.; A.N.S.
Nuclear interests: Metallic and ceramic fuels; Radiation effects; High-temperature materials; Solid state reaction mechanisms.
Address: 33 Forest Road, Burnt Hills, New York, U.S.A.

BURKE, Kevin Charles Antony, B.Sc., Ph.D. Born 1929. Educ.: London Univ. Sen. Geologist, Atomic Energy Div., Geological Survey of Great Britain, 1956-60; I.A.E.A. Adviser on nucl. raw materials to South Korea, 1960; Head, Geology Dept., Univ. of West Indies, 1961-.
Nuclear interest: Geology of nuclear raw materials.
Address: Geology Department, University of the West Indies, Mona, Kingston 7, Jamaica.

BURKHARD, Donald George, A.B., M.S., Ph.D. Born 1918. Educ.: California Inst. of Technol., California (Berkeley) and Michigan Univs. Prof., Phys., Colorado Univ., 1950-65; Consultant, Theoretical Phys., N.B.S., Colorado, 1955-58; Res. Director, PEC-Res. Assocs., 1956-67; Prof., and Head, Dept. of Phys. and Astronomy, Georgia Univ., 1965-. Societies: A.P.S.; A.A.A.S.
Nuclear interests: Nuclear physics; Nuclear engineering.

Address: Department of Physics and Astronomy, Georgia University, Athens, Georgia 30601, U.S.A.

BURKHARD, Eldred Lee, B.A., M.S. Born 1926. Educ.: Nebraska Wesleyan and Lehigh Univs. Test Eng., Babcock and Wilcox Co., 1951-53; Asst. Project Eng., Gen. Dynamics, Fort Worth, 1953-. Societies: A.N.S.; A.P.S. Nuclear interests: Radiation damage; Neutron activation; Isotopes and radiation; Management; Systems analysis.
Address: 4028 Eldridge Street, Fort Worth, Texas 76107, U.S.A.

BURKHARDT, Prof., Dr. Staatliche Zentrale für Strahlenschutz (Central Board for Radiol. Protection).
Address: Staatliche Zentrale fur Strahlenschutz, 336 Muggelseedamm, 1162 Berlin-Friedrichshagen, German Democratic Republic.

BURKHARDT, Gerd, Dr. phil., o. Prof. Born 1911. Educ.: Munich Univ. Director, Inst. f. theor. Physik, T.H. Hanover. Book: Grundlagen und Anwendung der Kerntechnik (Düsseldorf, VDI-Verlag, 1959). Society: Deutsche Physikalische Gesellschaft.
Nuclear interests: Shielding problems.
Address: Technische Hochschule, Hanover, Germany.

BURKITT, Robert William, M.A., A.M.I. Mech. E. Born 1908. Educ.: Cambridge Univ. Formerly Asst. Sec., Board of Trade; Principal Officer, Commercial Policy, U.K.A.E.A.
Address: United Kingdom Atomic Energy Authority, 11 Charles II Street, London, S.W.1, England.

BURKLE, Joseph Stewart, B.A., M.D. Born 1919. Educ.: Pennsylvania Univ. Director, Rad. Exposure Evaluation Lab., Bethesda, 1960-63; Director, Armed Forces Radiobiol. Res. Inst., Bethesda, 1966-67; Chairman, Dept. of Nucl. Medicine, York Hospital, Pennsylvania. Societies: Rad. Res. Soc., Soc. of Nucl. Medicine; Transplantation Soc.; American Coll. of Physicians.
Nuclear interest: Nuclear medicine.
Address: York Hospital, York, Pennsylvania 17403, U.S.A.

BURLAKOV, V. D., B.Sc., Atomic Energy Utilisation Board, U.S.S.R.; Deleg. to 2nd I.C.P.U.A.E., Geneva, Sept. 1958; Deleg. to Conférence sur la Physique des Plasmas et la Recherche concernant la Fusion Nucléaire Controlée, Salzbourg, Sept. 1961. Papers: On the Carrying of Residual Gas by a Stream of Vapour escaping into a Vacuum (letter to the Editor, Atomnaya Energiya, vol. 3, No. 9, 1957); co-author, Refining Beryllium and other Metals by Condensation on Heated Surfaces (2nd I.C.P.U.A.E., Geneva, Sept. 1958).
Address: Atomic Energy Utilisation Board, Council of Ministers of the U.S.S.R., Moscow, U.S.S.R.

BURLIN, Colonel Robert B., U.S. Army; B.S., M.S. Born 1920. Educ.: U.S. Military Acad. and M.I.T. Director, Army Nucl. Power Programme; Special Asst. to Chief of Engs., U.S. Army; In charge of Fort Belvoir Nucl. Power Field Office, 1960-62. Asst. Director, Reactor Development Div., U.S.A.E.C., 1962-65; Commanding Officer, 555th Eng. Combat Group. Societies: A.N.S.; Soc. American Military Eng.
Nuclear interests: Research, engineering, management, operation and technical support of military nuclear reactors and training of operators for military nuclear power plants (land based).
Address: 555 Engineer Combat Group, A.P.O. New York 09164, U.S.A.

BURMAN, Lawrence Cromer, B.S. (Ch.E.). Born 1912. Educ.: C.C.N.Y. Manager, Nucl. Service Dept., Engelhard Mineral and Chem. Corp., Newark, New Jersey; Formerly Director, Licensing Div., U.S.A.E.C., New York. Societies: A.I.Ch.E.; A.N.S.; Atomic Ind. Forum.
Nuclear interests: Management; Nuclear fuels manufacture.
Address: Engelhard Mineral and Chemical Corporation, 113 Astor Street, Newark, New Jersey 07114, U.S.A.

BURNAY, F. Ing., Inter-Faculty Centre of Nucl. Sci., Liège Univ.
Address: Inter-Faculty Centre of Nuclear Sciences, Liège University, Liège, Belgium.

BURNETT, James Stark Greig, M.D., D.P.H. Born 1904. Educ.: Glasgow and Manchester Univs. Medical Officer of Health, County Borough of Preston, 1949-; Past Chairman of Council, Chairman, Tech. Com., Nat. Soc. for Clean Air; Member of Council, Soc. of Medical Officers of Health; Director, Public Health Inspectors, Education Board; Member, Central Midwives Board. Societies: British Medical Assoc.; Royal Soc. of Health.
Nuclear interests: Medical and epidemiological aspects.
Address: Municipal Building, Preston, Lancs., England.

BURNETT, R. W. Director, Burnett and Lewis Ltd.
Address: Burnett and Lewis Ltd., Technical Sales Department, Redhouse Industrial Estate, Aldridge, Staffs., England.

BURNETT, Thomas H. Joseph, B.A. Born 1915. Educ.: Tennessee, Ohio State and Louisville Univs. At O.R.N.L., Health Phys. Div., 1946-. Societies: H.P.S.; A.I.H.A.
Nuclear interests: Health physics; Reactor safety; Nuclear standards.

Address: Health Physics Division, Oak Ridge National Laboratory, P.O. Box X, Oak Ridge, Tennessee, U.S.A.

BURNHAM, Donald C. Formerly Chief Eng., American Nucl. Div., American Electronics Inc., Formerly Vice-Pres., Ind. Group, now Pres., and Exec. Officer, Westinghouse Elec. Corp., 1963-.
Address: Westinghouse Electric Corp., P.O. Box 2278, Pittsburgh 30, Pennsylvania, U.S.A.

BURNOD, L. Sect. Accélérateur Linéaire et Techniques Annexes, Lab. de l'Accélérateur Linéaire, Ecole Normale Supérieure, Paris Univ.
Address: Ecole Normale Supérieure, Laboratoire de l'Accélérateur Linéaire, B.P. No. 5, Orsay, France.

BURNS, Charles F.W. Chairman of the Board, Burns Brothers and Denton Ltd. Director, Canadian Nucl. Assoc.
Address: Burns Brothers and Denton Ltd., Toronto, Ontario, Canada.

BURNS, John Francis, A.B., M.A., Ph.D. Born 1901. Phys., Gaseous Diffusion Plant, Oak Ridge, 1945-56; Sen. Phys., O.R.N.L., 1956-. Societies: A.P.S.; Albertus Magnus Guild.
Address: 131 Georgia Avenue, Oak Ridge, Tennessee, U.S.A.

BURNS, Ronald Hosendoff, B.Sc., F.R.I.C. Born 1905. Educ.: London Univ. Chief Ind. Chem., U.K.A.E.R.E., Harwell, 1948-. Books: Co-author, Radioactive Wastes; their Treatment and Disposal (E. and F.N. Spon Ltd., 1960); co-author, Atomic Energy Wastes – Its Nature, Use and Disposal (Butterworths). Societies: Soc. of Chem. Industry; Soc. of Water Treatment and Examination; Inst. Water Pollution Control.
Nuclear interests: General industrial chemistry with special reference to waste treatment and disposal; Water treatment and decontamination procedures.
Address: Atomic Energy Research Establishment, Harwell, Berkshire, England.

BURR, Arthur Albert, B.S., M.S., Ph.D. (phys.). Born 1913. Educ.: Saskatchewan and Pennsylvania State Univs. Prof. Phys. Metal., Dean Eng., Rensselaer Polytechnic Inst.; Visiting Com., O.R.N.L., 1961-63. Consultant, Espey Corp., Saratoga, N.Y., 1960-. Com. member, Eng. Council for Professional Development as representative of American Inst. of Mining, Metal. and Petroleum Eng. Societies: A.P.S.; A.S.M.; American Inst. of Mining, Metallurgical and Petroleum Eng.; American Soc. for Non-destructive Testing; Electrochem. Soc. Inc.; New York Acad. of Sci.
Nuclear interests: Metallurgy; Metals and alloys of interest to the nuclear industry (e.g. zirconium, rare earth alloys); Remote handling of radioactive metals; Tracer work; Radiation damage studies.
Address: Rensselaer Polytechnic Institute, Ricketts Building, Troy, New York, U.S.A.

BURR, John Green, B.S. (M.I.T.), M.S. (M.I.T.), Ph.D. (Northwestern). Born 1918. Educ.: M.I.T. and Northwestern Univ. Sen. Sci., O.R.N.L., 1948-57; Supv.,Rad. Chem. Unit (1957), Sect. Leader Rad. Chem. and Chem. Kinetics (1961-62), Atomics Internat., Canoga Park, California; Group Leader, North American Aviation Sci. Centre, 1962-. Book: Tracer Applications for the Study of Organic Reactions (New York, Interscience, 1957). Societies: Rad. Res. Soc.; Cosmos Club; A.C.S.; A.N.S.
Nuclear interests: Radiation chemistry, application of isotopic tracers in chemical research; Pyrolysis, and mass spectrometry; Nucleic acid photochemistry.
Address: Biosciences Group, North American Rockwell Corporation, Science Centre, 1049 Camino Dos Rios, Thousand Oaks, California, U.S.A.

BURRILL, Ernest Alfred, Jr., B.Sc. Born 1917. Educ.: M.I.T. Phys. (1947-54), Director Tech. Sales (1954-57), Vice-Pres. and Sales Manager (1957-60), Vice-Pres. and Director, Marketing (1960-67), Vice-Pres. and Manager, Western Region (1967-), High Voltage Eng. Corp. Societies: A.P.S.; A.A.A.S.; A.N.S.
Nuclear interests: Van de Graaff and electron linear accelerator design and applications; Characteristics and applications of X-rays, electrons, neutrons, positive ions from accelerators; Radiation shielding.
Address: 145 Golden Oak Drive, Portola Valley, California 94025, U.S.A.

BURRIS, Robert H., B.S., M.S., Ph.D. Born 1914. Educ.: South Dakota State Coll. and Wisconsin Univ. Prof. Biochem. (1951-), Chairman, Biochem. Dept. (1958-), Wisconsin Univ. Pres., American Soc. Plant Physiologists. Societies: A.C.S.; American Soc. Plant Physiologists: Biochem. Soc.; Soc. American Bacteriologists; N.A.S.
Nuclear interest: Tracer work in biochemistry, particularly biological nitrogen fixation.
Address: Biochemistry Department, Wisconsin University, Madison, Wisconsin 53706, U.S.A.

BURROWS, Belton A., M.D. Dept. Medicine, Boston Univ.
Address: Department of Medicine, Boston University, 80 East Concord Street, Boston 18, Massachusetts, U.S.A.

BURROWS, Don Steven, B.A. (Public Administration). Born 1909. Educ.: California Univ. (Berkeley). Comptroller, Nat. Security Resources Board, Executive Office of the President, 1950-51; Controller, Reconstruction Finance Corp., 1951-52; Controller, A.E.C., 1952-62; Martin Marietta Corp., Denver, 1962-.
Nuclear interest: Financial management.
Address: 1328 East Layton Avenue, Englewood, Colorado, U.S.A.

BURRUS, Walter Ross, Ph.D. Born 1931. Educ.: Ohio State Univ. and Georgia Inst. Technol. U.S.A.F. Nucl. Instrumentation and Rad. Damage, 1952-57; Consultant, Nucl. Environments, 1957-60; Phys., O.R.N.L., 1960-67; Pres., Tennecomp, Inc., 1967-. Societies: I.E.R.E.; A.P.S.
Nuclear interests: Low-energy nuclear physics; Nuclear instrumentation; Data acquisition and reduction; Radiation damage; Reactor and cosmic radiation shielding; Experimental mathematics; Domestication of computers.
Address: Tennecomp, Incorporated, Box J, Oak Ridge, Tennessee 37830, U.S.A.

BURSH, Talmadge. Com. Member, Nucl. Dept., Southern Univ.
Address: Southern University, Southern Branch Post Office, Baton Rouge, Louisiana, U.S.A.

BURSHTEIN, E. L. Papers: Co-author, Use of Autocorrelation Principle for the Magnetic Field in Cyclic Accelerators Designed to Produce Very High Energies (Atomnaya Energiya, vol. 12, No. 2, 1962); co-author, On the Effect of Beam Loading on the Parameters of a Linear Electron Accelerator (letter to the Editor, ibid, vol. 13, No. 5); co-author, Development of Super-High Energy (1000 Gev) Proton Synchrotron (ibid, vol. 23, No. 5, 1967).
Address: Academy of Sciences of the U.S.S.R., 14 Leninsky Prospekt, Moscow V-71, U.S.S.R.

BURSTEIN, Elias, A.B. (Brooklyn Coll.); A.M. (Kansas Univ.). Born 1917. Educ.: Brooklyn Coll., Kansas Univ., M.I.T. and Catholic Univ. Head, Phys. Sect., Crystal Branch (1948-58), Head, Semiconductor Branch (1958), U.S.N.R.L.; Prof. Phys., Pennsylvania Univ., 1958-; Consultant, Ford Sci. Lab., Dearborn, Michigan, 1966-; Univ. Advisor, Texas Instruments, Inc., Dallas, Texas, 1962-. Sec., Board of Editors, Solid State Communications, Pergamon Press, 1963-. Societies: Fellow, A.P.S.; Fellow, American Optical Soc.
Nuclear interest: Solid state.
Address: Physics Department, Pennsylvania University, Philadelphia, Pennsylvania 19104, U.S.A.

BURSUKOVA, M. I. Paper: Co-author, The Efficiency of H^3/Zr Sources in Nondispersive X-Ray Spectrometric Analysis (letter to the Editor, Atomnaya Energiya, vol. 19, No. 3, 1965).
Address: Academy of Sciences of the U.S.S.R., 14 Leninsky Prospekt, Moscow V-71, U.S.S.R.

BURT, John Corwin, A.B. Born 1914. Educ.: Colgate Univ., Hamilton, N.Y. and the City of New York Univ. Formerly Director, Nucl. Energy Writers Assoc., Inc.
Nuclear interests: Administrative liaison in national and international government, and public information programmes.
Address: International Atomic Energy Agency, United Nations, New York 17, N.Y., U.S.A.

BURT, P. Allister. Supt., Nine Mile Point Station, Niagara Mohawk Corp., 1964-.
Address: Niagara Mohawk Power Corp., Electric Building, Buffalo 3, New York, U.S.A.

BURTON, James Dennis, B.Sc. (London). Ph.D. (Liverpool). Born 1931. Educ.: Liverpool Univ. (post-graduate only). Sci. Officer, A.W.R.E., U.K.A.E.A., Aldermaston, 1955-58; S.S.O., Agricultural Res. Council Radiobiol. Lab., Letcombe Regis, Wantage, 1958-64; Lecturer in Chem. Oceanography, Southampton Univ., 1964-. Societies: A.R.I.C.; Soc. for Analytical Chem.; Chem. Soc.; Challenger Soc.
Nuclear interests: Application of radiochemical methods (tracer methods, neutron activation analysis, isotopic dilution techniques) in the field of marine geochemistry; Natural and artificial radioactive nuclides in the environment, particularly in the hydrosphere.
Address: Department of Oceanography, Southampton University, SO9 5NH, England.

BURTON, Jean. Inspector Gen. des Finances, Ministere des Affaires Economiques et de l'Energie.
Address: Ministere des Affaires Economiques et de l'Energie, 34 boulevard Pacheco (3e etage), Brussels 1, Belgium.

BURTON, John E., Dr. Vice-Pres., Business, Cornell Univ. Member, Gen. Advisory Com., New York State Office of Atomic Development.
Address: 1010 Triphammer, Ithaca, New York, U.S.A.

BURTON, John H. Supervisor, Remote Handling Unit, Space Systems Department, Atomics Internat. Member, Sec. (1966-), Remote Systems Technol. Div., A.N.S.
Address: 20718 Roscoe Boulevard, Canoga Park, California 91304, U.S.A.

BURTON, John Stewart, B.Sc. (London), M.I. Biol. Born 1932. Chief Chem., Thorn Electronics Ltd., 1962-. Book: Radioisotope Powered Battery Using Kr-85 hydroquinone Source in Batteries 2 (Editor, Collins) (Pergamon, 1965).
Nuclear interests: Scintillators; Nuclear instrumentation and nuclear battery systems.
Address: Research Division, Thorn Electronics Limited, Hook Rise South, Tolworth, Surbiton, Surrey, England.

BURTON, Joseph Ashby, Ph.D., B.S. Born 1914. Educ.: Washington and Lee, and Johns Hopkins Univs. Tech. Staff (1938-), Director, Chem. Phys. Res. (1958-), Bell Telephone Labs. Society: Fellow, A.P.S.
Nuclear interest: Nuclear physics; Radiation damage.
Address: Bell Telephone Laboratories, Murray Hill, New Jersey, U.S.A.

BURTON, Leonard Kenneth, B.Sc., Ph.D. Born 1928. Educ.: London Univ. M.O.S., R.A.E., Farnborough, 1951-53; Inst. Cancer Res., London, 1953-61; C.E.G.B., Berkeley

Nucl. Lab., Glos., 1961-. Societies: Inst. Phys. and Phys. Soc.; Soc. for Radiol. Protection. Nuclear interests: Nuclear instrumentation; Health physics.
Address: Central Electricity Generating Board, Berkeley Nuclear Laboratories, Berkeley, Glos., England.

BURTON, Milton, B.S. (Chem. Eng.), M.S., Ph.D. Born 1902. Educ.: N.Y.U. Prof. Chem., (1945-), Director, Rad. Lab. (1947-), Notre Dame Univ. Council, Rad. Res. Soc.; Advisory Board on Isotopes and Rad. Applications, U.S.A.E.C., 1967-. Societies: A.C.S.; Rad. Res. Soc.; A.P.S.; Faraday Soc.; Soc. de Chim. Physique.
Nuclear interests: Primarily in radiation chemistry and the fundamentals of the effects of radiation on matter.
Address: 730 Indiana Avenue, Mishawaka, Indiana, U.S.A.

BURTSCHER, Alfons, Dipl. Ing. Director, ASTRA Reactor, Osterreichische Studiengesellschaft für Atomenergie G.m.b.H.; Deleg., I.A.E.A. Symposium on Pile Neutron Res. in Phys., 17-21 October 1960, Vienna; I.C.P.U.A.E., Geneva, Sept. 1964.
Address: Osterreichische Studiengesellschaft für Atomenergie G.m.b.H., Reaktorzentrum, Seibersdorf, Austria.

BURZYNSKI, Zygmunt, Inzynier. Dr. Born 1911. Educ.: Polytechnic School, Warsaw and Liège Univ. Independent Sci. Worker, Inst. of Nucl. Res. Society: Polish Soc. Elec. Eng. Nuclear interests: Magnetic properties of particles; Physical quantity regulators; Dynamic stability of stabilisers.
Address: 23 m. 8 ul. Mickiewicza, Warsaw 32, Poland.

BUSBY, Jack. Member, Exec. Com., High Temperature Reactor Development Associates.
Address: High Temperature Reactor Development Associates, Inc., 89 East Avenue, Rochester 4, New York, U.S.A.

BUSCH, Alfred. Director, Tracerlab-Keleket.
Address: Tracerlab-Keleket, 1601 Trapelo Road, Waltham 54, Massachusetts, U.S.A.

BUSCH, Georg, Prof. Dr. Head, Lab. für Festkörperphysik, E.T.H., Zürich.
Address: 35 Gloriastrasse, Zürich 8006, Switzerland.

BUSCH, Joseph Sherman, B.S. (Hons. Zoology), B.S. (Chem. Eng.), M.S. (Chem. Eng.), Ph.D. (Chem. Eng.). Born 1927. Educ.: Northwestern and Johns Hopkins Univs. and Carnegie 'Inst. of Technol. Columbia University Res. Asst. studying heat transfer to liquid metals, Columbia Univ., 1950-53; Group Leader directing light water reactor core and primary system heat transfer, fluid flow, systems and hazards analyses,Westinghouse (Bettis), 1953-59; Head, A.P.D.A. core analysis and test sect. for Enrico Fermi reactor, 1959-62; Project Eng., Nucl. Projects Div., Kaiser Engineers, 1962-. Member, Subcom. 5.23 (developing standards for post operational energy in uranium dioxide fueled nucl. reactors), A.N.S. Book: Fast Reactor Technology: Plant Design (editor: J.G. Yevick) (M.I.T. Press, 1966). Societies: A.N.S.; A.I.Ch.E.
Nuclear interests: Heat transfer, fluid flow, systems response, shielding, containment, and general hazards work associated with light water, gas cooled, and liquid metal cooled nuclear power plants, and with large scale test facilities to develop data for those nuclear power plants.
Address: 1598 Hawthorne Terrace, Berkeley, California 94708, U.S.A.

BUSCHBECK, Frank Martin, Dipl. Ing. Born 1930. Educ.: Moscow Inst. for Telecommunication Technics. Sci. co-worker, Österreichische Studiengesellschaft für Atomenergie, Vienna. Nuclear interest: Nuclear electronics.
Address: Osterreichische Studiengesellschaft für Atomenergie, 10 Lenaugasse, Vienna 8, Austria.

BUSH, George L., B.S., M.A., Ed. D. Born 1897. Educ.: Ohio State and Columbia Univs. Formerly Prof. Chem., now Chairman of Gen. Chem., Kent State Univ. Societies: A.C.S.; A.A.A.S.
Nuclear interests: General.
Address: 7802 Birchwood Drive, Kent, Ohio, U.S.A.

BUSH, Harry Derrick, B.Sc. (Hons.), Ph.D. Born 1919. Educ.: Leeds Univ. Lecturer in Phys., Sheffield, 1947-52; Sen. Lecturer in Phys., Bradford Inst. of Technol., 1952-. Society: A.Inst.P.
Address: Bradford Institute of Technology, Bradford 7, Yorkshire, England.

BUSH, Philip David, B.S. (Electrochem. Eng.), M.S. (Metal. Eng.). Born 1917. Educ.: Stanford Univ. and M.I.T. Vice-Pres. Advanced Technol., Kaiser Eng. Div., Kaiser Industries Corp., Oakland, California. Societies: A.N.S.; Atomic Industrial Forum; A.I.Ch.E.; A.I.A.A.
Nuclear interest: Management.
Address: Kaiser Engineers, Division of Kaiser Industries Corporation, 300 Lakeside Drive, Oakland, California 94604, U.S.A.

BUSH, Spencer H. Born 1920. Educ.: Ohio State and Michigan Univs. Instructor, Michigan Univ., 1951-53; Sen. Sci. (1953-54), Supv., Phys. Metal. (1954-57), Supv., Fuels Fabrication Development (1957-60), Metal. Specialist (1960-62), Consulting Metal. (1962-63), Reactor and Fuels Lab. Consulting Metal. (1963-65), Hanford Labs., G.E.C., Richland, Washington; Consultant to Director, Battelle Memorial Inst. Pacific Northwest Labs., 1965-; Affiliate Prof. Metal. Eng., Graduate Study Centre, Washington Univ. U.S.A.E.C. Advisory Com. on Reactor Safeguards 1966-; Trustee, A.S.M.

1967-69; Annual Lecture Com. (1966-69), Chairman (1967-68), A.I.M.E.; Advisory Editorial Board, Nucl. Applications, A.N.S.; A.S.M.E.-Nucl. Code Com. 1966-; Com. on Uranium Standards and Com. A-10 A.S.T.M.; Com. N-6, A.S.A. Books: Chapter on Aluminium, Magnesium, Irrad. Effects, in Reactor Handbook; Effects of Irrad. on Cladding and Structural Materials (A.S.M.-A.E.C.). Societies: British Inst. Metals; A.N.S.; A.S.M.; A.S.T.M.; A.I.M.E.
Nuclear interests: Properties of metals and materials for nuclear applications; Structural components; Metallic and ceramic fuels and cladding materials; Fundamental mechanisms of radiation damage; Reactor safety and reactor failure mechanisms.
Address: Battelle-Northwest, Post Office Box 999, Richland, Washington 99352, U.S.A.

BUSH, Vannevar, B.S., M.S., D. Eng., 21 Hon. Doctor's Degrees. Born 1890. Educ.: Tufts Coll., Harvard and M.I.T. Pres. Carnegie Inst. of Washington, 1938-55; Nat. Sci. Foundation Advisory Com. on Govt.-Univ. Relationships, 1953-55; Director, American Telephone and Telegraph Co., 1947-62; Chairman, Merck and Co., 1957-62; Director, Graphic Arts Res. Foundation, Inc. Books: Operational Circuit Analysis; co-author, Principles of Elec. Eng.; Endless Horizons; Modern Arms and Free Men; Science, the Endless Frontier. Societies: Hon. Member, Franklin Inst.; Hon. Member, I.E.E.E.; Hon. Member, Soc. of Naval Architects and Marine Eng.; Hon. Member, A.S.M.E.; Hon. Fellow, American Coll. of Surgeons; Fellow, A.P.S.; American Philosophical Soc.; American Acad. Arts and Sci.; N.A.S.; Hon. Member, A.S.E.E.; Life Member, and Hon. Chairman (1959-), M.I.T. Corp.
Address: Massachusetts Institute of Technology, Cambridge, Massachusetts 02139, U.S.A.

BUSH, William R. Vice-Pres., Marketing, United Nucl. Corp., 1962-.
Address: United Nuclear Corporation, 1730 K Street, Washington 6, D.C., U.S.A.

BUSHKOV, A. P. Paper: Co-author, On Determination of Fluorite by Activation Analysis (letter to the Editor, Atomnaya Energiya, vol. 21, No. 3, 1966).
Address: Academy of Sciences of the U.S.S.R., 14 Leninsky Prospekt, Moscow V-71, U.S.S.R.

BUSHLAND, Raymond Cecil, B.S., M.S., Ph.D. Born 1910. Educ.: South Dakota State Coll. and Kansas State Univ. Entomology Res. Div., U.S. Dept. Agriculture, 1935-. Societies: Rad. Res. Soc.; Entomological Soc. America.
Nuclear interest: Insect sterilisation by radiation.
Address: Metabolism and Radiation Research Laboratory, State University Station, Fargo, N. Dak. 58102, U.S.A.

BUSHUEV, A. V. Papers: Co-author, Res. of ^{239}Pu Growth by ^{239}U γ-Rad. (letter to the Editor, Atomnaya Energiya, vol. 16, No. 6, 1964); co-author, Measurement of Ratio Capture Fissions in U^{238} by Gamma-Spectrometry Method (letter to the Editor, ibid, vol. 20, No. 1, 1966); co-author, Determination of Some Parameters of Reactors with Germanium Gamma-Ray Spectrometer (letter to the Editor, ibid, vol. 22, No. 6, 1967); co-author, Comparison of Measurement Methods of Fission Rates (letter to the Editor, ibid, vol. 24, No. 1, 1968).
Address: Academy of Sciences of the U.S.S.R., 14 Leninsky Prospekt, Moscow V-71, U.S.S.R.

BUSINARO, Ugo Lucio, Dr. Phys. Born 1929. Educ.: Milan Univ. Manager for Res., Fiat, Sezione Energia Nucleare, 1957-. Society: A.N.S.
Nuclear interests: Reactor physics and materials development.
Address: Fiat, Sezione Energia Nucleare, 235 Via Settembrini, Turin, Italy.

BUSO, Roberto, B.S. (Puerto Rico), M.D. (Paris). Born 1912. Educ.: Puerto Rico and Paris Univs. Clinical Assoc. in Medicine (Ad Honorem), School of Medicine and Tropical Medicine, Puerto Rico Univ., 1952-; Director, Radioisotopes Unit-Fundación Investigaciones Clinicas and Mimiya Hospital, Santurce P.R., 1952-. Member-Supervisor, Educational Council, Puerto Rico Univ.; Chairman Isotopes Com., and Sec., Fundación Investigaciones Clinicas. Papers: Blood Volume Studies in Healthy Puerto Ricans using Radiochromium (P.R. Medical Assoc. Bulletin 48: 156, 1956); Radiochromium Uptake in Vitro as a Measure of Red-cell Survival in Vivo (Proc. VIth Congress Internat. Soc. of Hematology, Aug.-Sept. 1956); Studies on the Pathogenesis of the Anemia of Hypothyroidism (J. Clinical Endocrinology and Metabolism, vol. xviii, No. 5, May 1958, pp. 501-505); Distribution of Abnormal Hemoglobins in Puerto Rico and Survival Studies of Red Blood Cells Using ^{51}Cr (to be published in Blood). Societies: A.N.S.; Puerto Rico Medical Soc.; American Medical Assoc.
Address: Box 9001, Santurce, Puerto Rico.

BUSOL, F. I. Papers: On the Refining of Zirconium by Iodide Method (letter to the Editor, Atomnaya Energiya, vol. 3, No. 10, 1957); On the Iodide Method of purifying Zirconium (ibid, vol. 4, No. 2, 1958); co-author, Cryogenic Magnetic Mirror Machine WGL-2 (letter to the Editor, ibid, vol. 21, No. 2, 1966).
Address: Academy of Sciences of the U.S.S.R., 14 Leninsky Prospekt, Moscow V-71, U.S.S.R.

BUSSAC, Jean, Diplôme Ecole Polytechnique, Paris. Born 1929. Educ.: Ecole Polytechnique and Paris Univ. Chef du Service de physique mathématique, C.E.A., France, 1951-. Expert, Third I.C.P.U.A.E., Geneva, Sept. 1964; Member, Computer Programme Library Com., O.E.C.D., E.N.E.A.
Nuclear interest: Reactor design.

Address: C.E.N. Saclay, Dept. des Etudes de Piles, B.P. No. 2, Gif-sur-Yvette, Seine-et-Oise, France.

BUSSARD, Robert William, B.S. (Eng.), M.S. (Eng.), A.M. (Phys.), Ph.D. (Phys.). Born 1928. Educ.: California (Los Angeles) and Princeton Univs. Hughes Aircraft Co., 1949-51; O.R.N.L., 1952-55; Los Alamos Sci. Lab., 1955-62; TWR/ Space Technol. Labs., 1962-64; Electro-Optical Systems, Inc. 1964-. Consultant, AGARD/NATO, 1960, 1962, 1964. Vice-Pres. (Tech.) and Member, Board of Directors, A.I.A.A., 1965-66. Books: Co-author, Nucl. Rocket Propulsion (New York, McGraw-Hill Publishing Co., 1958); co-author, Fundamentals of Nucl. Flight (New York, McGraw-Hill Publishing Co., 1965). Societies: Fellow, A.I.A.A.; A.P.S.; A.G.U.; British Interplanetary Soc.
Nuclear interests: Nuclear space power research and development, components, systems, and systems applications; High altitude effects of nuclear weapons.
Address: Electro-Optical Systems, Inc., 300 North Halstead Street, Pasadena, California 91107, U.S.A.

BUSSE, Ernst Hermann, Dr. phil. nat. Born 1902. Educ.: Karlsruhe T.H. and Jena Univ.; Zwangsaufenthalt Sowietunion, 1945-54; Arbeitsgemeinschaft für Kerntechnik, 1956-. Sekretär des Deutschen Atomforums; Formerly Geschäftsführer der Arbeitsgemeinschaft für Kerntechnik; Formerly Member, Exec. Com., Foratom. Societies: V.D.I.; Deutsche Physikalische Gesellschaft.
Nuclear interests: Management; Nuclear physics.
Address: 240 Koblenzer Str., 5300 Bonn, Germany.

BUSSEY, Brian William, B.A. (Hons., German). Born 1938. Educ.: Southampton Univ. Advertisement Manager, Nucl. Eng.
Address: 33-39 Bowling Green Lane, London, E.C.1, England.

BUSSIERE, Paul, L. ès Sc., Ing. E.S.C.I.L.— Docteur. Born 1927. Educ.: Lyon Univ. Maître de Recherches, C.N.R.S.; Chef du Service de Radiochimie, Institut de Recherches sur la Catalyse. Societies: Chimique de France; de Chimie Physique.
Nuclear interests: Utilisation des radioisotopes en analyse et en étude de mécanismes réactionnels; Utilisation des rayonnements pour irradiation, dans les systèmes catalytiques.
Address: Institut de Recherches sur la Catalyse, Centre National de la Recherche Scientifique, 39 boulevard du Onze-Novembre 1918, Villeurbanne, Rhône, France.

BUSSY, E. Member, Conseil d'Administration, Energie Nucléaire S.A.
Address: Energie Nucléaire S.A., 10 avenue de la Gare, Lausanne, Switzerland.

BUSTAD, Leo K., B.S., M.S., D.V.M., Ph.D. Born 1920. Educ.: Washington State and Washington Univs. Formerly Manager, Exptl. Animal Farm, Biol. Operation, Hanford Labs., G. E. C. and Battelle Memorial Inst., Pacific Northwest Lab.; now Director, Radiobiol. Lab. and Prof. of Rad. Biol., School of Veterinary Medicine, California Univ - Davis; N.S.F. postdoctoral fellow, Washington Univ. School of Medicine; Consultant, Veterans Administration, Livermore Hospital, Livermore, Calif., Lovelace Foundation and G.E.C., Vallecitos Atomic Lab.; Member, Sci. Advisory Com., Washington Regional Primate Centre; Member, Subcom. on Pathological Effects of Thyroid Irradiation and Inst. of Lab. Animal Resources, Com. on Professional Education of N.A.S.-N.R.C.; Guest Lecturer, Washington State Univ. Societies: American Veterinary Medical Assoc.; American Physiological Soc.; Rad. Res. Soc.; Soc. of Exptl. Biol. and Medicine; A.I.B.S.; Soc. of Nucl. Medicine.
Nuclear interests: Physiopathologic effects of irradiation; Metabolism and toxicity of radionuclides, especially bone-seeking isotopes and radioiodines; radiation life shortening and neoplasia.
Address: Rt. 1, 42 Walnut Lane, Davis, California 95616, U.S.A.

BUSTAMANTE, Amalio SAIZ DE. See SAIZ DE BUSTAMANTE, Amalio.

BUSTARRET, Jean. Member, Conseil d'Enseignement de l' I.N.S.T.N., C.E.A.
Address: Commissariat a l'Energie Atomique, 29-33 rue de la Federation, Paris 15, France.

BUSTRAAN, M., Drs. Head, Phys. Dept., Reactor Centrum Nederland, Petten. Deleg. I.A.E.A. Seminar on Codes for Reactor Computations, Vienna, 25-29 April 1960; Member, Neutron Data Compilation Centre Com., O.E.C.D., E.N.E.A.
Address: Netherlands Reactor Centre, 112, Scheveningseweg, The Hague, Netherlands.

BUTEMENT, Francis Dudley Stewart, B.Sc., Ph.D. (London). Born 1916. Educ.: London Univ. Sen. Harwell Res. Fellow, 1948-51; Principal Sci. Officer, A.E.R.E., Harwell, 1951-55; Sen. Lecturer in Radiochem., Liverpool Univ., 1955-. Society: Chem. Soc.
Nuclear interest: Research on radioisotopes.
Address: Department of Inorganic Physical Chemistry, Liverpool University, Liverpool, England.

BUTIN, H. Chief Eng., Sté. d'Exploitation et de Recherches Electroniques S.A.
Address: Société d'Exploitation et de Recherches Electroniques S.A., boulevard de Mantes, Aubergenville, (S. et O.), France.

BUTLER, Arthur P. Geological Survey, U.S. Dept. of the Interior.
Address: Geological Survey, U.S. Department of

BUT

the Interior, Federal Centre, Building 25, Denver, Colorado 80225, U.S.A.

BUTLER, Frank, C. Eng. Born 1909. Chief Eng., Fuel Element Plant Design, U.K.A.E.A., Risley. Societies: A.M.I.Struct.E.; Nucl. Eng. Soc.
Nuclear interest: Engineering.
Address: Woodstock, Clamhunger Lane, Mere, Cheshire, England.

BUTLER, Gordon Cecil, B.A., Ph.D. Born 1913. Educ.: Toronto and London Univs. Prof. Biochem., Toronto Univ., 1947-57; Director, Biol. and Health Phys. Div., A.E.C.L., 1957-65; Director, Div. of Rad. Biol., N.R.C., 1965-. Member, U.N. Sci. Com. on the Effects of Atomic Radiation; Com. 4, Internat. Commission on Radiol. Protection; Expert Advisory Panel on Rad., W.H.O. Societies: Fellow, Roy. Soc. Canada; Fellow, A.A.A.S.; American Soc. Biological Chemists; Canadian Biochem. Soc.; H.P.S.; Canadian Phys. Soc.
Nuclear interests: Radiation biology and health physics.
Address: National Research Council of Canada, Ottawa, Ontario, Canada.

BUTLER, James Wilford, B.S. (Chem. Eng., Georgia Inst. Technol.), M.A. (Rice), Ph.D. (Rice), Prof. Phys. Born 1924. Educ.: Georgia Inst. Technol., Baylor Univ., Waco, Texas and Rice Univ., Houston, Texas. At U.S. Naval Res. Lab., Washington, D.C. Nucleonics Div. (Head, 2-Mv Van de Graaff Sect.; Consultant, Van de Graaff Branch), 1951-61; Prof. Phys., Michigan State Univ., 1961-64. Consultant, Van de Graaff Branch, U.S. Naval Res. Lab., 1964-. Society: A.P.S.
Nuclear interests: Nuclear physics; Nuclear reactions in the low energy and medium energy regions, nuclear spectroscopy, radioactive decay schemes, neutron physics, particle accelerators, nuclear instrumentation, and electromagnetic theory.
Address: Nuclear Physics Division, U.S. Naval Research Laboratory, Washington, D.C. 20390, U.S.A.

BUTLER, M. Member, Study Group on Digital Techniques, O.E.C.D., E.N.E.A.
Address: O.E.C.D. European Nuclear Energy Agency, 38 boulevard Suchet, Paris 16, France.

BUTLER, Mrs. Margaret K., A.B. (Math.). Born 1924. Educ.: Indiana Univ. Mathematician, Argonne Nat. Lab., 1951-. Exec. Com., Maths. and Computation Div., Publications Com., A.N.S. Societies: A.N.S.; Assoc. for Computing Machinery.
Nuclear interests: Reactor mathematics; Digital computer programmes for the solution of reactor design and nuclear physics problems.
Address: Applied Mathematics Division, Argonne National Laboratory, 9700 South Cass Avenue, Argonne, Illinois 60439, U.S.A.

BUTLER, Stuart Thomas, M.Sc. (Adelaide), Ph.D. (Birmingham), D.Sc. (Australian National Univ.). Born 1926. Educ.: Adelaide Univ. Prof. Theoretical Phys., Sydney Univ., 1959-. Books: Nucl. Stripping Reactions (Horwitz, Wiley and Pitman, 1947); editor and part author, From Nucleus to Universe (Shakespeare Head Press, 1960); editor and part author, Space and the Atom (Shakespeare Head Press, 1961); co-author, A Modern Introduction to Phys., Volumes 1, 2, 3 (Horwitz Publications, 1960, 1961).
Nuclear interests: Nuclear physics, particularly direct nuclear reactions, and interactions of high energy (relativistic) nucleons with nuclei.
Address: 6 The Grove, Mosman, New South Wales, Australia.

BUTLER, Thomas Arthur, M.S. Born 1919. Educ.: Iowa State Univ. Dept. Supervisor, O.R.N.L., 1951-. Society: A.C.S.
Nuclear interests: Separation, purification and packaging of radioisotopes; Characteristics of radioactive sources.
Address: Isotopes Division, Oak Ridge National Laboratory, P.O. Box X, Oak Ridge, Tenn., U.S.A.

BUTRA, F. P. Papers: Co-author, The Diffuse Scattering of X-rays in Irradiated Diamond, Corundum, Silicon and Germanium Crystals (letter to the Editor, Atomnaya Energiya, vol. 5, No, 5, 1958); co-author, Temperature and Neutron Irradiation Effect on Plastic Deformation of Alpha Uranium Single Crystal (ibid, vol. 19, No. 4, 1965).
Address: Academy of Sciences of the U.S.S.R. 14 Leninsky Prospekt, Moscow V-71, U.S.S.R.

BUTT, David Keith, B.Sc., Ph.D., A.R.C.S., F.Inst.P. Born 1925. Educ.: London Univ. I.C.I. Res. Fellow, Edinburgh, 1949-50; D.S.I.R. Asst., Birkbeck Coll., London, 1951-53; Res. Fellowship, Exeter Univ. 1953-54; Lecturer in Phys. (1955-65), Reader in Phys. (1965-), Birkbeck Coll., London.
Nuclear interests: Nuclear physics; β and γ spectrometry.
Address: Physics Department, Birkbeck College, London, W.C.1, England.

BUTTERFIELD, William John Hughes, B.M. (Oxon), M.A. (Oxon), M.D. (Johns Hopkins), F.R.C.P. (London). Born 1920. Educ.: Oxford and Johns Hopkins Univs. Prof. of Medicine, Guy's Hospital Medical School, London Univ. 1963-. Chairman, Exec. Council, British Diabetic Assoc.; Ministry of Health Long Term Study Group; Chairman, Medical Planning Liaison Com. for new town of Thamesmead; Chairman, Sci. Advisory Panel,Army Personnel Res. Com., Medical Res. Council.
Nuclear interest: Biological effects of nuclear explosions.
Address: 3 Maids of Honour Row, The Green, Richmond, Surrey, England.

BUTTLAR, Haro VON. See VON-BUTTLAR, Haro.

BUTTLAR, Rudolph O., Dr. Asst. Prof. Chem., Kent State Univ.
Address: Kent State University, Kent, Ohio, U.S.A.

BÜTTNER, Heinz-Jürgen Franz Hermann, Dipl.-Phys. Born 1937. Educ.: Dresden T.H. Sci. co-worker, Inst. für Elektrochemie und physikalische Chemie Dresden T.U. Society: Member, Working-Association "Mass-spectrometry", Deutschen Akademie der Wissenschaften zu Berlin.
Nuclear interest: Mass-spectrometry.
Address: 10 Oederaner Strasse, Dresden A 28, German Democratic Republic.

BUTTON, James Charles Ezekial, B.Sc. (Hons.). Born 1924. Educ.: Birmingham Univ. Health Phys. and Safety Group Manager, Dept. of Atomic Energy, Springfields Works, Ministry of Supply, 1953-55; Health and Safety Manager, Operations, Dounreay Exptl. Reactor Establishment, U.K.A.E.A., 1956-60; Tech. Assistance Expert, Health Phys., attached to Thai A.E.C., I.A.E.A., 1960-61; Head, Safety Sect., Health and Safety Div., A.A.E.C., 1962-. Member, New South Wales Radiol. Advisory Council. Societies: H.P.S.; Fellow, Inst. of Phys. and Phys. Soc.; Fellow, Australian Inst. of Phys.
Nuclear interest: All aspects of operational health physics and safety.
Address: 19 Church Street, Woolooware, New South Wales 2230, Australia.

BUZINA, Ratko, M.D., Sc.D., Assoc. Prof. Born 1920. Educ.: Zagreb Univ. Adviser in Nutrition to W.H.O. Regional Office for S.E. Asia in New Delhi, India, 1959-61; Head, Div. Physiology of Nutrition, Inst. Public Health, Zagreb.
Nuclear interests: Metabolic studies particularly related to development of endemic goitre.
Address: Institute of Public Health, Zagreb, Yugoslavia.

BUZZATI-TRAVERSO, Adriano A., Dr. Natural Sci. (Milan). Born 1913. Educ.: Milan Univ. and Iowa State Coll. Prof. Genetics, Pavia Univ., 1948-; Res. Assoc., California Univ., 1952-61. Sci. Director, Biol. Div., C.N.R.N., Rome 1958; Council Member, Assoc. Genetica Italiana; Formerly Member, Assoc. Radiobiologistes des Pays de l'Euratom; Council Member, Biometric Soc.; Fellow, A.A.A.S. Books: Co-author, Teoria dell'urto ed unità biologiche elementari (1948, Longanesi); Editor, Perspective in Marine Biol. (California Univ. Press, 1958); Editor, Immediate and Low Level Effects of Ionising Rad. (London, Taylor and Francis, Ltd., 1960). Societies: Genetics Soc. of America; Soc. for the Study of Evolution.
Nuclear interests: Biological effects of ionising radiations (general interest); Radioactive contamination of the sea.
Address: Instituto di Genetica, Via S. Epifanio 14, Pavia, Italy.

BUZZI, Ettore, Dr. (Mech. Eng.). Born 1905. Educ.: Milan Polytech. Mech. Eng. Director, Franco Tosi S.p.A., Legnano Com. Impianti Nucl. Assoc. Termotecnica Italiana.
Nuclear interest: Conventional parts.
Address: Franco Tosi S.p.A., 27 Corso Italia, Legnano, Italy.

BYCHKOV, A. P. Paper: Co-author, Determination of Average Energy β-Rad. of Small Activity Samples (letter to the Editor, Atomnaya Energiya, vol. 16, No. 6, 1964).
Address: Academy of Sciences of the U.S.S.R., 14 Leninsky Prospekt, Moscow V-71, U.S.S.R.

BYCK, Harold T., Ph.D. Educ.: New York, Columbia and Johns Hopkins Univs. Manager of Administration Services, Exploration and Production Res. Div., Asst. Sec., Shell Oil Co., -1962; Chairman, Information and Exhibits Div., O.R.I.N.S. (now Oak Ridge Assoc. Univs.), 1962-67. Societies: A.C.S.; A.P.S.; A.A.A.S.
Nuclear interest: Directs nationwide programme of public information on peaceful uses of atomic energy for U.S.A.E.C.
Address: Oak Ridge Associated Universities, P.O. Box 117, Oak Ridge, Tennessee, U.S.A.

BYELDRE, V. Ya. Deputy Director, Phys. Inst., Dept. of Tech. Sci., Acad. Sci. of the Latvian S.S.R.
Address: Physics Institute, Riga, Latvian S.S.R., U.S.S.R.

BYHOVSKII, A. V. Papers: Co-author, Rad. Safeguarding of Personnel during Uranium Ore Yield (Atomnaya Energiya, vol. 19, No. 2, 1965); co-author, Use of Ampule with Hydraulic Block in large Gamma-Rad. Installations for Chemical Purposes (letter to the Editor, ibid, vol. 21, No.1, 1966).
Address: Academy of Sciences of the U.S.S.R., 14 Leninsky Prospekt, Moscow V-71, U.S.S.R.

BYKOV, G. S. Paper: Co-author, Potassium Boiling Heat Transfer in Tube at Moderate Steam Content Values (letter to the Editor, Atomnaya Energiya, vol. 21, No. 1, 1966).
Address: Academy of Sciences of the U.S.S.R., 14 Leninsky Prospekt, Moscow V-71, U.S.S.R.

BYKOV, V. N. Papers: Co-author, A Study of U-Ge System (letter to the Editor, Atomnaya Energiya, vol. 8, No. 2, 1960); co-author, Diffusion of Uranium in Molybdenum, Niobium, Zirconium and Titanium (ibid, vol. 19, No. 6, 1965); co-author, Vaporization Studies of Uranium Dioxide and Carbides (ibid,vol. 22, No. 6, 1967).
Address: Academy of Sciences of the U.S.S.R., 14 Leninsky Prospekt, Moscow V-71, U.S.S.R.

BYKOV, Valerii Nikolayevich. Eng., Nucl. Power Station, Inst. Phys. and Power. Member, Russian deleg. visiting Britain July 18-28, 1961,

to discuss solid state physics research.
Address: Obninsk Physico-Technical Institute, Obninsk, near Maloyaroslavets, Moscow, U.S.S.R.

BYKOV, Vladimir. At. Inst. of Phys. Problems, Moscow.
Address: Institute of Physical Problems, Academy of Sciences of the U.S.S.R., 2 Vorobevskoye Chaussée, Moscow, U.S.S.R.

BYKOVSKII, N. N. Paper: Co-author, Determination of Uranium and Plutonium in Organic Solutions (letter to the Editor, Atomnaya Energiya, vol. 24, No. 1, 1968).
Address: Academy of Sciences of the U.S.S.R., 14 Leninsky Prospekt, Moscow V-71, U.S.S.R.

BYKOVSKII, Ya. A. Paper: Co-author, The Twenty Fifth Anniversary of the M.I.F.I. (Atomnaya Energiya, vol. 23, No. 5, 1967).
Address: Academy of Sciences of the U.S.S.R., 14 Leninsky Prospekt, Moscow V-71, U.S.S.R.

BYLUND, Don Marvin, B.Sc. (Chem. Eng.). Born 1932. Educ.: Nebraska Univ. Sen. Eng., Thermal Analysis Sect., Atomic Energy Div., Babcock and Wilcox Co., Lynchburg, Virginia. Society: A.N.S.
Nuclear interests: Thermal and hydraulic research on nuclear reactor systems and components.
Address: 505 Hayes Drive, Lynchburg, Virginia, U.S.A.

BYRNE, James, B.Sc., M.Sc., Ph.D. Born 1933. Educ.: Nat. Univ. of Ireland, Michigan and Edinburgh Univs. Asst. Lecturer, Nat. Univ. of Ireland, 1956-57; Teach. Fellow and Res. Asst., Michigan Univ., 1957-58; Imperial Chemical Industries Res. Fellow (1958-61), Phys. Lecturer (1961-65), Edinburgh Univ.; Phys. Lecturer, Sussex Univ., 1965-.
Nuclear interests: Electromagnetic and weak decay processes in nuclei and effects associated with the interaction of nucleus and atomic electrons; Determination of electron and gamma-ray spin polarization; Radioactive decay of neutrons.
Address: Mathematical and Physical Sciences School, Sussex University, Falmer, Brighton, Sussex, England.

BYRNE, John Patrick, C. Eng., F.I.Mech.E. Born 1916. Head, Production and Works Div. A.E.R.E., Harwell, Berks.; Formerly Chief Eng., U.K.A.E.A. Tadley, Hants.; Asst. Director, Tech. Facilities, Nat. Gas Turbine Establishment, Pyestock. Chairman, Southern Branch Com., I.Mech.E. Member of Council, I.Mech.E.
Nuclear interests: Design, procurement, manufacture and construction of research facilities with particular reference to plant, equipment and services.
Address: A.E.R.E., Harwell, Berks., England.

BYRNES, James J., B.S.Ch.E. Born 1922. Educ.: Columbia Univ. With Assoc. Nucleonics Inc.; Project Manager, U.S. Army Nat. Rad. Lab., 1961-63; Project Manager, External Beam Facility for Princeton-Pennsylvania Accelerator, 1962-. Societies: Instr. Soc. America; Inst. Food Technol.
Nuclear interests: Project management on the design and supervision of construction of nuclear facilities; Engineering studies on applications of radiation in processes and food preservation.
Address: Associated Nucleonics, Inc., 975 Stewart Avenue, Garden City, New York, U.S.A.

BYURGANOVSKAYA, G. V. Paper: Co-author, Dosimeters Based on Glasses with Optimal Density Alternating by Irradiation (Atomnaya Energiya, vol. 21, No. 1, 1966).
Address: Academy of Sciences of the U.S.S.R., 14 Leninsky Prospekt, Moscow V-71, U.S.S.R.

CABANE, Gérard, Ing. Dr. Born 1923. Educ.: Ecole Supérieure de Physique et Chimie Industrielles de la ville de Paris. Dept. de Métallurgie et Chimie Appliquée, C.E.A., 1949-. (Head of Lab. since 1953.).
Nuclear interests: Deformation and recrystallisation of pure metals and alloys for fuel elements and cans – e.g., U, Be, FeAl; Purification and decontamination by process metallurgy; Solid state diffusion of metals and gaseous fission products.
Address: 24 avenue Lombart, Fontenay-aux-Roses, Seine, France.

CABANIUS, Jean. Born 1911. Educ.: Ecole Polytechnique. Directeur de l'Equipement, Electricité de France. Commission Consultative pour la Production d'Electricité d'Origine Nucléaire.
Address: Electricité de France, Direction de l'Equipement, 3 rue de Messine, Paris 8, France.

CABANYES, Cayetano, Dr. (Architecture). Born 1911. Educ.: Madrid Univ. Pres., Caltecnica S.A.; Adviser, I.N.T.A. (Aerospacial Spanish Inst.).
Nuclear interest: Buildings.
Address: 32 Jose Antonio, Madrid 13, Spain.

CABLE, Joe Wood, A.B., Ph.D. Born 1931. Educ.: Murray State Coll. (Ky.) and Florida State Univ. Chem., O.R.N.L., 1955-. Society: A.P.S.
Nuclear interest: Neutron diffraction.
Address: Oak Ridge National Laboratory, Post Office Box X, Oak Ridge, Tenn., U.S.A.

CABRAL, Joao Manuel PEIXOTO. See PEIXOTO CABRAL, Joao Manuel.

CABRAL, Joaquim L. Rocha, Elec. Eng. (Genie Atomique). Born 1934. Educ.: Inst. Superior Tecnico, Lisbon and Inst. Nat. Sci. Tech. Nucléaires, Saclay.

Nuclear interest: Reactor design.
Address: 16-6-D Av. Joao XXI, Lisbon 1, Portugal.

CABRERA, Oscar J., Manager, Gerencia de Servicios Generales, and Head, Organisation and Methods Dept., Comision Nacional de la Energie Atomica, Argentina.
Address: 8250 Avda. del Libertador Gral San Martin, Buenos Aires, Argentina.

CABRERA FELIPE, José. Director, Centrales Nucleares S.A.
Address: Centrales Nucleares S.A., c/o 1 Hermosilla, Madrid 1, Spain.

CABRERA LA ROSA, Augusto, Ing. de Minas, Dr. en Ciencias Fisicas y Geológicas. Born 1889. Educ.: Univ. Mayor Nacional de San Marcos and Escuela de Ingenieros de Lima. Former Director, Inst. Nacional de Investigación y Fomento Mineros; Deleg. del Ministerio de Fomento ante la Junta de Control de la Energia Atómica, 1954. Societies: Miembro vitalicio de la Soc. de Ingenieros de Perú; Soc. del Inst. de Ingenieros de Minas del Perú; Soc. de la Soc. Geográfica de Lima; Soc. fundador de la Soc. Geológica del Perú.
Nuclear interests: Geology and mining of radioactive minerals.
Address: 145 Pasaje Tello, Miraflores, Lima, Perú.

CABRERA LORENTE, Jose Ignacio, Director, Tecnatom S.A.; Director, Cenusa; Society: Atomic Ind. Forum.
Address: 1 Barquillo, Madrid, Spain.

CABRERO, Vinicio SERMENT-. See SERMENT-CABRERO, Vinicio.

CACCIA-DOMINIONI, Fabrizio, Dr. in Law (Internat. Public Law). Born 1934. Educ.: Milan Univ. Div. Internat. Relations, C.N.R.N., 1958-60; Head, Euratom Office of C.N.E.N., 1961-. Member, Advisory Com. on Nucl. Res. Euratom; Member, Group for the Atomic Affairs Euratom; Member, Budgetary and Financial Com., Euratom.
Nuclear interests: Political, economical and juridical aspects of international collaboration in nuclear field.
Address: C.N.E.N., 15 Via Belisario, Rome, Italy.

CACCIAPUOTI, Nestore Bernardo, Dr. in Phys. Born 1913. Educ.: Pisa and Rome Univs. Prof. Higher Phys. (1949-62), Dean, Faculty of Sci. (1950-51), Trieste Univ.; Deputy Director, Dept. of Natural Sci., U.N.E.S.C.O. Paris, 1951-58; Formerly Member, Sci. and Tech. Com., Euratom; Director, Inst. of Phys., and Dean, Faculty of Sci., Pisa Univ. Papers: 40 papers on nuclear physics, cosmic radiation, thermal diffusion. Societies: Soc. Italiana di Fisica; A.P.S.
Address: Istituto di Fisica dell'Universita, 2 Piazza Torricelli, Pisa, Italy.

CACCIARI, Alberto, Eng. in Ind. Chem. Born 1923. Educ.: Milan Polytech. High School. Prof. Chem. Metal. of Uranium and Thorium, Finishing School of Appl. Nucl. Phys., (1951), Teacher of Technol. of Nucl. Fuels, (1957), Univ. Asst. of Chem. Ind. Eng., Inst. of Chem. Industrial Plants, (1957), Milan Polytech. qualified teacher of Chem. Metal., 1958; At present Special Asst. for Nucl. Fuel Technol., C.N.E.N.; Formerly Director, C.N.E.N. Activities at Saluggia; Tech. Director, Italatom S.p.A., Saluggia (Vercelli), Italy.
Nuclear interests: Metallurgy and metallurgical chemistry.
Address: Comitato Nazionale per l'Energia Nucleare, 15 Via Belisario, Rome, Italy.

CACERES, Augusto. Fellow, Peruvian Soc. of Radiol.
Address: Sociedad Peruana de Radiologia, Casilla No. 2306, Lima, Peru.

CACERES, Eduardo, Dr.med. Born 1915. Educ.: San Marcos Univ. Director, Inst. Nacional de Enfermedades Neoplasicas, 1952-. Societies: American Radium Soc.; Fellow, Peruvian Soc. of Radiol.
Address: 825 Avenida Alfonso Ugarte, Lima, Peru.

CACHET, D. LION-. See LION-CACHET, D.

CACHO, Carlos Ferreira Madeira, Licence in Phys. and Chem. Sci. Born 1919. Educ.: Lisbon Univ. Asst. Phys., Faculty of Sci., Lisbon Univ.; Director-Gen., Laboratorio de Fisica e Engenharia Nucleares, Portuguese A.E.C. (Junta de Energia Nucl.). Portuguese Rep., Group on Cooperation in fields of Reactors; Portuguese Rep., E.N.E.A. Steering Com.; Portuguese Rep., 1st I.C.P.U.A.E.; Portuguese Deleg., 2nd I.C.P.U.A.E.; Portuguese Deleg., several I.A.E.A. Gen. Conferences; Vice-Pres., European Atomic Energy Soc., 1968-.
Nuclear interest: Nuclear physics.
Address: Laboratório de Fisica e Engenharia Nucleares, Sacavém, Portugal.

CADEAU, Jean. Eng. of Arts et Métiers, Sen. of P.R.A.I.E.N. Born 1914. Educ.: Arts et Métiers (Chalons-sur-Marne) and P.R.A.I.E.N. (Paris). Chief Eng. (in charge of co-ordination of nucl. activities), Bureau Veritas, Paris, 1961-. Society: Club of Arts et Métiers Seniors (Member, Nucl. Energy Com.).
Nuclear interests: Metallurgy and welded constructions (reactor vessels, containment building vessels, etc.).
Address: 31 rue Henri Rochefort, Paris 17, France.

CADOFF, Irving, Prof. Metal. Eng., School of Eng. and Sci. New York Univ.
Address: New York University, School of Engineering and Science, University Heights, New York, New York 10453, U.S.A.

CADWELL, Howard J. Vice-Pres., Yankee Atomic Elec. Co. Chairman of the Board and Chief Exec. Officer, Western Massachusetts Elec. Co.
Address: Yankee Atomic Electric Co., 441 Stuart Street, Boston 16, Massachusetts, U.S.A.

CADWELL, J. J. Member, Editorial Advisory Board, J. Nucl. Materials.
Address: c/o Journal of Nuclear Materials, North-Holland Publishing Co., P.O. Box 103, Amsterdam, Netherlands.

CADY, K. Bingham, S.B., Ph.D. Born 1936. Educ.: Wisconsin Univ. and M.I.T. Assoc. Prof., Eng. Phys., Cornell Univ., 1962-; Consultant, K.A.P.L., 1966-. Society: A.N.S.
Nuclear interests: Experimental and theoretical reactor physics; Nuclear engineering and reactor design.
Address: Nuclear Reactor Laboratory, Cornell University, Ithaca, New York 14850, U.S.A.

CAEMMERER, Ernst VON. See VON CAEMMERER, Ernst.

CAGLE, Charles D., B.S. Born 1920. Educ.: Tennessee Polytech. Inst. and O.R.S.O.R.T. Tech. Supervisor, Tennessee Eastman Corp., Y-12 Area, Oak Ridge, 1943-47; Supervisor (1947-50), Dept. Supervisor (1950-57), Development Eng. (1957-59), Dept. Superintendent (1959-), Operations Div., O.R.N.L.
Nuclear interests: Reactor design; Nuclear physics; Reactor and experiment safety; Radiation control; Personnel training; Management.
Address: Oak Ridge National Laboratory, Post Office Box X, Oak Ridge, Tenn., U.S.A.

CAGLIOTI, Giuseppe, Dr. in phys. Born 1931. Educ.: Rome Univ. Res. Asst., Rome Univ., 1953-56; Phys., Comitato Nazionale per l'Energia Nucleare, Rome, 1955-. Resident res. assoc., Phys. Div., Argonne Nat. Lab., 1956-57; Director, C.N.E.N. Activities at Ispra, 1964-. Society: Società Italiana di Fisica.
Nuclear interest: Spectrometry of low energy neutrons.
Address: Gruppo Diffrazione e Spettroscopia dei Neutroni del C.N.E.N., Ispra (Varese), Italy.

CAGNO, Vitantonio DI. See DI CAGNO, Vitantonio.

CAHEN, Gilbert. Born 1904. Educ.: Ecole Polytechnique, Ecole Nationale Supérieure de Génie Maritime. Ing. Gén. du Génie Maritime (Ret.); Directeur, Constructions et Armes Navales de Cherbourg, 1953-55; Directeur Scientifique, Compagnie des Compteurs à Montrouge; Directeur des Etudes au Centre de Préparation aux Applications Industrielles de l'Energie Nucléaire. Book: Co-author, Précis d'Energie Nucléaire, 3rd edition (Dunod, 1963). Societies: Assoc. Technique Maritime et Aéronautique; Sté. Française des Electriciens; Sté. Française des Electroniciens et Radioelectriciens; Assoc. Française de Régulation et d'Automatisme; British Nucl. Energy Soc.; Sen. Member, I.E.E.E.; Sen. Member, Instrument Soc. of America.
Nuclear interests: Control and equipment.
Address: 12 avenue Franklin Roosevelt, Paris 8, France.

CAHILL, William Joseph, Jr., Bachelor of Mech. Eng. Born 1923. Educ.: Brooklyn Polytech. Inst. Asst. Vice Pres., Consolidated Edison Co. of New York Inc., 1968-. Member, Tech. Com.; Empire State Atomic Development Assocs. Inc., Member, Reactor Safety Com., Atomic Ind. Forum; Member, Atomic Power Subcom. of Prime Movers, Edison Elec. Inst. Societies: A.S.M.E.; A.N.S.
Address: 4 Irving Place, New York, New York 10003, U.S.A.

CAHN, Robert Wolfgang, B.A., Ph.D. Born 1924. Educ.: Cambridge Univ. A.E.R.E., Harwell, -1951; With Phys. Metal. Dept. (1951-62), Reader (1958-), Birmingham Univ. Prof. Materials Technol., Wales Univ., 1962-64; Prof. Materials Sci., Sussex Univ., 1965-.Coeditor, J. Nucl. Materials and J. Materials Sci. Societies: Inst. Metals; Inst. Phys.; A.I.M.M.E.
Nuclear interest: Teaching of nuclear metallurgy.
Address: School of Applied Sciences, Sussex University, Brighton, England.

CAIANIELLO, Eduardo R., Dott. in Fisica (Naples), Ph.D. (Rochester). Born 1921. Educ.: Naples Univ. Prof. Theoretical Phys., Naples Univ. 1956-; Director, Inst. Theoretical Phys., and Director, Advanced School Theoretical and Nucl. Phys., Naples Univ.; previously various appointments in Italy, U.S.A., Denmark. Societies: Soc. Italiana di Fisica; A.P.S.; Accademia Pontaniana; New York Acad. of Sci.
Nuclear interests: Theoretical nuclear physics; Accelerator theory.
Address: Istituto di Fisica Theorica, Mostra d'Oltremare, Naples, Italy.

CAILLAT, Roger, Ing., D. ès Sc. Physiques (Paris). Born 1921. Educ.: Sorbonne. Chef de Sect., uranium et réfractaires (1948-55), Chief du Service de Chimie des Solides (1956-), Adjoint au Chef du Dept. de Métallurgie (1960-), C.E.A.; Prof., Inst. Nat. des Sci. et Techniques Nucl., 1962-. Book: Uranium et composés, nouveau traité de chimie minérale de Pascal. Societies: American Ceramic Soc.; Sté. Chimique de France; Sté. de Chimie Physique; Faraday Soc.; Sté. Française de Métallurgie; Inst. of Metals.
Nuclear interests: Nuclear metallurgy; Chemistry of solids.
Address: DM/CS, Centre d'Etudes Nucléaires de Saclay, B.P. No. 2, Gif-sur-Yvette 91, France.

CAIN, Francis M., Jr., B.S. Born 1919. Educ.: Waynesburg Coll. Fellow Eng. i/c Metallurgical Services, U.S.A.E.C. Bettis Plant, Westinghouse Elec. Corp., 1949-57; Manager, Metal. and Plant Services, Nuclear Materials and Equipment Corp., Apollo, Pa., 1957-. Societies: A.S.M. (served on Metallography Com. for A.S.M. Metals Handbook Supplement, 1954); A.S.T.M. (consulting member on E-4 Com. on Metallography).
Nuclear interests: Metallurgy; Ceramics; Electronics; Electron and light microscopy.
Address: Nuclear Materials and Equipment Corporation, P.O. Box 436, Apollo, Pennsylvania 15613, U.S.A.

CAIRNIE, Alan B., B.Sc., Ph.D. Asst. Res. Biophys., Lab. of Radiobiol., California Univ. Medical Centre.
Address: California University Medical Centre, Radiobiology Laboratory, San Francisco, California 94122, U.S.A.

CAIRNS, Robert Charles Phillip, A.S.T.C., B.Sc. (Chem. Eng.), Ph.D. Born 1927. Educ.: New South Wales Univ. Head, Chem. Eng. Sect., (1961), Acting Assoc. Director, Operations (1967), A.A.E.C. Societies: Inst. of Chem. Eng.; Inst. of Eng., Australia.
Nuclear interests: Fuel cycles and chemical reprocessing; Management.
Address: Australian Atomic Energy Commission, Research Establishment, Private Mail Bag, Sutherland, New South Wales 2232, Australia.

CAKENBERGHE, J. VAN See VAN CAKENBERGHE, J.

CALABRETTA, Peter Trent, B.M.E. Born 1918. Educ.: C.C.N.Y. and Brooklyn Polytech. Inst. Design Eng. (1942-50), Project Eng. (1950-55), Dept. Manager (1956-), AMF Atomics. Society: A.N.S.
Nuclear interests: Management of engineering services charged with the design of reactors, hot cells, manipulators, refuelling machinery, liquid metal loops, and special environmental machinery.
Address: 711 E. Hillcrest Road, York, Pa., U.S.A.

CALABRIA, Gerolamo, Dottore in Ingegneria. Born 1908. Educ.: Turin Politecnico. Manager, S.T.E.I.; Vice Pres. S.I.M.E.A.; Manager, Div. Agip Nucleare SNAM. Societies: Assoc. Termotecnica Italiana; A.E.I.; F.I.E.N.
Nuclear interests: Management, plant and reactor design and construction.
Address: c/o E.N.E.L., 3 Via G.B. Martini, Rome 00198, Italy.

CALAME, Gerald Paul, B.A. (Wooster), M.A., Ph.D. (Harvard). Born 1930. Educ.: Wooster Coll. and Harvard Univ. Theoretical Phys., K.A.P.L., 1959-61; Asst. Prof. Nucl. Eng. and Sci. (1961-63), assoc. Prof. (1963-66), Rensselaer Polytech. Inst.; Assoc. Prof., U.S.

Naval Acad., 1966-. Societies: A.N.S.; A.P.S.; A.A.P.T.
Nuclear interests: Application of neutron transport theory to reactor design problems; Nuclear structure physics.
Address: Department of Physics, U.S. Naval Academy, Annapolis, Md., U.S.A.

CALAMUR, Calamur Mahadevan, M.A., D.Sc., F.A.Sc., M.A.I.E., F.N.I., D.Sc. (Madras). Born 1901. Educ.: Madras Univ. Prof. and Head, Dept. of Geology, 1945-; Principal, Andhra Univ. Colleges, 1957-; Member of Council, Indian Acad. of Sci.; Nat. Inst. of Sci. of India; Geological Soc. of India; Board of Res. in Nucl. Sci. Societies: Indian Acad. of Sci.; Nat. Inst. of Sci. of India; Geological, Mining and Metal. Soc. of India; A.I.M.M.E.
Nuclear interests: Nuclear raw materials; distribution pattern of uranium and thorium in terrestrial materials; Prospecting for and processing of uranium and thorium ores.
Address: Andhra University College, Waltair, India.

CALAN, Pierre de. See de CALAN, Pierre.

CALAY, F., M.D. Thyroid Res. Unit, Lab. des Isotopes Radioactifs, Clinique Médicale, Hôpital St. Pierre.
Address: Laboratoire des Isotopes Radioactifs, Clinique Médicale, Hôpital St. Pierre, 115 boulevard de Waterloo, Brussels, Belgium.

CALCAGNO, Giorgio, Dr. Chem. Born 1902. Educ.: Padua Univ. Adviser to Central Office for Internat. Railway Transport.
Nuclear interest: Transportation of nuclear materials.
Address: 30 Gryphenhübeliweg, Berne, Switzerland.

CALDAS, Luiz Renato Carneiro da Silva, M.D., Assoc. Prof. Born 1929. Educ.: Rio de Janeiro Federal Univ. Rio de Janeiro Federal Univ., 1948-; Hospital dos Servidores do Estado, 1953-; U.N. Rad. Com., New York, 1960-61, 1968-. Brazilian Nucl. Energy Com., 1964-65. Co-opted Member, Comité Internat. de Photobiologie; Sec., Brazilian Soc. of Biophys.; Member, Molecular Biol. Com., Brazilian Soc. of Genetics. Books: Co-editor, Mammalian Cytogenetics and Specific Topics in Radiobiology (Pergamon Press, 1964).
Nuclear interests: Radiation biology at the molecular and cellular level.
Address: Instituto de Biofisica, 458 Avenida Pasteur, Rio de Janeiro, Guanabara, Brazil.

CALDECOTT, Richard S. Member, Subcom. on Radiobiol., N.A.S. Nat. Res. Council; Formerly Assoc. Editor, Rad. Res.; Delegate, I.A.E.A. Symposium on the Effects of Ionising Rad. on Seeds and their Significance for Crop Improvement, Karlsruhe, 8-12 August, 1960.
Address: Minnesota University, Minneapolis 14, Minnesota, U.S.A.

CALDERON, Derick, Ingeniere Electricista; Member, Inst. Nacional de Energia Nuclear, Guatemala.
Address: Institute Nacional de Energia Nuclear, Apartade Postal 1421, Guatemala, C.A.

CALDIROLA, Piero Carlo Augusto, Dr. Phys. Born 1914. Educ.: Pavia Univ. Fellow, Theoretical Physics (1941-47), Extraord. Prof. Theoretical Phys., (1947-50), Pavia Univ.; Ord. Prof. Gen. Phys., Milan Univ., 1950-. Director, Sezione di Milan, I.N.F.N., 1951-60; Director, Istituto di Sienze Fisica, Milan Univ., 1961-. Sci. Director, Camen (S. Piero a Grado Pisa); Consiglio Nazionale delle Ricerche (Rome); Pres., Lombard Phys. Soc.; Pres., Italian Radiol. Protection Soc.; Member of Directory, F.I.E.N.; Member, Sci. and Tech. Com., Euratom, 1968-. Societies: Accademia Nazionale des Lincei; A.P.S.
Nuclear interests: Reactor physics and isotopic separation.
Address: 33 Via Nino Bixio, Milan, Italy.

CALDWELL, Colin S. Formerly Project Eng., now Manager, Plutonium Process Chem., Nucl. Materials and Equipment Corp.
Address: Nuclear Materials and Equipment Corporation, Apollo, Pennsylvania 15613, U.S.A.

CALDWELL, Frank W., Member, Connecticut Advisory Com. on Atomic Energy.
Address: 54 Pilgrim Road, West Hartford, Connecticut, U.S.A.

CALDWELL, Harmon, Dr. Education Com., State of Georgia Nucl. Advisory Commission.
Address: State of Georgia Nuclear Advisory Commission, Office of Publications, Georgia Institute of Technology, Atlanta 13, Georgia, U.S.A.

CALDWELL, Richard L., Ph.D., M.S., B.S. Born 1917. Educ.: Missouri and Louisiana State Univs. Sen. Res. Phys. (1949-56), Res. Assoc. (1956-58), Sect. Supv., Nucl. Phys. and Well Logging (1958-), Mobil R. and D. Corp. Chairman, Special Weapons Sect., Dallas County Civil Defence and Disaster Commission. Societies: A.P.S.; A.N.S.; RESA.
Nuclear interests: Neutron physics; Applied nuclear geophysics.
Address: Field Research Laboratory, Mobil Research and Development Corporation, P.O. Box 900, Dallas, Texas 75221, U.S.A.

CALDWELL, Robert Tate. Born 1882. Educ.: Centre Coll. and Centre Coll. Law School. Private law practice, Ashland, Kentucky, 1919-. Member, Kentucky Advisory Com. on Nucl. Energy; Chairman, K.A.C.N.E. Subcom. on Workmen's Compensation.
Address: Caldwell and Robinson, 918 Second National Bank Building, Ashland, Kentucky, U.S.A.

CALDWELL, Roger Dale, B.S., B.A., M.S. Born 1933. Educ.: Ottenbein Coll., and Kansas Univ. Health Phys. Brookhaven Nat. Lab., 1960-63; Manager, Health and Safety Dept. Nucl. Materials and Equipment Corp., 1963-. Sec., Western Pennsylvania Chapter, H.P.S.; Advisory Com., Transportation of Radioactive Materials, Commonwealth of Pennsylvania. Societies: Conference on Bioassay and Analytical Chem.; American Ind. Hygiene Soc.
Nuclear interests: Radioaerosol measurement; Evaluation of internal radioactivity; Internal dosimetry; Gamma and neutron dosimetry; Environmental monitoring for radioactivity; Criticality control; Nuclear fuel health physics.
Address: Nuclear Materials and Equipment Corporation, Apollo, Pennsylvania 15613, U.S.A.

CALEEL, George Thomas, B.S., F.A.C.O.I., Dr. of Osteopathy. Born 1929. Educ.: Wayne Univ., Chicago Coll. of Osteopathy. Director, Nucl. Medicine, Chicago Coll. of Osteopathy, and Chicago Osteopathic Hospital; Prof. of Medicine, Chicago Coll. of Osteopathy; Medical Consultant, Director of Clinics, Chicago Osteopathic Hospital. Diplomate Fellow, American Coll. of Osteopathic Internists. Societies: American Osteopathic Assoc.; N.Y. Acad. Sci.; Soc. of Nucl. Medicine.
Nuclear interests: Medical application, diagnostic and therapeutic.
Address: 5200 S. Ellis, Chicago 15, Illinois, U.S.A.

CALEMBERT, L. Ing., Inter-Faculty Centre of Nucl. Sci., Liège Univ.
Address: Inter-Faculty Centre of Nuclear Sciences, Liège University, Liège, Belgium.

CALETKA, R. Paper: Co-author, Chem. Properties of Element 104 (Atomnaya Energiya, vol. 21, No. 2, 1966).
Address: Academy of Sciences of the U.S.S.R., 14 Leninsky Prospekt, Moscow V-71, U.S.S.R.

CALKINS, G. D., Member, Special Administrative Com. on Nucl. Problems, A.S.T.M.
Address: Atomics International Division, North American Aviation Inc., Box 309, Canoga Park, California, U.S.A.

CALKINS, Vincent Paul, B.S. (Chem. and Philosophy), M.S. (Gen. Chem.), Ph.D. (Phys. Chem. and Biochem.). Born 1916. Educ.: St. Ambrose Coll. and Iowa Univ. Head, Chem. Dept., N.E.P.A. Project, Oak Ridge, Tennessee, 1947-51; Supervisor, Materials Res. and Rad. Studies (1951-56), Manager, Appl. Materials Res. (1956-61), Gen. Elec. - Aircraft Nucl. Propulsion Dept., Evendale, Ohio; Exec. Eng. and Sci., Gen. Elec. Nucl. Materials and Propulsion Operation, Evendale, Ohio, 1961-. Com. Chairman, A.S.T.M. Books: Chapter on Metal. Principles in Powder Metal. (A.S.M. 1958); sect. on Rad. Damage to Liquids and Organic Materials in Nucl. Eng. Handbook (New York, McGraw-Hill, 1958); editor, sect.

on Shielding Materials in U.S.A.E.C. Reactor Handbook, 3rd edition, vol. I, Materials (New York and London, Intersci. Publishers, 1968). Socieites: A.C.S.; A.A.A.S.; A.N.S.; New York Acad. of Sci.
Nuclear interests: Engineering; Metallurgy; Chemistry; Ceramics; Management. Selection of materials and components for nuclear power plants; Novel means of utilising nuclear energy; Basic phenomena of nuclear physics; Oceanography; Nuclear safety; Nuclear standards; Utilising nuclear energy for peace and determining its role in resolving worldwide problems: mass starvation, pollution, desalination, massive irrigation.
Address: 570 Reilly Road, Wyoming, Ohio 45215, U.S.A.

CALLAHAN, J. Cal, A.B., B.S. (C.E.). Born 1910. Educ.: Xavier and Michigan Univs. Director, City Planning Div., Morris Knowles Inc.; Consultant to Governor of West Virginia; Formerly Chairman, City Planning Div., A.S.C.E.; Chairman, Com. A.S.C.E. Land Use Aspects of Nucl. Uses; A.S.C.E. Rep. Bureau of Standards Com. on Site Selection for Nucl. Uses; Sec. Internat. Com. on L'urbanisme Souterrain. Societies: A.S.C.E.; American Inst. Planners; A.N.S.; Soc. Professional Eng.
Nuclear interest: Planning for land use aspects of nuclear uses.
Address: 115 Parker Avenue, Easton, Pennsylvania, U.S.A.

CALLAHAN, M. Sec., Pacific Inst. of Advanced Studies.
Address: Pacific Institute of Advanced Studies, 448 N. Avenue 56, Los Angeles 42, California, U.S.A.

CALLAWAY, Howard. Industry Com., State of Georgia Nucl. Advisory Commission.
Address: State of Georgia Nuclear Advisory Commission, Office of Publications, Georgia Institute of Technology, Atlanta 13, Georgia, U.S.A.

CALLEJA y GONZALEZ-CAMINO, J. Vocal, Consejo, Junta de Energia Nucl., Ministerio de Industria.
Address: Departamento de Electricidad, Empresa Auxiliar de la Industria (Auxini), 8 Plaza de Salamanca, Madrid, Spain.

CALLEN, Herbert B., B.S., M.A. (Temple), Ph.D. (M.I.T.). Born 1919. Prof. Phys., Pennsylvania Univ., 1948-. Book: Thermodynamics (John Wiley, 1960). Society: Fellow, A.P.S.
Nuclear interests: Theoretical atomic physics; Statistical mechanics.
Address: Department of Physics, University of Pennsylvania, Philadelphia,Pa. 19104, U.S.A.

CALLENS, Robert. Member, Conseil d'Administration, Centre et Sud.
Address: Centre et Sud, 20 avenue de la Toison, Brussels 5, Belgium.

CALLERI, Giacomo. Italian Alternate, Board of Directors, Eurochemic, Mol. Adviser, Third I.C.P.U.A.E., Geneva, Sept. 1964.
Address: Comitato Nazionale per l'Energia Nucleare, 15 Via Belisario, Rome, Italy.

CALLIHAN, Dixon, A.B., M.A., Ph.D. Born 1908. Educ.: Marshall Coll., Duke Univ. and N.Y.U. Phys., Union Carbide Corp., 1945-. Editor, Nucl. Sci. and Eng., 1965-; Member, Panel to Provide Boards on Atomic Safety and Licensing, U.S.A.E.C.; Chairman, U.S.A. Standards Com. N-16, Nuclear Criticality Safety. Book: Progress in Nucl. Energy, Series 1, Phys. and Maths., Chapter 8, Homogeneous Critical Assemblies. Societies: Fellow, A.P.S.; Fellow, A.N.S.
Nuclear interests: Critical experiments: Nuclear safety.
Address: Oak Ridge National Laboratory, Box Y, Oak Ridge, Tennessee 37830, U.S.A.

CALLOW, J. H., B.Sc. (Eng.), A.M.I.E.E. Council Member, Inst. Nucl. Eng.
Address: Institution of Nuclear Engineers, 147 Victoria Street, Westminster, London, S.W.1, England.

CALMET, Jorge SARMIENTO. See SARMIENTO CALMET, Jorge.

CALNAN, Edward Arthur, B.Sc., Ph.D. Born 1923. Educ.: London Univ. At Nat. Phys. Lab., 1946-55; Fulmer Res. Inst., 1955-59; Director of Res., Gillette Industries Ltd., 1960-. Societies: Inst. Phys. and Phys. Soc.; Inst. Metals; Roy. Inst. of Great Britain.
Nuclear interests: Metallurgy; Irradiation of materials.
Address: Gillette Research Laboratory, 454 Basingstoke Road, Reading, Berkshire, England.

CALORI, Arturo, Elec. Eng. Degree; Nucl. Eng. (after graduation). Born 1932. Educ.: Rome Univ. Siemens Schuckertwerke A.G., Erlangen, Germany, 1956; C.N.E.N., Rome, 1956-. (Present position: Deputy Director of Safety and Control Div.). Member, Comitato Nazionale, and Sec. Gen., Assoc. Nazionale di Ingegneria Nucl.. Societies: Assoc. Nazionale Ingegneri Nucl.; Assoc. Elettrotecnica Italiana.
Nuclear interests: Reactor design, with particular reference to heat transfer and fluid mechanics problems; Safety problems specially connected with reactor design and operation.
Address: 15 Via M. Quadrio, Rome, Italy.

CALVETE, Alfredo DELGADO. See DELGADO CALVETE, Alfredo.

CALVI, G., Gruppo Van de Graaff, Centro Siciliano de Fisica Nucleare.
Address: Centro Siciliano di Fisica Nucleare, Instituto di Fisica, Universita de Catania, 57 Corso Italia, Catania, Italy.

CALVIN, Melvin, Ph.D. (Minnesota), Hon. D.Sc. (Nottingham), Hon. D.Sc. (Oxford), D.Sc. (Northwestern Univ.). Nobel Prize for Chem., 1961. Born 1911. Educ.: Michigan Tech. and Minnesota Univs. Prof. Chem., Dept. of Chem., California Univ., Berkeley; Director, Bio-Organic Chem. Group, Lawrence Rad. Lab., California Univ., Berkeley. Books: Co-author, Isotopic Carbon (John Wiley, 1948); co-author, Path of Carbon in Photosynthesis (Prentice-Hall, 1957). Societies: Royal Soc. (foreign member); N.A.S. (U.S.); Royal Netherlands Acad. of Sci.; Royal Soc. of Edinburgh (foreign member); Royal Irish Acad. of Sci.; British Chem. Soc.; A.C.S.
Nuclear interests: Photosynthesis; Use of isotopes in biology and biochemistry; Chemical evolution; Molecular biology and biophysics.
Address: California University, Lawrence Radiation Laboratory, Berkeley 4, California, U.S.A.

CALVO, H. E. Alberto Ullastres. Head, Spanish Mission Accredited to the European Atomic Energy Community.
Address: 25 rue de la Loi, Brussels, Belgium.

CAMARA, Santiago NORENA de la. See NORENA de la CAMARA, Santiago.

CAMBEL, Miss Perihan, M.D. Born 1915. Educ.: Istanbul Univ. Pathologist, Gureba Hospital, Ankara Nümune Hospital;Assoc. Prof. Cancer Res., Florida Univ., 1949-52. Book: Ojenik hakkinda (Eugenics) (1936). Societies: American Cancer Res. Assoc.; H.P.S.; Founder, Turkish Assoc. for Cancer Res. and Control.
Address: 1/6 Bilge Sokak, Bakanliklar, Ankara, Turkey.

CAMBORNAC, Michel Jean, Ing. (Ecole Centrale des Arts et Manufactures). Born 1923. Office Nat. d'Etudes et Recherches Aeronautiques, 1946-51; Sté. Servomecanismes Electroniques, 1951-58; C.G.E.I. Lepaute 1958-64; Grt. Automatisme de la C.I.T.E.C. 1964-. Societies: Sté. Française des Electriciens; Sté. Française des Électroniciens et Radio Electriciens; Assoc. Française de Regulation et d'Automatisme.
Nuclear interests: Servomécanismes appliqués aux techniques nucléaires (régulations de vitesse (laminoirs) de position (barres de réglages de reacteurs, chargement de reacteurs, d'intensité) alimentation d'électro aimants).
Address: 7 Rue Livingstone, Paris 18, France.

CAMBOU, Francis. D. ès Sc. Phys. Born 1930. Educ.: Paris Univ. Prof. Physique Spatiale, Faculté des Sci. de Toulouse. Societies: French Phys. Soc.; A.G.U.; Sté. Française des Electroniciens et Radioélectriciens.
Nuclear interests: Radiation in space and cosmic rays.
Address: Centre d'Etude Spatiale des Rayonnements, Faculté des Sciences de Toulouse, Toulouse, France.

CAMELIN. Director, Sous Direction des Essais, Direction des Applications Militaires, C.E.A.
Address: Commissariat a l'Energie Atomique, 29-33 rue de la Federation, Paris 15, France.

CAMENSON, Charles E., Ph.G. (Pharmacy Graduate). Born 1909. Educ.: California Univ., San Francisco. Res. Dept. (New Product Development), Microchem. Specialties Co.
Address: 2305 Stone Valley Road, Alamo, California 94507, U.S.A.

CAMERON, Angus Ewan, B.A., Ph.D. Born 1906. Educ.: Oberlin Coll. and Minnesota Univ. Asst. Director, Analytical Chem. Div., O.R.N.L. Book: Isotopic Composition of Uranium (National Nucl. Energy Series, 1-13, Tech. Information Service, U.S.A.E.C.). Society: A.S.T.M., Com. E-14 (Mass Spectrometry).
Nuclear interests: Mass spectrometry, isotope separation, variation of isotopic composition in nature, laboratory instrumental analysis.
Address: 114 W. Malta Road, Oak Ridge, Tennessee, U.S.A.

CAMERON, James, B.Sc. (Eng.). Born 1925. Educ.: Aberdeen Univ. Personal Asst. to Chairman and Managing Director (1951-53), Asst. to Chief Commercial Manager (1953-55), Commercial Officer, Atomic Power Div. (1955-63), Chief Commercial Manager, (1963-), English Elec. Co. Ltd.
Nuclear interests: Commercial applications of power reactors.
Address: 95 Bilton Road, Rugby, England.

CAMERON, John Alexander, B.A., Ph.D. Born 1936. Educ.: Toronto and McMaster Univs. Asst. Prof. of Phys., McMaster Univ. Society: Canadian Assoc. of Phys.
Nuclear interests: Hyperfine interactions in free atoms and solids. Atomic beam magnetic resonances. Perturbed angular correlations using internal fields in ferromagnetic metals.
Address: Department of Physics, McMaster University, Hamilton, Ontario, Canada.

CAMERON, John R., B.S. (Maths.), M.S., Ph.D. (Phys.). Born 1922. Educ.: Superior State Coll., Chicago and Wisconsin Univs. Asst. Prof. of Phys., Pittsburgh Univ., 1955-58; Assoc. Prof. of Radiol. and Phys., (1958-), Chairman, Rad. Safety Com., Wisconsin Univ. Board of Trustees, Soc. of Nucl. Medicine. Societies: A.P.S.; American Assoc. of Phys. in Medicine; A.A.A.S.; Illinois Medical Phys. Soc.; Biophys. Soc.; Rad. Res. Soc.
Nuclear interests: Medical applications of radioisotopes; Dosimetry of nuclear radiations.
Address: Radioisotope Service, University Hospitals, 1300 University Avenue, Madison, Wisconsin 53706, U.S.A.

CAMERON, Reid Anderson, B.Sc. (Mech. Eng.). Born 1921. Educ.: Illinois Inst. Technol. Eng., Argonne Nat. Lab., 1946-55; Sec.-Treas., Harper Eng. Co., 1955-57; Asst. Div. Manager (1957), now Manager Fuel Element

Manufacturing Div., Aerojet-General Nucleonics. Society: A.N.S.
Address: 428 Marion Lane, Danville, California, U.S.A.

CAMINO, J. CALLEJA y GONZALEZ-. See CALLEJA y GONZALEZ-CAMINO, J.

CAMONA, Gianangelo, Dr. in Ind. Chem., Prof. in Corrosion and Protection of Materials. Born 1936. Educ.: Univ. degli Studi, Milan. Researcher (1961-), Head, Metallographic Lab., (1967-), C.I.S.E. Voluntary Asst., Univ. Metallography Course. Society: Soc. of Electron Microscopy.
Nuclear interests: Metallographic studies on both irradiated and non irradiated constructive (steels, zirconium alloys, etc.) and nuclear (uranium, uranium dioxide, etc.) materials. Studies on the properties of the above materials also in function of thermal treatments and workings.
Address: c/o C.I.S.E., Casella Postale 3986, 20100 Milan, Italy.

CAMPANA, Robert Joseph, B.S. (Eng. Phys.). Born 1922. Educ.: Maine Univ. Eng., MSA Res. Corp., 1951; Nucl. Eng., Douglas Aircraft Co., 1956; Res. Staff Member, Gulf Gen. Atomics, 1959-. Vice Chairman, San Diego Sect., A.N.S. Societies: A.P.S.; A.I.A.A.
Nuclear interest: Design of systems for the generation of electric power from nuclear reactor and radioisotopic energy sources for various applications from medical research and prostheses to utility central stations at power ratings from microwatts to thousands of megawatts and operating in the environments of outer space, oceanography and the human body.
Address: 609 Ridgeline Place, Solana Beach, California 92075, U.S.A.

CAMPANINI, Mario, Dr. Eng. Born 1904. Educ.: Politecnico, Milan. Head, Tech. Dept., Capo del Servizio Tecnico S.N.A.M. S.p.A., 1950-55; Gen. Manager, Direttore Generale, Saipem, 1956-58; Gen. Manager, Direttore Generale Agip Nucleare, 1959-63; Amministratore, Delegato, S.N.A.M., 1964-65; Pres., Snam Progetti S.p.A., 1965-; Pres. Somiren S.p.A. (Società Minerale Radioattivi Energia Nucleare), 1966-; Consigliere E.N.I. - Ente Nazionale Idrocarburi. Membro Comitato. Consultivo Agenzia di Approvvigionamento dell'Euratom; Membro Giunta, F.I.E.N.
Nuclear interest: Management.
Address: c/o Snam Progetti S.p.A. Casella Postale 4169, Milan, Italy.

CAMPBELL, C. W. Vice-Pres., Administration, Sandia Lab.
Address: Sandia Laboratory, P.O. Box 5800, Albuquerque, New Mexico, U.S.A.

CAMPBELL, Charles C., Manager, Sandia (N. Mexico) Area (-1962), Manager, Los Alamos Area Office (1962-67), now Deputy Asst. Manager, Administration, Albuquerque Operations Office, U.S.A.E.C.
Address: Albuquerque Operations Office, U.S. Atomic Energy Commission, Post Office Box 5400, Albuquerque, New Mexico, U.S.A.

CAMPBELL, Clifford Graham, M.A., B.Sc. (Hons. Phys.), Ph.D. Born 1922. Educ.: Edinburgh Univ. Asst., Edinburgh Univ., 1950; I.C.I. Res. Fellow, 1952; Sci. Staff, Harwell (1953-59), Winfrith (1959), Head, Fast Water Reactor Phys. Div., Winfrith (including the Shielding Group, Harwell) (1962-), U.K.A.E.A. Society: Fellow, Inst. Phys.
Nuclear interest: Experimental and theoretical studies of reactor physics and shielding and their application to reactor design.
Address: Moor Lodge, Moor Road, Broadstone, Dorset, England.

CAMPBELL, Colin Blaine, A.R.S.M., B.Sc. (Mining Geology). Born 1919. Educ.: London Univ. Principal Geologist, Atomic Energy Div., Geological Survey of Great Britain, 1947-56; Mining Geologist, U.K.A.E.A., 1956-68. Society: I.M.M.
Nuclear interests: Raw materials of atomic energy; Prospecting methods.
Address: Contracts (Metals) Branch, United Kingdom Atomic Energy Authority, 11 Charles II Street, London, S.W.1, England.

CAMPBELL, Herbert, C. Eng. Born 1911. Eng. Clerk of Works, Ministry of Works, Capenhurst, 1950-54; Mech. Eng. (1954-57), Resident Eng., Springfields (1957-62), Resident Eng., Windscale and Calder Works (1962-66), Resident Eng., Capenhurst (1966-), U.K.A.E.A. Societies: Nucl. Eng. Soc.; A.M.I.Mech.E.
Nuclear interest: Construction.
Address: Woodlands, Vicarage Lane, Little Budworth, Cheshire, England.

CAMPBELL, Jack M. Formerly Governor of New Mexico; Partner, Law Firm, Stephenson, Campbell and Olmsted, Santa Fe, New Mexico. Member, Atomic Safety and Licensing Board Panel, U.S.A.E.C., 1967-.
Address: Stephenson, Campbell and Olsted, Santa Fe, New Mexico, U.S.A.

CAMPBELL, James H. Vice-Pres., Power Reactor Development Co.; Director, Atomic Power Development Associates, Inc.; Pres., Consumers Power Co. Director (1962-), Vice-Pres. (1965-), Member, Exec. Com., (1965-), Chairman, Finance and Budget Com. (1965-), Chairman, Reactor Regulation Com., Atomic Ind. Forum.
Address: Power Reactor Development Co., 1911 First Street, Detroit 26, Michigan, U.S.A.

CAMPBELL, James Philander, Jr., Bachelor of Sci. and Agriculture(Georgia). Born 1917. Educ.: George Washington, Virginia, and Georgia Univs. Elected 3 terms Georgia State Legislature from Oconee County, 1949-54; Elected Georgia State Commissioner of Agriculture,

1954, 1958, 1962 and 1966. Georgia Sci. and Technol. Commission.
Nuclear interests: Practical applications in agriculture; State laws pertaining to nuclear energy and its uses.
Address: Agriculture Building, 19 Hunter Street, S.W., Atlanta, Georgia, U.S.A.

CAMPBELL, John, Dipl. of Roy. Tech. Coll., Glasgow (Mech. Eng.). Assoc., of Manchester Coll. of Technol. (Chem. Eng.). Born 1919. Educ.: Roy. Tech. Coll., Glasgow and Manchester Coll. of Technol. Sen. Draughtsman, Design Eng. II, Design Eng. I, Asst. Chief Eng., Chem. Plant Design Office, Eng. Group, U.K.A.E.A., 1950-67. Society: Nucl. Eng. Soc.
Nuclear interests: Design and construction of irradiated fuel; Examination and reprocessing facilities and ex-reactor fuel handling and storage plants.
Address: 10 Beechmill Drive, Culcheth, Nr. Warrington, Lancs., England.

CAMPBELL, John Alexander, B.Sc., M.B., M.D. Born 1914. Educ.: Cincinnati Univ. Prof. and Chairman, Dept. of Radiol., Indiana Univ. 1st Vice Pres., Radiol. Soc. of North America. Societies: American Roentgen Ray Soc.; American Medical Soc.; Assoc. of Univ. Radiol.
Address: Indiana University, 1100 W. Michigan Street, Indianapolis, Indiana, U.S.A.

CAMPBELL, Jorge RICKARDS. See RICKARDS CAMPBELL, Jorge.

CAMPBELL, Keith George, M.Sc., B.Sc. (For.) (Sydney), Dip. For. (Canberra). Born 1922. Educ.: Sydney Univ. Silvicultural Res. Forester, 1951-53; Sen. Forester, Entomological Res., 1954-. Sec., Inst. Foresters Australia, N.S.W. Div.
Nuclear interests: Use of radioactive isotopes in tagging insects.
Address: Entomological Research Section, Division of Forest Management, Forestry Commission of N.S.W., 44 Margaret Street, Sydney, Australia.

CAMPBELL, Kenneth H., B.S. (Chem. Eng.). Born 1928. Educ.: California Univ. Berkeley. Supv., O.M.R.E. Operation, 1956-59; U.S. A.E.C.-A.E.C.L. Liaison, O.C.D.R. Programme, 1959-61; Organic Reactors Dept., Atomics Internat., 1961-62; Projects Manager, Whiteshell Reactor No. 1 (1962-63), Manager, Nucl. Plant Eng. (1963-67), Canadian G.E.C.; Projects Manager, N.U.S. Corp., 1967-. Society: A.N.S.
Nuclear interest: Nuclear business management.
Address: 1730 M. Street, Washington D.C. 20036, U.S.A.

CAMPBELL, Milton Hugh, B.S. (Chem.), M.S. (Nucl. Eng.). Born 1928. Educ.: Montana State Coll. and Washington Univ. Manager, Separations Chem. Lab., Atlantic Richfield Hanford Co., 1967-. Society: A.C.S.

Nuclear interest: Reactor fuel reprocessing. Of particular interest: solvent extraction separations for the transuranium elements as well as specific fission products; instrumental techniques for control or isotopic detection; radio and analytical chemistry; Remote handling techniques.
Address: 2119 Beech, Richland, Washington, U.S.A.

CAMPBELL, Hon. Phil. Agriculture, State of Georgia Nucl. Advisory Commission.
Address: State of Georgia Nuclear Advisory Commission, Office of Publications, Georgia Institute of Technology, Atlanta 13, Georgia, U.S.A.

CAMPBELL, Robert James, B.S. (M.E.). Born 1921. Educ.: Illinois Univ. Design Eng., Standard Oil Co. (Ind.), 1946-51; Plant Supervisor, Argonne Nat. Lab., 1951-56; Operations Supv., E.B.W.R., Argonne, 1956-60; Nucl. Plant Manager, Rural Co-operative Power Assoc., 1960-67; Reactor Eng., Reactor Licensing Div., U.S.A.E.C., 1967-. Society: A.N.S.
Nuclear interests: Operation of nuclear power plants, including related subjects such as general design, radiological physics, instrumentation, maintenance, chemical control, nuclear physics, fuel and plant management, operator licensing and regulating aspects.
Address: 700 Fletcer Place, Rockville, Maryland 20851, U.S.A.

CAMPBELL, Ronald Hugh, B.Sc. Born 1924. Educ.: Glasgow Univ. Equipment Eng., Sperry Gyroscope Co. Ltd., 1950-57; Chief Eng., Savage and Parsons, Watford, 1957-59; Asst. Chief Eng., A.G.R. (1959-61), Deputy Chief Eng., High Temperature Reactor (1962), Deputy Chief Eng., Steam Graphite Heavy Water Reactor (1962-65), Deputy Chief Eng., Fast Reactor (1965-), U.K.A.E.A. Society: I.E.E.
Nuclear interest: Reactor design.
Address: Larchwood, Blueberry Road, Bowdon, Cheshire, England.

CAMPBELL, Walter Edward, S.B., M. of Architecture. Born 1901. Educ.: M.I.T. Member, Building Res. Advisory Board, N.A.S. Com. on Res., American Inst. Architects. Society: Fellow, A.I.A.
Nuclear interests: Committee on Research of A.I.A. has occasionally been involved in design studies for nuclear facilities.
Address: 711 Boylston Street, Boston 16, Mass., U.S.A.

CAMPBELL, William Munro, B.A.Sc., M.S., Ph.D. Born 1915. Educ.: Toronto Univ., Case Inst. Technol., and Illinois Univ. Director, Chem. and Metal. Div., A.E.C.L., Ontario 1956-65; Director of Res., Ontario Res. Foundation, 1965-. Societies: A.I.Ch.E.; Chem. Inst. of Canada; A.P.E.O.; Canadian Nucl. Assoc.
Nuclear interests: Metallurgy and chemistry.

Address: Ontario Research Foundation, Sheridan Park, Ontario, Canada.

CAMPEN, J. van. See van CAMPEN, J.

CAMPION, Jean, B.Sc. Dept. of Phys. Applied to Medicine, Middlesex Hospital Medical School.
Address: Middlesex Hospital Medical School, Department of Physics Applied to Medicine, Cleveland Street, London W.1, England.

CAMPION, Peter James, B.A., M.A., D.Phil. Born 1926. Educ.: Oxford Univ. Nuffield Fellow, Oxford Univ. 1953-54; Post-doctoral Fellow (1954-55), Res. Officer (1955-60), A.E.C.L.; Head, Radiol. Sect. (1960-64), Supt., Appl. Phys. Div. (1964-66), Supt., Rad. Sci. Div. (1966-), Nat. Phys. Lab. Societies: F.Inst. P.; Hospital Phys. Assoc.
Nuclear interest: Radiation physics.
Address: The Hawthornes, Fee Farm Road, Claygate, Surrey, England.

CAMPO, Armando LOPEZ M. del. See LOPEZ M. del CAMPO, Armando.

CAMPO, César GOMEZ-. See GOMEZ-CAMPO, César.

CAMPOS, Paulo C., M.D. Born 1921. Educ.: Univ. of the Philippines. Resident Instructor in Medicine and Pathology (1949-50), Resident Instructor in Medicine (1950-52), Part-time Instructor in Medicine (1953-54), Asst. Prof. Medicine (1954-60), Assoc. Prof. Medicine (1960-62), Prof. Med. (1962-), Univ. Philippines; Physician in Charge: Thyroid Clinic (1954-59), Diabetic Clinic (1954-59), Radio-isotope Lab. (1956-), Metabolic Unit (1959-), Philippine Gen. Hospital; Consultant in Medicine, Veterans Memorial Hospital, 1958-; Consultant in Medicine, Rizal Provincial Hospital, 1953-; Member, Editorial Staff, J. Philippine Assoc. for the Advancement of Sci., 1959-; Gen. Lecturer, Radioisotope Course, Philippine A.E.C., 1959-; Director, Regional Training Course on Medical Uses of Radioisotopes, Manila, Philippines 1964; Principal Investigator, Endemic Goiter Survey, N.S.B.D. --I.A.E.A. project, 1959-; Attending Physician and Head, Medicine Dept., Philippine Gen. Hospital, 1960-; Head, Medicine Dept., College of Medicine, Univ. Philippines 1960-. Societies: A.N.S.; N.R.C.; Fellow, Philippine Assoc. for the Advancement of Sci.; Philippine Medical Assoc.; Vice-Pres. and Fellow, Philippine Coll. of Physicians; Philippine Public Health; Fellow, Philippine Heart Assoc.; Philippine Medico-Pharmaceutical Soc.; Manila Medical Soc.; Vice-Pres. and Director, Philippine Diabetes Assoc.
Nuclear interest: Nuclear medicine.
Address: 245 Villaruel, Pasay City, Philippines.

CAMPOS COSTA, Manuel José de. See de CAMPOS COSTA, Manuel José.

CAMUS, Guy. Member, Comité de Biologie, Commissariat à l'Energie Atomique.
Address: Commissariat à l'Energie Atomique, 29-33 rue de la Federation, Paris 15, France.

CANANZI, Renato, Dottore in Fisica. Born 1922. Educ.: Messina Univ., Sicily; Istituto di Fisica, Padua Univ. Ente Nazionale Energia Elettrica, Centro Progettazioni e Costruzioni Termiche, Venice, 1953-; Marine and Terrestrial Steam Power Plants Consultant. Societies: Assoc. Naz. di Fisica Sanitaria-Torino; Assoc. Naz. di Ingegneria Nucleare, Rome; Assoc. Eletrotecnica Italiana, Milan.
Nuclear interests: Radiochemistry and radiation dosimetry; Health physics; Environment survey; Reactor engineering; Corrosion; Metallurgy; Water chemistry, treatment and control; Steam-water cycles conditioning; Design features of large-scale water-treatment plants for power stations; Laboratory processes; Waste-disposal.
Address: 13/B Via Monte Grappa, 30171 Mestre, Venice, Italy.

CANAVAN, Frederick Louis, M.A., Ph.D. Born 1919. Educ.: Catholic Univ. Res. Assoc., Rad. Lab., California Univ., 1955-56; Asst. Prof. Phys., Fordham Univ., 1956-. Societies: A.P.S.; A.A.A.S.; A.A.P.T.
Nuclear interests: Nuclear spectroscopy; High-energy nuclear physics.
Address: Physics Department, Fordham University, New York, N.Y. 10458, U.S.A.

CANCELLARIO d'ALENA, Franco, Dr. of Laws. Born 1914. Educ.: Rome Univ. Deputy Chief, Italian Deleg. to the E.C.D.O. in Paris, 1960-63; Gen. Director for External Relations, European Atomic Energy Community (Euratom), 1963-.
Address: 51-53 rue Belliard, Brussels, Belgium.

CANCIO, Gregorio V. Sen. Sci. (Medical Officer), Philippine Atomic Res. Centre.
Address: Philippine Atomic Research Centre, Diliman,Quezon City, Philippines.

CANDEIRA, Daniel SUAREZ. See SUAREZ CANDEIRA, Daniel.

CANDES, Pierre. Ing. de l'Ecole Supérieure d'Electricité. Born 1923. Educ.: Ecole Navale and Ecole Supérieure d'Electricité. Chef du Service d'Etudes de Sûreté Radiologique, C.E.A. Societies: Sté. Française des Electriciens; Sté. Française des Electroniciens et Radioélectriciens; Sté. Française de Radioprotection; A.N.S.; H.P.S.
Nuclear interests: Nuclear safety, health physics, instrumentation and dosimetry.
Address: 7 rue d'Arcole, 75-Paris 4, France.

CANER, Abdullah. Treas., Turkish Cancer and Radiobiol Soc.
Address: Turkish Cancer and Radiobiological Society, 6 Tura Apt., 1 Bilge Sokak, Ankara-Bakanliklar, Turkey.

CANFIELD, Wright. Exec. Vice Pres., Southwest Atomic Energy Associates Rep., Public Service Co. of Oklahoma.
Address: Public Service Co. of Oklahoma, 600 S. Main Street, Tulsa, Oklahoma 74102, U.S.A.

CANNOBIO, Enrique Curti, Eng. Member, Permanent Com. on Internat. Affairs Relating to Atomic Energy.
Address: Permanent Committee on International Affairs Relating to Atomic Energy, c/o Ministerio de Relaciones Exteriores, Plaza de la Moneda, Santiago, Chile.

CANO, Rafael MARQUEZ. See MARQUEZ CANO, Rafael.

CANOY, Nestor R., M.D. Born 1928. Educ.: Silliman and Philippines Univs. Formerly Chief, Dept. of Radiol. and Nucl. Medicine, Chong Hua Hospital; Prof., Radiol., Southwestern Univ. Coll. of Medicine, Cebu City, Vice-Pres., Cebu Medical Soc.; Vice-Pres. for the Visayas, Philippine Medical Assoc. Societies: A.C.R.; Philippine Soc. of Radiol.
Nuclear interests: Purely in medical isotopes − practically interested in liver studies with the frequency of carcinoma of the liver in Cebu City.
Address: Cebu Medical Centre, Osmena Boulevard, Cebu City, Philippines.

CANTADA, Miss Erlinda. Chem., Radioisotope Lab., Philippine Gen. Hospital.
Address: Philippine General Hospital, Manila, Philippines.

CANTONE, Biagio, Dr. Phys., Lib. Docente in Struttura della Materia. Born 1926. Educ.: Catania Univ. Asst. Prof. Exptl. Phys. (1954-), i/c Phys. Medical School (1958-), Ottica Elettronica (1966-), Catania Univ.
Nuclear interests: Atomic and molecular physics; Mass spectrometry.
Address: Istituto Fisica Università,57 Corso Italia, Catania, Italy.

CANTONE, S. Gruppo Stato Solido e Positronio, Centro Siciliano di Fisica Nucl., Catania Univ.
Address: Catania University, Centro Siciliano di Fisica Nucleare, 57 Corso Italia, Catania, Sicily, Italy.

CANTRELL, Clifford M., B.Sc. (Mech. Eng.). Born 1915. Educ.: Tennessee Univ. (also attended Kentucky Univ., Cumberland Coll., and refresher studies at Cincinnati Univ.). R. and D. Eng., Fairchild Engine and Airplane Corp. (N.E.P.A. Project), 1949-51; Vice-Pres., Atlas Eng. Corp., 1951-53; Project Eng. (munitions plants, chemical and food products plants), 1953-57; Project Manager (nucl. projects), 1957-58; Asst. Director, Nucl. Div., 1958-59; Director, Nucl. Div., H. K. Ferguson Co., 1953-61. Society: Atomic Ind. Forum.
Nuclear interests: Administrative management and development, design and construction of facilities for conventional and advanced nuclear power, process steam, fuel processing and reprocessing, research, nuclear propulsion.
Address: 111310 Flower Avenue, Cleveland 11, Ohio, U.S.A.

CANTU, C., Dr. Director, Fratelli Koristka S.p.A.
Address: Fratelli Koristka S.p.A., 71 Via Ampère, Milan, Italy.

CAORSI, Juan H., Ing. Member, Comision Nacional de Energia Atomica, Uruguay.
Address: Comision Nacional de Energia Atomica, 565 J. Herrera and Reissig, Montevideo, Uruguay.

CAP, Ferdinand, Ph.D. Born 1924. Educ.: Vienna Univ. Prof. Theoretical Phys., Innsbruck Univ. Book: Physik und Technik der Atomreaktoren (Vienna, Springer, 1957). Societies: Austrian, Italian, German, U.S.A. Phys. Socs.; Indian Acad. Sci.; N.Y. Acad. Sci.
Nuclear interests: Reactory theory; Plasma physics.
Address: 103d Schneeburggasse, Innsbruck A6020, Austria.

CAPACCIOLI, Jorge Hugo, Dr. in Chem. Born 1925. Educ.: Buenos Aires Univ. Adviser, Raw Materials Direction, Argentine A.E.C., 1953; Assoc. Prof., Analytical Chem., Buenos Aires Univ. 1958. Society: Asociacion Quimica Argentina.
Nuclear interests: Analytical chemistry of reactor materials.
Address: 8250 Avda. Lib. Gral. San Martin, Buenos Aires, Argentina.

CAPDEVILLE, Pierre Germain Francois. Born 1908. Educ.: Ecole Polytechnique and Ecole Nat. Superieure d'Aeronautique. Ing. Conseil, 1951-58; Directeur Adjoint au Directeur Gen. (1958-66), Directeur (1966-), Sud-Aviation.
Address: Sud-Aviation S.A., 37 boulevard de Montmorency, Paris 16, France.

CAPLE, Donald, M.I.Mech.E. Born 1924. Educ.: Ipswich School of Eng. With Bristol Aeroplane Co. Ltd., 1939-55; Orenda Engines Ltd., 1955-62; Eng. Manager, Atomics, Hawker Siddeley Eng., 1962; Now Manager, Lab. Services and Nucl. Products, Orenda Ltd., Ontario. Society: A.P.E.O.
Nuclear interests: Mechanical design of nuclear power plants and components; Economic power programmes; Research and development in nuclear power reactors; Compilation and dissemination of design and manufacturing standards for nuclear power plants.
Address: Orenda Limited, Box 6001, Toronto International Airport, Ontario, Canada.

CAPPADONA, Carmelo, Prof. Chimica del reattore e Radioisotopi, Facolta d'Ing., Istituto di Applicazioni e Impianti Nucl., Palermo Univ.
Address: Palermo University, Viale delle Scienze, Palermo, Sicily, Italy.

CAPPELLER, Ulrich, Dr. rer. nat. habil. Born 1917. Educ.: Königsberg Univ. Pr. Kustos am Physikal. Inst., Marburg/L. Univ. Book: Beitrag, Some Relationships between Binding Energies and the Shell Model of Atomic Nuclei (Conference on Nucl. Masses, ed. H. Hintenberger Pergamon Press, 1957). Societies: Deutsche Physikal Ges.; A.P.S.
Nuclear interests: Angular correlations; Binding energies; Nuclear structure.
Address: Physik Institut, 5 Renthof, Marburg/L., Germany.

CAPPELLO, Alberto. Sen. Officer, Carlo Gavazzi S.P.A.
Address: Carlo Gavazzi S.P.A., 9 Via G. Cardi, Milan, Italy.

CAPPS, Richard Huntley, B.A. (Phys.), M.A. (Phys.), Ph.D. (Phys.). Born 1928. Educ.: Kansas and Wisconsin Univs. Consultant, Gen. Atomic Div., Gen. Dynamics Corp., La Jolla, California, 1958; Acting Asst. Prof. Phys., Washington Univ., Seattle, 1957-58; Res. Assoc., Lab. Nucl. Studies, Cornell Univ., 1958-60; Asst. Prof. Phys. (1960-62), Assoc. Prof. Phys. (1962-), Northwestern Univ. Society: A.P.S.
Nuclear interests: Primary interest: Theoretical research concerning the interactions between elementary particles. Secondary interests: High-energy particle accelerators and plasma physics.
Address: Physics Department, Northwestern University, Evanston, Illinois, U.S.A.

CAPRIO, Igino, Dr. Italian Member, Specialised Nucl. Sect. for Social Health and Development Problems, Economic and Social Com. Euratom.
Address: 67 Via Due Macelli, 00 187 Rome, Italy.

CAPRIOGLIO, Pietro, Dr. in Ind. Chem. Born 1929. Educ.: Milan Univ. C.I.S.E. Milan, 1952-55; Abbott Laboratories, Rome, 1955-57; Agip Nucleare, Milan, 1957-59; Euratom, Brussels, 1959-; Director, Petten Establishment of Joint Res. Centre; Member, Board of Management, O.E.C.D. High Temperature Reactor Project, Dragon.
Nuclear interests: Metallurgy and ceramics; Fuel elements development; High-temperature gas-cooled reactors.
Address: Euratom, 51 rue Belliard, Brussels, Belgium.

CAPRON, Paul Corneille, Dr. Sc. Born 1905. Educ.: Louvain Univ. Prof., Conseiller Scientifique, and Membre, Conseil d'administration, Univ. Catholique de Louvain. Member, Commission Sci. Inst. Interuniv. des Sci. Nucleaires; Membre, Comité de Rédaction, Industrie Chimique Belge. Society: Sté. de chimie physique de France.
Nuclear interests: Radiocarbon dating; Chemical consequences of nuclear transformations; Radiation chemistry.
Address: Centre de Physique Nucléaire, Parc d'Arenberg, Kardinaal Mercierlaan, Heverlee, Louvain, Belgium.

CAPURRO, Federico GARCIA. See GARCIA CAPURRO, Federico.

CARACH, Jozef, Dr. in Chem. Born 1920. Educ.: Bratislava Univ. Chief, Dept. of Rad. Hygiene, State Regional Hygienic Station, Bratislava. Societies: Czechoslovak Chem. Soc., Medical Soc. of J.E. Purkyne.
Nuclear interests: Biosphere contamination monitoring; Radiation hygiene in general; Atomic energy power plants.
Address: 60 Trnavská, Bratislava, Czechoslovakia.

CARAMELLI, Fabrizio, Ing. Director, Italelettronica.
Address: Italelettronica, 72 Via Ignazio Pettinengo, Rome, Italy.

CARASALES, Julio C., Dr. Minister Plenipotentiary, Embassy of Argentina, Austria; Resident Rep. of Argentina to I.A.E.A.
Address: 2/II/6/26 Golsdorfgasse, Vienna 1, Austria.

CARATI, Filippo, Ingegnere Elettrotechnico. Born 1901. Educ.: Politecnico Milan. Consigliere Amministrazione, Ente Nazionale Energia Elettrica.
Nuclear interest: Centrali termonucleari.
Address: E.N.E.L., 3 Via G. B. Martini, Rome, Italy.

CARBALLO FERNANDEZ, J. M. Spanish Member, Group of Experts on Third Party Liability, O.E.C.D., E.N.E.A.
Address: O.E.C.D. European Nuclear Energy Agency, 38 boulevard Suchet, Paris 16, France.

CARBIENER, Wayne A. Res. Sci.-Eng., Appl. Nucl. Phys. Div., Battelle-Columbus Labs.
Address: Battelle-Columbus Laboratories, 505 King Avenue, Columbus, Ohio, 43201, U.S.A.

CARBON, Max W., B.S. (M.E.), M.S. (M.E.), Ph.D. Born 1922. Educ.: Purdue Univ. Pile Eng. (1949-51), Head, Heat Transfer Unit (1951-55),Head, Contract Eng. Unit (1955), G.E. Co.; Chief, Thermodynamics Sect., AVCO Mfg. Co., 1955-58; Chairman, Nucl. Eng. Dept. and Prof. Nucl. Eng., Member, Rad. Safety Com., Wisconsin Univ., 1958-. Societies: A.N.S.; A.S.M.E.; A.S.E.E.; A.A.A.S.; N.S.P.E.
Nuclear interests: Heat transfer; Reactor design; Economics.
Address: Singapore Polytechnic, P.O. Box 2023, Singapore 2.

CARBONELL, Luis M., Dr. Deputy Director, Inst. Venezolano de Investigaciones Cientificas.
Address: Ministerio de Sanidad y Asistencia Social, Instituto Venezolano de Investigaciones Cientificas, Apartado 1827, Caracas, Venezuela.

CARDA APARICI, Pedro. Vocale, Comisiones Asesoras de Medicine y Biologia Animal, Junta de Energia Nucl.
Address: Junta de Energia Nuclear, Ciudad Universitaria, Madrid 3, Spain.

CARDENAS, Juan F., Fis. Encargado de la Sección de Radiación Cósmica, Instituto de Fisica, Universidad Autonoma de San Luis Potosi.
Address: Universidad Autonoma de San Luis Potosi, No. 64 Alvaro Obregón, San Luis Potosi, S.L.P., Mexico.

CARDENAS, René F., Medical Dr. Born 1932. Educ.: Univ. de La Habana. Head, Nucl. Medicine Dept., Oncological and Radiobiol. Inst. 1962-.
Nuclear interests: Nuclear medicine and radiobiology.
Address: Nuclear Medicine Department, Instituto de Oncologia y Radiobiologia, 29 y F. Vedado, Havana 4, Cuba.

CARDONA, Salvador, Attorney at Law (abogado). Born 1896. Educ.: Nat. Univ., Mexico. Prof. School of Social and Political Sci., Nat. Univ. Mexico, 1953; Tech. Consultant, Foreign Affairs Ministry of Mexico, 1954; Gen. Sec., Nat. Commission for Nucl. Energy, 1959-.
Books: The Law which creates the Nat. Commission for Nucl. Energy (Nat. Commission for Nucl. Energy, 1957); The Nucl. Energy and the Law (Nat. Coll. of Attorneys at Law of Mexico, 1960). Other Publications: Atomic Energy: Internat. Political Implications (School of Social and Political Sci. Revue, 1960).
Society: Mexican Soc. for the Study of Radioisotopes.
Address: 1079 Insurgentes Sur, México 18, D.F., Mexico.

CARDUS, Santiago CASTRO. See CASTRO CARDUS, Santiago.

CARERI, Giorgio. Born 1922. Educ.: Rome Univ. Prof. Phys., Padua Univ., 1958-61; Sec. Commission of Thermodynamics, I.U.P.A.P. 1955-.
Nuclear interests: Mass spectrometry; Isotopic reaction kinetics; Nuclear resonance in liquid ^{3}He.
Address: Istituto di Fisica, Università, Rome, Italy.

CAREW, Thomas Edward, M.S., M.N.E. Born 1918. Educ.: Catholic Univ. of America and U.S. Naval Postgraduate School. Former Director, U.S. Army Nucl. Defence Lab.; Chief, Nucl. Effects Res. Branch, Dept. of Defence; Nucl. Safety Officer, U.S. Army, Europe; Lieutenant Colonel U.S. Army Corps. of Eng., 1950-64; Res. Assoc., Nucl. Sci. (1965-67), Asst. Prof., Nucl. Sci. and Eng. (1967-), The Catholic Univ. of America. Society: A.N.S.
Nuclear interests: Nuclear engineering education; Radiation shielding; Isotopic neutron source design and use.
Address: Nuclear Science and Engineering Department, The Catholic University of America, Washington, D.C. 20017, U.S.A.

CAREY, Z. E. Exec. Asst. Computer Sci. Corp., Northwest Operations, Hanford Facilities, U.S.A.E.C.
Address: U.S.A.E.C., Computer Sciences Corporation, Northwest Operations, Richland, Washington, U.S.A.

CARFTON, William H. Lecturer in Radiol. Sci., Rad. Sci. Centre, Rutgers State Univ.
Address: Rutgers State University, New Brunswick, New Jersey, U.S.A.

CARIO, Günther, Prof. Dr. phil. Director, Physikalisch Inst., Braunschweig T.H.
Address: Technische Hochschule, Braunschweig, Germany.

CARLBOM, Lars Erik, Ph.D. Born 1917. Educ.: Uppsala and Stockholm Univs. At (1952-), Head, Safety and Operations Dept. (1957-), Deleg. for Radiobiol., Swedish Atomic Res. Council. Societies: Swedish Atomic Res. Council; Swedish Nat. Com. for Rad. Protection Res.
Nuclear interests: Health physics; Reactor safety.
Address: AB Atomenergi, Studsvik, Nyköping, Sweden.

CARLE, Remy Louis. Born 1930. Educ.: Ecole Polytechnique, Paris and Ecole Nationale Supérieure des Mines de Paris. At (1957), now Chef du Département de Construction des Piles, C.E.A. Expert, Third I.C.P.U.A.E., Geneva, Sept. 1964.
Nuclear interests: Design and engineering problems of heavy water reactors.
Address: 9 boulevard Soult, Paris 12, France.

CARLIER, Aimé Pierre, L. ès Sc., Diplome d'Etudes Supérieure, D. ès Sc. Born 1924. Educ.: Faculté des Sciences de Paris.
Nuclear interests: Geology of radioactive mineral deposits; Geostatistics.
Address: Commissariat à l'Energie Atomique, Direction de Productions, C.E.N.-F.A.R., Fontenay-aux-Roses, Seine, France.

CARLOS ROMERO, José E. DE. See DE CARLOS ROMERO, José E.

CARLSON, Bengt. G., Deleg. to 2nd I.C.P.U.A.E., Geneva, Sept. 1958. Member, Programme Com. and Member, Mathematics and Computations Div., A.N.S.; Member, Maths. and Computer Sci. Res. Advisory Com., U.S.A.E.C.

Address: Los Alamos Scientific Laboratory, Los Alamos, New Mexico, U.S.A.

CARLSON, Oscar Norman, B.A., Ph.D. Born 1920. Educ.: Yankton and Iowa State Colls. Prof., Dept of Metal., Iowa State Univ. Societies: A.S.M.; A.I.M.E.; A.S.E.E.; A.C.S. Nuclear interests: Metallurgy: preparation of rare nuclear metals, study of their mechanical properties and alloying behaviour.
Address: Metals Development Building, Iowa State University, Ames, Iowa, U.S.A.

CARLSON, Robert Werner Shelley, B.E. (1st Class Hons.) (Mech. Eng.), A.S.T.C. Mech. Eng. Diploma. Born 1929. Educ.: Sydney Univ. and Sydney Tech. Coll. Cadet Eng. and Eng., Dept. of Supply, 1951-56; Res. Officer (1956-59), Deputy Reactor Operations Eng. (1959-63), Acting Reactor Operations Eng. (1963-), now Head, Reactor Operations Sect., Australian A.E.C. Societies: A.M.I.Mech.E. (London); Assoc. Member, Inst. Eng. (Australia).
Nuclear interests: Reactor operation, design and management.
Address: 5 Homer Place, Caringbah, N.S.W., Australia.

CARLSON, Roger Dean, B.S. (Phys.). Born 1932. Educ.: North Central Coll., Naperville, Ill. Assoc. Eng., Argonne Nat. Lab., 1954-. Society: A.N.S.
Nuclear interests: Experimental nuclear engineering with special interest in: reactor fuels and coolants systems which employ liquid metals; homogeneous reactor fuels; special application of reactors; irradiation testing of potential reactor fuels and components.
Address: 116 South Madison Avenue, LaGrange, Ill., U.S.A.

CARLSON, Walter B. Asst. Manager for Administration, Nevada Operations Office (1962-64), Deputy Manager, Idaho Operations Office (1964-), U.S.A.E.C.
Address: United States Atomic Energy Commission, Idaho Operations Office, P.O. Box 2108, Idaho Falls, Idaho, U.S.A.

CARLSON, Walter Berndt, D.Sc. Born 1909. Educ.: Witwatersrand Univ. Johannesburg. Asst. to Chief Res. Eng. (1948-56), Chief Res. Eng. (1956-), Babcock and Wilcox, Ltd. Societies: M.I.Mech.E.; M.Inst.F.; B.N.E.S. Nuclear interests: All materials and engineering aspects of nuclear reactor design, including heat exchanger, steam raising and pressure containment plant which may be linked therewith.
Address: Babcock and Wilcox Ltd., Renfrew, Scotland.

CARLVIK, T. Ingvar, Tekn. Dr. Born 1923. Educ.: Stockholm Roy. Inst. Technol. S.A.A.B., 1953-55; A.B. Atomenergi, 1955-.
Nuclear interest: Reactor theory.
Address: A.B. Atomenergi, Stockholm, Sweden.

CARMAGO, Pedro BENTO de. See BENTO de CARMAGO, Pedro.

CARMICHAEL, Donald Macaulay, M.A. (1st Class Hons. in Philosophy) (Edin.). Born 1908. Educ.: Edinburgh and Cambridge Univs. Ministry of Works. Gen. Sec. D.E.R.E.
Nuclear interest: Administration.
Address: Dorrery Lodge, by Thurso, Caith., Scotland.

CARMICHAEL, Hugh, B.Sc. (Edinburgh), Ph.D. (Cambridge), M.A. (Cambridge), Sometime Fellow of St. John's Coll., Cambridge, F.R.S.C. Born 1906. Educ.: Edinburgh and Cambridge Univs. Sen. P.S.O., M.O.S., Atomic Energy Mission to Canada, 1944-50; Principal Res. Officer, Head, Chalk River Gen. Phys. Branch (1950-), A.E.C.L. Society: F.R.S.C. Nuclear interests: Reactor instrumentation; Radiation monitoring; Cosmic rays; Solar terrestrial physics.
Address: 9 Beach Avenue, Deep River, Ontario, Canada.

CARNEIRO, Octavio Augusto, DIAS. See DIAS CARNEIRO, Octavio Augusto.

CARO, David Edmund, M.Sc., Ph.D. Born 1922. Educ.: Melbourne and Birmingham Univs. Lecturer (1952), Sen. Lecturer (1954), Reader (1958), Prof. (1961-), Melbourne Univ. Societies: Fellow, Inst. Phys.; Fellow, Australian Inst. Phys.
Nuclear interests: Nuclear physics; accelerators; Elementary particle physics.
Address: 11 Fulham Avenue, South Yarra, Victoria, Australia.

CARO, Rafael, Dr. Phys. Born 1935. Educ.: Madrid Univ. On leave from Junta de Energia Nucl., Spain at: C.E.A., Saclay, 1959; U.K.A.E.A., Winfrith, 1961-62; Atomic Power Div., Westinghouse Elec. Corp., 1966-67. Editorial Board Member, Energia Nucl. Society: Real Sociedad Espanola de Fisica y Quimica.
Nuclear interests: Reactor design and operation analyses.
Address: 118 C/Embajadores, Madrid, Spain.

CARON, H. Marcel, Master in Commerce; Licentiate in Accountancy; Chartered Accountant. Born 1919. Educ.: Montreal Univ. Auditor, Bank of Canada, 1965-66; Auditor, Quebec Hydro Electric Commission, 1967; Director, Atomic Energy of Canada Ltd., 1967.
Nuclear interest: Management.
Address: 500 Saint James Street W., Montreal, Quebec, Canada.

CAROTHERS, James Edward, B.A., Ph.D. Born 1923. Educ.: California Univ. Phys., Lawrence Rad. Lab., 1952-. Societies: A.P.S.; A.N.S.
Nuclear interests: Reactor physics, criticality control.

Address: Box 808, Livermore, California, U.S.A.

CARPENDER, James W. J., M.D. Formerly Vice-Pres., now Member, American Board of Radiology.
Address: American Board of Radiology, Kahler Centre Building, Rochester, Minnesota 55901, U.S.A.

CARPENTER, J. M. Formerly Asst. Prof., now Assoc. Prof., Dept. of Nucl. Eng., Coll. of Eng., Michigan Univ.
Address: Michigan University, College of Engineering, Ann Arbor, Michigan, U.S.A.

CARPENTER, L. P. Nucl. Project Eng., Kaiser Engineers.
Address: Kaiser Engineers, Division of Henry J. Kaiser Co., 300 Lakeside Drive, Oakland 12, California, U.S.A.

CARPENTER, M. Scott, Commander. Formerly with N.A.S.A.; Asst. for Aquanaut Operations, Deep Submergence System Project, U.S. Navy; Member, Advisory Com. on Isotopes and Rad. Development, U.S.A.E.C., 1967-.
Address: U.S. Navy, Deep Submergence System Project, Chevy Chase, Maryland, U.S.A.

CARPENTER, Robert Thomas, B.S. (Chem. Eng.), M.S. (Nucl. Eng.). Born 1931. Educ.: Maryland Univ. and U.S.A.F. Inst. of Technol. Chief, Isotope Power Systems Branch, U.S.A.E.C., 1964-.
Nuclear interests: Isotopic space power systems and related subjects: isotopes, high temperature materials, power conversion techniques and aerospace nuclear safety.
Address: Space Nuclear Systems Division, United States Atomic Energy Commission, Washington, D.C. 20545, U.S.A.

CARR, Arthur John, A.C.W.A., M.P.O.A. Born 1918. Stores Superintendent, Merseyside and N. Wales Elec. Board, 1951; Area Stores Office, St. Helens (1955), Area Purchasing and Stores Manager, Ashington (1958), N.C.B. Chief Stores Officer (1959), Asst. Chief Purchasing and Stores Officer (1964), U.K.A.E.A. Hon. Treas., Nucl. Eng. Soc., Risley. Societies: A.C.W.A.; M.P.O.A.
Nuclear interest: Management.
Address: 38 Hardy Road, Lymm, Cheshire, England.

CARR, Donald R. Director and Vice-Pres., Isotopes Inc., 1955-.
Address: 32 Villa Road, Pearl River, New York, U.S.A.

CARR, Francis Herbert, M.E. Born 1921. Educ.: Western Australia Univ. Sen. Res. Officer, Australian A.E.C., Harwell, 1955-57; Principal Res. Sci., Australian A.E.C., Sydney, 1957-. Society: Assoc. Member, Inst. of Eng., Australia.
Nuclear interests: Reactor design; Nuclear power studies; Engineering research and design.
Address: 11 Harbour Street, Cronulla, N.S.W., Australia.

CARR, Fred. Chief Technician, Detectron Div., Computer Measurements Corp.
Address: Computer Measurements Corporation, 12970 Bradley Avenue, Sylmar, California, U.S.A.

CARR, Frederick, A.R.C.S., B.Sc., D.I.C. Born 1915. Educ.: London Univ. Project Eng. and Field Eng., Polymer Corp., Sarnia, Ontario, 1948-52; Director and Chief Eng., Process Development Ltd., Canada, 1952-54; i/c Boiler Div. (1954-56), Chief Boiler Eng. (Design) (1956-59), Nucl. Power Plant Co. Ltd. Development Manager, Head Wrightson Processes, 1959-60; Development Manager (1960-65), Tech. Director (1965-67), Servotomic Ltd.
Nuclear interests: Heat transfer and thermodynamics; Engineering design of graphite reactor cores.
Address: White Lodge, Blackpond Lane, Farnham Common, Buckinghamshire, England.

CARR, Howard Earl, B.S. (Alabama Polytech. Inst.), M.A. and Ph.D. (Virginia). Born 1915. Educ.: Alabama Polytech. Inst. and Virginia Univ. Assoc. Prof. Phys., Alabama Polytech. Inst., 1948-53; Consultant to Air Force at Holloman Air Force Base, 1957-60; Head Prof. Phys., Auburn Univ., 1953-. Sec., Southeastern Sect., A.P.S. Societies: A.P.S.; A.A.P.T.
Nuclear interest: Reactor physics.
Address: 342 Payne Street, Auburn, Alabama, U.S.A.

CARR, James Bryan, B.Ch.E., M.Ch.E. Born 1930. Educ.: Clarkson Coll. Tech. and Rensselaer Polytech. Inst. Societies: Northeastern New York Chapter, A.I.Ch.E.; Northeastern New York Chapter, A.N.S.
Nuclear interests: Reactor thermal and hydraulic design, with particular interest in the application of experimental results to design criteria. Other interests include experimental work in thermal, hydraulic and physics areas, and in dynamic analysis.
Address: 3313 Maryvale Drive, Schenectady, N.Y., U.S.A.

CARR, John Stewart, B.Sc., B.E., M.S., S.M., M.Sc. Born 1921. Educ.: Otago and Missouri Univs. and M.I.T. S.S.O., A.E.R.E., Harwell, 1951-52; Sen. Lecturer, Mineral Processing (1953-), Member, Board of Studies in Nucl. Sci. and Eng., Melbourne Univ. Book: Co-editor, Some Problems in the Development of Atomic Power (Melbourne Univ. Press, 1953). Society: A.I.M.E.
Nuclear interests: Mineral processing; Metal processing.
Address: 60 Durham Street, Heidelberg, Victoria 3084, Australia.

CARR, Robert Joseph, B.S., Ph.D. Born 1931. Educ.: Chicago and California (Los Angeles, Berkeley) Univs. Chem., Lawrence Rad. Lab., California Univ., 1951-55; Instructor and Asst. Prof. Chem., Washington State Univ., 1955-61; Chem., Shell Development Company, 1961-67; Prof. Chem., Merritt Coll., 1967-. Societies: A.C.S.; A.P.S.
Nuclear interests: Nuclear chemistry (charged-particle-induced nuclear reactions; compound nucleus theory; nuclear decay schemes); Radiochemistry (Activation analysis; Application of nuclear radiations and radioactive materials to analytical problems).
Address: Merritt College, Oakland, California, U.S.A.

CARR, Terence E. F., B.Sc. Educ.: St. Andrews Univ., Scotland, Windscale Works U.K.A.E.A.; Medical Res. Council, Biological res. at Harwell, 1953-. Sci. Sec., 2nd I.C.P.U.A.E.
Nuclear interests: Biological problems associated with atomic energy development, emphasis on mineral metabolism.
Address: Medical Research Council, Radiobiological Research Unit, Harwell, Didcot, Berks., England.

CARRASSE, Jean. Ing. Principal, Services d'Etudes Nucléaires, "Alsthom".
Address: Services d'Etudes Nucléaires, Société Générale de Constructions Electriques et Mécaniques "Alsthom", 20 rue d'Athènes, Paris 9, France.

CARRAT. Henri Germain. D. ès Sc., diplôme d'Ing. géologue de l'Ecole Nat. Supérieure de Géologie de Nancy. Born 1915. Educ.: Paris and Nancy Univs. Ancien Chef du Service des Recherches, Div. de Grury (Saône et Loire); Missions d'assistance technique, Péru and Argentina; Ing. en Chef, Centre de Recherches Radiogéologiques de Nancy. Societies: Sté. Géologique de France; Ste. Française de Minéralogie et de Cristallographie; Geochemical Soc.
Nuclear interests: Geologie et géochimie des roches cristallines; Géochimie de l'uranium et du thorium; Applications à la prospection des matériaux nucléaires.
Address: 16 rue Santerre, Paris 12, France.

CARREIRA PICH, Henrique Joao, Degree in Eng. Chem. Born 1926. Educ.: Inst. of Eng., Lisbon Univ. Chief Appl. Chem. Group, Nucl. Res. Centre, Junta de Energia Nucl. Society: Ordem dos Engenheiros.
Nuclear interest: Applied chemistry; Nuclear fuel cycle.
Address: Junta de Energia Nuclear, Laboratório de Fisica e Engenharia Nucleares, 10 Estrada Nacional, Sacavém, Portugal.

CARRELLI, Antonio. Born 1900. Educ.: Naples Univ. Vice-Pres., S.M.E.; Prof. Ordinario Fisica Sperimentale e Direttore Istituto Fisica, Naples Univ., 1932-. Societies: Socio Ordinario Accademia Nazionale Lincei; Socio Stà. Scienze Lettere ed Arti Napoli; Socio Stà. dei Quarante. Consiglio Nazionale Ricerche.
Address: 3 Via Antonio Tari, Naples, Italy.

CARRERAS MEJIA, Galo. Vocale, Comision Interministerial de Conservacion de Alimentos por Irradiacion, Junta de Energia Nucl.
Address: Junta de Energia Nuclear, Ciudad Universitaria, Madrid 3, Spain.

CARRICO, Paul E., Chief, Agreements and Regulations Branch, Maritime Administration and U.S.A.E.C.
Address: U.S. Department of Commerce, Maritime Administration, Washington 25, D.C., U.S.A,

CARRIER, Jean. At Batignolles-Chatillon (Mécanique Générale) S.A.
Address: Batignolles-Chatillon (Mécanique Générale) S.A., 5 rue de Monttessuy, Paris 7, France.

CARRIERE, Raymond. Directeur Gen. Adjoint Cie. de Construction Mécanique Procédès Sulzer.
Address: Cie. de Construction Mécanique Procédès Sulzer, 19 rue Cognacq-Jay, Paris 7, France.

CARRINGTON SIMOES DA COSTA, Joao, D.Sc., Dipl. Prof. Born 1891. Educ.: Lisbon and Porto Univs. Prof., Oporto Univ., 1938-; Pres., Junta de Investigacoes do Ultramar, 1953-. Vice-Pres., Comissao de Estudos da Energia Nucl., Inst. de Alta Cultura; Member, Exec. Com., Junta de Energia Nucl.; Director, Mines Dept., Inst. Superior Tecnico. Books: Several geological publications, with special interest for the prospection of radioactive minerals. Societies: Acad. de Ciencias de Lisboa (Effective Member); Real Acad. de Ciencias de Madrid (Correspondent Member); Soc. Geológica de Portugal (Pres.).
Nuclear interests: Mineralogy and geology of radioactive minerals.
Address: Praceta Santa Isabel lote 2, Parede, Portugal.

CARROL, David I. Electronic Eng., Franklin Systems Inc.
Address: Franklin Systems Inc., P.O. Box 3250, West Palm Beach, Florida 33402, U.S.A.

CARROLL, Clayton C., A.B., A.M. Born 1905. Educ.: Missouri Univ. Manager, New Products, Nucl. Dept., U.S. Radium Corp. Society: A.C.S.
Nuclear interests: Radioisotopes and their applications in commercial fields; Use in light and radiation sources; Radiochemistry, research and management.
Address: 9 Sage Drive, Plainfield, New Jersey, U.S.A.

CARROLL, David Shields, B.S., M.D. Born 1917. Educ.: Tennessee Univ. Prof. and Chairman, (-1964), Dept. of Radiol., Tennessee Univ. Coll. of Medicine; Chairman, Dept. of

Radiol., City of Memphis Hospital; Consultant in Radiol., Kennedy Veterans Hospital; Consultant in Radiol., Oak Ridge, Tennessee. Vice-Pres., Radiol. Soc. of North America. Societies: Radiol. Soc. of North America; American Coll. of Radiol.; Soc. of Nucl. Medicine; American Radium Soc.
Nuclear interest: Radioisotope teletherapy.
Address: 860 Madison Avenue, Memphis 3, Tennessee, U.S.A.

CARROLL, James G., B.A. (Economics), B.S. (Mech. Eng.), professional degree of Mech. Eng. Born 1918. Educ.: Wisconsin Univ. Research Eng., Chevron Res. Co., Richmond, California, 1948-. Book: Rad. Effects on Organic Materials (1963). Societies: A.I.A.A.; A.C.S.
Nuclear interest: Radiation effects on organic materials.
Address: 88 Valley Avenue, Martinez, California, U.S.A.

CARROLL, P. E. Sen. Nucl. Eng., Bechtel Nucl. Corp.
Address: Bechtel Nuclear Corporation, 220 Bush Street, San Francisco 4, California, U.S.A.

CARROLL, R. D. Geophys., Geological Survey, U.S. Dept. of the Interior.
Address: U.S. Department of the Interior, Geological Survey, Washington D.C. 20242, U.S.A.

CARRON, Lord William, M.A. (Oxon.) D.Sc. (Honoris Causa, Loughborough and Salford). Born 1902. Director, Bank of England; Part-time member U.K.A.E.A. 1967-71; Part-time member, Midlands Gas Board. Chairman, Foundation on Automation and Employment.
Address: U.K. Atomic Energy Authority, 11 Charles II Street, London, S.W.1, England.

CARRUTHERS, George Harry, LL.M., M.Sc., A.R.C.S., Barrister-at-Law. Born 1900. Educ.: London Univ. Past Asst. Sec., Board of Trade, Insurance and Companies Dept., London; now with British Insurance (Atomic Energy) Com. Member, O.E.E.C. Group of Experts on 3rd Party Liability; Member, Internat. Atomic Energy Agency Panel on Civil Liability.
Nuclear interest: Insurance and legal questions relating to reactors.
Address: Aldermary House, Queen Street, London, E.C.4, England.

CARRUTHERS, Robert, B.Sc. (Eng.), A.C.G.I. Born 1921. Educ.: London Univ. P.S.O., later Sen. Principal Sci., A.E.R.E., 1950-61; Sen. Principal Sci., Head of Technol. Div., U.K.A.E.A., A.E.R.E., Culham Lab., 1961-. Society: Fellow, I.E.E..
Nuclear interests: Technological problems associated with devices used in plasma physics and controlled thermonuclear reactions (Storage and switching of electrical energy, electrical insulation, and materials problems associated with the discharge vessel).
Address: 32 Norman Avenue, Abingdon, Berkshire, England.

CARSTENS, Karl, Master of Laws (Yale). Dr. jur., Prof. Born 1914. Educ.: Frankfurt/Main, Dijon, München, Königsberg, Hamburg and Yale Univs. Plenipotentiary of the Free Hanseatic City of Bremen to Federal Govt., Bonn, 1949-54; Minister, permanent rep. of Federal Republic of Germany to Council of Europe, Strasbourg, 1954-55; Ministerialdirigent at Foreign Office, Bonn, Deputy Chief of Political Dept., 1955-58; Head, Political Dept. "West I" (1958-60), State Sec. (1960-66), Foreign Office, Bonn; State Sec., Federal Ministry of Defence, Bonn, 1966-; Prof. of Constitutional and Internat. Law, Cologne Univ. Books: Grundgedanken der amerikanischen Verfassung und ihre Verwirklichung (1954); Das Recht des Europarats (1956).
Address: Bundes Ministerium des Verteidigung, Bonn, Germany.

CARSWELL, Douglas John, M.Sc.,Ph.D., Dip. Ed. Born 1929. Educ.: Sydney Univ. Res. Officer, Australian A.E.C., 1955-59 (stationed at Harwell, England, 1955-57); Sen Lecturer, Nucl. and Rad. Chem. Dept. (1959-66), Head of Dept. (1966-), New South Wales Univ. Books: Co-author, Introduction to Nucl. Sci. (A.A.E.C., 1959); Introduction to Nucl. Chem. (Elsevier, 1967).
Nuclear interest: Nuclear chemistry, especially the chemistry of the lower actinides.
Address: New South Wales University, P.O. Box 1, Kensington, New South Wales 2033, Australia.

CARTELLIERI, Wolfgang, Dr. jur., Staatssekretär a.D. Born 1901. Educ.: Jena and Munich Univs. Vorsitzer, Aufsichtsrat der Gesellschaft für Kernforschung m.b.H., Karlsruhe. Books: Grundsätzliche Rechtsprobleme des Atomgesetzes, Sonderdruck aus Veröffentlichungen des Instituts für Energierecht an der Universität Bonn, 1/2, (1960); Die Grundlinien der deutschen Gesetzgebungspolitik auf dem Gebiet des Strahlenschutzes (EUR 2791, 1960).
Address: 17 Bergstrasse, 532 Bad Godesberg, Germany.

CARTER, Alan, M.A. Born 1925. Educ.: Cambridge Univ. Res. Metal. Hard Metal Tools, Ltd., Coventry, 1944-54; Tech. Officer, Sect. Leader, Res. Dept., I.C.I. (Metals Div.), Ltd., 1954-57; Sen. Development Officer, Nucl. Eng. Products, Development and Tech. Service Dept., I.C.I. (Metals Div.), Ltd., Birmingham, 1957-63: Home Sales and Tech. Service Manager,Nucl. and Special Products, New Metals Division, Imperial Metal Industries (Kynoch) Ltd., 1964-. Book: Co-author, The Chem. and Metal.of Titanium Production (Monograph) (Roy. Inst. Chem., 1958). Societies: Fellow, Inst. of Metal.; Inst. of Metals.
Nuclear interests: Fuel canning materials and other reactor components; Superconducting

materials.
Address: Imperial Metal Industries (Kynoch) Limited, Kynoch Works, Witton, Birmingham 6, England.

CARTER, Bridget, B.Sc. Dept. of Phys. Applied to Medicine, Middlesex Hospital Medical School.
Address: Middlesex Hospital Medical School, Department of Physics Applied to Medicine, Cleveland Street, London W.1, England.

CARTER, G. A., Lt. Cdr. R.N., M.A. Sen. Lecturer, Dept. of Nucl. Sci. and Technol., Roy. Naval Coll.
Address: Royal Naval College, Department of Nuclear Science and Technology, Greenwich, London S.E.10, England.

CARTER, John Lemuel, Jr., B.A., M.Sc., Ph.D. Born 1920. Educ.: Baylor, Brown, and Cornell Univs. Measurements Lab., Instrument Dept., G.E.C., Lynn, Mass., 1952-55; Hanford Labs., G.E.C., Richland, Wash., 1955-65; Battelle-Northwest, Richland, Wash., 1965-. Societies: A.P.S.; A.N.S.
Nuclear interests: Reactory theory; Neutron transport theory; Digital computer codes; Radiation damage.
Address: 78 McMurray, Richland, Washington 99352, U.S.A.

CARTER, Melvin W., Dr. Officer-in-Charge, Southeastern Radiol. Health Lab.
Address: Southeastern Radiological Health Laboratory, Montgomery, Alabama, U.S.A.

CARTER, P. W., Assoc. Prof. Nucl. Dept., Miami Univ.
Address: Nuclear Department, University of Miami, Coral Gables, Florida 33124, U.S.A.

CARTER, Paul Evan, B.Sc., A.R.C.S. Born 1918. Educ.: London Univ. Radiochem. Centre, Amersham, 1950-56; Chief Planning Officer, Commercial Branch, Production Group, H.Q., U.K.A.E.A., Risley, 1956-.
Address: Commercial Branch, Production Group, H.Q., U.K.A.E.A., Risley, near Warrington, Lancashire, England.

CARTER, Randall Wayne, Sr., B.S. (Maths. and Chem.), M.S. (Phys.). Born 1930. Educ.: Vanderbilt Univ., Certified by American Board Health Phys. Res. Chem., Phillips Petroleum Co., 1956; Special Fellow in Radiol. Phys., O.R.I.N.S., U.S.A.E.C., 1956-58; Radiol. Safety Officer (1958-61), Asst. Res. Chem. (1959-61), Georgia Inst. Technol.; Assoc. in Radiol. (Phys.) (1961-63), Asst. Prof. Radiol. (Phys.) (1963-), Rad. Safety Officer (1961-), Emory University; Consulting Phys., Grady Memorial Hospital, Atlanta, 1961-; Consulting Phys., U.S.V.A. Hospital, Atlanta, 1962-. Georgia State Rad. Control Council, 1964-; Treas., Southeastern Sect., Soc. Nucl. Med. Societies: H.P.S.; A.P.H.A.; American Assoc. Phys. in Medicine; Soc. Nucl. Medicine; Southeastern Section, A.P.S.; Georgia Acad. Sci.
Nuclear interests: Low-energy x-ray; Nuclear medicine; Nuclear physics.
Address: Department of Radiology, Emory University, Atlanta, Georgia 30322, U.S.A.

CARTER, Robert Emerson, B.S., M.S. Born 1920. Educ.: Washington Coll. and Illinois Univ. Staff member, California Univ. Book: Co-author, Fast Neutron Phys. (Editors, Fowler and Marion) (Intersci. Publishers). Societies: A.P.S.; A.N.S.
Nuclear interests: Nuclear physics; Reactor physics; Radiobiology. Principal activities have been use of reactor radiations, both neutrons and gamma rays, to study their interactions with matter. Such experiments as non-elastic scattering cross sections of neutrons, thermal neutron capture gamma ray spectra and thermal neutron induced fission parameters have been studied. Present interests also include radiation induced biological processes.
Address: Armed Forces Radiobiology Institute, Defence Atomic Support Agency, Bethesda, Maryland, U.S.A.

CARTER, Commander Terence, Roy. Navy, B.Sc. (Hons. Phys.). Born 1929. Educ.: Birmingham Univ. and Roy. Naval Coll., Greenwich. Meteorologist, Roy. Naval Air Station, Halfar, Malta; Lecturer, Roy. Naval Elec. School; Lecturer, Roy. Naval Eng. Coll., Manadon, Plymouth; Sen. Lecturer, Roy. Naval Coll., Greenwich; Health Phys. and Lecturer, Nucl. Propulsion School, H.M.S. Sultan, Gosport; Sen. Instructor Officer, Britannia Roy. Naval Coll., Dartmouth; Principal, Gibraltar and Dockyard Tech. Coll., Gibraltar. Society: Inst. Phys.
Nuclear interests: Reactor design; Nuclear physics; Health physics.
Address: H.M.S. Rooke, Gibraltar.

CARTER, William Lloyd, B.S., M.S., Ph.D. Born 1920. Educ.: Georgia Inst. Technol., Texas Univ. At Union Carbide Nucl. Co. 1950-. Societies: A.N.S.; RESA.
Nuclear interests: Chemical reprocessing irradiated fuels; Reactor design.
Address: Oak Ridge National Laboratory, P.O. Box X, Oak Ridge, Tennessee, U.S.A.

CARTERON, Jean. Born 1926. Educ.: Ecole Polytechnique, Paris and Ecole Nationale Supérieure des Télécommunications. Directeur-Gén. de la SACS; Prof., Ecole Supérieure d'Electricité. Societies: Sté. Française des Electriciens; Sté. Française des Electroniciens et des Radioelectriciens; A.F.I.R.O.
Nuclear interests: System analysis and system design.
Address: 35 boulevard Brune, Paris 14, France.

CARTLEDGE, Groves Howard, A.B., A.M., Ph.D.,Sc.D. Born 1891. Educ.: Davidson Coll., Chicago Univ. Group Leader, Chem. Div., O.R.N.L., 1951-61; Consultant, 1961-. Society: A.C.S.

Nuclear interests: Electrochemistry and corrosion inhibitors, particularly studies of mechanism using the pertechnetate ion.
Address: Oak Ridge National Laboratory, P.O. Box X, Oak Ridge, Tennessee, U.S.A.

CARTWRIGHT, David Kendall, B.Sc. (Hons.), Phys.). Born 1928. Educ.: Liverpool Univ. P.S.O., U.K.A.E.A., 1952-.
Nuclear interests: Reactor safety and nuclear instrumentation of irradiation loops in reactors; Burst cartridge detection for reactors.
Address: United Kingdom Atomic Energy Authority, Reactor Engineering Laboratories, Risley, Warrington, Lancs., England.

CARTWRIGHT, Harry, M.A. Born 1919.
Educ.: Cambridge Univ. Director, Water Reactors, Reactor Group, U.K.A.E.A. Member, Nucl. Safety Advisory Com., M.O.P., 1962-; Vice-Pres., Risley Branch, Nucl. Eng. Soc. Societies: A.M.I.Mech.E.; A.M.I.E.E.
Nuclear interest: Reactor design.
Address: Norton Lodge Cottage, Norton, near Runcorn, Cheshire, England.

CARUSO, Lawrence R. Legal Counsel, Office of Res. Administration, Princeton Univ. Member, Board of Contract Appeals, U.S.A.E.C.
Address: Princeton University, Office of Research Administration, Princeton, New Jersey, U.S.A.

CARVAJAL y URQUIJO, Juan. Director, Tecnicas Nucleares S.A.
Address: Tecnicas Nucleares S.A., 46 Serrano, Madrid 1, Spain.

CARVALHO, Antonio Carlos, M.D. Res. Asst., Radioisotope Clinical Unit, Instituto de Biofisica, Brazil Univ.
Address: Brazil University, 458 Avenida Pasteur, Rio de Janeiro, Gb., Brazil.

CARVALHO, Antonio Herculano GUIMARAES CHAVES de. See GUIMARAES CHAVES de CARVALHO, Antonio Herculano.

CARVALHO, Armando de. See de CARVALHO, Armando.

CARVALHO, Hervasio GUIMARAES de. See GUIMARAES de CARVALHO, Hervasio.

CARVALHO, José Alberto FERNANDES de. See FERNANDES de CARVALHO, José Alberto.

CARVALHO, Rodrigo Alberto GUEDES DE. See GUEDES DE CARVALHO, Rodrigo Alberto.

CARVALHO e SOUZA, H. E. Madame Odette de. See de CARVALHO e SOUZA, H. E. Madame Odette.

CARVALHO FRANCO, Paulo de. See de CARVALHO FRANCO, Paulo.

CARVAO GOMES, F., Dr. Physician, Radioisotopes Res. Lab., Overseas Res. Council.
Address: Overseas Research Council, Av. da Ilha da Madeira (Encosta do Restelo), Lisbon 3, Portugal.

CASAGRANDE, Ivano, Dr. (Elec. Eng.), Libero Docente (Nucl. Eng.). Born 1929. Educ.: Bologna Univ. Manager, CIRENE Prototype Design Programme, Nucl. Plant Div., C.I.S.E., Milan, 1957; Lecturer on Nucl. Reactor Eng., Milan Politecnico. Editorial Board, Energia Nucleare.
Nuclear interest: Reactor design and evaluation.
Address: Centro Informazioni Studi Esperienze, Casella Postale 3986, Milan 20100, Italy.

CASALE, Renato, Dr. Born 1921. Educ.: Rome Univ. Pres., Ital Elettronica S.p.A., Rome; Pres., Soc. Italiana Ricerche Industriali, Terni; Member, Board of Directors, Ammonia Casale S.A., Lugano, and of Panammonia S.A., Panama City.
Nuclear interest: Nuclear instrumentation.
Address: 13 Viale Gorizia, Rome, Italy.

CASANOVAS, J. Asst., Centre de Physique Nucl., Faculte des Sci., Toulouse Univ.
Address: Toulouse University, 118 route de Narbonne, 31-Toulouse, (Haute-Garonne), France.

CASARES LOPEZ, Roman. Vocale, Comision Interministerial de Conservacion de Alimentos por Irradiacion, Junta de Energia Nucl.
Address: Junta de Energia Nuclear, Ciudad Universitaria, Madrid 3, Spain.

CASARETT, George William, Ph.D. Born 1920. Educ.: Rochester Univ. Prof. Rad. Biol., Rochester Univ. School of Medicine and Dentistry, 1959-. Societies: Rad. Res. Soc.; American Soc. for Experimental Pathology; Fellow, Gerontological Soc.; Fellow, New York Acad. of Sci.
Nuclear interests: Radiation pathology; Effects of ionising radiations on normal and diseased tissues.
Address: Department of Radiation Biology, Rochester University School of Medicine and Dentistry, P.O. Box 287, Station 3, Rochester 20, New York, U.S.A.

CASAS, Prof. Head, Group constructing mass spectrometers and studying isotopes, Zaragoza Univ.
Address: Zaragoza University, Zaragoza, Spain.

CASE, Edson Gardner, B.Sc. (Marine Eng.), Master of Naval Architecture and Marine Eng. Born 1924. Educ.: U.S. Naval Acad. and M.I.T. Officer, U.S. Navy, 1950-57; Nucl. Eng. (1957-59), Chief, Res. and Power Reactor Safety Branch (1959-61), Asst. Director, Div. of Reactor Licensing 1961-65), Deputy Director, Div. of Reactor Licensing, (1965-66), Director, Div. of Reactor Standards (1967-), U.S.A.E.C.
Nuclear interest: Nuclear safety criteria and standards for the location, design, construction,

and operation of nuclear reactors.
Address: United States Atomic Energy Commission, Division of Reactor Standards (BETH-010), Washington, D.C. 20545, U.S.A.

CASEY, A. T., Dr. Member, Board of Studies in Nucl. Sci. and Eng., Melbourne Univ.
Address: Melbourne University, Parkville N.2, Victoria, Australia.

CASEY, Daniel Joseph, B.Sc. (Civil Eng.). Born 1910. Educ.: Notre Dame, Princeton, and De Paul Univs. Director of Eng. and Construction, U.S.A.E.C., Chicago Operations Office. Society: Western Soc. of Eng.
Nuclear interests: Management; Design, and construction.
Address: 9800 South Cass Avenue, Argonne, Illinois 60439, U.S.A.

CASHMAN, Thomas J., B.S.Ch.E. Born 1920. Educ.: Rhode Island Univ. Manager, Reactor Component Test, then Manager, Reactor Assembly, then Manager, Reactor Operations, then Manager, Reactor Field Installation, Knolls Atomic Power Lab. (1950-56), Manager, Nucl. Projects, Installation and Service Eng. Dept. (1957-67), G.E.C.; Asst. Director, Planning, Office of Atomic and Space Development, 1967-; Registered Professional Eng., New York State. Societies: A.N.S.; A.I.Ch.E.
Nuclear interest: Management.
Address: Office of Atomic and Space Development, Alfred E. Smith State Office Building, Albany, New York 12225, U.S.A.

CASILLI D'ARAGONA, Massimo. Law, Economics and Political Sciences. Born 1913. Educ.: Naples Univ. Head of the Atomic Office, Ministry of Foreign Affairs. Career diplomat attached to O.C.E.D.; Chairman Co-ordinating Com. for East-West Trade Policy, Paris. Book: Organizzazione internazionale per lo sviluppo dell'energia atomica (International Organisation for the Development of Atomic Energy).
Nuclear interest: International field.
Address: Ministry of Foreign Affairs, Rome, Italy.

CASIMIR, Hendrik Brugt Gerhard, D.Sc., D.Sc. (Hon., Copenhagen, Louvain, Aachen and Edinburgh Univs.). Born 1909. Educ.: Leyden and Copenhagen Univs. Member, Board of Management responsible for Res., Philips Elec. Eindhoven; Prof. Extraordinary, Leyden Univ. Vice Pres., Roy. Acad. of Sci., Amsterdam; Pres., European Ind. Res. Management Assoc., Paris. Book: Interaction between Atomic Nuclei and Electrons (Haarlem 1936, reprinted by W. H. Freeman, 1963).
Nuclear interests: Nuclear instrumentation (cascade generators, cyclotrons, counting equipment) and radioactive isotopes; Theoretical physics.
Address: Philips Research Laboratories, Eindhoven, Netherlands.

CASO MONTANER, Alberto. Member, Forum Atomico Espanol.
Address: Forum Atomico Espanol, 38 General Goded, Madrid 4, Spain.

CASPARI, Max Edward, A.B., Ph.D. Born 1923. Educ.: Wesleyan Univ., Connecticut, and M.I.T. Res. Asst., M.I.T., 1948-54; Instructor (1954-55), Asst. Prof. (1955-56), Assoc. Prof. (1956-64), Prof. (1964-), Pennsylvania Univ. Societies: A.I.P.; A.P.S.
Nuclear interests: Nuclear and solid state physics; Measurement of hyperfine interactions by angular correlation techniques and determination of nuclear magnetic moments of short-lived excited states.
Address: Department of Physics, Pennsylvania University, Philadelphia, Pennsylvania 19104, U.S.A.

CASPER, Karl Joseph, B.Sc., Ph.D. Born 1932. Educ.: Ohio State Univ. Res. Assoc., Ohio State Univ., 1958-60; Instructor (1960-62), Asst. Prof. (1962-), Western Reserve Univ. Societies: A.P.S.; I.E.E.E. Professional Group on Nucl. Science.
Nuclear interest: Experimental low energy nuclear physics.
Address: Department of Physics, Western Reserve University, Cleveland 6, Ohio, U.S.A.

CASS, N. M., M.B., B.S., D.A., F.F.A.R.C.S. Radiobiol. Res. Anaesthetist, Cancer Inst. Board.
Address: Cancer Institute Board, 278 William Street, Melbourne, C.1, Australia.

CASSATT, Wayne A., Jr., A.B., M.S., Ph.D. Born 1925. Educ.: Michigan Univ. Res. Chem., Phillips Petroleum Co., 1954-59; Tech. Adviser, U.S. Internat. Co-operation Administration, Karachi, 1959-61; Assoc. Prof., Chem., Nucl. Reactor Project, Washington State Univ., 1961-. Society: A.C.S.
Nuclear interests: Neutron activation analysis; Perturbed angular correlations; Nuclear decay schemes.
Address: Washington State University, Pullman, Washington, U.S.A.

CASSELS, James Macdonald, M.A., Ph.D. Born 1924. Educ.: Cambridge Univ. Fellow, subsequently P.S.O., A.E.R.E., Harwell, 1949-53; Lecturer, subsequently Prof. Exptl. Phys. (1953-59), Lyon Jones Prof. Phys. (1960-), Liverpool Univ.; Visiting Prof., Cornell Univ., 1959-60. Societies: F.R.S.; Fellow, Phys. Soc.; A.P.S.
Nuclear interests: Nuclear and fundamental particle physics.
Address: Chadwick Physics Laboratory, Liverpool University, Liverpool 3, England.

CASSEN, Benedict, M.S., Ph.D., A.R.C.S. Born 1902. Educ.: California Inst. Technol. Prof. Biophys., California Univ., Los Angeles, 1947-. Book: Co-author, Practice of Nuclear Medicine (Charles C. Thomas, 1958).

Societies: Soc. of Nucl. Medicine; A.P.S.
Nuclear interest: Nuclear and radiation physics and applications to biology and medicine.
Address: Laboratory of Nuclear Medicine and Radiation Biology, 900 Veteran Avenue, Los Angeles, California 90024, U.S.A.

CASSIDY, Bernard F., B.E.E., M.S. (E.E.). Born 1927. Educ.: Manhattan Coll. and M.I.T. Manager, DIG Elec. Systems (1958-), Manager, D2 Test (1960-), Manager S5G Elec. (1962-), G.E.C., K.A.P.L. Computing Devices Com., I.E.E.E. Society: I.E.E.E.
Nuclear interests: Control and instrumentation for naval nuclear propulsion plants.
Address: 2155 Orchard Park Drive, Schenectady, New York, U.S.A.

CASTAGNA, A. Prof. Termotecnica del Reattore, Univ. degli Studi, Rome.
Address: Universita degli Studi di Roma, 18 Via Eudossiana, Rome, Italy.

CASTELLANI, Renato, Ing. Director, Italelettronica.
Address: Italelettronica, 72 Via Ignazio Pettinengo, Rome, Italy.

CASTELLANO, Enrique, Dr. Sec., Comision Especial de Fisica Atomica y Radioisotopes, Univ. Nacional de La Plata.
Address: Universidad Nacional de La Plata, La Plata, Argentina.

CASTELLI, Franco, Graduate Eng. Born 1901. Educ.: Milan Polytech. Central Director, Ente Nazionale Energia Elettrica. Vice-Pres., Assoc. Nazionale di Ingegneria Nucleare; Vice-Pres., F.I.E.N. Societies: Atomic Ind. Forum; Fellow, A.S.M.E.; A.S.T.M.; Assoc. Member, A.I.E.E.; A.N.S.
Nuclear interests: Management; Reactor design.
Address: 3 Via G. B. Martini, 00198 Rome, Italy.

CASTERA, Pierre, Ing. des Arts et Manufactures. Born 1928. Educ.: Bordeaux and Angers Univs. Asst. Gen. Manager, C.O.C.E.I. Societies: A.T.E.N.; Sté. Ing. Civils de France.
Nuclear interests: Reactor control; Nuclear power plants.
Address: c/o C.O.C.E.I., 22 rue de Clichy, Paris 9, France.

CASTILLO, Guillermo, Director, Dept. de Fisica, Universidad Nacional de Colombia.
Address: Departamento de Fisica, Universidad Nacional de Colombia, Bogota, Colombia.

CASTILLO DOIG, Contador Público Augusto. Departamento de Contabilidad, Instituto Superior de Energia Nucl., Junta de Control de Energia Atomica del Peru.
Address: Junta de Control de Energia Atomica del Peru, 3420 Avenida Arequipa, Apartado 914, Lima, Peru.

CASTILLO LEDON, Amalia G. C. de. See de CASTILLO LEDON, Amalia G. C.

CASTLE, Richard Thomas, B.S., M.S., Ph.D. Born 1934. Educ.: Otterbein Coll., Westerville and Ohio Univ. Societies: A.P.S.; A.A.A.S.
Nuclear interests: Nuclear structure; Decay schemes.
Address: Battelle Memorial Institute, 505 King Avenue, Columbus, Ohio 43201, U.S.A.

CASTLEMAN, Louis Samuel, B.S. (Metal.), D.Sc. (Metal.). Born 1918. Educ.: M.I.T. Sen. sci. and supervisory sci., Westinghouse Atomic Power Div., Pittsburgh, Pennsylvania, 1950-54; Metal. specialist and sect. head, Gen. Telephone and Electronics Labs., Inc. (formerly Sylvania Res. Labs.), Bayside, N.Y., 1954-64; Instructor and Adjunct Prof. (1955-64), Prof. Phys. Metal. (1964-), Brooklyn Polytechnic Inst.; Assoc. School of Mines, Columbia Univ., 1957-58. Societies: A.P.S.; A.I.M.M.E.; N.Y. Metal. Sci. Club; A.S.M.; A.S.E.E.
Nuclear interests: Effects of nuclear reactor environment on the physical and mechanical properties of, and on solid-state reactions within, fissionable and structural metals and alloys.
Address: 15 Oak Street, Lynbrook, New York 11563, U.S.A.

CASTORINA, Thomas C., B.A., M.S. Born 1918. Educ.: Brooklyn Coll. and Stevens Inst. Technol. Sen. Sci., Explosives Lab., Picatinny Arsenal. Societies: A.C.S.; Rad. Res. Soc.; RESA.
Nuclear interests: Radiation chemistry of heterogeneous systems; Preparative radiation chemistry.
Address: Explosives Laboratory, Picatinny Arsenal, Dover, New Jersey, U.S.A.

CASTRACANE, Nicola. Exec. Pres., Internat. Exhibition of Electronics, Nucl. Energy, Wireless, T.V. and Cine.
Address: International Exhibition of Electronics, Nuclear Energy, Wireless, T.V. and Cine, 9 Via Crescenzio, Rome, Italy.

CASTRO, Humberto RUIZ. See RUIZ CASTRO, Humberto.

CASTRO, Irene Emygdio de. See de CASTRO, Irene Emygdio.

CASTRO, J. L. SILVA e. See SILVA e CASTRO, J. L.

CASTRO, Jose Sarmento de VASCONCELLOS e. See de VASCONCELLOS e CASTRO, Jose Sarmento.

CASTRO, P. F. SAMPAIO e. See SAMPAIO e CASTRO, P. F.

CASTRO CARDUS, Santiago. Director, Forum Atomico Espanol.
Address: Forum Atomico Espanol. 38 General Goded, Madrid 4, Spain.

CASTRO de FARIA, Nelson V. de. See de CASTRO de FARIA, Nelson V.

CASTRO FARIA, Hugo. Prof., Physiological Chem., F.C.M., Dept. of Res. Radioisotope Lab., Ministerio da Saude D.N.S.-Servico Nacional de Cancer.
Address: Ministerio da Saude D.N.S.-Servico Nacional de Cancer, 23 Praca de Cruz Vermelha, Rio de Janeiro, Brazil.

CASTRO y BRAVO, F. de. See de CASTRO y BRAVO, F.

CASWELL, Randall Smith, S.B., Ph.D. Born 1924. Educ.: Oregon Univ. and M.I.T. Assoc. Prof. Phys., Kentucky Univ., 1950-52; Phys. (1952-), Chief, Neutron Phys. Sect. (1957-), Nat. Bureau of Standards, Washington, D.C.; Adjunct Prof. Phys., American Univ., Washington, D.C., 1958-. Member, Neutron Planning Board, I.C.R.U.; Member, Com. on Neutrons and Heavy Particles, Nat. Council on Rad. Protection and Measurements; Member, Neutron Working Group, Internat. Bureau of Weights and Measures, Sèvres, France; Member, Nucl. Cross Sections Advisory Com., U.S.A.E.C. Society: Fellow, A.P.S.
Nuclear interests: Neutron physics, average neutron cross sections, neutron standards, neutron dosimetry.
Address: 2209 Salisbury Road, Silver Spring, Maryland, U.S.A.

CATACOSINOS, William James, B.S., M.A., Ph.D. Born 1930. Educ.: New York Univ. Member, Ramey Procurement Study Panel, U.S. A.E.C., 1967; Consultant, Nucl. Res. Management, Democritos Nucl. Res. Centre, Greek A.E.C., 1967. With Assoc. Univs. Inc., Brookhaven Nat. Lab. Member, Exec. Com., Suffolk Div., American Cancer Soc.; Member, Exec. Com., Long Island Univ. School of Sci. and Eng.; Member, Industrial Commission, Town of Brookhaven.
Nuclear interest: Management of nuclear research facilities.
Address: 34 Landing Lane, Port Jefferson, New York 11777, U.S.A.

CATALA de ALEMANY, Joaquin, Dr. Phys. Born 1911. Educ.: Barcelona and Madrid Univs. Dean, Faculty of Sci., Univ. Valencia, (1954-) now Director, Particle Phys. Inst., Valencia Univ. Member, Acad. of Sci., Barcelona; Meteorologist; Director, Centro de Fisica Fotocorpuscular, 1955-58. Head, Exchange Unit, I.A.E.A. Books: Fisica general (1958); Introduccion a la Fisica Nucl.: David Halliday (translated into Spanish). Societies: Real Soc. Espanola de Fisica y Chimica; Soc. Italiana de Fisica.
Nuclear interests: Nuclear photographic emulsion techniques.
Address: Faculty of Science, Valencia University, Valencia, Spain.

CATANIUS, Jean. Member, Commission Consultative pour la Production d'Electricite d'Origine, C.E.A.
Address: Commissariat a l'Energie Atomique, 29-33 rue de la Federation, Paris 15, France.

CATCH, John Reynolds, B.Sc., Ph.D. Born 1918. Educ.: London Univ. Manager, Organic Dept., Radiochem. Centre, U.K.A.E.A. Book: Carbon-14 Compounds (Butterworth, 1961).
Nuclear interest: Preparation and use of compounds labelled with radioactive isotopes.
Address: Radiochemical Centre, United Kingdom Atomic Energy Authority, Amersham, Bucks., England.

CATCHPOLE, Arthur George, B.Sc., Ph.D., F.R.I.C. Born 1917. Educ.: London Univ. Vice Principal, Kingston Coll. of Technol., 1965-. Chairman, Midlands Assoc. for Qualitative Analysis.
Address: Kingston College of Technology, Penrhyn Road, Kingston upon Thames, Surrey, England.

CATE, James L., Dr. Prof. of History, Chicago Univ. Member, Historical Advisory Com., U.S.A.E.C.
Address: Chicago University, Chicago, Illinois 60637, U.S.A.

CATHERS, George Ivan, Ph.D. (Phys. Chem.). Born 1915. Educ.: West Virginia and Yale Univs. Problem Leader, Fluoride Volatility Group, O.R.N.L., 1952-. Society: A.C.S.
Address: 5312 Maywood Road, Knoxville, Tennessee, U.S.A.

CATHERWOOD, Martin P., Dr. Member, Co-ordinating Council, New York Office of Atomic Development.
Address: New York Office of Atomic Development, The Alfred E. Smith State Office Building, Albany, New York 12225, U.S.A.

CATHROW, William René. Born 1915. Chief Quantity Surveyor, Eng. Group H.Q., U.K.A.E.A., Risley. Society: F.R.I.C.S.
Address: Heathfield House, 732 Warrington Road, Risley, Warrington, Lancs., England.

CATLIN, Robert James, A.B. (Biol.). Born 1925. Educ.: Harvard Coll., and Princeton Univ. Sen. Supervisor, Works Tech. Dept., Health Phys. Sect., E. I. Du Pont de Nemours and Co., Savannah River Plant, 1951-56; Manager, Health Phys., Westinghouse Testing Reactor, 1956-. Chairman, W.T.R. Safeguards Com.; Certified Health Phys. (American Board of Health Phys.). Societies: H.P.S. (Pres., Western Pennsylvania Chapter); A.N.S.

CAT

Nuclear interests: Health physics, including facilities engineering for hazards control; Waste disposal, including spent reactor fuel and fueled experiments; Reactor and facilities safeguards; Licensing of facilities and materials.
Address: Westinghouse Testing Reactor, P.O. Box 1075, Pittsburgh 30, Pennsylvania, U.S.A.

CATSCH, Alexander, Prof. Dr. med. Born 1913. Educ.: Berlin, Freiburg im Breisgau and Karlsruhe Univs. and Inst. of Radiobiol., Nucl. Res. Centre, Karlsruhe.
Nuclear interests: Removal of internally deposited radionuclides; Metabolism of radionuclides.
Address: P. O. B. 947, Karlsruhe, Germany.

CATTANEO, Gianfranco. Operations Manager, S.G.S. Fairchild - Sta. Gen. Semiconduttori S.p.A.
Address: S.G.S. Fairchild - Societa Generale Semiconduttori S.p.A., 1 Via Olivetti, Agrate B, Milan, Italy.

CATTIN, Pierre BENOIT-. See BENOIT-CATTIN, Pierre.

CATTRALL, Robert Walter, B.Sc. (Hons.), Ph.D. Born 1936. Educ.: Adelaide Univ. Res. Chem., Australian Mineral Development Labs; Post-doctoral Scholar, London Univ.; Sen. Demonstrator Inorg. Chem., Melbourne Univ. Now Lecturer Inorg. Chem., La Trobe Univ. Co-editor, Reviews of Pure and Appl. Chem. Society: Roy. Australian Chem. Inst.
Nuclear interest: Liquid-liquid extraction systems involving high molecular weight amines and phosphorus compounds.
Address: Chemistry Department, La Trobe University, Bundoora 3083, Victoria, Australia.

CATTRELL, Victor Gordon, M.A. (Cantab.). Born 1924. Educ.: Cambridge Univ. Sci. Officer, Ministry of Supply, 1949-55; Principal Phys., Western Regional Hospital Board, Glasgow, 1955-64; Lecturer, Medical Phys., Edinburgh Univ., 1964-. Societies: Assoc. Inst. of Phys.; Hospital Phys. Assoc.; Soc. of Univ. Rad. Protection Officers.
Nuclear interest: Radiation protection and the application of radiation techniques to medical research.
Address: Department of Medical Physics, Edinburgh University, Royal Infirmary, Edinburgh, Scotland.

CATUDAL, Frank W., B.S. (Eng.), P.E. Born 1919. Educ.: Rhode Island Univ. Supt., Eng. Div., Travelers Insurance Co., 1948-; Eng. Consultant, K.A.P.L., 1954-55. Chairman, Eng. Com., Nucl. Energy Liability Insurance Assoc.; Consultant, Nucl. Energy Property Insurance Assoc.; Chairman, Subgroup on Gen. Requirements, Code Sect. III, Member, Task Group on Prestressed Reactor Vessels, A.S.M.E.; Member, Com. on Nucl. Power, Member, Nucl. Standards Board, Member, N-45 Nucl. Com., U.S.A.S.I.; Member, Eng. Com., Nat. Bureau

248

of Casualty and Surety Companies. Societies: A.N.S.; N.S.P.E.
Nuclear interests: Design and safety of nuclear reactors - power, research, testing and training; Design and safety of nuclear fuel reprocessing, uranium scrap recovery; Food and materials irradiation facilities.
Address: 23 Springbrook Lane, Simsbury, Connecticut 06070, U.S.A.

CAUSSE, Jean-Pierre, Agrégé des Sciences Physiques (France). Born 1926. Educ.: Ecole Normale Superieure, Paris. Head, Electron Phys. Sect., Schlumberger Well Surveying Corp., Ridgefield, Conn., 1955-60; Manager, Photoelec. Dept., ASCOP Div., Electro-Mech. Res., Inc., Princeton, N.J., 1960-; Societies: A.P.S.; Optical Soc. America; Inst. Radio Eng.; Sté. Française de Physique.
Nuclear interests: Photomultipliers; Neutron generators; Radioactivity well logging.
Address: 52 Hartley Avenue, Princeton, N.J., U.S.A.

CAUTIUS, Werner, Dipl. Ing. Geschäftsführer, Arbeitsgemeinschaft Versuchs-Reaktor G.m.b.H. Papers: Co-author, Tasks and work of the groups for the building and operations of reactors (Atomwirtschaft, vol. 5, No. 7/8 1960); co-author, The A.V.R. reactor in the Jülich atomic research station (ibid.).
Address: A.V.R., 105 Luisenstrasse, Düsseldorf, Germany.

CAVACA, Rogério A., Mining Eng. Born 1903. Educ.: Inst. Superior Técnico, Lisbon. Superior Mining Inspector, Overseas Ministry, 1944-55; Gen. Director, Prospecting and Mining Exploration Services, Junta de Energia Nuclear.
Nuclear interests: Uranium geology and prospecting.
Address: 36-2⁰ Av. Defensores de Chaves, Lisbon, Portugal.

CAVAGGIONI, Mario, Dott. Ing. Born 1925. Educ.: Naples Univ. Officer, Italian Military Navy, 1950-58; Head, Naval Propulsion Group, C.N.E.N., 1959-64; Member, for Italy, Nucl. Ship Propulsion Com., Euratom.
Nuclear interest: Nuclear naval propulsion.
Address: 7/A Via Ravenna, Rome, Italy.

CAVALIERI, Ralph R., M.D. Born 1932. Educ.: New York Univ. Chief, Radioisotope Service, Veterans Administration Hospital, San Francisco, 1963-; Asst. Clinical Prof., Medicine and Radiol., California Univ., San Francisco Medical Centre, 1963-. Societies: Soc. of Nucl. Medicine; Endocrine Soc.; American Federation for Clinical Res.
Nuclear interests: Nuclear medicine (application of radioisotopes to diagnosis and treatment of disease); Thyroid physiology.
Address: Veterans Administration Hospital, 42nd Avenue and Clement, San Francisco, California 94121, U.S.A.

CAVALLARO, S. Member, Gruppo Van De Graaff, Centro Siciliano di Fisica Nucleare, Istituto di Fisica, Catania Univ.
Address: Catania University, 57 Corso Italia, Catania, Sicily, Italy.

CAVALLERI, Giancarlo, Dr. in Eng. Born 1932. Educ.: Politecnico, Milan, Engaged in Res. at C.I.S.E. Lab.
Nuclear interests: Gaseous state physics and electromagnetism.
Address: 4 Via Randaccio, Milan, Italy.

CAVALLI-SFORZA, Luigi Luca, M.D., (Pavia) M.A. (Cambridge). Born 1922. Educ.: Pavia Univ. Istituto Sieroterapico Milanese, Milan, 1950-57; Prof. Genetics, Parma Univ., 1958-63; Prof. Genetics, Pavia Univ., 1963-. Book: La teoria dell'urto ed unita biologiche elementari (Target theory and biological elementary units) (Milan, Longanesi, 1948); Analisi Statistica per Medici e Biologi (Turin, Boringhieri.).
Nuclear interest: Human genetics.
Address: Istituto di Genetica, Università di Pavia, 14 Via Sant'Epifanio, Pavia, Italy.

CAVOLLORO, Raffaele, Dr. in Agricultural Sci. (magna cum laude). Born 1925. Educ.: Perugia Univ. Asst. Agriculturalist Entomology Lab., 1949-52; Exptl. Director, and Chief Entomology Lab., Istituto Scientifico Sperimentale per i Tabacchi, Scafati (Salerno), 1953-63; Saclay C.E.A., Radiobiol. Scholarship I.A.E.A., 1959-60; Prof. Appl. Nucl. Energy in Agriculture, Bari Univ. 1962; C.E.E.A., radioisotopes and labelled molecules, Brussels, 1962; Chief, Sect. Entomologie Service de Biologie, Ispra (Varese), C.C.R., C.E.E.A., 1963-. Deleg. to Internat. Congress in France (1955), Austria (1960), Greece (1963), Great Britain (1964). Societies: Italian Entomological Soc.; Italian Zoological Union; Drs. in Agricultural Sci. Soc.
Nuclear interests: Applications of nuclear energy in agriculture; Radioisotopes and ionising radiations in entomology.
Address: European Atomic Energy Community, Service de Biologie, Section Entomologie, Centre Commun de Recherche, Ispra, (Varese), Italy.

CAVANAGH, Patrick Edward, B.Sc., Ph.D., A.K.C. Born 1916. Educ.: London Univ. At Harwell (1946-) Sen. Principal Sci., Group Leader of Proton Phys. Group, A.E.R.E., (1958-), U.K.A.E.A. Society: A.P.S.
Nuclear interests: Nuclear physics, particularly nuclear reactions in the intermediate energy range; Elementary particle physics; Data processing.
Address: 2 Chaucer Crescent, Newbury, Berks., England.

CAVANAUGH, Richard G. Director, Security Div., Chicago Operations Office, U.S.A.E.C.
Address: Chicago Operations Office, 9800 South Cass Avenue, Argonne, Illinois 60439, U.S.A.

CAVEN, W. E. Vice Pres., Eng. and R. and D., Atlantic City Elec. Co.
Address: Atlantic City Electric Co., 1600 Pacific Avenue, Atlantic City, New Jersey, U.S.A.

CAVENDISH-BENTINCK, V. F. W. With Rio Tinto Zinc Corp.; Council Member, British Nucl. Forum, 1963-.
Address: Rio Tinto-Zinc Corp. Ltd., 6 St. James's Square, London S.W.1, England.

CAVERS, David Farquhar, B.S. (Econ.), LL.B. Born 1902. Educ.: Pennsylvania and Harvard Univs. Prof. Law, Harvard Univ., 1945-; Assoc. Dean, Harvard Law School, 1951-58; Pres., Walter E. Meyer Res. Inst. Law.
Nuclear interests: Research and instruction in laws and international agreements relating to the development and regulation of the use of nuclear energy for power and radiation, and also in legal aspects of international control of atomic weapons.
Address: 21 Buckingham Street, Cambridge, Massachusetts, U.S.A.

CAVILES, Alendry, Dr. Physician, Radioisotope Lab., Philippine General Hospital.
Address: Philippine General Hospital, Manila, Philippines.

CAWLEY, Sir Charles Mills, A.R.C.S., B.Sc. (1st class Hons. in Chem.), D.I.C., M.Sc., Ph.D., F.R.I.C., F.Inst.Pet., D.Sc. (London), F.Inst.F. Born 1907. Educ.: Imperial Coll. of Sci. and Technol. (Roy. Coll. of Sci.). Director, Headquarters, D.S.I.R., of Divs. concerned with the Department's Labs. and Grants to Univs., 1953-59; Chief Sci., M.O.P., 1959-67; Civil Service Commissioner (Sci. and Eng.) 1967-. Fellow, Imperial Coll. of Sci. and Technol.
Nuclear interest: General interest in nuclear power production.
Address: Springfield, Longlands, Worthing, Sussex, England.

CAWLEY, John H., M.S. (Elec. Eng.). Born 1925. Educ.: Notre Dame and California Univs. At Scripps Inst. Oceanography, California Univ., 1949-57; Gen. Atomic Div., Gen. Dynamics Corp., 1957-. Consultant to U.S. Navy, 1959-60. Societies: Sen. Member, I.S.A.; Sen. Member, I.R.E.
Nuclear interests: Nuclear reactor and radiation monitoring instrumentation, as well as special instrumentation suited to reactor requirements.
Address: General Atomic Division, General Dynamics Corp., P.O. Box 608, San Diego 12, Calif., U.S.A.

CAZAUX, Jean Maurice, B.L., B.Sc. Dr. en Droit, L. es lettres, Dipl., Ecole des Sci. Politiques. Born 1914. Educ.: Bordeaux and Paris Univs. Gen. Sec., G.I.I.N., Groupe Intersyndical de l'Industrie Nucléaire.
Address: G.I.I.N., 15 rue Beaujon, Paris 8, France.

CAZIER, Gail Allan, M.S. (Maths.). Born 1931. Educ.: Idaho Univ. Mathematician, Phillips Petroleum Co., 1955-. Society: Assoc. for Computing Machinery.
Nuclear interests: Theoretical physics and computer programming.
Address: Phillips Petroleum Company, Atomic Energy Division, P. O. Box 2067, Idaho Falls, Idaho, U.S.A.

CAZORLA TALLERI, Alberto. Delegado del Ministerio de Salud Publica ante la Junta de Control de Energia Atomica. Deleg., Third I.C.P.U.A.E., Geneva, Sept. 1964.
Address: Junta de Control de Energia Atomica del Peru, 3420 Avenida Arequipa, Apartado 914, Lima, Peru.

CEBESOY, I., Dr. Staff Member, Faculty of Medicine, Inst. and Clinic of Radiotherapy, Istanbul Univ.
Address: Istanbul University, Faculty of Medicine, Institute and Clinic of Radiotherapy, Topkapi, Istanbul, Turkey.

CECH, B., Doc. Ing. Member, Editorial Board, Jaderna Energie.
Address: Jaderna Energie, SNTL, Spalena 51, Prague 1, Czechoslovakia.

CEDERWALL, Gustav F. E., Licenciate of Philisophy. Born 1913. Educ.: Stockholm Univ. Adviser, Inst. for Business Res., 1946- and 1951-53; Economic Adviser, O.E.E.C., 1949-51; Head, Nat. Budget Div., Ministry of Finance, 1948-49 and 1953-56; Under-Sec., Ministry of Commerce, 1956-; Chairman, Atomic Energy Board, 1956-; Member, Nordic Contract Com. for Atomic Energy.
Address: Handelsdepartementet, Stockholm 2, Sweden.

CEFIS, Eugenio, Dr. Pres., S.N.A.M.
Address: S.N.A.M., S.p.A. Casella Postale 4159, Milan, Italy.

CELADA, Francesco. Director, R. F. Celada S.R.L.
Address: R. F. Celada S. R. L., 4 Viale Tunisia, Milan, Italy.

CELADA, Giovanni. Sen. Officer, R. F. Celada S.R.L.
Address: R. F. Celada S.R.L., 4 Viale Tunisia, Milan, Italy.

CELEDA, Jiri, Ing. (M.Sc.Eng.), doc. (Prof. Asst.), C.Sc. (Ph.D.). Born 1916. Educ.: Inst. Chem. Technol., Prague. Head, Metallograph. Dept., Res. Inst. CKD - Works, Prague, 1945-50; Head, Res. Dept., Ministry of Chem. Industry, 1951-54; Head, Dept. of Technol. of Nucl. Fuel and Radiochem. (1955-), Dean, Faculty of Inorganic Technol. (1959-61), Vice-Dean (1966-) Inst. Chem. Technol. Prague; Director, Nat. Tech. Museum, Prague, 1964-66; Chief editor, sci. journal Chemicky prumysl (Chem. Ind.), 1951-. Books: Co-author, Atom a jeho energie (The Atom and its Energy) (Orbis, 1954); Contribution, The Philosophical Problems of Modern Phys. included in collective publication Philosophy and Natural Sci. (Prague, SNPL, 1962-); co-author, Cesta do nitra hmoty (The Way to the Interior of Matter) (Prague, SNTL, 1961). Society: Czechoslovak Sci. and Tech. Soc. (Cs. vedeckotechnicka spolecnost), Sect. for Radiochem.; Czechoslovak Chemical Soc. (Cs. chemicka spolecnost), Sect. for Inorg. Chem. and Sect. for Chem. Education.
Nuclear interests: Chemistry of radioactive and trace elements especially chemistry of their complexes in highly concentrated solutions of background electrolytes, including the development of physical methods based on an extended theory of highly concentrated electrolytic solutions (density, activity coefficients, conductance, salting-out action).
Address: 21 Thamova, Prague 8, Czechoslovakia.

CELEMIN, Alfio VIDAL. See VIDAL CELEMIN, Alfio.

CELINSKI, Zdzisław Nikodem, D.Sc., Dipl. in Elec. Eng. Born 1932. Educ.: Warsaw Tech. Univ. Chief of Res. Group, Inst. of Nucl. Res., Swierk near Otwock, 1959-.
Nuclear interests: MHD Generators.
Address: 88 m. 35 ul. Woloska, Warsaw, Poland.

CELIS, René, Degree in Phys. Born 1925. Educ.: Louvain Univ. Manager, R. and D. Dept., Cimenteries C B R (in charge of nuclear problems).
Address: 34 boulevard de Waterloo, Brussels, Belgium.

CELLINI, Ricardo FERNANDEZ. See FERNANDEZ CELLINI, Ricardo.

CEMBER, Herman, B.S. (C.C.N.Y.), M.S., Ph.D. (Pittsburgh). Born 1924. Educ.: C.C.N.Y. and Pittsburgh Univ. Asst. Prof., then Assoc. Prof., Graduate School of Public Health, Pittsburgh Univ., 1950-60; Assoc. Prof., Univ. Cincinnati Coll. Medicine, 1960-64; Health Phys., Internat. Labour Office, Short-term temporary appointment, 1961; Prof., Northwestern Univ., 1964-. Societies: H.P.S.; Rad. Res. Soc.; A.I.H.A.; New York Acad. of Sci.; A.A.A.S.; A.A.U.P.; A.P.H.A.
Nuclear interests: Health physics; Research interest in radiobiology, with emphasis on pulmonary effects from inhaled radioactive aerosols.
Address: 626 Noyes Street, Evanston, Illinois 60201, U.S.A.

CENTER, Clark E. Formerly Gen. Manager, A.E.C. operations, Union Carbide Corp., Oak Ridge; Vice-Pres., and formerly Manager of Production, Union Carbide Nuclear Co. Director, Atomic Ind. Forum, Inc.

Address: Union Carbide Nuclear Co., Oak Ridge, Tennessee, U.S.A.

CEPEDA, Carlos, OQUENDO. See OQUENDO CEPEDA, Carlos.

CERMAK, Vladimir, RNDr., Can. Chem. Sci. Born 1920. Educ.: Charles Univ., Prague, Mass Spectrometry Sect., Inst. Phys. Chem., Czechoslovak Acad. Sci.
Nuclear interest: Mass spectrometry.
Address: Institute of Physical Chemistry, Czechoslovak Academy of Sciences, 7 Machova, Prague 2, Czechoslovakia.

CERNY, Pavel Petr, Dr. of Chem. Born 1927. Educ.: J. E. Purkyne Univ., Brno. Asst. Lecturer, J. E. Purkyne Univ., Brno, 1950-51; Asst. Lecturer, J. A. Komensky Univ., Bratislava, 1951-59; Director of Res., Centre for the Application of Radioactive Isotopes in the Textile Industry, Brno, 1959-. Societies: County Com., Czechoslovak Sci. and Tech. Soc.; Chairman, State Commission for Rad. Sterilisation.
Nuclear interests: Radiation chemistry; Radiation sterilisation of surgical and pharmaceutical products; Modification of polymers by radiation; Tracing methods; Beta gauging and electrostatic elimination as applied in the textile industry.
Address: State Textile Research Institute, Centre for the Application of Radioactive Isotopes in the Textile Industry, 6 Vaclavska, Brno, Czechoslovakia.

CERRAI, Enrico, Dr. Chem., Prof. Nucl. Chem. Born 1924. Educ.: Pisa Univ. C.I.S.E. (Milan), 1949- (now Gen. Director); Montecatini Co., Settore Progetti e Studi, 1950-55; Milan Politecnico, 1950; Steering Com., Energia Nucleare; Editorial Board, Acqua Industriale; Advisory Com., Talanta. Book: Co-author, Materiali Nucleari Speciali-Acqua Pesante e Deuterio. Societies: Soc. Chimica Italiana; Soc. Analytical Chem. (England).
Nuclear interests: Chemistry and technology of nuclear materials; Reactor technology; Radioactive waste decontamination; New separation procedures of stable and radioactive nuclides by ion exchange, selective chromatography and liquid-liquid extraction; Activation analysis.
Address: 8 Via Griziotti, Milan, Italy.

CERRE, Pierre Robert, Certificat d'Etudes Supérieures de Physique Générale. Born 1904. Educ.: Sorbonne. Manager, Mestre et Blatge firm; Gen. Manager, Sté. Plastiques-Ignifuges-Traitements; Chef des Etudes Techniques au Service de Contrôle des Radiations et de Génie Radioactif, C.E.A., Saclay, 1956-. Books: Processing and Pre-treatment of Solid Radioactive Wastes (I.A.E.A., Vienna); Traitement et Stockage des Matériels Contaminés – Centre d'Etudes Nucléaires de Saclay (Rapport C.E.A. No. 1197). Societies: H.P.S.; A.N.S.; Sté. Française de Radioprotection.
Nuclear interests: Health physics, and specially wastes disposal.
Address: 3 avenue Jean Perrin, Fontenay aux Roses, Seine, France.

CERRI, Brunello, M.D. Born 1932. Educ.: Pisa Univ. Res. Asst., Medical Clinic, Pisa Univ., 1957-. Society: Italian Soc. for Nucl. Biology and Medicine.
Nuclear interest: Experimental and clinical uses of radioisotopes in cardiology.
Address: Centro di Medicina Nucleare, Università di Pisa, Italy.

CERROLAZA ASENJO, Jose Angel, Licenciado (M.Sc.) (Chem.),Eng. in Naval Ordnance. Born 1926. Educ.: Madrid Univ. and Madrid School of Naval Ordnance Eng. Head, Tech. Information Sect., Junta de Energia Nucl., 1957; Prof., Statistics and Quality Control, Madrid School of Naval Ordnance Eng.
Nuclear interests: General and technical information; Quality of nuclear products and components.
Address: 106 Avenida del Generalisimo, Madrid 16, Spain.

CERTAINE, Jeremiah. With United Nucl. Corp. Member, Exec. Com., Mathematics and Computation Div., A.N.S.
Address: United Nuclear Corporation, 5, New Street, White Plains, N.Y., U.S.A.

CERVASEK, Jiri, D.Ph. Born 1918. Educ.: Prague Tech. Univ. Res. worker, Res. Inst. for Material and Technol., 1950-56; Chief of Phys. Metal. Dept., Nucl. Res. Inst., Czechoslovak Acad. Sci., 1956-.
Nuclear interest: Metallurgy.
Address: 68 Belohorska Street, Prague 6, Czechoslovakia.

CERVELLATI, Andrea, Dr. Ind. Chem. Born 1928. Educ.: Bologna Univ. Biol. and Health Phys. Div. Dosimetry Lab. C.N.E.N.; Asst. Nucl. Eng. Specialisation School, Bologna Univ., 1958-.
Nuclear interests: Radiochemistry; Health safety (general); Radioactivity environments dosimetry.
Address: 18/3 Via Ghirardacci, Bologna, Italy.

CERVELLINI, Admar, Livre-Docente. Born 1920. Educ.: Sao Paulo Univ. Prof. Catedrático, Phys. Dept., Escola Superior de Agricultura "Luiz de Queiroz", Sao Paulo Univ., 1953-; Director, Centro de Energia Nucl. na Agricultura.
Nuclear interests: Use of stable and radioactive isotopes in agriculture, mass spectrometry.
Address: Escola Superior de Agricultura "Luiz de Queiroz", Piracicaba, Sao Paulo, Brazil.

CERVIGNI, Tommaso, Docent in Biochem. Born 1921. Educ.: Rome Univ. Istituto Scientifico Sperimentale per i Tabacchi, 1950-53; Lega Italiana per la lotta contro i tumori,

1953-58; Laboratorio per le Applicazioni in Agricoltura del C.N.E.N., 1959-. Societies: Soc. Italiana di Fisiologia Vegetale; Assoc. des Radiobiologistes des pays de l'Euratom. Nuclear interests: Radiobiology; Biochemical and biophysical effects of radiations on plants.
Address: 5 Via Borgorne, Rome, Italy.

CERVINO, Jose, Prof. Dr. Sub-Director, Inst. Endocrinologia, Hospital Pasteur.
Address: Instituto de Endocrinologia, Hospital Pasteur, Montevideo, Uruguay.

CESARANO, Catello, Dr. in Ind. Chem. Born 1928. Educ.: Milan Univ. Director, Hot Operation Lab., Casaccia Nucl. Studies Centre, C.N.E.N., Italy. Society: Remote System Div., A.N.S.
Nuclear interests: Hot laboratory design and development; Hot laboratory management; Post irradiation analysis of fuel elements either by non-destructive or by destructive techniques; Fission products migration analysis; Activation analysis.
Address: Nuclear Studies Centre, Casaccia, S. Maria di Galeria, Rome, Italy.

CESARE, Mario de. See de CESARE, Mario.

CESONI, Giulio, Dr. Ing. (Industriale Chimica). Born 1916. Educ.: Politecnico, Milan and Politecnico, Turin. Consultant for Atomic Power Plants, Azienda Elettrica Municipale, Milan, 1955-; Consultant for Atomic Energy Industrial Uses, Fiat-Torino, 1955-; Manager, Sezione Energia Nucleare, Fiat, 1956-; Gen. Manager, Sorin-Società Ricerche Impianti Nucleari, Saluggia (Vercelli), 1956-; Prof., Nucl. Plants design, Politecnico, Turin. Sci. and Tech. Com., C.I.S.E.; Sci. and Tech. Com., C.E.E.A. Societies: Soc. Italiana per il Progresso delle Scienze; A.N.S.; Assoc. Italiana di Ingegneria Nucleare; Assoc. Italiana di Metallurgia; Atomic Ind. Forum.
Nuclear interests: Reactor design; Metallurgy and metallurgical chemistry.
Address: Sorin, Saluggia (Vercelli) Fiat, Sezione Energia Nucleari, 235 Corso Settembrini, Turin, Italy.

CESSION, J. P. Gen. Manager, Cie. Générale des Conduites d'Eau S.A.
Address: Cie. Générale des Conduites d'Eau S.A., 1 Quai des Vennes, Liège, Belgium.

CESSOT, Pierre, Ing. Chargé du Département Nucléaire, Sté. d'Etudes Industrielles de Villejuif, Brunet Pramaggiore et Cie.
Address: Société d'Etudes Industrielles de Villejuif, Brunet Pramaggiore et Cie., 21 rue Georges le Bigot, Villejuif, (Seine), France.

CESTEROS, Miguel A., Member, Comision Nacional de Investigaciones Atomicas, Dominica.
Address: Comision Nacional Investigaciones Atomicas, Ciudad Trujillo, Dominican Republic.

CHABOSEAU, Jacques Pierre. Born 1926. Educ.: Ecole Centrale des Arts et Manufactures, Paris. Chef de Departement, Div. Energie Nucléaire, Soc. Rateau. Societies: Assoc. des Ing. en Genie Atomique; B.N.E.S.
Nuclear interests: Nuclear reactors and components. Advanced nuclear system for power and propulsion. Rotating machinery for nuclear and advanced applications.
Address: Division Energie Nucléaire, Societe Rateau, 141 rue Rateau, La Courneuve, Seine, France.

CHABROL, Yves. Hon. Pres., Syndicat des Pharmaciens de France. French Member, Specialised Nucl. Sect. for Social Health and Development Problems, Economic and Social Com., Euratom, 1966-.
Address: Syndicat des Pharmaciens de France, 29 avenue du Général Lederc, 75 Paris 14, France.

CHACON, Eduardo BREGNI. See BREGNI CHACON, Eduardo.

CHADHA, Mohindra Singh, B.Sc. (Hons.), M.Sc. (Punjab), Ph.D. (Cornell). Born 1928. Educ.: Punjab and Cornell Univs. Jun. Res. Chem., Dept. Chem., California Univ., Berkeley, 1955-57; Sci. Officer (1958-), Biol. Div., Bhabha Atomic Res. Centre, Trombay, Bombay; Guest Chem., Lawrence Rad. Lab., Berkeley, 1961. Societies: Indian Chem. Soc.; Chem. Soc., London.
Nuclear interests: Synthesis of ^{14}C and ^{3}H labelled compounds; Metabolism of labelled chemotherapeutics; Biosynthesis using labelled precursors; Synthesis of potential anti-radiation compounds.
Address: Biology Division, Bhabha Atomic Research Centre, Modular Laboratories, Trombay, Bombay, India.

CHADWICK, Donald Roger, M.D. Born 1925. Educ.: Harvard Coll. and Harvard Medical School. Special Training in Radiol. Health, Reed Coll., Portland, Oregon, Sandia Base, Albuquerque, New Mexico, and O.R.N.L.; 1954-55; Occupational Health Field Headquarters, Cincinnati, Ohio, 1955-56; Chief, Programme Services, Radiol. Health Medical Programme, Div. of Special Health Services, 1956-57; Liaison Officer for Rad., Office of the Surgeon Gen., P.H.S., 1957-58; Acting Chief, Div. of Radiol. Health, 1958; Chief, Programme Operations Branch, Div. of Radiol. Health, 1958-59; Exec. Sec., Nat. Advisory Com. on Rad., 1958-59; Sec., Federal Rad. Council, 1959-61; Chief, Div. of Radiol. Health, 1961-66; Director, Nat. Centre for Chronic Disease Control, P.H.S. Bureau of Disease Prevention and Environmental Control, 1967-. Member, World Health Organisation, Expert Advisory Panel on Rad., 1962-; Member, Nat. Com. on Rad. Protection and Measurements, 1963-; Member, Atomic Ind. Forum, Com. on Public Understanding, Tech. Pension Group. Societies A.P.H.A., Sect. on Radiol. Health;

Conference on Radiol. Health; A.M.A.; American Thoracic Soc.; American Assoc. of Public Health Physicians; H.P.S.; Commissioned Officers Assoc., U.S. P.H.S.
Nuclear interest: Public health control of radiation exposure.
Address: R.F.D., Box 4515, Upper Marlboro, Maryland, 20870, U.S.A.

CHADWICK, Frank, B.A. Born 1922. Educ.: Cambridge Univ. Inspector of Taxes, 1949-52; Administrative Officer, Board of Inland Revenue, 1952-60; Administrative Officer (1960-67), Head, Finance Branch, London Office (1967-), U.K.A.E.A.
Nuclear interests: Financial policy and budgetary control; Economics and programming.
Address: United Kingdom Atomic Energy Authority, 11 Charles II Street, London, S.W.1, England.

CHADWICK, George B., B.A., M.A., Ph.D. Born 1931. Educ.: British Columbia and Cambridge Univs. Res. Assoc. and Asst. Phys., Brookhaven Nat. Lab.; Res. Assoc. and Asst. Phys., Oxford Univ.; Res. Assoc. and Asst. Phys., Stanford Linear Accelerator Centre, Stanford Univ. Society: A.P.S.
Nuclear interests: Study of high energy particle reactions in bubble chambers, principal interest at present is in photoproduction at high energies.
Address: Stanford University, Stanford Linear Accelerator Centre, P.O. Box 4349, Stanford, California 94305, U.S.A.

CHADWICK, Sir James, F.R.S., M.Sc. (Vict.), Ph.D. (Cantab.); Hon. DiSc. (Reading, Dublin, Leeds, Oxford, Birmingham and McGill Univs.), Hon. LL.D. (Liverpool and Edinburgh Univs.). Nobel Laureate (Phys.), 1935. Born 1891. Educ.: Manchester and Cambridge Univs. Master of Gonville and Caius Coll., Cambridge, 1948-58; Member (part-time), U.K.A.E.A., 1957-62. Societies: F.R.S.; F.R.S.E.; Corresponding member Sachs. Akad. der Wissenschaften (Leipzig); Assoc. Acad. Roy., Belgium; For. Member K. Ned. Akad. Wet. Amsterdam; K. Danske Vid. Selskab; Hon. Member, A.P.S.; Hon. Member, American Philosophical Soc., Philadelphia; Pontificia Academia Scientiarum.
Nuclear interests: All aspects of nuclear energy and its applications, but particularly nuclear physics.
Address: Wynne's Parc, Denbigh, Wales.

CHADWICK, Kenneth Helme, B.Sc., M.Sc. Born 1937. Educ.: Liverpool and London Univs. Asst. Health Phys., Berkeley Nucl. Labs., C.E.G.B., 1961-64; Rad. Dosimetrist, Assoc. Euratom – I.T.A.L. (Instituut voor Toepassing van Atoomenergie in de Landbouw), 1964-66, Euratom, 1966-. Societies: Soc. for Radiol. Protection; Inst. of Phys.
Nuclear interests: Radiation dosimetry; Radiobiology; Health physics.
Address: Association Euratom – I.T.A.L., Postbus 48, Wageningen, Netherlands.

CHAFER, Francisco Bosch. See BOSCH CHAFER, Francisco.

CHAFFIOTTE, L. Sen. Officer, Ets. C. Masson.
Address: Ets. C. Masson, 16 rue du Moulin-des-Bruyères, Courbevoie, (Seine), France.

CHAGAS, Carlos, Dr. en Médecine (Brazil Univ.), Diplomé par l'Inst. Oswaldo Cruz, Dr. Sciences (Paris Univ.), Prof. de la Faculté Nationale de Médecine, Brazil Univ.; Member, Pontifical Acad. of Sci. (Vatican), 1961; Membre de l'Acadèmie des XL (Rome); Membre de l'Acadèmie Bresilienne des Sciences; Doctor Honoris Causa, Mexico, Paris, Recife, Coimbra and Toronto Univs. Born 1910. Educ.: Brazil and Paris Univs. Prof. Biophys. School of Medicine, Brazil Univ.; Member (Director, 1952-53), N.R.C., Rio de Janeiro. Directeur de l'Inst. de Biophysique, Brazil Univ.; Prof. d'Echange de l'Inst. Franco-Brésilien de Haute Culture; Membre Associé de l'Academie Nationale de Médecine, France; Membre du Conseil d'Administration de l'Assoc. Internationale des Universités; etc. Délégué du Brézil, 1ère et 2ème Conf. Cén. de l'Unesco, etc. Délégué du Brésil au Comité de l'O.N.U. pour l'étude des effets des radiations atomiques, New York (1956), Geneva (1957), New York (1958); Vice-Prés. de l'Association Latin-Américaine de Sciences Physiologiques; Représentant dans l'Etranger au Congrès Internat. de Recherches sur la Radiation, Vermont, 1958; Sec. gen. to U.N. Conference on Sci. and Technol. (Geneva, 1963). Représentant du Brésil à la 1ère Conférence de l'Organisation des Etats Américains, Washington, June 1958. Book: Homens e Coisas de Ciência (1957).
Address: 38 Francisco Otaviano, Rio de Janeiro, Brazil.

CHAIKA, Taisiya Vasilyevna, Prof. Dr. Head, Pathological-Anatomic Lab., Leningrad. Prof. A.L. Polenov Neurosurgical Res. Inst., Ministry of Health.
Address: Leningrad Professor A. L. Polenov Neurosurgical Research Institute, 12 Mayakovskii Street, Leningrad D104, U.S.S.R.

CHAITCHIK, Samario, M.D. Born 1932. Educ.: Rio de Janeiro Univ. Asst., Dept. of Radiotherapy and Isotopes, Tel-Aviv Univ. Medical School. Society: Soc. of Nucl. Medicine.
Nuclear interest: Nuclear medicine.
Address: 14 Faivelet Street, Tel Aviv, Israel.

CHAJSON, Leon (Lee), B.S. (Chem. Eng., Cooper Union). Born 1928. Educ.: Cooper Union, and Idaho and Pittsburgh Univs. Eng., Atomic Energy Div., American Cyanamid Co. 1950-53; Eng., Atomic Energy Div., Phillips Petroleum Co., 1953-55; Res. Assoc., Eng. Res. Inst., Michigan Univ., 1955-57; Manager, Water Reactor Development, Atomic Power Div., Westinghouse Elec. Corp., 1957-.

Societies: A.N.S.; A.I.Ch.E.
Nuclear interests: Management, nuclear fuels processing, water reactors research and development.
Address: 1105 McCully Drive, Pittsburgh 35, Pennsylvania, U.S.A.

CHAKHOTIN, Sergei, Prof. Micro-surgery, Biophys. Lab., Atomic Energy Dept., Acad. Sci. of the U.S.S.R.
Address: Academy of Sciences of the U.S.S.R., 14 Leninsky Prospekt, Moscow V-71, U.S.S.R.

CHAKRABORTY, K. P., Dr. Sen. Officer, Inst. of Nucl. Medicine and Allied Sci.
Address: Institute of Nuclear Medicine and Allied Sciences, Metcalfe House, Delhi-6, India.

CHAKRAVARTI, Jyotsna, Dr. Lecturer, Saha Inst. of Nucl. Phys., Calcutta Univ.
Address: Calcutta University, Saha Institute of Nuclear Physics, 92 Acharya Prafulla Chandra Road, Calcutta-9, India.

CHAKRAVARTI, M. N. Project Administrator, Tarapur Atomic Power Project, A.E.C.
Address: Tarapur Atomic Power Project, Atomic Energy Commission, Chhatrapati Shivaji Maharaj Marg, Bombay 1, India.

CHALDER, Geoffrey Hay, B.Sc. (Hons., Metal.). Born 1930. Educ.: Durham Univ. U.K.A.E.A., Culcheth Labs., 1951-55; A.E.C.L., Chalk River, 1955-. Society: A.P.E.O.
Nuclear interests: Development, fabrication and evaluation of nuclear fuel materials.
Address: Fuel Materials Branch, Atomic Energy of Canada Ltd., Chalk River Nuclear Laboratories, Chalk River, Ontario, Canada.

CHALIKIAS, D., Dr. Sci. Director, Managing Com., Greek A.E.C.
Address: Greek Atomic Energy Commission, 5 Merlin Street, Athens (134), Greece.

CHALKE, Herbert Davis, M.A. (Cantab.), M.R.C.P. (London), D.P.H. (R.C.P.S. London). Born 1897. Educ.: Cambridge and Wales Univs. and St. Bartholomew's Hospital, London. Late Medical Officer of Health, Metropolitan Borough of Camberwell. Books: Co-editor, Rad. and Health; Hygiene and Public Health.
Nuclear interests: Public health and genetic aspects.
Address: 5 Egerton Gardens, Hendon, London, N.W.4, England.

CHALKER, Ralph Giles, B.S. (Civil Eng.). Born 1918. Educ.: Utah Univ. Designer, Atomic Res. Div. (1949-51), Supv. Atomic Res. Div. (1951-52), Group Leader (1952-61), Assoc. Director, Atomics Internat. Div., now Director, Eng. Systems Management Dept., Atomics Internat. Div., North American Rockwell Corp. Chairman, Standards Com., A.N.S.; Exec. com., Nucl. Standards Board, U.S.A.

Standards Inst. Societies: A.S.M.E.; A.N.S.; U.S.A. Standards Inst.
Nuclear interest: Technical management of personnel engaged in engineering design of nuclear reactors for space application, including facilities and associated equipment.
Address: 8046 Altavan, Los Angeles, California, U.S.A.

CHALLENS, Wallace John, B.Sc. (Special Phys.). Born 1915. Supt., Weapon Electronics (1950-57), Sen. Supt., Weapon Electronics (1957-59), Chief of Warhead Development (1959-), Asst. Director (1966-), A.W.R.E., U.K.A.E.A. Society: F. Inst. P.
Nuclear interests: Management of research, development and design, in fields of physics, electronics, engineering, and explosives, as applied to problems of nuclear weapons and their effects.
Address: U.K.A.E.A., Atomic Weapons Research Establishment, Aldermaston, Berkshire, England.

CHALMIN, Jean. Manager, Ets. Garczynski et Traploir.
Address: Ets. Garczynski et Traploir, 34 rue du Pavé, Le Mans, (Sarthe), France.

CHALOV, P. I. Papers: Co-author, U^{234} and U^{238} Isotopic Displacement in Secondary Uranium Minerals of Some Hydrothermal Deposits (letter to the Editor, Atomnaya Energiya, vol. 19, No. 1, 1965); co-author, On Relative Levels of Stratosphere Fallout (letter to the Editor, ibid, No. 5); co-author, Two-Canal Scheme of Synchronous Registration of Nuclei Fission Fragments of Standard and Analysed Sample (letter to the Editor, ibid, vol. 21, No.3, 1966).
Address: Academy of Sciences of the U.S.S.R., 14 Leninsky Prospekt, Moscow V-71, U.S.S.R.

CHALUPA, Bohumil, Ing. Born 1928. Educ.: Czechoslovak Electrotech. Univ., Prague. Nucl. Res. Inst., Czechoslovak Acad. of Sci., Rez. Society: Czech Sci. Tech. Soc.
Nuclear interests: Nuclear physics; Radioactive capture; Neutron spectrometry and diffractometry.
Address: Nuclear Research Institute, Czechoslovak Academy of Sciences, Rez, Czechoslovakia.

CHALUPA, Z., Ing. Member, Editorial Board, Jaderna Energie.
Address: Jaderna Energie, SNTL, 51 Spálená, Prague 1, Czechoslovakia.

CHAMBALOUX, Pierre, Ing., Electro-Entreprise, S.A.
Address: Electro-Entreprise, S.A., 32 rue de Mogador, Paris 9, France.

CHAMBARD, L. J. Formerly Member, now Chairman and Liaison Exec. (1967-), Standing Com. on Nucl. Energy in the Elec. Industry, Nat. Assoc. of Railroad and Utilities

Commissioners.
Address: National Association of Railroad and Utilities Commissioners, Interstate Commerce Commission Building, P.O. Box 684, Washington, D.C. 20044, U.S.A.

CHAMBERLAIN, Arthur Cyril, M.A. Born 1920. Educ.: Cambridge Univ. Group Leader, Aerosol Group (Health Physics), A.E.R.E., 1959-. Societies: Biometrics Soc.; Geol. Assoc.; Roy. Meteorological Soc.
Nuclear interests: Radioactive gases and aerosols; Meteorological aspects of nuclear energy; Containment of reactors; Clean-up of fission products from gases; Atmospheric physics.
Address: A.E.R.E., Harwell, Didcot, Berks., England.

CHAMBERLAIN, Owen, A.B. (Dartmouth Coll.), Ph.D. (Chicago). Shared Nobel Prize for Phys., 1959. Born 1920. Educ.: Dartmouth Coll. and Chicago Univ. Asst. Prof. Phys. (1950-54), Assoc. Prof. (1954-58), Prof. (1958-), California Univ. Berkeley. Guggenheim Fellowship, 1957-58; Loeb Lecturer at Harvard Univ., 1959. Societies: Fellow, A.P.S.; N.A.S.
Nuclear interests: High-energy nuclear physics.
Address: Department of Physics, California University, Berkeley 4, California, U.S.A.

CHAMBERLAIN, Richard Hall, A.B., M.D. Born 1915. Educ.: Centre Coll.,Danville, Kentucky, and Louisville Univ. Asst. Prof. Radiol. (1948-50), Assoc. Prof. Radiol. (1950-52), Prof. Radiol. (1952-), Chairman, Radiol. Dept. (1961-), School of Medicine, Pennsylvania Univ.; Asst. Prof. Radiol. (1948-50), Assoc. Prof. Radiol. (1950-51), Clinical Prof. Radiol. (1951-52), Prof. Radiol. (1952-), Graduate School of Medicine, Pennsylvania Univ. Consultant, Nat. Advisory Council on Rad., U.S.P.H.S., 1960-; Tech. Adviser, 2nd I.C.P.U.A.E., 1958; Official U.S. Deleg. to U.N. Sci. Commission of the Effects of Atomic Rad., 1963-; Board of Trustees, Assoc. Univs., Inc., 1963-; I.C.R.U.; Nat. Council on Rad. Protection; Commission on Radiol. Units, Standards and Protection, American Coll. Radiol., 1961-. Societies: Atomic Ind. Forum; American Radium Soc.; Radiol. Soc. of North America; A.M.A.; American Roentgen Ray Soc.; Rad. Res. Soc.; American Assoc. for Cancer Res.; A.A.A.S.
Nuclear interests: Clinical use of radioactive isitopes and high energy radiation in medicine, radioisotope scanning methods, radiation protection in medicine and from the standpoint of public health.
Address: 8327 Germantown Avenue, Philadelphia, Pennsylvania 19118, U.S.A.

CHAMBERS, F. W., Jr. Capt. U.S.N. Ret.; Vice-Pres., Head of R. and D., Dynatomics Inc., 1965-.
Address: Dynatomics,Inc., P.O. Box 9923, Atlanta, Georgia 30319, U.S.A.

CHAMBERS, Frank G. Chairman, Hazleton Nucl. Sci. Corp.; Board Member, Nucl. Sci. and Eng. Corp.,1963-.
Address: Hazleton Nuclear Science Corporation, 4062 Fabian Way, Palo Alto, California, U.S.A.

CHAMBOISIER, Jean. Vice-Pres., Assoc. les Amis de l'Atome.
Address: Association les Amis de l'Atome, 11 Square Moncey, Paris 9, France.

CHAMBON, Peter, Baccalauréat ès Sciences, B.Sc. Born 1921. Educ.: London Univ. (External). Deputy Chief Eng., and Chief Project Eng., Atomic Energy Dept., Babcock and Wilcox Ltd., London. Society: A.M.I.Mech.E.
Nuclear interests: Technical and commercial aspects of design engineering of components for nuclear applications; Project and contract management.
Address: Babcock and Wilcox (Operations) Ltd., Atomic Energy Department, 209 Euston Road, London, N.W.1, England.

CHAMP, Bruce Richard, B.Agr.Sc., Ph.D. Born 1930. Educ.: Queensland and London Univs. Res. Entomologist, Dept. Agriculture and Stock, Queensland; Gowrie Post-graduate Res. Travelling Scholarship, 1955-58.
Nuclear interests: Radio-isotopes as a tool in insecticide studies.
Address: Department of Primary Industries, Meirs Road, Indooroopilly, Brisbane, Queensland, Australia.

CHAMPEIX, Louis, Ing. I.C.C., L. ès Sc. Born 1927. Educ.: Clermont-Ferrand Univ. (Inst. de Chimie et de Technologie Industrielle). Ing. de recherches, Société Cadum, 1951-55; Ing. de recherches, C.E.A. (Sect. d'Etudes de la Çorrosion par Gaz et Métaux Liquides), 1955-. High Temperatures Commission, Internat. Union of Pure and Appl. Chem.
Nuclear interests: Nuclear metallurgy; Gas analysis (chromatography); Corrosion by liquid metals.
Address: Commissariat à l'Energie Atomique, C.E.N. Saclay, Département de Métallurgie B.P. No. 2, Gif-sur-Yvette, S. et O., France.

CHAMPETIER, Georges Hippolyte, D. ès Sc. Born 1905. Educ.: Paris Univ. Directeur-adjoint du Centre Nat. de la Recherche Scientifique, 1951; Prof. à la Faculté des Sciences de Paris, chaire de chimie macromoléculaire, 1953; Directeur honoraire du Centre Nat. de la Recherche Scientifique, 1956; Conseil Scientifique de C.E.A. Societies: Académie des Sciences; Sté. Chimique de France.
Nuclear interests: Applications scientifiques et techniques.
Address: 10 Rue Vauquelin, Paris 5, France.

CHAMPION, Frank Clive, M.A., Ph.D. Born 1907. Educ.: Cambridge, Nottingham, Bristol and London Univs. Reader in Phys. (1946-60), Prof. Exptl. Phys. (1960-), King's Coll., London Univ. Books: Co-author, Properties of Matter;

University Phys., vols. I-V; Electronic Properties of Diamonds.
Nuclear interests: Effects of nuclear radiations on fundamental solids (e.g. diamond, sulphur, etc.), and on liquids; Scintillation counters, semiconductor counters; General properties, particularly optical and electrical, of interaction of ionizing radiations on the condensed state.
Address: Flat 279, Chiswick Village, London, W.4, England.

CHAN, Wen-Oi. Papers: Co-author, On the possibility of Λhyperon creation through isobars in π-p interactions at 7 - 8 GeV (Nukleonika, vol. 9, No. 2 - 3, 1964); co-author, On the possibility of establishing of isobar state systems and their decay modes (ibid.).
Address: Joint Institute for Nuclear Research, Dubna, Nr. Moscow, U.S.S.R.

CHANCE, Henry M., II. Director, Atomic Power Development Assoc., Inc.
Address: Atomic Power Development Associates, Inc., 1911 First Street, Detroit 26, Michigan, U.S.A.

CHAND, Ramesh, B.Sc. (Hons.), M.Sc., Ph.D. (Phys.). Born 1934. Educ.: Delhi and Chicago Univs. Instructor, Phys. Dept., Syracuse Univ., 1962-64; Asst. Prof. (1964-66), Assoc. Prof. (1966-), Phys. Dept., Wayne State Univ. Societies: A.P.S.; Italian Phys. Soc.
Nuclear interests: Theoretical high energy nuclear physics involving interactions of Kaons with nucleons, deuterons and heavier nuclei, and the role played by the various pion, and hyperon resonances in these reactions; The use of the quark model within the framework of SU(3) symmetry, for the study of strong, electromagnetic and weak interactions of the elementary particles.
Address: Department of Physics, Wayne State University, Detroit, Michigan, U.S.A.

CHANDARASATIT, Insri. Member, Thai Atomic Energy Commission for Peace (Agriculture); Rector, Kasetsart Univ.
Address: Department of Science, Rama VI Road, Bangkok, Thailand.

CHANDOS, Viscount Oliver Lyttelton. Born 1893. Educ.: Cambridge Univ. Chairman, A.E.I., 1945-51, 1954-63; Director, Imperial Chem. Industries, 1955-; Chairman, Board of Partners, The Nucl. Power Group, 1960-; Pres., Royal Marsden Hospital.
Address: 9 Carlton Tower Place, Sloane Street, London, S.W.1, England.

CHANDRASEKHAR, B. S., B.Sc. (Hons.), M.Sc., D. Phil. Born 1928. Educ.: Mysore, Delhi and Oxford Univs. Res. Assoc., Illinois Univ., 1951-54; Res. Phys. (1954-59), Fellow Phys. (1959-61), Manager, Cryophys. Sect. (1961-63), Westinghouse Res. Labs.; Sen. Visiting Res. Fellow, Imperial Coll. of Sci. and Technol., 1961; Prof. Phys., Western Reserve Univ., 1963-67; Perkins Prof. of Phys., Case Western Reserve Univ., 1967-. Societies: Fellow, The Phys. Soc.; F. Inst. P.; Fellow, A.P.S.
Nuclear interests: Superconductivity and other electronic properties of uranium and its alloys.
Address: Department of Physics, Case Western Reserve University, Cleveland, Ohio 44106, U.S.A.

CHANDRASEKHAR, S. Member, Editorial Council, Annals of Phys.
Address: Yerkes Observatory, Williams Bay, Wisconsin, U.S.A.

CHANG, Chia-hua, Atomic Energy Inst., Acad. of Sci., Peking. Paper: The use of isotopes in industries today (Atomic Energy (China), vol. 4, No. 1, 1959, pp. 3-7).
Address: Academia Sinica, 3 Wen Tsin Chien, Peking, China.

CHANG, F. C., Ph.D. Born 1928. Educ.: Michigan Univ. Assoc. Prof., Phys., St. John's Univ., 1959. Society: A.P.S.
Nuclear interest: Nuclear physics.
Address: St. John's University, Jamaica, New York 11432, U.S.A.

CHANG, Jen-hua, Asst. Soil Technologist, Radioisotope Lab., Taiwan Sugar Experiment Station.
Address: Taiwan Sugar Experiment Station, 1 Sheng Chaan Road, Tainan, Formosa.

CHANG, Li-ning. Paper: Co-author, The Radiative Capture of μ-Meson by Nucleus (Scientia Sinica, vol. 10, No. 4, 1961).
Address: Academia Sinica, 3 Wen Tsin Chien, Peking, China.

CHANG, Min Chueh, B.Sc., (Tsing Hua), Dip. Agr. Sci. (Edinburgh), Ph.D. (Cambridge). Born 1908. Educ.: Tsing Hua, Edinburgh and Cambridge Univs. Res. Prof., Dept. of Biol., Boston Univ., 1961; Sen. Sci., Worcester Foundation for Exptl. Biol. Societies: American Physiol. Soc.; American Assoc. of Anatomists; American Acad. of Arts and Sci.
Nuclear interest: Irradiation of mammalian germ cells and early embryos.
Address: Worcester Foundation for Experimental Biology, Shrewsbury, Mass., U.S.A.

CHANG, Run-Hwa. Paper: Co-author, Angular distribution of neutrons due to μ-capture in calcium for various energy thresholds (Nukleonika, vol. 9, No. 2 - 3, 1964).
Address: Joint Institute for Nuclear Research, Laboratory of High Energy, Dubna, U.S.S.R.

CHANG, T. S. Paper: Analyticity of Perturbation Expansions (Scientia Sinica, vol. 10, No. 1, 1961).
Address: Academia Sinica, 3 Wen Tsin Chien, Peking, China.

CHANG, Tsung-Yeh. Paper: Co-author, A Fermi System Giving the Equally-Spaced Spectrum (Scientia Sinica, vol. 13, No. 12, 1964).
Address: Academia Sinica, 3 Wen Tsin Chien, Peking, China.

CHANG, Wen-yu, Ph.D. Educ.: Cambridge Univ. Teacher and research worker in atomic phys., U.S.A., 1943; Lecturer in atomic and theoretical phys., Purdue Univ., 1949; at present Deputy Director, Cosmic Ray Res. Lab., Academia Sinica, Peking.
Address: Cosmic Ray Research Laboratory, Academia Sinica, Peking, China.

CHANG, Won Pyo. Born 1921. Educ.: Seoul Nat. Univ. Analyst, Taejon Mineral Lab., U.N. Korean Reconstruction Agency, 1953-56; Lecturer in Analytical Chem., Chungnam Univ., 1955-; Chief Chem. and Chief Metal., Taejon Mineral Lab., Geological Survey of Korea, 1956-67; Chief Chem., Nat. Mining Res. Centre, 1967-. Societies: Korean Chem. Soc.; A.C.S.; The Japan Soc. for Analytical Chem.
Nuclear interests: Metallurgy of nuclear materials; Analysis of raw materials and refined products.
Address: 326 Taehung-dong, Taejon, Korea.

CHANGSOON, Koh, M.D., Ph.D. Born 1935. Educ.: Showa Coll. of Medicine, Tokyo. Clinical Asst. Prof., Coll.of Medicine, Seoul Nat. Univ.; Staff, Korean Soc. of Nucl. Medicine. Societies: Korean Soc. of Nucl. Medicine; Japanese Soc. of Nucl. Medicine; Soc. of Korean Internal Medicine; Soc. of Japanese Internal Medicine.
Nuclear interests: The application of radioisotopes in the field of diagnosis and treatment in addition to research work in the medical field.
Address: 44-3 Suhkyo-dong, Mapo-ku, Seoul, Korea.

CHANTAL, Sister Frances de. See de CHANTAL, Sister Frances.

CHANTEUR, Jean, M.D. Born 1924. Educ.: Paris Univ. Prof. Medical Phys., Medical School Paris; Chief Adjoint, Central Protection Service against Ionising Rads., Ministry of Health. Societies: French Soc. Radioprotection; H.P.S.
Nuclear interests: Biophysics; Health physics; Radioprotection; Radiobiology; Radiopathology.
Address: 39 boulevard Saint Michel, 75 Paris 5, France.

CHANTURIYA, V. M. Papers: Co-author, The Model of a Radiative Indium-Gallium Loop of the Tbilisi IRT-2000 Reactor (Atomnaya Energiya, vol. 13, No. 4, 1962); co-author, Indium-Helium Rad. Circuit for Pile-Type Reactor (Abstract, ibid, vol. 19, No. 2, 1965).
Address: Academy of Sciences of the U.S.S.R., 14 Leninsky Prospekt, Moscow V-71, U.S.S.R.

CHAO, Bates T., B.Sc. (E.E.), Ph.D. (Mech. Eng.). Born 1918. Educ.: Nat. Chiao-Tung, Shanghai and Manchester Univs. Prof., Mech. Eng. (1956-), Prof., Nucl. Eng. (1961-), Illinois Univ. Societies: A.S.M.E.; Soc. Eng. Sci.
Nuclear interest: Reactor heat transfer.
Address: 268 Mechanical Engineering Building, Illinois University, Urbana, Illinois 61801, U.S.A.

CHAO, Chung-yao, Prof. Teacher, Tsinghua Univ. China; Rep., People's Republic of China, Conference of the Joint Inst. for Nucl. Research, Moscow, Sept. 23, 1956.
Address: Academia Sinica, 3 Wen Tsin Chien, Peking, China.

CHAO, Tse-Hung, B.Sc. Born 1914. Educ.: Chiao Tung Univ. Assoc. Prof., Tainan Eng. Coll.. Acting Chief, Metal. Res. Lab., Union Ind. Res. Inst. Paper: Separation of Monazite from Taiwan Heavy Sands. Societies: Chinese Inst. Eng.; A.S.M.
Address: P.O. Box 100, Hsin-Chu, Formosa.

CHAPIN, Douglas S., B.S., M.S., Ph.D. Born 1922. Educ.: Kansas State Univ., Illinois Inst. Technol., Ohio State Univ. Asst. Prof. (1954-59), Assoc. Prof. (1959-66), Chem. Dept., Arizona Univ.; Staff Member, M.I.T. Lincoln Lab., 1963-64; Assoc. Programme Director, Graduate Fellowships and Traineeships, Nat. Sci. Foundation, Washington D.C., 1966-. Societies: A.C.S.; Faraday Soc.
Nuclear interests: Catalysis (ortho-parahydrogen conversion at low temperatures); Cryogenics (selective adsorption of orthohydrogen); Microcalorimetry; Thermoelectric properties of oxides.
Address: National Science Foundation, Washington D.C. 20550, U.S.A.

CHAPIRO, Adolphe, Ing. Dr. Born 1924. Educ.: Paris Univ. Directeur de Recherches, C.N.R.S., 1947-. Book: Rad. Chem. of Polymeric Systems (Interscience, 1962). Societies: Faraday Soc.; Sté. de Chimie physique; A.C.S.; Société Chimique de France.
Nuclear interest: Radiation chemistry.
Address: Laboratoire de Chimie des Radiations, C.N.R.S., Bellevue, S. et O., France.

CHAPMAN, Clifford. Formerly Gen. Supt., now Director of Manufacturing, Nat. Lead Company of Ohio.
Address: National Lead Company of Ohio, P.O. Box 39158, Cincinnati, Ohio, U.S.A.

CHAPMAN, Earle MacArthur, B.S., M.D. Born 1903. Educ.: Michigan State Coll. and Johns Hopkins Med. School. Assoc. Clinical Prof. Medicine, Harvard Med. School. Chairman, Isotopes Com., Massachusetts Gen. Hospital. Societies: American Thyroid Assoc.; Soc. Nucl. Medicine; Americal Coll. of

Physicians.
Nuclear interests: Therapy of thyroid disease with I^{131}. Studies on physiology of thyroid function.
Address: 275 Charles Street, Boston, Massachusetts 02014, U.S.A.

CHAPMAN, Noel George, M.Sc., Ph.D. Born 1928. Educ.: Victoria Univ. Lieutenant, R.N.Z.N., 1953-59; Sen. Lecturer in Phys. (1959-), now Prof., Nucl. Phys., Phys. Dept., Wellington Univ. Society: A. Inst. P.
Nuclear interest: Nuclear physics.
Address: Department of Physics, Wellington University, P.O. Box 196, Wellington, New Zealand.

CHAPMAN, Thomas Edwin. Born 1908. Formerly Gen. Manager, now Exec. Director, Fuel Element Div. and Works Manager, Fairey Eng., Ltd.
Nuclear interests: Management of development and production facility in connection with fuel element containers for U.K.A.E.A., 1950-64.
Address: 390 Hanworth Road, Hounslow, Middlesex, England.

CHAPMAN, Thomas Shelby, Ph.D. Born 1905. Educ.: Rice Inst. Director, Health Phys. and Medicine, Dow Chem. Co., Rocky Flats Plant, 1951-66; Chief, Regulation and Appraisal, Div. of Tech. Extension, U.S.A.E.C., 1967-. Society: A.C.S.
Nuclear interest: Management.
Address: 436-438 E. Tennessee Avenue, Oak Ridge, Tennessee 37830, U.S.A.

CHAPMAN, William Moore, B.S.E.E. Born 1906. Educ.: Colorado Univ. Vice-Pres., Ind. Products, Westinghouse Electric Internat. Co. Society: Atomic Ind. Forum.
Address: Westinghouse Electric International Company, 200 Park Avenue, New York, New York 10017, U.S.A.

CHAPPUIS, Pierre. Tech. Manager, Lausanne, Ed. Zublin et Cie. A. G.
Address: Ed. Zublin et Cie. A. G., 6 Okenstrasse, 8031 Zürich, Switzerland.

CHAPUIS, Ann Marie, L. ès Sc. Physiques. Born 1934. Educ.: Faculté des Sciences, Paris. Eng., C.E.A., 1956-.
Nuclear interests: Measurement and spectrometry of various samples for monitoring aims; fall-out, contamination, soils. Monitoring of tritium. Neutron detectors for personal dosimetry. Liquid scintillation. Gamma-spectrometry.
Address: 80 chemin Argoulets, Toulouse, Haute-Garonne, France.

CHARAP, John Michael, M.A., Ph.D. Born 1935. Educ.: Cambridge Univ. Member, Inst. for Advanced Study, Princeton, 1962-63; Sen. Sci. Officer, N.I.R.N.S., 1963-64; Lecturer, Imperial Coll., London, 1964-65; Visiting Sci.,
Brookhaven Nat. Lab., 1966; Visiting Sci., Internat. Centre for Theoretical Phys., 1967; Reader, Theoretical Phys., Queen Mary Coll., London. Society: A.P.S.
Nuclear interest: Elementary particle physics.
Address: Queen Mary College, Mile End Road, London, E.1, England.

CHARDON, P., Chef du Div. Minière du Forez, Direction des Productions, C.E.A.
Address: Commissariat à l'Energie Atomique, Direction des Productions, Division Minière du Forez, Saint Priest la Prugne, (Loire), France.

CHARLAND, Telesphore Lawrence, P.E. (Professional Eng.), B.S. (Ceramic Eng.), M.S. (Ceramic Eng.). Born 1921. Educ.: Iowa State Coll. and Alfred Univ., New York. Ceramic Eng., Westinghouse Elec. Corp., Fairmont, West Virginia, 1950-52; Res. Assoc., Alfred Univ., 1952-54; Ceramic Eng., Phillips Petroleum Co., of Oklahoma City, 1954-56; Ceramic Eng., Westinghouse Elec. Corp., Pittsburgh, 1956-. Societies: A.N.S.; American Ceramic Soc.; National Inst. Ceramic Eng.; Ceramic Assoc. New York.
Nuclear interests: Development and use of ceramic materials for nuclear fuels, control rods, moderators, shielding and refractory purposes in a nuclear reactor; also, development of thermoelectric materials to achieve direct conversion of nuclear heat energy to electrical energy.
Address: 316 Elias Drive, Pittsburgh 35, Pennsylvania, U.S.A.

CHARLES, Daniel Raymond, D ès Sc. Born 1912. Educ.: Paris Univ. Head, Special Tubes Dept., C.S.F.; Prof. Phys., Ecole Spéciale de Mécanique et d'Electricité. Societies: Sté. Française des Radioélectriciens et électroniciens.
Nuclear interests: Detection (P.M., image tube).
Address: Compagnie Générale de Télégraphie Sans Fil, Centre de Physique Electronique et Corpusculaire, Domaine de Corbeville, B.P. No. 10, Orsay, (S. et O.), France.

CHARLES, Jack. Born 1923. Formerly Deputy Chief Labour Officer and Establishments Officer, now Authority Personnel Officer (1965-), U.K.A.E.A. Society: Inst. Personnel Management.
Nuclear interest: Personnel management.
Address: United Kingdom Atomic Energy, 11 Charles II Street, London, S.W.1, England.

CHARLESBY, Arthur, D.Sc., Ph.D., D.I.C. Born 1915. Educ.: London Univ. Head, res. group dealing with rad. effects, Harwell, 1949-55; Head, Rad. Dept., T.I. Res. Labs., Cambridge, 1955-57; Prof. Phys., R.M.C.S., 1957-. Pres., British Soc. Rheology. Books: Atomic Rad. and Polymers (Pergamon, 1959); co-author, Effects of Rad. on Materials (Chapman and Hall, 1958); co-author, Chemische Reaktionen Ionisierender Strahlea (Sauerlander,

1958); co-author, Effects Chimiques et Biologiques de Radiation, vol. iii (Masson, 1958).
Societies: Fellow, Inst. Phys. and Phys. Soc.; Roy. Inst.; Assoc. for Rad. Res.; British Soc. Rheology.
Nuclear interests: Radiation effects in materials, particularly polymers; Radiation protection; Radiation damage; Conductivity; Luminescence; Radiation damage and protection mechanisms in biological molecules.
Address: Royal Military College of Science, Shrivenham, Swindon, Wiltshire, England.

CHARNOFF, Gerald, B.A., M.S., L.L.B. Born 1930. Educ.: Brooklyn Coll., Columbia Univ. and N.Y.U. Law School. In Office of Gen. Counsel (1957-60), Asst. to Director, Div. of Licensing and Regulation (1960), U.S.A.E.C.; Staff Counsel, Atomic Ind. Forum, Inc., 1960-66; Partner, Shaw, Pittman, Potts, Trowbridge and Madden, 1966-; Lecturer, Law of Atomic Energy, School of Law, N.Y.U., 1963-66.
Societies: Chairman, Com. on Atomic Energy, Assoc. of the Bar of the City of New York; Member, Atomic Energy Com., New York County Lawyers Assoc.
Nuclear interests: Legal aspects of atomic energy; insurance and indemnification protection against nuclear liability; Contracting with A.E.C.; A.E.C. patent policy; Compensation to workmen and members of public for radiation injuries.
Address: 910 17 Street N.W., Washington D.C., U.S.A.

CHARON, Jean Emile Octave, Phys. Eng., Dr. Born 1920. Educ.: Ecole Supérieure de Physique et Chimie, Paris. Sci. expert, French Embassy, Washington D.C., -1955; C.E.A., 1955-61. Pres. Assoc. pour la Cooperation de la Jeunesse Mondiale; Director, Centre Internat. Culturel d'Aigremont. Books: La Connaissance de l'Univers (Seuil, 1962); Du temps, de l'espace et des Hommes (Seuil, 1963); L'Homme à sa découverte (Seuil, 1964); Eléments d'une théorie unitaire d'Univers (Geneva, René Kister, 1962); Relativite Générale (Geneva, Kister, 1963). Society:Sté. Française de Physique.
Nuclear interests: Controlled thermonuclear fusion; Fundamental theoretical physics; General relativity.
Address: 30 avenue Saint-Laurent, Orsay, (S. et O.), France.

CHARPIE, Robert Alan, B.S., M.S., D.Sc. (Theoretical Phys.). Born 1925. Educ.: Carnegie Inst. Technol. At O.R.N.L. (1950-), Asst. Director, (1955-), Director, Reactor Div., (1958-61), Gen. Manager, Development Dept., then Director of Technol. (1964-), formerly Pres., Electronics Div., (1966-), Corporate Headquarters Exec. (-1968), Union Carbide Corp.; Pres., Bell and Howell Co., 1968-.
Scientific Sec., 1st I.C.P.U.A.E., Geneva, 1955; Editor-in-Chief, Proc. 1st I.C.P.U.A.E.; Coordinator, U.S. Fusion Research Exhibit, 2nd I.C.P.U.A.E., Geneva, 1958. Co-editor, Journal of Nucl. Energy; Formerly Advisory Editor, Nucl. Safety; Formerly Sec., Gen. Advisory Com., U.S.A.E.C. Books: Gen. Editor, International Monograph Series in Nuclear Energy; an editor of Progress Series in Nuclear Energy.
Societies: Fellow, American Phys. Soc.; Formerly Director, American Nucl. Soc.
Nuclear interests: Reactor physics; Administration.
Address: Bell and Howell Company, 7100 McCormick Road, Chicago 45, Illinois, U.S.A.

CHARRON, A. Ing. Welding, Bureau Veritas.
Address: Bureau Veritas, 31 rue Henri Rochefort, Paris 17, France.

CHARTOIRE, L. Prés. Directeur Gén., Glacières et Entrepôts Frigorifiques d'Auvergne S.A.
Address: Glacières et Entrepôts Frigorifiques d'Auvergne S.A., Chemin des Papeteries, Chamalières, (P. de D.), France.

CHASE, Carl T. Sen. Faculty Member, Villanova Univ.
Address: Villanova University, Villanova, Pennsylvania, U.S.A.

CHASE, E. J. A.M.I.E.E. Resident Site Eng., Oldbury Nucl. Power Station.
Address: Central Electricity Generating Board, Oldbury Nuclear Power Station, Oldbury, Glos., England.

CHASE, Grafton D., B.Sc., M.A., Ph.D. Born 1921. Educ.: Philadelphia Coll. of Pharmacy and Sci., Temple Univ.; Organised clinical lab. in Maracaibo, Venezuela (Clinica Quintero), 1955-56; Instructor (1948-54), Asst. Prof. (1956-57), Assoc. Prof. (1958-64), Prof. (1965-), Director, Radiochem. Labs. (1959-), Faculty of the School of Chem., Philadelphia Coll. of Pharmacy and Sci. Books: Principles of Radioisotope Methodology, 3rd edition (Burgess, 1967); Editor, Remington's Practice of Pharmacy (Mack); Co-author, Experiments in Nucl. Sci. (Burgess, 1964). Societies: A.A.A.S.; A.C.S.; A.N.S.
Nuclear interests: Nuclear education; Isotope effects; Nuclear instrumentation; Tracer applications of isotopes in biochemistry.
Address: School of Chemistry, Philadelphia College of Pharmacy and Science, Philadelphia 4, Pennsylvania, U.S.A.

CHASE, Robert Lloyd, B.S. (Elec.Eng.), M.E.E. Born 1926. Educ.: Columbia and Cornell Univs. Assoc. Head, Instrumentation Div., Brookhaven Nat. Lab. Member, Subcom. on Instr. and Techniques, N.R.C. Book: Nucl. Pulse Spectrometry (McGraw-Hill, 1961). Society: Inst. Radio. Eng.
Nuclear interest: Electronic instrumentation for nuclear research.
Address: 42 Namkee Road, Blue Point, New York, U.S.A.

CHASSAGNY, M. Pres. Directeur Gén., Sté. des Engins Matra.
Address: Société des Engins Matra, 49 rue de Lisbonne, Paris 8, France.

CHASSAGNY, Ph. Directeur Commercial, Sté. des Engins Matra.
Address: Société des Engins Matra, 49 rue de Lisbonne, Paris 8, France.

CHASSANG, R. Directeur du Service Technique Central, Groupement des Associations de Proprietaires d'Appareils a Vapeur et Electriques.
Address: 5 rue Blanche, Paris 9, France.

CHASSON, Robert L., A.B., M.A., Ph.D. Born 1919. Educ.: California Univ. Chairman, Dept. of Phys., Nebraska Univ. 1956-59; Prof. Phys., California Univ., 1959-62; Chairman, Dept. of Phys. and Director of Phys. Res., Denver Res. Inst., Denver Univ., 1962-. Societies: A.P.S.; A.A.U.P.; A.A.A.S.; A.G.U. Nuclear interests: Cosmic rays and high energy particle physics; Radiation detection devices, fields and particles in space.
Address: Department of Physics, Denver University, Denver, Colorado 80210, U.S.A.

CHASTAIN, Joel William, M.S. (Phys.). Born 1921. Educ.: Case Inst. Technol. and Kent State Univ. Chief, Reactor Phys. Div., Battelle Memorial Inst., Columbus, Ohio. Exec. Com., Reactor Operations Div., A.N.S.; Subcom. on Res. Reactors, Nat. Res. Council. Book: U.S. Res. Reactor Operation and Use (Addison-Wesley, 1958). Societies: A.N.S.; A.I.P. Nuclear interest: Reactor design.
Address: 4062 Fairfax Drive, Columbus, Ohio 43221, U.S.A.

CHASTEL, R., Prof., Lab. de Physique Nucléaire, Faculte des Sci., Bordeaux.
Address: Faculte des Sciences de Bordeaux, Laboratoire de Physique Nucleaire, 351 Cours de la Libération, Talence, (Gironde), France.

CHATBURN, Gerald, B.Sc. (Hons.). Born 1929. Educ.: Manchester Univ. Manager, Nucl. Safety, Windscale Works, U.K.A.E.A. Vice Chairman, Windscale and Calder Branch, Nucl. Eng. Soc. Society: Assoc. Member, Inst. of Phys.
Nuclear interest: Criticality safety.
Address: United Kingdom Atomic Energy Authority, Windscale Works, Sellafield, Calderbridge, Cumberland, England.

CHATELAIN, Jack E., Assoc. Prof., Phys. Dept., Utah State Univ.
Address: Physics Department, Utah State University, Logan, Utah, U.S.A.

CHATELIER, Jean Marie Joseph LE. See LE CHATELIER, Jean Marie Joseph.

CHATRATH, Manmohan Singh, B.Sc. (Hons. School, Botany), M.Sc. (Hons. School, Botany), Ph.D. Born 1931. Educ.: Panjab Univ. and Indian Agricultural Res.Inst. Res. Asst. (1955-59), Asst. Plant Pathologist (1959-64), Plant Pathologist in Scheme for use of Atomic Energy in Plant Pathological Res. (1964-), Indian Agricultural Res. Inst., New Delhi. Society: Indian Phytopathological Soc., New Delhi.
Nuclear interests: Radiation biology; Radiation induced mutation in plant pathogens to understand intricacies of host parasite relationship; Radiation therapy of plant diseases and physiological effects of ionising radiation on microorganisms.
Address: Division of Mycology and Plant Pathology, Indian Agricultural Research Institute, New Delhi, India.

CHATTERJEE, Aparesh, B.Sc. (Hons.), M.Sc., D.Phil.(Sc.). Born 1924. Educ.: Calcutta Univ. Postdoctorate Fellow, McMaster Univ., Hamilton, 1958-60; Postdoctorate Fellow, McGill Univ., Montreal, 1960-61; Consultant, A.E.R.E., Harwell, 1961-62; Lecturer (1963), Reader (1964), Saha Inst.; Sen. Reader in Nucl. Phys., Calcutta Univ., 1964-. Societies: A.P.S.; Canadian Assoc. of Phys.; Indian Phys. Soc.
Nuclear interests: Experimental and theoretical studies in nuclear reactions and nuclear structure; β-γ spectroscopy; Fast neutron physics and fast counting techniques; Reaction systematics and nuclear structure effects; Medium energy charged particle reactions; Nuclear fission phenomena.
Address: Saha Institute of Nuclear Physics, 92 Acharya Prafulla Chandra Road, Calcutta 9, India.

CHATTERJEE, S. M., Dr. Entomologist, Radioisotopes in Entomological Investigations, Div. of Entomology, Indian Agricultural Res. Inst.
Address: Indian Agricultural Research Institute, New Delhi 12, India.

CHATTERJEE, Santimay, M.Sc., D. Phil. Born 1924. Educ.: Calcutta Univ. Assoc. Prof. Nucl. Phys. and Head, Teaching Div., Saha Inst. Nucl. Phys., Calcutta. Society: Fellow, Indian Phys. Soc.
Nuclear interests: Low energy nuclear physics; Nuclear instrumentation; Variable energy A.V.F. cyclotron.
Address: Saha Institute of Nuclear Physics, 92 Acharyya Prafulla Chandra Road, Calcutta 9, India.

CHATTERJI, Snehamoy, M.Sc. (Zoology), Assoc., Indian Agricultural Res. Inst., Dr. Natural Philosophy. Born 1922. Educ.: Delhi and Agra Univs. Res. Asst. to Head of Div. of Entomology, Indian Agricultural Research Inst., New Delhi, 1946-51; Asst. Director, Ministry of Food and Agriculture (Food), Govt. of India, 1952-53; at Div. of Entomology (1953-59), Entomologist (radio tracer studies), in charge of entomological investigations with

radioactive isotopes (1960-), Indian Agricultural Research Inst., New Delhi. Managing Editor (1959-60), Branch Pres., Delhi (1961), Entomological Soc. of India.
Nuclear interest: Research in uses of nuclear energy in agriculture, particularly in field of entomology.
Address: Division of Entomology, Indian Agricultural Research Institute, New Delhi 12, India.

CHAUDHRI, Rafi Mohammed, Ph.D. (Phys. Cambridge). Born 1903. Educ.: Cambridge and Muslim (Aligarh, India) Univs. Prof., Phys., Head of Dept., Dean, Faculty of Sci., and Vice-Principal (1948-58), Director, High Tension, and Nucl. Res. Lab., Educ. Dept. (1958-65), Govt. Coll., Lahore. Hon. Univ. Prof., Panjab Univ., Lahore, 1960-. Societies: F. Inst. P., London; Fellow, N.A.S., Allahabad; Fellow, Panjab Univ.; Member, Syndicate and Selection Board of West Pakistan Univ. of Eng. and Technol., Lahore, 1961-65.
Nuclear interests: Scattering of fast neutrons from elements and their velocity determination by time of flight method; Study of nuclear isomerism; Half life of induced radioactive elements; Emission of soft electromagnetic radiation and ions from elements under the impact of high energy positive ions; Separation of isotopes by ion emission due to positive ions; Theory and operation of Geiger Müller counters; Ion sources; Study of the impact of high velocity neutral atoms on metals; Gas discharge problems and plasma physics; Study of fission fragments.
Address: Post Box 111, Lahore, West Pakistan.

CHAUDHURY, A. L., Dr. Lecturer, Nucl. Phys., Dept. of Phys., Dacca Univ.
Address: Dacca University, Dacca 2, Pakistan.

CHAUDHURY, A. M., M.Sc., Ph.D. Prof., Nucl. Phys. in the Dept. of Phys., Dacca Univ.
Address: Dacca University, Dacca 2, Pakistan.

CHAUDRON, Georges, Prof. Member, Editorial Advisory Board, J. Nucl. Materials; Member, Conseil Scientifique, C.E.A.
Address: Centre d'Etudes de Chimie Metallurgique, 15 rue Georges Urbain, Vitry-sur-Seine, France.

CHAUVEZ, Claude Emile Jean-Baptiste.
Born 1921. Educ.: Ecole Polytechnique, Paris Univ.; Ecole Nationale Supérieure des Mines, Paris; Ecole Supérieure d'Electricité. Head, Res. Reactors Div., Direction des Piles Atomiques, Centre d'Etudes Nucléaires de Saclay. Societies: Société Française des Electriciens; Société des Ingénieurs Civils de France.
Nuclear interests: Neutron physics; Electricity.
Address: 38 rue du Parc, 91 Orsay, France.

CHAUVIN, Rene. Head, Nucl. Dept., Sté. Française d'Etudes et de Constructions Electroniques, S.A.
Address: Société Française d'Etudes et de Constructions Electroniques, S.A., 11 rue Saint-Florentin, Paris 8, France.

CHAVANNE, André. Born 1916. Educ.: Geneva Univ. Sent by Unesco to Ecuador as an expert for instruction of engineers and technicians; since return to Europe appointed Prof. Phys., Technical High School, Geneva; Conseiller d'Etat, Chef du Département de l'Instruction publique; Délégué, Confédération suisse, C.E.R.N. Formerly Chief editor, Industries Atomiques.
Address: European Organisation for Nuclear Research (C.E.R.N.), Meyrin, Geneva 23, Switzerland.

CHAVE, Charles Trudeau, A.B., B.S., M.E. Born 1905. Educ.: Columbia Univ. Chief Nucl. Eng., Stone and Webster Eng. Corp. Member, Subcom. No. 2, Containment, A.S.A. Sectional Com, N6, Reactor Safety Standards. Societies: A.S.M.E.; A.I.Ch.E.; A.N.S.
Nuclear interests: Management of engineering of nuclear power plants.
Address: 225 Franklin Street, Boston, Massachusetts 02107, U.S.A.

CHAVES de CARVALHO, Antonio Herculano GUIMARAES. See GUIMARAES CHAVES de CARVALHO, Antonio Herculano.

CHAVEZ, Miss Carlita. Medical Technician, Radioisotope Lab., Philippine Gen. Hospital.
Address: Philippine General Hospital, Manila, Philippines.

CHAVEZ, Juan CONTRERAS. See CONTRE- RAS CHAVEZ, Juan.

CHAVEZ ORBEGOZO, Pedro. Secretario Técnico, Departamento de Reactores, Instituto Superior de Energia Nucl., Junta de Control de Energia Atomica del Peru.
Address: Junta de Control de Energia Atomica del Peru, 3420 Avenida Arequipa, Apartado 914, Lima, Peru.

CHAVIGNER, André Charles, Ancien Ing. des P.T.T. Born 1920. Educ.: Ecole Polytech., Paris. Directeur de Production, Sté. Alsacienne de Constructions Atomiques, de Telecommunication et d'Electronique (ALCATEL), Paris.
Address: ALCATEL, Société Alsacienne de Constructions Atomiques, de Telecommunication et d'Electronique, 32 rue de Lisbonne, Paris, France.

CHEATHAM, Hon. J. M. Industry Com., State of Georgia Nucl. Advisory Commission.
Address: State of Georgia Nuclear Advisory Commission, Office of Publications, Georgia Institute of Technology, Atlanta 13, Georgia, U.S.A.

CHEBAEVSKII, V. F. Paper: Co-author, Two-Cascade Power Mass-Analyser (letter to the

Editor, Atomnaya Energiya, vol. 24, No. 3, 1968).
Address: Academy of Sciences of the U.S.S.R., 14 Leninsky Prospekt, Moscow V-71, U.S.S.R.

CHEBOTAREV, N. T. Nuclear interest: Metallurgy of nuclear reactors.
Address: Nuclear Energy Institute, Academy of Sciences of the U.S.S.R., 14 Leninsky Prospekt, Moscow V-71, U.S.S.R.

CHEBOTAREVA, L. D. Paper: Co-author, On Expediency of Radiometric Uranium Ore Dressing and on Optimum Value of Separation during Dressing (letter to the Editor, Atomnaya Energiya, vol. 19, No. 1, 1965).
Address: Academy of Sciences of the U.S.S.R., 14 Leninsky Prospekt, Moscow V-71, U.S.S.R.

CHECHETKIN, Yu. V. Paper: Co-author, Radioactivity Carry-over in Boiling Water of Reactor WK-50 (Atomnaya Energiya, vol. 23, No. 4, 1967).
Address: Academy of Sciences of the U.S.S.R., 14 Leninsky Prospekt, Moscow V-71, U.S.S.R.

CHECHKIN, V. V. On Exchange Visit to Culham Laboratory, March - October 1965. Paper: Co-author, Acceleration of Large Current Pulses in Linear Electron Accelerators (Atomnaya Energiya, vol. 11, No. 1, 1961); co-author, a High-current Electron Accelerator (ibid, vol. 11, No. 1, 1961).
Address: Ukranian Physico-Technical Institute, 2 Yumovskii Tupik, Kharkov, U.S.S.R.

CHEESMAN, G. H., Dr. Radiochem., Dept. of Chem., Tasmania Univ. Councillor, Australian Inst. Nucl. Sci. and Eng.
Address: School of Chemistry, University of Tasmania, Hobart, Tasmania.

CHEESMAN, W. J. Gen. Manager, Atomic, Defence and Aerospace Group, Canadian Westinghouse Co. Ltd. Director (1966-67), Vice Pres. (1968-), Canadian Nucl. Assoc.
Address: Canadian Westinghouse Co. Ltd., Defence and Aerospace Group, P.O. Box 510, Hamilton, Ontario, Canada.

CHEIFETZ, Alan Sidney, B.S. (Nucl.), M.S. (Radiol. Health). Born 1941. Educ.: New York State and Rutgers Univs. Radiol. Health Fellow, U.S.P.H.S. (1962-63), Asst. Instructor, Rad. Sci. (1963-64), Lecturer, Rad. Sci., Univ. Health Phys. (1964-66), Radiol. Health Fellow, U.S.P.H.S. (1966-67), Air Pollution Special Fellow, Div. of Air Pollution, U.S.P.H.S. (1967-), Rutgers Univ. Societies: H.P.S.; Conference on Radiol. Health; American Public Health Assoc.
Nuclear interests: Effects of ionising radiation of liquid aerosols; Environmental radioactivity; Health physics; Nuclear engineering.
Address: 168 Throop Avenue, New Brunswick, New Jersey 08901, U.S.A.

CHEKUNOV, V. V. Papers: Co-author, Calculation of Neutron Fluxes in Cell of Intermediate Reactor with Control Rod (letter to the Editor, Atomnaya Energiya, vol. 24, No. 3, 1968); co-author, Probability of Neutron Absorption in the Block in the Unresolved Resonances Energy Range (letter to the Editor, ibid, vol. 25, No. 3, 1968).
Address: Academy of Sciences of the U.S.S.R., 14 Leninsky Prospekt, Moscow V-71, U.S.S.R.

CHELF, Max T. Born 1902. Vice Pres., Canon Gas Service Co., 1945-55; Member, Colorado House of Reps. 1955-56; Director, Fremont County Nat. Bank, 1956-; Pres. and Chairman, Board, Metrix, Inc., 1960-; Member, Canon City, Colorado City Council, 1967-.
Nuclear interests: Design, construction and installation of whole body counters; Consultant in procedures, testing and operation of low levels counting and health physics installations; Technical co-ordination and supervision of facilities in Europe, North and South America.
Address: P.O. Box 127, Canon City, Colorado, U.S.A.

CHELINTSEV, N. G. Paper: Approximative Solution of Equations of Reactor Dynamic (Atomnaya Energiya, vol. 21, No. 5, 1966).
Address: Academy of Sciences of the U.S.S.R., 14 Leninsky Prospekt, Moscow V-71, U.S.S.R.

CHELLEW, Norman R. B.S. Born 1917. Educ.: Minnesota Univ. Assoc. Chem., Argonne Nat. Lab. Societies: A.N.S.; A.C.S.; RESA.
Nuclear interest: Processing of reactor fuels by non-aqueous techniques.
Address: 811 Winthrop Avenue, Joliet, Illinois, U.S.A.

CHELNOKOV, L. P. Papers: Co-author, On Decay Characteristics of 102^{254} Isotope (Atomnaya Energiya, vol. 20, No. 3, 1966); co-author, On Nucl. Properties of Isotopes of Element 102 with Mass Numbers 255 and 256 (letter to the Editor, ibid., vol. 22, No. 2, 1967).
Address: Academy of Sciences of the U.S.S.R., 14 Leninsky Prospekt, Moscow V-71, U.S.S.R.

CHEMLA, Marius, Ing. Chimiste, D. ès Sc. (Physiques). Born 1927. Educ.: Paris Univ. Prof., Faculté des Sciences de Paris, 1963. Society: Sté. Chimiques de France.
Nuclear interests: Isotope effects and isotope separation; Mass spectrography; Molten salts.
Address: Laboratoire de Physique Nucléaire, Collège de France, Paris 5, France.

CHEN, Francis F., A.B., M.A., Ph.D. Born 1929. Educ.: Harvard Univ. Res. Phys., Plasma Phys. Lab., Princeton Univ. Society: A.P.S.
Nuclear interests: Plasma physics; Low-frequency instabilities; Langmuir probes; Anomalous diffusion; Quiescent plasmas.
Address: Princeton Plasma Physics Laboratory, P.O. Box 451, Princeton, New Jersey U.S.A.

CHEN, James T. T., M.D. Radiological Centre, Taipeih Provincial Hospital.
Address: Taipeih Provincial Hospital, 145 Cheng Chow Road, Taipeih, Formosa.

CHEN, Joseph S. Formerly Prof. and Head, Dept. of Biochem., now Deputy Director, Nat. Defence Medical Centre, Formosa.
Address: National Defence Medical Centre, P.O. Box 7432, Taipei, Formosa.

CHEN, Ko-Chung, Dr. Pres., Nat. Tsing-Hua Univ.; Member, Exec. Yuan, Atomic Energy Council, Taipeih.
Address: National Tsing-Hua University, Institute of Nuclear Science, Kuang Fu Road, Hsinchu, Formosa.

CHEN, Lan Kao, B.Sc., M.Sc. Born 1914. Educ.: Tokyo Nihon Univ. Chief, Dispatch Div. (1946-50), Director, Sun-Moon Lake Power Plants Administrative Office (1950-51), Formerly Vice-Pres. and Chief Eng., Formerly Chairman, now Convenor, Com. on Nucl. Power and Pres., Taiwan Power Co.; Board Chairman, Taiwan Rain Stimulation Res. Inst.; Director, Water Resources Planning Commission; Member, Atomic Energy Council, Formosa. Books: Electrotechnical Experiments; D.C. Machine Experiments. Societies: Director, Chinese Inst. of Eng.; Executive Director, Chinese Inst. of Elec. Eng.
Address: Taiwan Power Company, 39 East Hoping Road, Sec. 1, Taipei, Formosa.

CHENG, Chu-ying. Paper: Co-author, Notes on Dynamic Processes of Thermoregulation in White Mice (Scientia Sinica, vol. 12, No. 10, 1963).
Address: Institute of Biophysics, Academia Sinica, 3 Wen Tsin Chien, Peking, China.

CHENG, Chung-Fu, B.S. (Fukien), M.S. (Southern California), Ph.D. (Minnesota). Born 1912. Educ.: Fukien Christian, Southern California and Minnesota Univs. Sen. Plant Breeder and Head, Plant Breeding Dept., Taiwan Sugar Res. Inst., 1946-55; Sen. Specialist, Plant Industry Div., Sino-American Joint Commission on Rural Reconstruction, 1955-. Head, Res. and Sci. Lectures Com., Agricultural Assoc. of China; Member, Metal. Res. Lab., Union Industrial Res. Inst.
Nuclear interest: The uses of radiation for the improvement of agricultural crops.
Address: Joint Commission on Rural Reconstruction, 25 Nan Hai Road, Taipeih, Taiwan, Formosa.

CHENG, Kenneth T. C., Director, Taiwan Weather Bureau, Section for the Study of Radioactivity in Atmosphere.
Address: Taiwan Weather Bureau, 64 Park Road, Taipeih, Formosa.

CHENG, Victor Chen-Hwa, B.S. Born 1921. Educ.: Utopia Univ., Shanghai and I.S.N.S.E., Argonne Nat. Lab. Director, Inst. of Nucl. Sci., Nat. Tsing Hua Univ., 1961; Commissioner and Sec. Gen., Atomic Energy Council, 1966. Director, Chinese Inst. of Eng. Society: Atomic Energy Soc. of Japan.
Nuclear interests: Reactor design; Nuclear power plant design and management.
Address: Institute of Nuclear Science, National Tsing Hua University, 397 Kuang-Fu Road, Hsin-Chu, Formosa.

CHEOSAKUL, Pradisth, B.S., M.S., Ph.D. Born 1913. Educ.: Chulalongkorn (Bangkok), Philippines (Manila) and Cornell Univs. Asst. Prof., Albany Medical Coll., N.Y., 1951-55; Tech. Adviser, Dept. of Sci., Ministry of Industry, Bangkok, 1955-61; Deputy Sec.-Gen., Nat. Res. Council, Office of Prime Minister, Thailand, 1961-; Governor, Appl. Sci. Res. Corp., Thailand, 1964-. Advisory Member, Thai A.E.C.; Sec.-Gen., Sci. Soc. of Thailand; Advisory Member, Board of Regents, Kasetsart Univ.; Member, Nat. F.A.O.; Member, Nat. Education Council; Member, Nat. Com. for U.N.E.S.C.O. Societies: A.C.S.; Fellow, A.A.A.S.
Nuclear interest: Tracers in animal metabolism.
Address: National Research Council, Bangkhen, Thailand.

CHEPURENKO, I. A. Paper: Co-author, Gas Supply System for Ion Source of Electrostatic Generator (letter to the Editor, Atomnaya Energiya, vol. 20, No. 2, 1966).
Address: Academy of Sciences of the U.S.S.R., 14 Leninsky Prospekt, Moscow V-71, U.S.S.R.

CHERBULIEZ, Emile, Dr. phil. (Munich), Dr. rer. nat. (E.T.H., Zurich). Born 1891. Educ.: E.T.H. Zurich, Munich and Geneva Univs. Full Prof. of organic and pharmaceutical chem., Geneva Univ.; Pres., Commission of Atomic Sci., Berne, -1962; Vice-Pres., Nat. Council of Sci. Res., Berne; Pres., Board of Editors, Helvetica Chimica Acta (J. Swiss Chem. Soc.); Pres., Comité Suisse de la Chimie (adhering body to Internat. Union of Pure and Appl. Chem.).
Nuclear interest: Tracer studies.
Address: Ecole de Chimie, Geneva, Switzerland.

CHERDYNTSEV, Victor Victorovitch, Candidate of Chem. Sci. (Kazan State Univ.), Doctor of Physico-mathematical Sci. (Acad. Sci., U.S.S.R.), Prof. Born 1912. Educ.: Moscow Univ. Prof., Kasakhstan State Univ., Alma-Ata; Chief of Lab. of the determination of the absolute age, Geological Inst., Acad. Sci. of the U.S.S.R., Moscow, 1960-. Book: Abundance of Chem. Elements (Moscow, State Publishing House, 1956). English translation (second edition revised) (Univ. Chicago Press, 1961).
Nuclear interests: Nuclear processes in nature; Determination of geological age; Origin of nuclei.
Address: Geological Institute, Academy of

Sciences of the U.S.S.R., 7 Pyzevski pereulok, Moscow V-17, U.S.S.R.

CHEREMNYKH, P. A. Papers: Co-author, The Exptl. Plasma Facility C-1 with Screw Magnetic Fields (Atomnaya Energiya, vol. 14, Nos. 2, 1963); co-author, Effect of Helical Magnetic Field on Plasma Ohmic Heating in S-1 Machine (ibid, vol. 20, No. 4, 1966).
Address: Academy of Sciences of the U.S.S.R., 14 Leninsky Prospekt, Moscow V-71, U.S.S.R.

CHERENKOV, Pavel Alekseevich, Nobel Prize for Phys., 1958; Stalin Prize (1st Class), 1946. Born 1904. Educ.: Voronezh Univ. Worker of Phys. Inst., Acad. of Sci. of the U.S.S.R., 1930-; Prof., Moscow Phys. Eng. Inst.
Nuclear interests: Nuclear physics and instrumentation.
Address: Moscow Physical Engineering Institute, Moscow, U.S.S.R.

CHEREVATENKO, A. P. Paper: Co-author, Back-Yield of Neutrons from out Shielding Bombarded with 660 MeV Protons (letter to the Editor, Atomnaya Energiya, vol. 24, No. 4, 1968).
Address: Academy of Sciences of the U.S.S.R., 14 Leninsky Prospekt, Moscow V-71, U.S.S.R.

CHEREVATENKO, E. P. Papers: Co-author, Attenuation of High-Energy Neutron Fluxes in Heterogeneous Shielding (letter to the Editor, Atomnaya Energiya, vol. 21, No. 1, 1966); co-author, Attenuation of Stray Radiation from Synchrocyclotron in Shielding (ibid, vol. 24, No. 2, 1968); co-author, Measurement of Dose Field of Mixed Radiation behind Steel Shielding (letter to the Editor, ibid, vol. 24, No. 2, 1968); co-author, Measurements of the Quality Factor for High Energy Protons in the Water Phantom (Nukleonika, vol. 13, No. 2, 1968).
Address: Academy of Sciences of the U.S.S.R., 14 Leninsky Prospekt, Moscow V-71, U.S.S.R.

CHEREVATENKO, G. A. Paper: Co-author, Determination of Absolute and Relative Yield of Bremsstrahlung (letter to the Editor, Atomnaya Energiya, vol. 23, No.1, 1967).
Address: Academy of Sciences of the U.S.S.R., 14 Leninsky Prospekt, Moscow V-71, U.S.S.R.

CHEREZOV, N. K. Paper: Co-author, Possibility of Rd^{83} Use of Mossbauer's Effect Investigations with Kr^{83} (letter to the Editor, Atomnaya Energiya, vol. 23, No. 1, 1967).
Address: Academy of Sciences of the U.S.S.R., 14 Leninsky Prospekt, Moscow V-71, U.S.S.R.

CHERKASOV, A. S. Paper: Co-author, Sand Moisture Determination by Fast Neutron Flux Attenuation (letter to the Editor, Atomnaya Energiya, vol. 16, No. 4, 1964).
Address: Academy of Sciences of the U.S.S.R., 14 Leninsky Prospekt, Moscow V-71, U.S.S.R.

CHERMAK, I. Paper: Calculation of Efficiency of Partially Inserted Absorbing Rods (Atomnaya Energiya, vol. 22, No. 6, 1967).
Address: Academy of Sciences of the U.S.S.R., 14 Leninsky Prospekt, Moscow V-71, U.S.S.R.

CHERNAVSKY, D. S. Papers: Co-author, Strong Interactions Theory and Cosmic Rad. (Nukleonika, vol. 9, No. 2 - 3, 1964); On the Role of Fireballs in Recent Theory of Strong Inelastic Interactions at High Energy (ibid, vol. 13, No. 1, 1968).
Address: Academy of Sciences of the U.S.S.R., 14 Leninsky Prospekt, Moscow V-71, U.S.S.R.

CHERNICK, Jack, B.S. (Chicago), M.S. (Maths., Brooklyn Coll.). Born 1911. Phys., Assoc. Div. Head, Reactor Phys. Div., Brookhaven Nat. Lab. Societies: American Nucl. Soc. Fellow, A.N.S.; A.P.S.
Nuclear interests: Reactor physics, reactor design, reactor stability and safety; Design of natural and enriched uranium-graphite research reactors, liquid metal fuelled breeders, and the Brookhaven High Flux Beam Reactor.
Address: Brookhaven National Laboratory, Upton, Long Island, New York, U.S.A.

CHERNILIN, Yu. F. Deputy Director, Kurchatov Inst. Atomic Energy. Paper: Co-author, Production of Pure Fluxes of Fast Neutrons for Radiobiologic Res. on IRT- 1000 Reactor (letter to the Editor, Atomnaya Energiya, vol. 12, No. 6, 1962).
Address: Kurchatov Institute of Atomic Energy, 46 Ulitsa Kurchatova, P.O. Box 3402, Moscow, U.S.S.R.

CHERNOCK, Warren Philip, M.S. (Metal.), B.S. (Metal. Eng.). Born 1926. Educ.: Columbia and New York Univs. Manager, Nucl. Labs., Nucl. Div. Combustion Eng., Inc., Windsor, Conn., 1956-; Adjunct Assoc. Prof. Metal., Hartford Graduate Centre, Rensselaer Polytech. Inst.; Visiting Assoc. Prof. Nucl. Eng., M.I.T. Member, Com. E-10, Com. on Radioisotopes and Rad. Damage, A.S.T.M. Societies: A.S.M.; A.I.M.M.E.; A.N.S.
Nuclear interests: Metallurgy and ceramics; fuels, cladding, control and structural materials; Research and development; Fuel cycle economics.
Address: Nuclear Division, Combustion Engineering, Inc., Windsor, Connecticut, U.S.A.

CHERNOGOROVA, V. A. Paper: Co-author, Angular Distribution of Neutrons due to μ^--Capture in Calcium for Various Energy Thresholds (Nukleonika, vol. 9, No. 2 - 3, 1964).
Address: Joint Institute for Nuclear Research, Laboratory of High Energy Physics, Dubna, U.S.S.R.

CHERNOMORDIK, E. N. Paper: Co-author, Sodium Technol. and Equipment of BN-350 Reactor (Atomnaya Energiya, vol. 22, No. 1, 1967).

Address: Academy of Sciences of the U.S.S.R., 14 Leninsky Prospekt, Moscow V-71, U.S.S.R.

CHERNOPLEKOV, Nikolai Alekseyevich. Paper: Co-author, Investigation of Colb Neutron Inelastic Scattering in Certain Hydrogen-Bearing Materials (Atomnaya Energiya, vol. 14, No. 3, 1963).
Address: Kurchatov Institute of Atomic Energy, 46, Ulitsa Kurchatova, Post Box 3402, Moscow, U.S.S.R.

CHERNYAEV, V. A. Papers: Co-author, Method of Calculation of Water and Power Costs for Nucl. Desalination Plants (Atomnaya Energiya, vol. 19, No. 2, 1965); co-author, Some Features of Diphenyl Thermal Turbines and Their Maximum Power (ibid, No. 3).
Address: Academy of Sciences of the U.S.S.R., 14 Leninsky Prospekt, Moscow V-71, U.S.S.R.

CHERNYAHOVSKII, V. V. Paper: Co-author, Scattering of 0.5 MeV Gamma-Rays by Shielding Barriers (Atomnaya Energiya, vol. 22, No. 5, 1967).
Address: Academy of Sciences of the U.S.S.R., 14 Leninsky Prospekt, Moscow V-71, U.S.S.R.

CHERNYAYEV, I. I. Academician of the Acad. of Sci. of the U.S.S.R.; Deleg.to 2nd I.C.P.U.A.E., Geneva, Sept. 1958. Director, Kurnakov Inst. of Gen. and Inorganic Chem. Paper: Co-author, The Complexing Reaction in Chem. Technol. of Uranium (Atomnaya Energiya, vol. 14, No. 4, 1963).
Address: Kurnakov Institute of General and Inorganic Chemistry, 31 Leninsky Prospekt, Moscow, U.S.S.R.

CHERNYSHOV, A. K. Paper: How to Find Prandtle Number of Some Liquid Metal Nomograph (letter to the Editor, Atomnaya Energiya, vol. 20, No. 1, 1966).
Address: Academy of Sciences of the U.S.S.R., 14 Leninsky Prospekt, Moscow V-71, U.S.S.R.

CHERRY, Robin David, M.Sc., Ph.D. Born 1933. Educ.: Cape Town Univ. Sen. Lecturer, Phys., Cape Town Univ.
Nuclear interest: Natural radioactivity and its detection.
Address: Physics Department, Cape Town University, Rondebosch, Cape Town, South Africa.

CHERVETSOVA, I. N. Paper: Co-author, Determination of Spatial Distribution of Co^{60} Gamma-Rad. Absorbed Dose with Cellulose Triacetate Films (Atomnaya Energiya, vol. 20, No. 6, 1966).
Address: Academy of Sciences of the U.S.S.R., 14 Leninsky Prospekt, Moscow V-71, U.S.S.R.

CHERY, Roland. Prof., Group Spectrométrie, Inst. de Physique Nucléaire, Faculté des Sci., Lyon Univ.
Address: Lyon University, Institut de Physique Nucléaire, 43 boulevard du 11 Novembre 1918, 69 Villeurbanne, France.

CHESLEY, Frank G. Pres., Central Res. Labs. Inc.
Address: Central Research Laboratories Inc., Red Wing, Minnesota, U.S.A.

CHESNE, André. Head, Chem. and Radioactive Studies Sect., C.E.N., Fontenay-aux-Roses. Member for France, Group of Experts on Production of Energy from Radioisotopes, O.E.C.D., E.N.E.A.
Address: Commissariat a l'Energie Atomique, C.E.N., B.P. No. 6, Fontenay-aux-Roses, (Seine), France.

CHESNOKOV, N.I. Paper: Co-author, Rad. Safeguarding of Personnel during Uranium Ore Yield (Atomnaya Energiya, vol. 19, No. 2, 1965).
Address: Academy of Sciences of the U.S.S.R., 14 Leninsky Prospekt, Moscow V-71, U.S.S.R.

CHESTER, C. V., B.Sc. (Eng.), G.I.Mech.E. Chief Eng., and Local Director, Eng. Group, Vickers Ltd.
Address: Vickers-Armstrongs South Marston Works, Swindon, Wiltshire, England.

CHESTER, Peter Francis, Ph.D., B.Sc. Born 1929. Educ.: London Univ. Post Doctoral Fellowship, N.R.C., Ottawa, 1953-54; Advisory Phys., Westinghouse Res.Labs., Pittsburgh, 1954-60; Head, Solid State Phys. Sect. (1960-65), Head, Fundamental Studies Sect. (1965-66), Central Elec. Res. Labs., Leatherhead; Res. Manager, Sci., Elec. Council Res. Centre, Capenhurst, 1966-. Member, Council, Member Editorial Board, Proceedings of the Phys. Soc. and Fellow, Inst. of Phys. and the Phys. Soc. Nuclear interest: High magnetic fields.
Address: Electricity Council Research Centre, Capenhurst, Chester, England.

CHESWORTH, Robert Hadden, B.S. (Chem. Eng.); Nucl. Eng. Born 1929. Educ.: California Univ., Berkeley. Manager, Nucl. Eng. and Manufacturing Operations, Nucl. Products and Services Group, Aerojet-Gen. Corp. Societies: A.N.S.; A.I.Ch.E.; American Management Assoc.
Nuclear interests: Reactor fuel element design, development, and fabrication; Development of radioisotope power systems for terrestrial and underwater applications; Space power liquid metal boiler development; Measurement of thermophysical properties of alkali metals for nuclear application; Systems management.
Address: Aerojet-General Corporation, P.O. Box 77, San Ramon, California 94583, U.S.A.

CHEVALIER, Louis, Ing. Born 1903. Educ.: Ecole Polytechnique, Paris and Ecole du Génie Maritime. Pres., Sté. des Forges et Chantiers de la Mediterranée; Member, Council of Administration, A.T.E.N.
Address: Société des Forges et Chantiers de la Mediterranée, 6-8 rue Camou, Paris 7, France.

CHEVALLIER, J. Chef du Département de Propulsion Nucléaire, Direction des Piles Atomiques, C.E.A. Expert, Third I.C.P.U.A.E., Geneva, Sept. 1964.
Address: Commissariat à l'Energie Atomique, 29-33 rue de la Federation, Paris 15, France.

CHEVALLIER, Pierre, D. ès Sc. Born 1930. Educ.: Strasbourg Univ. Prof., Faculté Sci., Strasbourg Univ.
Nuclear interest: Nuclear physics.
Address: Centre de Recherches Nucléaires, Département de Physique Nucléaires, rue du Loess, Strasbourg, Cronenbourg, France.

CHEVEREV, N. S. At Kurchatov Inst. of Atomic Energy. Adviser, Third I.C.P.U.A.E., Geneva, Sept. 1964. Paper: Co-author, Paramagnetic High-Frequency Effect and Electron Paramagnetic Resonance in Plasma (Atomnaya Energiya, vol. 20, No. 5, 1966).
Address: Kurchatov Institute of Atomic Energy, 46 Ulitsa Kurchatova, Post Box 3402, Moscow, U.S.S.R.

CHEVREMONT, Maurice J. J., Agrégé de l'Enseignement Supérieur; Dr. (honoris causa). Born 1908. Educ.: Liège Univ. Formerly Prof. Ordinaire, Liège Univ.; Membre Correspondant, Académie Royale de Médicine de Belgique; Sec.-Treas., Internat. Soc. for Cell Biol.; Sec., Centre Anticancéreux, Liège Univ. Nuclear interest: Histoautoradiography.
Address: Institut d'Histologie, 20 rue de Pitteurs, Liège, Belgium.

CHEVREMONT-COMHAIRE, Madame S. Nucl. research work, Inst. d'Histologie, Liège.
Address: Institut d'Histologie, 20 rue de Pitteurs, Liège, Belgium.

CHEVRIER, Charles. Directeur du gaz et d'électricité, Ministère de l'Industrie. Adviser, Third I.C.P.U.A.E.; Geneva, Sept. 1964. Member, Comite des Mines, C.E.A.
Address: Director du Gaz et de l'Electricité, Ministere de l'Industrie et du Commerce, 24 rue de l'Université, Paris 7, France.

CHEW, B. B. Chief Project Eng., (-1962), Assoc. Director, Systems Development Dept., (1962-), Compact Systems Div., Div. Director, Marketing, Atomics International.
Address: Compact Systems Division, Atomics International, P.O. Box 309, Canoga Park, California, U.S.A.

CHEW, Geoffrey Foucar, B.S., Ph.D. Born 1924. Educ.: George Washington and Chicago Univs. Prof. Phys., Illinois Univ., 1950-57; Prof. Phys., California Univ., Berkeley, 1957-. Book: S-Matrix Theory of Strong Interactions (W. A. Benjamin, Inc., 1961); The Analytic S-Matrix (W.A. Benjamin, Inc., 1966). Societies: Fellow, A.P.S.; Nat. Acad. Sci.; American Acad. Arts and Sci.
Nuclear interest: The theory of the strong interactions between nuclear particles (e.g., nuclear forces).
Address: Department of Physics, University of California, Berkeley 4, California, U.S.A.

CHEW, H., Ph.D. Born 1935. Educ.: Chicago Univ. Res. Assoc., Pennsylvania Univ., 1961-62; Res. Sci., Columbia Univ., 1962-63; Instructor, Princeton Univ., 1963-64; Asst. Prof., Western Reserve Univ., 1964-67; Assoc. Prof., Clarkson Coll., 1967-. Society: A.P.S.
Nuclear interest: Theoretical particle physics.
Address: Department of Physics, Clarkson College, Potsdam, N. Y., 13676, U.S.A.

CHEW, Woodrow Wilson, B.S. (Chem. Eng.), M.S. (Chem. Eng.). Born 1913. Educ.: New Mexico and Oklahoma State Univs. Assoc. Prof. and Prof. and Head, (1945-), Prof. and Head, (1952-), Chem. Eng. Dept., Louisiana Polytech. Inst. Societies: A.I.Ch.E.; A.C.S.; A.S.E.E.; American Soc. Professional Eng.
Nuclear interest: Training and education of undergraduate and graduate students in chemical engineering in the area of industrial utilisation of radioactive isotopes.
Address: Louisiana Polytechnic Institute, Tech. Station, Ruston, Louisiana, U.S.A.

CHEZEM, Curtis Gordon, B.A., M.A., D.Ph. Born 1924. Educ.: Oregon Univ., Oregon State Coll. Phys. Sci. (Chief, System Studies Branch), Office of Safeguards and Materials Management.
Nuclear interests: Coupled reactor kinetics, critical assemblies; Nuclear propulsion; Nuclear materials management.
Address: United States Atomic Energy Commission, Washington D.C. 20545, U.S.A.

CHI, Benjamin Exner, B.S. (Antioch Coll.), Ph.D. (Rensselaer Polytech. Inst.). Born 1933. Educ.: Rensselaer Polytech. Inst. Postdoctoral Fellow, Lecturer, Instructor in Phys., Western Reserve Univ.; now Asst. Prof., New York State Univ., Albany. Book: Numerical Table of Clebsch-Gordan Coefficients (Rensselaer Polytech. Inst., 1962). Societies: A.P.S.; A.A.P.T.; Assoc. for Computing Machinery. Nuclear interest: Nuclear structure (experiment and theory).
Address: Department of Physics, New York State University, Albany, New York 12203, U.S.A.

CHIARINI, Arnaldo, Dr. Phys. Born 1930. Educ.: Bologna Univ. Researcher of I.N.F.N., 1955; Asst., Reactor Design Dept., Scuola di Specializzazione in Ingegneria Nucleare, 1957; Lecturer in Reactor Theory, 1959-61; Head, Computing Service, Centro di Calcolo del C.N.E.N., Bologna, 1961-. Lecturer on Statistics, Eng. Faculty, Bologna Univ., 1963-. Italian rep., Steering Com., E.N.E.A. Computer Programme Library. Societies: Assoc. for Computing Machinery; Soc. Italiana di Fisica; A.N.S.; Assoc. Italiana per il Calcolo Automatico.

Nuclear interests: Reactor physics and computation.
Address: Via Finelli, 7 Bologna 40126, Italy.

CHIAROTTI, Gianfranco, Prof. Prof. of Advanced Phys., Messina Univ. Member, Sicilian Nucl. Res. Com.; Director, Sicily Sect. (Messina), Nucl. Phys. Nat. Inst.
Address: Messina University, Via Tommaso Cannizzaro, Messina, Sicily, Italy.

CHIARUGI, A. Member, Editorial Com., Minerva Nucleare.
Address: 83-85 Corso Bramante, Turin, Italy.

CHICK, Douglas Richard, Prof. D.Sc. (Eng.), M.Sc. (Eng., London), M.Sc. (Phys., Reading). Born 1916. Educ.: London and Reading Univs. Nucl. Sci. Res. (1946-63), Leader Nucl. Sci. Group (1958-63), A.E.I. Res. Lab., Aldermaston Court; Vickers Res. Ltd., 1963-65; Head, Dept. Elect. and Control Eng., Surrey Univ., 1965-. Res. Com. I.E.R.E.; Council, Elec. Res. Assoc. Societies: Fellow, I.E.E.; I.E.R.E.; F. Inst. Phys.
Nuclear interests: Reactor design and management; Nuclear physics; Plasma physics; Ion implantation.
Address: Wisteria Lodge, Brewery Common, Mortimer, Berkshire, England.

CHIEFETZ, Alan S. Lecturer in Radiol. Sci., Rad. Sci. Centre, Rutgers State Univ.
Address: Rutgers State University, New Brunswick, New Jersey, U.S.A.

CHIEN, Hsueh-shen, Dr. Born 1912. Educ.: M.I.T. and California Inst. of Technol. Worked at M.I.T. with Prof. von Karman; During war worked with U.S. scientific commission on national defence (in charge of rocket sect.); Studied German rocket installations after war; Returned to China in 1955; Opened Inst. Dynamics, Peking, 1955; Now directs China's nucl. rocket programme.
Address: Academia Sinica, 3 Wen Tsin Chien, Peking, China.

CHIEN, Ji-Peng, Ph.D. (Nucl. Eng.). Born 1924. Educ.: Michigan Univ. Director, Inst. of Nucl. Energy Res., Atomic Energy Council, Formosa.
Nuclear interest: Neutron and nuclear physics, reactor theory.
Address: P.O. Box No. 3, Lung-Tan, Taiwan, Formosa.

CHIEN, San-chiang, Ph.D. (Paris). Pupil of Joliot Curie. Director of Atomic Energy, Chinese Acad. of Sci., Peking. Head, China's Nucl. Programme, and directing work on nucl. propelled rocket.
Address: Academia Sinica, 3 Wen Tsin Chien, Peking, China.

CHIEN, Shao-Chun. Papers: Co-author, On the possibility of Λhyperon creation through isobars in π^--p interactions at 7 -8 GeV (Nukleonika, vol. 9, No. 2 -3, 1964); co-author, On the possibility of establishing of isobar state systems and their decay modes (ibid.).
Address: Joint Institute for Nuclear Research, Dubna, Nr. Moscow, U.S.S.R.

CHIEN, Shih-liang. Member, Exec. Yuan, Atomic Energy Council, Taipei.
Address: Atomic Energy Council, 1-1 Lane 20, Sin-Yi Road, Section 1, Tapiei, Formosa.

CHILD, H. C. Res. Manager, Jessop-Saville Ltd.
Address: Jessop-Saville Ltd., Brightside Works, Sheffield 9, England.

CHILDS, Donald S., Jr., B.A., M.D. Born 1916. Educ.: Haverford Coll. and Yale Univ. Head, Sect. of Therapeutic Radiol., Mayo Clinic, Rochester, Minnesota, 1953-; Member, Medical Advisory Com., U.S.A.E.C., 1956-. Societies: Radiol. Soc. of North America; Soc. of Nucl. Medicine; American Radium Soc.
Nuclear interests: Medical application of radioactive isotopes; Radiation protection.
Address: Mayo Clinic, 200 First Street Southwest, Rochester, Minnesota, U. S. A.

CHILENSKAS, Albert Andrew, B.S. (Chem. Eng.). Born 1927. Educ.: Illinois and Chicago Univs. and Illinois Inst. Technol. Assoc. Chem. Eng., Argonne Nat. Lab. Societies: A.N.S.; A I.Ch.E.
Nuclear interests: Nuclear fuel processing specifically; Pyrometallurgical processes; Process and plant design and development regarding EBR-II fuel reprocessing; Corrosion studies and bench-scale investigations in the fluoride volatility fuel process.
Address: Argonne National Laboratory, 9700 South Cass Avenue, Building 205, Argonne, Illinois, U.S.A.

CHILTON, Arthur Bounds, B.S., B.C.E., M.C.E., M.Sc., Ph.D. Born 1918. Educ.: U.S. Naval Acad., Rensselaer Polytech. Inst. and Ohio State Univ. Manager, Atomic Energy Branch, Res. Div. (1954-58), Director, R. and D. Div. (1958-59), Bureau of Yards and Docks, U.S. Navy Dept.; C.O. and Director, U.S. Naval Civil Eng. Lab., 1959-62; Prof. Civil and Nucl. Eng., Illinois Univ., 1962-. Societies: A.P.S.; A.N.S.; Biophysics Soc.; H.P.S.
Nuclear interests: Interaction of radiation with matter; Radiation shielding.
Address: University of Illinois, Urbana, Illinois, U.S.A.

CHIN, S. C., M.D. Radiological Centre, Taipeih Provincial Hospital.
Address: Taipeih Provincial Hospital, 145 Cheng Chow Road, Taipeih, Formosa.

CHINAGLIA, Leopoldo, Chem. Eng., Aeronautical Eng. Born 1929. Educ.: Politecnico, Turin. Project Eng., FIAT Nucl. Energy Dept., 1953-.
Nuclear interest: Overall reactor plant design.
Address: 7 bis Via Villa della Regina, Turin, Italy.

CHINE, Mustapha. Adjoint du Commissaire, C.E.A.
Address: Commissariat à l'Energie Atomique, Secretariat d'Etat au Plan et aux Finances, Tunis, Tunisia.

CHIOTTI, Premo, B.S. (Chem. Eng.), Ph.D. (Phys. Chem.). Born 1911. Educ.: Illinois and Iowa State Univs. Asst. Prof. (1950-54), Assoc. Prof. (1954-61), Prof. (1961-), Iowa State Univ. Societies: A.C.S.; A.S.M.; A.A.A.S.; A.A.U.P.
Nuclear interests: High temperature X-ray diffraction, and thermal properties of materials; Phase relations and thermodynamic properties of alloys; Fused salt and liquid metal chemistry and applications to pyrometallurgical reprocessing of nuclear fuel materials.
Address: Metallurgy Building, Iowa State University, Ames, Iowa, U.S.A.

CHIPPERFIELD, F. P. Sec., State Elec. Commission of Victoria; nominated to receive correspondence relating to the Commission's membership of the Commonwealth States Com. on Exchange of Atomic Information.
Address: State Electricity Commission of Victoria, 15 William Street, Melbourne 3000, Australia.

CHIPPINDALE, Peter, B.Sc., Ph.D. Born 1925. Educ.: Manchester Univ. Tootal Broadhurst Lee Co., Ltd., Manchester, 1951-58; Lecturer, Phys., Salford Univ. Society: A.Inst.P.
Nuclear interest: Teaching of nuclear physics, including reactor engineering and radiological health and safety courses.
Address: 45 Woodfield Avenue, Accrington, Lancs., England.

CHIRIKOV, B. V. Papers: Resonance Processes in Magnetic Traps (Atomnaya Energiya, vol. 6, No. 6, 1959); Stability of Partly Neutralized Electron Beam (ibid, vol. 19, No. 3, 1965); co-author, Coherent Instability of Beam inside Chamber with Non-Conducting Walls (Abstract, ibid, vol.20, No. 4, 1966).
Address: Academy of Sciences of the U.S.S.R., 14 Leninsky Prospekt, Moscow V-71,U.S.S.R.

CHIRKIN, A. V. Papers: Co-author, Effect of Vanadium on Phase Constitution and Structure of High Boron Steel (Abstract, Atomnaya Energiya, vol. 20, No. 2, 1966); co-author, Effect of Titanium on Phase Constitution and Ductivity of Chrom - Nickel - Steel with High Content to Bore (ibid, vol. 22, No. 5, 1967).
Address: Academy of Sciences of the U.S.S.R., 14 Leninsky Prospekt, Moscow V-71, U.S.S.R.

CHIRKIN, V. S. Paper: Temperature Conductivity and Thermal Conductivity of Metallic Beryllium (letter to the Editor, Atomnaya Energiya, vol. 20, No. 1, 1966).
Address: Academy of Sciences of the U.S.S.R., 14 Leninsky Prospekt, Moscow V-71, U.S.S.R.

CHISHOLM, A. Lecturer, Phys. Dept., Auckland Univ.
Address: Auckland University, P. O. Box 2175, Auckland, New Zealand.

CHISHOLM, Roderick George, B.A., M.A., M.S. Dr. in Philosophy, Dr. de Filosofia en Sciencias Matematicas. Born 1913. Educ.: De Paul (Chicago) and Santo Tomas (Manila) Univs. Dean of Graduate Studies, St. Mary's Coll., Winona, Minnesota, 1958-64; Radiol. Defence Officer, Sect. Chief, Mobile Support Area I, Minnesota. Societies: I.E.R.E.; A.I.P.; U.S. Naval Inst.; A.N.S.
Nuclear interests: Radioisotope techniques: Agriculture applications; Nuclear physics; Radiological defence.
Address: St. Mary's College, Winona, Minnesota, U.S.A.

CHISTOV, E. D. Papers: Co-author, Penetration of Gamma-Rays through Spherical and Cylindrical Shields (letter to the Editor, Atomnaya Energiya, vol. 21, No. 5, 1966); co-author, Spectral and Angular Distributions of Cs^{137} Gamma-Rays Scattered in Spherical and Slab Shields (letter to the Editor, ibid, vol. 23, No. 2, 1967).
Address: Academy of Sciences of the U.S.S.R., 14 Leninsky Prospekt, Moscow V-71, U.S.S.R.

CHISTYAKOVA, G. A. Paper: Co-author, Determination of Uranium and Plutonium in Organic Solutions (letter to the Editor, Atomnaya Energiya, vol. 24, No. 1, 1968).
Address: Academy of Sciences of the U.S.S.R., 14 Leninsky Prospekt, Moscow V-71, U.S.S.R.

CHISTYAKOVA, G. I. Paper: Co-author, Investigation of Electrophoresis Filters with Graphite Anodes (letter to the Editor, Atomnaya Energiya, vol. 23, No. 4, 1967).
Address: Academy of Sciences of the U.S.S.R., 14 Leninsky Prospekt, Moscow V-71, U.S.S.R.

CHISWIK, Haim H., A.B., Sc.D. Born 1915. Educ.: Harvard Univ. Assoc. Director, Metal. Div., Argonne Nat. Lab. (U.S.A.E.C.). Transactions Com., A.S.M.; Editorial Advisory Board, Nucl. Materials. Societies: A.I.M.E., A.S.M.
Nuclear interest: Nuclear physical metallurgy.
Address: 1525 Thornwood Drive, Downers Grove, Illinois 60515, U.S.A.

CHITNIS, E. V. Assoc. Prof., Phys. Res. Lab.
Address: Physical Research Laboratory, Navrangpura, Ahmedabad 9, India.

CHITTENDEN, William Arthur, B.Sc. (Gen. Eng.). Born 1927. Educ.: Illinois and Northwestern Univs. Mech. Eng. (1952-55), Nucl. Power Eng. (1955-56), Head, Nucl. Power Staff (1956-), Sargent and Lundy. Societies: A.S.M.E.; A.N.S.
Address: 143 South West Avenue, Elmhurst, Illinois, U.S.A.

CHITTY, John Albert, M.A. (Hons.) (Phys.). Born 1924. Educ.: Oxford Univ. Asst. Tech. Manager, Capenhurst Works (1958-60), Asst. Works Manager, Chapel Cross Works (1960-61), With Economics and Programming Branch, London Office (1961-), U.K.A.E.A. Nuclear interest: Operations management and nuclear power economics.
Address: 9 Lakeside, Enfield, Middlesex, England.

CHMIELEWSKI, Jerzy I. Z., M.Sc. (Elec. Eng.). Born 1930. Educ.: Polytech. Warsaw. Sen. Asst., Instytut Badan Jadrowych, Zaklad XV, Warsaw, 1954-. Nuclear interests: Electronic instrumentation; Multi-channel pulse-height analysers; Industrial applications.
Address: 3a, m27 Szymanowskiego, Warsaw 52, Poland.

CHO, Baek Hyun. Commissioner, Office of Atomic Energy and A.E.C., Seoul.
Address: Office of Atomic Energy and Atomic Energy Commission, 2-1 Chung-dong, Sudaimoon-ku, Seoul, Korea.

CHOCHLOVSKY, Igor, Dipl. Ing. Born 1928. Educ.: Prague Tech. Univ. Eng. in Chief, Nucl. Div., Chemoprojekt, Design, Eng. and Consulting Corp., Prague.
Nuclear interest: Design of nuclear reactors and hot laboratories.
Address: 15 Stepanska, Prague 2, Czechoslovakia.

CHODOROW, Marvin, B.A., Ph.D. Born 1913. Educ.: M.I.T. Asst. Prof. (1947), Assoc. Prof. (1955), Prof. (1959-), At present Director Microwave Lab. and Exec. Head Appl. Phys. Div., Stanford Univ. Societies: A.P.S.; A.A.U.P.; A.A.P.T.; A.A.A.S.; Nat. Acad. Eng.; I.E.R.E.
Nuclear interest: Theory and design of linear electron accelerator.
Address: Microwave Laboratory, Stanford University, Stanford, California, U.S.A.

CHOI, Hyung Sup, B.S. (Mining and Metal.), M.S. (Phys. Metal.), Ph.D. (Metal. Eng.). Born 1920. Educ.: Waseda, Notre Dame and Minnesota Univs. Commanding Officer, 180 Air Maintenance Depot, ROKAF, 1951- 53; Res. Metal., U.S. Department of Interior, 1958-59; Vice-Pres., Kuksan Motor's Co. Ltd., Korea, 1959-61; Director, Bureau of Mines, Ministry of Commerce and Industry, Korea 1961-62; Director, Atomic Energy Res. Inst., Korea, 1962-. Director, Board of Directors, Res. Inst. of Mining and Metal., Seoul, Korea. Societies: A.I.M.M.E.; Canadian Inst. of Mining and Metal.; Korean Chem. Soc.; Korean Inst. of Metals.
Nuclear interest: Metallurgy.
Address: Atomic Energy Research Institute, P.O. Box 7, Chungryangri, Seoul, Korea.

CHOISY, E., Dr. Pres., Swiss Assoc. for Atomic Energy; Member, Steering Com., Forum Atomique Européen; Member, Swiss Federal Commission for Atomic Energy. Societies: Assoc. Univ. Prof.; American Assoc. Phys.
Address: Swiss Association for Atomic Energy, 21 Schauplatzgasse, 3001 Bern, Switzerland.

CHOKSI, R., Prof. Member, Governing Board, Tata Memorial Hospital.
Address: Tata Memorial Hospital, Parel, Bombay 12, India.

CHONE, Bruno Karl Ferdinand, Dr. med., Priv. Doz. Born 1925. Educ.: Heidelberg Univ. Wissenschaftliche Asst., Dept.of Clinical-Exptl. Radiol., Universitats Strahlenklinik. Societies: Soc. of Radiol.; Soc. Nucl. Medicine. Nuclear interest: Autoradiography.
Address: Universitäts-Strahlenklinik, 69 Heidelberg, Germany.

CHOONHAVAN, Chatichai, Brigadier Gen. Formerly Thai Alternate, now Thai Resident Rep., I.A.E.A.
Address: 48 Cottagegasse, Vienna 18, Austria.

CHOPE, Henry Roy, B.E.E. (Elec. Eng.), M.S. (Meteorology), S.M. (Eng. Sci.). Born 1921. Educ.: Ohio State, Louisville and Harvard Univs. and California Inst. Technol. Exec. Vice-Pres. and Director, Ind. Nucleonics Corp., Columbus, Ohio; Vice-Pres. and Director, AccuRay of Canada Ltd., Montreal; Vice-Pres. and Director, AccuRay (U.K.) Ltd., London; Chairman, Com. on Sci. and Technol., Chamber of Commerce of the United States; Member, Atomic Energy Labour-Management Advisory Com., and Member, Advisory Com. on Isotopes and Rad. Development, U.S.A.E.C.
Nuclear interests: Development and design of nuclear industrial process measuring and control systems.
Address: 3885 Woodbridge Road, Columbus, Ohio 43221, U.S.A.

CHOPE, Wilbert E., B.S. (E.E.) (Ohio State), Master's Degree in Business and Eng. Administration (M.I.T.). Born 1923. Educ.: Vanderbilt and Ohio State Univs. and M.I.T. Pres. and Director, Ind. Nucleonics Corp., Columbus, Ohio; Pres. and Director, AccuRay (U.K.), Ltd.; Pres. and Director, AccuRay of Canada, Ltd.; Pres., AccuRay Corp.; Chairman, World Neighbors, Inc.; Trustee for Foundation of Instrumentation Education and Res., Inc.; Young President's Organization. Societies: N.S.P.E.; I.E.E.E.; Instrument Soc.of America; A.I.A.A. Nuclear interests: Application of radioisotopes to industrial process control and military/space electronics; Research and development in newer fields of use of nuclear energy and advanced electronics.
Address: 650 Ackerman Road, Columbus 2, Ohio, U.S.A.

CHOPOROV, D. Ya. Paper: Co-author, Determination of Oxygen in Germanium and

Silicon Using Activation by Helium-3 (Atomnaya Energiya, vol. 23, No. 2, 1967).
Address: Academy of Sciences of the U.S.S.R., 14 Leninsky Prospekt, Moscow V-71, U.S.S.R.

CHOPPIN, Gregory R., B.S., Ph.D. Born 1927. Educ.: Texas Univ. Prof., Dept. of Chem., Florida State Univ. Books: Exptl. Nucl. Chem. (Prentice-Hall Co., 1961); Nuclei and Radioactivity (W.A. Benjamin Co., 1964); Chem. (Silver Burdett Co.). Societies: A.C.S.; A.A.A.S.
Nuclear interests: Lanthanide and actinide elements; Nuclear reactions and fission.
Address: Department of Chemistry, Florida State University, Tallahassee, Florida, U.S.A.

CHOU, Ken-Chuan. Head, Sect. for the Study of Radioactivity in Atmosphere, Taiwan Weather Bureau.
Address: Taiwan Weather Bureau, 64 Park Road, Taipeih, Formosa.

CHOU, Kuang-chao, working on T. alphamesons and tau-mesons, Joint Inst. for Nucl. Research, Dubna. Paper: Co-author, A Remark on an Exptl. Method to Detect the p-Resonance in π - π Scattering Process (Scientia Sinica, vol. 10, No. 4, 1961).
Address: Laboratory of Theoretical Physics, Joint Institute for Nuclear Research, Dubna, near Moscow, U.S.S.R.

CHOU, Liang-han, Dr. rer. nat. Born 1903. National Peking and Tübingen Univs. Head, Chem. Div. (1950-56), Director, Radiochem. Lab. (1957-63), Director (1964-66), Ordnance Res.Inst.; Dean, School of Eng., Chung Cheng Inst. of Technol., 1966-. Member, Com. of Sci. Educ., Ministry of Educ.; Member, Nat. Com. of Crystallography; Editor, J. of Chinese Chem. Soc. Book: Editor, Nucl. Sci. (Chinese).
Nuclear interest: Radiochemistry, especially studies on radioactive minerals and investigations of radioactive fall-out.
Address: 16 Lane 42, Chungshan N. Rd. Sec. 2, Taipei, Formosa.

CHOUTEAU, René Eugéne, Eng. Born 1915. Educ.: Ecole Polytechnique, Paris. Works Manager, Sté d'Electrochimie d'Ugine, Grenoble, Isère; Director, Sté. de Production et de Vente d'Aimants Allevard Ugine S.A. Society: Inst. of Metals.
Address: 60 rue Ampère, Grenoble, Isère, France.

CHOVE, Louis, Ing. Born 1892. Educ.: Ecole Polytechnique, Génie Maritime. Pres., Soc. Anonyme d'Etudes et Realisations Nucléaires SODERN; Administrateur, Sté. La Radiotechnique.
Nuclear interests: Nuclear physics and electronics.
Address: Societe Anonyme d'Etudes et Realisations Nucléaires SODERN, 10 rue de la Passerelle, Suresnes, Seine, France.

CHOW, K. T. Gen. Manager, Electro-Nucl. Labs. Inc.
Address: Electro-Nuclear Laboratories, Inc., 2433 Leghorn Street, Mountain View, California, U.S.A.

CHOWDHURY, K. P., Dr. Medical Officer, Saha Inst. of Nucl. Phys., Calcutta Univ.
Address: Calcutta University, Saha Institute of Nuclear Physics, 92 Acharya Prafulla Chandra Road, Calcutta-9, India.

CHOWDHURY, Sneha, M.Sc., D.Phil. Born 1924. Educ.: Calcutta Univ. Lecturer, Phys., Undergraduate Coll., 1951-59; Jun. then Sen. Res. Asst. (1959-64), Lecturer, Nucl. Electronics (1964-), Saha Inst. of Nucl. Phys.
Nuclear interest: Nuclear instrumentation.
Address: Saha Institute of Nuclear Physics, 92 Acharya Prafulla Chandra Road, Calcutta 9, India.

CHRETEN, Max, Ph.D. (Basel). Born 1924. Educ.: Basel Univ. Swiss Foundation "Pro Helvetia" Postdoctoral Fellowship, Birmingham Univ., 1951-53; Boese Postdoctoral Fellow (1953-54), Instructor (1954-56), Columbia Univ.; Asst. Prof (1956-58), Assoc. Prof. (1959-), Chairman, Dept. of Phys. (1961-63), Brandeis Univ. Society: A.P.S.
Nuclear interests: Elementary particle physics, experimental and theoretical (Decay of strange particles, π-p scattering). Co-operated with bubble chamber group at Harvard and M.I.T. Also in process of establishing data processing centre at Brandeis.
Address: Department of Physics, Brandeis University, Waltham 54, Massachusetts, U.S.A.

CHRIEN, Robert Edward, B.S. (Rensselaer Polytech. Inst.), M.S., Ph.D. (Case Inst. of Technol.). Born 1930. Educ.: Rensselaer Polytech. Inst. and Case Inst. of Technol. (now Case-Western Univ. Cleveland Ohio). Staff Res. Phys. Brookhaven Nat. Lab.; Member, Nuclear Cross Sect. Advisory Com., U.S.A.E.C. Societies: A.P.S.; A.A.A.S.
Nuclear interests: Neutron cross sections, neutron induced reactions, especially capture followed by γ-emission applications of computers to nuclear physics; Design of neutron time-of-flight apparatus.
Address: 51 South Country Road, Bellport, N.Y. 11713, U.S.A.

CHRISTEN, Hans Anton, Eng. Born 1933. Educ.: School of Eng., Winterthur. Irradiator sales, Sulzer Brothers Ltd., Winterthur.
Nuclear interests: Research irradiators; pilot plants; Large industrial irradiation facilities; Design and irradiation studies; Large isotope sources.
Address: 15 Im Langacher, CH-8606 Greifensee, Switzerland.

CHRISTENSEN, Borge Christian, M.D., D.Sc. Born 1911. Educ.: Copenhagen Univ. Physician in Chief, Medical Radioisotope Dept., Finsen

Inst., Copenhagen, 1948-. Adviser in radiobiol. to Danish Nat. Health Service and Civil Defence Directorate. Danish Rad. Protection Com.; W.H.O. Expert Advisory Panel on Rad.; Health and Safety Com. of E.N.E.A.; I.C.R.P. Societies: Rad. Res. Soc.; H.P.S.; N.Y.Acad. Sci.
Nuclear interests: Radiobiology and radiation protection.
Address: Finsen Institute, Strandboulevarden 49, Copenhagen Ø, Denmark.

CHRISTENSEN, Hans Peter, Civil eng., D.Sc. (tech.) h.c. Born 1886. Educ.: Roy. Danish Marine School for Eng. Vice Chairman, Danish A.E.C. Societies: M.I.N.A.; Soc. Naval Architects and Marine Eng.
Address: 42B Sdr. Strandvej, Elsinore, Denmark.

CHRISTENSEN, Helge, M.Sc. (Eng. Phys.), Ph.D. (Nucl. Eng.). Born 1929. Educ.: Denmark Roy. Tech. Univ. and M.I.T. Res Eng., Chr. Michelsen Inst., Bergen, Norway, 1954-57; Res. Asst., M.I.T., 1958-60; Res. Fellow, Argonne Nat. Lab., 1960-61; Sect. Head Dynamics (1961-66), Director Nucl. Eng. Div. (1966-), Inst. for Atomenergi, Kjeller, Norway. Societies: Norweigian Phys. Soc.; A.N.S.
Nuclear interests: Reactor design, especially reactor kinetics, the dynamics of BWR's, and reactor control.
Address: Institutt for Atomenergi, P.O. Box 40, Kjeller, Norway.

CHRISTENSEN, Knud. Born 1895. Educ.: Copenhagen Univ. Manager, Copenhagen Fire Insurance; Pres., Danish Insurance Com.; Formerly Chairman, Danish Atomic Insurance Com.
Address: 17 Ved Stranden, Copenhagen, Denmark.

CHRISTENSEN, Odd B. Res. Asst., Isotope Lab., Agricultural Coll of Norway.
Address: Agricultural College of Norway, Vollebekk, Norway.

CHRISTENSEN, Walter Clyde, B. Mech. Eng. Born 1933. Educ.: Georgia Inst. Technol., Nebraska and George Washington Univs. and Internat. Inst. Nucl. Sci. Nucl. Eng., Nucl. Power Field Office, Fort Belvoir, Va. 1958-59; Tech. Asst., Army Reactors, U.S. A.E.C. 1959-61; Exec., Army Nucl. Power Programme, U.S. Army, 1961-66; Acting Director, Tech. Information, Office of the Sec. of Defence, 1966-. Sec., Professional Eng. in Government; District of Columbia Soc. of Professional Eng. Society: Nat. Soc. of Professional Eng.
Nuclear interests: Management and technical direction of the development and utilisation of small nuclear power plants and research reactors utilised by the United States military.
Address: Office of the Director of Defence Research and Engineering, Office of the Director of Technical Information, Pentagon, Washington D.C. 20301, U.S.A.

CHRISTENSEN, Wayne John, Captain, Civil Eng. Corps, U.S.N.; B.S., B.C.E., M.C.E., M.S. Born 1920. Educ.: Utah State amd Michigan Univs., Rensselaer Polytech. and Inst. Naval Postgraduate School, California Univ. Staff, Armed Forces Special Weapons Project, 1953-56; Staff, Commander-in-Chief Allied Forces Northern Europe, 1956-58; Manager, Atomic Energy and Appl. Sci. Branch (1958-61), Director, Office of Res. (1961-62), Bureau of Yards and Docks; Commanding Officer and Director, U.S. Naval Civil Eng. Lab., 1962-66; Deputy officer i/c Construction, Thailand, 1966-67; Asst. Deputy Director, Sci. Defence Atomic Support Agency, 1967-. A.S.C.E. Com. for Construction of Nucl. Facilities N.A.S. Nat. Res. Council Subcom. on Protective Structures; Ex-Member, N.A.T.O. Ad Hoc Com. on Protective Structures. Societies: A.S.C.E.; American Soc. of Military Eng.; RESA.
Nuclear interests: Management of research and development activities involving portable and mobile nuclear power plants, utilisation of radioisotopes, radioactive waste disposal, shielding against nuclear radiations, nuclear weapons effects and protective construction.
Address: DDST, Defence Atomic Support Agency, Washington D.C. 20305, U.S.A.

CHRISTIAN, John Edward, B.S. (Pharmacy), Ph.D. Born 1917. Educ.: Kalamazoo Coll. and Purdue Univ. Co-ordinator of Bio-Nucleonics Res. (1948-), Prof., Pharmaceutical Chem. (1950-), Head, Dept. of Radiol. Control (1956-59), Head, Bionucleonics Dept. (1959-), Purdue Univ. Numerous committees, A.Ph.A.; Vice-Chairman and Chairman, Purdue Section, A.C.S.; Member, U.S. Pharmaceutical Rev. Com.; Editorial Adviser, J. American Pharmaceutical Assoc.; Member, Rad. and Chem. Defence Section, Ind. Dept. of Civil Defence; Advisory Com. on Isotope Distribution, U.S.A.E.C.; Sec., Pharmaceutical Sect.; A.A.A.S.; Chairman, Sci. Sect., A.Ph.A. Book: Quantitative Pharmaceutical Chem. (5th Ed., 1957). Societies: American Soc. of Bacteriology; Fellow, A.A.A.S.; A.A.U.P.; Fellow, Ind. Acad. of Sci.; H.P.S.; A.N.S.
Nuclear interests: Isotope tracer application; Measurements of natural levels of radioactivity; Activation analyses; Biophysics; Health physics.
Address: Bionucleonics Department, Purdue University, Lafayette, Indiana, U.S.A.

CHRISTIANSEN, Aksel, Graduate in law. Born 1906. Educ.: Copenhagen Univ. Danish Foreign Ministry, 1947-54; Consul-Gen. and Commercial Counsellor, London, 1954-59; Ambassador to Mexico and Minister to Costa Rica, Cuba, Dominican Republic, Salvador, Guatemala, Haiti, Honduras, Nicaragua and Panama 1959-65; Ambassador to Austria, 1965-; Ambassador to Hungary, 1966-. Denmark Permanent Resident Rep. to the I.A.E.A.; Denmark Alternate Deleg., I.A.E.A. Gen. Conferences, Vienna, Sept. 1966 and Sept. 1967.

Address: 6 Führichgasse, 1015 Vienna, Austria.

CHRISTIANSEN, Herman. Member, Board of Representatives, A/S Noratom.
Address: A/S Noratom, 20 Holmenveien, Oslo, Norway.

CHRISTIANSEN, R. Member of Board, Voimayhdistys Ydin (Nuclear Power Assoc.).
Address: Ekono, 14 S. Espl., Helsinki, Finland.

CHRISTIE, Jack. Editor, Electronic Week, -1962; Washington Editor, Electronic Design, -1962; News Editor, Nucleonics and Nucleonics Week, 1962-; Editor, Nucl. Industry, 1966-.
Address: Atomic Industrial Forum, Incorporated, 850 Third Avenue, New York, New York 10022, U.S.A.

CHRISTMAN, David Robert, B.Sc., (Ohio State Univ.), M.Sc., D.Sc. (Carnegie Inst. Technol). Born 1923. Educ.: Ohio State Univ. and Carnegie Inst. Technol. Assoc. Chem. (1951-64), Chem. (1964-), Brookhaven Nat. Lab. Lecturer in Chem., Columbia Univ., 1958-64. Societies: A.C.S.; Chem. Soc., London.
Nuclear interests: Organic radioactivity analysis; Organic radiation; Hot atom chemistry; Biosynthesis of nicotine (using tracer methods).
Address: Chemistry Department, Brookhaven National Laboratory, Upton, New York 11973, U.S.A.

CHRISTOPHERS, A. J. Chief Ind. Hygiene Officer, Victoria. Member, Rad. Health Standing Com., Nat. Health and Medical Res. Council. Deleg., A.A.E.C. Conference on the Technol. Use of Rad., May 1960.
Address: Department of Health, Corner Spring and Latrobe Streets, Melbourne, Victoria, Australia.

CHRISTOPHORIDOU, Miss Noria. Born 1939. Librarian, Democritus Nucl. Res. Centre, 1959-62; C.E.R.N. Library, Geneva, 1963; Librarian, Democritus Nucl. Res. Centre, 1964-.
Address: Nuclear Research Centre, Democritus, Library, Aghia Paraskevi Attikis, Greece.

CHRISTOV, Ch., Prof. Chief, Cosmic Ray Sect., Head, Field Theory and Elementary Particles Sect., Deputy Director, Inst. of Phys. and Atomic Res. Centre of the Bulgarian Acad. of Sci.
Address: Institute of Physics and Atomic Research Centre of the Bulgarian Academy of Sciences, 152 Lenin Street, Sofia 13, Bulgaria.

CHRISTOV, V. Head, Reactor Phys. Measurements Lab., Inst. of Phys. and Atomic Res. Centre, Bulgarian Acad. of Sci.
Address: Bulgarian Academy of Sciences, 152 Lenin Street, Sofia 13, Bulgaria.

CHRISTY, Robert Frederick, B.A., M.A., Ph.D. Born 1916. Educ.: British Columbia and California (Berkeley) Univs. Prof., Calif. Inst. Technol., 1946-64. Society: A.P.S.
Nuclear interests: Reactors; Bombs; Theory; Elementary particles; Cosmic rays; Nuclear physics; Astrophysics.
Address: California Institute of Technology, Pasadena, California, U.S.A.

CHRUSCIEL, Edward, M.Sc. Born 1937. Educ.: Jagiellonian Univ., Cracow. II Dept. of Phys., Acad. of Mining and Metal., 1959-.
Nuclear interests: Applied nuclear physics and industrial use of radioisotopes.
Address: Academy of Mining and Metallurgy, II Department of Physics, 30 A1. Mickiewicza, Cracow, Poland.

CHRYSOCHOIDES, Nicholas, B.Sc. (Phys.), M.D. (Electronics), Ph.D. (Reactor Phys.). Born 1931. Educ.: Athens Univ. Phys. 1949-53; Electronics, 1953-54; I.S.N.S.E., Argonne Nat. Lab., 1957; Phoenix Memorial Lab., Ann Arbor, Michigan, 1958; Director, Reactor Dept., Democritus Nucl. Res. Centre, Aghia Paraskevi, Attikis, Athens. Societies: Greek Phys. Soc.; A.N.S.
Nuclear interests: Experimental reactor physics; Reactor operation.
Address: 38 Leoforas Kifissias, Athens 608, Greece.

CHU, Ju-Chin, B.Sc. (Chem.), Sc.D. (Chem. Eng.). Born 1919. Educ.: Tsing Hua Univ. and M.I.T. Assoc. Prof. (1949-55), Prof. Chem. Eng. (1955-), Polytechnic Inst. Brooklyn; Tech. Director for Res. Development, Process Eng. and Patent, Chem. Construction Corp., 1956-57; Consulting Expert, U.S.D.A., 1956-; Consultant, Argonne Nat. Lab., 1955-; Space Technol. Labs., 1959; Aircraft Nucl. Propulsion Dept. 1958; Union Carbide Nucl. Co., 1955-58; Curtiss-Wright, 1955-57; Sen. Consultants, Inc. 1949-; Redstone Arsenal-Rohm Haas, 1955-; Sun Oil Co., 1952; Columbia Southern Chem., 1952; American Cynamid, 1953; International Bank, 1949-; Rocketdyne (1960-62), S.I.S.D. Div. (1962-63) North American Aviation; A.R.O., Inc., 1962-64; Huyck Corp., 1960-; Aerospace Corp., 1962; Grumann Aircraft Eng. Co., 1964; Gen. Elec., 1962; Books: 2 books on distillation; contributor to several chapters in Encyclopaedia Chem. Tech. and Fluidisation in Practice. Societies: Sen. Member, A.I.Ch.E.; A.C.S.; A.N.S.; A.S.E.E.; A.R.S.; S.C.I. (London); Fellow, A.A.A.S.
Nuclear interests: Processing and reprocessing of nuclear fuels; Reactor design; Heat and mass transfer, fluid dynamics; Materials; Utilisation of spent fuels; Fluidation; Gamma-ray application; Extraction,etc.
Address: 34 Linden Street, Garden City, New York, U.S.A.

CHU, K. Y. Educ.: California Inst. of Technol. At Chinese Nucl. Res. Bureau.
Address: Academia Sinica, 3 Wen Tsin Chien, Peking, China.

CHU, Pao-Hsi, B.S. (Phys.), Diploma in Nucl. Eng. Born 1920. Educ.: Yenching Univ., China and O.R.N.L. Exec. Sec., Nucl. Power Com., Taiwan Power Co., 1966-.
Nuclear interests: Nuclear reactor engineering; Nuclear power plant design; Nuclear physics; Nuclear fuel management; Nuclear power costs analysis.
Address: Taiwan Power Company, 39 East Hoping Road, Section 1, Taipei, Formosa.

CHUBURKOV, Yu. T. Paper: Co-author, Chem. Properties of Element 104 (Atomnaya Energiya, vol. 21, No. 2, 1966).
Address: Academy of Sciences of the U.S.S.R., 14 Leninsky Prospekt, Moscow V-71, U.S.S.R.

CHUCHALIN, I. P. Papers: Co-author, 1,5 GeV Electron Synchrotron at the Polytechnic Inst. of Tomsk (Atomnaya Energiya, vol. 21, No. 6, 1966); co-author, 300 MeV Electron Synchrotron at the Polytechnic Inst. of Tomsk (letter to the Editor, ibid, vol. 21, No. 6, 1966).
Address: Academy of Sciences of the U.S.S.R., 14 Leninsky Prospekt, Moscow V-71, U.S.S.R.

CHUECA, Carmelo, Prof. Nucl. Consultant, Gen. Electrica Espanola.
Address: General Electrica Espanola, 4 Plaza de Federico Moyüa, Bilbao, Spain.

CHUGANOV, V. A. Paper: Co-author, On Possibility of Suppression of Inhomogeneous Plasma Drift Instability by Feedback (Atomnaya Energiya, vol. 24, No. 4, 1968).
Address: Academy of Sciences of the U.S.S.R., 14 Leninsky Prospekt, Moscow V-71, U.S.S.R.

CHUJO, Kimbe, Dr. Director, Res. Lab., Nihon Cement Co. Ltd.
Address: Nihon Cement Co. Ltd., 8, 1-chome Fukagawa-Kiyosumicho, Koto-ku, Tokyo, Japan.

CHULKIN, V. L. Paper: Co-author, Back-Scattering of Gamma-Rays Isotropic Sources from Cylindrical Barriers (Atomnaya Energiya, vol. 23, No. 3, 1967).
Address: Academy of Sciences of the U.S.S.R., 14 Leninsky Prospekt, Moscow V-71, U.S.S.R.

CHULKOV, B. A. Paper: Co-author, Investigation of Heat Transfer in Alkali Metals Condensation (letter to the Editor, Atomnaya Energiya, vol. 22, No. 5, 1967).
Address: Academy of Sciences of the U.S.S.R., 14 Leninsky Prospekt, Moscow V-71, U.S.S.R.

CHUMBAROV, Yu. K. Papers: Co-author, Personnel Neutron Film Dosimetry Using In^{115} (Abstract, Atomnaya Energiya, vol. 20, No. 1, 1966); co-author, Neutrons Glass Track Detector (letter to the Editor, ibid, vol. 24, No. 5, 1968).
Address: Academy of Sciences of the U.S.S.R., 14 Leninsky Prospekt, Moscow V-71, U.S.S.R.

CHUMIN, Victor. At Joint Inst. for Nucl. Res., Dubna.
Address: Joint Institute for Nuclear Research, Dubna, Near Moscow, U.S.S.R.

CHUNG, Hack-Pil, B.Sc. Born 1936. Educ.: Dong-A Univ. Researcher, Sci. Res. Inst., Seoul, Korea, 1961-. Society: Korean Phys. Soc.
Nuclear interests: Nuclear physics and health physics.
Address: Radiochemistry Section, Scientific Research Institute, Ministry of National Defence, Seoul, Korea.

CHUNG, Man Young. Acting Chief, Electronics Div., Atomic Energy Res.Inst.
Address: Atomic Energy Research Institute, Kongneung-dong, Sungbook-ku, Seoul, Korea.

CHUNG, Myung Cho. Chief, Management Sect., Bureau of Administration, Office of Atomic Energy and A.E.C.
Address: Office of Atomic Energy and Atomic Energy Commission, 2-1 Chung-dong, Sudaimoon-ku, Seoul, Korea.

CHUPKA, Sh. Paper: Co-author, Content of Sr^{90} and Cs^{137} in Agricultural Products during 1963 and 1964 in West Slovakia (Atomnaya Energiya, vol. 21, No. 3, 1966).
Address: Academy of Sciences of the U.S.S.R., 14 Leninsky Prospekt, Moscow V-71, U.S.S.R.

CHUPRUNOV, D. L. Papers: Co-author, Utilisation of Fast Charged Particles Inelastic Scattering for Analysis of Composition of Materials (letter to the Editor, Atomnaya Energiya, vol. 20, No. 5, 1966); co-author, Scintillation Spectrometer for Simultaneous Investigation of Competing Nucl. Reactions (letter to the Editor, ibid, vol. 21, No. 1, 1966).
Address: Academy of Sciences of the U.S.S.R., 14 Leninsky Prospekt, Moscow V-71, U.S.S.R.

CHURCH, Thomas G., Graduate in Metal. Eng. Educ.: British Columbia Univ. Formerly Asst. to Vice-Pres., R. and D., A.E.C.L.;Sci. Sec., 2nd I.C.P.U.A.E.;Sec., 3rd I.C.P.U.A.E., Geneva, Sept. 1964.
Address: Atomic Energy of Canada Ltd., Chalk River, Ontario, Canada.

CHURIN, S. A. Paper: Co-author, Backscattering of Low-Energy Gamma Rays from Finite Barriers (letter to the Editor, Atomnaya Energiya, vol. 20, No. 4, 1966); co-author, Backscattering of Gamma-Rays from Aluminium Barriers (ibid, vol.22, No. 4, 1967).
Address: Academy of Sciences of the U.S.S.R., 14 Leninsky Prospekt, Moscow V-71, U.S.S.R.

CHURKIN, V. N. Paper: Co-author, Radioactive Fallouts on the U.S.S.R. Area in 1963 Year (Atomnaya Energiya, vol. 19, No. 1, 1965).
Address: Academy of Sciences of the U.S.S.R., 14 Leninsky Prospekt, Moscow V-71, U.S.S.R.

CHURSIN, G. N. Paper: Co-author, Sand Moisture Determination by Fast Neutron Flux Attenuation (letter to the Editor, Atomnaya Energiya, vol. 16, No. 4, 1964).
Address: Academy of Sciences of the U.S.S.R., 14 Leninsky Prospekt, Moscow V-71, U.S.S.R.

CHUSHKIN, Yu. V. Papers: Co-author, Gas-Release from Irradiated Beryllium Oxide and Boron Carbide During Isothermal Annealing (letter to the Editor, Atomnaya Energiya, vol. 22, No. 6, 1967); co-author, Irradiation Stability of Arbus Nucl. Reactor Fuels (letter to the Editor, ibid, vol. 24, No. 4, 1967); co-author, Reactor SM-2 Plate Fuel Elements Rad. Stability (ibid, vol. 24, No. 5, 1968); co-author, On Possible Character of Fuel Composition Volume Change under Solid Swelling (letter to the Editor, ibid, vol. 24, No. 5, 1968).
Address: Academy of Sciences of the U.S.S.R., 14 Leninsky Prospekt, Moscow V-71, U.S.S.R.

CHUTKIN, O. A. Papers: Co-author, Quick Measurement of Airborne Long-lived Alpha-Isotopes Concentrations of Order of $10^{-16} - 1 - 10^{-17}$ C/1 and of Thoron Decay Products Aerosols (letter to the Editor, Atomnaya Energiya, vol. 21, No. 2, 1966); co-author, Spectrometric Measurement of Long-Lived Alpha Activity Aerosols (ibid, vol. 24, No. 3, 1968).
Address: Academy of Sciences of the U.S.S.R., 14 Leninsky Prospekt, Moscow V-71, U.S.S.R.

CHUVILO, Ivan V., Candidate of Phys. mathematical Sci. Born 1924. Educ.: Moscow Lomonosov Univ. Lebedev Phys. Inst., U.S.S.R. Acad. of Sci., 1948-54; Elec. Phys. Lab., U.S.S.R. Acad. Sci., 1954-56; Research worker, Deputy Director of the Lab. of High Energy Phys., (1956-67), Director, High Energy Lab. (1967-), Joint Inst. for Nucl. Res. (Dubna, Moscow Region).
Nuclear interests: High energy and elementary particle physics.
Address: Joint Institute for Nuclear Research, Dubna, Moscow Region, U.S.S.R.

CHUYANOV, V. A. Member, Editorial Board, Atomnaya Technika za Rubezhom. Papers: A Multigroup Calculation Carried out on "Strela" for the Power Station Reactor (letter to the Editor, Atomnaya Energiya, vol. 7, No. 1, 1959); co-author, On Possibility of Plasma Current Column Stabilisation by Feedbacks (letter to the Editor, ibid, vol. 25, No. 2, 1968).
Address: Academy of Sciences of the U.S.S.R., 14 Leninsky Prospekt, Moscow V-71, U.S.S.R.

CHUZHINOV, V. A. Papers: Co-author, Separation of Isotopes of Neon in Mercury Mass-Diffusion Column (letter to the Editor, Atomnaya Energiya, vol.23, No. 1, 1967); co-author, Isotope Separation in Three Step Mercury Mass-Diffusion Column (letter to the Editor, ibid, vol. 24, No. 5, 1968).
Address: Academy of Sciences of the U.S.S.R., 14 Leninsky Prospekt, Moscow V-71, U.S.S.R.

CHVOJKA, Ludvik, M.Sc. (Agriculture), Ph.D. (Biol.). Born 1928. Educ.: Bratislava and Leningrad Agricultural Studies Univs. Asst. Univ. of Agricultural Studies, Brno; Sci. worker, Czechoslovak Acad. Sci. Coordinator for the exploitation of radioisotopes in agriculture, Ministry of Agriculture; Member, working group, Czechoslovak Atomic Commission.
Nuclear interests: Incorporation of radioisotopes in organic compounds of plants; Molecular biology.
Address: 7 Narodni obrany, Prague 6, Czechoslovakia.

CHWASZCZEWSKI, Stefan, Dr. Born 1935. Educ.: Moscow Univ. Inst. Nucl. Res., 1958.
Nuclear interest: Physics of nuclear reactors.
Address: bl. 9/16 ul. Wyspiańskiego, Otwock, Poland.

CHYRIN, S. A. Paper: Co-author, Method of Transfer Matrix for Calculation of Spectral-Angular Distribution of Gamma Rays in Slab Geometry (Atomnaya Energiya, vol. 24, No. 3, 1968).
Address: Academy of Sciences of the U.S.S.R., 14 Leninsky Prospekt, Moscow V-71, U.S.S.R.

CHYZEWSKI, Adam, M.Sc. Born 1936. Educ.: Lodz Tech. Univ. Dept. of Phys. Chem., Lodz Tech. Univ. Society: Polish Chem. Soc.
Nuclear interests: Radiochemistry and radiation chemistry of solids.
Address: 22a m14 ul.Krzemieniecka, Lodz, Poland.

CIBOROWSKI, Stanislaw, Ph.D., Assoc. Prof. Born 1925. Educ.: Lodz Tech. Univ. Head, Radiochem. Dept., Inst. of Gen. Chem., Warsaw. Books: Radiation Chemistry of Inorganic Compounds (Warsaw, Pánstwowe Wydawnictwo Naukowe, 1962); Radiation Chemistry of Organic Compounds (Warsaw, Panstwowe Wydawnictwo Naukowe, 1966). Societies: Polish Chem. Soc.; Polish Soc. for Rad. Res.
Nuclear interests: Radiation chemistry of organic compounds; Irradiation effects in solid catalysts; Application of radioisotopes in research of mechanisms of chemical reactions.
Address: Institute of General Chemistry, 8 Rydygiera Street, Warsaw 86, Poland.

CIECHANOWICZ, Wieslaw, Dr. Born 1926. Educ.: Technical Univ., Gdańsk. Sci. worker, Inst. Nucl. Res., Warsaw.
Nuclear interests: Transient state of thermal processes in nuclear plant and control of nuclear plant.
Address: 36 m.28 ul. Marchlewskiego, Warsaw, Poland.

CIFKA, Jiri, R.N.Dr. Born 1933. Educ.: Charles Univ., Prague. Inst. Nucl. Res., Czechoslovak Acad. Sci., 1956-. Society: Czechoslovak Chem. Soc.
Nuclear interests: Radiochemistry, especially chemical effects of nuclear transformations and production of radioisotopes and

radiopharmaceuticals.
Address: Institute of Nuclear Research,
Czechoslovak Academy of Sciences, Rez by
Prague, Czechoslovakia.

CIFUENTES, D. Manuel GOLMAYO. See
GOLMAYO CIFUENTES, D. Manuel.

CIGNA, Arrigo A., Dr. (Phys.), Libero docente
(Health Phys.). Born 1932. Educ.: Milan Univ.
i/c Radioactive Fallout Measurements, Osservatorio Astronomico di Brera, Milan, 1951-54;
i/c Environmental Monitoring, Health Phys.
Dept., C.N.R.N. Ispra, 1959; Director, Lab.
for the Study of Environmental Radioactivity,
Casaccia Nucl. Centre, C.N.E.N., 1961-. Member, Board of Directors, Associazione Italiana
di Fisica Sanitaria e Protezione contro le
Radiazioni. Societies: H.P.S.; Comitato Elettrotecnico Italiano; Società Astronomica
Italiana.
Nuclear interests: Health physics problems,
radiation protection ecology applied to
environmental radioactivity.
Address: c/o Lab. Radioattività Ambientale,
C.N.E.N., CSN Casaccia, C.P. 2400, 00100
Rome, Italy.

CIGRANG, René. Substitute du Procureur
d'Etat; Com. Member, Assoc. Luxembourgeoise pour l'Utilisation Pacifique de l'Energie
Atomique.
Address: Association Luxembourgeoise pour
l'Utilisation Pacifique de l'Energie Atomique,
9 rue Marie-Adelaide, Luxembourg, Luxembourg.

CILESIZ, Ayhan, D.Sc. Born 1928. Educ.:
Istanbul Univ. Reactor Supv., TR-1 Reactor,
Istanbul, 1961-. Society: Turkish Phys. Soc.
Nuclear interest: Reactor operations.
Address: P.K.1, Havaalane, Istanbul, Turkey.

CILLIERS, A. C., Prof. Member, Atomic
Energy Board.
Address: Atomic Energy Board, Private Bag
256, Pretoria, South Africa.

CIMBLERIS, Borisas. M.S. (Nucl. Eng.), Sc.D.
(Thermodynamics). Born 1923. Educ.: School
of Mines of Ouro Preto, N.C.S.C., School of
Nucl. Science and Eng., (Argonne), IIT., and
Michigan Univ. Instructor, Braz. Centre for
Phys. Res. 1950-51; Assoc. Phys., Armour Res.
Foundation, 1955-56; Teaching Fellow, Michigan Univ., 1956-57; Tech. Adviser, National
Nucl. Energy Commission of Brazil, 1957-60;
Prof., N. E. Course, Nat. School of Eng., Brazil
Univ.; Assoc. Prof. Phys., School of Mines of
Ouro Preto; Prof. Reactor Eng., Co-ordinator
Nucl. Eng. Course and Acting Prof. Thermodynamics, Head, Dept. of Mech. Eng., School
of Eng., Minas Gerais Univ., 1964-; Member,
Thorium Group, and Technical Adviser, Nat.
Nucl. Energy Commission. Societies: A.P.S.;
A.N.S.; Soc. Brasileira Para o Progresso da
Ciencia; Assoc. Brasileira de Energia Nucl.;
Assoc. Brasileira de Transferencia de Calor e
Mecanica dos Fluidos.
Nuclear interests: Heat transmission in reactors;
Thermodynamic cycles for nuclear power
stations; Reactor dynamics, oscillator techniques; Research reactors; Uses of radiation
for preservation of food; Nuclear engineering
education.
Address: Escola de Engenharia da UMG, Rua
Espirito Santo 35, Belo Horizonte, Minas
Gerais, Brazil.

CINI, Marcello, Prof. Member, Inst. Nazionale
di Fisica Nucleare, Italy. Papers: Co-author,
Theory of Low-Energy Scattering in Field
Theory (Annals of Physics, vol. 10, No. 3,
1960).
Address: Istituto di Fisica dell'Universita Roma
Istituto Nazionale di Fisica Nucleare, Sezione
di Roma, Rome, Italy.

CINNAMON, Carl A., B.A., M.A., Ph.D. Born
1901. Educ.: Wyoming, Chicago and Iowa
Univs. Head,Phys. Dept., (1946-66), Prof.
Phys. (1966-), Wyoming Univ.; Sen. Phys.,
U.S. Naval Operations, 1955-56. Director,
Summer Inst. in Rad. Biol., Wyoming Univ.,
1958-67. Radiol. Safety Com. Societies:
A.P.S.; A.A.U.P.; A.A.P.T.
Nuclear interests: Teaching radiation physics,
and directing institutes in radiation biology.
Address: Department of Physics, P. O. Box
3295, University Station, Laramie, Wyoming,
U.S.A.

CINTRA DO PRADO, Luiz, Civil Eng., Phys.,
Ph.D. (Phys.). Born 1904. Educ.: Sao Paulo
and Paris Univs. Full Prof., Exptl. Phys., Escola
Politecnica (1938-64), Lecturer, Phys. in
Buildings, School of Architecture (1950-),
Vice-Rector and Acting Rector (1953-54), Sao
Paulo Univ.; Prof., Pontif. Catholic Univ.,
1938-54. Member, Nat. Research Council, 1951-64; Chief, Div. of Nucl. Eng. and Director, Inst.
de Energia Atomica, 1961-63; Vice-Pres.,
Brazilian Acad. Sci. 1959-64; Chairman,
Brazilian Nucl. Energy Commission, 1964-65;
Member, Sci. Advisory Com., and Brazilian
Alternate to the Governor, I.A.E.A.; Member,
Internat. Com. Weights and Measures, 1967-;
Consultant, Centrais Electricas de Sao Paulo.
Societies: Eng. Assoc. of Sao Paulo; A.N.S.;
Atomic Ind. Forum.
Nuclear interests: Management; Power reactors;
Economics of nuclear power; Nuclear engineering.
Address: (apt. 51) 1347 Rua Bela Cintra, Sao
Paulo 5, S.P. – Brazil.

CIORASCO, Florin, Sci. Deputy Director,
Inst. of Atomic Phys., Academy of the Socialist Republic Roumania; Deleg. to 2nd
I.C.P.U.A.E., Geneva, Sept. 1958. Adviser,
Third I.C.P.U.A.E., Geneva, Sept. 1964.
Address: Institute of Atomic Physics, Casuta
Post 35, Bucharest, Roumania.

CIRAK, Julius, Ing. Dr. (Slovak Tech. Univ.).
Born 1924. Educ.: Slovak Tech. Univ.,

Bratislavia, and Polytech. Inst., Leningrad. Manager, Nucl. Phys., and Technics Chair, Slovak Tech. Univ., Bratislava, 1961-.
Nuclear interests: Experimental nuclear physics – namely, accelerators of particles and instruments and methods of nuclear radiation detection.
Address: 5 Teplická, Bratislava, Czechoslovakia.

CIRANNI, Elena, Laurea in Pharmacy. Born 1936. Educ.: Rome Univ. Radiochem., Centro di Chimica Nucl. del C.N.R., Istituto di Chimica Farmaceutica, Rome Univ., 1959-.
Nuclear interests: Organic radiochemistry; e.g., radiation chemistry, labelling techniques with C^{14} and H^3, analytical procedures to count C^{14} and H^3, recoil chemistry of hot H^3 and C^{14} atoms.
Address: Centro di Chimica Nucleare del C.N.R., Istituto di Chimica Farmaceutica, University of Rome, Rome, Italy.

CIRANNI, Giovanna, Laurea in Pharmacy. Born 1932. Educ.: Rome Univ. Radiochem., Centro di Chimica Nucleare del C.N.R., Istituto di Chimica Farmaceutica, Rome Univ., 1957-.
Nuclear interests: Organic radiochemistry; e.g., radiation chemistry, labelling techniques with C^{14} and H^3, analytical procedures to count C^{14} and H^3, recoil chemistry of hot H^3 and C^{14} atoms.
Address: Centro di Chimica Nucleare del C.N.R., Istituto di Chimica Farmaceutica, University of Rome, Rome, Italy.

CIRCAUD, Paul. French Member, Economic and Social Com. Specialised Nucl. Sect. for Economic Problems, and Member, Specialised Nucl. Sect. for Social Health and Development Problems, Euratom.
Address: 58 avenue Maréchal Foch, Lyon 6, (Rhône), France.

CISHEK, Edmundo. Tech. Adviser, Comision Nacional de Investigaciones Atomicas, Dominica; Jefe de las plantas Electricas de la Corporacion Dominicana de Electricidad.
Address: Comision Nacional de Investigaciones Atomicas, Ciudad Trujillo, Dominican Republic.

CISLER, Walker Lee, Dr. Eng. (Michigan Univ., Stevens Inst. Technol., South Dakota School of Mines and Technol.), Dr. Laws (Detroit and Wayne State Univs. Marietta Coll., Akron, Northern Michigan, Michigan State Univs.), Dr. Sci. (Toledo Univ., Indiana Tech. Coll., Michigan Technol. Univ.)Dr. Public Service (Detroit Inst. of Tech.). Born 1897. Educ.: Cornell Univ. Director: American Airlines, Assoc. of Edison Illuminating Cos., Atomic Ind. Forum, Brazilian Light and Power Co. Ltd., Burroughs Corp., Cornell Aeronautical Lab., Detroit Bank and Trust, Eaton Yale and Towne, Inc.,Equitable Life Assurance Soc. of the United States, Chemical Bank N.Y. Trust Co.- Advisory Board on Internat. Business; Holley Carburetor Co. Chairman of Board: Fruehauf Corp., Economic Club of Detroit, Michigan Com. of the Newcomen Soc. in North America, Detroit Edison Co.; Pres.: Atomic Power Development Assocs. Inc.; Power Reactor Development Co.; Fund for Peaceful Atomic Development Inc.; Thomas Alva Edison Foundation, Inc. Rep. for U.S.A., O.E.E.C., Energy Advisory Commission. Society: Newcomen Soc.
Nuclear interests: Confined mostly to the field of management.
Address: The Detroit Edison Company, 2000 Second Avenue, Detroit, Michigan 48226, U.S.A.

CITARELLA, Luigi. Direzione, Diritto ed Economia Nucleare.
Address: Diritto ed Economia Nucleare, 68 Viale Bruno Buozzi, Rome, Italy.

CITRON, A. Formerly Member, Planning Board, Rad., - Exposure, Kerma and Fluence, now Member, Planning Board, Rad. - Heavy Particles, I.C.R.U.
Address: International Commission on Radiological Units and Measurements, 4000 Brandywine Street, N.W., Washington D.C. 20016, U.S.A.

CITRON, Joachim Jörk Anselm, Dr. rer. nat., Doz. für Phys., ordentl. Prof. Born 1923. Educ.: Freiburg and Basel Univs. C.E.R.N., Geneva, 1953; Karlsruhe Univ., 1965. Society: Physikalische Gesellschaft Baden-Württemberg-Pfalz.
Nuclear interests: High energy physics; Accelerator and beam transport design shielding.
Address: 2 Erasmusstrasse, 75 Karlsruhe-Waldstadt, Germany.

CITTERIO, Rag. Ernesto. Director, Administrative Div., C.N.E.N.
Address: National Committee for Nuclear Energy, 15 Via Belisario,Rome, Italy.

CIUFFOLOTTI, Luigi, Dr. Born 1931. Educ.: Genoa Univ. Phys. Inst., Genoa Univ., 1958-59; C.E.S.N.E.F. – Enrico Fermi Centre for Nucl. Studies, Milan, 1960; S.O.R.I.N., Centre for Nucl. Researches, Saluggia (Vercelli), 1961.
Nuclear interest: Nuclear physics.
Address: 19 Via Ponte Rocca, Saluggia, Vercelli, Italy.

CIULLI, Sorin, Physicist. Born 1933. Educ.: Bucharest Univ. Asst. Prof., Bucharest Univ., 1955-56; Joint Inst. for Nucl. Res. (Lab. for Theoretical Physics), Dubna, U.S.S.R., 1957-.
Nuclear interests: Theoretical physics of high energy particles (field theory); Theoretical problems of controlled thermonuclear reactions.
Address: Joint Institute for Nuclear Research, Laboratory for Theoretical Physics, P.O. Box 79, Moscow, U.S.S.R.

CIVAOGLU, Ilhami, Prof. Member, Advisory Board, Turkish A.E.C.; Sen. Officer, Istanbul Tech. Univ.

Address: Istanbul Technical University,
Gümüssuyu, Istanbul, Turkey.

CIZERON, Georges Rene, Doc. Sc., Doc. Ing.,
Ing. E.C.P. and E.N.S.P.M. Born 1927. Educ.:
Paris Univ. Lecturer, Faculty of Sci. Orsay and
I.N.S.T.N. Saclay.
Nuclear interests: Studies on zirconium and
zirconium alloys. Sintering of U and UO_2
powders. Dilatometric investigations of U
single and polycrystals. Texture determinations on U sheets and rods. Purification of U
by zone melting. Creep behaviour of U and
U-Mo alloys. Recovery and recrystallization of
pure U. Mechanical testing of pure Nb.
Address: 131 rue Nationale, Paris 13, France.

CKHEIDZE, I. I. Paper: Co-author, Some
Remarks on Rad. Stability Low Melting Coolants in Liquid and Solid State (Atomnaya
Energiya, vol. 22, No. 1, 1967).
Address: Academy of Sciences of the U.S.S.R.,
14 Leninsky Prospekt, Moscow V-71,U.S.S.R.

CLACK, Robert Wynandus, B.S. Born 1921.
Educ.: U.S. Naval Acad. Asst. Prof. Nucl.
Eng. and Reactor Supervisor, Kansas State
Univ., 1959-. Societies: A.N.S.; H.P.S.;
A.S.E.E.; American Soc. of Naval Eng.; Registered Professional Engineer (Kansas), Licensed
Senior Reactor Operator (A.E.C.); Diplomate,
American Board of Health Phys.
Nuclear interests: Reactor operations; Radiation
safety; Reactor safety; Activation analysis.
Address: Kansas State University, Department
of Nuclear Engineering, Manhattan, Kansas,
U.S.A.

CLAESSON, Arne Lennart, F.D. Born 1925.
Educ.: Lund Univ. A.B. Atomenergi, 1960-66;
Umea Univ., 1966-.
Nuclear interest: Solid state physics.
Address: Umea University, Institute of Physics,
Department of Theoretical Physics, Axtorpsvagen, Umea 4, Sweden.

CLAIBORNE, Harry Clyde, B.S., M.S. Born
1921. Educ.: Louisiana State and Tennessee
Univs. Dev. Eng., O.R.N.L., 1949-52; Production Eng., U.S.A.E.C., 1952-53; Dev. Eng.,
O.R.N.L., 1953-. Societies: A.N.S.; RESA.
Nuclear interests: Reactor, space and weapons
effects; Radiation shielding; Thermal aspects
of reactor design.
Address: Oak Ridge National Laboratory, Post
Office X, Oak Ridge, Tennessee, U.S.A.

CLAIRE, Alan De La Mare LE. See LE
CLAIRE, Alan De La Mare.

CLANCY, Brian Edward, B.Sc., Dip. Ed.,
M.Sc. Born 1930. Educ.: Sydney and New
South Wales Univs. Lecturer, Maths., New
South Wales Univ., 1954-61: Sen. Res. Sci.
and Principal Res. Sci. (1962-), and Head,
Theoretical Phys. Sect., Phys. Div., Res. Establishment, A.A.E.C. Societies: Roy. Soc. of
New South Wales; Australian Mathematical

Soc.; A.N.S.
Nuclear interest: Reactor physics and mathematics, particularly neutron transport theory.
Address: Australian Atomic Energy Commission, Research Establishment, Sutherland,
New South Wales, Australia.

CLARK, A. J. Formerly Supervisor, Aerospace
Nucl. Safety Div. II, now Manager, Isotope
Power Dept., (1966-), Sandia Lab., U.S.A.E.C.
Address: U.S.A.E.C., Sandia Laboratory, Isotope Power Department, P.O. Box 5800,
Albuquerque, New Mexico, U.S.A.

CLARK, Arthur Colvin Lindesay, Prof., M.D.,
B.S., F.R.A.C.P. Born 1928. Educ.: Melbourne
Univ. Second and First Asst., Dept. of Paediatrics, Melbourne Univ. Societies: Anti-Cancer
Council of Victoria; Nat. Health and M.R.C.
Radioisotopes Standing Com.
Nuclear interest: Use of radioisotopes in
clinical medicine.
Address: Monash University, Department of
Paediatrics, Queen Victoria Hospital, 172
Lonsdale Street, Melbourne, Victoria 3000,
Australia.

CLARK, Avon Maxwell, M.Sc., Ph.D. Born
1918. Educ.: Melbourne and Cambridge Univs.
Reader in Genetics (1955-59), Member of
Board of Studies in Nucl. Sci. (1957-59),
Melbourne Univ.; Member, Nat. Rad. Advisory
Com., 1959-; Prof. Zoology, Tasmania Univ.,
1959-63; Prof. Biol., Flinders Univ. 1964-.
Alternate Australian delegate, U.N. Sci. Com.
on the Effects of Atomic Rad. Society:
Australian and New Zealand Assoc. for Advancement of Sci.
Nuclear interest: General radiation biology.
Address: School of Biological Sciences,
Flinders University, Adelaide, Australia.

CLARK, Claude L., B.S.E., M.S.E., Ph.D. Born
1903. Educ.: Michigan Univ. Staff Metal. Eng.,
Timken Roller Bearing Co., 1957-. Chairman,
A.S.T.M.Subcom. on Tubular Products, Producer-Vice-Chairman, Steel Com. Societies:
A.S.T.M.; A.S.M.E.; A.I.M.E.; N.A.C.E.
Nuclear interest: Metallurgy.
Address: The Timken Roller Bearing Company,
Steel and Tube Division, Canton 6, Ohio,
U.S.A.

CLARK, Colin Douglas, B.Sc. (Gen.), B.Sc.
(Special Phys.), Ph.D. Born 1929. Educ.: Reading Univ. Res. Phys., English Elec. Co., Stafford;
Res. Phys., Industrial Distributors (1946) Ltd.;
Lecturer, Reading Univ. Societies: Inst. Phys.;
Phys. Soc.
Nuclear interest: Radiation effects in solids.
Address: 52 The Crescent, Earley, Reading,
Berks., England.

CLARK, David Delano, A.B., Ph.D. (Phys.).
Born 1924. Educ.: California Univ., Berkeley.
Assoc. Prof. Eng. Phys., Cornell Univ., 1958-.
(Asst. Prof., 1955-58); Consultant, Gen.
Atomic, summer, 1959; Visiting Phys.,

Reactor Phys. Group, Brookhaven Nat. Lab., summers 1957 and 1958. Director, Nucl. Reactor Lab., 1960-. Societies: A.P.S.; A.N.S. Nuclear interests: Nuclear and reactor physics, primarily experimental; Nuclear instrumentation; Applications of isotopes.
Address: Nuclear Reactor Laboratory, Cornell University, Ithaca, New York, U.S.A.

CLARK, Donald, B.Sc. (Eng.), A.C.G.I. Born 1913. Educ.: City and Guilds (Eng.) Coll., London. Station Development Eng., British Elec. Authority and Central Elec. Authority, 1948-58; Chief Planning Eng., C.E.G.B., 1958-. Society: I.E.E.
Nuclear interests: Civil nuclear power stations; planning and siting, including safety aspects.
Address: c/o Central Electricity Generating Board, Sudbury House, 15 Newgate Street, London, E.C.1, England.

CLARK, H. Murray, B.S., Ph.D. (Illinois). Born 1916. Educ.: Illinois Univ. Prof. Chem. (1946-), Asst. to Dean, Div. of Sci. and Appl. Arts (1956-), San Jose State Coll., San Jose, California. Societies: A.C.S.; A.A.A.S.
Nuclear interest: Teaching (graduate and undergraduate courses).
Address: Division of Sciences and Applied Arts, San Jose State College, San Jose, California, U.S.A.

CLARK, Herbert Mottram, B.S., Ph.D. Born 1918. Educ.: Yale Univ. Assoc. Prof. (1949-51), Prof. Phys. and Nucl. Chem. (1951-), Rensselaer Polytech. Inst. Books: Co-author, Radioisotope Techniques (McGraw-Hill, 1960); co-author, Principles of Chem. (Prentice-Hall, 1966). Societies: Fellow, A.A.A.S.; A.C.S.; A.P.S.; A.S.E.E.; Fellow, N.Y. Acad. Sci.
Nuclear interests: Radiochemistry; Chemical processing of reactor fuels; Utilisation of radio-isotopes.
Address: Chemistry Department, Rensselaer Polytechnic Institute, Troy, New York 12181, U.S.A.

CLARK, Hugh Kidder, A.B., Ph.D. Born 1918. Educ.: Oberlin Coll. and Cornell Univ. Res. Assoc., E. I. du Pont de Nemours and Co., Inc. Societies: A.C.S.; Fellow, A.N.S.; A.A.A.S.
Nuclear interests: Nuclear safety (criticality); Reactor physics.
Address: Savannah River Laboratory, E. I. du Pont de Nemours and Company, Incorporated, Aiken, South Carolina, U.S.A.

CLARK, J. McADAM. See McADAM CLARK, J.

CLARK, John. Director, Automotive Products and Equipment Res., Southwest Res. Inst.
Address: Southwest Research Institute, 8500 Culebra Road, San Antonio 6, Texas, U.S.A.

CLARK, John Walter, B.S., M.A. (Texas), Ph.D. (Washington). Born 1935. Educ.: Texas and Washington (St. Louis) Univs. Res. Assoc., Washington Univ., 1959; Nat. Sci. Foundation Post-doctoral Fellow, Princeton Univ., 1959-61; Assoc. Res. Sci., Martin Co., Denver, Colorado, 1961; N.A.T.O. Postdoctoral Fellow, Birmingham Univ. and Saclay, 1962-63; Asst. Prof. (1963-66), Alfred P. Sloan Foundation Fellow (1965-67), Assoc. Prof. (1966-) Washington Univ. Societies: A.P.S.; Phys. Soc. Japan.
Nuclear interests: Nuclear many-body problem; Nuclear models; Spin-orbit splitting; Photonuclear effect.
Address: Department of Physics, Washington University, St. Louis, Missouri, U.S.A.

CLARK, John Wesley, B.A. (Phys., Maths., Montana), Ph.D. (Phys., Illinois). Born 1915. Educ.: Montana State and Illinois Univs. Vice-Pres., Eng., Deep Ocean Technol., Inc. Societies: A.A.A.S.; A.N.S.; A.P.S.; A.I.A.A.; Fellow, I.E.E.E.; Marine Technol. Soc.; N.Y. Acad. Sci.
Nuclear interests: Design and development of remote handling systems; Application of nuclear power to undersea installations.
Address: 422 23rd Street, Santa Monica, California 90402, U.S.A.

CLARK, Kenneth Courtright, Ph.D., A.M., B.A. Born 1919. Educ.: Texas and Harvard Univs. Assoc. Prof. (1955-61), Prof. (1962-), Washington Univ., Seattle; Res. Assoc. Prof., Geophys. Inst. Coll., Alaska, 1957-58; Consultant, Phys., Aerospace Corp. and Boeing Co., 1964-65; Consultant Prof., U.S.A.I.D. Programme, India, 1964-66. Societies: Fellow, A.P.S.; Fellow, Optical Soc. of America; A.G.U.
Nuclear interests: Atomic collision processes involving charge transfer; Upper atmosphere physics; Excitation of radiation; Optical spectroscopy; Geophysics.
Address: Department of Physics, Washington University, Seattle, Washington 98105, U.S.A.

CLARK, Melville, Jun., S.B., A.M., Ph.D. Born 1921. Educ.: M.I.T., New Mexico, Princeton and Harvard Univs. Brookhaven Nat. Lab., 1949-53; California Univ. Rad. Lab., 1953-55; M.I.T. 1955-62; Appl. Res. Lab., Sylvania Elec. Products, 1962-64; Avco Corp., 1964-67; Electronics Res. Centre, N.A.S.A., 1967-. Clark Music Co., 1940-60; Raytheon Manufacturing Co., 1955-58; United Shoe Machinery, 1956; Arthur D. Little, 1957-58; Edgerton, Germeshausen, and Grier, 1961-; 416 South Salina Street Corp., 1957-61; Meldor Corp., 1960-66; Melville Clark Assocs., 1955-. Books: Co-translator, Introduction to the Theory of Ionized Gases by J. L. Delcroix (Interscience, 1960); co-author, Plasmas and Controlled Fusion (M.I.T. Press, 1961); co-author, Numerical Methods of Reactor Analysis (Academic Press, 1964). Societies: A.I.P.; A.P.S.; A.A.A.S.; N.Y. Acad. Sci.; A.S.A.; American Symphony Orchestra League; American Musicological Soc.
Nuclear interests: Neutron and gamma ray transport theory; Reactor design; Numerical

analysis; Plasma physics; Hydromagnetics; Controlled fusion; Mathematical physics, atomic and nuclear physics; Shielding.
Address: 8 Richard Road, Cochituate, Mass., U.S.A.

CLARK, Michael David, M.Sc. (Maths.). B.Sc. (Gen.). Born 1928. Educ.: London Univ. Tech. Asst. (1952-55), Asst. Project Eng. (1955-57), Rockets Div., de Havilland Engine Co. Ltd.; Head, Shielding Sect. (1957-59), Head, Rad. Phys. Group (1959-63), Atomic Power Constructions Ltd.; Mathematician (1963-65), Nat. Account Manager (1965-), Sci. Computing Service.
Nuclear interests: Radiation shielding; Effects of irradiation on materials.
Address: 16 Plaitford Close, Rickmansworth, Herts., England.

CLARK, R. Lee. Director, M. D. Anderson Hospital and Tumor Inst., Texas Univ.; Member, Nat. Advisory Cancer Council, U.S. P.H.S. 1961-.
Address: M. D. Anderson Hospital and Tumor Institute, Texas University, Texas Medical Centre, Houston 25, Texas, U.S.A.

CLARK, Robert Kenley, Ph.D., M.S., B.A. Born 1918. Educ.: Illinois and Montana State Univs. Assoc. Phys., Argonne Nat. Lab., 1950-54; Phys., VA Res. Hospital, Chicago, 1954-63; Pres., Rad. Control, Inc., 1959-. Sen. Vice-Pres. and Treas., The Human Systems Inst., Chicago, 1963-. I.R.E. Professional Group on Bio-Medical Electronics. Societies: A.P.S.; Rad. Res. Soc.; A.A.P.T.; A.A.A.S.; H.P.S.; American Assoc. of Physicists in Medicine; Rad. and Medical Phys. Soc. of Illinois; New York Acad. Sci.; I.R.E.
Nuclear interests: Physical research; Dosimetry; Environmental monitoring; Isotope technology.
Address: The Human Systems Institute, 6 W. Ontario Street, Chicago 10, Illinois, U.S.A.

CLARK, Ulray, B.S. Born 1923. Educ.: Florida Univ. At O.R.I.N.S. Pres.-Elect, Florida Chapter, H.P.S.
Nuclear interest: Applied health physics in governmental regulatory agencies.
Address: 840 Shady Lane, Bartow, Florida 33830, U.S.A.

CLARK, Walter Ernest, B.S., M.A., Ph.D. Born 1916. Educ.: Virginia Military Inst., George Washington and Wisconsin Univs. Asst. Prof. Chem. Eng., Missouri Univ., School of Mines and Metal., 1949-51; Res. Chem., O.R.N.L. 1951-; Lecturer in Chem. Eng. (Electrochem.), Tennessee Univ., 1963-64; Visiting Prof., Villa Madonna Coll., 1966-67; Fulbright Lecturer, Tribhuvan Univ., Nepal, 1966-67. Societies: A.C.S.; Electrochem. Soc.; American Inst. of Chemists.
Nuclear interests: Radioactive waste disposal; Radiation corrosion; Reprocessing of nuclear fuels; Materials of construction for radio-chemical processing equipment; Radiation electrochemistry.
Address: 386 East Drive, Oak Ridge, Tennessee 37830, U.S.A.

CLARKE, Edgar John Southwell, M.A. Born 1913. Educ.: Oxford Univ. Asst. Sec., Lord President's Office, 1953; Principal Metals Officer (1954-62), Principal Finance Officer (1962-), U.K.A.E.A.
Address: 39 Lyford Road, London, S.W.18, England.

CLARKE, Eric Thacher, S.B., Ph.D. Born 1916. Educ.: Harvard Univ. and M.I.T. Vice Pres., Res. Div., Sec. and Director, Tech. Operations, Inc., 1951-. Member, Exec. Com., Shielding Div., A.N.S. Society: Soc. for Nondestructive Testing.
Nuclear interests: Nuclear instrument design; Isotope radiography; Gamma-ray scattering; Radiation shielding.
Address: Technical Operations, Incorporated, Burlington, Massachusetts 01803, U.S.A.

CLARKE, F. J. P. Member, Editorial Advisory Board, J. of Nucl. Materials.
Address: Journal of Nuclear Materials, North-Holland Publishing Co., P.O. Box 103, Amsterdam, Netherlands.

CLARKE, J. H. Atomic Energy Sect., Hebron and Medlock.
Address: Hebron and Medlock, St. George's Lodge, 33 Oldfield Road, Bath, Somerset, England.

CLARKE, Kenneth Henry, M.Sc. (Melbourne), A.R.C.S., F. Inst. P., F.A.I.P. Born 1923. Educ.: London and Melbourne Univs. Physicist-in-Charge, Cancer Inst., Melbourne, 1960-. Hon. Sec.-Treas., Biophys. Group, Australian Inst. Phys.; Hon. Sec.-Treas., Australian Regional Group, Hospital Phys. Assoc.; Editor, Australasian Newsletter of Medical Phys. Society: British Inst. Radiol.
Nuclear interests: Medical and biological physics including nuclear medicine.
Address: Cancer Institute, 278 William Street, Melbourne C.1., Victoria, Australia.

CLARKE, Norman, B.Sc. Born 1916. Educ.: Manchester Univ. Deputy Sec. (1945-65), and Fellow, The Inst. of Phys. (and Phys. Soc. after 1960); Sec. and Registrar, The Inst. of Mathematics and its Applications 1965-. Societies: British Acad. of Forensic Sci.; Mathematical Assoc.
Nuclear interest: Educational.
Address: The Institute of Mathematics and its Applications, Maitland House, Warrior Square, Southend-on-Sea, Essex, England.

CLARKSON, Stuart Winston, M.A. Born 1920. Educ.: Mount Allison and Toronto Univs. Economist, Joint Intelligence Bureau Defence Res. Board, 1950-54; Economist, Energy Branch, Dept. of Trade and Commerce,

1954-57; Economist, A.E.C.L., 1957-60; Deputy Minister, Ontario Dept. of Energy Resources, 1960-62; Deputy Minister, Ontario Dept. of Economics and Development, 1962-. Member, Foreign Liaison Com., Canadian Nucl. Assoc. Book: Uranium in Western World (A.E.C.L., 1959). Society: Canadian Nucl. Assoc.
Nuclear interests: Administration of Provincial Government policy with regard to nuclear manufacturing development.
Address: Deputy Minister of Economics and Development, Room 201, 454 University Avenue, Toronto 2, Ontario, Canada.

CLATWORTHY, Willard Hubert, B.A. (Mathematics), M.A. (Mathematics), Ph.D. (Mathematical Statistics). Born 1915. Educ.: Berea College, Kentucky and North Carolina Univs. Mathematician, U.S. N.B.S., 1952-55; Staff Statistician (1955-58), Manager, Statistical Methods (1958-62), Westinghouse Elec. Corp., Bettis Atomic Power Div., Pittsburgh, Pa.; Prof. of Statistics (1962-), Director (1963-65), Div. of Mathematical Statistics,New York State Univ. Societies: Fellow, A.S.A.; Fellow, Roy. Statistical Soc.; Fellow, A.A.A.S.; Inst. Mathematical Statistics; A.A.U.P.
Nuclear interests: Statistical design and analysis of experiments and testing programmes in metallurgy, chemistry; Nuclear physics; Reactor engineering; Instrumentation.
Address: 253 Sherbrooke Drive, Buffalo 21, New York, U.S.A.

CLAUDE, Albert, Prof., Univ. Libre de Bruxelles. Society: Member, Commission Scientifique, Inst. Interuniversitaire des Sciences Nucléaires.
Address: 62 rue des Champs-Elysées, Ixelles, Brussels,Belgium.

CLAUMARCHIRANT, D. DOMINGO PANIAGUA. See DOMINGO PANIAGUA CLAUMARCHIRANT, D.

CLAUS, Walter D., A.B., M.S., Ph.D. Born 1903. Educ.: Washington Univ., St. Louis. With U.S.A.E.C., 1949-67. Book: Rad. Biol. and Medicine. Societies: A.P.S.; H.P.S.; Radiation Res. Soc.; A.C.S.
Nuclear interests: Health physics; Radiation biology.
Address: 1040 N.W. Fourth Street, Boca Raton, Florida 33432, U.S.A.

CLAUSEN, Andreas Walter, Oberbaurat, Diplom-Ing. Born 1919. Educ.: Danzig T.H. Dozent an der staatlichen Ingenieurschule Flensburg. Schiffbautechn. Gesellschaft; Studiengesellschaft zur Förderung der Verwendung der Kernenergie in Schiffbau und Schiffahrt, Hamburg.
Nuclear interests: Metallurgy; Reactor-operating; Reactor design.
Address: 1 Grossestrasse, Flensburg, Germany.

CLAUSS, Christian Richard Alexander, Dr. rer. nat. Born 1929. Educ.: Darmstadt T.H. Asst.,
Darmstadt T.H., 1954; Asst., Inorganic Chem. Div., European Res. Assoc., Brussels, 1956; Head, Nucl. Div. (1959-), Sci. Adviser (1964-), Sci. Sub-Manager (1966-), S.E.R.A.I., Brussels.
Nuclear interests: Chemistry and diffusion of fission products; Chemistry of reactor materials, especially graphite; Testing of fuel elements.
Address: S.C. Société d'Etudes de Recherches et d'Applications pour l'Industrie, 1091 Chaussée d'Alsemberg, Brussels 18, Belgium.

CLAUSS, Julius, Dipl. Ing. Born 1921. Educ.: Stuttgart T.H. Kohlenscheidungsgesellschaft m.b.H., Stuttgart, 1954-61; Energieversorgung Schwaben A.G., Stuttgart, 1961-. Society: V.D.I.
Nuclear interests: Nuclear engineering (reactor design, applied thermodynamics and hydrodynamics, corrosion); Nuclear physics; Economic and legal aspects of nuclear energy.
Address: Energieversorgung Schwaben A.G., 12 Goethestrasse, Stuttgart N, Germany.

CLAUSSEN, Nils E., Dipl.-Ing., M.Sc. (Metal.). Born 1937. Educ.: Stuttgart Univ. and Georgia Inst. of Technol. Max-Planck-Inst. für Metallforschung, 1966-.
Nuclear interests: Reactor materials, cermets and ceramics; Mechanical properties of these materials and testing methods.
Address: Max-Planck-Institut für Metallforschung, Institut für Sondermetalle, 92 Seestrasse, 7 Stuttgart-1, Germany.

CLAVE, Marcel. Ing.-Constructeur, Dept. Nucleaire-Observation, Ets. S.R. Clave - Optique et Mecanique de Precision.
Address: Ets. S.R. Clave - Optique et Mecanique de Precision, 9 rue Olivier Metra, Paris 20, France.

CLAVE, Serge-Rene. Ing.-Constructeur, Dept. Nucleaire-Observation, Ets. S.R. Clave - Optique et Mecanique de Precision.
Address: Ets. S.R. Clave - Optique et Mecanique de Precision, 9 rue Olivier Metra, Paris 20, France.

CLAVIJO, J., L.G. Adviser to the Gen. Manager, Dept. of Sci. Instr., Philips Colombiana S.A.
Address: Philips Colombiana S.A., Calle 1351-03, Bogota, Colombia.

CLAYTON, Eugene Duane, B.A., M.Sc., Ph.D. (Phys.). Born 1923. Educ.: Oregon Univ. Teaching Fellow (1949-50), Res. Fellow (1950-51), Oregon Univ.; Phys., Appl. Res. (1951-54), Sen. Sci., Appl. Res. (1954-56), Supervisor, Reactor Lattice Phys. (1956-63), Supervisor (1963-64), Manager (1964-), Critical Mass Phys., G.E.C.; Faculty, (Critical Mass Phys.), Faculty of the Nucl. Eng. Dept., Washington Univ. Societies: A.N.S.; A.P.S.
Nuclear interests: Reactor physics; Nuclear safety and criticality control; Critical experiments: Nuclear physics.

Address: 2041 Davison, Richland, Washington, U.S.A.

CLAYTON, G. W. Vice Pres., Acres Res. and Planning. Chairman, Economic Planning Com., Canadian Nucl. Assoc.
Address: Acres Research and Planning, 74 Victoria Street, Toronto 1, Ontario, Canada.

CLAYTON, Henry Hubert, B.A., M.A. Born 1906. Educ.: British Columbia and Purdue Univs. Head, Theoretical Phys. Branch, A.E.C.L., Chalk River, Ontario, 1950-58, 1960-. Overseas Rep., A.E.C.L., 1959-60. Society: Canadian Assoc. of Physicists.
Nuclear interests: Nuclear and reactor physics; Computation.
Address: Theoretical Physics Branch, Atomic Energy of Canada Ltd., Chalk River, Ontario, Canada.

CLAYTON, John Charles, B.S. (St. Joseph's Coll., Philadelphia), M.S., Ph.D. (Pennsylvania). Born 1924. Educ.: Pennsylvania Univ. Post-doctoral Fellow, Pennsylvania Univ., 1953-54; Sen. Sci., Westinghouse Electric Corp., Bettis Atomic Power Lab., Pittsburgh, Pa., 1955-. Societies: A.C.S.; A.A.A.S.
Nuclear interests: Inorganic chemistry of nuclear materials.
Address: 5453 Youngridge Drive, Apt. 24, Pittsburgh 36, Pennsylvania, U.S.A.

CLEARE, Henry Murray, B.S. Born 1928. Educ.: Georgia Inst. of Technol. Phys. (1951-58), Res. Phys. (1958-62), Res. Assoc.(1962-), Head, Radiography Lab. (1964-), Res. Labs., Eastman Kodak Co. Society: H.P.S.
Nuclear interests: Photographic radiation dosimetry; Photographic action of charged particles; Photographic effects of radiations in space.
Address: Research Laboratories, Eastman Kodak Company, Rochester, New York 14650, U.S.A.

CLEARMAN, Harold, Dr. Sci. concerned with Nucl. Work, Hofstra Univ.
Address: Hofstra University, Hempstead, New York, U.S.A.

CLEBSCH, Alfred, Jr. Geological Survey, U.S. Dept. of the Interior.
Address: Geological Survey, U.S. Department of the Interior, Washington 25, D.C., U.S.A.

CLEEMPOEL, Henri, M.D. Born 1924. Educ.: Brussels Free Univ. Inst. de Recherches Cardiologiques, Centre de Médecine Nucléaire, Brussels Univ.; Contract Euratom, Brussels Univ. Societies: Belgian Soc. Cardiology; Belgian Soc. Internal Medicine.
Nuclear interests: Tracers in cardiology (cardiac output, shunts, renogramme, etc.).
Address: Hôpital Saint-Pierre, Institut de Recherches Cardiologiques, 322 rue Haute, Brussels, Belgium.

CLEERE, Roy Leon, B.S., M.D., Master of Public Health, Fellow, American Board of Preventative Medicine and Public Health. Born 1905. Educ.: A. and M. Coll. Texas; Texas Univ. School of Medicine and Johns Hopkins Univ. School of Hygiene and Public Health. Exec. Director, Colorado State Dept. of Public Health, 1935-; Asst. Prof. in Public Health and Lab. Diagnosis, 1947-; U.S. Deleg. to 4th World Health Assembly, Geneva, 1951; U.S.A.E.C. Advisory Com. State Officials, 1955-; Surgeon General's (U.S.P.H.S.) Advisory Com. on Occupational Health, 1957-67. Societies: A.M.A.: Colorado State Medical Soc.; Assoc. of State and Territorial Health Officers; A.P.H.A.; Western Branch, A.P.H.A.; Colorado Public Health Assoc.
Nuclear interests: Management; Administration radiological health programme, including supervision of use of Whole Body Counter.
Address: Colorado State Department of Public Health, 4210, E 11th, Denver 20, Colorado, U.S.A.

CLEEREMANS, Richard William, B.S. Born 1923. Educ.: Wisconsin Univ. Sec., Com. on Rad. Control, Mayo Clinic and Mayo Foundation.
Address: Mayo Clinic and Mayo Foundation, Rochester, Minnesota 55901, U.S.A.

CLEGG, Arthur B., M.A., Ph.D. Born 1929. Educ.: Cambridge Univ. Res. Fellow, California Inst. Technol. 1955-58; Sen. Res. Officer, Oxford Univ., 1958-66; Fellow, Jesus Coll., Oxford Univ., 1964-66; Prof. Nucl. Phys., Lancaster Univ., 1966-. Societies: F. Inst. P.; A.P.S.
Nuclear interests: Interactions of 100-200 MeV nucleons with nuclei, with related interests in nuclear structure; Elementary particle physics; Reaction mechanisms; Quark structure of elementary particles.
Address: Strawberry Bank, Cannon Hill, Westbourne Road, Lancaster, Lancashire, England.

CLEGG, John William, B.S. (Chem. Eng.), Ph.D. (Chem. Eng.). Born 1916. Educ.: Georgia Inst. Technol. and Minnesota Univ. Manager, Chem. Eng. Dept., Battelle Memorial Inst. -1963; Pres., North Star R. and D. Inst., Minnesota, 1963-. Book: Editor, Uranium Ore Processing (Addison Wesley, 1958). Societies: A.I.Ch.E.; A.C.S.; A.I.M.M.E.
Nuclear interests: Extractive metallurgy; Feed materials processing; Reprocessing of nuclear fuels; Nuclear ceramics.
Address: North Star Research and Development Institute, 3100 38th Avenue South, Minneapolis, Minnesota 55406, U.S.A.

CLELAND, Marshall R., B.A. (South Dakota), Ph.D.(Washington Univ., St. Louis). Born 1926. Educ.: Washington University, St. Louis. Sci. Staff, N.B.S., Washington, D.C., 1951-52; Development Staff, Nucl. Corp. of America, 1952-53; Pres. Teleray Corp., 1953-58;

Vice-Pres., Rad. Dynamics, 1958-. Societies: A.P.S.; A.R.S.
Nuclear interests: Nuclear physics, beta and gamma spectroscopy, Particle accelerator development, industrial applications of radiation.
Address: Radiation Dynamics, Inc., Westbury Industrial Park, Westbury, Long Island, New York, U.S.A.

CLELLAND, David Watson, Ph.D., B.Sc., A.R.T.C., F.R.I.C., A.M.I.Chem.E. Born 1925. Educ.: Glasgow Univ. and Roy. Coll. Technol., Glasgow, Manager of chemical plants producing detergent compounds, Thos. Hedley and Co., Ltd., 1951-54; Chem. Eng. Group and Manager, Works Development Group, R. and D. Branch, Windscale Works, Sellafield, Cumberland (1954-58), Chem. Adviser, Eng. Group, Risley, Lancs. (1958-.), U.K.A.E.A.
Nuclear interest: Chemical processing.
Address: 19 Park Crescent, Appleton, Cheshire, England.

CLEMEDSON, C. J., Prof. Member, Nat. Rad. Protection Inst., Sweden.
Address: National Radiation Protection Institute, Karolinska sjukhuset, Stockholm 60, Sweden.

CLEMENTE, Mrs. Alicia O. Medical Technician, Radioisitope Lab., Philippine Gen. Hospital.
Address: Philippine General Hospital, Radioisotope Laboratory, Manila, Philippines.

CLEMENTE, Amando, A.B., M.S., Ph.D. Born 1889. Educ.: Philippines and Chicago Univs. Pres., Univ. Paper Mills, 1958; Pres., Aclem Paper Mills, Inc., 1959; Pres. and Manager, Chem. Corp. of the Philippines, 1958. Chairman, Div. of Ind. and Eng. Res., N.R.C. of the Philippines; Member, Board of Directors, Chem. Soc. of the Philippines; Emeritus Prof., Chem., Philippines Univ. Societies: Philippines Assoc. for Advancement of Sci.; A.A.A.S.; New York Acad. Sci.
Nuclear interests: Nuclear chemistry and nuclear physics.
Address: 86 Times Street, West Triangle, Quezon City, Philippines.

CLEMENTEL, Ezio, Ph.D. (Theoretical Phys.). Born 1918. Educ.: Padua Univ. Prof. Theoretical Phys., Ferrara Univ. Sci. Director, Studies and Res. Div., C.N.R.N.; Director, Computing Centre, C.N.E.N.; Director, C.N.E.N. activities at Bologna; Prof. Nucl. Phys., Bologna Univ.; Member, Group of Specialists on Neutron Data Compilation, and Member, Computer Programme Library Com., E.N.E.A., O.E.C.D.
Nuclear interests: Nuclear physics and fast reactor physics.
Address: Centro di Calcolo del C.N.E.N., 2 Via Mazzini, Bologna, Italy.

CLEMOW, Brigadier John, M.A. Born 1911. Educ.: Cambridge Univ. Director, Guided Weapons, Ministry of Supply, 1950-57; Chief Eng. (Weapons), Vickers Aircraft, 1957-61; Director, G.E.C. (Electronics) Ltd., 1961-66; Head, Nucl. Phys. Div., Sci. Res. Council, 1966-. Societies: Fellow, Inst. of Elec. Eng.; Fellow, Inst. of Mathematics.
Nuclear interest: Nuclear and high energy physics.
Address: Science Research Council, State House, High Holborn, London, W.C.1, England.

CLERC, Blaise, notaire. Depute au Conseil des Etats. Member, Swiss Federal Commission for Atomic Energy.
Address: Conseil des Etats, 2 rue Pourtalis 2, 2000 Neuchatel, Switzerland.

CLERC, R. At Stephan S.A.
Address: Stephan S.A., Fribourg, Switzerland.

CLERMONT, Y GOLDSCHMIDT-. See GOLDSCHMIDT-CLERMONT, Y.

CLESS, Traude BERNERT-. See BERNERT-CLESS, Traude.

CLEWETT, Glen H., A.B. Born 1911. Educ.: California Univ., Los Angeles. Asst. Res. Director, Union Carbide Nucl. Co., Sterling Forest Lab., Tuxedo, New York. Book: Chapter on Isotope Separation in Nucl. Eng. Handbook (Etherington; Mc-Graw-Hill, 1958). Societies: A.C.S.; RESA.
Nuclear interests: Chemical separation of isotopes; Milling of uranium ores.
Address: 71 Jean Street, Ramsey, New Jersey, U.S.A.

CLIFFORD, Charles E., Ph.D., M.Sc., B.Sc. Born 1930. Educ.: Carleton and McGill Univs. Sci. Service Officer, Defence Res. Board, Ottawa, Canada, 1955-; Sessional Lecturer, Carleton Univ. Book: Contributor to Eng. Compendium on Rad. Shielding (I.A.E.A.). Societies: Canadian Assoc. Phys.; A.N.S.
Nuclear interest: Radiation shielding and dosimetry.
Address: Defence Research Board, Shirley Bay, Ottawa, Canada.

CLIFFORD, Daniel J. Manager, Material Dept., AMF Atomics.
Address: AMF Atomics, Whiteford Road, York, Pennsylvania 17402, U.S.A.

CLIKEMAN, Franklyn Miles, Ph.D., B.S. Born 1933. Educ.: Montana State Coll. and Iowa State Univ. Post Doctoral Res. Appointment, Ames Lab., Iowa State Univ., 1962-63; Ford Post Doctoral Appointment (1963-65), Asst. Prof. (1963-), M.I.T. Society: A.P.S.
Nuclear interests: Neutron and gamma ray spectroscopy; Neutron physics; Reactor physics; Neutron cross sections.
Address: Department of Nuclear Engineering, Building NW12, Massachusetts Institute of Technology, Cambridge, Massachusetts 02139, U.S.A.

CLINE, James Edward, B.S.E., M.S., Ph.D. Born 1931. Educ.: Michigan Univ. Teaching Fellow (1953-56), Res. Assoc. (1954-57), Michigan Univ.; Instructor, Idaho Univ., N.R.T.S. Campus, 1958-; Res. Phys., Atomic Energy Div., Phillips Petroleum Co., 1957-. Society: A.P.S.
Nuclear interests: Low energy nuclear spectroscopy – beta and gamma spectroscopy.
Address: Phillips Petroleum Company, Atomic Energy Division, P.O. Box 2067, Idaho Falls, Idaho, U.S.A.

CLINE, James G., B.E.E. M.Nucl.E. Born 1933. Educ.: Marquette and New York Univs. Reactor Eng., U.S.A.E.C., 1959-62; Principal Nucl. Eng., Office of Atomic and Space Development, New York State, 1962-65; Programme Manager, Atomic and Space Development Authority, New York State, 1965-; Registered Professional Eng., State of New York. Chairman, Subcom. on Isotope Availability, Nucl. Space Programme Com., Atomic Ind. Forum; Director, Metropolitan Sect., A.N.S. Society: New York Acad. of Sci. Book: Co-author, Atomic Power Waste Handling (Atomic Ind. Forum, Inc., 1965).
Nuclear interests: Reactor development; Reactor siting and systems integration; Isotopes; Reprocessing and nuclear fuel cycle development; Radiation applications.
Address: New York State Atomic and Space Development Authority, 230 Park Avenue, New York, N.Y. 10017, U.S.A.

CLINTON, Thomas Gerard John, A.R.I.C. Born 1910. Educ.: Roy. Salford Tech. Coll. Analytical Services Manager, Springfields Works, U.K.A.E.A., 1955-.
Nuclear interests: Chemistry with special reference to analysis; Laboratory management.
Address: 235 Black Bull Lane, Fulwood, Preston, Lancs., England.

CLIQUOT, Rene. Tech. Manager, Sté. d'Etudes Techniques et d'Entreprises Generales.
Address: 9 avenue Reaumur, Le Plessis-Robinson, (Seine), France.

CLOETE, Francis Louis Dirk, B.Sc. (Eng.), M.Sc. (Chem. Tech.) Born 1934. Educ.: Cape Town and Natal Univs. Chem. Eng., Distillers Co. Ltd., Hull, 1958-59; Sci. Officer, U.K.A.E.A., Aldermaston, 1959-61; Chem. Eng.. Nucl. Chem. Plant Ltd., 1961-62; Lecturer, Nucl. Technol., Imperial Coll. Societies: B.N.E.S.; Roy. Inst. Chem.; Inst. Chem. Eng.; Inst. Fuel.
Nuclear interests: Fuel processing techniques with particular reference to processing of solids; Ion exchange techniques.
Address: Nuclear Technology Laboratory, Department of Chemical Engineering and Chemical Technology, Imperial College, London S.W.7, England.

CLOSON, Auguste Georges, Ing. Civil. Born 1898. Educ.: Brussels Univ. Director, C.E.N.; Prés., Syndicat Vulcain; Vice-Prés., Belgo-nucléaire; Administrateur, Sté. Anonyme Anglo Belge Vulcain.
Address: 23 avenue Jeanne, Ixelles – Brussels 5, Belgium.

CLOSS, Hans, Dr. rer. nat. Born 1907. Educ.: Tübingen, Berlin and Vienna Univs. Leitender Direktor u. Professor at Bundesanstalt für Bodenforschung. Deutsche Atomkommission, Vorsitzender Arbeitskreis III/4.
Nuclear interest: Prospecting for raw materials.
Address: 3 Hannover-Buchholz, Alfred-Bentz-Haus, Postfach 54, Germany.

CLOTHIER, Robert Frederic, B.Eng., M.Sc., Docteur de l'Université en Sci., D.Sc. (phys.). Born 1925. Educ.: Southern California and Montpellier Univs. Consultant with Westinghouse, U.S.A.E.C.; Assoc. Prof. Eng., Alabama Polytech. Inst. 1950-55; Sen. Eng. and Administrator (Staff) Westinghouse Atomic Power Div., 1955-57; Prof. Eng., i/c Nucl. Eng. and Chairman, Mech. Eng. Dept., San Jose State Coll., 1957-. Society: A.S.E.E.
Nuclear interests: Reactor physics and core thermal-hydraulic design; Health physics.
Address: San Jose State College, San Jose, California, U.S.A.

CLOUET D'ORVAL, Christian Michel, Eng. Born 1931. Educ.: Ecole Superieure d'Electricite, Paris and Internat. Inst. of Nucl. Sci. and Eng., Argonne Nat. Lab. Society: Soc. Française des Electriciens.
Nuclear interests: Reactor physics and experiments.
Address: 41 boulevard Raspail, Paris 7, France.

CLUA DOMINGUEZ, Jose Javier. Vocale, Comisiones Asesoras de Biologia Vegetal y Aplicaciones Industriales, Junta de Energia Nucl.
Address: Junta de Energia Nuclear, Ciudad Universitaria, Madrid 3, Spain.

COADY, A. W. B., C.M.G., B.A., B.Ec. Member, Advisory Com., A.A.E.C.
Address: Australian Atomic Energy Commission, 45 Beach Street, Coogee, New South Wales, Australia.

COAKLEY, Sister Mary Peter, R. S. M., A.B., M.S., Ph.D. Born 1915. Educ.: Georgian Court Coll. and Notre Dame Univ. Chairman, Dept. of Chem., Georgian Court Coll., 1955-; Principal Investigator, Contracts with U.S.A.E.C., 1957-63. Societies: N. Y. Acad. Sci.; A.C.S.; Middle Atlantic Sci. Teachers Assoc.; Albertus Magnus Guild.
Nuclear interests: Chemical structure and bonding.
Address: Georgian Court College, Lakewood, N.J., U.S.A.

COATES, Fredrick William, M.A. Born 1926. Educ.: Cambridge Univ. Resident Eng.,

Ffestiniog Power Station, 1959-65; Resident Eng., Trawsfynydd Power Station, 1959-66; Project Eng., C.E.G.B., 1966-. Societies: A.M.I.Mech.E.; A.M.I.C.E.; A.M.I.Water E. Nuclear interests: Management; Reactor design. Address: Central Electricity Generating Board, Northern Project Group, Agecroft Road, Pendlebury, Swinton, Manchester, England.

COATES, G. P. Member, Group of Experts on Production of Energy from Radioisotopes, O.E.C.D., E.N.E.A.
Address: O.E.C.D. European Nuclear Energy Agency, 38 boulevard Suchet, Paris 16, France.

COB, Manuel DEL VAL. See DEL VAL COB, Manuel.

COBAS, Amador, B.A. (Puerto Rico), M.A., Ph.D. (Columbia). Born 1910. Educ.: Puerto Rico and Columbia Univs. Dean, Faculty Gen. Studies, Puerto Rico Univ., 1949-50; Visiting Prof. Phys., New York Univ., 1952-53; Director, Radioisotope Techniques Training Centre (1956-58), now Deputy Director, Puerto Rico Nucl. Centre, Puerto Rico Univ. Societies: A.P.S.; N.Y. Acad. Sci., Nat. Education Assoc. of U.S.; American Inst. Phys.; A.A.P.T.; Fellow, A.A.A.S.
Nuclear interest: Radioisotope applications; Solid state physics.
Address: 561 Ensenada Street, Santurce, Puerto Rico, U.S.A.

COBB, Carolus Melville, S.B., Ph.D. Born 1922. Educ.: M.I.T. Teach. Fellow, M.I.T., 1948-50; Ionics Inc., 1951-55; Allied Res. Assoc., 1955-60; American Sci. and Eng., Inc., 1960-. Societies: A.C.S.; A.P.S.
Nuclear interests: Scintillating chemicals and solid state particle detectors; Radiation safety; Weapons effects.
Address: 15 Elsie Road, Lynn, Massachusetts, U.S.A.

COBB, Edgar E., B.S. (Elec. Eng.). Born 1904. Educ.: Oklahoma State Univ. Societies: A.N.S.; I.E.E.E.
Nuclear interests: Application of nuclear reactors to central stations for the electric light and power industries.
Address: Gibbs and Hill, Incorporated, 393 Seventh Avenue, New York, N.Y. 10001, U.S.A.

COBB, George Hamilton, B.S. (Mech. Eng.). Born 1911. Educ.: Oklahoma State, Nebraska and Kansas Univs. Production Supt. and Production Manager (1939-53), Asst. to Pres. (1953-56), Kerr-McGee Corp., Oklahoma; Exec. Vice-Pres., Kermac Nucl. Fuels, 1956-64; Vice-Pres., Exploration (1959-64), Vice-Pres., Exploration and Res. (1964-67), Sen. Vice-Pres. (Kerr-McGee) (1967-), Kerr-McGee Corp. Societies: A.I.M.M.E.; A.A.P.G.; Registered Professional Eng. in States of Oklahoma and Texas.
Nuclear interests: Metallurgy; Production of U_3O_8 and additional refinement of atomic fuels.
Address: Kerr-McGee Corporation, Kerr-McGee Building, Oklahoma City, Oklahoma 73102, U.S.A.

COBBLE, James W., B.A., M.S., Ph.D. Born 1926. Educ.: Northern Arizona, Southern California and Tennessee Univs. Chem., O.R.N.L., 1949-52; Chem., Rad. Lab., 1953; Instructor, Chem. Dept., California Univ., Berkeley, 1954; Asst. Prof. (1955-58), Assoc. Prof. (1958-61), Prof. (1961-), Chem. Dept., Purdue Univ. Societies: A.C.S.; A.P.S.; A.A.U.P.
Nuclear interests: Nuclear chemistry; Radiochemistry; Chemistry of the heavy elements.
Address: Department of Chemistry, Purdue University, Lafayette, Indiana, U.S.A.

COBURN, Don Brandley, B.S. (Eng. Phys.). Born 1927. Educ.: Utah State Univ. and O.R.S.O.R.T. Instructor, Mathematics, Utah State Univ., 1952-53; Fellowship, O.R.S.O.R.T., 1953-54; Eng., Atomics Internat., Gen. Elec., 1954-57; Eng. Manager, Nucl. Res., Chrysler Corp.; Chief, Tech. Evaluation and High Temperature Reactor Branch, U.S.A.E.C., 1957-60; Assoc. Manager, Gas Cooled Fast Reactor Programmes, Gulf Gen. Atomic, 1960-. Various Offices, Local Sect., A.N.S. Societies: A.A.A.S.; American Management Assoc.
Nuclear interests: Technical management of reactor development programmes, and the design and development of the gas cooled fast breeder reactor concept.
Address: 3645 Brandywine Street, San Diego, California 92117, U.S.A.

COCCHI, Umberto Dante, Dr. med. (Berlin, Milan and Zurich). Born 1909. Educ.: Berlin, Zurich and Milan Univs. Docent Prof., Zurich Univ.; Docent Prof., Bologna Univ. Head Asst. Inst. of Roentgendiagnostics and Radiotherapeutic Clinic, Zurich Univ., 1937-63; Now Chief Radiol. (diagnostics and radiotherapy), Hospital Uster, Zürich, Sec. of the C.O.R.D. (Collegium Orbis Radiologiae Docentium) with seat at Zurich. Societies: Swiss, German and Italian Socs. of Radiol. and Nucl. Medicine.
Nuclear interest: Nuclear therapy.
Address: 62 Scheidegg Str., 8002 Zürich, Switzerland.

COCCONI, G., Prof. Director, Phys. 1 Dept., C.E.R.N., 1967-70.
Address: C.E.R.N., Physics 1 Department, Meyrin, 1211 Geneva 23, Switzerland.

COCHE, André, D.Sc. Born 1921. Educ.: Lyon and Paris Univs. Prof. Faculté des Sci. de Strasbourg. Societies: Sté. Française de Physique; Sté. de Chimie Physique; Sté. Française des Electroniciens et des Radioélectriciens.
Nuclear interests: Experimental nuclear physics (detection of nuclear radiation with liquid, plastic, gaseous scintillators and solid state counters).

Address: Centre de Recherches Nucléaires, rue du Loess, Strasbourg-Cronenbourg, France.

COCHRAN, James Charles, B.S., M.S. (Naval Architecture), M.S. (Nucl. Eng.). Born 1917. Educ.: U.S. Naval Acad. and M.I.T. Officer, U.S. Navy, 1940-55; U.S.A.E.C., 1953-55; Atomics Internat., 1955-64. Societies: A.S.M.E.; A.N.S.
Nuclear interests: Reactor design, development, operation; Primarily management.
Address: Atomics International, P.O.Box 309, Canoga Park, California, U.S.A.

COCHRAN, Joseph S. On-Site Supervisor and Tech. Director, Quehanna Nucl. Lab., Martin Co., 1962-.
Address: Quehanna Nuclear Laboratory, Martin Co., Quehanna, Pennsylvania, U.S.A.

COCHRAN, Lewis W., B.S., M.S., Ph.D. Born 1915. Educ.: Kentucky Univ. Prof. Phys., Dean Graduate School, Vice-Pres. for Res., and Member, Nucl. Phys. Res. Group, Dept. Phys., Kentucky Univ., 1946-; O.R.N.L., Summers 1950, 1953, and June 1958-Sept. 1960. Societies: A.P.S.; H.P.S.
Nuclear interests: Nuclear reaction physics; Interaction of radiation with matter.
Address: Kentucky University, Lexington, Kentucky 40506, U.S.A.

COCHRAN, Robert Glenn, B.A. (Phys.), M.S., Ph.D. (Nucl. Phys.). Born 1919. Educ.: Indiana and Pennsylvania State Univs. Res. Asst., Indiana Univ., 1947-50; Development Eng., Sarkes Tarzain Construction Eng., 1947-50; Nucl. Phys. and Group Leader, Oak Ridge Nat. Lab., 1959-54; Director, Res. Reactor, and Assoc. Prof. Nucl. Eng., Pennsylvania State Univ., 1954-59; Head, Nucl. Eng., Texas A. and M. Coll., 1959-. Member, Subcom. on Res. Reactor, N.R.C. 1957. Societies: A.N.S.; A.P.S.
Nuclear interests: Research reactor design; Neutron detector devices; Neutron and gamma ray spectroscopy; Nuclear reactions and decay schemes; Reactor shielding; Nuclear instrumentation and control.
Address: R.R.4, Bryan, Texas, U.S.A.

COCHRANE, Gordon S. Born 1915. Sen. Eng., G.E.C., Hanford Atomic Products Operation; Member, Public Relations Com., A.S.M.E., Columbia Basin Section; Acting Co-ordinator of Atomic Development Activities for the State of Washington; Society: A.S.M.E.
Nuclear interests: Reactor and fuel research and development; Facilities engineering.
Address: 1636 Howell Avenue, Richland, Washington, U.S.A.

COCKBAINE, David Robinson, Chartered Eng. Born 1925. Educ.: Oxford Coll. Technol. A.E.R.E. (1948-), now Electronic Eng.; Seconded to Cento Inst. Nucl. Sci., Teheran, Iran, 1959-62; Now Seconded to Turkish A.E.C. as Adviser in Electronics. Societies: A.N.S.; Assoc. Member, Inst. Radio and Electronic Eng.; A.M.I.E.E.
Nuclear interests: Reactor instrumentation; Nuclear electronics; Electronics in the field of isotope applications, especially in nuclear medicine and hydrology.
Address: c/o Ministry of Overseas Development, Eland House, Stag Place, London, S.W.1, England.

COCKERAM, Donald James, B.S. (E.E.). Born 1924. Educ.: Oregon State Univ., O.R.S.O.R.T. Bonneville Power Administration, 1949-54; Atomics Internat. (North American Rockwell), 1954- (at present Director), Analysis and Design Dept., Compact Systems Div. Societies: A.N.S.; A.I.A.A.; I.E.E.E.
Nuclear interests: Development of reactors for nuclear power plants, especially liquid metal cooled fast breeder reactors for central station power; Unmanned, lightweight systems for remote locations: for example, space vehicles and earth satellites.
Address: 10518 DeSoto Avenue, Chatsworth, California, U.S.A.

COCO, F. In Charge, Gruppo Stato Solido e Positronio, Centro Siciliano di Fisica Nucleare, Catania Univ.
Address: Catania University, Centro Siciliano di Fisica Nucleare, 57 Corso Italia, Catania, Sicily, Italy.

CODD, James, B.Sc. (Maths.). Born 1918. Educ.: Liverpool Univ.; A.E.R.E., Harwell 1947-60; A.E.E., Winfrith, 1960-.
Nuclear interest: Theoretical reactor physics.
Address: Atomic Energy Establishment, Winfrith, Dorchester, Dorset, England.

CODDING, James William, Jr., B. Chem. Eng. Born 1925. Educ.: Rensselaer Polytech. Inst., Troy, New York. Res. Chem., K.A.P.L., Gen. Elec. Co., Schenectady, New York, 1949-57; Separation Res. Chem., N.R.T.S., Atomic Energy Div., Phillips Petroleum Co., Idaho Falls, Idaho, 1957-66; Supervisor, Radioactive Sample Preparation, Idaho Nucl. Corp., Idaho Falls, Idaho, 1966-. Society: A.C.S.
Nuclear interests: Fuel reprocessing; Nuclear chemistry; Heavy element chemistry; Time-of-flight neutron spectroscopy.
Address: Route 4, Box 42A, Idaho Falls, Idaho, U.S.A.

COE, Roger J., E.E. Born 1901. Educ.: Cornell Univ. Vice Pres., Vermont Yankee Nucl. Power Corp.; Vice Pres., Maine Yankee Atomic Power Co.; Director and Vice Pres., Connecticut Yankee Atomic Power Co.; Vice Pres., Yankee Atomic Elec. Co. Industry Member, Labour-Management Advisory Com., U.S.A.E.C.; Member, Com. on Commercial Uses of Atomic Energy, Chamber of Commerce of the U.S. Society: A.N.S.
Address: New England Power Company, Boston, Massachusetts, U.S.A.

COEKELBERGS, R. F. R., Prof. Dr. Asst. Director and Head, Nucl. Chem. Dept., Centre de l'Ecole Royale Militaire; Sci. Adviser, Soc. Belge de l'Azote et des Produits Chimiques du Marly; Member, Management Com., Bureau Belge des Radioisotopes; Member, Commission Scientifique, Inst. Interuniv. des Sci. Nucléaires.
Address: Institute Interuniversitaire des Sciences Nucléaires, Centre de l'Ecole Royale Militaire, 30 avenue de la Renaissance, Brussels 4, Belgium.

COELHO, Aristides PINTO-. See PINTO-COELHO, Aristides.

COELHO, Fernando PINTO. See PINTO COELHO, Fernando.

COELHO SALGUEIRO, Lidia, Ph.D. Born 1917. Educ.: Lisbon Univ. Lecturer, Sci. Faculty, Lisbon Univ. Society: Soc. Portuguesa de Quimica e Fisica.
Nuclear interest: Nuclear spectroscopy.
Address: Laboratorio de Fisica, Faculdade de Ciencias, Rua da Escola Politecnica, Lisbon, Portugal.

COESSENS, A. Belgian Member, Special Group of Steering Com., and Member, Board of Directors, Eurochemic, E.N.E.A., O.E.C.D.
Address: O.E.C.D. European Nuclear Energy Agency, 38 boulevard Suchet, Paris 16, France.

COEURE, Marcel, Ing. (E.C.P.). Born 1915. Educ.: Paris Univ. Directeur, Div. Procédés et Exploitation, Sté. Saint-Gobain Techniques Nouvelles; Administrateur, Sté. de Raffinage de l'Uranium.
Address: 23 boulevard Georges Clemenceau, Courbevoie, (Seine), France.

COFER, Mrs. Agnes C., Prof. Member, Nucl. Advisory Com., Louisiana Polytech. Inst.
Address: Louisiana Polytechnic Institute, Nuclear Centre, Tech Station, Ruston, Louisiana, U.S.A.

COFFINBERRY, Arthur Shotter, B. Eng. Phys., M.Sc., D.Sc. Born 1902. Educ.: Case Inst. Technol., Ohio State and Harvard Univs. Group Leader for Plutonium Metal. (1947-58), Staff Member (1958-), Los Alamos Sci. Lab. Member, Editorial Advisory Com., J. of Nucl. Materials. Book: Editor, The Metal Plutonium (A.S.M.). Societies: A.N.S.; American Crystallographic Assoc.; A.S.M.; A.I.M.M.E.
Nuclear interest: Metallurgy.
Address: 6137 W. Highland Avenue, Phoenix, Arizona, U.S.A.

COFFMAN, Moody L., Ph.D. (Phys.), M.S. (Phys.), M.A. (Maths.), B.A. (Maths.). Born 1925. Educ.: Abilene Christian Coll., and Oklahoma and Texas A. and M. Univs. Instructor in Phys. and Maths., Texas A. and M. Univ., 1951-54; Sen. Nucl. Eng., Theory Group, Convair Div., Gen. Dynamics Corp., Fort Worth, Texas and Adjunct Prof. Phys., Texas Christian Univ., 1955-56; Assoc. Prof. Phys. Abilene Christian Coll., Abilene, Texas, and Consultant in Theoretical Nucl. Phys., Convair Div., Gen. Dynamics Corp., 1955-60; Sen. Phys., Space Phys. Sect., Space Systems Dept., Hamilton Standard Div., United Aircraft Corp., Windsor Locks, Conn. 2nd Adjunct Assoc. Prof. Phys., Rensselaer Polytech. Inst. Hartford Graduate Centre, 1960-61; Prof. and Head, Dept. of Phys., Oklahoma City Univ. and Consultant in Nucl. Phys. to Shiflet Bros., Abilene, Texas, 1961-. Book: Major contributor to Shielding Handbook for Aircraft Designers (Secret Restricted Data) (Published for Air Force by Convair Div., Gen. Dynamics Corp., Fort Worth, 1957). Societies: A.P.S.; American Mathematical Soc.; Mathematical Assoc. of America; A.G.U.; A.A.P.T.
Nuclear interest: Active interest in consulting on nuclear problems; Teaching courses in nuclear physics. Management of research and development of nuclear devices.
Address: 1832 N.W. 17th Street, Oklahoma City, Oklahoma 73106, U.S.A.

COGAN, D. H. Pres., Victoreen Instrument Co.
Address: Victoreen Instrument Co., 10101 Woodland Avenue, Cleveland 4, Ohio, U.S.A.

COGGAN, Bernard F., B.S. (Elec. Eng.), Masters work, U.S. Signal Corp. (Electronics). Born 1918. Educ.: Michigan State Univ. and U.S. Signal Corp. Recent Vice-Pres. and Gen. Manager, Convair; Pres., Southwest Capital Corp.; Corporate Vice-Pres., Operations, Douglas Aircraft Co.; Pres., San Diego Internat. Development Corp.; Consultant, General Dynamics Corp.; Consultant, Bendix Corp.; Director, Video Corp.; Director, Fluidgenics Corp.; Pres., B. F. Coggan and Associates; Vice-Pres., West Coast Financial Corp.; Currently: Corporate Vice-Pres., North American Rockwell Corp.; Special Consultant, U.S. Defence Dept.; Sen. Inter-Regional Adviser, U.N. Secretariate. Naval Weapons Advisory Board. Society: A.R.S.
Nuclear interests: Management of companies involved in the design and manufacture of nuclear reactors; Management of companies involved in the application of nuclear products to commercial uses, medical applications, space radiation effects, etc.
Address: 6436 Camino de la Costa, La Jolla, California, U.S.A.

COGHEN, Tomir, M.Sc., Ph.D. Born 1928. Educ.: Jagellonian Univ., Cracow. Inst. of Nucl. Res., Cracow Branch. Society: Polish Phys. Soc.
Nuclear interests: High energy physics; Nuclear physics.
Address: Institute of Nuclear Research, 2 Kategdra Fizyki, 30 Mickiewicza, AGH Cracow, Poland.

COHEN, Bernard L., B.S., M.S., Sc.D. Born 1924. Educ.: Case Inst. Technol., Pittsburgh Univ. and Carnegie Inst. Technol. Group Leader, O.R.N.L., 1950-58; Assoc. Prof. (1958-61), Pittsburgh Univ. Society: Fellow, A.P.S. Nuclear interests: Nuclear reactions; Nuclear structure; Accelerators; Fission; General nuclear physics.
Address: Physics Department, University of Pittsburgh, Pittsburgh 13, Pennsylvania, U.S.A.

COHEN, E. Richard, A.B. (Phys., Pennsylvania), M.S. and Ph.D. (California Inst. Technol.). E.O. Lawrence Memorial Award, 1968. Born 1922. Educ.: Pennsylvania Univ. and California Inst. Technol. Deleg. to 1st and 2nd I.C.P.U.A.E., Geneva, 1955 and 1958; Assoc. Director, North American Rockwell Sci. Centre, Thousand Oaks, California. Editorial Review Board, Nucl. Sci. and Eng.; Advisory Com. Reactor Phys., U.S.A.E.C. Book: Co-author, Fundamental Constants of Phys. (Interscience Publishers, 1957). Societies: A.I.P.; A.P.S.; A.N.S.
Nuclear interests: Reactor theory and nuclear physics.
Address: North American Rockwell Science Centre, 1049 Camino Dos Rios, Thousand Oaks, California, U.S.A.

COHEN, Jacob Antonie, Prof. Dr. Born 1915. Educ.: Leyden and Cambridge Univs. Director, Medical Biological Lab., Nat. Defence Res. Organization, T.N.O., 1947-; Prof. Applied Enzymology and Radiobiol., (1956-63), Prof. Physiology, Chem., Appl. Enzymology and Radiobiol., (1963-), Leyden Univ.; Member, U.N. Sci. Com. on the Effects of Atomic Rad.; Formerly Member, Nucl. Energy Commission, Netherlands. Formerly Director, Inst. of Radiopathology and Rad. Protection.
Address: 13 Lange Kleiweg, Rijswijk (Z.H.), Netherlands.

COHEN, James Alan, B.S.C., J.D. Born 1937. Educ.: Iowa Univ. Atorney, Office of the Gen. Counsel (1962-64), Legal Asst., Board of Contract Appeals (1964-), U.S.A.E.C.
Address: 5117 Brentford Drive, Rockville, Maryland, U.S.A.

COHEN, Karl Paley, B.A.(Columbia Coll.), M.A. (Columbia Univ.), Ph.D. (Columbia Univ.). Born 1913. Educ.: Columbia Coll., Columbia and Paris Univs. Tech. Director, H. K. Ferguson Co., 1948-52; Vice-Pres., Walter Kidde Nucl. Lab., 1952-55; Consultant to A.E.C., Columbia Univ., 1955-56; Manager of Advance Eng., Atomic Power Equipment Dept., (1956-65), Manager Advanced Products Sect. (1964-65), Gen. Manager, Advanced Products Operation, Atomic Div., (1965-), G.E.C. Fellow, and Vice Pres. (1967), Pres. (1968-), A.N.S. Book: The Theory of Isotope Separation as applied to Large-scale Production of U-235 (McGraw-Hill, 1951). Society: A.P.S.
Nuclear interests: Reactor design, analyses, safeguards; Fast reactors; Isotope separation, ultra-centrifuges.
Address: General Electric Co., Advanced Products Operation, 175 Curtner Avenue, San Jose, California 94566, U.S.A.

COHEN, Lionel, Ph.D., M.B., B.Ch. Educ.: Witwatersrand Univ. Chief Radiotherapist, Rad. Therapy Dept., Johannesburg Hospital.
Nuclear interest: Radiation biology.
Address: Radiation Therapy Department, General Hospital,Smit Street, Johannesburg, South Africa.

COHEN, Martin J. Vice-Pres., Franklin Systems Inc.
Address: Franklin Systems Inc., P.O. Box 3250, West Palm Beach, Florida 33402, U.S.A.

COHEN, Michael, Assoc. Prof., Thermodynamics and Statistical Mechanics, Phys. Dept., Pennsylvania Univ.
Address: Physics Department, Pennsylvania University, Philadelphia, Pennsylvania 19104, U.S.A.

COHEN, Morris, B.Sc., D.Sc. Educ.: M.I.T. Prof. Phys. Metal. (1946-62), Ford Prof. Materials Sci. and Eng. (1962-), M.I.T. Member, several Govt. panels and advisory committees. Societies: A.I.M.M.E.; A.S.M.; Indian Inst. Metals; British Inst. Metals; British Iron and Steel Inst.; Internat. Soc. Stereology; A.A.A.S.; Fellow, American Acad. Arts and Sci.; N.Y. Acad. Sci.; A.I.M.E.; Fellow, Metal. Soc., A.I.M.E.
Nuclear interests: Radioactive tracers; Materials of reactors.
Address: Massachusetts Institute of Technology, Cambridge 39, Massachusetts, U.S.A.

COHEN, Paul, B.Sc., M.Sc. Born 1912. Educ.: City Coll. of New York and Carnegie Inst. Technol. Sen. Sci., Bettis Plant, Westinghouse Elec. Corp., 1950; Manager, Chem. Process Section, 1951-53; Manager, P.W.R. Chemistry Sect., 1954-57; Manager, Chem. Development Activity, 1957-59; Manager, Chemistry Development, Westinghouse Atomic Power Dept., 1959-66; Manager, Advanced Concepts Analysis, Westinghouse Advanced Reactors Div., 1966-. Societies: A.C.S.; A.N.S.; A.S.M.E.
Nuclear interest: Nuclear power concepts.
Address: 3024 Beechwood Boulevard, Pittsburgh 17, Pennsylvania, U.S.A.

COHEN, Philip P., Dr. Member, Advisory Com. for Biol. and Medicine, U.S.A.E.C.
Address: Wisconsin University, School of Medicine, Department of Physiological Chemistry, Madison, Wisconsin, U.S.A.

COHEN, Pierre, Lic. Sci. Phys. Born 1925. Educ.: Faculté des Sci. à Paris.
Nuclear interest: General and applied radiochemistry.
Address: 37 boulevard du Cdt. Charcot, Neuilly-sur-Seine, Seine, France.

COHEN, Robert, M.Sc. (Elect. Eng.). Born 1920. Educ.: Paris Univ., M.I.T. European Manager for Tracerlab, Inc., 1949-57; European Manager for Texas Instrument, Inc., 1958-59; Commercial Director, Intertechnique, S.A., 1959-. Societies: Sté. d'Electroniciens et Radioelectriciens; I.E.E.E.
Nuclear interests: Nuclear instruments applied to physics, chemistry, biology, medicine and industry.
Address: 106 rue du Point du Jour, 92 Boulogne, France.

COHEN, Robert M., B.S.Ch.E., M.S.Ch.E. Born 1930. Educ.: Iowa State Univ. and Clarkson Coll. of Technol. Manager, Nucl. Eng., Aircraft Nucl. Propulsion Dept. (1953-60), Manager, Superheat Phys., Nucl. Energy Div. (1960-62), Consultant, Nucl. Technol., Missile and Space Div. (1962-68), G.E.C.
Nuclear interests: Space nuclear power; Isotope power; Isotope applications; Nuclear technology management; Direct energy conversion.
Address: General Electric Company, Missile and Space Division, P.O. Box 8661, Philadelphia, Pennsylvania 19101, U.S.A.

COHEN, Solly Gabriel. Born 1920. Educ.: Cambridge Univ. Staff Member, Dept. Phys. (1950-61), Prof. Phys. (1961-), Hebrew Univ., Jerusalem. Society: A.P.S.
Nuclear interests: Nuclear physics; Nuclear spectroscopy and applications to solid state physics.
Address: Department of Physics, Hebrew University, Jerusalem, Israel.

COHEN, Yves Yvan Charles, D. ès. Sci. (Etat) Born 1927. Educ.: Paris Univ.: Faculty of Sciences, Faculty of Pharmacy – Hôpital Cochin. Attaché de recherches au C.N.R.S., 1951-54; Professeur à l'Ecole Nationale de Médecine et de Pharmacie de Rouen, 1955-61, puis à Paris, 1962-; Chef du laboratoire de contrôle pharmaceutique des Radioéléments, Centre d'Etudes Nucleaires de Saclay, 1955-. Chef de la rubrique Sciences Pharmacologiques du Bulletin signalétique du C.N.R.S.; Membre correspondant de la Commission Permanente de la Pharmacopée Française; Expert pharmacologue auprès du Ministére de la Santé Publique et de la Population. Societies: Assoc. des Physiologistes de Langue Française; Sté. Française de Therapeutique et de Pharmacodynamie; Sté. chimique de France; Sté. de Biologie; Assoc. Européenne pour l'étude de la toxicité des médicaments.
Nuclear interests: Radiopharmacology; Pharmacological controls, including physical, chemical and biological studies on the radioactive isotopes and labelled compounds used in medicine; The use of radio isotopes and labelled compounds in pharmacological research; Distribution, fate and mode of action of drugs.
Address: 4 Avenue de l'Observatoire, Paris, France.

COHEUR, P., Prof. Administrateur Gérant, Centre Nat. de Recherches Métallurgiques.
Address: Centre National de Recherches Métallurgiques, Abbaye du Val-Benoît, Liège, Belgium.

COHN, Byron E., B.S., M.S., Ph.D. Born 1901. Educ.: Denver and Chicago Univs. Prof. Phys., Chairman, Denver Univ., 1943-62. Chairman, Inter-Univ. High Altitude Labs. Societies: A.P.S.; A.A.P.T.; Fellow, A.A.A.S.
Nuclear interests: Special interests in cosmic rays, luminescence, and upper atmospheric physics.
Address: Department of Physics, Denver University, Denver, Colorado 80210, U.S.A.

COHN, Charles Erwin, A.B., S.M., Ph.D. (Chicago). Born 1931. Educ.: Chicago Univ. At Argonne Nat. Lab., 1956-. Societies: A.N.S.; A.P.S.
Nuclear interests: Reactor noise and computer-aided experimentation applied to reactor physics.
Address: 445 Ridge Avenue, Clarendon Hills, Illinois 60514, U.S.A.

COHN, Stanton H., B.S., M.S., Ph.D. Born 1920. Educ.: Chicago and California (Berkeley) Univs. Head, Internal Toxicity Branch, U.S. Naval Radiol. Defence Lab., San Francisco, 1950-58; Sci., Medical Res. Centre, Brookhaven Nat. Lab., 1958-. Assoc. Editor, Rad. Res.; Editor, J. Lab. and Clinical Medicine; Member, Nat. Com. on Rad. Protection. Societies: Am. Physiology Soc.; Rad. Res. Soc.; A.A.A.S.; H.P.S..
Nuclear interests: Calcium and strontium kinetics; Compartmental analysis; Whole-body counting; Use of radioisotopes in clinical medicine.
Address: Medical Research Centre, Brookhaven National Laboratory, Upton, L.I., New York 11973, U.S.A.

COIFFARD, Jacques. Directeur Gén. Adjoint, Ing. des Travaux Publics, Omnium Technique "OTH" S.A.
Address: Omnium Technique "OTH" S.A., 18 boulevard de la Bastille, Paris 12, France.

COKER, Samuel Terry, B.S. (Auburn), M.S. (Purdue), Ph.D. (Purdue). Born 1926. Educ.: Auburn and Purdue Univs. Assoc. Prof. Pharmacology, Univ. Kansas City, 1956-59; Dean and Prof. Pharmacology, Auburn Univ. School of Pharmacy, 1959-.
Nuclear interests: Tracer studies in pharmacology and biochemistry.
Address: School of Pharmacy, Auburn University, Auburn, Alabama, U.S.A.

COLACO, Isabel DE MAGALHAES. See DE MAGALHAES COLACO, Isabel.

COLBY, Malcolm Y., Jr., B.A., M.D., M.S. Born 1916. Educ.: Texas and Minnesota Univs. Asst. Prof., Radiol., Mayo Graduate

School of Medicine, Minnesota Univ. Societies: Radiol. Soc. of North America; American Radium Soc.; American Soc. of Therapeutic Radiol.; A.C.R.; Diplomate, American Board of Radiol.
Nuclear interest: Nuclear medicine.
Address: Minnesota University, Mayo Graduate School of Medicine, Rochester, Minnesota, U.S.A.

COLDEWEY, T. C. Member, Florida Nucl. Commission.
Address: Florida Nuclear Commission, Room 113, 725 South Bronough Street, Tallahassee, Florida 32304, U.S.A.

COLE, Arthur, Ph.D. Phys., Phys. Dept., M.D. Anderson Hospital and Tumour Inst.
Address: Texas University, M.D. Anderson Hospital and Tumour Institute, Houston, Texas 77027, U.S.A.

COLE, Donovan Dennett Wilding. Born 1900. Dept. of Atomic Energy, 1946-; Deputy Director, Operations Branch, U.K.A.E.A. Ind. Group, 1959; Seconded to Chief Inspector Nucl. Installations, M.O.P., 1959-66. Society: Inst. Mech. Eng.
Address: Thellwall, Winchelsea, Sussex, England.

COLE, G. H. A., Prof., B.Sc., Ph.D., F.R.A.S., D.Sc. Formerly Theoretical Phys., now Consultant, Clarke, Chapman and Co. Ltd.
Address: Clarke, Chapman and Co. Ltd., Victoria Works, Gateshead 8, Co. Durham, England.

COLE, Leonard Jay, B.S., M.S. Born 1916. Educ.: C.C.N.Y.; New York. and California, Berkeley Univs. Asst. Specialist in Experiment Station, California Univ, Berkeley, 1948-50; Biochem., U.S. Veterans Hospital, Oakland, Calif., 1950-51; Head, Exptl. Pathology Branch, U.S. Naval Radiol. Defence Lab., San Francisco, 1951-. Assoc. Editor, Rad. Res. (Rad. Res. Soc.). Societies: A.C.S.; Rad. Res. Soc.; American Physiological Soc.; Fellow, A.A.A.S.; New York Acad. Sci.; American Soc. for Microbiol.
Nuclear interests: Modification of radiation injury; Radiation immunology, tissue and marrow-cell homotransplantation; Radiation carcinogenesis; Effects of radiation on nucleoproteins and on nucleic acid biosynthesis; Late effects of ionizing radiation; Fast neutron versus X-ray effects.
Address: 1569 Fernside Street, Redwood City, Calif., U.S.A.

COLE, Sterling, A.B. (Colgate), LL.B. (Albany Law School), Hon. LL.D. (Colgate), Hon. D.Sc. (Union Coll.). Born 1904. Educ.: Colgate Univ. and Albany Law School. U.S. Rep. in Congress, 1934-57; Chairman, Joint Com. Atomic Energy, and member, Armed Services Com.; Director-Gen., I.A.E.A., 1957-61; Former trustee: Colgate Univ., Elmira (N.Y.) College, and Woodlawn Foundation, Inc.; Consultant, atomic energy; Partner, Cole and Norris, attorneys, Washington, D.C. Vice-Pres., Nucl. Products and Services; Member, Board of Directors, Iso/Serve Inc., 1964-; Vice-Pres., Director, Chemtree Corp., 1964-.
Address: 2201 South Knoll Road, Arlington 2, Virginia, U.S.A.

COLE, Thomas Earle, B.S. (Phys.). Born 1922. Educ.: Rollins Coll., Miami, Columbia, and Tennessee Univs. Sen. Phys., O.R.N.L., 1946-. Societies: A.P.S.; A.N.S.; RESA.
Nuclear interests: Design, construction and operation of research reactors; reactor safety and control problems.
Address: 103 Disston Road, Oak Ridge, Tennessee, U.S.A.

COLE, Victor Harold Ernest. Born 1919. At (1958-), now Deputy Head, Commercial Policy, U.K.A.E.A., London.
Nuclear interests: Overseas exploitation, licensing and collaborative arrangements.
Address: United Kingdom Atomic Energy Authority, 11 Charles II Street, London S.W.1, England.

COLEMAN, John Howard, B.E.E. Born 1925. Educ.: Virginia and Princeton Univs. Pres., Rad. Res. Corp., New York, 1955-. Societies: A.N.S.; A.P.S.; Assoc. for Appl. Solar Energy; Electrochem. Soc.
Nuclear interests: Nuclear power sources and effects of radiation on dielectrics.
Address: 226½ East 62nd Street, New York 21, New York, U.S.A.

COLEMAN, Joseph Johnston, A.B. (Chem.), M.A. (Phys.), Ph.D. (Chem.). Born 1907. Educ.: Colorado Univ. Chief Eng. (1940-55), Vice-Pres., Eng. and Res. (1955-67), Burgess Battery Div., Clevite Corp.; Independent Consultant, 1968-. Societies: A.P.S.; Electrochem. Soc.; I.E.E.E.; Audio Soc. of America.
Nuclear interest: Portable power sources.
Address: 1547 West Harrison Street, Freeport, Illinois, U.S.A.

COLEMAN, Lamar William, B.S., M.S., Ph.D. Born 1934. Educ.: Oregon State Univ. Res. Assoc., Phys. Dept., Oregon State Univ., 1963-64; Phys., Lawrence Rad. Lab., Livermore, 1964-. Society: A.P.S.
Nuclear interests: Low energy nuclear physics; Beta and gamma ray spectroscopy.
Address: Lawrence Radiation Laboratory, P.O. Box 808, L-24, Livermore California 94550, U.S.A.

COLEMAN, Lester F., B.E. (Chem. Eng.). Born 1922. Educ.: Tulane Univ. Assoc. Eng., Argonne Nat. Lab. 1946-. Book: Contributor to Eng. Compendium on Rad. Shielding (I.A.E.A.). Society: Instrument Soc. of America.
Nuclear interests: Fuel processing; Fuel transportation; Shielding; Processing instrumentation; Hot cell design; Remote handling; Inert

gas purification; Gaseous dielectric breakdown.
Address: Chemical Engineering Division, Argonne National Laboratory, 9700 South Cass Avenue, Argonne, Illinois 60439, U.S.A.

COLEMAN, Robert Griffin, B.S., M.S. (Geology), Ph.D. (Geology). Born 1923. Educ.: Oregon State and Stanford Univs. Res. Geologist (1954-64), Branch Chief, Isotope Geology Branch, (1964-67), U.S. Geological Survey. Councillor, Mineralogical Soc. of America. Society: Geochem. Soc.
Nuclear interests: Natural materials and measurement of radioactive daughter products; Management of such research.
Address: 2025 Camino A1 Lago, Menlo Park, California, U.S.A.

COLEMAN, Sidney Richard, Ph.D., B.S., M.A. (Hon., Harvard Univ.). Born 1937. Educ.: California Inst. of Technol. and I.I.T. Res. Fellow (1961-66), Asst. Prof. (1963-66), Assoc. Prof. (1966-), Harvard Univ. Society: A.P.S.
Nuclear interest: Theoretical high energy physics.
Address: Lyman Laboratory, Harvard University, Cambridge, Massachusetts 02138, U.S.A.

COLES, Harold John. Born 1918. Sen. Designer, Power Station Div. (1946-55), Chief Eng., Nucl. Power Div. (1955-65), Mitchell Eng., Ltd.; Group Eng., Reactor Development Div., Atomic Power Constructions Ltd., 1965-. Societies: A.M.I.Mech.E.; A.M.I.Mar.E.
Nuclear interests: Steam cooled and boiling water reactor design for power generation and marine propulsion; Steam generating heavy reactors; Advanced gas cooled reactors and fast reactors for power generation.
Address: 44 Kenton Gardens, Kenton, Harrow, Middlesex, England.

COLGATE, Stirling A., Ph.D. Born 1925. Educ.: Cornell Univ. Rad.Lab., Berkeley (1951-52), Livermore Lab. (1952-55), Sherwood Project (1955-), Lecturer on Plasma Phys., Elec. Eng. Dept., California Univ., Berkeley; Pres., New Mexico Inst. of Mining and Technol. Deleg. to 2nd I.C.P.U.A.E., Geneva, Sept. 1958.
Nuclear interests: Astrophysics; Supernova origin of cosmic rays; Neutron star origin of supernova; Stable pinch confinement of high temperature plasmas.
Address: New Mexico Institute of Mining and Technology, Socorro, New Mexico 87801, U.S.A.

COLIN, F., M.D. Cardiac Res.Unit, Lab. des Isotopes Radioactifs, Clinique Médicale, Hôpital St. Pierre.
Address: Laboratoire des Isotopes Radioactifs, Clinique Médicale, Hôpital St. Pierre, 115 boulevard de Waterloo, Brussels, Belgium.

COLIN, Leopoldo GARCIA. See GARCIA COLIN, Leopoldo.

COLINO LOPEZ, D. Antonio, Member of Consejo, Formerly Vocale, Comisione Asesora de Reactores Industriales, now Pres., Comisione Asesora de Centrales Nucleares, Presidente en Junciones, Comisione Asesora de Biologia Vegetal y Aplicaciones Industriales, and formerly Pres., de Equipo Industrial, Junta·de Energie Nucl.; Director, Forum Atomico Espanol.
Address: c/o Junta de Energia Nuclear, Ciudad Universitaria, Madrid, Spain.

COLLATZ, Siegwart, Dr. rer. nat. Born 1936. Educ.: Humboldt-Univ., Berlin, and Dresden T.H.
Nuclear interest: Reactor theory.
Address: Deutsche Akademie der Wissenschaften zu Berlin Forschungsgemeinschaft, Zentralinstitut für Kernforschung, 8051 Dresden, PF 19, German Democratic Republic.

COLLEE, Robert F. J. M. J., Ing. Civil des Mines, Ing. Civil Metal., D. en Sci. Appliquées (Nucl. Appl. Sci.). Born 1924. Educ.: Liège Univ. Prof., Liège Univ. (non-ferrous metal., ore dressing, materials for nucl. reactors). Societies: A.I.Lg.; Sté. Roy. Belge Ing. et Ind.; Métallurgie Spéciale, Liège Univ.
Nuclear interests: Metallurgy; Ore-dressing; Materials for nuclear reactors; Nuclear chemistry; Electrochemistry; Reprocessing of irradiated fuels; Analytical chemistry; Industrial applications of radioisotopes and radiations; Reactor design; Management; Financial and economics aspects of electronuclear plants and atomic energy.
Address: 19 rue A. Ponson, Jupille, Liège, Belgium.

COLLET, M. Deleg. I.A.E.A. Symposium on Selected Topics in Rad. Dosimetry, Vienna, 7-11 June 1960; Member, Health and Safety Sub-com., E.N.E.A.
Address: Euratom, 103 avenue Franklin Roosevelt, Brussels, Belgium.

COLLEY, John Reginald, M.Sc. (Eng.), B.Sc. (Hons., Eng.). Born 1928. Educ.: Manchester Univ. Asst. Director Gen., South African Atomic Energy Board, 1957-. Society: A.M.I.Mech.E.
Nuclear interest: Reactor design and operation.
Address: South African Atomic Energy Board, Private Bag 256, Pretoria, South Africa.

COLLIE, Carl Howard, M.A., B.Sc. Born 1903. Educ.: Oxford Univ. Formerly Demonstrator, now Lecturer, Nucl. Phys. Dept., Oxford Univ. Societies: Phys. Soc.; A.P.S.
Nuclear interests: Low energy nuclear physics; Neutron physics; Radiochemistry.
Address: Nuclear Physics Laboratory, 21 Banbury Road, Oxford, England.

COLLIN, Frank F., Prof. Member, Rad. Safety Com., Wisconsin Univ.
Address: Wisconsin University, Madison 6, Wisconsin, U.S.A.

COLLINGE, Brian, B.Sc., Ph.D. Born 1921. Educ: Liverpool Univ. Reader, Nucl. Phys. Res. Lab., Liverpool Univ. 1949-; Consultant: C.E.R.N., Geneva; A.W.R.E., Aldermaston; A.E.R.E., Harwell. Nuclear interest: Instrumentation.
Address: Nuclear Physics Research Laboratory, Liverpool University, Mount Pleasant, Liverpool 3, England.

COLLINS, Clair Joseph, B.Chem. Eng., M.S. (Chem.), Ph.D. (Chem.). Born 1915. Educ.: Minnesota and Northwestern Univs. Group Leader, O.R.N.L., 1947; Prof. of Chem., Tennessee Univ., 1964-. Society: A.C.S.
Nuclear interest: Organic reaction mechanism elucidation with isotopes.
Address: Chemistry Division, Oak Ridge National Laboratory, Oak Ridge, Tennessee, U.S.A.

COLLINS, Edwin Richard, M.Sc., Ph.D. Born 1916. Educ.: Auckland and Birmingham Univs. P.S.O., U.K.A.E.A., Harwell, 1952-58; Assoc. Prof. Phys., Auckland Univ., N.Z., 1958-. N.Z. Univ. Grants Com. Society: Inst. Phys. and Phys. Soc.
Nuclear interests: Nuclear physics, particularly nucleon and nuclear polarisation phenomena.
Address: Physics Department, Auckland University, P.O. Box 2175, Auckland, New Zealand.

COLLINS, Gordon D., B.S. (E.E). Born 1924. Educ.: Washington Univ., Seattle. Elec. Eng., Stanford Univ.; Nucl. Eng. Manager, Sodium Technol., Advanced Products Operation, G.E.C.; one year assignment to C.E.A., France. Societies: A.N.S.; Atomic Ind. Forum.
Nuclear interests: Advanced experiments and testing; Sodium technology and sodium components; Fuel cycle and nuclear power plant costs; Refueling systems; Steam generator development and sodium water reactions; Development of electromagnetic pumps; Flow meters, level gauges, cold traps and sodium purity control.
Address: 1635 Kensington Avenue, Los Altos, California, U.S.A.

COLLINS, H. E., M. Eng., M. Inst. M. E., A.M.I.E.E. Member, Advisory Council on R. and D., Ministry of Power.
Address: Ministry of Power, Advisory Council on Research and Development, Thames House South, Millbank, London, S.W.1, England.

COLLINS, J. H. Formerly NRX Reactor, Operations Div., now Special Projects, Chalk River Nucl. Res. Labs., A.E.C.L.
Address: Chalk River Nuclear Research Laboratories, Chalk River, Ontario, Canada.

COLLINS, John Charles, B.Sc. (Eng.), M.S.E. Born 1930. Educ.: London and Johns Hopkins Univs. Health and Safety Branch, U.K.A.E.A., 1956-59; Lecturer in Civil Eng., Manchester Univ.; Rad. Protection Officer, Manchester Univ. Rad. Protection Service; Partner, Collins and Mason, Consulting Eng. Book: Radioactive Wastes: Their Treatment and Disposal (Spon, 1960). Societies: Inst. Civil Eng.; A.S.C.E.; Roy. Soc. Health; H.P.S.
Nuclear interests: Health physics and radiological protection, particularly public health aspects of radioactive pollution; Contamination of air, water and food and the disposal of radioactive wastes; Training courses in the above subjects.
Address: White Lodge, Leycester Road, Knutsford, Cheshire, England.

COLLIN, L. W. Gen. Manager, James Girdler and Co. Ltd.
Address: James Girdler and Co. Ltd., Mansell Works, Mansell Road, Acton Vale, London, W.3, England.

COLLINS, Otis G. Member, Illinois Commission on Atomic Energy.
Address: 3757 West 16th Street, Chicago, Illinois 60623, U.S.A.

COLLINS, Robert Dorrell. Res. Manager, Reactor Safety Group, Windscale, U.K.A.E.A.
Nuclear interests: Reactor technology; Safety analysis; Containment and filtration.
Address: 28 Hallsenna Road, Seascale, Cumberland, England.

COLLINS, T. Leo, Jun., B.S., M.S. Born 1925. Educ.: Siena Coll. Manager, Mass spectrometry, K.A.P.L. 1948-. Societies: A.N.S.; A.S.T.M., E-14.
Nuclear interests: The application of mass spectrometry to nuclear reactor technology; Design of special purpose multistage mass spectrometers with high sensitivity.
Address: Knolls Atomic Power Laboratory, P.O. Box 1072, Schenectady, N.Y., U.S.A.

COLLINS, Thomas L., Dr. Asst. Director, Cambridge Electron Accelerator, operated for U.S.A.E.C. by M.I.T. and Harvard Univ.
Address: U.S.A.E.C., Cambridge Electron Accelerator, Cambridge, Massachusetts, U.S.A.

COLLINS, William Leighton, B.S., M.S. Born 1906. Educ.: Illinois Univ. Prof. Theoretical and Appl. Mechanics, Illinois Univ.; Sec., A.S.E.E., Illinois Univ. Societies: A.S.E.E.; A.S.T.M.; A.S.C.E.
Nuclear interests: Director of the summer school programmes for engineering teachers jointly sponsored by the American Society for Engineering Education and the Atomic Energy Commission.
Address: American Society for Engineering Education, 2100 Pennsylvania Avenue, N.W., Washington, D.C. 20037, U.S.A.

COLLINSON, A. J. L., B.Sc. (Hons., Phys., London). Born 1918. Educ.: Nottingham Univ. Coll. Principal Lecturer Phys. (1947), now Head, Phys. Dept., Borough Polytechnic. Societies: Fellow, Inst. Phys. and Phys. Soc.; Fellow, Inst. Elec. Eng.

COL

Nuclear interests: Nuclear radiation detection and measurements.
Address: Borough Polytechnic, Borough Road, London, S.E.1, England.

COLLINSON, Edgar, M.A., Ph.D. Born 1921. Educ.: Cambridge Univ. Lecturer in Phys. Chem. (1952-61), Rad. Protection Officer (1959-64), Sen. Lecturer in Phys. Chem. (1961-65), Reader (1965-), Leeds Univ. Societies: Faraday Soc.; Chem. Soc.
Nuclear interests: Radiation chemistry; Health hazards from ionising radiations.
Address: Department of Physical Chemistry, Leeds University, Leeds, England.

COLMER, Frederick Charles William, M.A. Born 1920. Educ.: Cambridge Univ. Sci. Officer, A.E.R.E., Reactor Physics Div., 1947; P.S.O., R. and D. Branch, Windscale Works, U.K.A.E.A., 1950; Development Eng., Nucl. Power Branch, Central Electricity Authority, 1955; Principal Director, R. and D., (1960), Chief Res. Officer, (1963) C.E.G.B. Societies: F. Inst. P.; M. Inst. F.
Nuclear interest: Research and development in the nuclear power station field.
Address: Beverley,Burnt Common, Nr. Ripley, Surrey, England.

COLOMB, Henri. Prés. Directeur Gén., Sté. d'Etudes de Constructions et d'Equipements Industriels Modernes S.A.
Address: Société d'Etudes de Constructions et d'Equipements Industriels Modernes S.A., 27 boulevard Berthier, Paris 17, France.

COLOMBIER, Louis BUREAU du. See BUREAU du COLOMBIER, Louis.

COLOMBO, Roberto Lelio, Laurea Chem. Eng. (equivalent to M.D.). Born 1926. Educ.: Rome and Pittsburgh Univs., Polytech. School of Turin. Jun. Sci. (1950-53), Sen. Sci. (1953-55), Supv. (1955-56), Material Testing Lab., Fiat Aircraft Factories; Head, Solid State Group (1956-61), Manager, Material Development Lab. (1961-65), Fiat Nucl. Dept.; Head, Metal Phys. Dept., Centro Sperimentale Metallurgico (Exptl. Metal. Centre), 1965-67; Prof., Nucl. Materials Technology, Rome Univ., 1966-67; Guest Sci., Nat. Phys. Lab., Teddington, Middlesex (U.K.) 1966-67. Society: Associazione Italiana di Metallurgia (Italian Soc. of Metal.).
Nuclear interests: Metallurgy: Fuel elements fabrication, SS. properties; Ceramics:UO_2 sintering; Solid state: Theory of sintering and hot pressing, surface diffusion, grain boundaries, theory of recrystallisation, radiation damage in ferrous materials, fission gas release in sintered bodies.
Address: 19 Via G. Lorenzoni, 00143 Rome, Italy.

COLOMBO, Sergio, Dott. Technical Director, Soc. Elettronica Lombarda.
Address: Societa Elettronica Lombarda, 12 Via Monfalcone, Milan, Italy.

COLONNA, Marc. Member, Comite de l'Equipement Industriel (1965-), Member, Commission Consultative pour la Production d'Electricite d'Origine Nucleaire (1965-), C.E.A.
Address: Commissariat a l'Energie Atomique, 29-33 rue de la Federation, Paris 15, France.

COLONNA di PALIANO, Guido, Dr. jur. Born 1908. Educ.: Naples Univ. Member, Commission of the E.E.C., 1964-67; Member, Commission of the European Communities, 1967-.
Address: Commission of the European Communities, 23 avenue de la Joyeuse Entree, Brussels 4, Belgium.

COLONOMOS, Mentech, B.Sc. (Chem. Eng.), Dipl. Eng. (Chem. Eng.), Nucl. Eng. (I.N.S.T.N.). Born 1929. Educ.: Israel Inst. of Technol. and I.N.S.T.N. Nucl. Eng., Reactor Dept. (1959), i/c Nucl. Chem. (1964), Head, Nucl. Chem. Dept. (1966), I.V.I.C. Society: Assoc. des Ing. en Genie Atomique, Paris.
Nuclear interests: Activation analysis in biological media; Standardisation of radionuclides; Research in special production techniques of radioisotopes and instrument development.
Address: Venezuelan Institute for Scientific Research, Apartado 1827, Caracas, Venezuela.

COLSMANN, Paul, B.S. (Business and Eng. Administration; M.I.T.). Born 1922. Educ.: M.I.T. Society: A.N.S.
Nuclear interests: Management and engineering aspects of nuclear reactor operation and radioisotope production and distribution.
Address: P.O. Box 385, Brookhaven, Long Island, New York, U.S.A.

COLSON, Robert. Born 1917. Educ.: Ecole Spéciale Militaire de Saint-Cyr. Attaché Administratif et Adjoint au Directeur du Centre Technique de l'Aluminium à Paris, 1941-51. Adjoint au Directeur du Centre de Production de Plutonium de Marcoule (Gard, France).
Nuclear interest: Management of nuclear establishments.
Address: Centre de Marcoule, B.P. 106, 30 Bagnols-sur-Cèze, France.

COLTAS, Edward Kenneth, B.Sc. (Hons., Chem.). Born 1917. Educ.: Mount Allison Univ. Director, Production and Development Div., Commercial Products, A.E.C.L., 1964-.
Nuclear interests: Commercial applications of radioactivity.
Address: P.O. Box 93, Ottawa, Ontario, Canada.

COLTMAN, Ralph Read, Jr., B.S. Born 1924. Educ.: Carnegie Inst. Technol. and Tennessee Univ. Phys., O.R.N.L. Societies: A.P.S.; Sci. Res. Soc. of America.
Nuclear interest: Radiation damage.

Address: Solid State Division, Oak Ridge National Laboratory, P.O. Box X, Oak Ridge, Tennessee, U.S.A.

COLTON, Ervin, B.S., M.S., Ph.D. (Inorg. Chem.). Born 1927. Educ.: Georgia Inst. Technol., Kansas and Illinois Univs. Asst. Prof. Inorg. Chem., Georgia Inst. Technol., Atlanta, 1954-56; Principal Chem., Internat. Minerals and Chem. Corp., Chicago, 1956-58; Project Leader, High Temperature Materials Res., Allis-Chalmers Res. Div., Milwaukee, 1958-62; Manager, Cerac Sect., Allis-Chalmers New Products Div., 1962-64; Pres., Cerac, Inc., Milwaukee, 1964-; Pres., Cerac Hot-Pressing, Inc., 1967-. Societies: A.C.S.; American Ceramic Soc.
Nuclear interests: Non-metallic materials for high-temperature applications (inventor of Boron Silicide synthesis for commercial production); Corrosion resistance to high-temperature water; New borides for control devices, shielding structures, and burnable poisons; Ceramic nuclear fuels; Protective coatings for graphite.
Address: Cerac, Incorporated, Box 597, Butler, Wisconsin 53007, U.S.A.

COLVIN, D. W. Head of Neutron Data Compilation Centre, O.E.C.D., E.N.E.A.
Address: E.N.E.A. Neutron Data Compilation Centre, Saclay, France.

COLVIN, Ted Howard, B.S., M.S. Born 1920. Educ.: U.S. Naval Acad., M.I.T. Now Manager, Nucl. Systems and Space Power Dept., Bendix Aerospace Systems Div.
Nuclear interests: Reactor and weapon physics; Reactor safety; Management; Nuclear weapons effects; Radiographic, radioisotope camera (cancer diagnostics) and night vision equipment; Advanced nuclear instrumentation.
Address: Bendix Aerospace Systems Division, 3300 Plymouth Road, Ann Arbor, Michigan 48107, U.S.A.

COMAR, Cyril Lewis, B.S. (Chem., California, Berkeley), Ph.D. (Biochem., Purdue). Born 1914. Educ.: California (Berkeley) and Purdue Univs. Prof. and Head, Dept. of Phys., Biol., and Director, Lab. of Rad. Biol., Cornell Univ. Consultant to: U.N. F.A.O.; U.N. Sci. Com. on the Effects of Atomic Rad.; American Inst.of Biol. Sci.; Health and Safety Labs., U.S.A.E.C., U.S. Public Health Service, Cincinnati; O.R.I.N.S.; and Stanford Res.Inst. Books: Radioisotopes in Biology and Agriculture (McGraw-Hill, 1955); Editor, Atomic Energy and Agriculture (A.A.A.S. symposium); Editor, Mineral Metabolism – An Advanced Treatise. Societies: A.C.S.; Rad. Res. Soc.; Soc. Exptl. Biol. Med.; American Soc. Biol. Chemists; American Inst. Nutrition; A.A.A.S.
Nuclear interests: Research and educational aspects of atomic energy in biology.
Address: Department of Physical Biology, Cornell University, Ithaca, New York, U.S.A.

COMAS, Manuel M. URGELL-. See URGELL-COMAS, Manuel M.

COMBET, Georges, Ing. (E.P.), Ponts et chaussées (Ecole Supérieure d'Electricité). Born 1895. Educ.: Paris Univ. Directeur Gén. Honoraire, Gaz de France; Chairman, G.I.I.N. (Groupe Intersyndical de l'Industrie Nucléaire); Chairman, Comité Français, World Power Conference; Chairman, Table ronde nucléaire, Council of European Ind. Federations (C.I.F.E.).
Address: G.I.I.N., 15 rue Beaujon, Paris 8, France.

COMBI, John Vincent, B.S. Born 1933. Educ.: C.C.N.Y. and Texas Univ. Radiol. Hygienist, Div. of Occupational Health and Rad. Control, City of Houston Health Dept., 1961-. Sec.-Treas., Gulf Coast Sect., A.I.H.A. Societies: H.P.S.; American Conference of Governmental Ind. Hygienists.
Nuclear interest: Health physics or radiation safety.
Address: City of Houston Health Department, Division of Occupational Health and Radiation Control, 1115 N. MacGregor, Houston, Texas 77025, U.S.A.

COMBO, John X., B.S. (Geolological Eng.), LL.B., Professional Eng. Born 1922. Educ.: Montana School of Mines and Georgetown Law School. Geologist (1947-50), Attorney, (1950-52), U.S. Geological Survey, Washington, D.C.; Attorney, Grand Junction Office (1951-61), Chief Counsel, Grand Junction Office (1961-62), Chief Counsel, Idaho Operations Office (1962-), U.S.A.E.C. Chapter Organizer, Idaho Chapter, Federal Bar Assoc. Book: Co-author, American Law of Mining, (Matthew Bender and Co., Inc., 1960). Society: Geological Soc. of Washington.
Nuclear interests: Atomic energy law; Government procurement matters; Mining law.
Address: U.S. Atomic Energy Commission, Idaho Operations Office, P.O. Box 2108, Idaho Falls 83401, Idaho,U.S.A.

COMBRISSON, Jean. Born 1924. Formerly Adjoint au chef, Departement de Physique Nucléaire et de Physique du Solide, Centre d'Etudes Nucléaires, Saclay.
Address: Direction de la Physique, C.E.N.S., Saclay, Boite Postale No. 2, Gif-sur-Yvette, (Essonne), France.

COMBS, Cecil Edward, B.S., M.A. Hon. Ed. D. (Dayton Univ.). Born 1912. Educ.: Chicago Univ., U.S. Military Acad. and George Washington Univ. Commandant, Inst. of Technol., U.S.A.F., -1965; Assoc. Provost. Rochester Univ., 1965-. Societies: A.S.E.E.; Nat. Education Assoc.
Nuclear interests: Graduate education in astronautical and nuclear engineering; Special programmes in continuing scientific, engineering and management education.

COM

Address: Rochester University, Rochester, New York, U.S.A.

COMEFORO, Jay Eugene, B.S., M.S., Ph.D. (Ceramics). Born 1922. Educ.: Rutgers and Illinois Univs. Sect. Head, Synthetic Minerals Sect., U.S. Bureau of Mines, 1949-54; Eng.-in-Charge, Ceramics, Sylvania Electric Products, Woburn, Mass., 1954-55; Manager, Marketing and Eng., Frenchtown Porcelain Co., 1955-59; Consultant in Ceramics, 1959-61; Pres., Consolidated Ceramics and Metalizing Corp., 1961-. Societies: American Ceramic Soc.; Mineralogical Soc. of America; A.S.T.M.; Keramos. Officer of several sections and committees in these organisations.
Nuclear interest: Applications of ceramic materials in reactors, primarily on precise dimensioned BeO and ceramic-to-metal assemblies.
Address: Foxgrape Road, Flemington, New Jersey, U.S.A.

COMES, Salvatore, Dr. Gen. Director, Higher Education, Ministry of Education, Italy, 1965-. Member, Steering Commission, C.N.E.N., 1965-.
Address: Direttore Generale dell Istruzione Superiore, Ministero della Pubblica Istruzione, 76/A Viale Trastevere, 00100 Rome, Italy.

COMETTO, Mario, Degree in Phys. (specialisation in nucl. energy). Born 1920. Educ.: Turin Univ. and Milan Politecnico. With Soc. P.C.E. (S.I.P. holding), 1934-; posted at Soc. Elettronucleare Nasionale, Rome, 1957-; E.N.E.L., 1963-. Societies: Assoc. Elettrotecnica Italiana; Assoc. Fisica Sanitaria; Assoc. Nazionale Ingegneria Nucleare; F.I.E.N. Nuclear interests: Operation of nuclear power plants; Economic in nuclear energy.
Address: E.N.E.L.-D.P.T., 2 via G.B. Martini, Rome, Italy.

COMHAIRE, Mrs. F. BOUILLENNE-. See BOUILLENNE-COMHAIRE, Mrs. F.

COMHAIRE, Madame S. CHEVREMONT-. See CHEVREMONT-COMHAIRE, Madame S.

COMINGS, Edward Walter, B.S. (Chem. Eng.), Sc.D. (Chem. Eng.). Born 1908. Educ.: Illinois Univ. and M.I.T. Prof., Chem. Eng., Illinois Univ., 1947-51; Head, School of Chem. and Metal. Eng., Purdue Univ., 1951-59; Dean of Eng., Coll. of Eng., Delaware Univ., 1959-. Societies: A.C.S.; Fellow, A.A.A.S.; A.S.E.E.
Address: Delaware University, College of Engineering, Newark, Delaware, U.S.A.

COMMANDER, Robert E., B.S. (Process Eng.). Born 1928. Educ.: California Univ. Chem. Eng., Shell Oil Co., 1951-58; Production Eng., Operations Branch, Chem. Processing Plant, Phillips Petroleum Co., Atomic Energy Div., Idaho Falls, 1958-. Society: Idaho Section, A.I.Ch.E.

294

Nuclear interests: Spent fuel reprocessing and processes for conversion of waste liquids to solids.
Address: 847 Saturn Avenue, Idaho Falls, Idaho, U.S.A.

COMMELIN, Jean. Chairman, Cie. pour l'Etude d'Equipements Industriels S.A.
Address: Cie. pour l'Etude d'Equipements Industriels S.A., 12 rue de la Pierre Levée, Paris, France.

COMMERMAN, Marcel. Member, Comité de Gestion, Syndicat Belge d'Assurances Nucléaires.
Address: Syndicat Belge d'Assurances Nucléaires, 4 Place de Louvain, Brussels, Belgium.

COMMON, D. K., M.A. Mech. Design Eng., Nucl. Plant Design Branch, C.E.G.B.
Address: Central Electricity Generating Board, Nuclear Plant Design Branch, Walden House, 24 Cathedral Place, London E.C.4, England.

COMPERE, Edgar Lattimore, A.B., M.S., Ph.D. Born 1917. Educ.: Quachita Coll. and Louisiana State Univ. Asst. Prof. Chem. (1946-51), Assoc. Prof. Chem. (1951), Louisiana State Univ.; Sen. Chem, (1951-), Chief, Slurry Materials Sect., Reactor Exptl. Eng. Div. (1958), Heterogeneous Systems Sect., Reactor Chem. Div. (1960), O.R.N.L. Societies: A.N.S.; A.C.S.; N.A.C.E.; Fellow, A.A.A.S.
Nuclear interests: Integrity of reactor fuels and materials; Reactor chemistry; Alkali metals chemistry; Irradiation studies of molten salt fuel chemistry.
Address: Oak Ridge National Laboratory, Oak Ridge, Tennessee, U.S.A.

COMTE, Pierre. Vice-Chairman and Director, Compagnie des Ateliers et Forges de la Loire. Nuclear interests: General.
Address: Compagnie des Ateliers et Forges de la Loire, C.A.F.L., Département Energie Nucléaire, 12 rue de la Rochefoucauld, Paris 9, France.

CON, Ton That. Dept. of Electronics, Dalat Nucl. Res. Centre, Atomic Energy Office of the Republic of Vietnam.
Address: Atomic Energy Office of the Republic of Vietnam, Dalat Nuclear Research Centre, Dalat, Vietnam.

CONARD, Robert Allen, B.S., M.D. Born 1913. Educ.: South Carolina Univ. and South Carolina Medical Coll. Sen. Sci., Medical Res. Centre, and Chief, Medical Surveys Marshallese People Exposed to Fallout Radiation, Brookhaven Nat. Lab. Societies: Rad. Res. Soc.; A.A.A.S.; N.Y. Acad. Sci.; Soc. of Nucl. Medicine; Tissue Culture Soc.
Nuclear interests: Radiation effects on animals and human beings; Gastrointestinal effects of radiation; Effects of radioactive fallout on Marshall Islanders; Action of radioactive phytohaemagglutinin on lymphocyte

transformation in culture.
Address: Medical Research Centre, Brookhaven National Laboratory, Upton, Long Island, New York 11973, U.S.A.

CONDIT, Gomer. Formerly Resident Eng., now Supt. of Power Plant Construction, Idaho Power Co.
Address: c/o Idaho Power Company, P.O. Box 770, Boise, Idaho 83701, U.S.A.

CONDON, Edward Uhler, A.B., Ph.D., D.Sc. Born 1902. Educ.: California Univ. Prof. Phys., Colorado Univ., Boulder, Colorado. Editor, Reviews of Modern Phys. (American Phys. Soc.). Books: Co-author, The Theory of Atomic Spectra (Cambridge University Press, 1935); co-editor, Handbook of Phys. (McGraw-Hill, 1958). Societies:Past Pres., A.P.S.; Past Pres., A.A.A.S.
Nuclear interest: Fundamental nuclear theory.
Address: 761 Cascade Avenue, Boulder, Colorado 80302, U.S.A.

CONGDON, Charles C., A.B., M.S. (Pathology), M.D. Born 1920. Educ.: Michigan Univ. Exptl. Pathologist, Biol. Div., O.R.N.L., 1955-. Societies: Soc. of Exptl. Biol. and Medicine; American Assoc. of Pathologists and Bacteriologists; Rad. Res. Soc.
Nuclear interest: Treatment of radiation injury.
Address: Biology Division, Oak Ridge National Laboratory, Oak Ridge, Tennessee, U.S.A.

CONGER, Alan Douglas, B.A. (Hons.), M.A., Ph.D. (Harvard). Born 1917. Educ.: Harvard Coll. and Harvard Univ. Sen. Res. Biol., O.R.N.L., 1947-52, 1953-58; Fulbright Sen. Res. Scholar, London, 1952-53; Prof. Rad. Biol., Florida Univ., 1958-; Director, Radiobiol. Lab., School of Medicine and Hospital, Temple Univ. Editor, Rad. Res.; Hon. Editor, Rad. Biol. Societies: Rad. Res. Soc.; Biophys. Soc.
Nuclear interests: Radiation biology; Modification of biological radiosensitivity; Free radicals and post-radiation biological damage.
Address: Radiobiology Laboratory, School of Medicine and Hospital, Temple University, Philadelphia, Pennsylvania 19122, U.S.A.

CONKIE, William Rodgers, B.A.Sc., M.Sc., Ph.D. Born 1932. Educ.: Toronto, McGill and Saskatchewan Univs. Asst. Res. Officer, A.E.C.L., Chalk River, 1956-60; Asst. Prof., Queen's Univ., Kingston, Ontario, 1960-. Societies: A.P.S.; Canadian Assoc. Phys.
Nuclear interest: Theoretical physics.
Address: Department of Physics, Queen's University, Kingston, Ontario, Canada.

CONKLIN, Mrs. Marie Eckhardt, B.A., M.S., Ph.D. Born 1908. Educ.: Wellesley Coll., Wisconsin and Columbia Univs. Assoc. Director, A.E.C.-N.S.F. Summer Inst. 1958, 1959, 1960, 1961 and 1963. Res. Collaborator, Brookhaven Nat. Lab.; Formerly Prof., Biol., Dept. of Biol., Adelphi Univ. Societies: A.A.A.S.; A.I.B.S.; Botanical Soc.; Genetical Soc.
Nuclear interests: Cytogenetic and physiological effects of radiation on cells in plant tumor studies.
Address: 18 First Street, Garden City, New York 11530, U.S.A.

CONN, G. K. T., Prof. M.A., Ph.D. Head, Phys. Dept., Univ. Exeter.
Address: Department of Physics, Exeter University, The Washington Singer Labs., Prince of Wales Road, Exeter, England.

CONNELL, Richard P. Business Manager, E.O. Lawrence Rad. Lab., California Univ.
Address: E. O. Lawrence Radiation Laboratory, California University, Berkeley 4, California, U.S.A.

CONNELLY, T. F. Director, Res. Materials Information Centre, O.R.N.L., operated for U.S.A.E.C. by Union Carbide Corp.
Address: Oak Ridge National Laboratory, Post Office Box X, Oak Ridge, Tennessee, U.S.A.

CONNER, Willard P., Jr., B.S. (Colorado), Ph.D. (Wisconsin). Born 1915. Manager, Hercules Res. Centre, 1948-55, 1956-; leave of absence to serve as Tech. Director, U.S.A.E.C. Materials Testing Reactor, Arco, Idaho, 1955; U.S.A.E.C. Advisory Com. on Reactor Safeguards, 1951-; Tech. Asst. to Director (1960-), Res. Assoc. (1963-), Hercules Res. Centre. Society: A.N.S.
Nuclear interests: Reactor safety; Radiation chemistry.
Address: Atwater Road, R.D.1, Chadds Ford, Pennsylvania, U.S.A.

CONNERT, Winfried, Dr.-Ing. Born 1913. Educ.: Freiberg Bergakademie. Deutsche Edelstahlwerke A.G., -1964; Member, Directorate, Röchling'sche Eisen- und Stahlwerke G.m.b.H., 1964-65; Tech. Board Member, Deutsche Edelstahlwerke A.G., 1965-. Member, Directorate, Deutscher Normenausschuss; Member, Board, Verein Deutscher Eisenhüttenleute; Member, Administrative Board, Bundesverband der Deutschen Luft- u. Raumfahrtindustrie e.V.; Member, Administrative Board, A.E.G. -Elotherm G.m.b.H.; Member, Administrative Board, Continental Titanium Metals Corp. G.m.b.H.
Nuclear interest: Metallurgy.
Address: 20 Finkenweg, 415 Krefeld, Germany.

CONNOLLY, Thomas J., B.Ch.E., M.S., Ph.D. Born 1923. Educ.: Syracuse Univ. and Carnegie and California Technol. Insts. Prof. Mech. Eng., and Director Nucl. Div. Stanford Univ. Society: A.N.S.
Nuclear interest: Nuclear reactor theory and design.
Address: Nuclear Division, Department of Mechanical Engineering, Stanford University, Stanford, California, U.S.A.

CONNOLLY, W. H. Chairman, Elec. Commission, Victoria. Adviser, Third I.C.P.U.A.E. Conference, 1964. Member, Advisory Com., A.A.E.C.
Address: Electricity Commission of Victoria, 22, William Street, Melbourne, Victoria, Australia.

CONNOR, Gerald B., B.S. (U.S. Naval Acad.), M.S. (U.S.A.F. Inst. Technol.). Born 1927. Educ.: U.S. Naval Acad. and U.S. Air Force Inst. Technol. Major, U.S.A.F., 1951-. Aerospace Nucl. Safety Analyst, Div. Reactor Development, U.S.A.E.C., Washington 25, D.C., 1959-.
Nuclear interest: Nuclear safety of aerospace applications of nuclear energy.
Address: Division of Reactor Development Headquarters, U.S.A.E.C., Washington 25, D.C., U.S.A.

CONNOR, Robert Dickson, B.Sc., Ph.D. Born 1922. Educ.: Edinburgh Univ. Lecturer in Phys., Edinburgh Univ., 1949-57; Assoc. Prof. (1957-60), Prof. (1960-), Assoc. Dean, Arts and Sci. (1963-), Manitoba Univ. Society: A. Inst. P.
Nuclear interests: Alpha, beta and gamma ray spectroscopy; Low energy nuclear physics.
Address: 638 Elm Street, Winnipeg 9, Manitoba, Canada.

CONNORS, Philip I., Ph.D., M.S., B.S. Born 1937. Educ.: Pennsylvania State and Notre Dame Univs. Graduate Asst. (1959-63), Instructor (1963), Pennsylvania State Univ.; Jun. Res. Assoc., Brookhaven Nat. Lab., 1963-65; Res. Assoc., Phys., Maryland Univ., 1965-. Society: Nucl. Div., A.P.S.
Nuclear interest: Nuclear physics and instrumentation.
Address: Department of Physics and Astronomy, Maryland University, College Park, Maryland 20742, U.S.A.

CONRAD, E. E. Vice Chairman, Administrative Com., Nucl. Sci. Group, Inst. of Elec. and Electronic Eng.
Address: Institute of Electrical and Electronic Engineers, Nuclear Science Group, 345 East 47th Street, New York, New York 10017, U.S.A.

CONRAD, James W., A.B., M.S. Born 1925. Educ.: St. John's Coll., Annapolis and Carnegie Inst. Technol. Manager, Field Services, Artisan Industries, Inc., Boston, 1953-60; Vice-Pres. and Director, Mt. Pleasant Foundries, Inc., Pa., 1949-53; Director, Cleveland Instrument Co., Cleveland, Ohio, 1952-59; Director, Henolite Products, Inc., Phila., Pa., 1961-. Director, Terra-jets Corp., Phila., Pa., 1963-. Consulting Eng., 1960-. Membership Chairman, A.N.S., 1960-62; Chairman, Shell-Mold Process Com., American Foundry Assoc., 1952. Societies: A.N.S.; A.I.Ch.E.; American Ordnance Assoc.
Nuclear interests: Consulting in use of radiation for plastics and polymerisation; Engineering and production of process equipment for fuel reprocessing; Evaluation of high temp. fibre fuel elements; Acting as a consultant to two nuclear oriented firms.
Address: 1234 Denniston Avenue, Pittsburgh 17, Pa., U.S.A.

CONRATH, Joseph John, B. Eng. (Mech.), M. Eng. (Mech.). Born 1926. Educ.: McGill Univ. At A.E.C.L.
Nuclear interests: Design and development of reactor components.
Address: Sheridan Park, Ontario, Canada.

CONSOLO, F., Director, Industrial and Economic Div., U.S.A.-Euratom Dept. Formerly Special Adviser to the Commission, Euratom.
Address: 51-53 rue Belliard, Brussels, Belgium.

CONSTANT, René Joseph Pierre, Dr. en Sc. Chimiques. Born 1926. Educ.: Brussels Free Univ. Chief, Radioisotope Dept., Centre d'Etude de l'Energie Nucléaire, Mol, Belgium, 1955. Appointed lecturer of Radio chemistry, Univ. du Travail de Charleroi, Charleroi, Belgium. Appointed lecturer for the application of Radioisotopes, Brussels Free Univ.
Nuclear interests: Radioisotopes: production, research, applications. Hot chemistry.
Address: Centre d'Etude de l'Energie Nucléaire (C.E.N.), Mol, Belgium.

CONTANT, Rudolf Bernard, Lanbouwk. Ir. (M.Sc.) Born 1935. Educ.: Wageningen Univ. Asst., Coffee Div., Inst. Nat. pour l'Etude Agronomique du Congo, 1958-59; Head, Breeding and Genetics, Pyrethrum Res. Station, Ministry of Agriculture and Animal Husbandry, Molo, Kenya, 1961-63; Radiogeneticist, Assoc. Euratom-Inst. voor Toepassing van Atoomenergie in de Landbouw, Wageningen, 1963-; Adviser, to Netherlands and Kenya Govts. on Bilateral Development Aid Projects, 1966-. Societies: Founder Member, East African Acad.; Netherlands Radiobiol. Soc.; European Soc. for Rad. Biol.
Nuclear interests: Research management; Radiogenetics of plants (mainly tomato and arabidopsis thaliana); Mutation breeding; Radiation sterilisation of insects.
Address: Institute for Atomic Sciences in Agriculture, P. O. Box 48, Wageningen, Netherlands.

CONTE, Fernand, Ing. Arts et Métiers. Born 1923. Educ.: E.N.S.I.A.M. Aix-en-Provence. Ing. Service Construction des Piles, Chef, Ensemble Industriel G.3, Chef Service Ensembles Industriels G.2/G.3 à Marcoule, C.E.A.
Nuclear interest: Construction et exploitation de réacteurs.
Address: Commissariat à l'Energie Atomique, Centre de Marcoule, Service G2/G3, B.P. No. 106, Bagnols sur Cèze, Gard, France.

CONTREIRAS, J. CORREIRA. See CORREIRA CONTREIRAS, J.

CONTRERAS CHAVEZ, Juan, Dr. Salvador Resident Representative, I.A.E.A. Address: 8/11 Opernring, Vienna 1, Austria.

CONTZEN, Jean-Pierre, Ing. Civil Electricien Mécanicien. Born 1935. Educ.: Brussels Univ. Head of Sect., C.E.N., Mol, 1958-64; Responsible for Advanced Systems, Directorate Gen. for Studies and Exptl. Work, European Space Vehicle Launcher Development Organisation, 1964-; Consultant to E.N.E.A. Societies: A.N.S.; Deutsche Gesellschaft für Raketentecknik und Raumfahrt.
Nuclear interests: Space applications of nuclear energy; Nuclear safety; Land, space and undersea applications of isotopic power; Conversion systems.
Address: Résidence Bellevue, 36 rue Ernest Renan, 92 Meudon, France.

CONVERSI Marcello. Born 1917. Prof. Phys., Rome Univ. Socio corrispond. Accademia Naz. dei Lincei; Vice-Pres., I.N.F.N.; Fellow, A.P.S.; Member, Editorial Boards of "Il Nuovo Cimento" and "Acta Physica Austriace". Papers: Coauthor, Experimental proof that cosmic ray mesons are not Yukawa particles (1947); co-author, Development of first pulsed track detector, spark-chamber like (1955); co-author, Study of time-like electromagnetic structure of the nucleon (1963).
Address: Istituto di Fisica, Rome University, 00100 Rome, Italy.

CONVEY, John, B.Sc. (Phys.), M.Sc. (Phys.), Ph.D. (Atomic Phys.), D.Sc. Born 1910. Educ.: Alberta and Toronto Univs. Assoc. Prof. Phys., Toronto Univ., 1946-48; Chief, Phys. Metal. Div. (1948-51), Director (1951-), Mines Branch, Ottawa. Past Chairman, Electrochem. Soc., Canada; Member, Administrative Board, Canadian Welding Bureau; Member, Tech. Council, Canadian Standards Assoc.; Member of Board, A.S.M.; Past Pres., A.S.M.; Board of Directors, Canadian Standards Assoc.; Chairman, Commonwealth Com. on Mineral Processing. Societies: Fellow, Inst. Phys. G.B.; Faraday Soc.; Canadian Inst. Mining and Metal.; Electrochem. Soc. America; A.P.S.; A.I.M.E.; Inst. Mining and Metal., G.B.
Nuclear interests: Metallurgy and nuclear explosives. Actively engaged in the development of materials for nuclear reactors and associated research.
Address: Mines Branch, 555 Booth Street, Ottawa, Canada.

CONWAY, John T., B.S. (Eng.), Ll.B. Born 1924. Educ.: Tufts Coll., and Columbia School of Law. Special Agent, Federal Bureau of Investigation, 1950-56; Staff Member (1956-58), Asst. Director (1958-62), Exec. Director (1962-68), Joint Com. on Atomic Energy; Exec. Asst. to Chairman of the Board, Consolidated Edison Co. of New York, 1968-.
Address: Consolidated Edison Company of New York, 4 Irving Place, New York 3, New York, U.S.A.

CONWELL, William Anthony, B.S.(Civil Eng.), M.S. (Civil Eng.). Born 1908. Educ.: Carnegie Inst. Technol., Pittsburgh. Gen. Eng. (1950-54), Structural Eng. (1954-58), Vice-Pres. (1958-), Duquesne Light Co.; Instructor, Carnegie Inst. Technol., 1950-58. Societies: A.S.C.E.; A.S.E.E.; Soc.of American Military Eng.; Eng. Soc. of Western Pennsylvania; Pennsylvania Soc. of Professional Eng.; Eng. Council for Professional Development.
Nuclear interests: Management of engineering and construction of nuclear power stations.
Address: Duquesne Light Company, 435 Sixth Avenue, Pittsburgh 19, Pennsylvania 15219, U.S.A.

COOBS, John Henry, B.Sc. Born 1923. Educ.: Iowa State Univ. Ceramic Eng., NEPA Project, Oak Ridge, 1949-51; Metall., 1951- (assigned to Dragon Project, A.E.R.E., Winfrith, Dorchester, England, (1962-64), now Asst. Director, A.G.R. Programme, O.R.N.L. Societies: A.S.M.; A.I.M.M.E.; A.N.S.; A.A.A.S.
Nuclear interests: Materials problems pertaining to gas-cooled reactors, especially those problems concerning fuel elements, control rods, and moderating materials.
Address: Oak Ridge National Laboratory, Post Office Box X, Oak Ridge, Tennessee 37830, U.S.A.

COOK, B. B. Manager, Fed. and Marine Sales, De Laval Turbine Inc.
Address: De Laval Turbine Inc., 853 Nottingham Way, Trenton, New Jersey 08602, U.S.A.

COOK, Charles Falk, Ph.D. Born 1928. Educ.: Rice Inst., Houston, Texas. Asst. Prof. Phys., Florida Univ., 1954-55; Supervisor, Nucl. Programme, Convair, Fort Worth, 1955-58; Manager, Nucl. Phys. Sect. (1958-63), Manager, Phys. Branch (1963-), Phillips Petroleum Co. Societies: A.P.S.; A.A.A.S.
Nuclear interests: Management; Nuclear physics research, applied and fundamental.
Address: Radiation Laboratory, Phillips Petroleum Company, Bartlesville, Oklahoma, U.S.A.

COOK, Charles Jacob, B.S. (Elec. Eng.), M.A. (Phys.), Ph.D. (Phys.). Born 1924. Educ.: Nebraska Univ. Employed (1955-), Director, Chem. Phys. Div. (1962-) Exec. Director, Phys. and Chem. Phys. Labs. (1966-), Stanford Res. Inst.; Sen. Res. Sci., Queen's Univ. Belfast, 1962-63. Societies: A.I.P.; A.A.P.T.; A.I.A.A.; American Ordnance Assoc.
Nuclear interests: Study of atomic and molecular collision processes.
Address: Stanford Research Institute, Menlo Park, California, U.S.A.

COOK, Clarence Sharp, A.B. (De Pauw). M.A. (Indiana), Ph.D. (Indiana). Born 1918.

Educ.: De Pauw and Indiana Univs. Asst. Prof. Phys., Washington Univ., St. Louis, 1948-53; Head, Nucl. Rad. Branch (1953-60), Head, Nucleonics Div. (1960-61), Phys. Consultant to Sci. Director (1962-65), Head, Rad. Phys. Div. (1965-), U.S. Naval Radiol. Defence Lab., San Francisco; Fulbright Fellow, Aarhus Univ., Denmark, 1961-62. Societies: Fellow, A.P.S.; American Geophys. Union; Fellow, California Acad. Sci.; A.A.P.T.
Nuclear interests: Beta and gamma ray spectroscopy Radiations from nuclear weapons.
Address: United States Naval Radiological Defence Laboratory, San Francisco, California 94135, U.S.A.

COOK, David C., B.S. (E.E.). Born 1912. Educ.: Washington Univ. Head, Nucl. Systems Branch, N.R.L., Subcom. Instruments and Techniques, Com. Nucl. Sci., N.A.S. – Nat. Res. Council; Sec., U.S.A.S.I. Sect. Com. N42; U.S. Rep., Internat. Electro-tech. Commission; Rad. Detectors Com., Standards Com., Group on Nucl. Sci., I.E.E.E.; Scintillation and Semiconductor Counter Symposium Com. Societies: I.E.E.E.; RESA.
Nuclear interests: Detectors for nuclear radiation, particularly low-level detectors for neutron and gamma rays; Neutron activation and gamma ray fluoroscopy.
Address: Naval Research Laboratory, Washington D.C.20390, U.S.A.

COOK, Eugene. Attorney General, State of Georgia Nucl. Advisory Commission.
Address: State of Georgia Nuclear Advisory Commission, Office of Publications, Georgia Institute of Technology, Atlanta 13, Georgia, U.S.A.

COOK, F. A. Eng. Manager, Power Div., United Eng. and Constructors Inc.
Address: United Engineers and Constructors Inc., 1401 Arch Street, Philadelphia, U.S.A.

COOK, Gerald B., Ph.D. Educ.: Birmingham Univ. Chemistry Div. (1947-54), Head, Chem. Group, Isotope Res. Div. (1954-60), U.K.A.E.A., Harwell; Chief Chem., I.A.E.A., 1960-. Book: Co-author, Modern Radiochem. Practice.
Address: International Atomic Energy Agency, Seibersdorf, Nr. Vienna, Austria.

COOK, H. L. Jr., Formerly Chief Eng., now Vice Pres., Ohmart Corp.
Address: The Ohmart Corporation, 4241 Allendorf Drive, Cincinnati, Ohio 45209, U.S.A.

COOK, Harold Frederick, B.Sc., Ph.D. Born 1913. Educ.: London Univ. Reader in Phys., London Univ., at Middlesex Hospital Medical School. Societies: British Inst. of Radiol.; Hospital Phys. Assoc.
Nuclear interests: Radiation biology and protection; Radiological physics.
Address: 56 Twickenham Road, Teddington, Middlesex, England.

COOK, Leslie Gladstone, B.A. (Toronto), Dr. rer. nat. (Berlin). Born 1914. Educ.: Toronto and Berlin Univs. Director, Chem. and Metal. Div., A.E.C.L. -1956; Project Analyst, G.E. Res. Lab., Schenectady, 1956-59; Manager, Project Analysis Sect., G.E., 1959-. Sec., Northeastern New York Sect., A.N.S. Societies: A.C.S.; Chem. Inst. of Canada; A.N.S.
Address: 16 Cedar Lane, Scotia 2, New York, U.S.A.

COOK, Maurice, D.Sc., Ph.D., Hon. A.C.T. (Birm.), Hon. C.G.I.A. Born 1897. Educ.: Manchester and Cambridge Univs. Chairman, Metals Div., Imperial Chem. Industries, Ltd., 1957-59, Chairman, British Non-Ferrous Metals Research Assoc., 1955-59; Pres., Aluminium Development Assoc., 1957-58. Societies: F.I.M.; Fellow, Inst. Metal.; Inst. of Welding; Distinguished Life Member, A.S.M.; Hon. Life Member, A.I.M.M.E. Nuclear interest: Metallurgy.
Address: 48 Rocky Lane, Birmingham 22b, England.

COOK, P. D., Ph.D. Dept. of Phys. applied to Medicine, Middlesex Hospital Medical School.
Address: The Middlesex Hospital Medical School, Cleveland Street, London W.1, England.

COOK, R., M.Sc. Member, Joint Nucl. Marine Propulsion Propulsion Panel.
Address: Institute of Marine Engineers, 76 Mark Lane, London, E.C.3, England.

COOK, Richard W., B.S. (Civil Eng.). Born 1907. Educ.: Michigan State Univ. Manager, Oak Ridge Operations (1949-51), Director of Production (1951-54), Asst. Gen. Manager, Manufacturing (1954-55), Deputy Gen. Manager (1955-58), U.S.A.E.C.; Vice-Pres., Deputy Group Exec. and Titan Programme Director, Director of Administration and Atomics, Special Projects and Group Exec. A.M.F. Atomics, A.M.F., 1958-64; Asst. to Deputy Director, Tech. (1964-), Asst. Director, R. and D. Operations, Marshall Space Flight Centre. Societies: Registered Professional Eng.; A.S.C.E. Nuclear interest: Management.
Address: 1317 Forbes Drive, S.E. Huntsville, Alabama, U.S.A.

COOK, Thomas B., Jr. Director, Phys. and Mathematics Res. (-1967), Vice Pres., i/c, Res. Programmes (1967-), Sandia Labs., Sandia Corp.
Address: Sandia Corporation, Sandia Laboratories, P.O. Box 969, Livermore, California 94551, U.S.A.

COOK, Sir William Richard Joseph, M.Sc., F.R.S. Born 1905. Educ.: Bristol. Chief, Royal Naval Scientific Service, Admiralty, 1950-54; Deputy Director, A.W.R.E., 1954-57; Member for Engineering and Production (1957-59), Member for Development and Eng. (1959-61), Member for Reactors (1961-64), U.K.A.E.A. Deputy Chief Sci. Adviser (1964-67), Chief Adviser, Projects (1967-), Ministry of Defence.

Address: Adbury Springs, Newbury, Berks., England.

COOKE, C. H. Tech. Director, Electro-Mechanical Div., M.B. Metals Ltd.
Address: M.B. Metals Ltd., Electro-Mechanical Division, Victoria Road, Portslade, Sussex, England.

COOKE, L. F. Director, Electronic Instruments Ltd.
Address: Electronic Instruments Ltd., Richmond, Surrey, England.

COOKE-YARBOROUGH, Edmund Harry, M.A. Born 1918. Educ.: Oxford Univ. Head, Electronics and Appl. Phys. Div., A.E.R.E., Harwell, 1957-. Book: An Introduction to Transistor Circuits (Oliver and Boyd, 1960). Societies: Fellow, I.E.E.; I.E.E.E.; F. Inst. P. Nuclear interests: Nuclear instrumentation; Information processing; Non-linear electronic circuits; Direct conversion of nuclear heat to electricity.
Address: Electronics and Applied Physics Division, Atomic Energy Research Establishment, Harwell, near Didcot, Berks., England.

COOL, Rodney L. Assoc. Director, High Energy Phys., Brookhaven Nat. Lab., 1966-. Member, High Energy Phys. Advisory Panel, U.S.A.E.C., 1967-.
Address: Brookhaven National Laboratory, Upton, Long Island, New York, U.S.A.

COOLEY, H. J. Chief Eng., Nucl. Div., Nucl. Corp. of America, Inc.
Address: Nuclear Corporation of America, Inc., Nuclear Division, 2 Richwood Place, Denville, New Jersey, U.S.A.

COOLIDGE, Charles A. Director, Yankee Atomic Elec. Co.
Address: Yankee Atomic Electric Co., 441 Stuart Street, Boston 16, Massachusetts, U.S.A.

COOMBER, Denys Irvine, B.Sc., Ph.D., A.R.I.C. Born 1914. Educ.: London Univ. P.S.O. (1954-61), S.P.S.O. (1961-), Lab. of the Government Chem. (Div. Head of Radiochem. Div.). Societies: A.R.I.C.; Fellow, Chem. Soc.; Faraday Soc. Nuclear interests: Health physics; Low level environmental radioactivity.
Address: Laboratory of the Government Chemist, Cornwall House, Stamford Street, London, S.E.1, England.

COOMBS, Douglas McLeod. Member, Patent Compensation Board, U.S.A.E.C.
Address: Simmonds Precision Products, Inc., Tarrytown, New York, U.S.A.

COOMBS, F. L., LL.B., A.M.I.E.E. Formerly Junior Vice-Pres., now Vice Pres., Inst. Nucl. Eng.
Address: Institution of Nuclear Engineers, 147 Victoria Street, London, S.W.1, England.

COOMBS, Frederick William, B.Sc. (Eng.). Born 1915. Educ.: London Univ. Chief Eng. (London), Head Wrightson Processes, Ltd. Society: Inst. of Nucl. Eng. Nuclear interest: Research reactor design and construction.
Address: 16/26 Baltic Street, London, E.C.1, England.

COON, James H., A.B., Ph.D. Born 1914. Educ.: Indiana and Chicago Univs. Member of Staff Los Alamos Sci. Lab. Societies: A.P.S.; A.G.U. Nuclear interests: Nuclear physics; Solar and cosmic radiation; Nuclear explosions.
Address: Los Alamos Scientific Laboratory, Los Alamos, New Mexico, U.S.A.

COOPER, A. C. Sec., Head Office, A.A.E.C.
Address: Australian Atomic Energy Commission, 45 Beach Street, Coogee, New South Wales, Australia.

COOPER, Alfred Ronald, M.A. Born 1932. Educ.: Cambridge Univ. Sci. Officer (1954-58), Sen. Sci. Officer (1958-60), R. and D., Springfields Factory, U.K.A.E.A.; Lecturer in Chem. Eng. (1960-62), Sen. Lecturer (1962-), Coll. Advanced Technol., Birmingham. Societies: A.M.I.Chem.E.; Inst. Nucl. Eng. Nuclear interests: The chemical extraction and processing of nuclear fuels and of canning, moderator, control rod and coolant materials.
Address: Department of Chemistry, College of Advanced Technology, Gosta Green, Birmingham 4, England.

COOPER, Andrew Ramsden, M. Eng., C. Eng., F.I.E.E., F. Inst. F., M.R.I. Born 1902. Educ.: Sheffield Univ. Controller, Merseyside and N. Wales Div. (1948-51), N.W. Div. (1951-54), N.W. Merseyside and N. Wales Div. (1954-57), Director, Eastern, London and S. Eastern Region (1957), Central Elec. Authority; Member and Regional Director (1957-59), Member for Operations and Personnel, (1959-66), C.E.G.B. Society: Assoc. Mining, Elec. and Mech. Eng.
Address: 21 Bloomsbury Street, London, W.C.1, England.

COOPER, Arthur George Stening, M.B., Ch.M. (Sydney), D.M.R. (London), F.C.R.A., F.F.R. Born 1899. Educ.: Sydney and London Univs. Director, Queensland Radium Inst., -1965. Society: Internat. Club of Radiotherapists. Nuclear interest: Medical investigation and treatment by radioactive isotopes.
Address: 19 Apex Street, Clayfield, Brisbane, Australia.

COOPER, D. C., Assoc.Prof. Chairman, Radiol. Safety Com., Dept. of Elec. Eng., Nucl. Reactor Facility, Delaware Univ.
Address: Delaware University, Newark, Delaware, U.S.A.

COOPER, Daniel I., B.S., Ph.D. Born 1926. Educ.: M.I.T. Managing Editor, Nucleonics Magazine, McGraw-Hill Publishing Co., New

York, 1954-61; Exec. Editor, Internat. Sci. and Technol., Conover-Mast Publications, Inc., New York, 1961-. Societies: A.N.S.; A.P.S.; I.E.E.E.
Nuclear interests: Nuclear instrumentation; Thermonuclear power; Editorial and information.
Address: 48 Harding Drive, South Orange, New Jersey, U.S.A.

COOPER, Douglas Elhoff, Ph.D. Born 1912. Educ.: Purdue Univ. Res. Assoc., Ethyl Corp. Res. Labs. Society: A.C.S.
Nuclear interests: Use of isotopes as tracers; Radiochemistry; Activation analysis; Engineering uses of radioactivity.
Address: Ethyl Corporation Research Laboratories, 1600 W.8 Mile Road, Ferndale, Detroit, Michigan 48220, U.S.A.

COOPER, Eugene Perry, B.S. (Phys.), Ph.D. (Theoretical Phys.). Born 1915. Educ.: M.I.T. and California Univ. Assoc. Head, Underwater Ordnance Dept., Naval Ordnance Test Station, (1953-60), Tech. Director (1960-), Naval Radiol. Defence Lab. Societies: Radiol. Defence Lab. Societies: A.P.S.; A.A.A.S.
Nuclear interests: Nuclear physics; Radioactivity; Radiobiology; Effects of nuclear weapons and countermeasures therefor.
Address: 48 Lakemont Drive, Daly City, California, U.S.A.

COOPER, George, Jr., B.A., M.D. Educ.: Virginia Univ. Prof. Radiol., Virginia Univ. Hospital, 1954-64; Chairman, Radiol. Dept., Coll. of Medicine, Tennessee Univ., 1964-. Chancellor, American College of Radiol.
Address: Tennessee University, College of Medicine, Knoxville, Tennessee, U.S.A.

COOPER, Herbert J. Formerly Exec. Vice-Pres., now Pres., and Director, Nucl. Dept., Cooper Alloy Corp.
Address: Cooper Alloy Corporation, Hillside, New Jersey, U.S.A.

COOPER, John A. D., B.S., Ph.D.,M.B., M.D., Dr.h.c. Born 1918. Educ.: New Mexico State and Northwestern Univs. Prof. Biochem., (1957-), Asst. Dean (1956-59), Assoc. Dean, (1959-63), Dean of Sci. (1963-), Northwestern Univ.; Visiting Prof., Dept. of Biophys.: Brazil Univ., summer 1956; Buenos Aires Univ., 1958. Consulting Staff, Veterans Administration Res. Hospital, 1954-; Consultant, U.S.A.E.C., 1956-67; Consultant, Argonne Nat. Lab., 1957-63; Vice-Pres., Board of Trustees, Argonne Univ. Assoc.; Chairman, Illinois Sci. Advisory Council; Member: Illinois Legislative Com. on Atomic Energy, Illinois Board of Higher Education; Nat. Advisory Council on Health Res. Facilties, N.I.H.; N.A.S. Advisory Com. on Sci. Personnel; Editor, J. of Medical Education; Treas., Pan-American Federation of Nat. Assocs. of Medical Schools. Societies: American Soc. of Biol. Chemists; Central Soc. for Clinical Res.; Hon. member, Asoc. Venezolana Para el Avance de la Ciencia; Inst. of Medicine, Chicago; A.M.A.; Assoc. of American Medical Colleges; A.N.S.
Nuclear interests: Radiobiology and use of radioisotopes in biology and medicine.
Address: Northwestern University, 619 Clark Street, Evanston, Illinois, U.S.A.

COOPER, John Niessink, A.B. (Kalamazoo Coll.), Ph.D.(Cornell). Born 1914. Educ.: Cornell Univ. Asst. Prof., Assoc. Prof. and Prof. Phys., Ohio State Univ., 1946-56; Prof. Phys., U.S. Naval Postgraduate School, 1956-. Societies: Fellow, A.P.S.; A.A.P.T.; Sen. Member, I.E.E.E.
Nuclear interest: Penetration of charged particles in matter.
Address: Department of Physics, U.S. Naval Postgraduate School, Monterey, California, U.S.A.

COOPER, John Norman, A.M.I.Mech.E. Born 1926. N.C.B., 1952; U.K.A.E.A., 1952-. Society: Life Member, Nucl. Eng. Soc.
Nuclear interest: Process plant design and installation.
Address: 4 Pewterspear Lane, Appleton, Warrington, Lancs., England.

COOPER, Junius H. Controller, Vitro Corp. of America.
Address: Vitro Corporation of America, 90 Park Avenue, New York, New York 10016, U.S.A.

COOPER, Col. Kenneth B. Educ.: United States Military Acad., Naval Postgraduate School and M.I.T. Div. of Military Application, U.S.A.E.C., 1951-55, Principal Staff Officer, Plans and Policy Div., S.H.A.P.E., Paris, 1955-59; Project Manager, Tech. Operations Div., Office of the Director of Defence Res. and Eng., Advanced Res. Projects Agency, 1959-63; Chief, Army Eng. Nucl. Power Div., and Asst. Director, Army Reactors, Div. of Reactor Development and Technol., U.S.A.E.C., 1965-; Director, Army Nucl. Power Programme, Dept. of the Army.
Address: Division of Reactor Development and Technology, U.S.A.E.C., Washington D.C. 20545, U.S.A.

COOPER, Leon N, A.B., A.M., Ph.D. Born 1930. Educ.: Columbia Univ. Prof. Phys., Brown Univ.
Nuclear interests: Theoretical physics; Many body problems; Low temperature physics; Superconductivity; Elementary particle physics.
Address: Physics Department, Brown University, Providence, Rhode Island, U.S.A.

COOPER, R., Dr. Member, Board of Studies in Nucl. Sci. and Eng., Melbourne Univ.
Address: Melbourne University, Board of Studies in Nuclear Science and Engineering, Parkville N.2., Victoria, Australia.

COOPER, Raymond David, B.S., M.S., Ph.D. Born 1927. Educ.: Illinois and Iowa State Univs. and M.I.T. Research Asst., U.S.A.E.C., 1951-54; Chief, Accelerator Branch, U.S. Army, Natick Labs., Natick, Mass., 1954-67; Lecturer in Phys., Tufts Univ., 1957-62. Societies: A.P.S.; A.A.A.S.; RESA.
Nuclear interests: Photodisintegration cross-sections; Interaction and scattering of electrons; Research applications of electron accelerators; Activation analysis.
Address: 81 West Union Street, Ashland, Massachusetts, U.S.A.

COOPER, Vance Reece, A.B. (Chem.), M.S. (Chem. Eng.). Born 1913. Educ.: U.C.L.A. and Michigan Univ. Manager, Chem. R. and D., Hanford Labs. Operation, 1956-57; Manager, Res. and Eng., Chem. Processing Dept., Hanford Atomic Products Operation (1957-61), Manager, Chem. and Materials Eng. Lab., Advanced Technol. Labs., Schenectady, N.Y. (1961-65), Manager, Chem. and Systems Lab., R. and D. Centre (1965-), G.E.C. Chairman, Nucl. Technol. Subdiv., A.C.S. Societies: A.C.S.; A.I.Ch.E.; A.N.S.; A.A.A.S.
Nuclear interests: Separations processing; Fuel systems.
Address: General Electric Company, Research and Development Centre, Schenectady, New York, U.S.A.

COOPER, W. M., B.Sc., A.Inst.P. Reactor Supervisor, Jason Reactor, Greenwich, 1962-; Sen. Lecturer, Nucl. Sci. and Technol. Dept., Royal Naval Coll.
Address: Royal Naval College, Greenwich, London, S.E.10, England.

COOPER, Warwick Alan, B.Sc., Ph.D. Born 1932. Educ.: Manchester Univ. C.E.R.N. (Geneva), 1955-56; A.E.R.E. Harwell, 1957-59; C.E.R.N. (Geneva), 1960-65; A.N.L. (U.S.A.), 1965-.
Nuclear interests: Uses of bubble chamber in high-energy physics; Automation in data analysis; Elementary particles.
Address: H.E.P., A.N.L., Illinois 60439, U.S.A.

COOPER, William Eugene, B.S., M.S., Ph.D. Born 1924. Educ.: Purdue Univ. and Oregon State Coll. K.A.P.L., G.E.C., Schenectady, New York, 1952-63; Vice Pres. and Manager, Eng., Teledyne Materials Res. Co., 1963-. Member, Pressure Vessel Research Com.; Chairman, Subcom. on Design, Vice Chairman, Subcom. on Nucl. Power, A.S.M.E. Boiler and Pressure Vessel Code. Societies: A.S.M.; Soc. for Exptl. Stress Analysis.
Nuclear interests: Structural evaluation of reactor system components, particularly pressure retaining members.
Address: Teledyne Materials Research Company, 303 Bear Hill Road, Waltham, Massachusetts 02154, U.S.A.

COOPER, William Henry Bernard. Born 1910. Deputy Head, Electronics Div., A.E.R.E., Harwell. Society: A.M.I.E.E.
Nuclear interest: Electronics applied to atomic energy research problems.
Address: The Thatched Cottage, Bath Road, Hungerford, Berkshire, England.

COOPER, William J. Chairman, and Chief Exec. Officer, United Illuminating Co.
Address: United Illuminating Co., 80 Temple Street, New Haven, Connecticut 06506, U.S.A.

COOPER, Wilson Reid, Elec. Eng. Born 1905. Educ.: Rensselaer Polytech. Inst. Asst. to Director, Power Operations (-1953), Nucl. Development Eng. (1953-56), Head, Nucl. Development Sect. (1956-60), Manager, Exptl. Gas-Cooled Reactor Project (1960-66), Chief, Nucl. Power Staff (1966-), Tennessee Valley Authority. Society: I.E.E.E.
Nuclear interests: Management – nuclear power development, operations, safety analysis, and technical and economic evaluation.
Address: Tennessee Valley Authority, 309 Power Building, Chattanooga, Tennessee 37401, U.S.A.

COOR, Thomas, B.A., Ph.D. Born 1922. Educ.: Rice Inst. and Princeton Univ. Sci. Liaison Officer, U.S. Embassy, London, 1948-50; Res. Phys., Brookhaven Nat. Lab., 1950-53; Res. Assoc. (1954-59), Sen. Res. Assoc. (1959-63), Princeton Univ.; Vice-Pres., Princeton Appl. Res. Corp., 1962-. Scientific Sec., 2nd I.C.P.U.A.E., 1958. Papers: On cosmic-ray research, elementary particle physics, electronic instrumentation, plasma physics. Society: A.P.S.
Address: 60 Pheasant Hill Road, Princeton, New Jersey 08540, U.S.A.

COPE, David F., A.B., M.S. (West Virginia), Ph.D. (Phys., Virginia). Born 1912. Educ.: West Virginia and Virginia Univs. Chief, Res. Branch (1953-56), Deputy Director, R. and D. Div. (1956-59), Director, Reactor Div. (1959-66), Oak Ridge Operations Office, U.S.A.E.C.; Senior A.E.C.-R.D.T. Rep. at O.R.N.L., 1966-, U.S.A.E.C. Societies: A.N.S.; A.P.S.; A.A.U.P.; A.A.P.T.
Nuclear interests: Evaluation of technical and management aspects of nuclear reactor programmes at the Oak Ridge National Laboratory; Use of nuclear energy for large dual purpose power and water plants; General scientific administration and management of physical and life science programmes.
Address: United States Atomic Energy Commission, R.D.T. Site Representative Office, Post Office Box X, Oak Ridge, Tennessee 37830, U.S.A.

COPE, Oliver, B.A., M.D., Dr. h.c. (Toulouse). Born 1902. Educ.: Harvard Univ. Prof. Surgery, Harvard Medical School, 1963-; Chairman, Res. Com., Massachusetts Gen. Hospital, 1961-64; Chief of Staff, Shriners Burns Inst., Boston Unit, 1964-. Pres., American Surgical Assoc.
Nuclear interests: Biologic and medical aspects.
Address: Massachusetts General Hospital, Boston 14, Massachusetts, U.S.A.

COPELAND, Joseph, Dr. Project Eng., Advanced Technol. Div., Atlantic Res. Corp.
Address: Atlantic Research Corporation, Henry G. Shirley Memorial Highway at Edsall Road, Alexandria, Virginia, U.S.A.

COPENHAVER, Carl Monroe, M.S. (Mech. Eng.). Born 1929. Educ.: Columbia Univ. and O.R.S.O.R.T. Development Eng., O.R.N.L., 1955-58; Tech. Staff Member, Mitre Corp., Bedford, Mass., 1959-. Societies: A.N.S.; A.P.S.
Nuclear interests: Nuclear physics and reactor design.
Address: 31 North Avenue, Weston 93, Massachusetts, U.S.A.

COPIC, Milan, D.S. Born 1925. Educ.: Ljubljana Univ. Sci. Coll., Inst. Josef Stefan, Ljubljana, until 1959; Res. Phys., I.S.N.S.E., Argonne Nat. Lab., 1959-60; Head, Reactor Dept., Nucl. Inst. Josef Stefan, Ljubljana, 1961; Visiting Prof., Kansas State Univ., 1965-66. Book: Skrivnosti Atoma (Presernova Druzba, 1953). Societies: A.N.S.; Drustvo Matematikov in Fizikov L.R.S.
Nuclear interest: Reactor physics.
Address: 15 Pletersnikova, Ljubljana, Yugoslavia.

COPP, D. Harold, M.D. (Toronto), Ph.D. (California). Born 1915. Educ.: Toronto and California Univs. Prof. and Head, Dept. of Physiology, British Columbia Univ. Sci. Sec., 2nd I.C.P.U.A.E., Geneva, 1958; Member, Panel on Rad. Protection, D.R.B., Canada. Societies: American Physiological Soc.; Canadian Physiological Soc.; Roy. Soc. of Canada; Soc. for Exptl. Biol. and Medicine.
Nuclear interests: Kinetics studies with radiocalcium and radiostrontium in animals and man; Biological decontamination radiostrontium.
Address: Department of Physiology, British Columbia University, Vancouver 8, British Columbia, Canada.

COPPA, Rafael Carlos, Dr. Jefe a/c del Departamento de Plantas, Comisión Nacional de Energia Atomica, Argentina.
Address: Comisión Nacional de Energia Atomica, 8250 Av. del Libertador, Buenos Aires, Argentina.

COPPE, Albert, Dr. of Economics, B. Com. and Finance, Bachelor of Political and Social Sci. Member, Commission of the European Communities, 1967-.
Address: Commission of the European Communities, 23 avenue de la Joyeuse Entree, Brussels 4, Belgium.

COPPEE, Baron Evence Arnold Dieudonné Marie Joseph Ghislain. Mining Civil Eng. (Louvain). Born 1929. Educ.: Univ. Catholique de Louvain and M.I.T. Chairman, Board of Governors, Company Evence Coppée; Chairman of Board, Sté. d'Etude et de Construction Evence Coppée; Managing Director, Allegheny-Longdoz S.A.; Director, S.A. Métallurgique d'Espérance-Longdoz; Vice-Chairman, Bureau d'Etudes Industrielles Fernand Courtoy; Vice-Chairman, Constructions et Entreprises Industrielles; Director, Charbonnage André Dumont; Managing Director, Hydrometals S.A.; Chairman of Board, Sté. Belge d'Etudes Générales et Développement; Vice-Chairman, Sté. Belge de Chimie Nucléaire. Ministère des Affaires Economiques Belge; Member of working group on Ardennes atomic energy plant (1960); Member, Board of Management, Syndicat Vulcain, Belgonucleaire.
Address: 33b avenue Franklin Roosevelt, Brussels 5, Belgium.

COPPENS, Benignus Joseph, J.D. Born 1917. Educ.: Nijmegen Univ. Director of Administration, Ministry of Finance; Member, Netherlands deleg. to Paris Convention on Third-party Liability in Field of Nucl. Energy; Idem. to Diplomatic Conference on Maritime Law, 1961 (third-party liability of operators of nucl. ships); Idem. (and Vice-Pres.) to the Brussels Supplementary Convention; Idem. to Internat. I.A.E.A. Conference on Civil Liability for Nucl. Damage.
Nuclear interest: Legal problems involved in the utilisation of nuclear energy (third-party liability, insurance, State intervention).
Address: Ministry of Finance, 20 Kneuterdijk, The Hague, Netherlands.

COPPENS, René, D. ès Sci. Physiques. Born 1910. Educ.: Caen, Rennes, Paris. Prof. Faculté des Sciences de Nancy; Directeur-Adjoint du Centre de Recherches Radiogéologiques de l'Université de Nancy (liaison étroite avec le C.E.A.). Books: La Radioactivité des Roches (P.U.F., 1958); Co-author, Prédis de Géologie (P.U.F., 1966).
Nuclear interest: Géochimie de l'uranium et du thorium; Géologie nucléaire; Géochronologie U/Pb et ^{14}C.
Address: 35 boulevard Jean Jaurès, 54 Nancy, France.

COPPINI, Mario Alberto. With Previdenza. Italian Member, Specialised Nucl. Sect. for Social Health and Development Problems, Economic and Social Com., Euratom, 1966-.
Address: 34 Via Proba Petronia, 00 136 Rome, Italy.

COPPOCK, W. H., Ph.D. Prof. and Head, Dept. Chem. and Member, Radioisotopes Com.,

Drake Univ., Des Moines.
Address: Drake University, Des Moines, 11,
Iowa, U.S.A.

CORBASCIO, Aldo Nicola, M.D., D.Sc. Born
1928. Educ.: Bari and Pennsylvania Univs.
Fulbright Scholar, 1954-56; Instructor in
Pharmacology, Pennsylvania Univ. Medical
School, 1955-59; Asst. Res. Pharmacologist,
California Univ. Medical Centre, San Francisco,
1959-63; Assoc. Prof. Pharmacology, Pacific
Univ. School of Dentistry, 1963-. Sci. Director,
Berkeley Biomedical Res. Inc. Societies:
American Soc. for Pharmacology and Exptl.
Therapeutics; Ballistocardiographic Res. Soc.;
A.A.A.S.; A.M.A.; Western Pharmacological
Soc.
Address: 9451 Florio Street, Berkeley,
California 94618, U.S.A.

CORBEN, Herbert Charles, M.A., M.Sc., Ph.D.
Born 1914. Educ.: Melbourne, Cambridge
and California Univs. Prof. Phys., Carnegie Inst.
Technol., 1946-56; Assoc. Director, Res. Lab.
(1956-61), Director, Quantum Phys. Lab.
(1961-), T.R.W. Systems. Books: Co-author,
Classical Mechanics (J. Wiley, 1960); Spinning
Particles (Holden-Day, 1967). Societies: A.P.S.;
A.N.S.; A.A.A.S.
Nuclear interests: Reactor kinetics; Quantum
theory of elementary particles.
Address: 247-34th Street, Hermosa Beach,
California, U.S.A.

CORBETT, James William, Ph.D. Born 1928.
Educ.: Yale Univ. Res. Assoc., G.E.C. Res.
Lab., 1955-. Societies: Fellow, A.P.S.; A.A.P.T.
Nuclear interest: Radiation damage.
Address: General Electric Research and
Development Centre, P.O. Box 8, The Knolls,
Schenectady, New York 12301, U.S.A.

CORBETT, Mack Colvin, B.A. (Journalism).
Born 1910. Educ.: Utah Univ. Regional
Information Officer, Region 4, Salt Lake City,
Bureau of Reclamation, U.S. Dept. of the
Interior, 1945-53; Director of Information,
Bureau of Land Management, U.S. Dept. of
the Interior, Washington, D.C., 1954-57;
Director of Information, Idaho Operations
Office, U.S.A.E.C., Idaho Falls, Idaho, 1957-.
Address: 912 Eleventh Street, Idaho Falls,
Idaho,U.S.A.

CORBIAU, Paul-Emile, Ing. Civil Métal.,
Licencié Sci. économiques, politiques et
sociales, Advanced Management Programme
(Harvard). Born 1913. Educ.: Louvain and
Harvard Univs. Exec. Director, Fansteel-Hoboken, S.A., 1962; now Managing Director,
Metallurgie Hoboken, S.A.
Nuclear interests: Metallurgy; Chemistry.
Address: 4 Dennenlaan, Antwerp, Belgium.

CORBIN, Lawrence Tapp, B.S. Born 1918.
Educ.: Kentucky Univ. Asst. Director, Analytical Chem. Div., O.R.N.L. Membership Com.,
Analytical Chem. Sect., A.C.S. Societies:
A.N.S.; A.A.A.S.
Nuclear interest: Analytical chemistry.
Address: 121 Outer Drive, Oak Ridge, Tennessee, U.S.A.

CORDAY, Eliot, M.D. Member, Consulting
Editorial Board, J. Nucl. Medicine.
Address: 333 N. Michigan Avenue, Chicago 1,
Illinois,U.S.A.

CORDELLE, Pierre - Michel, Ing. Born 1922.
Educ.: Ecole Navale. Directeur du Lab. d'Electronique et de Technol. de l'Informatique au
Centre d'Etudes Nucléaires de Grenoble.
Nuclear interests: Nuclear instrumentation
and electronics; Research on electronic components and devices.
Address: Centre d'Etudes Nucléaires de Grenoble,B.P. 269, France.

CORDFUNKE, E. H. P., Dr., Ir. Member,
Editorial Com., Atoomenergie en Haar Toepassingen.
Address: Atoomenergie en Haar Toepassingen,
112 Scheveningseweg, The Hague, Netherlands.

CORDOVI, Marcel A., A.B., B.M.E., M.M.E.,
M.E. Born 1915. Educ.: Polytech. Inst. Brooklyn and Michigan Univ. In charge, Power Application, Internat. Nickel Co., 1958-; Consultant, Brookhaven Nat. Lab., A.E.C., 1950-59;
Manager, Materials and Testing Dept., Atomic
Energy Div., Babcock and Wilcox Co., 1953-58;
Adjunct Prof. Metallurgical Eng., Polytech.
Inst. Brooklyn (1950-), and New York Univ.
(1958-), Chairman, Coordinating Com. on
Materials Specifications for Nucl. Service,
A.S.T.M. Chairman, Com. on Pressure Vessels
for Nucl. Applications, Welding Res. Council;
Boiler and Pressure Vessel Code Com., Subcom.
on Nucl. Power, A.S.M.E.; Res. Com. on
Effects of Rad. on Materials for Nucl. Installation, A.S.M.E.; A.S.T.M. Rep. on Nucl. Standards Board, A.S.A.; A.E.C. Advisory Com.,
American Welding Soc. Societies: A.N.S.;
A.S.M.E.; A.S.M.; A.S.T.M.; Soc. of Naval
Architects and Marine Eng.
Nuclear interests: Primary interest in nuclear
metallurgy - both materials application and
teaching.
Address: 67-71 Yellowstone Boulevard, Forest
Hills 75, New York, U.S.A.

CORETTE, John E., LL.B. (Virginia), LL.D.
(Hon.) (College of Great Falls). Born 1908.
Educ.: Montana and Virginia Univs. Pres., Gen.
Manager and Director, The Montana Power Co.;
Pres. and Chief Exec. Officer, Canadian-Montana
Pipe Line Co.; Pres. and Chief Exec. Officer,
Canadian-Montana Gas Co., Ltd.; Pres., Atlanta
Exploration Co.; Director, Pacific Gas Transmission Co.; Director, First Bank Stock Corp.;
Member, Exec. Com., High Temperature
Reactor Development Associates Inc.; Trustee,
Endowment Foundation, Montana State Univ.;
Director, Edison Elec. Inst. (Past Pres. E.E.I.,
1958-59); Director, Stanford Res. Inst.;
Member, The Business Council; Member, Nat.

Association of Elec. Companies (Past Director); Member, Nat. Association of Manufacturers (Past Director); Member, U.S. Chamber of Commerce (Past Director); U.S. Deleg. to Com. on Flec. Power, subsidiary of United Nations Economic Commission for Europe, 1957, 1958. Society: Atomic Ind. Forum. Nuclear interest: Research and development of various types of reactors to produce lowcost power.
Address: 1245 West Platinum St., Butte, Montana, U.S.A.

COREY, Clark L., Ph.D. (Michigan). Born 1921. Prof., Wayne State Univ. Chairman, A.I.M.E., Detroit Sect. Societies: A.I.M.E.; A.S.M. Nuclear interests: Metallurgy; Reactor design.
Address: Department of Chemical and Metallurgical Eng., Wayne State University, Detroit 2, Michigan, U.S.A.

CORIN, C., Prof. Member, Comite d'Application des Methodes Isotopiques aux Recherches Agronomiques.
Address: Faculte des Sciences Agronomiques de l'Etat, Gembloux, Belgium.

CORIOU, Alain Henri Marcel, Ing. A.M., E.N.S.P.M.-Ing. en Génie Atomique. Born 1931. Educ.: E.N.S.I.A.M., E.N.S.P.M., I.N.S.T.N. Society: A.T.E.N. Nuclear interests: Control and monitoring of nuclear reactors; Nuclear power plants; Irradiated fuel treating plants; Isotopic separation plants.
Address: c/o COCEI, 22 rue de Clichy, Paris 9, France.

CORIOU, Henri, L. ès Sc. Physique, Born 1925. Educ.: Paris Univ. Chef du Service d'Etude de la Corrosion Aqueuse et d'Electrochimie, C.E.A., C.E.N., Saclay; Prés., Groupe de Travail "Corrosion Nucléaire", Fédération Européenne de la Corrosion; Conseiller Scientifique, Commission des Etudes Fondamentales et Applications (C.E.E.A.), CEBELCOR. Book: Electrochimie, Electrométallurgie, Corrosion. Societes: Comité International de Thermodynamique et de Cinétique Electrochimiques (Rapporteur de la Commission des Hautes Températures, 1961-65); Sté. de Chimie Industrielles; Sté. Française de Métallurgie; CEFRACOR. Nuclear interests: Métallurgie et electrométallurgie.
Address: Department de Chimie, Commissariat à l'Energie Atomique, B.P. No. 2, Gif-sur-Yvette, Seine et Oise, France.

CORLEO, G. Gruppo van de Graaff, Istituto di Fisica, Catania Univ.
Address: Catania University, Istituto di Fisica, 57 Corso Italia, Catania, Sicily, Italy.

CORLETTE, Philip Manning Christian, M.B., B.S., F.C.R.A., D.M.R.T. Born 1919. Educ.: Sydney Univ. Sen. Hon. Radiotherapist, Royal Prince Alfred Hospital and Sydney Hospital, N.S.W., Australia. Member, Radiol. Advisory Council, Public Health Dept., N.S.W. Societies: College of Radiologists of Australia; Faculty of Radiologists (England). Nuclear interests: Therapeutic and diagnostic uses of isotopes in medicine.
Address: 135 Macquarie Street, Sydney, N.S.W., Australia.

CORLEW, R. P. Vice Pres., Operations, Atlantic Richfield Hanford Co., Hanford Facilities, U.S.A.E.C.
Address: U.S.A.E.C., Atlantic Richfield Hanford Company, Richland, Washington, U.S.A.

CORLISS, William Roger, B.S. (Phys.), M.S. Born 1926. Educ.: Rensselaer Polytech. Inst. and Colorado Univ. Eng. Supervisor, Heat Transfer and Hydrodynamics Unit, Pratt and Whitney Aircraft Co., 1954-56; Space Propulsion System and Nucl. Power System Specialist, Gen. Elec. Co., 1956-59; Director, Advanced Programmes, Nucl. Div., Martin Co., 1959-63; Consultant and writer, 1963-. Books: Propulsion Systems for Space Flight (McGraw-Hill, 1960); Radioisotopic Power Generation (Prentice-Hall, 1964). Societies: American Rocket Soc.; British Interplanetary Soc.; A.A.A.S. Nuclear interest: Research and development management of nuclear auxiliary powerplants.
Address: Box 311, Manor Road, Glenarm, Md., U.S.A.

CORMACK, Douglas Villy, B.Sc.(Alberta), M.Sc. (Alberta), Ph.D. (Saskatchewan). Born 1926. Educ.: Alberta and Saskatchewan Univs. Asst. Prof. (1956-62), Assoc. Prof. (1962-), Saskatchewan Univ.; Phys., Manitoba Cancer Treatment and Res. Foundation, 1967-. Planning Board and Task Groups, I.C.R.U.; Assoc. Editor, Rad. Res. Societies: Canadian Assoc. Phys.; Rad. Res. Soc.; British Inst. Radiol. Nuclear interests: X-ray and gamma-ray dosimetry; Radiation biology; Radiological health.
Address: Manitoba Cancer Foundation, 700 Bannatyne Avenue, Winnipeg 3, Canada.

CORNAGLiA, L. Board Member, Centre Scientifique de Monaco.
Address: Centre Scientifique de Monaco, 16 boulevard de Suisse, Monte Carlo, Monaco.

CORNATZER, William Eugene, B.S., M.S., Ph.D., M.D. Born 1918. Educ.: Wake Forest and North Carolina Univs. and Bowman Gray School of Medicine. Asst. Prof. of Biochem., Bowman Gray School of Medicine, 1946-51; Prof. and Head, Dept. of Biochem., Director, Ireland Res. Lab., North Dakota Univ., School of Medicine, Grand Forks, North Dakota, 1951-. Societies: Rad. Res. Soc.; A.C.S.; American Soc. Biol. Chem.; American Inst. of Nutrition; American Assoc. Cancer Res.; Central Soc. Chem. Res.; Soc. Exptl. Biol. Medicine; Fellow, American Inst. of Chem.; Fellow,

Coll. of Physicians; Fellow, A.A.A.S.; Fellow, New York Acad. of Sci.; Affiliate, Roy, Soc. of Medicine.
Nuclear interests: Effects of whole body irradiation of gamma and beta rays, use of isotopes in biochemical reactions of the cell, phospholipid metabolism.
Address: Biochemistry Department, University of North Dakota Medical School, Grand Forks, North Dakota 58201, U.S.A.

CORNELIS, J. Charles. Netherlands Member, Health and Safety Sub-Com., O.E.C.D., E.N.E.A.
Address: O.E.C.D. European Nuclear Energy Agency, 38 boulevard Suchet, Paris 16, France.

CORNER, J., Dr. Chief, Mathematical Phys., Weapons Group, U.K.A.E.A.
Address: United Kingdom Atomic Energy Authority, Weapons Group, Atomic Weapons Research Establishment, Aldermaston, Berks., England.

CORNEY, George M., B.A. Born 1916. Educ.: Cornell Univ. Phys., Radiographic Dept., Kodak Res. Lab., 1937-. Societies: Soc. Photographic Sci. and Eng.; A.S.T.M.; Soc. Nondestructive Testing.
Nuclear interests: Photographic monitoring of radiation; Application of gamma-rays to nondestructive testing.
Address: Eastman Kodak Company, Rochester, New York 14650, U.S.A.

CORNGOLD, Noel Robert David, A.B. (Columbia), A.M., Ph.D. (Harvard). Born 1929. Educ.: Columbia and Harvard Univs. Assoc. Phys. (1954-59), Phys. (1959-66), Brookhaven Nat. Lab; Prof. Appl. Sci., Calif. Inst. of Technol., 1966-. Societies: A.P.S.; A.N.S.
Nuclear interests: Neutron transport theory; Statistical physics.
Address: California Institute of Technology, Pasadena, California, U.S.A.

CORNISH-BOWDEN, Athelstan Claude Muir, B.A. (Hons., Chem.). Born 1910. Educ.: Oxford Univ. Manager, Ind. Development Corp. of S.A., Ltd., to 1953; Chairman, Cape Asbestos South Africa Pty., Ltd., 1954-59; Atomic Energy Board, South Africa 1958-64; Consultant to Gen. Manager, I.D.C. of Rhodesia, 1964-67; Consultant to Gen. Manager, Malawi Development Corp., 1967-. Pres., Exporters Assoc. of South Africa. Society: I.M.M. of S. Africa.
Address: P.O. Box 2105, Johannesburg, South Africa.

CORNS, Harry, M.I.Mech.E. Born 1919. Educ.: St. Helens Municipal Tech. Coll. Chief Eng., Eng. Group H.Q., U.K.A.E.A., 1957-. Society: Nucl. Eng. Soc.
Nuclear interest: Reactor fuel processing.
Address: 33 Knowsley Road, Rainhill, Lancashire, England.

CORNUET, Robert Louis, Eng. (Ecole Nationale Supérieure d'Electrochimie et d'Electrométallurgie de Grenoble). Born 1925. Educ.: Grenoble Univ. Head, Sect. Applications of Radioisotopes, C.E.A., Paris. Societies: A.N.S.; Sté. Française des Electroniciens et des Radio-électriciens.
Nuclear interests: Industrial use of isotopes; Activation analysis; Irradiation sources of high intensity.
Address: 71 Cours Jean-Jaurès, Grenoble 38, France.

CORONADO, Abel. Consulting Eng., Tecnicas Nucleares S.A.
Address: Tecnicas Nucleares S.A., 46 Serrano, Madrid 1, Spain.

CORPUS, Guillermo C. Supervising Sci. (Head, Reactor Operations Dept.), Philippine Atomic Res. Centre.
Address: Philippine Atomic Research Centre, Diliman, Quezon City, Philippines.

CORRADI, Rear-Admiral Peter. Formerly Chief, Bureau of Yards and Docks, U.S. Navy, Vice-Pres., Gen. Manager, Dravo Corp. Subsidiary (1965-), Pres. (1967-), Gibbs and Hill Inc.
Address: Gibbs and Hill, Inc., 393 Seventh Avenue, New York 1, New York, U.S.A.

CORREA MILLER, Javier, Dr. Formerly Delegado, now Pres., Junta de Control de Energia Atomica del Peru; Governor for Peru to I.A.E.A.
Address: Junta de Control de Energia Atomica del Peru, 3420 Avenida Arequipa, San Isidro, Apartado 914, Lima, Peru.

CORREIRA CONTREIRAS, J. Portuguese Member, Study Group on Food Irradiation, O.E.C.D., E.N.E.A.
Address: O.E.C.D. European Nuclear Energy Agency, 38 boulevard Suchet, Paris 16, France.

CORRELL, William A. Educ.: McMaster Univ. Formerly with Steel Co. of Canada, Ltd.; Member, Royal Commission on Govt. Organization, 1961; Director, Personnel, A.E.C.L., 1967-.
Address: Atomic Energy of Canada Limited, Head Office, 275 Slater Street, Ottawa 4, Ontario, Canada.

CORSINI, Giuseppe, M.D. Born 1926. Educ.: Pisa Univ. Res. Asst., Inst. of Medical Pathology (1951-55), Res. Asst., Medical Clinic (1955-), Pisa Univ. Societies: Italian Soc. for Nucl. Biol. and Medicine; Italian Soc. of Cardiology; Italian Soc. of Metabolic Diseases.
Nuclear interests: Application of radioisotopes to medicine, for research and diagnosis, mainly in the field of metabolism. Previous work concerns turnover of plasma phospholipids, absorption of I-131 and Br-82 labelled fats.
Address: Centro di Medicina Nucleare, Università di Pisa, Italy.

CORTAZAR, Jaime, M.D. Born 1923. Educ.: Colombia Nat. Univ. Auxiliary Physician, Radiotherapy, Nat. Cancer Inst., 1951-53; Head Endocrinologist, Colombian Inst. of Social Security, 1953-55; Head, Endocrine and Radioisotopes Sect. (1954-55, 1957-66), Sci. Director (1955-57), Head, Endocrine Sect. (1967-), Nat. Cancer Inst. Societies: Colombian Endocrine Sect.; Colombian Cancer Soc.; American Thyroid Assoc.; American Endocrine Soc.; A.A.A.S.
Nuclear interest: Nuclear medicine.
Address: 5100 Apartado Aéreo, Bogotá, D.E., Colombia.

CORTE-REAL, Manuel, Dr. Vice Pres., Forum Atomico Portugues.
Address: Forum Atomico Portugues, 16 Trav Abarracamento de Peniche, Lisbon 2, Portugal.

CORTELEZZI, Cesar, Dr. Member, Comision Especial de Fisica Atomica y Radioisotopes, La Plata National Univ.
Address: La Plata National University, Comision Especial de Fisica Atomica y Radioisotopos, La Plata, Argentina.

CORTELLESSA, G., Prof. Head, High Energy Phys. Sect., Laboratori di Fisica, Istituto Superiore di Sanita'.
Address: Istituto Superiore di Sanita', 299 Viale Regina Elena, Rome, Italy.

CORTES LACUESTA, José Maria. Director, Forum Atomico Espanol.
Address: Forum Atomico Espanol, 38 General Goded, Madrid 4, Spain.

CORTINES, Manuel G. Ind. Eng. Born 1901. Educ.: Tech. School of Ind. Eng., Madrid. Managing Director, Electra de Viesgo, S.A., 1950; Chairman, Compañia de Contadores, 1957; Exec. Vice-Pres., Centrales Nucleares del Norte, S.A., Nuclenor, 1957. Chairman, Advisory Com., Atomic Branch, Bilbao Res. Labs.; Pres., Ind. Branch, Chamber of Commerce and Industry of Santander; Spanish Advisory Com. on Ind. Reactors; Pres., Assoc. of Ind. Eng. of Santander; Vice-Pres., Spanish Atomic Ind. Forum 1962-. Society: Atomic Industrial Forum, Inc.
Nuclear interest: Management.
Address: Nuclenor, Medio 12, Santander, Spain.

CORTINI, Giulio, laurea in fisica. Born 1918. Educ.: Rome Univ. Gen. Phys. Teacher, Naples Univ. Societies: Soc. Italiana di Fisica; A.P.S.
Nuclear interest: Nuclear physics.
Address: Istituto di Fisica Sperimentale, 3 Via Antonio Tari, Naples 80138, Italy.

CORYELL, Charles D(uBois), B.S., Ph.D. Born 1912. Educ.: California Inst. Technol. and Munich T.H. Prof. Chem., M.I.T., 1946-; Louis Lipsky Visiting Prof., Weizmann Inst. Sci., Rehovoth Israel, 1953-54; Fulbright Lecturer, Inst. du Radium, Paris, 1963. Board of Directors, A.N.S. 1962-65; Board of Trustees, Windham Coll., Putney, Vt., 1963-; Mark Hopkins Coll. Brattleboro, Vt. 1968-. Book: Co-editor, Radiochem. Studies: the Fission Products (Nat. Nucl. Energy Series, Div. IV. Vol. 9, pp. 2251. New York, McGraw-Hill Publishing Co., 1951). Societies: A.P.S.; F.A.S.; A.N.S.; A.C.S.
Nuclear interests: Radiochemistry; The fission process and charge distribution in fission; Basic chemistry of separations, especially ion exchange; The rare earths.
Address: Room NW-13-200, Department of Chemistry, Massachusetts Institute of Technology, Cambridge 39, Mass., U.S.A.

COSANDEY, Maurice, Prof., Ecole Polytechnique, Lausanne Univ. Directeur-adjoint, Zwahlen and Mayr, S.A.; Vice-Pres., Therm-Atom A.G.
Address: Zwahlen and Mayr, S.A., Constructions Metalliques, Lausanne (Vaud), Switzerland.

COSENTINO, Jorge, Ing. Jefe, Div. de Reactores Nucleares, Comision Nacional de Energia Atomica; Participant, 3rd Inter-American Symposium on the Peaceful Application of Nucl. Energy, Rio de Janeiro, July, 1960.
Address: Division de Reactores Nucleares, Comision Nacional de Energia Atomica, 8250 Avenida del Libertador General San Martin, Buenos Aires, Argentina.

COSGAREA, Andrew, Jr., B.S.Ch.E., M.S. (Nucl. Eng.), Ph.D. Born 1934. Educ.: Michigan Univ. Director,Nucl. Reactor Lab. (1960-64), Assoc. Prof. of Chem. Eng. and of Metal. Eng. (1962-64), Oklahoma Univ. Societies: Nucl. Div., A.I.Ch.E.; American Inst. of Met. Engs.; A.S.E.E.
Nuclear interests: Thermodynamics; Metallurgy of nuclear materials; Fuels reprocessing.
Address: Oklahoma University, Norman, Oklahoma, U.S.A.

COSTA, Adelino Augusto NOGUEIRA DA.
See NOGUEIRA DA COSTA, Adelino Augusto.

COSTA, Angelo, Cav.Lav.Dott. Member, Comitato Economica e Sociale delle Comunita Europee; Member, Comitato permanente per i Problemi dell'Impiego Pacifico dell'Energia Nucleare e per i Rapporti con Euratom; Member, Editorial Com., Minerva Nucleare.
Address: 2 Via G. D'Annunzio, Genoa, Italy.

COSTA, Giovanni, Ph.D. (M.I.T.). Born 1930. Educ.: Padua Univ. Res. Asst., Lab. for Nucl. Sci., M.I.T., 1957-59; Asst. Prof., Bari Univ., 1959-60; Asst. Prof., Padua Univ., 1960-61; Res. Assoc. C.E.R.N., Geneva, 1962-63; Assoc. Prof., Padua Univ., 1963-.
Nuclear interests: Theoretical nuclear physics; Nuclear reactions, elementary particle interactions.

Address: 37 Porta Adige, Rovigo, Italy.

COSTA, Joao CARRINGTON SIMOES DA. See CARRINGTON SIMOES DA COSTA, Joao.

COSTA, Joaquin ORTEGA-. See ORTEGA-COSTA, Joaquin.

COSTA, Manuel Jose de CAMPOS. See DE CAMPOS COSTA, Manuel Jose.

COSTA, Ricardo Gomez. Secretario de Presidencia, Comision Nacional de Energia Atomica, Argentina.
Address: Comision Nacional de Energia Atomica, 8250 Av. del Libertador, Buenos Aires, Argentina.

COSTA RIBEIRO, Carlos. Res. Asst., Radioisotope Lab., Inst. de Biofisica, Brazil Univ.
Address: Radioisotope Laboratory, Istituto de Biofisica, Brazil University, 458 Avenida Pasteur, Rio de Janeiro, Gb., Brazil.

COSTA RIBEIRO, Uriel da. See da COSTA RIBEIRO, Uriel.

COSTACHEL, Octav, M.D., D.Sc., Prof. Born 1911. Educ.: Iasi Univ., Roumania. Director, Oncological Inst., Bucharest. Vice-Pres., Internat. Union Against Cancer (U.I.C.C.); Pres., Oncological Soc., Roumania. Books: Editor, Gen. Oncology (1961); Early Diagnosis of Cancer (1963); Complex Treatment of Cancer (1965); Editura Medicala, Bucharest. Societies: Roy. Soc. Medicine, England; N. York Acad. Sci.; Internat. Soc. Chemotherapy; Soc. de Pathologie Comparée, France.
Nuclear interests: Oncology; Radiobiology; Biochemistry; Medicine.
Address: Oncological Institute, No. 11 Bd. 1 Mai, P.B.5916, Bucharest 62, Roumania.

COSTAIN, N. L., B.Sc., A.M.I.C.E. Member, Council, Inst. of Nucl. Eng.
Address: Institute of Nuclear Engineers, 147 Victoria Street, Westminster, London S.W.1, England.

COSTALES, Manuel. Sen. Officer, Compania Euskalduna de Construccion y Reparacion de Buques, S.A.
Address: Compania Euskalduna de Construccion y Reparacion de Buques, S.A., 3 Plaza Sagrado Corazon, Bilbao, Spain.

COSTANTINI, Baldassarre. With Sindacalista. Italian Member, Specialised Nucl. Sect. for Economic Problems, Economic and Social Com., Euratom, 1966-.
Address: 81 Via Giovanni Verga Palazzina, 10 137 Rome, Italy.

COSTET, Francois. Born 1897. Educ.: Ecole Navale. Directeur, Div. Nucléaire, Cie. Francais Thomson-Houston, 1958-63; Pres. Directeur Gen., Sté. d'Etudes Techniques et d'Entreprises Generales (SODETEG).
Nuclear interests: Reactor design; Management.
Address: 9 avenue Reaumur, Le Plessis-Robinson, 92, France.

COSTIKAS, Athanassios, Dipl. Eng., M.S. (Elec. Eng.), Ph.D. Born 1933. Educ.: Nat. Tech. Univ. of Athens, Greece, Purdue Univ. and I.I.T. Instructor, Dept. of Elec. Eng., I.I.T., 1958-61; Formerly Res. Phys., Head of Solid State Phys. Group, Nucl. Res. Centre Democritus, Athens, Greece, 1962-. Society: A.P.S.
Nuclear interests: Solid state physics using reactor neutrons.
Address: Nuclear Research Centre Democritus, Aghia Paraskevi, Athens, Greece.

COSTRELL, Louis, B.S., M.S. Born 1915. Educ.: Maine, Pittsburgh and Maryland Univs. Chief, Rad. Phys. Instrumentation Sect., Nat. Bureau of Standards, Washington, D.C. Chairman, A.E.C. Com. on Nucl. Instrument Modules; Chairman, U.S.A.S.I. Com. Rad. Instruments; Administrative Com., I.E.E.E. Nucl. Sci. Group. U.S. Delegate, Internat. Electrotechnical Commission. Societies: I.E.E.E.; A.P.S.; Washington Acad. Sci.; Philosophical Soc.
Nuclear interest: Nuclear instrumentation.
Address: 10614 Cavalier Drive, Silver Spring, Maryland, U.S.A.

COT, Pierre. Tech. Manager, Basel, Ed. Zublin et Cie. A.G.
Address: Ed. Zublin et Cie. A.G., 6 Okenstrasse, 8031 Zurich, Switzerland.

COTECCHIA, Vincenzo, Prof. Director, Lab. for the Application of Radioisotopes to Hydrogeology, C.N.E.N.
Address: C.N.E.N., 83 Via Bellomo, Bari, Italy.

COTELO NEIVA, J. M., Prof. Director, Centro de Estudos de Mineralogia e Geologia (Coimbra), Minsterio da Educacao Nacional.
Address: Minsterio da Educacao Nacional, Avenida Rovisco Pais, Lisbon, Portugal.

COTTER, Francis P., Jr. Formerly Manager, Operations Analysis, Bettis Atomic Power Div., now Vice Pres., Westinghouse Electric Corp. Member, Ad Hoc Advisory Panel on Safeguarding Special Nucl. Materials and Member, Advisory Com. on Nucl. Materials Safeguards (1967-), U.S.A.E.C.
Address: Westinghouse Electric Corp., 1625 K Street, N.W., Washington, D.C., U.S.A.

COTTINI, Carlo, Radio Designer. Born 1925. Educ.: Istituto Radiotecnico Milan (Technical School). Nucl. Instrumentation Designer, C.I.S.E., 1949-. Society: Assoc. Elettrotecnica Italiana.
Nuclear interests: Nuclear instrumentation for physical researches.
Address: 10 Via Bellincione, Milan, Italy.

COTTON, Clare M. European Editor, Business Atomic Reports.
Address: Business Atomic Reports, 315 Riverside Drive, New York 25, New York, U.S.A.

COTTRELL, Alan Howard, B.Sc., M.A., Ph.D., F.R.S. Born 1919. Educ.: Birmingham Univ. Prof. Phys. Metal., Birmingham Univ., 1955; Deputy Head, Metal. Div., A.E.R.E., Harwell, 1955-58; Goldsmiths Prof. Metal., Cambridge Univ., 1958-. Societies: Inst. Metals; Iron and Steel Inst., Inst. Metallurgists.
Nuclear interests: Metallurgy; Radiation damage in solids.
Address: Department of Metallurgy, Pembroke Street, Cambridge, England.

COTTRELL, Stanley Arthur, B.Sc., Ph.D. Born 1925. Educ.: Birmingham Univ. Group Leader, Fabrication, U.K.A.E.A., Culcheth, 1952-56; Res. Manager, Radioactive Examination, U.K.A.E.A., D.E.R.E., 1956-64; Head of Technol., U.K.A.E.A., Atomic Energy Establishment, Winfrith, 1964-.
Nuclear interests: Fuel element development; Post-irradiation techniques; High burn-up performance.
Address: 35 Herringston Road, Dorchester, Dorset, England.

COTTRELL, William Barber, B.A. (cum laude, Maths. and Phys.). Born 1924. Educ.: Alfred Univ. Tech. Asst. to ANP Project Director (1950-53), Supervisor, Operation Aircraft Reactor Experiment (1953-54), Group Leader, Reactor Safety (1954-), Editor, Nucl. Safety (1957-), Director, Nucl. Safety Information Centre (1963-), Director, Nucl. Safety R. and D. Programme (1964-), O.R.N.L. Chairman, Special Subcom. Nucl. Standards Board, U.S.A.S.I. Society: A.N.S.
Nuclear interests: Nuclear safety as related to the location, design, construction, and operation of nuclear facilities; Fission product release and transport behaviour under accident conditions; Containment systems and other engineered safeguards.
Address: Oak Ridge National Laboratory, P.O. Box Y, Oak Ridge, Tennessee, U.S.A.

COUCHMAN, Donald L., B.S. (Mech. Eng.). Born 1931. Educ.: California Univ. Eng.-in-Charge, Jet Engine Testing, G.E.C., Cincinnati, 1953-55; Sen. Eng. on reactor components for naval nucl. power programme, U.S. Navy, assigned to U.S.A.E.C., Washington, D.C., 1955-58; Project Eng., Shippingport PWR, Naval Reactors Branch, Div. of Reactor Development, Atomic Energy Commission, 1958-60; Asst. Manager, Utility Programmes, (1960-65), Manager, Eng. (1965-), Nucl. Utility Services, Inc. Societies: A.N.S.; A.S.M.E.
Nuclear interests: Reactor safeguards and reactor operation; Reactor mechanical design and problems associated with handling of spent fuel.
Address: Nuclear Utility Services, Inc., 1730 M Street, N.W., Washington, D.C. 20036, U.S.A.

COUGHLIN, F. H. Trustee, Southwest Atomic Energy Associates.
Address: Southwest Atomic Energy Associates, 306 Pyramid Building, Little Rock, Arkansas, U.S.A.

COULLIETTE, James Horace, A.B., A.M., Ph.D. Born 1899. Educ.: Birmingham Southern Coll. and Columbia Univ. Director, Ind. Res. Inst., Chattanooga Univ., 1945-60; Prof., and Chairman, Phys. Dept., Miami Univ., Florida, 1960-65. Societies: A.P.S.; Optical Soc. of America.
Nuclear interest: Spectroscopy.
Address: 63 Lake Morton Drive, Lakeland, Florida 33801, U.S.A.

COULOMB, R. D.-es-Sc. Ing., Service de Minéralogie, C.E.A.
Address: Commissariat à l'Energie Atomique, 29-33 rue de la Federation, Paris 15, France.

COULON, J. L. de. See de COULON, J. L.

COULOT, Robert, Ing. de l'Ecole Polytechnique de Paris. Born 1924. Ing. Principal du Génie Maritime. Directeur du Bureau Nucl. de la Compagnie Gén. d'Electricité, Paris. Societies: Sté. Française de Radioélectriciens et Electroniciens; I.E.E.
Nuclear interest: Electric and electronic equipment for reactors.
Address: 205 boulevard Péreire, Paris 17, France.

COULTER, Henry. Geologist, Geological Survey, U.S. Dept. of the Interior.
Address: U.S. Department of the Interior, Geological Survey, Washington D.C. 20242, U.S.A.

COULTER, Thomas. Born 1892. Educ.: Edinburgh Univ. Formerly Director, Anglo-American Corp. of South Africa, Ltd.; Director, Vereeniging Estates Ltd.; Director, Vereeniging Brick and Tile Co. Ltd.; Deputy Chairman, South African Board Mills Ltd. Member, Commission of Inquiry into the Use of Nucl. Power in South Africa. Societies: S.A. Inst. of Mining and Metal.; S.A. Inst. of Mech. Eng.; S.A. Inst. of Mining Eng.; A.I.Ch.E.
Address: "Fairways", 12 Federation Road, Parktown, Johannesburg, South Africa.

COUNE, A., Dr. Service des Medecine, Inst. Jules Bordet (Centre des Tumeurs).
Address: Service des Medecine, Institut Jules Bordet (Centre des Tumeurs), 1 rue Heger-Bordet, Brussels, Belgium.

COURANT, Ernest D., B.A., M.S., Ph.D. Born 1920. Educ.: Swarthmore Coll. and Rochester Univ. Sen. Phys., Brookhaven Nat. Lab., 1948-; Prof., (part time), Yale Univ., 1961-67; Prof., (part time), State University of New York, Stony Brook, 1967-. Book: Co-author, Article on Proton Synchrotrons in Handbuch der Physik. Volume 44. (Springer, 1959). Societies: A.P.S.;

A.A.A.S.
Nuclear interest: Design and orbit theory of particle accelerators.
Address: Brookhaven National Laboratory, Upton, New York 11973, U.S.A.

COURBON, M. Director Pres., Tech. Com., Sté. d'Etudes et d'Equipements d'Entreprises.
Address: Société d'Etudes et d'Equipements d'Entreprises, 25 rue de Courcelles, Paris 8, France.

COUROUBLE, J. M. Member for France, Group of Experts on Production of Energy from Radioisotopes, O.E.C.D., E.N.E.A.
Address: O.E.C.D. European Nuclear Energy Agency, 38 boulevard Suchet, Paris 16, France.

COURRIER, Robert. Sec. Perpetuel, Acad. Sci. Member, Conseil Scientifique, Pres., Comité de Biologie, et Member, Conseil d'Enseignement de l'I.N.S.T.N., C.E.A., France.
Address: Commissariat a l'Energie Atomique, 29-33 rue de la Federation, Paris 15, France.

COURSAGET, Jacques. Member, European Soc. for Rad. Biol.
Address: European Society for Radiation Biology, c/o Professor E. H. Betz, Institut de Pathologie, 1 rue des Bonnes Villes, Liege, Belgium.

COURSAGET, Jean, Prof. Dr. Chef du Service de Biologie, C.E.A.; Co-Editor, Internat. J. of Appl. Rad. and Isotopes, Deleg. to 2nd I.C.P.U.A.E., Geneva, Sept. 1958.
Address: Commissariat à l'Energie Atomique, Centre d'Etudes Nucléaires de Saclay, B.P. 2, Gif-sur-Yvette, Seine et Oise, France.

COURT BROWN, William Michael, M.B., B.Sc., Ch.B., M.R.C.P.E., F.F.R. Born 1918. Educ.: St. Andrews Univ. Director, M.R.C., Clinical Effects of Rad. Res. Unit, Western Gen. Hospital, Edinburgh; Hon. Prof. in Cytogenetics, Faculty of Medicine, Edinburgh Univ. Society: British Inst. Radiol.
Nuclear interest: In vivo studies of chromosome damage in man, particularly clone formation in lymphocytes.
Address: Medical Research Council, Clinical Effects of Radiation Research Unit, Western General Hospital, Edinburgh 4, Scotland.

COURTIAL, Jean, M.D. Born 1903. Educ.: Paris Univ. Director, Fondation Curie, Paris. Vice-Pres., Ligue Nat. Française contre le Cancer. Societies: Sté. de Biologie; Sté. Française d'Electro-Radiologie Médicale; Royal Soc. of Medicine (Affiliate).
Nuclear interests: Medical application of radioactivity, specially to diagnosis and treatment of cancer and allied diseases.
Address: Fondation Curie, 26 rue d'Ulm, Paris 5, France.

COURTIER, Geoffrey Bernard, B.Sc., D.I.C. Born 1906. Educ.: London Univ. Asst. Sci. Adviser, Greater London Council. Societies: F.R.I.C.; Fellow, Chem. Soc.; Fellow, Royal Soc. of Health; Inst. of Fuel.
Nuclear interest: Environmental safety.
Address: County Hall, London, S.E.1, England.

COURTOIS, Guy, Ingénieur. Born 1931. Educ.: Ecole Centrale de Paris. Ing., Sect. des Applications des Radioélements, C.E.A.
Nuclear interest: Applications des radioéléments.
Address: Section Applications des Radioéléments, Centre d'Etudes Nucléaires de Saclay, B.P. No. 2, Gif-sur-Yvette, S. et O., France.

COURVOISIER, Peter L., Dr. Phil. Born 1915. Educ.: Munich Univ. Chief, Sect. on Safety of Nucl. Installations; Sec., Federal Commission on Safety of Nucl. Installations. Treas., Internat. Rad. Protection Assoc. Societies: Fachverband für Strahlenschutz; Schweizerische Gesellschaft von Fachleuten der Kerntechnik; Schweizerische Gesellschaft für Strahlenbiol.
Address: Section on Safety of Nuclear Installations, 5303 Würenlingen, Aargau, Switzerland.

COUTEUR, Kenneth James LE. See LE COUTEUR, Kenneth James.

COUTINHO BRAGA, Carlos AZEVEDO. See AZEVEDO COUTINHO BRAGA, Carlos.

COUTRIS, André. Prés. Directeur Gén., Electro-Entreprise, S.A.
Address: Electro-Entreprise, S.A., 32 rue de Mogador, Paris 9, France.

COUTURE, Jean. Born 1913. Sec. Gén. de l'Energie, Ministère de l'Industrie, 1963-; Chairman, Commission Consultative pour la Production d'Electricité d'Origine Nucléaire, 1967-. Chairman, Inst. Française des Combustibles et de l'Energie.
Address: 99 rue de Grenelle, Paris 7, France.

COUTURE, Pierre. Born 1909. Educ.: Ecole Polytech. et Ecole Nationale Supérieure des Mines de Paris. Administrateur Gén., Délégué du Gouvernement, C.E.A.
Address: Commissariat à l'Energie Atomique, 29-33 rue de la Federation, Paris 15, France.

COVA, Carlo, Dr. Eng. Born 1932. Educ.: Politecnico di Milan, Pennsylvania State Univ. and Argonne Nat. Lab.; Pres., Cova Associates. Society: F.I.E.N.
Nuclear interest: Industrial applications of nuclear energy.
Address: Cova Associates, 10 Corso Matteotti, Milan, Italy.

COVERT, Roger A., B.S. (Purdue), S.M. (M.I.T.), Sc.D. (M.I.T.). Born 1929. Educ.: Purdue Univ. and M.I.T. i/c Corrosion Eng. Product Development, Internat. Nickel Co., Inc., New York. Societies: A.I.M.E.;

Electrochem. Soc.; N.A.C.E.
Nuclear interests: Application of radioactive tracers to physical metallurgical research, aqueous and high temperature corrosion of metals, and quantitative chemical analysis of trace elements in metals.
Address: 94 Deepdale Drive, Middleton, New Jersey, U.S.A.

COVEYOU, Robert Reginald, B.S., M.A. Born 1916. Educ.: Chicago and Tennessee Univs. Mathematician, O.R.N.L. Societies: A.N.S.; A.C.S.; American Mathematical Soc.; Soc. Ind. and Appl. Maths.
Nuclear interests: Applied mathematics, in particular reactor theory and Monte Carlo methods.
Address: Oak Ridge National Laboratory, Post Office Box X, Oak Ridge, Tennessee 37830, U.S.A.

COVINO, Mario, Dr. Ing. Gen. Manager, SENN; Deputy Director in charge of Construction of Thermal and Nucl. Power Plants,Ente Nazionale per l'Energia Elettrica. Member, Comitato Nazionale, Assoc. Nazionale di Ingegneria Nucleare; Director, C.I.S.E.
Address: C.I.S.E., Casella Postale 3986, Milan, Italy.

COWAN, Clyde Lorrain, Jr., B.S., M.S., Ph.D., Sc.D., Guggenheim Fellow, 1957. Born 1919. Educ.: The School of Mines and Metal., Missouri Univ. Ordinary Prof. Phys., The Catholic Univ. of America, Washington, D.C. and Washington Univ., St. Louis, Mo. Consultant to: Smithsonian Inst.; Naval Ordnance Lab.; U.S. A.E.C.; Gen. Atomics Div., Gen. Dynamics Corp. Societies: Fellow, A.P.S.; Fellow, A.A.A.S.; Washington Philosophical Soc.
Nuclear interests: Nuclear physics; Cosmic ray physics; Physics of the elementary particles and physics at high energies.
Address: 11108 Waycroft Way, Wickford, Rockville P.O., Maryland, U.S.A.

COWAN, Frederick P., B.A., M.A., Ph.D. Born 1906. Educ.: Bowdoin Coll. and Harvard Univ. Head, Health Phys. Div., Brookhaven Nat. Lab., New York, 1947-. Expert Advisory Panel on Rad., W.H.O.; Subcom. 4, Heavy Particles, N.C.R.P.; I.C.R.U.; Nat. Council on Rad. Protection; Com. 3, External Rad., I.C.R.P. Societies: H.P.S.; A.P.S.; Rad. Res. Soc.; I.E.E.E.; A.A.A.S.
Nuclear interests: Health physics administration and research; Dosimetry and applied health physics for GeV accelerators; Gamma-ray dosimetry; Neutron dosimetry.
Address: Brookhaven National Laboratory, Upton, Long Island, New York, U.S.A.

COWAN, George Arthur, B.S., D.Sc. E.O. Lawrence Memorial Award, 1965. Born 1920. Educ.: Worcester Polytech. Inst. and Carnegie Inst. Technol. Assoc. Div. Leader, Los Alamos Sci. Lab. Societies: Fellow, A.P.S.; A.C.S.;

Fellow, A.A.A.S.
Nuclear interests: Nuclear chemistry; The fission process; Neutron cross sections.
Address: 721 42nd Street, Los Alamos, New Mexico, U.S.A.

COWART, William S., Jr. Vice Pres., Marketing and Customer Relations, Atlantic City Elec. Co.; Director and Chairman, Nucl. Energy Com., New Jersey Council for R. and D., 1964-.
Address: Atlantic City Electric Co., 1600 Pacific Avenue, Atlantic City, New Jersey, U.S.A.

COWHERD, Sidney John, B.Sc. (Eng.). Born 1928. Educ.: London Univ. Sturtevant Eng. Co., 1952; At (1956-), now Chief Eng., Nucl. Power Dept. (1965-), Kennedy and Donkin, Consulting Eng.
Nuclear interests: The integration of nuclear reactors into power systems; Reactor design and the management of nuclear power projects.
Address: Kennedy and Donkin, Premier House, Woking, Surrey, England.

COWING, Russell Francis. Born 1907. Educ.: Wentworth Inst., Boston. Consultant to Commonwealth of Mass. on radiol. problems; Consultant Radiol. Phys. for Veterans Administration. Book: Protection in Diagnostic Radiol. (Rutgers Univ. Press, 1959, ed. Sonnenblick); Rad. Exposure Survey of X-ray and Isotope Personnel. Societies: H.P.S.; N.E. Roentgen-Ray Soc.; American Coll. Radiol.; A.N.S.
Nuclear interests: Radiological physics and health physics as they pertain to the use of X-rays, gamma rays and isotopic material in medicine and in industrial non-destructive inspection.
Address: Cancer Research Institute, 194 Pilgrim Road, Boston, Massachusetts 02215, U.S.A.

COWLEY, William. Born 1910. Chief Accountant, Directorate of Contracts, Ministry of Works, 1950-54; Chief Accountant, Internal Audit, U.K.A.E.A., London, 1954-57; Chief Accountant, U.K.A.E.A., Atomic Energy Establishment, Winfrith, 1957-.
Nuclear interest: Financial administration.
Address: Atomic Energy Establishment, Winfrith, Dorset, England.

COWLISHAW, Brian, B.Sc. (London), A.R.C.S., Ph.D. (Sydney). Born 1926. Educ.: London Univ. Post-doctoral Fellow, Australian Inst. Nucl. Sci. and Eng., 1963-66; Biochem. Dept., Faculty of Medicine, Singapore, 1966-.
Nuclear interest: Effects of atomic radiations on biological macromolecules, especially D.N.A.
Address: Biochemistry Department, Faculty of Medicine, Singapore 3.

COWPER, George, B.Sc. Born 1921. Educ.: Durham Univ. Atomic Energy Div., N.R.C., Chalk River, Ontario, 1948-52; Electronics Branch (1948-57), Head, Rad. Dosimetry Branch (1958-67), Head, Health Phys. Branch

(1967-), A.E.C.L., Chalk River, Ontario. Societies: A.Inst.P.; I.E.E.; H.P.S.; Canadian Assoc. Phys.
Nuclear interests: Design and development of instruments for radiation protection; Radiation dosimetry; Health physics.
Address: 4 Cartier Circle, Deep River, Ontario, Canada.

COX, Ellis Thurman, B.Sc. Born 1919. Educ.: New Hampshire and Houston (Texas) Univs. Project Eng., Advanced Submarine Fleet Reactor, 1955; Manager, Submarine Fleet Reactor Control Sect., 1955; Manager, Submarine Fleet Reactor Systems Design Sect., 1955-57; Manager, Submarine Fleet Reactor Power Plant Subdiv., 1957-58; Manager, AIW Power Plant Subdiv., 1958-59; Manager, Tech. Dept., Surface Ship Project, 1959-60; Manager, Surface Ship Project, 1960-.
Address: 242 Oakcrest Lane, Pittsburgh 36, Pennsylvania, U.S.A.

COX, James Albert, B.S. Born 1916. Educ.: Washington State Univ. Manager, Radioisotope Distribution (1946-50), Supervision of Reactor Operations (1950-57), Supt. of Reactor Operations and Radioactive Waste Disposal (1957-62), Supt. of Operations Div.-Reactor and Hot Cell Operation and Radioactive Waste (1962-64), Operations Div., O.R.N.L. Chairman, Reactor Operations Div., and Ex Officio Member, Professional Divisions Com., (1966-), A.N.S. Book: Manual for the Operation of Res. Reactors (I.A.E.A., 1965).
Nuclear interests: Reactor operation, hot-cells operation, and waste disposal.
Address: 112 Everest Circle, Oak Ridge, Tennessee, U.S.A.

COX, Jean A.M., Born 1923. Educ.: Utrecht and Leiden Univs. Prof., Leiden State Univ.
Nuclear interest: Nuclear physics - high energy physics.
Address: Institute Lorentz, 18 Nieuwsteeg, Leiden, Netherlands.

COX, Raymond John, B.Sc., A.M.I.E.R.E. Born 1921. Educ.: London Univ. S.S.O., Electronics Div., A.E.R.E., Harwell (1949), Sen. Principal Sci., Leader of Reactor Instrumentation Group, Electronics Div. (1956), Head, Control and Instrumentation Branch, Atomic Energy Establishment, Winfrith (1959), Deputy Chem. Sci., Control and Instrumentation Div., Atomic Energy Establishment, Winfrith (1960), U.K.A.E.A. Papers and Programmes Com., and Assoc. Member, I.E.R.E.; I.E.R.E. Rep., B.N.E.S.
Nuclear interest: Reactor design.
Address: Sherbrooke, High Park Road, Broadstone, Dorset, England.

COZZA, A. Member, Editorial Board, Minerva Nucleare.
Address: Minerva Nucleare, 83-85 Corso Bramante, Turin, Italy.

CRAIG, Harmon, Ph.D., M.S. Born 1926. Educ.: Chicago Univ. Prof., Geochem., Chairman, Div. of Earth Sci., Scripps Inst. of Oceanography, California Univ., San Diego. Book: Editor, Isotopic and Cosmic Chem. (North-Holland, 1964). Societies: A.G.U.; A.A.A.S.
Nuclear interests: Nuclear geochemistry; Isotope chemistry; Mass spectrometry; Atmospheric and marine chemistry.
Address: Scripps Institution of Oceanography, California University, San Diego, La Jolla, California, U.S.A.

CRAIG, James Morrison, A.B., M.A. Born 1916. Educ.: San Jose State Coll. and Stanford Univ. At San Jose State Coll. (1948-), now Assoc. Prof. Microbiology. Societies: American Inst. Biol. Sci.; Soc. American Microbiol.
Nuclear interests: Radiation biology; Isotope technology.
Address: 2524 Gerald Way, San Jose, California, U.S.A.

CRAM, S. Winston, Ph.D., B.A. Born 1907. Educ.: Wisconsin Univ. and St. Olaf Coll. Head, Phys. Sci., Kansas State Teachers Coll. Director and Assoc. Director, Nat. Sci. Foundation Insts.; Exec. Board Member, A.A.P.T. Societies: A.A.A.S.; A.Inst.P.
Nuclear interests: Instructor in two courses in nuclear physics.
Address: Kansas State Teachers College, Emporia, Kansas, U.S.A.

CRAMER, Thomas L., A.B., M.S. Born 1927. Educ.: Colgate Univ. and Carnegie Inst. Technol. Asst. Editor (1952-57), Assoc. Editor (1957-67), Nucleonics magazine; Sen. Tech. Editor, Chem. Eng. Div., Argonne Nat. Lab. Society: A.N.S.
Nuclear interests: Editorial; Information; Materials; Fuel; Fabrication; Fuel processing.
Address: Chemical Engineering Division, Argonne National Laboratory, Argonne, Illinois 60439, U.S.A.

CRANBERG, Lawrence, Ph.D. Born 1917. Educ.: C.C.N.Y., Harvard and Pennsylvania Univs. Consultant, Texas Nucl. Corp., Japan Atomic Energy Res. Inst., and Swedish Atomic Energy Co. Los Alamos Sci. Lab., Los Alamos, 1950-63; Prof. of Phys., (1963-), Director, Phys. Accelerator Lab. (1965-), Virginia Univ. Member, Nucl. Cross Section Advisory Group, U.S.A.E.C.; Ethics Com., Eng. Council for Professional Development; Participating Sci., Visiting Sci. Programme, A.I.P.; Guggenheim Fellow, 1961. Books: Chapters in Nucl. Spectroscopy (ed. Ajzenberg-Selove; Academic Press, 1960) and in Progress in Nucl. Energy (Series 1, Pergamon Press, 1956). Societies: Fellow, A.P.S.; A.A.A.S.
Nuclear interest: Neutron spectrometry, nuclear reactions, nuclear forces, few nucleon problems.
Address: 1934 Blue Ridge Road, Charlottesville, Virginia, U.S.A.

CRANDALL, J. U.S.A. Member, European American Com. on Reactor Phys., O.E.C.D., E.N.E.A.
Address: O.E.C.D. European Nuclear Energy Agency, 38 boulevard Suchet, Paris 16, France.

CRANDALL, John Lou, B.S.(Chem. Eng.), Ph.D. (Phys. Chem.). Born 1920. Educ.: M.I.T. Res. Manager, Exptl. Phys. Div., Savannah River Lab. (1953-67), Asst. Director, Advanced Operational Planning, Atomic Energy Div. (1967-), E.I. du Pont de Nemours and Co. Societies: Fellow, A.N.S.; A.C.S.
Nuclear interests: Reactor physics experimentation; Heavy water reactors; Operations planning.
Address: Savannah River Laboratory, E. I. du Pont de Nemours and Company, Aiken, South Carolina 29801, U.S.A.

CRANDALL, Walter Ellis, B.S., Ph.D. Born 1916. Educ.: Worcester Polytech. Inst., Stanford and California, Berkeley, Univs. Rad. Lab., California Univ., 1953; Lawrence Rad. Lab., Livermore, 1953-61; Consultant, Boeing Airplane Co., 1957-61; Northrop Corp., 1961-.
Nuclear interests: Nuclear weapon design; Nuclear weapon's effects; High energy nuclear physics; Nuclear instrumentation.
Address: 21930 Carbon Mesa Road, Malibu, California, U.S.A.

CRANSHAW, Thomas Edwin, M.A., Ph.D. Born 1922. Educ.: Cambridge Univ. At A.E.R.E., Harwell. Book: Cosmic Rays (O.U.P., 1963).
Nuclear interests: Cosmic rays, Mössbauer effect.
Address: Atomic Energy Research Establishment, Harwell, Berks., England.

CRANSTON, Frederick Pitkin, Jr., B.A., M.S., Ph.D. Born 1922. Educ.: Colgate and Stanford Univs. Staff Member, Los Alamos Sci. Lab., 1953-. Chairman, Los Alamos Chapter, Federation American Sci. Societies: A.P.S.; A.A.P.T.
Nuclear interests: Beta- and gamma-ray spectroscopy.
Address: P.O. Box 1663, Los Alamos, New Mexico, U.S.A.

CRASEMANN, Bernd, Ph.D. Born 1922. Educ.: California (Berkeley) Univ. Prof. Phys., Oregon. Consultant, California Univ. Rad. Lab. Societies: A.P.S.; A.A.P.T.
Nuclear interests: Nuclear spectroscopy; Nuclear models.
Address: Department of Physics, Oregon University, Eugene, Oregon, U.S.A.

CRAWFORD, Cyril G., M.Sc. Radona Irradiations A.B. Society: Assoc., Roy. Inst. of Chem.
Nuclear interest: Radiation sterilisation of medical products and irradiation of food.
Address: Radona Irradiations A.B., Skärhamn, Sweden.

CRAWFORD, George Wolf, B.S., M.A., Ph.D. Born 1922. Educ.: Trinity (San Antonia, Texas) and Texas Univs. Assoc. Prof. Phys., Clemson Coll., South Carolina, 1951-55; Asst. Director i/c Univ. Div. of Texas Petroleum Res. Com., Texas Univ., 1955-59; Chief, Rad. Phys. Sect., Radiobiol. Branch School of Aerospace Medicine, Brooks A.F.B., Texas, 1959-63; Prof. of Phys., Southern Methodist Univ., Dallas, Texas, 1963-. Societies: A.P.S.; Texas Acad. Sci.; A.N.S.
Nuclear interests: Design of charge mass energy spectrometers; Stopping power measurements; Monte Carlo stopping power and range calculations which include nuclear interactions; Effect of radiation on various detectors and physical objects as function of type and energy of all ionising radiations and neutrons; Dose measurements - i.e., energy absorbed by irradiated specimen.
Address: 9440 Brentgate, Dallas, Texas 75238, U.S.A.

CRAWFORD, James Homer, Jr., B.S., Ph.D. Born 1922. Educ.: North Carolina Univ. Assoc. Director, Solid State Div., O.R.N.L., 1951-67; Chairman, Phys. Dept., North Carolina Univ., 1967-. Editor, J. Appl. Phys., 1960-64. Societies: A.P.S.; A.A.A.S.
Nuclear interest: Radiation effects in solids.
Address: Department of Physics, North Carolina University, Chapel Hill, North Carolina 27514, U.S.A.

CRAWFORD, John E., B.A., M.A., Ph.D. Born 1933. Educ.: Toronto and McGill Univs. Asst. Prof., Phys., McGill Univ. Society: Canadian Assoc. of Phys.
Nuclear interests: Nuclear physics; Accelerator design; Nuclear instruments.
Address: Foster Radiation Laboratory, McGill University, Montreal, Quebec, Canada.

CRAWFORD, John E., B.S. Born 1924. Educ.: Johns Hopkins Univ. Formerly Head, Marine Minerals Tech. Centre, and Chief Nucl. Eng., U.S. Bureau of Mines; Now Founder and Pres., Crawford Marine Specialists, Inc. Societies: A.N.S.; Marine Technol. Soc.; American Soc. Oceanography.
Nuclear interests: Radioisotope application; Nuclear power in marine science and engineering.
Address: Crawford Marine Specialists, Inc.; Piers 38-40 The Embarcadero, San Francisco, California 94107, U.S.A.

CRAWFORD, Thomas Henry, Dr. Assoc. Prof. of Chem. Coll. of Arts and Sci., Louisville Univ.
Nuclear interest: Nuclear magnetic resonance.
Address: 2706 Woodmere Drive, Louisville, 16, Kentucky, U.S.A.

CRAWLEY, Ralph Hugh Alfred, B.Sc., A.R.C.S. Born 1925. Educ.: London Univ. Analytical Chem. Group Leader, Res. Dept., A.E.I., Rugby, 1950-58; Chem. Sect. Leader,

Atomic Power Div., English Elec. Co., 1958-66; Head Chem. Office, Nucl. Design and Construction Ltd., 1966-67. Societies: F.R.I.C.; Soc. Analytical Chem.
Nuclear interest: Chemistry.
Address: 136 Chester Road, Blaby, Leicester, LE8 3HA, England.

CRAYA, Antoine Edouard, D. ès Sc. Born 1911. Educ.: Paris Ecole Polytechnique and Grenoble Univ. Prof. à la Faculté des Sci. Grenoble Univ.; Conseiller Scientifique au Centre d'Etudes Nucléaires de Grenoble; Conseiller Scientifique Establissements S.O.G.R.E.A.H. (Grenoble).
Nuclear interest: Heat transfer in nuclear reactors.
Address: 9 Rue Charles Péguy, Grenoble, France.

CREAGAN, Robert Joseph, B.S. (Eng., Illinois Inst. Technol.), M.S., Ph.D. (Phys., Yale). Born 1919. Educ.: Illinois Inst. Technol. and Yale Univ. Eng.Manager, Westinghouse Commercial Atomic Power; Director, Nucl. Programme, Bendix Aviation; now Sen. Consultant, Westinghouse Atomic Power Div. Societies: A.P.S.; A.N.S.
Nuclear interests: Reactor design and engineering.
Address: 2305 Haymaker Road, Monroeville, Pennsylvania, U.S.A.

CREAMER, Willis B. Manager, Dayton (Miamisburg, Ohio) Area, Albuquerque Operations Office, U.S.A.E.C.
Address: Albuquerque Operations Office, U.S. Atomic Energy Commission, Post Office Box 5400, Albuquerque, New Mexico, U.S.A.

CREE, Albert A. Chief Exec. Officer (-1968), and Chairman, Central Vermont Public Service Corp.; Director, Yankee Atomic Elec. Co.; Director, Connecticut Yankee Atomic Power Co.
Address: Central Vermont Public Service Co., 77 Grove Street, Rutland, Vermont, U.S.A.

CREMER, Erika Hildegard, Dr. Phil., Dr. phil. habil. Educ.: Berlin, Freiburg i.Br., and Munich Univs. and M.I.T., Cambridge, U.S.A. Asst. of Georg von Hevesy, Freiburg i.Br. (1928-31), of M. Polany, K.W.I. f. physikal. Chemie, Berlin-Dahlem (1931-33), of Otto Hahn, K.W.I. f. Chemie, Berlin-Dahlem (1937-38), of M. Bodenstein, Berlin Univ. (1938-39), of O. Diebner, K.W.I. f. Physik, Berlin-Dahlem (1940). Asst. Prof. (1940-51), a.o. Prof. (1951-59) and Head, Inst. of Phys. Chem., o. Prof. (1959-), Innsbruck Univ. Books: Homogeneous Catalysis of the o-p-Hydrogen Conversion; Heterogeneous Catalysis of the o-p-Hydrogen Conversion (in Handbook on Catalysis, vol. i. and vi 1940); co-author, Kinetic der Gasreaktionen (Verlag W. de Gruyter, 1961). Societies: Deutsche Bunsengesellschaft; Verein Österreichischer Chemiker; Mathematisch-Physikalische Gesellschaft Innsbruck; Internationale Planseegesellschaft für Pulvermetallurgie; Österreichische Studiengesellschaft für Atomenergie; Österreichisches College-Collegegemeinschaft Innsbruck; I.F.U.W. Korrespondierendes Mitglied der österreichischen Akademie der Wissenschaften.
Nuclear interests: Kinetics of gases, reactions in the solid state, catalysis, adhesion, adsorption, gaschromatography, separation of isotopes, thin layers of metals.
Address: 1a/II Peter Mayrstrasse, Innsbruck, Austria.

CREMIERS, Roger Jacques AUGIER de. See AUGIER de CRÉMIERS, Roger Jacques.

CREN, Thomas Bernard LE. See LE CREN, Thomas Bernard.

CREPELLE, Jacques. Manager, Crépelle et Cie.
Address: Crépelle et Cie., Porte de Valenciennes, Lille, (Nord), France.

CREPELLE, Pierre J. Managing Director, Crépelle et Cie.
Address: Crépelle et Cie., Porte de Valenciennes, Lille, (Nord), France.

CRESPI, Martin B. A., M.Sc., Ph.D. (Chem.). Born 1921. Educ.: Buenos Aires Univ. Instructor, Gen. and Inorg. Chem. (1948-53), Asst. Prof. Inorg. Chem. (1955-58), Prof. Inorg. Chem. (1958-60), Buenos Aires Univ.; Postdoctoral fellowships in radiochem., Durham and Cambridge Univs., 1953-54; Res. Officer, Chem. Group (1952-55), Head, Inorg. Chem. Div. (1955-58), Head, Chem. Dept. (1958-60), Director of Res. (1963-), Argentine A.E.C.; Professional Officer, Dept. of Res. and Isotopes, Seibersdorf Lab., I.A.E.A., Vienna (1960-63). Titular Member, Commission on Analytical Radiochem. and Nucl. Materials, Anal. Chem. Div., I.U.P.A.C., 1965-. Societies: Argentine Chem. Soc.; Argentine Phys. Soc.
Nuclear interests: Inorganic radiochemistry, with emphasis on analytical and tracer chemistry and on the chemistry of nuclear materials; Research programming.
Address: Comision Nacional de Energia Atomica, 8250 Avenida del Libertador, Buenos Aires, Argentina.

CRESPI GONZALEZ, Maria Alicia, Dr. Chem. Sci. Born 1928. Educ.: Central Univ., Madrid. Coll. Inst. Especial de la Grasa y sus Derivados (Patronato Juan de la Cierva, C.S.I.C.) 1950-52; Jefe Sección Piritas Españolas I.N.I., 1952-57; Catedrático de Mercancias Escuelas de Comercio 1957-; Investigador División Ingenieria Junta de Energia Nucl. 1957-; Societies: Real Sociedad Española de Fisica y Quimica; Colegio Oficial de Quimicos; Sté. Française de Radioprotection.
Nuclear interests: Design and construction of nuclear reactors; Nuclear safety and engineering related with nuclear safety: nuclear filtration and ventilation systems, nuclear containments, radiological protection systems, radioactive

waste storage, transport and disposal; Inspection of reactors; construction and tests.
Address: 19 - 5º C, Avenida General Perón, Madrid 20, Spain.

CRESPO, Godofredo GOMEZ. See GOMEZ CRESPO, Godofredo.

CRESPO, V. PEREIRA. See PEREIRA CRESPO, V.

CRESTI, Marcello, Dr. Phys. Born 1928. Educ.: Pisa Univ. Scuola Normale Superiore, Pisa. Prof., Padua Univ. Bubble Chamber Group, I.N.F.N., Padua Sect., Padua Univ. Society: Soc. Italiana di Fisica.
Nuclear interests: Elementary particle physics; Automatic measuring of bubble chamber pictures; Transport and separation of beams from high energy accelerators.
Address: Istituto di Fisica, Padua University, 8 Via F. Marzolo, Padua, Italy.

CRETTE, M. Bureau d'Etudes, Chaudronnerie Industrielle et Constructions Spéciales S.A.
Address: Chaudronnerie Industrielle et Constructions Spéciales S.A., Vieux Chemin de Saint-Denis, Noisy-le-Sec, (Seine), France.

CRETTEZ, Jean Pierre. Directeur-Adjoint, Lab. Maurice de Broglie, Ecole Pratique des Hautes Etudes.
Address: Ecole Pratique des Hautes Etudes, Laboratoire Maurice de Broglie, 63 rue Alfred Leblance, Brətigny-sur-Orge, Essonne, France.

CREUTZ, Edward (Chester), B.A.,Mendenhall Fellow, Ph.D. (Phys.). Born 1913. Educ.: Wisconsin Univ. Prof. and Head, Dept. of Phys., and Director of Nucl. Res. Centre, Carnegie Inst. Technol., 1949-55; Director, John Jay Hopkins Lab. for Pure and Appl. Sci., and Vice-Pres., R. and D. Gulf Gen. Atomic, 1955-. Consultant, California Univ. Rad. Lab., Livermore, 1956; Nat. Sci. Foundation, 1950-. Sci. at Large, Res. Div., U.S.A.E.C., 1955-56. Visiting Sci., A.I.P., 1958. Societies: A.S.E.E.; Fellow, A.P.S.; A.A.P.T.; Phys. Soc. of Pittsburgh; A.A.U.P.; Pennsylvania Acad. of Sci.; N.Y. Acad. Sci.; Fellow, A.N.S.; Inst. of Radio Eng.
Nuclear interests: Nuclear physics and metallurgy.
Address: Paseo Delicias and Valle Plateada, Rancho Santa Fe, California, U.S.A.

CREW, William Henry, B.S., M.A., Ph.D. Born 1899. Educ.: U.S. Naval Acad. and Johns Hopkins Univ. Dean, Coll. Eng. Sci., U.S.A.F. Inst. Technol., 1948-50; Asst. Director for Sci. Personnel, Los Alamos Sci. Lab., 1950-65. Societies: Fellow, A.P.S.; Fellow, A.A.A.S. Nuclear interest: Management.
Address: 815 Camino de Fray Marcos, Tucson, Arizona 85718, U.S.A.

CREWE, Albert Victor, B.Sc. and Ph.D. (Phys.). Born 1927. Educ.: Liverpool Univ. Assoc. Prof.

Phys., Chicago Univ., 1956-58; Director, Particle Accelerator Div. (1958-61), Director (1961-67), Argonne Nat. Lab.; Prof. Phys., Chicago Univ., 1963-. Societies: Fellow, A.P.S.; Fellow, A.N.S.; RESA; E.M.S.A.
Nuclear interests: High energy physics; Particle accelerators; Electron microscopy.
Address: Chicago University, Enrico Fermi Institute for Nuclear Studies, 5630 Ellis Ave., Chicago, Illinois 60637, U.S.A.

CRISP, E. R., Dr. Formerly Pres., now Convenor, Fellowship Board. Coll. of Radiologists of Australasia; Formerly Member, Medical Rad. Com., Nat. Health and Medical Res. Council.
Address: College of Radiologists of Australasia, 12 Collins Street, Melbourne, Victoria 3000, Australia.

CRISTOBAL, Napoleon F., Bachelor of Laws. Born 1920. Educ.: Manila Law Coll. Practiced Law in Courts of Manila and its suburbs, 1950-54; Special Prosecutor, Office of the Provincial Fiscal, Provincial Govt. of Cebu, Philippines, 1954-59; Sen. Legal Officer (1959-64), Chief Legal Officer (1964-), Philippine A.E.C. Sec., Nucl. Indemnity Board, Philippines; Sec., Nat. Com. on the Safe Transport of Radioactive Materials in the Philippines.
Nuclear interests: Legal aspects of atomic energy. In particular: Licensing and regulation concerning the acquisition, possession and use of radioactive materials; Legislation on the licensing and regulation of the construction and operation of nuclear power plants; Insurance and third party liability aspects involved in reactors, especially power reactors;Rules of international law concerning the peaceful uses of atomic energy.
Address: Philippine Atomic Energy Commission, 727 Herran Street, Malate, Manila, Philippines.

CRISTOFORI, Franco, Ing. navale e meccanico (naval architect). Born 1912. Educ.: Genoa Univ. Vice Gen. Manager, Italcantieri, Trieste.
Address: Italcantieri, 1 Corso Cavour, Trieste, Italy.

CRITCHFIELD, Charles L., B.S., M.A., Ph.D. Born 1910. Educ.: George Washington Univ. Prof. Phys., Minnesota Univ., 1947-55; Director, Sci. Res., Div. Gen. Dynamics, Convair, 1955-60; Vice-Pres., Res, Telecomputing Corp., 1960-61; Assoc. Div. Leader, Los Alamos Sci. Lab., 1961-. Book: Co-author, Atomic Nucleus and Nucl. Energy Sources (Oxford, 1949). Societies: Fellow, A.P.S.; A.R.S.; A.A.U.P.; A.A.P.T.
Nuclear interests: Basic research; Scattering; Polarization; Cosmic rays; Field theory.
Address: 391 El Conejo, Los Alamos, N. Mex. 87544, U.S.A.

CRITOPH, Eugene, B.A.Sc., M.A.Sc. Born 1929. Educ.: British Columbia Univ. Sen. Res. Officer, Reactor Phys. Branch (1953), now

Head, Appl. Reactor Phys. Branch, A.E.C.L. Chairman, European American Reactor Phys. Com., E.N.E.A. Society: Canadian Assoc. Phys.
Nuclear interests: Reactor physics and reactor design.
Address: 4 Darwin Crescent, Deep River, Ontario, Canada.

CRITTENDEN, J. R. Product Specialist, Severe Environments, Gen. Elec. Co.
Address: 316 E. Ninth Street, Owensbro, Kentucky, U.S.A.

CRIVELLI, Roger Leo, B.A., B.Com. Born 1907. Educ.: Melbourne Univ. Commercial Manager (1955), Exec. Asst. (1958-), Director, Information Services (1960), Australian A.E.C. I.A.E.A., Vienna, 1963-66. Society: Fellow, Roy. Economic Soc.
Nuclear interests: Industrial and economic aspects.
Address: Australian Atomic Energy Commission, Box 41, Coogee, New South Wales, Australia.

CROACH, Jesse W., Jr., A.B. Born 1918. Educ.: Harvard Univ. Chicago Liaison Office (1951-53), Manager, Theoretical Phys. Div., Savannah River Lab. (1953-54), Director, Pile Eng. and Materials Sect. (1954-59), Gen. Supt., Works Tech. Dept., Savannah River Plant (1960-61), Asst. Director Savannah River Lab. (1962-63), Asst. Tech. Director, Wilmington (1964-67), Tech. Director (1967-), E. I. du Pont de Nemours and Co., Explosives Dept., Atomic Energy Div.
Nuclear interests: Nuclear reactor physics; Reactor engineering; Fuel element development; Reactor safety; Large-scale isotope production.
Address: Church Hill Road, RD 2, Landenberg, Pennsylvania 19350, U.S.A.

CROATTO, Ugo, Dr. Chem. (Padua). Born 1914. Educ.: Padua Univ. Prof. Gen. and Inorganic Chem., Padua Univ.; Director, Centre of Nucl. Chem., Italian Nat. Res. Council.; Director, School of Nucl. Chem., Padua Univ.
Nuclear interest: Nuclear chemistry.
Address: 6 Via Loredan, Padua, Italy.

CROCCHI, Peter Andrews, M.D. (Buenos Aires). Born 1901. Argentine Nat. A.E.C., 1954-. Inter-American Congress of Radiol., 1943; Nat. Congress of Radiol., 1945, 1952, 1957. Prof. Radiol. and Radiotherapy, La Plata Univ. (Medical School). Specialist in Radiol., Argentine Navy. Chief of Radiology Services, Buenos Aires. Book: Co-author, Medical Uses of Radioisotopes (in: Radiology and Physiotherapy, 5th ed.). Societies: Argentine Medical Assoc.; Argentine Radiol. Assoc.; Inter-American Coll. of Radiol.; Radiol. Assoc. Panama; Municipal Medical Assoc.
Nuclear interests: Application of radioisotopes in medicine, with special reference to ^{60}C (teletherapy) and ^{32}P in therapy.
Address: 1002 Billinghurst Street, Buenos Aires, Argentina.

CROCKER, Victor Simeon, B.Sc., Ph.D. Born 1927. Educ.: Birmingham Univ. Sci. Officer (1951), -Head, Phys. Dept., Res. Reactors Div. (1960-), U.K.A.E.A., A.E.R.E., Harwell. Society: Inst. Phys.
Nuclear interests: Design and use of research reactors.
Address: United Kingdom Atomic Energy Authority, Atomic Energy Research Establishment, Harwell, Berkshire, England.

CROCKET, James Harvie, B.Sc. (New Brunswick), B.Sc. (Oxford), Ph.D. (M.I.T.). Born 1932. Educ.: New Brunswick and Oxford Univs., and M.I.T. Assoc. Prof. Geology, McMaster Univ. Societies: Geochem. Soc.; A.G.U.
Nuclear interests: Neutron activation analysis; Radioactive decay of natural radionuclides and applications in geochronology.
Address: Room 301, Senior Science Building, McMaster University, Hamilton, Ontario, Canada.

CROCKETT, J. R. Gen. Manager, Radiol. Sci. Dept., Nevada Test Site, operated for U.S.A.E.C. by Reynolds Elec. and Eng. Co., Inc.
Address: Nevada Test Site, Radiological Sciences Department, A.E.C. Contract Office, P.O. Box 1360, Las Vegas, Nevada 89101, U.S.A.

CROFT, Huber O. Chairman, Res. Reactor Com., Missouri Univ.
Address: Missouri University, College of Engineering, Columbia, Missouri, U.S.A.

CROFT, John Frederick, B.Sc., Ph.D. Born 1925. Educ.: London Univ. Lecturer in Phys., Royal Military Coll. of Sci., Shrivenham, 1952-58; Health Phys., U.K.A.E.A., 1958-64; Dept. Head, Sci. and Maths., Rugby Coll.of Eng. Technol., 1964-. (At present Health Phys. Branch, A.E.E., Winfrith).
Nuclear interests: Reactor materials and engineering applications of radioisotopes; Experimental study of radiological problems in nuclear reactor operation, especially radiotoxic aerosols; Evolution of general safety design concepts.
Address: Rugby College of Engineering Technology, Eastlands, Rugby, Warwickshire, England.

CROLL, Millard Norval, A.B., M.D. Born 1922. Educ.: Pennsylvania Univ. and Jefferson Medical Coll. Assoc. Prof. of Radiol., Hahnemann Medical Coll. and Attending Radiol., Hahnemann Hospital, Philadelphia; Attending Radiol., Embreeville State Hospital; Director, Clinical Isotope Sect., Dept. Radiol., Hahnemann Hospital, Philadelphia. Board of Trustees, and Treasurer, Soc. of Nucl. Medicine; Consultant in Radioisotopes, Newcomb Hospital, Vineland,

New Jersey; Consultant in Radioisotopes, Veterans Administration Hospitals. Book: Recent Advances in Nucl. Medicine (Appleton-Century-Crofts, 1966). Societies: Soc. of Nucl. Medicine; Radiol. Soc. of North America; A.M.A.;Pennsylvania State Medical Soc.; American Radium Soc.; Assoc. of Univ. Radiol. Nuclear interest: Clinical nuclear medicine.
Address: The Hahnemann Medical College and Hospital, 230 North Broad Street, Philadelphia 2, Pennsylvania, U.S.A.

CROMER, Alan Herbert, B.S., Ph.D. Born 1935. Educ.: Wisconsin and Cornell Univs. Res. Fellow, Harvard Univ.; Assoc. Prof., Northeastern Univ., Boston. Societies: A.P.S.; A.A.A.S. Nuclear interests: Theoretical nuclear physics; High energy nucleon-nucleon and nucleon-nucleus scattering and polarization; Nucleon-nucleon interaction and nuclear structure.
Address: Northeastern University, Boston, Mass., U.S.A.

CROMER, Sylvan J., M.S. (Eng.), B.S. (Mech. Eng.). Born 1906. Educ.: Oklahoma Univ. Formerly Vice-Pres., Union Carbide Corp. Nucl. Div., 1958. Societies: A.S.M.E.; A.I.M.M.E. Nuclear interests: Reactors; Production of fissionable material; Mining and milling of uranium ore.
Address: Union Carbide Corporation, Nuclear Division, Post Office Box P, Oak Ridge, Tennessee 37830, U.S.A.

CROMPTON, Charles Edward, Ph.D. (Phys. Chem.), B.S. (Chem.). Born 1922. Educ.: California, Washington and Tennessee Univs. Director, Radioactivity Div., United States Testing Co., N.J., 1951-52; Deputy Director, A.E.C. Isotopes Div., Oak Ridge, 1949-51, 1953-55; Assoc. Tech. Director, Nat. Lead Co. of Ohio, 1956-60; Director, Advanced Development Sect., Auxiliary Power Systems Department (1960-61), Director, Nucl. Chem. Dept. (1961-62), The Martin Co., Baltimore, Maryland; Director of Res., Glidden Co. Chemicals Group, Baltimore, Maryland, 1962-. A.S.T.M. Representative to American Standards Association; A.S.T.M. Special Advisory Com. on Nucl. Problems; Consultant to U.S.A.E.C., 1951; Consultant to A.M.A., 1955-; Chairman, A.S.T.M. Committee E-10, Radioisotopes and Rad. Effects; Chairman, American Standards Assoc. Sub-committee on Chem. Eng. in Nucl. Fuel Manufacture and Fabrication. Books: Co-author, Radioactivity applied to Chem.; co-author, Nucl. Eng. Handbook (McGraw-Hill, 1958). Societies: A.N.S.; A.S.T.M.; Fellow, American Inst. Chem.; Armed Forces Chem. Assoc.; A.C.S. Nuclear interests: Management of chemical, metallurgical and production activities concerned with nuclear fuel cycle, hot cell operations, transplutonium chemistry, industrial applications of radioisotopes, application of isotopic power to direct conversion of heat to electricity.
Address: Glidden Co., Chemicals Group, Baltimore, Maryland, U.S.A.

CRONKHITE, Leonard W., Ph.B. (Brown Univ.), B.Sc. (Oxford). Born 1882. Educ.: M.I.T., Brown, Oxford and Harvard Univs. (Graduate School, International Law). Trustee World Peace Foundation. Formerly Chairman, Treasurer, Director of Atomium Corp. Director, Atomic Ind. Forum; Pres., Atomic Instrument Co., 1949-56; Vice-Pres., Baird-Atomic, Inc., 1956. Nuclear interests: Research and application of ultra low level radiation, particularly with reference to medical applications. For research, environmental aggregates of shielded sensitive detection instruments; blood volume measurement with accuracy, rapidity, and automated simplicity, transistorized gamma counting and computing automatically; and some industrial process applications.
Address: 5 Concord Avenue, Cambridge, Mass., U.S.A.

CRONKITE, Eugene P., M.D. Born 1914. Educ.: Stanford Univ. Chairman, Medical Dept., Medical Res. Centre, Brookhaven Nat. Lab.; Chairman, N.C.R.P., Com. on Hazards of Radioactive Nucleic Acids. Books: Rad. Injury in Man (Springfield, Illinois, Chas. C. Thomas, 1959); co-author, chapters in Atomic Medicine, 3rd edition (Baltimore, Maryland, Williams and Wilkins Co., 1959); The Hematology of Ionizing Rad.; Fallout Rad.; Effects on the Skin; Acute Whole-Body Rad. Injury; Pathogenesis, Pre- and Post-Rad. Protection; The Diagnosis and Therapy of Acute Rad. Injury. Societies: Rad. Res. Soc.; Internat. Soc. Hematology; American Soc. Hematology; American Soc. for Clinical Investigation; American Federation of Clinical Res.; Assoc. American Physicians. Nuclear interest: Radiation injury in man.
Address: Medical Department, Brookhaven National Laboratory, Upton, L.I., New York 11973, U.S.A.

CROOK, Delma Lomax, B.S. Born 1924. Educ.: Mars Hill Coll., U.S. Merchant Marine Acad. and Virginia Univ. Navy Dept., Norfolk Naval Shipyard, Portsmouth, Virginia; Various Eng. positions to Head Eng., Design Div., Supervisor of Shipbuilding (1950-63), now Navy Supv., Shipbuilding, Newport News, Virginia; Manager, A.E.C.-Maritime Administration Joint Group - Nucl. Ship Savannah Demonstration Programme for Peaceful Uses of Atomic Energy (1963-65), Manager, Nucl. Ship R. and D. Programme (1965-), Maritime Administration, Dept. of Commerce. Book: Technical, Operational and Economic Report on the N.S. Savannah First Year of Experimental Commercial Operation 1965-66 (The Federal Clearinghouse, 1966). Society: Soc. of Naval Architects and Marine Eng.
Nuclear interests: To improve the U.S. Merchant Marine by the use of commercial nuclear power

in all ships handling foreign and domestic cargo, then apply the commercial nuclear technology to other segments of transportation systems -i.e., trains, trucks, tugs and planes
Address: Nuclear Programme Maritime Administration, United States Department of Commerce, Washington, D.C. 20545, U.S.A.

CROOK, Guy Harman, A.B., M.D. Born 1914. Educ.: Rensselear Polytech. Inst., Cornell Univ. and Nebraska Univ. Coll. of Medicine. Assistant Medical Director, Hanford Environmental Health Foundation, March 1967-. Board of Directors, Northwest Association of Occupational Medicine. Society: Columbia Chapter, H.P.S.
Nuclear interests: Biological effects and treatment of radiation syndrome and deposited radionuclides.
Address: Hanford Environmental Health Foundation, P.O. Box 100, Richland, Washington 99352, U.S.A.

CROOKALL, John Ormand, A.R.I.C. Born 1928. Educ.: Burnley Municipal Coll. Inst. Cancer Res., London, 1952-.
Nuclear interests: Fall-out radionuclides, both natural and artificial in air, rain, soil and biological materials; Relationship between uptake by human tissues - e.g., of strontium 90 and cesium 137 - and the levels in the food chain and the pattern of fall-out.
Address: The Physics Department, Institute of Cancer Research, Royal Cancer Hospital, Clifton Avenue, Belmont, Surrey.

CROOME, John Lewis, B.Sc. (Economics). Born 1907. Educ.: London Univ. H.M. Treasury (Central Economic Planning Staff), 1948-51; Ministry of Food, 1951-54; U.K. Delegation to O.E.E.C., Paris, 1954-57; Chief Overseas Relations Officer, U.K.A.E.A., 1958-. Nuclear interest: Management.
Address: Pearmain, Ruxley, Claygate, Surrey, England.

CROSS, Wilfred. Member, Nat. Energy Commission.
Address: Comisión Nacional de Energia, Ministerio de Fomento y Obras Públicas, Managua, D.N., Nicaragua.

CROSSLEY, DeRyee Ashton, Jr., B.A., M.S., Ph.D. Born 1927. Educ.: Texas Tech. Coll. and Kansas Univ. Prof. Entomology and Inst. Ecology, Georgia Univ.
Nuclear interests: Accumulation and turnover of radioisotopes in insect food chains; Applications of radioisotopes in measurement of food consumption, energy flow, and material movement in food chains; Effects of radiation upon insect populations in field environments.
Address: Entomology Department, Georgia University, Athens, Georgia 30601, U.S.A.

CROSSLEY, Edward, C. Eng., M.I.Mech.E. Born 1905. Managing Eng., Non-Marine Dept., Lloyd's Register of Shipping; Council, British Nucl. Forum. Societies: B.N.E.S.; Inst. Mech. Eng.
Nuclear interest: Survey and inspection of nuclear plant.
Address: Lloyd's Register of Shipping, Non-Marine Department, Norfolk House, Croydon, Surrey, England.

CROSSLEY, H. E., Ph.D., M.Sc., B.Sc. (Tech.), A.R.I.C., F. Inst. F. Chief Fuel Technol., R. and D. Branch, C.E.G.B.
Address: Central Electricity Generating Board, Research and Development Branch, Grindall House, 25 Newgate Street, London E.C.1, England.

CROUSE, David J., B.S. (Chem. Eng.). Born 1920. Educ.: Pennsylvania State Univ. Chem. Eng., O.R.N.L. 1948-. Society: A.C.S.
Nuclear interests: Processing of nuclear raw materials; Nuclear fuels reprocessing; Fission product recovery.
Address: 249 Iroquois Road, Oak Ridge, Tennessee, U.S.A.

CROUT, J. Sen. Managerial Officer, F.H. Gottfeld G.m.b.H.
Address: F. H. Gottfeld G.m.b.H., 26 Robert-Koch-Strasse, Cologne-Lindenthal, Germany.

CROUTHAMEL, Carl E., Dr. Formerly at Argonne Nat. Lab.; Director of Res., Isotopes Inc., 1966-.
Address: Isotopes Inc., 123 Woodland Avenue, Westwood, New Jersey, U.S.A.

CROUZIER, M. Ing. Responsable, Dept. Nucl. Entrepose S.A.
Address: Entrepose S.A., 75 rue de Tocqueville, Paris 17, France.

CROVETTO, H. E. Arthur. Ministre Plénipotentiare; Prés., Centre Scientifique de Monaco. Deleg., I.A.E.A. Sci. Conference on the Disposal of Radioactive Wastes, Monaco, November 1959; Deleg., I.A.E.A. General Conference every year since 1957; Deleg., U.N. I.C.P.U.A.E., Geneva, 1958-64.
Address: Centre Scientifique de Monaco, Villa Girasole, 16 boulevard de Suisse, Monte-Carlo, Monaco.

CROW, James Franklin, A.B., Ph.D. Born 1916. Educ.: Friends (Wichita, Kansas) and Texas Univs. Prof. Medical Genetics, Wisconsin Univ. Pres., American Soc. for Human Genetics. Society: N.A.S.
Nuclear interest: Genetif effects of radiations.
Address: Genetics Building, Wisconsin University, Madison, Wisconsin, U.S.A.

CROWLEY, John Harold, B.S.M.E. Born 1924. Educ.: Purdue Univ. and O.R.S.O.R.T. U.S. Navy Bureau of Ships, 1948-56; Gen. Elec. Co., 1956-61; Jackson and Moreland, 1961-. Lecturer and instructor for A.S.M.E. and I.E.E.E. Society: A.N.S.

Nuclear interests: Technical and economic aspects of applying nuclear power to electric utility or special purpose military propulsion or electric generation systems. Previous experience in nuclear reactor design with special emphasis on nuclear superheater fuel design.
Address: 17 Louis Road, Framingham, Massachusetts, U.S.A.

CROWLEY-MILLING, Michael Crowley, M.A. Born 1917. Educ.: Cambridge Univ. Res. Eng., A.E.I. (Manchester) Ltd., 1939-63; Nucl. Phys. Lab., Sci. Res. Council, Daresbury, 1963-. Society: Fellow, I.E.E.
Nuclear interests: Design, development and applications of particle accelerators, particularly Microwave Electron Linear Accelerators and Electron Synchrotrons; The effects of radiation on the properties of materials.
Address: 210 Rusholme Gardens, Manchester 14, England.

CROWSON, Brigadier-General Delmar L., B.A., M.S. Born 1917. Educ.: California Univ. and California Inst. Technol. U.S. Air Force. Asst. to Asst. Sec. of Defence for Atomic Energy, 1955-59; Deputy Chief of Staff, R. and D., Field Command, Defence Atomic Support Agency, Albuquerque, 1960-62; Deputy Director (1962-64), Director (1964-67), Div. of Military Application, U.S.A.E.C.; Director, Office of Safeguards and Materials Management (1967-), Vice Chairman, Advisory Com. on Nucl. Materials Safeguards (1967-), U.S.A.E.C.
Address: Office of Safeguards and Materials Management, United States Atomic Energy Commission, Washington D.C. 20545, U.S.A.

CROZEMARIE, J. Administrateur, Inst. de Recherches Scientifiques sur le Cancer and Centre de Recherches Physiologiques sur la Cellule Normale et Cancereuse, Centre Nat. de la Recherche Scientifique.
Address: 16 bis Avenue Vaillant-Couturier, B.P. No. 8, Villejuif, (Seine), France.

CRUDELE, Joseph S., B.S. (Phys.), M.A. (Phys.). Born 1922. Educ.: Brown and Wesleyan Univs. Reactor Eng., Pratt and Whitney Aircraft, Fox Project, 1952-55; Reactor Phys., Combustion Eng., Inc., 1956-63; Reactor Phys., Ames Lab. Res. Reactor, 1963-. Operations Supv. of Flexible Critical Experiment, 1956-57; Operations Supv. of Critical Facilities, 1957-59; Operations Supv. SL-1 Idaho, 1959-60; Reactor Phys. at Critical Facilities, 1960-63; Reactor Measurements and Analysis and Reactor Dynamics, 1963-. Society: A.N.S.
Nuclear interests: Reactor measurements, kinetics and control; Operation of critical facilities of research reactors.
Address: Ames Laboratory, Reactor Division, Iowa State University, Ames, Iowa 50010, U.S.A.

CRUIKSHANK, Alexander John, B.A., M.A. Born 1921. Educ.: Toronto Univ. Defence Res. Board of Canada. Society: Chem. Inst. of Canada.
Nuclear interests: Radiation protection and measurement; Effects of nuclear weapons.
Address: Defence Research Board, Department of National Defence, Ottawa 4, Ontario, Canada.

CRUSSARD, Jean, D. ès Sc., Ph.D. (Faculté des Sciences de Paris). Born 1911. Educ.: Ecole Polytechnique, Paris. Lab. Leprince-Ringuet, Ecole Polytechnique, Paris, 1950-59; Dept. Phys., Rochester Univ., (N.Y.), 1953-54; Max Planck Inst., Göttingen, 1956-57; French C.E.A., 1959-.
Nuclear interests: Main interest: high energy corpuscular physics; production, interaction and decay of mesons and hyperones; High energy accelerators. Also, plasma physics and controlled fusion.
Address: 6 place du Pantheon, Paris 5, France.

CRUTCHFIELD, E. B. Formerly Vice Pres., now Sen. Vice Pres., (1967-), Virginia Elec. and Power Co.
Address: Virginia Electric and Power Co., Richmond, Virginia 23209, U.S.A.

CRUZ, A., Dr. Asst., Catedra de Fisica Matematica, Facultad de Ciencias, Barcelona Univ.
Address: Barcelona University, Barcelona, Spain.

CRUZ, Benjamin DE LA. See DE LA CRUZ, Benjamin.

CRUZ, Felipe DE LA. See DE LA CRUZ, Felipe.

CRUZ, Francisco GONZALEZ. See GONZALEZ CRUZ, Francisco.

CRUZ, Manuel de la PARRA y de la. See de la PARRA y de la CRUZ, Manuel.

CSUPKA, Stefan, Eng. (Chem.). Born 1927. Educ.: Chem. Eng. Univ. With Dept. of Rad. Hygiene, Regional Station of Hygiene. Society: Slovak Chem. Soc., Slovak Acad. of Sci.
Nuclear interest: Environmental monitoring of biosphere contamination.
Address: K.H.S., 60 Trnavská, Bratislava, Czechoslovakia.

CUBA, Eduardo GOYBURU de la. See GOVBURU de la CUBA, Eduardo.

CUCULLU, Lionel J., B.S. Born 1906. Educ.: Louisiana State Univ. Pres. and Chief Eng., New Orleans Public Service Inc. Trustee, Southwest Atomic Energy Assocs.; Trustee, High Temperature Reactor Development Assocs. Societies: Edison Elec. Inst.; Fellow, A.S.M.E.; I.E.E.E.; Co. Rep., Atomic Ind. Forum.
Nuclear interest: Management.
Address: 317 Baronne Street, P.O. Box 60340, New Orleans, Louisiana 70160, U.S.A.

CUCURON, Claude. Directeur Technique, Sté. d'Applications Mécaniques et de Robinetterie Industrielle.
Address: Société d'Applications Mécaniques et de Robinetterie Industrielle, 65 avenue Marcel-Cachin, Chatillon-sous-Bagneux, (Seine), France.

CUE, Ernesto MEDINA. See MEDINA CUE, Ernesto.

CUENOD, Michel, Ing. dipl. E.P.F. de Zürich, D. ès Sc. Tech. Born 1918. Educ.: Ecole Polytech. Fédérale de Zürich. Eng., Sté. Générale pour l'Industrie. Sec., Swiss Assoc. Automatic Control.
Nuclear interest: Reactor design.
Address: 7 Place Claparède, Geneva, Switzerland.

CÜER, Pierre, Dr. of Sci. Born 1921. Educ.: Lyon and Paris Univs. Full Prof. Gen. and Corpuscular Phys., Faculty of Sci., Strasbourg Univ. Director, Lab. of Corpuscular Phys., Strasbourg Univ., 1950; Director, Dept. of Corpuscular Physics, Centre Nucléaire, C.N.R.S. de Strasbourg, 1959; Prés., Internat. Com. of Corpuscular Photography, 1957; Pres., Internat. School of Elementary Particles, Herceg-Noñ (Yugoslavia), 1955-. Book: Comptes-Rendus du Ier Colloque Internat. de Photographie Corpusculaire (Strasbourg, C.N.R.S., 1959); Compte-rendu de la réunion de travail sur l'enregistrement des traces des particules chargées dans les cristaux (Strasbourg, C.N.R.S., 1963). Societies: Sté. Française de Physique; A.P.S.; British Phys. Soc.; Deutsche Gesellschaft für Wiss. Photographie.
Nuclear interests: Fundamental research in nuclear physics (light nuclei) low and high energy and corpuscular physics.
Address: Département de Physique Corpusculaire, Centre de Recherches Nucléaires, Strasbourg-Cronenbourg, Bas-Rhin, France.

CUEVA, J. Adan, Dr. Voter, Honduras A.E.C.
Address: Honduras Atomic Energy Commission, Tegucigalpa, D.C., Honduras.

CUKIERSZTAJN, Wlodzimierz, M.D. Formerly Head, Dept. for Radiobiology and Radiological Protection, Office of Govt. High Commissioner for Atomic Energy, Poland.
Address: Office of Government High Commissioner for Atomic Energy, Palace of Culture and Science, 18th floor, Warsaw, Poland.

CULLEN, Thomas G., B.S. Born 1912. Educ.: Rutgers and California Univs. Nucl. Projects Manager for Carboline Co., St. Louis, Mo., 1957-. Societies: N.A.C.E.; A.N.S.; A.I.Ch.E.
Nuclear interests: Corrosion, contamination and the prevention or reduction of both.
Address: 12 La Rancheria, Carmel Valley, California, U.S.A.

CULLER, Floyd LeRoy, Jr., B.S. (Chem. Eng.). E.O. Lawrence Memorial Award, 1965. Born 1923. Educ.: Johns Hopkins Univ. Director, Chem. Technol. Div., now Asst. Director, O.R.N.L. National Research Council Subcom. to Study Disposal and Dispersal of Radioactive Wastes, 1955-64; Societies: Fellow, A.N.S.; A.C.S.; A.I.Ch.E.; N.A.S.; Res. Eng. and Sci. of America.
Nuclear interests: Radiochemical processing technology; Process design engineering; Management; Nuclear engineering and economics; Reactor design; Fuel cycle development; Nuclear chemical development; Radioactive waste treatment and disposal.
Address: 109 Oneida Lane, Oak Ridge, Tennessee 37830, U.S.A.

CULLEY, Frederick John Spencer, LL.B., B.Sc. Born 1916. Educ.: London Univ. Principal, Ministry of Agriculture, Fisheries and Food, 1950-61; Principal, A.R.C., 1961-. Joint Sec., A.R.C./M.R.C. Com. on the monitoring of radioactivity from fallout.
Address: Agricultural Research Council, Cunard Building, 15 Regent Street, London, S.W.1, England.

CULLIGAN, Gerard, B.Sc., Ph.D. Born 1934. Educ.: Liverpool Univ. D.S.I.R. Res. Fellow, Liverpool Univ., 1958-59; Res. Assoc. C.E.R.N., 1959-61; Res. Assoc. Chicago Univ., 1961-64; Sen. Res. Officer, Oxford Univ., 1964-.
Nuclear interests: Research on the nature and interactions of fundamental particles using high energy accelerators.
Address: Nuclear Physics Laboratory, University of Oxford, 21 Banbury Road, Oxford, England.

CULP, Archie William, Jr., B.S. (Mech. Eng.), M.S. (Mech.Eng.). Born 1931. Educ.: Missouri School of Mines and Metal. Res. Asst. under a Naval Rocket Res. Grant, 1953; Graduate Asst., Missouri School of Mines and Metal., 1952-54; Student, O.R.S.O.R.T., 1954-55; Nucl. Eng., Convair, Texas, 1955-56; Nucl. Eng., ASTRA, Inc., 1956-61; Asst. Prof. Mech. Eng., Missouri School of Mines and Metal., 1960-. Society: A.N.S.
Nuclear interests: Reactor physics; Activation; Reactor engineering; Digital computers.
Address: 2 Curtis Drive, Rolla, Missouri, U.S.A.

CULTRERA, R. Italian Member, Study Group on Food Irradiation, E.N.E.A., O.E.C.D.
Address: O.E.C.D. European Nuclear Energy Agency, 38 boulevard Suchet, Paris 16, France.

CULVER, Donald H. Manager, Tech. Information, Nucl. Materials and Propulsion Operation, Gen. Elec. Co. Member, Tech. Information Panel, U.S.A.E.C.
Address: General Electric Company, Nuclear Materials and Propulsion Operation, Cincinnati, Ohio 45215, U.S.A.

CUMELLA PAU, Antonio. Vocal, Comision Asesora de Equipo Ind., Junta de Energia

Nucl., Ministerio de Industria.
Address: Junta de Energia Nuclear, Ciudad Universitaria, Madrid 3, Spain.

CUMMINGS, Robert E. Asst. Director, Cambridge Electron Accelerator, operated for U.S.A.E.C. by M.I.T. and Harvard Univ.
Address: U.S.A.E.C., Cambridge Electron Accelerator, Cambridge, Massachusetts, U.S.A.

CUMMINS, Cyril Joseph, M.B., B.S., D.P.H. Born 1914. Educ.: Sydney Univ. Director, Div. Ind. Hygiene (1950), Deputy Director Gen. of Public Health (1952), Director Gen. Public Health and Chief Medical Officer (1959), New South Wales Govt. Safety Review Com., A.E.C.; Chairman, Rad. Advisory Com., Clean Air Act, 1961.
Nuclear interests: Administration of legislation to conserve the public security from harmful radiation arising from either irradiating apparatus or sources.
Address: Department of Public Health, Winchcombe House, 52 Bridge Street, Sydney, Australia.

CUMMINS, John Edward, B.Sc., M.S. Born 1902. Educ.: Western Australia and Wisconsin Univs. Chief Sci. Liaison Officer, London, 1948-54, U.S.A. 1955-58; Director, Div. of Sci. and Tech. Information, I.A.E.A., 1958-60; C.S.I.R.O., 1961-62; Sec.-Exec. Officer, Ian Clunies Ross Memorial Foundation, 1963-. Societies: F.R.I.C. Great Britain and Northern Ireland; F.R.A.C.I.
Nuclear interest: Scientific and technical information.
Address: I.C.R.M.F., 191 Royal Parade, Melbourne 3052, Australia.

CUNHA, David MESQUITA da. See MESQUITA da CUNHA, David.

CUNNINGHAM, Burris Bell, B.Sc., Ph.D. Born 1912. Educ.: California Univ. Prof. Chem., California Univ., 1953-. Editorial Advisory Board, J. Inorganic and Nucl. Chem. Societies: A.A.A.S.; A.C.S.
Nuclear interest: Chemistry and metallurgy of the actinide elements.
Address: Department of Chemistry, California University, Berkeley, Calif., U.S.A.

CUNNINGHAM, Sir Charles Craik, K.C.B., K.B.E. M.A., B.Litt., Ll.D. Born 1906. Educ. : St. Andrews Univ. Sec., Scottish Home Dept., 1948-57; Under Sec. of State, Home Office, 1957-66; Deputy Chairman, U.K.A.E.A., 1966-; Chairman, Board, The Amersham/Searle Corp., 1968-.
Address: U.K.A.E.A., 11 Charles II Street, London, S.W.1, England.

CUNNINGHAM, Harold David, B.S. (Mech. Eng.). Born 1926. Educ.: New Mexico State Univ. Mech. Eng., Connally Stevens Co., 1948-51; Mech. Supt., Brown and Olds Plumbing and Heating Co., 1951; Mech. Supt., Reynolds Elec. and Eng. Co. Inc., 1953; Mech. Supt., Brown and Olds Plumbing and Heating Co., 1953-54; various posts including Manager, Mech. Dept., Area Supt., Mech. Supt. (1958-62), Asst. Gen. Manager, Eng. and Maintenance Div. (1962-65), Manager, Operations Div. (1965-), Reynolds Elec. and Eng. Co. Inc.
Nuclear interests: Weapons testing with emphasis on the management of certain support functions, including drilling, electrical, mining, construction, and other activities in support of our weapons testing programme.
Address: 2104 Kirkland Avenue, Las Vegas, Nevada 89102, U.S.A.

CUNNINGHAM, J. B. W. Asst. Director, Southern Project Group, C.E.G.B.
Address: Southern Project Group, Central Electricity Generating Board, Squires Lane, Finchley, London, N.3, England.

CUNNINGHAM, John Edward, B.Sc., M.Sc. Born 1920. Educ.: Illinois and Tennessee Univs. Tech. staff (1945) became Head, Metal. group that developed aluminium-base fuel elements for res. reactors and stainless steeluranium dioxide dispersion fuel for the Army Water Reactors Programme, became Asst. Director, Metals and Ceramics Div. (1955), engaged in tech. management of materials sci. and technol. activities since that time, O.R.N.L. Programme Com. Chairman, Materials Sci. and Technol. Div., A.N.S. Societies: A.N.S.; A.S.M.; A.I.M.,M.P.E.; A.S.T.M.
Nuclear interests: Metallurgy; Reactor design; Management; Fuel technology; Fabrication; Cladding and structural materials; and control materials.
Address: Metals and Ceramics Division, Oak Ridge National Laboratory, P.O. Box X, Oak Ridge, Tennessee 37830, U.S.A.

CUNNINGHAM, Malcolm Terence, B.Sc. (Hons.). Born 1927. Educ.: Leeds Univ. Eng. Manager, Dewrance and Co., Ltd., valve manufacturers and engineers to nucl. power industry. Societies: A.M.I.Mech.E.; A.M.Inst.F.
Nuclear interests: Management board member; Design and development of valves, fittings and instruments for gas and pressurised water reactor plant.
Address: c/o Dewrance and Co.,Ltd., 165 Great Dover Street, London, S.E.1, England.

CUNNINGHAM, Richard G., B.S., M.S., Ph.D., Registered Professional Eng. Born 1921. Educ.: Northwestern Univ. Res. Eng., Pure Oil Co., 1950-51; Assoc. Prof., Eng. Res., Pennsylvania State Univ., 1951-54; Res. Group Leader, Sen. Res. Eng., Shell Oil Co. (Word River, Illinois, Res. Lab. and Manufacturing Res. Dept., New York), 1954-61; Prof., Mech. Eng., Pennsylvania State Univ., 1961-. Chairman, Univ. Senate; Board of Accreditation Visitors, Eng. Council for Professional Development; Policy Board on Educ., A.S.M.E. Societies: A.I.A.A.; American Soc. of Lubricating Eng.

Nuclear interest: Two phase, two component flow, two phase jet pump.
Address: Pennsylvania State University, 207 Mechanical Engineering, University Park, Pennsylvania 16802, U.S.A.

CUNNINGHAM, Robert Leonard, B.Sc., M.Sc., Ph.D. Born 1915. Educ.: Dalhousie and McGill Univs. Head, Metal Phys. Sect. (1942-57), Principal Sci. (1957-), Phys. Métal. Div., Dept. of Mines and Tech. Surveys, Ottawa. Member, Editorial Advisory Board, J. of Nucl. Materials; Chairman, Ottawa Valley Chapter, and Member, Nat. Seminar Com., A.S.M. Societies: Canadian Inst. of Mining and Metal.; Mineralogical Assoc. of Canada; Inst. of Metals.
Nuclear interests: Kinetics of gas reactions, phase transformations in steels, applications of X-ray diffraction, alloy constitution, thermal analysis techniques, formation of graphite in cast irons.
Address: 309 Fairmont Avenue, Ottawa, Ontario, Canada.

CUOCOLO, Gastone, Dr. in Ind. Eng. Born 1928. Educ.: Rome and Milan Univs. Reactor Group, C.I.S.E., 1956; Vice-Chief Eng., Ispra I Reactor Construction, 1957; Vice-Head, Ispra I Reactor Operation, 1959; Asst. Chief Eng., Dragon Project, 1960; Chief, Programmes Coordination Service (1962), Chief, Organisation and Methods Service (1965), C.N.E.N. Tech. Correspondent for Italy, Dragon Project. Society: British Nucl. Soc.
Nuclear interests: Organisation and management; High temperature gas cooled reactors development.
Address: Via Raffaele Issel, 3 Rome, Italy.

CUPAK, Miloslav, B.Sc. Born 1932. Educ.: J. E. Purkyne Univ., Brno. Mosilana, wool mills, Brno, 1950-56; Military Acad. of A. Západocký, Brno, 1956-61; Res. Inst. of Macromolecular Chem., Brno, 1961-. Society: Czechoslovak Sci. and Tech. Soc.
Nuclear interests: Through the measurement of low energy β-emitters ^{14}C, ^{3}H, ^{35}S (proportional and liquid scintillation counters) the structural and quantitative analysis of polymers. Radiochromatography on columns and thin layers.
Address: 131 Skorkovského, Brno 15, Czechoslovakia.

CURAMI, Antonio, Dr. of Elec. Eng. Born 1913. Educ.: Inst. of Technol. (Politecnico), Milan. Vice-Gen. Manager, Soc. Orobia, Milan, 1948-55; Gen. Manager, Soc. Idroelettrica Subalpina, Como, 1956-; Director, Societa Elettronucleare Italiana. Societies: Assoc. of Italian Elec. Eng.; Assoc. of Power Generating and Power Self-Using Companies; Italian Electrotech. Com.
Nuclear interests: Nuclear power plant operation; Management.
Address: 11 Via Morigi, 20123 Milan, Italy.

CURET, Juan Daniel, B.S., M.S., Ph.D. Born 1914. Educ.: Puerto Rico and Michigan Univs. Prof. Chem. (1953), Chairman, Dept. of Chem. (1959-60), Dean, Coll. of Natural Sciences (1960-), Puerto Rico Univ.; Lecturer, Puerto Rico Nucl. Centre (part time), 1958-59. Society: A.C.S.
Nuclear interests: Absorption of beta rays by weakly paramagnetic substances; Use of radioisotopes to demonstrate chemical principles.
Address: University of Puerto Rico, Rio Piedras, Puerto Rico.

CURILLON, Robert, Ing. Arts et Métiers. Born 1913. Directeur, Eng. Div., Sté. Saint-Gobain Techniques Nouvelles; Vice-Prés., Directeur Gén. Sté. Conservatome; Administrateur, Sté. de Raffinage d'Uranium; Administrateur, Compagnie pour l'Etude et la Realisation de Combustibles Atomiques; Administrateur, Etablissements Briard; Directeur Technique, Sté. d'Etudes et de Travaux pour l'Uranium.
Address: Saint-Gobain Techniques Nouvelles, 23 Boulevard Georges Clemenceau, Courbevoie 92, France.

CURIONE, Charles, B.S.C.E., M.S.C.E., LL.B. Born 1922. Educ.: Omaha, Nebraska, Illinois, Lasalle and California (Los Angeles) Univs. Manager, Public Works Systems, Stanford Res. Inst., Menlo Park; Lecturer, Stanford Univ., 1967. Past Pres., Soc. of American Military Eng. Societies: RESA: A.S.C.E.; Soc. of Professional Cost Eng.
Nuclear interest: Management and cost research as related to protective shelter design and construction.
Address: Stanford Research Institute, Menlo Park, California, U.S.A.

CURMI, H. E. George T. Head, Malta Mission to the E.E.C. and Euratom, 1968-.
Address: Malta Mission, c/o Euratom, 51-53 rue Belliard, Brussels, Belgium.

CURRAN, Samuel Crowe, M.A., B.Sc., Ph.D., D.Sc. (Glasgow), Ph.D. (Cantab.). F.R.S.E., F.R.S. Born 1912. Educ.: Glasgow and Cambridge Univs. Chief Sci., U.K.A.E.A., A.W.R.E., Aldermaston, 1955-59; Principal and Vice-Chancellor, Strathclyde Univ., 1959-. Sci. Res. Council; Advisory Council on Technol. Books: Counting Tubes (London, Butterworths, 1949); Luminescence and the Scintillation Counter (London, Butterworths, 1953). Societies: Roy. Soc.; Roy. Soc. of Edinburgh; Inst. Phys. and Phys. Soc.
Nuclear interests: Nuclear physics; Controlled thermonuclear research; Nuclear instrumentation; Management of research projects; Education in science.
Address: Principal's Residence, Livingstone Tower, Strathclyde University, Glasgow C.1, Scotland.

CURRIE, D. A. Staff Eng., Sales, Eldorado Mining and Refining Ltd. Chairman, Nucl. Safety Com., Canadian Nucl. Assoc.
Address: Eldorado Mining and Refining Ltd., Port Hope, Ontario, Canada.

CURRIE, Lauchlin M., B.A., Ph.D., D.Sc. (Hon., Clarkson Coll. Technol. and Davidson Coll.). Born 1898. Educ.: Davidson Coll. and Cornell Univ. Vice-Pres., Nat. Carbon Co., 1945-55; Vice-Pres., Union Carbide Nucl. Co., 1955-58; Vice Pres., Babcock and Wilcox Co., 1958-62; Chairman, Advisory Com. on Development of Isotopes (1963-64), Consultant (1964-), U.S.A.E.C. Director, Nat. Assoc. Manufacturers; Chairman, Nucl. Energy Com., Nat. Assoc. Manufacturers; Nucl. Energy Com., U.S. Chamber of Commerce; Chairman, Nucl. Congress, Com., Eng. Joint Council. Societies: A.I.Ch.E.; A.N.S.; A.C.S.
Nuclear interests: Reactor design, uses and management.
Address: 574 Alda Road, Marmaroneck, New York, U.S.A.

CURRIE, William Masterton. Born 1910. Pres., Bank of Commerce, 1964-. Director, Canadian Nucl. Assoc., 1966-.
Address: Canadian Imperial Bank of Commerce, 25 King Street West, Toronto 1, Ontario, Canada.

CURRIER, Edwin L., B.S. (Chem.), B.S. (Ch.E.). Born 1920. Educ.: Creighton Univ. and Newark Coll. of Eng. At Argonne Nat. Lab., 1950-56; A.C.F. Industries, Inc., 1956-59; Bechtel Corp., 1959-. Societies: A.N.S.; A.S.M.
Nuclear interests: The design and manufacturing of nuclear fuel elements; The economics of the fuel cycle; Irradiation materials testing; Nuclear safety; Site safety analysis; Design and construction of nuclear power plants; Spent fuel reprocessing plants.
Address: 727 Butternut Drive, San Rafael, California, U.S.A.

CURRY, La Verne L., Ph.D. Prof. and Head, Biol. Dept., Central Michigan Univ.
Address: Department of Biology, Box 103, Central Michigan University, Mt. Pleasant, Michigan 48858, U.S.A.

CURTIS, Allan Raymond, M.A. Born 1922. Educ.: Cambridge Univ. Lecturer, Appl. Maths., Sheffield Univ., 1952-56; Theoretical Phys. Div., A.E.R.E., Harwell, 1957-, (now Head of Maths. Branch).
Nuclear interest: Mathematics, including computation.
Address: 257 Woodstock Road, Oxford, England.

CURTIS, Carl T. Born 1905. Educ.: Nebraska Wesleyan Univ. Senator. Member, Joint Com. on Atomic Energy, U.S.A.E.C., 1963-.
Address: Minden, Nebraska, U.S.A.

CURTIS, Howard James, B.S., M.S., Ph.D. Born 1906. Educ.: Michigan Univ., Swarthmore Coll. and Yale Univ. Prof. and Head Dept., Sch. Med., Vanderbilt, 1947-50; Chairman, Dept. Biol. (1950-65), Sen. Biol. (1965-), Brookhaven Nat. Lab. Com. on Growth, Nat. Res. Council, 1949-54; Com. on Radiobiol., 1949-; Com. on Rad. Studies, 1951-; Chairman, Subcom. on Radiobiol., Nat. Res. Council; Chairman, Editorial Board, Physiological Reviews; Advisory Board, Saratoga Spa; Chairman, Rad. Study Sect., P.H.S.; Director, Long Island Biol. Assoc.; Sci. Advisory Council, American Cancer Soc.; Board of Sci. Counsellors, Nat. Inst. of Neurological Diseases and Blindness, P.H.S.; Chairman, Sect. on Biophys. Com. on Growth, Nat. Res. Council. Societies: Phys. Soc.; Physiological Soc.; Soc. Gen. Physiologists: Rad. Res. Soc.; N.Y. Acad. Sci.
Nuclear interests: Basic radiobiology; Radiation-induced ageing; Biological effects of cosmic rays.
Address: Biology Department, Brookhaven National Laboratory, Upton, Long Island, New York, U.S.A.

CURTIS, Orlie Lindsey, Jr., B.A., M.S., Ph.D. (Tennessee). Born 1934. Educ.: Purdue Univ. Member, Sen. Tech. Staff, Northrop Corporate Labs. Society: A.P.S.
Nuclear interests: Radiation effects on solids, especially electrical properties of semiconductors; Recombination and trapping behaviour associated with radiation-induced defects; Disordered regions introduced by neutron and heavy charged-particle irradiation.
Address: Northrop Corporate Laboratories, 3401 West Broadway, Hawthorne, California 90250, U.S.A.

CURTISS, Leon Francis, A.B., Ph.D. Born 1895. Educ.: Cornell Univ. Consultant to Director, N.B.S. Washington, D.C., 1953-61. Chairman, Com. on Nucl. Sci., Nat. Res. Council, Washington, D.C. Book: Introduction to Neutron Phys. (Van Nostrand, 1959). Societies: A.P.S.; A.A.A.S.; Washington Acad. Sci.
Nuclear interest: Neutron physics.
Address: 1690 Bayshore Drive, Englewood, Florida, U.S.A.

CUSACK, John Hunter, Ph.D. (Chem. Eng.). Born 1926. Educ.: Pennsylvania State Univ. Catalytic processing of oil, radiation-induced polymerisation of olefins, Shell Development Co., 1954-59; Chemonucl. R. and D., Aerojet-Gen. Nucleonics, San Ramon, 1959-66; Head, High Intensity Rad. Development Lab., Brookhaven Nat. Lab., 1966-. Societies: A.I.Ch.E.; A.C.S.; A.N.S.
Nuclear interests: Design, testing and characterisation of package and flow irradiators; Development of standardised, high intensity ^{60}Co, ^{137}Cs and ^{90}Sr sources; Development of dose prediction and measurement techniques.
Address: Brookhaven National Laboratory, Upton, New York, U.S.A.

CUSH, J. M. Nucl. Valve Eng. and Works Manager, Scotts Eng. (Newport) Ltd. Address: Scotts Engineering (Newport) Ltd., Stafford Road, Newport, Monmouthshire, England.

CUSTERS, Jan Frans Henri, D.Sc. Born 1904. Educ.: Utrecht Univ. Director of Res., Diamond Res. Lab.,1953-63, and Adamant Res. Lab., 1955-63; Res. Consultant to Ind. Distributors (1946) Ltd., Johannesburg, 1963-. Societies: Physics Club, Univ. Witwatersrand; South-African Inst. of Physics. Nuclear interests: Effects of neutron, electron and gamma radiation on electrical, optical and mechanical properties of industrial and gem diamonds. Address: Industrial Distributors Ltd., P.O. Box 104, Crown Mines, Johannesburg, South Africa.

CUTHBERT, F. Leicester, B.A., M.A. (Geology) (Buffalo), Ph.D. (Geology and Soils) (Iowa State Coll.). Born 1913. Educ.: Buffalo Univ. and Iowa State Coll. Manager, R. and D. Lab., Baroid Sales Div., National Lead Co., 1945-51; Tech. Director, National Lead Co. of Ohio, 1951-62; Tech. Director, Hightstown Labs., National Lead Co., 1962-. Book: Production of Thorium Metal (1958). Societies: Geological Soc. of America; American Mineralogical Soc.; Foundryman's Soc.; American Petroleum Inst.; Ind. Res. Inst. Nuclear interests: Uranium extraction and purification for production of uranium metal; Fabrication of fuel elements; Research and development. Address: 529 Prospect Avenue, Princeton, New Jersey, U.S.A.

CUTHBERTSON, James Davidson. Born 1920. Com. Member, Risley Branch, Nucl. Eng. Soc. Society: Inst. of Mech. Eng. Nuclear interest: Irradiation rig design. Address: 30 Villiers Crescent, Eccleston, St. Helens, Lancs., England.

CUTLER, J. A., Dr. Sen. Sci., Markite Corp. Address: Markite Corporation, 155 Waverly Place, New York 14, New York, U.S.A.

CUTTITTA, Frank. Chem., Geological Survey, U.S. Dept. of the Interior. Address: U.S. Department of the Interior, Geological Survey, Washington D.C. 20242, U.S.A.

CUTTLER, Alan Howard, B.Sc., Ph.D. Born 1933. Educ.: Liverpool Univ. S.S.O., U.K.A.E.A., 1958-66; Sen. Lecturer, Plymouth Technol. Coll., 1967-. Nuclear interests: Radiation detection and instrumentation; Radiation protection; β/γ-ray spectroscopy; X-ray fluorescence; Neutron absorption/activation analysis. Address: E live Mhor, 24 Manor Park, Dousland, Yelverton, Devon, England.

CUTTS, Burton, B. E. Born 1923. Educ.: Sheffield Univ. Tech. Manager, Central Tech. Services, Reactor Group, U.K.A.E.A. Nuclear interests: Technical work associated with the design of nuclear reactors. Primarily reactor physics, kinetics and control system studies. Address: 6 Roughlea Avenue, Culcheth, near Warrington, Lancs., England.

CUYKENDALL, T. R., Dr. At Cornell Univ.; Head, A.S.E.E. Com. on Relations with U.S.A.E.C. Address: Cornell University, Ithaca, New York, U.S.A.

CUYPERS, Marc Yves Robert, Licencié en Sci. Chimiques, D. ès. Sc. Born 1936. Educ.: Liège and Paris Univs. Chercheur Agréé, Inst. Interuniversitaire des Sci. Nucléaires, Liège Univ., 1960-64; Asst. Res. Chem., Texas A and M Univ., 1964-66; Chercheur Agréé, Inst. Interuniversitaire des Sci. Nucléaires, Liège Univ., 1966-. Nuclear interests: Development of targets for neutron production; Activation analysis with charged particles and neutrons. Address: Laboratoire d'Application des Radio-éléments, Place du XX Août, 9 Liege, Belgium.

CVETKOVIC, Dragan, Sen. Officer, Energoinvest. Address: Energoinvest, Istrazivacko Razvojni Centar za Termotehniku i Nuklearnu Tehniku, Sarajevo, Stup-Yugoslavia.

CYFER, Dr. At Pacific Inst. Advanced Studies. Address: Pacific Institute of Advanced Studies, 448 N. Avenue 56, Los Angeles 42, California, U.S.A.

CYPHERS, Howard E., M.E., M.M.E. Born 1921. Educ.: Rensselaer Polytechnic Inst. K.A.P.L., Schenectady, New York, 1953-54; Riso Res. Establishment, Danish Atomic Energy Commission, 1959-60; Assoc. Prof. Mech. Eng., Rensselaer Polytechnic Inst. Nuclear interests: Heat transfer and fluid flow problems in reactors and component equipment. Address: Rensselaer Polytechnic Institute, Troy, New York, U.S.A.

CYPIN, S. G., Dr. Book: Contributor to Eng. Compendium on Rad. Shielding (I.A.E.A.). Papers: Co-author, The Investigation of Hydrogen - less Shielding of Nucl. Reactors (Jaderna Ernegie, vol. 11, No. 6, 1965); co-author, Empirical Methods of Calculation of Fast Neutron Transit through Hydrogen containing Medium (ibid, vol. 13, No. 5, 1967). Address: Fyzikalne energeticky institut, Obninsk, U.S.S.R.

CZARNECKI, Roman, Eng. Born 1930. Educ.: Eng. School Warsaw. Inst. Mechaniki Precyzyjinej Warszawa Duchnicka 3, 1950-56; Inst. Badań Jadrowych Swierk k/Otwock Polska

Akademia Nauk, 1956-.
Nuclear interest: Metallurgy (vacuum technique).
Address: Bl.19/4 ul.Andriollego, Otwock, Poland.

CZAUDERNA, Krzysztof (Christopher), M.Eng. Born 1936. Educ.: Cracow Acad. of Mining and Metal. Hydrological State Enterprise. Society: Suprema Tech. Organisation.
Nuclear interest: Use of isotopes in hydrogeology and engineering geology as well as in hydrotechnic building works.
Address: 2/12 ul.Kryniczna, Cracow, Poland.

CZEIJA, Carl Maria Conrad, Diploma Eng. (Chem.) Dr. of Laws. Born 1917. Educ.: Vienna Univ. and Vienna Techn. Univ. Licensed Eng., Ind. Consultant; Consultant to Österreichische Studiengesellschaft für Atomenergie; Austrian Deleg. to the "Working Party: Chemical Process Automation" in the European Federation of Chem. Eng. Societies: Verein Österreichischer Chemiker; Gesellschaft Deutscher Chemiker; A.C.S.
Nuclear interests: Applications and economics of radioisotopes in the chemical industry; Food conservation.
Address: 12 Gusshausstrasse, Vienna 4, Austria.

CZERNIAK, Pinchas A., M.D., H.D. Born 1909. Educ.: Montpellier and Wilno Univs. Asst., Radium and Tumours Inst., Hadassah Medical School, Jerusalem, 1950-; Head, Radium and Isotope Inst., Tel-Hashomer Govt. Hospital, Israel, 1954-; Medical adviser, Res. Nucl. Labs., 1953-; Tel-Aviv Univ. Medical School, 1965-. Societies: Israel Radiol. Assoc.; Soc. of Nucl. Medicine; Assoc. des Hygienistes Atomiques.
Nuclear interests: Radiotherapy - X-ray and radium for therapy; Medical uses of radioisotopes for diagnosis, therapy, and research; Hygiene and protection against ionising radiations.
Address: Tel-Hashomer Government Hospital, Israel.

CZERNICHOWSKI, Albin, Dr. eng. Born 1937. Educ.: Wroclaw Tech. Univ. Asst. (1960-), Adjunct (1966-), Inst. of Inorganic Chem. and Metal.. of Rare Elements, Wroclaw Tech. Univ.
Nuclear interests: Plasmochemistry; Metallurgy; Plasma spectroscopy; Molecular and atomic spectroscopy.
Address: 56 m. 8 ul Piastowska, Wroclaw, Poland.

CZERWIEC-POTE, Mrs. J. Reactor Study, State Tech. Coll. for Nucl. Industries, Board of Nat. Education and Culture.
Address: Institut Technique Supérieur de l'Etat pour les Industries Nucléaires, Ministère de l'Education Nationale et de la Culture, 150-152 rue Royale, Brussels, Belgium.

CZERWIK, Zbigniew, Dr. Phil. At. Inst. of Appl. Rad. Chem., Tech. Univ., Lodz.
Address: Technical University, Institute of Applied Radiation Chemistry, 15 Wroblewskiego, Lodz, Poland.

CZESCHIN, Charles. Trustee, Southwest Atomic Energy Associates.
Address: Southwest Atomic Energy Associates, 306 Pyramid Building, Little Rock, Arkansas, U.S.A.

CZIRR, John B. Formerly Res. Faculty, Phys. Dept., Washington Univ.
Nuclear interest: High energy physics.
Address: Lawrence Radiation Laboratory, Livermore, California 94550, U.S.A.

CZOSNOWSKA, Wanda Krystyna, M.Sc. (Chem. Eng.). Born 1925. Educ.: Lodz High Polytech. School. Adjunct, Inst. of Occupational Medicine, Lodz, 1957-62; Adjunct, Central Lab. for Radiol. Protection, Warsaw, 1962-. Polish Com. of Nucl. Technol. Society: Polish Chem. Soc.
Nuclear interests: Radiological protection; Radioactive contamination; Mechanism of transfer of radionuclides from the environment through food to man; Metabolism of radionuclides in man's body; Influence of dietary composition on radionuclide intake; Monitoring programme in food.
Address: 15/12 Karmelicka, Warsaw, Poland.

CZOSNOWSKI, Stanislaw, M.Sc. (Elec. Eng.). Born 1921. Educ.: Lodz and Warsaw High Polytech. Schools. Lecturer, Thermal and Nucl. Power Stations, Lodz High Polytech. School, 1947-66; Main Designer, Thermal Power Stations, Bureau of Power Designs and Studies "Energoprojekt", Warsaw, 1949-61; Chief Specialist on Power and Nucl. Energy (1961-), Sec., Main Power Commission, Vice Chairman, Direct Energy Conversion Group, Com. of Sci. and Technol., Warsaw. Member, Tech. and Economic Council, Ministry of Mining and Power; Member, Sci. Council, Power Inst., Warsaw; Member, Sci. Council, Inst. of Rad. Technol., Lodz; State Council for Peaceful Uses of Nucl. Energy.
Nuclear interests: Research and development in the field of nuclear power; Nuclear power stations - design, construction and exploitation.
Address: 15/12, Karmelicka, Warsaw, Poland.

CZYZ, Wieslaw, Sc.D. Formerly Sen. Officer, Cracow Centre of Nucl. Phys., now Head, Theoretical Phys. Dept., Inst. for Nucl. Phys. in Cracow.
Address: Cracow Institute for Nuclear Physics, Radzikowskiego, Krakow 23, Bronowice, Poland.

DA COSTA, Adelino Augusto NOGUEIRA.
See NOGUEIRA DA COSTA, Adelino Augusto.

DA COSTA, Joao CARRINGTON SIMOES.
See CARRINGTON SIMOES DA COSTA, Joao.

da COSTA RIBEIRO, Uriel. Chairman, Brazilian Nat. Nucl. Energy Commission; Brazilian Governor, I.A.E.A.
Address: National Nuclear Energy Commission, 90 (ZC-82), Rua General Severiano, Rio de Janeiro - GB, Brazil.

da CUNHA, David MESQUITA. See MESQUITA da CUNHA, David.

da PROVIDENCIA, Joao, Ph.D. (Mathematical Phys., Birmingham). Born 1933. Educ.: Coimbra and Birmingham Univs. Lecturer (1955-66), Assoc. Prof. (1967-), Coimbra Univ.; Res. Assoc., M.I.T., 1966-67.
Nuclear interests: Theoretical nuclear physics and many body problems in quantum mechanics.
Address: Laboratorio de Fisica, Universidade, Coimbra, Portugal.

da ROCHA, Manuel MENDES. See MENDES da ROCHA, Manuel.

da SILVA, Arthur Gerbasi, B.S. (Chem.), M.S. (Chem.). Born 1929. Educ.: Louisiana State Univ. Teaching Asst., Louisiana State Univ., 1953-55; Asst. Prof., centro Brasileiro de Pesquisas Fisicas, 1957-64; Head, Nucl. Phys. Div., Inst. de Engenharia Nucl., 1965-.
Nuclear interests: Nuclear physics; nuclear fission, nuclear spectroscopy, geochronology.
Address: Rua Leopoldo Miguez 14 apt. 201, Copacabana, Rio de Janeiro, GB., Brazil.

da SILVA, Denise Helena. Lecturer, Corpuscular Phys. Dept., Centro Brasileiro de Pesquisas Fisicas.
Address: Centro Brasileiro de Pesquisas Fisicas, 71 Avenida Wenceslau Braz, Rio de Janeiro, Brazil.

da SILVA, J. M. MACHADO. See MACHADO da SILVA, J. M.

DA SILVA, Mario Estevao, Ing., Bureau d'Etudes, Amoniaco Portugues S.A.R.L.
Address: Amoniaco Portugues S.A.R.L., 113-1 Rua do Poco dos Negros, Lisbon, Portugal.

da SILVA, Miss Monique MONTEAUX. See MONTEAUX da SILVA, Miss Monique.

DAANE, Adrian Hill, B.S., Ph.D. Born 1919. Educ.: Florida and Iowa State Univs. At Iowa State Univ., 1950-63; Head, Chem. Dept., Kansas State Univ., 1963-. Societies: A.C.S.; A.S.M.E.; A.I.M.M.E.; American Vacuum Soc.
Nuclear interests: Uranium metallurgy; High cross-section rare earths; Beryllium.
Address: Department of Chemistry, Kansas State University, Manhattan, Kansas, U.S.A.

DAATSELAAR, Cornelis Johan VAN. See VAN DAATSELAAR, Cornelis Johan.

DAAVETTILA, Donald Axel, B.S., M.S. (Eng. Phys.). Born 1934. Educ.: Michigan Technol. Univ. Asst. Phys., Argonne Nat. Lab., 1957-64 (including 9 month break to teach chem. at Kingswood School, Detroit); Exptl. Phys., Atomic Power Development Assocs.; Asst. Prof., Nucl. Eng., Michigan Technol. Univ. Faculty Adviser, Student Chapter, A.N.S.
Nuclear interests: Reactor engineering design and reactor parameter measurements. Also developing an interest in activation analysis.
Address: Michigan Technological University, Houghton, Michigan 49931, U.S.A.

DABBS, John Wilson Thomas, B.S., Ph.D. Born 1921. Educ.: Tennessee Univ. Phys., O.R.N.L., 1946-; Fulbright Lecturer, Inst. de Fisica, San Carlos de Bariloche, Argentina, 1961; Visiting Sci., Centre d'Etudes Nucleares, Saclay, France, 1967-68. Book: Co-editor, Semiconductor, Nuclear Particle Detectors (N.A.S., NSS-32, 1961). Society: Fellow, A.P.S.
Nuclear interests: Nuclear orientation at low temperatures; Semiconductor charged-particle detectors.
Address: P.O. Box X, Oak Ridge, Tenn., U.S.A.

DABEK, Tadeusz Jan, B.M. Born 1933. Educ.: Acad. of Mining and Metal., Cracow. Res. Worker, Inst. of Nucl. Techniques, Lab. of Nucl. Instr.
Nuclear interest: Nuclear electronics.
Address: 12/72 Sliska str., Cracow, Poland.

DABROWSKI, Andrzej Jan, M.Sc. Born 1939. Educ.: Warsaw Univ. Central Lab. for Radiol. Protection, Warsaw-Zeran, 1963.
Nuclear interest: Semiconductor nuclear particle detectors.
Address: 11 m. 177 ul. Szanajcy, Warsaw, Poland.

DABROWSKI, Janusz, Dr., Prof. Born 1927. Educ.: Warsaw Univ. Head, Atomic Nucleus Theory Dept., Soltan Nucl. Res. Centre, Inst. for Nucl. Res., Warsaw; Head, Theoretical Div., Inst. of Theoretical Phys., Warsaw Univ.; Assoc. Prof., Mathematical Phys. Dept., Warsaw Univ. Societies: Polish Phys. Soc.; A.P.S.
Nuclear interests: Theory of nuclear structure and nuclear reaction.
Address: 5/7 Hoza, Warsaw, Poland.

DACRE, B., Dr. Health Officer and Officer in Charge of Nucl. Chem. Teaching, Rutherford Lab., Royal Military Coll. of Sci.
Address: Royal Military College of Science, Shrivenham, Swindon, Wiltshire, England.

d'ADLER-RACZ, Joseph H., Eng. Born 1925. Educ.: Brussels Univ. Director, Ind. X-Ray Dept., Usines Balteau, Liège. Societies: American Soc. for Nondestructive Testing; Deutsche Gesellschaft für Zerstörungsfreie Prüfverfahren.
Nuclear interest: Quality control by X-rays.
Address: Beyne-Heusay-lez- Liège, Belgium.

DAGNINO, Virgilio. Italian Member, Specialised Nucl. Sect. for Economic Problems, and Member, Specialised Nucl. Sect. for Social

Health and Development Problems, Economic and Social Com., Euratom.
Address: 2 Via Ghislieri, 20123 Milan, Italy.

d'AGOSTINO, Vincent. Formerly Tech. Asst. to the Pres., now Vice Pres., Res., (1966-), Rad. Applications, Inc.
Address: Radiation Applications, Inc., 36-40 37th Street, Long Island City, New York 11101, U.S.A.

DAHANAYAKE, Charles, B.Sc. (1st class Hons.), Ph.D. (Elementary Particles). Born 1928. Educ.: Ceylon and Bristol Univs. Asst. Lecturer (1951-56), Lecturer, Grade 1 and Grade 2 (1956-67), Ceylon Univ.; Post Doctoral Fulbright Fellowship, Rochester Univ., New York, 1962-63; Prof., Phys., Vidyalankara Univ. of Ceylon, 1967-.
Nuclear interests: Nuclear physics, both medium energy and extreme high energy; Cosmic rays, their origin and composition; Elementary particles; Nuclear emulsion technique.
Address: Department of Physics, Vidyalankara University of Ceylon, Kelaniya, Ceylon.

DAHL, Adrian Hilman, B.A., Ph.D. (Biophys.). Born 1919. Educ.: St. Olaf Coll., Rochester Univ. Head, Rad. Phys. Sect., Atomic Energy Project, Rochester Univ., 1950-59; Consultant to Federal Civil Defence Administration in establishing its nation-wide programme in Radiol. Defence, Battle Creek, Michigan, 1956; Fulbright Lecturer in Health Phys., Argentina, 1958; Principal Sci., responsible for course on Ind. Applications of Radioisotopes, O.R.I.N.S., Tennessee; U.S.A.E.C. expert in Radioisotope Techniques on loan to I.A.E.A. programme in Indonesia, July 1961; Prof. of Phys. and Rad. Biol., Colorado State Univ., 1961. Director, Radiol. Health Specialists training programme. Societies: H.P.S.; A.P.S.; I.E.E.E.
Nuclear interests: Radiation dosimetry; Health physics; Biological effects of radiation; Radioisotope applications in general; Uptake of fall-out materials by cattle, wild-life, and humans through milk and meat; Radon and daughter exposure to lungs of uranium minerals.
Address: Department of Radiology and Radiation Biology, Colorado State University, Fort Collins, Col., U.S.A.

DAHL, John Blessum, M.Sc. Born 1921. Educ.: Oslo Univ. Inst. for Atomenergi, Kjeller, 1955-60; I.A.E.A., Vienna, 1960-62; Inst. for Atomenergi, Kjeller, 1962-. Society: Norwegian Chem. Soc.
Nuclear interest: Industrial applications of radioisotopes.
Address: Institutt for Atomenergi, Kjeller, Norway.

DAHLBERG, G. Member, Rad. Protection Board, Sweden.
Address: Radiation Protection Board, Karolinska sjukhuset, Stockholm 60, Sweden.

DAHLEN, Goran S., M.Sc. Born 1925. Educ.: Royal Inst. Technol., Sweden. Director of Radiol. Defence, Swedish Res. Inst. of National Defence; Member, Reactor Safety and Location Com., Atomic Energy Board of Sweden. Societies: Swedish Assoc. Eng. and Architects; Svenska Teknologföreningen.
Nuclear interests: Nuclear instruments and electronics; Reactor safety and locating problems.
Address: FOA 4, Stockholm 80, Sweden.

DAHLINGER, Alexander, B.Sc. (Elec. Eng.). Born 1928. Educ.: Manitoba Univ. At (1950-), now Supt., Control, Instrumentation and Elec. Branch, BLW (Boiling Light Water) Div., Power Projects, A.E.C.L.
Nuclear interest: Design of nuclear power plant control, instrumentation and electrical systems.
Address: Atomic Energy of Canada Limited, Sheridan Park, Ontario, Canada.

DAI, Yuan-ben. Papers: Co-author, The Radiative Capture of μ-Meson by Nucleus (Scientia Sinica, vol. 10, No. 4, 1961); The Probability of the $\pi^- \to \pi^0 + e + \bar{\nu}$ Decay in the Case of Nonconserved Weak Current (ibid., No. 6).
Address: Academia Sinica, 3 Wen Tsin Chien, Peking, China.

DAINTON, Frederick Sydney, M.A., B.Sc. (Oxon), Ph.D., Sc.D. (Cantab.). Born 1914. Educ.: Oxford and Cambridge Univs. Fellow and Praelector, St. Catharine's Coll., Cambridge, and H.O. Jones Lecturer in Phys. Chem., Cambridge, 1945-50; Prof. Phys. Chem., Leeds Univ., 1950-65; Vice Chancellor, Nottingham Univ., 1965-. Past Pres., and Associate for Rad. Res., Faraday Soc.; Council Member, Chem. Soc.; Assoc. Editor, Rad. Res.; Assoc. Editor, J. Polymer Res. Societies: F.R.S.; Fellow, Chem. Soc.; Faraday Soc.; Rad. Res. Soc.
Nuclear interests: The effects of ionizing radiations on materials, including water, ammonia and organic compounds; Isotopic exchange reactions in gas and solution.
Address: High Energy Radiation Research Centre, Leeds University, Leeds, Yorks., England.

DAITCH, Paul B., Prof., Ph.D. Member, Faculty associated with nucl. sci. and eng. activities, Rensselaer Polytech. Inst.; Honours and Awards Com. Chairman, Northeastern New York Sect., A.N.S.
Address: Rensselaer Polytechnic Institute, Troy, New York, U.S.A.

DAJCAROVA, Yvona, Ing. Radiochem., Div. of Isotopes, School of Medicine, Safarik Univ.
Address: Safarik University, School of Medicine, 53 Rastislavova, Kosice, Czechoslovakia.

DAKSHINAMURTI, Chirravuri, M.Sc., Ph.D. (London), D.Sc. (Banaras). Born 1914. Educ.: Andhra Univ. Waltair, Banaras Hindu Univ.

and Rothamsted Exptl. Station, London Univ. Sen. Phys. and Radiol. Safety Officer (1956), Head, Div. of Soil Sci. and Agricultural Chem. (1961) Head, Div. of Agricultural Phys. (1962), now Prof. Phys., Post-graduate School, Indian Agricultural Res. Inst., New Delhi. i/c Gamma Rad. Field, Indian Agricultural Res. Inst.; Principal Investigator in F.A.O./I.A.E.A. Scheme on Ionic Migration through Soils.
Nuclear interests: Health physics and nuclear physics; Health hazard control measures to be adopted for different types of experiments in radiation biology; Measurement of the radiation fields; Ionic migration by diffusion and mass flow through soils using isotope techniques; Root study in relation to soil structure using isotopes.
Address: Division of Agricultural Physics, Indian Agricultural Research Institute, New Delhi 12, India.

DAKUBU, S., B.Sc. Res. Fellow, Phys. Dept., Radioisotope and Health Phys. Unit, Ghana Univ.
Address: Ghana University, Physics Department, Radioisotope and Health Physics Unit, P.O. Box 25, Legon, Ghana.

DALDRUP, Heribert, Mech. Eng. (Dipl. Ing.). Born 1929. Educ.: Karlsruhe and Hanover T.H. and Inst. Marine Eng. At Hanover T.H., 1956-58; Gesellschaft für Kernenergieverwertung in Schiffbau und Schifffahrt (G.K.S.S.), Hamburg, 1958-60; Group for Reactor Safety, Vereinigung der Tech. Überwachungsvereine, Essen, 1960-. Society: Kernenergie Studiengesellschaft, Hamburg.
Nuclear interests: Nuclear marine engineering - projecting the primary and secondary systems; Nuclear engineering safety - examining safety reports of nuclear installations, checking safety provisions in nuclear installations and studying the possibilities of setting up rules for the atomic industry.
Address: 1 Norbertstrasse, Essen, Germany.

DALE, Geoffrey Carrington, B.Sc. (Phys.). Born 1921. Educ.: London Univ. Supv., A.W.R.E., Aldermaston, 1949-59; Deputy Chief Nucl. Health and Safety Officer (Safeguards), C.E.G.B., London. Society: H.P.S.
Nuclear interest: Radiological protection.
Address: Central Electricity Generating Board, Laud House, 20 Newgate Street, London, E.C.1, England.

DALE, Walter Bernard, B.Sc. (1st class Hons., Eng., External London). Born 1924. Educ.: Roy. Tech. Coll., Salford. At (1956-), Asst. Chief Eng., Gas Cooled Reactors Directorate, Risley, (1960-), U.K.A.E.A. Society: F.I.E.E.
Nuclear interest: Reactor design.
Address: United Kingdom Atomic Energy Authority, Gas Cooled Reactors Directorate, Reactor Group, Risley, Lancs., England.

DALEN, Helge, Cand. real. Lab. for Tissue Culture, Norsk Hydro's Inst. for Cancer Res.
Address: Norsk Hydro's Institute for Cancer Research, Norwegian Radium Hospital, Oslo, Norway.

d'ALENA, Franco CANCELLARIO. See CANCELLARIO d'ALENA, Franco.

DALFES, Abdi, M.S.(Elec. Eng.). M.S. (Nucl. Eng.). Born 1921. Educ.: Istanbul Tech. Univ. and California (Berkeley) Univ. Assoc. Prof., (1955-62), Prof. (1962-), Istanbul Tech. Univ.; (Leave of absence as Chief Reactor Supervisor, Cekmece Nucl. Res. Centre, (1961). Book: Solutions of Reactor Kinetics Equations (Istanbul, Tech. Univ., 1962). Society: A.N.S.
Nuclear interests: Reactor dynamics and stochastical studies of reactor noise.
Address: Technical University of Istanbul, Turkey.

DALGLEISH, James White, B.Sc. Born 1903. Educ.: Glasgow Univ. Exec. Director and Chief Eng., Pye Labs. Ltd., Cambridge, 1945-. Book: Co-author, Elec. Power Transmission and Interconnection. Societies: M.I.E.E.; Sen. M.I.E.E.; Sen.M.I.R.E. (U.S.A.).
Address: 37 Porson Road, Cambridge, England.

DALITZ, Richard Henry, B.A. (Hons.), B.Sc. (Melbourne); Ph.D. (Cantab.). Born 1925. Educ.: Melbourne and Cambridge Univs. Lecturer and Reader in Mathematical Phys., Birmingham Univ., 1951-56; Prof. Phys., Chicago Univ., 1956-63; Roy. Soc. Res. Prof., Oxford Univ., 1963-. Societies: F.R.S.; Fellow, Phys. Soc.; A.P.S.
Nuclear interests: Elementary particle physics; Hypernuclear physics.
Address: Department of Theoretical Physics, Oxford, England.

DALLA PORTA, Niccolò, Prof. Head, Nucl. emulsion group, Istituto di Fisica, Padua Univ.
Address: Istituto di Fisica, Universita di Padua, 8 Via F. Marzolo, Padua, Italy.

d'ALLEST, Jean. Prés. Directeur Gén., Sté. des Chantiers et Ateliers de Provence S.A.
Address: Société des Chantiers et Ateliers de Provence S.A., 130 Chemin de la Madrague, Marseille 15, (B.D.R.), France.

DALLEY, Paul, B.Sc. (Hons.). Born 1922. Educ.: Leeds Univ. Works Manager, Oxide Fuels, U.K.A.E.A., Springfields Works. Society: A.M.I.Ch.E.
Nuclear interest: Production management.
Address: 1 Stanagate, Clifton, near Preston, Lancs., England.

DALLYN, Stewart L., B.Sc., M.S., Ph.D. Born 1924. Educ.: Alberta and Cornell Univs. At Louisiana State Univ., 1950-52; Prof., Agricultural Expt. Station, Cornell Univ., 1952-. Societies: A.I.B.S.; American Soc. Horticultural Sci.
Nuclear interest: Effect of nuclear energy on biological systems - primarily physiology

and genetics.
Address: Long Island Vegetable Research Farm, Sound Avenue, Riverhead, Long Island, New York, U.S.A.

DALTON, G. R. B.S. (Mech. Eng.), Ph.D. (Nucl. Eng., Michigan). Born 1932. Educ.: Mich. Univ., O.R.S.O.R.T. Atomic Power Development Assoc., 1954, 1956; Teaching Fellow, Graduate School, Mich. Univ., 1955-58; Bendix Aviation Systems Div., 1959, 1963; Assoc. Prof. Nucl. Eng., Florida Univ. 1960-65; Westinghouse Atomic Power Div., 1965-66; Acting Chairman, Nucl. Eng. Sci., Florida Univ., 1966-67. Society: A.N.S. Nuclear interests: Reactor analysis; Numerical analysis; Thermonuclear theory.
Address: Department of Nuclear Engineering, University of Florida, Gainesville, Fla., U.S.A.

DALZELL, Robert Carson, B.E., M.Sc., Sc.D. Born 1906. Educ.: Johns Hopkins and Harvard Univs. Chief, Metal. (1950-56), Chief Eng. Development Branch, (1956-58), Special Asst. to the Director (1958-61), Asst. Director for Foreign Activities (1961-), U.S.A.E.C. Div. of Reactor Development. U.S. Sc. Sec., I.C.P.U.A.E., Geneva, 1955 and 1964; Nucl. Standards Board, U.S.A.S.I.; American deleg. to TC-85, I.S.O.; Chairman, Nucl. Eng. Div. Exec. Com., Chairman, Meetings Com., Member, Board on Technol., Chairman, Westinghouse Medal Com., and Res. Com. on Effect of Rad. on Materials, A.S.M.E.; Com. on Nucl. Problems, A.S.T.M.; Hon. Editorial Advisory Board, J. of Nucl. Energy; Exec. Com., and Com. on Internat. Standards, Nucl. Standards Board. Societies: A.N.S.; A.S.M.; A.S.M.E.; Washington Acad. Sci.
Nuclear interest: Research and development management.
Address: 2548 North Vermont Street, Arlington 7, Virginia, U.S.A.

D'AMATO, Francesco, Dr. nat. sci. Born 1916. Educ.: Pisa Univ. Prof. Genetics, Pisa Univ.; Sci. Consultant for Plant Genetics to C.N.E.N., Rome. Societies: International Soc. for Cell Biology; Gesellschaft für Genetik; American Genetic Assoc.
Nuclear interests: Radiation-induced genetic changed in Triticum durum; Embryo-endosperm relations in irradiated seeds; Resistance to chronic gamma radiation in higher plants with special reference to morphogenetic and chromosomal alterations.
Address: Istituto di Genetica della Università di Pisa, Italy.

DAMEWOOD, Glenn. Director, Industrial Phys., Southwest Research Inst.
Address: Southwest Research Institute, 8500 Culebra Road, San Antonio 6, Texas, U.S.A.

DAMKOHLER, Wilhelm Ludwig, Dr. Phil. (Maths.), Dr. phil. habil. (Maths.). Born 1906. Educ.: Munich Univ. Prof. contratado para Mecánica Racional, Tucumán Univ., Argentina, 1949-52; Prof. contratado para Matemáticas y Mecánica Rac., Potosi Univ., Bolivia, 1952-55; Prof. contratado de Matemáticas, La Paz Univ. Bolivia, 1955-. Societies: Deutsche Mathematiker-Vereinigung and Gesellschaft f. Angew. Math. und Mechanik; Soc. cientifica Argentina, Buenos Aires; Asoc. Boliviana para el progreso de las ciencias.
Address: Laboratorio de Fisica Cosmica, Universidad Mayor de San Andres, La Paz, Bolivia.

DAMME, K. VAN. See VAN DAMME, K.

DAMODARAN, Kochuparambil Kunjippennu, B.Sc., M.A., Ph.D. Born 1921. Educ.: Travancore, Madras and London Univs. Sen. Lecturer in Phys., Shree Narayana Coll., Quilon, Kerala, India, 1948-52; Res. Fellow, Phys. Dept., Birmingham Univ., 1954-56; Jun. Res. Officer (1956-59), Res. Officer (1959-62), Formerly Sci. Officer (1962-), Atomic Energy Establishment, Trombay, Bombay; Formerly Sci. Officer, Nucl. Phys. Div., and Sec., Training School Co-ordination Com., Atomic Energy Establishment, Trombay; Head, Training Div., Bhabha Atomic Res. Centre, A.E.C., India. Societies: Fellow, Phys. Soc., London; Assoc. Inst. of Phys., London; Indian Sci. Congress Assoc.
Nuclear interests: Nuclear physics; Training in nuclear science and technology.
Address: Atomic Energy Commission, Bhabha Atomic Research Centre, Trombay, Bombay 74, India.

DAMONTI, Franco. Born 1932. Instrument Eng., Sta. Edison, 1952-56; Market Manager, Sales for Carlo Gavazzi S.p.A., 1956-64; Pres., Reactor Controls, Inc., 1965-. Society: Instrument Soc. of America.
Nuclear interest: Instrumentation and control systems.
Address: c/o Reactor Controls, Incorporated, 1261 Lincoln Avenue, San Jose, California 95125, U.S.A.

DANBARA, Hiroshi. Chief, Lab. of Poultry Physiology, Nat. Inst. of Animal Industry.
Address: National Institute of Animal Industry, Chiba-shi, Japan.

DANCE, William E., B.S., M.S., Ph.D. Born 1930. Educ.: Carson-Newman Coll. and Louisiana State Univ. A.E.C. Contract Fellow (1958-59), Graduate Teaching and Res. Asst. (1953-58), Louisiana State Univ; Res. Sci. (1959-63), Sen. Sci. (1963-), Ling-Temco-Vought Res. Centre. Society: A.P.S.
Nuclear interests: Experimental low energy neutron scattering (below 4 MeV). Nuclear interactions. Bremsstrahlung (experimental) from electrons in the energy range .025-3.50 MeV.
Address: 3306 Whitehall Drive, Dallas, Texas 75229, U.S.A.

DANCEWICZ, Antoni M., Ph.D. Born 1922. Educ.: Gdansk Inst. Technol. Head, Rad.

Biochem. Lab., Dept. Radiobiol. and Health Protection, Inst. Nucl. Res., Warsaw, 1956-. Nuclear interests: Radiation biochemistry; Radiation effects on the mechanism of the biosynthesis of haemoglobine; Early biochemical changes in subcellular fractions after irradiation; Chemical protection.
Address: Department of Radiobiology and Health Protection, Institutę of Nuclear Research, ul. Dorodna 16, Warszawa 91, Poland.

DANDI, Raphael A., Deleg. to 2nd I.C.P.U.A.E., Geneva, Sept. 1958. Member, Subcom. on Instruments and Techniques, Com. on Nucl. Sci., N.A.S.-N.R.C.
Address: Union Carbide Nuclear Company, Oak Ridge, Tennessee, U.S.A.

d'ANDLAU, Charles. Principal collaborateur, Lab. de Physique, Ecole Polytechnique, Paris.
Address: Ecole Polytechnique, Laboratoire de Physique, 17 rue Descartes, Paris, France.

DANDURAND, Paul, Ing. Ecole Polytechnique, M.A., Ph.D. (Phys., Harvard). Born 1923. Educ.: Ecole Polytechnique and Harvard Univ. Director, Groupement d'Industries Atomiques, 1960-; formerly Head, Reactor Control Group, Cie. Française Thomson-Houston.
Nuclear interest: Nuclear electronics.
Address: 8 rue Darcel, 92 Boulogne, France.

DANE, Eric Humbert. Born 1907. Director and Gen. Manager, R.A. Stephen and Co. Ltd.
Nuclear interest: Nucleonic instruments.
Address: Howells Gill, Friday Street, Rusper, Horsham, Sussex, England.

DANEAU, Jacques, Dr. Asst., Inst. du Radium, Canada.
Address: Institut du Radium, 4120 Ontario Est, Quebec, Canada.

DANELYAN, L. S. Papers: Co-author, Radiative Capture Cross-Sections for Te-Isotopes in Relation Neutron Energies up to 1.5 kev (Atomnaya Energiya, vol.14, No. 3, 1963); co-author, Radiative Capture Cross Sections for Disprosium Isotopes in the Energy Region of 0.023-Iev (ibid, vol. 16, No. 1, 1964).
Address: Academy of Sciences of the U.S.S.R., 14 Leninsky Prospekt, Moscow V-71, U.S.S.R.

DANFORTH, John Loring, S.B., S.M. Born 1917. Educ.: M.I.T. Vice-Pres. and Director Mech. Eng., High Voltage Eng. Corp., Burlington, Mass., 1947-.
Nuclear interest: Accelerator engineering.
Address: 35 Farm Lane, Westwood, Massachusetts 02090, U.S.A.

DANIEC, Leszek, M.Sc. Formerly Head, Mining Dept., Office of Govt. High Commissioner for Atomic Energy, Poland.
Address: Office of Government High Commissioner for Atomic Energy, Palace of Culture and Science 18th floor, Warsaw, Poland.

DANIEL, Herbert Gustav Karl, Dr. rer. nat. Born 1926. Educ.: Heidelberg Univ. Res. Asst. (1954), Group Leader (1959-65), Div. Leader (1965-), Max-Planck-Inst. für Kernphysik, Heidelberg; Res. Assoc., Iowa State Univ., 1958-59; Privat-Dozent, Heidelberg Univ., 1961-; Group Leader and Visiting Sci., C.E,R.N., Geneva, 1966-. Society: Deutsche Physikalische Gesellschaft (German Phys. Soc.).
Nuclear interests: Beta- and gamma-ray spectroscopy; Nuclear models; Weak interaction; Magnetic spectrometers; Magnetic field; Muonic and pionic atoms; Neutrinos.
Address: Conseil Européen pour la Recherche Nucléaire, Geneva, Switzerland.

DANIEL, Ranjan Roy, M.Sc. (Banares), Ph.D. (Bristol). Born 1923. Educ.: Madras, Banares and Bristol Univs. Prof., Tata Inst. of Fundamental Res., Bombay. Society: Fellow, Indian Acad.Sci.
Nuclear interests: High energy nuclear physics; Cosmic radiation.
Address: Tata Institute of Fundamental Research, Homi Bhabha Road, Bombay-5, India.

DANIELOPOULOS, Stylianos Daniel, Dipl. Phys. Sc. Born 1930. Educ.: Athens and North Carolina Univs., North Carolina State Coll. and Argonne Nat. Lab. Instructor and Res. Asst., Phys. Lab., Athens Univ., 1953-56; Phys., Greek A.E.C., 1956-59; Acting Director, Phys. Div., Democritus Nucl. Res. Centre, 1959-61; Ph.D. candidate, North Carolina Univ., 1961-.
Address: Greek Atomic Energy Commission, Democritus Nuclear Research Centre, Aghia Paraskevi, Nr. Athens, Greece.

DANIELS, Edward William, B.A., M.S. (Illinois), Ph.D. (Illinois). Born 1917. Educ.: Cornell Coll., Iowa and Illinois Univ. Instructor (1950-53), Asst. Prof. (1953-54), Dept. Physiology, Chicago Univ.; Staff, U.S.A.F. Rad. Lab., Chicago Univ., 1952-54; Assoc. Biol., Div. Biol. and Medical Res., Argonne Nat. Lab., 1954-. Exec. Com., Soc. of Protozoologists. Societies: Soc. Protozoologists; American Soc. Zoologists; A.A.A.S.; Illinois Soc. for Medical Res.; Rad. Res. Soc.; E.M.S.A.; American Soc. for Cell Biol.; A.I.B.S.
Nuclear interests: Effects of ionising and ultraviolet radiations on living systems, particularly at the cellular and subcellular levels; Requirements for survival of animal cells following lethal radiation injury.
Address Division of Biological and Medical Research, Argonne National Laboratory, 9700 South Cass Avenue, Argonne, Illinois, U.S.A.

DANIELS, Farrington, B.S. (Minnesota), Ph.D. (Harvard), D.Sc. (Hon.) (Rhode Island, Minnesota, Dakar, Louisville and Wisconsin Univs.). Born 1889. Educ.: Minnesota and Harvard Univs.; Asst., Assoc. Prof., Prof. Chem. (1920-59), Prof. Emeritus (1959-), Wisconsin Univ.;

Advisory Policy Council, Argonne Nat. Lab., 1957-62. Books: Co-author: Challenge of Our Times; Phys. Chem., Exptl. Phys. Chem., Solar Energy Res.; Author: Chem. Kinetics, Mathematical Preparation for Phys. Chem., Direct Use of the Sun's Energy. Societies: Nat. Acad. Sci.; A.C.S.; American Philosophical Soc.; A.A.A.S.; American Acad. Arts and Sci.; Geochem. Soc.; Solar Energy Soc.
Nuclear interests: High-temperature gas-cooled nuclear reactors; Thermoluminescence; Radiation dosimetry; Effect of radiation on crystals.
Address: Solar Energy Laboratory, Wisconsin University, Madison 6, Wisconsin, U.S.A.

DANIELS, Irvine D., B.A. Born 1910. Educ.: Kentucky Wesleyan Coll. Gen. Manager, Receiving Tube Dept., Electronic Components Div., Gen. Elec. Co., Owensboro., 1956-. Member, Kentucky Advisory Com. on Nucl. Energy. Society: I.E.E.E.
Address: General Electric Co., 316 East Ninth Street, Owensboro, Kentucky, U.S.A.

DANIELS, James Maurice, B.A., M.A., D.Phil. Born 1924. Educ.: Oxford Univ. Res. Fellow, Clarendon Lab., Oxford, 1951-53; Asst. Prof. (1953-56), Assoc. Prof. (1956-60), Prof. (1960-61), British Columbia Univ.; U.N.E.S.C.O. Expert in Exptl. Phys., Buenos Aires Univ., 1958-59; Visiting Prof., Inst. de Fisica José A. Balseiro, Argentina, 1960-61; Prof., Toronto Univ., 1961-. Book: Oriented Nuclei: Polarised Targets and Beams (Academic Press, 1965). Societies: Fellow, Phys. Soc., London; Socio Activo Asociación Fisica Argentina.
Nuclear interest: Oriented nuclei.
Address: Department of Physics, Toronto University, Toronto 5, Ontario, Canada.

DANIELS, John T., B.Sc. (Eng.). Born 1926. Educ.: London Univ. Chief Criticality Inspector, Health and Safety Branch, U.K.A.E.A., Risley.
Nuclear interests: Nuclear safety; Operational management in engineering and processing industries.
Address: U.K.A.E.A., A.H.S.B., Risley, Warrington, Lancashire, England.

DANIELS, Raphael Sanford, B.S. (Civil Eng.), M.S. (Sanitary Eng. and Nucl. Eng.). Born 1931. Educ.: C.C.N.Y. and M.I.T. Res. Asst., Solar Energy, Conversion of Saline Waters, M.I.T., 1953-54; U.S.A.F. Sanitary and Ind. Hygiene Eng., Ramey A.F.B., Puerto Rico, 1954-57; Res. Asst., Radioactive Waste Disposal Project, Sanitary Eng. Dept., M.I.T., 1957-58; Staff Eng., Div. of Radiol. Health, U.S.P.H.S., 1959-60; Consultant, Waste Disposal and Environmental Surveillance, N.R.T.S., Idaho, 1960-61; Nucl. Facilities Consultant, Washington, D.C., 1961-62; Sen. Tech. Assoc., Environmental Safeguards Div., Nucl. Utility Services Corp., 1962-. Chairman, Washington Sect., A.N.S., 1967-68. Societies: A.S.C.E.; Water Pollution Control Federation; A.A.A.S.; H.P.S.; Air Pollution Control Assoc.
Nuclear interests: Site selection; Waste disposal; Environmental safety aspects of nuclear facilities; Environmental surveillance and meteorological diffusion programmes; Aerospace nuclear safety.
Address: Nuclear Utility Services Corporation, 1730 M Street, N.W., Washington, D.C. 20036, U.S.A.

DANIELSON, Gordon Charles, M.A., Ph.D. Born 1912. Educ.: British Columbia and Purdue Univs. Prof. Phys., Iowa State Univ., 1948-; Sen. Phys., Ames Lab. of U.S.A.E.C., 1948-. Guggenheim Fellowship, Roy. Soc. Mond Lab. (1958-59), Distinguished Prof. Sci. and Humanities (1964-), Cambridge; Editorial Advisory Board, Journal of Phys. and Chem. of Solids, 1958-; Chairman, O.N.R. Com. on Thermoelectric Conversion, 1958; Visiting Sci., American Inst. of Phys., 1960-; Review Com. for Solid State Sci. and Phys. Metal., Argonne Nat. Lab., 1960-63; Solid State Sci. Panel, Nat. Res. Council, 1961-; Consultant to Nat. Sci. Foundation, 1963-66. Societies: Fellow, A.P.S.; Fellow, Iowa Acad. Sci.
Nuclear interests: Electronic properties of solids; Thermal properties at high temperatures; Semiconductors; Tungsten bronzes.
Address: Department of Physics, Iowa State University, Ames, Iowa 50010, U.S.A.

DANIELSON, Paul K. Director, Nucl. Dept., Marion Laboratories Inc.
Address: Marion Laboratories Inc., 4500 East 75th Terrace, Kansas City 32, Missouri, U.S.A.

DANILA, Nicolae, B.Sc., M.Sc., Ph.D., Dr. of Tech. Sci. Born 1925. Educ.: Bucharest Tech. Univ. Prof., Bucharest Tech. Univ. Books: Centrale atomoelectrice (Nuclear power plants), (Editura Tehnica Bucuresti, 1956); Analiza Termodinamica a schemelor centralelor electrice nucleare (Thermodynamic analysis of nuclear power plant cycles) (Editura Academiei RSR, 1967).(Academy Publishing House).
Nuclear interests: Nuclear power plant design; Reactor design.
Address: 10-22, ap. 170 Sc.4 Bd. Duca, Bucharest, Roumania.

DANIL'CHENKO, I. D. Paper: Co-author, On Some Rapidity Criteria of Neutron-Activation Analysis (letter to the Editor, Atomnaya Energiya, vol. 22, No. 1, 1967).
Address: Academy of Sciences of the U.S.S.R., 14 Leninsky Prospekt, Moscow V-71, U.S.S.R.

DANILCHIK, Walter. Geologist, Geological Survey, U.S. Dept. of the Interior.
Address: U.S. Department of the Interior, Geological Survey, Building 25, Federal Centre, Denver, Colorado, U.S.A.

DANILIN, L. D. Paper: Co-author, Radioactive Source of Soft X-Ray Rad. for Phys. Investigation, Technique and Medicine (Atomnaya Energiya, vol. 21, No. 2, 1966).

Address: Academy of Sciences of the U.S.S.R., 14 Leninsky Prospekt, Moscow V-71, U.S.S.R.

DANILIN, V. S. Paper: Co-author, Some Features of Diphenyl Thermal Turbines and Their Maximum Power (Atomnaya Energiya, vol. 19, No. 3, 1965).
Address: Academy of Sciences of the U.S.S.R., 14 Leninsky Prospekt, Moscow, V-71, U.S.S.R.

DANILOV, L. L. Paper: Co-author, Injection and Accumulation of Positron (Atomnaya Energiya, vol. 22, No. 3, 1967).
Address: Academy of Sciences of the U.S.S.R., 14 Leninsky Prospekt, Moscow V-71, U.S.S.R.

DANILOV, V. I. Head, Synchrocyclotron Dept., Lab. of Nucl. Problems, Joint Nucl. Res. Inst., Dubna. Papers: Co-author, A. Cyclotron with a Spatially Varying Magnetic Field (Atomnaya Energiya, vol. 8, No. 3, 1960); co-author, Proton Beam Level Increase in 6-Metre Phasotron of the Joint Inst. for Nucl. Res. (ibid, vol. 16, No. 1, 1964); co-author, Duty Cycle Improvement of Particle Beams at a 680 Mev J.I.N.R. Synchrocyclotron (letter to the Editor, ibid, vol. 19, No. 3, 1965); co-author, Beam Current Dependence on Accelerating Voltage in 680 MeV Synchrocyclotron (letter to the Editor, ibid, vol. 21, No. 5, 1966).
Address: Laboratory of Nuclear Problems, Joint Nuclear Research Institute, Dubna, Nr. Moscow, U.S.S.R.

DANIN, D. Book: Dobroi atom (The Good Atom) (M. Molodaya Gvardiya, 1958).
Address: Academy of Sciences of the U.S.S.R., 14 Leninsky Prospekt, Moscow V-71, U.S.S.R.

DANKO, Joseph C., B.S. (Metal. Eng.) (Carnegie Inst. Technol.), M.S. (Metal. Eng.), Ph.D. (Metal. Eng.) (Lehigh). Born 1927. Educ.: Carnegie Inst. Technol. and Lehigh Univ. Instructor, Metal. Eng., Lehigh Univ., 1952-55; Metal. Eng., Gen. Elec. Co., 1955-56; Supervisor, Atomic Power Dept., Materials and Processes (1956-59), Manager, Atomic Power Dept., Materials and Processes (1959-60), Manager, Astronucl.Lab., Thermoelectric-Thermionic Development (1960-), Westinghouse Elec. Corp. Societies: British Inst. for Metals; A.S.M.
Nuclear interests: Technical direction of development programmes of reactor materials, in particular, those associated with direct conversion devices - i.e., thermo-electrics and thermionics in which a nuclear fission or decay heat of an isotope is used as heat source.
Address: Thermoelectric-Thermionic Development, Westinghouse Electric Corporation, Astronuclear Laboratory, Forest Hills Site, P.O. Box 355, Pittsburgh 30, U.S.A.

DANNEBERG, Wolfgang Gottfried Erich, Dr. rer. nat. Born 1928. Educ.: Berlin Freie Univ. Osram G.m.b.H., Studiengesellschaft, Augsburg, 1957-. Society: Deutsche Physikalische Gesellschaft (Fachausschuss Strahlenschutz und Strahlenwirkung).
Nuclear interests: Metallurgy; Radiochemical analysis.
Address: 11 Louis Braille-Strasse, Augsburg, Germany.

DANNEEL, Rolf, Dr. phil. Born 1901. Prof., Bonn Univ.; Leiter der Arbeitsgruppe, Inst. für Zoologie, Kernforschungsanlage Jülich des Landes Nordrhein-Westfalen e.V.
Nuclear interests: Forschung; Lehre; Strahlencytologie.
Address: Poppelsdorfer Schloss, 53 Bonn, Germany.

DANON, Jacques, Chem. Born 1926. Educ.: Brazil and Sorbonne Univs. Prof. (1960), Head, Molecular Phys. and Solid State Dept., Centro Brasileiro de Pesquisas Fisicas; Visiting Prof. of I.A.E.A. at Comision Nacional de Energia Nucl. de Mexico, 1962 and 1965. Books: Bases Fisico-Quimicas de la Radiochimiqua (Comision Nacional de Energia Nucl. de Mexico 1963); Topics on the Mössbauer Effect (Gordon and Breach, 1967); chapter on Mössbauer Spectroscopy in Phys. Methods in Advanced Inorganic Chem. (Intersci. Publishers, 1967); chapter on ^{57}Fe, Metal, Alloys and Inorganic Compounds in Chem. Applications of the Mössbauer Spectroscopy (Academic Press, 1968). Society: Brazilian Acad. Sci.
Nuclear interests: Nuclear chemistry - use of isotopes in physical chemistry; Physical chemistry - Mössbauer effect and applications to molecular structure; Electron spin resonance of irradiated transition metal complexes.
Address: Centro Brasileiro de Pesquisas Fisicas, 71 Avenida Wenceslau Bras, Rio de Janeiro, Brazil.

DANON, Roger Joseph, Phys. Eng. Born 1934. Educ.: E.S.P.C.I. Nucl. Dept., Sté. d'Exploitation des Matériels Hispano-Suiza, 1957-67; At Sogev.
Nuclear interests: Reactor safety; Reactor equipment and experimental facilities; Applications of vacuum and ultra-high vacuum technology; Environment facilities for nuclear and spatial applications; Management.
Address: Résidence du Parc Pierre, 91 St. Geneviève des Bois, France.

DANTZIG, R. van. See van DANTZIG, R.

DANYSZ, Marian, M.Sc. (Phys.) (Warsaw Univ.), M.Sc. (Electronics) (Warsaw Tech. Univ.). Born 1909. Educ.: Warsaw Univ. Prof., Warsaw Univ., 1954-; Dep. Director, Inst. for Nucl. Res., Dubna, 1956-59; Dep. Director Phys., Inst. for Nuclear Res., 1960-. Member, Polish Acad. of Sci. Societies: Polish Phys. Soc.; Italian Phys. Soc.; A.P.S.; Phys. Soc., London.
Nuclear interests: High energy physics and elementary particle physics.
Address: Physics Department, Warsaw University, 69 Hoza, Warsaw, Poland.

DAR, Kunwar Krishna, B.Sc. (Phys. and Geology), B.Sc. (Mining Eng.). Born 1914. Educ.: Punjab and Banaras Hindu Univs. Mining Geologist (1949-59), Formerly Superintending Geologist (1959-), now Regional Director, Atomic Minerals Div., Dept. of Atomic Energy, Govt. of India. Societies: Fellow, Geological, Mining and Metal. Soc. of India; Mining, Geological and Metal. Inst. of India; Hon. Sec. Rajasthan Branch, Mining, Geological and Metal. Inst. of India.
Address: Atomic Minerals Division, Department of Atomic Energy, West Block VII, R.K. Puram, New Delhi, India.

DARA, A. D. Paper: Co-author, X-Ray Diffraction and Thermal Studies of Uranium Molybdates (Atomnaya Energiya, vol.23, No. 2, 1967).
Address: Academy of Sciences of the U.S.S.R., 14 Leninsky Prospekt, Moscow V-71, U.S.S.R.

d'ARAGONA, Massimo CASILLI. See CASILLI D'ARAGONA, Massimo.

DARCEY, W. J. Res. Assoc., Nucl. Phys. Lab., Dept. of Nucl. Phys., Oxford Univ.
Address: Oxford University, Nuclear Physics Laboratory, 21 Banbury Road, Oxford, England.

DARD, Philippe. Born 1929. Educ.: Paris Univ. Sales Manager, Sci. Div., Cie. française Thomson Houston Hotchkiss-Brandt. Society: A.T.E.N.
Nuclear interests: Mass spectrometry; Isotopes separation; Nuclear physics.
Address: Compagnie Française Thomson Houston, D.T.N., P.O. Box 17, Chatou, (S.-et-O.), France.

DARDANO, Ruben, Dr. Sec., Comision de Energia Atomica de El Salvador.
Address: Comision de Energia Atomica de El Salvador, c/o Ministerio del Economia, San Salvador, El Salvador, C.A.

DARDEL, Guy Fredrik VON. See VON DARDEL, Guy Fredrik.

DARDEN, Sperry E., B.S., M.S., Ph.D. Born 1928. Educ.: Iowa State Coll., and Wisconsin Univ. At. Wisconsin Univ., 1956-57; Notre Dame Univ., 1957-. Society: A.P.S.
Nuclear interest: Interaction of fast neutrons with nuclei.
Address: Department of Physics, University of Notre Dame, Notre Dame, Indiana, U.S.A.

D'ARDENNE, Walter Herbert, B.S. (Mech. Eng.), Ph.D. (Nucl. Eng.). Born 1932. Educ.: Pennsylvania State Univ. and M.I.T. Asst. Prof., Nucl. Eng., Pennsylvania State Univ., 1964-. Societies: A.N.S.; A.S.M.E.; A.S.E.E.
Nuclear interests: Reactor design; Nuclear safety; Reactor physics.
Address: Nuclear Engineering Department, Pennsylvania State University, 231 Sackett Building, University Park, Pennsylvania 16802, U.S.A.

DARKER, Harold Alfred. Born 1919. Educ.: South-east London Tech. Coll. Control Eng., English Electric Co., Luton, 1950-55; Computer Eng., English Electric Co., Whetstone, 1955-66; Analogue Office Head, N.D.C. Ltd., 1966-. Society: C. Eng., I.E.E.
Nuclear interests: To obtain analogue computer solutions for problems arising in the nuclear field, consisting of the development of such techniques as nuclear reactor simulations appropriate to particular studies in their dynamic behaviour, and the development of the "Saturn" Analogue Computer specifically for this purpose.
Address: 35 The Fairway, Blaby, Leics., England.

DARLING, Byron Thorwell, B.S., M.S., Ph.D. Born 1912. Educ.: Illinois and Michigan Univs. Asst. Prof., Phys. (1947-51), Assoc. Prof., Phys. (1951-53), Ohio State Univ.; Prof. agrégé, Phys. (1955-58), Prof. titulaire, Phys. (1958-67), Laval Univ. Societies: A.P.S.; Canadian Assoc. of Phys.; Assoc. Canadienne Française pour l'Avancement des Sci.
Nuclear interest: Nuclear theory.
Address: Laval Université, Cité Universitaire, Québec 10, Quebec, Canada.

DARLING, George Bapst, S.B., Dr.P.H., M.A. Born 1905. Educ.: M.I.T., Michigan and Yale Univs. Director, Atomic Bomb Casualty Commission, Hiroshima, Japan, 1957-; Yale Univ. (now on leave of absence), 1943-.
Address: Atomic Bomb Casualty Commission, Hiroshima, Japan.

DARLÖF, Ivan. Formerly Head, Design Div., Thermal Power Dept. now Chief of Design Sect., Thermal Eng. Div., Swedish State Power Board.
Address: Swedish State Power Board, Fack, Vällingby 1, Sweden.

d'ARMINIO-MONFORTE, Giovanni, Degree in Elec. Eng. (Specialisation in Ind. Economics, Specialisation in Nucl. Eng.). Born 1927. Educ.: Naples Univ.; Economic and Commercial Univ., Bocconi, Milan, and I.S.N.S.E., Argonne Nat. Lab. Tech. Study Staff, Gen. Management, Edisonvolta S.p.A. 1950-57; Project Manager, (1957-63), D. Gen. Manager (1963-65), SELNI;Asst. to Sec. Gen. Montecatini Edison S.p.A. Societies: F.I.E.N.; Member, several tech. and econ. committees in Italy and abroad.
Nuclear interest: Nuclear energy management.
Address: Montecatini Edison S.p.A., 31 Foro Buonaparte, Milan 20121, Italy.

DARMON, Gilbert, Ing. en Chef du Génie Maritime. Born 1923. Educ.: Ecole Polytechnique. Ing. du Génie Maritime, Marine Nationale, 1951-59; Sec. Général du Groupement Atomique Alsacienne Atlantique (G.A.A.A.),

1960-. Rep. of Ste. Alsacienne de Constructions Atomiques, de Telecommunications et d'Electronique, Conseil d'Administration, and Treas., Bureau, A.T.E.N. Society: Assoc. Tech. Maritime et Aeronautique.
Nuclear interest: Management.
Address: Groupement Atomique Alsacienne Atlantique, 20 Avenue Edouard-Herriot, Le Plessis-Robinson, Hauts de Seine, France.

DARNELL, D. W. B.S. (M.E.). Born 1900. Educ.: California Inst. Technol. Director, The Fluor Corp., Ltd. Society: American Inst. Mech.Eng.
Nuclear interests: Reactor design; Management; Nuclear power installations.
Address: P.O. Box 7030, E. L. A. Branch, Los Angeles 90022, California, U.S.A.

DARNLEY, Arthur George, M.A., Ph.D. Born 1930. Educ.: Cambridge Univ. Geologist in Rhodesian Copperbelt, Rhodesian Selection Trust (Services) Ltd., 1952-54; Geologist (1957-59), Sen. Geologist (1960-63), Principal Geologist (1963-66), Atomic Energy Div., Geological Survey of Great Britain; Head, Remote Sensing Methods Exploration Geophys. Div., Geological Survey of Canada, 1966-. Society: Assoc. Member, I.M.M.
Nuclear interests: Use of gamma-ray spectrometry and radioisotope instrumentation for mineral exploration and geological mapping; Isotope geology; Nuclear raw materials.
Address: Geological Survey of Canada, 601 Booth Street, Ottawa 4, Ontario, Canada.

DARRAH, James Gore, B. Met. E., M. Met. E., Ph.D. Born 1928. Educ.: Rennselaer Polytech. Inst. and Lehigh Univ. Project Eng., Central Eng., Chrysler Corp., Highland Park, Michigan, 1955-56; Sen. Metal. Eng., Res. Staff, Gen. Motors, 1956-58; Supervisor, Nucl. Materials R. and D., Pratt and Whitney Aircraft, CANEL, Middletown, Conn., 1958-62; Manager, Thermionic Converter Group, Pratt and Whitney Aircraft, East Hartford, Conn., 1962-64; Manager, Process and Material Labs. (1964-65) Manager, Special Products Div. (1965-67), Eitel-McCullough, Inc., San Carlos, Calif.; Pres. and Chairman of the Board, Consolidated Gen. Corp., (merged with Teledyne, Inc.), 1967; Gen. Manager, Monolith, A Teledyne Co., Mt. View, California 1967-. Societies: A.N.S.; A.S.M.
Nuclear interests: Materials research and development - Nuclear fuel element simulators, electron beam heaters, and ceramics; Radioisotopes.
Address: 885 Maude Avenue, Mountain View, California, U.S.A.

DARRAS, Raymond, Dr.-Ing., Ing. E.N.S.E.E.G. Born 1927. Educ.: Paris, Grenoble Univs. Ing. de recherches, C.E.A., 1952-59; Chef de la Section d'Etude de la Corrosion par Gaz et Métaux Liquides, C.E.A., France, 1959-. Book: Chapter in Génie Atomique, tome IV, vol. II (Paris, Presses Universitaires de France, 1965). Societies: Sté. Française de Métallurgie; Centre Française de la Corrosion; A.N.S.
Nuclear interests: Nuclear metallurgy; Corrosion; Gas analysis; Thermodynamics.
Address: Commissariat à l'Energie Atomique, Centre d'Etudes Nucléaires de Saclay, Département de Métallurgie, B.P. No. 2, 91 Gif-sur-Yvette, France.

DARUGA, V. K. Papers: Co-author, The Investigation of Hydrogen - less Shielding of Nucl. reactors (Jaderna Energie, vol. 11, No. 6, 1965); co-author, Reaction $Li + He_2^4$, $Li + H^2$, and $Li + H_1^1$ as Cyclotron Sources of Fast Neutrons (letter to the Editor, Atomnaya Energiya, vol. 24, No. 1, 1968); co-author, Space-Energy and Angular Distributions of Neutrons in Lithium (ibid, vol. 24, No. 4, 1968).
Address: Fyzikalne energeticky institut, Obninsk, U.S.S.R.

DARUSMAN, B. Indonesian Resident Rep., I.A.E.A., 1962-.
Address: 39 Lannerstrasse, Vienna 1, Austria.

DARWIN, George Erasmus, M.A. Born 1927. Educ.: Cambridge Univ., Res. Dept., Metropolitan Vickers, 1954-57; attached A.E.R.E., Harwell, 1954-56; Babcock and Wilcox, Ltd., 1957-. (now Marketing Manager). Book: co-author, Beryllium (Butterworths, 1960). Societies: Inst. of Metals; Inst. Phys. and Phys. Soc.; B.N.E.S.
Nuclear interests: Metallurgy; Reactor design; Economic and commercial matters.
Address: 7 Tor Gardens, London, W.8, England.

DAS, Ranjit Kumar, M.Sc., D. Phil. (Cal.). Born 1923. Educ.: Calcutta Univ., Prof., Saha Inst. Nucl. Phys., Calcutta. Society: Indian Phys. Soc.
Nuclear interests: Design of nuclear instruments, especially various kinds of cloud chambers; High-intensity electro-magnets, geiger counters, ionisation chambers, etc.; Nuclear physics, especially theories of elementary particles and nuclear structure and cosmic radiator.
Address: Saha Institute of Nuclear Physics, 92 Acharya Prafulla Ch. Road, Calcutta 9, India.

DAS, S., Shri. Lecturer, Saha Inst. of Nucl. Phys., Calcutta Univ.
Address: Calcutta University, 92 Acharya Prafulla Chandra Road, Calcutta-9, India.

DAS, Tara Prasad, Dr. Phil. (Sci.) (Calcutta). Born 1932. Educ.: Patna and Calcutta Univs. Res. Assoc., Dept. Chem., Cornell Univ., 1955-56; Res. Assoc., Dept. Phys., California Univ., Berkeley, 1956-57; Reader of Nucl. Phys., Saha Inst. of Nucl. Phys., Calcutta, 1957-58; Res. Asst.Prof., Dept. Phys., Illinois Univ., 1958-59; Res. Assoc., Dept. Chem., Columbia Univ., 1959-60; Sen. Res. Officer, Atomic Energy Establishment, Bombay, Govt. of India,

DAS
1960-61; Assoc. Prof., (1961-65), Prof. (1965-), Phys. Dept., California Univ., Riverside. Member, Board of Studies, Utkal Univ., Cuttack, Orissa, India. Books: Co-author, Nucl. Induction (Saha Inst. of Nucl. Phys., Calcutta, 1957); co-author, Nucl. Quadrupole Resonance Spectroscopy (Academic Press, Inc., 1958). Society: A.P.S.
Nuclear interests: Theory of nuclear moments and hyperfine interactions in optical, microwave and atomic beam spectra; Theory of the applications of electron and nuclear magnetic resonance spectroscopy in the study of electron distributions in molecules and solids.
Address: Department of Physics, California University, Riverside, Calif., U.S.A.

DAS, Y.S., M.A. Born 1929. Educ.: Allahabad Univ. Asst. Comptroller and Auditor Gen., (1959-61) Deputy Sec. to the Govt. of India, Dept. of Atomic Energy, 1961-67; Sec., A.E.C. and Director, Dept. of Atomic Energy, 1967-.
Address: Atomic Energy Commission, Chhatrapati Shivaji Maharaj Marg, Bombay 1, India.

DAS GUPTA, Niraj Nath, M.Sc. (Calcutta), Ph.D. (London), F.N.I. Born 1909. Educ.: Calcutta, London and Stanford Univs. Reader in Biophys. (1946-52), Prof. and Head, Biophys. Div., Saha Inst. (1953-), Calcutta Univ.; Consultant Biophys., Chittaranjan Nat. Cancer Res. Centre, Calcutta. Sec., Biophys. Res. Com., Council of Sci. and Ind. Res., 1956-58; Com. for Biophys., Govt. of India; Exec. Com. of Internat. Federation of Socs. for Electron Microscopy. Societies: A.P.S.; Biophys. Soc., U.S.A.; F.N.I.;.Fellow, Indian Phys. Soc.; Electron Microscope Soc., India.
Nuclear interests: Biological and medical applications of nuclear energy and radioactive isotopes; teaching and research in this field; Radio-iodine and radiophosphorus uptake by normal and malignant tissues; Survival time of leukocycles in leukaemic patients; Incorporation of radioactivity in cell constituents.
Address: Biophysics Laboratory, Saha Institute, Calcutta University, 37 Belgachia Road, Calcutta 37, India.

DASANANDA, Sala, B.S., Ph.D. Born 1916. Educ.: Chulalongkorn and Cornell Univs. Lecturer (1950), Sen. Lecturer (1950), Asst. Prof. (1956), Chulalongkorn Univ.; Asst. Director-Gen. (1954), Deputy Director-Gen., Director-Gen. (1961), Rice Dept., Ministry of Agriculture; Member, Thai A.E.C. Societies: Science Soc. of Thailand; National Res. Council (Thailand).
Nuclear interests: Application of radioisotopes and radiation in agriculture and biology.
Address: Rice Department, Ministry of Agriculture, Bangkok, Thailand.

DASH, J. Gregory, Ph.D. Born 1923. Educ.: C.C.N.Y. and Columbia Univ. Staff. Member, Los Alamos Sci. Lab., 1951-60; Assoc. Prof., Phys. (1960-64), Prof., Phys. (1964-), Washington Univ., Seattle; Consultant, Boeing Co., 1961-64; Consultant, Los Alamos Sci. Lab., 1961-67. Societies: Fellow, A.P.S.; F.A.S.; Peace Res. Soc. (Internat.).
Nuclear interest: Mössbauer effect applied to solid state physics.
Address: Washington University, Physics Department, Seattle 5, Washington, U.S.A.

DASPIT, Woodson B., M.S. Born 1926. Educ.: Louisiana State Univ. Process Supervisor, Savannah River Plant, E. I. de Pont de Nemours and Co., Aiken, South Carolina, 1952-. Treas., Savannah River, South Carolina Sect., A.N.S. Society: A.N.S.
Nuclear interests: Reactor physics and reactor engineering from an operating viewpoint.
Address: 806 Oleander Drive, S.E., Aiken, South Carolina 29801, U.S.A.

DASTEUR, Darab K., M.D., M.Sc., Member, Coll. of Pathologists, (London). Born 1924. Educ.: Bombay Univ. Asst. Res. Officer, 1950; Res. Officer, 1958; Officer-in-Charge, Neuropathology Unit (1961), Neuropathologist (1964-), Indian Council of Medical Res. J.J. Hospital, Bombay. Societies: Neurological Soc. of India; Indian Assoc. of Pathologists; Indian Assoc. of Physiologists and Pharmacologists; Indian Assoc. of Leprologists. Corresponding Member, Problem Commission on Neuropathology, Neuro Muscular Disorders, Neurochemistry and Tropical Neurology, World Federation of Neurology.
Nuclear interest: Use of isotopes in biological and medical research.
Address: Neuropathology Unit, Indian Council of Medical Research, Post-graduate Research Laboratories, J.J. Group of Hospitals, Bombay 8, India.

DATH, Marie-Therese REMACLE. See REMACLE-DATH, Marie-Therese.

DATTA, Narayan Pada, M.Sc., D.Sc. Prof., P.G. School, Indian Agricultural Res. Inst. Born 1913. Educ.: Calcutta and California Univs. Agronomist (Soils), Soil Sci., Radiotracer Lab., Special Officer, and Head, Div.of Soil Sci. and Agricultural Chem., Indian Agricultural Res. Inst., Ministry of Agriculture, Govt. of India. Pres., Indian Soc. of Soil Sci., Member, Indian Council of Agricultural Res. Societies: Internat. Soc. of Soil Sci.; Indian Chem. Soc.
Nuclear interests: Use of radioisotopes in soil, plant-fertiliser use investigations.
Address: Indian Agricultural Research Institute, New Delhi 12, India.

DAUER, Maxwell, A.B., Sc.M., LL.B., Ph.D. Born 1913. Educ.: New York, George Washington and Chicago Univs. Res. Assoc., Toxicity Lab., Pharmacology Dept., Chicago Univ., 1949-51; Chief, Nucl. Energy Special Projects Branch, Office of the Surgeon-Gen., U.S. Army R. and D. Div., 1951-54; Sec. (1954-55), Chief,

Pharmacology Dept. (1955), Army Medical Service Graduate School, Walter Reed Army Medical Centre; Chief, Biophys. Dept., Exec. Officer and Rad. Protection Officer, U.S. Army, Japan, 1955-58; Chief, Radiol. Hygiene Div., U.S. Army Environmental Hygiene Agency, 1958-61; Assoc. Prof. Rad. Phys. (1961), Prof. Rad. Phys. (1962), Radiol. Dept., Miami Univ.; Res. Assoc. and Prof. Pharmacology, Miami Univ., 1961-; Rad. Phys., Radiol. Dept., Jackson Memorial Hospital, 1961-62. Rep. for Miami Univ., O.R.I.N.S. Council. Societies: H.P.S.; Rad. Res. Soc.; Fellow, A.P.H.A.; A.N.S.; A.A.A.S.; American Ind. Hygiene Assoc.; A.A.U.P.; American Conference of Govt. Ind. Hygienist; Radiol. Soc.of N. America; Assoc. Fellow, American Coll. Radiol.
Nuclear interests: Radiation dosimetry, health physics, radiological health and protection; Radioisotope applications, techniques and instrumentation; Radiation equipment and devices; Radiobiology: Nuclear weapons effects, radiation hazards evaluations.
Address: 6660 Montgomery Drive, Miami, Florida, U.S.A.

DAUM, C. Studietoelagen, physici, Institute for Nucl. Phys. Res. (I.K.O.).
Address: 18 Ooster Ringdijk, Amsterdam-O, Netherlands.

DAUTRAY, Robert, Eng., Ecole Polytech., Ing. en Chef des Mines. Born 1928. Direction des Piles Atomiques, C.E.A., 1955-; Maitre de Conférences, Ecole Polytech., 1958-; Member, Study Group on High Flux Reactor, E.N.E.A.; Project Head, German-French High Flux Reactor, Inst. Max von Laue-Paul Langevin. Society: A.N.S.
Nuclear interests: Reactor design and reactor physics.
Address: 7 Villa Madrid, (92) Neuilly-sur-Seine, France.

DAUTREPPE, Daniel Henri, Ing. E.S.E., D. es Sc. Born 1921. Educ.: Paris and Grenoble Univs. Ing., Electricité de France, 1948; Attaché de Recherches Centre Nat. de la Recherche Scientifique, 1952; Chef du Service de Physique du Solide et de Resonance Magnetique (Centre Nucléaire de Grenoble), 1956. Societies: Sté. française de Physique; Sté. française de Metallurgie.
Nuclear interests: Metallurgy; Radiation damage; Nuclear magnetic resonance and electron spin resonance; Organic chemistry.
Address: 33 boulevard Maréchal Foch, Grenoble, Isère, France.

d'AUZAC DE LA MARTINIE, Gerard. Chief Eng., Nucl. Dept., Lab. Industriel d'Electronique Belin.
Address: Laboratoire Industriel d'Electronique Belin, 296 avenue Napoléon Bonaparte, Rueil-Malmaison, (Seine et Oise), France.

DAVENPORT, Donald Emerson, B.S., M.S., Ph.D. (Phys. Chem.). Born 1920. Educ.: Washington Univ. and M.I.T. Societies: A.C.S.; A.N.S.; RESA.
Nuclear interests: Reactor design; Reactor physics; Effects of reactor runaway on structure; High pressure failure of components; Explosively actuated controls.
Address: Advanced Products Division, Link Group, General Precision Systems Inc., Sunnyvale, California, U.S.A.

DAVENPORT, J. F. Vice Pres., High Temperature Reactor Development Associates, Inc.; Formerly Exec. Vice Pres., Southern California Edison Co.; Member, Board of Directors, Atomic Ind. Forum, 1966-69.
Address: High Temperature Reactor Development Associates, Inc., 89 East Avenue, Rochester 4, New York, U.S.A.

DAVENPORT, Lee Losee, B.S. (Union College), M.S. and Ph.D. (Pittsburgh). Born 1915. Educ.: Pittsburgh Univ. Director, Nucl. Lab., Harvard Univ., 1946-50; Exec. Vice-Pres., Perkin-Elmer Corp., 1950-57; Pres., Sylvania-Corning Nucl. Corp., 1957-. Formerly Member, Gen. Advisory Com., N.Y. State Office of Atomic Development. Societies: A.P.S.; Optical Soc.; A.A.A.S.
Nuclear interests: Corporate management; Research and development; Fabrication of nuclear reactor fuel elements and componenets.
Address: Forbell Drive, Norwalk, Connecticut, U.S.A.

DAVEY, G. Res. Assoc., Nucl. Phys. Lab., Dept. of Nucl. Phys., Oxford Univ.
Address: Oxford University, Department of Nuclear Physics, 21 Banbury Road, Oxford, England.

DAVEY, Martin Geoffrey, M.D., M.B., B.S. Born 1936. Educ.: Sydney and Adelaide Univs. Res. Fellow, Nat. Heart Foundation of Australia: Adelaide Univ., 1962-64, Theodor Kocher Inst., Bern, 1964-66; Res. Fellow, The Wellcome Trust, Radcliffe Infirmary, Oxford, 1966-67; Director, Red Cross Blood Transfusion Service, Perth, Australia, 1967-. Book: The Survival and Destruction of Human Platelets (Basel, S. Karger, 1966).
Nuclear interests: Use of radioisotopes for the study of cell turnover in vivo; Applications of radioisotopes in studies of cell metabolism and function - e.g., divalent ion binding by cell membranes.
Address: Red Cross Blood Transfusion Service, 290 Wellington Street, Perth, Western Australia 6000, Australia.

DAVEY, Squire. Born 1913. At (1948-), Present position, Chief Coordinating Eng., Steam graphite heavy water reactor, Reactor Group, U.K.A.E.A. Societies: Life Member, Nucl. Eng. Soc.; Inst. of Production Eng.; B.N.E.S.
Nuclear interest: Reactor design and management.

DAVID (DAVIS), Gerald BEN-. See BEN-DAVID (DAVIS), Gerald.

DAVID, Milan. Civil Eng. Born 1931. Educ.: Prague Tech. Univ. Chief Specialist for special civil eng. structures (prestressed pressure concrete vessels), Energoprojekt, Prague. Society: Czechoslovak Sci.-Tech. Organisation.
Nuclear interests: Prestressed concrete pressure vessels for nuclear reactors.
Address: 7 Ke Krci, Prague 4, Czechoslovakia.

DAVID, Mrs. Vivian Mary, Ph.D. (Reading), M.Sc. (Manchester). Born 1933. Educ.: Manchester Univ. A.E.I. Res. Lab., Aldermaston, 1956-63; Berks. County Council, 1963-. Societies: F.R.A.S.; A.Inst.P.
Nuclear interest: Nuclear physics.
Address: Willow House, Reading Road, Finchampstead, Berkshire, England.

DAVIDGE, Peter Clifford, B.Sc. Born 1924. Educ.: Liverpool Univ. Res. Manager, Culcheth Labs., U.K.A.E.A., 1958-59; Chief Development Chem. (1959-61), Sen. Tech. Officer (1961-63), Production Group, U.K.A.E.A. Sen. U.K.A.E.A. Representative, A.E.C.L., Chalk River, Canada, 1963-. Society: F.R.I.C.
Nuclear interests: Chemical aspects of reactor design; Development of chemical plant for manufacturing reactor fuel and for processing irradiated fuel.
Address: Atomic Energy of Canada Ltd., Chalk River, Ontario, Canada.

DAVIDOVSKII, V. G. Papers: Oscillations of a Spatially Inhomogeneous Plasma in a Magnetic Field (letter to the Editor, Atomnaya Energiya, vol. 15, No. 1, 1963); co-author, About Damping of Betatron Oscillations in Weak Focussing Synchrotron and Storage Rings with Single Turn Injection (ibid, vol. 23, No. 4, 1967).
Address: Academy of Sciences of the U.S.S.R., 14 Leninsky Prospekt, Moscow V-71, U.S.S.R.

DAVIDS, Jan, Biol. Drs. Born 1928. Educ.: Leyden Univ. Radiobiol., Nat. Defence Res. Council T.N.O., 1953-55; Instructor Biochem., Leyden Univ. 1955-57; Radiobiol., Reactor Centrum Nederland, 1957-.
Nuclear interests: Radiobiology; Biological aspects of health physics.
Address: Reactor Centrum Nederland, Petten, N.H., Netherlands.

DAVIDSON, Donald S. Chief Eng., Technical Measurement Corp.
Address: Technical Measurement Corporation, 441 Washington Avenue, North Haven, Connecticut, U.S.A.

DAVIDSON, Jack Dougan, A.B., M.D. Born 1918. Educ.: Princeton and Columbia Univs. Asst. Prof. Medicine, Coll. of Physicians and Surgeons, Columbia Univ., 1953-57; Sen. Investigator, Nat. Cancer Inst. (1957-66), Chief, Dept. Nucl. Medicine (1966-), N.I.H.; Co-Chairman N.I.H. Rad. Com., 1962-. Societies: American Assoc. for Cancer Res.; American Soc. for Pharmacology and Exptl. Therapeutics; A.A.A.S.; Soc. Nucl. Medicine.
Nuclear interests: Biochemical and medical uses of radioisotopes; Techniques and instrumentation of liquid scintillation counting.
Address: Nuclear Medicine Department, Clinical Centre, National Institutes of Health, Bethesda, Maryland 20014, U.S.A.

DAVIDSON, James Taylor, M.Sc. Born 1911. Educ.: Rhodes Univ., South Africa. Prof., Director, Phys. Dept., Univ. Coll. of Fore Hare.
Nuclear interest: Training of physics students.
Address: University College of Fort Hare, Alice, Cape Province, South Africa.

DAVIDSON, John Keith, B. Chem. Eng., M.S. (Phys.). Born 1925. Educ.: Cornell Univ. and Union Coll.. Manager, Chem. Eng. Unit, K.A.P.L., G.E.C., 1954-57. Manager, Eng. Analysis Sect., Gas-cooled Reactor Project, (1957-60), Manager Thorium Fuel Cycle Project (1960-62), Manager, Fuel Manufacturing (1963-), Allis-Chalmers Mfg. Co., Washington, D.C. Societies: A.N.S.; A.I.Ch.E.
Nuclear interests: Nuclear engineering; Fuel element design; Fuel fabrication; Fuel reprocessing.
Address: 216 Hillsboro Drive, Silver Spring, Maryland, U.S.A.

DAVIDSON, John Mackinnon, M.D., Ch.B., D.P.H. Born 1903. Educ.: Edinburgh Univ. Principal Medical Officer, Ministry of Pensions and Nat. Insurance, 1948-56; Medical Adviser, Ministry of Power; Principal Medical Inspector of Mines, Quarries and Nucl. Installations, 1956-. Society: M.I.Min.E.
Nuclear interests: Safety and health in industrial applications of nuclear energy.
Address: Ministry of Power, Thames House South, Millbank, London, S.W.1,England.

DAVIDSON, R. P. Directeur Gén. Adjoint, Internat. Gen. Elec. France, S.A.
Address: International General Electric France, S.A., 42 avenue Montaigne, Paris 8, France.

DAVIDSSON, D., Prof. Chief, Radioisotope Lab., Dept. of Chem. Pathology, State Hospital, Reykjavik.
Address: State Hospital, Postbox 1036, Reykjavik, Iceland.

DAVIES, A., B.Sc., M.I.Biol. Radiol. Health and Safety, Roy. Coll. of Advanced Technol., Salford.
Address: Royal College of Advanced Technology, Salford 5, Lancashire, England.

DAVIES, A. G., Ph.D. Phys. Chem., Tracer Lab., Marine Biol. Assoc. of the United Kingdom.

Address: Marine Biological Association of the United Kingdom, Tracer Laboratory, Citadel Hill, Plymouth, Devon, England.

DAVIES, David Arthur, M.Sc. (1st class Hons. Maths., 1st Class Hons. Phys.). Born 1913. Educ.: Wales Univ. Director, East African Meteorological Dept., Nairobi, 1949-55; Sec.-Gen., World Meteorological Organisation, Geneva, 1955-. Societies: Fellow, Inst. Phys.; Fellow, Roy. Meteorological Soc.; Roy. Inst. of Public Administration.
Nuclear interests: Atmospheric radioactivity; Use of radioisotopes for meteorological purposes.
Address: World Meteorological Organisation, Geneva, Switzerland.

DAVIES, David Kaufman, B.S. (M.E.). Born 1925. Educ.: Lehigh Univ. Process Equipment Design Eng. (1948-53), Asst. Head, Structural Eng. Sect., Atomic Energy Div. (1953-55) Sales Liaison Eng., Atomic Energy Div. (1955-56), Chief Nucl. Component Specialist (1956-), Babcock and Wilcox Co. Chairman, Subcom. on Uniform Notation, Design Div., Pressure Vessel Res. Com., Welding Res. Council. Sect. Editor, Reactor Handbook. Societies: A.S.M.E.; A.N.S.; Atomic Ind. Forum.
Nuclear interests: Management, related particularly to nuclear plant components; Marketing of nuclear components and nuclear fuels.
Address: 562 Orchard Avenue, Barberton, Ohio, U.S.A.

DAVIES, Douglas Mackenzie, B.A. (Hon. Sc.), Ph.D. (Toronto). Born 1919. Educ.: Toronto Univ. Asst. Prof. Zoology (1951-57), Assoc. Prof. Zoology (1957-63), Prof. Zoology (1963-), McMaster Univ. Former Pres. and member of Editorial Board, Entomological Soc. of Ontario. Societies: Entomological Soc. of America; Royal Entomological Soc. of London; A.I.B.S.; Canadian Soc. Zoologists; Entomological Soc. of Canada.
Nuclear interest: Using radioactive isotopes as a tool in entomological research on nutritional physiology.
Address: Department of Biology, Hamilton College, McMaster University, Hamilton, Ontario, Canada.

DAVIES, Eric, B.Sc. (Tech.), A.M.I.Mech.E. Born 1925. Manchester Univ. Group Eng., Light Chem. and Pile Group, U.K.A.E.A., Seascale. Chairman, Eng. Equipment Users Assoc. Panel M/38 Sealing. Society: Nucl. Eng. Soc. Nuclear interest: Works engineering management.
Address: 19 Wastwater Rise, Seascale, Cumberland, England.

DAVIES, James Brian Meredith, M.D., M.B., B.S., D.P.H. Born 1920. Educ.: London Univ. Deputy Medical Officer of Health and Deputy School Medical Officer, Oxford, 1950-53; Deputy Medical Officer of Health and Deputy Port Medical Officer, Liverpool, 1953-;

Sen. Lecturer, Liverpool Univ., 1953-. Hon. Sec., Teaching Group, Soc. of Medical Officers of Health; Public Health Lab. Service Board, 1967-. Societies: Soc. of Medical Officers of Health, Liverpool Medical Inst.; Paediatric Club; British Medical Assoc.
Nuclear interests: Preventive public health aspects of health hazards associated with ionising radiation.
Address: 201 Thomas Lane, Broadgreen, Liverpool 14, England.

DAVIES, John Heddwyn, B.Sc. (Wales). Born 1926. Educ.: Wales and Bristol Univs. George Wills Res. Assoc. in Phys., Bristol Univ.
Nuclear interest: Cosmic rays.
Address: Bristol University, H.H. Wills Physics Laboratory, Royal Fort, Tyndall Avenue, Bristol 2, England.

DAVIES, Peter Trevor, M.A., Ph.D. Born 1925. Educ.: Cambridge Univ. Employed (1949-), now Principal Sci. Thornton Res. Centre Shell Res. Ltd. Societies: Assoc. Inst. Phys. and Phys. Soc.; Soc. for Analytical Chem.; Inst. of Maths. and its Applications.
Nuclear interests: Physical methods of chemical analysis; Applications of radioisotopes in research; Statistical and computing aspects of counting, especially of liquid scintillation counting.
Address: 17 St. James's Avenue, Upton Heath, Chester, England.

DAVIES, Sherwood, B.C.E., M.P.H. Born 1917. Educ.: Rensselaer Polytech. Inst. and Minnesota Univ. Director, Bureau of Radiol. Health Services, New York State Dept. of Health. Societies: American Public Health Assoc.; Conference of State Sanitary Eng.; A.A.A.S. Nuclear interests: Public health aspects of ionising radiation and the handling and disposal of radioactive wastes.
Address: 84 Holland Avenue, Albany, New York, U.S.A.

DAVIES, William John Morswyn. Born 1926. Resident Eng., Eng. Group, U.K.A.E.A. Societies: Inst. of Mech. Eng.; A.M.I.C.E.
Nuclear interest: Reactor construction.
Address: 34 Rothesay Road, Dorchester, Dorset, England.

DAVILA BURGA, Jorge, Ing. Geologo, Departamento de Control de Sustancias Radiactivas, Inst. Superior de Energia Nucl., Junta de Control de Energia Atomica del Peru.
Address: Junta de Control de Energia Atomica del Peru, 3420 Avenida Arequipa, Apartado 914, Lima, Peru.

DAVIS, Allan William, B.Sc., D.Sc. (Glasgow) Born 1912. Educ.: Glasgow Univ. (Formerly Vice-Chairman, Fairfield-Rowan Ltd., Deputy Managing Director, Fairfield Shipbuilding and Eng. Co. Ltd.; Manager, Marine Mech. Dept., Westinghouse Elec. Corp. 1966. Member, American Com., Lloyd's Register of Shipping.

Societies: Inst. Mech. Eng.; I.Mar.E.; Soc. Naval Architects and Marine Engs.; American Soc. of Naval Eng.; Inst. Eng. and Shipbuilders in Scotland; North East Coast Inst. Eng. Shipbuilders.
Nuclear interest: Coordination of the design of nuclear machinery submitted to the British Ministry of Transport for a 65,000 D.W.T. Tanker. Current U.S. Naval Mach.
Address: Westinghouse Electric Corp., Sunnyvale, Calif., U.S.A.

DAVIS, Charles. Keleket Div. Manager, Tracerlab-Keleket.
Address: Tracerlab-Keleket, 1601 Trapelo Road, Waltham 54, Massachusetts, U.S.A.

DAVIS, D. C. Chief, Eng., Ind. Eng. Div., Nucl. Eng. Dept., E.G. Irwin and Partners Ltd.
Address: E. G. Irwin and Partners Ltd., Nuclear Engineering Department, Kibworth Street, Dorset Road, London S.W.8, England.

DAVIS, Dent C., Jr., B.S., M.S. (Chem. Eng.). Born 1918. Educ.: Kansas State Univ. Asst. Prof. Chem. Eng. and Project Supervisor, Denver Res.Inst., Denver Univ., 1947-52; Project Eng. (1952-54), Asst. Chief, Reactor Branch (1955-58), Chief, Res. Branch (1958-), Oak Ridge Operations, U.S.A.E.C., Oak Ridge, Tennessee. Chairman, U.S.A.E.C. Special Com., U.S. Fusion Res. Exhibit, I.C.P.U.A.E., Geneva, 1958. Societies: A.C.S.; A.N.S.; A.I.Ch.E.
Nuclear interests: General.
Address: Research Branch, Oak Ridge Operations, U.S.A.E.C., Oak Ridge, Tennessee, U.S.A.

DAVIS, Donald D. Asst. Managing Editor, Nucl. Sci. Abstracts.
Address: Nuclear Science Abstracts, Division of Technical Information Extension, U.S. Atomic Energy Commission, P.O. Box 62, Oak Ridge, Tennessee, U.S.A.

DAVIS, F. F. Vice Pres., Projects, (Australasia), Kaiser Eng.
Address: Kaiser Engineers, Division of Kaiser Industries Corp., 300 Lakeside Drive, Oakland 12, California, U.S.A.

DAVIS, F. W. Vice-Pres. and Manager, Nuclear Facility, Fort Worth, Convair.
Address: Convair, Fort Worth, Texas, U.S.A.

DAVIS, Frank Wilson, B.S. (Metal.), M.S. (Mech. Eng.). Born 1889. Educ.: Lehigh and Nebraska Univs. Metal. Eng., Div. of Reactor Development, U.S.A.E.C., 1950-60; Chairman, A.E.C. Welding Forum, 1951-61; Staff Consultant, Southwest Res. Inst., 1960-; Consultant to U.S.A.E.C., 1960-61. Hon. Member, Main Boiler and Pressure Vessel Code Com. Societies: A.S.M.; A.S.M.E.; Hon. Member, American Welding Soc.
Nuclear interests: Reactor pressure vessel design, with particular reference to fatigue resistance as related to design; Materials engineering both as related to pressure vessel service and as reactor fuel and control materials.
Address: 133 East Mistletoe, San Antonio, Texas 78212, U.S.A.

DAVIS, George Kelso, Ph.D. Born 1910. Educ.: Pennsylvania State and Cornell Univs. Prof. Nutrition and Animal Nutritionist, Agricultural Experiment Station, Florida Univ., 1942-. Director of Nucl. Studies, 1960. Nat. Res. Council Com. on Animal Nutrition; Florida Nucl. and Space Commission; F.A.O. Expert Panel on Animal Nutrition; N.A.S. ad hoc Com. Internat. Biol. Programme; Inter-Institutional Com. on Nucl. Sci. and Res. Societies: A.C.S.; American Soc. of Biolog. Chem.; American Inst. of Nutrition; Soc. of Exptl. Biol. and Medicine.
Nuclear interests: Radiation safety; Isotopes; Metabolic research.
Address: Nuclear Sciences Building, Florida University, Gainesville, Florida, U.S.A.

(DAVIS), Gerald BEN-DAVID. See BEN-DAVID (DAVIS), Gerald.

DAVIS, H. Willard, Dr. Educ.: Cincinnati Univ. Head, Chem. Dept., South Carolina Univ., 1941-59; Principal Sci., Univ. Relations Div., O.R.I.N.S., 1959-60; Dean, Coll. Arts and Sci., (1960-66), Vice-Pres. for Academic Affairs (1966-), South Carolina Univ. Member, Visiting Scientists Panel, Div. Chem. Educ., A.C.S.; Counsellor, South Carolina Sect.,. A.C.S.; Formerly Member for South Carolina Univ., O.R.I.N.S. Council.
Address: College of Arts and Science, South Carolina University, Columbia, S.C., U.S.A.

DAVIS, Harold Shelley, B.S. (C.E.), M.S. (C.E.), Ph.D. Born 1919. Educ.: Idaho and Northwestern Univs. Asst.Prof. Civil Eng., Washington State Univ., 1949-51; Sen. Eng., Hanford Products Operation, G.E.C., 1951-67; Principal Civil Eng., Douglas United Nucl. Corp., 1967-. Com. on Structural Materials in Reactor Design, A.S.C.E.; Subcom. II-H, Concrete for Radiation Shielding, A.S.T.M.; Com. 349, A.C.I. Societies: A.N.S.; A.C.I.; A.S.C.E.
Nuclear interests: Design of plant and structures incorporating nuclear power facilities and reactors; Development, design and construction of structures for shielding nuclear radiations; Development and design of high-density concrete for use in shielding nuclear reactors and radioactive materials.
Address: Building 1101-N, Douglas United Nuclear Corporation, Richland, Washington, U.S.A.

DAVIS, Harold W. Born 1922. Educ.: California Univ., Berkeley. Chief Cost Accountant, Aerojet-Gen. Corp., 1948-56; Vice Pres. and Manager, Administration Div., Aerojet-Gen. Nucleonics, 1956-66; Manager, San Ramon Plant, Aerojet-Gen. Corp., 1966-. Society:

A.I.A.A.
Nuclear interests: Management; Reactor design; Terrestrial and hydrospace propulsion.
Address: Aerojet-General Corp., San Ramon Plant, P.O. Box 77, San Ramon, California 94583, U.S.A.

DAVIS, Howard F., Assoc. Prof. of Phys., Phys. Dept., Oregon State Univ.
Address: Oregon State University, Corvallis, Oregon, U.S.A.

DAVIS, Jared James, B.S., M.S. Born 1920. Educ.: Washington and Washington State Univs. Manager, Radioecology, Hanford Atomic Products Operation, G.E.C., 1948-62; Vice Pres. and Director, Biol. Sci. Div., Hazleton-Nucl. Sci. Corp., 1963; Aquatic Ecologist, Div. of Biol. and Medicine, U.S.A.E.C., 1963-; Tech. Asst., Exec. Office of the Pres., President's Office of Sci. and Technol., 1965-66 (on leave from U.S.A.E.C.). Books: Chapter in Radioecology (Editors, V. Schultz and A.W. Klement) (Reinhold Publishing Corp., 1963); Chapter in Radioactivity and Human Diet (Editor, R. Scott Russell) (Pergamon Press, 1966). Societies: A.A.A.S.; American Inst. of Biol. Sci.; Ecological Soc. of America; American Soc. of Limnology and Oceanography.
Nuclear interests: Technical direction; Administration and research in the field of radioecology; Effects of radioactivity upon ecological systems and the accumulations, cycling and fate of radionuclides in the natural environment and controlled ecological systems.
Address: Division of Biology and Medicine, United States Atomic Energy Commission, Washington, D.C. 20545, U.S.A.

DAVIS, John E. Sen. Administrative Asst., Dept. of Materials Eng., Battelle Memorial Inst., Columbus. Member, Tech. Information Panel, U.S.A.E.C.
Address: Battelle Memorial Institute, Department of Materials Engineering, Columbus, Ohio, U.S.A.

DAVIS, John Kennerly, B.S. (Elec. Eng.). Born 1906. Educ.: Virginia Military Inst. Pres., Toledo Edison Co., 1959-. Exec. Vice Pres. and Vice Chairman, Exec. Com., Atomic Power Development Associates; Member, Exec. Com., Power Reactor Development Co. Societies: I.E.E.E.; N.S.P.E.; Ohio Soc. of Professional Eng.; Eng. Soc. of Toledo.
Nuclear interest: Management.
Address: Toledo Edison Co., 420 Madison Avenue, Toledo, Ohio 43601, U.S.A.

DAVIS, John L., A.B., LL.B., Born 1913. Educ.: Kentucky Univ. Member, Kentucky Advisory Com. on Nucl. Energy.
Address: 310 First National Bank Building, Lexington, Kentucky, U.S.A.

DAVIS, John Milton, A.B. Born 1918. Educ.: California (Berkeley) Univ. California Univ. Rad. Lab., 1947-52; Self-employed, 1952-54; Atomics Internat, 1954-63; Consultant, 1963-. Exec. Com., Remote Systems Technol. Div., A.N.S. Society: A.N.S.
Nuclear interests: Hot laboratory design, equipment, and management; Project engineering management.
Address: Kernforschungsanlage Jülich, 517 Jülich, Postfach 365, Germany.

DAVIS, Miss Joyce P., S.B. (M.I.T.), M.S. (Rochester). Born 1934. Educ.: M.I.T. Nucl. Eng., Burns and Roe, Inc. Sec. N.Y. Metropolitan Sect., and Member, Membership Com., A.N.S. Societies: H.P.S.; A.P.S.; A.I.A.A.
Nuclear interests: Plant safety and hazards analysis; Shielding; Health physics; Reactor economics; Space environment; Advanced power systems.
Address: Burns and Roe Inc., 160 W. Broadway, New York City, U.S.A.

DAVIS, Michael, B.Sc. (Hons., London), Ph.D. (Bristol), F.Inst.P. Born 1923. Educ.: Exeter and Bristol Univs. Services Electronics Lab., 1951; R. and D., Culcheth Labs. (1956), Res. Manager (1957), Deputy Head of Labs. (1958), Special Asst. to Sir Leonard Owen, Member of U.K.A.E.A. for Production and Managing Director of the Production Group (1959), Chief Inspector of Fuel Elements (1960), Comptroller of Fuel Elements (1961-63), Commercial Director, Production Group (1963), Tech. Adviser, London Office (1965), U.K.A.E.A. Societies: Inst. of Metals.; Manchester Literary and Philosophical Soc.; B.N.E.S.
Nuclear interests: Management; Effectiveness of research; Uranium resources and their discovery.
Address: United Kingdom Atomic Energy Authority, 11 Charles II Street, London, S.W.1, England.

DAVIS, Monte V., B.A., M.A., Ph.D. Born 1923. Educ.: Linfield Coll. and Oregon State Univ. Prof. Nucl. Eng. and Director of Nucl. Reactor Lab., Arizona Univ. Societies: A.N.S.; A.P.S.; A.I.A.A.
Nuclear interests: Reactor physics and design, space power systems and high temperature materials technology; Direct energy conversion.
Address: Department of Nuclear Engineering, Arizona University, Tucson, Arizona 85721, U.S.A.

DAVIS, Peter Stanislaus, M.Sc., A.S.T.C., A.R.A.C.I. Born 1927. Educ.: Sydney Tech. Coll. and New South Wales Univ. C.S.I.R.O., Div. Fisheries and Oceanography, 1954-57; Officer in charge of Low Level Radiochem. Lab., Lucas Heights, Australian A.E.C., 1957-63; Lecturer (Clinical Biologist), Adelaide Univ., South Australia, 1964-. Societies: Assoc. Roy. Australian Chem. Inst.; Australian Rad. Soc.; Australian Soc. for Microbiol.

Nuclear interests: Measurement of trace amounts of radioactive nuclides in biological materials; Clinical applications of radioisotopes.
Address: Department of Medicine, Adelaide University, North Terrace, Adelaide, South Australia.

DAVIS, R. H. Manager, Marine and Nucl. Energy Div., Worthington Corp.
Address: Worthington Corporation, 401 Worthington Avenue, Harrison, New Jersey, U.S.A.

DAVIS, Raymond, Jr. Chairman, Subcom. on Low Level Contamination of Materials and Reagents, Com. on Nucl. Sci. and Member, Com. on Nucl. Sci., N.A.S.-N.R.C.
Address: National Academy of Sciences-National Research Council, 2101 Constitution Avenue, N.W., Washington 25, D.C., U.S.A.

DAVIS, Robert E. Geologist, Geological Survey, U.S. Dept. of the Interior.
Address: U.S. Department of the Interior, Geological Survey, Building 25, Federal Centre, Denver, Colorado, U.S.A.

DAVIS, Robert Gordon. Born 1925. Sales Manager, Nucl. Enterprises Ltd. Societies: I.E.E.E. (Australia); A.N.S.; British Inst. of Radiol.; Hospital Physicists' Assoc.
Nuclear interests: Management, particularly in connexion with the design, production and sale of nucleonic instruments for nuclear medicine, research in the life sciences and industry.
Address: Nuclear Enterprises Ltd., Sighthill, Edinburgh 11, Scotland.

DAVIS, Robert Houser, B.S. (Nebraska), M.S. and Ph.D. (Wisconsin). Born 1926. Educ.: Nebraska and Wisconsin Univs. Res. Assoc., Rice Inst., Houston, Texas, 1955-57; Asst. Prof. Phys. (1957-61), Assoc. Prof. (1961-63), Prof. and Principal Sci., Tandem Van de Graaff Lab. (1963-), Florida State Univ. Society: A.P.S.
Nuclear interests: Elastic and inelastic scattering of charged particles; Models for nuclear reaction mechanisms; Molecular collision physics.
Address: Physics Department, Florida State University, Tallahassee, Florida, U.S.A.

DAVIS, Thomas W., B.Sc., M.S., Ph.D. Born 1905. Educ.: N.Y.U. Prof. Chem. (1951-), Chairman, Dept. at Univ. Heights Centre (1956-64), N.Y.U.; Res. Chem. at U.S. Naval Ordnance Test Station, summer 1952; Res. Chem., O.R.N.L., summer 1954; Res. Chem., summers 1955 and 1956; and Res. Assoc., Argonne Nat. Lab.; Visiting Prof., Leeds Univ., 1964-65. Societies: Rad. Res. Soc.; Federation of Atomic Scientists.
Nuclear interests: Chemical effects of radiation in aqueous systems and in solids; Use of radioactive substances as tracers.
Address: New York University, University Heights, New York 10453, U.S.A.

DAVIS, W. A. Director, Fleming Instruments Ltd.
Address: Fleming Instruments Ltd., Caxton Way, Stevenage, Herts., England.

DAVIS, W. Kenneth, B.S. (Chem. Eng.) (M.I.T.); M.S. (Chem. Eng.) (M.I.T.). Born 1918. Educ.: California Univ. and M.I.T. Prof. of Eng., California Univ., 1949-53; Manager, Res. Div., California R. and D. Co., 1950-54; Asst. Director, Deputy Director (1954-55), Director (1955-58), Reactor Development Div., U.S.A.E.C.; Vice Pres., Bechtel Corp., 1958-; Vice-Pres., Bechtel Nucl. Corp., 1958-; Regents Lecturer, California Univ., 1961. Director Pres. (1965-66), Chairman, Exec. Com. (1965-), Atomic Ind. Forum. Societies: A.N.S.; A.C.S.; A.I.Ch.E.; A.S.E.E.; A.S.M.E.; A.A.A.S.; B.N.E.S.
Nuclear interests: Management of nuclear energy projects, especially research, development, and engineering; Economics and applications of nuclear energy.
Address: Bechtel Corporation, 220 Bush Street, San Francisco 4, California, U.S.A.

DAVIS, William D. Ph.D. (Phys. Chem.). Born 1921. Educ.: Pittsburgh Univ. Phys., Plasma and Vacuum Phys. Branch, Gen. Phys. Lab., Gen. Elec. R. and D. Centre, Schenectady, 1959-. Society: A.P.S.
Nuclear interest: Solid state counters.
Address: General Electric Research and Development Centre, P.O. Box 8, Schenectady, New York 12301, U.S.A.

DAVISON, Malcolm, M.A., D.Phil. Born 1932. Educ.: Oxford Univ. Lecturer, Phys. Dept., Imperial Coll., London, 1961-63; Sen. Phys. (1963-65), Asst. Regional Phys. (1965-), Western Regional Hospital Board. Societies: Inst. of Phys. and the Phys. Soc.; Hospital Phys. Assoc.
Nuclear interest: Medical applications of X-rays and isotopes.
Address: Regional Physics Department, 9 West Graham Street, Glasgow, C.4, Scotland.

DAVISON, Perry Woodruff, B.A., M.A., Ph.D. Born 1919. Educ.: Amherst Coll., Rice and Yale Univs. Res. Phys., R.C.A. Labs., Princeton, 1948-53; Phys., Sci. and Tech. Unit, U.S. Navy Dept., Frankfurt/M, Germany, 1953-56; Manager, Reactor Evaluation Sect., Westinghouse Atomic Power Dept., Pittsburgh, 1956-62; Advisory Scientist, Westinghouse Astronucl. Lab., assigned to Nucl. Rocket Development Station (N.R.D.S.) Jackass Flats, Nevada, 1962-63; Manager, Reactor Test Operations (1963-65), Deputy Manager, (1964-65), NERVA Test Operations, N.R.D.S. Deputy Manager, Test Systems and Operations, 1965-; Exptl. Rocket Engine Project Manager, 1966-. Societies: A.N.S.; A.P.S.; A.A.A.S.

Nuclear interests: Management and direct supervision of work in the field of experimental reactor physics, including the performance of critical experiments to obtain data to assist with the nuclear design of nuclear power reactors; Direct supervision of the preparations for and the performance of full power testing of nuclear reactors for rocket engines.
Address: Westinghouse Electric Corporation, Astronuclear Laboratory, P.O. Box 10864, Pittsburgh, Pennsylvania 15236, U.S.A.

DAVISON, William Henry Thomas, Born 1916. Educ.: Cambridge Univ. Dunlop Res. Centre, 1946-55; T.I. Res. Labs., 1955-. (now Asst. Director). Societies: F.R.I.C.; British Interplanetary Soc.
Nuclear interests: Radiation chemistry, particularly of organic and polymeric materials; Industrial applications of electron accelerators, particularly for curing coatings.
Address: 18 Landscape View, Saffron Walden, Essex, England.

DAVLETSHIN, A. N. Paper: Co-author, Two-Layer Shielding Optimisation Against Isotopic Fast Neutron Sources (Atomnaya Energiya, vol. 23, No. 6, 1967).
Address: Academy of Sciences of the U.S.S.R., 14 Leninsky Prospekt, Moscow V-71, U.S.S.R.

DAVYDOV, A. V. Paper: Co-author, Oxalic Complexes of Protactinium (V) (Atomnaya Energiya, vol. 22, No. 5, 1967).
Address: Academy of Sciences of the U.S.S.R., 14 Leninsky Prospekt, Moscow V-71, U.S.S.R.

DAVYDOV, E. F. Papers: Co-author, Irradiation Stability of Arbus Nucl. Reactor Fuels (letter to the Editor, Atomnaya Energiya, vol. 24, No. 4, 1968); co-author, Reactor SM-2 Plate Fuel Elements Rad. Stability (ibid, vol. 24, No. 5, 1968); co-author, On Possible Character of Fuel Composition Volume Change under Solid Swelling (letter to the Editor, ibid, vol. 24, No. 5, 1968); co-author, Study of Gas Behaviour Under Annealing of Irradiated Hot Pressurised Beryllium (letter to the Editor, ibid, vol. 25, No. 2, 1968).
Address: Academy of Sciences of the U.S.S.R., 14 Leninsky Prospekt, Moscow V-71, U.S.S.R.

DAVYDOV, V. I. Papers: Co-author, Analytical Method of Solution of Critical Problems (Atomnaya Energiya, vol. 22, No. 5, 1967); co-author, Heat Transfer Studies in High-Temperature Radiant Spray Drier (ibid, vol. 24, No. 3, 1968).
Address: Academy of Sciences of the U.S.S.R., 14 Leninsky Prospekt, Moscow V-71, U.S.S.R.

DAWALT, Brigadier Gen. Kenneth F. Member, Military Liaison Com., U.S.A.E.C.
Address: U.S. Atomic Energy Commission, Military Liaison Committee, Washington D.C. 20545, U.S.A.

DAWSON, Frank G., Jr. B.E.E. Born 1925. Educ.: N.C.S.C. and O.R.S.O.R.T. Eng. (1950-53), Sen. Eng. (Critical Experiments) (1953-56), Principal Eng. (Reactor Analysis) (1956-60), Tech. Specialist (Appl. Phys.) (1960), Manager (Appl. Phys. Operation) (1960-63), Manager, (Reactor Physics Operation) (1963-65), G.E.C.; Manager, Reactor Phys., Battelle Memorial Inst., 1965-. Acting Asst. Prof. Nucl. Eng., Cincinnati Univ. Graduate School, 1958-60. Society: A.N.S.
Nuclear interests: Management in applied reactor physics.
Address: 2012 W. Hood, Kennewick, Washington, U.S.A.

DAWSON, J. W. Director of Nucl. Dept., Stillite Products Ltd.
Address: Stillite Products Ltd., 15 Whitehall, London, S.W.1, England.

DAWSON, John Keith, B.Sc., Ph.D., D.Sc. Born 1923. Educ.: Birmingham Univ. M.O.S., later U.K.A.E.A., in Chem. Div., A.E.R.E., Harwell, 1947-, Group Leader, 1957-. Books: Chem. of Nucl. Power (Newnes, 1959); Chem. Aspects of Nucl. Reactors (Butterworths, 1963). Societies: Fellow, Chem. Soc. (London); Fellow, Roy. Inst. Chem.
Nuclear interests: Chemical and metallurgical problems of nuclear reactors with particular emphasis on the oxidation of metals and on ceramic fuels.
Address: United Kingdom Atomic Energy Authority, Atomic Energy Research Establishment, Harwell, Berks., England.

DAWSON, W. K. Member, nucl. physics group, Physics Dept., Alberta Univ.
Address: Physics Department, Alberta University, Edmonton, Alberta, Canada.

DAY, James Meikle, B.S., M.S., Ph.D. Born 1924. Educ.: New Hampshire, Illinois, Arkansas, and Wisconsin Univs. Anderson Phys. Labs., 1948-51; Phillips Petroleum Co., Bartlesville, Oklahoma, 1954-56; Phillips Petroleum Co. (Phillips' rep. with Rocky Mountain Nucl. Study Group), Idaho Falls, Idaho, 1956-58; Phillips Petroleum Co., Atomic Energy Div., Idaho Falls, Idaho, 1958-59; American Tobacco Co., Industrial Reactor Labs., Plainsboro, N.J., 1959-62; Whirlpool Corp., Rad. Chem. Lab., St. Joseph, Michigan, 1962-. Societies: A.C.S.; A.N.S.
Nuclear interests: Reactor physics; Radiation chemistry; Radioactivation analysis.
Address: Box 382, St. Joseph, Michigan, U.S.A.

DAY, Melvin Sherman, B.S. Born 1923. Educ.: Bates Coll. Deputy Chief (1950-55), Chief (1956-58), Tech. Information Service Extension, U.S.A.E.C., Transfer to U.S.A.E.C. Headquarters, Asst. Director, Tech. Information Service (1958-59), Director, Office of Tech. Information (1960), U.S.A.E.C.; Deputy Director, Office of Tech. Information and

Educational Programmes (1960), Director, Sci. and Tech. Information Div. (1962-65), Deputy Asst. Administration for Technol. Utilisation (1965-), N.A.S.A. Com. on Sci. and Tech. Information of the Federal Council for Sci. and Technol., 1962-. Advisory Board for Chem. Abstracts Service; Tech. Information and Documentation Com. of A.G.A.R.D.-N.A.T.O., 1960-; Sci. Information Exchange Advisory Board, 1960-. Societies: A.C.S.; A.A.A.S.; A.I.A.A.; American Space Club; N.Y. Acad. Sci.; Soc. Tech. Writers and Publishers; American Documentation Inst.; Special Libraries Assoc.
Nuclear interests: Documentation and dissemination of scientific and technical information; Management; Metallurgy.
Address: 7805 Beech Tree Road, Bethesda 14, Maryland, U.S.A.

DAY, R. A., Dr. Nucl. Res., Chem. Dept., Coll. of Arts and Sci., Emory Univ.
Address: College of Arts and Sciences, Emory University, Atlanta 22, Georgia, U.S.A.

DAY, Robert Briggs, A.B., M.S., Ph.D. Born 1923. Educ.: Haverford Coll. and Calif. Inst. Technol. Res. Fellow, Calif. Inst. Technol., 1951; Staff Member, Los Alamos Sci. Lab., 1952-, Fulbright Res. Scholar and Guggenheim Fellow, Inst. Theoretical Phys., Copenhagen, 1957-58. Society: A.P.S.
Nuclear interests: Studies in nuclear spectroscopy and nuclear reactions at low and medium energies.
Address: Los Alamos Scientific Laboratory, Los Alamos, N. Mex., U.S.A.

DAYAL, Maheshwar, B.Sc. (Hons.) (Phys.) (Delhi), Mech. Sciences Tripos (Cambridge). Born 1932. Asst. Design Eng. (1955-57), Res. Officer and Design Eng. (1957), Atomic Energy Establishment, Trombay, India; Consultant to I.A.E.A., 1959; Project Eng., Atomic Power Project, Atomic Energy Dept., India, 1959; i/c, Tarapur Atomic Power Project, 1961-. Indian deleg. 1st and 2nd I.C.P.U.A.E., Geneva, 1955 and 1958; Commonwealth Nucl. Scientists Conference, U.K., 1958; Tech. adviser to Indian deleg. to: Preparatory Commission of I.A.E.A., New York, 1957; Asian Regional Nucl. Centre Conference, Washington, 1957; expert for power and energy problems on 1st Preliminary Assistance Mission of I.A.E.A., to Burma, Thailand, Indonesia and Ceylon, Jan-March, 1959; consultant on Safeguards to I.A.E.A., Vienna, March-July 1959. Member, I.A.E.A. panel on Nucl. Power Costing.
Nuclear interests: Nuclear power stations (design and construction); Fuel cycles; Reactor engineering; Reactor physics; Power economics.
Address: Atomic Energy Commission, Tarapur Atomic Power Project, Chhatrapati Shivaji Maharaj Marg, Bombay 1, India.

DAYAL, Narendar, M.Sc. (Phys.). Born 1927. Educ.: Delhi Univ. Asst. Phys. (1952-56), Phys. (1956-61), Sen. Phys., Sci. Officer (SD-2) (1961-), Dept. of Atomic Energy, Govt. of India.
Nuclear interests: Nuclear instrumentation and measurements.
Address: Department of Atomic Energy, 32 Khasmahal, Tatanagar, India.

DAYTON, Irving E., B.A., Ph.D. Born 1927. Educ.: Swarthmore Coll. and Cornell Univ. Res. Asst., Lab. of Nucl. Studies, Cornell Univ., 1948-52; Instructor, Phys. Dept., Princeton Univ., 1952-54; Group Leader, Exptl. Phys., Atomic Energy Div., Babcock and Wilcox Co., 1954-57; Asst. Prof. Phys., Swarthmore Coll., 1957-61; Prof. Phys. (1961-), Head Phys. Dept. (1961-66), Vice-Pres. for Academic Affairs (1966-), Montana State Univ. Exec. Board (1964-), Treas. (1966-), A.A.P.T. Societies: History of Sci. Soc.; A.A.P.T.; A.P.S.; A.N.S.
Nuclear interests: Excitation by atomic and ionic collisions; Experimental nuclear and reactor physics; Education in nuclear science and engineering.
Address: Montana State University, Bozeman, Montana 59715, U.S.A.

DAYTON, Russell W., Ch. Eng., M.S., Ph.D. Born 1910. Educ.: Rensselaer Polytech. Inst. At Battelle Memorial Inst., 1934- (Asst. Tech. Director, 1953-64, Asst. Director 1964-). Book: Chapter on Zirconium and its Alloys in The Reactor Handbook, vol. iii, (Tech. Information Service, U.S.A.E.C., 1955-). Societies: A.S.M.; A.S.M.E.; A.A.A.S.; A.N.S.
Nuclear interests: Fabrication and properties, especially radiation damage resistance, of metallic and ceramic nuclear materials; High temperature nuclear fuels.
Address: Battelle Memorial Institute, 505 King Avenue, Columbus 1, Ohio, U.S.A.

DAZAI, T. Asst. Prof. Reactor Physics Div., Res. Lab. of Nucl. Reactors, Tokyo Inst. of Technol.
Address: Research Laboratory of Nuclear Reactors, Tokyo Institute of Technology, 1 Oh-okayama, Keguro-ku, Tokyo, Japan.

DE, Mihir Lal, B.Sc. (Hons.), M.Sc., P.R.S. Born 1918. Educ.: Calcutta Univ. Reader in Biophys., Saha Inst. of Nucl. Phys., Calcutta Univ., 1952-. Society: Sec., Electron Microscope Soc. of India.
Nuclear interests: Nuclear instrumentation and radiation dosimetry; Radiation effects on macromolecules; Health hazards due to nuclear radiation.
Address: Saha Institute of Nuclear Physics, 92 Acharya Prafulla Chandra Road, Calcutta-9, India.

de ABAIGAR, H.E. Joaquin Buxe-Dulce. Ambassador for Spain in Vienna. Resident Rep. for Spain to the I.A.E.A.
Address: 34 Argentinierstrasse, 1040 Vienna, Austria.

de ABREU, Maria Helena. Measurement of radioactivity, Electronics and Elec. Measurements Div., Inst. Nacional de Tecnologia, Brazil.
Address: Instituto Nacional de Tecnologia, 82 Avenida Venezuela, Rio de Janeiro, Brazil.

de ABREU FARO, Manuel José, Dipl. Eng. Born 1923. Educ.: Lisbon Tech. Univ., Prof. Telecommunications, Inst. Superior Tecnico, Lisbon, 1956; Fellow, Comissao de Estudos de Energia Nucl., Lisbon.
Nuclear interest: Application of electronics to nuclear energy.
Address: Instituto Superior Técnico, Avenida Rovisco Pais, Lisbon, Portugal.

de ALEMANY, Joaquin CATALA. See CATALA de ALEMANY, Joaquin.

de ALMEIDA, Yvone G., Asst. Prof. Dept. of Nucl. Chem., Centro Brasileiro de Pesquisas Fisicas.
Address: Centro Brasileiro de Pesquisas Fisicas, 71 Avenida Venceslau Bras, Rio de Janeiro, Brazil.

DE ANDRADE, Ayres Cunha, Commandante. Comissao Nacional de Energia Nucl.; Participant, 3rd Inter-American Symposium on the Peaceful Application of Nucl. Energy, Rio de Janeiro, July, 1960; Brazilian adviser, I.A.E.A. Gen. Conference, Vienna, September, 1960.
Address: Comissao Nacional de Energia Nuclear, 81 Avenida Almirante Barroso, Rio de Janeiro, Brazil.

de ARAUJO PENNA, Monica. Lecturer, Dept. of Radioactivity, Centro Brasileiro de Pesquisas Fisicas.
Address: Centro Brasileiro de Pesquisas Fisicas, 71 Avenida Wenceslau Braz, Rio de Janeiro, Brazil.

de ARIETA ARAUNABENA, Pedro. Director, Tecnicas Nucleares S.A.
Address: Tecnicas Nucleares S.A., 46 Serrano, Madrid 1, Spain.

de ARRIAGA, Kaulza. Governor for Portugal to I.A.E.A.
Address: Nuclear Energy Board, 79 rue de S. Pedro de Alcantara, Lisbon, Portugal.

de ARRUDA, Paulo Ribeiro. Born 1909. Prof. of Appl. Elec. and Lecturer on Appl. Electronics, Polytech. School, Sao Paulo Univ.; Director of Res., Electronic Microscopy Dept., Sao Paulo Univ.; Director, Dept. of Training and Technical Information, Div. of Exchange and Training, I.A.E.A., 1959-60; Member, Brazilian Nat. Nucl. Energy Commission; Alternate to Brazilian Governor, I.A.E.A. Societies: Nat. Res. Council of Brazil; Brazilian Acad. of Sci.; Inst. Radio Eng. (U.S.A.).
Address: Brazilian National Nuclear Energy Commission, 90 (ZC-82) Rua General Saveriano, Rio de Janeiro, Brazil.

de ARTIGAS SANZ, D. Jose Antonio. Formerly Member, Comisiones Asesoras de Reactors Industriales, now Member, Comisiones Asesoras de Centrales Nucleares, Junta de Energia Nucl., Spain.
Address: c/o Junta de Energia Nuclear, Ciudad Universitaria, Madrid 3, Spain.

DE AVILA, Aureo Fernandez. Formerly Vocal Comisiones Asesoras de Reactores Industriales, Junta de Energia Nuclear, Madrid; Director, Forum Atomico Espanol.
Address: Forum Atomico Espanol, 38 General Goded, Madrid 4, Spain.

de BAERDEMAEKER, Adolphe J. L., D. en Droit. Born 1920. Educ.: Catholique Univ. Louvain. Sec., N.A.T.O. Council, London, 1951; Adviser to Minister for Economic Affairs, 1952-54; Member, permanent deleg. to O.E.E.C., 1954-58; Director, Directorate-Gen. of External Radiations, E.E.C., 1958; European Community's Permanent Delegate to O.E.C.D., 1966-.
Address: 129 Avenue Malakoff, Paris 16, France.

de BARROS, Jose Antonio QUEIROZ. See QUEIROZ DE BARROS, Jose Antonio.

DE BARROS, L. A. Sen. Director, Nucl. Measurements Corp.
Address: Nuclear Measurements Corporation, International Division, 13 East 40th Street, New York 16, New York, U.S.A.

DE BARROS, Manuel Gaspar, Ing. Director, Bureau d'Etudes, Amoniaco Portugues S.A.R.L.
Address: Amoniaco Portugues S.A.R.L., 113-1⁰ Rua do Poco dos Negros, Lisbon, Portugal.

de BARROS, Solange M. C., Asst. Prof. Dept. of Radioactivity, Centro Brasileiro de Pesquisas Fisicas.
Address: Centro Brasileiro de Pesquisas Fisicas, 71 Avenida Wenceslau Braz, Rio de Janeiro, Brazil.

de BASTIANI, M. J. Mine Manager, Denison Mines Ltd.
Address: Denison Mines Ltd., 4 King Street West, Toronto 1, Ontario, Canada.

de BASTOS MARTINS, E. C. Lic. ex. Math. Sci. Born 1912. Educ.: Lisbon Univ. Actuary (1936-), Manager (1946-), Fidelidade Insurance Co.; Actuary of social insurance institutions, 1937-. Board Member, Permanent Com., Internat. Congress of Actuaries, Brussels; Fire Branch Permanent Com., Life Branch Permanent Com., Portuguese Atomic Risk Com., Gremio dos Seguradores, Lisbon. Societies: Inst. dos Actuarios Portugueses, Lisbon; Forum Atomico Portugues, Lisbon.
Nuclear interests: Insurance; Nuclear physics.
Address: Largo do Corpo Santo 13, Lisbon, Portugal.

de BEER, J. J. Director, Baird-Atomic
(Europe) N.V.
Address: Baird-Atomic (Europe) N.V., 26-27
Veenkade, The Hague, Netherlands.

de BEER, Johannes Frederik, D.Sc. Born 1930.
Educ.: Potchefstroom Univ. Lecturer, Potchefstroom Univ., 1952-55; Res. Fellow, A.E.R.E., Harwell, 1956-58; Lecturer (1959-68), Prof. Phys. (1968-), Potchefstroom Univ. Society: South African Inst. Phys.
Nuclear interest: Cosmic radiation.
Address: University Physics Department, Potchefstroom, Republic of South Africa.

DE BENAVENT, Ferdinand Leopold F.J., Bachelor in Humanities, Agric. Eng., Radioisotopes Tech., Special Studies on Nucl. Phys. and Nucl. Chem. Born 1932. Educ.: Madrid Univ., Nat. School of Agriculture, Peru,Inst. Nucl. Studies, Peru, and Inst. of Nucl. Studies, Tennessee Univ. Prof. of Radioisotopes applied to the Agricultural and Biol. Sci.; Agricultural member and Sci., Peruvian Atomic Energy Board. N.A.S.-N.R.C. and U.S.A.E.C.; Ind. Chem. Dept., Lima; William Crosby and Sons, North Carolina Univ. Mission at Peru; Gamma Liaison Officer Between U.S.A.E.C. and Peruvian Atomic Energy Board; Assisting the I.A.E.A., Vienna. Book: Synthetic Polyelectrolites' Effects on Soil Aggregation.
Nuclear interests: Research on radioisotopes applied to agriculture and nuclear physics.
Address: Junta de Control de Energia Atomica, 3420 Avenida Arequipa, Apt. 914, Lima, Peru.

de BENNETOT, Michel, D. ès Sc. Phys. Born 1927. Educ.: Paris Univ. Compagnie générale de télégraphie Sans Fil, 1952-57. Prés. Directeur Gén., Sté. le Materiel Magnetique. Societies: A.T.E.N.; Ste. Francaise des Electroniciens et des Radioelectriciens.
Nuclear interest: Magnetic devices used in nuclear field especially couplings and bearings.
Address: 13 rue Victor Hugo, 92 Puteaux, France.

DE BERGH, Anton Dirk Johan, Ir. Born 1907.
Educ.: Delft Univ. Director-Pres. N.V. Provinciaal en Gemeentelijk Utrechts Stroomleveringsbedrijf, 1967-. Member, Study Com. Regarding the Building of the First Dutch Nucl. Power Station, Samenwerkende Electriciteits Productiebedrijven.
Nuclear interest: Management.
Address: N.V. P.E.G.U.S., 189 Keulsekade, Utrecht, Netherlands.

de BOECK, Pedro, Administrateur Directeur-Gen. de la Sofina; Administrateur, Sté. Nucl. Franco-Belge des Ardennes S.E.N.A.; Member, Conseil d'Administration, Centre et Sud.
Address: 38 rue de Naples, Brussels, Belgium.

DE BOER, Abraham Adolf, Dr. nat. sc. Born 1928. Educ.: Utrecht State Univ. Chem., Joint Establishment for Nucl. Energy Res., 1952-54; Sci., Reactor Centrum Nederland, 1954-58; Eng., Industrial Dept., Euratom, 1958-61. Chef-adjoint of the Cabinet of the Dutch Member of the Euratom-Commission, Brussels, 1961-67; idem, Commission of the European Communities, 1967-68; Director for Nucl. Energy and Elec., Commission of the European Communities, 1968-. Books: Economische Aspecten van de ontwikkeling der kernenergie (Leiden, 1962); co-author, Kernenergie, een inleiding tot de reactorkunde (Groningen, 1954); Weten en Regeren (Amsterdam, 1967).
Nuclear interests: Economics of nuclear energy.
Address: 41 avenue des Eperviers, Brussels 15, Belgium.

de BOER, Jan Hendrik, Dr., Dr. h.c. (T.H. Hanover). Born 1899. Educ.: Groningen Univ. Extraord. Prof., Technol. Univ. Delft. Chairman, Central Council for Nucl. Affairs and Sci. Council for Nucl. Affairs, Netherlands; Comité Scientifique Inst. International de Chimie Solvay. Societies: Roy. Netherlands Acad. Sci.; Hon. Member, Roy. Netherlands Chem. Soc.; Hon. Member, Flemish Chem. Soc.; Faraday Soc.
Nuclear interests: Management; Radiation chemistry; Power reactors; Waste disposal.
Address: Aan de Heer Voorzitter van de Wetenschappelijke Raad voor de Kernenergie, Postbus 5086, Duinweg 24, 's-Gravenhage, Netherlands.

de BOERO, M. T., Asst. Investigator, Radioisotopes Lab., Dept. of Phys., Faculty of Mathematical Phys., Univ. Nacional de La Plata.
Address: Universidad Nacional de La Plata, La Plata, Argentina.

de BONAFFOS, J. Commercial-Tech. Eng., Ateliers de Montages Electriques S.A.
Address: Ateliers de Montages Electriques S.A., 77 rue Saint-Charles, Paris 15, France.

de BORDE, Anthony Hartley, B.Sc., Ph.D., F. Inst. P. Born 1922. Educ.: London Univ. Nuffield Res. Fellow, 1953-55; Lecturer in Natural Philosophy, Glasgow Univ., 1955-58; Theoretical Phys., English Elec. Co. Ltd., 1958-63; Chief Mathematician, Nelson Res. Labs., 1963-. Book: Co-author, Introduction to Statistical Mechanics (Pergamon Press, 1958). Society: Inst. Phys. and Phys. Soc.
Nuclear interests: Theoretical physics of controlled thermonuclear reactions and direct power generation methods.
Address: English Electric Co. Ltd., Nelson Research Laboratories, Blackheath Lane, Stafford, England.

de BREIDENBACH, Ernest C. G. STUECKELBERG. See STUECKELBERG DE BREIDENBACH, Ernest C. G.

de BRITO MARIANO, Maria Helena, B.Sc. (Hons. Phys. and Chem.). Born 1926. Educ.: Lisbon Univ. Asst. of Chem., Lisbon Univ., 1952-58; Asst., Lab. Junta de Energia Nucl., 1959-. Society: Forum Atómico Português.

Nuclear interest: Radiation chemistry.
Address: Av. Visconde Valmor, 9 -r/c, Esq.,
Lisbon 1, Portugal.

de BROGLIE, Louis. Formerly Member, now Pres. Conseil Scientifique, C.E.A.
Address: Commissariat à l'Energie Atomique, 29-33 rue de la Federation, Paris 15, France.

de BROUWER, Louis, Ing. civil des Mines. Born 1908. Educ.: Univ. Catholique de Louvain. Chef de la Division "Administration" au Centre d'Etude de l'Energie Nucléaire, Brussels.
Nuclear interest: Management.
Address: 20 avenue du Maréchal, Uccle, Belgium.

DE BUSTAMANTE, Amalio SAIZ. See SAIZ DE BUSTAMANTE, Amalio.

de CALAN, Pierre. With Babcock et Wilcox. Vice Pres., Sté. pour l'Industrie Atomique.
Address: Babcock et Wilcox, 48 rue La Boetie, Paris 8, France.

de CAMPOS COSTA, Manuel José, degree in Elec. Eng. Born 1924. Educ.: Porto Univ. Chief, Electronics Group, Nucl. Res. Centre, Junta de Energia Nucl. Sacavem.
Nuclear interest: Electrostatic accelerator.
Address: Nuclear Research Centre, 10 Estrada Nacional, Sacavem, Portugal.

DE CARLOS ROMERO, José E. Dr., Phys. Born 1932. Educ.: Madrid Univ. Sci. Staff, Junta de Energia Nucl., 1954-. Society: Real Sociedad Espanola de Fisica y Quimica.
Nuclear interests: Design and supervision of nuclear electronic systems for research; Reactor control instrumentation; Spectrometric instrumentation.
Address: 64 Guzman el Bueno, Madrid-15, Spain.

de CARMAGO, Pedro BENTO. See BENTO de CARMAGO, Pedro.

de CARVALHO, Antonio Herculano, GUIMARAES CHAVES. See GUIMARAES CHAVES DE CARVALHO, Antonio Herculano.

de CARVALHO, Hervasio GUIMARAES. See GUIMARAES DE CARVALHO, Hervasio.

de CARVALHO, José Alberto FERNANDES. See FERNANDES de CARVALHO, José Alberto.

DE CARVALHO, Rodrigo Alberto GUEDES. See GUEDES DE CARVALHO, Rodrigo Alberto.

de CARVALHO e SOUZA, H. E. Madame Odette. Head, Brazilian Mission to Euratom, 1964-.
Address: c/o Euratom, 51-53 rue Belliard, Brussels, Belgium.

de CARVALHO FRANCO, Paulo, Dr. Director, Inst. de Radium "Arnaldo Vieira de Carvalho", Sao Paulo Univ.
Address: Instituto de Radium "Arnaldo Vieira de Carvalho", 112 Rua Cesário Mota, Sao Paulo, Brazil.

de CASTILLO LEDON, Amalia G. C. Mexican Resident Rep., I.A.E.A.
Address: 12/II/10/67 Parkring, Vienna 1, Austria.

de CASTRO, Irene Emygdio. Measurement of Radioactivity, Electronics and Elec. Measurements Div., Inst. Nacional de Tecnologia, Brazil.
Address: Instituto Nacional de Tecnologia, 82 Avenida Venezuela, Rio de Janeiro, Brazil.

de CASTRO de FARIA, Nelson V. Asst. Prof., Corpuscular Phys. Dept., Centro Brasileiro de Pesquisas Fisicas.
Address: Centre Brasileiro de Pesquisas Fisicas, 71 Avenida Wenceslau Braz, Rio de Janeiro, Brazil.

de CASTRO y BRAVO, F. Spanish Member, European Nucl. Energy Tribunal, O.E.C.D., E.N.E.A.
Address: O.E.C.D. European Nuclear Energy Agency, 38 boulevard Suchet, Paris 16, France.

de CESARE, Mario. Member, du Bureau, Economic and Social Com., Euratom.
Address: 7 Piazzale Medaglie d'Oro, Rome, Italy.

de COULON, Jean-Louis, Legal Attorney. Born 1907. Educ.: Neuchâtel Univ. Directeur de la S.A. des Cableries et Trefileries de Cossonay; Member, Conseil d'administration, Energie Nucléaire S.A.
Nuclear interest: Metallurgy.
Address: S.A. des Cableries et Trefileries de Cossonay, Cossonay-Gare, Switzerland.

de CREMIERS, Roger Jacques AUGIER. See AUGIER de CREMIERS, Roger Jacques.

DE ENQUIN, Maria Cristina PALCOS. See PALCOS DE ENQUIN, Maria Cristina.

de FARIA, Nelson V. de CASTRO. See de CASTRO de FARIA, Nelson V.

DE FAZIO, Peter G., B.S. (Civil Eng.). Born 1911. Educ.: Alabama Univ. Director of Eng., Nat. Lead Co. of Ohio. Society: Nat. Soc. Prof. Eng.
Nuclear interests: Design, construction and management.
Address: 5933 Bellmeadows Drive, Cincinnati 24, Ohio, U.S.A.

de FEBER, K. Sen. Officer engaged in Nucl. Work, Zuid-Hollandsche Pletterijen v/h D.A. Hamburger N.V.

DE FIG

Address: Zuid-Hollandsche Pletterijen v/h D.A. Hamburger N.V., 73 Delftweg, Postbus 19, Delft, Netherlands.

DE FIGUEIREDO, R. J. PACHECO. See PACHECO DE FIGUEIREDO, R. J.

DE FRANCISCIS, Pietro, M.D. Born 1919. Educ.: Naples Univ. Asst. Prof., Human Physiology (1950-53), Assoc. Prof., Human Physiology (1953-), Full Prof. and Chairman, 2nd Chair of Human Physiology (1960), Naples Medical School. Co-editor, Rassegna di Medicina Sperimentale. Society: Soc. Italiana di Radiologia Medica e Medicina Nucleare. Nuclear interests: Radiation biology, and specifically the relationship between spleen and radiation mortality.
Address: Istituto di Fisiologia Umana, 8 via S. Andrea Delle Dame, Naples, Italy.

DE FREIGAS, Orlita GOMES. See GOMES DE FREITAS, Orlita.

de GALLIER de SAINT SAUVEUR, Henri, Born 1913. Educ.: Ecole Polytechnique. Directeur, Département des Activites Nucléaires, Sté. des Forges et Ateliers du Creusot. Address: 15 rue Pasquier, Paris 8, France.

DE GARAY, Alfonso L., Medical Dr. Born 1920. Educ.: Puebla Univ. Director of Genetics Programme, Comisión Nacional de Energia Nucl., Mexico Univ. (U.N.A.M.).
Nuclear interests: Human genetics; Genetical effects of radiations; Tissue culture and cytogenetics. Actually research on chromosomal aberrations in human beings.
Address: Medicina #56 Fraccionamiento Copilco-Universidad, México 20, D.F., Mexico.

de GAUDEMARIS, Gabriel Pierre, D. ès Sc. Physiques. Born 1926. Educ.: Grenoble Univ., 1947-56; C.E.A., 1956-57; Manager, Radiochem. Lab., Grenoble Centre of Nucl. Studies, Inst. Française du Pétrole, 1957-. Society: Sté. chimique de France.
Nuclear interests: General radiochemistry, reaction of hydrocarbons under radiation; Reactor design, especially organic moderated reactor, and chemistry problems associated therewith.
Address: 1 boulevard Joseph Vallier, Grenoble, Isère, France.

de GAULLE, Bernard Xavier, D. en Droit, Economie politique. Born 1923. Educ.: Grenoble and Paris Univs. Sec.-Gen., L'Electronique Appliquée, Paris, 1954; Prés. d'Honneur, Commission Intersyndicale de l'Instrumentation et de la Mesure Nucléaire Française. Address: 96-100 rue Maurice Arnoux, Montrouge, (Seine), France.

de GEFFRIER, G. Commercial Manager, Instr. Dept. Ribet-Desjardins.
Address: Ribet-Desjardins, 13-17 rue Périer, Montrouge, (Seine), France.

de GORTAZAR Y LANDECHO, Manuel M., Industrial Eng. Born 1909. Educ.: Escuela Técnica Superior de Ingenieros Industriales, Bilbao. Chairman, Española S.A.; Director: Centrales Nucleares del Norte de España, Centrales Nucleares S.A.
Nuclear interests: Reactor design; Management; Reactor operation; Nuclear physics; Power reactor economics.
Address: Génova 26, 1 Madrid, Spain.

de GRAAF, Jacob E., Masters degree in Eng. Born 1910. Educ.: Delft Technol. Univ. Prof. Metal. Technol. Delft Univ.; Member, Sci. Advisory Council, Reactor Centrum Nederland. Nuclear interest: Metallurgy.
Address: Technological University, Delft, Netherlands.

de GROOT, J. P. Managing Director, Wambesco.
Address: Wambesco, P.O. Box 1439, Rotterdam, Netherlands.

de GROOT, Sybren Ruurds, Dr. Born 1916. Educ.: Amsterdam Univ. Prof. Theoretical Phys., Utrecht Univ., 1948-53; Prof. Theoretical Phys., Leyden Univ., 1953-64; Prof. Theoretical Phys., Amsterdam Univ., 1964-. Society: Roy. Netherlands Acad. of Sci.
Nuclear interest: Nuclear physics.
Address: Instituut voor theoretische fysika, 65 Valckenierstraat, Amsterdam C, Netherlands.

de GROOTE, Paul Hubert, Graduate in Commercial Eng. Born 1905. Educ.: Brussels Free Univ. Pres., Governing Body, Brussels Free Univ., 1952; Pres., Central Economic Council, 1955; Vice-Pres., Nat. Com. for Studying Peaceful Uses of Atomic Power, 1956; Pres. d'Honneur, Central Economic Council (1956), Vice-Pres., Economic Expansion Study Centre (1957), Pres. (1957), Member d'Honneur, Belgian Assoc. for the Peaceful Development of Atomic Power; Member, European Commission, European Atomic Energy Community, Euratom, 1958-67.
Address: 294 Dieweg, Uccle (Brussels), Belgium.

DE GUZMAN, Arturo R., B.S.C.E., M.I.M. Born 1912. Educ.: Philippines Univ. Chief Sci., Philippine A.E.C., 1962; Instructor in Mathematics, Feati Univ. Society: Japan Radioisotope Assoc.
Nuclear interests: Management; Laboratory design; Application of radioisotopes in civil engineering.
Address: Philippine Atomic Energy Commission, 727 Herran, Manila, Philippines.

DE HAAN, Abel, Jr., B.S. (Chem., California). Born 1921. Educ.: California Univ. Gen. Manager and Vice-Pres., Tracerlab, Inc., Western Div., Richmond, California, 1948-. Society: A.N.S.
Nuclear interests: Manager of division which develops and manufactures reactor monitoring

equipment and provides specialised radiochemical services.
Address: Tracerlab, A division of L.F.E., Inc., 2030 Wright Avenue, Richmond 3, California, U.S.A.

de HAAS, Willem Alexander, Chem. Born 1906. Educ.: Delft Tech. Univ. Patent Attorney (1930-57), Sec. (1957-), Philips Gloeilampenfabrieken, Eindhoven; Director, Reactor Centrum in the Netherlands. Pres., Forum Atomique Européen; Member and Advisor, Nucl. Energy Commission, Netherlands; Member, Industrial Council for Nucl. Energy; Foundation for Nucl. Propulsion of Merchant Ships; Chairman, Netherlands Atomforum; Chairman, Nucl. Energy Div., Royal Inst. Eng. Address: 4 Treeswijklaan, Waalre N.B., Netherlands.

de HAAS van DORSSER, A. H., Ir. Asst. Director, Central Tech. Inst. T.N.O., Organisation for Ind. Res. T.N.O.
Address: Central Technical Institute T.N.O., 5 Koningskade, The Hague, Netherlands.

de HALAS, Don Richard, B.S. (Chem.), M.S. (Phys. Chem.), Ph.D. (Phys. Chem.). Born 1930. Educ.: California, Idaho and Oregon State Univs. Manager, Chem. and Chem. Eng. Dept., Battelle-Northwest Lab., Richland, Washington. Member, Exec. Com., Materials Sci. and Technol. Div., Member, Publications Com., A.N.S. Book: Co-author, Nuclear Graphite (Academic Press, 1962). Society: A.A.A.S.
Nuclear interests: Research management; Chemistry; Materials; Nuclear fuels.
Address: Battelle-Northwest, P.O. Box 999, Richland, Washington 99352, U.S.A.

de HALLER, Pierre, Dr. h.c. Swiss Member, Group on Co-operation in the Field of Reactors, E.N.E.A. Director, Sulzers; Vice-Pres., Arbeitsgemeinschaft Lucens; Director and Business Manager, Therm-Atom A.G., Zurich. Address: Sulzer Bros., Ltd., Winterthur, Switzerland.

DE HEEM, Louis Marie Paul, Civil Eng. Born 1904. Educ.: State Univ., Ghent. Gen. Manager, Belgian Atomic Centre, 1953-63; Chairman, Sté. pour la Co-ordination de la Production et du Transport de l'Energie Electrique; Pres., Assoc. Vinçotte (control and res.); Pres., Lab. Belge de l'Industrie Electrique (public utilities res. centre); Member, Board of Directors: Union pour la Coordination de la production et du transport de l'électricité; UNIPEDE; Assoc. Belge pour le Développement Pacifique de l'Energie Atomique; GECOLI; Sté. d'Energie Nucl. Franco-Belge des Ardennes. Societies: A.N.S.; Atomic Ind. Forum; Nat. Geographical Soc.; I.E.E.E.
Nuclear interests: Management and economic aspects of nuclear energy.
Address: C.P.T.E., 31 rue Belliard, Brussels 4, Belgium.

DE HEER, Frederik Jacques, Dr. Born 1926. Educ.: Amsterdam Univ. Inst. for Atomic and Molecular Phys., Amsterdam.
Nuclear interests: Electron and ionic impact; Excitation and ionisation of gases.
Address: Institute for Atomic and Molecular Physics, 407 Kruislaan, Amsterdam, Netherlands.

de HEMPTINNE, Marc, Comte, Ing. chim., Dr. en Sc. phys. et mathématiques. Educ.: Louvain Univ. Prof., Louvain Univ.; Directeur du Centre de Physique Nucl. de Louvain; Membre de l'Acad. Roy. des Sci. de Belgique.
Nuclear interests: Physique des interactions faibles; Spectroscopie nucléaire; Spectroscopie moléculaire spécialement Raman, Infra rouge et microondes.
Address: Université de Louvain, Centre de Physique Nucléaire, Parc d'Arenberg, Avenue Cardinal Mercier, Heverle, Louvain, Belgium.

de HOFFMANN, Frederic, B.S., M.A., Ph.D. Born 1924. Educ.: Harvard Univ. Los Alamos Sci. Lab., 1944-55; Phys. Dept., Cornell Univ., 1952-53; Pres., Gen. Atomic, 1955-67; Vice-Pres., Gen. Dynamics Corp., 1955-67; Pres., Gulf Gen. Atomic Co., 1967-; Vice-Pres., Gulf Oil Corp., 1967-; Pres., Gulf Gen. Atomic Europe, 1967-. Formerly Member, various coms., U.S.A.E.C.; formerly consultant, Joint Congressional Com. on Atomic Energy; U.S. Sci. deleg., third I.C.P.U.A.E., Geneva, 1964. Books: Co-author, Mesons and Fields 2 vols. (Evanston, Row, Peterson and Co., 1955); Science and Eng. of Nucl. Power. Vol. II. (Cambridge, Mass., Addison-Wesley Press Inc., 1949); co-author, Introduction to Neutron Diffusion Theory, Vol. I. (U.S. Govt. Printing Office, 1953); Chapter on High Conversion High-temperature Reactors and Neutron Thermalization in Perspectives in Modern Phys. (Editor, Robert E. Marshak), (New York, John Wiley and Sons, Inc., 1966). Societies: Fellow, A.P.S.; Fellow and Director, A.N.S.; Director, Atomic Ind. Forum.
Nuclear interests: Management of nuclear research development and construction. Professional research interests: determination of reactor constants, theory of chain reactor fluctuations, nuclear economics, high energy nuclear physics.
Address: Gulf General Atomic Inc., P.O. Box 608, San Diego, California 92112, U.S.A.

de JESUS MONZON, Maria, Q.F.B. Res. Worker, Dept. Sci. Investigation, Guanajuato Univ.
Address: Departamento de Investigaciones Fisico-Quimicas, Universidad de Guanajuato, No. 5 L. de Retana, Guanajuato, Mexico.

de JONG, Klaas Jozef, qualified Mech. Eng. Born 1924. Educ.: Delft Technol. Univ. Kon. Machinefabriek Gebr. Stork and Co., N.V., Hengelo (O), the Netherlands, 1952; Turbine Design Dept., N.V. Kema, Arnhem, 1956; Homogeneous Reactor Development Team,

Kon. Machinefabriek Gebr. Stork and Co., 1957 (Formerly Head, Nucl. Dept.); Asst. Gen. Manager (1963), Gen. Manager (1967-), Neratoom N.V. Formerly Plaatsvervanger, Com. No. 102-Nucl. Energy, Nederlands Normalisatie-Instituut. Society: Royal Inst. of Eng.
Nuclear interests: Management; Nuclear power station design, fabrication, manufacture and building. Research and development. Sodium technology.
Address: 15 Krimkade, Voorschoten, Netherlands.

de JONG, W. E. van RIJSWIJK. See van RIJSWIJK de JONG, W. E.

de JONGE, C. W., Ir. Manager, Contracts Dept., Neratoom N.V.
Address: Neratoom N.V., 13 Lange Voorhout, The Hague, Netherlands.

DE JUREN, James A., Ph.D. Born 1918. Educ.: California (Berkeley) Univ. At California Univ. Rad. Lab., 1942-45 and (research phys.) 1946-51; Nat. Bureau of Standards, 1951-55; Bettis Atomic Power Div. (Westinghouse), 1955-61; Atomics Internat., NAA 1961-. Sec., Com. on Nucl. Sci., Nat. Res. Council. Societies: A.P.S.; A.N.S.
Nuclear interests: Nuclear physics; Basic reactor measurements.
Address: 23061 Oxnard Street, Woodlands Hill, California, U.S.A.

DE KANY, John P., M.S.E. (Chem. Eng.). Born 1935. Educ.: Princeton Univ. and Polytech. Inst. of Brooklyn. Chem. Eng., Argonne Nat. Lab., 1957-61; U.S.A.E.C., 1961-66; Westinghouse Elec. Corp., 1966-67; Gulf Oil Corp., 1967-. Society: A.N.S.
Nuclear interests: Fuel cycle development and sodium systems engineering.
Address: Gulf Oil Corp., P.O. Box 1166, Pittsburgh, Pa. 15230, U.S.A.

de KERK, Jacob van. See van de KERK, Jacob.

DE KLUIVER, Hans, Ph.D. Born 1928. Educ.: Amsterdam Municipal Univ. Sci. Member, Foundation for Fundamental Res. on Matter, Netherlands.
Nuclear interests: Plasma physics; Controlled thermonuclear research.
Address: Foundation for Fundamental Research on Matter, Instituut voor Plasma-Fysica, Rijnhuizen, Jutphaas, Netherlands.

de la CAMARA, Santiago NORENA. See NORENA de la CAMARA, Santiago.

DE LA CRUZ, Benjamin L., Dr. of Medicine. Born 1921. Educ.: Santo Tomas Univ. Supervising Sci. and Head, Dept. Biol. and Medical Res., Philippine A.E.C. Societies: Philippine Medical Assoc.; Ochsner Fellows Assoc.
Nuclear interests: Radiation biology; Health physics; Uses of radioisotopes in medicine.
Address: Philippine Atomic Energy Commission, Diliman, Quezon City, Philippines.

DE LA CRUZ, Felipe, Licenciado en Ciencias Quimicas (M.Sc. Chem.). Born 1923. Educ.: Madrid Univ. Assoc. Prof. Phys. Chem., Madrid Univ., 1950-52; Res. Chem. (1952-56), Chief, Chem. Res. Sect. (1956-60), Asst. Head, Chem. Div. (1960-), Junta de Energia Nucl., Madrid; Member for Spain Study Group on Food Irradiation, O.E.C.D., E.N.E.A. Societies: Real Soc. Española de Fisica y Quimica, Spain; A.C.S.
Nuclear interests: Nuclear chemical research programme management; Production of radioisotopes; Research in radiochemistry.
Address: Junta de Energia Nuclear, Centro Nacional de Energia Nuclear "Juan Vigón", Av. Complutense, Madrid 3, Spain.

de la CRUZ, Manuel de la PARRA y. See de la PARRA y de la CRUZ, Manuel.

de la CUBA, Eduardo GOYBURU. See GOYBURU de la CUBA, Eduardo.

de la GENIERE, Renaud. Member, Commission Consultative des Marches, Member, Comite Financier, Member, Comite de l'Energie Atomique, Member, Commission Consultative pour la Production d'Electricite d'Origine Nucleaire, C.E.A.
Address: Commissariat a l'Energie Atomique, 29-33 rue de la Federation, Paris 15, France.

de la MARRE, Robert HUARD. See HUARD de la MARRE, Robert.

DE LA MARTINIE, Gerard d'AUZAC. See d'AUZAC DE LA MARTINIE, Gerard.

de la PARRA y de la CRUZ, Manuel. Member, Forum Atomico Espanol.
Address: Forum Atomico Espanol, 38 General Goded, Madrid 4, Spain.

de la PEZUELA PINTO, Pedro, B.S. (Chem.), M.S. (Chem. Eng.). Born 1932. Educ.: Valladolid and Madrid Univs. Res. Asst., Control and Instrumentation Sect., Reactor Div., Junta de Energia Nucl., 1958; Resident Res. Assoc., Reactor Eng. Div. Control Sect., Argonne Nat. Lab., 1959; Res. Asst., Electronic Sect., Phys. Div., Junta de Energia Nucl., 1960-61; Mech. Eng. (1962-64), Head Mech. Eng. Dept. (1965-), Lummus Española.
Nuclear interests: Nuclear reactor control systems design and testing; Reactor dynamics simulation and measurements; Servosystems and analog computing components design.
Address: Lummus Española, 14 Arapiles, Madrid 15, Spain.

de la ROCHETTE, Guy. Pres., Dept. Entreprises, Sté. Parisienne pour l'Industrie Electrique.

Address: Societe Parisienne pour l'Industrie Electrique, Departement Entreprises, 85 boulevard Haussmann, Paris, France.

de la SIERRA TORRES, Manuel. Pres., Tecnicas Nucleares S.A.
Address: Tecnicas Nucleares S.A., 46 Serrano, Madrid 1, Spain.

de la TORRE, Juan Antonio RODRIGUEZ. See RODRIGUEZ de la TORRE, Juan Antonio.

de LABOULAYE, Hubert, Diploma of Ing. de l'Ecole Centrale de Paris. Born 1921. Nucl. res. phys. (1947-54), Tech. Adviser for Foreign Relations, and Sen. Sci. (1954-58), C.E.A.; Deputy Director Gen., I.A.E.A. (1958-61; Special Asst. in High Commissionner's Cabinet, (1961-63), Head, Dept. of Programmes, (1963), C.E.A.
Nuclear interests: Experimental nuclear physics; Reactor physics and technology; Development of nuclear power; Economics and management; International problems.
Address: 5 rue Dufrenoy, Paris 16, France.

DE LANDSHEER, P. Ing. Principal, Cie. Maritime Belge S.A. (Lloyd Royal).
Address: Cie. Maritime Belge S.A. (Lloyd Royal), 61 Rempart Ste. Cathérine, Antwerp, Belgium.

DE LANGE, Pieter Willem, D.Sc. (Pretoria), Transvaal Dip. Ed. Born 1928. Educ.: Pretoria and Witwatersrand Univs. N.P.R.L., Pretoria, 1953-59; South African Atomic Energy Board, 1959-; Harwell Reactor School, 1960; Visitor, C.E.N. Saclay, 1960-61, and Inst. for Atomenergi, Kjeller, 1961; N.R.C. Post-doctorate Res. Fellow at Chalk River Nucl. Lab. 1964-65. Society: South African Inst. Phys.
Nuclear interests: Experimental reactor physics; Activation analysis; High resolution gamma-spectroscopy.
Address: Atomic Energy Board, Private Bag 256, Pretoria, South Africa.

de LANGEN, L. H., Prof. Member, Board of Governors, Reactor Centrum Nederland; Temporary Pres., Energiecommissie der drei verbonden, Verbond van Nederlandsche Werkgevers; Director, Algemene Kunstzijde Unie N.V.
Address: Reactor Centrum Nederland, Institute for the Development of Nuclear Science for Peaceful Purposes, 112 Scheyeningseweg, s'Gravenhage, Netherlands.

de LASTEYRIE, Bernard. Director, Cie. pour l'Etude et la Realisation de Combustibles Atomiques.
Nuclear interest: Economy and technology of nuclear energy, especially in the field of nuclear fuels.
Address: Cie. pour l'Etude et la Realisation de Combustibles Atomiques, 41 Avenue Montaigne, Paris 8, France.

DE LEENER, Marcel, Ing. civil. Born 1908. Educ.: Brussels Free Univ. Délégué Gén., Union des Centrales Electriques Linalux-Hainaut S.A.; Administrateur-Délégué, Assoc. des Centrales Electriques Industrielles de Belgique; Administrateur, du Groupement Professionnel de l'Industrie Nucléaire; Administrateur, Centre d'Etudes Nucléaires; Administrateur, Fondation Nucléaire.
Address: 49 square Marie-Louise, Brussels 4, Belgium.

de LEOBARDY, Member, Conseil d'Administration representing Caisse des Dépôts et Consignations, Sté Industrielle des Minerais de l'Ouest (S.I.M.O.).
Address: Societe Industrielle des Minerais de l'Ouest, 25 boulevard de l'Amiral-Bruix, Paris 16, France.

DE LEON, Julio MONTES PONCE. See MONTES PONCE DE LEON, Julio.

de LEON, Julio TORIELLO. See TORIELLO de LEON, Julio.

De LEON, Luis, Ing. Agr. Member, Comision Nacional de Energia Atomica, Uruguay.
Address: Comision Nacional de Energia Atomica, 565 J. Herrera y Reissig, P.2., Montevideo, Uruguay.

de LIMA, Ralph. Secretario, Liga Panamena Contra el Cancer.
Address: Liga Panamena Contra el Cancer, Apartado 7358, Panama, Panama.

de LIMA e SANTOS, Manuel RODRIGUES. See RODRIGUES de LIMA e SANTOS, Manuel.

DE LOOSE, Roger, Dr. in Agricultural Sci. Born 1928. Educ.: State Agricultural Univ., Ghent. Res. Fellow, I.R.S.I.A.; Asst., I.R.S.I.A. (1953-54), Asst. (1955-57), Chair for Analytical and Phys. Chem. State Agricultural Univ., Ghent. Asst., I.R.S.I.A. Government Plant Breeding Station, Lemberge, 1957-65; Res. Leader, I.R.S.I.A. Irrad. Lab., Melle-Ghent, 1965-.
Nuclear interests: Nuclear radiation in agriculture; Uptake of tagged fertilizers by plants; Use of ionising radiations in plant mutation work.
Address: Government Ornamental Plant Breeding Station, 17 Caritasstraat, Melle, Belgium.

DE LOS SANTOS LASURTEGUI, Alfonso, Dr. in law (diploma in Internat. Studies). Born 1913. Educ.: Valladolid and Madrid Univs.; Inst. Political Studies. Legal adviser to Junta de Energia Nucl.
Nuclear interests: All legal questions relating to nuclear energy.
Address: Junta de Energia Nuclear, Ciudad Universitaria, Madrid, Spain.

de LUC
de LUCCIA, E. Robert, S.B., C.E. Born 1904. Educ.: M.I.T. Sen. Vice-Pres. and Chief Eng., Pacific Power and Light Co. Firm's Rep. on atomic power study teams; Advisory Council, Atomic Ind. Forum. Society: A.N.S.
Address: Pacific Power and Light Company, 920 S.W. Sixth Avenue, Portland 4, Oregon, U.S.A.

DE MAGALHAES COLACO, Isabel, Doutora, Tech. Assessor, Comissao Do Risco Atomico, Portugal.
Address: Comissao Do Risco Atomico, 16 Largo Rafael Bordalo Pinheiro, Lisbon, Portugal.

de MAGNEE, Ivan, Ing. des Mines, Ing. Géologue. Born 1905. Educ.: Liège Univ. Prof. Brussels Univ., 1936-. Member, Commission Consultative de l'Agence des Approvisionnements de l'Euratom. Societies: Acad. Roy. des Sci. d'Outre-Mer; Inst. Mining and Metal. (London).
Nuclear interests: Geology; Prospecting; Ore-dressing.
Address: 72 avenue de l'Hippodrome, Brussels 5, Belgium.

de MARANON, Isidro Heredia MARTINEZ. See MARTINEZ de MARANON, Isidro Heredia.

DE MARCHI, Sante Bruno, M. Economic Sci. Born 1906. Educ.: Turin Univ. Chairman, Compagnia Lombarda di Assicurazione, 1963-; Director and Gen. Manager, Compagnia di Assicurazione di Milano; Vice-Chairman, Ticino, Stà. di Assicurazioni sulla Vita; Director: S.I.A.C., Uniorias, Coge and Consorziale. Steering Com., Pool Italiano per l'Assicurazione dei Rischi Atomici.
Address: 7 Via del Lauro, Milan, Italy.

DE MASTRY, John Andrew, B.Sc. (Chem.). Born 1930. Educ.: Ohio State Univ. General Motors Corp., 1953-56; Battelle Memorial Inst., 1956-; Societies: A.S.M.; A.R.S.; A.S.T.M.; A.N.S.
Nuclear interests: Reactor and high-temperature materials generally, with specific interest in irradiation damage to structural materials, fatigue, anf fracture mechanics studies of pressure vessel materials; Particular interests in mechanical metallurgy of materials for structural and cladding use in reactor applications.
Address: 1903 Brandywine Drive, Columbus, Ohio 43201, U.S.A.

DE MATTEIS, Artenio, D. Eng. Born 1930. Educ.: Milan Polytech. Centro di Ricerche Nucleari di Ispra, 1959-61. Centro di Calcolo del C.N.E.N., 1961-.
Nuclear interests: Reactor physics with special interest in Monte Carlo calculations.
Address: Centro di Calcolo del C.N.E.N., 2 Via Mazzini, Bologna, Italy.

de MEERSMAN, R. Belgian Member, Computer Programme Library Com., E.N.E.A.
Address: E.N.E.A. Computer Programme Library, Ispra, Varese, Italy.

de MEESTER de TILBOURG, Philippe Jacques, Chimiste. Born 1926. Educ.: Paris Univ. French Member, Social, Biology, and Safety Problems, Euratom. Society: Sté. de Cybernétique.
Nuclear interests: Biology and safety problems.
Address: Euratom, 51-53 rue Belliard, Brussels, Belgium.

DE MERRE, Marcel Albert, Ing. civil des Mines. Born 1899. Educ.: Louvain Univ. Administrateur, Centre d'Etude de l'Energie Nucl., 1952; Directeur, Sté. Gén. de Belgique, 1957; Chairman, Sté. Belge pour l'Industrie Nucl.; Administrateur, Sté. Belge de Chimie Nucl., 1958; Prés. and Administrateur-délégué de Métallurgie and Mécanique Nucl. (M.M.N.), 1958; Prés. and administrateur-délégué de la Sté. Gen. Métallurgique de Hoboken, 1961; Administrateur, Groupement Professional de l'Industrie Nucl.; Fondation Nucl.; Member, Sci. and Tech. Com., Euratom. Societies: Sté. de Chimie Industrielle de Paris; Fonds National de la Recherche Sci.; A.C.S.; A.I.M.M.E.; Inst. of Metals; Soc. of Chem. Industry; Australasian Inst. of Mining and Metal.
Nuclear interests: Metallurgy; Chemistry; Reactor design.
Address: 7 avenue Louise, Hoboken-lez-Anvers, Belgium.

de METZ, Joseph. Born 1918. Asst. Chief Quantity Surveyor, U.K.A.E.A. Society: Assoc., Roy. Inst. Chartered Surveyors.
Nuclear interest: Construction and contracting for works services.
Address: c/o United Kingdom Atomic Energy Authority, Risley, Nr. Warrington, Lancs., England.

de MEVERGNIES, Marcel J. B. NEVE. See NEVE de MEVERGNIES, Marcel J. B.

de MONCHAUX, C. F., Dr. Member, Dept. Public Health Radiol. Advisory Council.
Address: 189 Macquarie Street, Sydney, New South Wales, Australia.

DE MORI, Bruno. Prof., Rome Univ. Pres., Italian Pool for Insurance against Atomic Risks; Member of the Board, Unione Italiana di Riassicurazione; Member, F.I.E.N.
Address: 2 Via Petrolini, 0019 Rome, Italy.

DE MOURA SRIVASTAVA, Lêda, Bacharel and Licenciado (Phys.). Born 1932. Educ.: Rio de Janeiro Federal Univ. Asst. Prof., Centro Brasileiro de Pesquisas Fisicas, 1955-. Scholarship, N.R.C., Brazil.
Nuclear interest: Nuclear physics and high energy particle physics (mainly experimental aspects).

Address: C.E.R.N., 1211 Geneva 23, Switzerland.

DE MOYA, Emil BOYRIE. See BOYRIE DE MOYA, Emil.

de MUELENAERE, Félix A., L.C.R., Dr. en Droit. Born 1915. Educ.: Gent Rijksuniv. Service Juridique, Union Minière du Haut-Katanga; Conseiller Juridique, BelgoNucléaire; Sec. Gén., Sté. Anglo-Belge Vulcain. Nuclear interests: Legal aspects nuclear energy; Industrial property; Civilian nuclear responsibility.
Address: 20 avenue Jules César, Brussels 15, Belgium.

de NEMITZ, Serge, Dr. Asst., Demierre et Cie. S.A.
Address: Demierre et Cie. S.A., 5 Corraterie, Geneva, Switzerland.

de NERVO, Jacques. Prés. Directeur Gén., Denain Nord-Est Longwy, S.A.
Address: Denain Nord-Est Longwy, S.A. 25 rue de Clichy, Paris 9, France.

de NETTANCOURT, Dreux, M.Sc., Ph.D. Born 1933. Kentucky and McGill Univs. Graduate Res. Asst.,McGill Univ., 1960-63; Asst. Prof., Genetics, Loyola Coll., Montreal, 1964-65; Cytogeneticist, Biol. Div., Euratom, Assoc. Euratom-Instituut voor Toepassing van Atoomenergie in de Landbouw, Wageningen. Societies: European Soc. of Radiobiol.; Canadian Soc. of Genetics; American Genetic Assoc.
Nuclear interest: Radiobiology of higher plants and mutation breeding with tomatoes.
Address: Association Euratom - Instituut voor Toepassing van Atoomenergie in de Landbouw, Wageningen, Netherlands.

de NORRE, H. Asst. Manager, Nucl. Dept., Mercantile Marine Eng. and Graving Docks Co. S.A.
Address: Mercantile Marine Engineering and Graving Docks Co. S.A., Nuclear Department, 403 Hansadok, Antwerp, Belgium.

de OLIVEIRA, C. G., Asst. Prof. Theoretical Phys. Dept., Centro Brasileiro de Pasquisas Fisicas.
Address: Centro Brasileiro de Pasquisas Fisicas, 71 Avenida Weneceslau Braz, Rio de Janeiro, Brazil.

de OLIVEIRA GUTMAN, Maria Carmen. Chem., Radioisotope Lab., Inst. de Cancer, Ministerio da Saude D.N.S.-Servico Nacional de Cancer.
Address: Ministerio da Suade D.N.S.-Servico Nacional de Cancer, 23 Praca de Cruz Vermelha, Rio de Janeiro, Brazil.

de ORIOL y URQUIJO, José Maria, Dr. in ind. eng. Born 1906. Educ.: Escuela de Ingenieros Industriales, Madrid. Pres., Hidroelectrica Española; Pres., Sociedad Minera Guindos Guindos; Pres., Centrales Nucleares S.A. (CENUSA); Pres., Forum Atomico Español; Member of the Board, Junta de Energia Nucl.; Formerly Vice-Pres., Pres. (1966-67), and Member, Steering Com., Foratom.
Nuclear interest: Nuclear reactors for production of nuclear energy.
Address: 14 Montalbán, Madrid, Spain.

de OYARZABAL-ORVETA, Juan, Fisico. Born 1913. Educ.: Univ. Nacional Autónoma de México. Prof., Mexico Univ., 1945-; Prof., Inst. Politécnico, 1945-; Investigador, Mexico Univ., 1944-; Investigador, Inst. Nacional de la Investigación Cientifica, 1951-; Asesor Tecnico, Comisión Nacional de Energia Nucl., 1962-. Consejero consultivo de la Soc. Mexicana de Fisica. Societies: A.P.S.; A.I.P.; Acad. Nacional de la Investigación Ciéntifica (México).
Nuclear interests: High-energy nuclear physics; Reactor design.
Address: Calle C. Manzana V, Lote 141, Colonia Educación, Mexico 21 D.F., Mexico.

DE PARAVICINI, Thomas Pitt, M.A. Born 1909. Educ.: Cambridge Univ. Aircraft Industry (Rolls Royce, Armstrong Siddeley Motors, Bristol Aircraft, Normalair), to 1959; A.E.E. Winfrith, 1959-. Societies: A.M.I.Mech. E.; F.R.Ae.S.
Nuclear interests: Thermo-dynamics and engineering of power reactors with emphasis on heat transfer problems.
Address: The Old Manor, Abbotts Ann, Andover, Hants., England.

de PARDO, Carmen RUIZ. See RUIZ de PARDO, Carmen.

de PATER, Cornelis, qualified mech. eng. Born 1923. Educ.: Delft Tech. Univ. Kon. Machinefabriek Gebr. Stork and Co. N.V., Hengelo (O), Netherlands, 1949-65; Prof. Theoretical and Appl. Mech. Twente Tech. Univ., Netherlands. Society: Roy. Inst. of Eng.
Nuclear interest: Stress analysis of nuclear components.
Address: 50 Lansinkweg, Hengelo (O), Netherlands.

de PINEDO y ANGULO, Ignacio. Director, Centrales Nucleares S.A.
Address: Centrales Nucleares S.A., c/o 1 Hermosilla, Madrid 1, Spain.

de POLL, W. N. van. See van de POLL, W. N.

DE PRECIGOUT, Jean. Membre du Bureau and Vice Chairman (1968-70), Economic and Social Com., Euratom.
Address: 55 rue de la Boëtie, Paris 8, France.

de PROOST, M., Dr. Head, Irradiation Dept., Belgian Centre d'Etude de l'Energie Nucléaire. Vice-Chairman, Project Com., Internat. Food Irradiation Project, Seibersdorf.

DE QUE

Address: Centre d'Etude de l'Energie Nucléaire, Mol-Donk, Belgium.

DE QUEVEDO, José Luis GARCIA. See GARCIA DE QUEVEDO, José Luis.

de RAAF, H., Ing. Sen. Officer engaged in Nucl. Work, Leidsche Apparatenfabriek.
Address: Leidsche Apparatenfabriek N.V., 41-43 Os en Paardenlaan, Leiden, Netherlands.

de'REGUARDATI, Fausto MARINUCCI. See MARINUCCI de'REGUARDATI, Fausto.

de RHAM, E. Member, Conseil d'Administration, Energie Nucléaire S.A.
Address: Energie Nucléaire S.A., 10 avenue de la Gare, Lausanne, Switzerland.

de ROCHEMONT, R. du MESNIL. See du MESNIL de ROCHEMONT, R.

de ROUVILLE, Maurice A., Ecole Polytechnique, Ing. en Chef des Ponts et Chaussées. Born 1920. Naval Eng., 1944-52; Highway Dept. Eng., 1952-55; Directeur Centre Production Plutonium Marcoule, C.E.A., 1955.
Nuclear interests: Plutonium producing reactors graphite-moderated and gas-cooled; Plutonium extraction; Waste-disposal problems.
Address: Centre de Marcoule, Boite Postale 106, Bagnols-sur-Ceze, (Gard), France.

de RUYTER, A. W. Radiochem., Inst. for Nucl. Phys. Res., Netherlands.
Address: Institute for Nuclear Physics Research, 18 Oosterringdijk, Amsterdam O, Netherlands.

de S. MUTUCUMARANA, T., Dr. Member, Phys. Dept., Ceylon Univ.
Address: Ceylon University, Thurstan Road, Colombo 5, Ceylon.

de SAINT MAURICE, Arthur Bérault, B.S., M.S. Born 1932. Educ.: Georgetown and Tufts Univ. Sen. Eng. Ind. Reactor Labs., Rutgers Univ. Society: A.N.S.
Nuclear interests: Nuclear reactor design and operation; Nuclear applications.
Address: Industrial Reactor Laboratories, Incorporated, Plainsboro, New Jersey, U.S.A.

de SAINT SAUVEUR, Henri de GALLIER. See de GALLIER de SAINT SAUVEUR, Henri.

DE SALDANHA, Jose Luis da Camara, Dr Director of Central Services, Junta de Energia Nucl., Portugal.
Address: Junta de Energia Nuclear, 79 Rua de S. Pedro de Alcantara, Lisbon, Portugal.

de SALVO BRITO, Sergio. Civil Eng. Nucl. Eng. Born 1935. Educ.: Escola Politécnica, Univ. de Sao Paulo (E.P.U.S.P.), Inst. Militar de Engenharia (I.M.E.), Rio de Janeiro, I.N.S.T.N., Saclay and Ecole Pratique des Hautes Etudes, Paris Univ. Eng., Tech. Adviser to the Pres., Comissao Nacional de Energia

Nucl., 1960-; In charge of the Economic Branch, Power Reactors Work Group, 1962-.
Societies: Assoc. Brasileira de Engenheiros Eletricistas; Assoc. Brasileira de Energia Nucl.; Assoc. Internat. des Ing. au Génie Atomique; Assoc. Française d' Informatique et de Recherche Operationnelle.
Nuclear interests: Nuclear engineering; Economic aspects of nuclear power.
Address: Comissao Nacional de Energia Nuclear, Av. Almirante Barroso, 81 3^0 and., Rio de Janeiro, GB-Brazil.

DE SAUSSURE, Gerard, Diploma E.T.H., Ph.D. (M.I.T.). Born 1924. Educ.: Swiss Federal Inst. Technol., Zürich, and M.I.T. Res. Asst., Mass. Inst. Technol., 1952-54; Phys., O.R.N.L., 1955-. Societies: A.I.P.; A.N.S.; Società Italiana di Fisica.
Nuclear interests: Nuclear and reactor physics.
Address: 209 Louisana Avenue, Oak Ridge, Tennessee, U.S.A.

DE SCHRYNMAKERS, Thierry, Ing. Civil des Mines (Louvain Univ.). Educ.: Louvain Univ. Director, Nucl. activities, Bureau d'Etudes Industrielles Fernand Courtoy.
Address: Bureau d'Etudes Industrielles Fernand Courtoy, 43 rue des Colonies, Brussels 1, Belgium.

DE SENARCLENS, A., Dr. phil. Born 1905. Educ.: Ecole Polytechnique Fédérale, Zurich and Breslau Univ. Administrator Delegate, Energie Nucléaire S.A.; Vice-Pres., Sté. Nat. pour l'Encouragement de la Technique Atomique Industrielle (Berne).
Address: Chèserex, Nyon, Switzerland.

de-SHALIT, Amos, M.Sc., Dr.Sc.Nat. Born 1926. Educ.: Hebrew Univ. Jerusalem, and E.T.H., Zurich. Visiting Prof., Hebrew Univ., 1956-. Head, Dept. Phys. (1961-63), Sci. Director (1963-66), Director Gen. (1966-), Weizmann Inst. of Sci. Societies: A.P.S.; Israel Phys. Soc.; Israel Acad. of Sci.
Nuclear interest: Nuclear physics.
Address: The Weizmann Institute, Rehovoth, Israel.

De SHONG, James Anson, Jr., B.S. (I.E., Gen. Motors Inst.), B.S. (Elect. Eng., Purdue), M.S.E. (Purdue). Born 1917. Educ.: Washington Univ., Gen. Motors Inst., Purdue Univ., Illinois Inst. Technol. Sen. Elec. Eng., Argonne Nat. Lab. Consultant to Sweden, Switzerland, Halden (Norway), 1958-59. Societies: A.N.S.; RESA; A.A.A.S.; I.E.E.E.; A.P.S.
Nuclear interests: Reactor and high energy physics; Use of hydrogen bubble chambers, spark chambers, etc., together with data processing to read and measure tracks of high energy particles in such instruments; Measurement of primary particle mass and charge with balloon-borne spark chamber spectrometers; Reactor dynamics and control.

Address: 590 Park Avenue, Elmhurst, Illinois, U.S.A.

DE SILVA, Ginige Richard Walter, B.Sc. (Phys., London), M.A. (Phys. and Maths., Cambridge). Born 1911. Educ.: London and Cambridge Univs. Director of Commerce, (1961-63), Permanent Sec. (1963-65), Ministry of Commerce and Trade; Director, Div. of Conference and Gen. Services, I.A.E.A., 1966-. Address: International Atomic Energy Agency, 11 Kaerntnerring, Vienna 1, Austria.

de SILVA, W. A. Chairman, Atomic Energy Com., Nat. Planning Council of Ceylon. Address: Atomic Energy Committee, National Planning Council of Ceylon, 5 Galle Buck Road, Colombo 1, Ceylon.

de SOUSA TORRES, Branca Edmee MARQUES. See MARQUES de SOUSA TORRES, Branca Edmee.

de SOUZA SANTOS, Marcello Damy, B.Sc., Ph.D. Born 1914. Educ.: Sao Paulo and Cambridge Univs. Director, Inst. de Energia Atomica; Prof. Phys., Univ. Sao Paulo. Societies: A.N.S.; Atomic Forum; Academia Brasileira Ciências. Nuclear interests: Reactor physics and reactor design. Address: Caixa Postal 11049, Pinheiros, Sao Paulo, Brazil.

de SOUZA SANTOS, Tharcisio Damy, Metal. Eng., Civil Eng. Born 1912. Educ.: Sao Paulo Univ. Dean, "Livre-Docente", and Full Prof. of Non Ferrous Metal., Escola Politécnica, Univ. de Sao Paulo; Head, Nucl. Metal. Div. Inst. de Energia Atômica. Societies: A.I.M.M.E.; Associacao Brasileira de Metals; Inst. de Engenharia de Sao Paulo. Nuclear interests: Nuclear metallurgy and fuel element fabrication. Address: Instituto de Energia Atômica, Caixa Postal 11049 (Pinheiros), Sao Paulo, Brazil.

DE STAEBLER, H., Jr., Dr. Book: Contributor to Eng. Compendium on Rad. Shielding (I.A.E.A.). Address: Stanford University, P.O. Box 4349, Stanford, California, U.S.A.

de STEENE, J. van. See van de STEENE, J.

de STORDEUR, Arnold N., Ing. Civil Métallurgiste. Born 1928. Educ.: Liège Univ. Head, Fast Reactor Div., Gen. Directorate for Res. and Training, Euratom Headquarters, Brussels. Societies: Assoc. des Ingénieurs Diplômés de l'Université de Liège; A.N.S. Nuclear interests: Fast reactor core design (including heat transfer, hydraulics and metallurgy); Fast reactor economics. Address: 43 avenue G. Lebon, Bruxelles (16), Belgium.

de STTAU MONTEIRO, Miguel, Law graduate Born 1930. Educ.: Lisbon Univ. Director,
Mundial Insurance Co., 1954; Managing Director, Soc. Central de Cervejas, 1955; Member, Nucl. Risks Commission, Grémio dos Seguradores; Member, E.F.T.A. Consultive Com.; Director, Accident Prevention Centre; Director, Portuguese Chamber of Commerce; Pres., Insurance Sect., Ind. Assoc. Society: Portuguese Atomic Forum. Nuclear interests: All legal aspects concerning liability for accidents relating to nuclear risks, and all insurance and reinsurance aspects concerning cover of nuclear risks. Address: Gremio dos Seguradores, 16 Largo Rafael Bordalo Pinheiro, Lisbon, Portugal.

DE SWART, Johan J. Ir., Ph.D. Born 1931. Educ.: Delft Technol. Univ. and Rochester Univ. Rochester Univ., 1958-60; Chicago Univ., 1960-62; C.E.R.N., Geneva, 1962-63; Nymegen Univ., 1962; Pittsburgh Univ., 1966-67. Societies: Dutch Phys. Soc.; A.P.S. Nuclear interests: Nuclear physics; High energy physics. Address: 9,1 Driehuizerweg, Nymegen, Netherlands.

de TILBOURG, Philippe Jacques de MEESTER. See de MEESTER de TILBOURG, Philippe Jacques.

DE TOLEDO, Paul SARAIVA. See SARAIVO DE TOLEDO, Paulo.

de TORRONTEGUI, Leandro Jose. Formerly Member, Comisiones Asesoras de la Junta de Energia Nuclear De Reactores Industriales now Member, Comisiones Asesoras de Centrales Nucleares, Spain; Formerly Deputy Chairman, Managing Director, now Chairman of the Board, S. E. de C. Babcock and Wilcox C.A.; Pres., Construcciones Nucleares; Pres., Laboratorios de Ensayos e Investigaciones Industriales, Tech. High School for Industrial Engs., Bilbao, Director, Forum Atomico Espanol. Address: Technical High School for Industrial Engineers, Bilbao, Spain.

DE TROYER, André, D.Sc. Born 1924. Educ.: Louvain Univ. Asst. Kamerlingh Onnes Lab., Leiden Univ. 1950; Head, Working Group, Fundementeel Onderzoek der Materie; At (1952), at present, Chef de service principal, Union Minière du Haut-Katanga, Brussels. Nuclear interests: Neutron physics and neutron sources; Radioactivity measurements and calibrations; Research and development on ^{227}Ac, ^{228}Th and their production; Research and development on radioisotopic heat sources - e.g. for direct conversion. Address: c/o Union Minière du Haut-Katanga, 6 rue Montagne du Parc, Brussels, Belgium.

de URIBE, Mrs Esmeralda ARBOLEDA. See ARBOLEDA de URIBE, Mrs Esmeralda.

de USAOLA BARRENENGOA, Emilio. Director, Centrales Nucleares S.A.

de USE
Address: Centrales Nucleares S.A., c/o 1 Hermosilla, Madrid 1, Spain.

de USERA, Gabriel. Member, Atomic Risk Commission - Nat. Insurers Assoc.
Address: Atomic Risk Commission – National Insurers Association, 4 Avenida Calvo Sotelo, Madrid, Spain.

de VABRES, Jean Donnedieu. See DONNEDIEU de VABRES, Jean.

de VAISSIERE, Alfred. Prés. Directeur Gén., Saint-Gobain Techniques Nouvelles.
Address: Saint-Gobain Techniques Nouvelles, 23 Boulevard Georges Clemenceau, Courbevoie 92, France.

de VALENCE, Edgar Louis Philippe POUPINEL. See POUPINEL de VALENCE, Edgar Louis Philippe.

de VASCONCELLOS e CASTRO, Jose Sarmento, Prof. Senior Officer, Lab. di Fisica, Lisbon Univ.
Address: Laboratorio de Fisica, Faculdade de Ciencias, Rua da Escola Politecnica, Lisbon, Portugal.

DE VATHAIRE, Francois, Ing. Born 1930. Educ.: Ecole Polytechnique, Ecole Nationale Supérieure du Génie Maritime. Reactor shielding studies, 1955-59; Head Reactor Safety Group, 1960.
Nuclear interest: Reactor safety.
Address: Commissariat a l'Energie Atomique, B.P. No. 2, 91 Gif-sur-Yvette, France.

de VETTOR, Giorgio BOMBASSEI FRASCANI. See BOMBASSEI FRASCANI de VETTOR, Giorgio.

DE VILLE, Georges. Director, Chaudronneries Brouhon, S.A.
Address: c/o Chaudronneries Brouhon, S.A., 41 rue de la Station, Awans-Bierset, (Liège), Belgium.

DE VILLEPIN, Francois. Formerly Chef du Département, Déparement du Vide, Cie. Française Thomson-Houston. Directeur Gén. Adjoint, Sté. Gén. du Vide S.A.
Address: Société Générale du Vide S.A., 186 rue du Faubourg Saint Honoré, Paris 8, France.

de VILLIERS, G. S. Manager, Hartebeestfontein Gold Mining Co. Ltd.
Address: Hartebeestfontein Gold Mining Co. Ltd., Private Bag, Stilfontein, Transvaal, South Africa.

DE VILLIERS, Isak Frederick Albert, B.A. (Law). Born 1918. Educ.: Cape Town Univ. Formerly in the South African Foreign Service: Rome, 1945-51; Pretoria, 1952-56; London, 1957-60; U.N. General Assembly, 1960; Paris, 1961-65. Manager, Nucl. Fuels Corp. of South Africa (Pty) Ltd., and Uranium Adviser to the Chamber of Mines of South Africa.
Address: NUFCOR, P. O. Box 1162, Johannesburg, South Africa.

DE VILLIERS, J. Wynand L., B.Sc., M.Sc., D.Sc. (Stellenbosch). Born 1929. Educ.: Stellenbosch Univ. Head, Mass Spectrometry Div., Nat. Phys. Res. Lab., C.S.I.R., South Africa, 1956-58; Fellowship, Atomic Energy Board, 1958-59. Head, Reactor Phys. Subdiv. (1959-67), Director, Reactor Development Div. (1967-), South African Atomic Energy Board. Societies: S.A. Inst. for Phys. (Foundation Member); A.N.S.
Nuclear interests: Nuclear physics; Reactor physics; All aspects of nuclear engineering.
Address: S.A. Atomic Energy Board, Private 256, Pretoria, South Africa.

de VITO, Edgardo, Dr. Naval and Mech. Eng. Born 1905. Educ.: Genoa Univ. Managing Director, Ansaldo Meccanico-Nucleare S.p.A., Genoa, 1966; Pres., Progettazioni Meccaniche Nucleari S.p.A., Genoa, 1966; Pres., Italian Tech. Subcom., Lloyd's Register; Member, Italian Tech. Com., American Bureau of Shipping; Member, Tech. Com., Italian Register (RINa); Member, Tech. Com. CE.TE.NA. (Shipbuilding Technique Centre); Member, Board of Directors, C.I.S.E., Milan.
Nuclear interest: Management.
Address: Ansaldo Meccanico-Nucleare S.p.A., 2 Piazza Carignano, Genoa, Italy.

DE VOORDE, Norbert Leopold Carlos VAN. See VAN DE VOORDE, Norbert Leopold Carlos.

De VRIES, Adolf Eduard, Ph.D. Born 1924. Educ.: Amsterdam Municipal Univ. Head, Chem. Dept., F.O.M. Inst. for Atomic and Molecular Phys., Amsterdam, 1951-; Brookhaven Nat. Upton, L.I., 1956-57; Weizmann Memorial Fellow, Weizmann Inst., Rehovoth, Israel, 1962.
Nuclear interests: Physical chemistry of isotopes; Study of chemical reactions with atomic beams.
Address: F.O.M. Institute for Atomic and Molecular Physics, 407 Kruislaan, Amsterdam, Netherlands.

de VRIES, Coenraad, Ph.D. (Phys.). Born 1927. Educ.: Amsterdam Univ. Assoc. Sci. Director, Inst. for Nucl. Phys., Amsterdam. Nuclear interests: Nuclear physics; Electromagnetic interactions with nuclei and nucleons.
Address: Institute for Nuclear Research, 18 Oosterringdijk, Amsterdam, Netherlands.

de VRIES, Daniel Alexander, Ph.D. (Phys.). Born 1915. Educ.: Leyden Univ. Principal Res. Officer, C.S.I.R.O., Deniliquin, N.S.W., 1955-58; Prof. Phys. (Heat and mass transfer), Technolog. Univ., Eindhoven, 1958-. Society: Netherlands Phys. Soc. (Nederlandse

Natuurkundige Vereniging).
Nuclear interests: Reactor physics, especially heat transfer.
Address: Technological University, Postbus 513, Eindhoven, Netherlands.

de VRIES, M. J., Dr. Staff Member, Radiobiol. Inst., Organisation for Health Res. T.N.O.
Address: Organisation for Health Research T.N.O., 151 Lange Kleiweg, Rijswijk (ZH), Netherlands.

de VRIES, W. L., Director-General, Directorate-General for Shipping, Ministry of Transport and Inland Waterways, Netherlands.
Address: 400 Van Alkemadelaan, The Hague, Netherlands.

DE WAARD, Hendrik, Dr. of Phys. Born 1922. Educ.: Groningen Univ. Lecturer in Phys. (1957-58), Full Prof. Phys. (1958-), Groningen Univ. Books: Electronica (De Haan, 1959); co-author, Modern Electronics (Addison-Wesley, 1966). Society: Netherlands Phys. Soc.
Nuclear interests: Nuclear spectroscopy; Nuclear electronics; Mössbauer effect.
Address: 8 Werfstraat, Groningen, Netherlands.

de WAARD, Reinier Herman, D.Sc., Medicine doctorandus, arts. Born 1898. Educ.: Utrecht and Amsterdam. Chief Röntgendepartment, Medical Univ. Clinic, Utrecht, 1929-49; Prof. Radiology, Utrecht, 1949-. Commission of Dosimetry, Nederlandse Vereniging voor Radiologie; Board of Radiobiologisch Inst. der Gezondheidsorganisatie T.N.O. Book: Ferromagnetisme en kristalstructuur (1924).
Nuclear interests: Scattered X-radiation in diagnostics and in therapy.
Address: 2 Pieter Breughelstraat, Utrecht, Netherlands.

de WAEGH, Francois Albert, Ing. civil élec., M.Sc. (Phys.). Born 1935. Educ.: Louvain Catholique and Maryland Univs. At (1960-), at present Dept. Head, Belgonucléaire S.A. Societies: A.N.S.; Soc. Roy. des Ing. et Ind. belges.
Nuclear interests: Core design and performance studies (neutronic and thermohydraulic); Plutonium recycle in thermal reactors; Market surveys for small amd medium power nuclear reactors.
Address: 12 rue Eugène Denis, Brussels 16, Belgium.

de WALLE, Remy T. VAN. See VAN de WALLE, Remy T.

de WALQUE, Francois Antoine Félix Marie, Ing. civil des Mines, Ing. électricien. Born 1903. Educ.: Louvain and Liège Univs. Manager (1954-), and Director (1967-), Electrobel, S.A. holding co. of elec. generation and supply utilities; Member, Exec. Com., Trabel (Assoc. of the Eng. Depts. of the companies Electrobel and Traction and Electricité, S.A., for the eng. of the two large capacity nucl.

plants to be erected in Belgium), 1965; Director, Groupement Professionnel de l'Industrie Nucléaire, 1961; Director, Fondation Nucléaire, 1961. Society: Sté. Royale belge des Ing. et Industriels.
Nuclear interest: Generation and transmission of electricity, namely nuclear energy.
Address: 197 rue Belliard, Brussels 4, Belgium.

DE WOESTIJNE, W. J. VAN. See VAN DE WOESTIJNE, W. J.

DE YOUNG, H. George. Formerly Director, Formerly Exec. Vice Pres., now Pres., Rio Algom Mines Ltd. Director, Canadian Nucl. Assoc.
Address: Rio Algom Mines Ltd., 335 Bay Street, Toronto, Ontario, Canada.

de ZAFRA, Robert Lee, A.B., Ph.D. Born 1932. Educ.: Princeton and Maryland Univs. Instructor, Pennsylvania Univ., 1958-61; Asst. Prof. (1961-64), Assoc. Prof. (1964-), New York State Univ. Societies: A.P.S.; A.I.P.; A.A.P.T.
Nuclear interests: Atomic and nuclear physics; Quantum electronics.
Address: Physics Department, New York State University, Stony Brook, Long Island, New York, U.S.A.

DE ZEEUW, Dick, Dr. Agricultural Sci., Born 1924. Educ.: Agric. Univ., Wageningen, Netherlands. Asst. Agricultural Univ., Wageningen, 1950-54; Postdoctorate Fellowship, Purdue Univ., 1954-56; Sen. Sci. Officer, Horticultural Dept., Agricultural Univ., Wageningen, 1956-57; Principal Sci. Officer (1957-60), Director (1960-), Inst. voor Toepassing van Atoomenergie in de Landbouw, Wageningen.
Nuclear interest: Application of atomic sciences in agriculture.
Address: Association Euratom-Ital, 6 Keyenbergseweg (Postbox 48), Wageningen, Netherlands.

DEAGLIO, Romolo, Prof. Director, Istituto di Fisica Superiore; Preside, Facoltà di Scienze, Turin Univ.
Address: Istituto di Fisica Università, 1 Via Pietro Giuria, 10125 Turin, Italy.

DEAL, William Edgar, Jr., B.S., M.A., Ph.D. Born 1925. Educ.: Texas Univ. Staff Member (1950-60), Group Leader (1960-65), Asst. Div. Leader (1965-), Los Alamos Sci. Lab. Society: A.P.S.
Nuclear interests: Management; Nuclear weapons.
Address: 159 Monte Rey Drive South, Los Alamos, New Mexico 87544, U.S.A.

DEALE, Valentine B. Attorney, Washington, D.C. Member, Atomic Safety and Licensing Board Panel, and Member, Board of Contract Appeals, U.S.A.E.C.

Address: U.S. Atomic Energy Commission, Washington, D.C. 20545, U.S.A.

DEALLER, John Francis B., B.Sc. M.Sc. Born 1926. Educ.: Sheffield and London Univs. Asst. Phys., Regional Radium Inst., Bradford, 1951-55; i/c External Rad. Measurement Services, Radiol. Protection Service (M.R.C.), 1955-62; Sen. Phys., Smith and Nephew Res. Ltd., i/c Electron irradiation plant, 1962-. Societies: British Inst. Radiol.; Hospital Phys. Assoc.; A. Inst. P.; Assoc. Rad. Res. Nuclear interests: Radiation physics. Address: Botley House, School Lane, Hints, Nr. Tamworth, Staffs., England.

DEAN, Orlen Camp, B.A., M.S., Ph.D. Born 1907. Educ.: Wisconsin and Iowa Univs. Teaching, Oshkosh, Wisconsin State Coll., 1946-52 (Chairman, Dept. of Chem., 1949-52); Development Eng., O.R.N.L., 1952-. Societies: A.C.S.; A.N.S.; RESA. Nuclear interests: Preparation and processing of ceramic nuclear fuel; Pyrochemical processing of metals; Reduction and fabrication of refractory metals. Address: 32 Outer Drive, Oak Ridge, Tennessee, U.S.A.

DEAN, William Ernest, Jr., B.S. (Elec. Eng.). Born 1913. Educ.: Tennessee Univ. Chief, Power Economics Branch (1949-52), Asst. Director Power Supply (1952-58), Formerly Chief of Res. Staff, Office of Power (1958-), Tennessee Valley Authority; Consultant, I.A.E.A., 1961; Consultant, F.P.C., 1964. Nuclear interest: Application of nuclear power in the production of electricity on a commercial basis. Address: 3236 N. Lockwood Drive NW, Chattanooga 5, Tennessee, U.S.A.

DEAR, B. D., Instructor R.N., B.Sc. Sen. Lecturer, Dept. of Nucl. Sci. and Technol., Roy. Naval Coll. Address: Royal Naval College, Department of Nuclear Science and Technology, Greenwich, London S.E.10, England.

DEATHERAGE, Fred E., A.B., A.M., Ph.D., D.Sc.(Hon.). Born 1913. Educ.: Illinois Coll., Illinois and Iowa State Univs. Prof. and Chairman, Dept. of Biochem. (1951-64), Prof. (1964-), Ohio State Univ.; Ohio State Univ. - USAID Mission to Brazil, Sao Paulo Univ., School of Agriculture 'Luiz de Queiroz', 1964-68. Societies: A.C.S.; American Soc. Biol. Chem.; American Inst. Nutrition; Inst. of Food Technols.; American Oil Chem. Soc.; A.A.A.S.; American Soc. for Animal Sci. Nuclear interests: Use of radioactive isotopes as biological tracers: Radiation preservation of foods. Address: Biochemistry Department, Ohio State University, 2121 Fyffe Road, Columbus, Ohio 43210, U.S.A.

DeBAKEY, M. E. Member, Board of Directors, Oak Ridge Assoc. Univs. Address: Oak Ridge Associated Universities, P.O. Box 117, Oak Ridge, Tennessee, U.S.A.

DEBATISSE, Michel. Pres., Confédération Française de l'Aviculture; Sec.-Gen. Adjoint, F.N.S.E.A. Member, Specialised Nucl. Sect. for Economic Problems, Economic and Social Com., Euratom, 1966-. Address: c/o F.N.S.E.A., 8 avenue Marceau, 75 Paris 8, France.

DEBEAUVAIS, Monique, Dr. es Sci. Born 1925. Educ.: Strasbourg Univ. Nuclear interests: Nuclear photography and solid state nuclear tracks detectors. Address: Departement de Physique Corpusculaire du Centre de Recherches Nucleaires, Strasbourg-Cronenbourg, France.

DeBENEDETTI, Sergio, Ph.D. Born 1912. Educ.: Florence Univ. Prof. Phys., Carnegie Inst. Technol., 1949-; Fulbright Fellow, Turin Univ. 1956-57; At Centro Brazileiro de Pesquisas Fisicas, Rio de Janeiro, Summers 1952, 1962; At Brookhaven Nat. Lab., Summers, 1956, 1958, 1959, 1960; California Univ., Summer 1963. Editorial Board, Nucl. Instruments and Methods and Reviews Modern Phys. Book: Nucl. Interactions (New York, John Wiley, 1964). Society: Fellow, A.P.S. Nuclear interests: Particle physics; Positrons; Mössbauer effect. Address: Carnegie Institute of Technology, Department of Physics, Pittsburgh 13, Pennsylvania, U.S.A.

DEBERNARDI, Enzo, Dr. Ing. Born 1925. Educ.: Escuela Superior de Ingenieria Politecnico di Turin. Administrador Gen., Administración Nacional de Electricidad, 1959-; Prof., Facultad de Ciencias Fisicas y Matemáticas, 1958-. Society: Miembros, Comisión Nacional de Energia Atómica. Nuclear interests: Producción de energia elétrica. Address: c/o ANDE, Estrella y Ayolas, Asunción, Paraguay.

DEBERTIN, K., Dr., Dipl.-Phys. Born 1933. Educ.: Johann Wolfgang Goethe Univ. Sci. Asst., Inst. für Kernphysik, Frankfurt am Main, Germany; Sci. Asst., Physikalisches Institut der Universität Freiburg im Breisgau, Germany, 1966-. Nuclear interests: Fast neutron physics; Nuclear reactions at van de Graaff accelerator energies. Address: Physikalische Institut der Universität Freiburg im Breisgau, Germany.

DEBIESSE, Jean, D. ès Sc., Agrégé de Physique, D. es Sc. Phys. Born 1907. Prof. puis Inspecteur d'Académie; Directeur adjoint de l'Enseignement Primaire; Inspecteur Gén. de l'Instruction Publique détaché au C.E.A.; Directeur du Centre d'Etudes Nucléaires de

Saclay; Directeur I.N.S.T.N. Papers: 20 comptes-rendus à l'Académie des Sciences; quelques rapports à l'Unesco sur l'enseignement public.
Address: 62 rue de la Tourelle, Boulogne 92, France.

deBOISBLANC, Deslonde Raymond, B.S.E. Born 1914. Educ.: Louisiana State Univ. Director, Reactor Phys. and Eng. Branch, (1956-64), Formerly Asst. Manager, Res. Activities (1964-), Atomic Energy Div. Phillips Petroleum Co., now Manager, Nucl. and Chem. Technol. Div., Idaho Nucl. Corp., Nat. Reactor Testing Station, Nr. Arco, Idaho. Formerly Member, Advisory Com. on Reactor Phys., U.S.A.E.C., 1957-; Member, Nucl. Com., A.S.E.E., 1961-; Member, Exec. Com., Reactor Operations Div., A.N.S. 1962-63; Formerly Member, Board of Directors, A.N.S. Book: Co-author, Nucl. Power Reactor Phys. (D. Van Nostrand, 1961). Societies: A.N.S.; A.A.A.S.; A.S.E.E.
Nuclear interests: Reactor physics research; Reactor design; Nuclear instrumentation; Technical management.
Address: Idaho Nuclear Corporation, P.O. Box 2908, Idaho Falls, Idaho, U.S.A.

DeBOLT, Harold E., B.S., M.S., Sc.D. (Elec. Eng.). Born 1922. Educ.: Carnegie Inst. Technol. Principal Sci., AVCO Mfr. Corp., Wilmington, Mass. Societies: American Inst. Elec. Eng.; Inst. Radio Eng.
Nuclear interest: Space power reactors.
Address: 201 Lowell Street, Wilmington, Mass., U.S.A.

DEBRAINE, M. Adjoint au Directeur, Centre d'Etudes Nucléaires de Saclay, C.E.A.
Address: Commissariat à l'Energie Atomique, Centre d'Etudes Nucléaires de Saclay, Boite Postale No. 2, Gif-sur-Yvette, (Seine-et-Oise), France.

DEBRECZENI, Béla, Dr. Sen. Officer, Res. Inst. for Irrigation and Rice Cultivation.
Address: Research Institute for Irrigation and Rice Cultivation, 2 Szabadsag ut., Szarvas, Hungary.

DEBRECZENI BELANE, Mrs., Dr. Sen. Officer, Res. Inst. for Irrigation and Rice Cultivation.
Address: Research Institute for Irrigation and Rice Cultivation, 2 Szabadsag ut., Szarvas, Hungary.

DEBRUNNER, Hermann, Dr. Phys. Born 1931. Educ.: Bern Univ. At Phys. Inst., Bern Univ.
Nuclear interests: Nuclear physics; Cosmic radiation.
Address: Physical Institute, 5 Sidlerstrasse, Bern, Switzerland.

DeBUCHANANNE, George D. Geologist, Geological Survey, U.S. Dept. of the Interior.
Address: Geological Survey, U.S. Department of the Interior, Washington 20242, D.C., U.S.A.

DECKEN, Claus-Benedict von der. See von der DECKEN, Claus-Benedict.

DECKER, Clarence F., B.S. (Western Michigan), M.S. (Michigan State), Ph.D. (Michigan State). Born 1925. Educ.: Western Michigan and Michigan State Univs. Post-doctoral, U.S. P.H.S., Biochem. Dept., East Lansing, 1954-56; Asst. Sci. (1956-59), Assoc. Sci. (1959-), Argonne Nat. Lab. Biochem., Presbyterian-St. Luke's Hospital, Asst. Prof. Biochem., Illinois Univ. Coll. Medicine, Chicago. Societies: A.C.S.; New York Acad. Sci.; A.A.A.S.; Rad. Res.
Nuclear interests: Skeletal retention and metabolism of the alkaline earth radionuclides; Toxicity and metabolism of radionuclides of cadmium, cerium and other metals; Trace element metabolism; In vivo and in vitro metabolism of I^{131} labeled thyroid hormones.
Address: Section of Endocrinology, Presbyterian-St. Luke's Hospital, Chicago, Illinois, U.S.A.

DECKER, Peter, Dipl., Dr. rer. nat., Privatdoz. (Physiologische Chem.). Born 1916. Educ.: Munich Univ. Wiss. Mitarb. II Mediz, Univ. Klinik, Munich, 1943-55; Stipendiar, Deutsche Forschungsgemeinschaft, 1951-54; Asst. (1955-58), Formerly Privatdoz. (1958-), Physiologisches Inst., Tierärzliche Hochschule, Hanover. Society: Gesellschaft Physiologische Chem.
Nuclear interest: Biochemical and analytical applications.
Address: 74A Stammestrasse, 3000 Hanover-Linden, Germany.

DECKER, Wolfgang, Dipl. Phys. Sen. Officer, Inst. of Special Metals, Max-Planck Inst. for Metal Res.
Address: Max-Planck Institute for Metal Research, 92 Seestrasse, 7000 Stuttgart 1, Germany.

DECONNINCK, Gaston Germain, D.Sc. Born 1924. Educ.: Louvain Univ. Prof., Faculté, Namur Univ.; i/c, Van de Graaff Dept., Centre de Physique Nucléaire, and Prof., Faculté, Louvain Univ.
Nuclear interest: Neutron scattering, neutron producing reactions namely (p,n) and (αn) reactions.
Address: Centre de Physique Nucléaire, Parc d'Arenberg, Heverlee, Louvain, Belgium.

DECROLY, Claude, Prof. Member, Editorial Advisory Board, J. Nucl. Materials.
Address: Universite Libre de Bruxelles, Brussels, Belgium.

DeCROSTA, Edward Anthony, B.E.E. Born 1921. Educ.: Rensselaer Polytech. Inst. Tracer lab., 1952-58; Sales Manager, Lab. for

Electronics, 1958-60; Vice-Pres., Atomic Div., Baird Atomic, 1960-64; Operations Manager (1964-), and Director, Hamner Electronics Co., Inc. Societies: A.N.S.; Soc. of Nucl. Medicine. Nuclear interests: General management, nuclear instrumentation field.
Address: 1019 Evergreen Road, Morrisville, Pennsylvania, U.S.A.

DEDDENS, James Carroll, B. Mech. Eng., M. Mech. Eng. Born 1928. Educ.: Louisville Univ. O.R.S.O.R.T. 1953-54; Reactor Phys. (1954-56), Project Supv., Reactor Eng. Indian Point Reactor Project (1956-59), Project Eng., Indian Point Reactor Project (1959-62), Project Manager, Indian Point Reactor Project (1962-64), Asst. Marketing Co-ordinator, Atomic Energy Div., (1964-66), Manager, Service Sect., Nuclear Power Generation Dept., Boiler Div. (1966-), Babcock and Wilcox Co. Chairman, North Carolina - Virginia Sect., and local Sects., and Com. member, A.N.S. Societies: A.N.S.; A.S.M.E.
Nuclear interests: Nuclear plant design; Nuclear power plant testing, start up and operations; Nuclear fuel management services; Nuclear plant performance analysis; Engineer and operator training.
Address: Babcock and Wilcox Company, 5061 Fort Avenue, Lynchburg, Virginia 24505, U.S.A.

DEDEK, Vladislav, Dipl. Chem. Eng. Born 1924. Educ.: Prague Tech. Univ. Tech. Director, Inst. for Res. Production and Uses of Radioisotopes, Czechoslovak A.E.C. Society: Czechoslovak Chem. Soc., Nucl. Chem. Sect.
Nuclear interests: Radioisotope applications in science and industry; Analytical chemistry of nuclear materials; Radioanalytical methods.
Address: 39 Velvarská, Prague 6, Czechoslovakia.

DEE, P. I., Prof., M.A., F.R.S. Prof. Natural Philosophy, Glasgow Univ.; Member, Nucl. Safety Advisory Com., Ministry of Power.
Address: Department of Natural Philosophy, Glasgow University, Glasgow W.2, Scotland.

DEE, Robert W. R. Doctoral degree in phys. and maths. Born 1923. Director, R.C.N.; Sec. Gen., Netherlands Atoomforum; Member, Exec. Com., Forum Atomique Européen; Member, Nucl. Public Relations Contact Group. Nuclear interest: Management.
Address: 15 Laan van Leeuwesteyn, Voorburg, Netherlands.

DEEV, V. I. Paper: Co-author, Exptl. Study of Sodium Pool Boiling Heat Transfer (letter to the Editor, Atomnaya Energiya, vol. 22, No. 1, 1967).
Address: Academy of Sciences of the U.S.S.R., 14 Leninsky Prospekt, Moscow V-71, U.S.S.R.

DEEV, Yu S. Papers: The Uses of CdS Photoresistors in Rad. Dosimetry (Atomnaya Energiya, vol. 6, No. 4, 1959); co-author,

Methods of Preparing Low-Activity Polonium α-Ray Sources (letter to the Editor, ibid, vol. 12, No. 4, 1962); co-author, Determination of Indium by (γ,γ) - Reaction with 5 MeV Linear Electron Accelerator (letter to the Editor, ibid, vol. 24, No. 3, 1968).
Address: Academy of Sciences of the U.S.S.R., 14 Leninsky Prospekt, Moscow V-71, U.S.S.R.

DEFAZIO, Peter G. Formerly Plant Eng., now Director of Eng., Nat. Lead Co. of Ohio.
Address: National Lead Company of Ohio, P.O. Box 39158, Cincinnati, Ohio, U.S.A.

DeFELICE, Joseph Anthony, M.E., M.S. Born 1927. Educ.: Stevens Inst. Technol., and Harvard Univ. Res. Eng., North American Aviation, Inc., A.E.C., 1950-52; Chief, Nucl. Eng. Sect., U.S.A.E.C., Savannah River Operations Office, 1952-54; Manager, Preliminary Design Dept., Nucl. Development Corp. of America, 1954-60; Pres., Nucl. Technol. Corp., 1960-. Chairman, New York Metropolitan Sect., A.N.S.
Nuclear interests: Reactor design; Reactor physics; Radiation shielding; Hazards evaluation; Economics of nuclear reactors; Management of nuclear activities.
Address: Nuclear Technology Corp., 55 Church Street, White Plains, New York, U.S.A.

DEFRENNE, Guy Amand Théodore Charles, Ing. Commercial (Bruxelles). Born 1912. Educ.: Brussels Univ. Directeur de la Division industrielle, S.A. Philips, Brussels. Societies: Sté. belge de Radiologie; Sté. belge de Physique; membre fondateur de l'Organisation Internationale pour la Science et la Technique du Vide.
Nuclear interests: Accélérateurs; Réacteurs; Mesures de Radiations; Isotopes; Data processing and computing; Direction et Organisation.
Address: 18 avenue du Chili, Brussels, Belgium.

DEGAIN, R. Asst. Managing Director, Progil. Rep. for Progil, Conseil d'Administration, A.T.E.N.
Address: Progil, 77-79 rue de Miromesnil, Paris 8, France.

DEGANI, Meir Hershtenkorn, B.S., M.Sc., Sc.D. Born 1909. Educ.: M.I.T. Asst. Prof. Geophysics, Pennsylvania State Univ. 1942; Chairman, Sci. Dept., New York State Univ., Maritime Coll., Fort Schuyler, 1946-. Societies: A.P.S.; American Meteorological Soc.; A.G.U.
Nuclear interests: In charge of a department that offers a major in Nuclear Science with particular emphasis on the application of nuclear energy to the Merchant Marine.
Address: State University of New York, Maritime College, Fort Schuyler, New York 65, New York, U.S.A.

DEGEILH, André, Dr. ès-Sci. Born 1929. Educ.: Toulouse Univ. Maitre de Conférences à la Faculté des Sci.

Nuclear interests: Nuclear physics; Ion Sources (in particular H.F. ion source); Plasmas; Accelerators.
Address: Laboratoire d'Optique Electronique du C.N.R.S., 29 rue Jeanne Marvig, Toulouse, France.

DEGEN, Joseph. Member, Isotope Com., Massachusetts Gen. Hospital.
Address: Massachusetts General Hospital, Boston 14, Massachusetts, U.S.A.

DEGLAIRE, Philippe. Directeur Gen. Adjoint, Sté. Parisienne pour l'Industrie Electrique.
Address: Societe Parisienne pour l'Industrie Electrique, Departement Entreprises, 85 boulevard Haussman, Paris, France.

DEGREMONT, G. Prés. Directeur Gén., Degremont S.A.
Address: Degremont S.A., 183 Route de St. Cloud, Rueil Malmaison, (Seine-et-Oise), France.

DEGTYAREV, S. F. Papers: Co-author, Transmission of Fast Neutrons through Cylindrical Duct Partially Penetrating Slab Shield (letter to the Editor, Atomnaya Energiya, vol. 24, No. 1, 1968); co-author, Space, Energy and Angular Distributions of Fast Neutrons in Lithium Hydride, Water, Tungsten and Boron Carbide (ibid, vol. 24, No. 4, 1968); co-author, Exponential Angular Neutron Co-efficients for Nonhydrogen Media (ibid, vol. 24, No. 4, 1968); co-author, Space-Energy and Angular Distributions of Neutrons in Lithium (ibid, vol. 24, No. 4, 1968).
Address: Academy of Sciences of the U.S.S.R., 14 Leninsky Prospekt, Moscow V-71, U.S.S.R.

DEGTYAREV, Yu. G. Papers: Co-author, Inelastic Interaction Cross-section Measurements of 13-20 MeV Neutrons for Some Isotopes (letter to the Editor, Atomnaya Energiya, vol. 11, No. 4, 1961); Nonelastic Neutron Cross Sections for Li^7, C^{12}, N^{14}, Al^{27}, Fe^{56}, Cu, Pb, U^{235}, U^{238} and Pu^{239} (letter to the Editor, ibid, vol. 19, No. 5, 1965); co-author, Excitation of Al^{27}, Cr^{52}, Fe^{56} and Bi^{209} Low Levels at Inelastic Scattering of 1-4 MeV Neutrons (letter to the Editor, ibid, vol. 23, No. 6, 1967).
Address: Academy of Sciences of the U.S.S.R., 14 Leninsky Prospekt, Moscow V-71, U.S.S.R.

DEGTYAREVA, E. V. Paper: Co-author, Toroidal Chambers for Investigation of Plasma-RF Fields Interaction (letter to the Editor, Atomnaya Energiya, vol.23, No. 1, 1967).
Address: Academy of Sciences of the U.S.S.R., 14 Leninsky Prospekt, Moscow V-71, U.S.S.R.

DeHART, Robert C., B.S., M.S., Ph.D. Born 1917. Educ.: Wyoming Univ. and Illinois Inst. Technol. Assoc. Prof., Montana State Coll., 1946-53; Structural Analyst, Armed Forces Special Weapons Project, Dept. of Defence, 1953-58; Director, Structural Res., Southwest Res. Inst., 1958-. Societies: A.S.C.E.; N.Y. Acad. Sci.
Nuclear interest: Effects of nuclear weapons.
Address: 7707 Broadway, San Antonio, Texas, U.S.A.

DEHLER, Thomas, Dr. jur. et. rer. Born 1897. Educ.: Munich, Freiburg and Würzburg Univs. Bundesminister der Justiz, 1949-53. Landesvorsitzender der FDP in Bayern, Mitglied der Bayerischen Verfassunggebenden Landesversammlung, 1946-56; Mitglied des Deutschen Bundestages, 1949-; Vorsitzender der bundestagsfraktion der FDP, 1953-57; Bundesvorsitzender der Freien Demokratischen Partei; Vorsitzender des Bundestagsausschusses für Atomkernenergie und Wasserwirtschaft, 1957-61. Rechtsanwalt. Vizepräsident des Deutschen Bundestages, 1960-. Chairman, Law and Administration Com., Deutsches Atomforum E.V. Society: Gesellschaft für Atomenergie e.V.
Address: 6 Schleichstrasse, Bonn, Germany.

DEHMELT, Hans G. Prof. Phys. Dept., Washington Univ.
Nuclear interest: Radiofrequency spectroscopy.
Address: Washington University, Physics Department, Seattle, Washington 98105, U.S.A.

DEHNE, Edward James, Colonel, B.S., M.D., M.P.H. (Master Public Health), Dr. P.H. (Doctor Public Health). Born 1911. Educ.: North Dakota State Coll., N. Dakota, Oregon and Johns Hopkins Univs. Director, U.S. Army Environmental Health Lab., 1955-60; Consultant in Occupational Medicine to Surgeon-Gen., U.S. Army, 1957; Consultant in Occupational Medicine and Preventive Medicine to Surgeon-Gen., U.S. Army, 1960; Prof. Preventive Medicine, U.S. Army Medical Field Service School, Fort Sam Houston, Texas, 1960-; Director, Public Health and Welfare Dept., Office of the High Commissioner, Ryukyu Island, Japan. Exec. Com., American Conference Governmental Ind. Hygienists; Com. on Sci. Programme, American Coll. of Preventive Medicine. Societies: Fellow, American Coll. Preventive Medicine (Charter); Fellow, Roy. Soc. of Health; Fellow, Ind. Medical Assoc. (U.S.A.); Fellow, A.P.H.A.;A.M.A.; American Assoc. Military Surgeons; American Conference of Government Ind. Hygienists (Exec.).
Nuclear interests: As Director, U.S. Army Environmental Health Laboratory provided inspection and consultive services for the Army's Radiological Hygiene Programme to minimize human exposure through systems for: maintaining current knowledge of users and uses of radiological material, evaluating qualifications and facilities of applicants for radiation sources and radiomaterials, conduct of periodic surveys to insure that safe operating procedures and adequate controls are used, providing monitoring and dosimetry capabilities, providing radiation physics consultation.

Address: Public Health and Welfare Department, Office of the High Commissioner, Ryukyu Islands, Japan.

DEHON, S. Lab.-work; nucl. phys., State Tech. Coll. for Nucl. Industries, Board of Nat. Education and Culture.
Address: Institut Technique Superieur de l'Etat pour les Industries Nucléaires, Ministère de l'Education Nationale et de la Culture, 150-52 rue Royale, Brussels, Belgium.

DEHTYAR, I. Ya. Paper: Co-author, Change of Resistivity by Annealing of Nickel Irradiated with Alpha-Particles (Atomnaya Energiya, vol. 21, No. 6, 1966).
Address: Academy of Sciences of the U.S.S.R., 14 Leninsky Prospekt, Moscow V-71, U.S.S.R.

DeHUFF, Philip G., B.S. (Metal. Eng.). Born 1918. Educ.: Lehigh Univ. Sect. Eng. and Eng. Manager, Westinghouse Elec. Corp., Steam and Aviation Gas Turbine Div., 1942-55; Asst. Project Manager, PWR Project, Westinghouse Atomic Power Div., Bettis, 1955-59; Project Manager, CVTR Project, Westinghouse Atomic Power Dept., 1960-62; Manager, Reactor Eng. and Materials Dept., 1962-; Member, S.A.E. Aeronautical Material Specification Com.; Member, S.A.E. Subcommittees on High Temperature Alloy, Heat and Corrosion Resistant Alloy Castings, Non-Destructive Test Methods. Book: Co-author, The Shippingport Pressurized Water Reactor (Addison Wesley, 1958). Societies: Soc. of Automotive Eng.; A.S.M.
Nuclear interests: Reactor design, physics nuclear design and experimental work.
Address: 5659 Kings School Road, Bethel Park, Pennsylvania, U.S.A.

DEIGHTON, Thomas, M.A. Born 1921. Educ.: Cambridge Univ. Eng., Design and Construction Dept., State Elect. Commission of Victoria, 1947-50; Sen. Eng., London office, State Elect. Commission of Victoria, Australia, 1950-55; Sen. Technician, Vickers-Armstrongs (Eng.) Ltd., Barrow-in-Furness, (attached to Naval Section, Harwell), 1955-58; Sen. Project Eng., Nucl. Energy Dept., Babcock and Wilcox, 1958-. Societies: A.M.I.Mech.E.; B.N.E.C.; A.N.S.
Nuclear interests: Reactor engineering concerned with small and medium-sized power projects suitable for marine or land-based units, with particular respect to water-cooled and moderated reactor systems.
Address: 209 Euston Road, London, N.W.1, England.

DEIKE, G. H. Jr. Sec., MSA Res. Corp.
Address: MSA Research Corporation, 201 North Braddock Avenue, Pittsburgh, Pennsylvania, U.S.A.

DEILY, George J., B.S. (Phys.). Born 1926. Educ.: Rensselaer Polytech. Inst. Phys. (1951-56), Res. Supervisor (1956-), E.I. du Pont de Nemours and Co. Society: A.N.S.
Nuclear interests: Hot laboratory operations and management; Radiation damage; Radioisotopes for power generation.
Address: E. I. du Pont de Nemours and Co., Savannah River Laboratory, Aiken, S. Carolina, U.S.A.

DEISS, William P., Jr., M.D. Born 1923. Educ.: Illinois Univ. Coll. of Medicine. Assoc. Prof. Medicine, Duke Univ.; Chief, Medical Service, Durham (N.C.) V.A. Hospital, 1956-58; Assoc. Prof. Medicine (1958-61), Prof. Medicine and Biochem. (1961-), Indiana Univ. School of Medicine;Prof. and Chairman Medicine, Texas Univ. Medical Branch, Texas, 1967-. Societies: Fellow, American Coll. of Physicians; American Soc. for Clinical Investigation; Southern Soc. for Clinical Research; Endocrine Soc.; Assoc. American Physicians.
Nuclear interest: Medical research.
Address: Texas University Medical Branch, Galveston, Texas 77550, U.S.A.

DEISSLER, Robert George, B.Sc., M.S. Born 1921. Educ.: Carnegie Tech. and Case Inst. Technol. Chief, Fundamental Heat Transfer Branch, Lewis Res. Centre, N.A.S.A. Societies: A.S.M.E.; A.P.S.; A.I.A.A.; Soc. for Natural Philosophy; A.A.A.S.
Nuclear interests: Heat transfer and fluid flow; Turbulence phenomena.
Address: Lewis Research Center, N.A.S.A., 21,000 Brookpart Road, Cleveland, Ohio, U.S.A.

DEJONGHE Paul André Jozef, Dr. Appl. Sci., Chem. Agricultural Eng. Born 1928. Educ.: Ghent Univ. Asst., Ghent Univ., 1949-55; Head of the Waste Disposal Service, C.E.N. Mol, Belgium, 1955-61; Head, Waste Disposal Div., Sté. Belge de Chimie Nucléaire, Mol, 1961-64, Head, Div. Appl. Res., C.E.N., Mol, 1964-. Societies: Vlaamse Chemische Vereniging; Internat. Soil Sci. Soc.; H.P.S.
Nuclear interests: Radioactive waste disposal; Nuclear fuel cycle; Isotopes application; Research project management.
Address: C.E.N., Villa 11, Mol, Belgium.

DEJOU, Alexis, Ing. Ecole Polytechnique and Ecole Supérieure d'Electricité. Born 1920. Director des Etudes et Recherches, Electricité de France. Chairman, Sté. Française des Electriciens; Member, Commission Consultative pour la Production d'Electricite d'Origine Nucleaire, C.E.A.
Nuclear interest: Production of electricity.
Address: Electricité de France, Direction des Etudes et Recherches, 12 place des Etats – Unis, Paris 16, France.

DEKEYSER, Willy Clément, Dr. Phys. Born 1910. Educ.: Ghent Univ., Prof. Crystallography, Ghent Univ., 1945-. Member, Sci. and Tech. Com., Euratom. Societies: Phys. Soc.; Sté. Belge de Physique; Sté. Française de Cristallographie et Minéralogie.

Nuclear interest: Solid state aspects.
Address: 57 Rijssenbergstraat, Ghent, Belgium.

DEKONINCK, André. Sen. Officer, Tech. Dept., Crépelle et Cie.
Address: Crépelle et Cie., Porte de Valenciennes, Lille, (Nord), France.

del CAMPO, Armando LOPEZ M. See LOPEZ M. del CAMPO, Armando.

DEL PONT, Alberto MARCO. See MARCO DEL PONT, Alberto.

del POZO, Florencio. Agrupacion de Trabajo C.A.L. Oficinas Tecnicas, (Consulting Architects and Engineers).
Address: Agrupacion de Trabajo C.A.L. Oficinas Tecnicas, 32 Avda. Jose Antonio, Madrid 13, Spain.

DEL PRADO, Enrique BAONZA. See BAONZA DEL PRADO, Enrique.

DEL REGATO, Juan A., M.D. Born 1901. Educ.: Paris Univ. Clinical Prof., Radiol., Medical School, Colorado Univ. Member, Nat. Advisory Cancer Council; Pres., American Radium Soc., 1968-69; Consultant, Puerto Rico Nucl. Centre, U.S.A.E.C.; Sec., American Soc. of Therapeutic Radiol.; U.S. Consular, Interamerican Coll. of Radiol. Book: Co-author, Cancer Diagnoses, Treatment and Prognoses (C.V. Mosby, 1964); Editor, Conference, Radiobiol. and Radiotherapy (Monograph No. 24). (N.C. Inst. 1957). Societies: A.C.R.; Radiol. Soc. of North America; American Roentgen-Ray Soc.; Rad. Res. Soc.
Nuclear interests: Therapeutic radiobiology; Radiation physics; Therapeutic isotopology.
Address: 2215 North Cascade Avenue, Colorado Springs, Colorado 80907, U.S.A.

DEL RIO, Carlos SANCHEZ. See SANCHEZ DEL RIO, Carlos.

del RIO, Jose Francisco SAIZ. See SAIZ del RIO, Jose Francisco.

DEL ROSARIO, Casimiro, M.S. (Yale), Ph.D. (Pennsylvania). Born 1896. Educ.: Philippines, Yale and Pennsylvania Univs. Director, Weather Bureau, 1945-58; Pres., R.A.V., and ex-officio Member, Exec. Com., W.M.O., 1954-57; Vice-Chairman, Nat. Science Development Board, 1958-; Chairman, Div. of Math., and Chairman, Nat. I.G.Y. and I.G.C. Coms. Societies: A.A.A.S.; A.P.S.; Philippines Phys. Soc.
Nuclear interests: Nuclear physics; Reactor design.
Address: National Science Development Board, Manila, The Philippines.

DEL VAL COB, Manuel, Dr. en Ciencias Quimicas (Ph.D. Chem.). Born 1932. Educ.: Faculty of Sci., Madrid Univ. Head, Isotope Sect., Junta de Energia Nucl. (Nucl. Energy Board of Spain).
Nuclear interests: Isotopes (production, distribution, legislation, transport and applications).
Address: Sección de Isótopos, Junta de Energia Nuclear, Madrid, Spain.

del VALLE, Martin Gonzalez. Director, Tecnicas Nucleares S.A.
Address: Tecnicas Nucleares S.A., 46 Serrano, Madrid 1, Spain.

DEL VALLE S., Bernardo, Dr. en Medicina. Member, Inst. Nacional de Energia Nuclear, Guatemala.
Address: Instituto Nacional de Energia Nuclear, Apartado Postal 1421, Guatemala, C.A.

del VISO, Rafael PEREZ. See PEREZ del VISO, Rafael.

DELACARTE, Louis. Directeur Commercial, S.N.C.F. Member, Specialised Nucl. Sect. for Economic Problems, and Member, Specialised Nucl. Sect. for Social Health and Development Problems, Economic and Social Com., Euratom, 1966-.
Address: S.N.C.F., 54 boulevard Hausmann, 75 Paris 9, France.

DELANEY, Cyril F. G., B.A., Ph.D. Born 1925. Educ.: Dublin Univ. A.E.R.E., Harwell, 1953-55; Prof. Exptl. Phys., Trinity Coll., 1966-. Society: Roy. Irish Acad.
Nuclear interests: Instrumentation in general; Fundamental properties of scintillation counters and secondary electron statistics.
Address: Physical Laboratory, Trinity College, Dublin University, Dublin 2, Ireland.

DELANCE, Maurice, Ing. E.N.S. Electrochimie Electrométallurgie Grenoble and L. es Sc. Born 1928. Educ.: Grenoble Univ. Development and Production Eng. Uranium Refining Plant, Le Bouchet, (1953-61), Chef du Service de Fabrication d'Uranium (1962), Chef des Services de Production du Centre de Traitement de Combustibles Irradiés de la Hague (1964-), C.E.A. Deleg. to 2nd I.C.P.U.A.E., Geneva, 1958 and 3rd I.C.P.U.A.E. 1964. Society: Sté. Chimie Industrielle.
Nuclear interests: Uranium and uranium dioxide production; Extraction and metallurgy of nuclear fuels; Reprocessing of irradiated fuels; Plutonium production.
Address: 90 rue Emmanuel Liais, 50 - Cherbourg, France.

DELAPALME, Bernard. Born 1923. Educ.: Ecole Polytechnique, Paris. Gen. Manager, Sté. Rhône-Alpes. Society: Sté. de Chimie Industrielle.
Nuclear interests: Management; Economics.
Address: Elf-Rhone-Alpes, 12 rue Jean Nicot, Paris 7, France.

DELARUE, Roger André, Ing. Ecole Polytechnique, Paris. Born 1906. Educ.: Paris Univ. Ing. en Chef, Sté. C.E.C. Societies: A.T.E.N.;

Sté. de Radioprotection; Sté. de Chimie Industrielle.
Nuclear interests: Photoluminescent dosimeters, waste treatment, waste storage; Decontamination; Fission products vitrification; CO_2 production and purification; Ceramics for thermal neutrons protection.
Address: Société C.E.C., B.P. 60, Montrouge, (Seine), France.

DELAURA, Emil, Directeur, Budget Div., Chicago Operations Office, U.S.A.E.C.
Address: U.S. Atomic Energy Commission, Chicago Operations Office, 9800 South Cass Avenue, Argonne, Illinois 60439, U.S.A.

DELAVA, R. Head, Nucl. Applications Dept., Atomic Energy Div., Ateliers de Constructions Electriques de Charleroi.
Address: Ateliers de Constructions Electriques de Charleroi, avenue E. Rousseau, Charleroi, Belgium.

DELAY, A. Member, Conseil d'Administration, Energie Nucléaire S.A.
Address: Energie Nucléaire S.A., 10 avenue de la Gare, Lausanne, Switzerland.

DELBIANCO, Walter, Dottore in Phisica, Ph.D. Born 1933. Educ.: Rome and Pennsylvania Univs. Res. Assoc., Pennsylvania Univ., 1961; Res. Sci., Max Planck Inst., Mainz, 1962; Asst. Prof. (1965), Assoc. Prof. (1967-), Montreal Univ. Societies: Canadian Assoc. of Phys.; A.P.S.
Nuclear interest: Photonuclear reactions.
Address: Physics Department, Montreal University, Montreal, Quebec, Canada.

DELCROIX, Jean, Eng. E.C.P. Born 1921. Manager, Persan Plant, (1947-57); Tech. Manager, Montbard Plant (1957-62), Asst. Gen. Manager (1962-), Cie. du Filage des Métaux et des Joints Curty.
Nuclear interests: Bars, shapes, tubes, and extrudes in stainless steels, refractory alloys, zirconium alloys.
Address: Cie. du Filage des Métaux et des Joints Curty, 30 avenue de Messine, Paris 8, France.

DELCROIX, Jean-Loup, D. ès Sc. Born 1924. Educ.: Ecole Normale Supérieure and Paris Univ. Sous-Directeur, Lab. des Hautes Energies, Orsay, 1957-59; Prof., I.N.S.T.N., 1957; Maitre de Conférences de Physique des Plasmas, Faculté des Sci. de Paris, 1960; Prof. de Physique des Plasmas, Faculte des Sci. de Paris, 1963. Books: Introduction à la théorie des gaz ionisés (Paris, Dunod, 1959); Théorie des ondes dans les plasmas (Paris, Dunod, 1961); Physique des Plasmas I (Paris, Dunod, 1963) Physique des Plasmas II (Paris, Dunod, 1966). Societies: Sté. française de physique; A.P.S.
Nuclear interests: Accélérateurs; Plasma physics.
Address: Faculté des Sciences, Orsay, Seine et Oise, France.

DELEVAUX, Maryse H. Chem., Geological Survey, U. S. Dept. of the Interior.
Address: U.S. Department of the Interior, Geological Survey, Building 25, Federal Centre, Denver, Colorado, U.S.A.

DELFINO, Riccardo PARODI-. See PARODI-DELFINO, Riccardo.

DELFOSSE, J. M., Prof., D.Sc. Head, Mass Spectroscopy Dept., Louvain Univ.
Address: Centre de Physique Nucléaire Université de Louvain, Parc d'Arenberg, avenue Cardinal Mercier, Heverle, Louvain, Belgium.

DELGADO, Jorge Ospina, Dr. Principal Member, Instituto de Asuntos Nucleares.
Address: Instituto de Asuntos Nucleares, Apartado Aereo 8595, Bogota D. E., Colombia.

DELGADO CALVETE, Alfredo. Vocale, Comision Interministerial de Conservacion de Alimentos por Irradiacion, Junta de Energia Nucl.
Address: Junta de Energia Nuclear, Ciudad Universitaria, Madrid 3, Spain.

DELHOVE. Director, Controlatom, Assoc. Vincotte.
Address: Association Vincotte, Controlatom, 27-29 avenue A. Drouart, Brussels 16, Belgium.

DELL, J. Maxey, Jr., M.D. Member, Florida Nucl. Commission.
Address: Florida Nuclear Commission, Room 113, 725 South Bronough Street, Tallahassee, Florida 32304, U.S.A.

della ROCCA, Carlo, M.Ch.E. Born 1934. Educ.: Naples Univ. Staff Eng., Heat Transfer Dept., Atomic Power Div., English Elec. Co. Ltd., 1959-61; Head, Chem. Group, Garigliano Power Station (1961-), then responsible for Reactor Chem. Design Group, Ente Nazionale per l'Energia Elettrica; Asst. to Prof. of Electrochem., Naples Univ.
Nuclear interest: Chemistry and radiochemistry applied to the operation of nuclear reactors. This includes liquid and gaseous waste processing and disposal.
Address: C.P.N., Ente Nazionale per l'Energia Elettrica, 6 Via Torino, Rome, Italy.

DELLA VALLE, Robert. Directeur des Travaux, Sté. d'Entreprise de Montages.
Address: Société d'Entreprise de Montages, 22 rue Ampere, 38 Grenoble (Isere), France.

DELL'AGLIO, M. S. Military Electronics Division, Daystrom Inc.
Address: Daystrom Inc., Military Electronics Division, Archbald, Pennsylvania 18403, U.S.A.

DELLER, D. J., Dr., Prof. With Dept. of Medicine, Adelaide Univ.
Nuclear interest: Metal chelates and man (Fe, Cu, Ca).
Address: Adelaide University, Adelaide, South Australia, Australia.

DELLO SIESTO, Edoardo. Nucl. Eng., Bombrini Parodi-Delfino S.p.A.
Address: Bombrini Parodi-Delfino S.p.A., 31 Via Lombardia, Rome, Italy.

DELMAS, M. Chef de la Sect. des Combustibles Ceramiques, C.E.N., Grenoble, C.E.A.
Address: Section des Combustibles Ceramiques, C.E.N. Grenoble, B.P. No. 269, Grenoble (Isere), France.

DELMAS, M. J. Prés. Directeur Gén., Département Nucléaire, Centre d'Etudes de Prevention S.A.
Address: Centre d'Etudes de Prevention S.A., 34 rue Rennequin, Paris 17, France.

DELONEY, H. L. Formerly Asst. Chief Eng., then Chief Eng., now Vice Pres. Eng., Louisiana Power and Light Co.
Address: Louisiana Power and Light Co., 142 Delaronde Street, New Orleans, Louisiana 70114, U.S.A.

DELOURME, Alfred. Sec. Nat., Fédération Gén. du Travail de Belgique. Member for Belgium, Specialised Nucl. Sect. for Social Health and Development Problems, Economic and Social Com., Euratom, 1966-.
Address: Fédération Générale du Travail de Belgique, 42 rue Haute, Brussels, Belgium.

DELPECH, Jean-Laurens. At Batignolles-Chatillon (Mécanique Générale) S.A.
Address: Batignolles-Chatillon (Mécanique Générale) S.A., 5 rue de Monttessuy, Paris 7, France.

DELPLA, Jean. Tech. Manager, Cie. pour l'Etude d'Equipements Industriels S.A.
Address: Cie. pour l'Etude d'Equipements Industriels S.A., 12 rue de la Pierre Levée, Paris, France.

DELPLA, Maurice Joseph Albert, Diplôme d'études Supérieures de Sciences physiques. Dr. en médecine, diplômé d'Electro-Radiologie. Born 1912. Educ.: Montpellier and Toulouse Univs. Prof. Sci. Phys., Lycées, 1937-55; Maitre de conférence, Faculté de Médecine, Nantes 1955-57; Chef du Service Général de Radioprotection, Electricité de France, Paris, 1957-. Societies: Sté. Francaise d'Electroradiologie; Sté. de radioprotection.
Nuclear interests: Reactor safety; Health physics; Radiobiology; Radiopathology.
Address: 73 boulevard Haussmann, Paris 8, France.

DELSEMME, Paul, Administrateur Directeur de l'Electrorail.
Address: 16 rue du Congres, Brussels, Belgium.

DELVAUX, Armand. Administrateur-Directeur, Gradel S.A. Sté. Luxembourgeoise pour l'Industrie Nucleaire.
Address: Gradel S.A. Societe Luxembourgeoise pour l'Industrie Nucleaire, Steinfort, Luxembourg.

DELVAUX, Arthur. Administrateur-Delegue, Gradel S.A. Sté. Luxembourgeoise pour l'Industrie Nucleaire.
Address: Gradel S.A. Societe Luxembourgeoise pour l'Industrie Nucleaire, Steinfort, Luxembourg.

DELVENNE, L. Radiochem., Inst. for Nucl. Phys. Res., Netherlands.
Address: Institute for Nuclear Physics Research, 18 Oosterringdijk, Amsterdam 0, Netherlands.

DELVILLE, Pierre Lucien Marie, Civil Mining Eng. Born 1905. Educ.: Louvain Univ. Director Gen., Sté. Evence Coppée et Cie; Chairman, Centre Coal Assoc., 1949; Chairman, Belgian Coal Federation, 1958; Member, Consulting Com. to High Authority of European Coal and Steel Community, 1953. Administrateur, Centre d'Etude pour les Applications de l'Energie Nucléaire; Administrateur, Sté. Belge pour l'Industrie Nucléaire; Formerly Administrateur, Groupement Professionnel de l'Industrie Nucléaire; Formerly Administrateur, Fondation Nucléaire; Membre, Comité de Direction du Syndicat d'Etude de l'Energie Nucléaire; Membre, Comité de Direction di Syndicat Uraninga.
Address: 145 avenue Fond'Roy, Uccle, Brussels 18, Belgium.

DELZELL, T. W. Chairman of the Board, Portland Gen. Elec. Co.
Address: Portland General Electric Co., Electric Building, Portland 5, Oregon, U.S.A.

DeMARCUS, Wendell C., B.S., M.S., Ph.D. Born 1924. Educ.: Kentucky Univ. Principal Phys., Union Carbide Nucl. Corp., -57; Assoc. Prof. (1957), Prof. Phys. (1958-), Kentucky Univ. Rep. on Council for Univ. of Kentucky, Oak Ridge Assoc. Universities. Societies: Fellow, Roy. Astronomical Soc. London; A.P.S.; American Astronomical Soc.
Nuclear interest: Neutron physics.
Address: Department of Physics, Kentucky University, Lexington, Kentucky, U.S.A.

DEMARET, Ernest Fritz Paul, Ing. civil des Mines, A.I.Lg. Born 1904. Educ.: Liège Univ. Directeur-Général, Metallurgie et Mécanique Nucléaires S.A., 1959-.
Nuclear interests: Management; Fuel elements production.
Address: Metallurgie et Mécanique Nucléaires S.A., (12 Europalaan, Dessel, Belgium.

DEMBINSKI, Wojciech Stefan, Magister of Chem. Born 1934. Educ.: Nicholas Copernicus Univ. Inst. of Nucl. Res., Warsaw, 1955-. Society: Polish Chem. Soc.
Nuclear interest: Technology of reactor fuels.
Address: Piekna 31/37, m.3, Warsaw, Poland.

DEME, Sándor, Dipl. Elec. Eng. Born 1936. Educ.: Moscow Inst. of Energetics. Sen. Researcher, and Group Leader, Neutron Dosimetry, Central Res. Inst. for Phys., Budapest, 1960-. Societies: Eötvös Loránd Phys. Soc., Budapest; Internat. Rad. Protection Assoc.
Nuclear interest: Use of semiconductor detectors for radiation dosimetry.
Address: Central Research Institute for Physics, P.O.B. 49, Budapest 114, Hungary.

DEMERS, Alfred Ernest Pierre, B.A., L. ès Sc. (Phys.). L. ès Sc. (Maths.), M. ès Sc. (Montreal), D. ès Sc. (Paris), Agrégé Univ. de France. Born 1914. Educ.: Montreal, Cornell and Paris Univs. Prof. titulaire Phys., Montreal Univ.; Lecturer, Milan Univ., 1950. Member, Conseil Arts Quebec; Pres., Commission Recherche Sci.; Pres.-fondateur, Cercle Canado-Russe, Monetreal. Books: Ionographie (1958), Photographie Corpusculaire II (1959), Photographie Corpusculaire III (1964) (Presses Universitaires de Montreal). Societies: A.P.S.; Assoc. Canadienne Phys.; Stà. Italiana Fisica; Sté. Roy. Canada; A.C.F.A.S.
Nuclear interests: Ionographie les emulsions nucléaires; Fission; Low energy reactions.
Address: Université de Montréal, Departement de Physique, C.P.6128, Montreal, P.Q., Canada.

DEMEUR, Marcel, Prof. Formerly Chef du Département Physique nucléaire théorique, Labs., C.E.N.; Sen. Officer, Service de Physique Nucléaire, Brussels Univ.; Member, Commission Scientifique, Inst. Interuniv. des Sci. Nucléaires.
Address: Brussels Free University, avenue F.D. Roosevelt, Brussels, Belgium.

DEMEURE, Jacques, Administrateur délégué, Prés. du Conseil d'Administration de la Compagnie Auxiliaire d'Electricité.
Address: 8 rue de la Presse, Brussels, Belgium.

DEMICHEV, V. F. Paper: Co-author, Plasma Jet Rotation in Magnetic Fields (Atomnaya Energiya, vol. 19, No. 4, 1965).
Address: Academy of Sciences of the U.S.S.R., 14 Leninsky Prospekt, Moscow V-71, U.S.S.R.

DEMIDECKI, Andrzej J., M.Sc. Born 1933. Faculty of Phys., Warsaw Univ.; Dept. Phys., Inst. Oncology, Warsaw, 1955; Isotopes Lab., Medical High School, Warsaw, 1958-. Societies: Hospital Phys. Assoc., London; Polish Soc. Phys.
Nuclear interests: Medical physics; Nuclear medicine; Radiology; Radiobiology.
Address: 23 m 12 ul. Odolańska, Warsaw 12, Poland.

DEMIDOV, Anatoly M. Delegate, I.A.E.A. Symposium on inelastic scattering of neutrons in solids and liquids, Vienna, Oct. 1960; Delegate, I.A.E.A. Symposium on pile neutron research in physics, Vienna, Oct. 1960.
Nuclear interests: γ-ray spectra and the investigation of highly excited states.
Address: Institute of Atomic Energy, Academy of Sciences of the U.S.S.R., 14 Leninsky Prospekt, Moscow V-71, U.S.S.R.

DEMIDOV, B. A. Paper: Co-author, Plasma Turbulence Evaluation According to Electromagnetic Rad. and Ranan Scattering in Microwave Band (letter to the Editor, Atomnaya Energiya, vol. 20, No. 6, 1966).
Address: Academy of Sciences of the U.S.S.R., 14 Leninsky Prospekt, Moscow V-71, U.S.S.R.

DEMIDOV, P. I. Paper: Co-author, Effect of Vanadium on Phase Constitution and Structure of High Boron Steel (Abstract, Atomnaya Energiya, vol. 20, No. 2, 1966).
Address: Academy of Sciences of the U.S.S.R., 14 Leninsky Prospekt, Moscow V-71, U.S.S.R.

DEMIDOVA, P. G. Paper: Co-author, Uranium Nucleus Fission Fragment Tracks Determination of Age of Muscovites (letter to the Editor, Atomnaya Energiya, vol. 21, No. 4, 1966).
Address: Academy of Sciences of the U.S.S.R., 14 Leninsky Prospekt, Moscow V-71, U.S.S.R.

DEMIDOVICH, V. N. Paper: Co-author, Ionisation-Mechanic Detector of Ionisation Particles (letter to the Editor, Atomnaya Energiya, vol. 20, No. 5, 1966).
Address: Academy of Sciences of the U.S.S.R., 14 Leninsky Prospekt, Moscow V-71, U.S.S.R.

DEMIERBE, Edgard Floris, Ing. Civil Electromécanicien A.I.Br. Born 1904. Educ.: Brussels Univ. Managing Director, Atomic Energy Div., Ateliers de Constructions Electriques, Charleroi. Prof. Elec. Equipment Design and Construction, Faculty of Appl. Sci., Brussels Univ.; Administrateur, Groupement Professionel de l'Industrie Nucléaire; Director, Sté. Belge pour l'Industrie Nucléaire; Administrateur, Fondation Nucléaire.
Address: 89 avenue Mascaux, Marcinelle, Belgium.

DEMIN, A. G. Papers: Co-author, Synthesis and Determination of Radioactive Properties of Some Isotopes of Fermium (Atomnaya Energiya, vol. 21, No. 4, 1966); co-author, On Nucl. Properties of Isotopes of Element 102 with Mass Numbers 255 and 256 (letter to the Editor, ibid, vol. 22, No. 2, 1967).
Address: Academy of Sciences of the U.S.S.R., 14 Leninsky Prospekt, Moscow V-71, U.S.S.R.

DEMIN, V. P. Paper: Co-author, Secondary Gamma-Rays Coefficients for Aluminium Copper and Tungsten (letter to the Editor, Atomnaya Energiya, vol. 19, No. 5, 1965).

Address: Academy of Sciences of the U.S.S.R.,
14 Leninsky Prospekt, Moscow V-71, U.S.S.R.

DEMING, Robert L., B.S. (Elec. Eng.), S.M.
(Elec. Eng.). Born 1921. Educ.: New Hampshire Univ. and M.I.T. Div. Manager, Manager, Eng. and Development, Tracerlab, Richmond, Calif.; Manager, Nucleonic Systems Lab., Gen. Dynamics/Electronics, Div. of Gen. Dynamics; Head, Nucleonics Res. Sect., Stromberg-Carlson, a div. of Gen. Dynamics Corp., Rochester, New York; Sen. Res. Member, Los Alamos Sci. Lab.; Asst. Prof. Elec. Eng., U.S. Naval Post-graduate School; Teaching and Res. Asst., M.I.T.; Radar and C.I.C. Officer, U.S. Navy. Com. on Nucl. Instrumentation, Atomic Ind. Forum; Com. on Stands and Nucleonics, A.I.E.E.E. Societies: H.P.S.; A.N.S.
Nuclear interests: Radiation monitoring; Industrial nucleonics instrumentation; Reactor control; Space radiation detection; Nuclear activities management.
Address: Tracerlab, a Division of Laboratory for Electronics, 2030 Wright Avenue, Richmond, California, U.S.A.

DEMIRKHANOV, R. A., D.Sc., Acad. of Sciences of the Georgian S.S.R.; Deleg. to 2nd I.C.P.U.A.E., Geneva, Sept. 1958; Deleg. to Conference sur la Physique des Plasmas et la Recherche concernant la Fusion Nucléaire Controlée, Salzbourg, Sept. 1961. Papers: The Mass of He³ (letter to the Editor, Atomnaya Energiya, vol. 2, No. 5, 1957); Atomic Masses of the Isotopes C13, N14, N15 (ibid, vol. 2, No. 6, 1957).
Address: Academy of Sciences of the U.S.S.R., 14 Leninsky Prospekt, Moscow V-71, U.S.S.R.

DEMOPOULOS, Brigadier General Panayiotis, M.E. Born 1910. Educ.: Military Acad. and Athens Tech. Univ. Brigadier Gen., 1955; Director-Advisor, Tech. Matters, British Petroleum Greece, 1957-67; Pres., Greek Atomic Energy Commission, 1967-.
Address: Greek Atomic Energy Commission, Aghia Paraskevi, Attikis, Athens, Greece.

DEMOS, Peter Theodore, Ph.D. Director, Lab. for Nucl. Sci., M.I.T. Member, Subcom. on Nucl. Structure, Com. on Nucl. Sci., N.A.S.-N.R.C.
Address: Massachusetts Institute of Technology, Cambridge 39, Massachusetts, U.S.A.

DEMOULLIN, Mathias, physician, doctor of medicine, radiologist. Born 1917. Educ.: Giessen, Vienna, Strasbourg and Paris Univs. Member, Conseil National de l'Energie Nucléaire. Societies: Sté. de Radiologie de l'Est – France; Deutsche Roentgengesellschaft.
Nuclear interests: Nuclear physics and X-ray treatment.
Address: Hôpital de la Ville, Esch-sur-Alzette, Luxembourg.

DEMPSEY, H. Eugene, B.Sc., A.C.T. (Birm.), A.Inst.P. Born 1928. Educ.: Coll. of Technol., Birmingham. Serck Radiators Ltd., 1952-54; Coll. of Technol., Wolverhampton, 1954-56; Canadair Ltd., 1956-59; Canadian Westinghouse Co. Ltd., 1959-60; Computing Devices of Canada Ltd., 1961-62: Staff Sci., Atomic Products Dept., Hawker Siddeley Eng., 1962-66; Staff Sci., Decca Radar Canada Ltd., 1966-. Societies: Inst. Phys. and Phys. Soc.; B.N.E.S.
Nuclear interests: Reactor design and technology, physics, heat transfer, fluid flow; Economics of large and small power reactors; Design of marine reactors.
Address: Decca Radar Canada (1967) Ltd., 23 Six Point Road, Toronto 18, Ontario, Canada.

DEMPSEY, John M., Jr., A.B., M.A. (Mathematical Phys.). Born 1928. Educ.: Boston Coll. Exec. Vice Pres., Baird-Atomic, Inc., 1967-. Director and Sec., Assoc. of Nucl. Instrument Manufacturers. Society: Optical Soc. of America.
Nuclear interests: Management of a nuclear instrument corporation primarily in the field of clinical and diagnostic medicine,health physics and biochemical research.
Address: Baird-Atomic, Incorporated, 33 University Road, Cambridge, Massachusetts 02138, U.S.A.

DEM'YANOV, A. V. Paper: Co-author, Investigation of Diffusion of Nucl. Reaction Recoil Products in Different Materials (Atomnaya Energiya, vol. 22, No. 4, 1967).
Address: Academy of Sciences of the U.S.S.R., 14 Leninsky Prospekt, Moscow V-71, U.S.S.R.

den BAKKER, E. K., Master of Economic Sci. Born 1919. Educ.: Netherlands School of Economics. Managing Board, Nationale-Nederlanden N. V.
Nuclear interests: Insurance problems concerning nuclear risks.
Address: 130 Schiekade, Rotterdam, Netherlands.

DEN BERG, L. C. H. VAN. See VAN DEN BERG, L. C. H.

den BOESTERD, Johannes, eng. Born 1902. Educ.: Delft Tech. Univ. Director, Provinciaal Electriciteitsbedrijf van Noord-Holland at Bloemendaal, Netherlands, 1951-; Director, N. V. Provinciale en Gemeentelijke Electriciteits Maatschappij at Amsterdam, 1951-. Board of Control, Stichting Reactor Centrum Nederland; Commissie Kernenergiecentrale N.V. S.E.P., Arnhem; Man. Com., Vereniging van Exploitanten van Electriciteitsbedrijven in Nederland; Commissie van overleg met de Vereniging van Fabrikanten op electrotechnisch gebied in Nederland (Foegin); Nederlands Comité van de Commission for Control and Regulation of Electrical Equipment; Comité d'Etudes des Câbles à Très Haute Tension (Cigré). Society: Koninklijk Instituut van Ingénieurs.

DEN BOL

Nuclear interest: Application of nuclear energy for production of electricity.
Address: c/o Provinciaal Electriciteitsbedrijf van Noord-Holland, 19 Ign. Bispincklaan, Bloemendaal, Netherlands.

DEN BOLD, Hermanus Johannes VAN. See VAN DEN BOLD, Hermanus Johannes.

DEN BOSCH, J. VAN. See VAN DEN BOSCH, J.

den BRANDE, J. van. See van den BRANDE, J.

den BRENK, Hendrik Athos Sydney van. See van den BRENK, Hendrik Athos Sydney.

den BRINK, J. N. BAKHUIZEN van. See BAKHUIZEN van den BRINK, J. N.

den HENDE, A. van. See van den HENDE, A.

den HEUVEL, Jan Adrianus van. See van den HEUVEL, Jan Adrianus.

DEN HOEK, J. VAN. See VAN DEN HOEK, J.

DENAMUR, Henri. Gen. Manager, Sté. Française d'Etudes et de Constructions Electroniques, S.A.
Address: Société Française d'Etudes et de Constructions Electroniques, S.A., 11 rue Saint-Florentin, Paris 8, France.

DENIAU, Jean-Francois, Bachelor of Laws, Master of Literature. Born 1928. French Deleg., negotiations on the Common Market and Euratom. Member, Commission of the European Communities, 1967-.
Address: Commission of the European Communities, 23 avenue de la Joyeuse Entree, Brussels 4, Belgium.

DENIELOU, Guy Pierre Marie, Eng. (Naval School), Eng. (Nucl. Eng.). Born 1923. Head, Reactor Design Group, Nucl. Res. Centre, Grenoble, 1959-64; Asst. Manager, Fast Reactor Project, Euratom-C.E.A., 1964-. Society: A.N.S.
Nuclear interests: Reactor design. Has been in charge of designing the SILOE reactor. Participated in design of OSIRIS and High Flux Grenoble reactor. Has been in charge of tests, start up and commissioning of RAPSODIE reactor.
Address:Département de Recherche Physique, Centre d'Etudes Nucléaires de Cadarache, B.P. No. 1, 13 St. Paul lez Durance, France.

DENIS, André, Ing. Ecole Centrale des Arts et Manufactures, Paris. Born 1923. Chef du Service d'Extraction du Plutonium, C.E.A. Marcoule, 1955-59; Dir., Sté. Provencale des Ateliers Terrin, 1959; Responsable de Departement Nucleaire, Terrin Technique Industrie S.A.
Nuclear interests: Transfer and handling devices, stainless steel and aluminium alloys, vessels and equipment "hot" maintenance.
Address: Sociéte Provençale des Ateliers Terrin, 287 Chemin de la Madrague, Marseille, Bouches du Rhône, France. ·

DENIS, Jean-Paul. Chef Ing. Achats, U.S.S.I.
Address: U.S.S.I., 104 avenue Edouard Herriot, Le Plessis-Robinson, Seine, France.

DENIS, M. Member, Commission Nationale pour l'Etude de l'Utilisation Pacifique de l'Energie Nucléaire; Sec. Gen., Syndicat des Entreprises Publiques pour l'Etude et la Construction des Centrales Nucléaires.
Address: 24 rue du Luxembourg, Brussels, Belgium.

DENIS, Pierre M., D. ès Sc. Born 1923. Educ.: Genève Univ. Privat Docent à l'Univ. de Genève. Chef de travaux à l'Inst. de Phys., Genève, 1955; chef du Département de Résonance magnétique nucl., I.V.N.I.N., Caracas, Vénézuéla, 1956-57; Directeur du Lab. de Recherches Nucléaires, Inst. de Phys. Genève, 1957-60; Chef du Lab. de Methodes Nouvelles de Mesures de Rayonnement, C.E.A., Grenoble, Isère, 1964-. Book: Co-author: La résonance paramagnetique nucléaire (Paris, C.N.R.S., 1955). Societies: Sté. Française de Physique; A.P.S.
Nuclear interest: Nuclear physics.
Address: Centre d'Etudes Nucléaires, Grenoble, B.P. 269, Isère, France.

DENISEN, Mrs. Virginia P. Formerly Sec., A.S.E.E.-A.N.S. Com. on Objective Criteria in Nucl. Eng. Education. Assoc. in Nucl. Eng., Iowa State Univ.
Address: 2137 Friley Road, Ames, Iowa 50012, U.S.A.

DENISENKO, A. N. Paper: Co-author, Correction of Energy Dependency of Scintillation Dosimetry Detectors (letter to the Editor, Atomnaya Energiya, vol. 23, No. 3, 1967).
Address: Academy of Sciences of the U.S.S.R., 14 Leninsky Prospekt, Moscow V-71, U.S.S.R.

DENISOV, F. P. Paper: Co-author, About Possibility of Positron Production and Acceleration in Microtron (letter to the Editor, Atomnaya Energiya, vol. 16, No. 4, 1964).
Address: Academy of Sciences of the U.S.S.R., 14 Leninsky Prospekt, Moscow V-71, U.S.S.R.

DENISOV, V. A. Paper: Co-author, Effect of Fuel Density on Stability of Circulating Fuel Nucl. Reactor (Atomnaya Energiya, vol. 24, No. 4, 1968).
Address: Academy of Sciences of the U.S.S.R., 14 Leninsky Prospekt, Moscow V-71, U.S.S.R.

DENIVELLE, Leon. Member, Comite de l'Equipement Industriel, C.E.A., France.
Address: Commissariat a l'Energie Atomique, 29-33 rue de la Federation, Paris 15, France.

DENNE, William H., Jr. Sec., Vitro Corp. of America.
Address: Vitro Corporation of America, 90 Park Avenue, New York, New York 10016, U.S.A.

DENNIS, Sidney, B.S. (Production Management). Born 1926. Educ.: Syracuse Univ. Subcontract Buyer (1952-53), Contract Administrator (1953-57), Chief Contract Administrator (1957-58), Contract Manager (1958-64), Asst. Gen. Manager (1964-), A.M.F. Atomics, a Div. of American Machine and Foundry Co., Greenwich, Conn. Society: Atomic Ind. Forum.
Nuclear interest: Management.
Address: 100 Hoyt Street, Stamford, Conn., U.S.A.

DENNISTON, Rollin Henry II, A.B., M.A., Ph.D. Born 1914. Educ.: Wisconsin, Chicago and Yale Univs. Prof. of Physiology, Wyoming Univ., 1952-60; Special Sen. Res. Fellow, N.I.H., 1960-61; Sen. Prof. Physiology (1961-), Director, Res. Development (1965-), Wyoming Univ. Societies: American Physiology Soc.; A.I.B.S.; American Psychological Assoc.; American Soc. Zoology; N.Y. Acad. Sci.; Federation Socs. Exptl. Biol. and Medicine.
Nuclear interests: Radiobiology; Natural radiation tracers.
Address: Box 3371, University Station, Laramie, Wyoming 82071, U.S.A.

DENSCHLAG, Hans Otto, Diplomchemiker, Dr. rer. nat. Born 1937. Educ.: Mainz Univ. Postdoctoral Fellow, Michigan Univ., 1965-66; Instructor, California Univ., Irvine, 1966-67; Asst., Mainz Univ., 1963-. (On leave of absence, 1965-67). Society: Gesellschaft Deutscher Chemiker.
Nuclear interests: Nuclear fission; Hot atom chemistry; Radiation chemistry.
Address: Institut für Anorganische Chemie und Kernchemie, Mainz Universität, 65 Mainz, Germany.

DENT, Kenneth H., A.M.I.Mech.E., M.I.E.E. Born 1923. Eng. Div., Harwell (1947-53), Design Eng., Risley (1953-58), now Chief Eng., Advanced Gas-Cooled Reactors, U.K.A.E.A.
Nuclear interest: Reactor engineering.
Address: White Lodge Orchard, Norley Road, Cuddington, Cheshire, England.

DENTI, Ennio, Dr. Chem. Born 1932. Educ.: Turin Univ. Scholarship from C.N.E.N. (1959-60) for a post-graduate course in nucl. sci., Sta. Ricerche Impianti Nucl., 1960-.
Nuclear interests: Chemical effects of radiations; Chemical dosimetry; Organic chemistry.
Address: Società Ricerche Impianti Nucleari, Centro Ricerche Nucleari, Saluggia, Vercelli, Italy.

DENUAULT, J. Concerned with Nucl. Work, Sté. des Condenseurs Delas.
Address: Société des Condenseurs Delas, 8 rue Bellini, Paris 16, France.

DENZEL, Paul, Dr. Ing. Prof., Rheinisch-Westfälische T.H., Aachen.
Nuclear interest: Electrical plants.
Address: Rheinisch-Westfälische Technische Hochschule Aachen, 55 Templergraben, Aachen, Germany.

dePACKH, David Calvert, B.S. (Mathematics). Born 1921. Educ.: Chicago Univ. Supervisory Res. Phys. (Nucl. Phys.), Nucleonics Div. U.S. Naval Res. Lab., Washington, D.C. Societies: A.P.S.; Washington Philosophical Soc.; Washington Acad. Sci.; Sci. Res. Soc. of America; New York Acad. Sci.
Nuclear interests: Generation and control of intense energetic electron beams as applied to the design of particle accelerators and the production of high-level electro-magnetic radiation.
Address: Code 7710 Nucleonics Division, U.S. Naval Research Laboratory, Washington, D.C. 20390, U.S.A.

DEPOMMIER, Pierre Henri, Agrégé de Phys., D. ès Sc. Born 1925. Educ.: Lille, Paris and Grenoble Univs. Prof., Faculté des Sci., Grenoble Univ.
Address: Institut des Sciences Nucléaires, B.P. 21, 38 Grenoble, France.

DEPOVERE, Raymond L. E., Ing. Born 1907. Educ.: Louvain Catholic Univ. Conseiller Chef de Service, Administration Industrie, Ministère Affaires Economiques; Chef de Cabinet Adjoint, Ministre des Affaires Européennes. Belgian Alternate, Board of Directors, Eurochemic.
Address: 20 rue Faider, Brussels 5, Belgium.

DEPRAZ, J., Maitre de Conferences: Techniques en Physique Nucléaire, and Prof., Inst. de Physique Nucléaire, Lyon Univ.
Address: Institut de Physique Nucléaire, Lyon Universite, 43 boulevard du 11 Novembre 1918, 69- Villeurbanne (Rhône), France.

DEPRIMOZ, Jacques, Dr. en Droit, Diplômé de l'Ecole des Hautes Etudes Commerciales, Ancien Elève de l'Ecole des Sciences Politiques. Born 1926. Educ.: Faculté de Droit de Paris. Sec. Gen. du Pool Française d'Assurance des Risques Atomiques.
Nuclear interests: Biological effects of ionising radiations, safety and control devices for reactors; Reactor incidents and hazards; Problems of liability and material damages insurance for nuclear operators and users of radioisotopes.
Address: 118 rue de Tocqueville, Paris 17, France.

DEPTULA, Andrzej Jozef, M.Sc. (Chem.).
Born 1935. Educ.: Poznan Univ. Inst. of Nucl.
Res., Warsaw, 1955-. Society: Polish Chem.
Soc.
Nuclear interests: Technology of reactor fuels;
Chemistry of complex uranium compounds.
Address: 12, m.20 Nowotki, Warsaw, Poland.

der BEN, W. R. van. See van der BEN, W. R.

der DECKEN, Claus-Benedict von. See von
der DECKEN, Claus-Benedict.

der GROEBEN, Hans von. See von der
GROEBEN, Hans.

**DER HAVE VAN SIRJANSLAND, Adriana
ENDA MARAH- VAN.** See ENDA MARAH-
VAN DER HAVE SIRJANSLAND, Adriana.

DER HOVEN, Isaac VAN. See VAN DER
HOVEN, Isaac.

der KAM, J. A. van. See van der KAM, J. A.

DER KRAAN, W. VAN. See VAN DER
KRAAN, W.

DER LAAN, Pieter Cornelis Tobias VAN. See
VAN DER LAAN, Pieter Cornelis Tobias.

DER LEEDEN, P. VAN. See VAN DER
LEEDEN, P.

der MEER, J. van. See van der MEER, J.

DER MEULEN, Joseph VAN. See VAN DER
MEULEN, Joseph.

DER MEULEN, Peter Andrew VAN. See
VAN DER MEULEN, Peter Andrew.

DER MUHLL, Jan VON. See VON DER
MUHLL, Jan.

der PLAS, Theo van. See van der PLAS, Theo.

der PUIJL, G. van. See van der PUIJL, G .

DER SPEK, Jean Hector VAN. See VAN DER
SPEK, Jean Hector.

DER SPUY, E. VAN. See VAN DER SPUY, E.

DER TAK, C. VAN. See VAN DER TAK, C.

DER TORREN, J. H. VAN. See VAN DER
TORREN, J. H.

der VALK, F. van. See van der VALK, F.

DER VENNE, Marcel VAN. See VAN DER
VENNE, Marcel.

der VOORT, K. L. F. M. ROUPPE van. See
ROUPPE van der VOORT, K. L. F. M.

der WAAY, D. van. See van der WAAY, D.

der WART, R. van. See van der WART, R.

der ZWAN, L. van. See van der ZWAN, L.

DERBENEV, Ya. S. Papers: Co-author, Some
Effects of Electromagnetic Interactions in
Particle Bunch Intersection (Atomnaya Ener-
giya, vol. 20, No. 3, 1966); co-author, On the
Theory of Transverse Coherent Stability of
the Charged Particles Bunch (ibid, vol. 22,
No. 3, 1967).
Address: Academy of Sciences of the U.S.S.R.,
14 Leninsky Prospekt, Moscow V-71, U.S.S.R.

**DERCLAYE, Raymond Albert Georges Nico-
las,** Ing. Civil Métallurgiste A.I.Lg. Born 1914.
Educ.: Liège Univ. Directeur Gén., Sté.
Métallurgique Hainaut-Sambre S.A.
Nuclear interest: All that concerns the use
of steel in nuclear field.
Address: Société Métallurgique Hainaut-
Sambre S.A., Couillet, Belgium.

DERGATCHEV, N. P. At U.S.S.R. State
Com. for Utilisation of Atomic Energy.
Adviser, Third I.C.P.U.A.E., Geneva, Sept.
1964.
Address: U.S.S.R. State Committee for the
Utilisation of Atomic Energy, 26 Staromonet-
nii Pereulok, Moscow, U.S.S.R.

DeRIENZO, Paul D. Formerly Supervising
Eng., now Chief Eng., Nucl. and Process
Eng. Div., Burns and Roe, Inc.
Address: Burns and Roe, Inc., 160 W. Broad-
way, New York, N.Y. 10013, U.S.A.

DERINER, Ibrahim. Sec. Gen., Turkish
A.E.C.
Address: Turkish Atomic Energy Commission,
12 Ziya Gokalp Caddesi, Ankara, Turkey.

DERMENDJIEV, E. Paper: Co-author, U^{235}
Fission Cross Section for Resonance Neutrons
(letter to the Editor, Atomnaya Energiya,
vol. 19, No. 1, 1965).
Address: Academy of Sciences of the U.S.S.R.,
14 Leninsky Prospekt, Moscow V-71, U.S.S.R.

DEROME, Guy. Ing. en Chef (Entreprise
Générale), Dept. Energie Nucléaire, Télé-
communications, Electroniques, Sté. Alsaci-
enne de Constructions Mecaniques; Formerly
Chief, Dépt. Entreprise Gén., Groupement
Atomique Alsacienne Atlantique.
Address: Société Alsacienne de Constructions
Mecanique, 69 rue de Monceau, Paris 8,
France.

DERRICK, Charles L. Formerly Vice-Pres.
Electric Operations, now Pres., Hartford
Electric Light Co.
Address: Hartford Electric Light Co., 176
Cumberland Avenue, Wethersfield, Conn.,
U.S.A.

DERRICK, Malcolm, B.Sc., Ph.D., M.A. Born
1933. Educ.: Birmingham Univ. Instructor,

Carnegie-Mellon Univ., 1957-60; Sen. Res. Officer, Oxford Univ., 1960-63; Sen. Phys., High Energy Phys. Div., Argonne Nat. Lab. Societies: Phys. Soc., London; A.P.S. Nuclear interests: Interactions and decays of elementary particles; Bubble chamber technology; Design of high energy secondary beams from accelerators; Organisation of research groups.
Address: High Energy Physics Division, Argonne National Laboratory, Argonne, Illinois, U.S.A.

DERRICK, Ronald. Born 1920. Educ.: South Shields Marine Tech. Coll. Sect. Eng., Nat. Boiler and Gen. Insurance Co., 1948-58; Deputy Chief Inspection Eng., Nucl. Power Group, 1958-. Society: I. Mar. E.; Inst. Eng. Inspection.
Nuclear interests: Works inspection of plant for nuclear power stations.
Address: 2 Burford Crescent, Wilmslow, Cheshire, England.

DERRON, L., Dr. Member, Federal Commission for Atomic Energy, Switzerland; Director, Zentralverband Schweiz. Arbeitgeber-Organisationen.
Address: 44 Florastrasse, Zürich, Switzerland.

DERRY, John Alvin, B.S. (E.E.). Born 1907. Educ.: Rose Polytech. Inst., Terre Haute, Indiana. Co-ordinator of Special Assignments, Div. of Production (1949-50), Exec. Officer Div. of Biology and Medicine (1950), Chief Project "A", and Manager, San Francisco Office, Divs. of Res. and Production (1950-52), Asst. Director, Div. of Production (1953), Director, Div. of Construction and Supply (1954-), U.S.A.E.C. Societies: Member, American Inst. of Elec. Eng.; Registered Professional Eng., D.C.
Nuclear interests: Management of construction, engineering, and logistics.
Address: 5723 Ogden Road (Springfield), Washington 16, D.C., U.S.A.

DERUYTTER, Achiel Jerome, Dr. Sc. Born 1933. Educ.: Ghent Univ. Central Bureau for Nucl. Measurements, C.E.E.A., Geel; 1956-. Society: Belgian Phys. Soc.
Nuclear interests: Basic nuclear physics; Neutron cross-section work by time-of flight; Nuclear fission; Neutron cross-section standards.
Address: Ispralaan 70, Mol (Millegem), Belgium.

DERWAE, J. Director, State Tech. Coll. for Nucl. Industries, Board of Nat. Education and Culture.
Address: Institut Technique Superieur de l'Etat pour les Industries Nucléaires, Ministère de l'Education Nationale et de la Culture, 150-52 rue Royale, Brussels, Belgium.

DeSANA, James A. Member, Board of Directors, Dynatomics Inc., 1963-.
Address: Dynatomics Inc., 180 Mills Street, N.W., Atlanta, Georgia 30313, U.S.A.

DESBORDES, J. Prés., Trefimetaux G.P.
Address: Trefimetaux GP, 28 rue de Madrid, Paris 8, France.

DESBRUSLAIS, Eric Leonard, B.Sc. (Eng.). Born 1908. Educ.: London Univ. Elec. Eng., H.M. Dockyard, Portsmouth; Plant Maintenance Eng., Capenhurst (1951-55), Deputy Works Eng., Windscale (1955-56), Works Manager, Chapelcross (1956-59), Works Manager, Calder (1959-65), Supt., Chapelcross (1965-), U.K.A.E.A. Societies: I.E.E.; M.I.Mech. E.; Inst. Plant Eng.
Nuclear interest: Operation of nuclear power plant.
Address: 62 North Street, Annan, Dumfriesshire, Scotland.

DESCAMPS, Monseigneur Albert-Louis, D. et Maitre en Théologie, L. en Philosophie, L. en Sci. Bibliques. Born 1916. Educ.: Catholique Univ. Louvain, and Institut Biblique Pontifical, Rome. Reactor Magnificus, Catholic Univ., Louvain. Membre, Conseil d'Administration, Institut Interuniversitaire des Sciences Nucleaires.
Address: 13 Oude Markt, Louvain, Belgium.

DESCAMPS, Claude Jules Antoine, D. en Sci. Chimiques, Agrégé de l'Enseignement Moyen – Degré Supérieur. Born 1926. Educ.: Brussels Free Univ. and M.I.T. Carbochimique; Chef de Département, BelgoNucléaire, 1957-. Societies: Assoc. Docteurs en Sci., Brussels Free Univ.; A.N.S.
Nuclear interests: Reactor design; Nuclear technology; Application of radioisotopes; Waste disposal.
Address: 34 rue de Saphir, Brussels 4, Belgium.

DESCHREIDER, A. R. Belgian Member, Study Group on Food Irradiation, O.E.C.D., E.N.E.A.
Address: O.E.C.D. European Nuclear Energy Agency, 38 Boulevard Suchet, Paris 16, France.

DESCOMBES, Jacques, Diplômé d'Etudes Supérieures de Droit; ancien élève, Ecole des Sci. Politiques. Born 1917. Educ.: Paris Univ. Commissaire en Chef de la Marine; Adjoint au Directeur, Centre d'Etudes Nucléaires de Saclay.
Nuclear interests: Management: Administration and finances of nuclear centres.
Address: 36 rue de Varenne, Paris 7, France.

DESER, Stanley, B.A., M.A., Ph.D. Born 1931. Educ.: Harvard Univ. Prof., Brandeis Univ. Society: A.P.S.
Nuclear interests: Elementary particles; Field theory; General relativity.
Address: Brandeis Univ., Waltham, Mass. 02154, U.S.A.

DESIPRI, A. I. Paper: Co-author, Indium-Helium Alloy as Gamma Carrier for Rad. Circuits (Abstract, Atomnaya Energiya, vol. 19, No. 2, 1965).
Address: Academy of Sciences of the U.S.S.R., 14 Leninsky Prospekt, Moscow V-71, U.S.S.R.

DESMARQUEST, J. M. Prés. Directeur Gén., Société L. Desmarquest et Cie.
Address: Société L. Desmarquest et Cie., 89 rue de Monceau, Paris 8, France.

DESNEIGES, Paul, Eng. Born 1922. Educ.: C.N.A.M., Paris. Societies: Prés. de la IIème Sect. d'Etudes, Electronique Nucl. et Corpusculaire, Sté Française des Radioelectriciens; Sté. Française des Electroniciens et Radioelectriciens.
Nuclear interests: Développement industriel des appareillages d'electronique nucléaire et de mesures.
Address: Centre D'Etudes Nucléaires (Chef du Service d'Electronique Industrielle), B.P. No. 2, Gif-sur-Yvette 91, France.

DESNUELLE, Pierre. Prof., Faculté des Sci. de Marseille. Member, Comité de Biologie, C.E.A.; Sec. Gen., Internat. Union of Biochem.
Address: Institut de Chimie Biologique, Faculte des Sciences, place Victor Hugo, Marseille 3, France.

DESOV, A. E., Dr. Book: Contributor to Eng. Compendium on Rad. Shielding (I.A.E.A.).
Address: c/o V. Kandaritski, Esq., Department of International Relations and Scientific-Technical Information, U.S.S.R. State Committee for the Utilisation of Atomic Energy, 26 Staromonetny Pereulok, Moscow, U.S.S.R.

DESPORTES. Member, Conseil d'Administration representing Ets. Kuhlmann, Sté. Industrielle des Minerais de l'Ouest (S.I.N.O.). Adviser, Third I.C.P.U.A.E., Geneva, Sept. 1964.
Address: Ets. Kuhlmann, 25 boulevard de l'Amiral-Bruix, Paris 16, France.

DESPRAIRIES, Pierre. Member, Comite des Mines, C.E.A., France.
Address: Commissariat a l'Energie Atomique, 29-33 rue de la Federation, Paris 15, France.

DESREUX, V. Nucl. Chem. Inter-Faculty Centre of Nucl. Sciences, Liège Univ.
Address: Laboratory of Physical Chemistry, University of Liège, Belgium.

DESROSIER, Norman W., B.S., M.S., Ph.D. Born 1921. Educ.: Massachusetts Univ. Prof. Food Technol., Purdue Univ., 1949-61; Vice-Pres., R. and D., Beech-Nut Life Savers, Inc., 1963-65; Director of Res., Nat. Biscuit Co., 1965-; Director, Radiation Div., Quartermaster Food and Container Inst. for U.S. Armed Forces, 1957-59. U.S. Atoms for Peace Mission to Europe, 1958, 1960; Sci. Consultant, E.P.A. and E.N.E.A., O.E.E.C., 1958, 1960. Books: Technol. of Food Preservation (Avi Publishing Co., 1959); Attack on Starvation (Avi Publishing Co., 1961); Co-author, Rad. Technol., in Food, Agriculture and Biol. (Avi Publishing Co., 1960). Societies: Inst. Food Technologists; A.I.B.S.
Nuclear interests: Radiation technology in fields of food, agriculture and biology; Radiation preservation of foods; Radiation microbiology; Radiation induced metabolic changes in organisms; Radiation effects: plant and animal tissues.
Address: 509 Fowler Avenue, Pelham Manor, New York, U.S.A.

DESSAUER, Gerhard, M.A., Ph.D. Born 1910. Educ.: Frankfurt, Berlin, Munich, California and Rochester Univs. Director, Phys. Sect., Savannah River Lab., E.I. du Pont de Nemours and Co., Aiken, South Carolina. Societies: A.N.S.; A.P.S.
Nuclear interests: Nuclear physics; Reactor physics.
Address: 517 Sumter Street, S.E., Aiken, South Carolina 29801, U.S.A.

DESSEVRE, Jean. Administrateur, Sté. d'Etudes et de Travaux pour l'Uranium (S.E.T.U.).
Address: Société d'Etudes et de Travaux pour l'Uranium (S.E.T.U.), 23 boulevard Georges Clemenceau, Courbevoie, (Seine), France.

DESSUS, Gabriel, Ancien Prés. de l'Assoc. technique pour la production et l'utilisation de l'energie nucléaire, Paris; Deleg. to 2nd I.C.P.U.A.E., Geneva, Sept.1958; Pres. Auxi-Atome; Director, Groupe Intersyndical de l'Industrie Nucleaire.
Address: Auxi-Atome, 139 boulevard Haussmann, Paris, France.

d'ESTAING, Philippe GISCARD. See GISCARD d'ESTAING, Philippe.

DESTIVAL, Claude. Member, Commission Consultative pour la Production d'Elec. d'Origine Nucl., C.E.A.
Address: Commissariat a l'Energie Atomique, 29-33 rue de la Federation, Paris 15, France.

DESYATKINA, E. A. Paper: Co-author, Determination of Oxygen in Cesium Metal by Vacuum Distillation Method (Atomnaya Energiya, vol. 24, No. 2, 1968).
Address: Academy of Sciences of the U.S.S.R., 14 Leninsky Prospekt, Moscow V-71, U.S.S.R.

DETENBECK, Robert W., Ph.D. Born 1933. Educ.: Princeton Univ. Asst. Prof. (1959-65), Assoc. Prof. (1965-67), Maryland Univ.; W.A.E., U.S. Naval Ordnance Lab., 1964-; Assoc. Prof., Vermont Univ., 1967-. Acting Editor (1966-67), Member, Board of Editors, The Phys. Teacher. Societies: A.P.S.; A.A.A.S.; A.A.P.T.

Nuclear interests: Instrumentation; Reaction theory; Nuclear structure (basic research).
Address: Department of Physics, Vermont University, Burlington, Vermont 05401, U.S.A.

DETERDING, John Henry, M.A. Born 1923.
Educ.: Cambridge Univ. Res. Phys., "Shell" Res., Ltd.
Nuclear interest: Application of radioisotopes in research on petroleum products.
Address: Thornton Research Centre, "Shell" Research, Ltd., P.O. Box 1, Chester, Cheshire, England.

DETERING, Hans-Werner. Born 1923. Frank und Schulte G.m.b.H., Essen; Detinox G.m.b.H., Essen.
Nuclear interests: Management and nuclear and medical shielding.
Address: 23 Am Scheidbusch, Essen-Haarzopf, Germany.

DETILLEUX, Emile, Dr. Chem. Born 1927.
Educ.: Liège Univ. Sect. Head, (1958), Res. Director, (1964), Tech. Director, (1967-), Eurochemic.
Nuclear interests: Reprocessing of irradiated fuel; Reactor chemistry.
Address: 32 Strasbourglaan, Mol, (Province of Antwerp), Belgium.

DEUBLEIN, Otmar, Dr. Ing. Born 1921.
Educ.: Munich and Karlsruhe T.H. Vereinigte Elektrizitätswerke Westfalen A.G., Dortmund, 1952-63; Technical Manager, Kernkraftwerk Lingen G.m.b.H., 1963-; Gruppe 3, Reaktore zur Energiegewinnung Standiges Seminar für Kerntechnik, Haus der Technik e.V. Essen.
Address: 80 Märkische Strasse, Dortmund, Germany.

DEUSS, H., Dr. Member, Steering Com., Deutsches Atomforum E.V.
Address: Deutsches Atomforum E.V., 240 Koblenzer Strasse, 53 Bonn, Germany.

DEUTSCH, David Ernest, B.S. (Chem.), M.S. (Metal). Born 1918. Educ.: Lafayette Coll. and Carnegie Inst. Tech. Staff Metal., Los Alamos Sci. Lab., 1949-51; R. and D. Administrator, U.S.A.E.C., Washington, D.C., 1952-56; Nucl. and Refractory Alloy R. and D., Fansteel Metal. Corp., Chicago, and Stanford Res. Inst., Palo Alto, California, 1957-58; Head, Metal. and Chem. Depts. (1958-61), Manager, Nucl. Technol. Div. (1961-64), Aerojet-Gen. Nucleonics; Special Asst. to Sen. Vice Pres., Nucl. Div. (1965-66), Manager, Rad. Effects, N.E.R.V.A. Nucl. Rocket Programme (1967-), Aerojet-Gen. Corp. Societies: A.S.M.; A.N.S.; American Inst. of Mining and Metal. Eng.
Nuclear interests: Management and administration of research, development, test and engineering for current and advanced nuclear power and propulsion systems, particularly in nuclear metallurgy and materials, radiation chemistry and radiation effects.
Address: 996 Bartlett Court, El Dorado Hills, California 95630, U.S.A.

DEUTSCH, Martin, B.S., Ph.D., Dr. h.c. Born 1917. Educ.: M.I.T. Assoc. Prof. Phys. (1949-53), Prof. Phys. (1953-), Chairman, Directing Com., Lab. for Nucl. Sci., M.I.T.; Guggenheim Fellow, Paris, 1953-54, Rome, 1960-61; Assoc. Editor, Nucl. Instruments. Societies: Fellow, U.S. Nat. Acad. Sci.; Fellow, American Acad. Arts and Sci.; Fellow, A.P.S.; Fellow, A.A.A.S.; Sté. Française de Phys.; Phys. Soc. Japan.
Address: Massachusetts Institute of Technology, Cambridge 39, Massachusetts, U.S.A.

DEUTSCH, Robert William, B.S. (M.I.T.), Ph.D. (California). Born 1924. Educ.: Queens Coll., Flushing, N.Y., M.I.T. and California Univ., Berkeley. Phys., Lawrence Rad. Lab., 1950-53; Nucl. Phys., K.A.P.L., 1953-57; Chief Project Phys., Gen. Nucl. Eng. Corp., 1957-61; Sci., Republic Aviation Corp., 1961-62; Phys. Consultant, Eng. Dept., Martin Co., 1962-64; Prof., Catholic Univ., Washington, 1964-. Societies: A.P.S.; A.N.S.; A.A.A.S.; A.I.A.A.; A.A.U.P.
Nuclear interests: Reactor physics; Radiation detectors; Radiation transport and shielding.
Address: Catholic University, Washington D.C. 20017, U.S.A.

DEUTSCHMANN, Martin, Dr.-Ing. Born 1917.
Asst., Max-Planck-Inst. für Physik, 1950-57; Prof. Phys. Rheinisch-Westfälische T.H. Aachen, 1959-. Book: Contribution to Kosmische Strahlung (edited by W. Heisenberg; Springer Verlag, 1953).
Nuclear interests: Elementary particle and nuclear physics.
Address: Physics Department, Technische Hochschule, 14 Charlottenstr., Aachen, Germany.

DEVAKUL, Kanjika, M.B., Ph.D. Educ.: Bangkok Univ. of Medical Sci. and School of Tropical Medicine, Liverpool. Lecturer, Dept. of Physiology (1950-61), Faculty of Tropical Medicine (1961-68) at present Dept. Head, Univ. of Medical Sci., Bangkok.
Nuclear interest: Application of radioisotopes in research in tropical medicine.
Address: Faculty of Tropical Medicine, Rajvithi Road, Bangkok, Thailand.

DeVAN, Jackson H., B.S. (Metal. Eng.), M.S. (Metal. Eng.). Born 1930. Educ.: Stanford, Ohio State and Tennessee Univs. Project Eng., Wright Air Development Centre, 1952-54; Assoc. Metal., O.R.N.L., 1954-. Societies: A.S.M.; A.N.S.
Nuclear interests: Development and testing of high temperature reactor materials; Corrosion studies of liquid metal and fused salt reactor coolant systems.
Address: P.O. Box X, Oak Ridge, Tennessee, U.S.A.

DEVELL, Lennart C. E., Civil eng. Born 1934. Educ.: Roy. Inst. Technol., Stockholm. Phys. Chem. and Nucl. Chem. Divs., Roy. Inst. Technol., – 1961; A.B. Atomenergi, Studsvik, 1961–.
Nuclear interests: Health physics; Radiochemical analysis; Whole body counting; Research on fission product release, transport and removal.
Address: A.B. Atomenergi, Studsvik, Nyköping, Sweden.

DEVIK, Finn, Doctor's degree. Born 1916. Educ.: Oslo Univ. Res. Assoc., Norwegian Defence Res. Establishment, 1948-51; Res. Fellowship, Norwegian Assoc. against Cancer, 1951-57; Chief Medical Officer, State Inst. of Rad. Hygiene, 1957-; Medical Officer, Rad. and Isotopes Univ. W.H.O., Geneva, 1961-63. Societies: Rad. Res. Soc.; N.Y. Acad. Sci., and several Norwegian professional societies.
Address: Medical Section, State Institute of Radiation Hygiene, Institute of Pathology, Rikshospitalet, Oslo, Norway.

DEVLIN, Leighton Hugh, B. Eng., M. Eng. Administration. Born 1932. Educ.: Johns Hopkins and George Washington Univs. Pratt and Whitney Aircraft Co., United Aircraft Corp., East Hartford, Conn., 1954-57; Nucl. Power Dept., Allis-Chalmers Manufacturing Co., Washington D.C., 1957-61; Appl. Phys. Lab., Johns Hopkins Univ., 1961-67; Chief Naval Operations, Navy Dept., 1967-. Society: A.S.M.E.
Nuclear interests: Power plant performance analysis; Fuel element heat transfer analysis; Design and analysis of reactor components.
Address: 5602 Atta Vista Road, Bethesda, Md. 20034, U.S.A.

DEVOE, James R. Member, Subcom. on Low Level Contamination of Materials and Reagents, N.A.S.-Nat. Res. Council.
Address: National Academy of Sciences-National Research Council, 2101 Constitution Avenue, N.W., Washington 25, D.C., U.S.A.

DEVOINO, A. N. Paper: Co-author, Investigation Progress in the Inst. of Nucl. Power of the Bylorussian Acad. of Sci. (Atomnaya Energiya, vol. 24, No. 4, 1968).
Address: Academy of Sciences of the U.S.S.R., 14 Leninsky Prospekt, Moscow V-71, U.S.S.R.

DEVONS, Samuel, M.A., Ph.D., M.Sc. Born 1914. Educ.: Cambridge Univ. Prof. Phys., Imperial Coll., London, 1950-55; Langworthy Prof. Phys. and Director, Phys. Lab., Manchester Univ., 1955-60; Prof. Phys. (1960-), Chairman, Dept. of Phys. (1963-67), Columbia Univ. Vice-Pres., Phys. Soc. Books: Excited States of Nuclei (1949); Contribution to Nucl. Spectroscopy (1959). Societies: F.R.S.; Fellow, Phys. Soc.; A.P.S.
Nuclear interest: Nuclear physics; Particle physics.
Address: Department of Physics, Columbia University, New York, New York 10027, U.S.A.

DEVOOGHT, Jacques, Ing. civil mécanicien-electricien A.I.Br., Ingénieur radioelectricien (Univ. Libre Brussels)., M.Sc. (Nucl. Eng., M.I.T.). Agrégé de l'Enseignement Supérieur. Born 1932. Educ.: Univ. Libre de Bruxelles and M.I.T. Professeur extraordinaire, Univ. Libre de Bruxelles.
Nuclear interests: Nuclear reactor physics and reactor theory; Mössbauer spectroscopy.
Address: 36 avenue E. Duray, Brussels 5, Belgium.

DEWAR, Donald James, M.Sc., Ph.D. Born 1915. Educ.: Queen's (Kingston, Ont.,) and (McGill Univs. Sci. Adviser, Atomic Energy Control Board, Ottawa. Sci. Sec., U.N.I.C.P.U.A.E., 1955.
Nuclear interests: Radioisotope and reactor licensing; Radiation hazards control; Reactor safety.
Address: 462 Island Park Drive, Ottawa 3, Ontario, Canada.

DEWEY, David Lewis, M.A., Ph.D. Born 1927. Educ.: Cambridge Univ. Academic Staff, University Coll., London, 1953-56; Res. Staff, British Empire Cancer Campaign Res. Unit in Radiobiol., 1956-. Societies: Assoc. for Rad. Res.; Biochem. Soc.; Soc. for Gen. Microbiol.
Nuclear interest: Radiobiology.
Address: British Empire Cancer Campaign for Research, Research Unit in Radiobiology, Mount Vernon Hospital, Northwood, Middlesex, England.

DEWEY, Davis R., II, B.S. (Chem.), Sc.D. (Chem. Eng.). Born 1917. Educ.: Harvard Univ. and M.I.T. Tech. Director, American Res. and Development Corp., 1946-51; Vice-Pres., High Voltage Eng. Corp., 1951-57; Pres., Baird-Atomic, Inc., 1957-64. Director: Baird-Atomic, Inc.; Chemtrac Corp.; Hampshire Chem. Corp.; Atomic Ind. Forum. Societies: A.C.S.; A.A.A.S.; American Inst. of Chem.; A.N.S.; Rad. Res. Soc.
Nuclear interests: Particle accelerator applications; Radioactivity instrumentation.
Address: Baird-Atomic, Inc., 33 University Road, Cambridge 38, Massachusetts, U.S.A.

DEWEY, William Cornet, B.S. (Phys.), Ph.D. (Rad. Biol.). Born 1929. Educ.: Washington and Rochester Univs. In A.E.C., Rochester Univ., 1954-58; Texas Univ., M. D. Anderson Hospital, 1958-; Prof. Rad. Biol., Dept. Radiol. and Rad. Biol., Colorado State Univ. Societies: Nucl. Medicine Soc.; Rad. Res. Soc.
Nuclear interests: Effects of radiation on cellular systems especially at the chromosomal level; Radioisotopes as used in medicine for diagnosis and therapy and as physiological tracers.

Address: Department of Radiology and Radiation Biology, Colorado State University, Fort Collins, Colorado 80521, U.S.A.

DEWEZ, Denis, Ing. civil élec. Born 1928. Educ.: Liège Univ. Inst. Interuniversitaire des Sci. Nucl., 1956; A.C.E.C., 1956-57; Nucl. Development Corp. of America, Europe, 1957-60; Ing. Soc. Gen. des Minerais, 1960-; Collaborateur, Nucl. Reactor Core Technol., Liège Univ., 1960-.
Nuclear interest: Industrial utilisation of nuclear energy.
Address: 31 rue du Marais, Brussels 1, Belgium.

DeWITT, Russell B. Formerly Asst. Supt., now Supt., Big Rock Point Nucl. Plant, Consumers Power Co.
Address: Consumers Power Company, Big Rock Point Nuclear Plant, Charlevoix, Michigan, U.S.A.

DEXTER, Ralph Warren, B.S., Ph.D. Born 1912. Educ.: Massachusetts and Illinois Univs. Biol. Sci. Dept., Kent State Univ., 1937-. Pres., Ohio Acad. Sci. Societies: Ecological Soc. America; American Soc. of Limnology and Oceanography.
Nuclear interest: Passage of radioactive materials through food chains.
Address: Department of Biological Sciences, Kent State University, Kent,Ohio, U.S.A.

DEY, M. L., Shri. Reader, Saha Inst. of Nucl. Phys., Calcutta Univ.
Address: Calcutta University, 92 Acharya Prafulla Chandra Road, Calcutta-9, India.

D'EYE, Royston Walter Mastin, B.Sc., M.Sc., Ph.D. Born 1926. Educ.: London Univ. Res. Manager, Reactor Group, Springfields, Preston, Lancs., U.K.A.E.A. Societies: A.R.I.C.; Fellow, Chem. Soc.
Nuclear interests: Non-metallic nuclear fuels; Corrosion and solid state chemistry.
Address: 64 Kingsway, Penwortham, Preston, Lancs., England.

D'HEUR, Jacques H. E. A., Ing. civil des Mines. Born 1909. Educ.: Liège Univ. Director, Sté. Belge de Chimie Nucléaire.
Address: Mean par Havelange, Belgium.

D'HONT, Maurice Docile Emile, Dr. in Phys. Chem. Born 1922. Educ.: Louvain and Cambridge Univs. M.I.T., Harvard Univ. Res. Assoc., Chem. Eng., M.I.T., 1949-50; Res. Assoc. Harvard Univ., 1950-51; Lecturer (1951-63), Prof. (1963-), Louvain Univ.; Research Eng., Metallurgie Hoboken and C.E.N., 1951-60; Res. Manager (1960-63), Adviser to the Pres. (1963-), C.E.N. Sci. and Tech. Com., Eurochemic; Special Com. on Radioprotection; Pres., Controle-Radioprotection; Consulting board, Inst. Interuniv. des Nucléaires. Societies: Corresponding Member of the Belgian Acad. Sci.; Vlaamse Chemische Vereniging; Société de Chimie Industrielle; A.N.S.; Roy. Inst. of Metals.
Nuclear interests: Nuclear chemistry and metallurgy; Fuel element fabrication; Plutonium, reprocessing, etc.
Address: "Dennetoppen", S.C.K., Mol, Belgium.

d'HOSPITAL, Gilbert MELESE-. See MELESE d'HOSPITAL, Gilbert.

DHOUAILLY, Chef, Dept. de Sûreté et de Protection du Secret, C.E.A.
Address: Commissariat à l'Energie Atomique, 29-33 rue de la Federation, Paris 15, France.

DHUMWAD, Ramachandra, M.Sc. Born 1933. Educ.: Bombay and Karnatak Univs. Sci. Asst. (1954-55), Sen. Sci. Asst. (1955-58), Jun. Sci. Officer (1958-61), Sci. Officer (1961-65), S.S.O. (1965-), Fuel Reprocessing Div., Bhabha Atomic Res. Centre, Bombay, India; Training in Spectrochem. Analysis, O.R.N.L., Tennessee; Res. work in molecular spectroscopy, N.B.S., Washington, D.C., 1961-62.
Nuclear interests: Spectroscopic problems associated with the analysis of high purity plutonium and uranium; Development of emission and atomic absorption spectroscopic methods for the assay of plutonium isotopes; Study of molecular energy states, rotational and vibrational constants of carbon; Molecular isotope shifts.
Address: Plutonium Plant, Atomic Energy Establishment Trombay, Bombay 74, India.

DI CAGNO, Vitantonio, Avv. Pres., Board of Directors, Ente Nazionale per l'Energia Elettrica. Adviser, Third I.C.P.U.A.E., Geneva, Sept. 1964.
Address: Ente Nazionale per l'Energia Elettrica, 181 Via del Tritone, Rome, Italy.

DI FELICE, L., Ing. Tech. Manager and working in Nucl. Field, Dott. Ing. Giuseppe Torno and C.S.p.A.
Address: Dott. Ing. Giuseppe Torno and C. S.p.A., 7 Via Albricci, Milan, Italy.

DI GIOVANNI, Hugo J., B.S. (E.E.). Born 1912. Educ.: Cooper Union Inst. Technol. Chief, Eng. Sect., Health and Safety Lab. (1948-52), Director, Tech. Liaison Div., New York Operations Office (1952-54), U.S.A.E.C.; Vice-Pres. and Chief Eng., Del Electronics Corp., 1954-.
Nuclear interests: Instrumentation – radiation detectors, aerosol collection and measuring devices; Health physics; Plasma physics; Equipment devices; Energy discharge systems.
Address: 250 E. Sandford Boulevard, Mount Vernon, N.Y., U.S.A.

DI IANNI, Elmo J., B.S. (Phys.). Born 1923. Educ.: C.C.N.Y., New York Univ. and Pratt Inst. Head, Nucl. Instrumentation Development Sect., Naval Material Lab., 11 years; now Manager, Nucl. Div., and Vice-Pres., Nucl.

DI LAN

Corp. of America.
Nuclear interests: Developments in nuclear physics and instrumentation.
Address: Nuclear Corporation of America, Nuclear Division, 2 Richwood Place, Denville, New Jersey 07834, U.S.A.

DI LANDRO, Osvaldo A., D.V.M. Born 1918. Educ.: Montevideo Univ. Prof. Biophys., Faculty Veterinary Medicine, Montevideo; Radiologist and Phys. Medicine, Clinics Inst. Hospital, Veterinary Faculty; World Fellowship "André Mayer", F.A.O., 1961; "Applications of radioisotopes in Veterinary Medicine" – Studies in Cornell, Illinois and Pennsylvania Univs., EE.UU., 1962-63; Radiologist - Technician Senior in Public Health Protection and Inspector in this Field of Social Security, Bank of Uruguay - Statal Organism - Montevideo. Books: Radiol. in Veterinary Medicine. Coxa-Vara in Dogs. Electrocardiography in Horses and Dogs. Neutron-activation analysis in Fish (silage). Absorption of Ca45 and Sr85 in Dogs. Public Health Protection and Social Security: National Legislation in this Field. Societies: Vice-Pres., Atomic Energy Commission, Uruguay; Radiol. and Phys. Medicine Soc. of Uruguay; Nat. Commission of Rad. Protection, Uruguay; New York Acad. Sci.
Nuclear interests: Applications of radioisotopes in veterinary medicine and atomic energy research in biology. Public health protection and safety in atomic energy field.
Address: 5885 Friburgo, Montevideo, Uruguay, South America.

di LORETO, A., Dr. Ing. Member, Comitato Nazionale, Assoc. Nazionale di Ingegneria Nucl.
Address: Associazione Nazionale di Ingegneria Nucleare, 24 Piazza Sallustio, Rome, Italy.

DI MARZIO, Salvador, Cap. Frag. (R). Electromech. Eng. Dept. Head, Electronics Dept., Comision Nacional de Energia Atomica, Argentina.
Address: Comision Nacional de Energia Atomica, Electronics Department, 8250 Avda. del Libertador Gral. San Martin, Buenos Aires, Argentina.

DI MENZA, Raffaele, Mech. Eng., M.S. (Nucl. Eng.). Born 1932. Educ.: Stevens Inst. Technol. and M.I.T. Chief, Project Sect., Reactor Div. (1960-61), Manager, Organic Reactor Programme (P.R.O.), Plants and Construction Div. (1961-), C.N.E.N. Societies: A.S.M.E.; A.N.S.
Nuclear interests: Nuclear engineering and reactor design; Management; Application of nuclear energy to desalting of sea water.
Address: Comitato Nazionale per l'Energia Nucleare, 15 Via Belisario, Rome, Italy.

DI NARDI, Giuseppe. Member, Com. Scientifico, Diritto ed Economia Nucleare.
Address: Diritto ed Economia Nucleare, 68 Viale Bruno Buozzi, Rome, Italy.

di PALIANO, Guido COLONNA. See COLONNA di PALIANO, Guido.

DI PRETORO, Walter, Dr. Director, Nucl. Dept., Industrial Controls, S.A.S.
Address: Industrial Controls S.A.S., 90 Viale Lombardia, Monza, Italy.

DI STEFANO, Guiliano. Member of Board, F.I.E.N.
Address: Forum Italiano dell'Energia Nucleare, 26-28 Via Paisiello, Rome, Italy.

DI TORO, M. Member, Regional Nucl. Res. Com., Member, Gruppo Teorico, Istituto di Fisica, Catania Univ.
Address: Catania University, Istituto di Fisica, 57 Corso Italia, Catania, Sicily, Italy.

di VIGNANO, Angelo TOMMASI. See TOMMASI di VIGNANO, Angelo.

DI VITO, Giovanni, Prof. Ing. Director, Centrale Commerciale S.p.A. Ercole Marelli. Member, Comitato Permanente per i Problemi dell'Impiego Pacifico dell'Energia Nucleare e per i Rapporti con Euratom.
Address: Marelli Ercole and C.S.p.A., Casella Postale 3695, Milan, Italy.

DIAMOND, Edwin, Ph.B., M.A. Born 1925. Educ.: Chicago Univ. Sci. Editor (1957-61), Sen. Editor (1961-), Newsweek Magazine; Board of Directors, Nucl. Energy Writers Assoc. Book: The Rise and Fall of the Space Age. Societies: Nucl. Energy Writers Assoc.; Nat. Assoc. Sci. Writers.
Nuclear interests: Coverage of peaceful and military aspects of nuclear energy; The atom in space; Fallout hazards to man; Arms race and arms control; Nature of thermonuclear war and nature of a workable disarmament system.
Address: Newsweek Magazine, 444 Madison Avenue, New York, U.S.A.

DIAMOND, Jack, M.Sc. (Cambridge and Manchester). Born 1912. Educ.: London and Cambridge Univs. Atomic Energy, Montreal and Harwell, 1944-53; Prof. Mech. Eng., Manchester Univ., 1953-. Council member (1958-), and Vice-Pres. (1967-), Inst. Mech. Eng., Governing Board, N.I.R.N.S., 1957-60; Member of Board of B.N.E.C., 1957-63; Univ. Grants Com., 1965-; Nat. Res. Development Council, 1966-. Societies: I. Mar. E.; Inst. Mech. Eng.
Nuclear interests: Reactor design; Education and training.
Address: Manchester University, Manchester, England.

DIAMOND, Richard M. Member, Subcom. on Radiochem., Com. on Nucl. Sci., N.A.S.-N.R.C.
Address: National Academy of Sciences – National Research Council, Committee on

Nuclear Sciences, 2101 Constitution Avenue, N.W., Washington 25, D.C., U.S.A.

DIAS CARNEIRO, Octavio Augusto, B.S. (Brazilian Naval Acad.), M.A. (Economics) (George Washington Univ.), Ph.D. (Economics) (M.I.T.). Born 1912. Educ.: Lycée Français, Rio de Janeiro; Brazilian Naval Acad., Rio de Janeiro; School Fine Arts, Paris Univ.; George Washington Univ.; M.I.T. Minister Plenipotentiary, Brazilian Foreign Office; Member, National Nucl. Energy Commission of Brazil; Member, National Petroleum Council of Brazil; Participant, 3rd Inter-American Symposium on the Peaceful Application of Nucl. Energy, Rio de Janeiro, July, 1960. Book: Aspetos Econômicos do Aproveitamento Industrial da Energia Atomica (Rio de Janeiro, 1955). Societies: American Economic Soc.; Econometric Soc.; Philosophical and Phenomenological Assoc.
Nuclear interests: Economics of nuclear power; Nuclear engineering; Isotope applications to industry; Medicine and agriculture.
Address: Comissao Nacional de Energia Nuclear, 350 Avenida Marechal Camara, 7⁰ Andar, Rio de Janeiro, Brazil.

DIAZ, Augustin, B.S. (Eng. Phys.), M.S. (Mech. Eng.). Born 1936. Educ.: Chattanooga, Duke and Maryland Univs. Manager, Madrid operations, Gen. Elec. Tech. Services Co. (for the construction (turnkey) of the Santa Maria de Garona nuclear power plant (460 MWe), boiling water reactor, direct cycle). Societies: A.N.S.; A.S.M.E.
Nuclear interests: Physics; Mechanical system construction and start-up and operation of nuclear power plants.
Address: G.E.T.S. CO., 2 Ferraz, Madrid, Spain.

DIAZ, Carlos JIMENEZ-. See JIMENEZ-DIAZ, Carlos.

DIAZ, Jorge SUAREZ. See SUAREZ DIAZ, Jorge.

DIAZ ACHECAR, Abelardo ELIAS. See ELIAS DIAZ ACHECAR, Abelardo.

DIAZ-DUQUE, Ricardo, Chem. Eng. Born 1932. Educ.: S. Carlos Univ., Guatemala. Deleg. of Guatemala to Conference on Statute of I.A.E.A., New York, 1956; and to 1st Gen. Conference of I.A.E.A., Vienna, 1957; Formerly Exec. Sec., Guatemalan Nucl. Energy Commission. Society: Guatemalan Soc. Chem. Eng.
Address: 7a Calle 1-38, Zona 1, Guatemala City, Guatemala, Central America.

DIAZ LOSADA, Eduardo, Ing.; Advisory Board Member, Comision Nacional de Energia Nuclear, Mexico.
Address: Comision Nacional de Energia Nuclear, Av. Insurgentes Sur 1079 Tercer piso, Mexico 18, D.F., Mexico.

DIAZGRANADOS, Alberto MEJIA-. See MEIJA-DIAZGRANADOS, Alberto.

DIB, George Theofiel, B.S. (Chem. Eng.), graduate hons. credit in Maths. Born 1923. Educ.: Connecticut Univ. Special Project Eng., Gen. Dynamics Corp., 1952-54; Sen. Nucl. Eng., Martin Co., 1954-55; Nucl. Eng., Blaw-Knox Co., 1955-56; Nucl. Consultant (clients: Arthur G. McKee and Co.; Nucl. Sci. and Eng. Corp.; Argonne Nat. Lab.; Westinghouse Elec. Corp.), 1956-60; First Professional licensed Nucl. Eng., Commonwealth of Pennsylvania, 1957; Nucl. Design Coordinator, Vitro Eng. Co., 1960-63; Gen. Manager, Mark III Div., Victoria Industries, Inc., 1963-; Director, Office of Civil Defence and Disaster Control, Verona, New Jersey. A.E.C. sponsored A.S.E.E. and A.N.S. Com. on Optimum Criteria in Nucl. Eng. Educ.; Com. on Gen. and Administrative Nucl. Codes, American Atomic Ind. Forum.
Nuclear interests: Management interests in nuclear design and space sciences. Project management responsibilities where nuclear aspects are of prime importance, i.e. power, propulsion, and research reactors and facilities, radiation applications to industry and phenomena analysis.
Address: 4 Cypress Avenue, Verona, New Jersey, U.S.A.

DIBBEN, Harold Edwyn, B.Sc. Born 1917. Educ.: Nottingham Univ. Chief Chem., Production Group, Risley, Warrington, Lancs. (1958-59) now Deputy Gen. Manager, Production Group, U.K.A.E.A., Springfields Works, Salwick, Nr. Preston, Lancs. Society: Roy. Inst. of Chem.
Nuclear interests: General and maintenance engineering; Chemistry; Chemical engineering; Engineering; Metallurgy; Administration; Health and safety.
Address: 2 Norfolk Road, Lytham, Lancs., England.

DIBDEN, Frederick Andrew, M.B.B.S., D.T.R., F.C.R.A. Born 1917. Educ.: Adelaide Univ. Visiting Radiotherapist, Repatriation Dept. (1957), Clinical Instructor (1965), Adelaide Univ. Hon. Radiotherapist, Roy. Adelaide Hospital; Hon. Radiotherapist, Queen Elizabeth Hospital; Hon. Radiotherapist, Adelaide Children's Hospital. Society: Faculty of Radiol., England.
Nuclear interest: Clinical radiotherapy.
Address: 168 Ward Street, North Adelaide, South Australia.

DICHARRY, Richard F., B.S. (Mech. Eng.). Born 1920. Educ.: Louisiana State Univ. Gen. Manager, Gamma Industries, Inc. Societies: Louisiana Eng. Soc.; Soc. of Nondestructive Testing; A.S.M.E.; A.S.T.M.; A.W.S.
Nuclear interest: Encapsulation of radioactive materials - industrial application.
Address: c/o Gamma Industries, Inc., P.O. Box 2543, Baton Rouge, Louisiana, U.S.A.

DICKE, Hans. Member, Board of Management, Blohm und Voss A.G. Mitglied Arbeitskreis III/2, Kernenergie fur Schiffe, Deutsche Atomkommission, Federal Ministry for Sci. Res.
Address: Blohm und Voss A.G., 56 Am Elbtunnel, Hamburg-Steinwerder, Germany.

DICKEMAN, Raymond Louis, B.S., M.S. Born 1922. Educ.: Wisconsin Univ. Phys. (1948-50), Supervisor, Exptl. Phys. (1950-52), Supervisor, Reactor Phys. (1952-56), Manager, Process and Reactor Development (1956-60), Gen. Manager, Fuels Preparation Dept. (1960-62), Gen. Manager, N Reactor Department (1962-67), Gen. Elec. Co., Richland, Washington; General Manager, Domestic Turnkey Projects, Gen. Elec. Co., San Jose, California, 1967-. Societies: A.I.P.; A.N.S.
Address: 15362 Via Palamino, Monte Sereno, California, U.S.A.

DICKENSON, Aubrey Fiennes TROTMAN-.
See TROTMAN-DICKENSON, Aubrey Fiennes.

DICKERSON, Ronald F., B.S. (Metal. Eng.), M.S. (Metal. Eng.). Born 1922. Educ.: Virginia Polytech. Inst. Staff Manager, Battelle-Northwest, 1965. Societies: Electrochem. Soc.; A.S.M.; A.I.M.E.; Atomic Ind. Forum.
Nuclear interests: Reactor and high-temperature materials, particularly melting and casting, metallography, and physical metallurgy; Refractory metals; Compounds for high-temperature applications and radiation effects.
Address: Battelle-Northwest, P.O. Box 999, Richland, Washington, U.S.A.

DICKEY, Richard K., B.A. Born 1928. Educ.: California Univ. Vice-Pres. and Gen. Manager, West Coast Div., Nucl. Consultants Corp. Board Directors, H.P.S. S. California Chapter. Societies: H.P.S.; A.I.H.A.
Nuclear interests: Radiopharmaceuticals; Biological effects of radiation (dosimetry); Beta dosimetry; Waste disposal.
Address: 1549 Ben Lomond Drive, Glendale, California, U.S.A.

DICKINSON, David Franklin, A.B., M.S., Ph.D. (Iowa State). Born 1914. Educ.: Iowa State Coll., and Illinois Univ., Kansas State Teachers Coll., Pittsburgh. At New Mexico Univ., 1956-57; Argonne Nat. Lab., 1957-58; Boeing Airplane Co., summer of 1959; Nevada Univ., 1958-. Pres., Nevada Astronomical Soc. Societies: A.I.Ch.E.; A.C.S.; A.S.M.; Phys. Honor Soc.
Nuclear interests: Radiation damage, fuel reprocessing, and radiation processing.
Address: 770 MacDonald Drive, Reno, Nevada, U.S.A.

DICKINSON, Frederick Samuel. Born 1923. Educ.: Roy. Tech. Coll., Salford and Manchester Coll. Technol. Welding Eng., Metropolitan Vickers Elec. Co., 1953; Eng. III (1953-56), Eng. II (1956-59) Eng. I (1959-) responsible for all fabrication consultancy for Northern Groups of U.K.A.E.A., Dept. of Atomic Energy, Ministry of Supply (P) and later U.K.A.E.A. Societies: A.M.I.Mech. E.; Inst. Welding; Nucl. Eng. Soc. (within U.K.A.E.A.).
Nuclear interests: Interested in the design and particularly construction of all reactor types and the chemical processing plants necessary for the production of uranium fuels and the subsequent extraction of plutonium, in relation to the problems of material selection and fabrication difficulties.
Address: 74 Culcheth Hall Drive, Culcheth, Warrington, Lancs., England.

DICKINSON, Robert W., A.B., M.S., S.M. Born 1920. Educ.: California Univ., U.S. Naval Post Graduate School, and M.I.T. Director, Liquid Metal Eng. Centre, North American Rockwell Corp. Book: Sodium Graphite Reactors (Addison-Wesley Press, 1958). Societies: A.N.S.; A.S.M.
Nuclear interests: Engineering; Management.
Address: Atomics International, Liquid Metal Engineering Centre, Santa Susana Mountains, Los Angeles, California, U.S.A.

DICKINSON, Wade, B.S. Born 1926. Educ.: Carnegie Inst. Technol. (Aeronautical Eng.), U.S.M.A. (Military Eng.), Tennessee Univ. (Nucl. Phys.), O.R.S.O.R.T. (Nucl. Eng.), Ohio State Univ. (Business Administration). H.Q., U.S.A.F. Wright Air Development Centre, loaned to O.R.N.L. and RAND Corp., 1949-54; Bechtel Corp., 1954-64; Tech. Consultant, Joint Com. on Atomic Energy, U.S. Congress, 1957-58; Pres., W. W. Dickinson Corp., 1960-. Societies: A.N.S.; A.P.S.; A.S.M.E.; A.S.A.; American Ordnance Soc.
Nuclear interests: Commercial and military power plants, particularly nuclear monitoring and inspection; Missile and satellite power supplies, particularly nuclear; Mechanical design and management of nuclear facilities; Non-destructive testing; Computer controlled agricultural and medical equipment.
Address: 2125 Broderick Street, San Francisco, California, U.S.A.

DICKS, John B., Ph.D. Director, Radioisotope Lab., Univ. of the South.
Address: Radioisotope Laboratory, University of the South, Sewanee, Tennessee, U.S.A.

DICKSON, George Kenneth, B.Sc., C. Eng., F.R.I.C., M.I.Chem.E. Born 1913. Educ.: London Univ. Chem. Eng. Advisor, Chem. Plant Design Office, M.O.S. (Dept. of Atomic Energy), 1949-51; Works Manager, Windscale (1951-52), Deputy Chief Sci., R. and D. Branch, Risley (1953-57), Manager (Chem. Eng.)– Central Tech. Services (Risley) (1957-61), Chem. - Chem. Eng. Adviser, Reactor Group, (1961-), U.K.A.E.A.
Nuclear interests: The production of reactor materials; Chemistry and chemical engineering

of reactor circuits; Containment and handling of fission products.
Address: "Lane End", Hall Lane, Stretton, Cheshire, Via Warrington, England.

DICKSON, John Lewis, M.A. (Cantab.). Born 1924. Educ.: Cambridge Univ. U.K.A.E.A., Capenhurst and A.E.R.E., 1950-58; Asst. Chief Eng., Eng. Services Div., A.E.E., Winfrith, 1958-. Society: A.M.I.Mech.E.
Nuclear interest: Reactor design.
Address: 9 Came View Close, Dorchester, Dorset, England.

DIDEIKIN, T. S. Papers: Co-author, Determination of Effective Multiplication Factor on Basis of Measuring of Differential Reactivity (Atomnaya Energiya, vol. 22, No. 2, 1967); co-author, Correlation of Similar Reactivities in Different Reactors (letter to the Editor, ibid, vol. 24, No. 4, 1968).
Address: Academy of Sciences of the U.S.S.R., 14 Leninsky Prospekt, Moscow V-71, U.S.S.R.

DIDENKO, A. N. Papers: Co-author, On the Use of the "Pin" Moderating Systems in Acceleration Technique (letter to the Editor, Atomnaya Energiya, vol. 12, No. 3, 1962); co-author, 1,5 GeV Electron Synchrotron at the Polytech. Inst. of Tomsk (ibid, vol. 21, No. 6, 1966); co-author, Investigation of Auto-Resonant Particle Acceleration by Electromagnetic Waves (ibid, vol. 22, No. 1, 1967); co-author, Exptl. Investigation of Variation of Electron Beam Cross Dimensions in the 1.5 GeV Synchrotron (ibid, vol. 24, No. 1, 1968).
Address: Academy of Sciences of the U.S.S.R., 14 Leninsky Prospekt, Moscow V-71, U.S.S.R.

DIDIOT, Jean. Senior officer engaged in nucl. work, Sté des Forges et Chantiers de la Méditerranée.
Address: Société des Forges et Chantiers de la Méditerranée, 6-8 rue Camou, Paris 7, France.

DIEBOLD, Doushan, Dipl. Ing. (E.T.H.). Res. Div., Contraves A.G.
Address: Contraves A.G., 580 Schaffhauserstrasse, Zurich 52, Switzerland.

DIECKAMP, Herman M., B.S. (Eng. Phys.). Educ.: Illinois Univ. With Atomics Internat., (1950-), Space Nucl. Work (1956-), Programme Director, S.N.A.P. Systems, (1962-66), Vice Pres., Nucl. Eng., (1966-), Atomics Internat. Adviser, Third I.C.P.U.A.E., Geneva, Sept. 1964.
Address: Atomics International, P.O. Box 309, Canoga Park, California,U.S.A.

DIEDERICHS, Manfred K., Dipl. Ing., Dr. Ing. Born 1927. Educ.: Karlsruhe Tech. Univ. Design Eng. with Daimler-Benz A.G. Stuttgart, 1951-54; Asst. Prof. (1954-58), Sen. Eng. (1958), Karlsruhe Tech Univ.; Supervisor for Operation and Maintenance of Karlsruhe-Reactor FR 2, 1958-. Society: A.N.S.
Nuclear interests: Reactor operation; Reactor engineering.
Address: 2 Kantstrasse, Eggenstein/Karlsruhe, Germany.

DIEDERICHS, Nicolaas, M.A. (South Africa), D.Litt et Phil. (Leiden). Born 1903. Former Prof. Orange Free State Univ.; Minister of Economic Affairs, 1958-67; Minister of Mines, 1961-64; Minister of Finance, 1967-.
Address: Union Buildings 36, Pretoria, South Africa.

DIEHL, E. J., Ir. Director, Inst. TNO for Mech. Constructions, Organisation for Ind. Res. TNO.
Address: Institute TNO for Mechanical Constructions, 2 Mekelweg, Delft, Netherlands.

DIEHL, Johannes Friedrich, M.Sc. (Chem.), Dipl. Chem., Dr. rer. nat. Born 1929. Educ.: Kentucky and Heidelberg Univs. Res. Asst., German Wool Res. Inst., 1957; Res. Assoc., Dept. of Biochem. (1957-58), Asst. Prof. (1959-63), Assoc. Prof. (1963-65), Arkansas Univ. Medical School; Director and Prof., Inst. of Rad. Technol., Federal Res. Centre for Food Preservation, 1965-. Societies: German Chem. Soc.; A.C.S.
Nuclear interests: Food preservation by irradiation; Fallout radioactivity in foods; Decontamination of foods; Application of isotopic tracer technique in food science, technology and nutrition.
Address: Bundesforschungsanstalt für Lebensmittelfrischhaltung, 20 Engesserstr., 75 Karlsruhe, Germany.

DIEHL, Jörg, Dr. rer. nat. Sen. Officer, Inst. of Special Metals, Max-Planck Inst. for Metal Res.
Address: Max-Planck Institute for Metal Research, 92 Seestrasse, 7000 Stuttgart-1, Germany.

DIEHL, Peter, Prof. Born 1931. Educ.: Basel Univ. Nat. Res. Council, Ottawa, 1964. Society: Schweizerische Physikalische Gesellschaft.
Nuclear interest: Nuclear magnetic resonance isotopes.
Address: Physics Department, 82 Klingelbergstrasse, Basel, Switzerland.

DIEHL, Walter, Dr. rer. publ. Born 1927. Educ.: Univ. Economics, St. Gall. Supt., Swiss Reinsurance Co., Zürich. Administration Officer, Swiss Pool for Insurance of Atomic Risks. Society: Swiss Assoc. for Atomic Energy.
Nuclear interests: Atomic risks, insurance and reinsurance.
Address: 2 Steinbrüchelstrasse, Zürich 7/53, Switzerland.

DIEMAIR, Willibald, Prof., Dr. Dr. Mitglied, Arbeitskreis II/2, Chemie, Federal Ministry for Sci. Res.

DIE

Address: Federal Ministry for Scientific Research, 46 Luisenstrasse, 532 Bad Godesberg, Germany.

DIENES, George Julian, B.Sc., M.Sc., M.A., D.Sc. Born 1918. Educ.: Carnegie Inst. Technol. and Columbia Univ. Sen. Phys., Brookhaven Nat. Lab. Member, Solid State Advisory Panel; Assoc. Editor, J. Phys. Chem. Solids. Books: Co-author, Nucl. Fuels (Van Nostrand, 1956); co-author, Rad. Effects in Solids (Interscience, 1957); Point Defects in Metals (Gordon and Breach, 1963). Societies: Fellow, A.P.S.; Soc. of Rheology.
Nuclear interest: Radiation effects in solids.
Address: Physics Department, Brookhaven National Laboratory, Upton, Long Island, New York, U.S.A.

DIENNER, John Astor, B.S. (E.E.), E.E., LL.B., M.L.D., Dr. Eng. (Hon.)., LL.D. (Hon.). Born 1886. Educ.: Purdue, George Washington and Georgetown Univs. Commerce Dept. Patent Office Advisory Com., 1933-53; Patent Advisory Panel, Atomic Energy Commission, 1946-. Book: The U.S. Patent System (1936). Societies: I.E.E.E.; American Inst. Chem.; Western Soc. Eng.; Newcomen Soc. of London; Roy. Soc. Arts, London.
Nuclear interests: Advisor re interpretation and amendment of Act and Agreements and Treaties under same.
Address: 53 West Jackson Boulevard, Chicago, Illinois 60604, U.S.A.

DIENSTBIER, Zdenek, M.D., Dr. Sc., Professor. Born 1926. Educ.: Medical Faculty, Charles Univ., Prague. Head, Biophys. Inst., Charles Univ.; Chairman, Czechoslovak Soc. of Nucl. Medicine; Member, Council of European Soc. of Rad. Biol. Books: Nemoc z ozareni (Radiation Sickness) (Prague, State Publishing House of Health, 1957); Zaklady nuklearni mediciny (Fundamentals of Nuclear Medicine) (Prague, State Pedagogic Publishing House, 1961). Experimental Postirradiation Syndrome (SZdM, Prague 1966). Societies: Czechoslovak Biophys. Soc.; Czechoslovak Soc. of Nucl. Medicine; European Soc. of Rad. Biol.
Nuclear interests: Nuclear physics, nuclear medicine especially; Radiobiology with a view to pathophysiology of the biological mechanism of post-irradiation changes and the clinical application of radioisotopes.
Address: Biophysical Institute, 3 Salmovská, Prague 2, Czechoslovakia.

DIERKENS, Ferdinand E. E. Born 1922. Ing. Civil Electricien-Mécanicien (Brussels), Ingénieur des Constructions Aeronautiques (Brussels), M.S. (Eng.) (Michigan), Candidat. D.Sc. (M.I.T.), special nucl. classes, Univ. Brussels 1955; Belgian American Educational Foundation Fellow, 1946 and 1947. Manager, Sté. Anonyme Centre et Sud, grouping the Belgian participants to the Centrale Nucléaire des Ardennes. Asst. Manager, Sté. d'Energie Nucléaire Franco-Belge des Ardennes. Member of Board: S.C.K. Studiecentrum voor Kernenergie, Mol.; Vianova S.A., Brussels; Antwerps Asphalt and Beton Bedrijf, Antwerp; S.A. Fours Lecocq et Ateliers de Trazegnies Réunis, Brussels; S.A. Awans François, Awans-Bierset; S.A. Betafra, Brussels; Chantier Naval de Rupelmonde S.A., Rupelmonde; S.A. Ateliers Germain-Anglo, La Croyère; Administrateur, Fondation Nucleaire. Societies: A.N.S.; A.I.M.M.E.; Soc. Automotive Eng.; Chambre de Commerce, Bruxelles; Alumni Univ. Foundation of Belgium; Sté. Royale belge des Ingénieurs et Industriels.
Nuclear interests: Electric power production; Economic problems; Automation.
Address: 10 Square Frère Orban, Brussels 4, Belgium.

DIETHELM, Lothar, Prof. Dr. med. Director, Mainz Univ. Clinic, Inst. für Klinische Strahlenkunde, Johannes Gutenberg Univ.
Address: Mainz University Clinic, Institut für Klinische Strahlenkunde, 1 Langenbeckstrasse, Mainz, Germany.

DIETHORN, Ward S., B.S. (Lake Forest Coll.), M.S., Ph.D. (Chem.) (Carnegie Inst. Technol.). Born 1927. Educ.: Lake Forest Coll. and Carnegie Inst. Technol. Asst. Div. Chief, Radioisotope and Rad. Div., Battelle Memorial Inst., 1956-60; Assoc. Prof., Nucl. Eng. Dept., Pennsylvania State Univ., 1960-. Societies: A.C.S.; A.A.A.S.; A.N.S.
Nuclear interests: Nuclear reactor materials; Radioisotope; Large radiation sources; Radiation damage; Radiation dosimetry.
Address: Nuclear Engineering Dept., 231 Sackett Building, Pennsylvania State University, University Park, Pa., U.S.A.

DIETRICH, Joseph Robert, B.S. (William and Mary), M.S. and Ph.D. (Virginia). Born 1914. Educ.: Coll. of William and Mary, and Virginia Univ. Assoc. Director, Reactor Eng. Div., Argonne Nat. Lab., 1954-56; Formerly Vice-Pres. and Director Phys. Div., Gen. Nucl. Eng. Corp., Dunedin, Florida.; Chief Sci., Nucl. Div., Combustion Eng. Inc. Editor, Power Reactor Technol.; Member, Honours and Awards Com., American Nucl. Soc. Books: Co-author, Solid Fuel Reactors (Addison-Wesley, 1958); co-author, Sect. 6, Reactor Physics, Nucl. Eng. Handbook (McGraw-Hill, 1958). Society: Fellow, A.N.S.
Nuclear interests: Design and development of nuclear reactors; Reactor physics.
Address: 165 Wood Pond Road, West Hartford, Connecticut 06107, U.S.A.

DIETRICH, Klaus M., Ausseplanmässiger. Born 1934. Educ.: Munich and Heidelberg Univs. Res. Asst. (1960-62), Res. Fellow (1962-64), Lawrence Rad. Lab., Berkeley; Res. Fellow (1964-65), Ausseplanmässiger (Supernumerary) Prof., Phys. (1965-), Heidelberg Univ. Society: German Phys. Soc.

Nuclear interest: Theoretical nuclear physics.
Address: Institute for Theoretical Physics, 16 Philosophenweg, 69 Heidelberg, Germany.

DIETSCH, G. Director, Sté. Franco-Américaine de Constructions Atomiques S.A.
Address: Société Franco-Américaine de Constructions Atomiques S.A., 12 rue Lincoln, Paris 8, France.

DIETZ, E., Dr. Vertreter des Rechtsdienst, Eidgenössiche Politisches Departements, Alarmausschuss, Eidgenössische Departement des Innern.
Address: Rechtsdienst, Eidgenössische Politisches Departement, Bern, Switzerland.

DIETZ, George Robert, B.S., M.S., N.S. Born 1931. Educ.: U.S. Military Acad., West Point and Georgia Inst. Technol. U.S. Army Quartermaster Corps Liaison Officer to Div. of Isotopes Development, U.S.A.E.C., 1960-.
Nuclear interests: Isotopic sources and industrial uses, especially in conjunction with radiation preservation of foods; Design and construction of U.S. Army's million curie cobalt-60 source and 24 MeV Linac constructed in Massachusetts, and large A.E.C. isotopic radiation facilities.
Address: 1023 Welsh Drive, Rockville, Maryland, U.S.A.

DIETZE, H., Prof., Dr. Inst. für Reaktorwerkstoffe, Wissenschaftlicher Rat, Kernforschungsanlage Jülich des Landes Nordrhein-Westfalen E.V.
Address: Kernforschungsanlage Jülich des Landes Nordrhein-Westfalen E.V, Postfach 365, Jülich, Germany.

DIEU, H. Nucl. Chem., Inter-Faculty Centre of Nucl. Sci., Liège Univ.
Address: Laboratory of Physical Chemistry, University of Liège, Belgium.

DIEVOET, Jean Paul VAN. See VAN DIEVOET, Jean Paul.

DIGNEFFE, Alfred. Manager, Head of Eng. Dept., Sté. Financière de Transports et d'Entreprises Industrielles, S.A.
Address: Société Financière de Transports et d'Entreprises Industrielles, S.A., 38 rue de Naples, Brussels 5, Belgium.

DIJCK, Willem Johannes Dominicus VAN. See VAN DIJCK, Willem Johannes Dominicus.

DIJKSTRA, Jan Hille, drs. Born 1925. Educ.: Amsterdam Vrije Univ. Physicus, Wetenschappelijk medewerker, Inst. voor Kernphysisch Onderzoek (Inst. for Nucl. Phys. Res.), 1955-65; Laboratorium voor Ruimteonderzoek (Lab. for Space Res.), 1965-. Society: Netherlands Phys. Soc.
Nuclear interests: Astrophysics; Optics in wavelengths 10 – 2500 Ångströms, in particular soft X-rays and required equipment (grazing incidence mirrors, gratings, Bragg crystals, detectors, etc.).
Address: Laboratorium voor Ruimteonderzoek, 121 Huizingalaan, Utrecht, Netherlands.

DIKANSKII, N. S. Papers: Co-author, Transverse Coherent Instability of Charged Particles Bunched Beam (Atomnaya Energiya, vol. 21, No. 3, 1966); co-author, Interaction of Coherent Betatron Oscillations with External Systems (ibid, vol.22, No. 3, 1967); co-author, On the Theory of Transverse Coherent Stability of the Charged Particles Bunch (ibid, vol. 22, No. 3, 1967); co-author, Study of Self-Excitation and Rapid Damping of Coherent Transverse Oscillations in VEPP-2 Storage Ring (ibid, vol. 22, No. 3, 1967).
Address: Academy of Sciences of the U.S.S.R. 14 Leninsky Prospekt, Moscow V-71, U.S.S.R.

DIKII, A. G. Paper: Co-author, On Possibility of Plasma Injection into Closed Magnetic Trap through Divertor (Atomnaya Energiya, vol. 24, No. 1, 1968).
Address: Academy of Sciences of the U.S.S.R., 14 Leninsky Prospekt, Moscow V-71, U.S.S.R.

DILLARD, Clyde Ruffin, B.S., M.S., Ph.D. Born 1920. Educ.: Virginia Union and Chicago Univs. Prof. Chem., Tennessee A. and I. Univ., 1948-53; Prof. Chem., Morgan State Coll., 1953-59; Assoc. Prof. Chem., Brooklyn Coll., 1959. Societies: A.C.S.; A.A.A.S.; N.Y. Acad. Sci.; Soc. of Appl. Spectroscopy.
Nuclear interests: Fission product separations; Industrial utilisation of fission product activities.
Address: Brooklyn College, Brooklyn 10, New York, U.S.A.

DILLENBURG, Darcy. Born 1930. Educ.: Rio Grande do Sul and Sao Paulo Univs. At Inst. de Energia Atômica, Sao Paulo, 1956-57; Faculdade de Filosofia e Inst. de Fisica da Univ. do Rio Grande do Sul, 1958-.
Nuclear interest: Nuclear physics.
Address: Instituto de Fisica, U.F.R.G.S., Avenida Luiz Englert, Pôrto Alegre, Rio Grande do Sul, Brazil.

DILLINGER, Pavel, R.N.Dr. Born 1937. Educ.: Comenius Univ., Bratislava and Lomonosov Univ., Moscow. Asst. (1961-64), Asst. Prof. (1964-) Dept. of Radiochem. and Rad. Chem., Chem. Techn. Faculty, Slovak Tech. Univ., Bratislava.
Nuclear interests: Nuclear chemistry; Radioanalytical chemistry.
Address: Department of Radiochemistry and Radiation Chemistry, Chemicotechnological Faculty, Slovak Technical University, 1 Jánska, Bratislava, Czechoslovakia.

DILLON, Ira G., B.S., M.S., Ph.D. Born 1919. Educ.: Oregon State Univ. and I.I.T. Assoc. Chem. Eng., Argonne Nat. Lab., 1949-65; Prof., Mech. Eng. (Nucl. Eng.), Tuskegee Inst., 1965-. Societies: A.N.S.; A.S.M.E.; A.I.Ch.E.

Nuclear interests: Fuel processing; Metallurgy.
Address: School of Engineering, Tuskegee Institute, Alabama 36088, U.S.A.

DILLON, John A., Jr. Dean, Graduate School, Louisville Univ. Louisville Univ. rep. on Council, Oak Ridge Assoc. Univs., 1966-.
Address: Louisville University, Louisville 8, Kentucky, U.S.A.

DILUZIO, Frank. Deputy Manager, Albuquerque Operations Office, U.S.A.E.C., 1957-61; Former Staff Director, Senate Space Com., Head, Office of Saline Water, 1965-; Asst. Sec., Office of the Asst. Sec. for Water Pollution Control, Dept. of the Interior, 1966-67; Vice Pres., Edgerton, Germeshausen and Grier, Inc., 1967-.
Address: Edgerton, Germeshausen and Grier, Inc., Box 98, Goleta, California, U.S.A.

DILWORTH, Constance Charlotte, B.Sc., M.Sc., Dr. h.c. (Berne). Born 1924. Educ.: London Univ. Prof., Milan Univ. Societies: Societa Italiana di Fisica; American Phys. Union.
Nuclear interests: Elementary particles and cosmic radiation.
Address: Istituto di Fisica dell' Università, 16 via Celoria, I20133 Milan, Italy.

DILWORTH, J. Richardson. Director, United Nucl. Corp.
Address: United Nuclear Corp., 1730 K Street, N.W., Washington 6, D.C., U.S.A.

DILWORTH, Robert H. III, B.S. (Eng. Phys.). Born 1930. Educ.: Tennessee Univ. Development Eng., O.R.N.L., 1956-62; Product Manager, Electronics, ORTEC, Inc., 1962-.
Societies: I.E.E.E.; A.N.S.; RESA.
Nuclear interest: Nuclear instrumentation for physics research and life sciences.
Address: ORTEC, Inc., 100 Midland Road, Oak Ridge, Tennessee 37830, U.S.A.

DIMEGLIO, A. Francis, A.E.C. Fellow in Radiol. Phys., B.S. (Phys.). Born 1931. Educ.: Providence Coll., Rochester and Connecticut Univs., O.R.S.O.R.T. U.S.N. Underwater Sound Lab., 1953-55; N.R.L., 1956-; Director of Operations, Rhode Island Nucl. Sci. Centre.
Society: A.N.S.
Nuclear interests: Reactor design; Nuclear physics.
Address: Rhode Island Nuclear Science Centre, Narragansett, Rhode Island, U.S.A.

DIMOTAKIS, Paul Nicholas, Diploma in Chemistry, Ph.D. (Nucl. Chem.). Born 1928. Educ.: Athens and Cambridge Univs. Sen. Res. Chem. (1956-60), Head of the Chem. Div. (1960-62), Chief, Nucl. Chem. Lab. (1964-), Sci. Director (1967), Democritus Nucl. Res. Centre, Greek A.E.C. Book: How to Prospect for Uranium (Hellenic A.E.C., 1956). Societies: Assoc. of Greek Chem.; A.N.S.
Nuclear interests: Hot atom chemistry (in solid systems); Nuclear chemistry (nuclear fission, reactions and spallation); Radiochemistry (natural radioactivity, separation of radionuclides); Radiation chemistry (radiation damage in solids); Analytical chemistry (ion exchange, chromatography; IR, UV spectroscopy).
Address: 74 Nikopoleos Street, Athens 218, Greece.

DIMOV, G. I. Papers: Co-author, Experiments of Charge Exchange Injection in Storage Ring (Atomnaya Energiya, vol. 19, No. 6, 1965); co-author, 1, 5 GeV Electron Synchrotron at the Polytechnic. Inst. of Tomsk (ibid, vol. 21, No. 6, 1966); co-author, Experiments on Production of Intense Proton Beam by Charge Exchange Injection Method (ibid, vol. 22, No. 5, 1967).
Address: Novosibirsk Science Centre, 20 Sovietskaya Ulitsa, Novosibirsk, Siberia, U.S.S.R.

DIN, E. Tawdy Saad EL. See EL DIN, El Tawdy Saad.

DINGEE, David Aaron, B.S., M.S. Born 1929. Educ.: Michigan State Univ. Div. Chief, Appl. Nucl. Phys. Div., Battelle - Columbus Labs. Papers: (Battelle Reports): Several including A Study of Core Fuel Systems for a Fast Breeder Reactor; Critical Assembly Studies for the Variable Moderator Reactor; GCRE and ML-1 Critical Assembly Studies; Effect of Fuel Parameters on Specific Weight of Reactor Space-Power Systems. Society: A.N.S.
Address: Battelle - Columbus Laboratories, 505 King Avenue, Columbus, Ohio 43201, U.S.A.

DINGLE, Leo S. Head Nucl. Eng., Portsmouth Naval Shipyard, New Hampshire.
Address: Portsmouth Naval Shipyard, Nuclear Power Division, Portsmouth, New Hampshire, U.S.A.

DINTENFASS, Leopold, Ph.D., M.Sc., Dipl.-Ing. (Chem.). Born 1921. Educ.: Polytech. Lvov (Poland and U.S.S.R.). and New South Wales Univ. Res. Chem., B.A.L.M. Paints Pty. Ltd. (Div.of I.C.I.), 1950-56; Sen. Res. Chem., Taubmans Industries Ltd. 1956-61; Res. Rheologist, Surgical Res. Unit, Prince of Wales Hospital, Sydney, 1961-62; Sen. Res. Officer, Nat. Health and Medical Res. Council, and Medical Res. Fellow, Dept. of Medicine, Sydney Univ., 1962-63; Sen. Res. Fellow, Nat. Heart Foundation and Dept. of Medicine, Sydney Univ., 1964-. Hon. Consulting Biorheologist, Sydney Hospital, 1962-; Hon. Consulting Rheologist, Children Medical Res. Foundation, 1966-. Societies: F.R.A.C.I.; Fellow, Chem. Soc., London; Fellow, Internat. Coll. of Angiology, New York; British Soc. Rheology; Assoc. Member, Inst. of Eng., Australia; Australian Soc. Hematology; Sydney Biophys. and Medical Electronics Soc.; Australian Soc. for Medical Res.; American

Federation for Clinical Res.
Nuclear interests: Radio-rheology; Heterogeneous radio-catalysis; Effect of gamma radiation on biological tissues and cells.
Address: 74 Gilgandra Road, North Bondi, New South Wales, Australia.

Di NUNNO, Joseph John, B.S. (Elec. Eng.), M.S. (Elec. Eng.). Born 1921. O.R.S.O.R.T., Pennsylvania State and Maryland Univs. Electronics Sci., U.S. Naval Ordnance Lab., 1946-56; Nucl. Eng., U.S. Navy, 1956-59; Asst. Director, Div. of Reactor Studies, U.S. A.E.C., 1959-67; U.S.A.E.C. Sci. Rep., Paris, 1967-. Society: A.N.S.
Nuclear interests: Reactors - Design and operation; Reactor safety; Reactor materials, metallurgy and applications; Instrumentation and controls; Pressure vessel design, inspection and tests.
Address: United States Embassy, 2 Avenue Gabriel, Paris, France.

DINWIDDIE, Joseph Gray, Jr., B.S., Ph.D. (Chem.). Born 1922. Educ.: Randolph-Macon Coll., Ashland, Virginia, and Virginia Univ. On staff, Chem. Dept. (1948-), now Prof. Chem., Clemson Univ. Societies: A.C.S.; A.A.A.S.
Nuclear interests: Teaching of radiochemistry; Use of tracers in organic research.
Address: 114 Poole Lane, Clemson, South Carolina, U.S.A.

DISMUKE, Stewart Edgar, B.S. Born 1920. Educ.: Texas Technol. Coll. Phys., O.R.N.L.
Nuclear interests: Design and operation of hot cells and remotely controlled equipment.
Address: Operations Division, Oak Ridge National Laboratory, Post Office Box X, Oak Ridge, Tennessee, U.S.A.

DISNEY, Harold Vernon. Born 1907. Educ.: Nottingham Univ. Coll. Chief Eng. (Design) Diffusion Plant (1950), Asst. Director Defence Projects (1955), Deputy Director Defence Projects (1957), Director of Eng.. (1958), Deputy Managing Director, Eng. Group (1961-62), Managing Director, Eng., Groups (1962-), U.K.A.E.A. Pres., Risley Branch, Nucl. Eng. Soc. Societies: Inst. of Mech. Eng.; Nucl. Eng. Soc.
Nuclear interests: Design and construction of chemical and process plants handling toxic and radioactive materials.
Address: Delph House, Delphfields Road, Appleton, Warrington, Lancs., England.

DISPA, U. Site Direction, Nucl. Dept., Citar S.A.
Address: Nuclear Department, Citar S.A., 162-164 Quai des Usines, Brussels, Belgium.

DISSANAIKE, George Alexander, B.Sc., Ph.D. Born 1927. Educ.: Ceylon and Cambridge Univs. Lecturer in Phys., Ceylon Univ., 1963-; Visiting Asst. Prof., South Carolina Univ., 1956-57. Societies: H.P.S.; Ceylon Assoc. for the Advancement of Sci.
Nuclear interests: Nuclear physics; Radio tracer techniques applied to biology and medicine.
Address: Department of Physics, Ceylon University, Thurstan Road, Colombo 5, Ceylon.

DITCHBURN, Robert William, B.Sc., M.A., Ph.D., F.R.S. Born 1903. Educ.: Liverpool and Cambridge Univs. Prof. Phys., Reading Univ., 1946-. Societies: Roy. Irish Acad.; Inst. Phys. and Phys. Soc.
Nuclear interest: Radiation damage.
Address: The University, Reading, Berks., England.

DITTMAR, Rupprecht, Difpl.-Kaufm., Dr. rer. pol. Member of Präsidium, Member of Fachkommission V-Wirtschaftliche, Finanzielle und Soziale Probleme, Deusche Atomkommission.
Address: Federal Ministry for Scientific Research, 46 Luisenstrasse, 532 Bad Godesberg, Germany.

DIVATIA, Ajay Shrinivas, B.Sc. (Bombay), M.S. (Wisconsin), Ph.D. (Wisconsin). Born 1927. Educ.: Bombay, Chicago and Wisconsin Univs. Res. Asst., Wisconsin Univ., 1951-55; formerly Jun. Sci. Officer and Res. Officer, now sci. Officer., Van de Graaff Lab., Nucl. Phys. Div., Bhabha Atomic Res. Centre. Trombay. Societies: A.P.S.; Indian Sci. Congress Assoc.
Nuclear interests: Nuclear physics; Nuclear electronics; Vacuum physics; Plasma physics; Accelerator physics.
Address: Van de Graaff Laboratory, Nuclear Physics Division, Bhabha Atomic Research Centre, Trombay, Bombay 74 AS, India.

DIVEN, Benjamin C., A.B., M.S., Ph.D. Born 1919. Educ.: California and Illinois Univs. Staff Member, Los Alamos Sci. Lab., California Univ. Society: A.P.S.
Nuclear interests: Experimental nuclear physics, particularly fission process, and neutron interactions.
Address: Los Alamos Scientific Laboratory, P.O. Box 1663, Los Alamos, New Mexico, U.S.A.

DIXON, Harold Evans, B.Sc., M.Sc., F.I.M. Born 1921. Educ.: Durham Univ. Chief Metal. (1957-59), Head, R. and D. (1959), Chief Eng. (1960-63), Gen. Manager, (1964), Director and Gen. Manager (1965-), Atomic Power Constructions Ltd. Societies: A.W.S.; Inst. Welding; Inst. Metal.; B.N.E.S.
Nuclear interest: Management.
Address: 7 Tormead Road, Guildford, Surrey, England.

DIXON, John A., M.A. Born 1911. Educ.: Oxford Univ. Production Div., Dept. of Atomic Energy, Ministry of Supply, 1948-54; Public Relations Officer, Development and Eng. Group (1954-61), P.R.O., Reactor Group (1961-), U.K.A.E.A., Risley.

Nuclear interests: Dissemination of information on atomic energy to both technical and non-technical public.
Address: Reactor Group, United Kingdom Atomic Energy Authority, Risley, Warrington, Lancs., England.

DIXON, P., M.Sc. (New Zealand). Formerly Asst. Director, now Deputy Director, Australian Mineral Development Labs.
Address: Australian Mineral Development Laboratories, Conyngham Street, Parkside, South Australia.

DIXON, William R., B.A., M.A., Ph.D. Born 1925. Educ.: Saskatchewan and Queen's (Kingston, Ontario) Univs. Assoc. Res. Officer, X-Rays and Nucl. Rad. Section, N.R.C., Ottawa, Canada, 1950-. Society: Canadian Assoc. Physicists.
Nuclear interests: Nuclear physics; X-rays and nuclear radiations.
Address: X-Rays and Nuclear Radiations Section, National Research Council, Montreal Road Laboratories, Building M-35, Ottawa, Ontario, Canada.

DIZDAR, Zdenko, Chem. Eng. Born 1911. Educ.: Belgrade Univ. Res. worker (1951), Head, Hot Lab. (1965-), Boris Kidrich Inst. of Nucl. Sci., Belgrade. Books: Co-author, Practical Radiochem. (Belgrade, 1959); Radioisotopes and Radiations (Belgrade, 1963); Radiochemical Laboratory (Belgrade, 1960). Societies: Chem. Soc., Belgrade; Croatian Chem. Soc.
Nuclear interest: Nuclear chemistry.
Address: Boris Kidrich Institute of Nuclear Sciences, Belgrade, P.O.B. 522, Yugoslavia.

DJAKOV, Emile, Corresponding Member, Bulgarian Acad. Sci. Born 1908. Educ.: Sofia Univ. Vice Pres., Join Inst. Nucl. Res., Dubna, 1959-61; Director, Inst. Electronics, Sofia, 1963-; Holder, Chair Appl. Phys., Faculty of Phys., Sofia Univ. Pres., Nat. Measurement Com., 1958-; Member, Com. for Peaceful Use of Atomic Energy, 1956-.
Nuclear interests: Nuclear electronics and accelerators.
Address: Bulgarian Academy of Sciences, Institute of Electronics, 152 Boulevard Lenin, Sofia 13, Bulgaria.

DJIE, H. R. E. TJIN A. See TJIN A. DJIE, H. R. E.

DJORDEVIC, L. Head, Electronics Dept., Dept. for Prospection of Nucl. and Other Raw Materials, Inst. Nucl. Raw Materials, Belgrade. Deleg., I.A.E.A. Conference on Nucl. Electronics, Belgrade, 15-20 May 1961.
Address: Institute of Nuclear Raw Material, 12 Rovinjska, Belgrade, Yugoslavia.

DJUANDA, H. Chairman, Council for Atomic Energy, Indonesia.
Address: Council for Atomic Energy, Djl.

Palatehan I No. 26, Blok K5, Kebajoran-Baru, Djakarta, Indonesia.

DJUKIC, Zoran, D. of Medicine. Born 1927. Educ.: Belgrade Univ. Head, Medical Protection Dept., Boris Kidric Inst., Belgrade, 1960-. Member, Management Board, Yugoslav Soc. for Radiol. Protection. Book: Radioactive Isotopes and Rad., Book II, Chapter XVII Medical Protection. (Belgrade Naucna Knjiga, 1963).
Nuclear interest: Radiation health protection.
Address: Boris Kidric Institute of Nuclear Sciences, Belgrade-Vinca, P.O.B. 522, Yugoslavia.

DLOUHY, Zdenek, M.S., Ph.D. Born 1932. Educ.: Charles Univ., Prague, and Czechoslovak Acad. Sci. Head, Res. Group, Nucl. Res. Inst., Czech Acad. Sci., 1962-. I.A.E.A. fellowship, Nucl. Res. Centre, Casaccia, Rome, 1964-66.
Nuclear interests: Disposal of radiation wastes into the ground; Migration of radionuclides in soils and geologic formations; Application of natural sorbents in radiochemical separation processes.
Address: Nuclear Research Institute, Czechoslovak Academy of Sciences, Rez n. Prague, Czechoslovakia.

DLOUHY, Zdenek, Ph.D. Born 1932. Educ.: Charles Univ., Prague. Lebedev Inst. for Phys., Moscow, 1955-59; Head, Res. Group, Nucl. Res. Inst., Czechoslovakian Acad. Sci., 1959-.
Nuclear interest: Spectrometry of fast neutrons; Applications for reactor shielding.
Address: Nuclear Research Institute of Czechoslovak Academy of Sciences, Rez, Czechoslovakia.

DMITRIC, Vojin. Legal Adviser, and Director, Training and Foreign Relations Dept., Federal Nucl. Energy Commission. Adviser, Third I.C.P.U.A.E., Geneva, Sept. 1964.
Address: Federal Nuclear Energy Commission, 29 Kosancicev Venac, P.O. Box 353, Belgrade, Yugoslavia.

DMITRIEV, A. B. Papers: Voltage-current Characteristics of Boron Ionisation Chambers (letter to the Editor, Atomnaya Energiya, vol. 4, No. 4, 1958); co-author, Non Burned-up Radiator for Fission Chamber (letter to the Editor, ibid, vol. 22, No. 4, 1967).
Address: Academy of Sciences of the U.S.S.R., 14 Leninsky Prospekt, Moscow V-71, U.S.S.R.

DMITRIEV, I. D. Papers: Co-author, BN-350 and BOR Fast Reactors (Atomnaya Energiya, vol. 21, No. 6, 1966); co-author, Sodium Technology and Equipment of BN-350 Reactor (ibid, vol. 22, No. 1, 1967).
Address: Academy of Sciences of the U.S.S.R., 14 Leninsky Prospekt, Moscow V-71, U.S.S.R.

DMITRIEV, P. P. Papers: Co-author, Excitation Functions for Reactions Ag^{109} (p,n) Cd^{109} Ag^{108} (d,2n) Cd^{109} and Ag^{107} (α, 2n + pn) Cd^{109} and Yield of Radioisotope Cd^{109} (letter to the Editor, Atomnaya Energiya, vol. 22, No. 4, 1967); co-author, Yields of Isotope Au^{195} in Nucl. Reactions on Cyclotron (letter to the Editor, ibid, vol. 23, No. 1, 1967); co-author, Yields of Ce^{139} in Nucl. Reactions La^{139} (p,n) and La^{139} (d,2n) (letter to the Editor, ibid, vol. 24, No. 3, 1968); co-author, Excitation Function of Cu^{65} (p,n) Zn^{65} Reaction (letter to the Editor, ibid, vol. 24, No. 3, 1968).
Address: Academy of Sciences of the U.S.S.R., 14 Leninsky Prospekt, Moscow V-71, U.S.S.R.

DMITRIEV, V. D. Papers: Co-author, Mechanical Parameters and Microstructure of Structural Materials Irradiated by Neutrons (Atomnaya Energiya, vol. 8, No. 6, 1960); co-author, Influence of Neutron Irradiation on Structure and Properties of Uranium Alloys with 0, 6 - 9, 0 w/o of Mo (ibid, vol. 22, No. 6, 1967).
Address: Academy of Sciences of the U.S.S.R., 14 Leninsky Prospekt, Moscow V-71, U.S.S.R.

DMITRIEV, V. N. Papers: Co-author, The Comparative Characteristics of Ternary Fission in Uranium and Plutonium (letter to the Editor, Atomnaya Energiya, vol. 14, No. 6, 1963); co-author, The Kinetic Energy of Fragments and α-Particles in ^{235}U Ternary Fission (ibid, vol. 15, No. 1, 1963); co-author, Use of Gamma-Gamma Coincidence Spectrometer with Summation of Pulse Amplitudes for Analysis of Mixture Radioactive Isotopes (ibid, vol. 19, No. 4, 1965); co-author, Air Radioactivity above the Atlantic Ocean in May - July 1964 Year (letter to the Editor, ibid, No. 5).
Address: Academy of Sciences of the U.S.S.R., 14 Leninsky Prospekt, Moscow V-71, U.S.S.R.

DMITRIEVA, I. B. Papers: Co-author, Determination of Oxygen in Cesium Metal by Vacuum Distillation Method (Atomnaya Energiya, vol. 24, No. 2, 1968); co-author, Solubility of Oxygen in Alloy of Potassium-Sodium (letter to the Editor, vol. 24, No. 5, 1968).
Address: Academy of Sciences of the U.S.S.R., 14 Leninsky Prospekt, Moscow V-71, U.S.S.R.

DMITRIEVSKII, Vitaly Petrovich. Consultant, Main Administration for Utilization of Atomic Energy, Council of Ministers, U.S.S.R.; Member, J.I.N.R.; Inst. Nucl. Problems, Acad. of Sciences, U.S.S.R. Papers: Co-author, A Cyclotron with a Spatially Varying Magnetic Field (Atomnaya Energiya, vol. 8, No. 3, 1960); co-author, Beam Loss at the Limiting Radius in a Phasotron (letter to the Editor, ibid, vol. 9, No. 4, 1960); co-author, The Effect of Space Charge on the Frequency of Free Oscillations in an Isochronous Cyclotron (ibid, vol. 15, No. 3, 1963); co-author, Strong Focusing Ring Cyclotron for Multi-Charged Ions (ibid, vol. 24, No. 4, 1968).
Address: Joint Institute of Nuclear Research, Dubna, near Moscow, U.S.S.R.

DNEPROVSKII, I. S. Papers: New Isotopes of Holmium and Erbium (letter to the Editor, Atomnaya Energiya, vol. 8, No. 1, 1960); co-author, Work of the Union Scientific Res. Inst. of Instrument Construction (ibid, vol. 23, No. 5, 1967).
Address: Academy of Sciences of the U.S.S.R., 14 Leninsky Prospekt, Moscow V-71, U.S.S.R.

do AMARAL, C. M., Asst. Prof. Theoretical Phys. Dept., Centro Brasileiro de Pesquisas Fisicas.
Address: Centro Brasileiro de Pesquisas Fisicas, 71 Avenida Wenceslau Braz, Rio de Janeiro, Brazil.

DO PRADO, Luiz CINTRA. See CINTRA DO PRADO, Luiz.

DOAN, Richard Lloyd, B.S. (Indiana), M.S. (Indiana), Ph.D. (Chicago), D.Sc. (Hon., College of Idaho). Born 1898. Educ.: Indiana and Chicago Univs. Assoc. Director of Res. (1945-50), Director of Res. (1950-51), Manager, Atomic Energy Div. (1951-63), Project Manager (1963-64), Phillips Petroleum Co., Idaho Falls, Idaho. Director, Div. of Reactor Licensing (1964-66), Consultant to Director of Regulation (1966-68), Asst. Director of Regulation for Special Projects (1968-), U.S.A.E.C. Societies: Nat. Advisory Com. on Reactor Safeguards; A.P.S.; A.C.S.; Fellow, A.A.A.S.; Director, A.N.S.; Assoc. Member, Atomic Ind. Forum.
Address: 1715 Calkins, Idaho Falls, Idaho, U.S.A.

DOANE, J. William, Ph.D. (Phys.). Born 1935. Educ.: Missouri Univ. Asst. Prof., Phys., Kent State Univ. Societies: A.P.S.; A.A.P.T.
Nuclear interest: Nuclear magnetic resonance.
Address: Department of Physics, Kent State University, Kent, Ohio 44240, U.S.A.

DOBBINS, T. Controller, Northrop Nortronics, Northrop Corporation.
Address: Northrop Corporation, Northrop Nortronics, Applied Research Department, 2323 Teller Road, Newbury Park, California 91320, U.S.A.

DOBO, Janos, Dr. Cand. Chem. Sci., Dipl. Eng. Chem. Born 1921. Educ.: Federal Tech. High-School (E.T.H.), Zürich. Sen. Sci. Co-worker, Res. Inst. Plastics Ind., Budapest; Lecturer, Rad. Chem. of Polymers and Monomers, Tech. Univ., Budapest. Society: Soc. Hungarian Chem.
Nuclear interests: Radiation chemistry; Radiation polymerization and grafting; Effect of radiation on materials; Industrial application of large radiation sources.

Address: Research Institute of the Plastics Industry, 114 Hungária krt., Budapest 14, Hungary.

DOBROKHOTOV, E. I., Sci. Asst., Acad. of Sci. of the U.S.S.R.; Deleg. to 2nd I.C.P.U.A.E., Geneva, Sept. 1958; Soviet Deleg., Convention on Thermonucl. Processes, Inst. of Elec. Eng., London, April 29-30, 1959.
Address: Academy of Sciences of the U.S.S.R., 14 Leninsky Prospekt, Moscow V-71, U.S.S.R.

DOBROLYUBSKAYA, T. S. Paper: Spectrophotometric Determination of Uranium (VI) in Carbonate Solutions in Short Ultra-Violet (letter to the Editor, Atomnaya Energiya, vol. 19, No. 6, 1965).
Address: Academy of Sciences of the U.S.S.R., 14 Leninsky Prospekt, Moscow V-71, U.S.S.R.

DOBROMYSLOV, V. A. Paper: Co-author, Rad. Characteristics of ^{145}Sm and Enriched ^{75}Se Gamma-Sources (Atomnaya Energiya, vol. 15, No. 6, 1963).
Address: Academy of Sciences of the U.S.S.R., 14 Leninsky Prospekt, Moscow V-71, U.S.S.R.

DOBROVOL'SKII, S. P. Paper: Co-author, Distribution of Field Dose Rate Gamma-Rad. from Used Fuel Rod (Atomnaya Energiya, vol. 19, No. 4, 1965).
Address: Academy of Sciences of the U.S.S.R., 14 Leninsky Prospekt, Moscow V-71, U.S.S.R.

DOBSON, Robert Lowry, A.B., M.D., Ph.D. Born 1919. Educ.: California Univ. Director, Medical Services, Rad. Lab. (1947-58), Medical Phys. Physician (1950-58), Assoc. Res. Medical Phys., California Univ.; Chief Medical Officer, Rad. and Isotopes, W.H.O., Geneva, 1958-67; Sen. Sci., Bio-Medical Res. Div., California Univ. Lawrence Rad. Lab., Livermore (1967-). Nuclear interests: Radiobiology; Nuclear medicine; Radiation health.
Address: California University, Lawrence Radiation Laboratory, Livermore, California, U.S.A.

DOBZHANSKY, Theodosius, D.Sc. (Hon.: Sao Paulo, Munster, Montreal, Chicago, Sydney, Oxford, Syracuse and Michigan Univs. and Wooster Coll., U.S.A.). Born 1900. Educ.: Kiev Univ. Prof. Genetics, The Rockefeller Univ. Book: Co-author, Radiations, Genes, and Man (New York, Henry Holt, 1959). Societies: Nat. Acad. Sci. (U.S.A.); American Philosophical Soc.; Genetics Soc. of America; American Soc. of Naturalists; Accademia dei Lincei; Roy. Soc., London.
Address: The Rockefeller University, New York City, New York 10021, U.S.A.

DOCTOROFF, Michael. Phys., Phys. Dept., Space Sci., Inc.
Address: Space Sciences, Inc., Physics Department, 301 Bear Hill Road, Waltham, Massachusetts, U.S.A.

DODD, C., Dr. Reader, nucl. physics res. Phys. Dept., University Coll., London.
Address: Physics Department, University College, London, Gower Street, London W.C.1, England.

DODD, John Newton, M.Sc., Ph.D. (Birmingham), F. Inst. P., F.R.S.N.Z. Born 1922. Educ.: Otago and Birmingham Univs. Lecturer, Phys. (1952-59), Reader, Phys. (1959-65), Prof., Phys. (1966-), Otago Univ.; Visiting Fellow, Joint Inst. of Lab. Astrophys., Boulder, Colorado, 1967. Nuclear interests: Optical and radiofrequency methods for nuclear moments; Interaction of radiation and matter.
Address: 7 Malvern Street, Dunedin, New Zealand.

DODEMENT, Georges. Chef du Departement Controles et Analyses, Compagnie Generale de Radiologie.
Address: Compagnie Generale de Radiologie, 48-50 boulevard de Gallieni, Issy-les-Moulineaux, France.

DÖDERLEIN, Jan M., Cand. Real. Born 1932. Educ.: Oslo Univ. Director, Phys. Div., Inst. for Atomic Energy. Societies: Norwegian Phys. Soc.; A.N.S. Nuclear interests: Nuclear marine propulsion; Reactor physics.
Address: I.F.A., Box 40, Kjeller, Norway.

DODGEN, Harold W., B.S., Ph.D. (Chem. and Phys.). Born 1921. Educ.: California Univ. Hanford, G.E.C., 1967; Director, Nucl. Reactor Lab. and Chairman, Chem. Phys. Programme, Washington State Univ. Societies: A.C.S.; A.N.S.; A.A.U.P. Nuclear interest: NMR and NQR studies of structure, motion and bonding in crystals, fast reactions.
Address: Washington State University, Nuclear Reactor Laboratory, Pullman, Washington 99163, U.S.A.

DODSON, Wallace J., B.S. (Phys.). Born 1927. Educ.: California Univ., Berkeley and O.R.S.O.R.T. Nucl. Specialist, Kaiser Engineers. Society: A.N.S. Nuclear interests: Radiation shielding; Reactor physics; Hazards evaluations.
Address: Kaiser Engineers, Division of Henry J. Kaiser Co., Kaiser Centre, Oakland 12, California, U.S.A.

DOE, Bruce Roger, B.S., B. Geol. E., M.S., Ph.D. Born 1931. Educ.: Minnesota Univ., Missouri School of Mines and California Inst. of Technol. Geophys. Lab., Carnegie Inst. of Washington, 1960-61; U.S. Geological Survey, Washington, D.C., 1961-63; U.S. Geological Survey, Denver, 1964-. Societies: A.A.A.S.; Charter member, Geochem. Soc.; A.G.U.; Fellow, Geological Soc. of America. Nuclear interests: Nuclear geology; Mass spectrometry.

Address: United States Geological Survey, Denver, Colorado 80225, U.S.A.

DOE, William B., B.S. (Chem. Eng.). Born 1916. Educ.: Georgia Inst. of Technol. and Florida Univ. Assoc. Chem. Eng., Argonne Nat. Lab. Vice Chairman, Remote Systems Technol. Div., A.N.S.
Nuclear interest: Remote handling of radioactivity.
Address: 441 Blackstone, LaGrange, Illinois 60525, U.S.A.

DOEBLER, Errol W. Vice Pres., Power Reactor Development Co.; Vice-Pres., Empire State Atomic Development Associates, Inc.; Chairman of the Board (-1968), Director (1968-), Long Island Lighting Co.
Address: Power Reactor Development Co., 1911 First Street, Detroit 26, Michigan, U.S.A.

DOEDE, Clinton Milford, B.S., Ph.D. Born 1910. Educ.: Chicago Univ. Pres., Quantum Inc.; Chairman, Windsor Nuclear. Trustee, Nat. Security Ind. Assoc. Societies: Fellow, A.A.A.S.; A.C.S.; A.P.S.; Faraday Soc.
Nuclear interests: Management; Industrial utilisation; Synthesis; Effect of radiation on materials.
Address: 26 Underhill Road, Hamden, Connecticut, U.S.A.

DOERING, Willis P. Phys., Geological Survey, U.S. Dept. of the Interior.
Address: U.S. Department of the Interior, Geological Survey, Building 25, Federal Centre, Denver, Colorado, U.S.A.

DOERNER, Robert Carl, B.S., M.S., Ph.D. Born 1926. Educ.: St. John's, (Minnesota) and St. Louis (Missouri) Univs. Assoc. Phys., Argonne Nat. Lab. Societies: A.P.S.; A.N.S.; Inst. of Radio Eng.
Nuclear interests: Experimental reactor physics; Neutron slowing down; Flux perturbation by detecting foils, Fuel elements and control rods.
Address: Argonne National Laboratory, 9700 South Cass Avenue, Argonne, Illinois, U.S.A.

DOGGETT, Wesley Osborne, B. Nucl. Eng., B.E.E. (North Carolina State College), M.A. (Phys.), Ph.D. (Nucl. Phys., California). Born 1931. Educ.: North Carolina State and California Univs. Res. Phys., California Univ. Rad. Lab., 1954-56; Chief, Reactor Operations Branch, U.S.A.F. Nucl. Eng. Test Facility, 1956-58; Asst. Dean, School Phys. Sci. and Appl. Maths, North Carolina State Univ.; Asst. Prof. Phys. (1958-60), Assoc. Prof. Phys. (1960-62), Prof. Phys. (1962-), Phys. Dept., North Carolina State Univ.; Consultant, Republic Aviation Corp., 1956-59, Res. Triangle Inst., 1960-, Office of Civil Defence, 1962-; Director, Troxler Electronics, Inc., 1962-. Discoverer of isomer ^{81m}Rb and co-discoverer of isotopes ^{109}Sn and ^{119}Sn. Societies: A.P.S.; A.N.S.; North Carolina Acad. Sci.; A. Inst. P.; A.A.A.S.; A.S.E.E.; A.A.P.T.
Nuclear interests: Nuclear and reactor physics; Heat transfer analysis; Nuclear radiation attenuation.
Address: Department of Physics, North Carolina State University, Raleigh, North Carolina, U.S.A.

DOGLIOTTI, G. C. Member, Editorial Com., Minerva Nucleare.
Address: 83-85 Corso Bramante, Turin, Italy.

DOHERTY, David George, B.S., M.S. (Organic Chem.), Ph.D. (Biochem.). Born 1915. Educ.: Wisconsin Univ. Biochem., O.R.N.L. Vice-Pres., Oak Ridge Chapter, Sci. Res. Soc. of America. Societies: American Soc. of Biol. Chemists; A.C.S.; Rad. Res. Soc.; A.A.A.S.
Nuclear interests: All aspects of chemical protection against ionising radiation.
Address: Biology Division, Oak Ridge National Laboratory, Oak Ridge, Tennessee, U.S.A.

DOHNALEK, Josef, M.U.Dr., C.Sc. Born 1912. Educ.: Charles' Univ., Prague and J. E. Purkyne Univ., Brno. Faculty of Medicine, J. E. Purkyne Univ., Brno; Society: Sect. of Nucl. Medicine and Rad. Hygiene (Member of the Staff), Czechoslovak Medical Soc. J. E. Purkyne, Prague.
Nuclear interests: Nuclear medicine: the use of radioisotopes in haematology, endocrinology, gastroenterology.
Address: Radiology and Nuclear Medicine Department, Medical Faculty, University J. E. Purkyne, Komenskeho n. 2, Brno, Czechoslovakia.

DOI, Masaharu, Bachelor's Degree of Law. Born 1894. Educ.: Tokyo Imperial Univ. Chairman, Sumitomo Chem. Co., Ltd. Member, Economic Council; Member, Social Development Council; Vice Chairman, Federation of Economic Organisation; Counsellor, Osaka Chamber of Commerce and Industry; Standing Director, Kansai Economic Federation; Exec. Counsellor, Japan Chem. Industry Assoc.; Member, Japan Atomic Industry Council; Chairman, Isotope Council.
Address: 21-2 Kumoi-cho, Nishinomiya, Hyogo, Japan.

DOI, S. Chairman, Radioisotope Com., and Prof. Physics, Eng. Faculty, Iwate Univ.
Address: Radioisotope Research Laboratory, Iwate University, Ueda, Morioka, Iwate Prefecture, Japan.

DOIG, Contador Publico Augusto Castillo. See CASTILLO DOIG, Contador Publico Augusto.

DOIREAU, Michel, Ingénieur en Chef du Génie Maritime, Ancien Elève de l'Ecole Polytechnique. Born 1916. Chef du Département d'Electronique Générale, C.E.A.
Nuclear interest: Instrumentation nucléaire.

Address: C.E.N. Saclay, B.P.2, Gif-sur-Yvette, France.

DOKO, Toshio. Formerly Pres., Atomic Energy Dept., Eng. Div., Ishikawajima Heavy Industries Co. Ltd.; Formerly Vice-Pres., now Director, Nippon Atomic Industry Group. Address: Nippon Atomic Industry Group Co. Ltd., No. 12-1 1 Yurakucho 1-chome, Chiyoda-ku, Tokyo, Japan.

DOLBILOV, G. V. Paper: Co-author, Collective Linear Acceleration of Ions (Atomnaya Energiya, vol. 24, No. 4, 1968). Address: Academy of Sciences of the U.S.S.R., 14 Leninsky Prospekt, Moscow V-71, U.S.S.R.

DOLGINOVA, N. A. Papers: Co-author, Automatic Start-up Dynamic of Nucl. Reactor (Atomnaya Energiya, vol. 23, No. 3, 1967); co-author, About Choice of Temperature Coolant Programme (ibid, vol.25, No. 2, 1968). Address: Academy of Sciences of the U.S.S.R., 14 Leninsky Prospekt, Moscow V-71, U.S.S.R.

DOLGOV, V. V. Papers: Co-author, Start-up Conditions in a Uranium-Graphite Power Reactor Providing Superheated Steam (Atomnaya Energiya, vol. 9, No. 1, 1960); co-author, Temperature Operating Conditions of Heating Tube at Pulsing of Flow (letter to the Editor, ibid, vol. 20, No. 1, 1966). Address: Academy of Sciences of the U.S.S.R., 14 Leninsky Prospekt, Moscow V-71, U.S.S.R.

DOLGOV-SAVELIEV, G. G. Dr. Deleg. to Conference sur la Physique des Plasmas et la Recherche concernant la Fusion Nucléaire Controlée, Salzbourg, Sept. 1961. Papers: Co-author, Investigations of the Stability and Heating of Plasmas in Toroidal Chambers (2nd I.C.P.U.A.E., Geneva, Sept. 1958); co-author, Microwave Emission of Quasi-Stationary Plasma (letter to the Editor, Atomnaya Energiya, vol. 21, No. 4, 1966). Address: Nuclear Energy Institute, Academy of Sciences of the U.S.S.R., 14 Leninsky Prospekt, Moscow V-71, U.S.S.R.

DOLL, William Richard Shaboe, D.Sc., M.D., F.R.C.P., F.R.S. Born 1912. Educ.: London Univ. Director, M.R.C., Statistical Res. Unit; Assoc. Physician, Central Middlesex Hospital, Acton. Book: Leukaemia and Aplastic Anaemia in Patients irradiated for Ankylosing Spondylitis. Societies: Roy. Statistical Soc.; Medical Res. Soc.; F.R.C.P.; F.R.S.
Nuclear interest: Medical effects of nuclear radiations.
Address: University College Hospital Medical School, 115 Gower Street, London, W.C.1, England.

DOLLEY, J. Development Branch Chief, Eng., Ateliers de Montages Electriques S.A. Address: Ateliers de Montages Electriques S.A., 77 rue Saint-Charles, Paris 15, France.

DOLLEZHAL', N. A., Dr. Tech. Sci.; Power and Construction Eng.; Corresponding Member, Acad. of Sci. of the U.S.S.R.; Deleg. to 2nd I.C.P.U.A.E., Geneva, Sept. 1958. Member, Editorial Board, Atomnaya Energiya.
Nuclear interest: Nuclear reactor engineering.
Address: Nuclear Energy Institute, Academy of Sciences of the U.S.S.R., 14 Leninsky Prospekt, Moscow V-71, U.S.S.R.

DOLPHIN, Geoffrey William, B.Sc., Ph.D. Born 1923. Educ.: Reading Univ. St. Bartholomew's Hospital Medical School, 1950-53; British Empire Cancer Campaign, 1953-58; Biophys. Dept., Yale Univ., 1958-59; Health and Safety Branch, U.K.A.E.A., 1959-. Societies: Hospital Phys. Assoc.; Assoc. for Rad. Res.
Nuclear interest: Interaction of radiation with biological material, especially at the cellular level. Biological aspects of radiobiological protection.
Address: 1, Barrow Road, Shippon, Abingdon, Berks., England.

DOMANIC, Fahri, Ph.D. Born 1915. Educ.: Göttingen Univ., I.S.N.S.E., Argonne Nat. Lab. Prof., Ankara Univ. Sci. Adviser, Turkish A.E.C., Ankara. Societies: A.N.S.; Soc. for Turkish Phys.
Nuclear interests: Reactor physics; Neutron physics.
Address: Fen Fakültesi, Fizik Enst., Ankara, Turkey.

DOMANUS, Jozef, M.S. (Elec. Eng.), Asst. Prof. Born 1919. Educ.: Warsaw Polytech. Head, Ind. Radiol. Dept. and Sec. of Sci. Council, Electrotech. Inst., Warsaw, 1947-. Chairman, Nucl. Energy Commission, Polish Standards Com.; Chairman, Subcom. Radioisotopes and Working Group, Packaging of Radioisotopes, Tech. Commission Nucl. Energy, I.S.O.; Vice-Chairman, Nat. Commission of Rad. Protection; Vice-Chairman, Sect. Non-destructive Testing, Commission of Tech. Applications of Radioisotopes; Council of Peaceful Applications of Nucl. Energy; Member of Presidium, Com. of Sci. and Technique; Commission of Non-destructive Testing Methods, Polish Acad. Sci.; Standardisation Commission on Ind. Radiol. Books: Co-author, X-ray Technique (Trzaska, Evert and Michalski, 1948), and Miniature Radiography (Lekarski Instytut Naukowo-Wydawniczy, 1949); X-ray Technique (Trzaska, Evert and Michalski, 1951); Construction and Equipment of X-ray Darkroom (Panstwowe Wydawnictwa Techniczne, 1951); New Trends in X-ray Technique (Panstwowe Wydawnictwo Naukowe, 1952); Radiology (Trzaska, Evert and Michalski, 1956); Tech. Problems of the Applications of Radioisotopes (Panstwowe Wydawnictwa Techniczne, 1959); co-author, Invisible Detective (Wiedza Powszechna, 1966). Societies: Polish Acad. Sci.; Assoc. of Polish Elec. Eng.

Nuclear interests: Industrial radiography; Radioisotopic gauging; Radiation dosimetry; Radiation metrology; Radiation protection.
Address: 44 m 5 Polna, Warsaw 10, Poland.

DOMANUS, Stefania, M.S., Chem. Eng. Born 1915. Educ.: Warsaw Tech. Univ. Sci. officer, Inst. of Gen. Chem., 1948-55; Sci. officer, Nucl. Res. Inst., 1955-.
Nuclear interest: Uranium ore processing; Graphite technology.
Address: 44 m 5 Polna, Warsaw 10, Poland.

DOMASIK, Lech. Born 1943. Educ.: Pedagogical Coll.
Nuclear interests: Electronics, solid state physics and reactor physics.
Address: 16 m.59 ul. Krzemieniecka, Lodz 3, Poland.

DOMBOVARI, János, Dr. Head, Isotope Lab., Res.Inst. for Irrigation and Rice Cultivation.
Address: Research Institute for Irrigation and Rice Cultivation, 2 Szabadsag ut., Szarvas, Hungary.

DOMEIJ, Bo. Asst., Div. of Phys. IV, Royal Inst. Technol., Sweden.
Address: Division of Physics IV, Royal Institute of Technology, Stockholm 70, Sweden.

DOMENICHINI, P. Tech. Director, Riva e Mariani S.A.S.
Address: Riva e Mariani S.A.S., 5 Via Fatebenefratelli, Milan, Italy.

DOMENICI Marcello, Degree in Phys. Born 1932. Educ.: Pisa Univ., Scuola Normale Superiore di Pisa. C.S.C.E. Electronic Computers Studies Centre, Pisa Univ., 1958; SO.R.I.N., Nucl. Res. Soc., Saluggia, Vercelli, 1959; Stagiaire, C.E.N. Saclay (France) Nucl. Studies Centre of Saclay, 1961-.
Nuclear interest: Neutron diffraction as a tool in investigating the nuclear properties of matter and more generally solid state physics.
Address: Case SO.R.I.N. Saluggia, Vercelli, Italy.

DOMENJOUD, Paul Marie, Ing. Civil des Mines. Born 1918. Educ.: Ecole des Mines de Paris Univ. Ing. Chef de Service, La Technique Electronique, Paris, 1950; Ing. attaché à la Direction Technique (1953), Ing. des Services d'Etudes Nucléaires (1958), Ing. en Chef des Services d'Etudes Nucléaires (1959), Soc. Alsthom. Society: A.N.S.
Nuclear interests: Management; Heat transfer laboratory – parts of equipments.
Address: 96 rue de Longchamp, Neuilly-sur-Seine, (Seine), France.

DOMINGO, Vicente, Ph.D. Born 1934. Educ.: Valencia and Madrid Univs. Asst. Prof., Instituto de Fisica Corpuscular, Valencia Univ., 1956-62; Res. Assoc. Lab. de Fisica Cosmica de Chacaltaya, La Paz Univ., Bolivia. Society: Real Sociedad Española de Fisica y Quimica.
Nuclear interests: Low energy nuclear reactions (ternary fission and photo-reactions) and cosmic rays (extensive air showers).
Address: Laboratorio de Fisica Cósmica de Chacaltaya, Universidad Mayor de San Andrés, La Paz, Bolivia.

DOMINGO PANIAGUA CLAUMARCHIRANT, D. Asesor periodistico, Energia Nuclear.
Address: Junta de Energia Nuclear, Avda. Complutense, 22, Ciudad Universitaria, Madrid 3, Spain.

DOMINGUEZ, German, M.S. Born 1929. Educ.: Barcelona Univ. With Junta de Energia Nucl. of Spain, 1954-. At present Chief, Section of Radiochem.
Nuclear interests: Radiochemistry; Radioisotopes production; Radiochemical separations; Activation analysis; Isotopic power generators.
Address: Junta de Energia Nuclear, Dirección de Quimica e Isótopos, Madrid 3, Spain.

DOMINGUEZ, Jose Javier CLUA. See CLUA DOMINGUEZ, Jose Javier.

DOMINIONI, F. CACCIA-. See CACCIA-DOMINIONI, F.

DOMKEN, Jack William, degree of candidat ing. civil. Born 1921. Educ.: Liège Univ. Gen. Sec., Sec. to the Management Com. and Sec. to the Board, Belgonucléaire; Sec., Management Com., Vulcain Syndicate; Sec. to the Board, Sté. Anglo-Belge Vulcain; Sec. to the Management Com., Assoc. Industrielle Belge pour Centrales de Puissance à Réacteurs Rapides.
Address: Société Anglo-Belge Vulcain, 33 rue des Colonies, Brussels, Belgium.

DONALDSON, W. Lyle, B.S. (Elec. Eng.). Born 1915. Educ.: Texas Technol. Coll., Harvard Univ. and M.I.T. Asst. Prof. Elec. Eng., Lehigh Univ., 1946-51; Active duty, U.S. Navy, 1951-53; Assoc. Prof. Elec. Eng., Lehigh Univ., 1953-54; Sen. Res. Eng., Dept. of Phys. (1954-55), Manager, Communications Sect., Dept. of Electronics and Elec. Eng. (1955-59), Dept. Director (1959-64), Vice-Pres., Electronics and Elec. Eng. (1964-), Southwest Res. Inst. Societies: I.E.E.E.; Soc. for Nondestructive Testing; American Optical Soc.; American Soc. of Professional Eng.
Nuclear interests: Nuclear instrumentation development and nondestructive testing of nuclear components and materials.
Address: Southwest Research Institute, P.O. Box 2296, San Antonio, Texas 78206, U.S.A.

DONATI, Enrico, Dott.Ing. Head, Servizio Ricerche e Laboratori Centrali, Dalmine S.p.A.
Address: Dalmine S.p.A., Dalmine (Bergamo), Italy.

DONATO, Luigi, M.D. Born 1929. Educ.: Pisa Univ. Assoc. Prof., Inst. of Medical Pathology (1952-55), Res. Asst., Medical Clinic (1955-57), Pisa Univ.; Sen. Res. Asst., Dept. Medical Res., McMaster Univ., Canada, 1956-57; Dept. of Medicine, Columbia Univ., New York, 1959-60; Vice-Director, Centre of Nucl. Medicine, Pisa Univ., 1955-; Director, C.E.E.E.A. Res. Group in Nucl. Medicine, Pisa Univ. Editor, Minerva Nucleare; Sec., Italian Soc. Nucl. Biol. and Medicine. Books: Lectures on Isotopic Methodology in Biol. Res. (1957); Radioisotopes in Clinical Investigation (1960); Papers: 52 papers on exptl. and clinical uses of radioisotopes in cardiology and endocrinology, including, co-author, Investigation of Central Haemodynamics by Means of Selective Quantitative Radiocardiography (2nd I.C.P.U.A.E., Geneva, Sept. 1958). Societies: Italian Soc. Nucl. Biol. and Medicine; Italian Soc. Exptl. Biol.; Italian Soc. Cardiology; Italian Soc. Metabolic Diseases; European Soc. Clinical Investigation.
Address: Centro di Medicina Nucleare, Università di Pisa, Italy.

DONDELINGER, René. Deputy Director-Gen., ARBED. Member, Sci. and Tech. Com., Euratom, 1964-.
Address: c/o Euratom, 51-53 rue Belliard, Brussels, Belgium.

DONDES, Seymour, Ph.D. (Chem.) (Rensselaer Polytech. Inst.). Born 1918. Educ.: Brooklyn Coll., N.Y. and Rensselaer Polytech. Inst. Sen. Res. Assoc., Dept. Chem., Rensselaer Polytech. Inst. Societies: A.C.S.; A.A.A.S.; A.P.S.
Nuclear interests: Transformation of nuclear energy into chemical energy; Radiation chemistry; Nuclear reactor technology; Radiochemistry; Nuclear fuel element design; Kinetics of gas reactions under ionising radiation.
Address: Chemistry Department, Rensselaer Polytechnic Institute, Troy, N.Y., U.S.A.

DONELSON, R. N. Senior officer, U.S. Nucl. Corp.
Address: U.S. Nuclear Corporation, P.O. Box 208, 801 N. Lake Street, Burbank, California, U.S.A.

DONETS, E. D. Papers: Co-author, Spontaneous Fission and Production of Far Transuranium Elements (Atomnaya Energiya, vol. 14, No. 1, 1963); co-author, Synthesis of Element 102 Isotope with Mass Number 256 (ibid, vol. 16, No. 3, 1964); co-author, Lw^{256} Fusion (ibid, vol. 19, No. 2, 1965); co-author, On Properties of 102^{254} Isotope (ibid, vol. 20, No. 3, 1966); co-author, Element 103 of the Periodic System (ibid, vol. 25, No. 2, 1968).
Address: Academy of Sciences of the U.S.S.R., 14 Leninsky Prospekt, Moscow V-71, U.S.S.R.

DONI, Israel S., B.Sc. (Chem.). Born 1920. Educ.: Regele Carol II Czernowitz, Rumania, and Hebrew Univ. Jerusalem. Manager, "Indicator" Ltd.
Nuclear interests: Apparatus in testing, discovery and rating. Supervision, stocking and disposing of meters, scintillators, teaching and instructing units.
Address: 40 Lilienblum Street, Tel-Aviv, Israel.

DONNAN, M. G. S., Dr. Consultant Radiol. to Army and Hon. Diagnostic Radiol., St. Vincent's Hospital, Melbourne. Member, Rad. Health Standing Com., Nat. Health and Medical Res. Council.
Address: St. Vincent's Hospital, 111 Collins Street, Melbourne, Victoria, Australia.

DONNEAUX, E., Ing. electricien, Institut de Physique et de Chimie Nucléaires, Univ. Liège.
Address: Institut de Physique et de Chimie Nucléaires, Université de Liège, 9 Place du XX Août, Liège, Belgium.

DONNEDIEU de VABRES, Jean. State Counsellor. Member, Comité de l'Energie Atomique, and Member, Comité Financier, C.E.A.
Address: Commissariat à l'Energie Atomique, 29-33 rue de la Federation, Paris 15, France.

DONNELL, Alton P., B.S. Educ.: West Texas State Univ. and U.S. Military Acad., West Point Gen. Manager (1955-), Vice Pres. (1966-), Atomic Power Development Associates, Inc. Societies: A.N.S.; American Ordnance Assoc.
Address: Atomic Power Development Associates, Inc., 1911 First Street, Detroit 26, Michigan, U.S.A.

DONNELLY, Lt. Gen. Harold C., B.S. Born 1910. Educ.: U.S. Military Acad. Asst. Deputy Chief of Staff, Plans and Programmes, Air Force Headquarters, Washington, 1954-60; Commander, Field Command, Defence Atomic Support Agency, 1960-63; Asst. Deputy Chief, Staff, R. and D., U.S. Air Force, 1963-64; Director, Defence Atomic Support Agency, Dept. of Defence, 1964-68; Manager, Albuquerque Operations Office, U.S.A.E.C., 1968-.
Address: United States Atomic Energy Commission, Albuquerque Operations Office, Post Office Box 5400, Albuquerque, New Mexico 87115, U.S.A.

DONNELLY, Michael J., B.S. (Chem., Canisius Coll.). Radiochem., Savannah River Plant, 1954-56; Radiochem., Westinghouse-Bettis, 1956-60; Manager, Chem. Operations Dept., Controls for Radiation Inc., 1960-.
Address: Controls for Radiation Inc., 130 Alewife Brook Parkway, Cambridge 40, Massachusetts, U.S.A.

DONNELLY, Robert. Member, Illinois Legislative Commission on Atomic Energy.
Address: Illinois Legislative Commission on Atomic Energy, c/o George E. Drach, Myers Building, Springfield, Illinois, U.S.A.

DONNERT, Hermann Jakob Anton, Dr. of Philosophy (Maths. and Theoretical Phys.) Born 1929. Educ.: Leopold Franzens Univ., Innsbruck. Res. Fellow, Cologne Univ., 1953-54; Dozentenanwärter, Freiburg im Breisgau Univ., 1955-57; Phys., Radiol. Div., U.S. Army Chem. Warfare Lab., Army Chem. Centre, Maryland, 1957-60; Chief Rad. Phys. Branch, (1960-61), Acting Sci. Director (1961-62), U.S. Army Chem. Corps. Nucl. Defence Lab., Army Chem. Centre, Maryland; Chief, Nucl. Phys. Div. (1962), Acting Chief Sci. (1962-63), Chief Sci. (1963-66), U.S. Army Nucl. Defence Lab., Edgewood Arsenal, Maryland; Lecturer (Asst. Prof.) for Maths. and Phys., Maryland Univ. (part-time), 1960-66; Assoc. Prof., Nucl. Eng., Kansas State Univ., 1966-. Societies: Österreichische Physikalische Gesellschaft; A.A.A.S.; A.P.S.; RESA; A.N.S.; American Ordnance Assoc.
Nuclear interests: Theoretical nuclear physics; Neutron physics; Radiation transport and shielding; Neutron spectroscopy; Radiation dosimetry; Radiation chemistry.
Address: Department of Nuclear Engineering, Kansas State University, Manhattan, Kansas 66502, U.S.A.

DONOHUE, Robert J., B.S., M.S. Born 1934. Educ.: DePaul and Wayne State Univs. Res. Asst., DePaul Univ., 1954-56; Nucl. Phys., Gen. Motors Res. Labs., 1956-. Societies: A.I.P.; American Inst. of Physics Teachers.
Nuclear interests: Nuclear physics; Energy converter (thermionic) devices and plasma diagnostics.
Address: Physics Department, Research Laboratories, General Motors Corporation, 12 Mile and Mound Roads, Warren, Michigan, U.S.A.

DONON, M. Chef de Fabrication, Chaudronnerie Industrielle et Constructions Spéciales S.A.
Address: Chaudronnerie Industrielle et Constructions Spéciales S.A., Vieux Chemin de Saint-Denis, Noisy-le-Sec, (Seine), France.

DONOVAN, Hugh William, Bachelor of Sci. and Commerce (cum laude). Born 1924. Educ.: Santa Clara Univ. Specialist, Plant and Appropriation Accounting (1956-58), Specialist, Eng. Cost Accounting (1958-59), Specialist, Nucl. Materials Management (1959-66), Specialist, Nucl. Economic Studies (1966-), Atomic Power Equipment Dept., G.E.C. Member, Ad Hoc Com. on Nucl. Materials Management, Atomic Ind. Forum. Society: I.N.M.M.
Nuclear interest: Nuclear materials management and fuel cycle economic studies and evaluation.
Address: c/o General Electric Company, Atomic Power Equipment Department, 175 Curtner Avenue, San Jose, California 95125, U.S.A.

DONOVAN, James Britt, B.A., LL.B. Born 1916. Educ.: Fordham Univ. and Harvard Law School. Pres., Pratt Inst.; former Pres., Board of Education, N.Y.C. Trustee, Brooklyn Inst. of Arts and Sci.; Member, Governing Com., Brooklyn Museum; Pres., Assoc. Municipal Art Commission, N.Y.C.; Chairman, Insurance Negligence and Compensation Law Sect., American Bar Assoc. Gen. Counsel, Nucl. Energy Liability Insurance Assoc. and Nucl. Energy Property Insurance Assoc. Paper: Insurance Problems created by the Peace-time Use of Atomic Energy (Insurance Law Journal, Oct. 1957). Societies: Fellow, American Soc. of International Law; Fellow, American Coll. of Trial Lawyers; Fellow, Pierpont Morgan Library.
Address: Suite 1900, 161 William Street, New York 38, New York, U.S.A.

DONOVAN, John Kenmore, M.B., B.S., D.M.R.T., M.C.R.A. Born 1929. Educ.: Sydney and London Univs. Radiotherapist, Roy. Prince Alfred Hospital, Sydney.
Nuclear interests: Clinical radiotherapy; Clinical diagnostic and therapeutic uses of radioactive isotopes.
Address: Radiotherapy Department, Royal Prince Alfred Hospital, Sydney, Australia.

DONOVAN, M. S. Rad. Safety Officer, Nucl. Lab., Texaco Inc.
Address: Research and Technical Department, Texaco Inc., Box 509, Beacon, New York, U.S.A.

DONTH, Hans H., Dr. rer. nat., Dipl. Phys. Born 1927. Educ.: Heidelberg and Stuttgart Tech. Univs. Vacuumschmelze A.G., Hanau, 1952-53; Metal. Lab., Felten and Guilleaume Carlswerk A.G., Köln-Mülheim, 1953-57; Phys., Bundesministerium für Atomkernenergie und Wasserwirtschaft and Bundesministerium für wissenschaftliche Forschung, Bad Godesberg, 1957-.
Nuclear interests: Radiation detectors; Automatic data handling; Nuclear electronics; Computers; Reactor control and instrumentation.
Address: Bundesministerium für wissenschaftliche Forschung, 2-10 Heussallee (Hochhaus), 53 Bonn 9, Germany.

DOOLEY, H. H., B.A. Born 1917. Educ.: Central Coll., Missouri Univ. Director of Tech. Services, American Atomics Corp. Societies: A.N.S.; A.C.S.; American Inst. Chem.
Nuclear interests: Handling of radioisotopes; Processing of sealed and unsealed radiation sources; Manufacture and use of selfluminous materials, sources and devices.
Address: American Atomics Corp., 2255 Aviation Highway, Tucson, Arizona, U.S.A.

DOOLEY, John Raymond, Jr., B.S. (Mathematics and Philosophy), M.S. (Mathematics, Phys. and Educ.). Born 1925. Educ.: Regis Coll., Denver and Denver Univ. Phys., Nucl. Geology, Dept. of the Interior, U.S. Geological Survey, 1953-. Societies: A.C.S.; Geochem. Soc.; H.P.S.; Colorado Sci. Soc.

DOO

Nuclear interests: Nuclear geology; Natural radioactivity and uranium geochemistry; Uranium and thorium series disequilibrium; Uranium-234 fractionation; Autoradiography and fission tracks.
Address: 6579 Lewis Street, Arvada, Colorado 80002, U.S.A.

DOOLEY, R. C. Manager of Chem. and Ind. Eng. Dept., A.P.V. Co. Ltd.
Address: A.P.V. Co. Ltd., Manor Royal, Crawley, Sussex, England.

DOOLITTLE, James H., A.B., M.S., D.Sc. Born 1896. Educ.: California Univ., Berkeley and M.I.T. Vice-Pres. (1946-58), Director (1946-67), Shell Oil Co.; Chairman (1959-62), Director, (1959-63), Space Technol. Labs., Inc.; Director, Mutual of Omaha Insurance Co., 1961-; Director, Thompson Ramo Wooldridge Inc., 1961-; Director, United Benefit Life Insurance Co., 1964-; Trustee, Aerospace Corp., 1963-; Director, Tele-Trip Co., Inc. 1966-.
Address: 5225 Wilshire Boulevard, Room 702, Los Angeles, California, U.S.A.

DOOREN, R. VAN. See VAN DOOREN, R.

DOORLEY, Paul A., A.B. Born 1919. Educ.: Rider Coll., New Jersey. Pres., Permali, Inc.
Nuclear interest: Reactor shielding materials.
Address: Mount Pleasant, Pennsylvania, U.S.A.

DOPCHIE, Henri A., Ing. Civil, M.Sc. Born 1925. Educ.: Louvain Univ. and California Inst. Technol. Chief Eng., Sté. Traction et d'Electricité.
Nuclear interest: Power station design and safety.
Address: 7 avenue G. Abeloos, Brussels, Belgium.

D'OR, L. Nucl. Chem., Inter-Faculty Centre of Nucl. Sci., Liege Univ.
Address: Inter-Faculty Centre of Nuclear Sciences, Liege University, Liege, Belgium.

DORABIALSKA, Alice, D.Ph., C.Chem.Sci. Born 1897. Educ.: Warsaw Univ. and Radium Inst., Paris. Prof. Dept. Phys. Chem., Tech. Univ., Lodz, 1945-. Pres., Polish Chem. Soc. Books: Natural Radioactivity (Warsaw, P.Z.W.S., 1952). Societies: Polish Chem. Soc.; Sci. Soc. of Lodz.
Nuclear interest: Microcalorimetric investigations of nuclear processes and chemical effects of radiation.
Address: Lodz Technical University, Department of Physical Chemistry, 155 Gdanska, Lodz, Poland.

DORAM, Colin Carrington, B.Sc. Born 1929. Educ.: Hull Univ. Head, Phys. and Mathematics Dept., Whitehaven Coll. of Further Education, 1961-. Societies: H.P.S.; Inst. Nucl. Eng.; Inst. Phys. and Phys. Soc.
Nuclear interests: Radiological protection; Nuclear and reactor physics training.

Address: High House Farm, Parkside, Cleator Moor, Cumberland, England.

DORFMAN, Leon M., (Manitoba), M.A., Ph.D. (Toronto). Born 1922. Educ.: Manitoba and Toronto Univs. Res. Assoc., K.A.P.L. and G.E.C. Res. Lab., 1950-57; Sen. Chem., Chem. Div., Argonne Nat. Lab., 1957-64; Prof. Chem., Ohio State Univ., 1964-. Assoc. Editor, Rad. Res., 1967-; Editorial Board, Internat. Journal Chem. Kinetics, 1968-. Societies: A.P.S.; A.C.S.; Rad. Res. Soc.
Nuclear interests: Radiation chemistry and photochemistry; Kinetics of isotope reactions; Fast reaction studies in radiation chemistry.
Address: Department of Chemistry, Ohio State University, Columbus, Ohio, U.S.A.

DORFMAN, Ralph I., Ph.D. (Physiological Chem. and Pharmacology). Born 1911. Educ.: Chicago Univ. Director of Labs., Worcester Foundation, 1956; Res. Prof. Biochem., Boston Univ., 1951-67; Prof. Chem. (Affiliate), Clark Univ., 1956-64; Director, Inst. Hormone Biol. and Sen. Vice-Pres., Syntex Research, 1964-. Council, Endocrine Soc.; Chairman, Subcom. on Biol. Activity, Endocrinology Panel, Cancer Chemotherapy Nat. Service Centre, U.S. P.H.S.; Res. Advisory Council, American Cancer Soc. Societies: Fellow, American Acad. Arts and Sci.; Fellow, A.A.A.S.; Fellow, N.Y. Acad. Sci.; American Soc. Biol. Chem.
Nuclear interest: Use of radioisotopes in biochemistry and medicine.
Address: Syntex Research, Stanford Industrial Park, Palo Alto, California 94304, U.S.A.

DORIAN, P. F. Sec., Board of Studies in Nucl. Science and Eng., Melbourne Univ.
Address: Melbourne University, Board of Studies in Nuclear Science and Engineering, Parkville N.2, Victoria, Australia.

DORN, Thomas Felder, B.S., Ph.D. Born 1933. Educ.: Duke and Washington (Seattle) Univs. Instructor in Chem., Univ. South Sewanee, Tennessee, 1958-. Societies: A.C.S.; A.A.U.P.
Nuclear interests: C^{14} dating, isotopic tracers in inorganic chemistry.
Address: Department of Chemistry, The University of the South, Sewanee, Tennessee, U.S.A.

DORNALETECHE y RODENAS, Placido. Member, Forum Atomico Español.
Address: Forum Atomico Español, 38 General Goded, Madrid 4, Spain.

DÖRNER, H., Dr. Fragen des Strahlenschutzes Innenministerium Baden-Wurttemberg.
Address: Innenministerium Baden-Wurttemberg 6 Dorotheenstrasse, 7 Stuttgart 1, Germany.

DORNIER, Olivier. Vice-Pres., Assoc. les Amis de l'Atome.
Address: Association les Amis de l'Atome, 11 Square Moncey, Paris 9, France.

DOROFEEV, G. A.
Nuclear interest: Neutron measurements with reference to fission cross-sections and resonance absorption integrals.
Address: Academy of Sciences of the U.S.S.R., 14 Leninsky Prospekt, Moscow V-71, U.S.S.R.

DOROSH, M. M. Papers: Co-author, Method for Determination of Oil-Water Interface Based on Delayed Neutrons Detection (Atomnaya Energiya, vol. 21, No. 1, 1966); co-author, On Possibility of Analysis of Some Metals for Oxygen by Counting of Delayed Neutrons Produced by O^{18} (γ,p) N^{17} Reaction (ibid, vol. 21, No. 3, 1966); co-author, On New Method of Determination of Fluorine Concentration in Metals and other Substances (letter to the Editor, ibid, vol. 24, No. 3, 1968).
Address: Academy of Sciences of the U.S.S.R., 14 Leninsky Prospekt, Moscow V-71, U.S.S.R.

DOROSHENKO, G. G. Papers: Co-author, The Reliability Analysis of Methods for Studying Fast Neutron and γ-Ray Continuous Spectra (Atomnaya Energiya, vol. 16, No. 3, 1964); co-author, On Matrix Treatment of Data obtained by Fast Neutron Single Crystal Scintillation Spectrometer (letter to the Editor, ibid, vol. 19, No. 1, 1965); co-author, Analyses of Systematic Errors Due to Differentiation of Apparatus Spectra Measured by Fast Neutron Single Crystal Scintillation Spectrometer (letter to the Editor, ibid); co-author, Changes of Fast Neutron Spectrum after Passing through Aluminium, Paraffin and Aqueous Medium (letter to the Editor, ibid, No. 5).
Address: Academy of Sciences of the U.S.S.R., 14 Leninsky Prospekt, Moscow V-71, U.S.S.R.

DORRENBACHER, C. J. Vice Pres., Advance Systems and Technol., Missile and Space Systems Div., Douglas Aircraft Co., Inc.; Member, Board of Directors, Atomic Ind. Forum, 1967-70.
Address: Douglas Aircraft Company, Inc., Missile and Space Systems Division, 3000 Ocean Park Boulevard, Santa Monica, California, U.S.A.

DORSSER, A. H. de HAAS van. See de HAAS van DORSSER, A. H.

D'ORVAL, Christian Michel CLOUET. See CLOUET D'ORVAL, Christian Michel.

dos REIS PIEDADE, Lucio. Phys., Radioisotope Lab., Inst. de Cancer, Ministerio da Saude D.N.S.-Servico Nacional de Cancer.
Address: Ministerio de Saude D.N.S.- Servico Nacional de Cancer, 23 Praca de Cruz Vermelha, Rio de Janeiro, Brazil.

dos SANTOS, Pedro Lopes, Dr. Radioisotope Lab., Inst. Biofisica, Brazil Univ.
Address: Radioisotope Laboratory, Instituto de Biofisica, Brazil University, 458 Avenida Pasteur, Rio de Janeiro, Brazil.

DOSE, Klaus, Dr.phil.nat., Dipl.Chem., Privatdozent. Born 1928. Educ.: Kiel, Darmstadt and Frankfurt Univs. Head, Chem. Div. Max Planck-Inst. of Biophys., Frankfurt am Main, 1955-64; Dozent for Chem. Biophys., Frankfurt am Main Univ., 1963-; (On leave since 1965). Assoc. Prof. Biophys. Dept., Michigan State Univ. Societies: Gesellschaft Deutscher Chemiker; European Soc. for Rad. Biol.; Deutsche Gesellschaft für Biophysik.
Nuclear interests: Radiation chemistry of biochemicals, especially proteins and nucleic acids; Photochemistry; Radiation biochemistry of enzymes, protein and nucleic acid synthesis; Radiation chemical synthesis of biochemicals.
Address: Biophysics Department, Michigan State University, East Lansing, Michigan, U.S.A.

DOSOUDIL, Frantisek, Ing. With Czechoslovak A.E.C. Adviser, Third I.C.P.U.A.E., Geneva, Sept. 1964; Member, Editorial Board, Jaderna Energie.
Address: Czechoslovak Atomic Energy Commission, 7 Slezká, Prague 2, Czechoslovakia.

d'OSTIANI, Adrea fe'. See fe' d'OSTIANI, Adrea.

DOSTROVSKY, Israel, B.Sc., Ph.D. (London). Born 1918. Educ.: London Univ. Head, Isotope Dept., Weizmann Inst. of Sci., 1948-. Member, Israel A.E.C., 1952. Director of Res. (1952-56), Sen. Sci., (1966-), Israel A.E.C. Societies: Chem. Soc., Great Britain; A.C.S.; A.I.P.; Israel Chem. Assoc.
Nuclear interests: Physical and physical organic chemistry.
Address: c/o Weizmann Institute of Science, Isotope Dept., Rehovoth, Israel.

DOTTERILL, C. F. Manager, Cee Vee Eng. Co. (Carter and Viner).
Address: Cee Vee Engineering Co. (Carter and Viner), Cooden Sea Road, Bexhill, Sussex, England.

DOTY, Delbert Malcolm, B.S., M.S., Ph.D. Born 1908. Educ.: Purdue Univ. Asst. Director, Res. and Education (1949-56), Assoc. Director, Res. and Education (1956-61), Director, Sci. Activities (1956-64), Director, Res. and Education (1961-64), American Meat Inst. Foundation, Chicago; Tech. Director, Fats and Proteins Res. Foundation, Chicago, 1964-. Nat. Res. Council Advisory Board on Military Personnel Supplies; Chairman, Animals Products Com., Member, Rad. Preservation of Food Com. and Food Gen. Com.; U.S. Dept. Agriculture Advisory Com. on Utilisation Res.; Exec. Vice-Pres., Res. and Development Associates. Societies: A.C.S.; Inst. of Food Technologists.
Nuclear interests: Ionising radiations for food preservation.
Address: Fats and Proteins Research Foundation, 30 La Salle Street, Chicago, Illinois 60602, U.S.A.

DOUB, Howard P., A.B., M.D. Born 1890. Educ.: Johns Hopkins Univ. Radiol.-in-Chief (1923-55), Consulting Radiol. (1955-), Henry Ford Hospital. Editor (1941-66), Emeritus Editor (1966-), Radiology. Formerly Historian, Radiol. Soc. of North America. Societies: American Roentgen Ray Soc.; American Radium Soc.; Radiol. Soc. of North America. Nuclear interest: Medical: all types of diagnostic work.
Address: Henry Ford Hospital, Detroit 2, Michigan, U.S.A.

DOUB, William B., Ph.D. Born 1924. Educ.: California (Los Angeles) Univ. Westinghouse Elec. Corp., 1957-. Societies: A.N.S.; A.P.S. Nuclear interests: Reactor design and nuclear physics.
Address: Bettis Atomic Power Laboratory, Westinghouse Electric Corporation, P.O. Box 1468, Pittsburgh 30, Pa., U.S.A.

DOUD, Alfred H. Vice Pres., Nucl. Eng. Dept., Rochester Gas and Elec. Corp.
Address: Rochester Gas and Electric Corporation, Nuclear Engineering Department, 89 East Avenue, Rochester 4, New York, U.S.A.

DOUGAL, Richard Crombie, B.Sc., Ph.D. Born 1937. Educ.: Edinburgh Univ. Lecturer, Dept. Natural Philosophy, Edinburgh Univ., 1962-. Society: Assoc., Inst. of Phys. Nuclear interest: Electron-nuclear scattering.
Address: Edinburgh University, Department of Natural Philosophy, Drummond Street, Edinburgh 8, Scotland.

DOUGHERTY, Ivan Noel, E.D., B.Ec. Born 1907. Educ.: Sydney Univ. Teacher, headmaster, District Inspector of Schools and Staff Inspector, New South Wales Education Dept., 1928-55; Director of Civil Defence for New South Wales, 1955-; Member of Senate (1954-), Deputy Chancellor (1958-66), Sydney Univ. Pres., United Service Inst. of New South Wales, 1957-58; President of Sydney Legacy, 1958-59.
Nuclear interests: Effects of nuclear weapons and in methods of protecting people from those effects; Co-ordinating civil defence operations in time of war or a possible accident caused by a nuclear reactor, nuclear powered ship, etc.
Address: 4 Leumeah Street, Cronulla, New South Wales 2230, Australia.

DOUGLAS, Donald George, B.A., M.A., Ph.D. Born 1912. Educ.: Saskatchewan and McGill Univs. Prof., Phys. Dept., Manitoba Univ. Society: A.P.S.
Nuclear interest: Low energy nuclear spectroscopy, chiefly by use of bent crystal spectrometer.
Address: Physics Department, Manitoba University, Winnipeg, Canada.

DOUGLAS, Henry T. Born 1909. Educ.: Dartmouth Coll. and Johns Hopkins Univ. Sales and Requirements Dept., Nucl. Div., The Martin Co. Chairman, Maryland Advisory Com. on Atomic Energy; Maryland Rep. on Southern Regional Advisory Council on Nucl. Energy. Society: A.N.S.
Nuclear interests: Market analysis; Economic evaluation and sale of nuclear reactor systems.
Address: 123 Croydon Road, Baltimore 12, Maryland, U.S.A.

DOUGLASS, David L., B.S., M.S. (Pennsylvania), Ph.D. (Ohio). Born 1931. Educ.: Pennsylvania State and Ohio State Univs. E.I. du Pont de Nemours Co., 1953-54; Battelle Memorial Inst., 1955-57; Ohio State Univ., 1956-58; K.A.P.L. (1958-60), Vallecitos Atomic Lab. (1960-), Gen. Elec. Co. Book: Editor, Columbium Metallurgy (Interscience, 1961). Societies: A.I.M.E.; A.S.M.; N.A.C.E.; American Ceramic Soc.
Nuclear interests: Physical metallurgy of Zr, U, and Cb; Corrosion and oxidation; Radiation damage.
Address: Vallecitos Atomic Laboratory, General Electric Company, Pleasanton, Calif., U.S.A.

DOUGLASS, Joseph F. Com. Member, Nucl. Dept., Southern Univ.
Address: Southern University, Southern Branch Post Office, Baton Rouge, Louisiana, U.S.A.

DOULTON, P. D., M.I.Mech.E., M.Inst.Pet.E., M.I.Chem.E. Chairman and Managing Director, Matthew Hall Eng. Ltd.
Address: Matthew Hall Engineering Ltd., Matthew Hall House, 101-108 Tottenham Court Road, London, W.1, England.

DOUMERC, Jean, Ing. E.S.L., Ing. en Génie Atomique (I.N.S.T.N.). Born 1927. At Inst. de Recherches de la Sidérurgie, 1951-54; Sté des Forges et Ateliers du Creusot, 1955-58; Manager, Trade and Development Div., Compagnie pour l'Etude et la Réalisation de Combustibles Atomiques, Paris, 1958-. Societies: Sté. Française de Métallurgie; S.F.I.T.V.; Cercle d'Étude des Métaux; A.N.S.
Nuclear interests: Nuclear fuel elements; fundamental studies and technology, metallurgy of fissile, fertile and structural materials, neutronics, economy.
Address: 5 rue Michelet, Choisy-le-Roi 94, France.

DOUX, John Carver, LE. See LE DOUX, John Carver.

DOVBENKO, A. G. Papers: Co-author, Fast Neutron Radiative Capture for Y^{89} (letter to the Editor, Atomnaya Energiya, vol. 21, No. 6, 1966); co-author, Radiative Capture of Cl^{37}, Rb^{87}, Ir^{193} for Fast Neutrons Cl^{37}, Rb^{87}, Ir^{193} (letter to the Editor, ibid, vol. 23, No. 2, 1967); co-author, Radioactive Capture Cross Sect. of Fast Neutrons for Ge^{74}, Ca^{138} and Os^{192} (letter to the Editor, ibid, vol.23, No. 6, 1967); co-author, Radiative Capture of

Fast Neutrons by Nuclei Sn^{122}, Sn^{124} and Sb^{121}, Sb^{123} (letter to the Editor, ibid, vol. 24, No. 6, 1968).
Address: Academy of Sciences of the U.S.S.R., 14 Leninsky Prospekt, Moscow V-71, U.S.S.R.

DOVBNA, A. N. Papers: Co-author, 100 MeV Neutron Collector (Jaderna Energie, vol. 13, No. 11, 1967); co-author, Storage Ring for 100 MeV Electrons (Atomnaya Energiya, vol. 23, No. 6, 1967).
Address: Physico-Technical Institute, 2 Polytechnicheskaya Ulitsa, Leningrad, U.S.S.R.

DOWNES, Kenneth William, B.Sc., M.Sc., Ph.D. Born 1910. Educ.: McGill Univ. Chief, Extraction Metal. Div., Mines Branch, Dept. of Mines and Tech. Surveys, 1960-. Societies: Chem. Inst. of Canada; Canadian Inst. of Mining and Metal.; A.I.M.M.E.
Nuclear interest: Extraction metallurgy of uranium, thorium and rare earths.
Address: 300 LeBreton Street, Ottawa 1, Ontario, Canada.

DOWNING, Arthur Covington, Jr., A.B. (Phys., Princeton), M.S. (Maths., Michigan), Ph.D. (Maths., Michigan). Born 1925. Educ.: Rensselaer Polytechnic Inst., Princeton, New York and Michigan Univs. Teaching Fellow (Maths.) (1949-50, 1952-), Fellow (1950-51), Res. Asst. (1951-53), Michigan Univ.; Mathematician (1953-60), Asst. Director, Mathematics Panel (1960-64), O.R.N.L.; Manager, Tech. Applications Development, Control Data Corp., 1964-. Societies: American Mathematical Soc.; Assoc. for Computing Machinery; Mathematical Assoc. of America; Soc. for Ind. and Applied Math.
Nuclear interests: Reactor analysis computations; Real time data reduction; Digital computer applications to control of experiments.
Address: Control Data Corp., 4201 N. Lexington Avenue, St. Paul, Minnesota, U.S.A.

DOYEN, Robert Emile, Graduate Mech. and Elec. Eng. Born 1923. Educ.: Faculté Polytech. de Mons, London Univ. and M.I.T. Thermal Power Stations, Private Power Utilities, Belgium, 1945-59; Asst. Prof., Faculté Polytech. de Mons, 1958-; Asst. Prof., Brussels Univ., 1958-; Manager, BR3 Nucl. Power Station, Belgium, 1959-62; Manager, Sté. d'Energie Nucléaire Franco-Belge des Ardennes Nucl. Power Station, Sté. Centre et Sud, 1962-.
Nuclear interest: Power reactors.
Address: 7 rue de la Bonté, Brussels 5, Belgium.

DOYLE, D. J. Manager, Digital Equipment of Canada Ltd.
Address: Digital Equipment of Canada Ltd., 150 Rosamond Street, Carleton Place, Ontario, Canada.

DOZINEL, Paul. Head of Department, Bureau d'Etudes Nucléaires. Adviser, Third I.C.P.U.A.E., Geneva, Sept. 1964.
Address: Bureau d'Etudes Nucléaires, (B.E.N.), S.A., 47 rue Montoyer, Brussels, Belgium.

DRAB, Frantisek, Ing. Elec. Eng. Born 1930. Educ.: Kharkov Polytech. Inst., U.S.S.R. Res. worker, Skoda Works, Pilsen. Societies: Czechoslovak Sci. and Tech. Soc.
Nuclear interest: Reactor control.
Address: Skoda Works, Pilsen, Czechoslovakia.

DRACOPOLI, J. L. Director, Nucl. Eng. Dept., Miles Group of Companies.
Address: Miles Hivolt Ltd., Old Shoreham Road, Shoreham-by-Sea, Sussex, England.

DRAGANIC, Ivan, D.Sc. (Paris). Born 1924. Educ.: Belgrade and Paris Univs. Phys. Chem. (1950-), Head, Rad. Chem. Group (1955-), Boris Kidrich Inst. of Nucl. Sciences. Books: Introduction to Radiochem. (1957); co-author, Radiochem. Practice (1959); Radioactive Isotopes and Radiations vol. I-III (1963).
Nuclear interests: Radiation chemistry of liquids and solids; Large doses chemical dosimetry.
Address: Boris Kidrich Institute of Nuclear Sciences, Post Box 522, Belgrade, Yugoslavia.

DRÄGER, Heinrich, Dr. Senior officer, Nucl. work, Drägerwerk, Lübeck.
Address: Drägerwerk, Heinr. and Bernh. Dräger, 53/55 Moislinger Allee, Lübeck, Germany.

DRAGOUMIS, Paul, B.S.E.E., Dipl. Nucl. Eng. Born 1934. Educ.: Polytech. Inst. of Brooklyn and I.S.N.S.E. R. and D. Manager, East Central Nucl. Group, Inc., 1963-; Chief Nucl. Eng., American Elec. Power Service Corp., 1967-. Chairman, and Member, Exec. Com., Power Div., A.N.S.; Member, Reactor Safety Com., Atomic Ind. Forum.
Nuclear interests: Siting and licensing; Fuel management and physics; Radiological health; Raw materials procurement and handling; Nuclear safety.
Address: 51 Karen Way, Summit, New Jersey 07901, U.S.A.

DRAHNY, Milos, C.Sc. (Scientiarum Candidatus), Ing. (Eng.). Born 1924. Educ.: Brno Tech. and Charles (Prague) Univs. Sci. (1950-58), Manager, Nucl. Power Res. Dept. (1958-66), Deputy Director (1966-), Power Res. Inst., Prague. Chairman, Nucl. Power Group, Czechoslovak Sci. and Tech. Soc. Books: Nucl. Power Development (in Czech) (Prague, UTEIN, 1962); Electricity Producing Costs in Nucl. Power Stations (in Czech) (Prague, UTEIN, 1962); Thermal Cycle Economic in Nucl. Power Stations (in Czech) (Prague, UTEIN, 1963).
Nuclear interests: Nuclear power programmes and economics; Integration of nuclear power stations in power networks; Research and development planning and management. Safeguards, I.A.E.A.

Address: Power Research Institute, 7a Partyzánská, Prague 7-Holesovice, Czechoslovakia.

DRAKE, Antony Elliot, B.A. Born 1907. Educ.: Oxford Univ. Treasury, 1947-57; Principal Finance Officer (1957-), Finance and Programmes Officer (1963-), U.K.A.E.A. Address: Burwood, West Hill, Oxted, Surrey, England.

DRAKE, Francis E., Jun., B.S. (Elec. Eng.). Born 1915. Educ.: Columbia Univ. Exec. Vice-Pres., then Pres. (-1968), now Chairman (1968-), and Chief Exec. Officer, Rochester Gas and Elec. Corp.; Sec.-Treas., High Temperature Reactor Development Assoc., Inc. Society: Atomic Ind. Forum.
Nuclear interests: Reactor design; Management.
Address: 89 East Avenue, Rochester, New York, U.S.A.

DRAKE, Richard Vernon, B.A. (Liberal Arts). Born 1915. Educ.: Grinnell Coll. Manager-Finance, Gen. Elec. Co.
Address: 20232 Blauer Drive, Saratoga, California, U.S.A.

DRAKE, W. W. Chempump Division, Nucl. Dept., Fostoria Corp.
Address: Chempump Division, Nuclear Department, Fostoria Corporation, Buck and County Line Roads, Huntingdon Valley, Pa., U.S.A.

DRALEY, Joseph Edward, B. Appl. Chem., Ph.D. Born 1919. Educ.: Catholic Univ. of America. Metal. Lab., Chicago Univ., 1942-45; Kellex Corp., New York (1945-46), Silver Spring, Maryland (1946-47); O.R.N.L., 1947-48; Argonne Nat. Lab., 1948-; Chairman, Corrosion Div., Electrochem. Soc.; Corrosion Div. editor, J. Electrochem. Soc. Society: A.C.S.
Nuclear interests: Metallurgy; Design and corrosion of nuclear fuel elements.
Address: 814 S. Bruner, Hinsdale, Illinois, U.S.A.

DRANITSYNA, G. F. Book: Co-author, Sistematika energii beta-raspada (Systematics of Energy of Beta-decay) (Fasc. 3 Svoistva Atomnykh Yader (Properties of Atomic Nuclei) (AN SSSR, M., 1960) (Pergamon, London).
Address: Radium Institute, Academy of Sciences of the U.S.S.R., 14 Leninsky Prospekt, Moscow V-71, U.S.S.R.

DRAPCHINSKII, L. V. Papers: Co-author, The Comparative Characteristics of Ternary Fission in Uranium and Plutonium (letter to the Editor, Atomnaya Energiya, vol. 14, No. 6, 1963); U^{235}-U^{238} Ternary Fission Probability Ratio for Neutrons of Different Energies (letter to the Editor, ibid, vol. 16, No. 2, 1964).

Address: Academy of Sciences of the U.S.S.R., 14 Leninsky Prospekt, Moscow V-71, U.S.S.R.

DRAPER, Laurence Rene, A.B., Ph.D. Born 1930. Educ.: Middlebury Coll., Middlebury, Vermont, and Chicago Univ. Res. Assoc., Dept. of Microbiol., Chicago Univ., 1957; Res. Assoc., Argonne Nat. Lab., 1957-60; Res. Biol., Nat. Cancer Inst., 1960-. Assoc. Editor, J. Nat. Cancer Inst. Societies: Washington Area Radiobiol. Assoc.; Rad. Res. Soc.
Nuclear interests: Effects of acute and chronic irradiation on antibody formation.
Address: National Cancer Institute, Laboratory of Physiology, Building 10, Room B2B-44, Bethesda, Maryland, U.S.A.

DRAWBAUGH, Donald W., Ph.D. Born 1923. Educ.: Kansas Univ. Manager, Reactor Phys. and Maths., Westinghouse Astronucl. Lab., 1959-. Member, Tech. Group for Reactor Phys., A.N.S. Societies: A.N.S.; A.P.S.
Nuclear interest: Transport theory.
Address: Astronuclear Laboratory, Westinghouse Electric Corporation, P.O. Box 10864, Pittsburgh 36, Pa., U.S.A.

DRAWE, Horst Georg Friedrich, Dr. rer. nat., Dipl.-Chem. Born 1932. Educ.: Humboldt Freie and Technische Univs., Berlin. Sci. Asst., Hahn-Meitner-Inst. für Kernforschung, Berlin, 1960-. Society: Gesellschaft Deutscher Chemiker, Fachgruppe Kern und Strahlenchemie.
Nuclear interests: Recoil and radiation chemistry, especially phosphorus and arsenic; Radiation chemistry of polymeric systems; Kinetic and preparative aspects of radiation chemistry.
Address: 80 Zobeltitzstr., 1 Berlin 52, Germany.

DREESZEN, W. Emmett, B.S., M.A. Born 1912. Educ.: Iowa State and Iowa Univs. Administrative Aide to the Director, 1948-64; Head, Information and Security, Inst. Atomic Res. and Ames Lab., U.S.A.E.C., Iowa State Univ., 1964-. Society: A.N.S.
Nuclear interests: Technical information; Public relations; Security.
Address: Ames Laboratory, I.S.U., Ames, Iowa 50010, U.S.A.

DREIHELLER, Wilhelm Hermann, Diplom. Physiker. Born 1927. Educ.: Mainz and Marburg Univs. Isotope Lab., Strahlen Institut, Marburg Univ., 1953-58; Frieseke und Hoepfner Erlangen-Bruck, 1958-.
Nuclear interests: Radiation measurement; Isotope techniques; Isotope la aboratories; Radiation protection; Industrial applications.
Address: 18 Hummelweg, 8520 Erlangen-Bruck, Germany.

DREJAK, Zdzislaw Adam, Master of Tech. Sci. (Elec. Eng.). Born 1923. Educ.: Gdańsk Polytech. School. Head, Electronic Dept. and Electronic Development Lab., Inst. Nucl. Phys., Cracow, 1958-64; Manager, Office of Nucl. Tech. Apparatus, Cracow

Nuclear interest: Nuclear instrumentation.
Address: 10/16 Smoluchowskiego, Cracow, Poland.

DRESCHER, William J. Geological Survey, U.S. Dept. of the Interior.
Address: Room 175 Science Hall, University of Wisconsin, Madison, Wisconsin, U.S.A.

DRESEL, Hans, Dr. Working Community for Rad. Protection, Germany.
Address: Radiologischen Institut, Universität Freiburg i. Br., Freiburg i. Br., Germany.

DRESNER, Lawrence, B.S. (C.C.N.Y.), M.A., Ph.D. (Princeton). Born 1929. Educ.: C.C.N.Y. and Princeton Univ. Phys., O.R.N.L., 1954-.
Book: Resonance Absorption in Nucl. Reactors (London, Pergamon Press, 1960). Society: RESA.
Nuclear interests: Reactor theory; Theory of nuclear reactions; Water desalination.
Address: 111 Stanton Lane, Oak Ridge, Tennessee, U.S.A.

DRESSEL, Ralph William, B.S. (Phys.), Ph.D. (Phys.). Born 1922. Educ.: Union Coll., Schenectady, New York and Illinois Univ. Prof. Phys., New Mexico State Univ., 1950-; Phys., White Sands Missile Range, 1961-. Regional Counsellor in Phys. for New Mexico (A.I.P.). Societies: A.P.S.; A.A.P.T.; A.A.A.S.; New Mexico Acad. Sci.
Nuclear interests: Nuclear physics; Electron interactions with nuclei and atomic electrons; Bremsstrahlung and interaction of photons with matter (experimental).
Address: Department of Physics, New Mexico State University, Box 3D, Las Cruces, New Mexico 88001, U.S.A.

DRESSLER, Jacques. Administrateur-Directeur Gén. Adjoint, Sté. Saint-Gobain Techniques Nouvelles; Prés. Directeur Gén., Sté. de Raffinage de l'Uranium; Vice-Prés., Directeur Gen., Compagnie pour l'Etude et la Réalisation de Combustibles Atomiques; Directeur Gén., Sté. d'Etudes et Travaux pour l'Uranium; Pres., Directeur Gen., Compagnie Industrielle des Combustibles Atomiques Frittés.
Address: Saint Gobain Techniques Nouvelles, 23 boulevard Georges Clemenceau, Courbevoie Hauts de Seine 92, France.

DREW, Howard R., B.S. (Eng.). Born 1920. Educ.: Texas Univ. Exec. Vice-Pres., Texas Atomic Energy Res. Foundation, 1959-. Pres., Texas Soc. of Professional Eng.; Texas Atomic Energy Advisory Com. Societies: A.N.S.; A.A.A.S.; Texas Soc. of Professional Eng.; A.S.C.E.; Fellow, Texas Acad. Sci.
Nuclear interests: Thermonuclear reactions and their control.
Address: Box 970, Fort Worth, Texas 76101 U.S.A.

DREXLER, G., Dipl.-Ing. Rep. At Rad. Protection Service.

Address: Institut für Strahlenschutz, 1 Ingolstädter Landstrasse, 8042 Neuherberg bei Munich, Germany.

DREYER, Howard Silbert, B.S. (Chem. Eng.). Born 1928. Educ.: Bucknell Univ. Project Leader, Rad. Effects, N.R.L., 1950; Chem. Eng., Atomic Power Div., G.E.C., 1956; Sen. Chem. Eng., Aerojet-Gen. Nucleonics, 1959; Staff Eng., Lockheed Missile and Space Co., 1963; Consulting Eng., Re-entry Systems, G.E.C., 1966. Societies: A.N.S.; A.C.S.; A.A.A.S.; I.E.E.E.
Nuclear interests: Radiation effects (semiconductors, materials and fuel); Fission gas release (fuel and fuel elements); Space power; Gas and water cooled reactors.
Address: 580 Gulph Road, Wayne, Pennsylvania 19087, U.S.A.

DREYER, William G., B.S. (Chem. Eng.). Born 1924. Manager, Policy Holders Service Div., Insurance Co. of North America. Member, Eng. Com.,Nucl. Energy Liability Insurance Assoc. Societies: A.I.H.A.; American Soc. of Safety Eng.
Nuclear interests: Reactor safety; Criticality safety; Health physics.
Address: Insurance Company of North America, 1600 Arch Street, Philadelphia 1, Pennsylvania, U.S.A.

DRINKWATER, Bryan John, D.I.C. Born 1925. Educ.: London Univ. Chief Eng., (1965), Tech. Director (1967), Permali Ltd.; Tech. Director, Hordern Richmond Ltd., 1965. Exec. Director, Transmission Developments Ltd., 1966.
Address: 230 Stroud Road, Gloucester, England.

DRION, R. arts. Chief Inspector, Medical Inspectorate of Public Health, Ministry of Social Affairs and Public Health.
Address: Medical Inspectorate of Public Health, 8 Dokter Reijersstraat, Leidschendam, Netherlands.

DRISCOLL, William G. Chairman, Phys. Dept., Villanova Univ.
Address: Villanova University, Villanova, Pennsylvania, U.S.A.

DROBINSKI, Joseph Charles, Jr., B.A. (Chem.). Born 1917. Educ.: Wesleyan (U.S.A.) and Queensland Univs., Restricted work for Chem. Corps, Metal Hydrides, Inc., 1950; Chem.-Group Leader (1951), Sen. Sci. (1957-), Tracerlab, Inc.
Nuclear interests: Reactor monitoring; Liquid and gas scintillation counting; Isotope geology.
Address: 1601 Trapelo Road, Waltham, Massachusetts, U.S.A.

DROBNIK, Jaroslav, C. Sc., Assoc. Prof. Born 1929. Educ.: Charles Univ., Prague. At Charles Univ.; Michigan State Univ., 1963-64. Society: Czechoslovak Biophys. Soc.

DRO

Nuclear interest: The biological effects of radiation and transmutation on molecular level, particularly in nucleic acids.
Address: Division of Biophysics, Faculty of Science, Charles University, Prague 2, Vinicna 5, Czechoslovakia.

DROEVEN, G. Director, Station de Chimie et Physique Agricoles, Ministère de l'Agriculture.
Address: Station de Chimie et Physique Agricoles, 13 Chaussée de Wavre, Ernage-Gembloux, Belgium.

DROSTE, G. von. See von DROSTE, G.

DROSTE zu VISCHERING-PADBERG, Freiherr Gottfried v. See v. DROSTE zu VISCHERING-PADBERG, Freiherr Gottfried.

DROULARD, Nelson R. With Budd Co., 1944-60; Formerly Tech. Director of Nucl. and Mech. Eng. Activities, Deputy Director (1966-), Res. Labs., Franklin Inst. of the State of Pennsylvania. Member, Exec. Com., (1964-65), now Sec., Delaware Valley Sect., A.N.S.
Address: Franklin Institute, Philadelphia 3, Pennsylvania, U.S.A.

DROZDOV, F. S. Paper: Co-author, On Determination of Negative Reactivity by Shot Source Method (letter to the Editor, Atomnaya Energiya, vol. 20, No. 1, 1966).
Address: Academy of Sciences of the U.S.S.R., 14 Leninsky Prospekt, Moscow V-71, U.S.S.R.

DROZDOV, V. E. Papers: Co-author, To Calculation of Gamma-Source Absorbed Dose Rate from Nucl. Reactor Spent Fuel Elements (Atomnaya Energiya, vol. 23, No. 2, 1967); co-author, Rad.-Chem. Stand Plant for Thermorad. Vulcanization of Shims with Emitter from Spent Fuel Elements (letter to the Editor, ibid, vol. 23, No. 2, 1967); co-author, On Calculation of Absorbed Dose Rates from Hollow Cylinder Irradiator with Non-Uniform Distribution of Activity (ibid, vol. 24, No. 5, 1968); co-author, The Universal Graphs for Calculation Absorbed Dosed from Plane Irradiators (ibid, vol. 25, No. 3, 1968).
Address: Academy of Sciences of the U.S.S.R., 14 Leninsky Prospekt, Moscow V-71, U.S.S.R.

DROZDOVSKAYA, A. A. Papers: Co-author, Some Thermodynamic Stability Data on Uraninites with Variable Composition in Supergene Conditions (Atomnaya Energiya, vol. 21, No. 6, 1966); co-author, Thermodynamic Analysis of Stability of Uranium Oxides in Low-Temperature Carbonaceous Waters (ibid, vol. 22, No. 5, 1967); co-author, Thermodynamic Analysis of Uranium Migration in Hypergene Sulphate Waters (ibid, vol. 22, No. 5, 1967).
Address: Academy of Sciences of the U.S.S.R., 14 Leninsky Prospekt, Moscow V-71, U.S.S.R.

DROZDOVSKII, B. A. Paper: Co-author, Evaluation of Brittle Fracture Behaviour of Thick Sheet Materials (letter to the Editor, Atomnaya Energiya, vol. 23, No. 6, 1967).
Address: Academy of Sciences of the U.S.S.R., 14 Leninsky Prospekt, Moscow V-71, U.S.S.R.

DRUCKER, Eugene E., S.B., S.M. Born 1924. Educ.: M.I.T. Instructor-Assoc. Prof. Mech. Eng., U.S. Naval Postgraduate School, 1950-56; Assoc. Prof. Mech. Eng., (1956-61), Prof. (1961-), Syracuse Univ.; Fulbright Lecturer Nucl. Eng., Technische Hoogeschool, Delft, 1965-66. Societies: A.N.S.; A.S.M.E.; A.I.Ch.E.; A.S.E.E.; A.A.U.P.; American Soc. of Heating, Refrigerating and Air Conditioning Eng.
Nuclear interests: Reactor heat transfer; Pressuriser design.
Address: Syracuse University, Syracuse, New York 13210, U.S.A.

DRUDE, Burkhard Carl. Born 1903. Educ.: Goettingen and Kiel Univs. Siemens-Reiniger-Werke A.G., Erlangen, 1927-59; Reactor Development Dept., Siemens Werke A.G., Erlangen, 1959-.
Nuclear interests: Management; Desalination.
Address: Siemens A.G., 1 Günther-Scharowsky-Strasse, Erlangen, Germany.

DRUIN, V. A. Papers: Co-author, Synthesis and Determination of Radioactive Properties of Some Isotopes of Fermium (Atomnaya Energiya, vol. 21, No. 4, 1966); co-author, On Nucl. Properties of Isotopes of Element 102 with Mass Numbers 255 and 256 (letter to the Editor, ibid, vol. 22, No. 2, 1967); co-author, Measurements of Cf^{246} and Cf^{248} Spontaneous Fission Periods (letter to the Editor, ibid, vol. 24, No. 1, 1968); co-author, Element 103 of the Periodic System (ibid, vol. 25, No. 2, 1968).
Address: Academy of Sciences of the U.S.S.R., 14 Leninsky Prospekt, Moscow V-71, U.S.S.R.

DRUMMOND, James Edgar, Ph.D. Born 1924. Educ.: Stanford Univ. Book: Editor, Plasma Physics (New York, McGraw-Hill, 1961). Society: A.P.S.
Nuclear interest: Thermonuclear research.
Address: Boeing Scientific Research Laboratories, P. O. Box 3981, Seattle, Washington, U.S.A.

DRUMMOND, William Eckel, B.S., Ph.D. Born 1927. Educ.: Stanford Univ. Phys., Gen. Atomic, San Diego, 1959-65; Prof., Phys., and Director, Centre for Plasma Phys. and Thermonucl. Res., Texas Univ., 1965-; Pres., Austin Res. Assocs. Inc., Austin, 1967-. Assoc. Editor, Physics of Fluids. Society: Div. of Plasma Phys., A.P.S.
Nuclear interest: Controlled thermonuclear research.
Address: 3206 Greenlee Drive, Austin, Texas 78703, U.S.A.

DRYDEN, Charles E., B.S., M.Sc., Ph.D. Born 1917. Educ.: Drexel Inst. Technol., Princeton and Ohio State Univs. Asst. Div. Chief, Battelle Memorial Inst., 1951-54; Assoc. Prof. Chem. Eng. (1954-61), Prof. (1961-), Ohio State Univ. Sec.-Treas., Nucl. Eng. Div., A.I.Ch.E.; Public Relations Com., A.N.S. Book: Co-author, Chem. Eng. Plant Design (McGraw-Hill, 1959); contains work on nucl. chem. plant design. Societies: A.I.Ch.E.; A.N.S.; A.C.S.; A.S.E.E.
Nuclear interests: Reactor design; Fuel cycle systems; Radiation chemistry research.
Address: Dept. of Chemical Engineering, Ohio State University, Columbus 10, Ohio, U.S.A.

DRYLAND, Peter William, A.F.R.Ae.S. Born 1920. Educ.: Medway Tech. Coll. Manager, Design and Development, Eng. Products Div., Goodyear Tyre and Rubber Co., Ltd., 1947-58; Asst. Chief Eng., Windscale Chem. Plants Design Office (1958-63), Reactor Design (1963-), U.K.A.E.A., Reactor Group, Risley.
Nuclear interests: Design and construction of active processing plants; Reactor engineering design, control and instrumentation.
Address: Romney House, Norley Road, Cuddington, Cheshire, England.

DRYZEK, Tadeusz, Mgr. inz. Born 1919. Educ.: Lodz Polytech. Head, Elec. Energy Equipment Design Bureau; Director-Gen., Ministry of Power Industry; Director-Gen., Ministry of Mining and Power Industry. Sec., Com. on Sci. and Technics. Societies: Pres., Polish Elec. Eng. Assoc.; Member of the Presidium, State Council for Peaceful Uses of Nucl. Energy.
Nuclear interest: Nuclear power industry.
Address: Komitet Nauki i Techniki, 1 ul. Krakowskie Przedmiescie, Warsaw, Poland.

DU, Yuan-Tsai. Papers: Co-author, On the possibility of Λ hyperon creation through isobars in π^--p interactions at 7 - 8 GeV (Nukleonika, vol. 9, No. 2 - 3, 1964); co-author, On the possibility of establishing of isobar state systems and their decay modes (ibid.).
Address: Joint Institute for Nuclear Research, Dubna, Nr. Moscow, U.S.S.R.

DU BOIS, Bernard BAILLY. See BAILLY DU BOIS, Bernard.

du BOIS, Marcel. Deputy Gen. Manager, Conrad Zschokke S.A.
Address: Conrad Zschokke Ltd., 42 rue du 31 Décembre, Ch-1211 Geneva 6, Switzerland.

du BOULAY, G. H., M.B., B.S., D.M.R.D., F.F.R. Hon. Sec., British Inst. of Radiol.
Address: British Institute of Radiology, 32 Welbeck Street, London W.1, England.

du COLOMBIER, Louis BUREAU. See BUREAU du COLOMBIER, Louis.

du MESNIL de ROCHEMONT, R. Member, Editorial Board, Atompraxis.
Address: Atompraxis, Verlag G. Braun, 14-18 Karl-Friedrich Strasse, Karlsruhe, Germany.

DU PONT, Frank Melton, B.S. (Mech. Eng.). Born 1922. Educ.: The Cooper Union and Polytech. Inst. of Brooklyn. Chief Nucl. Eng., The Kuljian Corp. Societies: A.N.S.; A.S.M.E.
Nuclear interests: Power plants and facilities; Safety and safeguards systems; Systems dynamics; Management.
Address: 1823 Ft. Washington Avenue, Maple Glen, Pa. 19002, U.S.A.

DU PUI, Pierre A. C., Electrotech. Eng. Born 1907. Educ.: Tech. Univ., Delft. Director, Elec. Supply, Ministry of Economic Affairs.
Nuclear interests: Nuclear energy for the production of electricity.
Address: Ministry of Economic Affairs, 15 Eerste Van den Boschstraat, The Hague, Netherlands.

du ROURE, Edmond. Member, Forum Atomico Español.
Address: Forum Atomico Español, 38 General Goded, Madrid 4, Spain.

DU SERT, Thaddée PERCIE. See PERCIE DU SERT, Thaddée.

DU TEMPLE, Octave Joseph, M.S. (Chem. Eng.), Master in Business Administration. Born 1920. Educ.: Michigan Technol. and Northwestern Univs. Exec. Sec., A.N.S., Inc., 1958-. Societies: A.N.S.; A.I.Ch.E.; A.C.S.; Res. Eng. Soc. America; A.A.A.S.; American Management Assoc.
Nuclear interests: Nuclear economics; Management; Archeological dating methods.
Address: 244 E. Ogden Avenue, Hinsdale, Illinois, U.S.A.

DU TOIT, Stefanus Jakobus, B.Sc. (South Africa), M.Sc., D.Sc. (Stellenbosch). Born 1919. Educ.: Potchefstroom and Stellenbosch Univs., and Nobelinstitutet för Fysik, Stockholm. Res. Officer, Sen. Res Officer, Principal Res. Officer, Chief Res. Officer, N.P.R.L., Pretoria 1947-59; Chief Phys. and Asst. Director-gen., South African Atomic Energy Board, Pretoria, 1959-. Societies: Suid-Afrikaanse Akademie vir Wetenskap en Kuns.
Nuclear interests: Cyclotron design and construction; Nuclear spectroscopy; Lifetimes of excited states; Neutron physics.
Address: Atomic Energy Board, Private Bag 256, Pretoria, Republic of South Africa.

DUBEDOUT, Hubert, Ingénieur E. N., L. ès Sc., Master of Sciences. Born 1922. Educ.: French Naval Acad., Toulouse Univ., and Carnegie Inst. Technol. Now at Centre d'Etudes Nucléaires de Grenoble, Chargé des Relations Extérieures, Programmes et Relations Industrielles.

Address: Centre d'Etudes Nucléaires de Grenoble, B.P. 269, Grenoble, France.

DUBININ, N. P. Director, Inst. Cytology and Genetics, S.I. Vavilor Inst. of Phys. Problems. Paper: Mechanism of Rad. Effect on Heredity and Problem of Radiosensitivity (2nd I.C.P.U.A.E., Geneva, Sept. 1958).
Address: Institute of Cytology and Genetics, Novosibirsk Science Centre, 20 Sovietskaya Ulitsa, Novosibirsk, Siberia, U.S.A.

DUBININA, A. N. At Inst. of Nucl. Phys., Siberian Div., U.S.S.R. Acad. of Sci. Adviser, Third I.C.P.U.A.E., Geneva, Sept. 1964.
Address: U.S.S.R. Academy of Sciences, Siberian Department, Novosibirsk Science Centre, 20 Sovietskaya Ulitsa, Novosibirsk, Siberia, U.S.S.R.

DUBOIS, F., Dr. Book: Contributor to Eng. Compendium on Rad. Shielding (I.A.E.A.).
Address: Centre d'Etudes Nucléaires de Saclay, Boite Postale No. 2, Gif-sur-Yvette, (S. et O.), France.

DUBOIS, Gerard. Prés.-Directeur Gén., Cie. d'Applications et Recherches Atomiques (Caratom).
Address: Cie. d'Applications et Recherches Atomiques, 25 boulevard de l'Amiral Bruix, Paris 16, France.

DUBOIS, Jacques. Member, Conseil Scientifique, C.E.A.
Address: Commissariat a l'Energie Atomique, 29-33 rue de la Federation, Paris 15, France.

DUBOIS, Léon Max, Mining Civil Eng. Born 1921. Educ.: Louvain Catholic Univ. Vice-Pres., (1959-63), Exec. Vice-Pres. (1963-), Evence Coppee et Cie.; Director, Etude et Construction Evence Coppee-Rust, 1962; Director, Bureau d'Etudes Nucléaires, 1963-; Member, Commission Industrielle, Centre d'Etudes Nucléaires, 1963-; Administrateur, Fondation Nucleaire; Administrateur, Groupement Professionnel de l'Industrie Nucleaire. Nuclear interests: General; Management.
Address: 103 boulevard de Waterloo, Brussels, Belgium.

DUBOIS, Marcel, Ing. Civil des Mines. Born 1900. Educ.: Louvain Univ. Gen. Manager of Eng. offices and Director Manager, Traction and Electricité; Administrateur, C.E.N.; Administrateur, Fondation Nucleaire; Administrateur, Groupement Professionnel de l'Industrie Nucleaire. Societies: Sté. Royale Belge des Electriciens; Sté. Royale des Ingénieurs et Industriels.
Nuclear interests: Nuclear power plants; General economics of nuclear energy.
Address: 31 rue de la Science, Brussels 4, Belgium.

DUBOVSKY, B. G. Adviser, Third I.C.P.U.A.E., Geneva, Sept. 1964.
Nuclear interest: Reactor physics.
Address: Nuclear Energy Institute, Academy of Sciences of the U.S.S.R., 14 Leninsky Prospekt, Moscow V-71, U.S.S.R.

DUBOZ, Jacques, Eng. Born 1927. Educ.: Ecole Supérieure de Physique et Chimie, Paris. Chief of Div., C.E.A., Paris, 1959-; entered Commissariat, 1955. Alternate to Director, Eurochemic.
Nuclear interests: Nuclear engineering and design.
Address: 69 Route des Gardes, Meudon, Seine et Oise, France.

DuBRIDGE, Lee Alvin, A.B., A.M., Ph.D. and 17 honorary degrees. Born 1901. Educ.: Wisconsin Univ. Pres., California Inst. of Technol., 1946-. Chairman, Education Panel President's Sci. Advisory Com., 1957; Nat. Sci. Board, 1958-; Vice-Chairman, Education Foundation for Nucl. Sci. Books: Co-author, Photoelectric Phenomena (McGraw-Hill, 1932); New Theories of the Photoelectric Effect (Paris, Hermann and Co., 1934); An Introduction to Space (Columbia University Press, 1960). Societies: Fellow, A.P.S.; Fellow, A.A.A.S.; Member, American Philosophical Soc.; N.A.S. Nuclear interest: Nuclear physics.
Address: 415 South Hill Avenue, Pasadena, California, U.S.A.

DUBROVIN, K. P. Papers: Co-author, On Some Physico-chemical Processes occurring in Fissionable Materials under Irradiation (2nd I.C.P.U.A.E., Geneva, Sept. 1958); Study of Structural Changes produced in Uranium-molybdenum Alloys by Neutron Irradiation (Atomnaya Energiya, vol. 4, No. 1, 1958); co-author, Effect of Reactor Irradiation on Constitution of Some Diluted Uranium Alloys (ibid, vol.22, No. 6, 1967); co-author, Effect of Reactor Irradiation on Materials of Thermo-Couples (letter to the Editor, ibid, vol. 25, No. 3, 1968).
Address: Nuclear Energy Institute, Academy of Sciences of the U.S.S.R., 14 Leninsky Prospekt, Moscow V-71, U.S.S.R.

DUBROVSKII, G. P. Paper: Co-author, Exptl. Study of Sodium Pool Boiling Heat Transfer (letter to the Editor, Atomnaya Energiya, vol. 22, No. 1, 1967).
Address: Academy of Sciences of the U.S.S.R., 14 Leninsky Prospekt, Moscow V-71, U.S.S.R.

DUBROVSKII, V. B. Papers: Co-author, Shielding Properties of Borated Fire-Proof Chromite Concretes (Atomnaya Energiya, vol. 22, No. 2, 1967); co-author, Heat Emission in Borated Concrete Shields (ibid, vol. 22, No. 2, 1967); co-author, Shielding Properties of Borated Concretes (ibid, vol. 23, No. 1, 1967); co-author, On Radiational Stability of Plain Concrete (ibid, vol. 23, No. 4, 1968).

Address: Academy of Sciences of the U.S.S.R., 14 Leninsky Prospekt, Moscow V-71, U.S.S.R.

DUBSEK, Frantisek, Dipl. Ing., C.Sc. Born 1929. Educ.: Prague Tech. Univ. Leading Res. Worker, Vyzkumny ustav energetickych zarizeni, Prvni brnenska strojirna n.p., Brno. Society: Czechoslovak Sci. and Tech. Soc. (Cs. vedecko - technika spolecnost).
Nuclear interests: Nuclear power plants; Components of liquid metal circuits.
Address: 35 Krasneho, Brno, Czechoslovakia.

DUBUISSON, Marcel Georges Valère Céline, Ph.D. Born 1903. Educ.: Ghent Univ. Prof. Gen. Biol. and Zoology, Liège Univ.; Rector, Liège Univ., and Pres., Board of Directors, Administrative Com. on the patrimony and of the Acad. Council of Liège, 1953; Member, Conseil d'Administration, Inst. Interuniversitaire des Sciences Nucléaires. Societies: Acad. Sci. of Belgium; Assoc. Member, Physiological Soc. of London; Assoc. Member, Acad. Sci., Bologna.
Address: Hovade, Tilff-sur-Ourthe, Belgium.

DUCCI, Roberto, Dr. of Law. Born 1914. Educ.: Rome Univ. Italian Deleg. to C.E.E.A. Com., Six-Powers Conference, Brussels, 1955-57; Italian Deleg. to E.N.E.A., Paris, 1957-58; Italian Ambassador to Finland, Yugoslavia and Austria; Permanent Rep. to I.A.E.A. Book: A Short History of Euratom (Rome, S.O.I., 1957).
Nuclear interest: Management.
Address: Italian Embassy, Vienna, Austria.

DUCHESNE, Jules Charles Gérard Léon, B.Sc., Ph.D., D.Sc. Born 1911. Educ.: Liège Univ. Prof., Liège Univ., 1957; Assoc. Prof., Paris Univ., 1964; Director, Chair of Structure of Matter, Liège Univ. Chairman, Belgian Nat. Com. of Biophys.; Member, Editorial Board, Intersci. Publishers for series Advances in Chem. Phys. and of Elsevier Publishing Co. for series J. of Molecular Structure; Member, Commission for Phys. of the Fonds Nat. Belge de la Recherche Sci. Societies: Roy. Acad. of Belgium; Hon. Member, Teatina Acad. Sci. of Chieti, Italy; Hon. Member, Acad. Sci. and Arts of Abruzzi, Italy; Tiberina Acad. of Rome.
Nuclear interest: Magnetic and electrical interactions between nuclei and electrons in atoms and molecules, using radiofrequency spectroscopy.
Address: 3 bd Frère Orban, Liège, Belgium.

DUCKWORTH, Henry Edmison, B.A. (Manitoba), B.Sc. (Manitoba), Ph.D. (Chicago). Born 1915. Educ.: Manitoba and Chicago Univs. Assoc. Prof. Phys., Wesleyan Univ., 1946-51; Prof. Phys. (1951-), Chairman, Dept. of Phys. (1955-), McMaster Univ.; Formerly Head, Dept. of Phys., now Prof. engaged in nucl. work, Manitoba Univ. Editor, Canadian J. Phys., 1955-. Book: Mass Spectroscopy (1958). Societies: F.R.S.C.; Fellow, A.P.S.; Member, Canadian Assoc. of Physicists.
Address: Manitoba University, Physics Department, Winnipeg 19, Manitoba, Canada.

DUCKWORTH, John Clifford, M.A. Born 1916. Educ.: Oxford Univ. N.R.C., Chalk River, Ontario, 1946-47; A.E.R.E., Harwell, 1947-50; Chief Eng., Wythenshawe Labs., Ferranti, Ltd., 1950-54; Nucl. Power Eng. (1954-56), Deputy Chief Eng., Nucl. Power (1957-58), Central Electricity Authority; Chief R. and D. Eng.,C.E.G.B., 1958-59; Managing Director, Nat. Res. Development Corp., 1959-. Vice-Pres., Parliamentary and Sci. Com., 1963-66; Part-time Member, U.K.A.E.A., 1965-. Societies: F. Inst. P.; F.I.E.E.; F. Inst. F. (Pres., 1963-64).
Nuclear interests: Management; Nuclear physics; Reactor design.
Address: Hawks Hill House, Guildford Road, Fetcham, Leatherhead, Surrey, England.

DUCRET, J. Member, Conseil d'Administration, Energie Nucléaire S.A.
Address: Energie Nucléaire S.A., 10 avenue de la Gare, Lausanne, Switzerland.

DUDA, Richard Frank, B.Ch.E. Born 1923. Educ.: Rensselaer Polytech. Inst. With Vitro Engineering Co., 1948-.; Project Eng., TBP Solvent Extraction Waste Recovery Plant (1949-50), Project Eng., Purex Reprocessing Plant (1954), Hanford Works; Project Eng., Uranium Recovery from Oil Field Sands, Oklahoma, 1956; Project Manager, Uranium Solvent Extraction Plant and Metal Recovery Plant, Vitro Rare Metals Co., 1956; Project Eng., Heavy Water-Fertilizer Complex, Nangal, India, 1957-59; Project Manager, Uranium Solvent Extraction Plant, Vitro Uranium Co., 1958; Project Eng., Liquid Metal Fuel Reprocessing Plant Study, Brookhaven Nat. Labs., 1959; Project Manager, Reprocessing Plant for Power Reactor Fuel Elements, Hanford, Washington, 1959-60; Project Manager, Fuel Handling and Cleaning, Enrico Fermi Fast Breeder Reactor, 1960; Project Manager, Hazards Analysis for Fuel Handling, Cleaning and Storage for the Enrico Fermi Facility, 1960; Project Manager, Reprocessing Fuel Elements, Aktiebolaget Atomenergi, Sweden, 1961; Project Manager, Reprocessing Fuel Elements, C.N.E.N., Italy, 1961-62. Societies: A.I.Ch.E.; A.I.M.E.; Armed Forces Chem. Assoc.
Nuclear interests: Reprocessing fuel elements, both aqueous and non-aqueous methods; Uranium mill problems involving solvent extraction.
Address: 211 West Midland Avenue, Paramus, N.J., U.S.A.

DUDEK, Walter, Dr. jur. Member, Fachkommission V—Wirtschaftliche, Finanzielle und Soziale Probleme, Formerly Stv. Vorsitzender, Arbeitskreis V/1—Staatliche Förderungsmassnahmen für die Atomtechnische Entwicklung, Deutsche Atomkommission, Federal Ministry for Sci. Res., Germany.

Address: Federal Ministry for Scientific Research, 46 Luisenstrasse, Bad Godesberg, Germany.

DUDKIN, V. E. Papers: Co-author, Secondary Protons Emitted from Thick Aluminium Targets During Bombarding by 340 Mev Protons (Atomnaya Energiya, vol. 22, No. 6, 1967); co-author, High Energy Nucleons Emitted from Thick Targets of Aluminium During Bombarding by 660 Mev Protons (ibid, vol. 23, No. 3, 1967).
Address: Academy of Sciences of the U.S.S.R., 14 Leninsky Prospekt, Moscow V-71, U.S.S.R.

DUDLEY, Horace Chester, A.B., Ph.D. Born 1909. Educ.: Missouri State Coll. and Georgetown Univ. Commander then Captain, Medical Service Corps., U.S. Navy, 1947-62 (Retired); Head, Radioisotope Lab., U.S.N. Hospital, 1952-62; Prof. and Chairman, Dept. of Phys., Southern Mississippi Univ., 1962-. Book: New Principles in Quantum Mechanics (New York Exposition Press Inc., 1959). Societies: A.P.S.; H.P.S.; A.A.P.T.
Nuclear interest: Neutrino theory involving flux having finite mass.
Address: Department of Physics, Southern Mississippi University, Hattiesburg, Mississippi, U.S.A.

DUDLEY, Robert Augur, B.A., Ph.D. Born 1926. Educ.: Oberlin College, Pennsylvania Univ. and M.I.T. Consultant to Egyptian A.E.C., 1955-56; Res. Assoc., Dept. Phys., M.I.T., 1956-60; Sen. Officer, Div. Isotopes, I.A.E.A., 1960-. Societies: A.P.S.; A.A.A.S.; Hospital Phys. Assoc.
Nuclear interests: Medical applications of radioisotopes; Toxicity of internally deposited radioisotopes; Radioisotope instrumentation techniques.
Address: Division of Research, International Atomic Energy Agency, 11-13 Kaerntnerring, Vienna I, Austria.

DUDNIK, S. S. Paper: Co-author, Decontaminating Low-Activity Effluents by Electrodialysis with Ion-Exchange Membranes (Atomnaya Energiya, vol. 22, No. 5, 1967).
Address: Academy of Sciences of the U.S.S.R., 14 Leninsky Prospekt, Moscow V-71, U.S.S.R.

DUDNIKOV, V. G. Paper: Co-author, Experiments on Production of Intense Proton Beam by Charge Exchange Injection Method (Atomnaya Energiya, vol. 22, No. 5, 1967).
Address: Academy of Sciences of the U.S.S.R., 14 Leninsky Prospekt, Moscow V-71, U.S.S.R.

DUE-PETERSEN, J. Managing Director, Frichs A/S.
Address: Frichs A/S, P.O. Box 115, Åarhus, Denmark.

DUERDEN, Francis, B.Sc. (Hons.). Born 1917. Educ.: Manchester Univ. Formerly Manager, Electronics Div., Bruce Peebles and Co., Ltd., Edinburgh; Formerly Chief Eng., Cossor Electronics; Now Marketing Manager, Automation Div., Marconi Co. Council Member, Industries Council for Educational and Training Technol. Society: M.I.E.E.
Nuclear interests: Real time computer reactor control and monitoring; Computer graphics.
Address: 127 Rye Street, Bishops Stortford, Herts., England.

DUESBERG, Jean-Marie. Member, Conseil d'Administration, Centre et Sud.
Address: Centre et Sud, 20 avenue de la Toison, Brussels 5, Belgium.

DUFF, Brian George, B.Sc., Ph.D. Born 1938. Educ.: London Univ. Lecturer in Phys., Univ. Coll., London. Society: A. Inst. P.
Nuclear interest: High energy physics.
Address: 10 Glade Close, Ditton Hill, Surrey, England.

DUFF, C. Gordon, B.A.Sc. (Mech. Eng.). Born 1928. Educ.: Toronto Univ. Res. Fellow, Ontario Res. Foundation, Toronto, 1951-54; Mech. Development Eng., A.E.C.L., Chalk River, 1954-55; Sen. Design Eng. (Mech.), Canadian Westinghouse Co., Hamilton, Ont., 1955-61; Project Manager, A.M.F. Atomics, Greenwich, Con., 1961-64; Sen. Design Eng. (Mech.), A.E.C.L., Toronto, 1964-.
Nuclear interests: Design and development of research and power reactors and associated systems, especially high-pressure high-temperature loops; Design of nuclear equipment, remote-handling apparatus and fuel reprocessing systems.
Address: 3241 Flynn Crescent, Erindale Woodlands, Cooksville, Ontario, Canada.

DUFFEY, Dick, B.S., M.S., Ph.D. Born 1917. Educ.: Purdue, Iowa, and Maryland Univs. Formerly Director of Nucl. Reactor, now Prof., Maryland Univ. Tech. Sec., Advisory Com. on Reactor Safeguards, U.S.A.E.C.; Governor's Com. on Nucl. Energy, Maryland. Societies: A.I.Ch.E.; A.N.S.
Nuclear interests: Nuclear reactors; Applied gamma and beta radiation.
Address: Nuclear Reactor, Maryland University, College Park, Maryland 20740, U.S.A.

DUFFIELD, Robert B., B.A., Ph.D. Born 1917. Educ.: Princeton and California Univs. Asst. and Assoc. Prof., Depts. of Chem. and Phys., Illinois Univ., 1946-56; John Jay Hopkins Lab. for Pure and Appl. Sci., Gen. Atomic Div., Gen. Dynamics Corp., 1956-67; Lab. Director, Argonne Nat. Lab., 1967-. Societies: A.P.S.; A.C.S.
Nuclear interest: Reactors.
Address: Argonne National Laboratory, 9700 South Cass Avenue, Argonne, Illinois 60439, U.S.A.

DUFFY, Edward C. Director, Atomic Power Development Associates, Inc.; Sen. Vice Pres. (-1968), Pres. (1968-), Long Island Lighting Co.

Address: Atomic Power Development Associates, Inc., 1911 First Street, Detroit 26, Michigan, U.S.A.

DUFFY, Kenneth C., B.S. (Maths.). Born 1924. Educ.: Franklin Coll., Franklin, Indiana. Special Materials Representative, Argonne Nat. Lab., 1950-57; Nucl. Materials Manager, Nucl. Materials Div., Metals and Controls Inc., Attleboro, Mass., 1957-63; Nucl. Materials Manager, Gen. Dynamics Corp. Gen. Atomic Div., 1963-.
Nuclear interest: Nuclear materials management.
Address: 712 Ocean Crest Road, Cardiff, California, U.S.A.

DUFLO, Marcelle, Ing., Ecole Nationale Supérieure de Chimie de Strasbourg, and Licence d'enseignement es Sci. Born 1930. Educ.: Paris and Strasbourg Univs. Maître-Asst., Faculté des sciences.
Nuclear interest: Radiochemistry.
Address: 68 boulevard Soult, Paris 12, France.

DUFLO, P. Directeur Gén. Adjoint, Degremont S.A.
Address: Degremont S.A., 183 route de St. Cloud, Rueil Malmaison, (Seine-et-Oise), France.

DUFOSSEZ, Jean. Sous-Directeur, Atomic Dept., Etablissements Neu, Lille.
Address: Etablissements Neu, Sac Postal 28, Lille (Nord), France.

DuFRESNE, Maurice, Dr. Hon. Sec.-Treas., Canadian Assoc. of Radiologists.
Address: Canadian Association of Radiologists, Suite 511, 1555 Summerhill Avenue, Montreal 25, Quebec, Canada.

DUFRESNE, Origene, Hon. Member, Colegio Inter-Americano de Radiologia, Buenos Aires; B.A., M.D., F.R.C.P. (c), F.A.C.R. Born 1899. Educ.: Montreal and Paris Univs. Radiotherapeutist (1929-), Medical and Sci. Director (1946-67), Radium Inst., Montreal. Asst. Sec.-General of the Tenth Internat. Congress of Radiol., Montreal, 1962. Pres., Administration of l'Union Medicale du Canada, 1962-; Prés. Assoc. des Diplômés de l'Univ. de Montréal; Prés. Sté. Canadienne Française d'Electro-Radiologie. Societies: Sté. de Biologie; Membre titulaire, Sté. Médicale de Montréal; Assoc. des Médicins de Langue Française du Canada; Sté. d'Electro-Radiologie Canadienne-Française; Sté. de Radiologie Médicale de France; Founder member, Canadian Assoc. of Radiol; Radiol. Soc. of North America; Assoc. des Radiologistes de la Province de Québec; American Radium Soc.; Soc. of Nucl. Medicine; American College of Radiol.
Nuclear interests: New developments in nuclear physics, although dealing particularly with diagnosis by I_{131} P_{32} and with treatment by I_{131} P_{32} and Gold 198.

Address: L-Hotel-Dieu le Québec, 11 Cote du Palais, Québec 4, P.Q., Canada.

DUFTSCHMID, Klaus Erwin, Ph.D. Born 1937. Educ.: Vienna Univ. Austrian Atomic Energy Co. (Österreichische Studiengesellschaft für Atomenergie), Reactorcentre, Seibersdorf, 1962-67; Gen. Nucleonics Corp., Claremont, California, 1967-. Societies: Austrian Phys. Soc.; Chem. Phys. Soc. of Vienna Univ.; Austrian Health Phys. Soc.
Nuclear interest: Industrial radioisotope application.
Address: General Nucleonics Corporation, 115 South Spring Street, Claremont, California 91711, U.S.A.

DUGGAN, Herbert G., B.S. (M.E.). Born 1919. Educ.: Tennessee Univ. Union Carbide Nucl. Co., O.R.N.L.,Regional Chairman Public Affairs Com., A.S.M.E.; Subcom. on Hot Labs. and Equipment, A.N.S. Societies: A.N.S.; A.S.M.E.
Nuclear interests: Administering the design of unique remotely controlled nuclear equipment, reactors, hot laboratories, reactor experiments, mechanisms, and related items.
Address: 400 Virginia Road, Oak Ridge, Tennessee, U.S.A.

DUGGAN, Jerome L., Ph.D. Born 1933. Educ.: North Texas State and Louisiana State Univs. Asst. Prof., Georgia Univ., 1961-63; Assoc. Prof., Georgia Univ., on leave to Oak Ridge Assoc. Univs., 1963-65; Sen. Staff, Oak Ridge Assoc. Univs., 1963-. Society: A.P.S.
Nuclear interest: Nuclear physics.
Address: P.O. Box 117, Oak Ridge, Tennessee 37830, U.S.A.

DUHAMEL, Albert Marcel Francis, Ecole Polytechnique, E.S.E. Born 1915. Educ.: Paris Univ. État-Major de l'Armée, 1949-55; Chef du Service de Contrôle des Radiations et de Génie Radioactif, C.E.A., 1955-64; Chargé de mission au Cabinet du Haut-Commissaire à l'Energie Atomique, 1964-. Normalisation du contrôle de la contamination radioactive (1961); Choix des sites nucléaires (1963); Protection collective (1964); Risques associés à la gestion des déchets nucléaires (1966). Societies: Groupement Ampère (Geneva); H.P.S.; Sté. Française de Radioprotection; Internat. Radioprotection Assoc.
Address: 52 avenue Jean-Jaurès, 91 Athis-Mons, France.

DUHAMEL, F. French Member, Health and Safety Sub-com., O.E.C.D., E.N.E.A.
Address: O.E.C.D. European Nuclear Energy Agency, 38 boulevard Suchet, Paris 16, France.

DUHM, Bernhard, Dr. Phil. (Phys.). Born 1909. Educ.: Göttingen and Munich Univs. Head, Phys. and Isotope Inst., Farbenfabriken Bayer A.G., Werk Elberfeld. Board member, German Soc. for Biophys.; Atomic Commission, Div.

Nucl. Chem.; German Joint Com. on Rad. Protection; Pres., Working Com. concerning Rad. Effects and Rad. Protection, German Phys. Soc. and German Soc. for Biophys.
Nuclear interests: Tracer methods; Studies of metabolism; Radiation chemistry.
Address: 93 Am Elisabethheim, 56 Wuppertal-Elberfeld, Germany.

DUJOLS, Pierre. Eng. Born 1922. Educ.: Paris Univ. Manager, S.A. d'Etudes et Réalisations Nucléaires, 1957-.
Nuclear interests: Metallurgy; In-pile thermometry.
Address: 10 rue de la Passerelle, 92 Suresnes, France.

DUKE, Dorothy M. Tech. Librarian, Atomic Energy Div., The Babcock and Wilcox Co. Member, Tech. Information Panel, U.S.A.E.C.
Address: Babcock and Wilcox Company, Atomic Energy Division, 1201 Kemper Street, Lynchburg, Virginia, U.S.A.

DUKE, Philip John, B.Sc. (Hons. Class I), Ph.D. Born 1932. Educ.: Birmingham Univ. D.S.I.R. Res. Fellow, Bristol Univ., 1957-59; Res. Phys., Sci. Res. Council, Rutherford Lab. 1959-.; Sessional Lecturer, Reading Univ. Society: A.P.S.
Nuclear interest: Elementary particle physics.
Address: Rutherford Laboratory, Chilton, Didcot, Berks., England.

DUKE, Thomas Wade, B.S., M.S., Ph.D. Born 1931. Educ.: Texas Agricultural and Mech. Univ. Res. Sci., Res. Foundation, Texas A.and M., 1960; Asst. Lab. Director, Radiobiol. Lab., Bureau of Commercial Fisheries, North Carolina, 1961-; Adjunct Assoc. Prof., North Carolina State Coll., 1964. Societies: A.A.A.S.; Soc. of Limnology and Oceanography; Ecological Soc. of America.
Nuclear interests: Radioecology - the cycling of radioactive elements in the estuarine environment.
Address: Radiobiological Laboratory, Bureau of Commercial Fisheries, Beaufort, North Carolina 28516, U.S.A.

DUKERT, Joseph Michael, B.A. (Notre Dame). Born 1929. Educ.: Notre Dame and Georgetown Univs., Johns Hopkins School of Advanced Internat. Studies, Bologna. Psychological Warfare Officer, U.S.A.F., 1951-53; Editorial Staff, Baltimore News Post, 1953-55; Information Services Staff, (1956-59), Director Public Relations, Nucl. Div. (1960-62), Martin Co.; Manager Public Relations, Res. Inst. for Advanced Studies, 1962-65; Freelance consultant, marketing and ind. communications, 1965-. Com. on Public Understanding of the Atom, Atomic Ind. Forum. Books: Atompower (New York, Coward-McCann, 1962); This Is Antarctica (New York, Coward-McCann, 1965). Societies: Atomic Ind. Forum; Nat. Assoc. Sci. Writers.

Nuclear interests: New developments in the nuclear field, and their interpretation to various specialised groups; Films, articles, scripts and booklets in this field, both for technical and lay audiences.
Address: 222 East 39th Street, Baltimore, Maryland 21218, U.S.A.

DUKES, John Alexander. Born 1912. Res. Manager, Irradiation Specimens Group, Reactor Fuel Element Labs., (Reactor Group), Springfields, U.K.A.E.A., 1960-. Societies: A.R.I.C.; Inst. Nucl. Eng.
Nuclear interests: Automatic continuous process development; Enriched uranium processing.
Address: 88 Liverpool Road, Penwortham, Preston, PR1 0HT, England.

DULIN, V. A. Co-author, Angular Energy Neutron Distribution at the Boundary of Two Media (letter to the Editor, Atomnaya Energiya, vol. 14, No. 5, 1963); co-author, Transformation of Integral Amplitude Distribution into Neutron Energy Spectra (Abstract, ibid, vol. 20, No. 2, 1966); co-author, Angular Distributions of Fast Neutrons behind Iron Barriers (ibid, vol. 20, No. 6, 1966); co-author, Reaction $Li + He_2^3$, $Li + H_1^2$, and $Li + H_1^1$ as Cyclotron Sources of Fast Neutrons (letter to the Editor, ibid, vol. 24, No. 1, 1968).
Address: Academy of Sciences of the U.S.S.R., 14 Leninsky Prospekt, Moscow V-71, U.S.S.R.

DULIS, Edward J., B.S. (Metal. Eng.), M.S. (Sci.). Born 1919. Educ.: Alabama Univ. and Stevens Inst. of Technol. Res. Metal., Fundamental Res. Lab. (1945-52), Supervising Technol., Appl. Res. Lab. (1952-55), U.S. Steel Corp.; Supv., Stainless and High Temperature Materials (1955-60), Manager, Product Res. (1960-63), Asst. Director, Product Res. (1963-64), Director, Product R. and D. (1964-65), Director, Res. (1965-), Crucible Steel Co. of America. Societies: Iron and Steel Inst.; A.I.M.E.; A.S.M.; A.S.T.M.
Nuclear interest: Metallurgy.
Address: Crucible Steel Company of America, Hunter Research Laboratories, P.O. Box 988, Pittsburgh, Pennsylvania 15230, U.S.A.

DULLEMOND, Cornelis, Dr. Phil. Born 1931. Educ.: Amsterdam State and Rochester Univs. Res. Faculty, Phys. Dept., Washington Univ., Seattle, 1962; Res. Faculty, Phys. Dept. (1963), Lector (Assoc. Prof.) (1966-), Nijmegen Univ. Society: A.P.S.
Nuclear interest: Theoretical high energy physics.
Address: Nijmegen University, Physics Department, 200 Driehuizerweg, Nijmegen, Netherlands.

DULMEN, Adrianus Antonius van. See van DULMEN, Adrianus Antonius.

DUMAS, Jean-Claude, Asst., Coll. de France, Lab. de Physique Atomique et Moléculaire.

Address: College de France, Laboratoire de Physique Atomique et Moléculaire, 11 place Marcelin Berthelot, Paris 5, France.

DUMINI, Walfredo, Dott. Chief, Metallographic Dept., Istituto di Ricerche Breda S.p.A. Address: Istituto di Ricerche Breda S.p.A., 336 Viale Sarca, Milan, Italy.

DUMMER, Walter, Ing. Consejero, Consejo Directivo, Comision Chilena de Energia Nucl., 1966-. Address: Comision Chilena de Energia Nuclear, 138 Calle Miraflores, Santiago, Chile.

DUMOND, Jesse William Monroe, B.S. (Calif. Inst. Technol.), M.S. (Union Coll.), Ph.D. (Calif. Inst. Technol.). Born 1892. Educ.: Calif. Inst. Technol. and Union Coll., Schenectady. Prof. Phys., California Inst. Technol., 1946-; Consultant, Argonne Nat. Lab., 1952-57. Consultant, Calif. Univ. Rad. Lab., Livermore Site, 1957-; Commission on Nuclidic Masses, I.U.P.A.P., 1960-; Constants Com., U.S. Nat. Res. Council, Div. of Chem. and Chem. Technol., 1944-. Book: Co-author, The Fundamental Constants of Phys. (Intersci., 1957). Societies: A.P.S.; Sté. Française de Physique; Italian Phys. Soc.; U.S. N.A.S. Nuclear interests: Nuclear gamma-spectroscopy by direct crystal diffraction; Nuclear (magnetic) beta-ray spectroscopy, and particularly instrumentation for the two above techniques; Also nuclear finite size effects and perturbations on X-ray emission line phenomena; Verifications of the Bohr-Mottelson theory of collective states of nuclear excitation; Fundamental atomic constants of physics and physical chemistry. Address: 530 South Greenwood Avenue, Pasadena, Calif. 91107, U.S.A.

DUMONT, Jacques Emile, Medical Degree. Born 1931. Educ.: Brussels Univ. Belgian Army, 1956-58; U.S. Public Health Res. Fellow, 1958-59; Fonds Nat. de Recherche Scientifique, 1959-61; Residency, Hospital Saint Pierre, 1961-62; Fonds Nat. de Recherche Scientifique, 1962-63; Coordinator, Nucl. Medicine Programme, Euratom. 1963-. Nuclear interests: Nuclear medicine and the application of biochemical radioisotopic methods to medicine. Address: 12 Avenue Forêt, Brussels, Belgium.

DUMONT, M. Directeur-Conseil, Poudreries Réunies de Belgique. Administrateur, Groupement Professionnel de l'Industrie Nucléaire. Address: Poudreries Réunies de Belgique, 12 avenue de Broqueville, Brussels 15, Belgium.

DUMONT, Marc, Dr. of Law, Eng. Diploma of Electrotech. Inst. of Grenoble (France); Diploma of Commercial Inst. of Grenoble Univ. Born 1915. Educ.: Brussels, Aix-en-Provence and Grenoble Univs. and Harwell Reactor School. Manager of company i/c consulting eng., exclusive distribution and installation of electrotech. equipment, Casablanca, Morocco, 1947-57; Consul of Luxembourg in Morocco, 1953-57; Consultant, then Counsellor and Head of Div., O.E.C.D., E.N.E.A., Paris, 1957-. Societies: American Soc. of Refrigerating Eng.; Assoc. des Ingénieurs-Electriciens, "La Houille Blanche", Paris. Nuclear interests: Nuclear education and training; Application of radioisotopes, in particular food irradiation. Address: European Nuclear Energy Agency, 38 boulevard Suchet, Paris 16, France.

DUMONTET, Pierre, Dr. ès. Sc. Born 1927. Educ.: Algiers Univ. Prof., Algiers Univ.; Director, Inst. d'Etudes Nucléaires, Universitair and Cultural French Office for Algeria; Sen. Officer, Sté. Algérienne de Documentation Atomique. Societies: Sté. française de Physique; Sté. des Radioélectriciens. Nuclear interest: Theoretical physics. Address: Laboratoire de Physique Théorique et de Physique Nucléaire de l'Université d'Alger, Algeria.

DUMUR, M. Sec. Tech. Officer, Giovanola Frères S.A. Address: Giovanola Frères S.A., Monthey (Valais), Switzerland.

DUNAEV, L. M. Paper: Co-author, Rad.-Chem. Stand Plant for Thermorad. Vulcanization of Shims with Emitter from Spent Fuel Elements (letter to the Editor, Atomnaya Energiya, vol. 23, No. 2, 1967). Address: Academy of Sciences of the U.S.S.R., 14 Leninsky Prospekt, Moscow V-71, U.S.S.R.

DUNAVANT, Billy G., B.S., M.A., Ph.D. Born 1925. Educ.: George Peabody Coll. and Purdue Univ. Instructor, Sci., Freed-Hardeman Coll., Tennessee, 1950-54; Tech. Review and Asst. to Director, Isotopes Div., U.S. A.E.C., 1954-56; Instructor, Radiobiol. (1956-59), Radiol. Control Officer (1956-60), Asst. Prof., Health Phys. (1959-60), Purdue Univ.; Assoc. Prof., Rad. Biol. (1960-67), Director and Prof., Nucl. Sci. (1967-), Florida Univ. Pres., Florida Chapter, H.P.S. Societies: Soc. of Nucl. Medicine; American Assoc. of Phys. in Medicine; Conference on Radiol. Health. Nuclear interests: Radiation biology and biophysics; Low-level measurements of gamma radiation in biological systems, particularly whole-body counting; Nuclear medicine; Science administration. Address: 317 Nuclear Sciences Building, Florida University, Gainesville, Florida 32601, U.S.A.

DUNBAR, Kenneth A., Manager, U.S.A.E.C., Chicago Operations Office. Address: U.S.A.E.C., Chicago Operations Office 9800 South Cass Avenue, Argonne, Illinois 60439, U.S.A.

DUNBAR, W. R. Reactor Supervisor, and Director of Planning, Michigan Memorial

Phoenix Project, Michigan Univ.
Address: Michigan University, College of Engineering, Ann Arbor, Michigan, U.S.A.

DUNCAN, Ebb. Member, Education Com., State of Georgia Nucl. Advisory Commission. Address: State of Georgia Nuclear Advisory Commission, Office of Publications, Georgia Institutte of Technology, Atlanta 13, Georgia, U.S.A.

DUNCAN, James Francis, M.A., B.Sc., D.Phil., D.Sc. (Oxon.), M.Sc. (Melb.). Born 1921. Educ.: Oxford Univ. Prof. Inorg. and Theoretical Chem., Victoria Univ. of Wellington. Books: Co-author, Modern Radiochem. Practice, (Oxford Univ. Press., 1952); Isotopes in Chemistry (Oxford Univ. Press, 1967). Societies: Professional Societies in U.K., Australia and New Zealand. Nuclear interests: Inorganic and theoretical chemistry. Mossbauer effect in chemistry. Solid state kinetics. Address: Chemistry Department, Victoria University of Wellington, New Zealand.

DUNCAN, James Moyer, B.Ch.E., M.S. (Chem. Eng.). Born 1921. Educ.: Florida and Wisconsin Univs. and O.R.S.O.R.T. Chief Nucl. Eng., Patchen and Zimmerman Eng., 1953-58; Assoc. Prof., Nucl. Eng., Florida Univ. 1958-59; Consultant, Phillips Petroleum Co. (N.R.T.S.), 1959; Staff Eng., Gen. Nucl. Eng. Corp., 1959-61; Head, Reactor Lab., Northrop Corp., 1961-63; Manager, Nucl. Eng., Holmes and Narver, Inc., 1963-. Societies: A.N.S.; A.I.Ch.E.; N.S.P.E.; A.C.S.
Nuclear interests: Reactor and hot laboratory design and management; Applications of radiation; Systems engineering.
Address: Holmes and Narver, Inc., 828 So. Figueroa Street, Los Angeles, Calif. 90017, U.S.A.

DUNCAN, Hon. John P., Jr. Member, Agriculture Com., State of Georgia Nucl. Advisory Commission.
Address: State of Georgia Nuclear Advisory Commission, Office of Publications, Georgia Institute of Technology, Atlanta 13, Georgia, U.S.A.

DUNCAN, Kenneth Playfair, B.Sc., M.B., Ch.B., D.I.H. Educ.: St. Andrews Univ. Area Medical Officer, British Railways (Southern), 1951-54; Chief Medical Officer, S.W. Gas Board, 1954-58; Group Medical Officer, Industrial Group (1958-59), Medical Adviser, Health and Safety, Branch (1959-64), Chief Medical Officer, (1964-), U.K.A.E.A.
Nuclear interest: Occupational medicine, and care of over-exposure to internal or external radiation.
Address: Westfield, Steeple Aston, Oxon., England.

DUNCAN, T. C. Vice Pres., Eng., Indian Point Nucl. Energy Power Plant, Consolidated Edison Co. of New York, Inc.
Address: Consolidated Edison Co. of New York, Inc., Indian Point Nuclear Energy Power Plant, Buchanan, New York,U.S.A.

DUNCAN, Val (John Norman Valette), M.A. (Law). Born 1913. Educ.: Oxford Univ. Managing Director, Rio Tinto Co. Ltd., 1951-61; Chairman, Nucl. Developments Ltd., 1961-; Chairman and Managing Director, Rio Tinto-Zinc C orp. Ltd., 1964.
Nuclear interests: Production of uranium and research into its uses.
Address: 6 St. James's Square, London, S.W.1, England.

DUNCAN, William. Formerly Manager, Atomic, Defence and Space Group, now Manager, Aerospace Programmes, Aerospace Div., Defence and Space Centre, (1965-), Westinghouse Elec. Corp.
Address: Westinghouse Electric Corp., P.O. Box 2278, Pittsburgh, Pennsylvania 15230, U.S.A.

DUNCOMBE, Wilfred George, B.Sc., Ph.D. Born 1921. Educ.: Reading Univ. Isotope Sect., Nat. Inst. for Res. in Dairying, Reading Univ., 1949-56; Head, Radiochem. Sect., Wellcome Res. Lab. (Wellcome Foundation), 1956-. Society: Biochem. Soc.
Nuclear interest: Isotopic tracers in biochemistry and pharmacology.
Address: 19 Pollyhaugh, Eynsford, Kent, England.

DUNCTON, Percy John, B.Sc., A.R.C.S., D.I.C., C. Eng. Born 1920. Educ.: London Univ. Chief Exec., Fairey Eng. Ltd., Director of Fairey Co. Ltd. (progressively Chief Technician, Asst. Chief Eng., Tech. Manager since 1948). Council of British Nucl. Forum; Board Member, B.N.E.S. Societies: F. Inst. P.; F.R.Ae.S.
Nuclear interests: Corporate planning, power reactor and experimental nuclear equipment; Fuel elements; Fuel handling, etc.
Address: Riffhams, Henbury Rise, Henbury, nr. Macclesfield, Cheshire, England.

DUNFORD, James M. Manager, Nucl. Power Dept., then Vice-Pres. (1962-), New York Shipbuilding Corp.
Address: New York Shipbuilding Corp., Camden 1, New Jersey, U.S.A.

DUNHAM, Charles Little, B.A., M.D. Born 1906. Educ.: Yale and Chicago Univs. Director, Div. Biol. and Medicine, U.S.A.E.C., 1955-67; Chairman, Div. Medical Sci., N.A.S., Nat. Res. Council, 1967-. Societies: Rad. Res. Soc.; H.P.S.; A.A.A.S.; A.N.S.; A.M.A.; Ind. Medical Assoc.
Nuclear interest: Bio-medical aspects.
Address: National Academy of Sciences, National Research Council, 2101 Constitution Avenue, N.W., Washington D.C. 20418, U.S.A.

DUNHAM, William H. Pres., Central Maine Power Co.; Director, Yankee Atomic Elec. Co.; Pres., Maine Yankee Atomic Power Co., 1966-. Address: Yankee Atomic Electric Co., 441 Stuart Street, Boston 16, Massachusetts, U.S.A.

DUNIGAN, Paul Francis Xavier, B.A., M. Ed. Born 1918. Educ.: Boston Coll. and Massachusetts State Teachers' Coll. Manager, Facilities Operation, as contractor for U.S.A.E.C., Richland, Washington (Pacific Northwest Labs., Battelle Memorial Inst., formerly Hanford Atomic Products Operation, Gen. Elec. Co.), 1950-. Vice-Chairman, Richland Sect., A.N.S., 1961-. Societies: A.N.S.; A.C.S.; A.I.Ch.E. Nuclear interests: Design and operation of nuclear laboratories; Specialised enclosures ranging from glove boxes to massive cells; Methods of disposal of radioactive wastes from laboratories.
Address: 1942 Davison Avenue, Richland, Washington, U.S.A.

DUNKIN, Henry Haughton, B. Metal. E. Born 1909. Educ.: Melbourne Univ. i/c,Joint Univ.-C.S.I.R.O. Ore Dressing Lab., 1941-67; Assoc. Prof. Mining Dept., Melbourne Univ. 1949-. Chairman, Sludge Abatement Board (Victoria); Member, Council, Australian Inst. Mining and Metal. and Chairman, Publications Subcom. Nuclear interest: Beneficiation of mineral raw materials.
Address: Mining Department, Melbourne University, Parkville N2, Victoria, Australia.

DUNLAP, Charles E., A.B., M.D. Born 1908. Educ.: Harvard Univ. Prof. and Chairman, Dept. of Pathology, Tulane Univ. School of Medicine, 1945-; Consultant, Armed Forces Inst. of Pathology, 1952-. Com. on Pathological Effects of Rad., Nat. Res. Council; Board of Directors, O.R.I.N.S.; Board of Editors, Archives of Pathology. Book: Chapter on Effects of Rad. (in: Pathology, St. Louis, C.V. Mosby Co., 1961). Societies: American Assoc. Pathologists and Bacteriologists; Rad. Res. Soc.; Founding Fellow, Coll. American Pathologists; Internat. Acad. Pathology; American Soc. for Exptl. Pathology.
Nuclear interests: Radiation biology; Effects of radiation on normal and neoplastic mammalian tissues.
Address: Tulane University School of Medicine, 1430 Tulane Avenue, New Orleans, Louisiana 70112, U.S.A.

DUNLAP, Julian Lee, B.E.E., Ph.D. (Phys.). Born 1932. Educ.: Georgia Inst. Technol. and Vanderbilt Univ. Sen. Res. Staff, Thermonucl. Div., O.R.N.L., 1959-. Society: A.P.S. Nuclear interests: Research in controlled thermonuclear reactions and plasma physics.
Address: Thermonuclear Division, Oak Ridge National Laboratory, Oak Ridge, Tennessee, U.S.A.

DUNLAP, Robert Hawkins, Registered Professional Eng. Born 1913. Educ.: School Sci. and Technol., Pratt Inst., Washington Square Coll., New York. Asst. Chief Tooling Eng., Chase Aircraft Co., 1950-56; Sen. Eng. and Asst. to Manager, Development Dept. Babcock and Wilcox Co., Lynchburg, Va, 1956-60; Exec. Director (State of Florida) Nucl. Commission, 1961-65. Asst. to Pres.,·Florida Inst. of Technol., Melbourne, Florida, 1966-. Societies: A.N.S.; Inst. of Nucl. Eng.; Atomic Ind. Forum. A.S.M.E.; Florida Eng. Soc.; N.S.P.E.; American Management Assoc.; American Ordnance Assoc.; A.A.A.S.; Soc. of Nucl. Medicine; Marine Technol. Soc.; A.I.A.A.; Governor's Representative to Regional Advisory Council on Nucl. Energy.
Address: Florida Institute of Technology, Melbourne, Florida 32901, U.S.A.

DUNLOP, Donald P., B.Sc. (Hons., Appl. Phys.). Born 1925. Educ.: Carleton Univ., Ottawa. NRX operations (1951-53), Analytical Design (1954-57), Fuel Eng. (1958), Nucl. Submarine Survey Team (Roy. Canadian Navy) (1959), Canadian Sec., U.S.A.E.C.-A.E.C.L. Joint Board and Tech. Advisory Com. (1960-63), A.E.C.L. Sen. Eng. and Project Coordinator, Atomic Products Dept., Hawker Siddeley Eng., 1963-65; Sen. Eng., Eng. Development D.W., A.E.C.L. Power Projects. Societies: Canadian Assoc. of Phys.; I.E.E.E. Nuclear interests: Shielding; Reactor safety; Fuelling systems; Experimental loops; Engineering development.
Address: Atomic Energy of Canada Limited, Power Projects, Sheridan Park, Ontario, Canada.

DUNN, A. W. Formerly Director and Manager, Heat Transfer, Fuel Elements, Reinforced Plastics and Special Purposes Machine Groups, now Managing Director, Marston Excelsior Ltd.
Address: Marston Excelsior Ltd., Fordhouses, Wolverhampton, Staffs., England.

DUNN, Albert A., M.D. Pres., New York Roentgen Soc.
Address: Roosevelt Hospital, 428 West 59th Street, New York City, U.S.A.

DUNN, Arthur Lovell, A.B., A.M., Ph.D. Born 1908. Educ.: Omaha and Nebraska (Lincoln) Univs. and Nebraska Coll. of Medicine (Omaha). Chief, Electronics Res. Unit, Veterans Administration Hospital, Omaha. Societies: New York Acad. of Sci.; A.N.S.; Inst. of Radio Eng.; Fellow, A.A.A.S.
Nuclear interests: Medical research utilising radioisotopes; Neutron activation analysis; Electronic instrumentation for measurement of radioactivity.
Address: 5101 Rees Street, Omaha 6, Nebraska, U.S.A.

DUNN, Colon H., B.S., M.E.E. Born 1921. Educ.: John Brown Univ. and Rensselaer Polytech. Inst. Prof. and Head Elec. Eng., Wichita Univ., Consultant U.S.A.E.C., 1958; Assoc.

Head Elec. Eng., Rensselaer Polytech. Inst., 1959; Consultant, Argonne Nat. Lab., 1959; Consultant Boeing Airplane Co. and Beech Aircraft, 1960-. Societies: American Inst. of Elec. Eng.; A.S.E.E.; A.N.S.
Nuclear interests: Nuclear instrumentation and control.
Address: Wichita University, Wichita 14, Kansas, U.S.A.

DUNN, K. Vice Pres., Treas., Rio Algom Mines Ltd.
Address: Rio Algom Mines Ltd., 335 Bay Street, Toronto 1, Ontario, Canada.

DUNN, Peter Douglas, B.Sc. (Hons., Nottingham), B.Sc. (London), Ph.D., C. Eng. Born 1927. Educ.: Nottingham Univ. Scientific Officer (1950-53), S.S.O. (1953-57), P.S.O. (1957-60), S.P.S.O. (1960-65), A.E.R.E., Harwell; Prof. Eng. Sci., Reading Univ., 1965-. Societies: A.M.I.Mech.E.; I.E.E.
Nuclear interests: Conversion of reactor generated heat to electricity; Particle accelerators.
Address: Downlands, Glebe Close, Moulsford, Berks., England.

DUNN, Richard P. Adviser, Atomics, Phys. and Sci. Fund Inc.
Address: Atomics, Physics and Science Fund Inc., 1033 30th Street, N.W., Washington 7, D.C., U.S.A.

DUNN, Roy E., B.S. (Eng. Phys.). Born 1927. Educ.: Montana State Univ. Reactor Phys. (1951-54), Specialist, Measurements and Administrative Procedures (1954-58), Supv., Reactor Operations (1958), Sen. Control Eng., Plutonium Recycle Test Reactor (1958-61), Manager, C Reactor Plant (1961-64), Manager, Production Planning and Schedule (1964-65), Manager, KE-KW Reactor Plant (1965-67), Manager, N Reactor Sec. (1967-), G.E.C., Hanford; A.E.C. Reactor Operator License Examiner, 1966-. Past Chairman, Richland Sect., A.N.S. Societies: A.N.S.; Atomic Ind. Forum.
Nuclear interests: Reactor management; Reactor instrumentation; Reactor control; Reactor operator training and certification.
Address: 2047 Hudson, Richland, Washington 99352, U.S.A.

DUNN, William A., Editor, Science Horizons (U.S.A.).
Address: U.S. Information Service, 24 Grosvenor Square, London, W.1., England.

DUNNING, Gordon M., M.S., Ed.D. (Sci. Educ.). Born 1910. Educ.: Cortland State Coll. (New York) and Syracuse Univ. Biophys. Res. Analyst (1951-52), Chief, Rad. Effects of Weapons Branch (1952-59), Asst. Director, Deputy Director, Div. of Operational Safety (1960-67), Tech. Advisor, Div. of Operation Safety (1967-), U.S.A.E.C.; Sci. Advisor, U.S.P.H.S., 1959-60. Societies: H.P.S.; Fellow, A.A.A.S.; A.A.P.T.
Nuclear interests: Health physics; Radiation protection standards.
Address: United States Atomic Energy Commission, Washington, D.C. 20545, U.S.A.

DUNNING, John Ray, A.B. (Nebraska Wesleyan), Ph.D. (Columbia), Hon.: Sc.D., LL.D., D.Sc.Ed. Born 1907. Educ.: Nebraska Wesleyan and Columbia Univs. Dean, School of Eng., Columbia Univ. Director: A.A.A.S., Oak Ridge Assoc. Universities, Fund for Peaceful Atomic Development, Vitro Corp., Basic Sci. Foundation, City Investing Corp., Armstrong Memorial Res. Foundation, Sci. Service, Empire State Atomic Development Associates; Chairman, Nucl. Energy Policy Board; Chairman, New York State Sci. Advisory Council to the Legislature. Book: Matter, Energy and Rad. (McGraw-Hill, 1941). Societies: A.S.M.E.; Eng. Joint Council; N.S.F.; Fellow, A.A.A.S.; Nat. Acad. of Sci.; Fellow, New York Acad. of Sci.; Fellow, A.P.S.
Nuclear interests: Uranium fission; Gas diffusion; Nuclear physics; Neutron physics, accelerators; U-235 separation systems; Nuclear power plants.
Address: Spring Lake Road, Sherman, Connecticut, U.S.A.

DUNNING, Kenneth L., B.E.E., M.S. (Phys.). Born 1914. Educ.: Minnesota, Maryland Univs. Head, Van de Graaff Branch, Nucl. Phys. N.R.L., Washington, D.C., 1952-. Societies: A.P.S.; RESA; Washington Philosophical Soc.; Washington Acad. Sci.
Nuclear interests: Experimental nuclear physics; Accelerator design; Research management.
Address: Code 7670, Naval Research Laboratory, Washington, D.C. 20390, U.S.A.

DUNNINGTON, Bruce Willard, B.Met.E., M.Sc., Ph.D., Met. Born 1924. Educ.: Iowa State and Ohio State Univs. Sen. Res. Supervisor, Savannah River Lab., E. I. du Pont de Nemours and Co., Inc., 1951-57; Consultant, Battelle Memorial Inst., Columbus, Ohio, 1957-. Societies: A.I.M.E.; A.S.M.; A.N.S.
Nuclear interests: Nuclear fuels and structural components; Principal studies have been in development, fabrication, and irradiation of fuel assemblies.
Address: E. I. du Pont de Nemours and Co., Eastern Laboratory, Gibbstown, New Jersey, U.S.A.

DUNNINGTON, Frank G., B.S., Ph.D. Born 1903. Educ.: California Univ. Prof. Phys. and Rad. Sci. (1946-), Director, Rad. Sci. Centre (1966-), Rutgers Univ. Chairman, New Jersey Rad. Protection Commission, 1958-. Societies: Fellow, A.P.S.; H.P.S.
Nuclear interests: Nuclear physics (low energy); Radiological physics.
Address: Rutgers, The State University, Radiation Science Centre, New Brunswick, New Jersey, U.S.A.

DUNSTER, Herbert John, B.Sc., A.R.C.S. Born 1922. Educ.: London Univ. R. and D. Branch, Windscale Works, Cumberland (1950-53); Health Phys. Div., Harwell (1953-57), Chief Health Phys., Health and Safety Branch, Ind. Group, Risley (1957-59), Health Phys. Adviser, Authority Health and Safety Branch, Harwell, (1959-67), Deputy Head, Radiological Protection Div. Authority, Health and Safety Branch, Harwell (1967-), U.K.A.E.A. Societies: British Occupational Hygiene Soc.; Soc. for Radiological Protection; Internat. Rad. Protection Assoc.
Nuclear interests: Radiological protection and occupational hygiene.
Address: Radiological Protection Division, Authority Health and Safety Branch, United Kingdom Atomic Energy Authority, Harwell, Didcot, Berks., England.

DUNWORTH, John Vernon, M.A., Ph.D. Born 1917. Educ.: Cambridge Univ. Sometime Fellow of Trinity Coll., Cambridge, and Univ. Demonstrator in Phys. at Cambridge. Head, Reactor Div. (1951-58), Asst. Director (Reactor Res. Policy) (1958-59), A.E.R.E., Harwell; Deputy Director, Atomic Energy Establishment, Winfrith, 1959-62; Deputy Director, (1962-64), Director, (1964-), Nat. Phys. Lab. Editor-in-Chief, J. Nucl. Energy; Vice-Pres., Inst. of Phys. and Phys. Soc.; Chairman of Board and Exec. Com., B.N.E.S. Societies: Fellow, A.N.S.; F.I.E.E.; Fellow, Inst. of Phys. and Phys. Soc.; Member, Cambridge Univ. Eng. Assoc.
Nuclear interest: Production of useful energy for electricity generation, ship propulsion, and special applications.
Address: National Physical Laboratory, Teddington, Middlesex, England.

DUPEN, Douglas William, A.B., B.S., M.S. Born 1928. Educ.: California and Stanford Univs. Head, Tech. and Public Information Dept., Stanford Linear Accelerator Centre, Stanford Univ., 1962-. Tech. Information Panel, U.S.A.E.C. Book: The Story of Stanford's Two-Mile Accelerator (Springfield, Virginia, Clearing House for Federal Sci. and Tech. Information, 1966). Society: I.E.E.E.
Nuclear interest: Science information and communication.
Address: 850 Monte Rosa Drive, Menlo Park, California, U.S.A.

DUPLAN, Jean-Francois, M.D. Born 1921. Chef. de Service, Sect. de Biologie, Inst. du Radium. Societies: Rad. Res. Soc.; Sté. Européenne de Radiobiologie.
Nuclear interest: Radiobiology of mammals.
Address: Section de Biologie, Institut du Radium, 26 rue d'Ulm, Paris 5, France.

DUPLIN, Victor Joseph, Jr., S.B., S.M., C.E. Born 1904. Educ.: Dartmouth and M.I.T. With Babcock and Wilcox Co., 1931-. Trustee, Basic Science Div., and Vice-Pres., American Ceramic Soc.; Societies: Fellow, American Ceramic Soc.; British Ceramic Soc.; American Mineralogical Soc.; A.N.S.
Nuclear interests: Nuclear ceramics; Fuels, moderators, controls; Structures, irradiation effects, stability and corrosion resistance.
Address: P.O. Box 785, Lynchburg, Virginia, U.S.A.

DUPRE LA TOUR, Francois, Jesuit D. ès. Sc. physiques, Dr. em médecine. Born 1900. Educ.: Paris Univ. Prof. Biophys. S. Joseph Univ. Medical School, Beirut, Lebanon. Society: Sté. française de Physique.
Nuclear interests: Biological effects of nuclear radiation; Radiation protection.
Address: Faculté française de Médecine et de Pharmacie, Beyrouth, Lebanon.

DUPREE, A. Hunter, Dr. Prof., History, California Univ., Berkeley. Member, Historical Advisory Com., U.S.A.E.C.
Address: California University, Berkeley, California, U.S.A.

DUPREE, Thomas Henderson, Ph.D. Asst. Prof. Nucl. Eng., M.I.T.
Address: Massachusetts Institute of Technology, Cambridge 39, Massachusetts, U.S.A.

DUPUIS, Paul. Ing. attaché a la Direction, Forges et Acieries du Saut du Tarn, S.A.
Address: Forges et Acieries du Saut du Tarn, S.A., 6 avenue de Messine, Paris 8, France.

DUPUY, Jean-Charles. Born 1912. Educ.: Ecole National Supérieure des Mines (Nancy), Directeur, Indatom, 1961-.
Nuclear interest: Realisation de réacteurs de puissance, en particulier eau lourde et moderé en gaz.
Address: Indatom, 48 Rue la Boëtie, Paris 8, France.

DUPUY, P. Member for France, Study Group of Food Irradiation, O.E.C.D., E.N.E.A.
Address: O.E.C.D. European Nuclear Energy Agency, 38 boulevard Suchet, Paris 16, France.

DUQUE, Richardo DIAZ-. See DIAZ-DUQUE, Ricardo.

DUQUESNE, Henry F. C., Ing. civil electricien. Born 1930. Educ.: Liège Inst. Interuniv. des Sci. Nucléaires.
Nuclear interests: Accelerator engineering; Neutron physics.
Address: Laboratoire de Radioactivité et de Physique Nucléaire, Université de Liège, 9 Place du XX Août, Liège, Belgium.

DURAN, Armando. Born 1913. Educ.: Madrid Univ. Prof., Madrid Univ., 1945-; Counsellor, Nat. Council for Res., 1947-; Director, Torres Quevedo Inst., 1949-58; Dean, Faculty of Sci., Madrid Univ., 1958-63; Member of Board, Junta de Energia Nucl., 1951-; Formerly Vice-Pres., Junta de Energia Nuclear, 1958-. Board of Governors, I.A.E.A., 1959-61;

Pres., Royal Soc. of Phys. and Chem., 1962-; Director, Spanish Inst. of Nucl. Res., 1967-. Counsellor, Nat. Inst. for Tech. Aeronautics, 1955-; Nat. Council of Education, 1956-. Societies: Royal Soc. of Phys. and Chem.; Optical Soc. of America.
Nuclear interest: Nuclear physics.
Address: 1 Isaac Peral, Madrid 15, Spain.

DURAN, Efrain GOMEZ. See GOMEZ DURAN, Efrain.

DURAN MIRANDA, Armando, Prof. Vice Pres., Junta de Energia Nucl. Deleg., Third I.C.P.U.A.E., Geneva, Sept. 1964; Vice Chairman, Control Bureau, E.N.E.A., O.E.C.D.
Address: Junta de Energia Nuclear, Ciudad Universitaria, Madrid 3, Spain.

DURAND, E., Doyen, Centre de Physique Nucléaire, Faculte des Sci., Univ. Toulouse.
Address: Centre de Physique Nucléaire de la Faculte des Sciences, Université de Toulouse, 118 route de Narbonne, 31-Toulouse, France.

DURANDET, Jean, L. ès Sc. mathématiques; L. ès Sc. Physiques et chimiques; Eng., Ecole Nationale Supérieure du Pétrole. Born 1925. Educ.: Rennes and Paris Univs. Head, Appl. Phys. Chem. Dept., French Petroleum Inst. Maitre de Conférence, Louvain Univ. "Chemical Technology". Society: Sté de Chimie Industrielle (Sect. "Génie Nucléaire").
Nuclear interests: Fuel reprocessing; Liquid-liquid extraction.
Address: Institut Français du Petrole, 4 place Bir Hacheim, Rueil-Malmaison, (S. et O.), France.

DURANDIN, Roger. Directeur des Usines, Forges et Acieries du Saut du Tarn, S.A.
Address: Forges et Acieries du Saut du Tarn, S.A., 6 avenue de Messine, Paris 8, France.

DURANT, H. P. Chairman, Advisory Com., British Insurance (Atomic Energy) Com., 1967.
Address: Sun Alliance and London Group, 1 Bartholomew Lane, London, E.C.2, England.

DURANTE, Raymond Walter, M.E., M.S. (Ind. Eng.). Born 1928. Educ.: Stevens Inst. Technol. and Stevens Graduate School. Prohect Eng., Savannah River Project, 1950-53; Consultant Eng., K.A.P.L., 1953-55; Project Supervisor, American Machine and Foundry, 1955-57; Asst. Manager, Marketing, A.C.F. Industries, Inc., 1957-60, Manager of Operations, Aerojet-Gen. Nucleonics, 1960-63; Manager, Planning, Aerojet Gen. Nucleonics, San Ramon, California, 1963-65; on loan from Aerojet to the Dept. of the Interior, and subsequently assigned to Office of Saline Water as Project Manager, Metropolitan Water District dual-purpose nucl. power and desalting project, Chief Negotiator for the U.S.A.E.C. and Office of Saline Water, alternate Project Management Board member, and Project Eng., 1965-. Societies: Atomic Ind. Forum; A.N.S.; Washington (D.C.) Ind. Council; Atomic Energy Course for Management; A.R.S.; American Ordnance Assoc.
Nuclear interests: Supervision of marketing and planning activities, including the technical and business aspects of complete nuclear programme for military and peaceful applications.
Address: 8805 Daimler Court, Potomac, Maryland, U.S.A.

DURELLE, Jacques. Treas., Assoc. les Amis de l'Atome.
Address: Association les Amis de l'Atome, 11 Square Moncey, Paris 9, France.

DURET, Maurice Francis, B.Sc., M.Sc., Ph.D. Born 1922. Educ.: Queens (Kingston, Ontario) and Toronto Univs. Sen. Res. Officer, A.E.C.L. Society: A.N.S.
Nuclear interests: Mathematical aspects of reactor design and operation.
Address: 5 Kelvin Crescent, Deep River, Ontario, Canada.

DURFEE, Robert, Dr. Project Eng., Advanced Technol. Div:, Atlantic Res. Corp.
Address: Advanced Technology Division, Atlantic Research Corporation, Henry G. Shirley Memorial Highway at Edsall Road, Alexandria, Virginia, U.S.A.

DURGIN, Harold L. Pres. (-1968), Vice Chairman (1968-), Central Vermont Public Service Corp.
Address: Central Vermont Public Service Corp., 77 Grove Streeet, Rutland, Vermont, U.S.A.

DURHAM, Frank E., Dr. Prof. concerned with Nucl. Phys., Phys. Dept., Tulane Univ.
Address: Tulane University, New Orleans, Louisiana 70118, U.S.A.

DÜRIG, Günter, Prof. Dr. Mitglied, Fachkommission I Kernenergierecht, Deutsche Atomkommission, Federal Ministry for Sci. Res.
Address: 12 Viehweide, Tübingen, Germany.

DÜRR, Hans-Peter Emil, Diploma in Phys. (Stuttgart), Ph.D. (California). Born 1929. Educ.: Stuttgart Tech. and California Univs. Res. Asst., California Univ., 1956-57; Res. Asst. and Assoc., Max-Planck-Inst. für Phys. und Astrophys., Munich, 1958-62 and 1963-; Head, Theoretical Elementary Particle Group, 1961-; Privat-dozent, Munich Univ., 1961-; Res. Assoc., Rochester Univ., Consultant, California Univ., 1960; Visiting Assoc. Prof., California Univ., 1962-63; Visiting Prof. Matscience, Madras, India, 1963; Sci. Member, Max-Planck-Gesellschaft, 1963-.
Nuclear interests: Nuclear magnetic resonance; Experimental and theoretical nuclear physics; Field theory; Theory of elementary particles.

Address: Max-Planck-Institut für Physik und Astrophysik, 6 Föhringerring, Munich 23, Germany.

DURRILL, David C. B.S. (Chem. Eng.). Born 1920. Educ.: Denver Univ. Los Alamos Sci. Lab., 1945-51; Utah Univ. Dugway Project, 1951-54; A.N.P.D., (1954-60), Supv., Lab. Operations Nucl. Materials and Propulsion Operations, (1960-62), G.E.C.; Manager, R-MAD Operations, (1962-65), Deputy Manager, MAD Operations, N.E.R.V.A. Test Operations (1965-67), Westinghouse Astronucl. Lab.
Nuclear interests: Management in connection with assembly, maintenance, remote disassembly and material studies, radiochemistry and data reduction and evaluation of reactor and nuclear rocket engine systems.
Address: Westinghouse Astronuclear Laboratory, P.O. Box 2028, Jackass Flats, Nevada, U.S.A.

DURUISSEAU, J. C. Nucl. Lab., Serseg (Seguin-Sergot).
Address: Serseg (Seguin-Sergot), 4 place Felix Eboue, 75 Paris 12, France.

DURUM, Walton H. Chem., Geological Survey, U.S. Dept. of the Interior.
Address: U.S. Department of the Interior, Geological Survey, Washington D.C. 20242, U.S.A.

DUSABEK, Mark R., B.E.E. Born 1927. Educ.: Minnesota Univ. Nucl. Consultant, Fluor Corp. Ltd., 1953-. Society: A.N.S. Nuclear interest: Reactor design.
Address: 9340 S. Amelga Drive, Whittier, California, U.S.A.

DUSHIN, L. A. Paper: Co-author, Microwave Rad. of Electrodeless Induction Discharge (Atomnaya Energiya, vol. 16, No. 2, 1964).
Address: Academy of Sciences of the U.S.S.R., 14 Leninsky Prospekt, Moscow V-71, U.S.S.R.

DUSSART, Robert. Gen. Manager, Sté. Intercommunale Belge de Gaz et d'Electricité, S.A.; Member, Conseil d'Administration, Centre et Sud.
Address: Société Intercommunale Belge de Gaz et d'Electricité, S.A., 1 Place du Trône, Brussels, Belgium.

DUSSOIX, A. Member, Conseil d'Administration, Energie Nucléaire S.A.
Address: Energie Nucléaire S.A., 10 avenue de la Gare, Lausanne, Switzerland.

DUTILLIEUX, E. Belgian Member, Group of Experts on Third Party Liability, O.E.C.D., E.N.E.A.
Address: O.E.C.D. European Nuclear Energy Agency, 38 boulevard Suchet, Paris 16, France.

DUTREIX, Mme. A. Laboratoire de Radio-Physique, Département des Radiations, Inst. Gustave Roussy.
Address: Institut Gustave Roussy, 16 bis avenue P. Vaillant-Couturier, Villejuif (Seine), France.

DUTREIX, J.-M., Prof. Ag. Unité du Bétatron, Département des Radiations, Inst. Gustave Roussy. Member, Planning Board on Rad. - Medical Biol. Applications (Therapy), I.C.R.U.
Address: Institut Gustave Roussy, 16 bis avenue P. Vaillant-Couturier, Villejuif (Seine), France.

(DUTT), Mrs. Ila MUKHERJEE. See MUKHERJEE (DUTT), Mrs. Ila.

DUTTA-ROY, Binayak, M.Sc., Ph.D. Born 1936. Educ.: Allahabad and Princeton Univs. Assoc. Prof., Saha Inst. of Nucl. Phys., Calcutta, 1963-67; On leave of absence at Stevens Inst. of Technol., New Jersey.
Nuclear interest: Problems in high energy nuclear physics.
Address: Theoretical Physics Division, Saha Institute of Nuclear Physics, 92 Acharya Prafulla Chandra Road, Calcutta-9, India.

DUTTON, Arthur Morlan, B.S., Ph.D. (Iowa State Univ.). Born 1923. Instructor (1952-55), Asst. Prof. (1955-61), Assoc. Prof. (1961-), Rad. Biol. and Biophys., Rochester Univ., N.Y. Societies: A.S.A.; Biometric Soc.; Inst. Mathematical Statistics; A.A.A.S.
Nuclear interests: Application of statistical techniques to radiation biology experiments.
Address: P. O. Box 287, Station 3, Rochester, New York, U.S.A.

DUTTON, James J., B. Aeronautical Eng. Born 1929. Educ.: Catholic Univ. America; O.R.S.O.R.T., Southern Methodist, Tennessee and Columbia Univs. Asst. Director, Reactor Eng. Div., Argonne Nat. Lab. Society: A.N.S. Nuclear interest: Reactor development, design and performance.
Address: Argonne National Laboratory, 9700 South Cass Avenue, Argonne, Ill. 60439, U.S.A.

DUUREN, K. VAN. See VAN DUUREN, K.

DUURSMA, Egbert Klaas, D. Chem. Born 1929. Educ.: Free Univ. Amsterdam. Internat. Lab. Marine Radioactivity, I.A.E.A., Monaco.
Nuclear interests: Fall-out and neutron-induced radioisotope behaviour with suspended matter and bottom sediments of characteristic spots of the oceans; Kinetical processes of sorption and diffusion in sediments; Dissolved organic substances in the sea.
Address: International Laboratory of Marine Radioactivity, I.A.E.A., Musée Océanographique, Monaco.

DUVALL, George E., Ph.D. Born 1920. Educ.: Oregon State Coll and M.I.T. Phys., Head of Theoretical Group, Hanford Atomic Products Operation, 1948-53; Sen. Phys., Sci. Director, Director, Poulter Labs., Stanford Res. Inst., 1953-64; Prof. Phys., Washington State Univ., 1964-. Societies: A.P.S.; A.A.A.S.
Nuclear interests: Lattice theory; Neutron diffusion.
Address: Physics Department, Washington State University, Pullman, Washington, U.S.A.

DUVAUX, Yves. Adjoint au Chef, Service des Travaux et Installations, C.E.A. Expert, Third I.C.P.U.A.E., Geneva, Sept. 1964.
Address: Commissariat à l'Energie Atomique, C.E.N. Saclay, B.P. No. 2, Gif-sur-Yvette, (S. et O.), France.

DUVE, R. Sen. Managerial Officer, F. H. Gottfeld G.m.b.H.
Address: F. H. Gottfeld G.m.b.H., 26 Robert-Koch-Strasse, Cologne-Lindenthal, Germany.

DUYCKAERTS, G. Nucl. Chem. Inter-Faculty Centre of Nucl. Sci., Liège Univ. Deleg., I.A.E.A. Conference on the Use of Radioisotopes in the Phys. Sci. and Industry, Copenhagen, 6-17 September 1960.
Address: Inter-Faculty Centre of Nuclear Science, Liège University, Liège, Belgium.

DVORAK, Jan, Prof. Dr. Scientarium, Dipl. Eng. Born 1922. Educ.: Prague Tech. Univ. Res. worker, Nat. Res. Inst. Heat Eng., Prague. Society: Com. Czechoslovak Acad. Sci. for Problems of Czechoslovak Nucl. Power Plants; Technico-Sci. Soc., Czechoslovakia.
Nuclear interests: Calculations of shells, pressure vessels of nuclear reactors; Thermal stresses in nuclear reactor components.
Address: 19 Csl. armády, Praha 6-Bubenec, Czechoslovakia.

DVORETSKII, M. I. Paper: Co-author, 1,5 GeV Electron Synchrotron at the Polytech. Inst. of Tomsk (Atomnaya Energiya, vol. 21, No. 6, 1966).
Address: Academy of Sciences of the U.S.S.R., 14 Leninsky Prospekt, Moscow V-71, U.S.S.R.

DVORETSKII, V. N. Paper: Co-author, Determination of Some Parameters of Reactors with Germanium Gamma-Ray Spectrometer (letter to the Editor, Atomnaya Energiya, vol. 22, No. 6, 1967).
Address: Academy of Sciences of the U.S.S.R., 14 Leninsky Prospekt, Moscow V-71, U.S.S.R.

DVUHSHERSTNOV, V. G. Paper: Co-author, Reaction Li + He_2^4, Li + H^2, and Li + He_2^1 as Cyclotron Sources of Fast Neutrons (letter to the Editor, Atomnaya Energiya, vol. 24, No. 1, 1968).
Address: Academy of Sciences of the U.S.S.R., 14 Leninsky Prospekt, Moscow V-71, U.S.S.R.

DWIGHT, Austin Elbert, B.S., M.S. Born 1919. Educ.: Michigan State and Michigan Univs. Assoc. Metal. Eng., Argonne Nat. Lab., 1950-. Societies: A.I.M.M.E.; A.S.M.; American Crystallographic Assoc.; Inst. Metals (London).
Nuclear interests: Phase diagrams and transformation kinetics of uranium alloys; Alloy theory of transition elements, lanthanides and actinides.
Address: Argonne National Laboratory, 9700 South Cass Avenue, Argonne, Illinois 60439, U.S.A.

DWORK, Ralph Ephraim, B.S., L.R.C.P.S., M.P.H. Born 1915. Educ.: New York and Melbourne Univs., Anderson Coll. of Medicine, Glasgow and Columbia School of Public Health, N.Y. Chief, Div. of Tuberculosis (1950-54), Director (1954-63), Ohio Dept. of Health; Deputy Sec. of Health, Pennsylvania Dept. of Health, 1963-. Vice-Chairman, Residency Review Com. for Preventive Medicine, American Medication Assoc.; Member, Governing Council, American Public Health Assoc.; Member, Review Com. on Construction of Schools of Public Health, U.S. P.H.S.; Member, Nat. Advisory Com. on Rad.
Address: 4204 Cotswold Drive, Harrisburg, Pa. 17110, U.S.A.

DWYER, Orrington Embry, B.S., M.S., D.Eng. Born 1912. Educ.: Northeastern and Yale Univs. and M.I.T. Head, Heat Transfer Res. Group, Eng. Div., Brookhaven Nat. Lab. Societies: A.I.Ch.E.; Fellow, A.A.A.S.; Fellow, A.N.S.
Nuclear interest: Liquid-metal heat transfer.
Address: Brookhaven National Laboratory, Upton, New York, U.S.A.

D'YACHKOV, B. A. Paper: Co-author, Capture and Loss of Protons with Energy 5-40 KeV in Thick Streams of Lithium, Natrium, Magnesium and Zinc (Atomnaya Energiya, vol. 24, No. 1, 1968).
Address: Academy of Sciences of the U.S.S.R., 14 Leninsky Prospekt, Moscow V-71, U.S.S.R.

D'YACHKOVA, R. A. Paper: Co-author, Pa^{231} Concentrating from Uranium Production Wastes (Atomnaya Energiya, vol. 16, No. 2, 1964).
Address: Academy of Sciences of the U.S.S.R., 14 Leninsky Prospekt, Moscow V-71, U.S.S.R.

D'YAKOV, I. G. Paper: Co-author, Change of Beryllium Properties after Overaging Treatment (Atomnaya Energiya, vol. 19, No. 3, 1965).
Address: Academy of Sciences of the U.S.S.R., 14 Leninsky Prospekt, Moscow V-71, U.S.S.R.

DYBCZYNSKI, Rajmund, M.Sc., Ph.D. Born 1933. Educ.: Warsaw Univ. Inst. of Gen. Chem., Warsaw 1955-57; Inst. of Nucl. Res., Warsaw, 1957-63; I.A.E.A. post-doctoral fellow, Polytech. Inst. of Brooklyn, Brooklyn, N.Y., 1963-64; Head, Lab. of Radiometric Methods of

Analysis, Dept. of Analytical Chem., Inst. of Nucl. Res., Warsaw, 1964-. Society: Polish Chem. Soc.
Nuclear interests: Radiochemical separation techniques, ion exchange, activation analysis, application of tracers in chemical research, rare earth and noble metals radiochemistry.
Address: Department of Analytical Chemistry, Institute of Nuclear Research, 16 ul. Dorodna, Warsaw 91, Poland.

DYBOWSKI, Kazimierz, Magister ing. Born 1922. Educ.: Warsaw Tech. Univ. Inst. Nucl. Res., 1959.
Nuclear interest: Pulsed neutron sources (design) and pulse neutron technique.
Address: 4-6 Kowelska, Warsaw 4, Poland.

DYE, Captain Ira. Maritime Administration Member, Maritime Administration and U.S.A.E.C. Joint Liaison Com. on Nucl. Merchant Ships.
Address: U.S. Department of Commerce, Maritime Administration, Washington D.C. 20545, U.S.A.

DYE, Maynard Malden, B.S. (E.E.). Born 1924. Educ.: Colorado and Idaho Univs. Shift Supervisor, S1W Project, Nautilus Prototype, 1953-57; Shift Supervisor, PWR Project, 1957-58; Supervisor, Mechanical Test Section A1W (1958), Supervisor, Power Plant Test A1W (1959), Sen. Eng., Reactor Control Design, S5W core 4 (1960), Atomic Power Div., Westinghouse Elec. Corp. Chairman, Southern Idaho subsection, A.I.E.E.; Sec., Idaho section, A.I.E.E. Societies: A.N.S.; American Inst. of Elec. Eng.
Nuclear interests: Reactor control design; Nuclear power plant operation; Reactor plant test and analysis.
Address: 1470 Scorpius Drive, Idaho Falls, Idaho, U.S.A.

DYER, G. H. U.S.A. Member, Study Group on Nucl. Ship Propulsion, O.E.C.D., E.N.E.A.
Address: O.E.C.D. European Nuclear Energy Agency, 38 boulevard Suchet, Paris 16, France.

DYER, Henry Brooke, M.Sc., Ph.D. Born 1927. Educ.: South Africa and Cambridge Univs. Sen. Phys. (1953-59), Director Res. (1961-), Diamond Res. Lab.; Asst. Director Res., Adamant Res. Lab., 1959-61.
Nuclear interests: Irradiation damage in solids; Solid state nuclear particle counters.
Address: Box 916, Johannesburg, South Africa.

DYKEN, Alexander Robert VAN. See VAN DYKEN, Alexander Robert.

DYKES, Fred W., B.S. (Chem.). Born 1928. Educ.: Idaho State Univ. Chem., Phillips Petroleum Co., 1953-66; Idaho Nucl. Corp., 1966-. Societies A.C.S.; A.N.S.
Nuclear interest: Development of facilities, apparatus, and techniques for chemical analysis by remote control.
Address: 964 Wayne Avenue, Pocatello, Idaho 83201, U.S.A.

DYMKOV, Yu. M. Papers: Co-author, The Reddening of Minerals in Uranium-bearing Veins (Atomnaya Energiya, vol. 10, No. 1, 1961); co-author, Hard Bitumen in Uranium bearing Veins (ibid, vol. 16, No. 5, 1964); Contribution to Genesis of Spherolites of Nasturan (ibid, vol. 20, No, 3, 1966).
Address: Academy of Sciences of the U.S.S.R., 14 Leninsky Prospekt, Moscow V-71, U.S.S.R.

DYMOKE, Captain Lionel Dorian, R.N. Born 1921. Asst. Director, Marine Eng. for Nucl. Propulsion, Ministry of Defence (Navy), 1962-66; Commanding Officer, H.M.S. Sultan (Roy. Naval School of Marine Eng.), 1966-. Societies: Inst. if Mech. Eng.; I. Mar. E.
Nuclear interest: Management.
Address: H.M.S. Sultan, Gosport, Hants., England.

DYMOVICH, D. I. Paper: Co-author, Separation of Isotopes of Neon in Mercury Mass-Diffusion Column (letter to the Editor, Atomnaya Energiya, vol. 23, No. 1, 1967).
Address: Academy of Sciences of the U.S.S.R., 14 Leninsky Prospekt, Moscow V-71, U.S.S.R.

DYRSSEN, David Waldemar, D.Sc. Born 1922. Prof. Analytical Chem., Gothenburg Univ.
Nuclear interests: Process chemistry; Radiochemistry; Solution chemistry; Sea-water.
Address: 5 Lunnatorpsg, Göteborg 5, Sweden.

DYSON, Freeman John, B.A. Born 1923. Educ.: Cambridge and Cornell Univs. Prof., Phys., Cornell Univ., 1951-53; Prof., Phys., Inst. for Advanced Study, Princeton, 1953-. Book: Symmetry Groups in Nucl. and Particle Phys. (Benjamin, 1966). Societies: Roy. Soc. of London; N.A.S., U.S.A.
Nuclear interest: Orion nuclear space propulsion system.
Address: Institute for Advanced Study, Princeton, New Jersey 08540, U.S.A.

DYSON, James, B.A., M.A., Sc.D., F. Inst. P. Born 1914. Educ.: Cambridge Univ. Supt., Div. of Optical Metrology, Nat. Phys. Lab., Teddington.
Nuclear interest: Optical instrumentation especially interferometry and spectroscopy.
Address: 19 Hansler Grove, East Molesey, Surrey, England.

DYSON, Norman Allen, M.A., Ph.D. Phys. Born 1929. Educ.: Cambridge Univ. Sci. Staff, Medical Res. Council, Radiotherapeutic Res. Unit, Hammersmith Hospital, London, 1955-59; Lecturer in Phys., Birmingham Univ., 1959-.
Nuclear interests: Nuclear physics and its applications in medicine and biology; Nuclear

resonance fluorescence and the Mössbauer effect: Gamma ray physics.
Address: University of Birmingham, Birmingham, England.

DZAMPOV, B. V. Paper: Co-author, Thermal Properties of Hexafluorocarbon (letter to the Editor, Atomnaya Energiya, vol. 24, No. 3, 1968).
Address: Academy of Sciences of the U.S.S.R., 14 Leninsky Prospekt, Moscow V-71, U.S.S.R.

DZANTIEV, B. G. Deleg., I.A.E.A. Conference on the Use of Radioisotopes in the Phys. Sci. and Industry, Copenhagen, Sept. 6-17, 1960. Paper: Co-author, Lab. Installation for Investigation of Cheminucl. Synthesis (letter to the Editor, Atomnaya Energiya, vol. 20, No. 3, 1966).
Address: Academy of Sciences of the U.S.S.R., 14 Leninsky Prospekt, Moscow V-71, U.S.S.R.

DZHAGATSPANYAN, R. V. Papers: Co-author, Calculation of Efficiency of Rad. - Chem. Apparatus with Plain Beta-Sources (Atomnaya Energiya, vol. 20, No. 6, 1966); co-author, Calculation of Spherical Beta Source Efficiency (letter to the Editor, ibid, vol. 24, No. 6, 1968).
Address: Academy of Sciences of the U.S.S.R., 14 Leninsky Prospekt, Moscow V-71, U.S.S.R.

DZHELEPOV, Boris Sergeevich. Born 1910. Educ.: Leningrad Univ. Lecturer, Leningrad Univ., 1944; Radium Inst., Acad. of Sci. of the U.S.S.R., 1945; All-Union Inst. for Sci. Res. on Metrology, 1946; Member, Editorial Board, Nucl. Instruments and Methods: Deleg., 2nd I.C.P.U.A.E., Geneva, Sept. 1958. Books: Co-author, Effect of the Electric Field on the Atom on its Beta Decay (Moscow-Leningrad, 1956); co-author, Sistematika energii beta-raspada (Systematics of Beta-Decay Energies) (Fasc. 3. Svoistva Atomnykh Yader (Properties of Atomic Nuclei), AN SSSR, M, 1960: English Translation, Pergamon, London); co-author, Izobranye yadra s massovym chislom A = 140 (Isobaric Nuclei with the Mass Number A = 140) (Fasc. 5. Svoistva Atomnykh Yader (Properties of Atomic Nuclei) M.-L. AN SSSR, 1960; English Translation, Pergamon, London); Isobaric Nuclei with the Mass Number A = 74 (Pergamon, London); co-author, Isobaric Nuclei with the Mass Number A = 110 (Pergamon, London); co-author, Decay Schemes of Radioactive Nuclei (Pergamon, London). Society: Corr. Member, Acad. of Sci. of the U.S.S.R.
Nuclear interest: Physics of atom nucleus.
Address: Radium Institute, Academy of Sciences of the U.S.S.R., 14 Leninsky Prospekt, Moscow V-71, U.S.S.R.

DZHELEPOV, Venedict P., Prof.; Director, Lab. of Nucl. Problems, J.I.N.R. Papers: Co-author, A Cyclotron with a Spatially Varying Magnetic Field (Atomnaya Energiya, vol. 8, No. 3, 1960); Investigation of Hydrogen and Deuterium μ-Mesic Atom and μ-Mesic Molecule Properties at 680 Mev Synchrocyclotron in Dubna (ibid, vol. 14, No. 1, 1963); co-author, The Effect of Space Charge on the Frequency of Free Oscillations in an Isochronous Cyclotron (ibid, vol. 15, No. 3); co-author, Strong Focusing Ring Cyclotron for Multi-Charged Ions (ibid, vol. 24, No. 4, 1968).
Address: Joint Institute for Nuclear Research, Head Post Office, P.O.Box 79, Moscow, U.S.S.R.

DZIK, Wlodzimierz, Master of Mathematics. Born 1929. Educ.: Poznan Univ. Office for Projects of Machinery Industry.
Nuclear interests: The application of nuclear techniques in the investigation of ventilated objects and in flow medium.
Address: 12/25 Kosinskiego Street, Poznan, Poland.

DZIUNIKOWSKI, Bohdan, Dr. of Sci. Born 1930. Educ.: Jagiellonian Univ. and Acad. of Mining and Metal. Adjunct, Acad. of Mining and Metal., Cracow, 1955-.
Nuclear interests: Practical applications of nuclear physics and radioisotopes. Analytical method based on the emission and absorption of X or gamma rays.
Address: Institute of Nuclear Techniques, Academy of Mining and Metallurgy, 30 al. Mickiewicza, Cracow, Poland.

e CASTRO, J. L. SILVA. See SILVA e CASTRO, J. L.

e CASTRO, P. F. SAMPAIO. See SAMPAIO e CASTRO, P. F.

e SOUSA, J. BESSA MENEZES. See BESSA MENEZES e SOUSA, J.

e SOUZA, H. E. Madame Odette de CARVALHO. See de CARVALHO e SOUZA, H. E. Madame Odette.

EALES, Claude Sibthorpe. Born 1908. Chief Eng., Eng. Directorate, Production Group, U.K.A.E.A.
Nuclear interest: Engineering management.
Address: Southview, Guilden Sutton Lane, Chester, England.

EASON, Harold E. Asst. Treas., Atomic Power Development Assocs., Inc.
Address: Atomic Power Development Associates, Inc., 1911 First Street, Detroit 26, Michigan, U.S.A.

EASSON, Eric Craig, M.D., M.R.C.P. Born 1915. Educ.: Glasgow Univ. Director, Radiotherapy, Christie Hospital and Holt Radium Inst. Chairman, Commission of Cancer Control, Member: Exec. Com., Council, Com. on Internat. Congresses, Com. on Publications and Com. of Visiting Experts, U.I.C.C.; Member, Expert Com. on Cancer Treatment, W.H.O.; Fellow and Member, Fellowship Board,

Faculty of Radiol., London. Society: Fellow, British Inst. Radiol., Nuclear interest: Interested primarily in the medical applications of ionising radiations, whatever their sources.
Address: Christie Hospital and Holt Radium Institute, Wilmslow Road, Withington, Manchester 20, England.

EASTERDAY, Harry Tyson, A.B., Ph.D. (California, Berkeley). Born 1922. Educ.: California Univ., Berkeley. Prof., Oregon State Univ. Society: A.P.S.
Nuclear interest: Beta- and gamma-ray spectroscopy.
Address: Department of Physics, Oregon State University, Corvallis, Oregon, U.S.A.

EASTWOLD, Rear Admiral E. R. Asst. Chief for Field Support, Bureau of Naval Weapons, Dept. of the Navy.
Address: Department of the Navy, Bureau of Naval Weapons, Washington 25, D.C., U.S.A.

EASTWOOD, David Charlton. Born 1921. Eng. Officer, Royal Navy, 1939-56; Sen. Eng., Atomic Div., English Elec. Co., 1957-. Society: Assoc. Member, I. Mech. E.
Nuclear interests: Station economics; Overseas nuclear interests and sales possibilities.
Address: 3 Bilton Hall, Rugby, England.

EASTWOOD, E., Dr. Formerly Director, Eng. and Res., now Chief Sci., Marconi Co. Ltd.
Address: Marconi Co. Ltd., Marconi House, Chelmsford, Essex, England.

EASTWOOD, Thomas Alexander, B.A., M.A., Ph.D., D.Phil. Born 1920. Educ.: Western Ontario, McGill and Oxford Univs. Principal Res. Officer and Asst. Director, Chem. and Materials Div., A.E.C.L., Chalk River, Ontario. Societies: Chem. Inst. of Canada.; A.C.S.; A.P.S.
Nuclear interests: Neutron cross-section and flux determination by activation techniques; Radioactive decay; Low-energy nuclear reactions.
Address: 9 Tweedsmuir, Deep River, Ontario, Canada.

EASTWOOD, William Stuart, B.Sc. (Eng.). Born 1913. Educ.: London Univ. Isotope Div., A.E.R.E., Harwell, 1949-59; Head, Reactor Development Div., A.E.E., Winfrith, 1960-.
Nuclear interests: Water moderated reactor development and also heat transfer in the gas and steam/water fields.
Address: Atomic Energy Establishment, Winfrith, Dorchester, Dorset, England.

EATHERLY, Walter P., B.S., M.S. Born 1923. Educ.: California Inst. of Technol. and Illinois Univ. Consultant, O.R.N.L., 1967-. Book: Nucl. Graphite (Academic Press, 1962). Societies: A.P.S.; A.N.S.; A.A.A.S.
Nuclear interests: Radiation damage in structural materials; Design and fabrication of fuel elements; Reactor core design.
Address: Oak Ridge National Laboratory, P.O. Box X, Oak Ridge, Tenn. 37830, U.S.A.

EATON, James Robert, B.S. (E.E.), M.S. (E.E.), Ph.D. Born 1902. Educ.: Purdue and Wisconsin Univs. Prof. Elec. and Nucl. Eng., on staff of Purdue Univ., 1942-.
Nuclear interest: Instructional work in nuclear instrumentation and reactor behaviour.
Address: Electrical Engineering Building, Purdue University, Lafayette, Indiana, U.S.A.

EATON, Samuel Edward, B.S. Born 1915. Educ.: Wesleyan Univ. and M.I.T. Chairman, Adv. Com., Isotope Development, U.S.A.E.C., 1955; Head Radiochem. Group, Arthur D. Little Inc., 1937-64; At present Director, New Product Development, United States and Canada, United Shoe Machinery Corp. U.S. deleg. to 1st I.C.P.U.A.E., Geneva, 1955, and 1959 Warsaw Conference on Application of Large Rad. Sources in Industry; Board of Directors, Nucl. Metals, Inc., 1955-58. Books: Contributor to Ind. and Safety Problems of Nucl. Technol. (Shamos and Roth; Harper and Brothers, 1959); Rad: A Tool for Industry (U.S. Dept. of Commerce, 1959). Societies: Chem. Soc.; A.N.S.; Electrochem. Soc.; Soc. Testing Material.
Nuclear interests: Nuclear rocket hazards; Radiation processing; Radioisotope applications; Control rod drive mechanisms.
Address: United Shoe Machinery Corp., 140 Federal Street, Boston, Mass., U.S.A.

EAVES, George, B.Sc., Ph.D. Born 1921. Educ.: Manchester and Leeds Univs. Res. Asst., Dept. Biomolecular Structure (1949-51), Res. Asst., Dept. Cancer Res. (1951-54), Leeds Univ.; Sen. Lecturer in Phys., Roy. Coll. Advanced Technol., Salford, 1959-64; Lecturer in Electron Microscopy, Leeds Univ., 1964-. Society: A. Inst. P.
Nuclear interest: Health physics, cancer research.
Address: Department of Cancer Research, The University, Leeds 2, Yorks., England.

EBDON, H. G. Vice-Chairman, Combustion Eng.; Director, Atomic Ind. Forum, 1965-.
Address: Combustion Engineering, Inc., Windsor, Connecticut, U.S.A.

EBELING, Douglas Roy, B. Mech. E., Diploma Mech. and Elec. Eng. Born 1929. Educ.: Melbourne Univ. I.C.I.A.N.Z. – Nobels (Explosive Div.), Deer Park, Melbourne; I.C.I. – Blackley, Manchester (Dyestuffs Div.); A.A.E.C., Harwell; A.A.E.C., Lucas Heights, N.S.W., Australia. Head, Design and Development Sect., Eng. Res. Sec., Symposium on the Peaceful Uses of Atomic Energy in Australia, 1958 (Assoc. Techniques Sect.). Societies: A.M.I.Mech.E.; A.M.I.E.Aust.
Nuclear interests: Reactor design and feasibility studies, remote handling and design of associated nuclear equipment.

Address: 41 Smarts Crescent, Cronulla, New South Wales, Australia.

EBERHARDT, Peter, Dr. Born 1931. Educ.: Bern Univ. Assoc. Prof., Div. of Geological Sci., California Inst. Technol., 1966-67; Assoc. Prof., Phys. Dept., Bern Univ., 1967-. Nuclear interests: Mass spectrometry; Isotope research in meteorites.
Address: Bern University, Physikalisches Institut, 5 Sidlerstrasse, Bern, Switzerland.

EBERLEIN, G. Donald. Geological Survey, U.S. Dept. of the Interior.
Address: Geological Survey, U.S. Department of the Interior, 345 Middlefield Road, Menlo Park, California, U.S.A.

EBERLINE, Howard C. Pres., Eberline Instr. Corp.; Formerly Director, Atomic Ind. Forum, 1962; Formerly Member, Nat. Advisory Com. on Rad.
Address: Eberline Instrument Corp., 805 Early Street, P.O. Box 279, Santa Fé, New Mexico, U.S.A.

EBERSOLE, Captain John H., M.D. Born 1925. Educ.: Indiana Univ. Medical Officer, U.S.S. Nautilus, 1954-56; Medical Officer, U.S.S. Seawolf, 1956-59; Chief, Rad. Therapy Sect. (1962-), Acting Director, Rad. Exposure Evaluation Lab. (1962-), U.S. Naval Hospital, Nat. Naval Medical Centre; Director, Nucl. Medicine Training Div., Naval Medical School, 1962-. Societies: A.C.R.; Soc. of Nucl. Medicine; H.P.S.
Nuclear interest: Clinical radiation therapy.
Address: Naval Hospital, National Naval Medical Centre, Bethesda, Maryland 20014, U.S.A.

EBERT, Heinrich Georg, Diplomphysiker, Dr. rer. nat. Born 1931. Educ.: Göttingen Univ. Asst., Göttingen Univ., Inst. für Medizinische Phys. und Biophys., 1956-64. European Communities (Euratom), 1964-.
Nuclear interests: Biophysics; Radiation dosimetry; Radiation protection.
Address: European Communities - Euratom, 51 rue Belliard, Brussels, Belgium.

EBERT, Michael, Dipl. Ing. (Chem.), Dr. rer. nat. Born 1914. Educ.: Berlin and Mainz Univs. Asst. Chem. and Rad. Biol., Paterson Labs., Christie Hospital and Holt Radium Inst., Withington, Manchester. Member, Editorial Board, Internat. J. of Rad. Biol. Societies: Assoc. for Rad. Res.; British Inst. of Radiol.; Gesellschaft Deutscher Chemiker; Faraday Soc.
Address: Paterson Laboratories, Christie Hospital and Holt Radium Institute, Withington, Manchester 20, England.

EBINGER, Gerhard, Dipl. Ing. Born 1912. Educ.: Stuttgart T.H. Director, Heilmann and Littmann Bau A.G. Society: Studiengesellschaft zur Förderung der Kernenergieverwertung in Schiffbau und Schiffahrt e.V., Hamburg.

Nuclear interest: Rays shelter concrete.
Address: Heilmann and Littmann Bau A.G., 14 Auguststrasse, Hamburg 22, Germany.

EBLE, Charles E. Director, Empire State Atomic Development Assocs. Inc. Pres., Consolidated Edison Co. of N.Y. City. Member, Gen. Advisory Com., New York State Office of Atomic Development.
Address: New York State Office of Atomic Development, 1161 York Avenue, New York City, New York, U.S.A.

ECHEGARAY ROMEO, Miguel. Pres., Nat. Agricultural Inst. Member, Comisiones Asesoras de la Junta de Energia Nuclear de Biologia Vegetal y Applicaciones Industriales.
Address: Instituto Nacional de Investigaciones Agronomicas, Ministerio de Agricultura, Avda. Puerta de Hierro, Madrid, Spain.

ECHENIQUE, Miguel. Member, Permanent Com. on Internat. Affairs Relating to Atomic Energy, Chile.
Address: Permanent Committee on International Affairs Relating to Atomic Energy, c/o Ministerio de Relaciones Exteriores, Plaza de la Moneda, Santiago, Chile.

ECHOLS, T., Dr. Book: Contributor to Eng. Compendium on Rad. Shielding (I.A.E.A.).
Address: Argonne National Laboratory, 9700 South Cass Avenue, Argonne, Illinois, U.S.A.

ECK, John E., B.S. (Mech. Eng., Carnegie Inst. Technol.). Educ.: Carnegie Inst. Technol. Toolmaker, East Pittsburgh Works (1946-50), Mech. Eng., A.E.C. Bettis Plant (1950-57), Westinghouse Eléc. Corp.; Mech. Eng. (1957-67), Manager, Hafnium Facility (1967-), Nucl. Materials and Equipment Corp., Apollo, Pa. Society: A.S.M.E.
Nuclear interests: Special materials and forms of materials for the nuclear industry; Material process engineering.
Address: Nuclear Materials and Equipment Corporation, Apollo, Pennsylvania 15613, U.S.A.

ECKBERG, Herbert F., Bachelor of Naval Sci., M.S. (M.E.). Born 1906. Educ.: U.S. Naval Acad. and California Univ., Chairman, Dept. of Mech. Eng., and Dean of Eng., Bucknell Univ. Societies: A.S.M.E.; A.S.E.E.
Nuclear interest: Nuclear reactor design.
Address: Bucknell University, Lewisburg, Pennsylvania, U.S.A.

ECKEL, Paul, Dr. med. Facharzt für Röntgenologie.; Vorstand, Deutsche Gesellschaft für Atomenergie e.V.; Vorsitzender, Bundesarztekammer Atomkommission. Pres., and Member for Germany, Economic and Social Com. Specialised Nucl. Sect. for Social Health and Development Problems, Euratom.
Address: 8 Schwarzer Bär, 3 Hanover-Linden, Germany.

ECKERT, E., Dipl. Ing. Mitglied, Technisch-Wissenschaftlicher Beirat, Co. for the Utilisation of Nucl. Energy in Shipbuilding and Navigation.
Address: Company for the Utilisation of Nuclear Energy in Shipbuilding and Navigation, 2^V Gr. Reichenstrasse, 2 Hamburg 11, Germany.

ECKHARDT, F. W. Chief Eng., Power Plant Div., Borsig A.G.
Address: Borsig A.G., VK 26, 19-37 Berliner Strasse, 1 Berlin 27, Germany.

ECOCHARD, R. M., Dr. Staff Member, Cytogenetics, Inst. for Atomic Sci. in Agriculture.
Address: Institute for Atomic Sciences in Agriculture, 6 Keyenbergse Weg, Wageningen, Netherlands.

EDA, Bunzo. E.S.R. of Irradiated Organic Substances, Government Ind. Res. Inst., Ministry of Internat. Trade and Industry.
Address: Ministry of Interiational Trade and Industry, Government Industrial Research Institute, Hirate-machi, Kita-ku, Nagoya, Japan.

EDDLEMON, Joseph D., B.S. (Eng. Phys.). Born 1925. Educ.: Tennessee Univ. Assoc. Development Eng. (1953-55), Development Eng. (1955-60), O.R.N.L.; Sales Eng., Charles Walsh Assoc., Inc., 1960-63; District Sales Manager (1963-64), Field Eng. Manager (1964-65), Nucl. Data, Inc.; Sales Manager, Tullamore Div., Victoreen Instrument Co., 1966-67; Pres., Pulcir, Inc., 1967-. Society: RESA.
Nuclear interest: Nuclear instrumentation, multi-channel pulse height analysers and associated instrumentation for radiation spectrometry.
Address: 2711 Kingston Pike, Knoxville, Tennessee, 37919, U.S.A.

EDE, Allan John, M.A., B.Sc. Born 1915. Educ.: London Univ. Prof. of Mech. Eng., Aston Univ., Birmingham. Formerly Head, Heat Div., Nat. Eng. Lab. Societies: Fellow, Inst. Phys. and Phys. Soc.; Inst. Mech. Eng.; Fellow, Inst. Maths and Appl.
Nuclear interest: Heat transfer.
Address: Mechanical Engineering Department, Aston University, Gosta Green, Birmingham 4, England.

EDELMANN, Abraham, A.B., M.S., Ph.D. Born 1915. Educ.: Johns Hopkins, New York and Ohio State Univs. Sci. (T), Brookhaven Nat. Lab., 1947-55; Manager, Dept. of Biology and Medicine, Nucl. Sci. and Eng. Corp., 1955-. (Formerly Vice-Pres., 1959-); Lecturer, Pittsburgh Univ., 1962-; Sen. Lecturer, Carnegie Inst. of Technol., 1963-. Societies: American Physiological Soc.; Rad. Res. Soc.; Endocrine Soc. for Exptl. Biology and Medicine.
Nuclear interests: Radiobiology, effects of radiation on mammals, use of radioisotopes in biological problems.
Address: Nuclear Science and Engineering Corporation, P.O. Box 10901, Pittsburgh, Pennsylvania 15236, U.S.A.

EDELSACK, Edgar Allen, B.S. (Phys.). Born 1924. Educ.: New York and Southern Californian Univs. and Washington State Coll. Head, Nucl. Sci. Div., Office of Naval Res. Branch Office, San Francisco; Supervisory Nucl. Phys., U.S. Naval Radiol. Defence Lab., San Francisco; Phys., Emery Tumor Group, Los Angeles. Societies: A.P.S.; A.N.S.; A.A.A.S.; N.Y. Acad. of Sci.
Nuclear interests: Nuclear physics and nuclear medicine.
Address: Office of Naval Research, 1000 Geary Street, San Francisco, California 94109, U.S.A.

EDEN, Richard John, M.A., Ph.D. Born 1922. Educ.: Cambridge Univ. Smithson Res. Fellow, Roy. Soc., 1952-55; Sen. Lecturer in Phys. Manchester Univ., 1955-57; Member, Princeton Inst. Advanced Study, spring 1954, autumn 1959; Visiting Phys., Indiana Univ., 1954-55, California Univ., 1960; Lecturer in Mathematics, Cambridge Univ., 1957-; Fellow of Clare Coll., Cambridge, 1951-55, 1957-; Consultant to A.E.R.E., Harwell; Visiting Prof., Maryland Univ., 1961; Visiting Prof., Brandeis Univ., 1961; Columbia Univ. 1962; Maryland Univ. 1965; Visiting Sci., Lawrence Rad. Lab., 1967; Reader in Theoretical Phys., Cambridge Univ., 1964-. Books: Co-author, The Analytic S-Matrix (Cambridge Univ. Press, 1966); High Energy Collisions of Elementary Particles (Cambridge Univ. Press, 1967).
Nuclear interests: Theory of elementary particles; Analytic properties of collision amplitudes; Theoretical nuclear physics.
Address: Clare College, Cambridge, England.

EDENHOLM, Ingmar, M.Sc., M.E. Born 1932. Educ.: Chalmers Inst. of Technol. Manager, Nucl. Equipment, Svenska Masinverken, The Johnson Group.
Address: Svenska Maskinverken, Kallhall, Sweden.

EDER, Howard Abraham, B.A., M.D. Born 1917. Educ.: Wisconsin Univ. and Harvard Medical School. Res. Collaborator, Brookhaven Nat. Lab., 1956-; Assoc. Prof. Medicine (1957-60), Prof. Medicine (1961-), Albert Einstein Coll. of Medicine; Attending Physician, Bronx Municipal Hospital Centre, 1957-. Societies: American Soc. for Clin. Invest; Soc. of Biol. Chemist; Soc. for Exptl. Biol. and Medicine.
Nuclear interest: Use of radioactive tracers in medicine and biochemistry.
Address: Albet Einstein College of Medicine, Eastchester Road and Morris Park Avenue, New York 61, New York, U.S.A.

EDESKUTY, Frederick James, B.Ch.E., Ph.D. Born 1923. Educ.: Minnesota Univ. Assoc. Group Leader, Los Alamos Sci. Lab., California Univ., Los Alamos. Com. member on Planning Com. for Cryogenic Eng. Conference, 1958-61.
Nuclear interests: Cryogenic engineering as applied to nuclear rocket testing; involves large scale storage and transfer of liquid hydrogen both without and within a radiation field.
Address: Los Alamos Scientific Laboratory, P.O. Box 1663, Los Alamos, New Mexico, U.S.A.

EDGE, Ronald D., B.A., M.A., Ph.D. Born 1929. Educ.: Cambridge Univ. Res. Fellow, Australian Nat. Univ., 1954-58; Asst. Prof., South Carolina Univ., 1958-62; Res. Assoc., Yale Univ., 1962-63; Prof., South Carolina Univ., 1963-; Visiting Prof. Stanford Univ., 1961; Res. Assoc., California Inst. of Tech., 1962; Assoc. and Consultant, Oak Ridge. Society: Fellow, A.P.S.
Nuclear interests: High energy nuclear physics; Cosmic ray neutrons; Analog states; Low energy scattering.
Address: Physics Department, South Carolina University, Columbia, South Carolina 29208, U.S.A.

EDGE, Thomas, B.Sc. Tech. Born 1920. Educ.: Manchester Univ. Res. Manager, Windscale; Chief Metal., Operations Branch, Risley. Chief Tech. Officer, Tech. Directorate, Chief Metal., U.K.A.E.A. Production Group, Risley, Warrington, Lancs. Society: Inst. of Metal.
Nuclear interest: Metallurgy.
Address: U.K.A.E.A. Production Group, Risley, Warrington, Lancs., England.

EDHÄLL, Per Eric. Born 1913. Educ.: Chalmers Technol. Univ. Manager, Special Products Div., Nydqvist and Holm Aktiebolag.
Nuclear interests: Reactor design; Management.
Address: Nydqvist and Holm Aktiebolag, Special Products Division, Trollhättan, Sweden.

EDIS, Alfred Robert, B. Eng. (McGill). Born 1908. Educ.: London and McGill Univs. Res. Officer, A.E.C.L., Chalk River, Ontario (N.R.C., Canada), 1950; Assoc. Prof. Mech. Eng., McGill Univ. (responsible for nucl. eng. courses and undergraduate and graduate courses in mech. design), 1953-. Hon. Sec. Canadian Advisory Com., Inst. of Mech. Eng.; Internat. Editor (Canada), Nucl. Energy Eng. Societies: M.I.Mech.E.; Eng. Inst. Canada; Assoc. Fellow, Canadian Aero. Inst.; M.I.Nucl. E.; P. Eng. (Que.); P. Eng. (Ont.).
Nuclear interest: Nuclear engineering.
Address: 224 Clement Street, Dorval, Quebec, Canada.

EDISON, John G. Director, Rio Algom Mines Ltd.
Address: Rio Algom Mines Ltd., 335 Bay Street, Toronto, Ontario, Canada.

EDLING, Gustaf. Member, Sub-Com. on Reactor Location and Safeguards, Swedish Atomic Energy Board.
Address: Swedish Atomic Energy Board, Fack, Stockholm 2, Sweden.

EDLOW, Samuel. Born 1914. Educ.: Hebrew Union Coll., Cincinnati and Cincinnati Univ. Pres., Edlow Lead Co., Columbus, Ohio, 1953-61. Consultant, Samuel Edlow - Traffic Management of Nucl. Materials, 1961-. U.S. agent for A.B. Atomenergi, Stockholm, Sweden; European Atomic Energy Community, Brussels, Belgium; NUKEM, Wolfgang, Germany, and A.S.E.A., Vasteras, Sweden representing the interests of these companies in the international transportation of fissile and nuclear materials. Consultant to United Nucl. Corp. Industry Advisory Council to Div. of Licensing and Regulation, U.S.A.E.C. Societies: Atomic Ind. Forum.; Inst. Nucl. Materials Management.
Nuclear interests: Considered especially expert in the area of transportation of fissile materials.
Address: Suite 214, Port Columbus International Airport, P.O. Box 19827 Airport Station, Columbus, Ohio 43219, U.S.A.

EDLUND, Frans Erik Ronald. Univ. graduate. Born 1922. Educ.: Roy. Inst. of Technol., Stockholm. Group Leader, Design Dept., Karlstads Mekaniska Werkstad, Karlstad, Sweden. Societies: Atomic Ind. Forum; Svenska Teknologföreningen; Svetstekniska Föreningen.
Nuclear interest: Reactor design.
Address: Karlstads Mekaniska Werkstad, Karlstad, Sweden.

EDLUND, Milton Carl, B.S. and M.S. (Phys.), Ph.D. (Nucl. Sci.). E.O. Lawrence Award, 1965. Born 1924. Educ.: Michigan and Princeton Univs. Phys., MTR Design, Aircraft Nucl. Propulsion Project, Lecturer in Reactor Phys. (1949-51), Chief Phys., Aqueous Homogeneous Reactor Project (1953-55), O.R.N.L.; Manager, Phys. and Maths. Dept. (1955-60), Manager, Development Dept. (1960-62), Manager, Marketing Dept. (1962-65), Asst. Div. Manager, (1965-66), Atomic Energy Div., Babcock and Wilcox Co., Lynchburg, Virginia; Prof. Nucl. Eng., Michigan Univ., Ann Arbor, Mich. (1966-67); Consultant, Union Carbide Corp., Oak Ridge, Tennessee, (1967-), Member, Atomic Safety and Licensing Board Panel, U.S.A.E.C. Book: Co-author, The Elements of Nucl. Reactor Theory. Societies: A.P.S.; A.N.S.
Nuclear interest: Technical management - reactor design and development.
Address: 119 Balboa Circle, Oak Ridge, Tennessee, U.S.A.

EDMISON, Marvin Tipton, B.A. (Nebraska), M.Sc. (Nebraska), Ph.D. (Oklahoma State). Born 1912. Educ.: Nebraska and Oklahoma State Univs. Asst. Prof. to Assoc. Prof. and Asst. (Res.) to Vice-Pres., Arkansas Univ., 1951-55; Prof. Chem. and Director Res. Foundation, Oklahoma State Univ., 1955-. Exec. Com., Mid-America State Univs. Assoc.; Council Member, Assoc. Mid-West Univs. (Participating Univs. with Argonne Nat. Lab.). Societies: A.A.A.S.; A.C.S.
Nuclear interests: Research administration and radiological safety.
Address: 1107 N. Skyline Drive, Stillwater, Oklahoma, U.S.A.

EDMUNDS, Wade. Special Asst., Nucl. Power, Rural Electrification Administration.
Address: U.S. Department of Agriculture, Rural Electrification Administration, Washington 25, D.C., U.S.A.

EDSALL, John Tileston, B.A., M.D., Sc.D. (Hon., Chicago and Western Reserve Univs. and New York Medical Coll.). Born 1902. Educ.: Harvard and Cambridge Univs. Prof. of Biol. Chem., Harvard Univ.; Editor, Journal of Biol. Chem., 1958-68. Books: Co-author, Proteins, Amino Acids and Peptides (Academic Press, 1943); co-author, Biophysical Chemistry. Volume I. (Academic Press, 1958). Societies: N.A.S.; American Soc. Biol. Chem.; American Acad. of Arts and Sci.; A.C.S.; A.P.S.
Nuclear interests: Use of isotopes for labelling in the study of biochemical reactions; Social effects of nuclear energy, including control of nuclear weapons (Pugwash conferences) and effects of nuclear industry on the biological environment.
Address: The Biological Laboratories, Harvard University, Cambridge, Massachusetts 02138, U.S.A.

EDSTRÖM, John Olof, Dr. of Technol. Born 1926. Educ.: Roy. Inst. Technol., Stockholm and Minnesota Univ. Res. Metal., Jernkontoret; Docent, Roy. Inst. Technol.; Exec. Vice Pres. and Director, R. and D., Sandvik Steel Works, Sandviken. Societies: A.I.M.E.; Iron and Steel Inst. (England); Verein Deutscher Eisenhüttenleute (Germany).
Nuclear interests: Metallurgy; Management of research.
Address: 20 Strandvagen, Sandviken, Sweden.

EDUARDO, Alfonso Pajuelo, Dr. Staff Prof., Instituto Superior de Energia Nuclear, Peru.
Address: Junta de Control de Energia Atomica, Instituto Superior de Energia Nuclear, 611 Avenida Nicolas de Pierola, Apt. 914, Lima, Peru.

EDVARSON, Kay, D.Sc. Nucl. Dept., Res. Inst. Nat. Swedish Defence. Paper: Co-author, Transport of Long-lived Fission Products in Swedish Soils (2nd I.C.P.U.A.E., Geneva, Sept. 1958).
Address: Research Institute of National Defence, 10 Gyllenstiernsgatan, Stockholm 80, Sweden.

EDWARDS, A. J. Director, Thorium Ltd.
Address: Thorium Ltd., 6 St. James's Square, London S.W.1, England.

EDWARDS, C. F. Trustee, Southwest Atomic Energy Assoc.
Address: Southwest Atomic Energy Associates, 306 Pyramid Building, Little Rock, Arkansas, U.S.A.

EDWARDS, Doyle Ray, B.S., (Chem. Eng.), M.S. (Nucl. Eng.), Sc.D. (Nucl. Eng.). Born 1938. Educ.: Missouri School of Mines and Metal. and M.I.T. Asst. Prof., Nucl. Eng., and Director, Nucl. Reactor Facility (1963-64), Res. Phys. Prof., N.S.F.-A.E.C. Summer Inst. in Nucl. Sci. and Eng. (1964), Assoc. Prof., Nucl. Eng., and Director, Nucl. Reactor Facility (1964-), N.S.F. Res. Initiation Grant (1965-66), Missouri Univ., Rolla. Societies: A.N.S.; A.I.Ch.E.; Soc. of Ind. and Appl. Mathematics.
Nuclear interests: Computer simulation of radiation damage; Radiation induced polymerisation; Reactor physics; Computer techniques in nuclear engineering; Fast flux measurements.
Address: 54 Nancy Lane, Rolla, Missouri 65401, U.S.A.

EDWARDS, Edwin F., B.S. (Chem.)., B..S. (Chem. Eng.). Born 1918. Educ.: Louisiana State and Purdue Univs. Principal Development Eng. (Nucl. Systems Processes), Fluor Corp. Ltd.
Nuclear interests: Spent nuclear fuel processing; Waste disposal; Air cleaning.
Address: The Fluor Corporation Ltd., 2500 South Atlantic Boulevard, Los Angeles, California, U.S.A.

EDWARDS, Jack, B.Sc. (Eng.), A.C.G.I., Whitworth Scholar, M.I.Mech.E. Born 1919. Educ.: City and Guilds Coll., London Univ. Sen. Principal Sci. Officer, Roy. Naval Sci. Service, and Head of Naval Sect., A.E.R.E., Harwell, 1955-58; Prof. Nucl. Sci. and Technol., Royal Naval Coll., Greenwich, 1959-.
Nuclear interests: Marine reactor design and operation; Experimental and research programmes leading to small reactor systems; Training of reactor operators, maintenance and design staff.
Address: Maskells, Osgood Avenue, Farnborough, Orpington, Kent, England.

EDWARDS, John E., B.S. (Phys.), M.A., Ph.D. Born 1908. Educ.: Ohio and Ohio State Univs. Prof. Phys., Ohio Univ., 1951-. Pres. Appalachian Sect., A.A.P.T., 1963-64. Society: A.P.S.
Nuclear interests: Nuclear decay schemes, X-rays from nuclear processes. Neutron scattering. Also nuclear laboratory instruction.

Address: Physics Department, Ohio University, Athens, Ohio, U.S.A.

EDWARDS, Kenneth W. Chem., Geological Survey, U.S. Dept. of the Interior.
Address: U.S. Department of the Interior, Geological Survey, Building 25, Federal Centre, Denver, Colorado, U.S.A.

EDWARDS, Max N. Formerly Legislative Counsel, now Asst. Sec. of the Interior for Water Pollution Control and Desalination (1967-), U.S. Dept. of the Interior.
Address: U.S. Department of the Interior, Water Pollution Control and Desalination, Washington, D.C. 20242, U.S.A.

EDWARDS, Paul D. Staff Member, GMX Div., Los Alamos Sci. Lab. Chairman, Nucl. Components Com. Soc. for Nondestructive Testing. Paper: Leak, Thermal and Surface Test Methods (Symposium on Nondestructive Testing Trends in the A.E.C. Reactor programme, Germantown, U.S.A.E.C., 20 May 1960).
Address: Los Alamos Scientific Laboratory, California University, P.O. Box 1663, Los Alamos, New Mexico, U.S.A.

EDWARDS, R. D., A.M.I.W. Director, Nucl. Chem. Plant Ltd.
Address: Nuclear Chemical Plant Ltd., Chematom House, St. James Avenue, Ealing, London W.13, England.

EDWARDS, Raymond Richard, B.A., Ph.D. Born 1917. Educ.: Arkansas Univ. and M.I.T. Prof., Chairman, Arkansas Univ. Dept. of Chem., 1951-56; Asst. Director for Tech. Programmes, Div. of Internat. Affairs, U.S.A.E.C., 1956-57 (on leave); Director, Arkansas Univ. Graduate Inst. Technol., 1957-60; Guest Res. Staff, Lab. for Nucl. Sci., M.I.T., Jan. to Apr. 1961; Tech. Director, Nucl. Sci. and Eng. Corp., July 1961-65; Sen. Res. Chem., Carnegie Inst. of Technol. 1965-67; Lecturer, Carnegie-Mellon Univ., 1967-. Societies: A.C.S.; A.A.A.S.; A.N.S.; A.P.S.; A.G.U.
Nuclear interests: Chemical effects of nuclear transformations, inorganic reaction mechanisms, natural fission, isotope ratio variations, radioactivity; Application of tracers to biochemical problems.
Address: Department of Chemistry, Carnegie-Mellon University, Schenley Park, Pittsburgh, Pa., 15213, U.S.A.

EDWARDS, Rex, M.I.Nucl.Eng. Born 1935. Educ.: Reading Tech. Coll. Director, (1963), Joint Managing Director (1965), H.S. Marsh Ltd.
Nuclear interests: Design of irradiation plants of sizes varying from 500 curies to 2,000,000 curies.
Address: H.S. Marsh Ltd., 125 Southampton Street, Reading, Berks., England.

EDWARDS, Ronald. Director of Nucl. activities, Grand Rapids Jr. Coll.
Address: Grand Rapids Junior College, Nuclear Chemistry, 143 Bostwick Avenue, Grand Rapids, Michigan, U.S.A.

EDWARDS, William Elmer, B.M.E., M.S. Born 1922. Educ.: Ohio State Univ. Sen. Project Eng., G.E.C., 1951-. Past Chairman, Southwestern Ohio Sect., A.N.S. Society: N.S.P.E.
Nuclear interest: Reactor and shield physics.
Address: 10879 Reading Road, Cincinnati, Ohio 45241, U.S.A.

EDWARDS, William Farrell, B.S. (Utah), M.S., Ph.D. (Calif. Inst. Tech.). Born 1931. Educ.: Utah Univ. and California Inst. Technol. Asst. Prof., Utah State Univ., 1959-. Society: A.P.S.
Nuclear interests: Nuclear physics, specialising in high-precision spectroscopy of low-energy (to approx. 1 MeV) gamma radiation and beta radiation.
Address: Physics Department, Utah State University, Logan, Utah, U.S.A.

EEKELEN, M. van. See van EEKELEN, M.

EELES, Wilfred Trefor, M.Sc. Born 1930. Educ.: Univ. College, Cardiff. Sen. Sci. Officer, U.K.A.E.A., A.W.R.E., 1955-59; Res. Officer, C.E.G.B., Berkeley Nucl. Labs., 1959-. Society: Inst. of Phys.
Nuclear interests: Radiation damage in graphite and other materials; High-temperature shock phenomena.
Address: 10 Stanthill Drive, Dursley, Gloucester, England.

EERING, Johannes Theodorus. Born 1933. Lecturer, Rad. safety courses, Röntgen Technische Dienst N.V.; Deleg., I.A.E.A. Conference on the Use of Radioisotopes in the Phys. Sci. and Industry, Copenhagen, 6-17 Sept. 1960. Society: Nederlandse Vereniging voor Strahlingshygiëne.
Nuclear interests: Radiation safety; Industrial radiography; Nuclear physics; Radiation measurements; Legislative regulations; Industrial radiotracer applications.
Address: Röntgen Technische Dienst N.V., 144 Delftweg, Rotterdam 8, Netherlands.

EES, Ronald van. See van EES, Ronald.

EFANOV, A. I. Papers: Co-author, Measurement of IRT-1000 Reactor Frequency Response by Oscillator Method (letter to the Editor, Atomnaya Energiya, vol. 15, No. 4, 1963); co-author, Device for Oscillation measurements on Nucl. Reactor (letter to the Editor, ibid, vol. 20, No. 5, 1966).
Address: Academy of Sciences of the U.S.S.R., 14 Leninsky Prospekt, Moscow V-71, U.S.S.R.

EFFAT, Kamal Eldin Ahmed, B.Sc. (Elec. Eng.), B.Sc. (Hons., Phys.), Ph.D. (Nucl. Phys.). Born 1920. Educ.: Cairo and Birmingham Univs. Asst. Prof. (1958), Director Gen., Nucl. Power Project (1964), now Head, Training and Exchange of Experts Div., and Member, Board, United Arab Atomic Energy Establishment. Society: A. Inst. P., London. Nuclear interests: Reactors; Nuclear physics; Nuclear engineering.
Address: 25 Amin El Rafi Street, Dokki, Cairo, United Arab Republic.

EFFER, Franz. Manager, Hauptgemeinschaft des deutschen Einzelhandels, Cologne. Member, Economic and Social Com., Euratom, 1967-70.
Address: European Economic Community, Economic and Social Committee, 3 Boulevard de l'Empereur, Brussels, Belgium.

EFIMENKO, B. A. Papers: Co-author, Analyses of Systematic Errors Due to Differentiation of Apparatus Spectra Measured by Fast Neutron Single Crystal Scintillation Spectrometer (Atomnaya Energiya, vol. 19, No. 1, 1965); co-author, Analysis and Generalisation of Cross-Correlation Method for Measurement of Particles Life-Time Distribution in Physical System (ibid, vol. 20, No. 5, 1966); co-author, Differential Albedo for Gamma-Rays from Point Monodirectional Source (ibid, vol. 23, No. 3, 1967); co-author, Angular Distribution of Dose- Rate Gamma-Rays behind Carbon Barrier from Sources 4.45 MeV (letter to the Editor, ibid, vol. 24, No. 1, 1968).
Address: Academy of Sciences of the U.S.S.R., 14 Leninsky Prospekt, Moscow V-71, U.S.S.R.

EFIMOV, Aleksandr Nikitich, Eng. Graduate, North Caucasian Non-Ferrous Metals Inst.; Sci. Sec., 2nd I.C.P.U.A.E. Sen. Sci. Officer, Moscow Non-Ferrous Metals and Gold Inst., engaged on res. on technology of uranium, plutonium and radioactive metals production, 1946-.
Address: Moscow Non-Ferrous Metals and Gold Institute, Moscow, U.S.S.R.

EFIMOV, I. A. Paper: Co-author, BR-5 Reactor Operating Experience in 1959-66 (Atomnaya Energiya, vol. 23, No. 6, 1967).
Address: Academy of Sciences of the U.S.S.R., 14 Leninsky Prospekt, Moscow V-71, U.S.S.R.

EFSKIND, Jon, M.D. Pathologist, Norwegian Radium Hospital. Sen. Pathologist, Norsk Hydro's Inst. for Cancer Res.
Address: Norsk Hydro's Institute for Cancer Research, Norwegian Radium Hospital, Oslo, Norway.

EFSKIND, Leif Magnus. Born 1904. Educ.: Oslo Univ. Prof. Surgery, Oslo Univ.; Surgeon-in-Chief, Rikshopspitalet, Surg. Dept. A; Consulting Surgeon, Glittre Sanatorium. Societies: Norwegian Cancer Soc.; Norwegian Acad. of Sci.
Address: Rikshopspitalet, Oslo, Norway.

EGAMI, Nobuo, D.Sc. Born 1925. Educ.: Tokyo Univ. Head, Biol. Div., Nat. Inst. of Radiol. Sci. Society: Japan Rad. Res. Soc. Nuclear interest: Radiation biology.
Address: Division of Biology, National Institute of Radiological Sciences, 9-1, 4 chome, Anagawa, Chiba-City, Japan.

EGAMI, Tatsuhiko. Formerly Deputy Director, Atomic Energy Bureau, A.E.C. of Japan; Management Director, Hokuriku Elec. Power Co., Ltd.
Address: Hokuriku Electric Power Company, Limited, 10-3-3 Marunouchi, To yama-shi, Japan.

EGE, John Frederick, Jr., M.S. (Harvard Univ.), B.S. (Harvard Coll.). Born 1916. Educ.: Harvard Univ. and Harvard Coll. At Argonne Nat. Lab., 1949- (at present Director, Industrial Hygiene and Safety Div.). Health Phys. Fellowship Adviser for Argonne Nat. Lab. in Atomic Energy Commission's Special Fellowships Programme; Oak Ridge Associated Univs., Inc.; Member, Ind. Hygiene Fellowship Board; Nat. Safety Council rep. on N-12 Com., Nucl. Terminology, Units, Symbols, Identifications, and Signals, and N-12 ad hoc subcom. on Warning Means or Devices, U.S.A. S.I.; Alternate A.I.H.A. rep. on N-101 Com., Atomic Industry Faculty Design, Construction and Operating Criteria, U.S.A.S.I. Book: Chapter 43, entitled Ionising Rad. in Nat. Safety Council's Accident Prevention Manual for Ind. Operations (4th edition, 1959, Nat. Safety Council). Societies: A.A.A.S.; A.C.S.; A.I.H.A.; Atomic Ind. Forum, Inc.; American Management Assoc.; The Soc. of Harvard Eng. and Sci.; Assoc. of Harvard Chem.; Harvard Public Health Alumni Assoc.; H.P.S.; Sci. Res. Soc. of America; Certified by the American Board of Health Phys.; Certified by American Board of Ind. Hygiene.
Nuclear interests: Service function: Measure, record, and control employee exposures to radioactive materials; Review planned and existing equipment and facilities for adequacy of safeguards; Institute corrective actions; Provide consultation; Develop new instruments or techniques.
Address: Argonne National Laboratory, 9700 South Cass Avenue, Argonne, Illinois 60439, U.S.A.

EGELSTAFF, Peter A., B.Sc., Ph.D. Born 1925. Educ.: London Univ. (External). Sen. Principal Sci., U.K.A.E.A.; At present working in U.S.A.
Nuclear interests: Nuclear physics; Solid and liquid state physics.
Address: Atomic Energy Research Establishment, Harwell, Berks., England.

EGERT, Herwig, Dr. med., Facharzt für Innere Medizin. Born 1929. Educ.: Vienna Univ. Physician, 2nd Medical Univ. Clinic,

EGG

Vienna.
Nuclear interests: Diagnostic and therapeutic use of radioisotopes in the whole field of internal medicine, especially in thyroid disorders, and research work in this matter.
Address: Allgemeines Krankenhaus, Klinik Prof. Fellinger, Vienna 9, Austria.

EGGEBERT, Walter SEELMAN-. See SEELMANN-EGGEBERT, Walter.

EGGELING, Ernst, Bauing. Firmeninhaber und Geschäftsführer, Antitron-Gesellschaft für Strahlenschutz m.b.H. und Co.
Address: Antitron-Gesellschaft für Strahlenschutz m.b.H. und Co., 41-43 Castroper Strasse, 4600 Dortmund-Mengede, Germany.

EGGEN, Donald T., A.B. (Phys.), Ph.D. (Phys.). Born 1922. Educ.: Whittier Coll., Pennsylvania State and Ohio State Univs. Project Manager, Atomics Internat., a div. of North American Aviation, 1949-66; Formerly Programme Manager, Argonne Nat. Lab., 1966-; Prof., Nucl. Eng., Technol. Inst., Northwestern Univ. Societies: A.N.S.; A.P.S.
Nuclear interests: In the fields of reactor design, development and management as well as basic nuclear research, such as critical experiments, radiation damage and nuclear reactions.
Address: Northwestern University, Technological Institute, Evanston, Illinois, U.S.A.

EGGERS, F., Dipl.-Ing. Chief, Construction Eng., Howaldtswerke Hamburg, A.-G.
Address: Howaldtswerke Hamburg A.-G., Hamburg 11, Germany.

EGIAZAROV, B. G. Papers: Co-author, Apparatus for Neutron Activation Analysis (Atomnaya Energiya, vol. 20, No. 3, 1966); co-author, Choice of Analytical Methods for Instrumental Activation Analysis (ibid, vol. 24, No. 5, 1968).
Address: Academy of Sciences of the U.S.S.R., 14 Leninsky Prospekt, Moscow-V-71, U.S.S.R.

EGLI, J., Dir. Stellvertreter, Federal Commission for Atomic Energy.
Address: Federal Commission for Atomic Energy, 55 Effingerstrasse, Bern 3, Switzerland.

EGNELL, Gerard. Nucl. Dept. Business Manager, Sté. d'Etudes Techniques et d'Entreprises Generales.
Address: Societe d'Etudes Techniques et d'Entreprises Generales, 9 avenue Reaumur, Le Plessis-Robinson, (Seine), France.

EGOROV, A. I. Papers: Co-author, Pile-Neutron Burn-up Cross Section of Pm^{149} and Samarium Poisoning (letter to the Editor, Atomnaya Energiya, vol. 19, No. 2, 1965); co-author, Electrophoretic Filter for Purifying of Reactor Water (ibid, No. 4); co-author, Neutron Capture Cross-Sect. Measurement by Radioactive Nuclei Co^{58m}, Cu^{64} and Sc^{46} (ibid, vol. 24, No. 6, 1968).
Address: Academy of Sciences of the U.S.S.R., 14 Leninsky Prospekt, Moscow V-71, U.S.S.R.

EGOROV, N. N. BOBROV-. See BOBROV-EGOROV, N. N.

EGOROV, O. K. Paper: Co-author, Pulse Miniature Fission Chambers (letter to the Editor, Atomnaya Energiya, vol. 20, No. 3, 1966).
Address: Academy of Sciences of the U.S.S.R., 14 Leninsky Prospekt, Moscow V-71, U.S.S.R.

EGOROV, Yu. A. Papers: Co-author, Spectra for Reactor Fast Neutrons Passed through Graphite, Lead and Iron (Atomnaya Energiya, vol. 16, No. 2, 1964); co-author, <1.5 MeV Neutron Attenuation in Fe (letter to the Editor, ibid, No. 4); co-author, Attenuation of Reactor Radiations in Serpentine Sand (ibid, vol. 19, No. 4, 1965); co-author, Shielding Properties of Iron-Serpentine Concrete (letter to the Editor, ibid, vol. 20, No. 1, 1966).
Address: Academy of Sciences of the U.S.S.R., 14 Leninsky Prospekt, Moscow V-71, U.S.S.R.

EGOROV, Yu. V. Paper: Co-author, Use of Flotation Techniques for Clarifying and Cleaning of Radioactive Wastes (Atomnaya Energiya, vol. 16, No. 1, 1964).
Address: Academy of Sciences of the U.S.S.R., 14 Leninsky Prospekt, Moscow V -71, U.S.S.R.

EGRETEAU, Gaston. Manager, Nucl. Dept., Sté. Nessi Freres et Cie.
Address: Société Nessi Freres et Cie, 43 rue de la Vanne, Montrouge, (Seine), France.

EGUCHI, Wataru. Asst. Prof., Nucl. Chem. Eng., Dept. of Nucl. Eng., Kyoto Univ.
Address: Kyoto University, Department of Nuclear Engineering, Yoshida-Honmachi, Sakyo-ku, Kyoto, Japan.

EGUES, Antonio MORA. See MORA EGUES, Antonio.

EHLOTZKY, Fritz, Ph.D. Born 1929. Educ.: Innsbruck, Graz, Tübingen and Vienna Univs. Asst. (1958-), Sen. Asst. (1966-), Lecturer, Theoretical Phys. (1966-), Inst. for Theoretical Phys., Innsbruck Univ. Society: Austrian Phys. Soc.
Nuclear interest: Elementary particle physics, in particular quantum electrodynamics.
Address: Institut für Theoretische Physik, Innsbruck Universität, 52 Innrain, A-6020 Innsbruck, Austria.

EHMERT, A. On Editorial Board, Atompraxis.
Address: Atompraxis, Verlag G. Braun, 14-18 Karl-Friedrich-Strasse, Karlsruhe, Germany.

EHRENBERG, Hans, Prof. Dr. Born 1922. Educ.: Aachen, Göttingen and Bonn Univs.

Director, Inst. für Kernphysik, Mainz Univ.
Address: Institut für Kernphysik, Universität
Mainz, 21 Saarstrasse, 65 Mainz, Germany.

EHRENPREIS, Stanley Norvin, B.S. (Mech.
Eng.). Born 1929. Educ.: Pittsburgh Univ.
Mech. Eng., Westinghouse Elec. Corp., 1956-.
Society: A.N.S.
Nuclear interests: Nuclear instrumentation;
Radiation monitoring; Material activation
systems; Rod controls; Safeguard activities.
Address: 1023 Murray Hill Avenue, Pittsburgh,
Pennsylvania, U.S.A.

EHRENSPERGER, Fred Earl, B.M.E., M.S.
(Nucl. Eng.). Born 1933. Educ.: Georgia Inst.
Technol. Airborne Electronics Officer, U.S.
A.F., 1954-56; Aircraft Eng., Lockhead Aircraft Corp., Marietta, Georgia, 1956-60; Nucl.
Eng., Southern Services, Inc., Birmingham,
Alabama, 1960-.
Nuclear interests: Design, construction, and
operation of nuclear power reactors.
Address: 600 North 18 Street, Birmingham,
Alabama 35202, U.S.A.

EHRET, Adolf L., Dr. rer. nat., Dipl. Phys.
Born 1925. Educ.: Tübingen and Freiburg
Univs. Kernforschungsanlage Jülich des
Landes Nordrhein - Westfalen e.V., 1958-61;
Deutsche Babcock and Wilcox, 1962-66;
Gutehoffnungshütte Sterkrade, 1966-.
Nuclear interests: Reactor physics and technology; Sea water desalination.
Address: Gutehoffnungshütte Sterkrade A.G.,
42 Oberhausen - Sterkrade, Germany.

EHRET, Charles Frederick, M.S., Ph.D. Born
1923. Educ.: C.C.N.Y. U.S.P.H.S. Fellow, Inst.
de Biophysique, Genèva Univ., Geneva, Switzerland, 1960-61; Visiting Prof. in Zoology,
Indiana Univ., 1963; Fridley Lecturer in Chem.
Eng., Minnesota 1963; N.A.S.A. Fellow in
Theoretical Biol., Fort Collins, Colorado,
1966; Assoc. Biol. (1951-59), Biophysicist
(1964-66), Sen. Sci. (1966-), Argonne Nat.
Lab. Consultant, N.S.F. and U.S.A.E.C. Book:
Co-author, Mitogenesis (Chicago Univ. Press,
1958). Societies: Rad. Res. Soc.; American
Soc. of Zoologists; American Soc. of Naturaists; Soc. of Gen. Physiologists; Fellow,
A.A.A.S.
Nuclear interest: Radiation biology.
Address: Division of Biological and Medical
Research, Argonne National Laboratory,
Lemont, Illinois,U.S.A.

EHRINGER, Hermann J., Dr. rer. nat., Dipl.
Phys. Born 1921. Educ.: Tübingen and Stuttgart Univs. Asst., Stuttgart T.H., 1950-52;
Direktionsassistant Degussa Frankfurt/M.,
1952-58; Asst. Director, Euratom, Brussels,
1958-. U.S.-Euratom Joint R. and D. Board.
Nuclear interests: Metallurgy and management;
Materials testing reactors.
Address: 42 avenue Groelstveld, Brussels 18,
Belgium.

EHRLICH, Reinhart Paul, B.A.Sc. (Metal.
Hons.). Born 1922. Educ.: Toronto Univ.
Mill Superintendent, United Keno Hill Mines
Ltd., -1953; Chief Metal., Tech. Mine Consultants, 1953-56; Chief Metal. and Asst. Chief
Eng., Rio Tinto Mining Co. of Canada Ltd.,
1956-64; Principal, The Gen. Eng. Co. Ltd.,
1964-66; Consulting Metal. Eng., Western
Mining Divisions, Kennecott Copper Corp.
Societies: A.P.E.O.; Canadian Inst. of Mining
and Metal.; A.I.M.M.E.; I.M.M., London.
Nuclear interests: Production of uranium
raw materials including all necessary testing,
plant design, process development for production of nuclear grade raw materials.
Address: 1515 Mineral Square, Salt Lake
City, Utah 84111, P.O. Box 11299, U.S.A.

EHRLICH, Richard, B.A. (Phys.)., Ph.D.
(Theoretical Phys.). Born 1921. Educ.: Harvard and Cornell Univs. Manager, Advanced
Development Activity, K.A.P.L., Schenectady,
New York. First Chairman, Reactor Mathematics and Computations Div., A.N.S.; Chairman, Northeastern New York Sect., A.N.S.
Advisory Com. on Reactor Phys., U.S.A.E.C.
Chairman, Advisory Committee to Nat. Neutron Cross Section Centre, Brookhaven Nat.
Lab. Societies: Fellow, A.N.S.; A.A.A.S.;
A.P.S.; Assoc. for Computing Machinery.
Nuclear interests: Nuclear reactor physics;
Mathematics in nuclear reactor design;
Management.
Address: 2125 Lynn Plaza, Schenectady, New
York, U.S.A.

EHRMAN, Chester Samuel, B.Sc. (Marine
Eng.). Born 1931. Educ.: New York State
Univ. Reactor Plant Test Eng., Newport News
Shipbuilding and Dry Dock Co., 1953-62;
Project Eng., Atomics Div., American Machine
and Foundry Co., 1962-66; Allied Chem. Corp.,
1965-. Society: A.N.S.
Nuclear interests: Design of mechanical equipment for hot laboratories and spent nuclear
fuel reprocessing plants; Maintenance and
decontamination of equipment and cells by
remote and contact techniques; Radiation
effects on materials; Project management;
Marine reactor plant design and operation.
Address: 6 Warwick Lane, Basking Ridge, New
Jersey, U.S.A.

EICHELBERGER, John F., B.S. (Phys., Case
School of Appl. Sci.), M.S., Ph.D. (Phys.,
Ohio State). Born 1909. Educ.: Case School of
Appl. Sci. and Ohio State Univ. Chief
(1948-52), Sect. Supervisor (1952-54), Asst.
Res. Div. Director (1954-55), Res. Director
(1955-), Mound Lab.; Co-ordinating Organisation Director for Atomic Energy Commission
at Mound Lab., 1959-. Chairman, Southwestern
Ohio Div., A.N.S. Society: A.P.S.
Nuclear interests: Quantum mechanics; Heat
flow; Geometrical optics; Supersonic aerodynamics; Thermodynamics; Theory of
atomic spectra.

Address: 327 Jenny Lane, Dayton, Ohio, U.S.A.

EICHHOLZ, Geoffrey Gunther, B.Sc. (Hon.), Ph.D. (Leeds). Born 1920. Educ.: Bristol and Leeds Univs. Asst. Prof. Phys., British Columbia Univ., 1947-51; Head, Rad. Lab., Canadian Dept. Mines and Tech. Surveys, Ottawa, 1951-63; Prof., Nucl. Eng., Georgia Inst. of Technol., 1963-. Societies: Phys. Soc. (London); A.P.S.; A.N.S.; Canadian Assoc. Phys. Nuclear interests: Applications of radioactive isotopes in science and industry; Nuclear instrumentation.
Address: School of Nuclear Engineering, Georgia Institute of Technology, Atlanta, Ga. 30332, U.S.A.

EICHLER, Siegfried, Dr. Geschaftsfuhrer, Federation of German Industries.
Address: Federation of German Industries, 2-12 Habsburgerring, Postschliessfach 107, 5 Cologne 10, Germany.

EICHNER Madame C. Sec., I.N.S.T.N.
Address: Institut National des Sciences et Techniques Nucléaires, B.P. No. 6, Gif-sur-Yvette (Seine-et-Oise), France.

EIDENBERGER, Karl, Dipl.Ing. Head, Dept. of Measuring and Control Technique, Wiener Schwachstromwerke.
Address: Wiener Schwachstromwerke, 12 Apostelgasse, 1031 Vienna, Austria.

EIFLER, Friedrich Karl, Min. R. Member, Fachkommission IV-Strahlenschutz und Sicherheit, and Arbeitskreis I/2, Rechts- und Verwaltungsfragen des Strahlenschutzes, Deutsche Atomkommission, Federal Ministry for Scientific Research, Germany.
Address: Federal Ministry for Scientific Research, 46 Luisenstrasse, 532 Bad Godesberg, Germany.

EIGEN, Fritz, Dipl. Ing. Gen. Manager, Tech. Dept., Frieseke und Hoepfner.
Address: Frieseke und Hoepfner, G.m.b.H., Erlangen-Bruck, Germany.

EINNATZ, Alfred, Dr. jur. Mitglied, Arbeitskreis I/1 Haftung und Versicherung, Deutsche Atomkommission, Federal Ministry for Sci. Res.
Address: Federal Ministry for Scientific Research, 46 Luisenstrasse, Bad Godesberg, Germany.

EISBERG, Robert Martin, Ph.D., B.S. Fulbright Fellowship; Guggenheim Fellowship. Born 1928. Educ.: California (Berkeley), and Illinois Univs. Guest Res. Phys., Linear Accelerator Lab., Southern California Univ., 1961-; Visiting Res. Director, Cyclotron Lab., Buenos Aires, Argentina, June-Sept., 1962; Prof. California Univ., Santa Barbara, 1963-; Consultant, Rutherford Lab., Didcot, Berks., 1965. Book: Fundamentals of Modern Phys. (New York-London, John Wiley and Sons, Inc., 1961). Societies: A.P.S.; A.I.P.; A.A.P.T.; Federation of American Scientists.
Nuclear interests: Nuclear scattering, absorption and activation experiments at intermediate energies; Passage of particles through matter; Phenomenological theory of nuclear scattering and reactions.
Address: Department of Physics, California University, Santa Barbara, California, U.S.A.

EISENBERG, Yehuda, M.Sc. (Jerusalem), Ph.D. (Cornell) Born 1927. Educ.: Hebrew (Jerusalem) and Cornell Univs. Prof. of Phys., Phys. Dept., Weizmann Inst., Rehovoth, Israel; Visiting Sci.: Bern Univ. (Switzerland) M.I.T., Cambridge, Mass., SLAC, Stanford, California. Societies: A.I.P.; Italian Phys. Soc.; Israeli Phys. Soc.
Nuclear interests: High energy nuclear physics and cosmic rays and elementary particle physics.
Address: Weizmann Institute, Nuclear Physics Department, Rehovoth, Israel.

EISENBUD, Merril, B.S. (E.E.), Sc.D. (Hon.). Born 1915. Educ.: New York Univ. Director, Health and Safety Lab. (1949-57), Manager, New York Operations Office (1954-59), U.S.A.E.C.; Assoc. Prof. Environmental Medicine, Dept. of Environmental Medicine (1950-55), Adjunct Prof. Environmental Medicine, Postgraduate School of Medicine (1955-59), Prof. (1959-), New York Univ. Director, Env. Rad. Lab., Inst. Environmental Medicine, New York Univ., 1959-. Toxicology Com. (1952-59), Com. on Atmospheric and Ind. Hygiene (1952-), N.R.C.; Subcom. on Radium Dust and Radon Gas Z37 (1949-), Sectional Com. on the Use of X-rays Z54 (1951-), American Standards Assoc.; Alternate U.S. Rep., U.N. Sci. Com. on the Effects of Atomic Rad., 1956-60. Member, Com. on the Meteorological Aspects of the Effects of Atomic Rad., N.A.S., 1956-; Member, Mayor's Tech. Advisory Com. on Rad., New York City, 1958- (Chairman, 1961-). Societies: A.N.S. (Board of Directors, 1961-); Rad. Res. Soc.; A.A.A.S.; H.P.S. Nuclear interests: Reactor technology; Radiological hazards; Radioactive fallout; Beryllium poisoning.
Address: 550 First Avenue, New York, New York 10016, U.S.A.

EISENLOHR, Horst H., Dr. rer. nat., Dipl.-Phys. Born 1926. Educ.: Karlsruhe T.H. and Freiburg Univ. Res. Phys., Siemens and Halske 1953-56; Res. Asst., Freiburg Univ., 1956; Rad. Expert, Ministry of Defence, Germany, 1957-63; Professional Officer, Dosimetry Sect. I.A.E.A., Vienna, 1963. Consultant Member, Fachnormenausschuss Radiologie (Deutscher Fachnormenausschuss). Books: Coordinating Editor, Eng. Compendium on Rad. Shielding (Springer-Verlag and I.A.E.A., 1968); co-author, Dosimetrie und Strahlenschutz (Stuttgart, Thieme-Verlag). Society: Deutsche Physikalische Gesellschaft.

Nuclear interests: Elementary particles; Radiation protection and dosimetry; Radiation biophysics; Research management.
Address: International Atomic Energy Agency, 11 Karntnerring, A-1010 Vienna, Austria.

EISENMANN, W. Optical Problems Consultant, Optische Werke C.A. Steinheil Sohne G.m.b.H.
Address: Optische Werke C.A. Steinheil Sohne G.m.b.H., 10 Germersheimer Strasse, Munich 8, Germany.

EISMONT, V. P. Papers: The Angular Distributions of Light and Heavy Fragments from the 14 MeV Neutron Fission of U^{238} (letter to the Editor, Atomnaya Energiya, vol. 4, No. 2, 1958); co-author, Anisotropy in the 14 MeV Neutron-induced Fission of U^{238} (ibid, vol. 6, No. 6, 1959); co-author, Comparison of the Kinetic Energies of the Fission Fragments Produced from ^{238}U by Neutrons of Energies 3 and 15 Mev (letter to the Editor, ibid, vol. 12, No. 2, 1962); Fission Neutrons of Excited Nuclei (ibid, vol. 19, No, 2, 1965).
Address: Academy of Sviences of the U.S.S.R., 14 Leninsky Prospekt, Moscow V-71, U.S.S.R.

EITEL, Joseph Erwin, Dipl.-Ing. Born 1910. Educ.: Munich T.H. Vorstandsmitglied der Seereederei "Frigga", Hamburg. Societies: V.D.I., Düsseldorf; Schiffbautechnische Gesellschaft, Hamburg; Kernenergie-Studiengesellschaft, Hamburg; Deutsches Atomforum e.V., Bonn.
Nuclear interests: Reactor design, management, nuclear ship propulsion.
Address: 7 Am Pumpenkamp, 2 Hamburg 55, Germany.

EITZ, August-Wilhelm, Dr.-Ing. Born 1933. Educ.: Aachen T.H. Formerly Rhein.Westf. Elektrizitätswerk AG. (R.W.E.) Essen, Nucl. power station group; Geschäftsfuhrer, Gesellschaft für Kernforschung m.b.H. Versuchsanlagen, Leopoldshafen, 1968-. German Member, Study Group on the Long Term Role of Nucl. Energy in Western Europe, O.E.C.D., E.N.E.A.
Nuclear interests: Management in nuclear field (construction and operation of prototype reactors and reprocessing plant).
Address: Gesellschaft Kernforschung m.b.H., Versuchsanlagen, 7501 Leopoldshafen, Postfach, Germany.

EK, Corneille S. L., Chem. and Metal. Eng. Born 1926. Educ.: Liège Univ. Asst. Prof., Dept. of Non-ferrous Metal. and Oredressing, Liège Univ., 1951-. Societies: Assoc. Ing. sortis de l'Ecole de Liège; Sté. Roy. Belge des Ing. et Industriels; A.I.M.M.E.
Nuclear interests: Ore-dressing; Extractive metallurgy and fabrication of non-ferrous metals for atom industry.
Address: 14 avenue Saint-Michel, Esneux, Belgium.

EKEFALK, Sture Harald, Civil Eng. Born 1909. Educ.: Royal Inst. Technol., Stockholm. Exec. Vice-Pres., Souschief (1955-58), Head, Atomic Energy Dept. (1955-58), State Power Board; Deputy Managing Director, Atlas Copco AB, Stockholm, 1958-. Societies: Conférence Internationale des Grands Réseaux Electriques; A.A.A.S.; American Inst. Elec. Eng.; Union Internationale des Producteurs et Distributeurs d'Energie Electrique.
Nuclear interest: Nuclear reactor technology, especially concerning heavy-water moderated natural uranium combination.
Address: G:a Kyrkvägen 21, Danderyd, Sweden.

EKER, Bjarne Reidar, M.D., Ph.D. Born 1903. Educ.: Oslo Univ. Prosector (1939-), Director (1947-), Norwegian Radium Hospital, Director, Norsk Hydro's Inst. Cancer Res. Societies: Norwegian Cancer Soc.; Rad. Hygiene Advisory Council (Chairman).
Nuclear interests: Radiobiology and radiation protection.
Address: The Norwegian Radium Hospital, Oslo, Norway.

EKER, Per, cand. real. Born 1930. Educ.: Oslo Univ.
Address: Department of Biochemistry, Norsk Hydro's Institute for Cancer Research, Montebello, Oslo, Norway.

EKERT, B., Dr. Chef de Service, Sect. de Biologie, Inst. du Radium.
Address: Institut du Radium, Section de Biologie, 26 rue d'Ulm, Paris 5, France.

EKHOLM, Erik. Member, Washington State Advisory Council on Nucl. Energy and Rad.
Address: Washington State Advisory Council on Nuclear Energy and Radiation, General Administration Building, Olympia, Washington, U.S.A.

EKINS, Roger Phillip, M.A., Ph.D. Born 1926. Educ.: Cambridge Univ. Lecturer (1949-62), Sen. Lecturer, then Reader in Medical Nucleonics, Middlesex Hospital Medical School (Inst. of Nucl. Medicine). Societies: British Inst. Radiol.; Hospital Physicist's Assoc.
Nuclear interests: Use of isotopes in medicine, biology, biochemistry and physiology; Particular interest in radioisotopic methods of microanalytical assay of hormones and vitamins.
Address: Roffeys, Coldharbour, near Dorking, Surrey, England.

EKLUND, Arne Sigvard, D.Sc. Atoms for Peace Award, 1968. Born 1911. Educ.: Uppsala Univ. Sec. Gen., 2nd I.C.P.U.A.E., Geneva, Sept. 1958. Tech. Dir., A.B. Atomenergi, Sweden, 1957-61; Director-Gen., I.A.E.A., 1961-. Member, Sci. Council, Internat. Centre for Theoretical Phys., Trieste. Books: Studies in Nuclear Physics. Excitation by Means of X-rays. Activity of ^{87}Rb (Uppsala, Diss., 1946). Society: Roy. Swedish

Acad. of Eng. Sci.
Nuclear interests: Reactor design; Management.
Address: I.A.E.A., 11 Kaerntnerring, Vienna 1, Austria.

EKLUND, Karl Erik, B.S. (M.I.T.), M.A., Ph.D. (Columbia). Born 1929. Educ.: M.I.T. and Columbia Univ. Phys., U.S. Army Chem. Corps. 1951-54; Ind. Consulting Phys., 1955-60; Res. Assoc., Columbia Univ., 1959, 1962; Director of Res., Rad. Dynamics, Inc., 1960-61; Assoc. Prof. Phys., U.S. Merchant Marine Acad., 1962; Wanderjahr 1962; Res. Assoc. and Asst. Director, Nucl. Structure Lab., Yale Univ., 1963-65; Assoc. Director, Nucl. Structure Lab. and Director, Phys. Lab., New York State Univ. at Stony Brook, 1965-.
Societies: A.P.S.; A.A.A.S.
Nuclear interests: Accelerator design and use; Nuclear structure physics; Research management.
Address: The Physical Laboratories, State University of New York at Stony Brook, New York, U.S.A.

EKMAN, Gunnar. Member, Swedish Atomic Energy Board.
Address: Swedish Atomic Energy Board, Fack, Stockholm 2, Sweden.

EKSPONG, Anders Gosta, Ph.D. Born 1922. Educ.: Uppsala Univ., Sweden. Prof., Stockholm Univ., 1960-. Chairman, Phys. III Com., Nucl. Phys. Res. Com., Sci. Policy Com., C.E.R.N. Book: Kärnfysik (Almqvist and Wiksell, 1964).
Nuclear interest: Elementary particles.
Address: Institute of Physics, 9 Vanadisvägen, Stockholm Va, Sweden.

EL-BAZ, Edgard, D. es-Sc. Physiques. Born 1937. Educ.: Algiers Univ. Maitre de Conf., Faculté Sciences de Lyon, Inst. de Physique Nucléaire, Villeurbanne, France.
Nuclear interests: Theoretical nuclear physics at low energy and especially direct interaction processes; Graphical treatment of the algebra of angular momentum.
Address: 100 C Cours Lafayette, 69 Lyon, France.

EL-BEDEWI, F. Member, Editorial Board, Nuclear Physics.
Address: Atomic Energy Commission, Al-Tahrir Street, Dokki, Cairo, Egypt, United Arab Republic.

EL DIN, El Tawdy Saad, Dr. At Ein Shams Univ. Hospital, Egypt.
Address: Ein Shams University Hospital, Demerdache, Cairo, Egypt, United Arab Republic.

EL-FASSI, Nacer, Moroccan deleg., I.A.E.A. Third General Conference, Vienna, Sept. 1959; Now Moroccan Resident Rep., I.A.E.A.
Address: c/o Moroccan Embassy, 11 Marienstrasse, Berne, Switzerland.

EL-GARHY, Mohammed Ahmed, B.Sc. (Hons.), Ph.D. Born 1933. Educ.: Ain Shams (Cairo) and Vienna Univs. Demonstrator, Faculty of Sci., Alexandria Univ., 1954; Scholarship, Vienna Univ., 1956-59; Lecturer, Radiochem. (1960-64), Prof., Radiochem. (1964-), United Arab Atomic Energy Establishment; Prof., Radiochem., Middle Eastern Regional Radioisotope Centre, 1964-; Prof., Radiochem., Faculty of Sci., Ain Shams Univ., 1964-.
Nuclear interests: Inorganic nuclear chemistry; Radiochemical separations related to the production of radioactive isotopes and hot atom chemistry.
Address: United Arab Atomic Energy Establishment, Atomic Energy Post Office, Cairo, Egypt, United Arab Republic.

EL-GUBEILI, Mohammed Abdul-Mabood, Asst. Prof. Dr. Formerly Head, Nucl. Chem. Div., now Director Gen., Atomic Energy Establishment, United Arab Republic; United Arab Republic alternate deleg., Gen. Conference, Vienna, Sept. 1960, and Member, Sci. Advisory Com., (1966-), I.A.E.A.
Address: Atomic Energy Establishment, Inshas, Nr. Cairo, Egypt, United Arab Republic.

EL HADIDY, Ali Ramadan. Head, Sci. and Tech. Inf. Div., United Arab Atomic Energy Establishment.
Address: United Arab Atomic Energy Establishment, Inshas, Nr. Cairo, Egypt, United Arab Republic.

EL HALAWANI, Ahmed Abdel Salam, M.D., D.T.M., D.T.H. Born 1902. Educ.: Würzburg, London, and Liverpool Univs. Director, Res. Inst., Cairo; Undersec. of State for Ministry of Public Health (Egypt); Formerly Deputy Director W.H.O. Regional Office for Eastern Mediterranean; Prof. and Director, Centre of Medical Res., Coll. of Medicine, Baghdad. Editor, J. Egyptian Medical Assoc. Societies: Egyptian Soc. for Radioisotopes; Egyptian Medical Assoc.
Nuclear interests: Effects of ionising radiation and its role in public health.
Address: 2 Ismailia Street, Rushdi, Alexandria, United Arab Republic.

EL-KHALAFAWI, Miss T. United Arab Republic Deleg. to Conference sur la Physique des Plasmas et la Recherche concernant la Fusion Nucléaire Controlée, Salzburg, Sept. 1961.
Address: Institut für Plasmaphysik, Jülich, Germany.

EL-KHARADLY, M. E. A., Dr. U.A.R. Member, U.N. Sci. Com. on the Effects of Atomic Rad.

Address: United Nations Scientific Committee on the Effects of Atomic Radiation, United Nations, New York, U.S.A.

EL-KHASHAB, Sayed Amin, Eng. Sec. Gen., Egyptian A.E.C.
Address: Atomic Energy Post Office, Cairo, United Arab Republic.

EL-MOFTY, Osman Hassan, B.Sc. (Elec. Eng.), M.A. (Phys.), Ph.D. (Phys.). Born 1921. Educ.: Cairo and California (Berkeley) Univs. Instructor, Faculty of Eng., 1953-58; Assoc. Prof. (1958), Head, Reactor Div. (1959-), Atomic Energy Establishment, Cairo. Society: Mathematical Phys. Soc., United Arab Republic.
Nuclear interests: Reactor engineering; Nuclear physics.
Address: Atomic Energy Establishment, Dokki, Cairo, United Arab Republic.

EL-NADI, Mohamed, B.Sc. (Hons.), M.Sc., Ph.D. Born 1918. Educ.: Cairo and London Univs. Prof. Nucl. Phys. (1961-), Head, Phys. Dept. (1967-), Cairo Univ.; Acting Head, Phys. Dept., U.A.R. Atomic Energy Establishment, 1961-. Societies: Phys. Soc. of London; A.P.S.; Mathematical and Phys. Soc. of Egypt.
Nuclear interests: Nuclear physics (theory of low-energy nuclear reactions).
Address: Faculty of Science, Cairo University, Giza, Cairo, Egypt, U.A.R.

EL SAID, El Said Moustafa, Dr. Rector, Cairo Univ.
Address: Cairo University, Giza, Egypt, United Arab Republic.

EL SHAZLY, El Shazly Mohamed, B.Sc., D.I.C., Ph.D. Born 1923. Educ.: Cairo and London Univs. First Sci. Officer, I.A.E.A., Vienna, 1959-61; Director, Mineral Res. Labs., Cairo, -1962; Prof. of Geology, Head of Geology and Raw Materials Dept., Atomic Energy Establishment, Cairo, 1962-. Sec. Geological Soc. of Egypt; Editor, Journal of Geology of the U.A.R.; Rep., Internat. Mineralogical Assoc.; Organiser, Internat. Mineralogical Abstracts.
Nuclear interests: Nuclear materials, their exploration and processing. Reactor site studies. Ground disposal of radioactive wastes.
Address: Atomic Energy Establishment of the U.A.R., Atomic Energy Post Office, Cairo, United Arab Republic.

EL-SHERBINI, Mahmoud Ahmed, B.Sc. (Egypt), M.Sc. (Egypt), Ph.D. (London). Born 1909. Educ.: Cairo Univ., Dean, Faculty of Sci., Alexandria Univ.; Member, Egyptian Atomic Energy Establishment Board; Member, Com. of Measurements, Ministry of Industry; Member, Phys. Com., Nat. Res. Centre of Egypt. Member, Nat. Com. of Pure and Appl. Phys. Book: Our Place in Atomic Energy (Alexandria Univ. Press). Societies: Phys. Soc. of London; Acad. of Sci., Egypt; Sci. History Soc.; Mathematical and Phys. Soc. Nuclear interest: Nuclear physics.
Address: Faculty of Science, Moharrem Bay, Alexandria, United Arab Republic.

EL-SOLH, H. E. Abdel-Rahman. Lebanese Ambassador to Austria; Lebanese Resident Rep. to I.A.E.A.
Address: 50 Reisnerstrasse, Vienna 111, Austria.

EL-WAKIL, M. M., B.S., M.S., Ph.D. Born 1921. Educ.: Cairo and Wisconsin Univs. Lecturer in Mech. Eng., Alexandria Univ., 1950-52; Asst. Prof. of Mech. Eng., Minnesota Univ., 1954-55; Prof. Mech. and Nucl. Eng., Wisconsin Univ. 1955-. Book: Nucl. Power Eng. (New York, McGraw-Hill Publishing Co., 1962). Societies: A.N.S.; A.S.M.E.; A.S.E.E.; Combustion Inst.
Nuclear interests: Research and teaching; Nuclear power engineering, heat transfer, two-phase flow.
Address: Department of Mechanical Engineering, 1513 University Avenue, Wisconsin University, Madison, Wis. 53705, U.S.A.

ELBANIDZE, Yu. G. Paper: Co-author, Radiographic Study of Liquid Indium-Gallium Alloy with Eutectic Concentration (Atomnaya Energiya, vol. 22, No. 5, 1967).
Address: Academy of Sciences of the U.S.S.R., 14 Leninsky Prospekt, Moscow V-71, U.S.S.R.

ELBEK, B., Dr. Tandem Accelerator Dept., Niels Bohr Inst., Copenhagen Univ.
Address: Niels Bohr Institutet, 15-17 Blegdamsvej, Copenhagen Ø, Denmark.

ELDER, Glenn Earl, A.B., M.S. Born 1914. Educ.: Indiana and Tennessee Univs., Evansville Coll. Res. Asst., Tennessee Univ., 1950-54; Sen. Metallurgist, Oak Ridge Nat. Lab., 1954-55; Director of R. and D., Electro-Mechanical Labs. (1955-59); Chief, Nucl. Effects Directorate (1959-), White Sands Missile Range, New Mexico. Society: A.N.S. Nuclear interests: Direct operation of a laboratory with mission of testing and evaluating weapon system performance in a nuclear weapon environment. Laboratory operates a Fast Burst Reactor, of Godiva type; a 15 Mev linear electron accelerator for high intensity gamma pulses; and a pulsed neutron generator for high intensity 14 Mev neutron pulses.
Address: Army Missile Test and Evaluation, Nuclear Effects Directorate, White Sands Missile Range, New Mexico, U.S.A.

ELDJARN, Lorentz, M.D. Born 1920. Educ.: Oslo Univ. Prof. Clinical Biochem., Univ. Hospital, Rikshopspitalet, Oslo. Formerly head, Biochem. Labs., Norwegian Hydro's Inst. for Cancer Res., Norwegian Radium Hospital, Oslo. Member, Com. 1, The Internat. Commission on Radiol. Protection. Book: Chapter: Mechanisms of Action of Protective and Sensitising Agents (in: Mechanisms in Radiobiol.;

Academic Press, 1959). Society: Norwegian Acad. of Sci.
Nuclear interest: Basic mechanism in radiobiology.
Address: Institute of Clinical Biochemistry, University Clinical Laboratory, Rikshopspitalet, Oslo, Norway.

ELDRED, Norman Orville, B.S.E. (Chem. Eng., Michigan), M.S. (Chem. Eng., Michigan). Born 1916. Consulting Chem. Eng., Vicksburg, Michigan, 1948-52; Project Eng., The Lummus Co., N.Y., 1952-53; Feature Eng., Grinnell Co., Ohio, Gaseous Diffusion Plant, Portsmouth, Ohio, 1954-55; Project Eng., Leonard Construction Co., Chicago, Catalytic Acid Plants, 1955-57; Project and Process Eng., Sumner Sollitt Co., Chicago, 1957-59. Consulting Facilities and Elec. Eng., Titan Missile Project, Ellsworth A.F.B., Rapid City, S.D., 1961-62; Project Eng., Naphtha Catalytic Reformers and Steam Power Plants, Kaighin and Hughes Eng. Co., Toledo, Ohio, 1962; Project Eng. Plant Dept., McDonnell Aircraft Corp., Lambert Municipal Airport, St. Louis, Mo., engaged in the U.S. Navy Phantom 11 Fighter Plane Programme, 1963; Consulting Industrial Eng. in Naval Shipboard Atomic Power Reactor Refueling Machines, 1963. Member, Board of Directors, Illinois Assoc. of Consulting Eng.; Member, Nat. Naval Advisory Com., U.S.N.R., Reserve Officer Assoc. of the United States; Treas., Naval Armoury Chapter, Chicago, Reserve Officers Assoc. of the United States. Societies: Registered Professional Eng., State of Illinois, No. 16282; Illinois and Nat. Assoc. of Professional Eng.; Assoc. Member, A.I.Ch.E.; A.C.S.; American Petroleum Inst.; A.N.S.; American Ordnance Assoc.; A.S.T.M.; N.A.C.E.; A.S.M.E.; I.E.E.E.
Nuclear interests: Management, gaseous diffusion, reactor design.
Address: 5630 N. Sheridan Road, Apt. N-1021, Chicago 40, Illinois, U.S.A.

ELDRED, Vernon Walter, M.A., Ph.D. Born 1925. Educ.: Cambridge Univ. Metal., English Elec. Co. Ltd., Nelson Res. Lab., Stafford, 1953-55; Reactor Development Lab., U.K.A.E.A., Windscale Works, Cumberland, 1955-; Res. Manager (Metal.), 1959-. Societies: Inst. Metals; Iron and Steel Inst.; Assoc., Inst. Metal.
Nuclear interest: Metallurgy, with particular reference to behaviour of fuel elements under irradiation.
Address: Dunure, Seascale, Cumberland, England.

ELDREDGE, A. G. Vice Pres., Assoc. Eng. and Consultants, Inc.
Address: Associated Engineers and Consultants, Inc., 975 Stewart Avenue, Garden City, Long Island, New York, U.S.A.

ELDRIDGE, Hudson Bluford, B.S. (Eng. Phys.), M.S. (Phys.). Born 1933. Educ.: Tennessee Univ. Phys., O.R.N.L., 1958-61; Phys., Aerospace Corp., 1961-62; Student, California Univ., Los Angeles, 1962-; Member, Staff, Phys. Dept., Wyoming Univ. Societies: A.A.A.S.; A.P.S.
Nuclear interests: General experimental and theoretical nuclear physics; at present investigating proton damage in semiconductors.
Address: Wyoming University, Department of Physics, Box 3905, University Station, Laramie, Wyoming 82070, U.S.A.

ELDRIDGE, James Samuel, B.S. (Chem.), M.S. Born 1928. Educ.: Tennessee Univ. Asst. Sci., Medical Div., O.R.I.N.S., 1950-56; Chem., O.R.N.L., 1956-. Society: A.N.S.
Nuclear interests: Analytical radiochemistry; Scintillation spectrometry.
Address: Oak Ridge National Laboratory, Post Office Box X, Oak Ridge, Tennessee, U.S.A.

ELEKDAG, Sukru. Member, A.E.C., Turkey.
Address: Atomic Energy Commission, 12 Ziya Gokalp Caddesi, Ankara, Turkey.

ELIAS, Horit, Doz., Dr. Born 1932. Educ.: Darmstadt T.H. and Tübingen Univs. Society: Gesellschaft Deutscher Chemiker.
Nuclear interests: Nuclear chemistry; Radiochemistry; Radiation chemistry.
Address: Lehrstuhl für Kernchemie, Darmstadt Technische Hochschule, 61 Darmstadt, Germany.

ELIAS, K., Dr. Phys. Dept., Isotope Lab., Dept. of Clinical Therapeutics, Athens Univ.
Address: Alexandra Hospital, Isotope Laboratory, Vas. Sophias-K. Lourou Str., Athens, Greece.

ELIAS DIAZ ACHECAR, Abelardo, Member, Comision Nacional de Investigaciones Atomicas, Dominican Republic.
Address: Comision Nacional de Investigaciones Atomicas, Ciudad Trujillo, Dominican Republic.

ELIASSEN, Rolf, Dr. Formerly Member, Faculty: M.I.T., New York Univ. and I.I.T.; Prof., Civil Eng., Stanford Univ.; Consultant, President's Office of Sci. and Technol. Member, Panel of Experts to Provide Boards on Atomic Safety and Licensing, U.S.A.E.C., 1968-.
Address: Stanford University, Stanford, California, U.S.A.

ELINSON, M. I. Book: Co-author, Avtoelektronnaya emissiya (Auto-electron Emission) (M. Fizmatgiz, 1958).
Address: Academy of Sciences of the U.S.S.R., 14 Leninsky Prospekt, Moscow V-71, U.S.S.R.

ELISEEV, V. S. Paper: On β-Particle Bremsstrahlung and Protection (letter to the Editor,

Atomnaya Energiya, vol. 14, No. 4, 1963).
Address: Academy of Sciences of the U.S.S.R., 14 Leninsky Prospekt, Moscow V-71, U.S.S.R.

ELISEEVA, I. M. Paper: Co-author, Application of Harley-Hallden Method to Analysis of Integral Beta-Spectra (letter to the Editor, Atomnaya Energiya, vol. 21, No. 4, 1966).
Address: Academy of Sciences of the U.S.S.R., 14 Leninsky Prospekt, Moscow V-71, U.S.S.R.

ELKIND, Mortimer M., B.M.E., M.M.E., M.S., Ph.D. E. O. Lawrence Memorial Award, 1967. Born 1922. Educ.: Cooper Union School of Eng., Polytech. Inst., Brooklyn, and M.I.T. Phys. on assignment to M.I.T. (1949-53), Biophys. on assignment to Donner Lab., California Univ., Berkeley (1953), Biophys., Physiology Lab. (1954), Nat. Cancer Inst., Bethesda, Maryland. Societies: Biophys. Soc.; Rad. Res. Soc.
Nuclear interest: Radiation biology.
Address: 1705 Wilmart Street, Rockville, Maryland, U.S.A.

ELLEFSON, Bennett S., Dr. Vice Pres., Gen. Telephone and Electronics Labs., Inc. Member, Gen. Advisory Com., New York State Office of Atomic Development.
Address: 214-04 33rd Road, Bayside, New York, N.Y., U.S.A.

ELLENS, Siewert Hendrik. Born 1925. Educ.: Amsterdam Univ. Organisation for Appl. Sci. Res., Netherlands 1953-64; Reactor Centrum Nederland, 1964-.
Nuclear interest: General.
Address: 113 van Cranenburchlaan, Wassenaar, Netherlands.

ELLETT, D. Maxwell, B.M.E., M. Eng., D. Eng. Born 1922. Educ.: Virginia and Yale Univs. Asst. in Instruction, Yale Univ., 1951-52; Staff Member, Sandia Corp., 1952-. Exec. com., Trinity Sect., A.N.S.
Nuclear interests: Reactor operation; Mechanical design of pulsed reactors and auxiliary equipment; Radiation effects; Effects of nuclear weapons.
Address: 2000 Los Poblanos Pl., N.W., Albuquerque, N. Mex. 87107, U.S.A.

ELLETT, Julian D. Formerly at Oak Ridge Nat. Lab.; Formerly at Hanford Lab.; Manager, Savannah River Plant (1957-66), Explosives Dept. (1967-), E. I. Du Pont de Nemours and Co., Inc.
Address: E. I. Du Pont de Nemours and Co., Inc., Explosives Department, Wilmington, Delaware 19898, U.S.A.

ELLIOT, Joe Oliver, B.S., A.M., Ph.D. Born 1923. Educ.: Iowa State Coll., Columbia and Maryland Univs. Res. Phys., U.S. Naval Res. Lab., 1949-. Head, Reactors Branch, 1958-. Society: A.P.S.
Nuclear interests: Nuclear physics, neutron scattering, scintillation detectors, research reactors.
Address: Code 7460, Radiation Division, U.S. Naval Research Laboratory, Washington, D.C. 20390, U.S.A.

ELLIOTT, Lloyd George, B.Sc., M.Sc., Ph.D., D.Sc. Born 1919. Educ.: Dalhousie Univ. and M.I.T. Director, Phys. Div., A.E.C.L., 1952-67; Director, Chalk River Nucl. Lab. Res., 1967-. Societies: Fellow, Roy. Soc. of Canada; Fellow and Member of Council, A.P.S.; Canadian Assoc. of Phys.
Nuclear interest: Nuclear physics.
Address: Physics Division, Atomic Energy of Canada, Chalk River, Ontario, Canada.

ELLIOTT, R. B., Lecturer, Univ. Dublin, Trinity College.
Address: University of Dublin, Physical Laboratory, Trinity College, Dublin, Eire.

ELLIOTT, Vorras Alexander, B.S. (Mech. Eng.). Born 1912. Educ.: Kansas State Coll. Project Eng., Construction Eng. Dept. (1942-52), Manager, Mech. Eng., Construction Eng. Dept. (1952-54), Eng., Atomic Power Study (1954-55), Manager, Power Plant Design, Dresden Nucl. Power Station (1955-58), Manager, Power Reactor Design (1958-), G.E.C. Societies: A.S.M.E.; Airplane Owners and Pilots Assoc.
Nuclear interests: Management and reactor designs.
Address: 11550 Fairway Drive, Los Altos, California, U.S.A.

ELLIS, Frank, M.Sc., M.B., Ch.B., M.D., M.A. (Oxon.), D.M.R. (R.C.P. & S.), F.F.R., M.R.C.P. (Lond.). Born 1905. Educ.: Sheffield Univ. Director, Radiotherapy Dept., United Oxford Hospitals; Lecturer in Radiotherapy, Oxford Univ. Medical School. Sen. Adviser, I.C.R.U. Societies: Rad. Res. Soc.; Rad. Protection Soc.; Hon. Member, American Radium Soc.; British Commission on Radiol. Units and Measurements; British Inst. of Radiol.; Roy. Soc. of Medicine; Faculty of Radiol.
Nuclear interests: Radiotherapy in all forms; Radiobiology and radiation protection.
Address: Churchill Hospital, Oxford, England.

ELLIS, Geoffrey Courtnauld, B.Sc., M.Sc. Born 1918. Educ.: Wales Univ. Sen. Supt., Met. Div., Atomic Weapons Res. Establishment, Aldermaston, U.K.A.E.A.
Nuclear interest: Metallurgy.
Address: 8 Spencer Road, Newbury, Berkshire, England.

ELLIS, John L., M.A. Born 1912. Educ.: Brussels Univ. Pres. and Chief Eng., Sintercast Div. of Chromalloy American Corp.; Vice Pres., Chromalloy American Corp., 1959. Societies: American Soc. of Tool and Manufacturing Eng.; A.S.M.
Nuclear interest: Metallurgy.
Address: Sintercast, 169 Western Highway, West Nyack, New York, U.S.A.

ELLIS, Reed Hobart, Jr., A.B., A.M., Ph.D. Born 1918. Educ.: Bowdoin Coll., Maine and Columbia Univs. Res. phys., Radiol. Res. Labs., Columbia Univ., 1949-54; Assoc. Editor, (1956-61), Managing Editor (1961-62), Nucleonics; Adjunct Assoc. Prof., New York Univ., 1960-62. Editor, Nucl. Fusion, I.A.E.A., Vienna, 1962-64; Sen. Officer, I.A.E.A. New York Office, U.N. Building 1964-65; Exec. editor, then editor, Physics Today, 1965-. Book: Nucl. Technol. for Eng. (New York, McGraw-Hill Book Co., Inc., 1959). Societies: A.P.S.; A.N.S.; A.A.A.S. Nuclear interests: Radiotracers; Radiation; Reactor physics; Fusion.
Address: Physics Today, American Institute of Physics, 335 E. 45 Street, New York, N.Y. 10017, U.S.A.

ELLIS, Major Gen. Richard H. Member, Military Liaison Com., U.S.A.E.C.
Address: United States Atomic Energy Commission, Military Liaison Committee, Washington, D.C. 20545, U.S.A.

ELLIS, Roy Edwin, B.Sc., Ph.D. Born 1925. Educ.: London Univ. Sen. Lecturer, Dept. of Phys. Appl. to Medicine, Middlesex Hospital Medical School, London, 1954-64; Sci. Sec., U.N. Com. on Effects of Atomic Rad., New York, 1961-62; Reader, Phys. Dept., St. Bartholomew's Hospital Medical Coll., London, 1964-. Hon. Sec., Non-Medical, British Inst. of Radiol. Societies: Biol. Eng. Soc.; Hospital Phys. Assoc.; Soc. of Rad. Protection; Assoc. of Rad. Res.
Nuclear interests: Application of physics to medicine; Teaching of physics to undergraduate and postgraduate students; Research in radiation physics and radiobiology.
Address: Physics Department, Medical College, St. Bartholomew's Hospital, Charterhouse Square, London E.C.1, England.

ELLIS, Thomas Gordon, B.S. (Chem. Eng.), Ph.D. (Metal.). Born 1931. Educ.: Michigan Technol. and Iowa State Univs. Res. Assoc., and Post Doctoral Fellow, Ames Lab., U.S.A.E.C.; Assoc. Prof. Nucl. Eng., Director of Summer Sessions, Michigan Technol. Univ. Societies: A.N.S.; A.S.M.
Nuclear interests: Metallurgy; Isotope applications; Radiation chemistry; Activation analysis.
Address: Office of Summer Sessions, Michigan Technological University, Houghton, Michigan 49931, U.S.A.

ELLIS, Walter Roy, B.Sc. (Hons.). Born 1922. Educ.: Western Australia Univ. Sci. Officer, Dept.of Supply (Australia), 1950-57; Leader, Ind. Res. Group (1957-64), formerly Acting Head, Isotope Applications Res. Sect., Isotope Div., A.A.E.C.
Nuclear interests: Industrial and research applications of radioactive isotopes.
Address: 33 Stuart Street, Blakehurst, N.S.W., Australia.

ELLIS, William R., B.S. (Elec. Eng.). Born 1917. Educ.: Missouri School of Mines and Metal. Design and Project Eng. on Atomic Power Plants for U.S. Navy submarines and aircraft carrier, 1950-55; Manager, P.W.R. Power Plant Subdiv. – responsible for design of reactor plant for Shippingport Atomic Power Station, 1955-58; Manager, Westinghouse Special Atomic Project – responsible for procurement of power plant equipment for the British submarine Dreadnought, 1958-61; Gen. Manager, Westinghouse Atomic Fuel Div., manufacturing nucl. reactor cores, 1962; Tech. Asst. to Vice-Pres., Eng., 1963-; Manager, NERVA Materials and Fuel 1964-66; Tech. Asst. to NERVA Project Manager, 1967-. Societies: I.E.E.E.; A.N.S.
Nuclear interests: Overall design of reactor plants for naval and commercial power generation applications and nuclear rocket reactors.
Address: 1137 Gilchrest Drive, Pittsburgh 35, Pennsylvania, U.S.A.

ELLSWORTH, Melvin A. Exec. Vice Pres., Fluor Corp. Ltd.
Address: Fluor Corp. Ltd., 2500 South Atlantic Boulevard, Los Angeles 22, California, U.S.A.

ELLSWORTH, Robert W., B.S., Ph.D. Born 1937. Educ.: Yale and Rochester Univs. Res. Assoc., Washington Univ., Seattle.
Nuclear interest: Experimental high energy physics.
Address: Physics Department, Washington University, Seattle, Washington,U.S.A.

ELMANOV, N. M. Paper: Co-author, Exptl. Gamma-Rays Unit EKU-50 of the N. F. Gamalei Inst. of Epidemiology and Microbiology (letter to the Editor, Atomnaya Energiya, vol. 22, No. 4, 1967).
Address: Academy of Sciences of the U.S.S.R., 14 Leninsky Prospekt, Moscow V-71, U.S.S.R.

ELMESHAD, Yahia Amin, Ph.D. (Nucl. Eng.). Born 1932. Educ.: Alexandria Univ., Egypt and Energy Inst., Moscow, U.S.S.R.
Head, Nucl. Kinetics Sect., Reactor Eng. Dept. Atomic Energy Establishment, Cairo, 1962-66; Asst. Prof., Nucl. Eng. Dept., Faculty of Eng., Alexandria Univ.
Nuclear interests: Reactor physics and technology; Power reactors; Kinetics; Nuclear instrumentation; Education.
Address: Nuclear Engineering Department, Faculty of Engineering, Alexandria University, Alexandria, Egypt.

ELMING, Jorgen, M.Sc. Born 1915. Educ.: Tech. Univ. of Denmark. Head, Reactor Dept., Burmeister and Wain, Eng. and Shipbuilders Ltd., Copenhagen K. Society: Inst. Danish Civil Eng.
Nuclear interests: Reactor design and technology with reference to power stations and

nuclear ships.
Address: 2 Torvegade, 1400 Copenhagen K., Denmark.

ELORZA, Miguel GUINEA. See GUINEA ELORZA, Miguel.

ELSTON, Jean, Dr. of Phys. Sci. Born 1921. Chief, Sect. of Studies of Refractory Materials, C.E.A., C.E.N., Saclay. Book: Traité de Chimie Minérale du Pr. Pascal, Tome XV (Masson et Cie.).
Address: 26 Parc d'Ardenay, Palaiseau, (Seine-et-Oise), France.

ELSTON, Walter T., Elec. Eng. Born 1911. Educ.: Rensselaer Polytech. Inst. Res. Staff Coordinator, Nucl. Tech. Service Applications, Air Reduction Co. Inc., St. Marys, Pa. Com. C-5 (Standards for Nucl. Grade Graphite), A.S.T.M.
Nuclear interests: Nuclear moderators fuel and control materials such as purified graphite in bulk quantity; Graphite fuel element manufacture and boronated graphite materials for control and shielding purposes.
Address: 609 Sherry Road, St. Marys, Pa., U.S.A.

ELTEKOV, V. A. Papers: Co-author, On the Absorption of γ-Rays in Infinite Lattice Arrays (Atomnaya Energiya, vol. 13, No. 6, 1962); co-author, Calculation of γ-Rad. Energy Absorption in Heterogeneous Macrosystems (ibid., vol. 15, No. 5, 1963); co-author, Gamma Ray Spectrum and Partial Values of Absorbed Energy in Arbitrary Homogeneous Mixture (ibid, vol. 16, No. 4, 1964); co-author, About Absorption of Gamma Rays Energy of Point Source in Macrosystems (letter to the Editor, ibid, vol. 19, No. 2, 1965).
Address: Academy of Sciences of the U.S.S.R., 14 Leninsky Prospekt, Moscow V-71, U.S.S.R.

ELTHAM, Barry Edward, B.Sc. (Eng.). Born 1923. Educ.: Medway Tech. Coll. Chief Eng. (Design) (1955-57), Chief Eng. (Health and Safety Branch) (1957-59), Chief Tech. Eng., Production Group (1959-63), Deputy Tech. Director (1963-), U.K.A.E.A., Risley; Director-Gen., British Nucl. Export Exec., 1966-. Com. Member, Risley Branch, Nucl. Eng. Soc. Societies: M.I.Mech.E.; A.M.I.Chem.E.
Nuclear interest: Reactor and fuel element design.
Address: British Nuclear Export Executive, Dorland House, 14-16 Regent Street, London, S.W.1.

ELTON, Lewis Richard Benjamin, M.A. (Cantab.), B.Sc., D.Sc. (London). Born 1923. Educ.: Cambridge and London Univs. and Polytech., Regent Street. Lecturer, King's Coll., London, 1950-57; Res. Assoc., M.I.T. and Stanford Univ., 1955-56; Head, Phys. Dept., Battersea Coll. of Technol., 1958-66; Head, Phys. Dept. (1966-), Head, Inst. of Educational Technol., (1967-), Surrey Univ. Books: Introductory Nucl. Theory (Pitman, 1965); Nucl. Sizes (Oxford, 1961). Societies: Fellow, Inst. of Phys. and Phys. Soc.
Nuclear interest: Theoretical nuclear physics.
Address: Physics Department, Surrey University, Guildford, Surrey, England.

ELVAS-KESSE, M., Dr. Isotope Lab., Dept. of Clinical Therapeutics, Athens Univ.
Address: Alexandra Hospital, Isotope Laboratory, Vas. Sophias-K. Lourou Str., Athens, Greece.

ELWYN, Alexander J., A.B. (Grinnell Coll.), Ph.D. (Washington, St. Louis). Born 1927. Educ.: Grinnell Coll., Iowa and Washington Univ., St. Louis. Res. Assoc., Brookhaven Nat. Lab., 1956-59; Asst. Sci. (1959-62), Assoc. Sci. (1962-), Argonne Nat. Lab. Society: A.P.S.
Nuclear interests: Nuclear physics; Nuclear reactions, nuclear structure, polarisation of neutrons.
Address: Physics Division, Argonne National Laboratory, Argonne, Illinois, U.S.A.

ELY, Ralph Lawrence, Jr., B.S. (Washington and Jefferson Coll.), M.S. (Colorado), Ph.D. (Pittsburgh). Born 1917. Educ.: Washington and Jefferson Coll. and Colorado and Pittsburgh Univs., Sen. Sci., Westinghouse Atomic Power Div., Pittsburgh, 1951-54; Vice-Pres. and Tech. Director, Nucl. Sci. Eng. Corp., Pittsburgh, 1954-59; Head, Isotope Development Lab. (1959-65), Director, Office of Industry Services, (1965-), Res. Triangle Inst., Durham, North Carolina. Societies: A.P.S.; A.N.S.; Marine Technol. Soc.
Nuclear interests: Application of nuclear and radioisotope techniques to measurement problems; Development of special counting techniques and devices.
Address: P.O. Box 12194, Research Triangle Park, North Carolina, 27709, U.S.A.

ELY, Thomas Sharples, M.D. Born 1924. Educ.: Bethany Coll., Georgetown and Rochester Univs. Res. Project Officer, Naval Medical Res. Inst., Bethesda, Maryland, 1950-52 and 1954-56; Asst. Medical Officer, U.S.S. Kearsarge, F.P.O. San Francisco, 1953-54; Asst. Chief, Medical Branch, Div. of Biol. and Medicine (1956-57), Asst. Chief, Health Protection Branch (1957-60), Chief, Health Protection Branch, Office of Operational Safety (1960-61), U.S.A.E.C., Washington, D.C.; Fellow in Occupational Medicine (1961-63), Clinical Asst. Prof. Preventive Medicine (1963-), Rochester Univ.; Staff Physician, Eastman Kodak Co., Rochester, New York, 1963-. Societies: Ind. Medical Assoc.; American Acad. of Occupational Medicine; A.A.A.S.; A.I.H.A.
Nuclear interests: Administration and operation of a health protection programme in the nuclear industry in the fields of industrial hygiene, health physics, and particularly industrial medicine; Development of exposure standards for radiation protection.

Address: Laboratory of Industrial Medicine, Eastman Kodak Company, Rochester, New York 14650, U.S.A.

ELZA, Paul M. Educ.: Tennessee Univ. Employed (1948-51), Manager Administration (1951-61), Asst. Director for Administration, (1961-64, 1965-), O.R.I.N.S. Sen. Officer, Exchange and Training Div., I.A.E.A., Vienna, 1964-65.
Address: Oak Ridge Institute of Nuclear Studies, P.O. Box 117, Oak Ridge, Tennessee, U.S.A.

EMAMI, Jaffar SHARIF-. See SHARIF-EMAMI, Jaffar.

EMBEJE, Nestor P., Ing. Head, Civil Eng. Dept., Comision Nacional de Energia Atomica, Argentina.
Address: Comision Nacional de Energia Atomica, Electronics Department, 8250 Avda. del Libertador Gral. San Martin, Buenos Aires, Argentina.

EMELYANOV, I. Ya. Sci. Asst., Acad. of Sci. of the U.S.S.R.; Deleg. to 2nd I.C.P.U.A.E., Geneva, Sept. 1958. Papers: Co-author, A Uranium-graphite Reactor giving Superheated High-pressure Steam (Atomnaya Energiya, vol. 5, No. 3, 1958); co-author, The 50 MW Res. Reactor SM (ibid, vol. 8, No. 6, 1960); co-author, Dynamic Characteristic Res. of First Reactor on I.V Kurchatov Beloyarsk Nucl. Power Station (ibid, vol. 19, No. 2, 1965); co-author, Control Rod Operation in High Temperature Reactors (ibid, vol. 22, No. 5, 1967).
Address: Academy of Sciences of the U.S.S.R., 14 Leninsky Prospekt, Moscow V-71, U.S.S.R.

EMELYANOV, V. A. Candidate of Agricultural Sci. Lab., All-Union Inst. for Sci. Res. on Water Control and Land Improvement. Paper: Co-author, Capture, γ-rays from Cadmium as a Means of Slow-neutron Detection in Neutron Methods of Water Determination in Soil and Earth (letter to the Editor, Atomnaya Energiya, vol. 6, No. 5, 1959).
Nuclear interest: Land improvement.
Address: Laboratory for the All-Union Institute for Scientific Research on Water Control and Land Improvement, 19 ul Pryanishnikova, Moscow A-8, U.S.S.R.

EMELYANOV, Vasily Semenovich, Order of the Red Banner of Labour. Born 1901. Graduate, Moscow Mining Acad. Faculty of Metal. Lecturer, then Prof. Electro-metal., Moscow Steel Inst., 1938-; Chairman, Com. of Standards, U.S.S.R. Govt., 1940-41; Chief, Central Board on the Use of Atomic Energy of the U.S.S.R. Council of Ministers, till Feb. 1962; Corresponding Member, Soviet Acad. of Sci.; Member, Dept. of Eng. Sci. Academy of Sci.; Deputy Chairman, U.S.S.R. State Com. for the Utilisation of Atomic Energy; Formerly Pres., Council for Mutual Economic Assistance.

Formerly Member, Board of Governors, and formerly Member, Sci. Advisory Com., I.A.E.A., Vienna. Deleg. and Vice-Pres., 2nd I.C.P.U.A.E., Geneva, Sept. 1958. Formerly Member, Editorial Board, Atomnaya Energiya.
Nuclear interests: Metallurgy of reactor materials; Policy and administration.
Address: State Committee for the Utilization of Atomic Energy, 26 Staromonetnii Pereulok, Moscow, U.S.S.R.

EMENDÖRFER, Dieter, Dipl. Phys., Dr. rer. nat. Born 1927. Educ.: Stuttgart T.H. and N.C.S.C., Raleigh. Lecturer in Reactor Theory, Stuttgart Univ. Societies: German Phys. Soc.; A.N.S.
Nuclear interests: Nuclear physics; Reactor theory.
Address: Institut für Kernenergetik, Abteilung Reaktorphysik, 51 Herdweg, N. Stuttgart, Germany.

EMERSON, Charles Phillips, A.B. (Chem.), M.D. Born 1912. Educ.: Princeton and Harvard Univs. Member of Evans Memorial (1950-), Director, Clinical Labs. (1959-), Massachusetts Memorial Hospitals; Assoc. Prof. Medicine, Boston Univ. School of Medicine, 1952-; Consultant in Hematology, U.S. Public Health Hospital, Brighton, 1954-. Bio-Medical Advisory Com. for Nucl. Reactor, M.I.T.
Nuclear interest: Radioisotope applications in hematology.
Address: 65 East Newton Street, Boston 18, Massachusetts, U.S.A.

EMERY, Juel F., B.S., M.S. (Louisville). Born 1927. Educ.: Louisville Univ. Society: A.C.S.
Nuclear interests: Activation analysis; Nuclear analysis; Gamma-ray spectrometry.
Address: Rt. 1, Irwin Road, Powell, Tennessee, U.S.A.

EMLET, Logan B. Formerly Exec. Vice-Pres., Nucl. Div., Union Carbide Nucl. Co.; Vice-Pres., Mining and Metals Div., Union Carbide Corp., 1965-. Formerly Industry Member, Labour-Management Advisory Com., U.S.A.E.C.
Address: Mining and Metals Division, Union Carbide Corp., 270 Park Avenue, New York, New York 10017, U.S.A.

EMLEY, Edward Frederick, B.Sc., Ph.D. Born 1917. Educ.: Rutherford Tech. Coll., Newcastle. Deputy Chief Metal. (1951-53), Chief Metal. (1954-66), Magnesium Elektron, Ltd.; Head of Metal. Dept., British Aluminium Co. Res. Labs., Gerrards Cross, Bucks., 1966-. Societies: F.R.I.C.; Fellow, Inst. of Metal.; Assoc. Inst. of Mining and Metal.; Inst. Metals; Inst. Nucl. Eng.; Manchester Metal. Soc.
Nuclear interest: Magnesium fuel element components.
Address: Matfen, Mill Lane, Chalfont St. Giles, Bucks., England.

EMMA, V. Formerly Reparto Lastre, now Grupp Stato Solido e Positronio, Centro Siciliano de Fisica Nucleare, Catania Univ.
Address: Catania University, 57 Corso Italia, Catania, Sicily.

EMMER, Thomas Lee, B.S. (E.E.). Born 1930. Educ.: Wisconsin Univ. Society: Inst. of Radio Eng.
Nuclear interest: Nuclear instrumentation.
Address: 106 Carnegie Drive, Oak Ridge, Tennessee, U.S.A.

EMMERSON, Bruce William, A.M.I.E.E. Born 1929. Educ.: Regent Street Polytech., London. Reactor Supervisor, Jason res. reactor, Hawker Siddeley Nucl. Power Co., Ltd., 1959-61; Tech. Expert on Reactor Commissioning to Govt. of Thailand, I.A.E.A. appointment, 1961-62; Health Phys., C.E.G.B., Bradwell Nucl. Power Station. In charge of commissioning of L.F.R. res. reactor, R.C.N. Holland. Member, British Standards Inst., Coms. on Rad. Protection and Radioisotopes. Societies: Inst. of Elec. Eng.; Inst. of Radio Eng. (Australia); Member, American H.P.S.; Member, U.K. Soc. for Radiol. Protection.
Nuclear interests: Management; Reactor engineering; Research reactors.
Address: 69 Spital Road, Maldon, Essex, England.

EMMERSON, Peter Tunley, B.Sc. (Hons., Chem.). Born 1937. Educ.: King's Coll. Durham Univ.
Nuclear interests: Effect of ionising radiation on DNA and nucleoprotein; Electron spin resonance; General radiobiology.
Address: Department of Radiology, Yale University School of Medicine, 333 Cedar Street, New Haven, Connecticut, U.S.A.

EMMONS, A. H., Ph.D., M.S., B.S. Born 1924. Educ.: Dubuque and Michigan Univs. Lab. Supervisor, Phoenix Memorial Lab., Michigan Univ., 1954-60; Director, Res. Reactor Facilities and Prof. Nucl. Eng., Missouri Univ., 1960-. Societies: A.N.S.; A.A.A.S.; A.S.E.E.; H.P.S.; A.C.S.
Nuclear interests: Applications of radioisotopes; Reactor and radiation laboratory design and administration.
Address: Research Reactor Facility, Missouri University, Columbia, Missouri, U.S.A.

ENATSU, Hiroshi, Dr. of Sci. Born 1922. Educ.: Kyoto Univ. Asst., Kyoto Univ., 1946-57; Res. Asst., Columbia Univ., 1952-53; Visiting Member, Inst. for Theoretical Phys., Copenhagen, 1955-56; Asst. Prof. (1957), Prof., Phys. (1957-), Ritsumeikan Univ. Societies: Phys. Soc. of Japan; A.P.S.
Nuclear interests: Theoretical nuclear physics; Quantum theory of fields.
Address: Department of Physics, Ritsumeikan University, Toji-in, Kita-ku, Kyoto, Japan.

ENCHEVICH, Ivan B. At the Lab. of Nucl. Problems, Dubna.
Nuclear interest: Study of the movement of protons in the synchocyclotron.
Address: Joint Institute for Nuclear Research, Dubna, Nr. Moscow, U.S.S.R.

ENDA MARAH-VAN DER HAVE VAN SIRJANSLAND, Mrs. Adriana, Ahli Kimia (Chem. expert). Born 1912. Educ.: Delft and Indonesia Univs. Lector for Analytical Chem., Indonesia Univ., 1950-58; Appointment with I.A.E.A., Vienna, 1958-60; Appointment with Atomministerium Deutschland, 1960-64.
Nuclear interest: Environmental radiological health and safety.
Address: 3 Djalan Dajang Sumbi, Bandung, Indonesia.

ENDO, Nobuyuki. Born 1922. Educ.: Nippon Univ. Chief Eng., Rad. Measurement Service Dept., Nucl. Rad. Hazard Protection Div., Japan Safety Appliances Assoc. Books: Protective of X- and Gamma Ray (1955); The Measurement of Rad. (1960). Societies: The Japanese Soc. for Nondestructive Inspection.
Address: Japan Safety Appliances Association, Nagoya Branch of 2-1-chome Minamikuwanamachi, Naka-ku, Nagoya, Japan.

ENDO, Shinichiro, M.D. Prof. and Sen. Officer, Fukushima Radioisotope Res. Com., Fukushima Medical Coll.
Address: Fukushima Medical College, 14 Sugizuma-cho, Fukushima City, Japan.

ENDRES, H. Member for C.E.T.S., Group of Experts on the Production of Energy from Radioisotopes, O.E.C.D., E.N.E.A.
Address: O.E.C.D. European Nuclear Energy Agency, 38 boulevard Suchet, Paris 16, France.

ENDT, Pieter Maarten, Ph.Dr. Born 1916. Educ.: Utrecht State Univ. Prof. Exptl. Phys., Utrecht Univ., 1955-. Pres., Dutch Phys. Soc. Book: Nucl. Reactions, vol. i (Amsterdam, North-Holland Publ. Co., 1958). Societies: Royal Acad. Sci. of Amsterdam; A.P.S.
Nuclear interest: Nuclear physics.
Address: 9 Blauwkapelseweg, De Bilt, Netherlands.

ENENKL, Vladimir, Prof., Dr. habil., Dipl. Ing. Born 1920. Educ.: Coll. of Mech. Eng., Czech Tech. Univ., Brno. Dean, Mech. Eng., Coll. Tech. Univ., Brno, 1958-; Chairman, Czechoslovak Com. for the fuel elements of nucl. reactors, 1960-; Prof. Tech. Univ., Brno; Chief, cathedra of applied thermodynamics, Tech. Univ., Brno. Society: Czechoslovak Commission for Atomic Energy.
Nuclear interests: The design of fuel elements, heat transmission, and hydraulics of nuclear reactors.
Address: 1 Bayerova, Brno, Czechoslovakia.

ENGE, Harald Anton, Siviling (Norway Tech. Univ.), Dr. Philos. (Bergen). Born 1920. Educ.: Tech. Univ. Norway and Bergen Univ. Prof. Phys., M.I.T., 1963-. Book: Introduction to Nucl. Phys. (Reading, Addison-Wesley Publishing Co., 1966). Societies: A.P.S.; Norsk Fysisk Selskap.
Nuclear interests: Charged particle spectroscopy; Energy levels.
Address: 26 Prince Avenue, Winchester, Massachusetts, U.S.A.

ENGEL, David Walter, B.A., M.A. Born 1935. Educ.: Ohio Wesleyan and Duke Univs. Res. Asst., Duke Univ., 1960-61; Fishery Biol., Bureau of Commercial Fisheries, Radiobiol. Lab., Beaufort, North Carolina, 1961-. Societies: Rad. Res. Soc.; A.A.A.S.
Nuclear interests: Marine radiation biology. The effects of ionising radiations on the physiology and biochemistry of marine organisms.
Address: Bureau of Commercial Fisheries, Radiobiological Laboratory, Beaufort, North Carolina, U.S.A.

ENGEL, Robert, D. of Phys. and Mathematical Sci., Prof. Phys. and Maths. at the Athénee Grand-ducal of Luxembourg. Born 1912. Educ.: Sorbonne and Munich Univ. Societies: Luxembourg Nat. Council of Nucl. Energy; Administration Council, Assoc. Luxembourgeoise pour l'Utilisation Pacifique de l'Energie Nucléaire.
Address: c/o A.L.U.P.A., 9 rue Marie-Adelaide, Luxembourg, Luxembourg.

ENGEL, W., Pres. and Co-ordinating Director, Arbeitsgemeinschaft Deutscher Energieversorgungsunternehmen zur Vorbereitung der Errichtung eines Leistungsversuchs-Reaktors e.V.
Address: Arbeitsgemeinschaft Deutscher Energieversorgungsunternehmen zur Vorbereitung der Errichtung eines Leistungsversuchs-Reaktors e.V., 105 Luisenstrasse, Postfach 1411, 4 Düsseldorf, Germany.

ENGELAND, T. Member, Nucl. Phys. Lab., Oslo Univ.
Address: Oslo University, Nuclear Physics Laboratory, Institute of Physics, Blindern, Norway.

ENGELBERT, Lincoln E. I.A.E.A. Liaison Officer, F.A.O., 1964-.
Address: International Atomic Energy Agency, 11 Kaerntnerring, Vienna 1, Austria.

ENGELBERT, Ulrich, Dipl. Ing., Dr. Ing. Born 1920. Educ.: Berlin T.H. and Berlin Tech. Univ. Chief Metal., Clyde Eng. Co., New South Wales; Sen. Exptl. Officer, Metal. Sect., now Head, Radioisotope Production Sect., Res. Establishment, Australian A.E.C. Societies: Fellow, Inst. of Metal. (London); Australian Inst. Metals.
Nuclear interests: Radioisotopes production and application; Metallurgy.
Address: 9 Kethel Road, Cheltenham, New South Wales, Australia.

ENGELDER, Theodore Carl, B.S. (Michigan), M.S. (Yale), Ph.D. (Yale). Born 1927. Educ.: Michigan and Yale Univs. Lab. Manager, Babcock and Wilcox Co., Lynchburg, Va. Societies: A.P.S.; A.N.S.
Nuclear interest: Experimental reactor physics.
Address: 2236 Taylor Farm Road, Lynchburg, Virginia, U.S.A.

ENGELHARDT, H., Min. Dir. Dr. Leiter, Abteilung I, Zentralabteilung, Bundesministerium für Wissenschaftliche Forschung.
Address: Bundesministerium für Wissenschaftliche Forschung, 46 Luisenstrasse, 532 Bad Godesberg, Germany.

ENGELHARDT, Vladimir Aleksandrovich. Born 1894. Educ.: Moscow Univ. Prof., Kazan Univ. and Kazan Medical Inst., 1929-33; Prof., Leningrad Univ., 1934-40; Prof., Moscow Univ., 1936-; Acad. of Sci. of the U.S.S.R., 1933; Active Member (1944), Member-Correspondent (1946), Acad. of Sci. of the U.S.S.R.; Full Member, Acad. of Medical Sci. of the U.S.S.R., 1953; Inst. of Physiology, Acad. of Sci. of the U.S.S.R., 1944-50; Director, Biochem. Dept., Inst. of Exptl. Medicine, 1945-52; Director, Biochem. Lab., Inst. of Biochem., and Chairman, Commission on Radiobiol., Acad. of Sci. of the U.S.S.R., 1955-; Academician-Sec., Dept. of Biological Sci., Acad. of Sci., 1956; Chief Editor, Biochem. Deleg. to 2nd I.C.P.U.A.E., Geneva, Sept. 1958.
Address: Academy of Sciences of the U.S.S.R., 14 Leninsky Prospekt, Moscow V-71, U.S.S.R.

ENGELHARDT, Wolfgang v. See v. ENGELHARDT, Wolfgang.

ENGELKE, Charles Edward, B.S., M.A., Ph.D. Born 1930. Educ.: Culumbia Univ. Res. Asst., Columbia Univ., 1952-53; Commissioned in U.S.A.F., served as Phys. in R. and D. Command, Air Force Special Weapons Centre, Albuquerque, New Mexico, 1953-56; Res. Asst., Res. Assoc., Columbia Univ., 1956-63; Asst. Prof., Assoc. Prof., Phys., Graduate Faculty of City Univ. of New York, Hunter Coll. of N.Y.C., 1960-. Societies: A.P.S.; A.I.P.
Nuclear interests: Low energy nuclear physics, particularly two-body interactions; Precision neutron-proton cross-section measurements; Diffraction of recoil localised annihilation quanta.
Address: 4 Chemung Place, Jericho, New York, U.S.A.

ENGESET, Arnfinn, M.D. Born 1916. Educ.: Oslo Univ. Norsk Hydro's Inst. for Cancer Res., Norwegian Radiumhospital, Oslo. Book: Irradiation of lymph nodes and vessels. Experiments with rats with reference to cancer therapy. (Oslo, Oslo Univ. Press, 1954).

Nuclear interests: Irradiation effects on lymph nodes and lymphoid tissue.
Address: Radiumhospitalet, Montebello, Oslo, Norway.

ENGEVIK, Reidar, Mech. Eng. Born 1925. Educ.: Bergen Tekniske Skole. Bergens Mekaniske Verksteder A/S, Bergen, 1950-58; In charge of Design Office primarily engaged with reactor project and reactor components development for water cooled power and res. reactors (ship propulsion projects), Institutt for Atomenergi, Kjeller, 1958-. Society: Norwegian Soc. of Atomic Energy, Oslo.
Nuclear interests: Reactor design and development work.
Address: Institutt for Atomenergi, Kjeller, Norway.

ENGH, Howard Owen. Plant Manager, Rad. Instr. Devel. Lab., 1964-.
Address: Radiation Instrument Development Laboratory, Division of Nuclear-Chicago Corp., 4501 West North Avenue, Melrose Park, Illinois, U.S.A.

ENGLAND, Glyn, B.Sc. (Eng.). Born 1921. Educ.: London Univ. Appointments in Generation Design Branch, Chief Engineer's Dept., British Elec. Authority and Central Elec. Authority, 1951-57; Development Eng., Central Elec. Authority, 1957-58; Development Eng. (Policy) (1958-66), Chief Operations Eng., C.E.G.B.
Nuclear interests: Special interest in the integration of nuclear power stations in the electricity supply system, in the provision of fuel for and safety of operating reactors.
Address: 1 Ashley Close, Welwyn Garden City, Herts., England.

ENGLANDER, Alois G. Born 1907. Educ.: Prague Univ. Permanent Rep. and Deleg. of Republic of Honduras to Gen. Conferences of Atomic Energy Agency, Vienna. Honduran Consul for Austria.
Nuclear interests: Management; Prevention of radiation diseases and radioactive fallout.
Address: 28 Graben, Vienna I, Austria.

ENGLANDER, Marcel, Dr-Ing. Ing. E.P.C.I., E.N.S.P. Born 1913. Educ.: Paris Univ. and Ecole Supérieure de Physique et de Chimie. Chef du Service de Recherches Métallurgiques Appliquées C.E.A., Centre d'Etudes Nucléaires de Saclay. Books: Métallurgie Physique de l'Uranium (Cours de Génie Atomique, 1958, C.E.N. Saclay); Propriétés Physiques de l'Uranium (Nouveau Traité de Chimie Minerale). Societies: Inst. of Metals, London; Sté. française de Métallurgie, Paris; Sté. de Physique, Paris.
Nuclear interests: Nuclear metallurgy and reactor design.
Address: 68 rue Hallé, Paris 14, France.

ENGLER, O. Vice-Manager, Giovanola Frères S.A.
Address: Giovanola Frères S.A., Monthey, (Valais), Switzerland.

ENGLISH, Spofford Grady, B.S. (Chem.), M.S. (Chem.), Ph.D. (Chem.), Outstanding Service Award, U.S.A.E.C. Born 1915. Educ.: Chattanooga, Oklahoma and California (Berkeley) Univs. Chief, Chem. Branch, Res. Div. (1947-58), Deputy Director of Res. (1959-61), Asst. to Gen. Manager for Arms Control and Disarmament (1959-61), Asst. Gen. Manager for R. and D. (1961-), U.S.A.E.C.
Nuclear interests: Power reactor technology; Space atomic power systems; Basic physical and biological research in atomic energy; Application of radioisotopes; Peaceful applications of nuclear explosives; Education and training.
Address: 8204 Thoreau Drive, Bethesda, Md., U.S.A.

ENGVALL, Robert P. Member, Radioisotope Com., Worcester Foundation for Exptl. Biol., Inc.
Address: Worcester Foundation for Experimental Biology, Inc., 222 Maple Avenue, Shrewsbury, Massachusetts, U.S.A.

ENNERST, Karl, Dipl. Ing. Born 1931. Educ.: Bergakademie Clausthal-Zellerfeld. Deutsche Edelstahlwerke AG. Krefeld, 1957-58; Degussa, Wolfgang bei Hanau (Main), 1959-60; Nuklear-Chemie und -Metallurgie G.m.b.H., Wolfgang bei Hanau (Main), 1960-.
Nuclear interest: Metallurgy.
Address: Nuklear-Chemie und -Metallurgie G.m.b.H., Wolfgang bei Hanau (Main), Germany.

ENNOR, Howard R., B.A., M.A. Born 1914. Educ.: Willamette and Chicago Univs. Deputy Director, then Director, Budget Div., Hanford Operations Office, U.S.A.E.C., 1947-61; Acting Director (1961-63), Director (1963-), Budget and Finance Div., I.A.E.A.
Address: International Atomic Energy Agency, 11 Kaerntnerring, Vienna 1, Austria.

ENOMOTO, Shigemasa, Dr. Eng. Born 1926. Educ.: Nagoya Univ. Member, Div. of Rad. Applications, Government Ind. Res. Inst., Nagoya, 1953-64; Member, Div. of Rad. Applications, Japan Atomic Energy Res. Inst., 1964-. Societies: Japan Soc. Appl. Phys.; Japan Radioisotope Assoc.
Nuclear interest: Applications of radioisotopes to industrial and engineering problem, especially X-ray analysis using radioactive source.
Address: Division of Radiation Applications, Japan Atomic Energy Research Institute, Oarai-Cho, Ibaragi- Ken, Japan.

ENQUIN, Maria Cristina PALCOS DE. See PALCOS DE ENQUIN, Maria Cristina.

ENSIO, Peter J., D.Sc. Vice-Pres., Nucl. Shielding Supplies and Service Inc. Address: Nuclear Shielding Supplies and Service Inc., Northcourt Building 175 Main Street, White Plains, New York 10601, U.S.A.

ENTZIAN, Wolfgang K. R. Dr. med. Born 1931. Educ.: Göttingen, Marburg and Bonn Univs. Societies: Vereinigung Deutscher Strahlenschutzärzte; Arbeitsgemeinschaft der Strahlenschutzärzte des DRK. Nuclear interests: Applications of radioactive isotopes in medicine with special reference to the field of neuro-surgery. Address: Neurochirurg. Univ. Klinik, 31 Wilhelmstrasse, Bonn, Germany.

ENUSTUN, Bekir Vedad, Chem. Eng., Ph.D. Born 1921. Educ.: Istanbul and Leeds Univs. Res. Eng., Materials Dept., Turkish State Highways, Ankara, 1950-; Lecturer in Phys. Chem., Ankara Univ., 1953-. Formerly Member, Advisory Commission, Turkish A.E.C. Books: Estimation of Rubber Hydrocarbon (1944); A.A.S.H.O. Standards, Part II (1956). Societies: Turkish Chamber of Chem. Eng.; Turkish Chem. Soc. Address: Mithat Pasa Cad., 44/10, Ankara, Turkey.

ENY, Désiré Marc, (Chem. Physiol.), M.S. (Elec. Eng.), B.S. (Maths., Phys.), LL.B. Born 1915. Educ.: Brequet Inst., Paris; Algiers and Cornell Univs.; Blackstone Coll. of Law, Chicago, and Army War Coll., Washington, D.C. Principal Biochem., Nucl. Biol. Lab., U.S. Dept. of Agriculture, 1951-53; Chief, Biol. Warfare Br., 1953-54; Chief, CBR(ABC) Weapons Div., 1954-57; Consultant CBR(ABC) Warfare to Army Res. Office, Office of Civil Defence and Office of Naval Res., 1956-63; Chairman, Isotope Com., U.S. Army Chem. Corps. 1957-58; Director, Nucl. Directorate, and Sci. Adviser for Nucl. Activities to Commanding Officer, U.S. Army Chem. Corps. 1957-62; Assoc. Tech. Director, Glidden Res. Centre, 1963-. Member, Com. N5, A.S.A.; Nat. Councillor, American Inst. Chemists, Maryland Chapter; Programme Chairman, Govt.-Industry Council, Baltimore Assoc. of Commerce. Books: Manual of Health Phys. (U.S. Dept. of the Army, 1958); Manual for Nucl. Waste Disposal (U.S. Dept. of the Army, 1959). Societies: A.N.S.; H.P.S.; A.I.Ch.E.; American Ordnance Assoc.; A.S.T.M.; RESA; A.C.S.; American Management Assoc.; Maryland Acad. Sci.; Nat. Soc. Federal Eng.; N.S.P.E.; Fellow, American Inst. Chemists. Nuclear interests: Management of R. and D. and production activities related to dosimetry, biological effects of radiations, shielding, health physics, waste disposal, nuclear defence, semi-conductors. Address: Glidden Research Centre, 3901 Hawkins Point Road, Baltimore, Maryland 21226, U.S.A.

ENZMANN, Hans Ferdinand, Dipl. Masch. Ing. E.T.H., D.I.C. Born 1927. Educ.: Swiss Federal Inst. Technol., Zürich and City and Guilds Coll., London. With Oerlikon Eng. Co., Zürich, 1953-56; Swiss Federal Inst. Reactor Technol., Würenlingen, 1956; Sec., O.E.C.D. Project Dragon, 1959-. Societies: Swiss Nucl. Soc.; A.N.S. Nuclear interests: H.T.G. reactor design and technology; Gas turbine application. Address: 11 Clarendon Avenue, Weymouth, Dorset, England.

EPONESHNIKOV, V. N. Papers: Co-author, 1.5 GeV Electron Synchrotron at the Polytech. Inst. of Tomsk (Atomnaya Energiya, vol.21, No. 6, 1966); co-author, 300 MeV Electron Synchrotron at the Polytech. Inst. of Tomsk (letter to the Editor, ibid, vol. 21, No. 6, 1966). Address: Academy of Sciences of the U.S.S.R., 14 Leninsky Prospekt, Moscow V-71, U.S.S.R.

EPPELSHEIMER, Daniel Snell, B.S., D.Sc. Born 1909. Educ.: Harvard Univ. Prof., Metal. Eng. and Nucl. Eng., Missouri Univ. at Rolla, Missouri; Formerly Chairman, Metal. Eng. Dept. and Nucl. Eng. Com., M.S.M., Rolla; Missouri A.E.C.; Vice Chairman, Southern Interstate Nucl. Board. Societies: A.I.M.M.E.; A.C.S.; Inst. of Metals; Iron and Steel Inst.; Verein Deutscher Eisenhuttenleute; American Foundrymen's Soc. Nuclear interests: Application of materials in the field of nuclear engineering; Radiation damage studies; Extractive metallurgy of uranium and thorium. Address: Box 217, R. R. 1, Rolla, Missouri, U.S.A.

EPPERT, Ray R. Member, Board of Trustees, Power Reactor Development Co. Address: Power Reactor Development Co., 1911 First Street, Detroit 26, Michigan, U.S.A.

EPPLE, Robert Paul, B.S., Ph.D. Born 1916. Educ.: Juniata Coll. and M.I.T. Asst. Prof. Chem., Brown Univ., 1946-51; Head, Inorganic Radiochem. Dept., Tracerlab. Inc., 1951-53; Sen. Chem., Arthur D. Little, Inc., 1956-. Society: A.C.S. Nuclear interests: Radiation chemistry; Radiation effects in gases; High-frequency discharges. Address: Arthur D. Little, Inc., 15 Acorn Park, Cambridge, Massachusetts, U.S.A.

EPPRECHT, Willfried Th., Prof. Dr. Born 1918. Educ.: Swiss Federal Inst. of Technol. Head of X-ray Dept., Eidg. Materialprüfungs Anstalt, 1950-55; Member of res. staff, Sect. for Ind. Res. of the Inst. für Tech. Phys. (1955-66), Prof. for material sci. (1966-), Swiss Federal Inst. of Technol. Book: Werkstoffkunde der Kerntechnik (Basel/Stuttgart, Birkhäuser-Verlag, 1961). Society: Swiss Soc. of "Fachleute der Kerntechnik"; Schweizerische Vereinigung für Atomenergie. Nuclear interests: Crystallography; Metallurgy; Thermocycling and lattice distortion by radiations; Nuclear materials.

Address: Swiss Federal Institute of Technology, Aussenstation Honggerberg, 8049 Zürich, Switzerland.

EPPS, Eugene F. VAN. See VAN EPPS, Eugene F.

EPSTEIN, Harold M. Formerly Project Leader, Reactor Phys. Div., Battelle Memorial Inst., now Asst. Div. Chief, Appl. Nucl. Phys. Div., Battelle - Columbus Lab.
Address: Battelle - Columbus Laboratories, 505 King Avenue, Columbus, Ohio 43201, U.S.A.

EPSTEIN, Leo Francis, S.B. (Chem.), Ph.D. (Phys. Chem.). Born 1913. Educ.: M.I.T. K.A.P.L., G.E.C., Schenectady, 1947-57; Vallecitos, Nucl. Centre, G.E.C., Pleasanton, California, 1957-; Sci. Adviser to U.S. Deleg., Geneva Conference, 1955. Societies: A.A.A.S.; A.C.S.; A.P.S.
Nuclear interests: Liquids and gases, particularly liquid metals.
Address: Vallecitos Nuclear Centre, General Electric Company, Pleasanton, California, U.S.A.

EPSTEIN, Robert Joseph, B.E.E. Born 1928. Educ.: Rensselaer Polytech. Inst. Assoc. Director, Electronics Div., Argonne Nat. Lab., 1960-; Consultant to Packard Instrument Co., 1962-. Member, Electronics Com., Zero Gradient Synchrotron Users Group.
Nuclear interest: Nuclear electronics instrumentation.
Address: 520 63rd Street, Downers Grove, Illinois, U.S.A.

er–RAWI, Ihsan, Dr. Member, Iraqi A.E.C.
Address: Iraqi Atomic Energy Commission, Baghdad, Iraq.

ERAMETSA, Kurt Heikki Olavi, Dr. of Technol. Born 1906. Educ.: Tech. Univ. Prof. Inorg. Chem., Tech. Univ., 1946-. Societies: Phys. Soc. Finland; Finnish Acad. Sci. and Letters; Acad. Tech. Sci. Finland.
Address: c/o Technical University, Otaniemi, Finland.

ERB, Odo, Dipl. Ing. Born 1920. Educ.: Vienna T.H. With Wiener Schwachstromwerke Gesellschaft m.b.H.; Head of tech. Dept. for Process Control Equipment and Instrumentation, 1951-. Society: Österreichischer Verband für Ekeltrotechnik.
Nuclear interests: Reactor instrumentation; Reactor control; Reactor safety systems.
Address: c/o Wiener Schwachstromwerke G.m.b.H., 12 Apostelgasse, A-1030 Vienna, Austria.

ERBA, E., graduate. Born 1936. Educ.: Univ. degli studi di Milano. Lecturer, Dept. of Exptl. Phys., Milan Univ. Society: Lombard Phys. Soc.
Nuclear interest: Theoretical nuclear physics.

Address: 90 Corso di porta romana, Milan, Italy.

ERBACHER, Wilhelm. Dipl.-Ing., Dr.techn. Born 1916. Educ.: Vienna T.H. Authorised head of dept., Österreichische Elektrizitätswirtschafts A.G.
Nuclear interests: Projecting, building and running of an atomic power plant for the Austrian Power Industry.
Address: Österreichische Elektrizitätswirtschafts-Aktiengesellschaft (Verbundgesellschaft), 6a Am Hof, Vienna I, Austria.

ERBEN, M. TALAT-. See TALAT-ERBEN, M.

ERDIK, Enis, Ph.D. Born 1914. Educ.: Istanbul Univ. Prof. Phys., Sci. Faculty, Ankara Univ. Society: Türk Fizik Dernegi (Turkish Phys. Soc.).
Nuclear interests: Experimental reactor physics. Subcritical assembly.
Address: Ankara Fen Fakültesi, Ankara, Turkey.

ERDMAN, Howard E., Dr. With Battelle-Northwest. First Officer, Joint F.A.O. and I.A.E.A. Dept. of Res. and Isotopes, 1966-68.
Address: International Atomic Energy Agency, Department of Research and Isotopes, 11 Kaerntnerring, Vienna 1, Austria.

ERDMANNSDORFF, W. D. von. See von ERDMANNSDORFF, W. D.

ERGALANT, Jacques. Ing. en Chef, Installations, U.S.S.I.
Address: U.S.S.I., 104 avenue Edouard Herriot, Le Plessis-Robinson (Seine), France.

ERGANG, Richard, Dr. rer. nat. Born 1913. Educ.: Göttingen Univ. Personal Member, Study Assoc. for the use of Nucl. Energy in Navigation and Industry.
Nuclear interests: Nickel-containing materials in nuclear applications.
Address: c/o International Nickel Deutschland G.m.b.H., 34 Kreuzstrasse, 4000 Düsseldorf 1, Germany.

ERGEN, William Krasny, Ph.D. Born 1912. Educ.: Vienna Univ. Phys., N.E.P.A. Project, Fairchild Engine and Aircraft Corp., 1947-51; Phys., O.R.N.L., 1951-. Member, Advisory Com. on Reactor Safeguards, U.S.A.E.C., Washington, D.C., 1959-63. Societies: A.N.S.; A.P.S.; Mathematical Assoc. of America.
Nuclear interests: Reactor dynamics; Reactor physics; Nuclear safety.
Address: 103 Orkney Road, Oak Ridge, Tennessee, U.S.A.

ERGINSOY, Cavid, B.Sc. (Eng., London), Ph.D. (Eng., London). Born 1924. Educ.: Univ. Coll. and Queen Mary Coll., London Univ. Co-ordinating Eng., Saryar Hydroelectric Project, Etibank, 1952-54; Chief Eng. for

R. and D., Etibank, 1954-55; Chief Eng. for Atomic Energy, Etibank 1955-; Part-time Lecturer, Tech. Univ., Istanbul, 1957-; Member, Inst. of Nucl. Studies, Middle East Tech. Univ., Ankara, 1957-; Formerly Deputy Chairman, Advisory Board, Turkish A.E.C., 1957-; Member, Res. Reactor Project, Turkish A.E.C. Society: Turkish Phys. Soc.
Nuclear interests: Solid state physics and reactor theory.
Address: Etibank, Ankara, Turkey.

ERICHSEN, Lothar Carl Manuel von. See von ERICHSEN, Lothar Carl Manuel.

ERICKSON, Kenneth W., Ph.D. Born 1915. Educ.: Texas Univ. Manager, Weapons Systems Development Dept., Sandia Corp., 1949-56 (loaned to Weapons Systems Evaluation Group, Washington, D.C., 1954-55); Vice-Pres., Board of Directors, Kaman Aircraft Corp., 1957-; Pres. and Gen. Manager, Kaman Nucl. Societies: Phys. Soc.; Automatic Documentation Section (of European Atomic Energy Community Sci. Data Processing Centre).
Nuclear interest: Management of theoretical and experimental nuclear research.
Address: 6 Heather Circle, Colorado Springs, Colorado 80906, U.S.A.

ERICSON, Alfred, Ph.D. Nucl. chem., Kansas State Teachers Coll.
Address: Kansas State Teachers College, Emporia, Kansas, U.S.A.

ERICSON, Douglas. Site Rep., Idaho Falls, Chicago Operations Office, U.S.A.E.C.
Address: c/o Chicago Operations Office, U.S.A.E.C., 9800 South Cass Avenue, Argonne, Illinois 60439, U.S.A.

ERICSON, Magda Vera, Licence in Mathematics and Phys., Aggregation in Phys., Ph.D. Born 1929. Educ.: Ecole Normale Supérieure Paris. Prof. Phys., Lyon Univ., 1960-. Society: Sté. Française de Physique.
Nuclear interests: Slow neutron physics. Various problems of meson interactions with nuclei (pion optical model, pion absorption and radiative pion capture in nuclei.
Address: Institut de Physique Nucléaire, 43 boulevard du 11 Novembre 1918, 69-Villeurbanne (Rhône), France.

ERICSON, Malte, B.B.A. Born 1913. Educ.: C.C.N.Y. Comptroller, Nat. Lead Co., Ohio, 1951-.
Address: 2774 Breezyway, Cincinnati 39, Ohio, U.S.A.

ERICSON, Torleif Erik Oskar, Fil. Dr. (Lund). Born 1930. Educ.: Lund Univ., Sweden. Asst. in Phys., Lund Univ., 1956-57; Res. Fellow (1957-58), Res. Assoc. (1958-59), Nordita, Copenhagen; Instructor, M.I.T., 1959-60; Res. Assoc. (1960-62), Sen. Phys. (1962-), C.E.R.N., Geneva.
Nuclear interests: Theoretical nuclear physics; Intermediate energy physics, particularly the use of elementary particles for investigating nuclear structure.
Address: C.E.R.N., Geneva 23, Switzerland.

ERICSSON, Ulf, Fil. lic. Born 1923. Educ.: Stockholm Univ. Sen. Sci., Nucl. Dept., Res. Inst. of Nat. Defence, Sweden.
Nuclear interest: Effects of nuclear explosions.
Address: Nuclear Department, Research Institute of National Defence, Stockholm 80, Sweden.

ERIKSEN, Bjarne, Höyesterettsadvokat. Member, Statens Atomenergirad; Member, Board of Directors, Norsk Hydro's Inst. for Cancer Research. Formerly Chairman, Board of Directors, A/S Noratom.
Address: P.O. Box 40, Kjeller, Norway.

ERIKSEN, L. H. Chief, Explosives and Propellants Lab., Feltman Res. Labs., Picatinny Aresenal.
Address: Feltman Research Laboratories, Explosives and Propellants Laboratory, Picatinny Arsenal, Dover, New Jersey, U.S.A.

ERIKSEN, Th. Atomic Energy Matters, Ministry of Foreign Affairs, Oslo.
Address: Ministry of Foreign Affairs, 7 Juni pl. 1, Oslo, Norway.

ERIKSEN, Viking Olver, Cand. real. Born 1922. Educ.: Oslo Univ. Res. Asst., Inst. of Oceanography, Oslo Univ., 1949-51; Reactor Phys. (1951-56), Group Leader, Reactor Phys. (1956-59), Director of Phys. Dept. (1959-65), Asst. Director (1965-), Inst. for Atomenergi, Kjeller, Norway.
Nuclear interest: Reactor physics.
Address: Institutt for Atomenergi, Postboks 40, Kjeller, Norway.

ERIKSSON, Ake, Agr. Sen. Officer, Div. of Agricultural Radiobiol., Dept. of Soil Fertility and Management, Roy. Agricultural Coll.
Address: Royal Agricultural College, Uppsala 7, Sweden.

ERIKSSON, Erik Gustav Gösta, Fil. kand., Fil. lic. Born 1935. Educ.: Stockholm Univ. Asst., Roy. Coll. Forestry, 1963-.
Nuclear interest: Radiation biology.
Address: Department of Forest Genetics, Royal College of Forestry, Stockholm 50, Sweden.

ERIKSSON, Erik Ingemar, Civilingenjör. Born 1932. Educ.: Roy. Inst. Technol. Res. Asst., Div. of Phys. Chem., Roy. Inst. Technol., Stockholm, 1956-61; Sen. Res. Officer, Isotope Techniques Lab., Stockholm, 1961-66.
Nuclear interests: Isotope separation by gaseous diffusion; Technical isotope applications.
Address: Sandvikens Jernverks AB, Sandviken, Sweden.

ERIKSSON, S. Ind. X-Ray Dept., Svenska A.B. Philips.
Address: Svenska A.B. Philips, Box 6077, Stockholm 6, Sweden.

ERINMEZ, Nilüfer, Dr. Bacteriologist, Turkish Cancer Hospital.
Address: Turkish Cancer Hospital, 2 Imrahor Caddesi, Ankara, Turkey.

ERIVANSKII, Yu. A. Papers: Standardisation of Radiometric Separators (letter to the Editor, Atomnaya Energiya, vol. 11, No. 5, 1961); co-author, On Expediency of Radiometric Uranium Ore Dressing and on Optimum Value of Separation during Dressing (letter to the Editor, ibid, vol. 19, No. 1, 1965).
Address: Academy of Sciences of the U.S.S.R., 14 Leninsky Prospekt, Moscow V-71, U.S.S.R.

ERKES, Pierre A., Ing. Electricien UILv. Born 1922. Educ.: Louvain Univ. Ing. Sofina, 1947-51; Ing. détaché au Centre d'Etudes Nucléaires, 1951-55; Ing.,Syndicat de l'Energie Nucléaire, 1955-56; Directeur, Bureau d'Etudes Nucléaires, 1956-66; Directeur de la Construction, Sté d'Energie Nucleaire Franco-Belge des Ardennes, S.E.N.A., 1965-; Sous-Directeur, Sté. de Traction et d'Electricité, 1966-. Societies: A.N.S.; A.S.M.E.; American Elec. Inst.
Address: 283 avenue Defré, Brussels 18, Belgium.

ERLEBACH, Woodland Eustace, B.A.Sc., M.A.Sc. (British Columbia), Ph.D. (Cantab.). Born 1922. Educ.: British Columbia and Cambridge Univs. Asst. Res. Officer, Chalk River Nucl. Lab. (1955-59), Development Chem., Commercial Products (1959-64), A.E.C.L.; Sen. Sci., Isotope Inc., 1964-. Societies: The Soc. of Nucl. Medicine; Assoc. of Professional Eng. of the Province of British Columbia.
Nuclear interests: Fission product waste disposal; Radiation chemistry; Production and applications of radioisotopes.
Address: Isotopes Inc., 50 Van Buren Place, Westwood, N.J., U.S.A.

ERLER, Georg, Dr. jur. Born 1905. Educ.: Munster and Zürich Univs. Prof. of Law; Director, Inst. of Public Internat. Law, Göttingen Univ.; Judge, European Nucl. Energy Tribunal, O.E.C.D. Member, Deutsche Atomkommission, Fachkommission I-Kernenergierecht (German A.E.C. Dept. Atomic Energy Law); Internat. Com. on the Legal Aspects of Atomic Energy, Internat. Law Assoc. (I.L.A.); Chairman, German Com. on Atomic Energy, German Branch, I.L.A.; Chairman, Legal Com. Studiengesellchaft zur Förderung der Kernenergieverwerung in Schiffbau und Schiffahrt, Hamburg. Books: Grundproblem des internationalen Wirtchaftsrecht (Göttingen, 1956); Deutsches Atomenergierecht (Göttingen, 1955) (looseleaf collection; Die Rechtsentwicklung der iternationalen Zusammenarbeit im Atomereich (Göttingen, 1963); Formen und Ziele der internationalen Zusammenarbeit bei der friedlichen Kernenergienutzung (Düsseldorf, 1961); co-author, Deutsches Atomenergierecht und Strahlenschutzrecht (Baden-Baden, 1962); Chief editor, Beiträge zum internationelen Wirtschaftsrecht und Atomenergierecht; Chief editor, Göttinger Atomrechtskatalog; Chief editor of series Kernenergierecht (in cooperation with the Federal Ministry of Sci. Res.).
Nuclear interests: National and international atomic energy law. Collection of the laws and provisions regulating nuclear energy law in all states and of the respective bilateral and multilateral treaties; Particular interest in the unification of the national law as to nuclear liability and insurance and in the working of the international organisations and institutions in this field, furthermore in the public and private international law relating to nuclear propelled ships.
Address: 11 Schlegelweg, 34 Göttingen, Germany.

ERLEWINE, John Albert, B.A., LL.B. Born 1923. Educ.: Michigan Univ. and Columbia Law School. Deputy Asst. Gen. Counsel, Chicago Operations Office (1952-55), Director, Development Contracts Div., Chicago Operations Office (1955-58), Contract Negotiator and Asst. to Asst. Gen. Manager for Res. and Ind. Development, Headquarters (1958-61), Rep., Euratom, Brussels (1961-63), Director, Congressional Relations, Headquarters (1964), Asst. Gen. Manager, Operations, Headquarters (1964-), U.S.A.E.C.
Nuclear interests: Management: Field operations; Contracting; Labour relations; Operational safety; Construction; Workmen's compensation and community assistance.
Address: United States Atomic Energy Commission, Washington, D.C. 20545, U.S.A.

ERMAGAMBETOV, S. B. Paper: Co-author, Intergrated and Differential Cross-Sections of Fission Th^{232} by Neutrons near Threshold (Atomnaya Energiya, vol. 23, No. 1, 1967).
Address: Academy of Sciences of the U.S.S.R., 14 Leninsky Prospekt, Moscow V-71, U.S.S.R.

ERMAKOV, S. M. Papers: Co-author, On Penetration of Gamma-Ray Quantum through Shielding Barrier (letter to the Editor, Atomnaya Energiya, vol. 19, No. 1, 1965); co-author, About Gamma-Rays Angular and Energy Distribution on Surface of Volume Source (Abstract, ibid, No. 2); co-author, Angular Distributions of Fast Neutrons behind Iron Barriers (ibid, vol. 20, No. 6, 1966); co-author, Optimal Variant of Focusing by Accelerating Field (ibid, vol. 23, No. 3, 1967).
Address: Academy of Sciences of the U.S.S.R., 14 Leninsky Prospekt, Moscow V-71, U.S.S.R.

ERMAKOV, S. V. Paper: Co-author, Thermoelectronic Emission of Uranium Dodecarboride (letter to the Editor, Atomnaya Energiya, vol. 20, No. 5, 1966).

Address: Academy of Sciences of the U.S.S.R., 14 Leninsky Prospekt, Moscow V-71, U.S.S.R.

ERMAKOV, V. A. Papers: Co-author, Synthesis of Element 102 Isotope with Mass Number 256 (Atomnaya Energiya, vol. 16, No. 3, 1964); co-author, Lw^{256} Fusion (ibid, vol. 19, No. 2, 1965); co-author, On Decay Characteristics of 102^{254} Isotope (ibid, vol. 20, No. 3, 1966).
Address: Academy of Sciences of the U.S.S.R., 14 Leninsky Prospekt, Moscow V-71, U.S.S.R.

ERMANS, André-Marie, M.D., Agrégé de l'Enseignement Supérieur. Born 1923. Educ.: Brussels Univ. Res. Fellow, Inst. Interuniversitaire des Sci. Nucléaires, Brussels; Senior Asst., Medical Clinic, St. Peter Hospital, Brussels Univ.; Assoc. Prof., Directeur du Centre Euratom, Brussels Free Univ. (Nucl. Medicine). Book: Influence de l'Activité Thyroidienne sur la Vitesse de Phosphate Globulaire (Brussels, Editions Acta Medica Belgica.). Societies: Belgian Soc. of Endocrinology; Affiliate Member, Roy. Soc. of Medicine.
Nuclear interests: Thyroid physiology and biochemistry in man; Medical applications of radioisotopes (diagnostic aspects, visualisation problems).
Address: Hospital Saint Pierre, Service des Isotopes Radioactifs, 322 rue Haute,Brussels, Belgium.

ERNST, Eugene, Dr. med. Born 1895. Educ.: Pécs Univ. Prof. Biophys., Pécs Univ.,1947-. Societies: Hungarian Acad. Sci.; Council's member, Internat. Union of Pure and Appl. Biophys.; Hungarian Physiological Soc.; Internat. Physiological Soc.; Hungarian Phys. Soc.; Hungarian Math. Soc.; Hungarian Biol. Soc.; Hungarian Biophys. Soc.
Nuclear interest: Physiological effects of radiations; Biological ability to differentiate isotopes.
Address: Biophysical Institute, Pécs University, Pécs, Hungary.

ERNST, Peter C., B. Eng. Born 1939. Educ.: McMaster Univ. Reactor Supervisor, McMaster Univ., 1963-. Member Technol. Com., Canadian Nucl. Assoc.
Nuclear interests: Reactor operations and management; Nuclear physics; Reactor physics; Computer analysis associated with the above.
Address: Nuclear Research Building, McMaster University, Hamilton, Ontario, Canada.

ERNST, Robert Craig, B.S. (Chem. Eng.). (North Carolina State Coll.), M.S. (Chem. Eng.) (Minnesota), Ph.D. (Major in Chem. Eng.) (Minnesota). Dean, Speed Sci. School, (1947-), Pres., Inst. Ind. Res. (1960-), Louisville Univ. Consultant to various companies in U.S.A. and consultant (1951-55) to U.S.A.E.C. New York Operational Office. Societies: A.I.Ch.E.; A.S.E.E.; Engineers' Council for Professional Development; A.C.S.; O.R.I.N.S.; A.A.A.S.; Soc. of Chem. Industry; Southern Assoc. of Sci. and Industry; A.A.U.P. (former Pres., Local Chapter); Kentucky Acad. of Sci.; Chairman, Kentucky Advisory Com. on Nucl. Energy, 1957-60; Govt.'s Rep. and Member, Regional Advisory Council on Nucl. Energy, 1956-61, and numerous other committees and societies.
Nuclear interests: Radioisotope uses in research work is of confidential nature. Directing project for State of Kentucky through Institute of Industrial Research on effect of gamma radiation on kinetics of chemicals from coal.
Address: University of Louisville, Speed Scientific School, James Breckinridge Speed Foundation, Belknap Campus, Louisville 8, Kentucky, U.S.A.

ERNST, Theodor, Dr. phil. Born 1904. Educ.: Jena and Göttingen Univs. Prof. and Director, Mineralogical Inst., Munich Inst. of Technol., 1950; Prof. and Director, Mineralogical Inst., Erlangen Univ., 1950-; Editor, Fortschritte der Mineralogie der Deutschen Mineralogischen Gesellschaft; Member, Arbeitskreis III/4 – Beschaffung und Aufbereitung von Uranerzen, Federal Ministry for Sci. Society: Fellow, Mineralogical Soc. America, 1939.
Nuclear interests: Mineralogy of radioactive raw materials.
Address: Mineralogisches Institut, 5 Schlossgarten, 852 Erlangen, Germany.

ERO, János, Candidate of Phys. Sci. Born 1928. Educ.: Budapest Tech. Univ. Leader, Dept. of Atomic Phys., Central Res. Inst. for Phys.
Nuclear interests: Nuclear physics; Low-energy nuclear reactions, especially stripping reactions and polarisation phenomena.
Address: Nagysalló-u.7, Budapest 12, Hungary.

EROFEEV, D. B. V. Deleg. I.A.E.A. Symposium on the Chemical Effects of Nuclear Transformations, Prague, Oct. 24-27, 1960.
Address: Academy of Sciences, Minsk, U.S.S.R.

ERPERS ROYAARDS, Robbert VAN. See VAN ERPERS ROYAARDS, Robbert.

ERRERA, Jacques, Dr. spécial en Chimie Physique. Born 1896. Educ.: Brussels and Leipzig Univs.; Collège de France; Inst. Pasteur, Paris. Prof. (1926-), Brussels Univ. Commissaire à l'Energie Atomique en Belgique; Conseiller à la Délégation Permanente de la Belgique aux Nations Unies; Représentant Permanent (Vienna), Member, Board of Governors (1962-), I.A.E.A. Member, Steering Com., and Top Level Group on Co-operation in Research, E.N.E.A. O.E.C.D. Formerly Administrateur, Assoc. Belge pour le Developpement Pacifique l'Energie Atomique, A.S.B.L.; Asst., Conseil d'Administration, Inst. Interuniversitaire des Sci. Nucleaires. Books: Polarisation diélectriqu (1928); Actualitiés Scientifiques et Industrielles, No. 220; Le Moment Electrique en Chimie et en Physique (1935); No. 221:

Moment Electrique et Structure Moleculaire (1935); Chimie physique nucléaire appliquée (Sciences et Lettres Liège, 1955); co-author, Euratom, Analyse et Commentaires du Traité (1958).
Address: 14 rue Royale, Brussels, Belgium.

ERRERA, Maurice Leo, M.D., Agrégé in Biochem. Born 1914. Educ.: Brussels Free Univ. Sci. Sec. to Com. on the Effects of Atomic Radiations of United Nations, 1957; now Prof. ordinaire, Brussels Univ. (Biochem., Biophys., Radiobiol.). Books: Co-author, Mécanismes de l'action biologique des radiations (Desoer and Masson, 1952); Effets biologiques des Radiations - Aspects Biochimiques (Protoplasmatologia, vol. x, Vienna, Springer-Verlag, 1957); co-editor, Mechanisms in Radiobiol. (N.Y., Academic Press, 1960 and 1961).
Nuclear interests: Radiobiology: effects of radiation on nucleic acid and protein metabolism; mutagenesis; transfer of genetic information within the cells.
Address: Laboratoire de Biophysique et Radiobiologie, Université Libre de Bruxelles, 1850 chaussée de Wavre, Brussels 16, Belgium.

ERRINGTON, G. E. Tech. Director, Avimo Ltd.
Address: Avimo Ltd., Taunton, Somerset, England.

ERRINGTON, M. Canadian Member, Study Group on Food Irradiation, E.N.E.A., O.E.C.D.
Address: O.E.C.D. European Nuclear Energy Agency, 38 boulevard Suchet, Paris 16, France.

ERRINGTON, Roy Franklin, B.A. (Maths. and Phys.), M.A. (Phys.). Born 1915. Educ.: Toronto Univ. Sales Manager, Eldorado Mining and Refining, Ltd., 1946-52; Vice-Pres. formerly Manager, Commercial Products Div., A.E.C.L., 1952-. Director, Canadian Nucl. Assoc., 1964-. Society: Canadian Assoc. of Phys.
Nuclear interests: The entire field of radioisotope production, processing and usage. Particular emphasis is placed on cobalt 60, deep therapy machines and irradiation machines using radioisotopes. Interest and responsibility is in the planning and management of programmes mentioned above.
Address: P.O. Box 93, Ottawa, Ontario, Canada.

ERSHLER, L. V. Paper: Co-author, Mass-Spectroscopic Determination of Microquantities of Boron in Some Materials (Atomnaya Energiya, vol. 16, No. 5, 1964).
Address: Academy of Sciences of the U.S.S.R., 14 Leninsky Prospekt, Moscow V-71, U.S.S.R.

ERTAUD, André, Ing. Arts et Métiers, Ing. E.S.E., Ing. en Chef de Marine (C.R.), D. ès Sc. Born 1910. Educ.: Ecole Supérieure d'Electricité; Ecole des Ingénieurs de la Marine and Sorbonne. Chef du service de la Pile de Chatillon, C.E.A., 1946-54; Directeur Technique du Groupement Atomique Alsacienne Atlantique-Chateau d'Arny, Bruyère le Chatel. Seine et Oise, France. Societies: Ing. A. et M.; Ing. Civils de France; A.N.S.
Nuclear interests: Reactor design; Nuclear fuel; Nuclear propulsion; Instrumentation.
Address: 33 rue Lacepède, Paris 5, France.

ERTZ, Helmut, Dipl.-Ing. Born 1919. Educ.: Karlsruhe Univ. Formerly Chief Eng., GEA-Gesellschaft für Luftkondensation m.b.H., 1958-.
Nuclear interest: Heat exchangers for atomic industry.
Address: 14a Cranachstrasse, Bochum, Germany.

ERWALL, Lars-Gustaf, Tekn. lic. Born 1926. Educ.: Roy. Inst. Technol., Stockholm. Formerly Chief Asst. in Phys. Chem., Roy. Inst. Technol.; Head, Isotope Techniques Lab., Stockholm -1966; now Pres., Incentive R. and D. AB, Stockholm.
Nuclear interests: Technical applications of radioisotopes and radiation.
Address: Incentive Research and Development AB, 21 Ranhammarsvägen, Bromma, Sweden.

ERYKALOV, A. N. Papers: Estimation of Variation Method Accuracy (letter to the Editor, Atomnaya Energiya, vol. 19, No. 5, 1965); co-author, Phys. Res. Reactors Parameters (letter to the Editor, ibid, vol. 25, No. 1, 1968).
Address: Academy of Sciences of the U.S.S.R., 14 Leninsky Prospekt, Moscow V-71, U.S.S.R.

ESCARGUEIL, Georges. Formerly Planning, Sté. d'Etudes Techniques et d'Entreprises Générales.
Address: Société d'Etudes Techniques et d'Entreprises Générales, 9 avenue Reaumur, 92 Le Plessis Robinson, (Seine), France.

ESCHAUZIER, Henri F., Degree in Law., Born 1910. Educ.: Leyden Univ. Ambassador of The Netherlands in Vienna, 1958-64; Board of Governors (1959-61, 1964-66), Pres., 8th Internat. Conference (1964-), Resident Rep. of The Netherlands (1958-67), I.A.E.A.
Address: Netherlands Permanent Mission to the I.A.E.A., c/o Royal Netherlands Embassy, 10 Jacquingasse, Vienna 3, Austria.

ESCHBACH, Eugene Arment, B.S. Born 1923. Educ.: Washington State Univ. Formerly Member, Programme Com., A.N.S.
Nuclear interests: Plutonium recycle; Uranium utilisation; Overall optimisation.
Address: 1501 Hains, Richland, Washington, U.S.A.

ESCHER, Alfred. Swiss Resident Rep., I.A.E.A.
Address: 7 Prinz Eugenstrasse, Vienna 3, Austria.

ESCHER, G., Dipl. Ing. Asst., Inst. fur Kernverfahrenstechnik, Karlsruhe T.H.
Address: Technische Hochschule Karlsruhe, Reaktorstation Leopoldshafen/Karlsruhe, Germany.

ESCHLER, Max Paul, Dipl. Ing. Born 1927. Educ.: Federal Inst. of Technol., Zurich. Tech. Director, M. Eschler Urania - Accessoires, 1953-. Society: V.D.I.
Nuclear interests: Specialist for sealing problems for extreme conditions; Hydraulic actuations and controls; Special pumping problems.
Address: 288-296 Badenerstrasse, CH-8004 Zurich, Switzerland.

ESCHNAUER, Heinz, Dr. rer. nat., Dipl. Chem. Born 1930. Educ.: Mainz Univ. Degussa Nucl. Group, now Nukem, Nuclear-Chemie und -Metallurgie G.m.b.H., Wolfgang bei Hanau (Main), Germany. Societies: Gesellschaft Deutscher Chemiker; Gesellschaft für Kernenergieverwertung in Schiffbau und Schiffahrt m.b.H., Hamburg.
Nuclear interests: Fuel elements; Fuel cycle economy.
Address: Nukem, Nuklear- Chemie und —Metallurgie G.m.b.H., Wolfgang bei Hanau (Main), Germany.

ESCOBAR, José Wladimiro, Licenciado en Filosofia, Maestro en Ciencias y Matemáticas. Born 1922. Educ.: Javeriana (Bogota) and Nacional (Bogota) Univs. Director de la Estación Ionosférica, 1957; Miembro del Comité del Año Geofisico Internacional, 1957; Formerly Jefe de Laboratorios, Javeriana Univ., Dept. de Fisica Nuclear. Prof. de Fisica, Prof. de Matemáticas, Prof. de Fisica Nuclear y Afines, en la Fac. Eccl., Javeriana Univ. Formerly Decano, Facultad de Ingenieria Electrónica, and Prof. de Fisica en el Departamento de Fisica, Javeriana Univ., 1961-. Societies: Soc. Colombiana de Fisicos; Inst. of Radio Eng.
Nuclear interest: Nuclear physics.
Address: Carrera 7, No. 40-62, Bogota, Colombia, South America.

ESCOBAR V., Ismael, Dr. in Sci. Born 1918. Educ.: Valladolid, Barcelona and Toulouse Univs. and M.I.T. First and Gen. Director of Meteorology, 1945-52; Founder and Director, Cosmic Ray Lab., 1952-58; Head, Bolivian Deleg., I.G.Y. Western Hemisphere Conference, Rio de Janeiro, 1956; Head, Bolivian Deleg., I.C.P.U.A.E., Geneva, 1955; Director, Lab. de Fisica Cósmica de Chacaltaya; Prof., Centro Brasileiro de Pesquizas Fisicas, Brasil; Pres., Bolivian Nat. Com. for the I.G.Y., 1957-58; Invited Prof. Phys., New Mexico Univ., 1959; Res. Assoc., Enrico Fermi Inst. for Nucl. Studies, Chicago Univ., 1959-60; Visiting Prof., Summer Courses, Sci. Foundation, 1960; Member, Interamerican Com. for Space Res., 1961. Books: Elementos de Metereologia (1943); Estudio sobre el granizo (1948); Contribución al Estudio del Tiempo en Bolivia (1943); Regimen Pluviométrico (1948); La Radiación Solar y el Pronostico del Tiempo; Energias no Convencionales (1958). Societies: Pres., Bolivian Meteorological Soc. and Bolivian Assoc. for the Advancement of Science; A.P.S.; A.A.P.T.; Hon. Member, Mexican Astronomical Soc.; National Acad. of Sciences of Bolivia, 1960.
Address: Laboratorio de Fisica Cósmica, La Paz, Bolivia.

ESCOTO, Roberto MAASS-. See MAASS-ESCOTO, Roberto.

ESCUE, Richard Byrd, Jr., M.A., Ph.D. Born 1919. Educ.: California Inst. Technol. Prof., Chem., North Texas State Coll., 1945-. Councillor, O.R.I.N.S. (now Oak Ridge Assoc Univs.). Society: A.C.S.
Nuclear interest: Tracer techniques.
Address: North Texas State College, Denton, Texas, U.S.A.

ESCULIES, Don Jose, Dr. Member, Nat. A.E.C., Paraguay.
Address: National Atomic Energy Commission, Department of Organisations, Treaties and International Agreements, Ministry of Foreign Affairs, Asuncion, Paraguay.

ESGUERRA-GOMEZ, Gonzalo, M.D. Hon. Pres., Sociedad Colombiana de Radiologia.
Address: Sociedad Colombiana de Radiologia, Carrera 7ª No. 31-46 (401), Apartado aereo 58-04, Bogota D.E., Colombia, S.A.

ESHAYA, Allen M., B.S., M. Eng., D. Eng. Born 1922. Educ.: Rensselaer Polytech. Inst. and Yale Univ. Tech. Director for Europe and Near East (1950-54), Asst. Vice-Pres. (1954-56), Coca-Cola Export Corp.; Sci. Staff, Brookhaven Nat. Lab., 1956-61; Vice-Pres., Socsil S.A., Lausanne, 1961-; Consultant: Coca-Cola Export Corp., 1956-, U.S. Dept. of the Interior, 1957-60, Brookhaven Nat. Lab., 1961-, I.A.E.A., 1962-. Rep. of A.I.Ch.E. to Nucl. Terminology Subcom. (N2.4) of American Standards Assoc. Societies: A.I.Ch.E.; A.C.S.; A.N.S.
Nuclear interests: Radiative heat transfer, diffusion of fission products in nuclear field; Adsorption of fission gases at high temperatures; Nuclear energy in saline water conversion.
Address: Socsil S.A., Ecublens, Vd., Switzerland.

ESHKOL, Levi. Prime Minister, Israel. Head, Israel A.E.C., 1966-.
Address: Israel Atomic Energy Commission, P.O.B. 7056, Hakirya, Tel-Aviv, Israel.

ESIKOV, V. I. Papers: Co-author, On the Crisis of Heat Transfer in a Steam-Generation Tube (Atomnaya Energiya, vol.13, No. 4, 1962); co-author, Exptl. Study of Crisis Boiling Heat Transfer (ibid, vol. 16, No. 5, 1964);

co-author, Studies in Start-Up Conditions of I.V. Kurchatov Beloyarsk Nucl. Power Station Performed at Test Rig (ibid, vol. 19, No. 3, 1965).
Address: Academy of Sciences of the U.S.S.R., 14 Leninsky Prospekt, Moscow V-71, U.S.S.R.

ESKILDON, Hugo N., Jr. Assoc. Director, Div. of Reactor Development (-1962), and Formerly Manager, Idaho Operations Office (1962-), U.S.A.E.C.
Address: U.S. Atomic Energy Commission, Washington, D.C. 20545, U.S.A.

ESKREYS, Andrzej Wlodzimierz, Dr. of Phys. Born 1938. Educ.: Jagiellonian Univ., Cracow. Reader, Inst. of Nucl. Res., Cracow, 1966-. Nuclear interest: Elementary particle physics, at high energies.
Address: Institute of Nuclear Research, 30 al. Mickiewicza, Cracow, Poland.

ESNE, J. Member, Nucl. Phys. Lab., Oslo Univ.
Address: Oslo University, Nuclear Physics Laboratory, Institute of Physics, Blindern, Norway.

ESPINOSA, César Anibal, Dr. Prof. of Gen. Chem. and Vice-Rector, Univ. Central, Quito.
Address: Universidad Central, Quito, Ecuador.

ESPOSITO, Joseph John, B.A., M.D. Born 1912. Educ.: Yale and Columbia Univs. Director, Dept. of Radiol. and Chairman, Isotope Com., Bridgeport Hospital, Bridgeport, Conn. Member, Connecticut Advisory Com. on Atomic Energy; Pres., Bridgeport Medical Assoc.; Chairman, Medical Advisory Com., Conn. Div., American Cancer Soc. Societies: A.M.A.; Fellow, Amer. Coll. of Radiol.; Radiol. Soc. of North America; New York Roentgen Soc.
Nuclear interests: Medical uses of radioisotopes; Radiation therapy with cobalt and radium.
Address: 144 Golden Hill Street, Bridgeport, Conn., U.S.A.

ESSELMAN, Walter H., B.S., M.S., D.E.E. Born 1917. Educ.: Newark Coll. of Eng., Stevens Inst. Technol. and Polytech. Inst. of Brooklyn. Various technical positions at Westinghouse Bettis Atomic Power Lab., 1950-53. Tech. Asst. to Manager, Naval Reactor Facility, Prototype Testing of Nautilus Nucl. Power Plant, 1955-56; Manager, Advanced Development Dept., Westinghouse Bettis Atomic Power Dept., 1956-; Sen. Staff, Westinghouse Astronucl. Lab., 1959-61; Deputy Manager (1964-68), Project Manager (1968-), Nerva Project. Vice Chairman, Aerospace Div., A.N.S. Societies: A.N.S.; A.I.E.E.; A.A.S.; A.I.A.A. Nuclear interests: Applications of nuclear energy to space applications, nuclear rockets and auxiliary power units; Reactor design and analysis; Reactor kinetics and control.
Address: 208 Yarrow Lane, Pittsburgh 36, Pennsylvania, U.S.A.

ESSEN, Alexander Von. See Von ESSEN, Alexander.

ESSER, Josef A., Dr. iur. Born 1910. Educ.: Lausanne, Paris, Frankfurt-Main, Freiburg i.B., Greifswald, Innsbruck, Mainz and Tübingen Univs. Full Prof., 1943; Judge (Court of Appeals), 1956-61; Member, Deutsche Atomkommission; Director, Legal Div., I.A.E.A., Vienna, 1958-60. Books: Several works on law. Societies: Internat. Law Assoc.; Gesellschaft für Rechtsvergleichung; Int. Vereinigung für Rechtsphilosophie.
Address: Jur. Fakultät der Universität Tübingen, Germany.

ESSER, P., Dr. Stellv. Obmann, Arbeitsausschuss "Atomenergierecht", Vereinigung Deutscher Elektrizitätswerke.
Address: Stadtwerke Düsseldorf, Düsseldorf, Germany.

ESSEX, C. D. R. and D. Manager, Liebig Group Res. Centre, Liebig's Extract of Meat Co. Ltd.
Address: Liebig's Extract of Meat Co. Ltd., Liebig Group Research Centre, 208 Harlequin Avenue, Brentford, Middlesex, England.

ESTABROOK, Frank B., B.A., M.S., Ph.D. Born 1922. Educ.: Miami Univ. and California Inst. Technol. Asst. Prof., Assoc. Prof., Miami Univ., 1950-52; Reactor Phys., Atomics Internat., 1952-55; Phys., Office of Ordnance Res., U.S. Army, 1955-60; Staff Sci., Phys. Sect. Jet Propulsion Lab., California Inst. Technol., 1960-. Society: A.P.S.
Nuclear interests: Neutron physics; Reactor theory; Neutron transport theory; General relativity and gravitation.
Address: Jet Propulsion Laboratory, California Institute of Technology, 4800 Oak Grove Drive, Pasadena, California, U.S.A.

ESTADA GIRAUTA, Manuel. Vocale, Comision Interministerial de Conservacion de Alimentos Por Irradiacion, Junta de Energia Nucl.
Address: Junta de Energia Nuclear, Ciudad Universitaria, Madrid 3, Spain.

ESTAING, Philippe GISCARD d'. See GISCARD d'ESTAING, Philippe.

ESTEN, Michael John, B.Sc. (1st class Hons., Special Phys.), Ph.D. Born 1934. Educ.: London Univ. Fellow, C.E.R.N., 1960-61; Res. Asst. (1961-64), Lecturer (1964-), Phys. Dept., Univ. Coll., London.
Nuclear interests: Elementary particle physics, particularly using bubble chamber technique. On-line computer technique used in bubble chamber film measurement.
Address: 17 Highgate Spinney, Crescent Road, London, N.8, England.

ESTEP, Samuel D., A.B., J.D. Born 1919. Educ.: Kansas State Teachers Coll. and Michigan Univ. Asst. Prof. Law, 1948-51; Assoc. Prof. Law, 1951-54; Prof. Law, 1954-; Director, Legislative Res. Centre, Michigan Univ. Law School, 1951-57; Director, Atomic Energy Res. Project, Michigan Univ. Law School, 1961-. Societies: American Soc. of Internat. Law; American Bar Assoc.; State Bar of Michigan.
Nuclear interests: Legal and social policy problems created by use of nuclear energy, and particularly regulations of business, national and international, and compensation for radiation injuries.
Address: Michigan University Law School, Hutchins Hall, Ann Arbor, Michigan, U.S.A.

ESTERMANN, I., D.Sc. (Hamburg). Born 1900. Educ.: Hamburg Univ. Director, Material Sci. Div. and Res. Coordinator, U.S. Office of Naval Res., Washington, 1950-59; Sci. Director, U.S. Office of Naval Res., London Branch, 1959-. Books: Recent Res. in Molecular Beams (Academic Press, 1959); Methods of Exptl. Phys. (Academic Press, 1959). Societies: A.P.S.; A.A.A.S.; Washington Acad. Sci.; Philosophical Soc. of Washington.
Nuclear interests: Nuclear magnetic moments; Neutron effects on solids; Scattering cross-sections; Radiation damage; Instrumentation.
Address: Apartment 206, 2404 Fulton Street, Berkeley, California, U.S.A.

ESTES, William James, B.S.A. Born 1909. Educ.: Georgia Univ. Pres., Empire Pedigreed Seed Co., Haralson, Ga.; Director, C. and S. Newnan Bank; Director for 6 years, Nat. Cotton Council of America; Nat. Cotton Ginners Assoc. (Pres., 1955); Georgia Seedsmen's Assoc. (Pres., 1959-60); Pres., Esco Gas Co.; Owner, Esco Tractor Co.; Member, Co. Board Education for 10 years. Society: Nucl. Advisory Commission, Georgia, 4 years.
Nuclear interest: Agricultural developments coming from the nuclear research field.
Address: Haralson, Georgia, U.S.A.

ESTON, Tede, M.D., B.Sc. (Medicine). Born 1919. Educ.: Sao Paulo and Toronto Univs. Director, Centro de Medicina Nucl. (former Laboratório de Isótopos da Faculdade de Medicina da Universidade de Sao Paulo), 1949-. Pres., Brazilian Soc. of Nucl. Biol. and Medicine.; Vice-Pres., Latin-American Assoc. Socs. of Nucl. Biol. and Medicine. Societies: Assoc. Paulista de Medicina; Assoc. Médica Brasileira; Soc. Brasileira para o Progresso da Ciência; Reticuloendothelial Soc.; Brazilian Soc. of Nucl. Biol. and Medicine; Latin-American Assoc. Soc. of Nucl. Biol. and Medic.
Nuclear interests: Nucl. medicine; Biological application of radioisotopes; Health physics.
Address: Centro de Medicina Nuclear, anexo à Faculdade de Medicina, Caixa Postal 22022, Sao Paulo, Brazil.

ESTON, Veronica Rapp, M.D., B.S. (Med.), privat dozent of Physiological Chem. Born 1918. Educ.: Sao Paulo and Toronto Univs. Asst. Prof. Physiological Chem., Faculty of Medicine, Sao Paulo Univ.; Head of Training and Biological Res., Centre of Nucl. Medicine, Sao Paulo. Societies: Reticuloendothelial Soc.; Biochem. Soc.; Assoc. Paulista de Medicina; Assoc. Médica Brasileira; Soc. Brasileira para o Progresso da Ciência; Rad. Res. Soc.; Brazilian Coll. Radiol.
Nuclear interests: Nuclear medicine; Biological application of radioisotopes; Health physics.
Address: Centro de Medicina Nuclear, anexo à Faculdade de Medicina da University Sao Paulo, Caixa Postal 22022, Sao Paulo, Brazil.

ESTRIN, Gerald, Prof. Dept. of Eng., California Univ., Los Angeles. Member, Maths and Computer Sci. Res. Advisory Com., U.S.A.E.C.
Address: Department of Engineering, California University, Los Angeles, California, U.S.A.

ETEMAD, Akbar, Dr. (Reactor Phys.). Born 1930. Educ.: Ecole Polytechnique, Lausanne Univ. Chief, Shielding Group, Swiss Federal Inst. for Reactor Res., Wuerenlingen, 1964-65; Chief, Atomic Energy Bureau, Plan Organisation, Teheran, 1967-. Society: A.N.S.
Nuclear interests: General; Management; Nuclear materials; Economics; Radioisotopes; Reactors.
Address: Plan Organization, Teheran, Iran.

ETHERINGTON, Harold, A.R.S.M., B.Sc. Born 1900. Educ.: London Univ. Director, Naval Reactor and Reactor Eng. Div., Argonne Nat. Lab., 1948-53; Vice-Pres., Nucl. Products-Erco, Div. of A.C.F. Industries, Inc., Washington, D.C., 1953-59; Gen. Manager, Atomic Energy Div., Allis-Chalmers Manufacturing Co., 1959-63; Atomic Energy Consultant. Member, U.S. Advisory Com. on Reactor Safeguards. Books: Modern Furnace Technol.; Nucl. Eng. Handbook (Ed.). Societies: A.S.M.E.; A.N.S.; American Soc. for Metals.
Nuclear interests: Management and reactor design.
Address: 84 Lighthouse Drive, Jupiter, Florida, U.S.A.

ETO, A. Manager, Nucl. Reactor Design Sect., Hitachi Works, Hitachi Ltd.
Address: Hitachi Ltd., Hitachi Works, 1405 Sukegawa, Hitachi-shi, Ibaragi-ken, Japan.

ETO, Hideo, Ph.D. (Medical Sci.). Born 1911. Educ.: Tokyo Imperial Univ. Assoc. Prof., School of Medicine, Tokyo Univ., 1942-; Head, Div. Rad. Hazards (1957-), Sen. Sci. Res. Officer (1965-), Nat. Inst. Radiol. Sci., Japan. Books: Human and Rad. (Tokyo, Iwanami Publishing Co., 1958); Text Book of Radiol. (Tokyo, Igaku Shoin Ltd., 1959); Radiol. (Fundamentals). 5th edition. (Tokyo, Igaku Shoin Ltd., 1964); Rad. Protection (Tokyo, Maruzen Co. Ltd., 1965). Societies: Nippon Societas Radioligica; Japan Rad. Res. Soc.;

Atomic Energy Soc. of Japan.
Nuclear interests: Radiation hazards to humans;
Assessment, protection and treatment of radiation injury.
Address: National Institute of Radiological
Sciences, 4-9-1 Anagawa, Chiba, Japan.

ETOH, Kazuaki, B.Sc. Eng. Born 1928. Educ.:
Tokyo Univ. Nucl. Plant Eng. Dept., Hitachi
Works, Hitachi, Ltd., 1950-. Societies: Japan
Nucl. Soc.; Japan Soc. Mech. Eng.
Nuclear interests: Nuclear plant design and construction.
Address: 19-13, 3 Chome, Oose Machi, Hitachi
City, Ibaraki Ken, Japan.

ETTER, Hans, Dr. med. Educ.: Basel, Berlin
and Zürich Univs. Röntgeninstitut, Kantonsspital, Lucerne. Society: Sec. Gen., Schweizerische Gesellschaft für Radiologie und Nuklearmedizin.
Address: Röntgeninstitut, Kantonsspital, Lucerne, Switzerland.

ETTLINGER, Leopold, Prof. Dr. Born 1914.
Educ.: Swiss Federal Inst. Technol. Head,
Dept. of Microbiol., Swiss Federal Inst. of Technol.
Nuclear interests: Tracer experiments on microbial metabolism.
Address: Swiss Federal Institute of Technology, Buildings of Agriculture and Forestry, Department of Microbiology, 2 Universitätsstrasse, Zürich 8006, Switzerland.

EUBANK, Harold Porter, B.S., M.S., Ph.D.
Born 1924. Educ.: Coll. of William and Mary,
Syracuse and Brown Univs. Asst. Prof., Brown
Univ., 1954-59; Res. Staff, Project Matterhorn,
Princeton Univ., 1959-60. Society: A.P.S.
Nuclear interest: Neutron-induced reactions.
Address: Plasma Physics Laboratory, Princeton
University, Princeton, New Jersey, U.S.A.

EUDIER, M. Tech. Director, Res. Lab., Métallurgie Française des Poudres S.A. "Metafram".
Address: Métallurgie Française des Poudres
S.A. "Metafram", 53 Chaussée Jules-César,
Beauchamps, (S. et O.), France.

EUGENIO, Manuel R. Supervising Sci. (Head,
Dept. Nucl. Eng.), Philippine A.E.C.
Address: Philippine Atomic Research Centre,
Diliman, Quezon City, Republic of the Philippines.

EUKEL, Warren Wenzl, B.S. (Eng. Phys.).
Born 1921. Educ.: California (Berkeley) Univ.
Phys. Res. Eng., Lawrence Rad. Lab., 1950-54;
Departmental Manager, Appl. Rad. Corp.,
1955-65; At (1965-), Operations Manager, now
Vice Pres., Operations (1968-), William M.
Brobeck and Assocs., Berkeley. Com. on Radiation Dosimetry for Q.M. Corps. Societies:
American Men of Sci.; N.A.S.; Inst. Radio Eng.
Nuclear interests: Ion source; Radiation measurement; Machine-produced radiation and uses.
Address: 315 Crest Avenue, Walnut Creek, California, U.S.A.

EULA, Ernesto. Member, Com. Scientifico,
Diritto ed Economia Nucleare.
Address: Diritto ed Economia Nucleare, 68
Viale Bruno Buozzi, Rome, Italy.

EULER, Joachim Karl, Dipl.-Phys., Dr.-Ing.
Born 1922. Educ.: Hanover Tech. Univ. Varta
A.G., Frankfurt/Main, 1952-. Vice Pres.
Physikalischer Verein, Privatdozent appl. phys.,
Giessen Univ. Book: Neue Wege zur Stromerzeugung (New Paths to Generation of Electricity) (Frankfurt-am-Main, Akademische
Verlagsgesellschaft, 1963). Societies: Deutsche
Physikalische Gesellschaft; V.D.I.; Deutsches
Atomforum; Comité Internat. de Thermodynamique et Cinétique Electriques; Bunsengesellschaft für physikal. Chem.
Nuclear interests: Radiochemistry (tracer methods and autoradiography), nuclear batteries, thermal and photoelectric.
Address: 11 Pfungststrasse, D6 Frankfurt/Main, Germany.

EUROLA, Ahti Tapio, Ph.D., M.Sc., Dipl. Eng.
Born 1931. Educ.: Finland Inst. of Technol.
and Illinois Inst. of Technol. Res. Eng., City
Elec. Works of Helsinki, 1956-57; Resident
Student Assoc., Argonne Nat. Lab., 1960-61;
Res. Phys. (1961-64), Group Leader, Reactor
Dynamics Group (1965), O.E.C.D. Halden
Reactor Project; Project Eng., EKONO Association for Power and Fuel Economy, 1966-.
Treas., Finnish Nucl. Soc. Society: Finnish
Nucl. Soc.
Nuclear interests: Reactor dynamics; Safety of nuclear systems.
Address: EKONO, Association for Power and
Fuel Economy, 14 E.Esplanadikatu, Helsinki
13, Finland.

EVANI, Kondaiah, B.Sc., M.Sc., Fil. Lic., Fil.
Dr. (Stockholm) cum laude. Born 1923. Educ.:
Benares Hindu, Allahabad and Stockholm
Univs. Scholar, Australian Nat. Univ. Canberra,
1952-54; Res. Fellow (1954-57), Fellow (1957-),
Tata Inst. Fundamental Res., Bombay.
Societies: Indian Sci. Congress Assoc.; A.P.S.
Nuclear interests: Nuclear physics, research and teaching.
Address: Tata Institute of Fundamental Research, Colaba, Bombay-5, India.

EVANS, Alfred David, B.Sc. (Hons., Phys.).
Educ.: Liverpool Univ. Member, Calder Hall
Operations Staff, 1958-60; Works Manager,
A.G.R., 1960; P.F.R. Project Manager, Dounreay Establishment, 1967.
Nuclear interest: Nuclear reactor operation.
Address: 24 Miller Place, Scrabster, By Thurso, Caithness, Scotland.

EVANS, Charles C., B.Sc., Ph.D. Born 1918.
Educ.: London Univ. Technical Manager,
Amersham/Searle Corporation.
Nuclear interest: Radioisotope production.

Address: c/o Amersham/Searle Corporation, 2000 Nuclear Drive, Des Plaines, Illinois 60018, U.S.A.

EVANS, Ersel Arthur, B.A., Ph.D. Born 1922. Educ.: Reed Coll., Portland and Oregon State Univ. Directing R. and D. programmes at Hanford Site, G.E.C., 1951-67; Manager, Materials Dept., Battelle-Northwest, Richland. Ind. Coordinator, U.S.-U.K. Exchange on Ceramic (Uranium) Fuels. Societies: Fellow, A.N.S.; A.S.M.
Nuclear interests: Management of research and development programmes involving plutonium, uranium, and thorium alloys and compounds and reactor components for fast, thermal, special purpose reactors such as those for rocket propulsion and management of isotopic heat and radiation sources.
Address: Battelle-Northwest, P.O. Box 999, Richland, Washington 99352, U.S.A.

EVANS, G. R., Dr. Dept. Natural Philosophy, Edinburgh Univ.
Nuclear interest: High-pressure cloud chamber studies.
Address: Edinburgh University, Drummond Street, Edinburgh 8, Scotland.

EVANS, Gordon W. Born 1903. Educ.: Texas and Southern Calif. Univs. Pres., and Chairman of the Board, Kansas Gas and Elec. Co.; Vice-Pres. and Exec. Com. Member, Southwest Atomic Energy Assocs.; Board Member: Nat. Assoc. of Elec. Cos.; Elec. Cos. P.I.P.; E.C.A.P.
Address: P.O. Box 208, Wichita, Kansas 67201, U.S.A.

EVANS, Helen Harrington, Ph.D. Educ.: Western Reserve Univ. Asst. Prof. Biochem. and Radiol., Western Reserve Univ. Society: American Soc. Microbiologists.
Nuclear interests: Effects of X-irradiation on the biological and physical-chemical properties of D.N.A. and on the synthesis of messenger R.N.A.; Relation of the effects of X-irradiation on nucleic acids to the inhibition of cell division.
Address: Radiology Department, Wearn Research Building, Western Reserve University, 2064 Abington Road, Cleveland, Ohio 44106, U.S.A.

EVANS, Henry John, B.Sc., Ph.D. Born 1930. Educ.: Univ. Coll. of Wales, Aberystwyth. Res. Sci., M.R.C. Radiobiol. Res. Unit, Harwell, 1955; Prof., Genetics, Aberdeen Univ., 1966-. Sec., Assoc. of Rad. Res.; Editor, Mutation Res.; Editor, Rad. Botany. Book: Co-Editor, Human Radiation Cytogenetics (Amsterdam, North-Holland Publishing Co., 1967). Societies: Genetical Soc.; Soc. for Exptl. Biol. Nuclear interests: Radiation genetics and cellular radiobiology.
Address: Aberdeen University, Marischal College, Aberdeen, Scotland.

EVANS, Hugh Devereux, B.Sc., Ph.D. Born 1922. Educ.: Edinburgh Univ. Phys., Inst. of Cancer Res., Royal Cancer Hospital, 1949-55; Res. Fellow, Dept. of Phys. (1955-57), Lecturer in Phys. (1957-59), Lecturer in Rad. Hazards (1958-), Health Phys. (1957-), Imperial Coll. Societies: Assoc. Inst. Phys.; Soc. for Radiol. Protection.
Nuclear interests: Radiological protection (health physics); Design of apparatus for radioactive measurement.
Address: 10 Seymour Avenue, Ewell, Surrey, England.

EVANS, James Carey. Born 1900. Educ.: Yale Univ. Vice Pres. for Programme Development, Nucl. Res. Centre, Buffalo Univ. Societies: A.N.S.; Atomic Ind. Forum.
Nuclear interest: Management.
Address: WNY Nuclear Research Centre, Power Drive, State University of New York at Buffalo, Buffalo, New York 14214, U.S.A.

EVANS, John A., M.D. Formerly Member, then Vice Pres., now Pres., American Board of Radiol.
Address: American Board of Radiology, Kahler Centre Building, Rochester, Minnesota 55901, U.S.A.

EVANS, John Ellis, B.S. (Educ.), B.A. (Ohio State), M.A. (Maths., Ohio State), Ph.D. (Phys., Rice Inst.). Born 1914. Educ.: Rio Grande Coll., Ohio State Univ. and Rice Inst.; Staff member, Los Alamos Sci. Lab., 1948-52; Project Leader, Group Leader, and Sect. Head, Nucl. Cross-sections (1952-56), Director, Nucl. Phys. Branch (1956-61), Phillips Petroleum Co., Atomic Energy Div.; Consulting Sci., Lockheed Missile and Space Co., 1961-. Member, U.S.A.E.C. Nucl. Cross-section Advisory Group, 1956-61. Societies: Fellow, A.P.S.; A.N.S.; A.A.P.T.; A.A.A.S.; A.G.U.
Nuclear interests: Nuclear cross-sections; Fission process; Inelastic scattering of thermal neutrons; Energy spectra and flux-time behaviour of particles and electromagnetic radiations in space.
Address: Lockheed Missile and Space Co., 3251 Hanover Street, Palo Alto, California, U.S.A.

EVANS, Marjorie, B.A., Ph.D. Born 1921. Educ.: Colorado and California (Berkeley) Univs. Director, Poulter Labs., Stanford Res. Inst., 1965-. Programme Chairman, Twelfth Internat. Symposium on Combustion, Poitiers, France, 1968. Societies: A.P.S.; A.C.S.; A.I.A.A.
Nuclear interest: Management research on shock effects.
Address: Stanford Research Institute, Menlo Park, California, U.S.A.

EVANS, Robley Dunglison, B.S., M.S., Ph.D. Born 1907. Educ.: California Inst. Technol. Prof. Phys., M.I.T., 1945-; Staff Consultant, Biophys., Peter Bent Brigham Hospital,

Boston, 1945-; Vice-Chairman, Com. on Nucl. Sci., (1946-), Chairman, Subcom. on Shipment of Radioactive Substances, (1946-), N.A.S.-N.R.C.; Consultant, Phys., Massachusetts Gen. Hospital, 1948-; Consultant, Health Div., Los Alamos Sci. Lab., 1948-; Member, Isotope Com., Massachusetts Gen. Hospital, 1950-; Consultant, Div. of Biol. and Medicine, U.S.A.E.C., 1950-; Consultant, U.S. Naval Radiol. Defence Lab., 1952-; Member, Editorial Board, Internat. J. Appl. Rad. and Isotopes, 1955-; Member, Com. on Rad. Protection, M.I.T., 1955-; Member, Relative Hazard Factor Subcom., Nat. Com. Rad. Protection and Measurements, 1957-; Com. on Shipment of Radioactive Substances, American Standards Assoc., 1959-; Adviser on rad. protection problems, M.I.T., 1959; Member, Com. on Radioisotope Utilization, M.I.T., 1959-; Formerly Vice Pres., Rad. Res. Soc.; Phys. Editor, Rad. Res., 1959-; Consultant, U.S. P.H.S., 1961; Consultant to Surgeon Gen., Dept. of the Army, 1961; Book: The Atomic Nucleus (McGraw-Hill, 1955). Societies: Fellow, A.P.S.; Fellow, American Acad. of Arts and Sci.; Fellow, A.A.A.S.; A.A.P.T.; A.I.H.A.; American Assoc. of Physicists in Biol. and Medicine; A.N.S.; H.P.S.; Assoc. American Roentgen Ray Soc.; Hon. Foreign Member, Royal Sci. and Lit Soc., Goteborg; Rad. Res. Soc.
Nuclear interests: Nuclear physics; Biological effects of radiation.
Address: Massachusetts Institute of Technology, 77 Massachusetts Avenue, Cambridge 39, Massachusetts, U.S.A.

EVANS, Titus Carr, B.A., M.S., Ph.D. Born 1907. Educ.: Baylor and Iowa Univs. Prof. and Head, Radiation Research Lab., State Univ. Iowa, Coll. Medicine, 1948-; Medical Advisory Group and Life Sci. Working Group, U.S. Air Force and A.E.C. Managing Editor, Rad. Res. Assoc. Editor, J. of Nucl. Medicine. Pres., Soc. of Nucl. Medicine 1961; Member, Nat. Com. on Rad. Protection. Societies: Rad. Res. Soc.; Soc. Nucl. Medicine; American Cancer Res. Soc.; Radiol. Soc.; North American Roentgen Ray Soc.
Nuclear interests: Biological effects of radiation; Uses of radioisotopes in biology and medicine.
Address: Radiation Research Laboratory, College of Medicine, Iowa State University, Iowa City, Iowa, U.S.A.

EVANS MORGAN, Eilir, Geologist. Born 1911. Educ.: Univ. Nacional de Córdoba. Consulting Geologist and Tech. Manager, CURA—CO, SRL, 1946-56; Chief, Mining Div. (1956-60), Formerly Director (1960-), now Adviser, Comisión Nacional de Energia Atómica.
Nuclear interests: Prospecting, exploration and production of nuclear raw materials and concentrates.
Address: Comisión Nacional de Energia Atómica, 8250 Av. Libertador General San Martin, Buenos Aires, Argentina.

EVERETT, James LeGrand, III, B.S. (Mech. Eng.), M.S. (Mech. Eng.), M.S. (Ind. Management). Born 1926. Educ.: Penn. State Univ. and M.I.T. Eng. in Charge, Mech. Res. (1956-58), Staff Eng., R. and D. (1958-60), Director of Res. (1960-62), Manager, Eng. and Res., (1962-66), Vice Pres., Eng. and Res. (1966-), Philadelphia Electric Co. Societies: A.N.S.; A.S.M.E.
Nuclear interests: Management; Nuclear power technology; Reactor engineering; Fuel element design; Nuclear power economics and fuel utilisation; Nuclear power plant operations.
Address: Philadelphia Electric Company, 1000 Chestnut Street, Philadelphia, Pa. 19105, U.S.A.

EVERETT, Willis Lynn, B.S. (Eng. Phys.), M.S. (Nucl. Eng.), Ph.D. (Nucl. Eng.). Born 1923. Educ.: Michigan Univ. Teaching Fellow (1955-58), Res. Asst. (1958-60), Michigan Univ.; Project Eng., Bendix Systems Div., 1961-62; Asst. Prof., Phys. (1962-64), Assoc. Prof., Eng. (1964-65), Wyoming Univ.; Assoc. Prof., Nucl. Eng. (1965-), Acting Chairman, Nucl. Eng. Dept. (1966-67), New Mexico Univ. Book: Contribution to Third Internat. Rarefied Gas Dynamics Symposium (Ed. Laurmann, J. A. New York, Academic Press, 1963). Societies: A.N.S.; A.S.E.E.; A.P.S.
Nuclear interests: All aspects of nuclear science and engineering education; Research in the area of thermonuclear systems.
Address: 3225 Reina Drive, N.E., Albuquerque, New Mexico, U.S.A.

EVERS, Heinz, Dipl.-Ing., Dr. Born 1912. Educ.: Berlin and Braunschweig Tech. Univs. and Giessen Univ. Owner of the firm of Industrieplanung und Wassertechnik, Dipl-Ing. Dr. H. Evers G.m.b.H. and Co. Pres., Verband Selbständiger Ingenieure VSI. Society: Deutsche Kernenergiegesellschaft.
Nuclear interests: Nuclear research in the biochemical and food industry fields.
Address: 2 Slevogtstieg, Hamburg-Othmarschen, Germany.

EVERSOLE, Harold B. Asst. to Manager, Chicago Operations Office, U.S.A.E.C.
Address: U.S. Atomic Energy Commission, Chicago Operations Office, 9800 South Cass Avenue, Argonne, Illinois, U.S.A.

EVERTS, B. F. Deputy Member, Netherlands Pool for the Insurance of Atomic Risks.
Address: 75-79 Rokin, Amsterdam, Netherlands.

EVKINA, Z. V. Paper: Co-author, Temperature and Neutron Irradiation Effect on Plastic Deformation of Alpha Uranium Single Crystal (Atomnaya Energiya, vol. 19, No. 4, 1965).
Address: Academy of Sciences of the U.S.S.R., 14 Leninsky Prospekt, Moscow V-71, U.S.S.R.

EVOY, E. F. Gen. Supt., Gunnar Mining Ltd. Address: Gunnar Mining Ltd., Gunnar, Saskatchewan, Canada.

EVRARD, Charles. Directeur Général, Cie. Maritime Belge S.A. (Lloyd Royal). Director and Member, Board of Management, Syndicat Vulcain, Soc. Belge pour l'Industrie Nucléaire; Administrateur, Groupement Professionnel de l'Industrie Nucléaire; Belgian Member, Study Group on Nucl. Ship Propulsion, O.E.C.D., E.N.E.A.; Member for Belgium, Nucl. Ship Propulsion Com., Euratom; Director, Societe Anglo-Belge Vulcain S.A. Address: Cie. Maritime Belge S.A. (Lloyd Royal), 61 Rempart Ste. Cathérine, Antwerp, Belgium.

EVSEEV, V. S. Papers: Co-author, Angular Distribution of Neutrons due to μ^- Capture in Calcium for Various Energy Thresholds (Nukleonika, vol. 9, No. 2-3, 1964); co-author, On interaction constants in μ^- capture (ibid.). Address: Joint Institute for Nuclear Research, Laboratory of High Energy Physics, Dubna, U.S.S.R.

EWALD, Earl, B.E.E. Born 1908. Educ.: Minnesota Univ. Pres., Chief Exec. Officer and Member, Board of Directors, Northern States Power Co.; Chairman, Res. Div. Exec. Com., and Member, Board of Directors, Edison Elec. Inst.; Board of Directors: First Bank Stock Corp.; St. Paul Fire and Marine Insurance Co.; Upper Midwest R. and D. Council. Vice Pres., Minnesota Safety Council; Chairman, Mid-Continent Area Power Planners; Vice Chairman, Elec. Res. Council. Nuclear interest: Management. Address: 414 Nicollet Mall, Minneapolis, Minnesota 55401, U.S.A.

EWALD, Heinz Wilhelm, Dr. rer. nat. Born 1914. Educ.: Jena, Freiburg and Göttingen Univs. Kaiser Wilhelm Inst. (Max Planck Inst.) für Chemie, Berlin-Dahlem, Tailfingen, Mainz, 1939-51; apl. Prof. and Konservator, Physikalisches Inst., Munich T.H., 1952-62; o. Prof. und Direktor of II. Physikalisches Inst., Justus-Liebig-Univ. Giessen, 1963-. Book: Co-author, Methoden und Anwendungen der Massenspektroskopie (1953). Societies: Deutsche Physikalische Gesellschaft; A.P.S. Nuclear interests: Nuclear physics; Mass spectroscopy; Isotope geology; Fission process. Address: II. Physikalisches Institut der Justus-Liebig-Universität, 2 Arndstrasse, 63 Giessen, Germany.

EWAN, George Thomson, B.Sc. (Hons. Phys., Edinburgh), Ph.D. (Edinburgh). Born 1927. Educ.: Edinburgh Univ. Asst. Lecturer, Edinburgh Univ., 1950-52; Res. Assoc. (1952-53), N.R.C. Post-Doctoral Fellow (1953-55), McGill Univ.; Res. Officer, A.E.C.L., 1955-; Ford Foundation Fellow, Niels Bohr Inst., Copenhagen, 1961-62; Visiting Sci., Lawrence Rad. Lab., Berkeley, 1966. Societies: A.I.P.; Canadian Assoc. of Physicists. Nuclear interests: Structure of nuclei; Nuclear spectroscopy and disintegration schemes; High resolution β-ray spectrometers and high resolution solid state gamma-ray spectrometers. Address: Atomic Energy of Canada, Ltd., Chalk River, Ontario, Canada.

EWART, J. Norton, B.S.E.E. Born 1903. Educ.: Cornell Univ. Chief System Project Eng., Niagara Mohawk Power Corp. Societies: A.S.M.E.; I.E.E.E.; A.P.C.A.; A.N.S. Nuclear interests: Power generation design and construction for Niagara Mohawk Power Corp. including 500 MWe Nuclear Power Plant at Nine Mile Point Station and 750 MWe Nuclear Station at Easton. Address: Niagara Mohawk Power Corp., 535 Washington Street, Buffalo, N.Y. 14203, U.S.A.

EWBANK, Michael Henry, B.Sc. Born 1930. Educ.: London Univ. Tech. Eng., Ewbank and Partners, Ltd., 1953; seconded to U.K.A.E.A. for special training in nucl. power eng., 1954; Chief Tech. Eng., Nucl. Power (1955), Director (1957-67), Managing Director (1967-), Ewbank and Partners, Ltd. Societies: A.C.G.I.; M.I.Chem.E. Nuclear interest: Consultant engineering in nuclear power with particular reference to economics, tender assessment, design and project co-ordination. Address: Oakwood, Clayhill Road, Leigh, Nr. Reigate, Surrey, England.

EWELL, Raymond, B.S., M.S., Ph.D. Born 1908. Educ.: Toledo, Purdue and Princeton Univs. Vice-Pres., New York State Univ.; Consultant, Ford Foundation; Consultant, U.N.; Consultant, Agency for Internat. Development. Societies: A.I.Ch.E.; A.C.S. Nuclear interest: Administrative charge of Western New York Nuclear Research Centre, Inc. which is a wholly owned subsidiary of the State University of New York. Address: State University of New York at Buffalo, Buffalo 14, N.Y., U.S.A.

EWEN, Richard Lee, B.S. (M.E., Iowa), M.S. (M.E., Iowa), Ph.D. (M.E., Pittsburgh). Born 1930. Educ.: Iowa State Coll. and Pittsburgh Univ. Sen. Eng., Westinghouse Elec. Co., Bettis Plant, 1953-. Societies: A.N.S.; A.S.M.E. Nuclear interests: Thermal and nuclear design; Transient analysis, including space-time kinetics. Address: Westinghouse Electric Corp., Box 1468, Pittsburgh 30, Pennsylvania, U.S.A.

EWING, Ben B., Ph.D., B.S.C.E., M.S.C.E. Born 1924. Educ.: Texas and California Univs. Asst. Prof., Civil Eng., Texas Univ., Austin, 1952-58; Assoc., Civil Eng. (1955-56), Asst. Res. Eng. (1956-58), California Univ., Berkeley; Assoc. Prof. (1958-61), Prof. (1961-), Sanitary Eng., Prof., Nucl. Eng. (1966-), Director,

Water Res. Centre (1966-), Illinois Univ., Urbana. Societies: A.S.C.E.; American Water Works Assoc.; Water Pollution Control Federation; A.S.E.E.; American Assoc. Profs. in Sanitary Eng.
Nuclear interests: Radioactive waste disposal; Control of radioactive contamination of environment.
Address: 3220 Civil Engineering Building, Illinois University, Urbana, Illinois 61801, U.S.A.

EWING, James Houston, B.S. Born 1908. Educ.: Washington and Jefferson Univs., Pennsylvania State Coll. Vice-Pres., Andrews-Knapp Construction Co. Inc. Societies: A.N.S.; American Ordnance Assoc.; Atomic Ind. Forum; Assoc. of Purchasing Agents.; N.A.C.E.
Nuclear interests: Lead shielding and fabrications; Design and installation of shielding and hardware for the nuclear and process field.
Address: 59 West 10th Street, New York, N.Y., 10011, U.S.A.

EWING, John Arthur, B.S.A., M.S., D.P.A. Born 1912. Educ.: Tennessee and Harvard Univs. Asst. Director, Agricultural Experiment Station (1949-55), Sen. Vice Dean and Sen. Vice Director, Coll. of Agriculture, Teaching Res. Extension (1955-57), Director, Agricultural Experiment Station (1957-), Tennessee Univ. Member, State Air Pollution Control Board; Water Resource Res. Inst.
Nuclear interest: Research in the general field of radiation and radioisotopes in agriculture using important crops and farm animals.
Address: P.O. Box 1071, Knoxville, Tennessee 37901, U.S.A.

EWUSIE, Joseph Yanney, B.Sc. (London), Ph.D. (Cantab.). Born 1927. Educ.: Ghana and Cambridge Univs. Lecturer, Botany, Ghana Univ., 1957-62; Chief Exec. and Gen. Sec., Ghana Acad. of Sci. 1963-. Sec., Ghana Sci. Assoc., 1961-62; Editor, Ghana J. of Sci., 1961-; Council Member, Kumasi Sci. and Technol. Univ., 1963-66; Sec., Ghana Atomic Energy Commission, 1963-66.
Nuclear interest: Radiobiology.
Address: c/o Ghana Academy of Sciences, P.O. Box M.32, Accra, Ghana.

EXLEY, James, C. Eng., M.I.Mech.E., M.I.E.E., A.M.Inst.F., A.M.B.I.M. Born 1914. Educ.: Bradford Tech. Coll. Deputy Station Supt., Huddersfield Power Station, 1952-54; Deputy Station Supt., Kirkstall Power Station, 1954-56; Attachment to U.K.A.E.A., 1956-58; Station Supt., Berkeley Nucl. Power Station, 1958-62; Station Supt., Oldbury Nucl. Power Station, 1962-68.
Nuclear interests: Nuclear power station management.
Address: 26 Dryleaze, Merlin Haven, Wotton-under-Edge, Glos., England.

EXNER, Hans Eckart, Dipl. Ing., Dr. mont. Born 1938. Educ.: Leoben Inst. of Montanistic. Sci. Swedish Inst. for Metal Res., Stockholm, 1962-64; Max Planck Inst. for Special Metals, Stuttgart, 1965-; Brown Univ., Providence, 1967-68. Society: D.G.M.
Nuclear interests: Physical metallurgy; Phase diagrams for uranium systems; Power metallurgy of uranium compounds; Oxidation of uranium compounds.
Address: Max Planck Institut für Sondermetalle, Laboratorium für Pulvermetallurgie, D7000 Stuttgart, Germany.

EXNER, R. Austrian Member of Group of Experts on the Production of Energy from Radioisotopes, O.E.C.D., E.N.E.A.
Address: O.E.C.D. European Nuclear Energy Agency, 38 boulevard Suchet, Paris 16, France.

EYE, Royston Walter Mastin D'. See D'EYE, Royston Walter Mastin.

EYLER, William Ross, A.B., M.D. Born 1918. Educ.: Harvard Coll. and Harvard Medical School. Chairman, Dept. of Radiol., Henry Ford Hospital, 1955-. Editor, Radiology. Societies: Radiol. Soc. of North America; American Roentgen Ray Soc.
Nuclear interest: Application of radioactive materials to clinical diagnostic and therapeutic problems.
Address: Henry Ford Hospital, Detroit 2, Michigan, U.S.A.

EYSTER, Eugene, Dr. Alternate Leader GMX Div., Los Alamos Sci. Lab., U.S.A.E.C. Member, Com. of Senior Reviewers, U.S.A.E.C.
Address: Los Alamos Scientific Laboratory, U.S.A.E.C., Los Alamos, New Mexico, U.S.A.

EZELLE, Sam, LL.B. Born 1920. Educ.: Louisville Univ. Sec.-Treas., Kentucky State Federation of Labour, 1953-57; Exec. Sec.-Treas., Kentucky State AFL-CIO, 1958-. Member, AFL-CIO Deleg. to Nat. Labour Organisation of Israel, 1956; Formerly Member, Kentucky Atomic Energy and Space Authority.
Address: 312 Armory Place, Louisville 2, Kentucky, U.S.A.

FAAS, H. A., Ir. Member, Com. No. 102, Nucl. Energy, Nederlands Normalisatie-Instituut.
Address: Nederlands Normalisatie-Instituut, Postbus 70, 20-22 Duinweg, The Hague, Netherlands.

FABER, Mogens, M.D.; Director, Radiobiol. Lab., Radiumstationen. Medical Adviser, Danish A.E.C.; Member, Health and Safety Subcom., and Study Group on Food Irradiation, E.N.E.A.
Address: 10 Christiansborg Ridebane, Copenhagen K, Denmark.

FABIC, Stanislav, B. Mech.E., M.Mech.E., M.S., Ph.D. Born 1925. Educ.: Zagreb, Melbourne and California (Berkeley) Univs. At Inst. of Eng. Res., California Univ., 1958-63; Project Eng., Kaiser Eng., 1963-67; Fellow Eng., Atomic Power Div., Westinghouse Elec. Corp., 1967-. Society: A.N.S.
Nuclear interests: Fluid flow and heat transfer during reactor power transients or during blowdown.
Address: Westinghouse Electric Corporation, Pressurised Water Reactor Division, Pennsylvania Centre, Monroeville, Pennsylvania, U.S.A.

FABIOU, H. Nucl. Dept. Serseg (Seguin-Sergot).
Address: Serseg (Seguin-Sergot), 4 place Felix Eboue, 75 Paris 12, France.

FABOZZI, S., Ing. Tech. Manager, Dott. Ing. Giuseppe Torno and C.S.p.A.
Address: Dott. Ing. Giuseppe Torno and C. S.p.A., 7 Via Albricci, Milan, Italy.

FABREGUETTES, Jean. Ing., Coyne et Bellier.
Address: Coyne et Bellier, 19 rue Alphonse de Neuville, Paris 17, France.

FABRI, Gianni Bruno Giulio, Dr. in Phys. Born 1935. Educ.: Pavia Univ. Assoc. Researcher, C.I.S.E. labs., Segrate, Milan, 1960-; Teaching Asst. in Solid State Phys., Politecnico of Milan, 1960-.
Nuclear interests: Solid state detectors; Physics of semiconductor surfaces; Positron decay in organic materials.
Address: Centro Informazioni Studi Esperienze, via Redecesio 12, Segrate, Milan, Italy.

FACCHINI, Ugo, Prof. Phys. Born 1924. Head, Nucl. Phys. Dept., C.I.S.E., Milan, 1950-. Chair of Phys., Turin Univ., 1958-60; Milan Univ., 1960-. Papers: Papers on measurement of uranium fission cross sections (1949) and neutron source calibration (1950); on the behaviour of Geiger-counters (1951-52) and elementary processes in gas discharges (1953); on the detection of low light intensities from biologic substrate (1954); on the use and behaviour of ionisation chamber (1956-57); on nuclear reaction at intermediate energy (1957-58); on nuclear evaporation (1960-61).
Address: Lab. C.I.S.E., C.P. 3986, Milan, Italy.

FACUSSE, Jose, Ingro. Voter, Honduras A.E.C.
Address: Honduras Atomic Energy Commission, Tegucigalpa, D.C., Honduras.

FACY, Léopold, Ing. Ecole Supérieure d'Optique. Born 1909. Educ.: Paris Univ. Ing. Gén. Météorologie Nat. Member, Panel of Experts on Atomic Energy, World Meteorological Organisation; Rapporteur National de la France au Conseil de l'Europe sur le problème de pollution.
Nuclear interest: Atmospheric pollution-fallout.
Address: 8 place du Souvenir, Bures sur Yvette, Essonne, France.

FADER, Walter J., A.B., Ph.D. Born 1923. Educ.: Harvard Univ. and M.I.T. With United Aircraft Res. Lab., East Hartford. Societies: A.P.S.; A.N.S.
Address: 197 Foote Road, South Glastonbury, Connecticut 06073, U.S.A.

FAES, Maurice Herman, M.D. Born 1919. Educ.: Louvain Univ. Head, Medical Dept., C.E.N., Mol.; Lecturer, Faculty Medicine, Louvain Univ. Society: H.P.S.
Nuclear interests: Radiation dosimetry; Radiotherapy; Isotopes in medicine; Radiotoxicology; Health physics.
Address: Drève des Faisans, Keerbergen, Belgium.

FAGEN, William Federick, B.S., M.S. Born 1916. Educ.: Louisiana State Univ. Prof. Nucl. Eng., Prof. Elec. Eng., Florida Univ.
Nuclear interests: Education (nuclear engineering); Reactor simulation (research); Nuclear instrumentation (research); Linear accelerators (research); Fusion (research); Digital control (research); Quantum electronics.
Address: Department of Electrical Engineering, Florida University, Gainesville, Florida, U.S.A.

FAGERLUND, Gunnar. Design Office, Atomic Power Dept., Hydqvist and Holm A.B.
Address: Nydqvist and Holm A.B. (NOHAB), Atomic Power Department, Trollhättan, Sweden.

FAGGIANI, Sergio, Degree in Phys. Born 1932. Educ.: Genoa Univ. Istituto Fisica Tecnica, Rome Univ.
Nuclear interests: Reactor heat transfer; Heat transfer with boiling.
Address: Istituto di Fisica Tecnica, 18 Via Eudossiana, Rome, Italy.

FAGNANI, Massimo. Manager, Ing. C. Pavone S.A.S.
Address: Ing. C. Pavone S.A.S., 12 Via Natale Battaglia, Milan, Italy.

FAHLAND, J., Dr. Member, Nucl. Chem. Group, Inst. für Anorganische Chemie und Kernchemie, Johannes Gutenberg Univ.
Address: Johannes Gutenberg University, Institut fur Anorganische Chemie und Kernchemie, 21 Saarstrasse, 65 Mainz, Germany.

FÄHRMANN, Karlheinz, Dipl. Phys., Dr. rer. nat. Born 1931. Educ.: Dresden Tech. Hochschule.
Nuclear interest: Reactor physics.
Address: Deutsche Akademie der Wissenschaften zu Berlin, Forschungsgemeinschaft, Zentralinstitut für Kernforschung, 8051 Dresden, PF 19 German Democratic Republic.

FAHY, Edward Francis, B.Sc., M.Sc., Ph.D. Born 1922. Educ.: Ireland Nat. Univ. and Chicago Univ. Asst. Prof. Phys., Marquette Univ., Milwaukee, -1952; at present Prof. Exptl. Phys. and Head, Dept. Exptl. Phys. Univ. Coll., Cork. Pres. (1967-68), Irish Sci. Teachers Assoc. Societies: A.P.S.; Italian Phys. Soc.; Phys. Soc. of Japan.
Nuclear interests: Nuclear physics; Cosmic rays; Reactor physics; Relativity; Quantum mechanics.
Address: Physics Department, University College, Cork, Eire.

FAINBERG, Ya. B. Papers: Co-author, Study on Electromagnetic Rad. from a Plasma of a Linear Pinch Discharge (Atomnaya Energiya, vol. 14, No. 4, 1963); co-author, Interaction of Modulated Beam with Plasma (ibid, vol. 19, No. 4, 1965); co-author, Exptl. Res. of Fast Ion Fluxes, being Generated in Plasma-Beam Discharges (ibid, vol. 24, No. 5, 1968); co-author, High Intensity Electromagnetic Waves Propagation in Plasma Waveguides (ibid, vol. 15, No. 1, 1968).
Address: Academy of Sciences of the U.S.S.R., 14 Leninsky Prospekt, Moscow V-71, U.S.S.R.

FAIR, Donald Robert Russell, B.Sc., A.K.C., Born 1916. Educ.: London Univ. Deputy Regional Director, S.E. Region, C.E.G.B. Societies: F. Inst. P.; M. Inst. F.
Nuclear interests: Reactor design, operation and general management.
Address: 24 Bede House, Manor Fields, Putney, London, S.W. 15, England.

FAIRAND, Barry. Formerly Principal Phys., Reactor Phys. Div., Battelle Memorial Inst., now Res. Sci.-Eng., Appl. Nucl. Phys. Div., Battelle-Columbus Labs.
Address: Battelle-Columbus Laboratories, 505 King Avenue, Columbus, Ohio 43201, U.S.A.

FAIRBAIRN, Alan, M.Sc. Born 1913. Educ.: Sheffield Univ. Atomic Energy work, 1948-. Society: A.R.I.C.
Nuclear interest: Health and safety work associated with atomic energy.
Address: Arran, 19 Hillfield, Frodsham, via Warrington, Cheshire, England.

FAIRBAIRN, The Honourable D. E., D.F.C., M.P. Minister of State for Nat. Development, 1964-.
Address: c/o Australian Atomic Energy Commission, Post Office, Coogee, New South Wales, Australia.

FAIRBANK, Henry Alan, A.B., Ph.D., M.A. Born 1918. Educ.: Whitman Coll.; Yale Univ. Faculty, Yale Univ., 1945-62; Prof. and Chairman, Dept. of Phys., Duke Univ., 1962-. Societies: Fellow, A.P.S.; A.A.P.T.; A.A.A.S.
Nuclear interest: Low temperature research.
Address: Department of Physics, Duke University, Durham, North Carolina, U.S.A.

FAIRBANKS, George Charles. 1909. Educ.: Goldsmith's Coll. Director of the following companies: Elliott Brothers (London), Ltd. (Asst. Managing Director); Bristol's Instrument Co., Ltd.; Elliott Brothers Sales Agencies, Ltd.; Fisher Governor Co., Ltd.; Mech. Automation, Ltd.; Panellit, Ltd.; James Gordon and Co., Ltd.; James Gordon Valves, Ltd.; Reactor Control Div., Elliott Brothers (London), (also Chairman); Sauter Controls, Ltd.; The Swartwout Co., Ltd.; Isotope Developments, Ltd.; E-A Automation Services, Ltd.; E-A Tech. Services, Ltd.; E-A Automation Systems, Ltd.; Rotron Controls, Ltd.; Leybold-Elliott Ltd.; Satchwell Controls Ltd.; E-A Data Processing Ltd.; Elliott Process Automation (Chairman); Elliott Medical Automation Ltd.; Elliott Marine Automation Ltd.; Elliott Automation Ltd.; S.A.C.I.R.; De Var Controls Ltd.; Member of Council, Sci. Instrument Manufacturers' Assoc.; Member of Council, British Industrial Measuring and Control Apparatus Manufacturers' Assoc. Societies: M.I.Prod.E.; Inst. of Nucl. Eng.
Nuclear interest: Management.
Address: Elliott Brothers (London), Ltd., Century Works, Lewisham, London, S.E.13, England.

FAIRBROTHER, J. A. V., Prof. Head, Phys. Dept., Natal Univ.
Address: University of Natal, Pietermaritzburg, Natal, South Africa.

FAIRES, Ronald Arthur, F.R.I.C. Born 1910. Educ.: London Univ. Principal, Isotope School, Educ. and Training Dept., A.E.R.E., Harwell, U.K.A.E.A. Books: Co-author, Radioisotope Lab. Techniques (Newnes, 1958); co-author, Radioisotope Conference, (Butterworths, 1954). Society: Royal Inst. of Chem.
Nuclear interest: Radioisotope training.
Address: 42 Thesiger Road, Abingdon, Berks., England.

FAIRHALL, Arthur William, B.Sc., Ph.D. Born 1925. Educ.: Queen's Univ., Canada and M.I.T. Asst. Prof. (1954-58),Assoc. Prof. (1958-63), Prof. (1963-), Washington Univ. Societies: A.C.S.; A.P.S.
Nuclear interests: Nuclear fission; Nuclear reactions; Nuclear geochemistry.
Address: Department of Chemistry, Washington University, Seattle 5, Washington, U.S.A.

FAIRSTEIN, Edward, B.E.E. Born 1922. Educ.: C.C.N.Y. Basic Instrumentation Group, Instrumentation and Controls Div., O.R.N.L., 1946-59; Chief Eng., Fair Port Instruments, Inc., Oak Ridge, 1959-60; Pres., Tennelec Instrument Co., Inc., Oak Ridge, 1960-. Book: Contributing author, Nucl. Instruments and their Uses (John Wiley, 1962). Societies: RESA; I.E.E.E. (Sen. Member).
Nuclear interests: Design of advanced electronic instruments for use in nuclear and atomic research; Consultant in the electronic instrumentation field.

Address: 122 Westlook Circle, Oak Ridge, Tennessee, U.S.A.

FAIRWEATHER, Ian Lawson, B.Sc., Ph.D. Born 1933. Educ.: Edinburgh Univ. N.R.C. Postdoctorate Fellow, (1958-60), Asst. Prof., (1960-61), Manitoba Univ.; Asst. Prof. (1961-67), Assoc. Prof. (1967-), Ottawa Univ. Societies: Canadian Assoc. of Physicists; A.A.P.T.
Nuclear interest: Low energy experimental nuclear physics.
Address: Department of Physics, Ottawa University, Ottawa, Canada.

FAJNZILBERG, S. D. Papers: Co-author, 100 MeV Neutron Collector (Jaderna Energie, vol. 13, No. 11, 1967); co-author, Storage Ring for 100 MeV Electrons (Atomnaya Energiya, vol. 23, No. 6, 1967).
Address: Physico-Technical Institute, 2 Polytechnicheskaya Ulitsa, Leningrad, U.S.S.R.

FAKHRI, O., Ph.C., Diploma in Radiobiol. Physicist, Radioisotope Dept., Ministry of Health Republic Hospital.
Address: Ministry of Health Republic Hospital, Radioisotope Department, Baghdad, Iraq.

FALAHI, S. AL-. See AL-FALAHI, S.

FALER, Kenneth Turner, B.S., Ph.D. Born 1931. Educ.: Idaho State Coll., and California (Berkeley) Univ. Chem. Process Development Chem., Phillips Petroleum Co., 1953; U.S. Army, 1954-56; Civilian chem. res. supv., Brooke Army Medical Centre, 1956; Teaching Asst. and graduate student, California Univ., 1956-59; Group Leader, Nucl. Chem. Group, Chem. Sect., Nucl. Phys. Branch, Phillips Petroleum Co., Atomic Energy Div., 1959-66; Group Leader, Nucl. Chem. Group, Idaho Nucl. Corp., 1966-; Affiliate Asst. Prof. Chem., Idaho State Univ., 1967-. Societies: A.A.A.S.; A.P.S.; A.C.S.; Idaho Acad. Sci.
Nuclear interests: Nuclear spectroscopy; Slow neutron cross sections; Mechanism of fission; Resonance neutron capture gamma rays.
Address: Materials Testing Reactor, Idaho Nuclear Corporation, P.O. Box 1845, Idaho Falls, Idaho 83401, U.S.A.

FALK, Charles E., B.A., M.S., D.S. Born 1923. Educ.: New York Univ. and Carnegie Inst. Technol. Sci., Phys. Branch, Div. Res., U.S.A.E.C., Washington, D.C., 1956-58; Assistant to the Director (1958-61), Asst. Director (1961-62), Assoc. Director (1962-), Brookhaven Nat. Lab.; Planning Director, Nat. Sci. Foundation, 1966-. Societies: A.P.S.; New York Acad. Sci.
Nuclear interests: High energy physics; High energy accelerator technology; Research administration.
Address: 179 South Gillette Avenue, Bayport, New York, U.S.A.

FALK, Max. Born 1906. Managing Director, Nucl. Res. Applications, Ltd., Reigate, M. Falk and Co., Ltd., Reigate, Appl. Res. (Instruments) Ltd., Reigate. Book: Handbook on Gamma-radiographic Equipment (private publication, 1949; revised 1952 and 1958). Societies: Soc. of Non-Destructive Testing (American); Non-Destructive Testing Soc. of Great Britain.
Nuclear interests: Non-destructive testing of materials, nucleonic instrumentation, consultancy (technical and management).
Address: The White House, Croydon Road, Reigate, Surrey, England.

FALK-VAIRANT, Paul. Maitre de Conferences, Inst. du Radium, Faculte des Sci., Paris Univ.
Address: Institut du Radium, Laboratoire Curie, 11 rue Pierre Curie, Paris 5, France.

FALKENBERG, Dietrich, Dipl. Ing. Born 1934. Educ.: Dresden T.H. Abteilungsleiter, Zentralinst. für Kernforschung, Rossendorf.
Nuclear interest: Reactor theory.
Address: Zentralinstitut für Kernforschung Rossendorf, Rossendorf über Dresden-WH, Postfach 19, German Democratic Republic.

FALKENHEIM, Ernst Georg Paul. Born 1898. Educ.: Munich Univ. Formerly Managing Director, Deutsche Shell A.G., Hamburg; now Member, Board of Directors, Deutsche Shell A.G. and others. Member of the presidency, Bundesverband der Deutschen Industrie, Cologne; Member of the presidency, Economic and Social Com., European Economic Community and Euratom, Brussels; Member, Kuratorium of Deutsche Industrie Inst., Cologne. Nuclear interest: First line in legislature and management.
Address: 15 Hagrainerstrasse, Rottach-Egern Obb., Germany.

FALKOFF, David. Prof. Phys. Dept., Brandeis Univ.
Address: Brandeis University, Waltham 54, Massachusetts, U.S.A.

FALLS, Oswald Benjamin, Jr., B.S. (Phys.), S.B. (Elec. Eng.), S.M. (Elec. Eng.). Born 1913. Educ.: Richmond Univ. and M.I.T. Pres. and Director, Commonwealth Assocs. Inc.; Vice Pres. and Director, Commonwealth Services Inc.; Vice Pres. and Director, Commonwealth Services Internat. Inc.; Vice Pres. and Director, Arabian Development Assocs. Inc. Societies: I.E.E.E.; A.N.S.; A.A.A.S.
Nuclear interests: Reactor design; Nuclear power plant design and application; Fuel systems and economy; Management; Safety and licensing.
Address: 209 E. Washington Avenue, Jackson, Michigan 49203, U.S.A.

FAN, The Ngoc, Dr. In charge of general purpose large rockets, Space Res. Centre, Sin-Kiang.

Address: Space Research Centre, Sin-Kiang, China.

FANCHENKO, S. D. Paper: Co-author, Plasma Turbulence Evaluation According to Electromagnetic Rad. and Ranan Scattering in Microwave Band (letter to the Editor, Atomnaya Energiya, vol. 20, No. 6, 1966).
Address: Academy of Sciences of the U.S.S.R., 14 Leninsky Prospekt, Moscow V-71, U.S.S.R.

FANELLI, A. ROSSI-. See ROSSI-FANELLI, A.

FANJOY, George Robert, B.Sc. Born 1931. Educ.: Roy. Military Coll., Ontario and Queen's Univ., Ontario. Design Eng., Nucl. Fuel, Canadian Gen. Elec., Peterborough, 1958-64; Nucl. Materials Eng., Ontario Hydro, 1964-. Society: A.N.S.
Nuclear interest: Procurement, use and disposal of nuclear fuel and heavy water.
Address: Ontario Hydro, 620 University Avenue, Toronto 2, Ontario, Canada.

FANO, U., Sc.D. Born 1912. Educ.: Turin Univ. Prof. of Phys., Chicago Univ. Subcom. on Penetration of Charged Particles in Matter, Com. on Nucl. Sci., Nat. Acad. of Sci. Books: Irreducible Tensorial Sets (New York, Academic Press, 1959); Basic Phys. of Atoms and Molecules (New York, Wiley, 1959). Societies: A.P.S.; Rad. Res. Soc.
Nuclear interests: Radiation penetration; Various aspects of nuclear theory.
Address: Chicago University, Chicago, Illinois 60637, U.S.A.

FAORO, L., Ing. Working in Nucl. Field, Dott. Ing. Giuseppe Torno and C. S.p.A.
Address: Dott. Ing. Giuseppe Torno and C. S.p.A., 7 Via Albricci, Milan, Italy.

FARAGGI, Moshe, M.Sc. (Jerusalem), D.Sc. (Pàris). Born 1932. Educ.: Jerusalem and Paris Univs. Res. chem., A.E.C., Israel; Res. Assoc., A.E.C., France. (On a Sabbatical leave from the Israel A.E.C. Society: A.C.S.
Nuclear interests: Radiation chemistry, Radiochemistry; Radiation dosimetry.
Address: Service de Chimie Physique C.E.N. Saclay, B.P. 2, 91, Gif-sur-Yvette, France.

FARAGGI MATHIEU, Henriette, D. ès Sc. (Paris). Born 1915. Educ.: Paris Univ. Res. Phys., (1950-), Adjoint, Dept. de Physique Nucléairé, Direction de la Physique, C.E.A. Societies: Sté. Française de Physique; A.P.S.
Nuclear interests: Nuclear reactions with medium energy charged particles, either polarized or not, as a tool for nuclear spectroscopy; Structure of nuclei; Nuclear models; Solid state detectors and nuclear emulsions techniques.
Address: 53 rue Fontaine Grelot, Bourg-la-Reine, Seine, France.

FARAGO, Peter Stephen, M.A., Dr. phil. Born 1918. Educ.: Budapest Univ. Reader in Phys., Budapest Univ., 1950; Head, Dept. of Electromagnetic Waves (1950), Deputy Res. Director (1956), Central Res. Inst. of Phys., Hungarian Acad. Sci., Budapest; Sen. Lecturer in Phys., Edinburgh Univ., 1957; Visiting Res. Prof., Dept. of Phys., Maryland Univ., 1959-60; Reader in Phys., (1962), Prof. of Phys., (1967), Edinburgh Univ. Societies: F. Inst. P.; F.R.S.E.
Nuclear interests: Particle beam optics and polarization; Spin dependent phenomena in atomic and nuclear physics.
Address: Edinburgh University, Department of Natural Philosophy, Drummond Street, Edinburgh 8, Scotland.

FARBER, Sidney, M.D. Prof. Pathology, Harvard Medical School; Director of Res., Children's Cancer Res. Foundation. Formerly Pres., Worcester Foundation for Exptl. Biology.
Address: Harvard Medical School, Cambridge, Massachusetts 02138, U.S.A.

FARBMAN, Gerald H., B.M.E. Born 1931. Educ.: Polytech. Inst. of Brooklyn and O.R.S.O.R.T. Educational Dept. (1952-54), Atomic Power Div., Preliminary Plant Eng. (1954-63), Atomic Power Div., Manager, Plant Application (1963-64), Atomic Power Div., Manager, Application Eng. 1964-, Westinghouse Elec. Corp. Societies: A.N.S.; A.S.M.E.; American Management Assoc.
Nuclear interests: Preliminary design and analysis of nuclear power plants. Preparation of technical proposals to electric utilities. Study and evaluation of advanced water moderated and cooled reactors. Study and evaluation of new applications for present-day and advanced reactors.
Address: P.O. Box 355, Pittsburgh, Pennsylvania 15230, U.S.A.

FAREEDUDDIN, Syed, B.Sc. (Osmania), M.Sc. (Osmania), B.S.E. (Ch.E., Michigan), M.S.E. (Ch.E., Michigan). Born 1923. Educ.: Osmania and Michigan Univs. Jun. Sci. Officer, Dept. of Atomic Energy, 1953-55; Jun. Res. Officer (1955-57), Design Eng. (1957-), now Head, Chem. Eng. Div., Bhabha Atomic Res. Centre, A.E.C., India. Adviser, Third I.C.P.U.A.E., Geneva, Sept. 1964.
Address: 184 D'Monte Park Road, Bandra, Bombay-20, India.

FARHATAZIZ, B.Sc., B.Sc. (Hons.), M.Sc. (Hons.), Ph.D. (Phys. Chem. Cantab.). Born 1932. Educ.: Panjab (Lahore) and Cambridge Univs. Lecturer, Pharmaceutical Chem. (1954-56), Lecturer, Phys. Chem. (1959-60), Panjab Univ.; At (1960-) now Principal Sci., Pakistan A.E.C.; Trainee, Cento Inst. for Nucl. Sci., Tehran, 1960; Staff member, A.E.R.E., Harwell, 1961-62; Exchange Visitor, A.E.C.L., Chalk River, 1964; Postdoctoral Fellow, Rad. Lab., Notre Dame Univ., 1964-67. Book: A Practical Course in Rad. Chem. (Lahore, Atomic Energy Centre, 1963).

Nuclear interests: Primary processes in radiation chemistry; Applications of radiation chemistry.
Address: Atomic Energy Centre, Ferozepur Road, P.O. Box 658, Lahore, Pakistan.

FARIA, Hugo CASTRO. See CASTRO FARIA, Hugo.

FARIA, Nelson V. de CASTRO de. See de CASTRO de FARIA, Nelson V.

FARINELLI, Ugo, "Laurea" in Elec. Eng.; "Libera Docenza" in Nucl. Phys. Born 1930. Educ.: Turin Politecnico and Turin Univ. At Nat. Inst. of Nucl. Phys., Section of Turin, 1953-58; with Nat. Com. for Nucl. Energy, Lab. of Reactor Phys. and Calculations, 1958-. Nuclear interests: Nuclear physics; reactor physics.
Address: Via Fratelli Carle 19, Turin, Italy.

FARIS, Frank Edgar, B. Eng., Ph.D. (Phys.). Born 1919. Educ.: Southern California and California Univs. Pres., Faris, Jacobson, and Assoc., Inc. Societies: American Management Assoc.; A.N.S.; A.P.S.; W.G.L., Cologne. Nuclear interest: Consulting.
Address: 235 Bear Hill Road, Waltham, Mass. 02154, U.S.A.

FARKAS, Joseph, Dipl. Eng. Chem., Dr. tech., Ph.D. Born 1933. Educ.: Budapest Tech. Univ. Asst. researcher, Inst. for Res. in Canning, Meat-packing and Refrigeration, Budapest, 1957-59; I.A.E.A. Fellow at Federal Res. Inst. for Food Preservation, Karlsruhe, 1959; Res. Chem., Head of Res. Group for Rad. Preservation, Central Food Res. Inst., Budapest, 1959-. Societies: Hungarian Chem. Soc.; Hungarian Microbiol. Soc.
Nuclear interests: Radiation preservation of foods; Radiation pasteurisation of spices, and fruit juices; Investigating the effects of radiation in exterminating pests in stored products; Radiobiology: effect of radiation on the food spoilage micro-organisms; Application of radiotracers to microbiological problems.
Address: 6. II. 1 Elek utca, Budapest 11, Hungary.

FARLEY, Edward R., Jun., A.B., LL.B. Born 1918. Educ.: Princeton Univ. and Harvard Law School. Formerly Vice Pres., now Chairman of the Board and Pres., Nucl. Dept., Atlas Corp.; Formerly Vice Pres., Beryllium Resources Inc.
Nuclear interest: Management.
Address: Atlas Corporation, Nuclear Department, 345 Madison Avenue, New York, New York 10022, U.S.A.

FARMAKOVSKII, V. V. Paper: Co-author, About Cross-Section Averaging for Thermal Region in Media with Zirconium Hydride Content (letter to the Editor, Atomnaya Energiya, vol. 20, No. 5, 1966).
Address: Academy of Sciences of the U.S.S.R., 14 Leninsky Prospekt, Moscow V-71, U.S.S.R.

FARMER, Bobby Joe, B.A., M.S., Ph.D. Born 1932. Educ.: Texas Christian and Rice (Houston) Univs. Sen. Sci., Ling-Temco-Vought Res. Centre, Dallas. Societies: A.I.P.; A.N.S. Nuclear interests: Research in electron interactions with matter; Charged particle analysis; Development of space radiation spectrometers; Patent on real time spectrum to dose conversion system.
Address: Ling-Temco-Vought, Incorporated, Post Office Box 6144, Dallas, Texas 75222, U.S.A.

FARMER, Frank Reginald, B.A. Born 1914. Educ.: Cambridge Univ. Head, Safeguards Div., Risley, U.K.A.E.A. Health and Safety Branch. Chairman, Internat. Standards Organisation Tech. Com. 85, I.S.O. Sub-Com. 3 (Reactor Safety). Society: F.Inst.P.
Address: The Long Wood, Appleton, Nr. Warrington, Lancs., England.

FARMER, Frank Taylor, B.Sc. (Eng.) (London), Ph.D. (Cantab.). Born 1912. Educ.: London and Cambridge Univs. Chief Phys., Roy. Victoria Infirmary and Regional Hospital Board, Newcastle upon Tyne, 1945-; Prof., Medical Phys., Newcastle upon Tyne Univ., 1967-. Societies: F.I.E.E.; British Inst. of Radiol.; F. Inst. P.; Hospital Phys. Assoc. Nuclear interests: Medical applications; Nuclear disarmament.
Address: 81 Grosvenor Avenue, Newcastle upon Tyne 2, England.

FARMER, G. Head, Scintillator Dept., Thorn Electronics Ltd.
Address: Thorn Electronics Ltd., Scintillator Department, Hook Rise South, Tolworth, Surrey, England.

FARMER, Samuel J., A.B. (Political Sci.), LL.B. Born 1925. Educ.: Whitman Coll. and Stanford Univ. Attorney, U.S.A.E.C., 1950; Counsel, Elec. Boat Div. (1954), Vice Pres., and Counsel, Gen. Atomic Div. (1956), Gen. Dynamics Corp.; Vice Pres., and Counsel, Gulf Gen. Atomic Inc., 1967-. Societies: A.N.S.; Atomic Ind. Forum.
Nuclear interest: Management.
Address: Gulf General Atomic Incorporated, P.O. Box 608, San Diego 12, California, U.S.A.

FARMER, William S., Jr., B.S. (magna cum laude), M.S., Ph.D. Candidate. Born 1922. Educ.: Tufts Coll. and Tennessee Univs. Sen. Dev. Eng., O.R.N.L., 1945-53; Project Eng., Pratt and Whitney Aircraft Co., Hartford, Conn., 1953-57; Sect. Head, A.C.F. Industries, Washington, D.C., 1957-58; Project Manager, Manager Planning and Tech. Director, Allis-Chalmers Co., Washington, D.C., 1959-. Societies: A.C.S.; A.I.Ch.E.; A.N.S.
Nuclear interests: Reactor design and development; Heat transfer; Heat transfer; Hydrodynamics and nuclear engineering analysis; Nuclear plant design and construction.

Address: Allis-Chalmers Manufacturing Co., AED, 6935 Arlington Road, Bethesda, Maryland, U.S.A.

FARNES, W. R. Managing Director, Bristol Aeroplane Plastics Ltd.
Address: Bristol Aeroplane Plastics Ltd., Filton House, Bristol, England.

FARNSWORTH, P. T., Dr. Pres., Farnsworth Res. Corp.
Address: Farnsworth Research Corp., 3700 East Pontiac Street, Fort Wayne, Indiana, U.S.A.

FARO, Manuel Jose de ABREU. See de ABREU FARO, Manuel Jose.

FAROOQI, Wasim Ahmad, B.Sc. (Agriculture), M.Sc. (Horticulture). Born 1938. Educ.: Panjab and West Pakistan Agricultural Univs. Res. Asst., Fruit Sect., Agricultural Res. Inst. Lyallpur, 1958-64; Asst. Sci. Officer (1964-67), Sci. Officer (1967-), Pakistan Atomic Energy Centre. Member, Council, Horticultural Soc. of West Pakistan. Society: Fellow, Roy. Horticultural Soc., London.
Nuclear interest: Use of radiation and radioisotopes in tropical and sub-tropical fruits with special reference to fruit irradiation.
Address: Radiobiology Division, Atomic Energy Centre, P.O. Box 658, Lahore, West Pakistan, Pakistan.

FARR, Lee Edward, B.S., Dr. Med., Licentiate, American Board of Pediatrics, 1935. Born 1907. Educ.: Yale Univ. Prof. Environmental Health, Chairman, Dept. Environmental Health, Texas Univ. School of Public Health at Houston. Consultant: U.S.N. Medical Res. Mission to Formosa; Coms. of Nat. Res. Council, N.A.S.; Chairman, Advisory Com. Atomic Bomb Casualty Commission, Com. on Naval Medical Res.; Expert Advisory Panel on Rad., W.H.O.; Chairman, Com. on Environmental Hazards, American Acad. Pediatrics; Member, Board of Visitors, U.S. Naval Radiol. Defence Lab., San Francisco, California. Consultant, TRW Systems Inc., Houston, Texas. Societies: Soc. Nucl. Medicine; Soc. for Rad. Res.; Soc. for Exptl. Biol. and Medicine; Soc. for Exptl. Pathology; American Soc. for Clinical Investigation; Fellow, New York Acad. Sci.; Inst. Environmental Sci.; A.A.A.S.; Roy. Soc. Arts; A.M.A.; American Acad. Pediatrics; American Pediatric Soc.; Soc. Pediatric Res.; American Therapeutic Soc.
Nuclear interests: Application of radioactive isotopes and devices of nuclear physics to clinical and research medicine. Development and application of nuclear reactors as instruments of medical therapy and diagnosis; treatment of cancer with neutron capture therapy; Organisation of education and research programmes in nuclear medicine; Establishment of professional standards in nuclear medicine.
Address: 811 Lovett Boulevard, Houston, Texas 77006, U.S.A.

FARR, R. F., M.A., F. Inst. P. Hon. Director, Birmingham Regional Centre, Radiol. Protection Service M.R.C.
Address: Birmingham Regional Centre, Radiological Protection Service, Medical Research Council, Queen Elizabeth Hospital, Egbaston, Birmingham 15, England.

FARRAN, Miss H. E. A. Principal Phys., Isotope Lab., New End Hospital.
Address: New End Hospital, Isotope Laboratory, New End, Hampstead, London, N.W.3, England.

FARRELL, Michael Schiller, B.Sc. Born 1923. Educ.: Melbourne Univ. Sen. Res. Sci. Res. Establishment, Australian A.E.C., 1962-67; Atomic Energy Attache, Australian Embassy, Washington, D.C., 1967-. Societies: Assoc. Roy. Australian Chem. Soc.; A.N.S.; I.N.M.M. Nuclear interests: Specifically in inorganic and nuclear chemistry, nuclear materials and nuclear fuel cycles and generally in nuclear power and its economics.
Address: Australian Embassy, 1700 Massachusetts Avenue N.W., Washington, D.C. 20036, U.S.A.

FARRINGER, L. Dwight, Dr. Assoc., Phys. Dept., Manchester Coll.
Nuclear interest: Gamma ray spectroscopy.
Address: Physics Department, Manchester College, N. Manchester, Indiana, U.S.A.

FARROW, Joseph H., M.D. Formerly Pres. Elect. now Member., Exec. Com., American Radium Soc.
Address: American Radium Society, c/o John L. Pool, M.D., 444 East 68 Street, New York, New York 10021, U.S.A.

FARUQI, Azimuddin Ahmed, Ph.D. Born 1935. Educ.: Oregon State Coll. Sen. Sci. Officer, Pakistan A.E.C., 1961-.
Nuclear interest: Nuclear chemical engineering.
Address: Pakistan Atomic Energy Commission, Karachi, Pakistan.

FARWELL, George Wells, S.B. (Harvard); Ph.D. (Chicago). Born 1920. Educ.: Harvard, California and Chicago Univs. Assoc. Prof. Phys. (1955-59), Prof. Phys. (1959-), Assoc. Dean Graduate School (1959-65), Asst. Vice Pres. (1965-67), Vice Pres. for Res. (1967-), Washington Univ. Seattle. Societies: Fellow, A.P.S.; A.A.A.S.
Nuclear interests: Nuclear physics: fission process, especially spontaneous fission; nuclear reactions at low and medium energies; elastic and inelastic scattering of charged particles; collective states in nuclei.
Address: Department of Physics, Washington University, Seattle, Washington 98105, U.S.A.

FASSBENDER, Josef A., Dr. rer. nat., Dipl-Phys. Born 1920. Educ.: Bonn and Munich Univs. At Deutsche Akademie der Wissenschaften, Berlin, 1945-52; Bonn Univ., 1952-58; Formerly at Kernforschungsanlage Jülich, Leiter, Abt. Forschungsreaktoren, 1958-. Society: Nordwestdeutsche physikalische Gesellschaft e.V.
Nuclear interests: Reactor operations and management; Reactor physics.
Address: Jülich, 8 Frankenstrasse, Germany.

FASSI, Abdelkabir EL-. See EL-FASSI, Abdelkabir.

FASSIO, Ugo, Gr.Uff. Member, Steering Com., "Pool" Italiano per l'Assicurazione dei Rischi Atomici.
Address: "Pool" Italiano per l'Assicurazione dei Rischi Atomici, 2 Via Ettore Petrolini, Rome, Italy.

FAST, Edwin, A.B., M.S., Ph.D. Born 1914. Educ.: Tabor Coll. (Hillsboro, Kan), Friends Univ. (Wichita, Kansas), Oklahoma Univ. Sen. Phys., C-14 tracer studies in catalysts (1947-50), Health phys. (1951-52), Graphite damage studies (1952-54), Group leader i/c Reactivity Measurement Facility (MTR site) (1954-59), Sect. Head i/c Reactor Experiments (all zero power testing facilities), Phillips Petroleum Co., Atomic Energy Div.; Chief, Reactor Experiments Sect., Idaho Nucl. Corp., 1966-; Instructor, N.R.T.S. Educ. programme, Idaho Univ., 1954-. Societies: A.P.S.; A.N.S.; American Optical Soc.
Nuclear interests: Graphite irradiation damage; Radioactive tracer studies and measurement of reactor physics constants, resonance integrals, irradiation transients and long term reactivity effects, neutron reproduction constants in fiss. isotopes with critical facility.
Address: 1549 Beverly Road, Idaho Falls, Idaho, U.S.A.

FAST, Johan Diedrich, Dr. h.c. (Delft). Born 1905. Chief Metal., Philips, Eindhoven and Prof. extr., Tech. Univ., Eindhoven. Books: Entrophy (Centrex Publ. Co., 1962); Interaction of Metals and Gases (Centrex Publ. Co., 1965). Societies: Koninklijke Nederlandse Chemische Vereniging; Nederlandse Natuurkundige Vereniging; Deutsche Gesellschaft für Metallkunde.
Nuclear interest: Metallurgy.
Address: 29 Parklaan, Eindhoven, Netherlands.

FATEEV, A. P. Papers: Co-author, A Ringtype Accelerator with a Vertically Increasing Magnetic Field (letter to the Editor, Atomnaya Energiya, vol. 8, No. 6, 1960); On the Theory of the Ring Cyclotron (letter to the Editor, ibid., vol. 10, No. 4, 1961); co-author, New Accelerator-Symmetrical Ring Phasotron Lebedev Phys. Inst. AN U.S.S.R. Starting Up (letter to the Editor, ibid, vol. 20, No. 6, 1966); co-author, Acceleration of Electrons in FIAN Ring Phasotron (ibid, vol. 21, No. 3, 1966).
Address: Academy of Sciences of the U.S.S.R., 14 Leninsky Prospekt, Moscow V-71, U.S.S.R.

FATEEVA, Mrs. Margarita Nicolaevna, Prof., Dr. of Medicine. Born 1900. Educ.: 2nd Moscow Medical Inst.; Chief, Biophys. Lab., Inst. of Internal Medicine, Acad. of Medical Sci. of the U.S.S.R., 1946-56; Chief, Radiological Clinic, and Formerly Director, Inst. of Hygiene of Labour and Occupational Diseases, Acad. of Medical Sci. of the U.S.S.R., 1956-. Papers: 33 scientific articles, mostly in the field of medical radiology. Societies: All-Union Com. of Medical Radiology.
Address: Academy of Medical Sciences of the U.S.S.R., 14 Solyanka Street, Moscow, U.S.S.R.

FATKIN, N., Dr. Deleg. to Conference sur la Physique des Plasmas et la Recherche concernant la Fusion Nucléaire Controlée, Salzbourg, Sept. 1961. Adviser, Third I.C.P.U.A.E., Geneva, Sept. 1964.
Address: Academy of Sciences of the U.S.S.R., 14 Leninsky Prospekt, Moscow V-71, U.S.S.R.

FAUCETT, T. R., Dr. Chairman, Dept. of Mech. Eng., and Member, Nucl. Eng. Area Study Com., Missouri Univ. at Rolla.
Address: Missouri University at Rolla, Rolla, Missouri, U.S.A.

FAUCONNIER. Gen. Sec., Assoc. des Ing. en Genie Atomique, 1967-.
Address: Association des Ingenieurs en Genie Atomique, 6 rue de Castellane, Paris 8, France.

FAUL, Henry, S.B., M.S., Ph.D. Born 1920. Educ.: M.I.T. and Michigan State Univ. Geophys U.S. Geological Survey, 1947-63; Prof of Geophys., S.W. Centre for Advanced Studies, Dallas, Texas, 1963-66; Chairman, Dept. of Geology, Pennsylvania Univ., 1966-. Books: Nucl. Geology (New York, John Wiley and Son, 1954); Ages of Rocks, Planets, and Stars (New York, McGraw-Hill, 1966). Societies: A.G.U.; Geochem. Soc.; A.A.A.S.; Geological Soc. America.
Nuclear interests: Nuclear age determination of rocks.
Address: Department of Geology, Pennsylvania University, Philadelphia, Pa. 19104, U.S.A.

FAULKNER, J. B. L., B.Sc., A.M.I.C.E. Chief Project Eng., Hartlepool Nucl. Power Station, Babcock English Elec. Nucl. Ltd., 1969-. Board Member, British Nucl. Energy Soc.
Address: Babcock English Electric Nuclear Limited, Design and Commercial Offices, Whetstone, near Leicester, England.

FAULKNER, John Edward, B.A. (Oberlin Coll.), Ph.D. (Wisconsin). Born 1920. Educ.: Oberlin Coll. and Wisconsin Univ. Supervisor, Exptl. Nucl. Phys. (1952-57), Manager, Nucl. Phys. Res. (1957-63), Hanford Labs., G.E.C.; Consultant, Exptl. Phys., Westinghouse Astronucl. Lab. (1963-). Societies: A.P.S.; A.N.S.

Nuclear interest: Experimental reactor physics and nuclear safety.
Address: 1273 Gill Hall Road, Clairton, Pennsylvania, U.S.A.

FAULKNER, Rafford L., B.S. Born 1909. Educ.: M.I.T. Asst. to Director (1950-57), Deputy Director (1957-63), Director (1963-), Div.of Raw Materials, U.S.A.E.C. Society: D.C. Soc. of Professional Engs.
Address: 8624 Beech Tree Road, Bethesda, Maryland, U.S.A.

FAURE, René, Ing. des Arts and Manufactures. Born 1920. Educ.: Ecole Centrale des Arts et Manufactures. Directeur, Centre d'Etudes Nucléaires et Cadarache. Pres., Assoc. Provençale Enseignement Sci. Tech. et Economique.
Address: rue Philippe Solari, Aix en Provence, France.

FAUST, William R., B.S., M.S., Ph.D. Born 1918. Educ.: Oklahoma State and Maryland Univs. and I.I.T. Phys., U. S. Naval Res. Lab., 1941-. Societies: Fellow, A.P.S.; Washington Acad. of Sci.; Philosophical Soc. of Washington; American Astronautical Soc.
Nuclear interests: Nuclear physics; Plasma physics; Thermonuclear reactions.
Address: Code 5100, U.S. Naval Research Laboratory, Washington, D.C. 20390, U.S.A.

FAUVEL, René. Asst., Res. Dept., Mecaserto S.A.
Address: Mecaserto S.A., 126 boulevard d'Alsace-Lorraine, Le Perreux, (Seine), France.

FAVA, Giacomo, Dr. Chief Eng., Nuova San Giorgio S.p.A.
Address: Nuova San Giorgio S.p.A., 2 Via Luciano Manara, Genova-Sestri, Italy.

FAVIER, Andre. Member, Commission Consultative pour la Production d'Electricite d'Origine Nucleaire, C.E.A.
Address: Commissariat a l'Energie Atomique, 29-33 rue de la Federation, Paris 15, France.

FAVILLI, G. Member, Editorial Com., Minerva Nucleare.
Address: 83-85 Corso Bramante, Turin, Italy.

FAW, Richard E., B.S., Ph.D. Born 1936. Educ.: Cincinnati and Minnesota Univs. Prof. of Nucl. Eng. Kansas State Univ. Societies: A.N.S.; A.C.S.; A.S.E.E.
Nuclear interests: Research interests in radiation transport, radiation shielding, and radiation chemistry, in particular, the radiation chemistry of chain reactions.
Address: 1511 Leavenworth Avenue, Manhattan, Kansas, U.S.A.

FAWCETT, Sherwood L., B.Sc., M.S., Ph.D. Born 1919. Educ.: Ohio State Univ. and Case Inst. Technol. A.E.C. Predoctoral Fellowship, 1948-50; Res. Eng., Battelle Memorial Inst., (Columbus Labs.). 1950-51; Asst. Chief, 1951-54; Chief, Mathematical Phys. Div., 1954-; Asst. Manager, Dept. of Phys., 1957-60; Manager, Dept. of Phys., 1960-62; Manager, Dept. of Met. and Phys., 1963-65; Director, Pacific Northwest Labs., (1965-67), Exec. Vice Pres. (1967-), Battelle Memorial Inst. Societies: A.P.S.; A.N.S.; A.S.M.; Atomic Ind. Forum; A.A.A.S.
Nuclear interests: Management of nuclear reactor research in performance and reliability of fuel and structural components, evaluation of reactor concepts, engineering and design of reactor cooling and moderator systems, and nuclear physics.
Address: Battelle Memorial Institute, 505 King Avenue, Columbus, Ohio 43201, U.S.A.

FAWCETT, Sydney, M.A. Born 1914. Educ.: Cambridge Univ. Deputy Chief Eng., Permutit Co., 1946-56; Director, Ion Exchange (Canada) Ltd., 1955-56; Chief Eng. (1956), Asst. Director, Reactor Design (1958), Deputy Director, D. and E. Group (Design) (1959), Director, Reactor Group, Reactor Design (1960), Director, Fast Reactor (1964-65), U.K.A.E.A.; Director and Project Manager, Cammel Laird (Shipbuilding and Eng. Group), 1965-. Societies: A.M.I.C.E.; A.M.I.Mech.E.
Nuclear interest: Reactor design.
Address: Cheriton, Caldy Road, West Kirby, Wirral, Cheshire, England.

FAWKES, Rear Admiral E. E. Asst. Chief for Res., Development, Test and Evaluation, Bureau of Naval Weapons, Dept. of the Navy.
Address: Department of the Navy, Bureau of Naval Weapons, Washington 25, D.C., U.S.A.

FAWN, Geoffrey, B.A. Born 1920. Educ.: Oxford Univ. Gen. Manager, Rolls-Royce and Assocs., Ltd., 1959-. Society: A.F.R.Ae.S.
Nuclear interest: Nuclear propulsion for ships.
Address: Windrush, Edleston, Nantwich, Cheshire, England.

FAX, David Hirsch, B. Eng. Born 1919. Educ.: Johns Hopkins Univ. Asst. Prof. Mech. Eng., Johns Hopkins Univ., 1946-54; Advisory Eng. (1954-55), Asst. to Project Manager (Tech.), Pennsylvania Advanced Reactor Project (1955-58), Manager, Plant Development (1959), Manager, Advanced Development, Westinghouse Atomic Power Dept., Pittsburgh, 1960-61; Consultant, Westinghouse Astronucl. Lab., 1961-63; Tech. Asst. to Vice-Pres., Eng., (1963-67), Director, Eng. Consulting, Westinghouse Elec. Corp. Societies: A.S.M.E.; A.A.A.S.; A.I.A.A.
Nuclear interests: Reactor and plant design; Fluid mechanics; Heat transfer and thermodynamics; Technical management.
Address: Westinghouse Electric Corporation, P.O. Box 2278, Pittsburgh 30, Pennsylvania, U.S.A.

FAXER, Per O., M.Sc. Born 1933. Educ.: Uppsala Univ.
Address: Nuclear Power Department, Allmänna Svenska Elektriska Aktiebolaget, Västeras, Sweden.

FAYE, Jean, Membre, Comité des Mines, C.E.A. Deleg. to 2nd I.C.P.U.A.E., Geneva, Sept. 1958.
Address: Commissariat à l'Energie Atomique, 29-33 rue de la Federation, Paris 15, France.

FAYYAZUDDIN, M.Sc., D.I.C. (London), Ph.D. (London). Born 1931. Educ.: Panjab and London Univs. Sen. Sci. Officer, Atomic Energy Centre, Lahore, Pakistan, 1962-66; Res. Assoc., The Enrico Fermi Inst., Chicago Univ. 1966-68; Prof., Inst. of Phys., Islamabad Univ., Rawalpindi, Pakistan, 1968-.
Nuclear interest: Theoretical high energy nuclear physics (particle physics).
Address: Institute of Physics, Islamabad University, 77-E Satellite Town, Rawalpindi, Pakistan.

FAZIO, Michelangelo, Phys. degree. Born 1936. Educ.: Milan Univ. Fellowship at Milan Univ., 1960; Asst. in Exptl. Phys., 1961; Prof., 1962-; Researcher in Nucl. Phys., Milan Univ. Society: Stà. Lombarda di Fisica.
Nuclear interests: Nuclear reactions with fast neutrons; Now working on (n,d) reactions with 14 MeV neutrons; Study of nuclear reactions with charged particles (especially the (d, He^3) reactions).
Address: Istituto di Fisica, 16 Via Celoria, Milan, Italy.

FAZIO, Peter G. DE. See DE FAZIO, Peter G.

FEATHER, Norman, B.A., Ph.D. (Cantab.), B.Sc. (Lond.). Born 1904. Educ.: Cambridge Univ. Prof. of Natural Philosophy, Edinburgh Univ., 1945-. Pres., Roy. Soc. of Edinburgh.
Books: An Introduction to Nucl. Phys. (C.U.P., 1936); Lord Rutherford (Blackie, 1940); Nucl. Stability Rules (C.U.P., 1952). Societies: F.R.S., F.R.S.E.
Nuclear interests: Fission physics; Radioactivity and nuclear systematics.
Address: 9 Priestfield Road, Edinburgh 9, Scotland.

FEBER, K. de. See de FEBER, K.

FECHNER, Paul. Born 1921. Geschäftsführer, Oberingenieur.
Nuclear interest: Abschirmtecnologie.
Address: 345a Dorstener Str., 468 Wanne-Eickel, Germany.

FEDERENKO, N., Prof. Dr. Deleg.to Conference sur la Physique des Plasmas et la Recherche concernant la Fusion Nucléaire Controlée, Salzbourg, Sept. 1961.
Address: Physical Technical Institute, Academy of Sciences, Leningrad, U.S.S.R.

FEDERICI, Alfredo. Born 1908. Educ.: Trieste Univ. Manager for Italy, Insurance Co. of N. America. Member, Exec. Com., "Pool" Italiano per l'Assicurazione dei Rischi Atomici.
Address: 2 Via Verdi, Milan, Italy.

FEDERIGHI, Francis D., B.A. (Oberlin), M.A. (Harvard), Ph.D. (Harvard). Born 1931. Educ.: Oberlin Coll. and Harvard Univ. Assoc. Prof., N.Y. State Univ., Albany, N.Y. Societies: A.N.S.; Assoc. for Computing Machinery; Soc. for Ind. and Appl. Maths.
Nuclear interests: Reactor theory; Computational methods.
Address: 2109 Baker Avenue East, Schenectady, New York, U.S.A.

FEDORENKO, R. P. Paper: Co-author, Optimal Reactor Shutdown for Short-Term Work (letter to the Editor, Atomnaya Energiya, vol. 23, No. 2, 1967).
Address: Academy of Sciences of the U.S.S.R., 14 Leninsky Prospekt, Moscow V-71, U.S.S.R.

FEDOREYEV, Henrich, Dr. Born 1931. Oncologist working on cure of cancer of the blood vessels by deep X-ray.
Address: Leningrad Institute of Oncology, 2-ya Berezovaya Alleya 3-5, Leningrad, U.S.S.R.

FEDOROV, G. A. Papers: Co-author, Investigation of Fallouts of Cd^{109} in Some Places of the Soviet Union in 1964-65 (letter to the Editor, Atomnaya Energiya, vol.23, No. 3, 1967); co-author, Evaluation of Fallout Gamma-Rad. Attenuation by Snow Cover (letter to the Editor, ibid, vol. 25, No. 1, 1968); co-author, Investigation of Fallout Products at Suburb of Moscow in 1962-66 (ibid, vol. 25, No. 2, 1968).
Address: Academy of Sciences of the U.S.S.R., 14 Leninsky Prospekt, Moscow V-71, U.S.S.R.

FEDOROV, G. B. Papers: Co-author, Fe-Distribution in Microvolume of Zirconium Alloys (letter to the Editor, Atomnaya Energiya, vol. 8, No. 1, 1960); co-author, Thermo dynamic Properties of Gamma-Phase Uranium-Zirconium Alloys (ibid, vol. 21, No. 3, 1966).
Address: Academy of Sciences of the U.S.S.R., 14 Leninsky Prospekt, Moscow V-71, U.S.S.R.

FEDOROV, K. N. Sci. Asst., Inst. Oceanology, Moscow; Deleg., I.A.E.A. Sci. Conference on the Disposal of Radioactive Wastes, Monaco, Nov. 16-21, 1959.
Address: Institute of Oceanology, Moscow, U.S.S.R.

FEDOROV, M. I. Paper: Co-author, Destruction of Tubes by High-Frequency Pressure Oscillations Occuring during Heat-Transfer (letter to the Editor, Atomnaya Energiya, vol. 23, No. 2, 1967).
Address: Academy of Sciences of the U.S.S.R., 14 Leninsky Prospekt, Moscow V-71, U.S.S.R.

FEDOROV, V. A. Paper: Co-author, Changes of Fast Neutron Spectrum after Passing through Aluminium, Paraffin and Aqueous Medium (letter to the Editor, Atomnaya Energiya, vol. 19, No. 5, 1965).
Address: Academy of Sciences of the U.S.S.R., 14 Leninsky Prospekt, Moscow V-71, U.S.S.R.

FEDOROV, V. V. Paper: Co-author, On the Use of Ac-Be Neutron Sources in Trade Geophysics (letter to the Editor, Atomnaya Energiya, vol. 16, No. 3, 1964).
Address: Academy of Sciences of the U.S.S.R., 14 Leninsky Prospekt, Moscow V-71, U.S.S.R.

FEDOROVSKII, Yu. P. Paper; Co-author, Device for Continuous Registration of Low Levels of Salt Water Gamma-Radioactivity in Nucl. Geophys. (letter to the Editor, Atomnaya Energiya, vol. 23, No. 2, 1967).
Address: Academy of Sciences of the U.S.S.R., 14 Leninsky Prospekt, Moscow V-71, U.S.S.R.

FE' D'OSTIANI, Adrea, Comte. Member, Com. for the Problems of the Pacific Use of Nucl. Energy and for the Contacts with Euratom.
Address: Committee for the Problems of the Pacific Use of Nuclear Energy and for the Contacts with Euratom, 11 Piazza Venezia, Rome, Italy.

FEDRIGO, Gino. Director, Nucl. Dept., Carlo Gavazzi S.P.A.
Address: Carlo Gavazzi S.P.A., 9 Via G. Ciardi, Milan, Italy.

FEDULOV, M. V. Paper: Method for Calculation of Thermal Neutron Space-Energy Distribution in Heterogeneous Reactor (Atomnaya Energiya, vol. 22, No. 2, 1967).
Address: Academy of Sciences of the U.S.S.R., 14 Leninsky Prospekt, Moscow V-71, U.S.S.R.

FEDULOV, V. I. Paper: Co-author, Measurement of Pressure Distribution behind Strong Shock Wave (letter to the Editor, Atomnaya Energiya, vol. 19, No. 5, 1965).
Address: Academy of Sciences of the U.S.S.R., 14 Leninsky Prospekt, Moscow V-71, U.S.S.R.

FEDYUSHIN, B. K. Paper: Penetration of Neutron Capture Gamma-Rays through Spherical Homogeneous Shield (Atomnaya Energiya, vol. 21, No. 1, 1966).
Address: Academy of Sciences of the U.S.S.R., 14 Leninsky Prospekt, Moscow V-71, U.S.S.R.

FEFILOV, B. V. Paper: Co-author, On Nucl. Properties of Isotopes of Element 102 with Mass Numbers 255 and 256 (letter to the Editor, Atomnaya Energiya, vol. 22, No. 2, 1967).
Address: Academy of Sciences of the U.S.S.R., 14 Leninsky Prospekt, Moscow V-71, U.S.S.R.

FEHER, I. Deleg., I.A.E.A. Symposium on the Detection and use of Tritium in the Phys. and Biol. Sci., Vienna, 3-10 May, 1961.
Address: Central Research Institute for Physics, Kouholy Thege, Budapest 12, Hungary.

FEHRENTZ, Dieter, Dr. rer. nat., Dipl.-Phys. Born 1938. Educ.: Heidelberg and Saarbrücken Univs.
Nuclear interests: Biophysics; Health physics; Medical physics; Applied nuclear physics.
Address: 38 Wasserturmstrasse, 6900 Heidelberg-Eppelheim, Germany.

FEHRENTZ, E., Dipl-Phys. Wissenschaftliche Asst., Dept. of Clinical-Exptl. Radiol., Univ. Strahlenklinik (Czerny Krankenhaus).
Address: Universitäts Strahlenklinik (Czerny Krankenhaus), 3 Vosse Strasse, Heidelberg, Germany.

FEIGE, Yehuda, M.Sc., Ph.D. Born 1926. Educ.: Hebrew Univ., Jerusalem. Instructor, Dept. of Phys., Hebrew Univ., Jerusalem, 1951-54; Hospital Phys., Radium and Isotope Dept., Hadassah Univ. Hospital, Jerusalem, 1954-57; Head, Health Phys. Dept. Soreq Nucl. Res. Centre, Israel, A.E.C., 1957-65, 1967-. Sci. Consultant to the U.N. Sci. Com. on the effects of Atomic Rad., 1965-66; Res. Assoc., Radiol. Phys. Div., Argonne Nat. Lab., 1966-67. Societies: Israel Phys. Soc.; H.P.S.
Nuclear interests: Radiation dosimetry; Health physics; Radiation biology; Low level counting; Hospital physics; Environmental radiation.
Address: 36 Eisenberg Street, Rehovoth, Israel.

FEINBERG, Gerald, B.A., M.A., Ph.D. Born 1933. Educ.: Columbia Univ. Res. Assoc., Brookhaven Nat.Lab., 1957-59; Asst. Prof. (1959-61), Assoc. Prof. (1961-65), Prof. (1965-), Columbia Univ. Society: A.P.S.
Nuclear interest: Theory of elementary particles.
Address: Department of Physics, Columbia University, New York 27, N.Y., U.S.A.

FEINBERG, S. M., D.Sc., Acad. of Sci. of the U.S.S.R.; Deleg. to 2nd I.C.P.U.A.E., Geneva, Sept. 1958. Papers: Co-author, Burnup of Fuel in Water-moderated Water-cooled Power Reactors and Uranium-Water Lattice Experiments (2nd I.C.P.U.A.E., Geneva, Sept. 1958); co-author, The 50 MW Research Reactor SM (Atomnaya Energiya, vol. 8, No. 6, 1960).
Address: Nuclear Energy Institute, Academy of Sciences of the U.S.S.R., 14 Leninsky Prospekt, Moscow V-71, U.S.S.R.

FEINE, Ulrich, Dr. med. Doz. (Medizinische Strahlenkunde). Born 1925. Educ.: Tübingen Univ. Facharzt, Röntgenologie und Strahlenheilkunde. Röntgendiagnostisches Inst. und Radiotherapeutische Klinik, Zürich Univ., 1953-55; Max Planck Inst. für Biophysik, Frankfurt am Main, 1955-56; Oberarzt, Medizinischen Strahlen Inst., Tübingen Univ., 1956-. Societies: Assoc. Radiobiol. Pays Euratom; Gesellschaft Nuklearmedizin; Vereinigung Deutscher Strahlenschutzärzte; Deutsche

FEI

Röntgengesellschaft.
Nuclear interests: Nuclear medicine; Radiotherapy; Radiodiagnostics; Radiation protection.
Address: Medizinisches Strahleninstitut der Universität Tübingen, 74 Tübingen, Germany.

FEINER, Frank, A.B., M.S., Ph.D. Born 1928. Educ.: Princeton Univ. and Carnegie Inst. Technol. Manager, Exptl. Phys., (1958-64), Manager, Advanced Reactor Phys. (1964-), Advanced Development Activity, Knolls Atomic Power Lab.; Visiting Prof., Cornell Univ., 1967-68. Societies: A.P.S.; A.N.S.
Nuclear interests: Reactor physics; Nuclear physics; High-energy physics; Nuclear instrumentation; Reactor design.
Address: Knolls Atomic Power Laboratory, General Electric Company, Schenectady, New York, U.S.A.

FEINGOLD, Arnold M., A.B., A.M., Ph.D. Born 1920. Educ.: Brooklyn Coll. and Princeton Univ. Instructor (1950-53), Asst. Prof. (1953-55), Pennsylvania Univ.; Asst. Prof., Illinois Univ., 1955-57; Assoc. Prof., Utah Univ., 1957-60; Prof., State Univ. of N.Y. At Stony Brook, 1960-. Society: A.P.S.
Nuclear interests: Theoretical nuclear physics; Beta decay, angular correlations, tensor force effects, structure of light nuclei, origin and nature of the effective spin-orbit force.
Address: Physics Department, State University of New York, At Stony Brook, Stony Brook, New York, U.S.A.

FEINSTEIN, Robert N., Ph.D. (Physiological Chem.). Born 1915. Educ.: Wisconsin Univ. Assoc. Biochem. (1955-59), Sen. Biochem. (1959-), Argonne Nat. Lab.; Guggenheim Fellow, Inst. du Radium, Paris, 1959-60. Societies: Rad. Res. Soc.; A.C.S.; A.A.A.S.; American Soc. of Biol. Chem.
Nuclear interest: Radiation biochemistry.
Address: Argonne National Laboratory, Argonne, Illinois, U.S.A.

FEITKNECHT, Walter, Ph.D., D.I.C. Born 1899. Educ.: Berne Univ., Roy. School of Mines, and London Univ. Prof. Inorg. and Phys. Chem. and Head, Inst. of Inorg., Analytical and Phys. Chem., Berne Univ. Pres., Swiss Chem. Soc., 1960-62. Societies: Schweizerische Chemische Gesellschaft; Schweizensche Chemiker Verband; Deutsche Bunsengesellschaft; Gesellschaft Deutscher Chemiker.
Nuclear interests: Use of radioactive tracers in chemistry of solids, diffusion problems, solubility determination; Metallurgy, specially corrosion and corrosion products.
Address: Institut für anorganische, analytische und physikalische Chemie, 3 Freiestr., Berne, Switzerland.

FEITO, Jose SUAREZ-. See SUAREZ-FEITO, Jose.

FELD, Bernard Taub, B.S., Ph.D. Born 1919. Educ.: C.C.N.Y. and Columbia Univ. Prof. Phys., M.I.T., 1954-; Visiting Prof., Rome Univ., 1953-54; Visiting Sci., C.E.R.N., Geneva, 1960-61; John Simon Guggenheim Fellow, 1953-54, 1960-61. Book: The Neutron (Exptl. Nucl. Phys., Vol. II, E. Segre, Editor: New York, Wiley, 1953). Society: Fellow, A.P.S.
Nuclear interests: Neutron physics, nuclear theory, theory of elementary particles and their structure.
Address: Department of Physics, Massachusetts Institute of Technology, Cambridge 39, Massachusetts, U.S.A.

FELDEN, Marceau, D. ès Sc. Born 1932. Educ.: Nancy Univ. Prof., M.P.C., 1961-62; Prof. of Theoretical and Nucl. Phys., 1962-; Res. Contractor for Nucl. Res. Society: French Phys. Soc.
Nuclear interests: Plasma physics; Quantum mechanics; Weak interactions.
Address: 2 rue de la Citadelle, Nancy, France.

FELDER, Henri. Born 1927. Educ.: Paris Univ. Editor, Industries Atomiques. Societies: Assoc. Suisse pour l'Energie Atomique; A.T.E.N.
Nuclear interests: General.
Address: 34 chemin de Pont-Céard, Versoix-Geneva, Switzerland.

FELDES, John G. Health Phys. Supervisor, Enrico Fermi Atomic Power Plant.
Address: Enrico Fermi Atomic Power Plant, Post Office Box 725, Monroe, Michigan, U.S.A.

FELDMAN, Cyrus, B.S. and M.S. (Chem.). Born 1919. Educ.: Pennsylvania Univ. Supervisor, Spectrochem. and X-ray Analysis Labs., O.R.N.L., 1944-; Com. E-2 on Emission Spectroscopy, A.S.T.M. Societies: A.C.S.; Soc. for Applied Spectroscopy.
Nuclear interests: Spectrochemistry and X-ray analysis of materials used or produced in nuclear industries.
Address: Oak Ridge National Laboratory, Oak Ridge, Tennessee, U.S.A.

FELDMAN, David, B.S., M.S., Ph.D. Born 1921. Educ.: C.C.N.Y., New York and Harvard Univs. Asst. Prof. Phys., Rochester Univ., 1950-56; Assoc. Prof., Phys. (1956-59), Prof., Phys. (1959-), Brown Univ. Book: Editor, Proceedings of the Fifth Annual Eastern Theoretical Phys. Conference (New York, W.A. Benjamin, Inc., 1967). Societies: Fellow, A.P.S.; Italian Phys. Soc.
Nuclear interests: Quantum theory of fields; Theory of elementary particles.
Address: Brown University, Department of Physics, Providence, Rhode Island 02912, U.S.A.

FELDMAN, Henry R., A.B., A.M., Ph.D. Born 1932. Educ.: Harvard and Columbia Univs. Res. Asst., Dept. of Phys., Columbia Univ., 1958-63; Res. Asst. Prof., Dept. of Phys. (1963-65); Sen. Phys., Appl. Phys. Lab. (1966-)

Washington Univ. Society: A.P.S.
Nuclear interest: Nuclear physics.
Address: Applied Physics Laboratory, Washington University, 1013 N.E. 40 Street, Seattle, Washington 98105, U.S.A.

FELDMAN, Isaac, B.S., Ph.D. Born 1918. Educ.: George Washington and Illinois Univs. Instructor in Phys. Chem. (Pharmacology) (1947-51), Asst. Prof. (1951-56), Assoc. Prof. of Rad. Biol. (Phys. Chem.) (1956-65), Prof. Rad. Biol. and Biophys. (1965-), Rochester Univ. School of Medicine; Jun. Sci. (1949-51), Sci. (1951-), Rochester Univ. Atomic Energy Project. Special U.S.P.H.S. Res. Fellow, Univ. Coll. London, 1962-63. Societies: A.C.S.; Chem. Soc., London; Rad. Res. Soc.; Biophys. Soc.
Nuclear interests: Co-ordination chemistry; Chemistry of toxic metals, including molecular structure and chemical reactions of physiological interest; Luminescence of macromolecules; Nuclear magnetic resonance.
Address: P.O. Box 287, Station 3, Rochester 20, New York, U.S.A.

FELDMAN, L. I. Papers: Co-author, The Model of a Radiative Indium-Gallium Loop of the Tbilisi IRT-2000 Reactor (Atomnaya Energiya, vol. 13, No. 4, 1962); co-author, Indium-Helium Rad. Circuit for Pile-Type Reactor (Abstract, ibid, vol. 19, No. 2, 1965).
Address: Academy of Sciences of the U.S.S.R., 14 Leninsky Prospekt, Moscow V-71, U.S.S.R.

FELDMAN, Lawrence, B.A. (Brooklyn Coll.) M.S. (North Carolina, Ph.D. (Columbia). Born 1922. Educ.: North Carolina and Columbia Univs. Res. Assoc., Columbia Univ., 1952-. Society: A.P.S.
Nuclear interests: C.W. cyclotron; Low energy nuclear physics; Beta-ray spectroscopy.
Address: Department of Physics, Columbia University, New York City 27, New York, U.S.A.

FELDMAN, Melvin J., B.S. (Metal. Eng.), M.S. (Metal.). Born 1926. Educ.: Purdue and Tennessee Univs. O.R.N.L., 1950-56; Westinghouse Atomic Power Div., 1956-60; Operations Manager of the Fuel Cycle Facility, Argonne Nat. Lab., 1960-. Nat. Programme Com. and past Chairman, Remote Systems and Technol. Div., A.N.S. Society: American Metals Soc.
Nuclear interests: Radiation damage studies on metals and recently, studies and operations on the feasibility, efficiency, and methods for the remote refining fabrications of reactor fuel.
Address: Argonne National Laboratory, P.O. Box 2528, Idaho Falls, Idaho, U.S.A.

FELDMAN, Michael, Prof., M.Sc., Ph.D. Born 1926. Educ.:Hebrew Univ., Jerusalem. Head, Dept. of Cell Biol., Weizmann Inst. of Sci. Societies: Member, Council, Internat. Soc. for Cell Biol. (Internat. Cell Res. Organisation U.N.E.S.C.O.); Fellow, New York Acad. of Sci.; Internat. Inst. of Embryology.
Nuclear interest: Radiobiology.
Address: Weizmann Institute of Science, Rehovoth, Israel.

FELDMAN, Myer, B.S. (Economics), LL.B. Born 1917. Educ.: Pennsylvania Univ. Special Counsel, Securities and Exchange Com., 1950-57; Legislative Asst. to Senator John F. Kennedy, 1957-61; Deputy Special Counsel to the Pres., 1961-64; Special Counsel to the Pres., 1964-65; Counsel to C.E.A. and to other foreign governments in matters dealing with Nucl. Energy. Society: American Law Inst.
Address: 1700 Pennsylvania Avenue, Washington, D.C. 20006, U.S.A.

FELDMEIER, Joseph Robert, B.S. (Phys.), M.S. and Ph.D. (Nucl. Phys.). Born 1916. Educ.: Carnegie Tech. and Notre Dame Univ. Prof. and Chairman, Phys. Dept., St. Thomas Coll. (St. Paul, Minn.), 1948-52; Advisory Sci., Phys. Dept. Manager (1952-56), Asst. Manager, Central Phys. and Math. Dept. (1956-58), Asst. Manager, Div. Phys. (1958-60), Westinghouse Bettis Atomic Power Div.; Assoc. Director of Res., Philco Corp., 1960-62; Director, Philco Sci. Lab., 1962-64; Vice Pres. and Director, Franklin Inst. Res. Labs., 1964-. Book: Microwave Magnetrons (contributing author; M.I.T. Rad. Lab. Technical Series, 1948). Societies: A.P.S.; A.A.A.S.; I.R.E.; A.N.S.
Nuclear interests: Nuclear physics and reactor design.
Address: Midway Lane, Blue Bell, Philadelphia, Pennsylvania, U.S.A.

FELDT, Werner, Diplom Physiker. Born 1926. Educ.: Hamburg Univ. Inst. of Appl. Physics. In Dept. of Nucl. Instruments, Leybold, Köln, 1957; Federal Res. Board of Fisheries, Inst. of Fish Processing, Isotope Lab., Hamburg, 1959. Society: Phys. Soc. of Germany.
Nuclear interests: Radioactive contamination of marine organisms; Disposal of radioactive wastes; Nuclear radiation detection.
Address: Isotope Laboratory, Federal Research Board of Fisheries, Hamburg, Germany.

FELICE, Anthony John, B.A.Sc. (Eng. Physics). Born 1922. Educ.: Toronto Univ. Special Assignments, Res. and Ind. Products Branch, Sales Div., Commercial Products, A.E.C.L., 1958-. Society: A.P.E.O.
Nuclear interest: Radioisotopes.
Address: 2094 Black Friars Road, Ottawa 13, Ontario, Canada.

FELICE, L. DI. See DI FELICE, L.

FELICI, Noël Joseph, D. ès Sc. Born 1916. Educ.: École Normale Supérieure, Paris. Prof. Grenoble Univ. (Electrostatics); Consultant to C.E.A. Books: Elektrostatische Hochspannungsgeneratoren (Karlsruhe, Braun, 1957); Les accelerateurs de particules (Paris, Dunod, 1959). Societies: Sté. Francaise des Electriciens; Sté. Française des Radioélectriciens.

Nuclear interests: Electrostatic low-energy accelerators; Electrostatic high current generators; Electron irradiation; Activation analysis.
Address: 13 rue Charles Péguy, Grenoble, Isère, France.

FELIPE, Jose CABRERA. See CABRERA FELIPE, Jose.

FELIX, Fremont, L. ès Sc., and Elec. Eng. Degrees. Born 1903. Educ.: Paris Univ. Rep., Field Operations, Paris, for Internat. Gen. Elec. Co., U.S.A., 1953-59; Eng.-Consultant, Power and Industry, Paris, for Bechtel Internat. Corp., San Francisco, 1959-60; European Rep., Paris, for Stone and Webster Eng. Corp., Boston, 1960-63; Planning and Development Consultant, Gibbs and Hill, Inc., New York, 1964-. Member, U.S. Council, Internat. Chamber of Commerce's Commissions on European Affairs and on Asian and Far-Eastern Affairs. Societies: Life Member, I.E.E.E.; Atomic Ind. Forum, Inc.
Nuclear interests: Development and progress of nuclear power as a fully competitive source of electrical energy in Europe, Africa, the Middle East, Asia, Caribbean and South America, within the frame of optimum development of energy resources towards meeting forecasted growth of demand.
Address: 21 boulevard Lannes, Paris 16, France.

FELKER, V. M. Paper: Co-author, On Activation Detector of Thermal Neutrons (letter to the Editor, Atomnaya Energiya, vol. 20, No. 3, 1966).
Address: Academy of Sciences of the U.S.S.R., 14 Leninsky Prospekt, Moscow V-71, U.S.S.R.

FELL, Jesse, M.A. (1st class Hons., Phys. and Maths.). Born 1923. Educ.: Oxford Univ. With Sir W. G. Armstrong-Whitworth Aircraft, Ltd., Coventry, 1952-59; U.K.A.E.A., Atomic Energy Establishment, Winfrith, 1959-.
Nuclear interests: Reactor assessment and design, with particular reference to core performance problems; Numerical methods for digital computers.
Address: 1 Mellstock Avenue, Dorchester, Dorset, England.

FELL, L. R. Managing Director, MegaVolt Ltd.
Address: MegaVolt Ltd., Ilminster, Somerset, England.

FELLER, Robert R. Mech. Eng. Born 1923. Educ.: New York Univ. Formerly Project Eng., Nucl. Equipment Sect., A.M.F. Atomics, Greenwich, Connecticut; now Project Eng., Chem. Construction Corp., New York City, Subsidiary of Ebasco. Society: A.N.S.
Nuclear interests: Reactor component and refuelling equipment engineering and design.
Address: 16-32 202nd Street, Bayside, New York, U.S.A.

FELLINGER, Karl, Dr. med. (Vienna), Dr. med. h.c. (Thessaloniki). Born 1904. Educ.: Vienna Univ. Director, 2nd Med. Univ. Clinic of Vienna; Full Prof. (internal medicine), Vienna Univ. Pres., Austrian Supreme Council of Health; Co-editor, Nucl. Medicine; Pres., American Medical Soc. of Vienna; Pres., Vienna Ges. Int. Med. Books: Co-author, Radioaktive Isotope in Klinik und Forschung (3 vols., 1954, 1956, 1958).
Nuclear interest: Medical application is isotopes (particularly internal medicine).
Address: Allgemeines Krankenhaus, 13 Garnisongasse, Vienna 9, Austria.

FELLOWS, John Albert, A.B. (Phys., Williams Coll.), M.S. (Phys., M.I.T.), Sc.D. (Metal., M.I.T.). Born 1906. Res. Metal., American Brake Shoe Co., 1948-53; Manager, Metal. Development Dept., Mallinckrodt, 1953-58; Asst. Tech. Director, Mallinckrodt Chem. Works, 1959-. Member various committees of A.S.M., A.I.M.M.E. and A.S.T.M.; Vice-Pres. (1963-64), Pres. (1964-65), A.S.M. Book: Co-author: Uranium Production Technol. (Van Nostrand, 1959). Societies: A.I.M.E.; A.S.M.; Inst. of Metals (London); A.S.T.M.; A.N.S.
Nuclear interests: Metallurgy of uranium; bomb reduction, alloy additions, melting, extrusion, forging, rolling, heat treatment, mechanical and physical properties, grain size and orientation. Production and properties of UO_2: denitration, reduction of UO_3, conpaction and sintering, particle size, surface area, density.
Address: 5139 Westminster Place, St. Louis, Missouri 63108, U.S.A.

FELLOWS, Walter Scott, Jr., B.M.E., M.S.E. Born 1918. Educ.: Ohio State and Purdue Univs. Colonel (rank on retirement), U.S.A.F., 1941-65; Chief, Guided Missile Div., Air R. and D. Command, Dept. of the Air Force, 1951-53; O.R.S.O.R.T., 1953-54; Oak Ridge Operations Office, Air Force Special Projects Office, 1954-55; Chief, Direct Cycle Nucl. Propulsion Project, Reactor Development Div., U.S.A.E.C., 1955-58; Asst. Manager, Tech. Programmes, Lockland Aircraft Reactors Operations Office (1858-61), Chief, Nucl. Vehicle Projects Office (1961-64), Chief, Cost Reduction and Value Eng. Office (1964-65), Director, Operations Management Office (1965-), N.A.S.A. Marshal Space Flight Centre. Society: Assoc. Fellow, A.I.A.A.
Address: 8803 Bridlewood Drive, S.E., Huntsville, Alabama, U.S.A.

FELSENTHAL, Peter, B.S. Born 1932. Educ.: M.I.T. Phys., Arthur D. Little, Inc. Societies: A.I.P.; A.G.U.
Nuclear interest: Atmospheric and geophysical effects of bombs.
Address: Arthur D. Little Incorporated, 20 Acorn Park, Cambridge, Massachusetts, U.S.A.

FELT, Norris Elliott, Jr., B.S.E.E., M.S.E.E. Born 1924. Educ.: Maryland Univ. and Georgia Inst. Technol. Electromech. Eng. (1951-54),

Sales Manager, Ballistic Rocket Projects (1954-55), Operations Manager, Project Vanguard (1955-58), Business Manager, Space Flight Div. (1958-59), Gen. Manager, Corporate Advanced Programmes (1959-62), Gen. Manager, Programme Review (1962-63), Gen. Manager, Nucl. Div. (1963-), Vice Pres., Nucl. Div. (1968-), Martin Marietta Corp. Societies: A.I.A.A.; Atomic Ind. Forum.
Nuclear interests: Isotopic powered generators for space, terrestrial, and underwater uses; Small power reactors and offshore oil field equipment.
Address: 15-2B Warren Lodge Court, Cockeysville, Maryland, 21030, U.S.A.

FENDLER, Heinz Gerhard, Dr. rer. nat. Born 1921. Educ.: Marburg/Lahn Univ. and Hanover T.H. Phys., Labor Prof. Berthold, Wildbad, 1951-53; Sci. res., Hanover T.H., with Prof. Stuart, 1953-56; Phys., Vereinigte Glanzstoffwerke, Obernburg, 1956-57; Group leader with A.E.G., Abt. KEA, Frankfurt am Main 1957-65; T.U.V., Baden, 1966-. Society: German Phys. Soc.
Nuclear interests: Health physics; Shielding of reactors and accelerators; Nuclear physics; Management; Nuclear instrumentation; Radiation effects on materials.
Address: 25 Liebersbacherstr., 6941 Reisen, Germany.

FENECH, Henri, Ing. Arts et Métiers, S.M., Sc.D. (Nucl. Eng. (M.I.T.)). Born 1925. Educ.: Ecole Nat. des Ingénieurs Arts et Métiers and M.I.T. Head of Steam Div., Sté. Foster-Wheeler Français (Paris), 1952-55; Res. Staff Member, Reactor Phys. Sect., Gen. Atomic, San Diego, 1959-60; Asst. Prof. (1960-63), Assoc. Prof. (1963-), Dept. Nucl. Eng., M.I.T. Societies: A.N.S.; A.S.M.E.
Nuclear interests: Nuclear reactor physics (theoretical analysis and computational methds); Nuclear reactor design (fuel management, plant optimisation, burnable poisons); Heat transfer (conductance of surfaces in contact; liquid metal gas cooling).
Address: Room 24-204, Massachusetts Institute of Technology, Cambridge 39, Massachusetts, U.S.A.

FENEVES, Edvin. Deputy-Director, Joint Inst. for Nucl. Res., Dubna, 1964-.
Address: Joint Institute for Nuclear Research, Head Post Office, P.O. Box 79, Moscow, U.S.S.R.

FENG, Paul Y., Ph.D., B.S. Born 1926. Educ.: Washington Univ. and Catholic Univ. Chem. (1954), Chief Chem. (1954), Manu-Mine R. and D. Co.; Assoc. Phys. (1955), Res. Phys. (1956), Group Leader (1957), Asst. Supervisor (1958), Supervisor, Chem. Phys. (1959), Armour Res. Foundation; Assoc. Prof., Marquette Univ., 1967; Tech. Adviser, U.S. Deleg. 2nd I.C.P.U.A.E., Geneva, Sept. 1958; Visiting Prof., Tsing-Hua Inst. Nucl. Sci., Formosa, 1958; Fulbright Lecturer, Taiwan, 1965.

Societies: A.C.S.; Rad. Res. Soc.; Faraday Soc.; A.A.A.S.
Nuclear interests: Radiation chemistry of organic systems, particularly halogen compounds; Hot atom chemistry; Nuclear chemistry.
Address: 88 Robsart Road, Kenilworth, Illinois 60043, U.S.A.

FENNING, Frederick William, B.A. Born 1919. Educ.: Cambridge Univ. Group Leader (Thermal Reactors), Reactor Div. Harwell (1955), Chief Phys., R. and D. Branch, Risley (1958), Director, Reactor Technol. Branch, Risley (1960-66), Asst. Director, A.E.R.E. Harwell (1966-), U.K.A.E.A. Society: A.N.S.
Nuclear interest: Nuclear technology.
Address: 21 St. Peter's Hill, Caversham, Reading, Berks., England.

FENSKE, Merrill R., Dr. Head, Dept. Chem. Eng. and Prof. Chem. Eng. and Director, Petroleum Refining Lab., Pennsylvania State Univ.
Address: Pennsylvania State University, College of Chemistry and Physics, University Park, Pennsylvania, U.S.A.

FENTON, Dr. At Pacific Inst. Advanced Studies.
Address: Pacific Institute of Advanced Studies, 448 N. Avenue 56, Los Angeles 42, California, U.S.A.

FENTON, Arthur Geoffrey, B.Sc. (Tasmania), Ph.D. (Birmingham). Born 1920. Educ.: Tasmania Univ. Member, Radiol. Advisory Com., Dept. of Public Health, Tasmania; Member, Australian Nat. Com. for Pure and Appl. Phys.; Convenor, Australian Cosmic Ray Subcom. for the International Geophysical Year, 1957-58; Member, Phys. Advisory Com., Australian Nat. Antarctic Res. Expedition; Member, Planning Com., Internat. Years of the Quiet Sun 1964-67; Member, Council of Australian Inst. of Nucl. Sci. and Eng. 1964-65. Societies: Fellow, Inst. of Phys.; Phys. Soc.; Fellow, Australian Inst. Phys.
Nuclear interests: Nuclear physics; Cosmic-ray observations; Nuclear particle detection and measurement; Galactic X-ray astronomy.
Address: Physics Department, University of Tasmania, Box 252C, G.P.O., Hobart, Tasmania, Australia.

FENTON, Keith Brian, B.Sc. (Tasmania), Ph.D. (Tasmania). Born 1925. Educ.: Tasmania Univ. Member, Australian Nat. Antarctic Res. Expedition, 1950-52; Post-doctoral Fellow, N.R.C. of Canada, 1953-55; Res. Assoc., Chicago Univ., 1955-59; Sen. Lecturer in Phys., Tasmania Univ. Society: A.P.S.
Nuclear interests: Cosmic ray observations, nuclear particle detectors.
Address: Tasmania University, G.P.O. Box 252 C, Hobart, Tasmania.

FERD, Gunvald, Civ. ing. Born 1918. Chief and Sen. Eng., now Project Manager, Reactor Development Div., Institutt for Atomenergi.

Nuclear interests: Management; Reactor design.
Address: Institutt for Atomenergi, P.O. Box 40, Kjeller, Norway.

FERENCE, Michael, Jun., B.S., M.S., Ph.D. Born 1911. Educ.: Chicago Univ. Tech. Director, Signal Corps. Eng. Labs., 1950-59; Exec. Director, Sci. Lab., Ford Motor Co., 1959-. Member of Board of Directors, Atomic Power Development Assoc.; Chairman, Com. on Atmospheric Sci., N.A.S.; Chairman, Sci.-Eng. Activity Com., Society of Automotive Engs. Book: Analytical and Exptl. Phys. (Univ. of Chicago Press, 1943). Societies: A.P.S.; Inst. Radio Eng.; Soc. Automotive Eng.; American Rocket Soc.
Nuclear interests: Management of nuclear activities, including the metallurgy of fuel element design and nuclear physics.
Address: Scientific Laboratory, Ford Motor Company, P.O. Box 2053, Dearborn, Michigan, U.S.A.

FERETIC, Danilo, Diploma in Elec. Eng., M.Sc. (Reactor Phys. and Technol.), Ph.D. (Reactor Heat Transfer). Born 1930. Educ.: Zagreb, Birmingham and Belgrade Univs. Head, Nucl. Power Group, Elec. Undertaking of Croatia, Zagreb, Jugoslavia.
Nuclear interest: Nuclear power plant evaluations, thermodynamic optimisations of reactor design and heat transfer transient analysis of nuclear reactors.
Address: Udruzena elektroprivreda SRH, 37 Proleterskih brigada, Zagreb, Yugoslavia.

FERGUSON, David. With United States Steel Corp. Formerly Vice-Chairman, now Sec., Illinois Commission on Atomic Energy.
Address: 208 South LaSalle Street, Chicago, Illinois 60690, U.S.A.

FERGUSON, Don Ernest, B.S., M.A. Born 1923. Educ.: Tennessee Polytech. Inst. and Tennessee Univ. Sect. Chief (1946-65), Director (1965-), Chemical Technol. Div., Oak Ridge Nat. Lab. Adviser, Third I.C.P.U.A.E., Geneva, Sept. 1964. Member, Exec. Com. (1963-64), now Vice-Chairman, Oak Ridge Sect., A.N.S. Society: A.N.S.
Address: 198 Northwestern Avenue, Oak Ridge, Tennessee, U.S.A.

FERGUSON, Edward H. Gen. Manager, Rad. Instr. Development Lab., 1962-. Former Gen. Manager, now Vice-Pres., Manufacturing (1965-), Nucl. Data, Inc.
Address: Nuclear Data, Inc., 100 W. Golf Road, Palatine, Illinois, U.S.A.

FERGUSON, Edwin Earle, B.S. (Education, Drake), LL.B. (Drake), J.S.D. (Yale). Born 1910. Educ.: Drake and Yale Univs. Chief Counsel, U.S. Bureau of Indian Affairs, 1950-52; Attorney, (1952-55), Deputy Gen. Counsel (1955-62), Assoc. Gen. Counsel (1962-), U.S.A.E.C. Books: Contributor to Legal and Administrative Problems of Protection in the Peaceful Uses of Atomic Energy (Euratom, 1960); contributor to Atomic Energy and Law (Inter-American Symposium (1960), Univ. Puerto Rico, School of Law); Liability of Nuclear Powered Vessels (Atomic Energy Law Journal, 1960). Societies: Federal Bar Assoc.; American Law Inst.
Nuclear interests: All legal aspects of the formulation and management of governmental programmes for the development, utilisation, and control of nuclear energy. Special interests: Implementing programmes through contractual devices; Development of radiation protection regulations; Third party liability for nuclear hazards.
Address: 5821 Osceola Road, Washington 16, D.C., U.S.A.

FERGUSON, James Oliver, B.A., C.A. Born 1912. Educ.: Toronto Univ. Manager, Accounting, A.E.C.L.
Nuclear interest: Financial management.
Address: 2055 Carling Avenue, Apt. 510A, Ottawa 13, Ontario, Canada.

FERGUSON, Kenneth. Assoc. Director, Remote Control Eng. Div., Argonne Nat. Lab. Ex-Officio Member, Professional Div. Com., Member, Publications Com., Ex-Officio Member, Remote Systems Technol. Div., A.N.S.
Address: Argonne National Laboratory 9700 South Cass Avenue, Argonne, Illinois 60440, U.S.A.

FERGUSON, Norman Napier, M.A. Born 1910. Educ.: Edinburgh Univ. Asst. Director, Personnel Control Commission for Germany (British Element), 1950; Principal, Ministry of Supply, 1951; Chief Administrative Officer, A.E.R.E., Harwell, 1956.
Address: 9 South Drive, Atomic Energy Research Establishment, Harwell, near Didcot, Berks., England.

FERGUSON, Robert L., B.S. (E.E.). Born 1924. Educ.: Univ. Illinois Inst. Technol. Formerly Projects Eng., now Asst. Manager, Projects and Maintenance (1967-), Chicago Operations Office, U.S.A.E.C. Society: A.N.S. Nuclear interests: Reactor design, nuclear instrumentation.
Address: Chicago Operations Office, U.S.A.E.C., 9800 South Cass Avenue, Argonne, Illinois 60439, U.S.A.

FERGUSON, Whitworth, B.Sc., Dr. Sci. Formerly Vice-Chairman, New York State Atomic R. and D. Authority, now Vice-Chairman, New York State Atomic and Space Development Authority. Member, Com. on Commercial Uses of Atomic Energy, and Member, Com. on Sci. and Technol., Chamber of Commerce of the U.S.
Address: Ferguson Electric Construction Co., Inc., 333 Ellicott Street, Buffalo, New York 14203, U.S.A.

FERGUSSON, Gordon John, M.Sc. (1st class Hons.). Born 1922. Educ.: Victoria Univ., Wellington, New Zealand. With D.S.I.R. New Zealand, 1949-59. At present at California Univ. Societies: A.M.I.E.E. (London); N.Z. Assoc. Scientists; Fellow, Roy. Soc. New Zealand.
Nuclear interests: Radiocarbon dating; Nuclear geology; Nuclear instrumentation.
Address: Institute of Geophysics, California University, Los Angeles 24, California, U.S.A.

FERNANDES, Abilio Augusto, Eng. Attached to Empresa Termoeléctrica Portuguesa; Sec., Forum Atomico Portugues. Adviser, Third I.C.P.U.A.E., Geneva, Sept. 1964; Member, Internat. Union of Producers and Distributors of Elec. Energy; Member, Exec. Com., Foratom.
Address: Forum Atomico Portugues, 16 Trav Abarracamento de Peniche, Lisbon 2, Portugal.

FERNANDES de CARVALHO, José Alberto, Ph.D. (Cambridge, England), D. em Matemáticas (Coimbra, Portugal). Born 1933. Educ.: Coimbra Univ. Lecturer, Coimbra Univ., 1957-.
Nuclear interest: Atomic structure.
Address: 12-A 1º Av. D. Afonso Henriques, Coimbra, Portugal.

FERNANDEZ, Alonso, B.Sc. (Electronics and Communication Eng.), Ph.D. (Elec. Eng.). Born 1927. Educ.: Escuela Superior de Ingenieria Mecánica y Eléctrica, México, and Manchester Univ. Res. Worker, Inst. of Phys., Mexico Univ., 1952-; Director, Programme on Instrumentation, Nat. Commission of Nucl. Energy, Mexico, 1958-. Societies: Sociedad Mexicana de Fisica; Inst. of Phys. and Phys. Soc., London; A.P.S.; Academia de la Investigación Cientifica, México.
Nuclear interests: Nuclear instrumentation; Scintillation and solid state detectors; Crystal growth; Radiation damage and electrical properties of crystals.
Address: No. 1229 Providencia, México 12, D.F., Mexico.

FERNANDEZ, Antonio, Dr.-Ing. Agrónomo; Técnico Bromatólogo. Born 1935. Educ.: Madrid Univ. Ing. Agregado, Sección de Energia Nucl., Nat. Inst. of Agricultural Res., Madrid; Vocal, Comisión Asesora de Conservación de Alimentos por Irradiación.
Nuclear interests: Food irradiation; Radiation microbiology.
Address: 12 Calle Majada Alta, Madrid 20, Spain.

FERNANDEZ, Carlos GRAEF-. See GRAEF-FERNANDEZ, Carlos.

FERNANDEZ, Francisco. Vice Pres., Sociedad Espanola de Electro-Radiologia y Medicina Nucl.
Address: c/o Colegio de Medicos, 11 Calle de Villanueva, Madrid, Spain.

FERNANDEZ, J. M. CARBALLO. See CARBALLO FERNANDEZ, J. M.

FERNANDEZ, Miguel SERRANO. See SERRANO FERNANDEZ, Miguel.

FERNANDEZ CELLINI, Ricardo, Ing. de Armas Navales, Licenciado en Ciencias Quimicas. Born 1919. Educ.: Escuela de Ingenieros de Armas Navales and Barcelona and Madrid Univs. Prof., Escuela Técnica Superior de Ingenieros de Armas Navales. Director, Centro de Energia Nucl. Juan Vigon; Director de Quimica e Isotopos. Comisión Nacional de Investigación del Espacio. Society: Real Sociedad Española de Fisica y Quimica.
Nuclear interests: Quimica analitica; Quimica nuclear; Beneficio de minerales y radioquimica.
Address: Junta de Energia Nuclear, Dirección de Quimica e Isótopos Av. Computense, Madrid 3, Spain.

FERNANDEZ-MORAN V., Humberto, M.D., (Munich), M.D. (Univ. Central de Venezuela, Caracas), Ph.D. (Stockholm). Born 1924. Fellow in Neurology, George Washington Univ.; Foreign Asst., Neurosurgery Clinic Serafimerlasaretet Stockholm; Res. Fellow, Nobel Inst. of Phys., Stockholm; Res. Fellow, Inst. Cell Res., Karolinska Inst.; Asst. Prof., Cell Res., Karolinska Inst.; Prof. Biophys., Univ. Central de Venezuela, Director, Venezuelan Inst. for Neurology and Brain Res., 1954; Minister of Education, Venezuela, 1958; Head, Venezuelan Com. to Atomic Energy Conference (Geneva, 1955); Head, Venezuelan Com. to 1st Inter-American Symposium on Nucl. Energy (Brookhaven, 1957). Societies: Acad. of Phys. and Mathematical Sci. (Venezuela); Hon. Corresponding Member, American Soc. of Neurology; Electron Microscope Soc. of America; Soc. of Surgery, Buenos Aires; Soc. of Neurology, Peru; Soc. of Neurology, Santiago de Chile.
Nuclear interests: Radiation biology, studies of effects of ionising radiations in hydrated biological systems by low-temperature electron microscopy, electron diffraction, and neutron diffraction.
Address: Department of Biophysics, Chicago University, 5640 South Ellis Avenue, Chicago, Illinois 60637, U.S.A.

FERNANDO, E. C. Member, Atomic Energy Com., Nat. Planning Council of Ceylon.
Address: Atomic Energy Committee, National Planning Council of Ceylon, 5 Galle Buck Road, Colombo 1, Ceylon.

FERNANDO, Habaraduwa Kudabokola Teslon, L.M.S. (Ceylon), D.M.R.D. (Eng.), D.M.R.T. (Eng.), F.F.R. (Ireland). Born 1917. Educ.: Ceylon Univ. and Postgraduate Medical School, London. Radiotherapist-in-Charge, General Hospital, Colombo, 1956; Radiotherapist-in-Charge, Government Cancer Inst., Maharagama, 1958; Consultant Radiotherapist, Colombo Group of Hospitals, Lecturer in Radiotherapy,

Faculty of Medicine, Ceylon Univ. Com. Management, Ceylon School Radiography; Advisory Com. Radiotherapy and Cancer Services. Societies: British Inst. Radiol.; Faculty of Radiologists, England.
Nuclear interests: Application of ionizing radiations in medicine; Radiation physics; Public health aspects of nuclear energy.
Address: 28 Anderson Road, Colombo 5, Ceylon.

FERNANDO, L. J. D. Government mineralogist, Ceylon; Rep., Commonwealth nucl. scientists conference;Member, Atomic Energy Com., National Planning Council of Ceylon.
Address: Hunupitiya Lake Road, Colombo 2, Ceylon.

FERNANDO, Panagodage Chandra Bertram, B.Sc., Ph.D. Born 1930. Educ.: Ceylon and Cambridge Univs. Lecturer in Phys., Ceylon Univ., 1958-67; Prof. of Phys., Vidyodaya Univ. of Ceylon, 1967-; Res. Assoc., California Univ., Berkeley, 1965-66.
Nuclear interests: Atomic beam resonance radiofrequency spectroscopy; High-energy cosmic-rays.
Address: Department of Physics, Vidyodaya University of Ceylon, Nugegoda, Ceylon.

FERNBACH, Howard N. Group Leader, Decontamination Compound Development (-1968), Manager, Nucl. Div. (1967-), Turco Products, Inc.
Address: Turco Products Incorporated, Wilmington, California 90746, U.S.A.

FERNBACH, Sydney, Computation Div., Lawrence Rad.Lab.; Member, Mathematics and Computer Sci. Res. Advisory Com., U.S.A.E.C. Deleg. to 2nd I.C.P.U.A.E., Geneva, Sept. 1958.
Address: California University, Lawrence Radiation Laboratory, Livermore, California, U.S.A.

FERNET, Paul, Ing. des Arts et Manufacture. Born 1927. Educ.: Ecole Centrale des Arts et Manufacture, Paris; Génie Atomique, Saclay; I.S.N.S.E., Argonne Nat. Lab. Compagnie Française des Métaux, 1951-56; Tréfileries et Laminoirs du Havre, 1956-59; Eng., Res. and Training Div., Euratom, 1959-.
Nuclear interests: Metallurgy and ceramics; Fuel and cladding materials.
Address: Euratom, 51 rue Belliard, Brussels, Belgium.

FERRADINI, Christiane, Dr. Born 1924. Educ.: Toulouse and Paris Univs. Attachée de recherches (1948-56), Chargée de recherches (1956-58), Chef de travaux en phys. nucl. (1958-62), Maitre-Assistante en radiochimie (1962-), Faculté des Sciences, Paris Univ.
Nuclear interests: Radiation chemistry and hot atoms chemistry.
Address: Université de Paris, Laboratoire Curie, 11 rue P. Curie, Paris 5, France.

FERRAND, Pierre. Directeur de la Div. du Gros Matériel, Sté. Gén. de Constructions Electriques et Mécaniques "Alsthom".
Address: Société Générale de Constructions Electriques et Mécaniques "Alsthom", 38 avenue Kléber, Paris 16, France.

FERRARI, Alberto, LL.D., M.A., Prof. of Economic and Financial Politics. Born 1914. Educ.: Pisa and Yale Univs. Gen. Manager, Banca Nazionale del Lavoro. Member, Board, F.I.E.N.
Address: 119 Via Vittorio Veneto, Rome, Italy.

FERRARI, Harry M., B.S., M.S., Ph.D. Born 1932. Educ.: Wayne State and Michigan Univs. Manager, Irradiation Design and Testing, Atomic Power Div. (1958-60, 1962-66), Advisory Eng., Nucl. Fuel Div. (1966-), Westinghouse Elec. Co.; Tech. Adviser to Fiat, Turin, for Joint A.E.C.-Euratom R. and D. Contracts, 1960-62. Societies: A.N.S.; A.S.M.; A.I.M.E.; Inst. of Metals (British).
Nuclear interest: Fuel element technology - design, materials, fabrication, and performance.
Address: Westinghouse Electric Corporation, Nuclear Fuel Division, P.O. Box 355, Pittsburgh, Pennsylvania 15230, U.S.A.

FERRARI, Sergio, Laurea in Chem. Born 1932. Educ.: Milan Univ.Manager, Ceramic Technol. Lab., C.N.E.N., 1966-. Societies: A.N.S.; Smithsonian Inst.
Nuclear interests: Metallurgy; Management.
Address: Laboratorio Tecnologie Ceramiche, Casaccia Nuclear Studies Centre, C.N.E.N., S.P. Anguillarese km 1 300, Casaccia, Rome, Italy.

FERRARINI, Richard, Dipl. Masch. Ing. Born 1927. Diesel Engine Dept. (1953-56), Delegated to Federal Inst. for Reactor Res., Würenlingen (1956-58), Reactor design (1958-), In charge of a test facilities group and mechanical tests on reactor components (1959-), Sulzer Bros. Ltd. Society: Swiss Nucl. Soc.
Nuclear interests: Design of facilities for metallurgical experiments; High vacuum techniques; Mechanical and thermodynamical testing of reactor components.
Address: Sulzer Brothers Ltd., Winterthur, Switzerland.

FERRARO, Eugene R., A.B. (Ind. Administration). Born 1922. Educ.: Union Coll. Hofstra Coll. and Johns Hopkins Univ. Rad. Advisory Council, Dept. of Health, State of Ohio, U.S.A.
Nuclear interests: Management and regulatory control of atomic energy.
Address: 4410 Clearbrook Court, Columbus, Ohio 43221, U.S.A.

FERREIRA, Erasmo Madureira, B.Sc., Ph.D. Born 1930. Educ.: Brazil and London Univs. Assoc. Prof., Theoretical Phys. Dept., Centro Brasileiro de Pesquisas Fisicas, 1953-; Universidade Central de Venezuela, 1963-64; Internat.

Centre for Theoretical Phys., -1967; Universidade Catolica do Rio de Janeiro, 1967-.
Society: Brazilian Phys. Soc.
Nuclear interest: Theoretical elementary particle physics.
Address: Centro Brasileiro de Pesquisas Fisicas, 71 Avenida Wenceslau, Bras, Rio de Janeiro, Brazil.

FERREIRA, Jose Francisco Vitorino GOMES. See GOMES FERREIRA, Jose Francisco Vitorino.

FERRELL, Richard Allan, B.S. (California Inst. Technol.), M.S. (California Inst. Technol.), Ph.D. (Princeton). Born 1926. Educ.: California Inst. Technol. and Princeton Univ. Asst. Prof. (1953-56), Assoc. Prof, (1956-59), Prof. (1959-), Maryland Univ. Societies: A.P.S.; Wash. Acad. Sci.
Nuclear interests: Theoretical nuclear and solid state physics.
Address: Maryland University, College Park, Maryland, U.S.A.

FERRER MONCADA, Carlos, Prof. Eng. Director, Centro de Estudos de Electronica, Comissao de Estudos da Energia Nuclear, Portugal; Director, Electronics Dept., Inst. Superior Tecnico.
Address: Instituto Superior Tecnico, Avenida Rovisco Pais, Lisbon, Portugal.

FERRER-MONGE, José A., Ph.D. Born 1923. Educ.: Puerto Rico and Missouri Univs. Louisiana State Univ. Prof. Biol. CAM 1960; Director, Health Phys. Div., Puerto Rico Nucl. Centre, Mayaguez, P. R., 1958. Societies: H.P.S.; Rad. Res.
Nuclear interests: Health physics; Dosimetry; Biological effects of radiation.
Address: Health Physics Division, Puerto Rico Nuclear Centre, Coll. Station, Mayaguez, Puerto Rico, U.S.A.

FERRETTI, Bruno, Phys. Degree. Born 1913. Educ.: Bologna Univ. Head, Theoretical Phys. Dept., Bologna Univ.
Address: Istituto di Fisica, 46 Via Irnerio, 40126 Bologna, Italy.

FERRI, Mário GUIMARAES. See GUIMARAES FERRI, Mário.

FERRIER, Malcolm David, B.Sc. Born 1930. Educ.: St. Andrews (Scotland) Univ. Tech. Information Officer, Vice-President's Office, A.E.C.L., Chalk River Nucl. Labs., 1956-62; Managing Editor, A.N.S., 1962-65. Societies: A.N.S.; A.I.A.A.
Nuclear interest: All aspects of the industry through publishing technical journals, books and magazines.
Address: 128 North Park, Hinsdale, Illinois, U.S.A.

FERRINI, Omero, Prof. Director, Centro Isotope Roiattivi, Clinica Medica, Genoa Univ.
Address: Centro Isotopi Radioattivi, Clinica Medica Universita, Viale Benedetto XV, Genoa, Italy.

FERRONI, Sergio. Prof. Associe, Sect. Phys. des Hautes Energies, Lab. de l'Accelerateur Linéaire, Ecole Normale Superieure, Paris Univ.
Address: Laboratoire de l'Accelerateur Linéaire, Ecole Normale Superieure, Batiment 200, Faculte des Sciences, Orsay (Essonne), France.

FERUSIC, Seid, Sen. Officer, Energoinvest.
Address: Energoinvest, Istrazivacko Razvojmi Centar za Termotehniku i Nuklearnu Tehniku, Sarajevo, Stup-Yugoslavia.

FERZIGER, Joel Henry, B.Ch.E., M.S.E., Ph.D. Born 1937. Educ.: Cooper Union and Michigan Univ. Asst. Prof. (N.E.), Stanford Univ., 1961-; Consultant, G.E.C., 1961-.
Societies: A.N.S.; A.P.S.
Nuclear interests: Reactor theory; Neutron scattering; Plasma physics.
Address: Nuclear Engineering Laboratory, Stanford University, Stanford, California, U.S.A.

FESHBACH, Herman, Ph.D. (M.I.T.). Born 1917. Educ.: C.C.N.Y. and M.I.T. Consultant, U.S.A.E.C., Lincoln Lab., M.I.T. 1941-; Director, Centre for Theoretical Phys., M.I.T. Book: Co-author, Methods of Theoretical Phys. (New York, McGraw-Hill, 1953). Societies: A.P.S.; American Acad. of Arts and Sciences.
Nuclear interest: Nuclear physics.
Address: Department of Physics, Massachusetts Institute of Technology, Cambridge, Mass., U.S.A.

FESSLER, Hans, Diplom-Physiker. Born 1928. Educ.: Karlsruhe T.H. At Kernreaktor Bau- und Betriebsgesellschaft m.b.H. Karlsruhe, 1958-.
Nuclear interest: Health physics.
Address: 23 Gellerstrasse, Karlsruhe, Germany.

FETT, Walter, Diplom-Meteorologist, Dr. rer. nat. Born 1927. Educ.: Berlin Univ. Leader, Sci. Dept., Radiosonde Service, Berlin; Sci. collaborator, Inst. for Meteorology and Geophys., Berlin Free Univ., 1958-; Sci. Asst. and Leader of Dept. Radioactivity, 1960-.
Nuclear interests: Nuclear meteorology (tracer method of near ground and upper air level); Counter physics (especially exoelectrons and radioactive recoil); Health physics.
Address: 62 Podbielskiallee, Berlin-Dahlem, Germany.

FETTWEIS, Paul Felix Maria, D. en Sc. Born 1930. Educ.: Louvain Univ. C.E.N., Mol, 1958-. Societies: Sté. Belge de Phys.; Sté. Française de Phys.
Nuclear interests: Nuclear physics and especially nuclear spectroscopy.
Address: Centre d'Etude de l'Energie Nucléaire, Mol, Belgium.

FETZ, Hans, apl. Prof., Dr. Ing. Born 1912. Educ.: Munich T.H. Würzburg Univ., 1947-; Argonne Nat. Lab., 1960-61.
Nuclear interest: Ionisation phenomenas, sputtering and plasma phenomenas.
Address: 31 Ebertsklinge, 87 Würzburg, Germany.

FEUER, Irving, B.S. (Chem.), M.S. (Chem.). Born 1922. Educ.: C.C.N.Y. and Polytechnic. Inst. of Brooklyn. Phys., Beth Israel Hospital, 1950-58; Phys., Bellvue Hospital, 1954-58; Phys. Chem., Air Force Cambridge Res. Centre, 1959-60; Director of Appl. Res., Canrad Precision Industries, New York City, 1960-. Chemical Abstractor for Chemical Abstracts, 1954-64. Societies: A.C.S.; A.A.A.S.; American Vacuum Soc.; New York Acad. of Sci.
Nuclear interests: Nuclear medicine; Instrumentation; Radiation damage to surfaces; Gas analysis utilising nuclear particles; Catalytic properties of materials; Electroluminescence; Image intensifiers; Beta ray devices for physical chemical measurements; Tritiated phosphors.
Address: 94-25 57th Avenue, Elmhurst 73, New York, U.S.A.

FEUILLEBOIS, L., Jr. Director, Sté. des Compresseurs à Membrane Corblin.
Address: Société des Compresseurs à Membrane Corblin, 78-80 boulevard Saint-Marcel, Paris 5, France.

FEUVRAIS, L. Prof., Inst. de Phys. Nucl., Lyon Univ.
Address: Lyon University, Institut de Physique Nucleaire, 43 boulevard du 11 Novembre 1918, 69-Villeurbanne, (Rhône), France.

FEUZ, Peter, Dr. rer. pol., lic. oec. HSG. Born 1935. Educ.: St. Gall and Fribourg Univs. Sec. Gen., Swiss Assoc. for Atomic Energy. Member, Swiss Federal Commission for Atomic Energy; Member, Swiss Federal Commission for Rad. Protection; Member, Exec. Com., Foratom and the Nucl. Public Relations Contact Group. Editor in Chief, Information-Bulletin and the Atom Press Service Swiss Assoc. for Atomic Energy, the Nucl. Newsletter from Switzerland, and the Manual Nucl. Energy and Rad. Protection in Switzerland.
Address: Swiss Association for Atomic Energy, 2 Bärenplatz, P.OB. 2713, CH - 3001 Bern, Switzerland.

FEY, John T. Pres., Nat. Life Insurance Co., Montpelier. Member, Board of Contract Appeals, U.S.A.E.C.
Address: U.S. Atomic Energy Commission, Board of Contract Appeals, Washington D.C. 20545, U.S.A.

FEYNBERG, E. L. Paper: Some Remarks on the Lifetime of Fireballs (Nukleonika, vol. 13, No. 1, 1968).
Address: Academy of Sciences of the U.S.S.R., 14 Leninsky Prospekt, Moscow V-71, U.S.S.R.

FEYNMAN, Richard P. B.S., Ph.D. Albert Einstein Award, 1954; E. O. Lawrence Award, 1962; Nobel Prize in Physics, 1965. Born 1918. Educ.: M.I.T. and Princeton Univ. Deleg. to 2nd I.C.P.U.A.E., Geneva, Sept. 1958; Prof. Theoretical Phys., (1950-59), Richard Chace Tolman Prof. Theoretical Phys. (1959-), Caltech. Societies: A.P.S.; A.A.A.S.; N.A.S.; Foreign Member, Roy. Soc., London.
Address: California Institute of Technology, Pasadena, California, U.S.A.

FIALHO, Gabriel Emiliano de Almeida, M.A., Ph.D. Born 1920. Educ.: Columbia Univ. Director, Latin American Centre of Phys.; Prof., Theoretical Phys., and Member, Exec. Board, Centro Brasileiro de Pesquisas Fisicas. Member, Exec. Board, Instituto Brasileiro de Educacao e Cultura. Societies: Sociedad Colombiana de Fisica; A.P.S.; Sté. Française de Physique; Sociedade Brasileira Para o Progresso da Ciência.
Nuclear interests: Nuclear physics; Statistical mechanics; Field theory; Management.
Address: Centro Brasileiro de Pesquisas Fisicas, 71 Av. Wenceslau Braz, Rio de Janeiro, Brazil.

FIALIOS, Rafael TORRES. See TORRES FIALIOS, Rafael.

FIALIOS SILVA, Miguel, Dr. Voter, Honduras A.E.C.
Address: Honduras Atomic Energy Commission, Tegucigalpa, D.C., Honduras.

FICARELLI, Leopoldo. Tech. Manager, Instr. Dept., Applicazioni Elettroniche S.P.A.
Address: Applicazioni Elettroniche S.P.A., 14 V. de Monte Grappa, Milan, Italy.

FICKES, P. A. Manager, Nucl. Div., Teleflex Inc.
Address: Teleflex Inc., Nuclear Division, P.O. Box 218, North Wales, Pennsylvania, U.S.A.

FICQ, Adrienne, Dr. in Biol. Sci. Born 1914. Educ.: Brussels Univ. Chargé de Recherches, Fonds National de la Recherche scientifique, 1954-56; Chercheur agréé, Inst. Interuniv. des Sciences nucléaires, 1957-60. Chargé de cours, Univ. libre de Bruxelles, 1960-61. Book: Contribution à l'étude du métabolisme cellulaire au moyen de la méthode autoradiographique (Monographie No. 9, Ed. Inst. Interuniv. des Sciences Nucléaires, Brussels, 1961). Societies: Internat. Inst. of Embryology; Sté. belge de Biochimie.
Nuclear interests: Application of a high resolution method of autoradiography to the study of nucleic acids and protein metabolisms to different biological materials. Results were obtained at cellular and subcellular levels (isolated cells, nuclei, nucleoli chromosomes).

Address: Laboratoire de Morphologie animale, Université libre de Bruxelles, 1850 chaussée de Wavre, Brussels 16, Belgium.

FIDECARO, Giuseppe, Prof. Director, Trieste Sub-Sect., Nucl. Phys. Nat. Inst.
Address: Nuclear Physics National Institute, Trieste Sub-Section, Universita degli Studi, 158 Via Fabio Severo, Trieste, Italy.

FIDELIS, Irena, Dr. of Chem. Born 1933. Educ.: Warsaw Univ. Dept. of Radiochem., Inst. of Nucl. Res., Warsaw, 1955-. Society: Polish Chem. Soc.
Nuclear interests: Radiochemistry, particularly studies of rare earth elements; Partition chromatography.
Address: Instytut Badan Jadrowych, Warsaw 91, Poland.

FIDLER, Harold Alvin, B.S. (C.E.), S.M. (C.E.), Sc.D. (C.E.). Born 1910. Educ.: Drexel Inst. Technol. and M.I.T. Area Eng., Manhattan Eng. Dist., Berkeley, California, 1942-45; Deputy Director, Res. Div. and Chief, Declassification and Publications Branch, Manhattan Eng. Dist., Oak Ridge, 1945-57; Chief of Declassification, H.Q., Washington (1947-49), Area Manager, Berkeley Area Office, Berkeley, California (1949-52), U.S.A.E.C.; Deputy Manager (1952-54), Manager (1954-58), San Francisco Operations Office, U.S.A.E.C., Oakland, California; Asst. Director (1958-61), Assoc. Director (1961-), Lawrence Rad.Lab., California Univ., Berkeley. Ind. Member, Atomic Energy Labour Management Advisory Com., U.S.A.E.C. Society: A.S.C.E., N.Y.
Nuclear interest: Management.
Address: Lawrence Radiation Laboratory, University of California, Berkeley, California, U.S.A.

FIEBIG, Karl Reinhard, Dr. rer. nat. Born 1935. Educ.: Christian Albrecht Univ. Kiel. Inst. für reine und angewandte Kernphysik, Kiel Univ., 1961-. Societies: Nordwestdeutsche phys. Gesellschaft; A.G.U.
Nuclear interests: Theoretical nuclear physics, geophysics.
Address: Gesellschaft für Kernenergieverwertung in Schiffbau und Schiffahrt Forschungsreaktor Geesthacht, 2057 Geesthacht-Tesperhude, Germany.

FIEBIGER, N., Dr. Leader, Experiment Group Reactor, Inst. für Kernphysik, Johann Wolfgang Goethe Univ., Frankfurt am Main.
Address: Institut für Kernphysik, Johann Wolfgang Goethe Universität, 31 Am Romerhof, Frankfurt a.M., Germany.

FIELD, Edwin Oscar, B.M., B.Ch., M.A., D.M. Born 1923. Educ.: Oxford Univ. Res. Asst. in Organic Nervous Diseases. Courtauld Inst. of Biochem., Middlesex Hospital Medical School, London, 1950-51; Graduate Asst. in Biochem., Dept. of Biochem. (Radcliffe Infirmary), Oxford Univ., 1952-56; Clinical Res. Radiobiol. i/c Radiobiol. Labs., subsequently Head, Radiotherapy Res. Unit, Dept. of Radiotherapy, Inst. of Cancer Res., Roy. Cancer Hospital. London, 1956-. Societies: British Medical Assoc.; British Inst. of Radiol.
Nuclear interests: Isotope tracer techniques applied to clinical problems; Laboratory studies on radiobiological effects.
Address: Foxholes, Stokesheath Road, Oxshott, Surrey, England.

FIELDER, George Antonio, Dipl. (Loughborough Coll.). Born 1932. Educ.: Loughborough Coll. Mech. Eng., Saunders-Roe and Nucl. Enterprises Ltd.
Address: Saunders-Roe and Nuclear Enterprises Ltd., North Hyde Road, Hayes, Middx, England.

FIELDS, Paul R., B.S. Born 1919. Educ.: Chicago Univ. Chem., Met. Lab., Chicago Univ., 1943-54; Group Leader, Chem. Div., Argonne Nat. Lab., 1946-.
Nuclear interests: Nuclear physics and chemistry; Chemistry of the transuranium elements.
Address: 9860 Calhoun Avenue, Chicago, Illinois, U.S.A.

FIELDS, Thomas H., B.S., Ph.D. Educ.: Carnegie Inst. Technol. Phys. Prof., Northwestern Univ., 1962-; Res. Assoc. (1960-62), Assoc. Phys. (1962-65), Director, High Energy Phys. Div. (1965-), Argonne Nat. Lab., U.S.A.E.C.
Address: Argonne National Laboratory, 9700 South Cass Avenue, Argonne, Illinois 60440, U.S.A.

FIENNES, Sir Maurice. Chairman and Managing Director, Power-Gas Corp. Ltd.
Address: Power-Gas Corp. Ltd., P.O. Box 21, Stockton-on-Tees, England.

FIERENS, J., Ir. Director, Manager Res. Dept., Gevaert - Agfa N.V.
Address: Gevaert - Agfa N.V., Mortsel, Antwerp, Belgium.

FIERRO-BENITEZ, Rodrigo, Dr. en Medicina, Especialista en Endocrinologia. Born 1930. Educ.: Central de Quito, Madrid, Genoa and Pisa Univs. and Nat. Naval Medical School (U.S.A.). Director, Departamento de Radioisótopos, Escuela Politécnica Nacional, Quito; Endocrinólogo del Hospital Eugenio Espejo, Quito; Professor de Endocrinologia Facultdad de Medicina, Univ. Central, Quito. Miembro de la Comisión Ecuatoriana de Energia Atómica. Book: La Funcion Tiroidea en el Bocio Endémico de la Provincia de Pichincha (Ed. Casa de la Cultura Ecuatoriana, 1961). Societies: Soc. Bolivariana de Endocrinologia; Soc. Ecuatoriana de Endocrinologia; Academia Ecuatoriana de Medicina; Casa de la Cultura Ecuatoriana.
Nuclear interests: Aplicaciones diagnósticas y terapéuticas de los radioisótopos; especial interés en todos los aspectos relacionados con

la fisiopatologia tiroidea.
Address: Casilla Postal 2698, Quito, Ecuador.

FIERZ, Markus, Dr. phil. Born 1912. Educ.: Göttingen and Zürich Univs. Prof. Theoretical Phys., Basle, 1946-59; Director Theoretical Dept., C.E.R.N., 1959-60; Prof. Theoretical Phys., E.T.H. Zürich, 1960-. Swiss deleg. to the Council of C.E.R.N.
Address: 60 Hochstrasse, Zürich, Switzerland.

FIESCHI, A. Member, Editorial Com., Minerva Nucleare.
Address: 83-85 Corso Bramante, Turin, Italy.

FIGUEIREDO, R. J. PACHECO DE. See PACHECO DE FIGUEIREDO, R. J.

FIGUERA, A. S. Gruppo van de Graaff, Istituto di Fisica, Catania Univ.
Address: Catania University, Istituto di Fisica, 57 Corso Italia, Catania, Sicily, Italy.

FIGUEROA, David, Member, Junta Directiva, Inst. Colombiano de Asuntos Nucleares.
Address: Edificio Diario Oficial, Bogota, Colombia.

FILBERT, Robert B., Jr., B.S. (Chem. Eng.), M.S. (Chem.). Born 1921. Educ.: Pennsylvania State Coll. Assoc. Manager, Chem. Eng., Battelle Memorial Inst., Columbus, 1960-. Vice Pres., Exec. Board, Nat. Programme Com., A.I.Ch.E. Trustee, Ohio Soc. of Professional Eng. Book: Progress in Nucl. Energy, Volume 2 (Pergamon Press, 1958). Society: A.C.S. Nuclear interests: Fuels processing; Homogeneous fuels; Extractive metallurgy.
Address: 505 King Avenue, Columbus, Ohio 43201, U.S.A.

FILHO, Alceu G. PINHO. See PINHO FILHO, Alceu G.

FILIMONOVA, E. A. Papers: Co-author, On Increasing of Ion Trapping into Magnetic Confinement of System by Neutron Atom Photoionization (Atomnaya Energiya, vol. 21, No. 4, 1966); co-author, Ion Stream Interaction with Target Plasma during Recharging (ibid, vol. 25, No. 2, 1968).
Address: Academy of Sciences of the U.S.S.R., 14 Leninsky Prospekt, Moscow V-71, U.S.S.R.

FILIPCZAK, Wieslaw, Ph.D., M.Sc. Born 1931. Educ.: Polytech. Inst. Leningrad. At Inst. Nucl. Res., Polish Acad. of Sci., Dept. of Electronics and Reactor Control.
Nuclear interests: Reactor control; Analogue computers.
Address: 18 m. 5 ul. Andriollego bl., Otwock, Poland.

FILIPOV, Alexander Milan, B.S. (Elec. Eng., Communications). Born 1931. Educ.: Queen's Univ. Eng. Canadian Aero Services Ltd., Ottawa, 1955-59; Eng., Tracerlab, Inc., 1959-63; Eng. Manager, Tracerlab Div. L.F.E., Inc., 1963-66; Product Manager, Lab. Instr., Tracerlab-Nucl. Instr. Div., L.F.E., Inc. 1966-. Chapter Chairman, I.E.E.E. Group on Nucl. Sci., Boston Sect. Societies: I.E.E.E.; Assoc. of Professional Eng. of the Province of Ontario. Nuclear interests: Application and design of nuclear instrumentation in the laboratory and in the life sciences, including automatic counting systems, survey instruments, training instruments, and monitoring instruments.
Address: 4 Minot Road, Concord, Massachusetts 01742, U.S.A.

FILIPPOV, A. G. Adviser, Third I.C.P.U.A.E., Geneva, Sept. 1964. Papers: Co-author, The 50 MW Res. Reactor SM (Atomnaya Energiya, vol. 8, No. 6, 1960); co-author, Res. and Training Reactor IR-100 (ibid, vol. 21, No. 5, 1966).
Address: Academy of Sciences of the U.S.S.R., 14 Leninsky Prospekt, Moscow V-71, U.S.S.R.

FILIPPOV, D. Deleg., I.A.E.A. Conference on the Use of Radioisotopes in the Phys. Sci. and Industry, Copenhagen, Sept. 6-17, 1960.
Address: Academy of Sciences of the U.S.S.R., 14 Leninsky Prospekt, Moscow V-71, U.S.S.R.

FILIPPOV, N. V. Deputy Director, Kurchatov Inst. Atomic Energy. Paper: Co-author, Measurement of the Plasma Electron Temperature in a Strong Shock Wave (Nucl. Fusion, vol. 1, No. 3, 1961).
Address: I. V. Kurchatov Institute of Atomic Energy, Academy of Sciences of the U.S.S.R., 14 Leninsky Prospekt, Moscow V-71, U.S.S.R.

FILIPPOV. V. V. Papers: Co-author, Effects of Resonances on the Motion of Fast Neutrons in Iron (letter to the Editor, Atomnaya Energiya, vol. 11, No. 5, 1961); co-author, Measurements of Total Cross-Section Resonance Structure Parameters for Some Elements in Energy Region of 0.3 - 2.7 MeV (ibid, vol. 15, No. 6, 1963).
Address: Academy of Sciences of the U.S.S.R., 14 Leninsky Prospekt, Moscow V-71, U.S.S.R.

FILIPPOVA, T. I. Deputy Director, Kurchatov Inst. Atomic Energy. Paper: Co-author, Measurement of the Plasma Electron Temperature in a Strong Shock Wave (Nucl. Fusion, vol. 1, No. 3, 1961).
Address: I. V. Kurchatov Institute of Atomic Energy, U.S.S.R. Academy of Sciences, 14 Leninsky Prospekt, Moscow V-71, U.S.S.R.

FILLIPOV, D. P. Deputy Head, Dept. Internat. Relations and Sci. Tech. Information, State Com. for the Utilization of Atomic Energy.
Address: State Committee for the Utilization of Atomic Energy, 26 Staromonetrii Pereulok, Moscow, U.S.S.R.

FILLNOW, Ronald Henry, B.S., M.S. (Metal.). Born 1921. Educ.: Wisconsin Univ. Society: A.S.M.

Nuclear interests: Effect of neutron exposure on physical and mechanical properties of materials for nuclear application, particularly the mechanism of inducing or retarding phase changes.
Address: Westinghouse Electric Corporation, Bettis Atomic Power Division, P.O. Box 79, West Mifflin, Pennsylvania 15122, U.S.A.

FILLON, René, Agrégé de Univ. Born 1904. Educ.: Paris Univ. Managing Director, Rothschild Bank, -1955; Chairman, Board, Cie. Française des Minerais d'Uranium, 1957-; Vice Chairman, Board, Sté. Minière et Métallurgique de Penarroya, 1958-; Member, Council for Economic and Social Affairs (Conseil Economique et Social),1959-. Member, Board, Cie. Electro-Mécanique; Member, Board, Sté. Ugine-Kuhlmann; Member, Board, Cie. des Mines d'Uranium de Franceville; Member, Board, Sté. Minière de l'Air.
Nuclear interests: Prospecting, mining and processing of uranium ore; Development of nuclear power stations.
Address: 12 Place Vendôme, Paris, France.

FILYUSHKIN, I. V. Paper: Co-author, Spectra of Po-Be-Source Fast Neutrons Passed through Water Shielding (letter to the Editor, Atomnaya Energiya, vol. 16, No. 2, 1964); Co-author, The Reliability Analysis of Methods for Studying Fast Neutron and γ-Ray Continuous Spectra (ibid, No. 3).
Address: Academy of Sciences of the U.S.S.R., 14 Leninsky Prospekt, Moscow V-71, U.S.S.R.

FINAN, John Marshall, B.S. (E.E.). Born 1927. Educ.: Lehigh Univ. and O.R.S.O.R.T. Nucl. Eng., Atomic Power Development Assocs., 1955-57; Application Eng., Leeds and Northrup Co., 1957-. Member, Nucleonics Com., American Inst. of Elec. Eng.; Chairman, Subcom. on Instrumentation, Reactor Standards Com., American Standards Assoc. Society: A.N.S.
Nuclear interests: Instrumentation and control of nuclear devices and systems.
Address: Roberts Road, Merrill Hills, Ambler, Pennsylvania, U.S.A.

FINAN, W. J. Supervisor of Project Eng., Foster Wheeler Corp.
Address: Foster Wheeler Corporation, 666 Fifth Avenue, New York 19, New York, U.S.A.

FINCH, Harold William Stuart, Diploma City and Guilds of London Inst. Born 1912. Educ.: Regent Street Polytech., London. Tech. Director, Isotope Developments Ltd.; Tech. Director, Baldwin Instrument Co. Ltd.
Nuclear interest: Design of nucleonic instrumentation for use in medicine and research; Design of industrial process control instrumentation based on the use of radionuclides.
Address: 6 Croft Lane, Speen, near Newbury, Berks., England.

FINCKH, Eberhard, Dr. rer. nat. (Heidelberg); Dipl. Phys. (Heidelberg). Born 1929. Educ.: Frankfurt am Main and Heidelberg Univs. Sci. Asst., I Physikal. Inst., Heidelberg Univ., 1960-62; Hahn-Meitner-Inst. für Kernforschung, Berlin, 1962-. Society: Deutsche Physikalische Gesellschaft.
Nuclear interests: Photonuclear reactions; Low energy nuclear physics.
Address: Hahn-Meitner-Institut für Kernforschung Berlin, 100 Glienicker Strasse. 1 Berlin 39, Germany.

FINE, Paul C., B.A., M.S., Ph.D. Born 1915. Educ.: Oklahoma Univ. and California Inst. Technol. Tech. Asst., Military Application Div. (1948-55), Director, Division of Operations Analysis and Forecasting (1956-), U.S.A.E.C.
Nuclear interests: Technology, economics, and growth of nuclear power and the production, distribution, and utilisation of nuclear materials.
Address: U. S. Atomic Energy Commission, Washington, D.C. 20545, U.S.A.

FINE, Samuel. Manager, Eng. Div., Isotopes Inc., 1963-.
Address: Isotopes Inc., 123 Woodland Avenue, Westwood, New Jersey, U.S.A.

FINGER, Harold B., Bachelor of Mech. Eng. (C.C.N.Y.), M.S. (Aeronautical Eng., Case Inst. Technol.). Born 1924. Educ.: C.C.N.Y. and Case Inst. Technol. Assoc. Administrator for Organisation and Management, N.A.S.A. Society: A.I.A.A.
Nuclear interests: Nuclear propulsion and power for space missions; Management, technical direction and organisation, and planning of research and development programmes.
Address: 6908 Millwood Road, Bethesda, Maryland 20034, U.S.A.

FINK, Richard W., B.S., M.S., Ph.D. Born 1928. Educ.: Michigan, California (Berkeley) and Rochester Univs. Los Alamos Sci. Lab., 1951; Visiting Prof., Gustaf Werner Inst. for Nucl. Chem., Uppsala Univ.; At Arkansas Univ., 1953-61; Prof. Phys., Marquette Univ., 1961-. Consultant, Phillips Petroleum Co., Bartlesville, Oklahoma, 1958-. Society: A.P.S.
Nuclear interests: Nuclear reactions and nuclear spectroscopy of radioactive nuclei; Accelerator physics; Inner shell ionisation phenomena.
Address: Department of Physics, Marquette University, Milwaukee 3, Wisconsin, U.S.A.

FINKE, Joseph, B.S. and M.S. (Chem. Eng.). Born 1926. Educ.: Columbia Univ. Development Eng., Gen. Elec. Co., 1947-50; Res. Eng., North American Aviation, 1950-51; Asst. Project Eng., U.S.A.E.C., 1951-55; Principal Eng., (1955-), Formerly Asst. Manager, Nucl. Development, now Nucl. Project Eng. and Specialist, Kaiser Engineers Chairman, Northern California (San Francisco) Sect., A.N.S.

Societies: A.C.S.; A.I.Ch.E.; A.N.S.
Address: 3156 Mars Court, Lafayette, California, U.S.A.

FINKE, Wolfgang A., Dr. phil. Dipl. Volkswirt. Born 1925. Educ.: Göttingen, Tübingen and Harvard Univs. Economic Asst., Federal Ministry for Economic Affairs, Bonn, 1956-57; Economic Adviser, Federal Ministry for Sci. Res., Bad Godesberg, 1957-.
Nuclear interests: Economics of nuclear energy, relationship between Govt. and private industry, financing of reactor development and nuclear power stations; International cooperation.
Address: 24 Baumschulallee, Bonn, Germany.

FINKEL, Asher Joseph, Ph.D., M.D. Born 1915. Educ.: Chicago Univ. Director, Health Div., Argonne Nat. Lab., 1955-. Member, Expert Com. on Rad., W.H.O., Geneva, Sept. 1959. Sec.-Treas. (1962-64), Pres. (1964-65), Central Chapter, Soc. of Nucl. Medicine, U.S.A. Societies: A.M.A.; Rad. Res. Soc.; Soc. of Nucl. Medicine; American Soc. of Zoologists; Soc. for Exptl. Biol. and Medicine; Ecological Soc. of America; Fellow, American Acad. of Occupational Medicine.
Nuclear interests: Medical aspects of nuclear energy; Medical supervision of radiation workers; Effects of radium in man; Radium metabolism in mice; Biological effects of deuterium.
Address: Health Division, Argonne National Laboratory, 9700 South Cass Avenue, Argonne, Illinois 60440, U.S.A.

FINKEL, E. E. Paper: Co-author, Rad. Cross-Linking of Polyethylene Insulation of Cable Articles on Large Scale (letter to the Editor, Atomnaya Energiya, vol. 21, No. 1, 1966).
Address: Academy of Sciences of the U.S.S.R., 14 Leninsky Prospekt, Moscow V-71, U.S.S.R.

FINKEL, Miriam P., B.S., Ph.D. (Chicago). Born 1916. Educ.: Chicago Univ. Sen. Sci., Argonne Nat. Lab. Societies: Rad. Res. Soc.; American Assoc. for Cancer Res.; American Soc. Zoologists; American Soc. for Exptl. Pathology; Soc. for Exptl. Biol. and Medicine; Radiol. Soc. North America; A.A.A.S.
Nuclear interest: Toxicity of radioisotopes and radiocarcinogenesis.
Address: Argonne National Laboratory, Argonne, Illinois, U.S.A.

FINKEL, V. A. Paper: Co-author, Deformation and Fracture of Rolled Beryllium of Various Purity (Atomnaya Energiya, vol. 16, No. 5, 1964).
Address: Academy of Sciences of the U.S.S.R., 14 Leninsky Prospekt, Moscow V-71, U.S.S.R.

FINKELSTEIN, Abraham, Dr. Acting Chairman, Eng. Sci. Dept., Pratt Inst.
Address: Pratt Institute, Brooklyn 5, New York, U.S.A.

FINKELSTEIN, André, Ing. Chimiste, Ph.D. (Phys. Chem.). Born 1923. Educ.: Ecole de Chimie Industrielle de Lyon; Faculté des Sci., Paris. Attaché de Recherches, C.N.R.S. 1947-53; Post-Doctoral Fellow, Rochester Univ., 1950-52; Ing. (1953-), then Deputy Director, Office of the High Commissioner (1967-69), C.E.A.; Deputy Director, Gen., Dept. of Res. and Isotopes, I.A.E.A., 1969-. Sci. Sec., I.C.P.U.A.E., Geneva, 1955; Sec., French deleg., I.C.P.U.A.E., Geneva, 1958 and 1964. Societies: Sté. de Chimie Physique; Faraday Soc.; A.C.S.
Nuclear interest: International aspects of nuclear energy.
Address: International Atomic Energy Agency, Department of Research and Isotopes, 11 Kaerntnerring, Vienna 1, Austria.

FINKENSTAEDT, Kimball L., B.A. Born 1896. Educ.: Yale Univ. Vice-Pres. and General Direction of Nucl. Dept., W. F. and John Barnes Co.
Nuclear interests: Reactor design; Nuclear physics.
Address: W. F. and John Barnes Co., 301 South Water Street, Rockford, Illinois, U.S.A.

FINLAY, Roger William, A.B., Ph.D. Born 1935. Educ.: Johns Hopkins Univ. Asst. Prof. (1962-65), Assoc. Prof. (1965-), Ohio Univ.; Visiting Staff Member, Los Alamos Sci. Lab., 1966. Societies: Div. of Nucl. Phys. and Ohio Sect., A.P.S.; A.A.P.T.
Nuclear interests: Nuclear structure; Direct interactions; Neutron polarisation; Nuclear instrumentation.
Address: Ohio University, Athens, Ohio, U.S.A.

FINN, T. P. Chief Eng., W. W. Brown and Partners Ltd.
Address: W. W. Brown and Partners Ltd., Nuclear Department, Haddon House, Fitzroy Street, London, W.1, England.

FINNEY, D. J., Prof., Sc.D., F.R.S. Member, Managing Subcom. on Monitoring, M.R.C.
Address: c/o Medical Research Council, 20 Park Crescent, London W.1, England.

FINNILÄ, Pehr Olof, Dipl. Ing. Born 1911. Educ.: Tech. Univ. Finland. Asst. Managing Director, Atomivoima Oy, Helsinki, 1957-58; Managing Director, Oy Termodyn Ab, Helsinki, 1958-. Societies: Tekniska Föreningen i Finland; Svenska Teknologföreningen.
Nuclear interest: Design of nuclear plants for power and process heat generation.
Address: 22-24 Holländarvägen, Helsinki 33, Finland.

FINOGENOV, Yakov. Vice-Chairman, State Com. for Power Eng. and Electrification.
Address: State Committee for Power Engineering and Electrification, Moscow, U.S.S.R.

FINSTON, Harmon Leo, B.S.A.S. (I.I.T.),
Ph.D. (Ohio). Born 1922. Educ.: I.I.T. and
Ohio State Univ. Leader, Radiochem. Analysis
Group, Brookhaven Nat. Lab.; Prof. Chem.,
Brooklyn Coll., New York City Univ. Member,
Subcoms. on Radiochem. and Use of Radioactivity Standards, Nat. Res. Council, U.S.A.
Societies: A.C.S.; A.N.S.; N.Y. Acad. of Sci.
Nuclear interests: Radio, nuclear, and analytical chemistry.
Address: Brooklyn College, City University of
New York, Brooklyn 10, New York, U.S.A.

FINZI, Dante, Dr. Ing. Head, Energy Div.,
Montecatini. Member, Com. for the Problems
of the Pacific Use of Nucl. Energy and for the
Contacts with Euratom; Member, Board,
F.I.E.N.; Member, Sci. and Tech. Com.,
Euratom, 1968-.
Address: Committee for the Problems of the
Pacific Use of Nuclear Energy and for the
Contacts with Euratom, 11 Piazza Venezia,
Rome, Italy.

FINZI, Sergio, Elec. eng. Born 1928. Educ.:
Bologna Univ. Research man, C.I.S.E.,
Milan, 1952-58; Research man, AGIP Nucleare, Milan,
1959; Head, Eng. Dept., Euratom C.C.R.-
Ispra, 1959-.
Nuclear interest: Reactor technology and
design; Heat transfer; Electronics.
Address: Euratom-C.C.R., Ispra (Varese),
Italy.

FIREMAN, Edward L., B.S., M.S., Ph.D. Born
1922. Educ.: Princeton Univ. and Carnegie
Inst. Technol. Phys., Brookhaven Nat. Lab.,
1950-56; Phys., Smithsonian Astronomical
Observatory, 1956-. Res. Assoc., Harvard
Univ.; Guest Sci., Brookhaven Nat. Lab.
Societies: A.P.S.; A.G.U.; American Astronomical Soc.; Meteoritical Soc.
Nuclear interests: Low-level counting techniques; Neutron activation methods; Radioactive age determination.
Address: 60 Garden Street, Cambridge 38,
Massachusetts, U.S.A.

FIRESTONE, Richard, Prof. Chem. Dept.,
Coll. of Arts and Sci., Ohio State Univ.
Address: Ohio State University, Columbus
10, Ohio, U.S.A.

FIRING, Jorgen Andrew, B.Sc. (Mech. Eng.).
Born 1932. Educ.: Colorado Univ. Sec.,
Halden Branch, Norwegian Eng. Soc.
Nuclear interests: Reactor design and operation; Fuel technology; In-core instrumentation
development and application.
Address: Institutt for Atomenergi, Halden,
Norway.

FIRSOV, E. I. Paper: Co-author, Gamma Rad.
from AN BSSR Reactor IRT-2000 (letter to the
Editor, Atomnaya Energiya, vol. 16, No. 4,
1964).
Address: Academy of Sciences of the U.S.S.R.,
14 Leninsky Prospekt, Moscow V-71, U.S.S.R.

FIRSOVA, E. V. Papers: Co-author, Heat
Transfer for Liquid Sodium Flow Running
Lengthwise round a Bundle of Tubes (letter
to the Editor, Atomnaya Energiya, vol. 14,
No. 6, 1963); co-author, Heat Transfer in
Moning off Pencil Rods with Metallic Sodium
Flowing around Longitudinally (letter to the
Editor, ibid, vol. 16, No. 5, 1964); co-author,
Oxide Effect on Heat Transfer to Sodium
Flow across Staggered Tube Bank (letter to
the Editor, ibid, vol. 22, No. 4, 1967); co-author, Heat Transfer to Longitudinal-Flowing
Liquid Metal in Triangular Tube Banks (letter
to the Editor, ibid, vol. 22, No. 4, 1967).
Address: Academy of Sciences of the U.S.S.R.,
14 Leninsky Prospekt, Moscow V-71, U.S.S.R.

FISCHER, Claus, Dipl. chem. Born 1930.
Educ.: Greifswald Univ. and Dresden Tech.
Univ. Inst. für Metallphysik und Reinstmetalle,
Dresden, 1959-.
Address: 14 August-Röckel-Str., 8046 Dresden,
German Democratic Republic.

FISCHER, David Anthony Valdemar, B.A.,
Post-graduate qualification in Law, History,
Diplomacy. Born 1920. Educ.: Cape Town
and Pretoria. South African Foreign Service
(1st Sec.), 1945-57; Director of External
Liaison, I.A.E.A., 1958. Society: Economic
Soc. of South Africa.
Nuclear interests: General.
Address: 34 Pötzleinsdorferstrasse, Vienna 18,
Austria.

FISCHER, E. Mitglied, Alarmausschuss,
Eidgenössische Departement des Innern.
Address: Bundesamt für Zivilschutz, Bern,
Switzerland.

FISCHER, Erich G. K., Dr. rer. nat., Dipl.-
Chem. Born 1920. Educ.: Heidelberg Univ.
Head, Dept. Isotopen- und Kerntechnik,
Battelle-Institut e.V., 1957-66. Head, Dept.
Tracertechnik und Kontamination, Bundesforschungsanstalt für Lebensmittelfrischhaltung, Inst. für Strahlentechnologie, 1966-.
Society: Gesellschaft Deutscher Chemiker.
Nuclear interests: Application of radionuclides,
especially in food research; Radiochemistry;
Contamination and decontamination of foods;
Radiation chemistry and technology.
Address: 11 Posseltstrasse, Karlsruhe, Germany.

FISCHER, Erich H., Prof., Dr. Phil. Born
1910. Berlin (Dr. Docent), Tübingen (Docent.
Prof.), Ankara (Prof.), Hamburg (Prof.). Staff
member, Max Planck Inst. Phys., Berlin and
Hechingen (1937-51); Prof. and Chairman
of Phys., Ankara Univ., 1951-56; Phys. Res.
Director in Gesellschaft für Kernenergieverwertung, Res. Reactor Geesthacht, near Hamburg. Book: Lectures on General Atomic,
Nucl. and Molecular Phys. (1957).
Nuclear interests: Nuclear and reactor physics;
Radiation damage.
Address: Forschungsreaktor, Geesthacht,
near Hamburg, Germany.

FISCHER, Friedrich. Sen. Officer, Nucl. Div., Caliqua Warmegesellschaft Dr. Kloiber und Co.
Address: Caliqua Warmegesellschaft Dr. Kloiber und Co., 14 Mariannengasse, Vienna 9, Austria.

FISCHER, George James, B.S. and M.S. (Chicago), Ph.D. (Iowa). Born 1918. Educ.: Chicago and Iowa State Univs. Assoc. Phys., 1956-, Presently Head of Safety Analysis Sect., Argonne Nat. Lab. Societies: A.P.S.; A.N.S. Nuclear interests: Theoretical fast reactor studies, including: reactor concepts, with fuel cycle studies; major accident analysis; design of zero power fast reactor experiments. Doppler effect measurements.
Address: c/o Argonne National Laboratory, 9700 South Cass Avenue, Argonne, Illinois, U.S.A.

FISCHER, Richard, Dr. Ing. Formerly Mitglied, Präsidium, now Mitglied, Fachkommission 1, Recht und Verwaltung, Deutsche Atomkommission. Formerly Pres., Hamburgische Electricitats-Werke A.G.
Address: 48 Gerhart Hauptmann-Platz, 2 Hamburg 1, Germany.

FISCHER, Richard Philip, B.A., M.A., Ph.D. Born 1910. Educ.: Ohio Wesleyan and Princeton Univs. Geologist, U.S. Geological Survey, 1937-. Societies: Geological Soc. of America; American Assoc. of Petroleum Geologists; Colorado Sci. Soc.; Soc. of Economic Geologists.
Nuclear interest: Geology of uranium deposits.
Address: U.S. Geological Survey, Building 25, Denver Federal Centre, Denver, Colorado 80225, U.S.A.

FISCHER, Werner, Dr. Ing. Born 1902. Educ.: Hanover T.H. Ord. Prof. für anorganische Chemie an der Hanover T.H. Member, Arbeitskreis III/5-Aufarbeitung von Kernbrennstoffen, Deutsche Atomkommission.
Nuclear interest: Nuclear chemistry.
Address: 18 Appelstr., Hanover, Germany.

FISCHER, Wolfgang, Dr. phil. Born 1929. Educ.: Marburg Univ. Akad. Oberrat, Marburg Univ. Society: Deutsche Physikalische Gesellschaft.
Nuclear interests: Isotope separation; Hyperfinestructure.
Address: Physikalisches Institut, Universität Marburg, 5 Renthof, Marburg, Germany.

FISCHER-HJALMARS, Inga M., Fil. lic., Fil. Dr. Born 1918. Educ.: Stockholm Univ. Asst. Prof. (1953-59), Full Prof. (1963-), Stockholm Univ.; Assoc. Prof., Roy. Inst. of Technol., Stockholm, 1959-63. Stockholm Deleg. for Univ. Computers. Societies: Svenska Nationalkommittén för fysik (Swedish Nat. Com. of Phys.); A.P.S.
Nuclear interest: Fundamentals of solid state theory, many-body problems.
Address: Institute of Theoretical Physics, Stockholm University, 9 Vanadisvägen, Stockholm VA, Sweden.

FISCHERHOF, Hans, Prof., Dr. iur. Born 1908. Educ.: Munich, Paris, Berlin, Freiburg i.B. and Frankfurt a.M. Univs. Attorney and Counsellor at Law; Prof., Frankfurt a.M. Univ. (Energy Law, Atomic Energy Law). Chairman, A.E.C., German Bar Assoc. Books: Rechtsfragen der Energiewirtschaft I (Frankfurt a.M., 1956); II (Munich, 1966); Deutsches Atomgesetz und Strahlenschutzrecht I, II (Baden-Baden, 1962, 1966). Societies: Deutsches Atomforum; Internat. Law Assoc. (A.E.C.); Internat. Bar Assoc. (A.E.C.).
Nuclear interest: Atomic energy law.
Address: 41 Grillparzerstrasse, Frankfurt a.M., Germany.

FISCHESSER, Raymond. Member, Comite des Mines, C.E.A.
Address: Commissariat a l'Energie Atomique, 29-33 rue de la Federation, Paris 15, France.

FISH, Victor. J., M.D. Born 1919. Educ.: Breslau Univ. Cedars of Lebanon Hospital, 1960-62; Div. of Rad. Therapy and Nucl. Medicine, Los Angeles Palo Alto Medical Clinic, 1962-. Societies: Radiol. Soc. of North America; Soc. of Nucl. Medicine.
Nuclear interest: Nuclear medicine.
Address: 300 Homer Avenue, Palo Alto, California, U.S.A.

FISHEL, Derry Lee, Ph.D., A.B. Born 1929. Educ.: Ohio State Univ. Asst. Prof., South Dakota Univ., 1958-60; Asst. Prof. (1960-64), Assoc. Prof. (1964-), Kent State Univ. Society: A.C.S.
Nuclear interests: Tracer studies; Organic reaction mechanisms.
Address: Kent State University, Kent, Ohio, U.S.A.

FISHENDEN, Richard Martin, M.A. Born 1917. Educ.: Cambridge Univ. Telecommunications Res. Establishment, 1940-45; Head of Sci. Administration Office, A.E.R.E., Harwell, U.K.A.E.A., 1956-.
Nuclear interest: Management, including the information and publication fields.
Address: 8 Severn Road, Chilton, Berks., England.

FISHER, Beeman. Director, Texas Atomic Energy Res. Foundation.
Address: Texas Atomic Energy Research Foundation, P.O. Box 970, Fort Worth, Texas, U.S.A.

FISHER, Charlie, Ing. E.N.S.P.C.I. Born 1925. Educ.: Paris Univ. Head, Radioisotope Div., C.E.A.; Sec., Joint Commission on Appl. Radioactivity, I.C.S.U. Societies: Ste. Chimique de France; A.C.S.; A.N.S.
Nuclear interest: Production and utilisation of radioisotopes.

Address: 48 avenue de Clos Toutain, 92 Vaucresson, France.

FISHER, Dale John, Ph.D. Born 1925. Educ.: Wisconsin State Univ., Oshkosh and Indiana Univ. Group Leader, Analytical Instrumentation, in Analytical Chem. Div., O.R.N.L., 1951-. Societies: A.C.S.; Polarographic Soc.; A.N.S.; Tenn. Archaeological Soc.
Nuclear interests: Design and application of instruments and instrumental methods for the bench top or the remotely controlled analysis of chemical solutions.
Address: 22 Outer Drive, Oak Ridge, Tennessee 37830, U.S.A.

FISHER, E. B. Pres., Nucl. Dept., United States Radium Corp.
Address: United States Radium Corporation, P.O. Box 246, Morristown, New Jersey, U.S.A.

FISHER, Sir George. Member, Council, Sci. Foundation for Phys., Sydney Univ.
Address: Sydney University, Sydney, New South Wales, Australia.

FISHER, Joseph V. Pres., Semi-Elements, Inc.
Address: Semi-Elements, Inc., Saxonburg Boulevard, Saxonburg, Pennsylvania, U.S.A.

FISHER, L., A.M.I.E.D. Chief Eng. and Local Director, Vickers Res. Establishment.
Address: Vickers Research Establishment, Sunninghill, nr. Ascot, Berks., England.

FISHER, Leo. Production Manager, Isotope Accessories Co.
Address: Isotope Accessories Co., 5312 Westminster Avenue, Philadelphia 31, Pennsylvania, U.S.A.

FISHER, Norman Henry, Dr. Sci. Born 1909. Educ.: Queensland Univ. Chief Geologist, Bureau of Mineral Resources, Geology and Geophys.; Dept. of Nat. Development, Canberra; U.N. Adviser, Israel, on Mineral Development, 1963-64 (3 months). Member, Nat. Coms. on Geological Sci. and Hydrology; Assoc. Editor, Economic Geology, Mineralium Deposita. Societies: Geological Soc. of Australia; Soc. of Economic Geologists; Soc. for Geology Appl. to Mineral Deposits; Australasian Inst. of Mining and Metal. Nuclear interest: Geology and exploration for uranium deposits.
Address: 68 National Circuit, Deakin, Canberra, A.C.T., Australia.

FISHER, Peter Stevens, B.Sc. (London), Ph.D. (Birmingham), M.A. (Oxon.). Born 1933. Educ.: London and Birmingham Univs. Res. Assoc., Princeton Univ., U.S.A., 1957-59; Sen. Res. Officer, Dept. of Nucl. Phys., (1959-66), Univ. Lecturer in Phys. (1966-), Oxford Univ.; Official Fellow in Phys., Trinity Coll., Oxford, 1964-.
Nuclear interest: Nuclear reaction mechanisms at intermediate energies.
Address: Nuclear Physics Laboratory, Keble Road, Oxford, England.

FISHER, Philip C., B.S., Ph.D. Born 1926. Educ.: Rochester and Illinois Univs. Res. Asst., Illinois Univ., 1949-53; Staff Member, Los Alamos Sci. Lab., 1953-59; Consulting Sci., Lockheed Missiles and Space Co., 1959-. Societies: A.P.S.; A.G.U.; American Astronomical Soc.
Nuclear interests: Criticality safety, fission and nuclear physics, X-ray astronomy and instrumentation.
Address: Lockheed Missiles and Space Co., Palo Alto, California, U.S.A.

FISHER, Raoul Conrad, Dipl. Ing., Dr. Ing., Chartered Eng. Born 1900. Educ.: Hanover Tech. Univ. Formerly Managing Director, Head, Wrightson Processes, Ltd.; Industrial consultant, independent, Petroleum and chem. eng., nucl. eng. Societies: M.I.Mech.E.; M.I.Chem.E.; Inst. Fuel.; Fellow, Inst. Petroleum.
Nuclear interests: Petro-chemical industry, nuclear reactors and fuels.
Address: 6 Hayes Road, Bromley, Kent, England.

FISHER, Ray Wm., B.S. (Chem. Technol.). Born 1921. Educ.: Iowa State Univ. Ames Lab. U.S.A.E.C., I.S.U., 1947-. Books: The Metal Thorium, Chapter 5, The Preparation of Thorium Oxide and Thorium Fluoride from Thorium Nitrate (A.S.M., 1958); Progress in Nucl. Energy, vol. 11 of Process Chem. Series III, Chapter on The Preparation of Thorium Oxide and Thorium Fluoride from Thorium Nitrate (Pergamon Press, 1958). Society: A.C.S.
Nuclear interest: Metallurgy - high temp. molten metal corrosion.
Address: 3612 Mary Circle, Ames, Iowa, U.S.A.

FISHER, Russell Arden, M.A., Ph.D. Born 1904. Educ.: Michigan Univ. Prof. Phys. (1946-), Chairman, Dept.Phys. (1950-57), Northwestern Univ., Illinois. Societies: Fellow, A.P.S.; Optical Soc. America; A.A.P.T.
Nuclear interests: Nuclear spins and magnetic moments from optical hyperfine structure; Nuclear energy levels by methods of nuclear spectroscopy employing electrostatic accelerator.
Address: 810 Edgewood Lane, Glenview, Illinois, U.S.A.

FISK, W. B. Asst. Vice Pres., Eng., Indian Point Nucl. Energy Power Plant, Consolidated Edison Co. of New York, Inc.
Address: Consolidated Edison Co. of New York, Inc.,Indian Point Nuclear Power Plant, Buchanan, New York, U.S.A.

FISS, E. C. Director of Eng., Carolinas Virginia Nuclear Power Associates, Inc. Nucl.

Eng., Duke Power Co.
Address: Duke Power Co., P.O. Box 2178, Charlotte, North Carolina, U.S.A.

FISSORE, André, Dr. Service d'Electroradiologie Cobalt Therapy Unit, Polyclinique Princess Grace. Member, Board, Centre Scientific de Monaco.
Address: Polyclinique Princesse Grace, Monaco.

FOSSORE, Odette, Dr. Service d'Electroradiologie Cobalt Therapy Unit, Polyclinique Princesse Grace. Member of Board, Centre Scientifique de Monaco.
Address: Polyclinique Princesse Grace, Monaco.

FISZER, Wlodzimierz Wladyslaw, Dr. of Tech. Sci. Born 1930. Educ.: Poznan Agriculture Univ. Res. Asst., Univ. of Agriculture, Poznan, 1962-. Society: Polish Com. of Nucl. Sci.
Nuclear interests: The application of radioisotope methods in food technology, embracing the tracers, which have been used in studies on the precursors of smell and flavour in raw sausages. The application of a new technique for radiation preservation of food.
Address: 41/9 Szydlowska Street, Poznan, Poland.

FITCH, L. W. N. Member, N. Z. Atomic Energy Com.
Address: New Zealand Atomic Energy Committee, c/o D.S.I.R., Private Bag, Lower Hutt, New Zealand.

FITCH, Val L., Dr. E. O. Lawrence Memorial Award, 1968. Born 1923. Educ.: McGill and Columbia Univs. Instructor (1954-56), Asst. Prof. (1956-59), Assoc. Prof. (1959-60), Prof., Phys. (1960-), Princeton Univ. Formerly Consultant to U.S.A.E.C.; Trustee, Assoc. Univs., Inc., 1961-. Societies: A.P.S.; Italian Phys. Soc.; N.A.S.; American Acad. of Arts and Sci.
Address: 292 Hartley Avenue, Princeton, New Jersey, U.S.A.

FITOUSSI, Lucien, L. ès Sc. Physiques. Born 1929. Educ.: Paris Univ. Societies: Internat. Rad. Protection Assoc.; A.N.S.; Sté. Française de Radioprotection.
Nuclear interests: Health physics; Radiation dosimetry; Reactor safety.
Address: Centre d'Etudes Nucléaires de Saclay-S.P.R., B.P.No. 2, Gif-sur-Yvette, 91 Essonne, France.

FITZER, Erich, o. Prof., Dr. techn., Dipl. Ing. Born 1921. Educ.: Karlsruhe Univ. Director, Inst. für Chemische Technik.
Societies: Gesellschaft Deutscher Chemiker; Dechema; A.C.S.; Gesellschaft für Metallkunde; Verein Österreichische Chemiker, Eisenhütte Österreich.
Nuclear interests: Metallurgy, graphite, radiation damage, fuel elements.
Address: 14 Weberstrasse, 75 Karlsruhe, Germany.

FITZGERALD, Joseph J., B.S. (Phys.), M.S. (Phys.). Born 1919. Educ.: Boston Coll. and Rochester Univ. Consultant, Health Phys., K.A.P.L., G.E., 1958; Supervisor, Radiol. Phys. and Chem. (1957), Chairman, Reactor Safeguards Com. (1958), Chairman, Radioactive Wastes Com., (1958), K.A.P.L.; Consultant, Rad. Protection, Massachusetts Gen. Hospital; Consultant, Rad. Protection, Peter Bent Brigham Hospital; Asst. Prof. Phys., School of Public Health, Harvard Univ.; Consultant, Radiol. Services, University Health Services, Harvard Univ., 1961; Pres. and Director, S Sanders Nucl. Corp. and Cambridge Nucl. Corp. First Pres., New England Sect., H.P.S., 1962; Commissioner of Atomic Energy, Massachusetts; Consultant, U.S.P.H.S., U.S.A.E.C., Los Alamos Sci. Lab. and Gen. Atomic Director of Rad., Protection and Control, Massachusetts Gen. Hosp., 1962; Chairman, Special Fellowship Board in Health Phys., U.S.A.E.C.; Chairman, A.S.A. N 7.5 Com. on Rad. Protection Standards for Nucl. Reactors; A.S.A. N 6.7 Com. on Failure Probabilities (Reactor Safety), 1958; A.I.H.S., Rad. Sessions Com. Books: Reactor Safeguards Properties of Dangerous Materials (Sax, Rheingold Publishing Co.); co-author, Mathematical Book I - Theory of Rad. Dosimetry (New York, London and Paris, Gordon and Breach). Societies: A.N.S.; Ind. Hygiene Assoc.; Atomic Ind. Forum; H.P.S.
Nuclear interests: Reactor hazard analyses; Health physics; Radiological engineering; Isotope production and applications.
Address: 7 Squire Road, Winchester, Massachusetts, U.S.A.

FITZGERALD, Joseph M., Jr., B.S. (Seton Hall), M.A. (Montclair). Born 1920. Educ.: Seton Hall Univ. and Montclair State Coll. Director Summer Training Programme for Secondary School Sci. Students, 1960, 1961, 1962, Newark Coll. Eng. (Nat. Sci. Foundation sponsored); Exec. Assoc., Assoc. Prof., Dept. Chem., Newark Coll. Eng. Societies: A.C.S.; American Ceramic Soc.; A.S.E.E.; A.A.A.S.
Nuclear interests: Nuclear metallurgy; Chemical processes.
Address: 15 Madison Street, Glen Ridge, New Jersey, U.S.A.

FITZGERALD, P. R. Sec., Conzinc Riotinto of Australia Ltd.
Address: Conzinc Riotinto of Australia Ltd., 95 Collins Street, Melbourne, Victoria, Australia.

FITZGIBBONS, John David, Bachelor of Nucl. Eng., Graduate of O.R.S.O.R.T. Born 1922. Educ.: North Carolina State Coll. Nucl. Eng. (1953-57), Project Eng., Res. Reactors (1957-59), Project Manager, Res. Reactors (1959-61), Sen. Eng., Design and Erection,

Res. Reactors, Schedule of Power Reactors (1961-63), Sen. Eng., Nucl. Fuel Management (1963-67), Nucl. Fuel Specialist, Naval Reactors (1967-), Babcock and Wilcox, Boiler Div. Nuclear interests: Reactor engineering and management; Computer programming and scheduling.
Address: 3639 East Woodside Avenue, Lynchburg, Virginia, U.S.A.

FITZPATRICK, John Patrick, B.S. (Chem. Eng., Notre Dame). Born 1912. Educ.: Notre Dame Univ. and M.I.T. Formerly Assoc. Director, Particle Accelerator Div., Argonne Nat. Lab. Societies: A.I.Ch.E.; A.C.S.; A.A.A.S.; A.N.S.
Nuclear interest: Construction and operation of accelerators.
Address: 6121 N. Kedvale Avenue, Chicago 46, Illinois, U.S.A.

FIX, Richard Conrad, B.S., Ph.D. Born 1930. Educ.: Wisconsin Univ. and M.I.T., Res. Asst., Nucl. Chem. Group, M.I.T., 1952-56; Project Leader, investigative work in radiochem., Tracerlab, Inc., Waltham, Mass., 1956-57; Head, Nucl. Chem. Dept. (1957-59), Asst. Tech. Director (1959-63), Vice-Pres. (1963-65), Director, R. and D. Div. (1963-65), Controls for Rad., Inc., Cambridge, Mass. Sen. Staff Sci., Tracerlab Div. of LFE, Inc., Waltham, Mass., 1965-. Societies: A.C.S.; H.P.S.; A.N.S.; Rad. Res. Soc.
Nuclear interests: Radiation dosimetry; Radiation detection; Radiation safety; Nuclear chemistry.
Address: Tracerlab Div. of LFE, Inc., 1601 Trapelo Road, Waltham, Mass. 02154, U.S.A.

FIZZOTTI, Carlo, Dr. in Industrial Chem. Born 1927. Educ.: Milan Univ. At C.I.S.E., Milan, 1953-57; at C.N.E.N. 1957-; Tecnologia dei Materiali Impiegati Negli Impianti Nucleari, Facolta di Ingegneria, Rome Inst. of Technol.
Nuclear interest: Nuclear metallurgy.
Address: Rome Institute of Technology, 18 Via Eudossiana, Rome, Italy.

FLACHOWSKY, Paul Johannes, Dr. rer. nat., Dipl.-chem. Educ.: Leipzig Univ. Inst. of Appl. Radioactivity, German Acad. Sci., 1958-64; Isotopic Lab., Petroleum Refinery, Schwedt/ Oder, 1964-66; German Acad. Sci., 1966-. Society: Sect. Nucl. Chem., Soc. of German Chemists.
Nuclear interest: Application of radioisotopes in industry, especially in the petroleum industry; Radiochemical methods of analysis.
Address: 17 Louis Fürnberg Str., (705) Leipzig, German Democratic Republic.

FLACK, Frederick Colin, B.Sc., Ph.D. Born 1926. Educ.: Liverpool Univ. Res. Fellow, British Columbia Univ., 1949-52; I.C.I. Res. Fellow, Glasgow Univ., 1952-54; Sen. Lecturer, Exeter Univ.
Nuclear interests: Nuclear reactions; Gamma ray spectroscopy, especially applied to chemical analysis.
Address: Department of Physics, Exeter University, The Queens Drive, Exeter, Devon, England.

FLACK, J. C. Chief Nucl. Eng., Nucl. Labs., Lockheed Georgia Div., Lockheed Aircraft Corp. Vice Chairman, Huntsville, Alabama Sect., A.N.S.
Address: Lockheed Georgia Division, Marietta, Georgia, U.S.A.

FLAHERTY, John J., B.A. Born 1913. Educ.: Michigan Univ. Manager, Chicago Operations Office, U.S.A.E.C., 1953; Asst. to Gen. Manager (1957), Exec. Vice Pres. (1961), Pres. (1966), Atomics Internat. (a Div. of North American Rockwell Corp.). Society: A.I.A.A.
Nuclear interests: Management; Compact nuclear power plants.
Address: Atomics International, P.O. Box 309, Canoga Park, California, U.S.A.

FLAIG, Wolfgang, Prof. Dr. phil. nat. habil. Born 1912. Educ.: Freiburg, Würzburg, Frankfurt, and Halle Univs. and Max-Planck Inst., Mülheim-Ruhr. Director, Inst. Biochem. des Bodens, Forschungsanstalt für Landwirtschaft, Brunswick. Member, Arbeitskreis II/3 – Biol. und Medizin, Deutsche Atomkommission. Active member, New York Acad. Sci.; Conserjero hon., Consejo Superior de Investigaciones Sci., Madrid (Oberster Spanischer Wissenschaftsrat). Societies: Gesellschaft Deutscher Chem.; Deutsche Bodenkundliche Gesellschaft; Pres., Commission II, Internat. Soil Sci. Soc.; A.C.S.
Nuclear interests: Chemistry of soil organic matter; Plant metabolism.
Address: Institut für Biochemie des Bodens der Forschungsanstalt für Landwirtschaft, 50 Bundesallee, Brunswick-Völkenrode, Germany.

FLAJSHANZ, Milan, Mech. Eng. Born 1924. Educ.: Prague Tech. Univ. Chief Eng. of Reactor Project Zvil-Plzen; Lecturer, Reactor Design, Plzen Tech. Univ. Society: Czechoslovak Sci. and Tech. Soc.
Nuclear interests: Pressure vessel, reactor design, management of reactor project; Reactor safety.
Address: 108 Kyjevska, Plzen, Czechoslovakia.

FLAKUS, F. N. Member for Germany, Group of Experts on Production of Energy from Radioisotopes, O.E.C.D., E.N.E.A.
Address: O.E.C.D. European Nuclear Energy Agency, 38 boulevard Suchet, Paris 16, France.

FLAMANT, Pierre. Chief of Dept., Ets. Le Bourget, Cie. Electro-Mécanique.
Address: Cie. Electro-Mecanique, Ets. Le Bourget, 55 avenue Jean-Jaures, Le Bourget, (Seine), France.

FLAMIG, Gerhard. Member, Steering Com., Deutsches Atomforum e.V.
Address: Deutsches Atomforum e.V., Bundeshaus, 5300 Bonn, Germany.

FLAMMERSFELD, Arnold Rudolf Karl, Dr. phil., Dr. rer. nat. habil. Born 1913. Privatdozent, Tübingen, 1948; Ord. Prof., Göttingen, 1954. Ord. Prof. für Physik, Direktor des II. Physikalischen Inst., Göttingen Univ. Books: Co-author, Isotopenbericht (Tübingen, 1949); co-author, Atomphysik (Berlin, 1963).
Nuclear interest: Nuclear physics.
Address: 9 Bunsenstrasse, Göttingen, Germany.

FLAMMIA, Nicola, Political Sci. Born 1932. Educ.: Naples Univ. Chief, Congress and Exhibitions Office, Information Service, C.N.E.N., 1959-.
Nuclear interest: Public information.
Address: C.N.E.N., 15 Via Belisario, Rome, Italy.

FLANAGAN, Terence Patrick, M.Sc., B.Sc. Born 1924. Educ.: London Univ. Elec. Design Dept. (1950-56) Atomic Power Sect., (1956-57), Marconi Instruments, Ltd.; Electronics Dept. (1957), Head of Nucleonics Dept. (1958-64), Head of Ind. Measurement and Control Group, (1964-), British Sci. Instr. Res. Assoc. Societies: Institute of Physics; I.E.E.
Nuclear interests: Nuclear instrumentation and detection; Research work in the field of nuclear radiation detectors, and the application of radioisotope techniques to general instrumentation and research.
Address: 2 Homemead Road, Bromley, Kent, England.

FLANDERS, Robert Bernard, B.S. (Ch.E.), S.M. (Nucl. Eng.). Born 1915. Educ.: Tufts Univ. and M.I.T. Sen. Project Eng., Nucl. Materials Development Lab., Metals and Controls, Inc., Div. of Texas Instruments, Inc. Member, Programme Com., New England Sect., Treas., Northeastern (Boston) Massachusetts Sect., A.N.S.; Plant Visitations Com. for 1962, Boston, Mass., Annual Meeting (Nat.), Member, Executive Com. (1962-63), Treasurer (1963-64 and 1964-65), A.N.S. Societies: A.N.S.; A.A.A.S.; Registered Professional Eng.
Nuclear interest: Power (energy) generation and direct conversion and the associated development of materials, devices and systems for advanced concepts with improved efficiency and reliability. Applications include space, maritime and commercial use.
Address: Nuclear Materials Development Laboratory, Metals and Controls, Inc., Div. of Texas Instruments, Inc., P.O. Box 898, Attleboro, Massachusetts, U.S.A.

FLATBY, Jon, M.Sc. Div. Chief (Rad. Diagnosis and Dosimetry), State Inst. of Rad. Hygiene, Oslo.
Address: State Institute of Radiation Hygiene, Oslo, Montebello, Norway.

FLATSCHER, Josef Hermann. Born 1890. o. Prof., Univ. Agriculture, Vienna. Chairman, Forestry Sect., Osterreichische Studiengesellschaft für Atomenergie, Vienna.
Nuclear interests: Der Einfluss radioaktiver Strahlen (Isotopenbestrahlung) auf das Baumwachstum; Der Schutz vor schädlicher radioaktiver Bestrahlung durch den Wald; Holzbauweise ein Schutz gegen schädlicher Einwirkung radioaktiver Luft; Die Beeinflussung der Radioaktiven Strahlen durch auftretende Erdstrahlen.
Address: 13 Dittesgasse, Vienna 18, Austria.

FLATT, Horace P. Group Leader, Appl. Maths. Group, Atomics Internat. Chairman, Reactor Maths. and Computations and Ex-Officio Member, Professional Divs. Com., A.N.S., 1962-63.
Address: Manager, Problems Analysis Systems, International Business Machines, 2670 Hanover, Palo Alto, California, U.S.A.

FLECHSIG, W., Dr. Formerly Head, Lab. für Halbleiter, Abteilung für Atomphysik, Physikalische-Technische Bundesanstalt.
Address: Physikalische-Technische Bundesanstalt, 100 Bundesallee, Braunschweig, Germany.

FLECK, Lord, K.B.E., D.Sc., LL.D. Born 1889. Educ.: Glasgow Univ. Chairman, Imperial Chem. Industries, Ltd., 1953-60; Deputy Chairman, African Explosives and Chem. Industries, Ltd., 1953-60; Director, Imperial Chem. Industries of Australia and New Zealand, Ltd., 1953-60. Chairman, Internat. Res. and Development Co., Ltd., 1963-. Pres., Industrial Co-Partnership Assoc.; Chairman, Nucl. Safety Advisory Com.; Chairman, M.O.P. Advisory Com. on R. and D. Pres., Roy. Institution, 1963-. Societies: Treas., Roy. Soc.; Fellow, Chem. Soc.; Hon. Fellow, Roy. Soc. of Edinburgh; F.R.I.C.; Pres., Soc. of Chem. Industry, 1960-62.
Nuclear interests: Management; Chairman of 3 committees appointed following Report of Com. of Inquiry into accident at Windscale, 1957.
Address: 100 Roebuck House, Stag Place, London, S.W.1, England.

FLECK, Carl M., Dr. Phil. Born 1937. Educ.: Vienna Univ. Lecturer, Atominst. der Osterreichischen Hochschulen, Vienna, 1963-.
Book: Co-author, Neutronenphysik, Physik und Technik der Aktivierungssonden, Volumes I and II. (Mannheim, Bibliographisches Institut, 1967). Society: Austrian Phys. Soc.
Nuclear interests: Neutron physics: cold neutrons, resonance neutrons. Nuclear physics: neutron induced nuclear reactions, delayed neutrons. Low level counting.
Address: Atominstitut der Osterreichischen Hochschulen, 115 Schüttelstrasse, 1020 Wien 2, Austria.

FLECK, Rudolf, Dr. iur. Born 1908. Educ.: Leipzig Univ. Dozent, Ingenieurschule Kiel.

Book: Umgang mit radioaktiven Stoffen - Rechtsschutz und Haftung (Bonn, Hundesministerium für Wissenschaftliche Forschung 1965; Verlag Neue Wirtschafts-Briefe, Herne).
Nuclear interests: Atom law.
Address: 84 Forstweg, Kiel, Germany.

FLECK, William, M.Sc. Born 1923. Educ.: Queens Univ., Belfast Sci. Officer, Linen Ind. Res. Assoc., 1950-55; Sen. Development Eng., Short Brothers and Harland Ltd., 1955-59; Sen. Lecturer, Phys. Dept. (1960-65), Head, Dept. (1965-), Belfast Coll. of Technol. Society: Assoc., Inst. of Phys.
Nuclear interest: Courses of instruction for industrial personnel and teachers in the handling of isotopes.
Address: Belfast College of Technology, Belfast 1, Northern Ireland.

FLECKENSTEIN, Albrecht, Prof. Dr. med. Mitglied, Arbeitskreis II/3, Biologie und Medizin, Deutsche Atomkommission, Federal Ministry for Sci. Res.
Address: Federal Ministry for Scientific Research, 46 Luisenstrasse, 532 Bad Godesberg, Germany.

FLEGENHEIMER, Juan, Dr. (Chem.), Ph.D. Born 1927. Educ.: Buenos Aires and Cambridge Univs. In radiochem., Comisión Nacional Energia Atómica, Buenos Aires, 1952-; Head, Reprocessing Dept. Co-editor, Radiochimica Acta, 1962-.
Nuclear interest: Nuclear chemistry.
Address: Comisión Nacional de Energia Atómica, 8250 Avda. Libertador, Buenos Aires, Argentina.

FLEGONTOV, V. M. Paper: Co-author, Accumulation Artificial Radionuclides on Surface on Earth in Vicinity of Leningrad in 1954-65 (letter to the Editor, Atomnaya Energiya, vol. 23, No. 4, 1967).
Address: Academy of Sciences of the U.S.S.R., 14 Leninsky Prospekt, Moscow V-71, U.S.S.R.

FLEISCHER, Allan Arthur, B.S. (Phys.), M.S. (Phys.), Ph.D. (Phys.). Born 1931. Educ.: Yale Univ. Member, Tech. Staff, TRW, Inc., 1959-60; Dept. Manager, Edgerton, Germeshausen and Grier, Inc., 1960-63; Group Manager, William M. Brobeck and Assocs., 1963-65; Vice Pres., The Cyclotron Corp., 1965-. Societies: A.P.S.; A.N.S.; Nucl. Medical Soc.; I.E.E.E.
Nuclear interests: Design of nuclear particle accelerators; Development of applications for accelerators; Management of nuclear research-oriented company.
Address: 950 Gilman Street, Berkeley, California 94710, U.S.A.

FLEISCHMANN, Rudolf, Dr. rer. nat. Born 1903. Educ.: Erlangen, Münich and Göttingen Univs. o.Prof. of Phys., Hamburg Univ. and Director, Physikalisches Staatsinst., Hamburg, 1947-53; o.Prof. of Physics, and Head, Phys. Dept., Erlangen-Nurnberg Univ., 1953-.
Societies: Physikalische Gesellschaft, (Verband deutscher Physikalische Gesellschaften); Max Planck Gesellschaft. (Auswärtiges wissenschaftliche Mitglied); Bayerische Akademie der Wissenschaften (ordentlich Mitglied).
Nuclear interest: Nuclear physics.
Address: Physikal Institut, Erlangen Universität, 6 Glückstrasse, Erlangen, Germany.

FLEISHMAN, Morton Robert, B.S.E. (Aeronautical Eng., Michigan), M.S.E. (Nucl. Eng., Michigan). Born 1933. Educ.: Michigan and New York Univs. Asst., Michigan Univ., Aircraft Propulsion Lab., 1951-54; Powerplant Eng., Glenn L. Martin Co., summer 1954; Project Eng., Michigan Univ., Eng. Res. Inst., summer 1955; Student, O.R.S.O.R.T., 1955-56; Sen. Sci., United Nucl. Corp., 1956-63; Nucl. Phys., Space Nucl. Propulsion Office, 1963-. Book: Co-author on 4 chapters, Revised Edition of Reactor Handbook, (U.S.A.E.C.). Societies: A.N.S.; A.I.A.A.
Nuclear interest: Reactor and shielding physics.
Address: 5247 Evergreen Drive, North Olmsted, Ohio 44070, U.S.A.

FLEMES, R., Ministerialdirigent Dr. Mitglied, Aufsichtsrat, Gesellschaft für Kernenergieverwertung in Schiffbau und Schiffahrt m.b.H.
Address: Company for the Utilization of Nuclear Energy in Shipbuilding and Navigation, 2^V Gr. Reichenstrasse, 2, Hamburg 11, Germany.

FLEMING, Robben Wright, B.A., LL.B. Born 1916. Educ.: Beloit Coll. and Wisconsin Univ. Prof. Law, Illinois Univ.; Provost, Wisconsin Univ., 1964-. Member, Atomic Energy Labour Management Relations Panel, U.S.A.E.C.
Address: 708 La Sell Drive, Champaign, Illinois, U.S.A.

FLEMING, Robert W. Director, Atomic Phys. and Sci. Fund, Inc.
Address: Atomic Physics and Science Fund, Inc., 1033 30th Street, N.W., Washington 7, D.C., U.S.A.

FLEMING, Thomas B. Vice-Pres. and Gen. Manager, Reuter-Stokes Electronic Components, Inc.
Address: Reuter-Stokes Electronic Components, Inc., 18530 South Miles Parkway, Warrensville Heights, Cleveland, Ohio 44128, U.S.A.

FLEMING, William Herbert, M.Sc., Ph.D. (McMaster). Born 1925. Educ.: McMaster Univ. Postdoctorate Fellow in Phys. (1954), Reactor Supt. (1958), Assoc. Prof. (Appl. Mathematics) (1966), McMaster Univ. Societies: Canadian Assoc. Phys.; A.N.S.
Nuclear interests: Fission yield studies and reactor operation.
Address: McMaster University, Hamilton, Ontario, Canada.

FLENDER, Hans Walter, Dipl.-Phys. At Inst. für Strahlen- und Kernphysik, Bonn Univ.
Address: Institut für Strahlen- und Kernphysik, Bonn University, 16 Nussallee, Bonn, Germany.

FLEROV, Georgii Nikolaevich, Prof. Lenin Prize, 1967. Born 1913. Educ.: Industrial Inst., Leningrad. Labs. of the Acad. of Sci. of the U.S.S.R., 1938-; Director, Lab. of Nucl. Reactions, Joint Inst. for Nucl. Res., Dubna. Deleg. 2nd I.C.P.U.A.E., Geneva, Sept. 1958. Books: Co-author, Utilisation of radioactive emission in the exploration and exploitation of oil deposits in the U.S.S.R. (Moscow, 1955). Society: Corr. Member, Acad. of Sci. of the U.S.S.R.
Nuclear interests: Nuclear physics and activation analysis.
Address: Joint Institute for Nuclear Research, Head Post Office, P.O. Box 79, Dubna, Moscow, U.S.S.R.

FLETCHER, John, Board member, Atomkraftkonsortiet.
Address: Atomkraftkonsortiet, 19 Stureplan, Stockholm C., Sweden.

FLETCHER, John Molyneux, M.A., Ph.D., F.S.A. Born 1910. Educ.: Oxford and California (Berkeley) Univs. Consultant. Book: Process Chem. (Progress in Nucl. Energy, Series III), vol. 1, 1956; vol. 2, 1958; vol. 3, 1961 (Pergamon Press). Societies: Chem. Soc.; Soc. of Chem. Industry; I.M.M.
Nuclear interests: Chemical processing and extraction metallurgy.
Address: Bekynton House, Sutton Courtenay, Berks., England.

FLETCHER, Pablo, Dr. Medical Dept., Instituto de Estudios Nucleares, Panama Univ.
Address: Panama University, Faculty of Sciences, Instituto de Estudios Nucleares, Apartado 3277, Panama, Republic of Panama.

FLETCHER, Paul Thomas, B.Sc. (Eng.). Born 1912. Educ.: Medway Tech. Coll. Chief Eng. (M. and E.), Ministry of Works; Deputy Director of Eng. (1954-57), Director of Eng. (1957-58), Deputy Managing Director (1958), U.K.A.E.A., Ind. Group; Deputy Managing Director, Development and Eng. Group, U.K.A.E.A.; Director, United Power Co. Ltd. Managing Director, G.E.C. (Process Eng.) Ltd. Nucl. Safety Advisory Com., Ministry of Power. Societies: Inst. Mech. Eng.; Inst. Civil Eng.; I.E.E.
Nuclear interests: General management of design and development.
Address: G.E.C. (Process Engineering) Ltd., Birch Walk, Erith, Kent, England.

FLETCHER, Robert C., B.S., Ph.D. (Phys.). Born 1921. Educ.: M.I.T. Member, Tech. Staff (1949-56), Dept. Head, Solid State Devices Lab. (1956-58), Director, Solid State Devices Lab. (1958-62), Director, Electronic Devices Lab. (1962-64), Exec. Director, Military Systems Res. Div. (1967-), Bell Telephone Labs.; Vice Pres., Res., Sandia Corp., 1964-67; Societies: Fellow, A.P.S.; I.E.E.E.
Nuclear interest: Management.
Address: Bell Telephone Laboratories, Whippany, New Jersey 07981, U.S.A.

FLETCHER, Robert D., Jr., B.S. (Chem.). Born 1917. Educ.: Norwich Univ., Northfield, Vermont. With G.E.C., Hanford, 1950-52; American Cyanamid Co., Idaho Falls, 1952-53; Group Leader, Hot Lab. Equipment Development (1953-62), Group Leader, Coordination, Loft Project (1962-), Phillips Petroleum Co., Atomic Energy Div., Idaho Falls. Society: A.N.S.
Nuclear interests: Development and design of specialised equipment for research laboratories, radiochemical laboratories, and multicurie hot cells.
Address: 2829 Westmoreland Drive, Idaho Falls, Idaho, U.S.A.

FLETCHER, William, B.Sc. Born 1915. Educ.: London Univ. Sen. Sci. Officer, Springfields Works (1950), Sen. Sci. Officer, Capenhurst Works (1951), Principal Sci. Officer, Capenhurst Works (1952), M.O.S.; Chief Chem. (1954), Tech. Manager (1961-), Capenhurst Works, U.K.A.E.A. Society: Roy. Inst. of Chem.
Nuclear interests: Inorganic and analytical chemistry and the use of modern instrumental techniques; Laboratory management.
Address: 13 Green Way, Saughall, Chester, England.

FLEUR, J. J. C. LA. See LA FLEUR, J. J. C.

FLEURY, Georges, Pres., Assoc. Technique pour la Production et l'Utilisation de l'Energie Nucléaire; Pres., France-Atome; Pres., Editorial Com., Energie Nucléaire, Past-Pres., A.T.E.N.; Director, Groupe Intersyndical de l'Industrie Nucléaire.
Address: A.T.E.N., 26 rue de Clichy, Paris 9, France.

FLEURY, P., Prof. Sec., Internat. Union of Pure and Appl. Phys.
Address: International Union of Pure and Applied Physics, 3 boulevard Pasteur, Paris 15, France.

FLEXER, P. Manager, of Nucl. Dept., Cie. Générale Française d'Etudes Techniques.
Address: Cie. Générale Française d'Etudes Techniques, 3 rue Moncey, Paris 9, France.

FLINT, Reginald Arthur. Born 1929. Sect. Head, Atomic Power Div., English Elec. Society: I.E.E.
Nuclear interests: Reactor equipment design; Fuelling machinery.
Address: 12 Francis Avenue, Braunstone, Leicester, England.

FLINT, Robert Francis, A.L.A. Born 1922. Librarian, Fulmer Res. Inst. Ltd. Hon. Sec., Bucks., Berks., and Oxon Branch, Library Assoc. Society: Inst. of Metals.
Nuclear interests: Metals and materials.
Address: Fulmer Research Institute Limited, Stoke Poges, Bucks., England.

FLINTA, Jan Erland, Civ. Ing. (Dr.) (Roy. Inst. Technol., Stockholm). Born 1918. Educ.: Roy. Inst. Technol., Stockholm. In Dept. of Defence Res. Lab., F.O.A., 1947-50; Leader, Sect. for Heat Transfer and Mech. Lab., (1950-60), in Studsvik (1961-), AB Atomenergi.
Nuclear interest: Reactor design and development.
Address: 20 Rosenkällavägen, Nyköping, Sweden.

FLOBERG, John Forrest, A.B., LL.B. Born 1915. Educ.: Loyola and Harvard Univs. Asst. Sec. of the Navy for Air, 1949-53; Partner, Kirkland, Fleming, Green, Martin and Ellis, 1953-57. Member, U.S.A.E.C., 1957-60. Deleg. to 2nd I.C.P.U.A.E., Geneva, 1958. Advisor, 2nd Gen. Conference, I.A.E.A., 1958; Alternate Delegate, 3rd Gen. Conference, I.A.E.A., 1959; Dept. Member, Defence Dept.'s. R. and D. Board and Member, Air Co-ordinating Com., 1949-53; Consultant, 2nd Hoover Com.; Gen. Counsel (1960-), Sec. (1962-), Director (1964-), Vice Pres. (1965-), The Firestone Tire and Rubber Co.; Trustee, Res. Analysis Corp., 1961-; Director, Gen. Dynamics Corp. 1967-. America, Illinois, Chicago, District of Columbia, and Ohio Bar Assocs.
Nuclear interest: Management.
Address: 4240 Ira Road, Bath, Ohio, U.S.A.

FLOIRAT, S. Vice-Pres., Sté. des Engins Matra.
Address: Société des Engins Matra, 49 rue de Lisbonne, Paris 8, France.

FLOOK, George Alexander. Born 1921. Eng. with M.O.S. (Div. of Atomic Energy) and U.K.A.E.A., 1948-; Resident Eng., Springfields Works, Preston, for D. and E. Group (1955-59), Deputy Manager and Head of Construction, Southern Works Organisation, Eng. Group, (1959-65), Manager (Special Projects and Planning), Eng. Group, Risley (1965-), U.K.A.E.A. Societies: A.M.Inst.C.E.; M.I.Struct.E.; B.N.E.S.
Nuclear interests: Management of new works design and construction of buildings and plant for atomic energy sites.
Address: Room Y.522, Engineering Group, U.K.A.E.A., Risley, Nr. Warrington, Lancs., England.

FLOREEN, E. D. Eng. Manager, Marman Div., Aeroquip Corp.
Address: Aeroquip Corporation, Marman Division, 11214 Exposition Boulevard, Los Angeles 64, California, U.S.A.

FLORENT, Roger Charles, Ing. Born 1922. Educ.: Conservatoire Nat. des Arts et Métiers, Paris. Synchrotron à protons Saturne, Adjoint au Chef du Dépt. du Synchrotron Saturne, C.E.N., Saclay, C.E.A., 1955; Chef du Projet de la Grande Chambre Européenne, C.E.R.N., Genèva, 1967.
Nuclear interests: Etudes et réalisations d'équipements expérimentaux pour la physique des hautes énergies (transports de faisceaux, détecteurs, chambres à bulles, application des champs magnétiques supraconducteurs).
Address: C.E.R.N. CH-1211 Genèva 23, Switzerland.

FLORENTIN, P. Prés. Directeur Gén., Chaudronnerie Industrielle et Constructions Spéciales S.A.
Address: Chaudronnerie Industrielle et Constructions Spéciales S.A., Vieux Chemin de Saint-Denis, Noisy-le-Sec, (Seine), France.

FLORENTSEV, M. M. Paper: Co-author, Circulating Rate Effect on Rad.-Induced Conversion of Organic Coolants at Elevated Temperatures (Atomnaya Energiya, vol. 22, No. 5, 1967).
Address: Academy of Sciences of the U.S.S.R., 14 Leninsky Prospekt, Moscow V-71, U.S.S.R.

FLORES, Julio PIZARRO. See PIZARRO FLORES, Julio.

FLORES-MALDONADO, Victor, B.S. (E.E.), B.S. (Phys.), M.S. (Phys.), Ph.D. (Phys.). Born 1924. Educ.: E.S.I.M.E. Inst. Politecnico Nacional, Univ. Nacional Autonoma de Mexico, Internat. School of Nucl. Science and Eng., Argonne Nat. Lab., Lehigh and Cornell Univs. E. Eng., Nitrogeno S.A., 1947-54; Res. fellow, Univ. Nacional Autonoma de Mexico (Inst. of Phys.), 1954-61; Adviser, Comisión Nacional de Energia Nucl., 1959-; Dean, Escuela Superior de Fisica y Matematicas, Inst. Politécnico Nacional, 1961-64. Societies: Soc. Mexicana de Fisica; A.P.S.
Nuclear interests: Theoretical nuclear physics; Reactor physics; Nuclear engineering.
Address: 170 Oriente #64, Mexico 9, D.F., Mexico.

FLORKIN, M. Dr. in Medicine, Inter-Faculty Centre of Nucl. Sci., Liège Univ.
Address: Inter-Faculty Centre of Nuclear Sciences, Liège University, Liège, Belgium.

FLORKOWSKI, Tadeusz, Dr. phys., M. Eng. Elec. Born 1929. Educ.: Acad. of Mining and Metal., Cracow. At Inst. of Phys., Acad. of Mining and Metal., Cracow, 1950; Head of Dept., Inst. Nucl. Technol., 1962; Dept. Res. and Lab. I.A.E.A., 1964.
Nuclear interests: Application of radioisotopes in industry and geophysics; Applied X-and gamma ray spectrometry; Radioisotope X-ray fluorescence analysis; Low-level tritium and radiocarbon counting.

FLO

Address: 15/4/8 Hofzeile, Vienna 1190, Austria.

FLOUD, R. C. Chief Eng., Eng. and Nucl. Div., Fairey Eng. Ltd.
Address: Fairey Engineering Ltd., Heston, Middlesex, England.

FLOWERS, Brian Hilton, M.A. (Cantab.), M.A. (Oxon), M.Sc. (Manchester), F.Inst.P., D.Sc. (Birmingham), F.R.S. Born 1924. Educ.: Cambridge and Birmingham Univs. Jun. Sci. Officer, D.S.I.R. (Tube Alloys), Montreal and Chalk River, Ontario, 1944; Sci. Officer, Nucl. Phys. Div. (1946), transferred to Theoretical Phys. Div. (1948), Head, Theoretical Phys. Div. (1952), Chief Res. Sci. (1958), A.E.R.E., Harwell; Prof., Theoretical Phys. (1958-61), Langworthy Prof. Phys. (1961-), Manchester Univ.; Member, Council for Sci. Policy, 1965-67; Chairman, Computer Board for Universities and Res. Councils, 1966-; Chairman, Sci. Res. Council, 1967-. Societies: Sometime Vice-Pres. Inst. of Phys. and Phys. Soc.;
Nuclear interests: Nuclear physics and neutronics.
Address: Science Research Council, State House, High Holborn, London, W.C.1, England.

FLOYD, Lieutenant Commander E. D. M., R.N., B.A. Sen. Lecturer, Dept. of Nucl. Sci. and Technol., Roy. Naval Coll.
Address: Royal Naval College, Department of Nuclear Science and Technology, Greenwich, London S.E.10, England.

FLOYD, George C. Sen. Vice Pres., Vanadium Corp. of America.
Address: Vanadium Corp. of America, Pan Am Building, 200 Park Avenue, New York 17, New York, U.S.A.

FLOYD, John D., Born 1925. Educ.: Missouri School of Mines. Reactor Operations Supv., Gen. Elec., 1952-62; Site Rep. ERR, Elk River Reactor, La Crosse Boiling Water Reactor, U.S.A.E.C., 1962-. Society: A.N.S. Nuclear interests: Power reactor management and staff training.
Address: 401 N. 26th Place, La Crosse, Wisconsin 54601, U.S.A.

FLUHARTY, Rex Gilbert, B.S., Ph.D. Born 1918. Educ.: Idaho and Washington Univs. and M.I.T. Phys., O.R.I.N.S., 1949-52; Phys., Atomic Energy Div., Phillips Petroleum Co., 1952-66; Idaho Nucl. Corp., 1966-. Member, U.S.A.E.C. Neutron Cross-section Advisory Group. Societies: A.P.S.; A.N.S.
Nuclear interests: Reactor neutron cross sections; Low energy nuclear levels at neutron binding energy; Fission; Molecular and liquid-solid energy states; Reactor neutron physics; Radioactivity; Tracer techniques; Neutron source.
Address: Idaho Nuclear Corporation, P.O. Box 1845, Idaho Falls, Idaho 83401, U.S.A.

FLURSCHEIM, Cedric Harald, B.A. Born 1960. Educ.: Cambridge Univ. Director of Eng., A.E.I.
Nuclear interests: Special equipment such as ZETA. Generation, distribution and instrumentation equipment for nuclear power stations.
Address: Associated Electrical Industries Ltd., Power Group, Manchester 17, England.

FLYNN, Arthur William, B.Ch.E., M.Ch.E., D. Eng. Sc., Lic. Prof. Eng. Born 1919. Educ.: C.C.N.Y., N.Y.U. Vitro Corp. of America (Kellex), 1947-53; Bechtel Corp., 1953-55; Nucl. Development Corp. of America, 1955-61; Ebasco Services, Inc., 1961-. Societies: A.I.Ch.E.; A.N.S.; Nat. Soc. Professional Eng.; A.N.S. Standards Com. ANS-7.
Nuclear interests: Reactor systems design; In-pile experiments; Nuclear power plants.
Address: 1561 Unionport Road, Bronx 62, New York, New York, U.S.A.

FLYNN, Robert James, D.V.M. Born 1923. Educ.: Michigan State Univ. Veterinarian, Argonne Nat. Lab., 1948-. Sec.-Treas. (1953-62), Pres. (1964), Animal Care Panel. Sec.-Treas. (1957-62), Pres. (1963), American Coll. of Lab. Animal Medicine. Society: Rad. Res. Soc.
Nuclear interests: Radiobiology; Effects of external and internal emitters on living things; Particularly interested in the interference caused by the spontaneous diseases of the laboratory animals used.
Address: Argonne National Laboratory, Argonne, Illinois, U.S.A.

FOA. C. Co-editor, Minerva Nucleare.
Address: 83-85 Corso Bramante, Turin, Italy.

FOCH, René, Doctorate in law, lic. in history, degree of the French Colonial School. Born 1924. Educ.: Toulouse and Paris Univs. Head, Economic Sect., Secretariat, Council of Europe, 1951; liaison officer between Council of Europe and O.E.E.C., 1953-57; Director, External Relations Div., Euratom, 1958-; Formerly Member of Steering Com., now Member, Control Bureau, O.E.C.D., E.N.E.A. Book: La Haute Autorité de la Vallée du Tennessee.
Address: 27 avenue des Aubépines, Uccle, Brussels, Belgium.

FOCK, Cornelis Laurens Willem, LL.D. Born 1905. Educ.: Leiden Univ. Commissioner of H.M. the Queen in the Province of Groningen. Member, Board of Directors, Reactor Centre, Netherlands.
Address: 17 Marktstraat, Groningen, Netherlands.

FOCKE, Alberto TORRES-. See TORRES-FOCKE, Alberto.

FOCKE, Alfred B., B.S. (Phys.), Ph.D. (Phys.). Born 1906. Educ.: Case Inst. Technol. and California Inst. Technol. Chairman, Dept. of Phys., and Prof. Phys., Harvey Mudd Coll., Claremont, California. Weapons effects tests, Navy Electronics Lab., 1946-53; Director, California Univ. Marine Phys. Lab., 1954-58 (during this time also Sci. Director, Operation Wigwam, U.S. Dept. of Defence); Tech. Director, Naval Air Missile Test Centre; Tech. Director, Pacific Missile Range, 1958-59; Member, Mine Advisory Com., 1955 (Chairman, 1956-59, 1962-). Chairman, Navy Dept. of Civil Service Examiners, Pasadena, 1948. Societies: A.A.P.T.; A.S.A.; A.I.P.; A.G.U.; I.E.E.E.; Inst. for Aerospace Sci.; Seismological Soc. of America.
Nuclear interests: Effects of nuclear weapons; Under-water and air-borne acoustics.
Address: Harvey Mudd College, Claremont, California, U.S.A.

FOCKE, Arthur E., B. Met. E., M.S. (Phys. and Metal.), Ph.D. (Metal.). Born 1904. Educ.: Ohio State Univ. Chief Metal. Diamond Chain Co., Indianapolis, 1945-51; Manager, Materials Development, Aircraft Nucl. Propulsion Dept., Gen. Elec. Co., Evendale, Ohio, 1951-61. Nat. Trustee (1944-48), Vice-Pres. (1949), Pres. (1950), A.S.M.; Assoc. Prof. Met. Eng., Cincinnati Univ.; Pres., A. E. Focke Corp. Societies: A.S.M.; A.S.T.M.; A.M.I.E.E.; British Iron and Steel Inst.; A.N.S.; A.S.E.E.
Nuclear interests: Metallurgy; Reactor materials.
Address: 7420 Miami Hills Drive, Madeira, Cincinnati 43, Ohio 45243, U.S.A.

FODERARO, Anthony H., Ph.D. (Phys.). Born 1926. Educ.: Pittsburgh Univ. Westinghouse Atomic Power, 1954-56; General Motors Res. Labs., 1956-60; Pennsylvania State Univ., 1960-. Societies: A.N.S.; A.P.S.
Nuclear interests: Reactor shielding; Reactor hazards; Radiation damage research; Reactor theory; Nuclear weapons effects shielding; Teaching and research in all these fields.
Address: Nuclear Engineering Department, Sackett Building, Pennsylvania State University, University Park, Pa. 16802, U.S.A.

FODOR, Gyorgy. Member, Editorial Board, Energia es Atomtechnika.
Address: Muszaki Konyvkiado, Bajcsy-Zsilinszky ut 22, Budapest 5, Hungary.

FODOR, Joseph, Dipl. Chem. Ing., Dr.Sc.Tech. Born 1926. Educ.: Polytech. Univ., Budapest. At Inst. for Development of Vehicles, Chief of Radioisotopes Lab., Budapest. Book: Ind. Applications of Tracer Technique (Budapest, Müszaki Konyvkiadó, 1962). Society: Hungarian Chem. Soc.
Nuclear interest: Industrial uses of tracer technique.
Address: 5/b Naphegy tér, Budapest 1, Hungary.

FODOR, Zoltán. Born 1911. Different diplomatic posts, Ministry for Foreign Affairs of the Hungarian People's Republic. Hungarian Resident Rep., I.A.E.A.
Address: 4 Hollandstrasse, 1020 Vienna, Austria.

FOERSTNER, Peter. Pres., Abteilung Strahlenmessgerate, Totak Kom.-Ges. Foerstner und Co.
Address: Totak Kom.-Ges. Foerstner und Co., 6802 Ladeburg/Neckar, Germany.

FOGAGNOLO, Giorgio, Dr. Phys. Born 1932. Educ.: Bologna Univ. Chief, Group in charge of shielding studies and automatisation, Sorin, 1958-63; now Deputy Director, Fast Reactor Programme. C.N.E.N., P.R.V. Societies: A.N.S.; F.I.E.N.
Nuclear interests: Reactor design; Management.
Address: 125 Viale Regina Margherita, Rome, Italy.

FOGEL, Karl Jakob Gustav, Dr. phil. Born 1921. Educ.: Helsingfors and Lund Univs. Teacher Phys. (1950), Prof. Phys. (1954), Dean, Faculty Mathematics and Natural Sci. (1961-), Abo Akademi. Society: Soc. Sci. Fennica.
Nuclear interest: Theoretical nuclear physics.
Address: Department of Physics, Abo Akademi, Abo, Finland.

FOGLAR, Oswald, Dipl. Ing., Dr. Techn. Born 1925. Educ.: Graz T.H. At Metallwerk Plansee, 1949-54; Sté. d'Electro Chimie et Electro Metallurgie et Aciérie Electrique d'Ugine, 1954-.
Nuclear interest: Fabrication of borides, carbides, tungsten and tungsten alloys.
Address: 60 rue Ampère, Grenoble, France.

FOJTIK, Anton, Dipl. Ing. Born 1938. Educ.: Prague Tech. Univ. At Inst. of Rad. Hygiene, Prague, 1962-63; Res. worker, Inst. of Phys. Chem., Czechoslovak Acad. of Sci. Society: Czechoslovak Chem. Soc.
Nuclear interests: Radiation chemistry of organics; Radiation biology; Radiochemistry.
Address: 604 Invalidovna, Prague 8, Czechoslovakia.

FOKIN, A. V. Paper: Co-author, On Prospects of Decontaminating Liquid Radioactive Wastes by Method of "Oil" Flotation (Abstract, Atomnaya Energiya, vol. 20, No. 1, 1966).
Address: Academy of Sciences of the U.S.S.R., 14 Leninsky Prospekt, Moscow V-71, U.S.S.R.

FOKIN, G. N. Paper: Co-author, Absolute Reactivity Measurements and other Parameters Exptl. Determination with Method of Statistical Weights (Atomnaya Energiya, vol. 24, No. 1, 1968).
Address: Academy of Sciences of the U.S.S.R., 14 Leninsky Prospekt, Moscow V-71, U.S.S.R.

FOLDZINSKA, Aleksandra, M.Sc. Born 1932. Educ.: Poznan Univ.
Nuclear interests: Activation analysis; radiochemistry.
Address: 12m 258 ul. Stalingradzka, Warsaw, Poland.

FOLEY, H. Thomas. Sen. Investigator, Rad. Branch, Nat. Cancer Inst., U.S. Dept. of Health, Educ., and Welfare.
Address: U. S. Department of Health, Education, and Welfare, National Cancer Institute, 9000 Wisconsin Avenue, Bethesda 14, Maryland, U.S.A.

FOLEY, Henry M., Prof. Theoretical Phys. Res. (mainly fundamental particle research) (U.S.A.E.C. contract), Nevis Cyclotron Labs., Columbia Univ.
Address: Nevis Cyclotron Laboratories, Irvington, New York, U.S.A.

FOLOMESHKIN, V. N. Paper: Co-author, Calculation Spectra of Neutrino Produced by 70 GeV Protons (letter to the Editor, Atomnaya Energiya, vol. 23, No. 1, 1967).
Address: Academy of Sciences of the U.S.S.R., 14 Leninsky Prospekt, Moscow V-71, U.S.S.R.

FOLSOM, Francis Marion, Dr. Vatican Permanent Rep. (New York), I.A.E.A.; Vatican Deleg., I.A.E.A. Gen. Conference, Vienna, Sept. 1960.
Address: c/o Monsignor Gerolamo Prigione, Vatican Resident Representative, I.A.E.A., 31 Theresianumgasse, Vienna 4, Austria.

FOLSOM, Theodore Robert, B.S., M.S., Ph.D. Born 1908. Educ.: California Inst. Technol. and California Univ. (Scripps Inst. Oceanography). Societies: Geophys. Union; American Assoc. of Phys. in Medicine; A.A.A.S.; New York Acad. of Sci.; Human Ecological Soc.; Ecological Soc. of America.
Nuclear interests: Oceanographic use of nuclear science, especially nuclear tagged water masses, pollution from cities, and fallout, from waste disposal; Special nuclear instrumentation required for ship use.
Address: Scripps Institution of Oceanography, La Jolla, California 92037, U.S.A.

FOMENKO, G. P. Paper: Co-author, 1, 5 GeV Electron Synchrotron at the Polytech. Inst. of Tomsk (Atomnaya Energiya, vol. 21, No. 6, 1966).
Address: Academy of Sciences of the U.S.S.R., 14 Leninsky Prospekt, Moscow V-71, U.S.S.R.

FOMIN, G. S. Paper: Co-author, Fabrication of Thin Plates from Refractory Carbides (Atomnaya Energiya, vol. 20, No. 6, 1966).
Address: Academy of Sciences of the U.S.S.R., 14 Leninsky Prospekt, Moscow V-71, U.S.S.R.

FOMIN, Yu. K. Papers: Co-author, On Prospects of Decontaminating Liquid Radioactive Wastes by Method of "Oil" Flotation (Abstract, Atomnaya Energiya, vol. 20, No. 1, 1966); co-author, Concerning Influence of Surface Active Agents on Bituminisation Process of Radioactive Waste (ibid, vol. 22, No. 5, 1967).
Address: Academy of Sciences of the U.S.S.R., 14 Leninsky Prospekt, Moscow V-71, U.S.S.R.

FOMINYKH, V. I. Papers: Co-author, Absolute Measurement of Neutron Source Yield by Manganese Activation Method (letter to the Editor, Atomnaya Energiya, vol. 16, No. 3, 1964); co-author, Internat. Comparison of Neutron Sources (letter to the Editor, ibid, vol. 19, No. 1, 1965); co-author, Internat. Collation of Neutron Sources (ibid, No. 2).
Address: Academy of Sciences of the U.S.S.R., 14 Leninsky Prospekt, Moscow V-71, U.S.S.R.

FONDA, Luciano, Phys. graduate. Born 1931. Educ.: Trieste Univ. Asst. Prof., Trieste Univ., 1955-58; Res. Assoc., Indiana Univ., 1958-59; Member, Inst. for Advanced Study, Princeton, New Jersey, 1959-60; Full Prof. Quantum Mechanics, Palermo Univ., 1961; Parma Univ., 1962; Trieste Univ., 1963-. Societies: Soc. Italiana di Fisica; A.P.S.
Nuclear interests: Pion-nucleon and nucleon-nucleon interaction; Hyperon-nucleon interaction and hyperfragments; Theshold effects in nuclear (or other) collisions; Theory of resonance reactions.
Address: 19 Via Schiaparelli, Trieste, Italy.

FONG, James T., B.S., M.S. Born 1927. Educ.: M.I.T. and Columbia Univ. Formerly Project Eng., Res. Dept., Foster Wheeler Corp., New York; Tech. Staff of Vice-Pres. and Gen. Manager, (1957-66), Manager, Irradiated Components Eng. (1966-), Bettis Atomic Power Lab., Westinghouse Elec. Corp. Society: A.N.S.
Nuclear interests: Reactor engineering, metallurgy, economics, and power plant engineering.
Address: Westinghouse Electric Corporation, Bettis Atomic Power Laboratory, Box 1468, Pittsburgh 30, Pennsylvania, U.S.A.

FONRODONA, Jaime. Vocale, Comisiones Asesoras de Centrales Nucleares, Junta de Energia Nucl.
Address: Junta de Energia Nuclear, Ciudad Universitaria, Madrid 3, Spain.

FONTAINE, André Maurice, L. ès Sc., Ing. Chimiste E.N.S.C.M., D. ès Sc. Physiques. Born 1924. Educ.: Montpellier Univ. Attaché de recherches C.N.R.S., 1950-52; Asst. à la Faculté des Sci., Montpellier, 1952-56. Ing. chargé des Laboratoires au Service Extraction du Plutonium, Centre de Marcoule, Gard, France. Societies: Sté. Chimique de France; A.C.S.
Nuclear interests: Chimie analytique; Traitement des combustibles irradiés.
Address: Les Chardonnerets, Connaux, Gard, France.

FONTAINE, Maurice, D. ès Sc. Prof. Director, Inst. Océanographique and Muséum Nat. d'Histoire Naturelle, Paris. Ministry of Nat. Educ. French Alternate to Governor, I.A.E.A.; Deleg., I.A.E.A. Sci. Conference on the Disposal of Radioactive Wastes, Monaco, 16-21 Nov. 1959; French alternate deleg., I.A.E.A. Gen. Conference, Vienna, Sept. 1960. Society: Assoc. Internat. de Limnology.
Nuclear interest: Radioecology.
Address: Institut Océanographique, 195 rue St. Jacques, Paris 5, France.

FONTAINE, Richard Hector Jean Jacques, Civil Mine Eng. and Naval Architect. Born 1921. Educ.: Faculté Polytech. de Mons and Louvain Catholic Univ. Gen. Manager, Mercantile Marine Eng. and Graving Docks Co. S.A.
Address: 127 Baillet Latourlei, Brasschaat, Belgium.

FONTAN, Jacques Jean Paul, Dr. ès Sc. Born 1937. Maitre de Conférences, Nucl. and Atomic Phys., Toulouse Univ.
Nuclear interests: Radioactivity of the atmosphere.
Address: Centre de Physique Atomique et Nucléaire, Faculté des Sciences, Université de Toulouse, France.

FONTANA, Mars Guy, B.S. (Chem. Eng.), M.S., Ph.D. (Metal.). Born 1910. Educ.: Michigan Univ. Prof. and Chairman, Dept. Metal. Eng., Ohio State Univ.; Consultant to government and industry on corrosion. Many involve nucl. energy. Various committees, N.A.C.E., A.S.M., and A.I.Ch.E. Books: Corrosion: A Compilation (Columbus, Ohio, Hollenback Press, 1957); Corrosion Eng. (New York City, McGraw-Hill, 1967). Societies: A.S.M.; N.A.C.E.; A.I.Ch.E.; A.I.M.M.E.; Elchem. Soc.; A.S.E.E.; Nat. Acad. of Eng.
Nuclear interest: Corrosion.
Address: 116 West 19th Avenue, The Ohio State University, Columbus, Ohio 43210, U.S.A.

FONTELL, Nils Daniel, Dr. phil. Born 1901. Educ.: Helsinki Univ. Prof. Phys., 1942-. Editor, Annales Acad. Sci. Fennicae. Series A. VI. Phys. Societies: Finnish Acad. Sci. and Letters: Finnish Sci. Soc.
Address: Department of Physics, 20 Siltavuorenpenger, Helsinki, Finland.

FONTENAY, Fernand. Ing. Arts et Métiers. "Service Fabrication" (manufacturing department), Gamma Industrie.
Address: Gamma Industrie S.A.R.L., 52 rue de Dunkerque, Paris, France.

FOORD, F. F. Works Manager, and Local Director, Eng. Group, Vickers Ltd.
Address: Vickers-Armstrongs South Marston Works, Swindon, Wiltshire, England.

FOOTE, Frank G., A.B., M.A., Ph.D. Born 1906. Educ.: Ohio Wesleyan, Ohio State and Columbia Univs. Director, Metal. Div., (1948-65), Sen. Metal. (1965-), Argonne Nat. Lab. Books: Chapter, Phys. Metal. of Uranium (Nucl. Metal., vol I, A.I.M.E.); chapter, Phys. Metal. of Uranium (Metal. and Fuels). Societies: A.I.M.M.E.; A.S.M.; A.P.S.; A.N.S.; American Crystallographic Assoc.
Nuclear interest: Metallic and ceramic fuel materials, particularly those containing plutonium.
Address: 4822 Northcott Avenue, Downers Grove, Illinois, U.S.A.

FOOTE, James Harold, B.S. (Eng.), Dr. Sc. (Eng.) (Hon., Wayne State Univ.). Born 1891. Educ.: Michigan State Univ. Pres., Director (1949-58), Chief Eng. (1959-61), Commonwealth Assocs. Inc.; Vice-Pres., Director, Commonwealth Services Inc., 1949-61; Pres., Director, Commonwealth Buildings Inc., 1954-61; Director, Atomic Power Development Assocs., Inc., 1955-61; Director (1954-56), Vice-Pres. (1956-58), Pres. (1959-60), Board of Directors (1960-62), American Inst. Elec. Eng.; Director, American Standards Assoc., 1956-63. Societies: Fellow, American Inst. Elec. Eng.; A.S.T.M.; A.S.M.E.; American Inst. Consulting Eng.; Fellow, A.A.A.S.
Nuclear interests: Gross reactor design and containment, together with heat transfer and utilisation; Engineering management of atomic power plant design.
Address: Brethren, Mich. 49619, U.S.A.

FORBERG, Sevald, Tekn. lic. (Swedish). Born 1925. Educ.: Roy. Inst. Technol., Stockholm. Res. Asst., Swedish Atomic Res. Council, 1957-.
Nuclear interests: Chemical separation of stable isotopes; Threshold neutron reactions. Address: Division of Nuclear Chemistry, The Royal Institute of Technology, Stockholm 70, Sweden.

FORBES, Stuart Gordon, B.S. (Elec. Eng.), M.S. (Phys.), Ph.D. (Phys.). Born 1919. Educ.: Oregon State Coll. Res. Eng., Boeing Airplane Co., Seattle, Washington, 1951-52; Phys., Phillips Petroleum Co., Idaho Falls, Idaho, 1952-60; Phys., TRW Systems, Redondo Beach, California, 1960-67; Asst. Manager, Tech., Phillips Petroleum Co., A.E.D., Idaho Falls, 1967-. Societies: A.N.S.; A.P.S.; A.I.A.A.
Nuclear interests: Nuclear research and development, management, physics.
Address: Phillips Petroleum Company, Atomic Energy Division, P.O. Box 2067, Idaho Falls, Idaho 83401, U.S.A.

FORCELLA, Aldo, Hydraulic Eng. (Univ. Pisa), M.S. (Nucl. Eng., New York Univ.). Born 1923. European Rep., Atomic Energy Div., Allis-Chalmers Manufacturing Co.; Special Asst. to Sec. Gen., and Special Asst. for Reactors Development, (C.N.E.N.);

Tech. Correspondent, Dragon Project; Member, Consulting Com. for Nucl. Res. of Euratom; Former Deputy Director, E.N.S.I. Project Working Group; Member, Working Group of European Atomic Energy Soc.; Tech. Consultant to Euratom Div. of Res. and Training.
Nuclear interest: Reactor development.
Address: 2 Via Santa Costanza, Rome, Italy.

FORD, Charles Edmund, B.Sc., Ph.D., D.Sc., F.R.S. Born 1912. Educ.: London Univ. Member of Sci. Staff, Medical Res. Council, Radiobiol. Res. Unit, Harwell, 1949-. Societies: Linnean Soc.; Genetical Soc.; Soc. for Exptl. Biol.; Mammal Soc.; Soc. for Study of Human Biol.
Nuclear interest: Radiobiology, especially cell-population changes in irradiated mice as traced by cytogenetic methods.
Address: M.R.C. Radiobiological Research Unit, Harwell, Berks., England.

FORD, George William Kinvig, M.A. (Cantab.). Born 1920. Educ.: Cambridge Univ. Eng. Lab., A.E.R.E., Harwell (1946-55), Res. Manager, Exptl. Criticality Lab., Dounreay (1956), Sen. Principal Sci. Officer, Reactor Div., A.E.E., Winfrith (1959-64), U.K.A.E.A.; Chief, Eng. Res. Div., A.A.E.C. Res. Establishment, Lucas Heights, Sydney, 1965-.
Nuclear interests: Experimental engineering and physics.
Address: 133 Wattle Road, Jannali, N.S.W. 2226, Australia.

FORD, Hugh, F.R.S., D.Sc. (Eng.), Ph.D., B.Sc. (Eng.). Born 1913. Educ.: London Univ. Prof., Appl. Mechanics (1948-65), Head, Mech. Eng. Dept. (1966-), Imperial Coll. Book: Mechanics of Materials (Longmans, 1963). Societies: Inst. Mech. E.; Inst. Civil Eng.; Inst. of Metal.
Nuclear interests: Metallurgy; Stress analysis; Fracture mechanics.
Address: Imperial College, Mechanical Engineering Department, Exhibition Road, London S.W.7, England.

FORD, James W., Dr. Head, Appl. Phys. Dept., Cornell Aeronautical Lab. Inc.
Address: Cornell Aeronautical Laboratory Inc., P.O. Box 235, Buffalo 21, New York, U.S.A.

FORD, Kenneth W., Ph.D. Born 1926. Educ.: Harvard and Princeton Univs. At Indiana Univ., 1953-57; Los Alamos Sci. Lab., 1957-58; Brandeis Univ., 1958-64; California Univ., Irvine, 1964-. Society: A.P.S.
Nuclear interest: Nuclear physics.
Address: Physics Department, California University, Irvine, California 92664, U.S.A.

FOREMAN, Bruce Milburn, Jr., B.S., Ph.D. Born 1932. Educ.: California, Berkeley Univ. Res. Asst. (1954-58), Teaching Asst. (1955), California Univ.; Res. Assoc., Brookhaven Nat. Lab., 1958-60; Res. Sci. Columbia Univ., 1960-. Society: A.P.S.
Nuclear interest: Research in mechanisms of nuclear reactions, with application to nuclear structure.
Address: Department of Chemistry, Columbia University, New York 27, N.Y., U.S.A.

FOREMAN, Harry, M.D., Ph.D., B.Sc. Born 1915. Educ.: California Univ. Medical School, San Francisco. Assoc. Prof., School of Public Health, and Assoc. Dean, Office of Internat. Programmes, Minnesota Univ. Societies: A.A.A.S., Rad. Res. Soc.; Soc. for Exptl. Biol. and Medicine; Biophys. Soc.; American Physiological Soc.; Soc. for Exptl. Pharmacology and Therapeutics.
Nuclear interests: Toxicology of radioactive materials; Metabolism of metals and chelating agents; Treatment of radioactive metal poisoning.
Address: Office of International Programmes, 1224 Social Science Building, Minnesota University, Minneapolis, Minnesota 55414, U.S.A.

FORGO, Leslie, Dr. techn., conferred by the Tech. Univ., Budapest, and Candidate of Tech. Sci., conferred by the Hungarian Acad. Sci. Born 1907. Educ.: E.T.H., Zürich, Asst. Manager, Chief Eng., Res. Inst. Power Eng., 1950-52; Asst. Man., Chief Eng., Heat Economic and Design Inst., 1952-61. Com. for Energetics, Hungarian Acad. Sci.; Pres., Hungarian Soc. for Power Eng.; Hungarian Board for Atomic Energy.
Nuclear interests: Nuclear reactors, industrial heat supply for process heating; Heat transfer inside nuclear reactors moderated and cooled by organic liquids.
Address: 14 Néphadsereg u., Budapest 5, Hungary.

FORIEN, Michael, Ing. des Arts et manufactures. Born 1931. Educ.: Ecole Centrale des Arts et Manufactures. Manager, Nucl. Div., Ets. Pompes Guinard. Member, Board, A.T.E.N.
Nuclear interest: Nuclear applications for pumps.
Address: Ets. Pompes Guinard, avenue de Fouilleuse, 92 Saint Cloud, France.

FORMAN, A. Hennen, B.E. (Mech. and Elec.). Born 1906. Educ.: Tulane Univ. Pres. and Gen. Manager, Arizona Public Service Co., 1966-.
Address: Arizona Public Service Company, P.O. Box 21666, Phoenix, Arizona 85036, U.S.A.

FORNAGUERA, R. ORTIZ-. See ORTIZ-FORNAGUERA, R.

FORREST, Gerald, B.S.M.E. Born 1918. Educ.: Paris and California Univs. Nuclear Products Manager, Marman Div., Aeroquip Corp., 1960-. Societies: A.N.S.; Cryogenic Soc. of America.

Nuclear interest: Application engineering in the field of transfer of fluids and gases and remote handling capabilities.
Address: 16650 Oldham Street, Encino, California, U.S.A.

FORREST, John Samuel, M.A., D.Sc., F.R.S. Born 1907. Educ.: Glasgow Univ. Director, Central Electricity Res. Labs. Societies: F. Inst. P.; F.I.E.E.; F.R. Met. Soc.; Sec., Electricity Supply Res. Council.
Nuclear interests: Electrical power generation by fission, fusion, and magneto-hydrodynamic methods.
Address: Central Electricity Research Laboratories, Cleeve Road, Leatherhead, Surrey, England.

FORREST, Robert Neagle, B.S., M.S., Ph.D. Born 1925. Educ.: Oregon Univ. Lawrence Rad. Lab., Livermore, Calif., 1958-59; Boeing Sci. Res. Lab., Seattle, 1959; Formerly Asst. Prof., Phys. Dept., now Assoc. Prof. Phys., Oregon State Univ., 1959-. Societies: A.P.S.
Nuclear interest: Present interest is the investigation of ionic bombardment effects by radioactive tracer techniques.
Address: 700 North Third, Corvallis, Oregon, U.S.A.

FORSBERG, Hans Gosta. Civilingenjör, civilekonom. Born 1930. Educ.: Roy. Inst. Technol., Stockholm and Stockholm School of Economics. Res. Asst., Div. of Phys. Chem., Roy. Inst. Technol.,1955-59; Sen. Res. Officer, Isotope Techniques Lab., Stockholm, 1959-62; Isotope Survey Officer, Div. of Res. and Labs., I.A.E.A., Vienna, 1962-64; Head of Foreign Secretariat Roy. Swedish Acad. of Eng. Sci., 1965-.
Nuclear interest: Technical isotope applications.
Address: Royal Swedish Academy of Engineering Sciences, P. O. Box 5073, Stockholm 5, Sweden.

FORSBERG, Nils L., Civilingenjör (M.Sc. in Phys. Eng.). Born 1922. Educ.: Roy. Inst. Technol., Stockholm. Res. eng., A.B. Electrolux, Stockholm, 1950-52; Head, Lab., Tekniska Röntgencentralen, Stockholm, 1952-55; Sect. leader, A.B. Bofors, Bofors, 1955-56; Sect. leader, Res.Inst. Nat. Defence, Sundbyberg, 1957-. Societies: Swedish Assoc. Tech. Phys.; Sect. of Swedish Assoc. of Eng. and Architects; Svenska Fysikersamfundet; Swedish Welding Assoc.; Svenska Metallografförbundet.
Nuclear interest: Metallurgy.
Address: 85^{iii} Vasavägen, Jakobsberg, Sweden.

FORSCHER, Frederick, Ph.D. Born 1918. Educ.: Princeton and Columbia Univs. Manager, Advanced Fuels, Westinghouse Elec. Atomic Power Div. Societies: A.S.M.E.; A.I.M.E.; A.N.S.; A.S.M.
Nuclear interests: Nuclear materials; Metallurgy; Ceramics; Fuel cycle.

Address: N.F.D., Westinghouse, P.O. Box 355, Pittsburgh, Pennsylvania 15230, U.S.A.

FORSEN, Harold K., Dr. Phil., E.E. Born 1932. Educ.: California Inst. of Technol.and California Univ., Berkeley.
Member, Tech. Staff, Hughes Aircraft, Culver City, 1958-59; Res. Assoc., Gen. Atomic, San Diego, 1959-62; Res. Assoc., Dept. Elec. Eng., California Univ., 1962-65; Assoc. Prof., Dept. Nucl. Eng., Wisconsin Univ., 1965-. Societies: A.N.S.; A.P.S.; A.I.A.A.
Nuclear interest: Plasma research for electric propulsion and controlled thermonuclear fusion.
Address: Department of Nuclear Engineering, Wisconsin University, Madison, Wisconsin 53705, U.S.A.

FORSLING, Erik Wilhelm, Fil. dr. (Swedish). Born 1920. Educ.: Stockholm Univ. Assoc. Prof., Nucl. Chem. and Head, Nucl. Chem. Div., Res. Inst. for Phys., Stockholm; With government, 1966-. Books: Nuclear chemical investigations of alpha-radioactive neutron-deficient nuclides produced by heavy ion reactions, (Stockholm, Almqvist and Wiksell, 1961); Editor, Nuclides far off the stability line. Proceedings, Lysekil symposium, 1966, (Stockholm, Almqvist and Wiksell, 1967). Societies: Fysikersamfundet, Stockholm; Kemistsamfundet, Stockholm.
Nuclear interests: Nuclides far off the beta-stability line; Radiochemical applications in different fields; N.m.r. spectroscopy used in chemistry; Properties of heavy transuranium elements.
Address: Research Institute for Physics, Stockholm 50, Sweden.

FORSSBERG, A., Prof. Member, Nat. Rad. Protection Board Inst.
Address: National Radiation Protection Institute, Karolinska sjukhuset, Stockholm 60, Sweden.

FORT, Marc LE. See LE FORT, Marc.

FORTESCUE, Peter, B.Sc. (Eng.), Ph.D. (Eng.) (London External). Born 1913. Educ.: The College, Swindon (external London Univ. student). Deputy Chief Sci. Officer, U.K.A.E.A., Harwell, 1946-57; Chief R. and D. Eng., Gen. Atomic,San Diego, and Formerly Chief R. and D. Eng., Gen. Atomic Europe, Zurich.
Nuclear interest: Reactor design.
Address: Via de la Valle, Rancho Santa Fe, California, U.S.A.

FORTIER, Pierre, B.A.Sc., D.I.C. (Nucl. Power) Born 1932. Educ.: Montreal Univ. Manager, Nucl. and Thermal Dept., Surveyer, Nenninger and Chênevert Inc., Consultants, 1966-. Society: Canadian Nucl. Assoc.
Nuclear interest: Management of important nuclear projects.

FORTIER, Yves O., B.A. (Laval), B.Sc. (Queens), M.Sc. (McGill), Ph.D. (Stanford). Born 1914. Educ.: Laval, Queen's, McGill, and Stanford Univs. Sen. Geologist, Sect. Head, now Director, Geological Survey of Canada, 1943-.
Address: Geological Survey of Canada, Ottawa, Canada.

FORTIN, Luis BOGRAN. See BOGRAN FORTIN, Luis.

FORTIS, Sergio, Dr. Director, C.I.S.E.
Address: Centro Informazioni Studi Esperienze, 12 Via Redecesio, Segrate, Milan, Italy.

FORTSON, E. Norval. Asst. Prof., (Radiofrequency Spectroscopy), Dept. of Phys., Washington Univ., Seattle.
Address: Washington University, Department of Physics, Seattle, Washington 98105, U.S.A.

FOSSATI, Franco, Dr., Lecturer in Radiol. (Milan). Born 1911. Educ.: Milan Univ. Director, Inst. of Radiol. and Nucl. Medicine, Ospedale Magiore, Milan. Sec., Comitato per le Unità di Misura Radiologiche, Soc. Italiana di Radiologia Medica e di Medicina Nucleare; Member, Editorial Com., Minerva Nucleare. Books: Co-author, Norme per le protezioni radiologiche (Milan, Hoepli, 1956); Editor, Quantities, Units and measuring Methods of ionizing Radiations (Milan, Hoepli, 1959). Society: Soc. Italiana di Radiologia Medica e di Medicina Nucleare.
Nuclear interests: Radiological protection; Health physics; Diagnostic and therapeutic use of radioisotopes.
Address: Istituto di Radiologia e Medicina Nucleare, 3 Piazza Ospedale Maggiore, Milan, Italy.

FOSSOUL, Emile Auguste Joseph, Elec. and Mech. Eng. Born 1931. Educ.: Faculté Polytechnique de Mons. Eng. at Belgo-Nucléaire, 1955; Chef de Département, 1964; Adviser, Third I.C.P.U.A.E., Geneva, Sept. 1964. Societies: A.N.S.; Sté.Belgique de Physique. Nuclear interests: Reactor physics; Nuclear physics; Mathematics; Computation techniques.
Address: 221 avenue d'Huart, Crainhem, Belgium.

FOSTER, Bruce Parks, B.S., M.S. (Yale), Ph.D. (Yale). Born 1925. Educ.: Westminster Coll. (Pa.), Baldwin-Wallace Coll. and Yale Univ. Asst. Prof. Phys. (1953-56), Assoc. Prof. Phys. (1956-), North Texas State Univ.; Fulbright Lecturer in Phys., Univ. of Peshawar, Pakistan, 1960-61; Lecturer in Phys., Forman Christian Coll., Lahore, Pakistan, 1961-63. Societies: A.P.S.; A.A.P.T. Nuclear interests: Nuclear physics; X-ray scintillation spectrometry.
Address: Physics Department, North Texas State University, Denton, Texas, U.S.A.

FOSTER, Duncan Graham, Jr., B.A., Ph.D. Born 1929. Educ.: Swarthmore Coll and Cornell Univ. Phys., Hanford Labs., Richland, Washington, 1956-. Society: A.P.S.
Nuclear interest: Nuclear physics.
Address: 82 Thomas Street, Richland, Washington 99352, U.S.A.

FOSTER, J. G. Born 1904. Educ.: Oxford Univ. Parliamentary Under Sec., Commonwealth Relations, 1951-54. Chairman, Sir Isaicc Pitman and Sons; Chairman, One World Trust; Chairman, Federal Trust; Chairman, Exec. Com. Justice; Member, European Nucl. Energy Tribunal, O.E.C.D., E.N.E.A.
Address: 2 Hare Court, Temple, London, E.C.4, England.

FOSTER, James Lancelot, B.Sc. (Botany and Zoology), M.Sc. (Rad. Phys. and Rad. Biol.). Born 1940. Educ.: Durham and London Univs. Biol., British Empire Cancer Campaign for Res., Res. Unit in Radiobiol., Mount Vernon Hospital, Middlesex, 1964-. Society: British Assoc. for Rad. Res.
Nuclear interest: Radiobiology, chemical radiosensitisation in organised biological systems.
Address: British Empire Cancer Campaign for Research, Research Unit in Radiobiology, Mount Vernon Hospital, Northwood, Middlesex, England.

FOSTER, John Stanton, B.Eng., (Mech. Eng.), B. Eng. (Elec. Eng.), D. Eng. (Hon, Carleton Univ.), D. Eng. (Hon., Nova Scotia Tech. Coll.). Born 1921. Educ.: Dalhousie Univ. and Nova Scotia Tech. Coll. Manager, Douglas Point Project, -1960; Gen. Manager, (-1964), Vice-Pres. (-1966), Power Projects, A.E.C.L.
Address: 50 Farningham Crescent, Islington, Ontario, Canada.

FOSTER, John Stuart, Ph.D. Born 1890. Educ.: Acadia and Yale Univs. Macdonald Prof. Phys., McGill Univ., 1935-55; Rutherford Res.Prof. Phys., 1955-; Director, Rad. Lab., McGill Univ., 1947-. Chairman, Phys. Dept. McGill Univ., 1952-55. Societies: F.R.S.; F.R.S.C.
Address: 1065 Creston Road, Berkeley 8, California, U.S.A.

FOSTER, John Stuart, Jr., B.Sc. (Hons., Phys.), Ph.D. (Phys.). U.S.A.E.C. Ernest O. Lawrence Award, 1960, for contributions in field of nucl. weapons. Born 1922. Educ.: McGill and California (Berkeley) Univs. Director, Lawrence Rad. Lab., Livermore, 1961-65; Director, Res. and Eng., Defence Dept., 1965-. Air Force Sci. Advisory Board; Army Sci. Advisory Panel.
Nuclear interests: Nuclear and high explosives; hydrodynamics; gaseous discharge; radar.

Address: U. S. Department of Defence, Office of the Director of Research and Engineering, Washington 25, D.C., U.S.A.

FOSTER, Laurence Standley, A.B., A.M., Ph.D. Born 1901. Educ.: Clark and Brown Univs. Res. Metal., (1946-55), Chief Atomic Energy Div., Ordnance Materials Res. Office (1955-62), Watertown Arsenal, Mass. Chief, Tech. Information Centre, and Co-ordinator, Atomic Energy Res. Project, U.S. Army Materials Res. Agency, 1962-66; Special Asst. to Commanding Officer, Army Materials and Mechanics Res. Centre (A.M.M.R.C.), 1966-. Societies: A.N.S.; A.C.S.; A.A.A.S.; Acad. for Appl. Sci.; New England Assoc. of Chem. Teachers.
Nuclear interests: Was administratively responsible for installation of 5 megawatt tank type reactor at A.M.M.R.C.; serves as coordinator of atomic energy research programme for Nuclear Branch, U.S. Army Material Command.
Address: 16 West Street, Belmont, Massachusetts 02178, U.S.A.

FOTI, Erno. Member, Editorial Board, Energia es Atomtechnika.
Address: Muszaki Konyvkiado, 22 Bajcsy-Zsilinszky ut, Budapest 5, Hungary.

FOULDS, Ronald Bruce, B.S.E., M.S., M.B.A. Born 1931. Educ.: Michigan and Pittsburgh Univs. Sen. Eng., Reactor Core Design, Westinghouse Elec. Corp., 1953-. Society: A.N.S.
Nuclear interests: Design of advanced fuel cores for pressurised water nuclear reactors; Design of fuel elements and fuel element in-pile testing; Analysis of reactor fuel and reactor thermal and hydraulic performance throughout core life; Advanced reactor fuel component manufacture.
Address: 431 McClellan Drive, Pittsburgh, Pennsylvania 15236, U.S.A.

FOULKES, Frank Moulton, B.Sc. (1st class Hons., Eng.). Born 1923. Educ.: Birmingham Univ. Asst. Manager, Nucl. Div., Montreal Eng. Co. Ltd. Chairman, Ind. Participation Com., Canadian Nucl. Assoc. Society: I.E.E.
Nuclear interest: Management in nuclear consulting engineering, particularly associated with the Canadian CANDU reactor power stations, both pressurised heavy water, and boiling light water types.
Address: Montreal Engineering Company Limited, 276 St. James West, Montreal 1, Quebec, Canada.

FOULKES, Stanley Vernon. Assoc. of the Sydney Tech. Coll. Born 1913. Chief Inspector of Factories and Shops, Dept. of Labour and Industry, N.S.W., 1967-. Member, Radiol. Advisory Council, Dept. of Public Health, N.S.W.
Nuclear interest: Industrial application of radioisotopes, etc.

Address: Department of Labour and Industry, G.P.O. Box 21, Sydney 2001, N.S.W., Australia.

FOUQUET, Jean, L. ès Sc. Born 1915. Educ.: Ecole Polytechnique, Paris. Director of R. and D., Nobel-Bozel S.A., Paris. Societies: A.T.E.N.; Auxi-Atome.
Nuclear interests: Chemistry; Nuclear zirconium and hafnium; Electrocoating.
Address: 2 rue Mabillon, Paris 6, France.

FOURE, Claude. Director, Div. Atomique, Sté. Nat. d'Etude et de Construction de Moteurs d'Aviation (S.N.E.C.M.A.).
Address: S.N.E.C.M.A., Usine de Suresnes, 22 quai Gallieni, Suresnes, (Seine), France.

FOURIE, B. G. Sec. for Foreign Affairs, South Africa; Member, Atomic Energy Board, South Africa.
Address: Atomic Energy Board, Private Bag 256, Pretoria, South Africa.

FOURNIER, Jean H. Deputy Manager, Ets. Garczynski et Traploir.
Address: Ets. Garczynski et Traploir, 34 rue du Pavé, Le Mans, (Sarthe), France.

FOURNIER, Juan C. MUSSIO. See MUSSIO FOURNIER, Juan C.

FOURQUET, General Aerienne Michel. Member, Comite de l'Energie Atomique, C.E.A., 1966-.
Address: Commissariat a l'Energie Atomique, 29-33 rue de la Federation, Paris 15, France.

FOWLER, Earle Cabell, B.S., A.M., Ph.D. Born 1921. Educ.: Kentucky, Harvard and Chicago Univs. Asst. Prof. (1952), Assoc. Prof. (1958), Yale Univ.; Prof., Phys., Duke Univ., 1962. Consultant, Brookhaven Nat. Lab., 1952-; Member, Panel on Univ. Computer Facilities, N.S.F.; Member, High Energy Phys. Advisory Panel, U.S.A.E.C. Book; co-author, Strange Particles (New York, John Wiley and Sons, Intersci. Div., 1963). Society: Fellow, A.P.S.
Nuclear interests: Experimental techniques, especially visual; Analysis of photographs; Application of computers to data analysis; High energy nuclear physics, formation and decay of resonant states of sub-nuclear particles.
Address: 1821 S. Lakeshore Drive, Chapel Hill, North Carolina 27514, U.S.A.

FOWLER, Emil Eugene, B.S., M.S. Born 1923. Educ.: West Virginia Univ. Tech. Reviewer (1952-53), Asst. Chief, Radioisotopes Branch (1953-54), Chief, Allocations Branch, Washington, D.C. (1954-56), Chief, Radiations and Interagency Branch (1956-58), Deputy Director (1958-65), Director (1965-), Div. Isotopes Development, U.S.A.E.C.
Nuclear interests: Development of non-power uses of atomic energy.

Address: 6103 Wynnwood Road, Washington 16, D.C., U.S.A.

FOWLER, Gerald Nathan, B.Sc., Ph.D. Born 1921. Educ.: Manchester Univ. Prof. Theoretical Phys., Exeter, 1968.
Nuclear interest: Nuclear many body problem.
Address: Physics Department, Exeter University, Exeter, England.

FOWLER, Glenn A., Vice-Pres., Development, Sandia Corp. Adviser, Third I.C.P.U.A.E., Geneva, Sept. 1964.
Address: Sandia Corporation, Sandia Base, Albuquerque, New Mexico, U.S.A.

FOWLER, John Francis, Ph.D., M.Sc., F.Inst.P. Born 1925. Educ.: London Univ. Prof., Medical Phys., Postgraduate Medical School, Hammersmith Hospital, London; Principal Phys., King's Coll. Hospital, London, 1956-59. Societies: Hospital Physicists Assoc.; Assoc. for Rad. Res.
Nuclear interests: Research projects on cyclotron, (a) neutron beam dosimetry radiobiology and radiotherapy, (b) applications of cyclotron produced isotopes to medical research, (c) cell population kinetics.
Address: Postgraduate Medical School, Hammersmith Hospital, London, W.12, England.

FOWLER, Joseph Lee, Ph.D., M.S., A.B. Born 1913. Educ.: Princeton and Tennessee Univs. Director, High Voltage Lab., Phys. Div. (1951-57), Director, Phys. Div. (1957-), O.R.N.L. Chairman (1960-61), Southeastern Sect., Chairman, Div. of Nucl. Phys. (1967-69), A.P.S.; Chairman (1959-61), A.E.C. Cross Section Com. (Nuclear Cross Section Advisory Group). Book: Co-author and co-editor, Fast Neutron Physics (Interscience Publishers, Part I, 1960 and Part II, 1963). Society: Fellow, A.P S.
Nuclear interests: High energy neutrons, proton and deuteron interaction with nuclei; fission fragment mass and energy distribution.
Address: Physics Division, Oak Ridge National Laboratory, P. O. Box X, Oak Ridge, Tennessee, U.S.A.

FOWLER, P. H., Dr. Res. Fellow in Phys., H. H. Wills Phys. Lab. Bristol Univ.
Address: H. H. Wills Physics Laboratory, University of Bristol, Royal Fort, Bristol 8, England.

FOWLER, Richard Hindle, M. Ag. Sc. Born 1910. Educ.: Melbourne Univ. Asst. Director, Museum of Appl. Sci. of Victoria, 1952; Director, Inst. of Appl. Sci. of Victoria, 1962-. Society: F.R.S.A.
Nuclear interest: Radiocarbon dating laboratory, and museum displays on nuclear science (in executive capacity).
Address: Institute of Applied Science of Victoria, 304-328 Swanston Street, Melbourne 3000, Australia.

FOWLER, Robert Dudley, B.S. (California, Berkeley), M.A. (Michigan), Ph.D. (Michigan). Born 1905. Educ.: California (Berkeley) and Michigan Univs. Full Prof. (1943-52), Chairman, Chem. Dept. (1947-52), Johns Hopkins Univ.; Assoc. Div. Leader (1952-54), Alternate Div. Leader (1954-56), Div. Leader (1956-), Los Alamos Sci.Lab., California Univ. Societies: A.C.S.; A.P.S.
Address: 734 46th Street, Los Alamos, New Mexico, U.S.A.

FOWLER, William Alfred, B. Eng. Phys. (Ohio State), Ph.D. (California Inst. Technol.). Born 1911. Educ.: Ohio State Univ. and California Inst. Technol. Prof. Phys., California Inst. Technol. Books: Modern Phys. for the Eng., Chapter 6, vol i, Chapter 9, vol. ii (McGraw-Hill, 2nd edition, 1961); Nucl. Astrophys. (American Phil. Soc., 1967). Societies: A.A.U.P.; I.A.U.; A.P.S.; N.A.S.; American Phil. Soc.; American Acad. Arts and Sci.; A.A.A.S.; A.I.P.; American Astronomical Soc.; Membre correspondant, Soc. Roy. Sci. Liège.
Nuclear interests: Nuclear physics in astronomy; Energy levels and structure of light nuclei.
Address: 1565 San Pasqual, Pasadena, California 91109, U.S.A.

FOX, John M. Visiting Faculty, (Reactor Analysis and Design), Dept. of Nucl. Eng., Washington Univ., Seattle.
Address: Washington University, Department of Nuclear Engineering, Seattle, Washington 98105, U.S.A.

FOX, Marvin, B.S. (Carnegie Inst. Technol.). M.S., Ph.D. (Phys., Columbia). Born 1910. Educ.: Carnegie Inst. Technol. and Columbia Univ. Brookhaven Nat. Lab., 1946-52; Chairman, Reactor Dept., Brookhaven Nat. Lab., 1952-57; Vice-Pres., Internucl. Co., Rome, 1957-58; Director, Sodium Reactors Dept., Atomics Internat., 1959; Sen. Sci., Hughes Aircraft Co., 1960-; Assoc. Manager, Arms Control Programmes, Culver City, Calif. Societies: A.P.S.; A.A.A.S.; A.I.A.A.
Nuclear interests: Reactor design and construction; Management of R. and D.; Nuclear physics; Direct energy conversion; Arms control research.
Address: 3450 Scadlock Lane, Sherman Oaks, California 91403, U.S.A.

FOX, Thomas Allen, B.S., M.S. Born 1926. Educ.: Ohio Univ. Instructor, Phys., Ohio Univ., 1951-52; Aeronautical Res. Sci., N.A.C.A.-Lewis, 1952-56; Reactor Phys., O.R.N.L., (On loan from N.A.C.A.), 1956-57; Head, Exptl. Reactor Phys. Sect., N.A.S.A.-Lewis Res. Centre, 1958-. Society: A.N.S.
Nuclear interests: Experimental reactor physics, in particular critical mass studies, reactivity measurements, nuclear safety problems and reactor operations including licensing arrangements.
Address: 27522 Dunford Road, Westlake, Ohio 44145, U.S.A.

FOX, Thomas Bray. Born 1916. Managing Director, High Voltage Servicing Co. Ltd. Societies: British Inst. of Radiol.; Hospital Phys. Assoc.
Nuclear interests: Sales, service and installation of high voltage particle accelerators and instrumentation in the fields of nuclear physics, radiation chemistry, radiotherapy, radiography.
Address: High Voltage Servicing Company Limited, 8 Kildare Close, Eastcote, Ruislip, Middlesex, England.

FOX, W. W. Head, Eng. Dept. Group, Convair.
Address: Convair, Department of Scientific Research, 3165 Pacific Highway, San Diego, California, U.S.A.

FOY, Richard, D. ès Sc., L. en Droit. Born 1917. Educ.: École Polytechnique de Paris, Ecole Supérieure d'Aéronautique, Directeur Scientifique d'Auxi-Atome; Directeur du Dépt. "mesures" à l'A.O.I.P. Publications: Voyages interplanétaires et énergie atomique (Albin Michel, 1941); Compteurs de Geiger au brome (1953). Society: A.P.S.
Nuclear interest: Nuclear physics.
Address: 11 rue de la Pepinière, Paris 8, France.

FØYN, Johan Ernst Fredrik, Dr. philos. Born 1904. Educ.: Oslo Univ. Prof., Inst. of Marine Biology, Oslo Univ., 1948-. Book: Radioaktivitet, Moderne Naturvitenskap (Gyldendal, Norway, 1938). Societies: A.C.S.; Norsk Kjemisk Selskap; Board of Norske havforskeres forening.
Nuclear interest: Isotopes.
Address: Institutt for marin biologi, 3 Frederiksgt.,Oslo 1, Norway.

FRAAS, Arthur Paul, B.S. (M.E.; Case Inst. Technol.), M.S. (Aero. Eng.; N.Y.U.). Born 1915. Educ.: Case Inst. Technol. Assoc. Director, Reactor Div., O.R.N.L. Books: Aircraft Power Plants; Combustion Engines; Heat Exchanger Design (1964). Societies: A.S.M.E.; S.A.E.; A.I.A.A.; RESA; A.N.S.
Nuclear interests: High temperature nuclear power plant design and development, heat exchangers, stress analysis.
Address: 1040 Scenic Drive, Knoxville Tennessee, U.S.A.

FRACZEK, Edward, M.Sc. (Geology). Born 1931. Educ.: Wroclaw Univ. Radiometry Div., Hydrogeological Enterprise, Warsaw. Society: Polish Geological Assoc.
Nuclear interest: Radioisotope techniques in hydrogeology.
Address: 13 15-57 Nowolipie str., Warsaw, Poland.

FRADIN, J. Member for France, Group of Experts on Production of Energy from Radioisotopes, O.E.C.D., E.N.E.A.
Address: Service des Radioéléments Artificiels, Commissariat à l'Energie Atomique, 24-33 rue de la Federation, Paris 15, France.

FRADIS, M. Director, "Isratom" Israel Nucl. Eng. Co. Ltd.
Address: "Isratom" Israel Nuclear Engineering Co. Ltd., 6 Ahuzat Bayit Street, Tel-Aviv, Israel.

FRADKIN, G. M. Papers: Co-author, On the Use of Ac-Be Neutron Sources in Trade Geophys. (letter to the Editor, Atomnaya Energiya, vol. 16, No. 3, 1964); co-author, Some Characteristics of Back-Scattering Gamma-Rad. Field into Work Rooms (letter to the Editor, ibid, vol. 21, No. 4, 1966); co-author, Scattering of Collimated Co^{60}, Cs^{137} and Au^{198} Gamma Rad. near Interface between Two Media (ibid, vol. 23, No. 2, 1967); co-author, Some Aspects of Radioisotope Generator Fuel Production by Neutron Irradiation Methods (ibid, vol. 24, No. 4, 1968).
Address: Academy of Sciences of the U.S.S.R., 14 Leninsky Prospekt, Moscow V-71, U.S.S.R.

FRAENKEL, Stephen J., Ph.D. Born 1917. Educ.: Nebraska Univ. and I.I.T. Gen. Manager, R. and D., Container Div., Container Corp. of America, Chicago. Director, Chicago Sect., A.R.S. Societies: Tech. Assoc. of the Pulp and Paper Industry; Soc. for Exptl. Stress Analysis.
Nuclear interests: Disposition of spent nuclear fuel; Economics of transportation of radioactive materials.
Address: 1252 Spruce Street, Winnetka, Illinois, U.S.A.

FRAENKEL, Zeev, D.Sc. (Columbia), M.S. (Columbia), B.S. (Technion). Born 1925. Educ.: Technion (Israel Inst. Technol.) and Columbia Univ. Asst. Res. Phys., Rad. Lab. Columbia Univ., 1954-56; Res. Assoc. (1956-63), Sen. Sci. (1963-67), Assoc. Prof. (1967-), Weizmann Inst. Sci. Societies: Israel Phys. Soc.; A.P.S.
Nuclear interests: Nuclear physics; Elementary particle physics.
Address: Nuclear Physics Department, Weizmann Institute of Science, Rehovoth,Israel.

FRAENZ, Kurt Otto Max, Dr. phil., Dr. Ing. habil. Born 1912. Educ.: Berlin Univ. Prof., Buenos Aires Univ., 1960. Director, Telefunken Res. Inst., Ulm. Book: Antennen und Ausbreitung (1956). Societies: Fellow, Phys. Soc., London; Sen. Member, Inst. Radio Eng., New York.
Nuclear interest: Nuclear instruments.
Address: Telefunken GmbH, 100 Söflingerstrasse, Ulm-Donau, Germany.

FRAGOSO, Julio HERNANDEZ-. See HERNANDEZ-FRAGOSO, Julio.

FRAHN, Wilhelm Eberhard, Dr. rer. nat., Dipl. Phys. Born 1926. Educ.: Aachen Tech. Univ. Res. Officer (1955), Sen. Res. Officer

(1957), Principal Res. Officer (1959), Nucl. Phys. Div., Nat. Phys. Res. Lab., South African Council for Sci. and Ind. Res., Pretoria. Prof. Phys., Stellenbosch Univ., 1960-63; Prof. Theoretical Phys., Cape Town Univ., 1964-. Deleg., 2nd I.C.P.U.A.E., Geneva, 1958. Societies: A.P.S.; German Phys. Soc.; South African Inst. Phys.
Nuclear interest: Nuclear physics.
Address: Physics Department, Cape Town University, Rondebosch, South Africa.

FRALEY, R. F. Exec. Sec., Advisory Com. on Reactor Safeguards, U.S.A.E.C.
Address: U.S.A.E.C., Advisory Committee on Reactor Safeguards, Washington 20545, D.C., U.S.A.

FRAME, Alexander Gilchrist, B.A. (Cantab.), B.Sc. (Eng., Glasgow), A.M.I.Mech.E. Educ.: Cambridge and Glasgow Univs. and Roy. Coll. of Sci. and Technol., Glasgow. Chief Eng., Fast Reactor Design Office, U.K.A.E.A., Risley; Supv. construction of new aluminium smelter, Rio Tinto Zinc Corp., 1968-.
Nuclear interest: Reactor design.
Address: 6 Gorsey Road, Wilmslow,Cheshire, England.

FRAME, Frank Riddell, M.A., LL.B. Born 1930. Educ.: Glasgow Univ. Private legal practice, 1951-55; North of Scotland Hydro-Elec. Board 1955-60; Production Group, U.K.A.E.A., 1960-. Book: Co-author, Law Relating to Nuclear Energy (Butterworths, 1966).
Nuclear interest: Legal aspects.
Address: "Grasmere", Enville Road, Bowdon, Cheshire, England.

FRAME, John M. Formerly Gen. Manager, Vitro Italiana S.p.A.; Project Manager, Architect-Eng. Service, Vitro Eng. Co., 1963-.
Address: Vitro Architect-Engineering Service, Hanford Works, P.O. Box 296, Richland, Washington, U.S.A.

FRANCA, E. P. Member, Editorial Board, Atompraxis.
Address: c/o Atompraxis, Verlag G. Braun, 14-18 Karl Friedrich Strasse, Karlsruhe, Germany.

FRANCA, Eduardo PENNA-. See PENNA-FRANCA, Eduardo.

FRANCHETTI, Simone, Dottore in Fisica. Born 1907. Educ.: Florence and Pavia Univs. Prof., Gen. Phys., Florence Univ., 1949-.
Address: Istituto di Fisica, Florence Università, 2 Largo E. Fermi, Florence, Italy.

FRANCIS, J. David, A.B., LL.B. Born 1919. Educ.: Kentucky Univ. Law Coll. Chairman, Public Service Commission. Member, Kentucky Advisory Com. on Nucl. Energy.
Address: Public Service Commission, Old Capitol Annex, Frankfort, Kentucky, U.S.A.

FRANCIS, Owen. Born 1912. Asst. Sec. (1943-54), Under-Sec. (1954-62), Ministry of Power; Member, Finance and Administration (1962-64), Deputy Chairman (1965-), C.E.G.B. Society: Companion, I.E.E.
Address: Central Electricity Generating Board, Sudbury House, 15, Newgate Street, London, E.C.1, England.

FRANCIS, Warren Charles, B.S. (Chem. Eng.). Born 1918. Educ.: M.I.T., Post-graduate work at Purdue and Idaho Univs. Manager, Quality Control and Development, Cambridge Tile Mfg. Co., Cincinnati, 1950-53; Phillips Petroleum Co., Atomic Energy Div., 1953-66; Manager Reactor Eng. Branch, Idaho Nucl. Corp., 1966-. Societies: A.N.S.; Soc. for Nondestructive Testing.
Nuclear interests: Research and development in reactor fuels (engineering and metallurgy) heat transfer, hydraulics, corrosion, radiation effects; Conceptual design studies on reactors (A.T.R.).
Address: Idaho Nuclear Corporation, Atomic Energy Division, P.O. Box 1845, Idaho Falls, Idaho, U.S.A.

FRANCISCIS, Pietro DE. See DE FRANCISCIS, Pietro.

FRANCK, René, Commissaire au Controle des Banques, Luxembourg; Member, Conseil Nat. de l'Energie Nucléaire, Luxembourg.
Address: Ministère des Transports et de l'Electricité, 4 boulevard Roosevelt, Luxembourg, Luxembourg.

FRANCO, Gianfranco, Dr. Eng. (Mech. and Naval Eng.). Born 1922. Educ.: Naples Univ. Successively Vice-Head, Uranium Metal. Lab., Head, Special Technologies Lab., Vice-Head, Reactor Eng. Group, Supervisor, Design of C.N.R.N.'s Ispra I Reactor, C.I.S.E., 1950-57; Director of construction works of C.N.R.N.'s Ispra I Reactor and Head of Reactors Operation Group at C.N.R.N.'s Nucl. Res. Establishment at Ispra, 1957-60; Director, C.N.R.N.'s Nucl. Res. Establishment at Ispra, 1959-60; Chief Eng., Dragon Project, 1960-64; Director, Casaccia Nucl. Studies Centre, C.N.E.N., 1964-. Italian Tech. Com. for Nucl. Plants. Society: F.I.E.N.
Nuclear interests: Management; Reactor design; Fuel elements; Metallurgy and technology.
Address: 126 Via degli Orti della Farnesina, Rome, Italy.

FRANCO, Paulo de CARVALHO. See de CARVALHO FRANCO, Paulo.

FRANCOIS, Etienne. Asst., Station de Chimie et Physique Agricoles; Ministère de l'Agriculture.

Address: Station de Chimie et Physique Agricoles, 13 chaussée de Wavre, Ernage-Gembloux, Belgium.

FRANDSEN, Johannes, M.D. Born 1891. Educ.: Copenhagen Univ. Director Gen., National Health Service of Denmark, 1928-61. Pres., Danish Red Cross; Chairman, Danish Government Advisory Com. on Atomic Radiation.
Address: 16 Esperance Allee, Copenhagen, Charlottenlund, Denmark.

FRANEAU, Jacques, Prof., Polytech. Faculty of Mons. Formerly Vice Pres., Conseil d'Administration, formerly Pres., now Member, Commission Sci., Inst. Interuniv. des Sci. Nucléaires.
Address: 8 Rampe St-Waudru, Mons, Belgium.

FRANGELLA, Alphonse C. Medicine Dr., Prof. Emeritus of Radiology. Born 1900. Educ.: Medicine Faculty of Montevideo. Director, Inst. of Oncology, M.S.P.; Chairman, National Atomic Energie Commision. Book: Radioisotopos en Biologia y Medicina (Montevideo, Graficos Unidos, S.A., 1950). Societies: Radiologic Soc. of Uruguay; Interamerican Coll. of Radiol.; Biologic and Nucl. Medicine Soc. of Uruguay; American Coll. of Radiol.
Nuclear interests: Radioisotopes in biology and medicine; Nuclear radiations.
Address: Institute of Oncology, Avenida 8 de Octubre, 3265 - Montevideo, Uruguay.

FRANK, Abbott, B.C.E., M.C.E. Born 1926. Educ.: C.C.N.Y. and Polytech. Inst. Brooklyn. Instructor, Polytech. Inst. Brooklyn, 1947-51; Project Eng., Catalytic Construction Co., 1951-55; Consulting Eng., Gen. Elec. Nucl. Propulsion, and Defence Systems, 1955-63; Manager, Orbital Astronomical Observation Applications, Gen. Elec. Spacecraft Dept., 1963-65; Manager, Special Interplanetary Projects, Gen. Elec. Space Div., 1965-. Society: A.I.A.A.
Nuclear interest: Nuclear space power and propulsion.
Address: 1636 N. Santa Anita, Arcadia, California, U.S.A.

FRANK, Colonel Donald H. Deputy Director, Directorate for Nucl. Safety, U.S. Air Force.
Address: Directorate for Nuclear Safety, Kirtland Air Force Base, New Mexico, U.S.A.

FRANK, Helmut, Dipl.-Phys., Dr. rer. nat. Born 1921. Educ.: Göttingen Univ. Asst., II Physikalisches Inst., Göttingen Univ.; Wissenschaftliche Rat, Inst. für Technische Kernphysik, Darmstadt T.H., 1958-. Society: Deutsche Physikalische Gesellschaft.
Nuclear interests: Experimental nuclear physics; Electron scattering and nuclear structure; Education in nuclear physics; Radiation protection.

Address: Institut für Technische Kernphysik, 1 Hochschulstrasse, Darmstadt, Germany.

FRANK, Il'ya Mikhailovich, Nobel Prize for Phys., 1958; Order of Lenin. Born 1908. Educ.: Moscow Univ. Phys. Inst., Acad. of Sci. of the U.S.S.R., 1934; Prof. Moscow Univ., 1944-; Director, Neutron Phys. Lab., Joint Inst. for Nucl. Research, Dubna. Member, Working Group on Low Energy Nucl. Phys., Internat. Union of Pure and Appl. Phys.; Member, Academy of Sci. of the U.S.S.R., 1968-. Book: Co-author, Multiplication of neutrons in uranium-graphite systems (in book: Meeting of the Academy of Sciences, devoted to peaceful utilisation of atomic energy, held on 1-5 July, 1955, Moscow). Society: Corr. Member, Acad. of Sci. of the U.S.S.R.
Nuclear interests: Nuclear physics; Physical Optics.
Address: Moscow University, Moscow, U.S.S.R.

FRANK-KAMENETSKII, A. D. Paper: Application of Monte-Carlo Method for Reactor Multigroup Calculation (Atomnaya Energiya, vol. 16, No. 2, 1964).
Address: Academy of Sciences of the U.S.S.R., 14 Leninsky Prospekt, Moscow V-71, U.S.S.R.

FRANKE, Herbert Werner, Dr. phil. Born 1927. Educ.: Vienna Univ. Awarded scholarship of Inst. für Starkstromtechnik, Vienna T.H., 1952. German deleg. for Speleochronology for the Internat. Congresses of Speleology.
Nuclear interests: Dating by radioactivity; Paleoclimatic aspects of isotope rates; Didactic problems of nuclear physics.
Address: Jagdhaus Kreuzpullach, 8024 Deisenhofen, near Munich, Germany.

FRANKEL, Sherman, B.A., M.S., Ph.D. Born 1922. Educ.: Illinois Univ. Prof., Pennsylvania Univ. Society: A.P.S.
Nuclear interests: High energy and nuclear physics.
Address: Pennsylvania University, Physics Department, Philadelphia, Pennsylvania 19104, U.S.A.

FRANKHOUSER, William L., B.S., M.S., M.B.A. Born 1924. Educ.: Pennsylvanis State and Pittsburgh Univs. Supervisory Eng., Nucl. R. and D. Fabrication, Bettis Atomic Power Lab., Westinghouse Elec. Corp., 1954-67; Manager, Plutonium Manufacturing, Nucl. Materials and Equipment Corp., 1967-. Societies: A.S.M.; A.W.S.; American Vacuum Soc.
Nuclear interests: Management in nuclear fabrication, both research and development and production.
Address: Nuclear Materials and Equipment Corporation, Apollo, Pennsylvania 15613, U.S.A.

FRANKLIN, Jerrold, B.E.E., M.S. (Phys.), Ph.D. (Phys.). Born 1930. Educ.: Cooper Union and Illinois Univ. Instructor, Columbia Univ., 1956-59; Formerly Asst. Prof., Brown Univ., 1959-. Societies: A.P.S.; Federation American Sci.
Nuclear interest: Elementary particle theory.
Address: Lawrence Radiation Laboratory, California University, Livermore, California, U.S.A.

FRANKLIN, Norman Laurence, B.Sc., M.Sc., Ph.D., M.I.Chem.E. Born 1924. Educ.: Leeds Univ. Lecturer, Mech. Eng., Leeds Univ., 1948-55; Tech. Policy Branch, Ind. Group Headquarters (1955-58), Asst. Director, Tech. Planning, Ind. Group Headquarters (1958-59), Tech. Manager, Chapelcross (1959), Tech. Director, Production Group (1961-63), Nucl. Fuels Director (1962), Asst. Managing Director, Production Group (1963-64), Deputy Managing Director, Production Group (1964-), Member for Production (1969-), U.K.A.E.A. Member, Administrative Board representing U.K.A.E.A., Nukleardienst G.m.b.H.
Address: 9 Fir Tree Avenue, Knutsford, Cheshire, England.

FRANZ, H. Formerly Member, Now Sen. Adviser, and formerly Commission Sponsor, Planning Board, Rad. - Fundamental Phys. Parameters, I.C.R.U.
Address: International Commission on Radiation Units and Measurements, 4201 Connecticut Avenue, N.W., Washington, D.C. 20008, U.S.A.

FRANZ, H., Dipl. Ing. Spectrographic Equipment Consultant, Optische Werke C.A. Steinheil Sohne G.m.b.H.
Address: Optische Werke C.A. Steinheil Sohne G.m.b.H., 10 Germersheimer Strasse, 8 Munich 8, Germany.

FRANZ, Hans, Dr. phil. Born 1899. Educ.: Berlín Univ. Physikalisch Technische Bundesanstalt Braunschweig, 1949 (Director, Atomic Physics Div., 1952); Honorarprof., Braunschweig T.H., 1956. Books: Contributions to Kohlrausch, Praktische Physik, 17 Aufl. 1935. 19 Aufl. 1944, 20 Aufl. 1956; Geiger-Scheel, Handbuch d. Physik Bd. XXII, 1933; H. Ebert, Physikalische Taschenbuch, 1 Aufl. 1951, 2 Aufl. 1957. Angerer-Ebert, Techn. Kunstgriffe b. phys. Untersuchungen 1952 u. 1956.
Nuclear interests: Radioactivity and nuclear reactions.
Address: 50 Sulzbacher Str., Braunschweig, Germany.

FRANZ, Johannes, Prof. Dr. phil. Member, Fachkommission I – Recht und Verwaltung, Fachkommission IV – Strahlenschutz und Sicherheit, Arbeitskreis IV/3 – Strahlenschutz beim Umgang mit Radioaktiven Stoffen, Arbeitskreis I/2 – Rechts- und Verwaltungsfragen des Strahlenschutzes, Deutsches Atomkommission Federal Ministry for Scientific Research, Germany.
Address: Federal Ministry for Scientific Research, 46 Luisenstrasse, 532 Bad Godesberg, Germany.

FRANZ, K. Member, Editorial Board, Atompraxis.
Address: c/o Atompraxis, Verlag G. Braun, 14-18 Karl-Friedrich Strasse, Karlsruhe, Germany.

FRANZ, Kurt, Prof. Dr. phil., Dr.-Ing. habil. Mitglied, Arbeitskreis III/6 Mess- und Regelstechnik Datenverarbeitung, Federal Ministry for Sci. Res.
Address: Federal Ministry for Scientific Research, 46 Luisenstrasse, 532 Bad Godesberg, Germany.

FRANZINETTI, Carlo, Laurea in Fisica (Rome). Born 1923. Educ.: Rome Univ. Prof. Exptl. Phys., Trieste Univ., 1958; Prof. Exptl. Phys., Pisa Univ., 1959-. Society: Soc. Italiana di Fisica.
Nuclear interests: High energy nuclear physics; Interaction properties of mesons and baryons; Production of mesons and hyperons in very high energy collisions; High energy neutrino interactions.
Address: Istituto di Fisica dell'Università di Pisa, 2 Piazza Torricelli, Italy.

FRANZINI, Tito, Dr. Phys. (Genoa). Born 1902. Prof. Phys., Univs. Pavia (1935) and Florence (1938). Prof. Phys., Naval Acad. of Livorno; Formerly Member, Sci. Com., now Member for Italy, Nucl. Ship Propulsion Com., Euratom. Books: Lezioni di Fisica Generale (Florence Univ., 1939); Energia Atomica (Milan, Bompiani, 1946).
Nuclear interests: Nuclear physics, reactor physics and design.
Address: Marinaccad, Livorno, Italy.

FRASCANI de VETTOR, Giorgio BOMBASSEI. See BOMBASSEI FRASCANI de VETTOR, Giorgio.

FRASER, Anthony Robert NEWBY-. See NEWBY-FRASER, Anthony Robert.

FRASER, John Stiles, B.Sc., Ph.D. Born 1921. Educ.: Dalhousie and McGill Univs. Asst. Res. Officer (1951-53), Assoc. Res. Officer (1953-61), Sen. Res. Officer (1961-), A.E.C.L.; Visiting Sci., A.E.R.E., Harwell, 1959-60. Chairman, Secondary School Sci. Education Com., Canadian Assoc. of Phys., 1959-. Societies: Canadian Assoc. of Phys.; A.P.S.; New York Acad. of Sci.
Nuclear interests: Nuclear reactions; Physics of fission; Intense neutron generators.
Address: 7 Cartier Circle, Deep River, Ontario, Canada.

FRASURE, Carl. Dept. Political Sci., West Virginia Univ.; Member, Advisory Com. of State Officials, U.S.A.E.C.
Address: West Virginia University, Morgantown, W. Va., U.S.A.

FRATER, Richard James, M.A. (Cantab.). Born 1931. Educ.: Cambridge Univ. Production Manager (1966), Production Director (1968), Rolls-Royce and Associates Ltd., Derby. Societies: A.M.I.Mech.E.; M.I.E.E. Nuclear interests: Management, engineering and supply of nuclear steam raising plant for military propulsion.
Address: Rolls-Royce and Associates Ltd., P.O. Box 31, Derby, England.

FRATKIN, G. M. Paper: Co-author, Self-Shielding of Thermal Neutrons in Finite Cylinders and other Solids (Atomnaya Energiya, vol. 24, No. 4, 1968).
Address: Academy of Sciences of the U.S.S.R., 14 Leninsky Prospekt, Moscow V-71, U.S.S.R.

FRAZIER, Loring R. In charge of nucl. development work, Metal Chem. Div., Ventron Corp.
Address: Ventron Corporation, Metal Chemicals Division, Beverly, Massachusetts,U.S.A.

FRAZIER, T. V., Dr. Nucl. Physics, Nevada Univ.
Address: Nevada University, Physics Department, Reno, Nevada 89431, U.S.A.

FREA, Giovanni, Com. Member, Steering Com., "Pool" Italiano per l'Assicurazione dei Rischi Atomici.
Address: "Pool" Italiano per l'Assicurazione dei Rischi Atomici, 2 Via Ettore Petrolini, Rome, Italy.

FREDERICK, Bernard J., B.C.E. Born 1923. Educ.: Clarkson Coll. of Technol. Civil Eng. (hydraulics and hydrology), U.S. Geological Survey, Water Resources Div., 1951-. Chairman, Task Force on Discharge Measuring Techniques in Major Structures (1963-66), Pres., Tennessee Valley Sect., (1966), A.S.C.E. Nuclear interests: Application of radiotracer and/or activation analysis procedures in hydraulic and hydrologic investigations. Civilian uses of nuclear explosives. Power production, desalination, and mineral recovery using reactors and sea water.
Address: Room 236, 301 Cumberland Avenue, Knoxville, Tennessee 37902, U.S.A.

FREDERIKS, A. Formerly Deputy Member, now Member, Netherlands Pool for the Insurance of Atomic Risks.
Address: Netherlands Pool for the Insurance of Atomic Risks, Blaak 101, Rotterdam, Netherlands.

FREDERIKSEN, Per O., M.Sc. Born 1928. Educ.: Tech. Univ., Denmark. With Sadolin and Holmblad A/S, 1954-56; With (1965-), at present Chief Reactor Safety and Safeguards Officer, Danish Atomic Energy Commission.
Address: Atomic Energy Commission, Research Establishment Risö, Roskilde, Denmark.

FREDRIKSSON, Lars. Director, Div. of Agricultural Radiobiol., Roy. Agricultural Coll. U.N. Project Manager, Nucl. Res. Lab., Indian Agricultural Res. Inst., New Delhi.
Address: Royal Agricultural College, Uppsala 7, Sweden.

FREDRIKSSON, Tore. Manager, Dept. for Special Installations, Air Conditioning Div., Svenska Flaktfabriken, A.B.
Address: Svenska Flaktfabriken A.B., P.O. Box 20040, Stockholm 20, Sweden.

FREED, John Howard, B.S. (Nebraska), M.D. (Nebraska), D.Sc. (Medicine, Pennsylvania). Born 1915. Educ.: Nebraska and Pennsylvania Univs. Assoc. Prof. Radiol. at Univ. Colorado Medical Centre. At Univ. Colorado Medical Centre, 1952-. Formerly Sec.-Treas., formerly Pres. Elect., Rocky Mountain Radiol. Soc. Societies: Soc. of Nucl. Medicine; A.M.A.; American Coll. of Radiol.; American Radium Soc.; American Roentgen Ray Soc.; Radiol. Soc. of North America.
Nuclear interest: Diagnostic and therapeutic use of radioisotopes in medicine.
Address: 4200 East 9th Avenue, Denver, Colorado 80220, U.S.A.

FREEDBERG, A. Stone, A.B., M.D. Born 1908. Educ.: Harvard Univ. and Rush Medical Coll. of Chicago Univ. Director, Cardiology Unit, Beth Israel Hospital, 1964-; Assoc. Prof. Medicine, Harvard Medical School, 1958-. Societies: American Heart Assoc.; Assoc. of American Physicians; American Soc. for Clinical Investigation; American Physiological Soc.; American Thyroid Assoc.; Roy. Soc. of Medicine, London,
Nuclear interest: Application of radioactive isotopes to the elucidation of our understanding of thyroid function in health and disease.
Address: 330 Brookline Avenue, Boston, Massachusetts 02215, U.S.A.

FREEMAN, Gordon M. Pres., Internat. Brotherhood of Elec. Workers. Labour Member, Labour-Management Advisory Com., U.S.A.E.C.
Address: International Brotherhood of Electrical Workers, Washington, D.C., U.S.A.

FREEMAN, Harry, A.B., M.D. Born 1903. Educ.: Harvard Univ. Director of Res., Medfield State Hospital; Assoc. in Res., Worcester Foundation for Exptl. Biol. Member, Radioisotope Com.
Nuclear interest: Isotopes in medicine.
Address: Medfield State Hospital, Harding, Mass., U.S.A.

FREEMAN, Ira M., B.S., M.S., Ph.D. Born 1905. Educ.: Chicago Univ. and Johann

Wolfgang Goethe Univ., Frankfurt am Main. Prof. Phys., Rutgers Univ. Books: Collaborator, Theoretical Phys. (3d ed., Glasgow, Blackie and Son, 1958); Modern Introductory Phys. (1957); New World of Phys. (1963); Phys.: Principles and Insights (1968). Societies: Fellow, A.A.A.S.; A.A.P.T.
Nuclear interests: Theoretical particle physics and reactor waste disposal problems; Public education in science.
Address: Department of Physics, Rutgers University, New Brunswick, N.J., U.S.A.

FREEMAN, M. I. Director, Thorium Ltd.
Address: Thorium Ltd., 6 St. James's Square, London S.W.1, England.

FREEMAN, The Honourable O. L. Sec., Dept. of Agriculture, Member, Federal Rad. Council.
Address: U.S. Department of Agriculture, Washington D.C. 20250, U.S.A.

FREGEAU, T. A. Manager, Special Products Div., Farrel Corp.
Address: Farrel Corporation, Special Products Division, 25 Main Street, Ansonia, Connecticut, U.S.A.

FREIBERGER, Heinrich, Dr.-Ing. Born 1900. Educ.: Munich and Berlin T.H. Ehrensenator, Karlsruhe T.H. Co-editor, Die Atomwirtschaft.
Address: 90 Hellabrunner Strasse, Munich 1, Germany.

FREIDEN, Anatoli S., Bachelor of Tech. Sci. Born 1927. Educ.: Lensovet Technol. Inst. (Leningrad). Res. Inst. of Mech. Treatment of Wood, -1958; Sen. Sci. Worker, Central Inst. of Building Constructions, Moscow, 1958-. Book: Influence of Ionizing Rad. on Wood and Its Components (Moscow, Goslebumizdat, 1961, pp. 118). Society: Mendeleev Chem. Soc.
Nuclear interest: Radiation chemistry of polymers.
Address: Central Institute of Building Constructions, Moscow, U.S.S.R.

FREIRE, Arturo Anibal, Prof. of Phys. Born 1917. Educ.: Central Univ., Ecuador; Ohio State Univ., U.S.A.; Argonne Nat. Lab. Internat. School of Nucl. Sci. and Eng., Illinois, U.S.A. Prof. Central Univ. of Ecuador; Prof. Nucl. Sci., Polytech. School, Quito, Ecuador. Member, Nat. Commission of Nucl. Energy; Exec., Soc. of Astronomy. Society: Pres., Nat. Soc. Sci. Teachers.
Nuclear interest: Nuclear physics.
Address: P.O. Box 2110, Quito, Ecuador.

FREITAS, Orlita GOMES DE. See GOMES DE FREITAS, Orlita.

FREJACQUES, Claude Jacques, Polytechnicien, D. ès Sc. Born 1924. Educ.: Ecole Polytech., Paris Univ. Head, Dept. of Chem. Phys., C.E.A. Asst. Prof. Chem., École Polytech. Societies: Sté. Chimique de France; Sté. Chimie physique.
Nuclear interests: Isotope separation, reactor materials.
Address: Département de Physico-Chimie, Centre d'Etudes Nucléaires de Saclay, B.P. No. 2, Gif-sur-Yvette, Seine et Oise, France.

FREKSA, Hans FRIEDRICH-. See FRIEDRICH-FREKSA, Hans.

FREMEREY, G., Dr. Präsident, Bundesamt für Gewerbliche Wirtschaft.
Address: Bundesamt für Gewerbliche Wirtschaft, 38-40 Bockenheimer Landstrasse, 6 Frankfurt am Main, Germany.

FREMLIN, John Heaver, M.A., Ph.D. (Cantab.), D.Sc. (Birmingham). Born 1913. Educ.: Cambridge Univ. Lecturer (1948-58), Reader in Phys. (1958-65), Prof. of Appl. Radioactivity (1966-), Phys. Dept., Birmingham Univ. Books: Heavy Ions (section of Nucl. Reactions ed. Endt and Demeur); Applications of Nucl. Phys. (E.U.P., 1964). Societies: Fellow, Inst. of Phys.; Fellow, Phys. Soc.; A.M.I.E.E.; A.P.S.
Nuclear interests: Nuclear research with heavy ions and helium 3; Applications of radioactive tracers to chemical analysis and to dental research; Thermoluminescent dating.
Address: 53 Richmond Hill Road, Edgbaston, Birmingham 15, England.

FRENCH, Anthony Philip, B.A. (Hons.), M.A., Ph.D. Born 1920. Educ.: Cambridge Univ. Demonstrator and Lecturer, Cambridge Univ., Director of Studies in Natural Sci., Pembroke Coll., Cambridge, Fellow of Pembroke Coll., Cambridge, 1950-55; Res. Assoc., California Inst. Technol., summer 1951; Res. Assoc., Michigan Univ., summer 1954; Prof. Phys., South Carolina Univ., 1955-62 (Head of Dept., 1956-62); Visiting Prof., Wisconsin Univ., summer 1961; Visiting Prof. (1962-64), Prof. (1964-), M.I.T. Book: Principles of Modern Physics (John Wiley, 1958). Societies: A.P.S.; A.A.P.T.
Nuclear interest: Experimental and theoretical studies of nuclear reactions at low (Mev) energies.
Address: Physics Department, Massachusetts Institute of Technology, Cambridge, Mass., U.S.A.

FRENCH, Norman Roger, A.B., M.S., Ph.D. Born 1927. Educ.: Illinois, Colorado and Utah Univs. Chief, Ecology Branch, Idaho Operations Office, Idaho Falls, U.S.A.E.C., 1955-59; Assoc. Res. Ecologist, Lab. Nucl. Medicine and Rad. Biol., California Univ., 1959-. Societies: A.A.A.S.; Ecological Soc. America; A.I.B.S.
Nuclear interests: Radioecology; Environmental radiation studies.
Address: Environmental Radiation Division, Laboratory of Nuclear Medicine and Radiation Biology, California University, 900 Veteran Avenue, Los Angeles 24, California, U.S.A.

FRENCH, Phillip D., B.A. Born 1917. Educ.:
Albion Coll. Asst. Treas., Detroit Edison Co.,
1957-.
Address: Detroit Edison Company, 2000
Second Avenue, Detroit 26, Michigan, U.S.A.

FRENCH, Robert Lewis, A.A., B.S., M.S.
Born 1928. Educ.: Santa Ana Coll., Murray
State Coll. and Vanderbilt Univ. Nucl. Group
Eng., Gen. Dynamics/Fort Worth Div. of Gen.
Dynamics Corp., 1955-63; Staff Phys., Rad.
Res. Associates, Inc., Fort Worth, 1963-.
Books: Reactor Handbook, vol. iii, Chapter
15, 2nd Edition (Interscience); Eng. Compendium on Rad. Shielding, vol. i, Section 5.4.2
(Springer-Verlag). Societies: A.N.S.; H.P.S.
Nuclear interests: Analysis of radiation
environments from nuclear reactors and
nuclear weapons and development of theoretical models for predicting radiation spectra
and intensities; Development of methods for
design of radiation shields; Management of
research and development projects.
Address: 5667 Worrell Drive, Fort Worth,
Texas 76133, U.S.A.

FREREJACQUE, D. Maitre Asst., Sect.
Physique des Hautes Energies, Lab. de l'Accélérateur Linéaire, Ecole Normale Supérieure,
Paris Univ.
Address: Ecole Normale Supérieure, Laboratoire de l'Accélérateur Linéaire, Batiment 200,
Faculté des Sciences, Orsay, (Seine et Oise),
France.

FRERK, L. C. W., B.Sc. Member, Council,
Inst. of Nucl. Eng.
Address: Institution of Nuclear Engineers,
147 Victoria Street, London S.W.1, England.

FRESCO, James, A.B., M.S., Ph.D. Born 1926.
Educ.: Arizona Univ. Chem., Health and
Safety Labs., New York Operations Office,
U.S.A.E.C., 1949-56; Asst. Prof., Chem.,
Nevada Univ., 1962-64; Asst. Prof., Chem.,
Texas Technol. Coll., 1964-. Society: A.C.S.
Nuclear interest: Radiochemistry.
Address: Chemistry Department, McGill
University, Montreal, Quebec, Canada.

FRESON, Maximilien Jean, D. en Sc. Born
1904. Educ.: Univ. Libre de Bruxelles. Sec.
Gén. du Fonds Nat. de la Recherche Scientifique; Sec. Gén. de l'Inst. Interuniv. des
Sci. Nucléaires.
Nuclear interest: Nuclear physics.
Address: 11 rue d'Egmont, Brussels 5,
Belgium.

FRETTER, William Bache, Ph.D. (California,
Berkeley). Born 1916. Educ.: California
(Berkeley) Univ. Prof. Phys., California
(Berkeley) Univ. Book: Introduction to Exptl.
Phys. (Prentice-Hall, New York, 1954).
Society: Fellow, A.P.S.
Nuclear interest: High-energy nuclear interactions.

Address: Department of Physics, California
University, Berkeley 4, California, U.S.A.

FREUDENTHAL, Alfred Martin, D.Sc. (Eng.).
Born 1906. Educ.: German Tech. and Charles
(Prague) Univs. Prof., Civil Eng., Columbia
Univ., 1948-. Books: Sectional Editor, Nucl.
Eng. and Design; Chapter 11 on Stress
Analysis and Design in Nucl. Eng. (Editor, C.
Bonilla) (McGraw-Hill, 1957); Mechanics
of Solids (J. Wiley and Sons, 1966). Societies:
A.S.C.E.; Soc. Rheology.
Nuclear interest: Mechanical design of reactors,
containment, life estimation and reliability.
Address: Columbia University, 624 S. W.
Mudd Building, New York 27, New York,
U.S.A.

FREUND, George A., S.B., S.M. Born 1927.
Educ.: M.I.T. At Argonne Nat. Lab., 1950-61;
Pres., Western Nucl. Corp., 1961-66; Manager,
Idaho Office, N.U.S. Corp., 1966-. Books: Contributor, Reactor Handbook, vol. 1 (Interscience Publishers, 1959); Fast Reactor Handbook (1967). Societies: A.N.S.; A.I.C.E.;
H.P.S.; N.S.P.E.
Nuclear interests: Reactor design studies;
Irradiation testing and design of in-pile
facilities; Nuclear chemical engineering;
Reactor safety; Food irradiation.
Address: 576 N. Capital Avenue, Idaho Falls,
Idaho, U.S.A.

FREUND, Hugo, Dr. med., Ministerial-direktor
a.D. Born 1890. Educ.: Heidelberg, Münich,
Berlin and Tübingen Univs. Member, Deutsche
Atomkommission.
Nuclear interests: Protection; Mental-health
aspects; Public relations; Training.
Address: 34 Hiltenspergerstrasse, Munich 13,
Germany.

FREUNDORFER, Annaliese, Dr. med. Member, Fachkommission IV—Strahlenschutz und
Sicherheit, Member, Arbeitskreis 1/2—Rechtsund Verwaltungsfragen des Strahlenschutzes,
Deutsche Atomkommission.
Address: Federal Ministry for Scientific
Research, 46 Luisenstrasse, Bad Godesberg,
Germany.

FREWING, Joseph John, M.A., B.Sc. Born
1912. Educ.: Oxford Univ. Tech. Adviser on
Nucl. Power Projects, Shell Internat. Petroleum
Co., Ltd. Society: F.R.I.C.
Nuclear interests: Uses for petroleum products
in the nuclear industry and application to the
petroleum industry.
Address: 16 Finsbury Circus, London, E.C.2,
England.

FREY, Emil, Hon. Prof. Born 1904. Educ.:
Kiel and Munich Univs. Vorsitzer des Vorstandes, Mannheimer Versicherungsgesellschaft,
1950-; Vorsitzer des Vorstandes Mannheimer
Lebensversicherungs-Gesellschaft A.G., 1965-;
Mitglied des Praesidiums, Gesamtverband der
Versicherungswirtschaft, 1962-. Member,

Atomic Commission, Deutscher Transport Versicherungs Verband e.V.
Nuclear interest: Insurance.
Address: Mannheimer Versicherungsgesellschaft, 65 Augusta-Anlage, Mannheim, Germany.

FREY, Kurt August Gustav, 1 and 2 juristisches Staatsexamen. Born 1913. Educ.: Munich and Kiel Univs. Amtsgerichtsrat until 1951; Sec., Studentwerk, Kiel, 1946-50; Sec., Deutsches Studentenwerk, Bonn, 1950-52; Sec.-Gen., German Nat. Commission for Unesco, 1953-54; Sec.-Gen., Permanent Conference of Ministers of Culture and Education, 1955-. Deutsche Atom-Kommission-Fachkommission II Kernforschung.
Address: 8 Nassestr., Bonn, Germany.

FREYRE VILLAFANE, Alejandro, Bachiller en Ciencias, Ingeniero Minas. Born 1907. Educ.: Univ. Nacional de Ingenieria de Lima, Univ. Nacional de San Marcos, Peru. Prof. Director, Dept. de control de Sustancias Radioactivas, Inst. Superior de Energia Nucl., Junta de control de Energia Atomica del Peru. Congresos nacionales (del Peru) de Ingenieria y de Quimica. Soc. Quimica, Soc. Geografica, Inst. Minas, Soc. de Ingenieros, Soc. Geológia y otras. Societies: Divesas Soc. de Sud America y del Peru.
Nuclear interest: De metalurgia y geologia nuclear.
Address: 793 Sanches Carrion, apartado 2658, Lima, Peru.

FRIC, Claude. Prof., Universitair and Cultural French Office for Algeria, Inst. d'Etudes Nucléaires.
Address: Universitair and Cultural French Office for Algeria Institut d'Etudes Nucléaires, B.P. 1386, Algiers, Algeria.

FRICK, Leon Frederick, B.S. (Nucl. Eng.). Born 1930. Educ.: Kansas State Univ. Nucl. Safety Officer, PM-1 Nucl. Reactor, 1965-67; Chief, Nucl. Power Branch, Tenth Air Force, 1967-. Society: A.N.S.
Nuclear interest: Power reactor operation and management.
Address: Richards-Gebaur Air Force Base, 5227 Clark, Missouri 64030, U.S.A.

FRICKE, Edwin Francis, B.S., M.A., Ph.D. Born 1910. Educ.: Idaho and California (Los Angeles) Univs. Sen. Sci., Argonne Nat. Lab., 1950-56; Sen. Phys., A.C.F. Industries, 1956-59; Sen. Development Eng., Republic Aviation Corp., 1959-64; Sen. Staff Sci., Sanders Assocs., 1964-. Societies: A.P.S.; A.C.S.; A.N.S.; A.S.A.; A.A.A.S.; A.I.A.A.; Illinois Acad. of Science; RESA.; New York Acad. of Sci.
Nuclear interests: Nuclear powered space ships; Nuclear and atomic physics.
Address: RFD 2, Reeds Ferry, N.H. 03078, U.S.A.

FRICKE, Gerhard, Dr. Born 1921. Educ.: Göttingen Univ. Asst., Heidelberg Univ., 1953-57; Oberasst., Darmstadt T.H., 1957-63; Lehrstuhl Exptl. Kernphysik., Mainz Univ., 1964-. Society: Phys. Gesellschaft.
Nuclear interests: Nuclear physics; High energy electron-scattering; H.F.S. – measurements with atomic beam resonance techniques.
Address: Institut für Experimentelle Kernphysik der Universität Mainz, Saarstrasse, 65 Mainz, Germany.

FRID, E. S. Paper: Co-author, Neutron Chamber on Basis of "Long" Counter (letter to the Editor, Atomnaya Energiya, vol. 16, No. 4, 1964).
Address: Academy of Sciences of the U.S.S.R., 14 Leninsky Prospekt, Moscow V 71, U.S.S.R.

FRIDMAN, Jacques, M.D. Born 1932. Educ.: Rio de Janeiro Univ. Academical Auxiliary of the State of Guanabara, 1957; Res. Asst., Biophys. Inst., 1959; Fellowship of the Nat. Commission of Nucl. Energy, 1959. Asst. of the 3rd Chair of the Medical Clinic, Brazil Univ. Society: Brazilian Soc. of Endocrinology.
Nuclear interests: Radioactivity applications in medicine, mainly in the field of research on thyroid physiology and biochemistry.
Address: 610 Avenida Ataulfo de Paiva, Apartment 802 Leblon, Rio de Janeiro, Brazil.

FRIDMAN, Sh. D. Paper: Co-author, Experience of Spectral Gamma-Log with Multichannel Spectrometers (letter to the Editor, Atomnaya Energiya, vol. 23, No. 1, 1967).
Address: Academy of Sciences of the U.S.S.R., 14 Leninsky Prospekt, Moscow V-71, U.S.S.R.

FRIED, Burton David, B.S., M.S., Ph.D. Born 1925. Educ.: Chicago Univ. and Illinois Inst. Technol. Instructor, Illinois Inst. Technol., 1947-52; Res. Phys., California Univ. Rad. Lab., Berkeley, 1952-54; Member, Sen. Staff (1954-60). Director, Res. Lab. (1960-63), Ramo-Wooldridge Corp. and Space Technol. Labs., Inc.; Assoc. Prof. Phys. (1963-64), Prof. Phys. (1964-), California Univ., Los Angeles. Society: A.P.S.
Nuclear interests: Controlled fusion research; Plasma physics.
Address: Physics Department, California University, Los Angeles, Calif., U.S.A.

FRIED, Maurice, B.S., M.S., Ph.D. Born 1920. Educ.: Cornell and Purdue Univs. U.S. Dept. of Agriculture, -1960; Director, Joint F.A.O./-I.A.E.A. Div. of Atomic Energy in Food and Agriculture, I.A.E.A. Societies: A.C.S.; American Soc. of Agriculture; Soil Sci. Soc. of America; Internat. Soc. of Soil Sci.; American Soc. of Plant Physiology.
Nuclear interests: Application of isotopes and radiation to soil-plant relationships; Entomology, plant breeding, pesticide residues, animal science and food irradiation.

Address: Joint Food and Agricultural Organisation/International Atomic Energy Agency Division of Atomic Energy in Food and Agriculture, International Atomic Energy Agency, 11-13 Kärntner Ring, A-1010 Vienna, Austria.

FRIEDEL, Jacques, D. ès Sc. (Paris), Ph.D. (Bristol). Born 1921. Educ.: Ecole Polytech., Ecole des Mines, Faculté des Sci., Paris. Ing. (1950-55), Ing. en Chef (1955-56), Corps des Mines; Maitre de Conférences (1956-61), Prof. (1961-), Faculté des Sci. de Paris; Consultant for I.R.S.I.D. at St. Germain; Atomic Energy Centre at Saclay; Centre Métallurgique Ecole des Mines, Paris; Direction des Recherches et Moyens d'Essaie. Assoc. Editor, Phys. and Chem. of Solids; J. of Fracture.
Nuclear interest: Physics of metals.
Address: Physique des Solides, Faculté des Sciences de Paris, Centre d'Orsay, B.P. 11, Orsay, Essonne, France.

FRIEDELL, Hymer L., M.D., Ph.D. Born 1911. Educ.: Minnesota Univ. Prof. and Head, Dept. of Radiol., Western Reserve Univ., 1946-; Director, Dept. of Radiol., Univ. Hospitals of Cleveland, 1946-. Director, Atomic Energy Medical Res. Project, Western Reserve Univ., 1946-. Member, Rad. Study Section, Public Health Service; Member, Com. on Radiobiol., N.R.C.; Member, Plowshare Advisory Com., U.S.A.E.C. Societies: Rad. Res. Soc.; A.A.A.S.; American Roentgen Ray Soc.; Radiol. Soc. of North America.
Nuclear interests: The biological effects of ionising radiation and the application of radioisotopes to problems in medicine.
Address: 2065 Adelbert Road, Cleveland 6, Ohio, U.S.A.

FRIEDHEIM, K., Dr. Managing Director, Abteilung Strahlenmessgerate, Total Kom.-Ges. Foerstner und Co.
Address: Total Kom.-Ges. Foerstner und Co., 6802 Ladenburg/Neckar, Germany.

FRIEDLAND, Erich, D.Sc., M.Sc., B.Sc. Born 1933. Educ.: Pretoria Univ. Lecturer, Dept. of Phys., Pretoria Univ., 1959-65; Visiting Sci., Kernreaktor, Karlsruhe, 1966; Sen. Lecturer, Dept. of Phys., Pretoria Univ. Society: South African Inst. of Phys.
Nuclear interest: Low energy nuclear reactions and nuclear structure.
Address: Pretoria University, Pretoria, South Africa.

FRIEDLAND, Sidney, B.Sc., M.Sc. Born 1932. Educ.: Wisconsin Univ. Liaison Officer, Washington (1959-61), Vice-Consul, U.S. Consulate Gen., Toronto, Canada (1961-63), Third Sec. and Adviser, U.S. Mission to I.A.E.A. and Adviser, U.S. Deleg. to 8th I.A.E.A. Gen. Conference (1964-), U.S. Dept. of State.
Nuclear interests: Diplomacy, with special regard to nuclear energy matters.

Address: U.S. Mission to International Atomic Energy Agency, 14 Schmidgasse, Vienna 8, Austria.

FRIEDLANDER, Gerhart, B.S. (Chem.), Ph.D. (Chem.). Born 1916. Educ.: California Univ., Berkeley. Chem. (1948-52), Sen. Chem. (1952-), Brookhaven Nat. Lab. Assoc. Editor, Annual Review of Nucl. Sci., 1958-67. Chairman, Div. of Nucl. Chem. and Technol., A.C.S., 1967. Books: Co-author, Introduction to Radiochem. (John Wiley and Sons, 1949); co-author, Nucl. and Radiochem. (John Wiley and Sons, 1955; 2nd Ed., 1964) (Translations in French, German, Polish, Japanese, Italian, Serbo-Croatian, Russian, Korean). Societies: A.C.S.; A.P.S.; A.A.A.S.
Nuclear interests: Nuclear reactions, especially reactions induced by high-energy particles; Nuclear spectroscopy; Radiochemistry.
Address: Brookhaven National Laboratory, Upton, N.Y., U.S.A.

FRIEDMAN, Abraham Solomon, B.A., Ph.D. Born 1921. Educ.: Brooklyn Coll. and Ohio State Univ. Post-Doctoral Res. Assoc., Ohio State Univ., 1950-51; Fulbright Res. Fellow, Amsterdam Univ., 1951-52; N.B.S., 1952-56; Sen. Chem., U.S.A.E.C., 1956-62; A.E.C. Sci. Rep. in Paris, 1962-65; Deputy Director, Division of Internat. Affairs, U.S.A.E.C., 1965-. Book: Co-author, Ideal Gas Thermodynamic Functions and Isotope Exchange Functions N.B.S. Monograph 20, Washington U.S. Govt. Printing Office, 1961). Societies: A.C.S.; A.P.S.; Netherlands Phys. Soc.; Washington Acad. Sci.
Nuclear interests: Isotope effects and separation; Nuclear chemistry; Transplutonium elements; Thermodynamics; Chemical physics; International affairs.
Address: U.S. Atomic Energy Commission, Washington 20545, D.C., U.S.A.

FRIEDMAN, Helen, Dr. Sci. concerned with Nucl. Work, Hofstra Univ.
Address: Hofstra University, Hempstead, New York, U.S.A.

FRIEDMAN, Horace Allen, B.A., M.A. Born 1927. Educ.: Vanderbilt Univ. Chem., O.R.N.L., 1953-. Societies: American Ceramic Soc.; A.N.S.; A.C.S.
Nuclear interest: Study of the chemistry involved in nuclear reactors.
Address: Oak Ridge National Laboratory, Post Office Box X, Oak Ridge, Tennessee, U.S.A.

FRIEDMAN, Irving, B.S., M.S., Ph.D. Born 1920. Educ.: Chicago Univ. Geochem., U.S. Geological Survey, 1952. Societies: Geochem. Soc.; A.C.S.; A.A.A.S.
Nuclear interests: Abundance of stable isotopes in nature; Isotopic fractionation processes; Isotopic geochemistry.

FRIEDMAN, Marvin Harold, Ph.D., M.S., B.S. Born 1923. Educ.: Illinois Univ. Prof., Phys., Phys. Dept., Northeastern Univ., 1965-. Society: A.P.S.
Nuclear interests: The theory of elementary particles and their interactions; Many body problems.
Address: Northeastern University, Physics Department, 360 Huntington Avenue, Boston, Massachusetts 02115, U.S.A.

FRIEDMAN, Milton, M.D. Born 1903. Educ.: George Washington Univ. Prof., Clinical Radiol. (Rad. Therapy), New York Univ. Schools of Medicine, 1958-; Director, Rad. Therapy, New York Univ. Hospital and Hospital for Joint Diseases; Consultant, Rad.: Walter Reed Gen. Hospital, U.S.A.E.C., N.I.H., Veterans Administration Hospital. Pres., American Radium Soc.; Ex-Pres., New York Cancer Soc.; Treas., New York Roentgen Soc.; Medical Director, Lila Motley Cancer Foundation; Member, I.C.R.U.; Member, Com. on Internat. Affairs, A.C.R.; Member, American Joint Com. for Classification of Tumours, and Com. on Human Applications of Radioactive Isotopes, New York City. Books: Editor, American Lectures in Rad. Therapy. 6 Volumes; Roentgen Rads and Riddles (U.S.A.E.C., 1959); Contributing author: Clinical Therapeutic Radiol. (Nelson, 1958); Res. in Radiol. (N.A.S.-N.R.C., 1958); Treatment of Cancer and Allied Diseases (Paul B. Hoeber, 1960); Textbook of Radiotherapy (Lea and Febiger, 1966); Radiol. in World War II (U.S. Govt. Printing Office, 1966). Societies: Soc. of Nucl. Medicine; Radiol. Soc. of North America; American Roentgen Ray Soc.; Soc. of Therapeutic Radiol.; American Soc. of Cancer Res.
Nuclear interests: Treatment of cancer and allied diseases with radium; X-ray and radioactive isotopes; Development of supervoltage irradiation techniques; Radiation dosimetry; Histologic effects of irradiation on human cancer and normal tissues; Recovery from irradiation.
Address: University Hospital, 566 First Avenue, New York, N.Y. 10016, U.S.A.

FRIEDMAN, William Jack, A.B. (Biol. Sci. and Chem., San Francisco State Coll.). Born 1921. Educ.: Roosevelt Univ., Chicago; San Francisco State Coll.; California (Berkeley) Univ. Health Phys., U.S. Naval Radiol. Defence Lab., 1952-; Health Phys. Consultant, Diablo Labs., Berkeley, California, 1955-. Member, Tech. Rad. Com., A.I.H.A., 1957-61. Book: Contributing author, Rad. and Contamination Control (Washington, D.C., U.S. Navy Bureau of Ships, 1959). Societies: H.P.S.; A.I.H.A.; Nucl. Eng. Section, A.S.Ch.E.
Nuclear interests: Health physics, decontamination, radioactive waste disposal, bioassay, radiological safety training programmes.
Address: 1141 Southgate Avenue, Daly City, California, U.S.A.

FRIEDMANN, Efrain, Diplomate Civil Eng. Born 1926. Educ.: Chile Univ. Head, Mathematics Dept., Chile Univ., 1964; Exec. Director, Chilean A.E.C., 1966. Counselor, Nat. Commission of Sci. and Technol. Res., 1967; Pres., Chilean Computer and Information Processing Assoc., 1967; Counselor, Commission of Nat. Electronic Data Processing for the Public Administration, depending of the Republic President, 1967. Societies: Chilean Eng. Inst.; A.N.S.
Nuclear interests: Nuclear power economics; Nuclear research management.
Address: Comision Nacional de Energia Atomica, Miraflores 138 - 2º Piso, Santiago, Chile.

FRIEDRICH, Kurt. Director, Rad. Meter Dept., Abteilung Strahlenmessgerate, Total Kom.-Ges. Foerstner und Co.
Address: Total Kom-Ges. Foerstner und Co., 6802 Ladenburg/Neckar, Germany.

FRIEDRICH, Martin, Dr.jur. Born 1903. Educ.: Leipzig and Munich Univs. Ministerialdirigent a.D. Direktor, Allianz Versicherungs-A.G., Munich. Mitglied des Arbeitskreises, Haftung und Versicherung, Deutsche Atomkommission.
Address: 22/I Prösslstrasse, Munich 90, Germany.

FRIEDRICH, Otto A., Dr. h.c. Born 1902. Educ.: Marburg, Königsberg, Frankfurt, Heidelberg, Berlin and Vienna Univs. Exec. Pres., Phoenix Gummiwerke A.G., Hamburg-Harburg, 1949-; Raw Materials consultant to Federal Government in Bonn. Vice-Pres., Federal Assoc. of German Industry, Cologne; Board of Governors, Kulturkreis d. Bundesverbandes d. Deutschen Industrie; Exec. Board, Founders' Assoc. for German Sci.; Advisory Board, Bundesvereinigung der Deutschen Arbeitgeberverbände, Cologne; Fachkommission V-Wirtschaftliche, Finanzielle und Soziale Probleme, Deutsche Atomkommission; Personal Member, Study Assoc. for the use of Nucl. Energy in Navigation and Industry.
Nuclear interest: Economic problems.
Address: 248 Bremer Strasse, 2100 Hamburg-Harburg, Germany.

FRIEDRICH-FREKSA, Hans, Prof. Dr. rer. nat. Born 1906. Educ.: Tübingen Univ. Director, Abteilung für physikalische Biologie, Max-Planck-Inst. für Virusforschung. Member, Fachkommission IV — Strahlenschutz und Sicherheit, Deutsche Atomkommission; Vereinigung Deutscher Wissenschaftler e.V.
Address: Abteilung für physikalische Biologie, Max-Planck-Institut für Virusforschung, 35/I Spemannstrasse, 74 Tübingen, Germany.

FRIEDRICHS, Günter H. R., Dr., Dipl-Volkswirt. Born 1928. Educ.: Frankfurt am Main and Minnesota State Univs. Head, Nucl. Energy and Automation Dept., Metal Industry Trade Union for Federal Republic of Germany. Board Member, German Nat. Productivity Centre (RKW); Member Com. for Ship Propulsion, German Atomic Energy Commission; Tech. adviser to Economic and Social Council of Euratom and Common Market. Societies: Kernenergie Studiengesellschaft, Hamburg; Listgesellschaft, Düsseldorf; Rationalisierungs Kuratorium der Deutschen Wirtschaft, Frankfurt am Main.
Nuclear interests: Social and economic effects of nuclear energy; Ship propulsion; Isotopes; Radiation; Safety.
Address: 70-76 Untermainkai, Frankfurt am Main, Germany.

FRIEL, J. V. Pres., Internat. Atomic Exposition, Inc.
Address: International Atomic Exposition, Inc., Architects Building, 117 South 17th Street, Philadelphia 3, Pennsylvania, U.S.A.

FRIEMAN, Edward Allan, B.S. (Columbia), M.S. and Ph.D. (Brooklyn Polytech. Inst.). Born 1926. Educ.: Columbia Univ. and Brooklyn Polytech. Inst. Head, Theoretical Div., Plasma Phys. Lab. (formerly Project Matterhorn) (1953-), Prof., Dept. of Astrophys. Sci. (1961), Princeton Univ.; Consultant: Los Alamos Sci. Lab., 1953-, Aeronautical Res. Assoc. Princeton, 1959-, Inst. for Defence Analyses, 1960-; Member, N.A.S.A. Res. Advisory Com. on Nucl. Energy Processes. Societies: A.P.S.; American Astronomical Soc.
Nuclear interests: Controlled thermonuclear reactors and nuclear rockets.
Address: Plasma Physics Laboratory, Princeton University, Forrestal Research Centre, Box 451, Princeton, New Jersey, U.S.A.

FRIEND, Albert Gallatin, B.S., M.S., Ph.D. Born 1913. Educ.: Virginia Polytech. Inst. and Harvard Univ. Sanitary Eng. Consultant, Pan American Sanitary Bureau, W.H.O., 1956-58; Sanitary Eng., U.S.P.H.S., Nat. Centre for Radiol. Health, 1958-.
Nuclear interest: Research on radioactive waste disposal to fresh water environment.
Address: 3 Wachusett Circle, Lexington, Massachusetts 02173, U.S.A.

FRIEND, W. F., Nucl. Eng. Consultant, Ebasco Services Inc. Formerly Vice-Chairman, Atomic Energy Panel, Engineers Joint Council.
Address: Ebasco Services Inc., Two Rector Street, New York 6, New York, U.S.A.

FRIES, Bernard Albert, B.S., Ph.D. Born 1917. Educ.: California Univ. Sen. Res. Assoc., Chevron Res. Co., Richmond, California, 1945-. Societies: A.C.S.; H.P.S.
Nuclear interest: Industrial applications of radioisotopes, particularly in petroleum chemistry, refining and production.
Address: Chevron Research Company, 576 Standard Avenue, Richmond, California 94802, U.S.A.

FRIESE, Thomas, Dipl. Elec. Eng. Born 1929. Educ.: Berlin Tech. Univ. Shipbuilding Exptl. Tank in Berlin-West, 1957-58; Hahn-Meitner Inst. for Nucl. Res., Berlin-West, 1958-.
Nuclear interests: Nuclear engineering and equipment; Instrumentation; Electronics.
Address: 78 Glienicker Strasse, Berlin-Wannsee, Germany.

FRIESEN, Sten VON. See VON FRIESEN, Sten.

FRIGERIO, Norman Alfred, B.S. (M.I.T.), Ph.D. (Yale). Born 1929. Educ.: M.I.T. and Yale Univ. Chief Eng., Usher Aviation Radio, New Haven, Conn., 1953-57; Fellow, N.S.F. (U.S.A.), 1954-57; Asst. Sci. (1957-61), Assoc. Sci. (1961-), Argonne Nat. Lab.; Prof. Chem., St. Procopius Coll., Lisle, Ill., 1958-65. Abstractor and Translator, Chem. Abstracts Service, A.C.S. Societies: A.C.S.; I.E.E.E.; Rad. Res. Soc.; RESA; N.Y. Acad. Sci.; Ill. Acad. Sci.
Nuclear interests: Interaction of neutrons and charged particles with matter, particularly with organisms and with compounds and nuclides found in living organisms.
Address: D-202, Argonne National Laboratory, Argonne, Illinois, U.S.A.

FRILUND, Harald, Prof. h.c., Dipl.Ing. (Finland). Born 1889. Educ.: Finland's Inst. Technol. Managing Director, EKONO, 1946-63; Managing Director, A.B. Konsulterande Ingeniörbyrån Consulting, 1943-63; Member, Finnish A.E.C., 1958-64; Chairman, Energy Sec., Finnish Fuel Commission, and member various Govt. committees. Societies: Roy. Swedish Acad. of Eng. Sci.; Swedish Acad. of Eng. Sci. in Finland.
Nuclear interest: Power economy.
Address: c/o Ekono, 14 S. Esplanadgatan, Helsinki, Finland.

FRISANCHO-PINEDA, Ignacio, Bachelor in Maths., Dipl. Civil Eng. Born 1922. Educ.: Univ. San Andrés, La Paz, Bolivia; Cuzco Univ. and Eng. Univ., Lima, Peru; Pennsylvania State Univ.; Argonne Nat. Lab. Prof. Industrial Uses of Radioisotopes, Inst. Sup. de Energia Nucl., 1959; Prof. Nucl. Phys., 1961. Director, Reactor Dept., Junta de Control de Energia Atomica del Perú, 1956. Book: Industrial Uses of Radioisotopes (Junta de Control de Energia Atómica del Perú, 1960). Society: Nat. Civil Eng. Assoc.
Nuclear interests: Nuclear physics; Industrial uses of radioisotopes; Nuclear engineering.
Address: Junta de Control de Energia Atomica, Pierola 611-0f.21, Lima, Peru.

FRISCH, Otto Robert, Dr. Phil. (Vienna), D.Sc. (hon. (Birmingham), M.A. (Cantab.). Born 1904. Educ.: Vienna Univ.; Jacksonian

Prof. Natural Philosophy, Cambridge Univ., 1947-. Books: Meet the Atoms (New York, A.A. Wyn, 1947); Atomic Phys. Today (New York, Basic Books, 1961); Working with Atoms (Leicester, Brockhampton Press, 1965). Societies: F.R.S.; Phys. Soc.; A.P.S. Nuclear interest: Fundamental particles. Address: Trinity College, Cambridge, England.

FRISIUS, F., Dipl. Phys. Member, Fachausschuss für Strahlenschutz, (joint group of Deutsches Atomforum and Kernenergieverwertung Gesellschaft). Address: Gesellschaft für Kernenergieverwertung in Schiffbau und Schiffahrt m.b.H., Institut für Reaktorphysik, 2057 Geesthacht-Tesperhude, Postfach, Germany.

FRISKEN, William Ross, B.Sc., M.Sc., Ph.D. Born 1933. Educ.: Queen's (Canada) and Birmingham Univs. Teaching Fellow, McGill Univ., 1960-62; Asst. Prof., Dept. Phys., Montreal Univ., 1962-64; Assoc. Phys., Brookhaven Nat. Lab., 1964-66; Assoc. Prof., Case-Western Reserve Univ., Cleveland, 1966-. Member, Tech. Advisory Panel, Zero Gradient Synchrotron, Argonne Nat. Lab. Society: A.P.S.
Nuclear interest: Experimental high energy and particle physics and also design of experimental facilities for same. Address: Case Western Reserve University, Cleveland, Ohio, U.S.A.

FRISSEL, Martinus Johannes, Dr. Born 1929. Educ.: Rijksuniversiteit Utrecht, Netherlands. At Lab. of Soils and Fertilisers, State Agricultural Univ., Wageningen, 1956-60; Inst. for Atomic Sci. in Agriculture, Wageningen, 1960-. Nuclear interests: Activation; Transport of fall out through soils; Tracer methodology in soil-plant relationship. Address: 3 Torenlaan, Heelsum, Netherlands.

FRITSCH, Arnold Rudolph, B.S., Ph.D. Born 1932. Educ.: Rochester and California Univs. Sen. Sci., Westinghouse Elec. Corp., 1956-59; Br. Chief, Div. of Internat. Affairs (1959-61), Tech. Asst. to Chairman (1961-64), Special Asst. to Chairman (1964-), U.S.A.E.C. Societies: A.A.A.S.; A.C.S.; A.N.S.; A.P.S. Nuclear interests: Nuclear spectroscopy; Surface chemistry; Radiochemistry; Safeguards research and development. Address: 26723 Ridge Road, Damascus, Maryland 20750, U.S.A.

FRITSCHE, V. Manager, Münchener Rückversicherungs-Gesellschaft. Member, Atomic Commission, Deutscher Transport Versicherungs Verband e.V. Address: Münchener Rückversicherungs-Gesellschaft, 107 Königinstr., Munich 23, Germany.

FRITZ-NIGGLI, Mrs. H., Prof. Dr. Strahlenbiologisches Inst., Universität Zürich. Member, Council of Management, Swiss Assoc. for Atomic Energy; Council Member, Sté. Europeenne de Radiobiol. Address: Swiss Association for Atomic Energy, 21 Schauplatzgasse, 3001 Bern, Switzerland.

FRITZE, Klaus U. K. A., Dipl. Chem., Dr. rer. nat. Born 1926. Educ.: Mainz Univ. Res. Chem., McMaster Univ., 1957-. Nuclear interests: Nuclear and radiochemistry, particularly decay properties of short-lived fission products and independent and cumulative fission product yields; Neutron activation analysis. Address: McMaster University, Hamilton, Ontario, Canada.

FRITZSCHE, A. F., Dr. Tech. Director, Swiss Federal Inst. for Reactor Res. Com. Member, Swiss Nucl. Soc.; Member, Gen. Purposes Com., O.E.C.D. High Temperature Reactor Project (Dragon). Address: Swiss Federal Institute for Reactor Research, 5303 Würenlingen, Switzerland.

FRITZSCHE, Hans, Dr.-Ing., M.Sc. Born 1911. Educ.: Berlin T.H. and Montana School of Mines. Member, Board of Exec. Directors Stolberger Zink A.G. für Bergbau und Huttenbetrieb, 1958-. Chairman, Gesellschaft Deutscher Metallhütten-und Bergleute, Clausthal-Zellerfeld. Society: Deutsche Gesellschaft für Metallkunde e.V. Address: 10 Im Brockenfeld, Aachen, Germany.

FRITZSON, Per, Dr. phil. Born 1922. Educ.: Oslo Univ. Res. Fellow, Norwegian Cancer Soc., Norsk Hydro's Inst. for Cancer Res., 1955-. Nuclear interest: Nucleic acid metabolism. Address: Norsk Hydro's Institute for Cancer Research, Montebello, Oslo 3, Norway.

FRIZ, Carlos J. T. Jefe del Departamento de Exploración, Comision Nacional de Energia Atomica, Argentina. Address: Comision Nacional de Energia Atomica, 8250 Av. del Libertador, Buenos Aires, Argentina.

FRIZZI, D., Ing. Working in Nucl. Field, Dott. Ing. Giuseppe Torno and C.S.p.A. Address: Dott.Ing. Giuseppe Torno and C. S.p.A., 7 Via Albricci, Milan, Italy.

FRODESEN, A. G., Cand. real. Member, Res. on Elementary Particles and Cosmic Rad., Nucl. Phys. Lab., Oslo Univ. Address: Oslo University, Nuclear Physics Laboratory, Institute of Physics, Blindern, Norway.

FROHLINDE, Dietrich SCHULTE-. See SCHULTE-FROHLINDE, Dietrich.

FROHLY, Leon. Responsables du Departement Energie Nucl., Cie. Francaise d'Entreprises Metalliques.

Address: Cie. Francaise d'Entreprises Metalliques, 37 boulevard de Montmorency, Paris 16, France.

FROLOV, V. V. At Moscow Eng. and Phys. Institute. Adviser, Third I.C.P.U.A.E., Geneva, Sept. 1964.
Address: Moscow Engineering and Physics Institute, 31 Kirova Ulitsa, Moscow, U.S.S.R.

FROMAN, Darol Kenneth, B.Sc. (Alberta), M.Sc. (Alberta), Ph.D. (Chicago). Born 1906. Los Alamos Sci. Lab., 1943-62. Retired, Private Consultant. Member, Douglas Aircraft Co. Sci. Directorate. Member, Gen. Advisory Com. to U.S.A.E.C., 1964-66. Societies: A.P.S.; A.N.S.
Nuclear interests: Nuclear physics; Reactor design.
Address: Pajarito Village, Box 428, Route 1, Espanola, New Mexico 87532, U.S.A.

FROMM, Eckehard, Dr. rer. nat. Sen. Officer, Inst. of Special Metals, Max-Planck Inst. for Metal Res.
Address: Max-Planck Institute for Metal Research, 92 Seestrasse, 7000 Stuttgart 1, Germany.

FROMM, Leonard William, Jr., B.Ch.E. Born 1924. Educ.: Polytech. Inst. of Brooklyn. Development on S.T.R. for U.S.S. Nautilus (1949-54), Co-ordinator, Naval Reactor Development Programme (1954-55), Project Eng., E.B.W.R. Reactor and Containment Building (1955-57), Project Manager, ARBOR (1957-58), Head, Water Reactors Sect., Reactor Eng. Div. (1955-59), U.S. Representative on staff of O.E.E.C. Halden Reactor Project - Gen. Consultant to Project Management (1959-60), U.S. Representative on Halden Tech. Group (1961), Member, Organic Reactor Programme Task Group (1961), Manager, Argonne Advanced Res. Reactor Project (1962-), Argonne Nat. Lab. A.N.S. Standards Com., Reactor Components Project, 1958-; American Standards Assoc. B31 Code for Pressure Piping (Advisory Com. on Nucl. Piping); Chairman, Task Force on Reactor Plant Survey (1957-60), Com. on Nucl. Piping (1960-). Society: A.N.S.
Nuclear interests: Reactor design; Reactor technology development; Materials testing and development of material testing facilities; Management.
Address: 630 North Avenue, Glen Ellyn, Illinois, U.S.A.

FROMMHOLD, Ernst Alfred. Born 1924. Leader, Rad. measurement Lab., VEB Transformatoren-u. Röntgenwerk, Dresden, 1950-55; Leader, Elektrometer Lab., VEB Vakutronik, Dresden, 1956-64; Elektronik Lab., E. A. Frommhold, Dresden, 1964-.
Nuclear interests: Electronic instrumentation for reactor and X-ray equipments, especially radiation dosimeter, electrometer design.

Address: 17 Caspar David Friedrich Str., 8020 Dresden, German Democratic Republic.

FROMMHOLD, W. Editor, Fortschritte auf dem Gebiete der Rontgenstrahlen und der Nuklearmedizin.
Address: Georg Thieme Verlag, 63 Herdweg, Stuttgart N, Germany.

FROST, Brian Reginald Thomas, B.Sc. (Hons.) (Metal.), Ph.D. Born 1926. Educ.: Birmingham Univ. Sect. Leader, Liquid Metals, Metal. Div. (1955-59), Group Leader, Advanced Fuels Irradiation (1959-), A.E.R.E., Harwell. Book: Co-author, Nucl. Reactor Materials (Temple Press, 1959). Societies: Inst. Metals; Inst. Metallurgists; British Ceramic Soc.
Nuclear interests: Fast reactor technology; In the field of metallurgy irradiation studies of ceramic fuels and cermets, and of stainless steel canning materials. Corrosion and mass transfer in liquid metal circuits.
Address: 3 Mandeville Close, Abingdon, Berks., England.

FROST, Dietrich, Dipl.-Ing., Dr.-Ing. Born 1923. Educ.: Berlin-Charlottenburg Tech. Univ. Phys., Res. Lab., Auer-Gesellschaft AG, Berlin 1951-55; Chief Phys., Abteilung für Strahlungsphysik des Rudolf-Virchow-Krankenhauses, Berlin, 1955-; Prov.-Doz. Berlin-Charlottenburg Tech. Univ., 1964-. Books: Praktischer Strahlenschutz (Berlin, 1960); co-author, Radioactive Isotope in der Chirurgie (1961). Societies: Deutsche Röntgen-Gesellschaft; Physikalische Gesellschaft; H.P.S.; Internat. Rad. Protection Assoc.
Nuclear interests: High energy electron-accelerators; Radiation protection; Health physics; Radiation dosimetry; Radioisotopes in radiological research; Neutron physics.
Address: 79 Reichshofer Str., Berlin-Dahlem, Germany.

FROTA-PESSOA, E., Formerly Assoc. Prof. Exptl. Phys. Dept., now Assoc. Prof., Corpuscular Phys. Dept., Centro Brasileiro de Pesquisas Fisicas.
Address: Centro Brasileiro de Pesquisas Fisicas, 71 Avenida Venceslau Bras, Rio de Janeiro, Brazil.

FRUCHT, Robert, Ph.D. Prof. Born 1906. Educ.: Berlin Univ. Head, Mathematics and Phys. Dept., Univ. Tecnica Federico Santa Maria.
Address: Universidad Tecnica Federico Santa Maria, Casilla 110 V, Valparaiso, Chile.

FRUMERMAN, Robert, B.S., M.S. (Chem. Eng.). Born 1924. Educ.: Pittsburgh and Carnegie-Mellon Univs. Process Eng., Asst. Project Eng., Chemical Plants Div., Blaw-Knox Co., 1950-53; Process Sect. Manager, Dev. Dept., Elliott Co., Jeannette, Pa., 1953-56; Project Eng., Koppers Co. Inc., Pittsburgh, Pa., 1956-59; Chem. Processing Production Manager and Project Eng., Nucl. Materials and Equipment Corp. 1959-62, Consulting Eng., 1962-68;

Pres., Frumerman Associates, Inc., 1967-68. Society: A.I.Ch.E.
Nuclear interests: Chemical process designer, Savannah River Plant; Designed, built, and directed operations of plant converting enriched UF_6 to UO_2; Managed uranium scrap recovery facility; Criticality control in plant design and operations; Accountability officer; Consultant in development and design of nuclear process systems.
Address: 2323 Tilbury Avenue, Pittsburgh, Pa., U.S.A.

FRUNSE, V. V. Paper: Co-author, Experience in Operating the SM-2 Exptl. Reactor (Kernenergie, vol. 9, No. 10; 1966).
Address: Academy of Sciences of the U.S.S.R., 14 Leninsky Prospekt, Moscow V-71, U.S.S.R.

FRY, Donald William, B.Sc. (1st class Hons.), M.Sc., Fellow, King's Coll. London. Born 1910. Educ.: London Univ. Chartered Eng. Head, Gen. Phys. Div. (1950), Chief Phys. (1954), Deputy Director (1958), A.E.R.E., Harwell; Director, Atomic Energy Establishment, Winfrith (1959), U.K.A.E.A. Member, Board of Management, O.E.C.D. Halden Reactor Project and Dragon High Temperature Reactor Project. Societies: Fellow, Inst. of Phys. and Phys. Soc.; Fellow, I.E.E..; Fellow, I.E.E.E., 1960.
Nuclear interests: Reactor physics and technology, and management.
Address: Coveway Lodge, Overcombe, Weymouth, Dorset, England.

FRY, Kenneth, B.S., M.S., Ph.D. Educ.: Kansas State Teachers Coll. and Purdue Univ. Chairman, Biol. Dept., Chattanooga Univ.; Head, Fellowship Office Univ. Relations Div. (1964), Head, Fellowship Office (1964-), Oak Ridge Associated Univs.
Address: Oak Ridge Associated Universities, P.O. Box 117, Oak Ridge, Tennessee, U.S.A.

FRY, Richard M., A.B., M.S.P.H. Born 1937. Educ.: Asbury Coll., Rochester (New York) and North Carolina Univs. Director, Radiol. Health Programme, Kentucky State Dept. of Health. Exec. Council, Blue Grass Chapter, H.P.S.; Exec. Council, Rad. Health Soc., American Public Health Assoc. Society: Conference on Rad. Health.
Nuclear interest: Health physics.
Address: 220 Juniper Drive, Frankfort, Kentucky 40601, U.S.A.

FRY, Robert Mason, B.Sc. (Hons.). Born 1928. Educ.: Adelaide Univ. Res. Officer, C.S.I.R.O. Meteorological Phys. Sect., 1949-51; Lecturer, Cancer Res. Dept., Adelaide Univ., 1951-57; Res. Officer, Australian A.E.C., Health Phys. Sect. (attached to A.E.R.E., Harwell, England), 1957-59; Health Phys. Sect., A.A.E.C., Lucas Heights. Sydney, 1959-; Formerly Head, Health Phys. Res. Section, A.A.E.C., 1960-. Now Atomic Energy Advisor, Canberra House, London.
Nuclear interests: Health physics; Radiation dosimetry (especially neutrons), low energy particle physics, cosmic ray produced isotopes and neutrons, low level counting techniques, total body counting; Atmospheric diffusion and aerosol removal mechanisms.
Address: Canberra House, 10-16 Maltravers Street, Strand, London, W.C.2, England.

FRYAR, Robert Marshall, B.S., M.S., Ph.D., Nucl. Eng. Born 1921. Educ.: Idaho and Purdue Univs. Lecturer, Washington Univ. Graduate Centre, 1950-60; Manager, Eng. Development, Hanford Lab., G.E.C., 1950-61; Assoc. Manager, High Temperature Gas-Cooled Reactor Div., Gen. Atomic, 1961-65; Vice Pres., Nucl. Div., Kerr-McGee Corp., 1966-67. Societies: A.N.S.; Atomic Ind. Forum.
Address: Kerr-McGee Corporation, Kerr-McGee Building, Oklahoma City, Oklahoma 73102, U.S.A.

FRYE, John H., Jr., A.B. (Hons. Eng. Lit.), M.S. (Metal. Eng.), D.Phil. (Phys. Sci.). Born 1908. Educ.: Howard Coll., Lehigh and Oxford Univs. Director, Metals and Ceramics Div., O.R.N.L. 1948-; Lecturer, Graduate School, Tennessee Univ., 1950-. Member, Editorial Advisory Board, J. of the Less-Common Metals, 1962-; Member, Editorial Advisory Board, J. of Nucl. Applications, 1965-; Hon. Adjunct Prof., College of Eng., Alabama Univ., 1964-. Societies: A.I.M.M.E.; A.N.S.; Inst. of Metals; A.S.M.; Oak Ridge Rotary Club; A.A.A.S.
Nuclear interests: Metallurgy; Directs research and development in the field of metals and ceramics for nuclear applications.
Address: 210 Outer Drive, Oak Ridge, Tennessee, U.S.A.

FRYE, Robert. Deputy Director, Feltman Res. Labs., Picatinny Arsenal.
Address: Feltman Research Laboratories, Picatinny Arsenal, Dover, New Jersey, U.S.A.

FRYKLUND, Verne C., Jr. Geologist, Geological Survey, U.S. Dept. of the Interior.
Address: U.S. Department of the Interior, Geological Survey, Washington D.C. 20242, U.S.A.

FRYMANN, H. Pres., Suisatom A.G.
Address: Suisatom A.G., 1 Bahnhofplatz, Zurich, Switzerland.

FRYNTA, Zdenek, Eng. Born 1925. Educ.: Prague Tech. Univ. and Charles Univ., Prague. At Nat. Res. Inst. of Material, Prague. Societies: Czechoslovak Sci. and Tech. Soc.; Nucl. Eng. Commission, Prague.
Nuclear interests: Radioisotopes applications in metallurgy and mechanical engineering as sources (nondestructive testing, thickness measurement) and traces (especially wear measurement, diffusion, microautoradiography); respective radiometric measurement technique, low level counting installations, nuclear spectrometry.

Address: 4 Na Zavadilce, Dejvice, Prague 6, Czechoslovakia.

FU, Cheng-yi. Member, Editorial Board, Scientia Sinica.
Address: Scientia Sinica, Academia Sinica, Peking, China.

FU, Ying. Member, Editorial Com., Scientia Sinica.
Address: Scientia Sinica, Academia Sinica, Peking, China.

FUBINI, Alessandro, Dr. in Phys. Bcrn 1936. Educ.: Turin Univ. Res. Assoc., Italian Com. for Atomic Energy, 1960-.
Nuclear interest: Experimental reactor physics.
Address: 37 Corso Re Umberto, Turin, Italy.

FUBINI, Sergio Piero. Born 1928. Educ.: Turin Univ. Sen. Phys., C.E.R.N., 1957-60; Prof. Phys., Padua Univ., 1960-61; Prof. Theoretical Phys., Turin Univ., 1961; Visiting Prof., C.E.R.N.
Nuclear interests: Meson physics; Quantum field theory; Low-energy nuclear physics.
Address: Istituto di Fisica, 1 V. Giuria, Turin (10125), Italy.

FUCHS, Dr. Director, Electronic Dept., Mettler, Analytical and Precision Balances.
Address: Mettler, Analytical and Precision Balances, CH-8606 Greifensee, Switzerland.

FUCHS, Klaus, Ph.D., D.Sc. Born 1911. Educ.: Leipzig, Kiel, Berlin, Bristol and Edinburgh Univs. Tube Alloys, D.S.I.R., Birmingham, 1941-43; Member, British Atomic Energy Diffusion Mission, New York, 1943-44; Member, British Atomic Team, Los Alamos, 1944-46; Head, Theoretical Phys. Div., Harwell, 1946-49; Deputy Director, Central Inst. for Nucl. Res., Dresden, 1959-.
Nuclear interests: Nuclear physics; Reactor theory; Reactor design.
Address: Zentralinstitut für Kernforschung, Rossendorf near Dresden, Dresden WH, P.O. Box 19, German Democratic Republic.

FUCHSMANN, G. Chef du Bureau d'Etudes Chaudronnerie, Constructions Metalliques, Zwahlen and Mayr S.A.
Address: Zwahlen and Mayr S.A., Constructions Metalliques, Lausanne, (Vaud), Switzerland.

FUCKS, Wilhelm, o. Prof., Dr.-Ing. Born 1902. Educ.: Munich Univ. Director, Phys. Inst., Univ. Technol., Aachen. Pres., Sci. Council, Nucl. Res. Centre, Jülich, 1958-60; Director of the Plasma Phys. Inst., Prof., Wissenschaftlicher Rat, Kernforschungsanlage Jülich; Editor, Atomkernenergie. Book: Energiegewinnung aus Atomkernen (Essen, Verlag W. Giraret, 1948). Society: Fachkommission II Kernforschung, Deutschen Atomkommission.
Nuclear interests: High-temperature plasma physics and controlled nuclear fusion.
Address: Physics Institute, University of Technology, 55 Templergraben, Aachen, Germany.

FUENTE, Eddie S., B.S. Born 1934. Educ.: Southern Mississippi Univ. Supv., Radiol. Health Unit, Mississippi State Board of Health.
Address: Mississippi State Board of Health, P.O. Box 1700, Jackson, Mississippi, U.S.A.

FUENTES, General de Brigada Aerea Hugo, Ing. Vice Pres., Consejo Directivo, Comision Chilena de Energia Nuclear, 1966-.
Address: Comision Chilena de Energia Nuclear, 138 Calle Miraflores, Santiago, Chile.

FUERST, Robert, B.S., M.A., Ph.D. Born 1921. Educ.: Vienna Coll. for Ind. Arts, Houston and Texas Univs. Asst. Biol. and Head, Microbiol. Sect., M.D. Anderson Hospital and Tumour Inst., Texas Univ., 1955-57; Asst. Prof., Postgraduate School of Medicine, Texas Univ., 1956-57; Head, Microbiol. Res. Lab. (1957-), Prof., Biol. (1964-), Texas Woman's Univ. Societies: Genetics Soc. America; American Soc. for Microbiol.; New York Acad. Sci.; Human Genetics Soc. America; A.C.S.; Soc. Ind. Microbiol.; A.A.A.S.
Nuclear interests: Biological effects of radiation of different origins; Radiation protective effects by gases and other chemical agents; The use of nuclides in tracer studies of biochemical pathways in animals, fungi and bacteria.
Address: Box 2757, T.W.U. Station, Texas Woman's University, Denton, Texas 76204, U.S.A.

FUFAEVA, O. L. Paper: Co-author, Temperature and Neutron Irradiation Effect on Plastic Deformation of Alpha Uranium Single Crystal (Atomnaya Energiya, vol. 19, No. 4, 1965).
Address: Academy of Sciences of the U.S.S.R., 14 Leninsky Prospekt, Moscow V-71, U.S.S.R.

FUGIMOTO, Koichi. Manager, Atomic Energy Tech. Div., and Atomic Energy Manufacturing Div., Kobe Industries Corp.
Address: Kobe Industries Corp., Ohkubo-cho, Akashi, Japan.

FUJII, Akihiko, B.S. (Tokyo), Ph.D. (Rochester, N.Y.). Born 1927. Instructor (1958), Asst. Prof. (1959), Purdue Univ.; Res. Assoc., Imperial Coll., 1960; Res. Assoc., Frascati Nat. Lab., 1961; Sen. Res. Assoc., Inst. for Nucl. Study, Tokyo Univ., 1963; Assoc. Prof., Sophia Univ., Tokyo, 1963-. Society: Phys. Soc. of Japan.
Nuclear interests: Theoretical nuclear physics, in particular the study of nuclear structure and reaction mechanism by the use of mu-, pi- and kay-mesons; High energy nuclear physics.
Address: Department of Physics, Sophia University, 7 Kioi-cho, Chiyoda-ku, Tokyo, Japan.

FUJII, Saburo, Dr. of Sci. Born 1931.
Educ.: Kyoto Univ. Lecturer, Dept. of Phys.,
Nihon Univ., 1965-. Society: Phys. Soc. of
Japan.
Nuclear interest: Low energy nuclear physics
(theory).
Address: 1-8 Kanda-Surugadai, Chiyoda-ku,
Tokyo, Japan.

FUJII, Shigetada. Polymerisation, Govt. Ind.
Res. Inst., Ministry of Internat. Trade and
Ind.
Address: Ministry of International Trade and
Industry, Government Industrial Research
Institute, Hirate-machi, Kita-ku, Nagoya,
Japan.

FUJII, Shinzo. Chairman, Board of Directors,
Mitsubishi Heavy Industries, Ltd.
Address: Mitsubishi Heavy Industries, Ltd.,
10, 2-chome, Marunouchi, Chiyoda-ku,
Tokyo, Japan.

FUJII, Taro., D. Agr. Born 1927. Educ.:
Kyoto Coll. of Textile Fibres. Researcher,
Dept. of Induced Mutation, Nat. Inst. of
Genetics, 1950-. Societies: Genetics Soc. of
Japan; Japanese Soc. of Breeding; Japan Rad.
Res. Soc.
Nuclear interest: Radiation genetics in plants
by use of radiations.
Address: National Institute of Genetics,
Yata 1111, Misima, Sizuoka-ken 411, Japan.

FUJII, Yasunori, Ph.D. Born 1931. Educ.:
Nagoya Univ. Asst., High Energy Nucl. Phys.
(Theory), Atomic Energy Res. Inst., Nihon
Univ.; Assoc. Prof., Tokyo Univ., 1963-.
Society: Japan Phys. Soc.
Nuclear interests: Field theory; High energy
nuclear physics.
Address: Institute of Physics, College of
General Education, Tokyo University, Komaba,
Meguro-ku, Tokyo, Japan.

FUJIKUBO, S. Director, Tokyo Atomic
Industrial Consortium.
Address: Tokyo Atomic Industrial Consortium,
Hitachi Building, 6-4-chome, Surugadai,
Kanda, Chiyoda-ku, Tokyo, Japan.

FUJIMAKI, Shigeo, Prof. Member from
Medicine Dept., Radioisotope Res. Com.,
Niigata Univ. School of Medicine.
Address: Niigata University School of Medicine,
757, 1-Bancho Asahimachi-Street, Niigata,
Japan.

FUJIMOTO, Koichi. Born 1918. Educ.:
Tokyo Univ. of Technol. Manager, Atomic
Energy Div. (1963), Manager, Electronic Eng.
Div., (1966), Kobe Industries Corp.
Nuclear interest: Measuring and monitoring
equipment.
Address: Kobe Industries Corporation, Okubo-
cho, Akashi, Hyogo, Japan.

FUJIMOTO, Yoichi, Prof. Dr. Div. of Nucl.
Phys. and Eng., Sci. and Eng. Res. Lab.,
Waseda Univ.
Address: Waseda University, 17 Kikui-cho,
Shinjuku-ku, Tokyo, Japan.

FUJIMURA, Ryoichiro. Inorg. Phys., Government Ind. Res. Inst. Nagoya, Agency of Ind.
Sci. and Technol., Ministry of Internat. Trade
and Industry.
Address: Government Industrial Research
Institute, Ministry of International Trade and
Industry, Hirate-machi, Kita-ku, Nagoya,
Japan.

FUJINO, Akira, Chief, Rad. Chem. Div., Res.
Inst. for Atomic Energy, Osaka City Univ.
Address: Research Institute for Atomic
Energy, Osaka City University, 12 Minamioga-
machi, Kita-ku, Osaka, Japan.

FUJIOKA, Shingo, Bachelor of Laws. Born
1901. Educ.: Keio Univ., Tokyo. Pres., Mitsubishi Oil Co., Ltd., 1961-. Vice-Pres., Japan
Petroleum Assoc.; Com., World Petroleum
Congress, etc. Society: Japan Atomic Power
Industry Conference.
Nuclear interests: Use of isotopes in the oil
industry.
Address: Mitsubishi Oil Co., No. 1 Shiba
Kotohiracho, Minatoku, Tokyo, Japan.

FUJIOKA, Yoshio, Prof. Emeritus, Tokyo
Univ. of Education; Ph.D. (Tokyo). Born
1903. Educ.: Tokyo Univ. Prof. Tokyo Univ.
of Literature and Sci., 1941; Chairman, Atomic
Energy Problem Com. in Sci. Council of Japan,
1954; Member, A.E.C. of Japan, 1956. Director, I.A.E.A. (Div. of Isotopes), 1958-59. Pres.
of Saitama Univ., 1960-. Society: Phys. Soc.
of Japan.
Nuclear interests: Spectroscopy, quantum
mechanics and nuclear science.
Address: 10 Nishikata-machi, Komagome,
Bunkyo-ku, Tokyo, Japan.

FUJITA, Jun-Ichi, B.Sc., Ph.D. Born 1929.
Educ.: Tokyo Univ. Res. Assoc., Stanford
Univ., Calif., 1959-60; Res. Assoc., Indiana
Univ., 1960-61. Formerly Asst. Prof., Dept.
Phys., Coll. Sci. and Eng., Nihon Univ., Tokyo.
Societies: Japan Phys. Soc.; A.P.S.
Nuclear interest: Theoretical nuclear physics,
especially beta-decay and nuclear photo-effect.
Address: Institute for Nuclear Study, University of Tokyo, Tanashimachi, Tokyo, Japan.

FUKAI, Rinnosuke, Dr.Sc. Born 1925 Educ.:
Tokyo Univ. Radiochem., Isotope Res. Lab.,
Tokai Regional Fisheries Res. Lab., Japan
-1962; Sen. Sci., Internat. Lab. of Marine
Radioactivity, I.A.E.A., Monaco, 1962-.
Nuclear interests: Radiochemistry; Behaviour
of radionuclides and related trace elements in
marine environment; Application of neutron
activation analysis to the stated subject.

Address: International Laboratory of Marine Radioactivity, International Atomic Energy Agency, Musée Océanographique, Monaco.

FUKAI, Yuzo, Ph.D., B.S. Born 1928. Educ.: Waseda Univ. With Tokyo Shibaura Elec. Co., Ltd., 1952-58; Nippon Atomic Industry Group Co., Ltd., 1958-59, 1961-; Brookhaven Nat. Lab., 1960. Book: Nucl. Reactor (Tokyo, Corona Co., 1958). Society: Atomic Energy Soc. of Japan.
Nuclear interests: Reactor nuclear design and reactory theory. (Collision probability method in reactor calculations, resonance integral calculation in lattices and water-moderated reactor physics including void effect).
Address: #12-19 1-Chome, Yakumo, Meguro-ku, Tokyo, Japan.

FUKATA, Ryojiro. Chief, Fuel Sect., Atomic Energy Res. Group, Fujinagata Shipbuilding and Eng. Co. Ltd.
Address: Fujinagata Shipbuilding and Engineering Co. Ltd., Atomic Energy Research Group, 2-9 Shibatanicho, Sumiyoshi-ku, Osaka, Japan.

FUKSA, Josef, Dipl. Ing. Born 1921. Educ.: Brno Tech. Univ. Ministry of Chem. Industry, 1956; Ministry of Power, 1959-62; Deputy Director, Foreign Relations, State Commission for Technol.; Czechoslovak A.E.C.; Governor of Czechoslovak Socialist Republic at the I.A.E.A., 1961-62; Chairman, Editorial Board, Jaderná energie. Book: Rad. Measurements Instruments (1959). Societies: Czech. Sci. Assoc.— Commission for Nucl. Energy.
Nuclear interests: Nuclear chemistry; Application of radioisotopes.
Address: 9 Slezská, Prague 2, Czechoslovakia.

FUKUDA, Katsuji, B. Eng. Born 1900. Educ.: Tokyo Univ. Counsellor, Tokyo Elec. Power Co. Society: Atomic Energy Soc. of Japan.
Nuclear interest: Management.
Address: Tokyo Electric Power Company, No. 5-1, 1-chome, Uchisaiwai-cho, Chiyoda-ku, Tokyo, Japan.

FUKUDA, Masaomi, Dr. Medical and agricultural studies using isotope atoms as tracer, Kagoshima Univ.
Address: Kagoshima University, Kagoshima, Japan.

FUKUDA, Nobuyuki. Formerly Prof., Plasma Phys. (Theory), Nihon Univ. Chairman, Phys. Dept., Tokyo Univ. of Education.
Address: Tokyo University of Education, 3-29-1, Otsuka, Bunkyo-ku, Tokyo, Japan.

FUKUDA, Tamotsu, Dr. Technol. (Osaka). Born 1923. Educ.: Faculty of Technol., Osaka Univ. Chief Researcher, Instrumentation Sect., Third Div., Govt. Industrial Res.Inst., Osaka, 1950-. Sec., Colour Sci. Assoc. of Japan; Member, Japanese Industrial Standards Com. Society: Soc. of Appl. Phys., Japan.
Nuclear interests: Radiation damage and dosimetry.
Address: Government Industrial Research Institute, Osaka Saita-cho, Ikeda City, Osaka, Japan.

FUKUI, Shuji, Ph.D. (Phys.). Born 1923. Educ.: Osaka Univ. Prof., Phys., Nagoya Univ., 1962-; Ford Fellow, C.E.R.N., 1962-63; Res. Assoc., Chicago Univ., 1963-66, 1967. Society: Phys. Soc. of Japan.
Nuclear interest: Experimental studies of high energy (elementary particle) physics with the accelerator.
Address: Department of Physics, Nagoya University, Furo-Cho, Chikusa-Ku, Nagoya, Japan.

FUKUNAGA, Hiroshi, M. Eng. Born 1925. Educ.: Tokyo Univ. First Sec. (Sci. Attaché), Japanese Embassy in France.
Nuclear interest: Reactor engineer.
Address: 24 rue Greuze, Paris 16, France.

FUKUNAGA, Kiyoji, Dr. Asst., Div. 1, Lab. of Nucl. Study, Inst. for Chem. Res.,Kyoto Univ.
Address: Kyoto University, Institute for Chemical Research, Yoshida-Honmachi, Sakyo-ku, Kyoto, Japan.

FUKUTOMI, Hiroshi, Ph.D. Born 1926. Educ.: Tokyo Inst. Technol. Assoc. Prof. Nucl. Chem., Tokyo Inst. Technol., 1959-. Book: Modern Aspects of Ion Exchange (Tokyo, Hirokawa Shoten, 1960); Inorg. Chem. (Tokyo,Hirokawa Shoten, 1965). Societies: Chem. Soc. Japan; Atomic Energy Soc. Japan; Japan Soc. for Analytical Chem.; Japan Radioisotope Assoc.; Geochem. Soc. Japan; A.C.S.
Nuclear interests: Nuclear chemistry. Isotope separation. Tracer study in exchange reaction. Electron exchange reaction between uranium compounds. Solvent extraction chemistry. Chemical relaxation in uranium solution.
Address: Research Laboratory of Nuclear Reactor, Tokyo Institute of Technology, 2-12-1 Oh-okayama, Meguro-ku, Tokyo, Japan.

FUKUYAMA, Kazumasa. Director, Sumitomo Atomic Energy Industries Ltd.
Address: Sumitomo Atomic Energy Industries Ltd., Sumitomo Building, 22 5-chome, Kitahama, Higashi-ku, Osaka, Japan.

FUKUZAWA, Fumio. Asst. Prof., Nucl. Reaction, Faculty of Eng., Kyoto Univ.
Address: Kyoto University, Faculty of Engineering, Yoshida-Honmachi, Sakyo-ku, Kyoto, Japan.

FUKUZAWA, Heihachiro. Born 1904. Educ.: Tokyo Univ. Managing Director, Asahi Glass Co. Ltd., Tokyo.
Nuclear interests: X-ray masking glass; Radioactive waste disposal by vitrification method; Radioactive waste disposal by ion-exchange

resin membranes.
Address: 14 2-chome, Marunouchi, Chiyoda-ku, Tokyo, Japan.

FULLER, Harold Q., A.B., A.M., Ph.D. Born 1907. Educ.: Wabash Coll. and Illinois Univ. Employed (1947-), Prof. Phys. and Chairman of the Dept. (1948-), Missouri Univ., Rolla. Societies: A.P.S.; A.A.P.T.
Nuclear interest: Physics.
Address: Department of Physics, Missouri University, Rolla, Missouri 65401, U.S.A.

FULLMER, George C., B.S. (Phys.). Born 1922. Educ.: Washington State Coll. and Washington Univ. At (1947-), Manager, Operational Phys. (1956-65), Consulting Phys. (1966-67), G.E.C.; Battelle Memorial Inst., 1965-66; Manager, Instrumentation, Pacific Northwest Lab.; Manager, Phys., Douglas United Nucl., Inc., 1967-. Book: Reactor Operating Phys. (training manual) (G.E.C., 1963). Societies: A.N.S.; A.P.S.
Nuclear interests: Principal interests in physics of operating reactors, including integral experiments, flux distribution control, nuclear instrumentation and safety control and nuclear safeguards.
Address: Douglas United Nuclear, Incorporated, Richland, Washington 99352, U.S.A.

FULLWOOD, Ralph R., B.S., A.M. (Phys.). Born 1928. Educ.: Texas Technol. Coll., Harvard and Pennsylvania Univs. At Pennsylvania Univ., 1957-60; Formerly at Rensselaer Polytech. Inst., 1960-. Now at Los Alamos Sci. Lab. Societies: A.P.S.; I.E.E.E.; A.N.S.
Nuclear interests: Operation and development of electron linac for nuclear physics. Electronic instrumentation of nuclear experiments. Time dependent and asymptotic neutron spectre in moderators and subcritical assemblies.
Address: Los Alamos Scientific Laboratory, Los Alamos, New Mexico, U.S.A.

FULTON, David C. Formerly Asst. to Pres., (1964-), now Div. Director, Power Systems, Atomics Internat., Div. of North American Aviation Inc.
Address: Atomics International, North American Aviation Inc., P.O. Box 309, Canoga Park, California, U.S.A.

FULTON, Thomas, A.B., M.A., Ph.D. Born 1927. Educ.: Harvard Univ. Member, Inst. for Advanced Study, Princeton, N.J., 1954-55, 1955-56; Asst. Prof. (1956-59), Assoc. Prof. (1959-64), Prof. (1964-), Johns Hopkins Univ. Society: A.P.S.
Nuclear interests: Scattering theory, nucleon-nucleon interactions, dispersion relations, high energy particle physics.
Address: Johns Hopkins University, Baltimore 18, Maryland, U.S.A.

FUMI, Fausto, D.Sc. (Phys.). Born 1924. Educ.: Genoa Univ.; Lecturer in Statistical Mechanics, Milan Univ., 1951-54; Prof., Theoretical Phys., Pavia Univ., 1957-61; Sen. Phys., Argonne Nat. Lab., 1961-; Prof. Phys., Northwestern Univ., 1962-. Member, Editorial Board, Il Nuovo Cimento; Member, Editorial Advisory Board, J. of the Phys. and Chem. of Solids. Societies: Italian Phys. Soc.; A.P.S.
Nuclear interests: Solid state physics; Diffusion and radiation damage in solids; Lattice dynamics.
Address: Physics Department, Northwestern University, Evanston, Illinois, U.S.A.

FUNCK, Walter, Dr. Born 1904. Educ.: Tübingen and Vienna Univs. Director of Administration, Wirtschaftministerium des Landes Baden-Wuerttemberg, Germany; Bundesatomministerium, Bonn, Germany. Formerly Director General of Administration and Personnel of the Community of Atomic Energy.
Address: 621 Steenweg Terhulpen, Overijse-Malaise, Belgium.

FÜNFER, Ewald, Prof. Dr. Born 1908. Educ.: Tübingen and Berlin Univs. Lab. Tech. Phys., Munich T.H.; Inst. Plasmaphys., Garching. Book: Zählrohre und Szintillationszähler (Karlsruhe, G. Braun, 1959). Society: Deutsche Phys. Gesellschaft.
Nuclear interest: Plasmaphysik.
Address: 43 Gerstäckerstrasse, Munich 59, Germany.

FUNK, Emerson G., Jr., Dr. Nucl. Spectroscopy, Rare Earths, Notre Dame Univ.
Address: University of Notre Dame, Department of Physics, Notre Dame, Indiana 46556, U.S.A.

FUNK, Wesley C., B.S. Born 1910. Educ.: Montana State Coll. Deputy Manager, Idaho Operation Office, U.S.A.E.C., Nat. Reactor Testing Station, Idaho Falls, 1949-. Societies: Montana Soc. of Eng.; Idaho Soc. of Professional Eng.
Nuclear interest: Management of national reactor testing station.
Address: U.S. Embassy, Katmandu, Nepal.

FUNKE, Gösta Werner, Ph.D. Born 1906. Educ.: Stockholm Univ. and Tech. Univ., Darmstadt.; Sec.-Gen., Atomkommitten (1945-59), Statens rad för atomforskning (1959-), Statens naturvetenskapliga forskningsrad (1946-). Joint Com. for Nordic Res. Councils, 1948-; Swedish Nat. Com. for Unesco, 1951-60; Swedish Rep. (1953-), Pres. (1967-), Council, European Organisation for Nucl. Res.; Pres., Council, European Southern Observatory, 1966-; Swedish Rep. in coms. concerning Nordic or internat. collaboration in different fields of science.
Nuclear interests: Nuclear physics; Management; International collaboration.

Address: Statens rad för atomforskning, Wenner Gren Centre, 166 Sveavägen, Box 23136, Stockholm 23, Sweden.

FUNKE, Lothar, Dipl.-Ing., Dr. rer. nat. Born 1936. Educ.: Dresden Tech. Univ.
Nuclear interest: Nuclear physics.
Address: Zentralinstitut für Kernforschung Rossendorf, Dresden/Bad Weisser Hirsch, Postfach 19, German Democratic Republic.

FUNNELL, John E., Director, Mineral Technol., Southwest Res. Inst., Texas.
Address: Southwest Research Institute, 8500 Culebra Road, San Antonio 6, Texas, U.S.A.

FUNT, B. Lionel, B.Sc., M.Sc., Ph.D. Born 1924. Educ.: Dalhousie and McGill Univs. Asst. Prof. Chem. (1950-54), Assoc. Prof. (1954), Manitoba Univ.; Director, Nucl. Enterprises Ltd., 1954-. Society: Fellow, Chem. Inst. of Canada.
Address: 293 Carpathia Road, Winnipeg 9, Manitoba, Canada.

FUQUA, Philip Andrew, A.B., M.D. Born 1913. Educ.: Park Coll., Nebraska. Coll. of Medicine, Nebraska Univ.; Chicago Univ. Metal. Lab.; Clinton Labs., Oak Ridge, Tenn.; Pres. and Medical Director, Hanford Environmental Health Foundation, Richland, Washington. Vice-Pres., Northwest Assoc. of Occupational Medicine; Chairman, Washington State Medical Assoc., Com. on Occupational Health. Society: H.P.S. (Columbia Chapter).
Nuclear interests: Body contamination and deposition of plutonium. External radiation exposure. Internal radiation exposure. Wound contamination with radionuclides. Genetic and somatic effects of radiation exposures.
Address: P.O. Box 100, Richland, Washington 99352, U.S.A.

FURCHTGOTT, Ernest, Ph.D. Born 1922. Educ.: U.C.L.A. Asst. Prof. then Prof., Tennessee Univ., 1949-. Member, Exec. Com., Southeastern Psychological Assoc. Societies: Rad. Res. Soc.; A.A.A.S.
Nuclear interest: Behavioural effects of ionising radiation.
Address: Department of Psychology, Tennessee University, Knoxville, Tennessee 37916, U.S.A.

FURKEL, Th., Dipl.-Ing. Director, Technischer Uberwachungs-Verein Saarland e.V.
Address: Technischer Uberwachungs-Verein Saarland e.V., 12 Karcherstrasse, 66 Saarbrücken 3, Germany.

FURMAN, Sydney C., B.S., Ph.D. (Phys. Chem.). Born 1923. Educ.: California Univ., Los Angeles. Manager, Nucl. Materials Processing, Vallecitos Nucl. Centre, G.E.C., 1957-. Societies: A.C.S.; A.N.S.
Nuclear interests: Physical and inorganic chemistry as applied to nuclear fuel examination of irradiated fuel and other reactor materials, membrane technology, medical isotopes.
Address: General Electric Company, P.O. Box 846, Pleasanton, California 94566, U.S.A.

FURNAS, Clifford C., B.S., Ph.D. Dr. Eng. (Hon.), LL.D. (Hon.), D. Sci. (Hon.), Degree Honoris Causa (Univ. Nat. de Asuncion). Born 1900. Educ.: Purdue and Michigan Univs. Director and Exec. Vice-Pres., Cornell Aeronautical Lab., Inc. 1946-54. Chancellor, Buffalo Univ. 1954-62; Pres., State Univ. of New York at Buffalo, 1962-66; Pres. Emeritus 1966-; Pres., Western New York Nucl.Centre, 1966-. Defence Sci. Board, Naval Res. Advisory Com., Army Sci. Advisory Panel; Chairman, Advisory Council for Advancement of Ind. R. and D. Societies: A.C.S.; A.I.Ch.E.; Nat. Acad. of Eng.; A.A.A.S.
Nuclear interest: Management.
Address: State University of Buffalo at New York, Buffalo 14, New York,U.S.A.

FURR, Aaron Keith, A.B., M.S., Ph.D. Born 1932. Educ.: Catawba Coll., Emory and Duke Univs. Virginia Polytech. Inst., 1960-; Savannah River Plant, U.S.A.E.C., 1962. Society: A.N.S.
Nuclear interests: Fast neutron physics (specifically capture cross-sections); Neutron activation analysis.
Address: Physics Department, Virginia Polytechnic Institute, Blacksburg, Virginia 24061, U.S.A.

FURRY, Wendell Hinkle, A.B., A.M., Ph.D. Born 1907. Educ.: De Pauw and Illinois Univs. Assoc. Prof., Phys. (1945-62), Prof., Phys. (1962-), formerly Chairman, Phys. Dept. (1965-), Harvard Univ.; Lecturer, Summer Symposium on Theoretical Phys., Colorado Univ., 1962 and 1965. Society: A.P.S.
Nuclear interests: Quantum theory; Elementary particles.
Address: Harvard University, Physics Department, Jefferson Physical Laboratory, Cambridge 38, Massachusetts, U.S.A.

FURSOV, V. S. Member, Editorial Board, Atomnaya Energiya.
Address: 18 Kirov Ulitsa, Moscow, U.S.S.R.

FURST, Rudolf, Dipl. Ing. Born 1902. Educ.: Vienna Tech. Univ. Departmental Director, Austrian Federal Ministry for Communications and Elec. Power Economy. Vice-Pres., Österreichische Studiengesellschaft fur Atomenergie G.m.b.H.; Deleg. in the Elec. Com., O.E.E.C.; member of supervisory board in various companies. Society: Österreichischer Ingenieur- und Architektenverein.
Nuclear interest: Reactor economy.
Address: 1 Schwarzenbergplatz, Vienna 1, Austria.

FURTH, F. R. Rear Admiral U.S.N., Retd. Vice-Chairman, Farnsworth Res. Corp.

Address: Farnsworth Research Corp., 3700 East Pontiac Street, Fort Wayne, Indiana, U.S.A.

FURTH, Harold Paul, Ph.D. (Harvard). Born 1930. Educ.: Harvard and Cornell Univs. Nuclear interests: Controlled thermonuclear reactors and high-temperature plasma physics; High-energy nuclear physics.
Address: Plasma Physics Laboratory, Princeton University, P.O. Box 451, Princeton, New Jersey, U.S.A.

FURUICHI, Jiro, Prof. Member, Com. for Nucl. Study, Hokkaido Univ.
Address: Committee for Nuclear Study, Hokkaido University, North 13, West 8, Sapporo, Japan.

FURUKAWA, Y. Manager, Nucl. Dept., Kawasaki Dockyard Co. Ltd.
Address: Kawasaki Dockyard Co. Ltd., 14, 2-chome, Higashikawasaki-cho, Ikuta-ku, Kobe, Japan.

FURUTA, June Ichiro. Chief, Div. of Rad. Applications, Rad. Centre of Osaka Prefecture.
Address: Radiation Centre of Osaka Prefecture, Shinke-cho, Sakai, Osaka, Japan.

FURUYA, K. Gen. Manager, Tech. Dept., Maruzen Oil Co. Ltd.
Address: Maruzen Oil Co. Ltd., 3, 1-chome, Nagahoribashi-Suji, Minami-ku, Osaka, Japan.

FUSE, Takayoshi. Chief, Power Plant Sect., Ship Res. Inst., Ministry of Transportation.
Address: Ministry of Transportation, Ship Research Institute, 1-1057 Mejiro-cho, Toshima-ku, Tokyo, Japan.

FUSHIMI, Kodi. See HUSHIMI, Kodi.

FUSTEL, Emilio, Ing. Industrial. Born 1929. Educ.: Escuela Tec. Sup. de Ingenieros Industriales and Argonne Nat. Lab. Course. Eng., Tecnatom S.A., 1957-; Prof. Nucl. Eng., Escuela Tec. Sup. de Ingenieros Industriales, 1956-; Collaborated with Junta de Energia Nucl. for 30 MW Reactor at Westinghouse A.P.D., Pittsburgh, Pa., 1959; at Madrid for the construction of the Exponential Pile, 1961-62; Working for Zorita Nucl. Power Plant, 1963-. Society: Colegio de Ingenieros Industriales.
Nuclear interest: Nuclear power stations.
Address: 57 Fernan Gonzalez, Madrid 9, Spain.

FUTAMI, Shigeru, Dr. Technol. Born 1933. Educ.: Kyoto Univ. Sen. Staff, Teijin Ltd., Matsuyama Plant, 1964-. Society: Soc. of Polymer Sci. of Japan.
Nuclear interest: Radiation chemistry of organic compounds, especially radiation-induced reactions of polymerisable substances and radiation effects of polymersystems.

Address: Matsuyama Plant, Teijin Limited, 77 Kitayoshida, Matsuyama-shi, Japan.

FUTTERER, Edward, Jr., B. Eng. Born 1923. Educ.: McGill Univ. Mine Supt., Johnson's Co., Ltd., 1952-55; Chief Eng. (1955-61), Asst. Mine Manager (1961-64), Denison Mines, Ltd. Tech. Asst. to Pres. (1964-67), Vice Pres., Gen. Manager and Director (1967-), Kerr Addison Mines Ltd.; Vice Pres., Director and Gen. Manager of Joutel Copper Mines Ltd. and Agnew Lake Mines Ltd.; Director of Quemont Mining Corp. Ltd., Normetal Mining Corp. Ltd., Amalgamated Larder Mines Ltd. Societies: A.I.M.E.; C.I.M.M.; Professional Eng. of Ontario.
Nuclear interest: Mining and production of U_3O_8.
Address: Suite 1600, 44 King St. West, Toronto, Ontario, Canada.

FUZAK, William G., Director, Organisation and Personnel Div., Chicago Operations Office, U.S.A.E.C.
Address: U.S. Atomic Energy Commission, Chicago Operations Office, 9800 South Cass Avenue, Argonne, Illinois 60439, U.S.A.

FYODOROV, Evgeny K. Academician. Member, Soviet Pugwash Com.; Soviet adviser and Head of Deleg. of Working Group, Geneva Conference on the Discontinuance of Nucl. Weapons Tests.
Address: Academy of Sciences of the U.S.S.R., 14 Leninsky Prospekt, Moscow V-71, U.S.S.R.

GABANAC, J., Dr. Head, Czechoslovakian group of physicists and engineers engaged on development and design of new annular synchrocyclotron, Dubna; Czech leader of team (Soviet and Czech) jointly responsible for building new high-intensity cyclotron, Dubna, 1960.
Address: Joint Institute for Nuclear Research, Dubna, near Moscow, U.S.S.R.

GABAY, André. Directeur Gen., Pompes Wauquier.
Address: Pompes Wauquier, 69 rue de Wazemmes, Lille, (Nord), France.

GABBARD, Fletcher, B.S., M.A., Ph.D. Born 1930. Educ.: Kentucky Univ., Eastern Kentucky State Coll. and Rice Inst. Asst. Prof. Phys., Kentucky Univ., 1959-. Society: A.P.S.
Nuclear interests: Nuclear physics; Neutron-induced reactions; Reaction mechanisms.
Address: Department of Physics, Kentucky University, Lexington, Kentucky, U.S.A.

GABEAUD, M. Senior officer engaged in nucl. work, Bignier Schmid-Laurent.
Address: Bignier Schmid-Laurent, 25 Quai Marcel-Boyer, 94 Ivry-S/Seine (Seine), France.

GABORNE, Mrs. KOVACS. See KOVACS GABORNE, Mrs.

GABRIEL, David S., B.M.E. Born 1919. Educ.: Ohio Northern and Akron Univs. and Case Inst. Technol. Head, Sect. A, Altitude Test Chambers Branch (1949-51), Asst. Chief, Altitude Wind Tunnel Branch (1951), Chief, Engines Branch 1, (1951-53), Asst. Chief, Engine Res. Div. (1953-55), Assoc. Chief, Engine Res. Div. (1955-57), Assoc. Chief, Propulsion Systems Div. (1957-58), Lewis Flight Propulsion Lab., N.A.C.A.; Chief, Propulsion Systems Div. (1958-61), Chief, Nucl. Systems Div. (1961-63), Manager, Centaur Project (1963-65) Lewis Res. Centre, N.A.S.A.; Project Director, Rocket Engine Programmes, Bell Aerosystems Co., 1965-67; Deputy Manager, Space Nucl. Propulsion Office, N.A.S.A., 1967-.
Nuclear interest: Management.
Address: 768 Azalea Drive, Rockville, Maryland 20850, U.S.A.

GABRUSENKO, I. A. Papers: Co-author, 1, 5 GeV Electron Synchrotron at the Polytech. Inst. of Tomsk (Atomnaya Energiya, vol. 21, No. 6, 1966); co-author, 300 MeV Electron Synchrotron at the Polytech. Inst. of Tomsk (letter to the Editor, ibid, vol. 21, No. 6, 1966).
Address: Academy of Sciences of the U.S.S.R., 14 Leninsky Prospekt, Moscow V-71, U.S.S.R.

GABSATAROVA, S. A. Paper: Co-author, Determination of Dose Value for Products of Nucl. Reaction B^{10} (n,α) Li^7 and Temperature in Reaction Zone by Action of Thermal Neutrons on Borate Glasses (Atomnaya Energiya, vol. 21, No. 6, 1966).
Address: Academy of Sciences of the U.S.S.R., 14 Leninsky Prospekt, Moscow V-71, U.S.S.R.

GADAUD, Antoine, Diplom d'Ing. de l'Ecole Superieure d'Electricite. Born 1924. Educ.: Clermont Ferrand and Paris Univs. Centre d'Etudes Nucleaires de Saclay, 1952; Nardeux, 1953-.
Nuclear interests: Nuclear physics - health. Portable apparatus alpha beta gamma monitor; Radioprotection.
Address: Ets. Nardeux, B.P. 36, Loches, Indre et Loire, France.

GADDA, Ferdinando, Dr. in Chem. Born 1932. Educ.: Pavia Univ. Res., Chem. and Radiochem. Dept., (1957-), Head, Technol. Chem. Sect., Nucl. Plant Div., (1967-), C.I.S.E., Milan.
Nuclear interests: Radiochemistry; Corrosion of nuclear materials; Radiolysis and water chemistry.
Address: 51 Via B. d'Alviano, Milan 20146, Italy.

GADDIS, Paul O., B.S. (Annapolis), M.S. (Eng., Rensselaer), M.S. (Ind. Management, M.I.T.); Alfred P. Sloan Fellow in Exec. Development, M.I.T. School of Ind. Management, 1960-61. Born 1924. Educ.: U.S. Naval. Acad., Annapolis; Rensselaer Polytech. Inst., Troy, N.Y. Director of Management Systems, in charge of corporate management information systems and sci. eng. computer centre, Westinghouse Elec. Corp. Books: Contributor, Handbook of Ind. Res. Management; Corporate Accountability (N.Y., Harper and Row, 1964). Societies: A.S.M.E.; Board of Directors, American Management Assoc. and Management Sci. Centre, Wharton School.
Nuclear interests: Management of nuclear engineering and manufacturing enterprise.
Address: 2283 Country Club Drive, Pittsburgh, Pa. 15241, U.S.A.

GADE, Harald C., Chief Eng. Born 1926. Educ.: Architectural School, Hamburg. Design Eng. (Private Firm), 1952-56; Head, Building Dept., Co. for the Utilisation of Nucl. Energy in Shipbuilding and Navigation, Reactor Station, Geesthacht, Germany, 1957-. Society: Study Assoc. for the Use of Nucl. Energy in Navigation and Ind.
Nuclear interests: Civil engineering and building construction in the nuclear sphere.
Address: 4 Matthias-Scheits-Weg, Hamburg 33, Germany.

GADEV, Bogomil. Member, Com. for the Peaceful uses of Atomic Energy, Council of Ministers of the Bulgarian People's Republic.
Address: Council of Ministers, Sofia, Bulgaria.

GAEDE, Karl August Ernst, Dr. med. Born 1913. Educ.: Munich and Hamburg Univs. Head, Biochem. Dept., I.V.I.C., Caracas, 1958-; Prof., Physiopathology, and Lecturer, Biochem., Caracas Central Univ. Pres., Venezuelan Assoc. of Biochem.; Pres., Zone IV (Venezuela, Colombia, Ecuador and Trinidad and Tobago), Latin American Com. of Biochem., 1968-69.
Nuclear interest: Use of isotopes in biochemistry.
Address: Venezuela Institute for Scientific Research, Apartado 1827, Caracas, Venezuela.

GAERTNER, Henriette KNOERR-. See KNOERR-GAERTNER, Henriette.

GAERTTNER, Erwin Rudolf, E.E. and Ph.D. (Phys.). Born 1911. Educ.: Denver and Michigan Univs. Res. Assoc., G.E. Res. Lab., 1946; Res. Assoc. (1952), Manager, Exptl. Nucl. Phys. (1954), Manager, Phys. (1956), Manager, Exptl. Equipment Development (1957), G.E. K.A.P.L.; Dept. Head, Nucl. Eng. and Sci., and Director, Linac, Rensselaer Polytech. Inst.; Member, Advisory Com. on Reactor Phys., U.S.A.E.C. Society: Fellow, A.P.S.
Nuclear interests: Nuclear physics and nuclear instrument development.
Address: Rensselaer Polytechnic Institute, Troy, New York, U.S.A.

GAETA, Rafael, Chem. Born 1927. Educ.: Valencia Univ. Investigator of Junta Energia Nuclear of Spain. Society: Real Sociedad Española Fisica y Quimica. Nuclear interests: Scintillation counters; Gamma-spectrometry; Slow neutron detectors.
Address: Junta de Energia Nuclear, Division de Fisica, Ciudad Universitaria, Madrid-3, Spain.

GAGARINSKY, Yu. V. Papers: Application of the 'Gaseous' Fluorination in the Uranium Tetrafluoride Production (Atomnaya Energiya, vol. 6, No. 2, 1959); co-author, Radioactivity and the Physio-chem. Properties of a Material (ibid., vol. 10, No. 3, 1961); co-author, Binary System UF_4-UCl_4 (ibid, vol. 19, No. 5, 1965); co-author, DTA Study of Phase Transformations in Tetraiodides of Natural Uranium and Uranium - 233 (letter to the Editor, ibid, vol. 23, No. 2, 1967).
Address: Academy of Sciences of the U.S.S.R., 14 Leninsky Prospekt, Moscow V-71, U.S.S.R.

GAGELOT, P. C., Prof., Dr. Phys., Inst. for Nucl. Phys. Res., Amsterdam.
Address: Institute for Nuclear Physics Research, 18 Oosterringdijk, Amsterdam O, Netherlands.

GAGLIARDI, Georges. Directeur Technique, Bureau d'Etudes, Dept. Nucl., Siersatom.
Address: Siersatom, Department Nucleaire, 108 avenue du Maine, Paris 14, France.

GAGNE, C. Nucl. Dept., Jouan.
Address: Nuclear Department, Jouan, 1 avenue Carnot, Massy, (S. et O.), France.

GAGNON, Paul Edouard, D.I.C., Ph.D., D. ès Sc., D.Sc., LL.D., F.R.S.C. Born 1901. Educ.: Laval, Paris and London Univs. Former Director, Div. of Exchange and Training of Sci. and Experts, I.A.E.A., Past Pres., Chem. Inst. of Canada; Past Pres., Section III, Roy. Soc. of Canada. Societies: Chem. Soc.; Soc. of Chem. Industry; Chem. Inst. of Canada; Roy. Soc. of Canada. Nuclear interests: Synthetic organic compounds containing radioactive isotopes.
Address: National Research Council, 100 Sussex Drive, Ottawa, Canada.

GAIANI, Umberto. Sen. Officer, Carlo Gavazzi S.p.A.
Address: Carlo Gavazzi S.p.A., 9 Via G. Ciardi, Milan, Italy.

GAIDUCHIK, V. O. Paper: Co-author, Surface Relief Determination by Means of Back-Scattering Gamma-Rad. (Abstract, Atomnaya Energiya, vol. 19, No. 6, 1965).
Address: Academy of Sciences of the U.S.S.R., 14 Leninsky Prospekt, Moscow V-71, U.S.S.R.

GAILAR, Owen, B.S., M.S., Ph.D. Born 1925. Educ.: Rochester and Purdue Univs. Assoc. Prof., Dept. Nucl. Eng., Purdue Univ. Society: A.N.S.
Address: Purdue University, Lafayette, Indiana.

GAILLARD, Alphonse, Eng.-Arts et Métiers. Born 1931. Educ.: Paris Univ. Chief Eng. COCEI. Lecturer, Inst. Français du Pétrole. Society: A.T.E.N. Nuclear interests: Reactor control and safety design.
Address: c/o COCEI, 22 rue de Clichy, Paris 9, France.

GAILLARD, M. Maitre Asst., Inst. de Phys. Nucl., Lyon Univ.
Address: Institut de Physique Nucleaire, Lyon University, 43 boulevard du 11 Novembre 1918, 69-Villeurbanne, (Rhone), France.

GAILLARD, Pieter Johannes, M.D. Born 1907. Educ.: Leiden Univ. Lab. for Histology and Microscopical Anatomy, Leiden Univ.; Prof. Exptl. Histology (1947-51), Prof. Exptl. Histology, Cytology and Cytological Genetics (1951-58), Prof. Histology, Exptl. Histology, Cytology and Cytological Genetics (1958-64), Prof., Cell Biol. and Histology (1964-), Leiden Univ. Books: Hormones regulating Growth and Differentiation in Embryonic Explants (1942); Algemene Celleer (1962). Societies: Roy. Netherlands Acad. of Sci., Amsterdam; Consultants Com. on cell biol.; Vice-Pres., Internat. Soc. for Cell Biol.; Pres., I.U.B.S., Paris; Affiliate, Roy. Soc. of Medicine, London. Nuclear interests: Autoradiography and tracer studies in the field of cell biology.
Address: Academic Hospital, 10 Rijnsburgerweg, Leiden, Netherlands.

GAILLOUD, Maurice, D. ès Sci. Born 1927. Educ.: Lausanne Univ. Chargé de Cours, Lab. Physique Nucléaire, Ecole Polytechnique Lausanne. Society: Sté. Suisse de Phys. Nuclear interest: Nuclear physics.
Address: Institut de Physique Nucléaire de l'Université, 19 rue César Roux, Lausanne, Switzerland.

GAINES, Albert Lowery, B.M.E., M.S. (Mech. Eng.). Born 1920. Educ.: Auburn and Missouri Univs. Supervisor, Proposition Eng. Sect., Combustion Eng., Inc.; Supvr., Special Products Eng. Sect., Supvr., Thermodynamics Group, Nucl. Product Eng. Dept., CE, Inc.; Development Eng., O.R.N.L. (Y-12). Member, Tech. Com. on Unfired Heat Transfer Equipment (K-10), A.S.M.E.; Member, Res. Com. on Effect of Rad. on Materials, A.S.M.E. Societies: Charter Member, A.N.S.; A.S.M.E.; A.I.Ch.E. Nuclear interests: Nuclear power plant design, including thermal, structural, and economic evaluation, especially the heat exchange equipment and pressure containment components; Computer solution of parametric studies of system variables for minimum cost selection of over-all plant, including maintenance

evaluation.
Address: Combustion Engineering, Inc., 911 West Main Street, Chattanooga, Tennessee, U.S.A.

GAINES, Martin N. Project Manager, Uranium Ore and Concentrate Servicing Centre, operated for U.S.A.E.C. by Lucius Pitkin, Inc.
Address: U.S.A.E.C., Uranium Ore and Concentrate Servicing Centre, Grand Junction, Colorado, U.S.A.

GAINSBOROUGH, George F., Ph.D., F.I.E.E. Sec., Inst. Elec. Eng., 1962-; Member for Inst. Elec. Eng., Board, B.N.E.S., 1963-.
Address: Institution of Electrical Engineers, Savoy Place, London, W.C.2, England.

GAITAN, Mario, M.D. Born 1919. Educ.: National Univ. Colombia. Member, Board of Directors, Inst. de Asuntos Nucleares de Colombia, 1959; Member, Board of Directors of former Comisión de Energia Atómica, 1953; Director (1957-), Head of Radiotherapy Dept. (1949-), Nat. Cancer Inst. of Colombia. Societies: American Radium Soc.; Soc. Colombiana de Radiologia; H.P.S.; A.N.S.; Soc. de Radiologia de Panama, de Cuba, de México, del Perú y de la Argentina. Nuclear interests: Medicine, biology and health physics.
Address: Carrera 12 # 20-69, Bogotá, Colombia.

GAJA, Roberto, Law degree. Born 1912. Educ.: Turin Univ. Envoy Extraordinary and Minister Plenipotentiary, Italian Legation, Sophia, 1958; Deputy Director Gen. for Political Affairs, 1963; Director General for Political Affairs, Ministry for Foreign Affairs, Rome, 1964; Ambassador, 1967. Books: Foreign Policy and Nuclear Armaments (Bologna, Cappelli, 1964); Political Effects of Nuclear Weapons (Florence, 1959). Society: Inst. for Strategic Studies (London).
Address: Ministry for Foreign Affairs, Rome, Italy.

GAJARDO, Mallén, Elec. Eng. Born 1931. Educ.: Chile and Wisconsin Univs. Prof., Phys. Electronic Lab., Dept. de Fisica, Chile Univ., 1955-.
Nuclear interest: Nuclear electronics.
Address: Departamento de Fisica, Casilla 5487, Santiago, Chile.

GAJEWSKI, Ryszard, M.S. (Radio Eng.), Ph.D. (Phys.). Born 1930. Educ.: Warsaw Inst. Technol. (Politechnika Warszawska) and Inst. of Phys., Polish Acad. Sci. Sen. Staff Member, Theoretical Dept., Inst. for Nucl. Res., Warsaw; Asst. Prof. Phys., Warsaw Inst. Technol., 1959-60; Alfred P. Sloan Post-Doctoral Fellow, School for Advanced Studies, M.I.T. Member, Nat. Com. for Peaceful Uses of Atomic Energy, Warsaw.
Nuclear interests: Theoretical plasma physics.
Address: Institute for Nuclear Research, 69 Hoza, Warsaw, Poland.

GAJEWSKI, Walter Michael, B.S. (E.E.), M.S. (E.E.). Born 1923. Educ.: Connecticut Univ. Graduate Asst., School of Eng., Connecticut Univ., 1949-50; Westinghouse Elec. Corp., 1950-51; Positions from Jun. Eng. to Reactor Plant Manager, Bettis Atomic Power Lab., Westinghouse Elec. Corp., 1951-. Societies: A.N.S.; A.S.M.E.; American Inst. Elec. Eng.
Nuclear interests: Engineering management in the area of nuclear power plant design, construction, operator training and operation. Have performed studies of nuclear propulsion plant designs for various type reactors.
Address: 1178 Kojave Street, Idaho Falls, Idaho, U.S.A.

GAL, Dezso, Ph.D., Cand. of Chem., D.Sc. Born 1926. Educ.: Eötvös Loránd Univ., Budapest. Inst. for Phys. Chem., Debrecen Univ., 1949-52; Inst. for Inorganic and Analytical Chem., Szeged Univ., 1952-55; Agrochem. Res. Inst., Budapest, 1955-59; Prof., Inst. for Radiochem., Szeged Univ., 1959-; Fellowship, N.R.C., 1962-63. Visiting Prof., Inst. Chem. Phys., Acad. of Sci., Moscow, U.S.S.R., 1966-67. Societies: Hungarian Section, Combustion Inst.; Hungarian Chem. Soc.
Nuclear interests: Nuclear chemistry; Application of radioisotopes in investigation of mechanisms of chain reactions (especially of hydrocarbon oxidation) and of surface phenomena.
Address: Institute for Radiochemistry, Szeged University, Hungary.

GAL, Ivan, Dr. in Chem. (Belgrade), Chem. Eng. (Faculty of Technol., Belgrade). Born 1924. Educ.: Belgrade Univ. Doz. in Phys. Chem., Belgrade Univ., 1956-60; Sen. Sci., Hot Lab. Dept., Inst. of Nucl. Sci. Boris Kidrich, Belgrade, 1962. Book: Co-author, Radiochemiski praktikum (Belgrade, Naucna knjiga, 1959). Societies: Chem. Soc., Belgrade; Croatian Chem. Soc.
Nuclear interests: Actinides and lantanides chemistry; Radiochemistry; Reprocessing of nuclear fuel; Co-ordination chemistry.
Address: Boris Kidrich Institute of Nuclear Sciences, P.O.B. 522, Belgrade, Yugoslavia.

GALAHMATOVA, B. S. Paper: Co-author, Effect of Fission Product Migration upon Fuel Element Burn up Measurements (Atomnaya Energiya, vol. 23, No. 2, 1967).
Address: Academy of Sciences of the U.S.S.R., 14 Leninsky Prospekt, Moscow V-71, U.S.S.R.

GALAMBOS, John T., B.S., M.D. Born 1921. Educ.: Munich, Georgia and Emory Univs. Prof. Medicine, Emory Univ., 1957-. Societies: American Federation for Clinical Res.; Soc. Nucl. Medicine.
Nuclear interests: Metabolism of S-35 sulfomucopolysaccharides and glycoproteins;

GAL

Hepatic metabolism of various isotope-labeled compounds (I-131, Zn-65, Fe-59, Se-75).
Address: 69 Butler Street, S.E., Atlanta 3, Georgia, U.S.A.

GALANIN, A. D., D.Sc., Acad. of Sci. of the U.S.S.R. Deleg. to 2nd I.C.P.U.A.E., Geneva, Sept. 1958. Book: Teoriya yadernykh reaktorov no teplovykh neitronakh (Theory of Thermal Neutron Reactors) (Supplement Nos. 2-3, Atomnaya Energiya, Atomic Press, Moscow, 1957; translation, Consultants Bureau, Inc., 1958; 2nd Russian edition 1959. Papers: Co-author, Critical Test conducted on a Heavy-water Exptl. Reactor (2nd I.C.P.U.A.E., Geneva, Sept. 1958); Theory of a Heterogeneous Reactor Having Cylindrical Blocks of Finite Radius (Atomnaya Energiya, vol. 9, No. 2, 1960); co-author, On the Theory of Heterogeneous Reactors with Blocks of Finite Radius (ibid., vol. 15, No. 2, 1963).
Address: Nuclear Energy Institute, Academy of Sciences of the U.S.S.R., 14 Leninsky Prospekt, Moscow V-71, U.S.S.R.

GALANTIN, Rear Admiral I. J. Director, Special Projects Office, Bureau of Naval Weapons, Dept. of the Navy.
Address: Department of the Navy, Bureau of Naval Weapons, Washington 25, D.C., U.S.A.

GALBRAITH, Thomas Galloway Dunlop, M.P., M.A., LL.B. Born 1917. Educ.: Oxford and Glasgow Univs. M.P., Civil Lord of the Admiralty, 1957-59. Chairman Galbraith Com. (investigating the possibility of a nucl. powered merchant ship), 1957-59. Joint Parliamentary Under-Sec. of State for Scotland, 1959-62. Joint Parliamentary Secretary, Ministry of Transport, 1963-64.
Address: 2 Cowley Street, London, S.W.1, England.

GALBRAITH, William, B.Sc. (Gen.), B.Sc. (Spec.), Ph.D. (London). Born 1925. Educ.: London Univ. Harkness Fellow of Commonwealth Fund of New York, Rad. Lab., California Univ., Berkeley, 1957-58; Visiting Phys., Brookhaven Nat. Lab., New York, 1963-64; Prof. Nucl. Phys., Sheffield Univ., 1966-. Book: Extensive Air Showers (Cosmic Radiation) (Butterworths, London, 1958). Nuclear interests: High energy nuclear physics; cosmic radiation.
Address: Brackenhurst, 636 Abbey Lane, Sheffield, SI1 9NA, England.

GALE, Alfred John, (Special Honors) (Phys.). Born 1915. Educ.: London Univ. Formerly Vice-Pres. and Formerly Director Appl. Phys., High Voltage Engineering Corp., 1950-; Pres., Ion Phys. Corp., 1959-. Pres. (New England Sect.) American Rocket Soc. Societies: American Rocket Soc.; American Vacuum Soc.; A.N.S.; A.P.S.; A.A.A.S.

Nuclear interests: Particle accelerator development and applications.
Address: 10 Diana Lane, Lexington, Massachusetts, U.S.A.

GALE, Richard Harris, B.S. (Chem.), M.S., Ph.D. (Phys. Chem.). Born 1918. Educ.: Norwich, Vermont, Cornell and Delaware Univs. Director, R. and D., American Gas Assoc.; Manager, Materials Development, Nucl. Div., and Manager, Eng. and Materials Development Labs., Combustion Eng. Inc. Nucl. Technol. Study Com., State of Connecticut. Society: A.N.S.
Nuclear interest: Nuclear fuel element and control rods.
Address: 183 Ridgewood Road, West Hartford, Connecticut 06107, U.S.A.

GALEEV, A. A. Paper: Co-author, A Theory for Instability and Anomalous Diffusion in an Inhomogeneous Plasma (Atomnaya Energiya, vol. 16, No. 6, 1963).
Address: U.S.S.R. Academy of Sciences, 14 Leninsky Prospekt, Moscow V-71, U.S.S.R.

GALISHEV, V. S. Papers: On the Theory of γ-Ray Passage through Matter (Atomnaya Energiya, vol. 14, No. 5, 1963); On Theory of Gamma-Quantum Transmission through Layers of Finite Thickness (Abstract, ibid., vol. 20, No. 3, 1966); On Theory of Multiple Scattering of Electrons (ibid, vol.24, No. 2, 1968).
Address: Academy of Sciences of the U.S.S.R., 14 Leninsky Prospekt, Moscow V-71, U.S.S.R.

GALITSKII, V. M. Paper: Co-author, Production, of Photons and Positrons from Bombardment of Thick Foil by Fast Electrons (Atomnaya Energiya, vol. 21, No. 2, 1966).
Address: Academy of Sciences of the U.S.S.R., 14 Leninsky Prospekt, Moscow V-71, U.S.S.R.

GALL, Joseph le. See le GALL, Joseph.

GALL, William Rolfe, B.S. (Mech. Eng.), M. Eng. Born 1913. Educ.: Tennessee and Yale Univs. Asst. Chief, Design Dept., Reactor Div. (1947-64); Eng. Consultant, Gen. Eng. and Construction Div. (1964-), O.R.N.L. Chairman, U.S.A.S.I., B31.7, Com. for Nucl. Piping; Societies: Fellow, A.S.M.E.; A.N.S. Nuclear interests: Reactor design; Design of reactor systems and components.
Address: Oak Ridge National Laboratory, P.O. Box X, Oak Ridge, Tennessee, 37831, U.S.A.

GALLAGHAR, R. G., A.B. Educ.: Stanford and Cincinnati Univs., O.R.I.N.S. and Western Maryland and Hahnemann Medical Colls. Health Phys., U.S.P.H.S., 1949-54; Lecturer, Radiol. Health, Harvard Graduate School of Public Health, 1956-; Health Phys. Service, Liberty Mutual Insurance Co.; Asst. Manager, Dept. Biol. and Medicine, Nucl. Sci. and Eng. Corp. Vice-Pres. and Asst. to Pres., Pa. Elec. Co.

Societies: H.P.S.; A.C.S.; A.A.A.S.; American Conference Govt. Industrial Hygienists; New York Acad. Sci.;A.N.S.; A.I.H.A.
Address: Nuclear Science and Engineering Corp., Box 10901, Pittsburgh 36, Pa., U.S.A.

GALLAGHER, Michael John, B.Sc., Ph.D. Born 1934. Educ.: Univ. College, Durham Univ. Geologist, Inst. of Geological Sci., 1957-. Societies: Mineralogical Soc. of London; Geological Soc. of London.
Nuclear interests: Applications to identification of minerals and field surveys for radioelements and other elements detectable by techniques involving nuclear reactions.
Address: Institiute of Geological Sciences, Geochemical Division, Young Street, London, W.8, England.

GALLEY, Robert, Eng. Born 1921. Educ.: Ecole Centrale des Arts et Manufactures. Entered C.E.A., 1955; Formerly Chief of Dept., C.E.A., Paris, 1958-; Actuellement Director, Isotope Separation Plant, Pierrelatte, C.E.A., 1966-67; Ministre Delegue aupres du Premier Ministre, Charge de la Recherche Sci. et des Questions Atomiques et Spatiales.
Nuclear interest: Industrial problems.
Address: 2 rue Royale, Paris 8, France.

GALLIC, Yves LE. See LE GALLIC, Yves.

GALLIER de SAINT SAUVEUR, Henri de. See de GALLIER de SAINT SAUVEUR, Henri.

GALLIGAN, Thomas Joseph, Jr., A.B., M.B.A., C.P.A. Born 1919. Educ.: Boston Coll. and Harvard Graduate School of Business Administration. Pres., Boston Edison Co.
Nuclear interest: Management - Boston Edison Pilgrim Station and Boston Edison interests in Yankee Atomic and Connecticut Yankee.
Address: Boston Edison Company, 800 Boylston Street, Boston, Ma. 02199, U.S.A.

GALLIOT, J. Ing., Sen. Sci., Centre Scientifique de Monaco.
Address: Centre Scientifique de Monaco, 16 boulevard de Suisse, Monte Carlo, Monaco.

GALLMANN, André, Ph.D. Born 1926. Educ.: Strasbourg Univ. Res. Assoc., Brookhaven Nat. Lab., 1958-59; Maitre de Conférences (1959), Prof. (1962), Strasbourg Univ.
Nuclear interest: Nuclear physics.
Address: I.R.N., Boite Postale No. 16, Strasbourg-Cronenbourg, Bas-Rhin, France.

GALLONE, Patrizio, Dr. of Phys. Director of Res., Oronzio de Nora Impianti Elettrochimici.
Address: Oronzio de Nora Impianti Elettrochimici, 35 Via Bistolfi, Milan, Italy.

GALLONE, Sergio, Dr. Libero Docente Nucl. Phys. Born 1921. Educ.: Geneva and Milan Univs. Manager, Study and Res. Service, Agip Nucleare, 1957-. i/c Spectroscopy Course, Phys. Dept., Milan Univ. Member, Editorial Board, Energia Nucleare.
Nuclear interests: Nuclear models; Reactor physics.
Address: 2 Corso di Porta Nuova, Milan, Italy.

GALLONI, Ernesto Enrique, Civil Eng. Born 1906. Educ.: Buenos Aires Univ. Prof. Phys., Facultad de Ciencias (1944-54), Facultad de Ingenieria (1955-). Member, Academia Nacional de Ciencias Exactas Fisicas y Naturales, Buenos Aires; Correspondent Member, Real Academia de Ciencias Exactas Fisicas y Naturales of Madrid and Academia Nacional de Ciencias of Cordoba (Argentina); Member of Board, Argentine Nat. Atomic Energy Commission, 1955-58; Chief, Phys. Dept., Argentine Nat. Atomic Energy Commission, 1954-55; Pres., Comité Nacional de Cristalografia. Books: (in collaboration) Fisica Elemental (1939); Trabajos Prácticos de Fisica (1943); Fisico Quimica Elemental (1952); Fisica-Mecánica (1958). Societies: Assoc. Fisica Argentina (Sec.); A.P.S.; American Mineralogical Soc.; Assoc. Argentina para el Progreso de las Ciencias.
Nuclear interests: Crystal structure of uranium minerals, metallurgical problems.
Address: Yerbal 1763, Buenos Aires, Argentina.

GALLOWAY, Robert Bennett, B.Sc., Ph.D. Born 1934. Educ.: Edinburgh Univ. Lecturer, Dept. of Natural Philosophy, Edinburgh Univ., 1962-. Society: Fellow, Phys. Soc.
Nuclear interests: Low energy nuclear physics; Gamma-ray spectroscopy; Neutron scattering and polarisation; Nuclear instrumentation.
Address: Department of Natural Philosophy, Edinburgh University, Drummond Street, Edinburgh 8, Scotland.

GALOCI, Stephan, Eng. Born 1928. Educ.: Prague High Tech. Univ. Eng., Czechoslovak A.E.C. Society: Sci. and Tech. Soc.
Nuclear interests: Instruments and dosimetry.
Address: 7 Slezská, Prague 2, Czechoslovakia.

GALONSKY, Aaron I., A.B., M.S., Ph.D. Born 1929. Educ.: Wisconsin Univ. Phys., O.R.N.L., 1954-59; Group Leader, Midwestern Univs. Res. Assoc., 1959-64; Assoc. Prof. (1964-66), Prof. Phys. (1966-), Director, Cyclotron Lab. (1967-), Michigan State Univ.; Member, Board of Directors, Midwestern Univs. Res. Assoc. Society: A.P.S.
Nuclear interests: Nuclear physics; Shielding; Radiation damage; High-energy accelerators.
Address: Cyclotron Laboratory, Michigan State University, East Lansing, Michigan 48823, U.S.A.

GALPER, Milton Jerome, A.B. (Phys.), M.A. (Phys.). Born 1926. Educ.: California Univ., Los Angeles. Supervisor, PWR Core 1 Nucl.

Design, 1955-57; Manager, PWR Nucl. Design, 1957-61; Manager, D1W Nucl. Design and Analysis, 1961-62; Manager, LPR Nucl. Design and Analysis, 1962-66; Manager, PWR Nuclear and Thermal Design, 1966-67; Manager, PWR (Shippingport) Project, 1967-. Book: The Shippingport Pressurised Water Reactor, Chapter 3 (Addison-Wesley Publishing Co., Inc., 1958). Society: A.N.S.
Nuclear interests: The nuclear, thermal, and mechanical design of advanced power reactors and systems; also, the management of technical activities involved in the design and operational follow of reactor power plants employing such advanced systems.
Address: Bettis Atomic Power Laboratory, Westinghouse Electric Corporation, Box 79, West Mifflin, Pennsylvania 15122, U.S.A.

GALPIN, Floyd Lee, B.S. (Chem. Eng.), M.C.E. Born 1935. Educ.: Oklahoma and Michigan Univs. At Nat. Centre for Radiol. Health, U.S.P.H.S. Ionising Rad. Com., American Conference of Governmental Ind. Hygienists; Radiol. Health Sect., A.P.H.A. Societies: A.P.H.A.; H.P.S.; Conference on Radiol. Health; American Conference of Governmental Ind. Hygienists.
Nuclear interests: Training and manpower development activities related to radiation protection; Health physics.
Address: 12008 Rockinghorse Road, Rockville, Maryland 20853, U.S.A.

GALSTER, Siegfried, Dipl. Phys., Dr. Born 1933. Educ.: Erlangen and Mainz Univs. Ministerium für Atomkernenergie, Inst. für exp. Kernphysik, Karlsruhe T.H. Society: Physikalische Gesellschaft.
Nuclear interests: Polarization of bremsstrahlung weak and strong interactions, high energy electron scattering.
Address: Inst. für exp. Kernphysik, Karlsruhe, Germany.

GALTAROSSA, Giacomo, Dott. Pres., Industrie Siderurgiche; Member, Comitato Permanente per i Problemi dell'Impiego Pacifico dell'Energia Nucleare e per i Rapporti con Euratom.
Address: 37 Via Senato, Milan, Italy.

GALUSTASHVILI, V. V. Paper: Co-author, Behaviour of Borates and Boric Acid in Boiling Reactors (letter to the Editor, Atomnaya Energiya, vol. 16, No. 1, 1964).
Address: U.S.S.R. Academy of Sciences, 14 Leninsky Prospekt, Moscow V-71, U.S.S.R.

GALVAO, Júlio Pistacchini, Lic. in Phys. and Chem. Born 1925. Educ.: Lisbon Univ. Head, Rad. Protection Div., Junta de Energia Nucl. Tech. Assessor, Comissao do Risco Atomico.
Nuclear interest: Health physics.
Address: 79 Rua de S. Pedro de Alcântara, Lisbon, Portugal.

GALVEZ, Delfin, Prof. Chem. Dept., Instituto de Estudios Nucleares, Panama Univ.
Address: Panama University, Faculty of Sciences, Instituto de Estudios Nucleares, Apartado 3277, Panama, Panama.

GALVEZ ARMENGAUD, D. Diego. Director Administrativo, Junta de Energia Nucl., Spain.
Address: c/o Junta de Energia Nuclear, Ciudad Universitaria, Madrid 3, Spain.

GALVEZ ARMENGAUD, Francisco. Vocale, Comisiones Asesoras de Medicina y Biologia Animal, Junta de Energia Nucl.
Address: Junta de Energia Nuclear, Ciudad Universitaria, Madrid 3, Spain.

GAMBA, Dante Bautista Walter. Born 1920. Educ.: National Technician graduated in the Superior Industrial School of the Nation, annexed to the Univ. of Mathematics, Phys. and Natural Sci. of Rosario (Rep. Argentina). Tech. Director, Tecnitrón, Technological Applications of Atoms and Electrons, Commercial, Industrial and Financial, S.A. (Private); Vocal, Nat. Commission of Space Res. (Government). Pres., Professional Groups, Buenos Aires Section, I.R.E. Book: Bulletin Tecnitrón; Instrumentation for Nucl. Phys. (1953 to 1957). Societies: Sen. Member, Inst. of Radio Eng. (I.R.E.); Member, Argentine Assoc. of Electrotechnicians.
Nuclear interests: Technological applications of radioactive isotopes especially industrial automation; Works on cosmic radiation; Physics of the plasma; Electrical field; Magnetic field.
Address: 1337 Charcas, Buenos Aires, Argentina.

GAMBA, Otto O. M., Chem. Eng. Born 1917. Educ.: Nacional del Litoral (Argentina), Grenoble, Paris, Illinois and London Univs. Formerly Prof., Cuyo Univ., Argentina, and Prof., Buenos Aires Univ.; formerly Head, Nucl. Reactor Dept., Argentine A.E.C.; At Atomics Internat, 1960-.
Nuclear interests: Metallurgical and engineering problems associated with fuel elements.
Address: Atomics Internationa, P.O. Box 309, Canoga Park, California, U.S.A.

GAMBARYAN, R. G. Papers: Co-author, On Some Rapidity Criteria of Neutron-Activation Analysis (letter to the Editor, Atomnaya Energiya, vol. 22, No. 1, 1967); co-author, Neutron-Resonance Analysis of Substance Elemental Composition (ibid, vol.25, No. 2, 1968); co-author, Spectral-Resonance Analysis by Neutron Slowing Down Time (ibid, vol. 25, No. 2, 1968); co-author, Apparatus for Neutron-Resonance Analysis of Materials Elemental Composition (letter to the Editor, ibid, vol. 25, No. 3, 1968).
Address: Academy of Sciences of the U.S.S.R., 14 Leninsky Prospekt, Moscow V-71, U.S.S.R.

GAMBARYAN, V. G. Papers: Co-author, Indium-Helium Rad. Circuit for Pile-Type Reactor (Abstract, Atomnaya Energiya, vol. 19, No. 2, 1965); co-author, On Problem of Industrial Rad. Loop Construction (letter to the Editor, ibid, vol. 23, No. 6, 1967).
Address: Academy of Sciences of the U.S.S.R., 14 Leninsky Prospekt, Moscow V-71, U.S.S.R.

GAMBILL, Wallace Russell, B.S. (Chem. Eng.). Born 1931. Educ.: Georgia Inst. Technol. Design Eng., Union Carbide Chemicals Co., South Charleston, West Virginia, 1952-56; Res. Eng., O.R.N.L., 1956-. Book: 19-page sect. in 4th edition of McGraw-Hill's Chem. Eng. Handbook (1963).
Nuclear interests: Heat transfer, with emphasis on boiling, burnout, and swirl flow; Fluid dynamics; Estimation of thermophysical properties.
Address: 104 Mead Lane, Oak Ridge, Tennessee 37830, U.S.A.

GAMBOA LOYARTE, José Miguel, Dr. Sci., Prof. Inorg. Chem. Born 1919. Educ.: Saragossa and Cambridge Univs. Head, Radiochem. Sect., Inst. Phys. Chem., (Consejo Superior de Investigaciones Cientificas), Madrid; Prof. Radiochem., Faculty of Sci., Madrid Univ.; Head, Isotopes Sect., Junta de Energia Nucl., Madrid; Editor, Anales de la Real Sociedad Española de Fisica y Quimica. Society: Chem. Soc.
Nuclear interests: Application of radioisotopes in research and industry.
Address: 49-6^0 -dcha. Avenida de América, Madrid 2, Spain.

GAMMILL, Adrian Monroe, B.S., M.S., Ph.D. Born 1922. Educ.: Wisconsin, Illinois, Tulane and Mississippi Univs. Asst. Prof. Chem., Mississippi Univ., 1951-53; Supervisor, Nucl. Res., Davison Chem. Co., 1953-56; Director, Rare Earths and Thorium Div., Michigan Chem. Corp., 1956-. Now Tech. Director, W. R. Grace and Co., Davison Chem. Co. Societies: A.N.S.; A.C.S.; American Rocket Soc.; American Inst. of Metal. and Petroleum Eng.; Atomic Ind. Forum.
Nuclear interests: Reactor materials; Shielding; Process heat reactor technology; Rare earths as control rod materials.
Address: W. R. Grace and Co., Davison Chemical Co., Erwin, Tennessee, U.S.A.

GAMMON, Joseph A. Born 1921. Chairman, Southern Tank Lines; Member, Kentucky Sci. and Technol. Council; Vice-Pres., Nat. Industries Inc., Louisville.
Nuclear interests: Interstate and intra-state transportation of nuclear materials.
Address: Southern Tank Lines, Inc., 4107 Bells Lane, Louisville, Kentucky, U.S.A.

GAMREKELI, M. N. Paper: Co-author, Heat Transfer Studies in High-Temperature Radiant Spray Drier (Atomnaya Energiya, vol. 24, No. 3, 1968).
Address: Academy of Sciences of the U.S.S.R., 14 Leninsky Prospekt, Moscow V-71, U.S.S.R.

GANDINI, Augusto, Dr. in Phys. Born 1931. Educ.: Milan Univ. Head, Fast Reactor Phys. Sect., Casaccia Nucl. Studies Centre, C.N.E.N., Italy, 1962-. Society: A.N.S.
Nuclear interests: Calculation methods for the physics design of fast reactors and on the models of interpretation and correlation of integral experiments on fast critical facilities.
Address: Comitato Nazionale per l'Energie Nucleare, Nuclear Studies Centre, Casaccia, Rome, Italy.

GANEV, Gancho, Member, Com. for the Peaceful Uses of Atomic Energy, Bulgaria; Deputy Minister of Education and Culture. Deleg. to 2nd I.C.P.U.A.E., Geneva, Sept. 1958.
Address: Committee for the Peaceful Uses of Atomic Energy in the Bulgarian People's Republic, Council of Ministers, Sofia, Bulgaria.

GANGAS, Nicholas, Ph.D. Born 1937. Educ.: Vienna Univ. Forschungszentrum Seibersdorf, Vienna, 1960-61; Democritus Nucl. Res. Centre, Athens, 1962-. Society: Oesterreichische Phys. Gesellschaft, Vienna.
Nuclear interest: Low energy nuclear physics.
Address: No. 108 Kifissias, Athens 607, Greece.

GANGLOFF, Antoine, Eng. geologist. Born 1922. Educ.: Grenoble, Clermont-Ferrand and Strasbourg Univs. and Ecole Nationale Supérieure de Géologie de Nancy. C.E.A. (National authority), (1947-) now Chef du Département des Prospections et des Recherches Minières. Society: Sté. Géologique de France.
Nuclear interests: Uranium prospecting and mining.
Address: Commissariat à l'Energie Atomique, B.P. 4, 92 Chatillon-sous-Bagneux, (Hauts de Seine), France.

GANGULY, Anil Kumar, M.Sc., D.Sc. Born 1918. Educ.: Calcutta Univ. Post-Doctorate Res. Assoc., Notre Dame Univ., Indiana, 1953-55; now Head, Health Phys. Div., Electronics Group, Bhabha Atomic Res. Centre, A.E.C., India; Hon. Lecturer, Saha Inst. Nucl. Phys. Societies: Indian Science Congress; Indian Chem. Soc.
Nuclear interests: Health physics and nuclear chemistry; Reactor hazards evaluation; neutron and gamma-ray shielding; High field dosimetry; Build-up of elements in nuclear operations and radiochemical separation.
Address: Health Physics Division, Bhabha Atomic Research Centre, Trombay, Bombay 74, India.

GANOZA, Luis PINILLES. See PINILLES GANOZA, Luis.

GANTERT, Franz. Ministerialdirektor, Kultusministerium Baden-Württemberg.
Address: Kultusministerium Baden-Württemberg, 4 Schlossplatz, 7000 Stuttgart, Germany.

GANTT, Paul Hawkins, J.U.D., B.C.L. Born 1907. Educ.: Vienna Univ. and William and Mary Coll., Virginia. Procurement Counsel (1956-59), Chairman, Board of Contract Appeals (1959-64), U.S. Dept. of the Interior; Chairman, Board of Contract Appeals, U.S.A.E.C., 1964-. Pres., Federal Bar Assoc.; Pres., U.S. Sect., U.N. League of Lawyers; Rapporteur, Atomic Energy Law Com., World Peace Through Law Centre, Geneva.
Nuclear interest: Management.
Address: United States Atomic Energy Commission, Washington, D.C. 20545, U.S.A.

GANTY, Prabhakar Rao, B.Sc., M.Sc., D.Sc. (Geology). Born 1923. Educ.: Andhra Univ., Waltair, India. Sci. Officer (Sen. Geologist), Atomic Minerals Div., Dept. of Atomic Energy, Govt. of India, 1950-. Society: Mining, Metal. and Geological Soc. of India.
Nuclear interests: Nuclear geology, mineralogy and petrology and applied aspects.
Address: Department of Atomic Energy, 66 Upper Circular Road, Calcutta 9, India.

GANZHA, V. D. Papers: Co-author, Ion Exchange Unit for Production of Reactor Feed Water (letter to the Editor, Atomnaya Energiya, vol. 16, No. 5, 1964); co-author, Electrophoretic Filter for Purifying of Reactor Water (ibid, vol. 19, No. 4, 1965).
Address: Academy of Sciences of the U.S.S.R., 14 Leninsky Prospekt, Moscow V-71, U.S.S.R.

GAODU, A. N. Paper: Co-author, Toroidal Chambers for Investigation of Plasma-RF Fields Interaction (letter to the Editor, Atomnaya Energiya, vol. 23, No. 1, 1967).
Address: Academy of Sciences of the U.S.S.R., 14 Leninsky Prospekt, Moscow V-71, U.S.S.R.

GAPONOV, V. A. Papers: Co-author, Transformer Type High-Current Accelerator, (Atomnaya Energiya, vol. 20, No. 5, 1966); co-author, Vacuum Tubes for High-Current Accelerators (ibid, vol. 22, No. 1, 1967); co-author, High Intensity Electron Beams in Strong Focusing Acceleration Tubes (ibid, vol. 25, No. 2, 1968).
Address: Academy of Sciences of the U.S.S.R., 14 Leninsky Prospekt, Moscow V-71, U.S.S.R.

GARABEDIAN, Henry Leslie, B.S., M.A., Ph.D. Born 1901. Educ.: Tufts, Harvard and Princeton Univs. Consulting Phys., Atomic Power Div., Westinghouse Elec. Corp., 1949-56; Asst. Head, Nucl. Power Eng. Dept., Res. Lab. (1956-59), Head, Maths. Dept. (1959-66), Gen. Motors Corp.; Prof., Maths. and Energy Eng., Illinois Univ., Chicago Circle, 1967-. Societies: American Math. Soc.; A.N.S.
Nuclear interests: Theoretical nuclear reactor physics; Design of nuclear power reactors.
Address: 201 E. Chestnut Street, Chicago, Illinois, U.S.A.

GARAVAGLIA, C., Dr. in Medicine. Born 1924. Educ.: Milan Univ. Sci. Sec., 2nd I.C.P.U.A.E., Geneva, 1958; Chief, Radiol. Dept., Gen. Hospital, Vimercate (Milan), 1963-. Societies: Italian Soc. of Radiol. and Nucl. Medicine; Internat. H.P.S.
Nuclear interests: Nuclear medicine; Health physics.
Address: 7 Via de Castillia, Vimercate (Milan), Italy.

GARAY, Alfonso L. DE. See DE GARAY, Alfonso L.

GARBALYAUSKAS, Ch. A. Paper: Co-author, Radioactive Fallout in Vilnius in 1962 and 1964 (Atomnaya Energiya, vol. 22, No. 4, 1967).
Address: Academy of Sciences of the U.S.S.R., 14 Leninsky Prospekt, Moscow V-71, U.S.S.R.

GARBER, Harold Jerome, Ch.E. Born 1913. Educ.: Cincinnati Univ. Prof. Chem. Eng., Tennessee Univ., 1947-55; Consultant, O.R.N.L.; Manager, Chem. Development, Westinghouse Atomic Power Dept., 1955-59; Director of Plutonium Lab. and Director of Advanced Projects, Nucl. Materials and Equipment Corp., 1959-. Book: Slurry Eng. Handbook (WCAP-1150, Westinghouse Atomic Power Department, 1959). Societies: A.I.Ch.E.; A.N.S.; A.A.A.S.
Nuclear interests: Development and production of nuclear materials and equipment, particularly plutonium, ceramic and metallic fuels; Management of programmes in metallurgical and ceramic fuel development and production, and of alpha and neutron sources; Design of gamma irradiators and radioisotope fueled power generators.
Address: 5515 Darlington Road, Pittsburgh, Pa. 15217, U.S.A.

GARBER, Marion Harriett, A.B., M.S. (L.S.). Born 1921. Educ.: Temple Univ. and Drexel Inst. Technol. Librarian Tech. Information Div., U.S.A.E.C., Oak Ridge, 1949-50; Librarian (1950-), Assoc. Library Supervisor (1956-), Oak Ridge Assoc. Univs. Societies: American Library Assoc.; Special Libraries Assoc.; Tennessee Library Assoc.
Nuclear interest: Literature.
Address: 157 North Seneca Road, Oak Ridge, Tennessee, U.S.A.

GARCIA, Antonio M., M.D. Born 1930. Educ.: Univ. of the Philippines. Resident, Philippine Gen. Hospital, 1955-57; Fellowship in the Use of Radioisotopes in Medicine and Medical Res., by Internat. Cooperation Administration and N.R.C. at O.R.I.N.S. and Clinical Radioisotope Unit, Univ. Michigan

Hospital, 1957-58; Instructor, Univ. of the Philippines - Philippine Gen. Hospital Medical Centre, 1959-; Asst. Sci., Nat. Inst. of Sci. and Technol., and Lecturer for Basic Isotope Course, Philippine A.E.C., 1959-61; Res. Fellow in Medicine, Harvard Medical School and Peter Bent Brigham Hospital, supported by the U.S. N.A.S.-N.R.C., then by Nat. Institutes of Health, U.S.P.H.S., 1961-64. Sen. Sci., Nat. Inst. of Sci. and Technol.; Asst. Prof. Medicine, Univ. of the East School of Medicine, 1964-. Society: Soc. of Nucl. Medicine (U.S.).
Nuclear interests: Application of radioisotopes in medical diagnosis, treatment, and biomedical investigation.
Address: National Institute of Science and Technology, Herran Street, Manila, Philippines.

GARCIA, Paulino J., M.D. Born 1907. Educ.: Univ. Sto. Tomas, Manila. Head, Dept. of Radiol., Univ. Santo Tomas; Sec. of Health, 1953-58; Chairman, Nat. Sci. Development Board, 1958-. Chairman, Interdepartmental Com. on Atomic Energy, 1956-58. Chairman, U.N.I.C.E.F. Programmes of the Philippines, 1955-59; Chairman, Board of Governors, Philippine Nat. Red Cross, 1954-. Book: The Fundamentals of Radiol. (Community Publishing Co., 1951). Societies: Hon. Fellow, American Public Health Assoc.; Philippine Radiol. Soc. (former Pres.); Philippine Islands Medical Assoc.
Nuclear interests: The establishment of the Philippine Nuclear Research Centre with an open-pool type nuclear reactor.
Address: 3544 Saluysuy, Sta. Mesa, Manila, Republic of the Philippines.

GARCIA, R. VALDECANTOS. See VALDECANTOS GARCIA, R.

GARCIA AROSEMENA, Jorge, Dr. Vice-Pres., Liga Panamena Contra el Cancer.
Address: Liga Panamena Contra el Cancer, Apartado 7358, Panama, Panama.

GARCIA BLASQUEZ L., Teodoro, Ing. Geologo, Departamento de Control de Sustancias Radiactivas, Inst. Superior de Energia Nucl., Junta de Control de Energia Atomica del Peru.
Address: Junta de Control de Energia Atomica del Peru, 3420 Avenida Arequipa, Apartado 914, Lima, Peru.

GARCIA BONNELLY, Juan Ulises, Civil Eng. Born 1898. Educ.: S. Domingo Univ. Sec. of State for Agriculture, 1955-56; Sec. of State for Hydraulic Resources, 1956-57; Vice-Rector, S. Domingo Univ. Assessor, Dominican Commission for Atomic Investigations. Societies: Inst. for Geographical and Geologic Investigations of the Dominican Republic; Inter-American Inst. of Geography and History.
Nuclear interests: Studies on the neutron and nuclear physics in general.
Address: 14 Avenida Dr. Troncoso de la Concha, Ciudad Trujillo, Dominican Republic.

GARCIA CAPURRO, Federico, Medical Dr., Adjunt-Prof. Radiol. Born 1907. Educ.: State Univ. of Uruguay. Minister of Public Health; Chief Radiol., Radiol. Dept., Post Graduate Inst., Health Ministry of Uruguay. Member of the board of several health and civic associations. Society: Soc. de Radiol. del Uruguay.
Nuclear interests: Medical therapy and diagnosis.
Address: 2587 Avda. Italia, Montevideo, Uruguay, S.A.

GARCIA COLIN, Leopoldo, Dr. Programme of Statistical Mechanics, Comision Nacional de Energia Nuclear.
Address: Comision Nacional de Energia Nuclear, Av. Insurgentes Sur 1079, Tercer piso, Mexico 18, D.F., Mexico.

GARCIA DE QUEVEDO, José Luis, B.S. (E.E.), M.E.E., Ph.D. Born 1920. Educ.: Puerto Rico Univ., Rensselaer Polytech. Inst., and Duke Univ. Assoc. Prof. E.E. and Head of Dept., 1949-53; Prof. E.E. and Head of Dept., 1953-57; Acting Director (1958-60), Assoc. Director (1960-63), Head Nucl. Eng. Dept. (1963-), Puerto Rico Nucl. Centre. Societies: A.I.E.E.; A.P.S.; I.R.E.; Colegio de Ingenieros de P.R.
Nuclear interest: Nuclear physics.
Address: College Station, Mayaguez, Puerto Rico, U.S.A.

GARCIA-ONTIVEROS, Angel, Dipl. Ing. Born 1940. Educ.: Univ. Mayor de San Andrés, La Paz, Bolivia. Asst. Lecturer, Rational Mechanics, Univ. of San Andrés and Acad. of Military Eng., 1960-61; Asst. Lecturer, Inorganic Technol., Univ. of San Andrés, 1961; Asst. Res., Bolivian Air Showers Joint Experiment (B.A.S.J.E.) (1961), Assoc. Res., Space Phys. Group (Cosmic Rays, Time Variation) (1966), Laboratorio de Fisica Cósmica de Chacaltaya. Under Chief, Eng. Div. (1966), Head, Div. of Phys. and Eng. (1967-), Comisión Boliviana de Energia Nucl.; Lecturer on Introductory Phys., Inst. of Basic Sci. (1967), and Lecturer of Ind. Electronics, Faculty of Ind. Eng. Univ. Mayor de San Andrés; Lecturer on Introductory Atomic Energy, Tech. Faculty, Univ. Mayor de San Andrés; Lecturer on Applications of Atomic Energy, Short training courses held at the Comisión Boliviana de Energia Nucl.; Bolivian Deleg. at various Atomic Energy Events. Societies: Bolivian Phys. Soc.; Ind. Eng. Assoc.; Soc. of Bolivian Eng.
Nuclear interests: Applications of atomic energy on various features.
Address: Comisión Boliviana de Energia Nuclear, Casilla 802, La Paz, Bolivia.

GARCIA ORCOYEN, Jesús. Pres., Comisión Asesor de Medicina y Biologia Animal, y Vocale Ministerio de Industria, Junta de Energia Nucl. Madrid.
Address: Ministerio de Industria, Junta de Energia Nuclear, Ciudad Universitaria, Madrid 3, Spain.

GARCIA-ORCOYEN, Sergio. Vice-Sec., Comisiones Asesoras de Medicina y Biologia Animal, Junta de Energia Nucl.
Address: Junta de Energia Nuclear, Ciudad Universitaria, Madrid 3, Spain.

GARCIA PEREZ, Vicente. Vocale, Comisiones Asesoras de Biologia Vegetal y Aplicaciones Industriales, Junta de Energia Nucl.
Address: Junta de Energia Nuclear, Ciudad Universitaria, Madrid 3, Spain.

GARCIA RAMAL, Enrique. Member, Steering Com., Forum Atomico Espanol.
Address: Altos Hornos de Vizcaya Co., 27 Alameda Recalde, Bilbao, Spain.

GARCILLAN, Mario ALVAREZ. See ALVAREZ GARCILLAN, Mario.

GARD, L. M., Jr. Geologist, Geological Survey, U.S. Dept. of the Interior.
Address: U.S. Department of the Interior, Geological Survey, Building 25, Federal Centre, Denver, Colorado, U.S.A.

GARDEL, André, B.Sc. (Lausanne), Civil Eng. (Inst. Technol., Lausanne Univ.), Dr. Tech. Science (Inst. Technol., Lausanne Univ.). Born 1922. Educ.: Lausanne Univ. and Inst. Technol., Lausanne Univ. Consulting Eng., 1954-. Sec., Energia Nucléaire S.A. Lausanne, 1957-; Prof. Inst. Technol., Lausanne Univ.; Sen. Officer, Bonnard and Gardel, Ingenieurs-Conseil. Societies: Swiss Soc. of Eng. and Architects; A.S.C.E.; A.N.S.
Nuclear interests: Economical aspects of the production of energy, general problems concerning the lay-out of nuclear power plants, in particular civil engineering, safety, operation.
Address: 10 avenue de la Gare, Lausanne, Switzerland.

GARDELLA, Joseph W., M.D. Assoc. Dean, Harvard Medical School. Member, Isotope Com., Massachusetts General Hospital.
Address: Massachusetts General Hospital, Boston 14, Massachusetts, U.S.A.

GARDINER, Donald M. Director, Health and Safety Div., Chicago Operations Office, U.S.A.E.C.
Address: U.S. Atomic Energy Commission, Chicago Operations Office, 9800 South Cass Avenue, Argonne, Illinois 60439, U.S.A.

GARDNER, Andrew Leroy, B.S. (Radio Eng.), Ph.D. (Phys.). Born 1919. Educ.: Utah State Agricultural Coll. and California (Berkeley) Univ. Asst. Res. Eng., California Univ., Eng. Res. Lab., 1946-54; Phys., Lawrence Rad. Lab., California Univ., 1954-. Society: A.P.S.
Nuclear interest: Controlled thermonuclear reactor research.
Address: 2379 Palm Avenue, Livermore, California, U.S.A.

GARDNER, Annesta R., B.S., M.A. Educ.: Chicago Univ. Ind. Editor, Modern Industry, 1944-57; Sen. Assoc. Editor, Product Eng., 1957-. Board of Directors, Nucl. Energy Writers Assoc. Societies: A.C.S.; A.S.M.E.; Nat. Assoc. of Sci. Writers.
Nuclear interests: Materials; Reactor components; Radioisotope applications.
Address: Product Engineering, 330 West 42nd Street, New York, N.Y. 10036, U.S.A.

GARDNER, Donald Glenn, B.S. (Illinois), M.S. (Michigan), Ph.D. (Michigan). Born 1931. Educ.: Illinois and Michigan Univs. Chem., Westinghouse Elec. Corp., Pittsburgh, Pa., 1957-59; Asst. Prof., Arkansas Univ., 1959-61; Asst. Prof. (1961-64), Assoc. Prof. (1964-), Illinois Inst. Technol.; Chem., Lawrence Rad. Lab., 1965-67. Societies: A.C.S.; A.P.S.
Nuclear interests: Theory and measurement of cross-sections of reactions induced by fast neutrons; Beta- and gamma-ray scintillation spectroscopy; Nuclear decay scheme measurements.
Address: Department of Chemistry, Illinois Institute of Technology, Chicago, Illinois, U.S.A.

GARDNER, Herbert. Liaison, N.R.C. Medical Sci., Subcom. on Radiobiol., Nat. Acad.of Sci.- N.R.C.
Address: National Academy of Sciences - National Research Council, 2101 Constitution Avenue, N.W., Washington 25, D.C., U.S.A.

GARDNER, John William, B.Sc., Ph.D. Born 1919. Educ.: Southampton, London and Birmingham Univs., and Dublin Inst. for Advanced Studies. Various research fellowships in Europe, N. America and Australia, 1946-56; Sen. Phys., English Elec. Co., Ltd., 1956-. Book: Electricity Without Dynamos (Penguin, 1963). Societies: Phys. Soc.; Nucl. Eng. Soc. Nuclear interest: Reactor and nuclear physics.
Address: English Electric Co., Ltd., Whetstone, Leicestershire, England.

GARFINKEL, Samuel B. Member, Subcom. on Use of Radioactivity Standards, Com. on Nucl. Sci., N.A.S. - N.R.C.
Address: National Academy of Sciences - National Research Council, Committee on Nuclear Sciences, 2101 Constitution Avenue, N.W., Washington 25, D.C., U.S.A.

GARGAROS, Mrs. S. Sec., Centre de Physique Atomique et Nucléaire de la Faculté des Sciences, Toulouse Univ.
Address: Toulouse University, 118 route de Narbonne, 31-Toulouse, France.

GARHY, Mohammed Ahmed EL-. See EL-GARHY, Mohammed Ahmed.

GARLID, Kermit L., B.S., B.Ch.E., Ph.D. Born 1929. Educ.: Minnesota Univ. Assoc. Prof., Washington Univ.; Consultant, Battelle Memorial Inst. Societies: A.N.S.; A.I.Ch.E.; A.A.A.S.
Nuclear interests: Nuclear reactor theory; the determination of dynamic parameters of nuclear and chemical systems by pulse analysis; dynamics of two-phase (gas-liquid) flow systems.
Address: Department of Chemical Engineering, Washington University, Seattle, Washington 98105, U.S.A.

GARLOCK, Donald H. 2nd Vice Pres., Nucl. Insurance of Property, Travelers Indemnity Co.
Address: Travelers Indemnity Co., 1 Tower Square, Hartford, Connecticut 06115, U.S.A.

GARNAULT, Andre. Responsable du Departement Energie Nucl., Cie. Francaise d'Entreprises Metalliques.
Address: Cie. Francaise d'Entreprises Metalliques, 37 boulevard de Montmorency, Paris 16, France.

GARNER, Clifford Symes, B.S., Ph.D. (California Inst. Technol.). Born 1912. Educ.: Pasadena Jun. Coll. and California Inst. Technol. Assoc. Prof. (1947-53), Prof. (1953), Chem., California Univ., Los Angeles. Books: Co-author, The Rare-Earth Elements and their Compounds (New York, John Wiley, 1947); Radioactivity Appl. to the Discovery and Investigation of the Newer Elements (chapter 7 of Radioactivity Appl. to Chem.) (N.Y., John Wiley, 1951); Isotopic Tracers in Chem. (chapter 5 of Modern Chem. for the Eng. and Sci. (N.Y., McGraw-Hill, 1957). Societies: A.C.S.; A.P.S.; A.A.A.S.
Nuclear interests: Application of radioisotopes and enriched stable isotopic tracers in research in physical and inorganic chemistry, consulting in nuclear chemistry.
Address: Department of Chemistry, California University, Los Angeles, California 90024, U.S.A.

GARNETT, John Lyndon, Ph.D. (Chicago), M.Sc. (N.S.W.), A.S.T.C. (Sydney). Born 1929. Educ.: Chicago and N.S.W. Univs. Sen. Lecturer, New South Wales Univ., 1957-. Societies: Assoc. Roy. Australian Chem. Inst.; Univ. Sydney Chem. Soc.; Univ. N.S.W. Chem. Soc.
Nuclear interests: Radiation chemistry; Radiation induced reactions; Isotopic hydrogen exchange reactions; Radiation catalysis.
Address: Radiation Laboratory, Department of Physical Chemistry, New South Wales University, P.O. Box 1, Kensington, New South Wales, Australia.

GARNIER, J. Gen. Manager, Ateliers et Chantiers de Nantes (Bretagne-Loire), S.A.
Address: Ateliers et Chantiers de Nantes (Bretagne-Loire) S.A., Prairie au Duc, Nantes, (Loire-Atlantique), France.

GARR, Carl R., B.S., M.S., Ph.D. Born 1927. Educ.: Cast Inst. Technol., Cleveland, Ohio. Director, R. and D., A.C.F. Industries, Inc. Societies: A.I.M.M.; A.S.M.; Newcomen Soc. in North America.
Nuclear interest: Reactor metallurgy.
Address: A.C.F. Industries, Inc., 750 Third Avenue, New York, New York, U.S.A.

GARRALDA, D. Ignacio HERRERO. See HERRERO GARRALDA, D. Ignacio.

GARRETT, Alfred Benjamin, B.S., M.S., Ph.D. Born 1906. Educ.: Muskingum Coll. and Ohio State Univ. Prof. Chem., (1944-), Chairman, Dept. Chem. (1960-62), Vice-Pres. for Res. (1962-), Ohio State Univ. Chairman, Columbus Sect., and Chairman, Div. of Chem. Educ., A.C.S.; Chairman, Cooperative Com., A.A.A.S., 1963-64.
Nuclear interests: Neutron activation and radiotracers.
Address: Ohio State University, 190 North Oval Drive, Columbus 10, Ohio, U.S.A.

GARRETT, Cyril, B.Sc., M.Sc. Born 1921. Educ.: Saskatchewan Univ. Head, X-rays and Nucl. Rad., 1953-. Member, Rad. -Exposure, Kerma and Fluence Planning Board, I.C.R.U. Society: Canadian Assoc. of Phys.
Nuclear interests: Radiation physics; Nuclear physics.
Address: National Research Council of Canada, Montreal Road, Ottawa, Ontario, Canada.

GARRETT, George Alvin, A.B., A.M., Ph.D. Born 1910. Educ.: Mississippi and Rice Univs. Director of Information Processing, Lockheed Missiles and Space Co., Sunnyvale, Calif., 1962-. Societies: A.P.S.; American Mathematical Soc.; A.A.A.S.; RESA; Assoc. for Computing Machinery.
Nuclear interest: Isotope separation.
Address: 440 Marlowe Street, Palo Alto, California, U.S.A.

GARRETT, R. Lecturer, Phys. Dept., Auckland Univ.
Address: Auckland University, P.O. Box 2175, Auckland, New Zealand.

GARRETT, Robert A., A.B., M.D. Born 1919. Educ.: Miami and Indiana Univs. Prof. and Chairman, Dept. Urology, Indiana Univ. Medical Centre, 1954-. Member, Radioisotope Com., Indiana Univ.
Nuclear interest: Medical use of isotopes.
Address: Indiana University, Medical Centre, 1100 West Michigan Street, Indianapolis, Indiana 46207, U.S.A.

GARRICK, B. John. B.S., M.S., Ph.D. Born 1930. Educ.: California (Los Angeles) and Brigham Young Univs. and O.R.S.O.R.T.; Phys., Phillips Petroleum Co., Nat. Reactor Testing Station, Idaho, 1952-54; Phys., U.S.A.E.C., Washington, D.C., 1955-57;

Vice Pres., Nucl. Div., Holmes and Narver, Inc., Los Angeles, 1957-. Society: A.N.S. Nuclear interests: Reactor and nuclear safety; Systems analysis.
Address: Holmes and Narver, Inc., 828 So. Figueroa Street, Los Angeles, California, U.S.A.

GARRIDO, Luis M., M.A., M.Sc., Ph.D. Born 1930. Educ.: Harvard and Madrid Univs. Prof. Mathematical Phys., Barcelona Univ. Book: Quantum Mechanics (Madrid, Ed. Rialp, 1964).
Nuclear interests: Nuclear physics - theoretical; Plasma physics - theoretical; Solid state.
Address: Science Department, Barcelona University, Barcelona, Spain.

GARRISON, John Dresser, Ph.D. Born 1922. Educ.: California Univ. Instructor, Yale Univ., 1953-56; Asst. Prof. (1956-59), Assoc. Prof. (1959-62), Prof. (1962-), and Chairman, Phys. Dept., San Diego State Coll. Society: A.P.S. Nuclear interests: Neutron cross-sections; Nuclear physics.
Address: 9261 Virginian Lane, La Mesa, Calif., U.S.A.

GARRONE, L. Sen. Tech. Officer, Giovanola Frères S.A.
Address: Giovanola Frères S.A., Monthey, (Valais), Switzerland.

GARRONE, Silvio. Director, Ing. Silvio Garrone s.r.l.; Director, EKCO Nucleare Italiana s.r.l.
Address: Ing. Silvio Garrone s.r.l., 40 Via Marco Besso, Rome, Italy.

GARSOU, Julien Lambert, Dr. in Sci. (Liège), cand. and lic. in chem. sci., cand. in pharmaceutical sci., cand. in geological and mineralogical sci.; various certificates in phys., phys. chem., biochem., natural and medical sci., nucl. sci. Born 1930. Educ.: Liège and Michigan Univs. Researcher, Inst. Interuniv. des Sci. Nucléaires, Brussels, 1952-; Res. Assoc., Michigan Univ. (U.S. A.E.C. project), 1957-59. Nuclear interests: Radiation chemistry and biology; radiation effectiveness, kinetics of the effects, sensitivity and recovery; Dosimetry, radiation therapy (rotating source).
Address: 60 rue Auguste Ponson, Jupille, Belgium.

GARTNER, Henriette KNOOR-. See KNOOR-GARTNER, Henriette.

GÄRTNER, Walter, Ing., Direktor. Born 1892. Nuclear interests: Gamma irradiation equipment for kilocurie sources for research and industry; Medical equipment; Gammaradiographic equipment.
Address: 16 Florastrasse, Düsseldorf, Germany.

GARTON, William Reginald Stephen, D.Sc., A.R.C.S. Born 1912. Educ.: London Univ. Lecturer in Phys. (1946-54), Sen. Lecturer (1954-57), Reader in Phys. (1957-64), Prof. Spectroscopy (1964-), Imperial Coll., London. Society: Phys. Soc.
Nuclear interest: Thermal fusion; Astrophysics.
Address: 482 Upper Richmond Road, Richmond, Surrey, England.

GARUSOV, E. A. Papers: Co-author, Albedo of γ-rays from Co^{60} and Au^{198} from Different Materials (Atomnaya Energiya, vol. 5, No. 6, 1958); co-author, Ratio for Thermal Neutron Flux in Water to Strength of a Point Source (letter to the Editor, ibid, vol. 15, No. 1, 1963); co-author, Neutron Transport in Moving Medium (ibid, vol. 21, No. 2, 1966).
Address: Academy of Sciences of the U.S.S.R., 14 Leninsky Prospekt, Moscow V-71, U.S.S.R.

GARWIN, Richard L., Prof. Principal Investigator, U.S.A.E.C. contract at (30-1) (1932), Nevis Cyclotron Labs., Columbia Univ. Member, Tech. Working Group 2, Conference on the Discontinuance of Nucl. Weapons Tests, Geneva, Nov.-Dec. 1959.
Address: Nevis Cyclotron Laboratories, Irvington, New York, U.S.A.

GASALLA VALES, Manuel. Vocale, Comision Interministerial de Conservacion de Alimentos por Irradiacion, Junta de Energia Nucl.
Address: Junta de Energia Nuclear, Ciudad Universitaria, Madrid 3, Spain.

GASH, Donald, B.Sc. (Hons., Mech. Eng.). Born 1934. Educ.: Durham Univ. Society: Windscale and Calder Branch, Nucl. Eng. Soc.
Address: United Kingdom Atomic Energy Authority, Seascale, Cumberland, England.

GASKILL, James R., B.S., M.S. (Chem. Eng., Columbia). Born 1910. Educ.: Columbia, Princeton and California Univs. With California R. and D. Co. (U.S.A.E.C. contract), 1951-54; California Univ. Lawrence Rad. Lab., 1954-; City of Livermore, California, Dept. of Civil Defence, 1956-69. Societies: A.C.S.; A.I.Ch.E.; United States Civil Defence Council.
Nuclear interests: Standards and research in fire retardant coatings for 'hot' laboratories; Standards and research in floor coverings in 'hot' laboratories; Research and development in emergency protective equipment for radioactive operations and incidents, particularly in regard to communications devices; Research in fire extinction methods and smoke tests; Development of nuclear shipping containers.
Address: 875 Estates Street, Livermore, California, U.S.A.

GAST, Paul Frederick, B.S., Ph.D. Born 1916. Educ.: Ohio State and Washington Univs. Formerly Consulting Phys., Hanford Atomic Products Operation, G.E.C.; Manager, Phys. and Instrument R. and D., Hanford Labs., G.E.C., 1956-. Now Assoc. Director, Liquid Metal Fast Breeder Reactor Programme Office, Argonne Nat. Lab. Societies: Fellow, A.P.S.; Fellow, A.A.A.S.; Fellow, A.N.S.

Nuclear interests: Reactor physics; Research and development management.
Address: Argonne National Laboratory, Argonne, Illinois, U.S.A.

GASTEIGER, Edgar L., Jr., A.B., M.S., Ph.D. Born 1919. Educ.: Allegheny Coll., Illinois and Minnesota Univs. Res. Assoc., Neurosurgery, Massachusetts Gen. Hospital, 1951-57; Instructor, Assoc., Asst. Prof., Harvard Medical School, 1951-57; Asst. Prof., Physiology, School of Medicine, Rochester Univ., 1957-61; Prof., Phys. Biol., New York State Veterinary Coll., Cornell Univ., 1961-. Societies: American Physiological Soc.; A.A.A.S.; Biophys. Soc.; Eng. in Medicine and Biol. Group, I.E.E.E.
Nuclear interests: Action of ionising radiation on central and peripheral system; X-ray detection in rats.
Address: Cornell University, New York State Veterinary College, Department of Physical Biology, Ithaca, New York 14850, U.S.A.

GASTELBLUM, Camille. Sen. Officer Antwerpse Buizen Maatschappij "Anbuma" S.A.
Address: Antwerpse Buizen Maatschappij "Anbuma" S.A., 147 rue de Breda, Antwerp, Belgium.

GASTON, Ernest Charles, B.S. (Mech. Eng.), D.Sc. (Hon.). Born 1902. Educ.: Alabama Univ. Chief Design Eng. (1949-52), Vice-Pres. and Chief Eng. with responsibility for all eng. services (1952-57), Director (1952-67), Pres. (1957-67), Chairman of the Board (1967), Southern Services, Inc; Director, Mississippi Power Co. 1958-67; Director (1958-67), Pres. (1962-63), Southern Elec. Generating Co.; Director, Gulf Power Co. (1966-67); Director (1958-65), Member, Exec. Com. (1958-65), Member, Tech. and Eng. Com. (1954-65), Atomic Power Development Associates, Inc.; Member, Exec. Com. and Board of Trustees, High Temperature Reactor Development Associates, Inc. 1958-67; Member, Exec. Com. and Board of Trustees, Power Reactor Development Co., 1961-67; Director (1961-67), Vice-Pres. (1963-67), Southern Co.; Director, Atomic Ind. Forum, Inc. 1962-66; Director, Southeastern Elec. Exchange, 1962-63. Societies: Eng. Club, Birmingham, Alabama; A.S.M.E.; Alabama Society of Professional Eng.; Newcomen Soc.
Nuclear interests: Relates principally to reactor design and management for electric power production including water-cooled, fast breeder and gas-cooled designs.
Address: 91 Lucerne Boulevard, Birmingham, Alabama 35209, U.S.A.

GAT (GUTMANN), Joel Robert, Ph.D., M.Sc. (Jerusalem, Hebrew Univ.). Born 1926. Educ.: Jerusalem Hebrew Univ. Res. Chem., Govt. Labs. at Jerusalem and Rehovoth, 1950-55; Res. Assoc., Enrico Fermi Inst. for Nucl. Studies, Chicago, 1956; Res. Officer, Israel A.E.C. Labs., 1956-59; Sen. Sci., Isotope Dept. (1959-66), Assoc. Prof. (1966-), Weizmann Inst. of Sci., Rehovoth. Societies: Israel Chemists' Organisation; Israel Geophys. Soc.
Nuclear interests: Nuclear and isotope geochemistry; Radiochemistry, particularly environmental aspects of radiochemistry, fallout studies, etc.; Isotope chemistry, especially isotope separation processes.
Address: c/o Isotope Department, Weizmann Institute of Science, P.O.B. 26, Rehovoth, Israel.

GATES, George F., B.S.M.E. Born 1918. Educ.: Illinois Univ. Project Eng., Safety Test Exptl. Project, Project Eng., Exptl. Gas Cooled Reactor, Kaiser Eng. Society: A.N.S.
Nuclear interests: Technical management; Reactor plant test and operating procedures; Mechanical systems.
Address: Kaiser Engineers, Kaiser Centre, Oakland, California, U.S.A.

GATHERCOLE, George Edward, B.A. (McMaster), M.A. (Economics, Toronto), LL.D. (Hon. York). Born 1909. Educ.: McMaster and Toronto Univs. Provincial Economist of Ontario, 1951; Provincial Economist and Asst. Controller of Finances, 1954; Deputy Minister of Economics, 1956; Deputy Minister of Economics and Development 1961-62; First Vice-Chairman, (1961), Chairman, (1966), The Hydro-Electric Power Commission of Ontario; Director, and Pres. (1968-), Canadian Nucl. Assoc.
Nuclear interests: Broad interest in the operation of nuclear power plants, and particularly in the current construction of 2,160,00 kw. Pickering Generating Station near Toronto.
Address: 620 University Avenue, Toronto 2, Ontario, Canada.

GATTI, Emilio, Dr. Eng. Born 1922. Educ.: Padua Univ. Full Prof. Electronics and Nucl. Instrumentation, Politecnico di Milan; Consultant, C.I.S.E. Labs.; Editor, Alta Frequenza; Instrumentation Sect., Nuovo Cimento. Societies: Sen. Member, I.E.E.E.; Assoc. Elettrotecnica Italiana; Soc. Italiana di Fisica.
Nuclear interests: Nuclear physics and nuclear electronics; Multiplier photo tubes; Statistical problems regarding extraction of amplitude and time information from detection of nuclear events.
Address: Politecnico di Milano, 32 Piazza Leonardo da Vinci, Milan, Italy.

GATTY, Bernard, Agrégé de l'Université. Born 1938. Educ.: Paris Univ. Maitre Asst., Faculté des Sci. d'Orsay, Paris Univ.
Nuclear interests: Nuclear physics; Nuclear reactions induced by protons and heavy ions.
Address: Institut de Physique Nucléaire, 91 Orsay, France.

GATTYS, F. J. Proprietor, F. J. Gattys-Ingenieurbüro.
Address: F. J. Gattys-Ingenieurbüro, 36

GAU

Vilbeler Strasse, P. O. Box 3505, Frankfurt-am-Main, Germany.

GAUBAN, Daniel. Prés. Directeur Gén., Ets. G.B.G.
Address: Ets. G.B.G., 30 rue Eugène-Caron, Courbevoie, (Seine), France.

GAUCHER, Donald H., B.A., B.S., LL.B. Born 1931. Educ.: Rice Inst., Houston, Texas and Houston Univ. With Standard Oil Co., (N.J.), New York. Societies: A.I.M.M.; American Petroleum Inst.
Nuclear interests: Reactor design; Industrial radiation processes.
Address: 30 Rockefeller Plaza, New York, N.Y., U.S.A.

GAUDEFROY, Henri, B.A.Sc., B.S. (Elec. Eng.), D.Sc. (Hon., Laval, Sherbrooke and Western Ontario Univs.), LL.D. (Hon., Toronto Univ.). Born 1909. Educ.: Ecole Polytechnique, Montreal Univ. and M.I.T. Dean of Eng., Ecole Polytechnique, Montreal Univ., 1953-. Member, Reactor Safety Advisory Com., Canada, 1956-; Member, N.R.C., Canada, 1954-60; Member, Atomic Energy Control Board, Canada, 1961-. Active member of eng. soc. as A.S.E.E., E.I.C.
Address: 712 Rockland Avenue, Montreal 8, Canada.

GAUDEMARIS, Gabriel Pierre de. See de GAUDEMARIS, Gabriel Pierre.

GAUDERNACK, B. G. Sen. Chem., Chem. Div., Institutt for Atomenergi.
Address: Institutt for Atomenergi, P.O. Box 40, Kjeller, Norway.

GAUDRY, Andre. Administrateur, Sté. de Raffinage d'Uranium (S.R.U.).
Address: Société de Raffinage d'Uranium (S.R.U.), 23 boulevard Georges Clemenceau, Courbevoie, (Seine), France.

GAUL, Horst, Dr. Born 1919. Educ.: Göttingen Univ. Head, Mutation Lab., Max-Planck-Inst. für Züchtungsforschung, Köln Vogelsang, 1952-. Regional editor, Radiation Botany; Docent, Hanover Tech. Univ.
Nuclear interest: Radiation biology.
Address: Max-Planck-Institut für Züchtungsforschung, Köln-Vogelsang, Germany.

GAULARD, Marie-Louise. Travaux Pratiques, Maitre Asst., Faculté des Sciences, Ecole Nationale Supérieure d'Electricité et de Mécanique, Nancy Univ.
Address: Nancy University, 2 rue de la Citadelle, Nancy, (Meurthe-et-Moselle), France.

GAULLE, Bernard Xavier de. See de GAULLE, Bernard Xavier.

GÄUMANN, Tino, Prof. Dr. Born 1925. Educ.: Federal School Technol., Zurich. Prof. Phys. Chem., Head of Inst. Chem. Phys., E.P.U.L., Lausanne.
Nuclear interests: Radiation chemistry of organic compounds; Physical chemistry of reactor processes; Mass spectrometry.
Address: Institute of Chemical Physics, E.P.U.L., 31 avenue des Bains, Lausanne, CH-1007, Switzerland.

GAUME, Colonel. Attaché Militaire, C.E.A.
Address: Commissariat à l'Energie Atomique, 29-33 rue de la Federation, Paris 15, France.

GAUNT, Ian Alexander Butler, B.Sc. (Eng.). Born 1924. Educ.: London Univ. Head, Projects Sect. Atomic Energy Div., Gen. Elec. Co. Ltd., 1958-59; Chief Eng., British G.E.C. of Japan Ltd., 1959-65; Project Manager, Nucl. Eng., G.E.C. (Process Eng.) Ltd., 1965-66; Sen. Reactor Design Eng. Shipbuilding Group, Vickers Ltd., 1966-. Societies: Inst. Mech. Eng.; Assoc. Fellow, Roy. Aeronautical Soc.
Nuclear interests: Reactor design, construction and management.
Address: Vickers Ltd., Shipbuilding Group, Vickers-Armstrongs Barrow Shipyard, Barrow-in-Furness, Lancs., England.

GAUNT, John, B.Sc., Ph.D. Born 1920. Educ.: Leeds Univ. Spectroscopy Group, A.E.R.E., Harwell, -1955; Sci. Sec., First I.C.P.U.A.E., 1955; Group Leader, Spectroscopy Group, A.E.R.E., Harwell, 1956; Nucl. Attaché, British Embassy, Washington, D.C., 1957-60; Special Asst. to Deputy Chairman, U.K.A.E.A., London, 1961-63; Special Assistant to Chairman, U.K.A.E.A., London, and Head of U.K. Geneva Conference Secretariat, 1964; Head, Industrial Collaboration Office, A.E.R.E., Harwell, 1965.
Nuclear interests: All aspects of nuclear energy.
Address: United Kingdom Atomic Energy Authority, 11 Charles II Street, London, S.W.1, England.

GAUSSENS, Jacques Etienne, Ing. des Arts et Manufactures; Diplomè d'Etudes Superieures de Sci. economiques. Born 1920. Educ.: Ecole Centrale des Arts et Manufactures, Paris. Chargé des Etudes Economiques Générales, Adjoint au Chef du Département des Programmes, C.E.A.; Chargé de Mission au Secretariat gen. du l'Energie.
Nuclear interests: Economic studies and operation research.
Address: 12 rue A. Theuriet, Bourg la Reine, Seine, France.

GAUSTER, Wilhelm F., Dipl. Ing., Dr. Techn. Habil. Born 1901. Educ.: Vienna T.H. Ord. Prof., Vienna T.H. 1945-50; Prof., N.C. State Coll., Raleigh, N.C., 1950-57; Director, Magn. Lab., O.R.N.L. 1957-. Societies: Fellow, A.I.E.E.; Austrian Soc. of Elec. Eng.; A.N.S.; A.P.S.

Nuclear interests: Thermonuclear research; Applied electrophysics.
Address: 104 Seymour Lane, Oak Ridge, Tennessee, U.S.A.

GAUTHIER, Jean, Eng. Born 1919. Educ.: Conservatoire Nat. des Arts and Metiers, Paris. At Ste. Sciaky. Societies: Ste. des Ing. Soudeurs; Expert, Internat. Welding Soc.
Nuclear interest: Welding technology.
Address: Societe Sciaky, 119 Quai Jules Guesde, Vitry sur Seine 94, France.

GAUVENET, Andre Jean, L.es Sc., Diplome d'Etudes Superieures. Born 1920. Educ.: Ecole Normale Superieure de St. Cloud; Attache Sci. a l'Ambassade de France a New York, 1954; Ing. au C.E.A., 1956; Chef du Service des Programmes, 1961; Directeur de Cabinet du Haut Commissaire, 1963-67; Directeur de la Protection et de la Surete Radiologique, 1967-. Societes: A.P.S.; A.N.S.; Inst. of Radio Eng.; Ste. Francaise des Radioelectriciens.
Nuclear interests: Nuclear physics (electronics, ionised gases); Scientific management.
Address: 18 Residence du Petit Chambord, Bourg la Reine, 92, France.

GAUVIN, J. N. Laurie, B.Sc.A. (Phys.), D.Phil. Born 1929. Educ.: Laval and Oxford Univs. Asst. Prof. (1957-63), Assoc. Prof. (1963-), formerly Prof., Laval Univ. Pres., L'Association des Professeurs de l'Universite Laval, 1963-64; Third Vice-Pres., Canadian Assoc. Univ. Teachers, 1964-65. Societies: Canadian Assoc. of Phys.; A.P.S.; Inst. of Phys.
Nuclear interests: Theory of nuclear structure and of nuclear reactions.
Address: Faculte des Sciences, Universite Laval, Quebec 10, Quebec, Canada.

GAUWERKY, Friedrich, Prof., Dr.med. Vorsitzender, Fachnormenausschusses Radiol., Deutscher Normenausschuss; Chefarzt, Inst. für Strahlentherapie und Nuklearmedizin, Allgemeine Krankenhaus St. Georg.
Address: Institut fur Strahlentherapie und Nuklearmedizin, Allgemeine Krankenhaus St. Georg, 5 Lohmuhlenstrasse, Hamburg 1, Germany.

GAUZIT, Maurice, Dr. (Phys.). Born 1927. Educ.: Lyons and Paris Univs. C.N.R.S., 1950-55; Sté. Etudes Hispano Suiza, 1955-62; Co. Gen. Elec., 1962-. Gen. Sec., Assoc. Développement Applications Ind. des Rayonnements. Books: Physique et Calcul des Réacteurs Nucléaires (Paris, Dunod, 1957; Translated into Russian); Contrôle et Protection des Réacteurs Nucléaires (Paris, Dunod, 1957; Translated into Russian).
Nuclear interests: Reactor design; Control.; Shielding; Irradiation; Desalination.
Address: Bureau Nucleaire, 54 rue la Boetie, Paris, France.

GAVIAN, Peter W. Vice Pres., Finance, Nucl. Utility Services, Inc.
Address: Nuclear Utility Services, Inc., 1730 M Street, N.W., Washington, D.C. 20036, U.S.A.

GAVIN, James M. Former Pres., now Chairman (1964-), and Chief Exec. Officer, Arthur D. Little, Inc.
Address: Arthur D. Little, Inc., 30 Memorial Drive, Cambridge, Massachusetts, U.S.A.

GAVIN, Paul Henri, M.S., Ph.D. Born 1935. Educ.: Lausanne Univ. Lausanne Univ., 1960-66; Combustion Eng. Inc., Windsor, Connecticut. Society: A.N.S.
Nuclear interests: Reactor design; Nuclear engineering.
Address: 28 Country Lane, Simsbury, Connecticut 06070, U.S.A.

GAVLEFORS, Sture Harald, M.Sc. Born 1934. Educ.: Chalmers Tech. Univ., Gothenburg. Nucl. Power Dept., A.S.E.A.
Nuclear interest: Mechanicazl design in the nuclear field.
Address: Allmänna Svenska Elektriska Aktiebolaget, Nuclear Power Department, Vasteras, Sweden.

GAVRILOV, Boris Ivanovitch. Member, (Phys.), Thermonucl. Lab., Kurchatov Inst. of Atomic Energy. Papers: Co-author, Thermal-neutron Capture γ-rays (Atomnaya Energiya, vol. 6, No. 3, 1959); co-author, The Exptl. Plasma Facility C-1 with Screw Magnetic Fields (ibid, vol. 14, No. 2, 1963); Effect of Helical Magnetic Field on Plasma Ohmic Heating in S-1 Machine (ibid, vol. 20, No. 4, 1966).
Address: Kurchatov Institute of Atomic Energy, 46 Ulitsa Kurchatova, Post Box 3402, Moscow, U.S.S.R.

GAVRILOV, K. A. Paper: Co-author, Studies of the Extraction of Metal Ions using Radioactive Nuclides (Kernenergie, vol. 9, No. 9, 1966).
Address: Joint Institute for Nuclear Research, Dubna, U.S.S.R.

GAVRILOV, P. A. Papers: Co-author, On the Problem of Nucl. Power Plant Dynamics (Atomnaya Energiya, vol. 15, No. 2, 1963); co-author, Dynamic Characteristic Research of First Reactor on I.V. Kurchatov Beloyarsk Nucl. Power Station (ibid, vol. 19, No. 2, 1965).
Address: Academy of Sciences of the U.S.S.R., 14 Leninsky Prospekt, Moscow V-71, U.S.S.R.

GAW, James D. Pres., Rad. Counter Labs., Inc.
Address: Radiation Counter Laboratories, Inc., 5121 West Grove Street, Skokie, Illinois, U.S.A.

GAY, John Anderson, B.Sc. (Special Phys.). Born 1922. Educ.: London Univ. Commercial Dept., Ind. Group Headquarters, Risley (1956-58), Patents Licensing Officer, U.K.A.E.A. Society: Member, I.E.E.
Nuclear interests: Metallurgy, nuclear physics, nucleonic instruments, electronic circuits, measuring instruments, etc.
Address: United Kingdom Atomic Energy Authority, 11 Charles II Street, London, S.W.1, England.

GAZITH, Moshe, M.Sc., A.M., Ph.D. Born 1918. Educ.: Hebrew Univ., Jerusalem, and Columbia Univ., N.Y. Res. and Teaching Asst., Columbia Univ., N.Y., U.S.A., 1951-54; Postdoctoral Res. Assoc., Cornell and N.Y. State Univ., 1955-56; Weizmann Inst., Israel, 1950-51, 1958-60; Israel A.E.C., Labs., 1960-. Societies: Israel Chem. Soc.; A.C.S.; A.A.A.S.
Nuclear interests: Isotope separation, radiochemistry; Reaction kinetics, radiation chemistry; Nuclear information retrieval.
Address: Atomic Energy Commission, Labs., P.O. Box 527, Rehovoth, Israel.

GAZTELU Y JACOME, Jose Maria, Ind. Dr. Eng. Born 1907. Educ.: Escuela Tecnica Superior Ing. Ind., Barcelona. Tech. Director, Nat. Inst. Ind.; Vice-Pres., Forum Atomico Espanol; Member Com. Direction, Union Internat. Producteurs et Distributeurs Energie Elec. (UNIPEDE); Pres., Empresa Nacional Elec. Cordoba; Pres., Auxini Ingenieria Espanola, S.A.; Consejero de Potasas Navarra, S.A.; Managing Director, Gas y Elec., S.A.
Address: 1 Zurbano, Madrid 4, Spain.

GAZZABIN, Filippo, Chem. Eng., Bombrini Parodi-Delfino S.p.A.
Address: Bombrini Parodi-Delfino S.p.A., 31 Via Lombardia, Rome, Italy.

GEANKOPLIS, C. J., Prof. Dept. of Chem. Eng., Coll. of Eng., Ohio State Univ.
Address: College of Engineering, Ohio State University, Columbus 10, Ohio, U.S.A.

GEARY, Neville Rex, B.Sc. Born 1928. Educ.: Queen's Univ., Belfast. R. and D. Group (1950-58), Asst. Chief Chem., R. and D. Branch, Risley (1958-60), Sen. Rep. in Canada (1960-63), Sen. Fuel Technol., Production Group, Risley (1963-64), Special Asst. to the Member for Production (1964-), U.K.A.E.A.
Address: U.K.A.E.A., Production Group, Risley, England.

GEBALLE, Ronald, Prof. Chairman, Phys. Dept., Faculty, Washington Univ.
Nuclear interest: Atomic and molecular collisions.
Address: Washington University, Physics Department, Seattle, Washington 98105, U.S.A.

GEBAUER, Helwig, Dipl.-Phys. Born 1927. Educ.: Marburg/Lahn Univ. With Firma Siemens-Schuckert A.G., Bad Neustadt/Saale, 1954-55; Firma Frieseke and Hoepfner G.m:b.H., Erlangen-Bruck, 1955-. Vorsitzender des Technischen Ausschusses 18/2-Strahlungsmessgerate, im Zentralverband der Elektrotechnischen Industrie e.V., 1958-62; Obmann des Arbeitsausschusses FNE 414-Strahlungsmessgerate, im Deutschen Normenausschuss, 1959-.
Nuclear interests: Strahlungsmesstechnik, Zählrohre, Gasentladungsphysik, Vakuumtechnik.
Address: 7 Geisslerstr., 852 Erlangen-Bruck, Germany.

GEBAUER, Rudolf, Dr.rer.nat., o.ö. Prof. Born 1904. Educ.: Prague, Kiel and Darmstadt Univs. Darmstadt T.H., 1941-55; Graz T.H., 1955-; Dean of the Fakultät für Naturwissenschaften, 1957-58. Societies: Osterreichische Physikalische Gesellschaft; Verband Deutscher Physikalische Gesellschaften.
Address: 192 Körösistrasse, Graz, Austria.

GEBAUHR, Werner, Diplom-Chem., Dr.rer.nat., Privat-Dozent. Born 1920. Educ.: Hanover T.H. and Mainz Univ. Res. Lab., Siemens A.G., Erlangen. Society: Gesellschaft Deutscher Chemiker (Fachgruppe Kern- und Strahlenchemie).
Nuclear interests: Radiochemistry and application of chemistry in the field of nuclear power.
Address: Forschungslab. Erlangen der Siemens A.G., Germany.

GEBHARDT, Erich, Prof., Dr.-Ing.habil. Director, Inst. für Sondermetalle, Max-Planck-Inst. für Metallforschung. Member, Editorial Advisory Board, J. of Nucl. Materials. Formerly Mitglied, Technische Forschung, Ak II/5, Deutsche Atomkommission.
Address: Max-Planck-Institut für Metallforschung, 92 Seestrasse, 7000 Stuttgart 1, Germany.

GEBHARDT, Paul S., B.A., M.S. (Rad. Health Phys.). Born 1925. Educ.: Youngstown and Temple Univs. Instructor, Hahnemann Medical Coll. and Hospital.
Nuclear interests: Teaching bionucleonics; Management of Isotope Laboratories; Labelling of compounds.
Address: Hahnemann Medical College and Hospital, 235 North 15th Street, Philadelphia 2, Pennsylvania, U.S.A.

GEDEONOV, Lev I. Sen. Sci., Khlopin Radium Inst.
Nuclear interest: Health physics.
Address: Khlopin Radium Institute, 1 Roentgen Ulitsa, Leningrad, U.S.S.R.

GEFFRIER, G. de. See GEFFRIER, G.

GEGAUFF, Christiane WINTER-. See WINTER-GEGAUFF, Christiane.

GEHMAN, Samuel D., A.B., A.M., Ph.D. Born 1903. Educ.: Pennsylvania Univ. Manager, Phys. Res., Goodyear Tire and Rubber Co., 1956-. Societies: A.P.S.; A.C.S.
Nuclear interests: Research in effects of radiation on materials, especially polymers; Applications of tracers in the rubber industry.
Address: Goodyear Tire and Rubber Co., 142 Goodyear Boulevard, Akron, Ohio 44316, U.S.A.

GEHRCKENS, Heinrich Martin. Born 1890. Pres., Schutzverein Deutscher Rheder; Director: Verein Hamburger Rheder, Deutscher Rhederei-Verein; Baltic and Internat. Maritime Conference.
Address: 2 Beim Neuen Krahn, Hamburg 11, Germany.

GEHRHARDT, Heinz Willy, Dr.rer.pol. Born 1905. Educ.: Nürnberg and Frankfurt am Main Univs. Vorstand, Verband der Lebensversicherungsunternehmen e.V.; Member, Präsidialausschuss des Gesamtverbandes der Versicherungswirtschaft; Pres., Steuerausschuss des Gesamtverbandes der Versicherungswirtschaft; des Steuerausschusses des Verbandes der Lebensversicherungsunternehmen; Member, Versicherungsbeirat beim Bundesaufsichtsamt für das Versicherungs- und Bausparwesen, Berlin; Member, Vollversammlung der Industrie- und Handelskammer Frankfurt am Main; Vizepräsident der Industrie- und Handelskammer Frankfurt; Vorsitzender der Vorstandes der Alte Leipziger Lebensversicherungs gesellschaft auf Gegenseitigkeit und der Leipziger Feuer-Versicherungs-Anstalt.
Nuclear interests: Insurance; Economic; Fiscal.
Address: 42 Bockenheimer Landstrasse, Frankfurt am Main, Germany.

GEIER, Harald, Dr.rer.nat., Dipl. Phys. Born 1931. Educ.: Munich and Heidelberg Univs. Head, Nucl. Eng. Dept., Lurgi Gesellschaft für Wärme- und Chemotechnik m.b.H., Frankfurt am Main. Society: Deutsche Physikalische Gesellschaft (D.P.G.) (German Phys. Soc.).
Nuclear interests: Nuclear chemical processing; Criticality in nuclear chemical plant; Gas purification.
Address: Lurgi Gesellschaft für Wärme- und Chemotechnik m.b.H., Postfach 9181, 6 Frankfurt am Main, Germany.

GEIGER, Karl Martin, Dipl.-Meteorologist. Born 1920. Educ.: Prague and Leipzig Univs. Regierungsgewerberat (1962-), Oberregierungsgewerberat (1965-), Ministry of Labour and Social Affairs, Baden-Württemberg. Society: Deutsches Atomforum.
Nuclear interest: Problems of nuclear energy, but predominantly in safety problems of nuclear installations.
Address: Ministry of Labour and Social Affairs, Baden-Württemberg, 30 Rotebühlplatz, 7-Stuttgart-1, Germany.

GEIGER, Klaus Wilhelm, Dr.rer.nat. Born 1921. Educ.: Tübingen and Mainz Univs. With Appl. Phys. Div., X-rays and Nucl. Radiations Sec., N.R.C., Ottawa, 1957-. Society: Canadian Assoc. of Physicists.
Nuclear interests: Nuclear disintegration schemes and standardisation; Cosmic rays; Neutron spectroscopy.
Address: Division of Applied Physics, National Research Council, Ottawa, Ontario, Canada.

GEIGER, Lawton D. Manager, Pittsburgh (Pa.) Naval Reactors Operations Office, U.S.-A.E.C.
Address: Pittsburgh Naval Reactors Operations Office, U.S. Atomic Energy Commission, P.O. Box 109, West Mifflin, Pennsylvania 15122, U.S.A.

GEISENDORFER, Ingeborg, MdB. Born 1907. Verwaltungsrat, Deutsches Atom-Forum e.V.; Stellvertretende Vorsitzende (Vice-Pres.), Bundestagsausschuss für Wissenschaft, Kulturpolitik und Publizistik.
Nuclear interest: Gesetzgebung (legislation); Förderung von Forschung und technischer Entwicklung für friedliche Zwecke.
Address: Deutsches Atom-Forum e.V., 240 Koblenzer Strasse, 53 Bonn, Germany.

GEISER, Kenneth R. Chairman, Arizona A.E.C.
Address: Arizona Atomic Energy Commission, 40 East Thomas Road, Suite 107, Phoenix, Arizona 85012, U.S.A.

GEISER, P. Director, Bernische Kraftwerke A.G.
Address: Bernische Kraftwerke A.G., 2 Viktoriaplatz, 3000 Bern 25, Switzerland.

GEISLER, George Charles, B.S.E.E. Born 1927. Educ.: Bucknell, Pennsylvania State and Pittsburgh Univs. Asst. Director, Pennsylvania State Nucl. Reactor Facility; Supv., Plant Operations, Westinghouse Testing Reactor Control Eng., Honeywell, Inc. Society: Reactor Operations, A.N.S.
Nuclear interest: Application of operations research techniques to reactor operation and management.
Address: Nuclear Reactor Facility, University Park, Pennsylvania 16802, U.S.A.

GEISLER, Jan, Ph.D. Born 1925. Educ.: Warsaw Univ. Warsaw Univ., 1949-60; Inst. of Nucl. Res., Warsaw, 1960-.
Nuclear interest: Metabolism of bone-seeking radioisotopes.
Address: 10 m. 26 Ul. J. Bruna, Warsaw 12, Poland.

GEISLER, Martin, Dr.rer.nat., Dipl. Phys. Born 1938. Educ.: Karl-Marx-Univ., Leipzig. German Acad. of Sci., Inst. for Appl. Radioactivity, Leipzig, 1961-. Book: Co-author, Analyse von Mineral- und Syntheseölen mit radiometrischen Methoden (Berlin, Akademie-Verlag, 1968).
Nuclear interests: Gamma ray scintillation spectrometry; Instrumental activation analysis; X-ray fluorescence analysis with radioactive sources; Gamma ray backscattering for analytical purposes.
Address: Deutsche Akademie der Wissenschaften zu Berlin, Institut für angewandte Radioaktivität, 15 Permoserstrasse, 705 Leipzig, German Democratic Republic.

GEISS, Johannes, Ph.D., Prof. Phys. Born 1926. Educ.: Goettingen Univ. Books: Author and co-author, of about 60 publications. Societies: Schweizerische Phys. Gesellschaft; A.G.U.
Nuclear interests: Mass spectrometry; Natural radioactivity; Cosmic-ray produced isotopes in the atmosphere and meteorites.
Address: Physikalisches Institut der Universität Bern, 5 Sidlerstrasse, Bern, Switzerland.

GEITHOFF, Dietrich, Dipl.-Chem., Dr.rer.nat. Born 1929. Educ.: Hochschule Bamberg and Karlsruhe/Baden T.H. Society: Gesellschaft Deutscher Chemiker.
Nuclear interests: Fabrication technology and irradiation performance of fast reactor fuels, post-irradiation examination methods for ceramic matrices.
Address: Kernforschungszentrum Karlsruhe, Institut für Angewandte Reaktorphysik, Reaktorstation 7501 Leopoldshafen, Germany.

GEITTMAN, Frederick J., B.S. (Mech. Eng.). Born 1907. Educ.: Wisconsin Univ. Chief Eng., R. and D. Dept., Eddystone Div., Baldwin-Lima-Hamilton Corp. Society: A.N.S.
Nuclear interests: Reactor vessels; Shipping casks.
Address: 117 Shawnee Road, Ardmore, Pa., U.S.A.

GEKLER, Willard C., Pet. Ref. Eng. Born 1932. Educ.: Colorado School of Mines and Calif. Univ. Nucl. Eng., 1960-. Society: A.N.S.
Nuclear interests: Reactor protective systems reliability analysis; Reactor safety analysis; Reactor operations and operating experience; Nuclear facility siting.
Address: Holmes and Narver, Inc., 828 So. Figueroa Street, Los Angeles, California 90017, U.S.A.

GELBARD, Ely Meyer, M.S., Ph.D. Born 1924. Educ.: Chicago Univ. Phys., Westinghouse Elec. Corp. Advisory Sci., Bettis Atomic Power Lab., Westinghouse Elec. Corp., 1959-. Society: A.N.S.
Nuclear interest: Reactor physics computational methods.
Address: Westinghouse Electric Corporation, Bettis Plant, Bettis Site, P.O. Box 1468, Pittsburgh 30, Pennsylvania, U.S.A.

GELDERN, Eduard R. VON. See VON GELDERN, Eduard R.

GELIN, Olov Emil Vilhelm, Fil.dr. Born 1906. Educ.: Roy. Univ. Stockholm and Lund Univ. Director, Plant Breeding Inst., Weibullsholm. Member of Board, Swedish State Plant Protection Inst.; Member, Roy. Acad. of Agriculture and Forestry. Society: Eucarpia, group for theoretical and applied mutation research.
Nuclear interest: Applications of irradiation in plants.
Address: Plant Breeding Institute Weibullsholm, Landskrona, Sweden.

GELIN, Ragnar, Chem.Eng., M.Sc. Born 1925. Educ.: Roy. Inst. Technol., Stockholm. At (1952-), Head, Chem. Eng. Sect. (1957-65), Head, Fuel Elements Sect. (1966-), A.B. Atomenergi. Society: Swedish Assoc. Eng. and Architects.
Nuclear interests: Metallurgy; Nuclear fuel technology.
Address: A.B. Atomenergi, 32 Liljeholmsvägen, Stockholm 43, Sweden.

GELINSKY, P., Dr. Wissenschaftliche Asst., Dept. of Clinical-Exptl. Radiol., Univ. Strahlenklinik.
Address: Universitäts Strahlenklinik (Czerny Krankenhaus), 3 Voss Strasse, Heidelberg, Germany.

GELISSEN, Henri Caspar Joseph Hubert, Chem. Eng., Dr. Techn. Sci., Dr.Ing.h.c. (T.H. Aachen), Dr.h.c. (Leuven Univ.), Dr.h.c. (Gent Univ.), Dr.h.c. (Grenoble Univ.). Senator h.c. Cologne Univ. and Economics High School, Mannheim. Prof. Chem. Technol. Born 1895. Educ.: Tech. Univ., Delft. Hon. Pres., Assoc. of Directors of Netherlands Elec. Undertakings; Pres., Board Netherlands Railways; former Minister of Economic Affairs; Former Director, Prov. Limburg Elec. Co.; Former Minister plenipotentiary; Hon. Pres., Internat. Elec. Heat Assoc. Society: Chem. Soc.
Nuclear interest: Production nuclear energy.
Address: Breukelen, Netherlands.

GELL-MANN, Murray, B.S., Ph.D. E.O. Lawrence Memorial Award, 1966. Born 1929. Educ.: Yale Univ. and M.I.T. Member, Inst. for Advanced Study, Princeton, 1951-52; Instructor (1952-53), Asst. Prof. (1953-54), Assoc. Prof. (1954-55), Chicago Univ.; Assoc. Prof. Phys. (1955-56), Prof. (1956-), California Inst. Technol.; Visiting Prof., Coll. de France and Paris Univ., 1959-60. Societies: N.A.S.; American Acad. Sci.; Fellow, A.P.S.
Address: 3637 Canyon Crest Road, Altadena, California, U.S.A.

GELLER, Kenneth N., Ph.D., B.S. Born 1930. Educ.: Pennsylvania Univ. Asst. Prof., Dept. of Phys., Pennsylvania Univ., 1960-66; Assoc. Prof., Dept. of Phys., Drexel Inst. of Technol., 1966-. Society: A.P.S.
Nuclear interests: Low energy nuclear physics; Nuclear structure and reaction mechanisms.
Address: Drexel Institute of Technology, Department of Physics, Philadelphia, Pennsylvania 19104, U.S.A.

GELLER, Richard. Dr. Sci. Born 1927. Educ.: Conservatoire Nat. des Arts et Metiers and Paris Univ. C.E.A., Saclay, France, 1950-62, and 1963-64; Stanford Univ., Stanford, California, 1962-63.
Nuclear interests: Plasma physics research; Ion sources; Vacuum physics.
Address: 11 bis avenue de Sceaux, Antony, (Seine), France.

GELPERIN, B. B. Papers: A Description of Some of the Betatrons Produced by the Moscow Transformer Works (Atomnaya Energiya, vol. 7, No. 6, 1959); Influence of Insulating Current between Magnet Sheets on Work of Betatron and Synchrotron (ibid, vol. 22, No. 6, 1967).
Address: Academy of Sciences of the U.S.S.R., 14 Leninsky Prospekt, Moscow V-71, U.S.S.R.

GELSDORF, Karl, Dipl. Ing. Prokurist, Elektrizitätswerk Minden-Ravensberg G.m.b.H.
Address: Elektrizitätswerk Minden-Ravensberg G.m.b.H., 3b Bielefelder Strasse, Herford/-Westf., Germany.

GEMAHLING, Jean, L. ès Sc., Ing. Chimiste. Born 1912. Educ.: Strasbourg Univ. Adjoint au Directeur des Matériaux et Combustibles Nucléaires, C.E.A.
Address: Direction des Matériaux et Combustibles Nucléaires, B.P. No. 6, Fontenay-aux-Roses, Seine, France.

GEMMELL, Henry Maurice, B.Sc. (Hons.). Born 1906. Educ.: Durham Univ. Chief Tech. Eng., Yarrow-Admiralty Res. Dept., Yarrow and Co., Ltd., Glasgow, 1948-. Societies: M.I.Mech.E.; M.I.Mar.E.; Inst. Eng. and Shipbuilders in Scotland; North East Coast Inst.
Nuclear interests: Application of nuclear power in naval and mercantile machinery installations.
Address: 1 West Chapelton Crescent, Bearsden, Glasgow, Scotland.

GEMMELL, W., B.Sc., A.Inst.P. Head, Exptl. Phys. Sect., Phys. Div., A.A.E.C.
Address: Research Establishment, Australian Atomic Energy Commission, Private Mail Bag, Sutherland, New South Wales, Australia.

GENACHTE, Paul Frank, B.S. (Brussels), M.S. (M.I.T.), Ph.D. (California Inst. Technol.). Born 1909. Special Asst. to Pres. and Treas., Mexican Light and Power Co., Mexico City, 1948-52; Tech. Manager, Indussa Corp., New York, 1952-54; Vice-Pres., X-ray Equipment, Balteau Elec. Corp., Stamford, Conn., 1954-; Power Consultant to Ceylon, World Bank, Washington, D.C., 1954-; Power and Chem. Consultant (1954-), Director of Atomic Energy Div. (1954-63), Chase Manhattan Bank, New York. Chairman, Nucl. Industry Com., Investment Bankers Assoc.; Director, Homestake Mining Co.; Director, Rad. Dynamics, Inc.; Director, Nucl. Science and Eng. Co. Publications: Progress in Industrial Atomic Energy (1956, N.Y. Federationist); Moving Ahead with the Atom (Jan., 1957); Uranium and Atomic Energy Outlook (June, 1959, Business Atomics Report). Societies: M.I.T. Club of New York; American Inst. of Elec. Eng.; Assoc. of Appl. Solar Energy; Atomic Ind. Forum; A.N.S.
Address: 88 Ryder Road, Manhasset, New York, U.S.A.

GENCAY, Sarman, Asst. Prof. Mech. Eng. Faculty, Dept. of Nucl. Energy and Heat Transfer, Istanbul Tech. Univ.
Address: Istanbul Technical University, Gümüssuyu, Istanbul, Turkey.

GENDEREN, W. VAN. See VAN GENDEREN, W.

GENESIO, Corrado, Dott. Prof. (Industrial Eng.). Born 1912. Educ.: Genoa Univ. Tech. Director, Soc. OEG, Genoa, 1955; Vice Gen. Director (1959), Gen. Director (1962-64), Soc. Dinamo, Novara; Managing Director, C.I.S.E. (Centro Informazioni Studi Esperienze), Segrate (Milan), 1964; Vice Director (1965), Director (1966-67), Compartment of Firenze, E.N.E.L. (Ente Nazionale Energia Elettrica); Prof. incaricato, Official Course of Elec. Plants, Eng. Faculty, Genoa Univ., 1966. Society: A.E.I. (Italian Elec. Assoc.).
Nuclear interest: Management.
Address: E.N.E.L., 40 Via Bertola, 20122 Turin, Italy.

GENESLAY, Raymond. Director, Compagnie Générale de Geophysique.
Address: Compagnie Générale de Geophysique, S.E.R.C.E.L., 96 avenue Verdier, Montrouge, Seine, France.

GENIERE, Renaud de la. See de la GENIERE, Renaud.

GENTLES, John. Sen. Nucl. Eng., Bechtel Nucl. Corp.
Address: Bechtel Nuclear Corporation, 220 Bush Street, San Francisco 4, California, U.S.A.

GENTNER, Wolfgang, Dr. phil. nat. Born 1906. Educ.: Frankfurt-am-Main Univ. o. Prof. and Director, Inst. Phys., Freiburg Univ., 1946; Director, S.C. Div., C.E.R.N., Geneva, 1954; Director, Max Planck Inst. für

Kernphysik, Heidelberg, 1958; Chairman, Sci. Policy Com., C.E.R.N. Book: Atlas typischer Nebelkammerbilder (Berlin, Springer 1940; London, Pergamon Press). Societies: Leopoldina Akad. der Naturforschung; Heidelberger Akad. der Wissenschaften; Bayrische Akad. der Wissenschaften.
Nuclear interest: Nuclear physics.
Address: P.O.Box 1248, 6900 Heidelberg 1, Germany.

GENTON, André. Ing. en Chef, Chantier, Sté. de Construction d'Une Usine de Séparation Isotopique.
Address: Société de Construction d'Une Usine de Séparation Isotopique, 104 Avenue Edouard Herriot, Le Plessis-Robinson (Seine), France.

GENTRY, W. A. Director, Atomic Power Development Associates, Inc.
Address: Atomic Power Development Associates Inc., 1911 First Street, Detroit 26, Michigan, U.S.A.

GEOFFRION, Claude, B.Sc.A., M.Sc., D.Sc. Born 1918. Educ.: Laval Univ., Quebec City. Prof., Laval Univ., 1956; Chairman, Phys. Dept., 1960-61; Dean, Faculty of Sci., 1961-. Director, A.E.C.L. Societies: Canadian Assoc. of Physicists; A.C.F.A.S.
Nuclear interests: Beta and gamma spectroscopy; Design of magnetic spectrometers.
Address: Laval University, Quebec City 10, Canada.

GEOCHEGAN, Gerald Robert Holme, B.A., M.A. Born 1914. Educ.: Cambridge Univ. Chief Tech. Manager, Capenhurst Works, Production Group, U.K.A.E.A., 1958-. Previously in Tech. Sect., Ind. Group Headquarters, U.K.A.E.A. Society: British Computer Soc.
Nuclear interests: Theory and operation of isotope separation plants, particularly the gaseous diffusion and centrifuge processes; Planning and administration of research; General interest in technical and economic problems and progress of nuclear technology.
Address: Hunter's Lodge, Townfield Lane, Great Mollington, Chester, England.

GEORGE, Douglas E., B.C.S. Educ.: Benjamin Franklin Univ. Born 1913. Asst. Chief, Source and Special Nucl. Materials Accountability Branch (1954), Deputy Director, Div. of Source and Special Nucl. Materials Accountability (1954-59), Acting Director, Div. Nucl. Materials Management (1959-61), Director, Div. Nucl. Materials Management (1961-), U.S.A.E.C.
Address: 975 No. Rochester Street, Arlington, Virginia 22205, U.S.A.

GEORGE, Eric Paul, B.Sc., Ph.D., D.Sc. Born 1914. Educ.: London Univ. Reader in Phys., Sydney Univ., 1953-56; Director of Phys., St. Vincent's Hospital, Sydney, 1956-64; Prof. Phys., N.S.W. Univ., 1964-. Societies: F.Inst.P.; A.P.S.; Roy. Inst. of Great Britain; Soc. for Rad. Biol., Sydney.
Nuclear interests: Isotopic methods of investigation in medical and biological research.
Address: School of Physics, New South Wales University, Sydney, Australia.

GEORGE, Kenneth Dudley, M.Sc. (1st Class Hons.). Born 1916. Educ.: Auckland Univ. Phys., Atomic Energy Project, Chalk River, Canada, 1949-50; Phys., Picatinny Arsenal, Dover, New Jersey, 1951-64; Reactor Supervisor, Union Carbide Corp. Res. Centre, Tuxedo, New York, 1964-. Societies: A.P.S.; A.N.S.
Nuclear interests: Research reactor design, operation and utilisation.
Address: Kingsland Road, R.D. 3, Boonton, New Jersey, U.S.A.

GEORGI, M., Dr. Wissenschaftliche Asst., Dept. of Clinical-Exptl. Radiol., Univ. Strahlenklinik. (Czerny Krankenhaus).
Address: Universitats Strahlenklinik (Czerny Krankenhaus), 3 Voss Strasse, Heidelberg, Germany.

GEORGIEVSKII, A. V. Papers: Co-author, Longitudinal Magnetic Field of the Sirius Stellarator Reecetrack (Atomnaya Energiya, vol. 23, No. 2, 1967); co-author, Shielding of Thermonucl. Device Magnetic Systems from Accidental Overloadings (ibid, vol. 23, No. 2, 1967); co-author, Divertor of the Sirius Stellarator (ibid, vol. 23, No. 2, 1967); co-author, On Possibility of Plasma Injection into Closed Magnetic Trap through Divertor (ibid, vol. 24, No. 1, 1968).
Address: Academy of Sciences of the U.S.S.R., 14 Leninsky Prospekt, Moscow V-71, U.S.S.R.

GEOURS, Jean SAINT-. See SAINT-GEOURS, Jean.

GERARD, Albert J. A., Ing. Civil des Mines (A.I.Lg.), Ing. Civil Mecanicien, Degree in Business Administration (Ecole des Travaux Publics, Paris). Born 1901. Educ.: Liège Univ. Directeur Gen. Adjoint, S.A. des Nouveaux Ateliers Lebrun. Administrateur, Assoc. des Ing. de Liège; Vice-Pres., Sté. Belge des Mecaniciens, 1964-66; Pres., Res. Com., Sté. Belge des Mecaniciens; Member, Commission III, Internat. Inst. of Refrigeration.
Nuclear interests: High pressures. Low temperatures. Heat exchangers; Beam-ports (BR2); Air conditioning.
Address: Place, Casteau, Belgium.

GERARD, Charles J., Ing. Civil mécanicien et électricien. Born 1925. Educ.: Brussels Free Univ. Div. Electronique (1949-52), Lab. de Recherches Phys. (1952-57), Ing. en Chef, Div. Nucl. (1957-), A.C.E.C. Society: Belgicatom.
Address: c/o Ateliers de Constructions Electriques de Charleroi, B.P. 254, Charleroi, Belgium.

GERARD, Francis. Tech. Editor, Industries Atomiques.
Address: Industries Atomiques, Editions Sadesi, 39 rue Peillonnex, Geneva Chêne-Bourg, Switzerland.

GERARD, George, Dr. Formerly Director Eng. Sci., Aracon Labs., A Div. of Allied Res. Assocs., now Vice Pres., Allied Res. Assoc. Inc.
Address: Allied Research Associates, Inc., Virginia Road, Concord, Massachusetts, U.S.A.

GERARD, Jean. Tech. Manager, Ets. Geosyl S.A. Deleg., I.A.E.A. Conference on the Use of Radioisotopes in the Phys. Sci. and Industry, Copenhagen, 6-17 September 1960.
Address: Etablissements Geosyl S.A., 38 rue Auguste-Poullain, Saint-Denis, (Seine), France.

GERARD-KUMLEBEN, Francis, B.Sc. Born 1901. Educ.: Göttingen and Paris Univs. Directeur de la Rédaction technique, Revue Industries Atomiques and Sci. Director, Sadesi, Geneva; Pres., Exec. Bureau, World Assoc. of World Federalists; Asst. Editor in Chief, Space Res. and Eng. Nuclear interests: Peaceful applications of nuclear energy.
Address: 4 rue Banès, 92 Meudon, France.

GERASIMOV, V. V. Papers: Co-author, The Effects of Rad. on the Electro-chem. Behaviour of IXI8H9T Steel (letter to the Editor, Atomnaya Energiya, vol. 10, No. 2, 1961); co-author, Corrosion Resistance of Structural Materials in Bor Contained Solutions (letter to the Editor, ibid, vol. 19, No. 6, 1965); co-author, Effect of Gamma-Rad. on Deposition (letter to the Editor, ibid, vol. 20, No. 5, 1966).
Address: Academy of Sciences of the U.S.S.R., 14 Leninsky Prospekt, Moscow V-71, U.S.S.R.

GERBAULET, Karl, Dr. Asst., Inst. für Med. Isotopenforschung, Cologne Univ.
Address: Institut für med. Isotopenforschung, Cologne University, 15 Kerpener Str., Cologne-Lindenthal, Germany.

GERDENITSCH, Heinrich, Dr., Dipl. chem. Born 1916. Educ.: Dresden, Leipzig and Darmstadt Univs. Marketing Director, Balzers A.G. Society: Gesellschaft Deutscher Chemiker. Nuclear interest: High vacuum technique.
Address: 499 Sonnenhalde, Vaduz, Liechtenstein.

GERDES, Robert H. Exec. Vice-Pres. (-1963), Chairman of the Board (1963-), Pacific Gas and Elec. Co.
Address: Pacific Gas and Electric Co., 245 Market Street, San Francisco, California, U.S.A.

GERHART, James Basil, B.S., M.A., Ph.D. Born 1928. Educ.: California Inst. Technol. and Princeton Univ. Instructor, Princeton Univ., 1954-56; Asst. Prof., (1956-), Assoc. Prof. (1961-65), Prof. (1965-), Washington Univ., Seattle. Societies: A.P.S.; A.A.A.S.; A.A.P.T. Nuclear interests: Nuclear structure; Elementary particle interactions; Beta- and Gamma-ray spectroscopy; Nuclear scattering and angular correlations.
Address: Department of Physics, Washington University, Seattle, Washington 98105, U.S.A.

GERHART, James M. Irradiation Eng. (1959-64); Manager, Radioactive Materials Lab. Operations (1964-), Vallecitos Atomic Lab., Gen. Elec. Co.
Address: General Electric Co., Vallecitos Atomic Laboratory, P.O. Box 846, Pleasanton, California, U.S.A.

GERHOLM, Tor Ragnar, Ph.D. Born 1925. Educ.: Stockholm and Uppsala Univs. Asst. Prof., Phys. Dept., Uppsala Univ., 1956-61; Assoc. Prof., Phys. Dept., Stockholm Univ., 1961-. Contribution to Gamma and Beta Ray Spectroscopy (Editor, K. Siegbahn) (North Holland Publishing Co., 1955); Contribution to Encyclopaedia of Physics. Volume 33. (Editor, S. Flügge) (Berlin, Springer Verlag, 1956); Contribution to Methods of Experimental Physics. Volume V. (Editors, L. Yuan, C.S. Wu and L. Marton) (New York, Academic Press Inc., 1961); Contribution to Alpha, Beta- and Gamma-Ray Spectroscopy (Editor, K. Siegbahn) (North Holland Publishing Co., 1966). Society: A.P.S.
Nuclear interest: Nuclear spectroscopy.
Address: Physics Department, Stockholm University, 9 Vanadisvagen, Stockholm, Sweden.

GERICKE, William E., Dipl.-Phys., Dr. rer. nat. Born 1931. Educ.: Kiel Univs. Strahlenschutzreferent im Ministerium für Arbeit des Landes Schleswig-Holstein, 1959-62; Dozent, Staatlichen Ingenieur-Schule Kiel, 1962-. Nuclear interests: Reaktortechnik; Plasmaphysik.
Address: 35 Legienstr., Kiel, Germany.

GERIN-ROZE. Chef, Service des Affaires Budgetaires et du Contrôle, C.E.A.
Address: Commissariat à l'Energie Atomique, 29-33 rue de la Federation, Paris 15, France.

GERLACH, Albert August, B.Sc., M.Sc. (Elec. Eng.), M.Sc. (Math.), Ph.D. Born 1920. Educ.: Ohio State Univ. and Illinois Inst. Technol. Res. Sci., Armour Res. Foundation, 1948-53; Tech. Director, Instrumentation (1953-58), Manager, Res. Sect. (1958-61), Asst. Director of Res. (1961-), Cook Technol. Centre. Past Chairman, Chicago Sect., I.R.E.; Past Chairman, Chicago Sect., Prof. Group on Circuit Theory; Board of Counselors and Planners, Chicago Sect., I.R.E. Societies: Sen. Member, I.R.E.; A.I.E.E.; Acous. Soc. America; A.A.A.S.
Nuclear interests: Sensors, instrumentation, and control.
Address: 1211 East Euclid Street, Arlington Heights, Illinois, U.S.A.

GERLACH, Eckehart, M.D. Born 1927. Educ.: Göttingen and Heidelberg Univs. Prof. Physiology, Chairman, Dept. of Physiology, Med. Faculty, Aachen Tech. Univ.; Societies: Deutsche Physiologische Gesellschaft; Gesellschaft für Biologische Chem.; Gesellschaft für Nephrologie; New York Acad. of Sci. Nuclear interest: Application of stable isotopes and radioisotopes in the study of cell metabolism.
Address: Department of Physiology, Aachen Technical University, 7 Herm.-Herder-Str., 78 Freiburg i.Br., Germany.

GERMAI, G., Licencié en Sci. Chimiques. Lab. Application Radioéléments, Liège Univ.
Address: Liège University, Laboratoire Application Radioéléments, 9 Place de XX Aout, Liège, Belgium.

GERMAIN, Francis. Educ.: Paris Univ. Ministère de l'Industrie et du Commerce, 1952; Ville de Paris, 1958; Grands Travaux d'Electrification et de Canalisations, 1962; Exec. Vice Pres., Sté. d'Etudes et d'Equipements d'Entreprises. Society: A.T.E.N. Nuclear interest: Reactor design.
Address: Société d'Etudes et d'Equipements d'Entreprises, 25 rue de Courcelles, Paris 8, France.

GERMAIN, Paul. Prof., Faculté des Sci., Paris; Directeur, l'Office Nat. d'Etudes et de Recherches Aerospatiales. Formerly Member, Conseil Scientifique, C.E.A.
Address: O.N.E.R.A., 25 avenue de la Division-Leclere, Chatillon-sous-Bagneux, Seine, France.

GERMAIN, Pierre Sylvain, Dr. in Maths. Born 1922. Educ.: Brussels Univ. Chargé de cours on Particle Accelerators, Brussels Univ.; Director, Proton Synchrotron Dept., European Organisation for Nucl. Res. (C.E.R.N.). Nuclear interest: Particle accelerators.
Address: C.E.R.N., 1211 Geneva 23, Switzerland.

GERMAR, Herminio J., M.D. Educ.: Univ. of the Philippines. Instructor, Dept. of Medicine, Coll. of Medicine (1958-63); Assoc. Lecturer, Coll. of Dentistry (1961-); Asst. Prof., Dept. of Medicine, Coll. of Medicine (1964-), Univ. of the Philippines. Society: Radioisotope Soc. of the Philippines, 1959-. Nuclear interests: Medical applications of radioisotopes in diagnosis and therapy particularly in thyroid gland disorders.
Address: 1448 Taft Avenue, Manila, Philippines.

GERNERT, William E., B.A., B.S., M.B.A. Born 1917. Educ.: Bowling Green State and Ohio State Univs. and U.S. M.A. Brigadier Gen., U.S. A.F., 1967; Deputy Commander, Field Command, Defence Atomic Support Agency, Sandia Base, Albuquerque. Society: A.N.S. Nuclear interests: Nuclear reactors; Nuclear research; Nuclear physics; Nuclear safety; Research and development.
Address: 2206 22d Drive, Sandia Base, Albuquerque, New Mexico 87116, U.S.A.

GERSHTEIN, S. S. Paper: Co-author, Calculating Spectra of Neutrino Produced by 70 GeV Protons (letter to the Editor, Atomnaya Energiya, vol. 23, No. 1, 1967).
Address: Academy of Sciences of the U.S.S.R., 14 Leninsky Prospekt, Moscow V-71, U.S.S.R.

GERSHTEIN, Semyon, Academician, Nucl. Problems Lab., Joint Inst. for Nucl. Res., Dubna.
Address: Joint Institute for Nuclear Research, Dubna, near Moscow, U.S.S.R.

GERSTEIN, B. Nucl. Rad. Lab., Admiral Corp.
Address: Admiral Corp., 3800 W. Cortland Street, Chicago 47, Illinois, U.S.A.

GERSTEIN, Ira S. Asst. Prof., Elementary Particle Theory, Dept. of Phys., Pennsylvania Univ.
Address: Pennsylvania University, Department of Physics, Philadelphia, Pennsylvania 19104, U.S.A.

GERSTEN, Wolfgang, Dr. rer. nat., Dipl. Phys. Born 1925. Educ.: Hanover T.H., N.C. State Coll., Raleigh and Internat. School of Nucl. Sci. and Eng., Argonne Nat. Lab. U.S.A. With Preussische Elektrizitäts-A.G., Hanover, 1956-; Manager, Studiengesellschaft für Kernkraftwerke G.m.b.H. Nuclear interests: Design, operation and economics of nuclear power plants.
Address: Studiengesellschaft für Kernkraftwerke G.m.b.H., 10-12 Papenstieg, Hanover, Germany.

GERSTL, Siegfried Adolf Wilhelm, Dr. rer. nat., Dipl. Phys. Born 1939. Educ.: Stuttgart Univ., 1958-65; Karlsruhe Univ., 1965-67. Society: Fachausschuss Strahlenschutz in der Kernenergie-Studiengesellschaft. Nuclear interests: Shielding against several radiations; Gamma ray and neutron transport theory; Reactor physics.
Address: 6 Bahnhofstrasse, 7303 Neuhausen, Germany.

GERTSMAN, S. L. Chief, Phys. Metal. Div., Mines Branch, Dept. of Energy, Mines and Resources.
Address: Department of Energy, Mines and Resources, 555 Booth Street, Ottawa, Ontario, Canada.

GERTY, Francis J., Dr. Member, Illinois Legislative Commission on Atomic Energy.
Address: Illinois Legislative Commission on Atomic Energy, c/o George E. Drach, Myers Building, Springfield, Illinois, U.S.A.

GESKE, Gerhard Walter Georg, Dr. rer. nat., Dipl.-Phys. Born 1929. Educ.: Friedrich Schiller Univ., Jena. Wissenschaftlicher Asst. (1960-63), Wissenschaftlicher Ober asst. (1964-), Technisch-Physikalisches Inst., Friedrich Schiller Univ., Jena.
Nuclear interest: Particle accelerators, especially betatrons.
Address: 20 Ebertstrasse, Jena, German Democratic Republic.

GESSINESS, Bernard, B.A. (Chem.), M.S. (Chem.). Born 1924. Educ.: Cincinnati Univ. Instructor, Analytical Chem., Cincinnati Univ., 1951-52; Positions include Chief, Analytical Development, Sect. Leader, Ores and Special Samples Labs., Supv., Nucl. Materials Control Dept., and Nucl. Materials Rep., Nat. Lead Co. of Ohio, 1952-. Chairman, I.N.M.M., 1969.
Nuclear interest: All phases of nuclear materials management with emphasis on technical measurements, administration and the analytical chemistry of uranium and its compounds.
Address: National Lead Company of Ohio, P.O. Box 39158, Cincinnati, Ohio 45239, U.S.A.

GETLER, Morris W., B. Mech. Eng. Born 1918. Educ.: New York Univ. and Rensselaer Polytech. Inst. Asst. Chief Eng., S. and S. Machinery Corp., 1946-50; Pres., Alloy Flange and Fitting Corp. (reactor metallurgy and piping), 1950-. Societies: A.S.M.; A.S.T.M.; A.S.M.E.
Nuclear interests: Metallurgy and design.
Address: Bryant Avenue, Roslyn Harbor, L.I., New York, U.S.A.

GETOFF, Nikola, Dr. tech., Dipl. Ing. Born 1922. Educ.: Vienna Tech. Univ. Res. Assoc., Inst. of Unit Operation and Technol. of Fuels, Tech. Univ., 1952-56; Dept. of Inorganic Chem. Res. (1956-58), Head, Radioisotope Lab. (1958-59), Austrian Nitrogen Works, Ltd., Linz-Donau; Res. Assoc., Inst. of Radium Res. and Nucl. Phys., Austrian Acad. of Sci., Vienna, 1959-; Res. Assoc., Rad. Chem. Dept., Newcastle-upon-Tyne Univ., 1959-60; Res. Assoc., Cambridge Univ., 1960; Consultant, Austrian Reactor Centre, Seibersdorf, 1962-. Books: Kurzes radiochemisches Praktikum (Vienna Deuticke, 1961); Contributions in Isotope in der Landwirtschaft (Hamberg-Berlin, P. Parey, 1960); Strahlenchemie (Heidelberg, Dr. A. Hüthig, 1967). Societies: Austrian Chem. Soc.; Faraday Soc.
Nuclear interests: Radiation chemistry; Hot atom chemistry; Radiochemistry.
Address: Institut für Radiumforschung und Kernphysik, 3 Boltzmanngasse, A-1090 Vienna 9, Austria.

GETTINGS, J., C. Eng., M.I.Struct. E., M. Soc. E., M.R.S.H., F.R.S.A. Partner, John Liversedge and Associates, Consulting Engineers.
Address: 42 Portland Place, London W.1, England.

GETTNER, Marvin W., B.S., Ph.D. Born 1934. Educ.: Rochester and Pennsylvania Univs. Res. Assoc., Pennsylvania Univ., 1960-61; Asst. Prof. (1961-65), Assoc. Prof. (1965-), Northeastern Univ. Society: A.I.P.
Nuclear interest: Experimental elementary particle physics.
Address: Northeastern University, Physics Department, Boston, Massachusetts, U.S.A.

GEUGE, Horst, Dr. Sen. Officer, Reactor Station Garching, Munich T.H.
Address: Technische Hochschule Munich, 21 Arcisstrasse, Munich 2, Germany.

GEVANTMAN, Lewis Herman, B.E., Ph.D. Born 1921. Educ.: Johns Hopkins and Notre Dame Univs. Group Leader, Radioisotope Contamination-Decontamination Phenomena (1951-55); Acting Head, Appl. Res. Branch (1955-57), Head, Rad. Chem. Branch (1957-61), Head, Chem. Technol. Div. (1961-64), U.S. Naval Radiol. Defence Lab.; Sen. Sci. Advisor, U.S. Mission to the I.A.E.A., Vienna, Austria, 1964-67. Societies: A.C.S.; Rad. Res. Soc.; N.Y. Acad. of Sci.; A.A.A.S.
Nuclear interests: Radiochemistry, applied use of radioisotopes, effects of radiation on materials, radiation chemistry, radiation dosimetry and scientific management. Nuclear materials safeguards.
Address: Programme Manager, Office of Standard Reference Data, National Bureau of Standards, Washington, D.C. 20234, U.S.A.

GEVERS, R. Chef du Dépt. Phys. de l'Etat Solide, Labs., C.E.N.
Address: C.E.N., Mol-Donk, Belgium.

GEY, Albert, Ing. E.S.E., Ing. en Génie Atomique. Born 1920. C.E.A., France. Societies: Assoc. des Ingenieurs en Génie Atomique; Sté. Française de Radioprotection.
Nuclear interest: Nuclear physics and health physics.
Address: 138 rue Houdan, Sceaux, Hauts de Seine, France.

GEYER, Gerhard, Generaldirektor; Member, Studiengesellschaft zur Förderung der Kernenergiewertung in Schiffbau und Schiffahrt E.V. Mitglied, Präsidium, Mitglied, Fk V, Deutsche Atomkommission; Vorsitzer, Aufsichtsrat, Gesellschaft für Kernenergieverwertung in Schiffbau und Schiffahrt m.b.H.
Address: Esso A.G., Hamburg, Germany.

GEYER, John Charles, B.S.C.E., M.S.E., Dr. Eng. Born 1906. Educ.: Michigan, Harvard and Johns Hopkins Univs. Prof. of Environmental Eng. Sci., The Johns Hopkins Univ.; Consultant, Philadelphia Elec. Co., 1959-; Consultant, Baltimore Gas and Elec. Co., 1960-; Consultant, Potomac Elec. Co., 1960-; Consultant, Hanford Operations Office, (1966-),

Member, Advisory Com. on Reactor Safeguards (1961-64), Member, Atomic Safety and Licensing Examiners Panel (1966-), U.S.A.E.C.; Member, Maryland Advisory Com. on Atomic Energy, 1966-.
Nuclear interests: Environmental effects, safety, monitoring, waste handling, transportation and disposal.
Address: 513 Ames Hall, The Johns Hopkins University, Baltimore, Maryland 21218, U.S.A.

GHALIB, Selchouk Ahmed, B.Sc. (Eng.). Born 1913. Educ.: London Univ. Responsible for design of control mechanisms in Calder reactors, 1950; Group Leader (1954), Chief Eng. (1956), Gen. Manager (1959), A.E.I.-John Thompson Nucl. Energy Co. Ltd. Deputy Gen. Manager, (1960), Managing Director, (1966), The Nucl. Power Group Ltd.; Vice Pres., Inter Nucl. S.A. Societies: Inst. Elec. Eng.; Inst. Mech. Eng.; Soc. Inst. Technol.
Nuclear interests: Development of reactor systems; Technical executive.
Address: 238 Washway Road, Sale, Cheshire, England.

GHANEM, Nadim A., B.Sc., M.Sc. (Alexandria), Ph.D. (Birmingham). Born 1927. Educ.: Alexandria and Birmingham Univs. Demonstrator in Chem., Alexandria Univ., 1948-53; Univ. Res. Fellow, Birmingham Univ., 1955-56; Lecturer in Chem., Alexandria Univ., 1956-58; Res. Chem., Paints and Pigments Inst., Stuttgart, 1958-59; Res. Fellow, I.A.E.A., Nucl. Chem. Div., Roy. Inst. Technol., Stockholm, 1959-60; Res. Chem., Glasurit-Werke, A.G. in Hiltrup/Westf., Germany, 1961-62; Asst. Res. Prof., (1962-66), Res. Prof. (1966-), Nat. Res. Centre, Cairo. Society: Soc. of Chem. Industry.
Nuclear interests: Tracer techniques for study of high polymer reaction mechanisms; Industrial applications of radioisotopes; Preparation by accelerated techniques of labelled organic compounds for use as tracers in the medical, agricultural, biological and chemical fields, and the methods used for their purification; Effect of high energy radiation on high polymers, grafting.
Address: Department of Polymers and Paints, The National Research Centre, Dokki, Cairo, United Arab Republic.

GHANI, Abdul, M.Sc., Ph.D. Born 1925. Educ.: Panjab and London Univs. Res. Asst., Univ. Coll., London, 1959-60; S.S.O., Pakistan A.E.C., 1960; Res. Fellow, C.E.R.N., Geneva, 1961-63; Director, Nucl. Phys. Div. (1965-66), Director (1966-67), Pakistan Inst. of Nucl. Sci. and Technol., Islamabad.
Nuclear interest: Experimental nuclear physics.
Address: Pakistan Institute of Nuclear Science and Technology, Nilore, Rawalpindi, Pakistan.

GHERSINI, Federico, D. Econ. Born 1909. Educ.: Rome Univ. Deputy Manager, Unione Italiana di Riassicurazione; Gen. Sec., Italian Insurance Pool for Atomic Risks. Society: (F.I.E.N.).
Nuclear interest: Insurance of nuclear risks.
Address: 2 Via Ettore Petrolini, 00197 Rome, Italy.

GHERSINI, Giovanni, Dr. of Chem. Born 1937. Educ.: Rome Univ. Head, Chem.Lab., C.I.S.E., Milan, 1962-. Society: F.I.E.N.
Nuclear interests: Analytical chemistry of nuclear materials; Separation procedures by liquid-liquid extraction and chromatography; Fuel chemistry; Boiler water technology.
Address: 2 Via Cicognara, Milan, Italy.

GHIGLIA, Fernando, Dr. in Economy and Trade. Born 1905. Director Gen., Ministry of Merchant Marine, Rome.
Nuclear interests: Shipping, sea transport, nuclear coastal plants.
Address: Ministero della Marina Mercantile, Rome, Italy.

GHIGO, Giorgio. Physicist. Born 1929. Res. Fellow, Sezione di Torino, I.N.F.N., 1953; Member of Staff, Lab. di Frascati, Frascati, Rome, 1954-.
Nuclear interests: Management of accelerating machines; Researches on the storage rings.
Address: 25 Via Carlo Lucidi, 00044 Frascati, Rome, Italy.

GHISLANDI, Enrico, Medical Dr. Born 1928. Educ.: Milan Univ. Asst., Clinica del Lavoro (Occupational health), Milan Univ. 1956-62; C.I.S.E., Segrate-Milan, 1962-67.
Nuclear interests: Medical and haematological supervision in radiation work.
Address: 3 Via G. Ferraris 20025, Legnano, Milan, Italy.

GHOSH, Dilip Kumar, B.Sc., M.Sc., D.Phil. Born 1928. Educ.: Calcutta and Birmingham (England) Univs. Res. Scholar, Govt. of India, 1952-57; Post-doctoral Res. Fellow, Birmingham Univ., England, 1957-59; Sci. Pool Officer, Govt. of India, 1960-62; Lecturer (1962-63), Reader (1963-66), Assoc. Prof. (1966-), Saha Inst. Nucl. Phys. Society: Assoc. Member, Inst. Phys. and the Phys. Soc., London.
Nuclear interests: Nuclear information by electron spin resonance and microwave spectroscopy technique.
Address: 49 Badridas Temple Street, Calcutta 4, India.

GHOSH ROY, S. K., Dr. Lecturer, Saha Inst. of Nucl. Phys., Calcutta Univ.
Address: Calcutta University, 92 Acharya Prafulla Chandra Road, Calcutta 9, India.

GIACOMELLO, Giordano, Dr. in Chem. Born 1910. Educ.: Rome and Padua Univs. Chem. Prof.; Director, Ist. di Chimica Farmaceutica e Tossicologica, Rome Univ.; Director, Centro

di Studio del Chimica Nucleare del Consiglio Nazionale delle Ricerche, and Centro di studio delle Structuristica Chimica del C.N.R. Books: Quaderni di Chimica farmaceutica e tossicologica; Steroidi, chemioterapici, inorganici-Chimica farmaceutica generale. Societies: Accademia Nazionale dei Lincei; Pontificia Academia Scientiarum.
Nuclear interests: Radiochemistry; Nuclear chemistry.
Address: 93 Viale Ipprocrate, Rome, Italy.

GIAMBUSSO, Angelo, B.S. (Mech. Eng.). Born 1923. Educ.: M.I.T. Asst. Director, Project Management, Div. of Reactor Development and Technol., U.S.A.E.C. Society: A.N.S.
Nuclear interests: Nuclear power plant engineering; Project management.
Address: United States Atomic Energy Commission, Division of Reactor Development and Technology, Washington, D.C. 20545, U.S.A.

GIANNAZZO, E. Gruppo Elettronico, Centro Siciliano di Fisica Nucleare, Catania Univ.
Address: Catania University, Centro Siciliano di Fisica Nucleare, 57 Corso Italia, Catania, Sicily, Italy.

GIANNONE, L. Member for Italy, Study Group on Food Irradiation, O.E.C.D., E.N.E.A.
Address: O.E.C.D. European Nuclear Energy Agency, 38 boulevard Suchet, Paris 16, France.

GIAVI, Carlos, Ing. Vocales, Comision Nacional de Energia Atomica.
Address: Comision Nacional de Energia Atomica, 565 J. Herrera y Reissig, P.2, Montevideo, Uruguay.

GIBB, Donald, B.Sc. (Mech. Eng.). Born 1920. Educ.: Durham Univ. Steam Generation Design Dept. (1950-54), Res. Dept. (1954-55), Sen. Clarke Chapman Eng. with Nucl. Power Plant Co. Ltd. (1955-58), Manager, Thermal R. and D. Dept. (1958-66), Gen. Manager, Power Plant Div. (1966-), Clarke Chapman and Co. Ltd. Societies: Assoc. Member, Inst. Mech. Eng.; North East Coast Inst. of Eng. and Shipbuilders.
Nuclear interests: Research and development associated with present and future nuclear steam generators.
Address: Power Plant Division, Clarke Chapman and Co. Ltd., Victoria Works, Gateshead 8, Co. Durham, England.

GIBB, John Michel, M.A. (Glasgow), B.A. (Oxon.). Born 1930. Educ.: Glasgow and Oxford Univs. Managing Director, John Thompson S.A. (Belgium), 1960; Editor, Euratom Bulletin, 1961-.
Nuclear interests: All aspects of nuclear science and technology.
Address: 51 rue Belliard, Brussels 4, Belgium.

GIBB, Robert, M.B., Ch.B., D.M.R.T., F.F.R. Born 1920. Educ.: Aberdeen Univ. Deputy Director, Radiotherapy, Christie Hospital and Holt Radium Inst. Member, Gen. Assembly, European Assoc. of Radiol.; Member, Council, Faculty of Radiol., British Inst. of Radiol.
Nuclear interest: Clinical radiotherapy.
Address: 18 Carrwood Avenue, Bramhall, Cheshire, England.

GIBBONS, H. B. Head, Res. Centre, Ling-Temco-Vought, Inc.
Address: Ling-Temco-Vought, Inc., Post Office Box 6144, Dallas, Texas 75222, U.S.A.

GIBBONS, John J., Dr. Formerly Acting Head, Phys. Dept. and Prof. Phys., Coll. of Chem. and Phys., Pennsylvania State Univ.
Address: Department of Physics, Osmond Laboratory, Pennsylvania State University, University Park, Pennsylvania 16802, U.S.A.

GIBBONS, N. E., Dr. Sec. Gen., Internat. Assoc. of Microbiol. Socs.
Address: Division of Bioscience, National Research Council, Ottawa 7, Canada.

GIBBS, Dennis C. C., B.Sc., M.Sc. Born 1936. Educ.: London Univ. Sen. Lecturer, Dept. Nucl. Sci. and Technol., Roy. Naval Coll., Greenwich. Society: A. Inst. P.
Nuclear interests: Reactor physics; Design methods.
Address: Royal Naval College, Greenwich, London, S.E.10, England.

GIBBS, Harold. Born 1919. Head, Non-Destructive Testing, Ship Dept., Ministry of Defence (Naval). Member, Exec. Com., Soc. of Non Destructive Examination; Chairman, B.S.I.-WEE 34. Member, Advisory Com. on Ind. Radiography, City and Guilds of London Inst.
Nuclear interest: Non-destructive examination of materials and components.
Address: M.O.D. (N), Foxhill, Bath, Somerset, England.

GIBBS, Kenneth Peter, B.Sc. (Eng.). Born 1923. Educ.: Bristol Univ. Nucl. Plant Design Eng., C.E.G.B., 1964-.
Nuclear interest: Reactor design.
Address: Walden House, 24 Cathedral Place, London, E.C.4, England.

GIBBS, R. D. Exptl. Officer, Dept. of Nucl. Sci. and Technol., Roy. Naval Coll.
Address: Royal Naval College, Department of Nuclear Science and Technology, Greenwich, London, S.E.10, England.

GIBBS, Thomas Leonard, B.Sc.Eng. (Mining and Metal.). Born 1911. Educ.: Witwatersrand Univ. Govt. Mining Eng., 1960; Chairman, Marketing Advisory Com., Atomic Energy Board, Republic of South Africa. Society: Hon. Vice-Pres. and Life Member, South Africa Inst. of Mining and Metal.

Nuclear interests: Occurrence and mining of ores, metallurgy and health hazards; Marketing of products.
Address: Office of the Government Mining Engineer, P.O. Box 1132, Johannesburg, South Africa.

GIBBS, William James, M.Sc. (Sydney), S.M. (M.I.T.). Born 1916. Educ.: Sydney Univ. Commonwealth Director of Meteorology.
Nuclear interests: Meteorology: dispersal of products of nuclear explosions; radioactive material as atmospheric tracers.
Address: Bureau of Meteorology, Box 1289K, Melbourne, Australia.

GIBERT, Armando Carlos, Ph.D. (Zürich), D.Sc. Born 1914. Educ.: Lisbon Univ. and Federal Polytechnic School, Zürich. Asst. Prof., Lisbon Univ., 1939; Phys., Portuguese Cancer Hospital, Inst. Português de Oncologia, 1947; Phys., Civil Eng. Res. Inst., Lab. Nacional de Engenharia Civil, 1952. Phys., Building Phys. Sect., Lab. Nacional de Engenharia Civil. Societies: Soc. Portuguesa de Quimica e Fisica; Swiss Physical Soc.
Nuclear interests: Reactor design, management, nuclear power economics, applications of radioactive isotopes.
Address: c/o Cassel-Industrias Electronicas e Mecânicas, Estrada Do Outeiro, Amadora, Portugal.

GIBRAT, Robert Pierre Louis, Ecole Polytechnique (Corps des Mines), École Supérieure des Mines. Born 1904. Educ.: Paris Univ. Consulting Eng., Electricité de France, 1942-; Directeur Gén., Indatom, 1955-. Prés. Directeur Gén., Sté. pour l'Industrie Atomique (S.O.C.I.A.); Former Pres., Sté. de Statistique de Paris, 1966; Former Pres., Sté. des Ingénieurs Civils de France, 1966; Comité Scientifique et Technique, Euratom (Pres., 1962-63); Comité Scientifique, C.E.A.; Former Pres., Sté. française des Electriciens; Former Pres., Tech. Com., Sté. Hydrotechnique de France; Former Pres., Assoc. Technique pour le Développement de l'Energie Nucléaire, 1957. Societies: Fellow, International Soc. of Econometry; International Inst. of Statistics.
Nuclear interests: Nuclear physics; Reactor design; Management.
Address: 336-340 rue Saint Honore, Paris 1, France.

GIBSON, Henry C. Pres., and Director, Franklin Gno. Corp.
Address: Franklin Gno Corporation, P.O. Box 3250, West Palm Beach, Florida 33402, U.S.A.

GIBSON, Henry C., Jr. Pres., Franklin Systems Inc.
Address: Franklin Systems Inc., P.O. Box 3250, West Palm Beach, Florida, U.S.A.

GIBSON, Raymond A. Director, Yankee Atomic Elec. Co.
Address: Yankee Atomic Electric Co., 441 Stuart Street, Boston 16, Massachusetts, U.S.A.

GIBSON, William Martin, M.A., Ph.D. Born 1926. Educ.: Cambridge Univ. Lecturer, Phys., Queen's Univ., Belfast, 1953-57; Phys., C.E.R.N., Geneva, 1957-60; Sen. Lecturer (1960-66), Reader, Phys. (1967-), Bristol Univ.
Nuclear interest: Physics of elementary particles, as studied in experiments with high-energy accelerators.
Address: H. H. Wills Physics Laboratory, Bristol University, Royal Fort, Bristol 2, England.

GIDEL, Jean. Directeur, Sté. de Chauffage, Tuyauteries et Chaudronnerie d'Audincourt.
Address: Société de Chauffage, Tuyauteries et Chaudronnerie d'Audincourt, Audincourt, (Doubs), France.

GIDEZ, Lewis I., Dr. Sen. Officer, Dept. Medicine-Radiol., Albert Einstein Coll. of Medicine, Yeshiva Univ.
Address: Albert Einstein College of Medicine, Yeshiva University, Eastchester Road and Morris Park Avenue, New York 61, New York, U.S.A.

GIERDALSKI, Jerzy. Papers: Fast decade counting attachment type LE-2 (Communication, Nukleonika, vol. 4, No. 1, 1959); Some New Ideas in Application of Radioisotopes for Measuring Thickness and Density (ibid, vol. 10, No. 8, 1965).
Address: Zaklad Doswiadczalny, Biuro Urzadzen Techniki Jadrowej, Warsaw, Zeran, Poland.

GIERKE, Gerhart Otto Julius VON. See VON GIERKE, Gerhart Otto Julius.

GIERULA, Jerzy, Ph.D. Born 1917. Educ.: Jagiellonian Univ., Cracow. Prof. of Phys., Inst. of Nucl. Res., Cracow, 1960-; Prof. of Phys., Acad. of Mining and Metal., Cracow, 1960-. Chairman, Sci. Council, Inst. of Nucl. Phys., Cracow, 1966-. Societies: Polish Phys. Soc.; A.P.S.
Nuclear interest: High energy nuclear physics.
Address: 19 św Krzyża, Craców, Poland.

GIESE, K. Deleg., I.A.E.A. Conference on the Use of Radioisotopes in the Phys. Sci. and Industry, Copenhagen, 6-17 September 1960. Member, Study Group on Food Irradiation, O.E.C.D., E.N.E.A.; Alternate, Board of Directors, Member, Technical Com., Eurochemic, Mol, Belgium.
Address: Bundesministerium für Wissenschaftliche Forschung, 46 Luisenstrasse, Bad Godesberg, Germany.

GIESE, Werner Walter, Dr. med. vet. Born 1936. Educ.: Humboldt Univ. and Berlin Tierärztliche Hochschule. N.A.T.O. Res. Fellow, Dept. of Phys. Biol., Cornell Univ., Ithaca, 1962-64; At Physiologisches Inst.,

Tierärztliche Hochschule, Hanover. Society: Deutsche Gesellschaft für Biophysik.
Nuclear interests: Radiobiology studies on the metabolism of radiocesium; Neutron activation analysis for trace elements.
Address: Tierärztliche Hochschule, Physiologisches Institut, 15 Bischofsholer Damm, Hanover 3, Germany.

GIESEKE, Paul, Full Prof. Law, Dr. iur Dr.-Ing.E.h. Born 1888. Formerly Member, German A.E.C., Sub-commission IV (Protection against Ionising Radiation), now Vorsitzender Arbeitskreis I/2, Rechts-und Verwaltungsfragen des Strahlenschutzes. Book: Rechtsfragen der Atom-energie (Karlsruhe, Verlag C.F. Müller, 1956). Societies: Internat. Law Assoc., German branch; Deutsches Atomforum.
Nuclear interests: Legal questions of atomic energy.
Address: 9 Nachtigallenstrasse, Bad Godesberg, Germany.

GIESEN, K., Dr. Ing. habil. Prof., Standiges Seminar für Kerntechnik, Haus der Technik E.V., Essen. Member, Steering Com., Deutsches Atomforum e.V.
Address: Haus der Technik E.V. Essen, Postfach 767, la Hollestr., Essen, Germany.

GIESSEN, E. A., Met. E. Born 1900. Educ.: Munich and Aachen Univs. Owner of Firm E.A. Giessen K.G., Sole Agency of Superior Tube Co., Norristown, Pa., and Nat. Beryllia Corp., Haskell, N.J., U.S.A. Represent: Ugine - Kuhlmann, Paris; Cie. du Filage des Metaux et des Joints Curty, CEFILAC, Paris.
Nuclear interests: Applications of beryllium oxide in all fields as moderator and reflector; also varied canning tubing in all zirconium alloys.
Address: Werksvertretungen, Import Techn. Grosshandel, Postfach 9, Munich-Allach, Germany.

GIESSER, Walter, Diplom-Physiker. Born 1932. Educ.: Stuttgart T.H. and West Berlin Freie Univ. At Brown Boveri/Krupp Reaktorbau G.m.b.H., 1959-.
Nuclear interests: Reactor design; Nuclear reactor theory; Quantum mechanics.
Address: 3 Kniebisstrasse, Mannheim, Germany.

GIETZELT, Georg Fritz, Dr. med., Dr. Med. Habil., Full Prof. Radiol. Born 1903. Educ.: Leipzig, Graz and Düsseldorf Univs. Full Prof. of Radiol. held by Government of German Democratic Republic,1957. Member, Res. Council, G.D.R.; Member, Sci. Council for Peaceful Uses of Atomic Energy; Chairman, Commission for Rad. Protection. Societies: Med-Wissensch. Gesellschaft für Roentgenologie in der D.D.R.; Deut. Roentgengesellschaft.
Nuclear interests: Radiobiology; Telecurietherapy; Betatron.
Address: Direktor der Geschwulstklinik der Charité, 20 Schumanstrasse, Berlin, German Democratic Republic.

GIFFON, Maurice, D. ès Sc. Born 1931. Educ.: Lyon Univ. Maitre Asst., Faculté des Sci., Lyon.
Nuclear interests: Study of the analytic properties in the angular momentum for singular potentials; Inversion problem for regular and singular potentials in the complex angular momentum plane; Study of the differential properties of the Feynman graphs for polygonal diagrams.
Address: Institut de Physique Nucléaire, 43 boulevard du 11 novembre 1918, 69 Villeurbanne, France.

GIFFORD, Franklin A., Jr., B.S., M.S., Ph.D. Born 1922. Educ.: Pennsylvanis State Univ. Director, Atmospheric Turbulence and Diffusion Lab., Environmental Sci. Services Administration, U.S. Dept. of Commerce. Member, Chairman (1962), Member, Advisory Com. on Reactor Safeguards, U.S.A.E.C.; Chairman, Com. on Atmospheric Turbulence, American Meteorological Soc. Book: Contributor, Meteorology and Atomic Energy (U.S.A.E.C., 1967).
Nuclear interest: Nuclear reactor safety.
Address: A.T.D.L., E.S.S.A., Box E, Oak Ridge, Tennessee, U.S.A.

GIFFORD, Robert Henry. Exec. Director, Southern Interstate Nucl. Board. Member, Advisory Com. of State Officials, U.S.A.E.C.
Address: Southern Interstate Nuclear Board, Suite 664, 800 Peachtree Street, N.E., Atlanta 8, Georgia, U.S.A.

GIGAREL, M. Chef du Departement Nucléaire, Div. de la Ste. Alsthom, Ets. Neyrpic.
Address: Ets. Neyrpic, B.P. 52, 38 Grenoble, France.

GIGLI-BERZOLARI, Alberto, Ph.D. Born 1921. Educ.: Pavia and Rome Univs. Prof., Gen. Phys., Pavia Univ.; Member, Internat. Board of Editors, and Vice Director, Il Nuovo Cimento. Societies: Italian Phys. Soc.; A.P.S.
Nuclear interests: Nuclear physics; Nuclear instrumentation.
Address: Department of Physics, Pavia University, Pavia, Italy.

GIGUET, R. L. Pres., Westinghouse Internat. Atomic Power Co. Ltd., -1962; Chairman, Coordinating Board, Sena Project, 1962-.
Address: S.E.N.A., 68 Faubourg Saint-Honore, Paris 8, France.

GIJN, S. H. van. See van GIJN, S. H.

GIL, Lorenzo Horna. Fellow, Peruvian Soc. of Radiol.
Address: Sociedad Peruana de Radiologia, Casilla No. 2306, Lima, Peru.

GIL y GIL, D. Carlos, Dr.; Member, Comisión Asesor de Medicina y Biologia Animal de la Junta de Energia Nucl.
Address: c/o Junta de Energia Nuclear, Ciudad Universitaria, Madrid 3, Spain.

GILAR, Oldrich, M.S. (Eng.). Born 1928. Educ.: Brno Polytech. Tech. Director, Tesla Res. Inst. for Nucl. Instrumentation, 1960-. Member, Editorial Board, Jaderná Energie. Society: Czechoslovak Soc. for Sci. and Techniques.
Nuclear interests: Instruments for the measurement and detection of nuclear radiation; Industrial application of radioisotopes.
Address: Tesla Research Institute for Nuclear Instrumentation, Premysleni u Prahy, Czechoslovakia.

GILARDONI, Arturo, Dottore Ingegnere. Born 1905. Educ.: Politecnico di Milan. Founder and Unic Director, Gilardoni S.p.A., Apparecchiature Radiologiche e Nucleari, 1957; Founder and Unic Director, APEL Applicazioni Elettroniche - Fabbrica per la costruzione dei tubi a Raggi X e tubi elettronici, 1958-; Member, UNEL (Unificazione Elettronica) Commissione apparecchi Radiologici ed Elettromedicali, 1957-; Member, C.E.I. (Comitato Elettrotecnico) Sottocomitato no. 11 - Impianti elettrici degli apparecchi radiologici ed elettromedicali, 1957-; Member, C.E.I. (Comitato Elettrotecnico Italiano) Sottocomitato no. 108 - Apparecchi per radiologia ed elettromedicina, 1960-. Book: Co-author, Nondestructive Testing Handbook, Sez. 14 - Electronic Rad. Sources (di McMaster, Ed., N.Y., Ronald Press.). Societies: Stà. Italiana di Radiologia (Medica e Nucleare); Nondestructive Testing Soc.; Soc. Italiana di Fisica Sanitaria.
Nuclear interests: Nuclear equipment and techniques in nondestructive testing and medicine.
Address: Gilardoni S.p.A., Raggi X e Nucleari, 2 Via E. Fermi, Mandello Lario Como 22054, Italy.

GILAT, Eliezer, M.Sc., Ph.D. Born 1922. Educ.: Hebrew Univ., Jerusalem. I.A.E.A., Internat. Lab. of Marine Radioactivity, Musée Océanographique, Monaco, 1962-64; Sea Fisheries Res. Station, Dept. of Fisheries, Ministry of Agriculture, Haifa.
Nuclear interests: Radioactive contamination of marine environment (water, sediments and biota); Radioactive waste disposal into oceans.
Address: Sea Fisheries Research Station, P.O. Box 699, Haifa, Israel.

GILBERT, C. W. Dr. Director, Formerly Res. Phys., Exptl. Phys. Dept., Christie Hospital and Holt Radium Inst.; Formerly Sec., Assoc. for Rad. Res.
Address: Christie Hospital and Holt Radium Institute, Radium Institute, Wilmslow Road, Withington, Manchester 20, England.

GILBERT, Ernest. Born 1918. Sect. Leader, Marine Engine Drawing Office (1953-56), Asst. Chief Draughtsman, Marine Engine Drawing Office (1956-60), Chief Draughtsman, Dounreay Project, Nucl. Power Dept. (1960-64), Vickers Ltd. Eng. Group; Tech. Manager, Marine Eng., Shipbuilding Group Vickers Ltd., 1964-. Society: Inst. of Marine Eng.
Nuclear interests: Reactor design and installation in association with propulsion machinery. Have been associated with nuclear submarine propulsion plant from DS/MP Project, H.M.S. Dreadnought, Warspite, Resolution, Repulse, SSN.06 and currently involved in a new class of nuclear submarine propulsion machinery. Also associated with proposals for application of nuclear propulsion to surface ships.
Address: Vickers Limited, Shipbuilding Group, Barrow-in-Furness, Lancs., England.

GILBERT, Gareth E., Dr. Formerly Project Leader, Dept. Botany, Ohio Agricultural R. and D. Centre; Project Leader, Dept. of Botany and Plant Pathology, College of Agriculture and Home Economics, Ohio State Univ.
Address: Ohio State University, College of Agriculture and Home Economics, Columbus 10, Ohio, U.S.A.

GILBERT, Harris D., A.B., Diploma, Assoc. Mech. Eng. Born 1927. Educ.: Newark Coll. of Eng., Upsala Coll., Columbia and Rutgers Univs. Manager of Contracts, Thiokol Chem. Corp., 1951-61; Part and Full-time Management Consultant in Nucl. and Aerospace Management Field, 1955-; Manager, Products Contracting Div., Curtiss Wright Corp., 1961-62; Director of Contracts and Planning, Nucl. Div. (1962-63), Director Contracts (1963-) Martin Co.; Vice-Pres., Isochem Inc. (subsidiary of Martin Marietta Corp. and UniRoyal, Inc.), 1965-67; Vice-Pres., The Rocket Co., 1967-. Societies: American Management Assoc.; Atomic Ind. Forum; American Rocket Soc.
Address: 1365 Rimer Drive, Moraga, California 94456, U.S.A.

GILBERT, W., Formerly Assoc. Prof., now Prof., Phys. Dept., Harvard Univ.
Address: Harvard University, Jefferson Physical Laboratory, Cambridge 38, Massachusetts, U.S.A.

GILBERT, Walter D., M.S. (Chem. Eng.), B.S. (Chem. Eng.). Born 1921. Educ.: Washington State Univ. and Univ. of Washington. Eng., Gen. Elec. Co., Richland, Washington, 1950-55; Project Eng., U.S.A.E.C., Washington, D.C., 1955-57; Sen. Eng., Gen. Elec. Co., Richland, Washington, 1957-. Societies: A.N.S.; Columbia Basin Sect., A.I.Ch.E.
Nuclear interests: Research and development activities and nuclear safety studies associated with the Hanford Production Reactors.
Address: General Electric Company, Richland, Washington, U.S.A.

GILBERTSON, Stanley E. Member, Standing Com. on Nucl. Energy in the Elec. Industry, Nat. Assoc. of Railroad and Utilities Commissioners, 1967-.
Address: National Association of Railroad and Utilities Commissioners, Interstate Commerce Commission Building, P.O. Box 684, Washington, D.C. 20044, U.S.A.

GILCHRIST, William McKenzie, B.Sc. (Mining and Metal., Queen's Univ.). Born 1909. Educ.: Manitoba and Queen's (Ontario) Univs., Trans American Mining Co., Ltd. (exploration and development work in the Northwest Territories), 1946-50; Chief Eng. and Underground Supt., Giant Yellowknife Gold Mines, Ltd., 1950-51; Vice-Pres. and Manager of Mines, Transcontinental Resources, Ltd., 1951-52; Asst. Manager of Beaverlodge Operation (1952-55), Manager, Beaverlodge Operation (1955-58), Vice-Pres. i/c Western Operations (1958), Vice-Pres. i/c Operations (1958), Pres. (1958-), Eldorado Mining and Refining, Ltd.; Pres., Northern Transportation Co. Ltd.; Director: Eldorado Mining and Refining; Northern Transportation; Eldorado Aviation; Atomic Energy Control Board; Canadian Metal Mining Assoc.; Director, Canadian Nucl. Assoc.
Address: R.R. No. 3, Manotick, Ontario, Canada.

GILCREAS, Frank Wellington, A.B. Born 1893. Educ.: Harvard Univ. Asst. Director, Div.of Labs., New York State Dept. of Health, 1934-55; Prof., Dept. of Civil Eng., Florida Univ., 1955-. Hon. Member, Federation of Sewage and Industrial Wastes Assoc. Societies: A.P.H.A.; Federation of Sewage and Industrial Wastes Assoc.; American Water Works Assoc.; Roy. Soc. of Health.
Nuclear interests: Sanitation and public health.
Address: Department of Civil Engineering, Florida University, Gainesville, Florida, U.S.A.

GILCREST, John. Project Eng., Construction of A.E.C. Eng. Test Reactor, Idaho Falls, Manager, Space and Defence Projects Dept., (1965-), Kaiser Eng., Henry J. Kaiser Co.
Address: Kaiser Engineers, Division of Henry J. Kaiser Company, 300 Lakeside Drive, Oakland 12, California, U.S.A.

GILEADI, Aviva Eva, Ph.D. (Phys. (Budapest), M.Sc. (Nucl. Sci.) (Michigan). Born 1917. Educ.: Budapest and Michigan Univs. Lecturer (1955-58), Sen. Lecturer, Dept. of Nucl. Sci., Technion-Israel Inst. Tech.; Reactor Phys. at start-up of Bonus Reactor, Puerto Rico Water Resources Authority, Rincón, 1964. Society: A.N.S.
Nuclear interest: Reactor physics.
Address: 13 Adam Hacohen Street, Nave Sha'anan, Haifa, Israel.

GILES, E. W. Manager, Burlington (Iowa) Area, Albuquerque Operations Office, U.S. A.E.C.
Address: United States Atomic Energy Commission, Burlington (Iowa) Area Office, P.O. Box 561, Burlington, Iowa 52601, U.S.A.

GILKESON, Robert F. Vice-Pres. and Director, Atomic Power Development Assocs. Inc.; Exec. Vice-Pres. and Member, Exec. Com., High Temperature Reactor Development Assocs. Inc.; Pres., Philadelphia Elec. Co.; Member, Board of Directors, Atomic Ind. Forum, 1966-69.
Address: Atomic Power Development Associates, Inc., 1911 First Street, Detroit 26, Michigan, U.S.A.

GILL, Neal F. Director, Atomic Power Development Assocs. Inc.
Address: Atomic Power Development Associates, Inc., 1911 First Street, Detroit 26, Michigan, U.S.A.

GILL, Percy Roland. Born 1919. Asst. Shift Charge Eng. (1950-53), Maintenance Eng. (Mech.) (1953-56), C.E.G.B.; Mech. Eng. (1956-59), Sen. Mech. Eng. (1959-65), Eng. Manager (1965-), Calder Works, U.K.A.E.A. Societies: I.Mar.E.; B.N.E.S.; Windscale and Calder Branch, Nucl. Eng. Soc.
Nuclear interests: Management; Maintenace; Operation.
Address: United Kingdom Atomic Energy Authority, Calder Works, Sellafield, Calderbridge, Cumberland, England.

GILL, Philip, B.S. (Chem.) (Alaska), M.S. (Chem.) (Oregon). Born 1910. Educ.: Alaska Univ. and Oregon State Coll.
Nuclear interest: Management.
Address: U.S. Nuclear Corporation, 801 North Lake Street, Burbank, California, U.S.A.

GILL, Piara Singh, M.S. (Southern California), Ph.D. (Chicago). Born 1911. Educ.: Southern California and Chicago Univs. Prof. Phys., Aligarh Univ., 1949-. Director, Gulmarg Res. Observatory, Kashmir. Pres., Nat. Acad. of Sci., India, 1957 and 1958; Pres., Indian Phys. Soc., 1957 and 1958; Foreign Sec., Council, Nat. Inst. of Sci., India; Hon. Prof. Phys., Jammu and Kashmir Univs.; Hon. Prof. Phys., Punjab Univ.; Sci. advisor Punjab Govt.; Formerly Member, Radioisotopes and Ionizing Rad. Com., Indian Council of Agricultural Res. Societies: Fellow, A.P.S.; Fellow, Nat. Inst. of Sci., India; Fellow, Nat. Acad. of Sci., India; Fellow, Indian Phys. Soc.
Nuclear interests: Nuclear physics and cosmic rays.
Address: Central Scientific Instruments Organisation, Chandigarh, India.

GILL, Robert George SUMMERS-. See SUMMERS-GILL, Robert George.

GILLAMS, John Langford, M.A. Born 1922. Educ.: Oxford Univ. Lecturer in Maths., Christ Church, Oxford, 1948-52; Phys., and later

Asst. Director, Tech. Policy Branch, Risley, 1952-58; Economics and Programming Branch, A.E.A., London, 1958-65; Sen. Planner, British Petroleum Co. Ltd., 1965-.
Nuclear interest: Economic policy.
Address: 20 Hurst Way, South Croydon, Surrey, England.

GILLANDERS, Earl Burdette, B.A., M.A. (British Columbia), Ph.D. (Princeton). Born 1903. Educ.: British Columbia and Princeton Univs. Vice-Pres. and Manager, Rio Canadian Exploration, Ltd., 1953-54; Vice-Pres. in charge of Operations, Rio Tinto Mining Co. of Canada, Ltd. (1955), Executive Vice-Pres. (1957-64), Rio Tinto Mining Co. of Canada, Ltd. Consultant, Rio Algom Mines Ltd., 1964-. Part time consulting practice; Director, Spectroair Explorations Ltd., Vancouver; Director and Vice-Pres., Larnex Mining Corp. Ltd.; Director and Vice-Pres., New Imperial Mines Ltd. Pres., Canadian Metal Mining Assoc., 1961-62. Society: Canadian Nucl. Assoc.
Nuclear interests: Disposal of raw or more highly processed nuclear material.
Address: Box 85, Crescent Beach, British Columbia, Canada.

GILLEDO, Manuel F. GRAN. See GRAN GILLEDO, Manuel F.

GILLER, Brigadier Gen. Edward B., B.S., M.S., Ph.D. (Chem. Eng.). Born 1918. Educ.: Illinois Univ. Chief, Rad. Branch, Effects Div., Armed Forces Special Weapons Project, 1950-54; Director, Res. Directorate, Air Force Special Weapons Centre, 1954-58; Special Asst. to Commander, Office of Sci. Res., 1959-61; Special Asst. to Commander, Office of Aerospace Res., 1961-65; Director, Sci. and Technol., Office of Deputy Chief of Staff for R. and D., Dept. of the Air Force, 1964-67; Director, Div. of Military Application (1967); Asst. Gen. Manager, Military Application (1967-), U.S.A.E.C.
Address: United States Atomic Energy Commission, Division of Military Application, Washington D.C. 20545, U.S.A.

GILLESPIE, Arthur Blackley, B.Sc. (Hons., Elec. Eng.). Dipl. R.T.C. Born 1921. Educ.: Roy. Tech. Coll. and Glasgow Univ. S.O., S.S.O., P.S.O. and Group Leader, Nuclear Instrument Development, Electronics Div., A.E.R.E., Harwell 1946-63; Div. Head, Electronics Div., Culham Lab., 1963-66; Group Leader Analogue Circuits, Noise and Sub-mm Wave Techniques. Elect. and Appl. Phys. Div., A.E.R.E., Harwell, 1966-. Book: Signal, Noise and Resolution in Nucl. Counter Amplifiers (London, Pergamon Press, 1953).
Nuclear interests: Instrumentation and measurement techniques in the fields of nuclear physics, reactors, controlled fusion and materials science.
Address: 6 Kingston Close, Abingdon, Berks., England.

GILLESPIE, Arthur Samuel, Jr., B.S., M.A. (Phys. Chem.). Born 1931. Educ.: Mars Hill Coll., Wake Forest Coll., and Duke Univ.; Isotope Training Programme, O.R.I.N.S. Staff Member, Sandia Corp., Albuquerque, N.M., 1955-56; Res. Eng., Radiochem. Lab., Phys. Chem. Div., Alcoa Res.Labs., New Kensington, Pa., 1956-61; Radiochem., Res. Triangle Inst., Durham, N.C., 1961-66; Sen. Chem., Texas Gulf Sulphur Co., Aurora, N.C. 1966-67; Res. Chem., Lithium Corp. of America, Bessemer City, N.C., 1967-. Society: A.C.S.
Nuclear interests: Analytical radiochemistry, tritium tracer studies of hydrogen in metals, radiometric methods in analytical chemistry, autoradiography, neutron activation analysis, counter tube development, and miscellaneous applications of radioisotopes in electrochemical, metallurgical, water tracing, and oceanography.
Address: 317 East Second Avenue, Gastonia, N.C. 28052, U.S.A.

GILLESPIE, Peter James, B.Sc., B.E. Born 1921. Educ.: Sydney Univ. R.R.E. Malvern, Worcs., 1948-54; School of Elec. Eng., N.S.W. Univ. Technol., 1955-58; Cancer Inst. Board, Melbourne,1958-59; Amalgamated Wireless (Australasia) Pty., Ltd., 1959-64; Philips Elec. Ind. Pty. Ltd., 1964-66; Automatic Totalisators Ltd., 1966-.
Nuclear interests: Reactor simulation, nuclear instrumentation and control; Industrial uses of isotopes.
Address: 8 Alexander Avenue, Mosman 2088, New South Wales, Australia.

GILLETT, R. E. Administration Div., Radiol. Sci. Dept., Nevada Test Site operated for U.S.A.E.C. by Reynolds Elec. and Eng. Co., Inc.
Address: U.S.A.E.C., Nevada Test Site, Radiological Sciences Department, Administration Division, A.E.C. Contract Office, P.O. Box 1360, Las Vegas, Nevada 89101,U.S.A.

GILLETTE, John Harley, B.S. (Phys., Chem.). Born 1916. Educ.: St. Lawrence Univ. Supt., Radioisotope Distribution Dept., (1956-58). Supt., Isotopes Developments Centre (1958-), O.R.N.L. Society: A.N.S.
Nuclear interests: Research and development concerning uses of radio- and stable isotopes; Application of radioisotopes in medical and industrial fields, and in space programmes; Production of radioisotopes and stable isotopes.
Address: P.O. Box X, Oak Ridge National Laboratory, Oak Ridge, Tennessee 37830, U.S.A.

GILLI, Paul Viktor, Doz. Dipl.-Ing., Dr. techn. Born 1924. Educ.: Graz T.H. and I.N.S.N.E., Argonne Nat. Lab. At (1952-), Head, Nucl. Power Dept. (-1968), Waagner-Biro A.G.; Project Manager for 5 (12) MW Reactor ASTRA of Oesterreichische Studiengesellschaft fuer Atomenergie Ges. m.b.H.

(S.G.A.E.) at Seibersdorf, near Vienna, 1958-59; assigned to Eng. Div. O.E.C.D. High Temperature Reactor Project (Dragon) at A.E.E., Winfrith, Dorset, 1960-63; Lecturer, Vienna T.H., 1966-; Prof., Heat and Power Eng., and Head, Inst. of Heat and Power Eng., Faculty of Mech. Eng., Univ. of Technol., Graz, 1968-. Societies: Oesterreichischer Ingenieur- und Architektenverein; A.S.M.E.; A.N.S.; A.M.I.Mech.E.; M.I.Nucl.E.; B.N.E.S.; Commission Technique Générale, Comité Européen de la Chaudronnerie et de la Tolerie.
Nuclear interests: Layout and design of nuclear power plant; Thermodynamics; Heat transfer and fluid flow; Magnetohydrodynamics; Reactor theory.
Address: University of Technology, Faculty of Mechanical Engineering, 5 Universitatsplatz, Graz, Austria.

GILLMORE, Donald Wood, A.B., Ph.D. Born 1919. Educ.: Williams Coll. and Pennsylvania State Univ. Supv., Activated Carbon Res., Pittsburgh Coke and Chem. Co., 1953-60; Development Manager, (1960-61), Tech. Branch Manager (1961-67), Plant Manager, Niagara Falls, New York Plant (1967-), Electro Minerals Div., Carborundum Co. Societies: A.C.S.; American Ceramic Soc.; Abrasive Grain Assoc.
Nuclear interest: Boron carbide.
Address: 183 Amherstdale Road, Buffalo 14226, New York, U.S.A.

GILLON, Luc Pierre A., D.Sc. (Phys.). Born 1920. Educ.: Louvain Univ. Pres., Lovanium Univ., Kinshasa, Congo. Sec. Gen., Commissariat des Sci. Nucléaires du Congo; Member, C.C.T.A./Conseil Scientifique Africain (Nucl. Phys.). Societies: Academie Royale des Sci. d'Outre-Mer; A.P.S.
Nuclear interests: Reactor management; General organization of nuclear research; Nuclear spectroscopy.
Address: B.P. 121, Kinshasa 11, Congo.

GILLON, Philippe J. J., Elec. and Mech. Eng. Born 1909. Educ.: Louvain Univ. Deputy Manager, Sté. Financière de Transports et d'Entreprises Industrielles (SOFINA), S.A., 1958; Director, Sté. Belge de Chimie Nucléaire (BELCHIM); Member, Tech. Com., Sté. d'Energie Nucléaire Franco-Belge des Ardennes (SENA). Society: Atomic Ind. Forum.
Nuclear interests: Engineering and design of conventional parts of nuclear reactor installations and of nuclear power plants.
Address: SOFINA, 38 rue de Naples, Brussels 5, Belgium.

GILLOT, R. H. Member for Euratom, Group of Experts on Production of Energy from Radioisotopes, O.E.C.D., E.N.E.A.
Address:O.E.C.D. European Nuclear Energy Agency, 38 boulevard Suchet, Paris 16, France.

GILLOT-STOKKINK, A. J., L. ès Sc., D.Sc. Born 1922. Educ.: Brussels and Louvain Univs. Member, Res. Staff, Phys. Res. Lab. s.a. A.C.E.C., Charleroi, 1948-55; Res. Staff Member i/c Phys. Res., Lab. Central s.a. Union des Verreries Mécaniques Belges, Gilly, Charleroi, 1955-60; Euratom, C.C.R., Ispra, Italy, 1960-.
Nuclear interests: Diffusion, corrision, interface reactions, phase-separation, luminescence; Tracer applications; Nuclear batteries; research on solid radioactive waste treatment.
Address: Euratom, Ispra, Italy.

GILMORE, John Hamilton, B.S., M.D. Born 1907. Educ.: Illinois Univ. Director, Radiol., Illinois Masonic Hospital, 1935-61; Director, Radiol., West Suburban Hospital, Oak Park Illinois, 1961-. Asst. Prof., Radiol., Northwestern Univ.; Formerly Pres.-Elect, Radiol. Soc. of North America, Inc. Societies: Past Pres., Chicago Roentgen Ray Soc.; American Roentgen Ray Soc.; Fellow, A.C.R.; Soc. of Nucl. Medicine.
Nuclear interests: Practice of radiology and the practical applications of nuclear medicine in the diagnosis and treatment of disease.
Address: West Suburban Hospital, 518 North Austin Boulevard, Oak Park, Illinois, U.S.A.

GILROY, John, A.B., M.A. (Harvard), Ph.D. (Harvard). Born 1915. Educ.: Harvard Coll. and Harvard Univ. Phys., etc., Air Reduction Co., Inc., 1942-56; Assoc. Director, Electronics Div. (1956-61), Director (1961-), Argonne Nat. Lab. Society: Instrument Soc. of America.
Nuclear interests: Management; Instrumentation.
Address: Argonne National Laboratory, 9700 South Cass Avenue, Argonne, Illinois 60440, U.S.A.

GILSON, Albert J., B.A., M.D. Born 1928. Educ.: Cornell Univ. Asst. Prof., School of Medicine, Emory Univ., Atlanta,1961-63; Assoc. Prof. (1963-), Director, Div. of Nucl. Medicine (1963-), School of Medicine, Miami Univ. Sec. (1962-63), Pres. (1963-64), Southeastern Chapter, and Trustee, Soc. of Nucl. Medicine.
Nuclear interests: Research and clinical application of radioisotopes as applied to medicine and biology; Nuclear instrumentation and physics.
Address: Division of Nuclear Medicine, Miami University, School of Medicine, Miami, Florida 33136, U.S.A.

GIMENEZ-RAMOS, Gustavo, Ing. de Construcción y Electricidad; Licenciado en Ciencias Exactas; M.Sc. (Elec. and Nucl. Eng.). Born 1915. Educ.: Univ. Central de Madrid, Escuela Politécnica del Ejército, Madrid, and Stanford Univ. (California). Prof., Escuela Politécnico, Madrid; Eng., Tecnatom, S.A., Madrid, working in the Zorita Nucl. Power Plant Project. Books: Co-author, Coyuntura de la energia nuclear en España, presente y futuro (May, 1960);

Programa nuclear para la Zona Centro-Levante-Sur (Study for CENUSA, 1960). Societies: Asociación de Ingenieros de Construccion y Electricidad; A.N.S.
Nuclear interest: Nuclear power stations.
Address: 53 Velázquez, Madrid, Spain.

GIMENO, Carlos MENDOZA. See MENDOZA GIMENO, Carlos.

GIMERA, Ralph J., B.S.M.E., M.S.M.E., O.R.S.O.R.T. Graduate. Born 1927. Educ.: Pittsburgh, Connecticut and California (Los Angeles) Univs. Nat. Membership Com., A.N.S. Society: A.I.A.A.
Nuclear interests: Engineering management and nuclear reactor design of central station, marine, and space nuclear power plants.
Address: Atomics International, Department 731, C002, P.O. Box 309, Canoga Park, California 91304, U.S.A.

GIMSTEDT, Olle G., M.Sc. (Eng.). Born 1914. Educ.: Roy. Inst. Technol., Stockholm. Tech. Director, Sentab, Rio de Janeiro, 1950-55; Managing Director, Swedish Power Producers Assoc., 1955-59; Managing Director, Atomkraftkonsortiet (AKK Atomic Power Group) and Oskarshamnsverkets Kraftgrupp A.B., 1959-. Society: A.N.S.
Nuclear interests: Management; Plant construction.
Address: Atomkraftkonsortiet, 19 Stureplan, Stockholm, Sweden.

GINELL, William Seaman, B.S. (Chem.), Ph.D. (Chem.). Born 1923. Educ.: Polytechnic Inst. of Brooklyn and Wisconsin Univ. Assoc. Chem. (1949-52), Chem. (1952-58), Brookhaven Nat. Lab.; Sen. Tech. Specialist, Atomics Internat., 1958-61; Head, Inorganic Chem., Aerospace Corp., 1961-63; Sect. Chief, Douglas Aircraft Co. Societies: A.C.S.; A.N.S.; Fellow, A.A.A.S.; New York Acad. Sci.
Nuclear interests: Nuclear reactor chemistry; Chemical processing; High-temperature reactor materials; Nuclear propulsion; Liquid metal chemistry; Radiation effects.
Address: 16856 Escalon Drive, Encino, California, U.S.A.

GINGELL, C. E. L. Formerly Chief Instrumentation Eng., Linear Accelerator Groups, Yale Univ.; Manager, A.E.C.-Specified Instrumentation, Nucl. Data Inc., 1965-.
Address: Nuclear Data Inc., 911 Regent Street, Madison, Wisconsin, U.S.A.

GINGER, Donald Albert, M.A. (Cantab.), F.R.I.C. Born 1928. Educ.: Cambridge Univ. Manager, Res. Div., Thorn Bendix Ltd.
Nuclear interests: Manufacture of crystals and plastics for scintillation counting; Nuclear batteries; Clathrate compounds containing radioactive gases.
Address: c/o Thorn Bendix Ltd., Hook Rise South, Tolworth, Surrey, England.

GINKEL, William Louis, B.S. (Chem. Eng.), B.S. (Business Administration). Born 1920. Educ.: Rochester Univ., New York. Chief, Materials Control Branch, Div. of Production (1947-50), Chief, Chem. Processing Branch, Idaho Operations Office (1950-54), Director, Tech. Administration Divs., Schenectady Operations Office (1955-56), Asst. Director, Tech. Operations Div., Idaho Operations Office (1956-60), Asst. Manager for Tech. Operations (1961-62), Deputy Manager (1962-64), Manager (1964-), Idaho Operations Office, U.S.A.E.C.
Nuclear interests: Reactor engineering; Chemical processing; Management.
Address: 2825 W. Morningside Drive, Idaho Falls, Idaho, U.S.A.

GINNA, Robert E. Starting as Departmental Manager (1934-), Pres. (-1967), Chairman of the Board (-1968), Chief Exec. Officer, Director and Member, Exec. Com., Rochester Gas and Elec. Corp.; Pres., High Temperature Reactor Development Assoc. Inc.; Director: Security Trust Co. of Rochester, Atomic Power Development Assoc. Inc., Public Utilities Information Programme, Empire State Utilities Power Resources Assoc., Empire State Atomic Development Assoc. Inc.; Trustee, Power Reactor Development Co.; Director: Edison Electric Inst., Assoc. Edison Illuminating Cos., Nat. Assoc. of Elec. Cos.; Director and Member, Exec. Com., American Gas Assoc.; Director, Atomic Ind. Forum, 1965-; Member: Atomic Power Com., Edison Elec. Inst., U.S. Armed Forces Advisory Com., Nucl. Energy Com., Nat. Assoc. of Manufacturers, Com. on Commercial Uses of Atomic Energy, U.S. Chamber of Commerce. Alternate Member, U.S. Exec. Com., World Power Conference; Member, Gen. Advisory Com., New York State Office of Atomic Development.
Address: 12 San Rafael Drive, Pittsford, Rochester, New York 14618, U.S.A.

GINNIFF, Maurice Edmund, B.Sc. Born 1925. Educ.: Queen's Univ., Belfast. Gas Turbine Res. Eng. (1949-54), Eng. in charge of Gas Turbine Design and Res. (1954-57), Bristol Aeroplane Co.; Engineer I in Reactor Technol. (1957-59), Res. Manager, Reactor Eng. Group (1959-), U.K.A.E.A., Windscale. Societies: A.M.I.Mech.E.; Soc. of Instrument Technol.
Nuclear interests: Reactor design and development engineering, particularly associated with reactor experiments and irradiation studies.
Address: Lakenhow, Seascale, Cumberland, England.

GINOCCHIO, Roger, Ing. en chef des Ponts et Chaussées, Ancien élève de l'Ecole Polytech. Born 1918. Educ.: Ecole Polytech. Directeur à le Direction générale d'Electricité de France. Societies: Sté. française des Electriciens; Sté. française de Radioprotection.

Nuclear interest: Exploitation des centrales nucléaires.
Address: 3 rue de Messine, Paris 8, France.

GINOT, Paul Marie Camille, Ing., Bachelor Lettres-Mathématiques. Born 1931. Educ.: Ecole Polytechnique. Cours de Génie Atomique, Saclay, 1957-58; French Air Force (flying staff, 1952-57; atomic student, 1957-59); Thermonucl. Res. Lab., C.E.A. Fontenay-aux-Roses, 1959.
Nuclear interests: Thermonuclear fusion; Stability experiments on striction.
Address: 17 rue Neuve Notre-Dame, Versailles, Seine-et-Oise, France.

GINSBURG, David, B.A., LL.B. Born 1912. Educ.: West Virginia and Harvard Univs. Partner, Ginsburg and Feldman.
Nuclear interests: Legal aspects of U.S. atomic energy administration.
Address: Ginsburg and Feldman, 1700 Pennsylvania Avenue, N.W., Washington, D.C. 20006, U.S.A.

GINTHER, Robert J., B.S. Born 1917. Educ.: Northeastern Univ., Boston. Chem., U.S. Naval Res. Lab., Washington, D.C., 1946-. Societies: A.P.S.; Electrochem. Soc.; American Ceramic Soc.
Nuclear interests: Development of glass scintillators and materials for solid-state radiation dosimeters.
Address: 5507 Myrtle Avenue, Washington 22, D.C., U.S.A.

GIOK, Oey Hin. Lecturer, Nucl. Chem. Dept., Bandung Inst. of Technol.
Address: Bandung Institute of Technology, 10 Djl. Ganeca, Bandung, Indonesia.

GIORDANI, Roberto, Dr. Centro di Medicina Nucleare, Univ. Pisa.
Address: Universita di Pisa, Clinica Medica Generale, Centro di Medicina Nucleare, Pisa, Italy.

GIORDANO, Ch. Board Member, Centre Scientifique de Monaco.
Address: Centre Scientifique de Monaco, 16 boulevard de Suisse, Monte Carlo, Monaco.

GIORDANO, Roberto Luigi, Dottore in Fisica. Born 1934. Educ.: Catania Univ. Prof. of theoretical Phys. (1961-), Member, Regional Nucl. Res. Com., Catania Univ. Society: Italian Phys. Soc. (S.I.F.).
Nuclear interests: Nuclear physics; Theory of nuclear reactions and nuclear structure.
Address: Instituto Fisica Universita', 57 Corso Italia, Catania, Italy.

GIOVANNELLI, Giorgio, Prof., degree in Medicine. Born 1930. Educ.: Parma Univ. Formerly Asst. Ordinario, now Aiuto, Centro di Medicina Nucleare, Instituto di Clinica Pediatrica, Parma Univ. Society: Soc. di Medicina Nucleare.
Nuclear interest: Isotopes in childhood.
Address: Istituto di Clinica Pediatrica, Universita Parma, Italy.

GIOVANNI, Hugo J. DI. See DI GIOVANNI, Hugo J.

GIOVANNOZZI-SERMANNI, Giovanni, Dr. in Agricultural Sci.; Prof. in Appl. Biochem. Born 1931. Educ.: Perugia Univ. Head, Biochem. Lab., Istituto Sperimentale per i Tabacchi, 1957-60; Researcher, Consiglio Nazionale delle Ricerche and Head of Radiobiochem. Lab., 1961-. Societies: American Soc. Plant Physiologists; European Soc. of Radiobiol.
Nuclear interests: Studies of the ionization radiations on micro-organisms and plants; Biochemistry of irradiated cells; Use of tracers in normal and altered metabolism.
Address: Laboratory of Radiobiochemistry of Institute of Pharmaceutical Chemistry, University of Rome, Rome, Italy.

GIOVANOLA, M. Gen. Manager, Giovanola Frères S.A. Member, Conseil d'Administration, Energie Nucléaire S.A.
Address: Giovanola Frères S.A., Monthey, (Valais), Switzerland.

GIOVANOLA, R. Asst. to Manager, Giovanola Frères S.A.
Address: Giovanola Frères S.A., Monthey, (Valais), Switzerland.

GIRARDOT, Peter. Dean, Sci., Arlington State Coll., Texas A. and M. Univ.; Arlington Campus Rep., Oak Ridge Assoc. Univs. Council, 1967-.
Address: Arlington State College, Texas A. and M. University, Arlington, Texas, U.S.A.

GIRAUD, Francois., Ing. E.P.Z. et M.S.Sc., Dr. (M.I.T.). Directeur Sci., Bertin et Cie., S.A.
Address: Bertin et Cie., S.A., B.P. No. 3, Plaisir, (Seine et Oise), France.

GIRAUTA, Manuel ESTADA. See ESTADA GIRAUTA, Manuel.

GISCARD d'ESTAING, Philippe. Formerly Dept. Manager, Sté. d'Etudes Techniques et d'Entreprises Générales, Manager, Div. Recherches Instrumentation Nucl. et Sci., Cie. Francaise Thomson Houston Hotchkiss Brandt.
Address: 51 Boulevard de la Republique, P.O. Box 17, 78 Chatou, France.

GISLASON, Jakob, M.Sc., Icelandic elec. eng. Born 1902. Educ.: Copenhagen Univ. Director-Gen., State Elec. Authority, 1947-67; Director-Gen., Nat. Energy Authority, 1967-. Chairman, Icelandic Nat. Commission of World Power Conference; Chairman, Icelandic Nat. Commission of Internat. Commission on Large Dams; Hon. Sec. Icelandic Light-tech. Soc.; Vice-Chairman, Icelandic Nucl. Sci. Commission; Pres. Icelandic Management Assoc. Societies:

Icelandic Assoc. of Chartered Eng.; Icelandic Sci. Soc.; Icelandic Nat. Com. of C.I.G.R.E.
Nuclear interest: Nuclear power.
Address: 22 Barmahlid, Reykjavik, Iceland.

GITTERMAN, Henry. Chief Eng., Consultant to the U.S. Government (1956-66), Asst. Director, Nucl., Process and Aerospace Eng. Div., (1966-67), Deputy Director, Eng. Div. (1967-), Burns and Roe, Inc.
Address: Burns and Roe, Inc., Nuclear, Process and Aerospace Engineering Division, 160 West Broadway, New York 13, New York, U.S.A.

GITTUS, John Henry, B.Sc., A.I.M. Res. Manager, U.K.A.E.A., Springfields.
Nuclear interest: Metallurgy.
Address: Thatch Cottage, Wrea Green, near Preston, Lancs., England.

GIUFFRIDA, Anna Maria, Dottore in Scienze Biologiche e Naturali. Born 1932. Educ.: Catania and Messina Univs. Prof. di Chimica Biologica nella Facoltà di Scienze.
Nuclear interests: Researches with radioisotopes in biochemistry.
Address: Istituto Chimica Biologica, Via Androne 83, Catania, Italy.

GIULIANI, Alfred, Ing.; membre du Conseil Nat. de l'Energie Nucléaire. Ing.-électricien au Service d'Electricité de l'Etat, Luxembourg. Deleg. to 2nd I.C.P.U.A.E., Geneva, Sept. 1958.
Address: Ministère de l'Energie Nucléaire, 4 boulevard Roosevelt, Luxembourg, Luxembourg.

GIULOTTO, Luigi, Prof., Deleg. to 2nd I.C.P.U.A.E., Geneva, Sept. 1958. Chairman, Steering Com. Lab. Energia Nucl. Applicata, Pavia Univ. Director, Volta Exptl. Phys. Inst.
Address: Volta Experimental Physics Institute, Pavia University, Pavia, Italy.

GIUNTI, Torello, Dr. Sec., Com. for the Problems of the Pacific Use of Nucl. Energy and for the Contacts with Euratom. Italian member, Specialized Nucl. Section for Economic Problems, Economic and Social Com., Euratom.
Address: 11 Via Civinini, 00 187 Rome, Italy.

GIUNTINI, Carlo, M.D. Born 1932. Educ.: Pisa Univ. Res. Asst. Medical Clinic, Univ. Pisa, 1956-. Papers: On experimental and clinical uses of radioisotopes in endocrinology; the turnover of plasma phospholipids studied with P^{32}; the use of Cr^{51} for estimating blood red cells survival; the metabolism of plasma albumin labelled with I^{131}; the oxygen consumption by the lung tissue studied with Kr^{85}; the determination of cardiac output, right ventricular and pulmonary blood volumes by external counting; the statistics and the techniques of clinical uses of radioisotopes. Societies: Italian Soc. for Nucl. Biology and Medicine; New York Acad. Sci.
Address: Centro di Medicina Nucleare, Universita di Pisa, Italy.

GIUSTA, L. Directeur Gen., Sud-Aviation S.A.
Address: Sud-Aviation S.A., 37 boulevard de Montmorency, Paris 16, France.

GIUSTINA, Gianni, Ph.D. (Human Physiology), Ph.D. (Biochem.), Ph.D. (Medical Pathology). Born 1926. Educ.: Milan Univ. Asst., Dept. of Physiology (1950-55), Asst., Dept.of Medical Pathology (1955-), Milan Univ.
Nuclear interests: Nuclear medicine (thyroid function, haemodynamics, and haemoglobin biosynthesis).
Address: Istituto di Patologia Medica, Universita, Via F. Sforza 35, Milan, Italy.

GIUSTINIANI, Piero, Univ. degree in Eng. Born 1900. Educ.: Naples Univ. Managing Director, Montecatini Co., 1949-. Pres., Rhodiatoce Co.; Pres., Compagnie Néerlandaise de l'Azote; Sole Manager, Polymer Co.; Vice-Pres., SORIN, creators of Nucl. Res. Centre at Saluggia, Italy, Pres., Social and Economic Com., C.E.E. and C.E.E.A., 1964-; Member, Board of Directors, Lombard Industry Assoc.; Member for Italy, Economic and Social Com., Specialised Nucl. Sect. for Economic Problems, and Member, Specialised Nucl. Sect. for Social Health and Development Problems, Euratom. Society: Soc. of Chem. Industry, London.
Address: 30 Via Visconti di Modrone, 20 112 Milan, Italy.

GIVENS, James Wallace, Jr. B.S., M.S., Ph.D., D.Sc. Born 1910. Lynchburg Coll., Kentucky, Virginia and Princeton Univs. Prof., Tennessee Univ., 1947-56; Chairman, Mathematics Dept., Wayne State Univ., 1955-60; Prof., Northwestern Univ., 1960-; Assoc. Director (1962-64), and Director (1964-), Appl. Mathematics Div., Argonne Nat. Lab. Sect. Sec., American Assoc. for the Advancement of Science; Member of Council, Conference Board of Mathematical Sciences; Pres. Elect, Soc. for Ind. and Appl. Mathematics. Societies: Assoc. for Computing Machinery; Soc. for Ind. and Appl. Mathematics.
Nuclear interests: Concerned with provision of computing service for reactor design and management of theoretical research in this area.
Address: Applied Mathematics Division, Argonne National Laboratory, Argonne, Illinois 60439, U.S.A.

GJÖRUP, Henry Lund, M.Sc. (Elec. Eng.). Born 1921. Educ.: Denmark Tech. Univ. Head, Health Phys. Dept., Danish Atomic Energy Commission. Societies: H.P.S.; Inst. of Danish Civil Eng.
Nuclear interest: Health physics.
Address: Risö Huse, Pr. Roskilde, Denmark.

GJØTTERUD, Ole Kristoffer, Cand. real. Born 1931. Educ.: Oslo Univ. Fellow at Nordita, Copenhagen, 1959-61; Univ. Lecturer in Phys., Oslo Univ., 1962-.
Nuclear interest: Theoretical nuclear physics.
Address: 52 Hosleveien, Bekkestua, Norway.

GLADE, Edward H. Director, Div. of Headquarters Services, U.S.A.E.C.
Address: United States Atomic Energy Commission, Washington 20545, D.C., U.S.A.

GLADKOV, K. Book: Energiya atoma (Atomic Energy) (M. Detgiz, 1958).
Address: Academy of Sciences of the U.S.S.R., 14 Leninsky Prospekt, Moscow V-71, U.S.S.R.

GLADSTONES, John Sylvester, B.Sc. (Hons., Agriculture), Ph.D. Born 1932. Educ.: Western Australia Univ. Res. Officer, Western Australia Univ., 1955-58; Post-doctorate Fellow, Canadian N.R.C., Div. of Horticulture, Dept. of Agriculture, Central Exptl. Farm, Ottawa, 1958-59; Lecturer in Agronomy (1959-65), Sen. Lecturer (1966-), Western Australia Univ. Nuffield Foundation Dominion Travelling Fellow, 1968. Societies: Roy. Soc. of Western Australia; Australian Inst. of Agricultural Sci.
Nuclear interest: Use of artificially induced mutations in plant-breeding.
Address: Department of Agronomy, Institute of Agriculture, Western Australia University, Nedlands, Western Australia, Australia.

GLADWELL, G. E. Director and General Manager, Lintott Eng. Ltd.
Address: Lintott Engineering Ltd., Engineering Works, Foundry Lane, Horsham, Sussex, England.

GLADYSHEV, D. A. Papers: Co-author, Slow Neutron Spectrum in Horizontal Channel of VVR-S Reactor (letter to the Editor, Atomnaya Energiya, vol. 14, No. 5, 1963); co-author, Interaction Cross Section of Neutrons with ^{149}Sm and ^{115}In Nuclei (letter to the Editor, ibid, vol. 16, No. 6, 1964).
Address: Academy of Sciences of the U.S.S.R., 14 Leninsky Prospekt, Moscow V-71, U.S.S.R.

GLADYSHEV, V. A. Papers: Co-author, Design of 300 kev Sector-Focused Cyclotron with External Ion Injection (Abstract, Atomnaya Energiya, vol. 19, No. 5, 1965); co-author, Magnetic Field of 300 kev Sector-Focused Cyclotron with External Ion Injection (Abstract, ibid.).
Address: Academy of Sciences of the U.S.S.R., 14 Leninsky Prospekt, Moscow V-71, U.S.S.R.

GLAGOLEV, V. I. Paper: Co-author, The Reliability Analysis of Methods for Studying Fast Neutron and γ-Ray Continuous Spectra (Atomnaya Energiya, vol. 16, No. 3, 1964).
Address: Academy of Sciences of the U.S.S.R., 14 Leninsky Prospekt, Moscow V-71, U.S.S.R.

GLAGOLEV, V. M. Sci. Asst., Acad. of Sci. of the U.S.S.R.; Deleg. to 2nd I.C.P.U.A.E., Geneva, Sept. 1958; Deleg. to Conference sur la Physique des Plasmas et la Recherche concernant la Fusion Nucléaire Controlée, Salzbourg, Sept. 1961. Papers: Dispersion and Adsorption of Plasma Waves (Abstract, Atomnaya Energiya, vol. 20, No. 4, 1966); co-author, Paramagnetic High-Frequency Effect and Electron Paramagnetic Resonance in Plasma (ibid, vol. 20, No. 5, 1966).
Address: Academy of Sciences of the U.S.S.R., 14 Leninsky Prospekt, Moscow V-71, U.S.S.R.

GLASCOCK, Raymond Frederick, B.Sc. (1st class Hons. Chem.), Ph.D., D.Sc. (London). Born 1918. Educ.: London Univ. Head, Dept. of Radiobiochem., Nat. Inst. for Res. in Dairying, Reading Univ., 1949-. Member, M.R.C. Com. on Biol. (non-medical) Problems of Nucl. Phys. (jointly with the A.R.C. and the Natural Environment Res. Council). Books: Labelled Atoms (London, Sigma Books, Ltd., 1951); Isotopic Gas Analysis for Biochemists (New York, Academic Press, Inc., 1954). Society: Biochem. Soc.
Nuclear interest: Application of isotope techniques to biochemical research and especially in the use of tritium.
Address: Little Grazings, Church Lane, Aborfield, Berks., England.

GLASER, Arnold, Dr. Formerly Director Geophys., Aracon Labs., now Vice Pres., Aracon Div., Allied Res. Assoc., Inc.
Address: Allied Research Associates, Inc., Virginia Road, Concord, Massachusetts, U.S.A.

GLASER, Donald Arthur, B.S., Ph.D., Sc.D. (Hon., Case Inst. Technol.). Nobel Prize for Physics, 1960. Born 1926. Educ.: Case Inst. Technol. and California Inst. Technol. Prof., Michigan Univ., 1949-59; Prof. Phys. and Molecular Biol., California Univ., 1960-. Societies: Fellow, A.P.S.; Fellow, A.A.A.S.; N.Y. Acad. Sci.
Address: 229 Molecular Biology - Virus Laboratory, California University, Berkeley, California 94720, U.S.A.

GLASER, Herman, Dr. Sci. concerned with Nucl. Work, Hofstra Univ.
Address: Hofstra University, Hempstead, New York, U.S.A.

GLASGOW, Lyle E., B.S. (Mech. Eng.). Born 1920. Educ.: California Univ. Group Leader and Project Eng. for Sodium Reactor Experiment, 1956-63; Project Manager, Sodium-Cooled Fast Reactor Work, 1963-.
Nuclear interests: Reactor development, management, operations, engineering, physics, evaluation and design.
Address: P.O. Box 309, Canoga Park, California, U.S.A.

GLASKOV, Yu. Yu. Paper: Co-author, Measurements of Slow Neutron Spectra on Reactor Physical Stand at Beloyarskaya I.V. Kurchatov GRES (Atomnaya Energiya, vol. 15, No. 6, 1963).
Address: U.S.S.R. Academy of Sciences, 14 Leninsky Prospekt, Moscow V-71, U.S.S.R.

GLASOE, G. Norris, B.A., M.A., Ph.D. Born 1902. Educ.: St. Olaf Coll., Northfield, Minnesota and Wisconsin Univ. Assoc. Chairman, Phys. Dept., (1952-62), Asst. Director (1962-68), Sen. Sci., Phys. Dept. (1968-), Brookhaven Nat. Lab. Society: Fellow, A.P.S. Nuclear interests: Nuclear physics, neutron physics, particle accelerators, research and management.
Address: Brookhaven National Laboratory, Upton, New York 11973, U.S.A.

GLASOV, B. V. Paper: Co-author, Cryogenic Magnetic Mirror Machine WGL-2 (letter to the Editor, Atomnaya Energiya, vol. 21, No. 2, 1966).
Address: Academy of Sciences of the U.S.S.R., 14 Leninsky Prospekt, Moscow V-71, U.S.S.R.

GLASS, H. Bentley, Formerly Member and Chairman, Advisory Com. for Biol. and Medicine, U.S.A.E.C. Prof. Biol., Johns Hopkins Univ.; Deleg. to 2nd I.C.P.U.A.E., Geneva, Sept. 1958; Member, Governor's Com. on Nucl. Energy and Rad. Control Advisory Board, Exec. Dept., State of Maryland.
Address: Johns Hopkins University, Baltimore, Maryland, U.S.A.

GLASS, Harold I., B.A. Born 1933. Educ.: Trinity Coll., Dublin. Sen. Phys., Western Regional Hospital Board, Glasgow, 1959-63; Principal Phys., King's Coll. Hospital, 1963-64; Principal Phys., and Lecturer, Medical Phys., Roy. Postgraduate Medical School, Hammersmith Hospital, London. Society: Assoc., Inst. of Phys.
Nuclear interest: Application of radioactive isotopes in medicine.
Address: Department of Medical Physics, Hammersmith Hospital, Du Cane Road, London, W.12, England.

GLASSER, Georges. Prés.-Directeur Gén., Sté. Gén. de Constructions Electriques et Mécaniques "Alsthom". Member, Commission Consultative pour la Production d'Electricite d'Origine Nucléaire, C.E.A.
Address: Société Génerale de Constructions Electriques et Mécaniques "Alsthom", 38 avenue Kléber, Paris 16, France.

GLASSER, Herman. Born 1924. Pres., Nucl. Associates, Westbury, N.Y., 1966. Societies: H.P.S.; Soc. for Nucl. Medicine; Instrument Soc. America, A.N.S.; I.E.E.
Nuclear interests: Development of instruments for medical and research purposes.
Address: 4 Foxhurst Court, New Hyde Park, N.Y., U.S.A.

GLASSFORD, Hugh A. Formerly Pres., now Gen. Manager, Technical Assocs.
Address: Technical Associates, 140 West Providencia Avenue, Burbank, California, U.S.A.

GLASSON, Roy Acton. Born 1900. Educ.: London Univ. Formerly Managing Director, Johnsons Ethical Plastics Ltd. (now absorbed in Johnson and Johnson (Great Britain) Ltd.). Founder Member, U.K. Panel of Gamma and Electron Irradiation; Member, House Com., Canadian Red Cross Memorial Hospital, Taplow.
Address: Downham Lodge, Iver, Bucks., England.

GLATTER, Jacob, B.S., M.S. (Ceramic Eng.) (Illinois). Born 1923. Manager, Oxide Fuel Eng., Shippingport Cores Dept., Bettis Atomic Power Lab., Westinghouse Elec. Corp. Books: Contributor to: Corrosion and Wear Handbook for Water-cooled Reactors (U.S.A.E.C., 1957); The Shippingport Pressurized Water Reactor (Addison-Wesley, 1958); Uranium Dioxide: Properties and Nucl. Applications (U.S.A.E.C., 1961). Societies: American Ceramic Soc.; A.N.S.; A.S.M.E.
Address: 1043 Prospect Road, Pittsburgh 27, Pennsylvania, U.S.A.

GLATZ, G. Tech. Manager, Département Energie Nucléaire, S.A. des Ateliers de Sécheron; Sen. Officer, Seca.
Address: Département Energie Nucléaire, S.A. des Ateliers de Sécheron, Boite Postale, Geneva 21, Switzerland.

GLAUBER, Roy J., B.S., M.A., Ph.D. Born 1925. Educ.: Harvard Coll. and Harvard Univ. Postdoctoral Fellow, U.S.A.E.C., 1949-50; Frank B. Jewett Fellow, 1950-51; Lecturer, California Inst. Technol., 1951-52; Lecturer and Res. Fellow in Phys. (1952-54), Asst. Prof. Phys. (1954-56), Assoc. Prof. Phys. (1956-62), Prof.Phys. (1962-), Harvard Univ. Books: High Energy Collision Theory, Lectures in Theoretical Phys., Vol. I (New York, Interscience Publishers, 1959); Scattering of Neutrons by Statistical Media, Lectures in Theoretical Phys., Vol. IV (New York, Interscience Publishers, 1962). Societies: A.P.S.; American Acad. Arts and Sci.
Nuclear interests: Nuclear and elementary particle theory. Neutron physics, quantum electrodynamics, statistical mechanics, and solid state physics.
Address: Lyman Laboratory of Physics, Harvard University, Cambridge 38, Massachusetts, U.S.A.

GLAUBMAN, Michael Juda, M.S. (Hebrew Univ.), M.A., Ph.D. (Illinois). Born 1924. Educ.: Hebrew and Illinois Univs. At Princeton Univ., 1953-55; Columbia Univ., 1955-56; Atomics Internat., 1956-59; Northeastern Univ., 1959-. Society: A.P.S.

Nuclear interest: Nuclear structure.
Address: 9 Blueberry Lane, Lexington, Mass., U.S.A.

GLAUNER, Rolf. Born 1905. Educ.: Tübingen, Vienna, Berlin, and Munich Univs. Apl. Prof. für Radiologie, Tübingen Univ.; Chefarzt der Röntgen- und Strahlenabt., Marienhospital, Stuttgart; Editor, Fortschritte auf dem Gebiete der Röntgenstrahlen und der Nuklearmedizin. Society: Deutsche Röntgengesellschaft.
Nuclear interest: Medizinischer strahlenschutz.
Address: Röntgenabteilung, Marienhospital, Stuttgart, Germany.

GLAZKO, Anthony J., Ph.D. (Biochemistry), A.B. Born 1914. Educ.: California Univ., Berkeley. Lab. Director, Chem. Pharmacology, Res. Dept., and Chairman, Radioisotopes Committee, Parke, Davis and Co., Ann Arbor. Societies: A.C.S.; American Physiological Soc.; American Soc. for Pharmacology and Exptl. Therapeutics; Soc. for Exptl. Biol. and Medicine; American Therapeutic Soc.
Nuclear interest: Drug metabolism studies with isotope-labelled compounds.
Address: Research Laboratories, Parke, Davis and Co., Ann Arbor, Michigan, U.S.A.

GLAZKOV, O. M. Eng., U.S.S.R. State Com. for the Utilisation of Atomic Energy.
Address: U.S.S.R. State Committee for the Utilisation of Atomic Energy, 26 Staromonetnii Pereulok, Moscow, U.S.S.R.

GLAZKOV, Yu. Yu.
Nuclear interest: Reactor physics.
Address: Nuclear Energy Institute, Academy of Sciences of the U.S.S.R., 14 Leninsky Prospekt, Moscow V-71, U.S.S.R.

GLAZOV, A. A. Papers: Co-author, A Cyclotron with a Spatially Varying Magnetic Field (Atomnaya Energiya, vol. 8, No. 3, 1960); co-author, High Frequency System for the Proton Accelerator Constructed as a Cavity Resonator (Nukleonika, vol. 8, No. 2, 1963); co-author, The Effect of Space Charge on the Frequency of Free Oscillations in an Isochronous Cyclotron (Atomnaya Energiya, vol. 15, No. 3, 1963); co-author, Programming the Accelerating Voltage Amplitude of a Model of the Circular Proton Phasotron R.F. System (Nukleonika, vol. 12, No. 7-8, 1967).
Address: Academy of Sciences of the U.S.S.R., 14 Leninsky Prospekt, Moscow V-71, U.S.S.R.

GLAZUNOV, P. Ya. Paper: Co-author, Apparatus for Rad. - Chem. Process with Reactor, Provided with Uniform Temperature (Atomnaya Energiya, vol. 20, No. 5, 1966).
Address: Academy of Sciences of the U.S.S.R., 14 Leninsky Prospekt, Moscow V-71, U.S.S.R.

GLEASON, Geoffrey I. Nucl. Sci., Abbott Labs., -1964; Res. Sci., Special Training Div., O.R.I.N.S., 1964-.
Nuclear interests: Analytical problems of radioactive materials, application of radioactive materials to analysis and quality control, production of radionuclides, and measurement of nuclear properties.
Address: Oak Ridge Institute of Nuclear Studies, P.O. Box 117, Oak Ridge, Tenneseee, U.S.A.

GLEASON, James P. Private Law Practice, Washington; Formerly Chairman, Washington Suburban Transit Commission; Formerly Asst. Administrator, N.A.S.A.; Member, Panel of Experts to Provide Boards on Atomic Safety and Licensing, U.S.A.E.C., 1967-.
Address: United States Atomic Energy Commission, Panel of Experts to Provide Boards on Atomic Safety and Licensing, Washington, D.C. 20545, U.S.A.

GLEESON, Austin M., Ph.D., B.Sc., M.Sc. Born 1938. Educ.: Drexel Inst. of Technol. and Pennsylvania Univ. Res. Asst., Pennsylvania Univ., 1963-65; Res. Assoc. (1965-67), Asst. Prof. (1967-), Syracuse Univ. Society: A.P.S.
Address: Physics Department, Syracuse University, Syracuse, New York 13210, U.S.A.

GLEICH, David. Sen. Eng., Nucl. Div., Arde Inc.
Address: Nuclear Division, Arde Incorporated, 580 Winters Avenue, Paramus, New Jersey, U.S.A.

GLEIT, Chester Eugene, A.B., M.S., Ph.D. Born 1933. Educ.: Chicago Univ. and M.I.T. Union Carbide Fellow, M.I.T., 1957; Sen. Sci., Bettis Atomic Power Lab., Pittsburgh, Pa., 1958-60; Head, Advanced R. and D. Dept., Tracerlab., Div. of Lab. for Electronics, 1961-64; Assoc. Prof., North Carolina State Univ., Raleigh, 1964-. Societies: Analytical Div. A.C.S.; A.A.A.S.; American Soc. for Contamination Control; North Carolina Acad.of Sci.; American Microchem. Soc.
Nuclear interests: Application of nuclear techniques for the measurement of chemical and physical properties; Nuclear chemistry; Plasma chemistry.
Address: Withers Hall, North Carolina State University, P.O. Box 5247, Raleigh, North Carolina 27607, U.S.A.

GLEN, Herbert Moore, C.E. Born 1905. Educ.: South Carolina Univ. Supervisor of Structural Design, O.R.N.L., Oak Ridge, Tennessee, 1946-. Pres., Oak Ridge Branch, A.S.C.E.; Chairman, A.S.C.E.; Structural Div. Com. on Nucl. Structures and Materials; Editor, Radioisotope Facilities, Nucl. Structural Eng. Journal (North Holland Publishing Co.).
Nuclear interests: Radioisotope structures and materials.
Address: 111 East Magnolia Lane, Oak Ridge, Tennessee, U.S.A.

GLENNAN, T. Keith, B.S. (E.E.), D.Sc. (Hon.), D. Eng. Born 1908. Educ.: Yale Univ.

Commissioner, U.S.A.E.C., 1950-52; Member, Gen. Advisory Com., U.S.A.E.C., 1956; Director, Nat. Sci. Foundation, 1956; Administrator, N.A.S.A., 1958-61. Former Pres., Case Inst. Technol., Cleveland, Ohio. Pres., Assoc. Univs., Inc., 1965-. Society: American Acad. Arts and Sci.
Nuclear interests: Nuclear propulsion in missiles and space rockets, nuclear energy used to provide auxiliary power for satellite operations.
Address: Associated Universities Inc., c/o Brookhaven National Laboratory, Upton, Long Island, New York, U.S.A.

GLICK, Arnold Julian, B.A., Ph.D. Born 1931. Educ.: Brooklyn Coll. and Maryland Univ. N.S.F. Postdoctoral Fellow, Weizmann Inst. of Sci., 1959-61; Asst. Prof. (1961-67); Assoc. Prof. (1967-), Maryland Univ.; N.S.F. Postdoctoral Fellow, Paris Univ., 1967-68. Societies: A.P.S.; F.A.S.
Nuclear interests: Nuclear physics and the many-body problem.
Address: Physics Department, Maryland University, College Park, Maryland, U.S.A.

GLIMSTEDT, Ulf, M.Sc. (Eng.), Board Member, Atomkraftkonsortiet. Vice Chairman, Board of Directors, Oskarshamnsverkets Kraftgrupp A.B.
Address: Atomkraftkonsortiet, 19 Stureplan, Stockholm C, Sweden.

GLINSKII, Konstantin V. Sen. Eng., U.S.S.R. State Com. for the Utilisation of Atomic Energy.
Address: U.S.S.R. State Committee for the Utilisation of Atomic Energy, 26 Staromonetnii Pereulok, Moscow, U.S.S.R.

GLOCKER, Richard, Dr. phil. (Munich), Dr. med. h.c. (Tübingen). Born 1890. Educ.: Munich Univ. Prof., Stuttgart T.H.; Direktor des Röntgeninstitutes der T.H., Stuttgart (retired). Mitherausgeber der Fortschritte auf dem Gebiet der Röntgenstrahlen und der Nuklearmedizin, Verlag G. Thieme, Stuttgart. Book: Röntgen- und Kernphysik für Mediziner (2 Aufl. Stuttgart, Verlag G. Thieme, 1965). Societies: Hon. Member, Deutsche Röntgengesellschaft; Hon. Member, Gesellschaft für Metallkunde.
Nuclear interests: Dosimetry and protection.
Address: 10 Robert Boschstrasse, Stuttgart-N, Germany.

GLODEN, Raoul-Francois, L. ès Sc., Diplôme d'études supérieures ès Sciences math. Doctorat 3e cycle, spécialité: Physique des accélérateurs de particules. Born 1933. Educ.: Paris Univ. and I.N.S.T.N., Saclay. Nommé prof. à l'Athénée de Luxembourg en 1958; actuellement fonctionnaire scientifique du CETIS au Centre Commun de Recherche d'Ispra (Euratom). Societies: Sté. Mathématique de Belgique; Association Luxembourgeoise pour l'Utilisation Pacifique de l'Energie Atomique.
Nuclear interests: Théorie et technique des accélérateurs de particules; Etude des plasmas au point de vue théorique; Physique du réacteur, neutronique.
Address: 22 Vîa Pasubio, Varese, Italy.

GLOOR, Bruno R., M.S.E.E., Dr. sc. tech. Born 1921. Educ.: Swiss Federal Inst. Technol. Res. Assoc., Inst. for h.f. electronics, Swiss Federal Inst. Technol., Zurich, 1946-54; Testing Lab. rotating machinery (1954-58), Transformer Testing Lab. (1958-61), Magnet Design Dept. (1961-64), now Vice-Pres., Central Res., Oerlikon Eng. Co., Zurich. Book: Studien über einkreisige Schwingungssysteme mit zeitlich veränderlichen Elementen (Zürich, Verlag Lehmann, 1955). Societies: Schweiz. Elektrotechn. Verein (Swiss Assoc. of Elec. Eng.); Schweiz. Phys. Gesellschaft (Swiss Assoc. of Phys.); Zürcher. Phys. Gesellschaft (Zurich Assoc. of Phys.); I.E.E.E.; Deutsche Phys. Gesellschaft.
Nuclear interests: Nuclear physics (theoretical and experimental); Nuclear reactors.
Address: 10 Bohlstrasse, 8355 Aadorf TG, Switzerland.

GLOS, Margaret Beach, B.A. Born 1936. Educ.: Smith Coll., Columbia Univ. Sci. Editor, McGraw-Hill Book Co., 1958-60; Asst. Editor, Nucleonics Magazine, 1960-67; Asst. Managing Editor, Sci. Res. Mag., 1966-67; Administrator, Soc. of Nucl. Medicine, and Managing Editor, J. Nucl. Medicine, 1967-. Societies: American Astronomical Soc.; A.A.A.S.; A.N.S.; Soc. Nucl. Medicine. Nuclear interests: Nuclear medicine and biology, radiation applications, radiotracers, fusion.
Address: 211 East 43rd Street, New York, New York 10017, U.S.A.

GLOWER, Donald Duane, Sen., Ph.D. (Nucl. Eng.), M.S. (Mech. Eng.), B.S. (Sci. and Eng.), B.S. (Marine Eng.). Born 1926. Educ.: Iowa State Univ., Antioch Coll., Ohio and U.S. Merchant Marine Acad. Asst. Prof. and Member, Graduate Faculty, Iowa State Univ., 1954-60; Basic Res., Sandia Corp., 1960-63; Sen. Exec., Gen. Motors Corp., 1963-64; Chairman, Nucl. Eng., Ohio State Univ., 1964-. Tech. Consultant, 1964-. Chairman, A.S.E.E. Relations with A.E.C. (Subcom.of Com. on Relations with Federal Govt.); Chairman, Educ. Com., S.W. Ohio Sect., Member, Programme Com., A.N.S.; Member, Nucl. Eng. Educ. Com., Assoc. Midwest Univs.- Argonne Nat. Lab. Book: Experimental Reactor Analysis and Radiation Measurements (McGraw-Hill, 1965). Societies: A.P.S.; A.S.E.E.; I.E.E.E.; Ohio Acad. of Sci.; A.A.A.S.
Nuclear interests: Actively consulting with industry and government in the areas of: nuclear instrumentation, radiation effects in materials with emphasis on electronic devices, nuclear power, reactor design and safety.

Address: 2338 Kensington Drive, Columbus 21, Ohio, U.S.A.

GLOYNA, Earnest F., B.S., M.S., Dr. Eng. Born 1921. Educ.: Texas Technol. Coll., Texas and Johns Hopkins Univs. Prof. Environmental Health Eng. (Civil Eng.), Director Environmental Health Eng. Res. Lab., Director, Centre for Res. in Water Resources, Chairman, Radioisotope Com., Texas Univ.; Consultant, Air Force Directorate of Nucl. Safety; Special Consultant, U.S.P.H.S.; Consultant, U.S. Senate, Water Resources Select Com.; Consultant, Nucl. Div., Gen. Dynamics; Consultant, Los Alamos Sci. Lab.; Consultant, U.S. Army Chem. Centre Nucl. Defence Lab.; Res. participant, O.R.N.L.; Com. Chairman, Sanitary Eng. Aspects of Nucl. Energy, A.S.C.E.; Honorary member, Southwest Soc. of Nucl. Medicine; Chairman, Tech. Advisory Com. to Water Pollution Control Board; Com. Chairman, Advisory Com. to Texas State Health Dept. on Design, Construction, Operation of Sewerage Systems; Director, Radiol. Health Conference, A.P.H.A. Societies: A.S.C.E.; American Water Works Assoc.; American Assoc. of Profs. of Sanitary Eng.; Water Pollution Control Fed.; A.I.Ch.E.; Diplomate, American Acad. of Sanitary Eng.; American Environmental Eng. Intersociety Board.
Nuclear interests: Radiological health; Radioactive waste disposal; Air pollution control; Water treatment; Radioactivity transport in the streams and estuaries; High-level waste disposal in underground formations.
Address: 305 Engineering Laboratories Building, Department of Civil Engineering, Texas University, Austin 12, Texas, U.S.A.

GLUBRECHT, Hellmut. Born 1917. Educ.: Hanover T.H. and Göttingen Univ. Formerly Obering., Phys. Inst., Hanover T.H. and Docent of Biophys. Director, Inst. of Radiobiol., Hanover T.H., 1959-. Director, Ausseninst. der Hannoverschen Hochschulen. Societies: Deutsche Physikalische Gesellschaft; Deutsche Gesellschaft für Biophysik; Assoc. for Rad. Res.
Nuclear interests: Fundamental problems of radiobiology; Use of radioisotopes in biology, agriculture and horticulture; Shielding problems.
Address: 2 Herrenhäuserstrasse, Hanover, Germany.

GLUCKSTERN, Robert L., B.E.E., Ph.D. Born 1924. Educ.: M.I.T. Res. Assoc., Asst. Prof., Assoc. Prof., Yale Univ., 1950-64; Prof. and Head, Dept. of Phys. and Astronomy, Massachusetts Univ., 1964-. Society: Fellow, A.P.S.
Nuclear interests: Nuclear physics; Theory of nuclear accelerators.
Address: Hasbrouck Physics Laboratory, Massachusetts University, Amherst, Massachusetts 01002, U.S.A.

GLUECKAUF, Eugen, Dr.-Ing., M.Sc., D.Sc. Born 1906. Educ.: Berlin T.H. and London Univ. Group Leader, Phys. Chem. Group (1948-53), Group Leader, Fission Product Technol. Group (1953-58), Branch Head, Radiochem. Branch, (1958-), A.E.R.E., Harwell. Book: Atomic Energy Waste, its Nature, Use and Disposal (Butterworths and Interscience Publ., 1961).
Nuclear interests: Physical chemistry: ion exchange, solvent extraction, solution chemistry, adsorption, chromatography, isotope separation, high-temperature diffusion phenomena, high-temperature reactor systems, waste disposal processes; Desalination of sea water.
Address: Bankside, Chilton, Didcot, Berks., England.

GLUKHIKH, V., Ing. Deleg. to Conference sur la Physique des Plasmas et la Recherche concernant la Fusion Nucléaire Controlée, Salzbourg, Sept. 1961.
Address: Institute of the Electro-Physical Apparatus, Leningrad, U.S.S.R.

GLUSHKOV, E. S. Papers: Co-author, Methods of Neutron-phys. Calculation in the Phys. Design of Power Reactors (Atomnaya Energiya, vol. 11, No. 1, 1961); co-author, On Phys. Profiling Heat Production in Heterogeneous Power Reactors (letter to the Editor, ibid., vol. 12, No. 5, 1962); co-author, Solution of the Reactor Equations with Allowance for the Variable Density of the Moderator (letter to the Editor, ibid., vol. 12, No. 5, 1962); co-author, Exptl. Investigation for Shaping of Heat Rate by Non uniform Fuel Distribution (ibid., vol. 20, No.6, 1966).
Address: Academy of Sciences of the U.S.S.R., 14 Leninsky Prospekt, Moscow V-71, U.S.S.R.

GLUSHNEV, V. E. Paper: Co-author, Apparatus for Rad. - Chem. Process with Reactor, Provided with Uniform Temperature Field (Atomnaya Energiya, vol. 20, No. 5, 1966).
Address: Academy of Sciences of the U.S.S.R., 14 Leninsky Prospekt, Moscow V-71, U.S.S.R.

GNAM, Erich O. A., Dipl. Ing., Dr.Ing. Born 1908. Educ.: Munich T.H. Head, Nucl. Dept., M.A.N. Maschinenfabrik Augsburg-Nürnberg A.G. Society: V.D.I.
Nuclear interests: Management. Reactor design; Heat transfer.
Address: Maschinenfabrik Augsburg-Nürnberg A.G., Werk Nürnberg, 85 Nürnberg 2, Postfach, Germany.

GNANAOLIVU, Arjunan. Deputy Sec., Dept. of Atomic Energy, A.E.C., India.
Address: Atomic Energy Commission, Department of Atomic Energy, Chhatrapati Shivaji Maharaj Marg, Bombay 1, India.

GNEDENKO, B. V. Prof., Kiev Univ. Book: Co-author, Introduction à la théorie des

GOA

probabilities (Paris, Editions Dunod).
Address: Kiev University, 58 Vladimirskaya, Kiev, U.S.S.R.

GOAD, Colin, B.A. Born 1914. Educ.: Cambridge Univ. Deputy Sec. Gen., (1963-), now Sec. Gen., Intergovernmental Maritime Consultative Organisation.
Nuclear interest: Safety of nuclear ships
Address: 23A Burgh Heath Road, Epsom, Surrey, England.

GOBBATO, Bianca, Ing. Asst. to Director, FIAT, Sezione Energia Nucleare.
Address: 10/20 Corso Marconi, Turin, Italy.

GOBLE, Alfred Theodore, B.A., Ph.D. Born 1909. Educ.: Wisconsin Univ. Prof. Phys. (1945-), Acting Chairman, Dept. of Phys. (1956-57, 1962-63), Chairman (1966-), Union Coll.; Consultant, Revere Copper and Brass, Inc., 1948-; Consultant, Ramo-Wooldridge Corp., 1955-58; Consultant, Space Technol. Lab., 1958-60; Consultant, Aerospace Corp., 1960-64. Book: Co-author, Elements of Modern Phys. (The Ronald Press Co., 1962). Societies: A.P.S.; A.A.P.T.; Optical Soc. of America.
Nuclear interests: Educational aspects in general; Nuclear physics; Isotope shifts.
Address: Department of Physics, Union College, Schenectady, New York, U.S.A.

GODAR, Serge, L. en Sc. Chimiques. Born 1928. Educ.: Bruxelles Libre Univ. With Belgonucléaire S.A., 1956-59; Radioisotopes Group, Res. and Training; Euratom, 1960-66; Euratom Expert, 1966-67; Applied Research S.P.R.L., 1967-.
Nuclear interests: Industrial uses of radioisotopes, production of radioisotopes, recovery of fission products; Production of inorganic ion exchangers and chemical products for nuclear industry.
Address: 161 avenue Croix du Feu, Brussels 2, Belgium.

GODARD, Pierre. Gérant, Pompes Wauquier.
Address: Pompes Wauquier, 69 rue de Wazemmes, Lille, (Nord), France.

GODART, J. R., Ing. Civil, AI.Ms.Electromécanicien. Born 1924. Educ.: Faculté Polytechnique de Mons. Ing. en Chef, Ateliers J. Hanrez, Monceau-sur-Sambre. Societies: Sté. Belge de Physique; Sté. Française de Physique; Inst. Belge de Régulation et d'Automatisme.
Nuclear interests: Hot-cells, charging machines, manipulators, grabs, airlocks, transfer mechanisms.
Address: 203 rue de Marchiennes, Montignies le Tilleul, Belgium.

GODBOLD, Brian Craig, B.Sc. (Hons.) (Special, Phys.), C. Eng. Born 1914. Educ.: London Univ. Communications Eng., S.E. and E. areas, C.E.B., 1940-54; Sen. Asst. Eng., Nucl. Power Branch (1954-55), Sen. Asst. Eng. (Services, Health Phys., etc.), Nucl. Power Branch (1955-57), Reactor Design, Instrumentation and Control Eng., Nucl. Generation Sub. Dept. (1957-59), C.E.A.; Principal Health Phys., Nucl. Health and Safety Dept., C.E.G.B., 1959-. Societies: M.I.E.E.; A.Inst.P.; M.I.E.E.E.
Nuclear interests: Radiation protection aspects of station layout, design and operation; Environmental survey and communications; Health physics, instrumentation. Reactor control and instrumentation.
Address: Burnsands, 47 Main Road, Gidea Park, Romford, Essex, England.

GODDEN, Bernard, B.Sc. (Eng.). Born 1928. Educ.: Bristol Univ. Process Systems Design Branch, Chalk River Nucl. Labs., 1961-.
Nuclear interest: Design of irradiation equipment, particularly reactor loops.
Address: Box 651, Deep River, Ontario, Canada.

GODED, Federico, D.of C.C.P., Graduated at the Argonne I.S.N.S.E. Born 1917. Educ.: Escuela Tecnica Superior de Ingenieros de Caminos, Canales y Puertos (E.T.S. de I.C.C. y P.), Madrid and Ecole Politechnique de Lausanne. Prof. Nucl. Phys. and Nucl. Eng., E.T.S. de I.C.C. y P., 1957; Consulting Eng., Gen. Elec. Espanola, 1955; Vice Director, Gabinete de A.N. a las O.P., 1961; Prof. Nucl. Eng., E.T.S. de I.I., 1966-. Book: Teoria de Reactores y Elementos de Ingenieria Nuclear (Reactor Theory and Principles of Nucl. Reactor Eng. (Junta de Energia Nuclear, 1958 and 1964). Society: Assoc. of Spanish Civil Engs.
Nuclear interests: Reactor design and nuclear physics.
Address: 7 Fortuny, Madrid, 4, Spain.

GODEL, Dieter, Dipl.-Phys. Born 1937. Educ.: Stuttgart Univ. Standard Elektrik, Stuttgart; Standard Telephones and Cables Ltd., London; Max-Planck-Inst., Stuttgart.
Nuclear interests: Mass spectrometry; Microprobe-analysis; X-ray- and electron-diffraction of gas-metal systems e.g., refractory metals and o rare earth metals systems; Electrical resistivity of liquid metals and alloys.
Address: 92 Seestrasse, 7000 Stuttgart-N, Germany.

GODELLE, Maurice, Ing. E.S.E. Born 1930. Educ.: E.S.E., Malakoff (Seine), France. At C.C.R. Euratom, Ispra (Varese), Italy. Book: Chapter 18, vol. 2, of Génie Atomique: Mécanismes de contrôle (INSTN, France, 1960). Nuclear interests: Reactor engineering and reactor construction.
Address: 25 via Cavour, Malgesso (Varese), Italy.

GODFRAIND, Theophile, D. in Medicine, Agrégé de l'Enseignement Supérieur. Born 1931. Educ.: Louvain Univ. Prof. of Pharmacology; Medical School, Lovanium, Leopoldville Univ., Congo, 1958-63; Medical School, Louvain Univ., Belgium, 1964-.

Nuclear interests: Application of radioisotopes to the study of the mode of action of drugs acting on cell membrane, namely chemical transmitters and cardiac clycosides.
Address: Laboratoire de Pharmacodynamie Générale, University of Louvain, Belgium.

GODFREY, Robert K., Dr. Biol. Sci., Florida State Univ.
Address: Florida State University, Tallahassee, Florida, U.S.A.

GODFROI, Edmond Edgard André Jules, Dr. en Médecine. Born 1927. Educ.: Liège State Univ. Res. Medical Officer, Centre d'Etudes de l'Energie Nucléaire, 1957-59; Medical Consultant, Health Phys. Sect., Eurochemic, 1959-; Part-time Collaborator, Dept. Nucl. Phys. and Chem., Liège Univ., 1960-65; i/c., Dept. of Ind. Medicine and Health Phys., Assoc. Intercommunale de Services et Etablissements Medico - Sociaux du Bassin de Seraing. Society: H.P.S.
Nuclear interests: Medicine and biology; Health physics.
Address: 12 avenue de l'Europe, Seraing, Liège, Belgium.

GODSIN, Walter Woodrow, B.S. (Chem.). Born 1923. Educ.: Gannon Coll., Erie, Pennsylvania. With Westinghouse Elec. Corp., Bettis Atomic Power Div., 1951-56; Gen. Dynamics, Convair, 1956-57; Gen. Dynamics, Gen. Atomic Div., 1957-.
Nuclear interests: Design and operation of engineering and metallurgical irradiation experiments for the development of reactor fuels and reactor systems; Development of thermionic and thermo electric devices for space power systems. Study of transient radiation effects in structural materials.
Address: Gulf General Atomic, P.O. Box 608, San Diego 12, California, U.S.A.

GODWIN, John Thomas, B.S., M.D. Born 1917. Educ.: Emory Univ. Asst., Sloan-Kettering Inst., Radioautographic Div., Phys. Dept., 1949-50; Head, Pathology Div., Brookhaven Nat. Lab., 1951-55; Asst., Pathology Div., Sloan-Kettering Inst., 1954-55; Pathologist, Brookhaven Nat. Lab. Hospital, 1951-55; Res. Collaborator, Brookhaven Nat. Lab., 1955-58; Pathologist and Director of Radioisotope Labs., St. Joseph's Infirmary, Atlanta, 1955-63; With Dynatomics, Inc., 1963-. Chairman, Cytology Com. and Member, Board of Directors, American Cancer Soc., Georgia Div.; Medical Advisory Group, Georgia Tech. Res. Reactor; Member, Governors Georgia Nucl. Advisory Com., Head, Nucl. Reactor Biomedical Facility and Special Res. Sci., Georgia Inst. Technol. Societies: Fellow, College of American Pathologists; Fellow, American Soc. Clinical Patholigists; American Assoc. Pathologists and Bacteriologists; American Soc. for Exptl. Pathology; A.M.A.; Soc. Nucl. Medicine and many others.
Nuclear interests: Thermal neutron capture therapy; Application of short half-life isotopes in human therapy; General radiobiology.
Address: 1164 Springdale Road, N.E., Atlanta 6, Georgia, U.S.A.

GODWIN, Richard Philip, Bachelor of Eng. (Yale). Born 1922. Educ.: Ohio State and Yale Univs. Asst. Director, Div. Reactor Development (Maritime), U.S.A.E.C. and Chief, Office of R. and D., Maritime Administration, Dept. of Commerce, Washington, D.C., 1957-61; Manager, Corporate Planning and Development, Bechtel Corp., San Francisco, California, 1961-. Societies: Soc. of Naval Architects and Marine Eng.; Yale Eng. Club; A.N.S.
Nuclear interests: Reactor engineering and construction; Management of U.S. programme to build nuclear-powered merchant ships, principally the N.S. Savannah; Fast breeder reactor design.
Address: 130 Highland Avenue, San Rafael, California 94901, U.S.A.

GOECKERMANN, Robert H., B.S. (Chem.), Ph.D. (Chem.). Educ.: Wisconsin and California Univs. Res. Assoc., Chem. and Phys., Princeton Univ., 1949-52; Res. Chem., California R. and D. Co., 1952-53; Radiochem. Div. Leader (1953-64), Assoc. Director (1964-67), Lawrence Rad. Lab.; Sci. Rep. in Buenos Aires, U.S.A.E.C., 1967-.
Address: U.S. Embassy, Buenos Aires, Argentina.

GOEDECKE, Fritz, Ing. Born 1904. Educ.: Wuppertal Ingenieurschule. Geschäftsführer und technischer Direktor, Stahl- und Röhrenwerk Reisholz G.m.b.H., 1959-; Vorstandsmitglied, Thyssen Röhrenwerke A.G., Düsseldorf, 1966-. Vorstandsmitglied, Verein Deutscher Eisenhüttenleute, Düsseldorf; Aufsichtsratsmitglied: Baugesellschaft Reisholz m.b.H., Düsseldorf-Reisholz; Wuragrohr G.m.b.H., Wickede (Ruhr); Aufsichtsratsvorsitzender, Präzisrohrwerk Holzhausen G.m.b.H., Holzhausen.
Nuclear interests: Reactor vessels; Reactor cooling jackets; Pressure vessels; Support rings; flanged rings; Closures; Thick-walled tubes; Plating (build-up welding); Heat exchangers; Pipe coils; Pipe coil systems; Pipeline components; Large workpieces for pipelines; Components for turbines, such as shafts, discs, rotors etc.; Generator shafts; Boiler and supercharger tubes; Precision steel tubes.
Address: 7 Spohrstr., Düsseldorf-Benrath, Germany.

GOEDKOOP, Jacob A., Ph.D. (Amsterdam). Born 1921. Educ.: Amsterdam Univ. and Pennsylvania State Coll. At Joint Establishment for Nucl. Energy Research, Kjeller, Norway (1951-59), lastly as Director of Phys., with half a year's interruption in 1955, during which Scientific Sec. with U.N. for 1955 Geneva Conference; Director of Phys. (1959-61), Managing Director for Res. (1961-), Reactor Centrum Nederland, Petten, Netherlands; Extramural

**Prof., Leyden Univ., 1958-. Vice Pres. (1966-68), Exec. Vice Pres. (1968-), European Atomic Energy Soc. Societies: Netherlands and Norwegian Phys. Socs.; Royal Netherlands' Chem. Soc.; American Crystallographic Assoc.
Nuclear interest: Neutron beam research, especially neutron diffraction.
Address: R.C.N., Petten, Netherlands.

GOELER, Eberhard VON. See VON GOELER, Eberhard.

GOENS, Julien Raymond Jean, Ing. civil (A.I.A.-Gn.), L. en Sc. phys. (U.L.B.). Born 1916. Educ.: Ecole Roy. Militaire and Libre Univ. de Bruxelles. Asst. Prof. Phys., Ecole Royale Militaire, 1946-51; Sci. Attache, Belgian Embassy, Washington, 1951-55; Asst. Gen. Manager, C.E.N. Res. Centre, Mol, 1955-63; Gen. Manager, C.E.N., 1963-. Societies: Ste. Belge de Phys.; A.N.S.; A.P.S.; F. Inst. P.
Nuclear interest: Nuclear management.
Address: C.E.N., Mol-Donk, Belgium.

GOERTZ, Raymond C., B.S. (Eng. Phys.). Born 1915. Educ.: Montana State Coll., Polytech. Inst. Brooklyn and Illinois Inst. Technol. from Group Leader to Director, Remote Control Eng. Div., Argonne Nat. Lab., 1947-. Societies: A.I.E.E.; I.R.E.; Argonne Branch, RESA; Nucl. Eng. Div., A.I.Ch.E.
Nuclear interests: Development of remote handling systems such as general-purpose manipulators, viewing systems, slave robots, fuel process systems and remote handling aspects of reactor systems.
Address: 5510 Fairmont Avenue, Downers Grove, Illinois, U.S.A.

GOESCHEL, Heinz, Dr.-Ing., Dr.-Ing. E.h., Dr.-Ing. h.c., Honorarprof. und Ehrensenator Braunschweig T.H. Born 1906. Educ.: Munich and Braunschweig T.H. Vorstandsmitglied, Siemens-Schuckertwerke A.G. Mitglied, Prasidium, Deutsche Atomkommission; Mitglied des Präsidiums und Vorsitzender des Arbeitskreises I-Wissenschaft und Technik, Deutsches Atomforum; Mitglied, Wissenschaftsrat.
Address: 101 Burgbergstrasse, 852 Erlangen, Germany.

GOETHE, O. Norwegian Member, Nordic Contact Com. for Atomic Energy.
Address: c/o Dr. Hans Hakansson, Kgl. Handelsdepartementet, Stockholm 2, Sweden.

GOETHEM, A. VAN. See VAN GOETHEM, A.

GOFF, Lucien LE. See LE GOFF, Lucien.

GOFFART, J. Formerly Gen. Manager, Etude et Construction Evence Coppee-Rust, S.A.
Address: Evence Coppee et Cie., 103 boulevard de Waterloo, Brussels 1, Belgium.

GOFFIN, Jules. Gen. Director, Crépelle et Cie.
Address: Crépelle et Cie., Porte de Valenciennes, Lille, (Nord), France.

GOFMAN, John William, A.B. (Chem.). Ph.D. (Chem.), M.D. Born 1918. Educ.: Oberlin Coll. and California Univ. Prof. Medical Phys., California Univ., Berkeley, 1954; Director of Biol. and Medicine, (1963-), Formerly Assoc. Director (1963-), now Assoc. Director, Biomedical Res. California Univ. Lawrence Rad. Lab.
Nuclear interests: All aspects of the implications of radionuclide release, from whatever source, upon their biosphere.
Address: Lawrence Radiation Laboratory, P.O. Box 808, Livermore, California, U.S.A.

GOGLIA, Mario J., M.E., M.S., Ph.D. Born 1916. Educ.: Stevens Inst. of Technol. and Purdue Univ. Vice Chancellor for Res., Board of Regents, Univ. System of Georgia. Member, Nat. Advisory Com. Nat. Defence Education Act. Title IV Fellowship Programme, Office of Educ., and Member, Eng. Panel, Facilities Branch, Div. of Graduate Programmes, Bureau of Higher Educ., Office of Educ., U.S. Dept. of Health, Educ. and Welfare; Member, Advisory Panel, Div.of Nucl. Educ. and Training, U.S.A.E.C.; Chairman, Graduate Studies Div., A.S.E.E., 1967-68; Consultant, A.E.C., Office of Educ., N.S.F.; Consultant, Council of Graduate Schools in the U.S.; Consultant, Southern Assoc. of Colls. and Schools; Consultant, Southern Regional Educ. Board; Consultant, Graduate Record Examination Board, Educ. Testing Service, Princeton.
Book: Thermodynamics (Ronald Press, 1955). Societies: Fellow, A.A.A.S.; A.S.M.E.
Nuclear interest: Administration of research and education programmes.
Address: 244 Washington Street S.W., Atlanta, Georgia 30334, U.S.A.

GOHSHTEIN, L. E. Papers: Co-author, Backscattering of Gamma-Rays from Aluminium Barriers (Atomnaya Energiya, vol. 22, No. 4, 1967); co-author, Method of Transfer Matrix for Calculation of Spectral-Angular Distribution of Gamma Rays in Slab Geometry (ibid, vol. 24, No. 3, 1968).
Address: Academy of Sciences of the U.S.S.R., 14 Leninsky Prospekt, Moscow V-71, U.S.S.R.

GOICOECHEA, Jose Maria AGUILA. See AGUILA GOICOECHEA, Jose Maria.

GOICOLEA ZALA, Francisco Javier, Civil Eng. Born 1921. Educ.: Madrid Civil Eng. School and Escuela Ténica Superior de Ing. de Caminos, Canales y Puertos. Junta de Energia Nucl., 1956-. Gabinete de Aplicaciones Nucleares a las Obras Públicas, 1961-; Thermodynamics Prof. Escuela Tecnica Superior de Ing. de Caminos, Canales y Puertos. Societies: Asociación de Ingenieros de Caminos; A.N.S. Nuclear interests: Nuclear power stations;

Pressure vessels; Containment buildings.
Address: Centro Nacional de Energia Nuclear,
Juan Vigón, Madrid 3, Spain.

GOKHALE, Gajanan Shridhar, M.Sc., Ph.D.
Born 1920. Educ.: Benares and Bombay
Univs. Res. Asst. (1947-52), Res. Fellow
(1950-), then Reader, now Assoc. Prof., Tata
Inst. of Fundamental Res.
Address: Tata Institute of Fundamental
Research, Homi Bhabha Road, Bombay - 5,
India.

GOKSEL, S., Dr. Sen. Officer, Inst. for Nucl.
Energy, Istanbul Tech. Univ.
Address: Istanbul Technical University,
Gumussuyu, Istanbul, Turkey.

GOLAND, Martin, Pres. and Director, Southwest Res. Inst., Texas.
Address: Southwest Research Institute, 8500
Culebra Road, San Antonio 6, Texas, U.S.A.

GOLANSKI, Henryk, M.Sc. Member, State
Council for the Peaceful Use of Nucl. Energy.
Address: State Council for the Peaceful Use
of Nuclear Energy, Room 1819, Palace of
Culture and Science, Warsaw, Poland.

GOLAY, J. F. Dean, Graduate School, West
Virginia Univ.; West Virginia Univ.; Rep. on
Council, Oak Ridge Assoc. Universities, 1961-.
Address: West Virginia University, Morgantown, West Virginia, U.S.A.

GOLD, Raymond, B.A., M.S., Ph.D. Born
1927. Educ.: New York Univ., Illinois Inst.
Technol. Phys., Pupin Lab., Columbia Univ.,
1952-54; Res. Phys., Armour Res. Foundation,
Chicago, 1955-58; Consultant to Tech. Operations, Inc., Burlington, Mass., 1960-; Phys.,
U.S. Naval Radiol. Defence Lab., San Francisco,
summer 1959; Phys., California Univ., Lawrence Rad. Lab., summer 1961; Prof. and
Head, Dept. of Nucl. Sci. and Eng., Lowell
Technol. Inst., Lowell, Mass., 1958-62. Head,
Exptl. Reactor Phys. Sect., Reactor Phys.
Div., Argonne Nat. Lab., Argonne, Ill., 1962-.
Societies: A.A.A.S.; A.S.E.E.; A.P.S.; A.N.S.;
A.A.U.P.
Nuclear interests: Experimental nuclear
physics; Reactor physics; Numerical analysis;
Applied mathematics.
Address: 6402 Bradley Drive, Downers Grove,
Illinois, U.S.A.

GOLDAMMER, Rudolf, Dr. phil. Born 1907.
Educ.: Leipzig Univ. Sci. Sales Consultant,
Blaupunkt-Werke, Berlin, 1950-52; Tech.
Exec., Kadus-Werk Ludwig Kegel K.G., 1952-
54; Tech. Exec., Phywe, A.G., Göttingen,
1954-. Societies: Deutsche Physikalische
Gesellschaft; Deutsches Atomforum; Deutsche
Gesellschaft für den mathematischen und
naturwissenschaftlichen Unterricht.
Nuclear interests: Nuclear physics.
Address: P.O.B. 33, Göttingen, Germany.

GOL'DANSKII, B. I. Book: Co-author,
Statistika otschetov pri registratsii yadernykh
chastits (Statistics of Reading during the
Registration of Nucl. Particles) (M. Fizmatgiz,
1959).
Address: Academy of Sciences of the U.S.S.R.,
14 Leninsky Prospekt, Moscow V-71, U.S.S.R.

GOLDANSKII, Vitalii, Prof., Head of Lab. of
Nucl. and Rad. Chem., Inst. of Chem. Phys.,
U.S.S.R. Acad. of Sci.
Address: Laboratory of Nuclear and Radiation
Chemistry, Institute of Chemical Physics, 2-b
Vorobyevskoye Chaussée, Moscow V-334,
U.S.S.R.

GOLDBERG, Edward David, B.S., Ph.D. Born
1921. Educ.: California (Berkeley) and
Chicago Univs. Prof. Chem., California Univ.,
San Diego, La Jolla, 1949-.
Nuclear interests: Application of nuclear techniques to geochronology; Activation analysis;
Isotope dilution analysis; chemical composition
of sea water as determined by nuclear techniques; Low level radioactivity measurements.
Address: University of California at San Diego,
La Jolla, California, U.S.A.

GOLDBERG, Irving, B.S. (Chem.), M.S. (Soil
Sci.). Born 1927. Educ.: California Univ.
Berkeley and Los Angeles. Res. Eng., California Univ., Berkeley, 1950-58; Soils Specialist,
State of California, Dept. of Water Resources,
1958-. Societies: A.C.S.; A.N.S.; Western Soc.
Soil Sci.
Nuclear interests: Application of radioisotopic
tracing and gauging techniques to water
resources development and related projects.
Reactor technology, directed toward:
Research; Electric power production; Desalination; and control of effluents released to
the environment.
Address: 5900 Wymore Way, Sacramento,
California 95822, U.S.A.

GOLDBERG, John Edward, B.S. (C.E.), C.E.,
Ph.D. Born 1909. Educ.: Northwestern Univ.
and Illinois Inst. Technol. Assoc. Editor, Appl.
Mechanics Reviews, 1947-50; Asst. Prof.
Mechanics, Illinois Inst. Technol., 1947-50;
Prof. Civil Eng., Purdue Univ., 1950-. Member,
Res. Coms. Pressure Vessel Res. Com.; Welding
Res. Council, Column Res. Council; Chairman,
Task Com. on Nucl. Structures, A.S.C.E.
Societies: Internat. Assoc. for Bridge and
Structural Eng.; A.S.C.E.
Nuclear interests: Reactor design, fuel elements, controls, appurtenances; Stress analysis,
thermal stresses, stability, dynamic response;
Shells.
Address: School of Civil Engineering, Purdue
University, Lafayette, Indiana, U.S.A.

GOLDBERG, Marvin, Ph.D. Educ.: C.C.N.Y.
and Syracuse Univ. Asst. Prof., Phys., Syracuse
Univ., 1966-. Society: A.P.S.
Nuclear interest: Elementary particle physics.

GOLDBERG, Murrey D. Member, Subcom. on Nucl. Structure, Nat. Academy of Sci. - N.R.C.
Address: National Academy of Sciences - National Research Council, 2101 Constitution Avenue, N.W., Washington 25, D.C., U.S.A.

GOLDEMBERG, José, B.Sc., Ph.D., Privat-Dozent. Born 1928. Educ.: Sao Paulo, Saskatchewan and Illinois Univs. Res. Assoc. Illinois Univ., 1954 and 1957; Res. Assoc. Prof., Stanford Univ., 1962-63; Advisor, Brazilian A.E.C., 1956; Assoc. and Acting Prof. of Exptl. Phys., Sao Paulo Univ., 1957-. Society: A.P.S.
Nuclear interest: Nuclear physics.
Address: Departamento di Fisica, Universidade de Sao Paulo, 294 Rua Maria Antonia, Sao Paulo, Brazil.

GOLDEN, D. A., LL.B. Director, Atomic Energy of Canada Ltd.
Address: Atomic Energy of Canada Ltd., Head Office, 150 Kent Street, Ottawa 4, Ontario, Canada.

GOLDER, Jack Alexander. Born 1916. Educ.: Cranwell Elec. and Wireless School, R.A.F. Development Eng., I.C.I. Ltd.; Chief Instrument Eng., Reed Paper Group, 1953; Asst. Chief Instrument Eng. (1954-66), Deputy Chief Instrument Eng. (1966-), U.K.A.E.A.
Nuclear interests: Instrumentation and control of nuclear reactors; Radioactive chemical plant and chemical processes.
Address: "The Retreat", 5 Rutland Avenue, Lower Walton, Nr. Warrington, Lancs., England.

GOLDHABER, Gertrude SCHARFF-. See SCHARFF-GOLDHABER, Gertrude.

GOLDHABER, Maurice, Ph.D. (Cantab.). Born 1911. Educ.: Berlin and Cambridge Univs. Sen. Sci. (1950-60), Chairman, Dept. of Phys. (1960-61), Director (1961-), Brookhaven Nat. Lab. Societies: Fellow, A.P.S.; N.A.S.
Nuclear interests: Neutron physics; Radioactivity; Nuclear isomers; Nuclear photoelectric effect; Nuclear models; Fundamental particles.
Address: Brookhaven National Laboratory, Upton, New York 11973, U.S.A.

GOLDHAMMER, Paul, B.A., Ph.D. Born 1929. Prof., Nebraska Univ., 1957-64; Kansas Univ., 1964-. Societies: A.P.S.; A.I.P.
Nuclear interests: Nuclear structure and nuclear forces; Spin-orbit splitting in nuclei due to tensor interactions.
Address: Department of Physics, Kansas University, Lawrence, Kansas, U.S.A.

GOLDICH, Samuel S., B.A., M.A., Ph.D. Born 1909. Educ.: Minnesota Univ. Prof., Minnesota Univ., 1948-59; Geologist, U.S. Geological Survey, 1959-64; Prof., Pennsylvania State Univ., 1964-65; Prof., State Univ. of New York, Stony Brook, 1965-. Societies: A.C.S.; A.G.U.; Geochem. Soc.
Nuclear interests: Natural variations in isotopic abundance ratios and their geological implications; Geochronology - dating of rocks with emphasis on Early Precambrian history.
Address: Department of Earth and Space Sciences, State University of New York, Stony Brook, New York, New York 11790, U.S.A.

GOLDIN, Abraham Samuel, A.B., A.M., Ph.D. Born 1917. Educ.: Columbia and Tennessee Univs. With Union Carbide Corp., Oak Ridge 1946-50; U.S. Public Health Service, Robt. A. Sanitary Eng. Centre, Cincinnati, 1951-60; National Lead Co., Inc., Winchester, Mass., 1960-61; New York Univ. Medical Centre, 1961-62; U.S. P.H.S., Northeastern Radiol. Health Lab., Winchester, Mass., 1962-68; School of Public Health, Harvard Univ., 1968-. Societies: A.C.S.; A.A.A.S.; A.P.H.A.; H.P.S.
Nuclear interests: Radiochemical and instrumental assay; Environmental radioactivity; Radioactive waste treatment and disposal.
Address: Kresge Centre for Environmental Health, Harvard University, 665 Huntington Avenue, Boston, Massachusetts 02115, U.S.A.

GOLDIN, M. L. Papers: Co-author, An Approximate Calculation of the Mean Energy of the Electrons Ejected by γ-rays in an Ionization Chamber (Atomnaya Energiya, vol. 9, No. 2, 1960); A Method of Isotope Selection for Gamma-Relay (ibid, vol. 15, No. 6, 1963); Choice of Gamma-Sources used to Control Density of Pulp with Atomic Numbers $\geqslant 30$ (ibid, vol. 16, No. 1, 1964); co-author, Limits of Use of Co^{60} and Cs^{137} for Control of Levels (ibid, vol. 22, No. 6, 1967).
Address: Academy of Sciences of the U.S.S.R., 14 Leninsky Prospekt, Moscow V-71, U.S.S.R.

GOLDMAN, Allan E., Born 1926. Educ.: Texas Univ., El Paso. Metal. Eng., Union Carbide Nucl. Corp., O.R.N.L. Societies: A.S.M.; B.N.E.S.
Nuclear interests: Nuclear fuel element design, fabrication, irradiation testing, economics. Automatic data processing of fuel technology, especially fabrication and evaluation parameters.
Address: Oak Ridge National Laboratory, P.O. Box X, Oak Ridge, Tennessee 37830, U.S.A.

GOLDMAN, Arthur J., B.Ch.E., M.Ch.E. Born 1934. Educ.: C.C.N.Y. and New York Univ. Project Eng. for Feasibility Study of U^{233}-Th Fast Breeder Reactors, United Nucl. Corp., 1960-. Society: A.N.S.
Nuclear interests: Nuclear reactor design and analysis; Fast reactor design and safety

problems.
Address: United Nuclear Corporation, Development Division – NDA, 5 New Street, White Plains, New York, U.S.A.

GOLDMAN, David Tobias, A.B., M.S., Ph.D. Born 1933. Educ.: Brooklyn Coll., Vanderbilt and Maryland Univs. Phys., Evans Signal Lab., 1952-53; Res. Asst., Maryland Univ., 1954-58; Phys., O.R.N.L., 1957, 1958; Adjunct Asst. Prof. (1960-63), Adjunct Assoc. Prof. (1963-65), Rensselaer Polytech. Inst. (Nucl. Sci.); Res. Assoc., Pennsylvania Univ., 1958-59; Theoretical Phys., K.A.P.L. (1959-63), Supervising Phys., Nucl. Methods Development, (1963-65), K.A.P.L.; Chief, Theoretical Phys., Reactor Div., Programme Manager, Nucl. Data, Office of Standard Reference Data, N.B.S., 1965-; Lecturer, Maryland Univ. (Nucl. Eng.) 1965-. Societies: A.P.S.; New York Acad. of Sci.; A.N.S.
Nuclear interests: Theoretical nuclear physics; Nuclear reactions and structure; Thermal energy neutron scattering; Transport theory; Properties of chemical systems.
Address: National Bureau of Standards, Washington, D.C., 20234, U.S.A.

GOLDMAN, J. E. Director Sci. Lab., Ford Motor Co.
Address; Ford Motor Co., 20000 Rotunda Drive, P.O. Box 2053, Dearborn, Michigan, U.S.A.

GOLDMAN, Kenneth Marvin, B.S. (Met. Eng.), D.Sc. Born 1922. Educ.: Carnegie Inst. Technol. Graduate Student, Dept. Met.Eng. Carnegie Inst. Technol., 1949-51; Manager, Fuel Element Development, Bettis Atomic Power Lab., Westinghouse Elec. Corp., 1951- Societies: A.I.M.E.; A.S.M.; Inst. of Metals.
Nuclear interest: Metallurgy.
Address: 2223 Shady Avenue, Pittsburgh 17, Pennsylvania, U.S.A.

GOLDMAN, Morton I., B.S. (Civil Eng.), M.S. (Sanitary Eng.), M.S. (Nucl. Eng.), Sc.D. Born 1926. Educ.: New York Univ. and M.I.T. Res. Asst., M.I.T., 1950; U.S.P.H.S., 1950-61; Lecturer on Rad. Safety and Waste Disposal, 1950-54; Chief, Soils and Eng. Section, P.H.S., Waste Disposal Res. Activities, O.R.N.L., 1954-56; Project Leader, Radioactive Waste Disposal Project, Sanitary Eng. Dept., Sec., M.I.T. Reactor Safeguards Com., 1956-59; Nucl. Installations Consultant, Div. of Radiol. Health, Washington, D.C., 1959-61; Vice-Pres., Environmental Safeguards Div., N.U.S. Corp., 1961-. Societies: A.S.C.E.; Water Pollution Control Federation; A.A.A.S.; A.N.S.; American Acad.of Environmental Eng.; Air Pollution Control Assoc.
Nuclear interests: Site selection; Waste disposal; Environmental safety aspects of nuclear facilities; Radiochemistry; Environmental surveillance and meteorological programmes; Aerospace nuclear safety.
Address: N.U.S. Corp., 1730 M Street, N.W., Washington 20036, D.C., U.S.A.

GOLDRING, Benjamin Gvirol, M.Sc. (Jerusalem), Ph.D., D.I.C. (London). Born 1926. Educ.: Hebrew (Jerusalem) and London Univs. Prof., Weizmann Inst., Rehovoth,Israel (Dept. of Phys.). Societies: A.P.S.; Israel Phys. Soc.
Nuclear interests: Nuclear physics, nuclear reactions.
Address: Weizmann Institute, Rehovoth, Israel.

GOLDRING, Lionel S., A.B. (Chem., California), Dr. Phys. Chem. (M.I.T.). Educ.: California Univ. and M.I.T. Formerly Graduate Student and Res. Fellow, M.I.T.; Formerly Res. Chem. and Project Sci., Brookhaven Nat. Lab.; Sen. Res. Specialist, Res. Div., American Machine and Foundry Co., 1960-. Societies: A.A.A.S.; A.C.S.; A.N.S.
Address: American Machine and Foundry Company, 689 Hope Street, Springdale, Connecticut, U.S.A.

GOLDRING, Mary Sheila, P.T.E. (Modern Greats). Educ.: Oxford Univ. Sci. Correspondent, Economist, 1949-; London Correspondent, Nucl. Industry. Books: Economics of Atomic Energy (Butterworths, 1957); Economics of Atomic Energy (Lectures, E.N.I., Milan, 1959).
Nuclear interest: General.
Address: 16 Stanhope Row, London, W.1, England.

GOLDSCHMIDT, Bertrand Leopold, D. es Sc. Atoms for Peace Award, 1967. Born 1912. Educ.: Ecole de Physique et de Chimie, Faculte des Sciences de Paris. Head, Chemistry Div. (1946-59), Head for External Relations and Planning (1959-), C.E.A.; Governor for France (1957-), Member, Sci. Com. (1959-), I.A.E.A.; Member, Sci. Advisory Com., U.N., 1954-; Exec. Vice-Pres., European Atomic Energy Soc., 1955-58. Books: L'Aventure Atomique (1962); Les Rivalites Atomiques (1967).
Nuclear interests: Nuclear chemistry, and specially the technique of plutonium production; History of nuclear energy.
Address: 29-33 rue de la Federation, Paris 15, France.

GOLDSMITH, George Jason, B.S., M.S., Ph.D. Born 1923. Educ.: Vermont and Purdue Univs. Instructor in Phys., Purdue Univ., 1948-54; Res. Phys., R.C.A. Labs., Princeton, N.J. Book: Co-author, Exptl. Nucleonics (New York, Rinehart, 1952). Society: A.P.S.
Nuclear interests: Nuclear physics; Nuclear reactions; Detection of radiations; Nuclear physics as related to the physics of solids.
Address: R.C.A. Laboratories, Princeton, New Jersey, U.S.A.

GOLDSTEIN, Allen M., B.S. (Chem.). Born 1919. Educ.: M.I.T. Gen. Manager, Isotopes Specialities Co., 1952-59; Pres., U.S. Nucl. Corp., 1959-65; Pres., Tech. Assocs., 1965-. Radiol. Advisory Com., Los Angeles City Health Dept., Governor's Radiol. Defence Advisory Com., California; Chief, Radiol. Services, California Mutual Aid, Region I. Societies: A.C.S.; A.N.S.; A.A.A.S. Nuclear interest: Industrial application of radioactivity; Radiological defence.
Address: Technical Associates, 140 W. Providencia Avenue, Burbank, California 91502, U.S.A.

GOLDSTEIN, Carl. Asst. Editor, Nucleonics Week; Asst. Editor, Nucleonics, -1967.
Address: Nucleonics Week, McGraw-Hill Inc., 330 West 42nd Street, New York, New York 10036, U.S.A.

GOLDSTEIN, Herbert, B.S., Ph.D., E. O. Lawrence Award, 1962. Born 1922. Educ.: C.C.N.Y., Columbia Univ., and M.I.T. Instructor, Harvard Univ., 1946-50; Sen. Phys., N.D.A., 1950-61; Prof. Nucl. Eng. Sci., Columbia Univ., 1961-. Formerly Chairman, U.S.A.E.C. Nucl. Cross-sections Advisory Group; Sec., European-American Nucl. Data. Books: Classical Mechanics (1950); Fundamentals of Reactor Shielding (1958). Societies: A.N.S.; A.P.S.
Nuclear interests: Nuclear physics; Reactor theory; Shielding; Neutron cross sections.
Address: Division of Nuclear Science and Engineering, Columbia University, New York 27, New York, N.Y. 10027, U.S.A.

GOLDSTEIN, Jack S., B.S., M.S., Ph.D. Born 1925. Educ.: C.C.N.Y., Oklahoma and Cornell Univs. Prof. Phys., Chairman, Phys. Dept. and Director, Astrophys. Inst., Brandeis Univ. Societies: A.P.S.; American Astronomical Soc.
Nuclear interests: Physics of very dense systems; Physical properties of matter under conditions of white dwarf condensation; Nuclear reactions in dense systems; Plasma physics; Shock wave propagation in ionised gases.
Address: Physics Department, Brandeis University, Waltham 54, Massachusetts, U.S.A.

GOLDTHWAITE, William Harlow, A.B., M.S. Born 1921. Educ.: Middlebury Coll. and New Hampshire Univ. Principal Phys. (1951-55), Project Leader (1955-56), Asst. Chief (1956-57), Chief (1957-), Eng. Mechanics Div., Assoc. Manager Phys. Dept., (1965-), Battelle Memorial Inst. Societies: A.N.S.; Fellow, A.A.A.S. Papers: Numerous U.S.A.E.C. Topical Reports; Research in Bearings (Battelle Tech. Rev., Nov., 1954); Nondestructive Bond Inspection Test by Elec. Resistance Measurement for Complete Flat-Plate Fuel Element Subassemblies (A.S.T.M. XIV-143 MISC).
Address: 505 King Avenue, Columbus 1, Ohio, U.S.A.

GOLDWASSER, Edwin L., B.A., Ph.D. Born 1919. Educ.: Harvard and California (Berkeley) Univs. Asst. Prof., Phys., Illinois Univ., 1951-59; Fullbright and Guggenheim Fellow, Rome Univ., Italy, 1957-58; Professor, Phys., Illinois Univ., 1959-67; Member, Gen. Advisory Com., U.S.A.E.C.; Chairman, Division of Phys. Sci., Nat. Res. Council. Book: Optics, Waves, Atoms and Nuclei (W. A. Benjamin and Co.). Society: Fellow, A.P.S.
Nuclear interests: Elementary particle physics; Cosmic rays; Electron and proton interactions.
Address: National Accelerator Laboratory, P.O. Box 500, Batavia, Illinois, U.S.A.

GOLEB, Joseph Anthony, B.S.(Chem.). Born 1920. Educ.: Lewis Coll., Lockport, and Illinois Univ. R. and D. work in Spectroscopy, I.A.E.A., Vienna, 1967-69. Societies: Applied Spectroscopy; Optical Soc.of America; A.C.S. Nuclear interests: Atomic emission and absorption, and molecular emission as related to elements and their related isotopes.
Address: Argonne National Laboratory, Chemistry Division, Argonne, Illinois, U.S.A.

GOLINELLI, Giuseppe, Dott. Ing. Vice Pres. Nucl. Div., Sogene.
Address: Sogene, Divisione Nucleare, 24 Piazzale dell'Agricoltura, 00144 Rome, Italy.

GOLINSKI, Marek Juliusz, Dipl. Chem. Eng. Born 1934. Educ.: Gdansk Polytech. Res. worker, Dept. of Chem. Technol. (1957-63), Dept. of Application of Radioisotopes in Chem. and in Chem. Technol. (1964-), Inst. Nucl. Res., Warsaw.
Nuclear interests: Physical chemistry of solutions; Solvent extraction of inorganic compounds; Activation analysis.
Address: 6 m 58 ul.A.Sokolicz, Warsaw, Poland.

GOLLINGS, John Frederick. Assoc., Manchester Coll. of Technol. Born 1921. Educ.: Manchester Coll. of Technol. Chief Eng., Nucl. (1960), Company Director (-1966), now Overseas and Nucl. Director, W.H. Smith and Co. Elec. Eng. Ltd. Elec. Ind. Advisor, Union of Lancashire and Cheshire Insts. Society: I.E.E.E.
Nuclear interest: Complete installations and design of electrical systems on nuclear power stations.
Address: White Lodge, South Park Drive, Poynton, Cheshire, England.

GOLLNICK, Klaus, Dr. rer. nat. Born 1930. Educ.: Gottingen Univ. Sci. Asst. (1959-62), Sen. Officer (1962-), Div. of Rad. Chem., Max-Planck-Inst. of Coal Res., Muelheim-Ruhr; Visiting Assoc. Prof., Chem., Dept. of Chem., Arizona Univ., 1966-68. Book: Co-author, Chapter 10 on Oxygen as a Dienophile in 1, 4 - Cyclo addition Reactions (Editor, J. Hamer) (New York, Academic Press, 1967).

Societies: Gesellschaft Deutscher Chemiker; A.C.S.
Nuclear interest: Organic photochemistry and organic radiation chemistry.
Address: Max-Planck-Institute for Coal Research, Division of Radiation Chemistry, 34-36 Stiftstrasse, Mülheim-Ruhr, Germany.

GOLMAYO, Manuel, Ph.D. (Elec. Eng.). Born 1915. Educ.: Stanford Univ. Managing Director, Auxini Ingenieria Espanola S.A. (Auxiesa). Gen. Sec., Spanish Nat. Com., World Power Conference; Tech. Adviser, Forum Atomico Espanol. Society: I.E.E.E. Nuclear interests: Design work for the Spanish Nuclear Power Plants of Santa Maria de Garona (440 MWe) and Vandellos (500 MWe).
Address: 37 Don Ranion de la Cruz, Madrid 1, Spain.

GOLOIAN, Igor N. Nucl. Phys.; Dr., Physico-Mathematical Sci., Member, Deleg. of Soviet atomic scientists visiting the U.S., Nov. 1959.
Address: Academy of Sciences of the U.S.S.R., 14 Leninsky Prospekt, Moscow V-71, U.S.S.R.

GOLOVANOVA, V. N. Paper: Co-author, Study of System $UF_4 - CaF_2$ (Atomnaya Energiya, vol. 22, No. 4, 1967).
Address: Academy of Sciences of the U.S.S.R., 14 Leninsky Prospekt, Moscow V-71, U.S.S.R.

GOLOVIN, I. N., Dr., B.Sc., Acad. of Sci. of the U.S.S.R.; Deleg. to 2nd I.C.P.U.A.E., Geneva, Sept. 1958; Soviet deleg., Convention on Thermonucl. Processes, Inst. of Elec. Eng., London, April 29-30, 1959; Deleg. to Conference sur la Physique des Plasmas et la Recherche concernant la Fusion Nucleaire Controlee, Salzbourg, Sept. 1961. Member, Editorial Board, Atomnaya Energiya. Paper: Co-author, Stable Plasma Column in a Longitudinal Magnetic Field (2nd I.C.P.U.A.E., Geneva, Sept. 1958).
Address: Kurchatov Institut of Atomic Energy, Academy of Sciences of the U.S.S.R., 14 Leninsky Prospekt, Moscow V-71, U.S.S.R.

GOLOVLIN, I. S. Paper: Co-author, BN-350 and BOR Fast Reactors (Atomnaya Energiya, vol. 21, No. 6, 1966).
Address: Academy of Sciences of the U.S.S.R., 14 Leninsky Prospekt, Moscow V-71, U.S.S.R.

GOL'TSEV, V. Yu. Paper: Co-author, Evaluation of Brittle Fracture Behaviour of Thick Sheet Materials (letter to the Editor, Atomnaya Energiya, vol. 23, No. 6, 1967).
Address: Academy of Sciences of the U.S.S.R., 14 Leninsky Prospekt, Moscow V-71, U.S.S.R.

GOLUB, V. V. Paper: Co-author, Energy Build-up Factors in Barrier Geometry (letter to the Editor, Atomnaya Energiya, vol. 24, No. 1, 1968).
Address: Academy of Sciences of the U.S.S.R., 14 Leninsky Prospekt, Moscow V-71, U.S.S.R.

GOLUBEV, V. I. Papers: Co-author, Use of Resonance Absorbers for Measuring Neutron Spectra in Fast Reactors (Atomnaya Energiya, vol. 11, No. 6, 1961); co-author, The Effect of Reflectors Made from Different Materials on Neutron Capture Number Increase in Uranium Shield of a Fast Reactor (letter to the Editor, ibid, vol. 15, No. 3, 1963); co-author, The Effect of Reflectors Made from Different Materials on Neutron Capture Number Increase in a Uranium Carbide Screen of a Fast Reactor (letter to the Editor, ibid, vol. 15, No. 4, 1963); co-author, Measurement of Neutron Spectrum in Medium by Resonance Filters (ibid, vol. 23, No. 2, 1967).
Address: Academy of Sciences of the U.S.S.R., 14 Leninsky Prospekt, Moscow V-71, U.S.S.R.

GOLUBEV, V. S. Paper: Co-author, Exptl. Investigation of Self-Sustaining Elec. Discharge in Supersonic Gas Flow Across Transversal Magnetic Field (Atomnaya Energiya, vol. 23, No. 4, 1967).
Address: Academy of Sciences of the U.S.S.R., 14 Leninsky Prospekt, Moscow V-71, U.S.S.R.

GOLUBEV, Yu. M. Paper: Co-author, Measurements of Activity of Radioactive Gases by Means of Spherical Ionisation Chamber (letter to the Editor, Atomnaya Energiya, vol. 21, No. 2, 1966).
Address: Academy of Sciences of the U.S.S.R., 14 Leninsky Prospekt, Moscow V-71, U.S.S.R.

GOLUBKOV, A. I. Paper: Co-author, Penetration of Neutrons in Air (Atomnaya Energiya, vol.21, No. 4, 1966).
Address: Academy of Sciences of the U.S.S.R., 14 Leninsky Prospekt, Moscow V-71, U.S.S.R.

GOLUBNICHII, P. I. Paper: Co-author, Double Bremsstrahlung Emission in Electron-Electron Collisions at Energy of 2 X 160 MeV (Atomnaya Energiya, vol. 22, No. 3, 1967).
Address: Academy of Sciences of the U.S.S.R., 14 Leninsky Prospekt, Moscow V-71, U.S.S.R.

GOLUCKE, Karl, Dr.-Ing. E.h. Member Arbeitskreis III/4 Versorgung mit Brennstoffen, Deutsche Atomkommission.
Address: Federal Ministry for Scientific Research, 46 Luisenstrasse, 532 Bad Godesberg, Germany.

GOLVIN, I. N. Deputy Director, Kurchatov Inst. of Atomic Energy.
Address: Kurchatov Institute of Atomic Energy, 46 Ulitsa Kurchatova, Post Box 3402, Moscow, U.S.S.R.

GOMARD, Bernhard, cand. jur., Dr. jur. Born 1926. Educ.: Copenhagen Univ. At Ministry of Justice, 1950-58; Asst. Prof. (1951-58), Prof. (1958-), Copenhagen Univ.; Legal Adviser to Danish A.E.C., 1957-; Legal Adviser to Nat. Assoc. Insurers, 1958-. Member, exec. com., Danish branch of C.M.I. and Internat. Law Assoc., 1960-. Pres., Legal

GOM

Aid Soc., Copenhagen, 1961-.
Nuclear interest: Atomic law.
Address: 18 Grumslrupsvej, Hellerup, Denmark.

GOMBAS, Paul, Ph.D. Born 1909. Educ.: Budapest Univ. Full Prof., Univ. for Tech. Sciences in Budapest, 1944. Editor, Acta Phys. Hungarica. Books: Die statistische Theorie des Atoms (Vienna, Springer, 1949); Theorie u. Lösungsmethoden d. Mehrteilchenproblems d. Wellenmechanik (Basel, Birkhäuser, 1950); Statische Behandlung d. Atoms (Handbuch d. Physik, vol. 36, Berlin - Göttingen-Heidelberg, Springer, 1956); Pseudopotentiale (Vienna, Springer, 1967).
Society: Hungarian Acad. Sciences.
Nuclear interests: Atomic and nuclear physics, especially atomic structure, solid state physics, structure of matter and quantum chemistry.
Address: Physical Institute, University for Technical Sciences, 8 Budafoki ut, Budapest 112, Hungary.

GOMBERG, Henry Jacob, Ph.D., M.S., B.S. Born 1918. Educ.: Michigan Univ. Res. Assoc., Lab. Supervisor, Asst. Director, then Director, Phoenix Project, Michigan Univ., 1946-61; Asst. Prof., Assoc. Prof., then Prof. and Chairman, Nucl. Eng. Dept., Michigan Univ., 1946-61; Prof. Phys., Puerto Rico Univ., 1961-; Deputy Director (1961-66), Director (1966-), Puerto Rico Nucl. Centre; Consultant to Industries and Govt. Member, Education Com., and Fellow, A.N.S. Books: In: Recent Advances in the Eng. Sci. (McGraw-Hill, 1958); in: Reactors for Res. and Industry (Instrument Publishing Co., 1954). Societies: A.P.S.; I.E.E.E.; Nucl. Eng. Soc. (England).
Nuclear interests: Theory and practice in use of radiation; Nuclear engineering education; International development of nuclear energy.
Address: Puerto Rico Nuclear Centre, Caparra Heights Station, San Juan, Puerto Rico 00935, U.S.A.

GOMER, Robert. Member, Editorial Board, Bulletin of the Atomic Scientists. Member, Board of Directors, Educ. Foundation for Nucl. Sci.
Address: 935 East 60th Street, Chicago 37, Illinois, U.S.A.

GOMES, F. A. MAGALHAES. See MAGALHAES GOMES, F. A.

GOMES, F. CARVAO. See CARVAO GOMES, F.

GOMES, Harry, Prof. Inst. Res. in Radioactivity, Minas Gerais Univ.
Address: Instituto de Pesquisas Radioativas, Minas Gerais University, Belo Horizonte City, Minas Gerais, Brazil.

GOMES DE FREITAS, Miss Orlita. Radioisotope Lab., Brazil Univ.
Address: Radioisotope Laboratory, 458 Avenida Pasteur, Rio de Janeiro, Brazil.

GOMES FERREIRA, Jose Francisco Vitorino, Ph.D. Born 1923. Educ.: Lisbon Univ. Prof., Faculdade de Ciencias, Lisbon Univ. Societies: Phys. Soc. (London); Soc. Portuguesa de Quimica e Fisica.
Nuclear interest: Nuclear spectroscopy.
Address: Laboratorio de Fisica, Faculdade de Ciencias, Rua da Escola Politecnica, Lisbon, Portugal.

GOMES, Felix GOMEZ y. See GOMEZ y GOMEZ, Felix.

GOMEZ, Gonzalo ESCUERRA-. See ESCUERRA-GOMEZ, Gonzalo.

GOMEZ-CAMPO, Cesar, Dr. Ing. Agronomo. Born 1933. Educ.: Escuela Tecnica Superior de Ingenieros Agronomos, Madrid. At Inst. Nacional de Investigaciones Agronomicas, i/c the gamma rad. field res.programme, 1958-; Prof. of Plant Physiology, 1965-. Society: Rad. Res. Soc., U.S.A.
Nuclear interests: Radiosensitivity in plants; Radiation in plant genetics and plant development; Radiation sources for agricultural uses.
Address: Instituto Nacional de Investigaciones Agronomicas, Avda. Puerta de Hierro, Madrid, Spain.

GOMEZ-CRESPO, Godofredo, M.D. Born 1918. Educ.: Valencia and Salamanca Univs. Fellow, Inst. Cajal (Madrid) Radioisotopes, 1954-55; Fellow of J.E.N., Western Reserve Univ., Cleveland, U.S.A., 1955-56; Res. Assoc. in Radiol. (Atomic Energy Medical Res. Project) (1956), Damon Runyon Memorial Fellowship for Cancer Res. (1957), Western Reserve Univ.; Res. Assoc. in Radiol. (Radioisotopes) at Univ. Hospitals and Cuyahoga County Hospital, Cleveland, Ohio, 1958-60. First Officer - Medical Sect. I.A.E.A., 1960-65; I.A.E.A. Tech. Liaison Officer with W.H.O., Geneva 1965-66; Regional Adviser on Rad.and Isotopes, World Health Organisation, Regional Office for the Eastern Mediterranean, Alexandria, U.A.R., 1966-. Societies: Society of Nucl. Medicine; H.P.S.
Nuclear interests: Clinical research with radioisotopes; Radioisotope scanning of internal organs; International calibration and standardisation of thyroid radioiodine uptake measurements; Training on X-ray and medical uses of radioisotopes; Planning of X-ray and radioisotope facilities; Public health aspects of radiation use.
Address: World Health Organisation, Regional Office for the Eastern Mediterranean, P.O. Box 1517, Alexandria, U.A.R.

GOMEZ DURAN, Efrain, Q.F.B. Investigador, Dept.Investigaciones Fisico-Quimicas, Guanajuato Univ.

Address: Departamento de Investigaciones Fisico-Quimicas, Universidad de Guanajuato, L. de Retana No. 5, Guanajuato, Mexico.

GOMEZ LOPEZ, Juan. Vocale, Comisiones Asesoras de Medicina y Biologia Animal, Junta de Energia Nucl.
Address: Junta de Energia Nuclear, Ciudad Universitaria, Madrid 3, Spain.

GOMEZ-NUNEZ, Juan C., B.S. (Eng.), M.P.H., M.S. (Hygiene (Ecology)). Born 1924. Educ.: Pennsylvania Military Coll. and John Hopkins Univ. Chief, Vector Control Sect., Ministry of Health, Venezuela; Chief, Dept. of Ecology, I.V.I.C., Venezuela; Consultant, W.H.O., Geneva. Society: American Soc. of Tropical Medicine and Hygiene.
Nuclear interest: Radiation biology.
Address: Instituto Venezolano de Investigaciones Cientificas, Apartado 1827, Caracas, Venezuela.

GOMEZ y GOMEZ, Felix, Ingro. Voter, Honduras A.E.C.
Address: Honduras Atomic Energy Commission, Tegucigalpa, D.C., Honduras.

GONCALVES RAMALHO, António Joaquim. Degree in Phys. Born 1929. Educ.: Lisbon Univ., I.N.S.T.N., Saclay, France, and Michigan Univ. Portuguese Corresponding Member, European American Com. on Reactor Phys., O.E.C.D., E.N.E.A. Society: A.N.S.
Nuclear interest: Reactor physics.
Address: Junta de Energia Nuclear, Laboratório de Fisica e Engenharia Nucleares, Estrada Nacional No. 10, Sacavém, Portugal.

GONCHAROV, V. V., B.Sc., Acad. of Sci. of the U.S.S.R.; Deputy Director, Kurchatov Inst. Atomic Energy. Deleg. to 2nd I.C.P.U.A.E., Geneva, Sept. 1958. Papers: Graphite in Reactor Construction (Atomnaya Energiya, vol. 3, No. 11, 1957); co-author, Some New and Reconstructed Research Thermal Reactors (2nd I.C.P.U.A.E., Geneva, Sept. 1958); I. V. Kurchatov and Nucl. Reactors (ibid, vol. 14, No. 1, 1963).
Address: Kurchatov Institute of Atomic Energy, 46 Ulitsa Kurchatova, Post Box 3402, Moscow, U.S.S.R.

GONON, René, Eng. (Ecole Polytechnique), Eng. (Ecole Nationale des Ponts et Chaussées). Born 1906. Educ.: Ecole Polytechnique and Ecole Nationale des Ponts et Chaussées. Pres. Gen. Manager, Sté. des Grands Travaux de Marseille. Rep. of Sté. des Grands Travaux de Marseille, Conseil d'Administration, and Vice Pres., Bureau, A.T.E.N.
Nuclear interest: Prestressed concrete vessel for reactor design.
Address: 85 rue du Ranelagh, Paris 16, France.

GONZALEZ, C. GRANADOZ. See GRANADOS GONZALEZ, C.

GONZALEZ, Maria Alicia CRESPI. See CRESPI GONZALEZ, Maria Alicia.

GONZALEZ, Modesto, Prof. El Decano, Facultad de Ciencias, Cuyo Nat. Univ.
Address: Cuyo National University, Calle Chacabuco y Pedernera, San Luis, Argentina.

GONZALEZ, Otto, Dr. Member, Comision Nacional de Investigaciones Atomicas.
Address: Comision Nacional de Investigaciones Atomicas, Ciudad Trujillo, Dominican Republic.

GONZALEZ BAYLIN, Teodomiro. Vocale, Comisiones Asesoras de Centrales Nucleares, Junta de Energia Nucl.
Address: Junta de Energia Nuclear, Ciudad Universitaria, Madrid 3, Spain.

GONZALEZ-CAMINO, J. CALLEJA y. See CALLEJA y GONZALEZ-CAMINO, J.

GONZALEZ CRUZ, Francisco, Dr., Coronel, E.N. Member, Comision Nacional de Investigaciones Atomicas; Director del Cuerpo Medico y Sanidad Militar.
Address: Comision Nacional de Investigaciones Atomicas, Ciudad Trujillo, Dominican Republic.

GONZALEZ MASSENET, Rafael B., Dr. Sec.-Gen., Comision Nacional de Investigaciones Atomicas.
Address: Comision Nacional de Investigaciones Atomicas, Ciudad Trujillo, Dominican Republic.

GONZALEZ MICHELENA, Felipe. El titulo de Usuario de Isótopos Radiactivos mediante cursillo en la Junta de Energia Nucl. Born 1911. Técnico en Electrónica y mediciones, Técnicas Nucleares, 1958-.
Nuclear interests: Aparatos de medicion, como spectrómetros, ratémeters, escales, etc.
Address: Técnicas Nucleares, 46 Serrano, Madrid, Spain.

GONZALEZ SOLDEVILLA, Fausto, B.S. (Mech. Eng.). Born 1931. Educ.: Escuela de Peritos Industriales de Madrid. At Junta de Energia Nucl. Electronic Sect., 1957-.
Nuclear interest: Electronic and electromechanical components test and design for servo and nuclear reactor control systems.
Address: Junta de Energia Nuclear, Ciudad Universitaria, Madrid 3, Spain.

GONZALO, Jose Maria AGUIRRE. See AGUIRRE GONZALO, Jose Maria.

GONZENBACH, R., Ing. Stellvertreter, Federal Commission for Atomic Energy.
Address: Federal Commission for Atomic Energy, 55 Effingerstrasse, Bern 3, Switzerland.

GOOCH, Peter William, M.A.Sc. Born 1915.
Educ.: Toronto Univ. Pres., Canadian Vickers
Industries Ltd., 1964-. Council, A.S.M.E.
Society: Canadian Nucl. Assoc.
Nuclear interest: Manufacture of nuclear
energy equipment such as reactor components,
heat exchangers, etc.
Address: Canadian Vickers Industries Ltd.,
P.O. Box 7550, Montreal 3, Quebec, Canada.

GOOD, C. Allen, M.D. Formerly Member,
then Pres., now Sec., American Board of
Radiology.
Address: American Board of Radiology,
Kahler Centre Building, Rochester, Minnesota
55901, U.S.A.

GOOD, Edward John Gregory, B.A. (Oxon.).
Born 1932. Educ.: Oxford Univ. Sec., Nucl.
Chemical Plant, Ltd., 1958-.
Nuclear interests: Legal and insurance
aspects of the nuclear industry and transport.
Address: Nuclear Chemical Plant, Ltd.,
Chematom House, St. James Avenue, West
Ealing, London, W.13, England.

GOOD, John Lister, M.A. Born 1919. Educ.:
Oxford Univ. Sec., British Chem. Plant Manufacturers Assoc., Food Machinery Assoc.,
Bakery Equipment Manufacturers Soc.,
Brewery and Bottling Eng. Assoc., Dairy Eng.
Assoc.
Nuclear interest: Chemical plant for nuclear
engineering applications.
Address: 14 Suffolk Street, London, S.W.1,
England.

GOOD, Roland Hamilton, Jr., Ph.D. Born
1923. Educ.: Michigan Univ. Instructor, California Univ., 1951-53; Asst. Prof., Pennsylvania State Univ., 1953-56; Assoc. Prof. and
Prof. Phys., Iowa State Univ., 1956-. Member,
Inst. for Advanced Study, 1960-61. Society:
Fellow, A.P.S.
Nuclear interest: Theoretical physics.
Address: Physics Department, Iowa State
University, Ames, Iowa, U.S.A.

GOOD, Wilfred Manley, A.B. (Kansas), M.A.
(Kansas), Ph.D. (M.I.T.). Born 1913. Educ.:
Kansas Univ. and M.I.T. O.R.N.L.1946-52;
Co-director (1952-58), Director (1958-), High
Voltage Lab.; Head, Nucl. Data Univ., I.A.E.A.,
1966-. Society: Fellow, A.P.S.
Nuclear interest: Nuclear physics.
Address: 113 Taylor Road, Oak Ridge,
Tennessee, U.S.A.

GOODALL, R. D. Vice-Pres., Davison
Chemical Co.; Chairman of the Board, Nucl.
Fuel Services, Inc.
Address: Nuclear Fuel Services, Inc., 101
North Charles Street, Baltimore, Maryland
21201, U.S.A.

GOODELL, Warren F., Jr., Ph.D. Born 1924.
Educ.: Illinois and Columbia Univs. Vice Pres.
for Administration, Columbia Univ. Society:
A.P.S.
Nuclear interests: Laboratory management;
High energy particle physics.
Address: Columbia University, 211 Low
Memorial Library, Columbia University, New
York, N.Y. 10027, U.S.A.

GOODJOHN, Albert John, B.Sc., M.Sc., Ph.D.
Born 1928. Educ.: Alberta and Queen's Univs.
Lecturer in Phys. (1952-53, 1955, 1956); Phys.
Nucl. Div., Canadair, Ltd., 1956-; R. and D.
Staff Member, Gulf General Atomic. Vice
Chairman, San Diego Sect., A.N.S., 1966-.
Society: Canadian Assoc. of Physicists.
Address: Gulf General Atomic, P.O. Box 608,
San Diego, California 92112, U.S.A.

GOODLIFFE, A.W. Tech. Director, Flight
Refuelling Ltd.
Address: Flight Refuelling Limited, Nuclear
Engineering Department, Leigh Park, Wimborne, Dorset, England.

GOODMAN, Clark Drouillard, B.S., Ph.D.
Born 1909. Educ.: California Inst. of Technol.
and M.I.T. Assoc. Prof. of Phys., M.I.T., 1947-
58; Director of Res., Schlumberger Well
Surveying Corp., 1958-59; Vice Pres. - Technique, Schlumberger Ltd., 1959-62; Prof. and
Cha irman, Phys. Dept., Houston Univ., 1963-;
Vice Pres., Houston Res. Inst., 1962-. Consultant, Planetology Sub-Com. N.A.S.A.; Consultant for Lunar and Planetary Missions Board,
N.A.S.A.; Member, Atomic Safety and Licensing
Board Panel, U.S.A.E.C. Books: Sci. and Eng. of
Nucl. Power Vols. I and II (Addison Wesley,
1947-48); Nucl. Eng. Book IV Reactors (Inst. of
Sci. Studies, 1955). Societies: A.N.S.; A.I.A.A.;
A.P.S.; Geological Soc. of America.
Nuclear interests: Nuclear structure; Mossbauer
spectroscopy; Nuclear propulsion and energy
sources in space.
Address: 12511 Old Oaks Drive, Houston,
Texas 77042, U.S.A.

GOODMAN, Cyril William, B.E. Born 1893.
Educ.: Adelaide Univ. Director, English Elec.
Co. of Australia Pty. Ltd. Chairman, Australian
Computers Pty. Ltd. Societies: F.I.E.E.; Inst.
of Eng., Australia.
Nuclear interest: Atomic power development.
Address: c/o The English Company of
Australia Proprietary Limited, 365 Sussex
Street, Sydney, Australia.

GOODMAN, Eli I., S.B., S.M. Born 1929.
Educ.: M.I.T., Assoc. Eng., Brookhaven Nat.
Lab., 1951-55, Sen. Eng., Nucl. Sci. and Eng.
Corp., 1955-59; Sen. Eng., Westinghouse Elec.
Corp., 1959-65; Analyst, Div. of Operations
Analysis, (1965-67), Chief, Plans and Forecasts Branch, (1967-), U.S.A.E.C. Societies:
A.N.S.; A.I.Ch.E.
Nuclear interests: Predicting present and future
fuel cycles; Forecasting enriched uranium
non-weapon requirements to be met by U.S.
gaseous diffusion plants; Estimating of non-weapon requirements for products from

nuclear reactors and their production rates.
Address: 3304 Rittenhouse Street, N.W.,
Washington, D.C. 20015, U.S.A.

GOODRICH, Jack Knight, M.D. Born 1929.
Educ.: Tennessee Univ. Assoc. Prof. Radiol.
and Director, Div. of Nucl. Medicine, Dept.
of Radiol., Duke Univ. Medical Centre. Pres.
Elect, Southeastern Chapter, Soc. of Nucl.
Medicine. Societies: Southern Medical Assoc.;
A.M.A.; American Coll. of Radiol.; Radiol.
Soc. of North America; American Board of
Radiol. Diplomate.
Address: Box 3223, Duke Medical Centre,
Durham, North Carolina, U.S.A.

GOODWIN, Aubrey John Hutchinson, Cdr.
R.N. (Retd.). Born 1904. Educ.: R.N.C.
Keyham and Greenwich Advanced Eng. Course.
At Yarrow-Admiralty Res. Dept. Societies:
M.I.Mech.E.; M.I.Mar.E.
Nuclear interest: Management.
Address: Garemount, Shandon, Dunbarton-
shire, Scotland.

GOOGIN, John M., B.S. (Chem.), Ph.D. (Phys.
Chem., Tennessee). E. O. Lawrence Award,
1967. Born 1922. Educ.: Bates Coll. and
Tennessee Univ. Chem., Y-12 Plant, Tennessee
Eastman Corp., 1944; Tech. Asst. to Y-12
Plant Supt., Union Carbide Nucl. Div., Oak
Ridge. Societies: A.A.A.S.; A.C.S.; RESA.
Address: 111 Orkney Road, Oak Ridge,
Tennessee, U.S.A.

GOPAL-AYENGAR, A. R., Dr., U.N. Sci.
Com. on Effects of Atomic Rad., alternate
delegate to attend 4th session; Member, U.N.
Sci. Com.on the Effects of Atomic Rad.;
Formerly Member, Radioisotopes and Ionizing
Rad. Com., Indian Council of Agricultural
Res. Formerly Deputy Chief Sci. Officer now
Head, Biol. Div., and Director, Bio-Medical
Group, Bhabha Atomic Res. Centre, A.E.C.,
India.
Address: Atomic Energy Commission, Bhabha
Atomic Research Centre, Trombay, Bombay
74, India.

GOPEZ, Artemio Adolfo Castro, B.S. (Chem.
Eng.), M.S. (Chem.). Born 1921. Educ.:
Santo Tomas (Manila) and Madrid Central
Univs. Trained in radioisotope techniques,
Isotopes School, Harwell, England, summer
1955, and at O.R.I.N.S. and O.R.N.L., Oak
Ridge, Tennessee, 1963. Lecturer, Chem. and
Maths., Coll. of Eng., Santo Tomas Univ.,
1946-59; Acting Manager, Chem. Dept.,
Getz Bros. and Co., Manila Branch, 1959-60;
Sen. Sci., Res. Development Div., Philippine
A.E.C., 1961-. Fellow, I.A.E.A. Study Tour on
the Ind. Uses of Radioisotopes in Russia,
United Kingdom, France and Czechoslovakia,
1966. Societies: Philippine Assoc.of Chem.
Teachers; Philippine Inst. Chem. Eng.; Radio-
isotope Soc. of the Philippines.
Nuclear interests: Applications of radioisotopes
in industry; radiography, process tracing,
density and thickness gauging, activation
analysis, soil density and moisture gauging
and level gauging.
Address: 25 Tabayoc Street, Sta. Mesa Heights,
Quezon City, Philippines.

GORAN, Morris, B.S., M.S., Ph.D. Born 1916.
Educ.: Chicago. Oak Ridge, 1943-45; George
Williams Coll., 1953-54; Roosevelt Univ., 1946-
58. Book: Outline of Physical Sci. Societies:
A.C.S.; A.A.A.S.; A.P.S.; A.A.U.P.
Nuclear interests: Nuclear chemistry; Manage-
ment.
Address: Roosevelt University, 430 S. Michi-
gan Avenue, Chicago 5, Illinois, U.S.A.

GORBACHEV, S. K. Paper: Co-author, Pro-
duction of Multiply Charged Ions of Argon,
Krypton, Xenon and Tungsten from Arc
Discharged Ion Source on Stand (Atomnaya
Energiya, vol.24, No. 1, 1968).
Address: Academy of Sciences of the U.S.S.R.,
14 Leninsky Prospekt, Moscow V-71, U.S.S.R.

GORBAN', Yu. A. Paper: Co-author, Vapori-
sation Studies of Uranium Dioxide and Carbides
(Atomnaya Energiya, vol. 22, No. 6, 1967).
Address: Academy of Sciences of the U.S.S.R.,
14 Leninsky Prospekt, Moscow V-71, U.S.S.R.

GORBUNOV, L. M. Papers: Co-author, About
Cross-Section Averaging for Thermal Region
in Media with Zirconium Hydride Content
(letter to the Editor, Atomnaya Energiya,
vol. 20, No. 5, 1966); co-author, On Doppler
Temperature Coefficient Evaluation for Homogeneous Reactors (letter to the Editor, ibid,
vol. 21, No. 5, 1966).
Address: Academy of Sciences of the U.S.S.R.,
14 Leninsky Prospekt, Moscow V-71, U.S.S.R.

GORBUSHINA, L. V. Papers: Co-author,
Exact Measurements of the Concentrations
of Radioisotopes of Lead and Bismuth in the
Air of Underground Workings (letter to the
Editor, Atomnaya Energiya, vol. 9, No. 1,
1960); co-author, Measurement of Low
Radium Activity with Scintillation Electrostatic Cloud Chamber (letter to the Editor,
ibid, vol. 19, No. 1, 1965); co-author, Rise
of Sensitivity of Alpha-Scintillation Counter
(Abstract, ibid, No. 5).
Address: Academy of Sciences of the U.S.S.R.,
14 Leninsky Prospekt, Moscow V-71, U.S.S.R.

GORDEEV, V. V. Paper: Co-author, Investi-
gation of Diffusion of Nucl. Reaction Recoil
Products in Different Materials (Atomnaya
Energiya, vol. 22, No. 4, 1967).
Address: Academy of Sciences of the U.S.S.R.,
14 Leninsky Prospekt, Moscow V-71, U.S.S.R.

GORDEYEV, V. F., Eng., U.S.S.R. State Com.
for the Utilisation of Atomic Energy; Eng.,
Glavatom, Moscow; Deleg. to 2nd I.C.P.U.A.E.,
Geneva, Sept. 1958; Deleg. I.A.E.A. Conference
on the Disposal of Radioactive Wastes, Monaco,
Nov. 1959.

Address: U.S.S.R. State Com. for the Utilisation of Atomic Energy, 26 Staromonetnii Pereulok, Moscow, U.S.S.R.

GORDIYENKO, Ya. I. Paper: Co-author, Dependence of Hardened Uranium Texture from Heating Nature and Other Parameters of Heat Treatment (Atomnaya Energiya, vol. 16, No. 4, 1964).
Address: Academy of Sciences of the U.S.S.R., 14 Leninsky Prospekt, Moscow V-71, U.S.S.R.

GORDON, Andrew Robertson, M.A., Ph.D. Born 1896. Educ.: Toronto Univ. Head, Dept. Chem. (1944-60), Dean, School of Graduate Studies (1953-64), Toronto Univ. Formerly Director, A.E.C.L., 1952-; Member, Exec. Com., Board of Directors, 1953-. Societies: A.C.S.; Faraday Soc.; Fellow, Roy. Soc. of Canada; Fellow, Canadian Inst. Chem. Nuclear interests: Policy and administration of A.E.C.L.
Address: Chemistry Department, Toronto University, Toronto, Ontario, Canada.

GORDON, Angus N., Jr. Pres., United Illuminating Co.
Address: United Illuminating Co., 80 Temple Street, New Haven, Connecticut 06506, U.S.A.

GORDON, Emanuel, B.S. (Chem. Eng.). Born 1919. Educ.: Northeastern Univ. and Carnegie Inst. Technol. Chief Met. Eng., Metal Hydrides, Inc., Beverly, Mass., 1952-55; Supervising Metal., Nucl. Power Div., Combustion Eng., Inc., New York City, 1955-56; Production Manager, Nucl. Fuel Div. (1956-57), Manager, Nucl. Fuel Res. Labs., Metal. Labs. (1957-59), Tech. Manager, Fuels Div. (1959-62), Manager, Newhaven Res. Labs., Member, Operating Com. Development Div. (1962-), United Nucl. Corp. (formerly Olin), New Haven, Connecticut. Societies: A.I.M.M.E.; A.S.M.; Atomic Ind. Forum; Member, Com. on Reactor Materials; A.N.S.
Nuclear interests: Fuels, materials; Research, development, manufacture; Management.
Address: 90 Brooklawn Circle, New Haven, Connecticut, U.S.A.

GORDON, Hayden S(amuel), B.S. (Mech. Eng.), M.S. (Mech. Eng.), Ph.D. (Mech. Eng.). Born 1910. Educ.: California Univ., Berkeley. Sen. Eng., Berkeley (1946-56), Div. Leader, Nucl. Propulsion Div., Livermore (1956-58), Chief Mech. Eng., Berkeley (1958-66), Lawrence Radiation Lab.; Consulting Eng., W. M. Brobeck and Assoc., Berkeley, 1966-; Co-Founder, Director, Radiation Corp., 1953-. Co-Founder, Director, Board Chairman, Berkeley Sci. Capital Corp., 1962-. Societies: A.A.A.S.; A.S.M.E.
Nuclear interests: Management. Design of electro-nuclear machines, accelerators, research devices and appurtenances.
Address: 17 Culver Court, Orinda, California, U.S.A.

GORDON, Lois. Assoc. Editor, Bulletin of the Atomic Sci.
Address: Bulletin of the Atomic Scientists, 935 East 60th Street, Chicago 37, Illinois, U.S.A.

GORDON, M. M. Cyclotron Design, Nucl. Staff, Phys. and Astronomy Dept., Michigan State Univ.
Address: Michigan State University, East Lansing, Michigan, U.S.A.

GORDON, N. E. Manager, Tech. Services Labs., Waltz Mill Site, Westinghouse Elec. Corp.
Address: Westinghouse Electric Corporation, Waltz Mill Site, P. O. Box 158, Madison, Pennsylvania 15663, U.S.A.

GORDON, Robert, B.S., M.S., Ph.D. Born 1917. Educ.: Cooper Union Inst. Tech. and California Univ., Berkeley. Asst. Chief Eng., Aerojet-Gen. Corp., 1945-56; Tech. Specialist, 1956-57; Director, Army Reactor Programme, (1957-58), Tech. Specialist, (1959-60), Scientific Adviser, (1961), Vice Pres.-Tech. Director, (1961-65), Aerojet-Gen. Nucleonics; Manager, Power Systems Div., Aerojet-Gen. Corp., 1965-; Elec. Power Systems Com., A.I.A.A. Societies: American Nucl. Soc.; A.I.A.A.
Nuclear interests: Reactor design; Fuel element development; Specialised applications; Management; Direct conversion; Fissiochemistry; Research and development.
Address: 660 W. Sierra Madre Boulevard, Sierra Madre, California, U.S.A.

GORDON, Robert B., Dr., B.S. Educ.: Rochester Univ. and M.I.T. Metal. R. and D., Westinghouse Elec. Co., 1939-57; Deputy Director, Fuels and Material, then Asst. to Exec. Vice-Pres. (1957-62), Assoc. Manager, Gen. Development Div. (1962-63), Director, Product Operations (1963-), Atomics Internat.
Address: Atomics International, Box 309, Canoga Park, California, U.S.A.

GORDON, Sheffield, S.B., Ph.D. Born 1916. Educ.: Chicago and Notre Dame Univs. Assoc., Metal. Lab., 1942-46; Res. Assoc., Notre Dame Univ., 1946-49; Argonne Nat. Lab., 1950-. Book: Chapter Nucl. Reactor Experiments (by Hoag; Van Nostrand, 1958). Societies: A.C.S.; Faraday Soc.; Rad. Res. Soc.; A.A.A.S.; A.P.S.
Nuclear interest: Radiation chemistry.
Address: Chemistry Division, Argonne National Laboratory, Argonne, Illinois, U.S.A.

GORDON, Solon Albert, B.S., M.S., Ph.D. Born 1916. Educ.: Pennsylvania State and Michigan Univs. Sen. Biol., Argonne Nat. Lab. Societies: Rad. Res. Soc.; Soc. of Gen. Physiologists; Scandinavian Soc. Plant Physiologists; American Soc. Plant Physiologists.
Nuclear interests: Action of radiation on biomedical and physiological systems; Photobiology.

GOR

Address: Division of Biological and Medical Research, Argonne National Laboratory, Argonne, Illinois 60439, U.S.A.

GORDON, Vladimir. High Energy Div., Inst. of Radiography and Radiol.
Address: Institute of Radiography and Radiology, 7 Solyanka, Moscow, U.S.S.R.

GORDUS, Adon Alden, B.S., Ph.D. Born 1932. Educ.: I.I.T. and Wisconsin Univ. Assoc. Prof. Chem., Michigan Univ. Societies: A.C.S.; A.P.S.; A.A.A.S.
Nuclear interests: Chemical effects of nuclear transformations; Isotope separations; Activation analysis applied to archaeological specimens, ancient and medieval coins, and works of art; Geochemistry.
Address: Department of Chemistry, Michigan University, Ann Arbor, Michigan 48104, U.S.A.

GORE, Albert, B.S., LL.B. Born 1907. Member, Finance Com., U.S. Senate; Public Works Com., U.S. Senate; Joint Com. on Atomic Energy, U.S. Congressman, elected to 76th and successive Congresses until election to U.S. Senate, Nov. 4, 1952.
Address: Carthage, Tennessee, U.S.A.

GORESLINE, H. E. F.A.O. Member, Study Group on Food Irradiation, E.N.E.A., O.E.C.D. Deleg., Third I.C.P.U.A.E., Geneva, Sept. 1964.
Address: United Nations Food and Agriculture Organization (F.A.O.), Viale delle Terme di Caracalla, Rome, Italy.

GORETZKI, Hans, Dr. phil. Sen. Officer, Inst. of Special Metals, Max-Planck Inst. for Metal Res.
Address: Max-Planck Institute for Metal Research, 92 Seestrasse, 7000 Stuttgart 1, Germany.

GÖRISCH, Volker, Doz. Dr. med. habil. Born 1928. Educ.: Leipzig Univ. At Inst. für Pharmakoligie und Toxikologie, Karl-Marx-Univ., Leipzig, 1953-. Schriftführer der Arbeitsgemeinschaft der Industrie und Hochschulpharmakologen der D.D.R. Societies: Deutsche Pharmakologische Gesellschaft; Gesellschaft für experimentelle Medizin der D.D.R.; Medizinisch-biologische Gesellschaft Leipzig.
Nuclear interests: Tracers in biology and medicine, especially the stable isotopes ^{18}O, ^{15}N, D; Isotope exchange under biological conditions; Problems of measurement with mass-spectrometers.
Address: 14 Philipp-Rosenthal-Str., Leipzig C 1, German Democratic Republic.

GORJACEV, I. V. Paper: Co-author, The Measurement of Angular and Energy Distribution of Neutrons Recoiled from Different Media (Jaderna Energie, vol. 13, No. 8, 1967).

Address: Fyzikalne energeticky institut, Obninsk, near Maloyaroslavets, U.S.S.R.

GÖRLICH, Paul Robert, Dr.-Ing. habil. Born 1905. Educ.: Dresden Tech. Univ. Director for R. and D., VEB Carl Zeiss, Jena, 1952; Prof., Friedrich Schiller Univ., Jena, 1954; Director, Inst. for Optics and Spectroscopy, Berlin-Adlershof, 1959; Member, Council of Res., German Democratic Republic; Member, Exec. Com., Phys. Soc., German Democratic Republic; Member, Sci. Advisory Council, Phys. Berichte, Braunschweig. Societies: German Acad. Sci. in Berlin; German Acad. of Naturalists Leopoldina in Halle on the Saale. Nuclear interests: Measuring technique of nuclear physics; Employment of isotopes; Nuclear bombardment of crystals.
Address: 26 Humboldtstrasse, Jena, German Democratic Republic.

GORMAN, Arthur Ellsworth, B.S. (Civil Eng.). Born 1892. Educ.: Worcester Polytech. Inst. Chief Sanitary Eng., U.S.A.E.C., 1947-56. Consultant, U.S.A.E.C.
Nuclear interests: Plant site selection; Waste disposal and environmental sanitation.
Address: 133 South Halifax Drive, Ormond Beach, Florida, U.S.A.

GORMAN, Harry Hart, B.S. (Commerce and Finance). Born 1914. Educ.: St. Louis Univ. Deputy Asst. Director, Div. of Reactor Development, U.S.A.E.C., Washington, D.C., 1957-58; Manager, Lockland Aircraft Reactors Operations Office, U.S.A.E.C., 1958-60; Assoc. Deputy Director, George C. Marshall Space Flight Centre, Huntsville, Alabama, 1960-.
Nuclear interest: General management related to space research and development, including nuclear propulsion.
Address: 8512 Valley View Drive, Huntsville, Alabama, U.S.A.

GORN, L. S. Papers: Co-author, Determination of Near-Energy α-Emitters in Isotope Mixture (letter to the Editor, Atomnaya Energiya, vol. 14, No. 4, 1963); co-author, On the Apparatus Error in Measurements of Overlapping Spectral Line Intensity (ibid, vol. 16, No. 1, 1964); co-author, About Optimum Selection of Alpha-Emitter Thickness for Low-Level Radioactivity Probe Spectrometry (letter to the Editor, ibid, vol. 16, No. 5, 1964).
Address: Academy of Sciences of the U.S.S.R., 14 Leninsky Prospekt, Moscow V-71, U.S.S.R.

GORODETZKY, Serge, D. ès Sc. Born 1907. Educ.: Univ. Strasbourg and Ecole Polytechnique, Paris, Prof. à la Faculté des Sci., Strasbourg Univ., 1948-; Directeur de l'Inst. de Recherches Nucléaires de Strasbourg, 1948-; Examinateur des Elèves à l'Ecole Polytechnique, Paris, 1949-. Membre, Commission de Physique Nucléaire, C.N.R.S., Paris. Societies: Sté. française de Physique; A.P.S.

Nuclear interests: Nuclear physics.
Address: 46 boulevard Clémenceau, Strasbourg, Bas-Rhin, France.

GOROHOV, N. A. Papers: Co-author, Microwave Emission for Quasi-Stationary Plasma, (letter to the Editor, vol. 21, No. 4, 1966); Experimental Investigation of Microwave Emission of Quasi-Stationary Plasma (ibid, vol. 24,No. 2,1968).
Address: Academy of Sciences of the U.S.S.R., 14 Leninsky Prospekt, Moscow V-71, U.S.S.R.

GORSHKOV, G. V. Book; Gamma-Izluchenie radioaktivnykh tei i elementy rascheta zashchity ot izlucheniya (Gamma-rad. of Radioactive Bodies and Elements of Calculating Protection against them) (M.-LAN S.S.S.R., 1959). Paper: Co-author, Neutron yield for Nitrogen, Oxygen, Air and Water under the Action of Radium C α-Particles (letter to the Editor, Atomnaya Energiya, vol. 13, No. 5,1962).
Nuclear interest: Radiological physics.
Address: Academy of Sciences of the U.S.S.R., 14 Leninsky Prospekt, Moscow V-71, U.S.S.R.

GORSHKOV, S. I. Papers: Co-author, Electrohydraulic System for Safety Rods for Reactor SM-2 (letter to the Editor, Atomnaya Energiya, vol. 22, No. 3, 1967).
Address: Academy of Sciences of the U.S.S.R., 14 Leninsky Prospekt, Moscow V-71, U.S.S.R.

GORSHKOV, V. K. Papers: Yields of Certain Heavy Fragments in the Fission of U^{233} (letter to the Editor, Atomnaya Energiya, vol. 3,No. 12, 1957); co-author, The Fine Structure in the Fission Yield Curve for U^{233} (ibid, vol. 7, No. 2, 1959); co-author, Investigation of Thin Non-Metallic Samples Evaporation Produced by Fission Fragments (ibid, vol. 20, No. 4, 1966); co-author, Investigation of Zones Damaged by Fragments of Heavy Nuclei Fission (ibid, vol.22, No. 1, 1967).
Address: Academy of Sciences of the U.S.S.R., 14 Leninsky Prospekt, Moscow V-71, U.S.S.R.

GORSKI Ludwik Louis, Dr. Eng. (Chem.). Born 1924. Educ.: Jagiellonian Univ. and Faculty of Chem., Cracow. At Acad. of Medicine, Cracow, 1946-53; Zootech. Inst., Cracow, 1954-56; II Dept. of Phys., 1956-62; Lecturer, Nucl. Chem., Jagiellonian Univ., Cracow, 1962-; Chief, Radioanalytical and Radioactivation Lab., Inst. Radioisotopes Techniques, Acad. of Min. and Metall., 1963-.
Society: Polish Chem. Soc.
Nuclear interests: Application of radioisotopes and radiation in analytical chemistry; Applied gamma-ray spectrometry; Application of radioisotopes in chemistry and industry.
Address: 14 al. Krasińskiego, Cracow, Poland.

GORTAZAR Y LANDECHO, Manuel M. de. See de GORTAZAR Y LANDECHO, Manuel M.

GORTER, Cornelus Jacobus, Ph.D. (Leiden), Hon. D.Sc. (Grenoble, Paris, Nancy); LL.D. (Halifax). Born 1907. Educ.: Leiden Univ. Prof. Phys., Leiden, 1946-; Director, Kamerlingh Onnes Lab., 1946-. Pres., Roy. Netherlands Acad. Sci. 1960-66; Hon. Pres., Internat. Inst. Refrigeration 1955-; Vice-Pres., Internat. Union of Pure and Appl. Phys. 1960-66. Books: Paramagnetic Relaxation (Elsevier, 1947); Progress in Low Temperature Phys. I, II, III, IV, V (Noord. Holl. Uitg. My, 1955, 1957, 1961, 1964, 1967). Societies: Foreign Hon. Member, Roy. Swedish Acad. Sci.; Roy. Flemish Acad. Sci.; Finnish Acad. Sci.; Roy. Sci. Soc., Trondheim and Liège; American Acad. Arts and Sci., Boston; N.A.S., Washington.
Nuclear interest: Research on oriented nuclei.
Address: Kamerlingh Onnes Laboratorium, Leiden, Netherlands.

GORUM, Alvin Eugene, B.S., M.S., Ph.D. Born 1925. Educ.: Arizona and California Univs. Res. Eng., California Univ., 1956-59; Sect. Head, Los Alamos Sci. Lab., 1959-60; Head, Materials Sect., Rheem Semiconductor Corp., 1960-61; Director, Material Sci. Div., Stanford Res. Inst., 1961-. Societies: A.S.M.; American Ceramic Soc.
Nuclear interests: Metallurgy. Metallurgy of reactor fuel materials, particularly plutonium and its alloys. Essentially fabrication of fuel elements and associated problems.
Address: 2787 Alexis Drive, Palo Alto, California, U.S.A.

GORYACHENKO, V. D. Papers: On Stability of Coupled Coupled Core Reactor as Time Lag System (Atomnaya Energiya, vol. 23, No. 6, 1967); Acoustic Instability of Nucl. Reactor (ibid, vol.24, No. 4, 1968); co-author, Acoustic Oscillations in Circulating Gaseous Fuel Nucl. Reactor (ibid, vol. 24, No. 4, 1968); co-author, Effect of Fuel Density on Stability of Circulating Fuel Nucl. Reactor (ibid, vol. 24, No. 4, 1968).
Address: Academy of Sciences of the U.S.S.R., 14 Leninsky Prospekt, Moscow V-71, U.S.S.R.

GORYACHEV, I. V. Papers: Co-author, Measurement of Fast Neutron Dose Albedo for Various Barriers (Atomnaya Energiya, vol. 21, No. 4, 1966); co-author, Spectral and Angular Distributions of Gamma-Radiations from Nucl. Burst (ibid, vol. 25, No. 1, 1968).
Address: Academy of Sciences of the U.S.S.R., 14 Leninsky Prospekt, Moscow V-71, U.S.S.R.

GORYACHEV, S. B. Paper: Co-author, Impulse Electron Injector (Atomnaya Energiya, vol. 21, No. 1, 1966).
Address: Academy of Sciences of the U.S.S.R., 14 Leninsky Prospekt, Moscow V-71, U.S.S.R.

GORYACHEV, V. V. Papers: Angular Distributions of Neutron Doses in Atmosphere Near by Ground (letter to the Editor, Atomnaya Energiya, vol. 19, No. 4, 1965);

co-author, Angular Distributions of Fast Neutrons behind Iron Barriers (ibid, vol. 20, No. 6, 1966).
Address: Academy of Sciences of the U.S.S.R., 14 Leninsky Prospekt, Moscow V-71, U.S.S.R.

GORYANINA, E. N. Paper: Co-author, Decrease of Capture Gamma-Rad. and of Rad. Heat Release in Reactor Vessel with Some Screening and Boron Adding to Thermal Shielding (Abstract, Atomnaya Energiya, vol. 19, No. 4, 1965).
Address: Academy of Sciences of the U.S.S.R., 14 Leninsky Prospekt, Moscow V-71, U.S.S.R.

GOSLAWSKI, Eugeniusz, M.Sc., Formerly Head, Reactor Eng. Dept., Government High Commissioner for Atomic Energy.
Address: Government High Commissioner for Atomic Energy, Palace of Culture and Science, 18th floor, Warsaw, Poland.

GOSS, L. J. U.K. Member, Group of Experts on Third Party Liability, O.E.C.D., E.N.E.A.
Address: O.E.C.D. European Nuclear Energy Agency, 38 boulevard Suchet, Paris 16, France.

GOSS, S. H. Managing Director, Elliott (Treforest) Ltd.
Address: Elliott (Treforest) Ltd., Treforest, Pontypridd, Glamorgan, Wales.

GOSSELAIN, P. A. Asst. Centre des Sciences Nucléaires Ecole Royale Militaire, Brussels; Département Récherches Nucléaires, Société Belge de l'Azote et des Produits Chimiques du Marly S.A.; Deleg. to 2nd I.C.P.U.A.E., Geneva, Sept. 1958.
Address: Société Belge de l'Azote et des Produits Chimiques du Marly S.A., Departement Recherches Nucleaires, 4 Boulevard Piercot, Liege, Belgium.

GOSSELIN, Pierre, Ing. civil des Mines. Born 1902. Educ.: Brussels Free Univ. Directeur adjoint (1949), Directeur (1950), Directeur Gén. (1954), Administrateur-délégué (1958), Interbrabant. Prés., Union des Exploitations Electriques en Belgique, 1955-61; Prés., Comité de Gestion des Entreprises d'Electricité, 1955-59, 1963-; Adm. dél., Centre et Sud; Vice-Prés., Sté. d'Energie Nucléaire Franco-belge des Ardennes (S.E.N.A.); Administrateur dans diverses sociétés.
Nuclear interests: Nuclear power stations; Energy economics.
Address: Centre et Sud, 7 rue de la Bonté, Brussels 5, Belgium.

GÖSSWALD, Karl, Dr. phil. Born 1907. Educ.: Würzburg Univ. Prof. and Director, Inst. for Appl. Zoology, Würzburg Univ., 1948; Head, Official Testing Station for Pests of Raw Materials, 1951. Pres., German Sect., Internat. Union for the Study of Social Insects.
Nuclear interests: Application of radioisotopes for studies on metabolism and behaviour of insects, specially social insects, and on the working mechanisms of insecticides.
Address: Institute for Applied Zoology, 10 Röntgenring, 87 Würzburg, Germany.

GOSWAMI, Amit, B.Sc., M.Sc., D.phil. Born 1936. Educ.: Calcutta Univ. Formerly Instructor in Phys., now Asst. Prof., Phys., Western Reserve Univ., Cleveland, Ohio.
Nuclear interest: Nuclear physics.
Address: Physics Department, Western Reserve University, Cleveland 6, Ohio, U.S.A.

GOSWAMI, Upendra Lal, B.A. (Rangoon). Born 1912. Educ.: Rangoon and Cambridge Univs. Deputy Sec. to Govt. of India, 1948-53; Sec., Community Projects Administration and Joint Sec. to Govt. of India, in Ministry of Commerce and Industry, 1953-58; Deputy Director Gen., Dept. of Tech. Assistance, I.A.E.A.
Nuclear interest: Technical assistance in peaceful applications of atomic energy to underdeveloped countries.
Address: International Atomic Energy Agency, Kaerntnerring, Vienna 1, Austria.

GØTHE, Odd, Cand. econ. Born 1919. Educ.: Oslo Univ. Chief of Sect., Dept. of Industry, 1958. Member, Nat. Atomic Energy Council, Kjeller.
Address: Det Kgl. Departement for Industri og Handverk, Oslo, Norway.

GOTO, G., Prof. Dr. Sci. Dept. and Member, Isotope Res. Com., Kumamoto Univ.
Address: Department of Sciences, Kumamoto University, Kurokami-cho, Kumamoto, Kumamoto Prefecture, Japan.

GOTO, Tetsuo. Formerly Asst., now Asst. Prof., High Energy Nucl. Phys. (Theory), Atomic Energy Res. Inst., Nihon Univ.
Address: Atomic Energy Research Institute, Nihon University, 1 - 8 Kanda-Surugadai, Chiyoda-ku, Tokyo, Japan.

GOTOH, Shiroh, Dr. of Medicine. Born 1926. Educ.: Kanazawa Medical Coll., Japan. Asst. Kanazawa Medical Coll., 1953-55; Chief, Sect. of Preventive Medicine, Health Centre, Saitama Pref., Japan, 1956-59; Sect. Chief, Nat. Inst. of Nutrition, Japan, 1959-63; Chief, Div. Nutrition, Metropolitan Tokyo Labs. for Hygiene Sci., Japan, 1963-. Society: Nutrition Soc. of Japan.
Nuclear interests: Tracer experiment for mineral metabolism.
Address: 1106-3 Shinden, Ichikawa, Chiba Pref., Japan.

GOTOVSKII, M. A. Paper: Co-author, Heat Transfer to Longitudinal-Flowing Liquid Metal in Triangular Tube Banks (letter to the Editor, Atomnaya Energiya, vol. 22, No. 4, 1967).

Address: Academy of Sciences of the U.S.S.R., 14 Leninsky Prospekt, Moscow V-71, U.S.S.R.

GOTT, Brian, B.Sc. (Eng.). (London External). Born 1931. Eng., Hawker Siddeley, Canada, 1953; Res. Officer, British Ship Res. Assoc., 1959; Design Eng. U.K.A.E.A., 1964-. Hon. Sec., Nucl. Eng. Soc., Risley. Society: A.M.I. Mech.E.
Nuclear interest: Reactor design.
Address: U.K.A.E.A., Risley, Warrington, Lancs., England.

GOTT, Harold Howard, M.A., C. Eng. Born 1917. U.K.A.E.A., 1946-58; Nucl. Plant Design Eng. then Director, Southern Project Group, C.E.G.B., 1958-68; Managing Director, Associated Nucl. Services, 1968-.
Nuclear interest: Design of all types of reactors. General management of large projects.
Address: 123 Victoria Street, London, S.W.1, England.

GÖTTE, Hans, Assoc. Prof. Frankfurt a. M. Univ. (apl. Prof.), Dr. habil. rer. nat. Born 1912. Educ.: Berlin Univ. At Kaiser Wilhelm Inst. für Chemie, Berlin, and Max Planck Inst. für Chemie, Mainz, 1936-55. Head, Radiochem. Lab., Farbwerke Hoechst A.G., Frankfurt (Main)-Hoechst; Member, Board of Directors, Eurochemic, Mol. Member, Editorial Board, Die Atomwirtschaft.; Member, Arbeitskreis II/2, Chemie, and Member, Arbeitskreis I/2, Rechts- und Verwaltungsfragen des Strahlenschutzes, Deutsche Atomkommission. Nuclear interests: Pure and applied radiochemistry, including labelling techniques, hot atom chemistry, hot chemistry and side branches.
Address: 2 Mozartstrasse, Kelkheim-Münster (Taunus), Germany.

GOTTFELD, Fritz-Hermann. Joint Gen. Manager, F. H. Gottfeld G.m.b.H.
Address: F. H. Gottfeld G.m.b.H., 21 Wilhelm-Mauser-Strasse, Cologne-Bickendorf, Germany.

GOTTFELD, Mrs. M. Joint Gen. Manager, F. H. Gottfeld G.m.b.H.
Address: F. H. Gottfeld G.m.b.H., 21 Wilhelm-Mauser-Strasse, Cologne-Bickendorf, Germany.

GOTTFRIED, Kurt, B. Eng., Ph.D. Born 1929. Educ.: McGill Univ. and M.I.T. Assoc. Prof., Phys., Cornell Univ. Book: Quantum Mechanics (Benjamin Inc., 1966). Society: A.P.S.
Nuclear interests: Nuclear structure; High energy physics.
Address: Cornell University, Newman Laboratory, Ithaca, New York, U.S.A.

GOTTLIEB, Howard Lyle, B.S., M.S., Ph.D. (Wisconsin). Born 1918. Educ.: Wisconsin Univ. Fee Instructor, U.S. Armed Forces Inst. (1945-53), Lecturer in Nucl. Eng. and Technol., Extension Div., Mech. Eng. Dept. (1954), Wisconsin Univ.; Res. Sci., Oscar Mayer and Co., 1945-50; Res. Sci., Bjorksten Res. Labs. and Foundation, 1950-57; Radiological Defence Planner, Wisconsin Civil Defence, State Board of Health, 1958; Biochem. to Madison Gen. Hospital and Foundation, 1959-61; Asst. Prof. Chem., Wisconsin Univ. (i/c Chem. Dept. at the Kenosha Centre), 1961-. Editor, A.C.S., Wisconsin Sect. Book: Study Guide for Gen. Chem. (U.S. Armed Forces Inst.). Societies: A.N.S.; A.C.S.; American Inst. of Chemists; Inst. of Food Technologists; Non-destructive Testing Soc.; A.A.A.S.; American Soc. for the Study of Arteriosclerosis.
Nuclear interests: Radioisotopic tracers and radiation reactions; Teaching utilisation of nuclear radiation effects and application in industry, medicine and research; Radiological defenses from nuclear warfare.
Address: Wisconsin University, Madison 6, Wisconsin, U.S.A.

GOTTLIEB, Melvin B., B.S., Ph.D. Born 1917. Educ.: Chicago Univ. Asst. Prof. Phys. Iowa State Univ., 1950-54; Asst. Director and Head Exptl. Phys. (1954-57), Sen. Res. Assoc. and Assoc. Director (1957-61), Project Matterhorn, Princeton Univ.; Director, Plasma Phys. Lab., Princeton Univ., 1961-. Member, Board of Editors, Nucl. Fusion and Phys. of Fluids. Society: A.P.S.
Nuclear interests: Plasma physics and controlled thermonuclear reactors.
Address: James Forrestal Campus, Princeton University, P.O. Box 451, Princeton, New Jersey, U.S.A.

GOTTSCHALK, Alexander. Assoc. Prof., Radiol., and Chief, Sect. of Nucl. Medicine, Chicago Univ.; Director, Argonne Cancer Res. Hospital, 1967-.
Address: Argonne Cancer Research Hospital, 950 East 59th Street, Chicago, Illinois 60637, U.S.A.

GOTTSCHALK, Werner, Dr. Born 1920.Educ.: Freiburg Univ. Asst. Prof., Giessen Univ., 1951-55; Assoc. Prof., Bonn Univ., 1956-. Book: The Effects of Mutated Genes on Morphology and Function of Plant Organs (Fischer/Jena, 1964).
Nuclear interests: Experimental production of mutations by means of X-rays, neutrons and mutagenic chemical; Mutation breeding.
Address: Department of Agricultural Botany, Bonn University, Germany.

GÖTTSCHE, Hans, Dr. rer. nat., Dipl.-Phys. Born 1927. Educ.: Hamburg Univ. Staatliche Ingenieurschule Kiel, 1957-. Society: Deutsche Physikalische Gesellschaft.
Nuclear interests: Nuclear physics; Applications of radioisotopes in industry.
Address: 26 Allensteiner Weg, 23 Stift über Kiel-Holtenau, Germany.

GOTTSTEIN, Klaus Leo Ferdinand, Dipl.-Phys. (Göttingen), Dr. rer. nat. (Göttingen).

Born 1924. Educ.: Berlin, London and Göttingen Univs. Max-Planck-Inst. für Physik, Göttingen, 1953-58; Max-Planck-Inst. für Physik und Astrophysik, Munich, 1958-. Member, Sci. Council, German Electron Synchrotron (D.E.S.Y.), Hamburg, 1960; Privatdozent, (1961-67), Prof., (1967-), Munich Univ. Societies: Deutsche Physikalische Gesellschaft; Societa Italiana di Fisica; A.P.S.
Nuclear interests: High-energy nuclear physics; Elementary particles; Cosmic radiation.
Address: Max-Planck-Institut für Physik und Astrophysik, 6 Föhringer Ring, Munich 23, Germany.

GOUA, André. Born 1907. Educ.: Ecole Navale. Chef de Mission in Africa Equatoriale (1952-55), Chef de Mission at Madagascar (1955-62), Chef du Groupement Afrique, Madagascar (1962-), C.E.A.
Nuclear interests: Raw materials; Prospection; Geology.
Address: Commissariat à l'Energie Atomique, 92 avenue de Montredon, Marseille 8, France.

GOUBEAU, Josef, Dr. phil, Dr. rer. nat. h.c. Born 1901. Educ.: Munich Univ.; a.o. Prof., Göttingen Univ., 1937-51; Ord. Prof., Stuttgart T.H., 1951-. Mitglied der Heidelberger Akademie d. Wissenschaften; Mitglied der Leopoldina, Halle. Kor. Mitglied der Göttingen Akademie der Wissenschaften. Societies: Gesellschaft Deutscher Chemiker; Bunsengesellschaft; A.C.S.
Nuclear interest: Hot chemistry.
Address: 26 Schellingstr., Stuttgart N., Germany.

GOUDET, Georges, Dr. Sci. Born 1912. Educ.: Paris Univ. Prof., Nancy Univ.; Chairman of the Board, Lab. Central de Télécommunications; Managing Director, Cie. Gén. de Constructions Téléphoniques. Societies: Sté. Française des Electroniciens et des Radioélectriciens; I.E.E.E.; Sté. Française de Phys.
Nuclear interest: Particle detection.
Address: Compagnie Générale de Constructions Téléphoniques, 251 rue de Vaugirard, Paris 15, France.

GOUDIME, Paul, M.A. (Hons., Natural Sci.), Cantab.). Born 1910. Educ.: Cambridge Univ. Managing Director, Electronic Instr. Ltd., 1946-; Tech. Director, Cambridge Instrument Co. Ltd., 1965-.
Nuclear interest: Development of instrumentation for nuclear measurements.
Address: Orchard House, St. Ann's Hill, Chertsey, Surrey, England.

GOUDSMIT, Pieter F.A., M.S. Born 1937. Educ.: Amsterdam Univ. Phys., Inst. for Nucl. Phys. Res., Amsterdam. Society: Nederlandse Natuurkundige Vereniging.
Nuclear interest: Nuclear physics.
Address: Institute for Nuclear Physics Research, 18 Oosterringdijk, Amsterdam O, Netherlands.

GOUDSMIT, Samuel Abraham, Ph.D. (Phys.). Born 1902. Educ.: Leiden Univ. Deputy Chairman, Dept. of Phys., Brookhaven Nat. Lab. Visiting Prof., Rockefeller Univ., New York. Managing Editor, A.P.S. Books: Co-author, The Structure of Line Spectra (1930); co-author Atomic Energy States (1932). Societies: A.P.S.; N.A.S.; American Philosophical Soc.; A.N.S.
Nuclear interest: Research.
Address: Department of Physics, Brookhaven National Laboratory, Upton, New York, U.S.A.

GOUGH, Cecil Leonard Morris. Born 1925. Educ.: Witwatersrand Tech. Coll., Uranium Plant Supt. and Flotation Plant Supt., Vogelstruisbult G.M.A., Ltd., Springs, 1955-64; Reduction Manager, West Driefontein G.M. Co. Ltd., Carletonville, 1965-67; Zinc Plant Supt., Zinc Corp. of South Africa Ltd., Springs, 1968-. Society: Assoc. Member, South African Inst. Mining and Metal.
Nuclear interest: Metallurgy.
Address: 13 Commonwealth Road, Selcourt, Springs, Transvaal, South Africa.

GOUGH, William Cabot, B.S.E., M.S.E. (Elec. Eng.). Born 1930. Educ.: Princeton Univ. and Princeton Univ. Graduate School. Post Graduate Studies in Sci. and Public Policy, Harvard Univ. 1966-67. Asst. in Res., Plasma Phys. Lab., Princeton Univ., 1952-53; Administrative Eng., Civilian Power Reactors Branch U.S.A.E.C., 1954-55; Project Eng., Aircraft Reactors Branch, U.S.N. - U.S.A.E.C., 1955-58; Tech. Information Officer, Sci. Eng. Book Programme, U.S.A.E.C., 1958-60; Elec. Eng., Controlled Thermonuclear Research Programme, U.S.A.E.C., 1960-. Societies: A.N.S.; A.A.A.S.; Princeton Eng. Assoc.
Nuclear interest: The use of nuclear energy for power generation.
Address: U.S. Atomic Energy Commission, Division of Research, Controlled Thermonuclear Research Programme, Washington, D.C. 20545, U.S.A.

GOULD, A. R. Educ.: London Univ. Exptl. res. at A.R.D.E. Woolwich and A.W.R.E. Aldermaston, 1948-56; Health Phys. Officer, M.O.S., 1956-58; Sen. Exptl. Officer, R.M.C.S. Shrivenham, 1958-. Societies: Fellow, Phys. Soc.; Rad. Res. Assoc.
Nuclear interests: Radiological safety, radiation hygiene and control, health physics, radiobiology; Radiation effects (large molecules and L.E.T.), irradiation techniques by electrons, gamma rays, alpha particles, neutrons; Cosmic rays; Radiological backgrounds; Radiological instrumentation and nucleonics; Linear accelerators, thermonuclear devices, neutron generators.
Address: Physics Branch, Royal Military College of Science, Shrivenham, Wilts., England.

GOULD, Roy W., B.S., M.S., Ph.D. Born 1927. Educ.: California Inst. Technol. and Stanford Univ. Prof. Elec. Eng. and Phys.,

California Inst. Technol. Societies: A.P.S.;
Div. Plasma Phys., A.P.S.; Fellow, I.E.E.E.
Nuclear interests: Plasma physics; Nuclear
fusion.
Address: 808 Linda Vista Avenue, Pasadena,
California, U.S.A.

GOULD, William R. Vice Pres. i/c of Eng.
and Construction (-1968), Sen. Vice Pres.
(1968-), Southern California Edison Co.
Address: Southern California Edison Co., 601
W. Fifth Street, Los Angeles, California
90017, U.S.A.

GOULDING, Frederick Sydney, B.Sc. (Hons.).
Born 1925. Educ.: Birmingham Univ. S.O.
and S.S.O., A.E.R.E., Harwell, 1946-52; Asst.
Res. Officer, Sen. Res. Officer, 1952-58; Head,
Electronics Branch, A.E.C.L., 1958-61; Head,
Counting Development Group, Rad. Lab.,
California Univ., 1961-. Society: S.M.l.R.E.
Nuclear interests: All aspects of nuclear
instrumentation, particularly as applied to
nuclear physics research.
Address: 3731 Sundale Road, Lafayette,
California, U.S.A.

GOULDTHORPE, Hubert W. Vice-Pres. and
Gen. Manager, Power Transmission Div.
(-1967), Vice Pres., Power Generating Group
(1968-), G.E.C.
Address: General Electric Company, 175
Curtner Avenue, San Jose, California 95125,
U.S.A.

GOUPIL, J. Director, Studies Div., Sté.
d'Electronique Industrielle et Nucl.
Address: Société d'Electronique Industrielle
et Nucleaire, 57 rue Eugene Carriere, Paris 18,
France.

GOURE, Francois. Born 1919. Educ.: Ecole
Polytechnique; Ecole National Supérieure
des Télécommunications. Member, Commission
Consultatives des Marchés, and Chef du Service,
Relations Industrielles, C.E.A.
Nuclear Industry.
Address: Commissariat à l'Energie Atomique
8 avenue Kléber, Paris 16, France.

GOURSAUD, Robert. Dept. Manager, Sté.
d'Etudes Techniques et d'Entreprises Generales.
Address: 9 avenue Reaumur, Le Plessis-Robinson, (Seine), France.

GOUSHKOVA, M. Paper: I^{131} in Milk, Grass,
Bovine Thyroids in Czechoslovakia Following
Nucl. Tests (1964-66) (letter to the Editor,
Atomnaya Energiya, vol. 24, No. 2, 1968).
Address: Academy of Sciences of the U.S.S.R.,
14 Leninsky Prospekt, Moscow V-71, U.S.S.R.

GOUTELLE, J. Ing. Commercial Director,
C.R.C. Constructions Radioelectriques et
Electroniques du Centre.
Address: C.R.C. 5 rue Draguerre, St. Etienne,
(Loire), France.

GOUTTEFANGEAS, Maurice, Ing. Born
1926. Educ.: Ecole Polytechnique, Ecole
National Supérieure des Télécommunications.
Ing. Electronicien C.E.A. - Construction du
Synchrotron à Protons (spécialité: Haute-
Fréquence), 1957-. Chef de groupe, Dépt.
du Synchrotron à Protons 'Saturne.'
Nuclear interest: Developments in high energy
accelerators.
Address: Centre d'Etudes Nucleaires de
Saclay B.P. No. 2, Gif-sur-Yvette, Seine et
Oise, France.

GOUVEA PORTELA, António, Mech. Eng.
Born 1918. Educ.: Portuguese Tech. Univ.
Prof., Portuguese Tech. Univ.; Res. Director,
Companhia Uniao Fabril: Counsellor to Companhia Portuguesa de Indústrias Nucleares.
Vice-Pres., Forum Atomico Portugues; Chairman, Comissao do Risco Atomico. Societies:
A.S.M.E.; Nucl. Eng., Div., A.I.C.H.E.
Nuclear interest: Reactor design and management.
Address: 3 Rua Jao Campo-Grande, Lisbon,
Portugal.

GOVAERTS, Jean Marie Lambert Théo, D.Sc.,
Agrége de l'ensiegnement supérieure. Born
1913. Educ.: Liège Univ. Prof., Liège Univ.
Societies: Sté. Royal des Sci. de Liège;
Universitas Belgica; Sté. chimique de Belgique;
A.A.A.S.
Nuclear interests: Applied nucl. chem. and
dosimetry.
Address: 7 quai de la Boverie, Liège, Belgium.

GOVE, Harry Edmund, B.Sc., Ph.D. Born
1922. Educ. Queen's Univ., Ontario and M.I.T.
Res. Assoc., M.I.T., 1950-52; Asst. Res. Officer
(1952-53), Assoc. Res. Officer (1953-57),
Sen. Res. Officer (1957-), Branch Head, Nucl.
Phys. II (1956-63), Atomic Energy of
Canada, Ltd.; Prof. of Phys. and Director of
Lab. for Nucl. Structure Res., Rochester Univ.,
1963-. Societies: A.P.S.; Canadian Assoc. Phys.
Nuclear interests: Nuclear physics, in particular nuclear structure.
Address: Department of Physics and Astronomy, Rochester University, Rochester
N.Y. 14627, U.S.A.

GOVEAS, Cyril Maurice, B.A., B.E., C.E.
(Hons.). Born 1912. Educ.: Madras Univ.
Chief Eng. (Civil), Bhabha Atomic Res.
Centre, Trombay, Bombay, India. Society:
Inst. Eng. (India).
Nuclear interests: Structural and construction
engineering with special reference to atomic
research and power development.
Address: "Anand Bhavan", Warden Road,
Bombay 26, India.

GOVORKOV, A. B. Papers: On the Statistical
Straggling of Pulse Amplitudes in the Fast
Neutron Burst Reactor (Atomnaya Energiya,
Vol. 13, No. 2, 1962); co-author, On Statistics
of Amplitudes of Bursts from Pulsed Fast
Reactor (Abstract, ibid, vol. 20, No. 4, 1966).

Address: Academy of Sciences of the U.S.S.R., 14 Leninsky Prospekt, Moscow V-71, U.S.S.R.

GOWING, Mrs. Margaret Mary, B.Sc. (Economics). Born 1921. Educ.: London Univ. Historical Sect., Cabinet Office, 1945-59; Historian and Archivist, U.K.A.E.A., 1959-66; Currently Reader, Contemporary History, Kent Univ.; U.K.A.E.A. consultant-historian, writing further volume of history of U.K. atomic energy project. Book: Britain and Atomic Energy 1939-45 (1964).
Address: United Kingdom Atomic Energy Authority, 11 Charles II Street, London, S.W.1, England.

GOYBURU de la CUBA, Eduardo, Ing. Geologo, Departamento de Control de Sustancias Radiactivas, Inst. Superior de Energia Nucl., Junta de Control de Energia Atomica del Peru.
Address: Junta de Control de Energia Atomica del Peru, 3420 Avenida Arequipa, Apartado 914, Lima, Peru.

GRAAF, J. E. de. See de GRAAF, J. E.

GRABER, Heinz Erich, Dipl.ing. Born 1931. Educ.: Dresden Tech. Univ.
Nuclear interest: Nuclear physics.
Address: Zentralinstitut für Kernforschung, Bereich Kernphysik, Rossendorf bei Dresden, German Democratic Republic.

GRABOWSKI, Zbigniew Wojciech, M.S., Fil. lic., Ph.D. Born 1931. Educ.: Jagiellonian Univ., Cracow and Uppsala Univ., Sweden. Teaching Asst., Gliwice Tech. Univ., Poland, 1954-55; Res. Asst., Nucl. Res. Inst., Cracow, 1955-58; Fellowship (1958-61), Res. Asst. (1961-62), Uppsala Univ.; Project Manager, Kiruna Geophys. Observatory, Sweden, 1962-63; Res. Assoc., (1963-65), Asst. Prof., (1965-), Dept. of Phys. Purdue Univ., Lafayette. Society: A.P.S.
Nuclear interest: Nuclear spectroscopy.
Address: 108 N. Salisbury, W. Lafayette, Indiana 47906, U.S.A.

GRACE, C. S. Lab. Administrator and Officer in Charge of Nucl. Phys. Teaching, Rutherford Lab., Roy. Military Coll. of Sci.
Address: Royal Military College of Science, Shrivenham, Swindon, Wiltshire, England.

GRACE, Michael Anthony, M.A., D.Phil. Born 1920. Educ.: Oxford Univ. Univ. Senior Res. Officer, 1955-; Student of Christ Church, 1959-. Societies: Phys. Soc.; A.P.S.
Nuclear interest: Nuclear physics.
Address: 13 Blandford Avenue, Oxford, England.

GRACE, William E. Pres., Chief Exec. Officer, Fruehauf Corp.
Address: Fruehauf Corporation, Detroit, Michigan 48232, U.S.A.

GRACHEVA, L. M. Paper: Co-author, Pile-Neutron Burn-up Cross Section of Pm^{149} and Samarium Poisoning (letter to the Editor, Atomnaya Energiya, vol. 19, No. 2, 1965).
Address: Academy of Sciences of the U.S.S.R., 14 Leninsky Prospekt, Moscow V-71, U.S.S.R.

GRACHEVA, M. A. Book: Co-author, Radioaktivnye izotopy zolota - Au^{198} i Au^{199} (Radioactive Isotopes of Gold Au^{198} and Au^{199}) (M. Atomizdat, 1960).
Address: Academy of Sciences of the U.S.S.R., 14 Leninsky Prospekt, Moscow V-71, U.S.S.R.

GRACIA, A. J. Manager R. and D., Goodyear Tire and Rubber Co.
Address: The Goodyear Tire and Rubber Company, 1144 East Market Street, Akron 16, Ohio, U.S.A.

GRAD, Harold, B.E.E., M.S., Ph.D. Born 1923. Educ.: New York Univ. Prof. Mathematics, New York Univ.; Director, Magneto-Fluid Dynamics Div., Courant Inst. Math. Sci., New York Univ. Societies: American Mathematical Soc.; A.P.S.
Nuclear interests: Plasma physics; Magnetohydrodynamics; Controlled fusion.
Address: Courant Institute of Mathematical Sciences, New York University, 251 Mercer Street, New York, N.Y. 10012, U.S.A.

GRADY, Paul. Consultant, Price-Waterhouse Co., New York City. Member, Ad Hoc Advisory Panel on Safeguarding Special Nucl. Materials, and Member, Advisory Com. on Nucl. Materials Safeguards (1967-), U.S.A.E.C.
Address: U.S. Atomic Energy Commission, Advisory Committee on Nuclear Materials Safeguards, Washington, D.C. 20545, U.S.A.

GRAEF-FERNANDEZ, Carlos, Dr. Member, Advisory Board, Comision Nacional de Energia Nucl.; Member, Exptl. Nucl. Phys. Dept., Inst. de Fisica, Universidad Nacional Autonoma de Mexico. Pres., Soc. Mexicana de Estudios de Radioisotopos.
Address: Av. Insurgentes Sur 1079 Tercer piso, Mexico 18, D.F., Mexico.

GRAEFFE, Thor Gunnar, Ph.D., Ph.Lic., Ph. mag. Born 1935. Educ.: Helsinki Univ. Lecturer and Res. Assoc. Helsinki Univ., 1959-65. Res. Assoc., M.I.T., U.S.A., 1965-66. Assoc. Prof., Inst. of Technol., Helsinki, Finland, 1967-.
Nuclear interests: Nuclear spectroscopy.
Address: Institute of Technology, Laboratory of Physics, Otaniemi, Finland.

GRAENACHER, Ch., Prof. Dr. Member, Federal Commission for Atomic Energy, Switzerland; Member, Board of Directors, Eurochemic, Mol, Belgium.
Address: Federal Commission for Atomic Energy, 55 Effingerstrasse, Bern 3, Switzerland.

GRAF, Guenter, Diplom-Mathematiker. Born 1932. Educ.: Friedrich-Schiller-Univ., Jena. At Allgemeine Elektricitäts-Gesellschaft, Frankfurt am Main, 1957-.
Nuclear interests: Nuclear physics and reactor design; especially mathematical methods for solving problems in nuclear physics and reactor design with digital computers.
Address: 65 Ritterstrasse, Bad Vibel (Hess), near Frankfurt, Germany.

GRAF, Herbert, Dipl.-Ing. Born 1897. Educ.: Hanover T.H. an Darmstadt T.H. Zentrale Leitung der Mitarbeit (1951-63), Leiter des Archivs (1964-), Wernerwerks für Medizinische Technik der Siemens A.G. (früher Siemens-Reiniger-Werke A.G.) in Ausschüssen des Deutscher Normenausschuss and V.D.E. Vorsitzender, Kommission Elektromedizinische Geräte, V.D.E.; Stellv. Vorsitzender (1951-60), Vorsitzender (1960-63), Fachnormenausschuss Radiol., Mitglied, Präsidiums (1963-65), Deutscher Normenausschuss. Society: Deutsche Röntgen-Gesellschaft.
Nuclear interest: Entwicklungsgeschichte der radiologischen technik.
Address: 14 v. Bezzel-Strasse, 852 Erlangen, Germany.

GRAF, Peter, Dr. phil. nat. Nucl. Chem. Born 1929. Educ.: Berne Univ., Eidg. Inst. für Reaktorforschung, Wurenlingen. Society: Schweizerische Gesellschaft von Fachleuten der Kerntechnik.
Nuclear interests: Radio and nuclear chemistry and fuel metallurgy; Installations and equipment for handling highly radioactive materials; Nuclear power plants, design and operation; Radioactive waste management; Fuel procurement and fuel cycle management.
Address: Motor Columbus, Electrical Management Company Ltd., Consultant Engineers, Baden, Switzerland.

GRAFF, William John, Jr., B.S. (Mech. Eng.) (Texas A. and M.), M.S. (Mech. Eng.) (Texas A. and M.), Ph.D. (Purdue). Born 1923. Educ.: Texas A. and M. and Purdue Univs. Sen. Propulsion Eng. (1951-54), Nucl. Group Eng. (1954-56), Convair, Fort Worth, Texas; Res. Participant, O.R.N.L., summer 1959; Prof. and Chairman, Mech. Eng. Dept. (including Nucl. Eng.), Southern Methodist Univ., Dallas, Texas, 1956-61; Dean of Instruction, Texas A. and M., 1961-. Treas., Sec., Vice-Chairman, North Texas Sect., A.S.M.E.; Member, Nucl. Development Com., A.S.M.E. Societies: A.N.S.; A.S.M.E.; A.S.E.E.
Nuclear interests: Application of thermoelectrics and thermionics to nuclear energy sources for electric power; Heat transfer; Plasma dynamics; Thermal stresses.
Address: Dean of Instruction, Texas A. and M. University, College Station, Texas, U.S.A.

GRAFFSTEIN, Andrzej Wlodzimierz, Dr., Eng. Born 1926. Educ.: Gdańsk Tech. Univ. Inst. of Nucl. Res., Poland, 1956-.
Nuclear interest: Construction of nuclear instruments (choppers).
Address: 19 Kartonowa Street, Warsaw 92, Poland.

GRAFSTROM, Director Gen. E. Member, Swedish Atomic Res. Council.
Address: Swedish Atomic Research Council, Box 23 136, Stockholm 23, Sweden.

GRAGEROV, I. P. At Pisarzhevskij Inst. for Phys. Chem., Kiev. Paper: Co-author, Studies on the Participation of Phenyle Radicals in Reactions in Solution by using Deuterium (Kernenergie, vol. 5, No. 4/5, 1962).
Address: Pisarzhevskij Institute of Physics and Chemistry, Academy of Sciences of the Uranian S.S.R., 15 Ulitsa Lenina, Kiev, Ukranian S.S.R., U.S.S.R.

GRAHAM, Beardsley, B.S. Born 1914. Educ.: California Univ. Tech. Consultant to Vice-Pres., Bendix Aviation Corp., 1946-51; Tech. Consultant to Vice-Pres. (1949-51), Asst. Director (1951-56), Stanford Res. Inst.; Exec. Vice-Pres., Sequoia Process Corp., 1956-57; Lockheed Missiles and Space Co., 1957-62; Manager, Satellite Res. Planning (1958-60), Special Asst., Communications Satellites (1960-62), Pres. (1962-67), Spindletop Res. Inc.; Res. Prof. of Elec. Eng., Kentucky Univ. 1963-67; Consultant. Director, Communications Satellite Corp. 1963-64; Member, Kentucky Atomic Energy Authority; Director, Assoc. for Applied Solar Energy. Societies: Fellow, I.E.E.E.; Sen. Member, A.I.A.A.; Solar Energy Soc.
Address: P.O. Box 54, Noroton, Connecticut 06820, U.S.A.

GRAHAM, Clifton Brown, B.S. (M.E.). Born 1914. Educ.: O.R.S.O.R.T., Rochester Polytech. Inst. and Tennessee, Rochester, Illinois and Bradley Univs. Sect. Chief in charge of Eng. Development, Union Carbide Nucl. Corp., Oak Ridge, 1949-56; Chief Eng. (1956-61), Manager (1961-63), Nucl. Power Dept., Gen. Manager's Staff, Atomic Energy Div. (1963-), Allis-Chalmers Manufacturing Co. Society: A.N.S.
Nuclear interests: Nuclear power plant design and management.
Address: 1535 South East 15 Street, Fort Lauderdale, Florida 33316, U.S.A.

GRAHAM, James B. Formerly Exec. Sec., Advisory Com. on Reactor Safeguards, Formerly Tech. Advisor, Joint Com. on Atomic Energy, U.S.A.E.C. Member, Washington Staff, Gen. Elec. Co., 1967-.
Address: General Electric Company, Richland, Washington 99352, U.S.A.

GRAHAM, John. Special Tech. Editor, Transactions, Sen. Tech. Editor, Proceedings of the Hot Lab. Div. Conference, Ex Officio Member, Public Information Com., (1966-67), Editor, Nucl. News, (1967-), A.N.S.

Address: Transactions of the American Nuclear Society, 244 East Ogden Avenue, Hinsdale, Illinois, U.S.A.

GRAHAM, Richard Hugh, B.S. (Phys.). Born 1921. Educ.: Washington Univ. Sect. Leader, Phys. Group, Allis Chalmers Manufacturing Co., Milwaukee, 1948-51; Lead Phys., California R. and D. Corp., Livermore, Calif., 1951-54; Chief, Reactor Eng. Sect., Div. of Reactor Development,U.S.A.E.C., Washington, D.C., 1954-56; Head, Nucl. Eng. Sect., Lockheed Aircraft Corp., Palo Alto, Calif., 1956-59; Consultant, Product Planning, Manager, Energy Analysis, Manager, Uranium Development Operation, Nucl. Energy Div., G.E.C., San Jose, Calif., 1959-67; Development Director, Nucl. Fuels Div., Gulf Oil Corp., Pittsburgh, Pa., 1967-. Societies: A.N.S.; A.P.S.; Tech. Sec., Advisory Com. on Reactor Safeguards, U.S.A.E.C.
Nuclear interests: Fuel cycle technology and economics; Reactor design, control and kinetics; Nuclear raw materials and fuel processing; Recycle of uranium and plutonium.
Address: 101 Yorkshire Drive, Pittsburgh, Pa. 15238, U.S.A.

GRAHAM, Robert L. Member, Subcom. on Nucl. Structure, Nat. Academy of Sci. - N.R.C.
Address: National Academy of Sciences - National Research Council, 2101 Constitution Avenue, N.W., Washington 25, D.C., U.S.A.

GRAHAM, Thomas Edward, B.Sc., M.B., Ch.B. Born 1917. Educ.: Glasgow Univ. Sen. Medical Officer, Production Group, U.K.A.E.A. Nuclear interests: Health and safety.
Address: Tower Lodge, Sharoe Green Lane, Fulwood, Preston, Lancs., England.

GRAHN, Douglas, M.S., Ph.D. Born 1923. Educ.: Rutgers Univ. and Iowa State Coll. Assoc. Sci., Div. of Biol. and Medical Res., Argonne Nat. Lab., 1953-58; Geneticist, Div. of Biol. and Medicine, U.S.A.E.C., 1958-61; Assoc. Sci., Div. of Biol. and Medical Res. (1961-66), Assoc. Div. Director, Div. of Biol. and Medical Res. (1962-66), Sen. Sci. (1966-), Argonne Nat. Lab.; Consultant, Webb Assocs., Yellow Springs, 1962-64; Consultant, McDonnell Aircraft Corp., 1963-64; Consultant, Space Medicine, Office of Manned Space Flight, N.A.S.A., 1964-. Sustaining Membership Com., Genetics Soc. of America; Radiobiol. Advisory Panel, Space Sci. Board, U.S. N.A.S.-N.R.C. Society: Rad. Res. Soc.
Nuclear interests: Research in radiation genetics, toxicity of external radiations, epidemiology; Scientific programme management, United States Atomic Energy Commission foreign exhibit on peaceful uses of nuclear energy.
Address: Division of Biological and Medical Research, Argonne National Laboratory, 9700 S. Cass Avenue, Argonne, Illinois 60439, U.S.A.

GRALL, Lucien, Eng. from Nat. Superior School of Chem. of Paris, L. ès Sc. Born 1925. Educ.: Paris Univ. Service d'Etude de la Corrosion Aqueuse et d'Electrochimie, C.E.A., 1949-.
Nuclear interests: Metallurgy and aqueous corrosion of canning or structural materials; Electrochemistry of these materials; Fabrication of special alloys for nuclear field.
Address: 4 rue Thimonnier, Paris 9, France.

GRAM, Niels, M.Sc. (Eng.). Born 1925. Educ.: Denmark Tech. Univ. Development Manager, Paul Bergsoe and Son, 1963.
Nuclear interest: Lead shielding.
Address: Paul Bergsoe and Son, 2600 Glostrup, Denmark.

GRAMENITSKY, Igor M. High Energy Lab. Joint Inst. for Nucl. Research, Dubna. Papers: Interaction of 9 GeV Protons with Nuclei in Emulsions (letter to the Editor, Atomnaya Energiya, vol. 4, No. 3, 1958); co-author, Quasi-Elastic π^- N interactions at 9 GeV (Nukleonika, vol. 9, No. 2 - 3, 1964); co-author, Investigation of $\pi^- + Xe \rightarrow \pi^- + \pi^\circ + Xe$ interactions at 9 GeV/c primary momentum (ibid.).
Address: Joint Institute for Nuclear Research Dubna, Nr. Moscow, U.S.S.R.

GRAMMAKOV, A. G. Book: Co-author, Rukovodstvo op gamma-oprobovaniyu radio-aktivnykh rud v estestvennom zaleganii (Manual for Gamma-testing of Radioactive Ores in Natural Storage) (M. Atomizdat, 1959). Paper: Co-author, Effects of the Density of Uranium Ore and of the Thickness of Iron Absorber on the γ-ray Spectrum of the Ore as recorded by a Scintillation Counter (letter to the Editor, Atomnaya Energiya, vol. 11, No. 1, 1961).
Address: Academy of Sciences of the U.S.S.R., 14 Leninsky Prospekt, Moscow V-71, U.S.S.R.

GRAMMATER, R. D., Vice-Pres. and Director, Bechtel Nucl. Corp.
Address: Bechtel Nuclear Corp., 220 Bush Street, San Francisco 4, California, U.S.A.

GRAMMATIKATI, V. S. Papers: Co-author, The Dose Field of a Line Source (letter to the Editor, Atomnaya Energiya, vol. 8, No. 2, 1960); co-author, Exptl. Gamma-Rays Unit EKU-50 of the N. F. Gamalei Inst. of Epidemiology and Microbiology (letter to the Editor, ibid, vol. 22, No. 4, 1967).
Address: Academy of Sciences of the U.S.S.R., 14 Leninsky Prospekt, Moscow V-71, U.S.S.R.

GRANADOS GONZALEZ, C. Member, Editorial Board, Energia Nuclear.
Address: Junta de Energia Nuclear, 22 Avda. Complutense, Ciudad Universitaria, Madrid 3, Spain.

GRANATKIN, B. V. Papers: Co-author, Pulse Measurements on Neutron Diffusion and Thermalisation in Water and Ice over a Wide Range of Temperatures (Atomnaya Energiya, vol. 12, No. 1, 1962); co-author, The Investigation of Neutron Diffusion for Water and Ice at Temperatures near 0°C and -80°C (letter to the Editor, ibid, vol. 13, No. 4); co-author, Temperature Dependence of Neutron Diffusion Parameters in Water and Ice (letter to the Editor, ibid, vol. 20, No. 2, 1966).
Address: Academy of Sciences of the U.S.S.R., 14 Leninsky Prospekt, Moscow V-71, U.S.S.R.

GRANDBESANCON, Pierre. Directeur, Sté. des Chantiers et Ateliers de Provence S.A.
Address: Société des Chantiers et Ateliers de Provence S.A., 130 Chemin de la Madrague, Marseille 15 (B.D.R.), France.

GRANDGEORGE, René Jean, Ing., L.-ès Sc. Born 1899. Educ.: Ecole Centrale de Paris, Faculté des Sciences de Paris. Administrateur Directeur général honoraire, Compagnie de St.-Gobain,;ancien Membre du Comité Scientifique et Technique d'Euratom; Administrateur, Banco di Roma (France); Administrateur, La Pierre Synthétique; Pres., France-Investissement; Pres., Sté. de Placements Internat.; Vice-Pres., Seric-Industrie; Membre du Comité de l'Equipement Industriel, C.E.A.
Address: 8 rue de l'Abbaye, Paris 6, France.

GRANDJEAN, Pierre Jean Emile, Eng. Born 1914. Educ.: Saint-Brieuc's Lyceum. French Navy Captain (1955-), Adjoint au Directeur, Direction des Piles Atomiques (1962-), C.E.A. Sec., Navy - C.E.A. Nucl. Propulsion Liaison Coms., 1955-.
Address: 20 rue Moreau-Vauthier, 92 Boulogne sur Seine, France.

GRANDY, George L., B.S. (Chem.). Born 1933. Educ.: California State Coll., California, Pennsylvania. Society: A.N.S. Nuclear interests: Solid state diffusion studies on nuclear rocket fuels. Source term measurements during ground testing of nuclear rocket engines. Investigation of chemical and physical reactions affecting performance of fuel materials. Design, installation, and operation of radiochemistry laboratory equipment.
Address: 5734 Willow Terrace Drive, Bethel Park, Pennsylvania, U.S.A.

GRANGE, Alphonse. Director and Gen. Manager, Sté. Générale d'Exploitations Industrielles S.A.
Address: Société Générale d'Exploitations Industrielles S.A., 4 rue d'Aguesseau, Paris 8, France.

GRANIEWSKI, Marian, General. Member, Polish State Council for the Peaceful Use of Nucl. Energy.
Address: State Council for the Peaceful Use of Nuclear Energy, Room 1819, Palace of Culture and Science, Warsaw, Poland.

GRANJON, Denis. Directeur Gén., Cie. Française des Minerais d'Uranium S.A. French Member, Consultative Com. of the Supply Agency, Euratom.
Address: Cie. Française des Minerais d'Uranium S.A., 10 place Vendôme, Paris 1, France.

GRANOVSKY, Yakov, M.Sc. Nuclear interest: Theory of nuclear particles.
Address: Institute of Nuclear Physics Research, Alma-Ata, Kazakhstan, U.S.S.R.

GRANT, Albert Edgar, B.Sc. (1st class Hons.). Born 1913. Educ.: Liverpool Univ. Joined Ministry of Supply (Dept. of Atomic Energy), 1950; Res. Manager, R. and D. Branch, (1956), Sen. Tech. Officer, Production Group, (1961-65), Tech. Manager, Springfields Works (1965-), U.K.A.E.A. Nuclear interests: Design of large-scale plants for production of nuclear fuels and uranium hexafluoride; Economic assessments of possible future processes; Research and development work on potential new processes for natural and highly enriched uranium fuels and fuel canning materials.
Address: The Meades, Wrea Green, Preston, PR4 2WQ, England.

GRANT, Ian Seafield, M.A., Ph.D. Born 1931. Educ.: Cambridge Univ. With U.K.A.E.A., 1956-59; Manchester Univ., 1959-. Nuclear interests: Reactor physics; Particle physics.
Address: The Physical Laboratories, The University, Manchester 13, England.

GRANT, John W. Formerly R. and D. Director, R. R. Donnelley and Sons Co.; Pres., Rad. Counter Labs., Inc., 1964-.
Address: Radiation Counter Laboratories, Inc., 5121 West Grove Street, Skokie, Illinois, U.S.A.

GRANT, Peter John, M.A., Ph.D., A.M.I.Mech. E., F. Inst. P. Born 1926. Educ.: Cambridge Univ. Univ. Lecturer, Natural Philosophy, Glasgow Univ., 1950-55; Chief Phys., Atomic Energy Div., G.E.C., 1956-59; Univ. Reader in Eng. Sci., (1959-66), Prof., Nucl. Power (1966-), Imperial Coll. Sci. and Technol., London. Nuclear interests: Reactor physics and reactor performance.
Address: 49 Manor Road South, Esher, Surrey, England.

GRANT, S. P. Formerly Chief, Metal. and Welding, Res. Centre, Alliance, now Materials Application Eng., Atomic Energy Div., Babcock and Wilcox.
Address: Babcock and Wilcox Co., Mount Athos, Lynchburg, Virginia, U.S.A.

GRANT, Walter Lawrence, D.Sc. (Eng.), M.Sc. (Appl. Math.). Born 1922. Educ.: Witwatersrand

and Pretoria Univs. Head, Thermodynamic Div. (1952-57), Director (1957-59), Nat. Mech. Eng. Res. Inst., South Africa Council for Sci. and Ind. Res.; Chief Eng. (1959-64), Deputy Director Gen., Chief Eng. (1964-67), Director Gen., (1967-), Atomic Energy Board. Societies: A.N.S.; A.M.I. Mech. E. (London); M.S.A.I. Mech. E.; A.F. R. Ae. S.; S.A. Akademie vir Wetenskap en Kuns.
Nuclear interests: Reactor physics; Reactor design; Reactor economics.
Address: Private Bag 256, Pretoria, South Africa.

GRANTHAM, W. J., Dr. Book: Contributor to Eng. Compendium on Rad. Shielding (I.A.E.A.).
Address: Florida University, College of Engineering, Gainesville, Florida, U.S.A.

GRAS, Michel, L. ès Sc., Ingenieur, Frigoriste. Born 1925. Educ.: Sorbonne and Caen Univ. Direction des Etudes et Fabrications d'Armement, 1950-55; Directeur de Laboratoire, 1955-58; C.E.A., Prof. au Centre Associé de Saclay du C.N.A.M., 1958-. Sec., Sect. Francaise, H.P.S. Society: Sté. de Radioprotection.
Nuclear interests: Management and information.
Address: 5 rue Armand Gauthier, Paris 18, France.

GRASS, F., Dr. phil. Sen. Officer Atominstitut der Osterreichischen Hochschulen.
Nuclear interest: Radiochemistry.
Address: Atominstitut der Osterreischen Hochschulen, 115 Schüttelstrasse, Vienna 2, Austria.

GRASS, Günther, Dr.-phil., Physicien. Born 1924. Educ.: Tübingen, Cologne and Göttingen Univ's. Head, Heat Transfer Div., Euratom Res. Centre, Ispra, Italy. Member, Com. on Reactor Safety Technol., O.E.C.D., E.N.E.A.
Nuclear interest: Heat transfer.
Address: 50 Via Matteotti, Varese-Cascago, Italy.

GRASSAM, Norman Sydney John, Ph.D., B.Sc. (Eng.), C. Eng. Born 1922. Educ.: London Univ. Ind. Consultant on nucl. matters, 1951-. Sen. Lecturer in Mech. Eng., Southampton Univ. Societies: M.I.C.E.; M.I.Mech.E.
Nuclear interests: Reactor physics; Heat transfer.
Address: Engineering Department, The University, Southampton, England.

GRASSINI, Luigi, Dr. Vice-Pres., Board of Directors, Ente Nazionale per l'Energia Elettrica.
Address: Ente Nazionale per l'Energia Elettrica, 181 Via del Tritone, Rome, Italy.

GRASSO, Francesco, Dottore in Fisica. Born 1940. Educ.: Catania Univ. Ministero, Pubblica Istruzione, Centro Siciliano di Fisica Nucleare e Struttura della Materia, Catania Univ., 1962-. Gruppo Nazionale di Struttura della Materia. Society: Sta. Italiana di Fisica.
Nuclear interests: Electron impact processes; Study of electron impact ionisation of atoms and molecules; Determinations of threshold energies and cross sections under impact with electrons of low and well defined energy; Dynamics of impact processes and fragmentation.
Address: Catania University, 57 Corso Italia, Catania, Italy.

GRATTON, Joseph George, B.S., M.S. (Chem.). Born 1924. Educ.: Siena Coll. Chem.; GE-K.A.P.L., 1948-57; Tech. Information Officer, 1957-60; Chief Sci. Publications Branch, (1960-64), Asst. Director, Tech. Information Div., (1964-), U.S.A.E.C. Society: A.N.S.
Nuclear interests: Chemical processing, technical book preparation, reactor coolant technology.
Address: 11503 Monticello Avenue, Silver Spring, Maryland, U.S.A.

GRAU, Pierre Alphonse, L. ès Sc., E.S.E., G.A. (Saclay). Born 1921. Educ.: Toulouse and Paris Univs. and I.S.N.S.E. (Argonne). Deputy Gen. Manager, Hifrensa (Hisprano Francesa de Energia Nuclear, S.A.). Book: Introduction à l'Electronique. Societies: Sté. française des Electriciens; Sté. française des Radioélectriciens; A.N.S.; Assoc. Francaise d'Informatique et de Recherche Opérationnelle.
Nuclear interests: Reactor design; Management.
Address: Hifrensa, 20-24 Tuset, Barcelona, Spain.

GRAUE, Arnfinn, Dr. philos. Born 1926. Educ.: Bergen Univ. Assoc. Prof., Bergen Univ., 1961-.
Nuclear interests: Nuclear structure physics.
Address: Physics Department, Bergen University, Bergen, Norway.

GRAUL, Emil Heinz, Dr. rer. nat., Dr. med. Born 1920. Prof., Univ. Inst. of Radiol., Münster/Westf., 1946-53; Director, Dept. of Radiobiol. and Radioisotopes, Rad. Inst., Marburg/Lahn Univ., 1954-63; Direktor, Inst. of Biophys. and Nucl. Medicine; Editor-in-Chief, Atompraxis.
Nuclear interests: Nuclear medicine; Radiobiology; Biophysics; Space research.
Address: 4a Lahnstrasse, Marburg/Lahn, Germany.

GRAVELLE, Paul H. Formerly Vice-Pres., Power Group, Toledo Edison Co. Member, Operating Com., Power Reactor Development Co.
Address: 2143 Central Grove, Toledo, Ohio 43614, U.S.A.

GRAVES, Charles Carleton, B.E., M.E., D.Eng. (Mech. Eng.). Born 1922. Educ.: Yale Univ. Nat. Advisory Com. for Aeronautics, 1947-56;

United Nucl. Corp., White Plains, N.Y. 1956-65; Prof. of Nucl. Sci. and Eng., Catholic Univ., Washington, D.C. 1965-. Societies: A.N.S.; A.S.M.E.; A.A.U.P.
Nuclear interests: Reactor design and safety; Heat transfer.
Address: Department of Nuclear Science and Engineering, Catholic University of America, Washington, D.C. 20017, U.S.A.

GRAY, A. L., Dr. State Health Officer and Director, Radiol. Health Unit, Mississippi State Board of Health.
Address: Mississippi State Board of Health, Radiological Health Unit, Box 1700, Jackson 5, Mississippi, U.S.A.

GRAY, Alan Lyle, B.Sc. (Phys.), A. Inst. P. Born 1923. Educ.: London Univ. Caswell Res. Lab., Plessey Co. Ltd., 1950-56; Plessey Nucleonics Ltd., 1956-63 (Tech. Manager 1961-63); Chief Eng. (Nucl.) Plessey Automation Group, 1963-66; Chief Phys., Elliott Process Automation Ltd., 1966-.
Nuclear interests: Nuclear instrumentation; Applications of nuclear techniques in control, measurement and analysis.
Address: Elliott Process Automation Ltd., Century Works, Lewisham London, S.E.13, England.

GRAY, Allen G., B.S., M.S., Ph.D. Educ.: Vanderbilt and Wisconsin Univs. During World War II on loan from DuPont Co. to Metallurgical Lab., Chicago Univ. (Manhattan Atomic Bomb Project); Atomic Energy Div., DuPont Co. (special assignment, K.A.P.L., Gen. Elec. Co.), 1951-52; Editor, Metal Progress, 1958-; Member, Advisory Com. on Tech. Information, U.S.A.E.C.; In charge, U.S.A.E.C.-A.S.M. cooperative programme to develop series of monographs on nucl. metallurgy.
Address: 23499 Shelburne Road, Shaker Heights, Ohio, U.S.A.

GRAY, Chester Marmion, M.Sc. (Hons.), E.D. Born 1907. Educ.: Canterbury Coll., New Zealand Univ. C.S.I.R.O., 1946-50;Deputy Director, Div. of Ind. Development, 1951-53; Australian Tech. and Ind. Rep., London, 1953-56; Dept. of Trade, 1956-60; Consultant, Promotion of Use of Radioisotopes (1960-61), Commercial Manager, Isotopes, (1961), Leader, Production Group, Isotopes Div. (1962), Head, Services Sect., and Deputy to Chief, Isotopes Div. (1963-64), Australian Atomic Energy Adviser, London (1964-), Counsellor, Australian Mission to the Euratom Commission (1964-), A.A.E.C. Societies: Assoc. Member, Inst. Eng. Australia (Chartered Eng.); Assoc. Roy. Australian Chem. Inst.; Hon. Fellow, A.I.M.; Fellow, Roy. Soc. of Arts.
Address: Australia House, Strand, London, W.C.2, England.

GRAY, Dwight Elder, A.B., B.S., Ph.D. (Phys.). Born 1903. Educ.: Ohio State Univ. Chief, Tech. Information Div., Lib. of Congress, 1950-55; Office of Sci. Information Service, N.S.F., 1955-63; Chief, Sci. and Tech. Div., Library of Congress, 1963-65; Washington Rep., A.I.P., 1965-. Books: Co-author, Rad. Monitoring in Atomic Defence (Van Nostrand, 1951); co-author, Man and his Phys. World (Van Nostrand, 1966); co-ordinating editor, A.I.P. Handbook (McGraw-Hill, 1963). Societies: A.P.S.; A.A.A.S.; N.A.S.W.; Nat. Assoc. Sci. Writers.
Nuclear interests: In scientific information field.
Address: 117 4th Street, N.E., Washington, D.C. 20002, U.S.A.

GRAY, James Lorne, B. Eng., M.Sc. (Mech. Eng., Saskatchewan), LL.D. (Hon.), D.Sc. (Hon.). Born 1913. Educ.: Saskatchewan Univ. Gen. Manager (1952), Vice-Pres., Administration and Operations (1954), Pres. (1958-), Atomic Energy of Canada, Ltd. Societies: Eng. Inst. of Canada; Professional Eng. of Ontario.
Address: 275 Slater Street, Ottawa 4, Ontario, Canada.

GRAY, John E., B.S. (Chem. Eng., Rhode Island). Born 1922. Educ.: Rhode Island Univ. In Naval Reactors Branch, U.S.A.E.C. and U.S. Navy, 1949-50; Director, Tech. and Production Div., Savannah River Operations Office, U.S.A.E.C., 1950-54; Shippingport Project Manager, Duquesne Light Co., 1954-60; Pres. and Chief Exec. Officer, NUS Corp. (formerly Nucl. Utility Services, Inc.), 1960-. Vice Chairman, Nucl. Services Internat. Ltd., Zurich, Switzerland, 1962-; Chairman of the Board, Consultec, Inc., 1967-; Director, Cyrus Wm. Rice and Co., 1967-. Books: Editor, Shippingport Operations, 1960; contributor, Mark's 7th Edition, Standard Handbook for Mechanical Engineers, Nucl. Power Section. Societies: A.N.S.; A.I.Ch.E.; Atomic Ind. Forum; A.A.A.S.
Nuclear interests: Nuclear power development, design, and operation; Economics of nuclear power; Nuclear materials; Radiation processing; Peacetime uses of nuclear devices.
Address: NUS Corporation, 1730 M Street, N.W., Washington, D.C. 20036, U.S.A.

GRAY, Oscar S., B.A., LL.B. Born 1926. Educ.: Yale Univ. (Yale Coll. 1948, Yale Law School 1951). Attorney-adviser, economic and mutual security affairs, Legal Adviser's Office, U.S. Dept. of State, 1951-57; Sec. and Treas., (1957-67) Vice Pres. (1967-), Nucl. Materials and Equipment Corp.; Sec. and Director, Nucl. Decontamination Corp., 1959-; Sec., Treas. and Director, NUMEC Instruments and Controls Corp., 1962-; Pres. and Director, Apollo (Pa.) Chamber of Commerce, 1960-62; Member, Department of State Univ, Nat. Defence Executive Reserve, 1960-61; Sec. and Treas., Isotopes and Rad. Enterprise (ISORAD) Ltd., 1966-. Society: Atomic Ind. Forum.
Nuclear interests: Management; Public regulation; Legislation; Financing; International

trade; Procurement regulations and contract practices.
Address: 7551 Spring Lake Drive, Bethesda, Md. 20034, U.S.A.

GRAY, William, C. Eng., F.I.E.E., A.M.I.Mech. E. Born 1906. Director and Manager, Eng. Div., A. Reyrolle and Co. Ltd.
Nuclear interest: The development of control rod systems and faulty fuel can detection systems.
Address: A. Reyrolle and Co. Ltd., Hebburn, Co. Durham, England.

GREA, J. Maitre Asst., Inst. de Physique Nucléaire, Lyon Univ.
Address: Lyon University, 43 boulevard du 11 Novembre 1918, 69- Villeurbanne (Rhône), France.

GREAGER, Oswald Herman, B.S. (Maryland), M.S. and Ph.D. (Michigan). Born 1905. Educ.: Maryland and Michigan Univs. Manager, Res. and Eng., Irradiation Processing Dept., Hanford Atomic Products Operation, Richland, Washington, then Consultant, Atomic Power Equipment Dept., San Jose, California (-1968), now Manager, Div. Planning Operation, Nucl. Energy Div. (1968-), G.E.C. Societies: A.C.S.; A.N.S.; Atomic Ind. Forum.
Nuclear interests: Technical management; Reactor operation and design.
Address: 19311 Valle Vista Drive, Sarataga, California, U.S.A.

GREATHOUSE, Glenn A., B.S., M.S., Ph.D. Born 1903. Educ.: Illinois and Duke Univs. Consultant, N.A.S., U.S.A., 1945-; Pres., Nucl. Res. Chemicals, Inc.; Head Nucl. Consultant, Commonwealth of Mass. Book: Deterioration of Materials - Preventive Techniques (1954). Societies: A.N.S.; A.P.S.; A.C.S.
Nuclear interests: Nuclear chemistry and physics; Radiation research; Radioactive synthesis of organic chemicals.
Address: Special Consultant-Nuclear, P.O. Box 5395, Daytona Beach, Fla 32020, U.S.A.

GREBE, John Josef, B.S., M.S., Hon. D.Sc. Born 1900. Educ.: Case Inst. Technol. Consultant to Army Chem. Corps; Consultant A.E.C. Advisory Com. on Isotope and Rad. Development; Member, Com. on Automation and Manpower Development for U.S. Labour Dept. Societies: A.N.S.; N.Y. Acad. Sci.; A.P.S.; A.I. Ch. E.; and many others.
Nuclear interests: Reactors, peace-time uses of atomic energy (Project Plowshare), structure of the nucleus (nuclear physics).
Address: 12430 W. St. Andrew, Sun City, Arizona 85351, U.S.A.

GREBENNIKOV, K. V. Paper: Co-author, Control Rod Operation in High Temperature Reactors (Atomnaya Energiya, vol. 22, No. 5, 1967).
Address: Academy of Sciences of the U.S.S.R., 14 Leninsky Prospekt, Moscow V-71, U.S.S.R.

GREBENNIKOV, R. V. Papers: Co-author, Mechanical Properties and Corrosion Resistance of Hafnium-Zirconium Alloys in Steam-Water Media (Atomnaya Energiya, vol. 14, No. 3, 1963); co-author, Effect of Vanadium on Phase Constitution and Structure of High Boron Steel (Abstract, ibid, vol. 20, No. 2, 1966); co-author, Effect of Titanium on Phase Constitution and Ductivity of Chrome-Nickel-Steel with High Content to Bore (ibid, vol. 22, No. 5, 1967); co-author, Corrosion Resistance of Some Binary Hafnium Alloys (ibid, vol. 22, No. 6, 1967).
Address: Nuclear Energy Institute, Academy of Sciences of the U.S.S.R., 14 Leninsky Prospekt, Moscow V-71, U.S.S.R.

GREBOT, Daniel, Dr.-Vétérinaire. Lab. Directeur, J. Morey et Fils S.A.
Address: J. Morey et Fils S.A., B.P. No. 1, Cuiseaux, (Saône et Loire), France.

GRECHUHIN, D. P. Paper: Photoexcitation and Photoionisation of Energetic Hydrogen Atoms (Atomnaya Energiya, vol. 20, No. 5, 1966).
Address: Academy of Sciences of the U.S.S.R., 14 Leninsky Prospekt,Moscow, V-71, U.S.S.R.

GRECO, G., Ing. Strumenti e Misure nucl., Facolta d'Ing., Ist. di Applicazioni e Impianti Nucl., Palermo Univ.
Address: Palermo University, Viale delle Scienze, Palermo, Sicily, Italy.

GREEBLER, Paul, B.S., M.S., Ph.D. Born 1922. Educ.: Colorado and Rutgers Univs. Phys., Johns-Manville Res. Centre, 1946-55; Reactor Phys., K.A.P.L., 1955-56; Sen. Phys. (1956-60), Manager, Advance Reactor Phys., Atomic Power Equipment Dept. (1960-65), Manager, Phys. Subsect., Advanced Products Operation (1966-), G.E.C. Societies: A.N.S.; A.P.S.
Nuclear interests: Reactor physics and nuclear engineering with emphasis on development of fast breeder and other advance reactor concepts. Engaged in planning and conducting fast reactor physics development programmes in the areas of nuclear data, neutronics computational methods, dynamics and reactor safety, and in the nuclear design of advanced power reactors.
Address: 1461 Cherry Garden Lane, San Jose, California, U.S.A.

GREEN, Alex Edward Samuel, M.S., Ph.D. (Phys.). Born 1919. Educ.: California Inst. Technol. and Cincinnati Univ. Prof., Phys., Sci. Director (1953-59), Assoc. Prof., Phys., Acting Chairman, Tandem Van de Graaff Programme (1957), Florida State Univ.; Chief of Phys., Convair, a Div. of Gen. Dynamics, San Diego, 1959-63; Graduate Res. Prof., Florida Univ. Books: Nucl. Phys. (Internat. Series of Phys., McGraw-Hill, 1955); Editor, Proc. Internat. Conference on the Nucl. Optical Model (Florida State Univ. Studies, April 1959). Societies: Fellow, A.P.S.;

A.A.P.T.; Fellow, Optical Soc. of America.
Nuclear interests: Nuclear physics; Nuclear propulsion; Nuclear cross-sections; Radiation effects; Nuclear weapons effects; Radiation in atmosphere.
Address: 2900 N.W. 14th Place, Gainesville, Florida, U.S.A.

GREEN, Anthony Maurice, Ph.D., B.Sc. Born 1938. Educ.: Birmingham Univ. Res. Assoc., California Univ., San Diego, 1961-62; Res. Assoc., Maryland Univ., 1962-63; D.S.I.R. N.A.T.O. Fellow, Niels Bohr Inst., Copenhagen, 1963-64; Member, Inst. for Advanced Study (1964-66), Lecturer, Palmer Lab. (1966-68), Princeton Univ.; Res. Fellow, Res. Inst. for Theoretical Phys., Helsinki Univ., 1968-69.
Nuclear interests: Nuclear structure theory; Many body problems.
Address: Princeton University, Palmer Physical Laboratory, Princeton, New Jersey 08540, U.S.A.

GREEN, Ben Arthur, B.S., M.S., Ph.D. Asst. Prof., Phys. Dept., Western Reserve Univ.
Address: Western Reserve University, Physics Department, Cleveland 6, Ohio, U.S.A.

GREEN, Donald Thomas, B.Sc. (Eng. Phys.). Born 1920. Educ.: Saskatchewan Univ. Manager, Special Products Div., Picker X-ray Corp.; Gen. Manager, Picker Ind. Products.
Nuclear interest: Management.
Address: 1688 Arabella Road, Cleveland 12, Ohio, U.S.A.

GREEN, Earl E., Dr. Member, Advisory Com. for Biol. and Medicine, U.S.A.E.C.
Address: Jackson Laboratory, Bar Harbor, Maine, U.S.A.

GREEN, G. Kenneth, Dr. Chairman, Accelerator Dept., Brookhaven Nat. Lab. Member, Board of Contract Appeals, U.S.A.E.C.
Address: Brookhaven National Laboratory, Accelerator Department, Upton, Long Island, New York, U.S.A.

GREEN, Herbert Sydney, A.R.C.S., Ph.D., D.Sc. Born 1920. Educ.: Royal Coll.of Sci., London Univ. and Edinburgh Univ. Prof. Mathematical Phys., Adelaide Univ., 1951-; Visiting Prof., Dublin Inst. for Advanced Studies, 1950-51 and 1958. Societies: Fellow, Australian Acad. of Science; Australian Mathematical Soc.; Australian Inst. of Phys.
Nuclear interests: Theoretical nuclear physics and plasma physics.
Address: Adelaide University, South Australia, Australia.

GREEN, J. R., B.E., D. Phil., A.F.R. Ae.S Chief Sci. and Director, Vickers Res. Establishment.
Address: Vickers Research Establishment, Sunninghill, nr. Ascot, Berks., England.

GREEN, James Henry, M.Sc., Ph.D. Born 1922. Educ.: Queensland and Cambridge Univs. Formerly Assoc. Prof. Nucl. and Rad. Chem. (1960-), and formerly Member, Com. of the Inst. of Nucl. Eng., New South Wales Univ., 1960. Adviser to J. Inorg. and Nucl. Chem. Books: Co-author, Detection and Measurement of Radioactive Contamination (1960); co-author, Positronium Chem. (1964).
Societies: Fellow, Royal Australian Chem. Inst.; Australian Rad. Soc.; British Interplanetary Soc.
Nuclear interests: Head of dept. of nuclear and radiation chemistry, undergraduate and graduate teaching. Research. Positronium, mass spectrometry of solids and gases and correlation with radiation chemistry, cosmochemistry, radiocarbon dating, low-level analyses, radiation biology, ion-molecule reactions.
Address: New England Institute, P.O. Box 308, Ridgefield, Connecticut 06877, U.S.A.

GREEN, Leslie Leonard, M.A., Ph.D. (Cantab.). Born 1925. Educ.: Cambridge Univ. Lecturer (1948-57), Sen. Lecturer (1957-60), Reader (1960-64), Prof. Exptl. Phys. (1964-), Liverpool Univ. Papers: Papers on fission, nuclear reactions and γ-ray spectroscopy.
Society: Phys. Soc.
Address: Chadwick Laboratory, Liverpool University, Liverpool 3, Lancs., England.

GREEN, Ralph Ellis, B.Sc. (Dalhousie), M.Sc. (Dalhousie), Ph.D. (Phys., McGill). Born 1931. Educ.: Memorial (Newfoundland), Dalhousie (Halifax) and McGill Univs. Res. Officer, A.E.C.L. Society: Canadian Assoc. of Phys.
Nuclear interests: Early work in positron annihilation, both angular correlation and lifetime studies; Eleven years experience in experimental reactor lattice physics; Now working on accelerator physics research.
Address: 21 Frontenac Crescent, Deep River, Ontario, Canada.

GREEN, Richard C., B.S. (Business Administration). Born 1925. Educ.: Michigan State and Missouri Univs. Purchasing Agent (1950-51), Vice-Pres. and Purchasing Agent (1951-53), Exec. Vice-Pres. (1953-58), Pres. (1958-63), Pres. and Board Chairman (1963-), Missouri Public Service Co. Trustee, Southwest Atomic Energy Assoc.
Address: 10700 East Highway 50, Kansas City, Missouri 64138, U.S.A.

GREEN, Roy M., B.Sc. (Liverpool), M.Sc., Ph.D. (Toronto). Born 1935. Educ.: Liverpool and Toronto Univs. Res. Officer, Australian Atomic Energy Commission, 1961-64; Sen. Member, Sci. Staff, RCA Victor Co. Ltd., Montreal, 1964-67. Societies: Sen. Member, I.E.E.E. (Nucl. Sci. Group); Assoc. Member, Inst. of Phys. and Australian Inst. of Phys.; Canadian Assoc. of Physicists.
Nuclear interests: Gamma-ray spectrometry, with particular emphasis on scintillators and Ge(Li) detectors. Application to low-activity

measurements — e.g., whole body counting, Health physics applications.
Address: Research Laboratories, RCA Victor Co. Ltd., 1001 Lenoir Street, Montreal, P.Q., Canada.

GREEN, Stanley J., B.Ch.E., M.S.Ch.E. Born 1920. Educ.: C.C.N.Y. and Drexel Inst. Technol. Chem. Process Eng., ACME Coppersmithing and Machine Co., Oreland, Pa., 1945-54. Manager, Thermal and Hydraulics Eng. Sect., Fuel Element Development Dept., Bettis Atomic Power Lab., Westinghouse Elec. Corp., Pittsburgh, 1954-. Member, A.S.M.E. Standing Com. K-8 on Theory and Fundamental Res. Exec. Com. A.S.M.E. Heat Transfer Div. Societies: A.S.M.E.; A.I.Ch.E. Nuclear interests: Supervise heat transfer and fluid flow research relative to the design of pressurised water reactors.
Address: 5423 Northumberland Street, Pittsburgh 17, Pennsylvania, U.S.A.

GREENAWAY, David Stuart, B.Sc. (London). Born 1934. Educ.: London Univ. Mathematician, Fairey Aviation Co., Ltd., Hayes, Middx., 1955-57; Mathematician, Atomic Power Constructions, Ltd., London, 1957-.
Nuclear interest: Nuclear physics, particularly the fundamentals (e.g., quantum mechanics); All mathematical aspects of nuclear power station design, especially in the fields of optimisation and performance.
Address: 239 Kingston Road, London, S.W.19, England.

GREENBAUM, Leonard, Ph.D. Born 1930. Educ.: Michigan and California (Berkeley) Univs. Asst. Prof., English, Coll. of Eng. (1963-), Asst. Director, Michigan Memorial - Phoenix Project (1964-), Michigan Univ. Nuclear interest: Research administration and education.
Address: Michigan Memorial-Phoenix Project, Michigan University, Ann Arbor, Michigan, U.S.A.

GREENBERG, Elliott, B.S., M.S., Ph.D. Born 1927. Educ.: C.C.N.Y. and Michigan Univ. At Argonne Nat. Lab., 1954-. Societies: A.C.S.; Scientific Res. Soc. of America. Nuclear interests: The use of fluorine bomb combustion calorimetry as a technique in studying the thermochemistry of refractory type materials which are of interest in high-temperature chemistry.
Address: Argonne National Laboratory, Chemical Engineering Division, 9700 South Cass Avenue, Argonne, Illinois, U.S.A.

GREENBURG, Joseph, B.A., M.D. Born 1926. Educ.: New York Univ. Consultant, Nucl. Medicine, U.S. Naval Hospital, St. Albans, 1956-; Chairman, Com. on Educ. and Res., The Soc. of Nucl. Medicine, 1962-64; Consultant, Nucl. Medicine, U.S. Naval Hospital, Bethesda, 1962-65; Medical Advisor, Leukemia Soc. 1959-65; Prof. and Director, Dept. Nucl. Medicine, New York Polyclinic Medical School and Hospital, 1960-62; Chief, Div. of Nucl. Medicine, Long Island Jewish Hospital, 1956-60; Chief, Radioisotope Lab., U.S. Naval Hospital, St. Albans, 1954-56. Fellow, Nat. Cancer Inst., U.S.P.H.S., 1952-54. Pres., New York Metropolitan Chapter and Member, Nat. Exec. Com. and Board of Trustees, Soc. of Nucl. Medicine. Societies: Fellow, A.A.A.S.; Soc. of Nucl. Medicine; A.N.S.; American Physiological Soc.; American Assoc. for Cancer Res., New York Acad. Sci.; New York Cancer Soc.; American Medical Assoc. Nuclear interest: Nuclear medicine.
Address: 106 Clover Drive, Great Neck, Long Island, New York, U.S.A.

GREENE, David, M.Sc., Ph.D. Born 1923. Educ.: Queen's Univ., Belfast. Phys.-Eng., Betatron Res. Unit, M.R.C., 1953-. Societies: British Inst. Radiol.; Inst. Phys.; Hospital Physicists' Assoc.; Faculty of Radiologists (Honorary Assoc. Member).
Nuclear interest: Application of high-energy radiations to radiotherapy.
Address: Christie Hospital and Holt Radium Institute, Withington, Manchester 20, England.

GREENE, Margaret Willard, A.B., M.S. Born 1919. Educ.: Elmira Coll., N.Y. and Syracuse, New York and Columbia Univs. Asst. Prof. Chem., Elmira Coll. 1946-56; Member, Radio Isotope Development Group, Hot Lab., Brookhaven Nat. Lab., 1956-. Co-author, U.S. Patent No. 2942943 for invention titled Process and Apparatus for Separating Iodine-132 from Fission Products. Society: A.C.S. Nuclear interest: New radioisotope development.
Address: The Hot Laboratory, Brookhaven National Laboratory, Upton, New York, U.S.A.

GREENFIELD, Moses A., B.S., M.S. (Phys.), Ph.D. (Phys.), Certified by American Board of Radiol. (Radiol. Phys.), 1967. Born 1915. Educ.: New York and George Washington Univs. Phys., Atomic Energy Project, California Univ. at Los Angeles, 1948-51; Assoc. Prof. Radiol. (1951-56), Prof. Radiol. (1956-), Medical Centre, California Univ. at Los Angeles. American Specialist in South America, and in Asia (1964-65), for U.S. Dept. of State. Societies: Fellow, A.P.S.; New York Acad. Sci.; Rad. Res. Soc.; A.N.S.; Soc. Nucl. Medicine; American Coll. of Radiol. Nuclear interests: Application of radioisotope to medical problems; Reactor safety.
Address: Medical Centre, University of California at Los Angeles, Los Angeles 24, California, U.S.A.

GREENHALGH, Geoffrey Harvey. Born 1920. Educ.: London Univ. A.E.R.E. Harwell, 1948; Sci. Attaché, British Embassy, Stockholm, 1956; Nucl. Energy Attaché, U.K. Deleg. to the European Communities, Brussels, 1961; Sec.-Gen., British Nucl. Forum, 1964;

Member, Exec. Com., Foratom; Member for British Nucl. Forum, Nucl. Public Relations Contact Group.
Address: 21 Tothill Street, London, S.W.1, England.

GREENHAM, E. T. K., A.M.I.Mech.E. Project Eng. for Dungeness, Southern Project Group, C.E.G.B.
Address: Central Electricity Generating Board, Southern Project Group, Squires Lane, Finchley, London, N.3, England.

GREENHOW, Harold, B.Sc. Born 1923. Educ.: Durham Univ. Eng. II, U.K.A.E.A., -1955. Head, Procurement Office and Deputy Chief Construction Eng., English Elec. Co., Atomic Power Div., 1955-62; Chief Exec., Hinkley Point Group 1962-64; Works Manager, Rugby Steam Turbine Works, 1964-66; Chief Project Eng., Wylfa Nucl. Power Station, 1966-.
Nuclear interests: Off-site procurement of all components for nuclear power stations; All on-site operations.
Address: 84 Station Road, Wigston, Leicester, England.

GREENLAW, Robert Hiram, B.S., M.D. Born 1927. Educ.: Tufts Coll. and Rochester Univ. Radiol., Rochester Univ. Medical Centre, 1954-55; Radiol.-in-charge, 2500th U.S.A.F. Hospital, 1955-57; Instructor in Radiol. (1958-60), Sen. Instructor in Radiol. (1960), Rochester Univ.; Asst. Prof. Radiol., (1961-63), Assoc. Prof. Radiol., (1963-67), Prof. Radiol., (1967-), Kentucky Univ. Formerly Member, Kentucky Atomic Energy Authority; Member, Medical Advisory Com., U.S.A.E.C. Societies: A.C.R.; Assoc. Univ. Radiol.; Soc. Nucl. Medicine; Kentucky Radiol. Soc.
Address: College of Medicine, Department of Radiology, Kentucky University, Lexington, Kentucky, U.S.A.

GREENLEES, Frank Millar, B.Sc. (Eng.). Born 1909. Educ.: Glasgow Univ. Chief Eng., Eng. Services, Dept. of Atomic Energy, M.O.S. (later U.K.A.E.A.), 1950-56; Consultant (Ind. Reactors) U.K.A.E.A., Ind. Group (now Reactor Group), 1956-. Societies: F.I.E.E.; Inst. of Mech. Eng.; I.C.E.
Nuclear interests: Reactor design and nuclear engineering; Nuclear power stations (design, construction and economics).
Address: Oakfield, Hill Top Road, Grappenhall, Cheshire, England.

GREENOUGH, Geoffrey Blakeley, M.A., Ph.D. Born 1921. Educ.: Cambridge Univ. Sen. Lecturer in Phys. Metal., Sheffield Univ., 1951-56; Deputy Head of Lab., Windscale, 1956-61; Deputy Head of Lab., Springfields, 1961-. Societies: Fellow, Inst. Phys. and Phys. Soc.; Fellow, Inst. Metal.; F.I.M.
Nuclear interest: Behaviour of materials, particularly graphite and fuel elements, during irradiation.
Address: Springfield Laboratories, United Kingdom Atomic Energy Authority, Salwick, Preston, Lancashire, England.

GREENWOOD, G. W. Member, Editorial Advisory Board, J. of Nucl. Materials.
Address: Journal of Nuclear Materials, North-Holland Publishing Co., P.O. Box 103, Amsterdam, Netherlands.

GREENWOOD, J. Ward, B.Sc., M.A. Born 1925. Educ.: Manitoba and Minnesota Univs. With A.E.C.L., 1954-67; Head, Office of Internat. Affairs, A.E.C.L., 1959-67; Sci. Counsellor, Canadian Embassy, Washington D.C., 1967-.
Nuclear interest: International affairs.
Address: 1746 Massachusetts Avenue, N.W., Washington D.C. 20036, U.S.A.

GREGG, Earle C., Dr. Res. Assoc., Rad. Phys., Wearn Lab. for Medical Res., Western Reserve Univ.
Address: Wearn Laboratory for Medical Research, Western Reserve University, 2064 Abington Road, Cleveland 6, Ohio, U.S.A.

GREGOIRE, Georges, Ing. civil. Born 1915. Educ.: Ecole Royale Militaire. Directeur, groupe, Fabrimétal. Conseiller, Groupement Professionnel de l'Industrie Nucléaire.
Address: Fabrimétal, 21 rue des Drapiers, Brussels 5, Belgium.

GREGORIC, Teodor, Dipl. Eng. Born 1911. Educ.: Belgrade Univ. Chief, Development Dept., Energoinvest, 1950-60; Prof. Thermodynamics, Mech. Faculty, Sarajevo Univ. 1960-67. Society: Yugoslav Soc. Heat Eng. Nuclear interests: Heat problems in nuclear reactors and nuclear power plants.
Address: 12/V Obala Vojvode Stepe, Sarajevo, Yugoslavia.

GREGORY, Bernard Paul, Ph.D. Born 1919. Educ.: Paris Univ. and M.I.T. Prof., Phys., Ecole des Mines, Paris, 1953-58; Prof., Phys. (1959-), Vice Director, Lab. of Phys. (1959-65), Ecole Polytechnique, Paris; Director Gen., C.E.R.N., 1966-.
Nuclear interest: High energy physics.
Address: C.E.R.N., 1211 Geneva 23, Switzerland.

GREGORY, Charles O., A.B., LL.B. Born 1902. Educ.: Yale Univ. Law Prof., Virginia Univ. Member, Atomic Energy Labour-Management Relations Panel, U.S.A.E.C.
Address: Law School, University of Virginia, Charlottesville, Va., U.S.A.

GREGORY, Jack Norman, B.Sc., M.Sc., D.Sc. Born 1920. Educ.: Melbourne Univ. C.S.I.R.O. seconded to A.E.R.E. Harwell, 1948-53; Chief, Isotopes Div., Australian A.E.C., 1953-; Member, Radiol. Advisory Council, New South Wales, 1957-64. Book: The World of

Radioisotopes (Angus and Robertson, 1966). Society: F.R.A.C.I.
Nuclear interests: Radioisotope research, applications and production.
Address: 13 Nellella Street, Blakehurst, New South Wales, Australia.

GREGORY, Lloyd Patrick, B.Sc. Born 1924. Educ.: Canterbury Univ., New Zealand. Sen. Rad. Officer, Nat. Rad. Lab., New Zealand. Nuclear interest: Low level environmental radiochemical analyses.
Address: National Radiation Laboratory, 108 Victoria Street, Christchurch, New Zealand.

GREGORY, William Cyril Roy. Born 1913. Chief Planning Eng., Production Dept. (1947-55), Workshop Manager A.E.R.E., Harwell (1955-59), Manager, Eng. Dept., Radiochem. Centre, Amersham (1959-), U.K.A.E.A. Society: A.M.I.Mech.E.
Nuclear interest: Management of services and equipment provision for Atomic Energy applications.
Address: Lydstep, Clifton Road, Chesham Bois, Amersham, Bucks., England.

GREGSON, John, A.M.C.T. Born 1924. Educ.: Manchester Inst. of Sci. and Technol. Eng. Director, Fairey Eng. Ltd., 1965-.
Address: Fairey Engineering Limited, Heaton Chapel, Stockport, Cheshire, England.

GREGSON, Thomas C., B.S. (Ch.E.), M.S. (Ch.E.). Born 1927. Educ.: Purdue Univ. and O.R.S.O.R.T. With Goodyear Tire and Rubber Co., 1952-. Societies: A.N.S.; Isotopes and Rad. Div., A.N.S.; A.I.Ch.E.; A.C.S.; Rubber Chemistry Div., A.C.S.; Michigan Nucleonic Soc.
Nuclear interests: Radiation effects; Radiation chemistry; Tracer technology.
Address: 3380 Pendleton Street, Cuyahoga Falls, Ohio, U.S.A.

GREIDER, Kenneth R., B.S., M.S., Ph.D. Born 1929. Educ.: California (Berkeley) Univ. Phys., Lawrence Rad. Lab., 1958-59; Asst. Res. Phys., California, San Diego Univ., 1959-61; Asst. Prof., Yale Univ., 1961-. Society: A.P.S.
Nuclear interests: Scattering theory; Rearrangement collisions.
Address: Physics Department, Yale University, New Haven, Conn., U.S.A.

GREIFELD, Helmut Rudolf, Dr. jur. Born 1911. Educ.: Munich, Kiel and Leipzig Univs. Managing Director, Gesellschaft für Kernforschung m.b.H. Karlsruhe.
Address: 5 Weberstrasse, Karlsruhe, Germany.

GREIFER, Bernard, Ph.D. Born 1921. Educ.: Carnegie-Mellon Univ. Supervisory Chem., Phys. Res. Sect., U.S. Bureau of Mines, Pittsburgh, 1951-57; Phys. Chem., Kinetics and Combustion Group (1957-), now Head, Special Projects Sect., Advanced Technol. Dept., Res. Div., Atlantic Res. Corp., Alexandria, Virginia. Fundamental and applied research on gaseous and solid combustion. Studies on flammability limits of complex mixtures of gases. Chairman, Hospitality Com., Employment Com. and Membership Com., Chem. Soc. of Washington. Societies: A.C.S.; Sci. Res. Soc. of America.
Nuclear interests: Activation analysis, for determination of trace impurities in macro samples, and total analysis of micro samples of all types; Research in nuclear chemistry for eventual commercial application.
Address: Atlantic Research Corp., Shirley Memorial Highway at Edsall Road, Alexandria, Virginia, U.S.A.

GREIG, Pablo Willstätter, Ing. Staff Prof., Inst. Superior de Energia Nucl., Peru.
Address: Junta de Control de Energia Atomica, Instituto Superior de Energia Nuclear, 3420 Avenida Arequipa, Apt. 914, Lima, Peru.

GREINACHER, Ekkehard M., Dr. Born 1927. Educ.: Freiburg Univ. Plant Manager, Th. Goldschmidt A.G. Book: Winnacker-Küchler, Chemische Technologie, Sect.: Rare Earth (Munich, Carl Hanser Verlag, 1959).
Nuclear interests: Rare earth and zirconium as construction materials and for neutron absorbing. Rare earth metals and compounds.
Address: 41 Ahornstrasse, 43 Essen-Stadtwald, Germany.

GREINER, Walter, Ph.D. Born 1935. Educ.: Freiburg im Breisgau Univ. Wissenschaftlicher Asst., Freiburg Univ., 1960-62; Asst. Prof., Maryland Univ., 1963-64; Prof. Phys., Frankfurt am Main Univ.; Guest Prof. of Phys., Virginia Univ., 1967.
Nuclear interests: Nuclear structure; Nuclear models; Nuclear reactions; μ-mesic atoms; Nuclear spectroscopy; Bremsstrahlung.
Address: Institute for Theoretical Nuclear Physics, Frankfurt am Main University, Germany.

GREKOV, A. S. Paper: Co-author, Non Burned-up Radiator for Fission Chamber (letter to the Editor, Atomnaya Energiya, vol. 22, No. 4, 1967).
Address: Academy of Sciences of the U.S.S.R., 14 Leninsky Prospekt, Moscow V-71, U.S.S.R.

GRELL, Heinrich, Prof. Dr. Director, Mathematisches Inst. I, Humboldt Univ.
Address: Humboldt University, Mathematisches Institut I, 6 Unter den Linden, Berlin W 8, German Democratic Republic.

GRELL, Philippe L. G., Civil Eng. (Elec. and Mechanics), Naval Architect and Marine Eng., Eng. (Nucl. Eng.). Born 1931. Educ.: Liège Univ., Nat. High School of Marine Eng., Paris and Nat. Inst. for Nucl. Sci. and Technics, Saclay. Deputy Sen. Eng., Belgonucléaire S.A.

Society: Assoc. des Ing. en Génie Atomique.
Nuclear interest: Reactor design, construction, site erection, commissioning and operation.
Address: 80 avenue Robert Dalechamp, Brussels 15, Belgium.

GRENDON, Alexander, B.S., M.E., M. Bioradiol. Born 1899. Educ.: Columbia and California (Berkeley) Univs. Colonel, U.S. Army, 1950-54; Deputy Commander, Res. and Eng. Command, Army Chem. Corps., 1952-54; Coordinator of Atomic Energy Development and Rad. Protection, State of Calif., 1959-63; Biophys., Donner Lab., Calif. Univ., Berkeley, 1963-. Member, Nat.-Radiol. Health Advisory Council, U.S.P.H.S.; Member, Radiol. Health Advisory Com., State of California. Societies: Conference on Radiological Health; H.P.S.
Nuclear interests: Radiological health; Bioradiology; Governmental regulatory activities in the field of nuclear energy and radiation protection.
Address: Donner Laboratory, California University, Berkeley, California 94720, U.S.A.

GRENINGER, A. B., Dr. Formerly Manager, Irradiation Processing Dept., Hanford Atomic Products Operation, then Head, Nucl. Technol. Dept. (1966-), now Manager, Vallecitos Atomic Lab. and Deputy Div. Gen. Manager, Specialty Nucl. Operations, Gen. Elec. Co.
Address: General Electric Co., Nuclear Technology Department, Vallecitos Atomic Laboratory, P.O. Box 846, Pleasanton, California 94566, U.S.A.

GRENON, Michel Alain, L. ès Sc. Born 1928. Educ.: Faculté des Sciences, Paris. Chef du Groupe Réacteurs Homogènes, C.E.A., Saclay, 1955-60; On lease from C.E.A. to Euratom as Chief of the Reactor Project Group, Kema Project of the Euratom-RCN-Kema Reactor Development Group (Nederland), 1960-; Consultant to Euratom for Homgeneous and Liquid Fuel Reactors, 1960-; Deputy Director and Head of Technol. Dept., Euratom Petten Establishment, 1962-. Euratom co-ordinator, Euratom-U.S.A.E.C. Cooperation Agreement on Molten Salt Reactors.
Nuclear interests: Development of liquid fuel reactors (some patents taken for a boiling aqueous suspension reactor); Co-ordination of research programme in liquid fuel reactor field; Irradiations technology.
Address: 83 rue Charles Laffitte, Neuilly, Seine, France.

GREPPI, E. Member, Editorial Com., Minerva Nucleare.
Address: 83-85 Corso Bramante, Turin, Italy.

GRES, H., Dr. Ing. Vizepräsident, Kernforschungsanlage Jülich des Landes Nordrhein-Westfalen E.V.
Address: Postfach 365, Jülich, Germany.

GRES, Willi Hans, Dipl. Ing., Dr. Ing. Born 1916. Educ.: Hanover Tech. Univ. Member, Board of Directors, Fried. Krupp Essen, 1966-.
Society: Member, Supervisory Board, German Nucl. Energy Forum.
Address: 100 Altendorfer Strasse, 43 Essen, Germany.

GRESKY, Alan T., B.S. (Chem.). Born 1917. Educ.: Alabama Coll., Alabama Univ., Alabama Medical School and Tennessee Univ. Chem., Chem. Technol. Div. (On special assignment to Div. Director), O.R.N.L.; Lecturer, O.R.S.O.R.T. Editor, Process Chemistry (Pergamon Press). Society: A.C.S.
Nuclear interests: Reactor fuel reprocessing chemistry; Fuel cycle, long-range planning; Fundamental theory; Solvent extraction chemistry; Radioactive waste disposa; Processes for fission product recovery; Evaluation of advanced chemical and reactor concepts; Hazards analysis; etc.
Address: 113 Kingsley Road, Oak Ridge, Tennessee, U.S.A.

GRETHER, P. Engaged on Nucl. Work, Geilinger and Co.
Address: Geilinger and Co., 20 Werkstrasse, P.O. Box 112, 8401 Winterthur, Switzerland.

GRETZ, Joachim, Dipl.-Ing., Master of Eng. (U.S.A.). Born 1928. Educ.: Darmstadt and Hanover Tech. Univs. and Pennsylvania State Univ. Nuclear eng. working on the development of D_2O reactors with Euratom.
Address: C.C.R.-Euratom, Ispra, Varese, Italy.

GREULING, Eugene. Member, Panel of Experts to Provide Boards on Atomic Safety and Licensing, U.S.A.E.C.
Address: Department of Physics, Duke University, Durham, North Carolina, U.S.A.

GREVE, Lawyer Gunnar, Jr. Director, A/S Reactor.
Address: A/S Reactor, 4 Tollbugata, Oslo, Norway.

GREYBE, Grant, D. Admin. (Hons.). Born 1923. Educ.: Pretoria Univ. Head, Personnel Administration, Nat. Nucl. Res. Centre, Pelindaba, Atomic Energy Board. Society: South Africa Inst. of Public Administration. Nuclear interest: Personnel.
Address: 51 De Beer Street, Pretoria North, South Africa.

GRIBANOV, Yu. I. Papers: Co-author, Investigation of Nucl. Reactor ThreeDimensional Temperature Field by means of an Analogue Technics (Atomnaya Energiya, vol. 22, No. 4, 1967); co-author, Determination of Nucl. Core Temperatures with Heat Generation Varying in Longitudinal Direction (ibid, vol. 22, No. 5, 1967); co-author, Temperature Calculation in Nucl. Fuel Elements with Defects by Elec. Analog Method (ibid, vol. 25, No. 3, 1968).

Address: Academy of Sciences of the U.S.S.R., 14 Leninsky Prospekt, Moscow V-71, U.S.S.R.

GRIBANOVA, V. M. Paper: Co-author, NaJ(Tl) Scintillation Response of External Dimensions (letter to the Editor, Atomnaya Energiya, vol. 24, No. 3, 1968).
Address: Academy of Sciences of the U.S.S.R., 14 Leninsky Prospekt, Moscow V-71, U.S.S.R.

GRIBIBOV, B. S. Paper: Co-author, Neutron Detector with Alternating Thickness of Moderator (Atomnaya Energiya, vol. 22, No. 2, 1967).
Address: Academy of Sciences of the U.S.S.R., 14 Leninsky Prospekt, Moscow V-71, U.S.S.R.

GRIFFIN, Ralph G., Jr. Formerly Supervisor, Licensing and Regulation, now Supervisor, Interagency Relations,Div. of Occupational Health and Rad. Control, Texas State Dept. of Health.
Address: Division of Occupational Health and Radiation Control, Texas State Department of Health, 1100 West 49th Street, Austin, Texas 78756, U.S.A.

GRIFFIN, Miss Vera. Born 1917. Deputy Establishment Officer, U.K.A.E.A., London.
Nuclear interest: Personnel policy for scientific and technical staffs.
Address: 8 Broadlands Close, Highgate, London, N.6, England.

GRIFFITH, D. E., Prof. Chairman, Nucl. Eng. Com., Houston Univ.
Address: Houston University, Houston, Texas, U.S.A.

GRIFFITH, Frank Wells, B.S. (Gen. Eng.). Born 1921. Educ.: Iowa State Univ. Cadet Eng. (1948), Asst. Production Superintendent (1953), Gen. Superintendent Production (1955), Production Manager (1956), System Production Manager (1959), Asst. to the Pres. (1961), Vice Pres. Operations (1963), Exec. Vice Pres. (1965), Pres. Res. and Gen. Manager (1966), Sioux City Gas and Elec. Co. Societies: A.S.M.E.; American Gas Assoc.; American Public Works Assoc.
Address: 4455 Perry Way, Sioux City, Iowa 51104, U.S.A.

GRIFFITH, John William, B.A.Sc. Born 1924. Educ.: Toronto Univ. Geologist, Tech. Mines Consultants, Ltd., 1950-52; Geologist, Rio Tinto Exploration, Ltd., 1955; Sci. Officer, Dept. Mines and Tech. Surveys, Ottawa, 1955-.
Nuclear interests: Nuclear raw materials, uranium and thorium supply, demand, uses.
Address: Mineral Resources Division, Dept. of Mines and Technical Surveys, Ottawa, Canada.

GRIFFITH, Thomas Ceiri, B.Sc., Ph.D. (Wales). Born 1925. Educ.: Wales Univ. Lecturer (1950-65), Reader (1965-), Phys. Dept., Univ. College, London.
Nuclear interests: Low energy (20 - 150 MeV) nuclear physics involving two nucleon and few nucleon systems.
Address: Physics Department, University College, Gower Street, London, W.C.1, England.

GRIFFITH, W. C. Res. Director, Lockheed Missiles and Space Co., Lockheed Aircraft Corp.
Address: Lockheed Missiles and Space Co., P.O. Box 504, Sunnyvale, California, U.S.A.

GRIFFITHS, David Robert, B.E. Born 1918. Educ.: Adelaide Univ. Chief, Eng. Res. Div. (1961-63), Chief, Special Projects Div. (1964-), Formerly Head, Reactor Eng. Div., Australian A.E.C.
Nuclear interests: Investigations relating to application of nuclear power in Australia including assessment of reactor systems, siting and safety, overall planning of integrated nuclear power industry.
Address: Australian Atomic Energy Commission, 45 Beach Street, Coogee, New South Wales, Australia.

GRIFFITHS, George Motley, B.A.Sc. (Toronto), M.A. (British Columbia), Ph.D. (British Columbia). Born 1923. Educ.: Toronto and British Columbia Univs. Rutherford Memorial Fellow, Cavendish Lab., Cambridge, 1953-55. Asst. Prof. (1955-59), Assoc. Prof. (1959-63), Prof. (1963-), British Columbia Univ.; Res. Fellow, Calif. Inst. of Technol., 1962-63.
Societies: Inst. of Physics and Phys. Soc., London; Canadian Assoc. of Phys.; A.P.S.
Nuclear interests: Low energy nuclear physics; Nucleosynthesis in stars; Direct radiative capture reactions.
Address: The University of British Columbia, Physics Department, Vancouver 8, B.C., Canada.

GRIFFITHS, Peter John Felix, M.Sc. Born 1929. Educ.: Wales Univ. Lecturer, Chem., and Rad. Safety Officer, Inst. of Sci. and Technol., Wales Univ.; Regional Adviser, Use of Radioisotopes, Regional Sci. Training Officer, Scientific Advisers Branch, Home Office. Societies: F.R.I.C.; Chem. Soc.; Assoc. of Univ. Rad. Protection Officers; British Radiol. Protection Assoc.
Nuclear interests: Application of radioisotope techniques and irradiation in teaching and research; Radiological protection.
Address: Chemistry Department, Institute of Science and Technology, Wales University, Cathays Park, Cardiff CF1 3NU, Wales.

GRIFFITHS, S. H., A.I.M., M.I.W. Member, Advisory Com. on Ind. Radiography, City and Guilds of London Inst.
Address: Institute of Welding, 54 Princes Gate, London, S.W.7, England.

GRIFFITHS, Trevor, B.Sc. (Eng.), C. Eng. Born 1913. Educ.: London Univ. At Dounreay

Reactor Establishment (1955), Asst. Chief Eng., Production Reactor Design Office, Risley (1958), U.K.A.E.A.; Deputy Chief Inspector, (1960), Chief Inspector, (1964), Inspectorate of Nucl. Installations, Ministry of Power. Member, Nucl. Safety Advisory Com. Society: Fellow, I.E.E.
Nuclear interest: Nuclear installations.
Address: Ministry of Power, Thames House South, Millbank, London, S.W.1, England.

GRIFONE, Luigi, Dr. Manager, Amel - Apparecchiature di Misura Elettroniche.
Address: Amel - Apparecchiature di Misura Elettroniche, 15 Via F. Morandi, Milan, Italy.

GRIGOR'EV, E. P. Book: Izobranye yadra s massovym chislom A = 73 (Selected Nuclei with Matter Coefficient A = 73) (Fasc. 6 Svoistva Atomnykh Yader (Properties of Atomic Nuclei), M.-L. AN SSSR, 1960) (Pergamon, London).
Address: Leningrad State University, Leningrad, U.S.S.R.

GRIGOREV, I. S. Co-author, A Method of Predicting the Critical Mass and Neutron Flux Distribution of a Reactor by Use of a Physical Model (Atomnaya Energiya, vol. 7, No. 1, 1959); co-author, An Exptl. Reactor Using Gaseous Fissile Material (UF_6) (ibid, vol. 5, No. 3, 1958); Influence of Empty Channels on Slowing Down Length of Neutrons (letter to the Editor, ibid, vol. 23, No. 4, 1967).
Address: Academy of Sciences of the U.S.S.R., 14 Leninsky Prospekt, Moscow V-71, U.S.S.R.

GRIGOREV, K. B. Paper: Co-author, Sodium Technol. and Equipment of BN-350 Reactor (Atomnaya Energiya, vol. 22, No. 1, 1967).
Address: Academy of Sciences of the U.S.S.R., 14 Leninsky Prospekt, Moscow V-71, U.S.S.R.

GRIGOR'EV, V. N. Papers: Co-author, A Microcolumn for Separating and Analysing Mixtures of Hydrogen Isotopes (letter to the Editor, Atomnaya Energiya, vol. 12, No. 5, 1962); co-author, Investigation of Separation of Neon Isotopes in Rectification Film Column (ibid, vol. 20, No. 6, 1966).
Address: Academy of Sciences of the U.S.S.R., 14 Leninsky Prospekt, Moscow V-71, U.S.S.R.

GRIGOR'EV, Yu. N. Paper: Co-author, Storage Ring for 100 MeV Electrons (Atomnaya Energiya, vol. 23, No. 6, 1967).
Address: Academy of Sciences of the U.S.S.R., 14 Leninsky Prospekt, Moscow V-71, U.S.S.R.

GRIGORIEFF, Wladimir W., B.S. (Swiss Federal Univ., Zurich), Ph.D. (Chicago). Born 1908. Director, ORDARK Project, Arkansas, 1946-53; Director, Inst. of Sci. and Technol., Arkansas Univ., 1947-53; Chairman, Univ. Relations Div. (1953-64), Asst. to Exec. Director for Special Projects (1964-), O.R.I.N.S.; Exec. Sec., A.N.S., 1956-.
Societies: A.N.S.; A.C.S.; A.S.E.E.; A.A.U.P.; A.A.A.S.
Address: 104 Ogden Circle, Ridge, Tennessee, U.S.A.

GRIGORJEV, J. N. Paper: Co-author, 100 MeV Neutron Collector (Jaderna Energie, vol. 13, No. 11, 1967).
Address: Physico-Technical Institute, 2 Polytechnicheskaya Ulitsa, Leningrad, U.S.S.R.

GRIGOROV, Naum L. Head, Cosmic Ray Lab., Inst. Nucl. Phys. Papers: Co-author, Integral Spectrum of Ionisation Pulses Caused by Nucl. Active Particles of Cosmic Rad. at Mountain Altitudes (Nukleonika, vol. 7, No. 2, 1962); co-author, Investigations of High Energy Particles Interactions with Atomic Nuclei at the Mountain Altitudes (ibid, No. 12); co-author, Investigation of Nucl. Interactions of Particles with Energies $\sim 10^{13}$ eV by means of Controlled Photographic Emulsions (ibid, vol. 9, No. 4 - 5, 1964); co-author, Investigation of "young" E.A.S. at the Altitude of 3260m a.s.l. (ibid.).
Address: Institute of Nuclear Physics, University of Moscow, Moscow, U.S.S.R.

GRIGOROV, V. P. Papers: Method of Measurement of Active Concentration of Long-Live Alpha-Active Aerosols by Using Scintillation Spectrometer (Atomnaya Energiya, vol. 21, No. 6, 1966); Matrix Method of Calculations of Alpha-Ray Spectrum of Thick Sources (letter to the Editor, ibid, vol. 21 No. 6, 1966); co-author, Spectrometric Method for Measurement of Long-Lived Alpha Activity Aerosols (ibid, vol. 24, No. 3, 1968).
Address: Academy of Sciences of the U.S.S.R., 14 Leninsky Prospekt, Moscow V-71, U.S.S.R.

GRIGOROVICH, Yu. F. Paper: Co-author, 200 mA Helium Ion Beam at 70 keV Energy (letter to the Editor, Atomnaya Energiya, vol. 22, No. 2, 1967).
Address: Academy of Sciences of the U.S.S.R., 14 Leninsky Prospekt, Moscow V-71, U.S.S.R.

GRIGORYAN, S. V. Paper: Co-author, To Problem of Moving of Hydrothermal Solutions (Atomnaya Energiya, vol. 20, No. 6, 1966).
Address: Academy of Sciences of the U.S.S.R., 14 Leninsky Prospekt, Moscow V-71, U.S.S.R.

GRILLO, Gene P., B.S., M.S., Ph.D. Born 1927. Educ.: Boston Coll. and Boston Univ. Sen. Tech. Staff, Western Elec. Co., North Andover, Mass.; Chairman, Rad. Com., A.I.H.A. Nuclear interests: Leak testing crystal filter units by means of radioactive gases.
Address: 6 So. Maple Avenue, Bradford, Mass., U.S.A.

GRIMALDI, Frank S., B.S. (Eng.), M.Ch.E., Ph.D. Born 1915. Educ.: C.C.N.Y. and Maryland Univ. Adjunct Prof., The American Univ., 1954-62 (Part-time employment); U.S. Geological Survey, 1940-; At present, Chief, Branch

of Analytical Labs., U.S.G.S. Alternate Councillor (1960), Awards Com., Washington (1959, 1960), Budget Com., Washington (1961, 1962, 1963), A.C.S. N.I.H. Symposium Com. on Res. and Instrumentation, 1960, 1961. Book: Chapter on The Analytical Chemistry of Thorium in Treatise on Analytical Chemistry (Edited by I. M. Kolthoff and P. J. Elving; Interscience Publishers). Societies: A..C.S.; Geological Soc. of Washington.
Nuclear interests: Radiochemistry and radioactivation analysis.
Address: 3101 N. Toronto Street, Arlington, Virginia, U.S.A.

GRIMBERT, Arnold, Ing. géologue, L. ès Sc. Born 1919. Educ.: Sorbonne et Ecole National Supérieure de Géologie Appliquée et Prospection Minière de Nancy. Chef de la Section de Géochimie, C.E.A.; Conférencier à la Faculté des Sciences de Paris (géologie appliquée). Societies: Sté. géologique de France; Sté. géochimique de France; Geochem. Soc. (U.S.); Assoc. Internat. des hydrogeologues.
Nuclear interests: Géologie et prospection des gites uranifères.
Address: 106, rue de l'Abbé Groult, Paris 15, France.

GRIMELAND, Bertel, Cand. real. Dr. philos. (Oslo Univ.). Born 1920. Educ.: Oslo Univ. Phys., Inst. for Atomenergi, Kjeller, Lillestrom, 1951-59; Phys., Fysisk Inst., Blindern (Oslo Univ.), 1959-. Society: Norsk Fysisk Sekskap (Norwegian Phys. Soc.).
Nuclear interests: Nuclear physics and nuclear energy.
Address: Fysisk Institutt, Blindern, Norway.

GRIMES, Brian Kern, B.S. (Chem. Eng.), M.S. (Eng.). Born 1940. Educ.: Washington Univ., Seattle. Res. Assoc., Eng. Experiment Station, Washington Univ., Seattle, 1962-63; Nucl. Eng., Div. of Reactor Licensing, U.S.A.E.C., 1963-. Society: A.N.S.
Nuclear interests: Reactor safety, with particular reference to water-cooled power reactors; Regulatory aspects of atomic energy.
Address: United States Atomic Energy Commission, Mail Stop 010 Bethesda, Washington, D.C. 20545, U.S.A.

GRIMES, Warren Randall, B.A., M.S. Born 1919. Educ.: Wabash Coll. and Purdue Univ. Union Carbide, O.R.N.L., 1947-; Asst. Director, Chem. Div., 1956-58; Director, Reactor Chem. Div., 1958-. Books: Iron, Cobalt and Nickel (Chapter 16 of Nat. Nucl. Energy Series, VIII-1, 1950); Fused Salt Systems (Section 6, The Reactor Handbook, vol. 2, A.E.C.D. -3646); Chem. Aspects of Molten-Fluoride-Salt Reactor Fuels (Chapter 12 of Fluid Fuel Reactors, 1958). Societies: A.C.S.; A.N.S.
Nuclear interests: Chemistry of nuclear reactor systems; Molten salt behaviour; High temperature chemistry.
Address: 204 Englewood Lane, Oak Ridge, Tennessee, U.S.A.

GRIMMETT, Earl Shepherd, B.S. (Chem. Eng.), M.S. (Chem. Eng.). Born 1920. Educ.: Utah and Idaho Univs. The Galigher Co., 1949-52; American Cyanimid Co., 1952-53; Phillips Petroleum Co., Atomic Energy Div., 1953-. Society: A.I.Ch.E.
Nuclear interests: Chemical processing of nuclear fuels; Fluidised bed calcination of radioactive wastes.
Address: 1085 Syringa Drive, Idaho Falls, Idaho, U.S.A.

GRINBERG, Boris, Ing. E.P.C.I., D. ès Sc. Born 1909. Educ.: Paris Univ. C.E.A., 1949-67; Chef du Lab. de Mesures, Centre d'Etudes Nucléaires, Saclay; Formerly Prof., Conservatoire Nat. des Arts et Métiers; Formerly Prof., I.N.S.T.N.; Director, Div. of Res. and Labs., I.A.E.A., 1967-. Societies: Sté. de Physique; Membre du Comité Consultatif des Radiations Ionisantes du Bureau Internat. des Poids et Mesures; Sté. Française de Physique.
Nuclear interests: Metrology and standardisation of radionuclides.
Address: International Atomic Energy Agency, Division of Research and Laboratories, 11 Kaerntnerring, Vienna 1, Austria.

GRINDA, L. Sci. Director, Centre Sci. de Monaco.
Address: Centre Scientifique de Monaco, 16 boulevard de Suisse, Monte Carlo, Monaco.

GRISAK, Fred Richard, B.S., M.S. (Elec. Eng.). Born 1925. Educ.: Illinois Univ. and O.R.S.O.R.T. Sen. Group Eng., Corporate Planning, Temco Aircraft Corp., Dallas, Texas; industrial consultant on radiation problems and equipment design. Societies: American Inst. of Elec. Eng.; A.N.S.; Texas Soc. of Professional Eng.
Nuclear interests: Application of radiation to industrial processes.
Address: P.O. Box 517, Richardson, Texas, U.S.A.

GRISHAEV, I. A. Team leader, 2BeV linear accelerator, Physico-Tech. Inst., Kharkov. Papers: The Rationale of High-energy Linear-electron Accelerators Design (Atomnaya Energiya, vol. 4, No. 5, 1958); co-author, Storage Ring for 100 MeV Electrons (ibid, vol. 23, No. 6, 1967); co-author, 2 GeV Electron Linear Accelerator of the Phys. and Tech. Inst. of Acad. of Sci. of Ukrainian S.S.R. (ibid, vol. 24, No. 6, 1968).
Address: Academy of Sciences of the Ukrainian S.S.R., 2 Yumovskiy Tupik, Kharkov, U.S.S.R.

GRISHANIN, E. I. Papers: Co-author, The Analytical Method of Calculation: Irregular Burn out Fuel in Reactors (Atomnaya Energiya, vol. 16, No. 6, 1964); co-author, A Calculation of Efficiency for Control Rods Containing Moderator (ibid); co-author, Measurement of

Absorption Cross Section of Gd^{156} (letter to the Editor, ibid, vol. 19, No. 5, 1965); co-author, Measurements of Isotopes Gd^{154} and Gd^{156} Reactor Absorption Cross Section (letter to the Editor, ibid, vol. 22, No. 2, 1967).
Address: Academy of Sciences of the U.S.S.R., 14 Leninsky Prospekt, Moscow V-71, U.S.S.R.

GRISHKIN, O. E. Paper: Co-author, Investigation of Scintillation Characteristics of Lithium Glasses by Detection of Neutron, Beta and Gamma Rad. (Atomnaya Energiya, vol. 24, No. 2, 1968).
Address: Academy of Sciences of the U.S.S.R., 14 Leninsky Prospekt, Moscow V-71, U.S.S.R.

GRISMORE, Roger, B.S., M.S., Ph.D. Born 1924. Educ.: Michigan Univ. Asst. Phys. (1956-61), Assoc. Phys. (1961-62), Argonne Nat. Lab.; Assoc. Prof. Phys., Lehigh Univ., Bethlehem, Pennsylvania, 1962-67; Specialist in Phys., Scripps Inst. Oceanography, California Univ., San Diego, La Jolla, California, 1967-. Societies: A.P.S.; A.A.P.T.
Nuclear interests: High-energy particle physics, neutron spectrometry, nuclear structure and reactions, marine radioactivity.
Address: Marine Radioactivity Group, Oceanic Research Division, Scripps Institution of Oceanography, La Jolla, California 92037, U.S.A.

GRISON, Emmanuel, D. ès Sc. Born 1919. Educ.: Ecole Polytech., Paris. Ing. des Poudres, Lab. Central des Poudres, 1941-54; Prof., Ecole Polytech., Paris, 1964-.
Nuclear interests: Plutonium (métallurgie, chimie, fabrication et traitement des combustibles).
Address: 28 avenue de Verdun, Limours, Seine et Oise, France.

GRITCHENKO, Z. G. Papers: Co-author, Ra^{226} Half-life (Atomnaya Energiya, vol. 7, No. 5, 1959); co-author, The ^7Be Concentration in Ground-Level Air and in Precipitation (letter to the Editor, ibid, vol. 12, No. 1, 1962); co-author, Radioactive Fallout at the Crimea in 1960-61 (letter to the Editor, ibid., vol. 15, No. 3, 1963); co-author, Accumulation Artificial Radionuclides on Surface on Earth in Vicinity of Leningrad in 1954-65 (letter to the Editor, ibid, vol. 23, No. 4, 1967).
Address: Academy of Sciences of the U.S.S.R., 14 Leninsky Prospekt, Moscow V-71, U.S.S.R.

GRITSKOV, V. I. Papers: Co-author, Exptl. Res. of Boiling Water Reactor Stability (Atomnaya Energiya, vol. 24, No. 4, 1968); co-author, Exptl. Determination of the Boiling Reactor BK-50 Frequency Characteristics (ibid, vol. 25, No. 3, 1968).
Address: Academy of Sciences of the U.S.S.R., 14 Leninsky Prospekt, Moscow V-71, U.S.S.R.

GRIVET, P., Prof. Adjoint Director, Sect. de Besancon, Lab. de l'Horloge Atomique, C.N.R.S.
Address: C.N.R.S., Section de Besancon, 32 avenue de l'Observatoire, Besancon, (Doubs), France.

GRMELA, Miroslav, Candidate of Phys. and Math. Sci. Born 1939. Educ.: Prague Tech. Univ.
Nuclear interests: Neutron thermalisation theory, gas dynamics, non-equilibrium statistical mechanics.
Address: Nuclear Research Institute, Rez near Prague, Czechoslovakia.

GROBMAN TVERSQUI, Alexander, Dr. Delegado, Junta de Control de Energia Atomica del Peru.
Address: Junta de Control de Energia Atomica del Peru, 3420 Avenida Arequipa, San Isidro, Apartado 914, Lima, Peru.

GRODZINS, Lee, Ph.D. Born 1926. Educ.: New Hampshire and Purdue Univs. and Union Coll., Instructor, Purdue Univ.; Phys., Brookhaven Nat. Lab.; Assoc. Prof. now Prof. Phys., M.I.T. Society: A.P.S.
Nuclear interest: Nuclear spectroscopy.
Address: Physics Department, Massachusetts Institute of Technology, Cambridge 39, Massachusetts, U.S.A.

GROEBEN, Hans von der. See von der GROEBEN, Hans.

GROEN, Nicolaas Jan Alexander, Dr. Born 1916. Educ.: Amsterdam Municipal Univ. Private Practitioner Practitioner as Pharmaceutical Chem., -1955; Regional Public Health Officer (drugs, foods, environmental hygiene), 1955-62; Chief Public Health Officer (environmental health), 1962-. Society: Nederlandse Vereniging voor Strahlenbescherming (Dutch Assoc.for protection against Radioactive Rad.).
Nuclear interest: Protection of the population against nuclear radiation.
Address: 8 Dokter Reijersstraat, Leidschendam, Netherlands.

GROESEMAN, A. Responsable des Activités Nucléaires, Solvay et Cie.
Address: Solvay et Cie., 310 rue de Ransbeek, Brussels 12, Belgium.

GROMOV, A. M. Paper: External Injection of Bunched Electron Beam into Microtron (letter to the Editor, Atomnaya Energiya, vol. 23, No. 1, 1967).
Address: Academy of Sciences of the U.S.S.R., 14 Leninsky Prospekt, Moscow V-71, U.S.S.R.

GROMOV, B. F. Paper: Co-author, About Gamma Rays Angular and Energy Distribution on Surface of Volume Source (Abstract, Atomnaya Energiya, vol. 19, No. 2, 1965).

Address: Academy of Sciences of the U.S.S.R., 14 Leninsky Prospekt, Moscow V-71, U.S.S.R.

GROMOV, L. F. Paper: Co-author, Electrohydraulic System of Safety Rods for Reactor SM-2 (letter to the Editor, Atomnaya Energiya, vol. 22, No. 3, 1967).
Address: Academy of Sciences of the U.S.S.R., 14 Leninsky Prospekt, Moscow V-71, U.S.S.R.

GROMOVA, A. I. Papers: Co-author, Corrosion Resistance of Structural Materials in Bor Contained Solutions (letter to the Editor, Atomnaya Energiya, vol. 19, No. 6, 1965) co-author, Effect of Heat Treatment on Corrosion Resistance of Zirconium Alloys (ibid, vol. 20, No. 4, 1966); co-author, Effect of Cold Work on Corrosion Resistance of Zr - 2.5% Nb Alloy (letter to the Editor, ibid, vol. 24, No. 1, 1968).
Address: Academy of Sciences of the U.S.S.R., 14 Leninsky Prospekt, Moscow V-71, U.S.S.R.

GRÖNBLOM, Sven Edgar, Univ. degree of Civil Eng. Born 1913. Educ.: Helsinki Univ. Managing Director, Atomivoima Oy, Importers, Helsinki.
Address: P.O. Box 10370, Helsinki 10, Finland.

GRONEMANN, J. Manager, Forsikringsaktieselskabet Genatom.
Address: Forsikringsaktieselskabet Genatom, 12 Stormgade, Copenhagen K, Denmark.

GROOM, Alan C., B.Sc., Ph.D. (London) F. Inst.P. Born 1926. Educ.: London Univ. Sen. Physicist, St. Mary's Hospital, London, U.K., 1950; Leverhulme Res. Fellow (1957), Lecturer in Phys., (1958), St. Mary's Hospital Medical School, London Univ.; Res. Ass. in Physiology, Univ. of Buffalo, N.Y., U.S.A., 1962; Assoc. Prof. of Biophys., Western Ontario Univ., London, Canada, 1966. Societies: Biophys. Soc., Inst. Phys.; Hospital Physicists' Assoc.
Nuclear interests: Use of radioisotopes in studies of blood flow and capillary exchange in animals.
Address: Department of Biophysics, University of Western Ontario, London, Canada.

GROOT, J. P. de. See de GROOT, J. P.

GROOT, S. H., Jr., A.C.I.I. Sen. Officer in Charge of Nucl. Activities, Langeveldt de Vos de Waal.
Address: Langeveldt de Vos de Waal, Gebouw "de Walvis", 74 Grote Bickersstr., P.O. Box 357, Amsterdam C., Netherlands.

GROOT, Simon Hendrikus, Sr. Born 1904. Director, D. Hudig en Co.
Nuclear interest: All kinds of insurance.
Address: 61 Wijnhaven, Rotterdam, Netherlands.

GROOT, Sybren Ruurds de. See de GROOT, Sybren Ruurds.

GROOTE, Paul Hubert de. See de GROOTE, Paul Hubert.

GROS, Charles, L. ès Sc. (Mathématiques et phys.), Dipl. d'Etudes Supérieures de mathématiques et phys.-chimiques, Dr. en Médecine. Born 1910. Educ.: Montpellier Univ. Prof. titulaire de Chaire d'Electro-radiologie médicale, Strasbourg Univ.; Membre, Comité de Rédaction, J. de Radiologie et d'Electrologie et de Médicine Nucléaire, (Masson, Paris). Sec. Gen., Assoc. Européenne de Radiol. Societies: Sté. pour l'Avancement des Sci.; Sté. de Médecine Physique; Assoc. Française pour l'Etude du Cancer.
Nuclear interests: Physique nucléaire; Médecine nucléaire; Médecine physique.
Address: Faculté de Médecine, Hopital Civil de Strasbourg, Strasbourg, France.

GROS, Francois. Directeur de Recherche, C.N.R.S. Member, Comite de Biologie, C.E.A.
Address: Centre National de la Recherche Scientifique, 15 quai Anatole France, Paris 7, France.

GROSHEV, L. V. D.Sc., Acad. of Sci. of the U.S.S.R.; Deleg. to 2nd I.C.P.U.A.E., Geneva, Sept. 1958.
Nuclear interests: Physics of reactors and excited states.
Address: Academy of Sciences of the U.S.S.R., 14 Leninsky Prospekt, Moscow V-71, U.S.S.R.

GROSJEAN, Carl Clement, Dr. Sc., Ph.D. (Group Phys.), Agrégé de l'Enseignement Supérieur. Born 1926. Educ.: Ghent and Columbia Univs. Res. Phys., Interuniv. Inst. of Nucl. Sci., Brussels, 1949-58; Prof. Appl. Maths. (1958-), Director, Computing Lab. (1960-), Prof. higher Mathematical Analysis (1965-), Ghent Univ. Books: Formal Theory of Scattering Phenomena (Monograph No. 7, Brussels, Interuniv. Inst. of Nucl. Sci., 1960); Table of absolute Detection Efficiencies of cylindrical scintillation Gamma-ray Detectors (Computing Lab., Ghent Univ., 1965). Societies: Sté. Belge de Physique; Sté. Mathématique de Belgique.
Nuclear interests: Reactor theory; Transport theory; Quantum mechanics.
Address: Rekenlaboratorium van de Rijksuniversiteit te Gent, 6 Rozier, Ghent, Belgium.

GROSPIRON, A. F. Pres., Oil, Chem. and Atomic Workers Internat. Union, AFL-CIO-CLC.
Address: Oil, Chemical and Atomic Workers International Union, AFL-CIO-CLC, 1840 California Street, Denver 2, Colorado, U.S.A.

GROSS, Bernhard, Dipl. Ing. (Tech. Phys.), Dr. rer. nat. (Stuttgart). Born 1905. Educ.: Stuttgart T.H. and Berlin Univ. Director, Div. of Elec., Nat. Inst. Technol., Rio de Janeiro, 1946-67; Prof. Elec. Measurements, Catholic Univ., Rio de Janeiro, 1954-60; Prof. Elec. Measurements, School of Eng., Niteroy, Rio

de Janeiro, 1955-; Director, Div. of Phys., Brazilian Nat. Res. Council, 1951-54; Member, U.N.S.C.E.A.R., 1958-60; Brazilian Rep., U.N. Sci. Advisory Com. 1958-60; Member, Sci. Advisory Com., (1958-60), Director, Div. of Sci. and Tech. Information (1961-67), I.A.E.A.; Director, Dept. Sci. and Technol. Res., Nat. Nucl. Energy Commission of Brazil, Rio de Janeiro, 1967-. Books: Mathematical Structure of Theories of Viscoelasticity (Paris, Herrmann, 1953); Singularities of Linear System Functions (Amsterdam, Elsevier, 1961); Charge Storage in Solid Dielectrics (Amsterdam, Elsevier, 1964). Societies: A.P.S.; Deutsche Physikalische Gesellschaft; Brazilian Acad. of Sci.; Deutsche Rheologische Gesellschaft.
Nuclear interests: Irradiation effects; Radiation dosimetry; Fallout.
Address: Comissao Nacional de Energia Nuclear, 90 (ZC-82) Rua General Severiano, Rio de Janeiro, Brazil.

GROSS, George Roy Frederick, B.Appl.Sci. Born 1910. Educ.: Toronto Univ. Various positions, Roy. Canadian Air Force, 1935-61 (Last position, Naval, Military and Air Attache, Canadian Embassy, Ankara, Turkey); Gen. Manager, Canadian Nucl. Assoc. 1961-. Council Member, Eng. Alumni Assoc. Societies: Assoc. of Professional Eng., Province of Ontario; Eng. Inst. of Canada; A.N.S.;Atomic Ind. Forum.
Nuclear interests: To promote and foster an environment favourable to the healthy growth of the uses of nuclear energy and radioisotopes. To encourage co-operation between various industries, utilities, educational institutions, government departments, and agencies, and other authoritative bodies which have a common interest in the development of economic nuclear power and uses of radioisotopes.
Address: Canadian Nuclear Association, Board of Trade Building, Suite 1002, 11 Adelaide West, Toronto 1, Ontario, Canada.

GROSS, Joseph, Diplom Kaufmann. Born 1909. Educ.: Economic Univ. in Mannheim and Berlin. Chief, Public Authority of Security and Police, Karlsruhe.
Nuclear interest: Management.
Address: 11 Hansjakobstr., Karlsruhe, Germany.

GROSS, M. Member, Conseil d'Administration, Energie Nucléaire S.A.
Address: Energie Nucléaire S.A., 10 avenue de la Gare, Lausanne, Switzerland.

GROSS, Paul Magnus, B.S., M.S., Ph.D. Born 1895. Educ.: C.C.N.Y. and Columbia Univ. William Howell Pegram Prof. Chem. (1920-), Dean (1952-58), Dean, Graduate School of Arts and Sci. (1947-52), Vice-Pres., Educ. Div. (1949-60), Duke Univ.; Pres. and Member, Board of Directors, Oak Ridge Associated Univs., 1949-. Pres. (1962), Chairman, Board of Directors (1963), A.A.A.S.; Member, N. Carolina Atomic Energy Advisory Com., 1959-. Societies: Fellow, N.Y. Acad. Sci.; Fellow, A.P.S.; A.C.S.; A.A.A.S.
Address: Duke University, Durham, N. Carolina, U.S.A.

GROSS, Philipp, D. phil. (Vienna). Born 1899. Educ.: Vienna and Vienna Tech. Univs. Principal Sci., Fulmer Res. Inst., 1946-.
Nuclear interest: Physical chemistry involved in nuclear power production.
Address: 41 Long Drive, Burnham, Bucks., England.

GROSSE, Hans, Dr. -Ing. Born 1901. Educ.: Berlin T.H. Leiter der Arbeitsgruppe für Reaktorbauelemente der Kernforschungsanlage Jülich des Landes Nordrhein-Westfalen e.V.; Wissenschaftlicher Beirat der Deutschen Versuchsanstalt für Luftfahrt, Essen-Mülheim.
Nuclear interest: Reactor engineering.
Address: 13 Goldbachstr., Aachen, Germany.

GROSSE-BROCKHOFF, Franz, Prof., Dr. med. Born 1907. Educ.: Würzburg, Leipzig, Berlin, Kiel, Cologne, Graz and Bonn Univs. 1. Oberarzt, Med. Universitätsklinik, Bonn, 1945-54; Prof. (1954-), Rektor (1962-), Medizinischen Akademie, Düsseldorf, Verwaltungsratsmitglied der Kernforschungsanlage Jülich des Landes Nordrhein-Westfalen e.V.
Address: Med. Akademie Düsseldorf, 4 Strümppelstr., Düsseldorf, Germany.

GROSSMAN, Andrzej, Dr. Born 1908. Educ.: Lwow Polytech. and Paris Univ. Prof., Water and Sewage Technol., Silesian Tech. Univ., Gliwice, 1960-.
Nuclear interests: Nuclear graphite; Decontamination of radioactive water.
Address: 366 ul. Wolnosci, Zabrze, Poland.

GROSSMAN, Ely H., B.S. (M.E.), B.S. (I.E.). Educ.: Florida and Alabama Univs. Supv., Plant and Structural Quality Eng., Westinghouse Elec. Corp., Pgh., Pa.
Nuclear interest: Reactor design.
Address: 5142 Leona Drive, Pittsburgh, Pennsylvania 15227, U.S.A.

GROSSMAN, Ernesto, Dr. Member, Comision Ecuatoriana de Energia Atomica.
Address: Comision Ecuatoriana de Energia Atomica, Escuela Politecnica Nacional, Quito, Ecuador.

GROSSMAN, Lawrence M., B.Ch.E., M.Sc., Ph.D. Born 1922. Educ.: California Univ., Berkeley. Prof. Nucl. Eng., Dept. of Nucl. Eng., California Univ., Berkeley, 1954-. Societies: A.P.S.; A.N.S.
Nuclear interest: Nuclear reactor theory.
Address: Department of Nuclear Engineering, California University, Berkeley, California 94720, U.S.A.

GROSSMAN, Nicholas, M.S., B.S. Born 1920. Educ.: Case Inst. Technol. and M.I.T. Asst.

Prof. Mech. Eng., M.I.T., 1949-52; Head, Mech. Development Sect., Sylvania-Corning Nucl. Corp., Bayside, N.Y., 1952-57; Div. of Licensing and Regulation (1957-60), Div. of Compliance (1960-62), Chief, Eng. Development Branch, Div. of Reactor Development (1962-), U.S.A.E.C., Washington, D.C.; Licensed Professional Eng., Commonwealth of Mass., U.S.A. Book: Contributor, Materials for Nucl. Reactors (ed. B. Kopelman, N.Y., McGraw-Hill, 1959). Societies: A.S.M.E.; Inst. Mech. Eng., London; New York Acad. Sci. Nuclear interests: Materials for nuclear applications; Reactor safety; Economics of nuclear power.
Address: 2300 McAuliffe Drive, Rockville, Maryland, U.S.A.

GROSSMANN, Vojtech, Prof. of Pharmacology, M.D. Born 1922. Educ.: Charles Univ., Prague. Head, Dept. of Pharmacology, Faculty of Medicine, Charles Univ., Hradec Kralove, 1952-. Society: Sect. of Nucl. Medicine and Rad. Hygiene, J.E.P. Medical Assoc., Prague; European Soc. for Rad. Biol.; European Soc. for Res. of Drug Toxicity.
Nuclear interests: Changes of the metabolism of drugs in irradiated organisms and alterations in the reactivity of irradiated organisms on applied drugs.
Address: Charles University-Hradec Kralove, Simkova 870, Czechoslovakia.

GROSZKOWSKI, Janusz, Dr. Tech. Sci. Born 1898. Educ.: Warsaw Tech. Univ. Prof., Warsaw Tech. Univ., 1929-. Vice-Près. (1957-63), Pres. (1963-), Polish Acad. Sci.; Pres., Corp. for Planning and Co-ordination Sci. Res.; Member, State Council for Peaceful Use of Nucl. Energy. Books; Vacuum Technol. (1953-55; PWT, Warsaw, in Polish); Vacuum Technology (1955; Innostr. Literat., Moscow, in Russian); Frequency of Self-oscillations (Pergamon Press, 1964).
Nuclear interest: High vacuum techniques.
Address: 22 Nowowiejska Street, Warsaw, Poland.

GROTDAL, T. Senior Officer, Inst. Nucl. Phys., Bergen Univ.
Address: Institute of Nuclear Physics, Bergen University, Bergen, Norway.

GROTENHUIS, G. A. Chief Eng., Mech. Dept., Tebodin Advies en- Constructiebureau N.V.
Address: Tebodin Advies en- Constructiebureau N.V., 98-100 Koninginnegracht, P.O. Box 1029, The Hague, Netherlands.

GROTENHUIS, Marshall, B.S., M.S. Born 1918. Educ.: Wisconsin Univ. at Milwaukee and Marquette Univ. Supt. of Shops, Central Shops Dept., Argonne Nat. Lab. Societies: A.N.S. (Chairman, Shielding Div., 1960-62); RESA.
Nuclear interest: Reactor shielding.

Address: Argonne National Laboratory, 9700 South Cass Avenue, Argonne, Illinois, U.S.A.

GROTH, Donald P., Dr. Nucl. Res., Dept. Pharmacology, School of Medicine, Emory Univ.
Address: School of Medicine, Emory University, Atlanta 22, Georgia, U.S.A,

GROTH, Wilhelm, Dr. Born 1904. Educ.: Munich T.H., Munich and Tübingen Univs. Prof., Bonn Univ.; Director, Inst. of Phys. Chem.; Kernforschungsanlage Jülich. German A.E.C.: Fk III, Ak II/4. Society: Gesellschaft Deutscher Chemiker, Fachgruppe Kernchemie. Nuclear interests: Isotope separation; Nuclear chemistry.
Address: 38 Melbweg, Bonn, Germany.

GROTOWSKI, Kasimir, Magister of Philosophy, Dr. Sci. Habil., Docent Phys. Born 1930. Educ.: Jagellonian Univ., Cracow. Asst. Phys. (1950-57), Lecturer Phys. (1957-63), Docent Phys. (1963-), Jagellonian Univ.; Sen. Officer (1957-59), Head, Nucl. Reactions Lab. (1959), Head, Nucl. Reactions Lab. A (1962-), Inst. Nucl. Phys., Cracow. Society: Polish Phys. Soc.
Nuclear interests: Experimental nuclear physics, especially medium energy scattering and reactions, mechanism of nuclear reactions and nuclear structure.
Address: Institute of Nuclear Physics, ul. Radzikowskiego, Cracow, Poland.

GROUT, Henry James, B.Sc., A.C.G.I. (Hons.). Born 1915. Educ.: London Univ. Head, Reactor Eng. Div., Harwell, and Deputy Chief Eng. of the Establishment (1947-57), Chief Eng., Reactor Station, Winfrith, Dorset (1958-62), U.K.A.E.A. Director, W.S. Atkins and Partners, Consulting Eng., 1962-; Director, Associated Nucl. Services Ltd.; Council Member, British Nucl. Forum.
Nuclear interests: Nuclear engineering with particular reference to research reactors and associated research facilities.
Address: W. S. Atkins and Partners, Woodcote Grove, Ashley Road, Epsom, Surrey, England.

GROVE, Don J., Ph.D. Born 1919. Educ.: Coll. of Wooster and Carnegie Inst. Technol. Advisory Phys., Westinghouse Elec. Corp.; on loan to Plasma Phys. Lab., Princeton Univ., New Jersey, 1954-. Society: A.P.S.
Nuclear interests: Development and operation of experimental devices for the attainment of a controlled thermonuclear reactor.
Address: Princeton Plasma Physics Laboratory, P.O. Box 451, Princeton, N.J., U.S.A.

GROVE, George Richard, B.S., M.S., Ph.D. (Phys.). Born 1925. Educ.: Ohio State Univ. Phys., N.B.S., 1950-51; Res. Phys. (1951-53), Group Leader (1953-59), Sect. Manager (1959-62), Director, Res. Dept. (1962-67), Director, Nucl. Operations Dept. (1967-), Monsanto Res. Corp.; Consultant, Miami Valley Hospital, 1953-.

Societies: Fellow, Phys. Soc.; Soc. of Nucl. Medicine; Assoc. of Phys. in Medicine.
Nuclear interests: Radioactive sources and applications; Radiological physics; Isotope separation; Thermal diffusion; Nuclear medicine.
Address: Mound Laboratory, Mound Road, Miamisburg, Ohio 45342, U.S.A.

GROVE, Walter Patrick, Ph.D., F.R.I.C. Born 1914. Director, Radiochem. Centre, U.K.A.E.A., Amersham; Director, The Amersham/Searle Corp., 1965-. Societies: Chem. Soc.; Soc. Chem. Industry; Faraday Soc.; British Inst. Radiol.
Nuclear interest: Production and uses of radioactive isotopes.
Address: The Radiochemical Centre, United Kingdom Atomic Energy Authority, Amersham, Bucks., England.

GROVEN, Louis Jean Henri, Ph.D. (Phys.), (Brussels). Born 1910. Educ.: Brussels Univ. and Sorbonne, Maitre de Conférences, Brussels Univ., Sci. Counsellor, Belgian Embassy in Washington, D.C., U.S.A. Societies: Sté. Française de Physique; A.P.S.; A.N.S.
Nuclear interest: Consultant.
Address: 3330 Garfield Street, N.W., Washington 8, D.C., U.S.A.

GROVES, Leslie Richard, B.S., LL.D. (hon.), D.Sc. (hon.) Born 1896. Educ.: Washington (Seattle) Univ., M.I.T., U.S. Military Acad., West Point and Army Eng. School. Comm. and Gen. Staff Sch., 1936; Army War Coll., 1939; Cmdg. Gen., Manhattan Project, 1942-47; Vice-Pres., Remington Rand Div., Sperry Rand Corp., 1948-61; Lt.-Gen., U.S. Army (Ret.). Societies: Soc. Military Eng.; A.M. Soc. Civil Eng.
Nuclear interests: Management, executive decision-making; Broad general knowledge of nuclear science and engineering.
Address: 2101 Connecticut Avenue, Washington, D.C., U.S.A.

GRUBBS, K. R., Prof., Dr. Formerly Member, Nucl. Sci. and Eng. Development Com., now Member, Nucl. Advisory Com., Louisiana Polytech. Inst.
Address: Louisiana Polytechnic Institute, Nuclear Centre, Tech. Station, Ruston, Louisiana, U.S.A.

GRUBER, Alan Richard, S.B., S.M. Born 1927. Educ.: M.I.T. and Harvard Univ. Manager, Eng. Dept. and Member, Board of Directors, Nucl. Development Corp. of America, 1948-57; Asst. Chief Eng. (Nucl. Res.), Marquardt Aircraft Co., 1958-59; Director, Nucl. Systems Div., Marquardt Corp., Van Nuys, California, 1959-61; Vice-Pres. E. H. Plesset Assocs., Inc., Los Angeles, 1961-; Director, Solid State Radiations, Inc., Los Angeles, 1960-. Member, Tech. Com. on Nucl. Propulsion, American Rocket Soc.; Member, Exec. Com., Nucl. Eng. Div., A.S.M.E. Societies: A.N.S.; Atomic Ind. Forum; American Rocket Soc.; A.S.M.E.
Nuclear interests: Reactor design; Nuclear propulsion of aircraft, missiles, and spacecraft; Economics.
Address: 5831 Jed Smith Road, Calabasas, California, U.S.A.

GRUBER, Alvin V. Vice Pres., Condenser Service and Eng. Co. Inc.
Address: Condenser Service and Engineering Co., Inc., 150 Observer Highway, Hoboken, New Jersey, U.S.A.

GRUBER, Helmut, Dipl.-Phys. Born 1919. Educ.: Stuttgart Tech. Univ. Geschäftsführer, Leybold-Heraeus G.m.b.H. and Co. Vice-Chairman, Deutsche Arbeitsgemeinschaft für Vacuum (D.A.G.V.). Societies: Deutsche Physikalische Gesellschaft; Deutsche Gesellschaft für Metallkunde.
Nuclear interests: Metallurgy; Reactor design; All kinds of vacuum equipment used in nuclear energy techniques and nuclear research.
Address: Leybold-Heraeus G.m.b.H. and Co., 645 Hanau, Germany.

GRUEN, Dieter M., B.S., M.S., Ph.D. (Phys. Chem.). Born 1922. Educ.: Chicago Univ. Sen. Sci., Argonne Nat. Lab. Society: A.C.S.
Nuclear interests: Chemistry of the actinides; Chemistry and structure of fused salts; Electronic structure of open shell elements, including the 4f and 5f elements. High temperature spectroscopy of liquid and gaseous systems. Matrix isolation of high temperature molecules. Ligand field energy level calculations. The chemistry and structural properties of fused salts and concentrated electrolyte systems.
Address: Argonne National Laboratory, Argonne, Illinois, U.S.A.

GRUHN, C. Exptl. Nucl. Phys. Cyclotron, Dept. of Phys. and Astronomy, Michigan State Univ.
Address: Michigan State University, East Lansing, Michigan, U.S.A.

GRUMBKOV, A. P. Paper: Co-author, Background of Constructive Materials of Gamma-Ray Detectors Caused by Natural Radioelements (letter to the Editor, Atomnaya Energiya, vol. 23, No. 2, 1967).
Address: Academy of Sciences of the U.S.S.R., 14 Leninsky Prospekt, Moscow V-71, U.S.S.R.

GRÜMM, Hans Josef, Phil. Dr. (Vienna), M.I.Nucl.E. Born 1919. Educ.: Vienna Univ. Head, Inst. for Reactor Eng., Österreichische Studiengesellschaft für Atomenergie (Vienna). Books: Co-author, Lineare Reaktorkinetik und -Störungstheorie (Ergebn. ex. Naturwiss., vol. xxx, Springer, 1958); Kernreaktortheorie (Springer, 1962); Kernenergie (Oldenbourg, 1964). Society: A.N.S.
Nuclear interests: Reactor physics; Reactor design.

Address: Österreichische Studiengesellschaft für Atomenergie Ges.m.b.H., 10 Lenaugasse, Vienna 8, Austria.

GRUMMITT, William Edmund, B.Sc., M.Sc., Ph.D. Born 1917. Educ.: Alberta and McGill Univs. Sen. Res. Officer, A.E.C.L. Alternate Canadian Deleg., U.N. Sci. Com. on the Effects of Atomic Rad. Society: Fellow, Chem. Inst. of Canada.
Nuclear interests: Nuclear physics and chemistry; Dispersal of radioactive wastes in the environment.
Address: 3 Macdonald Street, Deep River, Ontario, Canada.

GRUNBERG, Leander, D.Sc., M.Sc. Born 1911. Educ.: Birmingham Univ. Chief Res. Chem., Dr. Rosin's Ind. Res. Co., Wembley, 1942-51; Principal Sci. Officer, Mech. Eng. Res. Lab. D.S.I.R., 1951-58; Sen. Principal Sci. Officer, 1958-59; Head, Lubrication and Wear Div. (1959-61), Dep. Chief Sci. Officer, (1961), Supt. Fluids Group, (1961-), Nat. Eng. Lab. (formerly M.E.R.L., D.S.I.R.); Visiting Prof., Strathclyde Univ. 1963-. Societies: F.R.I.C.; M.I.Mech.E.; M.I.Chem.E.; F.Inst.Pet.; Faraday Soc.
Nuclear interests: Application of radioactive materials to scientific and industrial research, particularly to the study of the properties of fluids and the solution of problems in fluid flow and heat transfer.
Address: Fluids Group, National Engineering Laboratory, East Kilbride, Glasgow, Scotland.

GRÜNBERGER, Dezider, Candidate of Chem. Sci. Born 1922. Educ.: Chem. Univ. Now at Czechoslovak Acad. of Sci., Inst. of Organic Chem. and Biochem., Prague.
Nuclear interests: In the field of radiobiochemistry the application of radioactive isotopes in biosynthetic processes and the methods of measuring radioactivity of weaker β-emitters.
Address: Inst. of Organic Chemistry and Biochemistry, 2 Na cvicisti, Prague 6, Czechoslovakia.

GRUND, W. Staatssekretär. Mitglied, Aufsichtsrat, Gesellschaft für Kernforschung m.b.H.; Vizeprasident, Kernforschungsanlage Jülich des Landes Nordrhein-Westfalen E.V.
Address: c/o Gesellschaft für Kernforschung m.b.H., 5 Weberstrasse, 75 Karlsruhe, Germany.

GRUNDER, E., Ing. Sen. Officer, Electronic Dept., Mettler, Analytical and Precision Balances.
Address: Mettler, Analytical and Precision Balances, 52 Militarstrasse, Zürich, Switzerland.

GRUNDY, Fred, M.D., M.R.C.P., D.P.H., Barrister-at-Law, Inner Temple. Born 1905. Educ.: Leeds Univ. Formerly Mansel Talbot Prof. Preventive Medicine, Welsh Nat. School of Medicine; now Asst. Director-Gen., W.H.O., Geneva.
Nuclear interests: International health aspects of radiation, isotopes and human genetics.
Address: Palais des Nations, Geneva, Switzerland.

GRÜNENFELDER, Marc, Asst. Prof. Dr. Competent for determination of mineral age, Dept. of Crystallography and Petrology, Swiss Federal Inst. of Technol.
Address: Department of Crystallography and Petrology, Swiss Federal Institute of Technology, 5 Sonnenggstrasse, Zürich 8006, Switzerland.

GRÜNEWALD, Theo, Dr. Ing., Dipl. Phys. Born 1922. Educ.: Dresden T.H. and Darmstadt T.H. Private firm, 1950-55; Bundesforschungsanstalt für Lebensmittelfrischhaltung, Karlsruhe, 1955-.
Nuclear interest: Food radiation technique.
Address: 4 Hoffstrasse, Karlsruhe, Germany.

GRUNWALD, Gerhard, Dipl. Ing. Born 1935. Educ.: Dresden Tech. Univ. Studies in eng. for power-engines and hydraulics, -1962; Eng., Zentralinstitut für Kernforschung, Rossendorf, 1962-.
Nuclear interest: Heat transfer and hydraulics for reactor calculating.
Address: Deutsche Akademie der Wissenschaften zu Berlin, Forschungsgemeinschaft, Zentralinstitut für Kernforschung, 8051 Dresden, PF 19, German Democratic Republic.

GRUSAJEV, I. A. Paper: Co-author, 100 MeV Neutron Collector (Jaderna Energie, vol. 13, No. 11, 1967).
Address: Physico-Technical Institute, 2 Polytechnicheskaya Ulitsa, Leningrad, U.S.S.R.

GRUSE, Erich Max Paul, Dr. rer. pol. Born 1897. Educ.: Berlin Univ. Member, 'Fachkommission IV' and 'Arbeitskreis Haftung und Versicherung', Deutsche Atomkommission. Aufsichtsrat, Kernreaktor-Finanzierungs-Gesellschaft m.b.H.
Address: 8161 Hammer, Fischbachau, Germany.

GRUSIN, P. L. Adviser, Third I.C.P.U.A.E., Geneva, Sept. 1964. Paper: Co-author, Fe-Distribution in Microvolume of Zirconium Alloys (letter to the Editor, Atomnaya Energiya, vol. 8, No. 1, 1968).
Address: Academy of Sciences of the U.S.S.R., 14 Leninsky Prospekt, Moscow V-71, U.S.S.R.

GRÜTTER, Fritz, Dipl. Phys. Born 1915. Educ.: Swiss Federal Inst. Technol. (E.T.H.). Brown, Boveri Co., Ltd., Baden, Switzerland; C.E.R.N., Geneva, 1954-63; Advanced Accelerator Study Group, Lawrence Rad. Lab., California Univ., Berkeley (on leave of absence from C.E.R.N.), 1963-. Society: Schweizerische Naturforschende Gesellschaft (Swiss Acad. of Natural Sci.).

GRU

Nuclear interests: Planning and design of high energy particle accelerators and experimental equipment.
Address: C.E.R.N., Meyrin, Geneva, Switzerland.

GRUVERMAN, Irwin Jerry, B.S. (Chem. Eng., Cooper Union), M.S. (Nucl. Eng., M.I.T.). Born 1933. Educ.: Cooper Union and M.I.T.; Sen. Sci. (1954-62), Manager, Radioactive Materials Dept. (1962-63), Nucl. Sci. and Eng. Corp.; part-time instructor in Radiochem., Carnegie Inst. Technol., 1957-58; Head, Special Sources Dept., New England Nuclear Corp, 1963-. Organises and publishes annual Mössbauer Methodology Symposia. Societies: A.N.S.; A.C.S.; A.I.Ch.E.
Nuclear interests: Radionuclide and radioactive device preparation and application; Industrial applications; Mössbauer methodology; Technical management; Radiochemistry laboratory management.
Address: 16 Tanglewood Road, Needham, Massachusetts 02194, U.S.A.

GRUZIN, P. L., Prof., D.Sc., Acad. of Sci. of the U.S.S.R.; Deleg. to 2nd I.C.P.U.A.E., Geneva, Sept. 1958. Papers: Determination of Spectral Characteristics of Isotopic Neutron Sources (letter to the Editor, Atomnaya Energiya, vol. 19, No. 5, 1965); co-author, Surface Relief Determination by Means of Back-Scattered Gamma-Rad. (Abstract, ibid, No. 6); co-author, Irradiation Effect on Magnetic Properties in Order-Disorder Alloy Ni_3Fe (ibid, vol. 24, No. 1, 1968); co-author, Investigation of Scintillation Characteristics of Lithium Glasses by Detection of Neutron, Beta and Gamma Rad. (ibid, vol. 24, No. 2, 1968).
Address: Nuclear Energy Institute, Academy of Sciences of the U.S.S.R., 14 Leninsky Prospekt, Moscow V-71, U.S.S.R.

GRYZINSKI, Michal Witold, Ph.D. (Phys.), Diploma in Telecommunication Eng. Born 1930. Educ.: Warsaw Tech. Univ. Chief, Thermonucl. Res. Group, Inst. of Nucl. Res., Swierk k. Otwocka.
Nuclear interests: Plasma physics and controlled thermonuclear reactions; theory of atomic collisions (with special interest to classical methods).
Address: 39 m.24 ul. Waszyngtona, Warsaw, Poland.

GRZYMALSKI, Zdzislaw Jerzy, M.Sc. (Chem.). Born 1933. Educ.: Gliwice Polytech. Univ. Asst. Sci. Worker, Gliwice Univ., 1957; Sen. Asst. Sci. Worker, Wroclaw Univ., 1964-.
Nuclear interest: Application of heterogeneous radioisotopic exchange for chemical structural studies.
Address: Wroclaw University, Department of Inorganic Chemistry, 27 Wyspianskiego, Wroclaw, Poland.

GSCHNEIDNER, Karl Albert, Jr., B.S. (Chem.), Ph.D. (Phys. Chem.). Educ.: Iowa State and Detroit Univs. Staff Member (1957-63), Sect. Leader (1961-63), Los Alamos Sci. Lab.; Visiting Asst. Prof., Phys., Illinois Univ., 1962-63; Assoc. Prof. and Metal. (1963-67), Prof. and Sen. Metal. (1967-), Director, Rare Earth Information Centre (1966-), Dept. of Metal., Ames Lab., Iowa State Univ. Chairman, Com. on Alloy Phases, Metal. Soc. of America; Inst. of Mining, Metal. and Petroleum Eng., 1965-67. Book: Rare Earth Alloys (D. Van Nostrand, 1961). Societies: A.C.S.; A.S.M.; American Crystallographic Assoc.; A.A.A.S.
Nuclear interest: Physical metallurgy or rare-earth metals and alloys.
Address: Ames Laboratory, Iowa State University, Ames, Iowa 50010, U.S.A.

GSPANN, Jürgen, Dr.-Ing., Dipl. Phys. Born 1937. Educ.: Karlsruhe Univ. Asst., Inst. fur Kernverfahrenstechnik, Karlsruhe Univ.
Nuclear interest: Nuclear physics.
Address: Kernforschungszentrum Karlsruhe, Institut für Kernverfahrenstechnik, Leopoldshafen bei Karlsruhe, Germany.

GUALANDI, Dante, Degree in Ind. Chem., Libera Docenza in Metal. and Metallography. Born 1921. Educ.: Bologna Univ. At Istituto Sperimentale dei Metalli Leggeri, Novara, Italy. Societies: Italian Metal. Soc.; Nucl. Metal. Centre.
Nuclear interests: Metallurgy of aluminium, aluminium alloys and sintered aluminium for applications in the nuclear field; Zirconium alloys.
Address: c/o Istituto Sperimentale Metalli Leggeri, c.p. 129, 28100 Novara, Italy.

GUALDONI, Oreste, Dr. Director, C.I.S.E.
Address: Centro Informazioni Studi Esperienze, 12 Via Redecesio, Segrate, Milan, Italy.

GUALTIERI, Giordano, Dr. Eng.; Chief of Operation, Soc. Italiana Meridionale Energia Atomica, Nucl. Reactor Div. Adviser, Third I.C.P.U.A.E., Geneva, Sept. 1964.
Address: 35 Via S. Teresa, Rome, Italy.

GUANES SERRADO, Captain Don Benito. Member, Nat. A.E.C., Paraguay.
Address: National Atomic Energy Commission, Department of Organisations, Treaties and International Agreements, Ministry of Foreign Affairs, Asuncion, Paraguay.

GUARD, Robin F. W., B.Sc. (Eng.). Born 1924. Educ.: London Univ. Manager, Nucl. Dept., Shawinigan Eng. Co. Ltd., Montreal, Canada; Vice Pres. and Manager, Special Projects, Canatom Ltd. Chairman, Internat. Affairs Com., Canadian Nucl. Assoc. Societies: M.I.Mech.E.; F.I.E.E.; A.N.S.
Nuclear interest: Consulting engineering services in nuclear power and particle accelerators.
Address: Shawinigan Engineering Co. Ltd., 620 Dorchester Boulevard W., Montreal, Canada.

GUARDIA, Simon QUIROS. See QUIROS GUARDIA, Simon.

GUARINO, Angelo, Laurea in chem. Born 1932. Educ.: Rome Univ. Radiochem., Centro di Chimica Nucleare del C.N.R., Istituto di Chimica Farmaceutica, Rome Univ., 1958-.
Nuclear interests: Organic radiochemistry – e.g., radiation chemistry, labelling techniques with C^{14} and H^3, analytical procedures to count C^{14} and H^3, recoil chemistry of hot H^3 anc C^{14} atoms.
Address: Centro di Chimica Nucleare del C.N.R., Istituto di Chimica Farmaceutica, Rome University, Rome, Italy.

GUAZZONI, Silvio. Nucl. Dept., Off. Ing. de Michelis and Co.
Address: Nuclear Department, Off. Ing. de Michelis and Co., 31 Via Bistolfi, Milan, Italy.

GUAZZUGLI MARINI, Giulio, Dr. phil. Born 1914. Educ.: Rome Univ. Counsellor, Council of Europe; Director of Gen. Affairs, Council of Ministers, European Coal and Steel Community; Director, Secretariat, Conference for Establishment of Treaties of E.E.C. and Euratom; Gen. Director, Joint Nucl. Res. Centres, Euratom.
Address: 341 avenue Louise, Brussels, Belgium.

GUBATOVA, D. Ya. Paper: Co-author, Fast Neutron Flux Measurement for IRT-200 Reactor (letter to the Editor, Atomnaya Energiya, vol. 20, No. 2, 1966).
Address: Academy of Sciences of the U.S.S.R., 14 Leninsky Prospekt, Moscow V-71, U.S.S.R.

GUBEILI, Mohammed Abdul-Mabood EL-. See EL-GUBEILI, Mohammed-Abdul-Mabood.

GUBENKO, V. V. Paper: Co-author, Heat Economy of Nucl. Power Station with High Production Desalination Plants (Atomnaya Energiya, vol. 23, No. 1, 1967).
Address: Academy of Sciences of the U.S.S.R., 14 Leninsky Prospekt, Moscow V-71, U.S.S.R.

GUBY, C. Chief Eng., Giovanola Frères S.A.
Address: Giovanola Frères S.A., Monthey, (Valais), Switzerland.

GUCZI, László, Dr. rer. nat. Born 1932. Educ.: Szeged Univ. At Isotope Lab., Agrochem. Res. Inst., 1955-59; Dept. of Radiol., Central Food Res Inst., 1959-61; Inst. of Isotopes, Group of Phys. Chem., Budapest, 1961-.
Nuclear interests: Application of radioisotopes in the investigation of chemical reaction mechanism, and radiation chemistry; Radiation effects on heterogeneous catalytic reactions, adsorption, etc.
Address: Institute of Isotopes, Konkoly Th. u., Budapest 12, Hungary.

GUDDEN, Frdr. Oberasst., Inst. für Tech. Kernphysik, Darmstadt T.H.
Address: Institut für Technische Kernphysik, 9 Schlossgartenstrasse, Darmstadt, Germany.

GUDEHUS, Herbert, Dipl.-Ing. Born 1907. Educ.: Munich and Berlin T.H. Employee of Behorde für Wirtschaft und Verkehr, Hamburg, 1946-.
Nuclear interests: Economics of nuclear energy and nuclear ships.
Address: 53 Strandweg, Hamburg-Blankenese, Germany.

GUDKOVA, L. Ya. Papers: Co-author, Differential Albedo of Fast Neutrons Thin Ray for Semi-Infinite Water Scatter (Atomnaya Energiya, vol. 22, No. 2, 1967); co-author, Differential Albedo of Fast Neutron Thin Ray for Semi-infinite Iron Scatterer (ibid, vol. 25, No. 3, 1968).
Address: Academy of Sciences of the U.S.S.R., 14 Leninsky Prospekt, Moscow V-71, U.S.S.R.

GUEBEN, Georges Charles Michel, Dr. in mathematical and phys. sci. Agrégé Liège Univ. Born 1897. Educ.: Liège Univ. Prof., Liège Univ. Book: Phénomènes radioactifs et introduction à la physique nucléaire (Liège, Desoer, 2e édition, 1956). Societies: Sté. Royale des Sci. de Liège; Sté. scientifique de Bruxelles; A.A.A.S.
Nuclear interests: Nuclear physics and chemistry.
Address: 2 rue Grétry, Liège, Belgium.

GUEDES DE CARVALHO, Rodrigo Alberto, Dr. Chem. Eng., Extraordinary Prof. in Chem. Eng. Born 1917. Educ.: Porto Univ. Stages at Inst. du Radium (Paris) in 1955 and 1957. Collaboration with Commission of Studies of Nucl. Energy, Lisbon, and Junta de Energia Nucl. Lisbon. Societies: A.C.S.; Soc. Portuguesa de Quimica e Fisica; Textile Inst.
Nuclear interests: Chemistry of rare earths; Nuclear chemistry.
Address: Faculdade de Engenharia, Porto, Portugal.

GUELLEC, Maurice le. See le GUELLEC, Maurice.

GUENTHER, Peter, Dipl. Ing. Born 1931. Educ.: Vienna T.H. Member, Siemens-Schuckertwerke G.m.b.H., Vienna, 1956-; delegated to Reaktor-Interessen-Gemeinschaft (R.I.G.), Vienna as Manager, 1964-65; with Siemens-Schuckertwerke, 1966-. Deputy Manager, Power Plant Dept., Siemens G.m.b.H., Vienna.
Nuclear interests: Nuclear power plant projects.
Address: Siemens G.m.b.H., 15 Nibelungengasse, 1010 Vienna, Austria.

GUERCHET, Jean-Marie. Directeur, Saint-Gobain Techniques Nouvelles.
Address: Saint-Gobain Techniques Nouvelles,

23 boulevard Georges Clemenceau, Courbevoie, (Seine), France.

GUERIN, General Maurice. Pres., Comite d'Action Scientifique de la Defense nationale. Member, Conseil Scientifique, C.E.A., France.
Address: Commissariat a l'Energie Atomique, 29-33 rue de la Federation, Paris 15, France.

GUERNSEY, Ernest William, B.S. (Illinois), M.S. (American Univ.), Ph.D. (George Washington Univ.). Born 1896. Educ.: Illinois, American (Washington, D.C.) and George Washington Univs. Director Res., Baltimore Gas and Electric Co. Governor's (Maryland) Advisory Com. on Atomic Energy, -1964, now Consultant. Societies: A.I.Ch.E.; Fellow, American Inst. Elec. Eng.; A.C.S.; Fellow, A.A.A.S.
Nuclear interests: Management; Continuing appraisal of potential for economic nuclear power.
Address: Baltimore Gas and Electric Company, Lexington and Liberty Streets, Baltimore 3, Maryland, U.S.A.

GUERON, Jules, D. ès Sc. phys. Born 1907. Educ.: Paris. Director, C.E.A., 1946-58; Formerly Director Gen. of Res., Euratom, 1958-; Prof., Conservatoire Nat. des Arts et Métiers, 1951-60. Member, Board of Management, O.E.C.D. Dragon Project. Societies: Sté. de Chimie Physique; A.C.S.; Sté. Chimique de France; Sté. Française de Physique; A.N.S.; British Nucl. Soc.
Nuclear interests: Research management; Reactors and materials; Physical chemistry.
Address: Euratom, 51 rue Belliard, Brussels, Belgium.

GUERRA, Pietro, Degree in Law. Born 1926. Educ.: Rome Univ. Legal Counsel of Società elettronucleare nazionale (E.N.E.L.); Lecturer of Trade Law, Rome Univ.
Nuclear interest: Nuclear liability.
Address: 6 Via Girolamo de Carpi, Rome, Italy.

GUERRA, Ranulfo LOBATO. See LOBATO GUERRA, Ranulfo.

GUERRERO, Jorge HALVAS. See HALVAS GUERRERO, Jorge.

GUERRERO AZULA, Coronel (r) Gerardo. Director, Departamento de Personal, Inst. Superior de Energia Nucl., Junta de Control de Energia Atomic del Peru.
Address: Junta de Control de Energia Atomica del Peru, 3420 Avenida Arequipa, Apartado 914, Lima, Peru.

GUERRIERI, G. Member for Italy, Study Group on Food Irradiation, O.E.C.D., E.N.E.A.
Address: O.E.C.D. European Nuclear Energy Agency, 38 boulevard Suchet, Paris 16, France.

GUERRIERO, L. Sen. Officer, Spark Chamber Group, Inst. di Fisica, Univ. Padua.
Nuclear interests: Study of strange particle and pion interactions.
Address: Istituto di Fisica, Universita de Padova, 8 Via F. Marzolo, Padua, Italy.

GUERROUE, M. LE. See LE GUERROUE, M.

GUERY, Arieh Yehuda, B.Sc. (Chem. Eng.). Born 1918. Educ.: Israel Inst. Technol., Haifa. At Consolidated Refineries, Ltd., Haifa, R. and D. Lab., 1951-54; Israel A.E.C.: Res. on extraction of uranium from low-grade ores, Head of exptl. unit for extraction of uranium from phosphates, 1955-59; C.E.A. Mol, Belgium, Training in radioactive waste disposal technol., 1959; Res. on waste disposal problems, 1959-. Societies: Israel Chem. Soc.; A.C.S.
Nuclear interests: Extraction of uranium from low-grade ores by leaching, solvent extraction and electrodeposition; Research on methods of radioactive waste disposal with emphasis on low activity waste.
Address: Israel Atomic Energy Commission, P.O.B. 7056, Hakirya, Tel-Aviv, Israel.

GUEUTAL, Pierrette BENOIST-. See BENOIST-GUEUTAL, Pierrette.

GUGELOT, Piet Cornelis, Phys. degree; Dr. Sc. Nat. Born 1918. Educ.: E.T.H., Zürich. Asst. Prof., Princeton Univ., 1949-56; Formerly Director, Inst. for Nucl. Phys. Res., Amsterdam, 1956-. Societies: Swiss Phys. Soc.; A.P.S.; Netherlands Phys. Soc.
Nuclear interest: Nuclear reactions.
Address: Virginia Associated Research Centre, 12070 Jefferson Avenue, Newport News, Virginia 23606, U.S.A.

GUHA, Arabinda, B.Sc. (Hons., Calcutta), M.Sc. (Calcutta), D. Phil. (Science, Calcutta). Born 1929. Educ.: Calcutta Univ.; also worked at King's Coll., London Univ. Previously Lecturer in Biophys., Saha Inst. of Nucl. Phys., Calcutta Univ. Awarded Sen. Res. Fellowship, Nat. Inst. of Sci. of India; Present position: Reader and in charge of Biophys. Lab., Coll. of Medical Sci., Banaras Hindu Univ.
Nuclear interests: Application of nuclear physics in biology.
Address: Biophysics Laboratory, College of Medical Sciences, Banaras Hindu University, Varanasi 5, India.

GUIDO, Hector Fernandez, Ing. Member. Comision Nacional de Energia Atomica, Uruguay.
Address: Comision Nacional de Energia Atomica, 565 J. Herrera y Reissig, P.2., Montevideo, Uruguay.

GUIDOTTI, Mario, Dr. in Chem. Eng., Diploma in Nucl.Eng. Born 1932. Educ.: Naples and Rome Univs. With Compagnia Tecnica Industrie Petroli (Rome), 1957-58; P.R.O. Organic Reactor Programme: design of

chemical auxiliary systems; Later in charge of R. and D. programme; Head, Chem. Eng. Office for ROVI Nucl. Desalination Programme; Head, R. and D. office for PCUT (Thorium Recycle) programme, C.N.E.N., 1959-.
Nuclear interests: Reactor design and chemical engineering problems.
Address: c/o C.N.E.N., 15 via Belisario, Rome, Italy.

GUILD, Walter R., B.S., M.A., Ph.D. Born 1923. Educ.: Swarthmore, Texas and Yale Univs. Instructor, Asst. Prof., Yale Univ., 1951-60; Assoc. Prof., Biophys. (1960-65), Prof. (1965-), Duke Univ. Assoc. Editor, Rad. Res., -1968. Book: Chapter in Fallout (Editor, John M. Fowler). (Basic Books, 1960).
Societies: Rad. Res. Soc.; Biophys. Soc.; American Soc. Biol. Chem.; American Soc. Cell Biol.; Genetics Soc. of America.
Nuclear interest: Radiobiology of D.N.A. and cells.
Address: Department of Biochemistry, Duke University, Durham, North Carolina 27705, U.S.A.

GUILL, James Harold, B.S., M.A., Ph.D. Born 1924. Educ.: California and Georgetown Univs. With American Potash and Chem. Corp., 1951-52; U.S.N., 1952-56; Supv., Applications Res. (Nucl. Flight Systems), Aircraft Nucl. Propulsion Dept., G.E.C., 1957-59; Advanced Projects Eng.-Nucl. Space Systems, Gen. Dynamics Astronautics, 1960-61; Manager, Advanced Systems Planning, Nucl. Space Programmes Div. (1962-65), Manager, Cryogenic and Nucl. Stage Programmes, (1965-), Lockheed Missiles and Space Co.
Societies: A.N.S.; U.S. Naval Inst.; Air Force Assoc.; American Ordnance Assoc.
Nuclear interests: Systems management and technical development of nuclear flight systems.
Address: Lockheed Missiles and Space Co., Sunnyvale, California, U.S.A.

GUILLAIN, Bernard. Prés. Directeur Gén., Cie. de Construction Mécanique Procédés Sulzer.
Address: Cie de Construction Mécanique Procédés Sulzer, 19 rue Cognacq-Jay, Paris 7, France.

GUILLARD, Jean. Pres. Directeur Gén., Le Nickel S.A. and Administrateur, Cie. Française des Minerais d'Uranium.
Address: Nickel, S.A., 92 rue de Courcelles, Paris 8, France.

GUILLAUMAT, Pierre Lucien Jean. Ing. Gén. des Mines. Born 1909. Ministre délégué chargé de l'Energie Atomique.
Address: 127 rue de Grenelle, Paris 7, France.

GUILLAUME, Marcel A. V., Dr. Born 1931. Educ.: Liège Univ. Chef de Travaux, Liège Univ. Society: Sté. Chimique Belge.
Nuclear interests: Activation analysis by 14 MeV neutrons; Tritium targets for high flux 14 MeV neutrons with little accelerators.
Address: Laboratoire d'Application Radioéléments, 9 place du XX Août, Liège, Belgium.

GUILLIEN, Robert, Agrégé de l'Univ. (Phys.), D. ès sc. physiques, ancien élève, E.N.S. Born 1909. Educ.: E.N.S., Paris and Sorbonne. Prof., European Univ. Sarrebruck (Germany), 1955; Prof., Faculty of Sci., Nancy Univ. (Electronics), 1955-; Director of the Inst. de Génie Biologique et Médical and of Electronic Lab. of the École Nat. Supérieure d'Electricité et de Mécanique in Nancy. Books: Physique Nucléaire Appliquée (Applied Nucl. Phys.) (Paris, Eyrolles, 1963); Electronics (Paris, Presses Universitaires, 3rd Edition, 1966).
Societies: Sté. française de physique; Sté. française des electriciens; Sté. française des electroniciens et radio electriciens.
Nuclear interests: Reactor design; Nuclear physics, and specially action of radiations on semiconductors.
Address: Institut de Génie Biologique et Médical, B.P. 31, Nancy 54, France.

GUILLOT, Marcel. Ing., Soc. des Chantiers Reunis Loire-Normandie; Membre du Comité de Biologie, C.E.A., Paris.
Address: Société des Chantiers Reunis Loire-Normandie, rue Jean Voruz, Nantes, (Loire-Atlantique), France.

GUILLOUX, Raymond. Born 1914. Educ.: E.N.S. de l'Enseignement Technique. Fonctionnaire, Ministère de l'Educ. Nat. Chef du Service Central de documentation, C.E.A. Deleg. to 2nd I.C.P.U.A.E., Geneva Sept. 1958.
Address: Commissariat à l'Energie Atomique, Centre d'Etudes Nucléaires de Saclay, B.P. No. 2 Gif sur Yvette, 91, France.

GUIMARAES CHAVES de CARVALHO, Antonio Herculano, Ind.-Chem. Eng. Born 1899. Educ.: Inst. Superior Técnico, Lisbon Univ. Prof., Inst. Superior Técnico, Lisbon Univ. Reactor, Univ. Técnico, Lisbon., Pres., Soc. Portuguesa de Quimica e Fisica. Book: A Guide to Water Analysis (Lisbon, 1961).
Societies: Acad. Sci. of Lisbon; Acad. Sci. of Madrid.
Nuclear interest: Radiochemistry.
Address: Instituto Superior Técnico, Lisbon, Portugal.

GUIMARAES de CARVALHO, Hervasio, Ind. Chem., Ph.D. (Nucl. Eng.), Dr. Phys. Born 1916. Educ.: Recife Univ., Brazil. Prof. (1950-), Chairman, Exptl. Phys. Dept. (1957), and Phys. Chem. Dept. (1958-60), now Sci. Director, Brazilian Centre for Phys. Res., Rio de Janeiro; Prof. (substitute) Phys., Nat. School of Chem. (1956-), Prof., Nucl. Eng. (1954-), Brazil Univ.; Asst., Tech. and Sci. Div., Nat. Res. Council of Brazil, 1951-54; Tech. Asst. (1958), Member (1967), Brazilian Nucl. Energy Commission, 1958; Res. Adviser, North Carolina State Coll., 1952-53;

Res. Assoc., Chicago Univ., 1953-54, 1955-56; Visiting Prof., Naples Sect., Nat. Inst. Nucl. Phys., 1960-61; Visiting Prof., Centre of Nucl. Res. of Casaccia, Rome, 1961; Director, First Itinerant Exposition on Phys. and Astronomy (U.N.E.S.C.O.), 1950; Societies: Brazilian Acad. Sci.; A.P.S.; Astronomy Soc. of Peru; Mathematical and Phys. Soc. of Cuba.
Nuclear interests: Nuclear physics; Nuclear fission; Nuclear engineering.
Address: Centro Brasileiro de Pesquisas Fisicas, Av, Wenceslau Braz 71–ZC–82, Rio de Janeiro, GB, Brazil.

GUIMARAES FERRI, Mário, Prof. Dr. Director, Faculdade de Filosofia, Ciencias e Letras, Sao Paulo Univ.
Address: Faculdade de Filosofia, Ciencias e Letras, Sao Paulo University, 310 Rua Maria Antonia, Sao Paulo, Brazil.

GUIMONT, Frederic. Commercial Director, Ateliers de Constructions Electroniques Pontier.
Address: Ateliers de Constructions Electroniques Pontier, 327-331 rue de la Garenne, Nanterre, (Seine), France.

GUINARD, P. Pres., Ets. Pompes Guinard S.A.
Address: Ets. Pompes Guinard S.A., 89 avenue de Fouilleuse, 92 St. Cloud, France.

GUINDON, (Rev.) William Gartland, S.J., A.B., A.M., Ph.L., Ph.D., S.T.L. Born 1916. Educ.: Boston Coll., Weston Coll., M.I.T. Vice Pres. and Dean, Coll. of the Holy Cross (Assoc. Prof. Phys.). Societies: A.P.S.; Phys. Soc. (London).
Nuclear interests: Nuclear forces; Angular correlation of successive gamma rays; Low energy nuclear reactions; Angular distribution of elastically scattered fast neutrons; Interactions of fundamental particles.
Address: College of the Holy Cross, Worcester, Massachusetts 01610, U.S.A.

GUINEA ELORZA, Miguel. Director, Tecnicas Nucleares S.A.
Address: Tecnicas Nucleares S.A., 46 Serrano, Madrid 1, Spain.

GUINIER, Andre Jean, D. ès Sc. Born 1911. Educ.: Paris Univ. Prof. à la Faculte de Sci., Orsay, et au Conservatoire des Arts et Metiers, 1949-. Books: Theorie et Technique de la Radiocristallographie (1956); Small Angle Scattering of X-rays (1955). Societies: Sté. de Physique française; Inst. Metals; Gesellschaft für Metallkunde.
Nuclear interests: Radiation damage; Atomic structure of crystals after neutron irradiation.
Address: 87 avenue Denfert Rochereau, Paris 5, France.

GUINN, Vincent Perry, A.B., M.S. (Chem.) (Southern California), Ph.D. (Phys. Chem.) (Harvard). Born 1917. Educ.: Southern California and Harvard Univs. Chem., Shell Development Co., Emeryville, Calif., 1949-61; Formerly Supervisor of Radiochem. Group, Shell Dev. Co.; Manager, Activation Analysis Programme, Gulf General Atomic Inc., San Diego, Calif., 1961-. Consultant and Lecturer, Oak Ridge Assoc. Univs., 1958-. Lecturer, activation analysis courses, Texas A and M Univ., 1961, and Glasgow Univ., 1964. Chairman, Isotopes and Rad. Div., A.N.S. Societies: A.N.S.; A.P.S.; A.C.S.; A.A.A.S.; Soc. for Appl. Spectroscopy.
Nuclear interests: Activation analysis; Radiochemistry; Nuclear instrumentation; Radiotracers; Low-energy accelerators; Low-energy nuclear reactions.
Address: Gulf General Atomic Inc., P.O. Box 608, San Diego, California 92112, U.S.A.

GUIRALDENQ, Pierre-Henri, D. es Sc. Phys., Diplomé de l'I.N.S.T.N. Born 1934. Educ.: Paris Univ. and C.E.A.
Nuclear interests: Studies on intermetallic diffusion by radio tracers, particularly in the grain boundaries and also on crystallographic problems of the allotropic phases in metals.
Address: Département des Recherches de la C.A.F.L., Unieux, Loire, France.

GUISAN, Francois, Civil Eng. (Inst. Technol., Lausanne Univ.). Born 1922. Assoc. of Bonnard et Gardel. Consulting Eng., Lausanne. Societies: Swiss Soc. of Eng. and Architects; Sté. suisse des spécialistes du génie nucléaire.
Nuclear interest: Underground power plant design and construction.
Address: 10 avenue de la Gare, Lausanne, Switzerland.

GULLATT, Sam P. Prof. Member, Nucl. Advisory Com., Louisiana Polytech. Inst.
Address: Louisiana Polytechnic Institute, Nuclear Centre, Tech Station, Ruston, Louisiana, U.S.A.

GULMANELLI, Paolo. Born 1928. Educ.: Pavia Univ. Prof. ordinario de Istituzioni di Fisica Teorica nell'Univ. di Pavia; doing res. also for I.N.F.N.
Nuclear interests: Low energy nuclear physics; Application of group theory to physics.
Address: Istituto di Fisica dell'Università, Pavia, Italy.

GUNCKEL, James E., B.S. Ed., M.A., Ph.D. Born 1914. Educ.: Miami and Harvard Univs. Instructor, Harvard Univ.; Assoc. Prof., Botany (1951-56), Chairman, Botany Dept. (1954-), Prof., Botany and Rad. Sci. (1957-), Rutgers Univ.; Res. Collaborator, Biol. Dept., Brookhaven Nat. Lab., 1951-57; Editor, Bulletin Torrey Botanical Club. Consultant, Biol. Dept., Brookhaven Nat. Lab., 1950-; Hon. Editorial Advisory Board, Rad. Botany. Society: Rad. Res. Soc.
Nuclear interest: Radiation botany. Effects of chronic and acute radiation on the cytology,

morphology and physiology of plants.
Address: Nelson Biological Laboratories, University Heights, Rutgers, The State University, New Brunswick, New Jersey 08903, U.S.A.

GUNDELACH, H. E. Finn Olav. Head, Danish Mission to Euratom, 1967-.
Address: Danish Mission to Euratom, 12 rue Belliard, Brussels 4, Belgium.

GUNN, J. Member, Sci. Staff, Nordic Inst. for Theoretical Atomic Phys.
Address: Nordic Institute for Theoretical Atomic Physics, 17 Blegdamsvej, Copenhagen Ø, Denmark.

GUNN, J. C., Prof., M.A. Cargill Prof. Theoretical Phys., Natural Philosophy Dept., Glasgow Univ.
Address: Department of Natural Philosophy, University of Glasgow, Glasgow W.2, Scotland.

GUNNELS, Lee O. Res. Sci.-Eng., Appl. Nucl. Phys. Div., Battelle-Columbus Labs.
Address: Battelle-Columbus Laboratories, 505 King Avenue, Columbus, Ohio 43201, U.S.A.

GUNST, Samuel Burton, A.B., M.S., Ph.D. Born 1917. Educ.: Olivet Coll. and Pittsburgh Univ. Gulf Res. and Development Co., 1941-50; Rad. Lab., Pittsburgh Univ., 1950-53; Fellow Sci., Bettis Atomic Power Lab., Westinghouse Elec. Corp., 1953-. Societies: A.N.S.; A.P.S.; A.A.A.S.
Nuclear interests: Experimental nuclear physics; Elementary particles; Neutron detectors; Reactivity of irradiated nuclear reactor fuels; Neutron interaction cross sections.
Address: Bettis Atomic Power Laboratory, Westinghouse Electric Corporation, P. O. Box 79, West Mifflin, Pennsylvania 15122, U.S.A.

GUNTEN, Hans R. Von. See Von GUNTEN, Hans R.

GÜNTHER, Alfred, Dr. rer. nat. Born 1921. Educ.: Göttingen Univ. Deputy Librarian, (1955-), Head, Sci. Information Service (1954), C.E.R.N.
Nuclear interest: Literature of nuclear science.
Address: C.E.R.N., 1211 Geneva 23, Switzerland.

GÜNTHER, Christian, Dr. Inst. für Strahlen und Kernphysik, Bonn Univ.
Address: Institut für Strahlen und Kernphysik, 14-16 Nussallee, Bonn, Germany.

GUO, Yong-huai. Member, Editorial Com., Scientia Sinica.
Address: Scientia Sinica, Academia Sinica, Peking, China.

GUPTA, K. K., Dr., Ph.D. (Bombay). Reader, Tata Inst. of Fundamental Res.
Address: Tata Institute of Fundamental Research, Homi Bhabha Road, Bombay 5, India.

GUPTA, Niraj Nath DAS. See DAS GUPTA, Niraj Nath.

GUR'EV, M. V. Paper: Co-author, A Study of Co^{60} γ-Ray Effect on Highly Basic Anionites AM-17 and AM (Atomnaya Energiya, vol. 16, No. 3, 1964).
Address: Academy of Sciences of the U.S.S.R., 14 Leninsky Prospekt, Moscow V-71, U.S.S.R.

GURIKOV, Yu. V. Paper: Co-author, About Cation Hydration in Heavy Water (Atomnaya Energiya, vol. 19, No. 5, 1965).
Address: Academy of Sciences of the U.S.S.R., 14 Leninsky Prospekt, Moscow V-17, U.S.S.R.

GURIN, V. N. Papers: To Calculation of Epithermal Neutron Capture by Infinite Lattice of Control Plates (Atomnaya Energiya, vol. 21, No. 5, 1966); Application of Method of Incomplete Division of Variables to Calculation of Epithermal Neutron Capture by Infinite Lattice of Control Plates (ibid, vol. 24, No. 3, 1968).
Address: Academy of Sciences of the U.S.S.R., 14 Leninsky Prospekt, Moscow V-71, U.S.S.R.

GURINSKY, David H., B.S., Ph.D. Born 1914. Educ.: New York Univ. Head, Metal. Div., Brookhaven Nat. Lab., 1947-67; Consultant, Gen. Atomic, 1958-59. Book: Co-editor, Nucl. Fuels (1956). Societies: A.S.M.; A.I.M.E.; A.N.S.
Nuclear interests: Metallurgy; Nuclear fuels; Graphite; Liquid metal technology; Gas-cooled reactors; Radiation effects.
Address: Brookhaven National Lab., Upton, Long Island, New York, U.S.A.

GURR, Edward, Ph.D., F.R.I.C., F.L.S., F.I.Biol., F.R.M.S., F.S.D.C. Born 1905. Educ.: Tech. Coll., Leicester; Chelsea Colleges of Sci. and Technol., London. Director, Edward Gurr, Ltd., Michrome Labs., London. Hon. Res. Assoc. in Anatomy, Univ. Coll., Cork. Books: Microscopic Staining Techniques, Nos. 1 and 2 (1950), No. 3 (2nd ed., 1958), No. 4, (1958) (Edward Gurr, London); Methods of Analytical Histology and Histochem. (Leonard Hill, 1958); A Practical Manual of Medical and Biol. Staining Techniques (Leonard Hill, 1956); Encyclopaedia of Microscopic Stains (Leonard Hill, 1960); Staining Animal Tissues (Leonard Hill, 1962); Rational Use of Dyes in Biol. (Leonard Hill, 1964). Societies: Linnean Soc.; Roy. Microscopical Soc.; Roy. Inst. Chem.; Inst. Biol.; Anatomical Soc.
Nuclear interest: Radiobiology.
Address: 19 Fife Road, East Sheen, London, S.W.14, England.

GURVICH, M. G. Papers: Co-author, The Mechanism of Negative Effect of Oxygen-Containing Uranium Compounds on the Course and Results of Metallothermic Reduction of Uranium Tetrafluoride (Atomnaya Energiya, vol. 13, No. 1, 1962); co-author, Preparation and Studies of some Plutonium Monocarbide Properties (ibid, vol. 22, No. 6, 1967).
Address: Academy of Sciences of the U.S.S.R., 14 Leninsky Prospekt, Moscow, V-71, U.S.S.R.

GURVICH, M. Yu. Papers: Co-author, Determination of Uranium Clarke Concentration in Ionic Crystals (letter to the Editor, Atomnaya Energiya, vol. 22, No. 6, 1967); co-author, Determination of Uranium Concentration and its Spatial Distribution in Minerals and Rocks (ibid, vol. 23, No. 6, 1967).
Address: Academy of Sciences of the U.S.S.R., 14 Leninsky Prospekt, Moscow V-71, U.S.S.R.

GUSAKOW, Mark, D. ès Sc. Phys. Born 1930. Educ.: Paris Univ. Prof. Lyon Univ. Societies: Sté. Française de Phys.; A.P.S.
Nuclear interest: Nuclear physics: nuclear reactions and structure of nuclei.
Address: Institut de Physique Nucleaire, Lyon Université, 43 boulevard du 11 novembre 1918, 69 Villeurbanne, France.

GUSEV, N. G. Book: Co-author, Gamma-izluchenie radioaktivnykh izotopov i produktov deleniya (Gamma-rad. of Radioactive Isotopes and Products of Fission) (M. Fizmatgiz, 1958). Papers: Measurement of Small α-emitter Concentrations in Water by Freezing-out (letter to the Editor, Atomnaya Energiya, vol. 3, No. 10, 1957); co-author, The Universal Tables for Calculation of γ-Ray Attenuation in Thin Filters (letter to the Editor, ibid., vol. 13, No. 5, 1962); co-author, Rad.Characteristics of Mixed Instantaneous Fission Products of V^{238} Generated by Neutrons of 14 MeV Energy (letter to the Editor, ibid, vol. 23, No. 1, 1967).
Address: Academy of Sciences of the U.S.S.R., 14 Leninsky Prospekt, Moscow V-71, U.S.S.R.

GUSHEE, Richard B. Sec. and Treas., Power Reactor Development Co.
Address:Power Reactor Development Co., 1911 First Street, Detroit 26, Michigan, U.S.A.

GUS'KOVA, V. N. Paper: Co-author, Rad. near Reactor VVR-M (letter to the Editor, Atomnaya Energiya, vol. 19, No. 1, 1965).
Address: Academy of Sciences of the U.S.S.R., 14 Leninsky Prospekt, Moscow V-71, U.S.S.R.

GUSSGARD, Knut, Cand. real. Born 1931. Educ.: Oslo Univ. Health Phys., State Inst. of Rad. Hygiene, Oslo, 1958-60; Health Phys. (1960-67), Head, Tech. Secretariat (1967-), Inst. for Atomenergi, Kjeller; Sec., Statens Atomenergirad, 1967-. Vice Chairman, Norwegian Atomic Energy Soc.; Sec., Nordic Soc. for Rad. Protection.
Nuclear interests: Radiation protection; Nuclear safety; Technical management.
Address: Institutt for Atomenergi, P.O. Box 40, Kjeller, Norway.

GUSTAFSON, John Kyle, A.B., A.M., Ph.D. Born 1906. Educ.: Washington and Harvard Univs. Consulting Geologist (1950-56), Director, Explorations (1956-), M. A. Hanna Co.; Vice-Pres., Hanna Mining Co. and predecessor Corp., 1953-61; Pres., Homestake Mining Co., 1961-. Member, Advisory Com. Raw Materials, U.S.A.E.C., 1950-59; Advisory Panel Mineral Exploration Res., N.S.F., 1952.
Papers: Several including Atomic Energy and the Mining Business: Uranium Supplies (1948); Impact of Atomic Energy on Economic Geology (1949); Uranium Resources (1949); How a Miner Sees It (Nucleonics, 1967). Societies: Fellow, Geological Soc. America (Member of Council, 1957-60); Member, Soc. Economic Geologists (Member council, Pres. 1966-67; American Inst. Mining and Metal. Eng.; Canadian Inst. Mining and Metal.; Mining, Metal. Soc. America.
Address: 41 Alvarado Road, Berkeley, California 94705, U.S.A.

GUSTAFSON, Torsten, Ph.D. Born 1904. Educ.: Lund and Copenhagen Univs. Member, Swedish Atomic Res. Council, 1945-; Member, Swedish Atomic Energy Deleg., 1956-. Chairman of Board, Nordita, Copenhagen, 1963-; Govt. Res. Board, 1963-. Societies: Swedish, Danish and Finnish Acads. Sci.
Nuclear interest: Nuclear physics.
Address: Institute of Theoretical Physics. University, Lund, Sweden.

GUSTAFSSON, C. Ake T., Born 1908. Educ.: Lund Univ. Prof., and Head, Dept. of Forest Genetics, Royal Coll.of Forestry, Stockholm. Societies: Roy. Swedish Acad. of Agriculture; Roy. Swedish Acad. Sci.; Roy. Physiographic Soc.; Kgl. Akademie der Naturforscher Leopoldina; Roy. Danish Acad. of Sci.; Indian Soc. of Genetics and Plant Breeding; N.A.S., Washington.
Nuclear interests: Radiation and mutations; Chemical mutagenesis.
Address: Royal College of Forestry, Stockholm 50, Sweden.

GUSTORF, Ernst KOERNER von. See KOERNER von GUSTORF, Ernst.

GUT, Marcel, Ph.D. Born 1922. Educ.: Basel Univ. Sen. Sci., Worcester Foundation for Exptl. Biol., Inc., Shrewsbury, Mass., 1948-. Societies: A.C.S.; Fed. of American Scientists; Swiss Chem. Soc.
Nuclear interests: Labelling of organic compounds.
Address: 147 Coolidge Road, Worcester 2, Massachusetts, U.S.A.

GUTBROD, Fritz, Dr. Formerly Wissenschaftlich Asst., Inst. für Theoretische Physik und

Mechanik, Heidelberg Univ.
Address: DESY (Deutsches Elektronen - Synchrotron), 149 Luruper Chaussee, 1000 Hamburg-Bahrenfeld, Germany.

GUTHRIE, Robert R. Chairman, Florida Nucl. Commission; Formerly Treas., Southern Interstate Nucl. Board.
Address: Florida Nuclear Commission, Room 113, 725 South Bronough Street, Tallahassee, Florida 32304, U.S.A.

GUTIERREZ-JODRA, Luis, B.Sc. (Chem.), Ph.D. (Chem,), Ph.D. (Ind. Chem.). Born 1922. Educ.: Madrid Univ. Prof., Valladolid Univ., 1955; Prof. Madrid Univ., 1958; Director, Plantas Piloto e Industriales, Junta Energia Nucl., 1958; Formerly Vice Chairman (1964-), now Chairman, Board of Directors, Eurochemic.
Nuclear interests: Reactor materials; Irradiated fuel reprocessing; Fuel elements.
Address: Junta de Energia Nuclear, Ciudad Universitaria, Madrid 3, Spain.

GUTIN, E. I. Paper: Co-author, Decontaminating Low-Activity Effluents by Electrodialysis with Ion-Exchange Membranes (Atomnaya Energiya, vol. 22, No. 5, 1967).
Address: Academy of Sciences of the U.S.S.R., 14 Leninsky Prospekt, Moscow V-71, U.S.S.R.

GUTKOWSKI, D. Gruppo Teorico, Istituto di Fisica, Catania Univ.
Address: Catania University, Istituto di Fisica, 57 Corso Italia, Catania, Sicily, Italy.

GUTMAN, Maria Carmen de OLIVEIRA.
See de OLIVEIRA GUTMAN, Maria Carmen.

GUTMAN, T. Nucl. Safety Eng., Nucl. Materials and Equipment Corp.
Address: Nuclear Materials and Equipment Corp., Apollo, Pennsylvania 15613, U.S.A.

GUTTINGER, Werner, Dr. rer. nat. Born 1925. Educ.: Tübingen Univ. Prof., Sao Paulo Univ. 1957-61; Max-Planck-Inst. für Phys., Munich, 1961-63; Munich Univ. 1963-.
Nuclear interests: Elementary particle physics (theory); Quantum field theory; Mathematical physics.
Address: Physics Department, Theory Division, University of Munich, 2 Schellingstr., Munich 13, Germany.

GUZINA, Vojin R. Pres., Federal Nucl. Energy Commission of the Socialist Federal Republic of Yugoslavia.
Address: Federal Nuclear Energy Commission, 29 Kosancicev Venac, P. O. Box 353, Belgrade, Yugoslavia.

GUZMAN, Arturo R. DE. See DE GUZMAN, Arturo R.

GUZMAN, Ferrer LOPEZ. See LOPEZ GUZMAN, Ferrer.

GUZMAN, Mario, Ing. Space Phys. and Aeronomy, Lab. de Fisica Cosmica, Univ. Mayor de San Andres.
Address: Laboratorio de Fisica Cosmica, Universidad Mayor de San Andres, La Paz, Bolivia.

GUZOVSKAYA, E. K. Paper: Co-author, Gamma-Ray Transmission through Oblique Barriers (Atomnaya Energiya, vol. 23, No. 1, 1967).
Address: Academy of Sciences of the U.S.S.R., 14 Leninsky Prospekt, Moscow V-71, U.S.S.R.

GVERDTSITELI, I. G., B.Sc., Acad. of Sci. of the Georgian S.S.R.; Deleg. to 2nd I.C.P.U.A.E., Geneva, Sept. 1958. Adviser, Third I.C.P.U.A.E., Geneva, Sept. 1964.
Papers: Relative Diference of $B^{11}F_3 - B^{10}F_3$ Steam Pressure (Atomnaya Energiya, vol. 19, No. 1, 1965); co-author, Investigation of Pressure Influence on Isotope Separation B^{11} and B^{10} Process by Distillation of BF_3 (ibid, vol. 23, No. 4, 1967).
Nuclear interest: Isotope enrichment.
Address: Academy of Sciences of the Georgian S.S.R., Tbilisi, Georgian S.S.R.

GVOZDEV, B. A. Paper: Co-author, Measurements of Cf^{246} and Cf^{248} Spontaneous Fission Periods (letter to the Editor, Atomnaya Energiya, vol. 24, No. 1, 1968).
Address: Academy of Sciences of the U.S.S.R., 14 Leninsky Prospekt, Moscow V-71, U.S.S.R.

GVOZDEV, E. G. Paper: Co-author, Dosimeters Based on Glasses with Optimal Density Alternating by Irradiation (Atomnaya Energiya, vol. 21, No. 1, 1966).
Address: Academy of Sciences of the U.S.S.R., 14 Leninsky Prospekt, Moscow V-71, U.S.S.R.

GWOZDZ, Rajmund. Head, Dept. of Radioisotopes and Tracer Compounds, Technol., Soltan Nucl. Res. Centre, Inst. of Nucl. Res.
Papers: Co-author, Separation of the Various Oxidation States of Plutonium by Reversed-Phase Partition Chromatography (Nukleonika, vol. 5, No. 10, 1960); co-author, Separation of Protactinium from Thorium by Reversed-Phase Partition Chromatography (ibid, vol. 8, No. 4, 1963); co-author, The Separation of Arsenic from Germanium by Reversed-Phase Chromatography (ibid, No. 5); co-author, Separation of Carrier-Free ^{199}Au from Platinum by Reversed-Phase Partition Chromatography (ibid.).
Address: Institute of Nuclear Research, Radiochemistry Department, Warsaw 9, Poland.

GYFTAKI, E., M.D. Director, Haematological Applications, Isotope Lab., Alexandra Hospital, Athens.
Address: Alexandra Hospital, Isotope Laboratory, Vas. Sophias-K. Lourou Str., Athens, Greece.

GYFTOPOULOS, Elias Panayiotis, Diploma in Mech. and Elec. Eng., Sc.D. (Elec. Eng.). Born 1927. Educ.: Athens Tech. Univ. and M.I.T. Instructor Elec. Eng. (1954-58), Asst. Prof. Elec. and Nucl. Eng. (1958-61), Assoc. Prof. Nucl. Eng. (1961-65), Prof. Nucl. Eng. (1965-), M.I.T.; Consultant to Brookhaven Nat. Lab., Phillips Petroleum Co. (Div. Atomic Energy) and several other industries. Nat. Programme Com., A.N.S.; Education Com. A.N.S.; A.N.S.-A.S.E.E. Com. on Objective Criteria in Nucl. Eng. Education; Board of Directors A.N.S.; Vice-Pres., American Hellenic Educational and Welfare Fund. Societies: Fellow, A.N.S.; A.P.S.; A.I.A.A.; A.A.A.S.; Fellow, American Acad. of Arts and Sci.
Nuclear interests: Dynamics of nuclear reactors and nuclear power plants; Fusion technology and direct conversion of nuclear energy into electricity (teaching and research).
Address: 18 Hastings Road, Lexington, Mass. 02173, U.S.A.

GYGI, Hans A., Dr. Sci. (Eng.). Born 1906. Educ.: Swiss Federal Inst. Technol., Zurich. Member, Therm-Atom A.G.; Managing Director, Escher Wyss Ltd. Society: Nationale Gesellschaft zur Förderung der industriellen Atomtechnik (N.G.A.).
Address: Escher Wyss Ltd., Escher Wyss-Platz, Post Office Box, CH-8023 Zurich, Switzerland.

GYI, Ku Aung, B.Sc. (Hons.), M.Sc. Born 1930. Educ.: Rangoon Univ. Phys., Rangoon Gen. Hospital, 1958-. Society: Inst. Nucl.Eng., London.
Nuclear interests: Radiation physics and nuclear physics.
Address: Radiotherapy Department, Rangoon General Hospital, Rangoon, Burma.

GYLES, Thomas BENSON. See BENSON GYLES, Thomas.

GYOREY, Geza Leslie, B.S.E., B.S., M.S.E., Ph.D. Born 1933. Educ.: Calvin Coll. and Michigan Univ. Asst. and Assoc. Prof. Nucl. Eng., Michigan Univ., 1960-66; Manager, Nucl. Methods, Nucl. Energy Div., G.E.C. 1966-. Society: A.N.S.
Nuclear interests: Analytical methods for reactor neutronics and fuel cycle analysis; Reactor core design; Reactor dynamics; Application of computers in nuclear engineering; Technical management; Nuclear engineering education.
Address: General Electric Co., Nuclear Energy Division, 175 Curtner Avenue, San Jose, California, 95125, U.S.A.

HA-VINH, Phuong, Dr. of Laws. Born 1926. Educ.: Toulouse Univ. Alternate Rep. of Vietnam, Board of Governors (1961-63), First Officer, Legal Div., Secretariat (1964-), I.A.E.A.
Nuclear interests: Nuclear legislation; Regulatory activities of international organisations in the nuclear field.

Address: International Atomic Energy Agency, 11 Kaerntnerring, A-1010 Vienna, Austria.

HAACK, Hans Werner. Chief Eng., Berkefeld-Filter G.m.b.H.
Address: Berkefeld-Filter G.m.b.H., Postfach 12,31 Celle, Germany.

HAAG, Fred G., M.E. (Stevens Inst.), M.S., Dr. Eng. Sci. (Rensselaer Polytech.). Born 1931. Educ.: Rensselaer Polytech. O.R.S.O.R.T. 1953-54; K.A.P.L., 1954-67; Union Coll., 1967-. Societies: A.S.M.E.; I.E.E.E.; A.A.A.S.
Nuclear interests: Design and analysis of reactor control systems; Safeguard and accident analysis and studies.
Address: Mechanical Engineering Department, Union College, Schenectady, New York 12308, U.S.A.

HAAGEN-SMIT, Arie J., Dr. Prof., Div. of Biol., California Inst. of Technol. Member, Advisory Com. for Biol. and Medicine, U.S.A.E.C.
Address: California Institute of Technology, Division of Biology, Pasadena, California, U.S.A.

HAAGER, Michael, Dr. phil. Born 1934. Educ.: Vienna Univ. II Physikalisches Inst., Vienna Univ.; Telephon- u. Telegraphen-Fabriks-A.G. Kapsch and Söhne, Vienna. Societies: Chemisch Physikalische Gesellschaft, Vienna; Österreichische Physikalische Gesellschaft.
Nuclear interests: Metallurgy and solid state physics; Management.
Address: 12 Auhofstrasse, 1130 Vienna, Austria.

HAAK, Enok, Agr. Sen. Officer, Div. of Agricultural Radiobiol., Dept. of Soil Fertility and Management, Roy. Agricultural Coll.
Address: Royal Agricultural College, Uppsala 7, Sweden.

HAALAND, John, Cand. real. Born 1923. Educ.: Oslo Univ. Group leader, Head, Spectrographic Lab., Inst. for Atomenergi, Kjeller. Society: Phys. Soc. of Norway, 1965.
Nuclear interests: Metallurgy, raw materials control and analysis of reactor materials for impurities; Nuclear physics, mass spectrometric burn up studies.
Address: Institutt for Atomenergi, P.O. Box 40, Kjeller, Norway.

HAAN, Abel, Jr., DE. See DE HAAN, Abel, Jr.

HAAS, Frederick Alfred, B.Sc., Ph.D. Born 1932. Educ.: Univ. Coll. of the South West, Exeter, and London Univ. English Elec. Co. 1958-60; A.W.R.E. (1960-62), A.E.R.E., Culham (1962-), U.K.A.E.A. Society: A.Inst. Phys.
Nuclear interests: Controlled thermonuclear reactions; Plasma physics.

Address: United Kingdom Atomic Energy
Authority, Culham Laboratory, Culham, Nr.
Abingdon, Berks., England.

HAAS, Paul Arnold, B.S., M.S., Ph.D. (Chem.
Eng.). Born 1929. Educ.: Missouri(Rolla) and
Tennessee Univs. Development Eng. (1952-
57), Group Leader (1957-), Chem. Technol.
Div., O.R.N.L. Societies: A.I.Ch.E.; A.C.S.
Nuclear interests: Nuclear fuel preparation
and processing; Solvent extraction; Phase
separation; Hydraulic cyclones; Waste disposal.
Address: Oak Ridge National Laboratory,
P.O. Box X, Building 4505, Oak Ridge,
Tennessee, U.S.A.

HAAS, Willem Alexander de. See de HAAS,
Willem Alexander.

HAAS van DORSSER, A. H. de. See de HAAS
van DORSSER, A. H.

HAASE, Alfred. Mitglied, Prasidium, Deutsche
Atomkommission, Federal Ministry for Sci.
Res.
Address: Federal Ministry for Scientific
Research, 2-10 Heussallee, P.O. Box 9124,
53 Bonn 9, Germany.

HAASE, Werner, Ministerialdirigent, Dipl. Ing.
Born 1902. Educ.: Tech. Univ. Danzig.
Bundesministerium für Wirtschaft, 1950-55;
Bundesministerium für Atomkernenergie, now
Bundesministerium für wissenschaftliche For-
schung, 1956-. Member and President, Supply
Agency of C.E.E.A.
Nuclear interests: Fuel elements; Metallurgy;
Uranium ore; Building material for reactors;
Separation of isotopes; Heavy water.
Address: Federal Ministry for Scientific
Research, 2-10 Heussallee, Bonn 53, Germany.

HABAHPASHEV, A. G. Paper: Co-author,
Status Report of Positron-Electron Storage
Ring VEPP-2 (Atomnaya Energiya, vol. 19, No.
6, 1965).
Address: Novosibirsk Science Centre, 20
Sovietskaya Ulitsa, Novosibirsk, Siberia,
U.S.S.R.

HABARU, J.-M. Asst., Exptl. Nucl. Phys.,
Inst. de Physique, Liège Univ.
Address: Institut de Physique, 1A Quai Roose-
velt, Liège, Belgium.

HABASHI, Fathi, Dr. techn. Born 1928.
Educ.: Cairo Univ. and T.H. Vienna.In Radio-
chem. Div., Chem. Inst., Vienna Univ., 1959-
60; Extraction Metal. Div., Dept. Mines and
Tech. Surveys, Ottawa, 1960-62; Formerly
Assoc. Prof., Metal. Dept., Montana School
of Mines. Now in Extractive Metal. Res. Div.,
The Anaconda Co., Arizona.
Nuclear interests: Geochemistry, analysis and
extraction metallurgy of uranium, thorium
and the rare earths; Distribution of fission
products in the biosphere.
Address: Extractive Metallurgical Research
Division, The Anaconda Company, Tucson,
Arizona, U.S.A.

HABERER, Gudrun, Dipl. Phys. Born 1939.
Educ.: Dresden Tech. Univ.
Nuclear interests: Strahlen biophysik; Dosi-
metrie.
Address: Deutsche Akademie der Wissenschaf-
ten zu Berlin, Forschungsgemeinschaft,
Zentralinstitut für Kernforschung, 8051
Dresden, P.F.19, German Democratic
Republic.

HABERLANDT, Reinhold Johannes Karl,
Dipl. Phys., Dr. rer. nat. Born 1936. Educ.:
Martin-Luther-Univ., Halle-Wittenberg. Wissen-
schaftlicher Asst., Inst. für physikalische
Stofftrennung, Leipzig (1959-60); Wissen-
schaftlicher Asst. (1961), Wissenschaftlicher
Oberassistent (1962-65), Wissenschaftlicher
Arbeitsleiter (1965-), Arbeitsstelle für Statis-
tische Physik, Leipzig, Deutsche Akademie
der Wissenschaften zu Berlin; Lecturer in
Statistical Phys. and Thermodynamics, Karl-
Marx-Univ., Leipzig. Arbeitsgemeinschaft
Analyse stabiler Isotope; Arbeitsgemeinschaft
Anwendung stabiler Isotope in Physik und
Chemie. Society: Phys. Soc., D.D.R.
Nuclear interests: Statistical mechanics and
quantum mechanics, especially theory of
isotopic systems; Molecular interaction.
Address: 9 Louis-Fürnberg-Strasse, Leipzig
705, German Democratic Republic.

HÄBERLI, Roland, Dr. Chem. Born 1922.
Educ.: Berne Univ. At Eidgenossische Inst.
für Reaktorforschung, 1959-. Society:
Schweiz. Gesellschaft von Fachleuten der
Kerntechnik.
Nuclear interests: Radiation and hot atom
chemistry.
Address: Eidgenossische Institut für Reaktor-
forschung, Würenlingen, Switzerland.

HABERMAN, Jerome, B.S., M.S. Born 1920.
Educ.: C.C.N.Y. and Stevens Inst. Technol.,
Isotopes Unit, Explosives Lab., Feltman Res.
Labs., Picatinny Arsenal, 1953-. Society:
A.C.S.
Nuclear interests: Tracer techniques in surface
studies. Radiation chemistry.
Address: Feltman Research Laboratories,
Isotopes Unit, Explosives Laboratory, Pica-
tinny Arsenal, Dover, New Jersey, U.S.A.

HABIBULLAH, A. K. M., M.Sc. Born 1932.
Educ.: Dacca Univ. Res. Asst. to Agricultural
Chem. (1955-60), Asst. Soil Chem., Soil
Fertility Survey and Popularisation of the use
of fertilisers in East Pakistan (1960-61), Govt.
of Pakistan; Asst. Sci. Officer (1961-62),
Sci. Officer (1964-), Pakistan A.E.C. Society:
Pakistan Assoc. for the Advancement of Sci.
Nuclear interests: Application of atomic energy
in agriculture. Soil chemistry and soil
fertility problems and soil-plant relation
studies using tracer technique. Tracing plant

nutrient through autoradiography and chromotography studies.
Address: Atomic Energy Centre, P. O. Box 164, Ramna, Dacca, East Pakistan.

HABRAKEN, L., Assoc. Prof. Chief Res. Metal., Div. of Phys. Metal. and Fundamental Res., Centre Nat. de Recherches Métallurgiques.
Address: Centre National de Recherches Métallurgiques, Abbaye du Val-Benoit, Liège, Belgium.

HACHATURYAN, M. N. Paper: Co-author, Pulse Height with Photography on Moving Film (letter to the Editor, Atomnaya Energiya, vol. 16, No. 5, 1964).
Address: Academy of Sciences of the U.S.S.R., 14 Leninsky Prospekt, Moscow V-71, U.S.S.R.

HACHEZ, Pierre. Formerly Administrateur Directeur, Metallurgique Hoboken; Formerly Director, now Administrateur-Delegue, Belgonucléaire.
Address: Societe Belge pour l'Industrie Nucleaire, 35 rue des Colonies, Brussels, Belgium.

HACKBARTH, Winston P., B.A., B.S., M.S., Ph.D. Born 1924. Educ.: Iowa State Univ. Augustana Coll., 1956-59; Member, Rad. Com., Louisiana Polytech. Inst., 1959-68. Nuclear interest: To review all uses, safety practices and procedures involving any kind of ionising radiation on the campus before its use will be allowed.
Address: Box 966, Tech Station, Ruston, Louisiana 71271, U.S.A.

HACKE, Jürgen, Dipl.-Ing., Dr. rer. nat. Born 1928. Educ.: Tech. Univ. Berlin and Giessen Univ. Wiss. Asst., Giessen Univ., 1958-59; Wiss. Rat, Hahn-Meitner-Inst. für Kernforschung Berlin, 1959-. Societies: Verband Deutscher Physikalischer Gesellschaften; Fachverband für Strahlenschutz.
Nuclear interests: Dosimetry; Spectroscopy; Health physics; Radiation effects in materials.
Address: Hahn-Meitner-Institut für Kernforschung Berlin, Abt. Strahlenphysik, Berlin 39, Germany.

HÄCKER, Otto, Dr. phil. Born 1898. Educ.: Tübingen Univ. Leader, Deutscher Forschungsdienst (agency for scientific news).
Nuclear interests: Applied nuclear power.
Address: Deutscher Forschungsdienst, 136 Südstrasse, Bad Godesberg, Germany.

HADDAD, Eugene, Dr. With Div. of Res., and Member, Nucl. Cross Sections Advisory Group, U.S.A.E.C.
Address: U.S. Atomic Energy Commission, Division of Research, Washington D.C. 20545, U.S.A.

HADDEN, G. R. Vice Pres., Construction, Indian Point Nucl. Energy Power Plant, Consolidated Edison Co. of New York, Inc.
Address: Consolidated Edison Co. of New York, Inc., Indian Point Nuclear Energy Power Plant, Buchanan, New York, U.S.A.

HADDOCK, Roy P. Prof., Nucl. Group, Phys. Dept., California Univ., Los Angeles.
Address: Department of Physics, California University, Los Angeles 24, California, U.S.A.

HADDOW, Alexander, M.D., D.Sc., Ph.D. Born 1907. Educ.: Edinburgh Univ. Director, Chester Beatty Res. Inst., and Prof. Exptl. Pathology, London Univ., 1946-. Book: Biological Hazards of Atomic Energy (Ed., 1952). Societies: Fellow, Royal Soc.; Manager, Royal Inst.; Member, Com. of Management, Inst. of Cancer Res.; Life Member, Royal Medical Soc. of Edinburgh; Member, Exec. Council, Ciba Foundation; Member, Grand Council and Sci. Advisory Com., British Empire Cancer Campaign. Member: Pathological Soc. of Great Britain and Ireland; Chemical Soc.; Genetical Soc.; Zoological Soc.; Soc. for Endocrinology; Soc. for Gen. Microbiol.; Soc. for Exptl. Biol.; Inst. of Biol.; Internat. Soc. of Haematology.
Address: Chester Beatty Research Institute, Institute of Cancer Research, Royal Cancer Hospital, Fulham Road, London, S.W.3, England.

HADIDY, Ali Ramadan EL. See EL HADIDY, Ali Ramadan.

HADLEY, L. N. Now Head and Formerly Reactor Administrator, Dept. Phys., Colorado State Univ.
Address: Department of Physics, Colorado State University, Fort Collins, Colorado, U.S.A.

HADRILL, Harold Frederick John, M. Eng. Educ.: Liverpool Univ. Jun. Eng., Metropolitan-Vickers Elec. Co., 1950-53; Kennedy and Donkin, 1953-58 (Head, Nucl. Dept., 1956-58); Staff Member (1958-60), Head, Eng. Div., Berkeley Nucl. Labs. (1960-65), C.E.G.B.; Asst. Director, Marchwood Eng. Labs., 1965-. Societies: M.I.Mech.E.; F.I.E.E.
Nuclear interests: Research and development applied to power reactors.
Address: Marchwood Engineering Laboratories, Marchwood, Southampton, England.

HADYANA, Pratiwi. Lecturer, Nucl. Chem. Dept., Bandung Inst. of Technol.
Address: Bandung Institute of Technology, 10 Djl. Ganeca, Bandung, Indonesia.

HAEFFNER, Erik Axel, tekn. lic. Born 1918. Educ.: Chalmers Univ. of Technol., Göteborg and Roy. Inst. Technol., Stockholm. Formerly Manager, Chem. Dept., now Head, Reactor Chem. Dept., AB Atomenergi, Stockholm, 1950-. Director, Eurochemic, Study and Res. Office, Mol, Belgium, 1958-59.

Nuclear interests: Reprocessing of reactor fuels, enrichment of isotopes.
Address: AB Atomenergi, P.O. Box 9042, Stockholm 9, Sweden.

HAEGERMARK, Bo. Asst., Div. of Nucl. Chem., Roy. Inst. of Technol.
Address: Royal Institute of Technology, Stockholm 70, Sweden.

HAELG, Walter, Dr. phil., Phys. Born 1917. Educ.: Basel Univ. Now Prof.Nucl. Technol., Federal Inst. Technol., Zürich; Formerly at A. G. Brown Boveri u. Cie. Societies: Swiss Phys. Soc.; A.N.S.
Nuclear interests: Reactor physics; Reactor design; Reactor technology.
Address: Villigen CH-5234, Switzerland.

HAENNY, Charles Bertrand, D. ès Sc. (Lausanne), D. ès Sc. (Paris), State Diplom. Born 1906. Educ.: Lausanne and Paris Univs. Prof.-Director, Nucl. Phys. Inst., Faculty of Sci., Lausanne Univ. Societies: Swiss Phys. Soc.; French Phys. Soc.; A.P.S.; Italian Phys. Soc.
Nuclear interests: Neutron reaction -pi physics; High-energy interactions.
Address: 21 Avenue Général-Guisan Pully, Lausanne 1009, Switzerland.

HAESSNER, Frank, Dr. rer. nat. Born 1927. Educ.: Göttingen Univ. Res. Assoc., Göttingen Univ., 1953-57; Exchange Visitor,North Carolina State Coll., and Argonne Nat. Lab., 1958; Res. Assoc., Max-Planck-Inst. für Metallforschung, Stuttgart, and Dozent, Stuttgart T.H., 1959-. Society: Deutsche Geseilschaft für Metallkunde.
Nuclear interest: Basic and applied research on reactor materials.
Address: 37 Traubergstrasse, Stuttgart, Germany.

HÄFELE, Wolf, Dr. rer. nat. Born 1927. Educ.: Munich and Göttingen Univs. Head, Fast Breeder Reactor Project, Karlsruhe, and Director, Inst. Appl. Phys., Kernforschungszentrum Karlsruhe, 1956-. Society: Fellow, A.N.S.
Nuclear interests: Reactor theory in general; Basic reactor design; Inherent safety problems; Fuel cycle; Physics of fast critical facilities and fast experimental reactors.
Address: Kernforschungszentrum Karlsruhe, Leopoldshafen bei Karlsruhe, Germany.

HAFERKAMP, Wilhelm. Born 1923. Member, Federal Exec., and Head, Economic Policy Dept., German Trade Unions Federation; Mitglied, Fachkommission V, Federal Ministry for Sci. Res.; Member, Commission of the European Communities, 1967-; Member, Economic and Social Com., Euratom, -1968.
Address: European Communities Commission, 23 avenue de la Joyeuse Entree, Brussels 4, Belgium.

HAFEZ, Mostafa Mahmoud, B.Sc. (1st class Hons.), Ph.D. (Chem.). Born 1909. Educ.: Higher Training Coll., Sheffield and London Univs. Sec. Gen., Nat. Res. Centre, 1950-61; Sec. of State,Ministry of Sci. Res., 1961-65; Vice Pres., Supreme Council, of Sci. Res., 1965-. Book: Plastics (in Arabic). Society: Egyptian Chem. Soc.
Nuclear interest: Atoms for peace.
Address: 3 Hasba Street, Heliopolis, Cairo, Egypt, United Arab Republic.

HAFSTAD, Lawrence R., B.S., Ph.D. Born 1904. Educ.: Minnesota and Johns Hopkins Univs. Director, Reactor Development Div., U.S.A.E.C., 1949-55; Director, Atomic Energy Div., Chase Manhattan Bank, 1955; Vice-Pres., Gen. Motors Corp.,1955-; Director, Atomic Power Development Assocs.; formerly Trustee, Power Reactor Development Corp. Member, Chairman (1964-), Gen. Advisory Com., U.S.A.E.C. Societies: Atomic Ind. Forum; Fellow, A.N.S.; A.P.S.; A.I.A.A.; Geophysical Union; I.E.E.E.; Nat. Acad. of Eng.
Address: 191 Marblehead Drive, Bloomfield Hills, Michigan, U.S.A.

HAFTKE, Jozef Jerzy, B.Sc. (Eng.). Born 1924. Educ.: London Univ. Design Eng., Chief Stress Analyst, Chief Special Design Eng., Process Plants Div., Foster Wheeler Ltd., 1950-55; Design Eng., Eng. i/c Pressure Vessel Design (1955-61), Chief Contract Eng. (1962-68), Chief Eng. (1968-), Atomic Energy Dept., Babcock and Wilcox (Operations), Ltd. Societies: A.M.I.Mech.E.; B.N.E.S.
Nuclear interests: Boiler and auxiliary plant, pressure vessel, materials design and contracting.
Address: Babcock and Wilcox (Operations) Limited, 209 Euston Road, London, N.W.1, England.

HAGA, Phillip Blair, B.S. (Elec. Eng.), Master in Letters in Maths. Born 1930. Educ.: Virginia Polytech. Inst., Pittsburgh Univ. and O.R.S.O.R.T. Manager, Reactor and Steam Systems Eng., Westinghouse Elec. Corp., 1956-. Society: A.N.S.
Nuclear interests: Plant design and development; Management.
Address: Atomic Power Division, Westinghouse Electric Corporation, P.O. Box 355, Pittsburgh 30, Pennsylvania, U.S.A.

HAGART, Richard Bein, Hon. LL.D.(Witwatersrand). Born 1894. Pres., Transvaal and Orange Free State Chamber of Mines, 1950-51. Joint Deputy Chairman, Anglo-American Corp.of South Africa, Ltd.; Director, Nat. Finance Corp. of South Africa, and in many other companies. Papers: Nat. Aspects of the Uranium Industry (Annual Meeting of South African Inst. of Mining and Metallurgy, 1957); co-author, Uranium Production in the Union of South Africa (4th Annual Conference of Atomic Ind. Forum, 1957).
Address: 43 Seventh Street, Lower Houghton, Johannesburg, South Africa.

HAGEBØ, Einar, Cand. real. Born 1933. Educ.: Oslo Univ. Res. Asst., Chem. Dept., Oslo Univ., 1959-60; Res. Asst., Gustaf Werner Inst. Nucl. Chem., Uppsala Univ. 1960-64; Res. Assoc. (1964), Staff Member (1965-), C.E.R.N.; Nucl. Chem. Div., Oslo Univ. Society: Norwegian Chem. Soc. Nuclear interest: Nuclear chemistry. Address: Oslo University, Nuclear Chemistry Division, Blindern, Oslo-3, Norway.

HAGEMAN, Roy C., B.S. (Mech. Eng.), M.E. Born 1904. Educ.: Denver and Illinois Univs. Tech. Adviser to Manager, Chicago (1950-56), Special Com. to Examine Organisation Management and Administrative Practices and to Recommend Improvements (1952-53), Regional Director (1956-), U.S.A.E.C. Member, Atomic Ind. Forum, Inc. Society: A.N.S. Nuclear interest: Direction of programmes for radiation, nuclear and reactor safety. Address: Division of Compliance, U.S. Atomic Energy Commission, Suite 410, Oakbrook Professional Building, Oak Brook, Illinois, U.S.A.

HAGEMANN, French T., Ph.D. Born 1909. Educ.: Washington Univ. Sen. Sci. and Assoc. Div. Director, Chem. Div., Argonne Nat. Lab. Society: A.C.S. Nuclear interests: Nuclear chemistry; Radiocarbon applications. Address: Argonne National Laboratory, 9700 South Cass Avenue, Argonne, Illinois, U.S.A.

HAGEN, Jack I., M.S. (Phys. and Maths.). Born 1914. Educ.: Oregon State and Johns Hopkins Univs. Res. Dept., Anaconda Copper Mining Co., 1950-51; Res. Staff Assoc., Johns Hopkins Univ., 1951-53; Sen. Sci., Westinghouse Elec. Corp., 1953-58; Assoc. Phys., Argonne Nat. Lab., 1958-65; Assoc. Prof. Elec. Eng., Idaho Univ., 1965-. Societies: A.P.S.; A.N.S. Nuclear interests: Mathematical aspects of reactor design; Theoretical nuclear and solid state physics. Address: R.F.D. 1, Viola, Idaho 83872, U.S.A.

HAGEN, Ulrich, Privat-Dozent, Dr. med., Dr. rer. nat. Born 1925. Educ.: Munich Univ. Wissenschaftlicher Assistent, Heiligenberg Inst., Heiligenberg (Baden), 1953-60; Wissenschaftlicher Assistent, Radiologisches Inst., Universität Freiburg i. Br., 1960-; Dozent, Universität Freiburg i. Br., 1962-; Abt.leiter, Kernforschungszentrum Karlsruhe, Germany, 1965-. Books: Co-author, Einfluss der ionisierenden und der UV-Strahlung auf die Atmung, Handbuch der Pflanzenphysiologie (Encyclopedia of Plant Physiology), Bd XII, (Berlin, Göttingen, Heidelberg, Springer-Verlag, 1960); Biochemie der biologischen Strahlenwirkung in Ergebnisse der Med. Strahlenforschung. Neue Folge, Bd 1 (George Thieme Verlag, 1964). Societies: Assoc. des Radiobiologistes des pays de l'Euratom; Deutsche Gesellschaft für Biophysik. Nuclear interests: Radiobiology: Biochemistry of irradiated cells; Radiosensitivity of deoxyribonucleic acid; Mechanism of biological protective substances. Address: Institut für Strahlenbiologie, Kernforschungszentrum Karlsruhe, Postfach 947, Karlsruhe 75, Germany.

HAGENDOORN, P. J. Jr. Managing Director, Philips N.V. Member, Board of Governors, Reactor Centrum Nederland. Address: Philips N.V., Eindhoven, Netherlands.

HAGGERTY, Wilburt E. Project Eng., Advanced Technol. Div., Atlantic Res. Corp. Address: Advanced Technology Division, Atlantic Research Corporation, Henry G. Shirley Memorial Highway at Edsall Road, Alexandria, Virginia, U.S.A.

HAGGLUND, Robert A. Formerly Sales Manager, now Gen. Manager (1965-), Rad. Instrument Devel. Labs. Div., Nucl.-Chicago Corp. Address: Nuclear-Chicago Corp., 333 East Howard Avenue, Des Plaines, Illinois, U.S.A.

HAGHIRI, Faz, B.S., M.S., Ph.D. Born 1930. Educ.: Nebraska Univ. Assoc. Prof., Ohio Agricultural Res. and Development Centre, and Ohio State Univ., 1959-. Society: American Soc. Agronomy. Nuclear interest: Movement of radioisotopes in soils and their translocation in plants. Address: Agronomy Department, Ohio Agricultural Research and Development Centre, Wooster, Ohio, U.S.A.

HAGINOYA, Tohru, B.E. Born 1924. Educ.: Tokyo Univ. Bureau of Mines, 1950-55; Japan A.E.C., 1956-60; Atomic Energy Attache, Japanese Embassy, Washington, D.C., 1961-64; Chief, Nucl. Fuel Div., Japan A.E.C., 1965-. Member, Steering Com., Japan Nucl. Soc. Society: A.N.S. Nuclear interests: Metallurgy; Fuel management; Accountability; International safeguards. Address: 3-2-2 Kasumigaseki, Chiyoda-ku, Tokyo, Japan.

HAGLER, Howard. Vice Pres. and Gen. Manager, Systems Eng. Div., Hittman Assocs., Inc., 1967-. Address: Hittman Associates Incorporated, Systems Engineering Division, 9190 Red Branch Road, P.O. Box 810, Columbia, Maryland 21043, U.S.A.

HAGLUND, Wilhelm. Gen. Manager, Sandvik Steel Works. Address: Sandvik Steel Works, Sandviken, Sweden.

HAGMAIER, Heinrich, Dr. jur. Member, Fachkommission I-Kernenergierecht, and Arbeitskreis I/1–Haftung und Versicherung der Deutschen Atomkommission; Stellv.

Vorsitzender, Deutsche Kernreaktor - Versicherungsgemeinschaft.
Address: 28 Königinstrasse, Munich 22, Germany.

HAGUE, John. Member, Advisory Com. for Standard Reference Materials and Methods of Measurement, U.S.A.E.C.
Address: National Bureau of Standards, Department of Commerce, Washington, D.C., U.S.A.

HAHN, Beat, Ph.D. Born 1921. Educ.: Basle Univ. Prof. Phys., Physikalisches Inst., Fribourg Univ. Societies: A.P.S.; Swiss Phys. Soc.
Nuclear interests: Nuclear physics; High energy physics.
Address: Physikalisches Institut, Universität Fribourg, Switzerland.

HAHN, Harold Thomas, Ph.D. (Phys., Inorg. Chem. Texas), B.S. (Ch. E., Columbia). Born 1924. Educ.: Texas and Columbia Univs. Graduate student, res. and teaching fellow, Texas Univ., 1950-53; Sen. Sci., Chem. Res., G.E.C., Hanford Lab. Operation, 1953-58; Sect. Chief, Chem. Res., Phillips Petroleum Co., Atomic Energy Div., 1958-. Society: A.C.S.
Nuclear interests: Research and development supervision; Chemical reprocessing of reactor fuel elements; Physical-inorganic chemistry including aqueous and pyrochemical dissolution processes, solvent extraction, ion exchange, isotope separation, transport processes, electrochemistry, radiation stability.
Address: 1920 Sequoia Drive, Idaho Falls, Idaho, U.S.A.

HAHN, Harro, Dr. rer. nat. (Aachen), Diplom-Physiker (Göttingen). Born 1932. Educ.: Göttingen, Vancouver (Canada) and Aachen Univs. Sci. Collaborator, Inst. für Reaktorwerkstoffe der Kernforschungsanlage Jülich, Germany; Privat doz., Aachen Inst. Technol.
Nuclear interests: Study of thermal motions in solids and liquids and of phase transitions by X-ray and neutron scattering (theory).
Address: Institut für Reaktorwerkstoffe, Kernforschungsanlage Jülich, Jülich, Germany.

HAHN, Richard B., B.S., M.S., Ph.D. Born 1913. Educ.: Wayne State and Michigan Univs. Res. Chem., O.R.N.L., 1950-51; Assoc. Prof. (1951-57), Prof. Chem. (1957-), Wayne State Univ. Books: Semi Micro Qualitative Analysis (1955); Inorganic Qualitative Analysis (1963). Societies: A.C.S.; Assoc. Analytical Chem.; American Public Health Assoc. (Com. on Radiochem.).
Nuclear interests: Radiochemistry; Radiochemical analysis; Determination of low-levels of radioactive substances in water, sewage, and foods; Analytical chemistry of zirconium and hafnium.
Address: Chemistry Department, Wayne State University, Detroit 2, Michigan, U.S.A.

HAHN, Thomas Marshall, Jr., B.S. (Phys., Kentucky), Ph.D. (Phys., M.I.T.). Born 1926. Educ.: Kentucky Univ. and M.I.T. Assoc. Prof. Phys. (1950-52), Prof. Phys., Director of Graduate Study in Phys., Director, Nucl. Accelerator Labs. (1952-54), Kentucky Univ.; Prof. and Head, Dept. of Phys., Virginia Polytech. Inst., 1954-59; Dean of Arts and Sci., Kansas State Univ., 1959-62; Pres., Virginia Polytech. Inst. 1962-. Societies: A.P.S.; A.A.P.T.
Nuclear interests: Low energy and accelerator nuclear physics; Energy levels of light nuclei; Reactor analysis, pulsed neutron systems, administration.
Address: Virginia Polytechnic Institute, Blacksburg, Virginia, U.S.A.

HAHNEL, Helmut, Dipl. Chem. At Berkefeld-Filter G.m.b.H.
Address: Berkefeld-Filter G.m.b.H., Postfach 12, 31 Celle, Germany.

HAIDER, Georg Heinrich Johannes Ernst. Born 1902. Formerly Vice Pres. and Director, AVR eV.; Pres., Stadtwerke Bremen, A.G.
Address: Stadtwerke Bremen A.G., Bremen, Germany.

HAIGH, Clement Percy, B.Sc. (Hons.), Ph.D. Born 1920. Educ.: Leeds and London Univs. Principal Phys., Barrow Hospital, Barrow Gurney, Bristol, 1949-56; Head, Phys. Sect., Nucl. Development Branch, London, (1956-59), Director, Berkeley Nucl. Labs., Glos. (1959-), C.E.G.B. Society: F. Inst. P.
Address: Berkeley Nuclear Laboratories, Central Electricity Generating Board, Berkeley, Gloucestershire, England.

HAILLEZ, Pierre. Director, Belgonucléaire.
Address: Belgonucléaire, 35 rue des Colonies, Brussels 1, Belgium.

HAIMANN, Otto. Sen. Lecturer, Dept. of Atomic Phys., Eotvos Univ., Budapest.
Address: Eotvos University, 5-7 Pushkin-u., Budapest 8, Hungary.

HAINE, Michael Edward, D.Sc. Born 1914. Educ.: Bristol Univ. Deputy Director, A.E.I. Central Res. Lab., Rugby. Book: The Electron Microscope (E. and F. N. Spon, 1961). Societies: M.I.E.E.; F. Inst. P.
Nuclear interests: Electron optics; Solid state nuclear particle detectors; Special devices and tubes for use in nuclear instrumentation.
Address: Central Research Laboratory, Associated Electrical Industries Ltd., Rugby, Warwickshire, England.

HAINES, E. B. Formerly Manager, Rad. Eng. Operation, now Manager, Relations and Auxiliary Operations, K.A.P.L.
Address: Knolls Atomic Power Laboratory, Schenectady, New York, U.S.A.

HAINES, John F., A.B., M.S. Born 1910. Educ.: Oberlin Coll. and M.I.T. Pres., Chief Eng., Haines Designed Products Corp., 1949-52; Consulting Eng. in private practice, 1949-55; Vice-Pres., Chief Eng., McCauley Ind. Corp., 1952-55; Chief Development Eng., Nucl. Power Eng. Dept., Alco Products Inc., 1955-. Societies: A.S.M.E.; A.N.S.
Nuclear interests: Management of design and development functions for nuclear reactors and associated control and power conversion equipment; includes mechanical and electrical engineering, development testing, metallurgy, chemical engineering and radiochemistry.
Address: 7400 Lakeview Drive, Apt. 304, Bethesda, Maryland 20034, U.S.A.

HAINSKI, Zvonimir, Ph.D. (D. en Sc.). Born 1922. Educ.: Louvain Univ. Centre Nat. de Recherches Métallurgiques (C.N.R.M.), Belgium, 1950-56; Sté. d'Etude de Recherches et d'Application pour l'Industrie (S.E.R.A.I.), Brussels, 1956-61; Euratom, Centre Commun de Recherches (C.C.R.), Ispra, Italy, 1961-. Editor for Spectroscopy, Sci. Journal, Energia Nucleare (C.I.S.E., Milan). Society: Groupement pour l'Avancement des Méthodes Spectrographiques (G.A.M.S.), Paris.
Nuclear interest: Emission spectrum analysis of atomic materials.
Address: Spectrographic Laboratories, C.C.R., Euratom, Ispra, Italy.

HAINZELIN, Jean Louis, Diplôme d'Etudes Supérieures (Mathématiques), Ing. civil du Génie Maritime. Born 1922. Educ.: Paris Univ. Formerly Ing. en Chef, Chef du Département Etudes Industrielles, Groupement Atomique Alsacienne Atlantique. Book: EL.3. (Presses Univ. de France, Paris, 1958). Societies: Assoc. Technique Maritime et Aéronautique, Paris; Sté. des Ingénieurs Soudeurs, Paris.
Nuclear interest: Reactor design.
Address: 15 quai Louis Blériot, Paris 16, France.

HAIRE, James Curtis, Jr. B.S. (Phys., St. Mary's Coll., California), M.S. (Phys., Notre Dame). Born 1927. Graduate teaching fellow, Dept. of Phys., Notre Dame Univ., 1950-53; At Kennecott Copper Corp., Res . Centre, Salt Lake City, 1953-56. Phillips Petroleum Co., Atomic Energy Div., Spert Project, Idaho Falls, 1956-. Chief, Spert Nucl. Test Sect. Member, Local Sections Com., A.N.S. Society: A.N.S.
Nuclear interests: Reactor safety investigations; Reactor kinetics; Experimental reactor testing.
Address: 586 Safstrom Place, Idaho Falls, Idaho, U.S.A.

HAIRE, T. P. Principal Inspector, Nucl. Health and Safety Dept., C.E.G.B.
Address: Nuclear Health and Safety Department, Central Electricity Generating Board, Laud House, 20 Newgate Street, London, E.C.1, England.

HAIRR, Graham M. Rad. Phys., Div. of Radiol. and Occupational Health, Florida State Board of Health.
Address: Florida State Board of Health, Division of Radiological and Occupational Health, Post Office Box 210, Jacksonville, Florida 32201, U.S.A.

HAISSINSKI, Jacques, M.A. (Stanford), Agrégé de l'Univ. (Paris). Born 1935. Educ.: E.N.S., Paris and Stanford Univ. Maitre de Conf., Lab. de l'Accélérateur Linéaire, Faculté des Sci. d'Orsay.
Nuclear interest: Storage rings in high energy physics.
Address: Laboratoire de l'Accélérateur Linéaire, Faculté des Sciences d'Orsay, (Essonne), France.

HAISSINSKY, Moise, D. ès. Sc., D. en Chimie. Born 1898. Educ.: Rome. Sci. Director, Curie Lab., Paris. Hon. Member, Acad. Sci. and Letters of Genoa. Book: La Chimie Nucléaire et ses Applications (Paris, Masson, 1957, English, Russian and other translations). Societies: Sté. de Physique; Faraday Soc.; A.A.C.S.; A.P.S.
Nuclear interests: Radiochemistry; Radiation chemistry; Indicators.
Address: Laboratoire Curie, Institut du Radium, 11 rue Pierre Curie, Paris 5, France.

HAJDU, Elemer. Member, Editorial Board, Energia es Atomtechnika.
Address: Muszaki Konyvkiado, Bajcsy-Zslinszky ut 22, Budapest 5, Hungary.

HAJDUKOVIC, Srdan, M.D., Ph.D. Born 1922.* Educ.: Belgrade and Liège Univs. Head, Radiobiol. Lab., Inst. Nucl. Sci. Boris Kidrich. Sci. Com. J. Nucl. Haematology. Societies: Soc. Nucl. Haematology; Roy. Soc. Medicine, London; N.Y. Acad. Sci.
Nuclear interests: Radiobiology; Nuclear haematology.
Address: Boris Kidrich Institute of Nuclear Sciences, P.O. Box 522, Belgrade, Yugoslavia.

HAJJAR, Aly, Eng. Sec., Lebanese A.E.C.
Address: Lebanese Atomic Energy Commission, Ministry of Public Works, Beirut, Lebanon.

HAKANSSON, Hans., Lic. Sci. Born 1922. Educ.: Uppsala Univ. Adviser, Ministry of Communications, Sweden, 1952-56; Sec., Atomic Energy Board, Sweden, 1956-. Head, Industry Div., Ministry of Commerce.
Address: Handelsdepartementet, Stockholm 2, Sweden.

HAKE, G. Canadian Member, Com. on Reactor Safety Technol., O.E.C.D., E.N.E.A.
Address: O.E.C.D. European Nuclear Energy Agency, 38 bouelvard Suchet, Paris 16, France.

HAKKARAINEN, Urho, Dipl. Eng. Born 1914. Educ.: Finland Inst. Technol. Sen. Officer, Imatran Voima Oy 1948-57; Chief of Power Bureau, Ministry of Commerce and Ind., 1957-. Vice-Chairman, Nat. Com., World Power Conference, 1961-; Energy Sect., State Advisory Ind. Commission 1957-; Nordic Power Contact Commission "Nordel", 1963-; Various Govt. coms.
Nuclear interest: Power economics.
Address: Ministry of Commerce and Industry, 10 Aleksanterinkatu, Helsinki, Finland.

HALAS, Don Richard de. See de HALAS, Don. Richard.

HALAWANI, Ahmed Abdel Salam EL. See EL HALAWANI, Ahmed Abdel Salam.

HALBERSTADT, Johan, Ph.D. (Chem.). Born 1916. Educ.: Utrecht Univ. Sen. Radiochem., Inst. for Nucl. Res., Isotope Div., Chem.-Pharmaceutical Industry Philips-Duphar, Amsterdam, 1950-59; Sen. Officer, I.A.E.A., Vienna, 1959-. Society: Roy.Netherlands Chem. Soc.
Nuclear interests: Production of isotopes and labelled organic compounds; Organic radiochemistry; Use of radioisotopes and radiation in the control of agricultural pests; Chemical effects of radiation.
Address: I.A.E.A., 11-13 Kärntnerring, Vienna 1, Austria.

HALBERT, Melvyn Leonard, A.B., Ph.D. Born 1929. Educ.: Cornell and Rochester Univs. Phys., O.R.N.L. Society: A.P.S.
Nuclear interest: Nuclear physics.
Address: Oak Ridge National Laboratory, Oak Ridge, Tennessee, U.S.A.

HALBRONN, G. Etablissements Neyrpic; Directeur du Departement des Applications Industrielles, SOGREAH, Soc. Grenobloise d'Etudes et d'Applications Hydrauliques.
Address: B.P. 145, Grenoble (Isère), France.

HALDI, John, Dr. Nucl. Res., Physiology Dept., School of Medicine, Emory Univ.
Address: School of Medicine, Emory University, Atlanta 22, Georgia, U.S.A.

HALE, William E. Geological Survey, U.S. Dept. of the Interior.
Address: U.S. Geological Survey, P.O. Box 4217, Room 279 Geology Building, University of New Mexico Campus, Albuquerque, New Mexico 87106, U.S.A.

HALE LULL, S. Vice Pres. in Charge of Eng., Western Massachusetts Elec. Co.
Address: Western Massachusetts Electric Co., 174 Brush Hill Avenue, West Springfield, Massachusetts 01089, U.S.A.

HALES, Ian Barnewall, M.B., B.S., M.R.C.P. Born 1926. Educ.: Sydney Univ. Res. Fellow, Endocrinology, now i/c, Thyroid Investigation Clinic, Unit of Clinical Investigation, Roy. North Shore Hospital, Sydney. Societies: Endocrine Soc. Australia; Endocrine Soc. America.
Nuclear interests: Medical application of isotopes.
Address: Unit of Clinical Investigation, Royal North Shore Hospital, Crows Nest, New South Wales, Australia.

HALES, Max Price, B.S. (Geology). Born 1924. Educ.: Idaho and Idaho State Univs. Exploration Geologist, J. R. Simplot Co., Boise, Idaho, 1951-52; Operations Foreman, American Cyanamid Co. (1953), Production Eng., Operations Foreman, Phillips Petroleum Co. (1954), Operations Shift Supt., Phillips Petroleum Co. (1955-66), Operations Shift Supt., Idaho Nucl. Corp. (1966-67), Chief, Production Eng., Idaho Nucl. Corp. (1967-), Chemical Process Plant N.R.T.S.
Nuclear interests: Operation and design of high enriched uranium reprocessing facilities; Production of kilo-curie quantities radioisotopes; Operation and design of fluidised bed waste calcination facilities; Nuclear plant decontamination; Radioactive waste management.
Address: 365 Holbrook Drive, Idaho Falls, Idaho, U.S.A.

HALEVY, Elkana, Ph.D. Born 1929. Educ.: California (Berkeley) and Hebrew (Jerusalem) Univs. At Weizmann Inst. Sci., 1956. Societies: Inst. Soil Sci. Soc.; Israel Chem. Soc.
Nuclear interests: Radiochemistry waste disposal; Application of radioisotopes.
Address: Weizmann Institute of Science, P.O. Box 26, Rehovoth, Israel.

HALEY, Sam. Member, Standing Com. on Nucl. Energy in the Elec. Industry, Nat. Assoc. of Railroad and Utilities Commissioners, 1967-.
Address: National Association of Railroad and Utilities Commissioners, Interstate Commerce Commission Building, P.O. Box 684, Washington, D.C. 20044, U.S.A.

HALEY, Thomas John, B.S. (Pharmacy, South California), M.S. (Chem., South California), Ph.D. (Pharmacology, Florida). Born 1913. Educ.: South California and Florida Univs. Chief, Pharmacology and Toxicology Div., Biophys. and Nucl. Medicine Dept., Nucl. Medicine and Rad. Biol., U.C.L.A. School of Medicine, 1947-66; Assoc. Clinical Prof. Medicine (Ind.) U.C.L.A., 1953-59; Prof. Pharmacology, Hawaii Univ., 1966-. Sec., Pharmacology and New Drug Sect., Pan American Medical Assoc. Societies: Fellow, A.A.A.S.; Rad. Res. Soc.; Fellow, Walter Reed Soc.; N.Y. Acad. Sci.; A.I.H.A.; Biometric Soc.; Pharmacology Soc.; Internat. Assoc. Forensic Toxicologists; European Soc. for Study of Drug Toxicity; Charter member, Soc. Toxicology; Pan American Medical Assoc.

Nuclear interests: Toxicology of materials used in atomic reactors; Effects of radiation on the nervous system.
Address: 3518 Alohea Avenue, Honolulu, Hawaii, U.S.A.

HALG, Walter. See HAELG, Walter.

HALKERSTON, Ian, Dr. Member, Radioisotope Com., Worcester Foundation for Exptl. Biol., Inc.
Address: Worcester Foundation for Experimental Biology, Inc., 222 Maple Avenue, Shrewsbury, Massachusetts, U.S.A.

HALKETT, James Alexander Elder, B.A., M.S., Ph.D. Born 1920. Educ.: Wooster Coll., M.I.T. and Boston Univ. Sen. Biol., Field of Radioisotopes, Boston Veterans Administration Hospital, 1952-. Asst. Medicine, Medicine Dept., Boston Univ.
Nuclear interests: Application of radiation and radioisotopes to medical research.
Address: Parks Drive, Sherborn, Mass., U.S.A.

HALL, Sir Arnold Alexander, M.A. Born 1915. Educ.: Cambridge Univ. Head, Dept. of Aeronautics, Imperial Coll. of Sci. and Technol., London Univ., 1945-51; Director, R.A.E. Farnborough, 1951-55; Director, Hawker Siddeley Group, Ltd., 1955-58; Managing Director, Bristol Siddeley Engines, Ltd., 1958-63; Vice-Chairman and Managing Director (1963-67), Chairman and Managing Director (1967-), Hawker Siddeley Group Ltd. Council, Internat. Aeronautical Sci.; Chairman, Board of Sci. and Ind. Studies for Coll. of Technol.; British Nat. Com. on Space Res. Societies: F.R.S.; Fellow, Roy. Aeronautical Soc.; Hon. A.C.G.I.
Nuclear interest: Management.
Address: Hawker Siddeley Group Limited, 18 St. James's Square, London, S.W.1, England.

HALL, Arthur, E., B.S. (Metal. Eng., Magna Cum Laude). Born 1938. Educ.: Washington Univ., Seattle. Reactor eng., Naval Reactors Branch, U.S.A.E.C.
Nuclear interest: Metallurgy of nuclear reactor materials, with an emphasis on solid state interactions.
Address: 1402 N. Beauregard Street, Alexandria, Virginia, U.S.A.

HALL, David Ballou, B.S. (Rutgers), M.S. (Chicago), Ph.D. (Chicago). Born 1915. Educ.: Rutgers and Chicago Univs. P-5 Group Leader (1945-51), Alt. Div. Leader, W-DO (1951-56), K-Div. Leader (1956-), Los Alamos Sci. Lab. Member, Panel of Experts to Provide Boards on Atomic Safety and Licensing, U.S.A.E.C. Societies: Fellow, A.P.S.; Fellow, A.N.S.
Address: Los Alamos Scientific Laboratory, P.O. Box 1663, Los Alamos, New Mexico, U.S.A.

HALL, Duane Charles, B.S., M.S. Born 1940. Educ.: Wisconsin and Vanderbilt Univs. Manager, Radiol. Phys. Dept., Rural Cooperative Power Assoc., Elk River, 1964-. Society: H.P.S.
Nuclear interest: Supervision of the health physics, waste disposal, radiochemistry and conventional chemistry programmes required for the operation of a nuclear power plant.
Address: Rural Cooperative Power Association, Elk River, Minnesota 55330, U.S.A.

HALL, E. O., M.Sc. (N.Z.), Ph.D. (Cantab.), F. Inst. P., F.I.M., M. Aus. I. M. M. Assoc. Prof., Newcastle Univ. Coll., New South Wales Univ.; Councillor, Australian Inst. of Nucl. Sci. and Eng.
Address: New South Wales University, Box 1 Post Office, Kensington, New South Wales, Australia.

HALL, G. VAN. See VAN HALL, G.

HALL, Geoffrey Ronald, B.Sc., F.R.I.C., M. Inst. F. Born 1928. Educ.: Manchester Univ. Formerly P.S.O., Chem. Div., Harwell and Colombo Plan Expert loaned to Indian A.E.C., Bombay. Now Prof. Nucl. Technol., Imperial Coll., London. Educ. Com., Inst. Fuel; Board member and Hon. Editor of J., B.N.E.S. Societies: Chem. Soc.; Roy. Inst. Chem.; B.N.E.S.; Assoc. Rad. Res.; Soc. Radiol. Protection; M. Inst. F.
Nuclear interests: Research in nuclear fuel reprocessing; Breakdown of ion exchange resins by heat and ionising radiations; Effect on mass transfer performance; Plutonium ion exchange; Radiation chemistry and engineering; Radiation chemistry and engineering Purification of water for nuclear reactors and boilers.
Address: Chemical Engineering and Chemical Technology Department, Imperial College, London, S.W.7, England.

HALL, Jane H., Dr.; Formerly Sec., now Member, (1966-), Gen. Advisory Com., and Member, Advisory Com.on Nucl. Materials Safeguards (1967-), U.S.A.E.C.; Formerly Asst. Director, now Assoc. Director and Phys., Los Alamos Sci. Lab. Deleg. to 2nd I.C.P.U.A.E., Geneva, Sept. 1958.
Address: Los Alamos Scientific Laboratory, P.O. Box 1663, Los Alamos, New Mexico, U.S.A.

HALL, John Allen, A.M., Ph.D. Born 1914. Educ.: Northwestern and Harvard Univs. Director, Office of Special Projects (1948-55), Director, Div. of Internat. Affairs (1955-58), Asst. Gen. Manager for Internat. Activities (1958-61, and 1964-67), U.S.A.E.C.; Deputy Director Gen. i/c Administration, I.A.E.A., Vienna (1961-64, and 1967-).
Address: 7 Hasenauerstrasse, Vienna 1180, Austria.

HALL, Kenneth Lynn, B.A., M.S., Ph.D. Born 1927. Educ.: Reed Coll., California and Michigan Univs. Res. Asst., California Univ., 1949-51; Res. Asst., Michigan Univ., 1951-55; Res. Chem., (1955-63), Sen. Res. Chem. (1964-), California Res. Corp. Society: A.C.S. Nuclear interests: Measurement of cross sections, half lives, branching ratios; Mass assignments; Absolute beta counting; Radiation chemistry of organic reactor coolants; Vacuum technology; Monitoring reactor radiation.
Address: 16 Sussex Court, San Rafael, California, U.S.A.

HALL, Nathan S., B.S., M.S., Ph.D. Born 1913. Educ.: Rhode Island State Coll., Missouri and North Carolina Univs. Assoc. Prof. to Prof. Agronomy, North Carolina State Coll., 1948-57; Soil Sci., Biol. and Medicine Div., U.S.A.E.C., 1954-55; Soil Sci., North Carolina Agricultural Mission to Peru, 1955-57; Director, Tennessee Univ. -U.S.A.E.C. Agricultural Res. Lab., 1957-.
Nuclear interests: Application of radioisotopes in agriculture.
Address: Agricultural Research Laboratory, Tennessee University, Oak Ridge, Tennessee, U.S.A.

HALL, Thomas Albert, B.Sc., Ph.D., D.I.C., A.R.C.S., C. Eng. Born 1916. Educ.: London Univ. Sen. Principal Sci., U.K.A.E.A., Harwell. Societies: A.R.I.C.; A.M.I.Chem.E.
Nuclear interests: Managerial aspects of safety, including nuclear and industrial safety.
Address: Maplewood, Picklers Hill, Abingdon, Berkshire, England.

HALL, Vincent C., Jr., Sc.B. (Eng.). Born 1924. Educ.: Brown Univ. Eng., Nat. Res. Corp., Cambridge, Mass., 1949-55; Eng. (1955-58), Asst. Div. Director, Reactor Eng. Div. (1959-60), Argonne Nat. Lab.; Sen. Eng. Combustion Eng., Inc., Windsor, Conn., 1960-. Society: A.N.S.
Nuclear interests: Nuclear power plant operational experiences which can be applied to the improvement of the design for future plants.
Address: 19 Cadwell Road, Bloomfield, Conn., U.S.A.

HALL, W. R. Formerly Director and Chief Eng., now Managing Director, Nucl. Eng. Dept., E. G. Irwin and Partners Ltd.
Address: E. G. Irwin and Partners Ltd., Kibworth Street, Dorset Road, London S.E.8, England.

HALL, William B., B.S. Born 1916. Educ.: Princeton Univ. Pres., Vitro Minerals and Chemical Co.; Vice Pres., Vitro Corp. of America. Societies: A.I.M.M.E.; Atomic Ind. Forum.
Nuclear interest: Mining and milling of uranium ores.
Address: Vitro Corporation of America, 90 Park Avenue, New York 16, New York, U.S.A.

HALL, William Bateman, B.Sc. (Eng.), M.Sc., M.I.Mech.E. Born 1923. Eng III M.O.S., Dept. of Atomic Energy, Risley, 1946; P.S.O., R. and D. Branch (1952), S.P.S.O., (1956), Deputy Chief Sci. Officer (1959), U.K.A.E.A. Prof. Nucl. Eng., Manchester Univ., 1959-.
Nuclear interests: Heat transfer and fluid mechanics research as applied to reactors; Reactor design.
Address: Manchester University, Manchester 13, England.

HALL, Willian Cornelius. Born 1912. Pres., Arboreal Associates, 1951-; Pres., Chemtree Corp., 1959-. Societies: A.A.A.S.; A.I.B.S.; A.I.M.M.E.; American Inst. of Powder Metal.; American Management Assoc.; American Sect., of Sté. des Ing. Civils de France; A.S.M.; I.E.E.E.; International Shade Tree Conference; N.Y. State Arborist Assoc.; North Eastern Weed Control Conference; American Ordnance Assoc.; Director, N.Y. Metropolitan Sect., A.N.S.
Nuclear interests: Materials for attenuation of nuclear radiations. Effects of nuclear radiations on materials and management.
Address: Chemtree Corp., Central Valley, New York 10917, U.S.A.

HALL, William Franklin, B.S., M.S. (Chem. Eng.). Born 1931. Educ.: Pennsylvania State and Iowa State Univ. Project Eng., Contract Administration, Foster-Wheeler Corp., New York City, 1956-58; Reactor Eng./Reactor Supv., Curtiss-Wright Corp., Quehanna, Pennsylvania, 1958-60; Deputy Director, Western New York Nucl. Res. Centre, Inc., 1960-. Chairman, Subcom. for Honours and Awards, Local Sects. Com., A.N.S. Society: A.N.S.
Nuclear interests: Reactor operation; Experiments involving reactor engineering education; Irradiation damage studies; Gamma and neutron dosimetry; Isotope production; Economics of the power reactor fuel cycle.
Address: Western New York Nuclear Research Centre, Incorporated, Power Drive, Buffalo, New York 14214, U.S.A.

HALL, William J. Nucl. Product Manager, Manning, Maxwell and Moore, Inc.
Address: Manning, Maxwell and Moore, Inc., Stratford, Connecticut, U.S.A.

HALLAMA, Jaakko. Member, A.E.C., Ministry of Commerce and Industry.
Address: Ministry of Commerce and Industry, 4 Hallituskatu, Helsinki, Finland.

HALLBERG, Olle, Med. lic. Born 1917. Educ.: Karolinska Institutet-Karolinska Sjukhuset Radiumhemmet (King Gustaf V's Jubilee Clinic) and Radiopathology Inst., Stockholm. Head Physician, Radiotherapy Dept., Central Hospital, Gävle, 1950-57; Head Physician, Radiotherapy Dept., Regional Hospital, Örebro, 1957-. Societies: Swedish Assoc. for Medical Radiol.; Cancer Assoc. in Stockholm; Northern Radiotherapists Club.

Nuclear interests: Medical use of radioactive isotopes in diagnosis and treatment of thyroid diseases and in cancer therapy.
Address: Radioterapeutiska kliniken, Regionsjukhuset, Örebro, Sweden.

HALLER, P. de. See de HALLER, P.

HALLET, Raymon W., Jr., B.S. (Mech. Eng.), M.S. (Eng.). Born 1920. Educ.: Purdue and California Univs. Director, R. and D., Missiles and Space System Div., Douglas Aircraft Co., 1942-66; Vice Pres., Douglas United Nucl., Inc., 1966-. Societies: A.I.A.A.; A.N.S.; Atomic Ind. Forum; American Astronautical Soc.; Inst. of Navigation.
Nuclear interests: Has responsibility for operating the U.S. Atomic Energy Commission's production reactors and dual purpose reactor at Hanford and guiding associated research and development programmes. Also directs Company's nuclear oriented commercial activities.
Address: Douglas United Nuclear Incorporated, P.O. Box 490, Richland, Washington, U.S.A.

HALLIDAY, John Stephen, B.Sc., Ph.D. Born 1927. Educ.: Reading Univ. Surface Phys. Sect. (1950-57), Electron Phys. Sect. (1957-61), Res. Lab., A.E.I., Aldermaston; Sect. Leader Electron Optics and X-ray Analysis (1962-63), Sect. Leader (1963), Eng. in Charge (1964-67), Manager (1967-), Mass Spectrometry, Sci. Apparatus Dept., A.E.I., Urmston, Manchester. Editorial Board, Proc. Phys. Soc. Book: Chapters on Electron Diffraction and Reflection Electron Microscopy in Techniques for Electron Microscopy, ed. D. Kay (Blackwell Sci. Publications, 1961). Society: F. Inst. P.
Nuclear interests: Electron scattering by atoms and solids; X-ray production; Ionisation processes; Crystallography; Electron and ion optics; Mass spectroscopy; Isotope ratio measurements.
Address: 16 Ferndale Road, Brooklands, Sale, Cheshire, England.

HALLMAN, Theodore M., B.S. (M.E.) (Washington), M.S. (M.E.) (Purdue), Ph.D. (Purdue). Born 1926. Educ.: Olympic Coll., Washington and Purdue Univs. Asst. Chief, Heat Transfer Res. Branch, Nucl. Reactor Res. Div., Lewis Flight Propulsion Lab., Nat. Advisory Com. for Aeronautics, Cleveland, Ohio, 1956-58; Chief, Plum Brook Reactor Facility, N.A.S.A., Sandusky, Ohio, 1958-61; Tech. Asst., Appl. Phys. Sect., Boeing Co., Seattle, Washington, 1961-; Chief, Nucl. Sci. Group, Northrop Ventura, 1962-65; Director, Atomic Phys., Northrop Nortronics A.R.D., 1965-67; Director, Advanced Eng. Technologies Lab., Northrop Corporate Labs., 1967-. Societies: A.N.S.; A.S.M.E.
Nuclear interests: Nuclear rockets; Nuclear electric space propulsion; Test reactor management; Reactor hazards; Heat transfer.
Address: Northrop Corporate Laboratories, Hawthorne, California, U.S.A.

HALLMARK, W. L. Manager, Operations, Willamette Iron and Steel Co.
Address: Willamette Iron and Steel Co., 2800 Northwest Front Avenue, Portland, Oregon 97210, U.S.A.

HALLOWELL, H. Thomas, Jr. Pres., Standard Pressed Steel Co.
Address: Standard Pressed Steel Co., Jenkintown, Pennsylvania, U.S.A.

HALMANN, Martin Mordehai, Ph.D., M.Sc., Born 1925. Educ.: Hebrew Univ., Jerusalem. At Weizmann Inst. Sci., Rehovoth, Israel. Societies: Israel Chem. Soc.; Chem. Soc. (London); A.C.S.
Nuclear interests: Isotope effects on rates of chemical reactions and on infra-red absorption spectra; Chemical effects of nuclear reactions (particularly organic phosphorus compounds.
Address: Isotope Department, Weizmann Institute of Science, Rehovoth, Israel.

HALNAN, Keith Edward, M.A., M.D. (Cantab.), F.F.R., D.M.R.T., M.R.C.P. Born 1920. Educ.: Cambridge Univ. and Univ. Coll. Hospital, London. Sci. staff, M.R.C., Dept. of Clinical Res., Univ. Coll. Hospital Medical School, London, 1954-58; Consultant Radiotherapist, Christie Hospital and Holt Radium Inst., and Hon. Lecturer in Radiotherapy, Manchester Univ., 1960-66; Director, Glasgow Inst. for Radiotherapy, and Hon. Clinical Lecturer, Glasgow Univ., 1966-. Book: Atomic Energy in Medicine (Butterworths, 1957). Societies: Medical Res. Soc.; Faculty of Radiologists; British Inst. Radiol.; Roy. Soc. of Medicine.
Nuclear interest: Medical applications of atomic energy.
Address: Institute for Radiotherapy, Western Infirmary, Glasgow W.1, Scotland.

HALPERIN, Joseph, Ph.D. Born 1923. Educ.: Chicago Univ. Chem., O.R.N.L. Societies: A.A.A.S.; A.C.S.; A.N.S.; A.P.S. Nuclear interests: Neutron cross-section measurements utilising activation techniques (particularly of heavy element and fission product nuclides produced during reactor operation); Nuclear properties and reactions; Nuclear chemistry; Radiochemistry.
Address: Oak Ridge National Laboratory, P.O. Box X, Oak Ridge, Tennessee, U.S.A.

HALPERN, Isaac, Ph.D. (M.I.T.). Born 1923. Educ.: C.C.N.Y. and M.I.T. Prof. Phys., Washington Univ. 1960-. Society: A.P.S.
Nuclear interests: Nuclear physics; Nuclear fission; Nuclear reactions.
Address: Physics Department, Washington University, Seattle, Washington 98105, U.S.A.

HALPERN, Julius, B.S. (Carnegie Inst. Technol.), M.S. (Carnegie Inst. Technol.), Sc.D. Born 1912. Prof. Phys., Pennsylvania Univ., 1952-. Societies: Fellow, A.P.S.; A.A.U.P.
Nuclear interests: Nuclear physics; Electronics; Low temperature physics; Scattering of slow neutron by liquid ortho- and parahydrogen.
Address: Pennsylvania University, Physics Department, Philadelphia, Pennsylvania 19104, U.S.A.

HALSBURY, Rt. Hon. John Anthony Hardinge, Earl of. B.Sc., C.E., D.Tech.(Hon.), F.R.I.C., F. Inst. P. Pres., Inst. of Nucl. Eng., 1963-65.
Address: Flat 2, 17 Stanhope Terrace, London, W.2, England.

HALSEY, George Hancock, B.S., M.S. Born 1914. Educ.: Oklahoma State Univ. Advanced Development Eng., Large Power Transformer, Eng. Dept., G.E.C., 1947-55; Manager, Submarine Advanced Reactor, Thermal and Hydraulic Design (1955-58), Manager, Natural Circulation Reactor Thermal and Hydraulic Design (1959-61), Consultant, Heat Transfer (1962 and 1967), K.A.P.L. Societies: A.S.M.E.; A.N.S.
Nuclear interest: Reactor mechanical, thermal and nuclear design.
Address: Knolls Atomic Power Laboratory, General Electric Company, Schenectady, New York, U.S.A.

HALSTEAD, T. H. Sales Manager, Nucl. Res., Honeywell Controls Ltd.
Address: Honeywell Controls Ltd., Ruislip Road East, Greenford, Middlesex, England.

HALTER, Josef, Phys. Diploma, Dr. sc. nat. Born 1918. Educ.: Swiss Federal Inst. Technol. Postdoctorate fellowship N.R.C., Ottawa, 1950-52; Asst., E.T.H. Zürich, 1952-57; Phys., Swiss Federal Commission for Radioactivity, Fribourg, 1957-. Societies: A.P.S.; Swiss Phys. Soc.
Nuclear interests: Nuclear physics (especially beta- and gamma-spectroscopy); Health physics.
Address: Swiss Federal Commission for Radioactivity, Pérolles, Fribourg, Switzerland.

HALTER, Samuel, Dr. in Medicine (Graduate in Hygiene). Born 1916. Educ.: Brussels Free Ministry of Health, Brussels; Prof., Faculty of Medicine, Brussels Free Univ. Sec., Conseil Supérieur d'Hygiène, Belgium; Expert in medical care of W.H.O., Geneva; Exper in hospital construction of Internat. Hospital Federation; Expert, group of 12, C.E.E.A. (Health protection and basic standards); Expert and chairman, Group for Health and Safety, E.N.E.A., O.E.C.D., Paris; Administrator, C.E.N. (Mol, Brussels).
Nuclear interests: Health protection and safety; Radiobiology.
Address: Quartier Vésale, Cité Administrative, Brussels, Belgium.

HALTRICH, Stefan, Eng. (Electronics). Born 1931. Educ.: Bucharest Polytech. Inst. Sen. Eng. Inst. for Atomic Phys., Bucharest, 1961-.
Nuclear interests: Circular electron accelerators; Magnetic circuits and measurements; Control engineering.
Address: Institute for Atomic Physics, P.O.B. 35, Bucharest, Roumania.

HALVAS GUERRERO, Jorge, Chemist, Chem. Eng., B.Sc., M.Sc. Born 1927. Educ.: Mexico Nat. Univ. Prof. Phys., Chem. and Maths., Mexico Nat. Univ., 1954-57; Director, Phys. Chem. seminars, Mexico Nat. Univ., 1958-61; Prof. Rad. Phys., Doctorate Div., School of Medicine, Mexico Nat. Univ., 1960-; Prof. Nucl. Eng., Nat. School of Chem. Sci., Mexico Nat. Univ., 1964-. Head, Phys. Dept., Oncological Hospital, Mexican Social Security Inst., 1956-62; Tech. Advisor on Rad. Phys., (1958-59), Director, Radiol. Protection Programme (1960-), Nat. Nucl. Energy Commission; Member (Voter), Consulting Council, Mexican Phys. Soc.; Treas., Mexican Soc. for Radioisotope Studies; Pres., Mexican Hospital Phys. Assoc.; Member, Editorial Board, H.P.S. Book: Co-author, Selected Topics on Rad. Dosimetry (I.A.E.A., 1961); co-author, Radiol. Health and Safety in Mining and Milling of Nucl. Materials (I.A.E.A., 1964). Societies: Mexican Phys. Soc.; Mexican Soc. for Radioisotope Studies; Mexican Hospital Phys. Assoc.; Mexican Seminar on Sci. Philosophical Studies; H.P.S.
Nuclear interests: Radiation physics; Physical aspects on radiation dosimetry; Radiological physics; Radioisotope research in medicine; Radiation biophysics; Radioactive contamination research; Radiation protection.
Address: National Nuclear Energy Commission, Apartado Postal 30190, Suc. 27, Mexico City, Mexico.

HALVERSON, J. W. Manager, Finance and Administration, Computer Sci. Corp., Northwest Operations, Hanford Facilities, U.S.A.E.C.
Address: U.S.A.E.C., Computer Sciences Corporation, Northwest Operations, Richland, Washington, U.S.A.

HALVORSEN, H. E. Jahn B. Ambassador of Norway; Head, Norwegian Mission Accredited to the European Atomic Energy Community.
Address: 16 Place Surlet de Chokier, Brussels, Belgium.

HALZL, Jozsef. Born 1933. Educ.: Budapest Univ. of Technol. Societies: Sci. Assoc. for Power Economy, Budapest.
Nuclear interest: Reactor design with special emphasis on the utilisation of nuclear energy for process heat production and saline water conversion.
Address: II. Garas-u. 12, Budapest, Hungary.

HAM, Harold John, M.B., B.S. (Melbourne), D.M.R.E. (Cambridge), F.F.R., F.C.R.A., F.R.A.C.P. Born 1902. Educ.: Melbourne Univ. Formerly Lecturer in Radiotherapy, and Honorary Assoc., Dept. Medicine, Sydney Univ.; Formerly Chairman, Therapeutic Trials Com. Radioisotopes, New South Wales; Prof., Radiotherapy, Utrecht Univ., Netherlands. Societies: British Medical Assoc.; Assoc. Member, Dermatological Assoc., Australia; Dutch Radiol. Soc.
Nuclear interest: Medical use of radioisotopes in diagnosis, therapy and research.
Address: Department of Radiotherapy, University Hospital, Utrecht, Netherlands.

HAM, William T., Jr., B.S. (Eng.), M.S. (Phys.), Ph.D. (Phys.). Born 1908. Educ.: Virginia Univ. Assoc. Prof. (1948-53), Prof. and Chairman, Dept. Biophys. and Biometry (1953-), Medical Coll., Virginia. Consultant, Health Phys. Div., O.R.N.L., 1950-; Rad. Cataract Com. (1954-57), Consultant, Atomic Bomb Casualty Com. (1956-), N.A.S. Controlled Background Facilities Com., A.I.B.S.-A.E.C., 1958-; Consultant, Army Medical Service Graduate School, 1952-; Consultant, Defence Atomic Support Agency, Weapons Effects Board; Dean's Advisory Com., Radioisotope Com. and Medical Curriculum Planning Com., Medical Coll., Virginia. Societies: Fellow, A.P.S.; Fellow, A.A.A.S.; A.N.S.; N.Y. Acad. Sci.; Biophys. Soc.; A.S.A.; Virginia Acad. Sci.
Nuclear interests: Nuclear physics; Separation uranium isotopes; Radiation dosimetry; Radiobiology; Thermal injury; Biophysics.
Address: Biophysics Department, Medical College of Virginia, Richmond 19, Virginia, U.S.A.

HAMA, George Major, B.S. (Chem. Eng.), M.S. (Chem. Eng., Minnesota), M.S. (Sanitory Eng., Harvard). Born 1910. Educ.: Minnesota and Harvard Univs. Sen. Assoc. Ind. Hygienist, Detroit City; Assoc. Prof., Div. Occupation Health and Medicine, Wayne State Univ. Exec. Com., American Conference Governmental Ind. Hygienists.
Nuclear interests: Industrial ventilation control of toxic materials; Beryllium radioactive isotopes.
Address: 13003 Santa Clara, Detroit, Michigan 48235, U.S.A.

HAMACHI, Tadao, B.S. Born 1922. Educ.: Tokyo Univ. With Geological Survey of Japan, 1943-. Formerly Chief Researcher, Radioactive Mineral Deposit Sect., Mineral Deposit Dept., Geological Survey of Japan. Societies: Mineralogical Soc. of America; Mineralogical Soc. of Japan; Geological Soc. of Japan.
Nuclear interest: Nuclear raw material, especially uranium minerals and their deposits.
Address: Mineral Deposit Dept., Geological Survey of Japan, Hisamoto-cho 135, Kawasaki City, Japan.

HAMADA, Hiroya, B.E. Born 1924. Educ.: Tokyo Univ. Member, Subcom., Power Reactor, 1966-67. Society: Atomic Energy Soc. of Japan.
Nuclear interests: Reactor design; Nuclear system engineering; Future feasibility of advanced thermal reactor.
Address: Meidensha Electric Manufacturing Company Limited, 2-1-17 Osaki, Shinagawa-ku, Tokyo, Japan.

HAMADA, Masayuki, Dr. Chief, Div. of Organic Chem., Dept.of Chem., Rad. Centre of Osaka Prefecture.
Address: Radiation Centre of Osaka Prefecture, Department of Chemistry, Shinkecho, Sakai, Osaka, Japan.

HAMADA, Shigeo. Born 1931. Educ.: Kyoto Univ. Asst. (1958), Lecturer (1961-), Dept. of Phys., Coll. of Sci. and Eng., Nihon Univ.; Formerly Asst., now Asst. Prof., Plasma Phys. (Experiment), Atomic Energy Res. Inst., Nihon Univ. Society: Phys. Soc. of Japan.
Nuclear interest: Nuclear fusion reaction.
Address: c/o Mr. Okasawa, 1-254 Amanuma, Suginami-ku, Tokyo, Japan.

HAMADA, Takeshi, Dr. Eng. Born 1907. Educ.: Tokyo Univ. Adviser, Mitsubishi Heavy-Ind., Ltd. Society: Soc. Naval Architects Japan.
Nuclear interests: Ship design; Collision-resisting construction of ships' sides.
Address: 474 Shinohara-cho, Kohoku-ku, Yokohama, Japan.

HAMADA, Toichi. Formerly Exec. Director, now Managing Director, Mitsubishi Estate Co. Ltd.
Address: Mitsubishi Estate Co. Ltd., Marunouchi Building, 2 Marunouchi 2-chome, Chiyoda-ku, Tokyo, Japan.

HAMAGUCHI, Hiroshi, D.Sc. Born 1915. Educ.: Tokyo Imperial Univ. Prof. Chem., Tokyo Univ. Book: Radioactivation Analysis (in Japanese). Society: Atomic Energy Soc. Japan.
Nuclear interest: Nuclear chemistry and radiochemistry, especially in the application to analytical chemistry.
Address: Department of Chemistry, Tokyo University, Hongo, Tokyo, Japan.

HAMAKER, Jacobus, Dr. Phil. Born 1911. Educ.: Utrecht Univ. Director, Central Tech. Inst. (1952), Gen. Res. Co-ordinator, Organisation for Ind. Res. (1963), Director Gen. Organisation for Ind. Res. (1965), T.N.O. Exec. Com. European Federation of Chem. Eng.
Nuclear interest: Heat transfer in reactors.
Address: 63 Thorbeckestraat, Delft, Netherlands.

HAMANN, Cecil Boyce, B.S., M.S., Ph.D. Born 1913. Educ.: Taylor and Purdue Univs. Chairman, Div. Sci. and Mathematics, Asbury Coll., 1946-. Pres. Elect, Kentucky Acad. Sci. Address: Asbury College, Wilmore, Kentucky, U.S.A.

HAMANO, Kazuo, Dr. Formerly Chief Eng., Reactor Sect., now Chief, Atomic Energy Res. Group, Fujinagata Shipbuilding and Eng. Co. Ltd. Address: Fujinagata Shipbuilding and Engineering Co. Ltd., 2-9 Shibatani-cho, Sumiyoshi-ku, Osaka, Japan.

HAMBLY, Colin Keith, M.B., B.S., D.T.R., F.C.R.A. Born 1918. Educ.: Sydney Univ. Director, Hallstrom Inst. Radiotherapy and Nucl. Medicine, Sydney Hospital, Sydney, Australia. Member, Federal Council, Coll. of Radiologists of Australasia; Member, Nat. Health and Medical Res. Council. Society: Coll. of Radiologists of Australasia.
Nuclear interest: Clinical nuclear medicine.
Address: Hallstrom Institute of Radiotherapy and Nuclear Medicine, Sydney Hospital, Macquarie Street, Sydney 2000, Australia.

HAMBURGER, Mrs. A. Director, Zuid-Hollandsche Pletterijen v/h D.A. Hamburger N.V.
Address: Zuid-Hollandsche Pletterijen v/h D.A., Hamburger N.V., 73 Delftweg, Postbus 19, Delft, Netherlands.

HAMBURGER, D. A. Director, Zuid-Hollandsche Pletterijen v/h D.A. Hamburger N.V.
Address: Zuid-Hollandsche Pletterijen v/h D.A. Hamburger N.V., 73 Delftweg, Postbus 19, Delft, Netherlands.

HAMEL, Paul, B.Sc.A. Born 1925. Educ.: Ecole Polytechnique, Montreal. Design N.R.U. (1954-57), Daniels-Boyd Nucl. Steam Generator (1957-58), C.D. Howe Co., Ltd.; Canadian Vickers, 1958-59; Canadian Pratt-Whitney Aircraft Co., Ltd., 1959; Orenda Engines, Ltd., 1959-62; Assoc. Sci. Adviser, Atomic Energy Control Board, 1962-. Society: A.P.E.O.
Nuclear interests: Reactor design and technology; Reactor safety; Reactor economics and applications; Isotopes separation and applications; Nuclear safety and criticality control; Safety of particle accelerators; Atomic energy regulations.
Address: Atomic Energy Control Board, P.O. Box 1046, Ottawa, Ontario, Canada.

HAMELIN, Raymond A. Agrégé de l'Université, Dr. ès-Sc. Born 1930. Educ.: Ecole Normale Supérieure, Paris. Sci. Attaché, Embassy of France, Washington D.C., 1962-65; Sci. Adviser Ugine Kuhlmann Co., Paris, 1965-; Prof. of Chem. Inst. Nat. des Sci. et Techniques Nucléaires, Saclay, 1965-; Editor-in-Chief, Energie Nucléaire, Paris, 1966-. Societies: New-York Acad. of Sci.; A.C.S.; Société de Chimie Physique; Société Chimique de France.
Nuclear interests: Nuclear management; Nuclear chemistry: ore treatment, fuel elaboration and reprocessing; Nuclear literature.
Address: 74bis, boulevard Maurice Barres, 92 Neuilly, France.

HAMER, Eberhard E. H., B.S. (Mech. Eng.), B.S. (Ind. Eng.). Born 1921. Educ.: Texas Technol. Coll. L.M.F.B.R. Programme Office (Components Sect.), Design Eng., Project Eng. and Construction Eng., Argonne Nat. Lab. Societies: A.N.S.; N.S.P.E.
Nuclear interests: Reactor design; Project management; Programme planning particularly R. and D. activities related to reactor components such as fuel handling, reactivity control devices, and vessels including appurtences and internals.
Address: 9701 S. Kenton Avenue, Oak Lawn, Illinois 60453, U.S.A.

HAMER, Pierre, Dr. en Droit, Ing. commercial. Born 1916. Educ.: Paris, Brussels and Frankfurt-am-Main Univs. Luxembourg Govt. Commissioner; Nat. Council for Nucl. Energy; Administrateur delégué, Sté. Electrique de l'OUR.
Address: 7 Boulevard Royal, Luxembourg, Luxembourg.

HAMERMESH, Morton, B.S., Ph.D. Born 1915. Educ.: C.C.N.Y. and New York Univ. Sen. Phys. (1948-59), Director, Phys. Div. (1959-63), Assoc. Lab. Director (1963-65), Argonne Nat. Lab.; Prof. and Head, School of Phys. and Astronomy, Minnesota Univ., 1965-. Com. on Russian Translations, A.I.P. Societies: A.P.S.; A.I.P.
Nuclear interest: Nuclear physics.
Address: School of Physics and Astronomy, Minnesota University, Minneapolis, Minnesota, U.S.A.

HAMILL, William H., B.S. (Notre Dame), Ph.D. (Columbia). Born 1908. Educ.: Notre Dame and Columbia Univs. At Notre Dame Univ., 1938-. Societies: A.C.S.; Faraday Soc.
Nuclear interests: Ion-molecule reactions and energetics by mass spectrometry; Ionic processes in irradiated organic solids; Physical and chemical effects produced by low energy electron impact on thin organic films.
Address: Radiation Laboratory, University of Notre Dame, Notre Dame, Indiana 46556, U.S.A.

HAMILTON, B. M. Director of Metal., Crucible Steel Co. of America.
Address: Crucible Steel Co. of America, Four Gateway Centre, Pittsburgh, Pennsylvania 15230, U.S.A.

HAMILTON, DeWitt Clinton, Jr., B.S., M.S., Ph.D. Born 1918. Educ.: Oklahoma, California (Berkeley) and Purdue Univs. Asst. Prof., Purdue Univ., 1949-51; Principal

Development Eng., Reactor Eng. Div. (1951-56), Lecturer, Eng. Sci., O.R.S.O.R.T. (1956-65), O.R.N.L. Prof., and Head, Mech. Eng. Dept., Tulane Univ., 1965-. Societies: RESA; A.N.S.; A.S.M.E.; A.S.E.E.; A.A.A.S.
Nuclear interest: Heat transfer, fluid mechanics, and the thermal aspects of nuclear reactors.
Address: Mechanical Engineering Department, Tulane University, New Orleans, Louisiana 70118, U.S.A.

HAMILTON, James, M.A. (Cambridge), M.Sc. (Belfast), Ph.D. (Manchester). Born 1918. Educ.: Queen's Univ. (Belfast), Inst. for Advanced Study (Dublin) and Manchester Univ. Lecturer in Maths., Cambridge Univ., 1950-60; Fellow, Christ's Coll., Cambridge, 1953-60; Res. Assoc., Nucl. Studies Lab., Cornell Univ., 1957-58; Prof. Phys., Univ. Coll., London, 1960-64; Prof. Phys., Nordic Inst. Theoretical Atomic Phys., Copenhagen, 1964-.
Book: The Theory of Elementary Particles (Oxford Univ. Press, 1959); editor, Phys. Letters. Societies: A.P.S.; Italian Phys. Soc.; Foreign Member, Roy. Danish Acad.
Nuclear interests: Theoretical aspects of nuclear physics and elementary particle studies; Quantum theory of fields.
Address: Nordita, Blegdams vej 17, Copenhagen, Denmark.

HAMILTON, Walter A., A.B., M.B.A. Born 1922. Educ.: Amhurst Coll. and Harvard Univ. Staff, U.S. Congressional Joint Com. on Atomic Energy, 1946-56; Exec. Sec., Panel on the Impact of the Peaceful Uses of Atomic Energy, U.S. Congress, 1955-56; Vice-Pres., Nucl. Development Corp. of America, 1956-61; Review of U.S. Internat. Atomic Energy Policies and Programmes, U.S.A.E.C. (consultant), 1960-61; Director of Public Affairs (1961-62), Vice-Pres., Administration (1962-), United Nucl. Corp.; Chairman, Price-Anderson Study Group, (-1966), Chairman, ad hoc Com. on Workmen's Compensation (1966-), Atomic Ind. Forum. Society: A.N.S.
Nuclear interest: Management.
Address: United Nuclear Corporation, 1730 K Street N.W., Washington 6, D.C., U.S.A.

HAMILTON, William H., B.S. (Maths. and Phys.), M.S. (Maths. and Eng.). Born 1918. Educ.: Washington and Jefferson Coll. and Pittsburgh Univ. Res. Eng., Res. Labs., Westinghouse Elec. Corp., 1945-50; Staff Asst. to Tech. Director (1950-51), Manager, Plant Analysis Sect. (1951-52), Manager, Control Systems Sect. (1952-54), Manager, S.F.R. Control Systems Sect. (1954-55), Manager, S.F.R. Power Plant Subdiv. (1955-56), Chief S.F.R. Test Eng. for U.S.S. Skate (1956-57), Manager, S3W/S4W Project (1957), Manager, Submarine Tech. Dept. (1958-59), Manager, PWR Project (1959-60), Manager, Materials Dept. (1960-61), Plant Technol. (1961-65), Manager, Operating Plants (1965-), Bettis Plant, Westinghouse Electric Corp. Papers: Reactor Plant Instruments and Controls (Pittsburgh, I.R.E. Professional Group on Nucl. Sci., 1956); Power and Temperature Control of Pressurised Water-cooled Reactors (Cleveland, Ohio, Nucl. Sci. Congress, 1955). Societies: Fellow, I.E.E.E.; A.N.S.
Address: 1128 Savannah Avenue, Pittsburgh, Pennsylvania 15218, U.S.A.

HAMILTON, William Maxwell, Master, Agricultural Sci., D.Sc., N.D.H. (New Zealand), F.R.S. (New Zealand). Born 1909. Educ.: Massey Agricultural Coll. and New Zealand Univ. Director-Gen., formerly Sec., D.S.I.R., 1953-.
Nuclear interest: Administration of nuclear research.
Address: Department of Scientific and Industrial Research, Private Bag, Wellington C.1, New Zealand.

HAMIPRODJO, N. Minister of Finance; Member, Council for Atomic Energy.
Address: c/o Institute for Atomic Energy, 7 Djl. Palatehan 1 No. 26, Blok 5k, Kebajoran-Baru, Djakarta, Indonesia.

HAMM, William Joseph, Ph.D. (Washington, St. Louis), M.S. (Catholic Univ., Washington, D.C.), B.S. (Dayton), Graduate of O.R.I.N.S. Born 1910. Educ.: Dayton, Washington (St. Louis) and Catholic Univs. Res. participant at O.R.N.L., summer 1960; Full Prof., St. Mary's Univ., San Antonio, Texas, 1946-. Fellow, Inst. of Radio. Eng.; Member, Professional Group on Nucl. Instrumentation, Inst. of Radio. Eng.
Nuclear interests: Instrumentation; Isotope technology and tracer applications.
Address: Department of Physics, St. Mary's University, 2800 Cincinatti Avenue, San Antonio 28, Texas, U.S.A.

HAMMAR, Lennart Harry, Teknologie Licentiat (Dr.). Born 1930. Educ.: Roy. Inst. of Technol., Stockholm. Aktiebolaget Atomenergie, Nyköping, 1957-65; Allmänna Svenska Elektriska Aktiebolaget (A.S.E.A.), (Västeras) 1965-.
Nuclear interests: Nuclear chemistry; Radiation chemistry; Reactor chemistry.
Address: 27 B Pettersbergsgatan, Västeras, Sweden.

HAMMER, Charles Lawrence, B.S., M.S., Ph.D. Born 1922. Educ.: Michigan Univ. Instructor, Michigan Univ., 1953-54; Res. Assoc. (1954-55), Asst. Prof. (1955-59), Assoc. Prof. (1959-61), Prof. (1961-), Iowa State Univ. Society: Fellow, A.P.S.
Nuclear interests: Nuclear physics; Elementary particles; Particle accelerators.
Address: Physics Department, Iowa State University, Ames, Iowa, U.S.A.

HAMMER, E. Walter, Jr., A.B. (cum laude), M.S. Born 1915. Educ.: Syracuse and Harvard Univs. Product Development Div., The Budd Co. Society: A.S.M.E.

Nuclear interests: Design of shield structures for aircraft nuclear propulsion; Metallurgy; Manufacture of core parts for submarine reactors.
Address: The Budd Company, 300 Commerce Drive, Washington, Pennsylvania 19034, U.S.A.

HAMMER, Gerhard, Priv-Dozent (Cologne). Born 1890. Educ.: Leipzig, Halle, Kiel, Berlin, and Munich Univs. Stadtobermedizinalrat. Ehemaliger Chefarzt des Strahleninstituts der Städt. Krankenanstalten Nuernberg, 1928-55. Societies: Deutsche Roentgen Gesellschaft; Bayr. Roentgen-Vereinigung.
Nuclear interest: Medizin.
Address: 4 Weidmannstrasse, Nürnberg, Germany.

HAMMERLING, F., Dr.-Ing. Vizepräsident, Kernforschungsanlage Jülich des Landes Nordrhein-Westfalen E.V.; Mitglied des Vorstandes, A.E.G.; Member, Steering Com., Deutsches Atomforum e.V.
Address: Allgemeine Elektricitats - Gesellschaft, AEG- Hochhaus, 6000 Frankfurt/Main-Sud 10, Germany.

HAMMERSCHMIDT, William Warner, B.A., M.A., Ph.D. Born 1916. Educ.: Ohio State and Cornell Univs. Chief, Atomic Energy Div., Office of Director of Defence Res. and Eng., and Exec. Sec., Defence Sci. Board, Office of Sec. of Defence. Societies: American Astronomical Soc.; American Philosophical Assoc.; Washington Acad. of Sci.; A.P.S.; N.Y. Acad. Sci.
Nuclear interest: Management.
Address: 7818 Holmes Run Drive, Falls Church, Virginia 22042, U.S.A.

HAMMERSLEY, Charles Arthur, B.Sc. (Hons. Metal.). Born 1933. Educ.: Birmingham Univ. Materials Eng., English Elec. Aviation Ltd., Warton, Lancashire, 1956-61; Tech. Manager, Fine Tubes Ltd., Plymouth, 1961-. Societies: Assoc. Inst. Metal.; Iron and Steel Inst.; Inst. Metals; A.S.M.
Nuclear interest: Manufacture of precision, high quality metallic tubing for nuclear application, e.g., fuel element tubes, instrumentation, heat exchangers, etc.
Address: Fine Tubes Limited, Estover Works, Crownhill, Plymouth, Devon, England.

HAMMING, Kenneth W., B.S. (Gen. Eng.). Born 1918. Educ.: Illinois Univ. Sen. Partner, Sargent and Lundy, Eng., Chicago. Director, Atomic Ind. Forum, Inc. Society: A.S.M.E.
Nuclear interest: Architect-engineering and consulting for design of thermo-nuclear power plants.
Address: 140 S. Dearborn Street, Chicago, Illinois 60603, U.S.A.

HAMMITT, Frederick Gnichtel, B.S. (Mech. Eng.), M.S. (Mech. Eng.), M.S. (Appl. Mech.), Ph.D. (Nucl. Eng.). Born 1923. Educ.: Princeton, Cornell, Michigan and Pennsylvania Univs. Power Generators Ltd., Trenton, New Jersey, 1948-50; Stevens Inst. Technol., Reaction Motors, Inc., 1950-53; Worthington Corp., 1953-56; Res. Assoc. (1956-57), Assoc. Res. Eng. (1957-58), Assoc. Prof. Nucl. and Mech. Eng. (1958-61), Formerly Prof. Nucl. Eng. (1961-), now Mech. Eng. Dept., Michigan Univ. Vice-Chairman. A.S.T.M. Com. G-2 on Cavitation and Impingement Erosion, 1965-. Societies: A.N.S.; A.S.M.E.; A.S.T.M.
Nuclear interests: Cavitation; Impingement; Heat transfer and fluid flow; Thermodynamic aspects of reactor design; Reactor power-plant economics; Reactor core design.
Address: Mechanical Engineering Department, 312 Automotive Engineering Laboratory, North Campus, Michigan University, Ann Arbor, Michigan 48105, U.S.A.

HAMMOND, F. J. Formerly Administration Div., Town, A.E.C.L.
Address: Atomic Energy of Canada Ltd., Whiteshell Nuclear Research Establishment, Pinawa, Manitoba, Canada.

HAMMOND, P. L., G.M.I.Mech.E., A.M.I.Nucl. E. Sen. Eng., W. S. Atkins and Partners.
Address: W. S. Atkins and Partners, Woodcote Grove, Ashley Road, Epsom, Surrey, England.

HAMMOND, R. Philip, B.S. (Chem. Eng.), Ph.D. (Inorg. Chem.). Born 1916. Educ.: Southern California and Chicago Univs. Staff member, Los Alamos Sci. Lab., 1947-62 (Assoc. Div. Leader, Reactor Development); Director, Nucl. Desalination Programme, O.R.N.L. Societies: A.C.S.; A.N.S.; American Assoc. of Cost Eng.
Nuclear interests: Reactor design; Reactor materials; Reactor economics; Desalination and agro-industrial complexes.
Address: 879 W. Outer Drive, Oak Ridge, Tennessee 37830, U.S.A.

HAMOLSKY, Milton W., A.B., M.D. Born 1921. Educ.: Harvard Coll. and Harvard Medical School. Lecturer on Medicine, Harvard Medical School; Physician-in-Chief, Dept. of Medicine, Rhode Island Hospital; Prof. Medicine, Brown Univ. Societies: American Soc. for Clinical Investigation; American Goitre Soc.; A.A.A.S.; American Physiological Soc.
Nuclear interest: Diagnosis and therapy of medical diseases.
Address: Rhode Island Hospital, Eddy Street, Providence 2, Rhode Island, U.S.A.

HAMOUDA, Ibrahim Fathi, B.Sc. (Hons.), Dr. sc. nat. Born 1924. Educ.: Cairo Univ. and Zurich E.T.H. Lecturer, Alexandria Univ., 1951; Head, Phys. Dept., Khartoum Univ., 1955-58; Prof., Neutron Phys., Head, Reactor and Neutron Phys. Dept., and Member, Board, United Arab Atomic Energy Establishment. Pres., Egyptian Soc. for Nucl. Sci. and

Applications.
Nuclear interest: Neutron and reactor physics.
Address: 29 Amin El-Rafi Street, Dokky, Giza, United Arab Republic.

HAMPTON, George Hughan, M.A. Born 1920. Educ.: Oxford Univ. Communications Eng., Ministry of Civil Aviation, 1945-54; Principal, Ministry of Transport and Civil Aviation, 1954-56; Administration Manager, Special Duties, Ind. Group (1956-59), Sec., Capenhurst Works, Production Group (1959-61), Deputy Director, Personnel and Administration, Production Group (1961-), U.K.A.E.A.; Now Director, Administration Dept., C.E.R.N.
Nuclear interests: Management in technological industry; Personnel matters.
Address: Conseil Européen pour la Recherche Nucléaire, Meyrin, Geneva, Switzerland.

HAMSTRA, Jan, Mech. Eng. Born 1923. Educ.: Delft Tech. Univ. with Koninklijke Maatschappij De Schelde N.V., Vlissingen, Netherlands, 1949-52; Bureau voor Scheepsbouw, Amsterdam and Bloemendaal, Netherlands, 1952-56; Head, Tech. Dept., Reactor Centrum Nederland, Petten (N-H.), Netherlands, 1956-.
Nuclear interest: Nuclear engineering.
Address: 16 Irenelaan, Bergen (N-H), Netherlands.

HAN, Kwon Shin. Acting Chief, Agricultural Div., Atomic Energy Res. Inst., Office of Atomic Energy and A.E.C.
Address: Atomic Energy Research Institute, Kongneung-dong, Sungbook-ku, Seoul, Korea.

HANAHAN, Thomas J., Jr. Member and Sec., Illinois A.E.C.
Address: 2012 Grandview, McHenry, Illinois 60050, U.S.A.

HANAUER, Stephen Henry, B.S.E.E., M.S.E.E., Ph.D. Born 1927. Educ.: Purdue and Tennessee Univs. Sen. Phys., O.R.N.L., 1950-65; Prof., Nucl. Eng., Tennessee Univ., 1964-. Member, Advisory Com. on Reactor Safeguards, U.S.-A.E.C., 1965-; Standards Com., A.N.S.; Chairman, Subcom. on Reactor Instrumentation, Internat. Electrotech. Commission; Member, Tech. Com. N42 Nucl. Instr., U.S.A. Standards Inst. Member, U.S.A.E.C. Com. on Nucl. Instr. Modules. Societies: A.P.S.; A.S.E.E.; A.N.S.
Nuclear interests: Teaching and research (primarily experimental) in nuclear reactor physics and instrumentation; Safety of nuclear power plants.
Address: Department of Nuclear Engineering, Tennessee University, Knoxville, Tennessee 37916, U.S.A.

HANBURY BROWN, Robert, D.Sc., D.I.C., B.Sc.Eng. Born 1916. Educ.: Brighton Tech. Coll. and City and Guilds Coll. Prof., Radio-Astronomy, Manchester Univ., 1960-64; Prof., Phys. (Astronomy), Sydney Univ., 1964-.
Societies: F.R.S.; Fellow, Roy. Astronomical Soc.; Fellow, Australian Acad. Sci.
Address: School of Physics, Sydney University, Sydney, Australia.

HANDWERK, Joseph Henry, B.S. (Ceramics), M.S. (Ceramics). Born 1919. Educ.: Alabama Univ. Asst. Prof. Ceramic Technol., Alabama Univ., 1950-54; Assoc. Ceramic Eng. and Group Leader, Fuels Properties, Metal. Div., Argonne Nat. Lab., 1954-. Vice-Chairman, Nucl. Div., American Ceramic Soc. Societies: Fellow, American Ceramic Soc.; A.S.T.M.; Nat. Inst. Ceramic Eng.
Nuclear interests: Ceramic nuclear fuels: oxides, carbides, sulfides, nitrides and phosphides of uranium and plutonium; Development of neutron-absorbing materials for control purposes.
Address: Metallurgy Division, Argonne National Laboratory, 9700 South Cass Avenue, Argonne, Illinois 60439, U.S.A.

HANDY, Lawrence J. Member, Standing Com. on Nucl. Energy in the Elec. Ind., Nat. Assoc. of Railroad and Utilities Commissioners.
Address: National Association of Railroad and Utilities Commissioners, Interstate Commerce Commission Building, P.O. Box 684, Washington D.C. 20044, U.S.A.

HANFORD, William Edward, B.S., M.S., Ph.D. Born 1908. Educ.: Philadelphia Coll. of Pharmacy and Sci. and Illinois Univ. Vice-Pres. for Res. and Member of Board, M.W. Kellogg Co., 1946-57; Vice-Pres. for R. and D., Olin Mathieson Chem. Corp., 1957-; Director, United Nucl. Corp. Societies: A.C.S.; A.A.A.S.; Soc. of Chemical Industry; Hon. Member, American Inst. of Chemists.
Address: 211 Central Park West, New York 24, New York, U.S.A.

HANG, Daniel F., B.S.E.E., M.S.E.E. Born 1918. Educ.: Miami and Illinois Univs. Director, A.M.U.-A.N.L. Summer Eng. Practice School, 1964. Assoc. Prof. Elec. and Nucl. Eng. Illinois Univ. Board of Direction, Illinois Professional Eng. Soc. Societies: A.S.E.E.; I.E.E.E.; A.N.S.
Nuclear interest: Nuclear engineering economy.
Address: 376 EEB, Illinois University, Urbana, Illinois 61801, U.S.A.

HANIGER, Ladislav, Ing., Elec. Eng. Born 1931. Educ.: Brno Polytech. Coll. Res. worker, Skoda Works. Society: Czechoslovak Sci. and Tech. Soc.
Nuclear interest: Use of the digital techniques for measurements and control of the power and period of nuclear reactors.
Address: Nuclear Power Division, Skoda Works, Pilsen, Czechoslovakia.

HANLE, Wilhelm, Dr. phil., o.Prof. Born 1901. Educ.: Heidelberg and Göttingen Univs. Full Prof. Exptl. Phys. and Director, Exptl. Phys. Inst., Giessen Univ. Head, Commission

on measuring techniques of nucl. rad., Branch of the Bundesatomministerium; Head, Diplomprüfungskommission des Verbandes deutscher Physikalischer Gesellschaften; Editor, Kerntechnik. Book: Künstliche Radioaktivitat (Stuttgart, Fischer-Verlag, 1952). Societies: Verband Deutscher Physikalischer Gesellschaften; A.P.S.; Societa fisica Italiana; Deutsche Biophysikalische Gesellschaft. Nuclear interests: Radiation damage; Measuring of nuclear radiation; Dosimetry; Polarisation and angular correlation of radiation; Neutron source; Linear accelerator.
Address: Physikalisches Institut der Universität Giessen, 104 Leihgesterner Weg, Germany.

HANLEN, Don Franklyn, A.B. (Psychology), M.S.(Chem.). Born 1921. Educ.: Denver Univ. Nucl. Eng.-Crew Chief, Reactor Operations Group, Convair-Fort Worth, Texas, 1955-57; Manager, Reactor Evaluation Centre, Westinghouse Atomic Power Dept., 1957-. Society: A.N.S.
Nuclear interest: Facility management, including conceptual design of controls, instrumentation, and general facility layout as well as organisational problems, procedures, training, and operations.
Address: 209 Woodbury Drive, Greensburg, Pennsylvania, U.S.A.

HANNA, Geoffrey Chalmers, M.A. Born 1920. Educ.: Cambridge Univ., N.R.C., A.E.C.L., 1945-.
Address: 5 Tweedsmuir Place, Deep River, Ontario, Canada.

HANNA, Stanley Sweet, A.B., Ph.D. Born 1920. Educ.: Denison and Johns Hopkins Univs. Asst. Prof., Johns Hopkins Univ.; Sen. Phys., Argonne Nat. Lab.; Prof., Stanford Univ. Society: A.P.S.
Nuclear interests: Nuclear physics; Study of the properties of nuclear energy levels, nuclear reactions, and nuclear structure. Mossbauer effect.
Address: 784 Mayfield Avenue, Stanford, California. U.S.A.

HANNERZ, Anders Kare, Tekn. lic. Born 1927. Educ.: Roy. Inst. Technol., Stockholm. Postgraduate res., Roy. Inst. Technol., 1949-52; Radiochem. Res., Swedish A.E.C., 1952-54; High polymer res., A.B. Casco, 1954-56; at present Head, Fast reactor development, Nucl. Power Dept., A.S.E.A.
Nuclear interests: Reactor design; Materials (metallurgy); Chemistry.
Address: Nuclear Power Department, Allmanna Svenska Electriska A.B., Vasteras, Sweden.

HANNOTHIAUX, André, Controleur Gen., de l'Equipement, Electricité de France. Adviser, Third I.C.P.U.A.E., Geneva, Sept. 1964.
Address: Electricite de France, 3 rue de Messine, Paris 8, France.

HANSCOME, Thomas Dixon, B.A., M.A. (Phys.). Born 1914. Educ.: Minnesota Univ. Head, Nucl. Instrumentation Branch, N.R.L., Washington D.C., 1942-58; Manager, Rad. Effects Res. Dept., Hughes Fullerton, 1958-. Societies: A.P.S.; RESA.
Nuclear interests: Radiation effects; Light particle reactions; Nuclear weapon phenomenology; Space radiation.
Address: 1782 Terry Lynn Drive, Santa Ana, California 92705, U.S.A.

HANSEN, Carl Ludwig, Jr., B.A., M.D., Ph.D. (Rad. Biol.). Born 1920. Educ.: American Internat. Coll., Tufts and Rochester Univs. Graduate work, Reed Coll., Portland, Oregon and O.R.N.L., 1952-53; Clinical Res. in Infectious Diseases, Fort Detrick, Maryland, 1953-54; Deputy Surgeon and Surgeon, Armed Forces Special Weapons Project, Washington, D.C., 1954-57; Graduate work in Rad. Biol. and Clinical Assoc. in Internat. Medicine, Rochester Univ., N.Y., 1957-60; Chief of Bio-nucleonics Sect., Office of Surgeon Gen., U.S.A.F., Washington, D.C., 1960-61; Deputy Director, Armed Forces Radiobiol. Res. Inst., Defence Atomic Support Agency, Bethesda, Maryland, 1961-. Societies: A.M.A.; American Physiological Assoc.; Aerospace Medical Assoc.; Rad. Res. Soc.
Nuclear interest: Mammalian radiation biology.
Address: 9807 Ashburton Lane, Bethesda, Maryland 20034, U.S.A.

HANSEN, Gordon E. Member, Advisory Com. on Reactor Physics, U.S.A.E.C. Deleg. to 2nd I.C.P.U.A.E., Geneva, Sept. 1958.
Address: California University, Los Alamos Scientific Laboratory, Los Alamos, New Mexico, U.S.A.

HANSEN, Henrik AGER-. See AGER-HANSEN, Henrik.

HANSEN, Ian Alfred, Ph.D. Born 1927. Educ.: Adelaide Univ. Lecturer, Biochem., Christian Medical Coll., Vellore, India, 1956-60; Lecturer, Biochem., Western Australia Univ., 1961-. Societies: Australian Biochem. Soc.; Australian and New Zealand Assoc. for the Advancement of Sci.
Nuclear interest: Use of isotopic tracers in lipid metabolism.
Address: Western Australia University, Nedlands, Western Australia 6009, Australia.

HANSEN, Joh., Prof. Dr. Mitglied, Technisch-Wissenschaftlicher Beirat, Study Assoc. for the use of Nucl. Energy in Navigation and Industry. Mitglied, Co. for the Utilisation of Nucl. Energy in Shipbuilding and Navigation; Vorsitzender, Arbeitskreis III/2, Kernenergie fur Schiffe, and Member, Fachkommission III, Kerntechnik, Deutsche Atomkommission, Federal Ministry for Sci. Res.

Address: 59 Cranachstrasse, Hamburg-Gr. Flottbek, Germany.

HANSEN, K. High Energy Phys. Dept., Niels Bohr. Inst., Copenhagen Univ.
Address: Niels Bohr Institutet, Copenhagen University, 15-17 Blegdamsvej, Copenhagen Ø, Denmark.

HANSEN, Kent Forrest, S.B., Sc.D. Born 1931. Educ.: M.I.T. Assoc. Prof., M.I.T. Book: Numerical Methods of Reactor Analysis (Academic Press, 1964). Society: A.N.S. Nuclear interests: Reactor physics; Reactor theory; Numerical analysis; Computer applications.
Address: 15 Paul Revere Road, Bedford, Massachusetts, U.S.A.

HANSEN, Niels, Chem. Eng., M.Sc. Born 1933. Educ.: Tech. Univ. Denmark. Head, Metal. Dept., A.E.C. Res. Establishment, Risoe. Danish Member, Halden Programme Group, O.E.C.D. Halden Reactor Project. Nuclear interest: Metallurgy.
Address: 131C Frederiksborgvej, Roskilde, Denmark.

HANSEN, Otto M. KOFOED-. See KOFOED-HANSEN, Otto M.

HANSEN, Paul Bjerre, M.D. Formerly Asst. Director, Radiumstationen for Jylland; Director, Radiumstationen, Odense.
Address: Radiumstationen, Odense Amts Og Bys, Sygehus, Odense, Denmark.

HANSEN, Robert Suttle, B.S., M.S., Ph.D. Born 1918. Educ.: Michigan Univ. Prof. of Chem. (1955-), Chairman, Chem. Dept. (1965-68), Distinguished Prof. of Sci. and Humanities (1967-), Iowa State Univ.; Sen. Chem. (1955-), Chief, Chem. Div. (1965-68), Director (1968-) Ames Lab., U.S.A.E.C. Societies: A.C.S.; A.P.S.
Nuclear interests: Reactions at metal surfaces, interface transport and dynamic properties, disperse systems.
Address: Ames Laboratory, U.S.A.E.C., Iowa State University, Ames, Iowa 50010, U.S.A.

HANSHAW, Bruce B., S.B., M.S., Ph.D. Born 1930. Educ.: M.I.T., Colorado and Harvard Univs. U.S.A.E.C., 1953-56; Petroleum Res. Corp., 1958-60; U.S. Geological Survey, 1961-. Treas., Geochem. Soc.; Member, U.S. Nat. Com. for Internat. Hydrological Decade; N.A.S.-N.R.C.; Member, Geochem. Hydrology Com., A.G.U. Society: A.C.S.
Nuclear interests: The use of natural light stable and radioactive isotopes in hydrologic and geologic investigations. Specialised research into the use of radiocarbon for the dating of ground water.
Address: United States Geological Survey, Washington, D.C. 20242, U.S.A.

HANSON, Alden Wade, B.S. Born 1910. Educ.: Alma Coll., Michigan. Consultant, California Univ. Rad. Lab.; Director, Nucl. and Basic Res. Lab., Dow Chemical Co., Midland, Michigan, U.S.A. Societies: A.S.T.M.; A.C.S.; A.A.A.S.; A.N.S.
Nuclear interests: Metallurgy; Reactor design; Management.
Address: 1605 St. Andrews Road, Midland, Michigan, U.S.A.

HANSON, Alfred Olaf, B.S., M.A. (North Dakota), Ph.D. (Wisconsin). Born 1914. Educ.: North Dakota and Wisconsin Univs. Prof. Phys., Illinois Univ., 1951-. Fulbright Scholar, Turin, Italy, 1955-56; Fulbright Lecturer, Sao Paulo, Brazil, 1960. Nat. Res. Council Advisory Com. to Nat. Bureau of Standards. Society: A.P.S.
Nuclear interests: Neutron physics with Van de Graaff, 1940-46; Photoneutron reactions with betatron electron scattering, photon scattering, 1946 to date.
Address: Department of Physics, Illinois University, Urbana, Illinois, U.S.A.

HANSON, Bertram Speakman, M.B., B.S. (Adelaide), F.C.R.A., Hon. F.F.R. Born 1905. Educ.: Adelaide Univ. In private medical practice. Hon. Consultant Radiotherapist, Roy. Adelaide Hospital; Chairman, Exec. Com., Anti-Cancer Foundation, Adelaide Univ.; Member, Council, Internat. Union Against Cancer. Society: Australian Cancer Soc.
Nuclear interest: Nuclear medicine as part of radiotherapy.
Address: 163 North Terrace, Adelaide, South Australia.

HANSON, George Henry, Ph.D., M.S., B.S.E. (Chem. Eng.) (Michigan). Born 1918. Educ.: Michigan Univ. Staff Member, Theoretical Phys. Sect., Atomic Energy Div. (1951-54), Rep in Rocky Mountain Nucl. Power Study Group (1954-56), Chief, M.T.R.-E.T.R. Reactor Eng. Sect., Atomic Energy Div. (1956-60), Chief, E.O.C.R., Organic Technol. Sect., Atomic Energy Div. (1960-62), Sen. Sci., Reactor Eng., Atomic Energy Div. (1962-64), Sen. Eng., Long Range Planning, Atomic Energy Div. (1964-), Phillips Petroleum Co. Societies: A.N.S.; A.I.Ch.E.; A.C.S.
Nuclear interests: Reactor engineering and physics; Reactor design; Fuel processing.
Address: 444 Seventh Street, Idaho Falls, Idaho, U.S.A.

HANSPETER, Marti. Verein Schweiz, Maschinen-Industrieller. Stellvertreter, Swiss Federal Commission for Atomic Energy.
Address: 4 Kirchenweg, 8032 Zurich, Switzerland.

HANSSEN, K. J., Dr. Lab. für Elektronenoptik, Atomic Phys. Div., Physikalisch-Technischen Bundesanstalt.

Address: Atomic Physics Division, Physikalisch-Technischen Bundesanstalt, 100 Bundesallee, 33 Braunschweig, Germany.

HANSSON, L. Member, Computer Programme Library Com., O.E.C.D., E.N.E.A.
Address: c/o E.N.E.A. Computer Programme Library, B.P. No. 15, Ispra (Varese), Italy.

HANSSON, Per M., B.A. in law. Born 1905. Educ.: Oslo Univ. Directorates in different banking, insurance and industrial enterprises. Member, Board of Reps., Noratom. Chairman, Norwegian Pool of Atomic Risk Insurers.
Address: Norwegian Pool of Atomic Risk Insurers, 19 V Bygdo Allé, Oslo 2, Norway.

HANSTEEN, Johannes Mathias, Cand. real. Born 1927. Educ.: Oslo Univ. Res. Assoc., Inst. Phys., Oslo Univ., 1959-63; Res. Assoc., Phys. Dept., Florida State Univ., 1963-. Societies: Norwegian Phys. Soc.; A.P.S.
Nuclear interest: Low-energy theoretical nuclear physics. At present working with the Cluster Model for light atomic nuclei.
Address: Physics Department, Florida State University, Tallahassee, Florida, U.S.A.

HANTZSCHE, Erhard, Dr. rer. nat. Born 1929. Educ.: Leipzig Univ. Phys.-Tech. Inst., German Academy of Sciences, 1956-. Society: Physikalische Gesellschaft in the German Democratic Republic.
Nuclear interests: Nuclear fusion, plasma physics and diagnostics, gas discharges.
Address: 24 Rossmässlerstr., Berlin-Karlshorst, German Democratic Republic.

HANZLIK, Jan, M.S. (Eng.). Born 1929. Educ.: Prague Tech. Coll. Energoprojekt Prague.
Nuclear interest: Atomic power plant design.
Address: 20 Belocerkevska, Prague 10, Czechoslovakia.

HAPPELL, John J., Jr., B.S. (Mech. Eng.), M.S. (Mech. Eng.). Born 1925. Educ.: Newark Coll. of Eng., Delaware Univ. and O.R.S.O.R.T. Project Eng., Liquid Metal Fuel Reactor Experiment and Spectral Shift Controlled Reactor, Babcock and Wilcox Co. Registered Prof. Eng., State of Virginia.
Nuclear interests: Reactor design; Fuel management; Component design and plant design.
Address: 2108 Woodcrest Drive, Lynchburg, Virginia, U.S.A.

HAQ, M. Shamsul. B.Sc. (Agriculture), Bachelor of Agriculture (Dacca), M.S. (Louisiana State), Ph.D. (Adelaide). Born 1921. Educ.: Dacca, Louisiana State and Adelaide Univs. District Agricultural Officer, 1944-52; Cotton Breeder, Pakistan Central Cotton Com.; Botanist, East Pakistan Agricultural Inst., 1952-58; Res. Officer, Pakistan A.E.C., 1958-60; Officer-in-Charge, Atomic Energy Agricultural Res. Centre, 1960-62; S.S.O. and Head, Agriculture Div., Atomic Energy Centre, Dacca, 1964-66; P.S.O., Head of Agriculture Div. and Director-designate, Inst. of Rad. Genetics and Plant Breeding, Mymensingh, 1967-.
Nuclear interest: Induced mutation breeding and radiation genetics. Concentrated mainly on the evolution of high yielding disease resistant rice strains through gamma irradiation and chemical mutagens under East Pakistan conditions. In recent years, programme relating to evolution of high yielding, sugarcane and mustard strains through gamma irradiation have been included.
Address: Atomic Energy Centre, P. O. Box No. 164, Ramna, Dacca, East Pakistan.

HAQUE, Ashraful, M.Sc. (Agri), Ph.D. Born 1919. Educ.: Benares, London and Oxford Univs. and Radioisotope School, Japan. Sen. Lecturer (1948-56), Reader (1956-61), Dacca Univ.; Sen. Sci. Officer, Pakistan A.E.C., 1961-; Sci. Officer, John Innes Inst., Hertford, U.K., 1952-54; Res. Asst., Oxford Univ.; Guest Biologist as Res. Scholar, Brookhaven Nat. Lab. Societies: Genetical Soc. of U.K.; Soc. of Agricultural and Biological Sci., Pakistan (Exec. member); Academic Council, Dacca Univ.
Nuclear interests: Cytological studies with ionising-radiation and radioactive isotopes.
Address: Department of Botany, Dacca University, Pakistan.

HAQUE, B., Dr. Reader, Nucl. Phys., Dept. of Phys., Dacca Univ.
Address: Dacca University, Dacca 2, Pakistan.

HAQUE, M. S., M.Sc., M.S. Nucl. Phys. in the Dept. of Phys., Dacca Univ.
Address: Department of Physics, Dacca University, Dacca 2, Pakistan.

HARA, Osamu, D.S. Born 1923. Educ.: Nagoya Univ. Prof. Theoretical Phys., Nihon Univ., Tokyo, 1956-. Societies: Phys. Soc. Japan; A.P.S.
Nuclear interest: Theoretical nuclear physics.
Address: Department of Physics, Faculty of Science and Engineering, Nihon University, Tokyo, Japan.

HARA, Saburo. Formerly Chief Surveyor, now Managing Director, Ship Classification Soc. of Japan.
Address: Ship Classification Society of Japan, 17-26, Akasaka 2 chome, Minato-ku, Tokyo, Japan.

HARA, Tamashige, Bachelor of Law. Born 1896. Educ.: Chuo Univ., Tokyo. Lawyer, 1921-; Formerly Vice Pres., Atomic Fuel Corp., 1956. Member, Economic Com., Japan Atomic Ind. Forum, Inc., 1958-.
Nuclear interests: Management; Prospecting and mining of uranium.
Address: 14 Wakamatsu-cho, Shinjuku-ku, Tokyo, Japan.

HARAGUCHI, Kuman, D.Sc. Prof. Sen. Officer, Res. Inst. for Radioisotopes, Yamagata Univ.
Address: Research Institute for Radioisotopes, Yamagata University, Koshirakawacho, Yamagata-shi, Japan.

HARAGUCHI, Miss Kyoko, Chief Officer, Radioisotope Inst., Radiological Dept., Kanto Communication Hospital.
Address: Radiological Department, Kanto Communication Hospital, 5-55 Gotanda, Shinagawa-ku, Tokyo, Japan.

HARAM, Sevrin A., M.E. Born 1922. Educ.: Stevens Inst. of Technol. Project Eng., Devenco, Inc., 1951-55; Asst. Dept. Manager and Staff Eng., Assoc. Nucleonics, Inc., 1955-62; Mech. Dept. Manager, Nucl. Materials and Equipment Corp., 1962-.
Nuclear interests: Facility design and construction for: nuclear power plants, fuel processing plants, process and food irradiation facilities, mobile and portable irradiators, hot cells. Management, administration, project engineering. Developed: source manufacturing facilities, gamma dose calculational procedures, irradiator patents.
Address: 1337 Foxwood Drive, Monroeville, Pennsylvania, U.S.A.

HARASAWA, Susumu, Dr. of Nucl. Eng., Master of Appl. Phys., Bachelor of Appl. Phys. Born 1931. Educ.: Tokyo Univ. Japan Atomic Energy Res. Inst., 1956-60; Rikkyo Univ., 1960-. Member, Subcom., Japan Soc. of Appl. Phys. Books: Fundamentals of Atomic Energy (Corona Co. Ltd., 1962); Introduction to Nucl. Reactors (Corona Co. Ltd., 1962). Society: A.N.S.
Nuclear interest: Interaction of radiation and matter, especially neutron irradiation effect of order-disordering in alloys.
Address: 715 Akiya, Yokosuka, Kanagawa, Japan.

HARBOE, Hans Christian, M.Sc. (Mech. Eng.). Born 1919. Educ.: Tech. Univ. Denmark. Eng., United Papermills, 1945-54; Chief, Tech. Dept., Federation of Danish Industries, 1954-57; Director, Danatom (Danish Assoc. for Ind. Development of Atomic Energy), 1957-63; Res. Director, Karl Kroyer Co., Aarhus, 1963-66; Director, Postgraduate Educ., Soc. Eng. 1966-; Board, Danish Inst. Computing Machinery; Isotopecentral; Danish Tech. Information Service, 1956-64. Chairman, Com. for Development of Ideas, Inventions and Res. Results, 1964-.
Address: 55 Lovgaardsvej, Sorgenfri (near Copenhagen), Denmark.

HARBOTTLE, Garman, B.S., Ph.D. Born 1923. Educ.: California Inst. Technol. and Columbia Univ. Brookhaven Nat. Lab., 1949-; Director, Res. Div. and Labs., I.A.E.A., Vienna, 1965-67.
Nuclear interests: Chemical effects of nuclear transformations; Nuclear chemistry.
Address: Chemistry Department, Brookhaven National Laboratory, Upton, Long Island, New York, U.S.A.

HARDE, Rudolf, Dr. rcr. nat., Dipl. Phys. Born 1922. Educ.: Bonn Univ. Gen. Tech. Manager, Interatom. German Member, Top Level Group on Co-operation in Res., O.E.C.D., E.N.E.A.
Nuclear interests: Development of nuclear power installations; Planning of research and development work.
Address: Interatom G.m.b.H., Bensberg/Cologne, Germany.

HARDER, Dietrich Hubert Wilhelm, Dr. phil. nat., Priv. Doz. (Exptl. Phys.). Born 1930. Educ.: Frankfurt am Main Univ. Max Planck Inst. für Biophys., 1955-60; Phys. Inst., Würzburg Univ., 1960-67. Second Chairman, Com. 1, Fachnormenausschuss Radiol.; Member, I.C.R.U. Task Group on Electron Beam Dosimetry. Societies: Deutsche Phys. Gesellschaft; Deutsche Gesellschaft für Biophys.; Deutsche Röntgengesellschaft.
Nuclear interests: Radiation physics; Biophysics.
Address: Physikalisches Institut, Würzburg Universität, Germany.

HARDER, Lewis B. Chairman, Board, Internat. Mining Corp.; Board Member, United Nucl. Corp., 1966-.
Address: United Nuclear Corporation, 1730 K Street, N.W., Washington 6, D.C., U.S.A.

HARDING, Geoffrey Norman, B.Sc. (Hons.). Born 1919. Educ.: Birmingham Univ. A.E.R.E., Harwell (1948-61), Culham Lab. (1961-), U.K.A.E.A. Society: Fellow, Phys. Soc.
Nuclear interests: Nuclear physics; Plasma physics; Controlled thermonuclear reactions.
Address: Culham Laboratory, Abingdon, Berkshire, England.

HARDING, Henry W. Director, Tracerlab-Keleket.
Address: Tracerlab-Keleket, 1601 Trapelo Road, Waltham 54, Massachusetts, U.S.A.

HARDING, M. Export Sec., Nucl. Dept., Anc. Ets. Aubert and Duval Acierie des Ancizes.
Address: Nuclear Department, Anc. Ets. Aubert and Duval Acierie des Ancizes, 41 rue de Villiers, Neuilly-sur-Seine, (Seine), France.

HARDMEIER, B., Dr. Stellvertreter, Federal Commission for Atomic Energy; Member, Swiss Assoc. for Atomic Energy.
Address: Federal Commission for Atomic Energy, 55 Effingerstrasse, Bern 3, Switzerland.

HARDT, Werner Eduard Karl, Dipl. Phys., Dr. rer. nat. Born 1923. Educ.: Bonn Univ. Inst. für Strahlen- und Kernphysik, Bonn Univ., 1952-56; Deutsches Elektronen-Synchrotron,

Hamburg, 1957-64; Sen. Phys., Accelerator Res. Div. (1964-66), Sen. Phys., 300 GeV Study Group, I.S.R. Div. (1966-), C.E.R.N., Geneva. Society: Deutsche Physikalische Gesellschaft e.V.
Nuclear interest: Accelerators, especially magnet design.
Address: 23 avenue St.-Cécile, Meyrin/ Geneva, Switzerland.

HARDTKE, Fred C., Jr., B.S., Ph.D. (Phys. Chem.). Born 1931. Educ.: Illinois Inst. Technol. and Oregon State Univ. At Chem. Dept., Missouri Univ.
Nuclear interest: Irradiation-induced defects in dielectric solids, especially the effect of these defects on the electrical properties of the dielectrics.
Address: Chemistry Department, Missouri University, Rolla, Missouri, U.S.A.

HARDUNG-HARDUNG, Heimo, Dr. Phys. Born 1926. Educ.: Vienna, Aarhus, Copenhagen and Stockholm Univs. Asst. Prof., Tech. Versuchsaustalt, Vienna, 1951; Deleg. to 1st I.C.P.U.A.E., Geneva, 1955; Tech. Sales Manager, Degussa (nucl. group); Head of Economics Office, E.N.E.A., 1960-; Tech. Director, Hans J. Zimmer A. G., 1964, and Vorstand Vickers-Zimmer A.G., 1965; Aufsichtsrat, High Polymer and Petrochem. Eng., Ltd., 1965-; Geschäftsführer, Syntex G.m.b.H., 1966. Vorstand Technischer Überwachungsverein 1967.Director, Vickers Res. Council, 1967. Books: Die industrielle Anwendung radioaktiver Isotopen (Vienna, 1951); Chancen in der Atomwirtschaft (Düsseldorf, 1957); Die Prokuristen (Gabler Verlag).
Nuclear interests: Economics of nuclear fuel cycles; Management.
Address: 139 Grosser Hasenpfad, Frankfurt/ Main-Sud, Germany.

HARDWICK, Thomas James, B.Sc., Ph.D. Born 1923. Educ.: McGill Univ. Member, N.R.C. of Canada, 1944-51; A.E.C.L., 1951-53; Leeds Univ., 1953-55; A.E.C.L., 1955-57; Head, Rad. Chem. Group, Gulf R. and D. Co., 1957-. Societies: Faraday Soc.; A.C.S.
Nuclear interests: Radiation chemistry of aqueous solutions, hydrocarbons and petroleum products.
Address: 214 Field Club Ridge Road, Pittsburgh 38, Pennsylvania, U.S.A.

HARDY, David Heugh, B.Sc. Eng. Born 1931. Educ.: The Polytechnic, Regent Street, London. Development Eng., Mullard Res. Labs., 1955-62; Chief Eng., Panax Equipment Ltd., 1962-.
Nuclear interests: Nuclear instruments and applications.
Address: Panax Equipment Ltd., Holmethorpe Industrial Estate, Redhill, Surrey, England.

HARDY, Harold Kenyon, D.Sc., Ph.D., A.R.S.M., F.I.M. Born 1919. Educ.: London Univ. Head, Phys. Metal. Sect., Fulmer Res. Inst., Stoke Poges, 1946-55; Deputy Head of Labs., Ind. Group, Springfields Works,near Preston (1955-58), Chief Metal., Ind. Group, R. and D. Branch, Risley (1958-60), Director of Fuel Element Development, Reactor Fuel Element Labs., Reactor Group, Springfields Works, near Preston (1960-66), Director, Reactor Fuels and Materials (1966-), U.K.A.E.A. Societies: Inst. Metal.; Inst. Metals.
Nuclear interests: The return from development.
Address: 15 Hastings Road, Hillside, Southport, England.

HARDY, Harriet Louise, A.B., M.D. Born 1906. Educ.: Wellesley Coll. and Cornell Univ. Asst. Physician (1950-51), Assoc. Physician (1951-), Isotope Com. (1949-63), Massachusetts Gen. Hospital; Assoc. Physician i/c Occupational Medical Service, Medical Dept. (1949-50), Asst. Medical Director i/c Occupational Medical Service, Medical Dept. (1950-), Lecturer, Civil and Sanitary Eng. Dept. (1951-62), Rad. Protection Com. (1950-), Nucl. Reactor Safeguard Com. (1958-), M.I.T.; Instructor in Ind. Hygiene, Ind. Hygiene Dept., Harvard School of Public Health, 1947-52; Clinical Assoc. Preventive Medicine (1952-55), Asst. Prof. Preventive Medicine (1955-58), Assoc. Prof. Preventive Medicine (1958-59), Lecturer on Medicine (1959-), Harvard Medical School; Consultant: Biol. and Medicine Div., U.S.A.E.C. (1949-64), Los Alamos Sci. Lab., U.S.A.E.C. (1949-), Mass. Div. of Occupational Hygiene (1949-54), Atomic Res. Centre, Ames, Iowa (1949-), Com. on Social and Occupational Health, W.H.O. (1951), I.L.O., and in Internal Medicine, Veterans Administration Hospital, Boston (1961). Book: Co-author, Sect. on Radiant Energy, in Ind. Toxicology (New York, Paul. B. Hoeber, 1949). Societies: A.M.A.; Assoc. American Physicians; Ind. Medical Assoc.; Fellow, American Coll. Physicians; American Conference of Governmental Ind. Hygienists.
Nuclear interest: Radiation protection.
Address: Massachusetts Institute of Technology, Cambridge, Massachusetts 02139, U.S.A.

HARDY, Judson, Jr., B.S., Ph.D. Born 1931. Educ.: North Carolina and Princeton Univs. Fellow Sci., Bettis Atomic Power Lab., Westinghouse Elec. Corp., Pittsburgh. Societies: A.N.S.; A.P.S.
Nuclear interest: Nuclear physics.
Address: Bettis Atomic Power Laboratory, Pittsburgh, Pennsylvania, U.S.A.

HARFORD, William B. Deputy Business Manager, Lawrence Rad. Lab.
Address: U.S.A.E.C., Lawrence Radiation Laboratory, Berkeley 4, California, U.S.A.

HARGÖ, B. Chief Eng., Nucl. Dept., Rosenblads Patenter A.B.
Address: Rosenblads Patenter A.B., P.O. Box 5088, Stockholm 5, Sweden.

HARGROVE, Clifford Kingston, B.A., B.Sc., M.Sc., Ph.D. Born 1928. Educ.: McGill and New Brunswick Univs. N.R.C. 1961-64. Society: Canadian Assoc. of Phys.
Nuclear interests: Nuclear physics; Elementary particle physics with emphasis on mesonic X-rays.
Address: Division of Pure Physics, National Research Council, Ottawa, Ontario, Canada.

HARIHAR, P., B.S., Ph.D. Dept. of Nucl. Sci. and Eng., Lowell Technol. Inst.
Address: Lowell Technological Institute, Department of Nuclear Science and Engineering, Lowell, Massachusetts 01854, U.S.A.

HARIN, V. P. Paper: Co-author, Medium Spectra of Neutrons in Double and Triple Fission of U^{235} by Thermal Neutrons (Abstract, Atomnaya Energiya, vol. 20, No. 4, 1966).
Address: Academy of Sceinces of the U.S.S.R., 14 Leninsky Prospekt, Moscow V-71, U.S.S.R.

HARINGTON, Sir Charles, K.B.E., Sc.D., F.R.S. Member, Joint M.R.C./U.K.A.E.C. Coordination Com. for Radiobiol. Res.
Address: Medical Research Council, 20 Park Crescent, London W.1, England.

HARISPE, Jean-Vincent, D. ès Sc. physiques, Pharmacien, Ingénieur-Chimiste E.N.S.C.P. Born 1904. Educ.: Paris Univ. Chef de Travaux, Chargé de Cours, Faculté de Pharmacie, Paris, 1950; Inspecteur Général des établissements dangereux et insalubres (Préfecture de Police, à Paris), 1960. Member, Consultative Com. of Dangerous Manufactories, French Ministère de l'Industrie; National Inspector of Manufactures using Radioactive Substances. Society: Sté. Chimique de France.
Nuclear interest: Protection of public health and safety.
Address: 194 Boulevard Saint Germain, Paris 7, France.

HARKE, Paul, Dipl.-Ing. Born 1904. Educ.: Darmstadt T.H. Member, Fachkommission III — Technisch-Wirtschaftliche Fragen bei Reaktoren, Deutschen Atomkommission; Mitglied, Verwaltungsrat, Deutsches Atomforum.
Addresss: A.E.G.—Zentralverwaltung, A.E.G. Hochhaus, 6 Frankfurt-am-Main-Süd, Germany.

HARKER, Wesley H., B.A. (Phys.). Born 1927. Educ.: California Univ. Manager, Rad. Effects Operation, Defence Systems Dept., G.E.C., Syracuse, New York. Societies: A.N.S.; A.P.S.; Assoc. for Comp. Mach.
Nuclear interests: Reactor physics; Reactor dynamics; Pulsed reactors; Radiaton effects.
Address: General Learning Corp., 5454 Wisconsin Avenue, Washington D.C. 20015, U.S.A.

HARKIN, Anthony, B.E., M.E. Born 1905. Educ.: Nat. Univ. Ireland. Deputy Chief Eng., Elec. Supply Board, 1950-. Chairman, Radioactivity Consultative Council to Minister of Health (Ireland). Deleg., Third I.C.P.U.A.E., Geneva, Sept. 1964. Societies: M.I.E.E.; M.I.Mech.E.; A.M.I.C.E.I.
Address: 52 Farney Park, Sandymount, Dublin, Ireland.

HARLEN, Frank, M.Sc. (Tech.). Born 1926. Educ.: Sheffield Univ. With Metropolitan-Vickers, 1946-56; Dowty Nucleonics, Ltd., 1957-. Society: Inst. of Phys.
Nuclear interest: Instrumentation.
Address: Dowty Technical Developments, Ltd., Andoversford, near Cheltenham, Glos., England.

HARLEY, John H., Sc.B., M.S., Ph.D. Born 1916. Educ.: Brown Univ. and Rensselaer Polytech. Inst. Member, U.S. Delegation U.N. Sci. Com. on Effects of Atomic Rad., 1957-; Director, Analytical Div. (1949-60), Director (1960-), Health and Safety Lab., U.S.A.E.C. Society: Bio-assay and Analytical Chem. Group.
Nuclear interests: Radiochemistry; Radioactive allout from weapons tests; General health protection in the field of atomic energy.
Address: 376 Hudson Street, New York, New York 10014, U.S.A.

HARMAN, Helen B. Born 1910. Educ.: Illinois and Northwestern Univs. Editor, Argonne News, 1954-. Sec., Assoc. of Nucl. Editors, and affiliate of Internat. Council of Ind. Editors.
Nuclear interest: Writing and editing nuclear subjects for the layman.
Address: Argonne National Laboratory, 9700 South Cass Avenue, Argonne, Illinois, U.S.A.

HARMER, David E., A.B., M.S., Ph.D. Born 1929. Educ.: Albion Coll. and Michigan Univ. Teaching Fellow (1950-51), Res. Asst. (1951-55), Michigan Univ.; Group Leader, The Dow Chem. Co., 1955-. Chairman, Subcom. on Process Rad., Advisory Com. on Rad. and Isotopes Development, U.S.A.E.C., 1966-. Societies: A.C.S.; A.I.Ch.E.
Nuclear interests: Industrial utilisation of radiation chemistry; Radiation polymerisation; Free radical chemical reactions under radiation; Solid state polymerisation by irradiation; Radiation graft copolymerisation; Dosimetry.
Address: 5908 Evergreen, Midland, Michigan 48640, U.S.A.

HARMEYER, Johein, Dr. med. vet. Born 1934. Educ.: Hanover, Munich and Vienna Univs. Wissenschaftliche Asst., Physiologisches Inst., Hanover Tieraerztliche Hochschule.
Nuclear interests: Tracer studies; Protein and amino acid metabolism (^{17}C, ^{3}H) comparative biochemistry of the urea cycle.
Address: Physiologisches Institut, Hanover Tieraerztliche Hochschule, 15 Bischofsholerdamm, Hanover, Germany.

HARMON, Dale Joseph, B.S. (Chem.), M.S. (Polymer Chem.). Born 1927. Educ.: Kent State (Ohio) and Akron Univs. Sen. Res. Chem., Plastics and Chem. Div., B. F. Goodrich Res. Centre. Society: A.C.S.
Nuclear interests: Effects of nuclear radiation on plastics and elastomers; Beneficial uses of radiation related to polymer chemistry such as polymerization and cross-linking.
Address: B. F. Goodrich Research Centre, Brecksville, Ohio, U.S.A.

HARMS, Jürgen, Dipl. Ing., Dr. tech. Born 1937 Educ.: Vienna Tech. Univ. Univ. Asst., Atominstitut der Osterreichischen Hochschulen, Vienna Tech. Univ., 1961-; On leave, Guest Researcher, Lawrence Rad. Lab. California Univ., Berkeley. Society: Austrian Phys. Soc.
Nuclear interests: Nuclear physics instrumentation; Nuclear electronics.
Address: 16 Schiffnerstrasse, A 4810 Gmunden, Austria.

HARMS, Keith L. Asst. to Vice-Pres., Planning, Atomics Internat., -1962; Assoc., Economics Res. Div., Planning Res. Corp., 1962-.
Address: Planning Research Corp., 1333 Westwood Boulevard, Los Angeles 24, California, U.S.A.

HARNWELL, Gaylord P., Dr. Pres., Pennsylvania Univ. Member, Undersea Warfare Advisory Panel, Joint Com. on Atomic Energy, U.S.A.E.C.
Address: Pennsylvania University, Philadelphia, Pennsylvania 19104, U.S.A.

HARPAZ, Yoav, M.Sc., Dip. Eng. Born 1927. Educ.: Technion and Israel Inst. Technol. Member, Israel A.E.C.; Dept. of Hydrological Res., Water Planning for Israel, Ltd.; Nat. Sec., Israel Soc. of Geodesy and Geophys. Societies: International Assoc. of Sci. Hydrology; I.U.G.G.
Nuclear interests: Siting and safety of nuclear plants; Radioactive waste disposal; Use of radioisotopes in hydrology.
Address: 48 Sokolov Street, Tel-Aviv, Israel.

HARPER, John, B.Sc. Educ.: London Univ. P.S.O., Roy. Aircraft Establishment, Farnborough, 1947-50; Res. Manager (1952-55), Head of Labs. (1955-), Reactor Fuel Element Labs., Reactor Group, Springfields, U.K.A.E.A. Papers: Co-author, Factors influencing the Magnesium Reduction of Uranium Tetrafluoride (Inst. Mining and Metal., 1956). Society: Assoc., Inst. of Metal.
Address: United Kingdom Atomic Energy Authority, Springfields Works, Salwick, Preston, Lancs., England.

HARPER, M. E. Vice-Pres., Treas. and Asst. Sec., Bechtel Nucl. Corp.
Address: Bechtel Nuclear Corp., 220 Bush Street, San Francisco 4, California, U.S.A.

HARPER, Mark, C. Eng., M.I.Mech.E., M.I.Mar.E. Born 1908. Manager, Marine Div., Vickers-Armstrongs (Engineers) Ltd., 1963-64; Marine Eng. Manager and Local Director, Vickers Ltd., Shipbuilding Group, 1964-. Chairman, Barrow-in-Furness Panel, Inst. Mech. Eng. (North Western Branch). Societies: North East Coast Inst. of Eng. and Shipbuilders; Foreign Affiliate Member, Soc. of Naval Architects and Marine Eng.
Nuclear interests: Marine nuclear installations for British naval submarines, surface warships and merchant cargo vessels.
Address: Vickers Ltd., Shipbuilding Group, Vickers-Armstrongs Barrow Shipyard, Barrow-in-Furness, Lancs., England.

HARPER, Paul Vincent, A.B., M.D. Born 1915. Educ.: Harvard Coll. and Harvard Medical School. Asst. Prof. (1953-55), Assoc. Prof. (1955-60), Prof. (1960-), Surgery Dept., Chicago Univ. Assoc. Director, Argonne Cancer Res. Hospital, 1963-69. Societies: Soc. Exptl. Biol. and Medicine; Rad. Res. Soc.; Soc. Nucl. Medicine.
Nuclear interests: Use of radioisotopes in clinical diagnosis and therapy; Study of biological processes.
Address: 950 East 59th Street, Chicago, Illinois 60637, U.S.A.

HARRELL, George T., Jr., M.D. (Duke). Born 1908. Educ.: Duke Univ. Prof. Medicine and Director, Internal Medicine Dept. (1947-52), Res. Prof. Medicine (1952-54), Bowman Gray School of Medicine, Wake Forest Univ.; Dean, Coll. of Medicine and Prof. Medicine, Florida Univ. Coll. of Medicine, 1954-64; Prof. Medicine, Dean Coll. of Medicine, Director, Milton S. Hershey Medical Centre, Pennsylvania State-Univ., 1964-. Editorial Board, Medicine, 1947-. Special Courses in isotopes at Oak Ridge and Brookhaven National Lab. Papers: Author and co-author, numerous contributions to medical journals. Societies: American Medical Assoc.; Fellow, American Coll. of Physicians; American Soc. Clinical Investigation; A.A.A.S.; Soc. Exptl. Biol. and Medicine; American Soc. Tropical Medicine; Fellow, Roy. Soc. Tropical Medicine.
Address: The Pennsylvania University College of Medicine. The M. S. Hershey Medical Centre, 500 University Drive, Hershey, Pennsylvania 17033, U.S.A.

HARREMOES, Poul, M. of Sci., M.S. Born 1934. Educ.: Tech. Univ. Denmark and M.I.T. Head, Fluid Mechanics Dept., Danish Isotope Centre; Assoc. Prof., Danish Inst. Technol.
Nuclear interest: The use of radioactive tracers as applied to flow problems of any kind.
Address: Danish Isotope Centre, 2 Skelbaekgade 1717, Copenhagen 5, Denmark.

HARRER, Joseph M., B.S. (E.E.) (Rensselaer Polytech. Inst.), M.S.E. (Illinois Inst. Technol.). Born 1913. Educ.: Rensselaer

Polytech. Inst. and Illinois Inst. Technol. At Argonne Nat. Lab., 1948-. Chairman, Chicago Sect., A.N.S.; Chairman, Reactor Instrumentation Subcom., A.I.E.E. Books: Chapter 8, Control Instrumentation and Drives in Reactor Handbook, vol. ii by H. Etherington (McGraw-Hill, 1958); Chapter 7.4, Nucl. Reactor Instrumentation in Reactor Handbook, vol. ii, Eng. (U.S.A.E.C., A.E.C.D.-3646); Nucl. Reactor Control Eng. (Van Nostrand, 1963). Societies: American Inst. of Elec. Eng.; A.N.S.; RESA.
Nuclear interests: Nuclear reactor design; Nuclear research and management.
Address: Argonne National Laboratory, 9700 S. Cass Avenue, Argonne, Illinois, U.S.A.

HARRINGTON, Charles Dana, B.S., M.A., Ph.D. Born 1910. Educ.: Harvard Coll. and Harvard Univ. Graduate School. Sen. Vice-Pres., United Nucl. Corp.;Gen. Manager and Pres., Douglas and United Nucl. joint venture at Hanford, 1965-; Ind. Member, Atomic Energy Labour-Management Advisory Com., U.S.A.E.C. Book: Uranium Production Technol. (Van Nostrand, 1959). Societies: A.N.S.; A.C.S.; Atomic Ind. Forum.
Nuclear interests: Fuel materials; Nuclear reactor cores; Chemical processing nuclear fuels.
Address: 303 St. Ives Drive, Severna Park, Maryland, U.S.A.

HARRIS, Carrol Fremont, B.S., M.S., Assoc. Prof. Born 1906. Educ.: Oklahoma State and Illinois Univs. Phys. Dept. (1942-), Radiol. Safety Officer (1956-), Oklahoma State Univ.; Res. Participant, Exptl. Measurements in Rad. Group, Health Phys. Div., O.R.N.L., 1952-53; Zone Commander, Off-Site Monitoring Programme, U.S.P.H.S. and U.S.A.E.C. (1955), Consultant, Off-Site Monitoring, U.S.P.H.S. (1957), Nucl. Test Series, Nevada; School in Theory and Operation of AG N-201 Nucl. Reactor, Aerojet Gen. Nucleonics, 1957; Member (1958-), Chairman (1965-), State Rad. Safety Advisory Com., Oklahoma State Board of Health; Participant, Inst. Nucl. Sci. and Technol. for Coll. Teachers, California Univ., Berkeley, 1960. Society: H.P.S.
Nuclear interests: Nuclear physics; Health physics (radiation safety).
Address: Oklahoma State University, Stillwater, Oklahoma, U.S.A.

HARRIS, Cecil Craig, B.S.E.E., M.S. Born 1925. Educ.: Tennessee Univ. Instructor, Mississippi State Univ., 1949-50; Res. Staff Member, O.R.N.L., 1950-67; Asst. Prof. Medical Electronics, Duke Univ. Medical Centre, Durham, North Carolina, 1967-. Pres., Soc. Nucl. Medicine. Books: Contributor to: Ch. 4 Scintillation Spectrometry in External Collimation Detection of Intracranial Neoplasms with Unstable Nuclides; Chs. 2 and 3, and co-editor, Progress in Medical Radioisotope Scanning (Tech. Information Dept. 7673, U.S.A.E.C., 1963); Ch. 1 in Scintillation Scanning in Clinical Medicine (Editor, J. L. Quinn, W. B. Saunders Co., 1964); Measurement of Radioactivity, in Principles of Nucl. Medicine (Editor, H. N. Wagner, Jr., W. B. Saunders Co., 1968).
Nuclear interests: Nuclear medical instrumentation, clinically and in research, particularly radionuclide imaging systems (scanning).
Address: Nuclear Medicine Division, Radiology Department, Duke University, Medical Centre, Durham, North Carolina, U.S.A.

HARRIS, David L. Chief Eng., Thermal and Nucl. Power Dept., Asselin, Benoit, Boucher, Ducharme, Lapointe.
Address: Asselin, Benoit, Boucher, Ducharme, Lapointe, 4200 Dorchester Boulevard West, Montreal 6, Canada.

HARRIS, Donald Rosswell, Jr., B.S., M.S. (Carnegie Inst. of Technol.), M.S. (Princeton). Born 1925. Educ.: Carnegie Inst. of Technol. and Princeton Univ. Fellow Sci., Bettis Atomic Power Lab., Westinghouse Elec. Corp. Treas., Reactor Phys. Div., A.N.S. Societies: A.N.S.; A.P.S.; A.A.A.S.
Nuclear interests: Reactor kinetics; Transport theory; Reactor fluctuations; Nuclear data for reactor studies.
Address: 5664 Valley View Drive, Bethel Park, Pennsylvania 15102, U.S.A.

HARRIS, Edward Grant, Ph.D. Born 1924. Educ.: Tennessee Univ. Theoretical Phys., N.R.L., Washington D.C., 1953-57; Asst. Prof. (1957-60), Assoc. Prof. (1960-63), Prof. (1963-), Tennessee Univ. Societies: A.P.S.; A.A.P.T.
Nuclear interests: Controlled thermonuclear reactions; Plasma instabilities.
Address: Physics Department, Tennessee University, Knoxville, Tennessee, U.S.A.

HARRIS, Emlyn. Chartered Accountant. Born 1905. With Air Ministry, Ministry of Aircraft Production and M.O.S. as professional accountant, 1937-53; Director of Accounts, U.K.A.E.A., 1954-67.
Nuclear interest: Finance and accounting techniques as applied in the atomic energy field.
Address: United Kingdom Atomic Energy Authority, 11 Charles II Street, London S.W.1, England.

HARRIS, F. A., F.C.A. Director, Electronic Instruments Ltd.
Address: Electronic Instruments Ltd., Richmond, Surrey, England.

HARRIS, Gerard G., Ph.D. (Phys.). Born 1926. Educ.: Princeton Univ. Asst. Prof. Phys., Columbia Univ., 1955-58; Member, Tech. Staff, Bell Telephone Labs., 1958-. Society: Acoustical Soc. of America.
Nuclear interests: High-energy nuclear physics; at present working in fields of psychoacoustics

and neurophysiology.
Address: Bell Telephone Laboratories, Murray Hill, New Jersey, U.S.A.

HARRIS, Gordon McLeod, B.Sc., M.Sc., A.M., Ph.D. Born 1913. Educ.: Saskatchewan and Harvard Univs. Sen. Lecturer Phys. Chem., Melbourne Univ., 1948-52; Res. Assoc., Wisconsin Univ., 1952-53; Assoc. Prof. (1953-56), Chairman, Chem. Dept., and Prof. Chem. (1956-), Buffalo Univ. Councillor, W.N.Y. Sect., A.C.S. Societies: A.C.S.; Chem. Soc. (London); A.A.A.S.; A.A.U.P.
Nuclear interest: Applications of isotopes to the study of the mechanism and kinetics of chemical reactions.
Address: Chemistry Department, New York State University at Buffalo, Buffalo 14, New York, U.S.A.

HARRIS, James T., Director, Supply and Services Div., Chicago Operations Office, U.S.A.E.C.
Address: U.S. Atomic Energy Commission, Chicago Operations Office, 9800 South Cass Avenue, Argonne, Illinois 60439, U.S.A.

HARRIS, John Albert, Jr., A.B. Born 1926. Educ.: Tulane and Tennessee Univs. Newsman, Nashville (1950-56), Gen. News Editor, Washington (1956-58), Assoc. Press; Public Information Officer (1958-66), at Oak Ridge (1964-66), Director, Div. of Public Information (1966-), U.S.A.E.C.
Address: Ridge Road, Damascus, Maryland, U.S.A.

HARRIS, L. H. Sec., Pacific Inst. of Earth Sci., Atomic Energy and Solar Rad.
Address: Pacific Institute of Earth Sciences, Atomic Energy and Solar Radiation, 448 North Avenue 56, Los Angeles 42, California, U.S.A.

HARRIS, Lieutenant Colonel Milford Douglas, Jr., D.V.M., Ph.D. (Radiobiol.). Born 1921. Educ.: Auburn and Rochester Univs. Chief, Appl. Radiobiol. Branch, U.S.A.F. School of Aerospace Medicine (1960-63), Tech. Director, Radiobiol., Headquarters, Aerospace Medical Div. (1963-66), Brooks Air Force Base, Texas; Deputy Chief, Biophys. Branch, Air Force Weapons Lab., Kirtland Air Force Base, 1966-.
Nuclear interests: Radiobiology; Research management.
Address: 2230 Stockton Loop, Kirtland Air Force Base, New Mexico 87118, U.S.A.

HARRIS, Milo Truman, M.D., M.S. (Radiol., Minnesota). Born 1903. Educ.: Texas Univ. Former Clinical Assoc. Prof. Radiol., Washington Univ. Societies: Soc. of Nucl. Medicine; American Roentgen Ray Soc.; Radiol. Soc. of North America; A.M.A.
Nuclear interest: Use of radioactive isotopes in diagnosis and therapy in the general practice of radiology.
Address: 252 Paulsen Building, Spokane 1, Washington, U.S.A.

HARRIS, Norman Allan, B.A. (Phys.). Born 1933. Educ.: Occidental Coll. and Vanderbilt Univ. Society: H.P.S.
Nuclear interests: Nuclear physics, space radiation shielding, astrobiomedicine, nuclear weapons effects, radiological safety.
Address: Planning Research Corporation, 1100 Glendon, Los Angeles, California, 90024, U.S.A.

HARRIS, Sam. Director, Rio Algom Mines Ltd.
Address: Rio Algom Mines Ltd., 335 Bay Street, Toronto, Ontario, Canada.

HARRIS, Saul Joseph, B.S. (Phys.), M.S. (Ind. and Management Eng.). Born 1923. Educ.: Queens Coll., New York and Columbia Univ. Regional Rep., Nat. Centre for Radiol. Health, P.H.S., Region II; Formerly Tech. Director, Atomic Accessories, Inc. and Asst. Manager, Tech. Services, Atomic Ind. Forum, Inc. Treas., H.P.S.; Health Phys. Member, New York State Board of X-ray Technician Examiners, 1965-68. Books: Nucl. Power Safety Economics (New York, Pilot Books, 1961); The Impact of the Peaceful Uses of Atomic Energy on State and Local Govt. (Atomic Ind. Forum, Inc., 1959); State Activities in Atomic Energy, 4th edition (Atomic Indust. Forum, Inc., 1958). Societies: American Assoc. of Phys. in Medicine; H.P.S.; Rad. Res. Soc.; A.N.S.; A.A.A.S.; Fellow, A.P.H.A.
Nuclear interests: Health physics; Codes and standards; Administrative law; Economic aspects; X-ray technician training.
Address: 14 Stillman Road, Glen Cove, Long Island, New York 11542, U.S.A.

HARRIS, Shearon. Vice-Pres. and Gen. Counsel (-1963), Pres. (1963-), Carolina Power and Light Co.
Address: Carolina Power and Light Co., 336 Fayetteville Street, Raleigh, North Carolina, U.S.A.

HARRIS, Thomas W. Rad. Phys., Div. of Radiol. and Occupational Health, Florida State Board of Health.
Address: Florida State Board of Health, Division of Radiological and Occupational Health, Post Office Box 210, Jacksonville, Florida 32201, U.S.A.

HARRIS, Walter E., B.Sc., M.Sc., Ph.D. Born 1915. Educ.: Alberta and Minnesota Univs. Now on academic staff, Alberta Univ. Societies: A.C.S.; Fellow, Chem. Inst. of Canada.
Nuclear interests: Nuclear chemistry, chemical effects of nuclear transformations, chromatographic separations.
Address: Department of Chemistry, Alberta University, Edmonton, Alberta, Canada.

HARRISON, Donald, Diploma of Faraday House, M.I.E.E. Born 1921. Educ.: Faraday House Elec. Eng. Coll., London. Sen. Principal Sci., Group Leader, Techniques Group, Control and Instr. Div., A.E.E., Winfrith. Nuclear interest: Reactor instrumentation. Address: Burchmore, Corfe Way, Broadstone, Dorset, England.

HARRISON, Edwin Davies, B.S., M.S., Ph.D. Born 1916. Educ.: U.S. Naval Acad., Virginia Polytech. Inst. and Purdue Univ. Res. Assoc., Purdue Univ., 1950-52; Asst. Dean of Eng., Assoc. Prof., Virginia Polytech. Inst., 1952-55; Dean of Eng., Toledo Univ. (Ohio), 1955-57; Pres., Georgia Inst. Technol., 1957-. Board of Visitors, U.S. Military Acad. Society: A.S.M.E. Nuclear interests: Use of high-intensity neutron beam in areas of materials and health; Use of short-lived radioisotopes. Address: Georgia Institute of Technology, 225 North Avenue, N.W., Atlanta, Georgia 30332, U.S.A.

HARRISON, James Merritt, B.Sc. (Manitoba), M.A. and Ph.D. (Queen's). Born 1915. Educ.: Manitoba and Queen's (Ontario) Univs. Asst. Deputy Minister, Mines and Geosci., 1964-. Pres., Internat. Council of Sci. Unions; Pres., Roy. Soc. Canada. Societies: Roy. Soc. Canada; Fellow, Geological Soc. America; Soc. Economic Geologists; Canadian Inst. of Mining and Metal.; Foreign Assoc., N.A.S. Address: Department of Energy, Mines and Resources, 588 Booth Street, Ottawa 4, Ontario, Canada.

HARRISON, John Raymond, M.A. (Oxon.). Born 1929. Educ.: Oxford Univ. Phys. Reactor Div., A.E.R.E., Harwell, 1953-59; Principal Reactor Phys., Nucl. Health and Safety Dept., C.E.G.B., London, 1959. Book: Nucl. Reactor Shielding (Temple Press, 1959). Nuclear interests: Reactor physics and safety. Address: 7 High Firs Crescent, Harpenden, Herts., England.

HARRISON, Kenneth Bond. Born 1907. Sen. Construction Eng. (1948-50), Generation Eng. (Construction) (1950-58), C.E.A.; Project Eng., Northern Project Group, C.E.G.B., 1958-. Societies: Assoc. M.C.T.; A.M.I.Mech.E. Address: 471 Liverpool Road, Ainsdale, Southport, Lancs., England.

HARRISON, L. W. Head, Sci. Equipment Div., Electronic Development and Applications Co. Ltd. Address: Electronic Development and Applications Co. Ltd., P.O. Box 6415, Wellington, New Zealand.

HARRISON, Richard Holmes, B.S. Born 1907. Educ.: Washington Univ. and U.S. Military Acad. 2nd Lieutenant (1931), Brigadier Gen. (1955), U.S. Army. Deputy Chief, Defence Atomic Support Agency, and Chief, Joint Atomic Information Exchange Group, 1958-61. Asst. to Vice-Pres. (New York City) (1961-62), Gen. Manager, Atomic Energy Div. (New York City) (1962-63), Vice-Pres. in charge of Atomic Energy Div. (Lynchburg, Virginia) (1963-68), Member, Chairman's and President's Staff handling Special Assignments, Power Generation Div. (1968-), Babcock and Wilcox Co. Director, Atomic Ind. Forum. Societies: Chamber of Commerce of the U.S.; Nat. Assoc. of Manufacturers; A.N.S. (local chapter - North Carolina/Virginia Sect.); Soc. of Naval Architects and Marine Eng.; American Management Assoc.; Newcomen Soc. of America. Nuclear interest: Management. Address: 3901 Peakland Place, Lynchburg, Virginia, U.S.A.

HARRISON, William Burr, III, B.S. (Chem. Eng.), M.S., Ph.D. (Chem. Eng. major, Maths. minor). Born 1922. Educ.: Tennessee Univ. Consultant, O.R.N.L., 1948-50; Instructor of Chem. Eng., Tennessee Univ., 1949-50; Development Eng., O.R.N.L., 1950-52; Formerly Prof. and Director, School of Nucl. Eng.; Chief, Nucl. Sci. Div., Georgia Inst. of Technol., 1953-66; Director, Eng. Experiment Station, Virginia Polytech. Inst., 1966-. Treas., Vice-Chairman, Chairman, Knoxville-Oak sect., A.I.Ch.E.; Director, Atlanta sect., A.I.Ch.E. Member, Professional Status Com., A.N.S. Society: A.N.S. Address: 1357 Paces Forest Drive, N.W., Atlanta, Georgia, U.S.A.

HARRISON-SMITH, Captain Sydney Alick. Born 1905. Educ.: R.N. Colleges Osborne, and Dartmouth. R.N. Eng. Coll., Keyham, R.N. Coll., Greenwich. Asst. Eng.-in-Chief, in charge of Dreadnought Project, Admiralty, 1954-58; Nucl. Liaison Officer, Swan, Hunter Group of Companies, 1958-; Head, Eng. Dept., Neptune Works, Swan, Hunter and Wigham Richardson Ltd., 1960-62; Director, Wallsend Slipway and Eng. Co., Ltd., 1962-. Societies: M.I.Mech.E.; M.I.Mar.E.; B.N.E.S. Nuclear interests: General in connection with ships. Address: Wallsend Slipway and Engineering Co. Ltd., Wallsend-on-Tyne, Northumberland, England.

HARRY, Ralph Lindsay, LL.B., B.A. (Jurisprudence). Born 1917. Educ.: Tasmania and Oxford Univs. Australian Ambassador to Belgium; Head, Australian Mission to E.E.C., E.C.S.C. and Euratom. Address: Australian Embassy, Brussels, Belgium.

HART, Edwin James, B.S., M.S., Ph.D. (Chem.). Born 1910. Educ.: Washington State Coll. and Brown Univ. Argonne Nat. Lab., 1948-. Assoc. Editor, Rad. Res. Societies: Faraday Soc.; Rad. Res. Soc.; A.A.A.S.; A.C.S. Nuclear interest: Radiation chemistry. Address: Chemistry Division, Argonne National Laboratory, Argonne, Illinois 60439, U.S.A.

HART, Hans Karl Richard, Dr. rer. nat., Dipl.-Phys. Born 1923. Educ.: Brandenburg State Univ., Potsdam. Asst., Inst. of Phys., Brandenburg State Univ., Potsdam; Sci. Contributor, Tech. Univ. for Chem., Merseburg; Head of Sci. Operations, Res. Inst. Potsdam of the Acad. Sci., Berlin; Dozent, Tech. Univ. for Chem., Merseburg. Books: Radioaktive Isotope in der Dickenmessung (Berlin, VEB Verlag Technik, 1958); Radioaktive Isotope in der Betriebsmesstechnik (Berlin, VEB Verlag Technik, 1962). Society: Phys. Soc. of D.D.R., Board of Technics.
Nuclear interest: Process control and automation by means of nuclear radiation.
Address: 6 Krumme Gehren, Kleinmachnow bei Berlin, German Democratic Republic.

HART, Hiram Emanuel, B.S., Ph.D. Born 1924. Educ.: C.C.N.Y. and New York Univ. Phys., Montefiore Hospital, 1952-; Instructor (1953-57), Asst. Prof. (1957-62), Assoc.Prof. (1962-), Dept. of Phys., C.C.N.Y. Societies: A.P.S.; American Assoc. Health Phys.; A.A.A.S.
Nuclear interests: Tracer analysis and perturbation-tracer analysis. Focusing collimator coincidence scanning. Electron and X-ray diffraction systems.
Address: Department of Physics, City College of N.Y., New York 31, N.Y., U.S.A.

HART, Robert J. Deputy Director, Div. of Contracts (-1968), Deputy Manager, Richland Operations Office (1968-), U.S.A.E.C.
Address: United States Atomic Energy Commission, Richland Operations Office, P.O. Box 550, Richland, Washington, U.S.A.

HARTECK, Paul, Ph.D. (Berlin). Born 1902. Educ.: Vienna and Berlin Univs. Prof. Phys. Chem., Hamburg, 1934-51; Rector, Hamburg Univ., 1948-50. Distinguished Res. Prof. Phys. Chem., Rensselaer Polytech. Inst., 1951-. Book: Photo-Chem. (1934). Societies: A.C.S.; German Chem. Soc., Foreign Member of Max Planck Soc.; Bunsen Soc.
Nuclear interests: Radiation chemistry: investigations on the transformation of nuclear into chemical energy in basic chemical reactions like the systems N_2-O_2; CO_2-N_2; CO, and so on; Photochemistry of the atmosphere of the earth and the planets.
Address: Brunswick Hills, Troy, New York, U.S.A.

HARTFIELD, Richard Abraham, B.M.E. Born 1931. Educ.: Rensselaer Polytech. Inst. Graduate Trainee (1953-55), Asst. Eng., Nucl. Power Dept. (1955-56), Allis-Chalmers Manufacturing Co.; Phys., Argonne Nat. Lab.,1956-57; Eng. (1957-61), Sen. Eng. (1961-63), Nucl. Power Dept., Allis-Chalmers Manufacturing Co.; Nucl. Safety Eng., U.S.A.E.C., 1963-. Society: A.N.S.
Nuclear interests: Reactor design and operation; Nuclear propulsion; Nuclear safety.
Address: 21420 Maplewood Avenue, Rocky River, Ohio 44116, U.S.A.

HARTH, Erich M., Ph,D. Born 1919. Educ.: Syracuse Univ. At G.S. 12 N.R.L., 1951-57; Res. Assoc., Duke Univ., 1954-57; Assoc. Prof. Phys. (1957-63), Prof. Phys. (1963-), Syracuse Univ. Societies: A.P.S.; A.I.P.; Biophys. Soc.; A.A.A.S.
Nuclear interests: High-energy nuclear physics; Elementary particles.
Address: Physics Department, Syracuse University, Syracuse, New York 13210, U.S.A.

HARTIG, Manfred, Ing.-Chem. Born 1930. Educ.: Fachschule für Chemie, Köthen. Stellv. Arbeitsgruppen-Leiter, Inst. für Ernährung, Deutschen Akademie der Wissenschaften zu Berlin.
Nuclear interests: Radiation protection; Irradiation sterilisation; Tracer employment.
Address: Institut für Ernährung, Arbeitsgruppe Isotopentechnik, 114/116 A.-Scheunert-Allee, Potsdam-Rehbrücke 1505, German Democratic Republic.

HARTILL, Eric Raymond, B.Sc. (Eng.). Born 1923. Educ.: Nottingham Univ. With Associated Elec. Industries Ltd., (now Chief Eng., Transformer Development), 1944-. Society: Inst.of Elec. Eng.
Nuclear interests: Nuclear fusion and particle accelerators.
Address: A.E.I. Transformer Division, Southmoor Road, Wythenshawe, Manchester 23, England.

HARTLAND, Stanley, M.A. Born 1932. Educ.: Cambridge Univ. With U.K.A.E.A., Risley, 1955-58; Imperial College, London, 1958-61 (at Harwell to Aug. 1959); Nottingham Univ., 1961. Society: Inst. Chem. Eng.
Nuclear interests: Chemical engineering problems in solvent extraction; Coalescence of drops; Optimum design and operation of stagewise and differential contactors; Effect of back mixing on their performance.
Address: 78 Moor Lane, Bramcote, Nottingham, England.

HARTLEY, Frank Ramsay, M.Sc. Born 1920. Educ.: Melbourne Univ. Res. Officer, C.S.I.R.O. Australia, 1946-52; Asst. Director, Australian Mineral Development Labs., 1959-. Societies: Roy. Australian Chem. Inst.; Australian Inst. Mining Metal.
Nuclear interest: Process engineering in extractive metallurgical field related to uranium, thorium, lanthanum and allied metals, particularly scandium.
Address: Australian Mineral Development Laboratories, Thebarton, South Australia.

HARTLEY, Hubert C., Lieutenant Colonel. Base Commander, 731st Radar Squadron, U.S. Air Force.
Address: U.S. Air Force, 731st Radar Squadron, Sundance Air Force Station, Wyoming, U.S.A.

HARTMAN, Americo, civil eng. Born 1926. Educ.: Montevideo Univ. Asst. Prof. Mathematic Analysis, Economics School, Montevideo Univ., 1955-56; Eng., Thermal Plants Dept., Administracion de las Usinas y los Telefonos del Estado; Prof. Construction Processes, Construction School, Trades Univ., Uruguay; Member, Comision Nacional de Energia Atomica, Uruguay; Eng., Dept. of Works for Interconnection of Transmission Systems.
Address: 8 de octubre Street No. 2297 bis. apto. 9, Montevideo, Uruguay.

HARTMANN, Arnold, Ing. Mitglied, Fachkommission 1, Recht und Verwaltung, Federal Ministry for Sci. Res.
Address: Federal Ministry for Scientific Research, 46 Luisenstrasse, 532 Bad Godesberg, Germany.

HARTMANN, Günther, Dr. rer. nat., Dipl. Phys. Inst. für angewandte Radioaktivitat Leitende Mitarbeiter, Deutsche Akademie der Wissenschaften zu Berlin.
Address: Institut für angewandte Radioaktivität, 15 Permoser Strasse, Leipzig 05, German Democratic Republic.

HARTMANN, Peter, Dr., Dipl.-Volksw. Born 1917. Educ.: Cologne Univ. Vorstandsmitglied, Kommunales Elektrizitätswerk Mark A.G., Hagen; Kfm. Geschäftsführer, Arbeitsgemeinschaft Versuchs-Reaktor G.m.b.H., Düsseldorf.
Address: 9 Schiefe Hardt, 5800 Hagen, Germany.

HARTMANN, Robert Frederick, B.S. (Metal. Eng., Missouri School of Mines and Metal.), M.S. (Automotive Eng., Chrysler Inst. of Eng.), Metal. Eng. (Missouri School of Mines and Metal.). Born 1922. Plant Metal., Gear and Axle Plant, Chrysler Corp., 1948-51; Supervisor Metal. Labs. (1951-54), Chief, Basic Metal. (1954-56), Western Brass Mills, Olin-Mathieson Chem. Corp.; Asst. Manager, Metal. Development Dept. (1956-59), Manager, Metal. Technol. Dept. (1959-), Uranium Div., Mallinckrodt Chem. Works. Member of Working Com. of Fuel Element Development Com., U.S.A.E.C. Society: A.S.M.
Nuclear interests: Uranium production and development of thermite reductions, melting, casting and alloying, extrusion, forging, machining, heat treatment, mechanical and physical properties, microstructure and orientation; Production and properties of UO_2; Fuel element technology and performance.
Address: 1532 Surf Side Drive, St. Louis 38, Missouri, U.S.A.

HARTMAN, Werner, Prof. Dr. Ing. habil. Born 1912. Educ.: Berlin-Charlottenburg T.H. Chief of Labs., Soviet Union, 1945-55; Managing Director and Director of Res., Vakutronik, Dresden, 1955-62; Prof., Dresden Tech. Univ. for Rad. Measurements, 1956-; Director, Inst. for Microelectronics, Dresden, 1961-. Books: Grundlagen und Arbeitsmethoden der Kernphysik (chapter: Nucl. Instruments; Berlin, Akademie-Verlag, 1957); co-author, Fotovervielfacher und ihre Anwendung in der Kernphysik (Berlin, Akademie-Verlag, 1957); Lehrbuch der Kernphysik (chapter: Measurements of Nucl. Particles; Leipzig, Verlag Teubner, 1966). Societies: Phys. Soc. of the German Democratic Republic; Session for Physics, Acad. of Sci., Berlin; A.N.S.
Nuclear interests: Radiation measurements; Management.
Address: 6 Klengelstrasse, 8054 Dresden, German Democratic Republic.

HARTRIDGE, Alfred L. Formerly Vice-Pres., Assoc. Nucleonics, Inc.; Vice Pres., Assoc. Eng. and Consultants, Inc. Director, Atomic Ind. Forum, Inc.
Address: Associated Engineers and Consultants, Inc., 975 Stewart Avenue, Garden City, Long Island, New York, U.S.A.

HARTSHORNE, Edward, B.S. Born 1906. Educ.: M.I.T. Gen. Manager (1956-58), Vice-Pres. (1958-61), Nucl. Fuels Operation, Olin Mathieson Chem. Corp.; Vice Pres., United Nucl. Corp., 1961-62; Olin Mathieson Chem. Corp., 1962-. Society: A.I.M.E.-A.S.M.
Nuclear interests: Metallurgy, reactor design and fabrication and management.
Address: Olin Mathieson Chemical Corp., 275 Winchester Avenue, New Haven 4, Connecticut, U.S.A.

HARTWEG, Helmut, M.D. Born 1920. Educ.: Munich and Tuebingen Univs. Prof. for Radiol., Tuebingen Univ. Pres., Roentgen Soc. of Southwest Germany. Society: German Röntgen Soc. Democratic Republic.
Address: Robert-Bosch-Hospital, Stuttgart, Germany.

HARTWELL, Robert Wellington, B.S. (Eng.), M.S. (Business Administration). Born 1917. Educ.: Michigan Univ. Naval Reactors Branch, Bureau of Ships, U.S.A.E.C., 1951-52; Assisted in design of reactor power plant for submarines Nautilius and Sea Wolf and Carrier Vessel Reactor and Pressurised Water Reactor; Director, Nucl. Power Development Dept., Detroit Edison Co., 1954-56; Asst. Gen. Manager (1956-57), Gen. Manager (1957-63), Power Reactor Development Co.; Asst. Manager Eng. (1964-66), Asst. Vice-Pres., and Manager of Eng. (1967-), Detroit Edison Co., A.N.S. Tech. Adviser, U.S., Deleg. Second I.C.P.U.A.E., Geneva, 1958; Participant: Conference on Operating Experience with Power Reactors, I.A.E.A., Vienna, 1963, and VIII Nucl. Congress, Rome, 1963. Societies: A.S.M.E.; A.N.S.
Nuclear interests: Reactor design, management, operating results, and economics.
Address: 30050 Bayview Drive, Grosse Ile, Michigan, U.S.A.

HARTWELL, Stephen, B.S. (Administrative Eng.). Born 1916. Educ.: Lafayette Coll., Easton, Pennsylvania. Chief, Construction Eng. Reports Branch, U.S.A.E.C., 1951-54; Exec. Vice-Pres., Atomics, Phys. and Sci. Fund, Inc., 1957-; Director, Ohmart Corp., 1963-. Nuclear interests: Development and management of all phases of the nuclear industry.
Address: River Road, Lorton, Virginia, U.S.A.

HARTWIG, Günther, Dipl. Phys., Dr. rer. nat. Born 1934. Educ.: Erlangen and Mainz Univs. Inst. für exptl. Kernphysik, Karlsruhe T.H.; Harvard Univ., U.S.A. Nuclear interests: Strong and electromagnetic interaction; Electromagnetic form factors of nucleons (especially neutron form factors); Quasielastic e-d scattering; Inelastic e-p scattering; Electro pion production.
Address: Institut für Experimental Kernphysik, Kernreaktor, Karlsruhe, Germany.

HARTY, Harold, B.S. (Mech. Eng.), M.S. (Mech.Eng.). Born 1924. Educ.: Washington Univ. and California Inst. of Technol. Manager, Heavy Water Reactor Programme Office, Pacific Northwest Lab., Battelle Memorial Inst., Richland. Society: A.N.S. Nuclear interest: Nuclear reactor design, development and operation.
Address: Pacific Northwest Laboratory, Battelle Memorial Institute, Box 999, Richland, Washington 99352, U.S.A.

HARTZ, Gustav Emil, M.Sc. (Eng.). Born 1888. Educ.: Roy. Danish Tech. Coll., Copenhagen. Managing Director (1947-60), Vice-Chairman of Board (1960), Thomas B. Thrige's Foundation and Works, Odense; Member, Roy. Elec. Council; Member of Board: Scandia Waggon Works, Ltd., Brandt Klaedefabrik, Ltd., and Thomas B. Thrige Copenhagen Ltd. Member, Board, Federation Danish Industries; Hon. Pres., Danish Nat. Com. for World Power Conference; Co-Founder and Member, Acad. Tech. Soc.; Formerly Member, Danish A.E.C.
Address: 25 Bredstedgade, Odense, Denmark.

HARTZELL, Harley Weikert, B. Mech. Eng., Registered Professional Eng. (Elec.), Michigan. Born 1911. Educ.: Rensselaer Polytech. Inst. and Ohio State Univ. Eng., Supervisor of Operations and Consulting Eng. (1949-61), Director of Office Operations (1961-), Sec. (1950-), formerly Administration Eng., Commonwealth Assoc. Inc.; Vice-Pres. and Director (1955-62). Pres. and Director (1962-), Commonwealth Buildings Inc.; Asst. Sec., Commonwealth Services, Inc., 1957-. Societies: A.N.S.; American Inst. of Elec. Eng.; A.S.E.E.; American Management Assoc.; N.S.P.E.; Michigan Soc. of Professional Eng.; Michigan Eng. Soc.; Michigan Assoc. of the Professions. Nuclear interests: Management of consulting and design engineering services for nuclear power plants,laboratories, and facilities. Projects include AE services for Enrico Fermi Atomic Power Plant, C-stellerator Laboratory and APDA Test Facility.
Address: 811 Union Street, Jackson, Michigan, U.S.A.

HARTZER, L. J. Pres., Rad. Instrument Development Lab. Inc.
Address: Radiation Instrument Development Laboratory Inc., 4501 West North Avenue, Melrose Park, Illinois, U.S.A.

HARTZER, Pieter Daniël, B.Sc., M.Sc. (cum laude). Born 1927. Educ.: Pretoria Univ. Rad. Phys., South African Atomic Energy Board, 1956-; Member, South African Provisional Council for Civilian Protection. Society: South African Inst. Physics. Nuclear interests: Production and uses of radioactive isotopes; Health physics; Civil defence.
Address: South African Atomic Energy Board, Private Bag 59, Pretoria, South Africa.

HARUN-ar-RASHID, A. M. B.Sc. (Hons.), M.Sc., Ph.D. (Theoretical Phys., Glasgow). Born 1933. Educ.: Dacca and Glasgow Univs. Principal Sci. Officer, Pakistan A.E.C., 1962-67; Prof. of Theoretical Phys., Univ. of Islamabad, 1967-; Visiting Prof., Internat. Centre for Theoretical Phys., Trieste, 1966 and 1968. Gen. Sec., Pakistan Phys. Soc., 1966-67. Book: Nucl. Structure, (North Holland, 1967). Society: A.P.S. Nuclear interests: Theoretical physics especially elementary particle physics.
Address: Department of Physics, University of Islamabad, Pakistan.

HARVEY, Douglas G. Former Manager, Space Auxiliary Power Systems, Nucl. Div., formerly Asst. Director, Small Power Systems, and formerly Programme Director, SNAP-29 Programme, Martin Co.; Vice Pres. and Manager, Nucl. Systems Development, Sanders Nucl. Corp., 1967-.
Address: Sanders Nuclear Corporation, Nashua, New Hampshire, U.S.A.

HARVEY, J. L. Chief Rad. Officer, Nucl. Eng. Co., Inc.
Address: Nuclear Engineering Co., Inc., P.O. Box 594, Walnut Creek, California,U.S.A.

HARVEY, J. R. Manager of Eng., Babcock and Wilcox Canada Ltd.
Address: Babcock and Wilcox Canada Limited, Coronation Boulevard, Galt, Ontario, Canada.

HARVEY, John Arthur, B.Sc., Ph.D. Born 1921. Educ.: Queen's Univ., Ontario and M.I.T. Phys., Brookhaven Nat. Lab., 1951-55; Phys., O.R.N.L., 1955-. Societies: Fellow, A.P.S.; RESA. Nuclear interests: Neutron and nuclear physics; Neutron cross-sections; Low-energy neutron spectroscopy.

Address: Oak Ridge National Laboratory, P.O. Box X, Oak Ridge, Tennessee 37830, U.S.A.

HARVEY, John Philip, B.Sc. (Hons.). Born 1911. Educ.: London Univ., Director, Harvey Electronics, Ltd., Farnborough, Hants, 1950-; Director, (1953-), and Elec. Designer, Phillips Control (G.B.), Ltd., Farnborough, Hants.; Director, Servo Units, Ltd., Farnborough, Hants, 1957-; Director, Materials Data, Ltd., Farnborough, Hants, 1957-. Convenor, Farnborough Area of Southern Centre, I.E.E.; Committee Member, Southern Centre, I.E.E. Society: M.I.E.E.
Nuclear interests: Design and manufacture of servo mechanisms and electrical solenoids.
Address: The Croft, Gough Road, Fleet, Hants, England.

HARVEY, N. D. M., M.B., Ch.B. (N.Z.), D.M.R.T., M.C.R.A. Sen. Staff Radiotherapist, Roy. Adelaide Hospital.
Address: Royal Adelaide Hospital, North Terrace, Adelaide, South Australia.

HARVEY, P. J. Chief of Instr. Div., Ekco Electronics Ltd.
Address: Ekco Electronics Ltd., Ekco Works, Southend-on-Sea, Essex, England.

HARWELL, W. L. Member, Tech. Information Panel, U.S.A.E.C.
Address: Legal and Information Control Department, Union Carbide Co., Oak Ridge, Tennessee, U.S.A.

HASAGAWA, Kunihiko, B.Sc. Concerned with Radiochem. and Nucl. Chem., Radiochem. Res. Lab., Shizuoka Univ.
Address: Shizuoka University, Radiochemistry Research Laboratory, 2 Oiwa-cho, Shizuoka City, Japan.

HASEGAWA, Haruo. Born 1926. Educ.: Osaka Univ. of Commerce. Sub-Manager, Atomic and Electronic Dept., Nissho-Iwai Co., Ltd., Tokyo. Planning Com. First Atomic Power Industry Group, Japan.
Nuclear interests: Planning and management of atomic power industry group; Importation of atomic power plant and related equipment, instrument and materials.
Address: B-307 Nissho Apt., 1-17 Ishikawacho, Ohta-ku, Tokyo, Japan.

HASEGAWA, Keishi, B.S. Born 1921. Educ.: Tokyo Inst. of Technol. Chief, 1st Dept. Toa Nenryo Kogyo K. K. Central Res. Lab. Societies: Chem. Soc. of Japan; Japan Petroleum Inst.; Tokyo Atomic Ind. Consortium.
Nuclear interests: Utilisation of radioisotopic technique for research on petroleum refining and petrochemical production.
Address: Toa Nenryo Kogyo K. K. Central Research Laboratory, Oi-machi, Iruma-gun, Saitama-Ken, Japan.

HASEGAWA, Osamu, Prof. Lab. of Appl. Nucl. Phys., Faculty of Eng., Kyushu Univ.
Address: Kyushu University, Fukuoka, Japan.

HASEK, Carl William, Jr., B.S. (Mech. Eng.), M.S. (Marine Eng.). Born 1915. Educ.: Pennsylvania State Univ. and M.I.T. Societies: Soc. Naval Architects and Marine Eng.; Soc. Naval Eng.; U.S. Naval Inst.
Nuclear interests: Thermal and breeder reactors for utilities; Nuclear propulsion systems for merchant ships; Design of cores, reactor vessels, and heat exchangers.
Address: The Babcock and Wilcox Company, Atomic Energy Division, P.O. Box 1260, Lynchburg, Virginia 24501, U.S.A.

HASELKORN, Robert. Member, Editorial Board, Bulletin of the Atomic Sci.
Address: Bulletin of the Atomic Scientists, 935 East 60th Street, Chicago 37, Illinois, U.S.A.

HASENCLEVER, Dieter Friedrich Walter, Dr.-Ing., Dipl. Phys. Born 1921. Educ.: Tübingen and Breslau Univs., Dresden T.H., Braunschweig T.H. and Aachen T.H. Director, Staubforschungsinst. des Hauptverbandes der gewerbl. Berufsgenossenschaften e.V., Bonn. Member, Reaktor-Sicherheitskomm, Bundesminister für wissenschaftliche Forschung; Counsellor, Fachgruppe Staubtechnik, V.D.I.; Societies: Deutsche physikalische Gesellschaft e.V.; Gesellschaft der Freunde und Förderer, Deutsche Röntgenmuseum e.V.
Address: Staubforschungsinstitut des Hauptverbandes der gewerbl. Berufsgenossenschaften e.V., 103 Langwartweg, Bonn, Germany.

HASENFUSS, W., Rechtsanwalt. Geschäftsführer, Kernkraftwerk Baden-Württemberg Planungsgesellschaft m.b.H. Director, Kernkraftwerk Obrigheim G.m.b.H. Member, Arbeitsausschuss Atomenergierecht Vereinigung Deutscher Elektrizitätswerke.
Address: Kernkraftwerk Obrigheim G.m.b.H., 6951 Obrigheim/Neckar, Germany.

HASENKAMP, Arthur, B.S. (Ch.E.), M.S. (Ch.E.); Graduate, O.R.S.O.R.T. Born 1920. Educ.: Denver Univ., O.R.S.O.R.T. and Denver Res. Inst. A.C.F. Industries, Albuquerque, N. M., 1952-54; Sandia Corp., Albuquerque, N. M., 1954-. Societies: A.I.Ch.E.; A.N.S.
Address: 804 Manzano NE, Albuquerque, N. M. 87110, U.S.A.

HASHIMI, H. E. Khalid M. EL-. See EL-HASHIMI, H. E. Khalid M.

HASHIMOTO, Ichiro. Prof., Nucl.Lab., Faculty of Sci. and Eng., Ritumeikan Univ.
Address: Nuclear Laboratory, Faculty of Science and Engineering, Ritumeikan University, 28 Toziin Kitamati, Kyoto, Japan.

HASHIMOTO, Seinosuke. Born 1894. Sen. Managing Director, J.A.I.F.; Managing Director, Inst. for Elec. Power Economy; Director, Fund for Peaceful Atomic Development of Japan; Member, Nucl. Public Relations Contact Group.
Address: Japan Atomic Industrial Forum, Incorporated, No. 1-1-13, Shimbashi, Minato-ku, Tokyo, Japan.

HASHIMOTO, Shozo. Chief, Lab. of Radioisotopes, Keio Univ.
Address: Laboratory of Radioisotopes, School of Medicine, Keio-Gijuku University, 35 Shinanomachi, Shinjuku-ku, Tokyo, Japan.

HASHIMY, S. T. AL-. See AL-HASHIMY, S. T.

HASHINO, Tomoyasu. Asst. Prof., Nucl. Fuel, Dept. of Nucl. Eng., Kyoto Univ.
Address: Kyoto University, Department of Nuclear Engineering, Yoshida-Honmachi, Sakyo-ku, Kyoto, Japan.

HASHISH, Salah el Din, B.Sc. (Hons.), Ph.D. Born 1926. Educ.: Cairo and London Univs. Lecturer in Physiology; Asst. Prof., Faculty of Medicine, Ein Shams Univ. Formerly Head, Biol. Sec., now Head, Radioisotopes Div., A.E.E., U.A.R. Book: Radioisotopes in Biol. (in Arabic) (Cairo, Ministry of Education, 1956). Society: Sec.-General, Egyptian Soc. for Radioisotopes.
Nuclear interest: Radiobiology.
Address: 11 El-Mahalawi Street, Dokki, Cairo, Egypt.

HASHMALL, M. L., M.E.E. Director, Radioisotope Lab., Brooklyn Coll. of Pharmacy, Long Island Univ.
Address: Long Island University, Brooklyn College of Pharmacy, 600 Lafayette Avenue, Brooklyn 16, New York, U.S.A.

HASIGUTI, Ryukiti Robert, D. Eng. (Tokyo). Born 1914. Educ.: Tokyo Univ. Prof. Metal Phys., Tokyo Univ., 1954-; Res. Official, Nat. Res. Inst. for Metals; Chief Researcher, Metal Phys. Div., Inst. of Phys. and Chem. Res. Chairman, Nucl. Reactor Materials Com., Foundation for the Promotion of Sci. Books: Rad. Damage (Chapter 3 of Lattice Defects, vol. 10 of Solid States Phys.) (Tokyo, Kyoritsu Publishing Co., 1959); Nucl. Materials (Chapter 3 of Reactor Eng., vol. 3 of Nucl. Eng.) (Tokyo, Kyoritsu Publishing Co., 1956). Societies: Phys. Soc. of Japan; Japan Inst. of Metals; Soc. of Appl. Phys., Japan; Iron and Steel Inst. of Japan.
Nuclear interests: Solid state physics, metallurgy.
Address: Department of Metallurgy, Tokyo University, Tokyo, Japan.

HASKELL, Donald F. Formerly with Martin Marietta; Formerly with N.A.S.A.; Chief, Appl. Mechanics Sect., Hittman Assocs., Inc., 1967.
Address: Hittman Associates Incorporated, Applied Mechanics Section, Columbia, Maryland, U.S.A.

HASLAM, Robert James, B.Sc. (Eng., Hons.), A.M.I.Mech.E. Born 1923. Educ.: London Univ. Asst. Chief Eng., U.K.A.E.A., Reactor Group, Risley, 1950-.
Nuclear interests: Reactor design and comparative studies of reactor systems.
Address: Reactor Group H.Q., U.K.A.E.A., Risley, Warrington, Lancs., England.

HASLETT, Arthur Woods, M.A. Born 1906. Educ.: Cambridge Univ. Sci. Correspondent, The Times, 1953-; Tech. Adviser, Sci. and Electronic Industries Trust Ltd., 1958-.
Nuclear interests: Technical journalism and advisory services.
Address: 104 Clifton Hill, London, N.W.8, England.

HASNAIN, S. Ahmed, M.Sc., Ph.D. Born 1931. Educ.: Punjab Univ. and North Carolina State Coll. Res. Asst., Brookhaven Nat. Lab., 1957: Post-doctoral Fellow as Resident Res. Assoc. in Reactor Eng. Div., Argonne Nat. Lab., 1959-60; P.S.O., Pakistan A.E.C.
Nuclear interest: Reactor physics.
Address: c/o Pakistan Atomic Energy Commission, P.O. Box 3112, Karachi 29, Pakistan.

HASSE, Robert A., B.S., M.S. (Chem.). Born 1932. Educ.: Michigan Tech. Univ. Student Assoc., Argonne Nat. Labs., 1956-57; Nucl. Chem., Alco Products Inc., 1958-61; Head, Radiochem. Labs. (1961-66), Head, Hot and Metal.Labs. (1966-), Reactor Div., Plum Brook Station, N.A.S.A. Societies: A.N.S.; Sen. Member, A.C.S.
Nuclear interests: Radiation damage to materials; Radiochemistry; Technical management; Remote systems technology; Build-up and transport of radioactive material in reactor systems.
Address: R.D. No. 1, Box 29, Milan, Ohio 44846, U.S.A.

HASSIALIS, Menelaos Dimitri, B.A. (Columbia Coll.), D.Sc. (hon., Bard Coll.). Born 1909. Educ.: Columbia Univ. Prof. Mineral Eng. (1951-), and Exec. Officer, School of Mines, Columbia Univ.; Pres., Pacific Uranium Mines Co., 1959-61; Director, Kermac Nucl. Fuels Corp., 1959-61; Consulting work: Socony Mobil Oil Co., 1945-; Vice-Pres. and Director of Technol. Investors Corp.; Adviser to Pakistan Govt. mineral development programme, 1957-58; Jones and Laughlin Steel Co.; Swedish Govt.'s A.E.C.; Haile Mines; investment houses in N.Y.; Director, Columbia Mineral Beneficiation Lab., 1951-60; Member, U.S. deleg. to 1st and 2nd I.C.P.U.A.E., Geneva, 1955 and 1958; U.S. Rep. at Unesco Meeting on Social and Moral Implications of the Peaceful Uses of Atomic Energy, 1958;

Member of Council, Atomic Age Studies.
Book: Contributor to Uranium Ore Processing (ed. Battelle Memorial Inst.). Societies: A.I.M.M.E.; A.C.S.; A.A.A.S.
Nuclear interests: Mining and beneficiation of uraniferous ores; Nuclear technology; Administration.
Address: 122 Phelps Road, Ridgewood, New Jersey, U.S.A.

HASSID, Andrea. Degree in Nucl. Eng. Born 1935. Educ.: Milan Polytech. C.E.S.N.E.F., Milan Polytech., 1960-61; Researcher (1961-63), i/c heat transfer and fluid dynamics group, (1964-), C.I.S.E., Segrate, Italy. Societies: Comitato Termotecnico Italiano Sottocomitato N. 13; Assoc. Termotecnica Italiana.
Nuclear interests: Heat transfer and fluid flow.
Address: C.I.S.E., Casella Postale 3986, 20100 Milan, Italy.

HAST, Paul Ferdinand, Dr.-Ing. E.h. Member, Fachkommission V—Wirtschaftliche Finanzielle und Soziale Probleme, Deutsche Atomkommission.
Address: Federal Ministry for Scientific Research, 46 Luisenstrasse, Bad Godesberg, Germany.

HASTE, Eric Leighton, B.Sc. Born 1916. Educ.: Leeds Univ. With Dowsett Eng. (Australia), Pty., Ltd., 1950-56; Asst. Chief Construction Eng., A.E.I. —John Thompson Nucl. Energy Co., Ltd., 1956-59; Asst. Chief Construction Eng., Nucl. Power Group, 1959-62; Group Resident Eng., Oldbury Nucl. Power Station, 1962-. Society: A.M.I.C.E.
Nuclear interests: Construction management; power stations;
Address: 1 Crantock Drive, Almondsbury, Bristol, England.

HASTE, Glenn Roy, Jr., B.A., M.S., M.S. Born 1934. Educ.: New Mexico and Virginia Univs. Thermonucl. Div., O.R.N.L., 1959-. Society: A.P.S., Plasma Phys. Sect.
Nuclear interest: Thermonuclear research.
Address: Building 9201-2, Y-12, P.O. Box Y, Oak Ridge, Tennessee, U.S.A.

HASTERLIK, Robert Joseph, S.B. M.D. Born 1915. Educ.: Chicago Univ. Sen. Sci. (1948-53), Director, Health Div. (1950-53), Argonne Nat. Lab., U.S.A.E.C.; Asst. Prof. Medicine (1948-53), Assoc. Prof. Medicine (1953-60), Prof. Medicine, (1960-), Chicago Univ.; Assoc. Director, Argonne Cancer Res. Hosp., 1952-62; Consultant, Argonne Nat. Lab., U.S.A.E.C. Sci. Member, Illinois State Rad. Protection Advisory Council and Legislative Commission on Atomic Energy; Subcom. 14, Nat. Com. on Rad. Protection. Societies: Rad. Res. Soc.; American Coll. Physicians; Radiol. Soc. North America; Central Soc. Clinical Res.
Nuclear interests: Long-term effects of deposit of radium in skeleton of human being; Metabolism of strontium in human being; Effects of ionising radiation on protein metabolism in human being and experimental animals; Mechanism of diuresis following total-body irradiation in rats; Means of modifying radiation effects; Health physics problems in general.
Address: 950 East 59th Street, Chicago 37, Illinois, U.S.A.

HASTINGS, Julius Mitchell, B.A., Ph.D. Born 1920. Educ.: New York and Cornell Univs. Chem., Brookhaven Nat. Lab., 1947-.
Nuclear interest: Neutron diffraction.
Address: Brookhaven National Laboratory, Upton, New York, U.S.A.

HATAYE, Itsuhachiro, D.Sc. Concerned with Radiochem. and Nucl. Chem., Radiochem. Res. Lab., Shizuoka Univ.
Address: Shizuoka University, Radiochemistry Research Laboratory, 2 Oiwa-cho, Shizuoka City, Japan.

HATCH, Colonel L. M. Chief, Plans and Requirements, Air Force Special Weapons Centre, Air Force Systems Command, U.S. Air Force.
Address: Air Force Special Weapons Centre, Kirtland Air Force Base, New Mexico, U.S.A.

HATHWAY, Charles W. Formerly with Atomic Power Equipment Dept., Gen. Elec. Co.; Formerly with Vallecitos B.W.R.; Manager, Nucl. Div., Todd Shipyards Corp., 1967-.
Address: Todd Shipyards Corporation, Nuclear Division, 1 Broadway, New York 4, New York, U.S.A.

HATINGUAIS, M. P. Deputy Gen. Manager, Soc. d'Assurances Mutualles de la Seine et de Seine-et-Oise contra l'Incendie les Accidents et Risques Vivers, 1957. Vice-Pres., Pool Français d'Assurances des Risques Atomiques.
Address: Pool Français d'Assurances des Risques Atomiques, 118 rue de Tocqueville, Paris 7, France.

HATTORI, Manabu. Asst. Prof., Inst. for Atomic Energy, St. Paul's Univ.
Address: Institute for Atomic Energy, St. Paul's University, Sajima, Yokosuka, Japan.

HATTORI, Takeshi, B.E. Eng., Atomic Power Div., Shimizu Construction Co. Ltd.
Address: Shimizu Construction Co. Ltd., Atomic Power Centre, No. 1 Takara-cho 2, Chuo-ku, Tokyo, Japan.

HATTORI, Yoshio, Dr. Eng. Born 1926. Educ.: Kyoto Univ. Asst. Prof. (1957), Prof. (1963-), Eng. Res. Inst., Kyoto Univ. Societies: Inst. Elec. Eng., Japan; Inst. Electronic and Communication Eng. of Japan; I.E.E.E.
Nuclear interests: Nuclear physics and reactor control; MHD power generator.
Address: Engineering Research Institute, Kyoto University, Uji, Kyoto-Fu, Japan.

HATZIKAKIDIS, Athanassios, B.S. (Chem.), Dr. of Phys. Sci., Diploma of Greek A.E.C., Diploma of O.R.I.N.S. Born 1918. Educ.: Nat. Univ. Athens and O.R.I.N.S. Head, Chem. Dept., Hydrobiol. Inst.,Nat. Acad. Athens, 1949-; Head, Solar Energy Dept., Greek A.E.C., 1962-. Editorial Director, Solar Energy; Collaborator, Roy. Foundation of Greece; Sec.-gen., Assoc. Greek Govt.'s Chem.; Sec.-gen., Hellenic Sci. Soc. of Solar and Wind Energy; Internat. Sec. and Council, Internat. Solar Energy Soc. Books: The Atom of Matter, Atomic Energy and the Atomic Bomb (1946); New Power - Solar Energy in the Service of Mankind (1957); New Natural Sources of Wealth in the Service of Humanity (1957); Inside the Atom (1957); Some Aspects of Aristotle's Ideas on Phys. and Chem. (1958); Relations between Air and Sea Temperatures (1963); Evaporation by Solar Energy (1964); L'Evaporation des Résidus Radioactifs par l'Energie Solaire (1966). Societes: Commissione Internationale per l'Explorazione Scientifica del Mediterraneo; American Soc. Limnology and Oceanography; American Assoc. Appl. Solar Energy; Greek Biol. Soc.; Archaeological Soc.; Roy. Foundation of Greece; Assoc. Greek Govt.'s Chem.; Hellenic Sci. Soc. of Solar and Wind Energy; Internat. Solar Energy Soc.
Nuclear interest: Radioactivity of sea waters.
Address: 30 Atlantos Str., Palaeon Phaliron, Athens, Greece.

HAUBENREICH, Paul N., B.S. (Mech. Eng.), M.S. (Mech. Eng.). Born 1925. Educ.: Tennessee Univ. Development Eng. (1951-), In charge of reactor analysis, Molten Salt Reactor Experiment (1962-). In charge of reactor analysis for MSRE (1962-64), In charge of MSRE operation (1964-), O.R.N.L. Lecturer (parttime), Nucl. Eng. Dept., Tennessee Univ., 1958-61.
Nuclear interests: Reactor design, experimental operation of reactors (planning programme and individual experiments and analysing results).
Address: 501 West Hills Road, Knoxville, Tennessee 37919, U.S.A.

HAUBROK, Heinz-Peter, Diplom-Volkswirt. Director, Käufmännische Leitung, Elektrizitätswerk Minden-Ravensberg G.m.b.H.
Address: Elektrizitätswerk Minden-Ravensberg G.m.b.H., 3b Bielefelder Strasse, Herford/Westf., Germany.

HAUCK, Alfred Werner, Dipl.-Ing., Dr.-Ing. Born 1927. Educ.: Karlsruhe T.H. Asst. of Prof. Dr. h.c. F. A. Henglein, Karlsruhe T.H., 1956-60; Reactorcentre, Karlsruhe, 1960-61; Gutehoffnungshütte Oberhausen-Sterkrade nucl. eng., 1962.
Nuclear interest: Reactor design and management.
Address: 31 Preussenstrasse, Oberhausen-Sterkrade, Germany.

HAUER, Josef, Mech. Eng. Born 1919. Educ.: Prague Tech. Univ. Director, Nucl. Power Plant Div., Skoda Concern, Pilsen. Society: Czechoslovak Sci. and Tech. Soc.
Nuclear interest: Management of reactor plants.
Address: 228 Radcice, Pilsen-sever, Czechoslovakia.

HAUFE, Siegfried, Dipl. Phys. Born 1936. Educ.: Dresden T.H. At Inst. für Chemieanlagen.
Nuclear interests: Nuclear physics; Isotopes.
Address: Institut für Chemieanlagen, 26 Mansfelder Str., Dresden A 19, German Democratic Republic.

HAUGE, Jens Christian. Born 1915. Educ.: Oslo Univ. Vice-Chairman, Norwegian A.E.C.; Norwegian Rep., Steering Com. for Atomic Energy, Member, Eurochemic Special Group of Steering Com., O.E.C.D.; legal adviser to Norwegian Inst. for Atomic Energy; Member, Nordic Contact Com. for Atomic Energy.
Nuclear interests: Management and atomic law.
Address: 2 Youngstorget, Oslo, Norway.

HAUGG, Werner. Born 1908. Educ.: Berlin Univ.Ministerialdirigent bei dem Ministerpräsidenten des Landes Nordrhein-Westfalen - Landesamt für Forschung 1961-; zuvor Ministerialdirigent im Kultusministerium des Landes Nordrhein-Westfalen. Societies: Kuratorium der Deutschen Gesellschaft für Flugwissenschaften e.V.; Kuratorium des Max-Planck-Inst. für Ernährungsphysiologie in Dortmund.
Address: 14 Wasserstr. Düsseldorf, Germany.

HAUGHEY, Francis James, B.S. (Phys.) (Hofstra Coll.), M.S. (Rutgers Univ.), Ph.D. (Rutgers Univ.), Certified Health Phys. (American Board of Health Phys.). Born 1930. Educ.: Hofstra Coll. and Rutgers Univ. Asst. Prof., Rutgers Environmental Sci. Dept. Societies: H.P.S.; A.G.U.; American Meteorological Soc.; A.A.A.S.; Air Pollution Control Assoc.; A.A.U.P.
Nuclear interests: Radioactive aerosols; Environmental radioactivity; Health physics.
Address: Environmental Sciences Department, Rutgers University, New Brunswick, New Jersey 08903, U.S.A.

HAUL, Robert, Prof., Dr.-Ing. Born 1912. Educ.: T.H. Brunswick, Graz, Danzig and Berlin. Direktor des Inst. für Physikalische Chemie und Elektrochemie der T.H., Hanover. Societies: Bunsen-Gesellschaft; Faraday Soc., England; Gesellschaft Deutscher Chemiker.
Nuclear interests: Physikalisch-chemisches verhalten; Stabiler isotope.
Address: 46 Callinstrasse, Hanover 3, Germany.

HAUMONT, S., Dr. Formerly Sen. Officer, Faculty of Medicine, now i/c of Radiobiol. Lab. and Related Matters, Trico Centre,

Lovanium Univ.
Address: Lovanium University, B.P. 868, Kinshasa 11, Congolese Republic.

HAUN, R. R., Ph.D. Prof. and Head, Dept. Phys. Sci. and Member, Radioisotopes Com. Drake Univ., Des Moines.
Address: Drake University, Des Moines 11, Iowa, U.S.A.

HAUNSCHILD, Hans-Hilger, First and second state law examination. Born 1928. Educ.: Berlin Univ. Law School. Federal Ministry for Economic Co-operation, Bonn, 1955-57; Head, C.E.E.A. Sect., Internat. Co-operation Dept., Federal Ministry for Nucl. Energy, Bonn, 1957-62; Director for Programmes, C.E.E.A., Brussels, 1962-67; Head, Co-operation Dept., Federal Ministry of Sci. Res., Bonn, 1967-. Vice Chairman, Board of Directors, Eurochemic.
Nuclear interests: Nuclear energy law; International co-operation in the field of nuclear energy.
Address: 1-10 Heussallee, Bonn 53, Germany.

HAUSER, Edouard. Director Gen. Manager, Ateliers et Chantiers de Nantes (Bretagne-Loire) S.A.
Address: Ateliers et Chantiers de Nantes (Bretagne-Loire) S.A., Prarie au Duc, Nantes, (Loire-Atlantique), France.

HAUSER, George, M.D. Member, Isotope Com., Massachusetts Gen. Hospital.
Address: Massachusetts General Hospital, Boston 14, Massachusetts, U.S.A.

HAUSER, Louis Glenn, B.S. (E.E.). Born 1919. Educ.: Carnegie Inst. Technol. Generation Consultant, Elec. Utilities Headquarters Dept., Westinghouse Elec. Corp. Societies: A.N.S.; Atomic Ind. Forum.
Nuclear interest: Management.
Address: Westinghouse Electric Corporation, East Pittsburgh, Pennsylvania, U.S.A.

HAUSER, O. Dr. rer. nat. habil., Head, Materials and Solids Div., Zentralinst. für Kernforschung.
Address: Zentralinstitut für Kernforschung, Bereich Werkstoffe und Festkörper, Rossendorf bei Dresden, German Democratic Republic.

HAUSER, Ulrich, Dr., Prof. Wiss, Oberräte II Physikalisches Inst., Heidelberg Univ.; Director, Erstes Phys. Inst., Cologne Univ.
Address: I Physikalisches Institut, 14 Universitätstrasse, Cologne University, Cologne, Germany.

HAUSER, Walter, B.S., Ph.D. Born 1924. Educ.: Brooklyn Coll. and M.I.T. Boston Univ., 1950-55; Lincoln Lab., M.I.T., 1955-60; Northeastern Univ., 1960-. Society: A.P.S.
Address: Northeastern University, Physics Department, 360 Huntington Avenue, Boston 02115, U.S.A.

HAUSHILD, William L. Eng., Geological Survey, U.S. Dept. of the Interior.
Address: U.S. Department of the Interior, Geological Survey, P.O. Box 3202, Room 416 Old Post Office Building, 511 N.W. Broadway, Portland, Oregon, U.S.A.

HAUSLER, Leland Meissner, B.S. (Mech. Eng.). Born 1922. Educ.: Iowa Univ. Field Elec. Supv., Gen. Construction Dept., Consumers Power Co., 1950-52; Gen. Eng. (1952), Asst. Generating Plant Supt. (1952-60), J. R. Whiting Plant, Erie, Michigan; Generating Plant Supt., Big Rock Point Nucl. Plant, Charlevoix, Michigan (1960-66), Generating Plant Supt., Palisades Plant (Nucl.), Covert, Michigan (1966-), Consumers Power Co. Societies: A.N.S.; A.S.M.E.
Nuclear interest: Management of nuclear power plants utilised for the economical production of electrical power.
Address: Palisades Plant, Consumers Power Co., P.O. Box 486, South Haven, Michigan 49013, U.S.A.

HAUSMAN, Eugene Arnold, B.A., M.S., Ph.D. Born 1923. Educ.: Brooklyn Coll., Iowa State and New York Univs. Fellowship, New York Univ., 1949-52; Res.Lab., Engelhard Industries, Inc., Newark, New Jersey, 1952-61. Societies: A.C.S.; A.A.A.S.
Nuclear interests: Research on the refining of enriched, cold uranium scrap; Decontamination of fission product containing precious metals.
Address: 44 Nomahegan Court, Cranford, New Jersey, U.S.A.

HAUSMAN, Robert F., B.S. (Chem. Eng.). Born 1921. Educ.: Cincinnati Univ. Staff Eng., Vehicle Flight Safety, Cryogenic and Nucl. Stage Programmes, R. and D. Lockheed Missiles and Space Co. Society: A.N.S.
Nuclear interests: Development of nuclear safety criteria and safeguards requirements for nuclear rocket flight vehicle applications. Launch and flight hazards analysis for advanced nuclear rocket vehicle systems. Nuclear reactor applications to space programmes for both power and propulsion. Technological aspects of isotopic heat source applications and utilisation in manned and unmanned space craft.
Address: Lockheed Missiles and Space Co., Dept. 30-60, Bldg. 534, P.O. Box 504, Sunnyvale, California 94088, U.S.A.

HAUSNER, Henry H., E. E., Dr. Eng. Born 1901. Educ.: Vienna T.H. Sylvania Elec. Products, Inc., 1948-56; Penn-Texas Corp., 1956-58; Consultant; Prof. Polytech. Inst., Brooklyn, 1952-; Visiting Prof., California Univ., 1963, 1964. Chairman, Powder Met. Com., A.I.M.E.; Pres., Int. Plansee Soc. Books: Powder Metal. (1947); Materials for Nucl. Power Reactor (1955); Powder Metal. in Nucl. Eng. (ed. 1958); Effects of Irradiation

on Materials (ed. 1958); Nucl. Fuel Elements (1959); Powder Metal.in Nucl. Reactor Construction (1961). Societies: Inst. Metals.; A.I.M.M.E.; A.S.M.; A.N.S.; Int. Plansee Soc. for Powder Met.; Deutsche Gesellschaft für Metallkunde; Japanese Powder Met. Soc.
Nuclear interests: Nuclear metallurgy; Fuel element fabrication; Development of control rods; High-temperature materials; Radiation effects on materials.
Address: 549 West 123rd Street, New York, New York 10027, U.S.A.

HAUSSER. Member, Conseil d'Administration, Sté. Industrielle des Minerais de l'Ouest (S.I.M.O.).
Address: Societe Industrielle des Minerais de l'Ouest (S.I.M.O.), 25 boulevard de l'Amiral Bruix, Paris 16, France.

HÄUSSERMANN, Wolfgang Friedrich, Dr. rer. nat., Dipl.-Chem. Born 1932. Educ.: Eberhard-Karls-Univ., Tübingen. Phys. Inst., Agricultural Univ., Hohenheim, 1959-65; Principal Administrator, European Nucl. Energy Agency, Paris, 1965-. Societies: Gesellschaft Deutscher Chemiker; Deutsches Atomforum.
Nuclear interests: Radiochemistry; Scientific and technical application of radioisotopes: Radioisotopic generators; Information exchange in nuclear field.
Address: European Nuclear Energy Agency, 38 boulevard Suchet, Paris 16, France.

HAUSSLER, Warren M., B.Sc., M.S. (Chem. Eng.). Born 1927. Educ.: Rochester Univ. Naval Reactors Branch, U.S.A.E.C. Washington D.C., 1953-57; Atomics Internat., Inc., 1958-62; Electro-Optical Systems, Pasadena, California, 1962-63; Product Manager, A.V.C.O. Ind. Products Subdiv., Lowell, Massachusetts, 1963-66; Marketing, Planning and Sales Management Services, Harbridge House, Inc., 1966-. Societies: A.N.S.; American Rocket Soc.; A.S.M.
Nuclear interests: Reactor design and operation; Project engineering and management; Product development.
Address: 12 Larchmont Lane, Lexington, Massachusetts, U.S.A.

HAUSSMANN, Alfred Carl, B.S., M.S.. Born 1924. Educ.: U.S. Military Acad., West Point, Pennsylvania State Univ., California Inst. Technol. and U. S. Naval Acad. Postgraduate School, Anapolis. Nucl. Supervisor, Armed Forces Special Weapons Centre, 1951-52; Phys. team member, Project Matterhorn, Princeton, 1952-53; Army Officer (1953-55), Phys. Group Leader (1955-59), Div. Leader (1959-62), Assoc. Director (1962-), Lawrence Rad. Lab.;Member, Strategic Systems Com., Dept. of Defence; Senior Reviewer, U.S.A.E.C.
Nuclear interest: The design and application of nuclear explosives.
Address: California University, Lawrence Radiation Laboratory, P.O. Box 808, Livermore, California, U.S.A.

HAVAS, Bela. Member, Editorial Board, Energia es Atomtechnika.
Address: Muszaki Konyvkiado, 22 Bajcsy-Zsilinszky ut, Budapest 5, Hungary.

HAVAS, Peter, Ph.D. Born 1916. Educ.: Vienna T.H. and Columbia Univ. Assoc. Prof. (1954), Prof., Phys. (1954-65), Lehigh Univ.; Visiting Res. Assoc., Argonne Nat. Lab., Summer 1958; Prof., Phys., Temple Univ., Philadelphia, 1965-. Society: A.P.S.
Nuclear interests: Nuclear theory; Elementary particle theory; Theory of radiation.
Address: Temple University, Department of Physics, Philadelphia, Pennsylvania 19122, U.S.A.

HAVE VAN SIRJANSLAND, Adriana ENDA MARAH- VAN DER. See ENDA MARAH-VAN DER HAVE VAN SIRJANSLAND, Adriana.

HAVEL, Stanislav, Metal. Eng. Born 1930. Educ.: Tech. Univ. Ostrava Res. Manager, Skoda Works, Pilsen, 1957-65; I.A.E.A. Officer, 1965-. Society: Czechoslovak Sci. and Tech. Soc.
Nuclear interests: Research of structural materials; Radiation damage and material testing; Reactor pressure vessels production technology; Power reactors in general.
Address: 11-13 Kärntnerring, Vienna I, Austria.

HAVELKA, Stanislav, Ing., C.Sc. Born 1929. Educ.: Tech. Univ., Prague, High School of Chem. Technol. Faculty of Chem. Technol., Dept. of Analytical Chem., Prague Tech. Univ., 1952-60; Nucl. Res. Inst., Czechoslovakian Akad. Sci., Rez, 1960-.
Nuclear interests: Extraction chemistry and technology of spent fuel reprocessing; Extraction chemistry of actinides.
Address: Nuclear Research Institute, Czechoslovakian Academy of Sciences, Rez, Czechoslovakia.

HAVENS, William Westerfield, Jr., B.S. (C.C.N.Y.), M.A. (Columbia), Ph.D. (Columbia). Born 1920. Prof. of Phys., Director, Pegram Nucl. Phys. Labs. and Div. of Nucl. Sci. and Eng., Columbia Univ.; Former Member and former Chairman, Nucl. Cross Sections Advisory Com., U.S.A.E.C.; Exec. Sec., European American Nucl. Data Committee; Member, Internat. Nucl. Data Com. of the I.A.E.A.; Exec. Sec., A.P.S.; Member, Exec. Com. and Governing Board, American Inst. of Phys.; Vice Pres., A.A.A.S.
Nuclear interest: The measurement of neutron cross sections and their use in research and the development of nuclear energy systems. Particle accelerators and their use in nuclear physics; The use of computers in the

acquisition, reduction and analysis of nuclear data. Nuclear standardisation. Neutron interactions used in chemical analysis. Nuclear radiation detectors and associated electronic systems.
Address: 219 Palisade Avenue, Dobbs Ferry, New York 10522, U.S.A.

HAVIE, Tore, Lic. tech. Born 1930. Educ.: Tech. Univ., Norway. Asst. Prof., Tech. Univ., Norway, 1953-57; Sen. Res. Mathematician, Kjeller Computer Installation, Norway, 1957-. Nuclear interests: Numerical analysis, digital computations and nuclear calculations.
Address: 6 Stubben, Skedsmokorset, Norway.

HAVIN, Edel, Cand. real. Member, Lab. for Genetics, Norsk Hydro's Inst. for Cancer Res.
Address: Norsk Hydro's Institute for Cancer Research, The Norwegian Radium Hospital, Oslo, Norway.

HAVLOVIC, Vratislav, Docent, MUDr., RNDr., C.Sc. Born 1927. Educ.: Charles Univ., Prague. Inst. of Phys., Medical Faculty, Charles Univ., Pilsen, 1950-60; Inst. of Phys., Medical Faculty, Charles Univ., Hradec Kralove, 1960-65; Faculty of Medical Hygiene, Charles Univ., Prague, 1965-.
Nuclear interests: Health physics, with special references to radioactive aerosols and environmental radioactivity.
Address: 32 Bendova, Pilsen, Czechoslovakia.

HAVRANEK, William Alfred, M.I.E.E., M.I.E.E.E. Born 1927. Educ.: Ingenieurschule Aussig and Coventry Tech. College. Test Lab. Technician, Armstrong Whitworth Aircraft Co., 1950-52; Electronic Eng., Tech. Eng. Div., Metal Box Co., 1952-53; Design Eng., Gen. Precision Systems, 1953-57;Sen. Nucleonics Applications Eng., Labgear, Ltd., 1957-58; Chief Project Eng., Reactor and Computer Div., Miles Electronics Ltd., 1958-61; Group Leader for Electronics and Simulation, Dornier Werke G.m.b.H., 1961-63; Eng. in Charge, Special Projects, Redifon Ltd., Simulator Div., 1963-66; Marketing Director, Appl. Dynamics Europe, 1966-.
Nuclear interests: Reactor and nuclear power station simulation; Analogue computers; On-line computer applications; Digital simulation; Hybrid systems for nuclear applications.
Address: Carinthia, Jersey Road, Ferring by Sea, Sussex, England.

HAW, Kum, M.P.H. Director, Isotope Lab., Physico-chem. Div., Nat. Chem. Labs., Seoul.
Address: Isotope Laboratory, Physicochemistry Division, National Chemistry Laboratories, Ministry of Health and Social Affairs, 79 Saichong- No, Chongno-ku, Seoul, Korea.

HAWES, Christine Ann, B.Sc. Born 1936. Educ.: London Univ. Res. Staff, British Empire Cancer Campaign for Res., Res. Unit in Radiobiol., Mount Vernon Hospital, Northwood, Middlesex, 1957-64. Society: Assoc. for Rad. Res.
Nuclear interest: Radiobiology.
Address: 245a College Street, Port Arthur, Ontario, Canada.

HAWKINS, Arthur Ernest, B.Sc. (Eng.). Born 1913. Educ.: London Univ. System Planning Eng., C.E.G.B., -1959; Formerly Chief Operations Eng., now Regional Director, Midlands Region, C.E.G.B. Chairman, A.E.A./ C.E.G.B. Nucl. Operations Com. Society: Corporate Member, Inst. Elec. Eng.
Nuclear interests: The operation of the nuclear power stations forming part of the civil programme and their integration into the Public Electricity Supply in England and Wales.
Address: Wraybank, 14 Coulsdon Rise, Coulsdon, Surrey, England.

HAWKINS, George Andrew, B.S. (M.E.), M.S. (M.E.), Ph.D. Born 1907. Educ.: Purdue Univ. Vice Pres. for Academic Affairs, Purdue Univ., 1967. Policy Advisory Board, and Res. Assoc., Argonne Nat. Lab. Books: Multilinear Analysis for Students in Eng. and Sci. (John Wiley, 1963); co-author, Eng. Thermodynamics (John Wiley, 1960); Elements of Heat Transfer, 3rd ed. (John Wiley, 1957). Societies: Fellow, A.S.M.E.; A.S.E.E.; A.P.S.; N.S.P.E.; A.I.Ch.E.; The Mathematical Assoc. of America; Nat. Acad. of Eng.
Address: President's Office Executive Building, Purdue University, Lafayette, Indiana, U.S.A.

HAWKINS, Vice Admiral Sir Raymond Shayle, K.C.B. Born 1909. Educ.: Roy. Navy. Eng. Coll. Naval Asst. to Third Sea Lord, 1953-57; Deputy Director, Marine Eng. (Rear Admiral Nucl. Propulsion), 1959-61; Director Marine Eng., 1961-63; Fourth Sea Lord 1963-64; Chief, Naval Supplies and Transport, and Vice-Controller, 1964-67.
Nuclear interest: Nuclear propulsion.
Address: Hodshill, Southstoke, Bath, England.

HAWKINS, William Madison, Jr., A.B. (Phys., Harvard), M.S.E. (Mech.Eng., Harvard), M.M.E. (Automotive Eng., Chrysler Inst. Eng.). Born 1909. Educ.: Harvard Univ. and Chrysler Inst. Eng. at A.C.F. Industries, Inc., 1948-59; Manager, R. and D., Nucl. Power Dept., Washington (1959-64), Manager, Application Eng., Atomic Energy Div. (1964-66), Manager, Special Projects Atomic Energy Div. (1966-67), Allis-Chalmers Manufacturing Co.; Advanced Projects, Southern Nucl. Eng., Inc., Bethesda, Maryland, 1967-; Space Nucl. Power Com., Atomic Ind. Forum. Societies: A.N.S.; A.S.M.E.; Atomic Ind. Forum.
Nuclear interest: Reactor engineering.
Address: 9414 Thrush Lane, Potomac, Maryland, U.S.A.

HAWKINS, William Marion. Born 1915. Educ.: Drury Coll. Springfield, Missouri and Missouri Univ. Mech. Design Eng. and Eng. Liaison

(M.T.R.), Union Carbide Nucl. Corp., O.R.N.L., 1947-51; Formerly Manager, Div. Eng. Branch, Phillips Petroleum Co., Atomic Energy Div., Idaho Falls, Idaho, 1951-. Address: 582 East 16th Street, Idaho Falls, Idaho, U.S.A.

HAWLEY, Ronald William. Born 1926. Res. Labs., Guest Keen Nettlefolds Group, 1951-54; Design and Development (1954-60), Managing Tech. and Production aspects (1960-), R. A. Stephen and Co. Ltd. Member, Com. NCE/2/1/2, Direct Reading Personal Dosimeters, B.S.I. Nuclear interest: Personal dosimetry. Address: R. A. Stephen and Company Limited, Miles Road, Mitcham, Surrey, England.

HAWN, Clinton V., M. D. Member, Isotopes Com., Mary Imogene Bassett Hospital. Address: The Mary Imogene Bassett Hospital, Cooperstown, New York, U.S.A.

HAWORTH, Leland John, A.B., A.M., Ph.D., D.Sc. (Hon.), D. Eng., D.C.L., Dr. of laws. Born 1904. Educ.: Indiana and Wisconsin Univs. Director, Brookhaven Nat. Lab. 1948-61; Vice-Pres. (1951-60), Pres. (1960-61), Assoc. Univs., Inc.; Member, U.S.A.E.C., 1961-63; Director, N.S.F., 1963-. Book: Ch. 7 Electronucl. Machines in Recent Advances in Sci.; Several chs. of Rad. Lab. Tech. series. Societies: A.N.S.; A.P.S.; N.Y. Acad. Sci.: A.A.A.S. Nuclear interests: Nuclear physics; High-energy accelerators; Management. Address: National Science Foundation, Washington D.C. 20550, U.S.A.

HAWTHORNE, Edward Peterson, M.A., M.I.Mech.E., A.F.R.Ae.S. Born 1920. Educ.: Cambridge Univ. Eng.-in-Charge, Power Plant Dept., C. A. Parsons and Co., Newcastle upon Tyne, 1947; A. V. Roe and Co., Manchester, 1954; Managing Director, Hawker Siddeley Nucl. Power Co., Ltd., Langley, Bucks.; Director, Tech. Development Capital Ltd., London, 1961-64. Managing Director, Urwick Technol. Management Ltd., 1965-. Address: Maloja, Woodside Avenue, Beaconsfield, Bucks., England.

HAXEL, Otto Philipp Leonhard, Dr. rer. nat. Born 1909. Educ.: München and Tübingen Univs., and Berlin T.H. Max Planck Inst. für Physik., 1945-50; Director des II Physik. Inst., Heidelberg Univ., 1950-. Prof. Phys., Heidelberg Univ.; Member, Deutsche Atomkommission; Member, Aufsichtsrat, Gesellschaft für Kernforschung m.b.H.; Formerly Member, Sci. and Tech. Com. of Euratom. Editor, Zeitschrift für Physik. Society: Akademie der Wissenschaften, Heidelberg. Nuclear interest: Nuclear physics. Address: 12 Philosophenweg, Heidelberg, Germany.

HAY, Charles C. Pres., British American Oil Co. Ltd. Director, Canadian Nucl. Assoc. Address: British American Oil Co. Ltd., Toronto, Ontario, Canada.

HAY, G. A., M.Sc. Sen. Lecturer, Dept. of Medical Phys., Leeds Univ. Address: The University of Leeds, Department of Medical Physics, The General Infirmary, Leeds 1, Yorkshire, England.

HAY, J., A.M.I.Mech.E. Works Manager and Local Director, Eng. Group, Vickers Ltd. Address: Vickers-Armstrongs Barrow Engineering Works, Barrow-in-Furness, Lancashire, England.

HAY, Willis. Formerly Deputy Director, now Acting Director (1965-), Div. of Labour Relations, U.S.A.E.C. Address: Division of Labour Relations, U.S.A.E.C., Washington D.C. 20545, U.S.A.

HAYAKAWA, Jun'ichi. Born 1908. Director and Sec. Gen., J.A.I.F.; Editor, Genshiryoku Kaiga Jijo (Development of Atomic Activities Abroad); Editor in Chief, Atoms in Japan; Editor, Nucl. Sci. Bibliography; Editor, Development of Atomic Energy in Japan; Editor, Atomic Industry Newspaper; Editor, Atomic Res. J. Address: Japan Atomic Industrial Forum, Incorporated, No. 1-1-13, Shimbashi, Minato-ku, Tokyo, Japan.

HAYAKAWA, Kiyoshi, Dr. of Eng. Born 1929. Educ.: Nagoya Univ., Faculty of Sci., Chem. Dept. Second Sect., Fourth Div., Govt. Ind. Res. Inst., Nagoya, Agency of Ind. Sci. and Technol., Ministry of Internat. Trade and Industry, Japan. Societies: Chem. Soc. of Japan; Soc. of Polymer Sci., Japan. Nuclear interests: Radiation chemistry, especially radiation-polymerisation and radiation-graft copolymerisation. Address: 10, 2-chome, Arata-machi, Gifu-shi, Japan.

HAYAKAWA, Masahiko. Asst. Chief, Nucl. Fuel Sect., Atomic Energy Bureau, A.E.C. of Japan. Address: Atomic Energy Commission of Japan, 3-4 Kasumigaseki, Chiyoda-ku, Tokyo, Japan.

HAYAKAWA, Satio, D.Sc. Born 1923. Educ.: Tokyo Univ. Asst. Prof., Osaka City Univ., 1950; Prof., Kyoto Univ., 1954; Prof., Nagoya Univ., 1959-. Book: Co-author, Nucl. Fusion (Iwanami Shoten, 1959). Societies: Phys. Soc. of Japan; A.P.S. Nuclear interests: Nuclear physics; High energy physics; Nuclear astrophysics; Plasma physics. Address: Nagoya University, Furo-cho, Chikusa-ku, Nagoya, Japan.

HAYAKAWA, Sohachiro, Dr. Sci.(Eng.). Born 1926. Educ.: Tokyo Univ. Asst. Prof., Tokyo Inst. Technol., 1959; Visiting Asst. Prof.,

New York Univ., 1960; Prof. Dept. of Appl. Phys., Tokyo Inst. Technol. Societies: Japanese Soc. Atomic Energy (Programming Com.); Japanese Soc. Phys.; Chem. Soc. in Japan.
Nuclear interests: Scintillators (organic, mainly). Radiation effects on solids.
Address: Toyotama-naka 1-1057, Nerimaku, Tokyo, Japan.

HAYAMI, Hiroshi, M.D. Born 1908. Educ.: Tokyo Imperial Univ. Chief, Food Chem. Div., Nat. Inst. of Nutrition. Society: Japan Rad. Res. Soc.
Nuclear interests: Tracer experiments on metabolic studies of mineral elements. Studies on prevention or decrease of radiation injury due to fall-out.
Address: The National Institute of Nutrition, 1 Toyamacho, Shinjuku-ku, Tokyo, Japan.

HAYAMI, Jun-ichi, D.Sc. Asst., Lab.of Radiochem., Inst. for Chem. Res., Kyoto Univ.
Address: Kyoto University, Institute for Chemical Research, Yoshida-Honmachi, Sakyo-ku, Kyoto, Japan.

HAYANO, Nobuo. Born 1925. Educ.: Kyushu Univ. Formerly Res. Assoc., Phys. Dept., Kyushu Inst. Technol.; Asst. Prof., Dept. of Mech. Eng., Faculty of Eng., Yamaguchi Univ.
Nuclear interest: Nuclear physics.
Address: Government House (East Campus), Faculty of Engineering, Tokiwadai, Ube-city, Japan.

HAYASHI, Chikara, D.Sc. Born 1922. Educ.: Tokyo Univ. Director: Japan Vacuum Eng. Co., 1959-; Japan Atomic Energy Res. Inst., 1957-; Chairman, Com. on Res. Programme, Vacuum Soc. of Japan. Book: Co-author, Particle Accelerator, in Progress in Phys., No. 4.
Society: Phys. Soc. of Japan.
Nuclear interests: Equipment for processing nuclear metals and particle accelerators.
Address: 146-5 Kikuna, Kohokuku, Yokohama, Japan.

HAYASHI, Chushiro, D.Sc. Born 1920. Educ.: Tokyo Univ. Prof.,Kyoto Univ., 1957-.
Societies: A.P.S.; Phys. Soc. of Japan.
Nuclear interest: Nuclear astrophysics.
Address: Department of Physics, Kyoto University, Kyoto, Japan.

HAYASHI, Hiroshi. Formerly Asst. Chief, Reactor Licensing and Regulation Sect., now Asst. Chief, Radioisotope Sect., Atomic Energy Bureau, A.E.C. of Japan.
Address: Atomic Energy Commission of Japan, 3-4 Kasumigaseki, Chiyoda-ku, Tokyo, Japan.

HAYASHI, Kiyozumi. Asst. Chief, Res. Sect., Atomic Energy Bureau, A.E.C. of Japan.
Address: Atomic Energy Commission of Japan, 3-4 Kasumigaseki, Chiyoda-ku, Tokyo, Japan.

HAYASHI, Mitsuaki, Dr. of Veterinary Medicine. Born 1928. Educ.: Hokkaido Univ. Chief Researcher, Radioisotope Lab. Societies: Nippin Societas Radilogica; Japan Endocrinological Soc.
Nuclear interest: Applied radiobiology.
Address: Radioisotope Laboratory, Third Research Division, National Institute of Animal Health, Kodaira, Tokyo, Japan.

HAYASHI, Shoichiro, Dr. Sci. Born 1920. Educ.: Tokyo Univ. Tech. Official (1944-68), Studying the nucl. raw materials in Japan (1956-68), Geological Survey of Japan.
Societies: Atomic Energy Soc. Japan; Mineralogical Soc. America; Mineralogical Soc. Japan.
Nuclear interests: Geology and mineralogy of nuclear raw materials, especially X-ray mineralogy of uranium and thorium minerals.
Address: Geological Survey of Japan, Tokyo Branch, 8 Kawada-cho, Shinjuku-ku, Tokyo, Japan.

HAYASHI, Takeo. Formerly Director, Taiyo Fire and Marine Insurance Co. Ltd.; Prof., Div. of Sci. Instr., Res. Reactor Inst., Kyoto Univ.
Address: Kyoto University Research Reactor Institute, Noda, Kumatori-cho, Sen-nan-gun, Asaka Prefecture, Japan.

HAYASI, Sigenori. Prof. in Charge of Reactor Instrumentation, Eng. Res. Inst., Kyoto Univ. Address:Engineering Research Institute, Kyoto University, Kyoto, Japan.

HAYBITTLE, J. L., M.A. Hon. Sec., British Inst. of Radiol.
Address: British Institute of Radiology, 32 Welbeck Street, London W.1, England.

HAYDAROV, T. Paper: Co-author, Interaction Cross Section of Neutrons with ^{149}Sm and ^{115}In Nuclei (letter to the Editor, Atomnaya Energiya, vol. 16, No. 6, 1964).
Address: Academy of Sciences of the U.S.S.R., 14 Leninsky Prospekt, Moscow V-71, U.S.S.R.

HAYDAY, F. Member, Nat. Union of Gen. and Municipal Workers; Member, Gen. Council, T.U.C.; Member, Nucl. Safety Advisory Com., Ministry of Power.
Address: Nuclear Safety Advisory Committee, Ministry of Power, Thames House South, Millbank, London, S.W.1, England.

HAYDEN, Owen, A.M.I.Mech.E., D.I.C. (Nucl. Power). Born 1929. Design Eng., Fast Reactor Design Office, U.K.A.E.A., Risley.
Nuclear interests: Various aspects of fast reactor design - feasibility studies, project work, operational rectifications and investigations into advanced concepts.
Address: 30 Belmont View, Harwood, Nr. Bolton, Lancs., England.

HAYDEN, Paul V. Formerly Exec. Vice Pres., now Pres. (1964-), Connecticut Light and Power Co.
Address: Connecticut Light and Power Co., Box 2010, Hartford, Connecticut, U.S.A.

HAYDEN, R. L. J. Director, Rolls-Royce and Assoc. Ltd.
Address: Rolls-Royce and Associates Ltd., P.O. Box 31, Derby, England.

HAYDEN, Robert E., B.A. (Phys., Minor in Math. and Chem., Colgate). Born 1908. Educ.: Colgate, New York, St. Lawrence and Tennessee Univs. Societies: H.P.S.; A.I.H.A.
Nuclear interest: Health physics.
Address: 195 Tautphaus Drive, Idaho Falls, Idaho, U.S.A.

HAYDEN, Vernon Hilary, B.S. (Elec. Eng.). Born 1914. Educ.: Pittsburgh Univ. Formerly Manager, Submarine Power Plant Eng., now Manager, Submarine Field Eng., Westinghouse Elec. Corp., Bettis Plant. Society: American Inst. Elec. Eng.
Nuclear interests: Management; Reactor plant design.
Address: Westinghouse Electric Corporation, Bettis Atomic Power Laboratory, Post Office Box 1526, Pittsburgh 30, Pennsylvania, U.S.A.

HAYDEN, W. Lee. Res. Sci.-Eng., Appl. Nucl. Phys. Div., Battelle-Columbus Labs.
Address: Battelle-Columbus Laboratories, 505 King Avenue, Columbus, Ohio 43201, U.S.A.

HAYES, Albert J. Born 1900. Educ.: Wisconsin Univ. Internat. Pres., Internat. Assoc. of Machinists. Vice Pres., AFL-CIO; Co-Chairman, American Foundation on Automation and Employment. Society: Atomic Energy Conference, Internat. Assoc. of Machinists.
Nuclear interests: Machinists Union represents 12,000 employees in atomic energy industry in the U.S.A.
Address: 1300 Connecticut Avenue, Washington, D.C., U.S.A.

HAYES, James Ernest, B.Sc., B.E. (Hons.). Born 1934. Educ.: Sydney Univ. Res. Sci. then Sen. Res. Sci., Australian A.E.C., 1957-67; Nucl. Investigations Eng., State Elec. Commission of Victoria, 1967-. Society: Inst. of Eng., Australia.
Nuclear interests: Reactor design; Nuclear power plant performance and economics; Nuclear education and training.
Address: 15 William Street, Melbourne, Victoria 3000, Australia.

HAYMAN, C., M.A. Sect. Leader, Phys. Chem., Fulmer Res. Inst., Ltd.
Address: Fulmer Research Institute, Ltd., Stoke Poges, Bucks., England.

HAYNES, Robert Hall, B.Sc., Ph.D. Born 1931. Educ.: Western Ontario Univ. Phys., Ontario Cancer Foundation, London (Ontario) Clinic, 1955-57; British Empire Cancer Campaign Exchange Fellow, Phys. Dept., St. Bartholomew's Hospital Medical Coll., London, 1957-58; Instructor in Biophys. (1959-61), Asst. Prof. Biophys. (1961-64), Chicago Univ.; Assoc. Prof.Biophys. and Med. Phys., California Univ., Berkeley, 1964-. N.A.S.-Nat. Res. Council Subcom. on Radiobiol., 1963-. Societies: Biophys. Soc.; Rad. Res. Soc.; American Soc. Cell Biol.; A.A.A.S.; Photobiol. Group, United Kingdom.
Nuclear interests: Cellular and molecular radiobiology; Repair of radiation damage to D.N.A.; Radiation physics; Photobiology.
Address: Lawrence Radiation Laboratory, California University, Berkeley, California 94720, U.S.A.

HAYNES, Sherwood K., A.B. Educ.: Williams Coll. Chairman, Phys. and Astronomy Dept., Michigan State Univ., 1957-. Societies: Fellow, A.P.S.; A.A.P.T.; Sté. Française Phys.; A.A.A.S. Nuclear interests: Beta ray spectroscopy; Auger effect.
Address: Department of Physics and Astronomy, Michigan State University, East Lansing, Michigan, U.S.A.

HAYS, Esther Fincher, A.B. (Cornell), M.D. (Cornell Univ. Medical Coll.). Born 1927. Educ.: Cornell Univ. and Cornell Univ. Medical Coll. Asst., Dept. of Medicine, Cornell Univ. Medical Coll., 1952-54; Instructor in Residence, Dept.of Medicine (1955-57), Asst. Prof., Dept. of Medicine (1957-66), Assoc. Prof., Depts. Medicine and Biophys. (1966-), California Univ. at Los Angeles; Intern in Medicine (1951-52), Asst. Resident in Medicine (1952-54), the New York Hospital; Jun. Res. Physician (1954-57), Asst. Res. Physician (1957-66), Assoc. Res. Physician (1966-), Chief, Medicine Div. (1958-63), Lab. of Nucl. Medicine and Rad. Biol., U.C.L.A. Papers: Co-author, several contributions to learned journals, including Induction of Mouse Leukemia with Purified Nucleic Acid Preparations (U.C.L.A., 410, Oct. 1957). Societies: N.Y.Acad. Sci.; American Assoc. for Cancer Res.; American Soc. of Hematology; Western Soc. for Clinical Res.
Address: U.C.L.A. Laboratory of Nuclear Medicine and Radiation Biology, 900 Veteran Avenue, Los Angeles, California 90024, U.S.A.

HAYWARD, Benjamin R., B.S. (Met. Eng.). Born 1921. Educ.: Rensselaer Polytech. Inst. Societies: A.S.M.; A.N.S.
Nuclear interests: Management of fuel material and fuel element component problems in all phases of fuel cycle, including radiation effects, processing of raw materials, fabrication and assembly of components, fuel element engineering, fuel cycle costs.
Address: 29149 Cliffside Drive, Malibu, California, U.S.A.

HAYWOOD, Frederick Wardle, B.Sc. (Hons.), Ph.D.(London), D.L.C. Born 1905. Educ.: Loughborough Coll. Tech. Director, Wild-Barfield Ltd., 1947-; Tech. Director, Refractory Mouldings and Castings, Ltd., 1951-. Books: Sen. author, Metal. Analysis by Means of the Spekker Photoelectric Absorptiometer; co-author, Steels in Modern Industry. Societies: F.R.I.C.; F.I.M.; Iron and Steel Inst.; Inst. of Metals; British Ceramic Soc.; Soc. of Chem. Industry; A.S.M.; Soc. of Analytical Chem. Nuclear interests: Supply of heat treatment and allied equipment for nuclear power sources. Address: 37 Oakroyd Avenue, Potters Bar, Hertfordshire, England.

HAYWOOD, Leslie Rupert, B.A., (Pre-Law, Phys. and Maths.). B.E., M.Sc. (Eng. Phys.). Born 1919. Educ.: Saskatchewan Univ. Supv., Instrumentation and Control,A.E.C.L., 1947-54; Manager, Instrumentation and Control (1955-57), Manager, Reactor Design (1957-59), Manager, Fuel and Materials (1959-61) Canadian Gen. Elec.; Manager, Reactor Development Projects (1961-63), Vice-Pres., Eng. (1963-67), Vice-Pres., Chalk River Nucl. Labs. (1967-), A.E.C.L. Nuclear interest: R. and D. Management. Address: Box 995, Deep River, Ontario, Canada.

HAZANOV, B. I. Paper: Co-author, About Optimum Selection of Alpha-Emitter Thickness for Low-Level Radioactivity Probe Spectrometry (letter to the Editor, Atomnaya Energiya, vol. 16, No. 5, 1964). Address: Academy of Sciences of the U.S.S.R., 14 Leninsky Prospekt, Moscow V-71, U.S.S.R.

HAZARD, Jacques. Expositions Publications, Direction des Relations Exterieures et des Programmes, C.E.A. Address: Commissariat à l'Energie Atomique, 29-33 rue de la Federation, Paris 15, France.

HAZELWOOD, R. M. Geophys., Geological Survey, U.S. Dept. of the Interior. Address: U.S. Department of the Interior, Geological Survey, Washington D.C. 20242, U.S.A.

HAZZAA, Ismail Bassyouni, Ph.D., M.Sc., B.Sc. (1st class hons.). Born 1919. Educ.: Cairo and Manchester Univs. Demonstrator (1950), Lecturer (1952), Faculty of Sci., Cairo Univ.; Head, Radioisotope Dept., U.A.R. Atomic Energy Establishment 1957; Director, Middle Eastern Regional Radioisotope Centre for the Arab Countries in Cairo, 1963. Books: Co-author, Radar in Peace (1956) (in Arabic); co-author, Atomic Energy (1956) (in Arabic); co-author, What you know about the Atom (1960) (in Arabic); History of the Atom (1960) (in Arabic). Societies: Egyptian Radioisotopes Soc.; Egyptian Mathematical and Phys. Soc. Nuclear interests: Applied nuclear and radiation physics. Address: Middle Eastern Regional Radioisotope Centre for the Arab Countries in Cairo, Sh. Malaeb El Gamaa, Dokki, Cairo, U.A.R.

HEAD, John Lawrence, B.Sc. (Eng., Hons.). Born 1932. Educ.: London Univ. Lecturer in Nucl. Power, Imperial Coll. of Sci. and Technol., London, 1963-. Society:A.M.I.Mech.E. Nuclear interest: Structural aspects of reactor design, particularly core design. Address: Boxgate Farm, St. John's Road, Crowborough, Sussex, England.

HEAD, Manuel A., B.S. (E.E.). Born 1922. Educ.: Tulane Univ. Society: A.I.E.E. Nuclear interests: Reactor core design; Reactor control and transient analysis. Address: Atomic Power Equipment Department, General Electric Company, San Jose, California, U.S.A.

HEADLAM-MORLEY, Kenneth Arthur Sonntag, B.A. Born 1901. Educ.: Oxford Univ. Sec., Iron and Steel Inst., 1933-67. Societies: Hon. Life Member, A.I.M.M.E.; A.S.M.; Hon. Member, l'Assoc. des Ing. sortis de l'Ecole de Liège; Verein deutscher Eisenhuttenleute; Sté. Française de Metallurgie. Address: The Iron and Steel Institute, 4 Grosvenor Gardens, London, S.W.1, England.

HEAL, Thomas John, M.Sc., B.Sc. Born 1915. Educ.: Wales Univ. Head Phys.Lab., Fulmer Res. Inst., 1950-55; Res. Manager, Culcheth Labs. (1955-59), Special Asst. to Member for Production (1959-60), Deputy Chief Eng. (Fuel Elements), Production Group, Springfields Works (1960-61), Chief Tech. Officer, Production Group, Springfields Works (1961-63), Chief Tech. Manager, Production Group, Springfields Works (1963-), U.K.A.E.A. Books: The Nucl. Handbook (Newnes, 1958); Materials for Nucl. Eng. (Temple Press, 1960). Societies: F. Inst. P.; Fellow, Inst. Metal. Nuclear interests: Metallurgy; Fuel element technology and design. Address: 24 Regent Avenue, Ansdell, Lytham St. Annes, Lancashire, England.

HEALD, Mark A., B.A., M.S. (Yale), Ph.D. (Yale). Born 1929. Educ.: Oberlin Coll. and Yale Univ. Assoc. Prof. of Phys., Swarthmore Coll., Pennsylvania, 1959-. Attached staff member, U.K.A.E.A., Culham Lab., Abingdon, Berks., 1963-64. Staff member, Experimental Phys. Div., Project Matterhorn, Princeton Univ., 1954-59. Book: Co-author, Plasma Diagnostics with Microwaves (New York, Wiley, 1965). Societies: A.P.S.; A.A.P.T. Nuclear interests: Controlled thermonuclear reactions; Microwave plasma physics; Microwave diagnostics. Address: Department of Physics, Swarthmore College, Swarthmore, Pennsylvania, U.S.A.

HEALY, John W., B.S. (Chem. Eng.). Born 1920. Educ.: Pennsylvania State Univ. G.E.C.,

1946-67. Nat. Com. Rad. Protection and Measurement; Inhalation Hazards Subcom., N.A.S.-N.R.C. Com. on Pathological Effects of Atomic Rad.; Chairman, American Board of Health Phys.; Chairman, Subcom. 6, Handling of Radioactive Isotopes and Fission Products, N.C.R.P.; Chairman, U.S.A.S.I. Com. N-13, Rad. Protection, Environmental Rad. Exposure Advisory Com., U.S.P.H.S. Societies: Rad. Res. Soc.; A.N.S.; A.I.H.A.; H.P.S.; A.C.S.; Atomic Ind. Forum.
Nuclear interests: Health physics; Reactor safety; Radiation limits; Environmental effects.
Address: Engineering Services, General Electric Company, 1 River Road, Schenectady 5, New York 12305, U.S.A.

HEALY, Paul William, B.A., B.S., M.A. (Ohio State), Ph.D. (Kentucky). Born 1915. Educ.: Rio Grande Coll. and Ohio State and Kentucky Univs. Assoc. Prof., Southwestern Coll., Winfield, Kansas, 1948-51; Asst. Prof., New Mexico Univ., 1952-56; Mathematician, Atomic Energy Div., Phillips Petroleum Co., Idaho Falls, Idaho, 1956-58; Sen. Phys., Aerojet-Gen. Nucleonics, 1958-60; Mathematician, U.S. Naval Weapons Station, Q.E.L., 1960-. Societies: Fellow, A.A.A.S.; Fellow and Treas., Meteoritical Soc.; A.N.S.; Mathematical Assoc. America.
Nuclear interests: Nuclear physics; Special weapons.
Address: 2448 Westcliffe Lane, Walnut Creek, Calif., U.S.A.

HEALY, R. M., Dr. Director, Chem. Sci. Res. Dept., and Director, Materials Sci. Res. Dept., Bunker-Ramo Corp.
Address: Bunker-Ramo Corporation, Corporate Research and Engineering, 2801 South 25th Avenue, Broadview, Illinois 60155, U.S.A.

HEAP, James C., Mech. Eng. B.S. (Mech. Eng.), M.S. (Mech. Eng.). Born 1921. Educ.: Illinois Inst. Technol. At Argonne Nat. Lab., 1956-. Societies: A.S.M.E.; A.N.S.
Nuclear interest: Reactor and particle accelerator design.
Address: Argonne National Laboratory, Argonne, Illinois, U.S.A.

HEARD, Manning Wright, LL.B. Born 1896. Educ.: Tulane Univ. and Coll. of Law, Tulane. Pres., Hartford Accident and Indemnity Co., 1961-; Chairman of the Board, Hartford Insurance Group, 1964-. Formerly Director, Connecticut Nucl. Centre for Res., Training and Education, Inc., Member, Hartford Redevelopment Agency; Member, Connecticut Advisory Com. on Atomic Energy.
Address: 1391 Asylum Avenue, Hartford, Conn. 06105, U.S.A.

HEARN, Richard Lankaster, B.A.Sc., D.Eng., LL.D., Consulting Eng. Born 1890. Educ.: Toronto Univ. Gen. Manager and Chief Eng. (1947-55), Chairman (1955-56), Ontario Hydro. Formerly Director, A.E.C.L. Societies: Hon. Member, Eng. Inst. of Canada; A.S.C.E.; Hon. Member, the Moles; Hon. Member, Inst. Elec. Eng., England.; I.E.E.E.; A.P.E.O.; Assoc. Professional Eng.of British Columbia; Newcomen Soc.
Address: Glen Nevis, Box 135, Queenston, Ontario, Canada.

HEATH, Charles Douglas, Chartered Eng., F.I.E.E. Born 1918. Deputy Supt., Hinkley Point Power Station (1959-62), Station Supt., Dungeness'A' Nucl. Power Station (1962-), Station Supt., Dungeness 'B' Nucl. Power Station (1967-), now No. 4 Group Manager, North Western Region, C.E.G.B.
Address: "The Spinney", Gannock Park, Deganwy, Caernarvonshire, Wales.

HEATH, Dennis Frederick, M.A., D.Phil. Born 1924. Educ.: Oxford Univ. Pest Control Ltd., finally as Chem. Res. Manager, 1949-54; Sci. Staff, Medical Res. Council, Toxicology Res. Unit, 1954-. Societies: Biochemical Soc.; Chem. Soc.; Faraday Soc.
Nuclear interest: Tracers in biochemical research.
Address: Toxicology Research Unit, Medical Research Council Laboratories, Carshalton, Surrey, England.

HEATH, Russell LaVerne, B.S. (Colorado State), M.S.(Vanderbilt). Born 1926. Educ.: Colorado State, Vanderbilt and Rutgers Univs. Graduate Asst. in Phys., Rutgers Univ., 1950-51; U.S.A.E.C. Graduate Fellow, O.R.N.L. and Vanderbilt Univ., 1951-53; Sen. Radiol. Phys., Atomic Energy Div., American Cyanamid Co., 1953-54; Leader, Decay Scheme Study Group (1954-58), Phys. Sect. Chief, Atomic Energy Div. (1959-66), Phillips Petroleum Co.; Idaho Nucl. Corp., 1966-. Societies: Fellow, A.P.S.; A.N.S.; Professional Group on Nucl. Sci., I.E.E.E.
Nuclear interests: Low-energy nuclear physics; Decay schemes; Radioactivity; Nuclear physics instrumentation; Gamma-ray scintillation spectrometry; Development of radiation detectors and systems for nuclear physics research; Reactor development activities; Management of scintillation spectrometry — gamma-ray spectrum cataloging programme.
Address: Idaho Nuclear Corporation, National Reactor Testing Station, P.O. Box 1845, Idaho Falls, Idaho, U.S.A.

HEATON, H., A.M.I.Struct. E. Chief Designer, Robert Watson and Co. (Constructional Engs.) Ltd.
Address: Robert Watson and Co. (Constructional Engineers) Ltd., Bolton, Lancs., England.

HEAVENS, Oliver Samuel, B.Sc., Ph.D., D.Sc. Born 1922. Educ.: London and Reading Univs. Reader in Phys., London Univ., 1964; Prof. of Phys., York Univ., 1964-. Societies: Fellow, Optical Soc. of America; Fellow, Inst. of

Phys. and the Phys. Soc.
Nuclear interest: Optical methods in nuclear instrumentation.
Address: Department of Physics, York University, Heslington, York, England.

HEBB, Malcolm H., B.A. (British Columbia), Ph.D. (Harvard). Born 1910. Educ.: British Columbia, Wisconsin and Harvard Univs. Manager, Gen. Phys. Res., Gen. Elec. Res. Lab., 1952-. Society: A.P.S.
Address: 2012 Lexington Parkway, Schenectady, New York, U.S.A.

HEBBELER, Brigadier General James A. Member, Military Liaison Com., U.S.A.E.C.
Address: U.S. Atomic Energy Commission, Military Liaison Committee, Washington D.C. 20545, U.S.A.

HEBEL, Y. Directeur Technique, Sté. des Engins Matra.
Address; Societe des Engins Matra, B.P. No. 1, 78 Velizy, France.

HECHT, G. Ch. C. Karl, Prof., Dr. Born 1903. Educ.: Göttingen Univ. Director, Inst. für die Pädagogik der Naturwissenschaften, Kiel Univ. Chairman, Com. for Educ.,German Atomforum; Member, Editorial Board, Die Atomwirtschaft. Book: German edition, Radioisotopes Lab. Techniques (by R. A. Faires and B. H. Parks) (Braunschweig, Friedr. Vieweg and Sohn, 1961).
Nuclear interest: Education in nuclear physics.
Address: Institut für die Pädagogoik der Naturwissenschaften, Christian-Albrechts-Universität Kiel, 40/60 Olshausenstrasse, 23 Kiel, Germany.

HECKER, Oskar, Dipl.-Phys. Born 1940. Educ.: Karl Marx Univ. Sci. Staff Member, German Board for Metrology and Goods Testing, of the German Democratic Republic.
Nuclear interest: Dosimetry of X- and γ-rays.
Address: Deutsches Amt für Messwesen und Warenprüfung der German Democratic Republic, Postfach 1542, Berlin, German Democratic Republic.

HECKLER, George E., Ph.D. Assoc. Prof. of Chem., Chairman, Dept. of Chem., Assoc. Director, I.S.C.-A.E.C. Res., Idaho State Univ.
Address: Idaho State University, Pocatello, Idaho, U.S.A.

HECKMAN, Thomas Paul, B.S.M.E., Prof. Eng. Born 1917. Educ.: Purdue Univ. Classification and Tech. Information Officer, Chicago Operations Office, U.S.A.E.C., 1965. Member, Nat. Membership Com., Chairman, Chicago Sect., Finance Com., A.N.S. Society: A.S.M.E.
Nuclear interest: Technical information management.
Address: United States Atomic Energy Commission, Chicago Operations Office, 9800 South Cass Avenue, Argonne, Illinois 60439, U.S.A.

HEDAYAT, H. E. Khorsrow. Chief of Iranian Mission to Euratom, 1963-.
Address: c/o Euratom, 51-53 rue Belliard, Brussels, Belgium.

HEDAYAT, Salah-El-Din. Formerly Director Gen., now Chairman, Atomic Energy Establishment, United Arab Republic; U.A.R. alternate deleg. I.A.E.A. Gen. Conference, Vienna, Sept. 1960.
Address: Atomic Energy Establishment, Inshas, Nr. Cairo, Egypt, United Arab Republic.

HEDBERG, Carl A. Head, Electronics Div., Denver Res. Inst., Denver Univ.
Address: Denver University, University Park, Denver 10, Colorado, U.S.A.

HEDDEN, Daniel Melvin, B.Sc. Born 1916. Educ.: McMaster Univ. Vice-Pres., Administration, McMaster Univ., 1963.
Nuclear interest: Management.
Address: McMaster University, Hamilton, Ontario, Canada.

HEDDEN, Daniel T., A.B., M.Sc. Born 1923. Educ.: Middlebury Coll.and New Hampshire Univ. Supervising Nucl. Test Facilities Group, Connecticut Aircraft Nucl. Eng. Lab., Pratt and Whitney Aircraft, Middletown, Conn.; Plant Eng., Snap 50 Programme, Idaho Test Labs., Pratt and Whitney Aircraft, 1962-; Nucl. Eng., Connecticut Yankee Atomic Power Co., 1964-; Now Chief, Environmental Sci. and Services, Northeast Utilities Service Co., Hartford, Connecticut. Radiol. Defence Advisory Com., Connecticut; Radiol. Rep., Governor's Civil Defence Advisory Com. Societies: A.N.S.; A.P.S.
Nuclear interest: Nuclear power station design and operation, with emphasis on environmental and thermal effects of nuclear power station operation.
Address: 114 Ridgewood Road, Glastonbury, Connecticut, U.S.A.

HEDGE, Carl E. Geologist, Geological Survey, U.S. Department of the Interior.
Address: U.S. Department of the Interior, Geological Survey, Building 25, Federal Centre, Denver, Colorado, U.S.A.

HEDGRAN, Arne, Ph.D. Born 1921. Educ.: Stockholm Univ. Div. Head (radiation safety and regulation as regards to radioactive substances and nuclear energy), Nat. Inst. Radiol. Protection.
Nuclear interest: Health physics.
Address: National Institute of Radiation Protection, Stockholm 60, Sweden.

HEDIN, Tore Knut Albert, Civil eng. Born 1900. Educ.: Kungl. Tekniska Högskolan, Stockholm. Pres. of the board, Gullspangs

Kraftaktiebolag and Svenska Kraftverksföreningen.
Address: Gullspangs-Kraftaktiebolag, Örebro, Box 472, Sweden.

HEDLUND, Roland G. M., M.S. (Elec. Eng.). Born 1921. Educ.: Chalmers Tech. Univ., Gothenburg. G.E.C., Pittsfield, 1950-55; Semko, Stockholm, 1956-59; Head, Sect. for Operation of Labs., A.B. Atomenergi, 1960-.
Nuclear interests: Operation of radioactive laboratories and plants; Waste management; Radioactive service.
Address: 6 Hagavägen, Studsvik, Tystberga, Sweden.

HEDVIG, Peter, Ph.D., Candidate of Phys. Sci. Born 1928. Educ.: Eötvös Univ., Budapest. Assoc. Prof., Dept. of Atomic Phys., Eotvos Univ., Budapest, 1957-; Phys., Central Res. Inst. of Phys., Budapest, 1953-.
Nuclear interests: Irradiation damage; Nuclear magnetic resonance.
Address: Department of Atomic Physics, Eötvös University, Budapest, Puskin u. 5-7, Hungary.

HEEGER, Alan Jay, Ph.D., B.S. Born 1936. Educ.: Nebraska and California Univs. Prof., Phys., Pennsylvania Univ. Society: A.P.S.
Nuclear interest: Solid state physics.
Address: Pennsylvania University, Physics Department, Philadelphia, Pennsylvania 19104, U.S.A.

HEELEY, Pierre George, Enseigne de Vaisseau, Ing. E.S.E. Born 1898. Educ.: Ecole Supérieure d'Electricité and Ecole Navale. Prés. Directeur Gen., Cie. des Compteurs.
Address: 12 Place des Etats-Unis, 92 Montrouge, (Hauts-de-Seine), France.

HEEM, Louis Marie DE. See DE HEEM, Louis Marie.

HEEMSTRA, Raymond Jacob, B.A., M.S. Born 1926. Educ.: Hope Coll., Michigan and Iowa State Univ. Phys. Chem., U.S. Bureau of Mines, Bartlesville, Oklahoma, 1951-. Society: A.C.S.
Nuclear interests: Use of radioactive isotopes for tracing the flow of subsurface oil field water between wells; Determination of oxygen contents in petroleum and petroleum fractions using neutron activation methods.
Address: 705 Yale Drive, Bartlesville, Oklahoma, U.S.A.

HEER, E., D. ès Sc. Born 1928. Educ.: Federal Inst. of Switzerland, Zurich. Prof. ordinaire, Lab. de Phys. Nucl. Expérimentale, Geneva Univ., 1961-.
Nuclear interest: Elementary particle physics.
Address: Ecole de Physique, 1211 Geneva 4, Switzerland.

HEER, Frederick Jacques DE. See DE HEER, Frederick Jacques.

HEERDEN, Izak Jacobus VAN. See VAN HEERDEN, Izak Jacobus.

HEERTJE, Isaäc, Dr. Born 1935. Educ.: Amsterdam Univ. At Unilever Research Laboratory, Vlaardingen, Netherlands. Society: Koninklijke Nederlandsche Chemische Vereniging.
Nuclear interests: Nuclear chemistry; Research with radioactive tracers.
Address: Unilever Research Laboratory, Vlaardingen, Netherlands.

HEESEN, Wilhelm von. See von HEESEN, Wilhelm.

HEFNER, Frank K. Minister, U.S. Resident Rep. to I.A.E.A. Adviser, Third I.C.P.U.A.E., Geneva, Sept. 1964.
Address: U.S. Mission to I.A.E.A., 14 Schmidgasse, Vienna 8, Austria.

HEGER, Adolf, Dipl.-Ing., Dr.-Ing. Born 1936. Educ.: Dresden T.H.
Nuclear interests: Application of radioisotopes in textile research and production. Irradiation treatment of textile fibres modification of natural and synthetic fibres by radiation chemistry - e.g., graft polymerisation.
Address: Institut für Technologie der Fasern der Deutschen Akademie der Wissenschaften zu Berlin, 6 Hohe Strasse, 801 Dresden, German Democratic Republic.

HEGGENHOUGEN, Rolv. Deputy Chairman, A/S Noratom.
Address: A/S Noratom, 20 Holmenveien, Oslo, Norway.

HEHN, Gerfried, Dipl. Phys., Dr. rer. nat. Born 1933. Educ.: Berlin Tech. and Stuttgart Univs. Sci., Inst. für Hochtemperaturforschung (1962), Group Leader, Inst. für Kernenergetik (1965), Stuttgart Univ. Chairman, Subcom. 1, Terminology, Definitions, Units and Symbols in Nucl. Energy, Fachnormenausschuss Kerntechnik (German Standardisation Organisation). Book: Das Strahlungsfeld des Reaktors (Münich, K. Thiemig K. G., 1964). Societies: Deutsche Physikalische Gesellschaft; Kernenergie-Studiengesellschaft Hamburg, Fachausschuss Strahlenschutz.
Nuclear interests: Reactor shielding; Neutron physics; Dosimetry; Reactor physics; Reactor technology.
Address: Institut für Kernenergetik, 51 Herdweg, 7 Stuttgart, Germany.

HEIDE, Harm ter. See ter HEIDE, Harm.

HEIDELBERGER, Charles, S.B., M.S., Ph.D. Born 1920. Educ.: Harvard Univ. Prof. Oncology, Wisconsin Univ., 1958-. Board of Directors, American Assoc. for Cancer Res.; Consultant to Cancer Chemotherapy Nat. Service Centre, of U.S.P.H.S. Book: Co-author,

Isotopic Carbon (New York, John Wiley and Sons, 1949). Societies: A.C.S.; American Soc. of Biol. Chem.; American Assoc. for Cancer Res.; A.A.A.S.; British Biochem. Soc. Nuclear interest: Tracer studies in the determination of the biochemical mechanism of action of substances of pharmacological interest, such as carcinogenic hydrocarbons and tumour inhibitory compounds.
Address: McArdle Memorial Laboratory, Wisconsin University, Madison, Wisconsin 53706, U.S.A.

HEIFETS, M. I. Paper: Co-author, On Limitation on Value Density of Interacting Currents in Colliding Ultrarelativistic Beams (letter to the Editor, Atomnaya Energiya, vol. 19, No. 4, 1965).
Address: Academy of Sciences of the U.S.S.R., 14 Leninsky Prospekt, Moscow V-71, U.S.S.R.

HEIKO, Adalbert. Inst. fur Theoretische Kernphysik, Bonn Univ.
Address: Institut fur Theoretische Kernphysik, 16 Nussallee, Bonn, Germany.

HEIL, A., Dr. Zentralsekretär, Christlichen Metallarbeiter-Verbandes der Schweiz. Member, Federal Commission for Atomic Energy, Switzerland.
Address: 56 Büelrainstr., Winterthur, Switzerland.

HEILIG, C. E. Design Eng.-Mechanical, R. and D., Baldwin-Lima-Hamilton Corp.
Address: Baldwin-Lima-Hamilton Corp., Eddystone Division, Philadelphia 42, Pa., U.S.A.

HEILMEYER, Ludwig, Dr. med., Dr. med. h.c. Born 1899. Rector, Ulm Univ.; Hon. Prof., Freiburg Univ. Vorsitzender, Vorstand, Soc. of Radioisotopes in the German Medical Assoc. Societies: Hon. Member, Roy. Soc., London; Hon. Member of many medical socs. throughout the world.
Address: Universität Ulm, 10 Parkstrasse, 79 Ulm/Donau, Germany.

HEIM, Roger. Member, Conseil Scientifique, C.E.A.
Address: Commissariat à l'Energie Atomique, 29-33 rue de la Federation, Paris 15, France.

HEIMAN, Warren Jonas, B.S. (Chem.), M.S. (Radiochem.), M.S. (Nucl. Eng.). Born 1923. Educ.: Tulane and California (San Francisco City) Univs, and Iowa State Coll. At Iowa State Coll. Atomic Res. Inst., 1948-50; California Univ. Rad. Lab., 1950-51; U.S. Naval Radiol. Defence Lab., 1951-56; Westinghouse Bettis Atomic Power Div., 1957-59; U.S. Naval Radiol. Defence Lab., 1959-60; Pres., Reactor Experiments, Inc., 1960-. Societies: A.N.S.; Nucl. Eng. Div., A.I.Ch.E.; H.P.S.
Nuclear interests: Nuclear engineering education; Experimentation in research and training reactors, nuclear-powered rockets and satellites, radiological safety, special shielding materials, reactor water chemistry, activation analysis.
Address: 40 Roxbury Lane, San Mateo, California, U.S.A.

HEIMBERG, Julius, Dipl.-Ing. Born 1897. Educ.: Berlin T.H. Mitglied, Vorstandsrat, Schiffbautechnischen Gesellschaft e.V., Hamburg. Member, Reaktor-Sicherheitskommission, Deutsche Atomkommission.
Address: 14 Göslingstr., 328 Bad Pyrmont, Germany.

HEIMBUCH, Alvin H., B.A., M.Sc., Ph.D. Born 1923. Educ.: Oregon and Portland Univs. and North Carolina State Coll. Res. Asst., N.C. State Coll., 1951-55; Res. Chem., G. Frederick Smith Chem. Co., 1955-58; Staff Chem., Tracerlab. Div. of LFE, 1958-. Societies: Sen. Member, C. S. (Div. of Analytical Chem. and Div. of Nucl. Chem. and Technol.); A.A.A.S. Nuclear interests: Analytical radiochemistry; Organic scintillators; Nuclear chemistry and applications; Activation analysis.
Address: 900 Pomona Avenue, El Cerrito, California, U.S.A.

HEIN, Hans-Jürgen, Diplom-Chemiker, Dr. rer. nat. Born 1920. Educ.: T.H. Hanover. DEGUSSA-Wolfgang, 1955-60; NUKEM, 1960-66; Gesellschaft zur Wiederaufarbeitung von Kernbrennstoffen m.b.H., Leopoldshafen, 1966-.
Nuclear interests: Chemical development and research; Reprocessing; Accountability of nuclear materials.
Address: Gesellschaft zur Wiederaufarbeitung von Kernbrennstoffen m.b.H., 7501 Leopoldshafen, Germany.

HEINDERTS, R. Eng. engaged in Nucl. Work, Koninklijke Maatschappij "de Schelde" N.V.
Address: Koninklijke Maatschappij "de Schelde" N.V., Vlissingen, Netherlands.

HEINDL, Clifford Joseph, B.S., M.S., A.M., Ph.D. Born 1926. Educ.: Northwestern and Columbia Univs. Sen. Phys., Bendix Aviation Res. Labs., 1952-54; Asst. Sect. Chief, Babcock and Wilcox Co., Atomic Energy Div., 1955-58; Group Supv., Nucl. Phys. (1959-65), Tech. Manager, Res. (1965-), Jet Propulsion Lab. Sec., Aerospace Div., A.I.A.A. Societies: A.N.S.; A.P.S.; H.P.S.; A.I.A.A. Nuclear interests: Nuclear physics; Reactor physics; Management.
Address: Jet Propulsion Laboratory, California Institute of Technology, 4800 Oak Grove Drive, Pasadena, California 91103, U.S.A.

HEINE, Kurt Egon, Dr. (Cologne). Born 1929. Educ.: Cologne Univ. and Max-Planck-Inst. für Chemie, Mainz. Asst. Inst. für Kernchemie, Cologne Univ., 1959-. Society: Gesellschaft Deutscher Chemiker.

Nuclear interest: Radiation chemistry.
Address: Farben Fabriken Bayer A.G., 509
Leverkusen, Ing. Abt. A.P.11, Germany.

HEINEMAN, Robert E., Ph.D. (Phys.), M.S.
(Phys.), B.S.E. (Phys. and Maths.). Born
1926. Educ.: Michigan Univ. Manager, High
Temperature Reactor Phys., Pacific Northwest Lab., Battelle Memorial Inst. Societies:
A.N.S.; A.P.S.
Nuclear interests: Reactor and nuclear physics;
Use of test reactors in reactor development,
nuclear technology, and neutron physics.
Address: Battelle-Northwest, Richland,
Washington 99352, U.S.A.

HEINEMANN, Hermann, Dr. Ing. Born 1901.
Educ.: Göttingen Univ. and Braunschweig
T.H. Chief of Div., Buchler and Co., Braunschweig. Society: Gesellschaft Deutscher
Chemiker, Fachgruppe Kern- und Strahlenchemie.
Nuclear interests: Technics of isotopes;
Chemistry of radioactive materials, Methods
of measurement.
Address: 6 Wilmerdingstrasse, Braunschweig,
Germany.

HEINTZ, W., Dr. Reactor Operation and
Reactor Properties, Nucl. Reactor Div.,
Physikalisch-Technischen Bundesanstalt.
Address: Physikalisch-Technischen Bundesanstalt, 100 Bundesallee, 33 Braunschweig,
Germany.

HEISE, Karl-Heinz, Dipl.-Ing. Born 1938.
Educ.: Martin-Luther-Univ., Halle, and Dresden
Tech. Univ.
Nuclear interests: Reinheitsprüfung und
Anwendung markierter Verbindungen (Purity
analysis and application of labelled compounds).
Address: Zentralinstitut für Kernforschung,
8051 Dresden, German Democratic Republic.

HEISENBERG, Werner Karl, Nobel Prize for
Physics, 1932; Pour le mérite für Wissenschaften und Künste, 1957; Dr.phil. Born 1901.
Educ.: Munich and Göttingen Univs. Director,
Max Planck Inst. for Phys., and Prof., Göttingen Univ., 1946-58; Director, Max Planck
Inst. for Phys. and Astrophys., and Prof.,
Munich Univ., 1958-; Member, Exec. Com.,
Member, Steering Com., Deutsches Atomforum E.V.; Formerly Mitglied, now Stv.
Vorsitzender, Prasidium, and Gast, Fachkommission II, Kernforschung, Member,
Fachkommission III, Kerntechnik, Deutsche
Atomkommission, Federal Ministry for Sci.
Res. Books: Die physikalischen Prinzipien
der Quantentheorie (1930); Wandlungen in
den Grundlagen der Naturwissenschaft
(1935); Die Physik der Atomkerne (1943);
Vorträge über die komische Strahlung (1943);
Das Naturbid der heutigen Physik (1955);
Phys. and Philosophy (1958). Societies: Royal
Soc. of London; Phys. Soc. of London;
American Philosophical Soc.; Academies of
Göttingen, Uppsala, Berlin, Bucharest, Oslo,
Halle, Leipzig, Munich, Madrid, Copenhagen,
Rome, Amsterdam, Helsinki, Pontificia
Academia Scientiarum. Pres., Alexander von
Humboldt-Stiftung.
Address: 6 Föhringer Ring, Munich 23,
Germany.

HEISKANEN, Paavo Kalevi, M.A. Born 1935.
Educ.: Turku Univ. Res. Asst., Turku Univ.,
1964-. Society: Phys. Soc. of Finland.
Nuclear interest: X-ray physics.
Address: Wihuri Physical Laboratory, Turku
University, 5 Vesilinnantie, Turku, Finland.

HEITLER, H., Dr. Res. Asst., Bristol Univ.
Address: Bristol University, H. H. Wills
Physics Laboratory, Royal Fort, Bristol 8,
England.

HEITMANN, Hans-Günter, Dr. rer. nat. Dipl.-Chemiker. Born 1923. Educ.: Humboldt Univ.,
Berlin. At Max-Planck Inst., Berlin-Dahlem,
1951; Märkische Kabelwerke, Berlin, 1952;
Siemens A.G., Erlangen, 1952-. Books: Contributions to: K. R. Schmidt, Nutzenergie aus
Atomkernen (vol. i, Berlin, 1959; vol. ii,
Berlin, 1960); E. Tödt, Korrosion und Korosionsschutz (Berlin, 1961); K. Schröder,
Grosse Dampfkraftwerke (Berlin, 1966).
Societies: Gesellschaft Deutscher Chemiker;
Vereinigung der Grosskesselbesitzer.
Nuclear interests: Chemistry of actinides;
Fuel reprocessing; Ore processing; Water
conditioning and treatment; Chemical problems in nuclear plants; Corrosion problems in
nuclear plants; Specific analysis problems
associated with radioactive waters; Liquid
effluent treatment and disposal; Application of
ion exchangers in primary loops.
Address: 10 Im Heidewinkel, Buckenhof, Kr.
Erlangen, Germany.

HEJTMANEK, Johann, Dr. phil., Doz. Born
1931. Educ.: Vienna and Ohio State Univs.
At Atominstitut der Österreichischen Hochschulen, 1960-. Society: Austrian Mathematical
Soc.
Nuclear interests: Statistical mechanics of
neutrons; Mathematical foundation of transport
theory; Ergodic theory.
Address: 19/5 Khevenhuellerstrasse, Vienna
18, Austria.

HEKHUIS, Gerrit L., A.B., M.D. Born 1917.
Educ.: Yale, Kansas, Chicago and California
Univs. Radiologist; Chief of Radiobiol.,
U.S.A.F. School of Aerospace Medicine.
Societies: Aerospace Medical Assoc.; American
Coll. of Radiol.; American Board of Radiol.
(Certified); A.A.A.S.; A.M.A.; Soc. of Nucl.
Medicine; Soc. AF Internists and Allied
Specialists.
Nuclear interests: Research in radiobiology,
clinical radiology, teaching. Special emphasis:
basic, applied, clinical radiobiology; nuclear
propulsion and power source problems of
human hazard, protection and treatment.

Address: School of Aerospace Medicine, U.S.A.F. Aerospace Medical Centre, Brooks Air Force Base, Texas, U.S.A.

HEKMAN, H., Drs. Born 1926. Educ.: Free Univ., Amsterdam. Dept. Phys., Free Univ., Amsterdam, 1947-60; Health Phys., Inst. for Atomic Sci. in Agriculture, Wageningen, 1960-67; Health Phys., Tech. Univ., Eindhoven, 1968-. Societies: Dutch Phys. Soc.; Christian Soc. for Phys. and Physicians in the Netherlands; Dutch Soc. for Rad. Hygiene. Nuclear interests: Nuclear physics; Health physics.
Address: Technische Hogeschool Eindhoven, 2 Insulindelaan, Eindhoven, Netherlands.

HELBICH, Jan, M.D. Born 1927. Educ.: Charles and Palacky Univs. Dept. of Ionizing Rad., Inst. of Industrial Hygiene and Occupational Diseases, Prague, 1957-60; Head, Information Centre, Inst. of Rad. Hygiene, Prague, 1960-; Editor, Index Radiohygienicus. Society: Sect. of Nucl. Medicine and Rad. Hygiene, Czechoslovak JE Purkyne Medical Assoc.
Nuclear interests: Scientific information in radiation hygiene and radiobiology.
Address: 48 Srobarova, Prague 10, Czechoslovakia.

HELBLING, Willy Anton, Dipl. Ing. E.T.H. Born 1923. Educ.: Swiss Federal Inst. Technol., Zürich and Internat. School of Nucl. Sci. and Eng. Mech. Eng., Sulzer Bros., Ltd., Winterthur, 1950-. Lectures in Reactor Design at the Technikum des Kantons Zürich, Winterthur, 1959-. Society: S.I.A. (Soc. of Swiss engineers and architects).
Nuclear interests: Reactor design; Economics of nuclear power plants.
Address: 81 Anton Graffstrasse, Winterthur, Switzerland.

HELD, Christian, Dipl. Ing. Formerly in Eng., Head, Reactor Plant Layout Div., Siemens-Schuckertwerke A.G.
Address: Siemens Schuckertwerke A.G., 1 Günther Scharowsky Strasse, Erlangen, Germany.

HELD, Kalman, B.A., M.A., Ph.D. Born 1919. Educ.: Brooklyn Coll., Columbia Univ. and Polytech. Inst. of Brooklyn. Sen. Sci., United Nucl. Corp., 1951-55; Vice Pres., Tech. Res. Group, 1955-62; Tech. Director, Radioptics, Inc., 1963-. Books: Nat. Catalogue of Patents. 10 Volumes. (Rowman and Littlefield, 1963); Internat. Index of Patents. 11 Volumes. (The Interdex Corp., 1965). Societies: A.P.S.; A.C.S.; Electrochem. Soc. Nuclear interests: Materials for reactors and shields; Effects of nuclear radiation on materials; Synthesis of materials using nuclear radiation.
Address: 10 Dupont Street, Plainview, Long Island, New York, U.S.A.

HELDACK, John Martin V., B.M.E. Born 1917. Educ.: Ohio State Univ. Various Exec. Management Positions to Manager, G.E.C., 1950-64; Corp. Director, North American Aviation, 1964-66; Pres. and Chairman, North American Aviation Internat., 1966-67; Exec. Vice Pres., Atomics Internat., 1967-. Societies: A.S.M.E.; A.I.A.A.; Nat. Security Ind. Assoc.; Assoc. of U.S. Army; U.S. Navy League. Nuclear interest: Management in reactor design and nuclear physics.
Address: Atomics International, P.O. Box 309, Canoga Park, California 91304, U.S.A.

HELF, Samuel, B.S. Born 1923. Educ.: Brooklyn Coll., Syracuse and Ohio State Univs. At Nucleonics Res. Sect. Picatinny Arsenal, Dover, N.J., 1950-; Internat. Inst. for Nucl. Sci. and Eng., Argonne Nat. Lab., 1960-61; Special assignment (on loan) to Health Phys. Div., O.R.N.L., 1961-62. Societies: A.C.S.; RESA.
Nuclear interest: Health physics; Environmental monitoring for radioactivity; Wholebody counting; Liquid scintillation counting; Internal and external radiation dosimetry.
Address: 6 Egbert Avenue, Morristown, New Jersey, U.S.A.

HELFRICH, Gerard. Director, Reactor Div., San Francisco Operations Office, U.S.A.E.C., -1963; U.S.A.E.C. Sci. Advisor, Euratom, 1963-. Tech. Asst. to Commissioner Johnson.
Address: U.S.A.E.C. Washington D.C. 20545, U.S.A.

HELG, M. R. Vice Pres., Energie Nucléaire S.A.
Address: Energie Nucléaire S.A., 10 avenue de la Gare, Lausanne, Switzerland.

HELGESON, B. P., B.S. Born 1920. Educ.: Columbia Univ. Industrial Staff Member from Reaction Motors to Project Rover, Los Alamos Sci. Lab.; Group Leader, Sequence Controls Group, Reaction Motors; Head, Nucl. Rocket Sect., Reaction Motors Div., Thiokol Chem. Corp.; Manager, Engine Technol., Nucl. Space Systems Div., Lockheed Missiles and Space Co.; Chief, Nevada Extension, Space Nucl. Propulsion Office; Deputy Manager, Richland Operations Office, U.S.A.E.C., 1967-68; Safety Director, N.A.S.A., 1968-. Societies: I.E.E.E.; A.I.A.A.
Nuclear interest: Development of nuclear rockets.
Address: Space Nuclear Propulsion Office, Nevada Extension, Nuclear Rocket Development Station, P.O. Box 1, Jackass Flats, Nevada, U.S.A.

HELLEN, E. Principal Member, Occupational Safety and Health Div., Internat. Labour Office; Deleg., I.A.E.A. Symposium on Nucl. Ship Propulsion with Special Reference to Nucl. Safety, Taormina, Nov. 14-18, 1960. Adviser, Third I.C.P.U.A.E., Geneva, Sept. 1964.

Address: Occupational Safety and Health Division, International Labour Office, Geneva, Switzerland.

HELLENS, Robert L., Sc.B., Ph.D. Born 1925. Educ.: Brown and Yale Univs. Group Leader, Brookhaven Nat. Lab., 1961-; Visiting Sci., U.K.A.E.A., Winfrith, 1964-65; Adjunct Prof., Columbia Univ., 1966. Societies: Fellow, A.N.S.; A.P.S.
Nuclear interest: Physics problems of thermal reactor design.
Address: Combustion Engineering Corporation, Nuclear Power Department, Windsor, Connecticut 06095, U.S.A.

HELLER, Edward L., B.S. (Eng.), M.B.A. (Transportation). Born 1912. Educ.: Lehigh Univ., Harvard Business School. Tech. Manager, Eastern Office, Gen. Atomic Div., Gen. Dynamics Corp.; Director, Nucl. Div., H. K. Ferguson Co. Societies: A.S.M.E.; A.N.S.
Nuclear interests: Reactors for universities, industries, utilities, merchant ships, underseas, and space applications; Direct conversion energy sources and activation analysis.
Address: 1783 East 11 Street, Cleveland 14, Ohio, U.S.A.

HELLER, László, Mech. Eng., Dr. sc. tech. (Eidgenössische T.H., Zürich). Born 1907. Educ.: Eidgenössische T.H., Zürich. Tech. Chief., Thermotech. Planning Bureau, 1949; Ord. Prof., Budapest Tech. Univ., 1951. Member, Hungarian Acad. of Sci.; Pres., Supreme Energetics Com., Hungarian Acad. of Sci. Societies: Sci. Soc. for Power Economy; Hungarian Electrotech. Soc.
Nuclear interests: Reactor design; Thermodynamic aspects of nuclear research in general.
Address: 45 Bimbó ut, Budapest 2, Hungary.

HELLER, Sheldon. Asst. Manager, Internat. Div., Nucl. Measurements Corp.
Address: Nuclear Measurements Corp., 13 East 40th Street, New York 16, New York, U.S.A.

HELLMAN, Stanley K., Bachelor Aeronautical Eng. Born 1928. Educ.: Southern California Univ. and Brooklyn Polytech. Inst. Advanced Eng. Studies Programme (1951-52), Guided Missile Dept. (1952-53), Knolls Atomic Power Dept. (1953-56), G.E.C.; Exec. Management Development Programme, Cornell Univ., 1952; Nucl. Eng. Dept., Vitro Eng. Co., 1956-60; Manager, Nucl. Programmes, 1963. Societies: A.S.M.E.; American Rocket Soc.
Nuclear interests: Nuclear design of powerplants, test facilities, industrial irradiators, and nuclear rocket test facilities; Consultant services and reactor design lecturing.
Address: Vitro Engineering Company, 100 Church Street, New York City, New York, U.S.A.

HELLMUTH, James G., B.S. (Industrial Administration), LL.B. Born 1923. Educ.: Yale and George Washington Univs.
Nuclear interests: Legal; Licensing; Safety; Health.
Address: 89 Beach Street, Boston 11, Massachusetts, U.S.A.

HELLRIEGEL, Werner, M.D., Prof. Born 1913. Educ.: Jena, Heidelberg, Frankfurt and Leipzig Univs. At Univ. Clinics Frankfurt/Main, Univ. Röntgeninstitut, 1941-; Director, Strahlenklinik, Stuttgart. Societies: Deutsche Röntgengesellschaft; Hon. member, Turkish Cancer and Radiobiol. Soc.
Nuclear interests: In medicine and biology (therapy and diagnosis).
Address: Strahlenklinik Katharinenhospital, 60 Kriegsbergstrasse, Stuttgart 7, Germany.

HELLSTEN, Aulis Alfred, M.Sc. Born 1933. Educ.: Helsinki Inst. Technol. Res. Eng., Imatran Voima Osakeyhtiö, Helsinki, 1958-60, 1962-63; Reactor Phys., O.E.C.D. Halden Reactor Project, Halden, Norway, 1960-62; Phys., C.C.R. Euratom, Ispra, Italy, 1963-. Societies: Phys. Soc. in Finland; A.N.S.
Nuclear interests: Reactor design, reactor physics.
Address: C.C.R. Euratom, Ispra (Varese), Italy.

HELLSTRAND, Eric, Dr.in Reactor Phys. Born 1923. Educ.: Roy. Inst. of Technol., Stockholm. Head, Sect. for Exptl. Reactor Phys., A. B. Atomenergi, Studsvik, Sweden. Swedish Member, European American Com. on Reactor Phys., O.E.C.D., E.N.E.A. Society: A.N.S.
Nuclear interest: Reactor physics.
Address: Aktiebolaget Atomenergi, Fack, S-611 01 Nyköping 1, Sweden.

HELLSTRÖM, P. Olof, M.S. (Metal.). Born 1927. Educ.: Roy. Polytech. Univ., Stockholm. Metal. (1950-57), Chief Metal. (1957-61), Works Manager (1961-65), Uddeholm Co.; Vice Pres. (1965-), Chairman, Exec. Council, and Board Member, Pressure Vessel Com.,Board Member, Swedish Steam Users' Assoc. Society: Teknologföreningen.
Address: Swedish Steam Users' Association, 7 Fleminggatan, P.O. Box 783, Stockholm 1, Sweden.

HELLWIG, Fritz, Dr. Phil., Dr. habil. Born 1912. Educ.: Marburg, Vienna and Berlin Univs. Economic Adviser and Director, Deutsches Industrieinst., Cologne, 1947-59; Substitute Delegate, Consultative Assembly, Council of Europe, 1953-56; Chairman, Economic Affairs Com. (1956-59), and Member (1953-59), Bundestag, Bonn; Member, European Parliament, 1959; Member, High Authority, European Coal and Steel Community, Luxembourg, 1959-67; Vice Pres., (Commissioner responsible for general and nucl. research), Commission of the European Communities, Brussels, 1967-.

Address: 23-27 Avenue de la Joyeuse Entrée, Brussels, Belgium.

HELMICK, P. S. Ph.D. Prof. and Head, Dept. Phys. and Member, Radioisotopes Com., Drake Univ., Des Moines.
Address: Drake University, Des Moines 11, Iowa, U.S.A.

HELSETH, S. Norwegian Member, Study Group on Long Term Role of Nucl. Energy in Western Europe, O.E.C.D., E.N.E.A..
Address: O.E.C.D. European Nuclear Energy Agency, 38 boulevard Suchet, Paris 16, France.

HEMBREE, Howard Gilbert, B.S. Born 1914. Educ.: Oklahoma State and Harvard Univs., M.I.T. and Argonne School of Nucl. Sci. and Eng. Asst. Regional Eng., R.E.A., 1946-52; Chief, Design Sect., Eng. and Const., (1952-56), Chief, Reactor Safety Branch, Reactor Development (1956-59). Sci. Rep. to A.E.C.L., Chalk River (1956-61), Chief, Eng. Test Branch, Reactor Development (1961-), U.S.A.E.C. Societies: A.N.S.; A.I.A.A.
Nuclear interests: Reactor and nuclear minetics; Nuclear aerospace; Reactor engineering, control, instrumentation and test.
Address: 9607 Byeford Road, Kensington, Maryland, U.S.A.

HEMILY, P. W. U.S. Member, Steering Com., Member, Group of Experts on Production of Energy from Radioisotopes, Member, Study Group on Food Irradiation, Member, Health and Safety Sub-Com., Member, Study Group on Long Term Role of Nucl. Energy in Western Europe, E.N.E.A., O.E.C.D.
Address:O.E.C.D. European Nuclear Energy Agency, 38 boulevard Suchet, Paris 16, France.

HEMMEL, L. Chairman, Press and Public Relations Com., Deutsches Atomforum E.V.
Address: Deutsches Atomforum E.V., 240 Koblenzer Strasse, 53 Bonn, Germany.

HEMMENDINGER, Arthur, B.A., Ph.D. Born 1912. Educ.: Cornell Univ. and California Inst. Technol. Phys., Los Alamos Sci. Lab., 1945-. Society: Fellow, A.P.S.
Nuclear interests: Reactions of the very light elements, neutron physics.
Address: 1442 47th Street, Los Alamos, New Mexico, U.S.A.

HEMMER, Norbert, Diploming. Born 1917. Educ.: Bonn, Prague and Stuttgart Univs. With Siemens and Halske, Wernerwerk für Messgeräte, Karlsruhe.
Nuclear interests: Monitoring systems for a high level of safety in operation of nuclear reactors.
Address: 79 Graf Ebersteinstrasse, Karlsruhe-Rüppurr, Germany.

HEMMERDINGER, Madame Perle. Born 1895. Présidente-Directeur Général, Manufacture Française des Vide-Touries Automatiques (en fonctions).
Address: 169 avenue Victor Hugo, Paris 16, France.

HEMMERLE, Elmer H., M.S. (Cincinnati). Born 1930. Educ.: Cincinnati Univ. Society: A.N.S.
Nuclear interests: Nuclear safety; Nuclear rocket development; Ground testing and flight safety.
Address: P.O. Box 10864, Pittsburgh, Pennsylvania 15236, U.S.A.

HEMPELMANN, L. H. Deleg., Joint I.A.E.A.-W.H.O. Sci. Meeting on Diagnosis and Treatment of Acute Rad. Injury, Geneva, Oct. 17-21, 1960; Member, Plowshare Advisory Com., U.S.A.E.C.
Address: Rochester University, Strong Memorial Hospital, Rochester, New York, U.S.A.

HEMPHILL, Houston Longino, B.S. Born 1900. Educ.: Louisiana Polytech. Inst., Mississippi Southern and Louisiana State Univs. Chem., O.R.N.L., 1944-65; Consultant, Chem. Separations Corp., Oak Ridge, 1965-. Societies: A.C.S.; A.S.M.; A.N.S.
Nuclear interests: Effect of radiation on corrosion by fused salt fuels: Measurement of high temperature in nuclear reactors; Dry boxes with ultra-pure atmosphere; Fluorimetric analysis of irradiated uranium.
Address: Chemical Separations Corporation, Oak Ridge, Tennessee 37830, U.S.A.

HEMPTINNE, Marc Comte de. See de HEMPTINNE, Marc Comte.

HEMS, Gordon, B.Sc., M.S., Ph.D. Born 1929. Educ.: Nottingham Univ. Res. Assoc., Brown Univ.,U.S.A., 1953-55; Visiting Res. Fellow, Sloan-Kettering Inst., New York City, 1955-56; Res. Asst., Middlesex Hospital, London, 1956-58; William Shepherd Fellow, Chester Beatty Res. Inst., Inst. of Cancer Res., London, 1958-. Society: Assoc. for Rad. Res.
Nuclear interests: Chemical effects of ionising radiation upon nucleic acid and its constituents, with particular reference to the biological action of ionising radiation.
Address: Chester Beatty Research Institute, Institute of Cancer Research, Fulham Road, London, S.W.3, England.

HENCHMAN, Michael John, M.A. (Cantab.), M.S. (Yale), Ph.D. (Yale). Born 1935. Educ.: Cambridge and Yale Univs. D.S.I.R. Res. Fellow, Cambridge, 1960-61; Asst. Lecturer (1961-63), Lecturer (1963-67), Leeds Univ.; Assoc. Prof., Brandeis Univ., Massachusetts.
Nuclear interests: Chemical effects of nuclear transformations; Hot atom chemistry; Application of isotopes to the study of

reaction kinetics and mechanisms; Radiation chemistry and mass spectrometric study of ion-molecule reactions.
Address: Physical Chemistry Department, Leeds University, Leeds 2, England.

HENCK, Frederick Seymour, B.S. Born 1923. Educ.: Tennessee Univ. Public Information Officer (1950-53), Asst. to Manager, Oak Ridge Operations (1953-55), A.E.C.; Pres., Henck Assocs., Atlanta, Ga., 1956-62; Director of Public Relations, Mead Packaging, Atlanta, 1963-67; Vice Pres., The Tiderock Corp., New York, 1967-; Manager, nucl. development study, 16 Southern states, U.S., 1956; Consultant, Florida Nucl. Development Commission, Tennessee Advisory Com. on Atomic Energy, 1956-59; Consultant in establishment and operation of Nucl. Inf. Div., State of Kentucky, 1956; Consultant, U.S.A.E.C., on public information programme, Nucl. Ship Savannah, 1959-61; Consultant, States Marine Lines, U.S., on public relations activities, Nucl. Ship Savannah, 1962; Exec. Vice-Pres., Communications Internat., Inc., Atlanta, 1961-62; Consultant, nucl. industrial development, States of South Carolina and Arkansas, 1961; Consultant, public information, Southern Interstate Nucl. Board, Atlanta, 1962-65. Societies: A.N.S.; Public Relations Soc.of America.
Nuclear interests: Industrial development, management, and public relations aspects of nuclear energy; Radioisotopes in industry; Development of process heat reactors; Power reactors.
Address: 301 E. 47th Street, New York, N.Y. 10017, U.S.A.

HENDE, A. van den. See van den HENDE, A.

HENDERSON, C., B.Sc., Ph.D. Born 1923. Educ.: London Univ. Lecturer, Phys. (1952-64), Reader, Phys. (1964-67), Univ. Coll., London; Sen. Lecturer, Aberdeen Univ., 1967-.
Nuclear interests: Particle physics; Bubble chamber technique.
Address: Department of Natural Philosophy, The University, Aberdeen, Scotland.

HENDERSON, Christopher L., B.S. (Ind. and Labour Relations). Born 1926. Educ.: Maryland and Cornell Univs. At (1948-), Asst. Director, Regulation for Administration (1963-), U.S.A.E.C. Society: Washington Chapter, A.N.S.
Nuclear interest: Assists Director of Regulation in day to day management of programme involving licensing and regulation of use of radioisotopes, source and special nuclear material, and licensing of nuclear reactors.
Address: 15104 Westbury Road, Rockville, Maryland 20853, U.S.A.

HENDERSON, James H. M. Educ.: Howard and Wisconsin Univs. Res. Fellow, Kerckkoff Biol. Lab., California Inst. Tech., 1948-50; Prof. and Head, Biol. Dept., Tuskegee Inst., Alabama; Tuskegee Inst. Rep. on Council, Oak Ridge Assoc. Univs., 1962-.
Address: Tuskegee Institute, Alabama, U.S.A.

HENDERSON, Joseph E., Prof. Director, Appl. Phys. Lab., Washington Univ., Seattle.
Nuclear interests: Field emission; Cosmic rays; Accelerators; Photonuclear disintegration; Cerenkov radiation.
Address: Washington University, Applied Physics Laboratory, Seattle, Washington 98102, U.S.A.

HENDERSON, R. W. Vice-Pres., Weapon Programmes, Sandia Lab., Albuquerque. Member, Sen. Reviewers Com., U.S.A.E.C., 1966-.
Address: Sandia Laboratory, P.O. Box 5800, Albuquerque, New Mexico, U.S.A.

HENDERSON, William Beatty, B.S. (Electronics), M.S. (Phys.). Born 1924. Educ.: U.S. Naval Academy and M.I.T. Tech. Eng., Reactor Phys., Aircraft Nucl. Propulsion Dept. (1955-61), Phys., Appl. Phys., Hanford Atomic Products Operation (1961-63), Eng., Nucl. Systems, Nucl. Materials and Propulsion Operation (1963-), Gen. Elec. Co. Society: A.N.S.
Nuclear interests: Nuclear physics; Reactor physics; Digital electronics; Machine programmes for reactor analysis; Machine programming methods.
Address: General Electric Co., Nuclear Materials and Propulsion Operation, Cincinatti 15, Ohio, U.S.A.

HENDERSON, William T. Chem., Geological Survey, U.S. Dept. of the Interior.
Address: U.S. Department of the Interior, Geological Survey, Building 25, Federal Centre, Denver, Colorado, U.S.A.

HENDEWERK, Hans J., Chief Eng., Nucl. Div., Ameray Corp., -1961; Formed own Nucl. Eng. Co., 1961. Pres., Monarch Eng.
Address: R.D.1, Box 317, Salida, Colorado, U.S.A.

HENDIN, J. H., M.I.Prod.E. Gen. Manager and Local Director, Eng. Group, Vickers Ltd.
Address: Vickers-Armstrongs South Marston Works, Swindon, Wiltshire, England.

HENDRICKS, Charles Durrell, B.S. (Phys.), M.S. (Phys.), Ph.D. (Phys.). Born 1926. Educ.: Utah State, Wisconsin and Utah Univs. Member, Tech. Staff, Lincoln Lab., M.I.T., 1955-56; Prof., Elec. Eng. (1956-67), Prof., Nucl. Eng. (1965-67), Illinois Univ.; Prof., Elec. Eng., M.I.T., 1967-68. Societies: A.P.S.; I.E.E.E.; A.I.A.A.
Nuclear interests: Energy conversion; Controlled thermonuclear fusion; Containment; Plasma physics; Propulsion.
Address: Illinois University, Department of Electrical Engineering, Charged Particle

Research Laboratory, Urbana, Illinois 61801, U.S.A.

HENDRICKSON, J. R., Prof. Com. Chairman Nucl. Sci. and Eng. Graduate Programme, Vanderbilt Univ.
Address: Vanderbilt University, Nashville 5, Tennessee, U.S.A.

HENDRICKSON, John Reese, Sen., B.S. (Chem. Eng.), M.S. (Chem. Eng.). Born 1910. Educ.: Washington (Seattle), Tennessee, Maryland and Pennsylvania Univs. Supervisory Phys. Sci., Radiol. R. and D., U.S. Army's Nucl. Defence Lab., Edgewood Arsenal, Maryland; Sen. Consulting Eng., (Chem. and Nucl.), 1949-56; Defence Electronics Plant, Radio Corp. of America, 1956-67. Societies: Nucl. Eng. Div., A.I.Ch.E.; A.N.S.; American Ordnance Assoc.; U.S.A.F. Assoc.; Franklin Inst.
Nuclear interests: Management; Nuclear radiation effects on materials; Radiation protection of man and electronics; Radiation curing of polymers; Cooling of electronics with fluorocarbon liquids; Nuclear hardening of electronic systems; Studies of nuclear vulnerability/survivability.
Address: Lou-Mar Estates, Abingdon, Maryland 21009, U.S.A.

HENDRICKSON, Waldemar Forrsel, M.S. (Chem. Eng.). Born 1934. Educ.: Idaho Univ. U.S. Naval Radiol. Defence Lab., 1956; G.E.C., Hanford, Washington, 1955, 1957-58; Nucl. Reactor Lab., Pullman, Washington, 1959-.
Nuclear interests: Shielding; Neutron scintillators; Neutron activation analysis.
Address: Nuclear Reactor Laboratory, Washington State University, Pullman, Washington, U.S.A.

HENDRIE, Joseph Mallam, Ph.D. (Columbia), B.S. (Case Inst. Technol.). Born 1925. Educ.: Case Inst. Technol. and Columbia Univ. Asst. Phys. (1955-57), Assoc. Phys. (1957-60), Phys. (1960-), Project Eng. and Chairman, Steering Com., Brookhaven High Flux Beam Reactor Project (1958-65), Acting Assoc. Head, Exptl. Reactor Phys. Div. (1966), Project Manager and Chairman, Steering Com., Brookhaven Pulsed Fast Reactor Project (1967-), Assoc. Head, Eng. Div., Nucl. Eng. Dept. (1967-), Brookhaven Nat. Lab. A.E.C. Advisory Com. on Reactor Safeguards, 1966-. Societies: A.N.S.; A.P.S.; A.S.M.E.
Nuclear interests: Reactor physics; Reactor design, construction and operation.
Address: Brookhaven National Laboratory, Upton, New York 11973, U.S.A.

HENDRIX, Velmar Van, B.S. (Chem. Eng.), M.S. (Chem. Eng.). Born 1916. Educ.: Oklahoma Univ. Chief, Source and Fissionable Materials Branch, Oak Ridge Operations Office (1950-56), Director, Reactor Eng. and Tech. Services Div., Schenectady Operations (1956-58), Director, Military Reactor Div., Idaho Operations Office (1958-61), Director, Tech. Services Div., Idaho Operations Office (1961-62), TAN (Test Area North) Site Manager, N.R.T.S., Idaho Operations Office (1962-66), Chief, Euratom Affairs Branch, Headquarters, (1966-), U.S.A.E.C. Societies: A.N.S.; A.C.S.
Nuclear interest: Coordination of design, testing and post-operative analysis of reactors.
Address: R.F.D. 2 Fox Chapel, Germantown, Maryland 20767, U.S.A.

HENGEHOLD, R. L., Ph.D., Formerly Instructor, now Asst. Prof., Air Force Inst. Technol., Air Univ., U.S.A.F.
Address: Air Force Institute of Technology, Air University, U.S. Air Force, Wright-Patterson Air Force Base, Ohio 45433, U.S.A.

HENGLEIN, Arnim, Dr. rer. nat. Born 1926. Educ.: T.H. Karlsruhe; Max Planck Inst. für Chemie, Mainz. Full Prof. for Rad. Chem., Tech. Univ., Berlin; Director at Hahn-Meitner-Inst. for Nucl. Res., Berlin-Wannsee. Societies: Gesellschaft Deutscher Chemiker; Bunsengesellschaft für physikalische Chemie.
Nuclear interests: Radiation chemistry; Mass spectrometry; Hot atom chemistry.
Address: Hahn-Meitner-Institut für Kernforschung, Berlin-Wannsee, Germany.

HENIN, Robert. Directeur Technique, Chargé des Activités Nucléaires, Sté. d'Etudes de Constructions et d'Equipements Industriels Modernes S.A.
Address: Société d'Etudes de Constructions et d'Equipements Industriels Modernes S.A., 27 boulevard Berthier, Paris 17, France.

HENK, H. V., Elec. Eng. Born 1902. Educ.: Denmark Tech. Univ. Tech. Director, Messrs. Laur. Knudsen, Copenhagen, 1956-. Pres., Danish Electrotech. Com. 1963-. Society: Member, Exec. Com., Danatom.
Nuclear interests: Application of electrical switchgear and control gear.
Address: 53 Haraldsgade, Copenhagen Ø, Denmark.

HENKEL, C. F., Jr. Manager, Market Development Optical Glass, Corning Glass Works.
Address: Corning Glass Works, Corning, New York, U.S.A.

HENLEY, Ernest J., B.S. (Delaware), M.S. (Columbia), Dr. Eng. Sci (Columbia). Born 1926. Educ.: Delaware and Columbia Univs. Board of Directors, Rad. Applications Inc., 1954-; Procedyne Assocs., Inc.; Asst. Prof., Columbia Univ., 1954-58; Prof., Stevens Inst. Technol., 1958-; Chief, Party AID Mission, Brazil Univ., 1964; Assoc. Plan, Eng., Houston Univ., 1966. Societies: A.N.S.; Rad. Res. Soc.; A.C.S.; American Soc. of Chem.Eng.
Nuclear interests: Radiation chemistry; Dosimetry; General reactor engineering.

Address: Houston University, Houston, Texas, U.S.A.

HENLEY, Ernest M., Ph.D. Born 1924. Educ.: California Univ., (Berkeley). Prof. Phys., Washington Univ., Seattle, 1961-; Sen. Postdoctoral Fellow, N.S.F., 1958-59; Guggenheim Fellow, 1967-68. Society: A.P.S. Nuclear interest: Theoretical nuclear physics. Address: Physics Department, Washington University, Seattle 98105, Washington, U.S.A.

HENNAUX, L., Prof. Member, Comité d'Application des Methodes Isotopiques aux Recherches Agronomiques (C.A.M.I.R.A.), Inst. pour l'Encouragement de la Recherche Scientifique dans l'Industrie et l'Agriculture (I.R.S.I.A.). Address: Faculte des Sciences Agromomiques de l'Etat Gembloux, Belgium.

HENNEBERKE, George, Phys. Eng. (Delft Tech. High School). Born 1917. Educ.: Delft Tech. High School. Dutch Army Tech. Staff, 1950-53; Inst. for Nucl. Studies, Amsterdam Univ., 1953-55; N. V. Philips Gloeilampen-fabrieken, Netherlands, 1955-. Nuclear interests: Accelerators; Reactor instrumentation and control; Radioactive isotope applications for level, thickness and density measurement; Radiation detectors and measuring instruments; Management. Address: 9 R. J. Schimmelpennickstraat, Almelo, Netherlands.

HENNEL, Jacek Witold, Ph.D. Born 1925. Educ.: Jagiellonian Univ., Cracow. At Inst. Nucl. Phys., Cracow, 1953-. Society: Polish Phys. Soc. Nuclear interest: Nuclear magnetic resonance. Address: 8 m. 8 Smoluchowskiego, Cracow, Poland.

HENNELLY, Edward Joseph, Ph.D. (Phys. Chem.), B.S. Born 1923. Educ.: Princeton Univ. and Union Coll. Sen. Res. Supv., Exptl. Phys. (1954-63), Sen. Res. Supv., Theoretical Phys. (1963-67), Res. Assoc. (1967-), Savannah River Lab. Chairman, Savannah River Sect., A.N.S., 1968. Society: A.C.S. Nuclear interests: Reactor design; Radioisotope production and application. Address: Savannah River Laboratory, Aiken, South Carolina 29801, U.S.A.

HENNESSEY, Joseph F., A.B., LL.B. Born 1910. Educ.: Holy Cross Coll. and Harvard Law School. Asst. Gen. Counsel (1956-60), Assoc. Gen. Counsel (1960-62), Gen. Counsel (1962-), U.S.A.E.C. Member, Nat. Council, Federal Bar Assoc. Nuclear interest: Atomic energy law. Address: United States Atomic Energy Commission, Washington D.C. 20545, U.S.A.

HENNESSY, James. Maitre de Recherches, Inst. du Radium, Faculte des Sci., Paris Univ. Address: Institut du Radium, Laboratoire Curie, 11 rue Pierre Curie, Paris 5, France.

HENNESSY, Thomas Gerard, B.S. (Chem.), M.D., Ph.D. (Med. Phys.). Born 1919. Educ.: St. John's, North Dakota and California Univs. Medical Officer, Radiological Safety Sec. Operation Crossroads, Bikini Operation, U.S. Navy, 1946-57; Instructor, Medical Aspects of Atomic Bomb, 1947; Lt. U.S.N. Naval R.O.T.C. Unit, California Univ., Student School of Medical Phys., 1947-50; Head, Clinical Radiobiol. Sect., U.S. Naval Rad. Defence Lab., 1950-52; i/c Radioisotope Lab., U.S. Naval Hospital, Philadelphia, 1952-54; Asst. Director, Assoc. Prof. in Residence, Lab. of Nucl. Medicine and Rad. Biol. and Dept. of Biophys., California Univ. at Los Angeles, 1954-. Society:California Univ., Berkeley Soc. of Nucl. Medicine. Nuclear interests: Research and teaching in radiation biology; Radioiron and Radiozinc metabolism studies; Radiation effects on bone marrow. Address: Laboratory of Nuclear Medicine and Radiation Biology, 900 Veteran Avenue, Los Angeles 24, California, U.S.A.

HENNIES, Hans Henning, Dr. rer. nat. Born 1935. Educ.: Tübingen and Göttingen Univs. Univ. Asst., 1959-61; now at Interatom, Internationale Atomreaktorbau G.m.b.H. Nuclear interests: Nuclear spectroscopy; Neutron physics in connection with reactor design and construction. Address: Interatom, International Atomreaktorbau G.m.b.H., Bensberg bei Köln, Germany.

HENNINGSEN, Eigil Juel, Medical Dr. Born 1906. Educ.: Copenhagen Univ. Deputy-Director Gen., Nat. Health Service of Denmark. Societies: H.P.S.; Nordic Soc. for Rad. Protection. Nuclear interests: Administration of radiation protection within public health administration. Address: Sundhedsstyrelsen (National Health Service), 1 Store Kongeensgade, Copenhagen K., Denmark.

HENNY, George Christian, B.A., M.S., M.D. Born 1899. Educ.: Reed Coll., Calif. Inst. Technol. Harvard Univ., Oregon Univ. Medical School and O.R.I.N.S. Emeritus Prof. Medical Phys., Temple Univ. Medical School. Books: Co-author, X-ray Diffraction Studies in Biol. and Medicine (Grune and Stratton). Societies: Radiol. Soc. North America; American Coll. Radiol.; American Roentgen Ray Soc.; Philadelphia County Medical Soc.; Delaware Valley Soc. Rad. Safety. Nuclear interests: Medical applications; Safety control. Address: 3400 North Broad Street, Philadelphia, Pennsylvania 19140, U.S.A.

HENOCH, William W., B.S. (Chem.), M.S. (Phys.). Born 1922. Educ.: Northwestern Univ. Staff Member, Los Alamos Sci. Lab., 1952-55; Project Eng., Div. Reactor Development, U.S.A.E.C., 1955-59; Project Eng., Nucl. Operations, Canoga Park, California (1959-), then Asst. Manager, Piqua Project (-1968), Atomics Internat.; Civilian Power Branch, U.S.A.E.C., 1968-. Gen. Chairman, A.N.S. Nat. Topical Meeting on Nucl. Power Reactor Siting; Formerly Sec., now Chairman, and Member, Exec. Com., Los Angeles Sect., A.N.S.
Nuclear interests: Project engineering or management of R. and D. programmes relating to nuclear power sources for terrestrial or space application. Management of projects for the development of equipment and techniques for remote maintenance, assembly, and disassembly of nuclear or other systems in space or on the lunar surface.
Address: 9931 Garden Grove Avenue, Northridge, California, U.S.A.

HENRICOT, Paul Emile Ernest Alfred André, Eng., A.I.Lg. Born 1931. Educ.: Liège Univ. Director, Usines Emile Henricot S.A. Societies: Belgonucléaire; Association Belge des Réacteurs rapides; Syndicat Vulcain.
Nuclear interests: Metallurgy - applications of stainless steel (castings, forging, tubes, bars) for nuclear plants.
Address: 8 rue de la Motte, Céroux-Mousty, Belgium.

HENRIE, Thomas A., B.S., Ph.D. Born 1923. Educ.: Brigham Young and Utah Univs. Sect. Leader, Titanium Metal Processing, Union Carbide, 1955-58; Project Coordinator, Electrowinning Rare Earths, Uranium, Thorium and other Reactive Metals (1958-66), Res. Director, Reno Metal. Res. Centre (1966-), U.S. Bureau of Mines. Member, and Chairman, Electroytic Processes Com. (1960-66), Member, Publications Com. (1963), Member, Exec. Com. (1965), Chairman, Publications Com. (1967), Chairman, 1968 Operating Conference, Extractive Metals Div., A.I.M.E.; Member, Publications Com., Metal. Soc., 1965. Societies: American Electrochem. Soc.; A.A.A.S.
Nuclear interest: Process metallurgy and chemical reactions for the preparation and production of rare earth, uranium, and thorium metals, alloys and compounds.
Address: Reno Metallurgy Research Centre, Bureau of Mines, Department of the Interior, Reno, Nevada 89505, U.S.A.

HENRIKSEN, Hans Chr., L.ès Sc. Economique et Commerciale. Born 1909. Educ.: Neuchâtel Univ. Member of Board of: A/S Meraker Smelteverk; Elec. Furnace Products Co., Ltd.; A/S Saudefaldene; Assuranceforeningen Skuld; Brague-Fram Livs- og Pensjonsforsikring A/S; Forsikringsaksjeselskapet Polaris-Norske Sjo; Akers Mek, Verksted; Christiania Bank og Kreditkasse; Den Norske Krigsforsikring for Skib; Norsk Marconikompani A/S; Pres. Norsk Selskab til Skibbrudnes Redning, 1956-61. Chairman, Oslo Stock Exchange.
Nuclear interest: Management.
Address: Den norske Amerikalinje A/S, Oslo, Norway.

HENRIKSEN, Tormod Egil, Cand. Real., Dr. Phil. Born 1928. Educ.: Oslo Univ. Biophys., Norsk Hydro's Inst. Cancer Res.; Biophys., Lawrence Rad. Lab., Berkeley, 1964-. Societies: Fellow, Norwegian Cancer Soc.; Norwegian Phys. Soc.
Nuclear interests: Effects of nuclear radiation in chemical and biological systems.
Address: Norsk Hydro's Institute for Cancer Research, Oslo, Norway.

HENRY, Allan F., B.S. (Chem.), M.S. (Phys.), Ph.D. (Phys.). 1950. E. O. Lawrence Award, 1967. Born 1925. Educ.: Yale Univ. Manager, Reactor Theory Sect., Reactor Development and Analysis Dept., Bettis Atomic Power Lab. Societies: A.P.S.; A.N.S.
Nuclear interests: Reactor physics; Reactor kinetics.
Address: Westinghouse Electric Corporation, Bettis Atomic Power Laboratoty, P.O. Box 79, West Mifflin, Pennsylvania 15122, U.S.A.

HENRY, Horace Lynford, Sc.B. Born 1916. Educ.: Brown Univ. Society: A.N.S.
Address: 75 McMurray, Richland, Washington, U.S.A.

HENRY, Hugh Fort, B.A., B.S. (Emory and Henry Coll.), Ph.D., M.S. (Virginia). Born 1916. Educ.: Emory and Henry Coll. and Virginia Univ. Head, Safety, Fire, and Rad. Control Dept., Oak Ridge Gaseous Diffusion Plant, Union Carbide Corp. 1949-61; Head, Dept. of Phys., De Pauw Univ., 1961-. Vice-Chairman, Com. N-7, Rad. Protection, Chairman Subcom. N-7.3, Rad. Protection Standards for Isotopic Separation Plants, Member, Subcoms. N-7.2, Safety Standards in Uranium and Thorium Refineries, and N-6.8, Fissionable Material outside Reactor, American Standards Assoc.; Consultant to Union Carbide Nucl. Co. and U.S.A.E.C. Books: Co-author, K-1019, Criticality Data and Nucl. Safety Guide applicable to the Oak Ridge Gaseous Diffusion Plant (5th revision, 1959, Union Carbide Nucl. Co.); editor and compiler, TID-7019, Guide to Shipment of U-235 Enriched Uranium Materials (U.S.A.E.C., Tech. Information Service, 1959); editor and compiler, K-1380, Studies in Nucl. Safety (Union Carbide Nucl. Co., 1958). Societies: A.P.S.; A.N.S.; H.P.S.; A.A.A.S.; A.A.P.T.
Nuclear interests: Radiation dosimetry; Longevity effects of radiation; Development of basic nuclear safety criteria; Interaction of fissile materials; Practical application of health physics and nuclear safety to plant operations.

Address: 404 Linwood Drive, Greencastle, Indiana, U.S.A.

HENRY, Kenneth James, B.Sc., A.M.I.Mech.E. Born 1922. Educ.: Univ. Coll., Cardiff. Operations Eng., B.E.P.O., GLEEP, 1951-56; Chief Reactor Operations Eng., Res. Reactor Div., Harwell, 1956-61; Project Manager, Dounreay Fast Reactor, 1961-.
Nuclear interests: Management and development of research and prototype reactors.
Address: 78 Duncan Street, Thurso, Caithness, Scotland.

HENRY, Robert, D. ès Sc. Born 1927. Educ.: Montpellier Univ. Sci., C.E.A., 1953-. Head, Service de préparation des Radioéléments, C.E.N. Saclay, 1959-. Societies: A.N.S.; Sté. Chimique de France.
Nuclear interests: Radioisotopes production and applications; Radiochemistry.
Address: Département des Radioéléments, Centre d'Etudes Nucléaire de Saclay, B.P. No. 2, Gif-sur-Yvette 91, France.

HENRY, W. H., Dr. Formerly Assoc. Res. Officer, now Sen. Res. Officer, X-rays and Nucl. Rad. Sect., Nat. Res. Council, Canada.
Address: X-rays and Nuclear Radiations Section, National Research Council, Montreal Road Laboratories, Building M-35, Ottawa 7, Canada.

HENSEL, F. R., Dr. Vice-Pres., Eng., Mallory Metallurgical Co.
Address: Mallory Metallurgical Co., Division of P. R. Mallory and Co. Inc., Corporate Research and Development Laboratories, 3029 East Washington Street, Indianapolis 6, Indiana, U.S.A.

HENSLER, J. Raymond, Dr. Formerly Dept. Head, Ceramics R. and D. Dept., now Director, Materials R. and D. Directorate, Bausch and Lomb Inc.
Address: Bausch and Lomb Inc., Rochester 2, New York, U.S.A.

HENSON, Arthur Harold, B.A. (Cantab.). Born 1916. Educ.: Cambridge Univ. Head, Dept. of Chem., Inst. of Sci. and Technol., Wales Univ. Societies: F.R.I.C.; Plastics Inst.
Address: Department of Chemistry, Wales University, Institute of Science and Technology, Cardiff, Wales.

HEPBURN, William Arthur, B. Mech.E. Born 1910. Educ.: Catholic Univ. of America. Exec. Eng., U.S.A.F. Air R. and D. Command, Development and Construction, Cape Kennedy, 1946-52; Project Chief, Reactor Development Div. (Army Reactors); Chief, Tech. Programme Branch (Maritime Reactors), Chief, Nucl. Res. Branch (Office of R. and D.), Maritime Administration. Sen. Project Officer (Gas-cooled Reactors), U.S.A.E.C., 1952-. Member, Study Group on Nucl. Ship Propulsion, E.N.E.A. Societies: A.N.S.; N.S.P.E.; District of Columbia Soc. Professional Eng.; Soc. Naval Eng.
Nuclear interests: Reactor development, design and construction.
Address: 10401 Grosvenor Place, Rockville, Maryland 20852, U.S.A.

HEPNER, I. L., Ph.D., A.M.I.Chem.E., A.I.R.I., F.N.C.R.T. Editor, Atomic World (incorporated in Chem. and Process Eng.).
Address: The Tower, Shepherds Bush Road, London, W.6, England.

HEPPE, Hans VON. See VON HEPPE, Hans.

HERAK, Marko, Dr. Born 1922. Educ.: Zagreb Univ. Inst. for Nucl. Res., Rudjer Boskovic, Zagreb, 1955-. Society: Croatian Chem. Soc.
Nuclear interests: Radiochemistry; Use of tracers in physical chemistry.
Address: Institute Rudjer Boskovic, 54 Bijenicka, Zagreb, Yugoslavia.

HERBER, Rolfe H., Ph.D. (Oregon State). Born 1927. Educ.: U.C.L.A., Oregon State Univ. and M.I.T. Prof. Phys. Chem., Rutgers Univ. Societies: A.S.C.; A.P.S.
Nuclear interests: Isotopic exchange kinetics; Mössbauer effect; Hot-atom chemistry; Radiochemistry.
Address: Chemistry Department, Rutgers University, New Brunswick, New Jersey, U.S.A.

HERBERT, John Ferguson, B.E. Born 1908. Educ.: Sydney Univ. Manager, Far East (1947-53), Export Area Manager, Australasia and Far East (1953-57), Manager, Overseas Atomic Projects (1957-63), Gen. Export Manager (1963-), English Elec. Co., Ltd.; Chairman, English Elec. Export and Trading Co., Ltd. Societies: M.I.E.E.; M.I.E.Aust.
Address: 14 Belsize Avenue, London, N.W.3, England.

HERBERT, Robert John, B.S. Born 1928. Educ.: Carnegie Inst. Technol. and New York Univ. With G.E.C., K.A.P.L., 1949-54; U.S. Army Chem. Corps, 1955-56; Nucl. Development Corp. America, 1957-60; Assoc. Nucleonics now Stone and Webster Eng. Corp., 1960. Societies: A.N.S.; Inst. of Management Sci.
Nuclear interests: Management systems development; Nuclear physics and engineering; Reactor design and engineering; Irradiation facilities design and engineering; Reprocessing facilities design.
Address: 105 Woodland Hills, White Plains, New York, U.S.A.

HERBST, John J. Asst. Prof., Nucl. Eng., School of Eng. and Sci., New York Univ., 1966-.
Address: New York University, School of Engineering and Science, University Heights, Bronx, New York 10453, U.S.A.

HERBST, Walter, Dr. phil. nat. Born 1907.
Educ.: Jena/Thür Univ. At Radiol. Inst.,
Freiburg/Breisgau Univ.
Nuclear interests: Health physics; Radiobiology.
Address: 23 Albertstrasse, Freiburg/Breisgau,
Germany.

HERCUS, Victor Macky, M.D., M.R.C.P.
(Edin.), D.A. (Lond.). Born 1917. Educ.:
Sydney Univ. Formerly Res. Anaesthetist,
Unit of Clinical Investigation, Royal North
Shore Hospital, Crows Nest, New South Wales;
Honorary Anaesthetist, Royal Alexandra
Hospital for Children, Sydney; Sen. Lecturer
in Surgery (Anaesthetics), New South Wales
University; Director of Respiratory and
Resuscitation Unit, Prince Henry Hospital,
Sydney; Honorary Anaesthetist, Royal
Hospital for Women, Sydney. Societies:
Australian Soc. of Anaesthetists; Surgical
Res. Soc. of Australia.
Nuclear interests: Clinical investigation in
various diseases and during induced
physiological abnormal states.
Address: Department of Surgery, Prince
Henry Hospital, Little Bay, N.S.W., Australia.

HERCZYNSKA, Elwira, Dr. Born 1922.
Educ.: Bucharest and Moscow Univs. Sci.,
Radiochem. Dept., Inst. Nucl. Res., Warsaw.
Nuclear interests: Radiochemistry; Surface
chemistry; Isotopic exchange.
Address: Instytut Badan Jadrowych, Warsaw,
Poland.

HEREDIA ZAVALA, Roberto, Staff Prof.,
Inst. Superior de Energia Nucl., Peru.
Address: Junta de Control de Energia
Atomica, 3420 Avenida Arequipa, Apt. 914,
Lima, Peru.

HERENGUEL, Jean Francoise Germain, D.
ès Sc. Born 1909. Educ.: Lille Univ. Directeur
du Dépt. des Recherches Métallurgiques de la
Sté. Tréfimétaux G.P.; Prof. de Métallurgie
Spéciale à l'I.N.S.T.N., Centre d'Etudes
Nucléaires de Saclay. Prés. de la Sté. Française de Métallurgie, Paris, 1961; Directeur
Gen., Ste. Industrielle du Zirconium. Book:
Cours de Métallurgie Spéciale de l'Inst.
National des Sciences et Techniques Nucléaires (Paris, Centre de Documentation Universitaire, 1958). Societies: Inst. Metals; Sté.
Française de Métallurgie; Sté. Ing. Civils de
France; Corresponding Member, Inst. Metals,
London.
Nuclear interest: Metallurgy: metallurgical
materials for nuclear fuels, their working and
fabricating practices.
Address: Tréfimétaux G.P., Département des
Recherches Métallurgiques, 141 rue Michel
Carré, Argenteuil (Val d'Oise), France.

HERFORTH, Lieselott, Prof., Dr.-Ing. habil.
Director, Bereich Anwendung radioaktiver
Isotope, Technische Univ. Dresden.
Address: Bereich Anwendung radioaktiver
Isotope, Technische Universität Dresden,
Zellescher Weg, 8027 Dresden, German Democratic Republic.

HERFURTH, Rolf. Director, Nucl. Dept.,
Herfurth G.m.b.H.
Address: Herfurth G.m.b.H., 6-8 Beerenweg,
2 Hamburg-Altona, Germany.

HERING, Hermann, D. ès Sc. Born 1904.
Educ.: Strasbourg Univ. Chef du Service de
Chimie physique, C.E.A., 1950. Societies:
Sté. chimique de France; Sté. de Chimie
physique; Sté. Française de Physique.
Nuclear interests: Radiation chemistry and
photochemistry of simple compounds; Chemistry and physical chemistry of graphite.
Address: 34 avenue de Suffren, Paris 15,
France.

HERMAN, Harry Hirsch, Jr., B.S. (Mech.
Eng.). Born 1930. Educ.: Colorado, California
(Los Angeles) and New Mexico Univs. Project
Officer, U.S.A.F., 1953-56; Tech. Staff Asst.,
Systems Eng. Res. and Advanced Development
Div., AVCO Manufacturing Corp.; Phys.
Dept. Convair Div., Gen. Dynamics Corp.;
Kaiser Industries; Asst. to Works Manager,
Willys Overland do Brazil, Acting Director,
Foreign Manufacturing, Kaiser Jeep Corp.,
Toledo, Ohio; Tech. Director, Kaiser Jeep
Overseas S.A. Teheran, Iran; Sen. Sci., Kaman
Nucl. Div., Kaman Aircraft Corp.; Served as
Staff Member, Instituto do Energia Atomica,
Sao Paulo, Brazil. Societies: American Soc.
of Aeronautics and Astronautics; A.N.S.
Nuclear interests: Operational techniques and
problems; Industrial applications of isotopes;
Hydro-magnetics.
Address: 4201 Cathedral Avenue, N.W.,
Washington, D.C., U.S.A.

HERMANS, Marie Egidius Antonius, Dr., Ir.
Born 1923. Educ.: Delft Tech. Univ. N.V.
KEMA, Reactor Development Group, Arnhem.
Group Leader, Homogeneous Reactor
Development Team, N. V. Gemeenschappelijke
Kernenergiecentrale Nederland. Societies: Roy.
Dutch Chem. Soc.; Dutch Ceramic Soc.;
A.C.S.
Nuclear interests: Fuel materials; Fuel suspensions; Suspension reactors.
Address: 4 Diependalstraat, Arnhem, Netherlands.

HERMANS, Willem Carel, Drs. Born 1929.
Educ.: Amsterdam Univ. Wetenschappelijk
medewerker, Inst. Nucl. Phys. Res., 1959-.
Society: Nederlandse Natuurkundige Vereniging.
Nuclear interest: Theoretical nuclear physics.
Address: Institute of Nuclear Physics Research
(IKO), 18 Ooster Ringdyk, Amsterdam O,
Netherlands.

HERMANSKY, Bedrich, M.Sc., Ph.D. Born
1933. Educ.: Prague Tech. and London Univs.

Lecturer, Dept. of Tech. and Nucl. Phys., Prague Tech. Univ., 1957-; External appointment, Pilsen-Skoda Nucl. Power Station Dept., 1965-66; External appointment, Power Res. Inst., 1966-67. Book: Nuclear Reactor Calculation (Prague, S.N.T.L., 1965). Nuclear interests: Power reactor design and calculation, particularly steady-state and transient temperature field in a fuel channel and active core temperature coefficients, transfer functions of a fuel channel and active core dynamics of the gas cooled heavy water moderated reactor.
Address: 10 Kremencova, Prague 1, Czechoslovakia.

HERME, Frederic REY-. See RAY-HERME, Frederic.

HERMINGHAUS, Dr. Inst. fur Kernphysik, Johannes Gutenberg Univ.
Address: Johannes Gutenberg University, Institut fur Kernphysik, 21 Saarstrasse, 65 Mainz, Germany.

HERNANDEZ, G. Prof. of Phys. Chem., Univ. Central, Quito.
Address: Universidad Central, Quito, Ecuador.

HERNANDEZ AQUIJE, Carlos, B.S. (Mech. and Elec. Eng.), M.S. (Mech. and Elec. Eng.), Diploma Nucl. Science and Eng. (U.S.A.). Born 1934. Educ.: Univ. Nacional de Ingenieria Lima-Perú. Consultant, Atomic Energy Control Board of Perú; Assoc. Prof., Univ. Nacional de Ingenieria Lima-Perú, 1958-. Now at Maryland Univ., studying and doing res. work on nucl. phys. Societies: Assoc. Electrotécnica Peruana (Peru); Inst. Peruano de Ingenieria Mecánica (Peru); A.S.M. Nuclear interests: Nuclear physics and peaceful applications of nuclear energy.
Address: Av. Brasil 2569, Dpto. 204, Lima, Peru.

HERNANDEZ-FRAGOSO, Julio, B.S.Ch.E., M.S.Ch.E. Born 1925. Educ.: Puerto Rico Univ., I.I.T. and O.R.S.O.R.T. Instructor, Chem. Eng. Dept., Puerto Rico Univ., 1950-53; Asst. Chem., Central Aguirre Sugar Co., 1953-56; Chem. Eng. (1952-62), Member, Nucl. Power Group (1958-62), Bonus Plant Supt. (1962-), Puerto Rico Water Resources Authority. Member, Tech. Com., Puerto Rico Nucl. Centre, 1963-. Societies: N.E.D., A.I.Ch.E.; A.N.S.
Nuclear interest: Management.
Address: Puerto Rico Water Resources Authority, P.O. Box 4267, San Juan, Puerto Rico 00905, U.S.A.

HERNANDEZ VARELA, José L., Ind. Eng. (Madrid), Ing. en Genié Atomique. Born 1930. Educ.: Madrid Univ., Eng. Nucl. Dept. Hidroeléctrica Espanola, S.A., Madrid.
Nuclear interests: Nuclear power stations: design and operation.

Address: 130 Doctor Esquerdo, Madrid 7, Spain.

HERNEGGER, Friedrich, Dr. phil. Born 1908. Educ.: Vienna Univ. Res. Assoc., Inst. für Radiumforschung u. Kernphysik, 1938-. Societies: Verein Osterreichischer Chemiker; Osterreichische Physikalische Gesellschaft. Nuclear interests: Radiochemical and geochemical problems and fluorimetric analysis methods for uranium.
Address: Institut für Radiumforschung u. Kernphysik, 3 Boltzmanngasse, Vienna 9, Austria.

HEROLD, Edward William, B.Sc., M.Sc., D.Sc. Born 1907. Educ.: Virginia Univ. and Polytech. Inst. of Brooklyn. Director, Electronic Res. Lab., R.C.A. Labs., 1953-59; Vice-Pres., Res., Varian Assocs., 1959-. U.S. Dept. of Defence Advisory Group on Electron Devices. Societies: A.P.S.; Fellow, Inst. E.E.E.; A.I.A.A. Nuclear interests: Controlled fusion; Application of radio-frequencies to gas plasmas.
Address: Varian Associates, Palo Alto, California, U.S.A.

HERPIN, André Georges, D. ès Sc. Born 1920. Educ.: Paris Univ. Ing. au service de Physique Mathématique, C.E.A., Saclay, 1951-; Adjoint au chef du Service de Physique du Solide, 1958-; chef du service de Physique du Solide, 1959. Sec. gén., Sté. de Phys. Societies: Sté. française de Physique; Union Internationale de Cristallographie.
Address: 40 avenue de Saxe, Paris 7, France.

HERR, Wilfrid, o.Prof., Dr. rer. nat. Born 1914. Educ.: Berlin-Charlottenburg T.H., Tubingen Univ. and Kaiser Wilhelm Inst. für Chemie, Berlin-Dahlem, later named Max-Planck-Inst. für Chemie (Otto-Hahn-Inst.), Mainz. At (1958-), now Director, Inst. for Nucl. Chem., Cologne Univ.; Head, Radiochem. Group, Kernforschungsanlage, Jülich. Societies: Gesellschaft Deutscher Chemiker; A.C.S.
Nuclear interest: Radiochemistry, Geo- and cosmochemistry.
Address: Institut für Kernchemie, Cologne University, 47 Zülpicher Strasse, 5 Cologne, Germany.

HERRERA, Diego MARTINEZ. See MARTINEZ HERRERA, Diego.

HERRERA, F. PEDRO. See PEDRO HERRERA, F.

HERRERA, Francisco Carlos, M.D. Born 1935. Educ.: Venezuela Central Univ. Res. Fellow, Physiology, Biophys. Lab., Harvard Medical School, 1959-61; Graduate Student (1959-61), Assoc.Investigator (1962-), I.V.I.C. Societies: A.A.A.S.; Asociacion Venezolana para el Avance de la Ciencia.
Nuclear interest: Use of isotopes in the study of active transport of electrolytes in biological

systems, especially in polar epithelia of amphibian bladder and skin.
Address: Institute Venezolano de Investigaciones Cientificas, P.O. Box 1827, Caracas, Venezuela.

HERRERA, Jose Maria ALBAREDA. See ALBAREDA HERRERA, Jose Maria.

HERRERO GARRALDA, D. Igancio. Formerly Member, Comisiones Asesoras De Reactors Industriales, now Member, Comisiones Asesoras de Centrales Nucleares, Junta de Energia Nuclear; Director, Forum Atómico Español.
Address: c/o Junta de Energia Nuclear, Ciudad Universitaria, Madrid 3, Spain.

HERRHAUSEN, Alfred, Dr. rer. pol. Born 1930. Educ.: Cologne Univ. Energiewirtschaftliches Inst., Cologne Univ., 1952; Ruhrgas Aktiengesellschaft, Essen, 1952-55; Vereinigte Elektrizitätswerke Westfalen, Aktiengesellschaft, Dortmund, 1955-67; Vorstandsmitglied der V.E.W. A.G., 1967-. Book: Aufsatz, Grundlagen für die Berechnung der Stromerzeugungskosten von Kernkraftwerken, (Frankfurt am Main, Verlags- und Wirtschaftsgesellschaft der Elektrizitätswerke m.b.H., 1960).
Address: i.Hs. Vereinigte Elektrizitätswerke Westfalen Aktiengesellschaft, 51 Ostwall, Dortmund, Germany.

HERRICK, H. T., LL.B., B.S. Educ.: Cornell Law School and Hamilton Coll. Gen. Counsel to Federal Mediation and Conciliation Service, 1963-65; Director, Div. of Labour Relations, (1965-), and Chairman, Atomic Energy Labour-Management Advisory Com., U.S.A.E.C.
Address: Division of Labour Relations, U.S.A.E.C., Washington D.C. 10545, U.S.A.

HERRING, William Conyers, A.B., Ph.D. Born 1914. Educ.: Kansas and Princeton Univs. Tech. Staff, Bell Telephone Labs., 1946-. Societies: Fellow, A.P.S.; A.A.A.S.; Fellow, American Acad. Arts and Sci.
Nuclear interest: Theoretical solid-state physics.
Address: Bell Telephone Laboratories, Murray Hill, New Jersey, U.S.A.

HERRLI, W., Ir. Chief Eng., Nucl. Dept., Machinefabriek "Breda" Voorheen Backer en Rueb N.V.
Address: Machinefabriek "Breda" Voorheen Backer en Rueb N.V., Postbus 260, Breda, Netherlands.

HERRMANN, Franz Karl Albert, Dipl.-Phys. Born 1920. Educ.: Rostock Univ. At Isotopen-Studiengesellschaft (1959-61), Gesellschaft für Kernforschung (1962-64), Kernforschungszentrum Karlsruhe; formerly at Physikalisches Institut der Universität Rostock.
Nuclear interests: Isotopen-Anwendungen in der Technik und Mess- und Regeltechnik in der Forschung.
Address: 12 Max-Planck-Strasse, Leopoldshafen 7501, Germany.

HERRMANN, Günter, Dr. rer. nat. Born 1925. Educ.: Mainz Univ. Asst. (1956), Privat-Dozent (1962), Assoc. Prof. (1967), Mainz Univ. Society: Gesellschaft Deutscher Chemiker.
Nuclear interests: Nuclear fission; Nuclear spectroscopy; Rapid chemical separations.
Address: Institut für Anorganische Chemie und Kernchemie, Universität Mainz, Postfach 606, Germany.

HERRMANN, Heinrich Walter, Dr. rer. nat. habil. Born 1910. Educ.: Halle Univ., Leipzig. At (1955-), Director (1966-), Inst. für angewandte Radioaktivität, Deutschen Akademie der Wissenschaften zu Berlin; Nebenamtlicher Prof., Karl Marx Univ., Leipzig, 1966-. Society: Fachnormenausschuss Kerntechnik, Deutschen Normenausschuss.
Nuclear interests: Kernphysikalische Messtechnik; Nutzanwendung von Radionukliden in Forschung und Technik.
Address: Institut für angewandte Radioaktivität, 15 Permoserstrasse, 705 Leipzig, German Democratic Republic.

HERRON, David Poston, B.S., M.S. (Chem. Eng.), M.B.A. Born 1919. Educ.: Wabash Coll., M.I.T. and Harvard Univ. Vice-Pres. and Director of Eng., Atlantic Res. Corp., 1949-51; Director of Operations Analysis, U.S.A.E.C., 1951-54; Director of Eng., Atomic Energy Div., American Standard, 1955-. Societies: A.I.Ch.E.; Operations Research Soc. of America.
Nuclear interests: Management of power reactor design projects; Nuclear power systems - economics, systems optimisation.
Address: 369 Whisman Road, Mountain View, California, U.S.A.

HERSH, Herman I. Member, Patent Compensation Board, U.S.A.E.C.
Address: Ooms, McDougall and Hersh, Chicago, Illinois, U.S.A.

HERTEL, Karl Wilhelm Gottfried, Assessor. Born 1925. Educ.: Mainz Univ. Geschäftsführer, Deutsche Kenreaktor-Versicherungs-Gemeinschaft. Deutsche Versicherungs-Akademie, Cologne. Society: Studiengesellschaft zur Förderung der Kernenergieverwertung in Schiffbau und Schiffahrt e.V., Hamburg.
Nuclear interest: Insurance.
Address: Deutsche Kernreaktor-Versicherungs-Gemeinschaft, 18 Theodor-Heuss-Ring, 5 Cologne 1, Germany.

HERTZ, Gustav, Dr. phil. Nobel Prize, 1926. Born 1887. Educ.: Göttingen, Munich and Berlin Univs. Leiter eines Forschungslaboratoriums in der Sowjet-union, 1945-54; Prof. und Direktor, Physikalischen Inst., Karl-Marx-

Univ., Leipzig, 1954-. Books: Lehrbuch der Kernphysik (1958); Arbeiten mit J. Franck über quantenhaften Energieaustausch zwischen Elektronen und Atomen (1914); Arbeiten über Isotopentrennung (1933). Societies: Deutschen Akademie der Wissenschaften zu Berlin; Ehrenmitglied, Ungarischen Akademie der Wissenschaften; Korrespondierendes Mitglied, Göttingen Akademie der Wissenschaften.
Address: 47 Lienhardweg, 117 Berlin-Kopenick, German Democratic Republic.

HERWIG, Lloyd Otto, B.A., M.S., Ph.D. Born 1921. Educ.: Luther Coll., Iowa and Iowa State Univs. Res. Asst. (1949-53), Res. Assoc. (1953), Iowa State Univ.; Westinghouse Elec. Corp., Bettis Atomic Power Lab.,1954-61; United Aircraft Res. Labs., 1961-64; Staff Assoc., Institutional Programmes, N.S.F., 1964-. Societies: A.P.S.; A.N.S.; A.A.A.S.
Nuclear interests: Nuclear physics; Atomic physics; Reactor physics, analysis and experiment; Nuclear rockets; Stimulated emission of optical radiation; Science administration.
Address: National Science Foundation, Institutional Programmes, 1800 G. Street N.W., Washington D.C. 20550, U.S.A.

HERZBERG, Gerhard, Dr. Ing. (Darmstadt Inst. Technol.), LL.D. (Saskatchewan), D.Sc. (McMaster), D.Sc. (Nat. Univ. Ireland), LL.D. (Toronto), D.Sc. (Oxford), LL.D. (Dalhousie), LL.D. (Alberta), Fil. Hed. Dr. (Stockholm), D.Sc. (Chicago). Born 1904. Educ.: Darmstadt Inst. Technol.; Goettingen and Bristol Univs. Director, Div. of Pure Phys., N.R.C., Ottawa, 1949-. Assoc. Editor, Canadian J. Phys. Books: Atomic Spectra and Atomic Structure (2nd ed., New York, Dover Publications, 1944); Molecular Spectra and Molecular Structure, vol. i: Spectra of Diatomic Molecules (2nd ed., New York, Van Nostrand, 1950); vol. ii: Infra-red and Raman Spectra of Polyatomic Molecules (New York, Van Nostrand, 1945) vol. iii: Electronic Spectra and Electronic structure of Polyatomic Molecules (Princeton, Van Nostrand, 1966). Societies: Roy. Soc. of London; Roy. Soc. of Canada; A.P.S.; American Astronomical Soc.; Hon. Foreign Member, American Acad. Arts and Sci.
Nuclear interests: Nuclear moments; Lamb shifts.
Address: Division of Pure Physics, National Research Council of Canada, Ottawa, Canada.

HERZENBERG, Caroline Stuart Littlejohn, S.B., S.M., Ph.D. Born 1932. Educ.: M.I.T. and Chicago Univ. Chicago Univ., 1958-59; Argonne Nat. Lab., 1959-61; I.I.T., 1961-67; I.I.T. Res. Inst., 1967-. Pres., I.I.T. Branch, Federation of American Sci. Society: A.P.S.
Nuclear interests: Experimental low energy nuclear spectroscopy; Mössbauer effect-resonant absorption of gamma rays.
Address: Nuclear Radiation Physics Department, I.I.T. Research Institute, Chicago, Illinois 60616, U.S.A.

HERZOG, Leonard Frederick, II, C.C.E., B.S., Ph.D. Born 1926. Educ.: Oregon State Coll., California Inst. Technol., M.I.T., and Harvard Univ. Visiting Sci., Dept. Terrestrial Magnetism, Carnegie Inst. Washington, 1951-52; Director, Nucl. Geophys. Lab., M.I.T., 1952-56; Asst. Prof. (1956), Assoc. Prof. (1961), Pennsylvania State Univ.; Pres., Nuclide Analysis Assocs., 1954; Board Chairman, Nuclide Corp., 1961. Chairman, Com. on Analysis of Solids, A.S.T.M. E-14 on Mass Spectroscopy; U.S.A.E.C. Com. on Uranium Isotope Measurements. Societies: A.S.T.M. E-14 (Mass Spectroscopy); Phys. Soc.; A.G.U.; Geochem. Soc.; Vacuum Soc.; Newcomen Soc.
Nuclear interests: Chemical and isotopic analysis by mass spectroscopy; X-ray fluorescence analysis; Design of nuclide analysis devices; Neutron activation analysis; Geological age determination; Cosmic-ray meteorite interactions; Geological environment - radiation site interaction problems.
Address: Box 752, State College, Pennsylvania, U.S.A.

HESBURGH, Father Theodore Martin, B. Phil., S.T.D. Born 1917. Educ.: Notre Dame and Gregorian Univs., and Catholic Univ. of America. Pres., Notre Dame Univ., 1952-; Permanent Rep. of Holy See to I.A.E.A., 1957-; Formerly Director, Midwest Univs. Res. Assoc.; Formerly Member, Policy Advisory Board, Argonne Nat. Lab.
Address: University of Notre Dame, Notre Dame, Indiana, U.S.A.

HESKETH, G. E., M.Sc., F.R.I.C. Radiochem. Inspector, Ministry of Housing and Local Govt.
Address: Ministry of Housing and Local Government, Whitehall, London, S.W.1, England.

HESLEP, John M., B.S., Ph.D. Born 1921. Educ.: California Univ. and Mississippi State Coll. Chief Radiol. Safety Eng. (for all campuses), California Univ., 1956-60; Chief, Bureau of Radiol. Health (1960-67), Chief, Div. of Environmental Sanitation (1967-), Now Chief, Div. of Environmental Health, State of California Dept. of Public Health. Societies: Fellow, A.P.H.A.; Fellow, A.A.A.S.; H.P.S.
Nuclear interest: Radiological health.
Address: State Department of Public Health, 2151 Berkeley Way, Berkeley, California 94805, U.S.A.

HESLOT, Henri, Ing. Agronome, D. ès Sc. Born 1921. Educ.: Paris and Cambridge Univs. Maitre de Conférences de Génétique à l'Inst. Agronomique, Paris. Society: Sté. Française de Génétique.
Nuclear interests: Induction of mutations in yeast and cultivated plants (barley, marigold, roses); Use of radiations and radiomimetic chemicals.

Address: Institut National Agronomique, 16 rue Claude Bernard, Paris 5, France.

HESS, E., Director, Lonza Ltd., Basle. Member, Federal Commission for Atomic Energy, Switzerland.
Address: Lonza Ltd., Basle 2, Switzerland.

HESS, Irmgard, Dipl. Ing. Sen. Officer, Inst. of Special Metals, Max-Planck Inst. for Metal Res.
Address: Max-Planck Institute for Metal Research, 92 Seestrasse, 7000 Stuttgart 1, Germany.

HESSABY, Mahmoud, B.A. (American Univ. Beirut), Elec. Eng. (E.S.E., Paris), D.Sc. (Phys., Paris Univ.). Born 1903. Educ.: American (Beirut) and Paris Univs. Prof. Phys. (1934-), Dean (1951-57), Faculty of Sci., Tehran; Minister of Educ., 1951-52; Senator for Tehran, 1949-51 and 1954-61; Iran A.E.C.; Pres., Iranian Astronautical Soc. Books: Etude de la sensibilité des cellules photoelectriques (1927); Essai d'Interpretation des Ondes de Broglie (1945); A Strain Theory of Matter (1946); A course of Phys. Optics (1962); Electromagnetic Theory (1965). Societies: American Phys. Soc.; Iranian Acad.
Nuclear interests: Elementary particles; Structure and spectrum of masses of elementary particles; Nature of nuclear forces.
Address: Char Rah Hessaby, Tajrish, Tehran, Iran.

HESSE, Walter J., B.S.M.E., M.S.M.E., Ph.D. Born 1923. Educ.: Valpariso and Purdue Univs. Chief Eng., Chief Academic Instructor and Project Eng., Test Pilot School, U.S. Naval Air Test Centre, Patuxent River, Maryland, 1949-55; Supervisor, Theoretical Propulsion (1956), Chief, Advanced Development Planning (1956), Manager, Advanced Systems Eng. (1959), Programme Director, Nucleonic Systems (1961), Programme Director, Advanced Missile Systems (1964), Vice Pres., Programme Director, V/STOL Programmes (1965-), LTV Aerospace Corp. and its predecessors. Member, Sci. Panel, Congressional House Com. on Sci. and Astronautics; Dept. of Defence Consultant to the Tech. Advisory Panel on Aeronautics; Advisory Board, Joint Task Force Two, Joint Chiefs of Staff; Texas Commission on Atomic Energy; Board Chairman, Aerospace Education Foundation. Books: Jet Propulsion (Pitman Publishing Corp., 1958); co-author, Jet Propulsion for Aerospace Application (Pitman Publishing Corp., 1964). Societies: A.N.S.; A.I.A.A.
Nuclear interests: Managed the Nuclear Powered Missile Weapons Systems called SLAM or PLUTO, including the design of the overall missile system and integration of the nuclear power plant as developed and tested by the Lawrence Radiation Laboratory. The nuclear aspects considered were shielding studies, radiation effects on the many subsystems from the high-powered reactor and special safety and hazards criteria for flight testing and disposal of the missile system after each flight.
Address: 4847 Allencrest Lane, Dallas, Texas 75234, U.S.A.

HESSEL, Philipp, Dr. Deputy of the Minister, Arbeitsministerium Baden-Württemberg.
Address: Arbeitsministerium Baden-Würtemberg, 30 Rotebühlplatz, 7 Stuttgart W, Germany.

HESSEN, Victor Boris, First Class Faraday House Diploma. Born 1909. Educ.: Faraday House Elec. Eng. Coll. Formerly Principal, Tech. Services, Ministry of Supply; at present Sales Manager, Pye Labs. Ltd. Society: A.M.I.E.E.
Nuclear interests: Remote handling; Remote viewing; Teaching and experimental reactors; Instrumentation.
Address: Pye Laboratories, Ltd., Radio Works, Cambridge, England.

HESSINGER, Philip S. Formerly Manager, Development Eng., now Manager, Tech. Operations, Nat. Beryllia Corp.
Address: National Beryllia Corp.,1st and Haskell Avenues, Haskell, New Jersey 07420, U.S.A.

HESSLER, Richard, Dipl.-Eng. Born 1901. Educ.: Darmstadt Tech. Univ. Director and Member, Board of Directors, Tech. Supervising Assoc. Stuttgart, Inc., 1939-; Chairman, Medical-Psychological Inst. for Traffic Safety and Safety in Factories. Pres., Low Pressure Boiler Com. and Regular Member, German Boiler Com.; Pres., Tech. Com. for Boilers, Assoc. of the Tech. Supervising Assocs. Societies: Soc. of German Eng.; Soc. of members of German Iron Works.
Nuclear interests: Metallurgy; Reactor design; Management; Nuclear physics.
Address: 48 Bebel-Strasse, 7 Stuttgart-W, Germany.

HESTIN, Francois, Ing. Chief Eng., Sechaud et Metz.
Address: Sechaud et Metz, 2 rue de la Baume, Paris 8, France.

HETHERINGTON, Cameron Hilliard, B.Sc., M.Sc. (Hons., Phys.). Born 1918. Educ.: Alberta Univ. Director of Marketing, A.E.C.L.
Nuclear interests: Sale of isotopes and associated equipment for industry, medicine and research.
Address: P.O. Box 93, Ottawa, Ontario, Canada.

HETHERINGTON, J. H. Nucl. Theory, Dept. of Phys. and Astronomy, Michigan State Univ.
Address: Michigan State University, East Lansing, Michigan, U.S.A.

HETHERINGTON, Robert M. Vice-Pres., Iowa-Illinois Gas and Elec. Co.; Member, Operating Com., Power Reactor Development Co.
Address: Iowa-Illinois Gas and Electric Co., Davenport, Iowa, U.S.A.

HETRICK, David L., B.S., M.S., Ph.D. (Phys.). Born 1927. Educ.: Rensselaer Polytech. Inst. and California (Los Angeles) Univ. Instructor of Phys., Rensselaer Polytech. Inst., 1947-50; Specialist, Eng. Res., Atomics Internat. Div. of North American Aviation, Inc., 1950-59; private consultant, 1959-60; member, tech. staff, Systems Labs. Div., Electronic Specialty Co., 1960; Assoc. Prof. Phys., San Fernando Valley State Coll., 1960-63; Prof., Nucl. Eng., Arizona Univ., 1963-. Societies: A.P.S.; A.N.S.; A.A.U.P.
Nuclear interests: Nuclear reactor theory; Neutron transport theory; Reactor dynamics.
Address: 2232 Rainbow Vista Drive, Tucson, Arizona, U.S.A.

HETSRONI, Gad, B.Sc. (Cum Laude), M.Sc., Ph.D. Born 1934. Educ.: Technion, Israel Inst. of Technol., and Michigan State Univ. Sen. Eng., Westinghouse Atomic Power Div., Pittsburgh, 1962-65; Sen. Lecturer, Nucl. Sci. Dept., Technion, Haifa, 1965. Member, Minister of Development's Com. on Nucl. Reactors. Society: A.S.M.E.
Nuclear interests: Reactor thermal and hydraulic advanced design and development; Research on heat transfer and fluid flow; Research on two-phase flow; Nuclear desalination.
Address: Nuclear Science Department, Technion, Haifa, Israel.

HETTCHE, Hans Otto, Prof., Dr. med., Dr. phil. Born 1902. Educ.: Frankfurt, Marburg, Freiburg, Munich and Greifswald Univs. Dozent, Königsberg Univ., 1934; Prof., Munich Univ., 1940; Prof. und Anstaltsleiter für Städtehygiene, Hygiene Inst., Hamburg, 1951-63; Direktor d. Landesanstalt f. Immissions-u. Bodennutzungsschutz d. Landes Nordrh.-Westfalen, 1963-. Wasserkommission Deutsche Forschungsgemeinschaft, Bad Godesberg; Kommission Reinhaltung der Luft im Verein Deutscher Ingenieure, Düsseldorf. Societies: Deutsche Gesellschaft für Hygiene und Mikrobiologie; Gesellschaft Deutscher Naturforscher und Arzte; Deutsche Gesellschaft für Arbeitsmedizin.
Nuclear interests: Reinhaltung der Gewässer, des Trinkwassers und Abwassers; Kontrolle durch eigene Messungen; Hygiene von Wasser, Boden und Luft.
Address: 6 Wallneyer Strasse, 43 Essen-Bredeney, (Landesanstalt), Germany.

HEUCK, F., Prof., Dr. Chief Medical Director, Radiologisches Zentrum, Katharinenhospital der Stadt Stuttgart.
Address: Radiologisches Zentrum, Katharinehospital der Stadt Stuttgart, 60 Kriegsbergstrasse, Stuttgart N., Germany.

HEUR, Jacques H. E. A. D'. See D'HEUR, Jacques, H. E. A.

HEURTEY, F. Prés. Directeur Gén., Heurtey S.A.
Address: Heurtey S.A., 30 rue Guersant, Paris 17, France.

HEUSCH, Clemens August, Dr. rer. nat., Dipl., Phys., Born 1932. Educ.: Bowdoin Coll., T.H. Aachen and Munich, Res. Asst., Labor.f. Techn. Physik, Munich, 1956-59; A.E.G. - Entwicklungsleitung, Frankfurt/Main, 1956-61; Staff Phys., Deutsches Elektronen-Synchrotron, Hamburg, 1961-; Assoc. Prof. Phys., California Inst. Technol.
Nuclear interests: High-energy physics (Photoproduction, etc.); Solid-state physics.
Address: Synchrotron Laboratory, California Institute of Technology, Pasadena, California, U.S.A.

HEUSLER, Konrad E., Dr., Privat dozent. Born 1931. Educ.: Göttingen Univ. Inst.für Sondermetalle, Max-Planck Inst. für Metallforschung (Inst. of Special Metals, Max-Planck Inst. for Metal Res.). Societies: D.G.M.; Deutsche Bunsen-Gesellschaft; Internat. Com. for Thermodynamics and Electrochem. Kinetics.
Nuclear interests: Kinetics of electrochemical reactions; Corrosion.
Address: Max Planck Institute for Metal Research, 92 Seestrasse, 7 Stuttgart-N, Germany.

HEUSON, J.-C., M.D. Born 1929. Educ.: Univ. Libre de Brussels. Attending Physician, Service of Internal Medicine, Inst. J. Bordet, Brussels Univ.
Nuclear interest: Medical applications of radioisotopes.
Address: Institut Jules Bordet, Centre des Tumeurs, Brussels University, 1 rue Héger-Bordet, Brussels 1, Belgium.

HEUSS, Kurt, Dipl. Phys., Dr. phil. nat. Born 1929. Educ.: Frankfurt am Main and Bern Univs. Formerly Asst., Max Planck-Inst. für Biophysik, Frankfurt am Main, 1955-. Societies: Gesellschaft Deutscher Naturforscher; Deutsche Gesellschaft für Biophysik; Deutsche Röntgengesellschaft.
Nuclear interests: Action of ionising radiation on biological materials, especially on mammals; X-ray technique and dosimetry.
Address: Strahleninstitut der A.O.K. Cologne, 19/27 Machabaerstrasse, Cologne, Germany.

HEUTCHY, Alvin E., B.A., M.A., LL.B. Born 1915. Educ.: Pennsylvania State and Pennsylvania Univs. Gen. Counsel, Sec., Director, and Member of the Board of Directors, F. and M. Schaefer Brewing Co., Brooklyn. Member, Gen. Advisory Com., New York State Office of Atomic Development.
Address: 430 Kent Avenue, Brooklyn, New York 11211, U.S.A.

HEUVEL, Jan Adrianus van den. See van den HEUVEL, Jan Adrianus.

HEWITT, Harold Barnett, M.D., B.S. Born 1915. Educ.: London Univ. Res. Fellow, British Empire Cancer Campaign for Res., 1947-. Societies: Fellow, Roy. Soc. of Medicine; British Inst. of Radiol.; Assoc. for Rad. Res.; British Medical Assoc.
Nuclear interest: Radiobiology of tumours with reference to the radiotherapy of human cancer.
Address: British Empire Cancer Campaign for Research, Research Unit in Radiobiology, Mount Vernon Hospital, Northwood, Middlesex, England.

HEWLETT, Richard Greening, M.A., Ph.D. Born 1923. Educ.: Dartmouth Coll. and Chicago Univ. Chief Historian (1957-), Member, Historical Advisory Com. (1968-), U.S.A.E.C. Societies: American Historical Assoc.; Organisation of American Historians; Soc. History of Technol.
Nuclear interests: History of the nuclear sciences and of the development of atomic energy in the United States.
Address: 7909 Deepwell Drive, Bethesda, Maryland 20034, U.S.A.

HEYBLOM, Tom. Born 1918. Coordinator, Marketing Services, Benzine and Petroleum Handel Maatschappij N.V., Amsterdam.
Address: 10 Park Leeuwensteyn, Voorburg, Netherlands.

HEYBOER, R. J. Member for Netherlands, Group of Experts on Production of Energy from Radioisotopes, E.N.E.A., O.E.C.D.
Address: O.E.C.D. European Nuclear Energy Agency, 38 boulevard Suchet, Paris 16, France.

HEYDORN, Kaj, M.Sc. (Chem. Eng.). Born 1931. Educ.: Tech. Univ. Denmark. With Danish Civil Defence Administration, 1954-56; Inst. for Theoretical Phys., 1955-56; Danish A.E.C., 1956-; I.A.E.A., 1960-; Gen. Atomic, San Diego, 1965-66.
Nuclear interest: Radioisotope production; Activation analysis.
Address: Isotope Division, Atomic Energy Commission, Research Establishment Riso, Roskilde, Denmark.

HEYM, A. Phys., Lab. de Recherches sur la Physique des Plasmas, Swiss Nat. Foundation for Sci. Res.
Address: Laboratoire de Recherches sur la Physique des Plasmas, 2 avenue Ruchonnet, Lausanne, Switzerland.

HEYMANN, Franz Ferdinand, B.Sc. (Eng.), Ph.D. Born 1924. Educ.: Cape Town and London Univs. Res. Eng., Metropolitan Vickers, Manchester, 1947-50; Asst. Lecturer, Phys. (1950-52), Lecturer, Phys. (1952-60), Reader, Phys. (1960-66), Prof. Phys. (1966-), University Coll., London.
Nuclear interest: Nuclear physics.
Address: Physics Department, University College, Gower Street, London, W.C.1, England.

HEYMANN, G., Dr. Res. Officer, Nucl. Phys. Div., Nat. Phys. Res. Lab., South African Council for Sci. and Ind. Res.
Address: National Physical Research Laboratory, South African Council for Scientific and Industrial Research, P. O. Box 395, Pretoria, South Africa.

HEYN, F. A., Prof., Dr., Ir. Electrotech. Lab., Dept. of Elec. Eng., Delft Tech. Univ.
Address: Electrotechnical Laboratory, 2b Kannaalweg, Delft, Netherlands.

HEYNE, Werner Alfred, Dipl.-Chem. Born 1931. Educ.: Dresden Tech. Univ. Sci., Zentralinst. für Kernforschung, Rossendorf near Dresden.
Nuclear interests: Nuclear fuels; Fuel reprocessing; Radiochemistry.
Address: 16 Oswald-Rentzsch-Strasse, 801 Dresden, German Democratic Republic.

HEYWOOD, Edgar Francis, B.Sc., Dip. Ed. Born 1922. Educ.: Melbourne Univ. Chem., G. Mowling and Son Pty. Ltd., Footscray, 1951; Chem., Newport Labs., Victorian Railways, Melbourne, 1952; Lecturer (1956), now Sen. Lecturer, Appl. Phys. Dept., Roy. Melbourne Inst. of Technol. Society: Assoc., Roy. Australian Chem. Inst.
Nuclear interest: Nuclear physics, especially the industrial applications of radioactivity and radioisotopes, including lecturing and consulting work in this field.
Address: 10/101 Gipps Street, East Melbourne, Victoria 3002, Australia.

HIBBS, John William, B.Sc. (Agriculture), M.Sc., Ph.D. Born 1917. Educ.: Ohio State Univ. At Ohio Agricultural R. and D. Centre. Societies: American Dairy Sci. Assoc.; American Soc. Animal Sci.; American Inst. Nutrition.
Nuclear interest: Isotopes used in animal nutrition and physiology.
Address: Ohio Agricultural Research and Development Centre, Wooster, Ohio, U.S.A.

HIBBS, Roger F. Superintendent, Y-12 Plant, Oak Ridge Nat. Lab., 1962-67; Manager, Production (1967), Vice Pres., Production (1967-), Nucl. Div., Union Carbide Corp.
Address: Union Carbide Nuclear Company, Y-12 Plant, P.O. Box Y, Oak Ridge, Tennessee 37830, U.S.A.

HICKCOX, Leigh H., B.S. (Elec. Eng.), M.B.A. Born 1932. Educ.: Worcester Polytech. Inst. and Harvard Univ. Packard Instrument Co., Inc., 1959-63; Philips Electronic Instr., 1963-65; Picker Nucl. Div., Picker X-Ray Corp., 1965-67. Marketing Manager, Asst. to the Pres., Sci. Accessories Corp., 1968-. Societies: I.E.E.E.; A.N.S.

Nuclear interest: Research and development and manufacturing in high energy nuclear physics.
Address: Beaver Brook Road, Weston, Connecticut 06880, U.S.A.

HICKENLOOPER, Bourke Blakemore, B.S. (Iowa State), J.D. (Law School, Iowa Univ.). Hon. degrees: LL.D. (Parsons Coll.), LL.D. (Loras Coll.), D.C.L. (Elmira Coll.). Born 1896. Educ.: Iowa State Coll. and the Law School, Iowa Univ. Elected to Iowa Legislature (House), 1934; re-elected 1936; elected Lieutenant-Governor of Iowa, 1938; re-elected 1940; elected Governor of Iowa, 1942; elected to the U.S. Senate, 1944; re-elected 1950 and 1956. Society: Member and former Chairman, Joint Com. on Atomic Energy, U.S. Congress.
Address: Cedar Rapids, Iowa, U.S.A.

HICKIE, John Bernard, M.B., B.S., M.R.C.P., F.R.A.C.P., F.A.C.C. Born 1926. Educ.: Sydney Univ. Part-time Lecturer in Medicine (1958-60), Sen. Lecturer in Medicine (1960-63), Assoc. Prof. in Medicine (1963-68), Sydney Univ.
Nuclear interests: Application of isotopic techniques to investigations and research in clinical medicine and cardiology.
Address: St. Vincent's Hospital, Sydney, Australia.

HICKMAN, Bernard Turner, M.D. Born 1925. Educ.: Tulane Univ., New Orleans. Assoc. Prof. Radiol., Mississippi Univ. Medical School, Jackson. Consultant Radiol. (in Radiotherapy) Veterans Administration Hospital, Jackson. Past-Pres., Mississippi Radiol. Soc. Societies: A.C.R.; Radiol. Soc. of North America; Mississippi Radiol. Soc.
Nuclear interests: Nuclear physics; Nuclear medicine.
Address: Mississippi University Medical Centre, Department of Radiology, 2500 North State Street, Jackson, Mississippi, U.S.A.

HICKMAN, Brian Stuart, M.Sc., F.I.M., F.A.I.P. Born 1932. Educ.: Melbourne Univ. Sen. Res. Officer, Metal. Sect. (1955-62), Head, Materials Phys. Sect. (1962-66), Australian A.E.C.; Tech. Staff, North American Rockwell Corp. Sci. Centre, 196
Nuclear interests: Irradiation damage in metals and ceramics; Structure and properties of metal oxides.
Address: 1049 Camino Dos Rios, Thousand Oaks, California 91360, U.S.A.

HICKOX, George H., B.E., M.S. (Iowa), Ph.D. (California). Born 1903. Educ.: Iowa and California Univs. Director, Eng. Experiment Station, Tennessee Univ., 1948-54; Programme Director for Eng. Sci., Nat. Sci. Foundation, Washington, D.C., 1954-56; Director of Res., Eng. R. and D. Labs., Ft. Belvoir, Virginia, 1956-62; Res. Management, The Boeing Co., Renton, Washington, 1962-; Chairman, Nat. Capital Sect., A.S.E.E.; Chairman, Com. on Eng. Education, Nat. Capital Sect., A.S.C.E. Com. on Fellowships, Scholarships, Grants and Bequests, A.S.C.E. Societies: A.A.A.S.; Washington Acad. of Sci.; A.S.C.E.;A.S.M.E.
Nuclear interest: Development and utilisation of nuclear power for military purposes.
Address: 417 S.W. 189th Place, Seattle, Washington, U.S.A.

HICKS, Donald A., A.B. (Phys.), M.A. (Phys.), Ph.D. (Nucl. Phys.). Born 1925. Educ.: California Univ., Berkeley. Teaching Asst. Phys. (1950-51), Nucl. Res., California R. and D. Corp. California Univ. (1951-56), Res. Phys. (1956-59), Chief of Nucl. Phys. Unit and Chief of Appl. Phys. Sect. (1960-61), Boeing Co.; Vice Pres. and Manager, Northrop Nortronics, Northrop Corp.; Tech. Consultant, Weapons Effects Board, Dept. of Defence. Society: A.P.S.
Nuclear interest: Management.
Address: Northrop Corporation-Nortronics Division, Applied Research Department, 2323 Teller Road, Newbury Park, California 91320, U.S.A.

HICKS, Samuel Pendleton, B.A., M.D. Born 1913. Educ.: Pennsylvania Univ. Prof. of Pathology, Michigan Univ. Medical Centre, Ann Arbor, Michigan. Societies: A.C.S.; American Acad. Neurology; Genetics Soc. America; American Soc. for Experimental Pathology.
Nuclear interest: Radiobiology, especially embryology and nervous system.
Address: Pathology Department, Michigan University Medical Centre, Ann Arbor, Michigan 48104, U.S.A.

HICKS, Thomas E., Dr. Prof. of Eng., California Univ., Los Angeles. Member, Advisory Com. on Isotopes and Rad. Development, U.S.A.E.C.
Address: California University, Los Angeles 24, California, U.S.A.

HICOCK, Russell. Chairman, Northeast Utilities Planning and Operating Com., Connecticut Light and Power Co.; Director, Yankee Atomic Elec. Co.
Address: Connecticut Light and Power Co., P.O. Box 2010, Hartford, Connecticut 06101, U.S.A.

HIDA, Noboru, Dr. Sci. Born 1920. Educ.: Hokkaido Imperial Univ. Study on beryllium ore deposit in the U.S.A., by the I.A.E.A. fellowship, 1962-63; Nucl. Material Resources Sect., Mineral Deposits Dept., Geological Survey of Japan. Book: Raw Materials of Newer Metals (Seibundo-shinkosha Publishing Co., 1963). Society: Atomic Energy Soc. of Japan.
Nuclear interest: Geology and mineralogy of nuclear materials, especially on the prospecting of beryllium minerals.
Address: Mineral Deposits Department, Geological Survey Tokyo Branch, 8 Kawada-cho, Shinjuku-ku, Tokyo, Japan.

HIDAKA, Takemichi. Director, Isotope Res. Lab., Tokai Regional Fisheries Res. Lab.
Address: Tokai Regional Fisheries Research Laboratory, Isotope Research Laboratory, Kachidoki 5-5-1, Chuo-ku, Tokyo, Japan.

HIDLE, Nils, Dipl. Ing. Born 1924. Educ.: Swiss Federal Inst. Technol., Zurich. Director of Eng., Joint Establishment for Nucl. Energy Res., Kjeller, Norway, 1950-57; Director of Eng., Noratom-Norcontrol A/S, Oslo, 1957-. Society: Norwegian Eng. Soc.
Nuclear interests: Reactor design; Design of research instruments and equipment.
Address: Noratom-Norcontrol A/S, 20 Holmenveien., Oslo 3, Norway.

HIEKMANN, Siegfried, Dipl.-Ing. Born 1925. Educ.: Dresden T.H.
Nuclear interest: Accelerator technics, particularly for the technic of cyclotron and electrostatic generator.
Address: 106 Bernhardstrasse, 8027 Dresden, Germany.

HIESINGER, Leopold Karl, Dr. phil. Born 1915. Educ.: Vienna Univ. and Inst. für Radiumforschung, Vienna. Manager, W. C. Heraeus, Hanau, 1950-57; Allgemeine Elektricitäts Gesellschaft, Kernenergieanlagen, Frankfurt, 1957-64. Societies: Deutsche Physikalische Gesellschaft; Österreichische Physikalische Gesellschaft; V.D.E.
Nuclear interests: Management of nuclear research establishment: Research reactors, particle-accelerator storage-banks, transport of irradiated fuel from nuclear power stations, nuclear physics, metallurgy.
Address: 19 Amselstrasse, Hohe Tanne, 645 Hanau, Germany.

HIGAKI, Bun-ichi, Bachelor of Laws (Hogakushi). Born 1900. Educ.: Tokyo Univ. Adviser, Yasuda Fire and Marine Insurance Co., Ltd. Society: Supv., Japan Nucl. Vessel Assoc.
Nuclear interest: Atomic energy insurance (liability and property damage).
Address: Yasuda Fire and Marine Insurance Company Limited, 6 Otemachi Itchome, Chiyoda-ku, Tokyo, Japan.

HIGASHIMURA, Takenobu, Dr. Asst. Prof., Div. of Sci. Instr., Res. Reactor Inst., Kyoto Univ.
Address: Research Reactor Institute, Noda, Kumatori-cho, Sen-nan-gun, Asaka Prefecture, Japan.

HIGATSBERGER, Michael J., Dr. phil., Hon. Prof. (Tech. Univ. Graz). Born 1924. Educ.: Vienna Univ. Res. Fellow, Minnesota Univ., 1952; Lecturer, Catholic Univ. Washington, 1954; now Prof., Vienna Univ.; and Tech. and Sci. Director, Austrian Atomic Energy Organisation; Director, Metallwerk Plansee A.G. Chairman, Internat. Halden Programme Group; Chairman, Internat. Dragon Gen. Purposes Com.; Board of Directors, Oesterreichisches Inst. fuer Energierecht; Council and Working Group, European Atomic Energy Soc.; Director, Soc. for Promotion of Technique in Plastic Industry; Member, Board of Management, O.E.C.D. High Temperature Reactor Project. Societies: Oesterr. Phys. Gesellschaft; Chem.-Phys. Gesellschaft; European Atomic Energy Soc.; A.N.S.; British Nucl. Soc.
Nuclear interests: Nuclear physics; Reactor physics; Isotope separation; Electron optics; Gamma-spectroscopy; Semiconductor-detector physics; Power reactor technology; Ion and plasma physics; High vacuum technology; Management.
Address: 6/9 Graf Starhemberggasse, Vienna 1040, Austria.

HIGGINS, Irwin R., B.S. (Biochem., Maine). Born 1919. Educ.: Maine Univ. Pres., Chem. Separations Corp. Book: Co-author, Chapter 16 in Ion Exchange Tech. (by Nachod and Schubert). Society: A.I.Ch.E.
Nuclear interests: Ion exchange process development; Construction of continuous ion exchange and solvent extraction equipment.
Address: 101 Midway Lane, Oak Ridge, Tennessee, U.S.A.

HIGGINS, Robert W., Ph.D. Formerly Director, now Chairman, Chem. and Phys. Dept., Texas Women's Univ.
Address: Texas Women's University, Chemistry and Physics Department, Denton, Texas, U.S.A.

HIGGINS, Thomas J., E.E. (Cornell), M.A. (Cornell), Ph.D. (Purdue). Born 1911. Educ.: Cornell and Purdue Univs. Prof. Elec. Eng., Wisconsin Univ., 1948-. Societies: Fellow, I.E.E.E.; Sen. Member, Instrument Soc. America; A.S.E.E.; American Mathematical Assoc.; Wisconsin Soc. Professional Eng.; Tensor Soc. of Great Britain.
Nuclear interests: Instrumentation; Automatic control theory of nuclear reactors with particular interests on stability theory.
Address: Electrical Engineering, Engineering Building, Wisconsin University, Madison, Wisconsin, U.S.A.

HIGGINSON, D. Reactor Operator, Dept. of Nucl. Sci. and Technol., Roy. Naval Coll.
Address: Royal Naval College, Department of Nuclear Science and Technology, Greenwich, London S.E.10, England.

HIGGS, Paul M. Assoc. Prof. Phys. Dept., Washington Univ.
Nuclear interest: Low temperature physics.
Address: Washington University, Physics Department, Seattle, Washington 98105, U.S.A.

HIGH, Edward Garfield, A.B., A.M., Ph.D. Born 1918. Educ.: Indiana, Purdue, Michigan and Columbia Univs. Prof. and Director, Biochem. Res., Prairie View A. amd M. Coll., 1949-53; Assoc. Prof., Biochem. (1953-59),

Prof. and Acting Head, (1959-67), Prof. and Head (1967-), Meharry Medical Coll.; Prof. (Part-time), Tennessee Agricultural and Ind. Univ.; Consultant, U.S.P.H.S., Guatemala and Paraguay, 1965. Member, Council, Oak Ridge Assoc. Univs.; Member, Board, Indiana Univ. Alumni Assoc., School of Arts and Sci. and Graduate School. Societies: Fellow, A.A.A.S.; A.M.Chem. Soc.; American Inst. of Nutrition. Nuclear interests: Radiation biology, in particular the use of isotopes in the elucidation of the mechanism of biochemical reactions. Also the effects of ionising radiation on animals.
Address: Meharry Medical College, Nashville, Tennessee 37208, U.S.A.

HIGH, James Millar, B.Sc. Born 1911. Educ.: St. Andrews Univ. Chief of Eng. Services, A.W.R.E., 1957-. Society: M.I.Mech.E. Nuclear interest: Engineering management.
Address "Treetops", Aldworth Road, Upper Basildon, Berks., England.

HIGHLAND, Virgil. Asst. Prof., High Energy Experimentalist, Dept. of Phys., Pennsylvania Univ.
Address: Pennsylvania University, Department of Physics, Philadelphia, Pennsylvania, U.S.A.

HIGHTON, Cecil Joseph, B.A., Barrister-at-Law. Born 1906. Educ.: Oxford Univ. Legal Adviser (1954-67), Consultant on Legal Matters (1967-), U.K.A.E.A. Nuclear interest: Legal aspects.
Address: The Dover House, Poling, Arundel, Sussex, England.

HIGHTOWER, Dan, D.V.M., M.S. Born 1925. Educ.: Agricultural and Mech. Coll. Texas, Reed Coll., and N.C.S.U. At present Assoc. Prof., Veterinary Physiology and Pharmacology Dept., Coll. of Veterinary Medicine, Texas A. and M. Univ. Societies: Rad. Res. Soc.; American Veterinary Medical Assoc.; A.N.S. Nuclear interest: Nuclear medicine; Biomedical engineering.
Address: Veterinary Physiology and Pharmacology Department, College of Veterinary Medicine, Texas A. and M. University, College Station, Texas 77843, U.S.A.

HIGHTOWER, R. E. Instructor, Nucl. Eng. Dept., Kansas State Univ.
Address: Kansas State University, Manhattan, Kansas, U.S.A.

HIGINBOTHAM, William Alfred, A.B., Hon. D.Sc. (Williams Coll.). Born 1910. Educ.: Williams Coll. (Williamstown, Mass.) and Cornell Univ. Assoc. Head, Electronics Divs. (1947-52), Head, Instrumentation Div. (1952-), Brookhaven Nat. Lab.; Chairman, Prof. Group on Nucl. Sci., I.E.E.E., 1962-63; Chairman and Member, Exec. Com., Federation of American Scientists, -1967. Societies: Fellow, A.P.S.; Fellow, A.I.E.E.E.; Fellow, A.N.S.

Address: 11 N. Howell's Point Road, Bellport, New York, U.S.A.

HIKOSAKA, Tadayoshi, Prof. Sci. Dept. Member, Radioisotope Res. Com., Niigata Univ. School of Medicine.
Address: Department of Science, Niigata University School of Medicine, 757, 1-Bancho Asahimachi-Street, Niigata, Japan.

HILBERRY, Norman, A.B., Ph.D., Hon. LL.D. (Elmhurst Coll. and Marquette Univ.), Hon. D.Sc., (Monmouth Coll.). Born 1899. Educ.: Oberlin Coll. and Chicago Univ. Deputy Director (1949-56), Director (1957-61), Sen. Sci. (1961-64), Argonne Nat. Lab., Prof., Nucl. Eng. Dept., Arizona Univ., 1964-. Director, Atomic Ind. Forum, 1962-. Societies: Fellow, A.N.S.; Fellow, A.P.S.; Fellow, A.A.A.S.; Fellow, New York Acad. Sci.; RESA; A.S.E.E.; American Inst. of Chem. Eng.; American Management Assoc. Nuclear interest: Research and development administration.
Address: Department of Nuclear Engineering, College of Engineering, Arizona University, Tucson, Arizona 85721, U.S.A.

HILBERT, Fritz, Dr. rer. nat. Born 1923. Educ.: Friedrich-Schiller-Univ., Jena. Abteilungsleiter, Zentralinstitut für Kernforschung, Rossendorf. Societies: Physikalische Gesellschaft in der Deutschen Demokratischen Republik; Deutsche Gesellschaft für Elektronenmikroskopie. Nuclear interests: Festkörperphysik (Strahlenwirkung, Elektronenmikroskopie).
Address: Deutsche Akademie der Wissenschaften zu Berlin, Zentralinstitut für Kernforschung Rossendorf, 8051 Dresden-Bad Weisser Hirsch, Postfach 19, German Democratic Republic.

HILBORN, J. W., Dr. Book: Contributor to Eng. Compendium on Rad. Shielding (I.A.E.A.).
Address: Atomic Energy of Canada Ltd., Reactor Research Division, Chalk River, Ontario, Canada.

HILDEBRAND, Roger Henry, A.B. (Chem.), Ph.D. (Phys.). Born 1922. Educ.: California Univ. Berkeley, Phys., Lawrence Rad. Lab., California Univ., Berkeley, 1942-51; Asst. Prof. Phys. (1953-55), Assoc. Prof. Phys. (1955-60), Prof. Phys. (1960-), Director (1965-), Enrico Fermi Inst. for Nucl. Studies, Chicago Univ.; Assoc. Lab. Director, High Energy Phys. (1958-64), Acting Director, High Energy Phys. Div. (1959-65), Argonne Nat. Lab. Society: Fellow, American Phys. Soc. Nuclear interests: Nuclear particle physics; Cosmic rays.
Address: 5722 S. Kimbark Avenue, Chicago 37, Illinois, U.S.A.

HILDGEN, Alphonse. Inspecteur principal, Service du Personnel, C.F.L.; Pres. f.f.,

Federation Nat. des Cheminots et Travailleurs du Transport Luxembourgeois. Member for Luxembourg, Specialised Nucl. Sect. for Social Health and Development Problems, Economic and Social Com., Euratom, 1966-.
Address: 9 rue Blochausen, Luxembourg, Luxembourg.

HILDREW, Bryan, M.Sc. (Eng.), D.I.C. Born 1920. Educ.: London Univ. Chief Eng. Surveyor, Lloyds Register of Shipping. Societies: Inst. Mech. Eng.; I.Mar.E.; B.N.E.S.
Nuclear interest: Nuclear plant design and manufacture for marine and land applications.
Address: Lloyds Register of Shipping, 71 Fenchurch Street, London E.C.3, England.

HILL, Albert James, B.Sc. Born 1904. Educ.: London Univ. Director (1940-51), Managing Director (1951-57), Chairman (1957-), Taylor Woodrow Construction Ltd.; Director, Nucl. Design and Construction Ltd., 1966-; Vice-Pres., Federation of Civil Eng. Contractors. Society: Inst. Civil Eng.
Nuclear interest: Design and construction of civil engineering works for nuclear power plants.
Address: Coggers Hall, Lamberhurst, near Tunbridge Wells, Kent, England.

HILL, Arthur M. Adviser, Atomics, Phys. and Sci. Fund Inc.
Address: Atomics, Physics and Science Fund Inc., 1033 30th Street, N.W., Washington 7, D.C., U.S.A.

HILL, Austin Bradford, D.Sc., Ph.D. Born 1897. Educ.: London Univ. Prof. Emeritus Medical Statistics, London School of Hygiene and Tropical Medicine, and lately Hon. Director, Statistical Res. Unit, M.R.C. Member, M.R.C. Com. on Hazards to Man of Nucl. and Allied Rad., and Adrian Com. on Radiol. Hazards to Patients. Societies: Roy. Soc.; Roy. Statistical Soc.; Roy. Soc. of Medicine.
Nuclear interest: Medical research into hazards, etc.
Address: Green Acres, Little Kingshill, Great Missenden, Buckinghamshire, England.

HILL, David A., B.S., Ph.D. Educ.: Princeton Univ. and M.I.T. Res. Phys., Nucl. Sci. Lab., M.I.T. 1957-64; Chief Phys., West Orange Lab., Vitro Corp., 1965-.
Address: 98 Beechwood Road, Summit, New Jersey, U.S.A.

HILL, Donald Harold, B.Sc. (Hons.). Born 1919. Educ.: London Univ. Ministry of Supply, 1951-56; Commercial Dept. Tech. Policy Directorate, Industrial Group, U.K.A.E.A., Risley, 1956-59; Nucl. Energy Attaché, United Kingdom Delegation to the European Communities, Brussels, 1959-61; Reactor Relations Dept., Reactor Group, U.K.A.E.A., Risley, 1961-67; First Sec. (Atomic Energy), United Kingdom Delegation to the European Communities, Brussels, and U.K.A.E.A. Rep. to Euratom, 1967-. Societies: F.R.I.C., London; Fellow, Inst. of Linguists, London; Inst. of Nucl. Eng., London; Société Royale Belge des Ingénieurs et des Industriels, Brussels.
Nuclear interests: International affairs.
Address: United Kingdom Delegation to the European Communities, 52 avenue des Arts, Brussels 4, Belgium.

HILL, Durwood W. Gen. Manager, Hallam Nucl. Power Facility, operated for U.S.A.E.C. by Consumers Public Power District.
Address: Hallam Nuclear Power Facility, Lincoln, Nebraska, U.S.A.

HILL, Ernest E., B.S.(Mech. Eng.), M.S. (Nuclear Eng.). Born 1922. Educ.: Univ. California. Production Supt., Federal Pacific Elec. Co. 1947-55; Nucl. Eng. (1955-59), Supervisor of Reactor Operations (1959-64), Lawrence Rad. Lab., Univ. California; Consultant to U.S.A.E.C., 1961-62. Society: A.N.S.
Nuclear interests: Reactor operations; Reactor physics; Nuclear engineering.
Address: 210 Montego Drive, Danville, California, U.S.A.

HILL, Hans Wilhelm, Dr. med. vet., o. Prof. Born 1916. Educ.: Tierärztliche Hochschule, Hanover. Direktor, Physiologisches Inst., Tierärztliche Hochschule, Hanover. Chairman, Com. of Rad. and Isotopes, German Soc. of Veterinary Medicine.
Nuclear interest: Fall-out problems in domestic animals and their products; Radioactive isotopes in physiology and biochemistry of farm animals.
Address: 15 Bischofsholer Damm, Hanover, Germany.

HILL, Jack W. Director, Dept., of Radiol. and Nucl. Medicine, Chong Hua Hospital.
Address: Department of Radiology and Nuclear Medicine, Chong Hua Hospital, Cebu City, Philippines.

HILL, James Fearnley, B.Sc. Born 1921. Educ.: Leeds Univ. Principal, Harwell Reactor School, A.E.R.E., 1958-60; Sen. Sci. and Dep. Head, Res. Reactors Div. (1960-63), Head, Post-Graduate Educ. Centre (1963-65), Head, Educ. and Training Dept. (1965-), A.E.R.E., Harwell. Book: Textbook of Reactor Phys – An Introduction (Allen and Unwin, 1961).
Nuclear interest: Reactor theory.
Address: 10 South Drive, Harwell, Didcot, Berkshire, England.

HILL, James Howard, B.S. (Chem.), M.S. (Nucl. Chem.). Born 1922. Educ.: Chattanooga and Ohio State Univs. Chief, Tech. Branch, Lockland Office (1951-54), Project Eng., Aircraft Nucl. Propulsion Programme (1954-58), Deputy Chief, Direct Cycle Sect., Aircraft Nucl. Propulsion Programme (1959-61),

Special Asst. to Commissioner (1961-64), Special Asst. to Chairman (1964-65), Deputy Director, Reactor Div., Oak Ridge Operations (1965), Asst. Director, Div. of Ind. Participation (1967-), U.S.A.E.C.; Vice Pres., Planning United Nucl. Corp., 1965-67. Societies: A.N.S.; Nat. Space Club.
Nuclear interest: Long range planning and management in the industrial nuclear power field.
Address: Route 3, Centreville, Maryland 21617, U.S.A.

HILL, James T., Jr. Director, United Nucl. Corp.
Address: United Nuclear Corporation, 1730 K Street, N.W., Washington 6, D.C., U.S.A.

HILL, John McGregor, B.Sc. (Phys.), Ph.D. Born 1921. Educ.: London and Cambridge Univs. At (1950-), Member for Production (1964-67), now Chairman, U.K.A.E.A. Chairman, representing U.K.A.E.A., Nukleardienst G.m.b.H.
Nuclear interest: All aspects of nuclear energy.
Address: United Kingdom Atomic Energy Authority, 11 Charles II Street, London, S.W.1, England.

HILL, Joseph A., B.A. (Biol.). Born 1926. Educ.: Clark Univ. Society: Analytical Chem. Soc.
Nuclear interest: Detection of radiation damage by means of analytical chemistry.
Address: General Electric Co., Schenectady, New York, U.S.A.

HILL, Joseph MacGlashan, B.S., M.D., D. Honoris Causa (Hon.), D.Sc. (Hon.). Born 1905. Educ.: Buffalo Univ., Buffalo, New York.
Prof. Pathology, School of Dentistry (1945-), Dean, Graduate Res. Inst. (1948-), Res. Consultant, Medical Centre (1959-), Baylor Univ.; Director, J. K. and Susie L. Wadley Res. Inst. and Blood Bank, 1951-; Clinical Prof. of Pathology, Southwestern Medical School (1957-), Sen. Consultant in Pathology, M. D. Anderson Hospital and Tumour Inst. (1965-), Texas Univ. Societies: A.A.A.S.; American Assoc. for Cancer Res.; A.M.A.; Fellow, Internat. Soc of Hematology; Leukemia Soc. of America; Soc. of Nucl. Medicine; Soc. of Exptl. Biol. and Medicine; Texas Soc. of Electron Microscopy.
Nuclear interests: Medical applications of radioisotopes in treatment of leukemia and other malignancies.
Address: 3600 Gaston Avenue, Dallas, Texas 75246, U.S.A.

HILL, Robert Sidney, M.A., D.Phil. Born 1926. Educ.: Oxford Univ. Res. Studentship, University Coll. London, 1954-55; Res. Phys., Clarke, Chapman and Co., Ltd., 1955-.
Nuclear interests: Physics of systems and techniques for heat removal from nuclear reactors; Reactor coolant technology; Nuclear physics; Reactor physics.
Address: Physics Division, Clarke, Chapman and Co., Ltd., Victoria Works, Gateshead 8, England.

HILL-RODRIGUEZ, Walter S., Ing. Born 1903. Educ.: Montevideo Univ. Director, Inst. de Fisica, and Prof. Phys., Facultad de Ingenieria de Montevideo; Director, Laboratorio MC^2, Montevideo (private). Member and Past Pres., Nat. Atomic Com. of Uruguay.
Nuclear interests: Nuclear physics; Isotopes; Non-destructive testing.
Address: Instituto de Fisica, Herrera y Reissig 565, Montevideo, Uruguay.

HILLAS, Anthony Michael, B.Sc., Ph.D., A. Inst. P. Born 1932. Educ.: Leeds Univ. Harwell Jun. Res. Fellow, A.E.R.E., 1956-59; Lecturer in Phys., Leeds Univ., 1959-. Society: A. Inst. P.
Nuclear interest: Cosmic rays: particularly extensive air showers and very high energy particle interactions.
Address: Physics Department, Leeds University, Leeds 2, England.

HILLEBOE, Herman E., B.S., M.B., M.D., M.P.H. Born 1906. Educ.: Minnesota Univ. Commissioner, New York State Dept. of Health, 1947-63; Prof., Public Health Practice, Columbia Univ. School of Public Health, New York City, 1963-. Book: Co-author, Mass Radiography of the Chest. Societies: A.M.A.; American Public Health Assoc.; American Assoc. of Public Health Physicians; Nat. Tuberculosis Assoc.
Nuclear interest: Public health.
Address: 600 West 168th Street, N.Y.C., New York 10032, U.S.A.

HILLEY, James Franklin, Geophys. Eng. Born 1923. Educ.: Colorado School of Mines. Eng., Nucl. Power Sect., Southern Services, Inc., 1960-.
Nuclear interests: Design, construction and operation of electric power reactors.
Address: Southern Services, Inc., 600 North 18 Street, Birmingham 2, Alabama, U.S.A.

HILLMAN, Peter, Assoc. Prof., B.A., M.A., Ph.D. Born 1928. Educ.: Harvard Univ. Sci. Officer, A.E.R.E., Harwell, 1954-55; Lecturer, Witwatersrand Univ., 1956; C.E.R.N. Fellow, Uppsala Univ., 1956-58; Ford Fellow, C.E.R.N., 1958-59; Res. Staff, Weizmann Inst., 1960-; Italian A.E.C. Fellow, Frascati, 1963-64; Head, Nucl. Phys. Dept., Weizmann Inst., 1964-67; Guest Investigator, Rockefeller Univ., 1967-68. Societies: Israel Phys. Soc.; A.P.S.
Nuclear interests: Low energy reactions; Mossbauer effect.
Address: Weizmann Institute, Rehovoth, Israel.

HILLS, E. S., Prof. Member, Board of Studies in Nucl. Sci. and Eng., Melbourne Univ.

HIL

Address: Geology Department, Melbourne University, Carlton N.3, Victoria, Australia.

HILLS, Peter Robert, B.Sc., M.Sc. (Dunelm), A.R.I.C. Born 1929. Educ.: Durham Univ. Sen. Sci. Officer, Wantage Rad. Lab., U.K.A.E.A. Treas., Miller Conference for Rad. Chem.; Treas., Assoc. for Rad. Res. Societies: Roy. Inst. of Chem.; Assoc. for Rad. Res. Nuclear interests: Radiation chemistry of organic systems.
Address: Wantage Radiation Laboratory, Isotope Research Division, Atomic Energy Research Establishment, Harwell, Didcot, Berks., England.

HILLS, William B., B.S. (Elec. Eng.). Born 1920. Educ.: Iowa State Univ. Supervising Eng., Control Design, SAR Project, K.A.P.L. (1955-58), Project Eng. SAR Test Planning, K.A.P.L. (1958-59), Manager, Elec. Systems, K.A.P.L. (1959-64), Project Eng., Test Site, K.A.P.L. (1965-67), Consulting Project Eng., Atomic Power Equipment Dept. (1967-), G.E.C. Societies: N.S.P.E.; I.E.E.E. Nuclear interests: Control systems for reactors and nuclear power plants; Electrical systems for naval service; Reactor safety.
Address: General Electric, Atomic Power Equipment Department, 175 Curtner Avenue, San Jose, California 95125, U.S.A.

HILMI, Abdul Karim, B.Sc. and M.Sc. (Chem. Eng.). Born 1928. Educ.: Texas Univ. Chem. Eng., Directorate Gen. of Ind. Planning, Baghdad. Treas., Iraqi Sci. Res. Soc. Society: Iraqi Sci. Res. Soc., Baghdad. Nuclear interest: Industrial applications of radioactive isotopes.
Address: 20A/78 Raghiba Khatoon, Adhamiya, Baghdad, Iraq.

HIMSWORTH, Sir Harold Percival, K.C.B., M.D., F.R.C.P., F.R.S. Born 1905. Prof. Medicine, London Univ., and Director Medical Unit, Univ. Coll. Hospital, London, 1939-49; Sec., Medical Res. Council, 1949-.
Address: Medical Research Council, 20 Park Crescent, London, W.1, England.

HINCHEY, John J., Commander, U.S.N. Nucl. Power Supt., Portsmouth Naval Shipyard.
Address: Portsmouth Naval Shipyard, Nuclear Power Division, Portsmouth, New Hampshire, U.S.A.

HINDAWI, Ali Yahya AL-. See AL-HINDAWI, Ali Yahya.

HINE, Alan Rowland, B.Sc. (Hons., Chem.). Born 1926. Educ.: Southampton Univ. Process Eng., Esso Petroleum Co. Ltd., 1950-56; Chem. Eng., R. and D. Branch, Tech. Directorate Production Group Headquarters, now Heavy Chem. Plant Development Group, Tech. Dept., Windscale Works, U.K.A.E.A., 1956-. Nuclear interest: Technical work on fuel reprocessing and assessment of future processes.
Address: Fair View, Cross Lane, Seascale, Cumberland, England.

HINE, gerald J., Ph.D. Born 1916. Educ.: Federal Inst. Techno Administration Hospital, Boston, Mass., 1952-64. I.A.E.A., Vienna, 1964-. Books: Co-editor, Rad. Dosimetry (1956); Editor, Instrumentation in Nucl. Medicine (1967). Societies: American Assoc. Phys. in Medicine; Rad. Res. Soc.; Soc. of Nucl. Medicine. Nuclear interests: Medical radiation physics. Instrumentation for the diagnostic applications of radioisotopes.
Address: Section of Nuclear Medicine, Division of Life Sciences, International Atomic Energy Agency, 1010 Vienna, Austria.

HINE, Gerald J., Ph.D. Born 1916. Educ.: Federal Inst. Technol., Zürich. Phys., Veterans Address: Department of the Navy, Bureau of Naval Weapons, Washington 25, D.C., U.S.A.

HINKLE, Norman E., B.S. Born 1931. Educ.: Grove City Coll. and Tennessee Univ. Development Eng., O.R.N.L. Society: A.S.M. Nuclear interest: Effect of neutron irradiation on reactor structural metals.
Address: Solid State Division, Oak Ridge National Laboratory, P.O. Box X, Oak Ridge, Tennessee, U.S.A.

HINKS, K. F. C. Resident Eng., Hartebeestfontein Gold Mining Co. Ltd.
Address: Hartebeestfontein Gold Mining Co. Ltd., Private Bag, Stilfontein, Transvaal, South Africa.

HINMAN, George Wheeler, B.S. (Phys.), B.S. (Maths.), M.S. (Phys.), D.Sc. (Phys.). Born 1927. Educ.: Carnegie Inst. Technol. Asst. Prof. (1953-59), Assoc. Prof. (1959-63), Carnegie Inst. Technol.; Chairman, Phys. Dept., Gulf Gen. Atomic, Inc. Societies: Fellow, A.P.S.; A.N.S.; Soc. Nucl. Medicine. Nuclear interests: Nuclear physics; Reactor physics; Solid state physics; Nuclear medicine.
Address: General Atomic, Box 608, San Diego, California, U.S.A.

HINNERS, Robert A., Captain, U.S.N.(ret.) Prof. of Naval Architecture, and Head of Luckenbach Graduate School, Webb Inst. of Naval Architecture.
Address: Webb Institute of Naval Architecture, Crescent Beach Road, Glen Cove, New York, U.S.A.

HINO, Jun, Ph.D. Born 1917. Educ.: Illinois Univ. Societies: Inst. of Metals; A.S.M. Nuclear interests: Physical and mechanical metallurgy of nuclear fuels, cladding alloys and control materials.

Address: Westinghouse Electric Corporation, P.O. Box 1468, Pittsburgh, Pennsylvania, U.S.A.

HINSLEY, John Frederick. Born 1902. Chief Phys., Edgar Allen and Co. Ltd., Sheffield, -1967; Director, J. H. Humphreys and Sons, Ltd., Oldham, 1961-67; Consultant, U.N.E.S.C.O., 1967-; U.N.E.S.C.O. Prof., Regional Eng. Coll., Durgapur, India, 1967-. Chairman, Com. NCE/2/2 (Shielding from Ionising Radiations), B.S.I. Book: Non-Destructive Testing (Macdonald and Evans, 1959). Societies: Fellow, Inst. of Metal.; Fellow, Inst. Phys. and Phys. Soc.; Iron and Steel Inst.; Inst. of Metals; Soc. for Analytical Chem.; Mathematical Assoc.; Soc. for Non-Destructive Examination.
Nuclear interests: Design and manufacture of magnetic devices – e.g., beam-bending magnets, particle accelerators, laboratory electromagnets.
Address: Regional Engineering College, Durgapur 9, West Bengal, India.

HINTENBERGER, Heinrich, Prof. Dr. phil. Born 1910. Educ.: Vienna Univ. Univ. Lecturer on Exptl. Phys. Mainz Univ., 1952; Head, Mass Spectroscopic Dept., Max Planck Inst., 1956; Member, Max Planck Soc.; Director, Max Planck Inst., Mainz, 1959. Books: Co-author, Methoden, und Anwendungen der Massenspektroskopie (Verlag Chemie, 1953); Editor, Nucl. Masses and their Determination (Pergamon Press, 1955). Societies: German Phys. Soc.; A.P.S.; A.G.U.
Nuclear interests: Cosmic nuclear abundances; Nuclear synthesis; Nuclear evaporation and spallation; Long lived radioactivities; Nuclear masses.
Address: Max Planck Institut für Chemie, Otto-Hahn-Institut, 23 Saarstrasse, Mainz, Germany.

HINTERMANN, Karl, Ph.D. (Mech. Eng.). Born 1924. Educ.: Swiss Federal Inst. Technol. and Berne Univ. Div. Head, O.E.C.D. High Temperaure Reactor Project Dragon, A.E.E., Winfrith. Societies: Schweiz. Ges. von Fachleuten der Kerntechnik; A.N.S.
Nuclear interests: Reactor physics and nuclear engineering.
Address: Organisation for Economic Cooperation and Development, High Temperature Reactor Project Dragon, Atomic Energy Establishment, Winfrith, Dorset, England.

HINTERMAYER, F., Dipl. Ing. Vorstandsmitglied, Oesterreichische Elektrizitätswirtschaft A.G. Vorsitzender der Mitgliederversammlung und des Verwaltungsausschusses, Arbeitsgemeinschaft Kernkraftwerk der Elektrizitätwirtschaft. Member, Comite d'Etudes de l'Energie Nucléaire, Union Internationale des Producteurs et Distributeurs de l'Energie Electrique; A.K.E.W. Rep., Koordinationsausschuss Osterreichisches Kernenergieprogramm; First Vice Pres., Osterreichisches Atomforum.
Address: 6A Am Hof, Vienna 1, Austria.

HINTON, The Lord Christopher, K.B.E., M.A., Hon. D. Eng. (Liverpool), Hon. D.Sc. (Eng., London), Hon. D.Sc. (Oxford), Hon. LL.D. (Edinburgh), Hon. Sc.D. (Cambridge), Hon. D.Sc. (Southampton, Durham and Bath). Born 1901. Educ.: Trinity Coll., Cambridge (Hon. Fellow, 1957). M.O.S., Dep. Controller of Production, Div. of Atomic Energy, 1946-54; Board, U.K.A.E.A. and Managing Director of Ind. Group, 1954-57; Chairman, C.E.G.B., 1958-64. Chairman, Internat. Exec. Council, World Power Conference. Societies: F.R.S.; M.I.Mech.E.; I.C.E.; M.I.E.E.; F.Inst.F.; F.R.S.A.; Hon. Mem. Inst. Metals; Hon. Mem., Inst. of Gas Eng.; Hon. Assoc., Manchester Coll. Sci. and Technology; Inst. Chem. Eng.; Fellow, British Inst. Management.
Nuclear interests: Design, construction and operation of nuclear power plants and of ancillary chemical, metallurgical and diffusion plants.
Address: Tiverton Lodge, Dulwich Common, London, S.E.21, England.

HINZNER, Fritz, Dr. rer. nat. Sen. Officer, Inst. for Special Metals, Max-Planck Inst. for Metal Res.
Address: Max-Planck Institute for Metal Research, 92 Seestrasse, 7000 Stuttgart 1, Germany.

HIRABAYASHI, Makoto, Prof. Div. of Nucl. Metal., Res. Inst. for Iron, Steel and other Metals, Tohoku Univ.
Address: Tohoku University, 75 Katahira-cho, Sendai, Japan.

HIRAKI, Hiromichi. Asst. Prof., Nucl. Phys. Div., Kyushu Inst. of Technol.
Address: Kyushu Institute of Technology, Nuclear Physics Division, Tobata, Kitakyushu, Fukuoka-ken, Japan.

HIRAMATSU, Hiroshi, M.D. Born 1909. Educ.: Kanazawa Coll. of Medicine, Prof., Faculty of Medicine, Kanazawa Univ., 1945-. Councillor, Japanese Medico-Radiol. Soc.; Chairman, Radioisotope Res. Com., Kanazawa Univ. Societies: Japanese Medico-Radiol. Soc.; Japanese Assoc. Balneoclimatology; Japanese Assoc. Hotspring Sci.; Japanese Assoc. Haematology.
Nuclear interest: Nuclear medicine.
Address: Kodatsuno 5-1-25, Kanazawa-city, Ishikawa Prefecture, Japan.

HIRAOKA, Masasuke. Asst. Manager Tech. Dept., Iino Shipbuilding and Eng. Co. Ltd.
Address: Iino Shipbuilding and Engineering Co. Ltd., 6 Marunouchi 3-chome, Chiyoda-ku, Tokyo, Japan.

HIRATSUKA, Masatoshi, B.A. Born 1900. Educ.: Tokyo Univ. Pres. Sumitomo Atomic Energy Industries Ltd.; Managing Director, Japan Atomic Ind. Forum; Councillor, Fund

for Peaceful Atomic Development of Japan.
Nuclear interest: Management.
Address: 485, Furudera, Sumiyoshi-cho,
Higashinada-ku, Kobe City, Hyogo Pref.,
Japan.

HIRAYAMA, Toru. Asst. Manager and Chief,
Radio- and Rad. Chem. Sect., Central Res.
Lab., Showa Denko K.K.
Address: Showa Denko K.K., 34 Shiba
Miyamoto-cho, Minato-ku, Tokyo, Japan.

HIRD, Albert Hamilton, B.Sc. (Eng.), A.C.G.I.
Born 1902. Educ.: London Univ. Director:
Vickers Ltd., Vickers-Armstrongs Ltd.,
English Steel Corp. Ltd., Metropolitan-
Cammell Carriage and Wagon Co. Ltd. Society:
M.I.Mech.E.
Nuclear interest: Management.
Address: 146 Marsham Court, Marsham
Street, London, S.W.1, England.

HIRD, Brian, M.A., Dr. Phil. Educ.: Oxford
Univ. Assoc. Prof., Ottawa Univ. Society:
Canadian Assoc. Phys.
Nuclear interests: Low energy nuclear physics.
Address: Physics Department, Ottawa University, Ottawa, Ontario, Canada.

HIRONE, Tokutaro, Ph.D. (Rigaku-hakushi).
Born 1906. Educ.: Tohoku Univ. Formerly
Res. Inst. for Iron, Steel and other Metals,
Tohoku Univ., Sendai. Society: Atomic Energy
Soc. Japan.
Nuclear interest: Nuclear materials.
Address: 6-10 Kunimi-1-chome, Sendai,
Japan.

HIROSE, Iwakichi. Manager, Tech. Res. Lab.,
Hokuriku Elec. Power Co. Inc.
Address: Hokuriku Electric Power Co. Inc.,
1 Sakurabashi-Dori, Toyama City, Japan.

HIROSIGE, Tetu. Lecturer, History of Sci.,
Atomic Energy Res. Inst., Nihon Univ.
Address: Atomic Energy Research Institute,
Nihon University, 1-8 Kanda-Surugadai,
Chiyoda-ku, Tomyo, Japan.

HIROTA, Juichi. Director, Japan Atomic
Power Co. Ltd. Japanese Member, European
American Com.on Reactor Phys., O.E.C.D.,
E.N.E.A.
Address: Japan Atomic Power Co. Ltd.,
Otemachi Building, Otemachi 1-chome,
Chiyoda-ku, Tokyo, Japan.

HIROTA, Kozo, D.Sc. (Tokyo). Educ.: Tokyo
Imperial Univ. (now Tokyo Univ.). Prof.
Phys. Chem., Sci. Faculty, Osaka Univ.; Res.
member, Osaka Labs., Japanese Assoc. Rad.
Res. on Polymers. Society: Soc. Isotopes and
Rad. Japan; Res. member, Inst. Phys. and
Chem. Res., Tokyo.
Nuclear interests: Radiation chemistry – e.g.,
telomerisation of olefins induced by radiation
and mechanism of primary process caused by
electron impact.

Address: Hyogo-Prefecture, Takarazuka-
City, Nakasuji, Higashi-iguchi 3, Japan.

HIROTA, Seiichiro. Director, Nippon Atomic
Industry Group Co. Ltd.
Address: Nippon Atomic Industry Group Co.
Ltd., 2-5 Kasumigaseki 3-chome, Chiyoda-ku,
Tokyo, Japan.

HIROTA, Shiro. Chief, Radioisotope Sect.,
(-1966), Chief, Technol. Promotion Sect.
(1966-), Atomic Energy Bureau, A.E.C. of
Japan.
Address: Atomic Energy Commission of Japan,
3-4 Kasumigaseki, Chiyoda-ku, Tokyo, Japan.

HIRSCH, Peter Bernhard, M.A., Ph.D. Born
1925. Educ.: Cambridge Univ. Reader, Phys.,
Cambridge Univ., 1964-66; Isaac Wolfson
Prof., Metal., Oxford Univ., 1966-. Societies:
F.R.S.; F.Inst.P.; F.I.M.
Nuclear interests: Metallurgy; Radiation
damage; Channelling of high energy particles.
Address: Metallurgy Department, Parks Road,
Oxford, England.

HIRSCH, Robert, Ancien élève de l'Ecole
Polytechnique. Born 1912. Educ.: Ecole Polytechnique. Préfet – Directeur Gén., Sûreté
Nationale, 1951-54; Préfet, Seine-Maritime,
1954-59; Inspecteur Gén. de l'Administration,
en mission extraordinaire pour la région du
Nord, 1959-63; Administrateur Gén., Délégué
du Gouvernement près le C.E.A., 1963-.
Address: 33 rue de la Fédération, Paris 15,
France.

HIRSCHBERG, C. F. G. Von. See Von
HIRSCHBERG, C. F. G.

HIRSCH, Merle Norman, B.S., Ph.D. Born
1931. Educ.: Pittsburgh and Johns Hopkins
Univ. Res. Phys., I.T.T. Federal Labs., Nutley,
New Jersey, 1958-61; Tech. Director, Space
Phys. Lab., G. C. Dewey Corp., New York,
1961-. Societies: A.P.S.; A.G.U.; American
Vacuum Soc.
Nuclear interest: Effect of nuclear radiation
on the properties of the ionosphere.
Address: The G. C. Dewey Corporation, 331
East Street, New York, New York 10016,
U.S.A.

HIRST, J. Production Manager, Reactor
Control Div., Elliott Bros. (London), Ltd.
Address: Elliott Bros. (London) Ltd.,
Reactor Control Division, Century Works,
Lewisham, London S.E.13, England.

HIRZEL, Oskar, Dr. Sc. nat. Born 1919.
Educ.: Federal Inst. Technol., Zürich. Formerly
Phys., now Nucl. Specialist, Nucl. Dept.,
Oerlikon Eng. Co., Zürich, 1949-. Societies:
A.P.S.; Schweizerische und Zuercher
Physikalische Gesellschaft.
Nuclear interests: Nuclear physics and reactor
technology.

Address: 14 Hoehenstrasse, Wetzikon, Zh., Switzerland.

HISATOMI, Yo. Born 1931. Educ.: Tokyo Inst. of Technol. Res. Div., Shimizu Construction Co., Ltd. Society: Architectural Inst. of Japan.
Nuclear interest: Construction (civil) nuclear power station.
Address: Shimizu Construction Company Limited, No. 1, 2-chome, Takara-cho, Chuo-ku, Tokyo, Japan.

HISKES, John Robert, A.B., M.A., Ph.D. Born 1928. Educ.: California Univ., Berkeley. Phys., Lawrence Rad. Lab., Livermore, California, 1954-; Phys. (Visiting), Culham Lab., Abingdon, Berks., 1963-64. Society: A.P.S.
Nuclear interest: Controlled fusion research.
Address: Lawrence Radiation Laboratory, P.O. Box 808, Livermore, California, U.S.A.

HITCHCOCK, John Alan, B.Sc. (Special, Mathematics, 1st-class Hons.). Born 1929. Educ.: London Univ. Asst. Head, Thermal Group, Mining Res. Establishment, N.C.B., Isleworth, Middx., 1953-58; Head, Heat Transfer Sect., Central Elec. Res. Labs., Leatherhead, Surrey, 1958-. Society: F.Inst.P.
Nuclear interests: Heat transfer aspects of nuclear plant, with particular reference to heat exchanger systems of drum or once-through type. Performance considerations in relation to the C.E.G.B.'s requirements for the satisfactory design and operation of nuclear stations. This involves the experimental determination of performance characteristics in the laboratory, plant measurement, fundamental studies of convection heat transfer, transient aspects in relation to operation and control, stability studies, vibration phenomena and theoretical modelling for computer simulation.
Address: 11 Carlton Green, Redhill, Surrey, England.

HITCHCOCK, John William, Dipl. of Coll. of Aeronautics. Born 1926. Educ.: Cranford Coll. of Aeronautics. Deputy Divisional Manager, Rad. and Nucl. Eng. Div., Vickers Ltd., South Marston Works. Societies: A.M.I.Mech.E.; B.N.E.S.
Nuclear interest: Experimental equipment for use with material test reactors.
Address: Vickers Limited, South Marston Works, Swindon, Wilts., England.

HITCHENS, John Gilbert KEITH-. See KEITH-HITCHENS, John Gilbert.

HITI, Thabit NAMAN AL-. See NAMAN AL-HITI, Thabit.

HITTMAIR, Otto, Dr. phil., Prof. Born 1924. Educ.: Innsbruck and Basle Univs., and M.I.T. Prof. Theoretical Phys., Vienna Tech. Univ. Book: Co-author, Nucl. Stripping Reactions (Wiley, 1957). Society: Corresponding member, Austrian Acad. Sci. and Austrian Phys. Soc.
Nuclear interest: Nuclear physics.
Address: Technical University, 13 Karlsplatz, Vienna, Austria.

HITTMAN, Fred, B.Sc. (Chem. Eng.) (Cum laude). Born 1929. Educ.: Michigan Univ. Process Eng., Hanford Works, 1951-53; Reactor Eng., Brookhaven Nat. Lab., 1953-55; Manager, Nucl. Powerplant Dept., Martin Co., 1955-62; Pres. (1962-), and Gen. Manager, (1962-67), Hittman Associates, Inc. Book: Co-author, Metals for Nucl. Reactors (A.S.M.). Societies: A.N.S.; Nucl. Eng. Sect., A.I.Ch.E.
Nuclear interests: Nuclear engineering, energy conversion, materials, space and oceanographic sciences.
Address: 2500 Ozark Circle, Baltimore 9, Maryland, U.S.A.

HIXSON, Orton F. Tech. Director, Rosner-Hixson Labs. Inc., Lab. of Vitamin Technol.
Address: Rosner-Hixson Laboratories Inc., 7737 South Chicago Avenue, Chicago 19, Illinois, U.S.A.

HJALMARS, Inga M. FISCHER-. See FISCHER-HJALMARS, Inga M.

HJERTBERG, Peter, G., M. Eng. (Stockholm), M.S. (M.I.T.). Born 1920. Educ.: Roy. Inst. Technol. Stockholm and M.I.T. Chief, Sect. for Control and Instrumentation, Nucl. Power Dept., A.S.E.A., Sweden. Societies: Svenska Teknologföreningen, Sweden; Tekniska Föreningen, Sweden.
Nuclear interests: Reactor instrumentation; Reactor kinetics.
Address: 41 Nordanbygatan, Västeras, Sweden.

HLADIK, Rudolf, Dr. rer. com. Born 1931. Educ.: Commerce Univ., Vienna. First Sec., Dept. for Atomic Energy Affairs, Federal Chancellery.
Nuclear interests: International cooperation and economic problems of nuclear energy, management and administration.
Address: Federal Chancellery, Department for Atomic Energy Affairs, 1-3 Hohenstaufengasse, 1010 Vienna, Austria.

HLAWATY, E., Dr. Stellv. Obmann, Arbeitsausschuss Atomenergierecht, Vereinigung Deutscher Elektrizitätswerke.
Address: Bayernwerk A.G., 6 Blutenburgstrasse, 8 Munich 2, Germany.

HLI, Freddy BA. See BA HLI, Freddy.

HO, D. C.
Address: P.O. Box 96, Peking, China.

HOAGLAND, Hudson, A.B. (Columbia), M.S. (M.I.T.), Ph.D. (Harvard), Sc.D. (hon.) (Colby Coll., Wesleyan and Clark Univs.). Born 1899. Educ.: Columbia and Harvard Univs. and M.I.T. Exec. Director, Worcester Foundation for Exptl. Biol., Inc.; Consultant,

Nat. Sci. Foundation. Pres., American Acad. Arts and Sci.; Trustee, Woods Hole Oceanographic Inst.; Trustee, Memorial Hospital, Worcester; Harvard Overseers Visiting Com. to Visit the Harvard Medical School and School of Dental Medicine; Prof. Biol. and Physiology in Psychiatry, Boston Univ.; Advisory Com., Federation of American Scientists. Society: American Physiological Soc.
Nuclear interests: Executive Director of Institution engaged in a variety of metabolic studies using radioactive tracers.
Address: Worcester Foundation for Experimental Biology, Inc., 222 Maple Avenue, Shrewsbury, Mass., U.S.A.

HOAGLAND, Mahlon Bush, M.D. Born 1921. Educ.: Williams and Harvard Colls., Harvard Medical School. Assoc. Biochem. (Medicine), Massachusetts Gen. Hospital, 1953-; Asst. Prof. Medicine (1958-60), Assoc. Prof. Bacteriology and Immunology (1960-67), Harvard Medical School; Prof. and Chairman, Biochem. Dept., Dartmouth Medical School, Hanover, New Hampshire, 1967-. Societies: American Soc. Biol. Chem.; American Acad. Arts and Sci.
Nuclear interest: Use of radioisotopes in the study of cellular biosynthetic and regulatory mechanisms.
Address: Box 62, Thetford Hill, Vermont 05074, U.S.A.

HOANG, Xuan Han, Prof. Mathematics. Ing. Principal, Service d'Études Nucléaires, Soc. Alsthom.
Address: Société Alsthom, 20 rue d'Athènes, Paris 9, France.

HOASHI, Komayoshi. Asst. Chief, Administrator of Atomic Energy Development Agencies, Atomic Energy Bureau, A.E.C. of Japan.
Address: Atomic Energy Commission of Japan, 3-4 Kasumigaseki, Chiyoda-ku, Tokyo, Japan.

HOBART, Lawrence, B.S., M.P.A. Born 1931. Educ.: Oregon and Michigan Univs. Director, American Public Power Assoc. Atomic Energy Service, 1962-.
Nuclear interest: Nuclear power reactor development.
Address: Watergate Office Building, 2600 Virginia Avenue, N.W. Washington D.C.20037, U.S.A.

HOBBIS, Leo Cyrus William, M.Sc., Ph.D. Born 1927. Educ.: Auckland Univ., New Zealand. Jun. Lecturer in Phys., Auckland Univ., 1948-53; A.E.R.E., Harwell, 1953-61; Sen. Commonwealth Fellow, 1953-56; Rutherford High Energy Lab., N.I.R.N.S., 1961-65; Sci. Res. Council, 1965-.
Nuclear interests: Particle accelerators and their utilisation.
Address: Coromandel, 122 Oxford Road, Abingdon, Berks., England.

HOBBS, Edwin J. Formerly Manager of Procurement, now Manager of Materials (1965-), Atomic Power Equipment Dept., Gen. Elec. Co.
Address: Atomic Power Equipment Department, General Electric Co., P.O. Box 254, 175 Curtner Street, San Jose, California, U.S.A.

HOBBS, W. R., B.Sc., A.R.C.S. Council Member, Inst. of Nucl. Eng.
Address: Institution of Nuclear Engineers, 147 Victoria Street, London S.W.1, England.

HOBLYN, Edward Henry Treffry, Ph.D., A.R.C.S., F.R.I.C., Inst. Chem. Eng. Born 1910. Educ.: Roy. Coll. Sci. Director, British Chem. Plant Manufacturers Assoc.; Director, Food Machinery Assoc.; Member of delegacy, Nat. Coll. Food Technol., Reading Univ. Societies: Inst. Chem. Eng.; Roy. Inst. Chem.; Soc. Chem. Ind.
Nuclear interest: Chemical processing plant.
Address: 14 Suffolk Street, London, S.W.1, England.

HOBSON, R. G., Mechanical Apparatus Supervisor, Bettis Atomic Power Div. Manager, Plant Eng., Westinghouse Elec. Corp. Member, Standards Com., A.N.S.
Address: Westinghouse Electric Corp., Atomic Power Division, P.O. Box 355, Pittsburgh, Pennsylvania 15230, U.S.A.

HOBSON, Robert Marshall, B.Sc., Ph.D. Born 1926. Educ.: Queen's Univ., Belfast. Lecturer, Phys. Dept., Queen's Univ., Belfast 1952-54; Roy. Soc. Warren Res. Fellow, University Coll. London, 1954-57; Consultant, Ministry of Aviation, 1956-; Formerly Phys., Formerly Res. Manager (1957-), now Consultant, Clarke, Chapman and Co. Ltd.
Nuclear interests: The experimental study of magneto-hydrodynamic interactions in fully ionised and partially ionised plasmas with application to thermonuclear reactions and the direct conversion of electrical power from thermal sources.
Address: Clarke, Chapman and Co. Ltd., Victoria Works, Gateshead 8, Co. Durham, England.

HOCHANADEL, Clarence Joseph, Ph.D. Born 1916. Educ.: Indiana and Chicago Univs. Metal. Lab., 1943-46; O.R.N.L., 1946-. Societies: A.C.S.; Rad. Res. Soc.
Nuclear interest: Radiation chemistry.
Address: Oak Ridge National Laboratory, Oak Ridge, Tennessee, U.S.A.

HOCHBERG, Samuel, B.Sc. (Hons. Phys.), B.Sc. (Hons. Maths.), Ph.D., A.R.C.S. Born 1920. Educ.: London Univ. Lecturer in Maths., Imperial Coll. of Sci. and Technol., London Univ., 1954-.

Nuclear interest: Theoretical nuclear physics; Elementary particles and light nuclei; Relevant symmetry, field, and dispersion theories, integral equations and numerical methods.
Address: Imperial College of Science and Technology, London University, South Kensington, London, S.W.7, England.

HOCHMUTH, M. S., Lieutenant Colonel. Commanding Officer, Harry Diamond Lab., Material Command, U.S. Army.
Address: U.S. Army, Material Command, Harry Diamond Laboratories, Walter Reed Army Medical Centre, Forest Glen Section, Forest Glen, Maryland, U.S.A.

HOCHNER, Royal M., Prof. Formerly Assoc. Prof. Mech. Eng., Fenn Coll.; Assoc. Prof. of Mech. Eng., Cleveland State Univ.
Address: Cleveland State University, Cleveland, Ohio 44115, U.S.A.

HOCHREUTINER, René, Eng. lic. jur. Born 1908. Educ.: Geneva Univ., Zurich Polytech. High School and M.I.T. Manager, Kraftwerk Laufenberg and Elektrizitäts-Gesellschaft Laufenberg A.G.; Vice-Pres., Board of Directors, Suisatom Ltd.; Nucl. Services Internat. Ltd. Books: Production d'énergie électronucléaire (Committee on Electric Power, ECE United Nations, 1957); L'économie électrique suisse et la production d'énergie nucléaire (Agence Economique et Financière, 1955). Societies: Nationale Gesellschaft zur Förderung der Industriellen Atomtechnik; Vereinigung Exportierender Elektrizitätsunternehmungen; Verband Schweizerischer Elektrizitätswerke, Zurich.
Address: 303 Baslerstrasse, Laufenburg A.G., Switzerland.

HOCHSTRASSER, Urs, Dr. of Maths., Diploma in Phys. Born 1926. Educ.: Swiss Federal Inst. Technol. Asst. (1950-51), Fellowship (1951-52), U.C.L.A. Appl. Mathematician with private company, 1952-54; Asst. Prof., American Univ. and Guest Worker, N.B.S., 1955-57; Assoc. Prof. and Director of Computation Centre, Kansas Univ., 1957-58; Sci. Counsellor, Swiss Embassy, Washington D.C., 1958-61. 1961 Deleg. for Atomic Energy Matters of Swiss Federal Council; Chairman, Swiss Federal A.E.C.; Formerly Chairman, Steering Com., formerly Vice Chairman, Study Group on Digital Techniques, now Chairman, Board of Management, High Temperature Reactor Project Dragon, E.N.E.A., O.E.C.D.; Pres., European Atomic Energy Soc., 1968-.
Nuclear interests: Reactor computations; Management.
Address: 3 Gürtengasse, Berne 3003, Switzerland.

HOCHWALT, Carroll Alonzo, B.Ch.E., D.S., D.Sc. (Hon., Washington Univ., 1962), D.Sc. (Hon., St. Louis Univ., 1964). Born 1899. Educ.: Dayton Univ. Monsanto Chem. Co.,

1945-64. Director, Central Transformer Corp.; Carboline Co.; Petrolite Corp.; Midwest Res. Inst.; Trustee: Catholic Univ. of America; Charles F. Kettering Foundation; Dayton Univ.; Saint Louis Univ.; Member, President's Council, St. Louis Univ. Societies: A.C.S.; A.I.Ch.E.; Soc. of Chem.Industry; American Inst. of Chemists.
Nuclear interests: Management of contracts from Atomic Energy Commission; Production of radioisotopes.
Address: 7 Upper Ladue Road, St. Louis 24, Missouri, U.S.A.

HOCKENBURY, Robert Wesley, B.S. (Phys. Major), M.S. (Phys.), Ph.D. (Nucl. Sci.). Born 1928. Educ.: Union Coll. and Rensselaer Polytech. Inst. Lab. Asst. (1953-57), Nucl. Analyst (1957-58), K.A.P.L.; Res. Asst. (1958), now Res. Assoc., Rensselaer Polytech. Inst.
Nuclear interest: Nuclear physics-neutron cross-sections.
Address: 8 Oak Drive, Albany 3, New York, U.S.A.

HÖCKER, Karl-Heinz Friedrich, Prof., Dr. rer. nat. Born 1915. Educ.: Marburg and Berlin Univs. Stuttgart Univ., 1948-; Now, Director, Institut für Kernenergetik, Stuttgart. Deutscher Normenausschuss, Vorsitzender des Fachnormenausschusses Kerntechnik. Book: Lexikon der Kern- und Reaktortechnik (Stuttgart, Franckh'sche Verlagshandlung, 1959). Societies: Deutsche Physikalische Gesellschaft; A.N.S.
Nuclear interests: Reactor theory and technology; Energy conversion; Nuclear reactors in space.
Address: Technische Hochschule Stuttgart, Institut für Kernenergetik, Allmandstr., 7 Stuttgart-Vaihingen, Germany.

HOCKER, W. H. Alexander, Ministerialdirigent a.D., Dr. iur. Born 1913. Educ.: Innsbruck, Hamburg and Leipzig Univs. Vorstandsmitglied der Kernforschungsanlage Jülich G.m.b.H., 1961-; Vice-Pres. (1964), Pres. (1965-67), European Space Res.Organisation; Member, Deutsche Kommission für Weltraumforschung, 1964-. Books: Co-editor, Taschenbuch für Atomfragen (1959, 1960/61, 1964, 1968); co-author, Kerntechnik (1958).
Nuclear interests: Research organisation, management; Atomic energy and space law.
Address: Kernforschungsanlage Jülich, Postfach 365, Jülich, Germany.

HODDINOTT, A. C. Mill Superintendent, Gunnar Mines Ltd.
Address: Gunnar Mines Ltd., Uranium City, Saskatchewan, Canada.

HODGE, Harold Carpenter, B.S., M.S., Ph.D. Born 1904. Educ.: Illinois Wesleyan and Iowa State Univs. Prof. Pharmacology, Rochester Univ. School of Medicine, 1946-. Book: Co-author, Pharmacology and Toxicology of Uranium Compounds, Nat. Nucl. Energy

Series VI-1. Parts 1, 2, 3 and 4 (McGraw Hill, 1953). Societies: A.C.S.; American Soc. Biol. Chem.; Soc. Exptl. Biol. and Medicine; American Soc. Pharmacology and Exptl. Therapeutics,Inc.; Soc. Toxicology.
Nuclear interests: Pharmacology and toxicology of uranium, thorium, beryllium, fluorine.
Address: Rochester University, School of Medicine, 260 Crittenden Boulevard, Rochester, New York 14620, U.S.A.

HODGE, Ronald Inglis, B.Sc. (Hons.). Born 1927. Educ.: Edinburgh Univ. S.S.O., Nat. Gas Turbine Establishment, M.O.S., 1949-57; Head, Eng. Res. Branch, Chalk River Nucl. Labs., A.E.C.L., 1957-. Society: A.M.I. Mech.E.
Nuclear interests: Management of engineering research; Applied mechanics; Instrumentation; Fluid dynamics and heat transfer.
Address: Applied Physics Division, Chalk River Nuclear Laboratories, Atomic Energy of Canada Limited, Chalk River, Ontario, Canada.

HODGE, Sir William Vallance Douglas, M.A., Sc.D., D.Sc. (Hon.), Dr. of Laws (Hon.). Born 1903. Educ.: Edinburgh and Cambridge Univs. Loundean Prof. Astronomy and Geometry, Cambridge, 1936-; Fellow (1935-58), Master (1958-), Pembroke Coll., Cambridge. Books: Theory and application of Harmonic Integrals (1952); co-author, Methods of Algebraic Geometry (Vol. I, 1947; Vol. II, 1952; Vol. III, 1954). Papers: Papers on geometry in various British and foreign mathematical journals. Societies: F.R.S.; London Mathematical Soc.; Cambridge Philosophical Soc.; Mathematical Assoc.; Hon.Member, Edinburgh Mathematical Soc.; Hon. Foreign Member, American Acad. Arts and Sci.; Foreign Assoc., N.A.S.; Corresponding Member, Akad. den Wissenschaft, Gottingen; Foreign Member, Roy. Danish Acad.
Address: The Master's Lodge, Pembroke College, Cambridge, England.

HODGES, Fred Jenner, B.S., M.D. Born 1895. At Medical Centre (1931-), Emeritus Prof. and Chairman, Radiol. Dept. (1966-), Michigan Univ. Advisory Com., Biol. and Medicine Div., U.S.A.E.C., 1960-66 (Chairman, 1964-66); Exec. Com., Michigan Memorial Phoenix Project. Societies: American Röntgen Ray Soc.; Radiol. Soc. North America.
Nuclear interests: Biological and medical research.
Address: Michigan University Medical Centre, Ann Arbor, Michigan, U,S.A.

HODGES, The Honourable L. H. Sec., Dept. of Commerce; Member, Federal Rad. Council.
Address: Federal Radiation Council, Executive Office Building, Washington 25, D.C., U.S.A.

HODGINS, John Willard, B.A.Sc., Ph.D. Born 1917. Educ.: Toronto Univ. Prof., Chem. Eng., Roy. Military Coll. of Canada, 1950-56; Dean, Eng., McMaster Univ., 1956-. Director, Professional Affairs, Chem. Inst. of Canada; Member, Nat. Advisory Com. on Appl. and Eng. Res., N.R.C.; Member, Advisory Com. on Sci. and Medicine, Expo '67; Board Member, Mohawk Coll. of Appl. Arts and Technol.; Chairman, Nat. Com. of Deans of Eng. and Appl. Sci.
Nuclear interests: Radiation initiated chemical reactions; Radiation initiated polymerisation.
Address: 25 Joanne Court, Ancaster, Ontario, Canada.

HODGSON, George G., Dr. Sen. Staff, Biophys., Lab. de Fisica Nucl., Inst. de Fisica y Matematicas, Univ. de Chile.
Address: Laboratorio de Fisica Nuclear, Universidad de Chile, 2008 Blanco Encalada, Casilla 2777, Santiago, Chile.

HODGSON, Peter Edward, Ph.D. (London), D.Sc. (London), B.Sc. (London), M.A. (Oxon), A.R.C.S., D.I.C., F. Inst. P. Born 1928. Educ.: London Univ. Lecturer in Phys., Reading Univ., 1957-58; Sen. Res. Officer, Nucl. Phys. Lab., Oxford, 1963-67; Lecturer in Theoretical Phys., Pembroke Coll., Oxford, 1960-62; Sen. Res. Fellow, Corpus Christi Coll., Oxford, 1962-; Univ. Lecturer in Nucl. Phys., Oxford, 1967-. Books: Nucl. Phys. in Peace and War (London, Burns Oates, and New York, Hawthorn Books, Inc., 1961); The Optical Model of Elastic Scattering (Oxford Univ. Press, 1963). Societies: Inst. of Phys. and Phys. Soc.; A.P.S.
Nuclear interests: Theoretical nuclear physics, especially medium energy scattering and reactions, nuclear models and nuclear forces.
Address: Nuclear Physics Laboratory, Oxford, England.

HODNETT, Ernest Matelle, B.S., M.S., Ph.D. Born 1914. Educ.: Florida and Purdue Univs. Prof. Chem., Oklahoma State Univ., Stillwater, Oklahoma, 1957-. Societies: A.C.S.; A.A.A.S.
Nuclear interests: In general, the use of radioisotopes in the study of chemical reactions. In particular, the investigation of reactions of organic compounds by means of the isotope effects of carbon-14 and of hydrogen-3.
Address: Department of Chemistry, Oklahoma State University, Stillwater, Oklahoma, U.S.A.

HODSON, Cecil John, Dr., F.R.C.P. Born 1915. Educ.: London Univ. Director, X-ray Diagnostic Dept., Univ. Coll. Hospital, London. Hon. Sec. and Fellow, Faculty of Radiol. Society: Radio-diagnostic Sect., Roy. Soc. of Medicine.
Address: 11 Perceval Avenue, London, N.W.3, England.

HOECK, Fernand VAN. See VAN HOECK, Fernand.

HOEK, J. VAN DEN. See VAN DEN HOEK, J.

HOEKSEMA, G. D. Deputy Gen. Manager, Tebodin Advies en-Constructiebureau N.V. Address: Tebodin Advies en-Constructiebureau N.V., 61 Laan van Nieuw Oost-Indie, P.O. Box 1029, The Hague, Netherlands.

HOEKSTRA, J. Netherlands Director, and Member, Tech. Com., Eurochemic, Mol. Address: Eurochemic, Mol, Belgium.

HOEKSTRA, R. Phys., Inst. for Nucl. Phys. Res., Amsterdam. Address: Institute for Nuclear Physics Research, 18 Oosterringdijk, Amsterdam O, Netherlands.

HOERNICKE, Heiko, Prof., Dr. vet. med. Born 1927. Educ.: School of Veterinary Medicine, Hanover, and Göttingen Univ. Abteilungsvorsteher, Dept. of Physiology, School of Veterinary Medicine, Hanover, 1954-. Society: Deutsche Physiologische Gesellschaft. Nuclear interests: Study of physiological and metabolic processes in domestic and farm animals with the aid of isotopic tracers. Address: Physiologisches Institut, Tierärztliche Hochschule, 15 Bischofsholer Damm, Hanover, Germany.

HOEVE, Abraham Jacob Cornelis, Chem. Dr. Born 1927. Educ.: Amsterdam Municipal Univ. Radiochem., Isotopecommission, Inst. for Nucl. Res., Amsterdam, 1956-58; Radiochem., Isotope-service, Central Lab. of T.N.O., Delft, 1958-61; Asst. and sci. worker, Lab. of Phys. and Inorganic Chem., Leiden Univ., 1961-. Society: Roy. Dutch Chem. Soc. Nuclear interests: Radiochemistry; Nuclear physics; Applications of radioisotopes in the scientific and technical sciences; Counting apparatuses and methods of all kinds for nuclear radiations. Address: Laboratory of Physical and Inorganic Chemistry, 27 Hugo de Grootstraat, Leiden, The Netherlands.

HÖFER, Rudolf, Universitätsdozent, medical dr. Born 1923. Educ.: Vienna Univ. Sen. Officer, Second Medical Clinic, and Head, Medical Isotopes Inst., Vienna Univ. Member, Board of Consultants, Gesellschaft für Nuclearmedizin. Nuclear interest: Medicine. Address: Second Medical Clinic, Medical Isotope Institute, 13 Garnisongasse, 1090 Vienna, Austria.

HOFF, Harry Summerfield, M.A., B.Sc. Born 1910. Educ.: Cambridge Univ. Civil Service Commission (full-time), 1945-58; Personnel Consultant, U.K.A.E.A. (part-time), 1958-; Personnel Consultant, C.E.G.B. (part-time), 1959-. Nuclear interest: Experience of appointment of scientists and engineers to research and development in nuclear field since 1941. Address: 14 Keswick Road, London, S.W.15, England.

HOFFMAN, Everett John, B.Ch.E., Ph.D. Educ.: Minnesota Univ. At (1950-), Chief, Chem. Sect., Sci. and Technol. Branch (1966-), Div. of Tech. Information Extension, U.S.A.E.C. Society: A.C.S. Nuclear interests: Radiation effects; Chemical separation processes; Metallurgy. Address: United States Atomic Energy Commission, Division of Technical Information Extension, Science and Technology Branch, Chemistry Section, P.O. Box 62, Oak Ridge, Tennessee 37830, U.S.A.

HOFFMAN, J. Member, Study Group on Long Term Role of Nucl. Energy in Western Europe, Corresponding Member, Com. on Reactor Safety Technol., O.E.C.D., E.N.E.A. Address: O.E.C.D. European Nuclear Energy Agency, 38 boulevard Suchet, Paris 16, France.

HOFFMAN, Joseph Gilbert, A.B. (Hons. Phys.), Ph.D. (Exptl. Phys.). Born 1909. Educ.: Cornell Univ. Director of Cancer Res., Roswell Park Inst., Buffalo, 1946-54; Res. Prof. Biophys., Buffalo Univ. School of Medicine, 1947-; Res. Prof. Biophys., Roswell Park Inst., Buffalo, 1955-; Prof. Biophys., Grad School, Buffalo Univ., 1957-; Staff Sci. (1944-46), Consultant (1946-), Los Alamos Sci. Lab.; Consultant to U.S.A.E.C., 1950-. Books: The Life and Death of Cells (1957); The Size and Growth of Tissue Cells (1953). Societies: Fellow, A.P.S.; American Assoc. Cancer Res.; A.A.A.S.; Soc. Exptl. Biol. and Medicine. Nuclear interests: Nuclear physical applications in biological processes; Solid and liquid state physics using nuclear methods. Address: Buffalo University, Department of Physics, Buffalo 14, New York, U.S.A.

HOFFMAN, William H. Manager, Nucl. Equipment Dept., AMF Atomics. Address: AMF Atomics, Whiteford Road, York, Pennsylvania, U.S.A.

HOFFMAN, Zbigniew, Magister inzynier (equivalent to M.Sc.). Born 1929. Educ.: Gdansk Tech. and Warsaw Tech. Univs. Inst. Tele- and Radio- Res., Warsaw, 1953-58; Inst. Nucl. Res., Warsaw, 1958-. Nuclear interests: Multiparameter analysers; Large capacity magnetic core memories; Content addressable systems. Address: 41/44 Miedzynarodowa, Warsaw, Poland.

HOFFMANN, Alexander Peter, B.Sc. (Mech. Eng.), M.Sc. (Nucl. Sci.). Born 1939. Educ.: Technion-Israel Inst. of Technol. Tech. Director, Res. Reactor IRR-1, Soreq Nucl. Res. Centre, Yavne, 1965-. Nuclear interest: Reactor design and operation.

Address: Soreq Nuclear Research Centre, Israel Atomic Energy Commission, Yavne, Israel.

HOFFMANN, Ewald A. H., Dr. of Sci. Born 1905. Educ.: Brussels Free Univ. Expert, Eurochemic, Mol, 1958; Sect. Head, Rad. Dept., Union Minière du Haut-Katanga, Brussels, 1958-59; Director Eng. Office, Belchim, Brussels, 1961-64; Director, Belgo-Nucléaire, Brussels, 1964-. Society: A.N.S. Nuclear interests: Reprocessing plant; Waste treatment and disposal; Nuclear chemical apparatus.
Address: 26 avenue des Hêtres Rouges, Wezembeek, Belgium.

HOFFMANN, F. de. See de HOFFMANN, F.

HOFFMANN, G., Prof. Dr. Sec., Gesellschaft für Nuklearmedizin.
Address: Medizinische Universitätsklinik, 78 Freiburg i. Br., Germany.

HOFFMANN, Günter. Born 1923. Educ.: Berlin, Strassburg, Göttingen and Freiburg Univs. Oberarzt, Medizinischen Univ. Klinik, Freiburg; Prof., Freiburg Univ. Sekretär, Gesellschaft für Nuklearmedizin. Books: Radioisotope in der Hämatologie (Schattauer Verlag, 1963); Radioisotope in der Endokrinologie (Schattauer Verlag, 1964). Nuclear interest: Anwendung von Radioisotopen in der Inneren Medizin.
Address: Medizinische Universität Klinik, Freiburg im Breisgau, Germany.

HOFFMANN, K., Dipl. Ing. Managing Director, Preussische Elektrizitäts A.G.; Member, Steering Com., Deutsches Atomforum e.V.
Address: Preussische Elektrizitäts A.G., 10-12 Papenstieg, Hanover 3000, Germany.

HOFFMANN, Karl-Heinz. Hauptgeschäftsführer, Sozialausschüsse der christlichdemokratischen Arbeitnehmerschaft. Member, Specialised Nucl. Sect. for Economic Problems, Economic and Social Com., Euratom, 1966-.
Address: 8 Brühler Platz, 5 Cologne-Raderthal, Germany.

HOFFMANN, Klaus, Ing. Member, Sen. Staff engaged in Nucl. Work, Abteilung Strahlungsmesstechnik, Reichert-Elektronik G.m.b.H. und Co. K. G.
Address: Reichert-Elektronik G.m.b.H. und Co. K.G., Postfach 743, Trier/Petrisberg, Germany.

HOFFMANN, L. C. Maritime Administration Member, Maritime Administration and U.S.A.E.C. Joint Liaison Com. on Nucl. Merchant Ships.
Address: U.S. Department of Commerce, Maritime Administration, Washington D.C. 20545, U.S.A.

HOFFMANN, Paul Otto, B.S. (Newark College of Eng.), A.M. (Columbia), Ph.D. (New York). Born 1909. Educ.: Newark Coll. of Eng., Columbia and New York Univs. Chairman, Dept. of Phys. and Mechanics, Newark Coll. of Eng. Societies: A.S.E.E.; A.S.P.T.; A.A.A.S.
Nuclear interests: As Chairman of Nuclear Engineering Committee at Newark College of Engineering, supervising the nuclear programme; Teaching nuclear engineering courses.
Address: 6 Park Avenue, Maplewood, New Jersey, U.S.A.

HOFFMAN, Tibor Andrews, Ph.D. (Budapest), Candidate Phys. Sci., Dr. Phys. Sci. Born 1922. Educ.: Budapest Univ. Head, Microwave Dept., Res. Inst. Telecommunication, Budapest, 1952-65; Head, Computer Centre, Chem. Industries, Budapest, 1965-; Nucl. Chem. Lab., Central Res. Inst. Phys., 1959-; Former Adviser for solid states, United Incandescent Lamp Co. Vice-Sec., Eötvös Loránd Phys. Soc., Hungary; Sub-group for Fundamental and Appl. Res., Hungarian A.E.C.; Biophys. Council and Atomic Shell Com., Hungarian Acad. Sci.
Nuclear interests: Nuclear solid state; Biophysics; Economic aspects of nuclear sciences. Biophysics; Economic aspects of nuclear sciences.
Address: 36 Maros utca, Budapest 12, Hungary.

HOFFMEISTER, Wolfgang, Dr. Diplomchemiker. Born 1930. Educ.: Cologne, Germany. Kernforschungsanlagen des Landes Nordrhein-Westfalen Jülich, 1958-61; I.B.M. Labs., Böblingen, Germany, 1961-.
Nuclear interests: Radiochemistry; Activation analysis; Geochemistry.
Address: I.B.M. Laboratories, 10 Schönaicher First, Böblingen Württbg., Germany.

HOFMAN, G. J. Member, Advisory Com., N.V. Gemeenschappelijke Kernenergie centrale Nederland.
Address: N.V. Gemeenschappelijke Kernenergie centrale Nederland, 310 Utrechtseweg, Arnhem, Netherlands.

HOFMANN, C. S. Manager - AIG Project, Knolls Atomic Power Lab., operated for U.S.A.E.C. by Gen. Elec. Co.
Address: Knolls Atomic Power Laboratory, Schenectady, New York, U.S.A.

HOFMANN, Ernst-Günter, Dr., Diplom-Physiker. Born 1923. Educ.: Stuttgart T.H. and Justus Liebig-Univ., Giessen. Now with Algemeine Elektrizitäts-Gesellschaft AEG - Telefunken, Hamburg. Societies: Deutsche Physikalische Gesellschaft e.V.; Deutscher Arbeitsgemeinschaft Vakuum (DAGV).
Nuclear interests: Development and construction of large radiation sources, especially high-power X-ray equipment; Applications of large radiation sources in industry – e.g.,

food preservation, radiation chemistry and sterilisation of pharmaceuticals and medical supplies; Radiation dosimetry.
Address: 37 Heidrehmen, 2 Hamburg 55, Germany.

HOFMANN, Peter L., B.E.E., M.S., D.Eng. Sci. Born 1925. Nucl. Eng. (1951-55), Manager, Nucl. Eng. Unit, Submarine Intermediate Reactor Project (1955-57), Manager, Nucl. Analysis Unit, Nucl. Destroyer Project (1957-59), Consulting Nucl. Eng. (1960-61), K.A.P.L.; Tech. Consultant (1961-63), Manager, Eng. Phys. and Nucl. Analysis Groups (1963-67), Manager, Fast Test Reactor Phys. (1967-), Battelle-Northwest.
Nuclear interests: Reactor theory and nuclear design.
Address: 2904 S. Benton Place, Kennewick, Wash., U.S.A.

HOFSTADTER, Robert, B.S. (magna cum laude), M.A., Ph.D. LL.D. (Hon.), D.Sc. (Hon.), Laurea h.c. Nobel Prize in Phys., 1961. Born 1915. Educ.: C.C.N.Y. and Princeton Univ. Assoc. Prof., Phys. (1950-54), Prof., Phys. (1954-), Director, High Energy Phys. Lab. (1967-), Stanford Univ. Assoc. Editor, Reviews of Modern Phys., 1958-61; Co-editor, Investigations in Phys., 1951-; Editorial Com., Nuovo Cimento, 1963-; Board of Governors, Weizmann Inst. of Sci., Rehovoth, 1967-. Books: Co-author, High Energy Electron Scattering Tables (Stanford Univ. Press, 1960); Editor, Electron Scattering and Nucl. and Nucleon Structure (Benjamin, 1963); co-editor, Nucleon Structure (Stanford, Stanford Univ. Press, 1964). Societies: Fellow, A.P.S.; Fellow, Phys. Soc. of London; Italian Phys. Soc.; Fellow, A.A.A.S.; N.A.S.
Nuclear interests: Nuclear physics; High energy physics.
Address: Department of Physics, Stanford University, Stanford, California, U.S.A.

HOFSTATTER, Anton, Dipl. Ing., Dr. techn. Dept. Head, Osterreichische Draukraftwerke A.G.
Address: Osterreichische Draukraftwerke A.G., 50 Anzengruberstrasse, Klagenfurt, Austria.

HOGAN, Aloysius Joseph, Jr., B.S. (Chem.), M.S. (Chem.). Born 1928. Educ.: Holy Cross Coll., Worcester, Mass. and Rensselaer Polytech. Inst. Teaching Asst. (1950-52, 1954), Res. Assoc. (1952-54), Dept. of Chem., Rensselaer Polytech. Inst.; Eng., Philadelphia Elec. Co., 1954-55; Eng., Atomic Power Development Assoc., Inc., 1955-63; Sen. Eng., Res. Div., Philadelphia Elec. Co., 1963-. Vice Chairman, Delaware Valley Sect., A.N.S. Society: A.C.S.
Nuclear interests: Research, development and design of nuclear reactor plants for production of electric power, especially the sodium cooled fast breeder type and the helium cooled, graphite-moderated high temperature type; nuclear and physical chemistry.
Address: 1221 Concord Avenue, Drexel Hill, Pennsylvania, U.S.A.

HOGAN, William Sanford, M.S. Born 1925. Educ.: Missouri School of Mines. Reactor Eng., Monsanto Chem. Co., 1952-56; Principal Phys., Battelle, 1956-58; Asst. Div. Consultant (1958-62), Formerly Res. Assoc. (1962-), Battelle Memorial Inst., now Div. Consultant, Battelle-Columbus Laboratories. Books: Contributing author to U.S. Res. Reactors (Addison-Wesley, 1958); Contributing author to Neutron Absorber Materials for Reactor Control (U.S.A.E.C., 1962). Society: A.N.S.
Nuclear interests: Reactor design, experimental reactor physics (critical assembly studies) and neutron physics in regard to reactor design problems.
Address: 3015 Mountview Road, Columbus, Ohio, U.S.A.

HOGARTH, Cyril Alfred, B.Sc., Ph.D. Born 1924. Educ.: Lond. Univ. Prof. and Head, Phys. Dept., Brunel Univ.; Chairman, Membership Sub-Com., Inst. Phys. Book: Co-author, Techniques of Non-destructive Testing (1960); Materials used in Semiconductor Devices (1965). Society: F. Inst. P.
Nuclear interests: Metallurgy; Non-destructive testing.
Address: Brunel University, Woodlands Avenue, Acton, London, W.3, England.

HØGDAHL, Ove Tormod, Cand. real. Born 1933. Educ.: Oslo Univ. Now at Central Inst. for Ind. Res., Oslo. Society: Norwegian Chem. Soc.
Nuclear interests: Use of radioisotopes in research and industry; especially neutron activation analysis and use of radioisotopes in analytical chemistry.
Address: Central Institute for Industrial Research, 1 Forskningsveien, Blindern, Oslo, Norway.

HOGG, Benjamin Gregory, B.Sc. (Hons.), M.A., Ph.D. Born 1924. Educ.: Manitoba, Wesleyan and McMaster Univs. Sci. Defence Res. Board, 1949-51; Assoc. Prof., Roy. Military Coll., 1954-57; Assoc. Prof., Manitoba Univ., 1957-. Councillor, Canadian Assoc. Phys., 1957-. Societies: A.P.S.; Canadian Assoc. Phys.
Nuclear interest: Positronium.
Address: 1587 Wolseley Avenue, Winnipeg, Manitoba, Canada.

HOGG, Ian Henry, B.Sc. (Hons.). Educ.: Glasgow Univ. Welding Eng., Div. of Atomic Energy, M.O.S., 1946-54; Head of Tech. Administration, R. and D. Branch and Reactor Eng. Lab., Risley (1954-63), Deputy Head, Tech. Secretariat (1963-66), Special Asst. to Managing Director (1966-), U.K.A.E.A. Societies: A.M.I.Mech.E.; A.M.I.E.E.
Nuclear interest: Management of research and development.

HOG

Address: Reactor Group, United Kingdom Atomic Energy Authority, Risley, near Warrington, Lancs., England.

HOGG, K. B. Director, Electronic Instruments Ltd.
Address: Electronic Instruments Ltd., Richmond, Surrey, England.

HOGLUND, Barton McMichael, B.S., M.S. Born 1928. Educ.: Kansas and Stanford Univs. Assoc. Mech. Eng., Argonne Nat. Lab. Societies: A.S.M.E.; RESA.
Nuclear interests: Heat transfer, particularly two-phase heat transfer; Fast reactor design and fast reactor safety.
Address: Argonne National Laboratory, 9700 South Cass Avenue, Argonne, Illinois, U.S.A.

HOGNESS, Thorfin R., B.S. and Chem. Eng. (Minnesota), Ph.D. (California). Born 1894. Educ.: Minnesota and California Univs. Prof., Dept. of Chem., Chicago Univ.; Director, Chicago Midway Labs., Chicago Univ.; Consultant, U.S.A.E.C. Society: A.C.S.
Nuclear interest: Industrial applications of isotopes.
Address: 5755 South Kenwood Avenue, Chicago 37, Illinois, U.S.A.

HOGREBE, Kurt, Dr. phil. Born 1912. Educ.: Münster and Leipzig Univs. Head, Isotopenlabor, Max-Planck-Inst. Göttingen; Isotopenlabor of Kernreaktor-Bau-u. Betriebs-G.m.b.H., Karlsruhe, 1957-64;Head, Health Phys. Gesellschaft für Kernforschung m.b.H.; Karlsruhe, 1964-.
Nuclear interests: Health physics; Radioactive isotopes; Production of isotopes and application.
Address: Abteilung Strahlenschutz und Dekontamination, Gesellschaft für Kernforschung m.b.H., Karlsruhe-Leopoldshafen, Reaktorgelände, Germany.

HÖHLE, Reinhard, Dr. rer. nat. Born 1924. Educ.: Rostock Univ. Rostock Univ., 1951-55; Inst. für angewandte Radioaktivität Leipzig, Deutsche Akademie der Wissenschaften zu Berlin, 1955-.
Nuclear interests: Dosimetry; Health physics.
Address: 10 Schwindstrasse, 705 Leipzig, German Democratic Republic.

HOHMUTH, Karl, Dipl. Phys., Dr. Ing. Born 1929. Educ.: Karl-Marx-Univ., Leipzig and Dresden Tech. Univ. Arbeitsgruppenleiter, Zentralinst. für Kernforschung, Rossendorf.
Nuclear interest: Nuclear physics.
Address: Deutsche Akademie der Wissenschaften zu Berlin, Forschungsgemeinschaft, Zentralinstitut für Kernforschung, 8051 Dresden, PF 19, German Democratic Republic.

HOHN, Hans, Dr. phil., o.ö. Prof. Born 1906. Educ.: Vienna, Heidelberg and Erlangen Univs. several appointments in Austrian chem. industry, Austrian ERP Office, Austrian Ministries of Industry and Economical Planning, 1944-50; o.ö. Prof., Vienna Tech. Univ., 1950; Chief Manager, Austrian chem. industry, 1958-. Director, Inst. Technol. and Inorganic Materials, Vienna Tech. Univ.; Chief Manager, Austrian Nitrogen Fertiliser Works, Linz; Pres., and Chairman of Control Board, Austro-Chematom; Vice-Dean, Faculty of Nature Sci., Vienna Tech. Univ. Book: Polarography (1936). Societies: M.N.Y.A.S.; Deutsche Bunsengesellschaft; Gesellschaft Deutscher Chemiker; Verein österreichischer Chemiker; Gesellschaft für Natur und Technik.
Nuclear interests: Metallurgy; Fuel elements; Radiation chemistry.
Address: 11 Zehenthofgasse, Vienna 19, Austria.

HØJERUP, Carl Frank, M.Sc. Born 1933. Educ.: Copenhagen Tech. Univ. Danish Atomic Energy Commission, 1958-.
Nuclear interests: Reactor development; Reactor physics.
Address: Danish Atomic Energy Commission, Risö Research Establishment, Risö, Roškilde, Denmark.

HOJMAN, Jolanda, Dr. Chem. Born 1901. Educ.: Berlin and Göttingen Univs.; Belgrade Univ. (extraord. prof.). Societies: Chem. Soc. Belgrade; Chem. Soc. Croatia.
Nuclear interests: Analytical chemistry of nuclear impurities; Radiochemistry.
Address: Faculty of Pharmacy, 35 Deligradska, Belgrade, Yugoslavia.

HOKE, George Robert, A.B., M.S., Ph.D. Born 1921. Educ.: Vanderbilt and North Carolina Univs. Phys., Savannah River Lab., E.I. du Pont de Nemours and Co., 1951-58; Prof. and Head, Nucl. Eng. Dept., Mississippi State Univ., 1958-64; Prof. and Asst. Dean, School of Technol., Southern Illinois Univ. Societies: A.N.S.; A.P.S.; A.A.P.T.
Nuclear interests: Nuclear and reactor physics; Radioisotopes techniques; Cosmic rays.
Address: School of Technology, Southern Illinois University, Carbondale, Illinois, U.S.A.

HOKR, Josef, Dipl. Chem. Eng. Born 1919. Educ.: Prague Tech. Univ. Director, Inst. for Res., Production and Application of Radioisotopes, Prague. Member, Czechoslovak Commission of Atomic Energy. Societies: A.C.S.; Gesellschaft Deutscher Chemiker.
Nuclear interest: Radioisotope and radiation applications in research and industry.
Address: 13 Pod vyhlidkou, Prague 5, Czechoslovakia.

HOLADAY, Duncan Asa, B.S., M.A. Born 1907. Educ.: Oregon State Coll. In U.S.P.H.S., 1943-; Chief, Occupational Health Field Station, 1953-. Director, A.I.H.A. Societies: A.I.H.A.; H.P.S.; American Conference of Governmental Industrial Hygienists; American Public Health Assoc.

Nuclear interest: Health physics.
Address: Occupational Health Field Station, U.S. Public Health Service, Box 2539, Fort Douglas Station, Salt Lake City, Utah 84113, U.S.A.

HOLAHAN, Frederick Stanley, Ph.D., B.A., M.S. Born 1929. Educ.: Rutgers Univ., Newark Colleges and Stevens Inst. Technol. Chem. (1951-56), Res. Chem. (1956-), Picatinny Arsenal. Society: A.C.S.
Nuclear interests: Use of radioactive isotopes as a radiation source and/or tracer in the study of the fundamental properties of explosives, propellants and polymers.
Address: Star Route, Morristown, New Jersey, U.S.A.

HOLBROOK, Philip. Former Station Supt., Trawsfynydd Nucl. Power Plant, now Station Supt., Wylfa Nucl. Power Station, C.E.G.B.
Address: C.E.G.B., Wylfa Nuclear Power Station, Wylfa, Anglesey, Wales.

HOLBROOK, R. K. Director, Gen. Services Div., Atomics Internat.
Address: Atomics International, Division of North American Aviation Inc., P.O. Box 309, Canoga Park, California, U.S.A.

HOLBROW, Charles H., B.A., A.M., M.S., Ph.D. Born 1935. Educ.: Wisconsin and Columbia Univs. Asst. Prof., Haverford Coll., 1962-65; Res. Investigator, Pennsylvania Univ., 1965-66; Assoc. Prof., Colgate Univ., 1967-. Assoc. Editor, Phys. Today. Societies: A.P.S.; A.A.P.T.; A.A.U.P.
Nuclear interests: Fast neutron spectroscopy; Charged particle spectroscopy with broad range magnetic spectrograph; Direct reactions.
Address: Colgate University, Hamilton, New York, U.S.A.

HOLDEN, Abe Noel, B.S. (Chem. Eng.), M.S. (Metal. Eng.). Born 1921. Educ.: Wisconsin Univ. Manager, Phys. Metal., K.A.P.L., 1954-56; Manager, Metallurgy and Ceramics, G. E. Co. Vallecitos Atomic Lab., 1956-67; Manager, Fuel and Materials, Westinghouse Astronucl. Lab., 1967-. Nucl. Metal. Com., A.I.M.E. Books: Phys. Metal. of Uranium (Reading, Massachusetts, Addison-Wesley, 1958); Dispersion Fuel Elements (New York, Gordon and Breach, 1968). Societies: A.S.T.M.; American Ceramic Soc.; American Electrochem. Soc.; A.S.M.; A.I.M.M.E.
Nuclear interests: Metallurgy; Reactor design; Radiation effects.
Address: 2225 Country Club Drive, Pittsburgh, Pennsylvania 15241, U.S.A.

HOLDEN, Robert B., Dr. With United Nucl. Corp. Book: Ceramic Fuel Elements (New York, Gordon and Breach Sci. Publishers, 1967).
Address: United Nuclear Corporation, 1730 K Street, N.W., Washington 6, D.C., U.S.A.

HOLE, Njal, Dr. tech. Born 1914. Educ.: Tech. Univ. Norway. Prof. Phys., 1954-.
Nuclear interests: Radioactivity; Nuclear reactions (low energy).
Address: Fysisk Instituut, NTH, Trondheim, Norway.

HOLFORD, Lord William Graham, M.A., D.C.L., Dr. of Letters, Dr. of Laws, Fellow, Roy. Inst. British Architects, Member, Town Planning Inst. Born 1907. Educ.: Liverpool Univ. Part-time Member, C.E.G.B.; Prof. Town Planning, London Univ. Societies: Fellow, Roy. Inst. British Architects; Town Planning Inst.
Address: 20 Eccleston Square, London, S.W.1, England.

HOLIFIELD, Chet. Born 1903. Congressman from State of California: elected to 78th - 87th Congresses. Member, Pres. Special Evaluation Commission on Atomic Tests at Bikini Atoll; Member, Joint Com. on Atomic Energy, 1946- (Chairman, 1961-62, 1965-); Member, Com. on Govt. Operations, Com. on Post Office and Civil Service; Member, Commission on Organisation of Exec.Branch of Govt.; Congressional Adviser on U.S. Deleg. to I.C.P.U.A.E., Geneva, 1955; rep., Joint Com. on Atomic Energy at 1st organisational meeting of I.A.E.A., Vienna, Oct. 1957.
Nuclear interests: Peacetime application.
Address: Montebello, California, U.S.A.

HOLLAENDER, Alexander, A.B., M.A., Ph.D., D.Sc. h.c. (Vermont, Leeds and Marquette). Born 1898. Educ.: Wisconsin Univ. Director (1946-66), Sen. Res. Adviser (1966-), Biol. Div., O.R.N.L. Hon. Pres., Comité International de Photobiologie, 1960-; Photobiology Com., Biol. and Agriculture Div., Nat. Res. Council, 1956-. Hon. Pres., Internat. Assoc. Rad. Res., 1964-. Books: Editor, Rad. Biol., vol. i (New York, McGraw-Hill, 1954); editor, Rad. Biol., vol. ii (1955); editor, Radiation Biology, vol. iii (1955); co-author, Effects of Rad. on Bacteria, pages 365-430 in Rad. Biol., vol. ii (New York, McGraw-Hill, 1955); editor, Rad. Protection and Recovery (Pergamon Press, Ltd., 1960). Societies: N.A.S.; Rad. Res. Soc.; Genetics Soc. America; American Soc. Bacteriologists; Internat. Assoc. Rad. Res.
Nuclear interests: Effects of radiation on mutation production; Long-term implications of exposure to atomic radiation.
Address: Biology Division, Oak Ridge National Laboratory, P.O. Box Y, Oak Ridge, Tennessee, U.S.A.

HOLLAND, Lieutenant Colonel Donald B., M.S.C. U.S. Army. Head, Phys. Sci. Dept., Armed Forces Radiobiol. Res. Inst., Nat. Naval Medical Centre.
Address: Armed Forces Radiobiology Research Institute, National Naval Medical Centre, Bethesda 14, Maryland, U.S.A.

HOLLAND, F. V. Maintenance and Construction, A.E.C.L.
Address: Atomic Energy of Canada, Ltd., Chalk River Project, Chalk River, Ontario, Canada.

HOLLAND, Joshua Zalman, B.S. (Maths.). Born 1921. Educ.: Chicago Univ. Graduate work in phys. and maths., Tennessee Univ., 1950-52; Graduate work in atmospheric sci., Washington Univ. (Seattle), 1964-65; Meteorologist in Charge, Oak Ridge Area Weather Bureau Office, U.S.A.E.C., Oak Ridge; Research meteorologist and civil defence coordinator, Weather Bureau Central Office, U.S.A.E.C., Washington D.C. Past Exec. Sec., U.S.A.E.C. Advisory Com. on Reactor Safeguards and Staff Consultant on problems of radioactive contamination of the air due to peaceful uses of atomic energy; Consultant on Meteorological Aspects of the Effects of Atomic Rad., N.A.S.
Nuclear interests: Micrometeorology; Turbulence; Atmospheric diffusion; Air pollution; Reactor hazards; Circulation of the atmosphere; Radioactive fallout; Atmospheric radioactivity preduction and measurement.
Societies: American Meteorological Soc.; A.G.U.; A.A.A.S.
Address: Fallout Studies Branch, Division of Biology and Medicine, United States Atomic Energy Commission, Washington D.C. 25, U.S.A.

HOLLAND, Leslie Arthur, D. ès Sc., F.Inst.P., F.S.G.T. Born 1921. Director, Res., Central Res. Lab., Edwards High Vacuum Internat. Ltd.; Assoc. Reader, Phys. Dept., Brunel Univ. Chairman, Vacuum Phys. Group, Inst. Phys. and Phys. Soc.
Nuclear interest: Vacuum systems and phenomena occurring in ultra-high vacuum.
Address: "Hazelwood", Balcombe Road, Pound Hill, Crawley, Sussex, England.

HOLLAND, Lester Kaye, B.S. (E.E.). Born 1926. Educ.: Utah Univ. Eng. (1952), Advanced Eng. Programme (1953-55), G.E.C. Society: I.E.E.E.
Nuclear interests: Boiling water reactor control, performance, and stability studies, including hydrodynamic consideration; Applications of digital and analog computers for systems analysis and for plant data logging and control; Research and development programme technical administration.
Address: General Electric Company, Atomic Power Equipment Department, 175 Curtner Avenue, San Jose, California, U.S.A.

HOLLANDER, Jack M., B.Sc., Ph.D. Born 1927. Educ.: Ohio State and California Univs. At Lawrence Rad. Lab., California Univ. 1948-. Society: A.P.S.
Nuclear interests: Nuclear physics; Nuclear spectroscopy; Radioactive isotopes; Nuclear data compilation.
Address: California University, Lawrence Radiation Laboratory, Berkeley, California, U.S.A.

HOLLANDS, George. Pres., Bar Ray Products Inc.
Address: Bar Ray Products Inc., 209 25th Street, Brooklyn 32, New York, U.S.A.

HOLLECK, Helmut, Dr. Phil. Born 1939. Educ.: Hamburg and Vienna Univs. Kernforschungszentrum Karlsruhe, 1965-. Societies: Planseegesellschaft für Pulvermetallurgie; Deutsche Gesellschaft für Metallkunde.
Nuclear interest: Metallurgy: constitution of systems containing fuel-compounds and fission product compounds. Thermodynamic properties of such compounds.
Address: Institut für Material und Festkörperforsch, Kernforschungszentrum, 75 Karlsruhe, Germany.

HÖLLER, Hugo, Dr. med. vet., Dipl. trop. med. Born 1929. Educ.: Munich Univ. and Veterinary Coll., Hanover. Res. Asst., Dept. of Physiology, Veterinary Coll. Hanover; seconded to Sen. Lecturer Physiol. and Biochem., Faculty Veterinary Sci., Khartoum Univ.
Nuclear interest: Application of radioisotopes in medicine and biology, especially metabolic studies.
Address: Khartoum University, P.O. Box 32, Khartoum-North, Sudan.

HOLLEY, George M., Jr. Board of Trustees, Power Reactor Development Co.
Address: Power Reactor Development Co., 1911 First Street, Detroit 26, Michigan, U.S.A.

HOLLIDAY, P. D. Manufacturing Div. B. Manager, Burlington A.E.C. Plant, operated for U.S.A.E.C. by Mason and Hanger-Silas Mason Co. Inc.
Address: U.S.A.E.C., Burlington A.E.C. Plant, Burlington, Iowa, U.S.A.

HOLLIDAY, W. C., B.Sc., A.M.I.Mech.E. Eng. Manager, Whessoe Ltd.
Address: Whessoe Ltd., Darlington, England.

HOLLINGSWORTH, Dorothy F., B.Sc., F.R.I.C., M.I.Biol. Member, Managing Subcom. on Monitoring, M.R.C.
Address: Medical Research Council, 20 Park Crescent, London W.1, England.

HOLLINGSWORTH, Guilford L., B.S. (Elec. Eng.), M.S. (Elec. Eng.), D.S. (Hon., Pacific Lutheran Univ.). Born 1918. Educ.: Oregon Univ. and Oregon State Coll. Director, Boeing Sci. Res. Labs., Seattle, 1959-. Member, Governor's Advisory Council on Nucl. Energy and Rad., State of Washington; Member, Com. on Sci. and Technol., United States Chamber of Commerce; Member, Advisory Board, Coll. of Eng., Washington State Univ.; Member, Visiting Com., Dept. of Aeronautics and Astronautics, Washington Univ.

Nuclear interests: Management; Industry-government relations.
Address: Boeing Scientific Research Laboratories, P.O. Box 3981, Seattle, Washington 98124, U.S.A.

HOLLINGSWORTH, Robert E. A.B. Born 1918. Educ.: Columbia Univ. Deputy Gen. Manager (1959-64), Manager (1964-), U.S.A.E.C.
Address: 4750 Chevy Chase Drive, Chevy Chase, Maryland, U.S.A.

HOLLIS, Roy F., B.S. (Mining Eng.). Born 1911. Educ.: Oregon State Coll. and Idaho Univ. Pres., Atlas Minerals Div. and Vice-Pres., Atlas Corp. Society: A.I.M.M.E.
Nuclear interests: Raw material processing; Nuclear reactor research.
Address: 910 Security Life Building, Denver, Colorado 80202, U.S.A.

HOLLOWAY, Bruce William, B.Sc. (Adelaide), Ph.D. (California Inst. Technol.), D.Sc. (Melbourne). Born 1928. Educ.: Adelaide Univ. Res. Fellow in Microbial Genetics, Australian Nat. Univ., 1953-57; Sen. Lecturer in Bacteriology (1957-60), Reader in Microbial Genetics (1961), Melbourne Univ. Prof. Genetics, Monash Univ., 1968. Societies: Australian Rad. Soc.; Australian Genetics Soc.; Australian Soc. of Biochem.; Australian Soc. Microbiology.
Nuclear interests: Biological effects of radiation, with particular reference to genetic effects and repair mechanisms; Genetic basis of radiation sensitivity in bacteria and bacteriophage; Relationship of radiation repair mechanisms to normal cellular functions.
Address: Faculty of Science, Monash University, Clayton, Victoria 3168, Australia.

HOLLOWAY, John Thomas, A.B., Ph.D. Born 1922. Educ.: Millikin Univ. (Illinois), M.I.T.; Maryland and Iowa State Univs. Nucl. Phys. Branch, Office of Naval Res., Washington, D.C., 1946-53; Ames Lab. of U.S.A.E.C., 1954-57; Office of Sci., Office of Director of Defence Res. and Eng., Dept. of Defence, Washington, D.C., 1958-61; Deputy Director, Grants and Res. Contracts (1961-67), Chief, Advanced Programmes and Technol., Space Applications (1967-), N.A.S.A., Washington, D.C. Societies: Phys. Soc.; A.A.A.S.;A.I.A.A.; A.P.S.
Nuclear interests: Management; Nuclear physics; Accelerators; Detectors; Space research.
Address: 2220 Cathedral Avenue, Washington D.C. 20008, U.S.A.

HOLLOWAY, Pierre E. Vice Pres., Iberian Gulf Oil Co., Madrid; Vice Pres., Gulf Gen. Atomic, 1967-.
Address: Gulf General Atomic, P.O. Box 608, San Diego 12, California, U.S.A.

HOLLSTEIN, Martin, Dr. rer. nat. Born 1933. Educ.: Johannes Gutenberg Univ. Mainz. At Kernforschungszentrum Karlsruhe, Inst. für Radiochemie, 1960-.
Nuclear interests: Nuclear reactions, especially fission of heavy nuclei: range-energy relations, mass- and charge-distribution of fission fragments.
Address: 28 Jahnstrasse, 7521 Neuthard, Germany.

HOLLUTA, Josef, Dr. tech., Dipl. Ing. Born 1895. Educ.: Brünn T.H. Ord. Prof., Lehrstuhl für Wasserchemie, Karlsruhe T.H.; Direktor, Abt. Wasserchemie, Inst. für Gastechnik, Feuerungstechnik und Wasserchemie, Karlsruhe T.H.; Strahlenschutzkommission beim Bundesministerium für Atomkernenergie und Wasserwirtschaft; DVGW/VGW/ATV/FW-Kommission Radioaktive Substanzen und Wasser; Deutsche Gesellschaft für Kerntechnik, Düsseldorf, Wissenschaftlicher Beirat; Strahlenschutzkommission Baden-Württemberg, Stuttgart, Wissenschaftlicher Beirat; Reaktor-Bau- und Betriebs-GmbH Karlsruhe, Wissenschaftlicher Beirat. Societies: Gesellschaft Deutscher Chemiker/Fachgruppe Wasserchemie; Deutscher Verein von Gas- und Wasserfachmännern; Abwassertechnische Vereinigung.
Nuclear interests: Dekontamination von wasser, wasserversorgung kerntechnischer betriebe.
Address: 21 Durlacher Strasse, Ettlingen bei Karlsruhe, Germany.

HOLM, Bror Herman, M.B.A., M.A. Born 1919. Educ.: Stockholm School of Economics and Stockholm Univ. Asst. Gen. Manager, Gen. Atomic Europe. Society: Roy. Acad. Eng. Sci., Stockholm.
Nuclear interests: Industrial, commercial, financial and organisational aspects.
Address: Gulf General Atomic Europe, Weinbergstrasse 109, Zurich 8006, Zwitserland.

HOLM, Lennart W., Ph.D. (Chem.). Born 1925. Educ.: Roy. Inst. Technol., Stockholm. Assoc. Res. Director (1957-63), Res. Director (1963-), Res. Inst. Nat. Defence, Sweden. Swedish Atomic Res. Council. Societies: Swedish Chem. Soc.; Chem. Assoc., Roy. Inst. Technol.; Swedish Assoc. Eng. and Architects.
Nuclear interests: Nuclear chemistry; Actinide chemistry; Nuclear metallurgy; Hot laboratory design and equipment; Analysis of nuclear fuels; Plutonium production.
Address: Research Institute of National Defence, Sundbyberg 4, Sweden.

HOLM, Niels Wilhelm, M.Sc. (Chem. Eng.). Born 1933. Educ.: Tech. Univ. Denmark. At Nordisk Insulinlaboratorium, 1958; At Danish A.E.C. Res. Establishment Risö, 1958-. (Head, Accelerator Dept., 1965-). Associateship from United States N.A.S. at Rad. Branch, Natick Labs., Massachusetts, 1964. Society: Inst. Danish Civil Eng.

Nuclear interests: Radiation chemistry; Chemical dosimetry; Industrial irradiation processing.
Address: Accelerator Group, Danish Atomic Energy Division, Research Establishment Risö, pr. Roskilde, Denmark.

HOLM, Warren M., B.S. Born 1913. Educ.: Long Island Univ. Vice-Pres., Radium Chem. Co. Inc. Society: A.S.T.M.-S.N.T.
Nuclear interests: Management; Isotopes research.
Address: Radium Chemical Co. Inc., 161 East 42nd Street, New York 17, New York, U.S.A.

HOLMAN, Colin B., M.D. Sect. of Diagnostic Roentgenology, Mayo Clinic and Mayo Foundation.
Address: Mayo Clinic and Mayo Foundation, Rochester, Minnesota, U.S.A.

HOLMAN, William Prout, M.B., B.S., F.R.A.C.P., F.F.R. (Lond.), F.C.R.A. Born 1899. Educ.: Melbourne Univ. Recently Medical Director, Cancer Inst. Board. Nat. Rad. Advisory Com. Australia.
Nuclear interest: Medical radiotherapy.
Address: 278 William Street, Melbourne 3000, Victoria, Australia.

HOLMES, Charles R., Dr. Sen. Geophys., Supervisor of Tritium Lab., New Mexico Inst. Mining and Technol.
Address: New Mexico Institute of Mining and Technology, Socorro, New Mexico, U.S.A.

HOLMES, David K., D.Sc., A.B. Born 1919. Educ.: Carnegie Inst. Technol. and Kansas Univ. Theoretical Phys., Solid State Div., O.R.N.L., 1949-. Book: Co-author, Reactor Analysis (McGraw-Hill, 1960). Society: A.P.S.
Nuclear interests: Reactor theory; Fast neutron spectra; Radiation damage to reactor materials.
Address: Solid State Division, Oak Ridge National Laboratory, P.O. Box X, Oak Ridge, Tennessee, U.S.A.

HOLMES, John Ernest Raymond, Ph.D., B.Sc. Born 1925. Educ.: Birmingham Univ. Chief Phys., U.K.A.E.A., Atomic Energy Establishment, Winfrith, 1966-.
Address: U.K. Atomic Energy Authority, A.E.E., Winfrith, Dorchester, Dorset, England.

HOLMES, John Winspere, B.Sc., M.Sc. Born 1921. Educ.: Sydney and Adelaide Univs. Appointed to C.S.I.R.O., Div. of Soils, 1950; studied at Cambridge for one year, 1955. Societies: A. Inst. P.; Assoc., Australian Inst. of Phys.
Nuclear interests: Application of isotope and radiation techniques in agricultural physics and hydrological research; Application of isotopes and radiation for measurements in agricultural physics – e.g., tracers for fluid and nutrient flow, water flow in acquifers, dating of underground waters by tritium and ^{14}C methods, and adsorption-desorption phenomena in leaching.
Address: Commonwealth Scientific and Industrial Research Organisation, Division of Soils, Private Bag No. 1, Glen Osmond, South Australia.

HOLMES, Raymond Leslie. Born 1926. Manager, Nucl. Eng. Dept., Clarke, Chapman and Co. Ltd., 1959-.
Nuclear interest: Steam generating plant for nuclear power stations.
Address: Clarke, Chapman and Company Limited, Victoria Works, Gateshead 8, County Durham, England.

HOLMES, Roger A. Formerly Assoc. Prof., Elec. Eng., Purdue Univ.; Director of Res., Rad. Dynamics Inc., 1964-.
Address: Radiation Dynamics Inc., Westbury Industrial Park, Westbury, Long Island, New York, U.S.A.

HOLMES, William Louis. Born 1917. Works Sec., Springfield Works (1951), Works Sec., Windscale Works (1954), Chief Establishment Officer, Ind. Group (1955), Asst. Director, Establishments, Ind. Group (1957), Deputy Director, Personnel, D. and E. Group (1959), Deputy Director, Personnel and Administration, Reactor Group, Risley (1961), Sec., A.E.E., Winfrith (1961-65), Deputy Director, Personnel and Administration, Reactor Group, Risley (1965-), U.K.A.E.A.
Address: 1 Park Crescent, Appleton, Warrington, Lancs., England.

HOLMGREN, Harry Dahl, Bachelor of Phys., M.A., Ph.D. Born 1928. Educ.: Minnesota Univ. U.S. Naval Res. Lab., 1954-61; Assoc. Prof. (1961-65), Prof., and Director, Cyclotron Project (1965-), Maryland Univ. Societies: A.P.S.; Washington Acad. Sci.
Nuclear interest: Experimental nuclear physics, nuclear reactions and structure.
Address: Maryland University, Department of Physics and Astronomy, College Park, Maryland, U.S.A.

HOLMQUIST, Carl-Eric Ragnar, Chem. Eng., M.Sc. Born 1921. Educ.: Roy. Inst. Technol., Stockholm and Harvard Univ. Head, Health and Safety Branch, Elec. and Thermal Eng. Div., Swedish State Power Board, 1956; Assoc. Prof. Ind. Hygiene, Roy. Inst. Technol., Stockholm, 1956. Societies: Ergonomics Res. Soc.; A.I.H.A.; Internat. Permanent Com. on Occupational Health.
Nuclear interests: Health and safety.
Address: Swedish State Power Board, Fack, Vällingby 1, Sweden.

HOLOWAYCHUK, Nicholas, B.Sc., M.Sc., Ph.D. Born 1907. Educ.: Alberta, California and Ohio State Univs. Prof., Ohio State Univ.; Formerly Prof., Ohio Agricultural Experiment

Station; Formerly Project Leader, Dept. of Agronomy, Ohio Agricultural R. and D. Centre. Societies: Soil Science Soc. of America; Indian Soil Science Soc. of America; Indian Soil Science Soc.; A.A.A.S. Nuclear interest: Disposition of fission materials found on land surface.
Address: Department of Agronomy, Ohio State University, Columbus 10, Ohio, U.S.A.

HOLROYD, Louis Vincent, B.A. (British Columbia), M.A. (British Columbia), Ph.D. (Notre Dame). Born 1925. Educ.: British Columbia and Notre Dame Univs. Prof. and Chairman, Phys. Dep., Missouri Univ., 1950-. Society: A.P.S.
Nuclear interests: Nuclear and electron resonance spectroscopy and radiation damage in solids.
Address: Physics Department, Missouri University, Columbia, Missouri, U.S.A.

HOLSTE, Werner, Prof., Dr.-Ing. Born 1927. Educ.: Aachen Tech. and Münster Univs. Managing Director, Demag A.G., Duisburg. Fachkommission III Kerntechnik, Deutsche Atomkommission. Society: Deutsches Atomforum e.V.
Nuclear interests: Reactor design; Management.
Address: Demag Aktiengesellschaft, Wolfgang-Reuter-Platz, 41 Duisburg, Germany.

HOLSTI, Lars R., Born 1926. Educ.: Helsinki Univ. Assoc. Prof., Radiotherapy, Radiotherapy Clinic, and Director, Radiotherapy Dept., Helsinki; Assoc. Prof., Helsinki Univ. Societies: Finnish Soc. of Nucl. Medicine; Scandinavian Soc. of Radiol.
Nuclear interest: Diagnosis and treatment of cancer.
Address: Radiotherapy Clinic, University Central Hospital, Helsinki 29, Finland.

HOLT, John Riley, Ph.D., F.R.S. Born 1918. Educ.: Liverpool Univ. Prof. of Exptl. Phys., Liverpool Univ. Society: Inst. Phys. and Phys. Soc.
Nuclear interest: Particle physics experiments with a 4 GeV electron synchrotron.
Address: Physics Department, Liverpool University, Liverpool, England.

HOLT, Norman, Cand. real. Born 1921. Educ.: Oslo Univ. Sen. Phys., Inst. for Atomenergi, Kjeller, 1956-; Principal, Netherlands-Norwegian Reactor School Kjeller, 1958-; Member, Neutron Data Compilation Centre Com., O.E.C.D., E.N.E.A. Societies: Norwegian Phys. Soc.; Norwegian Polytech. Soc.
Nuclear interest: Neutron physics.
Address: Institutt for Atomenergi, Kjeller, Norway.

HOLTE, Johan B. Born 1915. Educ.: Norges Tekniske Hoyskole, Trondheim. Pres., Norsk Hydro-Elektrisk Kvaelstofaktieselskab.
Address: Norsk Hydro, 2 Bygdoy Alle, Oslo 2, Norway.

HOLTE, Per Gunnar, Prof., Ph.D. (Phys.). Born 1920. Educ.: Uppsala Univ., Head, Dept. Reactor Phys. (1953-61), Director of Res. (1961-), A.B. Atomenergi.
Nuclear interest: Nuclear physics.
Address: A.B. Atomenergi, Studsvik, Nyköping, Sweden.

HOLTEBEKK, Trygve, Dr. philos. Born 1922. Educ.: Oslo Univ. Univ. lecturer. Society: Norsk Fysisk Selskap.
Nuclear interests: Nuclear physics; Low energy reactions.
Address: Physics Department, University of Oslo, Blindern, Norway.

HOLTER, Heinz, Ph.D. (Chem.). Born 1904. Educ.: Vienna Univ. Head, Cytochem. Dept. (1942-56), Prof. and Head, Physiological Dept. (1956-), Carlsberg Lab., Copenhagen. Treas., Roy. Danish Acad. of Sci. and Letters; Formerly Vice Pres., Internat. Soc. for Cell Biol. Societies: Histochem. Soc.; Biol. Soc., Copenhagen.
Nuclear interest: Biological tracer work.
Address: 10 Gamle Carlsbergvej, Copenhagen Valby, Denmark.

HOLTER, Norman J., M.A., M.S. Member, Consulting Editorial Board, J. Nucl. Medicine.
Address: 333 N. Michigan Avenue, Chicago 1, Illinois, U.S.A.

HOLTHUSEN, Hermann, Prof. Dr., Dr. rer. nat. h.c. Born 1886. Educ.: Heidelberg and Hamburg Univs. o. Prof. Hamburg, 1952. Books: Einführung in die Röntgenologie, Therapeutischer Teil (1933); Dosimetrie der Röntgenstrahlen (1934). Societies: I.C.R.P.; I.C.R.U.; German Atomic Commission.
Nuclear interest: Radiation protection.
Address: 25 Badestrasse, Hamburg 13, Germany.

HOLTON, Gerald, B.A., M.A., Ph.D. Born 1922. Educ.: Harvard Univ. Prof. Phys., Harvard Univ., 1958-. Chairman, N.E. Sect., A.P.S.; Council, History of Sci. Soc.; Council, Federation of American Scientists. Society: American Acad. of Arts and Sci.
Nuclear interest: Ultrasonic techniques for samples under high pressures.
Address: Jefferson Physical Laboratory, Harvard University, Cambridge 38, Mass., U.S.A.

HOLTSINGER, Clarence Eugene, B.S. (Mech. Eng.), M.S. (Mech. Eng.), M.S. (Aeronautical Eng.). Born 1921. Educ.: Florida and Connecticut Univs. and Rensselaer Polytech. Inst. Formerly Asst. Chief Eng., Connecticut Aircraft Nucl. Engine Lab., Pratt and Whitney Aircraft, Div. United Aircraft Corp., 1951-. Member, N.A.S.A. Com. on Nucl. Energy Processes, Nucl. Energy Com. of Nat. Assoc. of Manufacturers. Societies: Soc. of Automotive Eng.; Air Force Assoc.
Nuclear interests: Management; Reactor development.

Address: 10 Orchard Road, West Hartford, Connecticut, U.S.A.

HOLTSMARK, Johan Peter, Dr. philos. (Oslo). Born 1894. Educ.: Oslo, Würzburg, Leipzig, Göttingen and London Univs. Prof. Phys., Oslo Univ. Societies: Kgl. Norske Videnskabers Selskab; Videnskapsakademiet, Oslo.
Nuclear interest: Nuclear theory.
Address: Fysisk Institutt, Blindern, Norway.

HOLWAY, Lowell Hoyt, Jr., A.B. (Maths.), A.M. (Phys.), Ph.D. (Appl. Maths., Harvard). Born 1931. Educ.: Dartmouth and Harvard Univs. Societies: A.N.S.; A.P.S.
Nuclear interest: Neutron transport theory.
Address: Raytheon Co., Research Division, Waltham, Massachusetts, U.S.A.

HOLY, Z., Dipl. Ing. (Prague), M.Sc. (Birm.), M. Eng. Sci. (N.S.W.). At School of Nucl. Eng., New South Wales Univ.
Address: New South Wales University, Box 1, Post Office, Kensington, New South Wales 2203, Australia.

HOLZ, Guenter, Ing. Sen. Officer, Nucl. Dept., Philips Perunana S.A.
Address: Philips Perunana S.A., 1268 Alfonso Ugarte, Lima, Peru.

HOLZER, Franz Stephan, M.E.E. Born 1921. Educ.: Louisiana State Univ. and Polytech. Inst. Brooklyn. Control Eng., A.M.F.-Savannah River Project, 1952-54; Manager, Reactor Eng. Sect. (1954-60), R. and D. Co-ordinator (1960-), A.M.F. Atomics. Societies: A.N.S.; RESA.
Nuclear interests: Reactor development; Radiation effects on organisms and materials; Nuclear hazards analysis; Nuclear weapons effects; Control and instrumentation.
Address: 44 Saddle Hill Road, Stamford, Connecticut, U.S.A.

HOLZER, Helmut, Dipl. Chem., Dr. rer. nat. Born 1921. Educ.: Munich Univ. Dozent, Biochem., Hamburg Univ., 1953-57; Prof. of Biochem. and Director, Inst. of Biochem., Freiburg Univ., 1957-. Vorsitzender, Arbeitskreis Biologie und Medizin, Deutsches Atomkommission; Mitglied, Berufungsausschuss, Regensburg Univ. Book: Molekulare Biologie des malignen Wachstums, (17. Colloquium, Gesellschaft für Physiologische Chemie, 1966, Mosbach/Baden) (Heidelberg-New York, Springer-Verlag Berlin, 1966). Societies: Gesellschaft für Biologische Chemie; Gesellschaft Deutscher Chemiker; A.C.S.; Gesellschaft Deutscher Naturforscher und Ärzte; New York Acad. of Sci.; Roy. Soc. of Medicine; Biochem. Soc.
Address: Biochemisches Institut, 7 Hermann Herder Strasse, 78 Freiburg i.Br., Germany.

HOLZGRAF, Dean Bonner, B.S. (Chem.). Born 1919. Educ.: California (Berkeley) Univ.

Reactor Operations, Oak Ridge Nat. Lab., 1947-51; Supt., M.T.R. Operations, Phillips Petroleum Co., Idaho Falls, Idaho, 1951-56; Supervisor, Reactor Operations, Union Carbide Nucl. Co., New York 1957-. Member, Reactor Operations Div., A.N.S., 1966-. Societies: A.C.S.; A.N.S.
Address: 129 Morton Drive, Ramsey, New Jersey 07446, U.S.A.

HOMEISTER, Orville Ernest, B.S., M.S.E. Born 1923. Educ.: Wayne and Michigan Univs., and Illinois Inst. Technol. Nucl. Power Eng., Detroit Edison Co., Detroit. Societies: A.S.M.E.; A.N.S.
Nuclear interests: Fuel element design, fabrication inspection; Testing of nuclear reactor components; Nuclear power plant design and economic evaluation.
Address: 1787 Oakwood Drive, Trenton, Michigan, U.S.A.

HOMES, Marcel. Recteur, Brussels Free Univ. Formerly Member, Conseil d'Administration, Inst. Interuniv. des Sci. Nucl.
Address: Brussels Free University, 50 avenue F. D. Roosevelt, Brussels 5, Belgium.

HOMMEL, Carlton Olmsted, B.S. (Chem.), M.S. (Chem.). Born 1929. Educ.: Clarkson Coll. Technol. and Rensselaer Polytech. Inst. At Anton Electronic Labs., 1957-59; Tracerlab., a Div. of L.F.E., Inc., 1959-63; G.C.A. Corp., 1963-. Societies: New England Ch., Air Pollution Control Assoc.; A.C.S.
Nuclear interests: Research on gas-solid reactions involving their application to high sensitivity radiochemical release systems, and the application of radioactive rare-gas clathrates; Radiation protection.
Address: 134 Oak Crest Drive, Framingham Centre, Massachusetts, U.S.A.

HONCARIV, Robert, Univ. Doz., R.N.Dr., C.Sc. Born 1931. Educ.: Komensky's Univ., Bratislava. Botanical Garden, Slovak Acad. Sci., Kosice; Faculty of Natural Sci., P. J. Safarik's Univ., Kosice. Com. of Czechoslovak Sci. and Tech. Soc., Bratislava; Regional Com. Czech Tech. Sci. Soc., Kosice; Com., Biol. Soc. Societies: Czechoslovak Sci. and Tech. Soc.; Slovak Acad. Sci.; Czechoslovak Medical Soc. J. E. Purkyne.
Nuclear interests: Radiogenetics, especially chronic irradiation and low dose genetic effects.
Address: P. J. Safarik's University, Kosice, Czechoslovakia.

HONDA, Hidehisa. Asst. Chief, Reactor Regulation Sect., Atomic Energy Bureau, A.E.C. of Japan.
Address: Atomic Energy Commission of Japan, 3-4 Kasumigaseki, Chiyoda-ku, Tokyo, Japan.

HONDA, Takashi. Chief 2nd Dept., Isotope Dept., Cancer Res. Centre, Nat. Kanazawa Hospital.

Address: Isotope Department, Cancer Research Centre, National Kanazawa Hospital, 76 Shimoishibiki-cho, Kanazawa City, Ishikawa Prefecture, Japan.

HONDT, Henri D'. See D'HONDT, Henri.

HONECK, Henry Charles, B.S. (Phys.), Sc.D. (Nucl. Eng.). Born 1930. Educ.: Rensselaer Polytech. Inst. and M.I.T. With Pratt and Whitney Aircraft, 1952-56; Brookhaven Nat. Lab., 1959-. Society: A.N.S. Nuclear interest: Reactor theory. Address: Theoretical Reactor Physics, Brookhaven National Laboratory, Upton, L.I., New York, U.S.A.

HONG, Moon Wha, B.S., M.S., M.D., Ph.D. Born 1916. Educ.: Seoul Nat. Univ., Korea and Purdue Univ. Prof. and Dean, Coll. of Pharmacy, Seoul Nat. Univ., 1954-; Member, Tech. Consulting Board, Office of Atomic Energy, Republic of Korea, 1959-; Editor in Chief, J. of Korean Pharmaceutical Soc.; Vice-Pres., Pharmaceutical Soc. of Korea. Society: Pharmaceutical Soc. of Korea. Nuclear interests: Radioisotope tracer technique in pharmaceutical science; Activation analysis; Radio-pharmaceuticals. Address: 28 Yunkun-dong, Chong-ro-Ku College of Pharmacy Bdg., S.N.U., Seoul, Korea.

HONIG, Arnost, Doc. habil., Ing. (Civil Eng.), Dr.Sc.techn. Born 1928. Educ.: Brno Tech. Univ.; Course of nucl. phys. and technics, Brno J. E. Purkyne's Univ. Course of application of radioisotopes, Inst. Nucl. Res., Czechoslovak Acad. Sci. Asst., Sanitary Eng. Dept., Brno Tech. Coll., 1947-51; Sen. Asst. Dept. for Concrete and Bridges (1951-62), Asst. Prof., Building Materials and Testing Methods Dept. (1962-), Brno Tech. Univ.; Head, Central Defectoscopy Centre for Building Industry and Materials, Czechoslovak Ministry of Education, 1962-. Member, editorial board, tech. bulletin Inzenyrske stavby (Civil Eng.; Nucl. Industry Building), 1959-, and the periodical Veda a zivot (Science and Life), 1963-; co-editor, Eng. Compendium on Rad. Shielding (I.A.E.A.), 1963-; Member, Editorial Board, Nucl. Eng. and Design, formerly Nucl. Structural Eng., 1964-. Books: Radioisotopes and Rad. Sources in Building Industry and Res. Habilitation thesis (Brno Technical Univ. 1961); Nondestructive Testing (Prague, SNTL, 1966). Societies: Czechoslovak Sci. Assoc. (Cs. V.T.S.), sect. for Concrete and Masonry Structures, sect. for Application of Radioisotopes, section for Precasting and Prestressed Concrete; A.N.S.; A.S.T.M. Nuclear interests: Radioisotopes in building research, reactor and laboratory design, reactor containment buildings, prestressed concrete pressure vessels. Photon attenuation, cross sections and build-up factors. Nuclear shielding. Nuclear, physical and mechanical properties and technology of shielding materials. Concrete with ordinary, barite, magnetite, limonite iron punchings and other aggregates. Testing of reactor shields. Address: 176 Merhautova, Brno 14, Czechoslovakia.

HONIGMAN, Jacobo ZENDER. See ZENDER HONIGMAN, Jacobo.

HONKAJUURI, Paavo. Chairman, Administrative Board, Atomienergia Oy. Address: Atomienergia Oy, 13 Ristolantie, Helsinki, Finland.

HONKARANTA, H. E. Reino Ilmari. Head, Finnish Mission to the E.E.C. and Euratom, 1968-. Address: Finnish Mission, c/o Euratom, 51-53 rue Belliard, Brussels, Belgium.

HÖNNERSCHEID, Georg. Chief Eng., Gebr. Böhling. Address: Gebr. Böhling, 118 Grossmannstrasse,Hamburg 27, Germany.

HONSTEAD, John Frederick, B.S. (Chem. Eng.), M.S. (Chem. Eng.). Born 1922. Educ.: Washburn Univ., M.I.T., Kansas State and Texas A. and M. Univs. Radiol. Eng. (1950-55), Supv., Geochem. and Geophys. Res. (1955-62), Sen. Officer, Health, Safety and Waste Disposal Div., I.A.E.A., Vienna 1962-64; Tech. Specialist, Rad. Protection, Hanford Lab., 1964-65; Tech. Specialist, Rad. Protection (1965-67), Manager, Environmental Studies Sect. (1967-), Battelle-Northwest Lab. Publicity Com., H.P.S. Societies: A.I.Ch.E.; A.N.S.; H.P.S.; A.A.A.S.; Rad. Res. Soc.; Pacific N.W. Pollution Control Assoc. Nuclear interests: Treatment and disposal of radioactive wastes, particularly the release of low-level effluents to the environment; Environmental research and evaluation, radiation protection standards and controls. Address: Battelle-Northwest Laboratories, P.O. Box 999, Richland, Washington 99352, U.S.A.

HONT, Maurice Docile Emile D'. See D'HONT, Maurice Docile Emile.

HOOD, St. Clair Clinton, M.A. Born 1921. Educ.: Oxford Univ. Central Statistical Office, 1948; Board of Trade, 1950-56; Office of the Registrar of Restrictive Trading Agreements, 1957; London Office, U.K.A.E.A., 1958-. Nuclear interests: Commercial policy: diversification. Address: Flat 2, Quex Court, West End Lane, London, N.W.6, England.

HOOF, Jean VAN. See VAN HOOF, Jean.

HOOGWATER, Jan, M.A. (Economics). Born 1921. Educ.: Rotterdam Univ. Director-Gen. for Internat. Affairs, Ministry of Social

Affairs and Public Health.
Address: Ministry of Social Affairs and Public Health, The Hague, Netherlands.

HOOKE, Sir Lionel A. Managing Director, Amalgamated Wireless (A'sia) Ltd.; Formerly Member, Business Advisory Com., now Member, Advisory Com., A.E.C.
Address: Amalgamated Wireless (A'sia) Ltd., 47 York Street, Sydney, N.S.W., Australia.

HOOPER, Howard C. Deputy Director, Eng. and Construction Div., Chicago Operations Office (-1967), Area Manager, Palo Alto Area Office (1967-), U.S.A.E.C.
Address: U.S.A.E.C., San Francisco Operations Office, 2111 Bancroft Way, Berkeley 4, California, U.S.A.

HOOTON, David John, M.Sc. (N.Z.), Ph.D. (Edin.), F. Inst. P. Born 1926. Educ.: Auckland, Edinburgh and Munich Univs. Res. Asst., Max-Planck Inst., Göttingen, 1954-55; Lecturer: Aberdeen Univ., 1955-56; Liverpool Univ., 1956-59; Wellington Univ., 1959-62; Prof. of Theoretical Phys., and Head, Phys. Dept., Auckland Univ., 1963-. Societies: Royal Soc. of New Zealand; Inst. Phys.; Phys. Soc.
Nuclear interests: Theoretical nuclear physics, in particular direct interactions and deuteron polarisation.
Address: Department of Physics, Auckland University, P.O. Box 2175, Auckland, New Zealand.

HOOVER, John Irvin, A.B. (Phys.). Born 1911. Educ.: American Univ. Phys., U.S. N.R.L., 1939-67. Book: Co-author, Liquid Thermal Diffusion, (Washington, D.C., Dept. of Commerce, 1958). Societies: A.P.S.; Washington Acad. Sci.; Philosophical Soc. of Washington; A.A.A.S.; RESA; A.G.U.
Nuclear interests: Nuclear physics; Nuclear instrumentation; Application of nuclear techniques to oceanography.
Address: U.S. Naval Research Laboratory, Washington D.C. 20390, U.S.A.

HOOVER, Reynold Leonard, B.A. Born 1927. Educ.: Illinois Univ. Manager, Health and Safety, Nucl. Div., Combustion Eng., Inc., Windsor, Connecticut. Pres., Windsor Health Council, Windsor, Connecticut. Pres., New England Sect., A.I.H.A.; Rad. Sessions Arranger, American Ind. Hygiene Conference 1962; Certified Health Physicist, 1960; Councilman, Connecticut Chapter, H.P.S.; Founder, American Bioassay and Analytical Chem. Group. Societies: A.I.H.A.; U.S.A.S.I.; H.P.S.; N.E. Sect. A.I.H.A.; Connecticut Chapter, H.P.S.
Nuclear interests: Industrial hygiene; Health physics; Radiochemistry, bio-assay and safety.
Address: 74 Robin Road, Windsor, Connecticut, U.S.A.

HOPE, George Stanley, M.I.E.E. Born 1925. Asst. Chief Eng. (Instruments), Springfields, U.K.A.E.A.; previously at Capenhurst and Risley.
Nuclear interest: Management.
Address: 59 Singleton Avenue, St. Annes, Lytham-St.-Annes, Lancashire, England.

HOPE-JONES, Ronald Christopher, B.A. Born 1920. Educ.: Cambridge Univ. U.K. Resident Rep. to I.A.E.A., 1964-67; Head, Atomic Energy and Disarmament Dept., Foreign Office, 1967-.
Address: Foreign Office, London, S.W.1, England.

HOPE, Peter. Architect, Brodsky, Hopf, and Adler.
Address: Brodsky, Hopf, and Adler, 300 E 40 Street, New York, N.Y. 10016, U.S.A.

HOPKIN, Graham Llewellyn, B.Sc. Born 1909. Educ.: Univ. Coll. of South Wales and Monmouthshire. Asst. Director, and Chief, Materials, A.W.R.E., U.K.A.E.A. Society: F.I.M.
Nuclear interest: Nuclear materials.
Address: United Kingdom Atomic Energy Authority, Atomic Weapons Research Establishment, Aldermaston, Berks., England.

HOPKINS, Donald Walter, B.Sc., M.Sc. (Wales) (Hons. Metal.). Born 1913. Educ.: Univ. Coll., Swansea and Wales Univ. Lecturer, Univ. Coll. Swansea, 1943-51; Gen. Manager, Western Metal. Industries, Ltd., 1951-53; Head, Dept. Metal. and Chem., Swansea Coll. Technol. 1954-. Member, City and Guilds Metal. Panel; Member of Com. South Wales Sect., Inst. Metals; Past Pres., Swansea and District Metal. Soc.; Member, Central Advisory Council for Education, Wales. Societies: Fellow, Inst. Metallurgists; Assoc. Member, Inst. Mining and Metal.; Member, Inst. Metals.
Nuclear interests: Use of 5C Caesium 137 source for non-destructive testing; Radiochemistry.
Address: 23 Eaton Crescent, Uplands, Swansea, South Wales.

HOPKINS, Harold Horace, Ph.D., D.Sc., D. ès Sc. (Hon.). Born 1918. Educ.: London Univ. Prof., Appl. Optics, Reading Univ. Society: F. Inst. P.
Address: Department of Applied Physical Sciences, Reading University, Reading, Berks., England.

HOPKINS, Ivor, B.Sc. (Hons.). Born 1912. Educ.: Wales Univ. Group Personnel Officer, Reactor Group (1961-67), Head, Staffing Branch, Res. Group, Harwell, U.K.A.E.A.
Address: 'Mericona', Collaroy Road, Coldash, Newbury, Berks., England.

HOPKINS, Theodore Louis, B.S., M.S. (Oregon State), Ph.D. (Kansas State). Born 1929. Educ.: Oregon State and Kansas State

Univs. Entomologist, Entomology Res. Div., United States Dept. of Agriculture, 1953-57; Instructor of Entomology (1958-60), Asst. Prof. (1960-63), Assoc. Prof. (1963-67), Kansas State Univ. Societies: A.C.S.; Entomological Soc. of America; A.A.A.S.
Nuclear interests: Synthesis and use of radioisotope labelled molecules in insect biochemistry; Physiology and metabolism of insecticides.
Address: Entomology Department, Kansas State University, Manhattan, Kansas 66502, U.S.A.

HOPPER, Arthur F. Prof., Zoology, Rad. Sci. Centre, Rutgers State Univ.
Address: Rutgers State University, New Brunswick, New Jersey, U.S.A.

HOPPER, Victor David, D.Sc. (Melbourne). Born 1913. Educ.: Melbourne, Perth, and Birmingham Univs. Nuffield Fellow, Birmingham Univ., 1949-50; Sen. Lecturer (1944), Reader (1957), Melbourne Univ.; Prof. Phys. R.A.A.F. Acad., 1961-; Dean, Faculty of Sci., Melbourne Univ., 1963. Society: F. Inst. P. Nuclear interests: Nuclear physics with particular emphasis on ultra high energy interactions; Cosmic radiation and plasma fields.
Address: c/o Physics Department, R.A.A.F. Academy, Melbourne University, Parkville N.2, Australia.

HOPPMANN, William H. II, B.S. (Coll. of Charleston), M.A. (George Washington), Ph.D. (Columbia). Born 1908. Director of Res., Material Lab., New York Naval Shipyard, U.S. Navy, 1938-47; Assoc.Prof. and Consultant in Appl. Mechanics, Johns Hopkins Univ., 1947-57; Prof. Appl. Mechanics, Rensselaer Polytech. Inst., 1957-. Societies: A.S.M.E.; S.E.S.A.; Soc. of Rheology.
Nuclear interests: Thermal stresses; Reactor containment design; Explosive loading of structures; Rheological mechanics; Shock and vibrations.
Address: Rensselaer Polytechnic Institute, Troy, New York, U.S.A.

HORAK, Zdenek, D.Sc. Born 1898. Educ.: Charles Univ., Prague. Prof. and Head, Phys. Dept., Tech. Univ., Prague. Book: Introduction to Molecular and Atomic Phys. (Prague, Stat. Nakl. Tech. Lit., 1957). Societies: Commission of Tech. Nucleonics; Sté. Française de Physique; Gesellschaft für Angewandte Mathematik und Mechanik.
Nuclear interests: Nuclear physics; Isotopes.
Address: Technical University, 4 Technická, Prague 6, Czechoslovakia.

HORAN, John R., B.S. (Hons., Phys.). Born 1923. Educ.: Loyola and Vanderbilt Univs. Sen. Health Phys., American Cyanamid Co., Idaho Chem. Processing Plant, 1952-53; Rad. Eng., Test Eng., Supv. of Ind. Hygiene, Westinghouse Elec. Corp., Atomic Power Div., U.S.S. Natilus Prototype, Naval Reactor Facility, Idaho, 1953-57; Director, Health and Safety Div. (1957-67), Director, Operational Safety and Tech. Support Div. (1967-), Idaho Operations Office, U.S.A.E.C. Member, Exec. Council, Internat. Rad. Protection Assoc., 1966-74; Board of Directors (1962-68), Pres. (1967), H.P.S.
Nuclear interests: Emergency planning; Reactor siting; Applied health physics.
Address: United States Atomic Energy Commission, Idaho Operations Office, P.O. Box 2108, Idaho Falls, Idaho 83401, U.S.A.

HORAN, Richard F., B.A. (Chem.). Born 1931. Educ.: LaSalle Coll. Society: A.N.S. Nuclear interests: Radiation chemistry; Radiation polymerisation; Surface grafting; Radiation effects.
Address: Quantum, Inc., Lufbery Avenue, Wallingford, Conn., U.S.A.

HORI, Noriyoshi. Director, Mitsubishi Atomic Power Industries Inc.
Address: Mitsubishi Atomic Power Industries Inc., Otemachi Building, 4 Otemachi 1-chome, Chiyoda-ku, Tokyo, Japan.

HORI, Sanyo, Bachelor of Phys. Born 1912. Educ.: Hokkaido Univ. Prof., Phys. Lab., Toho Univ., Japan. Book: The Treatment of X-ray. (Kanda, Tokyo, Seibundo-Shinkosha, 1968). Societies: Japan Soc. of Appl. Phys.; Atomic Energy Soc. of Japan.
Nuclear interests: Radiation of polymers and living cells.
Address: Shimotakaido-4-chome, 870-50, Suginami-ku, Tokyo, Japan.

HORI, Shin. Vice Chairman, Japan Atomic Ind. Forum Inc.
Address: Japan Atomic Industrial Forum Inc., 1-13, 1-chome, Shimbashi, Minatoku, Tokyo, Japan.

HORI, Shiro, Dr. Chief, Div. of Agricultural Products, Dept. of Appl. Biol., Rad. Centre of Osaka Prefecture.
Address: Department of Applied Biology, Radiation Centre of Osaka Prefecture, Shinke-Cho, Sakai, Osaka, Japan.

HORI, Sumio, D.Sc. Born 1915. Educ.: Tokyo Imperial Univ. Director, Res. Dept., Agency of Ind. Sci. and Technol., 1950; Director, Atomic Energy Office, Ministry of Internat. Trade and Industry, 1955; Deputy Director, Atomic Energy Bureau, Agency of Sci. and Technol., 1956; Counsellor, Japanese Embassy in Paris, 1959; Director, Japan Nucl. Ship Development Agency, 1963-. Book: Atomic Energy (Sci. and Tech.. Publisher, 1957). Societies: I.M.M. of Japan; Japan Geological Soc.
Nuclear interests: Nuclear economics; Nuclear administration in each country; Nuclear development in connection with demand and supply of energy.

HOR

Address: 1-105, Kamiogikubo, Suginami, Tokyo, Japan.

HORI, Ushio, Dean, Economic Dept. and Formerly Sen. Officer, Atomic Energy Res. Assoc., Kanto Gakuin Univ.
Address: Kanto Gakuin University, Mutsuura, Kanazawa-ku, Yokohama, Japan.

HORII, Seisho. Born 1906. Educ.: Hikone Commercial Coll. Managing Director, Nippon Atomic Industry Group Co., Ltd., Tokyo; formerly Sub-Manager, Administrative Dept., Mitsui Bussan Kaisha, Ltd.
Nuclear interest: Management of nuclear industry.
Address: 482 Zaimokuza, Kamakura, Kanagawa Prefecture, Japan.

HORIKOSHI, Teizo. Formerly Managing Director, formerly Head, Com. on Development Programmes, now Vice Chairman, Japan Atomic Ind. Forum; Sec.-General, Federation of Economic Organisations; Formerly Director, now Councillor, Fund for Peaceful Atomic Development of Japan.
Address: Fund for Peaceful Atomic Development of Japan, No. 1-13, 1-chome, Shimbashi, Minato-ku, Tokyo, Japan.

HORIO, Masao, Prof. Director, Eng. Res. Inst., Dept. of Nucl. Eng., Kyoto Univ.
Address: Kyoto University, Department of Nuclear Engineering, Yoshida-Honmachi, Sakyo-ku, Kyoto, Japan.

HORION, P. Sciences Juridiques, Inter-Faculty Centre of Nucl. Sci., Liège Univ.
Address: Inter-Faculty Centre of Nuclear Sciences, Liège University, Liège, Belgium.

HORKSTRA, W. G., Prof., Member, Rad. Safety Com., Wisconsin Univ.
Address: Wisconsin University, Madison 6, Wisconsin, U.S.A.

HORLACHER, Hans Heinrich v. See v. HORLACHER, Hans Heinrich.

HORLITZ, Ferdinand Wilhelm Gerhard, Dr. rer. nat. Born 1926. Educ.: Halle and Bonn Univs. At Dept. Phys., Bonn Univ., 1953-57; Deutsches Elektronen Synchrotron, Hamburg, 1957-61; C.E.N., Saclay, 1961-64; Deutsches Elektronen Synchrotron, Hamburg, 1964-.
Nuclear interests: Accelerators; High energy physics; Bubble chambers; Mass-spectrometry.
Address: 6f Wientapperweg, 2 Hamburg 55, Germany.

HÖRMANDER, Olof, Civilingenjör (Chem. Eng.). Born 1923. Educ.: Royal Inst. Technol. Formerly Director of Production, A.B. Atomenergi, Sweden, 1962-. Managing Director, The Axel Johnson Inst. for Ind. Res., Rederi A.B. Nordstjernan.
Nuclear interests: Chemical engineering; Metallurgy; Management; Uranium-, plutonium-, D_2O- production, capacity and cost figures.
Address: Rederi A.B. Nordstjernan, The Axel Johnson Institute for Industrial Research, Nynäshamn, Sweden.

HORN, E. L. VAN. See VAN HORN, E. L.

HORN, George, Ph.D., B.Sc. Tech., M.Inst.F. Born 1928. Educ.: Sheffield Univ. Res. Eng., Atomic Energy Div., G.E.C., 1955-58; Head, Thermodynamics Sect., Marchwood Eng. Labs. (1958-65), Head, Eng. Div., Berkeley Nucl. Labs. (1965-), C.E.G.B.
Nuclear interests: Reactor engineering; Heat transfer.
Address: Central Electricity Generating Board, Berkeley Nuclear Laboratories, Berkeley, Gloucestershire, England.

HORN, Walter Gualtiero, degree in phys. Born 1922. Educ.: Padova Univ. Chief Eng. in charge of Electronic Dept., Sorin Nucl. Centre, Saluggia (Vercelli, Italy). Prof. Electronics and Electronic Measurements, Istituto Tecnico Industriale Statale A. Volta, Trieste, 1946-57. Society: Full member, Com. 45, Comitato Elettrotecnico Italiano (Ionised particles measurements).
Nuclear interests: Electronic control of nuclear reactors and power plants; Electronic instruments for nuclear physics measurements.
Address: Electronic Department, SORIN-Societa Richerche Impianti Nucleare, Saluggia, Vercelli, Italy.

HORNBECK, J. A. Pres., Sandia Lab.
Address: Sandia Laboratory, P.O. Box 5800, Albuquerque, New Mexico, U.S.A.

HORNBECK, Lyle W. Member, Board of Atomic and Space Development Authority, New York State, 1965-.
Address: Bond, Schoeneck and King, 1000 State Tower Building, Syracuse, New York, U.S.A.

HORNE, Gordon Stewart, B.Sc. Born 1924. Educ.: Strathclyde Univ. Combustion Eng., Federated Foundries, Glasgow, 1948-52; Tech. Officer, Water-Tube Boilermakers' Assoc., 1952-. Societies: A.M.I.Mech.E.; B.N.E.S.
Nuclear interest: Primarily steam raising units and associated equipment.
Address: c/o The Water-Tube Boilermakers' Association, 8 Waterloo Place, Pall Mall, London, S.W.1, England.

HORNE, John Ewen Troup, M.A. Born 1918. Educ.: Cambridge Univ. Principal Geologist, Geological Survey of Great Britain, 1953-. Societies: Mineralogical Soc. of Great Britain; Mineralogical Soc. of America; Geological Soc. of London; Inst. of Navigation.

Nuclear interests: Radioactive minerals, age determination; X-ray spectrographic analysis.
Address: Institute of Geological Sciences, 64, Gray's Inn Road, London W.C.1, England.

HORNER, Richard E. Sen. Vice Pres. and Gen. Manager, Northrop Space Labs., Northrop Corp.
Address: Northrop Space Laboratories, Hawthorne, California, U.S.A.

HORNIG, Donald F. Formerly Chairman, Dept. of Chem., Princeton Univ.; Director, Office of Sci. and Technol., Exec. Office of the President, and Special Asst. to the President for Sci. and Technol., 1964-.
Address: Office of Science and Technology, Executive Office of the President, Washington D.C., U.S.A.

HORNYAK, William F., B.S. (Elec. Eng., C.C.N.Y.), M.S. (Phys., California Inst. Technol.). Born 1922. Educ.: C.C.N.Y., and California Inst. Technol. Sci., Brookhaven Nat. fornia Inst. Technol. Sci., Bro Lab., 1949-55 (on leave of absence to Princeton Univ., 1953-55); Lecturer, Princeton Univ., 1955-56; Assoc. Prof. Phys. (1956-61), Prof. (1961-.), Maryland Univ. Societies: A.P.S.; A.A.U.P.; Washington Philosophical Soc.
Nuclear interests: Nuclear reactions in the light elements, particularly capture gamma rays, He^3 induced reactions, high resolution charged particle spectroscopy, and the use of neutron time of flight techniques.
Address: Physics Department, Maryland University, College Park, Maryland, U.S.A.

HOROWITZ, Jules, L. ès Sc., Ing. E.P. Born 1921. C.E.N., Saclay; Chef du Département des Etudes de Piles. Member, Commission Consultative pour la Production d'Electricité d'Origine Nucléaire, Member, Comite des Programmes, Member, Commission Scientifique du Centre d'Etudes Nucléaires de Grenoble, Member, Conseil d'Enseignement de l' I.N.S.T.N., Directeur, Direction des Piles Atomiques, C.E.A.; Formerly Special Adviser on Res. Problems, Euratom; Formerly Member, Study Group on the High Flux Reactor, Member, Board of Management, Dragon Project, E.N.E.A., O.E.C.D.; Member, Comité Consultatif de la Recherche Sci. et Technique. Society: Atomic Ind. Forum.
Nuclear interests: Reactor design; Reactor physics.
Address: 15 rue Philibert Delorme, Paris 17, France.

HORRILLENO, Emilio C., Dr. Physician, Radioisotope Lab., Philippine Gen. Hospital.
Address: Philippine General Hospital, University of the Philippines, Manila, Philippines.

HORSFALL, James Gordon, B.S. (Arkansas), Ph.D. (Cornell), Hon. D.Sc. (Vermont and Turin). Born 1905. Educ.: Arkansas and Cornell Univs. Director, Connecticut Agricultural Experiment Station, 1948-; Lecturer, Yale Univ. Advisory Com. on Biol. and Medicine, U.S.A.E.C., 1957-64; Board of Trustees, Biol. Abstracts. Books: Fungicides and their Action (1945); Principles of Fungicidal Action (1956). Societies: American Phytopathological Soc.; N.A.S.; American Acad. Arts and Sci.; Assoc. Appl. Biol., Great Britain; Hon. Member, Soc. Italiana di Fitoiatria; Botanical Soc. America.
Address: The Connecticut Agricultural Experiment Station, P.O. Box 1106, New Haven, Connecticut 06504, U.S.A.

HORSMAN, J. C. Reactor Loops, Operations Div., A.E.C.L. Expert, Third I.C.P.U.A.E., Geneva, Sept. 1964.
Address: Atomic Energy of Canada, Ltd., Chalk River Project, Chalk River, Ontario, Canada.

HORST, Wolfgang, Dr. med. Born 1920. Educ.: Hamburg Univ. Lecturer on Radiol. (1954-60), Director, Radioisotope Unit and Radiotherapeutic Dept. (1955-63), Prof. Radiol. (1960-63), Hamburg Univ.; Ord. Prof. Radiol, Zurich Univ., 1963-; Director, Univ. Clinic and Radiotherapy and Nucl. Medicine Polyclinic, Zurich, 1963-. Consultant: German Federal Govt., German Ministry of Health, German Ministry of Sci. and Res., I.A.E.A., Swiss Com. Isochroncyclotron, Swiss Federal Inst. Technol. Societies: German Assoc. Radiol.; Swiss Assoc. Radiol. and Nucl. Medicine; Soc. Nucl. Medicine, U.S.A.; American Thyroid Assoc.; Assoc. Nucl. Medicine; Roy. Soc. Medicine, London.
Nuclear interests: Radiotherapy - treatment with high energy photons and electrons; Isochroncyclotron for negative pi-mesons and neutrons under construction; Development of new methods of radioisotope scanning; Radioactive compounds in clinical diagnosis and treatment; Electronics for radiation measurements, human counting and radioisotope-scanning; Application of computer techniques for in vivo-turnover and clearance studies.
Address: University Hospital, 100 Rämistrasse, Zurich 8006, Switzerland.

HORSTMANN, Heinz, Dr. rer. nat., Dipl. Phys. Born 1932. Educ.: Mainz, Göttingen, Hamburg and Kiel Univs. At Inst. für Reine und Angewandte Kernphysik, Kiel Univ., 1958-59; Gesellschaft für Kernenergieverwertung in Schiffbau und Schiffahrt, Hamburg, 1959-60; Inst. für Reine und Angewandte Kernphysik, Kiel Univ., 1960-62; Head, Data handling and analysis group, Central Bureau for Nucl. Measurements, Euratom, Geel, 1962-. Societies: Deutsche Physikalische Gesellschaft; Gesellschaft Deutscher Naturforscher und Ärzte; Studiengesellschaft zur Förderung der Kernenergieverwertung in Schiffbau und Schiffahrt; Deutsches Atomforum.

HOR

Nuclear interests: Nuclear theory; Nuclear reactor theory.
Address: Central Bureau for Nuclear Measurements, Euratom, Geel, Belgium.

HORTIG, Günther, Dr. At Max-Planck Inst. für Kernphysik.
Address: Max-Planck-Institut für Kernphysik, Saupfercheckweg, Heidelberg, Germany.

HORTON, Charles Abell, A.B., M.S., Ph.D. Born 1918. Educ.: Cornell and Michigan Univs. Chem., Oak Ridge Gaseous Diffusion Plant, (Union Carbide Corp.,), 1949-56; Chem., O.R.N.L., Union Carbide Corp., 1956-; Sen. Officer, I.A.E.A., 1966-68. Books: Chapters in Fluorine Chem., volume II (editor, J. H. Simons); Advances in Analytical Chem. and Instrumentation, Treatise on Analytical Chem.; Sections in Pharmacology and Toxicology of Uranium (editors, Voegtlin and Hodge). Societies: A.C.S.; A.A.A.S.; American Inst. for Chem.; Soc. Public Analysts.
Nuclear interests: Analytical methods for materials used in or produced by nuclear reactors, or atomic energy projects, especially uranium and fluorine.
Address: 11 Karntnerring, Vienna 1, Austria.

HORTON, Clifford Charles, B.Sc. (1st Hons.). Born 1928. Educ.: Bristol Univ. At U.K.A.E.A. Harwell, 1949-57; Tech. Manager, Rolls-Royce and Assocs. Ltd. Book: Co-author, Rad. Shielding (Pergamon Press, 1957). Society: A.N.S.
Nuclear interests: Reactor physics; Propulsion reactor design.
Address: 24 Evans Avenue, Allestree, Derby, England.

HORTON, Jack K. Pres. (-1968), Chairman (1968-), and Chief Exec. Officer, Southern California Edison Co.
Address: Southern California Edison Co., 601 W. Fifth Street, Los Angeles, California 90017, U.S.A.

HORTON, James Henry, Jr., B.S., M.S., Ph.D. Born 1923. Educ.: Clemson Coll., North Carolina State Coll., Pennsylvania State Univ. Chem., E. I. du Pont de Nemours. Societies: H.P.S.; A.C.S.
Nuclear interest: Radioactive waste management.
Address: E. I. du Pont de Nemours, Savannah River Plant, Aiken, S.C., U.S.A.

HORTON, S. G. Supt., Nucl. Power Demonstration, Rolphton (1965-), Hydro-Elec. Power Commission of Ontario.
Address: Hydro-Electric Power Commission of Ontario, 620 University Avenue, Toronto 2, Canada.

HORUBALA, Zenon, M.Sc. Born 1936. Educ.: Warsaw Univ. of Technol. Warsaw Univ. of Technol., -1965; Now at Inst. of Nucl. Res.

Nuclear interests: Metallurgy, especially powder metallurgy and ceramics in nuclear application.
Address: 33c Walcownicza, Warsaw 90, Poland.

HORVIK, Einvind. Assoc. Prof., Phys. Dept., North Dakota State University of Agriculture and Appl. Sci.
Address: North Dakota State University of Agriculture and Applied Science, Physics Department, Fargo, North Dakota 58102, U.S.A.

HORWITZ, Nahmin, B.S., M.S., Ph.D., Prof. Born 1927. Educ.: Minnesota Univ. At Lawrence Rad. Lab., Berkeley, California, 1954-59; Syracuse Univ., 1959-. Societies: A.P.S.; A.A.P.T.
Nuclear interest: Elementary particles.
Address: Physics Department, Syracuse University, Syracuse 10, New York, U.S.A.

HORZ, Gerhard, Dr. rer. nat. Sen. Officer, Inst. of Special Metals, Max-Planck Inst. for Metal Res.
Address: Max-Planck Institute for Metal Research, 92 Seestrasse, 7000 Stuttgart 1, Germany.

HOSAKI, Takumi. Alternate Governor from Japan, Board of Governors, I.A.E.A.
Address: 1 Neuer Markt, Vienna 1, Austria.

HOSEGOOD, Samuel Brittan, B.Sc. (Eng.) (2nd (1st Div.), Mech. Eng.) C. Eng., M.I.Mech. E., A.M.I.Chem.E. Born 1923. Educ.: Bristol Univ. Univ. (Student), 1949-53; Tech. Dept. Admiralty Development Establishment Barrow, Vickers Armstrongs, Barrow-in-Furness, 1953-55; S.S.O./P.S.O., U.K.A.E.A., Harwell. 1956-59; Deputy Chief Eng., O.E.C.D. Dragon Project, Winfrith, Seconded from U.K.A.E.A. (U.K.A.E.A. Grade, Asst. Chief Eng.), 1959-.
Nuclear interests: Reactor design; Associated plant and component design.
Address: Shipstal Point, Arne, near Wareham, Dorset, England.

HOSHIAI, M., Ph.D. Gen. Manager, Central Res. Lab., Hitachi Ltd.
Address: Hitachi Ltd., Central Research Laboratory, 280 Koigakubo, Kokubunji, Kitatamagun, Tokyo, Japan.

HOSHINO, Fumihiko. Prof., Course of Medical Radiol., Faculty of Medical Sci., Tohoku Univ.
Address: Tohoku University, 75 Katahira-cho, Sendai, Japan.

HOSHINO, Ichiro, B.E. Eng., Atomic Power Div., Shimizu Construction Co. Ltd.
Address: Shimizu Construction Co. Ltd., Atomic Power Centre, No. 1 Takara-cho 2, Chuo-ku, Tokyo, Japan.

HOSKING, P. K. Officer in Charge, Metal., Australian Mineral Development Labs.

Address: Australian Mineral Development
Laboratories, Conyngham Street, Frewville,
South Australia 5063, Australia.

HOSMER, Craig, A.B., LL.B. Born 1915.
Educ.: California, Michigan and South California Univs. Ranking House Minority Member,
Joint Com. on Atomic Energy, U.S. Congress
(Congress Member from California).
Nuclear interest: Legislation.
Address: House Office Building, Washington
D.C. 20515, U.S.A.

HOSOE, Masanao. Prof., Phys. Div., Inst.
for Atomic Energy, St. Paul's Univ. (Rikkyo
Univ.).
Address: St. Paul's University, Institute for
Atomic Energy, Sajima, Yokosuka, Japan.

HOSOI, Tsuyoshi. Formerly Sen. Managing
Director, now Pres., Dowa Fire and Marine
Insurance Co. Ltd.
Address: Dowa Fire and Marine Insurance Co.
Ltd., 61 Shinmei-cho, Kita-ku, Osaka, Japan.

HOSS, Donald E., B.S., M.S. Born 1936.
Educ.: Missouri and North Carolina State
Univs. Fishery Biol., Res., Radiobiol. Lab.,
Bureau of Commercial Fisheries, 1958-.
Societies: American Fisheries Soc.; Assoc. of
Southeastern Biol.; Atlantic Estuarine Res.
Soc.
Nuclear interest: Passage of radionuclides
through food chains to fish and the effects of
radionuclides on fish.
Address: Fish and Wildlife Service, Bureau of
Commercial Fisheries, Radiobiological Laboratory, Beaufort, North Carolina 28516, U.S.A.

HOSSAIN, Anwar, M.Sc., Ph.D. Born 1931.
Educ.: Dacca and Bristol Univs. Lecturer,
Dacca Univ., 1955-57; Head, Sci. Dept., Army
Apprentice School, 1957-60; S.S.O., Pakistan
A.E.C., 1960-; Res. Fellow, C.E.R.N., Geneva,
1960-61; P.S.O., 1962-; Director, Atomic
Energy Centre, Dacca, 1964-67; Director Gen.,
Pakistan Inst. of Nucl. Sci. and Technol.,
Islamabad, 1967-.
Nuclear interest: Experimental nuclear physics.
Address: Pakistan Institute of Nuclear Science
and Technology, P.O. Nilore, Islamabad,
Pakistan.

HOSTE, Julien, Dr. in Sci. Born 1921. Educ.:
Ghent Univ. Prof. Analytical Chem., Ghent
Univ. Society: Vlaamse Chemische Vereniging.
Nuclear interest: Activation analysis.
Address: 22 J. Plateaustraat, Ghent, Belgium.

HOSTRUP-PEDERSEN, N., M.Sc. Member,
Executive Com., Danatom.
Address: Danatom, 7 Tingvej, DK-4690,
Haslev, Denmark.

HOTCHEN, Edwin Philip, M.Sc., M.I.Mech.E.,
F.I.E.E., F. Inst. P. Born 1924. Educ.: Manchester Univ. R. and D. Headquarters, Risley
(1955-58), Tech. Eng., Central Tech. Services
(1958-59), Sen. Phys., Chapelcross (1959-62),
Commissioning Supt. Phys., Hunterston (1962-65), A/Tech. Manager, Chapelcross (1965-66),
U.K.A.E.A.; First Sec., Atomic Energy,
British Embassy, Tokyo. Society: B.N.E.S.
Address: British Embassy, Ichiban-cho,
Chiyoda-ku, Tokyo, Japan.

HOUBART, Emile. Administrateur Directeur
des Unions de Centrales Electriques.
Address: 5 rue de la Bonté, Brussels 5,
Belgium.

HOUCKGEEST, Floris A. van BRAAM. See
van BRAAM HOUCKGEEST, Floris A.

HOUPILLART, Jean. Born 1912. Ing. de
l'Ecole Supérieure de Physique et Chimie de
la Ville de Paris. Directeur Scientifique, S.A.
des Filtres Philippe; Formerly Dept. Head,
Philippe Hilco Europe; Prof., Centre de Perfectionnement Technique.
Nuclear interests: Installations d'usines chimiques.
Address: Filtres Philippe S.A., 111 boulevard
Henri Barbusse, 78 Houilles, France.

HOURIET, A., Prof., Dr. Dept. Theoretical
Phys., Fribourg Univ.
Address: Department of Theoretical Physics,
Fribourg University, Fribourg, Switzerland.

HOURS, Richard, Ing. Diplomé de l'Ecole
Centrale des Arts et Manufactures, Paris.
Born 1922. Educ.: Paris Univ. With the French
A.E.C., 1950-. Deputy Head, Sect. des Applications de Radioéléments, Département des
Radioéléments, Saclay.
Nuclear interest: Applications (mainly industrial) of radioactive isotopes.
Address: Centre d'Etudes Nucléaires de
Saclay, B.P. No. 2, Gif-sur-Yvette 91, France.

HOUSEHOLDER, Alston Scott, B.S., M.A.,
Ph.D., Ehrendoktor (Munich). Born 1904.
Educ.: Northwestern, Cornell, and Chicago
Univs. Mathematician (1946-48), Director
Maths. Div. (1948-), O.R.N.L. Editorial Com.,
Assoc. Computing Machinery and Soc. Ind.
Appl. Maths; Editorial Staff Numerische
Mathematik. Societies: American Mathematical
Soc.; Mathematical Assoc. America; Soc. Ind.
Appl. Maths; Assoc. Computing Machinery;
A.A.A.S.
Nuclear interests: Numerical analysis; Computing.
Address: Oak Ridge National Laboratory, P.O.
Box X, Oak Ridge, Tennessee, U.S.A.

HOUSER, Frederick N. Geological Survey, U.S.
Dept. of the Interior.
Address: Geological Survey, U.S. Department
of the Interior, Washington D.C. 20242,
U.S.A.

HOUSTON, Robert Wayne, B. Eng., D. Eng.
Born 1924. Educ.: Yale Univ. Res. Assoc.,
K.A.P.L., 1949-51; Asst. Prof., Chem. Eng.,

New Hampshire Univ., 1955-56; Assoc. Prof., Chem. Eng., Pennsylvania Univ., 1956-60; Prof., Chem. Eng., Columbia Univ., 1960-67; Vice Pres. and Lab. Director, Ind. Reactor Labs., 1960-. Director, Reactor Operations Div., A.N.S., 1966-. Book: Co-author, Nucl. Eng. (McGraw-Hill, 1957). Societies: A.N.S.; A.C.S.; A.I.Ch.E.
Nuclear interest: Management of a nuclear and radiation research facility, including direction of research and research services.
Address: Industrial Reactor Laboratories, Plainsboro, New Jersey 08536, U.S.A.

HOUTEN, G. R. VAN. See VAN HOUTEN, G. R.

HOUTERMANS, Hans, Dr. rer. nat., Dipl.-Phys. Born 1919. Educ.: Göttingen Univ. Isotope Lab. (MFA), Max-Planck-Gesellschaft, Göttingen, 1951-58; Max-Planck-Inst. für Physik, München, 1958-60; Phys. Sect., I.A.E.A. Lab., Vienna, 1960-.
Nuclear interests: Low energy nuclear physics, calibration of radionuclides, in-pile dosimetry.
Address: International Atomic Energy Agency, 11 Kärntnerring, Vienna, Austria.

HOUTMAN, Johannes Paulus Willem, Degree in Chem. and Chem. Eng. Born 1917. Educ.: Tech. Univ. Delft. Prof. Chem. Eng. and Technol., Bandung, 1949-55; Sen. Chem., Homogeneous Nucl. Reactor Development Project, Dutch State Mines, Arnhem, 1956-59; Head, Chem. and Metal. Dept., Reactor Centrum Nederland, 1960-62; Prof. Nucl. Chem. and Nucl. Chem. Eng., Tech. Univ. Delft, 1959-. Societies: Royal Dutch Inst. Eng.; Royal Dutch Soc. Chem.
Nuclear interests: Radiochemistry; Activation analysis; Radiation chemistry; Radiation activation of catalysts; Nuclear chemical engineering.
Address: Reactor Institute, 15 Berlageweg, Delft, Netherlands.

HOUWELING, Jacobus Hermanus. Born 1907. At (1934-), Asst. Managing Director (1955-60), Gen. Director (1961-), Industrieele Maatschappij Activit N.V.
Nuclear interest: Manufacture of ion exchangers, among others for the application on water, contaminated with radioactivity (primary circuit, condensate circuit, waste treatment and water conditioning in general).
Address: Industrieele Maatschappij Activit N.V., P.O.B. 240C, 1-3 Nieuwendammerkade, Amsterdam, Netherlands.

HOUWINK, Roelof, Ph.D. Born 1897. Educ.: Delft Tech. Univ. Director, Euratom, Brussels. Societies: Fellow, Inst. of the Rubber Industry, London; Hon. Member, Sté. de Chimie Industrielle, Paris.
Nuclear interests: Management; Training, tuition.
Address: 5 Bremhorstlaan, Wassenaar, Netherlands.

HOVE, L. Ch. P. VAN. See VAN HOVE, L. Ch.P.

HOVEKE, George F., B.S. (Phys.). Born 1934. Educ.: Illinois Inst. Technol. and O.R.S.O.R.T. Nucl. Eng., Sargent and Lundy, Chicago, 1956-. Society: A.N.S.
Nuclear interest: Reactor design; Nuclear power plant design; Fuel cycles.
Address: 105 Perth Road, Cary, Illinois, U.S.A.

HOVEN, Isaac VAN DER. See VAN DER HOVEN, Isaac.

HOVI, Väinö Toivo, M.A., Ph.D. (Helsinki). Born 1913. Asst. of Phys. (1945-53), Docent of Phys. (1950-), Helsinki Univ.; Prof. Phys., Turku Univ., 1953-; Director Wihuri Phys. Lab., Turku Univ., 1957-; Sci. Com., Finnish A.E.C. Chairman (1954-55), Member of Board (1952-57, 1964-65), Phys. Soc. of Finland. Assoc. Editor for Finland of the internat. journal Acta Metallurgica, 1952-; Member, Editorial Board, internat. journal Phys. of Condensed Matter, 1963-. Societies: Finnish Sci. Acad. 1960-; Phys. Soc. of Finland; Japan Phys. Soc.; A.P.S.; American Inst. of Phys.
Nuclear interests: Cosmic rays; β-disintegration energy; Nuclear magnetic resonance in solids.
Address: Wihuri Physical Laboratory, Turku University, 5 Vesilinnantie, Turku, Finland.

HOVORKA, George, B.S. (Civil Eng.), Certificate in Agricultural Eng. Born 1919. Educ.: Wayne State and Iowa State Univs. Administrative Eng., Chief Eng. Office, Chicago Univ., 1948-50; Director of Eng., Savannah River Operations Office (1950-56), Specialist, Reactor Hazards Evaluation Staff, (1956-57), U.S.A.E.C.; Asst. Manager, Gas-cooled Reactor Project, A.C.F.Industries, Washington, D.C., 1957-; Director, Nucl. Services, Commonwealth Assoc. Inc. Societies: A.S.C.E.; A.N.S.
Address: 3376 Eastlane Drive, Jackson, Michigan 49203, U.S.A.

HOWARD, Albert W. Vice Pres., Montreal Eng. Co. Ltd. Director, Canadian Nucl. Assoc., 1965-.
Address: Canadian Nuclear Association, 11 Adelaide Street West, Toronto 1, Canada.

HOWARD, Alma, B.Sc., Ph.D. Born 1913. Educ.: McGill Univ. Medical Res. Council, Hammersmith Hospital, London, 1948-55; Res. Staff, British Empire Cancer Campaign Res. Unit in Radiobiol., 1955-62; Paterson Labs., Christie Hospital and Holt Radium Inst., Manchester, 1963-. Member, Editorial Board, Internat. J. of Rad. Biol. Societies: Assoc. for Rad. Res.; Soc. for Exptl. Biol.; Genetical Soc.; Rad. Res. Soc. of America; Inst. of Biology; Internat. Soc. for Cell Biol.
Nuclear interests: Radiobiology; survival, proliferation, and metabolic function in

irradiated cells.
Address: Paterson Laboratories, Christie Hospital and Holt Radium Institute, Manchester 20, England.

HOWARD, Charles P., B.S.M.E., M.S.M.E., Eng. Born 1923. Educ.: Texas A. and M. and Stanford Univs. Prof., Mech. Eng., Catholic Univ., 1966. Societies: A.N.S.; A.S.M.E.; A.S.E.E.
Nuclear interest: Reactor design (heat transfer and thermodynamics).
Address: Mechanical Engineering Department, Catholic University of America, Washington, D.C. 20017, U.S.A.

HOWARD, Frederick Thomas, B.Sc., M.A., Ph.D. Born 1909. Educ.: Parsons Coll., Colorado State Coll. of Education, and Columbia Univ. Phys., O.R.N.L., 1949-. Societies: A.P.S. (S.E.); RESA; A.A.A.S.
Nuclear interest: Accelerators; maintains directory of world's accelerators.
Address: Oak Ridge National Laboratory, Oak Ridge, Tennessee, U.S.A.

HOWARD, J. Don, B.S. (Elec. Eng.). Born 1903. Educ.: Iowa State Univ. Formerly Pres., now Director, Wisconsin Power and Light Co. Director and Member, Exec. and Finance Coms., Atomic Power Development Assoc.
Nuclear interest: Management: installation of a 527,000 kilowatt nuclear electric generating station.
Address: 122 W. Washington Avenue, Madison, Wisconsin 53701, U.S.A.

HOWARD, James R. Director, Organisation and Personnel Div., Idaho Operations Office, U.S.A.E.C.
Address: Idaho Operations Office, U.S. Atomic Energy Commission, P.O. Box 2108, Idaho Falls, Idaho 83401, U.S.A.

HOWARD, Norman James, B.A. Born 1925. Educ.: McMaster Univ. Manager, Finance, Civilian Atomic Power Dept., Canadian Gen. Elec. Co. Ltd., 1958-.
Nuclear interest: Financial management in the nuclear field.
Address: 1491 Sherwood Cr., Peterborough, Ontario, Canada.

HOWARD, Raydeen R., Dr. Manager, Electronics and Phys. Dept., Planning Res. Corp.
Address: Planning Research Corp., 1333 Westwood Boulevard, Los Angeles 24, California, U.S.A.

HOWARD, Robert Adrian, B.S., M.S., Ph.D. Born 1913. Educ.: California Inst. Technol. and Washington Univ. Ballistic Missile Div., Ramo-Wooldridge Corp., 1955-57; Consultant, Space Technol. Labs., 1959-63; Prof. Phys. (1947-), Director, Nucl. Reactor Lab. (1958-60), Oklahoma Univ.; Director, Instituto Central de Fisica, Universidad de Concepción, Concepción, Chile, 1962-64. Book: Nuclear Physics (Wadsworth Publishing Co., 1963). Societies: A.P.S.; Fellow, Oklahoma Acad. of Sci.
Nuclear interests: Nuclear physics; Fundamental particle physics; Cosmic radiation.
Address: University of Oklahoma, Norman, Oklahoma, U.S.A.

HOWARD, Robert Clark, B.S. (Eng., Maths.), M.S. (Nucl. Eng.). Born 1931. Educ.: Michigan Univ. and M.I.T. Res. Staff Member, M.I.T.; Asst. Eng., Argonne Nat. Lab.; R. and D. Staff Member and Head Eng. Sect., Direct Conversion Project, Gen. Atomic Div., Gen. Dynamics Corp.; Manager, Nucl. Thermionic Systems Thermo Electron Eng. Corp., Waltham, Massachusetts. Societies: A.N.S.; A.I.A.A.
Nuclear interests: Direct conversion of nuclear heat to electricity through use of thermionic vapour diodes or solid state thermoelectric couples; Primarily involved in engineering development of practical systems utilising these basic energy production and conversion principles.
Address: 63 Woodridge Road, Wayland, Massachusetts, U.S.A.

HOWARD, V. W., Dr. Vice Pres. and Manager, Systems Eng., Northrop Space Laboratories, Northrop Corp.
Address: Northrop Space Laboratories, Hawthorne, California, U.S.A.

HOWARD, William J. Formerly Director, Systems Devel., now Vice-Pres., Sandia Lab., Sandia Corp.
Address: Sandia Laboratory, P.O. Box 5800, Albuquerque, New Mexico, U.S.A.

HOWARTH, John Lee, M.A., M.Sc., Ph.D. Born 1924. Educ.: Cambridge and London Univs. Sen. Phys., Sheffield Nat. Centre for Radiotherapy, -1953; Rad. Phys., Lovelace Clinic and Lovelace Foundation, Albuquerque, New Mexico, 1953-64; Assoc. Prof., Phys., New Mexico Univ., 1964-; Consultant, Lovelace Clinic and Veterans Administration Hospital, Albuquerque, 1964-. Societies: British Inst. of Radiol.; Hospital Phys. Assoc.; Rad. Res. Soc.; American Assoc. of Phys. in Medicine.
Nuclear interests: Radiological physics; Radiobiology; Nuclear medicine.
Address: Department of Physics and Astronomy, New Mexico University, Albuquerque, New Mexico 87106, U.S.A.

HOWARTH, Ronald Matthews, M.A. (Cantab.). Born 1920. Educ.: Cambridge Univ. Chief Designer (1963-66), Chief Eng. (1966-), Bristol Aerojet Ltd. Societies: Fellow, Roy. Aeronautical Soc.; Fellow, British Interplanetary Soc.
Nuclear interest: In-pile and out-of-pile equipment for reactors.
Address: Elmdene, Sandford Road, Winscombe, Somerset, England.

HOWE, E. A. Managing Director, Film Cooling Towers (1925) Ltd.
Address: Film Cooling Towers (1925) Ltd., Chancery House, Parkshot, Richmond, Surrey, England.

HOWE, James Taburn, B.S. (M.E.). Born 1919. Educ.: Duke Univ. Sen. Development Eng., 1946-. Societies: A.N.S.; A.S.M.E.; A.S.M.
Nuclear interests: Design and development of in-pile facilities, especially low-temperature devices; also equipment for pre- and post-irradiation studies.
Address: 106 Columbia Drive, Oak Ridge, Tenneseee 37830, U.S.A.

HOWE, John Perry, B.S. cum laude (Chem.), Ph.D. (Phys. chem.). Born 1910. Educ.: Hobart Coll. and Brown Univ. Manager, Metal. Sect., K.A.P.L., G.E.C., 1945-52; Res. Assoc., Gen. Elec. Res. Lab., 1952-53; Sect. Chief, Reactor Materials (1953-57), Director, Res. Dept. (1957-61), Atomics Internat. Div., North American Aviation Inc., Downey and Canoga Park; Ford Prof. Eng. Director, Dept. Eng. Phys. and Materials Sci., Cornell Univ., 1961-65; on leave from Cornell Univ. as staff member, Inst. for Defence Analyses, 1967-68. Editorial Advisory Board, Reactor Handbook (1953-), Member, Advisory Com. Reactor Safeguards (1961), U.S.A.E.C.; Sci. Advisory Group, Office Aerospace Res.
Book: Co-editor, Progress in Nucl. Energy Series V, Metal. and Fuels, Vols. 1-4 (Pergamon Press Ltd., 1955-). Societies: Fellow, A.N.S.; Fellow, A.P.S.; Fellow, A.A.A.S.
Nuclear interests: Nuclear reactors for power; Materials for nuclear reactors; Physical and chemical effects of radiation; Metallurgy.
Address: 102 Woodcrest Terrace, Ithaca, New York, U.S.A.

HOWELL, Barbara F., B.A., M.S., Ph.D. Born 1924. Educ.: Minnesota, Kansas and Missouri Univs. Arlington State Coll., 1957-61; Kansas State Teachers Coll., 1964-. Society: A.C.S.
Nuclear interest: Studying reaction mechanisms using radioactive tracers.
Address: Chemistry Department, Kansas State Teachers College, Emporia, Kansas 66801, U.S.A.

HOWELL, E. Irl. Mississippi State Univ. Rep. on Council, Oak Ridge Assoc. Univs.
Address: Mississippi State University, P.O. Box 184, State College, Mississippi, U.S.A.

HOWELLS, Gordon Rushworth, B.Sc. Born 1914. Educ.: Wales Univ. Chem. Eng. Adviser, Risley (-1958), Gen. Manager, Production Group, Chapelcross Works (1958-64), Gen. Manager, Windscale, Calder and Chapelcross Works (1964-), U.K.A.E.A. Societies: Inst. Chem. Eng.; British Inst. Management.
Nuclear interests: Reactor operation; Chemical plant design and operation; General management.
Address: Bankfield, Wasdale Road, Gosforth, Cumberland, England.

HOWELLS, Huw, B.Sc. Born 1918. Educ.: Wales Univ. Health and Safety Manager, Windscale and Calder Works, U.K.A.E.A., 1955-.
Society: Inst. of Phys. and Phys. Soc.
Nuclear interests; Operational radiological and nuclear safety control with special interests in radioactive effluent disposal problems.
Address: 6 Wastwater Rise, Seascale, Cumberland, England.

HOWERTON, Robert James, B.S., M.S. Born 1923. Educ.: Northwestern Univ. Asst. Prof. Math. and Phys., Regis Coll., Denver, Colo., 1948-56; Reactor Core Phys. and Eng., MTR Idaho Falls, 1956-57; Group Leader, Theoretical Phys. Div., Lawrence Rad. Lab., Livermore, California, 1957-. Society: A.P.S.
Nuclear interest: Neutron physics, especially the determination of best values for nuclear constants needed for reactor computations.
Address: 11594 Ladera Court, Dublin, California, U.S.A.

HOWES, Edward Arthur, B.Sc. (Hons., Chem.), Dipl. Chem. Eng. (London Univ.). Born 1914. Chief Chem., London Div. (1948-55), Chem. Eng. Nucl. Power Dept. (1955-58), R. and D. Dept. (1958-63), R. and D. Officer, South Eastern Region (1963-), C.E.G.B. Societies: F.R.I.C.; Inst. Chem. Eng.; Assoc. Member, British Inst. Management.
Address: Hobdens, Five Ashes, Mayfield, Sussex, England.

HOWES, James E., Jr., B.S. Born 1935. Educ.: Otterbein Coll., Radioisotope Lab. Supervisor, 1959-.
Nuclear interests: In the fields of radiochemistry, nuclear fuels development, with regard to burn-up determination and fission product diffusion studies.
Address: Battelle Memorial Institute, 505 King Avenue, Columbus 1, Ohio, U.S.A.

HOWES, R. M. Formerly Administrative Manager, Gen. Elec. Co., Nat. Reactor Testing Station.
Address: Re-Entry Systems Department, General Electric Company, 3198 Chestnut Street, Philadelphia, Pennsylvania, U.S.A.

HOWEY, J. H., Dr. Director, School of Phys., Georgia Inst. Technol.
Address: Georgia Institute of Technolohy, Atlanta, Georgia, U.S.A.

HOWIESON, Joseph, B.Sc. (M.E.), D.C.Ae. Born 1927. Educ.: Edinburgh Univ. and Cranfield Coll. of Technol. A.E.C.L., 1953-63; Eng. Manager, Atomic Fuel Dept., Canadian Westinghouse, 1963-. Societies: A.M.I.Mech.E.; P.Eng. (A.P.E.O.); C.N.A.
Nuclear interests: Nuclear fuel design, development, fabrication and metallurgy.

Address: Atomic Fuel Department, Canadian Westinghouse, Port Hope, Ontario, Canada.

HOWLAND, George P., B.S. (Mech. Eng.). Born 1926. Educ.: Clarkson Coll. of Technol., New York. Manager, Navy and Ind. Nucl. Products, Metals and Controls Inc., Nucl. Products Group, Attleboro, Mass. Societies: A.S.M.E.; Atomic Ind. Forum; A.N.S.
Nuclear interests: Particularly sales, marketing and contract administration functions of fabrication of nuclear reactor fuels and components.
Address: 4 Hatch Road, Attleboro, Massachusetts, U.S.A.

HOWLAND, Joe Wiseman, B.S., M.Sc., Ph.D., M.D., D.Sc. Born 1908. Educ.: Denison, Ohio State and Rochester Univs. Prof. Rad. Biol. and Biophys. and Instructor in Medicine, Rochester Univ. School of Medicine; Consultant, Sci. and Technol. Office, President's Sci. Advisory Com.; Consultant, Surgeon Gen. and Quartermaster Gen. U.S. Army; Consultant, Rochester Gas and Elec. Co. on Reactor Safety. Steering Com., Public Information, Atomic Ind. Forum; Rad. Utilisation Com., New York Office of Atomic Development. Societies: A.N.S.; Rad. Res. Soc.; Ind. Hygiene Foundation of America; Ind. Medical Assoc. Inc.; Atomic Ind. Forum.
Nuclear interests: Health physics; Diagnosis and treatment of radiation injury; Reactor siting; Biological effects of ionizing radiation; Biologic effects of microwave radiation.
Address: 260 Crittenden Blvd., Rochester, New York, U.S.A.

HOWLES, Lawrence Reginald, D.F.H. Born 1922. Educ.: Faraday House. Tech. Sales Rep., Birmingham (1948), Deputy Commercial Officer, Atomic Power Div. (1955), Chief Sales Eng. and Deputy Commercial Officer, Atomic Power Div. (1957-), English Elec. Co., Ltd. Society: A.M.I.E.E.
Nuclear interests: Commercial aspects of complete nuclear power station design; Economics; Finance.
Address: The English Electric Company, Ltd., Cambridge Road, Whetstone, near Leicester, England.

HOWORTH, Muriel. M.O.S., Atomic Energy Dept., 1948. Pres., Inst. Atomic Information; Founder, Atomic Gardening Soc. for Laymen; Founder Director, Seed Mutation R. and D. Lab. for Lay Experimenters. Books: Atomic Transmutation (1953); Learning about the Atom (1957); Pioneer Res. on the Atom (1958); Biography of Frederick Soddy (1958); Atomic Gardening for the Laymen (1959); Impact of a Million Stars (1964). Societies: Fellow, Roy. Soc. Arts; Fellow, Roy. Commonwealth Soc.; Fellow, Roy. Astronomical Soc.; British Soc. for the History of Sci.
Nuclear interests: Authentic history of atomics.
Address: 14 Eversfield Road, Eastbourne, Sussex, England.

HOWTON, David R., B.S., Ph.D. Born 1920. Educ.: California Inst. Technol. Instructor Physiological Chem. (Clinical) (1951-52), Asst. Prof. Physiological Chem. (Clinical) (1953-54), Assoc. Prof. Physiological Chem. (Clinical) (1955-56), Resident Assoc. Prof. Physiological Chem. (1957-), Resident Assoc. Prof. Biophys. and Nucl. Medicine (1961-), Resident Prof. Biophys. and Biol. Chem. (1964-), Res. Biochem., Nucl. Medicine and Rad. Biol. Lab., Biophys. and Nucl. Medicine Dept., (1959-), California Univ. Medical Centre. Chief, Organic Chem. Sect., Biochem. Div., Atomic Energy Project, School of Medicine, U.C.L.A. (now Nucl. Medicine and Rad. Biol. Lab). Book: Radioisotope Studies of Fatty Acid Metabolism (Pergamon Press, 1960). Societies: A.C.S.; Chem. Soc. (London); Rad. Res. Soc.
Nuclear interests: Effects (direct and indirect) of ionising radiation on lipids and related substances; Use of radioisotopes in study of intermediary metabolism of lipids.
Address: California University Medical Centre, Biophysics and Nuclear Medicine Department, Nuclear Medicine and Radiation Biology Laboratory, 900 Weyburn Place, Los Angeles 24, California, U.S.A.

HOYAUX, Max, Ing. Civil Electromécanicien A.I.Ms., Dr. en Sci. Physiques (U.L.B.). Born 1919. Educ.: Faculté Polytechnique de Mons and Brussels Free Univ. Head, Development Dept., Nucl. Div., Ateliers de Constructions Electriques de Charleroi, 1957-. Prof. Nucl. Sci., Univ. de Travail de Charleroi; At Carnegie Inst. of Technol. Book: L'Atome et ses Applications (Brussels, Science et Technique, 1953). Societies: Assoc. des Ingénieurs de Mons; Assoc. des Ingénieurs de Montefiore; Sté. Belge de Physique.
Nuclear interests: Reactor design; Nuclear physics.
Address: 629 Mountain View Drive, Murrysville, Pennsylvania, U.S.A.

HOYLE, Dixon Barclay, B.S. (Meteorology), B.Chem.E. Born 1923. Educ.: New York and Cornell Univs. Chem. Eng., Standard Oil Co. (Indiana), 1948-52; Production Eng., Production Div. (1952-59), Chief, Materials Branch (1957-59), Chief, Euratom Affairs Branch (1959-61), Asst. Director, Tech. Implementation, Div. of Internat. Affairs (1961-66), U.S.A.E.C.; Sen. U.S.A.E.C. Rep., Brussels Office and Deputy for Euratom Affairs, U.S. Mission to the European Communities, 1966-.
Nuclear interests: Management; Materials supply; International relations and technical exchanges.
Address: United States Mission to the European Communities, 23 avenue des Arts, Brussels 4, Belgium.

HOYT, Frank C., Dr. Missiles System Div., Lockheed Aircraft Corp. Member, Com. of Sen. Reviewers, U.S.A.E.C.

Address: Lockheed Aircraft Corp., Palo Alto, California, U.S.A.

HOYT, Harlan K., B.S. and Professional in Mech. Eng. Born 1912. Educ.: Missouri School Mines and Metal. Efficiency Eng., Generating Stations, 1950-52; Asst. Chief Eng., Fisk Station, 1952-55; Asst. to Sup., Generating Stations, 1955-57; Sup., Dresden Nucl. Power Station, 1957-. Society: A.S.M.E.
Nuclear interests: Operating, maintenance, and development of large power reactors.
Address: Dresden Nuclear Power Station, R. R. #1 Morris, Illinois, U.S.A.

HOYT, Henry Reese, A.B. Born 1902. Educ.: Harvard Univ. Asst. Director for Administration, Los Alamos Sci. Lab., Los Alamos, New Mexico, 1946-.
Address: P.O. Box 1663, Los Alamos, New Mexico, U.S.A.

HOZO, H. Chief, Tech. Res. Dept., Kurashiki Rayon Co. Ltd.
Address: Kurashiki Rayon Co. Ltd., 8 Umeda, Kita-ku, Osaka, Japan.

HOZZA, Victor, Ing. Born 1931. Educ.: Moscow State Univ. Sen. Asst., Faculty of Elec. Eng., Slovak Tech. Univ., Bratislava, 1962-.
Nuclear interest: Nuclear physics.
Address: 48/61 Astrová, Bratislava, Czechoslovakia.

HRABOVCOVA, Alojzia, promovaný matematik. Born 1931. Educ.: J. A. Komenského Univ., Bratislava. Society: Sect. Rad. Protection, Czechoslovak Sci. and Tech. Soc. (C.S.V.T.S. - Czechoslovak Vedecko - Techniká Spolecnost).
Nuclear interests: Computer analysis of gamma ray spectra; Applied mathematics; Nuclear spectrometry; Radiation protection.
Address: 49 Ul. Cervenej armády, Bratislava, Czechoslovakia.

HRDLICKA, Jan, M.S. (Eng.). Born 1927. Educ.: Prague Tech. Coll. Energoprojekt Prague, Czechoslovak Atomic Energy Commission.
Nuclear interest: Atomic power plant design.
Address: 10 U Nikolajky, Prague 5, Czechoslovakia.

HRON, Miloslav, Ing. (Nucl. Eng. and Reactor Phys.). Born 1940. Educ.: Prague Tech. Univ. Nucl. Res. Inst., Czechoslovak Acad. of Sci., 1963-.
Nuclear interest: Reactor theory and calculation.
Address: Nuclear Research Institute, Czechoslovak Academy of Sciences, Rez near Prague, Czechoslovakia.

HRSTKA, Vladimir, Ing. (Mech. Eng.). Born 1933. Educ.: Moscow Tech. Univ. Res. Worker, Skoda Works. Society: Czechoslovak Sci. and Tech. Soc.

Nuclear interests: Reactor control; Computer control
Address: Nuclear Power Division, Skoda Works, Pilsen, Czechoslovakia.

HRUSKA, Karel J., M.V.Dr., C.Sc., Docent. Born 1935. Educ.: Faculty of Veterinary Sci., Brno. Biochem. Dept. Head, Faculty of Veterinary Sci., Brno. Societies: Czechoslovak Medical Soc.; Czechoslovak Biochem. Soc.
Nuclear interests: Use of labelled compounds for studying the metabolism and physiology of animals with special interest in biochemistry of reproduction.
Address: 1-3 Palackého, Brno, Czechoslovakia.

HRYNISZAK, Waldemar, D.Sc., (habil.), Dipl. Ing. Born 1910. Educ.: Vienna Technol. Univ. and Konsular Academi, Vienna. Docent, Gas Turbine Technol., Vienna Technol. Univ.; Director for Res., Development and Prototype Design Dept., Clarke, Chapman and Co., Ltd.; Consultant, Gen. Atomic Div., Gen Dynamic Corp., U.S.A., and other firms. Book: Heat Exchangers (Butterworths, 1958). Societies: M.I.Mech.E.; A.S.M.E.
Nuclear interests: All activities concerning the technology of nuclear power stations, with special reference to steam-raising, high temperature gas cooled reactors working in conjunction with gas turbines, compact heat exchangers etc.
Address: 48 St. George's Road, Cullercoats, North Shields, Northumberland, England.

HRYNKIEWICZ, Andrzej Z., Ph.D. Born 1925. Educ.: Jagellonian Univ., Cracow. Docent of Phys. (1954), Prof. Phys. (1961), Inst. of Phys., Jagellonian Univ., Cracow. Head, Nucl. Magnetic Resonance Lab. (1955-60), Head, Gamma Ray Spectroscopy Lab. (1960), now Vice Director, Inst. of Nucl. Phys., Cracow. Societies: Polish Soc.; A.P.S.
Nuclear interests: Nuclear spectroscopy; Nuclear magnetic resonance, Mössbauer effect.
Address: Pl. Wolności 4/14, Cracow, Poland.

HRYNKIEWICZ, Janus, M.Sc. Member, State Council for the Peaceful Use of Nucl. Energy.
Address: State Council for the Peaceful Use of Nuclear Energy, Room 1819, Palace of Culture and Science, Warsaw, Poland.

HRYNKIEWICZ, Marek SUDNIK-. See SUDNIK-HRYNKIEWICZ, Marek.

HSIAO, Sinju Pu, B.S. (Yale-in-China), M.A. (American Univ.), M.S. (Inst. of Advanced Res.). Born 1905. Educ.: Yale-in-China, Changsha; American Univ., Washington, D.C.; Inst. of Advanced Res., Chungking; Graduate School, Command and Gen. Staff College, Kansas; Yang-Ming-Shan-Chuan, Taipei; Sen. Officers' Training Corps, Yuan Shan, Taipei; Radiological Defence Course, U.S. Army, Washington; Management of Mass Casualties Course, U.S. Army Medical Service Graduate School, Walter Reed Army Medical Centre,

Washington; Staff College, U.S. Federal Civil Defence Administration, Battle Creek, Michigan; Special Training Class, U.S.A.E.C., Washington. Sen. Military Aide-de-Camp to Pres. of Republic of China, 1948-52; Military Attaché, Chinese Embassy, Washington, 1952-56; Military Counsellor to Pres. of the Republic of China, concurrently Member of Atomic Energy Council, 1957-. Spoken commentary and movie film on: Atomic Energy and Nucl. Structure of the Atom; Self-preservation under Atomic Attack; U.S. Civil Defence and Its Training Methods; Atom for Peace; An Historical Approach of Atomic Phys.
Nuclear interest: Nuclear physics.
Address: 2 Fu Shing Street, Yung Ho Cheng, Chung Ho Hsiang, Taipei, Formosa.

HSIEH, Hsi-tê, Prof. Book: Co-author, The Phys. of Semi-conductors (Peking Scientific Publishing Soc.).
Address: Academia Sinica, 3 Wen Tsin Chien, Peking, China.

HSING, Y. S. Res. Chem., Radiochem. Lab., Ordnance Res. Inst., Formosa.
Address: Radiochemical Laboratory, Ordnance Research Institute, 2 Tsinan Road, 1st Section, Taipei, Formosa.

HSU, Kuan-jen, Dr. Head, Lab. for Res. in the Application of Atomic Energy in Agriculture; formerly at Minnesota Univ.
Address: Chinese Academy of Agricultural Sciences, Peking, China.

HSU, Shoei-chyuan, Dr. Agr. Director, Taiwan Agricultural Res. Inst.
Address: Taiwan Agricultural Research Institute, Roosevelt Road, Tapieih, Formosa.

HU, Herbert Schih-tschang, B. Eng. Born 1921. Educ.: Nat. Tung-Chi Univ., Shanghai. Formosa's exchange scholar for German Alexander-von-Humboldt Foundation, 1955. Prof., Taipeih Inst. Technol., Formosa, 1953; Prof., Univ. Success, Formosa, 1957; Tech. expert, Atomic Energy Council, Formosa, 1960; Deleg. of Formosa at Symposium on Chem. Effects of Nucl. Transformations in Prague, 1960; Delegate of Formosa on Conference of Nucl. Electronics in Belgrade, 1961.
Nuclear interests: Applied physical metallurgy; Nuclear physics.
Address: Atomic Energy Council, Taipeih, Formosa.

HU, Ning. Papers: Up-down Asymmetries of Λ- and Σ-Decays (Scientia Sinica, vol. 10, No. 7, 1961); On the π-π-Interaction (ibid., vol. 11, No. 3, 1962); The π-π-Interaction in Three-Pion States (ibid.).
Address: Department of Physics, Peking University, Peking, China.

HU, Pung, Dr. Phys., Phys. Dept., Space Sciences, Inc.
Address: Space Sciences, Inc., Physics Department, 301 Bear Hill Road, Waltham, Massachusetts, U.S.A.

HU, Y. H., M.D. Member, Radiol. Centre, Taipeih Provincial Hospital.
Address: Radiological Centre, Taipeih Provincial Hospital, 145 Cheng Chow Road, Taipeih, Formosa.

HUANG, Frederick Chao-pang, B. Eng. (Electronics). Born 1932. Educ.: Nat. Taiwan Univ. Eng., Ministry of Communications, Formosa, 1956; Res. Officer, Union Ind. Res. Inst., Formosa, 1959; I.A.E.A. Fellow, Tech. Phys. Sect., A.A.E.C. Res. Establishment, Australia, 1960-62. Member, Formosan A.E.C.
Nuclear interests: Reactor instrumentation and operation – i.e., reactor simulation using analogue computers, counting devices, monitoring instruments, coincidence circuits using semiconductors, etc.
Address: Union Industrial Research Institute, P.O. Box 100, Hsin-chu, Formosa.

HUANG, Hui, B.S. (E.E.), M.E.E. Born 1903. Educ.: Nanyang (Shanghai) and Cornell Univs. Pres., Taiwan Power Co., 1950-; Metal. Res. Lab., Union Ind. Res. Inst.; Pres., Assoc. of Elec. Enterprises of Taiwan. Formerly Member, Atomic Energy Council. Societies: Chinese Inst. of Eng.; Chinese Inst. of Elec. Eng.
Nuclear interests: Construction, operation, cost analysis and management of nuclear power plants.
Address: Union Industrial Research Institute, P.O. Box 100, Sinchu, Formosa.

HUANG, Wen-xi. Member, Editorial Com., Scientia Sinica.
Address: Scientia Sinica, Academia Sinica, Peking, China.

HUARD de la MARRE, Robert. Prés. Directeur Gén., Sté. Générale d'Applications Electro-Thermiques Fours-Cyclop.
Address: Société Générale d'Applications Electro-Thermiques Fours-Cyclop, 24 rue de Meudon, Boulogne Billancourt, (Seine), France.

HUB, Wilfried, Dr. Born 1919. Educ.: Stuttgart Univ. Nucl. Div., Sigri Elektrographit Meitingen.
Nuclear interest: Nuclear materials.
Address: 10 Sudentenstrasse, 885 Donauwörth, Germany.

HUBBARD, Eugene T. Manager, Vallecitos Superheat Reactor, Gen. Elec. Co., 1963-.
Address: Vallecitos Atomic Laboratory, General Electric Co., P.O. Box 846, Vallecitos Road, Pleasanton, California, U.S.A.

HUBBELL, David W., B.S., M.S. (Eng.).
Born 1925. Educ.: Minnesota Univ. Res.
Hydrologist, Geological Survey, U.S. Dept. of
the Interior.
Nuclear interest: Movement of radionuclides
in the Columbia River Estuary.
Address: Geological Survey, United States
Department of the Interior, 830 N.E. Holladay,
Box 3202, Portland, Oregon, U.S.A.

HUBBELL, Harry Hopkins, Jr., B.A., M.S.,
Ph.D. Born 1914. Educ.: Williams Coll.,
Lafayette Coll. and Princeton Univ. Assoc.
Prof., Middlebury Coll., 1947-50; Sen. Phys.,
and Health Phys., O.R.N.L., 1950-. Societies:
A.P.S.; H.P.S.; A.A.P.T.; A.A.A.S.
Nuclear interests: Continuous spectra and
dosimetry of X-rays and beta rays; Health
physics.
Address: 248 Outer Drive, Oak Ridge,
Tennessee, U.S.A.

HUBBELL, John Howard, B.S. (Eng. Phys.),
M.S. (Phys.). Born 1925. Educ.: Michigan
Univ. Phys. (1950-), Rad. Theory Sect. (1951-),
N.B.S., Washington D.C.; Director, X-Ray
Attenuation Coefficient Information Centre
(Nat. Standard Reference Data System),
1963-. Societies: A.P.S.; Rad. Res. Soc.;
A.A.A.S.; Philosophical Soc. Washington.
Nuclear interests: Radiation penetration and
diffusion through materials; Geometrical
radiation problems; X- and gamma-ray
attenuation coefficients; Gamma-ray scintillation spectroscopy.
Address: 11830 Rocking Horse Road, Rockville, Maryland 20853, U.S.A.

HUBER, J. W., Dr. Eng. Works Div., Schweizerische Bundesbahnen; Vertreter des Eidg.
Verkehrs- und Energiewirtschaftsdepartements,
Alarmausschuss-Departement des Innern.
Address: Schweizerische Bundesbahnen, Bern,
Switzerland.

HUBER, Otto, Dr. sci. nat. Born 1916. Educ.:
Federal High School of Technol., Zürich.
Privat Doz. (1951), Titular Prof. (1952),
Federal High School of Technol., Zurich. o.
Prof., Fribourg Univ., 1953-. Eidgenössiche
Kommission zur Uberwachung der Radioaktivität. Societies: Schweizerische Physikalische
Gesellschaft; Stà. Italiana di Fisica; A.P.S.
Nuclear interests: Nuclear physics, especially
beta and gamma spectroscopy; Health physics.
Address: Physikalisches Institut Universität,
Fribourg, CH-1700, Switzerland.

HUBER, Paul, Dr. sc. nat., o. Prof. Born
1910. Educ.: Swiss Federal Inst. Technol.,
Zürich. Director, Inst. Phys., Basel Univ.
Pres., Swiss Com. of Phys., I.U.P.A.P.; Member, Swiss A.E.C.; Swiss Nat. Council of Sci.;
Alarmausschuss, Eidgenössische Departement
des Innern. Societies: Swiss Phys. Soc.;
A.P.S.
Nuclear interests: Neutron physics; Reactions
with polarized particles; Polarized particle
source; Absolute measurements of radioactive source strength.
Address: 13 Hungerbachweg, Riehen/BS,
Switzerland.

HUBER, Rudolf A. M., Diploma Elec. Ing.
Born 1907. Educ.: Swiss Federal Inst. Technol., M.I.T. and Cambridge Univ. Chairman,
Oerlikon Eng. Co.; Chairman, Thermatom
A.G.; Board, Nat. Soc. Development of Ind.
Nucl. Technol. Societies: I.E.E.; Schweizerische Ing.- und Architekten-Verein; Schweizerische Elektrotechnischer Verein; Schweizerische
Gesellschaft von Fachleuten der Kerntechnik.
Nuclear interest:Management.
Address: Oerlikon Engineering Company,
Zurich 50, Switzerland.

HUBER, Wolfgang, B.Sc., M.Sc., Ph.D. (Org.
Chem., Goettingen Univ.). Born 1910.
Educ.: Med. School, Freiburg, Kiel and
Berlin, Univs. Vice-pres. and Lab. Director,
Electronized Chemicals Corp., 1945-53;
Independent tech. consultant, 1953-; Consultant, Veterans Administration Hospital
1956-; Exec. Vice-pres. and Tech. Director,
Beta Labs. Inc., Palo Alto, California, 1963-;
Vice-pres. and Director, Diagnostic Data Inc.,
Palo Alto, California, 1964-; Lecturer, Biophys.,
Stanford Univ.; Chairman, Pacific Chem.
Exposition, A.C.S.; Member, Animal Nutritional Res. Council. Societies: Fellow,
A.A.A.S.; Soc. of Cosmetic Chemists; Soc.
for Microbiol.; Soc. for Promoting Internat.
Sci. Relations; Fellow, American Inst. of
Chemists; Solar Energy Soc.; Swiss Chem.
Soc.; A.C.S.; Rad. Res. Soc.; Atomic Ind.
Forum; Inst. Food Technologists; Soc. Chem.
Industry Gt. Brit.; N.Y. Acad. Sci.; A.N.S.
Nuclear interests: Mechanism of effects of
high-energy radiation on chemical and biological systems. Industrial uses of ionizing
radiation such as radiation preservation, polymerization, chemical synthesis, materials
testing. Space environment simulation.
Address: 66 Cleary Court, San Francisco,
California 94109, U.S.A.

HUBERLANT, M. R. and D. Manager,
Metallurgie et Mecanique Nucléaires (M. M.
N.), S.A.
Address: Metallurgie et Mecanique Nucléaires
(M.M.N.), S.A., Europalaan, Dessel, Belgium.

HUBERT, Emile-Herman, Civil Eng., Elec. and
Radio-elec. Eng., Bachelor in Phys. Born
1915. Educ.: Liège Univ. and Inst. Electrotechnique Montefiore, Liège. Head, Res.
Dept., U.C.E. Linalux, 1945-52; Administrative Director (1953-57). Director of Planning
(1958-60), Centre d'Etude de l'Energie
Nucléaire; Director of Gen. Affairs, Gen.
Direction Res. and Training, Euratom, 1960-;
Assoc. Prof., Reactor Technol., Liège Univ.,
1964-. Societies: Assoc. des Ingénieurs sortis
de l'Ecole de Liège; Sté. Royale Belge des
Ingénieurs et Industriels.

Nuclear interest: Administration of research organisations.
Address: Euratom, 51 rue Belliard, Brussels 4, Belgium.

HUBERT, Jerzy Zbigniew, magister fizyki. Born 1937. Educ.: Jagiellonian Univ., Cracow. Asst. and Sen. Asst. Inst. of Nucl. Techniques, Akademia Gorniczo-Hutnicza, Cracow, 1964-.
Nuclear interests: Application of nuclear techniques to the mass transport, particularly in porous media (filtration).
Address: 7/12 Warynskiego, Cracow, Poland.

HUBERT, Pierre Louis, D. ès Sc. Born 1923. Educ.: Ecole Supérieure de Physique et de Chimie, Paris. Ingénieur, C.E.A., 1953-. Chef du Service de Recherches sur la Fusion Contrôlée. Society: Sté. Française de Physique.
Nuclear interest: Fusion contrôlée.
Address: C.E.N.F.A.R., B.P. No. 6, Fontenay-aux-Roses, Seine, France.

HÜBERTY, Francois, Ing.-Directeur du Travail et des Mines, Luxembourg; Member, Conseil National de l'Energie Nucléaire.
Address: Ministère de l'Energie Nucléaire, 4 boulevard Roosevelt, Luxembourg, Luxembourg.

HUBIN, G., Ing.-Civil. Administrateur-délégué, Toleries Gantoises, S.A.
Address: Toleries Gantoises, S.A., Drongen, Ghent, Belgium.

HUBLEY, Reginald A., B.S. Born 1928. Educ.: Bucknell Univ. Publisher, Scientific Research Magazine and Nucleonics Week, McGraw-Hill, Inc.
Address: 330 West 42nd Street, New York, New York 10036, U.S.A.

HÜBNER, Gustav Paul Walter, Prof. Dr. Ing. Born 1906. Educ.: Braunschweig T.H. Leiter des Laboratoriums für Dosimetrie von Röntgen und Gammastrahlen Physikalisch-Technische Bundesanstalt, Braunschweig. Book: Co-author, Darstellung, Wahrung und Übertragung der Einheit der Dosis für Röntgen und Gammastrahlen mit Quantenenergien zwischen 3keV and 500keV (Braunschweig, Physikalisch-Technische Bundesanstalt, 1955). Society: Deutsche Röntgengesellschaft.
Nuclear interests: Measurement of ionising radiations; Dose measurements.
Address: 4 Elversbergerst., Brunswick, Germany.

HÜBNER, Heinrich, Dipl.-Chem., Dr. rer. nat. Born 1930. Educ.: Karl Marx Univ. At Inst. für Stabile Isotope, D.A.W.
Nuclear interests: Application of stable isotopes in chemistry; Isotopic effects in reaction rates, especially autoxidations.
Address: 4 Störmthalerstrasse, Leipzig 7027, German Democratic Republic.

HUBY, Ronald, M.A. (Cantab.), Ph.D., (Bristol). Born 1921. Educ.: Cambridge Univ. Lecturer (1948-58), Sen. Lecturer (1958-65), Reader (1965-), Liverpool Univ. Societies: Phys. Soc.; A.P.S.; Soc. Italiana di Fisica.
Nuclear interests: Theoretical low-energy nuclear physics; Nuclear reactions.
Address: 14 Marine Terrace, Magazines Promenade, New Brighton, Wallasey, Cheshire, England.

HUDDLE, R. A. U. Formerly High Temperature Reactor Div. U.K.A.E.A., Winfrith; now Deputy Head, R. and D. Div., Dragon Project.
Address: Dragon Project, Atomic Energy Establishment, Winfrith, Dorchester, Dorset, England.

HUDDLESTON, Charles M., B.S., M.S., Ph.D. Born 1925. Educ.: Texas, Northwestern and Indiana Univs. Assoc. Phys., Argonne Nat. Lab., 1953-61; Director, Phys. and Maths. Div., Appl. Sci. Dept., U.S. Naval Civil Eng. Lab., 1961-67; Phys. Sci. Administrator, U.S. Naval Radiol. Defence Lab., 1967-. Shielding Subcom. of Civil Defence Advisory Com., N.A.S. Societies: A.P.S.; RESA; A.A.A.S.; A.N.S.; N.A.S.
Nuclear interests: Neutron cross-section measurements; Reactor spectra; Nuclear spectroscopy; Gas scintillation counters; Cloud chambers; Radiation shielding; Fallout studies.
Address: 998 Corbett, San Francisco, California, U.S.A.

HUDDLESTON, Stanley Ernest, B.Sc., B.E., B.Ec. Born 1913. Educ.: Adelaide and Sydney Univs. Administration Manager, Elec. Trust of South Australia. Society: Inst. of Eng., Australia.
Nuclear interest: Nuclear power.
Address: 5 Rodney Street, Woodville, South Australia.

HUDIMOTO, Busuke. Prof. in Charge, Eng. Res. Inst., Dept. of Nucl. Eng.,Kyoto Univ.
Address: Kyoto University, Department of Nuclear Engineering, Yoshida-Honmachi, Sakyo-ku, Kyoto, Japan.

HUDIS, Jerome. Member, Subcom. on Radiochem., Com. on Nucl. Sci., N.A.S. N.R.C.
Address: National Academy of Sciences - National Research Council, Committee on Nuclear Sciences, 2101 Constitution Avenue, N.W., Washington 25, D.C., U.S.A.

HUDSON, Miller N., Jr., B.S. (Chem. Eng.), Registered Professional Eng. Born 1924. Educ.: New Mexico State Univ. Process and Reactor Eng., Phillips Petroleum Co., 1948-52; Chief,Nucl. Materials Management, Idaho Operations Office (1952-54), Chief, Mathematical Statistics Branch, Nucl. Materials Management Div. (1954-55), U.S.A.E.C.; Asst. to Chief Nucl. Reactor Operation and

Design for Reactor Operations, Convair, 1955-56; Project Director, Operations Res. Tech. Operations, Inc., 1956; Special Asst. to Director, Nucl. Materials Management Div., (1956-59), Chief, Tech. Branch, Internat. Affairs Div. (1959-60), Asst. Director for Safeguards Internat. Affairs Div. (1960-63), Sci. Rep. to Canada, (1963-67), U.S.A.E.C.; Sci. Attaché American Embassy, Rio de Janeiro, U.S. Dept. of State, 1967-.
Nuclear interests: Management; Engineering; Statistics; Nuclear aspects of foreign policy.
Address: American Embassy, Rio de Janeiro, A.P.O., New York, New York 09676, U.S.A.

HUDSPETH, Emmett L., M.A., Ph.D. Born 1916. Educ.: Rice Inst. Asst. Director, Bartol Res. Foundation, 1945-50; Prof., Phys., and Director, Nucl. Phys. Lab., Texas Univ., 1950-; Consultant, Texas Nucl. Corp., Radiobiol. Lab., and Southwest Res. Inst. Society: Fellow, A.P.S.
Nuclear interests: Nuclear physics; Nuclear disintegrations with Van de Graff machines; Nuclear decay schemes; Analysis by radiation.
Address: 6104 Janey Drive, Austin, Texas 78731, U.S.A.

HUDSWELL, Fred, M.A. Born 1909. Educ.: Cambridge Univ. Group Leader, Preparative Group, Chem. Div., A.E.R.E., Harwell, 1947-. Member, Sci. Secretariat for U.N. 2nd I.C.P.U.A.E., March-Sept. 1958.
Nuclear interests: Radiochemical methods applied to research in inorganic chemistry; Chemistry of actinides and rare earths; Trapping and safe disposal of fission products; Preparative inorganic chemistry.
Address: United Kingdom Atomic Energy Authority, Atomic Energy Research Establishment, Harwell, Didcot, Berkshire, England.

HUESTON, Frank Harris, B.A.Sc. Born 1930. Educ.: Toronto Univ. and Chalk River Reactor School. At Eldorado Mining and Refining Ltd., 1956-, (Asst. Supt. Green Salt Operations, 1958, Supt. Green Salt Operations, 1958, Development Eng.,1959, Project Supt., 1962). Societies: A.P.E.O.; Canadian Nucl. Assoc.; A.S.M.; American Vacuum Soc.; American Inst. Metal. Eng.
Nuclear interest: Uranium processing, with particular interest in metallurgy (vacuum)and fuel fabrication.
Address: Eldorado Mining and Refining, Ltd., 215 John Street, Port Hope, Ontario, Canada.

HUET, Henri René, Eng. E.S.P.C.I., L. ès. Sc. Born 1913. Educ.: E.S.P.C.I., Faculté des Sci. de Paris. Manager, Le Bouchet Centre, C.E.A. Book: Chapter on Uranium Manufacture, Nouveau Traité de Chimie (P. Pascal, 1960). Society: Sté. de Chimie Industrielle.
Nuclear interests: Chemistry, refining and metallurgy of uranium - development.
Address: Usine du Bouchet, B.P. No. 6, Ballancourt, (Seine et Oise), France.

HUET, Jean Jacques. Ing. civil métallurgiste. Born 1929. Educ.: Faculté polytechnique de Mons. Head, Metal. Dept., Centre d'Etudes de l'EnergieNucléaire. Societies: Inst. of Metals (London); Benelux Metallurgie (Brussels); A.S.M.
Nuclear interests: Nuclear metallurgy - principally fuel element fabrication studies and related physical and structural properties of metals and ceramics, pre- and post-irradiation testing.
Address: 260 Boeretang, Mol, Belgium.

HUET, Philippe. Directeur Gén. des Prix et des Enquêtes Economiques, Ministère des Finances et des Affaires Economiques. Member, Commission Consultative des Marchés, C.E.A.
Address: Commissariat à l'Energie Atomique, 29-33 rue de la Federation, Paris 15, France.

HUET, Pierre. Graduated magna cum laude at Law Faculty and at School of Political Sci., Paris. Born 1920. Educ.: Paris Univ. Sen. Lecturer, Political Studies Inst., Paris Univ., 1947. Legal Adviser, O.E.E.C., 1948; Sen. Member (Maitre des Requêtes), Conseil d'Etat, 1954; Gen. Counsel, O.E.E.C., 1956; Prof., Inst. for Higher International Studies, Law Faculty, Paris, 1956; Director Gen., E.N.E.A., O.E.C.D., Paris, 1958-64; Pres., A.T.E.N., 1965; Sec. Gen., Conseil d'Etat, 1966.
Nuclear interest: International co-operation in the field of nuclear energy.
Address: Conseil d'Etat, Palais Royal, Paris 1, France.

HUFF, George Albert, B.S. (Chem.). Born 1924. Educ.: Brigham Young Univ. Supv., Chem. Analysis Sect., Idaho Nucl. Corp. Society: A.C.S.
Nuclear interest: Chemistry of nuclear fuels.
Address: Idaho Nuclear Corporation, Box 1845, Idaho Falls, Idaho 83401, U.S.A.

HUFF, James B., B.S., M.B.A. Born 1919. Educ.: Indiana Univ. and Harvard Graduate School of Business. Treas. and Business Manager, Atomic and Space Development Authority, New York State, 1964-.
Nuclear interest: Financial.
Address: New York State Atomic and Space Development Authority, 230 Park Avenue, New York, N.Y. 10017, U.S.A.

HUFF, John B. Born 1910. Educ.: Washington Univ., Seattle (Chem. Eng.). Health Phys., Atomic Energy Div., Phillips Petroleum Co., Idaho Falls, 1954-62; Decontamination specialist, Lawrence Rad. Lab., Livermore, Calif., 1962-64; Health Phys., Bureau Commercial Fisheries, Gloucester, Mass., 1964-. Society: H.P.S.
Nuclear interests: Decontamination; Waste disposal; Food irradiation.
Address: 4 Paradis Circle, Rockport, Mass., U.S.A.

HUFFMAN, John Randolph, B.S., Ph.D. Born 1905. Educ.: Yale Univ. and Yale Univ. Graduate School. Project Eng., Argonne Nat. Lab., 1948-51; Tech. Consultant (1951-53), MTR Tech. Director (1953-54), Asst. Manager, Tech. (1954-65), Phillips Petroleum Co.; Nucl. Power Consultant, Jackson and Moreland Div., United Eng. and Constructors Inc., 1965-. Societies: A.C.S.; A.I.Ch.E.; A.A.A.S.; Licensed Professional Eng. in State of New York; Fellow, A.N.S. Nuclear interests: Direction of research and development, reactor design.
Address: United Engineers and Constructors, Inc., 1401 Arch Street, Philadelphia, Pa., U.S.A.

HUG, Otto Franz Josef, Dr. med., Prof. Rad. Biol. Born 1913. Educ.: Munich, Frankfurt/Main, Heidelberg and Freiburg Univs. Lehrstuhl für Biologie, Phil-Theol. Hochschule, Regensburg, 1956-57; Sen. Officer, Health and Safety Div., I.A.E.A., 1958-59; Prof. Radiobiol. and Director, Inst. Radiobiol., Munich Univ., 1959-; Director Inst. für Biologie, Ges. f. Strahlenforschung m.b.G., Neuherberg-Munich. Books: Strahlendosis und Strahlenwirkung, (1953); Wissenschaftliche Grundlagen des Strahlenschutzes (1957); co-author, Stochastik der Strahlenwirkung (1966). Papers: 80 papers on ultrasonics, electron microscopy, rad. biol. and pathology, radioisotopes and rad. protection. Societies: Auswärtiges wissenschaftliches Mitglied des Max Planck-Inst. für Biophysik, Frankfurt a. M.; Deutsche Röntgengesellschaft; Rad. Res. Soc.; Deutsche Gesellschaft für Pathologie.
Address: Strahlenbiologisches Institut, Munich University, 19 Bavariaring, 8 Munich 15, Germany.

HUGGARD, A. J., Dr. U.K.A.E.A. Liaison Officer, Overseas Offices, Canada.
Address: c/o Atomic Energy of Canada Ltd., Chalk River, Ontario, Canada.

HUGGETT, Clayton, Dr. Sen. Phys. Chem., Advanced Technol. Div., Atlantic Res. Corp.
Address: Atlantic Research Corporation, Henry G. Shirley Memorial Highway at Edsall Road, Alexandria, Virginia, U.S.A.

HUGH-JONES, Philip, M.A., M.D. (Camb.), F.R.C.P. (London). Born 1917. Educ.: Cambridge and London Univs. Sci. Staff, M.R.C., 1942-52; Sen. Lecturer in Medicine and Physician in Univ. Coll. of West Indies, Jamaica, 1952-55; Physician, Hammersmith Hospital, 1953-63; Physician and Part-time Director, Pulmonary Res. Unit, King's Coll. Hospital, 1963-. Societies: Assoc. of Physicians of Great Britain and Ireland; Medical Res. Soc.; Thoracic Soc. Nuclear interests: Use of radioactive gases for medical research and for the assessment of patients.
Address: King's College Hospital, Denmark Hill, London, S.E.5, England.

HUGHES, C. R., M.D. Formerly Director, Dept. of Rad. Therapy and Isotopes, Cleveland Clinic Foundation.
Address: Cleveland Clinic Foundation, Division of Radiology, 2020 East 93rd Street, Cleveland 6, Ohio, U.S.A.

HUGHES, Clifford Arthur, M.Sc., Ph.D., A.R.I.C. Born 1935. Educ.: Liverpool Coll. of Technol. and Natal Univ. Shell Refining and Marketing Co., 1952-55; U.K. Atomic Energy Authority, Capenhurst, 1955-59; Natal Univ., South Africa, 1959-63; Manchester Univ. 1963-64; Sen. Lecturer in Nucl. Chem., Liverpool Regional Coll. of Technol., 1964-. Society: Fellow, Chem. Soc.; Mossbauer Spectroscopy Discussion Group, Chem. Soc. Nuclear interests: Nuclear reaction studies including fission process. Activation analysis. Mossbauer spectroscopy.
Address: 4 Newton Park Road, West Kirby, Cheshire, England.

HUGHES, Donald, B.Sc., Ph.D., C. Eng. Born 1931. Educ.: Nottingham Univ. Res. Student under contract from A.E.R.E., Harwell, 1951-54; G.E.C., 1954-58; Roy. Marsden Hospital and Cancer Res. Inst., London, 1958-59; M.R.C. Environmental Rad. Res. Unit, Leeds Univ., Medical Phys. Dept., Leeds Gen. Infirmary, 1959-64. Societies: M.I.E.E.; A.Inst. P.; British Inst. Radiol.; Hospital Phys. Assoc. Nuclear interests: Radiation protection; Clinical application of radioactivity in man.
Address: Radiation Protection Officer, Leeds University, Leeds 2, Yorkshire, England.

HUGHES, Howard Arthur, B.Sc. (Special), A.R.C.S. Born 1923. Educ.: Roy. Coll. of Sci., Imperial Coll., London. Principal Phys., Medical Phys. Unit, Edinburgh, 1954-58; Asst. Tech. Manager, U.K.A.E.A., Windscale and Calder Works, 1958-. Society: F. Inst. P. Nuclear interest: Reactor technology.
Address: 11 Wastwater Rise, Seascale, Cumberland, England.

HUGHES, J. E. Vice-Pres. and Asst. Gen. Manager, Kaiser Engineers.
Address: Kaiser Engineers, Division of Henry J. Kaiser Co., Kaiser Centre, Oakland 12, California, U.S.A.

HUGHES, Lloyd. Manager, Cost Estimating, Nucl. Materials and Equipment Corp.
Address: Nuclear Materials and Equipment Corp., Apollo, Pennsylvania 15613, U.S.A.

HUGHES, Thomas Garfield, B.Sc., A.M.I. Chem.E. Born 1917. Educ.: London Univ. Manager, Separation Plants (1949-54), Asst. Works Manager (1954-57), Supt., Chem. Plants (1957-), U.K.A.E.A., Windscale Works, Sellafield, Seascale, Cumberland. Nuclear interest: Chemical reprocessing of reactor fuels; Manufacture of plutonium containing ceramic nuclear reactor fuels.

Address: 4 Wastwater Rise, Seascale, Cumberland, England.

HUGHES, Vernon Willard, B.S., M.S., Ph.D. (Phys.). Born 1921. Educ.: Columbia Coll., California Inst. Technol. and Columbia Univ. Consultant, Internat. Business Machines Corp., 1957-60; Consultant, O.R.N.L., 1960-; Asst., Assoc., and full Prof. Phys., Chairman, Dept. of Phys., Yale Univ.; Consultant, N.A.S.A.; Trustee Assoc. Univs., Inc. Books: Electron Magnetic Moment and Atomic Nagnetism in Recent Res. in Molecular Beams (N.Y., Academic Press, 1959); co-author, Atomic and Molecular Beam Spectroscopy in Handbuch der Physik, Vol. 37/1 (Springer-Verlag, 1959). Society: Fellow, A.P.S., 1957-. Nuclear interests: Radiofrequency and microwave spectroscopy of atoms and molecules; Particle physics.
Address: Physics Department, Yale University, New Haven, Conn. 06520, U.S.A.

HUGHES, Walter L. Formerly Head, Div. of Biochem., Brookhaven Nat. Lab.; Chairman, Dept. of Physiology, Tufts Univ., 1964-. Deleg. to 2nd I.C.P.U.A.E., Geneva, Sept. 1958.
Address: Department of Physiology, Tufts University, Medford, Massachusetts 02155, U.S.A.

HUGHEY, Robert W., B.S. (Chem.). Born 1921. Educ.: Stanford Univ., California. Director, Tech. Div., San Francisco Operations Office, U.S.A.E.C., 1951-. Chairman, Northern California Sect., A.N.S.
Nuclear interests: Technical administration in the fields of physical research, nuclear weapons, reactor development, radioisotopes, controlled thermonuclear reactions, health and safety.
Address: San Francisco Operations Office, United States Atomic Energy Commission, 2111 Bancroft Way, Berkeley, California 94704, U.S.A.

HUGHSON, Robert C. Asst. Prof., Inst. for Res., Seattle Pacific Coll.
Address: Institute for Research, Seattle Pacific College, Seattle 19, Washington, U.S.A.

HUGH, Jacques, Pierre, B.Sc. (Eng.), Ph.D., F.I.M. Born 1927. Educ.: Witwatersrand and Sheffield Univs. Head, Metal. Div., Council for Sci. and Ind. Res., Pretoria, South Africa, 1954-59; Chief, Phys. Metal. (1960-67), Deputy Director Gen. (1967-), South African Atomic Energy Board. Council, South African Inst. Mining and Metal. Society: B.N.E.S.
Nuclear interest: Nuclear metallurgy, with particular reference to nuclear fuels, cladding materials and irradiation testing.
Address: Atomic Energy Board, Private Bag 256, Pretoria, South Africa.

HUGUENIN, Pierre Louis, Prof., Dr. Phys. Born 1932. Educ.: Neuchâtel and Bonn Univs.
Nuclear interest: Theoretical nuclear physics.
Address: Institut de physique, Neuchâtel University, Breguet 1, Neuchâtel, Switzerland.

HUGULEY, Charles Mason, Jr., A.B., M.D. Born 1918. Educ.: Emory and Washington Univs. Assoc. Prof., and Chairman, Radioisotope Com. I (Human Use), School of Medicine, Emory Univ.
Nuclear interest: Investigational use of radioactive tracers in man.
Address: School of Medicine, Emory University, Atlanta 30322, Georgia, U.S.A.

HUISING, Wilhelmus Bernardus, Chem. Born 1929. Sen. Officer, Isotopes Div., N. V. Philips-Duphar, 1953-.
Nuclear interests: Isotopes production and use in medicine, industry and research, and its commercial aspects.
Address: 33 Goudenregenlaan, Castricum, Netherlands.

HUISKAMP, Willeminus Jan, Dr. Born 1925. Educ.: Utrecht Univ. Lector, Leiden Univ. Kamerlingh Onnes Lab., Leiden. Sec., Dutch Phys. Soc. Society: Dutch Phys. Soc.
Nuclear interests: Orientation of atomic nuclei at very low temperatures; Study of the radiations emitted from oriented radioactive nuclei.
Address: 35 Willem de Zwijgerlaan, Oegstgeest, Netherlands.

HUIZENGA, H., Dr. Sec. for the Health Council; Vice Pres., Central Council for Nucl. Energy; Adviser, Third I.C.P.U.A.E., Geneva, Sept. 1964; Formerly Member, Health and Safety Sub-Com., O.E.C.D., E.N.E.A.
Address: Health Council, 8 Dr. Kuyperstraat, The Hague, Netherlands.

HUIZENGA, John R., A.B., Ph.D. Ernest O. Lawrence Memorial Award, 1966. Born 1921. Educ.: Calvin Coll., Michigan, and Illinois Univ. Sen. Sci., Argonne Nat. Lab., 1949-67; Prof. of Chem. and Phys., Rochester Univ., 1967-. Books: Chapter in Nuclear Structure and Electromagnetic Interactions (London, Oliver and Boyd, 1965); Chapter II in Nuclear Reactions, Volume II, (Amsterdam, North Holland Publishing Co., 1962); Chapter in Nuclear Chemistry, (New York, Academic Press, 1967). Societies: A.P.S.; A.C.S.
Nuclear interests: Nuclear reactions and nuclear fission.
Address: Nuclear Structure Research Laboratory, Rochester University, Rochester, New York 14627, U.S.A.

HUKE, Frank B. Manager, Product Eng., Refractory Div., Norton Co.
Address: Norton Company, Worcester 6, Massachusetts, U.S.A.

HUKKINEN, Lars Johan, B.Sc. (Ch.E., diplomi-insinööri). Born 1928. Educ.: Finland Inst. Technol. Asst. teacher in inorganic chem.,

Finland Inst. Technol., 1955-. Societies: Finnish Chem. Soc.; Assoc. Finnish Eng.
Nuclear interests: Nuclear materials, their chemistry, analysis, spectrochemistry.
Address: Laboratory of Spectrochemistry, Finland Institute of Technology, Otaniemi, Finland.

HUKUNAGO, Hiroshi. Chief, Radioactivity Sect., Atomic Energy Bureau, A.E.C. of Japan.
Address: Atomic Energy Commission of Japan, 3-4 Kasumigaseki, Chiyoda-ku, Tokyo, Japan.

HUKUO, Ken-iti, Commissioner, Atomic Eng. Res. Com., Nagoya Inst. Technol.
Address: Nagoya Institute of Technology, Gokiso-cho, Showa-ku, Nagoya, Japan.

HULKA, Ivan, Mech. Eng. Born 1933.
Educ.: Pilsen Inst. of Technol. Nucl. Power Plant Div., Skoda Concern, Pilsen. Member, Editorial Council, Jaderná Energie, Journal of Czech A.E.C. Society: Nucl. Power Section, Czechoslovak Soc. for Sci. and Technol.
Nuclear interests: Reactor heat removal problems.
Address: tr. 109 Slovanská, Plzen, Czechoslovakia.

HULL, A. C. Sec., Joint Marine Atomic Energy Com.
Address: The Institute of London Underwriters, 40 Lime Street, London, E.C.3, England.

HULL, F. P. Director, F. Hull and Sons Ltd.
Address: F. Hull and Sons Ltd., 174 London Road, Mitcham, Surrey, England.

HULL, Harvard Leslie, A.B., Ph.D. (Phys.). Born 1906. Educ.: Nebraska Wesleyan and Columbia Univs. Director Remote Control Eng. Div., Argonne Nat. Lab., 1949-53; Vice-Pres., R. and D., Capehart-Farnsworth Div. (1953-54), Pres., Farnsworth Electronics Div. (1954-56), Internat. Telephone and Telegraph Corp.; Vice-Pres., Litton Industries, Inc., 1956-57; Pres., Hull Assoc., Nucleonics-Electronics Consultants, 1957-. Societies: A.P.S.; A.N.S.; Inst. Aerospace Sciences; Inst. Radio Eng.; American Rocket Soc.; A.A.A.S.; American Soc. Naval Eng.
Nuclear interests: Management and consulting in areas of hot laboratory design and operation; Viewing and handling of radioactive materials; Master-slave manipulators; Stereo television and remote control; Nondestructive testing and effect of radiation on electronic equipment.
Address: Hull Associates, 30 West Monroe Street, Chicago 3, Ill., U.S.A.

HULL, John Neville, B.Sc. (Eng.), (Hons., London), F.I.E.E. Born 1917. Educ.: Rugby Coll. Technol. Sen. Eng., Standards Div., Dominion Phys. Lab., New Zealand, 1950-53; Light Div., Nat. Phys. Lab., 1954; Reactor and Eng. Divs. (1954-60). Manager, Harwell Reactor School (1960-63), Post-graduate Education Centre (1963-), Education and Training Dept. (1965-), A.E.R.E., Harwell. Societies: I.E.E.; B.N.E.S.
Nuclear interests: Reactor physics and engineering. Post-graduate education courses.
Address: Post-Graduate Education Centre, Atomic Energy Research Establishment, Harwell, Didcot, Berks., England.

HULL, McAllister, Jr. Assoc. Prof., Phys. Dept., Yale Univ.
Address: Yale University, Sloane Physics Laboratory, 217 Prospect Street, New Haven, Connecticut 06520, U.S.A.

HULL, William Griffin, F.I.M., C.G.I.A., A.M.Inst. W. Born 1920. Principal Res. Officer, British Welding Res. Assoc., 1950-58; Head of Labs., R. and D. Div., Atomic Power Constructions Ltd., 1950-. Societies: F.I.M.; Assoc. Member, Inst. Welding.
Nuclear interests: Metallurgy with particular reference to welding and engineering materials; Laboratory management; Research administration.
Address: Atomic Power Constructions Ltd., Research and Development Division, Cranford Lane, Hounslow, Middlesex, England.

HULME, H. R., Dr. Chief of Nucl. Res., A.W.R.E., U.K.A.E.A.
Address: Atomic Weapons Research Establishment, U.K.A.E.A., Aldermaston, Berks., England.

HULOVEC, Jan, M.M.E. Born 1932. Educ.: Pilsen Polytech. Inst. and Coll. Mech. Eng. Head, Reactor Phys. and Computation Group, Nucl. Power Plant Div., Skoda Concern, Pilsen, Czechoslovakia. Society: Czechoslovak Tech. and Sci. Assoc.
Nuclear interests: Reactor kinetics and control; Nuclear station control and safety; Reactor design; Reactor calculation.
Address: 23 St. Vodicky, Pilsen, Czechoslovakia.

HULSMAN, Johannes Paulus, Eng. Born 1900. Educ.: Technische Hogeschool Delft. Director of Res., Netherlands Res. Centre for Shipbuilding and Navigation; Formerly Member of Board, Stichting Kernvoortstuwing Koopvarrdijschepen. Societies: K.I.v.I.; M.I.N.A.; M.I.Mar.Eng.
Nuclear interest: Application of nuclear engineering for ship propulsion.
Address: Flatgebouw Spiegelhoek Herenstraat-Bussum, Netherlands.

HULSTON, John Richards, B.Sc., M.Sc. (Hons.), Ph.D., A.Inst.P. Born 1932. Educ.: New Zealand and McMaster (Canada) Univs. In D.S.I.R. (New Zealand Govt. Service), Inst. of Nucl. Sci., New Zealand, 1955-. Societies: Roy. Soc. of New Zealand; Inst.

Phys. and Phys. Soc.; N.Z. Inst. of Chem.; A.G.U.
Nuclear interests: Mass spectrometry (instrumentation and vacuum techniques); Isotope geology (natural variations of carbon, sulphur, hydrogen and oxygen isotopic ratios in nature; potassium argon dating).
Address: Institute of Nuclear Sciences, Department of Scientific and Industrial Research, Private Bag, Lower Hutt, New Zealand.

HULSWIT, Charles L. Director, Empire State Atomic Development Assocs., Inc.
Address: Empire State Atomic Development Associates, Inc., 4 Irving Place, New York 3, New York, U.S.A.

HULTGREN, Ake Valdemar, Fil. Mag. (M.Sc.). Born 1925. Educ.: Stockholm Univ. Head, Plutonium Fuel Sect., A.B. Atomenergi, Stockholm.
Nuclear interests: Fuel fabrication and reprocessing.
Address: A.B. Atomenergi, Studsvik, Nyköping, Sweden.

HULTHEN, Lamek, Ph.D. Born 1909. Educ.: Stockholm Univ. Prof. Mathematical Phys., Roy. Inst. of Technol., Stockholm, 1949-. Member, Swedish Council for Nucl. Res., 1965; Member, 1966 Atomic Energy Study Com. (Swedish Government Com.); Chairman, Swedish Nat. Com. for Phys. Book: Co-author, The Two-Nucleon Problem. Handbuch der Physik. Volume 39 (Berlin, Springer Verlag, 1957). Societies: Fellow, Swedish Acad. of Eng. Sci.; Fellow, Roy. Swedish Acad. of Sci.; Member, Nobel Com. for Phys.
Nuclear interests: Theoretical nuclear physics. Scattering theory. Mathematical methods in nuclear physics and reactory theory. Planning of energy production with special regard to nuclear reactors.
Address: The Royal Institute of Technology, Stockholm 70, Sweden.

HULTIN, Sven Olof, Diplomingenjör (Chem. Eng.). Born 1920. Educ.: Abo Akademi. Vice Managing Director (1954-63), Managing Director (1964-), Ekono, Helsinki; Tech. Manager, Voimayhdistys Ydin (Nucl. Power Assoc.), Helsinki, 1956-67. Finnish Commission of Radiol. Protection, 1957-60; Tech. Adviser, Finnish Deleg. to 2nd I.C.P.U.A.E., Geneva, 1958; Finnish Deleg., Halden Tech. Group, Halden, Norway, 1959-64; Formerly Finnish Deleg., Halden Programme Group, 1964-. Societies: A.C.S.; T.A.P.P.I.; Tekniska Föreningen i Finland.
Nuclear interest: Power economics.
Address: Ekono, 14 S. Esplanadg., Helsinki, Finland.

HULUBEI, Horia, D. ès Sc. (Paris), Dr. hon. causa (Jassy). Born 1896. Educ.: Jassy (Roumania) and Paris Univs. Director, Inst. de Fizica Atomica, Acad. of the Roumanian People's Republic, 1949-; Now Pres., Com. for Nucl. Energy, Council of Ministers, Bucharest; Vice Chairman, Board of Governors, I.A.E.A., 1963-64; Member, Editorial Board, Atompraxis; Head, Roumanian Delegation to Standing Commission for Peaceful Uses of Atomic Energy, Council for Mutual Economic Aid; Member, titulaire, Roumanian Acad.; Member Correspondant, Acad. des Sci., Paris; Member Correspondant, Acad. de Lisbon; Societies: Sté. Française de Phys.; Soc. Suisse de Phys.; Soc. Alemande de Phys.; Life Member, Acad. of Sci., New York.
Address: Institute of Atomic Physics, Bucharest, Roumania.

HUMBACH, Walter Hans Joerg, Dipl. Phys., Dr. rer. nat. Born 1920. Educ.: München and Göttingen Univs. With Siemens-Reiniger Werke A.G. Erlangen (Betatron Labs.), 1951-53; Siemens-Schuckert Werke A.G. Erlangen (Research Labs.), 1954-61; Prof. Reactor Eng., Head, Inst. Reactor Eng., Darmstadt Tech. Univ., 1962-; Co-editor, Nukleonik. Books: 2 articles in: Kerntechnik (by Rietzler and Walcher, Teubner, 1958). Society: Deutsche physikalische Gesellschaft.
Nuclear interests: Reactor theory and experiments; Nuclear safety.
Address: Institut für Reaktortechnik der Technischen Hochschule, 1 Hochschulstrasse, 61 Darmstadt, Germany.

HUMBEL, J. H. Sen. Tech. Officer, Giovanola Freres S.A.
Address: Giovanola Freres S.A., Monthey, (Valais), Switzerland.

HUMBLE, LeRoy V., Nucl. Reactor Div., N.A.S.A., Lewis Res. Centre.
Address: Nuclear Reactor Division, National Aeronautics and Space Administration, Lewis Research Centre, 21000 Brookpark Road, Cleveland, Ohio, U.S.A.

HUMBLET, Jean E., Dr.Sc. and Agrégé. Born 1918. Educ.: Liège Univ. Prof., Liège Univ. 1959-. Centre pour l'Etude de l'Energie Nucléaire, Mol; Institut Interuniversitaire des Sciences Nucléaires, Brussels. Society: A.P.S.
Nuclear interest: Theoretical nuclear physics.
Address: University of Liège, Theoretical Nuclear Physics, 15 avenue des Tilleuls, Liège, Belgium.

HUME, F. C. Construction Manager, B. D. Bohna and Co. Inc.
Address: B. D. Bohna, Inc., 515 Market Street, San Francisco 5, California, U.S.A.

HUMMEL, Dieter, Prof., Dr. Mitglied, Arbeitskreis 11/2, Chemie, Federal Ministry for Sci. Res.
Address: Federal Ministry for Scientific Research, 46 Luisenstrasse, 532 Bad Godesberg, Germany.

HUMMEL, Dietrich Oskar, Diplomchemiker, Dr. rer. nat., Assoc. Prof. Born 1925. Educ.: T.H. Stuttgart. Lecturer, Phys. Chem., Cologne Univ., 1961. Visiting Asst. Prof., Cincinnati Univ., 1962-63. Societies: Gesellschaft Deutscher Chemiker; Bunsengesellschaft für Physikalische Chemie; Gesellschaft Deutscher Naturforscher und Aerzte.
Nuclear interests: Chemical effects of ionising radiation: kinetics of chain reactions, mechanism of radiation-induced polymerisations and copolymerisations.
Address: Institut für physikalische Chemie und Kolloidchemie, Cologne University, 34 Severinswall, Cologne, Germany.

HUMMEL, F. S. Sect. Head, Fuel Handling Power Projects, A.E.C.L.,1958-64; Sen. Assoc. in charge, Nucl. Eng. Dept., Dilworth, Secord, Meagher and Assocs. Ltd. Chairman, Technol. Com., Canadian Nucl. Assoc.
Address: Dilworth, Secord, Meagher and Associates Limited, 4195 Dundas Street West, Toronto 18, Ontario, Canada.

HUMMEL, Harry H., B.Ch.E., M.S.,Ph.D. Born 1917. Educ.: Louisville and Wisconsin Univs. Assoc. Phys. (1948-58), Sen. Phys. (1958-63), Assoc. Director, Reactor Phys. Div. (1963-), Argonne Nat. Lab. Societies: Fellow, A.N.S.; A.P.S.
Nuclear interests: Fast reactor physics, particularly with respect to reactivity coefficients, especially with Doppler and sodium void effects, and also with respect to group cross section evaluation and generation.
Address: Argonne National Laboratory, 9700 South Cass Avenue, Argonne, Illinois 60439, U.S.A.

HUMMEL, John Philip, B.S., Ph.D. Born 1931. Educ.: Rochester and California Univs. Instructor (1956-58), Asst. Prof. (1958-62), Assoc. Prof. (1962-67), Prof. (1967-), Chem. and Phys., Illinois Univ. Society: A.P.S.
Nuclear interests: Nuclear physics; Study of mechanisms of photonuclear reactions.
Address: Chemistry Department, Illinois University, Urbana, Illinois, U.S.A.

HUMMEL, Ulrich, Dipl.-Eng. Born 1931. Educ.: Leoben, Clausthal Univs. Sales Manager (processes and equipment), Nukem, Nuklear-Chemie und -Metallurgie G.m.b.H. Wolfgang bei Hanau/M., 1960-; Vickers-Zimmer A.G., Planung und Bau von Industrieanlagen, Frankfurt-am-Main.
Nuclear interests: Nuclear fuels, fuel cycle economics, management, isotopes separation.
Address: 22 Amselstrasse, 6450 Hanau-Hohe Tanne, Germany.

HUMPHREY, A. E., Dr. Sec., Sect. on Economic and Appl. Microbiol., Internat. Assoc. of Microbiol. Socs.
Address: International Association of Microbiological Societies, c/o Dr. N. E. Gibbons, Division of Applied Biology, National Research Council, Ottawa 2, Canada.

HUMPHREY, Ronald Mack, B.A. (Biol.), M.A. (Bacteriology), Ph.D. (Bacteriology). Born 1932. Educ.: Hardin-Simons (Abilene) and Texas Univs. Assoc. Prof. Biophys. and Rad. Biol., Texas Univ. Graduate School of Biomedical Sci., Houston, 1960-. Societies: A.A.A.S.; Rad. Res. Soc.
Nuclear interests: Radiation biology and radiation genetics.
Address: Department of Physics, University of Texas M. D. Anderson Hospital and Tumour Institute, Texas Medical Centre,Houston 25, Texas, U.S.A.

HUMPHREYS, George Baxter, B.S. (Process Eng., California), M.S. (Mech. Eng., California); Registered Eng., State of California. Born 1929. Educ.: California Univ. and O.R.S.O.R.T. Teaching Asst., California Univ., 1951; Student-employee, O.R.N.L., 1952-53; Asst. Eng. (1955-57), Sen. Eng. (1957-62), Assoc. Eng. (1962-), Kaiser Engs. Societies: A.N.S.; N.S.P.E.
Nuclear interests: Reactor hazards evaluation; Neutron and gamma heating; Reactor design; Process heat reactors; Thermal stress analysis; Fuel cycle analysis; Dual purpose reactors power-water production.
Address: 3433 Moraga Boulevard, Lafayette, California 94549, U.S.A.

HUMPHREYS, John Ross, Jr., B.S. (Chem. Eng.). Born 1922. Educ.: Illinois Inst. Technol. and Texas A. and M. Univ. Argonne Nat. Lab., 1948-59; Los Alamos Sci. Lab., 1959-65; Argonne Nat. Lab., 1965-. Societies: A.N.S.; RESA.
Nuclear interests: Reactor design; Liquid metal technology; Reactor materials.
Address: Argonne National Laboratory, 9700 South Cass Avenue, Argonne, Illinois 60439, U.S.A.

HUMPHREYS, Richard Franklin, B.A., M.A., Ph.D. Born 1911. Educ.: DePauw, Syracuse and Yale Univs. Asst. Manager, Phys. (1949-51), Manager, Phys. (1951-56), i/c installation of res. reactor facility (1956), Vice-Pres. (1956-61), Armour Res. Foundation; Pres., The Cooper Union, (1961-); Director, Assoc. Hospital Service of New York; Trustee, Josiah Macy, Jun., Foundation. Book: First Principles of Atomic Phys. (Harper Bros., 1950). Societies: A.A.A.S.; A.P.S.; American Ordnance Assoc.
Nuclear interests: Nuclear physics; Research reactors; Research management; Education.
Address: The Cooper Union, Cooper Square, New York 3, N.Y., U.S.A.

HUNDESHAGEN, Heinz, Priv. Doz., Dr. med. Born 1928. Educ.: Univ. Marburg/Lahn. Oberarzt, Dept. of Radiobiol. and Radioisotopes, Rad. Inst., Marburg/Lahn Univ., 1956-; Pres., German Acad. for Nucl. Medicine, Hanover, 1968-.

Nuclear interests: Nuclear medicine; Radiobiology.
Address: 8a Robert-Koch-Strasse, Marburg/Lahn, Germany.

HUNGERFORD, Herbert Eugene, Jr., B.S. (Phys.), M.S. (Phys.), Ph.D. (Nucl. Eng.). Born 1918. Educ.: Trinity Coll. (Connecticut), Alabama and Purdue Univs. Exptl. Phys., O.R.N.L., 1950-55; Phys., Shielding Eng. (1955-58), Head, Shielding and Health Phys. Sect. (1958-62), Atomic Power Development Assocs., Inc.; Res. Assoc., Nucl. Eng. Dept. (1963-64), Assoc. Prof. Nucl. Eng. (1964-), Purdue Univ. Societies: A.P.S.; A.N.S.; H.P.S.; A.A.P.T.; A.S.E.E.
Nuclear interests: Shielding; Shield materials and design; Nuclear radiation transport; Application of computer codes to shielding.
Address: Nuclear Engineering Department, Purdue University, Lafayette, Indiana, U.S.A.

HUNT, C. J., B.Sc., A. Inst. P. Sen. Lecturer, Radiol. Lab., City of Portsmouth Coll. of Technol.
Address: City of Portsmouth College of Technology, Portsmouth, Hants., England.

HUNT, Howard B., M.D. Member, Consulting Editorial Board, J. Nucl. Medicine.
Address: 333 N. Michigan Avenue, Chicago 1, Illinois, U.S.A.

HUNT, R. A. H.N.C. Chief Technician, Health Phys. and Radioisotope Unit, Nat. Res. Council of Ghana.
Address: National Research Council of Ghana, Physics Department, University of Ghana, Legon, Accra, Ghana.

HUNT, Stanley Ernest, B.Sc., Ph.D. Born 1922. Educ.: Liverpool and Reading Univs. Sect. Leader, Nucl. Phys. Sect., A.E.I. Res. Lab., Aldermaston Court, Berkshire, 1957-63; Visiting Prof. in Modern Phys., Algiers Univ., 1959; Head, Phys. Dept. (1963-), Dean, Faculty of Sci. (1967-), Univ. of Aston in Birmingham. Society: F.Inst.P.
Nuclear interests: Low energy nuclear physics; Absolute measurement of (p, γ) resonances, γ-ray spectroscopy; Neutron and charged particle activation, reactor physics, radiological protection.
Address: 159 Burfield Road, Solihull, Warwickshire, England.

HUNT, W. M. Manager, Purchasing and Stores, ITT Federal Support Services, Hanford Facilities, U.S.A.E.C.
Address: U.S.A.E.C., ITT Federal Support Services, Richland, Washington, U.S.A.

HUNTER, Colin Graeme, B.Sc. (London), M.D., M.B., Ch.B. (New Zealand), M.R.C.P. (London), D.P.H. (London), D.I.H. Born 1913. Educ.: Otago Univ., New Zealand. Surgeon Commander Roy. Navy (Retired), Sen. Specialist in Pathology, 1955; Assoc. Prof., Banting and Best Medical Res. Dept. and Prof. and Head Physiological Hygiene Dept., Toronto Univ., 1955-58; Director, Tunstall Lab., Shell Res., Ltd., Kent, 1958-. Societies: Rad. Res. Soc.; Pathological Soc. Great Britain and Ireland; Atomic Sci. Assoc.; Occupational Hygiene Soc.
Nuclear interests: Mammalian radiation injury; Toxicity of chemical agents.
Address: 193 The Gateway, Dover, Kent, England.

HUNTER, Francis A. Vice-Pres. and Director, Nucl. Dept., Chapman Valve Manufacturing Co.
Address: Chapman Valve Manufacturing Co., 203 Hampshire Street, Indian Orchard, Massachusetts, U.S.A.

HUNTER, Jack Allen, B.Sc. (Eng. Phys.), Ph.D. (Nucl. Major). Born 1920. Educ.: Ohio State Univ. Chairman, N2.4 Com., American Standards Assoc.; Pres., Maryland Acad. of Sci.; Chairman, Tech. Com. 85, Subcom. 1, Working Group, I.S.O. Societies: A.N.S.; A.I.A.A.
Nuclear interests: Management and reactor design, especially with respect to compact power reactors, space reactors for auxiliary power and reactors for space propulsion.
Address: 9210 Smith Avenue, Baltimore 34, Maryland, U.S.A.

HUNTER, Thomas Girvan, A.R.T.C. (Glasgow), B.Sc., Ph.D., D.Sc. (Birm.). Born 1903. Educ.: Roy. Tech. Coll., Glasgow. Prof. Chem. Eng., and Head, Chem. Eng. Dept., Sydney Univ.
Address: Chemical Engineering Department, Sydney University, Sydney, New South Wales, Australia.

HUNTER, Windsor H. Director, Transitron Electronic Corp.
Address: Transitron Electronic Corporation, 168 Albion Street, Wakefield, Massachusetts, U.S.A.

HUNTINGTON, Hillard B., A.B., M.A., Ph.D. Born 1910. Educ.: Princeton Univ. Prof. Phys. (1950), Assoc. Head, Phys. Dept. (1952), Chairman, Phys. Dept. (1961), Rensselaer Polytechnic Inst., Troy, N.Y.; Liaison Officer, Office of Naval Res., London, 1954-55; Visiting Prof., Metal. Dept., Yale Univ., 1960-61. Society: A.P.S.
Nuclear interest: Radiation damage.
Address: Department of Physics, Rensselaer Polytechnic Institute, Troy, New York, U.S.A.

HUNTLEY, Herbert Edwin, B.Sc. (Hons.), Ph.D. (Witwatersrand). Born 1892. Educ.: Bristol Univ. Chair of Phys., Ghana Univ., 1949-57. Book: Nucl. Species (Macmillan, 1954).
Nuclear interest: Nuclear physics.

Address: Nethercombe Cottage, Canada Combe, Hutton, Weston-super-Mare, England.

HUNZINGER, Werner, Ph.D. Born 1921. Educ.: Basel Univ. Eidg. Inst. for Reactor Res., 1956-62; Head, Health and Safety, Eurochemic, Mol, 1962-. Societies: H.P.S.; Schweizerische Physikalische Gesellschaft.
Nuclear interest: Health physics.
Address: Eurochemic, Mol, Belgium.

HUPKA, Stefan, M.D. Born 1920. Educ.: J. A. Komenský Univ., Bratislava. Head, Dept. of Radioisotopes, Cancer Res. Inst., Bratislava, 1951; Head, Inst. of Medical Phys. and Nucl. Medicine, Medical Faculty, Univ. of Komenský, Bratislava, 1963; Asst. Prof. of Nucl. Medicine, 1964-. Society: Medical Sci. Soc. of J. E. Purkynje, Prague.
Nuclear interests: Nuclear medicine; Biophysics; Influence of ionising radiation on living matter; Diagnostics and therapy with radio-isotopes.
Address: 4 Dohnany Street, Bratislava, Czechoslovakia.

HUPKES, Johannes Willem. Born 1908. Pres., N.V. Koninklijke Maatschappij "de Schelde", Vlissingen; Managing Director, Rijn-Schelde Machinefabrieken en Scheepswerven, Rotterdam. Member, Foundation for the Nucl. Propulsion of Merchant Ships.
Address: N.V. Koninklijke Maatschappij "de Schelde", Vlissingen, Netherlands.

HUPP, Eugene Wesley, B.S., M.S., Ph.D. Born 1933. Educ.: Nebraska and Michigan State Univs. Agricultural Res. Lab., Tennessee Univ. -U.S.A.E.C., Oak Ridge, 1958-62; Radiation Biol. Lab. and Biol. Dept., Texas A and M Univ., College Station, 1962-64; Biol. Dept., Texas Woman's Univ., 1964-. Societies: Rad. Res. Soc.; Soc. for the Study of Fertility; Soc. for the Study of Reproduction; Texas Acad. Sci.; American Soc. Animal Sci.; American Assoc. Lab. Animal Sci.
Nuclear interests: Radiation effects on mammalian systems, especially reproductive and hematopoietic systems and on embryos.
Address: Box 2923, TWU Station, Denton, Texas 76204, U.S.A.

HUQ, Mahfuzul, B.Sc. (Hons.), M.Sc., Ph.D. Born 1932. Educ.: Birmingham and Dacca Univs. Lecturer in Phys., Dacca Univ., 1953-54; Res. Scholar, Nucl. Phys. Lab., Birmingham Univ., 1955-58; Lecturer in Phys., Malaya Univ., 1958-.
Nuclear interests: High energy nuclear physics - study of nuclear structure by scattering; Cosmic rays - study of mu-mesons by Cerenkov detectors; Study of low-level activity - mainly due to fall-out.
Address: Physics Department, University of Malaya in Singapore, Singapore 10, Malaya.

HURDIS, Everett Cushing, Sc.B., M.A., Ph.D. Born 1918. Educ.: Brown and Princeton Univs. Res. Chem., U.S. Rubber Co., Wayne, 1942-58; Manager, Styrene Polymers Group, Koppers Co. Res. Centre, Monroeville, 1958-62; Asst. Prof., Chem., Texas Woman's Univ., 1962-.
Nuclear interests: Activation analysis; Radioactive tracers in organic reactions.
Address: 302 Texas Street, Denton, Texas 76201, U.S.A.

HURE, Daniel Gilbert, Civil Eng. in Aeronautics. Born 1925. Educ.: Econe Nationale Supérieure de l'Aéronautique, Paris. Mobil Oil, 1950-53; S.N.E.C.M.A., 1953-61; Asst. Prof., Strength of Materials, Ecole Nationale Supérieure de l'Aéronautique, Paris, 1960-. Eng., Exptl. Res., Bureau Veritas, Paris.
Nuclear interests: Metallurgy; Reactor design; Safety.
Address: Bureau Veritas S.A., 31 rue Henri Rochefort, Paris 17, France.

HURLEY, Frederick Ian, B.A., B.Sc. Born 1922. Educ.: Dublin Univ. R. and D. Lab., Capenhurst, Chester (1951), Head, Lab. (1956), Head of Reactor Eng. Lab., Risley (1961), Head of Reactor Group Planning (1967), U.K.A.E.A. Society: F. Inst. P.
Address: 'Mara Green', Waste Lane, Cuddington, Northwich, Cheshire, England.

HURLEY, Patrick Mason, B.A., B.A.Sc., Ph.D. Born 1912. Educ.: British Columbia Univ. and M.I.T. Prof. Geology, M.I.T.; Consultant to domestic and foreign companies in mining exploration and evaluation. Book: How Old is the Earth? (Doubleday and Co., Inc., 1959). Societies: Geochem. Soc.; Soc. Economic Geologists; A.I.M.M.E.
Nuclear interests: Nuclear geophysics; Natural radioactivity; Measurement of geologic age utilising abundances of radiogenic nuclides; Measurement of natural isotopic variations by mass spectrometry.
Address: Massachusetts Institute of Technology, Cambridge 39, Massachusetts, U.S.A.

HURSH, John Bachman, B.A., Ph.D. Born 1907. Educ.: Rochester Univ. School of Medicine and Western Reserve Univ. Assoc. Prof. (1948-59), Prof. (1959-), Rad. Biol., Rochester Univ. Societies: Rad. Res. Soc.; A.P.S.; H.P.S.
Nuclear interest: Research on assimilation and metabolism of radionuclides by man, particularly those associated with the thorium and uranium natural series.
Address: Medical Centre, Rochester University, Rochester, New York, U.S.A.

HURST, Charles Angas, B.A. (Hons.), B.Sc. (Melbourne), Ph.D. (Cantab.). Born 1923. Educ.: Melbourne and Cambridge Univs. Lecturer, Maths. Dept., Melbourne Univ., 1952-56; Lecturer, Mathematical Phys. Dept. (1957-63), Prof. (1964-) Adelaide Univ. Societies: A.I.P.; Australian Inst. Phys.

Nuclear interest: Elementary particle physics.
Address: Mathematical Physics Department, Adelaide University, Adelaide, South Australia.

HURST, Donald Geoffrey, B.Sc., M.Sc., Ph.D. Born 1911. Educ.: McGill Univ. Phys. Div. and Atomic Energy Div., N.R.C., 1939-52; A.E.C.L., 1952-. (Asst. Director Reactor R. and D. Div., 1955-61, Director Reactor R. and D. Div., 1961-67, Director Appl. R. and D., Chalk River, 1967-); On leave of absence as Director, Nucl. Power and Reactors Div., I.A.E.A., Vienna, 1965-67. Societies: F.R.S.C.; Canadian Assoc. Physicists; Fellow, A.N.S.
Nuclear interests:Reactor lattice theory; Reactor design; Nuclear physics.
Address: Atomic Energy of Canada, Ltd., Chalk River, Ontario, Canada.

HURST, George Samuel, B.A., M.S., Ph.D. Born 1927. Educ.: Berea Coll., Kentucky and Tennessee Univs. At Phys. and Astronomy Dept., Kentucky Univ.; Formerly at Health Phys. Div., O.R.N.L., Oak Ridge, Tennessee. Societies: Rad. Res. Soc.; A.P.S.; H.P.S.
Nuclear interests: Radiation physics; Dosimetry.
Address: Physics and Astronomy Department, Kentucky University, Lexington, Kentucky, U.S.A.

HURST, John, B.Sc. Born 1910. Educ.: Victoria Univ., Manchester. Sen. Lecturer, Pure and Appl. Phys., Univ. Rad. Protection Officer, Salford Univ. Society: A.Inst.P.
Nuclear interests: Reactor physics; Reactor instrumentation; Radiological health and safety.
Address: Pure and Applied Physics Department, Salford University, Salford, Lancs., England.

HURST, Lynn K. Director, Special Materials and Services Div., Argonne Nat. Lab. Formerly Vice Chairman, now Chairman (1967-), Inst. of Nucl. Materials Management.
Address: Argonne National Laboratory, 9700 South Cass Avenue, Argonne, Illinois 60440, U.S.A.

HURST, Robert, M.Sc. (N.Z.), Ph.D. (Cambridge). Born 1915. Educ.:Canterbury (New Zealand) and Cambridge Univs. Group leader, Heavy Elements Group (Chem. Div.), Group leader, Reactor Chem. Group, Project leader, Homogeneous Aqueous Reactor Project, A.E.R.E., Harwell (1948-57), Chief Chem., R. and D. Branch, Risley (1957-58), Director, Dounreay Exptl. Reactor Establishment (1958-63), U.K.A.E.A. Director of Res., British Ship Res. Assoc. Book: Progress in Nucl. Energy, Series IV. Vol. 1, Reactor Technol. (Pergamon Press). Societies: Chem. Soc.; Faraday Soc.
Nuclear interests: Fast breeder reactors; Chemical processing of nuclear fuels.

Address: Gorton Oaks, 17 Murdoch Road, Wokingham, Berkshire, England.

HURST, Thomas L. Formerly Asst. Res. Director, Inorganic Div., Monsanto Chem. Co., now Manager, Res., Kerr-McGee Oil Industries Inc., 1964-.
Address: Kerr-McGee Oil Industries Inc., Kerr-McGee Building, Oklahoma City, Oklahoma, U.S.A.

HURT, Charles, Chef du Service Electrique à la Sté. Hadir, Differdange. Member, Conseil National de l'Energie Nucléaire.
Address: Ministère des Transports et de l'Electricité, 4 boulevard Roosevelt, Luxembourg, Luxembourg.

HURTADO, J. LIARTE. See LIARTE HURTADO, J.

HURWIC, Józef, D.Sc. (Chem. Eng.). Born 1911. Educ.: Warsaw Inst. Technol. Prof. Phys. Chem. (1951-60), Prof. Phys. (1960-), Dean, Faculty of Chem. (1962-), Warsaw Inst. of Technol.; Editor, Problemy. Vice-Pres., (1953-64), Pres. (1964-), Polish Chem. Soc.; Praesidium, Polish State Council for Peaceful Use of Nucl. Energy; Praesidium, Com. for the Peaceful Use of Nucl. Energy, Polish Acad. Sci. Book: Budowa Materii (Warsaw, PWN, 1964). Societies: Polish Phys. Soc.; Sté. Chimique de France; Groupement Ampère.
Nuclear interest: Nuclear physical chemistry.
Address: Katedra Fizyki, Wydzial Chemiczny, Politechnika Warszawska, Warsaw, Poland.

HURWITZ, Henry, Jr., B.A. (Cornell), M.A. (Harvard), Ph.D. (Harvard). Born 1918. Worked at Los Alamos Sci. Lab. in group of Dr. E. Teller on problems related to hydrogen bomb, 1944-46; Manager, theoretical group and consulting phys. for Advanced Naval Reactor Physics Activity, Gen. Electric Co., K.A.P.L., 1946-56; Manager, Nucleonics and Rad., Gen. Elec. Co., Res. Lab., 1957-. E. O. Lawrence medalist, 1961. Book: Progress in Nucl. Energy, Series I (co-author, chapter, Theory and Experiment of Highly Enriched Intermediate and Thermal Assemblies, 1956). Societies: Fellow, A.P.S.; Fellow, A.N.S.; Fellow, New York Acad. of Sci.; Fellow, A.A.A.S.
Nuclear interests: Reactor theory; Reactor safety; Thermonuclear research.
Address: 827 Jamaica Road, Schenectady, New York, U.S.A.

HUSAIN, Syed Abid, M.Sc., Ph.D. Born 1928. Educ.: Muslin, Aligarh and Punjab Univs. Lecturer Phys., Karachi Univ., 1952-56; Sci. Officer, A.E.C., Govt. of Pakistan, 1956-58. Societies: Working Com., Pakistan Sci. Workers Assoc.; Pakistan Assoc. for the Advancement of Sci.
Address: Atomic Energy Centre, Ferozepur Road, P.O. Box No. 658, Lahore, Pakistan.

HUSAIN, Tahir, M.Sc., D.Phil. Sen. Prof. Phys., Dept. Head and Head, Nucl. Res. Lab., Govt. Coll., Lahore. Societies: F.P.S. (London); M.A.P.S. (U.S.A.); W.P.E.S. (India).
Address: Physics Department, Government College, P.O. Bag No. 1105, Lahore, West Pakistan.

HUSIMI, Kodi, D.Sc. Born 1909. Educ.: Tokyo Univ. Director, Inst. of Plasma Phys., Nagoya Univ., 1961-. Member, Advisory Com. on Reactor Safety, Japan A.E.C., 1956-. Societies: Phys. Soc. of Japan; Atomic Energy Soc. of Japan.
Nuclear interest: Reactor safety.
Address: Institute of Plasma Physics, Nagoya University, Furocho, Chikusaku, Nagoya, Japan.

HUSS, Alphonse, Master in Law. Born 1902. Educ.: Paris, Berlin and Louvain Univs. Former attorney gen.; Board, Int. Univ. Comp. Sci., Luxembourg. Vice-Pres., Luxembourg Soc. for Pacific Use of Nucl. Energy; Hon. Pres., Luxembourg U.N. Assoc., Vice-Pres., Soc. Luxembourg Naturalists. Societies: Luxembourg Soc. for Pacific Use of Nucl. Energy; Soc. Luxembourg Naturalists.
Nuclear interest: Law.
Address: 32 rue Albert I, Luxembourg, Grand Duchy of Luxembourg.

HUSSAIN, Dilwar, Dr. Lecturer, Nucl. Phys., Dept. of Phys., Dacca Univ.
Address: Dacca University, Dacca 2, Pakistan.

HUSSAIN, M., Dr. Reader, Nucl. Phys., Dept. of Phys., Dacca Univ.
Address: Dacca University, Dacca 2, Pakistan.

HUSSAIN, Syed Rafat, B.Sc., B.E. (Civil). Born 1928. Educ.: Agra (India) and Sind (Pakistan) Univs. Project Director (Pakistan A.E.C.), Construction and Installation of 5 MW Pool Type Res. Reactor, Islamabad, 1962-66; Resident Eng., Construction and Installation of 137 MW Natural Uranium Heavy Water Nucl. Power Station, (Pakistan A.E.C.), Karachi, 1966-. Society: Assoc. Member, Inst. of Structural Engs., London.
Nuclear interest: Structural engineering aspects of nuclear power station construction management.
Address: Karachi Nuclear Power Project, P.O. Box No. 3183, Karachi, Pakistan.

HUSSEINI, Jassem AL-. See AL-HUSSEINI, Jassem.

HUSSON, René. Faculté des sci., Ecole Nationale Supérieure d'Electricité et de Mécanique, Nancy Univ.
Address: Nancy University, 2 rue de la Citadelle, Nancy (Meurthe-et-Moselle), France.

HUSTER, Erich F. K., Dr. phil. Born 1910. Educ.: Marburg, Greifwald and Berlin Univs. Asst., Marburg Univ., 1950-56; Prof. Karlsruhe T.H., 1957-58; Prof., Marburg Univ., 1958-59; Prof., Münster/Westf. Univ., 1959-. Book: Contributor to Kerntechnik (Riezler-Walcher. Verlagsanstalt Stuttgart, B.G. Teubner, 1958). Societies: Deutsche Physikalische Gesellschaft; A.P.S.
Nuclear interests: Nuclear physics (β- and γ-spectroscopy at low energies; L/K-capture; polarisation of radiation; counter physics).
Address: Institut für Kernphysik der Universität, 7-15 Tibusstr., Münster i. W., Germany.

HUSTON, Norman Earl, A.B. (Phys.), Ph.D. (Phys.). Born 1919. Educ.: California (Berkeley) and Southern California Univs. Res. Eng. (1950-51), Sen. Res. Eng. (1951-55), Supervisor, Appl. Phys. (1955-57), Group Leader, Instrumentation and Control (1957-59), Sen. Tech. Specialist (1959-61), Asst. to Director, Reactor Phys. and Instrumentation (1961-62), Chief, Tech. Operations (1962-63), Director, Rad. Technol. and Instrumentation (1963-65), Atomics International; Sci. Adviser to Vice-Pres., Autonetics, 1965-66; Prof., Nucl. Eng., Director, Instrumentation Systems Centre, Director, Ocean Eng. Labs., Wisconsin Univ., 1966-68. Books: Summary of Reactor Designs – A.E.C. Reactor Handbook, Vols. I and II (A.E.C., 1955; 2nd Edition, McGraw-Hill); Reactor Safety Devices, Technol., Eng. and Safety, Progress in Nucl. Energy, Series IV, Vol. 3 (Pergamon Press, 1960). Societies: A.P.S.; A.N.S.; Sen. Member, I.E.E.E.; Sen. Member, I.S.A.; A.A.A.S.; A.S.E.E.
Nuclear interests: Reactor design and operations; Management; Reactor safety; Radiation technology and applications; Nuclear and radiation instrumentation; Utilisation and applications of radioisotopes.
Address: 4556 Winnequah Road, Monona, Wisconsin, U.S.A.

HUTCHEON, Ian Carrodus, M.A. (Oxon). Born 1923. Educ.: Oxford Univ. Chief Electronics Eng. (1951-66), Deputy Gen. Manager, Group Res. (1966-), G. Kent Ltd., Luton; Consultant, Tiltman Langley Ltd., Redhill, 1958-59; Director, Record Electrical Co., Altrincham, 1965-67; Societies: F.I.Mech.E.; F.I.E.E.
Nuclear interest: Instrumentation.
Address: 47 Fairford Avenue, Luton, Beds., England.

HUTCHEON, John Malcolm, B.Sc. Born 1915. Educ.: London Univ. Group Leader, Chem. Eng. Div., A.E.R.E., Harwell, 1954-58; Deputy Head of Labs., Head of Labs. (1959-), Reactor Materials Lab., U.K.A.E.A., Culcheth. Nuclear interests: Reactor materials, particularly graphite.
Address: 62 Brooklands Road, Sale, Cheshire, England.

HUTCHINSON, E. J., A.M.I.E.E. Director, Strachan and Henshaw Ltd.

HUT

Address: Strachan and Henshaw Ltd., Ashton Works, P.O. Box No. 103, Bristol 3, England.

HUTCHINSON, Franklin. Member, Subcom. on Radiobiol., Com. on Nucl. Sci., N.A.S.-N.R.C.
Address: National Academy of Sciences - National Research Council, Committee on Nuclear Sciences, 2101 Constitution Avenue, N.W., Washington 25, D.C., U.S.A.

HUTCHINSON, George W., M.A., Ph.D. Born 1921. Educ.: Cambridge Univ. Prof. Phys., Southampton Univ. Societies: Fellow, Phys. Soc.; F.R.A.S.; A.P.S.; Italian Phys. Soc.
Nuclear interests: High energy particle physics; Cosmic ray research with artificial satellites.
Address: Physics Department, Southampton University, England.

HÜTTEL, Günter, Dipl. Phys. Born 1930. Educ.: Bergakademie Freiberg. Arbeitsgruppenleiter, Zentralinstitut für Kernforschung, Rossendorf bei Dresden.
Nuclear interests: Reactor physics, especially reactor kinetics, pile oscillator, cross-section determination resonance integrals, Doppler coefficient.
Address: 8353 Langburkersdorf Nr. 152 b. Neustadt/Sa, German Democratic Republic.

HUTTERER, Leopold. Director, Hutterer und Lechner, Kommanditgesellschaft.
Address: Hutterer und Lechner, Kommanditgesellschaft, 5 Brauhausgasse, Himberg bei Wien, Austria.

HÜTTIG, Wilfried, Dipl.-Ing., Ing. Born 1938. Educ.: Ingenieurschule für Werkstofftechnik und Materialprüfung, Bergakademie Freiberg, Deutsche Akademie der Wissenschaften zu Berlin and Zentralinstitut für Kernforschung in Rossendorf bei Dresden. Book: Werkstoffe der Kerntechnik, Vols. I and II (Berlin, V.E.B. Deutscher Verlag der Wissenschaften, 1964).
Nuclear interests: Materials for fuel cannings; Metallurgy; Material testing; Irradiation effects on metals.
Address: 9 Hans-Beimler-Platz, Dresden 801, German Democratic Republic.

HUTTON, Stanley Peerman, D. Eng., Ph.D. Born 1921. Educ.: Liverpool and London Univs. Sen. Principal Sci. Officer, Head of Fluid Mechanics Div. (1949-60), Deputy Director (1960-61), Mech. Eng. Res. Lab., D.S.I.R.; Prof. Mech. Eng., University Coll., Cardiff, 1961-. Member, Hydraulics Group Com., I. Mech. E. Societies: M.I.Mech.E.; A.M.I.C.E.; A.F.R.Ae.S.
Nuclear interests: Control systems and fluid machinery.
Address: University College of South Wales, Cardiff, Wales.

HUUS, Torben, Dr. phil. Born 1919. Educ.: Copenhagen Univ. Res. Assoc., California Inst. Technol., 1951-52; Docent (1958-60), Prof. (1960-), Copenhagen Univ. Society: Roy. Danish Acad. Sci.
Nuclear interests: Nuclear structure; Van de Graaff accelerators.
Address: The Niels Bohr Institute, 17 Blegdamsvej, Copenhagen, Denmark.

HUVELIN, Paul. Member, Comitè de l'Equipement Industriel, C.E.A.
Address: Commissariat à l'Energie Atomique, 29-33 rue de la Federation, Paris 15, France.

HUYNH NGOC, L. Asst., Internat. Lab. of Marine Radioactivity.
Address: International Laboratory of Marine Radioactivity, Musée Océanographique, Monaco.

HVINDEN, Torleif, Director, Norwegian Defence Res. Establishment; Member, Norwegian Radiohygienic Board; Member, Rad. Advisory Council, Norway; Member, Health and Safety Subcom., Member, Group of Experts on Production of Energy from Radioisotopes, O.E.C.D.,E.N.E.A.
Address: Helsedirektoratat, Health Services of Norway, Royal Norwegian Ministry of Social Affairs, Oslo, Norway.

HYDE, Earl K., B.S., Ph.D. (Chem.). Born 1920. Educ.: Chicago Univ. Sen. staff, Nucl. Chem. Div., Lawrence Rad. Lab., Calif. Univ., Berkeley. Book: Nucl. Properties of the Heavy Elements (Prentice Hall Inc., 1964). Societies: A.P.S.; A.C.S.
Nuclear interests: Nuclear chemistry, radioactivity, chemical and nuclear properties of heavy elements.
Address: Lawrence Radiation Laboratory, California University, Berkeley, California, U.S.A.

HYDEN, Anders Lennart Fredrik, M. Eng. (Chem. Eng.). Born 1924. Educ.: Roy. Inst. Technol., Stockholm. At AB Svenska Metallverken, Vasteras, 1957-60; Project leader for the manufacture of fuel elements, AB Atomenergi, Stockholm, 1960-.
Address: AB Atomenergi, P.O. Box 9042, Stockholm 9, Sweden.

HYDER, Henry Richard McKenzie, M.A. Born 1928. Educ.: Cambridge Univ. At U.K.A.E.A., Harwell, 1954-63. Nat. Inst. for Res. in Nucl. Sci., 1963-. Book: Editor, Progress in Nucl. Energy, Series II, vol. ii (Pergamon Press).
Nuclear interests: Reactor physics; Operation of research reactors.
Address: National Institute for Research in Nuclear Science, Rutherford High Energy Laboratory, Chilton, Didcot, Berks., England.

HYER, Frank P. Director, Atomic Power Development Assocs., Inc.; Member, Board of Trustees, Power Reactor Development Co.

Address: Atomic Power Development Associates, Inc., 1911 First Street, Detroit 26, Michigan, U.S.A.

HYLLERAAS, E., Prof. Norwegian deleg. on the Council, C.E.R.N., Meyrin, Geneva. Member, Editorial Board, Nucl. Phys.
Address: Universitetet i Oslo, Blindern, Norway.

HYMAN, Herbert H., B.S., M.S., Ph.D. Born 1919. Educ.: C.C.N.Y., Polytech. Inst., Brooklyn, and Illinois Inst. Technol. Sen. Chem., Argonne Nat. Lab. Vice-Pres., Chicago Chapter, American Technion Soc. Books: Editor: Process Chemistry, vol. i (1956), vol. ii (1958), vol. iii (1960) (Series III, Progress in Nucl. Energy Series, Pergamon Press); Noble Gas Compounds (Univ. of Chicago Press, 1963). Societies: A.C.S.; Soc. for Applied Spectroscopy; N.Y. Acad. Sci.; A.N.S.; American Technion Soc.; RESA; Fellow, A.A.A.S.
Nuclear interests: Reactor concepts; Chemical processing; Fluorine and noble gas chemistry.
Address: 1347 Park Place, Chicago, Illinois, U.S.A.

HYODO, Tomonori, Dr. Eng. Born 1922. Educ.: Kyoto Imperial Univ. Lecturer (1951), Asst. Prof. (1955), Ritsumeikan Univ.; Asst. Prof. (1958-61), Prof. (1961-), Kyoto Univ. Societies: Phys. Soc. Japan; Atomic Energy Soc. Japan; A.P.S.; A.N.S.
Nuclear interests: Radiation shielding; Radiation physics.
Address: Nuclear Engineering Department, Kyoto University, Kyoto, Japan.

IACI, Giuseppe, Assoc. Prof. Born 1926. Educ.: Catania Univ. Reparto Elettronica, Centro Siciliano di Fisica Nucleare e di Struttura della Materià, Catania Univ., 1958-.
Nuclear interest: Nuclear electronics.
Address: Centro Siciliano di Fisica Nucleare, Istituto di Fisica dell'Universita, 57 Corso Italia, Catania, Italy.

IANEVA, N. Paper: Co-author, Measurements of U^{235} and Pu^{239} Neutron Capture-to-Fission Ratio for Resonance Incident Neutron Energies (Atomnaya Energiya, vol. 24, No. 4, 1968).
Address: Academy of Sciences of the U.S.S.R., 14 Leninsky Prospekt, Moscow V-71, U.S.S.R.

IANNI, Elmo J. DI. See DI IANNI, Elmo J.

IBATULIN, M. S. Paper: Co-author, Rad. near Reactor VVR-M (letter to the Editor, Atomnaya Energiya, vol. 19, No. 1, 1965).
Address: Academy of Sciences of the U.S.S.R., 14 Leninsky Prospekt, Moscow V-71, U.S.S.R.

IBE, Librado D., B.S. (E.E.), Ph.D. Born 1924. Educ.: Philippines and Purdue Univs. Supervising Sci., R. and D. Div. (1958-61), Deputy Director (1961-63), Director (1963-), Philippine Atomic Res. Centre; Chief, Power Planning Sect., Nat. Power Corporation (N.P.C.); Chief, Electrolytic Cell and Elec. Dept., Maria Cristina Fertiliser Plant, N.P.C. Society: A.N.S.
Nuclear interests: Reactor engineering and reactor physics.
Address: Philippine Atomic Research Centre, Diliman, Quezon City, Philippines.

IBRAGIMOV, M. Kh. Head of Hot Lab., Soviet Atomic Lab., Obninsk.
Nuclear interests: Heat transfer problems, especially those associated with liquid metal circuits.
Address: Soviet Atomic Laboratory, Obninsk, near Maloyaroslavets, Moscow, U.S.S.R.

IBRAGIMOV, Sh. Sh.
Nuclear interests: Metallurgy of neutron irradiated materials.
Address: Academy of Sciences of the U.S.S.R., 14 Leninsky Prospekt, Moscow V-71, U.S.S.R.

IBRAHIM, Abdel Aziz el Sayed. Rector Alexandria Univ.
Address: Alexandria University, Alexandria, Egypt, United Arab Republic.

IBRAHIM, Col. Khalil. Specialist Member, Iraq deleg. to Soviet Union to sign pact on peaceful uses of atomic energy, Aug. 1959. Member, Iraq A.E.C.
Address: c/o Iraq Atomic Energy Commission, Baghdad, Iraq.

IBRAHIMOV, Ahmed. Nucl. Phys. Inst., Acad. Sci. of the Uzbek S.S.R.
Address: Institute of Nuclear Physics (Uzbek), Kibrari, near Tashkent, U.S.S.R.

IBRAHIMOV, Shukur. Inst. of Genetics and Plant Physiology, Acad. Sci. of the Uzbek S.S.R.
Address: Institute of Genetics and Plant Physiology, 42 Ulitsa Lafarga, Akadgorodok, Tashkent, U.S.S.R.

ICE, C. H. Asst. Director, Savannah River Lab., operated for U.S.A.E.C. by E. I. du Pont de Nemours and Co.
Address: U.S.A.E.C., Savannah River Laboratory, Aiken, South Carolina 29801, U.S.A.

ICHIKAWA, Kazutaka. Manager, Administration Dept., Atomic Energy Dept., Ishikawajima-Harima Heavy Ind. Co. Ltd.
Address: Ishikawajima-Harima Heavy Industries Co. Ltd., Atomic Energy Department, 3-chome, Toyosu, Koto-ku, Tokyo, Japan.

ICHIKAWA, Shinobu, Master of commerce. Born 1897. Educ.: Kobe Coll. of Commerce. Chairman, Marubeni-Iida Co., Ltd., May 1964-; Pres., Fujikoshi, Ltd., May 1964-. Pres., Osaka Chamber of Commerce and Industry, July 1966-; Vice-Pres., Japan Chamber of Commerce and Industry, Oct.

1966-; Vice-Pres., Federation of Economic Organisations, May 1964-; Exec. Director, Japan Assoc. for the 1970 World Exposition, Aug. 1966-; Formerly Director, The Tokyo Atomic Ind. Consortium; Director, Japan Atomic Ind. Forum, Inc.
Nuclear interest: Everything concerning nuclear industry.
Address: 701, 2-chome, Tamagawa, Denenchofu, Setagaya-ku, Tokyo, Japan.

ICHIKAWA, Yoshi-hiko, Dr. of Sci. Born 1929. Educ.: Tohoku Univ. Res. Asst., Tohoku Univ., 1954-58; Lecturer (1958-63), Assoc. Prof. (1963-), Nihon Univ. Society: Phys. Soc. of Japan.
Nuclear interests: Plasma physics and nuclear physics (theory).
Address: Nihon University, 1-8 Kanda-Surugadai, Chiyoda-ku, Tokyo, Japan.

ICHIMURA, Munetake, Dr. Sc. Born 1938. Educ.: Tokyo Univ. Lecturer, Dept. of Phys. and Atomic Energy, Res. Inst., Nihon Univ., 1966-. Society: Phys. Soc. of Japan.
Nuclear interest: Nuclear physics.
Address: 3-25-53 Namiki-cho, Kokubunji-shi, Tokyo, Japan.

IDDINGS, Frank Allen, B.S., M.S., Ph.D. Born 1933. Educ.: Midwestern and Oklahoma Univs. Chem., Esso Res. Labs., 1959-64; Asst. Prof., Louisiana State Univ., 1964-. Societies: A.N.S.; A.C.S.
Nuclear interests: Education; Activation analysis; Industrial applications of radioisotopes; Liquid scintillation counting; Carbon age dating; Radiography.
Address: Nuclear Science Centre, Louisiana State University, Baton Rouge, Louisiana 70803, U.S.A.

IDDLES, Alfred, B.S., M.E. Born 1889. Educ.: Michigan State Univ. Vice-Pres. (1945-48), Pres. (1948-57), Babcock and Wilcox Co. Vice-Pres., Machinery and Allied Products Inst., 1955-58. Societies: A.S.M.E.; Nat.Ind. Conference Board; Franklin Inst.; Atomic Ind. Forum, Inc. (Pres., 1956-58).
Nuclear interests: Reactor design, management, manufacture.
Address: 303 Orchard Way, Wayne, Pennsylvania, U.S.A.

IDE, Yoshio. Formerly Asst. Chief, Rad. Safety Sect., now Asst. Chief, Internat. Co-operation Sect., Atomic Energy Bureau, A.E.C. of Japan.
Address: Atomic Energy Commission of Japan, 3-4 Kasumigaseki, Chiyoda-ku, Tokyo, Japan.

IDE, Yu, Dr. Chem. Sect., Third Res. Div., Nat. Inst. of Animal Health.
Address: National Institute of Animal Health, Kodaira-shi, Tokyo, Japan.

IDENO, Eikichi, D.Ph. Born 1930. Educ.: Tokyo Kyoiku Univ. Formerly Lecturer now Asst. Prof., Tokai Univ., 1959-68; Hitotsubashi Univ., 1968-. Societies: Atomic Energy Soc. Japan; Chem. Soc. Japan.
Nuclear interest: Radiochemistry.
Address: 1-31-12 Adachi, Adachi-ku, Tokyo, Japan.

IGNATENKO, Anatoly. Nucl. Problems Lab., Joint Inst. for Nucl. Res., Dubna.
Address: Joint Institute for Nuclear Research, Dubna, near Moscow, U.S.S.R.

IGNAT'EV, B. G. Paper: Co-author, Fabrication of Thin Plates from Refractory Carbides (Atomnaya Energiya, vol. 20, No. 6, 1966).
Address: Academy of Sciences of the U.S.S.R., 14 Leninsky Prospekt, Moscow V-71, U.S.S.R.

IGNAT'EV, K. G. Papers: Co-author, Variation of η and U^{235} and Pu^{239} Partial Cross-Sections for Neutrons of Resonance Energies (Atomnaya Energiya, vol. 16, No. 2, 1964); co-author, Interference Effects in Fission Cross-Sections (ibid, No. 3).
Address: Academy of Sciences of the U.S.S.R., 14 Leninsky Prospekt, Moscow V-71, U.S.S.R.

IGNATOV, Kiril, Dr. Member, Com. for the Peaceful Uses of Atomic Energy, Council of Ministers of the Bulgarian People's Republic.
Address: Council of Ministers, Sofia, Bulgaria.

IHARA, Yoshinori, B. Eng. Born 1924. Educ.: Tokyo Inst. Technol. Chief, Nucl. Fuels Sect., Atomic Energy Bureau, 1962-64; Sci. Attaché, London, 1964-67; Chief, Power Reactor Development Sect. Atomic Energy Bureau, Sci. and Technol. Agency, Govt. of Japan, 1968-. Book: Nucl. Energy (Tokyo, Keizai-Orai-sha, 1960). Societies: Atomic Energy Soc. of Japan; Inst. Elec. Eng. of Japan. Nuclear interest: Power reactor development.
Address: Atomic Energy Bureau, Science and Technology Agency, 3-2-2 Kasumigaseki, Chiyoda-ku, Tokyo, Japan.

IIDA, Hiroyoshi, B.Sc., Ph.D. Born 1916. Educ.: Tokyo Univ. Lecturer, School of Medicine, Nagoya Univ., 1956; Chief, Instruction, (1961), Head (1964), Div. of Training, Nat. Inst. of Radiol. Sci. Member, Japanese Ind. Standards Com.; Member, Com. of Nat. Examination for the Licence of X-ray Technicians. Books: Phys. of Radioisotopes; Rad. Protection. Societies: Councillor, Atomic Energy Soc. of Japan; Councillor, Japanese Soc. of Radiol.; Japan Health Phys. Soc.
Nuclear interests: From 1956-61, research and education in physical foundation of radiology particularly radiation therapy; Since 1961, research and professional education in nuclear medicine, radiology and health physics.
Address: 1453-13, 7-chome, Yazu-cho, Narashino-shi, Chiba-Pref., Japan.

IIDA, Shozo. Polymer Phys., Government Ind. Res. Inst. Nagoya, Agency of Ind. Sci. and Technol., Ministry of Internat. Trade and Industry.
Address: Government Industrial Research Institute, Ministry of International Trade and Industry, Hirate-machi, Kita-ku, Nagoya, Japan.

IIZUKA, Miki, Dr. Formerly in Chem. Sect., Third Res. Div., now in Radioisotope Lab., Third Res. Div., Nat. Inst. of Animal Health.
Address: National Institute of Animal Health, Kodaira-shi, Tokyo, Japan.

IKEDA, Nagao. Chem. Dept., Tokyo Univ. of Education.
Address: Tokyo University of Education, 3-29-1, Otsuka, Bunkyo-ku, Tokyo, Japan.

IKEDA, Toshio. Born 1917. Educ.: Tokyo Imperial Univ. Prof., Education Ministry, Shizuoka Univ., 1965; Res. Assoc. of Radiochem., Res. Lab., Shizuoka Univ. Society: Chem. Soc. Japan.
Nuclear interest: Radiation chemistry.
Address: Department of Chemistry, Shizuoka University, Ohya, Shizuoka, Japan.

IKEDA, Yuichi. Born 1926. Educ.: Tokyo Univ. Res. Asst., Kobayasi Inst. of Phys. Res., 1949-57; Fuji Spinning Co., Ltd., 1957-; Res. Assoc., Wayne State Univ., Dept. of Chem., 1960-62. Society: Phys. Soc. Japan.
Nuclear interest: Radiation chemistry.
Address: Fuji Spinning Co., Ltd., No. 1, 2-chome, Honcho, Nihonbashi, Chuo-ku, Tokyo, Japan.

ILAKOVIC, Ksenofont, Prof., Dr. Head, Dept. of Nucl. and Atomic Phys., Rudjer Boskovic Nucl. Inst.
Address: Rudjer Boskovic Nuclear Institute, 54 Bijenicka Cesta, P.O. Box 171, Zagreb, Yugoslavia.

ILBERY, Peter Leslie Thomas, M.D. (Sydney), B.S., M.C.R.A., D.M.R.T. Born 1923. Educ.: Shore School and Sydney Univ.
Nuclear interests: Cytogenetic changes in irradiated cells. Experimental studies in radiation induced leukaemia. Radiation therapy. Radiodiagnosis.
Address: Radiation Biology, School of Public Health and Tropical Medicine, Sydney University, Australia.

ILIFFE, Cedric Euan, M.A. Born 1915. Educ.: Cambridge Univ. Chief Tech. Eng., Risley (1953-58), Head, Central Tech. Services, Reactor Group (1958-), U.K.A.E.A. Societies: Inst. Mech. Eng.; B.N.E.S.
Nuclear interests: Economics of nuclear power; Reactor engineering; Reactor physics.
Address: 52 Styal Road, Wilmslow, Cheshire, England.

IL'IN, O. G. Paper: Co-author, Storage Ring for 100 MeV Electrons (Atomnaya Energiya, vol. 23, No. 6, 1967).
Address: Academy of Sciences of the U.S.S.R., 14 Leninsky Prospekt, Moscow V-71, U.S.S.R.

IL'IN, Yu. I. Papers: Co-author, A Two-dimensional 1024-channel Pulse-height Analyser (DMA-1024) (letter to the Editor, Atomnaya Energiya, vol. 11, No. 1, 1961); co-author, The Specific Ionization Distribution along Track as a Function of U^{235} Fission Fragments Initial Energy (ibid, vol. 19, No. 3, 1965).
Address: Academy of Sciences of the U.S.S.R., 14 Leninsky Prospekt, Moscow V-71, U.S.S.R.

ILJAS, Jasif. Formerly Head, Reactor Div., Inst. for Atomic Energy, Djakarta; Now Head, Personnel and Education Div., Nat. Atomic Energy Agency. Deleg., I.A.E.A. Conference on Small and Medium Power Reactors, Vienna, Sept. 5-9, 1960; Indonesia Alternate Deleg., I.A.E.A., Gen. Conference, Vienna, Sept. 1960.
Address: National Atomic Energy Agency, Djl. Palatehan I No. 26, Blok K 5, Kabajoran-Baru, Djakarta, Indonesia.

ILJICEV, B. I. Paper: Co-author, The Determination of Migration Length and Multiplication Coefficient in a Heavy-Water Natural Uranium Lattice (Jaderna Energie, vol. 13, No. 11, 1967).
Address: U.S.S.R. State Committee for Atomic Energy, 26 Staromonetnii Pereulok, Moscow, U.S.S.R.

ILJIN, O. G. Paper: Co-author, 100 MeV Neutron Collector (Jaderna Energie, vol. 13, No. 11, 1967).
Address: Physico-Technical Institute, 2 Polytechnicheskaya Ulitsa, Leningrad, U.S.S.R.

ILLEI, Vilmos. Member, Editorial Board, Energia es Atomtechnika.
Address: Muszaki Konyvkiado, 22 Bajcsy-Zsilinszky ut, Budapest 5, Hungary.

ILLIES, Kurt, Dr. Ing. Born 1906. Educ.: T.H. Munich. o. Prof., T.H. Hanover; Honorarprof., Hamburg Univ. Vorsitzender der Schiffbautechnische Gesellschaft in Hamburg, Leiter Fachausschusses für Schiffsmaschinenbau. Book: Schiffskessel 3 Bände Handbuch für Schiffsingenieure und Seemaschinisten. Societies: S.T.G.; V.D.I.
Nuclear interest: Kernenergie-Schiffsantriebsanlagen.
Address: Technische Hochschule Hanover, 1 Welfengarten, Hanover, Germany.

ILYASOV, V. M. Paper: Co-author, Investigation of PuO_2 Fuel Element Assembly of BR-5 Reactor (Atomnaya Energiya, vol. 24, No. 2, 1968).

Address: Academy of Sciences of the U.S.S.R., 14 Leninsky Prospekt, Moscow V-71, U.S.S.R.

ILYASOVA, G. A. Papers: Co-author, The Methods for Calculation of Slow Neutron Spectrum (Atomnaya Energiya, vol. 13, No. 6, 1962); co-author, Measurements of Slow Neutron Spectra on Reactor Physical Stand at Beloyarskaya I.V. Kurchatov GRES (ibid, vol. 15, No. 6, 1963); co-author, Effective Method of Solving Neutron Transport Equation in P_3- approximation for Hexagonal and Square Cells of Heterogeneous Reactor (ibid, vol. 24, No. 3, 1968).
Address: Academy of Sciences of the U.S.S.R., 14 Leninsky Prospekt, Moscow V-71, U.S.S.R.

ILYUSHCHENKO, V. I. Papers: Co-author, Synthesis of 102 Isotopes with Mass Numbers 254, 253 and 252 (Atomnaya Energiya, vol. 22, No. 2, 1967); co-author, Synthesis of Fm Isotopes With Mass Numbers 247 and 246 (ibid, vol. 22, No. 5, 1967).
Address: Academy of Sciences of the U.S.S.R., 14 Leninsky Prospekt, Moscow V-71, U.S.S.R.

IMAEDA, Kuni. Born 1918. Educ.: Osaka Univ. Asst. Prof., Yamanashi Univ., 1951-59; Asst. Prof. (1959-62), formerly Assoc. Prof. (1963-), now Prof., Dublin Inst. for Advanced Studies. Society: Phys. Soc. Japan.
Nuclear interests: High-energy nuclear interaction; Cosmic rays.
Address: Dublin Institute for Advanced Studies, 5 Merrion Square, Dublin, Ireland.

IMAEV, E. G. Paper: Co-author, Synthesis and Determination of Radioactive Properties of Some Isotopes of Fermium (Atomnaya Energiya, vol. 21, No. 4, 1966).
Address: Academy of Sciences of the U.S.S.R., 14 Leninsky Prospekt, Moscow V-71, U.S.S.R.

IMDI, Kiyokazu, Dr. Born 1926. Educ.: Kyoto Univ. Kurashiki Rayon Co., 1950-. Society: Chem. of High Polymers, Japan.
Nuclear interest: Application for chemistry of high polymers.
Address: Kurashiki Rayon Company Limited, 8 Umeda, Kita-ku, Osaka, Japan.

IMAI, Yoshiki, D.Sc. Born 1903. Educ.: Tokyo Imperial Univ. Head, Mining Res. Inst., Mitsubishi Metal Mining Co., Ltd., 1955-56; Formerly Director, Atomic Fuel Corp., 1956-; Now Vice Pres., Power Reactor and Nucl. Fuel Development Corp. Member, Atomic Power Com., Japan Atomic Energy Ind. Forum, 1959-; Member, Special Com. (Reprocessing), Japan Atomic Energy Ind. Forum, 1959-. Society: Chem. Soc. of Japan.
Nuclear interest: Chemistry.
Address: 2-842 Shimo-Ochiai, Shinjuku-ku, Tokyo, Japan.

IMAI, Yoshio, Formerly Deputy-Chief, Planning Sect., Technical Div., now Supt., Tokai Power Station, (1968-), Japan Atomic Power Co. Ltd. Adviser, Third I.C.P.U.A.E., Geneva, Sept. 1964.
Address: Japan Atomic Power Co., Otemachi Building, Otemachi 1-chome, Chiyoda-ku, Tokyo, Japan.

IMAMURA, Hiroshi, Dr. Eng. Born 1913. Educ.: Tokyo Univ. Manager By-Product Dept., Ube-Nitrogen Plant, 1947; Manager of Erection, Ube Caprolantan Plant, 1954; Manager, 7th Dept., Central Res. Lab., Ube Industries Ltd., 1957.
Nuclear interests: Utilisation of radioisotopes in industry.
Address: Miyanoo Magura, Kusunoki-machi, Yamaguchi-ken, Japan.

IMANISHI, Bunryu. Asst., Low Energy Nucl. Phys. (Theory), Atomic Energy Res. Inst., Nihon Univ.
Address: Nihon University, Atomic Energy Research Institute, 1-8 Kanda-Surugadai, Chiyoda-ku, Tokyo, Japan.

IMAZAKI, Masahide. Asst. Prof., Nucl. Phys. Div., Kyushu Inst. of Technol.
Address: Kyushu Institute of Technology, Nuclear Physics Division, Tobata, Kitakyushu, Fukuoka-ken, Japan.

IMHOFF, Donald H., B.S. (Chem. Eng.). Graduate work in Phys. and Maths. Born 1921. Educ.: Washington State Coll., S. California and California (Los Angeles) Univs. Res. Eng., Union Oil Co. of California, 1943; Manager, Nucl. Phys., California R. and D. Co., 1951; Manager, Eng. Phys. (1955), Manager, Development Eng. (1958-), Atomic Power Equipment Dept., G.E.C. Society: A.I.Ch.E.
Nuclear interests: Reactor physics, reactor design, reactor fuel, reactor dynamics, two-phase flow and heat transfer, reactor equipment development, nuclear superheat and boiling water reactors.
Address: 1577 Alisal Avenue, San Jose, California, U.S.A.

IMO, Saburo. Pres., Mitsubishi Atomic Power Industries, 1965-.
Address: Mitsubishi Atomic Power Industries, Inc., Otemachi Building, 4 Otemachi 1-chome, Chiyoda-ku, Tokyo, Japan.

IMOTO, Mitsuo, D.S. Born 1930. Educ.: Kyoto Univ. With Dept. Phys., Coll. Sci. and Eng., Nihon Univ., 1957-. Society: Japan Phys. Soc.
Nuclear interests: Elementary particle physics (especially symmetry theories and weak interactions).
Address: 3-10-310 Narashinodai-Danchi, Funabashi-shi, Chiba, Japan.

IMPE, Jean van. See van IMPE, Jean.

IMRE, Lajos, Dipl. Chem., Ph.D., D.Sc. (Chem.). Born 1900. Educ.: Pázmány P. Univ., Budapest. Prof. Phys. Chem., Cluj, Roumania,

1940-50; Prof. Phys. Chem., Debrecen Univ., Hungary, 1950-; Res. Member, Central Res. Inst. for Phys., Budapest, Csillebérc, 1957-62. Com. of Phys. Chem. and Com. Nucl. Chem., Hungarian Acad. Sci.; Sci. Board, Hungarian A.E.C., 1958-66. Books: Radiating Nuclei (Cluj, 1947); Gen. Chem. (Cluj, 1948). Society: Soc. Hungarian Chem. Nuclear interests: Tracer methodology concerning reaction mechanisms, especially in heterogeneous systems; Carrier-free separation of fission products (especially by electrolysis); Problems of decay rates; Standardisation of radioactive nuclides.
Address: Debrecen 10, Hungary.

INAGAKI, Katsuhiko, M.D. Born 1911. Educ.: Medical Faculty, Tokyo Univ. Chief Dept. of Clinical Pathology and Radiol., Tokyo Metropolitan Police Hospital, 1953-. Councillor, Japanese Soc. of Physiology. Societies: Japanese Soc. of Internal Medicine; Japanese Soc. of Radioisotopes; Royal Soc. of Medicine (affiliate). Nuclear interests: Biology and medicine.
Address: Tokyo Metropolitan Police Hospital, Fujimicho 3-chome, Chiyoda-ku, Tokyo, Japan.

INDECK, Bernard, Deg. of Assoc. in Electronic Eng. Born 1919. Educ.: N.U., Boston and Shrivenham Army Univ. Formerly Manager, Electron Beam Dept., Ethicon Inc. Nuclear interests: Engineering management of high-kilowatt electron particle accelerators and high kilocurie radioactive sources for use in industrial applications.
Address: 270 Rolling Knolls Way, Somerville, New Jersey, U.S.A.

INDIATI, Marcello, Laurea in Ingegneria Industriale. Born 1925. Educ.: Rome Univ. Univ. Asst. of Electrotechnique, Rome, 1948-50; Consultant, 1950-51; Chief, Transmission Project Dept., F.A.T.M.E., Rome, 1951-57; Chief, Elec. Dept., S.I.M.E.A., Rome, 1957-. Societies: Assoc. Elettrotecnica Italiana; Assoc. Nazionale di Ingegneria Nucleare.
Address: 11 Via Centuripe, Rome, Italy.

INDJOUDJIAN, Mardiros Dickran. Born 1920. Educ.: Ecole Polytechnique. Prof. de Statistique, Inst. de Statistique, Paris Univ.; Directeur adjoint, Banque de Paris et des Pays-Bas; Membre du Conseil, Sté. Indatom. Nuclear interests: Scientific, technical and industrial evolution; Economical and industrial aspects; Problems of financing.
Address: 3 rue d'Antin, Paris 2, France.

INDRIAS, Gheorghiu. Prof. Rumanian rep., presented paper to meeting to work out programme of res. on the new heavy ion accelerator, Joint Inst. for Nucl. Research, Dubna, Sept. 14-17, 1960.
Address: Joint Institute for Nuclear Research, Dubna, near Moscow, U.S.S.R.

INGEBRETSEN, F. Member, Nucl. Phys. Lab., Oslo Univ.
Address: Oslo University, Nuclear Physics Laboratory, Institute of Physics, Blindern, Norway.

INGHAM, George Kenneth, M.D., F.R.C.P. (C.). Born 1918. Educ.: Western Ontario Univ. Director, Dept. of Radioisotopes, St. Joseph's Hospital; Res. Assoc., McMaster Univ., Hamilton, Ontario. Societies: Soc. of Nucl. Medicine; American Soc. of Haematology. Nuclear interest: Nuclear medicine.
Address: Department of Medical Research, McMaster University, Hamilton, Ontario, Canada.

INGHRAM, Mark Gordon., B.A. (Olivet Coll.), Ph.D. (Chicago). Born 1919. Educ.: Chicago Univ. Asst. Prof. Phys. (1949-51), Assoc. Prof. Phys. (1951-57), Prof. Phys. (1957-59), Prof. and Chairman, Phys. Dept. (1959-64), Assoc. Dean., Phys. Sci. Div. (1964-), Chicago Univ. Com. on Sci. and Public, N.A.S., 1966-. Societies: Member, National Acad. of Sci.; Fellow, American Phys. Soc.
Address: Department of Physics, University of Chicago, Chicago, Illinois, U.S.A.

INGLES, John Cowan, B.A. (Hons., Chem.). Born 1920. Educ.: Western Ontario Univ. Head, Control Analysis Sect., Extraction Metal. Div., Mines Branch, Canada, Dept. of Mines and Tech. Surveys, 1954-. Book: Manual of Analytical Methods for the Uranium Concentrating Plant (Ottawa, the Queen's Printer, 1958). Societies: A.C.S.; Chem. Inst. of Canada. Nuclear interests: Analytical chemistry of radioactive ores, process materials and products; Chemistry of hydrometallurgical process for the treatment of radioactive ores.
Address: 300 Lebreton Street, Ottawa, Ontario, Canada.

INGLIMA, John N., B.A. (Natural Sci.). Born 1923. Educ.: Colgate Univ., Hamilton, New York. Tech. Information Officer, Division of Tech. Information, U.S.A.E.C., Washington, D.C. Society: A.N.S. Nuclear interests: Scientific publications, reactor safety analysis, research reactor operations, reactor licensing and regulations, report writing, reactor physics, radiation monitoring, administration.
Address: 7400 Lakeview Drive, Bethesda, Maryland, U.S.A.

INGLIS, David Rittenhouse, A.B., D.Sc., D.Sc. (Hon.). Born 1905. Educ.: Amherst Coll. and Michigan Univ. Theoretical Nucl. Phys., Argonne Nat. Lab.; Editorial Board, Bulletin of the Atomic Scientists. Society: Fellow, A.P.S. Nuclear interests: Shell structure of light nuclei; Collective dynamics of nuclei; Direct nuclear reactions.

INGLIS, George Harrison, B.Sc. Born 1927. Educ.: Durham Univ. Design Eng., C.A. Parsons and Co., Ltd., 1951; Chief Mech. Eng., Nucl. Power Plant Co., Ltd., 1956; Deputy Chief Eng. (Fuel Elements) (1960-62), Chief Eng., Fuel Element Design (1962-), U.K.A.E.A. Society: Inst. of Mech. Eng.
Address: Springfields Works, U. K. Atomic Energy Authority, Salwick, near Preston, Lancs., England.

INGOLFSRUD, Leif John, B.S. (Mech. Eng.), M.S. (Mech. Eng.). Born 1928. Educ.: Queen's Univ. and M.I.T. Superintendent, Reactor Design Branch, Power Projects, Atomic Energy of Canada Ltd., Sheridan Park, Ontario.
Nuclear interest: Reactor design.
Address: Reactor Design Branch, Power Projects, Atomic Energy of Canada Ltd., Sheridan Park, Ontario, Canada.

INGRAHAM, Hollis S., Dr. Commissioner of Health, Nucl. Dept., Health Dept., New York State. Member, Coordinating Council, New York State Office of Atomic Development.
Address: New York State Office of Atomic Development The Alfred E. Smith State Office Building, Albany, New York 12225, U.S.A.

INGRAHAM, Samuel C., II, B.A., M.D., M.P.H. Born 1913. Educ.: Pennsylvania, Johns Hopkins, Colorado and Denver Univs. Asst. Chief, Radiol. Health Programme (1949-54), Head, Field Studies Sect., Field Investigations and Demonstrations Branch, Nat. Cancer Inst. (1954), Head, Gen. Field Studies Sect., Field Investigations and Demonstrations Branch, Nat. Cancer Inst. (1955), Head, Cytology Sect., Field Investigations and Demonstrations Branch, Nat. Cancer Inst. (1956-59), Head, Diagnostic Development Activities, Field Investigations and Demonstrations Branch, Nat. Cancer Inst. and Asst. Chief for Operations, Field Investigations and Demonstrations Branch, Nat. Cancer Inst. (1959-60), Asst. Chief for Operations, Diagnostic Res. Branch, Nat. Cancer Inst. (1960-61), Chief, Training Branch, Radiol. Health Div. (1961-64), Project Director, Radium Res. Project, New Jersey State Dept. of Health (on detached duty from U.S.P.H.S.) (1964-67), U.S.P.H.S.; Director, Office of Comprehensive Health Planning, New Jersey State Dept. of Health, 1967-. Alternate member, New Jersey Commission on Rad. Protection, 1965-. Societies: A.M.A.; American Coll. of Preventive Medicine; American Assoc. of Public Health Physicians; A.P.H.A.; Assoc. of Military Surgeons; H.P.S.; Conference on Radiol. Health; Pan-American Medical Assoc.
Nuclear interest: Radiological public health practices.
Address: 2144 Central Avenue, Ocean City, New Jersey 08226, U.S.A.

INGRAM, Frank. Information Asst., Chicago Operations Office, U.S.A.E.C.
Address: U.S.A.E.C., Chicago Operations Office, 9800 South Cass Avenue, Argonne, Illinois 60439, U.S.A.

INGRAM, Maurice, M.A., Ph.D. Born 1912. Educ.: Cambridge Univ. Director, Meat Res. Inst., A.R.C.
Nuclear interest: Preservation of foodstuffs, especially meats, by means of ionising radiations.
Address: Agricultural Research Council, Meat Research Institute, Langford, near Bristol, England.

INK, Dwight A., B.S., M.A. (Minnesota) U.S.A.E.C. Distinguished Service Award, 1966. Born 1922. Educ.: Iowa State Coll. and Minnesota Univ. Office of Community Affairs, Oak Ridge Operations Office (1951-52), Savannah River Operations Office (1952-55), Management Asst. Office of Gen. Manager, Washington (1955-58), Special Asst. to Chairman McCone (1958-59), Asst. Gen. Manager (1959-66), U.S.A.E.C.; Asst. Sec. for Administration, Dept. of Housing and Urban Development, 1966-.
Nuclear interest: Management.
Address: 13025 Bluhill Road, Silver Spring, Maryland, U.S.A.

INMAN, Maurice Charles, B.Sc. (Hons., Chem.), Ph.D. (Metal.). Born 1923. Educ.: Leeds Univ. Prof., Dept. of Materials Sci., Pennsylvania State Univ. Societies: Inst. of Metals; A.I.M.E.; A.S.M.
Nuclear interests: Metallurgy and solid state generally; Use of radioisotopes in such fields as diffusion and solute segregation, at grain boundaries, surfaces, etc., and correlation with interface energies.
Address: 600 East Waring Avenue, State College, Pennsylvania 16802, U.S.A.

INO, Mitsuyoshi, Dr. Eng. Born 1892. Educ.: Tokyo Imperial Univ. Managing Director, Mitsubishi Heavy Industries, Ltd., 1945; Director and Counsellor, Mitsubishi Nippon Heavy Industries, Ltd., 1950; Managing Director (1958), Exec. Vice-Pres. (1960), Mitsubishi Atomic Power Industries, Inc. Chairman, Mitsubishi Atomic Power Research Com. Society: Hon member, Japan Soc. of Mech. Eng.
Nuclear interests: Administration of nuclear energy development policy and management of atomic power industry.
Address: 17-13 2-chome, Sanno, Otaku, Tokyo, Japan.

INO, Shigeru, M.D. Member, Fukushima Radioisotope Res. Com.
Address: Fukushima Radioisotope Research Committee, c/o Fukushima Medical College, 14 Sugizuma-cho, Fukushima City, Japan.

INOMOTO, Zenichiro, Prof. Agriculture Dept. Member, Radioisotope Res. Com., Niigata Univ. School of Medicine.
Address: Niigata University School of Medicine, Department of Agriculture, 757 1-Bancho Asahimachi-Street, Niigata, Japan.

INONU, Erdal Ismet, B.S., M.S., Ph.D. Born 1926. Educ.: Ankara Univ. and California Inst. Technol. Instructor (1952-53), Docent (1955-57), Ankara Univ.; Prof. Theoretical Phys., Middle East Tech. Univ., Ankara, 1960-.
Nuclear interests: Reactor theory; Transport theory; Nuclear physics.
Address: Ayten Sokak, No. 22, Mebus Evleri, Ankara, Turkey.

INOUE, Takeaki. Casualty Manager, Toa Fire and Marine Reinsurance Co. Ltd.
Address: Toa Fire and Marine Reinsurance Co. Ltd., 6, 3-chome, Kanda Surugadai, Chiyodaku, Tokyo, Japan.

INOUE, Yasushi, D.Sc. Born 1934. Educ.: Kyoto Univ. Res. Asst. (1959-65), Asst. Prof. (1965-), Tohoku Univ. Societies: Chem. Soc. of Japan; Japan Inst. of Metals; Japan Radioisotope Assoc.
Nuclear interest: Radiochemistry including chemistry of actinides and radiochemical separation.
Address: Research Institute for Iron, Steel and other Metals, Tohoku University, 75 Katahira-cho, Sendai, Japan.

INOUYE, Goro. Born 1899. Educ.: Tokyo Univ. Pres., Power Reactor and Nucl. Fuel Development Corp., 1967.
Address: 5-17-11 Daita-cho, Setagaya-ku, Tokyo, Japan.

INOUYE, Henry, B.S., M.S. Born 1920. Educ.: Colo. School of Mines and M.I.T. Air Force, Dept. of Defence, 1947-50; Graduate Student, 1950-52; Union Carbide Nucl., 1952-. Chairman, Oak Ridge Chapter, A.S.M.; Member, Materials Advisory Board, Nat. Acad. Sci. Books: Co-author, chapter in Fluid Fuel Reactors (Addison-Wesley, 1958); author, chapter in Reactor Handbook (Interscience Publishers, 1960). Society: A.S.M.
Nuclear interests: Metallurgy; High-temperature materials; Gas-metal reactions; Transformation kinetics; Alloy development; Cladding.
Address: Oak Ridge National Laboratory, Oak Ridge, Tennessee, U.S.A.

INOUYE, Hidehiro. Born 1899. Educ.: Tokyo Imperial Univ. Chairman, Board of Directors, Nihon Cement Co., Ltd., 1966-. Soc., Japan Com. for Economic Development, 1951; Regular Director, Employers' Assoc., 1953; Regular Director, Federation of Economic Organisation, 1954; Chairman, Cement Assoc. of Japan, 1967.
Address: Nihon Cement Company Limited, 4,1-chome, Ohtemachi, Chiyoda-ku, Tokyo, Japan.

INOUYE, Tatsuo. Born 1904. Educ.: Tokyo Imperial Univ. Chief, Chem. Eng. Dept. (1949), Director, Chief Eng. (1951), Managing Director (1955), Sen. Managing Director (1960), Pres. (1962), Nissan Chemetron Catalyst Co., Ltd. Japan Ind. Production Technique Council of Ministry of Internat. Trade and Ind.; Director, Japan Assoc. of Metals for Atomic Energy; Member, Com. for Chem. Industry and Eng. Development, Japan. Societies: Japan Catalyst Manufacturers' Assoc.; Japan Assoc. of Metals for Atomic Energy.
Nuclear interests: Chemical reaction under radioactive substance; Production of rare metals for the use of nuclear industry.
Address: 2nd Floor, Shinjo Building, No. 9 2-chome Ta-cho, Kanda, Chiyoda-ku, Tokyo, Japan.

INSCH, Gordon McConochie, B.Sc., Ph.D. Born 1927. Educ.: St. Andrews and Glasgow Univs. Head, Performance Office, (1957-59), Chief Eng., Commissioning Dept. (1959-66), Atomic Power Div. English Elec. Co., Ltd.; Manager of Eng., Nucl. Design and Construction Ltd., 1966-.
Nuclear interest: Reactor design and operation.
Address: 449 London Road, Leicester, Leics., England.

INSINGER, Francois Gerard, Physics Eng. Born 1929. Educ.: Delft Univ. Member, sci. staff, Stichting F.O.M., Holland.
Nuclear interest: Plasma physics.
Address: Laboratory of Atomic and Molecular Physics, 407 Kruislaan, Amsterdam, Netherlands.

in 't VELD, J. C. Born 1914. Head, Library and Documentation Dept., Reactor Centre, Netherlands, Petten, 1957-.
Address: Reactor Centrum Nederland, Petten, N.-H., Netherlands.

INTELMANN, Jürgen, Dr. Formerly Office Manager, now Managing Director, Krauss-Maffei-Imperial G.m.b.H.
Address: Krauss-Maffei-Imperial G.m.b.H., 4 Tanneweg, B.P. 20, Munich-Obermenzing, Germany.

INTOCCIA, Alfred Paul, B.S., M.S., Ph.D. Born 1928. Educ.: Scranton and Tennessee Univs. Physiologist, Biochem., Nucl. Sci. and Eng. Corp. Smith Kline and French Labs.
Nuclear interest: Radioisotopic uses and measurement techniques in mammalian physiology and drug metabolism studies.
Address: 1151 Thrush Lane, Audubon, Pennsylvania 19401, U.S.A.

INUGAI, Kanichi, D.Sc. Formerly Second Sect., Synthesis of Fluorine Compounds by Irradiation, now in Third Div., Chem., Ministry of Internat. Trade and Ind.

Address: Ministry of International Trade and Industry, Government Industrial Research Institute, Hirate-machi, Kita-ku, Nagoya, Japan.

INVERNIZZI, C. G. Pres., U.S. Radium Corp. (Europe).
Address: United States Radium Corporation (Europe), 36 Avenue Krieg, 1200 Geneva, Switzerland.

INYUTIN, E. I. Papers: Co-author, Physical Characteristics of Beryllium Moderated Reactor (2nd I.C.P.U.A.E., Geneva, Sept. 1958); Thermal Neutron Flux Shaping in Heterogeneous Nucl. Reactors by Fuel Charge Profiling (Abstract, Atomnaya Energiya, vol. 19, No. 1, 1965); Flattening of Volume Energy Release in Heterogenous Nucl. Reactors by Fuel Charge Profiling (ibid.); co-author, Thermal Neutron Flux Flattening in Uranium-Water Reactors (ibid.).
Address: Nuclear Energy Institute, Academy of Sciences of the U.S.S.R., 14 Leninsky Prospekt, Moscow V-71, U.S.S.R.

IOANID, George, Chem., Dr. Phil. Born 1911. Educ.: Bucharest Univ. Oncology Inst., Bucharest, 1951-54; Chem. Res. Inst., Bucharest, 1954-63; Ministry of Chem. Ind., Bucharest, 1963-67. Member, Permanent Commission of Roumania for Economic, Sci. and Tech. Cooperation for the use of Nucl. Power in Peaceful Purposes. Book: Radioactivity (Bucharest, Scientific Publishing House, 1964).
Nuclear interest: The applications of nuclear power in the chemical and oil industry field.
Address: Nr. 108 Calea Rahovei, Bucharest, Roumania.

IOKHEL'SON, S. V. Paper: Co-author, On the Content of Sb^{125} in Top-Soil and Plants (letter to the Editor, Atomnaya Energiya, vol. 16, No. 2, 1964).
Address: Academy of Sciences of the U.S.S.R., 14 Leninsky Prospekt, Moscow V-71, U.S.S.R.

IONAITIS, R. R. Papers: Co-author, Calculation of Transient States in a Hydraulic Loop Containing a Falling Body (letter to the Editor, Atomnaya Energiya, vol. 12, No. 5, 1962); The Calculation of Flow Rate Regulator with Flow Portion in Form of a Long Slot (letter to the Editor, ibid, vol. 15, No. 2, 1963); co-author, Experiments on Initial Travel of Safety Rod (letter to the Editor, ibid, vol. 22, No. 1, 1967); co-author, Electrohydraulic System of Safety Rods for SM-2 (letter to the Editor, ibid, vol. 22, No. 3, 1967).
Address: Academy of Sciences of the U.S.S.R., 14 Leninsky Prospekt, Moscow V-71, U.S.S.R.

IONOV, V. A. Papers: Co-author, The Exptl. Assessment of the Ra, Th and K Contents in Rocks from an Aircraft by Means of a NaJ (Tl) Pick-up (letter to the Editor, Atomnaya Energiya, vol. 13, No. 3, 1962); Spectral Distribution of Gamma-Rays from Co^{60} Point Source Screened with Aluminium Layer in Atmosphere Near by Ground (ibid, vol. 19, No. 4, 1965).
Address: Academy of Sciences of the U.S.S.R., 14 Leninsky Prospekt, Moscow V-71, U.S.S.R.

IORI, Ileana, Dr. in Phys.; Diploma in Appl. Nucl. Phys., Politecnic School of Milan; Libera Docenza in Nucl. Phys. Born 1931. Educ.: Modena Univ. With C.I.S.E., Milan. Collaborator, I.N.F.N., 1964-. Society: Italian Soc. of Phys.
Nuclear interest: Nuclear physics.
Address: 9 Via Bramante, Milan, Italy.

IOVNOVICH, M. L. Paper: Co-author, Collective Linear Acceleration of Ions (Atomnaya Energiya, vol. 24, No. 4, 1968).
Address: Academy of Sciences of the U.S.S.R., 14 Leninsky Prospekt, Moscow V-71, U.S.S.R.

IPPONMATSU, Tamaki, Dr. Eng. Born 1901. Educ.: Kyoto Univ. Managing Director (1951-57), Director (1957-60), Auditor (1960-), Kansai Elec. Power Co.; Director, Japan Land Development Co., 1952; Vice-Pres. (1957-62), Pres. (1962-), Japan Atomic Power Co.; Pres., Elec. Eng. Inst. of Japan, 1958-59; Vice Pres. (1959-60), Pres. (1965-67), Atomic Energy Soc. of Japan; Managing Director, Keizai Doyukai; Board of Council, Japan Atomic Energy Res. Inst.; Managing Director, Atomic Ind. Forum; Board of Council,Elec. Assoc. of Japan.
Nuclear interests: Management and engineering.
Address: 4-23, 3-chome, Jiyugaoka, Meguro-ku, Tokyo, Japan.

IRELAND, George Donald. Born 1909. Chief Eng., Windscale and Calder Works, U.K.A.E.A., 1959-. Society: M.I.Mech.E.
Nuclear interests: Engineering management and maintenance associated with reactors and chemical plants.
Address: Ashlea, Gosforth,Cumberland, England.

IRIARTE, Modesto, Jr., Ph.D., M.S.E.E., B.S.E.E., B.S.C.E. Born 1923. Educ.: Michigan and Puerto Rico Univs., Texas A. and M. Coll. and Argonne School Nucl. Sci. and Eng. Asst. Prof., Puerto Rico Univ., 1945-51; Design and Planning Eng., Puerto Rico Water Resources Authority, 1951-59; Staff Eng., General Nucl. Eng. Corp., Florida, 1959-60; Sen. Elec. and Nucl. Power Eng. (1960-65), Asst. Exec. Director for Elec. Planning, Res. and Construction (1965-), Puerto Rico Water Resources Authority. Societies: A.N.S.; I.E.E.E.
Nuclear interests: Nuclear reactor design; Plant instrumentation; Mathematical treatment of reactor kinetics; Associated dynamics of control systems; System synthesis.
Address: 1635 San Julian, Sagrado Corazón, Rio Piedras, Puerto Rico.

IRIBARREN, Ricardo PEREZ. See PEREZ IRIBARREN, Ricardo.

IRIZARRY, Sergio, M.D. Born 1925. Educ.: Buffalo Univ. Resident Medicine, Dept. of Health of Puerto Rico, 1952; Physician, (1953-56), Radiotherapy Resident, Dept. of Radiotherapy (1956-58), Puerto Rico League Against Cancer Hospital; Clinical Fellow, Internal Medicine, Francis Delafield Hospital, New York, 1958-60; Director, Clinical Applications Div., Puerto Rico Nucl. Centre, 1960-; Asst. Clinical Prof. of Medicine, School of Medicine, Puerto Rico Univ., 1963.
Nuclear interests: Nuclear medicine. Research, teaching and diagnostic work.
Address: Puerto Rico Nuclear Centre, Caparra Heights Station, San Juan, Puerto Rico 00935.

IRVINE, John W., Jr., B.A., Ph.D., Sc.D. (Hon.), D.Sc. (Hon.). Born 1913. Educ.: Missouri Valley Coll. and M.I.T. Prof. Chem., M.I.T., 1958-. Societies: A.C.S.; American Acad. of Arts and Sciences.
Nuclear interests: Radiochemical separations; Solvent extraction and ion exchange studies; Deuteron excitation functions.
Address: Department of Chemistry, Massachusetts Institute of Technology, Cambridge, Massachusetts 02139, U.S.A.

IRVING, John, M.A. (Glasgow), Ph.D. (Birmingham). Born 1920. Educ.: Glasgow Univ. Nuffield Res. Fellow, Glasgow, 1949-51; Sen. Lecturer in Appl. Maths., Southampton, 1951-59; Prof. Theoretical Phys., Cape Town, 1959; Freeland Prof. Natural Philosophy, Roy. Coll. Sci. and Technol., Glasgow, now Strathclyde Univ., 1961. Book: Maths. in Phys. and Eng. (Academic Press, Inc., 1959). Society: Inst. Phys. and Phys. Soc.
Nuclear interests: Thermonuclear problems; Three- and Four-body problems.
Address: Natural Philosophy Department, Strathclyde University, George Street, Glasgow, C.1, Scotland.

IRWIN, Glen W., Jr., M.D. Member, Radioisotope Com., School of Medicine, Indiana Univ.
Address: Indiana University, School of Medicine, 1100 West Michigan Street, Indianapolis, Illinois 46207, U.S.A.

IRWIN, Samuel Nelson, B.S.E. (E.E.), M.S.E. (N.). Born 1927. Educ.: Michigan Univ. Res. Assoc., Eng. Res. Inst., Coll. of Eng., Michigan Univ., 1951-56; Eng. in Charge, Advanced Eng. and Res. Dept., Holley Carbureter Co., Warren, Michigan, 1956-60; Manager, Inteledata Div., R. P. Scherer Corp., Detroit, 1960-61; Vice-Pres., Tech. Management Corp., Detroit, 1961-62; Pres., Data Systems Incorporated, subsidiary of Union Carbide Corp., Grosse Pointe Woods, 1962-. Societies: Inst. of Radio Eng.; A.N.S.
Nuclear interest: Reactor system kinetics.
Address: 128 Sunningdale Drive, Grosse Pointe Woods, Michigan, U.S.A.

ISAAC, Nadine VALENTIN-. See VALENTIN-ISAAC, Nadine.

ISAACSON, David Martin, B.S., M.S., Ph.D. Born 1924. Educ.: New York and Pittsburgh Univs. Graduate Student Asst., Pittsburgh Univ., 1948-50; Sanitarian (1950-51), Bacteriologist (1951-53), Chief Analyst (1953-55), Food Technologist (1955-56), Dept. of Public Health, City of Pittsburgh; Intermediate Sci. (1956-58), Advanced Sci. (1958-63), Nucl. Sci. and Eng. Corp.; Graduate Teach. Asst. (1959-61), Graduate Res. Asst. (1961-63), Pittsburgh Univ.; Sen. Res. Sci., E. R. Squibb and Sons, 1964-. Societies: American Soc. for Microbiology; Soc. of Gen. Microbiology.
Nuclear interests: Health physics; Radiobiology, with emphasis on mammalian and microbial membrane transport.
Address: 6 Thrush Drive, East Brunswick, New Jersey 08816, U.S.A.

ISABAEV, E. A. Paper: Co-author, Determination of Uranium and Plutonium Content in Samples of Mountain Rocks and Ores by means of Account of Fission Fragments (letter to the Editor, Atomnaya Energiya, vol. 23, No. 6, 1967).
Address: Academy of Sciences of the U.S.S.R., 14 Leninsky Prospekt, Moscow V-71, U.S.S.R.

ISABELLE, Didier Bernard Marie, D. ès Sc. (Ph.D.). Born 1934. Educ.: Paris Univ. C.E.A., 1957-61; C.N.R.S., 1961-62, 1964-65; N.B.S., U.S.A., 1963-64; Faculté des Sci. Clermont-Ferrand, 1965-. Societies: Sté. Française de Phys.; A.P.S.
Nuclear interest: Nuclear physics: electron scattering, experimental and theoretical. Apparatus for nuclear physics.
Address: 10 avenue R. Bergougnan, 63 Clermont-Ferrand, France.

ISAEV, B. M. Prof.; Deputy Director, Inst. Biophys., Acad. of Sci. Formerly Adviser, U.S.S.R. delegation to I.A.E.A. Papers: Co-author, Tissue dose Measurements for Hard X-rays (Atomnaya Energiya, vol. 6, No. 1, 1959); co-author, Ionization Methods of Measuring the Energy Absorbed from a Rad. Flux consiting of Neutrons Mixed with γ-rays (ibid, vol. 9, No. 2, 1960); co-author, Production of Pure Fluxes of Fast Neutrons for Radiobiologic Res. on IRT-100 Reactor (letter to the Editor, ibid, vol. 12, No. 6, 1962).
Address: Academy of Sciences of the U.S.S.R., 14 Leninsky Prospekt, Moscow V-71, U.S.S.R.

ISAKARI, H. Nucl. Specialist, Kaiser Eng.
Address: Kaiser Engineers, Kaiser Engineers International, 300 Lakeside Drive, Oakland 12, California, U.S.A.

ISAKOVA, L. Ya. Papers: Milnes Problem Solution for Multiplying Medium in Two-Group Approximation (Abstract, Atomnaya Energiya, vol. 20, No. 2, 1966); co-author, Method for Calculations of Neutron Flux and Control Rods Efficiency for Three-Dimensional Reactor (Abstract, ibid); co-author, The Effective Boundary Conditions at the Surface of the Black Rod (letter to the Editor, ibid, vol. 25, No. 3, 1968).
Address: Academy of Sciences of the U.S.S.R., 14 Leninsky Prospekt, Moscow V-71, U.S.S.R.

ISANO, M. Vice Chairman, First Atomic Power Ind. Group.
Address: First Atomic Power Industry Group, Tokyo Boeki Kaikan Building, 2, 1-chome Ohtemachi, Chiyoda-ku, Tokyo, Japan.

ISBERG, Pelle, Civil Ingenjör. Born 1926. Educ.: Stockholm Roy. Inst. of Technol. A.B. Atomenergi, 1950-53; North Carolina State Coll., 1952-53; Formerly at A.S.E.A., 1953-; Sci. Attaché, Swedish Embassy, Washington, 1960-64.
Nuclear interests: Reactor design; Nuclear plant design and safety.
Address: 17 Spannmalsgatan, Vasteras, Sweden.

ISBIN, Herbert S., B.S., M.S., Sc.D. Born 1919. Educ.: Washington (Seattle) and Minnesota Univs. and M.I.T. Member, Advisory Com. on Reactor Safeguards, and Staff Consultant, Argonne Univs. Assoc.; Prof., Dept. Chem. Eng., Minnesota Univ. Past Chairman, Nucl. Eng. Div., A.I.Ch.E.; Nucl. Eng. Educ. Com., Assoc. Midwest Univs. Book: Introductory Nuclear Reactor Theory (Reinhold Publishing Corp., 1963). Societies: A.C.S.; A.N.S.
Nuclear interests: Two-phase flow; Radiation chemistry; Nuclear reactor safety.
Address: Department of Chemical Engineering, Minnesota University, Minneapolis, Minnesota 55455, U.S.A.

ISELIN, Donald Grote, B.S., B.C.E., M.C.E. Born 1922. Educ.: Marquette Univ., U.S. Naval Acad. and Rensselaer Polytech. Inst. Eng.-construction Dept., Bureau of Yards and Docks, Washington, D.C. Past Pres., Oxnard-Ventura Post, Soc. of American Military Eng.; E.I.T., California. Book: Contributor to: Shippingport Pressurised Water Reactor (Reading, Mass., Addison-Wesley Pub. Co., 1958). Society: RESA.
Nuclear interest: Management of reactor plant design and construction.
Address: Bureau of Yards and Docks, Washington 25, D.C., U.S.A.

ISENBURGER, Herbert Rudolf. Born 1900. Educ.: Charlottenburg Univ. St. John X-ray Lab., 1927-. (Long Island City, N.Y., 1927-46; Califon, N.J. at present.). Governor's Advisory Com. on Rad. Protection, 1954-56. Books: Ind. Radiol. (1943); Bibliography on Ind. Radiol. (1943-); Bibliography on X-ray Stress Analysis (1953-). Societies: American Crystallographic Assoc.; A.S.M.E.; A.S.M.; A.S.T.M.; Soc. for Nondestructive Testing; H.P.S.; A.N.S.
Nuclear interests: Health physics; Industrial radiology.
Address: St. John X-ray Laboratory, Califon, New Jersey 07830, U.S.A.

ISHI, Eiichi. Sen. Officer, Lab. of Radiochem., Osaka Ind. Res. Inst.
Address: Laboratory of Radiochemistry, Osaka Industrial Research Institute, Daini Nishi-2-chome, Oyoda-ku, Osaka, Japan.

ISHIDA, Kyuichi, B. Law. Born 1901. Educ.: Tokyo Univ. Managing Director, Mitsubishi Atomic Power Industries, Inc., 1965-.
Address: Mitsubishi Atomic Power Industries, Incorporated, 4 Ohtemachi 1-chome, Chiyoda-ku, Tokyo, Japan.

ISHIDA, Masahiro R., D.Sc. Born 1926. Educ.: Kyoto Univ. Sen. Sci., Dept. of Zoology, Columbia Univ., 1961; Assoc. Prof., Res. Reactor Inst., Kyoto Univ., 1964-. Societies: Genetic Soc. of Japan; Japanese Biochem. Soc.
Nuclear interest: Radiation genetics and radiation protection.
Address: Research Reactor Institute, Kyoto University, Kumatori-cho, Sennan-gun, Osaka, Japan.

ISHIDA, Shoji. Born 1925. Educ.: Tokyo Univ. Literature and Sci. Prof., Nucl. Eng. Dept., Tokai Univ. Societies: Phys. Soc. of Japan; Japanese Soc. of Appl. Phys.; Japanese Soc. of Atomic Energy.
Nuclear interests: Reactor design; Education in reactor physics.
Address: Nuclear Engineering Department, Tokai University, 2-28 Tomigaya, Shibuya-ku, Tokyo, 158 Japan.

ISHIDA, Sueo. Formerly Asst. Chief, Rad. Safety Sect., now Asst. Chief, Res. Sect., Atomic Energy Bureau, A.E.C. of Japan.
Address: Atomic Energy Commission of Japan, 3-4 Kasumigaseki, Chiyoda-ku, Tokyo, Japan.

ISHIDA, Takanobu, B.S. (Chem.), M.S. (Nucl. Eng.), Ph.D. (Nucl. Eng.). Born 1931. Educ.: M.I.T., Kyoto and New York Univs. Res. Assoc., Brookhaven Nat. Lab., 1964-66; Res. Assoc., Belfer Graduate School of Sci., Yeshiva Univ., New York, 1966-. Society: A.C.S.
Nuclear interests: Separation of stable isotopes; Physics and chemistry of stable isotopes; Fuel cycle analysis.
Address: Yeshiva University, Belfer Graduate School of Science, New York, N.Y., U.S.A.

ISHIDA, Yoshio. Born 1906. Educ.: Tohoku Univ. Chief, Gen. Affairs Dept. (1952-57), Director (1957-65), Elec. Power Development

Co.; Managing Director, Japan Atomic Power Co., 1965-.
Address: 506, 15-1, 5-chome, Hibunya, Meguro-ku, Tokyo, Japan.

ISHIDA, Yukio. Member, Atomic Energy and Isotope Com., Tokushima Univ.
Address: c/o Medical Faculty, Tokushima University, Kuramoto-cho, Tokushima-shi, Japan.

ISHIGURO, Yoshitane. Member, Atomic Energy and Isotope Com., Tokushima Univ.
Address: c/o Medical Faculty, Tokushima University, Kuramoto-cho, Tokushima-shi, Japan.

ISHIHARA, Shunso, B.S., M.A., D.Sc. Born 1934. Educ.: Hiroshima and Columbia Univs. and Colorado School of Mines. Uranium Deposits Res. Group, Geological Survey of Japan.
Nuclear interest: Nuclear geology and geochemistry.
Address: Radioactive Mineral Resources Section, Geological Survey of Japan, 135 Hisamoto-cho, Kawasaki-shi, Japan.

ISHIHARA, Takehiko, M.S. Born 1924. Educ.: Tokyo Univ. and Internat. School of Nucl. Sci. and Technol. Asst. Prof., Gifu Pharmaceutical Coll. -1950; Res. Assoc., Tokyo Inst. of Technol., 1951-56; Researcher (1956-61), Chief, Chem. Eng. Lab. (later Reprocessing Lab.) (1961-65), Deputy Chief, Fuel R. and D. Div. (1966-67), Deputy Chief, Planning Office (1967-), Japan Atomic Energy Res. Inst. Books: Contributing author, Isotope Handbook (Maruzen, 1962); Radiochem. Handbook (Asakura-Shoten, 1962). Societies: Atomic Energy Soc., Japan; Japan Radioisotope Assoc.; Soc. of Chem. Eng., Japan; Chem. Soc. of Japan.
Nuclear interests: Nuclear chemical engineering, especially on reprocessing (radiation damage to solvents, non-aqueous reprocessing by chloride volatility) and radioactive waste management (chemical treatment, treatment plant design, cost analysis).
Address: Japan Atomic Energy Research Institute, 1-1 Shimbashi, Minato-ku, Tokyo, Japan.

ISHIHARA, Takeo. Director, Fund for Peaceful Atomic Development of Japan; Director, Japan Atomic Power Co. Ltd.
Address: Fund for Peaceful Atomic Development of Japan, No. 1-13, 1-chome, Shimbashi, Minato-ku, Tokyo, Japan.

ISHII, Ichiro, B.A. (Economics). Born 1905. Educ.: Tokyo Univ. Managing Director (1951), Vice Pres. (1962), Ind. Bank of Japan, Ltd.; Pres., Nissan Chem. Industries, Ltd., 1964; Director, Tokyo Atomic Ind. Consortium. Chairman, Japan Phosphatic and Compound Fertiliser Manufacturers Assoc.
Address: 2-191 Tamagawa-okusawa-cho, Setagaya, Tokyo, Japan.

ISHII, Shozo. Born 1914. Educ.: Kobe Univ. of Commerce. Director and Manager, Atomic and Electronic Dept., The Nissho Co. Ltd. Manager, Planning and Res. Com., First Atomic Power Industry Group; Auditor, F.A.P.I.G. Rad. Res. Labs., Ltd.
Nuclear interest: Management.
Address: 725-30 Tamagawa Seta-cho, Setagaya-ku, Tokyo, Japan.

ISHIKAWA, Ichiro. Born 1885. Educ.: Tokyo Imperial Univ. Pres., Japan Ind. Assoc., 1946-52; Pres., Federation of Economic Organisations, 1948-56; Chairman, Showa Denko Co., 1949-55; Chairman, Economy Council, 1952-67; Pres., Japanese Standard Assoc., 1955-; Commissioner, A.E.C. of Japan, 1956-63; Pres., Japan Nucl. Ship Development Agency, 1963-68. Societies: Japan Chem. Soc.; Japan Organic Chem. Soc.; Japan Chem. Industry Assoc.; Soc. of Economic Co-operation in Asia; Adviser, J.A.I.F., 1956-; Adviser, Atomic Energy Soc. of Japan, 1958-.
Address: 2-10-2 Tobitakyu, Chofushi, Tokyo, Japan.

ISHIKAWA, Kiyoshi. Born 1897. Educ.: Tokyo Univ. Sen. Managing Director (1958-67), Director, Nucl. Res. Lab. (1960-67), Adviser (1967-), Nippon Atomic Industry Group Co. Ltd. (NAIG Co.). Society: Nucl. Soc. of Japan.
Nuclear interests: General.
Address: Nippon Atomic Industry Group Company Limited, 2-5 Kasumigaseki 3-chome, Chiyoda-ku, Tokyo, Japan.

ISHIKAWA, Rokuro. Born 1925. Educ.: Civil Eng. Sect., Technol. Dept., Tokyo Univ. Director (1955), Managing Director (1957), Formerly Vice-Pres. (1959-), Kajima Construction Co., Ltd. Societies: Japanese Soc. of Civil Eng.; Japanese Atomic Ind. Forum.
Nuclear interest: Civil engineering pertaining to atomic power stations.
Address: 5-5 Mejirodai, 3-chome, Bunkyo-ku, Tokyo, Japan.

ISHIKAWA, Tachiomaru, Prof. Dr. Member (Pathology), Radioisotope Res. Com., Kanazawa Univ.
Address: Kanazawa University, Marunouchi 1-1, Kanazawa, Ishikawa Prefecture, Japan.

ISHIKAWA, Toshikatsu. Formerly Chief, Yokohama Factory, now Managing Director, Production Div., Nippon Carbon Co. Ltd.
Address: Nippon Carbon Company Limited, 2, 2-chome Nishi-Hachi-Chobori, Chuo-ku, Tokyo, Japan.

ISHIMATSU, Masakane. Pres., Sumitomo Coal Mining Co. Ltd.

ISH

Address: Sumitomo Coal Mining Co. Ltd., 2, 1-chome, Marunouchi, Chiyoda-ku, Tokyo, Japan.

ISHIMATSU, Toshiyuki. Asst. Prof., Nucl. Phys. Lab., Kyushu Univ.; Prof., Course of Appl. Nucl. Phys., Tohoku Univ.
Address: Tohoku University, 75 Katahira-cho, Sendai, Japan.

ISHIMATSU, Tsuyoshi. Director, R. and D. Div. (Nucl. Glass Technol.), Asahi Glass C. Ltd.
Address: Asahi Glass Co. Ltd., 14, 2-chome, Marunouchi, Chiyoda-ku, Tokyo, Japan.

ISHIYAMA, Shoji, Master of Law. Born 1924. Educ.: Hokkaido Univ. Director, Overseas Div., Rigaku Denki Co., Ltd.
Address: Rigaku Denki Company Limited, Overseas Division, 9-8, 2-chome, Sotokanda, Chiyoda-ku, Tokyo, Japan.

ISHIZAKA, Taizo, B.A. at law. LL.D. (hon., Loyola Univ.), LL.D. (hon., Hawaii Univ.). Born 1886. Educ.: Tokyo Imperial Univ. Chairman of the Board, Tokyo Shibaura Elec. Co., Ltd., 1957-; Director, Japan Atomic Power Generation Co. Ltd., 1957-; Pres., Federation of Economic Organisations, 1956-; Chairman of the Board, Nippon Atomic Industries Group Co., Ltd., 1958-; Vice-Pres., Japan Red Cross.
Nuclear interest: Management.
Address: 13-6 Shoto 1-chome, Shibuyaku, Tokyo, Japan.

ISHIZAWA, Schuichi, Dr. Chairman, Nucl. Com., Nat. Inst. of Agricultural Sci.
Address: National Institute of Agricultural Sciences, Nishigahara, Kita-ku, Tokyo, Japan.

ISHKOV, A. P. Paper: Co-author, Investigation of Auto-Resonant Particle Acceleration by Electromagnetic Waves (Atomnaya Energiya, vol. 22, No. 1, 1967).
Address: Academy of Sciences of the U.S.S.R., 14 Leninsky Prospekt, Moscow V-71, U.S.S.R.

ISKANDAR, M. Sen. Officer, X-Ray and Electromedical Div., N.E.E.A.-S.A.A.
Address: N.E.E.A.-S.A.A., X-Ray and Electromedical Division, 26 sh. Adly, Cairo, Egypt, United Arab Republic.

ISKENDERIAN, Haig P., B.S. (E.E.), M.A. and Ph.D. (Phys.). Born 1905. Educ.: Michigan and Columbia Univs. Federal Telecommunications Labs., 1943-50; Argonne Nat. Lab., 1950-. Societies: A.P.S.; A.N.S.
Nuclear interests: Nuclear physics; Reactor physics.
Address: 165 Fellows Court, Elmhurst, Illinois, U.S.A.

ISLA, Manuel, Dr. Ind. Eng. Born 1931. Educ.: Madrid Univ. Eng., Nucl. Dept., Hidroeléctrica Española, S.A. Society: A.N.S.

Nuclear interests: Power reactor stations design, construction and operation.
Address: Nuclear Department, Hidroelectrica Española S.A., 1 Hermosilla, Madrid 1, Spain.

ISLA, Ricardo, Ing. Agronomo. Consejero, Consejo Directivo, Comision Chilena de Energia Nucl., 1966-.
Address: Comision Chilena de Energia Nuclear, 138 Calle Miraflores, Santiago, Chile.

ISLAM, A. H. M. Habibul, M.Sc. (Chem.). Born 1924. Educ.: Dacca Univ. Lecturer, Sylhet Govt. Coll., 1950-56; Lecturer, Baghdad Pact Nucl. Centre, Baghdad, 1957-58; Res. Officer, A.E.C., Karachi, 1958-60; Sci. Officer (1961-63), Sen. Sci. Officer (1964-65), Atomic Energy Agricultural Res. Centre, Dacca; Sen. Sci. Officer, Atomic Energy Centre, Dacca, 1965-.
Nuclear interest: Tritium enrichment and counting and its application in hydrology. Measurement of velocity and direction of underground water by radioisotope tracer technique. Industrial uses of radioisotopes.
Address: Atomic Energy Centre, P.O. Box No. 164, Ramna, Dacca, East Pakistan.

ISLAM, Muhammad Munirul, B.Sc. (Hons.), M.Sc., Ph.D. Born 1936. Educ.: London Univ. Res. Assoc. (1962-65), Asst. Prof., Res. (1965-67), Brown Univ.; Asst. Prof., Connecticut Univ., 1967-. Society: A.P.S.
Nuclear interest: Particle physics.
Address: Department of Physics, Connecticut University, Storrs, Connecticut, U.S.A.

ISLAM, S., Dr. Reader, Nucl. Phys., Dept. of Phys., Dacca Univ.
Address: Dacca University, Dacca 2, Pakistan.

ISO, Y. Chief Architectural Eng., Construction Div., Japan Atomic Energy Res. Inst., Tokai Res. Establishment. Sect. Editor, Nucl. Structural Eng.
Address: Japan Atomic Energy Research Institute, Tokai Research Establishment, Tokaimura, Naka-gun, Ibaraki-ken, Japan.

ISOGAI, Makoto. Formerly Managing Director, Mitsubishi Nippon Heavy-Industries Ltd.; Managing Director and Gen. Manager of Tech. Headquarters, Mitsubishi Heavy Industries, Ltd.
Address: Mitsubishi Heavy-Industries Ltd., 10, 2-chome, Marunouchi, Chiyoda-ku, Tokyo, Japan.

ISOLA, Aulis Ales, Fil. lic. Born 1935. Educ.: Helsinki Univ. Asst., Nucl. Phys., Helsinki Univ.; Special Investigator, Geological Survey of Finland; Head, Dept. of Rad. Inspection, Inst. of Rad. Protection. Sec., Finnish Rad. Advisory Com. Societies: Nordic Soc. for Rad. Protection; Internat. Rad. Protection Assoc.; Finnish Nucl. Soc.
Nuclear interests: Siting of nuclear power plants; Radiological protection; Environmental

monitoring; Radiation protection standards.
Address: 184 Merikorttitie 1. D., Helsinki 96, Finland.

ISOLABELLA, Manuel, Dr. Comision Especial de Fisica Atomica y Radioisotopes, Univ. Nacional de La Plata.
Address: Universidad Nacional de La Plata, La Plata, Argentina.

ISOYA, Akira, D.Sc. Born 1921. Educ.: Tokyo Univ. Tokyo Univ., 1950-62; Pittsburgh Univ., 1959-61; Lawrence Rad. Lab., 1961-62; At (1962-), Prof., Dept. of Phys., Faculty of Sci., Director, Nucl. Phys. Lab., Kyushu Univ. Societies: Phys. Soc. of Japan; A.P.S. Nuclear interests: Nuclear spectroscopy; Nuclear instrumentation.
Address: Kyushu University, Fukuoka, Japan.

ISRAELACHVILI, Marcel, B.S. (Phys.), M.S. (Nucl. Eng.). Educ.: London and Stanford Univs. Gen. Manager, D.I.S.I. Nucl. Corp., Milan. Societies: Stà. Italiana di Medicina Nucl.; Assoc. Italiana Calcolo Automatico; F.I.E.N.
Nuclear interest: Management.
Address: D.I.S.I. Nuclear Corp., 14 Via Frua, 20146, Milan, Italy.

ISSEROW, Saul, B.A., M.S., Ph.D. Born 1922. Educ.: Brooklyn Coll. and Pennsylvania State Univ. Group Leader (1956-59), Project Manager (1959-64), Manager, Process Development (1964-66), Manager, Materials Res. (1966-), Nucl. Metals Div., Whittaker Corp. Books: Chapter in Rare Metals Handbook (Reinhold, 1961); Nucl. Reactor Fuel Elements (Wiley, 1962); The Technology of Nucl. Reactor Safety, Vol. II (M.I.T. Press, 1967). Societies: A.C.S.; A.S.M.
Nuclear interest: Development of improved fuels for nuclear reactors.
Address: 47 Stetson Street, Brookline, Mass. 02146, U.S.A.

ISSHIKI, Naotsugu, Dr. of Eng. Born 1922. Educ.: Tokyo Univ. and M.I.T. Ship Res. Inst., Japan, 1948-. Society: Atomic Energy Soc. of Japan.
Nuclear interest: Nuclear ship and marine reactors.
Address: 29-6 Kyodo 2 chome, Setagayaku, Tokyo, Japan.

ISTINYELI, H. E. Hasan. Turkish Ambassador to Austria; Turkish Governor and Resident Rep. to I.A.E.A.
Address: 40 Prinz Eugen-Strasse, Vienna 4, Austria.

ISUPOV, I. A. Papers: Co-author, Calculation of Shear Stresses on Wall and of Velocity Distribution for Turbulent flow of Liquid in Channels (Atomnaya Energiya, vol. 21, No. 2, 1966); co-author, Calculation of Hydraulic Resistance Coefficients for Turbulent Flow in Noncircular Cross-Section Channel (ibid, vol. 23, No. 4, 1967).
Address: Academy of Sciences of the U.S.S.R., 14 Leninsky Prospekt, Moscow V-71, U.S.S.R.

ISUPOV, V. K. Paper: Co-author, About Accumulation of Hydrogen Peroxide in Primary Cooling System of VVR-M Reactor (letter to the Editor, Atomnaya Energiya, vol. 20, No. 4, 1966).
Address: Academy of Sciences of the U.S.S.R., 14 Leninsky Prospekt, Moscow V-71, U.S.S.R.

ITAYA, Mitsuo. Res. Assoc., Nucl. Phys. Div., Kyushu Inst. of Technol.
Address: Kyushu Institute of Technology, Nuclear Physics Division, Tobata, Kitakyushu, Fukuoka-ken, Japan.

ITO, Chuko. Manager, Nucl. Fuel Res. Dept., Mitsubishi Metal Mining Co. Ltd.
Address: Mitsubishi Metal Mining Co. Ltd., Research Laboratory, No. 297, 1-chome, Kitabukuro, Omiya-city, Saitama Prefecture, Japan.

ITO, Hanjiro. Member, Atomic Energy and Isotope Com., Tokushima Univ.
Address: c/o Medical Faculty, Tokushima University, Kuramoto-cho, Tokushima-shi, Japan.

ITO, Hidehiko. Formerly Res. Assoc., Dept. Nucl. Sci., now Asst., High Energy Phys., Dept. of Phys. II, Kyoto Univ.
Address: Department of Physics II, Faculty of Science, Kyoto University, Sakyo-ku, Kyoto, Japan.

ITO, Norio. Sen. Officer, Onoda Cement Central Res. Lab.
Address: Onoda Cement Central Research Laboratory, 1-1-7, Toyosu, Koto-ku, Tokyo, Japan.

ITO, Otomasa, M.D. Prof., Dept. of Radiol., Yokohama Univ. School of Medicine.
Address: Department of Radiology, Yokohama University School of Medicine Hospital, Urafune-cho, Minami-ku, Yokohama, Japan.

ITO, Taiichi, Prof. Director, Radioisotope Res. Com., Niigata Univ.
Address: Radioisotope Research Committee, Niigata University, Asahimachi, Niigata, Japan.

ITO, Tatsuzi, D.M., M.M. Born 1904. Educ.: Niigata Univ. School of Medicine. Dean of Faculty of Medicine (1953-59), Pres. (1959-), Niigata Univ. Inspector of Medical Education (appointed by the Ministry of Education); Member, 7th Branch, Japan Sci. Council; Director, Assoc. Nat. Univs. of Japan; Councillor, Kanto Branch, Assoc. Univ. Professors; Director, Japan Pathological Soc.; Inspector of Public Univ. Education (appointed by the Ministry of Education); Councillor, Univ.

Standard Assoc.
Nuclear interest: Management.
Address: Niigata University, Asahi-Machi, Niigata City, Japan.

ITOH, Gakuro. Vice Director, Nucl. Rad. Hazard Protection Div., Japan Safety Appliances Assoc.
Address: Japan Safety Appliances Association, c/o Kyoiku Building, 4-6-16 Kohinata, Bunkyo-ku, Tokyo, Japan.

ITOH, Junkichi, Dc. Sci. Born 1914. Educ.: Osaka Univ. Prof. of Material Phys., Faculty of Eng. Sci., Osaka Univ., 1945-. Pres., Phys. Soc. of Japan, 1967-68. Book: Magnetic Resonance (Kyoritsu Shuppan, 1957, Iwanami Shoten, 1958). Society: Phys. Soc. of Japan.
Nuclear interest: Nuclear magnetic resonance.
Address: Faculty of Engineering Science, Osaka University, Toyonaka, Osaka, Japan.

IVANCHENKO, A. I. Paper: Co-author, Production of Megagauss Fields by Magnetodynamics Cumulation (Atomnaya Energiya, vol. 23, No. 6, 1967).
Address: Academy of Sciences of the U.S.S.R., 14 Leninsky Prospekt, Moscow V-71, U.S.S.R.

IVANCHEV, Neno. Born 1921. Educ.: Sofia State Univ. Prof., Civil Eng. Inst.; Exec. Sec., Com. for Peaceful Use of Atomic Energy. Books: Optic and Introduction in Atomic Phys. 2nd edition (Sofia Publishing House Technika, 1965); Application of Radioactive isotopes in Technics and Industry (Technika, 1962).
Nuclear interest: Nuclear physics.
Address: Council of Ministers, Sofia, Bulgaria.

IVANENKO, Dmitrii Dmitrievich. Born 1904. Educ.: Leningrad Univ. Prof., Moscow Univ., 1943; Inst. of the History of Natural Sci. and Eng., Acad. of Sci. of the U.S.S.R., 1949. Books: Co-author, Classical Field Theory: New Problems (2nd ed., Moscow-Leningrad, 1951); Quantum Field Theory (Moscow-Leningrad, 1952).
Nuclear interest: Physics of elementary particles.
Address: Academy of Sciences of the U.S.S.R., 14 Leninsky Prospekt, Moscow V-71, U.S.S.R.

IVANITSKII, P. G. Paper: Co-author, Study of Inelastic Scattering of Slow Neutrons by Polyethylene (Atomnaya Energiya, vol. 20, No. 1, 1966).
Address: Academy of Sciences of the U.S.S.R., 14 Leninsky Prospekt, Moscow V-71, U.S.S.R.

IVANOV, Boris. Chief Eng., Beloyarsk Atomic Power Station.
Address: Beloyarsk Atomic Power Station, Beloyarsk, Nr. Sverdlovsk, Urals, U.S.S.R.

IVANOV, G. A. Paper: Co-author, Collective Linear Acceleration of Ions (Atomnaya Energiya, vol. 24, No. 4, 1968).
Address: Academy of Sciences of the U.S.S.R., 14 Leninsky Prospekt, Moscow V-71, U.S.S.R.

IVANOV, G. K. Papers: On Scattering of Neutrons by Titanium and Calcium Nuclei (letter to the Editor, Atomnaya Energiya, vol. 12, No. 1, 1962); co-author, About Resonance Neutron-Molecule Interaction (letter to the Editor, ibid, vol. 19, No. 2, 1965).
Address: Academy of Sciences of the U.S.S.R., 14 Leninsky Prospekt, Moscow V-71, U.S.S.R.

IVANOV, I. N. Paper: Co-author, Collective Linear Acceleration of Ions (Atomnaya Energiya, vol. 24, No. 4, 1968).
Address: Academy of Sciences of the U.S.S.R., 14 Leninsky Prospekt, Moscow V-71, U.S.S.R.

IVANOV, Mari. Member, Committee for the Peaceful Uses of Atomic Energy, Council of Ministers of the Bulgarian People's Republic.
Address: Commiteee for the Peaceful Uses of Atomic Energy, Council of Ministers, Sofia, Bulgaria.

IVANOV, R. N.
Nuclear interest: Fission product yields.
Address: Academy of Sciences of the U.S.S.R., 14 Leninsky Prospekt, Moscow V-71, U.S.S.R.

IVANOV, S., Prof. Deputy Director, Inst. of Phys. and Atomic Res. Centre of the Bulgarian Acad. of Sci.
Address: Institute of Physics and Atomic Research Centre of the Bulgarian Academy of Science, 72 Lenin Street, Sofia 13, Bulgaria.

IVANOV, V. I.
Nuclear interests: Metallurgy of fuel elements and liquid metal coolant circuits.
Address: Nuclear Energy Institute, Academy of Sciences of the U.S.S.R., 14 Leninsky Prospekt, Moscow V-71, U.S.S.R.

IVANOV, V. N. Papers: Co-author, Study of Spectra and Doses Generated by Monoenergetic Neutron Source in Iron-Water Shielding (Atomnaya Energiya, vol. 21, No. 1, 1966); co-author, Two-Layer Shielding Optimisation Against Isotopic Fast Neutron Sources (ibid, vol. 23, No. 6, 1967).
Address: Academy of Sciences of the U.S.S.R., 14 Leninsky Prospekt, Moscow V-71, U.S.S.R.

IVANOV, Viktor Evgenyevich. Leader, Russian deleg. visiting Britain July 18-28, 1961, to discuss solid state phys. res. Papers: Co-author, Pin Fuel-element for Gas-cooled Heavy-water Power Reactor (Jaderná Energie, vol. 4, No. 11, 1958); co-author, Dependence of Hardened Uranium Texture from Heating Nature and Other Parameters of Heat Treatment (Atomnaya Energiya, vol. 16, No. 4, 1964); co-author, Deformation and Fracture of Rolled Beryllium of Various Purity (ibid, No. 5); co-author, Impurity State Effect on Deformation Texture of Low-Alloyed

Alpha-Uranium (letter to the Editor, ibid, vol. 24, No. 1, 1968).
Address: Physico-Technical Institute, Ukrainian Academy of Sciences, Kharkov, U.S.S.R.

IVANOVA, K. N. Paper: Co-author, Niobium-Base Alloys and Their Properties (Atomnaya Energiya, vol. 23, No. 1, 1967).
Address: Academy of Sciences of the U.S.S.R., 14 Leninsky Prospekt, Moscow V-71, U.S.S.R.

IVANOVSKII, M. N. Papers: Co-author, About Condensation Coefficient of Mercury Vapour (Atomnaya Energiya, vol. 24, No. 2, 1968); co-author, Quantitative Determination of Hydrogen Impurity in Liquid Sodium by Method of Hydride Thermal Dissociation (ibid, vol. 24, No. 3, 1968); co-author, Conductivity Indicator of Admixtures in Liquid Metal Coolants (letter to the Editor, ibid, vol. 24, No. 4, 1968); co-author, Dependence of Dropwise Condensation Heat Transfer Coefficient on Temperature Head (ibid, vol. 24, No. 6, 1968).
Address: Academy of Sciences of the U.S.S.R., 14 Leninsky Prospekt, Moscow V-71, U.S.S.R.

IVANOVSKII, N. N. Papers: Co-author, A Study of Heat Transfer to Liquid Sodium in Tubes (letter to the Editor, Atomnaya Energiya, vol. 13, No. 4, 1962); co-author, Natrium Decontamination from Interaction Products with Water in Circulation Loop (letter to the Editor, ibid, vol. 19. No. 3, 1965); co-author, Leakage Discovery in Sodium-Water Steam Generator (ibid, vol. 20, No. 6, 1966).
Address: Academy of Sciences of the U.S.S.R., 14 Leninsky Prospekt, Moscow V-71, U.S.S.R.

IVANTCHEV, Neno Petrov, Candidate of Tech. Sci. Born 1921. Educ.: Sofia State Univ. Reader in Phys., High School of Building-eng., Sofia. Book: Optics and Basis of Nucl. Phys. (third part of text-book in Phys. for tech. high schools in Bulgaria, edition 1961). Society: Soc. of Sci. Workers in Bulgaria. Nuclear interests: Nuclear physics and use of the positrons' annihilation for studying the physics of the solid body; Use of isotopes in industry (measuring of diffusion coefficient in different kinds of metals and alloys).
Address: 4 Dimtcho Debelijanov st., 13 Sofia, Bulgaria.

IVASHCHENKO, N. I. Papers: Co-author, The Effect of an Internal Heat Source on the Heat Transfer Coefficient (letter to the Editor, Atomnaya Energiya, vol. 7, No. 3, 1959); co-author, Calculation of Heat Transfer for Liquids Flowing Turbulently in Tubes at Values of Prandtl's Number Much Less than One (ibid, vol. 11, No. 5, 1961); co-author, On Heat Transfer to Liquid Metals Flowing in Tubes (letter to the Editor, ibid, vol. 14, No. 3, 1963); co-author, Heat Transfer With Metallic Sodium, Moving in Tube (letter to the Editor, ibid, vol. 16, No. 6, 1964).
Address: Academy of Sciences of the U.S.S.R., 14 Leninsky Prospekt, Moscow V-71, U.S.S.R.

IVASHIN, V. V. Paper: Co-author, 1, 5 GeV Electron Synchrotron at the Polytech. Inst. of Tomsk (Atomnaya Energiya, vol. 21, No. 6, 1966).
Address: Academy of Sciences of the U.S.S.R., 14 Leninsky Prospekt, Moscow V-71, U.S.S.R.

IVASHKEVICH, A. A. Papers: Burn-out Heat Flows during Forced Convection of Fluid in Channels (letter to the Editor, Atomnaya Energiya, vol. 8, No. 1, 1960); Simplified Model of Burnout at Forced Liquid Flow in Tubes (ibid, vol. 23, No. 1, 1967).
Address: Academy of Sciences of the U.S.S.R., 14 Leninsky Prospekt, Moscow V-71, U.S.S.R.

IVES, Michael Brian, B.Sc., Ph.D. Born 1934. Educ.: Bristol Univ. Res. Metal. Eng., Carnegie Inst. Technol., Pittsburgh, 1958-61; Assoc. Prof. of Metal. and Metal. Eng., McMaster Univ., Hamilton, 1961-. Councillor, Ontario Chapter, A.S.M.; Education Chairman and Councillor, Hamilton Branch, Canadian Inst. of Mining and Metal. Society: A.I.M.M.E. Nuclear interest: Metallurgy of corrosion inhibition, particularly liquid metal environments.
Address: Department of Metallurgy and Metallurgical Engineering, McMaster University, Hamilton, Ontario, Canada.

IVES, Patricia A. Chem., Geological Survey, U.S. Dept. of the Interior.
Address: U.S. Department of the Interior, Geological Survey, Washington D.C. 20242, U.S.A.

IVLEV, A. A. Papers: Co-author, Some Laws in the Thermodynamics of Isotope Exchange (Kernenergie, vol. 10, No. 4, 1967); co-author, The Partition Function Ratios of Isotopically Substituted Molecules Calculated by using the Urey-Bradley-Simanouti Potential Function (ibid, vol. 10, No. 5, 1967).
Address: Academy of Sciences of the U.S.S.R., 14 Leninsky Prospekt, Moscow V-71, U.S.S.R.

IWAI, Shigehisa. Prof. i/c, Eng. for Reactor Integrity, Dept. of Nucl. Eng., Kyoto Univ.
Address: Kyoto University, Department of Nuclear Engineering, Yoshida-Honmachi, Sakyo-ku, Kyoto, Japan.

IWAI, Tadashi. Sen. Res. Officer, Seventh Res. Dept., Central Res. Lab., Ube Industries Ltd.
Address: Ube Industries Ltd., Ogushi, Ubeshi, Yamaguchi Pref, Japan.

IWAMOTO, Tsuneji. Formerly Chairman and Vice Pres., Atomic Energy Com., now Pres., Hokkaido Elec. Power Co. Ltd.
Address: Hokkaido Electric Power Co. Ltd., 2, Higashi 1-chome, Odari, Sapporo, Hokkaido, Japan.

IWA

IWANAGA, Eiichi. Asst. Director, R. and D. Div. (Nucl. Glass Technol.), Asahi Glass Co. Ltd.
Address: Asahi Glass Co. Ltd., 14, 2-chome, Marunouchi, Chiyoda-ku, Tokyo, Japan.

IWANOWSKI, Rudolf, Dr.-Ing., Dipl. Ing., Chemie Rat. Born 1899. Educ.: Breslau Univ. Stadtwerke Wiesbaden A.G. in Hessen. Forschungsgemeinschaft Godesberg/Bonn/Rhein.
Nuclear interest: Decontamination by coagulation.
Address: 5 Adolfsallee, 62 Wiesbaden, Germany.

IWASAKI, Izuru. Asst. Chief, Res. Promotion Sect., Atomic Energy Bureau, A.E.C. of Japan.
Address: Atomic Energy Commission of Japan, 3-4 Kasumigaseki, Chiyoda-ku, Tokyo, Japan.

IWASAKI, Machio, D. Sc. Born 1924. Educ.: Nagoya Univ. Dept. of Chem., Oregon State Coll., 1958-60; Govt. Ind. Res. Inst., Nagoya, 1953-68. Societies: Japan Chem. Soc.; Soc. of Polymer Sci., Japan; Japan Radioisotope Assoc.
Nuclear interest: Electron spin resonance of organic free radicals produced by irradiation.
Address: 2-20 Kobai-cho, Showa-ku, Nagoya, Japan.

IWASAKI, Shoji. Res. Sect., Geophys. Dept., Geological Survey of Japan.
Address: Geophysics Department, Geological Survey of Japan, Hisamoto-cho 135, Kawasaki-chi, Japan.

IWASE, Y. Director, Tokyo Atomic Ind. Consortium.
Address: Tokyo Atomic Industrial Consortium, Hitachi Building, 6-4-chome, Surugadai Kanda, Chiyoda-ku, Tokyo, Japan.

IWATA, Shiro, Dr. Asst. Prof., Div. of Hot Lab., Res. Reactor Inst., Kyoto Univ.
Address: Kyoto University, Research Reactor Laboratory, Noda, Kumatori-cho, Sen-nan-gun, Asaka Prefecture, Japan.

IWATA, Yoshito, Dr. Member, Nucl. Com., Nat. Inst. of Agricultural Sci.
Address: National Institute of Agricultural Sciences, Nishigahara, Kita-ku, Tokyo, Japan.

IWATAKE, T. Exec. Director, First Atomic Power Industry Group.
Address: First Atomic Power Industry Group, Nissho Building, 10 Nihonbashi-Edobashi 1-chome, Chuo-ku, Tokyo, Japan.

IWATAKE, Teruhiko. Director-in-Charge, Res. Lab. and Atomic Power, Kobe Steel Ltd.
Address: Kobe Steel Ltd., 36-1, 1-chome, Wakinohama-cho, Fukiai-ku, Kobe, Japan.

IYA, Vasudeva Kilara, B.Sc. (Hons.) (Mysore), M.Sc., D. ès Sc. (Paris). Born 1927. Educ.: Mysore and Paris Univs. Formerly Attaché de Recherches, C.N.R.S. Paris; Head, Isotope Div., Metal. Group, Bhabha Atomic Res. Centre, A.E.C., India.
Nuclear interests: Nuclear and inorganic chemistry, with special reference to radio-isotopes.
Address: Atomic Energy Commission, Bhabha Atomic Research Centre, Trombay, Bombay 74, India.

IYENGAR, Padmanabha Krishnagopala, Ph.D. Born 1931. Educ.: Travancore and Bombay Univs. Res. Asst., Tata Inst. of Fundamental Res.; Sci. Officer, and Head, Nucl. Phys. Div., Bhabha Atomic Res. Centre, Trombay.
Nuclear interests: Interaction of neutrons with matter; Solid state physics; Mossbauer effect and nuclear techniques.
Address: Modular Laboratory, Bhabha Atomic Research Centre, Trombay, Bombay-74 AS, India.

IZATT, James Alexander, B.Sc. (Mech. Eng.). Born 1929. Educ.: Strathclyde Univ. Eng. U.K.A.E.A., Aldermaston, 1955-62; Deputy Director, Scottish Res. Reactor Centre, 1963.
Nuclear interests: Operation of research reactors. Design and operation of experiments for research reactors.
Address: Scottish Research Reactor Centre, East Kilbride, Glasgow, Scotland.

IZAWA, Masami, Dr. Eng. Born 1920. Educ.: Tokyo Univ. Div. of Antibiotics, Nat. Inst. of Health, -1957; Chief, Chem. Div., Nat. Inst. Radiol. Sci., 1957-. Sec., Japan Rad. Res. Soc.; Exec. Member, Japanese H.P.S.; Board, Japan Radioisotope Assoc. Societies: Chem. Soc. of Japan; Rad. Res. Soc.; Atomic Energy Soc. of Japan; H.P.S.; Japanese H.P.S.; Japan Radioisotope Assoc.
Nuclear interests: Nuclear chemistry; Environmental hygiene; Health physics.
Address: National Institute of Radiological Sciences, 9-4-1 Anagawa, Chiba-shi, Japan.

IZENSTARK, Joseph Louis, B.A., M.D., F.A.C.R. Born 1919. Educ.: California Univ. Asst. Prof., Radiol. (1960-62), Assoc. Prof. Radiol. (1962-63), Tulane Univ. School of Medicine; Assoc. Prof., Radiol., Emory Univ. School of Medicine, 1963-67. Consultant, U.S.P.H.S., 1961; Consultant, U.S. Army, 1965; Consultant, Veterans Administration, 1967. Societies: Soc. of Nucl. Medicine; American Roentgen Ray Soc.; Radiol. Soc. of North America.
Nuclear interests: Clinical nuclear medicine and nuclear education.
Address: William Beaumont Hospital, Royal Oak, Michigan, U.S.A.

IZRAEL', Yu. A. Papers: On the Calculation of the Dose Rate Inside a Spherical Source with Gaussian Distribution of a γ-Radiator (letter to the Editor, Atomnaya Energiya, vol. 14, No. 3, 1963); co-author, Gamma Rad.

Spectrum from Artificial Model of Radioactive Fallout (letter to the Editor, ibid, vol. 19, No. 2, 1965); co-author, Modulation of Radioactive Fallout Particles (letter to the Editor, ibid, vol. 24, No. 6, 1968).
Address: Academy of Sciences of the U.S.S.R., 14 Leninsky Prospekt, Moscow V-71, U.S.S.R.

IZZO, Theodore F., B.S. (Chem.). Born 1930. Educ.: Northeastern Univ. Mill Supt., Atlas Minerals Div. Chairman, Uranium Sect., A.I.M.E. Books: Section in Uranium Ore Processing (Addison-Wesley Publishing Co., 1958).
Nuclear interests: In complete charge of the metallurgical, research, and analytical sections of the Moab Mill of the Atlas Minerals Div. of Atlas Corp.
Address: 455 W Andrea Ct., Moab, Utah, U.S.A.

JAAG, Otto, Dr. rer. nat., Prof. (Swiss Federal Inst. Technol., Zürich). Director, Swiss Federal Inst. for Water Supply, Sewage Purification and River Protection. Pres., European Federation for Water Protection against Pollution.
Nuclear interests: Control of radioactivity in drinking water in surface and underground waters, rain and cisterns; Methods of decontamination of radioactive waters.
Address: 5 Physikstrasse, Zürich CH 8044, Switzerland.

JABLONSKI, Francis E. See JAMERSON, Frank E.

JACCARD, Pierre, Eng. Director, Service de l'Electricite of Geneva, Services Industriels de Geneve.
Address: Services Industriels de Geneve, Geneva, Switzerland.

JACCHIA, Enrico. Dr. in Economics, Lic. in Political Sci., Lic. in Social Sci., Degree in Statistics. Born 1923. Educ.: Geneva and Venice Univs. Graduate, Inst. of Internat. Studies, Geneva; Inst. Nat. de la Statistique, Paris. Attaché, Italian Deleg. to the Council of Europe, 1949-51; Deputy-Head, Economic Sect., Council of Europe, 1951-54; Head, Social Affairs Div., Western European Union, 1955-58; Director, Health and Safety Dept., (-1968), Head, Safeguards Div. (1968-), Euratom. Books: Il rischio da radiazioni nell'era nucleare (Milan, Giuffré, 1963); Atome et Securité (Paris, Dalloz, 1964); Atom Sicherheit und Rechtsordnung (Bonn-Freudenstadt, Lutzeyer, 1965).
Nuclear interests: Nuclear law; Regulatory activities in the field of radiation protection.
Address: 27 rue Edmond Picard, Brussels 18, Belgium.

JACH, Joseph, Dr. Phil. (Oxon.). Born 1929. Educ.: Cape Town and Oxford Univs. Postdoctoral Res. Fellow, Brookhaven Nat. Lab., 1956-57; Defence Dept., U.S.A., 1957-. Societies: Faraday Soc.; A.C.S.; American Assoc. Rhodes Scholars.
Nuclear interests: Radiation damage to solids and its effect on chemical reactivity (in the solid state).
Address: Materials Science Department, New York State University, Stoney Brook, New York 11790, U.S.A.

JACKE, Stanley Emil, B.S.E.E. Born 1925. Educ.: Purdue Univ. Vice-Pres., Branson Instruments, Inc., 1960; Pres., Branson Sonic Power Co., 1963. Societies: I.E.E.E.; Acoustical Soc. America; Soc. Plastics Eng.
Nuclear interests: Use of ultrasonics in nuclear research; Nuclear cleaning problems; Metallurgical research.
Address: Branson Sonic Power Company, Miry Brook Road, Danbury, Connecticut, U.S.A.

JACKLYN, Robin Mainwaring, B.Sc., Ph.D. Born 1922. Educ.: Tasmania Univ. Res. Officer, Antarctic Div., Dept. of External Affairs.
Nuclear interest: Variations of cosmic-ray intensity.
Address: Physics Department, Tasmania University, Hobart, Tasmania, Australia.

JACKMAN, J. H., A.C.I.S. Local Director and Commercial Manager, Shipbuilding Group, Vickers Ltd.
Address: Shipbuilding Group, Vickers Ltd., Barrow Shipyard, Barrow-in-Furness, Lancashire, England.

JACKSON, G. B., D.F.H., M.I.E.E. Chief Operations Eng., Nucl. Plant Design Branch, C.E.G.B. Headquarters, 1964-.
Address: Central Electricity Generating Board, Walden House, 24 Cathedral Place, London E.C.4, England.

JACKSON, George Gordon. Born 1909. Educ.: Glasgow Tech. Coll. Controller for Scotland, Ministry of Supply, 1950-53. Chairman and Managing Director (-1963), Director (1963-), A.M.F. Ltd.; Manager, Nucl. Power Div., Mitchell Engineering, Ltd. Societies: M. Inst. Pet.; M. Inst. Prod. Eng.
Address: A.M.F. Ltd., 4-6 Savile Row, London, W.1, England.

JACKSON, H. W. Formerly Asst. Managing Director, now Managing Director (1968-), Internat. Combustion (Holdings) Ltd.; Deputy Chairman and Chairman Designate, Atomic Power Constructions Ltd., 1968-; Chairman and Managing Director, Internat. Combustion Ltd., 1969-.
Address: International Combustion (Holdings) Limited, 19 Woburn Place, London, W.C.1, England.

JACKSON, Henry Martin, LL.B. (Washington), Dr. of Laws (Alaska). Born 1912. Educ.: Washington Univ., Seattle. U.S. Senator,

1952-; Chairman, Nat. Democratic Party Com., 1960-61. Armed Services Com.; Joint Com. on Atomic Energy and Chairman, Military Applications Subcom., U.S. Congress; Govt. Operations Com.; Chairman, Interior and Insular Affairs Com.
Address: Senate Office Building, Washington D.C. 25, U.S.A.

JACKSON, Herbert Lewis, Ph.D. Born 1921. Educ.: Wisconsin Univ. Asst. Prof. Phys., Nebraska Univ., 1954-. Society: A.P.S.
Nuclear interest: Emulsion techniques for high energy research or for studies of excited states of nuclei.
Address: Department of Physics, Nebraska University, Lincoln, Nebraska, U.S.A.

JACKSON, John Early, S.B. (Elec. Eng.). Born 1902. Educ.: Vanderbilt Univ. and M.I.T. Intelligence specialist, U.S. Army Dept., G-2, 1946-53; Sec., Co-ordinating Com. Atomic Energy (1954-58), Director, Office Atomic, Biol. and Chem. Warfare (1958-62), Director, Office Atomic Programmes, Office Director Defence Res. and Eng. (1962-), U.S. Defence Dept.
Nuclear interests: Management; Nuclear research and development.
Address: Room 3 E 1071, The Pentagon, Washington 25, D.C., U.S.A.

JACKSON, John Harry, Ch.E. Born 1916. Educ.: Rensselaer Polytech. Inst. Manager, Dept. of Metal. (1953-61), Asst. to Pres. (1961-), Battelle Memorial Inst.; member, Materials Panel to Aircraft Nucl. Propulsion Office, 1959-61, Consultant to Sci. Advisory Board; to Chief of Staff, U.S. Air Force, 1960-61; Chairman, Inst. Of Metals of Metallurgical Soc. of A.I.M.E. 1961-62; Director, Metallurgical Soc., 1961-62; Member, Tech. Council, A.S.M., 1960-63. Societies: A.I.M.M.E.; A.S.M.; A.S.M.E.; A.S.T.M.
Nuclear interests: Specific interests are in the use of metals for nuclear applications, also including a broader interest in the utilisation of non-metallic materials, especially ceramics.
Address: 160 E. Schreyer Place, Columbus, Ohio, U.S.A.

JACKSON, Kenneth L., A.B., Ph.D. Born 1926. Educ.: California Univ., Berkeley. Sen. Investigator, Biochem. Branch, U.S. Naval Radiol. Defence Lab., San Francisco, 1954-60; Head, Radiobiol. Group, Bioastronautics Sect., The Boeing Co., 1960-62; Asst. Prof. Radiol. and Head, Rad. Biol. Div., Radiol. Dept. and Chairman, Radiol. Sci. Group, Washington Univ., Seattle, 1963-. Societies: A.A.A.S.; American Physiological Soc.; H.P.S.; Rad. Res. Soc.
Nuclear interest: Radiation biology.
Address: Radiology Department, Washington University, Seattle, Washington, U.S.A.

JACKSON, Max E. Asst. Manager for Tech. Operations, Chicago Operations Office, U.S.A.E.C.
Address: U.S.A.E.C., Chicago Operations Office, 9800 South Cass Avenue, Argonne, Illinois 60439, U.S.A.

JACKSON, Ray W., Ph.D., B.A.Sc. Born 1921. Educ.: Toronto, McGill and Yale Univs. Res. Asst., Phys., McGill Univ., 1950-51; American Council of Learned Socs., 1951-52; Advanced Graduate Fellow (1951-52), Res. Asst., Phys. (1952-54), Yale Univ.; Sen. Eng., Sprague Elec. Co., North Adams, 1954-56; Director, Semiconductor Lab., RCA Victor Co. Ltd., Montreal, 1956-64; Visiting Prof., Phys., McMaster Univ., 1964-65; Sci. Adviser, Sci. Secretariat, Privy Council Office, Ottawa, 1966-. Member, Standards Com. on Nucl. Rad. Detectors, I.E.E.E.
Nuclear interests: Participated in the development of some of the earliest semiconductor nuclear radiation detectors; Doctoral thesis at McGill on aspects of design of synchrocyclotron.
Address: Science Secretariat, Privy Council Office, 110 Argyle Avenue, Ottawa, Ontario, Canada.

JACKSON, Reginald Arthur, M.Sc., B.Sc. Born 1932. Educ.: Bristol Univ. Head, Mech. and Production Eng., Norwich City Coll. Societies: A.M.I.Mech.E.; A.F.R.Ae.S.
Nuclear interests: Heat transfer; Mechanical design.
Address: Norwich City College, Norwich, England.

JACKSON, Robert F., M.A. Born 1921. Educ.: Cambridge Univ. Director Reactor Technol., Reactor Group, U.K.A.E.A., Risley. Member, Nucl. Safety Advisory Com., M.O.P. Societies: Inst. of Mech. Eng.; A.M.I.E.E.; B.N.E.S.
Nuclear interests: Reactor development and operation; Engineering management.
Address: United Kingdom Atomic Energy Authority, Risley, Warrington, Lancs., England.

JACKSON, Roland, B.Sc. (Manchester), M.Sc. (Manchester), Ph.D. (Sheffield). Born 1910. Educ.: Manchester and Sheffield Univs. Supt., Phys. Dept., B.C.U.R.A.,-1963; Deputy Sec., later Sec., Inst. of Fuel, 1963-. Board Member representing Inst. Fuel, B.N.E.S. Societies: Fellow, Inst. of Fuel; F.Inst.P.; Soc. of Instr. Technol.; F.I.E.E.
Nuclear interests: These have been in radioisotopy, but now are limited to general interests in nuclear energy as a form of energy; but particularly in the educational field concerned with entry qualifications into the Institute of Fuel.
Address: 9 Ashley Avenue, Epsom, Surrey, England.

JACKSON, Roy P. Vice Pres. and Asst. Gen. Manager, Northrop Space Laboratories, Northrop Corp.
Address: Northrop Space Laboratories, Hawthorne, California, U.S.A.

JACKSON, William Dalziel, B.Sc., Ph.D. Born 1921. Educ.: Tasmania Univ. Sen.

Demonstrator (1952-58), Lecturer (1958-61), Tasmania Univ. Society: Australian Rad. Soc. Nuclear interests: Biological and genetical effects of ionising radiations, especially states of dormancy, paramagnetic gas effect and hydration.
Address: Botany Department, University of Tasmania, Box 252C, G.P.O., Hobart, Tasmania, Australia.

JACKSON, William Morrison, B.S., M.S., Ph.D. (Chem.), M.S. (Ind. Management). Born 1926. Educ.: Alabama and Tennessee Univs. Chem., Goodyear Atomic Corp., 1953-55; Chem., Union Carbide Nucl. Co., 1955-61; Group Leader, Diamond Alkali Co., 1961-66; Chem., Sen. Staff, Oak Ridge Assoc. Univs., 1966-. Societies: A.C.S.; H.P.S.
Nuclear interests: Radiochemistry; Health physics; Nuclear safety; Nuclear fuels.
Address: 304 E. Faunce Lane, Oak Ridge, Tennessee 37830, U.S.A.

JACKSON, The Lord Willis, D.Ph., D.Sc., D. Eng. Dr. Sc. Tech., Dr. of Laws, F.R.S. Born 1904. Educ.: Manchester and Oxford Univs. Prof. Elec. Eng., Imperial Coll., London Univ., 1946-53 and 1961-; Director, Res. Educ., Metropolitan-Vickers Elec. Co., Ltd., Manchester and London, 1953-61. Univ. Grants Com.; Res. Council, D.S.I.R.; Governing Board, British Nucl. Energy Conference; Advisory Council on Sci. Policy; Chairman, Com. on Manpower Resources for Sci. and Technol.; Pres., British Assoc. for Advancement of Sci. Societies: M.I.E.E.; M.I.Mech.E.; F. Inst. P.; F.R.S.
Nuclear interests: Direction of nuclear research; Particle accelerators; Instrumentation; Metallurgy.
Address: Imperial College, London, S.W.7, England.

JACOB, Edouard. Chef du Service Ensembles Industriels, Chantiers de l'Atlantique Penhoet-Loire.
Address: Chantiers de l'Atlantique Penhoet-Loire, 7 rue Auber, Paris 9, France.

JACOB, F. E., Ph.D. Assoc. Prof. Chem., and Member, Radioisotopes Com., Drake Univ., Des Moines.
Address: Drake University, Des Moines 11, Iowa, U.S.A.

JACOB, Gerhard, Dipl. Phys., Dipl. Maths., Ph.D. (Phys.). Born 1930. Educ.: Rio Grande do Sul Univ. Lecturer Theoretical and Atomic Phys., Faculty of Sci., Rio Grande do Sul Univ., 1953-58; Six months' stay (1958-59), Visiting Prof. (six months, 1962-63), Inst. Theoretical Phys., Heidelberg Univ.; Prof. Theoretical and Atomic Phys., Catholic Univ., Rio Grande do Sul, 1955; Sci. Staff, Atomic Energy Inst., Sao Paulo, 1956-58; Prof. Theoretical and Atomic Phys. (1958-), Head, Phys. Dept. (1963-65), Faculty of Sci., Rio Grande do Sul Univ.; One year's stay, Inst. Theoretical Phys., Copenhagen Univ., 1961-62; Head, Education Div. (1958-62), Prof. (1958-), Deputy Director (1965-), Inst. of Phys., Rio Grande do Sul Univ. Brazilian Nat. Res. Council. Societies: Brazilian Soc. for Progress of Sci.; Brazilian Phys. Soc.
Nuclear interests: Theoretical nuclear physics (low and high energy, in particular nuclear reactions and nuclear structure); Quantum electrodynamics; Elementary particles.
Address: Rio Grande do Sul University, Instituto de Fisica, Pôrto Alegre, Rio Grande do Sul, Brazil.

JACOBI, William M., B.S. (Chem. Eng.), M.S. (Chem. Eng.), Ph.D. (Chem. Eng.). Born 1930. Educ.: Syracuse and Delaware Univs. With Westinghouse Corp., Bettis Atomic Power Lab., 1955-61; Sci. i/c S5W Nucl. Design, 1957-59; Supv., A1W Core 3 Nucl. Design, 1960; Supv., D1W Nucl. Design, 1961; Sen. Tech. Assoc., Nucl. Utility Services, Inc., 1961-63; Manager, Design and Analysis, Astronucl. Lab., Westinghouse Elec. Corp. Societies: A.I.Ch.E.; A.N.S.; A.R.S.; A.C.S.
Nuclear interests: Reactor design; Reactor physics; Heat transfer; Space nuclear power.
Address: 413 Manordale Road, Pittsburgh, Pennsylvania 15241, U.S.A.

JACOBI, Wolfgang Gerhard, Dipl.-Phys., Dr. rer. nat., Priv. Dozent (Tech. Univ., Berlin). Born 1928. Educ.: Frankfurt am Main Univ. Max Planck-Inst. for Biophys., Frankfurt am Main, 1953-57; Head, Rad. Phys. Div., Hahn-Meitner-Inst. für Kernforschung Berlin, 1957-. Book: Strahlenschutzpraxis, Teil I: Grundlagen. (Munich, Verlag K. Thiemig, 1962). Societies: H.P.S.; Deutsche Physikalische Gesellschaft; Deutsche Gesellschaft für Biophysik; Bunsen-Gesellschaft für physikalische Chemie.
Nuclear interests: Health physics; Radiation biophysics.
Address: Hahn-Meitner-Institut für Kernforschung Berlin, Abt. Strahlenphysik, 100 Glienicker Strasse, 1 Berlin-Wannsee, Germany.

JACOBS, Alan M., B. Eng. Phys., M.S., Ph.D. Born 1932. Educ.: Cornell and Pennsylvania State Univs. Assoc. Prof. (Nucl. Eng.), Pennsylvania State Univ. Book: Co-author, Basic Principles of Nucl. Sci. and Reactors (D. Van Nostrand, 1960). Society: A.N.S.
Nuclear interests: Reactor physics; Transport theory; Physics of many bodies.
Address: 231 Sackett Building, Pennsylvania State University, University Park, Pa., U.S.A.

JACOBS, John J., Jr., B.S.M.E. Born 1912. Educ.: New York Univ. Manager, Atomic Energy Sales Dept., New York (1955-62), Director, Atomic Power Div., New York (1962-64), Vice Pres. and Regional Director, Tokyo (1964-67), Vice Pres. and Regional Director, Sundbyberg, Sweden (1967-), Westinghouse Elec. Internat. S.A. Societies: A.N.S.; A.S.M.E.; Atomic Ind. Forum.

Nuclear interest: Developing nuclear activities for Westinghouse in Finland, Sweden, Norway and Denmark.
Address: 123 Albygatan, Sundbyberg, Sweden.

JACOBS, Robert M. Chairman, Com. on Nucl. Energy, Power Div., A.S.C.E.
Address: American Society of Civil Engineers, 345 East 47th Street, New York 17, New York, U.S.A.

JACOBS, Walter, Chem. Born 1932. Sen. Officer, Isotopes Div., N. V. Philips-Duphar, 1955-.
Nuclear interests: Isotopes production and use in medicine, industry and research, and its commercial aspects.
Address: 21 Wouwermanstraat, Amsterdam, Netherlands.

JACOBSEN, Cecil Felix, M.Sc., Ph.D. Born 1914. Educ.: Tech. Univ. Denmark. Carlsberg Labs., Chem. Dept., 1939-50; Nordic Insulin Labs., 1950-54; Consultant Eng., 1954-56; Formerly Head, Chem. Dept. (1956-), now Asst Director, Atomic Energy Res. Station, Risoe, Denmark.
Address: Chemistry Department, Atomic Energy Research Station, Risoe, Roskilde, Denmark.

JACOBSEN, Jacob Christian Georg, Ph.D. Born 1895. Educ.: Copenhagen Univ. Phys., Radiumstationen, Copenhagen, 1921-56; Asst. Prof. (1924-41), Prof. (1941-56), Copenhagen Univ.; Res. Director, Atomic Energy Res. Station, Risoe, Denmark, 1956-58; Prof., Copenhagen Univ., 1958-; Consultant to Danish A.E.C., 1958-. Societies: Royal Danish Acad. of Sci. and Letters; Kungliga Fysiografiska Sällskapet; Kungliga Vetenskapssocieteten; Acad. of Tech. Sci.
Nuclear interest: Nuclear physics.
Address: Kaervangen 41, Gentofte, Denmark.

JACOBSEN, Tor, Cand. real. Res. on Elementary Particles and Cosmic Rad., Nucl. Phys. Lab., Oslo Univ.
Address: Oslo University, Institute of Physics, Blindern, Norway.

JACOBSON, Jerome, B.S. (M.E.) (Illinois), M.S. (Maryland). Born 1929. Educ.: Illinois and Maryland Univs. Nucl. Eng., Dept. of Navy, Bureau of Ships, Nucl. Power Div., Washington, D.C., 1951-54; Eng., G.E.C., Atomic Power Equipment Dept., San Jose, California, 1954-. Societies: A.S.M.E.; A.N.S.
Nuclear interests: Reactor design; Nuclear engineering, and nuclear fuels.
Address: General Electric Company, Atomic Power Equipment Department, 2151 South First Street, San Jose, California, U.S.A.

JACOBSON, Leon Orris, B.S. (N.D. State Coll.), M.D. (Chicago). Born 1911. Educ.: N. Dakota State Coll. and Chicago Univ. Assoc. Dean, Div. Biological Sci. (1945-51), Prof. Medicine (1951-), Chairman, Dept. of Medicine (1961-), and Dean, Div. of Biol. Sci., Chicago Univ.; Director, Argonne Cancer Res. Hospital, Chicago Univ., 1951-67. Consultant, Argonne Nat. Lab., Biol. Div.; Member, Advisory Com. on Isotope Distribution, U.S.A.E.C.; Member, Com. on Rad. Studies, U.S.P.H.S., Hematology Study Sect.; Com.on Cancer diagnosis and therapy, Nat. Res. Council, Panel on Chemotherapy. Books: Rad. Injury (in Textbook of Medicine, 1955); Lymphosarcoma (in Current Therapy, 1955); Hematologic Effects of Ionizing Rad. (in Rad. Biol., vol. I, 1954); co-author, Industrial Medicine on Plutonium Project (1951); co-author, Biological Effects of External and Gamma Rad., Part I (1954). Societies: Fellow, A.C.P.; Assoc. American Physicians; A.A.A.S.; Rad. Res. Soc.
Nuclear interests: Effect of various radiations on blood-forming tissue;Studies on recovery from irradiation injury after transplantation of normal blood-forming tissue; Effects of irradiation on immune reactions; Tracer and therapeutic uses of isotopes.
Address: 1222 E. 56th Street, Chicago 37, U.S.A.

JACOBSON, Norman Harry, B.A. Born 1915. Educ.: Wisconsin Technol. Univ. Information Officer, later Chief, Ind.Information Services, U.S.A.E.C., Washington D.C., 1950-54; Exec. Business Editor, Elec. Light and Power (Haywood Pub. Co., Chicago), 1954-63; Managing Editor, Atomics and Editor, Business Atomics Report (Technical Pub. Co., Barrington, Illinois), 1963-66; Consultant (1958-66), now Asst. Director, L.M.F.B.R.Programme Office, Argonne Nat. Lab. Formerly Member, Advisory Com. on Tech. Information, U.S.A.E.C.; Vice-chairman, Nucl. Div., American Power Conference. Societies: A.N.S.; Nat. Assoc. Sci. Writers, Inc.
Nuclear interest: Business communication especially the interpretation of developments in science and technology for industrial management, with particular application to the nuclear field.
Address: 25 Logan Terrace, Golf, Illinois, U.S.A.

JACOBUS, David Penman, A.B., M.D. Born 1927. Educ.: Harvard Coll. and Pennsylvania Univ. School of Medicine. Hospital School of Medicine, Pennsylvania Univ., 1953-57; Chief, Dept. of Radiobiol., Walter Reed Inst. of Res., Walter Reed Army Medical Centre, Washington, D.C., 1958-. Member, N.A.S.-N.R.C. Com. on Modern Methods of Handling Chem. Information. Society: A.C.S.
Nuclear interests: The development of chemicals effective in protecting against ionising radiation injury, including studies on their mechanism of action.
Address: 7309 Holly Avenue, Takoma Park, Maryland, U.S.A.

JACOBY, Bruno, Dipl.-Ing. Born 1901. Educ.: Hanover T.H. Tech. Adviser, Losenhausen Maschinenbau A.G. Societies: Personal Member, Study Assoc. for use of Nucl. Energy in Navigation and Ind.; A.S.T.M.; German Standards for Materials Testing.
Address: 35 Burgmüllerstrasse, Düsseldorf-Grafenberg 4, Germany.

JACOME, D. Jose Ma. GAZTELU y. See GAZTELU y JACOME, D. Jose Ma.

JACQMARD, Miss A. Sci. concerned with Radioisotope Work, Centre de Recherches des Hormones Vegetales de l'I.R.S.I.A.
Address: Centre de Recherches des Hormones Vegetales de l'I.R.S.I.A., Institut de Botanique, 3 rue Fusche, Liège, Belgium.

JACQUES, Philippe G., B.A. Educ.: Assumption Coll. Deputy Director, Div. of Public Information, U.S.A.E.C., -1967; Sen. Adviser, U.S. Mission to the I.A.E.A., 1967-69. Adviser, Third I.C.P.U.A.E., Geneva, Sept. 1964.
Address: U.S. Mission to the I.A.E.A., 14 Schmidgasse, Vienna 8, Austria.

JACQUINOT, Pierre, Agrégation de Physique D. es Sc. (Ph.D.). Born 1910. Educ.: Nancy and Paris Univs. Director Gen., Centre Nat. de la Recherche Scientifique, 1962-. Societies: Société Française de Physique; A.P.S.
Nuclear interests: Nuclear physics: properties of nuclei revealed by optical spectroscopy (hyperfine structure, isotope shifts).
Address: C.N.R.S., 15 quai Anatole France, Paris 7, France.

JACROT, Bernard. Born 1927. Educ.: Ecole Polytechnique, Paris. Phys., French A.E.C., 1950-67; Deputy Director, Inst. Max Von Laue – Paul Langevin, 1967-. Societies: Sté. Française de Physique; A.P.S.
Nuclear interest: Neutron physics.
Address: Institut Max Von Laue – Paul Langevin, rue des Martyrs, B.P. 508, Grenoble 38, France.

JACROT, J. Deputy Director, Nucl. Reactor, Paul Langevin-Max von Laue Inst., 1967-.
Address: Paul Langevin-Max von Laue Institute, c/o Centre d'Etudes Nucléaires de Grenoble, B.P. 269, Chemin des Martyrs, Grenoble, (Isere), France.

JAE, Won Mok, M.S. Born 1930. Educ.: Internat. School of Nucl. Sci. and Eng., Lemont, Ill. and Seoul Nat. Univ., Korea. Chairman, Dept.of Nucl. Sci. and Eng., Han Yang Univ., Seoul, 1959-. Book: Principle of Radiochem. (1959). Society: Korean Chem. Soc.
Nuclear interest: Radiochemistry.
Address: Department of Nuclear Science and Eng., Han Yang University, Seoul, Korea.

JAEGER, Hendrik Elias, Naval Architect. Born 1903. Educ.: Technol. Univ. Prof. Naval Architecture, Delft Technol. Univ., 1946-. Societies: Inst. of Naval Architects, London; Assoc. Technique Maritime, Paris; Roy. Inst. of Eng., The Hague; Académie de Marine de France, Paris; Union Belge des Ing. Navals, Brussels; Assoc. of Astronomers of the Netherlands; Roy. Swedish Acad. of Eng. Sciences, Stockholm.
Nuclear interest: Nuclear propulsion of ships.
Address: 21 Maerten Trompstraat, Delft, Netherlands.

JAEGER, Robert Gottfried, Dr. phil. (Berlin), Honorarprof. (Mainz). Born 1893. Educ.: Marburg a.d.Lahn, Munich and Berlin Univs. Oberreg. Rat a.D. Physikal. Techn. Bundesanstalt-Braunschweig, T. H. Braunschweig. Mainz Univ. I.A.E.A., Vienna; Liaison Officer, I.A.E.A. - W.H.O., Geneva, 1965-. Membre de 12 experts du C.E.E.A., Brussels; Fachnormenausschuss Radiologie; Comité Consultatif pour les Etalons de Mesure des Radiations Ionisantes, Bureau Internat. des Poids et Mesures, Sèvres-France; Assoc. Member, British Hospital Phys. Assoc.; Co-ordinating editor, Eng. Compendium on Rad. Shielding. Book: Strahlenschutz und Dosimetrie (Stuttgart, Verlag G. Thieme, 1959). Societies: Deutsche Röntgengesellschaft; Deutsche Physikal. Gesellschaft.
Nuclear interests: Dosimetry of ionising radiations; Scientific foundation and instrumentation; Standardisation; Health physics; Radiation protection; Radiobiology; Medical physics.
Address: 10 Otto-Weiss-Strasse, Bad Nauheim 635, Germany.

JAEGER, Thomas A., Diploming., Dr.-Ing. Born 1929. Educ.: T.H. Dresden. Entwurfsbüro für Industriebau Rostock, 1956-57; Tech. Univ. Berlin, 1958-62. Kernforschungsanlage Jülich, 1962-63; Habilitation Scholarship Deutsche Forschungsgemeinschaft, 1963-; Lecturer, Nucl. Structural Eng., Tech. Univ. Berlin, 1964-; Co-editor, Nucl. Eng. and Design. Books: Technischer Strahlenschutz (Munich, Verlag Karl Thiemig, 1959); Grundzüge der Strahlenschutztechnik. (Berlin, Göttingen, Heidelberg, Springer-Verlag, 1960); co-author, Grenztragfähigkeits-Theorie der Platten (Berlin, Springer-Verlag, 1963); co-editor, Eng. Compendium on Rad. Shielding (I.A.E.A.). Societies: American Concrete Inst.; A.N.S.; B.N.E.S.
Nuclear interests: Structural engineering aspects of nuclear engineering; Radiation protection engineering.
Address: 26 Nordschleswig Strasse, Essen-Heisingen, Germany.

JAFFE, Aubrey Abram, B.Sc. (Hons.), Ph.D. Born 1926. Educ.: Manchester and Hebrew (Jerusalem) Univs. Demonstrator in Phys., Hebrew Univ., Jerusalem, 1949-53; I.C.I. Res. Fellow, Nucl. Phys. Res. Lab. Liverpool Univ., 1953-55; Lecturer in Phys., Manchester Univ., 1955-63; Visiting Assoc. Prof., Maryland Univ.,

1963-64; Assoc. Prof., Hebrew Univ., Jerusalem, 1964-. Book: Contribution to Fast Neutron Phys. (1959).
Address: Physics Department, Hebrew University, Jerusalem, Israel.

JAFFE, Harold, B.S. (Chem.), Ph.D. (Nucl. Chem.). Born 1930. Educ.: Illinois Univ. at Chicago and Urbana, and California Univ. Asst. Res. Chem., Union Oil Co., 1954-55; Sen. Chem., Tracerlab, Inc., 1955-56; Principal Nucl. Chem. (1956), Asst. Programme Manager, (1956-58), Programme Manager, (1958-59), Gas Cooled Reactor Experiment, Assoc. Manager, Advanced Projects Dept. (1959-60), Manager, Fuel Development Dept. (1960-62), Asst. Manager, Nucl. Technol. Div. (1962-64), Manager, Appl. Sci. Div. (1964-66), Corporate Nucl. Coordinator (1966-), Aerojet-Gen. Corp.; Assoc. Prof., Kennedy Univ., Martinez, California, 1966-. Societies: A.C.S.; A.N.S.
Nuclear interests: Nuclear fuels and materials; Special purpose nuclear power and propulsion systems; Radioisotope power systems; Nuclear research and development management.
Address: Aerojet-General Corporation, Post Office box 77, San Ramon, California, U.S.A.

JAFFE, Henry L., M.D. Director, Rad. Therapy and Nucl. Medicine Div., Cedars of Lebanon Hospital.
Address: Division of Radiation Therapy and Nuclear Medicine, Cedars of Lebanon Hospital, 4833 Fountain Avenue, Los Angeles 29, California, U.S.A.

JAFFEY, Arthur H., Dr. Member, Subcom. on Instr. and Techniques, Com. on Nucl. Sci., N.A.S.
Address: Argonne National Laboratory, P.O. Box 299, Lemont, Illinois, U.S.A.

JAFS, Daniel, Diplom. Ing. Born 1930. Educ.: Abo Akademi and Pennsylvania State Univ. Argonne Nat. Lab. Sales and Design Eng. (1955), i/c Nucl. Activities (1958), Head, Heat Transfer and Nucl. Dept. (1959), A. Ahlström Osakeyhtiö, Warkaus; Nucl. Adviser, A. Ahlström Osakeyhtiö, Helsingfors, 1966. Deputy Chairman, Finnish Nucl. Soc. Societies: A.N.S.; Tech. Soc. of Finland; Finnish Nucl. Soc.
Nuclear interests: Economical and technical aspects of nuclear heat and power.
Address: 31 Tegelbacken, Helsingfors 33, Finland.

JAGD, Per, M.Sc. Born 1935. Educ.: Denmark Tech. and Birmingham Univs. Sales Manager, Paul Bergsoe and Son, 1967-.
Nuclear interest: Lead shielding.
Address: Paul Bergsoe and Son, 2600 Glostrup, Denmark.

JAGERSBERGER, Josef, Dipl. Ing. Born 1922. Educ.: Montanische Hochschule, Leoben. Director of Production, Boehlerwerk, Gebr. Bohler and Co. A.G. Paper: High-alloy Steels in Nucl. Energy Plants I and II (Atomwirtschaft, vol. 6, Nos. 3 and 5, 1961).
Address: Gebr. Bohler and Co. A.G., Boehlerwerk an der Ybbs, Lower Austria.

JAGGIA, S. S. Project Administrator, Rajasthan Atomic Power Project, A.E.C., India.
Address: Atomic Energy Commission, Rajasthan Atomic Power Project, P.O. Anushakti, Rawatbhata Via Kota, Rajasthan, India.

JAHN, Walter, Dr. rer. nat., Dipl. Chem. Born 1921. Sci., Jenaer Glaswerk Schott und Gen. Nuclear interests: Protective windows against radiation; Radiation resistant optical glasses; Interaction between radioactive emission and glass; Dosimeter glasses (radiophotoluminescence glasses, high-dose glasses).
Address: 22 Alemannenstrasse, 6507 Ingelheim, Germany.

JAIN, H. K., Dr. Head, Genetics Div., and Controller, Gamma Garden, Indian Agricultural Res. Inst.
Address: Indian Agricultural Research Institute, New Delhi 12, India.

JAINE PEYRE, Juan. Licenciado. Prof. Fisica Nucl., Buenos Aires Univ.
Address: Buenos Aires University, Facultad de Ciencias Exactas y Naturales, 222 Calle Peru, Buenos Aires, Argentina.

JAKEMAN, Derek, B.Sc., Ph.D. Born 1925. Educ.: Manchester Univ. Post-doctoral Fellow, N.R.C., 1951-54; S.S.O. U.K.A.E.A. (Windscale), 1954-56; Lecturer in Reactor Phys., Birmingham Univ., 1956-60; P.S.O., U.K.A.E.A. (Winfrith), 1960-. Book: Phys. of Nucl. Reactors (E.U.P. and American Elsevier, 1966).
Nuclear interest: Neutron energy spectrum measurements in thermal and fast reactors.
Address: Atomic Energy Establishment, Winfrith, Dorchester, Dorset, England.

JAKES, Dusan, C.Sc., Chem. Eng. Born 1930. Educ.: Tech. Military Coll., Brno. Chem. Technol., 1955-56; Post-graduate course nucl. chem. and technol., Charles Univ. 1956-57; Dr. Thesis, Non-ferrous Metals Inst., Moscow, 1957-60; Nucl. Res. Inst., Rez, 1961-64. Head, Ceramic Branch, Nucl. Res. Inst., 1962-66. Editorial Board, Jaderná Energie; Nucl. Patent Com. Book: Nucl. Chem. Tables (Prague, S.N.T.L., 1964). Society: Czechoslovak Sci. and Tech. Soc.
Nuclear interest: Nuclear ceramic fuel.
Address: Nuclear Research Institute, Academy of Sciences, Rez, Czechoslovakia.

JAKESOVA, Ludmila, C.Sc. Born 1930. Educ.: Lomonosov's Inst. Fine Tech. Technol., Moscow. Head, Analytical Control, Metal. Plant, 1952-55; Sci. worker, Inst. Metals, 1955-58; C.Sc. Thesis, Inst.

Non-ferrous Metals, 1958-61; Nucl. Res. Inst., Rez, 1961-.
Nuclear interests: Nuclear fuel materials; Fuel element development.
Address: Nuclear Research Institute, Czechoslovak Academy of Sciences, Rez, Czechoslovakia.

JAKOB. Dipl.-Ing. Heat Exchangers Sect., Maschinenbau-A.G. Balcke.
Address: Maschinenbau-A.G. Balcke, Marienplatz, 463 Bochum, Germany.

JAKOBSEN, E. L., M.Sc. Member, Exec. Com., Danatom.
Address: Danatom, 7 Tingvej, DK-4690 Haslev, Denmark.

JAKOBSON, Mark John, A.B., M.A. (Montana Univ.), Ph.D. (California). Born 1923. Educ.: Montana and California Univs. Whiting Fellow, 1949-50; Phys. Rad. Lab., 1950-52; Instr. Phys., Washington Univ., 1952-53; Prof. Phys., Montana Univ., 1953-. Society: Phys. Soc.
Nuclear interests: Accelerator design; Photonuclear reactions; Electronics.
Address: Montana University, Missoula, Montana, U.S.A.

JAKUBKA, Karel, Ing. Head of Res. Group, Inst. of Plasma Phys., Czechoslovak Academy of Sci.
Address: Czechoslovak Academy of Sciences, Institute of Plasma Physics, 600 Nademlynska, Prague 9, Czechoslovakia.

JAKUBOWSKI, Janusz Lech, D. Tech. Sc., Dr. Habil. Born 1905. Educ.: Warsaw Tech. Univ. Ord. prof. Technol., Warsaw Univ.; Director, Central Electrotech. Inst., Poland, -1955; Director, Paris Sect., Polish Acad. Sci., 1960-61. Formerly Member, State Council for Peaceful Use of Nucl. Energy; Praesidium, and Pres. Electrotech. Com., Polish Acad. Sci.; U.N.E.S.C.O. Project Manager, Nat. Polytech. School, Alger-Algeria, 1967-.
Nuclear interests: Nuclear and electric energy.
Address: 9 Igańska, Warsaw, Poland.

JALICHAN, Nitipat, S.B., S.M. Born 1920. Educ.: M.I.T. Formerly Chief, Tech. Div., formerly Deputy Sec. Gen., now Sec. Gen., Thai Nat. Energy Authority.
Nuclear interests: Nuclear raw materials; Medium-size power reactors.
Address: National Energy Authority, Technical Division, Pibultham Villa, Krasatsuk Bridge, Bangkok, Thailand.

JAMEEL, Muhammad, M.Sc. (Phys.), Ph.D. (Theoretical Phys.). Born 1939. Educ.: Karachi and Edinburgh Univs. Sen. Sci. Officer (1963-68), Principal Sci. Officer (1968-), Pakistan A.E.C.; Res. Fellow, Sydney Univ. 1964-65. Editor, The Nucleus, quarterly journal of the Pakistan A.E.C.; Sec., Phys. Sci., Sci. Soc. of Pakistan Annual Conference, 1967.
Nuclear interests: Theoretical aspects of elementary particles physics; Nuclear reactions theory.
Address: Director, Scientific Information, Pakistan Atomic Energy Commission, P.O. Box No. 3112, Karachi, Pakistan.

JAMERSON, Frank E., S.B., Ph.D. Born 1927. Educ.: M.I.T. and Notre Dame Univ. Phys., U.S. Naval Res. Lab., Washington, D.C., 1951-52; Sen. Sci., Westinghouse Atomic Power Div., 1953; Neutron Phys. Sect. Head, U.S. Naval Res.Lab., 1954-57; Supervisory Res. Phys., Gen. Motors Res.Labs., Warren, Michigan, 1957-. Societies: RESA; A.P.S.; A.N.S.; A.I.A.A.; A.A.A.S.
Nuclear interests: Direct conversion of nuclear heat to electricity; Thermionic conversion; Plasma physics; Neutron physics; Reactor physics.
Address: 3251 Newgate Road, Birmingham, Michigan, U.S.A.

JAMES, Daniel Alfred Jules SAINT-. See SAINT-JAMES, Daniel Alfred Jules.

JAMES, G. M. Operations, A.E.C.L. Formerly Member, Reactor Safety Advisory Com., Atomic Energy Control Board. Deleg., Third I.C.P.U.A.E., Geneva, Sept. 1964.
Address: Atomic Energy of Canada, Ltd., Chalk River Project, Chalk River, Ontario, Canada.

JAMES, Peter Anthony, B.Sc., A.S.T.C. Born 1935. Educ.: N.S.W. Univ. Tech. Officer, Dept. of Nucl. and Rad. Chem., N.S.W. Univ., 1962-.
Nuclear interest: Industrial applications of radioisotopes.
Address: Department of Nuclear and Radiation Chemistry, New South Wales University, Australia.

JAMES, R. X. Agriculture and Radioactive (Isotopes) Exptl. Station, Pacific Inst. Earth Sci.
Address: Agricultural and Radioactive (Isotopes) Experimental Station, Pacific Institute of Earth Sciences, P.O. Box 998, Mojave, California, U.S.A.

JAMES, William Thomas. Born 1920. Field Sales Manager, Baldwin Instrument Co. Ltd.
Nuclear interest: Use of isotopes in industry.
Address: Baldwin Instrument Company, Bath Road, Beenham, Reading, Berks., England.

JAMET, Jean. Directeur à la production, Sté. Française Bitumastic.
Address: Societé Française Bitumastic, 8 rue Bayard, Paris 8, France.

JAMET, Y. Ing., Thermodynamics, Bureau Veritas.

JAMEZ, Jean. Born 1902. Educ.: Classical High School (technol. and ergological studies). Gen. Manager, Compagnie Belge d'Assurances Générals contre les Risques d'Incendie, A.G.; Director of Auxifina. Chairman, Tariffs Com. Ind. and Special Risks; Pres., A.N.P.I. (Fire Protection Association); Vice Pres., Comite de Gestion, Syndicat Belge d'Assurances Nucleaires.
Nuclear interest: Insurance of nuclear risks.
Address: 199 avenue Charles Woeste, Brussels 9, Belgium.

JAMIESON, Mrs. Dana Domnica, B.Sc. (Sydney), M.Sc. (Sydney). Born 1937. Educ.: Sydney Univ. Res. Asst., Pharmacology Dept., Sydney Univ., 1957-58; Res. Student (State Cancer Council Grant), 1959; Res. Worker, Radiobiol. Res. Unit, Cancer Inst. Board, Melbourne, 1960-.
Nuclear interests: Biochemistry, pharmacodynamics pertaining to radiation, effects and interaction of oxygen with chemical and biological systems.
Address: Radiobiological Research Unit, Cancer Institute Board, 278 William Street, Melbourne, Victoria, Australia.

JAMMET, Henri Paul, Officeri d'Académie; Dr. en médecine. Born 1920. Educ.: Paris Univ. Chef du Département de la Protection Sanitaire, C.E.A., 1950-; Chef du Service des radioisotopes à l'Hôpital Curie, 1948-. Book: Protection contre les Radiations. Societies: Sté. française d'Electro-Radiologie; Commission nationale de Protection radiologique; Commission internationale de Protection radiologique (Comité V); Conseil supérieur d'Hygiène de France.
Nuclear interests: Radiological protection; Health physics; Radioisotopes in medicine.
Address: 114 quai Louis Blériot, Paris 16, France.

JAMNE, Einar, M.Sc. Born 1921. Educ.: Tech. Univ. of Norway, Trondheim, Instituttingeniör, Tech. Univ. of Norway, 1950-55; Inst. for Atomenergi, Norway (1955-), O.E.C.D. Halden Reactor Project (1958-), Chief of Operation (1961-).
Nuclear interest: Reactor design and operation.
Address: O.E.C.D. Halden Reactor Project, Halden, Norway.

JANCAREK, Jan, Candidate of Tech. Sci. Born 1929. Educ.: Charles Univ., Prague. Ore Res., Prague, 1954-. Society: Czechoslovak Sci. and Tech. Assoc.
Nuclear interest: Processing of radioactive ores and minerals.
Address: 19 Pod Dékankou, Prague 4, Czechoslovakia.

JANGG, Gerhard, Dipl. Ing., Dr. techn. Born 1927. Educ.: Vienna T.H. Asst. Chem., Vienna T.H., 1955- and Consulting Chem., Austrochematom-Kernbrennstoff Gesellschaft m.b.H., Linz.
Nuclear interests: Chemical and metallurgical processing of nuclear fuels.
Address: 97 Hetzendorferstrasse, Vienna 12, Austria.

JANIK, Jerzy Antoni, Ph.D. Born 1927. Educ.: Jagiellonian Univ., Cracow. Prof. and Head, Chair of Structure Res., Jagiellonian Univ.; Head, Neutron Lab., Inst. Nucl. Phys., Cracow. Society: Polish Phys. Soc.
Nuclear interests: Scattering of neutrons by solids, liquids and molecules; Neutron physics.
Address: 8 ul. Sw. Marka, Cracow, Poland.

JANISCH, Duncan Basil Bailey, A.R.C.S., B.Sc. Born 1918. Educ.: London Univ. and R.M.C.S. Attached from War Office to A.E.R.E., Harwell (1954-58), Health Phys. and Safety Manager, Production Group, Capenhurst Works (1958-60), Group Health and Safety Officer, Production Group Headquarters (1960-64), Group Headquarters Staff (1965-), U.K.A.E.A.
Nuclear interests: Health physics; International safeguards.
Address: Himley House, Curzon Park North, Chester, England.

JANITSCHEK, Friedrich, Dipl. Ing., Dr. techn. Ing. en Génie Atomique, Ziviling. für Maschinenbau. Born 1929. Educ.: Polytech. Vienna and Saclay, France. Head of branch projecting thermal power plants, Österreichische Draukraftwerke, Klagenfurt, Austria.
Nuclear interests: Nuclear power plants; Metallurgy.
Address: 50 Anzengruberstrasse, Klagenfurt, Austria.

JANKOVIC, Dobrila, Mr.Ph. Born 1933. Educ.: Belgrade Univ. Medical Protection Dept., Boris Kidric Inst., Belgrade. Society: Yugoslav. Soc. for Radiol. Protection.
Nuclear interest: Radiotoxicology – human internal contamination.
Address: Boris Kidric Institute of Nuclear Sciences, Belgrade-Vinca, P.O.B. 522, Yugoslavia.

JANKOVIC, Slobodan, Ph.D. (Geology). Born 1925. Educ.: Belgrade Univ. Prof. Economic Geology, Belgrade Univ., 1950-. Member, Com. of Experts, Federal Nucl. Energy Commission, Yugoslavia. Book: Economic Geology.
Nuclear interests: Exploration of nuclear mineral deposits, and specially prospecting methods.
Address: 7 Djusina, Belgrade, Yugoslavia.

JANKOVIC, Zlatko. Born 1916. Educ.: Zagreb Univ. Prof., Faculty of Sci., Zagreb Univ.; Sci. Adviser, Phys. Dept., Inst. Rudjer Boskovic, Zagreb.

Nuclear interests: Nuclear theoretical physics, especially theory of nuclear reactions; Mathematical methods of theoretical physics, especially special functions.
Address: Institute Rudjer Boskovic, 54 Bijenicka, Zagreb, Yugoslavia.

JANKOWSKA, Mrs. Teresa, M.S. Born 1928. Educ.: Warsaw Polytech. At Inst. of Gen. Chem., Warsaw, 1952-54; Inst. of Nucl. Res., Polish Acad. of Sci., Warsaw, 1954. (Head, Lab. in Analytical Chem. Dept.). Member, Analytical Commission, Polish Acad. of Sci., Sci., Spectroscopic Section.
Nuclear interests: Analytical chemistry of nuclear pure materials; Spectrophotometric methods of analysis; Chemistry of complex compounds.
Address: 14 m 72 Nowolipki, Warsaw 47, Poland.

JANKUS, Vytautas Zachary, Ph.D. (Phys.). Born 1919. Educ.: Stanford Univ. Assoc. Phys., Argonne Nat. Lab., 1955-. Societies: A.P.S.; A.N.S.
Nuclear interests: Fast reactor safety; Thermalisation.
Address: Argonne National Laboratory, Lemont, Illinois, U.S.A.

JANNELLI, Sante. Prof. of Exptl. Phys., Messina Univ. Member, Sicilian Nucl. Res. Com.
Address: Messina University, Via Tommaso Cannizzaro, Messina, Sicily, Italy.

JANOSSY, Lajos, Dr. phil. (Berlin). Born 1912. Educ.: Vienna and Berlin Univs. Head, School of Cosmic Rays, Inst. for Advanced Studies, Dublin, 1947-50; Director, Central Res. Inst. Phys., Hungarian Acad. Sci., Budapest, 1956; Prof. Phys. (1950-), Head, Dept. of Atomic Phys. (1957-), L. Eötvös Univ., Budapest; Vice-Pres., Hungarian Acad. Sci., Budapest, 1961-. Vice-Pres., Hungarian A.E.C.; Member, Sci. Council, United Inst. Nucl. Res., Dubna, 1955-. Books: Cosmic Rays (Oxford, Clarendon Press, 2nd ed., 1950); Cosmic Rays and Nuclear Physics (London, Pilot Press, Ltd., 1948); Theory and Practice of the Evaluation of Measurements (Oxford, Clarendon Press, 1964). Societies: Roy. Irish Acad., 1948; Hungarian Acad. Sci., 1951; Corresp. Member, German Acad. Sci., 1955; Bulgarian Acad. Sci., 1958; Mongolian Acad. Sci., 1961.
Nuclear interest: Fundamental questions.
Address: Hungarian Academy of Sciences, 9 Roosevelt-tér, Budapest 5, Hungary.

JANOUCH, Frantisek, R.N.Dr., C.Sc. Born 1931. Educ.: Leningrad State and Moscow State Univs. Head, Theoretical Nucl. Phys. Dept., Nucl. Res. Inst., Czechoslovak Acad. Sci., Rez, 1959-67; Internat. Centre for Theoretical Phys., Trieste, 1964-65. Books: Co-author, Introduction to the Theory of Atomic Nuclei (Prague, S.N.T.L., 1963);
Editor, Selected Topics in Nucl. Theory (Vienna, I.A.E.A., 1963).
Nuclear interests: Theoretical nuclear physics; Elementary particle physics.
Address: 7 Prubezna, Prague 10, Czechoslovakia.

JANSEN, Cornelius Rudolph, M.B., Ch.B., M. Med. Rad. T. (Pretoria Univ.). Born 1928. Educ.: Pretoria Univ. Medical Practitioner, 1950-58; Dept. of Radiotherapy, Medical School, Pretoria Univ., 1958-60; Res. Collaborator, Brookhaven Nat. Labs., 1960-62; Head, Div. of Life Sci., Atomic Energy Board, South Africa, 1962-. Radioisotope Dept., and Sen. Lecturer, Medical School, Pretoria Univ. Societies: Soc. Nucl. Medicine; New York Acad. Sci.
Nuclear interests: Nuclear medicine; Radiobiology.
Address: Atomic Energy Board, Private Bag 256, Pretoria, South Africa.

JANSEN, Hendrikus Stephanus, M.Sc. (Hons.). Born 1927. Educ.: Victoria Univ. of Wellington, New Zealand. Head, Gen. Phys. Dept., Inst. of Nucl. Sci., D.S.I.R., Lower Hutt, New Zealand, 1957-. Member, Exec. Com., P.S.A. Book: Directory of New Zealand Sci. (New Zealand Assoc. of Sci., 1962). Society: New Zealand Assoc. of Sci.
Nuclear interests: Isotopic geology, meteorology and oceanography in particular ^{14}C and tritium; Cosmic radiation; Industrial applications of isotopes.
Address: Institute of Nuclear Sciences, D.S.I.R., Private Bag, Lower Hutt, New Zealand.

JANSEN, Johannes F. W. Drs. Phys. Born 1932. Educ.: Amsterdam Univ. Formerly Res. Phys., Inst. for Nucl. Phys. Res., Amsterdam, 1960-. Society: Netherlands Phys. Soc;
Nuclear interest: Nuclear physics.
Address: 168 Prins Bernhardplein, Amsterdam, Netherlands.

JANSSEN, H. J. H., Drs. Head of Office for Coordination of Nucl. Matters, and Director-Gen., Directorate-Gen. of Shipping.
Address: Ministry of Communications and Water Control,400 Van Alkemadelaan, The Hague, Netherlands.

JANSSEN, Roger. Vice Pres., Union Chimique - Chemische Bedrijven. Administrateur, Groupement Professionel de l'Industrie Nucléaire.
Address: Union Chimique-Chemische Bedrijven, 4 chaussee de Charleroi, Brussels 6, Belgium.

JANSSON, Erik, M.S. (Chem.Eng.). Born 1922. Educ.: Chalmers Univ. of Technol., Gothenburg. Nucl. Inspector and Sec., Reactor Safety Com., Swedish Atomic Energy Board. Society: B.N.E.S.
Nuclear interest: Nuclear safety.
Address: Atomic Energy Board, Fack, Stockholm 2, Sweden.

JANTSCH, Erich, Dr. Born 1929. Educ.: Vienna and Indiana Univs. Reactor phys., A. G. Brown, Boveri et Cie., Baden, Switzerland.
Nuclear interest: Reactor design.
Address: 2 St. Ursusstrasse, Baden, Switzerland.

JANZER, Victor J. Chem., Geological Survey, U.S. Dept. of the Interior.
Address: U.S. Department of the Interior, Geological Survey, Building 25, Federal Centre, Denver, Colorado, U.S.A.

JAQUISH, Richard Eugene, B.S., M.S. Born 1935. Educ.: Washington State and Harvard Univs. Asst. Chief, Environmental Radiol. Health Training Sect., Cincinnati, Ohio, 1961-63; Chief, Training Branch Activities (1963-66), Chief, Education and Information Programme (1966-), Southwestern Radiol. Health Lab., Las Vegas, Nevada. Societies: H.P.S.; A.P.H.A.; Diplomat, American Acad. Environmental Eng.; Conference on Radiol. Health.
Nuclear interests: Training in radiological health and radiological safety.
Address: Southwestern Radiological Health Laboratory, P.O. Box 15027, Las Vegas, Nevada 89114, U.S.A.

JAROSLOW, Bernard Norman, B.S., M.S., Ph.D. Born 1924. Educ.: C.C.N.Y., Iowa State and Chicago Univs. Now at Argonne Nat. Lab. Societies: Rad. Res. Soc.; American Soc. of Parasitologists; American Soc. of Tropical Medicine and Hygiene; American Inst. of Biological Sci.; American Assoc. of Immunologists.
Nucléar interest: Radiobiology.
Address: Argonne National Laboratory, Division of Biology and Medicine, Lemont, Illinois, U.S.A.

JAROSS, Robert Alan, B.S. Born 1927. Educ.: Illinois Univ. Assoc. Elec. Eng., Argonne Nat. Lab.
Nuclear interests: Liquid metal technology.
Address: 417 N. Main Street, Sandwich, Illinois, U.S.A.

JARVIS, Alan Maurice, B.Sc. Born 1920. Educ.: Durham Univ. Sen. Eng., R. and D. Dept., Merz and McLellan, 1948-55; Naval Nucl. Propulsion Project, Vickers Armstrongs (Eng.) and Rolls-Royce and Assocs., 1958-63; Director, Rolls-Royce Developments Ltd.; Director, Nucl. Developments Ltd.; Special Asst. to Chief Exec., Rolls Royce Ltd. Member, Council, British Nucl. Forum. Societies: I.E.E.; B.N.E.S.
Nuclear interest: Power generation and ship propulsion.

JARVIS, Guy Meredith, Barrister-at-Law. Born 1896. Educ.: Osgoode Hall. Legal Adviser and Sec. (1966-67), Legal Consultant (1967-), Atomic Energy Control Board. Gen. Counsel, Eldorado Mining and Refining Ltd., 1955-; Gen. Counsel, A.E.C.L., 1952-63.
Nuclear interests: Legal and administration.
Address: 127 Glen Road, Toronto 5, Ontario, Canada.

JASPERT, Jürgen, Dipl.-Chem. Born 1929. Educ.: Cologne Univ. Rad. Chem., Abt. Isotopen-und Kerntechnik, Battelle-Inst. e.V., Frankfurt a. M., 1957-61; Reactor Dosimetry, Service des Piles, C.E.N., Grenoble, 1961-64; Nucl. Chem., C.C.R. Euratom, Ispra, 1964-.
Nuclear interests: Health physics and dosimetry.
Address: C.C.R. Euratom, Ispra, Italy.

JASTRAM, Philip Sheldon, S.B., Ph.D. Born 1920. Educ.: Harvard and Michigan Univs. Asst. Prof., Washington Univ., St. Louis, 1949-55; Asst. Prof. (1955-57), Assoc. Prof. (1957-64), Prof. (1964-), Ohio State Univ.; Consultant, USAID, India, 1964-65. O.E.E.C. Fellow, Copenhagen, 1961; Formerly Member, Exec. Com., F.A.S. Societies: A.P.S.; A.A.A.S.; I.E.E.E.
Nuclear interests: Nuclear physics; Nuclear structure; Radiation processes; Nuclear orientation; Angular correlation.
Address: 115 West Royal Forest, Columbus, Ohio, U.S.A.

JAUERNICK, R., Dr-Ing. Director, Technischer Überwachungs-Verein Hamburg e.V.
Address: Technischer Überwachungs-Verein Norddeutschland e.V., 31 Gr. Bahnstrasse, 2 Hamburg 54, Germany.

JAUHO, Pekka Antti Olavi, Ph.D. Born 1923. Educ.: Helsinki Univ. Prof. Nucl. Phys. and Director Reactor Lab., Tech. Univ., Otaniemi, Finland. Society: Finnish Phys. Soc.
Nuclear interests: Nuclear physics; Reactor theory.
Address: Technical Physics Laboratory, Otaniemi, Finland.

JAUNEAU, Louis, Doctorat. Born 1924. Educ.: Paris Univ. Maitre de Recherche, C.N.R.S.; Lab. de Physique, Ecole Polytechnique, Paris.
Nuclear interest: Elementary particle physics.
Address: Ecole Polytechnique, Laboratoire de Physique, 17 rue Descartes, Paris, France.

JAWOROWSKI, Zbigniew, Ph.D. Born 1927. Educ.: Jagiellonian Univ., Cracow and Medical Acad., Cracow. Res. worker, Physiological Chem. Dept., Medical Acad., Cracow, 1951-52; Physician and Res. worker, Inst. Oncology, Gliwice, 1952-58; Res. worker i/c Internal Contamination Lab., Inst. Nucl. Res., Warsaw, 1958-. Books: Manual of Rad. Protection; Radioactivity and Human Health; Stable and Radioactive Lead in Environment and Human Body.
Nuclear interests: Health physics; Internal and environmental contaminations; Natural radioactivity of the body; Radiotoxicology.

Address; Nuclear Research Institute, Medical Department, 16 Ul. Dorodna, Warsaw, Poland.

JAYSON, Gerald Gert, B.Sc., Ph.D., F.R.I.C. Born 1928. Educ.: Queen's (Belfast) and Durham Univs. Res. Chem., Unilever, Ltd., 1955-59; Sen. Lecturer, Radio- and Rad. Chem. (1959-65), Principal Lecturer, Nucl. Studies, (1965-), Liverpool Regional Coll.of Technol. Societies: Rad. Res. Soc.; F.R.I.C. Nuclear interests: Radiation chemistry and biology; Radiochemistry; Health physics.
Address: Chemistry and Biology Department, Byrom Street, Liverpool 3, England.

JEAN, Maurice, D. ès Sc. Born 1921. Educ.: Paris Univ., Prof., Faculté des Sci. d'Orsay, Paris Univ.
Nuclear interest: Nuclear and particle theory.
Address: Institut de Physique Nucléaire, B.P. No. 1, Orsay 91, France.

JEANNET, Eric, Dr. Born 1932. Educ.: Neuchâtel Univ. Chef du Groupe, Emulsions Nucléaires, Inst. Physique (-1962), Chargé de cours (1962-67), Prof. extraordinaire (1967-), Neuchâtel Univ. Societies: Sté. suisse de Phys.; Societá Italiana di Fisica.
Nuclear interests: High energy physics; Nuclear emulsions; Bubble chamber.
Address: Institut de Physique, Université de Neuchâtel, 1 rue Breguet, Neuchâtel, Switzerland.

JEANPIERRE, Guy. Chef du Service de fabrication industrielle des éléments combustibles de matériaux de structure, C.E.A. Adviser, Third I.C.P.U.A.E., Geneva, Sept. 1964.
Address: Commissariat à l'Energie Atomique, 29-33 rue de la Federation, Paris 15, France.

JEANS, R. L. Vice Pres., Elec. Utility and Marine Products, Westinghouse Elec. Internat. Co.
Address: Westinghouse Electric International Co., 200 Park Avenue, New York, New York 10017, U.S.A.

JEBB, Alan, B.Sc. (Elec. Eng.), D.I.C. (Nucl. Power). Born 1933. Educ.: Manchester and London Univs. Elec. Eng., North Western Elec. Board, 1950-57; Res. Eng. Nucl. Power, Central Elec. Generating Board, 1957-60; Lecturer, Elec. Eng., City Univ., 1960-66; Lecturer, Nucl. Power, Imperial Coll., London Univ., 1966-. Societies:M.I.E.E.; A.M.I.Mech.E. Nuclear interests: Reactor kinetics and control. Research into reactor noise and control. Analogue and digital computation in nuclear field. Systems engineering and nuclear power.
Address: 10 Doves Close, Bromley, Kent, England.

JECH, Cestmir, R.N.Dr., C.Sc. Born 1925. Educ.: Charles Univ. Prague. Res. Worker, Inst. of Phys. Chem., Czechoslovak Acad. Sci., 1953-; Lecturer in Radiochem., Faculty of Natural Sci., Charles Univ., Prague, 1959-. Member, Editorial Board, journal, Chemické Listy. Book: Radioaktivni aerosoly (Radioactive aerosols) (Prague, NCSAV, 1955). Society: Czechoslovak Chem. Soc. Nuclear interests: Application of radioisotopes; Radiation physics of solids.
Address: Institute of Physical Chemistry, 7 Máchova, Prague 2, Czechoslovakia.

JEE, Webster Shew Shun, B.A., M.A., Ph.D. Born 1925. Educ.: California (Berkeley) and Utah Univs. Teach. Asst. Zoology, California Univ., 1949-51; Res. Asst. Anatomy (1956-59), Bone Group Leader, Radiobiol. (1956-), Instructor, Anatomy (1959-60), Asst. Res. Prof. Anatomy (1960-61), Asst. Prof. Anatomy (1961-63), Assoc. Prof. Anatomy (1963-), Assoc. Prof. Anatomy (1963-67), Prof. Anatomy (1967), Utah Univ. Societies: Rad. Res. Soc.; American Assoc. Anatomists; Orthopedic Res. Soc.; Gerontological Soc.; Internat. Soc. Stereology; Internat. Assoc. Dental Res.
Nuclear interests: Pathologic physiology, metabolism and pathology of osseous and dental tissues containing bone-seeking radionuclides.
Address: Radiobiology Division,, Anatomy Department, 2C105 Medical Centre, Utah University College of Medicine, Salt Lake City, Utah 84112, U.S.A.

JEFFERIES, H. V. Deputy Station Supt. (-1967), Station Supt. (1967-), Sizewell Nucl. Power Station, C.E.G.B.
Address: Central Electricity Generating Board, Sizewell Nuclear Power Station, Sizewell, Suffolk, England.

JEFFERSON, Merrill E., B.A. Born 1907. Educ.: George Washington Univ. Health Phys., U.S. Dept. of Agriculture, 1949-63. Radiol. Safety Officer, Certified American Board of Health Phys., 1960. Societies: A.N.S.; H.P.S.
Nuclear interests: Radiological health and safety; Applications of radioisotopes and radiation in agricultural research.
Address: 205 Dale Drive, Silver Spring, Maryland, U.S.A.

JEFFERSON, Robert M., B.S. (Mech. Eng.), M.B.A. Born 1932. Educ.: Michigan Coll. of Mining and Technol. and New Mexico Univ. Mech. Eng., Convair (Aircraft Co.), Fort Worth, 1953-54; Phys. Instructor, U.S. Air Force, Sandia Base, 1954-57; Supv., Reactor Applications Div., Sandia Lab., Sandia Corp., 1957-. Chairman, Trinity Sect., Member, Exec. Com., Remote Systems Technol. Div., and Member, Membership Com., A.N.S.; Chairman, Reactor Safeguards Advisory Com., New Mexico Univ.; Sec., Sandia Reactor Safeguards Advisory Com.

Nuclear interests: Reactor design and development; Reactor operations; Remote handling; Hot cell design; Shielding; Management.
Address: 10905 Elvin, N.E., Albuquerque, New Mexico 87112, U.S.A.

JEFFERSON, Sidney, B.Sc. (Eng.). Born 1907. Educ.: London Univ. Ind. applications of electronics, M.O.S., 1947-51; Ind. applications of radioisotopes (1951-55), Leader, Technol. Irradiation Group, Isotope Div. (1955-60), Branch Head, Rad. Branch, Isotope Res. Div. (1960-), U.K.A.E.A. Books: Radioisotopes – a New Tool for Industry (Newnes, 1957, 1960); Handbook of the Atomic Energy Industry (Newnes, 1958); Massive Rad. Techniques (Newnes, 1964). Nuclear interests: Research on the effects of radiation and the development of industrial radiation processing.
Address: Wantage Research Laboratory (A.E.R.E.), Wantage, Berks., England.

JEFFREY, Alan, Ph.D. (Maths.), M.Sc. (Maths.), B.Sc. (Maths.). Born 1929. Educ.: London Univ. Res. Mathematician, G.E.C., Ltd., Middlesex, 1953-56; Sen. Mathematician, Rolls-Royce, Ltd., Derby, 1956-; Temporary leave from Rolls-Royce, Ltd. as Exchange Visitor at Inst. Mathematical Sci., New York Univ., 1960-61; Prof. of Eng. Maths., Newcastle Univ., 1965-. Book: Co-author, Nonlinear Wave Propagation (Academic Press, 1964). Societies: F.R.A.S.; F.S.S.
Nuclear interests: Mathematical aspects of reactor theory – transport theory and nuclear codes.
Address: 22 Wilson Gardens, Newcastle Upon Tyne 3, England.

JEFFREY, B. S. Trustee, Southwest Atomic Energy Assoc.
Address: Southwest Atomic Energy Associates, 306 Pyramid Building, Little Rock, Arkansas, U.S.A.

JEFFRIES, Thomas Oliver, M.A., D.Phil. Born 1927. Educ.: Oxford Univ. Fereday Fellow, St. John's Coll., Oxford, 1949-52; U.K.A.E.A., Extra-mural Res. Phys., 1951-53; Lecturer in Phys., Birmingham Univ., 1953-56; Sect. Leader, Analogue Computer Sect. (1957), Head, Analogue Office (1958), Asst. Chief Development Eng., Phys. (1959), Head, Control and Instrumentation Design Group (1961), English Elec. Co., Atomic Power Div., Whetstone, Leicester.
Nuclear interests: Reactor physics and kinetics; Analogue computers and computing techniques; Reactor control systems; Reactor safety circuits; High voltage generators and particle accelerators.
Address: 52 Outwoods Road, Loughborough, Leicestershire, England.

JEGOUZO, Albert. At Batignolles-Chatillon (Mécanique Générale) S.A. Administrateur Directeur Générale, Forges et Acieries du Saut du Tarn S.A.
Address: Batignolles-Chatillon (Mécanique Générale) S.A., 5 rue de Monttessuy, Paris 7, France.

JEGU, Pierre. Born 1924. Educ.: Paris Univ. Chairman and Managing Director, Bureau Internat. de Relations Publiques.
Nuclear interests: Editor of Nuclelec bi-weekly bulletin on nuclear energy.
Address: 27 rue de Rome, Paris 8, France.

JEHN, Hermann, Dipl.-Phys. Born 1937. Educ.: Technische Hochschule, Stuttgart and Technische Hochscule, Vienna. Sci., Max-Planck-Inst. für Metallforschung,Inst. für Sondermetalle, (Max-Planck-Inst. for Metal Res., Inst. for Special Metals), Stuttgart, 1964-. Society: Deutsche Physikalische Gesellschaft (German Phys. Soc.).
Nuclear interests: Reactor and refractory metals; Gas-metal-reactions; Kinetics and equilibria; Thermodynamics.
Address:Max-Planck-Institut für Metallforschung, Institut für Sondermetalle, 92 Seestrasse, 7000 Stuttgart 1, Germany.

JEITNER, Franz, Dr. Ing.tech. Born 1912. Educ.: Prague T.H. Sigri Elektrographit G.m.b.H., Meitingen, 1950-.
Nuclear interests: Metallurgy, ceramic, special graphite for reactor purposes, including fuel-elements, further on impermeable graphite and silicon-carbide.
Address: 9 Peissenbergstrasse, 89 Augsburg, Germany.

JELATIS, Demetrius G. Vice-Pres., Central Res. Laboratories Inc. Member, Exec. Com., Remote Systems Technol. Div., A.N.S., 1966-.
Address: Central Research Laboratories Inc., Red Wing, Minnesota, U.S.A.

JEN, Chun-Hua, B.Sc. Born 1912. Educ.: Nat. South-west Univ. Assoc. Chem., Chinese Petroleum Corp., Chem., Inorganic Chem. Res. Lab., Union Industrial Res. Inst. Society: Chinese Chem. Soc.
Address: P.O. Box 100, Hsin-chu, Formosa.

JEN, Lang. Paper: Radiation from a Linear Monopole at the Tip of a Metallic Prolate Spheroid (Scientia Sinica, vol. 11, No. 10, 1962).
Address: Tangshan Railway College, Tangshan, China.

JENCKEL, Ludolf H. F., Dipl. Ing., Dr. Ing. Born 1912. Educ.: Göttingen, Munich, Berlin-Charlottenburg and Kiel Univs. Gen. Manager, Atlas Mat, 1948-; Gen. Manager, Fried. Krupp, Mat, 1966-; Gen. Manager, Varian Mat, G.m.b.H. 1967-. Society: Deutsche Physikalische Gesellschaft.
Address: Varian Mat G.m.b.H., 442-448, Woltmershauser Strasse, 28 Bremen 10, Germany.

JENKINS, A. E., B. Met. E., Ph.D., F.I.M. Born 1924. Educ.: Melbourne Univ. Prof., Chem. and Extraction Metal., New South Wales Univ. Technol.
Nuclear interests: Metallurgy, particularly reactions between metals and liquids and gases at high temperatures; Structure and properties of interstitial solid solutions.
Address: Box 1, Post Office, Kensington, New South Wales, Australia.

JENKINS, Dale W. Liaison (N.A.S.A.), Subcom. on Radiobiol., Nat. Acad. of Sci.-N.R.C.
Address: National Academy of Sciences - National Research Council, 2101 Constitution Avenue, N.W., Washington 25, D.C., U.S.A.

JENKINS, Ivor, B.Sc. (1st Class Hons., Metal.) (Univ. Coll., Swansea), M.Sc. (Wales), D.Sc. (Wales). Born 1913. Educ.: Univ. Coll., Swansea. Director of Res., Manganese Bronze Holdings Ltd., 1952-. Book: Controlled Atmospheres for the Heat-treatment of Metals. Societies: Inst. Metals; Inst. Metallurgists; Iron and Steel Inst.
Nuclear interest: Metallurgy.
Address: Godwins Place, Hoo, nr. Woodbridge, Suffolk, England.

JENKINSON, Horace Arnold. Born 1908. Managing Director, W. G. Jenkinson Ltd. Societies: Stainless Steel Fabricators Assoc.; Inst. Plumbing.
Nuclear interest: Lead shieldings.
Address: 156-160 Arundel Street, Sheffield 1, England.

JENKINSON, Ivan Searle, B.Sc. (First class Hons., Phys.), Dr. Phil. (Biophys.). Born 1934. Educ.: Queensland Univ. Res. Officer, Australian A.E.C., 1960-64; Phys., i/c Phys. Dept., St. Vincent's Hospital, Sydney, 1964-. Societies: Hospital Phys. Assoc. (Member of Australian Com.); Australian Inst. of Phys. (Biophys. Group); Australian Rad. Soc. (Sec. 1961-63); Assoc. Member, Coll. of Radiols. of Australasia.
Nuclear interests: Nuclear medicine, use of radioisotopes in diagnosis, treatment and research. Emphasis on use in cancer and endocrine research.
Address: Physics Department, St. Vincent's Hospital, Darlinghurst, Sydney, New South Wales 2010, Australia.

JENN, Jean-Tony Paul, Ing. en Chef des Mines. Born 1919. Educ.: Ecole Polytechnique and Ecole Nationale Superieure des Mines de Paris. Formerly Directeur Gen., Compagnie Française de l'Eau Lourde (Cette Compagnie est également Gérante de la Sté. d'Etudes pour l'Obtention du Deuterium); Directeur, l'Air Liquide; Pres.-Directeur Gen., Sté. d'Etudes et de Construction d'Appareillages pour tres basses Temperatures.
Nuclear interests: Heavy water and isotopes; Purification of nuclear reactor gases; Gas liquefaction and separation plants; General use of very low temperatures in the nuclear energy field.
Address: 15 boulevard Richard-Wallace, Neuilly-sur-Seine, France.

JENNE, Everett A. Soil Sci., Geological Survey, U.S. Dept. of the Interior.
Address: U.S. Department of the Interior, Geological Survey, Building 25, Federal Centre, Denver, Colorado, U.S.A.

JENNEKENS, John Hubert Felix, B.Sc. (Hons., Mech. Eng.), Dipl. Mech. Eng. (R.M.C.). Born 1932. Educ.: Roy. Military Coll. and Queen's Univ. Lieutenant, Roy. Canadian Elect. and Mech. Eng., 1954-58; Reactor Operations Eng., A.E.C.L., 1958-62; Assoc. Sci. Adviser, Atomic Energy Control Board, 1962-. Society: A.P.E.O.
Nuclear interests: Reactor safety; Nuclear material control (safeguards).
Address: Atomic Energy Control Board, P.O. Box 1046, Ottawa, Ontario, Canada.

JENNINGS, Harry. Analytical Lab. Supervisor, Elgin Softener Inc.
Address: Elgin Softener Inc., 440 South McLean Boulevard, Elgin, Illinois, U.S.A.

JENNINGS, R. E., Formerly Phys. Lecturer, Nucl. Phys. Res., University Coll., London.
Address: Physics Department, University College, London, Gower Street, London, W.C.1, England.

JENS, Wayne Henry, B.S. (Mech. Eng.), Ph.D. (Mech. Eng.). Born 1921. Educ.: Wisconsin and Purdue Univs. Societies: A.S.M.E.; A.N.S.
Nuclear interests: Core design; Engineering analysis materials; Safety; Fuel cycle.
Address: 1246 Balfour Road, Grosse Pointe Park 30, Michigan, U.S.A.

JENSCH, Samuel W. Educ.: Carleton Coll., Minnesota and Harvard Law School. Federal Power Commission, 1946; Central Elec. and Gas Co., Lincoln, Nebraska, 1952-55; Federal Power Commission, 1955-58; Hearing Examiner, Office of Hearing Examiner, (1958-), and Member, Panel of Experts to Provide Boards on Atomic Safety and Licensing, U.S.A.E.C.
Address: U.S. Atomic Energy Commission, Washington 20545, D.C., U.S.A.

JENSEN, Aage, M.Sc. (Mech. Eng.). Born 1928. Educ.: Tekniske Hojskole, Denmark. Danish Central Welding Inst., -1956; At (1956-), Head, Reactor Eng. Dept. (1964-), Danish Atomic Energy Commission.
Nuclear interest: Reactor design.
Address: Danish Atomic Energy Commission, Research Establishment Riso, Roskilde, Denmark.

JENSEN, Eiler Aage. Born 1894. Formerly Pres., Confederation of Danish Trade Unions; Member, Exec. Com., Danish A.E.C.

Address: 18 Aakjaersalle, Copenhagen Soborg, Denmark.

JENSEN, Erling. Member, Danish A.E.C.
Address: Danish Atomic Energy Commission, 29 Strandgade, Copenhagen K, Denmark.

JENSEN, H. F. Chief Res. Officer, Disa Elektronik.
Address: Disa Elektronik, 17 Herlev Hovedgade, Herlev, Denmark.

JENSEN, H.-W., Dr. Dozent, Nucl. Dept., Staatliche Ingenieurschule.
Address: Staatliche Ingenieurschule, Nuclear Department, 3 Munketoft, 2390 Flensburg, Germany.

JENSEN, Henning Hojgaard, Cand. mag. Born 1918. Educ.: Copenhagen Univ. Prof., Copenhagen Tech. Univ., 1952-60; Prof., Copenhagen Univ., 1960-. Com., Danish Atomic Energy Commission, 1960-.
Nuclear interests: Solid state and nuclear physics.
Address: 18A Rodtjornevej, Vanlose, Copenhagen, Denmark.

JENSEN, J. Hans D., Dr. rer. nat., Hon. Prof. (Hamburg Univ.). Nobel Prize for Phys., 1963. Born 1906. Educ.: Freiburg and Hamburg Univs. Prof. (1949-), Dean of Faculty (1953-), Senator (1958-), Heidelberg Univ.
Book: Co-author, Elementary Theory of Nucl. Shell Structure; co-editor, Zeitschrift für Physik. Societies: Max Planck Gesellschaft; Heidelberger Akademie d. Wissenschaften.
Address: 16 Philosophenweg, Heidelberg, Germany.

JENSEN, Jay O. Asst. Prof., Phys. Dept., Utah State Univ.
Address: Utah State University, Physics Department, Logan, Utah, U.S.A.

JENSEN, Niels ARNTH-. See ARNTH-JENSEN, Niels.

JENSEN, Nils Allan, B.S. Born 1917. Educ.: Arizona Univ. Project Manager, Paceco, 1956-. Societies: A.W.S.; A.N.S.; N.S.P.E.
Nuclear interests: Custom design, design analysis, and fabrication of nuclear reactor pressure vessels, pressurizers, steam generators. Reactor vessel structural internals and metallic shielding.
Address: Paceco, 2350 Blanding Avenue, Alameda, California 94501, U.S.A.

JENSEN, Sigurd TOVBORG-. See TOVBORG-JENSEN, Sigurd.

JENSEN, Tirkil Hesselberg, M.Sc. Born 1932. Educ.: Denmark Tech. Univ.
Nuclear interest: Plasma physics.
Address: Gulf General Atomic, Inc., P.O. Box 608, San Diego 12, Calif., U.S.A.

JENSSEN, Gudbrand, Cand. real. Born 1922. Educ.: Oslo Univ. Director, Health Phys., Inst. for Atomenergi, Norway; Member, Norwegian Govt's. Council on Rad. Hygiene. At present professional officer, Div. of Health, Safety and Waste Disposal, I.A.E.A., Vienna.
Nuclear interests: Health physics; Radiation measuring instruments.
Address: International Atomic Energy Agency, 11 Kaerntnerring, Vienna 1, Austria.

JENSSEN, Sverre Knut, M.Sc. (Tech. Eng.). Born 1911. Educ.: Roy. Inst. Technol., Stockholm. Chief Eng., Nucl. Dept., A.B. Rosenblads Patenter, -1962; Director, Thermal Tech. Div., Alfa-Laval A.B., 1963-.
Nuclear interests: Heat transfer; Heat exchanger and reactor design; Design of heat transferring components in nuclear reactors.
Address: Alfa-Laval A.B., Postfack, Tumba, Sweden.

JENTSCHKE, Willibald Karl, Prof., Dr. Born 1911. Educ.: Vienna Univ. Elec. Eng. Dept., Illinois Univ., 1948-50; Cyclotron Lab., Urbana, Illinois, 1950-56. Director, Phys. Dept., Hamburg Univ., 1956-; Guest Prof., C.E.R.N. Societies: Deutsche Physikalische Gesellschaft; Soc. Italiana di Fisica; A.P.S.
Nuclear interest: Elementary particle physics.
Address: II Institut für Experimentalphysik, 149 Luruper Chaussee, Hamburg-Bahrenfeld, Germany.

JENTZER, Albert, Dr. Prof. Pres., Geneva Commission on Radioactivity.
Address: Institut du Radium, 16 rue Alcide-Jentzer, Geneva, Switzerland.

JERCHEL, Dietrich, Dr. phil. nat. habil., Prof., Honorarprof. (Munich T.H.). Born 1913. Educ.: Munich and Freiburg/Breisgau Univs. Lecturer Organic Chem. and Biochem., Mainz Univ., 1946; appl. Prof., Mainz Univ., 1951; Res. Dept., Dr. Karl Thomae G.m.b.H., Biberach/ Riss, 1958; Res. Dept., Chemische Fabrik C. H. Boehringer Sohn, Ingelheim am Rhein, 1964. Societies: Gesellschaft Deutscher Chemiker; Gesellschaft für Physiologische Chemie.
Nuclear interests: Organic chemistry and biochemistry with isotopic-labelled substances.
Address: 1 Tiefenweg, Ingelheim am Rhein 6507, Germany.

JEREMIE, Hannes Arthur, Dipl. Phys. (Hamburg), D. es Sc. (Paris). Born 1933. Educ.: Hamburg Univ. Collège de France, 1959-64; Montreal Univ., 1965-.
Nuclear interest: Nuclear physics.
Address: Université de Montreal, Departement de Physique, B.P. 6128, Montreal, Canada.

JERRERO, F. Spanish Member, Group of Experts on Third Party Liability, O.E.C.D., E.N.E.A.
Address: O.E.C.D. European Nuclear Energy Agency, 38 boulevard Suchet, Paris 16, France.

JERTRUM, Paul. Officer in charge, Nautisch-Technische Abteilung, Hamburg-Sudamerikanische Dampschifffahrts-Gesellschaft.
Address: Hamburg-Sudamerikanische Dampschifffahrts-Gesellschaft, 59 Ost-West-Str., Hamburg 11, Germany.

JERVIS, Max William, M.Sc. Tech., B.Sc. Tech. (Hons.), F.I.E.E. Born 1925. Educ.: Manchester Univ. Sen. Asst. Eng., Nucl. Plant Design Branch, C.E.G.B., London, 1964; Formerly at the Nucl. Power Group, Knutsford, Cheshire, and Res. Lab., A.E.I., Aldermaston. Book: Nucl. Reactor Instrumentation, Nucl. Eng. Monograph 13 (Temple Press, Ltd., 1961). Society: F.I.E.E.
Nuclear interests: Instrumentation for safety and control of reactors, including neutron flux, temperature instruments and safety system equipment; Ergonomic aspects of control and application of digital computers in power stations.
Address: Walden House, 24 Cathedral Place, London, E.C.4, England.

JERVIS, Robert E., M.A., Ph.D. Born 1927. Educ.: Toronto Univ. Assoc. Res. Officer, A.E.C.L., Chalk River, 1952-58; Assoc. Prof. (1958-66), Prof. (1966-), Chem. Eng. and Appl. Chem. Dept., Toronto Univ.; Visiting Prof. (Radiochem.), Tokyo Univ., Japan, 1965-66. Books: Activation Analyses of Interest to Atomic Energy Programmes (Chapter in Analytical Chem., Series IX, vol. i, Progress in Nuclear Energy, ed. M. Kelley). Societies: Fellow, Chem. Inst. Canada; Can. Soc. Forensic Sci.; A.A.A.S.; American Sci. Affiliation; Canadian Youth Sci. Foundation.
Nuclear interests: Neutron activation analysis; Fast neutron reactions; Photo-fission; General radiochemistry research such as in the study of ion exchange and solvent extraction behaviour; Isotope exchange study of inorganic complexes; Activation analysis applied to forensic analyses.
Address: Chemical Engineering and Applied Chemistry Department, Toronto University, Toronto 5, Canada.

JESCHKI, Wolfgang, Dipl. Ing. Sen. Officer, Atominst. der Osterreichischen Hochschulen.
Address: Atominstitut der Osterreichischen Hochschulen, 115 Schuttelstrasse, Vienna 2, Austria.

JESPERSON, Richard E. Asst. to Director, Public Relations, Atomics Internat. Member, Public Information Com., A.N.S., 1966-68.
Address: Atomics International, P.O. Box 309, Canoga Park, California, U.S.A.

JESSE, William Polk, A.B., Ph., Sc.D. Born 1891. Educ.: Missouri, Chicago and Yale Univs. Argonne Nat. Lab., 1946-56; St. Procopius Coll., 1956-. Societies: A.A.A.S.; A.P.S.
Nuclear interests: Ionisation in gases from alpha, beta and gamma rays with special regard to W values; Isotope effect in ionisation in polyatomic gases.
Address: St. Procopius College, Lisle, Illinois, U.S.A.

JESUS MONZON, Ma. de. See de JESUS MONZON, Ma.

JETTER, L. K. Member, Editorial Advisory Board, J. Nucl. Materials.
Address: c/o Journal of Nuclear Materials, North-Holland Publishing Co., P.O. Box 103, Amsterdam, Netherlands.

JEWELL, R. B. Contract Manager (Vice Pres.), Burlington and Pantex Plants, U.S.A.E.C.
Address: Burlington A.E.C. Plant, Burlington, Iowa, U.S.A.

JEWETT, Frank B., Jr. Pres. and Chief Exec. Officer, Vitro Corp. of America.
Address: Vitro Corporation of America, 90 Park Avenue, New York, New York 10016, U.S.A.

JEWETT, John P., B.S., M.B.A. Born 1921. Educ.: Harvard Coll., and Harvard Graduate School of Business Administration. Director, Finance Div. (1957-58), Asst. Manager for Administration (1958-62), U.S.A.E.C., Chicago Operations Office; Asst. Chief for Administration (1962-67), Chief (1967-), Space Nucl. Propulsion Office, Nevada.
Nuclear interest: Management.
Address: 1820 Sweeney Avenue, Las Vegas, Nevada, U.S.A.

JEZDIC, Vojislav. Member, Sci. Advisory Com., Federal Nucl. Energy Commission.
Address: Boris Kidric Institute of Nuclear Sciences, P.O. Box 522, Belgrade, Yugoslavia.

JEZOWSKA-TRZEBIATOWSKA, Boguslawa, Dr. First Prize of Nat. Council for Peaceful Use of Atomic Energy, 1966. Born 1908. Educ.: Lvov Tech. Univ. Assoc. Prof., (1950-60), Full Prof. of Inorganic Chem., and Head, Inorganic Chem. Dept. (1960-), Wroclaw Univ. Vice-Pres., Polish Chem. Soc.; Tutular Member, Inorganic Chem. Div., I.U.P.A.C., 1963-67. Book: Chapter on Electronic Structure of Uranium in Fundamental Particles, Atomic Nucleus, Radioactivity (Warsaw, PWN, 1967). (In Polish). Societies: Com. for Rad. Chem., Nat. Council for Peaceful Use of Atomic Energy; Member, Polish Acad. Sci.
Nuclear interests: Nuclear chemistry: Chemistry, electronic structure, spectroscopy, magnetochemistry etc. of uranium, transuranium elements and lanthanides. Radiation chemistry of solutions. Application of radioisotopes in structural and reaction mechanisms studies.
Address: Department of Inorganic Chemistry, Wroclaw University, 1 Pl. Uniwersytecki, Wroclaw, Poland.

JIACOLETTI, Richard John. B.S. (Chem.), B.S. (Chem. Eng.), M.S. (Nucl. Eng.), Assoc. Prof.

Nucl. Eng. Born 1931. Educ.: Arizona and Wyoming Univ. Formerly Chief Reactor Supervisor, Wyoming Univ., 1958-; Reactor Analysis Consultant, Argonne Nat. Lab., 1963-64. Faculty Advisor, Student Branch, A.N.S., Wyoming Univ. Societies: A.N.S.; A.S.M.E.; N.S.P.E.
Nuclear interests: Nuclear analysis and engineering analysis of nuclear reactor systems; Nuclear engineering education.
Address: 2455 Park, Laramie, Wyoming 82070, U.S.A.

JIKHAREV, E. Deleg., I.A.E.A. Conference on the Use of Radioisotopes in the Phys. Sci. and Industry, Copenhagen, Sept. 6-17, 1960.
Address: Academy of Sciences of the Byelorussian S.S.R., Minsk, Byelorussian S.S.R.

JIMENEZ, Luis RUIZ. See RUIZ JIMENEZ, Luis.

JIMENEZ-DIAZ, Carlos, Prof. a. Dr. Med. Born 1898. Educ.: Madrid. Prof. Clinical Medicine, Madrid Univ., 1926; Director, Inst. for Medical Res., Madrid. President, Internat. Soc. Internal Medicine; President, Internat. Soc. Allergy.
Nuclear interests: Isotopic studies as applied in nutritious haematology, and internal medicine.
Address: 9 Avenida General Mola, Madrid, Spain.

JIRKOVSKY, Rudolf, Prof., R.N.Dr. Born 1902. Educ.: Charles Univ., Prague. Prof. and Head, Dept. Analytical and Radioactive Chem., Mining and Metal. Univ., Ostrava, Czechoslovakia. Com. for Utilisation of Radioactive Isotopes. Nucl. Energy Com., Prague; Chairman, Chem. Soc., Czechoslovak Acad. Sci.; Chairman, Radioactivity Sect., Region Com., Sci.-Tech. Soc., Ostrava. Books: The Application of Natural and Artificial Radioactivity in Mining, Geology and Preparation (Prague, Práce, 1961); The Utilisation of Radioactive Isotopes in Metal. and Mining (Prague, SNTL., 1967).
Nuclear interests: The application of radioactive isotopes in mining, metallurgy and geology; The utilisation of back-scattering of beta particles for chemical analysis.
Address: Metallurgy Faculty, Mining and Metallurgy University, 31 Osvoboditelu, Ostrava I, Czechoslovakia.

JOANOU, George David, B.S. (Chem.), M.S. (Phys.). Born 1931. Educ.: Arizona and Idaho Univs. Societies: A.N.S.; A.P.S.
Nuclear interests: Reactor physics; Management; Large scale scientific computing.
Address: c/o I.B.M. Scientific Centre, 2670 Hanover Street, Palo Alto, California, U.S.A.

JODRA, Luis GUTIERREZ. See GUTIERREZ JODRA, Luis.

JOEL, N., Dr. Chief Sci., Crystallography, Instituto de Fisica y Matematicas, Chile Univ.
Address: Instituto de Fisica y Matematicas, Casilla 2777, Santiago, Chile.

JOFEH, M. L., M.I.E.E., A.F.R. Ae.S. Managing Director, Sperry Gyroscope Co. Ltd.
Address: Sperry Gyroscope Co. Ltd., Great West Road, Brentford, Middlesex, England.

JOFFE, Mikhail, Dr. At Thermonucl. Lab., Kurchatov Inst. Atomic Energy. Deleg. to Conference sur la Physique des Plasmas et la Recherche concernant la Fusion Nucléaire Controlée, Salzbourg, Sept. 1961.
Address: Kurchatov Institute of Atomic Energy, 46 Ulitsa Kurchatova, Post Box 3402, Moscow, U.S.S.R.

JOHANSEN, H., M.D. Medical Director, Radium Centre, Copenhagen.
Address: Radium Centre, 49 Strandboulevard, Copenhagen Ø, Denmark.

JOHANSEN, Paul. Member, Board of Directors, Dansk Atomforsikrings Pool.
Address: Dansk Atomforsikrings Pool, 23 Gronningen, Copenhagen K, Denmark.

JOHANSSON, Erik Anders, Tekn. dr. Born 1929. Educ.: Roy. Inst. Technol., Stockholm. At A.B. Atomenergi, Stockholm, 1955-.
Nuclear interests: Experimental reactor physics with the emphasis on neutron spectra measurements and on physics experiments in power reactors.
Address: A.B. Atomenergi, 32 Liljeholmsvägen, Stockholm 43, Sweden.

JOHANSSON, Karl Evert Arne, Dr. Born 1930. Educ.: Uppsala Univ. Asst. Prof., Gustaf Werner Inst., Uppsala Univ.; Res. Fellow, Swedish Atomic Res. Council. Societies: A.P.S.; Fysikersamfundet, Sweden. Nuclear interests: Nuclear physics at medium energies; Nuclear reactions; Particle scattering.
Address: Gustaf Werner Institute, Uppsala University, Sweden.

JOHANSSON, S. A. E., Prof. Member, Nat. Inst. of Rad. Protection; Member, Swedish Atomic Res. Council.
Address: National Institute of Radiation Protection, Fack, S-104 01 Stockholm 60, Sweden.

JOHN, George, B.Sc., Ph.D. Born 1921. Educ.: Ohio State Univ. Res. Assoc., Ohio State Univ., 1952-53; Group Leader, Tracer Applications, Nucleonics Branch, Materials Lab. (1953-56), Assoc. Prof., Phys. Dept., Air Force Inst. of Technol. (1956-), Wright-Patterson A.F.B. Educ. Com., Southwest Ohio Chapter, A.N.S. Societies: A.C.S.; A.A.P.T. Nuclear interests: Nuclear radiation detectors and detection systems; Fluors and fluorescent

lifetimes; Solid state detectors.
Address: Air Force Institute of Technology-SEP, Wright-Patterson Air Force Base, Ohio 45433, U.S.A.

JOHN, Walter, B.S., Ph.D. (Phys.). Born 1924. Educ.: California Inst. Technol. and California Univ., Berkeley. Instructor, Phys. Dept., Illinois Univ., 1955-58; Res.Phys., Lawrence Rad. Lab.,Livermore, 1958-. Society: A.P.S.
Nuclear interests: Experimental nuclear physics; Nuclear reactions induced by charged particles; Photonuclear reactions; Nuclear spectroscopy; Nuclear fission.
Address: Lawrence Radiation Laboratory, California University, Livermore, California, U.S.A.

JOHNEN, Wilhelm, Born 1902. Educ.: Tübingen, Bonn, and Cologne Univs. Präsident, Landtag Nordrhein-Westfalen; Mitglied des Verwaltungsrats der Kernforschungsanlage Jülich des Landes Nordrhein-Westfalen e.V.
Address: 10 Dr. Weyer-Strasse, Jülich, Germany.

JOHNER, Werner Rodolphe, D. ès Sc. Born 1901. Educ.: Bern and Bruxelles Univs. and Tech. High School, Dresden. Chief Phys., Radiol. Centre, Inst. J. Bordet, Brussels. Societies: Swiss Phys. Soc.; Belgian Phys. Soc.
Nuclear interests: Dosimetry; Gamma-ray spectroscopy.
Address: Institut J. Bordet, rue Heger-Bordet, Brussels, Belgium.

JOHNS, Harold Elford, M.A., Ph.D., LL.D. Elected Fellow, Roy. Soc. of Canada; Roentgen Award, Br.Inst. of Radiol. (1953). Born 1915. Educ.: McMaster and Toronto Univs. Asst., Assoc., and Prof. Phys., Saskatchewan Univ.,and Phys., Saskatchewan Cancer Commission, 1945-56; Prof. Phys., Toronto Univ., and Head, Phys. Div., Princess Margaret Hospital (Ontario Cancer Inst.), Toronto, 1956-; Prof. and Head, Dept. Medical Biophys., Toronto Univ., 1962; Book: The Phys. of Rad. Therapy (1953); The Phys. of Radiol. (Springfield, Ill., Charles C. Thomas, Publisher, 1961). I.C.R.U., 1952; Board, Nat. Cancer Inst. of Canada, 1954; Pres., Canadian Assoc. of Medical Physicists, 1955; Assoc. Editor for Canada of British J. of Radiol.; Assoc. Editor of Rad. Res., 1955-; Member, Rad. Study Sect., Nat. Insts. of Health, Washington. Societies: British Inst. of Radiol.; Canadian Assoc. of Radiologists; Canadian Assoc. of Physicists; Canadian Assoc. of Medical Physicists; Royal Soc.; A.P.S.; American Radium Soc.; Rad. Res. Soc.; Biophys. Soc.
Nuclear interests: In field of ultraviolet photochemistry and photobiology.
Address: 24 Anderson Avenue, Toronto, Canada.

JOHNS, Kurt. Member, Fachausschuss Versicherung in der Kernenergie Studiengesellschaft.
Address: i/Fa, Kurt Johns Assekuranz, "Afrikahaus", Hamburg 11, Germany.

JOHNS, Martin Wesley, B.A., M.A., Ph.D. Born 1913. Educ.: McMaster and Toronto Univs. Prof., McMaster Univ., 1947-. Societies: A.P.S.; Canadian Assoc. of Physicists; A.A.P.T.; Roy. Soc. of Canada.
Nuclear interests: Nuclear physics; Beta and gamma-ray spectroscopy.
Address: McMaster University, Hamilton, Ontario, Canada.

JOHNS, Richard, Dr. Sen. Phys. Chem. Advanced Technol. Div., Atlantic Res. Corp.
Address: Atlantic Res. Corp.,Henry G. Shirley Memorial Highway at Edsall Road, Alexandria, Virginia, U.S.A.

JOHNS, S. Russell. Graduate of Officers School (Hospital Administration), Bethesda, Md. Born 1913. Lab. Manager, in charge of Dosimetry Interpretations, Rad. Detection Co., 1957-.
Nuclear interest: Film dosimetry for personnel and special tests.
Address: 1054 Robin, Sunnyvale, Calif. 94087, U.S.A.

JOHNSEN, Kjell, E.E., Dr. tech. Born 1921. Educ.: Norway Tech. Univ. Prof., Norwegian Inst. of Technol., 1957-59; Phys. (1952-57), Sen. Phys. (1959-65), Director, Intersecting Storage Rings Construction Dept. (1966-), C.E.R.N. Book: On the Theory of the Linear Accelerator (Bergen, Chr. Michelsens Inst., 1954). Societies: Norwegian Eng. Assoc.; Norwegian Phys. Soc.; Norwegian Acad. for Tech.Sci.
Nuclear interest: Theory and construction of particle accelerators and storage rings associated with such accelerators.
Address: Intersecting Storage Rings Division, C.E.R.N., 1211 Geneva 23, Switzerland.

JOHNSEN, Russell H. A., B.S., Ph.D. Born 1922. Educ.: Chicago and Wisconsin Univs. Prof. Phys. Chem., Florida State Univ., 1951-; O.R.N.L., summers 1954, 1957; U.S. Naval Radiol. Defence Lab., 1961; Rad. Lab., Berkeley,California, summer 1966. Societies: A.C.S.; Chem. Soc., London; Fellow, A.A.A.S.; A.P.S.; Rad. Res. Soc.
Nuclear interests: Radiation chemistry of organic systems; Reactions of atoms in solid substates at low temperatures; Optical and electron spin resonance spectroscopy of radiation-produced transients.
Address: Chemistry Department, Florida State University, Tallahassee, Florida, U.S.A.

JOHNSON, A. W. Director, Strachan and Henshaw Ltd.
Address: Strachan and Henshaw Ltd., Ashton Works, P.O. Box No. 103, Bristol 3, England.

JOHNSON, Alan, B.Sc. Born 1931. Educ.: Manchester Univ. Phys., Capenhurst (1953-56), Asst. Reactor Manager, Calder Hall (1956-59), Sen. Tech. Officer, Risley Production Group Headquarters (1959-63), Sen. Commercial Manager, Production Group London Office (1964-65), Sen. Tech. Officer, Production Group, Risley (1965-), U.K.A.E.A. Nuclear interests: Fuel cycles.
Address: Commercial Directorate, Production Group, United Kingdom Atomic Energy Authority, Risley, Warrington, Lancs., England.

JOHNSON, Alan D., B.S.M.E., M.S.M.E. Born 1919. Educ.: Purdue Univ. Chief, Reactor Operations Branch Plum Brook Reactor, now Director, Plum Brook Station, N.A.S.A. Society: Instrument Soc. of America.
Nuclear interest: Management.
Address: National Aeronautics and Space Administration, Plum Brook Station, Sandusky, Ohio 44870, U.S.A.

JOHNSON, Alfred A., Ph.D. (Phys. Chem.). Born 1916. Educ.: Illinois Univ. Reactor Technol. Supt., Savannah River Plant (1953-61), Sect. Director, Nucl. Eng. and Materials Sect., Savannah River Lab. (1961-64), Asst. Lab. Director, Savannah River Lab. (1964-67), Asst. Tech. Director, Atomic Energy Div., Wilmington, (1967-), E. I. du Pont de Nemours and Co. Society: A.N.S.
Nuclear interest: Management of research and development for reactor fuels and targets, reactor operation and separations chemistry and engineering.
Address: E. I. du Pont de Nemours and Company, Nemours Building, Wilmington, Delaware 19898, U.S.A.

JOHNSON, Bruce Connor, M.A., Ph.D. Born 1911. Educ.: McMaster and Wisconsin Univs. Prof., Animal Biochem., Illinois Univ., 1943-65; Prof. and Chairman, Dept. of Biochem., Oklahoma Univ. Medical School, 1965-; Sect. Head, Dept. of Biochem., Oklahoma Medical Res. Foundation, 1965-. Consultant, O.E.E.C. (Paris), on Peaceful Applications of Atomic Energy to Animal Science, 1958; Member of President's Second Atoms for Peace Mission to South America, 1956. I.A.E.A. Panel on the Application of Isotopes and Rad. to Agricultural Res. in Tropical Africa, I.A.E.A. Panel, Vienna, 1961. F.A.O./W.H.O./I.A.E.A. Tech. Meeting on the Evaluation of the Wholesomeness of Irradiated Foods, Brussels, 1961. Editorial Board, Proc. Soc. Exptl. Biol. Med. Societies: Central Soc. for Nucl. Medicine; Biochem. Soc. of Great Britain; A.C.S.; American Soc. of Biol. Chem.
Nuclear interest: Applications of radioactive tracers to research in the field of biochemistry.
Address: Department of Biochemistry, Oklahoma University Medical School, 800 Northeast Thirteenth Street, Oklahoma City, Oklahoma 73104, U.S.A.

JOHNSON, Christopher Michael Paley, Ph.D. Born 1931. Educ.: Cambridge Univ. 1851 Studentship (1956-58), I.C.I. Fellowship (1958-60), Cambridge Univ.; Res. Assoc., Nucl. Phys. Dept., Oxford Univ., 1960-63; Fellow, Bursar and Director, Studies in Phys., Selwyn Coll., Cambridge Univ.
Nuclear interest: Elementary particle physics and nuclear structure.
Address: Selwyn College, Cambridge, England.

JOHNSON, Cleland H., Ph.D. (Wisconsin). Born 1922. Educ.: Hastings Coll. and Wisconsin Univ. Phys., O.R.N.L., 1951-. Book: Contributor to Fast Neutron Physics (Marion and Fowler, Editors, Interscience Publishers, Inc., 1959-60). Society: A.P.S.
Nuclear interest: Low-energy nuclear physics.
Address: Building 5500 X-10 Site, Oak Ridge National Laboratory, Oak Ridge, Tennessee, U.S.A.

JOHNSON, D. M. Formerly Nucl. Specialist, Kaiser Engineers.
Address: Kaiser Engineers, Division of Kaiser Industries Corporation, Kaiser Centre, Oakland, California 94604, U.S.A.

JOHNSON, Dag G., Mech. Eng., Dr. Tech. (Tech. Univ. Norway). Born 1909. Educ.: Tech. Univ. Norway. Chief Mech. Eng. Norsk Hydro, Rjukan, 1949; Prof., Tech. Univ. Norway (steam technics, industrial heat technics, combustion), 1950. Tech. Director, Raufoss Ammo. Fact., Raufoss, 1953-56; Prof., Tech. Univ. Norway, 1956-60; Prof. Delft Technol. Univ., 1960-61; Prof. Tech. Univ. Norway, 1961-. Member, Board of Directors, Noratom, Oslo. Book: Turbulent Flow Friction and Heat Transfer in Tubes (Thesis, 1948). Societies: Det Kongelige Norske Videnskabers Selskab (Trondheim); Norges Tekniske Vitenskapsakademi (Trondheim); Inst. of Fuel (London); Combustion Inst. (Pittsburgh, U.S.A.).
Nuclear interests: Power reactors; Heat transfer; Fluid flow; Steam cycles.
Address: Laboratory of Thermodynamics, Institute of Steam Technics and Combustion, The Technical University of Norway, Trondheim, Norway.

JOHNSON, David L., Dr. Prof. and Head, Elec. Eng., Formerly Member, Nucl. Sci. and Eng. Dev. Com., now Member, Nucl. Advisory Com., Louisiana Polytechnic. Inst.
Address: Louisiana Polytechnic Institute, Tech. Station, Ruston, Louisiana, U.S.A.

JOHNSON, Douglas M. Manager, Finance, Nucl. Energy Div., Gen. Elec. Co., -1967; Vice Pres., Finance, and Treas., United Nucl. Corp., 1967-.
Address: United Nuclear Corporation, 1730 K Street, N.W., Washington 6, D.C., U.S.A.

JOHNSON, Edway Richard, B.S. (Chem.). Born 1927. Educ.: Bowling Green State Univ.

Asst. Res. Chem., Central Res. Div., Monsanto Chem. Co., Dayton, Ohio, 1951; Sect. Head, Chem. Development, Nat. Lead Co. of Ohio, U.S.A.E.C. Contract Operator, Cincinnati, Ohio, 1952-57; Tech. Director (1957-58), Asst. to Gen. Manager (1959-60), Gen. Sales Manager (1961-62), Davison Chem. Div., Nucl. Reactor Materials Plant, W. R. Grace and Co., Erwin, Tennessee; Vice-Pres., Nucl. Fuel Services, Inc., Wheaton, Maryland, 1963-67; Pres., E. R. Johnson Assocs., Fairfax, Virginia, 1967-. Nucl. Material Control Com., Atomic Ind. Forum; Exec. Com. and Chairman Safeguards Com., Inst. Nucl. Materials Management; Exec-Com., Sect. Com. N5, Methods of Nucl. Material Control, U.S.A.S.I. Societies: A.C.S.; A.I.Ch.E.; A.N.S.; A.S.M.; American Ordnance Assoc.; Atomic Ind. Forum; Inst. Nucl. Materials Management; U.S.A.S.I.
Nuclear interests: Management and technical consultancy.
Address: 9307 Convento Terrace, Fairfax, Virginia 22030, U.S.A.

JOHNSON, Gerald Woodrow, B.S., M.S., Ph.D. Born 1917. Educ.: Western Washington State Coll., Washington State and California Univs. Assoc. Phys., Brookhaven Nat. Lab., 1949-51; Captain, U.S.N., Washington, D.C., 1951-53; Special Asst. to Director of Res., U.S.A.E.C., 1953; Test Director and Assoc. Director (1953-61), Lawrence Rad. Lab., California Univ.; Asst. to Sec. of Defence (Atomic Energy), 1961-63; Chairman, Military Liaison Com., U.S.A.E.C., 1961-63; Assoc. Director for Plowshare, Lawrence Rad. Lab., California Univ., 1963-66; Director of Navy Labs., U.S.N., 1966-68; Director, Field Development, Gulf Gen. Atomic, 1968-. Societies: A.P.S.; A.G.U.; American Ordnance Association.
Nuclear interests: Phenomenology of nuclear explosions. Directing and participating in measurements and tests of nuclear explosives. Recent work in underground and cratering explosions with explorations and publications in peaceful uses (Project Plowshare). Direction and management of test and experimental programmes in these fields.
Address: 1220 Tanley Road, Silver Spring, Maryland 20904, U.S.A.

JOHNSON, Gifford K. Pres., Southwest Centre for Advanced Studies, Graduate Res. Centre of the Southwest.
Address: Graduate Research Centre of the Southwest, P.O. Box 30365, Dallas, Texas 75230, U.S.A.

JOHNSON, Harold Robert, B.Sc. (Eng.), F.I.E.E. Born 1913. Educ.: London Univ. Radar Res., Ministry of Supply, Air Ministry, 1939-53; Chief Electronic Eng., Vickers Armstrong Ltd., 1953-56; Asst. Director, Overseas Liaison, U.K.A.E.A., 1956-59; Deputy Director, R. and D. Dept. (1959-61), Director, Marchwood Eng. Labs. (1961-), C.E.G.B.
Address: Central Electricity Generating Board, Marchwood Engineering Laboratories, Marchwood, Southampton, Hants., SO4 4ZB, England.

JOHNSON, J. H., Dr. Head, Nucl. Sci. Group, Res. Centre, Ling-Temco-Vought, Inc.
Address: Ling-Temco-Vought, Inc., Post Office Box 6144, Dallas, Texas 75222, U.S.A.

JOHNSON, James R., B. Cer. E., M.Sc., Ph.D. (Ohio State). Born 1923. Educ.: Ohio State Univ. Tech. Adviser, Ceramics, O.R.N.L., 1951-56; at present Director, Phys. Sci., Central Res., 3M Co. Societies: American Ceramic Soc.; A.N.S.
Nuclear interests: Reactor materials (ceramics); Radiochemistry; Radiation.
Address: 3M Co., St. Paul, Minnesota, U.S.A.

JOHNSON, James Steven, Jr., B.S., Ph.D. Born 1921. Educ.: The Citadel, North Carolina Univ. Chem., O.R.N.L., 1949-. Societies: A.C.S.; RESA.
Nuclear interests: Aqueous solution chemistry of heavy elements and fission products; Hydrolytic polymerisation; Ultracentrifugation and light scattering; Ion exchange properties of hydrous oxides; Hyperfiltration.
Address: 918 W. Outer Drive, Oak Ridge, Tennessee, U.S.A.

JOHNSON, James T., B.S. (E.E.). Born 1927. Educ.: Illinois Univ. Sales Manager, Rad. Instrument Development Labs., Inc., Melrose Park, Illinois, 1960-. Societies: Inst. of Radio Eng.; A.N.S.
Nuclear interests: Management with regard to sales and establishment of requirements and specifications for new instruments.
Address: Radiation Instrument Development Labs., Inc., H501 W North Avenue, Melrose Park, Illinois, U.S.A.

JOHNSON, Jesse C., B.S. (Mining Eng. and Geology). U.S.A.E.C. Service Award, 1956. Born 1894. Educ.: Washington Univ. With U.S.A.E.C., 1948-July 1963 (Director, Div. of Raw Materials, 1949-63), Chem., Geological Survey, U.S. Dept. of the Interior. Chairman, internat. meeting on Uranium Resources, Buenos Aires, Nov. 1960. Societies: A.I.M.M.E.; Mining and Metal. Soc. of America.
Nuclear interests: Exploration for and production of uranium, thorium, and other raw materials used in nuclear energy programmes.
Address: 5900 Bradley Boulevard, Bethesda 14, Maryland, U.S.A.

JOHNSON, John Lowell, B.S., M.S., Ph.D. Born 1926. Educ.: Montana State Coll. and Yale Univ. Sen. Sci., Westinghouse Elec. Corp., Atomic Power Dept., 1954-64; On loan to Princeton Univ. Plasma Phys. Lab., 1955-; Fellow Eng., Westinghouse Elec. Corp., Res. Lab., 1964-. Societies: A.P.S.;A.A.P.T.
Nuclear interest: Thermonuclear power.

JOHNSON, K., Prof. Director, Bau der Speicherringe, C.E.R.N., 1966-.
Address: European Organisation for Nuclear Research (C.E.R.N.), Meyrin, 1211 Geneva 23, Switzerland.

JOHNSON, Keith David Bebb, M.A., B.Sc. Born 1923. Educ.: Oxford Univ. S.S.O., Chem. Div., A.E.R.E., Harwell (1950-51), Sen. Chem., R. and D. Branch, Capenhurst, (1951-58), Branch Head, Process Technol. Branch, Chem. Eng. and Process Technol. Div., A.E.R.E., Harwell (1958-), U.K.A.E.A. Nuclear interests: Separation of stable isotopes; Diffusion plants; Nuclear fuel reprocessing; Chemical engineering research; Management of radioactive wastes: Graphite technology.
Address: Chemical Engineering Division, Atomic Energy Research Establishment, Harwell, Didcot, Berkshire, England.

JOHNSON, Kenneth P. Maintenance Eng., Enrico Fermi Atomic Power Plant.
Address: Enrico Fermi Atomic Power Plant, Post Office Box 725, Monroe, Michigan, U.S.A.

JOHNSON, Leland P., B.S., M.S., Ph.D. Born 1910. Educ.: Iowa State Univ. Prof. Biol. (1947-), Dept. Chairman, (1956-), Co-ordinator Sci. Div. (1966-), Drake Univ.; Iowa Board Basic Sci. Examiners, 1955-. Pres., Assoc. Midwest Coll. Biol. Teachers. Societies: A.A.A.S.; Protozoology Acad.; American Microscopic Soc.; N.Y. Acad. Sci.
Nuclear interest: Tracer studies.
Address: Biology Department, Drake University, Des Moines 2, Iowa, U.S.A.

JOHNSON, Leslie E. Director, Neutronics Lab.
Address: Neutronics Laboratory, Tinley Park (Chicago Suburb), Illinois, U.S.A.

JOHNSON, Lyall E. Asst. Director for Facilities and Materials Licensing, Div. of Licensing and Regulation (-1964), Acting Director, Div. of Materials Licensing (1964-), U.S.A.E.C.
Address: Division of Materials Licensing, U.S. Atomic Energy Commission, Washington D.C. 20545, U.S.A.

JOHNSON, Max T., B.S. (E.E.). Born 1921. Educ.: Utah State Univ. Design Eng., Bureau of Reclamation, 1946-51; Operations Shift Supervisor, STR-Mark I, 1951-53; Operations Manager, SIW-ARCO, 1953-57; Operations Manager, Shippingport Atomic Power Station, 1957-58; Operations Manager, AIW Project, 1958-61; Gen. Manager, Westinghouse Testing Reactor, Waltz Mills, Pa., 1961-62; Test Operations Manager (1962-64), NRX Project Manager (1964-), NERVA Project, Westinghouse, Astronucl. Lab., Large, Pa. Society: A.N.S. (Charter Member).
Nuclear interests: Reactor plant operations and management.
Address: 935 Sunset Drive, Greensburg, Pa., U.S.A.

JOHNSON, Orlen N., B.S. (Minnesota), D.D.S. (Minnesota), M.S. (Radiol. Health, Wayne). Educ.: Minnesota and Wayne Univs. Director, Professional Educ., Dental X-Ray Programme, State Assistance Branch, Div. of Radiol. Health, U.S.P.H.S., Rockville, Maryland, 1964-65; Radiol. Health Specialist, Div. of Radiol. Health, Nebraska State Health Dept., 1965-.
Address: Division of Radiological Health, Nebraska State Health Department, Lincoln, Nebraska, U.S.A.

JOHNSON, Philip Carl, Jr., B.S., M.D. Born 1924. Educ.: Michigan Univ. Asst. Director, Radioisotope Unit, Instructor of Medicine, Univ. Hospital, Michigan Univ., 1954-55; Chief, Radioisotope Service and Asst. Director, Professional Services for Res., V. A. Hospital, Oklahoma City, 1955-60; Asst. Prof. Medicine (1955-58), Assoc. Prof. Medicine (1958-60), Oklahoma Univ. Medical School; Director, Radioisotope Lab., Methodist Hospital, Houston, 1960-; Assoc. Prof. Medicine (1960-67), Prof. Medicine (1967-), Baylor Univ., Houston.Rad. Health Officer, Baylor Univ. Coll. of Medicine, 1964-. Societies: Southern Soc. for Clinical Res.; American Federation for Clinical Res.; Fellow, American Coll. of Physicians; Soc. for Exptl. Biol. and Medicine; A.M.A.; Southern Medical Soc.; Central Soc. for Clinical Res.; Soc. of Nucl. Medicine; Fellow, American Coll. of Angiology; A.A.A.S. Nuclear interests: Biological tracer experiments; Clinical applications; Endocrinology and cardiovascular uses.
Address: The Methodist Hospital, Radioisotope Laboratory, 6516 Berther, Houston, Texas 77025, U.S.A.

JOHNSON, Ralph E., M.D. Chief, Rad. Branch, Nat. Cancer Inst., U.S. Dept. of Health, Educ., and Welfare.
Address: U.S. Department of Health, Education and Welfare, National Cancer Institute, 9000 Wisconsin Avenue, Bethesda 14, Maryland, U.S.A.

JOHNSON, Thomas Hope, A.B., M.S. (Amherst Coll.), Ph.D. (Yale). Born 1899. Educ.: Amherst Coll. and Yale Univ. Chairman, Phys. Dept., Brookhaven Nat. Lab., 1947-51; Director, Res. Div., U.S.A.E.C., 1951-57; Vice-Pres. and Res. Manager, Raytheon Co., 1957-65. Editorial Board, Phys. Rev., A.P.S. Book: Translation, Cosmic Radiation (ed. W. Heisenberg, Dover Publications, 1946). Societies: A.A.A.S.; A.P.S.; A.G.U.
Nuclear interests: Thermonuclear reactors; Nuclear physics research.
Address: Heads Corner, Denmark, Maine 04022, U.S.A.

JOHNSON, W. C., Dr. At Notre Dame Univ. Nuclear interest: Photonuclear reactions. Address: University of Notre Dame, Notre Dame, Indiana, U.S.A.

JOHNSON, W. P. Cyclotron Design, Nucl. Phys. Staff, Phys. and Astronomy Dept., Michigan State Univ. Address: Cyclotron Design, Nuclear Physics Staff, Physics and Astronomy Department, Michigan State University, East Lansing, Michigan, U.S.A.

JOHNSON, Warren C., B.S., M.A., Ph.D., D.Sc. (hon.). Born 1901. Educ.: Kalamazoo Coll., Clark and Brown Univs. Consultant, U.S.A.E.C., Union Carbide Nucl. Co.; Member, Board of Trustees, Inst. for Defence Analyses; Member, Board of Trustees, Mellon Inst.; Vice-Pres. and Prof. Chem., Chicago Univ. Member, Board of Directors: A.N.S., Atomic Ind. Forum, O.R.I.N.S. Chairman, Gen. Advisory Com., U.S.A.E.C.; 1956-60; Chairman, Com. of Sen. Reviewers, U.S.A.E.C., 1949-56; Dean of Phys. Sci. (1955-58), Vice-Pres., Special Sci. Programmes (1958-67), Vice-Pres. Emeritus (1967-), Chicago Univ. Societies: Fellow, A.N.S.; Fellow, A.A.A.S.; Consultant, N.S.F.; A.C.S. Nuclear interests: Management; Nuclear chemistry; Declassification of information; Chemical processing of nuclear fuel; Education in nuclear fields. Address: 5825 Dorchester Avenue, Chicago, Illinois 60637, U.S.A.

JOHNSON, Wendell P. Asst. Supt., then Plant Supt., Nucl. Power Plant, (1964-67), now Manager, Operations (1967-), Yankee Atomic Elec. Co. Member, Tech. Advisory Panel on Peaceful Use Safeguards, U.S.A.E.C. Address: Yankee Atomic Electric Co., 441 Stuart Street, Boston 16, Massachusetts, U.S.A.

JOHNSON, Wesley M., B.S. Educ.: Tennessee Univ. Director, Tech. Liaison Div. and Asst. Manager, Tech. Operations (1957-59), Deputy Manager, (1959-64), Manager (1964-), New York Operations Office, U.S.A.E.C. Address: U.S.A.E.C., New York Operations Office, 376 Hudson Street, New York, New York 10014, U.S.A.

JOHNSON, Wilfrid E., B.S. (Mech. Eng.), Prof. M.E., Sc.D. (hon.). Born 1905. Educ.: Oregon State Coll. Professional Eng., New York State; Manager, Design and Construction (1948-51), Asst. Gen. Manager (1951-52), Gen. Manager (1952-66), Hanford Atomic Products Operation, G.E.C.; Commissioner, U.S.A.E.C., 1966-. Societies: Fellow, A.S.M.E.; Assoc. Member, A.S.E.E.; A.N.S.; Newcomen Soc. Nuclear interest: General management, nuclear industry. Address: 12711 River Road, Potomac, Maryland 20854, U.S.A.

JOHNSON, William, B.Sc. (Hons.). Educ.: Western Australia. Geologist, Geological Survey of Western Australia, 1946-50; Geologist, Metropolitan Water, Sewerage and Drainage Board, Sydney, 1950-55; Geologist (1955-57), Sen. Geologist (1957-63), Geological Survey of South Australia; Consultant Geologist, W. Johnson and Assoc. Pty. Ltd., 1963-. Societies: Royal Soc., Western Australia; Roy. Soc., South Australia; Geological Soc., Australia; Australian Inst. Mining Metal.; A.A.P.G. Nuclear interests: Exploration, assessment and evaluation of radioactive mineral deposits. Address: 5 Horn Court, Walkerville, South Australia.

JOHNSON, Woodrow Eldred, B.S., M.Sc., Ph.D. (Major field, Phys.). Born 1917. Educ.: Hamline and Brown Univs. Project Manager, Pennsylvania Advanced Reactor (1949-), Director of Projects, Atomic Power Dept. (1958-), Formerly Gen. Manager, Atomic Power Div. (1961-), Gen. Manager, Astronucl. Labs. (1964-68), Vice Pres. (1967-), Head, Astronucl.-Underseas Divs. (1968-), Westinghouse Elec. Corp. Societies: A.N.S.; A.P.S.; Atomic Ind. Forum. Nuclear interest: Reactor design and manufacture. Address: Westinghouse Electric Corporation, Astronuclear Laboratory, P.O. Box 10864, Pittsburgh, Pennsylvania 15236, U.S.A.

JOHNSSON, Karl O., Jr., B.Ch.E., M.S. (Eng.). Born 1920. Educ.: Florida Univ. Chem. Eng., Carbide and Carbon Chem. Corp., 1947-50; Sen. Chem. Eng., O.R.N.L., 1950-. Societies: A.C.S.; A.I.Ch.E.; A.N.S.; RESA. Nuclear interests: The use of computer systems for technical information and data handling. Information storage and retrieval about processes for the desalting of seawater with nuclear reactor heat sources. Equipment for desalting seawater by the distillation process. Address: 110 Pomona Road, Oak Ridge, Tennessee 37830, U.S.A.

JOHNSTON, Bradley M., M.A. (Maths.). Born 1929. Educ.: California Univ., Berkeley. Mathematician, Lawrence Rad. Lab., Livermore, California. Nuclear interest: Applications of mathematical analysis to nuclear engineering. Address: Lawrence Radiation Laboratory, Livermore, California, U.S.A.

JOHNSTON, Gordon Basil, Assoc. of the Sydney Tech. Coll., B.Sc., M.Sc. Born 1929. Educ.: New South Wales Univ., Lysaghts Works Pty. Ltd., Newcastle; Tech. Officer then Lecturer, New South Wales Univ.; Lecturer, Newcastle Univ. Nuclear interests: Neutron diffraction investigations of the magnetic and chemical structures of intermediate phases containing transition metals. Radiation damage in

metallic materials.
Address: Newcastle University, Tighes Hill 2N, Newcastle, N.S.W., Australia.

JOHNSTON, T. R. Transport, Eng. Services Div., Chalk River Nucl. Res. Labs., A.E.C.L.
Address: Chalk River Nuclear Research Laboratories, Chalk River, Ontario, Canada.

JOHNSTONE, A. Com. Member, Risley Branch, Nucl. Eng. Soc.
Address: U.K.A.E.A., Risley, Warrington, Lancs., England.

JOLCHINE, Yves, D. ès Sc. Born 1922. Educ.: Montpellier Univ. Head, Etudes et Productions Spéciales Sect., Dépt. des Radioéléments, C.E.A. Societies: Sté. Chimique de France; A.N.S.
Nuclear interests: Radioisotopes production and applications; Radiochemistry.
Address: 14 rue Weber, Paris 16, France.

JOLY, René Louis, Ing. (E.P.), D. es Sc. Born 1924. Educ.: Ecole Polytechnique, Paris and E.T.H., Zürich. Head, Sect. des Mesures Neutroniques Fondamentales, French A.E.C. Societies: Sté. Française de Physique; A.P.S.
Nuclear interests: Nuclear physics at low energy; Experimental neutronics.
Address: S.M.N.F./D.R.P., C.E.N. Saclay, B.P. No. 2, Gif-sur-Yvette, (S.et. O.), France.

JONAS, Heinz, Dr.-Ing. Member, Arbeitskreis III/3 Werkstoffe Bauteile, Deutsche Atomkommission.
Address: Federal Ministry for Scientific Research, 46 Luisenstrasse, 532 Bad Godesberg, Germany.

JONCKHEERE, Charles Jean, Elec. Eng. Born 1933. Educ.: Ghent Univ. At Centre d'Etudes de l'Energie Nucléaire, 1956-.
Nuclear interest: Electronics applied to reactor control and nuclear measurements.
Address: Centre d'Etudes de l'Energie Nucléaire, 31 rue Belliard, Brussels 4, Belgium.

JONCKHEERE, Edouard, Metal. Eng. Born 1929. Educ.: Louvain Univ. Centre d'Etudes Nucl., Belgium, 1955-62; Chef de Département, Belgonucl., Brussels, 1962-. Society: Sté. Française de Metal.
Nuclear interests: Metallurgy; Processing of metals; Fuel fabrication; Fuel cycle.
Address: 261 Boeretang, Mol-Donk, Belgium.

JONES, A., A.R.I.C. Radiochem. Lab., Nucleonics Lab., Plymouth Coll. of Technol.
Address: Plymouth College of Technology, Nucleonics Laboratory, Portland Square, Plymouth, Devon, England.

JONES, A. H., M.I.Mech.E. Formerly Project Eng. for Sizewell, now Asst. Project Group Director, Midlands Project Group, C.E.G.B.
Address: Central Electricity Generating Board, Midlands Project Group, P.O. Box 314, 341 Bournville Lane, Birmingham 30, England.

JONES, A. W. Managing Director, Fleming Instruments Ltd.; Chairman, Nucleonics Group, Scientific Instrument Manufacturers Assoc. of Great Britain Ltd.
Address: Fleming Instruments Ltd., Caxton Way, Stevenage, Herts., England.

JONES, Allan C., Jr., M.S. (Eng. Sci. and Nucl. Eng.), B.S. (Architectural Eng.). Born 1927. Educ.: California (Berkeley) and Colorado Univs. Manager, Exptl. Beryllium Oxide Reactor Facility, N.R.T.S., Idaho, 1953-66; Asst. to the Pres., Idaho Nucl. Corp., 1966-. Societies: A.N.S.; American Soc. of Planning Officials.
Nuclear interests: Management; Organisation and planning.
Address: 2006 McKinzie Avenue, Idaho Falls, Idaho 83401, U.S.A.

JONES, Allan E. Born 1906. Educ.: Oregon State Coll. Manager, Grand Junction Office, U.S.A.E.C.
Nuclear interest: Administers programmes for the acquisition and production of uranium concentrates throughout the western U.S.A., including mining and appraisal of uranium and thorium resources.
Address: United States Atomic Energy Commission, Grand Junction Office, Grand Junction, Colorado 81501, U.S.A.

JONES, Alun Richard, B.Sc., M.Sc. Born 1928. Educ.: Bristol and McGill Univs. Jun. Res. Officer Electronics Branch (1954-56), Asst. Res. Officer Electronics Branch (1956-58), Asst. Res. Officer Rad. Dosimetry Branch (1958-61), Assoc. Res. Officer Rad. Dosimetry Branch (1961-), A.E.C.L. Societies: H.P.S.; Canadian Assoc. Phys.
Nuclear interest: Development of instruments and techniques for measurements in radiation dosimetry with special reference to health physics problems.
Address: 13 Highland Crescent, Deep River, Ontario, Canada.

JONES, Andrew Ross, B.E.E. (Elec. Eng.), M.S. (Elec. Eng.). Born 1921. Educ.: Clemson and Pittsburgh Univs. Student Eng. (1947), then res. in Elec. Utility Dept., and study group (1951), Application Eng., Ind. Atomic Power Study Group (1952), and later Supervisory Eng., Application Eng., Atomic Power Div., then Manager, Preliminary Plant Eng. (1958), Manager, Advanced Development and Planning (1964), Westinghouse Elec. Corp. Member, Industry Advisory Com., American Bureau Shipping; Member, Reactor Safety Com., Atomic Ind. Forum. Societies: I.E.E.E.; A.N.S.; Atomic Ind. Forum; Soc. of Naval Architects and Marine Eng.

Nuclear interests: Improved and advanced fast reactor design, total power plant design and economics for electric utilities, desalination, and other process heat, and marine propulsion.
Address: Atomic Power Department, Westinghouse Electric Corp., Pittsburgh, Pennsylvania, U.S.A.

JONES, Barclay George, B.Sc. (Mech. Eng.), M.S. (Nucl. Eng.), Ph.D. (Nucl. Eng.). Born 1931. Educ.: Illinois and Saskatchewan Univs. Eng., A.E.C.L., 1954; Athlone Fellow, English Elec. Co., Rugby and A.E.R.E., Harwell, 1954-57; Eng., Nucl. Div., Canadair Ltd., 1957-58; Eng., Atomic Power Dept., Westinghouse Elec. Corp., 1958; Graduate Asst. (1958-60), Graduate Coll. Fellow (1960-63), Instructor (1963-66), Asst. Prof., Nucl. and Mech. Eng. (1966-), now Member, Exec. Com., Nucl. Eng. Programme, Illinois Univ.; Res. Assoc., Argonne Nat. Lab., 1961; Member, Tech. Staff (1965), Consultant (1965-), T.R.W. Systems. Societies: A.N.S.; Eng. Inst. of Canada; Assoc. Member, A.S.M.E.; Graduate Member, Inst. Mech. Eng., London.
Nuclear interest: Reactor heat transfer and coolant systems two-phase flow.
Address: Nuclear Engineering Programme, Illinois University, Urbana, Illinois 61801, U.S.A.

JONES, Cecil E., Project Eng., (LACBWR), Chicago Operations Office, U.S.A.E.C.
Address: c/o Chicago Operations Office, U.S.A.E.C., 9800 South Cass Avenue, Argonne, Illinois 60439, U.S.A.

JONES, Charles F., B.S.M.E. Born 1927. Educ.: Worcester Polytech. Inst. Vice Pres. and Gen. Manager, Eng. Div., NUS Corp., 1967. Vice Chairman, Power Div., A.N.S. Society: A.S.M.E.
Nuclear interest: Primary interest is in the field of nuclear power, including the planning, procurement, design, construction and operation of nuclear-fueled, electric generating plants.
Address: NUS Corporation, 1730 M Street, N.W., Washington, D.C. 20036, U.S.A.

JONES, Daniel T., Jr., B.S. (Ch.E.). Born 1915. Educ.: Alabama Polytech. Inst. Eng., Parma Res. Centre, Union Carbide Corp., Cleveland, Ohio. Societies: RESA; A.N.S.
Nuclear interests: Reactor design and reactor hardware.
Address: 41 Oak Road, Rocky River 16, Ohio, U.S.A.

JONES, David Hunter, A.B., M.S. Born 1931. Educ.: Oberlin Coll. and Washington Univ. Sci., Bettis Atomic Power Lab., Westinghouse Elec. Corp. Society: A.N.S.
Nuclear interest: Nuclear design.
Address: 150 McClellan Drive, Pittsburgh 36, Pennsylvania, U.S.A.

JONES, Edward Hywel, Dipl. Eng. (Liverpool). Born 1902. Educ.: Liverpool Univ. Chief Generation Eng., British Elec. Authority, 1948; Deputy Eng. (1955-62), Formerly Chief Eng. (1962-), now Gen. Manager (Eng. Services), South of Scotland Elec. Board. Societies: Inst. Mech. Eng.; Inst. Elec. Eng.
Nuclear interests: Management and operation of reactors.
Address: South of Scotland Electricity Board, Inverlair Avenue, Glasgow, S.4, Scotland.

JONES, Emerson, B.Sc. (Elec. Eng.), M.Sc. (Phys.), Ph.D. (Phys.). Born 1924. Educ.: Nebraska Univ. Staff Member, Los Alamos Sci. Lab., 1948-54; Formerly Special Asst. to Gen. Manager, Consumers Public Power District, 1954-; Special Consultant to Gen. Manager, Hallam Nucl. Power Facility, operated for U.S.A.E.C. by Consumers Public Power District; Pres., Tech. Management, Inc.; Head, Nucl. Power and Advanced Systems Development, R. W. Beck and Associates, 1966-. Societies: A.P.S.; A.N.S.
Nuclear interest: Management.
Address: 3152 South 25 Street, Lincoln, Nebraska, U.S.A.

JONES, Frank Culver, B.A. (Rice Inst.), M.S. (Chicago), Ph.D. (Chicago). Born 1932. Educ.: Rice Inst. and Chicago Univ. Res. Assoc. (1960-61), Instructor (1961-63), Phys. Dept., Princeton Univ. Nat. Res. Council Resident Res. Assoc., N.A.S., 1963-65; Staff Phys., Goddard Space Flight Centre, 1965-. Societies: A.P.S.; A.A.P.T.
Nuclear interest: Cosmic-ray physics.
Address: Theoretical Studies Laboratory, National Aeronautics and Space Administration, Goddard Space Flight Centre, Greenbelt, Maryland, U.S.A.

JONES, Frank Richard Reece, B.Sc. (Eng.). Born 1921. Educ.: London Univ. Station Supt., Hinkley Point Nucl. Power Station, C.E.G.B., 1967. Societies: M.I.E.E.; M.I.Mech. E.; M.B.I.M.; B.N.E.S.
Nuclear interest: Nuclear power station management.
Address: Hinkley Point Power Station, near Bridgwater, Somerset, England.

JONES, Gwyn Owain, M.A., D.Sc. (Oxon.), Ph.D. (Sheffield). Born 1917. Educ.: Oxford and Sheffield Univs. Reader in Exptl. Phys. (1949-53), Prof. Phys. and Head Phys. Dept. (1953-), Queen Mary Coll., London Univ. Books: Co-author, Atoms and the Universe (1956); Glass (1956). Society: Inst. of Phys. and Phys. Soc.
Nuclear interests: Solid state physics, etc.
Address: Queen Mary College, London, E.1, England.

JONES, H. F., D.G.S., M.Inst.F., A.I.Mech.E. Overseas Member (Australia), Council Member, Inst. of Nucl. Eng.

Address: Institution of Nuclear Engineers, 147 Victoria Street, London, S.W.1, England.

JONES, Hardin Blair, Ph.D. Born 1914. Educ.: California Univ. Asst. Director, Donner Lab., 1948-; Prof. Medical Phys. and Physiology, 1954-; Consultant, U.S. Public Health Service, 1959-; Sci. Adviser, Kaiser Foundation Res. Inst., 1962-. Societies: Rad. Res. Soc.; Eugenics Soc.; A.A.A.S.; American Physiological Soc.; Gerontological Soc.
Nuclear interest: Effects of radiation.
Address;: Donner Laboratory, California University, Berkeley, California 94720, U.S.A.

JONES, Hugh, Former Deputy Supt., now Station Supt., Trawsfynydd Nucl. Power Plant, C.E.G.B.
Address: Trawsfynydd Nuclear Power Station, C.E.G.B., Trawsfynydd, Merionethshire, Wales.

JONES, I. R. Phys., Swiss Nat. Foundation for Sci. Res.
Address: Swiss National Foundation for Scientific Research, Laboratoire de Recherches sur la Physique des Plasmas, 2 avenue Ruchonnet, Lausanne, Switzerland.

JONES, Ian Puleston, B.Sc. Born 1922. Educ.: Wales Univ. Works Manager, Magnox Fuel, Production Dept., Springfields Works, U.K.A.E.A. Society: A.R.I.C.
Nuclear interest: Fuel element manufacture.
Address: 3 Woodland Avenue, Off Pope Lane, Penwortham, Preston, Lancs., England.

JONES, Ian Robertson, B.E., M.S., Ph.D., A.E.C. Special Fellowship in Nucl. Sci. and Eng., 1959-61. Born 1936. Educ.: Yale and California (Berkeley) Univs. Head, Thruster Development Sect. (1963-65), Asst. Manager, Nucl. Technol. Dept. (1965-67), Manager, Nucl. Technol. Dept. (1967-), TRW Systems. Society: Yale Eng. Associates.
Nuclear interests: Radioisotope power and propulsion, aerospace nuclear safety, and radioisotope measurement systems.
Address: TRW Systems, Bldg. 01, Room 1010, 1 Space Park, Redondo Beach, Calif. 90278, U.S.A.

JONES, J. T. Trustee, Southwest Atomic Energy Assoc.
Address: Southwest Atomic Energy Associates, 306 Pyramid Building, Little Rock, Arkansas, U.S.A.

JONES, Lieutenant Colonel Kenneth L. U.S. Air Force. Staff Member, Office of Atomic Programmes, Office of the Director of Defence Res. and Eng., U.S. Dept. of Defence.
Address: U.S. Department of Defence, Office of the Director of Defence Research and Engineering, Washington 25, D.C., U.S.A.

JONES, Kenneth Surridge, B.Sc. (Eng.). Born 1922. Educ.: Birmingham Univ. Member, Soc. of Instrument Technol. Society: A.M.I. Mech.E.
Nuclear interests: Design and application of all instruments for nuclear reactors and the logical extension of these to automatic control systems.
Address: 15 Malpas Drive, Pinner, Middlesex, England.

JONES, L. V. Director, Tech. Dept., Mound Lab.
Address: Mound Laboratory, Mound Road, Miamisburg, Ohio, U.S.A.

JONES, Lawrence W., B.S., M.S., Ph.D. Born 1925. Educ.: Northwestern and California (Berkeley) Univs. Prof. Phys., Michigan Univ.; Phys., Midwestern Univs. Res. Assoc.; Phys., Lawrence Rad. Lab., Berkeley; Consultant: Space Technol. Labs., Bendix Aerospace Div., Allis-Chalmers Corp.; Ford Foundation Fellow, C.E.R.N., Geneva; Guggenheim Foundation Fellow. Society: A.P.S.
Nuclear interests: Accelerator design; Luminescent chamber development; Spark chamber development; High-energy elementary particle experimental research; Cosmic ray physics.
Address: Physics Department, Michigan University, Ann Arbor, Michigan, U.S.A.

JONES, Lewis J., B. Met. Eng. Born 1924. Educ.: Fenn Coll., Cleveland. Supervisory Eng., P.W.R. Irradiation Effects Group, Bettis Atomic Power Lab., Westinghouse Elec. Corp., Pittsburgh; Supervisory Eng., Materials Fabrication and Evaluation, Advanced Materials Centre, Nucl. Materials and Equipment Corp.
Nuclear interests: Development of new and improved nuclear fuel materials, with particular emphasis on the high-temperature ceramic fuel materials, such as PuO_2-UO_2 and PuC-UC combinations. Specialty within field is the determination of irradiation effects on fuel materials.
Address: Nuclear Materials and Equipment Corporation, Advanced Materials Centre, Leechburg, U.S.A.

JONES, Norman S. Vice-Pres., Nucl. Data Inc.
Address: 100 W. Golf Road, Palatine, Illinois, U.S.A.

JONES, Peter Bernard, M.A., Ph.D. Born 1933. Educ.: Cambridge Univ. Sen. Student, Roy. Commission for the Exhibition of 1851, 1958-60; Res. Fellow, St. Catherine's Coll. (1960-65), Sen. Res. Officer, (1965-), Oxford Univ. Book: Optical Model in Nucl. and Particle Phys. (New York, John Wiley and Sons, 1963).
Nuclear interests: Elementary particle physics; High-energy accelerators; The production and decay of unstable particles studied by electronic techniques.

Address: Oxford University, Nuclear Physics Laboratory, Keble Road, Oxford, England.

JONES, Ralph John, B.S., M.B.A. Born 1923. Educ.: Kansas State and Rutgers Univs. Chief, Tech. Studies Branch, Safeguards and Materials Management Office, U.S.A.E.C., 1967-. Book: Selected Measurement Methods for Plutonium and Uranium in the Nucl. Fuel Cycle.
Nuclear interests: Nuclear materials safeguards; Inventory management.
Address: United States Atomic Energy Commission, Washington 25, D.C., U.S.A.

JONES, Reginald Victor, M.A., Dr. Phil. Born 1911. Educ.: Oxford Univ. Prof. Natural Philosophy, Aberdeen Univ. Chairman, Paul Instrument Fund Com., Roy. Soc.; Chairman, Electronics Res. Council, Ministry of Aviation; Sci. Advisor on Civil Defence, Scotland.
Societies: Roy. Soc., London; Inst. Phys. and Phys. Soc.; Roy. Soc., Edinburgh; Inst. Strategic Studies; Inst. Measurement and Control.
Nuclear interest: Crystal scintillators.
Address: Aberdeen University, Scotland.

JONES, Robert Murph, B.S., M.E. Born 1918. Educ.: Rice Univ., Houston Texas. Director, Nucl. Products, Monsanto Res. Corp.; Asst. Manager, Lockheed Nucl. Products, Lockheed Aircraft Corp.; Manager, Marketing, Nucl. Div.,A.C.F. Industries; Manager, Eng. Branch, Atomic Energy Div., Phillips Petroleum Co.; Sen. Eng., O.R.N.L.; Jun. Eng., Oak Ridge Y-12 Area. Society: A.N.S.
Nuclear interests: Management; Planning.
Address: P.O. Box 8, Sta. B, Dayton, Ohio 45407, U.S.A.

JONES, Ronald Christopher HOPE-. See HOPE-JONES, Ronald Christopher.

JONES, Royce J., B.S. Born 1920. Educ.: Baylor Univ. Phys., R. and D., electromagnetic isotope separation, Manhatten Eng. Project (O.R.N.L.) 1943-47; Res. Phys., azimuthally varing field cyclotrons, arc plasma and high magnetic fields, O.R.N.L., 1948-64; Prof., Cyclotron Inst., Texas A. and M. Univ. 1964-65; Asst. Director, Particle Accelerator Div. (1965-67), Director, High Energy Facilities Div. (1967-) Argonne Nat. Lab. Society: A.P.S.
Address: Argonne National Laboratory, 9700 South Cass Avenue, Argonne, Illinois 60439, U.S.A.

JONES, Stuart M., B.A.Sc. Born 1917. Educ.: Toronto Univ. Atomic Energy Div. (1956-58), Manager, Development Sect., Eng. Dept. (1958-59), Eng. Manager, Atomic Energy Div. (1959-61), Formerly Manager, Atomic Energy Dept. (1961-), now Asst. to Manager, Atomic Power Div., Canadian Westinghouse Co. Ltd. Society: A.N.S.
Nuclear interests: Development, design and manufacture of special components and systems for nuclear power reactors.
Address: Atomic Power Division, Canadian Westinghouse Company, Limited, Box 510, Hamilton, Ontario, Canada.

JONES, Thomas Alun, M.Sc. (Wales), F.R.I.C. Born 1919. Educ.: Wales Univ. Sen. Soil Chem., Sudan Govt. Service, 1945-53; Reader in Soil Sci. (1953-55), Head, Soil Res. and Survey Dept., Regional Res. Centre (1955-59), Imperial Coll. Tropical Agriculture, Trinidad; Director, Res., Coconut Industry Board, Kingston, Jamaica, 1959-62; Sen. Agronomist, Borax Consolidated Ltd., London, 1963-.
Societies: F.R.I.C.; British Soil Sci. Soc.; Internat. Soc. of Soil Sci.; Fertilizer Soc. of Great Britain; Soc. of Chem. Industry.
Nuclear interests: General application to problem of improved agricultural productivity.
Address: Borax House, Carlisle Place, London, S.W.1, England.

JONES, Thomas Francis, B.S.M.E. Born 1911. Educ.: Illinois Inst. Technol., O.R.S.O.R.T. and Maryland Univ. Bureau of Ships (1940-64), Head, Regulatory Branch, Nucl. Power Div., Naval Facilities Eng. Command (1964-), U.S.N. Programme Branch Manager, Nucl. Power Div., U.S. Navy Bureau of Yards and Docks, 1964-. Societies: American Soc. Naval Eng.; A.N.S.; N.S.P.E.
Nuclear interests: Safety matters associated with shore nuclear reactors and radioisotope power generators.
Address: 12217 Brookhaven Drive, Silver Spring, Maryland 20902, U.S.A.

JONES, Thomas V. Pres. and Chief Exec. Officer, Northrop Corp.
Address: Northrop Corp., 9744 Wiltshire Blvd., Beverly Hills, California, U.S.A.

JONES, William Morton, B.A., B.Sc., Ph.D. Born 1926. Educ.: Dublin and Reading Univs. Sub-Sect. Leader, A.E.I. Res. Lab., Aldermaston, Berkshire, 1959-63; Visiting Prof. Modern Phys., Algiers Univ., 1959; Phys. Centre d'Etudes Nucléaires, Fontenay-aux-Roses, 1964-67; Lecturer, Strathclyde Univ., 1967-.
Nuclear interest: Controlled thermonuclear fusion.
Address: Natural Philosophy Department, Strathclyde University, Glasgow C.1, Scotland.

JONG, Klaas Jozef de. See de JONG, Klaas Jozef.

JONG, W. E. van RIJSWIJK de. See van RIJSWIJK de JONG, W. E.

JONGE, C. W. de. See de JONGE, C. W.

JONKE, Albert A., B.Ch.E., M.S. Born 1920. Educ.: Fenn Coll. and Illinois Inst. Technol. At Argonne Nat. Lab., 1947-. Societies: A.I.Ch.E.; A.C.S.; A.N.S.; RESA.

Nuclear interests: Nuclear fuel reprocessing; Radioactive waste treatment; Applications of fluidisation in nuclear field.
Address: 264 Highview Avenue, Elmhurst, Illinois, U.S.A.

JONKER, Charles Christiaan, Ph.D. Born 1908. Educ.: Utrecht, Paris and Leyden Univs. Lecturer, Theoretical Phys. (1947-51), Prof. Theoretical Phys. (1951-), Director Phys. Lab., Free Univ., Amsterdam. Chairman, Nucl. Res. Dept., Organisation for Fundamental Res. of Matter; Chairman, Governing Board, Inst. Nucl. Res., Amsterdam; Member, Liaison Com. for C.E.R.N. Matters, and Member, Advisory Com. on Nucl. Affairs, Roy. Netherlands Acad. of Sci. and Letters.
Nuclear interest: Low-energy nuclear physics.
Address: 30 Watteaustraat, Amsterdam, Netherlands.

JONKER, Willem. Member, Specialised Sect. for Economic Questions in the Nucl. Field, Economic and Social Com., E.E.C. and Euratom.
Address: Euratom, Economic and Social Committee, 3 boulevard de l'Empereur, Brussels, Belgium.

JONNERBY, Erik, Civil eng. Born 1917. Educ.: Roy. Tech. Inst., Stockholm. Design Manager, Uddeholms A.B., Degersfors Järnverk. Pressure Vessel Commission IVA, Stockholm; Welding Commission IVA, Stockholm; ISO/TC85/SC3/WG3. Book: Nucl. Pressure Vessels (IIW XI A. Internat. Inst. of Welding).
Nuclear interest: Reactor design.
Address: Uddeholms A.B., Degerfors Järnverk, Degerfors, Sweden.

JONSSON, John Erik, M.E., D. Eng. (Hon., Rensselaer Polytech. Inst.), D.Sc. (Hon., Hobart and William Smith Colls. and Austin Coll.), Dr. Laws (Hon., Southern Methodist Univ.). Educ.: Rensselaer Polytech. Inst. Mayor, City of Dallas, Texas, 1964-69; Hon. Chairman, Board, Texas Instr. Inc.; Director, Republic Nat. Bank, Dallas; Director, Equitable Life Assurance Soc. of the U.S., New York; Director, Neiman-Marcus Co., Dallas; Tech. and Economic Consultant, Investors Management Co., Inc., Elizabeth, New Jersey; Chairman, Board of Governors, Southwest Centre for Advanced Studies. Member, Board of Trustees, Nat. Ind. Conference Board; Life Member, American Management Assoc.; Vice Chairman, Board of Trustees, Rensselaer Polytech. Inst., Troy; Member, Board of Trustees, Austin Coll.
Address: Southwest Centre for Advanced Studies, P.O. Box 30365, Dallas, Texas 75230, U.S.A.

JONSSON, Steingrimur, Master of Elec. Eng. Born 1890. Educ.: Polytech. Inst. Copenhagen. Director, Reykjavik Municipal Elec. System, 1921-61; Director, Sog Hydro Power Development, 1949-65. Hon. Pres., Assoc. Icelandic Power Works; Hon. Pres., Icelandic Light-tech. Soc. Papers: Many papers in Icelandic Eng. Proc. and in Yearbook of Assoc. of Elec. Power Works in Iceland. Societies: Icelandic Sci. Soc.; Assoc. Icelandic Elec. Eng.; Icelandic Light-tech. Soc.
Address: 73 Laufásvegur, Reykjavik, Iceland.

JORBA, Jean Paul PEREZ y. See PEREZ y JORBA, Jean Paul.

JORDAN, Denis Oswald, Ph.D., D.Sc. (London). Born 1914. Educ.: London, Nottingham, and Princeton Univs. Reader in Phys. Chem., Nottingham Univ., 1953; Angas Prof. Phys. and Inorganic Chem., Adelaide Univ., 1954-. Pres. (1959, 1961-62), now Councillor, Australian Inst. of Nucl. Sci. and Eng. Societies: Chem. Soc.; Faraday Soc.; Roy. Australian Chem. Inst.
Nuclear interests: Liquid metals; Irradiation of solids and polymerisation in solid state induced by radiation.
Address: 33 Craighill Road, St. George's, South Australia.

JORDAN, Edward Daniel, B.S., M.S., Ph.D. Born 1931. Educ.: Fairfield, New York and Maryland Univs. Instructor (part-time), New York Univ., 1955-56; Reactor Phys., Foster Wheeler Corp., 1955-57; Reactor Phys., U.S.A.E.C., 1957-59; Prof. Nucl. Eng. (1959-), Chairman, Nucl. Sci. and Eng. Dept. (1961-), Catholic Univ. Societies: I.E.E.E.; A.A.U.P.; A.N.S.; A.S.E.E.
Nuclear interest: Nuclear radiation engineering.
Address: 2915 Rittenhouse Street, N.W., Washington D.C. 20017, U.S.A.

JORDAN, Ernst Pascual, Dr. phil. Born 1902. Educ.: Göttingen Univ. Prof. Theoretical Phys., Hamburg Univ.; Member of Parliament, Bonn, 1957-61. Society: Akademie der Wissenschaften und der Literatur (Mainz).
Nuclear interests: Relations of nuclear physics to cosmology and geophysics.
Address: 123 Isestrasse, Hamburg 13, Germany.

JORDAN, Hermann L., Dr. rer. nat. Born 1922. Educ.: Munich T.H., California and Munich Univs., Aachen T.H. Direktor, Inst. für Plasmaphysik, K.F.A., Julich; On leave of absence at European Space Res. Inst., Frascati (Rome), Italy.
Nuclear interests: Plasma physics; Controlled thermonuclear fusion.
Address: Institut für Plasmaphysik der Kernforschungsanlage Jülich des Landes Nordrhein-Westfalen e.V., Jülich, Germany.

JORDAN, James Verdun, B.S., M.Sc., Ph.D. Born 1917. Educ.: British Columbia and Oregon State Univs. Asst. Prof. and Asst. Agricultural Chem. (1948-51), Assoc. Prof. and Assoc. Agricultural Chem. (1951-59), Prof. and Agricultural Chem. (1959-), Idaho Univ. Consultant, Concrete Chem. and Eng.,

Sydney, on leave from Idaho Univ. 1959-60. Societies: Idaho Acad. Sci.; Western Soc. of Soil Sci.
Nuclear interest: Using radiotracers in biological and soil systems.
Address: Unit 102 Glenhurst, 11 Yarranabbe Road, Darling Point, Sydney, N.S.W., Australia.

JORDAN, Kurt, Diplomingenieur, Phys. Born 1930. Educ.: Munich T.H. Rad. Measuring Instruments Dept. (1954-), Chief, development group in phys., electronics and nucl. medicine, (1955-), Frieseke and Hoepfner G.m.b.H.
Nuclear interests: Nuclear medicine, especially diagnostics; Electronics, especially all kinds of radiation measuring instruments.
Address: 6 Goethestrasse, Frauenaurach 8521, Germany.

JORDAN, Pierre A. Born 1921. Educ.: Swiss Federal Inst. Technol. Head, Nucl. Chem. Lab., Phys. Dept. (-1958), Head, Isotope Lab., Organic Chem. Dept. (1958-), Swiss Federal Inst. Technol.; Affiliate of the Argonne Nat. Lab.
Nuclear interests: Radiation and hot atom chemistry; Tracer techniques with radioactive and stable isotopes; Measurement of soft β-emitters (especially C 14 and T in organic substances).
Address: Organic Chemistry Institute, Eidgenössische Technische Hochschule, 6 Universitätstrasse, Zürich, Switzerland.

JORDAN, R. L. Manager, Relations and Auxiliary Operations, Atomic Power Equipment Dept., Gen. Elec. Co.
Address: General Electric Co., Atomic Power Equipment Department, P.O. Box 254, 175 Curtner Street, San Jose, California 94566, U.S.A.

JORDAN, Robert Gerald, B.S. Born 1920. Educ.: Ohio State Univ. Supt., Paducah Gaseous Diffusion Plant (1954), Supt., Y-12 Plant (1961), Supt., Oak Ridge Gaseous Diffusion Plant (1963), Nucl. Div., Union Carbide Corp. Society: A.C.S
Nuclear interests: Large-scale separation of isotopes; Manufacture of uranium hexafluoride; Fabrication of uranium materials.
Address: Union Carbide Corporation, Nuclear Division, P.O. Box P, Oak Ridge, Tennessee 37830, U.S.A.

JORDAN, Walter Harrison, A.B., M.S. (Oklahoma), Ph.D. (California Inst. Technol.). Born 1908. Educ.: Oklahoma Univ. and California Inst. Technol. Asst. Director, O.R.N.L., 1958-; Tennessee Univ., 1964-. Advisory Editor, Nucl. Safety. Societies: A.N.S.; A.P.S.
Nuclear interests: Nuclear reactors; Applications of nuclear energy to space exploration; Nuclear instrumentation.
Address: 881 W. Outer Drive, Oak Ridge, Tennessee, U.S.A.

JORDAN, Willard Clayton, B.A., M.S., Ph.D. Born 1922. Educ.: Miami and Michigan Univs. At Argonne Nat. Lab., 1951-54; Sen. Phys., Bendix Corp. Res. Labs. Div., 1954- (at present assigned to Lawrence Rad. Lab., Livermore, California). Societies: A.P.S.; A.A.P.T.; A.N.S.
Nuclear interests: Nuclear physics; Fission and fusion reactor development.
Address: Lawrence Radiation Laboratory, Building 131, P.O. Box 808, Livermore, California, U.S.A.

JORDANOV, Jordan POP-. See POP-JORDANOV, Jordan.

JORDY, George Y., M.B.A., B.S. Born 1932. Educ.; Carnegie Inst. of Technol., and Pennsylvania Univ. Asst. Director, Div. of Isotopes Development, U.S.A.E.C. Society: Inst. of Management Sci.
Nuclear interests: Programme planning and evaluation; Applications of operations analysis techniques to the solution of problems in the nuclear field.
Address: United States Atomic Energy Commission, Washington, D. C. 20545, U.S.A.

JORGENSEN, A. T. Vice Pres., Elec. Production, Long Island Lighting Co.
Address: Long Island Lighting Company, 250 Old Country Road, Mineola, New York, U.S.A.

JØRGENSEN, Frank Vernon, Elec. Eng. Born 1933. Educ.: Aarhus Tech. Coll., Denmark. Danish Isotope Centre, Div. Danish Acad. Tech. Sci., 1957-60; C.E.R.N., Geneva, 1960-62; Lawrence Rad. Lab., Berkeley, California, 1962-63; Danish Isotope Centre, Copenhagen, 1963-67; U.N. Development Programme, Bandoeng, Indonesia, 1962. Society: I.E.E.E.
Nuclear interests: Nuclear electronics; Applied nuclear physics; Semiconductor detectors.
Address: 5 Carinavoenget, Birkerod, Denmark.

JØRGENSEN, Jorgen Helms, Lic. agro. Born 1930. Educ.: Royal Veterinary and Agricultural Coll., Copenhagen. At Royal Veterinary and Agricultural Coll., Copenhagen, 1956-57; Atomic Energy Commission, Res. Establishment Riso, Roskilde, 1957-.
Nuclear interests: Application of ionising radiation in plant-breeding and genetics.
Address: Atomic Energy Commission, Research Establishment Riso, Roskilde, Denmark.

JORGENSEN, Theodore Prey, A.B., M.A., Ph.D. Born 1905. Educ.: Nebraska and Harvard Univs. Prof. Phys., Nebraska Univ., 1947-. Society: Fellow, A.P.S.
Nuclear interest: Slowing of ions in kilovolt range.
Address: Behlen Laboratory of Physics, University of Nebraska, Lincoln, Nebraska 68508, U.S.A.

JORTNER, Donald, Dr. Nucl. Co-ordinator, TRW Systems Group, TRW Inc.
Address: TRW Systems Group, One Space Park, Redondo Beach, California 90278, U.S.A.

JOSE, Rosario D. Sen. Teacher, Dept. Phys., Univ. Santo Tomas.
Address: University of Santo Tomas, Manila, Philippines.

JOSEPH, Claude. Chef de Groupe, Lab. de Recherches Nucléaires, Ecole Polytechnique, Lausanne Univ.
Address: Ecole Polytechnique, Lausanne University, 19 rue César-Roux, Lausanne, Switzerland.

JOSEPH, Joachim, Dr. Oberregierungsrat; Head, Oceanographic Sect., Deutsches Hydrographisches Inst., Hamburg; Sci. Director, Internat. Lab. of Marine Radioactivity, Monaco.
Address: International Laboratory of Marine Radioactivity, Musée Oceanographique, Monaco.

JOSEPHSON, Edward, Dr. Director, U.S. Army Rad. Lab., Natick Labs.
Address: U.S. Army Natick Laboratories, Radiation Laboratory, Natick, Massachusetts 01760, U.S.A.

JOSHEL, Lloyd Marvin, B.S., Ph.D (Chem.). Born 1914. Educ.: Illinois, Ohio State and Harvard Univs. Gen. Manager, Rocky Flats Div., The Dow Chem. Co., 1964-. Society: A.C.S.
Address: P.O. Box 888, Golden, Colorado 80401, U.S.A.

JOSLIN, Murray, B.S. (Elec. Eng.). Educ.: Iowa State Univ. Asst. to Exec. Vice Pres., Public Service Co. of Northern Illinois, 1951; Formerly Vice-Pres. in charge of nucl. activities, Commonwealth Edison Co. Member, Illinois Legislative Commission on Atomic Energy; Member, Illinois Atomic Energy Commission; Member-at-large, Board of Trustees, Argonne Univs. Assoc.; Chairman, Nucl. Standards Board N45 Com., United States of America Standards Inst. Societies: Fellow, I.E.E.E.; A.S.M.E.; A.N.S.
Address: 2121 78th Avenue, Elmwood Park, Illinois 60635, U.S.A.

JOSS, James Ormond, B.Sc. (Hons.). Born 1927. Educ.: Edinburgh Univ. Asst. Elec. Eng., British Electricity Authority and S.O.S.E.B. Edinburgh, 1950-55; Asst. Eng., Central Electricity Authority, Nucl. Design Branch, 1956-57; Asst. to Chief Elec Eng (1957), Senior Eng., Milan Office, (1958), N.P.P.C.; Asst. to Chief Elec. Eng., N.P.P.C. and Nucl. Power Group, 1959-. Society: Associate Member, I.E.E.
Nuclear interests: Design of reactor cores, fuel-handling equipment, burst cartridge detection equipment, and control devices reactor control.
Address: 8 Acacia Drive, Hale, Cheshire, England.

JOST, Res W., Dr. Phil.II. Born 1918. Educ.: Bern and Zurich Univs. Prof., Theoretical Phys., Swiss Federal Inst. of Technol.
Address: Swiss Federal Institute of Technology, Seminar of Theoretical Physics, 60 Hochstrasse, CH8044 Zurich, Switzerland.

JOUAN, René, D. ès Sc. Born 1911. Educ.: Paris Univ. Directeur, Adjoint à la Direction Générale, Saint-Gobain Techniques Nouvelles; Board, Sté. d'Etudes et de Travaux pour l'Uranium; Board, Centre Lyonnaise d'Applications Atomiques. Conseil de Surveillance, Indatom. Societies: A.N.S.; Deutsches Atomforum.
Nuclear interests: Nuclear chemistry; Nuclear chemical engineering.
Address: Saint-Gobain Techniques Nouvelles, 23 boulevard Georges Clemenceau, Courbevoie 92, France.

JOUANNAUD, Claude. Born 1921. Educ.: Inst. de Chimie Appliquée, Lille Univ. Directeur de l'Usine de la Rochelle de la Sté. des Terres Rares, 1948; Chef, Section Technologie Chimique, C.E.A., 1957; Chef Service Extraction du Plutonium de Marcoule, 1961.
Nuclear interests: Chemical engineering; Fuel processing; Plutonium production.
Address: Centre de Marcoule, Chusclan, Gard, France.

JOUGLET, Marcel. Sen. Officer, Tech. Dept., Crépelle et Cie.
Address: Crépelle et Cie, Porte de Valenciennes, Lille, (Nord), France.

JOUSSON, Henri. Editor, Industries Atomiques.
Address: Editions Sadesi, 39 rue Peillonnex, Geneva Chêne-Bourg, Switzerland.

JOUVEN, Pierre. Member, Commission Consultative pour la Production d'Electricite d'Origine Nucleaire, C.E.A.
Address: Commissariat a l'Energie Atomique, 29-33 rue de la Federation, Paris 15, France.

JOVANOVIC, Olga KOVACEVIC. See KOVACEVIC JOVANOVIC, Olga.

JOVER, Pierre. Born 1930. Educ.: Ecole Supérieure d'Electricité, Malakoff (Seine), France. At C.E.A., 1956-. Society: Sté. Française des Electriciens.
Nuclear interests: Nuclear reactors and power plants from a control point of view (radiation detectors, instrumentation, control mechanisms).
Address; Département d'Electronique, Centre d'Etudes Nucléaires de Saclay, B.P. No. 2, Gif-sur-Yvette, Seine-et-Oise, France.

JOVIC, Djordje, Ph.D. Born 1929. Educ.: Belgrade Univ. Solid State Phys. Lab., Boris Kidric Nucl. Sci. Inst., Belgrade. Society: Yugoslav Soc. Mathematicians and Phys. Nuclear interests: Neutron physics; Solid state physics.
Address: Boris Kidric Nuclear Sciences Institute, Belgrade, P.O. Box 522, Yugoslavia.

JOYCE, Barry, Editor, Atom.
Address: U.K. Atomic Energy Authority, 11 Charles II Street, London, S.W.1, England.

JOYET, Gustave. Dipl. Elec. Eng., Dr. Sci., Privat-Docent, Titular Prof. Born 1904. Educ.: Lausanne Univ. Head, Dosimetry and Protection Lab., Canton Hospital, Zürich, 1949-. Advisory Boards of Health Phys., Nucl. Medicine, Annales de Radiologie, and Minerva Nucleare.; Commission des Isotopes de l'Académie Suisse des Sciences Médicales. Societies: A.P.S.; Sté. Suisse de Physique; Physikalische Gesellschaft, Zürich. Nuclear interests: Dosimetry of beta- and gamma-rays; Health physics; Diagnostic in nuclear medicine; Whole-body counting.
Address: Dosimetry and Protection Laboratory, Kantonsspital, Zürich, Switzerland.

JOZEFOWICZ, Edward Tadeusz, D.Sc. Born 1936. Educ.: Tech. Univ., Lódz. Sci. worker, Phys. Chem. Dept., Tech. Univ., Lódz, 1956-58; Sci. worker, Reactor Phys. Dept., Inst. of Nucl. Res., Swierk, near Warsaw, 1958-. Nuclear interests: Reactor physics – static parameters; Neutron detectors and neutron activation detectors; Liquid scintillators; Radiochemistry.
Address: Anin 4B m.6 Ul. Zambrowska, Warsaw 89, Poland.

JOZEFOWICZ, Krystyna, M.Sc. (Chem.). Born 1936. Educ.: Tech. Univ., Lódz. Sci. worker, Dept. of Reactor Phys., Inst. of Nucl. Res., Swierk, near Warsaw, 1959-.
Nuclear interests: Reactor physics – static parameters; Activation and particle track detectors; Burst slug detection.
Address: Anin, 4 B m.6 Ul. Zambrowska, Warsaw 89, Poland.

JOZWIK, Boleslaw, Master's degree. Born 1916. Educ.: Warsaw Tech. Univ. In industrial design office,-1960; In nucl. technic design office, Proatom, Warsaw, 1960-. Books: 4 books on building.
Nuclear interests: Radiation protection, shielding design and building problems in nuclear technic, especially the research and design of shieldings from reinforced and prestressed concrete.
Address: 30B m.6 ui. Zorzy, Warsaw, Poland.

JU, Schobern, Dr. Resident Rep. for Formosa to I.A.E.A.
Address: 12/IV Marokannergasse, 1030 Vienna, Austria.

JUANG, Tzo-chuan, B.S. (Agricultural Chem.), M.S. (Soil Sci.). Born 1931. Educ.: Chung Hsin (Formosa) and Hawaii Univs. Asst. (1956-58), Asst. Soil Chem., Assoc. Soil Chem., Soil Chem. (1958-65), Chung Hsin Univ., Formosa; Acting Head, Dept. of Soils and Fertilisers, and Chief, Radioisotope Lab., Taiwan Sugar Experiment Station. Editor, Proc. of the Soc. of Soil Sci. and Fertiliser Technologists of Taiwan; Chairman, Subcom. for Soil Clay Mineralogy, Sect. of the Soc. of Soil Sci. and Fertiliser Technologists of Taiwan. Societies: Chinese Chem. Soc.; Chinese Agricultural Chem. Soc.
Nuclear interests: Soil science; Plant nutrition; Health physics.
Address: Taiwan Sugar Experiment Station, 1 Sheng Chaan Road, Tainan, Formosa.

JUDD, David, Dr. Assoc. Director, Phys. Div., Lawrence Rad. Lab.
Address: U.S.A.E.C., Lawrence Radiation Laboratory, Berkeley 4, California, U.S.A.

JUDY, J. Nelson. Formerly Chem. Production Manager, U.S. Rubber Co.; Formerly Pres., Isochem. Inc., 1965-; Manager, Chem. Res. and Analysis, Battelle-Northwest, 1967-.
Address: Battelle-Northwest, 3000 Stevens Drive, P.O. Box 999, Richland, Washington 99352, U.S.A.

JUILLARD, Madame Jacqueline, Chem. Eng. Born 1922. Educ.: Ecole Polytechnique de l'Univ. Lausanne. Free-lance sci. writer, specialised in atomic energy and space technol., for Gazette de Lausanne, Journal de Geneve, Neue Zurcher Zeitung, etc. Member of Board, Assoc. Suisse pour l'Energie Atomique. Book: L'Atome source d'Energie (Lausanne, Payot et Cie., in French,and Bern, Hallwag A.G., in German). Societies: Commission Nat. Suisse pour l'U.N.E.S.C.O. (Prés. Sect. du Sci. exactes, naturelles et appliquées). Sté. Suisse des Ingenieurs et des Architectes; Assoc. Suisses des Femmes universitaires; Union Suisse de la presse technique et professionelle; American Soc. of Women Eng.; British Interplanetary Soc.
Nuclear interests: Overall picture of nuclear energy, its technical evolution as well as its economic incidences in the world.
Address: Ch les Clys, 1293 Colovrex, Geneva, Switzerland.

JUKES, John Andrew, M.A. (Phys.), B.Sc. (Economics). Born 1917. Educ.: Cambridge Univ.; London School of Economics. Economic Adviser in Economic Sect., Her Majesty's Treasury and Cabinet Office, 1948-54; Economic Adviser (1954-64), Principal Economics and Programmes Officer (1957-64), U.K.A.E.A. Deputy Director Gen. (1964-67), Deputy Under Sec. of State (1967-), Dept. Economic Affairs.
Nuclear interests: Economics of nuclear power, including fuel and material requirements; Programmes for reactor development;

Programmes for introduction of nuclear power.
Address: 38 Albion Road, Sutton, Surrey, England.

JULIA, Roger. Prés.-Directeur Gén., Groupement Atomique Alsacienne Atlantique; Vice-Prés., Directeur Gén., Sté. Industrielle de Combustible Nucléaire; Formerly Prés.-Directeur Gén., Sté. Alsacienne de Constructions Atomiques, de Télécommunications et d'Electronique (Alcatel); Vice-Prés. Sté. Hispano-Alsacienne; Admin., Sté. Alsthom.
Address: Groupement Atomique Alsacienne Atlantique, 32 rue de Lisbonne, Paris 8, France.

JULIEN-LAFERRIERE Georges. Born 1897. Educ.: Ecole Centrale des Arts et Manufactures, Paris. Ing. Conseil Ste. pour Industrie Atomique. Societies: Sté. des Ing. Civils de France; Sté. de l'Industrie Minérale.
Nuclear interest: Nuclear engineering.
Address: Société pour l'Industrie Atomique, 48 rue la Boétie, Paris 8, France.

JULIEN-LAFERRIERE, Paul. Asst. Div. Energie Nucléaire, Sté. Rateau.
Address: Sociéte Rateau, 141 rue Rateau, La Courneuve (Seine), France.

JULIUS, Henry Willem, Prof. Dr. Pres., Central Organisation for Appl. Sci. Res. in the Netherlands (TNO). Board of Governors, Reactor Centrum Nederland; Member, Sci. Council for Nucl. Affairs.
Address: TNO Headquarters, 148 Juliana van Stolberglaan, P.O.B. 297, The Hague, Netherlands.

JUNA, Jaromir, R.N.Dr., C.Sc. Born 1928. Educ.: Masaryk Univ., Brno. Asst., Medical Phys. Inst., Masaryk Univ., Brno, 1951-54; Aspirant (1954-57), Sci. Worker (1957-), Nucl. Res. Inst., Czechoslovak Acad. Sci.
Nuclear interests: Fast neutron physics; Time-of-flight techniques; Mechanism of nuclear reactions; Direct interactions; Structure of light nuclei (n, γ) reactions.
Address: 3 Basta sv. Tomase, Prague 1, Czechoslovakia.

JUNCO, Fernando Sanchez. Formerly Sec., now Director, Tecnicas Nucleares S.A.
Address: Tecnicas Nucleares S.A., 46 Serrano, Madrid, Spain.

JUNCOSA, Mario, Dr. Chairman, Mathematics and Computer Sci. Res. Advisory Com., U.S.A.E.C.
Address: The Rand Corporation, 1700 Main Street, Santa Monica, California, U.S.A.

JUNG, Bo, Asst. Prof. Gustaf Werner Inst., Uppsala Univ.
Address: Uppsala University, Uppsala, Sweden.

JUNG, Richard. Formerly Principal Phys., Reactor Phys. Div., Battelle Memorial Inst., now Sen. Research Sci.-Eng., Battelle-Columbus Labs.
Address: Battelle-Columbus Laboratories, 505 King Avenue, Columbus, Ohio 43201, U.S.A.

JUNGE, Otto, Dr. jur. Director, Rheinisch-Westfalisches Elektrizitatswerk A.G.
Address: Rheinisch-Westfalisches Elektrizitatswerk A.G., 5 Kruppstrasse, 43 Essen, Germany.

JUNGERMAN, J. A., Prof. Sen. Investigator, Crocker Nucl. Lab., Phys. Dept., California Univ., Davis.
Address: Crocker Nuclear Laboratory, California University, Davis, California, U.S.A.

JUNGNELL, Dag Nils Olof, M.Sc. (Elec. Eng.). Born 1921. Educ.: Chalmers Tech. Univ. Head, Thermal and Nucl. Development Dept., Elec. and Thermal Eng., Swedish State Power Board.
Nuclear interests: Management; General development in the nuclear power field.
Address: Electrical and Thermal Engineering, Swedish State Power Board, S-162 87 Vällingby, Sweden.

JUNGNICK, Werner MEYER-. See MEYER-JUNGNICK, Werner.

JUNGST, Wolfgang, Dr. rer. nat., Dipl. Phys. Born 1931. Educ.: Göttingen, Erlangen and Mainz. Sci. Asst., Inst. Nucl. Studies,Mainz; Akad. Rat, Inst. Exptl. Nucl. Phys., Karlsruhe, 1959-. Society: Physikalische Gesellschaft Baden-Württemberg.
Nuclear interests: Nuclear physics; Low-energy physics and electronics in the field of nuclear physics; Particle accelerators; Superconducting RF Particle separators.
Address: Institut für Experimentelle Kernphysik, Karlsruhe Universität, Karlsruhe, Germany.

JUNKERMANN, Wolfgang, Dr.-Ing. Born 1907. Educ.: Munich and Charlottenburg T.H. Chief, Atomic Dept. (1955-66), Nucl. Consultant (1966-), Deutsche Babcock and Wilcox Dampfkesselwerke A.G., Oberhausen, 1955-. Co-editor, Atomkernenergie (Munich). Papers: Several contributions to German atomic journals. Societies: Verein Deutscher Ing.; Kernenergie Hamburg.
Address: 11 Graf-Bernadotte-Strasse, Mülheim-Ruhr, Germany.

JUNKINS, Philip D. Vice Pres., Schwarz Bioresearch Inc.
Address: Schwarz Bioresearch Inc., Mountain View Avenue, Orangeburg, New York, U.S.A.

JUNOD, Blaise. Gen. Manager, Conrad Zschokke S.A.
Address: Conrad Zschokke Ltd., 42 rue du 31 Décembre, Ch-1211 Geneva 6, Switzerland.

JUNOD, E. A. Gen. Manager, F. Hoffman-La Roche and Co. A.G., Basel. Member, Council of Management, Swiss Assoc. for Atomic Energy. Member, Federal Commission For Atomic Energy.
Address: Swiss Association for Atomic Energy, 21 Schauplatzgasse, 3001 Bern, Switzerland.

JUNQUERA, Emilio. Member, Atomic Risk Commission-Nat. Insurers Assoc.
Address: Atomic Risk Commission - National Insurers Association, 4 Avenida Calvo Sotelo, Madrid 1, Spain.

JUREN, James A. DE. See DE JUREN, James, A.

JURGENSEN, Jose E. SALLENT. See SALLENT JURGENSEN, Jose E.

JURI, René, Ing. Agronome. Born 1922. Educ.: Swiss Polytech. Univ. Director, Swiss Farmers Union, Brugg/AG. Exec. European Confederation of Agriculture; Exec. Internat. Federation of Agricultural Producers; Federal Commission for Atomic Energy.
Address: 10 Laurstrasse, Brugg CH - 5200, Switzerland.

JURIC, Mira, Dr. Phys. Born 1916. Educ.: Zagreb Univ. Inst. Nucl. Sci. Boris Kidric, Belgrade; Prof., Faculty of Natural Sci., Belgrade Univ.
Nuclear interests: Nuclear reactions; High energy physics.
Address: 16 Jevremova, Belgrade, Yugoslavia.

JURITZ, John Walter Faure, B.Sc. (Phys. and Chem.), M.Sc. (Physics). Born 1920. Educ.: Cape Town Univ. Lecturer in Phys. (1946-57), Sen. Lecturer in Phys. (1958-), Cape Town Univ. Society: British Astronomical Assoc.
Nuclear interests: Nuclear emulsion technique, hyperons and mesons.
Address: Physics Department, Cape Town University, South Africa.

JURKIEWICZ, Leopold, Ph.D. Born 1906. Educ.: Warsaw and Jagiellonian (Cracow) Univs. Extraordinary Prof. Phys., Acad. of Mining and Metal., Cracow; Sci. worker, Inst. of Nucl. Res.; Member, Rad. Com., journal Nukleonika, Warsaw. Pres. Commission of Radioactive Contamination of Atmo-, Hypro- and Geo-sphere, Polish Com. for Radiol. Protection; Vice-Pres., Polish Bio-Phys. Soc.; Member, Reaction Com. journal Haematologica Polonica (olim Cracoviensia). Societies: Polish Phys. Soc.; Polish Bio-Phys. Soc.
Nuclear interests: Cosmic radiation, application of nuclear physics methods in technical research, radioactive contamination by fission products of nuclear explosions.
Address: 6 ul. Mikolajska, Cracow, Poland.

JURKOVIC, Vilo, M.D., Sc.D. Born 1915. Educ.: Masaryk Univ., Brno. Prof. of Medicine, Charles Univ., Prague; Head, Dept. of Internal Medicine, Charles Univ. Medical School, Hradec Králové, 1954-. Books: Co-author, Rad. Disease (in Czech) (Prague, SZN, 1957); Monograph: Ventricular Tachycardia in Radiation Disease – Contribution to the Study of Heat Excitability (in Czech, with English and Russian summaries) (Hradec Králové, Sbornik Vlvdú, 1965). Societies: Cardiac Soc.; Biophys. Soc.
Nuclear interests: Radiation disease and nuclear medicine. Research on cardiovascular changes during the acute phase of radiation disease.
Address: Charles University Medical School, Department of Medicine, Hradec Králové, Czechoslovakia.

JURY, Stanley H., Ch.E., M.S., Ph.D. Born 1916. Educ.: Louisville and Cincinnati Univs. Prof., Tennessee Univ., 1949-. Director, Nucl. Eng. Div., A.I.Ch.E. Book: Convectional Heat and Momentum Transfer (Tennessee Univ., 1956). Societies: A.A.U.P.; A.S.E.E.
Nuclear interest: Unit operations of nuclear engineering.
Address: Department of Chemical and Metallurgical Engineering, Tennessee University, Knoxville, Tennessee 37916, U.S.A.

JUSTI, Eduard Wilhelm Leonhard, Dr. phil., Dr. phil. habil. Born 1904. Educ.: Marburg, Kiel and Berlin Univs. Prof. Appl. Phys., Braunschweig Tech. Univ. Consultant phys. of 5 leading firms in the field of German electrotechnics and chem. Past-Pres., Acad. of Sci. and Literature, Mainz. Past-Pres., Braunschweig Tech. Univ. Books: Spezifische Wärme, Enthalpie, Entropie und Dissociation technischer Gase (Berlin, 1937); Leitfähigkeit und Leitungsmechanismus fester Stoffe (Göttingen, 1947); High Drain Hydrogendiffusion Electrodes (Mainz, 1959); Cold Combustion-Fuel Cells (Mainz, 1962). Societies: Akademie der Wissenschaften und der Literatur, Mainz; Roy. Swedish Acad. of Eng. Sci., Stockholm; Kgl. Svenska Vitterhets- och Vetenskaps Samhället, Gothenburg; P.E.N. Club, Western Germany.
Nuclear interests: Conversion of nuclear heat into electricity by methods of direct energy conversion – e.g., thermo-elements and fuel cells.
Address: 14 Pockelsstrasse, Postfach 7050, 33 Braunschweig, Germany.

JUUL, Flemming, Dr. techn. Born 1911. Educ.: Tech. Univ. Denmark. Kemo-Scandia, Ltd., 1949-57; Danish A.E.C., 1957- (Deputy Director, 1965-). Gen. Purposes Com., High Temperature Reactor Project Dragon, O.E.C.D.
Address: Atomic Energy Commission, Research Establishment, Risö, Roskilde, Denmark.

KAAE, Sigvard, M.D. Chief Director, Radiumstationen for Jylland.
Address: Radiumstationen for Jylland, Kommunehospitalet, Aarhus, Denmark.

KAASA, Olav Robert. Member, Board of Representative, A/S Noratom.
Address: A/S Noratom, Holmenveien, Oslo, Norway

KABAEV, V. P. Paper: Co-author, Stopping Power of Ni for Proton and H_2^+-Ions of Energies 20-90 keV (letter to the Editor, Atomnaya Energiya, vol. 22, No. 2, 1967).
Address: Academy of Sciences of the U.S.S.R., 14 Leninsky Prospekt, Moscow V-71, U.S.S.R.

KABAKCHI, A. M. Dr. Head, Rad. Chem. Lab., Phys. Chem. Inst., Ukranian Acad. of Sci.
Address: Physical Chemistry Institute, Academy of Sciences of the Ukranian S.S.R., Kiev, Ukranian S.S.R., U.S.S.R.

KABAT, Miloslav, R.N.Dr. Born 1932. Educ.: Charles Univ. Prague. J.Heyrovský Polarographic Inst., Czechoslovak Acad. of Sci., 1952-53; Res. Inst. of Energetic, Prague, 1959-65; Nucl. Res. Inst., Czechoslovak Acad. of Sci., Rez.
Nuclear interest: Radiometry of technological circuits of nuclear reactors and exhausted active gaseous products.
Address: 9 Plynární, Prague 7, Czechoslovakia.

KABITZKE, Bruno, Obering. Direktor, Korrokunststoff Korrosionstechnik G.m.b.H.
Address: Korrokunststoff Korrosionstechnik G.m.b.H., 143 Frankfurter Strasse, Postfach 226, Giessen, Germany.

KACAR, B., Ph.D. Principal Sci. Officer, Cento Inst. of Nucl. Sci.
Address: Cento Institute of Nuclear Science, P.O. Box 1828, Tehran, Iran.

KACAREVIC, M. Head Protection Dept., Dept. for Processing of Nucl. Raw Materials, Inst. Nucl. Raw Materials.
Address: 86 Franse Deperea, Belgrade, Yugoslavia.

KACENA, Vladimir, Dr. rer. nat. Born 1924. Educ.: Charles Univ., Prague. At Chem. Inst. (1950-55), Nucl. Res. Inst. (1956-), Czechoslovak Acad. Sci.; Lecturer, Faculty Tech. and Nucl. Phys., Prague Tech. Univ., 1959-. Radioisotope Sect. Board, Czechoslovak Tech. Soc. Societies: Czechoslovak Tech. Soc.; Czechoslovak Chem. Soc.
Nuclear interest: Radiochemistry, especially chemical effects of nuclear transformations.
Address: Institute of Nuclear Research, Czechoslovak Academy of Sciences, Rez, Czechoslovakia.

KACHADOORIAN, Reuben. Geological Survey, U.S. Dept. of the Interior.
Address: Geological Survey, U.S. Department of the Interior, 345 Middlefield Road, Menlo Park, California, U.S.A.

KACHIKIN, V. I. Paper: Co-author, Miniature Instrument for Measurement of Mean Total Radon Concentration (letter to the Editor, Atomnaya Energiya, vol. 21, No. 3, 1966).
Address: Academy of Sciences of the U.S.S.R., 14 Leninsky Prospekt, Moscow V-71, U.S.S.R.

KACHUROV, B. P. Paper: Minimum Critical Mass with Bounded Fuel Density (Atomnaya Energiya, vol. 20, No.3, 1966).
Address: Academy of Sciences of the U.S.S.R., 14 Leninsky Prospekt, Moscow V-71, U.S.S.R.

KADLEZ, Karl, Dr. techn. Born 1918. Educ.: Vienna Technol. Univ. Head, Nucl. Energy Dept., Simmering-Graz-Pauker A.G. Sec., Reaktor-Interessengemeinschaft. Society: Austrian Welding Soc.
Nuclear interests: Metallurgy and engineering.
Address: Simmering-Graz-Pauker A.G., 32 Mariahilferstrasse, Vienna 7, Austria.

KADMOTSEV, B., Prof. Co-director, I.A.E.A. Theoretical Phys. Centre, Trieste.
Address: I.A.E.A. Theoretical Physics Centre, near Miramare Castle, Trieste, Italy.

KADOMSTEV, B.B., Dr. Deputy Director Kurchatov Inst. of Atomic Energy. Sci. Asst., Acad. of Sci. of the U.S.S.R.; Deleg. to 2nd I.C.P.U. A.E., Geneva, Sept. 1958; Deleg. to Conference sur la Physique des Plasmas et la Recherche concernant la Fusion Nucléaire Controlée, Salzburg, Sept. 1961. Papers: Co-author, Stabilisation of Plasma by Non-Uniform Magnetic Fields (2nd I.C.P.U.A.E., Geneva, Sept. 1958); Turbulent Convection of a Plasma in a Magnetic Field (Nucl. Fusion, vol. 1, No. 4, 1961); co-author, The Stabilsation of a Low-Pressure Plasma by a High-Frequency Field (Atomnaya Energiya, vol. 13, No. 5, 1962).
Address: Kurchatov Institute of Atomic Energy, 46 Ulitsa Kurchatova, Post Box 3402, Moscow, U.S.S.R.

KADOSCH, Marcel, Ing.E.M.P., D.es.Sc. Formerly Conseiller technique, now Conseiller Sci., Bertin et Cie., S.A.
Address: Bertin et Cie., S.A., B.P. No. 3, Plaisir, (Seine et Oise), France.

KADOWAKI, Yoshio, Sec., Inst. for Atomic Energy, St. Paul's Univ.
Address: The Institute for Atomic Energy, St. Paul's University, Ikebukuro, Toshima-ku, Tokyo, Japan.

KADRI, Louay, B.A. (Chem.), M.S. (Soils), Ph.D. (Soil Chem.). Born 1923. Educ.: Beirut American and Utah State Univs. and Texas A. and M. Coll. Acting Sec. Gen., (1963-64), now Member, Iraqi A.E.C. Societies: Iraqi Chem.

Soc.; Soil Sci. Soc. America.
Nuclear interest: Use of radioisotopes in the agricultural sciences and in chemistry.
Address: Agricultural Experimental Farm, Abughraib, Iraq.

KAEGI, W., Prof. Dr. Member, Council of Management, Swiss Assoc. for Atomic Energy.
Address: Swiss Association for Atomic Energy, 21 Schauplatzgasse, 3001 Bern, Switzerland.

KAEPPELER, H.J. Born 1926. Educ.: Tübingen and Stuttgart Univs. Phys. and Group Leader, Plasma and Thermodynamics Group, Forschunginst. für Physik der Strahlantriebe, Stuttgart, 1954-58; Chief, Plasma Theory Div., Inst. für Plasmaforschung, Stuttgart Univ., 1958-. Books: Collaborator, Encyclopedia of Reactor Phys. (Lexikon der Reaktorphysik) and Encyclopedia of Phys. (Lexikon der Physik) (Stuttgart, Frankh'sche Verlagsbuchhandlung 1959). Society: German Phys Soc.
Nuclear interests: Formerly engaged in nuclear rocket propulsion (theoretical work). At present engaged in basic research (plasma waves and shock waves) connected with nuclear fusion.
Address: 72 Böblingerstrasse, Stuttgart - S, Germany.

KAFENGAUZ, N. L. Paper: Co-author, Destruction of Tubes by High-Frequency Pressure Oscillations Occurring During Heat-Transfer (letter to the Editor, Atomnaya Energiya, vol. 23, No. 2, 1967).
Address: Academy of Sciences of the U.S.S.R., 14 Leninsky Prospekt, Moscow V-71, U.S.S.R.

KÄFER, Ernst. Director, Graetz Raytronik G.m.b.H.
Address: Graetz Raytronik G.m.b.H., P.O.B. 57, 50 Graetzstrasse, Altena/Westfalen, Germany.

KAGEYAMA, Seizaburo. Phys. Dept., Tokyo Univ. of Education.
Address: Tokyo University of Education, 24 Otsuka-Kubomachi, Bunkyo-ku, Tokyo, Japan.

KAHALAS, Sheldon L., A.B., M.S., Ph.D. Born 1933. Educ.: Harvard Coll. and Illinois and Boston Univs. Sylvania Elec. Products, Inc., 1955-58; Allied Res. Assoc., Inc., 1959-. Society: A.P.S.
Nuclear interests: Controlled thermonuclear reactors; Plasma physics; Nuclear physics.
Address: Allied Research Associates, Inc. 43 Leon Street, Boston 15, Mass., U.S.A.

KAHAN, Ralph Sidney, B.Sc., D.I.C., F.R.I.C., A.M.I.Chem.E., C.Eng. Born 1917. Educ.: London Univ. Chem. Eng., Heavy Water Pilot Plant, Rehovot (1955-59), Coordinator, Survey, Ind. Uses of Isotopes (1959), Citrus fruit irradiation (1960-), Head, Technol. Evaluation Unit (1963-64), Head, Agricultural Applications Group, Soreq Nucl. Res. Centre (1964-), Israel A.E.C. Books: Contribution, Ind.

Applications of Isotopes in: The Economic Potential of Nucl. Sci. in Israel (Weizmann Inst. of Sci., 1959); Radioisotopes and Chem. Eng. (Israel A.E.C., 1960).
Nuclear interests: Radiation preservation of fruits and vegetables. Development of commercial processes for irradiated foods, and design and development of large irradiators. Radiostimulation of plant growth and crop yield.
Address: 11 Uziel Street, Ramat-Gan, Israel.

KAHANE, Jean, L. ès Sc. Born 1930. Educ.: Toulouse Univ. Asst., Lab de Physique Atomque et Moleculaire, Coll. de France. Society: Sté. Française de Physique.
Nuclear interests: Physique nucléaire; Electronique; Physique du solide.
Address: 9 Alleé Apollinaire, Sarcelles, (S. et O.), France.

KAHANPÄÄ, Veikko, Prof. Chief, Gynecologic Units, Dept. of Radiotherapy, Univ. Central Hospital, Helsinki.
Address: Department of Radiotherapy, University Central Hospital,4 Haartmaninkatu, Helsinki 29, Finland.

KAHLSON, Donald Eric, B.Sc. Chem. Eng. Born 1931. Educ.: Case Inst. of Technol. Shift Supv., Du Pont Savannah River Project, 1953-58; Sen. Nucl. Eng., Lockhead Nucl. Lab., 1958-63; U.S.A.E.C., 1963-64; European Nucl. Power Sales Manager, G.E.C., 1965-67; Manager, Nucl. Systems, Europe and Mediterranean, Combustion Eng., U.S.A., 1967-. Director, Georgia Div., American Cancer Soc. Societies: A.N.S.; Soc. of Nucl. Medicine.
Nuclear interests: Commercial nuclear power and its application to utility practice in Europe and the Mediterranean countries; Advancement of nuclear medicine, particularly radiotherapy.
Address: 86 B rue de Florissant, Geneva, Switzerland.

KAHN, Bernd, Ph.D., B.S., M.S. Born 1928. Educ.: M.I.T., Vanderbilt Univ. and Newark Coll. of Eng. Chief, Nucl. Eng. Lab., Environmental Surveillance and Control Programme, Nat. Centre for Radiol. Health, Public Health Service, Cincinnati, Ohio. Chairman, Subcom. on Use of Radioactivity Standards, Nucl. Sci. Com., N.R.C. - N.A.S. Societies: A.C.S.; A.P.S.; RESA; A.S.T.M.
Nuclear interests: Radiochemistry; research in the movement of radionuclides in the environment.
Address: National Centre for Radiological Health, 4676 Columbia Parkway, Cincinnati, Ohio 45226, U.S.A.

KAHN, Erwin, C.E. Born 1927. Educ.: Brussels Univ. Gen. Manager, S.A. Graver N.V., 1961-. Pres. Comité Belge des Constructeurs d'Equipement Pétrolier; Pres., Federation Européenne des Constructions d'Equipment Petrolier. .

Nuclear interest: Steel construction in nuclear field.
Address: S.A. Graver N.V., 1 Oude Stationsplaats, Willebroek, Belgium.

KAINARSKII, N. S. Paper: Co-author, Toroidal Chambers for Investigation of Plasma-RF Fields Interaction (letter to the Editor, Atomnaya Energiya, vol. 23, No. 1, 1967).
Address: Academy of Sciences of the U.S.S.R., 14 Leninsky Prospekt, Moscow V-71, U.S.S.R.

KAINDL, Karl, Dr., Dozent. Born 1913. Educ.: Vienna Univ. Director, Austro-Chematom, Kernbrennstoffgesellschaft, m.b.H., Linz (Nucl. Fuel Ltd., Linz); Consultant of Österreichische Stickstoffwerke A.G., Linz.; Director, Agriculture and Biol. Inst., Österreichische Studiengesellschaft für Atomenergie m.b.H. Vienna; Leader, E.N.E.A. Food Irradiation Project, Seibersdorf; Formerly Co-editor of the scientific German journal, Atompraxis. Books: Quantenbiologie (Vienna Verlag Gebr. Hollinek, 1951); Anwendungsmethoden von Radioisotopen in der Landwirtschaft (Beitrag in: Fortschritte in der Radioisotopie und Grenzgebiete, Bd. I.) (Heidelberg Hüthig Verlag, 1951); Isotope in der Landwirtschaft (Hamburg, Parey-Verlag, 1959). Societies: Österr. Physikalische Gesellschaft; Österr. Biochemische Gesellschaft. Österr. Bodenkundliche Gesellschaft; Ing. und Architekteverein. Nuclear interests: Biophysics; Biology and agriculture; Food preservation; Management in fuels and moderators.
Address: c/o Österreichische Studiengesellschaft fur Atomenergie, G.m.b.H., 10 Lenaugasse, Vienna 8, Austria.

KAIPOV, R. L. Paper: Co-author, On the Use of Ac-Be Neutron Sources in Trade Geophys. (letter to the Editor, Atomnaya Energiya, vol. 16, No. 3, 1964).
Address: Academy of Sciences of the U.S.S.R., 14 Leninsky Proskept, Moscow V-71, U.S.S.R.

KAISEKI, Shinzo. Sen Managing Director, C.Itoh and Co., Ltd.
Address: 4, 2-chome, Hon-cho, Nihonbashi, Chuo-ku, Tokyo, Japan.

KAJAMAA, Jaakko Pekka, Tech. Licenciate. Born 1939. Educ.: Helsinki Tech. Univ. Res. Eng., Reactor Lab., Dept. of Tech. Phys., Helsinki Tech. Univ., 1964-. Societies: Finnish Soc. for Nucl. Eng.; Finnish Phys. Soc.; Eng. Soc., Finland.
Nuclear interests: Neutron diffraction; Physical metallurgy.
Address: A 3 Mäntypohja, Lohja as., Finland.

KAJFOSZ, Josef, R.N.Dr. Born 1933.Educ.: Masaryk Univ., Brno. Neutron Phys. Dept., Nucl. Res. Inst., Prague, 1956-.
Nuclear interests: Nuclear physics; Reactions with slow neutrons; Radiative capture; Gamma-ray spectroscopy; Linear and circular polarisation; Polarised neutrons; Time-reversal invariance..
Address: Nuclear Research Institute, Rez, Prague, Czechoslovakia.

KAKAR, Abdul Ghaffar, Dr. Dean, Faculty of Sci., Kabul Univ.; Pres., A.E.C. Afghanistan. Afghanistan Deleg., I.A.E.A. Gen. Conference, Vienna, September 1960.
Address: Kabul University, Faculty of Sciences, Kabul, Afghanistan.

KAKEHI, Hirotake, M.D. and Degree of Medical Sci. Born 1911. Educ.: Tokyo Univ. School of Medicine. Prof., Dept. of Radiol., Chiba Univ. School of Medicine, 1954-. Book: Co-author, Clinical Nucl. Medicine (Tokyo, Asakura Shoten, 1967). Societies: Nippon Soc. Radiologica; Japanese Assoc. of Nucl. Medicine.
Nuclear interest: Nuclear medicine (clinical applications of radioisotopes).
Address: Department of Radiology, Chiba University Hospital, Chiba, Japan.

KAKIGI, M. S. Shigeru, Asst., Div. 1, Lab. of Nucl. Study, Inst. for Chem. Res., Kyoto Univ.
Address: Kyoto University, Institute for Chemical Research, Yoshida-Honmachi, Sakyo-ku, Kyoto, Japan.

KAKIHANA, Hidetake, D. Sci. (Tokyo). Born 1920. Educ.: Tokyo Univ. Prof. Nucl. Chem., Tokyo Inst. Technol., 1958-. Japan Corp. of Atomic Fuels. Book: Ion Exchange Resins (Toyko, Hirokawa-Shoten, 1954). Societies: Chem. Soc. of Japan; Atomic Energy Soc. of Japan.
Nuclear interest: Nuclear chemistry.
Address: Research Laboratory of Nuclear Reactor, Tokyo Institute of Technology, 1 Oh-okayama, Meguro-ku, Tokyo, Japan.

KAKULIA, G. L. Paper: Co-author, Investigation of Pressure Influence on Isotope Separation B^{11} and B^{10} Process by Distillation of BF_3 (Atomnaya Energiya, vol. 23, No. 4, 1967).
Address: Academy of Sciences of the U.S.S.R., 14 Leninsky Prospekt, Moscow V-71, U.S.S.R.

KALAFATI, D. D. Papers: The Effect of Reactor Temperature Characteristics on the Choice of an Atomic Power Station Optimal Thermodynamic Cycle (Atomnaya Energiya, vol. 8, No. 1, 1960); Thermodynamic Cycle Parameters Analysis of Space Nucl. Power Plants (ibid, vol. 22, No. 6, 1967).
Address: Academy of Sciences of the U.S.S.R., 14 Leninsky Prospekt, Moscow V-71, U.S.S.R.

KALASHNIKOV, A. G. Paper: Co-author, On Self-Shielding Factor for Slab Array (letter to the Editor, Atomnaya Energiya, vol. 24, No. 3, 1968).
Address: Academy of Sciences of the U.S.S.R., 14 Leninsky Prospekt, Moscow V-71, U.S.S.R.

KALDOR, I., Dr. Sen. Lecturer in Physiology, Western Australia Univ.
Address: Western Australia University, Nedlands, Western Australia.

KALEBIN, S. M. Paper: Co-author, Total Neutron Cross Section of Th ^{230} at Thermal Region (Atomnaya Energiya, vol. 24, No. 3, 1968).
Address: Academy of Sciences of the U.S.S.R., 14 Leninsky Prospekt, Moscow V-71, U.S.S.R.

KALININ, B. N. Papers: Co-author, 1, 5 GeV Electron Synchrotron at the Polytech. Inst. of Tomsk (Atomnaya Energiya, vol. 21, No. 6 1966); co-author, 300 MeV Electron Synchrotron at the Polytech. Inst. of Tomsk (letter to the Editor, ibid, vol. 21, No. 6, 1966).
Address: Academy of Sciences of the U.S.S.R., 14 Leninsky Prospekt, Moscow V-71, U.S.S.R.

KALININ, V. F. Member, Editorial Board, Atomnaya Energiya.
Address: 18 Kirov Ulitsa, Moscow, U.S.S.R.

KALININA, M. D. Paper: Co-author, A Study of Co ^{60} γ -Ray Effect on Highly Basic Anionites AM-17 and AM (Atomnaya Energiya, vol. 16, No. 3, 1964).
Address: Academy of Sciences of the U.S.S.R. 14 Leninsky Prospekt, Moscow V-71, U.S.S.R.

KALISH, Herbert S., B.S. (Metal. Eng.), M.S. (Metal. Eng.), Professional degree of Metal. Eng. Born 1922. Educ.: Case Inst. Technol., Missouri Univ. School of Mines and Metal., and Pennsylvania Univ. Sen. Eng., Eng.-in-Charge, Eng. Manager (1955-57), Atomic Energy Div., Sylvania Elec. Products, Inc., and Sylvania-Corning Nucl. Corp., Bayside, New York, 1948-57; Sect. Chief, Nucl. Metal. (1957-62), Manager, Commercial Fuel Operations (1962-), United Nucl. Corp. (formerly Nucl. Fuels Operation, Olin Mathieson Chem. Corp.), New Haven, Connecticut. Exec. Com., A.S.M., New Haven, Connecticut; A.S.M.-A.E.C. Advisory Com. on Monographs; A.S.T.M. Com. A-10 on Iron-Chromium, Iron-Chromium-Nickel and Related Alloys, and Subcom. XIII, Specifications for Nucl. Reactor Structural Materials; A.I.M.E. I.M.D. Publications Com. (Key Reader); Advisory Com., Mech. Technol., State Univ. New York, Agricultural and Tech. Inst. Societies: A.I.M.M.E.; A.S.M.; A.S.T.M.; American Welding Soc.; Metal Powder Industries Federation; A.N.S.; Sci. Res. Soc. of America.
Nuclear interests: Metallurgy, fabrication, fuel elements, control rods and related components and materials for use in nuclear reactors and other atomic energy applications; interests are in both fields of research and production.
Address: United Nuclear Corporation, 365 Winchester Avenue, New Haven 4, Connecticut, U.S.A.

KALISZ, József Roman, Ph.D. (Eng.). Born 1935. Educ.: Gliwice Tech. Univ. Tech. Univ. 1957-60; Nucl. Ind.,Biuro Urzadzen Techniki Jadrowej (Nucl Apparatus Office), 1960-63; Inst. of Nucl. Res. Swierk, 1963-.
Nuclear interests: Nuclear electronics; Statistical problems involved in the analysis of coincidence circuits.
Address: 26/17 Zorzy, Warsaw, Poland.

KALITINSKY, Andrew, Diploma of Mech. Eng. Born 1914. Educ.: Swiss Federal Inst. Technol., Zürich. Nucl. Programme Manager, Convair Fort Worth Div., Gen. Dynamics Corp. Member, Texas Com. on Atomic Energy. Societies: A.N.S.; American Rocket Soc.; Inst. of the Aeronautical Sciences.
Nuclear interests: Nuclear-powered aircraft, shielding, radiation effects, reactor design.
Address: c/o Convair, Fort Worth,Texas, U.S.A.

KALIVODA, Radoslav, Ing. (Nucl. Eng.). Born 1935. Educ.: Prague Tech. Univ. Res. Worker, Skoda Works. Society: Czechoslovak Sci. and Tech. Soc.
Nuclear interests: Reactor control; Computer control.
Address: Nuclear Power Division, Skoda Works, Pilsen, Czechoslovakia.

KÄLLEN, Anders Olof Gunnar, Dr. Phil. Born 1926. Educ.: Lund Univ. Docent, Lund Univ., 1950-53; Theoretical Study Div., C.E.R.N., Copenhagen, 1953-57; Nordita, 1957-58; Prof., Theoretical Phys., Lund Univ., 1958-.
Books: Chapter on Quantum Electrodynamics in Handbuch der Physik. Volume 1. (Springer, 1958); Elementary Particle Phys. (translated to German and Russian) (Addison-Wesley, 1964). Societies: Danish and Swedish Academies of Sci., Fysiografiska Sällskapet.
Nuclear interest: High-energy physics.
Address: Department of Theoretical Physics, Lund University, Lund, Sweden.

KALLEN, Hans. Dr. rer. nat. h.c. Born 1901. Member, Exec. Board, Fried. Krupp; Member, Arbeitskreis für Atomfragen beim Bundesverband der Deutschen Industrie e.V.; Formerly Verwaltungsrat Deutsches Atomforum; Special Com. for Nucl. Power Problems with U.N.I.C.E., Brussels.
Address: 60 Elsassstrasse, Essen-Heisingen, Germany.

KALLENBACH, Reinhard, Dipl. Ing. Born 1917. Educ.: Darmstadt T.H. Geschäftsführer, Kernkraftwerk Baden-Württemberg Planungsgesellschaft m.b.H. 1960-; Direktor, Kernkraftwerk Obrigheim G.m.b.H., - 1968; Head, Plant Div., Energie - Versorgung Schwaben, 1968-. Gutachter für KNK-Reaktor im Programm fortgeschrittener mittlerer und kleinerer Reaktoren; Mitglied des Unterausschusses Kernkraftwerke im Deutschen Atomforum.
Nuclear interests: Stromerzeugung aus Kernenergie.
Address: 38 Schottstr., Stuttgart-N, Germany.

KALLMAN, Hartmut P. Member, Subcom. on Effects of Ionising Radiations, Com. on Nucl. Sci., N.A.S. - N.R.C. Prof. of Phys., Washington Square Coll. of Arts and Sci., New York Univ.
Address: National Academy of Sciences - National Research Council, Committee on Nuclear Sciences, 2101 Constitution Avenue, N.W., Washington 25, D.C., U.S.A.

KALMYKOV, V. D. Paper: Co-author, Experience of Spectral Gamma-Log with Multichannel Spectrometers (letter to the Editor, Atomnaya Energiya, vol. 23, No. 1, 1967).
Address: Academy of Sciences of the U.S.S.R., 14 Leninsky Prospekt, Moscow V-71, U.S,S.R.

KALOGEROPOULOS, Theodore E., Ph.D. Born 1931. Educ.: Athens and California (Berkeley). Univs. Instructor in Phys., Columbia Univ., 1959-62; Asst. Prof. (1962-65), Assoc. Prof. (1965-), Syracuse Univ. Assoc. Phys., Brookhaven Nat. Lab. Societies: A.P.S.; A.A.U.P.
Nuclear interests: Experimental high energy nuclear physics: Emulsion, bubble chamber, counter, and spark chamber techniques;. Nucleon-antinucleon interactions.
Address: Physics Department, Syracuse University, Syracuse 10, New York, U.S.A.

KALOS, M. H., Dr. Member, Group of Specialists on Neutron Data Compilation, E.N.E.A., O.E.C.D. Book: Contributor to Eng. Compendium on Rad. Shielding (I.A.E.A.).
Address: United Nuclear Corporation, 5 New Street, White Plains, New York, U.S.A.

KALSTAD, H. M. Gen. Manager, Nat. Beryllia Corp.
Address: National Beryllia Corporation, 1st and Haskell Avenues, Haskell, New Jersey 07420, U.S.A.

KALUS, Jürgen, Diplom-Physiker, Dr. rer. nat. Born 1935. Educ.: Munich T.H. Graduate student Munich T.H., 1955-62.
Nuclear interests: Van-de-Graaff tandem accelerator; Design of ion sources for negative ions; Nuclear resonance flourescence.
Address: Labor für Technische Physik, 21 Arcisstr., Munich, Germany.

KALUSZYNER, Louis. Director, Soc. d'Applications Industrielles de la Physique S.A.I.P. Formerly Representative to Syndicat des Industries de Materiel Professionnel Electronique et Radioelectrique.
Address: Societe d'Applications Industrielles de la Physique, S.A.I.P., 38 rue Gabriel Crié, Malakoff (Seine), France.

KALVEN, Harry, Jr. Member, Board of Directors, Educ. Foundation for Nucl. Sci.
Address: Educational Foundation for Nuclear Science, 935 E. 60th Street, Chicago 37, Illinois, U.S.A.

KALVIUS, Georg Michael, Dr. rer. nat., Dipl. Phys. Born 1933. Educ.: Munich T.H. Phys. Dept., Munich T.H.; Phys. Dept., Western Reserve Univ.; Assoc. Phys., Argonne Nat. Lab. Societies: A.P.S.; A.A.A.S.
Nuclear interests: Low energy nuclear physics; Mossbauer effect.
Address: S.S.S. - D212, Argonne National Laboratories, Argonne, Illinois 60439, U.S.A.

KAM, J. A. van der. See van der KAM, J. A.

KAMAEV, A. V. Papers: Co-author, Physical Characteristics of Beryllium-moderated Reactor (2nd I.C.P.U.A.E., Geneva, Sept. 1958); co-author, Intereactions in a Subscritical Reactor (Atomnaya Energiya, vol. 16, No. 1, 1964); co-author, Critical Parameters for Aqueous Solutions of $UO_2(NO_3)_2$ (ibid); co-author, Heterogeneous Absorbers Efficiency in Homogeneous Uranium-Water Reactors (ibid, vol. 19, No. 1, 1965).
Address: Nuclear Energy Institute, Academy of Sciences of the U.S.S.R., 14 Leninsky Prospekt, Moscow V-71, U.S.S.R.

KAMAL, Abdul Naim, M.Sc. Ph.D. Born 1935. Educ.: Dacca and Liverpool Univ. Lecturer, Educ.: Dacca and Liverpool Univs. Lecturer, Liverpool Univ. 1963; Postdoctoral Fellow, Theoretical Phys. Inst. Edmonton, 1963-64; Asst. Prof., Alberta Univ., Edmonton, 1964-.
Book: Problems in Particle Physics (London, McGraw-Hill, 1966).
Nuclear interest: Particle physics.
Address: Department of Physics, University of Alberta, Edmonton, Canada.

KAMAYEVA,, L. A. Paper: Co-author, Angular Distributions of Fragments from Fission of ^{235}U, ^{239}Pu by 0.08 - 1.25 Mev Neutrons (letter to the Editor, Atomnaya Energiya, vol. 16, No. 6, 1964).
Address: Academy of Sciences of the U.S.S.R., 14 Leninsky Prospekt, Moscow V-71, U.S.S.R.

KAMBARA, Tomihisa, D.Sc. Director, and concerned with Radiochem. and Nucl. Chem., Radiochem. Res. Lab., Shizuoka Univ.
Address: Shizuoka University, Radiochemistry Research Laboratory, 2 Oiwa-cho, Shizuoka City, Japan.

KAMBARA, Toyozo, Dr. Sci. Born 1913. Educ.: Tokyo Univ. Manager, Reactor Operation Div., Japan Atomic Energy Res. Inst. 1955-61; Manager, Ozenji Div. (1962-65), Gen. Manager, Hitachi Central Res. Lab. (1966-), Hitachi Ltd. Societies: A.N.S.; Atomic Energy Soc. Japan.
Nuclear interests: Reactor design; Operation management.
Address: Osawa, 2-7-21, Mitaka, Tokyo, Japan.

KAMBER, Franz Walter, Dipl. Eng. Born 1909. Educ.: E.T.H. Sales Eng. Manager, Steel Foundry. Society: Swiss Assoc. for Atomic Energy.
Address: George Fischer Limited, 8201 Schaffhausen, Switzerland.

KAMBER, R. S. Power Supply Manager, Hallam Nuel. Power Facility, operated for U.S.A.E.C. by Consumers Public Power District.
Address: Hallam Nuclear Power Facility, Consumers Public Power District, Lincoln, Nebraska, U.S.A.

KAMEDA, K., Prof. Dr. Attached Hospital Dept. of Medicine and Member, Isotope Res. Com., Kumamoto Univ.
Address: Department of Medicine, Kumamoto University, Honzyomati, Kumamoto, Japan.

KAMENETSKII, A.D. FRANK-. See FRANK-KAMENETSKII, A.D.

KAMETANI, Fujio. Member, Atomic Energy and Isotope Com., Tokushima Univ.
Address: c/o Medical Faculty, Tokushima University, Kuramoto-cho, Tokushima-shi, Japan.

KAMEYAMA, Koji, Keizai-Gakushi. Born 1904. Educ.: Kyoto Univ., Faculty of Economics. Managing Director, Nippon Fire and Marine Insurance Com., Ltd. Member, Managing Com., Japan Atomic Energy Insurance Pool.
Address: 127 1-chome, Narimune, Suginamiku, Toyko, Japan.

KAMIENSKI, E. Doc. Dr. Deleg., I.A.E.A. Symposium on Fuel Element Fabrication, with Special Emphasis on Cladding Material, Vienna, 10-13 May, 1960.
Address: Institute for Nuclear Research, 10 m. 8. ul. Nowotki, Warsaw, Poland.

KAMIMURA, Tadao, Prof. Medicine Dept. Member, Radioisotope Res. Com., Niigata Univ. School of Medicine.
Address: Department of Medicine, Niigata University School of Medicine, 757, 1-Bancho Asahimachi-Street, Niigata, Japan.

KAMINISHI, Tokishi. Born 1927. Educ.: Kyushyu Univ. Official, Ministry of Internat. Trade and Industry. Books: Co-author, Ind. Uses of Radioisotopes (Chem. Industry Co., 1958); co-author,Ind. Radioisotopes (Ind. Daily News Co., 1957). Societies: Japan Appl. Phys.; Japan Radioisotopes Assoc.
Nuclear interests: Activation analysis (specially by γ-ray); Radiation gauging; Radiation shielding and dosimetry.
Address: Hirate-machi, Kitaku, Nagoya, Japan.

KAMINKER, D. M. At Phys. -Tech. Inst., U.S.S.R. Acad. of Sci. Adviser, Third I.C.P.U. A.E., Geneva, Sept. 1964. Papers: Co-author, Electrophoretic Filter for Purifying of Reactor Water (Atomnaya Energiya, vol. 19, No. 4, 1965); co-author, Reduction of Radioactive Disposal into Atmosphere and Res. of Water Deaeration of WWR-M Reactor (ibid, vol. 19, No. 6, 1965); co-author, The Investigation of Radiolytic Products During Reactor WWR-M Operation without Degasation System (ibid, vol. 24, No. 4, 1968); co-author, Neutron Capture Cross-Sect. Measurement by Radioactive Nuclei Co^{58m}, Cu^{64} and Sc^{46} (ibid, vol. 24, No. 6, 1968).
Address: Physico-Technical Institute, 2 Polytechnicheskaya Ulitsa, Leningrad, U.S.S.R.

KAMINSKII, V.A. Papers: A Study on Mass Transfer in Seperation of Isotopes by Chem. Exchange and Distillation Methods (letter to the Editor, Atomnaya Energiya, vol. 14, No. 6, 1963); co-author, Elementary Separation Factor for Exchange Distillation of Compound $(CH_3)_2$ $O.BF_3$ (letter to the Editor, ibid, vol. 23, No. 3, 1967).
Address: Academy of Sciences of the U.S.S.R., 14 Leninsky Prospekt, Moscow V-71, U.S.S.R.

KAMKE, Detlef Gustav, Dr. Phil. Born 1922. Educ.: Tübingen, Göttingen and Marburg Univs. Sci. Asst. (1952), Head Asst. (1958-), Marburg Univ.; Prof. Exptl. Phys., Bochum Univ., 1963. Societies: Deutsche Physikalische Gesellschaft; A.P.S.
Nuclear interest: Nuclear reactions.
Address: 262 Markstrasse, Bochum 463, Germany.

KAMMASH, Terry, B.S., M.S. (Pennsylvania State), Ph.D. (Michigan). Born 1927. Educ.: Pennsylvania State and Michigan Univs. Asst. Prof., Nucl. Eng. and Eng. Mechanics, Michigan Univ., 1958-; Sci., Los Alamos, (summer 1959); Assoc. Prof., Nucl. Eng., Michigan Univ., 1960-; Sci. Lawrence Rad. Lab., Livermore, California, 1962-63. Societies: A.N.S.; A.P.S.; A.S.E.E.; Soc. of Eng. Sci.
Nuclear interests: Plasmas and controlled fusion; Interaction of radiation with matter; Nuclear physics; Thermal stresses in reactor components.
Address: Nuclear Engineering Department, Michigan University, Ann Arbor, Michigan, U.S.A.

KAMMERLOCHER, Louis. Born 1900 Educ.: Ecole spéciale des Travaux Publics, Paris.Ing. Conseil, Stés. Indatom et SOCIA. Societies: Sté Française des Electriciens; Assoc. des Ing. en Anticorrosion.
Nuclear interests: Reacteurs de nuisance eau lourde-gaz haute pression. Protection installations contre la corrosion marine.
Address: Indatom, 48 rue La Boëtie, Paris 8, France.

KAMOGAWA, Hiroshi, D.Sc. B.Sc. Born 1913. Educ.: Kyoto Univ. Sen. Manager, Development (Atomic Energy) Gen Eng. Group, Tokyo-Shibaura Elec. Co. Ltd. (Toshiba), 1961-. Societies: Atomic Energy Soc. of Japan; Atomic Ind. Forum of Japan; Phys. Soc. of Japan.
Nuclear interests: Management in nuclear materials, applied radiation and power reactor

development.
Address: 1-6, 1-chome Uchisaiwaicho, Chiyodaku, Tokyo, Japan.

KAMOUCHI, Masaichi. Born 1906. Educ.: Tokyo Univ. Mercantile Marine. Prof., Toyko Univ. Mercantile Marine. Society: Japan Soc. of Mech. Eng.
Nuclear interest: Ship's reactors.
Address: Tokyo University of Mercantile Marine, 2-I Fukagawa-Etchujima, Koto-Ku, Tokyo, Japan.

KAMP, A. W. Deputy Member, Netherlands Pool for the Insurance of Atomic Risks.
Address: Netherlands Pool for the Insurance of Atomic Risks, 101 Blaak, Rotterdam, Netherlands.

KAMPEN, C. F. A. van. See van KAMPEN, C. F. A.

KAMPEN, N. G. VAN. See Van Kampen, N. G.

KANAJI, Toru, Dr. Eng. Linear Accelerator, Government Ind. Res. Inst., Nagoya.
Address: Government Industrial Research Institute, Nagoya, Agency of Industrial Science and Technology, Ministry of International Trade and Industry, Hirate-machi, Kita-ku, Nagoya, Japan.

KANAREIKIN, V. A. Papers: Co-author, Relative Natural Radioactivity of Photomultipliers FEU-49 (letter to the Editor, Atomnaya Energiya, vol. 22, No. 1, 1967). co-author, About Contents of Cs^{137} in Human Body, Milk and Bread (letter to the Editor, ibid, vol. 24, No. 6, 1968).
Address: Academy of Sciences of the U.S.S.R., 14 Leninsky Prospekt, Moscow V-71, U.S.S.R.

KANCHANARAK, Sauwaros, B.S. Officer and Member Co-operating Staff, Atomic Energy Lab., Kasetsart Univ.
Address: Kasetsart University, Bangkhen, Bangkok, Thailand.

KANDARITSKI, V. S. Head, Dept. of Internat. Relations and Sci.-Tech. Information, U.S.S.R. State Com. for the Utilisation of Atomic Energy.
Address: U.S.S.R. State Committee for Utilisation of Atomic Energy, 26 Staromonetnii Pereulok, Moscow, U.S.S.R.

KANDIAH, Kathirkamathamby, M.A., B.Sc. Born 1916. Educ.: Univ. Coll., Ceylon and Cambridge Univ. Group Leader, Advanced Electronic Systems, Elec. and Appl. Phys. Div. A.E.R.E., Harwell.
Nuclear interests: Semiconductor device physics and applications; Ion implantation; Electronic instrumentation such as amplifiers, counters, pulse analysers, time spectrometers, signal and data processing systems; Magnetic and optoelectronic systems; Information storage.
Address: 13 Tullis Close, Sutton Courtenay, Berkshire, England.

KANDLE, Roscoe P., B.S., M.D., M.P.H. Born 1909. Educ.: Johns Hopkins Univ. and Jefferson Medical Coll. New Jersey State Commissioner of Health, 1959-. Member, New Jersey Commission on Rad. Protection; Member, Interstate Sanitation Commission (New Jersey, New York and Connecticut). Societies: Fellow A.M.A.; Fellow, A.P.H.A.
Nuclear interests: Radiological health.
Address: New Jersey State Department of Health, John Fitch Plaza, Trenton, New Jersey 08625, U.S.A.

KANE, D. E., Jr. Formerly Asst. Eng., Machinery Plant Design, now Chief Eng., Nucl. New Design, Newport News Shipbuilding and Dry Dock Co.
Address: Newport News Shipbuilding and Dry Dock Co., Newport News, Virginia, U.S.A.

KANE, Edward D., B.Sc. (Chem. Eng.), Licensed Professional Eng. Born 1927. Educ.: M.I.T., Buffalo and Hartford Univs. Development Eng., Ontario Paper Co., Ltd., Thorold, Ontario, 1947-50; Rocket Test Eng., Bell Aircraft Co., Niagara Falls, New York, 1950; Chem. Eng., Director Micro Klean Development, Manager Aircraft and Nucl. Sales, Div. Sales Manager, Cuno Engineering Corp., Meriden, Connecticut, 1950-60; Consultant, Westinghouse Elec. Corp., Atomic Power Dept., Pittsburgh, 1955-56; Vice-Pres. Sales, Kahn and Co., Wethersfield, Connecticut, 1960-61; Pres. and Gen. Manager, Nat. Sherardizing and Machine Co. Inc., Hartford, Connecticut, 1961-63; Founder and Pres.,E. D. Kane and Assocs., Ind. Consultants, W. Hartford, Connecticut, 1963-64; Sen. Product Analyst, Manager Desalination Programme, Manager Market Development, Combustion Eng., Windsor, Connecticut, 1964-. Chairman, A.I.Ch.E. Nat. Meeting, Seminar-Sales Eng.; S.A.E. Com. on Filter Test Methods Standardisation. Societies: A.I.Ch.E.; A.C.S.; S.A.E.; N.S.P.E.; Chem. Club, New York; T.A.P.P.I.
Address: 37 Magnolia Hill Road, West Hartford, Connecticut, U.S.A.

KANE, Brother Gabriel, F.S.C., Ph.D. Director, Phys. Dept. Nucl. Emulsion Res., and Chairman, Com. for Nucl. Studies, Manhattan Coll. Member, Gen. Advisory Com., New York State Office of Atomic Development.
Address: Physics Department, Manhattan College, Riverdale, New York 71, New York, U.S.A.

KANE, Miss J. Asst., Internat. Lab. of Marine Radioactivity.
Address: International Laboratory of Marine Radioactivity, Musee Oceanographique, Monaco.

KANE, Rajaram Purushottam, B.Sc., M.Sc., Ph.D. Born 1926. Educ.: Agra, Banaras, Bombay Univs. Fulbright and Smith-Mundt scholar, Chicago Univ., 1953-54, Assoc. Prof., Cosmic Ray Sect., Phys. Res. Lab., Navrangpura, Ahmedabad.
Nuclear interest: Research in time variation of cosmic rays.
Address: Physical Research Laboratory, Navrangpura, Ahmedabad 9, India.

KANEDA, Hiromu, Dr. Born 1906. Educ.: Kyoto Prefectural Medical Univ. Prof., Shinshu Univ., 1951-58; Director, Dept. of Radiol. and Prof., Kyoto Prefectural Medical Univ., 1958-. Societies: Counsellor, Japan Radiol. Soc.; Counsellor, Japanese Soc. Nucl. Medicine.
Nuclear interests: Radiation therapy (sieve irradiation); Sieve therapy of bronchogenic carcinoma.
Address: Department of Radiology, Kyoto Prefectural Medical University, Hirokoji, Kawara-machi, , Kyoto, Japan.

KANEKAR, Chandrakant Ramchandra, M.Sc. (Bombay), Ph.D. (Bombay), Ph.D. (London). Born 1919. Educ.: Bombay and London Univs. Res. Asst. in Chem., Inst. Sci., Bombay, 1948-55; Res. Officer, Atomic Energy Establishment, Trombay, Bombay, 1955-. Papers: 48 papers in magneto-chem., including nucl. magnetic resonance and Mössbauer effect. Societies: Fellow, Chem. Soc.; Fellow, Indian Chem. Soc.
Address: Modi Lodge, Vacha Ghandi Road, Gamdevi, Bombay 7, India.

KANEKO, Katsu, D. Sci. Born 1905. Educ.: Tokyo Univ. Director, Geological Survey of Japan, 1953. Vice-Pres., Tokyo Geographical Soc.; Exec. Member of Council, Geological Soc. of Japan; Member of Council, Japanese Assoc. of Petroleum Technologists; Member, Atomic Energy Special Com., Sci. Council of Japan. Publication: Natural Occurrences of Uranium in Japan (Geol. Survey of Japan, Jan. 1962).
Nuclear interests: Radioactive mineral resources, especially the deposits in sedimentary rocks.
Address: 2284, Kugenuma, Fujisawa City, Kanagawa Prefecture, Japan.

KANEKO, Osamu. Chief of Nucl. Res. Room, R. and D. Dept., Mitsui Mining and Smelting Co. Ltd.
Address: Mitsui Mining and Smelting Co. Ltd., 2-chome, Nihonbashi-Muromachi, Chuo-ku, Tokyo, Japan.

KANEKO, Toshi, Hon. Chairman, Fuji Bank Ltd. Director, Tokyo Atomic Ind. Consortium.
Address: The Fuji Bank Ltd., 6 1-chome, Otemachi, Chiyoda-ku, Tokyo, Japan.

KANELLOPOULOS, Th., Dr. Assoc. Prof. Sci. Director, and Formerly Director, now Member Sci. Com., "Democritos" Nucl. Res. Centre, Greek A.E.C. Member, Group of Experts on Nucl. Data, E.N.E.A., O.E.C.D.
Address: Greek Atomic Energy Commission, 5 Merlin Street, Athens (134), Greece.

KANEMATSU, Kazuo. Asst., Solid State Phys. (Experiment), Atomic Energy Res. Inst., Nihon Univ.
Address: Atomic Energy Research Institute, Nihon University, 1-8 Kanda Surugadai, Chiyodaku, Tokyo, Japan.

KANESTRØM, Ingolf, Dr.Phil. Born 1933. Educ.: Oslo Univ. NORDITA Fellowship, 1963-65; Res. Asst. (1962-63), Lecturer (1965-), Oslo Univ. Society: Norwegian Phys. Soc.
Nuclear interests: Theoretical nuclear physics: presently especially low energy nuclear structure of light nuclei; formerly problems concerning nuclear level densities at high excitation energies.
Address: Institute of Physics, Oslo University, Blindern, Norway.

KANG, Man-Sik, M.Sc. Born 1933. Educ.: Seoul Nat. Univ. and Kansas Univ. Full-time Instructor, Seoul Nat. Univ., 1959-60; Graduate Training in Rad. Biol. on Fellowship from Office of Atomic Energy, R.O.K., 1960-62; Res. Asst., Kansas Univ., 1961-62; Sen Res., Sci. Res. Inst., Seoul, 1963-. Societies: Zoological Soc. of Korea; Korean Chem. Soc.
Nuclear interests: Radiation biology and radiation biophysics.
Address: Radiochemistry Section, Scientific Research Institute, Ministry of National Defence, Seoul, Korea.

KANG, Yung Sun, Ph.D. Born 1917. Educ.: Hokkaido Imperial (Japan) and California (Berkeley) Univs. Asst. Prof., Zoology and Genetics (1946-51), Assoc. Prof., Zoology and Genetics (1951-53), Prof. (1953-), Liberal Arts and Sci. Coll., Head, Pre medical School (1957-60), Seoul Nat. Univ.; Res., Worcester Foundation for Exptl. Biol., Shrewsbury, Massachusetts, 1960-61. Member, N.A.S., Korea, 1954-; Com. member, Atomic Energy Res. Com., Seoul Nat. Univ., 1958-60; Member, Advisory Com., Office of Atomic Energy, Korea, 1959-; Vice-Pres. (1955-56), Pres. (1959-60, 1963-), Korean Zoological Soc.; Chairman, Korean Nat. Com. for Internat. Biol. Programme, 1965-.
Nuclear interests: Radiation genetics; Using X-ray and some kinds of isotopes to make any chromosome aberration and the other mutation with fruit flies (Drosophila), Chinese hamsters and human (cultured cells).
Address: 318 Buam-dong, Surdaemoon-Ku, Seoul, Korea.

KANIE, Matsuo, Dr. Medical and Agricultural Studies using Isotope Atoms as Tracer.
Address: Kagoshima University, Kagoshima, Japan.

KANNANGARA, Milindu Lakshaman Tennekoon, B.Sc. (Ceylon), Ph.D. (Manchester). Born 1915. Educ.: Ceylon Univ. Lecturer in Phys., Ceylon Univ.
Nuclear interests: Cosmic rays; Fundamental particles; Nuclear physics.
Address: Physical Laboratories, Ceylon University, Colombo 3, Ceylon.

KANNE, W. Rudolph, Ph.D. (Phys). Born 1913. Educ.: Johns Hopkins Univ. Formerly Manager, now Consulting Eng., Core and Fuel Eng., Atomic Power Equipment Dept., G.E.C. Societies: A.N.S.; A.P.S.
Nuclear interests: Nuclear physics; Reactor physics and reactor design; Nuclear instruments; Critical assemblies; Reactor tests and operations physics; Safeguards and hazard analysis.
Address: General Electric Company, Atomic Power Equipment Dept., 175 Curtner Avenue, San Jose, California, U.S.A.

KANNO, Akira. Chief Non-Destructive Testing Sect., Ship Res. Inst., Ministry of Transportation.
Address: Ministry of Transportation, Ship Research Institute, 1-1057 Mejiro-cho, Toshima-ku, Tokyo, Japan.

KANNO, Takiyi, D.Sc. Born 1920. Educ.: Tohoku Univ. Prof., Metal. of Radioelements Res. Div., Faculty of Medical Sci., Tohoku Univ. Res. Inst. of Mineral Dressing and Metal. Society: Atomic Energy Soc. of Japan.
Nuclear interest: Radiochemistry.
Address: 26 Fukusyuinmae, Nagamachi, Sendai, Japan.

KANNUNA, H. E. Abdul Karim, Dr. Born 1913. Educ.: Zurich Univ. Minister of Industry, Iraq. Chairman, Iraq A.E.C.
Address: Ministry of Industry, Baghdad, Iraq.

KANO, Shigeru. Asst. Prof. Rad. Measurement Dept., Graduate School of Nucl. Sci., Faculty of Sci., Tohoku Univ.
Address: Tohoku University, Katahiracho, Sendai, Japan.

KANTAN, Srinivas Kumar, B.Sc., B.Sc. (Tech). Born 1929. Educ.: Bombay Univ. Demonstrator in Chem. Eng., Bombay Univ., 1951-53; Head, Ceramic Sect., Atomic Fuels Div., Bhabha Atomic Res. Centre, Bombay, 1953-. at Centre de Recherches Metallurgiques de l'Ecole des Mines, Paris, 1959-60. Society: Indian Inst. of Metals.
Nuclear interests: Material science; Fabrication of ceramic nuclear fuel and fertile materials, involving general powder metallurgical techniques. Development of production methods for the manufacture of thorium, uranium carbide, etc. Actively associated with fabrication of fuel elements (zircaloy clad UO_2) for India's power reactors.
Address: Atomic Fuels Division, Bhabha Atomic Research Centre, Trombay, Bombay 74, India.

KANTEMIR, Izzet, Dr. Born 1903. Educ.: Istanbul Univ. Prof. Pharmacology, Medical Faculty, Ankara Univ. Chief, Pharmacological Sect., Oncological Inst. Medical Faculty; Gen. Sec., Turkish Assoc. Cancer Res. and Control; Radioisotope Com., Ahmet Andiçen Cancer Hospital. Society: Turkish Assoc. Cancer Res. and Control.
Nuclear interests: Thallium as metal; Hexachlorbenzene and alkylating agents as organic materials.
Address: Türk Kanser Arastirma ve Savas Kurumu, 73 Atatürk Bulvari, Ankara, Turkey.

KANTOLA, Martti Heikki, Ph.D. Born 1909. Educ.: Helsinki Univ. Res. Eng., State Inst. of Tech. Res., Helsinki, 1945-51; Prof. Phys. Turku Univ., 1950. Societies: Phys. Soc. of Finland; A.I.P.
Nuclear interest: Nuclear physics.
Address: 4a.A.10 Puolalanpuisto, Turku, Finland.

KANUNNIKOV, V. N. Papers: Co-author, Features of the 280-MeV Synchrotron at the Inst. of Phys., Acad. of Sci., U.S.S.R. (Supplement No. 4 of Atomnaya Energiya) (Moscow, Atomic Press, 1957). co-author, New Accelerator - Symmetrical Ring Phasotron Lebedev Phys. Inst., A U.S.S.R. Starting Up (letter to the Editor, Atomnaya Energiya, vol. 20, No. 6, 1966); co-author, Acceleration of Electrons in FIAN Ring Phasotron (ibid, vol. 21, No. 3, 1966).
Address: Academy of Sciences of the U.S.S.R., 14 Leninsky Prospekt, Moscow V-71, U.S.S.R.

KANY, John P. DE. See DE KANY, John P.

KAPCHIGASHEV, S. P. Papers: Co-author, Capture Cross Sections of some Structural Materials for Neutrons of Energies to 50 KeV (Atomnaya Energiya, vol. 15, No. 2, 1963); co-author, Capture Cross Sections of Energies to 50 KeV for Cr, Cr^{50}, Cr^{52} and Cr^{53} Nuclei (letter to the Editor, ibid, vol. 16, No. 3, 1964); 1 - 50,000 eV Neutron Capture Cross Sections by Nuclei of V, Zr, Zr^{90}, Zr^{91}, Zr^{94} (letter to the Editor, ibid, vol. 19, No. 3, 1965).
Address: Academy of Sciences of the U.S.S.R., 14 Leninsky Prospekt, Moscow V-71, U.S.S.R.

KAPCHINSKII, I. M. Papers: On the Realisation of Permissible Injection Currents in a Strong-Focussing Proton Synchroton (Atomnaya Energiya, vol. 13, No. 3, 1962); Longitudinal Coulomb Repulsion of Ions in Linear Accelerator in the Case of the Utmost High Value of the Beam Phase Space Density (ibid, vol. 25, No. 2, 1968).
Address: Academy of Sciences of the U.S.S.R., 14 Leninsky Prospekt, Moscow V-71, U.S.S.R.

KAPILA, Lieutenant Gen. C. C. Director, Medical Div., Atomic Energy Establishment Trombay.
Address: Atomic Energy Establishment Trombay, Apollo Pier Road, Bombay 1, India.

KAPITOLA, Jiri, M.D. Born 1928. Educ.: Charles Univ., Prague. Formerly Teaching Asst., IIIrd Medical Clinic, Faculty of Medicine; Now Res. Worker, Res. Lab. for Endocrinology and Metabolism, Charles Univ., Faculty of Medicine, Prague. Society: Sect. of Nucl. Medicine, Czechoslovak Medical Soc.
Nuclear interest: Nuclear medicine.
Address: 1 U nemocnice, Prague 2, Czechoslovakia.

KAPITSA, S. P. Paper: Co-author, 30 MeV Microtron-Injector for Pulsed Fast Reactor (Atomnaya Energiya, vol. 20, No. 2, 1966).
Address: Academy of Sciences of the U.S.S.R., 14 Leninsky Prospekt, Moscow V-71, U.S.S.R.

KAPITSA, Sergei. At Inst. of Phys. Problems, Moscow.
Address: Institute of Physical Problems, Academy of Sciences, of U.S.S.R., 2 Vorobevskoye Chaussée, Moscow, U.S.S.R.

KAPITZA, Peter, Ph.D. (Cantab.). Stalin Prize for Phys., 1941 and 1943; Order of Lenin, 1943, 1944 and 1945; Rutherford Medal and Prize, 1965. Educ.: Petrograd Polytech. Inst. (Faculty of Elec. Eng.) and Cambridge Univ. Lecturer, Petrograd Polytech. Inst., 1919-21; Clerk Maxwell Student, Cambridge Univ., 1923-26; Asst. Director, Magnetic Res.,Cavendish Lab., 1924-32; Royal Soc. Messel Res. Prof. and Director of Royal Soc. Mond. Lab., 1932-34; Director, Inst. of Phys. Problems, Acad. of Sci. of the U.S.S.R., 1935-. Fellow, Trinity Coll., Cambridge, 1925; Fellow, Royal Soc., 1929; Member, Acad. of Sci. of the U.S.S.R., 1939; Hon. Member, American Acad. of Sci. and Arts, 1968-.
Address: Institute of Physical Problems, Academy of Sciences of the U.S.S.R., 32 Kaluzhskaye Chaussee, Moscow, U.S.S.R.

KAPLAN, Bernard, B.S., Ph.D. Born 1921. Educ.: C.C.N.Y. and Ohio State Univ. Principal Eng., Aircraft Nucl. Propulsion Dept., G.E.C., 1953-61; Assoc. Prof. Phys., Air Force Inst. Technol., Wright-Patterson A.F.B., Ohio, 1961-. Societies: A.A.P.T.; A.N.S.
Nuclear interests: Reactor computations and mathematics; Education.
Address: 416 Rendale Place, Dayton 26, Ohio, U.S.A.

KAPLAN, Henry Seymour, B.S., D.M., M.S. (Radiol). Born 1918. Educ.: Rush Medical Coll., Chicago, Minnesota and Chicago Univs. Prof. and Exec., Radiol. Dept., Stanford Univ. School of Medicine, 1948-; Director, Biophys. Lab., Stanford Univ., 1957-62. Panel on Pathologic Effects of Atomic Rad., N.A.S. - Nat. Res. Council, 1957-. Societies: A.A.A.S.; American Assoc. Cancer Res.; American Cancer Soc.; A.C.R.; American Radium Soc.; Rad. Res. Soc.; Radiol. Soc. N. America; American Soc. Therapeutic Radiol.
Nuclear interests: Carcinogenesis by irradiation, especially leukemia induction by X-irradiation in mice; Biological and biochemical effects of radiation; Clinical and experimental radiation therapy of cancer.
Address: Department of Radiology, Stanford University School of Medicine, Palo Alto, California, U.S.A.

KAPLAN, Irving, A.B., Ph.D. Born 1912. Educ.: Columbia Univ. Sen. Phys., Brookhaven Nat. Lab., 1946-58; Prof. Nucl. Eng., M.I.T., 1958-. Book: Nucl. Phys. Papers: Various in Nucl. Reactor Phys., Reactor Design, Nucl. Phys., Isotope Separation. Societies: A.P.S.; A.A.A.S.; A.N.S.
Address: Nuclear Engineering Department, Massachusetts Institute of Technology, Cambridge, Massachusetts, U.S.A.

KAPLAN, Jerome I., Ph.D. Born 1926. Educ.: California Univ. Assoc. Prof., Res., Brown Univ., 1964-67. Fellow, Fundamental Physics Group., Battelle Memorial Inst. Society: A.P.S. Nuclear interest: Nuclear magnetic resonance in metals.
Address: Battelle Memorial Institute, Columbus, Ohio, U.S.A.

KAPLAN, Louis, Ph.D. Born 1917. Educ.: Stanford Univ. Sen. Chem., Argonne Nat. Lab., 1946-. Societies: A.C.S.; A.A.A.S.
Nuclear interests: Isotope effects; Radioactive tracers in organic chemistry; Organic radiation chemistry.
Address: Argonne National Laboratory, Argonne, Illinois, U.S.A.

KAPLAN, Reinhard Walter, Dr. phil. Born 1912. Educ.: Leipzig Univ. o. Prof. Microbiol., Director of Microbiol. Dept., Frankfurt-am Main Univ. Pres., Verband Deutscher Biologen. Societies: Deutsche Lichtkommitee; Deutsche Botanische Gesellschaft; Deutsche Gesellschaft für Hygiene und Mikrobiol.
Nuclear interest: Radiation genetics.
Address: 70 Siesmeyerstrasse, Frankfurt-am Main, Germany.

KAPLAN, S. A. Paper: Co-author, On the Theory of γ-Ray Non-Steady Multiple Compton Scattering (letter to the Editor, Atomnaya Energiya, Vol. 16, No. 2, 1964).
Address: Academy of Sciences of the U.S.S.R., 14 Leninsky Prospekt, Moscow V-71, U.S.S.R.

KAPLAN, Stanley, B.C.E., M.S., Ph.D. Born 1931. Educ.: C.C.N.Y., Pittsburgh Univ., and O.R.S.O.R.T. Sci. (1955-59), Sen. Sci. (1961-63), Fellow (1963-65), Advisory Sci. (1965-67), Bettis Atomic Power Lab.; Special Res. Fellow in Biomathematics, N.I.H.
Nuclear interests: Effective computational methods for realistic three dimensional

problems in reactor theory, especially in space-time dynamics, transport and diffusion theory. Approximation methods, synthesis techniques, calculus of variations, modern control theory. Inverse problems in nuclear medicine.
Address: Department of Electrical Engineering, Southern California University, Los Angeles, California 90007, U.S.A.

KAPLAN, Sylvan, J., B.A., M.A., Ph.D. Born 1919. Educ.: Texas and Stanford Univs. Director, Radiobiol. Lab., Texas Univ., U.S.A.F. School of Aviation Medicine, 1950-54; Prof. and Head, Dept. of Psychology, Texas Technol. Coll., 1954-62; Chairman, Board of Directors, Inst. of Internat. R. and D., 1960-; Deputy Director, Selection Div., U.S. Peace Corps, 1962-64; Chairman, Behavioural Sci. Dept., Armed Forces Radiobiol. Res. Inst. 1964-. Societies: Rad. Res. Soc.; American Psychological Assoc. Nuclear interest: Effects of radiation upon behaviour.
Address: 1421 Claremont Dr., Falls Church, Virginia, U.S.A.

KAPLON, Morton F., B.S. (Lehigh), M.S. (Lehigh), Ph.D. (Rochester). Born 1921. Educ.: Lehigh Univ. Assoc. Prof. (1955-61), Prof. Phys. (1961-63), Assoc. Dean, Coll. Arts and Sci. (1963-64), Chairman, Phys. and Astronomy Dept. (1964-), Rochester Univ.; N.S.F. Sen Fellow, 1960-61. Societies: Fellow, A.P.S.; Italian Phys. Soc.; A.A.A.S.; A.G.U. Nuclear interests: High energy particle physics; Primary cosmic radiation.
Address: Physics and Astronomy Department, Rochester University, Rochester 20, New York, U.S.A.

KAPPELMANN, Frederick Albert, B.S. (Chem. Eng.). Born 1921. Educ.: High Point Coll., North Carolina. With Union Carbide Nucl. Corp., O.R.N.L. 1944-. Society: A.C.S. Nuclear interests: Solvent extraction technology - such systems as: separation of the rare earths with TBP; separation and recovery of Np, Tc, and U from fluorination plant residues; plutonium extraction. The extraction behaviour of various elements.
Address: 106 Vernon Road, Oak Ridge, Tennessee, U.S.A.

KAPPER, R.S. Tech. Adviser, D. Hudig en Co.
Address: D. Hudig en Co., 23 Wijnhaven, Rotterdam, Netherlands.

KAPUSTIN, I. A. Paper: Co-author, Boron Isotope Separation with Phenetole-Boron Trifluoride Complex (Abstract, Atomnaya Energiya, vol. 20, No. 4, 1966).
Address: Academy of Sciences of the U.S.S.R., 14 Leninsky Prospekt, Moscow V-71, U.S.S.R.

KAPUSTIN, Yu. L. Paper: Accecorial Uraninite from Nepheline Syenites of Tuva (Atomnaya Energiya, vol. 20, No. 6, 1966).
Address: Academy of Sciences of the U.S.S.R., 14 Leninsky Prospekt, Moscow V-71, U.S.S.R.

KARABEKOV, I. P. Papers: Co-author, Production of Single-energy Beams of Accelerated Particles (letter to the Editor, Atomnaya Energiya, vol. 10, No. 3, 1961); co-author, Calculation and Designing Signal Electrodes for Accelerators (ibid, vol. 13, No. 4, 1962); Measurement of the Beam Aperture in an Accelerator Chamber without Beam Destruction (ibid, vol. 15, No. 6, 1963); Determination of Violation of Accelerator Parameters on Information about Lost Particles Distribution (letter to the Editor, ibid, vol. 21, No. 5, 1966).
Address: Academy of Sciences of the U.S.S.R., 14 Leninsky Prospekt, Moscow V-71, U.S.S.R.

KARADENIZ, Cemil, Dr. Nucl. Phys. Dept., Faculty of Sci., Istanbul Univ.
Address: Nuclear Physics Department, Faculty of Science, University of Istanbul, Beyazit-Istanbul, Turkey.

KARADY, Pal. Tech. Editor, Energia es Atomtechnika.
Address: Muszaki Konyvkiado, 22 Bajcsy-Zsilinszky ut, Budapest 5, Hungary.

KARALOVA, Z. K. Paper: Co-author, Total Neutron Cross Section of Th^{230} at Thermal Region (Atomnaya Energiya, vol. 24, No. 3, 1968).
Address: Academy of Sciences of the U.S.S.R., 14 Leninsky Prospekt, Moscow V-71, U.S.S.R.

KARAMIKHAILOVA, Elisaveta, Dr.; Member, Bulgarian Acad. Sci. Head Radioactivity Sect., Inst. of Phys., Sect. of Atom Phys.
Address: 1, 7th November Street, Sofia, Bulgaria.

KARAMYAN, A. T. Papers: Co-author, Relative Difference of $B^{11}F_3 - B^{10}F_3$ Steam Pressure (Atomnaya Energiya, vol. 19, No. 1, 1965); co-author, Elementary Separation Factor for Exchange Distillation of Compound $(CH_3)_2 O.BF_3$ (letter to the Editor, ibid, vol. 23, No. 3, 1967); co-author, Investigation of Pressure Influence on Isotope Separation B^{11} and B^{10} Process by Distillation of BF_3 (ibid, vol. 23, No. 4, 1967).
Address: Academy of Sciences of the U.S.S.R., 14 Leninsky Prospekt, Moscow V-71, U.S.S.R.

KARANDEEV, K. B. Director, Inst. of Automation and Electrometrics, Novosibirsk Sci. Centre, Acad. Sci. of the U.S.S.R.
Address: Novosibirsk Science Centre, 20 Sovietskaya Ulitsa, Novosibirsk, Siberia, U.S.S.R.

KARASEV, G. V. Paper: Co-author, Start-Stop Simulation Problems of Nucl. Reactors (letter to the Editor, Atomnaya Energiya, vol. 22, No. 6, 1967).

Address: Academy of Sciences of the U.S.S.R., 14 Leninsky Prospekt, Moscow V-71, U.S.S.R.

KARASEV, V. S. Papers: Co-author, Calorimetric Dosimetry at Nucl. Reactor (Atomnaya Energiya, vol. 21, No. 4, 1966); co-author Quasi-Stationary Calorimetric Method of Dosimetry of Powerful Fluxes of Ionizing Irradiations (letter to the Editor, ibid, vol. 21, No. 6, 1966); co-author, Exptl. Study of Irregularity of Flux Thermal Neutrons at Surface of Absorbed Specimen with Calorimeter Device (letter to the Editor, ibid, vol. 22, No. 4, 1967); co-author, Equipment for Temperature Control During In-Reactor Irradiation of Construction Materials Samples (ibid, vol. 22, No. 6, 1967).
Address: Academy of Sciences of the U.S.S.R., 14 Leninsky Prospekt, Moscow V-71, U.S.S.R.

KARCHER, K. H., Dr., Priv. Doz. Head, Dept. of Clinical-Exptl. Radiol., Univ. Strahlenklinik, (Czerny Krankenhaus).
Address: Universitäts Strahlenklinik (Czerny Krankenhaus), 3 Voss Strasse, Heidelberg, Germany.

KARCHER, Robert H., B.A. (Phys.). Born 1931. Educ.: Western Reserve Univ. Drexel Inst of Technol., Martin, 1956-58; Atomics Internat.,1958-61; Douglas, 1961-63; Project Sci., Holmes and Narver, Inc.,1963-. Societies: A.N.S.; A.S.A.; I.E.E.E.; American Ordnance Assoc.
Nuclear interests: Radiation transport and shielding; Applications of statistical methods and mathematical simulation; Nuclear weapons effects.
Address: Holmes and Narver, Incorporated, 828 So. Figueroa Street, Los Angeles 17, California, U.S.A.

KARCHES, Gerald J., B.S. (Phys.), M.S. (Hygiene). Born: 1934. Educ.: Xavier and Pittsburgh Univs. Asst. Chief, Radiol. Health Res. Activities, U.S. Dept. of Health, Educ., and Welfare. Societies: H.P.S.; Cincinnati Rad. Soc.
Nuclear interests: Environmental surveillance activities, especially around nuclear power reactors. Reducing unnecessary exposure to patients given diagnostic tracers and personnel in industrial uses of isotopes or machines producing ionising radiation.
Address: 4676 Columbia Parkway, Cincinnati, Ohio 45226, U.S.A.

KARGEROVA,Mrs. Vera. Born 1920. Editor Jaderná energie.
Nuclear interests: Nuclear physics and engineering.
Address: SNTL-Publishers of Technical Literature, 51 Spálená, Prague 1, Czechoslovakia.

KARHOV, A. N. Paper: Co-author, Dissociation of Fast H_2^+ Ions and Charge Exchange of Fast Protons in Lithium Arc (Abstract, Atomnaya Energiya, vol. 19, No. 4, 1965).

Address: Academy of Sciences of the U.S.S.R., 14 Leninsky Prospekt, Moscow V-71, U.S.S.R.

KARIM, Syed Mujtaba, B.Sc. (Hons.), M.Sc., Ph.D. (Alig.), Ph.D. (Toronto). Born 1908. Educ.: Muslim (Aligarh, India) and Toronto Univs. Head, Dept. of Phys. Karachi Univ. Societies: Fellow, Phys. Soc., London; Appl. Solar Energy Soc. Arizona, U.S.A.
Nuclear interests: Nuclear physics - isotopes, and mass spectroscopy.
Address: IIIC, 7/1A, Nazimabad, Karachi-18, Pakistan.

KARKHANAVALA, Minocher Dadabhoy, B.A. (Hons.), M.Sc., M.S., Ph.D. Born 1926. Educ.: Bombay, New York State and Pennsylvania State Univs. Sen. Res. Officer.
Nuclear interests: Reactor materials chemistry; Non-metallic fuel, and fertile materials; Fission products release and in-pile tests.
Address: Chemistry Division, Bhabha Atomic Research Centre, Trombay, Bombay 74 AS, India.

KARL, Clarence L. Born 1909. Educ.: Chicago and Washington Univs Manager, Cincinnati Area, U.S.A.E.C. Society: A.N.S.
Nuclear interest: Management.
Address: U.S. Atomic Energy Commission, Post Office Box 39188, Cincinnati 39, Ohio, U.S.A.

KARLIK, Berta, Dr. phil. Born 1904. Educ.: Vienna Univ. Director, Inst. for Radium Res. and Nucl. Phys., Vienna Acad. of Sci., 1945-. Pres. Chem. Phys. Soc. of Vienna, 1951-52. Societies: Acad. of Sci., Vienna; Acad. of Sci., Göteborg; Austrian Phys. Soc.; A.P.S.; Chem. Phys. Soc. of Vienna.
Address: Institut für Radiumforschung und Kernphysik, 3 Boltzmanngasse, Vienna 9, Austria.

KARLINER, M. M. Papers: Co-author, Phase Instability of Intensive Electron Beam in Storage Ring (Atomnaya Energiya, vol. 20, No. 3, 1966); co-author, Interaction of Coherent Betatron Oscillations with External Systems (ibid, vol. 22, No. 3, 1967); co-author, Investigation of Coherent Phase Self-Oscillations in Storage Rings (ibid, vol. 22, No. 3, 1967); co-author, Study of Self-Excitation and Rapid Damping of Coherent Transverse Oscillations in VEPP-2 Storage Ring (ibid, vol. 22, No. 3, 1967).
Address: Novosibirsk Science Centre, 20 Sovietskaya Ulitsa, Novosibirsk, Siberia, U.S.S.R.

KARLINER, Oskar, M.Sc. Head, Dept. for Foreign Relations, Office of Govt. High Commissioner for Peaceful Uses of Atomic Energy, Poland; Polish Adviser, I.A.E.A. Gen. Conference, Vienna, Sept. 1960, 1961, 1962, 1963, 1964, 1966, 1967.
Address: Office of Government High Commis-

sioner for Peaceful Uses of Atomic Energy, Palace of Culture and Science 18th floor, Warsaw, Poland.

KARMILOV, A. G. Paper: Co-author, Influence of Neutron Irradiation on Structure and Properties of Uranium Alloys with 0.6 - 9. 0 w/o of Mo (Atomnaya Energiya, vol. 22, No. 6, 1967). Address: Academy of Sciences of the U.S.S.R., 14 Leninsky Prospekt, Moscow V-71, U.S.S.R.

KARMOHAPATRO, Surjendu Bikas, D.Ph. (Phys.). Born 1924. Educ.: Calcutta Univ. Res. Asst. (1952-61), Reader (1961-66), Assoc. Prof. (1967-), Saha Inst. Nucl. Phys. Society: Fellow, Indian Phys. Soc.
Nuclear interests: Nuclear physics; Atomic collision processes in gases and solids; Spectrometry of charged particles; Isotope separation; Ion physics.
Address: Saha Institute of Nuclear Physics, 92 Acharya Prafulla Chandra Road, Calcutta 9, India.

KARNAUHOV, I. M. Paper: Co-author, Source of Polarized Ions with Current 1.2 mka (letter to the Editor, Atomnaya Energiya, vol. 21, No. 2, 1966).
Address: Academy of Sciences of the U.S.S.R., 14 Leninsky Prospekt, Moscow V-71, U.S.S.R.

KARNIKOVA, Eva, Dipl. Ing. Born 1928. Educ.: Prague High Tech. School. Státní výzkumný ústav ochrany materiálu G.V. Akimova, 1956-.
Nuclear interest: Corrosion in nuclear energy.
Address: Státní výzkumný ústav ochrany materiálu G.V.Akimova, 4 U mestanského pivovaru, Prague 7, Czechoslovakia.

KARP, William, A.B. (Chem.). Born 1924. Educ.: Boston Univ. Head, Health Phys. Dept. and Rad. Sources Dept., Tracerlab. Div., L.F.E. Inc., Waltham,Massachusetts.
Nuclear interests: Health physics; Industrial applications of radioisotopes; Design of radiation sources.
Address: 1601 Trapelo Road, Waltham, Massachusetts, U.S.A.

KARPEKOV, M. Soviet deleg., Convention on Thermonucl. Processes, Inst. of Elec. Eng., London, April 29-30, 1959.
Address: Academy of Sciences of the U.S.S.R., 14 Leninsky Prospekt, Moscow V-71, U.S.S.R.

KARPLUS, Robert, S.B., A.M., Ph.D. Born 1927. Educ.: Harvard Univ. Prof. Phys., Phys. Dept., California Univ. Society: A.P.S.
Nuclear interest: Theoretical physics.
Address: Physics Department, California University, Berkeley 4, California, U.S.A.

KARPOV, V. L. Papers: Co-author, Large Rad.-Chem. Unit with Irradiator Made from Spent Fuel Elements of a Nucl. Reactor (Atomnaya Energiya, vol. 15, No. 4, 1963); co-author, Rad. Cross-Linking of Polyethylene Insulation of Cable Articles on Large Scale (letter to the Editor, ibid, vol. 21, No. 1, 1966).
Address: Academy of Sciences of the U.S.S.R., 14 Leninsky Prospekt, Moscow V-71, U.S.S.R.

KARPUHIN, O. A. Paper: Co-author, Apparatus for Neutron Activation Analysis (Atomnaya Energiya, vol. 20, No. 3, 1966).
Address: Academy of Sciences of the U.S.S.R., 14 Leninsky Prospekt, Moscow V-71, U.S.S.R.

KARPUSHKINA, E. I. Paper: Photoexcitation and Photoionisation of Energetic Hydrogen Atoms (Atomnaya Energiya, vol. 20, No. 5, 1966).
Address: Academy of Sciences of the U.S.S.R., 14 Leninsky Prospekt, Moscow V-71, U.S.S.R.

KARR, H. German Member, Group of Experts on Third Party Liability, O.E.C.D., E.N.E.A.
Address: O.E.C.D. European Nuclear Energy Agency, 38 boulevard Suchet, Paris 16, France.

KARR, Hugh James, A.B., M.S., Ph.D. Born 1916. Educ.: Washington Univ., St. Louis. Staff member, Los Alamos Sci. Lab. Society: A.P.S.
Nuclear interests: Nuclear physics: Plasma physics; Magneto hydrodynamics; Reactors (experimental).
Address: 1509 42nd Street, Los Alamos, New Mexico, U.S.A.

KARRAS, Reino Matti, Ph.D. Born 1932. Educ.: Helsinki Univ. Res. Asst., Finnish A.E.C. at Helsinki Univ., 1959-60; Asst. Prof., Arkansas Univ., 1961-62; Assoc. Prof. Oulu Univ., 1962-65; Sen. Fellow, Nat. Res. Council for Tech. Sci., 1965-.
Nuclear interest: Nuclear spectroscopy.
Address: Physics Department, Oulu University, Oulu, Finland.

KARRER, Lawrence E. Sen. Vice Pres., Puget Sound Power and Light Co.
Address: Puget Sound Power and Light Co., Puget Power Building, Bellevue, Washington 98004, U.S.A.

KARTASHEV, E. P. Paper: Co-author, On Some Rapidity Criteria of Neutron-Activation Analysis (letter to the Editor, Atomnaya Energiya, vol. 22, No. 1, 1967).
Address: Academy of Sciences of the U.S.S.R., 14 Leninsky Prospekt, Moscow V-71, U.S.S.R.

KARTASHEV, K. B. Papers: Co-author, On Increasing of Ion Trapping into Magnetic Confinement of System by Neutral Atom Photoionization (Atomnaya Energiya, vol. 21, No. 4, 1966); co-author, Ion Stram Interaction with Target Plasma during Recharging (ibid, vol. 25, No. 2, 1968).
Address: Academy of Sciences of the U.S.S.R., 14 Leninsky Prospekt, Moscow V-71, U.S.S.R.

KARTASHOV, N. P. Papers: Co-author, Modification of the Equilibrium between Radon and its Decay Products in an Air Current (letter to the Editor, Atomnaya Energiya, vol. 6, No. 5, 1959); Determination of Traces of Uranium, Thorium and Potassium in Rocks by Means of γ-Ray Spectra (letter to the Editor, ibid, vol. 10, No. 5, 1961); co-author, Determination of the Concentration of an Aerosol Formed by the Short-Lived Decay Products of Radon (letter to the Editor, ibid, vol. 12, No. 4, 1962); Express - Method for Determination of Concentration of Aerosol RaA and 'Latent Energy' in Air (letter to the editor, ibid, vol. 20, No. 5, 1966).
Address: Academy of Sciences of the U.S.S.R., 14 Leninsky Prospekt, Moscow V-71, U.S.S.R.

KARTER, Peter, B.S., M.A. Born 1922. Educ.: U.S. Military Acad., West Point, and Harvard Univ. Captain, U.S. Army, - 1957; Supervisor, Waste Management Systems, Advanced Products Group, American Machine and Foundry, Atomics Div.
Address: Daleview Drive, York, Pa., U.S.A.

KARTOVITSKAYA, M. A. Paper: Co-author, Effectiveness of Borating of Metal-Water Shieldings (Atomnaya Energiya, vol. 21, No. 4, 1966).
Address: Academy of Sciences of the U.S.S.R., 14 Leninsky Prospekt, Moscow V-71, U.S.S.R.

KASATCHKOVSKY, O. D. Director Atomic Reactors Res. Inst. U.S.S.R. State Com. for Utilisation of Atomic Energy.
Address: Atomic Reactors Research Institute, U.S.S.R. State Committee for the Utilisation of Atomic Energy, 26 Staromonetnii Pereulok, Moscow, U.S.S.R.

KASBERG, Alvin H., B.S., M.S. Educ.: Wisconsin Univ. Supervisory Eng. i/c melting, welding and fabrication of reactor core components, and fabrication of major core assemblies, Westinghouse Electric Corp., Bettis Plant, 1950-57; Manager, Metal. and Ceramics, Nucl. Materials and Equipment Corp., Pennsylvania, 1957-.
Nuclear interests: Metallurgy; Ceramics; Fuel element development; Control element development.
Address: Nuclear Materials and Equipment Corporation, P.O. Box 436, Apollo, Pennsylvania 15613, U.S.A.

KASCHLUHN, Frank, Prof. mit Lehrstuhl, Dr. rer. nat. habil. Born 1927. Educ.: Jena Univ. Director, Inst. for Theoretical Phys., Humboldt-Univ., Berlin, 1960-. Member, Theoretical Council, Joint Inst. for Nucl. Res., Dubna. Society: Physikalische Gesellschaft, Deutschen Demokratischen Republik.
Nuclear interests: Theoretical high energy physics, in particular dynamical S-matrix theory.
Address: Institut für Theoretische Physik der Humboldt-Universität, 6 Unter den Linden, 108 Berlin, German Democratic Republic.

KASCHTOWICZKA, K., Gen. Dir. Dr. Rep. of Reaktor-Interessengemeinschaft, Koordinationsausschuss Österreichisches Kernenergieprogramm.
Address: Koordinationsausschuss Österreichisches Kernenergieprogramm, c/o Gen. Dir. Dipl. Ing. F. Hintermayer, p.A. Verbundgesellschaft, 6A Am Hof, Vienna 1, Austria.

KASEDA, Noboru. Asst. Chief, Reactor Licensing and Regulation Sect., Atomic Energy Bureau, A.E.C. of Japan.
Address: Atomic Energy Commission of Japan, 3-4 Kasumigaseki, Chiyoda-ku, Tokyo, Japan.

KASHUKEEV, Nicifor Todorov. Born 1917. Educ.: Sofia Univ. Inst. Phys. and Sofia Univ., 1949-; Joint Inst. Nucl Res., Dubna, U.S.S.R., 1957-59. Union of the Sci. Workers, Bulgaria. Society: Sect. Nucl. Phys. of Low Energies, Joint Inst. Nucl. Res., Dubna, U.S.S.R.
Nuclear interest: Nuclear physics.
Address: Neutron Physics Section, Institute of Physics and Atomic Research Centre of the Bulgarian Academy of Sciences, 152 Lenin Street, Sofia 13, Bulgaria.

KASHY, E. Exptl. Nucl. Phys. Cyclotron, Dept. of Phys. and Astronomy, Michigan State Univ.
Address: Michigan State University, East Lansing, Michigan, U.S.A.

KASKA, Nurettin, Ph.D. Born 1930. Educ.: Ankara Univ. Assoc. Prof., Faculty of Agriculture, Ankara Univ., 1954-67; P.S.O., Cento Inst. Nucl. Sci., Teheran, 1963-64.
Nuclear interest: Use of radioisotopes and radiations in horticulture, mainly nutrition of fruit trees, root growth of fruit trees, effects of radiation on pollen, seed and cuttings, and autoradiography.
Address: Faculty of Agriculture, Ankara University, Ankara, Turkey.

KASSCHAU, Kenneth, Mech. Eng., M.S. (Mech. Eng.). Born 1915. Educ.: Stevens Inst.of Technol and Harvard Univ. Mech. Eng., N.E.P.A. Project, 1948-50; Chief, Reactor Branch (1950-52), Director of Res. and Medicine Div. (1952-54), Oak Ridge Operations Office, U.S.A.E.C., Manager of Eng. Atomic Energy, Alco Products, Inc., Schenectady, N.Y., 1954-62; Manager, Advanced Projects, Westinghouse Astronucl. Lab., 1962-66; Manager,Product Eng., Westinghouse Marine Div., 1966-. Member, Board of Directors, North-east New York Sect., American Nucl. Soc. Societies: A.N.S.; Atomic Indu. Forum; Soc. of Automotive Eng.; American Soc. Naval Eng.; Soc. Naval Architects and Marine Eng.
Nuclear interests: Nuclear power for military and space applications.
Address: 248 Greenoaks Drive, Atherton, California 94025, U.S.A.

KASSEM, K. Abdel Khalik, M.B., Ch.B., D.M.R.E., M.D. (Rad.). Born 1914. Educ.: Edinburgh, Cairo and Cambridge Univs. Prof., Faculty of Medicine, Cairo Univ. Societies: Faculty of Radiologists, London; British Inst. of Radiol.
Nuclear interests: Application to diagnostic radiology; Radiation hazards.
Address: Cairo University Faculty of Medicine, Kasr el Aini, Cairo, Egypt, United Arab Republic.

KASTEN, Paul Rudolph, B.S. and M.S. (Missouri), Ph.D. (Minnesota). Born 1923. Educ.: Missouri Univ. School of Mines and Minnesota Univ. Staff Member (1950-55), Chief, Reactor Analysis Sect. (1955-60), Assoc. Head, Reactor Analysis Dept. (1960-63), Assoc. Director Molten Salt Reactor Programme (1964-), O.R.N.L.; Lecturer Chem. and Nucl. Eng. (1953-61), Prof., Nucl. Eng. Dept. (1964-), Tennessee Univ.; Guest Director, Reactor Development Inst., Kernforschungsanlage, Jülich, 1963-64.
Nuclear interests: Comparative economic evaluation of nuclear power reactors; Interplay of reactor statics, kinetics, and engineering factors in reactor design and development; Reactor development.
Address: Oak Ridge National Laboratory, P.O.Box Y, Oak Ridge, Tennessee, U.S.A.

KASTLER, Alfred, Prof. Nobel Prize for Physics, 1966. Born 1902. Director, Sect. de Besançon, Lab. de l'Horloge Atomique, C.N.R.S. At l'Ecole Normale Supérieure, Paris.
Address: Laboratoire de l'Horloge Atomique, 32 avenue de l'Observatoire, Besançon, (Doubs), France.

KASTNER, Jacob, B.Sc. (Hons.), M.A., Ph.D. Born 1919. Educ.: Manitoba and Toronto Univs. Res. Officer, N.R.C., Canada, 1950-52; Radio Phys., G.E.C., 1952-59; Director, Res., Picker X-Ray Corp., 1959-62; Assoc. Phys., Argonne Nat. Lab., 1962-. Pres., Midwest Chapter, American Assoc. of Medical Phys.; Director, Midwest Chapter, H.P.S. Societies: A.N.S.; Rad. Res. Soc.; Soc. Non-Destructive Testing.
Nuclear interests: Radiological physics; Interaction of radiation with matter; Radiation potential.
Address: 840 Oxford Street, Downers Grove, Illinois 60439, U.S.A.

KAT, Dirk, Mech. Eng. Born 1925. Educ.: High School for Tech. Sci., Amsterdam. Asst. of Head of Boilers and Pressure Vessels Dept., Werkspoor N.V., Amsterdam, 1950-56; Asst. Managing Director (1956-59), Managing Director (1959-68), Röntgen Technische Dienst N.V., Rotterdam; Managing Director, Röntgen Technischer Dienst G.m.b.H., Wels, Austria, 1968-. Societies: Industriële Vereniging tot Bevordering van de Stralingsveiligheid; Nondestructive Testing Soc.; Inst. Phys.; A.S.T.M.; Nederlands Instituut voor Lastechniek; Koninklijk Instituut van Ing.; Internat. Sci. Assoc. Eurotest, Brussels; A.I.M.M.E.
Nuclear interests: Metallurgy and reactor design with respect to non-destructive testing, in particular quality-control on boilers and pressure vessels and other reactor components; Nuclear physics in connection with the application of radioactive material in industry.
Address: Röntgen Technische Dienst N.V., 144 Delftweg, Rotterdam 8, Netherlands.

KATAGIRI, Masayuki, Dr., Prof. Member, Radioisotope Res. Com., Kanazawa Univ.
Address: Kanazawa University, Marunouchi 1-1, Kanazawa, Ishikawa Prefecture, Japan.

KATAGIRI, Michio, B.Sc. Born 1924. Educ.: Tokyo Univ. Now at Yokogawa Elec. Works, Ltd., Tokyo. Societies: Atomic Energy Soc. of Japan; Elec. Eng. Soc. of Japan.
Nuclear interests: Nuclear reactor control and instrumentation; Nuclear radiation applied instruments.
Address: Yokogawa Electric Works Ltd., 3000 Kichijoji Musashinoshi, Tokyo, Japan.

KATAKIS, Demetrius, Chem. Eng. Diploma, M.S., Ph.D. (Chem.). Born 1931. Educ.: Nat. Tech. Univ., Athens, Greece and Chicago Univ. Res. Assoc., Brookhaven Nat. Lab.; Res. Group Leader, Greek A.E.C.
Nuclear interest: Radiation chemistry.
Address: Nuclear Research Centre Democritos, Aghia Paraskevi Attikis, Athens, Greece.

KATAL'NIKOV, S. G. Papers: Co-author, Partition Coefficient for Lithium Isotopes in Vacuum Distillation (Atomnaya Energiya, vol. 11, No. 3, 1961); co-author, Determination of the Separation Factor for Lithium Isotopes in Ion Exchange (ibid, No. 6); co-author, Boron Isotope Separation with Phenetole-Boron Trifluoride Complex (Abstract, ibid, vol. 20, No. 4, 1966); co-author, Separation of Boron Isotopes on Basis of Isotopic Exchange Reaction between BF_3 and BF_3-phenetole Complex (ibid, vol. 22, No. 4, 1967).
Address: Academy of Sciences of the U.S.S.R., 14 Leninsky Prospekt, Moscow V-71, U.S.S.R.

KATAOKA, Iwao, Dr. Eng. Born 1930. Educ.: Tokyo Coll. of Sci. Researcher, 1954-61; Chief. Nucl. Ship Equipment Lab., Marine Engine Div., Transportation Tech. Res. Inst., 1961-63; Chief, Shielding Lab., Nucl. Ship Div., Ship Res. Inst., Ministry of Transport, 1963-; Lecturer, Faculty of Eng., Tokyo Univ. 1964-. Societies: Atomic Energy Soc. of Japan; Soc. of Naval Architects of Japan.
Nuclear interests: Radiation shielding; Marine reactor.
Address: 1-35-9 Sakuragaoka, Tama-machi, Minamitama-gun, Tokyo, Japan.

KATAOKA, Sentaro. Pres., Ando Elec. Co. Ltd.
Address: Ando Electric Co. Ltd., 4, 3-chome, Naka Kamata, Ota-ku, Tokyo, Japan.

KATASE, Akira. Born 1926. Educ.: Kyoto Univ. Asst. Prof., Nucl. Eng., Kyushu Univ., 1958-. Societies: Phys. Soc. of Japan; Atomic Energy Soc, of Japan; A.P.S.
Nuclear interests: Nuclear physics; Neutron physics; Reactor physics.
Address: Department of Nuclear Engineering, Faculty of Engineering, Kyushu University, Fukuoka, Japan.

KATAYAMA, Saburo, B.Sc. Born 1920. Educ.: Kyoto Imperial Univ. Vice-chief, Eng. Dept., Headquarters, Furukawa Elec. Co. Ltd. Societies: Atomic Energy Soc. of Japan; A.S.Q.C. (U.S.A.).
Nuclear interests: Metallurgy, fuel manufacturing.
Address: 27 Yoshikubo-cho, Megro, Tokyo, Japan.

KATCHUR, Lidia A., Dr. Sen. Sci. Worker, V.G. Khlopin Radium Inst.; U.S.S.R. expert, Advisory Panel on Rad., W.H.O.
Address: V.G. Khlopin Radium Institute, 1 Roentgen Ulitsa, Leningrad, U.S.S.R.

KATHREN, Ronald L., B.S., M.S. Born 1937. Educ.: California and Pittsburgh Univs. Industrial Hygienist, Los Angeles City Health Dept., 1957-59; Health Phys. (1959-60), Supervisory Health Phys. (1960-62), Mare Island Naval Shipyard; Health Phys., Lawrence Rad. Lab., California Univ., 1962-67; Instructor, Rad. Technol., Chabot Coll., California, 1966-67; Lecturer, Rad. Technol, Columbia Basin Coll., 1967-; Manager, External Dose Evaluation Unit, Battelle-Northwest, Richland, 1967-. Abstractor, Chem., Abstracts, 1962-; Member, Rad. Com., A.I.H.A., 1964-; Member, Interprofessional Relations Com., 1966-67. Societies: Charter Member, Conference on Radiol. Health; H.P.S.; A.A.A.S.; American Assoc. of Phys. in Medicine.
Nuclear interests: Health and radiological physics, with special interest in radiation dosimetry, instrumentation, radiological standardisation and the history of radiation protection and radiology; Education and training.
Address: Battelle-Northwest, P.O. Box 999, Richland, Washington 99352, U.S.A.

KATO, Goichi. Pres., Mitsui Ship Building and Eng. Co., Ltd.; Vice-Pres., Nippon Atomic Energy Industry Group.
Address: c/o Daiichi Bussan Kaisha, Ltd., 1-chome, Shibatamura-cho Minatoku, Tokyo, Japan.

KATO, H. T., Dr. Wissenschaftliche Asst., Dept. of Clinical-Exptl. Radiol., Univ. Strahlenklinik (Czerny Krankenhaus).
Address: Universitäts Strahlenklinik (Czerny Krankenhaus), 3 Voss Strasse, Heidelberg, Germany.

KATO, Masao, Prof., Dr. Eng. (Tokyo), Ph.D. Born 1916. Educ.: Tokyo Univ. At Tokyo Univ., 1940-. Books: Atomic Energy Eng., 4th volume, Ind. Application of Radioisotopes (Kyoritsu Publishing, Ltd., 1957); Exptl. Technique of Radioisotopes, 3rd volume Ind. Application of Radioisotopes (Nanko, Ltd., 1957); Studies on Corrosion of Aluminium Alloys for Water Cooled Reactor (Light Metal Soc., 1960). Societies: Japan Inst. of Metals; Japan Inst. of Light Metals; Atomic Energy Soc. of Japan; Japan Radioisotope Assoc.
Nuclear interests: Application of radioisotopes in industry; Metallurgy (e.g., metallic material for reactor).
Address: Institute of Industrial Science, Tokyo University, 22-1, Roppongi 7-chome, Minato-ku, Tokyo, Japan.

KATO, Ryuhei, Prof., Director and Chief Operator, Nucl. Lab., Faculty of Sci. and Eng., Ritumeikan Univ.
Address: Nuclear Laboratory, Faculty of Science and Engineering, Ritumeikan University, 28 Toziin Kitamati, Kyoto, Japan.

KATO, Walter Y., B.S., M.S., Ph.D. Born 1924. Educ.: Haverford Coll., Illinois and Pennsylvania State Univs. Res. Assoc., Ordnance Res. Lab., School of Eng., Pennsylvania State Univ., 1949-52; Jun. Res. Assoc., Phys. Dept., Brookhaven Nat. Lab., 1952-53; Assoc. Phys., Reactor Eng. Div. (1953-63), Head, Fast Reactor Experiments Sect., Reactor Phys. Div. (1963-67), Sen. Phys. (1967-), Argonne Nat. Lab. Fulbright Lecture and Res. Fellowship, 1958-59, affiliated with Japan Atomic Energy Res. Inst. Societies: A.P.S.; A.N.S.
Nuclear interests: Experimental reactor physics; Neutron physics; Critical assembly experiments; Nuclear physics; Fast reactor physics.
Address: Argonne National Laboratory, 9700 South Cass Avenue, Argonne, Illinois, U.S.A.

KATORI, Kazuo, D.Eng. Pres., Government Ind. Res. Inst. Nagoya, Agency of Ind. Sci. and Technol., Ministry of Internat. Trade and Industry.
Address: Government Industrial Research Institute, Ministry of International Trade and Industry, Hirate-machi, Kita-ku, Nagoya, Japan.

KATRICH, N. P. Papers: Co-author, H_1^+ Fast Ions Interaction with Metal Surface in Ultrahigh Vacuum (Atomnaya Energiya, vol. 21, No. 5, 1966); co-author, Determination of H_1^+ Fast Ions Inclusion by Condensation Methods (ibid, vol. 23, No. 2, 1967); Self Desorption of Hydrogen Included in Metals (ibid, vol. 23, No. 6, 1967); co-author, Desorption of Hydrogen Included in Nickel and and Titane (ibid, vol. 23, No. 6, 1967).
Address: Academy of Sciences of the U.S.S.R., 14 Leninsky Prospekt, Moscow V-71, U.S.S.R.

KATSAUROV, L. N.
Nuclear interest: Fusion reactions.
Address: Academy of Sciences of the U.S.S.R. 14 Leninsky Prospekt, Moscow V-71, U.S.S.R.

KATSOYANNIS, Panayotis G., Dr. Formerly with School of Medicine, Pittsburgh Univ.; Head. Div. of Biochem., Medical Dept., Brookhaven Nat. Lab., 1964-.
Address: Brookhaven National Laboratory, Upton, Long Island, New York, U.S.A.

KATSUME, Takuo, Dr. Medical and agricultural studies using isotope atoms as tracer.
Address: Kagoshima University, Kagoshima, Japan.

KATSUMORI, Hiroshi. Asst. Prof., Dept. of Nucl. Eng., Faculty of Eng., Kyoto Univ.
Address: Kyoto University, Department of Nuclear Engineering, Yoshida-Honmachi, Sakyo-ku, Kyoto, Japan.

KATSURAYAMA, Yukinori, Dr. Prof., Div. of Rad. Control, Res. Reactor Inst., Kyoto Univ.
Address: Kyoto University, Research Reactor Institute, Noda, Kumatori-cho, Sen-nan-gun, Asaka Prefecture , Japan.

KATZ, Donald LaVerne, B.S.E.,Ph.D. Born 1907. Educ.: Michigan Univ. Prof. Chem. Eng., Chairman of Dept., Michigan Univ., 1951-62. Pres., A.I.Ch.E., 1959. Book: Co-author, Heat Transfer and Fluid Dynamics. Societies: Fellow, A.N.S.; A.I.Ch.E.; A.S.M.E.; A.I.M.M.E. Nuclear interests: General, educational, reactor design, chemical processing of fuels.
Address: 2028 E. Engineering Bldg., Michigan University, Ann Arbor, Michigan, U.S.A.

KATZ, Elli.
Address: Joint Institute for Nuclear Research, Dubna, Nr. Moscow, U.S.S.R.

KATZ, Joseph J., Ph.D. Born 1912. Educ.: Chicago Univ. Sen. Chem., Argonne Nat. Lab., 1943-; Professorial Lecturer, Chicago Univ., 1964-. Books: Co-author, The Chem. of the Actinide Elements (Methuen, 1957); co-author, Res. U.S.A. (McGraw-Hill, 1964). Society: A.C.S.
Nuclear interest: Deuterium isotope effects in chemistry and biology.
Address: Argonne National Laboratory, 9700 South Cass Avenue, Argonne, Illinois 60439, U.S.A.

KATZ, Leon, B.Sc. and M.Sc. (Queen's, Ontario), Ph.D. (California Inst. Technol.). Born 1909. Educ.: Queen's Univ., Kingston, Ontario and California Inst. Technol. Prof. Phys. and Director, Linear Accelerator Lab., Head, Phys. Dept., (1965-), Saskatchewan Univ. Societies: Fellow, Roy. Soc. of Canada; Fellow, A.P.S.; Past Pres., Canadian Assoc. of Physicists; Phys. Soc.
Nuclear interest: Nuclear structure.
Address: Accelerator Laboratory, Saskatchewan University, Saskatoon, Sask., Canada.

KATZ-SUCHY, Julian, Member, State Council for the Peaceful Use of Nucl. Energy.
Address: State Council for the Peaceful Use of Nuclear Energy, Room 1819, Palace of Culture and Science, Warsaw, Poland.

KAUDERN, G. Swedish Member, Study Group on Nucl. Ship Propulsion, O.E.C.D., E.N.E.A.
Address: O.E.C.D. European Nuclear Energy Agency, 38 boulevard Suchet, Paris 16, France.

KAUFMAN, A. S., B.Sc., Ph.D., F. Inst. P. Born 1925. Educ.: Edinburgh Univ. Res. Lab., A.E.I. Ltd., Aldermaston, Berks. 1954-59; Sen. Lecturer, Hebrew Univ., Jerusalem, 1959-. Societies: Israel Phys. Soc.; Israel Astronomical Soc.; Inst. Phys. and Phys. Soc. (Great Britain). Nuclear interest: Thermonuclear reactions.
Address: Department of Physics, Hebrew University, Jerusalem, Israel.

KAUFMAN, Ernest E., Dr. Chem. Dept. Director, Saint Mary's Coll.
Address: Saint Mary's College, Terrace Heights, Winona, Minnesota, U.S.A.

KAUFMAN, J. V. Richard, B.S. (Chem.), Ph.D. (Inorganic Chem.). Born 1917. Educ.: Dickinson Coll., Carlisle, Pa. and M.I.T. Sec. of the Army Res. and Study Fellowship, Solid State Phys., Reading Univ., 1959-60. Societies: A.C.S.; A.P.S.; New York Acad. of Sci. Nuclear interests: Research in solid state physics and chemistry, using the nuclear reactor as a tool; Fundamental studies on explosives and related compounds.
Address: 53 Main Street, Sparta, New Jersey, U.S.A.

KAUFMAN, Raymond, B.S., M.S., Ph.D. Born 1917. Educ.: C.C.N.Y. and New York Univ. Director, Res., Del Electronics Corp. Nuclear interests: Instrumentation-radiation detectors; Aerosol collection and measuring devices; Health physics; Plasma physics; Equipment devices; Energy discharge systems.
Address: 250 E. Sandford Boulevard,Mount Vernon, New York, U.S.A.

KAUFMAN, Samuel J. Reactor Exptl. Projects Panel, N.A.S.
Address: National Aeronautics and Space Administration, Lewis Research Centre, 21000 Brookpark Road, Cleveland 35, Ohio, U.S.A.

KAUFMAN, W. J., Mr. i/c of Nucl. Dept., Koninklijke Machinefabriek Gebr. Stork en Co. N.V.
Address: Koninklijke Machinefabriek Gebr. Stork en Co. N.V., Hengelo, Netherlands.

KAUFMAN, Warren John, B.S., M.S., Sc.D. Born 1922. Educ.: California and Harvard Univs. and M.I.T. Prof. Sanitary Eng. 1950-61; various consulting assignments with U.S.A.E.C., I.A.E.A., etc. Societies: A.S.C.E.; A.W.W.A.; A.P.H.A.
Nuclear interest: Radioactive waste disposal and the application of radioisotopes to hydraulic and hydrologic problems.

Address: Room 112, Engineering Building, California University, Berkeley, California, U.S.A.

KAUFMANN, Albert R., B.S. (Mech. Eng.), Sc.D. (Metal.). Born 1911. Educ.: Lafayette Col. and M.I.T. Vice-Pres. and Tech. Director, Nucl. Metals, Inc., 1954; Sen. Sci.,Nucl. Metals Div., Whittaker Corp., 1966-. Societies: A.P.S.; A.S.M.; A.I.M.M.E.; A.N.S.
Nuclear interest: Metallurgy.
Address: Whittaker Corporation, Nuclear Metals Division, Concord, Massachusetts, U.S.A.

KAUFMANN, Ernst. Member, Federal Commission for Nucl. Safety Control.
Address: Federal Commission for Nuclear Safety Control, c/o Dr. Peter Courvoisier, Federal Institute for Research on Reactors, Würenlingen, Aargau, Switzerland.

KAUFMANN, Heinz, Attorney. Geschäftsführer, Kernreaktor-Finanzierungs-Gesellschaft m.b.H.
Address: Kernreaktor-Finanzierungs-Gesellschaft m.b.H., 45 Brüningstrasse, Frankfurt (M)-Hoechst, Germany.

KAUFMANN, John Francis, B.S.Ch.E. Born 1922. Educ.: Drexel Inst. of Technol., Brooklyn Polytech. and Tennessee and Idaho Univs. Project Eng., Civilian Power (1953-54), Sen. Project Eng., Sodium Cooled Reactors (1954-55), Chief, Education and Training (1955-57), Chief, Tech. Services Branch (1957-58), Chief, High Temperature Reactors Branch (1958-59), Deputy Asst. Director and Chief, Evaluation and Planning, Office of Civilian Power (1959-61), Deputy Asst. Director, Office of Advanced Concepts (1961-62), Reactor Development Div., U.S.A.E.C.; Asst. Manager, Tech. Operations,Idaho Operations Office, U.S.A.E.C., 1962-. Society: A.N.S.
Nuclear interests: Management of reactor and nuclear programmes and projects.
Address: 564 Safstrom Drive, Idaho Falls, Idaho, U.S.A.

KAUFMANN, Otto K., Dr. iur. (Zürich), LL.M. (Yale). Born 1914. Prof. Law. Member of Swiss Federal Court.
Nuclear interest: Atomic law.
Address: Pierrefleur 22 CH, Lausanne, Switzerland.

KAUL, P. Project Eng., Atomenergi AB; Tech. Sec., Dragon Project, 1963-.
Address: O.E.C.D. High Temperature Reactor Project, (Dragon), Atomic Energy Establishment, Winfrith, Dorchester, Dorset, England.

KAURANEN, Pentti Matti, Ph.D. Born 1930. Educ.: Helsinki Univ. Docent, Radiochem. Helsinki Univ.
Nuclear interest: Nuclear chemistry.
Address: Helsinki University, Department of Radiochemistry, Helsinki, Finland.

KAVANAGH, George M., B.S., Ph.D. Educ.: M.I.T. Supervisor, Chem. Res., Res. Div. (1952-56), Tech. Asst. to Commissioners (1956-59), Asst. Director, Nucl. Technol., Reactor Development Div. (1959-61), Deputy Asst. for Disarmament, Office of the Gen. Manager (1961-62), Acting Special Asst. for Disarmament (1962-64), Deputy Asst. Gen. Manager, R. and D. (1962-66), Asst. Gen. Manager for Reactors (1966-), and Member, Advisory Com. on Isotopes and Radiation Development, U.S.A.E.C.
Address: U.S.A.E.C., Washington D.C. 20545, U.S.A.

KAVANAGH, Michael Thomas, B.Sc., A.R.C.S., B.Sc. (Economics), F.S.S. Born 1915. Educ.: London Univ. At Roy. Ordnance Factories, explosives manufacture, 1936-51; Ind. Group, (1951), now Chief Commercial Officer, Production Group, U.K.A.E.A., Risley, Lancs.
Address: United Kingdom Atomic Energy Authority, Production Group, Risley, Warrington, Lancs., England.

KAVANAGH, Ralph William, B.A., M.A., Ph.D. Born 1924. Educ.: Reed Coll. and California Inst. Technol. Res. Fellow (1956-58), Sen. Res. Fellow (1958-60), Asst. Prof. Phys. (1960-65), Assoc. Prof. (1965-), Calif. Inst. Technol. Society: A.P.S.
Nuclear interest: Structure of nuclei, especially light nuclei.
Address: Kellogg Radiation Laboratory, California Institute of Technology, Pasadena, California, U.S.A.

KAVANAGH, Thomas Murray, B.A., M.A., Ph.D. Born 1932. Educ.: Saskatchewan and McGill Univs. Eng. Phys., A.E.R.E., Harwell, 1956-57; Lecturer (1960-62), Asst. Prof. (1962-64), Dept. of Phys., McGill Univ.; Res. Phys., Nucl.-Chicago Corp., 1964-66; Phys., Lawrence Rad. Lab., California Univ., 1966-. Societies: Canadian Assoc. of Phys.; A.P.S.
Nuclear interest: Nuclear physics.
Address: California University, Lawrence Radiation Laboratory, Livermore, California 94550, U.S.A.

KAWABATA, M., Dr. Director, Asst. Plant Manager, Kawasaki Plant, Nippon Yakin Kogyo Co. Ltd.
Address: Nippon Yakin Kogyo Co. Ltd., Ajinomoto Building, No. 7, 1-chome, Takaracho chuo-ku, Tokyo, Japan.

KAWAI, R. Manager, Sixth Dept., (Nucl. Technol.), Hitachi Res. Lab., Hitachi Ltd.
Address: Hitachi Ltd., Hitachi Research Laboratory, 4026 Omika, Kuji-machi, Hitachishi, Ibaragi-ken, Japan.

KAWAI, Toshio. Born 1932. Educ.: Tokyo Univ. Now with Hitachi, Ltd. Societies: Atomic Energy Soc. of Japan; Phys. Soc. of Japan.
Nuclear interests: Reactor core design;

Hydrodynamic stability; Optimal control-rod programming and fuel management of power reactors.
Address: Hitachi Central Research Laboratory, Kobubunji, Tokyo, Japan.

KAWAI, Yoshinari. Born 1886. Educ.: Tokyo Univ. Pres., Komatsu Manufacturing Co., Ltd.; Director, Japan Atomic Power Co. Ltd.
Address: 45 Yoshikubo-cho, Meguro-ku, Tokyo, Japan.

KAWAKAMI, Ichiro. Asst., Plasma Phys. (Theory), Atomic Energy Res. Inst., Nihon Univ.
Address: Nihon University, Atomic Energy Research Institute, 1-8 Kanda-Surugadai, Chiyoda-ku, Tokyo, Japan.

KAWAKAMI, Tatzuya, Dr. Eng. Born 1923. Educ.: Tokyo Inst. Technol. Government official eng. subordinated to the Ministry of Trade and Industry, 1951; Head of the process dynamics section, 1964.
Nuclear interests: Instrumentation studies on β-gauge, especially the studies on the bulk effect, the cover effect and the pass-line.
Address: Radioisotope Research Department Applied to Textile Industry, Textile Research Institute, 4 Sawatari, Kanagawa-ku, Yokohama, Japan.

KAWALA, Brunon O., M.D. Clinical Instructor of Medicine, Lab. of Radiobiol., California Univ. Medical Centre.
Address: California University Medical Centre, Laboratory of Radiobiology, San Francisco, California 94122, U.S.A.

KAWAMURA, Fumio. Member, Atomic Energy and Isotope Com., Tokushima Univ.
Address: c/o Medical Faculty, Tokushima University, Kuramoto-cho, Tokushima-shi, Japan.

KAWAMURA, Hazime, Prof. Lab. of Nucl. Studies, Faculty of Sci., Osaka Univ.
Address: Osaka University, Nakanoshima, Kita-ku, Osaka, Japan.

KAWAMURA, Satoshi. Managing Director, Atomic Energy Dept., Sumitomo Shoji Kaisha Ltd.
Address: Sumitomo Shoji Kaisha Ltd., 15, 5-chome, Kitahama, Higashi-ku, Osaka, Japan.

KAWAMURA, Tadao. Chief of Sect., Atomic Energy Manufacturing Div., Kobe Industries Corp.
Address: Kobe Industries Corp., Ohkubo-cho, Akashi, Japan.

KAWAMURA, Yoshio. Res. Eng., Atomic Power Plant Construction, Nucl. Sect. Eng. Div., Takenaka Komuten Co. Ltd.
Address: Nuclear Section, Engineering Division, Takenaka Komuten Co. Ltd., 2-30 Dojimanaka-machi, Kitaku, Osaka, Japan.

KAWANISHI, Masaharu, Asst. Prof. Rad. Lab., Osaka Univ.
Address: Osaka University, Kitahanada-cho, Sakai-si, Osaka, Japan.

KAWARA, Kiyoshi. Director, Inst. of Rad. Breeding.
Address: Institute of Radiation Breeding, Ohmiya-machi, Naka-gun, Ibaraki-ken, Japan.

KAWASAKI, Eiichi. Born 1920. Educ.: Tokyo Univ. of Literature and Science. Prof., Nihon Univ., 1956-. Sci. Council of Japan (special com. for nucl. fusion, special com. for atomic energy), 1957. Societies: Phys. Soc. Japan; Atomic Energy Soc, Japan.
Nuclear interests: Nuclear fusion reaction; Extremely low-energy nuclear cross-section for light element; Plasma physics (plasma oscillation).
Address: Department of Physics, College of Science and Engineering, Nihon University, Kanda-Surugadai, Tokyo, Japan.

KAWASE, Kinjiro, Prof. Agriculture. Member, Radioisotope Res. Com., Niigata Univ. School of Medicine.
Address: Department of Agriculture, Niigata University, 106 Koganecho, Niigata, Japan.

KAWASE, Osamu, M.D., D.M.Sc. Born 1909. Educ.: Kyoto Univ. Formerly Prof. Pathology, Osaka Medical Univ. for Women; Prof. Pathology, Inst. Constitutional Medicine, Kumamoto Univ., 1951-. Councillor: Japanese Pathological Soc., Japanese Soc. Electron-Microscopy; Managing Officer, Collagen Res. Soc., Japan. Societies: Japan Radioisotope Assoc.; Japan Polymer Soc.; Collagen Res. Soc., Japan; N.Y. Acad. Sci.; Japanese Pathological Soc.; Japanese Soc. Electron - Microscopy.
Nuclear interests: Tracer-chemical studies on the components of connective tissue, especially on elastin, collagen, mucopolysaccharides and lipids in relation to ageing and arteriosclerosis; Physico-chemical and chemical studies on the influence of γ- and neutron-radiation on elastin and collagen.
Address: Pathology Department, Institute of Constitutional Medicine, Kumamoto University, Honzyomati, Kumamoto, Japan.

KAWASHIMA, E. Member, Editorial Advisory Board, Nucl. Structural Eng.
Address: Section of Nuclear Ship, Kawasaki Dockyard, Kobe, Japan.

KAWASHIMA, Yoshio. Chief, Internat. Cooperation Sect., Atomic Energy Bureau, Sci. and Technol. Agency.
Address: Atomic Energy Bureau, 3-2-2 Kasumigaseki, Chiyoda-ku, Tokyo, Japan.

KAWASSIADES, Constantine Theodossius, B.Sc., D.Sc. Born 1904. Educ.: Athens, Liège and London Univs. Extraordinary Prof. Ind. Inorg. Chem., Athens Univ., 1953-; Ordinary Prof. Inorg. Chem., Aristotelian Univ., Salonika;

Rector, Salonika Univ., 1955-56. Books: Gen. and Inorg. Chem. (1952); Gen. and Ind. Chem. (1957); Electronic Theory of Valency (1958). Papers: Energy and Civilisation (1948); Atomic Energy, a Need of Humanity (1956). Societies: Greek A.E.C.; Assoc. Greek Chem.; Assoc. British Chem. Eng.; Ramsay Soc. Chem. Eng.; N.Y. Acad. Sci.
Address: 1 Anghelaki Street, Salonika, Greece.

KAY, Humphrey Edward Melville, M.D. (Lond.), M.R.C.P. (Lond.). Born 1923. Educ.: London Univ. Consultant Clinical Pathologist, Royal Marsden Hospital 1956-. Book: Bone Marrow Replacement: Principles and Applications (Cancer Progress, London, Butterworth and Co., 1960). Societies: British Soc. for Immunology; British Soc. of Haematology. Nuclear interest: Treatment of radiation injury by transplantation of bone marrow cells.
Address: 102 Lexham Gardens, London, W.8, England.

KAY, J. Lecturer in Geology, Western Australia Univ.
Address: Western Australia University, Nedlands, Western Australia.

KAY, J. M., M.A., Ph.D., M.I.Mech.E., Inst. Chem. Eng. Educ.: Cambridge Univ. Chem. Eng. Dept., Cambridge Univ., 1948-52; Chief Tech. Eng., Atomic Energy Production Dept. (later U.K.A.E.A.) Ind. Group, Risley, 1952-56; Prof. Nucl. Power, Imperial Coll. Sci. and Technol., 1956-61; Director of Eng. Development, Tube Investments Ltd., 1961-64; Chief Eng., Richard Thomas and Baldwins Ltd., 1965-. Rep., Inst. Chem. Eng., on Board of British Nucl. Energy Conference; Deleg. to 2nd. I.C.P.U.A.E., Geneva, 1958; Nucl. Safety Advisory Com., M.O.P. Societies: M.I.Mech.E.; Inst. Chem. Eng. Nuclear interests: Reactor design; Fuel element technology; Nuclear safety.
Address: R.T.B. House, 151 Gower Street, London, W.C.1, England.

KAYA, Seiji, D.Sc. Born 1902. Educ.: Tohoku Imperial Univ. Pres. (1959-65), now Prof. Emeritus, Tokyo Univ. Pres., Japan Radioisotope Assoc., 1951-; Chairman, Sci. Council of Japan, 1957-59; Chairman, Sci. Com. (Gakujutsu Shingikai), 1967-; Member, Japan Acad. of Sci.
Address: 28-45 Hon-Komagome 2-chome, Bunkyo-ku, Tokyo, Japan.

KAYAS, Georges. Principal collaborateur, Lab. de Physique, Ecole Polytechnique, Paris.
Address: Ecole Polytechnique, Laboratoire de Physique, 17 rue Descartes, Paris, France.

KAYE SCOTT, Rutherford, M.B., B.S., M.D., M.S., D.T.R.E. (Melbourne), F.R.A.C.S., F.F.R. (London), F.C.R.A. Born 1903. Educ.: Melbourne Univ. Hon. Consulting Radiotherapist, Royal Melbourne Hospital, Queen Victoria Hospital, Austin Hospital; Consultant Radiotherapist, Peter MacCallum Clinic, Melbourne. Warden of the Membership, and Past Pres., Coll. of Radiol. of Australasia.
Address: Strathnoon, 39 Auburn Road, Auburn, 3122, Victoria, Australia.

KAYLOR, John Daniel, B.Sc., M.Sc. Born 1910. Educ.: Massachusetts and Brown Univs. Supervisor, Marine Products Development Irradiator; 1963-.Society: Inst. of Food Technologists. Nuclear interests: Research to determine feasibility of commercial scale irradiation and shipment of seafoods using Marine Products Development irradiator of 250,000 curies 60 cobalt.
Address: Bureau of Commercial Fisheries, Technological Laboratory, Emerson Avenue, Gloucester, Massachusetts 01930, U.S.A.

KAYRA, Cahit. Born 1917. Councellor to Gen. Directorate of Finance, 1950-55; Advisor in private sector and in public or semi-public concerns such as postal services or Turkish airlines, 1955-59; Head, Foreign Trade Department, Ministry of Trade, 1959-60; Head, Turkish Permanent Delegation to G.A.T.T., 1960-63; Deputy Under-Sec. of State to Minister of Finance, 1963-64; Head, Turkish Delegation to O.E.C.D., 1964-.
Address: O.E.C.D., 38 boulevard Suchet, Paris 16, France.

KAYSER, P. Member for Luxembourg, Health and Safety Sub-com., O.E.C.D., E.N.E.A.
Address: O.E.C.D. European Nuclear Energy Agency, 38 boulevard Suchet, Paris 16, France.

KAZACHKOVSKII, Oleg Dmitrievich, Dr. of Maths. and Phys. Born 1915. Deputy Director, Phys. Tech. Inst. Obninsk. Sci. member, Fast Neutron Reactor, Moscow Inst. Phys.; Member, Soviet fast reactor team visiting U.K., Jan. 1960.
Nuclear interest: Physics of fast-neutron reactors.
Address: Physical Technical Institute, Obninsk, Near Moscow, U.S.S.R.

KAZAKEVICH, A. T. Paper: Co-author, Application of Autoradiography for Control of Irregularity of Actinide Element Layers (letter to the Editor, Atomnaya Energiya, vol. 21, No. 2, 1966).
Address: Academy of Sciences of the U.S.S.R., 14 Leninsky Prospekt, Moscow V-71, U.S.S.R.

KAZANSKII, L. N. Papers: Co-author, New Accelerator-Symmetrical Ring Phasotron Lebedev Phys. Inst. A U.S.S.R. Starting Up (letter to the Editor, Atomnaya Energiya, vol. 20, No. 6, 1966); co-author, Acceleration of Electrons in FIAN Ring Phasotron (ibid, vol. 21, No. 3, 1966).
Address: Academy of Sciences of the U.S.S.R., 14 Leninsky Prospekt, Moscow V-71, U.S.S.R.

KAZANSKY, Yu A.
Nuclear interests: Distribution and scattering of neutrons and γ-rays.
Address: Academy of Sciences of the U.S.S.R., 14 Leninsky Prospekt, Moscow V-71, U.S.S.R.

KAZARINOV, N. M. Papers: Co-author, Use of Semiconductor Spectrometric Counters to Measure the Energies of Fission Fragments (letter to the Editor, Atomnaya Energiya, vol. 12, No. 2, 1962); co-author, Surface Barrier Silicon Detectors for Measurements in Neutron and Fission Fragment Fluxes (letter to the Editor, ibid, vol. 16, No. 1, 1964).
Address: Academy of Sciences of the U.S.S.R., 14 Leninsky Prospekt, Moscow V-71, U.S.S.R.

KAZARNIKOVA, E. E. Paper: Co-author, About Gamma Rays Angular and Energy Distribution on Surface of Volume Source (Abstract, Atomnaya Energiya, vol. 19, No. 2, 1965).
Address: Academy of Sciences of the U.S.S.R., 14 Leninsky Prospekt, Moscow V-71, U.S.S.R.

KAZARNOVSKY, M. V. Papers: The Energy Spectrum of Neutrons from a Pulsed Source in a Heavy Moderator with Constant Path Length (Atomnaya Energiya, vol. 4, No. 6, 1958); co-author, Neutron Thermalisation and Diffusion in Heavy Media (2nd I.C.P.U.A.E., Geneva, Sept. 1958); Analytic Solution of Neutron Thermalisation Equation (Atomnaya Energiya, vol. 22, No. 2, 1967).
Address: Nuclear Energy Institute, Academy of Sciences of the U.S.S.R., 14 Leninsky Prospekt, Moscow V-71, U.S.S.R.

KAZATCHOVSKII, Oleg D. Director, Atomic Reactors Res. Inst. Member, State Com. for the Use of Nucl. Energy.
Address: Atomic Reactors Research Institute, Ulyanovsk Region, New Melekees, U.S.S.R.

KAZIMIERSKI, Adam, Dipl. Eng. (Electronics). Born 1930. Educ.: Tech. Univ. Warsaw. In Electronics Dept., Inst. Fundamental Tech. Problems, 1953; Head, Res. Group of Nucl. Detectors, Nucl Ind. Electronics Dept., Inst. Nucl. Res. 1955.
Nuclear interest: Nuclear detectors (especially neutron detectors): technology, physical phenomena and practical applications.
Address: Nowolipki 14 m 78, Warsaw 1, Poland.

KAZUNO, Miss Mitsuko, Ph.D. (Trinity Coll., Dublin). Born 1928. Educ.: Toho Univ., Tokyo. Research Scholarship (1960-63), Res. Asst. (1963-67), Asst. Prof. (1968-), Dublin Inst for Advanced Studies. Society: Phys. Soc. of Japan.
Nuclear interests: Nuclear physics, particularly ultra high energy nuclear interaction using cosmic rays. Inelastic collisions in the accelerator energy region.
Address: Dublin Institute for Advanced Studies, 5 Merrion Square, Dublin 2, Eire.

KEAGY, Walter Robert, Jr., B.S. Born 1922. Educ.: Yale Univ. Gen. Manager, Atomelectra, Ltd., Zürich, 1957-; Gen. Manager, Nucl. Services Internat. Ltd., Zurich, 1963-; European Representative, Nucl. Utility Services, Inc., Washington, D.C. Societies: A.N.S.; Atomic Ind. Forum.
Nuclear interests: Nuclear economics; Power plant design; Plant management; Fuel cycle analysis; Reactor safety.
Address: 48 Beethovenstrasse, Zürich, Switerland.

KEALY, W. A. Born 1912. Educ.: Seddon Tech. Coll., Auckland. Div. of Atomic Energy, Ministry of Supply, 1953; Anglo-American Corp. of South Africa, Zambia, 1953-60; Chief Instrument Eng., Eng. Group, U.K.A.E.A., 1960. Societies: Inst. of Mech. Eng.; Soc. Instr. Technol.; B.N.E.S.
Nuclear interest: Instrumentation and control.
Address "Cherry House", The Avenue, Lymm, Cheshire, England.

KEAM, D. W. Formerly C.S.I.R.O. Rep. Radioactive Isotopes Com., now Member, Radioisotopes Standing Com., Nat. Health and Medical Res. Council. Deleg., Australian A.E.C. Conference on the Technol. Use of Rad., May 1960.
Address: Commonwealth X-Ray and Radium Laboratory, 30, Lonsdale Street, Melbourne C.1, Victoria, Australia.

KEAM, R. F. Lecturer, Phys. Dept., Auckland Univ.
Address: Auckland University, P.O. Box 2175, Auckland, New Zealand.

KEANE, Austin, M.Sc., Ph.D. Born 1927. Educ.: Sydney and N.S.W. Univs. Lecturer in Maths. (1949-56), Sen. Lecturer in Maths. (1956-60), Assoc. Prof. Maths (1960-61), Member, Com. of the Inst. of Nucl. Eng., N.S.W. Univ.; Principal Res. Officer, A.A.E.C. 1961-; Visiting Prof. Nuc. Eng., 1961-. Societies: Fellow, Roy. Astronomical Soc.; Roy. Soc. of New South Wales; Australian Mathematical Soc.; A.N.S.
Nuclear interests: Reactor physics and reactor mathematics with specific interest in the resonance absorption and moderation of neutrons.
Address: A.A.E.C. Research Establishment, Private Mail Bag, Sutherland, New South Wales, Australia.

KEAR, David, B.Sc.(Eng.), B.Sc., Ph.D., A.R.S.M. Born 1923. Educ.: London Univ. District Geologist, Ngaruawahia 1949-58, and Auckland, 1958-65; Chief Economic Geologist (1963-67), Director (1967-), N.Z. Geological Survey. Societies: Geological Soc. of N.Z.; Geological Soc. of London.
Nuclear interests: Geology and exploitation of radioactive minerals.
Address: N.Z. Geological Survey, P.O. Box 30368, Lower Hutt, New Zealand.

KEARNEY, John J., B.S. (Mathematics), B.S. (Elec. Eng.). Born 1924. Educ.: Notre Dame Univ. Sec., Com. on Atomic Power, Edison Elec. Inst.
Address: Edison Electric Institute, 750 Third Avenue, New York City 10017, U.S.A.

KEARTON, Christopher Frank, M.A., B.Sc., F.R.S. Born 1911. Educ.: Oxford Univ. Chairman, Courtaulds Ltd.; Chairman, Industrial Reorganisation Corp. Member (1954-), Chairman (1960-), Elec. Supply Res. Council; Member, U.K.A.E.A., 1955-; Member Nat. Economic Development Council, 1965-; Member, Advisory Council on Technol., 1964-.
Address: Courtaulds Ltd., P.O. Box 2BB, 18 Hanover Square, London, W.1, England.

KEARY, Frank Vicent, B.Sc. M.Sc. Born 1934. Educ.: St. Bernardine of Siena, Loudonville, N.Y. and Rochester Univ. Director of Res., U.S.A. Medical Res. Unit, Europe, Landstuhl, Germany, 1959-61.
Nuclear interests: Radiation biology, nuclear medicine and biophysics.
Address: 2013 15th Street, Troy, New York, U.S.A.

KEAST, Asdruebal James, B.S.E.M., Dr. Eng. (hon.), D.Sc. (hon.). Born 1892. Educ.: Michigan Coll. of Mining and Technol. Formerly Managing Director, Mary Kathleen Uranium, Ltd.; Formerly Director, Rio Tinto Mining Co. (Australia), Ltd.; Consulting Mining Eng. Societies: Australasian Inst. of Mining and Metal.; American Inst. of Mining Eng.; Canadian Inst. of Mining and Metal.; Com. for Economic Development of Australia.
Nuclear interest: Brought into production Australia's largest uranium mine.
Address: 1 Monomeath Avenue, Canterbury, Victoria, Australia.

KEAST, Francis Henry, M.A. Born 1920. Educ.: Cambridge Univ. Asst. Chief Eng. (Tech.)(1952), Deputy Chief Eng. (1954), Chief Eng. (1957), Orenda Engines Ltd.; Tech. Director, A. V. Roe Canada Ltd., 1959; Director of Res., Hawker Siddeley Canada Ltd., 1962. Society: A.M.I.Mech.E.
Nuclear interests: Research and development in the field of power generation machinery, thermodynamic design, heat transfer.
Address: Box 4015, Terminal 'A', Toronto, Ontario, Canada.

KEATING, W. T. Eng. Dept. Head of Nucl. Components, Ford Instrument Co.
Address: Ford Instrument Company, 31-10 Thomson Avenue, Long Island City, 1, New York, U.S.A.

KEBADZE, B. V. Paper: Co-author, Exptl. Res. of Boiling Water Reactor Stability (Atomnaya Energiya, vol. 24, No. 4, 1968).
Address: Academy of Sciences of the U.S.S.R., 14 Leninsky Prospekt, Moscow V-71, U.S.S.R.

KECKES, Stjepan, Ph.D. (Biol.). Born 1932. Educ.: Zagreb Univ. Asst., Higher Asst., Res. collaborator (1957-62), Chief, Lab. of Marine Radiobiol. (1962-66), Inst. "Ruder Boskovic", Zagreb; First Officer and Chief, Biol. Sect., I.A.E.A., Lab. of Marine Radioactivity, Monaco, 1966-.
Nuclear interest: Marine radiobiology: influence of the physicochemical form of trace elements on their metabolism in marine organisms.
Address: International Atomic Energy Agency, Laboratory of Marine Radioactivity, Monaco.

KECKI, Zbigniew, Docent, Dr. (Warsaw). Born 1926. Educ.: Gdansk Polytech. High School, Poland. Docent, Dept. of Phys. Chem., Warsaw Univ.; Head, Spectroscopic Lab., Dept. of Rad. Chem., Inst. of Nucl. Res., Warsaw. Society: Polish Chem.Soc.
Nuclear interests: Molecular structure, intermolecular interactions, mechanism of radiolysis and chemical reactions studied by molecular spectroscopy.
Address: 62 m6 Wilcza, Warsaw, Poland.

KEDDAR, Ahmed, Dr. Resident Rep. for Algeria to I.A.E.A.
Address: Nuclear Research Institute, Boulevard Franz Fanon, Algiers, Algeria.

KEDROV-ZIKHMAN, O. K. VASKhNIL Academician. Director, Lab. of All-Union Inst. of Sci. Res. on Fertilisers and Soil Sci.
Nuclear interests: Soil science; Agricultural chemistry.
Address: Laboratory of the All-Union Institute of Scientific Research on Fertilisers and Soil Science, 31 ul. Pryanishnikova, Moscow A-8, U.S.S.R.

KEEDY, Curtis R., Ph.D. Born 1938. Educ.: Wisconsin Univ. and Argonne Nat. Lab. Asst. Prof. Chem., Reed Coll., Portland, Oregon. Asst. Director, Reed Reactor Project. Society: A.C.S. (Div. of Nucl. Chem.Technol.).
Nuclear interests: Activation analysis (for trace elements in geological materials). Neutron capture cross sections; Low energy nuclear reactions. Nuclear science education.
Address: Reed College, Portland, Oregon 97202, U.S.A.

KEEFE, Denis, B.Sc., M.Sc., Ph.D. Born 1930. Educ.: Dublin and Bristol Univs. Lecturer, University Coll., Dublin, 1952-59; Res. Phys., Lawrence Rad. Lab., California Univ., Berkeley, 1959-. Societies: Royal Irish Acad.; A.P.S.
Nuclear interests: Strong and weak interactions of heavy mesons and hyperons; High-energy pion and nucleon interactions; High energy accelerator design.
Address: Lawrence Radiation Laboratory, California University, Berkeley, California 94707, U.S.A.

KEELING, David H., M.A., M.B., B.Ch. Born 1934. Educ.: Cambridge Univ. Clinical Lecturer. Asst. Editor, British J. of Cancer. Societies:

Roy. Soc. of Medicine; Nucl. Medicine Soc. Nuclear interest: Diagnostic nuclear medicine. Address: The Institute of Nuclear Medicine, The Middlesex Hospital Medical School, The Middlesex Hospital, London, W.1, England.

KEEN, Richard C. Louisiana State Univ. rep. on Council, Oak Ridge Assoc. Universities. Address: Louisiana State University, Baton Rouge, Louisiana, U.S.A.

KEENAN, Charles Henry, A.B. Born 1914. Educ.: Coll. of the Holy Cross, Worcester. Brookhaven Nat. Lab., 1949-58; Yankee Atomic Elec. Co., 1958-. Labour Management Advisory Com., U.S.A.E.C.; Public Understanding Com., Workmens Compensation Com., Membership Com., Atomic Ind. Forum; Massachusetts Governor's Advisory Com. on Rad. Protection.
Nuclear interest: Management in non-technical side of nuclear industry.
Address: Yankee Atomic Electric Company, 441 Stuart Street, Boston 16, Massachusetts, U.S.A.

KEENAN, J. H., Asst. Gen. Manager, Nucl. Energy Property Insurance Assoc. Address: Nuclear Energy Property Insurance Association, 85 Woodland Street, Hartford, Connecticut 06102, U.S.A.

KEENAN, James T. Formerly Project Supv., N.E.R.V.A. Programme, Westinghouse Astronucl. Lab.; Formerly at Direct Conversion Project, R.C.A.; Chief, Isotopes Systems Eng. Sect., Hittman Assocs., Inc., 1967-. Address: Hittman Associates Incorporated, Isotopes Systems Engineering Section, Columbia, Maryland, U.S.A.

KEENAN, Thomas K., B.S. (South Dakota School of Mines and Technol.), M.S. (New Mexico), Ph.D. (New Mexico). Born 1924. Educ.: South Dakota School of Mines and Technol. and New Mexico Univ. Staff Member, Los Alamos Sci. Lab. Society: A.C.S. Nuclear interest: Inorganic chemistry of transplutonium elements.
Address: Box 1663, Los Alamos, New Mexico, U.S.A.

KEENE, Arthur Roy, B.A. (Bacteriology, Eng.). Born 1919. Educ.: California Univ. at Los Angeles. Director, H.P.S., 1961-64; Manager, Rad. Protection, Pacific Northwest Lab. Com., Chairman, Nat. Council on Rad. Protection and Measurements; Com. Chairman, American Standards Assoc. Societies: H.P.S.; A.I.H.A.; A.A.A.S.
Nuclear interests: Management of comprehensive radiation protection programmes; Environmental health.
Address: Pacific Northwest Laboratories, Battelle Memorial Institute, Richland, Washington, U.S.A.

KEEP, Anthony L. A.E.C. Accountability Rep., L. and S. Machine Co. Inc. Address: L. and S. Machine Co. Inc., Box 317, R.D. 2, Latrobe, Pennsylvania, U.S.A.

KEEP, Philip R., B.S., M.E. Born 1916. Educ.: Northeastern and Harvard Univs. Formerly Director, Power Systems Sales, Atomics Internat. Member, Tech. Group for Power, A.N.S. Societies: A.N.S.; I.E.E.E.; A.S.M.E. Nuclear interests: Management, nuclear power plants.
Address: 8900 DeSoto Street, Canoga Park, California, U.S.A.

KEEPIN, George Robert, Ph.B., B.S., M.S., Ph.D. Born 1923. Educ.: Chicago and Northwestern Univs. and M.I.T. U.S.A.E.C. Postdoctoral Fellow, California Univ., Berkeley, 1950; Consultant, Los Alamos Sci. Lab., summer, 1950, 1951; Phys. Dept. Staff, Minnesota Univ., 1951; Staff Member, Los Alamos Sci. Lab., 1952-63; Head, Phys. Group, Div. Res. and Labs., I.A.E.A., Vienna, 1963-65. Books: Delayed Neutrons (Progress in Nucl. Energy, Vol. 1, Series 1) (Oxford, Pergamon Press, 1956). Phys. of Nucl. Kinetics (U.S.A., Addison Wesley, 1964). Societies: Fellow, A.P.S.; A.N.S.
Nuclear interests: Neutron and reactor physics, nuclear kinetics, critical assemblies, nuclear propulsion reactors, pulsed neutron research. Address: Headquarters, International Atomic Energy Agency, Vienna I, Austria.

KEERY, Robert James, M.Sc., Ph.D. Born 1926. Educ.: Queens (Belfast) and Manchester Univs. Canadian Celanese Ltd., 1950-52; Manchester Univ., 1952-55; Sen. Development Eng., Short Bros. and Harland, 1955-56; Instrument and Elec. Eng., Chemstrand Ltd., 1956-59; U.K.A.E.A., 1959-. Society: Soc. of Instrument Technol.
Nuclear interest: Control and instrumentation of nuclear fuel plants and nuclear power stations.
Address: 24 Culcheth Hall Drive, Culcheth, Nr. Warrington, Lancs., England.

KEGEL, Gunter H. R., B.S., Ph.D. Born 1929. Educ.: Berlin T. H., Aachen T. H., Brazil Univ., Rio de Janeiro and M.I.T. Prof., Rio de Janeiro Catholic Univ., 1961-64; Prof., Lowell Tech. Inst., Lowell, 1964-.
Nuclear interest: Nuclear physics, low energy, experimental.
Address: Lowell Technological Institute, Lowell, Massachusetts 01854, U.S.A.

KEH, Zygmunt, M.Sc. (Eng.). Born 1904. Educ.: Gdansk Technol Univ. Managing Director, Group of Heavy Eng. Industries, -1952. Gen. Director (1952-55), Under Sec. of State (1955-), Ministry of Heavy Industries. Member, Exec. Board Com. for Economic Cooperation with Abroad to Council of Ministers; Member, Council for Peaceful Utilisation of Nucl. Energy, Warsaw; Chairman, Polish Delegation to Com.

for Radio and Electronic Ind. of the Comecon. Nuclear interest: Technical, engineering and economic matters concerning the equipment for the utilisation of nuclear energy and the exploitation of such equipment.
Address: 2/4 M.23 Aleja I Armii Wojska Polskiegu, Warsaw, Poland.

KEIDERLING, Walter, Dr. Born 1914. Educ.: Jena, Hamburg and Freilburg Univs. Prof. (1950-65), now Apl. Prof., Freiburg Univ.; Med. Leiter, Firma C.H.Boehringer Sohn, 1965-. Vorsitzender, Gesellschaft für Nuklearmedizin e.V. Books: Eisenstoffwechsel (Keiderling, 1959); Eisenstoffwechsel, and co-author, Blutkrankheiten in Künstliche Radioaktive Isotope in Physiologie, Diagnostik und Therapie (Schwiegk und Turba, 1961); co-author, Radioisotope in der Hämatologie (1963); co-author, Radioisotope in der Endokrinologie (1964). Societies: Gesellschaft für Innere Medizin; Gesellschaft deutscher Strahlenschutzärzte. Nuclear interests: Nuklearmedizin; Radiopharmaceuticals.
Address: Freiburg Universität, Med. Fakultat, 55 Hugstetter Strasse, 7800 Freiburg i Br., Germany.

KEILHOLTZ, Gerald W., A.B. M.A., Ph.D. Born 1912. Educ.: Lincoln, Utah and Oregon State Univs. Mass Analysis (1948-52), Rad. Damage (1952-58), Reactor Chem. (1958-), O.R.N.L. U.S. Patents: Cathode Coatings (2), Micro Light Filaments (1). Societies: A.C.S.; A.N.S.; A.A.A.S.
Nuclear interests: Reactor chemistry; Radiation damage; High temperature materials; Nuclear safety.
Address: Reactor Chemistry Division, Oak Ridge National Laboratory, P.O. Box X, Oak Ridge, Tennessee, U.S.A.

KEIM, Christopher Peter, Distinguished Alumni Award, Nebraska Wesleyan Univ.; A.B. (Nebraska Wesleyan), M.Sc., Ph.D. (Nebraska). Born 1906. Educ.: Pittsburgh, Nebraska and Tulsa Univs. and York Coll. Director, Stable Isotopes Div. (1947-57), Director, Tech. Information Div. (1957-), Oak Ridge Nat. Lab; Member, Tech. Information Panel, U.S.A.E.C. Books: Chapter contributor to Annual Review of Nucl. Sci., vol. 1, 1957; Electromagnetically Enriched Isotopes and Mass Spectrometry (Harwell Conference, 1955); Terms in Nucl. Sci. and Technol. (A.S.M.E., 1957); Nucl. Eng. Handbook (1958). Societies: A.C.S.; Fellow, A.P.S.; A.A.A.S.; A.N.S.
Address: 102 Orchard Lane, Oak Ridge, Tennessee 37830, U.S.A.

KEIRIM-MARKUS, I. B.
Nuclear interest: Health physics.
Address: Nuclear Energy Institute, Academy of Sciences of the U.S.S.R., 14 Leninsky Prospekt, Moscow V-71, U.S.S.R.

KEIRS, Russell J. Assoc. Dean, Graduate School, Assoc. Prof. Chem., Florida State Univ.; Florida State Univ. Rep. on Council, Oak Ridge Assoc. Universities, 1962-.
Address: Florida State University, Tallahassee, Florida, U.S.A.

KEISCH, Bernard. Member, Subcom. on Use of Radioactivity Standards, Nat. Acad. of Sci. N.R.C.
Address: National Academy of Sciences-National Research Council, 2101 Constitution Avenue, N.W., Washington 25, D.C., U.S.A.

KEISER, Henry Bruce, B.A. (Hons., Economics), LL.B. (Cum Laude). Born 1927. Educ.: Michigan Univ. and Harvard Law School. Member, Board of Contract Appeals, U.S.A.E.C., 1965.
Nuclear interest: Nuclear contract law.
Address: RCA Building, Washington, D.C. 20006, U.S.A.

KEITH-HITCHENS, John Gilbert. Born 1903. Educ.: Swindon Coll. Atomic Power Div., English Elec. Co., 1950-64. Formerly Member of the Board, B.N.E.S., London and Founder Member, East Midland Branch. Societies: M.I.Mech.E.; M.I.Loco.E.; F.I.Arb.
Nuclear interests: Reactors design with particular application to methods of fuelling and control. Occasional arbitration in differences nuclear inter alia.
Address: 33 Portland Road, Leicester, England.

KEKELIDSE, N. Deleg., I.A.E.A. Symposium on the Chemical Effects of Nuclear Transformations,Prague, Oct. 24-27, 1960.
Address: Academica Nauk U.S.S.R., 14 Leninsky Prospekt, Moscow V-71, U.S.S.R.

KELBER, Charles N., Ph.D., M.S., B.A. Born 1928. Educ.: Minnesota Univ. Assoc. Phys. (1955-63), Sen. Phys. (1963-), Argonne Nat. Lab. Sec., Reactor Phys. Div.,A.N.S., 1965-. Society: A.P.S.
Nuclear interest: Reactor physics, especially low energy nuclear physics.
Address: D-208, Argonne National Laboratory, 9700 S.Cass Avenue, Argonne, Illinois 60439, U.S.A.

KELDYSH, Academician Mstislav V. Pres., Acad. Sci. of the U.S.S.R.
Address: Academy of Sciences of the U.S.S.R., 14 Leninsky Prospekt, Moscow V-71, U.S.S.R.

KELLAWAY, F. W., B.Sc. Principal, Coll. of Technol., Letchworth.
Address: College of Technology, Broadway, Letchworth, Herts., England.

KELLER, Antoni, M.Sc. (Chem.). Born 1941. Educ.: Wroclaw Univ. Asst. Sci. Worker, Wroclaw Univ. 1965-.
Nuclear interest: Application of radioisotopic exchanges in chemical structural studies.

Address: Wroclaw University, Department of Inorganic Chemistry, 27 Wyspianskiego, Wroclaw, Poland.

KELLER, Arnold, Prof. Dr. Nucl. Phys., Univ. Tecnica Federico Santa Maria, Valparaiso. Address: Universidad Tecnica Federico Santa Maria, Casilla 110 V, Valparaiso, Chile.

KELLER, Curt, Dr.sc.tech. Prof. h.c. Brazil Univ.,Dr. ing. e.h. (Hanover T.H.). Born 1904. Educ.: Swiss Federal Inst. Technol., Zurich. Director, R. and D., Escher Wyss A.G., Zurich, 1935-. Co.-Editor, Atomkernenergie (Germany). Book: Theory and Performance of Axial Flow Fans (1935). Papers: Different articles on closed cycle gas turbines; Application of the closed cycle helium gasturbine to high temperature reactors; Aerodynamic tests method for hydraulic and caloric turbomachines. Societies: A.S.M.E.; S.J.A. (Swiss); Soc. Mech. Eng. (England).
Address: 200B Seestrasse, Küsnacht 8700, (ZH), Switzerland.

KELLER, Donald L., B.S. (Metal. Eng.). Born 1926. Educ.: Virginia Polytech. Inst. Asst. Chief, Advanced Materials Development Div., Battelle Memorial Inst. 1950-62. Books: Chapter, Reactor Handbook, 2nd edition, vol. 1, Materials. Societies: A.S.M.; American Ceramic Soc.
Nuclear interest: Reactor materials development.
Address: 505 King Avenue, Columbus 1, Ohio, U.S.A.

KELLER, Frederick Ralph, B.S. (Chem. Eng.). Born 1918. Educ.: Syracuse Univ. Supt. Operations, Materials Testing Reactor (1956-60), Supt. Operations, Eng. Test Reactor (1960-62), Formerly Supt. Operations, Advanced Test Reactor (1962-), Phillips Petroleum Co., Nat. Reactor Testing Station. Societies: A.I.Ch.E.; A.N.S.
Nuclear interests: Reactor operations.
Address: 2847 Westmoreland Drive, Idaho Falls, Idaho, U.S.A.

KELLER, J. C. Head, Insect Eradication and Pest Control Sect., Joint Div. of I.A.E.A. and F.A.O., U.N.
Address: International Atomic Energy Agency, 11 Kaerntnerring, Vienna 1, Austria.

KELLER, Joseph B., B.A., M.S., Ph.D. Born 1923. Educ.: New York Univ. Prof. Maths., New York Univ.; Head, Maths. Branch Naval Res. Office, 1953-54. Societies: American Mathematical Soc.; A.P.S.
Nuclear interest: Theoretical problems of scattering theory.
Address: Courant Institute of Mathematical Sciences, 251 Mercer Street, New York, New York 10012, U.S.A.

KELLER, Juliusz, Prof. Born 1911. Educ.: Warsaw Tech. Univ. Prof., Warsaw Tech. Univ., 1954-; Chief, Electronic Dept., Nucl. Res. Inst., 1955-59; Chief, Lab. of Nucl. Electronic Instr., Central Lab. for Radiol. Protection. Society: Nat. Rep., Internat. Federation for Medical Electronics and Biological Eng.
Nuclear interests: Measurements and instrumentation applied to health protection.
Address: 4 Szczuczynska, Warsaw 33, Poland.

KELLER, Karl Friedrich Wolfgang, Dr.-Ing., Dipl.-Phys. Born 1928. Educ.: Stuttgart T.H. Prokurist, Siemens A.G.
Nuclear interests: Reactor design; Management.
Address: Siemens A.G., RE 9, 50 Werner-v.-Siemens-Strasse, 852 Erlangen, Germany.

KELLER, Keaton Kent, B.S. (Phys.) (Southeast Missouri State Coll.), M.A. (with major in Phys.) (Washington), Ph.D. (Washington). Phys., Mallinckrodt Res. Inst., 1951; Asst. Prof. Phys., Alaska Univ., 1951-54; Phys. Teleray Corp., St. Louis, Missouri, 1955-56; Asst. Prof. Phys., Wyoming Univ., 1956-59; Assoc. Prof. Nucl. Eng.,Prof. Nucl. Eng. (1960-), Arizona Univ.
Nuclear interests: Thermonuclear reactions; Effects of radiation on materials; Nuclear theory and models.
Address: Nuclear Engineering Department, Arizona University, Tucson, Arizona, U.S.A.

KELLER, Robert, Ph.D. Born 1921. Educ.: Swiss Federal Inst. Technol., Zurich. Battelle Memorial Inst., Geneva, 1954; C.E.R.N., Geneva, 1955-60. Centre de Recherches sur la Physqiue des Plasmas, Lausanne, 1961-. Society: Swiss Phys. Soc.
Nuclear interests: Plasma physics;Controlled nuclear fusion.
Address: 21 avenue des Bains, Lausanne, Switzerland.

KELLERMANN, Ernst Walter, Dr. Phil., Ph.D. Born 1915. Educ.: Vienna and Edinburgh Univs. Sen. Lecturer Phys., Leeds Univ., 1949-. Societies: Fellow, Inst. Phys. and Phys. Soc.; A.P.S.
Nuclear interests: High energy physics; Cosmic rays.
Address: The University, Leeds 2, Yorkshire, England.

KELLERMANN, Otto A., Dipl.-Ing. Born 1925. Educ.: Aachen T.H. Design and Calculation of steam boilers, 1953-56; Technischer Überwachungs-Verein, Köln e.V., 1956; Chief Eng., Nucl. Group, Technischer Überwachungs-Verein, 1959-; Institutsleiter, Inst. for Reactor Safety of the Tech. Supervisory Societies; Member, Arbeitsgruppe Reaktorsicherheit VdTÜV.
Nuclear interests: Matters of safety in reactorplants; Reactor materials and reactor design.
Address: 27 von Görschenstrasse, Aachen, Germany.

KELLERSHOHN, C., Dr. en Médecine. Prof. agrége de physique médicale, Faculté de Médecine, Paris Univ. Member, Editorial Board, Atompraxis; Member, Medical and Biological Applications Planning Board, I.C.R.U., Washington.
Address: Service Frédéric Joliot, Centre d' Application Médicales du Service de Biologie, Commissariat à l'Energie Atomique, 29-33 rue de la Federation, Paris 15, France.

KELLEY, George G., B.S. Born 1920. Educ.: California Inst. Technol. At O.R.N.L., 1947-. Societies: A.P.S.; A.I.P.
Nuclear interests: Controlled thermonuclear power; Nuclear instrumentation.
Address: Route 1, Box 40B, Kingston, Tennessee, U.S.A.

KELLEY, Lois. Formerly Res. Asst., Forum Memo to Members. Now Res. Asst., Nucl. Industry.
Address: Atomic Industrial Forum, Inc., 850 Third Avenue, New York, New York 10022, U.S.A.

KELLEY, Wilbur Edrald, B.S. (Civil Eng.), C.E., D. Eng. Born 1908. Educ.: Louisville Univ. Manager, New York Operations Office, U.S.A. E.C., 1947-53; Vice-Pres. and Director, Catalytic Construction Co., Philadelphia, 1953-56; Pres. and Director, Walter Kidde Nucl. Labs., Inc., 1956-58; Pres. and Director, Assoc. Nucleonics, Inc., 1958-66; Vice-Pres. and Manager, Garden City Eng. Office, Stone and Webster Eng. Corp., 1966-. Societies: A.S.C.E.; A.I.Ch.E.; A.N.S.; Atomic Industrial Forum.
Nuclear interest: Design and construction of nuclear facilities.
Address: 35 Rhodes Street, New Hyde Park, New York, U.S.A.

KELLIHER, Maurice Gordon, B.A., M.Sc. Born 1920. Educ.: Dublin Univ. Head, Linear Accelerator Group, Mullard Res. Labs., -1954; Assoc. Chief Elec. Eng., High Voltage Eng. Corp., 1954-56; Special Director (Nucl. Projects), Vickers Res. Ltd., 1956-63; Local Director and Manager, Rad. and Nucl. Eng. Div., Vickers Ltd. Eng. Group, 1963-.
Nuclear interest: Design of particle accelerators in particular linear electron accelerators.
Address: Radiation and Nuclear Engineering Division, Vickers Limited Engineering Group, South Marston Works, Swindon, Wiltshire, England.

KELLY, Denis Thomas William ALIAGA-.
See ALIAGA-KELLY, Denis Thomas William.

KELLY, Elmer Lewis, B.S., Ph.D. (Phys.). Born 1915. Educ.: Virginia and California Univs. Res. Phys. (1946), Assoc. Director (1963), Lawrence Rad. Lab. Societies: A.P.S.; A.A.A.S.
Nuclear interests: Nuclear physics. Excitation functions of induced radioactivity; high energy neutron scattering; accelerator development and construction.
Address: Lawrence Radiation Laboratory, California University, Berkeley, California 94720, U.S.A.

KELLY, John C. R., Jr. Gen.Manager, Advanced Reactors Div., Atomic Power Div., Westinghouse Elec. Corp.
Address: Westinghouse Electric Corporation, P.O. Box 2278, Pittsburgh, Pennsylvania 15230, U.S.A.

KELLY, John Samuel, B.S. Born 1922. Educ.: Western Kentucky State Coll. and Ohio State Univ. Navy Dept., Bureau of Ships, 1948-56; Director, Div. of Peaceful Nucl. Explosives, U.S.A.E.C., 1956-.
Nuclear interest: Peaceful uses of nuclear explosives.
Address: 25101 Woodfield Road, Damascus, Maryland, U.S.A.

KELLY, Joseph A., Jr. Pres., Radium Chem. Co. Inc.
Address: Radium Chemical Co. Inc., 161 East 42nd Street, New York 17, New York, U.S.A.

KELLY, M. Supt., Berkeley Nucl. Power Station (-1968), Supt., Oldbury-on-Severn Nucl. Power Plant (1968-), C.E.G.B.
Address: Central Electricity Generating Board, Oldbury-on-Severn Nuclear Power Plant, Glos., England.

KELLY, Mervin J. Trustee, Atoms for Peace Awards.
Address: Atoms for Peace Awards Inc., 77 Massachusetts Avenue, Cambridge 39, Massachusetts, U.S.A.

KELLY, Peter John, B.A. Born 1922. Educ.: Oxford Univ. With U.K. Ministry of Technol. Book: Co-translator, Controlled Thermonuclear Reactions (L. A. Artsimovich: Oliver and Boyd, 1964).
Nuclear interests: Political and economic aspects of atomic energy; International relations.
Address: Ministry of Technology, Millbank Tower, Millbank, London, S.W.1, England.

KELLY, Robert Clyde, B.S. (Mech. Eng.), M.S. (Ind. Eng.). Born 1921. Educ.: Auburn and Tennessee Univs. Assoc. Editor, Eng., Earth Sci., and Metal. Sects. of Nucl. Sci. Abstracts.
Nuclear interests: Documentation and dissemination of technical information on nuclear science and technology.
Address: United States Atomic Energy Commission, Division of Technical Information Extension, P.O. Box 62, Oak Ridge, Tennessee 37830, U.S.A.

KELLY, William Harold, B.S.E. (Phys.), M.S., Ph.D. Born 1926. Educ.: Michigan Univ. Asst. Prof. Phys. and Astronomy (1955-61), Assoc. Prof. Phys. and Astronomy (1961-67), Prof. Phys. (1967-), Associate Chairman for Under-

graduate Affairs (1967-), Michigan State Univ. Society: A.P.S.
Nuclear interests: Natural radioactivity; Nuclear structure; Nuclear spectroscopy.
Address: Physics Department, Michigan State University, E. Lansing, Michigan, U.S.A.

KELSEY, Fremont Ellis, Ph.D. Born 1912. Educ.: Rochester Univ. Assoc. Prof., Chicago Univ, 1940-51; Director, Chem. Div., Nucl. Chicago, 1951-52; Prof. and Chairman, Physiology and Pharmacology, South Dakota Univ., 1952-60; Sci. Administrator, Nat. Institutes of Health, 1960-62; Special Asst. to Surgeon Gen., U.S.P.H.S., 1962-. Societies: American Soc. for Pharmacology and Exptl. Therapeutics; Soc. for Exptl. Biol. and Medicine.
Nuclear interests: Basic research on drugs, diagnostic applications of radioisotopes.
Address: 5811 Brookside Drive, Chevy Chase 15, Maryland, U.S.A.

KELSO, Albert F. B.S. (George Williams Coll., Chicago), M.S. (George Williams Coll., Chicago), Ph.D. (Loyola Univ. Graduate School, Chicago). Born 1917. Asst. Prof., Physiology and Pharmacology (1948-55), Assoc. Prof., Physiology and Pharmacology (1955-59), Chairman, Dept. of Physiology and Pharmacology (1956), Prof., Dept. of Physiology and Pharmacology (1959), Chicago Coll. of Osteopathy. Societies: American Physiological Soc.; A.A.A.S.; A.I.B.S.; I.E.E.E.
Nuclear interests: Teaching; Biological research; Medical instrumentation.
Address: Chicago College of Osteopathy, 1122 East 53rd Street, Chicago 15, Illinois, U.S.A.

KELTSCH, Ernst August Erhard, Dipl. Ing. Born 1912. Educ.: Braunschweig T.H. Vorstandsmitglied, Nordwestdeutsche Kraftwerke AG, Hamburg. Geschäftsführer, Studiengesellschaft für Kernkraftwerke G.m.b.H.
Nuclear interest: Power station application.
Address: 58 Up de Schanz, Hamburg-Hochkamp, Germany.

KEMBER, Norman Frank, B.Sc. (Phys.), Ph.D. (Biophys.). Born 1931. Educ.: Exeter Univ. Coll. Asst. Phys., Scunthorpe War Memorial Hospital, 1953-55; Phys., North Middlesex Hospital, 1955-57; Biophys. student, Inst. Cancer Res., 1957-60; Post-doctoral Fellowship, Biol. Dept., Brookhaven Nat. Lab., 1960-61; Lecturer (1961-65), Sen. Lecturer (1965-), Medical Phys. Dept., Royal Free Hospital School of Medicine, 1961-. Society: Hospital Phys. Assoc.
Nuclear interests: Microdosimetry of radiations within bone; Radiobiological studies on bone; Computer models of irradiated cell population kinetics.
Address: 20A Lower Road, Harrow, Middlesex, England.

KEMENY, Leslie George, B.S., A.M.I.E. Aust. Born 1931. Educ.: Sydney Univ. (Australia), Res. Eng., Metropolitan Vickers, 1955-56; U.K.A.E.A. Res. Fellow, (1957-58), Lecturer, Nucl. Eng. Lab. (1959-61), Queen Mary Coll. (London Univ.); School of Nucl. Eng., New South Wales Univ. Societies: A.N.S.; B.N.E.S.
Nuclear interest: Nuclear reactor theory - in particular, kinetics and control, and the stochastic theory of nuclear reactors.
Address: New South Wales University, School of Nuclear Engineering, Box 1, Post Office, Kensington, New South Wales, Australia.

KEMER, R. Ya. Paper: Co-author, Fast Neutron Flux Measurement for IRT-200 Reactor (letter to the Editor, Atomnaya Energiya, vol. 20, No. 2, 1965).
Address: Academy of Sciences of the U.S.S.R., 14 Leninsky Prospekt, Moscow V-71, U.S.S.R.

KEMMER, Nicholas, Dr. Phil. (Zurich), M.A. (Cantab.). Born 1911. Educ.: Göttingen,Zurich and London Univs. Univ. Lecturer, Cambridge, and Fellow, Trinity Coll., 1946-53; Tait Prof. Mathematical Phys., Edinburgh Univ., 1953-. Societies: F.R.S.; F.R.S.E.; A.P.S.; Phys. Soc.; Cambridge Philosophical Soc.; Edinburgh Mathematical Soc.
Nuclear interest: Elementary particle and nuclear physics.
Address: 35 Salisbury Road, Edinburgh 9, Scotland.

KEMMERLING, Carl, Dipl. Ing. Born 1927. Educ.: Aachen T.H. Allgemeine Elektricitäts Gesellschaft, 1957-59; Landesamt für Forschung, Düsseldorf, (delegated), 1959-61; Kernforschungsanlage Jülich des Landes Nordrhein - Westfalen e.V., 1959-.
Nuclear interests: Nuclear research; Reactor instrumentation.
Address: Kernforschungsanlage Jülich des Landes Nordrhein - Westfalen e.V., Postfach 365, 517 Jülich, Germany.

KEMP, D. M., D.Sc. Born 1928. Educ.: Stellenbosch Univ., South Africa. Lecturer, Stellenbosch Univ., 1954-58; Sen. Harwell Fellow, A.E.R.E., Harwell, 1958-61; At (1962-), Director, Chem. (1966-), South African Atomic Energy Board. Society: South African Chem. Inst.
Nuclear interests: Radiation chemistry; Nuclear chemistry; Isotope production; Instrumental activation analysis; Liquid sodium chemistry; Solvent extraction; Mass spectrometry; Analytical chemistry.
Address: National Nuclear Research Centre, Pelindaba, Private Bag 256, Pretoria, South Africa.

KEMP, Edwin Frank, B.Sc. Born 1921. Educ.: London Univ. With U.K.A.E.A., 1947- (present position, Sen. Chem., Eng. Directorate, Risley Headquarters). Societies: F.R.I.C.; Fellow, Chem. Soc.
Nuclear interests: Chemical plant management; Nuclear chemistry.
Address: 21 Oakways, Appleton Park, Warrington, Lancs., England.

KEMP, M. C., M.A., Ph.D. Member, Advisory Com., A.A.E.C.
Address: Australian Atomic Energy Commission, 45 Beach Street, Coogee, New South Wales, Australia.

KEMPE, Lloyd Lute, B.Ch.Eng., M.S., Ph.D. Registered Professional Eng., Michigan and Minnesota. Born 1911. Educ.: Minnesota Univ. Prof. Chem. Eng., Michigan Univ.; Consultant to Industry - Biochem. Eng. Member, Com. on Microbiol., Advisory Board on Quartermaster R. and D., Nat. Acad. Sci. - Nat. Res. Council; Member, Advisory Com. on Food Irradiation to Atomic Energy Commission, A.I.B.S.; Member, U.S. Advisory Com. on Botulism, Food and Drug Administration. Societies: A.C.S.; A.I.Ch.E.; Soc. American Microbiologists; Water Pollution Control Federation; Inst. of Food Technologists.
Nuclear interests: Radiation sterilisation of biologicals, radiation preservation of foods with particular reference to problems associated with Clostridium botulinum.
Address: Michigan University, Department of Chemical Engineering, Ann Arbor, Michigan, U.S.A.

KEMPF, Captain Hermann J. A. Born 1904. Educ.: Nautical Coll., Hamburg Univ. Managing Director, Poseidon Schiffahrt G.m.b.H., Hamburg. Society: Gesellschaft für Kernenergieverwertung in schiffbau und Schiffahrt m.b.H.
Address: 30 Jungfernstieg, Hamburg 36, Germany.

KEMPINSKI, Waldemar Stanislaw, Master, Ing. of Tech. Sci. Born 1936. Poznań Tech. Univ. Tech. Res. Asst., Coll. of Agriculture, Poznań, 1962-.
Nuclear interests: Nuclear electronics, electronic devices for the detection and measurement of radiation, electronic circuitry and advanced electronic systems used in nuclear research and applications.
Address: 86b/19 Pogodna Street, Poznań, Poland.

KENDALL, Ernest G., B.S. and M.S. (Metal. Eng.), Ph.D. (Eng.). Born 1926. Educ.: Brooklyn Polytechnic Inst., New York Univ., Stevens Inst. Technol. and Kentucky Univ. Nat. Lead Co., South Amboy, New Jersey, 1950-51; Titanium Metals Corp. of America, Henderson, Nevada, 1951-55; Metal. Eng. Dept., Kentucky Univ., 1955-57; Atomics Internat.,1957-61; Head, Metal. and Ceramics Dept., Materials Lab.-Aerospace Corp., El Segundo, California, 1961-. Societies: A.I.M.M.E.; A.S.M.; A.N.S.; A.I.A.A.; Ceramic Soc.
Nuclear interests: Nuclear metallurgy; Nuclear fuel element fabrication; Nuclear propulsion reactor materials; Thermoelectric and thermionic high-temperature materials.
Address: Aerospace Corporation, P.O. Box 95085, Los Angeles 45, California, U.S.A.

KENDALL, Frank Henry, B.Sc., Ph.D., F.R.I.C. Born 1924. Educ.: London Univ. Sci. Officer, Res. Assoc. British Rubber Manufacturers, 1950-51; Sen. Lecturer Radiochem., Sir John Cass Coll., and U.N. Radiochem. Advisor to Ceylon and Hong Kong Govts., 1951-67; Director, Radioisotope Unit, Hong Kong Univ., 1967-.
Nuclear interest: Use of radiation and radioisotopes in pure and applied research.
Address: Radioisotope Unit, Hong Kong University, Hong Kong.

KENDALL, Henry W., Ph.D. Born 1926. Educ.: M.I.T. Prof., Phys., M.I.T., 1966-. Society: A.P.S.
Nuclear interests: Elementary particle structure; Electromagnetic form factors.
Address: Physics Department, Massachusetts Institute of Technology, Cambridge, Massachusetts, U.S.A.

KENDALL, James Douglas, B.A.Sc., M.Sc. Born 1928. Educ.: Toronto Univ. and Stevens Inst. of Technol. General Precision Inc., Systems Eng. 1959-62; Sen. Systems Consultant (1962-66), Manager, Systems Consulting (1966-67), DCF Systems Ltd.
Nuclear interests: Nuclear reactor control systems, in particular the application of digital computers to control systems.
Address: 74 Victoria Street, Suite 725, Toronto 1, Ontario, Canada.

KENDON, Martin Honess, B.Sc. (Hons.), D.F.H. Born 1929. Educ.: Bromsgrove and Faraday House. Project Manager, Rolls-Royce and Assocs. Ltd., Derby; Formerly Elec. Design and Contracts Eng., Ewbank and Partners. Societies: M.I.E.E.; A.M.I.Mech.E.
Nuclear interests: Design, development and procurement of naval reactor propulsion plants.
Address: Fieldfare, 15 Hillcross Drive, Littleover, Derby DE3/7BW, England.

KENEY, Peggy M. Fishery Res. Biol., Radiobiol. Lab., Bureau of Commercial Fisheries, U.S. Dept. of the Interior.
Address: U.S. Department of the Interior, Bureau of Commercial Fisheries, Radiobiological Laboratory, Beaufort, North Carolina, U.S.A.

KENIS, Y., Dr. Service des Medecine, Inst. Jules Bordet (Centre des Tumeurs).
Address: Institut Jules Bordet (Centre des Tumeurs), 1 rue Heger-Bordet, Brussels, Belgium.

KENMOKU, Akitsugu, B.Pharm. Born 1934. Educ.: Tokyo Coll. of Pharmacy. Member, Isotope Lab., Nat. Inst. of Nutrition, Japan. Society: Pharmaceutical Soc. of Japan.
Nuclear interest: Radiation protection from the point of view of nutrition.
Address: Isotope Laboratory, National Institute of Nutrition, 1 Toyamacho, Shinjuku ku, Tokyo, Japan.

KENNEDY, Clyde Crawford, B.Sc. (McGill), Dipl. in Eng. (Carleton). B.Sc. Born 1917. Educ.: McGill and Carleton Univs. Public Relations Officer, A.E.C.L.; Member for A.E.C.L, Nucl. Public Relations Contact Group. Societies: Soc. for American Archeology; Soc. for Pennsylvania Archeology.
Address: P.O. Box M, Deep River, Ontario, Canada.

KENNEDY, D. S. Pres. and Board Chairman, Oklahoma Gas and Elec. Co. Trustee, Southwest Atomic Energy Associates.
Address: Oklahoma Gas and Electric Co., 321 North Harvey, Oklahoma City 1, Oklahoma, U.S.A.

KENNEDY, Geoffrey Farrer, M.A. Born 1908. Educ.: Cambridge Univ. Partner, Kennedy and Donkin. Societies: Inst. Civil Eng.; Inst. Mech. Eng.; I.E.E.; A.S.M.E.
Nuclear interests: Design and supervision of construction of complete nuclear power plants.
Address: Premier House, Woking, Surrey, England.

KENNEDY, Klem K., Graduated in Chem. Eng. Born 1922. Educ.: California Univ., Berkeley and Georgia School of Technol. Asst. Operations Supt., Idaho Chem. Processing Plant, American Cyanamid Co. and Phillips Petroleum Co., Idaho Falls, 1951-56; Sen. Res. Eng., Atomics Internat., Los Angeles, 1955-59; Chief, Chem. Processing and Materials Branch, Idaho Operations Office, U.S.A.E.C., 1959-. Societies: Nucl. Eng. Div., A.I.Ch.E.; A.I.Ch.E.
Nuclear interest: Technical administration of A.E.C. Idaho Operations Office programmes related to chemical processing of high enrichment reactor fuels and chemical and engineering research and development activities in fields of fuels processing, waste management and fission product isotopes separations.
Address: 820 Eighth Street, Idaho Falls, Idaho, 83401, U.S.A.

KENNEDY, R. H. Vice Pres., South Carolina Elec. and Gas. Co.
Address: South Carolina Electric and Gas Co., 328 Main Street, Columbia, South Carolina 29201, U.S.A.

KENNEDY, Vance C. Geological Survey, U.S. Dept. of the Interior.
Address: Geological Survey, U.S. Department of the Interior, Federal Centre, Building 25, Denver 25, Colorado, U.S.A.

KENNETT, Terence James, B.Sc., M.Sc., Ph.D. Born 1927. Educ.: McMaster Univ. Assoc. Phys., Argonne Nat. Lab., 1957-59; Asst. Prof. Phys. (1959-62), Assoc. Prof. Phys. (1962-66), Prof. Phys. (1966-), McMaster Univ. Society: A.P.S.
Nuclear interests: Nuclear physics: Decay scheme studies of fission products; Neutron capture gamma rays; Data handling and data reduction techniques; Development of counters.
Address: McMaster University, Hamilton, Ontario, Canada.

KENNEY, Edward Stephen, B.S. (Phys.), M.S. (Phys.), Ph.D. (Phys.). Born 1928. Educ.: St. Bonaventure, Rochester, Pittsburgh and Pennsylvania State Univs. O.R.I.N.S. Fellowship in Radiol. Phys., Rochester Univ., 1953-54; Health Phys., Westinghouse Atomic Power Plant, Bettis Field, Pittsburgh, 1954-55; Univ. Health Phys. (1955-59), Asst. Director Nucl. Reactor Facility (1959-), Acting Director Nucl. Reactor Facility (1960-61), Pennsylvania State Univ. Societies: Pennsylvania Acad. Sci.; A.N.S.; H.P.S.
Nuclear interests: Radiation-measuring equipment; Environs of the operating nuclear reactor core; Dynamic response of a reactor core to cyclic perturbations of reactivity; High-energy neutron spectral measurements.
Address: Nuclear Reactor Facility, College of Engineering and Architecture, University Park, Pennsylvania, U.S.A.

KENNY, A. W., M.A., B.Sc. F.R.I.C. Sen. Radiochem. Inspector, Ministry of Housing and Local Government. Member, Com. on the Biological Problems (Non-medical) of Nucl. Phys., M.R.C.
Address: Ministry of Housing and Local Government, Whitehall, London, S.W.1,. England.

KENSIT, Maurice Frank, M.A., A.M.I.C.E. Born 1908. Educ.: Cambridge Univ. Civil Eng., Ministry of Works, 1950-54; Sen. Civil Eng., U.K.A.E.A., 1954-.
Nuclear interest: Reactor construction.
Address: Santon Bridge, Holmrook, Cumberland, England.

KENSLER, Charles J., Ph.D., M.A., A.B. Born 1915. Educ.: Cornell and Columbia Univs. Asst. Prof. (1950-53), Assoc. Prof. (1953-54), Cornell Univ.; Head, Biol. Lab., Arthur D. Little, Inc., 1954-57; Prof. and Chairman, Dept. Pharmacology and Exptl. Therapeutics, Boston Univ. School Medicine, 1957-60; Sen. Vice Pres. i/c Life Sci. Div., Arthur D. Little, Inc., 1960-; Prof.. Pharmacology, Boston Univ. School Medicine, 1960-. Consultant, Nat. Cancer Inst., N.I.H.; Assoc. Editor, Editorial Board, Cancer Res.; Member, Board, Massachusetts Health Res. Inst., Inc.; Member, Council, Soc. of Toxicology. Societies: American Soc. for Pharmacology and Exptl. Therapeutics; American Assoc. for Cancer Res.
Nuclear interest: Drug metabolism and mode of action studies using radioactive isotopes (radiation effects on biological systems).
Address: Arthur D. Little Incorporated, 30 Memorial Drive, Cambridge, Massachusetts 02142, U.S.A.

KENT, James A., Ph.D. Born 1922. Educ.: West Virginia Univ. Res. Eng., Dow Chem. Co., 1950-52; Res. Group Leader, Monsanto Chem. Co., 1952-54; Director, Nucl. Eng. Programme (1954-63), Prof. Nucl. Eng. (1954-), Assoc. Director, Eng. Exptl. Station (1963-), West

Virginia Univ. Dean Eng., Michigan Technol. Univ., 1967-. Society: A.N.S.
Nuclear interests: Radiation effects; Radiochemical processing.
Address: Michigan Technological University, Houghton, Michigan, U.S.A.

KENT, William Charles Leonard, C.Eng., A.M.I. Mech.E. Born 1918. Educ.: Coventry Tech. Coll. Asst. Chief Eng., Fuel Element Plant Design Office, Eng. Group, Risley, 1960-. Society: Nucl. Eng. Soc.
Nuclear interests: Fuel fabrication facilities and chemical processing.
Address: 26 Hartley Road, Altrincham, Cheshire, England.

KENTON, John E., A.B., M.A. Born 1921. Educ.: Brown and Columbia Univs. Managing Editor, News, Nucleonics, 1955-66; Managing Editor (1960-66), Consulting Editor (1966-), Nucleonics Week; Exec. Editor, Sci. Res., 1966-. Society: Nat. Assoc. Sci. Writers.
Nuclear interests: All peaceful uses of nuclear energy, particularly naval and maritime reactors, central station power reactors, nuclear power economics.
Address: Scientific Research, 330 West 42nd Street, New York 36, New York, U.S.A.

KENTON, William B., B.S. (E.E.). Born 1913. Educ.: Union Coll. Schenectady, New York. With Turbine Div. (1940), Atomic Power Equipment Dept. (1956), Gen. Elec. Co.
Nuclear interest: Reactor design.
Address: 4922 Sandy Lane, San Jose, California, U.S.A.

KENWORTHY, Ray W. Assoc. Prof., Phys. Dept., Washington Univ.
Nuclear interest: Acoustics.
Address: Washington University, Physics Department, Seattle, Washington 98105, U.S.A.

KENYERES, Mihály, Dipl. Eng. Elec. Born 1922. Educ.: Moscow Energetical Inst. Head, External Relations,Hungarian Nat. A.E.C., 1962-. Hungarian Alternate Deleg., I.A.E.A. Gen. Conference, Vienna, 1964, 1966, 1967. Society: Hungarian Sci. Soc. for Measurement and Automation.
Nuclear interests: Management; Nuclear electronics.
Address: 32 Szemlohegy u., Budapest 2, Hungary.

KENZHEBAEV, Sh. Papers: Heavy Monoatomic Gas Model and Unsteady Thermalisation of Neutrons in Lead (letter to the Editor, Atomnaya Energiya, vol. 19, No. 3, 1965); co-author, Calculation of Neutron Thermalisation Parameters in Monoatomic Gases (letter to the Editor, ibid, vol. 24, No. 4, 1968).
Address: Academy of Sciences of the U.S.S.R., 14 Leninsky Prospekt, Moscow V-71, U.S.S.R.

KEOHANE, K. W. Prof., F.Inst.P. Head, Phys. Dept., Chelsea Coll. of Sci. and Technol.
Address: Physics Department, Chelsea College of Science and Technology, Manresa Road, London, S.W.3, England.

KEON, Edward Francis, B.A. (Chem. and Maths., St. Anselm's Coll.). Born 1928. Educ.: St. Anselm's Coll., Harvard Graduate School of Arts and Sci. and Northeastern Univ., Graduate School of Eng. Chem., Pratt and Whitney Aircraft Co., Middletown, Conn., 1952-54; Sen. Project Eng., Universal Match Co., Maynard, Mass. 1954; Administrative Asst. and Res. Chem., Arthur D. Little, Inc., Maynard, Mass., 1955; Evaluation Group Leader, Combustion Eng., Inc., Windsor, Conn., 1955-57; Res. Staff Member (1957-58), Sect. Manager, Nucl. Power Group (1958-59), Dept. Manager, High Temperature Materials Dept. (1959), Raytheon Manufacturing Co., Waltham, Mass.; Pres. and Director, High Temperature Materials Inc., Brighton, Mass., 1959-. Societies: A.I.A.A.; A.S.T.M.; Navy League.
Nuclear interests: Fuel and core materials development and manufacture.
Address: 780 E. Merrimack Street, Lowell, Massachusetts, U.S.A.

KEPAK, Frantisek, M.Sc. (Chem.), Ph.D. (Chem.). Born 1931. Educ.: Chem. Technol. Univ., Prague. Societies: Czechoslovak Soc. for Sci. and Technol.; Czechoslovak Chem. Soc.
Nuclear interests: Radiochemistry, sorption processes: study of sorption of radioisotopes from water solutions on inorganic sorbents (ionic precipitates, insoluble hydrated oxides and hydroxides), study of radiocolloids in aqueous solutions and their separation.
Address: Nuclear Research Institute, Czechoslovak Academy of Sciences, Rez near Prague, Czechoslovakia.

KEPP, Richard, M.D., ord. Prof. Obstetrics and Gynaecology. Born 1912. Educ.: Goettingen Univ. Director, Obstetrical and Gynaecological Clinic, Giessen Univ., 1956-. Member, German Reactor Safety Commission. Books: Grundlagen der Strahlentherapie (Stuttgart, G. Thieme-Verlag, 1952); Gynaekologische Strahlentherapie (Stuttgart, G. Thieme-Verlag, 1952). Societies: German Roentgen Soc.; German Central Com. of Fight against Cancer; Member, Special com., Protection against Irradiation, German Atom Commission; Hon. Member, Soc. of Gynaecology and Obstetrics, La Plata, Argentina.
Nuclear interests: Biological and medical radiation research.
Address: 28 Klinikstrasse, Giessen/Lahn, Germany.

KERESE, István, Dr. Biol. Sci. Born 1911. Educ.: Budapest Univ. Sci. and Pécs Univ. Sci., Leader, Lab. in Leather Factory, Pécs, 1948-56; Sci. worker (1956-58), Leader, Isotope Lab. (1958-), Leather Industry's Res. Inst., Budapest. Candidate of Tech. Sci., Hungarian Acad. of Sci.

Nuclear interests: Study of the protein-chemical changes during the hide-conservation, soaking and liming processes of leather manufacturing using labelled materials /^{22}NaCl, Na$_3$^{35}S/, partly study of collagen fibre renaturation, observed on ^{14}C labelled hides/rat hides/ and using ^{14}C labelled and other materials.
Address: Leather Industry's Research Institute, 43 Paksi J.-u., Budapest 4, Hungary.

KERJEAN, J. Director, Sté d'Etudes de Protections des Installations Atomiques.
Address: Societe d'Etudes de Protections des Installations Atomiques, 38 rue Jean de la Fontaine, Versailles, (Seine et Oise), France.

KERK, Jacob van de. See van de KERK, Jacob.

KERLEE, Donald D., B.S., Ph.D. Born 1926. Educ.: Seattle Pacific Coll. and Washington Univ. Director, Inst. for Res. 1960; Chairman Phys. Dept., Seattle Pacific Coll. 1961. Societies: A.P.S.; A.G.U.
Nuclear interests: Heavy ion elastic scattering; Heavy ion fission.
Address: Physics Department, Seattle Pacific College, Seattle 98119, Washington, U.S.A.

KERLEY, R. F. Manager, Industrial Sales Div., X-ray materials, Kodak Ltd.
Address: Kodak House, Kingsway, London, W.C.2, England.

KERMAN, Arthur Kent, B.S., Ph.D. Born 1929. Educ.: McGill Univ. and M.I.T. Asst. Prof. (1956-60), Assoc. Prof. (1960-64), Prof. (1964), M.I.T.; Exchange Prof. and Guggenheim Fellow, Paris Univ., 1961-62; Consultant: K.A.P.L., 1961-62, Los Alamos Sci. Lab., 1961-, Brookhaven Nat. Lab., 1963-, Argonne Nat. Lab. 1964-, Lawrence Rad. Lab., Livermore, 1964-. Societies: Canadian Assoc. Phys.; A.P.S.
Nuclear interest: Theoretical nuclear physics.
Address: Physics Department, Massachusetts Institute of Technology, Cambridge, Massachusetts, U.S.A.

KERN, Bernard Donald, M.S., Ph.D. Born 1919. Educ.: Indiana Univ. Sen. Phys., O.R.N.L., 1949-50; Prof. Phys., Kentucky Univ., 1950-; 1 year leave-of-absence as Phys., U.S. Naval Radiol. Defence Lab., 1957-58; A.I.D. Visiting Prof. (1961-62), Chairman (1967-), Inst. Teknologi Bandung, Indonesia. Societies: A.P.S.; A.I.P.
Nuclear interests: Nuclear structure; Particle-gamma and gamma-gamma angular correlations; Neutron-induced reactions; Beta - and gamma-ray spectroscopy. Electrostatic accelerators.
Address: Physics and Astronomy Department, Kentucky University, Lexington, Kentucky 40506, U.S.A.

KERN, Eberhard, Dr.rer.nat. (Phys.). Born 1924. Educ.: Tübingen,Göttingen and Munich Univs. Brown Boveri/Krupp Reaktorbau G.m.b.H., Mannheim, 1957-61; Vereinigung der Techn. Überwachungsvereine, Essen. Books: Nukleonik (Springer-Verlag, 1962); Atomkernenergie (Thiemig-Verlag, 1962). Society: Deutsche Physikalische Gesellschaft.
Nuclear interests: Nuclear physics; Neutron theory; Burn-up calculation; Mathematical methods. Turbulent gas flow in gas-cooled reactors; Reactor safety.
Address: Vereinigung der Techn. Überwachungsvereine, 17 Rottstrasse, Essen, Germany.

KERNAN, Anne, Ph.D. Born 1933. Educ.: Univ. Coll., Dublin. Asst. (1955-59), Asst. Lecturer (1959-), Phys. Dept., Univ. Coll. Dublin.
Nuclear interest: Interactions of elementary particles in bubble chambers.
Address: Lawrence Radiation Laboratory, Berkeley 4, California, U.S.A.

KERNOHAN, Robert H., A.B., M.A. Born 1913. Educ.: Oberlin Coll. and Columbia Univ. Phys., Solid State Div., O.R.N.L. Society: A.P.S.
Nuclear interests:Effects of radiation on metals and alloys; Superconductivity.
Address: Solid State Division, Building 3025, Oak Ridge National Laboratory, P.O. Box X, Oak Ridge, Tennessee, U.S.A.

KERNS, Quentin A. Member, Subcom. on Instruments and Techniques, Com. on Nucl. Sci., N.A.S. - N.R.C.
Address: National Academy of Sciences - National Research Council, Committee on Nuclear Sciences, 2101 Constitution Avenue, N.W., Washington 25, D.C., U.S.A.

KERNS, R. E. Sec., Southwest Atomic Energy Associates; Formerly Sec., Oklahoma Gas and Elec. Co.
Address: Southwest Atomic Energy Associates, 306 Pyramid Building, Little Rock, Arkansas, U.S.A.

KERR, Adrian George BRUCE-. See BRUCE-KERR, Adrian George.

KERR, H. Dabney, M.D. Sec., American Board of Radiol.
Address: American Board of Radiology, Office of the Secretary, Kahler Centre Building, Rochester, Minnesota 55901, U.S.A.

KERR, William, B.S. (E.E., Tennessee), M.S. (E.E., Tennessee), Ph.D. (E.E., Michigan). Born 1919. Educ.: Tennessee and Michigan Univs. Prof. Nucl. Eng. (1958-), Chairman, Dept. Nucl. Eng. (1961-), Director, Michigan Memorial-Phoenix Project (1965-), Michigan Univ. Project Supv., A.I.D. (I.C.A.) Nucl. Energy Project (1956-65); Consultant, Atomic Power Development Assocs.(1954-), Argonne Nat. Lab., Colorado Commission on Higher Education; Board of Directors (1965-), Pres., Board of Directors (1966-67), Assoc. Midwest Univs. Board of Trustees, Argonne Univs. Assoc., 1965-. Societies: A.N.S.; I.E.E.E.; I.E.R.E.

Nuclear interests: Reactor system dynamics; Reactor shielding; Reactor safety analysis; Nuclear radiation detectors.
Address: Nuclear Engineering Department, Michigan University, Ann Arbor, Michigan, U.S.A.

KERRIDGE, Maurice, M.A. Born 1922. Educ.: Cambridge Univ. Reactor Manager, A.E.I. Ltd., Aldermaston Court, 1960-63; Manager, London Univ. Nucl. Reactor, 1963-. Society: F.Inst.P.
Nuclear interests: Experimental work using research reactors; Radioisotope applications; Reactor operations and safety.
Address: London University Reactor, Silwood Park, Sunninghill, Ascot, Berkshire, England.

KERST, Donald W., B.A., Ph.D. (Phys., Wisconsin), Hon. Sc.D. (Lawrence Coll.), Hon. Dr. (Univ. Sao Paulo), Hon. Sc.D. (Wisconsin). Born 1911. Educ.: Wisconsin Univ.Tech. Director, Midwestern Univs. Res. Assoc. 1956-57. Societies: A.P.S.; A.A.A.S.; Nat. Acad. of Sci.
Nuclear interests: Thermonuclear reactor development; Development of spiral sector accelerators.
Address: 1506 Wood Lane, Madison, Wisconsin, U.S.A.

KERSTEN, J. A. H. Dr. Sen. Officer, Nucl. Work, Group Leader, Homogeneous Reactor Development Team, Tot Keuring van Electrotechnische Materialen N.V.; Member, Editorial Com., Atoomenergie en haar toepassingen.
Nuclear interests: Reactor physics and ion-accelerators.
Address: Tot Keuring van Electrotechnische Materialen N.V., 310 Utrechtseweg, Arnhem, Netherlands.

KERSTEN, Martin, Prof., Dr.-Ing. Born 1906. Educ.: Berlin T.H. Präsident, Physikalisch-Technischen Bundesanstalt; Mitglied des Wissenschaftlichen Rats der Kernforschungsanlage Jülich des Landes Nordrhein-Westfalen e.V.; Präsident, Deutschen Physikalischen Gesellschaft. Vorsitz d. Arbeitskreises "Brenn- und Werkstoffe, Bauteile" der Deutschen Atomkommission.
Nuclear interests: Festkörperphysik; Ferromagnetismus; Reaktorwerkstoffe.
Address: 8 Knappstrasse, Brunswick 33, Germany.

KERTES, Aviezer Stevan, Chem. Eng., Ph.D. Born 1922. Educ.: Hebrew Univ., Jerusalem. Res. Assoc., Inst. de Radium,Paris Univ. 1955; Sen. Lecturer, Hebrew Univ. 1960; Res. Assoc., Chem. Dept.,M.I.T. Cambridge, 1962; Visiting Sci., Ispra Establishment, Euratom, 1965; Visiting Sci., Dept. of Inorganic Chem., Roy. Inst. of Technol., Stockholm, 1966; Assoc. Prof., Dept. of Inorganic Chem., Hebrew Univ. Editorial Board, J. Inorg. Nucl. Chem.; Member, I.U.P.A.C. Com. on Equilibrium Data.
Books: Co-author, Ion-Exchange and Solvent Extraction of Metal Complexes (Wiley-Interscience, 1968); contribution to Recent Advances in Liquid-Liquid Extraction (Editor: C.Hanson) (Pergamon, 1968). Societies: Israel Chem. Soc.; British Chem. Soc.
Nuclear interests: Complex and solution chemistry; Solvent extraction and process chemistry; Radiochemical separations; Molten salts chemistry.
Address: Department of Inorganic and Analytical Chemistry, The Hebrew University of Jerusalem, Israel.

KERTESZ, François, Sc.D., Ch.E. Born 1911. Educ.: Stuttgart T.H. and Paris Univ. (Sorbonne). At O.R.N.L., Director's Div., 1951-64; Asst. Director, Tech. Information Div. (Coordinator of information centres), 1964-; previously with E. I. DuPont de Nemours and Co. and N.E.P.A. Project, Oak Ridge. Co-Chairman, A.N.S. Annual Meeting in Gatlinburg, Tennessee; Pres., Tech. Societies Joint Council, Oak Ridge. Societies: A.N.S.; A.C.S.; A.S.M.; New York Acad. of Sci.; Nat. Assoc. Corrosion Eng.
Nuclear interests: Reactor materials; Corrosion; Technical literature; Radiation safety.
Address: 236 Outer Drive, Oak Ridge, Tennessee, U.S.A.

KERTESZ, Z. I. Member for F.A.O., Study Group on Food Irradiation, O.E.C.D., E.N.E.A..
Address: O.E.C.D. European Nuclear Energy Agency, 38 boulevard Suchet, Paris 16, France.

KERWICK, Walter, B.E.E. Born 1921. Educ.: Detroit Univ. Tech. Asst., Michigan Con. Gas Co., 1949-52; Chairman, Industry Com., Michigan Sect., American Inst. of Elec. Eng.; Elec. Eng., McArthy Elec. Co., 1952-55; Chief Elec. Eng., A.F.Holden Co., 1955-56; Elec. Eng. (1956-), Sen. Project Eng. (1961-), Gen. Motors Res. Labs. Director, Michigan Sect. I.E.E.E., 1962-65. Society: American Inst. of Elec. Eng.
Nuclear interest: Industrial applications of radioisotopes.
Address: Research Laboratories, General Motors Corp., 12 Mile and Mound Roads, Warren, Michigan, U.S.A.

KERWIN, Larkin, Ph.D. Born 1924. Educ.: St. Francis Xavier Univ., Antigonish, Nova Scotia. Prof. (1956), Chairman, Dept. of Phys. (1961-), Laval Univ. Books: Atomic Phys. - An Introduction (Holt-Rinehard and Winston, Inc., 1963); Physique atomique - une introduction (Québec, Les Presses de l'Université Laval, 1964). Societies: Roy. Soc. of Canada; Canadian Assoc. of Phys.; A.P.S.; A.C.F.A.S.
Nuclear interest: Mass spectrometry.
Address: Department of Physics, Laval University, Québec 10, P.Q., Canada.

KERZE, Frank, Jr., B.S. (Ch.E.), M.A., M.Ch.E. Born 1908. Educ.: Case Inst. Technol., Columbia and New York Univs. Chief of Eng. Mat Materials Sect., Tech. Div. O.R.N.L., 1948-50; Materials Eng., Naval Reactors Branch (1950-55), Chief, Chem. Eng. Sect., Eng. Development

Branch (1955-59), Chief, Tech. and Economic Data Section (1959-62), Gas-cooled Reactors Branch (1962-), Div. of Reactor Development, U.S.A.E.C., Washington. Societies: A.S.M.; A.C.S.; A.N.S.
Nuclear interests: Development of processes, materials, components.
Address: Division of Reactor Development, U.S.Atomic Energy Commission, Washington 25, D.C., U.S.A.

KERZNER, Edward J., M. Chem. Eng., B. Chem. Eng. Born 1924. Educ.: Brooklyn Polytech Inst. and Pratt Inst. Production Project Eng. to Eng. Supervisor, Kollsman Instrument Corp., 1950-56; Div. Manager, Commercial Sales, Curtiss Wright Quehanna Div., 1956-59; Asst. to the Pres., Atomic Associates Inc., 1959-61; Pres., Rad. Equipment and Accessories Corp., 1961-66; Vice-Pres., Capintec Inc., and Pres., Milletron Inc., 1966-.
Nuclear interests: Radiation detection and measuring instruments; Non-contact temperature indicating and controlling systems.
Address: 1306 Barry Drive South, Valley Stream, New York, U.S.A.

KESHISHIAN, Vahe, B.S., M.S. Born 1929. Educ.: American (Beirut) and Kansas State Univs. Graduate Res. Asst., Kansas State Univ., 1952-54; Phys. Asst., Eng. R. and D. Labs., Fort Belvoir, Virginia, 1955-56; Nucl Specialist, Atomics Internat., Canoga Park, California, 1956-. Societies: A.P.S.; A.N.S.
Nuclear interests: Reactor design and shielding; Nuclear physics.
Address: Atomics International, Box 309, Canoga Park, California, U.S.A.

KESSE, M. ELVAS-. See **ELVAS-KESSE, M.**

KESSELRING, Kenneth A., B.E.E. (Cornell) Born 1917. Educ.: Syracuse and Cornell Univs. Project Manager, Submarine Intermediate Reactor Project, S.S.N. Seawolf (1955-57), Project Manager, Nucl Destroyer Project (1957-61), Manager (1961-), K.A.P.L., G.E.C., Schenectady, New York. Societies: Fellow A.N.S.; A.S.M.E.; American Soc. Naval Architects and Eng.; New York Acad. Sci.
Nuclear interests: Metallurgy; Core design; Management.
Address: 1951 Village Road, Schenectady, New York, U.S.A.

KESSLER, Gen. Chief Eng., Dept. Energie Nucléaire, Cie. des Ateliers et Forges de la Loire.
Address: Department Energie Nucléaire, Cie. des Ateliers et Forges de la Loire, 12 rue de la Rochefoucauld, Paris 9, France.

KESSLER, Oberst F. ABC-Sektion der Abteilung für Sanität des Eidgenossisches Militärdepartements. Mitglied, Alarmausschuss, Eidgenössische Departement des Innern.
Address: Abteilung für Sanität, Eidgenössische Militärdepartemente, Bern, Switzerland.

KESSLER, Paul, D. ès Sc. Physiques. Born 1926. Educ.: Marseille and Sorbonne Univs. Attaché de Recherches (1954-56), Chargé de Recherches (1956-61), Maître de Recherches (1961-), C.N.R.S.
Nuclear interest: Elementary particle physics.
Address: Laboratoire de Physique Atomique, Collège de France, Paris, France.

KESSLER, Wayne Vincent, B.S., M.S., Ph.D. Born 1933. Educ.: North Dakota State and Purdue Univs. Asst. Prof., Pharmaceutical Chem., North Dakota State Univ., 1959-60; Instructor (1959), Asst. Prof., Health Phys. (1960-64), Assoc. Prof., Health Phys. (1964-) Purdue Univ. Societies: American Pharmaceutical Assoc.; Acad. of Pharmaceutical Sci.; A.C.S.; A.A.A.S.
Nuclear interests: Applications of radioactive tracers in the biological and analytical areas; Radiological health studies; Body composition studies by ^{40}K; Use of large volume radioactivity counters.
Address: Bionucleonics Department, Purdue University, Lafayette, Indiana 47907, U.S.A.

KETCHUM, Bostwick Hawley, A.B. (Columbia), Ph.D. (Harvard). Born 1912. Educ.: Bard Coll., Columbia and Harvard Univs. Microbiol. (1945-53), Sen. Biol. (1953-54), Sen. Oceanographer (1955-), Woods Hole Oceanographic Inst. Ex-Pres., American Soc. of Limnology and Oceanography. Societies: A.A.A.S.; Ecological Soc. America; A.G.U.
Nuclear interest: Marine biological and oceanographic problems in radioactive contamination of the sea.
Address: Woods Hole Oceanographic Institution, Woods Hole, Massachusetts, U.S.A.

KETELAAR, Jan Arnold Albert, Ph.D. (Amsterdam). Born 1908. Educ.: Amsterdam Univ. and California Inst. Technol. Prof. Phys. Chem. (1941-60), Prof. Electrochemistry (1960-), Amsterdam Univ.; Prof. Chem., Brown Univ., Providence (R.I), 1958-59; Dean, Faculty of Sci., Amsterdam Univ., 1953-57; Co-ordinator of Res.,K.Z.K, Hengelo, 1960-. Member of Board, Reactor Centrum Nederland; Member of Board, F.O.M. (Foundation Fundamental Res. of Matter). Books: Chemical Constitution (2nd ed., Elsevier Publishing Co., 1958); Liaisons et Propriétés Chimiques (Paris,Dunod, 1960); Chemische Konstitution (Braunschweig, Vieweg, 1964). Societies: Koninklijke Nederlandse Chem. Vereniging; Nederlandse Natuurk. Vereniging; Roy. Netherlands Acad. of Sci.
Nuclear interests: Radiation chemistry and use of isotopes in chemical and electrochemical reaction; Fused salts.
Address: 91 Markeloseweg, Rijssen, Netherlands.

KETOLAINEN, Pertti Pekka Juhani, Ph. Lic. Born 1937. Educ.: Turku Univ. Formerly Asst. Phys. (1961-), now Res. Phys., Turku Univ. Society: Phys. Soc. of Finland.
Nuclear interest: Colour centres in solids.

Address: Wihuri Physical Laboratory, Turku University, 5 Vesilinnantie, Turku, Finland.

KETTNER, M. E. Prof. engaged in Nucl. Work, Phys. Dept., Manitoba Univ.
Address: Physics Department, Manitoba University, Winnipeg 19, Manitoba, Canada.

KETTNER, Robert, E., B.S., M.B.A. Born 1925. Educ.: U.S. Military Acad., West Point, Harvard Graduate School of Business Administration and New Mexico Univ. Regular Officer, U.S. Air Force. Electronic Specialist, Special Weapons A.F.S.W.P., 1947-53; Atomic warhead for Guided Missiles, Office Asst. Special Weapons, Air Material Command, 1953-54; Asst. Div. Manager, Westinghouse Elec. Corp., Bettis Atomic Power Div., 1955-56; Asst. to P.W.R. Project Manager, 1956-; Formerly Director, Nucl. Activities, Asst. to Pres., (1964-68), Consumers Power Co.
Address: 1848 N Walmont, Jackson, Michigan, U.S.A.

KETTUNEN, Pentti Olavi, Dipl. Eng., Licentiate of Technol. Dr. of Technol. Born 1932. Educ.: Inst. of Technol., Helsinki. Lecturer and Sen. Officer, Lab. of Phys. Metal., Inst. of Technol., Helsinki-Otaniemi, 1962-66; Res. Officer, Outokumpu-Copper Co., Helsinki, 1965-66; Docent, Phys. Metal., Inst. of Technol., Helsinki-Otaniemi; Res. Assoc., Argonne Nat. Lab., 1966-. Societies: A.I.M.E.; Scandinavian Soc. for Electron Microscopy; Vuorimiesyhdistys (Soc. for Finnish Mining and Metal. Eng.); S.T.S. (Finnish Technol. Soc.). Nuclear interest: Metallurgy: mechanical properties of materials, physical metallurgy.
Address: Argonne National Laboratory, Metallurgy Division, 9700 S. Cass Avenue, Argonne, Illinois 60439, U.S.A.

KETUDAT, Sippanondha, Ph.D. Born 1931. Educ.: U.C.L.A. and Harvard Univ. Lecturer, Chulalongkorn Univ., Bangkok. Society: Sci. Soc. of Thailand.
Nuclear interests: Reactor instrumentation; Nuclear physics.
Address: Department of Physics, Chulalongkorn University, Bangkok, Thailand.

KETZLACH, Norman, B.S. and M.S. (Chem. Eng.). Born 1921. Educ.: Washington Univ. Chem. Eng. and Chief Chem., Manganese Products, Inc., 1944-51; Chem. Eng. and Nucl. Safety Specialist, G.E.C., 1951-59; Criticality Safeguards Adviser, Atomics Internat., 1959-. A.N.S.-8, A.N.S.Standards Com. on Nucl. Safety of Fissionable Materials outside Reactors, 1963-; Exec. com., A.N.S. Tech. Group for Nucl. Criticality Safety. Societies: A.N.S.; A.I.Ch.E.; A.C.S.
Nuclear interest: Nuclear safety in storage, handling, processing and shipment of reactor fuels.
Address: 12966 La Maida Street, Sherman Oaks, California, U.S.A.

KEUSTERMANS, Jacques. Eng. and Manager, Nucl. organization, Etablissements Neu, Brussels. Commissaire, Fondation Nucleaire.
Address: Etablissements Neu, 497 avenue Louise, Brussels, Belgium.

KEYNES, Richard Darwin, M.A., Ph.D., Sc.D. (Cantab), F.R.S. Born 1919. Educ.: Cambridge Univ. Fellow, Trinity College (1948-52), Lecturer, Physiology (1949-60), Fellow, Peterhouse (1952-60), Fellow, Churchill Coll., Cambridge Univ.; Director, A.R.C. Inst. of Animal Physiology, Babraham, Cambridge; Fellow, Eton. Representing Internat. Union of Physiological Sci. and Internat. Union of Pure and Appl. Biophys.,Joint Commission on Appl. Radioactivity, I.C.S.U. Societies: Physiological Soc.; U.S. and British Biophys. Socs.
Nuclear interests: Passage of radioactive ions across excitable membranes; Determination of sodium and potassium by activation analysis.
Address: Agricultural Research Council, Institute of Animal Physiology, Babraham, Cambridge, England.

LEYS, David Arnold, M.A., Ph.D., D.Sc. (Hon.), LL.D. (Hon.), D.Humanities (Hon.), Hon. Fellow, Trinity Coll., Toronto. Born 1890. Educ.: Trinity Coll., Toronto. Sci. Vice-Pres., N.R.C., Ottawa, 1947-55; Chairman, Co-ordinating Com. (1952-53), Sci. Advisor to Pres. (1953-), Overseas Rep., London (1960-61), A.E.C.L.; Canadian Rep., N.A.T.O. Sci. Com., 1961-63. Societies: Roy. Soc. of Canada; Fellow, A.P.S.; Canadian Assoc. of Phys.; Fellow, A.A.A.S.; Hon. Member, Eng. Inst. of Canada; Hon. Fellowship, Chem. Inst. of Canada; Hon. Member, Roy. Canadian Inst. Nuclear interests: Administration, lecturing and writing; Interpreting the work to the general public and engineering societies, with emphasis on Canadian contributions.
Address: Atomic Energy of Canada Limited, Chalk River, Ontario, Canada.

KEYS, J. D. Dr. Sen. Sci. Officer, Dept. of Mines and Tech. Surveys, Canada.
Address: Department of Mines and Technical Surveys, 555 Booth Street, Ottawa, Ontario, Canada.

KHA, Maung Maung, B.Sc. (Hons., Rangoon), M.Sc. (London), Ph.D. (London), D.I.C. Born 1917. Educ.: Rangoon, London and California (Los Angeles) Univs. Prof. Phys., Rangoon Univ., 1946-. Chief Consultant, Union of Burma Atomic Energy Centre, 1955-59. Former Vice-Pres., Burma Sci. Assoc.; Member, Burma Res. Council, and Member, Exec. Res. Com. Societies: A.Inst.P. (Great Britain); Com. Member, Burma Res. Soc.; Burma Sci. Assoc.
Nuclear interest: Research using nuclear track plates.
Address: Rangoon University, Rangoon, Burma.

KHABAHPASHEV, A. G. Paper: Co-author, Beginning of Experiments on Positron-Electron Storage Ring VEPP-2 (Atomnaya Energiya, vol. 22, No. 3, 1967); co-author, Measurements of Luminosity in Colliding Beams by Small Angle Scattering (ibid, vol. 22, No. 3, 1967). Address: Academy of Sciences of the U.S.S.R., 14 Leninsky Prospekt, Moscow V-71, U.S.S.R.

KHACHATUROV, T. S. Director, Inst. of Complex Transportation Problems, Moscow. Address: Institute of Complex Transportation Problems, 39 Nizhnaya Krasnosel'skaya Ulitsa, Moscow, U.S.S.R.

KHADIVI, H., Eng. Member, Nat. Iranian A.E.C. Address: National Iranian Atomic Energy Commission, Ministry of Economy, Tehran, Iran.

KHAJAVI, Abolghassem, M.D., Diplomate of American Board of Radiol. Born 1932. Educ.: Tehran Univ. Asst. Prof. of Radiol., Pahlavi Univ., Shiraz, Iran, 1963-. Nuclear interests: Diagnostic and therapeutic radiology and nuclear physics. Address: Nemazi Hospital, Shiraz, Iran.

KHALAFAWI, Miss T. EL-. See EL-KHALA-FAWI, Miss T.

KHAL'CHITSKII, E. P. Paper: Co-author, Linear Induction Accelerator (Atomnaya Energiya, vol. 21, No. 6, 1966). Address: Academy of Sciences of the U.S.S.R., 14 Leninsky Prospekt, Moscow V-71, U.S.S.R.

KHALIFA, E. Sen. Officer, X-Ray and Electro Medical Div., N.E.E.A.-S.A.A. Address: N.E.E.A.-S.A.A., X-Ray and Electromedical Division, 26 sh. Adly, Cairo, Egypt, United Arab Republic.

KHAM'YANOV, L. P. Paper: Co-author, Analysis and Generalisation of Cross-Correlation Method for Measurement of Particles Life-Time Distribution in Physical System (Atomnaya Energiya, vol 20, No. 5, 1966). Address: Academy of Sciences of the U.S.S.R., 14 Leninsky Prospekt, Moscow V-71, U.S.S.R.

KHAN, Abdul Baten, B.Sc. (Agriculture), M.Ag. (Dacca), Ph.D., D.I.C. (London). Born 1924. Educ.: London Univ. Res. Asst. to Agricultural Chem., East Pakistan; Asst. Soil Survey Officer, East Pakistan, 1956-57; Agronomist, East Pakistan Agricultural Inst., 1957; S.S.O., Pakistan A.E.C., 1961; Officer-in-Charge, Atomic Energy Agricultural Res. Centre, Dacca, 1962-64; S.S.O. and Head, Plant Physiology Sect. (1964-66), P.S.O. (1967-), Atomic Energy Centre, Dacca. Post Doctoral Res. Work, Nat. Inst. of Agricultural Sci., Tokyo, 1966-67. Nuclear interest: Application of radioisotopes in the study of soil plant relationship with special reference to plant nutrition and plant physiology. Address: Atomic Energy Centre, P.O. Box No. 164, Ramna, Dacca, East Pakistan.

KHAN, Faruq Aziz, M.Sc., M.S., Ph.D. Born 1929. Educ.: Dacca and Western Ontario Univs. Post-doctoral res. fellow, O.R.N.L., 1960-61; Sen. Sci. Officer, Pakistan A.E.C., 1960-. Nuclear interest: Experimental nuclear physics. Address: Pakistan Atomic Energy Commission, Karachi, Pakistan.

KHAN, Naeem Ahmad, M.A., M.Sc., Ph.D. Born 1928. Educ.: Delhi, Karachi and Manchester Univs. Asst. Metrologist, India and Pakistan Metrological Depts., 1946-61; S.S.O., Pakistan A.E.C., 1961-66; P.S.O. and Director, Atomic Energy Centre, Lahore, 1967-. Res. Fellow, A.E.R.E., Harwell, 1961-62; Res. Fellow, Bartol Res. Foundation of the Franklin Inst., Swarthmore, U.S.A., 1964-65. Society: Inst. of Phys., London. Nuclear interest: Experimental nuclear physics. Address: Atomic Energy Centre, Ferozepur Road, Lahore, West Pakistan.

KHAN, Shams-ul Islam, M.Sc. (Hons.), Ph.D. Born 1920. Educ.: Panjab, Chekiang and Minnesota Univs. Lecturer Peshawar Univ., 1948-56; Nuffield Foundation Fellow, Cambridge Univ., 1957; Res. Botanist, Dept. of Agriculture, Canada, 1958-62; Principal Sci. Officer, Pakistan A.E.C., 1962-63; Principal Agricultural Coll. and Dean, Agricultural Faculty, Peshawar Univ. Nuclear interests: Agriculture. Cytogenetics: wheat. Address: College of Agriculture, Peshawar University, Peshawar, Pakistan.

KHANDAMIROV, Yu. E. Papers: Co-author, Radioactive Depositions on Equipment Surface at Kurchatov Nucl. Power Plant (Atomnaya Energiya, vol. 24, No. 3, 1968); co-author, Investigation of Long-Lived Radioisotopes in Coolant at Kurchatov Nucl. Power Plant (ibid., vol. 24, No. 3, 1968). Address: Academy of Sciences of the U.S.S.R., 14 Leninsky Prospekt, Moscow V-71, U.S.S.R.

KHANOLKAR, Vasant Ramji, B.Sc. (Lond.), M.D. (Lond.), Hon. F.R.C.P. (Edin.), Hon. LL.D. (Melbourne), Hon. D.M.S. (Perugia), F.N.I., F.A.Sc. Born 1895. Educ.: London and Bombay Univs. Director of Labs., (1941-51), Hon. Director of Labs. (1951-), Tata Memorial Hospital,Bombay; Director, Indian Cancer Res. Centre, Bombay, 1951-63; Nat. Res. Prof. in Medicine, 1963-. Vice-Chairman, U.N. Sci. Com. on the Effects of Atomic Rad., 1958-; Pres., Union Internat. Contre le Cancer, 1958-62; Vice-Chancellor, Bombay Univ. 1960-63; Chairman, Biological and Medical Advisory Com., Dept. of Atomic Energy, Govt. of India, 1955-59; Member, W.H.O. Expert Com. on Cancer, 1955-; Consultant W.H.O. Expert Com. on Rad. (Effect of Rad. on Human Heredity), 1958-. Books: Chapter on Cancer in India in Relation to

Habits and Customs, Cancer, vol. 3 (London, Butterworth and Co. (Publishers), Ltd., 1958); Pathology of Leprosy, Chapter 7 of book, Leprosy in Theory and Practice by Dr. R. G. Cochrane (1958); A Look at Cancer (Bombay Indian Cancer Res. Centre, 1958). Societies: Fellow, Roy. Soc. of Medicine, England; Hon. Fellow, Roy. Coll. of Physicians, Edinburgh; Hon. Member, Austrain Cancer Soc.; Member, Acad. of Medical Sci., U.S.S.R.; Pres., Indian Acad. of Medical Sci.
Address: Indian Cancer Research Centre, Parel, Bombay 12, India.

KHARADLY, M. E. A. EL-. See EL-KHARADLY, M. E. A.

KHARBANDA, Sant Raj Ram. Born 1917. Educ.: Cambridge Univ. Tech. Director, Labgear Ltd., 1946-. Com. Member, Panel 'K' - Aerials, Radio and Electronic Component Manufacturers' Federation. Societies: I.E.R.E.; Assoc. Member, I.E.E.
Nuclear interest: Nucleonic instrumentation.
Address: Ivett Lodge, London Road, Harston, Cambs., England.

KHARCHENKO, I. F., Dr. Deleg. to Conference sur la Physique des Plasmas et la Recherche concernant la Fusion Nucléaire Controlée, Salzburg, September, 1961. Paper: Coherent Electron Beam-Plasma Interaction (letter to the Editor, Atomnaya Energiya, vol. 14, No. 3, 1963).
Address: Physical-Technical Institute, Academy of Sciences of the U.S.S.R., Kharkov, Ukrainian S.S.R., U.S.S.R.

KHARITON, Yulii Borisovich. Born 1904. Educ.: Polytechnic Inst., Leningrad. Inst. of Chem. Phys., Acad. of Sci. of the U.S.S.R., 1931. Paper: Co-author, Chain-reaction Decay of Uranium, caused by Slow Neutrons (Zh. Eksper. Teor. Fiz., 1940, 10, No. 1). Society: Member, Acad. of Sci. of the U.S.S.R.
Nuclear interest: Nuclear physics.
Address: Academy of Sciences of the U.S.S.R., 14 Leninsky Prospekt, Moscow V-71, U.S.S.R.

KHARITONOV, Yu. P. Paper: Co-author, Synthesis of 102 Isotopes with Mass Numbers 254, 253 and 252 (Atomnaya Energiya, vol. 22, No. 2, 1967).
Address: Academy of Sciences of the U.S.S.R., 14 Leninsky Prospekt, Moscow V-71, U.S.S.R.

KHARLAMOV, A. G. Paper: Heat Conductivity of BeO for Temperature Range of 1000-2000°C (letter to the Editor, Atomnaya Energiya, vol. 15, No. 6, 1963).
Address: Academy of Sciences of the U.S.S.R., 14 Leninsky Prospekt, Moscow V-71, U.S.S.R.

KHARLAMOV, Yu. S. Paper: Co-author, Irradiation Effect on Magnetic Properties in Order-Disorder Alloy Ni_3Fe (Atomnaya Energiya, vol. 24, No. 1, 1968).
Address: Academy of Sciences of the U.S.S.R., 14 Leninsky Prospekt, Moscow V-71, U.S.S.R.

KHAR'YUZOV, R. V. Paper: Co-author, 30 MeV Microtron-Injector for Pulsed Fast Reactor (Atomnaya Energiya, vol. 20, No. 2, 1966).
Address: Academy of Sciences of the U.S.S.R., 14 Leninsky Prospekt, Moscow V-71, U.S.S.R.

KHASHAB, Sayed Amin EL- See EL-KHASHAB, Sayed Amin.

KHAZANOV, B. I. Papers Co-author, Determination of Near-Energy α-Emitters in Isotope Mixture (letter to the Editor, Atomnaya Energiya, vol. 14, No. 4, 1963). γ-Ray Rad. of Uranium and Thorium Series Elements in Low Energy Region (ibid, No. 5); co-author, On the Apparatus Error in Measurements of Overlapping Spectral Line Intensity (letter to the Editor, ibid, vol. 16, No. 1, 1964).
Address: Academy of Sciences of the U.S.S.R., 14 Leninsky Prospekt, Moscow V-71, U.S.S.R.

KHER, Y. G., Dr. Asst. Prof., Nucl. Chem. Sect., Chem. Dept., Saugar Univ.
Address: Saugar University, Department of Chemistry, Saugar, M.P., India.

KHIN, Aung. Deputy Res. Officer, Radioisotope Lab., Agricultural Res. Inst.
Address: Agricultural Research Institute, Gyogon, Insein P.O., Burma.

KHINTCHINE, A.L. Prof., Moscow Univ. Book: Co-author, Introduction à la théorie des probabilités (Paris Editions Dunod).
Address: Moscow University, Moscow, U.S.S.R.

KHLESTOV, I. J. Counsellor, Ministry of Foreign Affairs, U.S.S.R.; U.S.S.R. adviser, I.A.E.A. Gen. Conference, Vienna, Sept. 1960.
Address: Ministry of Foreign Affairs, Moscow, U.S.S.R.

KHOHLOV, S. F. Paper: Co-author, Crystallisation of Indium-Gallium Alloy with Eutectic Concentration (Atomnaya Energiya, vol. 22, No. 5, 1967).
Address: Academy of Sciences of the U.S.S.R., 14 Leninsky Prospekt, Moscow, V-71, U.S.S.R.

KHOL, Frantisek, Dr. of Phys. (RNDr.). Born 1915. Educ.: Charles Univ.,Prague. At Inst. of Radio Eng. and Electronics, Czechoslovak Acad. Sci. Prague. Com. for Radioisotopes Applications, Czechoslovak A.E.C., Prague. Societies: Czechoslovak Sci. and Tech. Soc.; Nucl. Eng. Commission.
Nuclear interests: Radioisotopes application in metallurgy, nuclear physics, health physics, radiation shielding.
Address: 834 Safarikova Street, Lysá n. Labem, Czechoslovakia.

KHOL'NOV, Yu. V. Book: Co-author, Isobaric Nuclei with the Mass Number A = 140 (London Pergamon).

Address: Radium Institute, Academy of Sciences of the U.S.S.R., 14 Leninsky Prospekt, Moscow V-71, U.S.S.R.

KHONIKEVICH, Aleksandr Aleksandrovich. Sen. Sci., Radium Inst., Leningrad.
Address: V.G. Khlopin Radium Institute, 1 Roentgen Ulitsa, Leningrad, U.S.S.R.

KHOVANOVICH, A. I. Papers: Co-author, Time Dependence of Neutron Yield for (Ra = MsTh) = Be Source (letter to the Editor, Atomnaya Energiya, vol. 20, No. 1, 1966); co-author, Dosimeters Based on Glasses with Optimal Density Alternating by Irradiation (ibid, vol. 21, No. 1, 1966); co-author, Ionisation Chamber with Silver Electrodes for Measuring of Thermal Neutron Fluxes under High Level Gamma - Rad. (letter to the Editor, ibid, vol. 21, No. 1, 1966); co-author, Propagation of Air of Biological Neutron Doses from Point Sources (letter to the Editor, ibid, vol. 23, No. 1, 1967).
Address: Academy of Sciences of the U.S.S.R., 14 Leninsky Prospekt, Moscow V-71, U.S.S.R.

KHRAMCHENKOV, V. A. Papers: Co-author, The Thermal Stability of an Organic Coolant-Monoisopropyl-diphenyl (Atomnaya Energiya, vol 13, No. 1, 1962); Formation of High Boiling Products of Radiolysis by Rad. Mixtures of Perfluorobenzene with Perfluorocyclohexane and Perfluorononane (ibid, vol. 21, No. 5, 1966); co-author, Some Remarks on Rad. Stability Low Melting Coolants in Liquid and Solid State (ibid, vol. 22, No. 1, 1967).
Address: Academy of Sciences of the U.S.S.R., 14 Leninsky Prospekt, Moscow V-71, U.S.S.R.

KHRIPIN, L. A. Papers: Co-author, Binary Systems UF_4 - UCl_4 (Atomnaya Energiya, vol. 19, No. 5, 1965). DTA Study of Phase Transformation in Tetraiodides of Natural Uranium and Uranium - 233 (letter to the Editor, ibid, vol. 23, No. 2, 1967).
Address: Academy of Sciences of the U.S.S.R., 14 Leninsky Prospekt, Moscow V-71, U.S.S.R.

KHRISTENKO, P. I. Papers: Thermodynamics of Using a Turbine with an Organic Liquid Heated in a Power Reactor (Atomnaya Energiya, vol. 8, No. 3, 1960); Some Ways of Increasing the Power Output of a Gas-Cooled Reactor (ibid, vol. 11, No. 6, 1961); On Recycling of Plutonium in Heavy Water Power Reactors (ibid, vol. 20, No. 1, 1966).
Address: Academy of Sciences of the U.S.S.R., 14 Leninsky Prospekt, Moscow V-71, U.S.S.R.

KHROMKOV, I. N. Paper: Co-author, Paramagnetic High-Frequency Effect and Electron Paramagnetic Resonance in Plasma (Atomnaya Energiya, vol. 20, No. 5, 1966).
Address: Academy of Sciences of the U.S.S.R., 14 Leninsky Prospekt, Moscow V-71, U.S.S.R.

KHROMOV, V. V. Papers: Co-author, Conventional Division of Dimensional and Angular Variable in Solution of Neutron Transport Equation (letter to the Editor, Atomnaya Energiya, vol 19, No. 6, 1965); co-author, Large Fast Plutonium Breeders with Nonuranium Dilutors of Plutonium (ibid, vol. 21, No. 2, 1966); co-author, Physical Problems in Developing Fast Power Reactors (Kernergie, vol. 9, No. 9, 1966). co-author, Multigroup Effective Method for Reactor Calculation (letter to the Editor, Atomnaya Energiya, vol. 21, No. 5, 1966).
Address: Physikalisch - Energetischen Institut, Obninsk, U.S.S.R.

KHRYASTOV, N. A. Paper: Co-author, Res. and Training Reactor IR-100 (Atomnaya Energiya, vol. 21, No. 5, 1966).
Address: Academy of Sciences of the U.S.S.R., 14 Leninsky Prospekt, Moscow V-71, U.S.S.R.

KHUDYAKOV, A. V. Papers: Co-author, Electron Microscopy Investigation of Beryllium Oxide Rad. Damage (Atomnaya Energiya, vol. 23, No. 3, 1967); co-author, Study of Rad. Stability of BeO at 100 C° (letter to the Editor, ibid, vol. 24, No. 5, 1968).
Address: Academy of Sciences of the U.S.S.R., 14 Leninsky Prospekt, Moscow V-71, U.S.S.R.

KHURI, Nicola N., B.A., M.A., Ph.D. Born 1933. Educ.: American Univ. Beirut and Princeton Univ. Asst. Prof. Phys., American Univ. Beirut, 1957-58; Member, Inst. for Advanced Study, Princeton, N.J., 1959-60; Assoc. Prof. Phys., American Univ. Beirut, 1960-.
Nuclear interests: Quantum field theory; Dispersion relations, and their application to strong and weak interactions and to potential scattering.
Address: Physics Department, American University of Beirut, Beirut, Lebanon.

KIACHIF, M. F. Manager, New Products Branch, High Temperature Materials Project, Carborundum R. and D. Division.
Address: Research and Development Division, Carborundum Co., Niagara Falls, New York, U.S.A.

KIBA, Toshiyasu, D.Sc. Born 1913. Educ.: Tohoku Univ. Prof., Analytical Chem. (1949), Director, Lab. of Sci. Res. for the Application of Radioisotopes (1957), Dean, Faculty of Sci. (1967), Kanazawa Univ. Vice Pres., Japan Soc. for Analytical Chem., 1967. Societies: A.C.S.; Chem. Soc. of Japan; Japan Rad. Res. Soc.
Nuclear interests: Separation of nuclides; Activation analysis of rare elements.
Address: Kanazawa University, No. 1 Marunouchi, Kanazawa, Japan.

KIBBE, Albert Payne, B.S. Born 1917. Educ.: Southwestern Louisiana Inst. Pres., A.P. Kibbe and Co., 1948-57; Sec.-Director, Utah Sand and Gravel Products, 1948-; Pres., Lisbon Uranium Corp., 1954-59; Director, Uranium Reduction Co., 1957-62; Pres., Hidden Splendor

Mining Co., 1957-62; Director, Beryllium Resources, Inc., 1960-63; Pres., Atlas Minerals Div. (1962-66), Pres. (1964-66), Atlas Corp.; Director, First Security Bank of Utah, 1962-; Director, Surety Life Insurance Co., 1964-67; Director and Vice-Pres., Atomic Ind. Forum, 1965-67; Pres., Kibbe and Co., 1966-; Pres., Kibbe and Assocs., 1967-. Director, Utah Mining Assoc.; Director Utah, Ch. Arthritis and Rheumatism Foundation.
Nuclear interests: Consultant and production interests in nuclear raw materials.
Address: Kibbe and Associates, 304 First Security Building, Salt Lake City, Utah 84111, U.S.A.

KIBBLE, Thomas Walter Bannerman, M.A., B.Sc., Ph.D. Born 1932. Educ.: Edinburgh Univ. Commonwealth Fund Fellow, California Inst. of Technol., 1958-59; N.A.T.O. Fellow (1959-60), Lecturer in Phys., (1961-65), Sen. Lecturer (1965-66), Reader in Theoretical Phys. (1966-), Imperial Coll., London Univ.
Nuclear interests: Theory of elementary particles; Quantum electrodynamics.
Address: Department of Physics, Imperial College, London S.W.7, England.

KICHEV, A. Z. Paper: Co-author, Determination of Spectral Characteristics of Isotopic Neutron Sources (letter to the Editor, Atomnaya Energiya, vol. 19, No. 5, 1965).
Address: Academy of Sciences of the U.S.S.R., 14 Leninsky Prospekt, Moscow V-71, U.S.S.R.

KICHIGIN, G. N. Papers: Co-author Production of Megagauss Fields by Magnetodynamics Cumulation (Atomnaya Energiya, vol. 23, No. 6, 1967); co-author Pulsed High Density Thermonuclear System (ibid, vol. 25, No. 1, 1968).
Address: Academy of Sciences of the U.S.S.R., 14 Leninsky Prospekt, Moscow V-71, U.S.S.R.

KICK, Hermann, Prof., Dr.agriculture. Born 1912. Educ.: Freiburg Univ. and Landw. Hochschule Stuttgart-Hohenheim. Direktor, Agrikulturchemischen Inst., Bonn Univ.; Leiter, Inst. für Landwirtschaft, der Kernforschungsanlage Jülich des Landes Nordrhein-Westfalen e.V.; Mitglied des Wissenschaftlichen Rats der Kernforschungsanlage Jülich des Landes Nordrhein-Westfalen e.V. Präsident des Verbandes Deutscher landwirtschaftlicher Untersuchungs- und Forschungsanstalten. Societies: Verbandes Deutscher landwirtschaftlicher Untersuchungs- und Forschungsanstalten.
Nuclear interest: Agrikulturchemie (Pflanzenernährung).
Address: 176 Meckenheimer Allee, Bonn, Germany.

KIDD, Donald Ewing, Ph.D. Lecturer, Strathclyde Univ., Glasgow.
Nuclear interest: Plasma physics.
Address: Strathclyde University, Glasgow, C.l, Scotland.

KIDD, R. M. Mech. Div. (Production and Design), Nucl. Enterprises (G.B.) Ltd.
Address: Nuclear Enterprises (G.B.) Ltd., Bankhead Crossway, Sighthill, Edinburgh 11, Scotland.

KIEFER, Hans, Dipl. Phys., Dr. rer. nat. Born 1923. Educ.: Karlsruhe T.H. Kernforschungszentrum Karlsruhe, 1956; Head, Nucl. Eng. School (1964), Privatdoz. (1965), Karlsruhe Univ. Sci. Council, Nucl. Res. Centre, Karlsruhe. Societies: Deutsche Atomkommission, Arbeitskreis III/6; Fachverband für Strahlenschutz e.V. (IRPA). Books: Monitoring the radioactivity in Liquid and Gaseous Radioactive Effluents (Teubner Verlag, 1967); co-author, Strahlenschutzmesstechnik (Braun-Verlag, 1964).
Nuclear interest: Health physics.
Address: Kernforschungszentrum Karlsruhe, Germany.

KIEHN, R. W. Deputy Gen. Manager, Radiol. Sci. Dept.. Nevada Test Site.
Address: U.S.A.E.C., Nevada Test Site, Radiological Sciences Department, A.E.C. Contract Office, P.O. Box 1360, Las Vegas, Nevada 89101, U.S.A.

KIELBASINSKI, Janusz, M.Sc. Born 1929. Educ.: Lódź Tech. Univ. Director, Polish Nucl. Energy Information Centre; Sec., Commission of Publications and Popularisation, State Council for Peaceful Uses of Nucl. Energy; Chief editor, Progress of Nucl. Eng.
Nuclear interests: Abundance of elements; Nuclear astrophysics; Meteoritic matter.
Address: Palace of Culture and Science, Warsaw, Poland.

KIELY, John Roche, B.A. (Civil Eng.). Born 1906. Educ.: Washington Univ. Exec. Vice-Pres. and Director, Bechtel Corp.; Vice-Pres. and Director, Eng. Joint Council. Society: Nat. Acad. Eng.
Address: Bechtel Corporation, 220 Bush Street, San Francisco 4, California, U.S.A.

KIELY, Joseph M., M.D. Hematologist, Com. on Rad. Control, also in Sect. of Internal Medicine, Mayo Clinic and Mayo Foundation.
Address: Mayo Clinic and Mayo Foundation, Rochester, Minnesota, U.S.A.

KIENBAUM, G., Dipl.-Ing. Minister für Wirtschaft, Mittelstand and Verkehr des Landes Nordrhein-Westfalen. Vizepräsident, Verwaltungsrat, Kernforschungsanlage Jülich des Landes Nordrhein-Westfalen E.V.
Address Kernforschungsanlage Jülich des Landes Nordrhein-Westfalen E.V., Postfach 365, Jülich, Germany.

KIENITZ, Hermann, Dr.rer.nat., Dipl. Chemiker. Born 1913. Educ.: Breslau Univ. At Badische Anilin and Sodafabrik A.G., Ludwigshafen/ Rhein. Managing Com., Fachgruppe Analytische Chemie der Gesellschaft Deutscher Chemiker;

Titular Member, Commission on Physico-chem. Data and Standards, Internat. Union of Pure and Appl. Chem.
Nuclear interests: Tracer methods with active and inactive isotopes for studying mechanism of organic reactions; Radiation chemistry.
Address: Ludwigshafen/Rhein, Wolframstrasse 4, Germany.

KIENLIN, Albrecht VON. See VON KIENLIN, Albrecht.

KIESSLING, Roland Richard, Prof. Ph.D. (Inorg. Chem.). Born 1921. Educ.: Uppsala Univ. Assoc. Prof., Uppsala Univ., 1950-51; Res. metal., Söderfors Bruk, 1951-53; Chief Metal., AB Atomenergi, Stockholm, 1953-60; Director, Swedish Inst. Metal Res., 1960-. Societies: Inst. of Metals; A.I.M.E.; Iron Steel Inst.; A.C.S.
Nuclear interest: Metallurgy.
Address: Swedish Institute of Metal Research, Drottn. Krist. V. 48, Stockholm Ö, Sweden.

KIGAR, Donald F. Pres., and Chief Operations Officer, Detroit Edison Co., 1964-.
Address: Detroit Edison Co., 2000 Second Avenue, Detroit 26, Michigan, U.S.A.

KIGER, Eugene Oliver, B.S. (Elec. Eng.), M.S. (Nucl. Eng.). Born 1924. Educ.: North Carolina State Univ. Chief Tech. Supv., Reactor Technol. Sect., Savannah River Plant (A.E.C.), E. I. de Pont de Nemours and Co., Inc. Local Sect. Com., and Member, Exec. Com., Savannah River Sect., A.N.S. Society: A.A.A.S.
Nuclear interests: Reactor engineering; Reactor operation. Radioisotope production and applications; Reactor fuel development; Power reactor technology.
Address: 384 Barnard Avenue, S.E., Aiken, South Carolina 29801, U.S.A.

KIGOSHI, Akiichi, Dr. Eng. Born 1926. Educ.: Tohoku Univ. Asst. (1951), Asst. Prof. (1958), Res. Inst. Mineral Dressing and Metal., Tohoku Univ. Societies: Atomic Energy Soc, of Japan; Japan Inst. Metals; Mining and Metal. Inst. Japan.
Nuclear interest: Fundamentals on metallurgical reprocessing of spent reactor fuels.
Address: 39-1 Sanjo Dori, Sendai, Japan.

KIHARA, Hitoshi, Dr. Sci., (Kyoto), Dr. Sci. h.c. (McGill), Dr. Agr. h.c. (College of Agr., Uppsala). Born 1893. Educ.: Hokkaido (Sapporo) Univ. Prof., Kyoto Univ., 1927-55; Director, Nat. Inst. of Genetics, Misima, 1955-. Commissioner, Japan A.E.C., 1961-64. Society: Genetics Soc. of Japan.
Nuclear interest: Radiation genetics.
Address: National Institute of Genetics, Yata 1,111, Misima, Sizuoka-ken, Japan.

KIHARA, Taro, D.Sc. Born 1917. Educ.: Tokyo Univ. Prof., Tokyo Univ. Societies: Phys. Soc. of Japan; A.P.S.
Nuclear interest: Theoretical aspects of controlled nuclear fusion.
Address: Department of Physics, Tokyo University, Tokyo, Japan.

KIHLBOM, J. O. Niklas, LL.B. Born 1921. Educ.: Uppsala Univ. Deputy Gen. Manager, Öresund Marine Insurance Co., 1958; Managing Director, Atlantica Insurance Co., 1959-; Vice-chairman, Swedish Atomic Insurance Pool; Chairman, Nucl. Com., Nordic Pool; Chairman, Nucl. Information Com., Internat. Union of Marine Insurance. Swedish Deleg. to O.E.E.C. and C.M.I. Nucl. Liability coms.; Tech. Consultant to Swedish legislative com. on nucl. liability. Book: Co-author, Nucl. Liability (Pergamon Press, 1959). Society: Swedish Assoc. Internat. Maritime Law.
Nuclear interests: Safety, liability and insurance problems, particularly regarding carriage of nuclear substances and the operation of nuclear ships.
Address: Atlantica Insurance Company, 5 S. Hamngatan, P.O. Box 2251, Gothenburg 2, Sweden.

KIHLMAN, A. Vice-Chairman, Voimayhdistys YDIN (Nucl. Power Assoc.).
Address: Ekono, 14 S. Espl., Helsinki, Finland.

KIJM, C. At Nederlandsche Dok en Scheepsbouw Mij. V.O.F.
Address: Nederlandsche Dok en Scheepsbouw Mij, V.O.F., Amsterdam, Netherlands.

KIKAWADA, Kazutaka. Councillor, Fund for Peaceful Atomic Development of Japan; Director, Japan Atomic Power Co. Ltd.
Address: Fund for Peaceful Atomic Development of Japan, No 1-13, 1-chome, Shimbashi, Minato-ku, Tokyo, Japan.

KIKKAWA, Akira. Pres., Nippon Carbon Co. Ltd.
Address: Nippon Carbon Co., Ltd., 2, 2-chome, Nishi - Hachi - Choboni, Chuoku, Tokyo, Japan.

KIKKAWA, Mitsuo, Dr. Manager, Third Res. Sect., Res. Dept., Hitachi Wire and Cable Ltd.
Address: Hitachi Wire and Cable Ltd., 2-16 Marunouchi, Chiyoda-ku, Tokyo, Japan.

KIKNADZE, G. I. Papers: Co-author, Indium-Helium Alloy as Gamma Carrier for Rad. Circuits (Atomnaya Energiya, vol. 19, No. 2, 1965); co-author, Radiographic Study of Liquid Indium - Gallium Alloy with Eutectic Concentration (ibid, vol. 22, No. 5, 1967); co-author, Crystallisation of Indium - Gallium Alloy with Eutectic Concentration (ibid, vol. 22, No. 5, 1967); co-author, On Problem of Industrial Rad. Loop Construction (letter to the Editor, ibid, vol. 23, No. 6, 1967).
Address: Academy of Sciences of the U.S.S.R., 14 Leninsky Prospekt, Moscow V-71, U.S.S.R.

KIKUCHI, Chihiro, B.S. (Phys.) (Washington); M.A. (Math.), (Cincinnati), Ph.D. (Phys.) (Washington). Born 1914. Educ.: Washington and Cincinnati Univs. Assoc. Prof., Michigan

State Univ., 1953; Visiting Phys., Brookhaven Nat. Lab., 1951-52; Res. Phys., Naval Res. Lab., 1953-55; Res. Phys. and Head, Solid-State Phys. Lab. Willow Run Labs. (1955-59), Prof. Nucl. Eng. (1959-), Michigan Univ.; Consultant, Gen. Motors Corp., 1960-; Consultant, Nat. Bureau of Standards (Boulder), 1961-. Society: A.P.S.
Nuclear interests: Radiation effects in solids; Neutron physics.
Address: Department of Nuclear Engineering, Michigan University, Ann Arbor, Michigan, U.S.A.

KIKUCHI, K., Asst. Prof. Sen. Officer, Dept. of Phys., Nucl. Res. Lab., Nagoya Univ.
Address: Nagoya University, Furo-cho, Chikusa-ku, Nagoya, Japan.

KIKUCHI, Minoru. Deputy Chairman, Managing Com., Japan Atomic Energy Insurance Pool.
Address: Japan Atomic Energy Insurance Pool, Ida Building, 1, Yaesu 2-chome, Chuo-ku, Tokyo, Japan.

KIKUCHI, Ryoichi, Ph.D. (Tokyo). Born 1919. Educ.: Tokyo (then Imperial) Univ. Asst. Prof., Inst. for the Study of Metals, Chicago Univ., 1953-55; Res. Phys., Armour Res. Foundation, Chicago, 1955-56; Assoc. Prof., Dept. of Phys., Wayne State Univ., 1956-58. Sen. Sci., Hughes Res. Labs., Malibu, California 1958-. Lecturer, Dept. of Eng., California Univ. at Los Angeles. Societies: A.P.S.; Phys. Soc. of Japan.
Nuclear interest: Statistical mechanics (equilibrium and irreversible).
Address: Hughes Research Laboratories, Malibu, California, U.S.A.

KIKUCHI, Seishi, D.Sc. Born 1902. Educ.: Tokyo Univ. Prof. (1953), Chief, Inst. for Nucl. Study (1955), Chief, Cosmic-Ray Lab. (1955), Tokyo Univ.; Commissioner, A.E.C. of Japan, 1958-; Board Chairman, Japan Atomic Energy Res. Inst., 1960-64.
Nuclear interests: Electron diffraction and atomic nucleus.
Address: 2926 Tanashimachi, Kitatama-gun, Tokyo, Japan.

KIKUCHI, Tohru, B.S. (Eng.). Born 1928. Educ.: Tokyo Univ. Atomic Energy Bureau, 1957-64; Japanese Atomic Energy Attaché stationed in Washington, D.C., 1964-68.
Nuclear interest: Administration.
Address: 2520 Massachusetts Avenue, N. W., Washington, D. C. 20008, U.S.A.

KILLELEA, Joseph R., B.S., Ph.D. Born 1917. Educ.: Manhattan Coll. and New York Univ. Director, Nucl. Sci. and Eng. Centre, Lowell Technol. Inst. Society: A.N.S.
Nuclear interest: Reactor operations and safety.
Address: Lowell Technological Institute, Nuclear Science and Engineering Centre, Lowell, Massachusetts, U.S.A.

KILLGORE, Charles A., B.S., M.S. (Chem. Eng.). Born 1934. Educ.: Louisiana Polytech. Inst. Officer, U.S.A.F., 1956-59; Graduate Student and Instructor, Chem. Eng., Louisiana Polytech. Inst., 1959-62; Director, Louisiana Tech. Nucl. Centre and Asst. Prof., Chem. Eng., 1962-. Vice Pres., Louisiana Branch, A.S.E.E. Societies: A.I.Ch.E.; A.N.S.; A.S.E.E.; Louisiana Eng. Soc.
Nuclear interests: Nuclear education and research in radioisotope applications in industry.
Address: Nuclear Centre, Louisiana Polytechnic Institute, Ruston, Louisiana 71270, U.S.A.

KILLIAN, James R., B.S. Sc.D. (Hon.), LL.D. (Hon.), Sc.D., D.Eng. (Hon.). Educ.: Duke Univ. and M.I.T. Pres., M.I.T. 1948-; Member, President's Advisory Com. on Management, 1950-52. Deleg. to 2nd I.C.P.U.A.E., Geneva, Sept., 1958. Member, Sci. Advisory Com., O.D.M., 1951; Chairman, Army Sci. Advisory Panel, 1951; Member, Board of Visitors, U.S. Naval Acad., 1953; Chairman, President's Board of Consultants on Foreign Intelligence Activities, 1956; Special Asst. to Pres. for Sci. and Technol., 1957; Member, Board of Directors, Educ. Foundation for Nucl. Sci. Societies: Fellow, American Acad. of Arts and Sci.; A.S.E.E.
Address: U.S. Atomic Energy Commission, Washington 25, D.C., U.S.A.

KILLMAR, Henry M., B.S. Born 1918. Educ.: Illinois Univ. Manager, New Products Branch, R. and D. Div., Carborundum Co. Societies: American Ceramic Soc.; Inst. of Ceramic Eng.; American Ordnance Assoc.
Nuclear interest: Non-metallic materials.
Address: The Carborundum Company, P.O. Box 337, Niagara Falls, New York, U.S.A.

KILPATRICK, James Nelson, B.B.A. Born 1907. Educ.: William and Mary Coll., Virginia. Sen. Civilian Eng., Tech. Div., U.S. Naval Supply Centre, Norfolk, Virginia, 1948-59; Eng. Editor, Westinghouse Atomic Power Divs., 1959-. Editorial Board, Soc. of Tech. Writers and Publishers. Societies: American Soc. of Naval Eng.; A.N.S.
Nuclear interests: Technical editing and writing in field of commercial atomic power.
Address: Westinghouse Atomic Power Divisions, P.O. Box 355, Pittsburgh 30, Pennsylvania, U.S.A.

KIM, Chi Eun. Chief, Gen. Affairs Sect., Bureau of Administration, Office of Atomic Energy and A.E.C.
Address: Office of Atomic Energy and Atomic Energy Commission, 2-1 Chung-dong, Sudaimoon-ku, Seoul, Korea.

KIM, Chung Han. Chief, Div. of Rad. Therapy, Radiol. Res. Inst., Office of Atomic Energy and A.E.C.
Address: Radiology Research Institute, 2-1 Chung-dong, Sudaimoon-ku, Seoul, Korea.

KIM, D. S. Member, Subcom. on Use of Radioactivity Standards, Nat. Acad. of Sci. - N.R.C.
Address: National Academy of Sciences-National Research Council, 2101 Constitution Avenue, N.W., Washington 25, D.C., U.S.A.

KIM, Hi In. Korean scientist, member of group obtaining for first time xi minus hyperons.
Address: Joint Institute for Nuclear Research, Dubna, Nr. Moscow, U.S.S.R.

KIM, J. M. Formerly Babcock and Wilcow Res. Fellow, now Res. Asst., Nucl. Eng. Dept., Queen Mary Coll.
Address: Queen Mary College, Nuclear Engineering Department, Mile End Road, London, E.1, England.

KIM, Jai Baik, Sen. Chem., Isotope Lab., Physico-chem. Div., Nat. Chem. Labs., Seoul.
Address: Isotope Laboratory, Physico-chemistry Div., National Chemistry Laboratories, Ministry of Health and Social Affairs, 79 Saichong-No, Chongno-ku, Seoul, Korea.

KIM, Ok Joon, Dr. Director, Geological Survey of Korea.
Address: Geological Survey of Korea, Namyoung Dong, Seoul, Korea.

KIM, Won Jo. Nucl. Material Prospecting Group, Geophysical Prospecting Sect., Geological Survey of Korea.
Address: Geological Survey of Korea, Namyoung Dong, Seoul, Korea.

KIM, You Sun, Ph.D., B.S. Born 1924. Educ.: Purdue and Seoul Univs. Head, Chem. Div., Atomic Energy Res. Inst., Seoul, Korea, 1961-. Societies: Korean Chem. Soc.; A.C.S.
Nuclear interest: Organic chemistry, especially the tracer application in organic reaction study and nuclear related preparative works.
Address: Division of Chemistry, Atomic Energy Research Institute, P.O. Box 7, Chyung Ryang Ri, Seoul, Korea.

KIM, Young Nok, B.Sc., M.Sc., Ph.D. Born 1921. Educ.: Seoul and Birmingham Univs. Res. fellow, Inst. of Theoretical Phys., Copenhagen, 1956-58; Asst. Prof. Seoul Nat. Univ., 1959; Chief, Fundamental Res. Div., Atomic Energy Res. Inst., Seoul,1960; Visiting Res. Sci., Washington Univ., Seattle, 1960-62; Res. Alberta Univ., 1962-63; Assoc. Prof., Newfoundland Univ., 1963-. Societies: A.P.S.; Italian Phys. Soc.
Nuclear interest: Theoretical nuclear physics.
Address: 5 Maple Street, St. John's, Newfoundland, Canada.

KIM, Yung Ill. Sen. Chem., Isotope Lab., Physico-chem. Div., Nat. Chem. Labs., Seoul.
Address: Isotope Laboratory, Physico-chemistry Div., National Chemistry Laboratories, Ministry of Health and Social Affairs, 79 Saichong-No, Chongno-ku, Seoul, Korea.

KIMBALL, Allyn W., B.S., Ph.D. Born 1921. Educ.: Buffalo Univ., M.I.T. and N.C.S.C. Chief, Statistics Sect. Maths. Panel, O.R.N.L., 1950-60; Prof. and Chairman, Biostatistics Dept. (1960-66), Dean, Faculty of Arts and Sci. (1966-), Johns Hopkins Univ.; Trustee, Assoc. Univs., Inc., 1961-. Treas., Biometric Soc. Societies: Biometric Soc.; A.A.A.S.; Inst. of Mathematical Statistics; A.S.A.
Nuclear interests: Statistics and mathematics applied to radiation biology and medicine.
Address: Office of the Dean,Faculty of Arts and Sciences, Johns Hopkins University, Baltimore, Maryland 21218, U.S.A.

KIMBALL, Dan Able, B.S. (Hon. Iowa Wesleyan College). Born 1896. Asst. Sec. of the Navy for Air, 1949; Undersec. of the Navy, 1949-51; Sec. of the Navy, 1951-53. Exec. Vice Pres. and Gen. Manager (1944-49), Pres. (1953-63), Chairman of the Board (1963-65), Chairman, Exec. Com. (1965-), Aerojet - Gen. Corp. Society: Inst of the Aeronautical Sci.
Address: Aerojet-General Corp., El Monte, California, U.S.A.

KIMBALL, Richard Fuller, A.B., Ph.D. Born 1915. Educ.: Johns Hopkins Univ.; Director, Biol. Div., O.R.N.L., 1967-. Societies: Rad. Res. Soc.; Genetics Soc. of America; Soc. of Protozoologists; American Soc. of Zoologists; American Soc. Cell Biol.
Nuclear interests: Radiation biology; Induction of mutation; Autoradiography with tritiated compounds; Effects of radiation on protozoa.
Address: Biology Division, Oak Ridge National Laboratory, P.O. Box Y, Oak Ridge, Tennessee 37830, U.S.A.

KIMEL', L. R. Papers: Co-author, Monte-Carlo Calculation for Scattered γ-Ray Spectral-Angular Distribution of Cs^{137} Point Monodirectional Source in Iron (letter to the Editor, Atomnaya Energiya, vol. 15, No. 4, 1963); co-author, Measurements of Stray Neutrons Spectra at Synchrocyclotron Hall (letter to the Editor, ibid, vol. 24, No. 1, 1968); co-author, Attenuation of Stray Rad., from Synchrocyclotron in Shielding (ibid, vol. 24, No. 2, 1968); co-author, Back - Yield of Neutrons from out Shielding Bombarded with 660 MeV Protons (letter to the Editor, ibid, vol. 24, No. 4, 1968).
Address: Academy of Sciences of the U.S.S.R., 14 Leninsky Prospekt, Moscow V-71, U.S.S.R.

KIMEL, William R., B.S., M.S. (Mech. Eng., Kansas State), Ph.D. (Eng. Mechanics, Wisconsin). Born 1922. Eudc.: Kansas State and Wisconsin Univs. Eng., U.S. Forest Products Labs., Madison, Wisconsin, 1955-56; Assoc. Prof. Mech. Eng. Dept., Kansas State Univ., 1956-58; Resident Res. Assoc., Argonne Nat. Lab., 1958; Prof. and Head, Dept. of Nucl. Eng., Kansas State Univ., 1958-. Societies: A.N.S.; A.S.M.E.
Nuclear interests: Reactor design, research and pedagogy in nuclear engineering.

Address: Department of Nuclear Engineering, Kansas State University, Manhattan, Kansas, U.S.A.

KIMMIG, Joseph, Prof., Dr. Mitglied, Arbeitskreis II/3 Biologie und Medizin, Deutsche Atomkommission, Federal Ministry for Sci. Res.
Address: Federal Ministry for Scientific Research, 46 Luisenstrasse, 532 Bad Godesberg, Germany.

KIMURA, Kenjiro, D.Sc. Born 1896. Educ.: Tokyo Univ. Prof. (1933-56), Dean, Faculty of Sci. (1953-55), Tokyo Univ.; Councillor and Director of Radioisotope School, Director of Isotope Applications, Japan Atomic Energy Res. Inst., 1956-64. Member, Rad. Effect Inquiry Special Com., Sci. Council of Japan. Books: Inorganic Qualitative Analysis (1947); Inorganic Quantitative Analysis (1949). Analytical Procedure of Inorganic Compounds (1956); Data Book of Isotopes.
Nuclear interests: Nuclear chemistry; Radioisotope production and applications.
Address: 1723 Higashiterao-Machi, Tsurumi-Ku, Yokohama, Japan.

KIMURA, Ki-ichi, D.Sc. Born 1904. Educ.: Kyoto Univ. Prof. Phys., Kyoto Univ., 1945; Director, Res. Reactor Inst.,and Prof., Div. of Reactor Accessory, Kyoto Univ. Societies: Phys. Soc. of Japan; A.P.S.
Nuclear interest: Nuclear physics (low energy).
Address: 5 Izumiden-cho, Yoshida, Sakyo-ku Kyoto, Japan.

KIMURA, Motoharu, Ph.D. Born 1908. Educ.: Tokyo Univ. Res. Phys., Phys. and Chem. Inst., Res., Inc., Tokyo, 1933-50; Prof. Phys., Tohoku Univ., 1950-. Societies: Phys. Soc. Japan; Japan Atomic Energy Soc.
Nuclear interests: Nuclear reactions and structure in low and medium energy region, especially photoreactions with Brems or monoenergetic gammas and electrons; Neutron physics and neutron diffraction; Design and management of accelerators.
Address: 13-20 Mukaiyama 3, Sendai, Japan.

KIMURA, Takayoshi. Eng. Surveyor, Ship Classification Soc. of Japan.
Address: Ship Classification Society of Japan, 1 Akasaka-Fukuyoshi-cho, Minato-ku, Tokyo, Japan.

KINARD, Frank Efird, Ph.D. (Phys.), M.S. (Phys.), A.B., B.S. Born 1924. Educ.: Newberry Coll. (South Carolina) and North Carolina Univ. Phys. (1953-63), Director, Univ. Relations Office (1963-67), Res. Assoc. with F. Reines, Antinuetrino Experiments (1964-67), Savannah River Lab., E.I. DuPont de Nemours and Co., Inc.; Exec. Director, South Carolina Commission on Higher Educ.. 1967-. Societies: A.N.S.; A.P.S.
Nuclear interest: Neutrino and neutron physics.
Address: South Carolina Commission on Higher Education, 2712 Millwood Avenue, Columbia, South Carolina, U.S.A.

KINCHIN, George Henry, M.A. Born 1923. Educ.: Cambridge Univ. Head, Gen. Reactor Phys. Div., A.E.E., Winfrith. Member, Internat. Nucl. Data Com., I.A.E.A.
Nuclear interest: Reactor physics.
Address: Cheriton House, Winfrith, Dorset, England.

KIND, Adolfo, Mech. Eng., Dr. in Tech. Sci.; Born 1916. Educ.: Polytech. School, Zürich. Res., Swiss Atomic Commission, Geneva, 1947-51; Lecturer, Padua Univ., 1951-55, 1956-63; Res., C.E.R.N. Theoretical Div., Copenhagen, 1955-56; Chief, Eng., Reactor Calculation Div., Agip Nucleare, Milan, 1957-59; Head, Reactor Theory and Analysis Div., Euratom C.C.R., Ispra 1959-; Lecturer, Bari Univ., 1964-. Societies: F.I.E.N.; Italian Phys. Soc.; Swiss Phys. Soc.; B.N.E.S.
Nuclear interests: Nuclear physics and reactor design.
Address: Euratom C.C.R., 21027 Ispra, Italy.

KINDERMAN, Edwin M., M.S., Ph.D. Born 1916. Educ.: Oberlin Coll. and Notre Dame Univ. Sen. Sci., G.E.C., 1949-56; Sen. Phys. (1956-57), Manager, Nucl. Phys. Dept. (1957-67), Director, Rad. Phys. Div. (1967-), Stanford Res. Inst. Societies: A.C.S.; A.N.S.; A.P.S.; RESA.; Atomic Ind. Forum; A.S.T.M.; I.N.M.M.
Nuclear interests: Radiation chemistry and radiation physics; Nuclear industry and applications; Nuclear materials management and control.
Address: Stanford Research Institute, Menlo Park, California 94025, U.S.A.

KINDSVATTER, Victor. Member, Rad. Com., A.I.H.A.
Address: American Industrial Hygiene Association, 14125 Prevost, Detroit 27, Michigan, U.S.A.

KINELL, Per-Olof, Dr. Sci. Born 1914. Educ.: Uppsala Univ. Asst. Prof., Uppsala Univ. 1953-; Res. leader, A.B. Atomenergi, Studsvik, Nyköping, 1960-63; Head, Swedish Res. Councils' Lab., Studsvik, Nyköping, 1963-.
Nuclear interests: Radiation chemistry; study of free radicals in heterogeneous systems and radiation and induced polymerisation; E.P.R. studies of free radicals.
Address: Swedish Research Councils' Laboratory, Studsvik, Nyköping, Sweden.

KINET, J. M. Sci. concerned with Radioisotope Work, Centre de Recherches des Hormones Vegetales de l' I.R.S.I.A.
Address: Centre de Recherches des Hormones Vegetales de l' I.R.S.I.A., Institut de Botanique, 3 rue Fusch, Liege, Belgium.

KING, A. J., Dr., F.Inst.P. Chairman, Acoustics Group, Inst. Phys. and the Phys. Soc.

Address: Institute of Physics and the Physical Society, Acoustics Group, 47 Belgrave Square, London, S.W.1, England.

KING, C.D.G., Assoc. Prof., Dept. Mech. Eng., U.S. Naval Postgraduate School.
Address: Department of Mechanical Engineering, U.S. Naval Postgraduate School, Monterey, California, U.S.A.

KING, Charles A. Supervision and Control of Nucl. Work, R. and D. Dept., Elcor, a Div. of Halliburton Co.
Address: Elcor, Research and Development Department, 2431 Linden Lane, Silver Spring, Maryland 20910, U.S.A.

KING, Daniel Clifford, Major, B.S. (E.E.), B.B.A., M.B.A. Born 1921. Educ.: Oklahoma Univ. and Air Force Inst. Technol. Major, U.S.A.F., 39476A, 1951-. Chief, Training Branch, Nucl. Power Field Office, Army Nucl. Reactor Programme, Ft. Belvoir, Virginia, 1956-59. Assisted in startup and operation of the APPR-1, Ft. Belvoir, and organisation of training programme for training nuclear power plant operators for the Army, Navy,and Air Force, Student Council Rep., Air Force Inst. Technol., Wright-Patterson A.F.B., Ohio. Presently a Project Officer in Reactor and Advanced Systems Div., Directorate of Nucl. Safety, Office of Deputy Inspector Gen. for Safety, Kirtland A.F.B., N.Mex. Society: American Inst. Elec. Eng.
Nuclear interests: Managing a nuclear project and inspecting nuclear reactor facilities for safety.
Address: 1930A Mercury Drive, Kirtland A.F.B., New Mexico, U.S.A.

KING, David C., B.S. (Mech. Eng.). Born 1926. Educ.: Baldwin Wallace Coll. and Case Inst. Technol. Eng., Savannah River Project, E.I. duPont de Nemours, 1951-52; Analytical Eng., Aircraft Nucl. Propulsion Project, Pratt and Whitney Aircraft Co., 1952-54; Project Eng., Materials Testing Reactor and Eng. Test Reactor, Phillips Petroleum Co., 1954-58; Operations Supt., Gas-Cooled Reactor Experiment, Aerojet-Gen. Nucleonics, 1958-63; Reactor Eng., Nucl. Safety Experiments, Loss-of-Fluid-Test, SNAPTRAN, Water Reactor Safety Programme Office, Idaho Operations Office, U.S.A.E.C., 1963-. Societies: Idaho Acad. Sci.; A.N.S.; A.A.A.S.
Nuclear interests: Reactor operations; Experimental operation; Heat transfer; Fluid flow; Design.
Address: 787 Sonja, Idaho Falls, Idaho, U.S.A.

KING, E. Richard, B.A., M.D. Born 1916. Educ.: Ohio State Univ. Chief of Radiol., U.S. Naval Hospital, 1958-61; Director, Dept. Nucl. Medicine, U.S. Naval Medical School, Nat. Naval Medical Centre, Bethesda, Maryland, 1954-61; Prof. of Radiol. and Chief, Div. of Radiotherapy, Medical College of Virginia, Richmond, Virginia, 1961-; Member Consulting Editorial Board, J. Nucl. Medicine, 1958-61. Member, U.S.A.E.C. Advisory Com. on Medical Uses of Isotopes, 1964-. Books: Contributor to Atomic Medicine, 1949, 1953, 1959, 1964; A Manual for Nucl. Medicine (1961). Societies: A.M.A.; A.A.A.S.; American Coll. Radiol.; Radiol. Soc. of North America; Soc. of Nucl. Medicine; American Roentgen Ray Soc.; American Radium Soc.; American Soc. of Univ. Radiologists; British Inst. of Radiol.
Nuclear interests: Clinical radioisotopes; Medical reactors; Medical radiobiology; Radiation therapy.
Address: 8928 Cherokee Road, Richmond 25, Virginia, U.S.A.

KING, George, Dr. Agriculture Com., State of Georgia Nucl. Advisory Commission.
Address: State of Georgia Nuclear Advisory Commission, Office of Publications, Georgia Institute of Technology, Atlanta 13, Georgia, U.S.A.

KING, John Callen, S.B. (M.I.T.), S.M. (Colorado). Born 1911. Educ.: M.I.T. Product Development Eng., The Master Builders Co. of Cleveland, Ohio; formerly Chief Eng. and Manager of Sales, Intrusion-Prepakt, Inc. and Prepakt Concrete Co. of Cleveland, Ohio; Sen. Eng., Internat. Eng. Co., San Francisco. Vice-Pres., Ohio Soc. of Prof. Eng.; Chairman, New Construction Techniques Com., A.S.C.E. Societies: National, Ohio and Cleveland Socs. of Professional Eng.; Fellow, A.S.C.E.; American Concrete Inst.; Internat. Soc. for Bridges and Structures.
Nuclear interests: Biological shielding and nuclear construction in concrete, normal and high density, placed by conventional and preplaced-aggregate methods; Non-shrink, metallic-aggregate grouts.
Address: 101 East 252nd Street, Cleveland 32, Ohio, U.S.A.

KING, John Swinton, Ph.D. (Phys.). Born 1920. Educ.: Michigan Univ. Res. Assoc. K.A.P.L. (1953-57), Manager, Phys. Sect., Submarine Reactor Project (1958-59), Gen. Elec. Co.; Assoc. Prof. (1959-62), Prof. (1963-), Nucl. Eng. Dept., Michigan Univ.; Low Energy Nucl. Phys., Dept. of Phys., Toronto Univ. Societies: A.P.S.; A.N.S.
Nuclear interests: Reactor physics and nuclear physics.
Address: 2311 Vinewood Boulevard, Ann Arbor, Michigan, U.S.A.

KING, L. D. Percival, B.S. (Mech. Eng.), Ph.D. (Phys.). Born 1906. Educ.: Rochester Univ. M.I.T. and Wisconsin Univ. Group Leader,Phys. and Reactor Div., Los Alamos Sci. Lab., 1944-57; Tech. Director, U.S.A.E.C., 1958; Atoms for Peace Conference, 1957-58; Asst. Div. Leader, Reactor Div., Los Alamos Sci. Lab., 1958-60; Chairman, Rover Flight Safety Office, 1960-; Societies: Fellow, A.N.S.; Fellow, A.P.S.; Fellow, A.A.A.S.
Nuclear interests: Design and development of

KING, Richard Wayne, Ph.D. Born 1924. Educ.: Washington Univ., St. Louis. Phys., Nucl. Development Assoc., White Plains, New York, 1952-53; Phys. Nat. Res. Council, Washington D.C., 1953-55; Asst. Prof. (1955-58), Assoc. Prof. Phys. (1958-61), Prof. Phys. (1961-), Head, Phys. Dept., (1966-), Purdue Univ.; Alfred P. Sloan Res. Foundation Fellow, 1959-63. Societies: A.P.S.; A.A.U.P.
Nuclear interests: Theoretical nuclear physics; Neutrino interactions; Nuclear structure; Radiations associated with fission.
Address: Physics Department, Purdue University, Lafayette, Indiana, U.S.A.

KING, William James, B.Sc., Ph.D. Born 1933. Educ.: Western Ontario and McMaster Univs. Post Doctorate Fellowship (1959), Asst. Res. Officer, X-Rays and Nucl. Rad. Sect. (1960), N.R.C. of Canada; Manager, Solid State Phys., Ion Phys. Corp., Burlington, Mass., U.S.A. (Subsidiary of High Voltage Eng. Corp.), 1961-64. Societies: A.P.S.; Canadian Assoc. of Phys.
Nuclear interests: Accelerators - primarily Van de Graaff for ion implantation semi-conductor doping, radiation simulation and radiation damage; Nuclear physics - low energy.
Address: 9 Putnam Road, Reading, Massachusetts 01867, U.S.A.

KINGSLAND, Lawrence Chappell, LL.B. Born 1884. Educ.: Washington Univ. U.S. Commissioner of Patents, 1947-50; Patent Compensation Board, U.S.A.E.C., 1957-. Papers: Articles on Patent Law in various legal journals. Societies: American State and Local Bar Assocs.
Address: 121 South Meramec Avenue, St.Louis, Missouri 63105, U.S.A.

KINLOUGH, R. L. Dr. Radiochromium Labelling of Red Cells and Platelets, Dept. of Medicine, Adelaide Univ.
Address: Adelaide University, Adelaide, South Australia, Australia.

KINNUNEN, Erkki Johannes, Tech. lic. Born 1908. Educ.: Finland Inst. Technol. Consulting Eng., Suomen Puunvieti Oy, and Managing Director, Valkon Laiva Oy, 1942-51; Acting Prof. Wood Technol., Finland Inst. Technol. Director, Rauma-Repola Oy, 1951-55; Director-Gen., Ministry of Commerce and Industry, 1955-; Member, Finnish A.E.C., 1956-; Member, Nordic Contact Com. for Atomic Energy, 1957-.
Nuclear interest: Management.
Address: 3A Katajanokankatu, Helsinki, Finland.

KINOSHITA, Jin, M.D. Member, Isotope Com., Massachusetts Gen. Hospital.
Address: Massachusetts General Hospital, Boston 14, Massachusetts, U.S.A.

KINOSHITA, Masao, Dr. Manager, Tech. Res. Inst., Hitachi Shipbuilding and Eng. Co. Ltd.
Address: Hitachi Shipbuilding and Engineering Co. Ltd., 60 Kitanocho, Sakurajima, Konohanaku, Osaka, Japan.

KINSEY, B. B., Prof. Phys. Dept., Texas Univ.
Nuclear interests: Nuclear reactions, gamma-rays produced by neutron capture.
Address: Texas University, Physics Department, Austin 12, Texas, U.S.A.

KINSEY, R. P. Born 1917. Educ.: Leeds Univ. Deputy Director, Reactor Technol., U.K.A.E.A., Risley; Joint Managing Director, Sté. Anglo-Belge Vulcain. Society: I.Mar.E.
Nuclear interests: Reactor design and development; Application of nuclear power to ship propulsion.
Address: Beech House, Hale Road, Hale Barns, Cheshire, England.

KINSMAN, George, B.S. (Mech. Eng.). Born 1902. Educ.: Armour Inst. Technol. Vice-Pres., Florida Power and Light Co. Southern Interstate Nucl. Board; Director, Florida Eng. Soc. and Nat. Soc. Prof. Eng.; Director, Atomic Ind. Forum. Societies: A.S.M.E.; A.N.S.; A.S.E.E.
Nuclear interest: Power reactors.
Address: Florida Power and Light Company, 4200 W Flagler Street, (P.O. Box 3100), Miami 33101, Florida, U.S.A.

KINTER, Lester Lee, B.S. (Mech. Eng.). M.S. (Mech. Eng.). Born 1925. Educ.: Illinois Univ. and Illinois Inst. of Chicago. At Argonne Nat. Lab., 1950-55; Atomic Power Development Assocs., Inc., Detroit, 1955-57; Allis-Chalmers Manufacturing Co., Atomic Energy Div., 1957-67; Reactor Licensing Div., U.S.A.E.C., 1967-. Book: Co-author, Chapters 1-7, Steady-State Heat Removal in Reactor Handbook, vol. 2, Engineering (A.E.C., 1955). Society: A.N.S.
Nuclear interests: Heat transfer; Fluid flow; Reactor instrumentation; Fuel element design; Power reactor performance. Dynamic analysis; Hazard analysis.
Address: 930 Randolph Road, Silver Spring, Maryland, U.S.A.

KINYON, Brice Whitman, B.S. (Mech. Eng.). Born 1911. Educ.: Purdue Univ. and O.R.S.O. R.T. Sen. Analyst, Nucl. Components Div., Combustion Eng., Inc. Societies: A.N.S.; A.S.M.E.; N.S.P.E.
Nuclear interest: Heat transfer and thermal stress in reactor vessels and related components.
Address: Combustion Engineering, Incorporated, 911 West Main Street, Chattanooga, Tennessee 37401, U.S.A.

KIOES, Camille, M.D. (Strasbourg). Born 1922. Educ.: Munich and Strasbourg Univs. Radiologist, Clinique St. François (Luxembourg); Director, Lab. des Radioisotopes, Maternité G.D. Charlotte, Luxembourg. Societies: Sté. Française d'Electroradiologie Médicale; Sté. des Sciences Medicales, G. D. de Luxembourg.

Nuclear interest: Medical use of radioisotopes.
Address: Bascharage, 105 rue de Luxembourg, Luxembourg.

KIÖNIG, Tore. Economic Manager, A/S Thunes Mekaniske Vaerksted.
Address: A/S Thunes Mekaniske Vaerksted, 130 Drammensvn., Postbox 225, Oslo, Norway.

KIPPENHAHN, Rudolf Johann, Dr.phil.nat., Dr.habil. Born 1926. Educ.: Halle/Saale and Erlangen Univs. Head, Div. of Astrophys., Max-Planck-Inst. für Physik und Astrophys.; Privatdozent, Munich Univ. Societies: Formerly Sci. Member, Max-Planck-Inst. für Physik und Astrophysik, Munich; Astronomische Gesellschaft.
Nuclear interests: Thermonuclear reactions in stars.
Address: 35 Ruemannstr., Munich 23, Germany.

KIRBY, Harold Wallace, M.S. Born 1919. Educ.: Iowa State Univ. Sen. Res. Specialist, Monsanto Res. Corp., Mound Lab., Miamisburg, Ohio. Society: A.C.S.
Nuclear interests: Radiochemical separations and analysis; Counting techniques; Bioassay, especially of the naturally occurring alpha-emitters.
Address: Mound Laboratory, Monsanto Research Corporation, Miamisburg, Ohio, U.S.A.

KIRBY, Sir James N. Member, Council, Nucl. Res. Foundation; Managing Director, J. N. Kirby Pty. Ltd.; Deleg. to 2nd I.C.P.U.A.E., Geneva, Sept. 1958.
Address: Nuclear Research Foundation, University of Sydney, Sydney, New South Wales, Australia.

KIRCH, P., Dipl. Phys. Asst., Inst. für Kernverfahrenstechnik, Karlsruhe T.H.
Address: Technische Hochschule Karlsruhe, Reaktorstation Leopoldshafen/Karlsruhe, Germany.

KIRCHENMAYER, Anton, Dr.Ing. Born 1925. Educ.: Belgrade Univ. Asst. and Lecturer, Belgrade Univ., 1952-57; Asst. and Lecturer, Stuttgart Tech. Univ., 1957-62; Head, Reactor R. and D. A.E.G., Frankfurt/Main, 1962-. Societies: Deutsches Atomforum; A.N.S.
Nuclear interests: Nuclear reactor theory and engineering.
Address: A.E.G. - Kernenergieversuchsanlage, 8752 Grosswelzheim, Germany.

KIRCHHEIMER, Franz, Dr. phil., o. Prof. Born 1911. Educ.: Giessen Univ. Präsident des Geologischen Landesamtes in Baden-Württemberg (1952); Honorarprofessor an den Universitäten Freiburg, Heidelberg und Stuttgart; Berater für Bergbau und Landesgeologie im Wirtschaftministerium Baden-Württemberg. Book: Das Uran und seine Geschichte (Stuttgart, 1963). Societies: Deutschen Akademer der Naturforscher; Hon. Fellow, Botanical Soc., Edinburgh; o. Mitglied der Heidelberger Akademie der Wissenschaften; korresp. Mitglied der Österreichischen Akademie der Wissenschaften.
Nuclear interests: Geologie und Mineralogie des Urans.
Address: 5 Albert-Strasse, Freiburg i.Br., Germany.

KIRCHMEYER, F. J. i/c of Radioactive Pharmaceutical Div., Abbott Labs.
Address: Abbott Laboratories, North Chicago, Illinois, U.S.A.

KIRCHNER, Fritz, Dr.phil. Born 1896. Educ.: Jena and Munich Univs. o. Prof. f. Experimentalphysik, Cologne Univ.
Nuclear interest: Nuclear physics.
Address: 5 Bardenheuerstr. Köln-Lindenthal, Germany.

KIRDIN, G. S. Dr. Member for U.S.S.R., Panel on Atomic Energy, World Meteorological Organisation.
Address: Panel on Atomic Energy, World Meteorological Organisation, Geneva, Switzerland.

KIRENSKII, L. V., Dr. Director, Phys. Inst., Acad. Sci. of the U.S.S.R.
Address: Physics Institute, 42 Karl Marx Ulitsa, Krasnoyarsk, U.S.S.R.

KIRICHENKO, V. N. Papers: Co-author, Trapping of Shortlived Radon Decay Daughter Products with Fibre Filters (Atomnaya Energiya, vol. 15, No. 3, 1963); co-author, Miniature Instrument for Measurement of Mean Total Radon Concentration (letter to the Editor, ibid, vol. 21, No. 3, 1966).
Address: Academy of Sciences of the U.S.S.R., 14 Leninsky Prospekt, Moscow V-71, U.S.S.R.

KIRILIN, Vladimir Alexeyevich, D.Eng. Sci. Lenin Prize, 1959. Born 1913. Educ.: Moscow Power Eng. Inst. Prof. (1943-), Res. and Teaching (1955-), Moscow Power Eng. Inst.; U.S.S.R. Deputy Minister of Higher Education, 1954-; Vice-Chairman, U.S.S.R. State Com. on New Techniques, 1955-; Director, High Temperature Lab., Dept. Tech. Sci., Chairman, U.S. S.R. Co-ordinated Com. on Properties of Steam; Chairman, U.S.S.R. State Com. for Sci. and Technol., 1965-; Vice-Chairman, U.S.S.R. Council of Ministers, 1965-. Deleg. to 2nd I.C.P.U.A.E., Geneva, Sept. 1958. Society: Vice-Pres., U.S.S.R. Acad, of Sci., 1963-65.
Address: Academy of Sciences of the U.S.S.R., 14 Leninsky Prospekt, Moscow V-71, U.S.S.R.

KIRILLOV, E. Y. Paper: Co-author, Value of Plutonium as Nucl. Fuel (Atomnaya Energiya, vol. 22, No. 6, 1967).
Address: Academy of Sciences of the U.S.S.R., 14 Leninsky Prospekt, Moscow V-71, U.S.S.R.

KIRILLOV, P. L.
Nuclear interest: Liquid metal coolants.
Address: Nuclear Energy Institute, Academy of Sciences of the U.S.S.R., 14 Leninsky Prospekt, Moscow V-71, U.S.S.R.

KIRILLOV-UGRYUMOV, V. G. Papers: Co-author, Angular and Energy Dispersions of π-mesons in the Leakage Field of a Six-metre Synchrocyclotron (letter to the Editor, Atomanay Energiya, vol. 11, No. 3, 1961); co-author, The Twenty Fifth Anniversary of the MIFI (ibid, vol. 23, No. 5, 1967).
Address: Academy of Sciences of the U.S.S.R., 14 Leninsky Prospekt, Moscow V-71, U.S.S.R.

KIRILYUK, A. L. Papers: Co-author, Resonance Levels of Erbium Isotopes (letter to the Editor, Atomnaya Energiya, vol. 15, No. 3, 1963); co-author, Total Neutron Cross-Sections of Re^{185} and Re^{187} (ibid, vol. 19, No. 3, 1965); co-author, Configuration of VVR-M Reactor Core and Neutron Spectrum from Horizontal Channel (letter to the Editor, ibid, No. 5); co-author, Observation of He^3 Behaviour in ZrT_2 and Zirconium (letter to the Editor, ibid, vol. 22, No. 3, 1967).
Address: Academy of Sciences of the U.S.S.R., 14 Leninsky Prospekt, Moscow V-71, U.S.S.R.

KIRK, Richard L. Deputy Director, Div. of Licensing and Regulation, U.S.A.E.C.; Member for U.S.A., Group of Experts on Production of Energy from Radioisotopes, O.E.C.D., E.N.E.A.
Address: Division of Licensing and Regulation, U.S. Atomic Energy Commission, Washington 25, D.C., U.S.A.

KIRKALDY, John Samuel, B.A.Sc., M.A.Sc., Ph.D. Born 1926. Educ.: British Columbia and McGill Univs. Chairman, Dept. of Metal. (1962), Prof. Metal. (1963), McMaster Univ. Societies: Canadian Assoc. of Phys.; C.I.M.; A.S.M.; A.I.M.E.
Nuclear interests: Application of nuclear techniques and materials in metallurgy.
Address: Engineering Building, McMaster University, Hamilton, Ontario, Canada.

KIRKBRIDE, Louis Dale, Ph.D.,M.S., B.S. Born 1932. Educ.: Carnegie Inst. of Technol. K.A.P.L., G.E.C., Schenectady, 1957-61; Los Alamos Sci. Lab., 1961-66; R. and D. Centre, G.E.C., 1966-. Exec. Com., Materials Div., A.N.S. Societies: A.I.M.E.; A.A.A.S.
Nuclear interest: Management of materials activities related to fuel, control rods and structurals.
Address: 2345 Knolls View Drive, Schenectady, New York 12309, U.S.A.

KIRKO, I. M. Director, Phys. Inst., Dept. of Tech. Sci., Acad. Sci. of the Latvian S.S.R.
Address: Physics Institute, Riga, Latvian S.S.R., U.S.S.R.

KIRKPATRICK, James Stanley, B.S., M.B.A. Born 1912. Educ.: Ohio State and Detroit Univs. Factory Supt. (1946-51), Vice-Pres., R. and D. (1952-60), Vice-Pres., Market Development (1960-62), Brooks and Perkins, Inc.; Exec. Vice-Pres. and Tech. Director, Magnesium Assoc., 1962-66; Exec. Director, Aluminium Smelters Res. Inst., 1966-. Societies: A.S.M.; Soc. American Military Eng.; American Soc. of Tool and Manufacturing Eng.; A.I.A.A.
Nuclear interests: Neutron shielding; Boral; Reactor components; Low capture light alloys; Aluminium components.
Address: 244 Columbia Avenue, Park Ridge, Illinois 60068, U.S.A.

KIRN, Frederick Shelly, M.A. Born 1922. Educ.: Illinois Univ. and North Central Coll. Nat. Bureau of Standards, 1949-54; Argonne Nat. Lab., 1954-. Society: A.N.S.
Nuclear interests: Reactor physics; Absolute fission rates and reactor power measurements; Reactor shielding; Fast reactor safety.
Address: Argonne National Laboratory, P.O. Box 2528, Idaho Falls, Idaho, U.S.A.

KIRPICHNIKOV, I. V. Papers: Co-author, A Property of the Resonance Levels of Fissile Nuclides (Atomnaya Energiya, vol. 7, No. 5, 1959); co-author, Variation of η and U^{235} and Pu^{239} Partial Cross-Sections for Neutrons of Resonance Energies (ibid, vol 16, No. 2, 1964); co-author, Interference Effects in Fission Cross-Sections (ibid, No. 3); Neutron Level Resonance Parameters of Fissionable Nucleus (ibid, vol. 23, No. 1, 1967).
Address: Academy of Sciences of the U.S.S.R., 14 Leninsky Prospekt, Moscow V-71, U.S.S.R.

KIRSANOV, V. V. Paper: Co-author, Electronic Computer Investigation Methods of Crystals Rad. Damage Dynamics (letter to the Editor, Atomnaya Energiya, vol. 23, No. 4, 1967).
Address: Academy of Sciences of the U.S.S.R., 14 Leninsky Prospekt, Moscow V-71, U.S.S.R.

KIRSCHBAUM, Albert J., Ph.D. (Phys.). Born 1923. Critical Assembly Group Leader (1952-56), Neutronics Div. Leader (1956-64), Lawrence Rad. Lab.; Vice Chairman, Dept. of Appl. Sci. (1963-66), Chairman, Dept. of Appl. Sci. (1966-), Assoc. Prof. and Assoc. Dean, Coll. of Eng. (1966-), California Univ., Davis. Member, Atomic Safety Licensing Panel, U.S.A.E.C., 1963-66. Societies: A.P.S.; A.A.A.S.; A.N.S.
Nuclear interests: Nuclear physics; Neutron physics.
Address: 1244 Madison Avenue, Livermore, California, U.S.A.

KIRSCHBAUM, Wolfgang von. See VON KIRSCHBAUM, Wolfgang.

KIRSCHNER, István. Docent. Born. 1934. Educ.: Roland Eötvös Univ., Budapest. Asst. (1957-62), Adjunct (1962-67), Leader, Low Temperature Lab.(1963-), Docent (1967-), Dept. of Atomic Phys., Budapest Univ.; Sci.

employee, Joint Inst. for Nucl Phys., 1958-60. Com. Member, Hungarian Phys. Soc.; Member, Phys. Com. Cultural Ministry. Nuclear interests: Low temperature methods in nuclear physics; Nuclear polarisation; Low temperature physics.
Address: Atomfizikai Tanszék, 5-7 Puskin utca, Budapest 8, Hungary.

KIRSHENBAUM, Maurice S., B.S. (Chem. Eng.), M.S. (Chem. Eng.). Born 1924. Educ.: Columbia Univ. Chem. Eng., Picatinny Arsenal, Dover, New Jersey, 1954-; Participant, Internat. Inst. of Nucl. Sci. and Eng., Argonne Nat. Lab., 1960-61; Res. Assoc., Res. Inst., Temple Univ., Philadelphia, 1949-54. Society: A.N.S. Nuclear interests: Plans, co-ordinates and conducts work pertaining to the establishment of an Ordnance Corps research reactor. Activities include evaluating the development of reactor design, the detailed site survey, hazards analyses and designing of supporting facilities.
Address: Clover Lane, R.F.D. #2, Dover, New Jersey, U.S.A.

KIRYLUK, W., I.E.E., Assoc. Member I.E.R.E., Diploma in Telecommunications Eng. (Northern Polytech.), Full Technol, Certificate of City and Guilds of London Inst., H.N.C. (Elec. Eng.). Born 1926. Educ.: Northern Polytech., London. Development Eng., Dawe Instruments Ltd., 1949-51; Sen. Development Eng., All Power Transformers, Ltd., 1951-52; Sect. Leader, Wayne Kerr Labs., Ltd., 1952-54; Sen. Development Eng., later Asst. Chief Eng., Electronics Div. (1954-57), Chief Eng., Nucleonics Lab. (1957), Manager, Ind. and Nucleonics Div. (1960), Burndept, Ltd.; Managing Director, Ind. Instruments Ltd., 1963-. Societies: I.E.E.; Inst. British Radio Eng.; Fellow, Phys, Soc.; Radio Inst. (U.S.A.).
Nuclear interest: Design and marketing of health physics instruments.
Address: 30A Whitecroft Way, Beckenham, Kent, England.

KISELEV, G. V. Paper: Co-author, The Nucl. Power Plant with BN-350 Reactor (Atomnaya Energiya, vol. 23, No. 5, 1967).
Address: Academy of Sciences of the U.S.S.R., 14 Leninsky Prospekt, Moscow V-71, U.S.S.R.

KISELEV, I. E. Paper: Co-author, Source of Polarised Ions with Current 1. 2 mka (letter to the Editor, Atomnaya Energiya, vol. 21, No. 2, 1966).
Address: Academy of Sciences of the U.S.S.R., 14 Leninsky Prospekt, Moscow V-71, U.S.S.R.

KISILEV, A. V. Papers: Co-author, Status Report of Positron-Electron Storage Ring VEPP-2 (Atomnaya Energiya, vol. 19, No. 6, 1965); co-author, Putting into Operation of B-3M Synchrotron as Injector for Electron-Positron Storage Ring (ibid, vol. 20, No. 3, 1966).
Address: Novosibirsk Science Centre, 20 Sovietskaya Ulitsa, Novosibirsk, Siberia, U.S.S.R.

KISILEVA, L. V. Paper: Co-author, Determination of Sum of Corrosion Products in Nucl. Reactors Waters (letter to the Editor, Atomnaya Energiya, vol. 20, No. 4, 1966).
Address: Academy of Sciences of the U.S.S.R., 14 Leninsky Prospekt, Moscow V-71, U.S.S.R.

KISK, I. Paper: Co-author, Study of Radiochem. Stability of High-Boiling Hydrocarbons in Reactor (Atomnaya Energiya, vol. 20, No. 1, 1966).
Address: Academy of Sciences of the U.S.S.R., 14 Leninsky Prospekt, Moscow V-71, U.S.S.R.

KISO, Yoshiyuki, Dr. Asst. Prof., Div., of Hot Lab., Res. Reactor Inst., Kyoto Univ.
Address: Kyoto University, Research Reactor Institute, Noda, Kumatori-cho, Sen-nan-gun, Asaka Prefecture, Japan.

KISS, Dezsö. Dr. of Phys. Sci. Born 1929. Educ.: Debrecen Univ., Hungary. Leader, Nucl. Phys. Lab., Central Res. Inst. for Phys., Budapest; Lecturer, Budapest Univ. Member, Com. for Educational Problems, Hungarian A.E.C.; Member, Com. for Phys., Hungarian Acad. of Sci. Society: Hungarian Phys. Soc., Budapest. Nuclear interests: Neutron-gamma reactions; Nuclear spectroscopy.
Address: Central Research Institute for Physics, Budapest 114, P.O.B. 49, Hungary.

KISS, Ibolya, Dr. Sen. Officer, Res. Inst. for Irrigation and Rice Cultivation.
Address: Research Institute for Irrigation and Rice Cultivation, 2 Szabadsag ut. Szarvas, Hungary.

KISS, István, Ph.D., Dipl. Chem., Cand. of Chem. Sci. Born 1923. Educ.: Eötvös Loránd (Budapest), and Szeged (Hungary) Univs. Head, Dept. for Nucl. Chem., Central Res. Inst. for Phys., 1955-. Member, Sci. Board, Hungarian A.E.C. Societies: Soc. of Hungarian Chemists; Hungarian Eötvös Loránd Phys. Soc. Nuclear interest: Nuclear and radiation chemistry.
Address: 6 Konkoly Thege ut, Budapest 12, Hungary.

KISSER, Josef Georg, Dr. phil. Born 1899. Educ.: Vienna Univ. Direktor des Botanischen Insts. der Hochschule für Bodenkultur, Vienna; Wissenschaftlicher Leiter des Österreichischen Holzforschungsinstitutes, Vienna. Societies: Deutsche Botanische Gesellschaft; Österreichische Gesellschaft für Ernährungsforschung; American Botanical Soc.; A.A.A.S. Nuclear interest: Biology.
Address: 33 Gregor Mendelstrasse, Vienna 18, Austria.

KISSLINGER, Leonard S., B.D., M.D., Ph.D. Born 1920. Educ.: St. Louis and Indiana Univs. Res. Assoc., Kentucky Univ., summer 1956; Res. Participant, O.R.N.L., summer 1957;

Post-doctoral Fellowship (Res. Corp.), Univ. Inst. for Theoretical Phys., Copenhagen, 1958-59; Asst. Prof., Phys. (1957-60), Assoc. Prof., (1960-63), Western Reserve Univ.; Fellowship Weizmann Inst. of Sci., Rehovoth, 1962-63; Prof., Case Western Reserve Univ., 1963-; Res. Assoc. M.I.T., 1966-67; Visiting Phys., Brookhaven Nat. Lab., summer 1954, 1964, 1966; Lawrence Rad. Lab., California Univ., summer 1961, 1965; Colorado Univ., summer 1960. Society: A.P.S.
Nuclear interests: Nuclear structure and intermediate energy theory; Particle theory.
Address: Physics Department, Case Western Reserve University, Cleveland 6, Ohio, U.S.A.

KISSOPOULOS, A. Special Applications Head, Greek A.E.C.
Address: Greek Atomic Energy Commission, 5 Merlin Street, Athens (134), Greece.

KISTEMAKER, Jacob, Dr. Phys. Born 1917. Educ.: Leyden Univ. Director, F.O.M. Atomic and Molecular Phys. Inst., Amsterdam, 1947-. Book: Co-author, Symposium on Isotope Separation (Amsterdam, North-Holland Publishing Co., 1957). Societies: Dutch Phys. Soc.; Svenska Fysikersamfundet; A.P.S.
Nuclear interests: Mass spectrometry; Isotope separation; Fusion; Ion-atom collisions.
Address: F.O.M. Atomic and Molecular Physics Institute, 407 Kruislaan, Amsterdam/Wgm., Netherlands.

KISTIAKOWSKY, Vera, B.A., Ph.D. Born 1928. Educ.: California Univ. (Berkeley). Res. Assoc. (1954-57), Instructor (1957-59), Columbia Univ.; Asst. Prof., Brandeis Univ., 1959-63; M.I.T., 1963-. Society: A.P.S.
Nuclear interest: Elementary particles.
Address: Laboratory for Nuclear Sciences, Massachusetts Institute of Technology, Cambridge, Massachusetts, U.S.A.

KISTLER, Ronald Wayne, B.A., Ph.D. Born 1931. Educ.: Johns Hopkins and California (Berkeley) Univs. Geologist, U.S. Geological Survey. Society: Geological Soc. of America.
Nuclear interests: Geochronology, K-Ar, Rb-Sr.
Address: United States Geological Survey, Branch of Isotope Geology, 345 Middlefield Road, Menlo Park, California 94025, U.S.A.

KISTNER, Georg, Dr.rer.nat. Born 1927. Educ.: Kiel Univ. Physiological Chem. and Physico-chem. Inst., Kiel, 1956-57; Federal Res. Inst. for Food Preservation, Karlsruhe, 1957-65; Rad. Hygiene Div., Federal Health Office, Berlin, 1965-.
Nuclear interests: Radiobiology; Fallout problems and fission-products in food; Transport mechanism in biological systems; Low-level-counting and tracer techniques; Radiobiology.
Address: 1 Correnspl., Berlin 33, Germany.

KITAEV, G. I. Papers: Comparison of Cascade Generator Circuits (letter to the Editor, Atomnaya Energiya, vol. 14, No. 2, 1963); Comparison of m - Phase Schemes of Cascades Generators (ibid, vol. 22, No. 3, 1967).
Address: Academy of Sciences of the U.S.S.R., 14 Leninsky Prospekt, Moscow V-71, U.S.S.R.

KITAEVSKII, L. Kh. Papers: Co-author, Longitudinal Magnetic Field of the Sirius Stellarator Reecetrack (Atomnaya Energiya, vol. 23, No. 2, 1967); co-author, Shielding of Thermonucl. Device Magnetic Systems from Accidental Overloadings (ibid, vol. 23, No. 2, 1967); co-author, Divertor of the Sirius Stellarator (ibid, vol. 23, No. 2, 1967).
Address: Academy of Sciences of the U.S.S.R., 14 Leninsky Prospekt, Moscow V-71, U.S.S.R.

KITAGAKI, Toshio, D.Sc. Born 1922. Educ.: Osaka Univ. Formerly Res. Assoc., Dept. of Phys., Osaka Univ.; Prof., Dept. of Phys., Tohoku Univ., 1952-; Res. Assoc. with rank of Visiting Prof., Princeton Univ., 1956-59. Societies: Phys. Soc. of Japan; A.P.S.
Nuclear interests: High energy nuclear physics; Accelerator physics.
Address: Physics Department, Tohoku University, Sendai, Japan.

KITAGAWA, Jiro. Pres., Nippon Light Metal Res. Lab. Ltd.,
Address: Nippon Light Metal Co. Ltd., 2, 7-chome, Ginza-Nishi, Chuo-ku, Tokyo, Japan.

KITAGAWA, M. Head, Nucl. Dept., Yamatake-Honeywell Keiki Co.
Address: Yamatake-Honeywell Keiki Co., Yaesu Building 2-6 Marunouchi, Chiyoda-ku, Tokyo, Japan.

KITAKA, Kunio. Sub Manager, Tokyo Tsusho Kaisha Ltd.
Address: Tokyo Tsusho Kaisha Ltd., 1-6, Marunouchi, Chiyoda-ku, Tokyo, Japan.

KITAMURA, Ryuichi. Born 1928. Educ.: Kyushu Univ. At Tokai Univ., 1959-. Society: Atomic Eng. Soc. Japan.
Nuclear interest: Reactor physics.
Address: 1616 Hoya-machi Kamihoya, Kitatama-gun, Tokyo, Japan.

KITANI, Kazuo, Dr. Ph. Born 1914. Educ.: Tokyo Univ. Director, Gen Manager, of Central Res. Lab., Mitsui Chem. Industry Co. Ltd.
Address: Mitsui Chemical Industry Co. Ltd., No. 1, 2-chome, Nihonbashi- Muromachi, Chuo-ku, Tokyo, Japan.

KITANISHI, Yasuhisa. Sci., Teijin Ltd.
Address: Teijin Ltd., Iino Building, 22, 2-chome, Uchisaiwai-cho, Chiyoda-ku, Tokyo, Japan.

KITAWAKI, Yasuyoshi. Sen. Managing Director, Sumitomo Joint Elec. Power Co. Ltd.
Address: Sumitomo Joint Electric Power Co. Ltd., 1840 Kaneko-Otsu, Niihama City, Ehime Prefecture, Japan.

KITCHEN, Sumner Wendell, A.B., Ph.D. Born 1921. Educ.: New York Univ. and Oberlin Coll. Manager, Nucl. Analysis, K.A.P.L.; Group Leader, Ernest O. Lawrence Rad. Lab., California Univ. 1950-54; Supervising Phys., K.A.P. L., 1954-65. Chairman, Northeastern New York Sect., A.N.S. Societies: A.P.S.; A.N.S.
Nuclear interests: Reactor physics; Reactor safeguards.
Address: Knolls Atomic Power Laboratory, P.O. Box 1072, Schenectady, New York 12301, U.S.A.

KITTEL, John Howard, B.S. (Metal. Eng.). Born 1919. Educ.: Washington State Univ. With Nat. Aeronautics and Space Agency, 1943-51; Argonne Nat. Lab., 1951-. Societies: A.S.M.; A.N.S.; RESA; A.I.M.M.E.
Nuclear interests: Metallurgy of reactor materials; Fuel element design; Irradiation effects on reactor materials.
Address: Argonne National Laboratory, 9700 South Cass Avenue, Argonne, Illinois, U.S.A.

KITZES, Arnold Stanley, Ph.D.,Professional Eng. Born 1917. Educ.: Minnesota Univ. and C.C.N.Y. Group Leader, O.R.N.L., 1948-57; Sect. Manager, Westinghouse Elec. Corp., 1957-. Book: Chapter on Thoria-Urania Slurries in Nucl. Energy Series (McGraw Hill).
Societies: A.N.S.; A.C.S.; Fellow, A.A.A.S.; A.I.Ch.E.
Nuclear interests: Reactor technology, especially homogeneous and pressurised water type reactors.
Address: Westinghouse Atomic Power Divisions, P.O. Box 355, Pittsburgh, Pennsylvania 15230, U.S.A.

KIUKKOLA, Kalevi Viljam, Dipl. Eng., M.Sc. (Metal. Eng.), Sc.D. Born 1925. Educ.: Helsinki Inst. Technol., Carnegie Inst. Technol. and M.I.T. Res. Eng., Helsinki Inst. Technol., Lab. of Metal. (Otaniemi, Finland). Societies: Vuorimiesyhdistys r.y. (Finland); Suomen Teknillinen Seura (Finland).
Nuclear interest: Metallurgy.
Address: Oy Fiskars Ab, Aminnefors, Finland.

KIVALO, Pekka, Dr. techn. Born 1919. Educ.: Abo Akademi, Finland and Illinois Univ. Prof. Phys. Chem. (1958-), Head, Chem. Dept. (1962-), Otaniemi Tech. Univ., Finland. Societies: Finnish Chem. Soc.; A.C.S.
Nuclear interests: Radiochemical applications; Application of radioisotopes in diffusion problems.
Address: Technical University,(Teknillinen Korkeakoulo), Otaniemi, Finland.

KIVEL, Joseph, A.B., M.S., Ph.D. Born 1934. Educ.: New York, Michigan and Iowa State Univs. Res. Asst., Eng. Res. Inst., Ann Arbor, Michigan,-1955; Res. Fellow, Ames Lab., U.S.A.E.C. Ames, Iowa 1956-60; Sen. Res. Sci. Honeywell Res. Centre, Hopkins, Minnesota, 1960-63; Head, Nucl. Lab., Atlantic Res. Corp. Alexandria,Virginia, 1963-. Societies: A.N.S.; A.C.S.
Nuclear interests: Isotope source fabrication and process radiation applications, radiation chemistry, dosimetry, radiochemistry.
Address: Atlantic Research Corporation, S Highway at Edsall Road, Alexandria, Virginia 22314, U.S.A.

KIYONARI, S. Director, Tokyo Atomic Ind. Consortium.
Address: Tokyo Atomic Industrial Consortium, Hitachi Building, 6-4-chome, Surugadai Kanda, Chiyoda-ku, Tokyo, Japan.

KJELBERG, Arve, Cand. real. Born 1928. Educ.: Oslo Univ. Res. Asst., Roy. Norwegian Council for Sci. and Ind. Res., 1955-57; Sci. Sec., U.N. Sci. Com. on Effects of Atomic Rad., 1958; Res. Fellow, McGill Univ., 1959-60; Res. Fellow, Roy. Norwegian Council for Sci. and Ind. Res., 1960-61; Univ. Lecturer, 1962-; Group Leader, Nucl. Chem. Group, C.E.R.N., 1965-. Society: Norwegian Chem. Soc.
Nuclear interests: Nuclear chemistry; Nuclear spectroscopy.
Address: N.P. Division, Conseil Européen pour la Recherche Nucléaire, Geneva, Switzerland.

KJELLGREN, Bengt R. F., Degree in Chem. Eng. Born 1894. Educ.: Roy. Tech. Inst., Stockholm, and M.I.T. Vice-Pres. and Director, Brush Labs. Co., Cleveland, Ohio, 1931-51; Pres. (1948-60), Chairman (1960-64), Brush Beryllium Co.; Consultant on crystal growth, Brush Electronics Co., 1930-54. Books: Contributor on Beryllium to Rare Metals Handbook and to Eng. Materials Handbook. Societies: A.A.A.S.; A.C.S.; Electro-chem. Soc.; Inst. Metals.
Nuclear interests: Applications for beryllium metal and beryllium oxide in nuclear reactors and the metallurgy involved in producing these materials of the purity specifications and physical properties required for nuclear use.
Address: 4355 Giles Road, Moreland Hills, Chagrin Falls, Ohio, U.S.A.

KJENSLI, O. Sen. Chem., Chem. Div., Inst. for Atomenergi.
Address: Institutt for Atomenergi, P.O. Box 40, Kjeller, Norway.

KLADNIK, Rudi, D.Sc. (Phys.). Born 1933. Educ.: Ljubljana Univ. Docent, Phys., Ljubljana Univ., 1964-; Visiting Prof., Nucl. Eng. Dept., Kansas State Univ., 1967-.
Nuclear interests: Reactor physics; Transport theory; Slow neutron scattering.
Address: Nuclear Engineering Department, Seaton Hall, Kansas State University, Manhattan, Kansas 66502, U.S.A.

KLADNITSKAYA, Eugenia. Soviet scientist, member of group obtaining for the first time xi minus hyperons.
Address: Joint Institute for Nuclear Research, Dubna, Nr. Moscow, U.S.S.R.

KLADNITSKII, V. S. Paper: Co-author, Focusing and Analysis of Secondary Beam with Magnetic Field of Proton-Synchrotron (Abstract, Atomnaya Energiya, vol. 20, No. 4, 1966).
Address: Academy of Sciences of the U.S.S.R., 14 Lensinky Prospekt, Moscow V-71, U.S.S.R.

KLAHR, Carl N., B.S., M.S. D.Sc. Born 1927. Educ.: Carnegie Inst. Technol. Phys., Nucl. Development Corp. of America, 1952-57; Formerly Sen. Phys., Tech. Res. Group, Inc., N.Y., 1957-. Societies: A.P.S.; A.N.S.; Operations Res. Soc.; Inst. of Management Sci. Nuclear interests: Reactor physics; Reactor design; Neutron distributions.
Address: 678 Cedar Lawn Avenue, Lawrence, Long Island, New York 11559, U.S.A.

KLAIBER, Edward, Dr. Member, Radioisotope Com., Worcester Foundation for Exptl. Biol., Inc.
Address: Worcester Foundation for Experimental Biology, Inc., 222 Maple Avenue, Shrewsbury, Massachusetts, U.S.A.

KLAINER, Stanley, R. and D. Manager, Tracerlab-Keleket.
Address: Tracerlab-Keleket, 1601 Trapelo Road, Waltham 54, Massachusetts, U.S.A.

KLAMUT, Carl J., B.S. (Mech. Eng.). Born 1926. Educ.: New York Univ. Metal., Brookhaven Nat. Lab., 1951-.
Nuclear interest: Metallurgy; Liquid metal corrosion.
Address: Brookhaven National Laboratory, Associated Universities, Incorporated, Upton, Long Island, New York, U.S.A.

KLANN, Paul Gerhardt, B.S., M.S. (Phys.). Born 1927. Educ.: Western Reserve Univ. and Case Inst. Technol. Phys., Combustion Eng. Co., Windsor, Conn., 1960-. Now Staff Phys., Gen. Nucl. Eng. Corp. Society: A.N.S. Nuclear interests: Reactor design, critical facilities.
Address: 1128 Burke Avenue, Dunedin, Florida 33528, U.S.A.

KLASEN, K., Dr. Mitglied, Aufsichtsrat, Gesellschaft für Kernenergieverwertung in Schiffbau und Schiffahrt m.b.H.; Mitglied Vorstand, Study Assoc. for the use of Nucl. Energy in Navigation and Industry; Member, Board of Managing Directors, Deutsche Bank A.G.
Address: Deutsche Bank A.G., Alter Wall, 2 Hamburg 11, Germany.

KLASSEN, C. W. Chief Sanitary Eng., Dept. of Public Health, Springfield, Ill.; Member, Advisory Com. of State Officials, U.S.A.E.C.
Address: Department of Public Health, Springfield, Illinois, U.S.A.

KLASTERSKY, J., Dr. Service des Medecine, Inst. Jules Bordet (Centre des Tumeurs).
Address: Service des Medecine, Institut Jules Bordet (Centre des Tumeurs), 1 rue Heger-Bordet, Brussels, Belgium.

KLAUDER, John Rider, B.S., Ph.D. (Theoretical Phys.). Born 1932. Educ.: California (Berkeley) and Princeton Univs. Theoretical Phys. Group, Bell Telephone Labs., Murray Hill, New Jersey, 1959-67; Visiting Assoc. Prof., Berne Univ., 1961-62; Adjunct Prof., Rutgers Univ., 1965; Prof., Syracuse Univ., 1967-. Society: A.P.S.
Nuclear interests: Elementary particle interactions; Description of elementary particles through unified field theories.
Address: Physics Department, Syracuse University, Syracuse, New York, U.S.A.

KLECHKOVSKY, V. M. Academician of the V.I. Lenin All-Union Acad. of Agricultural Sci., Moscow; VASKhNIL Academician; In charge of Chair of Agricultural Chem., K.A. Timiryazev Moscow Agricultural Acad. Deleg. to 2nd I.C.P.U.A.E., Geneva, Sept. 1958. Paper: Co-author, The Sorption of Microquantities of Strontium and Caesium in Soils (2nd I.C.P.U.A.E., Geneva, Sept. 1958).
Nuclear interest: Agricultural chemistry.
Address: K. A. Timiryazev Moscow Agricultural Academy, 51 Novoe Shosse, Moscow A-8, U.S.S.R.

KLECKER, Raymond William, B.S. (E.E.). Born 1920. Educ.: Illinois Inst. Technol., South California, Wayne, and Wisconsin Univs. and O.R.S.O.R.T. Eng., Nucl. Power Dept., Allis-Chalmers Manufacturing Co., Milwaukee, 1949-53; Sect. Head, Instrumentation and Control Sect. (on loan from Allis-Chalmers), Atomic Power Development Assocs., Detroit, 1953-56. Supervisory Eng., Reactor Design, Nucl. Power Div. (1956-58), Project Eng., Pathfinder Atomic Power Plant, Nucl. Power Dept., (1958-62), Manager of Eng. Nucl. Power Dept., Greendale (1962-63), Allis-Chalmers Manufacturing Co., Milwaukee; Manager Analysis and Development Dept., Atomic Energy Division, Allis-Chalmers Manufacturing Co., Bethesda, Maryland, 1963-. Societies: American Inst. of Elec. Eng.; A.N.S.; Sectional Com. on Reactor Safety Standards, American Standards Assoc.
Address: 9503 Hollins Court, Bethesda, Maryland, U.S.A.

KLEFFENS, A. See VAN KLEFFENS, A.

KLEIBEL, Francis, Dr. of Medicine. Born 1922. Educ.: Klausenburg, Giessen/Lahn, Würzburg Univs. Reader of Clinical Oncology and Radiotherapy. Member, Anglo-German Medical Assoc. Book: Co-author, Einführung in die klinisch experimentelle Radiologie (Berlin-München, Verlag Urban-Schwarzenberg, 1964).
Nuclear interest: Diagnostic and therapy with ^{32} phosphorus.
Address: 21 B Jahnstrasse, Heidelberg, Germany.

KLEIBER, Gottfried, Dipl. Ing. Director, Klein, Schanzlin und Becker A.G.
Address: Klein, Schanzlin und Becker A.G., Postfach 205/225, Frankenthal/Pfalz, Germany.

KLEIJN, Hans Robert, Dr. (Tech. Sci.). Ir. (Mech. Eng.). Born 1932. Educ.: Delft Technol. and Michigan Univs. Head, Reactor Phys. Dept., Reactor Inst., Delft; Lecturer, Reactor Phys., Delft Univ. Societies: Sec., Royal Inst. (England).; A.N.S.
Nuclear interests: Reactor physics; Engineering management.
Address: 17 Landsteinerbocht, Delft, Netherlands.

KLEIN, Abraham, B.A. (Brooklyn Coll), M.A., Ph.D. (Harvard). Born 1927. Educ.: Brooklyn Coll. and Harvard Univ. Instructor (1950-52), Lowell Fellow (1952-55), Harvard Univ.; Assoc. Prof. (1955-58), Prof. (1958-), Pennsylvania Univ. Societies: A.P.S.; A.A.A.S.; A.A.U.P.
Nuclear interests: Nuclear physics; Elementary particle physics.
Address: Physics Department, Pennsylvania University, Philadelphia, Pennsylvania 19104, U.S.A.

KLEIN, Bernhard K., Generaldirektor Dipl. Ing. Sen. Officer, Vereinigte Metallwerke Ranshofen-Berndorf A.G.
Address: Vereinigte Metallwerke Ranshofen-Berndorf A.G., Braunau am Inn, P.O.B. 94, Austria.

KLEIN, Daniel, Ph.D. Born 1927. Educ.: Rochester Univ. Now at Westinghouse Elec. Corp., Bettis. Societies: A.P.S.; A.N.S.
Nuclear interest: Reactor physics experiments.
Address: Bettis Atomic Power Div., P.O. Box 79, West Mifflin, Pennsylvania, U.S.A.

KLEIN, Fritz Shalom, M.Sc., Ph.D. Born 1920. Educ.: Hebrew Univ. Jerusalem. Sen. Res. Chem. Weizmann Inst. Sci., Israel, 1955-; Visiting Chem., Brookhaven Nat. Lab., 1956-57 and 1962-63. Nat. Bureau of Standards, 1963-64. Societies: Chem. Soc., London; Israel Chem. Soc.
Nuclear interests: Isotopic exchanges; Isotope effects; Mass spectroscopy.
Address: Weizmann Institute of Science, Rehovoth, Israel.

KLEIN, George F. Director of Nucl. Dept. and Formerly Vice-Pres. for Eng., Catalytic Construction Co.
Address: Catalytic Construction Co., 1528 Walnut Street, Philadelphia 2, Pennsylvania, U.S.A.

KLEIN, Georges. Chef du Départment des Etudes Atomiques, Centre de Recherches de la Compagnie Generale d'Electricité.
Address: Départment des Etudes Atomiques, Centre de Recherches de la Compagnie Generale d'Electricite, Marcoussis, Route de Nozay, (Seine et Oise), France.

KLEIN, J. C. Dr. Staff Member, Radiobiol. Inst., Organisation for Health Res. T.N.O.
Address: Organisation for Health Research T.N.O.; 151 Lange Kleiweg, Rijswijk (ZH), Netherlands.

KLEIN, J. Lester, B.S. Born 1919. Educ.: M.I.T. Vice-Pres. (1959-), Formerly Manager, R. and D., now Manager, Composites, Nucl. Metals Inc., Concord, Mass. Chairman, Nucl. Metal. Com., Inst. of Metals Div., Metal. Soc., A.I.M.M.E. Books: Chapter, Orientation and Grain Size, in The Metal. of Beryllium (A.S.M. Book); Chapter, Uranium and Its Alloys, in Nucl. Reactor Fuel Elements, Metal. and Fabrication. Societies: A.I.M.E.; A.S.M.; A.I.A.A.
Nuclear interests: Physical and fabrication metallurgy; Management.
Address: Nuclear Metals, Inc., Concord, Massachusetts, U.S.A.

KLEIN, Milton, B.S. (Chem. Eng.), M.B.A. Born 1924. Educ.: Washington (St. Louis), and Harvard Univs. Chem. Eng. (1950-51), Contact Administrator (1951-55), Director Res. Contracts Div. (1955-58), Asst. Manager for Tech. Operations (1958-60), Chicago Operations Office, U.S.A.E.C.; Deputy Manager (1960-67), Manager (1967-), A.E.C.-N.A.S.A. Space Nuclear Propulsion Office; Director, A.E.C. Div. of Space Nucl. Systems, 1967-. Societies: RESA; A.N.S.; A.I.A.A.; A.A.A.S.
Nuclear interests: Management; Research and development.
Address: 4810 Stonewall Avenue, Downers Grove, Illinois, U.S.A.

KLEIN, Oskar, Prof. Formerly Prefect, Inst. of Theoretical Phys. Stockholms Hogskola. Visiting Prof., Dept. of Phys., Brandeis Univ. Member, Editorial Board, Nucl. Phys., Amsterdam.
Address: Brandeis University, Department of Physics, Waltham 54, Massachusetts, U.S.A.

KLEIN, Robert L., Mech. Eng., M.S. (Metal.), Registered Professional Eng., Commonwealth of Massachusetts, Educ.: Stevens Inst. Technol. Pres., Windalume Corp., Kenvil, New Jersey, 1947-65; Sec.-Treas., Ameray Corp., Dover, New Jersey, 1950-. Societies: A.S.M.; A.I.M.M.E.; A.N.S.
Nuclear interests: Shielding for hot laboratories and elsewhere where radio materials are handled.; Design and development of tongs and other remote handling devices.
Address: 9 Woodlawn Drive, Morristown, New Jersey, U.S.A.

KLEINERT, H., Dr. med. Wissenschaftliche Asst. Dept. of Clinical-Exptl. Radiol., Univ. Strahlenklinik (Czerny Krankenhaus).
Address: Universitäts Strahlenklinik (Czerny Krankenhaus), 3 Voss Strasse, Heidelberg, Germany.

KLEINFELD, Morris, B.A., M.D. Born 1915. Educ.: Baylor Univ. Texas, Clinical Assoc. Prof.

Medicine, State Univ., Downstate Medical Centre; Consultant Occupational Health, New York City Health Dept.; Director, Ind. Hygiene Div., New York State Labour Dept.; Clinical Prof. Environmental Medicine, New York Univ. Medical Centre. N.C.R.P. Advisory Com.; Nucl. Standards Board, A.S.A. Book: Control of Rad. Hazards in Industry (New York State Labour Dept., 1957). Societies: American Conference of Govt. Ind. Hygienists; American Coll. Preventative Medicine; N.C.R.P.; American Coll. of Physicians.
Nuclear interests: Biological effects of ionising radiation; Administrative control; Physics as it relates to understanding mechanism of radiation effects; Medical use of radiositopes.
Address: 80 Centre Street, New York 13, New York, U.S.A.

KLEINHEINS, P., Dr. Tech. leader of accelerator group, Inst. für Kernphysik, Johann Wolfgang Goethe Univ., Frankfurt.
Address: Institut für Kernphysik, Johann Wolfgang Goethe Univeristät, 31 Am Römerhof, Frankfurt a.M., Germany.

KLEINMAN, Leonard, Ph.D. Born 1933. U.C.L.A. and California (Berkeley) Univ. Res. Assoc., Chicago Univ.,1960-61; Asst. Prof., Pennsylvania Univ., 1961-64; Assoc. Prof., Southern California Univ., 1964-67; Prof., Texas Univ., 1967-. Society: A.P.S.
Nuclear interest: Theoretical solid state physics.
Address: Physics Department, Texas University, Austin, Texas, U.S.A.

KLEINMANN, Alfred, Jurist. Born 1929. Educ.: Tübingen Univ. Oberregierungsrat, Verwaltungsberichterstatter für Kernenergie, Wirtschaftsministerium Baden - Württemberg, 1959-.
Address: 4 Theodor Heuss Strasse, Stuttgart-N, Germany.

KLEIST, B., Phys. Eng. Sen. Sci., Nucl. Dept., Res. Inst. of Nat. Defence, Sweden.
Address: Research Institute of National Defence, 10 Gyllenstiernsgatan, Stockholm 80, Sweden.

KLEITMAN, Daniel. Asst. Prof., Phys. Dept., Brandeis Univ.
Nuclear interests: High energy behaviour of deuteron cross sections. High energy behaviour of scattering amplitudes by many particle systems. Properties of analyticity domains of Wightman and retarded functions in field theory. Field theoretic formulation of strong interactions among elementary particles.
Address: Physics Department, Brandeis University, Waltham 54, Massachusetts, U.S.A.

KLEM, Per Gustav, S.Bc. Born 1930. Educ.: M.I.T. Inst. for Atomenergi, 1956-60; Hagb. Waage, Shipowner, 1960-.
Nuclear interests: Reactor design and reactor safety, with special reference to ship-propulsion.
Address: 47D Gjettum V, Gjettum, Norway.

KLEMA, Ernest Donald, A.B., M.A., Ph.D. Born 1920. Educ.: Kansas Univ. and Rice Inst. Prof. Nucl. and Sci. Eng. (1958-), Chairman, Dept. Eng. Sci. (1960-66), Prof., Nucl. and Sci. Eng. (1968-), Northwestern Univ.; summer Participant, O.R.N.L., 1958; Resident Res. Assoc., Argonne Nat. Lab., 1959 and 1966-67; Chairman, Subcom. on Neutron Measurements and Standards, Com. on Nucl. Sci.,N.R.C., 1958-63. Society: A.P.S.
Nuclear interests: Nuclear physics; Nuclear engineering education; Fission and nuclear reaction cross-section measurements; Neutron flux measurements; Nuclear model calculations; Semiconductor particle detector development; Gamma-Gamma angular correlation measurements.
Address: 2435 Pomona Lane, Wilmette, Illinois, U.S.A.

KLEMM, Alfred, Prof. Dr. Born 1913. Educ.: Munich Univ. Wissenschaftl. Mitglied, Max Planck Inst. für Chemie (Otto Hahn Inst.), Mainz. Societies: Deutsche Physikalische Gesellschaft; A.P.S.; A.C.S.
Nuclear interests: Isotope separation, mass spectrography.
Address: Max Planck Institut für Chemie, 23 Saarstrasse, Mainz, Germany.

KLENK, Ernst, Prof.,Dr.rer.nat., Dr.med. h.c. Born 1896. Educ.: Tübingen Univ. Prof., Cologne Univ., 1936-.
Nuelear interest: Fettstoffwechselversuche mit Isotopen.
Address: 18 Kermeter Strasse, Köln-Lindenthal, Germany.

KLEPPE, Kjell, Ph.D. Lab. for Biochem., Norsk Hydro's Inst. for Cancer Res.
Address: Norsk Hydro's Institute for Cancer Research, The Norwegian Radium Hospital, Oslo, Norway.

KLEPPNER, D., Asst. Prof. Phys. Dept., Harvard Univ.
Address: Harvard University, Jefferson Physical Laboratory, Cambridge 38, Massachusetts, U.S.A.

KLERKX, Liliane Marie Catherine Jeanne, Licencié en Sci. Chimiques, Agrégé de l'Enseignement secondaire supérieur. Born 1939. Educ.: Liège Univ. Asst., Lab. d'Utilisation des Radioéléments, Liège Univ., 1962-.
Nuclear interests: Nuclear chemistry; Radiocarbon dating with liquid scintillation; Labelling of photographic emulsions with marked compounds, such as S - 35, Au - 198, Ag - 110.
Address: Liège University, Laboratoire d'Utilisation des Radioéléments, 9 place du XX Août, Liège, Belgium.

KLESSMANN, Horst L., Diplom-Ing. Born 1928. Educ.: Berlin Tech. Univ. R. and D., A.E.G.-Forschungsinst., Berlin, 1954; R. and D., A.E.G., Industrie, Berlin, 1955-58; Sen. Eng., Nucl. Electronics, Hahn-Meitner Inst. für

Kernforschung, Berlin, 1959-. Societies: Verband Deutscher Elektrotechniker; und Nachrichtentechnische Gesellschaft.
Nuclear interests: Nuclear electronics; Reactor instrumentation.
Address: Hahn-Meitner Instut für Kernforschung, Glienicker Strasse, Berlin-Wannsee, Germany.

KLETT, Robert J. With Dynatomics Inc., 1963-.
Address: Dynatomics Inc., 180 Mills Street, N.W., Atlanta, Georgia 30313, U.S.A.

KLEVER, Horst Heinz, Dr.med. Born 1926. Educ.: Berlin Freie Univ. Hautklinik, Freie Univ. Berlin, Bürgerhaus Hospital Berlin-Charlottenburg, 1950-53; Hospital Berlin-Britz, 1954-55; Hospital Berlin-Wilmersdorf, 1955-57; Hospital Rudolf Virchow, Berlin-Wedding, 1957-64; Hospital St. Josef, Berlin-Tempelhof, 1964-.
Nuclear interests: All medical nuclear problems, especially in tests of thyroid gland, liver, kidney, heart and circulation of blood and tests for searching tumors.
Address: Rö.-Abteilung des St. Josef-Krankenhauses, 24 Bäumerplan, Berlin-Tempelhof, Germany.

KLEVIN, Paul D., (Phys.,New York). Born 1920. Educ.: New York Univ., Eng. Coll. of New York, Brooklyn Polytech. Inst. Certified by American Board of Health Phys.; Phys., Health and Safety Lab. (1947-55), Asst. to Director, Inspection Div. (1957-62), N.Y.O.O., U.S.A.E.C. Chief, Health and Safety, Brookhaven Office, U.S.A.E.C., 1962-. Member, Rad. Com., and Rad. Sessions Arranger, A.I.H.A.; Standards Com., and Meeting Placement Com., H.P.S. Guest Lecturer, Yeshiva and Rutgers Univs. Society: A.P.S.
Nuclear interests: Health physics, nuclear safety and radiation protection; Pre- and post-operational health and safety in medical research, industrial and agricultural uses of radio-material.
Address: 52 Remson Street, Valley Stream, Long Island, New York, U.S.A.

KLEY, Gisbert, Dr. jur. Born 1904. Educ.: Heidelberg, Munich and Berlin Univs. Member, Presidential Board, Bundesvereinigung der Deutschen Arbeitgeberverbände; German Member, Specialised Nucl. Sect. for Economic Problems and Specialised Nucl. Sect. for Social and Safety Problems, Economic and Social Com., Euratom.
Address: 2 Wittelsbacherplatz, 8 Munich 2, Germany.

KLICKA, Vladimír, Eng.,C.Sc. Born 1930. Educ.: Prague Tech. Univ. At Res. Inst. of Chem. Equipment. Society: Nuclear Energy Sect., Czechoslovak Sci. and Tech. Soc.
Nuclear interests: Research and design of chemical equipment for nuclear plants and uranium metallurgy.
Address: 51 Hradesínská, Prague 10, Vinohrady, Czechoslovakia.

KLICHMAN, Alton Earl, B.S., M.S. Born 1931. Educ.: Michigan Univ. Head, Nucl. Test Sect., Atomic Power Development Assocs. Vice-Chairman, Minhigan Sect. A.N.S. Society: A.N.S.
Nuclear interests: Reactor kinetics and fast reactor safety.
Address: 1911 First Street, Detroit, Michigan 48226, U.S.A.

KLIEFOTH, Werner, Prof., Dr. Born 1901. Educ.: Breslau Univ. Nucl. Phys., Kiel Univ. Chief Editor, Atomkernenergie; Sci. Councillor, Ministerium für Wirtschaft und Verkehr des Landes Schleswig-Holstein, Kiel; Director, Deutsches Atomforum, Bonn; Managing Com., Vereinigung Deutscher Wissenschaftler e.V. Books: Grundaufgaben des Physikalischen Praktikums; Sind wir bedroht?(1956); Atomkernreaktoren (1968); Vom Atom zum Kernkraftwerk (1963).
Nuclear interest: Reactor physics.
Address: Kiel,Hindenburufer 78/79 Haus Weltclub, Germany.

KLIK, Frantisek, Mech. Ing., C.Sc. Born 1930. Educ.: Prague Tech. Univ. Sen. Lecturer, Prague Tech. Univ., 1952-55; Sci. Worker, Sci. Sec., Nucl. Res. Inst., Czechoslovak Acad. Sci. Rez, 1955-; Professional staff member, Safeguards and Inspection Dept., I.A.E.A. 1967-. Society: Czechoslovak Sci. and Tech. Soc.
Nuclear interest: Reactor design.
Address: 81 Cimicka, Prague 8, Czechoslovakia.

KLIMANOV, V. A. Papers: Co-author, Gamma-Rays Distribution along Straight Empty Cylindrical Channels (Atomnaya Energiya, vol. 20, No. 2, 1966); co-author, Build-up Factors of Gamma-Rad. for Limited Mediums (ibid, vol. 22, No. 3, 1967); co-author, Differential Albedo for Gamma-Rays from Point Monodirectional Source (ibid, vol. 23, No. 3, 1967). Address: Academy of Sciences of the U.S.S.R., 14 Leninsky Prospekt, Moscow V-71, U.S.S.R.

KLIMENKOV, V. I. Papers: Co-author, Electron Microscopy Investigation of Beryllium Oxide Rad. Damage (Atomnaya Energiya, vol. 23, No. 3, 1967); co-author, Electronic Computer Investigation Methods of Crystals Rad. Damage Dynamics (letter to the Editor, ibid, vol. 23, No. 4, 1967); co-author, On Irradiation Stability of Zirconium - 1 wt % Niobium Alloy in SM - 2 Reactor Conditions (letter to the Editor, ibid, vol. 23, No. 5, 1968); co author, Study of Rad. Stability of BeO at $100°$C (letter to the Editor, ibid, vol. 24, No. 5, 1968).
Address: Nuclear Energy Institute, Academy of Sciences of the U.S.S.R., 14 Leninsky Prospekt, Moscow V-71, U.S.S.R.

KLIMENKOVA, N. A. Papers: Co-author, Investigation of Optimum Position Neutron Beams in SM-2 Reactor (letter to the Editor,

Atomnaya Energiya, vol. 22, No. 5, 1967); co-author, Influence on Reactivity with Moderator Concentration Change in Reactor SM-2 Flux Trap (letter to the Editor, ibid, vol. 24, No. 1, 1968).
Address: Academy of Sciences of the U.S.S.R., 14 Leninsky Prospekt, Moscow V-71, U.S.S.R.

KLIMFNTOV, V. B. Papers: Co-author, Reactor Phys. Exptl. Establishment of the Phys. Inst. of the Ukranian Acad. of Sci. (letter to the Editor, Atomnaya Energiya, vol 20, No. 1, 1966); co-author, Flux Disturbation Effects Caused by Indium Detectors in Reactor Spectra (letter to the Editor, ibid, vol. 24, No. 1, 1968); co-author, Study of Loop's Canal Cells with Highly Enriched Fuel (ibid, vol. 25, No. 1, 1968).
Address: Institute of Physics, Academy of Sciences of the Ukranian S.S.R., 42 Ulitsa Dobriy Put', Kiev, Ukranian S.S.R., U.S.S.R.

KLIMESCH, Dir. Dipl. Ing. Rep. of Arbeitsgemeinschaft Kernkraftwerk des Elektrizitätswirtschaft., Koordinationsausschuss Osterreichisches Kernenergieprogramm.
Address: Koordinationsausschuss Osterreichisches Kernenergieprogramm. c/o Gen. Dir. Dipl. Ing. F. Hintermayer, p.A. Verbundgesellschaft, 6a Am Hof, Vienna 1, Austria.

KLINE, Donald E., B.S., Ph.D. (Phys.). Born 1928. Educ.: Pennsylvania State Univ. Asst. Prof. Phys. (1954-56), Visiting Phys., Nucl. Reactor Facility (1956-57), Assoc. Prof. Nucl. Eng. (1961-65), Prof. Nucl. Eng. (1965-), Pennsylvania State Univ.; Staff Res. Phys., H.R.B.-Singer, Inc., 1957-61. Book: Co-author, Basic Principles of Nucl. Sci. and Reactors (New York, D. Van Nostrand Co., Inc., 1960.). Societies: A.N.S.; A.P.S.; A.S.E.E.
Nuclear interests: Radiation effects in materials, particularly polymers and carbons; Materials research; Radiation dosimetry; Teaching.
Address: 29 Hammond Building, Pennsylvania State University, University Park, Pennsylvania, U.S.A.

KLINE, Gordon M., A.B., M.S., Ph.D. Born 1903. Educ.: Colgate, George Washington and Maryland Univs. Tech. Editor, Modern Plastics, New York, 1936-; Chief, Polymers Div. (1951-63), Consultant (1964-), Nat. Bureau Standards. Director, A.S.T.M.; Chairman, Plastics and High Polymers Sect., Internat. Union Pure and Appl. Chem. Society: A.C.S.
Nuclear interests: Effects of nuclear energy on polymeric materials; Use of nuclear energy in polymerisation reactions and chemical modification of polymers.
Address: National Bureau of Standards, 331 South Palmway, Lake Worth, Florida 33460, U.S.A.

KLING, H., Dipl. - Phys. Öffentlichkeitsarbeit, Kernforschungsanlage Jülich des Landes Nordrhein-Westfalen e.V.
Address: Kernforschungsanlage Jülich des Landes Nordrhein-Westfalen e.V., Postfach 365, 517 Jülich, Germany.

KLING, Harry Pearce, Sc.D., S.M., B.S. Born 1917. Educ.: M.I.T. and California Inst. Technol. Sylcor Nuclear Corp., New York, 1949-59; Martin-Marietta Corp., Baltimore, 1959-63; Vice-Pres., Advanced Technology, Hittman Associates Inc., Baltimore, 1963-. Society: A.S.M.
Nuclear interests: Metallurgy; Fission fuel; Radioisotope heat auxiliary power systems.
Address: Hittman Associates Inc., 4715 East Wabash Avenue, P.O. Box 2685, Baltimore, Maryland 21215, U.S.A.

KLINGELHÖFER, Rolf, Dr. Phil. Born 1926. Educ.: Marburg am Lahn Univ. Wissenschaftliche Asst., Marburg am Lahn Univ. 1957-59; Wissenschaftliche Mitarbeiter. Gesellschaft für Kernforschung, Karlsruhe, 1959-65; Akademische und Wissenschaftliche Rat, Karlsruhe Univ., 1965-. Society: Deutsche Phys. Gesellschaft.
Nuclear interest: Plasma physics.
Address: 1 Max-Planck Strasse, 7501 Leopoldshafen, Germany.

KLINGENSMITH, R. W., A.B., M.A., Ph.D. Born 1931. Educ.: Miami and Ohio State Univs. Project Leader, Appl. Nucl. Phys. Div., Battelle-Columbus Labs.
Nuclear interest: Experimental nuclear physics.
Address: 2247 Woodstock Road, Columbus 21, Ohio, U.S.A.

KLINGSMÜLLER, Ernst, Dr. Phil., Habilitation. Born 1914. Educ.: Berlin Univ. and Karlsruhe Univ. (Karlsruhe T.H.). Full Prof. of Civil Law, Commerical Law and Insurance Law, Cologne Univ., 1961; Director Inst. of Insurance Law, Cologne Univ.; Judge, Court of Appeal, Coglenz. Co-editor of the Insurance Weekly: Versicherungsrecht. Chairman, Executive Com., Cologne Univ. with regard to the Volkswagen Foundation. Member, Com. I/1 on Liability and Insurance, German Atomic Commission.
Address: Institute of Insurance Law, University of Cologne, 11 Meister-Ekkehart Str., 5 Cologne-Lindenthal, Germany.

KLINGMÜLLER, Walter, Dipl.-Biol., Dr.rer. nat. Born 1929. Educ.: Frankfurt am Main, Giessen and Cologne Univs. Asst., Giessen and Cologne Univs., 1955-58; D.S.I.R. Exchange Fellow, England, 1958-59; Wissenschaftlicher Assistent, Bot. Dept., Giessen, 1959-61; N.A.T.O.-Exchange Fellow, Genetics Dept., Yale, 1961-62; Wissenschaftlicher Asst., Max-Planck-Inst. für Molekulargenetik, Berlin-Dahlem, 1962-. Societies: Deutsche Botanische Gesellschaft; Deutsche Gesellschaft für Biophysik.
Nuclear interests: Radiation biology; UV-Photolysis of plant hormones; Modification of X-ray damage in seeds and free radical studies; Interaction of chemical mutagens and radiations on survival and mutation rate

in Neurospora.
Address: Max-Planck-Institut für Molekulargenetik, 14 Ihnestrasse, Berlin 33, Germany.

KLINKE, J. D., Dr. Wissenschaftliche Asst., of Clinical-Exptl. Radiol., Univ. Strahlenklinik (Czerny Krankenhaus).
Address: Universitäts Strahlenklinik (Czerny Krankenhaus), 3 Voss Strasse, Heidelberg, Germany.

KLINOV, A. V. Paper: Co-author, Investigation of Optimum Position Neutron Beams in SM-2 Reactor (letter to the Editor, Atomnaya Energiya, vol. 22, No. 5, 1967).
Address: Academy of Sciences of the U.S.S.R., 14 Leninsky Prospekt, Moscow V-71, U.S.S.R.

KLIWER, James, Dr. Nucl. Phys., Phys. Dept., Nevada Univ.
Address: Nevada University, Reno, Nevada, U.S.A.

KLÖBER, Jürgen, Dr. rer.nat. Born 1934.
Educ.: Freiberg Bergakademie.
Nuclear interests: Neutron physics; Transport theory.
Address: 7 Obermarkt, 92 Freiberg, German Democratic Republic.

KLOECKNER, Erich. Born 1903. Societies: Deutsche Gesellschaft für Mineraloelwissenschaft und Kohlechemie e.V., Hannover. Deutsche Rheologische Gesellschaft, Berlin-Dahlem; Studiengesellschaft zur Foerderung der Kernenergieverwertung in Schiffbau u. Schiffahrt e.V., Hamburg.
Nuclear interests: Literature and documentation.
Address: Reuter und Kloeckner Buchhandlung, 7 Steinstrasse, Hamburg 1, Germany.

KLOETZLI, Commander M.A. Nurse Corps, U.S. Navy. Head, Nucl. Nursing, Nat. Naval Medical Centre.
Address: National Naval Medical Centre, Bethesda 14, Maryland, U.S.A.

KLOFT, Werner, Dr. rer. nat., Univ. Prof. Born 1925. Educ.: Berlin, Prague and Würzburg Univs. Wissenschaftlicher Asst. (1950-56), Privatdoz. Appl. Zoology (1956-63), Prof. (1963-65), Faculty for Natural Sci., Würzburg Univ.; Full Prof. and Director, Inst. Appl. Zoology, Faculty of Maths. and Natural Sci., Bonn Univ., 1965-.
Nuclear interests: Use of radiation and radioisotopes in entomology and plant pathology.
Address: Institute of Applied Zoology, Bonn University, 1 An de Immenburg, 53 Bonn, Germany.

KLOIBER, H., Dr. Wissenschaftliche Asst., Dept. of Clinical-Exptl. Radiol., Univ. Strahlenklinik (Czerny Krankenhaus).
Address: Universitäts Strahlenklinik (Czerny Krankenhaus), 3 Voss Strasse, Heidelberg, Germany.

KLOIBER, Herbert, Dr. Managing Director, Caliqua Wärmegesellschaft Dr. Kloiber und Co. Pres., Österreichisches Atomforum; Pres., Foratom (Forum Atomique Europeén), 1968-69.
Address: Caliqua Wärmegesellschaft Dr. Kloiber und Co., 14 Mariannengasse, Vienna 1095, Austria.

KLOKE, Adolf, Privatdozent, Dr. agr. Born 1921. Educ.: Göttingen Univ. Sci. Asst., Inst. for Agricultural Chem., Göttingen Univ. 1951-59; Agricultural Officer (Soil Sci. and Chem.), F.A.O. (U.N.), Rome, 1958; Wissenschaftlicher Oberrat, Head, Inst. Nonparasitic Plant Diseases, West-Berlin, 1959-. Societies: Verband Deutscher Landwirtschaftlicher Untersuchungs und Forschungsanstalten, Fachgruppe für Isotopenforschung; Deutsche Bodenkundliche Gesellschaft; Deutsche Phytomedizinische Gesellschaft.
Nuclear interests: Tracer-methods in agricultural chemistry, plant nutrition, plant diseases; Nuclear fallout, radioactivity in plant and soil.
Address: 6 Marinesteig, 1 Berlin 38, Germany.

KLOTZBACH, Robert James, B.S. Born 1922. Eudc.: Fordham and New York Univs. Design Problem Leader (1948-53), Chairman, Chem. Technol. Div., Long Range Planning Group (1953-55), O.R.N.L.; Eng. Project Supv. (1955-57), Eng. Dept. Manager (1957-), Mining and Metals Div., Union Carbide Corp. Societies: A.C.S.; A.I.Ch.E.; A.N.S.
Nuclear interests: Fuel reprocessing; Hot laboratory design; Radioisotope production and utilisation; Raw material mining and milling; Radioactive waste disposal.
Address: Union Carbide Corporation, Mining and Metals Division, 137 47th Street, Niagara Falls, New York 14304, U.S.A.

KLOUWEN, H. M., Dr. Staff Member, Radiobiol Inst. Organisation for Health Res. T.N.O.
Address: Organisation for Health Research T.N.O., 151 Lange Kleiweg, Rijswijk (ZH), Netherlands.

KLUCHAREV, A. P. Member, Soviet deleg. visiting U.S. low energy phys. labs., January 1966. Paper: Co-author, Isotope Effect in Elastic Proton-Nucleus Scattering (Atomnaya Energiya, vol. 14, No. 1, 1963).
Address: Ukrainian Physical Technical Institute, 2 Yumovskiy Tupik, Kharkov, U.S.S.R.

KLUGE, W. Prof. Dr. Director, Inst. für Gasentladungstechnik und Photoelektronik, and Director, Inst. für Plasmaforschung, Stuttgart Univ.
Address: Universität Sttuttgart, 72 Böblingerstrasse, 7 Stuttgart, Germany.

KLUIVER, Hans DE. See DE KLUIVER, Hans.

KLUMPAR, Josef, R.N. Dr.,C.Sc. Born 1909. Educ.: Charles Univ., Prague. Phys. Dept., Radiotherapeutical Inst., Prague, 1945-55; Radiol. Dosimetry Dept., Nucl. Res. Inst., Czechoslovak Acad. Sci., Prague, 1956-. Book: Co-author, Radiol. Phys. (Prague, S.N.T.L., 1958). Society: Sect. for Oncology, Sect. for Röntgenology, Czechoslovak Medical Soc. J.E. Purkyne.
Nuclear interests: Dosimetry; Metrology.
Address: Nuclear Research Institute, 2A Na Truhlárce, Prague 8, Czechoslovakia.

KLVANA, Miroslav, M.U.Dr., C.Sc. Born 1922. Educ.: Prague Univ. Sen. Reader Nucl. Medicine and Clinical Radiol. and Deputy-Chief Rad. Dept. Safárik Univ.,Kosice. Society: Sect. of Nucl. Medicine and Sect. of Oncology, Czechoslovak Medical Soc. J. E. Purkyne.
Nuclear interests: Nuclear medicine, especially cellular radiation pathology, chemical modifiers of sensitivity and tissue culture sensitivity tests; Radionuclides in cancer diagnosis; Combined radiation and chemotherapy of cancer.
Address: Radiation Department, Oncology Division, Safárik University School of Medicine, 53 Rastislavova, Kosice, Czechoslovakia.

KNACKE, Otmar, Prof., Dr. rer. nat. Born 1920. Educ.: Königsberg, Berlin, Danzig, Breslau and Rostock Univs. Prof., Inst. für chemische Technologie; Mitglied des Wissenschaftlichen Rats der Kernforschungsanlage Jülich des Landes Nordrhein-Westfalen e.V.
Address: 56 Schevenhüttener Str., Vicht bei Stollberg, Germany.

KNAPP, Adolf, Business Manager, Thermatom A.G. Com. Member, Arbeitsgemeinschaft Lucens (AGL).
Affress: Thermatom A.G. 9 Zürcher Strasse, Winterthur, Switzerland.

KNAPP, Alfred P. Director and Chairman, Nucl. Dept., Kanpp Mills Inc.
Address: Knapp Mills Inc., Nuclear Department, 23-15 Borden Avenue, Long Island City 1, New York, U.S.A.

KNAPP, Sherman R. Pres., Chief Exec. Officer and Trustee, Northeast Utilities; Pres., Chief Exec. Officer and Director, Northeast Utilities Service Co.; Chairman and Director, Connecticut Light and Power Co.; Pres. and Director, Connecticut Yankee Atomic Power Co.; Vice-Pres. and Director, Yankee Atomic Elec. Co.; Vice-Pres. and Director, Maine Yankee Atomic Power Co.; Director, Western Massachusetts Elec. Co.; Director,Millstone Point Co.; Director Vermont Yankee Nucl. Power Corp.; Director, Emhart Corp.; Director, Hartford Accident and Indemnity Co.; Director, Hartford Fire Insurance Co.; Director, Hartford Steam Boiler and Insurance Co.; Director, Scovill Manufacturing Co.; Director, Fafnir Bearing Co.; Trustee, Connecticut Coll.; Trustee, Inst. of Living;Vice-Pres. and Director, Atomic Ind. Forum, Inc.
Address: P.O. Box 270, Hartford, Connecticut 06101, U.S.A.

KNAPP, Vladimir, Ph.D. (Phys.). Born 1929. Educ.: Zagreb and Birmingham Univs. Prof. Phys., Zagreb Univ.; Head, Beta and gamma spectroscopy Lab., Inst. Rudjer Bošković, Zagreb.
Nuclear interests: Nuclear physics; Mössbauer effect; Science in world affairs.
Address: Institut Rudjer Bošković, Zagreb, Yugoslavia.

KNAPPWOST, Adolf, Dr.-Ing. Born 1913. Educ.: Hanover T.H., Karlsruhe T.H., Kaiser Wilhelm Inst. für Metallforschung Stuttgart and Tübingen Univ. Ruf auf das Ordinariat für Physikalische Chemie der Jena Univ. (abgelehnt), 1954; apl. Prof. für physikalische Chemie, Tübingen Univ., 1952; Auswärtiges Mitglied des wissenschaftlichen Rats der Kernforschungsanlage Jülich; Ordinarius und Direktor des Instituts für Physikalische Chemie, Hamburg Univ., 1960. Book: Mitverfasser von Ulich: Kurzes Lehrbuch der physikalischen Chemie (1941). Societies: Deutsche Bunsen-Gesellschaft für physikalische Chemie; D.G.M.; Gesellschaft Deutscher Chemiker e.V.; European Organisation for Res. on Fluorine and Dental Caries Prevention.
Nuclear interests: Metallurgy; Nuclear physics.
Address; Institut für Physikalische Chemie der Hamburg Universität, 9 Jungiusstrasse, Hamburg 36, Germany.

KNAPTON, Arthur George, B.Sc. (Wales), Ph.D. (Wales). Born 1927. Educ.: Wales Univ. Leader, Refractory Metals Sect., Power Group Res. Lab., A.E.I., Manchester. Papers: On uranium alloys and refractory metals. Societies: Inst. Metals; Inst. Metal.
Address: A.E.I. Ltd., Power Group Research Laboratory, Trafford Park, Manchester 17, England.

KNEBEL, Rudolf, Dr. med. Born 1910. Leiter der Kardiologischen Abteilung der Max-Planck Gesellschaft und Direktor der Kerckhoff-Klinik, Bad Nauheim. Wissenschaftliches Mitglied, Max-Planck Gesellschaft; Dozent des Deutsches Prague Univ., 1943; Münster Univ., 1947; apl. Prof., 1951, Univ. Giessen, 1956. Book: Mitherausgeber der Zeitschrift Atomkernenergie (Munich, Verlag Karl Thiemig F.K.G.).
Nuclear interest: Reactor design.
Address: Kerckhoff-Institut, Sprudelhof, Bad Nauheim, Germany.

KNECHT, Othmar, Dr. rer. nat. (Phys.). Born 1928. Educ.: Heidelberg Univ. Max-Planck-Institut, 1953-56; Kernreaktor G.m.b.H., Karlsruhe, 1956-58; Interatom, 1958-. Subcommission Reactor Safety Fachnormenausschuss.
Nuclear interests: Reactor development; Nuclear instrumentation and control; Reactor safety.

Address: Interatom, Internationale Atomreaktorbau G.m.b.H., Bensberg/Cologne, Germany.

KNELLER, Christian, Dipl.-Ing. Born 1903. Educ.: Stuttgart T.H. Vorsitzer des Vorstandes, Energie-Versorgung Schwaben A.G.; Vorsitzer des Ständigen Ausschusses, Kernkraftwerk Obrigheim G.m.b.H. Societies: Fachausschuss Atomenergie; Vereinigung Deutscher Elektrizitätswerke.
Nuclear interests: Stromerzeugung mit Kernenergie für die Elektrizitätsversorgung.
Address: Energie-Versorgung Schwaben A.G., 12 Goethestrasse, 7000 Stuttgart-N, Germany.

KNIGHT, Alan T. Pres., Catalytic Construction Co.
Address: Catalytic Construction Co., 1528 Walnut Street, Philadelphia 2, Pennsylvania, U.S.A.

KNIGHT, Henry de Boyne M.Sc. Born 1898 Educ.: Manchester Univ. Consultant. Societies: Fellow, I.E.E.; F.Inst.P.
Nuclear interests: Plasma physics and characteristics; Arc discharge devices for control of pulsed power supplies.
Address: Carloggas,Rugby Road, Swinford, Rugby, Warwickshire, England.

KNIGHT, James Albert, Jr., B.S., M.S., Ph.D. Born 1920. Educ.: Wofford Coll., Georgia Inst. Technol. and Pennsylvania State Univ. Asst. Prof. (1950-53), Assoc. Prof. (1953-57), Res. Assoc. Prof. (1957-62), Res. Prof. (1962-), Georgia Inst. Technol. Societies: A.C.S.; A.A.A.S.
Nuclear interests: Radiation chemistry of organic systems, including pure compounds and mixtures. Radiation-induced reactions of synthetic value.
Address: Radioisotopes Laboratory, Engineering Experiment Station, Georgia Institute of Technology, Atlanta, Georgia, U.S.A.

KNIGHT, Jere D. Member, Subcom. on Radiochem., Com. on Nucl. Sci., N.A.S.-N.R.C.
Address: National Academy of Sciences - National Research Council, Committee on Nuclear Sciences, 2101 Constitution Avenue, N.W., Washington 25, D.C., U.S.A.

KNIGHT, Ray. J. Chem., Geological Survey, U.S. Dept. of the Interior.
Address: U.S. Department of the Interior, Geological Survey, Building 25, Federal Centre, Denver, Colorado, U.S.A.

KNIPPER, Albert Charles, Dr. Phys. Sci. (French State degree). Born 1930. Educ.: Strasbourg Univ. Oxford Univ., 1954-56; C.E.A., 1954-56, 1958-60. C.N.R.S., 1951-.
Nuclear interest: Nuclear physics.
Address: 9 rue Curie, 67 Strasbourg 3, France.

KNIPPING, H. W., Prof. Dr. Dr. h.c. Born 1895. Educ.: Munich Univ. Direktor der Medizinischen Universitätsklinik, Cologne.
Member, Sheering Com., Deutsches Atomforum E.V. Editor, Atomkernenergie.
Nuclear interest: Nuclear medicine.
Address: Direktor der Medizinischen Universitätsklinik, Cologne, Germany.

KNOBEL, Werner. Sen. Officer, Losinger and Co. S.A.
Address: Losinger and Co. S.A., 9 Chemin de Longeraie, Lausanne, Switzerland.

KNOERR-GAERTNER, Henriette, Prof., Dr. Med. Born 1916. Educ.: Göttingen, Freiburg, Münster and Tübingen/N. Univs. Frauenklinik, Münster Univ., 1945-50; Frauenklinik, Tübingen Univ., 1950-55; Mitarbeit, Strahleninst. Tübingen Univ. neben eigener Fachpraxis, 1956-66; Umhabilitation an die Ulm Univ., 1967-. Books: Strahlentherapie und Strahlencytologische Hand- und Lehrbuchbeitrage. Societies: Deutsche Roentgen Gesellschaft; Deutsche Gesellschaft für Geburtshilfe und Gynaekologie, Div. Regionale Gesellschaft; Affiliate, Roy. Soc.
Nuclear interests: Radiobiologische Experimente an Gewebekulturen nach Einwirkung versch. Strahlenqualitäten, RBW-Faktoren, Radiocytologie.
Address: Universität Ulm. 16 Steinhoevelstrasse, Germany.

KNOLL, Glenn Frederick, B.S., M.S., Ph.D. Born 1935. Educ.: Case Inst. Technol., Stanford and Michigan Univs. Asst. Prof., Nucl. Eng., Michigan Univ., 1962-. Societies: A.P.S.; A.N.S.
Nuclear interests: Radiation detection and measurements, fast neutron spectroscopy, neutron dosimetry, plasma diagnostics.
Address: Department of Nuclear Engineering, Michigan University, Ann Arbor, Michigan, U.S.A.

KNOP, Gerhard H.H., Dr. Born 1923. Educ.: Göttingen and Bonn Univs. a.o. Prof. (1963), o. Prof. (1967), Bonn Univ.
Nuclear interests: High energy physics; Electron accelerators.
Address: Bonn University, 12 Nussallee, 53 Bonn, Germany.

KNORRING, H. von. See VON KNORRING, H.

KNOTT, Carl Michael, Dr. Ing., Dr.rer.nat. h.c. Born 1892. Educ.: Munich T.H. (Inst. Technol.). Former Member, Board of Management, Siemens-Schuckertwerke A.G. Societies: Board of Curators, Max-Planck Inst. for Chem., Mainz; Hon. Member of Senate and Hon. Citizen of Erlangen Univ.; Hon. Counsellor, Comité internat. de l'organisation scientifique.
Nuclear interest: Construction of nuclear reactors.
Address: 2 Faistenbergerstrasse, 8000 Munich 90, Germany.

KNOTTS, Joseph B., Jr. A.B., LL.B. Born 1938. Educ.: Princeton Univ. and Harvard Law School. Attorney, U.S.A.E.C., 1963-66; Legal Projects Manager, Atomic Ind. Forum, 1966-. Vice-

chairman, Federal Bar Association Com. on Atomic Energy Law. Society: Bar Association of the City of New York, Com. on Atomic Energy.
Nuclear interests: Legal and financial aspects of commerical nuclear activities, including regulatory, contractual, insurance-indemnification-liability matters.
Address: Atomic Industrial Forum, 850 Third Avenue, New York, N.Y., U.S.A.

KNOWLES, Harrold Brook, A.B., M.A., Ph.D. (Phys.). Born 1925. Educ.: California Univ., Berkeley. Res. Asst. (1951-57), Sen. Exptl. Phys., Livermore (1957-61), Lawrence Rad. Lab., California Univ.; Res. Assoc., Yale Univ., 1961-64; Assoc. Prof., Phys. (1964-), Assoc. Phys., Nucl. Reactor, i/c, Accelerator (Part-time) (1965-), Washington State Univ. Member, Review Board, Pacific Northwest Assoc. for Coll. Phys. Society: A.P.S.
Nuclear interests: Structure of light nuclei; Accelerator design.
Address: 121 Sloan Hall, Physics Department, Washington State University, Pullman, Washington 99163, U.S.A.

KNOWLES, Peter, M.A. Born 1921. Educ.: Oxford Univ. Head, Steam Raising Dept., Merz and McLellan, Consulting Eng. Societies: A.M.I.Mech.E.; I.E.E.
Nuclear interests: Design, operation and economics of nuclear reactors and nuclear generating stations.
Address: Merz and McLellan, Carliol House, Newcastle-upon-Tyne NE1 6UX, England.

KNOX, W. Eugene, Dr. Biochem., New England Deaconess Hospital, Cancer Res. Inst.
Address: Cancer Research Institute, New England Deaconess Hospital, 194 Pilgrim Road, Boston, Massachusetts 02215, U.S.A.

KNOX, William Jordan, B.S. (Chem.), Ph.D. (Phys.). Born 1921. Educ.: California (Berkeley) Univ. Phys., California Univ. Rad. Lab., Berkeley, 1950-51; Asst. Prof. of Phys., Yale Univ., 1951-53 and 1955-59; Phys., Div. of Res., U.S. A.E.C., Washington D.C., 1953-55; Res. Assoc., Yale Univ., 1959-60; Assoc. Prof. of Phys., (1960-), now Chairman, Dept. of Phys., California Univ., Davis. Society: Fellow, A.P.S.
Nuclear interests: Experimental nuclear physics - fission, nuclear reactions, reactions between complex nuclei.
Address: Department of Physics, California, University, Davis, California, U.S.A.

KNUDSEN, Erik Stenberg, M.Sc. Born 1924. Educ.: Copenhagen Univ. Phys., Danish Meat Research Inst.; Member, Study Group on Food Irradiation, E.N.E.A., O.E.C.D. Book: Co-author, Applications of Atomic Science in Agriculture and Food (Paris, E.P.A.-O.E.E.C., 1958).
Nuclear interests: Atomic radiations and radioisotopes.

Address: Slagteriernes Forskningsinstitut, 2 Maglegaardsvej, Roskilde, Denamrk.

KNUDSEN, K. D. Sen. Eng., Development Div., O.E.C.D. Halden Reactor Project.
Address: O.E.C.D. Halden Reactor Project, P.O. Box 173, Halden, Norway.

KNUDSEN, Per, M.Sc. (Chem. Eng.). Born 1932. Educ.: Tech. Univ., Denmark. Res. Establishment Risö, Danish Atomic Energy Commission. Societies: Danish Soc. of Eng.; A.S.M.
Nuclear interests: Metallurgy; Fuels.
Address: Danish Atomic Energy Commission, Risö Research Establishment, Risö, Roskilde, Denmark.

KNULST, B. J. Pres. and Gen. Manager, Tracerlab S.A.
Address: Tracerlab S.A., 277 Antwerpse Steenweg, Mechelen, Belgium.

KNUTSEN, Gunnar. Tech. Director, A/S Kvaerner Brug.
Address: A/S Kvaerner Brug, 65 Enebakkvn, Postbox 3610, Oslo, Norway.

KNYAZEV, D. A. Papers: Co-author, Some Laws in the Thermodynamics of Isotope Exchange (Kernenergie, vol. 10, No. 4, 1967); co-author, The Partition Function Ratios of Isotopically Substituted Molecules Calculated by using the Urey - Bradley - Simanouti Potential Function (ibid, vol. 10, No. 5, 1967).
Address: Academy of Sciences of the U.S.S.R., 14 Leninsky Prospekt, Moscow V-71, U.S.S.R.

KNYAZEVA, G. D. Paper: Co-author, I.V. Kurchatov Beloyarsk Nucl. Power Station (Atomnaya Energiya, vol. 16, No. 6, 1964).
Address: Kurchatov Beloyarsk Nuclear Power Station, Zarechny Settlement, Sverdlovsk Region, U.S.S.R.

KO, In Suk. Div. Chief, Isotope Lab., Physicochemistry Div., Nat. Chem. Labs., Seoul.
Address: Isotope Laboratory, Physico-chemistry Division, National Chemistry Laboratories, Ministry of Health and Social Affairs, 79 Saichong-No, Chongno-ku, Seoul, Korea.

KO, Mi Yang, Biomedical Res., Nucl. Inst., Peking.
Address: Academia Sinica, Nuclear Institute, 3 Wen Tsin Chien, Peking, China.

KO, Paoshu, L. ès Sc. (Mathématiques et Physique), Ing. Mécanicien-Electricien, Dr. en Phys. Nucl. Born 1920. Educ.: l'Aurore (Shanghai) and Paris Univs., North Carolina State Coll. and I.S.N.S.E. (Argonne Nat. Lab.). Eng., Taiwan Power Co., Taipei, 1948-59; Assoc. Prof., Inst. Nucl. Sci., Tsinghua Univ., Hsinchu, Taiwan, 1959-61; Res. Fellow, C.E.N. Cadarache, C.E.A., 1963-64; Asst. Prof., Manhattan Coll., 1964-. Societies: Assoc. des Ing. en Génie Atomique; A.N.S.; A.A.A.S.
Nuclear interests: Nuclear engineering; Reactor

physics; Nuclear instrumentation; Applied nuclear physics.
Address: Physics Department, Manhattan College, Bronx, N.Y. 10471, U.S.A.

KO, Ting-Sui. Born 1910. Educ.: Peking; California Inst. Technol. Associated with U.S. atomic project, June 1944-Nov. 1945; Returned to China, 1949; now distinguished nucl. specialist.
Address: Academia Sinica, Peking, China.

KOBA, I. I. Papers: Co-author, 100 MeV Neutron Collector (Jaderna Energie, vol. 13, No. 11, 1967); co-author, Storage Ring for 100 MeV Electrons (Atomnaya Energiya, vol. 23, No. 6, 1967).
Address: Physico-Technical Institute, 2 Polytechnicheskaya Ulitsa, Leningrad, U.S.S.R.

KOBASHI, Yozo. Exec. Director, Taiyo Fire and Marine Insurance Co. Ltd.
Address: Taiyo Fire and Marine Insurance Co. Ltd., No. 5, 3-chome, Nihonbashi-Tori, Chuo-ku, Tokyo, Japan.

KOBAYASHI, Hisao. Lecturer, Inst. for Atomic Energy, St. Paul's Univ. (Rikkyo Univ.).
Address: St. Paul's University, Ikebukuro, Toshima-ku, Tokyo, Japan.

KOBAYASHI, Koji, Dr. Chief, Div. of Gen. and Inorganic Chem., Dept. of Chem., Rad. Centre of Osaka Prefecture.
Address: Department of Chemistry, Radiation Centre of Osaka Prefecture, Shinke-Cho, Sakai, Osaka, Japan.

KOBAYASHI, Masaharu. Improvements of the Textile and Fibres by Rad. Chem., Radioisotope Res. Dept. Appl. to Textile Industry, Textile Res. Inst.
Address: Textile Research Institute, 4 Sawatari, Kanagawa-ku, Yokohama, Japan.

KOBAYASHI, Masatsugu, Dr. Eng. Born 1902. Educ.: Tokyo Univ. Exec. Director, Nippon Elec. Co., Ltd., Tokyo, 1956; Vice-Chairman, Sumitomo Atomic Industry Group, Osaka, 1958; Vice-Chairman Sumitomo A.E.C., Osaka.
Address: c/o Nippon Electric Co., Ltd., 2 Shibamita-Shikokumachi Minatoku, Tokyo, Japan.

KOBAYASHI, Osamu, Prof., Dr. Born 1910. Educ.: Kioto Univ. Medicine Dept., Member, Radioisotope Res. Com., Niigata Univ. School of Medicine.
Address: Department of Medicine, Niigata University School of Medicine, 757, 1-Bancho Asahimachi-Street, Niigata, Japan.

KOBAYASHI, Shigehiro, Prof., Phys. Dept., Kagawa Univ.
Address: Physics Department, College of Art and Sciences, Kagawa University, Miyawakicho, Takamatsu, Japan.

KOBAYASHI, Shoichi, Prof. Dept. of Medicine, Radioisotope Res. Com., Niigata Univ. School of Medicine.
Address: Niigata University School of Medicine, 757, 1-Bancho Asahimachi-Street, Niigata, Japan.

KOBAYASHI, Tadanori. Born 1903. Educ.: Waseda Univ. Director, Mitsui Mutual Life Insurance Co., 1947; Pres., Toyo Fibre Co., Ltd., 1951. Societies: Tokyo Chamber of Commerce; Japan Atomic Industrial Com.; Japan Atomic Soc.; Com. of Japan Industrial Productivity. Nuclear interest: Absorbent materials for fission product etc. disposal.
Address: Mitsui Seimei Building, 1 1-Chome, Ohtemachi, Chiyoda-ku, Tokyo, Japan.

KOBAYASHI, Takao, Master of Agriculture. Born 1935. Educ.: Tokyo Agriculture and Textile Industry and Tohoku Univs. Faculty of Medicine, Osaka Univ., 1967-.
Address: Department of Fundamental Radiology, Faculty of Medicine, Osaka University, Joancho-33, Kita-ku, Osaka, Japan.

KOBAYASHI, Tetsuro. Res. Asst., Div. Nucl. Phys. and Eng., Sci. and Eng. Res. Lab., Waseda Univ.
Address: Waseda University, 17 Kikui-cho, Shinjuku, Tokyo, Japan.

KOBAYASHI, Ugoro. Born 1909. Educ.: Tokyo Univ. of Literature and Sci. Prof. Eng. Dept., Niigata Univ., 1950-. Societies: Japan Radioisotope Assoc.; Japan Rad. Res. Soc.; Radioisotope Res. Com., Niigata Univ. Nuclear interests: Radioactive fallout.
Address: Department of Engineering, Niigata University, 1277, 1-chome, Gakkocho, Nagaoka, Japan.

KOBAYASHI, Yutaka, B.S. (Chem. Technol.), M.S. (Bio-organic Chem.), Ph.D. (Biochem.). Born 1924. Educ.: Iowa State Univ. Res. Fellow, Rheumatic Fever Res. Inst., Chicago, 1953-57; Sen. Sci., Worcester Foundation for Exptl. Biol., Shrewsbury, Mass., 1957-. Societies: A.C.S.; American Soc. of Biol. Chemists.
Nuclear interests: Assay of carbon 14 and tritium.
Address: Worcester Foundation for Experimental Biology, 222 Maple Avenue, Shrewsbury, Massachusetts, U.S.A.

KOBAYASI, Kiyosi, Ph.D. Born 1922. Educ.: Tohoku Imperial Univ. Prof. Reactor Eng., Tohoku Univ., 1960-. Societies: Atomic Energy Soc. Japan; Japan Soc. Mech. Eng. Nuclear interests: Heat problems in reactors; Thermal properties of nuclear fuels; Boiling; Sodium heat transfer.
Address: Tohoku University, Aramaki - Aoba, Sendai, Japan.

KOBAYASI, Minoru, D.Sc. Prof., Theoretical Nucl. Phys., Dept. of Phys. II, Faculty of Sci.,

Kyoto Univ.
Address: Kyoto University, Department of Physics II, Yoshida-Honmachi, Sakyo-ku, Kyoto, Japan.

KOBAYASI, R. Director, Tokyo Atomic Ind. Consortium.
Address: Tokyo Atomic Industrial Consortium, Hitachi Building, 6-4-chome, Surugadai Kanda, Chiyoda-ku, Tokyo, Japan.

KOBZAR, I. G. Paper: Co-author, Radioactivity Carry-over in Boiling Water of Reactor WK-50 (Atomnaya Energiya, vol. 23, No. 4, 1967).
Address: Academy of Sciences of the U.S.S.R. 14 Leninsky Prospekt, Moscow V-71, U.S.S.R.

KOBZAR, L. L. Papers: Co-author, Calculation of Shear Stresses on Wall and of Velocity Distribution for Turbulent Flow of Liquid in Channels (Atomnaya Energiya, vol. 21, No. 2, 1966); co-author, Calculation of Hydraulic Resistance Coefficients for Turbulent Flow in Noncircular Cross-Section Channel (ibid, vol. 23, No. 4, 1967).
Address: Academy of Sciences of the U.S.S.R., 14 Leninsky Prospekt, Moscow V-71, U.S.S.R.

KOCAK, Cevdet, Assoc. Prof. Born 1926. Educ.: Istanbul Univ. Sen. Officer, Inst. for Nucl. Energy, Istanbul Tech. Univ.
Address: Istanbul Technical University, Gümüssuyu, Istanbul, Turkey.

KOCH, Donald Fielding, B.S. Born 1927. Educ.: U.S. Naval Acad., Annapolis and Naval Advanced Nucl. Power School, New London, Connecticut. U.S. Naval Officer serving in various ships and commands, 1950; Reactor Officer, U.S.S. George Washington (SSB (N)-598), 1960; Operations Officer, U.S.S. George Washington (SSB (N)-598), 1961; Staff Member, Los Alamos Sci. Lab., Jackass Flats, Nevada, 1962; Exec. Director, State of Washington, Office of Nucl. Energy Development, 1964-. Exec. Sec., State of Washington Governor's Advisory Council on Nucl. Energy and Rad., 1964-; Coordinator and Administrator, Western Interstate Nucl. Compact Com., Western Governors' Conference, 1966-; Chairman, Nat. Conference of State Nucl. and Sci. Officials, 1966; Staff Director, State of Washington Legislative Joint Com. on Nucl. Energy, 1967-. Book: A Programme for the Development of the State of Washington in the Nucl. Age; a Western Interstate Nucl. Compact (Washington State Printer, 1964). Societies: Atomic Ind. Forum; A.N.S.; A.S.E.E.; A.A.A.S.
Nuclear interests: Executive and corporate management, nuclear economic promotion, marine propulsion, space applications, nuclear rocket propulsion, agricultural applications, underground nuclear engineering, ocean applications, regional - state - national - international nucl.development cooperation, nuclear - medical applications.
Address: Office of Nuclear Energy Development, Room 101 E = mc^2, General Administration Building, Olympia, Washington 98501, U.S.A.

KOCH, Edgar, Ing. Member, Sen. Staff, engaged in Nucl. Work, Abteilung Strahlungsmesstechnik, Reichert-Elektronik G.m.b.H. und Co. K.G.
Address: Reichert-Elektronik G.m.b.H. und Co. K.G., Postfach 743, Trier/Petrisberg, Germany.

KOCH, Francisc, Dr. Born 1925. Babes-Bolyai Univ., Cluj. At Babes-Bolyai Univ., 1949-. Nuclear interests: Nuclear physics; Nuclear magnetism; Isotope analysis; Experimental methods in nuclear physics.
Address: Universitatea Babes-Bolyai, Fac. Fizica, 1 Str. Kogalniceanu, Cluj, Roumania.

KOCH, George Karel, Dr. Born 1926. Educ.: Municipal Univ., Amsterdam. Sci., Inst. voor Kernfysisch Onderzoek, Amsterdam, -1963; Sect. Manager, Radiochem. Dept., Unilever Res. Lab.
Nuclear interest: Tracer applications in food research.
Address: Unilever Research Laboratory, 120 Olivier van Noortlaan, Vlaardingen, Netherlands.

KOCH, H. William, B.S. (Queens Coll., Flushing, N.Y.), Ph.D. and M.S. (Illinois). Born 1920. Educ.: Illinois Univ. Chief, High Energy Rad. Sect. (1949-63), Chief, Rad. Phys. Div. (1963-), Nat. Bureau of Standards, Washington, D.C. (Leave of absence from N.B.S.) Consultant, Rad. Lab., California Univ., Livermore, 1959. Society: Fellow, A.P.S.
Nuclear interest: Photonuclear research.
Address: National Bureau of Standards, Washington 25, D.C., U.S.A.

KOCH, Hans Henrik, M.A. (law). Born 1905. Educ.: Copenhagen Univ. Permanent Under-Sec. of State, Ministry of Social Affairs, 1942-56. Member (1955), Chairman, Exec. Com. (1956-), Danish A.E.C.; Govt. Deleg. to Internat. Labour Conference and Member of Governing Body, Internat. Labour Office, 1948-51; Chairman, Greenland Commission, 1949-50; Nordic Co-operation Com. on Peaceful Use of Atomic Energy, 1957; Danish Energy Council, 1957; Chairman,Danish Deleg.,to Internat. Atomic Energy Conference, 1957-; Board of Governors I.A.E.A., 1959, 1963; Vice-Chairman, Steering Com., E.N.E.A., 1958-; Govt. Disarmament Com., 1961-; Govt. Com. on Civil Service Organisation, 1960-.
Address: Atomic Energy Commission, 29 Strandgade, Copenhagen K, Denmark.

KOCH, Hartwig Walter Fritz, Prof., Dr. rer.nat. habil. Born 1921. Educ.: Karl-Marx-Univ., Leipzig and T.H. für Chemie, and Leuna-Merseburg. Inst. für Verfahrenstechnik, D.A.W. Leipzig, 1952; Inst. für angewandte Radioaktivität, Leipzig, 1955-; Prof. for Radiochem., Tech. High School. Book: Co-author, Radiophysikalisches und radiochemisches Grund-

praktikum, (Berlin Deutscher Verlag der Wissenschaften 1959). Society: Chemische Gesellschaft der D.D.R.
Nuclear interests: Applied radiochemistry; Radioactivity; Management.
Address: Institut für angewandte Radioaktivität, 15 Permoserstr., 705 Leipzig, German Demociatic Republic.

KOCH, Hermann Calderwood, B.A. (Rhodes), M.A., B.C.L. (Oxon). Born 1907. Educ.: Rhodes and Oxford Univs. Director and Manager, Anglo-American Corp. of South Africa, Ltd.; Chairman, Vaal Reefs Exploration and Mining Co., Ltd.; Western Reefs Exploration and Development Co. Ltd., Western Deep Levels, Ltd.; Atomic Energy Board, South Africa.
Nuclear interest: Management.
Address: Anglo-American Corporation of South Africa, Ltd., P.O. Box 4587, Johannesburg, South Africa.

KOCH, Jorgen, Dr.phil., Dr.ing. Born 1909. Educ.: Danzig and Berlin Tech. Univs. and Copenhagen Univ. Asst. Prof. (1954), Prof. Phys. and Director Biophys. Lab. (1957), Director Phys. Lab. II (1963), Copenhagen Univ. Counsellor, Danish Nat. Health Service on Rad. Protection. Book: Electromagnetic Isotope Separators and Applications of Electromagnetically Enriched Isotopes (Amsterdam, North Holland Publishing Co., 1957). Societies: N.Y. Acad. Sci.; A.P.S.; Nordic Soc. for Rad. Protection; H.P.S.; Danish Acad. Tech. Sci.; Roy. Physiografic Soc., Sweden.
Nuclear interests: Nuclear physics; Electromagnetic isotope separation; Particle accelerators.
Address: 20 Juliane Mariesvej, Copenhagen, Denmark.

KOCH, Leonard J., B.S. Educ.: I.I.T. Senior Mech. Eng., Argonne Nat. Lab., 1948; Assoc. Project Eng., Experimental Breeder Reactor I; Project Manager, Experimental Breeder Reactor II; Director, Reactor Eng. Div., 1963. Fellow, A.N.S.
Nuclear interests: Reactor design, development, and construction; Project management and reactor technology development management.
Address: 937 S. Adams Street, Hinsdale, Illinois, U.S.A.

KOCH, Lydie, Ing. (E.S.P.C.I.), Dr.-Ing. Born 1931. Educ.: Faculté des Sciences de Paris. Ing. de recherche, C.E.N., Saclay. Societies: Sté. des Radioelectriciens; Sté. de Physique.
Nuclear interests: Solid state detectors of nuclear particles; Radiation damage of semiconductors; Direct conversion of nuclear energy by plasma diodes and semi-conducting devices; Nuclear reactions in space cosmic ray physics.
Address: Service d'Electronique Physique, Centre d'Etudes de l'Energie Nucléaire de Saclay, B.P. No. 2, Gif-sur-Yvette, Seine et Oise, France.

KOCH, Robert C., Ph.D. Born 1927. Educ.: Chicago Univ. Sen. Eng., Westinghouse Elec. Corp., Bettis Plant, 1955-57; Asst. Manager, Chem. Dept. (1957-61), Manager, Phys. Sci. Dept. (1961-63), Manager of Operations (1963-), Nucl. Sci. and Eng. Corp., Pittsburgh. Book: Activation Analysis Handbook (New York, Academic Press, 1960). Society: A.N.S.
Nuclear interests: Applications of nuclear chemistry to reactor development and other industrial problems.
Address: Nuclear Science and Engineering Corporation, P.O. Box 10910, Pittsburgh, Pennsylvania 15236, U.S.A.

KOCH, Roland. Director Gen., Electro-Mécanique. Tresorier and Member, Conseil d'Administration, A.T.E.N.
Address: Cie. Electro-Mécanique, 12 rue Portalis, Paris 8, France.

KOCHAR, H. K. Joint Sec., Dept. of Atomic Energy, A.E.C., India.
Address: Atomic Commission, Department of Atomic Energy, Apollo Pier Road, Bombay 1, India.

KOCHAR, Mrs. S. Deputy Sec., Dept. of Atomic Energy, A.E.C., India.
Address: Atomic Energy Commission, Department of Atomic Energy, Apollo Pier Road, Bombay 1, India.

KOCHEGUROV, V. A. Papers: Co-author, 1,5 GeV Electron Synchrotron at the Polytech. Inst. of Tomsk (Atomnaya Energiya, vol. 21, No. 6, 1966); co-author, 300 MeV Electron Synchrotron at the Polytech. Inst. of Tomsk (letter to the Editor, ibid, vol. 21, No. 6, 1966).
Address: Academy of Sciences of the U.S.S.R., 14 Leninsky Prospekt, Moscow V-71, U.S.S.R.

KOCHENOV, A. S. Papers: The Stability of a Nucl. Power Plant (Atomnaya Energiya, vol. 7, No. 2, 1959); co-author, On Comparison of Theoretical and Exptl. Parameters of Homogeneous Uranium-Water Critical Assemblies (letter to the Editor, ibid, vol. 19, No. 5, 1965); Decrease of Thermal Neutron Flux due to Hollow Channel in Reflector (Abstract, ibid, No. 6). Influence of Research Reactor Parameters on Thermal Neutron Flux in Reflector and Fuel Inventory (ibid, vol. 21, No. 2, 1966).
Address: Academy of Sciences of the U.S.S.R., 14 Leninsky Prospekt, Moscow V-71, U.S.S.R.

KOCHENOV, I. S. Papers: Co-author, The Calculation and Analysis of Thermodynamic Cycle Parameters of a Nucl. Power Station (Atomnaya Energiya, vol. 13, No. 1, 1962); co-author, On Hydraulic Design of Nucl. Reactor Cooling System (ibid, vol. 23, No. 2, 1967).
Address: Academy of Sciences of the U.S.S.R., 14 Leninsky Prospekt, Moscow V-71, U.S.S.R.

KOCHERGIN, V. P. Papers: Co-author, Neutron Moderation Lengths (Atomnaya Energiya, vol. 6, No. 1, 1959); Thermal Neutron Flux Flattening in Uranium-Water Reactors (Abstract, ibid, vol. 19, No. 1, 1965).
Address: Academy of Sciences of the U.S.S.R., 14 Leninsky Prospekt, Moscow V-71, U.S.S.R.

KOCHETKOV, Levalekseyevich A. Lab. Chief, Inst. of Phys. and Power Eng.,Obninsk. Papers: Co-author, Start-up Conditions in a Uranium-Graphite Power Reactor Providing Superheated Steam (Atomnaya Energiya, vol. 9, No. 1, 1960); co-author, Chem. Stability of Deposits and Transport of Radioactive Material in Water and Steam in a Water-Steam Loop in the First Soviet Power Station (ibid, vol. 9, No. 2, 1960); co-author, Experience with the Operation of the First Nucl. Power Station (ibid, vol. 11, No. 1, 1961); co-author, Operational Experience of the First Nucl. Power Station (ibid, vol.16, No. 6, 1964).
Address: Institute of Physics and Power, Obninsk, Near Maloyaroslavets, Moscow, U.S.S.R.

KOCHEVANOV, V. A. Paper: Co-author, Photoactivation Determination of Zirconium in Alloys with Beryllium, Magnesium, Aluminium and Niobium (Atomnaya Energiya, vol. 24, No. 2, 1968).
Address: Academy of Sciences of the U.S.S.R., 14 Leninsky Prospekt, Moscow V-71, U.S.S.R.

KOCHHAR, Bal Raj, M.B.B.S., M.D., F.C.P.S. Born 1923. Educ.: Punjab and Bombay Univs. Pathologist in Indian Army, 1950-64. Nuclear interests: Nuclear medicine and radiation pathology. Interested in bone marrow transplantation in irradiated animals and development of isotopic techniques in diagnosis of haematological disorders. Teacher for radiation pathology and haematology in Diploma Radiation Medicine Course.
Address: Institute of Nuclear Medicine and Allied Sciences, Probyn Road, Delhi 6, India.

KOCHIN, A. E. Papers: Co-author, Absolute Measurement of Neutron Source Yield by Gold Foil Method (letter to the Editor, Atomnaya Energiya, vol. 16, No. 3, 1964); co-author, Internat. Comparison of P^{32}, Co^{60}, Tl^{200} Solution Specific Activity and Co^{60} Hard Source Activity (letter to the Editor, ibid, vol. 19, No. 1, 1965).
Address: Academy of Sciences of the U.S.S.R., 14 Leninsky Prospekt, Moscow V-71, U.S.S.R.

KOCHINA, Pelageya. Member, Inst. of Hydrodynamics, Novosibirsk Sci. Centre, Acad. Sci. of the U.S.S.R.
Address: Novosibirsk Science Centre, 20 Sovietskaya Ulitsa, Novosibirsk, Siberia, U.S.S.R.

KOCHUROV, B. P. Papers: Co-author, Critical Condition for Inhomogeneous Reactor with Rods of Finite Radius (Nukleonika, vol. 9, No. 6, 1964); Calculation of Dipole Momentum of Cylindrical Fuel Rod (Abstract, Atomnaya Energiya, vol. 19, No. 6, 1965); co-author, Maximum Reactor Power Problem (ibid, vol. 22, No. 1, 1967); co-author, The Determination of Migration Length and Multiplication Coefficient in a Heavy-Water Natural Uranium Lattice (Jaderna Energie, vol. 13, No. 11, 1967).
Address: U.S.S.R. State Committee for Atomic Energy, 26 Staromonetnii Pereulok, Moscow, U.S.S.R.

KOCK, Winston E., E.E., M.S. (Phys.), Ph.D., D.Sc. (Hon., Cincinnati Univ.). Born 1909. Educ.: Cincinnati and Berlin Univs. Director, Acoustics Res., Bell Telephone Labs., 1950; Chairman, Anti-Submarine Warfare Com., Nat. Security Ind. Assoc., 1964; First Director, N.A.S.A. Electronics Res. Centre, 1964; Vice Pres. and Chief Sci., Bendix Corp., 1966-. Chairman, Space Com., Member, Board of Directors, and Member, Exec. Com., Atomic Ind. Forum. Societies: A.P.S.; A.S.A.; I.E.E.E.; A.I.A.A.
Address: The Bendix Corporation, 1104 Fisher Building, Detroit, Michigan 48202, U.S.A.

KOCSIS, Ernest J. Member, Board of Trustees, Power Reactor Development Co.
Address: Power Reactor Development Co., 1911 First Street, Detroit 26, Michigan, U.S.A.

KODA, Shigeyasu, Dr.Eng. Born 1907. Educ.: Tokyo Univ. Prof., Res. Inst. for Iron, Steel and Other Metals, Tohoku Univ. Societies: Inst. Metals; Atomic Energy Soc. Japan; Phys. Soc. Japan; Japan Inst. Metals.
Nuclear interest: Metallurgy.
Address: 76 Eifuku-cho, Suginami-ku, Tokyo, Japan.

KODA, Yoshio, B.Sc. Born 1926. Educ.: Kyoto and Tokyo Univs. Tech. Official, Ministry of Internat. Trade and Industry, 1952. Book: Co-author, Industrial Uses of Radioisotopes (Chem. Industry Co., 1958). Society: Japan Radioisotope Assoc.
Nuclear interests: Radiochemistry; Chemical separation of fission products, especially separation of radioactive ruthenium from other elements.
Address: Government Industrial Research Institute, Nagoya, Hirate-machi, Kita-ku, Nagoya, Japan.

KODAMA, Keizo. Chairman, Atomic Energy and Isotope Com., Tokushima Univ.
Address: c/o Medical Faculty, Tokushima University, Kuramoto-cho, Tokushima-shi, Japan.

KODIREE, S. Paper: Co-author, Determination of Indium by (γ,γ)- Reaction with 5 MeV Linear Electron Accelerator (letter to the Editor, Atomnaya Energiya, vol. 24, No. 3, 1968).
Address: Academy of Sciences of the U.S.S.R., 14 Leninsky Prospekt, Moscow V-71, U.S.S.R.

KOECHLIN, André, Eng. E.P.F. Born 1899. Educ.: Ecole Polytechnique Fédérale, Zurich. Administrateur-délégué: Ofinco, Energie Electrique du Simplon S.A., and Sté. Générale pour l'Industrie. Societies: Sté. Suisse des Ingénieurs et Architectes; Sté. Française des Electriciens; Assoc. Suisse des Electriciens.
Nuclear interests: Reactor design; Management.
Address: Société Générale pour l'Industrie, 17 rue Bovy-Lysberg, Geneva, Switzerland.

KOECHLIN, Jean-Claude, Ing. E.S.E. Born 1927. Educ.: Paris Univ. Chef, Sect. des Relations avec Euratom, C.E.A.
Nuclear interests: Nuclear energy research and production; International cooperation.
Address: 29-33 rue de la Fédération, Paris 15, France.

KOECHLIN, Raymond, Dipl. civil eng. Born 1903. Educ.: Federal Inst. of Technol., Zurich. Managing Director, Conrad Zschokke Ltd., 1941-. Societies: Soc. suisse des Ing. et Architectes; Soc. des Ing. Civils de France; Assoc. des Anciens Eleves de l'Ecole Polytechnique Federale, Zurich.
Address: 9 avenue Bertrand, Geneva, Switzerland.

KOECK, Wolfgang, Dipl.-Chem., Dr. phil., Dr.jur.et rer.pol. Born 1905. Educ.: Munich, Berlin and Königsberg Univs. Employed since 1935 now Director, Verband der Chemischen Industrie e.V. Society: Member, Exec. Council, Deutsches Atomforum e.V.
Nuclear interests: Atomic industry; Application of radioisotopes; Radiation protection; Nuclear chemistry, radiochemistry and radiation chemistry; Fuel cycle; Special materials for nuclear engineering.
Address: Verband der Chemischen Industrie e.V., 21 Karlstrasse, 6 Frankfurt-am-Main, Germany.

KOEHLER, Harold P., B.A.Sc. Born 1924. Educ.: Toronto Univ. Chief Instrumentation Eng., Atomics, Hawker Siddeley Eng. Society: Soc. for Exptl. Stress Analysis.
Nuclear interests: Design and development of reactor instrumentation, in-pile experiments for measuring materials properties, creep, fuel expansion; In-pile heaters and temperature measurement and control systems; Analytical and experimental stress analysis of reactor systems including vibration, heat transfer and hydraulic problems.
Address: Atomic Products Department, Hawker Siddeley Engineering, Division of Hawker Siddeley Canada Ltd., P.O. Box 4015, Terminal 'A', Toronto, Ontario, Canada.

KOEHLER, J. S. Phys. Additional Graduate Teaching Faculty, Illinois Univ.
Address: Illinois University, 214 Nuclear Engineering Laboratory, Urbana, Illinois, U.S.A.

KOEHLER, John T. Attorney at Law, Butler, Keohler and Tausig, Washington. Member, Board of Contract Appeals, U.S.A.E.C.
Address: U.S. Atomic Energy Commission, Board of Contract Appeals, Washington D.C. 20545, U.S.A.

KOEN, Johannes Gerhardus, D.Sc., B.Ed. Born 1925. Educ.: Pretoria and Potchefstroom Univs. Sen. Lecturer, Phys. Dept., Pretoria Univ., 1961-. Sec. Mathematics and Natural Sci. Div., Acad. Arts and Sci., S. Africa. Society: M.Inst.P.
Nuclear interests: Reactor design and training; Pulsed neutron technique.
Address: Physics Department, Pretoria University, Pretoria, South Africa.

KOENIG, H. P. Member, Steering Com., O.E.C.D., E.N.E.A.
Address: European Nuclear Energy Agency, 38 boulevard Suchet, Paris 16, France.

KOENNE, Werner Georg, Dr. Techn., Dipl. Ing. Born 1933. Educ.:Vienna Tech. and Stuttgart Tech. Univs. With Austrian Elec. Board, Vienna, 1958-. Societies: Österr. Ingenieur und Architektenverein, Vienna; Abwassertechnische Vereinigung, Bonn; B.N.E.S.
Nuclear interests: Civil engineering work on nuclear field (containment, lay out); Waste disposal; Technical shielding problems; Use of radioactive isotopes in civil engineering; Management.
Address: 96 Landstr. Haupstr.,Vienna 3,Austria.

KOEPCKE, S. Asst. Gen. Manager for Finance and Administration, Vice Pres. (1967-), and Sec.-Treas., Douglas United Nucl., Inc. Joint Venture at Hanford.
Address: Douglas United Nuclear, Inc., P.O. Box 490, Richland, Washington 99352, U.S.A.

KOEPCKE, Werner, Dr.-Ing. Prof. Reinforced Concrete Construction. Director, Materials and Structures Testing Lab., Berlin Tech. Univ. Member, Editorial Advisory Board, Nucl. Structural Eng.
Nuclear interests: Research and consulting work concerning structural engineering aspects of nuclear power plants, especially prestressed concrete reactor pressure vessels.
Address: Lehrstuhl für Stahlbetonbau, Technische Universität Berlin, 35 Hardenbergstrasse, 1 Berlin-Charlottenburg 2, Germany.

KOEPPE, P., Wiss.Rat.Dr.Ing. Phys., Strahleninst.-Klinik, Berlin Free Univ.
Address: Strahleninstitut-Klinik, Berlin Free University, 130 Spandauer Damm, Berlin-Charlottenburg 19, Germany.

KOERNER VON GUSTORF, Ernst, Dr.rer.nat., Dipl. Chem. Born 1932. Educ.: Göttingen Univ. Organisch-Chemisches Inst., Göttingen Univ., 1956-57; Sci. Asst., Abteilung Strahlenchemie, Max-Planck-Inst. für Kohlenforschung, Mülheim/Ruhr, 1958-; Sen. Postdoctoral Resident Res. Assoc. at the U.S. Army Natick Labs., Natick, Mass. (Sponsor: Nat.Acad. Sci.-Nat. Res. Council), 1964-66; Assoc. Prof. Chem.

Boston Coll., Massachusetts, 1966-. Societies: Gesellschaft Deutscher Chemiken; A.C.S.
Nuclear interest: Radiation chemistry.
Address: Abteilung Strahlenchemie, Max-Planck-Institut für Kohlenforschung, 34-36 Stiftstrasse, Mülheim/Ruhr, Germany.

KOERTS, L. A. Ch., Dr. Physicist, Inst. for Nucl. Phys. Res., Amsterdam.
Address: Inst. for Nuclear Physics Research (I.K.O.), 18 Ooster Ringdijk, Amsterdam-O, Netherlands.

KOESTER, Lothar Ludwig, Diplom-Physiker, Dr. rer. nat. Born 1922. Educ.: Heidelberg Univ. At. Max-Planck-Inst. Heidelberg, 1950-55; Farbenfabriken Bayer, Isotopenlaboratory, 1955-58; Director of Operation,F.R.M., Res. Reactor, Munich, 1958-.
Nuclear interests: Nuclear physics; Isotopes reactor design; Management.
Address: Reaktorstation Garching der Technischen Hochschule München, 8046 Garching bei München, Germany.

KOFFMANN, Eugen. Born 1916. Educ.: Berlin Inst. Technol. and O.R.S.O.R.T. Sen. Nucl. Eng., Los Angeles Water and Power Dept. Societies: A.N.S.; A.S.M.E.
Nuclear interest: Power.
Address: 111 North Hope Street, Los Angeles, California, U.S.A.

KOFFOLT, J. H., Dr. Chairman, Dept. Chem. Eng., Coll. of Eng., Ohio State Univ.
Address: College of Engineering, Ohio State University, Columbus 10, Ohio, U.S.A.

KOFOED-HANSEN, Otto M., Dr. phil. Born 1921. Educ.: Copenhagen Univ. Res. Assoc., Columbia Univ., 1954-55; Lektor, Copenhagen Univ., 1955-56; Head, Phys. Dept., Danish A.E.C., 1956-; Prof. Reactor Phys., D.T.H. Copenhagen, 1960-.
Nuclear interests: Nuclear physics; Applied physics; Reactor physics; Plasma physics.
Address: Risö Research Laboratory, Physics Department, Risö, Roskilde, Denmark.

KOGA, Masami. Tech. Director, Atomic Energy Dept., Ishikawajima Harima Heavy Industries Co. Ltd.
Address: Ishikawajima Harima Heavy Industries Co. Ltd., 2-chome, Fukagawa-Toyosu, Kotoku, Tokyo, Japan.

KOGA, S. Chief, Atomic Power Sect.,Steam Power Dept., Kyushu Elec. Power Co. Inc.
Address: Kyushu Electric Power Co. Inc., 3/58 Higashiyakuin, Fukuoka City, Japan.

KOGA, Yuzo, Dr. Technol. Educ.: Tokyo Univ. of Tech. Inst. Director, Central Res. Lab., Toa Nenryo Kogyo K.K. Societies: Japan Atomic Industry Conference; Japan Chem. Soc.
Nuclear interests: Management; Radioisotope tracers; Radiation chemistry.
Address: Toa Nenryo Kogyo K.K., Ohi-mura, Ihara-gun, Saitama-ken, Japan.

KOGAN, V. I. Papers: Estimation of the Electron Temperature and Degree of Ionisation in the First Stage of a Powerful Pulsed Discharge (letter to the Editor, Atomnaya Energiya, vol. 4, No. 2, 1958); co-author, Optimum Parameters of Arc as Source of Ionising Rad. (letter to the Editor, ibid, vol. 23, No. 6, 1967).
Address: Academy of Sciences of the U.S.S.R., 14 Leninsky Prospekt, Moscow V-71, U.S.S.R.

KOGURE, Makita, Dr. Prof. Faculty of Textile Industry, Tokyo Univ. of Agriculture and Technol.
Nuclear interest: Radiation sensitivity of silkworm tests.
Address: Faculty of Textile Industry, Tokyo University of Agriculture and Technology, P.O.D. Koganei, Tokyo, Japan.

KOH, Kwang Jae, B.S.,M.S. Born 1932. Educ.: Seoul and Sung Kyun Kwan Univs. Nucl. Measurement Group, Geochem. Sect., Geological Survey of Korea; Lecturer in Analytical Chem., Sung Kyun Kwan Univ. Society: Korean Chem. Soc.
Nuclear interests: Ore analysis by emission, X-ray, absorption spectroscopy-investigation in minerals for atomic energy.
Address: Geological Survey of Korea, Namyoung Dong, Seoul, Korea.

KOHL, Jerome, B.S. (Hon., Appl. Chem.). Born 1918. Educ.: California Inst. Technol. Chem. Eng. (1948-51), Sect. Leader (1951-53), Chief Eng. (1953-58), Manager, Eng. and Development (1948-60), Tracerlab, Inc., Western Div., Richmond, California; In charge of Eng. Group designing and constructing rad. measuring equipment, shielding, radioisotope handling tools. Consulting on numerous field and lab. applications of radioisotopes; Coordinator of Special Products, General Atomic Div., General Dynamics Corp., San Diego, California, 1960-64; Marketing Services Manager, Oak Ridge Tech. Enterprises Corp., 1964-. Instructor, teaching Measurement of Nucl. Rad. and Properties and Applications of Radioisotopes, California Univ. Extension, Eng. and Sci., Berkeley, 1954-59; Instructor, on Radioactive Isotopes in Eng. (1962-63), and on Measurement and Uses of Rad. and Radioisotopes (1964), California Univ. Extension Div., San Diego. Books: Sen. author, Radioisotope Applications Eng. (D. Van Nostrand, 1961); chapter, Ind. Uses of Radioisotopes, in Modern Nucl. Tech nol. (Editors, Biehl, Mainhardt and Mills) (McGraw-Hill, 1959); Chapter 4, Rad. Techniques, in Advances in Petroleum Chem. and Refining. vol. II. (Editors, Kobe and McKetta) (New York, Interscience Publishers, Inc., 1959). Societies: A.I.Ch.E.; A.N.S.
Nuclear interests: Radiation monitoring instrumentation; Industrial applications of radioisotopes; Activation analysis.

Address: 113 Taylor Road, Oak Ridge, Tennessee, U.S.A.

KOHL, Richard J. Managing Director, Tech. Measurement Corp. G.m.b.H.; Managing Director, Tech. Measurement Corp. (Italia) S.R.L.; Managing Director, Tech. Measurement Corp. (U.K.) Ltd.
Address: Technical Measurement Corp. G.m.b.H., 51 Mainzer Landstrasse, Postfach 9090, Frankfurt-am-Main, Germany.

KÖHLER, Hilding Sigurd, Fil. Dr. Born 1928 Educ.: Uppsala Univ. With C.E.R.N., Geneva, 1957-59; Cornell Univ., 1959-60; U.C.L.A., 1960-61; Statens råd für atomforskning, Stockholm, 1961-63; California Univ., San Diego, 1963-65; Rice Univ., Houston, 1965-.
Nuclear interest: Nuclear many body problems; Scattering theory.
Address: Rice University, Physics Department, Houston, Texas 77001, U.S.A.

KOHLER, Max, Dr. phil. Born 1911. Educ.: Berlin Univ. Prof. für Theoretische Physik, Braunschweig T.H., 1949-. Society: Korrespondierendes Mitglied, Akademie der Wissenschaften und der Literatur, Mainz.
Nuclear interest: Feldtheoretische Methoden der Kernphysik.
Address: Technische Hochschule, Braunschweig, Germany.

KOHMAN, Lucja, M.Sc. Born 1923. Educ.: Warsaw Univ. At Inst. of Gen. Chem., Warsaw, -1955; At Inst. of Nucl. Res., 1955-.
Nuclear interest: Electrochemistry.
Address: Institute of Nuclear Research, 16 Dorodna, Warsaw 9, Poland.

KOHMAN, Truman Paul, A.B., Ph.D. Born 1916. Educ.: Harvard and Wisconsin Univs. Asst. Prof. (1948-52), Assoc. Prof. (1952-57), Prof. (1957-), Carnegie Inst. Technol. (now Carnegie - Mellon Univ.). Societies: A.A.A.S.; A.C.S.; A.P.S.; F.A.S.; Geochem. Soc.; A.G.U.; Meteoritical Soc.
Nuclear interests: Nuclear chemistry; Radioactivity measurement; Natural radioactivity; Nucleosynthesis; Nuclear processes in meteorites and the earth.
Address: Chemistry Department, Carnegie-Mellon University, Schenley Park, Pittsburgh, Pennsylvania 15213, U.S.A.

KOHN, André, Ing., Graduate of Ecole Nationale Superieure d'Electrométallurgie de Grenoble. Born 1922. Educ.: Grenoble Univ. Head, Use of Radioelements Dept., French Iron and Steel Res. Inst. (I.R.S.I.D.), St-Germain-en-Laye. Books: The Use of the Autoradiographic Method with Metals, The Use of Radioisotopes in the Iron and Steel Industry, in The Industrial Uses of Radioelements (A.N.R.T., 1961). Society: Sté. Française de Métallurgie.
Nuclear interests: All applications of radioisotopes in the iron and steel industry (researches in the laboratory or the plant, scale, gauges, control units etc.).
Address: 37 rue des Ursulines, Saint-Germain-en-Laye, Seine-et-Oise, France.

KOHN, Anthony, B.S., M.D. Born 1906. Educ.: C.C.N.Y., Tufts Univ. and O.R.I.N.S. Consulting Endocrinologist and Chief of Radioisotope Service, Central Islip State Hospital; Consultant, Internal Medicine (Endocrinology), Pilgrim State Hospital, W. Brentwood, N.Y.; Endocrinologist, Southside Hospital, Bay Shore, N.Y.; Medical Staff, Good Samaritan, Fordham, Lakeside Hospitals,etc.; Consultant in Atomic Medicine, Huntington Hospital. Sec., Southside Clinical Soc.; Chairman, Graduate Education and Atomic Medicine Coms., Suffolk County Medical Soc. Societies: A.N.S.; Soc. Nucl. Medicine; New York Acad. Sci.; A.A.A.S.; Professional Tech. Group Bio Medical Electronics; I.E.E.E.; Corresponding member, Biol. Eng. Soc. (England).
Nuclear interest: Radioisotopes in diagnosis and therapy, especially in thyroid disorders.
Address: 375 East Main Street, Bay Shore, New York, U.S.A.

KOHN, Harold William, B.S., Ph.D. Born 1920. Educ.: Michigan and Syracuse Univs. At Syracuse Univ., 1948-53; Chem., O.R.N.L., 1954-; Visiting Lecturer, California Univ., 1963-64. Societies: A.A.A.S.; A.C.S.
Nuclear interests: Effects of radiation on surfaces, especially catalysts; Radiation induced reactions between gases and solids; Effects of radiation on solids, minerals and molten salt colloids.
Address: 111 Carnegie Drive, Oak Ridge, Tennessee, U.S.A.

KOHN, Henry Irving, A.B. (Dartmouth), Ph.D. (Harvard), M.D. (Harvard). Born 1909. Educ.: Dartmouth and Harvard Univs. Fuller - American Cancer Soc. Prof. Radiol., Harvard Medical School, 1963-; Director, Shields Warren Rad. Lab., New England Deaconess Hospital, 1964-. Societies: Radiol. Soc. of North America; Rad. Res. Soc.; American Physiological Soc.
Nuclear interests: Radiobiology; Therapy.
Address: 50 Binney Street, Boston, Massachusetts 02115, U.S.A.

KOHN, Stéphane, Ing. diplômé E.S.E. Educ.: E.S.E., Paris. Directeur Technique, Jeumont-Schneider Sté. Conférencier, E.S.E.; Ancien Pres.,Section Gros Matériel Electrique, Sté. Française des Electriciens.
Nuclear interests: En relation avec la Sté. Framatome et Westinghouse, s'occupe de l' équipement des Centrales Atomiques.
Address: 5 Place de Rio-de-Janeiro, Paris 8, France.

KÖHNENKAMP, Johann D. Tech. Director, H.C. Stulcken Sohn.
Address: H. C. Stulcken Sohn, Steinwerder, Hamburg 11, Germany.

KOHONEN, Teuvo Kalevi, Ph.D. Born 1934. Educ.: Helsinki Tech. Univ. Prof., Helsinki Tech. Univ. Societies: Finnish Acad. Tech. Sci.; Phys. Club, Suomen Teknillinen Seura; Finnich Phys. Soc.
Nuclear interest: Fast electronics in nuclear physics.
Address: Helsinki Technical University, Otaniemi, Finland.

KOI, Pham Trong. Eng. Dept., of Radiobiol., Dalat Nucl. Res. Centre, Atomic Energy Office of the Republic of Vietnam.
Address: Atomic Energy Office of the Republic of Vietnam, Dalat Nuclear Research Centre, Dalat, Vietnam.

KOICKI, S. Member, Editorial Board, Bulletin of the Boris Kidric Inst. of Nucl. Sci.
Address: Bulletin of the Boris Kidric Institute of Nuclear Sciences, P.O. Box 522, Belgrade, Yugoslavia.

KOIDA, Akihiko, Director, Sumitomo Atomic Energy Industries Ltd.
Address: Sumitomo Atomic Energy Industries Ltd., Sumitomo Building, 22 5-chome, Kitahama, Higashi-ku, Osaka, Japan.

KOIDE, T., Dr. Formerly Managing Director, now Exec. Vice Pres., Yokohama Rubber Co. Ltd.
Address: Yokohama Rubber Co. Ltd., 36-11, 5-chome, Shinbashi, Minatoku, Tokyo, Japan.

KOIFMAN, O. S. Paper: Co-author, Calculation of Yield and Meansquare Deflection of Positrons in Case of Passing of Electrons through Thick Foil (Atomnaya Energiya, vol. 20, No. 5, 1966).
Address: Academy of Sciences of the U.S.S.R., 14 Leninsky Prospekt, Moscow V-71, U.S.S.R.

KOIZUMI, Buyo. Chief, Elec. Eng Dept., Federation of Elec. Power Cos., Japan.
Address: Federation of Electrical Power Companies, 1-3 Yuraku-cho, Chiyoda-ku, Tokyo, Japan.

KOJIMA, Gisei, Dr.Eng. Adviser, Sumitomo Metal Industries Ltd.
Address: Sumitomo Metal Industries Ltd., 15, 5-chome, Kitahama, Higashi-ku, Osaka, Japan.

KOJIMA, Kohei, Dr. of Eng. Born 1904. Educ.: Kyoto Univ. Director of Rad. Lab., 1961. Societies: Atomic Energy; Non-Destructive Testing; Applied Phys.
Nuclear interests: Metallurgy; Nuclear physics.
Address: The Institute of Scientific and Industrial Research, Osaka University, Suita-shi, Osaka, Japan.

KOJIMA, Tadanobu. Chief, Theoretical Phys. Div., Res. Inst. for Atomic Energy, Osaka City Univ.
Address: Research Institute for Atomic Energy, Osaka City University, 12 Minamiogimachi, Kita-ku, Osaka, Japan.

KOJO, Esko, Ph. Lic. Born 1934. Educ.: Turku Univ. Asst. Phys., Turku Univ., 1960-. Society: Phys. Soc. of Finland.
Nuclear interests: Solid state physics.
Address: Wihuri Physical Laboratory, Turku University, 5 Vesilinnantie, Turku, Finland.

KOKENY, Mihaly, Dipl. Eng.; Ing. mecanicien; Séc., Commission nationale de l'energie atomique, Hungary. Deleg. to 2nd I.C.P.U.A.E., Geneva, Sept. 1958.
Address: Hungarian Atomic Energy Committee, 1-3 Kossuth Lajos-ter, Budapest 5, Hungary.

KOKES, Antonín, C.Sc. Born 1931. Educ.: Charles Univ., Prague. Nucl. Res. Inst., Czechoslovak Acad. of Sci.
Nuclear interest: Nuclear spectroscopy.
Address: Ustav jaderného výzkumu, Rez, Prague, Czechoslovakia.

KOKIN, I. Acting Manager, Standing Com. for Peaceful Application of Nucl. Energy, Comecon Council for Mutual Economic Assistance.
Address: Council for Mutual Economic Assistance, Moscow, U.S.S.R.

KOKOREV, L. S. Paper: Co-author, Exptl. Study of Sodium Pool Boiling Heat Transfer (letter to the Editor, Atomnaya Energiya, vol. 22, No. 1, 1967).
Address: Academy of Sciences of the U.S.S.R., 14 Leninsky Prospekt, Moscow V-71, U.S.S.R.

KOKOVIHIN, V. F. Papers: Co-author, Time Dependence of Neutron Yield for (Ra + MsTh) = Be Source (letter to the Editor, Atomnaya Energiya, vol. 20, No. 1, 1966); co-author, Penetration of Neutrons in Air (ibid, vol. 21, No. 4, 1966); co-author, Propagation of Air of Biological Neutron Doses from Point Sources (letter to the Editor, ibid, vol. 23, No. 1, 1967); co-author, Distribution of Neutrons in Air Close to the Earth Surface (ibid, vol. 25, No. 2, 1968).
Address: Academy of Sciences of the U.S.S.R., 14 Leninsky Prospekt, Moscow V-71, U.S.S.R.

KOLAR, Miroslav, M.D. C.Sc. Born 1922. Educ.: Charles Univ., Prague. Prof., Asst. and Res. worker, Biophys. Inst., Charles Univ., Prague, 1955-. Society: Sect. Nucl. Medicine and Rad. Hygiene, Czech. Medical Soc. J.E. Purkyne.
Nuclear interest: Medical use of radioisotopes, especially in the field of nephrology and radiociroculography.
Address: Biophysics Institute, 3 Salmovska, Prague 2, Czechoslovakia.

KOLAR, Oscar Clinton, B.A., Ph.D. Born 1928. Educ.: California (Los Angeles) and California (Berkeley) Univs. Sen. Phys. (1955-67), Asst. Div. Leader (1967-),

Lawrence Rad. Lab. Societies: A.P.S.; A.N.S.; A.A.P.T.
Nuclear interests: Nuclear physics, especially nuclear reactions; Reactor physics, including criticality hazards evaluation; Design of scientific apparatus for educational purposes.
Address: 834 Estates Street, Livermore, California 94550, U.S.A.

KOLARIK, Zdenek, Dr., C.Sc. Born 1933. Educ.: Brno Univ. At the Nucl. Res. Inst., Czechoslovak Acad. of Sci., 1957-. Society: Czechoslovak Chem. Soc., Sections Analytical Chem. and Nucl.Chem.
Nuclear interests: Nuclear chemistry. The study of basic chemistry of solvent extractions; separations by the solvent extraction method.
Address: Nuclear Research Institute, Czechoslovak Academy of Sciences, Rez near Prague, Czechoslovakia.

KOLB, Alan Charles, B.S., M.S., Ph.D. (Phys.). Born 1928. Educ.: Georgia Inst. Technol. and Michigan Univ. Phys., Supt., Plasma Phys. Div., N.R.L. Prof. (Part-time), Dept. of Phys. and Astronomy, Maryland Univ. Society: A.P.S.
Nuclear interests: Controlled thermonuclear research.
Address: Naval Research Laboratory, Washington, D.C. 20390, U.S.A.

KOLB, K. E. Chem. Eng., Nucl. Design B. D. Bohna and Co. Inc.
Address: B.D. Bohna and Co. Inc., 515 Market Street, San Francisco 5, California, U.S.A.

KOLB, Walter, Dipl. Phys., Dr.rer.nat. Born 1926. Educ.: Karlsruhe T.H. Physikal. Tech. Bundesanstalt, 1954-; Chief, Lab. for Rad. Protection, Physikal.Tech. Bundesanstalt, 1958-. Society: Internat. Rad. Protection Assoc.
Nuclear interests: Nuclear physics, especially low level scintillation spectrometry; Transport of radioactive materials.
Address: 19 Wöhlerstrasse, 33 Braunschweig, Germany.

KOLB, William L. Born 1922.
Nuclear interests: Reactor design, test equipment, fabrication procurement and specifications; Test loops and in-pile loop design and field installation and testing.
Address: Argonne National Laboratory, 9700 S. Cass Avenue, Argonne, Ill., U.S.A.

KOLDE, Harry E., B.S. Born 1929. Educ.: Villa Madonna and Cincinnati Univ. Aircraft Nucl. Propulsion Dept., G.E.C., 1953-61; i/c, Airborne Radioactive Pollution, Nucl. Eng. Lab., Nat. Centre for Radiol. Health, U.S.P.H.S., 1961-. Sec., Cincinnati Rad. Soc. Society: H.P.S.
Nuclear interests: Health physics; Environmental radioactivity; Radionuclide standardisation; Whole-body monitoring.
Address: Robert A. Taft Sanitary Engineering Centre, 4676 Columbia Parkway, Cincinnati, Ohio 45226, U.S.A.

KOLEGANOV, Yu. F. Papers: Co-author, Neutron Spectra Measurements up to 3 KeV by Means of Resonance Detectors (letter to the Editor, Atomnaya Energiya, vol. 20, No. 6, 1966); co-author, Measurement of Neutron Spectrum in Medium by Resonance Filters (ibid, vol. 23, No. 2, 1967).
Address: Academy of Sciences of the U.S.S.R., 14 Leninsky Prospekt, Moscow V-71, U.S.S.R.

KOLESAR, Donald C. Formerly Res. Assoc., Nucl. Eng. Group Faculty, Eng. Coll., Washington Univ.
Nuclear interest: Reactor analysis.
Address: Washington University, College of Engineering, Seattle, Washington 98105, U.S.A.

KOLESNICHENKO, Ya. I. Paper: Co-author, On Instability of Inhomogeneous Plasma Caused by Thermonucl. Reaction Products (Atomnaya Energiya, vol. 23, No. 4, 1967).
Address: Academy of Sciences of the U.S.S.R., 14 Leninsky Prospekt, Moscow V-71, U.S.S.R.

KOLESOV, B. E. Papers: Co-author, Radiative Capture of Fast Neutrons for Cu^{63} (Atomnaya Energiya, vol. 21, No. 1, 1966); co-author, Fast Neutron Radiative Capture for Y^{89} (letter to the Editor, ibid, vol. 21, No. 6, 1966).
Address: Academy of Sciences of the U.S.S.R., 14 Leninsky Prospekt, Moscow V-71, U.S.S.R.

KOLESOV, B. M. Papers: Co-author, γ-Ray Spectra of Radioactive Ores in Natural Bedding Measured with Proportional Counters (letter to the Editor, Atomnaya Energiya, vol. 14, No. 5, 1963); co-author, On Effective Atomic Number of Infinite Homogeneous Media with Gamma-Ray Source (letter to the Editor, ibid, vol. 25, No. 1, 1968).
Address: Academy of Sciences of the U.S.S.R., 14 Leninsky Prospekt, Moscow V-71, U.S.S.R.

KOLESOV, I. V. Paper: Co-author, Synthesis and Determination of Radioactive Properties of Some Isotopes of Fermium (Atomnaya Energiya, vol. 21, No. 4, 1966).
Address: Academy of Sciences of the U.S.S.R., 14 Leninsky Prospekt, Moscow V-71, U.S.S.R.

KOLESOV, V. E. Papers: Co-author, Radiative Capture of Cl^{37}, Rb^{87}, Ir^{193} for Fast Neutrons Cl^{37}, Rb^{87}, Ir^{193} (letter to the Editor, Atomnaya Energiya, vol. 23, No. 2, 1967); co-author, Radioactive Capture Cross Section of Fast Neutrons for Ge^{74}, Cs^{138} and Os^{192} (letter to the Editor, ibid, vol. 23, No. 6, 1967); co-author, Isomeric Ratios for Radiative Capture of Fast Neutrons by Se^{80} Isotope (letter to the Editor, ibid, vol. 24, No. 2, 1968); co-author, Radiative Capture of Fast Neutrons by Nuclei Sn^{122}, Sn^{124} and Sb^{121}, Sb^{123} (letter to the Editor, ibid, vol. 24, No. 6, 1968).

Address: Academy of Sciences of the U.S.S.R., 14 Leninsky Prospekt, Moscow V-71, U.S.S.R.

KOLESOV, V. F. Papers: On the Dynamics of a Spherically Symmetric Fast Pulse Reactor (Atomnaya Energiya, vol. 14, No. 3, 1963); Some Problems of Impluse Reactor Dynamics (ibid. vol. 16, No. 4, 1964); Parametric Equations of Fast Pulse Reactor Dynamics (ibid, vol. 20, No. 3, 1966); Impluse on Delayed Neutrons in Fast Reactor (ibid, vol. 22, No. 4, 1967).
Address: Academy of Sciences of the U.S.S.R., 14 Leninsky Prospekt, Moscow V-71, U.S.S.R.

KOLESOV, V. V. Papers: Co-author, Irradiation Stability of Arbus Nucl. Reactor Fuels (letter to the Editor, Atomnaya Energiya, vol. 24, No. 4, 1968); co-author, Reactor SM-2 Plate Fuel Elements Rad. Stability (ibid, vol. 24, No. 5, 1968).
Address: Academy of Sciences of the U.S.S.R., 14 Leninsky Prospekt, Moscow V-71, U.S.S.R.

KOLGA, V. V. Papers: Co-author, A Cyclotron with a Spatially Varying Magnetic Field (Atomnaya Energiya, vol. 8, No. 3, 1960); co-author, Beam Loss at the Limiting Radius in a Phasotron (letter to the Editor, ibid, vol. 9, No. 4, 1960); co-author, The Effect of Space Charge on the Frequency of Free Oscillations in an Isochronous Cyclotron (ibid, vol. 15, No. 3, 1963); co-author, Strong Focusing Ring Cyclotron for Multi-Charged Ions (ibid, vol. 24, No. 4, 1968).
Address: Joint Institute of Nuclear Research, Dubna, near Moscow, U.S.S.R.

KOLGANOV, L. I. Paper: Co-author, Experiments on Initial Travel of Safety Rod (letter to the Editor, Atomnaya Energiya, vol. 22, No. 1, 1967).
Address: Academy of Sciences of the U.S.S.R., 14 Leninsky Prospekt, Moscow V-71, U.S.S.R.

KOLLAR, Jozef, graduate (Phys). Born 1941. Educ.: Komenský Univ., Bratislava. Res. Inst. of Ind. Hygiene and Occupational Medicine, Bratislava, 1962-.
Nuclear interests: Health physics; Spectroscopy; Dosimetry.
Address: 4 Druzicová, Bratislava, Czechoslovakia.

KOLLBRUNNER, Curt F., Dr. sc.techn., Dr. h.c., Senator h.c., grad. Civil Eng. (E.T.H., S.I.A.). Born 1907. Educ.: Federal Inst. Technol., Zurich. Societies: Internat. Assoc. for Bridge and Structural Eng.; Swiss Soc. of Eng. and Architects (S.I.A.); Soc. for Exptl. Stress Analyses.
Nuclear interest: Design and construction of all steel structures for nuclear subjects.
Address: 50 Witellikerstr., CH - 8702 Zollikon, Switzerland.

KOLLENBERGER, Dipl.-Ing. Manager, Nucl. Dept., Deutsche Werft A.G.

Address: Deutsche Werft A.G., Postfach 889, 2 Hamburg 1, Germany.

KOLODNEY, Morris, B.S., M.S., Ch.E., Ph.D. Born 1911. Educ.: C.C.N.Y., Columbia Univ. Prof. Chem. Eng., C.C.N.Y.; Consultant,U.N. Corp. White Plains, New York; Consultant, G.E.C. Aircraft Nucl. Propulsion Project, 1951-58. Societies: A.S.M.; A.I.M.M.E.; A.I.Ch.E. American Electrochem. Soc.
Nuclear interests: Metallurgy and materials of construction of nuclear reactors.
Address: City College, New York 31, New York, U.S.A.

KOLOKOLTSOV, N. A. Deputy Chief Editor, Atomnaya Energiya.
Address: Atomnaya Energiya, 18 Kirov Ulitsa, Moscow, U.S.S.R.

KOLOMENSKII, Andrey Aleksandrovich, Physico-Mathematical Sci. Member, Lebedev Phys. Inst., Acad. of Sci. of the U.S.S.R. Paper: Co-author, New Accelerator-Symmetrical Ring Phasotron Lebedev Physical Inst. AN U.S.S.R. Starting Up (letter to the Editor, Atomnaya Energiya, vol. 20, No. 6, 1966). Nuclear interest: Particle accelerators.
Address: P.N. Lebedev Institute of Physics, 53 Leninsky Prospekt, Moscow, U.S.S.R.

KOLOMIITSEV, M. A. Paper: Co-author, On Activation Detector of Thermal Neutrons (letter to the Editor, Atomnaya Energiya, vol. 20, No. 3, 1966).
Address: Academy of Sciences of the U.S.S.R., 14 Leninsky Prospekt, Moscow V-71, U.S.S.R.

KOLOS, Richard, Head, Hungarian Deleg. to Standing Com. for Peaceful Application of Nucl.Energy, Comecon Council for Mutual Economic Assistance. Deleg., Third I.C.P.-U.A.E., Geneva, Sept. 1964.
Address: Council for Mutual Economic Assistance, Moscow, U.S.S.R.

KOLOS, Wlodzimierz, Prof., Dr. Born 1928. Educ.: Poznan and Warsaw Univs. At Inst. Theoretical Phys., Warsaw Univ., 1951-54; Inst. Phys. Chem., Polish Acad. Sci., Warsaw, 1954-60; Phys. Dept., Chicago Univ., 1958-59; Inst. Nucl. Res., Polish Acad. Sci., Warsaw, 1959-65; Visiting Assoc. Prof. Phys., Chicago Univ. 1965-67; Now at Theoretical Chem. Dept., Warsaw Univ. Societies: Polish Chem. Soc.; Polish Phys. Soc.
Nuclear interests: Theory of radiation chemistry and of chemical effects of nuclear processes; Mesic atoms and molecules.
Address: Theoretical Chemistry Department, Warsaw University, 1 Pasteura, Warsaw 22, Poland.

KOLOSKOV, A. S. Paper: Co-author, Doubled Mass-Spectrometer with Nonuniform Magnetic Field (letter to the Editor, Atomnaya Energiya, vol. 22, No. 6, 1967).

Address: Academy of Sciences of the U.S.S.R., 14 Leninsky Prospekt, Moscow V-71, U.S.S.R.

KOLOSOV, B. I. Paper: Co-author, Application of Eigenfunction Expansions Methods to Multidimensional Reactor Computations (Atomnaya Energiya, vol. 23, No. 2, 1967).
Address: Academy of Sciences of the U.S.S.R., 14 Leninsky Prospekt, Moscow V-71, U.S.S.R.

KOLOTKOV, A. V. Paper: Co-author, Neutron Detector with Alternating Thickness of Moderator (Atomnaya Energiya, vol. 22, No. 2, 1967).
Address: Academy of Sciences of the U.S.S.R., 14 Leninsky Prospekt, Moscow V-71, U.S.S.R.

KOLOTYI, V. V. Papers: Co-author, Spectrum of the Slow Neutrons from a Horizontal Channel in the VVR-M Reactor (letter to the Editor, Atomnaya Energiya, vol. 12, No. 4, 1962); co-author, Total Neutron Cross-Sections of Re^{185} and Re^{187} (ibid, vol. 19, No. 3, 1965).
Address: Academy of Sciences of the U.S.S.R., 14 Leninsky Prospekt, Moscow V-71, U.S.S.R.

KOLSTAD, George A., Ph.D., B.S. (magna cum laude). Born 1919. Educ.: Yale Univ. Instructor at Yale; Visiting Staff Sci., Univ. Inst. f. Teor. Fysik, Copenhagen; Asst. Director of Res., Phys. and Mathematics, U.S.A.E.C., Washington 25, D.C. Society: Fellow, A.P.S. Nuclear interests: Nuclear physics research and management.
Address: Oak Hill, Laytonsville, Maryland 20760, U.S.A.

KOLSTAD, Per, Dr.med. Born 1925. Educ.: Oslo Univ. Resident in medicine, surgery, anaesthesia, radiology, obstetrics and gynaecology, Drammen Hospital, Vestfold Central Hospital, Rikshospitalet, Norwegian Radium Hospital 1950-60; Res. Fellow, Norwegian Cancer Soc., 1961-63; Res. Fellow, Norwegian and Swedish Cancer Soc., (working at Karolinska Sjukhuset and Radiumhemmet, Stockholm), 1963-64; Asst. Chief (1964-67), Chief (1967-), Gynaecological Dept., Norwegian Radium Hospital. Book: Vascularisation, Oxygen Tension and Radiocurability in Cancer of the Cervix (Oslo Univ. Press, 1964).
Nuclear interests: Radiotherapy of gynaecological cancer; Oxygen effect in clinical radiobiology.
Address: Norwegian Radium Hospital, Oslo 3, Norway.

KOLTZENBURG, G., Dr. Sen. Officer, Div. of Irradiation Chem., Max-Planck Inst. for Coal Res.
Address: Max-Planck Institute for Coal Research, 34-36 Stiftstrasse, Mülheim-Ruhr, Germany.

KOLYADA, V. M. Papers: Co-author, Calorimetric Dosimetry at Nucl. Reactor (Atomnaya Energiya, vol. 21, No. 4, 1966); co-author, Quasi- Stationary Calorimetric Method of Dosimetry of Powerful Fluxes of Ionising Irradiations (letter to the Editor, ibid, vol. 21, No. 6, 1966); co-author, Exptl. Study of Irregularity of Flux Thermal Neutrons at Surface of Absorbed Specimen with Calorimeter Device (letter to the Editor, ibid, vol. 22, No. 4, 1967); co-author, Some Features of Rad. Damage of Titanium Diboride (ibid, vol. 23, No. 4, 1967).
Address: Academy of Sciences of the U.S.S.R., 14 Leninsky Prospekt, Moscow V-71, U.S.S.R.

KOLYADIN, A. B. Paper, Co-author, Electrophoretic Filter for Purifying of Reactor Water (Atomnaya Energiya, vol. 19, No. 4, 1965).
Address: Academy of Sciences of the U.S.S.R. 14 Leninsky Prospekt, Moscow V-71, U.S.S.R.

KOLYCHEV, Boris S., B.Sc., Eng., Head of Sect., U.S.S.R. State Com. for the Utilisation of Atomic Energy; Deleg. to 2nd I.C.P.U.A.E., Geneva, Sept. 1958. Paper: Sorption and Extraction Processes in Uranium Hydrometallurgy (Atomnaya Energiya, vol. 6, No. 5, 1959).
Address: U.S.S.R. State Committee for the Utilisation of Atomic Energy, 26 Staromonetnü Pereulok, Moscow, U.S.S.R.

KOLYZHENKOVA, V. V. Paper: Co-author, Angular Distributions of Fast Neutrons behind Iron Barriers (Atomnaya Energiya, vol. 20, No. 6, 1966).
Address: Academy of Sciences of the U.S.S.R., 14 Leninsky Prospekt, Moscow V-71, U.S.S.R.

KOLZ, Helmuth, Dipl. Ing. Member, Sen. Staff engaged in Nucl. Work, Abteilung Strahlungsmesstechnik, Reichert-Elektronik G.m.b.H. und Co. K.G.
Address: Reichert-Elektronik G.m.b.H. und Co. K.G., Postfach 743, Trier/Petrisberg, Germany.

KOMAGATA, Sakuji, Dr. of Eng. Born 1904. Educ.: Tokyo Univ. Director, Electro-tech. Lab., 1945-52; Pres., Ind. Technol. Agency, 1952-56; Deleg. of Japan, Geneva Atoms for Peace Conference, 1955; Vice-Chairman and Director, Tokai Lab., Japan Atomic Energy Res. Inst., 1956-57 Commissioner, A.E.C. of Japan, -1966. Member, Sci. Council of Japan; Member, Japan Ind. Standards Inquiry Com.; Ministry of Internat. Trade and Ind. Book: Outline of Surface Electro-chem. (1948).
Address: 1027 Wadahon-machi, Suginami-Ku, Tokyo, Japan.

KOMAI, Kenichiro, B.Eng. Born 1900. Educ.: Tokyo Univ. Pres. Hitachi, Ltd. Member, Atomic Energy Council, Sci. and Technol. Agency; Exec. Director, Federation of Economic Organisations; Director, Japan Sci. Federation; Councillor, Atomic Energy Soc. of Japan; Managing Director, Japan Atomic Ind. Forum Inc., Councillor, Fund for Peaceful Atomic Development of Japan; Vice Pres.,

Japan Elec. Assoc.; Pres., Japan Electronic Ind. Development Assoc.; Director, Tokyo Atomic Ind. Consortium.
Address: Hitachi, Ltd., 1-4 Marunouchi, Chiyodaku, Tokyo, Japan.

KOMAI, Tomoyoshi, Ph.D. Born 1920. Chief, Lab. of Rad. Res., N.I.H., Japan. Nuclear interest: To product biological active substances labelled with radioisotope.
Address: 2-10-35 Kamiosaki, Shinagawaku, Tokyo, Japan.

KOMAR, Anton Panteleimonovich. Born 1904. Educ.: Polytechnic Inst., Kiev. Inst. of Tech. Phys., Acad. of Sci. of the U.S.S.R., 1950. Paper: Co-author, Use of Pulse Ionisation Chamber as Alpha Spectrometer (Izv. Akad. Nauk SSSR, Ser. Fiz., 1956, 20, 1455). Society: Member, Acad. of Sci. of the Ukrainian S.S.R. Nuclear interests: Accelerators; Photonuclear reactions.
Address: Academy of Sciences of the U.S.S.R., 14 Leninsky Prospekt, Moscow V-71, U.S.S.R.

KOMAR, E. G., D.Sc., Acad. of Sci. of the U.S.S.R.; Deleg. to 2nd I.C.P.U.A.E., Geneva, Sept. 1958. Director, Yefremov Sci.-Tech. Inst. for Electrophys. Apparatus. Papers: A Cyclotron with a Radially Travelling Magnetic Wave (Atomnaya Energiya, vol. 7, No. 1, 1959); co-author, Magnetic Characteristics of the 10-GeV Proton Synchrotron of the Joint Inst. for Nucl. Res. (Supplement No. 4 of Atomnaya Energiya, Atomic Press, Moscow, 1957); co-author, Designing the 7 BeV Proton Synchrotron (Atomnaya Energiya, vol. 12, No. 6, 1962).
Address: Yefremov Scientific Technical Institieta for Electrophysical Apparatus, P.O. Box 42, Leningrad, U.S.S.R.

KOMAREK, Arnost, Ing., C.Sc. Born 1926. Educ.: Prague Tech. Univ. and Moskovskij Energeticeskij Inst., Moscow. Chief Eng., Nucl. Power Div., Skoda Concern, Pilsen. Society: Czechoslovak Sci. Tech. Soc. Nuclear interests: Reactor design; Nuclear power economics; Heat transfer.
Address: 47 Cástkova, Pilsen, Czechoslovakia.

KOMARENKO, Yury Grigoryevich. Tech. Advisor, Council of Ministers, Ukrainian S.S.R.; Ukrainian S.S.R. alternate deleg., I.A.E.A. Gen. Conference, Vienna, Sept. 1960.
Address: Council of Ministers, Kiev, Ukrainian S.S.R.

KOMAROV, V. E. Papers: Co-author, Interaction of Uranium (IV) with Chloride-Fluoride Melt NaCl - KCl - NaF (Atomnaya Energiya, vol. 21, No. 6, 1966); co-author, Decomposition of Uranylchloride and its Interaction with Deoxide of Uranium in Melt NaCl - KCl (ibid, vol. 22, No. 1, 1967); co-author, Fluoride Ion Effect on Uranium Dioxide Deposition by Molten Salt Electrolyze (letter to the Editor, ibid, vol. 24, No. 4, 1968).
Address: Academy of Sciences of the U.S.S.R., 14 Leninsky Prospekt, Moscow V-71, U.S.S.R.

KOMAROV, V. N. Paper: Co-author, Calculation Method for Azeotropic Steam Fractionation, Applied to Tributylphosphate-Carbon Tetrachloride System (Atomnaya Energiya, vol. 20, No. 5, 1966).
Address: Academy of Sciences of the U.S.S.R., 14 Leninsky Prospekt, Moscow V-71, U.S.S.R.

KOMES, J. W. Vice-Pres. and Director, Bechtel Nucl. Corp.
Address: Bechtel Nuclear Corporation, 220 Bush Street, San Francisco 4, California, U.S.A.

KOMETIANI, Peter Antonovich, Prof., Dr. Biol. Sci. Born 1901. Educ.: State Univ. Tbilisi, Georgian S.S.R. Head, Biochem. Dept., Inst. of Physiology, Acad. of Sci., Georgian S.S.R. Papers: Co-author, 2nd I.C.P.U.A.E., A/Conf. 15/P.2318, 7/VII, 1958; Problems of Biochemistry of the Nervous System (Kiev, 1957). Societies: Biochemical Soc. of U.S.S.R.; Physiological Soc. of U.S.S.R. Nuclear interest: Radiobiochemistry.
Address: Institute of Physiology of the Academy of Sciences of the Georgian S.S.R., 62 Voenno Gruz. dor, Tbilisi, Georgian S.S.R.

KOMEYI, S. Manager, Atomic Power Div., Ataka and Co. Ltd.
Address: Atomic Power Division, Ataka and Co. Ltd., 14, 5-chome, Imabashi, Higashiku, Osaka, Japan.

KOMISSAROV, V. A. Paper: Co-author, Measurement of dT-Neutron Yield with Auxilary Annihilation Rad. Source (letter to the Editor, Atomnaya Energiya, vol. 22, No. 5, 1967).
Address: Academy of Sciences of the U.S.S.R., 14 Leninsky Prospekt, Moscow V-71, U.S.S.R.

KOMISSAROVA, L. N. Papers: Co-author, The Corroding Effect of Zirconium Tetrachloride Vapours on Steel IXI8H9T and Nickel at High Temperatures (Atomnaya Energiya, vol. 13, No. 1, 1962); co-author, Interactions between Zirconium Tetrachloride and Metallic Zirconium (ibid, vol. 24, No. 1, 1968).
Address: Academy of Sciences of the U.S.S.R., 14 Leninsky Prospekt, Moscow V-71, U.S.S.R.

KOMISSARZHEVSKAYA, G. F. Paper: Co-author, On Classification of Gamma-Methods in Nucl. Geophys. (letter to the Editor, Atomnaya Energiya, vol. 23, No. 3, 1967).
Address: Academy of Sciences of the U.S.S.R., 14 Leninsky Prospekt, Moscow V-71, U.S.S.R.

KOMOCHKOV, M. M. Papers: Co-author, Measurement of Dose Field of Mixed Rad. behind Steel Shielding (letter to the Editor, Atomnaya Energiya, vol. 24, No. 2, 1968); co-author, Measurements of the Quality Factor

for High Energy Protons in the Water Phantom (Nukleonika, vol. 13, No. 2, 1968), co-author, Attenuation of High Energy Neutrons Fluxes from Volume Source in Iron (letter to the Editor, ibid, vol. 24, No. 3, 1968); co-author, Back-Yield of Neutrons from out Shielding Bombarded with 660 MeV Protons (letter to the Editor, ibid, vol. 24, No. 4, 1968).
Address: Academy of Sciences of the U.S.S.R., 14 Leninsky Prospekt, Moscow V-71, U.S.S.R.

KOMPAORE, H. E. Michel. Head, Upper Volta Mission to Euratom, 1966-.
Address: c/o Euratom, 51-53 rue Belliard, Brussels, Belgium.

KOMURA, Kohjiro, D.Sc. Born 1926. Educ.: Kumamoto Univ. Chief, 2nd Lab. of Radioactive Mineral Resources Sect., Geological Survey of Japan. Book: Uran-Resources and Minerals (Tokyo, Asakura Book Co., 1960). Societies: Soc. of Mining Geologists of Japan; Assoc. for the Geological Collaboration of Japan; Japanese Assoc. of Mineralogists; Petrologists and Economic Geologists.
Nuclear interest: Mineralogical study of uraniferous minerals and the genesis of sedimentary deposit of uran.
Address: RH 17, Ikejirijutaku 537, Ikejiri-Machi, Setagaya-K Tokyo, Japan.

KOMURKA, Milos, Ing. Born 1928. Educ.: Tech. Univ., Faculty of Chem., Brno and Charles Univ., Faculty of Nucl. Phys. In Ministry of Chem. Industry, Dept. of Nucl. Chem. and Radiochem., 1956-59; Czechoslovak A.E.C, Prague, 1959-. Formerly at Res. Inst. of Air Technics. Member, Editorial Board, Jaderna Energie. Book: Application of Radioisotopes and Rad. in Nat. Economy (Prague, Utein, 1961). Societies: Czechoslovak Sci. Assoc.; Commission for Nucl. Energy.
Nuclear interests: Industrial application of radioisotopes; Radiation chemistry.
Address: Czechoslovak Atomic Energy Commission, 7 Slerská, Prague 2, Czechoslovakia.

KONCHINSKII, G. A. Papers: Co-author, Reactor Phys. Exptl. Establishment of the Phys. Inst. of the Ukranian Acad. of Sci. (Atomnaya Energiya, vol. 20, No. 1, 1966); co-author, Study of Loop's Canal Cells with Highly Enriched Fuel (ibid, vol. 25, No. 1, 1968).
Address: Institute of Physics, Ukranian Academy of Sciences, 42 Ulitsa Dobriy Put', Kiev, Ukranian S.S.R., U.S.S.R.

KONDAIAH, Evani, B.Sc., M.Sc., Fil. Lic., Fil. Dr. (cum laude, Stockholm). Educ.: Benares Hindu, Allahabad and Stockholm Univs. Scholar, Australian Nat. Univ., 1952-54; Reader, Tata Inst. of Fundamental Res., Bombay; Head, Nucl. Phys. Dept., Andhra Univ., 1966-; Sen. Foreign Sci., N.S.F. (U.S.A.), as Visiting Prof., School of Phys., Georgia Inst. Technol., 1967-68. Society: A.P.S.
Nuclear interests: Nuclear reactions and spectroscopy.
Address: Tata Institute of Fundamental Research, Colaba, Bombay 5, India.

KONDIC, Nenad, Dr.Sci. (Mech. Eng.). Born 1926. Educ.: Belgrade Univ. Constructor, Central Shipbuilding Office, Belgrade, 1951; Designer and Supervisor tool machines factory Ivo Lola Ribar, Belgrade-Zeleznik, 1952-54; Sen. Sci., Inst. of Nucl. Sci. Boris Kidric, Belgrade-Vinca, 1954-; Argonne Nat. Lab., 1960-61; Assoc. Prof. Thermodynamics, Univ., Nis, 1967-.
Nuclear interests: Intensification of heat and mass transport in two-phase and two-component systems (separation, superheating); Analysis and measurement of temperature, concentration, and velocity fields in multi-element reactor cores. Theory and development of advanced measuring methods, use of radiation and tracers.
Address: 54-I Cara Urosa, Belgrade, Yugoslavia.

KONDO, Masaharu, D.Sc. Born 1918. Educ.: Osaka Univ. Prof. Chem., Tokyo Metropolitan Univ., 1957-; Res. Assoc. (1959-61), Visiting Prof. (1963-64), Rad. Chem., Notre Dame Univ., U.S.A. Books: Chem. Dosimetry (Tokyo, Kyoritsu Shuppan Co., 1959); High Polymer Eng. (Chem. Action of Rad.) (Tokyo, Chijin Shokan, 1967). Societies: Chem. Soc. of Japan; Faraday Soc.
Nuclear interests: Radiation chemistry and radiochemistry.
Address: Department of Chemistry, Faculty of Science, Tokyo Metropolitan University, 950 Fukasawacho Setagayaku, Tokyo, Japan.

KONDO, Sohei, B.Sc., D.Sc. Born 1922. Educ.: Kyoto Univ. Head, Rad. Lab., Dept. of Induced Mutation, Nat. Inst. of Genetics, Japan, 1956-63; Prof., Fundamental Radiol., Faculty of Medicine, Osaka Univ., Japan, 1963-. Societies: Phys. Soc. of Japan; Japanese Soc. of Genetics; H.P.S. (U.S.A.); Biophys. Soc. of Japan; Rad. Res. Soc.
Nuclear interests: Radiation dosimetry and biophysical aspects of radiation biology and mutagenesis.
Address: Fundamental Radiology, Faculty of Medicine, Osaka University, Osaka, Japan.

KONDO, Sukenobu. Member, Atomic Energy and Isotope Com., Tokushima Univ.
Address: Atomic Energy and Isotope Committee, c/o Medical Faculty, Tokushima University, Kurmoto-cho, Tokushima-shi, Japan.

KONDO, Yoshio, Dr. Eng. Born 1924. Educ.: Kyoto Univ. Prof. Metal., Kyoto Univ., 1961-. Societies: A.I.M.M.E.; Japan Inst. of Metals; Mining and Metal. Inst. of Japan.
Nuclear interest: Reduction of uranium tetrafluoride with alkali earth metals.
Address: 57 Kitashirakawa Shimo-bettocho, Sakyo-ku, Kyoto, Japan.

KONDO, Yutaka, Dr. Eng. Born 1926. Educ.: Tokyo Univ. Chief, High Alloy Steel Sect., Central Res. Labs., Sumitomo Metal Industries Ltd. Society: Atomic Energy Soc. of Japan. Nuclear interests: Nuclear materials, especially zircalloy tubes for nuclear fuel cladding.
Address: Sumitomo Metal Industries Limited, Central Research Laboratories, 1-3 Nishinagasu Hondori, Amagasaki City, Japan.

KONDR, Miroslav, Ing. (Dipl.), Ing. Born 1919. Educ.: Prague High Tech. Coll. and Charles Univ., Prague. Chief, Very High Tension Exptl. Stand, Dept. of Electrotechniques (1952-55), Member, Group for Instrumentation and Counting Techniques, Dept. of Nucl. Power Stations (1958-), Skoda Works. Society: Czechoslovak Sci. Tech. Assoc. Nuclear interest: Automatic measurements in reactor design and counting techniques.
Address: 262 Hlavní, Tlucná, okres Pilzensever, Czechoslovakia.

KONDRASHOV, A. P. Papers: Co-author, The Effect of Layers with Boron Content on the Secondary Gamma-Rays Yield (letter to the Editor, Atomnaya Energiya, vol. 8, No. 1, 1960); co-author, Use of Multigroup Methods to Compute a Biological Shield (ibid, vol. 12, No. 2, 1962); co-author, Some Methods of Decrease of Secondary Penetrating Gamma-Radiation Fluxes (ibid, vol. 19, No. 5, 1965); co-author, Some Methods of Lowering Fluxes of Penetrating Secondary Gamma Rad.(Jaderna Energie, vol. 13, No. 11, 1967).
Address: Academy of Sciences of the U.S.S.R., 14 Leninsky Prospekt, Moscow V-71, U.S.S.R.

KONDRASHOVA, Z. S. Paper: Co-author, Application of Multigroup Methods for the Calculation of Neutron Flux Distributions in Inhomogeneous Shields (Kernenergie, vol. 11, No. 3, 1968).
Address: Academy of Sciences of the U.S.S.R., 14 Leninsky Prospekt, Moscow V-71, U.S.S.R.

KONDRATENKO, A. N. Paper: On Nonlinear Theory of Electronic Cyclotronic Resonance (Atomnaya Energiya, vol. 16, No. 5, 1964).
Address: Academy of Sciences of the U.S.S.R., 14 Leninsky Prospekt, Moscow V-71, U.S.S.R.

KONDRAT'EV, F. V. Papers: Co-author, Investigation of Xenon Operation of Caesium Thermionic Converter with Molybdenium Cathode (Atomnaya Energiya, vol. 23, No. 3, 1967); co-author, Investigation of Xenon on Operation of Caesium Thermionic Converter (ibid, vol. 23, No. 3, 1967); co-author, Exptl. Investigation of Cesium Thermionic Converter with Tungsten Cathode (ibid, vol. 24, No. 6, 1968).
Address: Academy of Sciences of the U.S.S.R., 14 Leninsky Prospekt, Moscow V-71, U.S.S.R.

KONDRAT'EV, L. N. Paper: Co-author, Thermal Neutron Activation Cross-Sections for Cd^{108} (letter to the Editor, Atomnaya Energiya, vol. 16, No. 2, 1964).
Address: Academy of Sciences of the U.S.S.R., 14 Leninsky Prospekt, Moscow V-71, U.S.S.R.

KONDRAT'EV, V. N., Academician. Asst. Director, Inst. of Chem. Phys. Acad. Sci. of the U.S.S.R. Member, Editorial Board, Internat. J. of Appl. Rad. and Isotopes. Book: Struktura atomov i molekul (Structure of Atoms and Molecules) (2nd revised edition, M. Fizmatgiz, 1959).
Address: Institute of Chemical Physics, Academy of Sciences of the U.S.S.R., 2-b Vorobyevskoye Chaussée, Moscow, V-334, U.S.S.R.

KONDRAT'KO, M. Ya. Papers: Co-author, Delayed Neutrons in ^{238}U Gamma-Fission (letter to the Editor, Atomnaya Energiya, vol. 15, No. 2, 1963); co-author, U^{235} Photofission Asymmetry as Function of Peak Bremsstrahlung Energy (letter to the Editor, ibid, vol. 20, No. 6, 1966); co-author, Photofission Yields for U^{235} (letter to the Editor, ibid, vol. 23, No. 6, 1967).
Address: Academy of Sciences of the U.S.S.R., 14 Leninsky Prospekt, Moscow V-71, U.S.S.R.

KONDUROV, I. A. Papers: Co-author, Pile-Neutron Burn-up Cross Section of Pm^{149} and Samarium Poisoning (letter to the Editor, Atomnaya Energiya, vol. 19, No. 2, 1965); co-author, Neutron Capture Cross-Section Measurement by Radioactive Nuclei Co^{58m}, Cu^{64} and Sc^{46} (ibid, vol. 24, No. 6, 1968).
Address: Academy of Sciences of the U.S.S.R., 14 Leninsky Prospekt, Moscow V-71, U.S.S.R.

KONDURUSHKIN, N. A. Papers: Co-author, Penetration of Neutrons in Air (Atomnaya Energiya, vol. 21, No. 4, 1966); co-author, Propagation of Air of Biological Neutron Doses from Point Sources (letter to the Editor, ibid, vol. 23, No. 1, 1967); co-author, Distribution of Neutrons in Air Close to the Earth Surface (ibid, vol. 25, No. 2, 1968).
Address: Academy of Sciences of the U.S.S.R., 14 Leninsky Prospekt, Moscow V-71, U.S.S.R.

KONEV, V. N. Paper: Co-author, Phase Diagrams of Plutonium with IIIA, IVA, VIII and IB Group-Metals (Atomnaya Energiya, vol. 23, No. 6, 1967).
Address: Academy of Sciences of the U.S.S.R., 14 Leninsky Prospekt, Moscow V-71, U.S.S.R.

KONGSHEM, Torstein Halvor. Born 1928. Director, A/S Reactor, 1959-. Nuclear interest: Ship reactors.
Address: A/S Reactor, 4 Tollbugata, Oslo, Norway.

KÖNIG, Lothar Alfons, Dr. rer. nat., Dipl.-Physiker. Born 1928. Educ.: Johannes Gutenberg Univ., Mainz. At Max-Planck-Inst. für Chemie (Otto Hahn Inst.), Mainz, 1953-62; Gesellschaft für Kernforschung m.b.H., Karlsruhe, 1962-. Books: Co-author, 1960 Nucl. Data Tables, Parts 1 and 2 (Washington

D.C., U.S. Govt. Printing Office); contributions to: Nucl. Masses and their Determination (ed. H. Hintenberger, Pergamon Press, 1957), Advances in Mass Spectrometry (ed. J. D. Waldron, Pergamon Press, 1958), Proc. of the Internat. Conference on Nuclidic Masses (ed. H. E. Duckworth, Toronto Univ. Press, 1960). Societies: Physikalische Gesellschaft; H.P.S.; Fachverband für Strahlenschutz; Gesellschaft Deutscher Naturforscher und Ärzte.
Nuclear interests: Health physics; Reactors; Reactor safety problems; Nuclidic masses; Ion optics.
Address: Gesellschaft für Kernforschung m.b.H., Kernforschungszentrum Karlsruhe, Abt. Strahlenschutz und Dekontamination/Strahlenschutzdienst, Germany.

KÖNIG, MEYER-. See MEYER-KÖNIG.

KONIJN, Joseph, Dr. Born 1931. Educ.: Delft Tech. Univ. Societies: A.P.S.; Koninklijk Inst. van Ingenieurs; Nederlandse Natuurkundige Vereniging.
Nuclear interest: Nuclear physics.
Address: Institute for Nuclear Physics Research, 18 Oosterringdijk, Amsterdam O, Netherlands.

KON'KOV, N. G. Paper: Co-author, Putting into Operation of B-3M Synchrotron as Injector for Electron-Positron Storage Ring (Atomnaya Energiya, vol. 20, No. 3, 1966).
Address: Academy of Sciences of the U.S.S.R., 14 Leninsky Prospekt, Moscow V-71, U.S.S.R.

KONN, W. H. Member, Rad. Com., American Ind. Hygiene Assoc.
Address: American Industrial Hygiene Association, 14125 Prevost, Detroit 27, Michigan, U.S.A.

KONNEKER, A. L. Treas., Nucl. Consultants Corp.
Address: Nuclear Consultants Corporation, Box 6172 Lamber Field, St. Louis, Missouri 63145, U.S.A.

KONNEKER, Wilfred R., B.S. (Ohio), M.S. (Ohio), Ph.D. (Washington Univ., St. Louis, Missouri). Born 1922. Educ.: Ohio and Washington (St. Louis) Univs. Vice Pres., Nucl. Consultants, Inc., 1950-55; Vice Pres., Nucl. Corp. of America, 1955-58; Pres., Nucl. Consultants Corp., 1958-66; Vice Pres. and Gen. Manager, Nucl. Div., Mallinckrodt Chem. Works, 1966. Member, Radioisotope Advisory Com., U.S.A.E.C. Societies: A.P.S.; Soc. of Nucl. Medicine; American Medical Phys. Soc.; Atomic Ind. Forum; H.P.S.; A.N.S.
Nuclear interests: The practical applications of nucleonics both to the medical diagnosis and therapy as well as industrial applications.
Address: Box 10172, Lamber Field, St. Louis, Missouri 63145, U.S.A.

KONO, Fumihiko. Formerly Pres., Mitsubishi Nippon Heavy-Industries Ltd.; Pres., Mitsubishi Heavy Industries, Ltd.
Address: Mitsubishi Heavy Industries Ltd., 10, 2-chome, Marunouchi, Chiyoda-ku, Tokyo, Japan.

KONO, Sachu, Prof. Medicine Dept. Member, Radioisotope Res. Com., Niigata Univ. School of Medicine.
Address: Department of Medicine, Niigata University School of Medicine, 757, 1-Bancho Asahimachi-Street, Niigata, Japan.

KONO, Y. Member, Editorial Board, Atompraxis.
Address: c/o Atompraxis, Verlag G. Braun, 14-18 Karl Friedrich Strasse, Karlsruhe, Germany.

KONOBEEVSKY, Sergei Tikhonovich. Born 1890. Educ.: Moscow Univ. At the Acad. of Sci. of the U.S.S.R. Member, Editorial Advisory Board, J. of Nucl. Materials. Book: Effect of Rad. on Structure and Properties of Fissile Materials (in book: Investigation in the fields of Geology, Chem. and Metal., Moscow, 1955). Society: Corr. Member, Acad. of Sci. of the U.S.S.R.
Nuclear interest: Effect of radiations on materials.
Address: Nuclear Energy Institute, Academy of Sciences of the U.S.S.R., 14 Leninsky Prospekt, Moscow V-71, U.S.S.R.

KONOCHKIN, V. G. Papers: Co-author, Experience with the Operation of the First Nucl. Power Station (Atomnaya Energiya, vol. 11, No. 1, 1961); co-author, Operational Experience of the First Nucl.Power Station (ibid, vol. 16, No. 6, 1964).
Address: Academy of Sciences of the U.S.S.R., 14 Leninsky Prospekt, Moscow V-71, U.S.S.R.

KONONENKO, V. I. Paper: Co-author, Microwave Rad. of Electrodeless Induction Discharge (Atomnaya Energiya, vol. 16, No. 2, 1964).
Address: Academy of Sciences of the U.S.S.R., 14 Leninsky Prospekt, Moscow V-71, U.S.S.R.

KONONOV, V. N. Papers: Co-author, Measurement of the Radiative-capture Cross-Section of ^{127}I for Fast Neutrons (letter to the Editor, Atomnaya Energiya, vol. 10, No. 2, 1961); co-author, Fast Neutrons Radioactive Capture Cross-Sections of Re and Ta (letter to the Editor, ibid, vol. 19, No. 5, 1965).
Address: Academy of Sciences of the U.S.S.R., 14 Leninsky Prospekt, Moscow V-71, U.S.S.R.

KONONOVICH, A. A. Paper: Co-author, On Calculation of Direct Nucl. Source (letter to the Editor, Atomnaya Energiya, vol. 16, No. 4, 1964).
Address: Academy of Sciences of the U.S.S.R., 14 Leninsky Prospekt, Moscow V-71, U.S.S.R.

KONOPINSKI, Emil Jan, B.A., M.A., Ph.D. Born 1911. Educ.: Michigan Univ. Consultant, Los Alamos Sci., Lab., 1946-; Visiting Board, Argonne Nat. Lab., 1959-61; Prof., Phys. Dept., Indiana Univ. Book: Chapter, Theory of β-Radioactivity in β- and γ-Ray Spectroscopy (editor: Siegbahn; Oxford, Clarendon Press).Society: A.P.S.
Nuclear interests: Nuclear theory and radioactivity.
Address: Physics Department, Indiana University, College of Arts and Sciences, Bloomington, Indiana, U.S.A.

KONOPLEV, K. A. Papers: Co-author, Ion Exchange Unit for Production of Reactor Feed Water (letter to the Editor, Atomnaya Energiya, vol. 16, No. 5, 1964); co-author, Electrophoretic Filter for Purifying of Reactor Water (ibid, vol. 19, No. 4, 1965); co-author, Reduction of Radioactive Disposal into Atmosphere and Res. of Water Deaeration of WWR-M Reactor (ibid, No. 6); co-author, The Investigation of Radiolytic Products During Reactor WWR-M Operation with Outcluding Degasation System (ibid, vol. 24, No. 4, 1968).
Address: Academy of Sciences of the U.S.S.R., 14 Leninsky Prospekt, Moscow V-71, U.S.S.R.

KONOPLEVA, R. F. Papers: Co-author, Semiconductor Devices for Recording the Relative Distribution of the Fast-Neutron Flux in the VVR-M Reactor (letter to the Editor, Atomnaya Energiya, vol. 11, No. 6, 1961); co-author, Low Temperature Canal of WWR-M Reactor of Phys. Tech. Inst. AS U.S.S.R. (letter to the Editor, ibid, vol. 20, No. 3, 1966).
Address: Physical-Technical Institute, Academy of Sciences of the U.S.S.R., 2 Polytechnicheskaya Ulitsa, Leningrad, U.S.S.R.

KONOVALOV, E. A. Papers: Co-author, Application of Polyethylene Pipes in Dosimetric Lines for Air Control (letter to the Editor, Atomnaya Energiya, vol. 19, No. 2, 1965); co-author, Tangential Channels and Reconstruction of Thermal Column of VVR-M Reactor (letter to the Editor, ibid, No. 5); co-author, Improvement of System of Stationary Dosimetry Control of VVR-M Reactor (ibid, vol. 21, No. 5, 1966).
Address: Academy of Sciences of the U.S.S.R., 14 Leninsky Prospekt, Moscow V-71, U.S.S.R.

KONOVALOV, E. E. Papers: Co-author, Determination of Oxygen in Cesium Metal by Vacuum Distillation Method (Atomnaya Energiya, vol. 24, No. 2, 1968); co-author, Quantitative Determination of Hydrogen Impurity in Liquid Sodium by Method of Hydride Thermal Dissociation (ibid, vol. 24, No. 3, 1968); co-author, Solubility of Oxygen in Alloy of Potassium-Sodium (letter to the Editor, ibid, vol. 24, No. 5, 1968).
Address: Academy of Sciences of the U.S.S.R., 14 Leninsky Prospekt, Moscow V-71, U.S.S.R.

KONRAD, Maksimilijan, Dr. Head, Electronics Dept., Rudjer Boskovic Nucl Inst. Deleg., I.A.E.A. Conference on Nucl. Electronics, Belgrade, 15-20 May 1961.
Address: Rudjer Boskovic Nuclear Institute, 54 Bijenicka Cesta, P. O. Box 171, Zagreb, Yugoslavia.

KONRATIEV, V. N. Member, Editorial Board, Internat. J. of Appl. Rad. and Isotopes (U.S.S.R.).
Address: Academy of Sciences of the U.S.S.R., 14 Leninsky Prospekt, Moscow V-71, U.S.S.R.

KON'SHIN, V. A. Papers: Co-author, Yield and Angular Distribution of Secondary Nucleons from Flat Shieldings Bombarded by 660 MeV Protons (Atomnaya Energiya, vol. 20, No. 2, 1966); co-author, Yield and Dose of Secondary Nucleons from Flat Ti Shield Layers Bombarded by Protons with 660 MeV Energy (letter to the Editor, ibid, vol. 24, No. 1, 1968).
Address: Academy of Sciences of the U.S.S.R., 14 Leninsky Prospekt, Moscow V-71, U.S.S.R.

KONSTANTINIDES, K., Dir. Isotope Lab., Dept. of Clinical Therapeutics, Athens Univ.
Address: Alexandra Hospital, Isotope Laboratory, Vas. Sophias-K. Lourou Str., Athens, Greece.

KONSTANTINOV, A. A. Papers: Co-author, Absolute Measurement of Neutron Source Yield by Manganese Activation Method (letter to the Editor, Atomnaya Energiya, vol. 16, No. 3, 1964); co-author Internat. Comparison of P^{32}, Co^{60}, Tl^{200} Solution Specific Activity and Co^{60} Hard Source Activity (letter to the Editor, ibid, vol. 19, No. 1, 1965).
Address: Academy of Sciences of the U.S.S.R., 14 Leninsky Prospekt, Moscow V-71, U.S.S.R.

KONSTANTINOV, B. Soviet deleg., Convention on Thermonucl. Processes, Inst. of Elec. Eng., London, April 29-30, 1959.
Address: Academy of Sciences of the U.S.S.R., 14 Leninsky Prospekt, Moscow V-71, U.S.S.R.

KONSTANTINOV, B. M. Paper: On Connection of Uranium Mineralisation with Extrusions of Acid Rocks (letter to the Editor, Atomnaya Energiya, vol. 22, No. 5, 1967).
Address: Academy of Sciences of the U.S.S.R., 14 Leninsky Prospekt, Moscow V-71, U.S.S.R.

KONSTANTINOV, Boris P., Prof. Director, Physico-Tech. Inst. Corresponding Member, and Vice Pres., Acad. of Sci. of the U.S.S.R.; Déleg. to 2nd I.C.P.U.A.E., Geneva, Sept. 1958.
Address: Physico-Technical Institute, Academy of Sciences of the U.S.S.R., 2 Polytechnicheskaya Ulitsa, Leningrad, U.S.S.R.

KONSTANTINOV, G. N. Eng., U.S.S.R. State Com. for the Utilisation of Atomic Energy;

Deleg. to 2nd I.C.P.U.A.E., Geneva, Sept. 1958.
Address: U.S.S.R. State Committee for the Utilisation of Atomic Energy, 26 Staromonetnii Pereulok, Moscow, U.S.S.R.

KONSTANTINOV, I. E. Papers: Co-author, Investigation of Fallouts of Cd^{109} in Some Places of the Soviet Union in 1964-65 (letter to the Editor, Atomnaya Energiya, vol. 23, No. 3, 1967); co-author, Evaluation of Fallout Gamma-Radiation Attenuation by Snow Cover (letter to the Editor, ibid, vol. 25, No. 1, 1968); co-author, Investigation of Fallout Products at Suburb of Moscow in 1962-66 (ibid, vol. 25, No. 2, 1968).
Address: Academy of Sciences of the U.S.S.R., 14 Leninsky Prospekt, Moscow V-71, U.S.S.R.

KONSTANTINOV, I. O. Papers: Co-author, Excitation Functions for Reactions Ag^{109} (p,n) Cd^{109}, Ag^{109} (d, 2n) Cd^{109} and Ag^{107} (α, 2n + pn) Cd^{109} and Yield of Radioisotope Cd^{109} (letter to the Editor, Atomnaya Energiya, vol. 22, No. 4, 1967); co-author, Yields of Isotope Au^{195} in Nucl. Reactions on Cyclotron (letter to the Editor, ibid, vol. 23, No. 1, 1967); co-author, Yields of Ce^{139} in Nucl. Reactions La^{139} (p,n) and La^{139} (d,2n) (letter to the Editor, ibid, vol. 24, No. 3, 1968); co-author, Excitation Function of Cu^{65} (p,n) Zn^{65} Reaction (letter to the Editor, ibid, vol. 24, No. 3, 1968).
Address: Academy of Sciences of the U.S.S.R., 14 Leninsky Prospekt, Moscow V-71, U.S.S.R.

KONSTANTINOV, L. V. Advisor, Third I.C.P.U.A.E., Geneva, Sept. 1964. Papers: Co-author, Pulse Miniature Fission Chambers (letter to the Editor, Atomnaya Energiya, vol. 20, No. 3, 1966); co-author, Device for Oscillation on Measurements on Nucl. Reactor (letter to the Editor, ibid, vol. 20, No. 5, 1966); co-author, Res. and Training Reactor IR-100 (ibid, vol. 21, No. 5, 1966); co-author, Sellection of Scatterer for Neutron Beam Extraction from Reactor through Tangential Channel (ibid, vol. 25, No. 1, 1968).
Address: Academy of Sciences of the U.S.S.R., 14 Leninsky Prospekt, Moscow V-71, U.S.S.R.

KONSTANTINOV, Yu. O. Paper: Co-author, Rad. near Reactor VVR-M (letter to the Editor, Atomnaya Energiya, vol. 19, No. 1, 1965).
Address: Academy of Sciences of the U.S.S.R., 14 Leninsky Prospekt, Moscow V-71, U.S.S.R.

KONSTANTINOVIC, Jovan, D.Sc. (Phys.). Born 1931. Educ.: Belgrade and Paris Univs. Sen. Assoc., Solid State Phys. Lab., Boris Kidric Inst. Nucl. Sci., Belgrade. Society: Yugoslav Soc. Mathematicians and Phys. Nuclear interests: Solid state physics; Magnetic properties; Neutron critical scattering.
Address: Boris Kidric Institute of Nuclear Sciences, Belgrade, P.O.Box 522, Yugoslavia.

KONYAEVA, G. P. Paper: Co-author, Thermodynamic Properties of UF_6 (Atomnaya Energiya, vol. 24, No. 2, 1968).
Address: Academy of Sciences of the U.S.S.R., 14 Leninsky Prospekt, Moscow V-71, U.S.S.R.

KÖNZ, Peider, L. en Droit (Geneva), LL.B. (Washington, D.C.)., Member, District of Columbia Bar. Born 1927. Educ.: Geneva, New York and George Washington Univs. Res. Assoc., Harvard Univ., 1957-58; Consultant; Sen. Legal Officer, I.A.E.A., Vienna, 1959-61; Deputy Sec., Gen. Brussels Diplomatic Conference on Maritime Law, 1961-62; Sec. of the Council and Head of Legal Service, O.E.C.D., Paris, 1961-. Book: Co-author, Financial Protection against Atomic Hazards - The Internat. Aspects (Atomic Ind. Forum, 1959). Societies: American Soc. Internat. Law; Maritime Law Assoc. United States. Nuclear interests: Liability; State responsibility; Legal aspects of health and safety.
Address: Organisation for Economic Cooperation and Development, 2 rue André Pascal, Paris 16, France.

KOOI, Jacob, Dr. Sc. Born 1922. Educ.: Amsterdam Municipal Univ. Res. chem. (until 1956 at Joint Establishment for Nucl. Energy Res., Kjeller, Norway, finally as Head Dept. Chem. and Materials), Foundation for Fundamental Res. of Matter and Reactor Centrum Nederland, 1952-59; Dept. Head, European Transuranium Inst., C.E.E.A., 1959-. Co-editor, Actinides Reviews. Society: Roy. Dutch Chem. Soc. Nuclear interests: Nuclear and radiochemistry; Chemistry heavy elements.
Address: Institute for Nuclear Physics Research, Ooster Ringdijk, Amsterdam, Netherlands.

KOOPS, Wilhelm K. A., Dipl.-Ing. Born 1913. Educ.: Hanover Tech. Univ. Personal Member, Study Assoc. for the use of Nucl. Energy in Navigation and Industry. Nuclear interests: Designing and production of testing machines for investigation of influence of radioactive rays on metallic and non-metallic material to be used in reactors installed in ships.
Address: c/o Hahn und Kolb Stuttgart, 10 Kajen, Hamburg 11, Germany.

KOOY, J. G. van. See van KOOY, J. G.

KOOYMAN, Eduard Cornelis, Ph.D. Born 1916. Educ.: Amsterdam Univ. Head, Dept. of Organic and Biol. Chem., Shell Labs., Amsterdam, -1958; Prof., Organic Chem. (1958-), Dean, Faculty of Sci. (1967-), Leyden Univ. Pres., Roy. Netherland's Chem. Soc., 1965-; Member, Advisory Board, J. of Labelled Compounds. Society: Chem. Soc., London.
Nuclear interest: Labelled compounds.
Address: Chemical Laboratories, 64 Wassenaarseweg, Leiden, Netherlands.

KOPACZ, Stanislaw Marian, magister chemii (M.Sc. in Chem.). Born 1938. Educ.: Wroclaw and Moscow Univs. Sen. Asst. Sci. Worker, 1963-. Society: Polish Chem. Soc.
Nuclear interest: Application of radioisotopes in extraction and separation of various metals.
Address: 26/94 ul. Legnicka, Wroclaw, Poland.

KOPCHINSKII, G. A. Paper: Co-author, Flux Disturbtion Affects Caused by Indium Detectors in Reactor Spectra (letter to the Editor, Atomnaya Energiya, vol. 24, No. 1, 1968).
Address: Academy of Sciences of the U.S.S.R., 14 Leninsky Prospekt, Moscow V-71, U.S.S.R.

KOPEC, Maria, Docent, Dr. (Medicine). Born 1919. Educ.: Warsaw Acad. of Medicine. Inst. of Haematology, Warsaw, 1951-61; At (1955-), now Head, Dept. of Radiobiol. and Health Protection (1967-), Inst. of Nucl. Res., Warsaw. Vice Pres., Polish Haematological Soc. Book: Co-author, Atom Cures (Państwowe Zaktady Wydawnictw Lekarskich, 1958). Societies: Polish Biochem. Soc.; Polish Rheumatological Soc.
Nuclear interest: Radiobiology, in particular radiation haematology.
Address: Department of Radiobiology and Health Protection, Institute of Nuclear Research, Warsaw 91, Poland.

KOPECKY, Jiri, R.N.Dr., C.Sc. Born 1932. Educ.: Charles Univ., Prague. Neutron Phys. Dept., Nucl. Res. Inst., Prague, 1956-.
Nuclear interests: Nuclear physics; Reactions with slow neutrons; Radiative capture; Gamma-ray spectroscopy; Linear and circular polarisation; Polarised neutrons; Time-reversal invariance.
Address: Nuclear Research Institute, Rez, Prague, Czechoslovakia.

KOPELMAN, Bernard, A.B., A.M., Ph.D. Born 1916. Educ.: Clark Univ. Manager, Tech. Co-ordination (1951-54), Chief Eng., Atomic Energy Div. (1954-57), Sylvania Elec. Products, Inc.; Director, Market Development, Sylvania-Corning Nucl. Corp., 1957-59; Tech. Director, Beryllium Corp., 1959-. Book: Materials for Nucl. Reactors (1958). Societies: A.N.S.; A.C.S.; A.S.M.; American Electrochem. Soc.; Cosmos Club (Washington, D.C.).
Nuclear interests: Utilisation of beryllium metal; Beryllium oxide in reactors.
Address: Messenger Lane, Sands Point, Long Island, New York, U.S.A.

KOPITZKI, Konrad, Dipl. Phys. Dr.rer.nat. Privatdozent Inst. für Strahlen- und Kernphysik, Bonn Univ.
Address: Institut Strahlen- und Kernphysik, Universität Bonn, 14-16 Nussallee, Bonn, Germany.

KOPP, Paul Joseph, B.A. (Chem.), M.A. (Chem.), LL.B. Born 1909. Educ.: Lehigh, Duke and George Washington Univs. Staff Asst., Office of the Sec. of Defence. Societies: A.A.A.S.; A.C.S.; B.N.E.S.; A.N.S.; American Bar Assoc.
Nuclear interests: Research and development on nuclear reactor power for generation of heat and electricity, and for propulsion.
Address: 5031 N. 33rd Street, Arlington 7, Virginia, U.S.A.

KOPP, Peter, B.Sc., Ph.D. Born 1932. Educ.: London Univ. Sci. Officer, Roy. Military Coll. of Sci., Shrivenham, 1960-65; Asst., Medizinisch - chemisches Inst., Bern Univ., 1965-. Societies: Fellow, Chem. Soc.; Assoc. for Rad. Res.
Nuclear interests: Fundamental and applied problems in radiation chemistry and radiation biology; Irradiation of natural and synthetic high molecular weight polymers; Problems of radiation sterilisation and radiation dosimetry; Design of radiation sources.
Address: Medizinisch - chemisches Institut, Bern Universität, 28 Bühlstrasse, 3000 Bern, Switzerland.

KOPPEL, Juan Ulrico, Eng. Born 1926. Educ.: Buenos Aires Univ. Res. Assoc., Brookhaven Nat. Lab., 1959-.
Nuclear interest: Reactor physics.
Address: General Atomic, Post Office Box 608, San Diego, California, U.S.A.

KOPS, Sheldon, B.Sc. (Commerce), Certified Public Accountant. Born 1923. Educ.: Roosevelt Univ. Chief, Nucl. Materials Branch, Chicago Operations Office, U.S.A.E.C., 1952-. Society: Inst. Nucl. Materials Management.
Nuclear interests: Chemical and isotopic measurements of uranium and plutonium; Management of uranium and plutonium.
Address: 2510 West Jarvis Avenue, Chicago 45, Ill., U.S.A.

KOPYTIN, N. S. Paper: Co-author, Medium Energy Neutron Scattering by Atomic Nuclei (letter to the Editor, Atomnaya Energiya, vol. 16, No. 3, 1964).
Address: Academy of Sciences of the U.S.S.R., 14 Leninsky Prospekt, Moscow V-71, U.S.S.R.

KORBEL, Kazimierz, Dr. Tech. Sc. Born 1927. Educ.: Acad. Mining and Metal., Cracow. Tech. Phys. Dept. (1952-60), Gen., Phys. Dept. (1960-61), Acad. Mining and Metal., Cracow; Inst. Nucl. Res., Polish Acad. Sci., Cracow Dept., 1961-.
Nuclear interests: Nuclear instrumentation; Application of radioactive isotopes in science and industry; Nuclear electronics.
Address: 30 Al. Mickiewicza, Institute of Nuclear Research, Cracow Department, Cracow, Poland.

KORBEL, Z. F. Paper: Co-author, Elastic scattering of 4 GeV/c π- mesons on protons (Nukleonika, vol. 9 No. 2 - 3 1964).
Address: Joint Institute for Nuclear Research, Dubna, Nr. Moscow, U.S.S.R.

KÖRBER, W., Dir.Dipl.Ing. Treas., Isotopen-Studiengesellschaft E.V.
Address: Isotopen-Studiengesellschaft E.V., 12 Am Hauptbahnhof, Frankfurt-am-Main, Germany.

KORCHYNSKY, Michael, Dipl. Ing. Born 1918. Educ.: Tech. Univ. Lviv (Lvov), Ukraine. Res. Metal. (1951-), Tech. Supervisor (1960-), Union Carbide Metals Co. Niagara Falls, New York; Res. Supervisor, (1961-), Asst. Res. Director, Product Res. (1965-), Jones and Laughlin Steel Corp. Societies: A.I.M.E.; A.S.M.; A.S.T.M.; Inst. of Metals (London).
Nuclear interests: Technology of nuclear fuel elements, synthesis and fabrication of refractory uranium compounds; Cladding materials for nuclear fuel elements and their compatability with fissionable materials.
Address: Jones and Laughlin Steel Corp., Graham Research Laboratory, 900 Agnew Road, Pittsburgh, Pa. 15230, U.S.A.

KOREN, Kristian Johannes, M.Sc. Born 1911. Educ.: Dresden Tech. Univ. Chief Phys. (1956), Director (1957), State Inst. of Rad. Hygiene. Govt. deleg. at E.N.E.A., O.E.C.D. and I.L.O. rad. coms.; Nat. Rad. Council; Com. 4, I.C.R.P. Book: Straler, liv og helse (Oslo, Aschehougs forlag, 1960). Societies: Norwegian Soc. of Medical Radiol.; H.P.S.; Nordic Soc. for Rad. Protection.
Nuclear interests: Public health in relation to nuclear installations and general use of radioactive products; Radiation safety on a national basis.
Address: State Institute of Radiation Hygiene, Oslo-Montebello, Norway.

KORENCHENKO, Spartak. Joint Inst. for Nucl. Research, Dubna, working on elastic and inelastic interaction of π-mesons with hydrogen.
Address: Joint Institute for Nuclear Research, Dubna, Nr. Moscow, U.S.S.R.

KORENEK, Jan. Born 1917. Skoda Nucl. Power Plant Div., Plzen. Society: Czechoslovak Sci. and Tech. Soc.
Nuclear interest: Fuel handling and transportation.
Address: 3 Ujezd, Plzen, Czechoslovakia.

KORGAONKAR, Kashinath Shamrao, M.Sc., Ph.D. Born 1917. Educ.: Bombay and Colorado Univs. Lecturer in Phys., Inst. Sci., Bombay, -1954; Asst. Res. Officer (1954-55), Res. Officer (1955-60), Sen. Res. Officer (1960-64), Asst. Director Biophys. (1964-), Indian Cancer Res. Centre, Bombay. Nat. Com. for Biophys., Govt. of India; Ad Hoc Com., Biophys. and Molecular Biol., Indian Council of Medical Res., Govt. of India; Com. for Standardisation of Food Colours, Govt. of India; Council, Indian Phys. Soc. Societies: Fellow, Indian Phys. Soc.; N.Y.Acad. Sci.; Biophys. Soc. (U.S.A.).
Nuclear interests: Dosimetry; Radiation effects on biomolecules by monolayer, spectral absorption and other physical tecniques; Radiation protection and sensitisation in animals and normal tumour cells; Spontaneous cancers and aging in relation to mechanisms of radiation injury.
Address: Biophysics Department, Indian Cancer Research Centre, Bombay 12, India.

KORISTKA, G. F., Dr. Director, S.p.A. Fratelli Koristka.
Address: Nuclear Department, S.p.A. Fratelli Koristka, 47 Via Ampère, Milan, Italy.

KORISTKA, Italo G. Pres., S.p.A. Fratelli Koristka.
Address: Nuclear Department, S.p.A. Fratelli Koristka, 47 Via Ampère, Milan, Italy.

KORMAN, R. Z. Res. Assoc., Rad. Biol. Lab., New York State Veterinary Coll., Cornell Univ.
Address: New York State Veterinary College, Cornell University, Ithaca, New York, U.S.A.

KORMICKI, Jan R., M.S. Born 1933. Educ.: Jagellonian Univ., Cracow. Sen. Res. Asst., Inst. of Nucl. Phys., Cracow; Sen. Res. Asst., Dept. of Phys., Medical Acad., Cracow. Society: Polish Phys. Soc.
Nuclear interests: Nuclear spectroscopy; Nuclear models.
Address: 1/97 ul. Sadowa, Craców, Poland.

KORMUSHKIN, Yu. P. Papers: Co-author, Experience in Operating the SM-2 Exptl. Reactor (Kernenergie, vol. 9, No. 10, 1966); co-author, Investigation of Optimum Position Neutron Beams in SM-2 Reactor (letter to the Editor, Atomnaya Energiya, vol. 22, No. 5, 1967); co-author, Influence on Reactivity with Moderator Concentration Change in Reactor SM-2 Flux Trap (letter to the Editor, ibid, vol. 24, No. 1, 1968).
Address: Academy of Sciences of the U.S.S.R., 14 Leninsky Prospekt, Moscow V-71, U.S.S.R.

KORNBERG, Harry Alexander, B.S. (Chem.), M.S. (Chem. Organic), Ph.D. (Chem. Bioorganic). Born 1914. Educ.: Illinois, Washington State and Texas Univs. Manager, Biol., Pacific Northwest Lab. Member, Group for Aerospace Biomedical Nucl. Medicine, U.S. A.E.C., 1964; Member, Pathological Effects of Atomic Rad. Com. and Chairman, Subcom. on Inhalation Hazards, N.A.S. 1957-63; U.S. Deleg., Tri-Partite Conference on Radiobiol., Harwell, England; U.S. Deleg. to Conference on Passage of Fission Products through Food Chains, Harwell, England, (1959); U.S. Deleg. to 4th Internat. Congress on Nucl. Energy Rome; Chairman,Northwest Sect., and Nat. Councilman, Soc. for Exptl. Biol. and Medicine. Societies: A.C.S.; H.P.S.; Rad. Res. Soc.; Soc. of Toxicology.
Nuclear interests: Radiation biology.
Address: Biology Department, Pacific Northwest Laboratory, Battelle Memorial Institute, Richland, Washington, U.S.A.

KORNBICHLER, Heinz, Dr. Ing. Born 1925. Educ.: Munich T.H. Dept. Manager, AEG Allgemeine Elektrizitäts-Gesellschaft, Kernenergieanlagen 1955-. Nuclear interest: Management.
Address: 10 Gerhardshainerstrasse, 6243 Falkenstein (Taunus), Germany.

KORNBLITH, Lester, Jr., S.B. (E.E.). Born 1917. Educ.: M.I.T. Chief Eng., Enrico Fermi Inst. for Nucl. Studies, Chicago Univ., 1947-55; Manager, Reactor Tech. Operation, Vallecitos Atomic Lab., G.E.C., 1956-63; Asst. Director for Reactors, Div. of Compliance, U.S.A.E.C., 1963-. Coms., I.E.E.E. and A.N.S. Societies: I.E.E.E.; A.N.S. Nuclear interests: Reactor design, construction and operation; Accelerator design, construction and operation.
Address: 6708 Tulip Hill Terrace, Washington, D.C. 20016, U.S.A.

KORNILENKO, I. I. Paper: Co-author, On Input and Output Concentration Ratio of Radioactive Gases Flowing through Dosimetric Camera DZ-70 (letter to the Editor, Atomnaya Energiya, vol. 21, No. 2, 1966).
Address: Academy of Sciences of the U.S.S.R., 14 Leninsky Prospekt, Moscow V-71, U.S.S.R.

KORNPROBST. Chef, Département Administratif et Financier, Direction des Applications Militaires, C.E.A.
Address: Commissariat à l'Energie Atomique, 29-33 rue de la Federation, Paris 15, France.

KOROBCHENKO, L. A. Paper: Co-author, Equipment for Temperature Control During In-Reactor Irradiation of Construction Materials Samples (Atomnaya Energiya, vol. 22, No. 6, 1967).
Address: Academy of Sciences of the U.S.S.R., 14 Leninsky Prospekt, Moscow V-71, U.S.S.R.

KOROBEINIKOV, I. A. Paper: Co-author, Temperature and Neutron Irradiation Effect on Plastic Deformation of Alpha Uranium Single Crystal (Atomnaya Energiya, vol. 19, No. 4, 1965).
Address: Academy of Sciences of the U.S.S.R., 14 Leninsky Prospekt, Moscow V-71, U.S.S.R.

KOROBEINIKOV, L. S. Papers: Co-author, System for Control and Measuring of Electron Beams Parameters in VEP-1 Electron-Electron Storage Ring (Atomnaya Energiya, vol. 20, No. 3, 1966); co-author, Non Linear Resonances of Betatron Oscillations in Storage Rings (ibid, vol. 22, No. 3, 1967).
Address: Academy of Sciences of the U.S.S.R., 14 Leninsky Prospekt, Moscow V-71, U.S.S.R.

KOROL', V. M. Paper: Co-author, Lithium Ion Source for Electrostatic Generator (letter to the Editor, Atomnaya Energiya, vol. 21, No. 3, 1966).
Address: Academy of Sciences of the U.S.S.R., 14 Leninsky Prospekt, Moscow V-71, U.S.S.R.

KOROLEFF, Folke, Ph.D. Head, Oceanographic Dept., Inst. of Marine Res.
Address: Institute of Marine Research, 2 Tähtitorninkatu, Helsinki, Finland.

KOROLEVA, V. P. Papers: Co-author, Radiative Capture of Cl^{37}, Rb^{87}, Ir^{193} for Fast Neutrons Cl^{37}, Rb^{87}, Ir^{193} (letter to the Editor, Atomnaya Energiya, vol. 23, No. 2, 1967); co-author, Radioactive Capture Cross Sect. of Fast Neutrons for Ge^{74}, Cs^{138} and Os^{192} (ibid, vol. 23, No. 6, 1967); co-author, Isomeric Ratios for Radiative Capture of Fast Neutrons by Se^{80} Isotope (letter to the Editor, ibid, vol. 24, No. 2, 1968); co-author, Radiative Capture of Fast Neutrons by Nuclei Sn^{122}, Sn^{124} and Sb^{121}, Sb^{123} (letter to the Editor, ibid, vol. 24, No. 6, 1968).
Address: Academy of Sciences of the U.S.S.R., 14 Leninsky Prospekt, Moscow V-71, U.S.S.R.

KOROTAEV, S. K. Paper: Co-author, Conductivity Indicator of Admixtures in Liquid Metal Coolants (letter to the Editor, Atomnaya Energiya, vol. 24, No. 4, 1968).
Address: Academy of Sciences of the U.S.S.R., 14 Leninsky Prospekt, Moscow V-71, U.S.S.R.

KOROTKOV, R. I. Paper: Co-author, Influence on Reactivity with Moderator Concentration Change in Reactor SM-2 Flux Trap (letter to the Editor, Atomnaya Energiya, vol. 24, No. 1, 1968).
Address: Academy of Sciences of the U.S.S.R., 14 Leninsky Prospekt, Moscow V-71, U.S.S.R.

KOROZA, V. I. Paper: Co-author, Radial Expansion of Beam in Linear Electron Accelerator Caused by Action of Nonsymmetric Mode (Atomnaya Energiya, vol. 20, No. 1, 1966).
Address: Academy of Sciences of the U.S.S.R., 14 Leninsky Prospekt, Moscow V-71, U.S.S.R.

KORP, Henry J., B.S. (Chem. Eng.), LL.B., Dr. jur. Educ.: Pittsburgh and St.Mary's Univs. Sen. Technologist, Socony Vacuum R. and D. Labs., 1949-53; Manager, Fuels Res. Sect., Dept. Engines, Fuels and Lubricants Res. (1953-56), Chairman, Automotive Dept., (1956-58), Vice-Pres. (1959-), Southwest Res. Inst. Diesel and Motor Vehicle Fuel, Lubricant, and Equipment Res. Groups, Co-ordinating Res. Council; Automotive Emissions and Air Pollution Com., Res. Advisory Com., Chairman Internat. Liaison Com., S.A.E.; D2 Coms., A.S.T.M.; U.S.Army Engine Oil and Gear Lubricants Reviewing Coms. Societies: Air Pollution Control Assoc.; A.C.S.; A.I.Ch.E.; American Ordnance Assoc.; American Petroleum Inst.; A.P.H.A.; Co-ordinating Res. Council; S.A.E.; RESA.; A.S.T.M.; American Bar Assoc.; Texas Bar Assoc.; San Antonio Bar Assoc.
Address: 7103 Oakridge Drive, San Antonio, Texas, U.S.A.

KORPAK, Wincenty Michal, Dr.Dipl. Eng. Born 1921. Educ.: Silesia Polytech.,Gliwice. Asst., Silesia Polytech., 1946-55; Res. worker, Inst. of Nucl. Res., Warsaw, 1955-.
Nuclear interests: Physical chemistry of solutions; Solvent extraction of inorganic compounds.
Address: Institute of Nuclear Research, Polish Academy of Sciences, 16 ul.Dorodna, Warsaw-Zeran, Poland.

KORRINGA, Pieter, Prof. Dr. (Amsterdam). Born 1913. Educ.: Amsterdam Univ. Director, Netherlands Inst. for Fishery Investigations; Extraordinary Prof. Appl. Hydrobiology,Amsterdam Univ. Responsible as Director of Fisheries Research for consequences for the Netherlands Fishing Industry of discharge of radioactive waste in sea. Member of several Government Commissions charged with advising various Ministries re the regulation of discharge of radioactive waste in sea, re the measuring of the degree of radioactivity in marine organisms and marine deposits in the Netherlands coastal waters, and re research to be carried out in this field.
Address: Netherlands Institute for Fishery Investigations, 1 Haringkade, Ijmuiden, Netherlands.

KORSHAKOV, A. I. Paper: Co-author, Intensification of Heat Transfer in Channel (Atomnaya Energiya, vol. 22, No. 6, 1967).
Address: Academy of Sciences of the U.S.S.R., 14 Leninsky Prospekt, Moscow V-71, U.S.S.R.

KORTUS, Josef, Dipl. Eng. Born 1926. Educ.: Charles Univ., Prague. Eng., Chemoprojekt, 1951-59; M.I.T., 1959-60; Specialist, Chemoprojekt, 1961-67; C.E.N., Cadarache, 1967-68. Society: Ceskoslovenska Spoleonost Chemicka. Nuclear interests: Radioactive waste; Treatment of uranium ores.
Address: 7 V Horni Stromce, Prague, Vinohrady, Czechoslovakia.

KORTUS, Jozef, Dipl. Eng. (Chem.), C.Sc. Born 1928. Educ.: Bratislava Tech. High School. Lab. of the Hygiene of Rad., Inst. of Hygiene, 1963-.
Nuclear interests: Radiochemical methods; Labelled pesticides.
Address: Research Institute of Hygiene, Bratislava, Czechoslovakia.

KORUR, Nuri Refet, D.Sc. Born 1899. Educ.: Göttingen, Instanbul and Ankara Univs. Chief, Chem. Sect. (1930-53), Head, Atomic, Biol. and Chem. Dept. (1954-56), Ministry of Defence; Gen. Sec., Turkish A.E.C., 1957-63. Books: War Economy; Chem. War-agents; Latest Discoveries in Chem.; Atomic Bomb. Societies: Turkish Soc. Chem.; German Soc. Chem.
Nuclear interests: Management; Radiation chemistry.
Address: Büklüm Sokak 35, Kavaklidere, Ankara, Turkey.

KORVENKONTIO, Osmo Olavi, Dipl. Eng. (M.Sc.). Born 1917. Educ.: Finland Inst. of Technol. Structural Designer (-1951), Head, Power Plant Designing (1951-55), Asst. Director, Civil Eng. Dept.(1955-61), Director, Civil Eng. Dept. (1961-), Imatran Voima Oskeyhtio. Member, Board of Directors Soil and Water; Member, Board of Directors, Killin Voima Oy (Killi Power Co.); Exec. Sec.,The Finnish Com. on Large Dams; Member, Board of Directors,Federation of Finnish Power Plant Builders; Member, Administrative Council, Foundation of the Delegation for the Development of the Building Line (Rakeva-säätiö).
Nuclear interests: Civil engineering and construction in connection with nuclear power plants.
Address: 4 F Tykkitie, Tapiola, Finland.

KORYAKIN, Yu. I. Papers: Co-author, Method of Calculation of Water and Power Casts for Nucl. Desalination Plants (Atomnaya Energiya, vol. 19, No. 2, 1965); co-author, Nucl. Power and Desalination (ibid, vol. 20, No. 3, 1966); co-author, Some Aspects of Nucl. Power Economics Incentive (ibid, vol. 20, No. 5, 1966); co-author, Nucl. Fuel Efficiency Criterion (ibid, vol. 21, No. 3, 1966).
Address: Academy of Sciences of the U.S.S.R., 14 Leninsky Prospekt, Moscow V-71, U.S.S.R.

KORYUSHKIN, A. P. Papers: Co-author, Interaction of Uranium (IV) with Chloride-Fluoride Melt NaCl - KCl - NaF (Atomnaya Energiya, vol. 21, No. 6, 1966); co-author, Decomposition of Uranylchloride and its Interaction with Deoxide of Uranium in Melt NaCl - KCl (ibid, vol. 22, No. 1, 1967); co-author, Fluoride Ion Effect on Uranium Dioxide Deposition by Molten Salt Electrolyze (letter to the Editor, ibid, vol. 24, No. 4, 1968).
Address: Academy of Sciences of the U.S.S.R., 14 Leninsky Prospekt, Moscow V-71, U.S.S.R.

KORZH, I. A. Papers: Co-author, Medium Energy Scattering by Atomic Nuclei (Atomnaya Energiya, vol. 16, No. 3, 1964); co-author, Medium Energy Neutron Scattering by Atomic Nuclei (letter to the Editor, ibid); co-author, Scattering of Medium Energy Neutrons (ibid, vol. 20, No. 1, 1966); co-author, Radiative Capture Cross Section of Isotopes Ti^{50} and V^{51} for Fast Neutrons (letter to the Editor, ibid, vol. 23, No. 1, 1967).
Address: Academy of Sciences of the U.S.S.R., 14 Leninsky Prospekt, Moscow V-71, U.S.S.R.

KOSA-SOMOGYI, Istvan, Grad. Eng., Candidate of Chem.-Sci. Born 1930. Educ.: Chem. Eng. School, Veszprem, Hungary, and Lensoviet, Inst. of Technol., Leningrad, U.S.S.R. Head, Rad. Chem. Group, Central Res. Inst. of Phys., 1958-. Society: Hungarian Chem. Soc.
Nuclear interests: Radiation chemistry;

Corrosion of reactor materials.
Address: Hungarian Academy of Sciences,
Central Research Institute for Physics, Budapest 114, P.O.B. 49, Hungary.

KOSACKI, Józef Stanislaw. Born 1909. Educ.:
Warsaw Tech. Univ. Chief, Transmission Div.,
State Inst. of Telecommunication, 1947-56;
Chief, Electronics Div., Inst. Nucl. Res., 1956-;
Prof., Military Tech. Acad., 1957-.
Nuclear interest: Nuclear electronics and
instrumentation for physical research.
Address: 7/9 m 5 Ratuszowa, Warsaw, Poland.

KOSAKA, Takao, Prof. Public Health. Member, Radioisotope Res. Com., Niigata Univ.
School of Medicine.
Address: Department of Public Health,
Niigata University School of Medicine,
Asahimachi, Niigata, Japan.

KOSAKI, Masahide, Dr.Sc. Manager, Nucl.
Fuel Res. Lab., Sumitomo Metal Industries
Ltd.
Address: Nuclear Fuel Research Laboratory,
Sumitomo Metal Indsutries Ltd., Ronowari,
Minato-ku, Nagoya, Japan.

KOSEK, Stanislaw Wojciech, Magister. Born
1935. Educ.: Warsaw Univ. Asst., Exptl.
Phys. Dept., Warsaw Univ., 1960; Asst., Rad.
Chem. Dept., Nucl. Res. Inst., Warsaw, 1967;
Adjunct, Light Technics Dept., Electrotech.
Res. Inst., Warsaw, 1967-.
Nuclear interests: The luminescence induced
by gamma rays; Radical structure studied by
e.p.r. spectroscopy; Optical spectrophotometry
and colorimetry.
Address: 32 m.6 Zorzy, Warsaw 89, Poland.

KOSEKI, Koji. Dr. Sci. Born 1917. Educ.:
Hokkaido Imperial Univ. Nucl. Material Sect.,
Mineral Deposits Dept., Geological Survey of
Japan. Society: Atomic Energy Soc. of Japan.
Nuclear interest: Geology and mineralogy of
nuclear materials, especially on the prospecting of uranium ore deposits.
Address: Mineral Deposit Department, Geological Survey Tokyo Branch, 8 Kawada-cho,
Shinjuku-ku, Tokyo, Japan.

KOSFELD, Robert, Dozent, Dr.rer.nat. Born
1925. Educ.: Bonn Univ. and Aachen T.H.
Industrietätigkeit, 1952-54; Batelle Stipendium, 1955; Wissenschaftliche Asst. (1956-60),
Oberingenieur Privatdozent (1961-67), Dozent
für Phys. Chemie (1967), Inst. für Phys.
Chem., Aachen T.H. Arbeitsgruppe Strahlenschutz, Hanover. Society: Fachausschuss
Strahlenschutz, Deutsche Physikalische
Gesellschaft.
Nuclear interests: Strahlenschutz; Kernmagnetische Resonanz.
Address: Aachen Technische Hochschule,
Institut für Physikalische Chemie, 51 Aachen,
Germany.

KOSHA-SHAMODI, I. Paper: Co-author,
Study of Radiochem. Stability of High-Boiling
Hydrocarbons in Reactor (Atomnaya Energiya,
vol. 20, No. 1, 1966).
Address: Academy of Sciences of the U.S.S.R.,
14 Leninsky Prospekt, Moscow V-71, U.S.S.R.

KOSHINO, Masao, D.Eng. Third Sect., Rad.
Chem. for Low Polymers, Govt. Ind. Res. Inst.,
Ministry of Internat. Trade and Industry.
Address: Ministry of International Trade and
Industry, Government Industrial Research
Institute, Hirate-machi, Kita-ku, Nagoya, Japan.

KOSHKIN, Yu. I. Paper: Co-author, The Nucl.
Power Plant with BN-350 Reactor (Atomnaya
Energiya, vol. 23, No. 5, 1967).
Address: Academy of Sciences of the U.S.S.R.,
14 Leninsky Prospekt, Moscow V-71, U.S.S.R.

KOSHKIN, Yu. N. Paper: Co-author, BN-350
and BOR Fast Reactors (Atomnaya Energiya,
vol. 21, No. 6, 1966).
Address: Academy of Sciences of the U.S.S.R.,
14 Leninsky Prospekt, Moscow V-71, U.S.S.R.

KOSIEK, Rolf, Dipl. Phys., Dr.rer.nat. Born
1934. Educ.: Göttingen and Heidelberg Univs.
I.Physikalisches Inst., Heidelberg Univ., 1963-.
Society: Deutsche Phys. Gesellschaft.
Nuclear interest: Photonuclear effect.
Address: 12 Ladenburger Strasse, 69 Heidelberg, Germany.

KOSITSYN, L. G. Paper: Co-author, 300 MeV
Electron Synchrotron at the Polytech. Inst.
of Tomsk (letter to the Editor, Atomnaya
Energiya, vol. 21, No. 6, 1966).
Address: Academy of Sciences of the U.S.S.R.,
14 Leninsky Prospekt, Moscow V-71, U.S.S.R.

KOSKELA, Urpo, B.S. (Analytical Chem.).
Born 1916. Educ.: Tufts and Minnesota Univs.
U.S. Army, Manhattan Project, Technician
3rd Grade, 1944-46; Chem., Clinton Labs.,
1946-48; Chem., M.I.T., 1948-51; Chem.,
American Cyanamid Co., 1951; Chem.,
O.R.N.L., 1951-. Societies: A.C.S.; A.N.S.
Nuclear interest: Analysis of "hot" samples
using remote techniques and equipment.
Address: 313 West Outer Drive, Oak Ridge,
Tennessee 37830, U.S.A.

KOSKINEN, Heikki Antero, Dipl. Eng. (M.S.
in Eng.), Lic. Tech. Born 1937. Educ.: Finland Inst. Technol. Asst., Finland Inst. Technol., 1961-65; Acting Prof. Appl. Maths.
Oulu Univ., Finland, 1965-. Societies: Finnish
Phys. Soc.; Finnish Soc. for Atomic Technol.
Nuclear interests: Reactor theory and analysis:
Transport theory; Computational aspects.
Address: Institute of Applied Mathematics,
Oulu University, Oulu, Finland.

KOSLOV, Samuel, A.B., M.A., Ph.D. (Phys.).
Born 1927. Educ.: Columbia Univ. Asst. Phys.,
Nevis Cyclotron Lab., Columbia Univ.,

1950-54; Member, Tech. Staff, Electron Tube Development, Bell Telephone Labs., N.J., 1954-56; Physicist, Vitro Corp. of America, Eng. Div., N.Y.C., 1956-57; Asst. Prof. (1957-60), Assoc. Prof. (1960-61), Stevens Inst. Technol., Hoboken, N.J.; Consultant, Republic Aviation Co., 1958-60; Asst. Head, Phys. and Space Sci. Dept., Vitro Labs., Div. of Vitro Corp. of America, West Orange, N.J. 1961- (Consultant, 1957-61); Vice-Pres., N.Y. Section, American Rocket Soc. Societies: A.P.S.; American Rocket Soc.; Inst. of Radio Eng.; A.G.U.
Nuclear interests: High energy nuclear physics; Mesonic atoms; Nuclear particle detectors; Nuclear weapons; Reactor physics; Reactor hazard evaluation; Controlled thermonuclear fusion; Plasma physics.
Address: 49 Bashkin Road, Lexington, Massachusetts, U.S.A.

KOSMACH, V. F. Papers: Co-author, Secondary Protons Emitted from Thick Aluminium Targets During Bombarding by 340 Mev Protons (Atomnaya Energiya, vol. 22, No. 6, 1967); co-author, High Energy Nucleons Emitted from Thick Targets of Aluminium During Bombarding by 660 Mev Protons (ibid, vol. 23, No. 3, 1967).
Address: Academy of Sciences of the U.S.S.R., 14 Leninsky Prospekt, Moscow V-71, U.S.S.R.

KOSODAEV, M. S., Prof. Member, Editorial Board, Nucl. Instruments and Methods.
Address: Academy of Sciences of the U.S.S.R., 14 Leninsky Prospekt, Moscow V-71, U.S.S.R.

KOSS, Peter, Ph.D., Univ. Doz. Born 1932. Educ.: Vienna Univ. Res. Asst., Inst. Phys., Vienna Univ., 1954-58; Head, Inst. Metal., Reactor Centre, Seibersdorf, 1958-. Societies: Österreichische Chemisch-Physikalische Gesellschaft; Österreichische Physikalische Gesellschaft. Nuclear interests: Metallurgy of reactor fuels, especially for high temperature application; Radiation damage in fuels and structural materials; Fission gas release; Fuel swelling.
Address: Österreichische Studiengesellschaft für Atomenergie, Ges.m.b.H., Vienna 8, Lenaugasse 10, Austria.

KOSS-ROSENQVIST, Agathe, Dr. phil. chem. Born 1915. Educ.: Vienna Univ. Österr. Stickstoffwerke, Linz; Austro-Chematom, Linz. Society: Verein Österreichischer Chemiker. Nuclear Interest: Metallurgy.
Address: 7 Lindaustr., Bad Ischl, Austria.

KOSSOVICH, Leone, Dr. of law. Born 1909. Educ.: Rome Univ. Chief. Transatom Service, Euratom, C.E.E.A., Brussels.
Address: 83 avenue du Pesage, Brussels,Belgium

KOST, Heinrich, Bergassessor, Dr.Ing.E.h. Born 1890. Educ.: Munich and Berlin Univs. Präsident, Wirtschaftsvereinigung Bergbau, Bad Godesberg; Fachkommission V, Federal Ministry for Sci. Res.
Address: Kapellen b. Moers, Agnetenhof, Germany.

KOSTA, Lado, Dipl. Eng., Ph.D. (Chem.). Born 1921. Educ.: Ljubljana Univ. In charge, Toxicological Lab., Inst. for Legal Medicine, Ljubljana, 1945-52; Sen. Sci. Officer, Head, Analytical Sect., Nucl. Inst. "J. Stefan" Ljubljana, 1952-63; Lecturer,Analytical chem. (1955-62), Prof. (1962-63), Ljubljana Univ. Analytical Chem., responsbile for analytical activities of the Chem. Programme, Div. of Res. and Labs., Lab. Seibersdorf, I.A.E.A., 1963-.
Nuclear interests: Analytical chemistry of nuclear materials; Activation analysis.
Address: International Atomic Energy Agency, 11 Kaerntnerring, Vienna 1, Austria.

KOSTALOS, John, Jr.,B.E.E., M.A. (Phys.). Born 1921. Educ.: Columbia Univ. Eng. Supervisory Eng., Advisory Eng. (1949-59), Advisory Eng., Advanced Systems Eng. (1960-67), Westinghouse Elec. Corp. Society: I.E.E.E.
Nuclear interests: Reactor and industrial plant instrumentation, control, and simulation.
Address: Westinghouse Electric Corporation, Research and Development Centre, Pittsburgh, Pennsylvania 15235, U.S.A.

KOSTAMIS, P., M.D. Director, Cardiovascular Applications, Whole Body Counting, Athens Univ.
Address: Athens University, Department of Clinical Therapeutics, Thyroid Clinic, Alexandra Hospital, Isotope Laboratory, Vas. Sophias-K. Lourou Str., Athens, Greece.

KOSTIC, C. Head, Metal. Dept., Dept. for Processing of Nucl. Raw Materials, Inst. Nucl. Raw Materials.
Address: 86 Franse Deperea, Belgrade, Yugoslavia.

KOSTIC, Emilija, Dr. of Tech. Sci. Born 1936. Educ.: Belgrade Univ. Asst. (1961), Sci. Res. Worker (1966), Boris Kidric Inst. of Nucl. Sci. Sec., Ceramics Sect., Serbian Chem. Soc., 1965.
Nuclear interests: Sintering of oxide powders; Kinetics and thermodynamics of the sintering of non-stoichiometric oxides, particulary uranium dioxide; Research in the oxide reactor fuel technology, and similar ceramic and metallo-ceramic materials.
Address: Laboratory 28, P.O. Box 522, Belgrade, Yugoslavia.

KOSTIC, Evgenije, Mining Eng. Born 1907. Educ.: Liège Univ. Director of Federal Geological Survey. Member, Com. of Experts of Federal Nucl. Energy Commission. Nuclear interests: Geology, mining metallurgy.
Address: 12 Roviljska, Federal Geological Survey, Belgrade, Yugoslavia.

KOSTIC, Ljiljana, M.Sc. Born 1930. Educ.: Univ. of Pharmacy, Belgrade. Now at Inst. of Nucl. Sci. Boris Kidrich, Dept. of Radiobiol. Nuclear interests: Radiobiology-microbiology (problems concerning effects of ionising radiation on living cells); Problems of biochemistry, genetics and radiological protection of radiated cells.
Address: Maja 94, Belgrade, Yugoslavia.

KOSTOCHKIN, O. I. Papers: Co-author, Yields of Delayed Neutrons from Fission in ^{239}Pu and ^{232}Th as Induced by Neutrons of Energy 14.5 MeV (letter to the Editor, Atomnaya Energiya, vol. 11, No. 6, 1961); co-author, Delayed Neutron Emission Probability from Halogen (letter to the Editor, ibid, vol. 16, No. 4, 1964).
Address: Academy of Sciences of the U.S.S.R., 14 Leninsky Prospekt, Moscow V-71, U.S.S.R.

KOSTRITSA, A. A. Papers: On the Neutron Diffusion in a Mobile Medium (letter to the Editor, Atomnaya Energiya, vol. 14, No. 2, 1963); co-author, Nonstationary Problems in Genetic Theory of Neutron Transport (ibid, vol. 16, No. 6, 1964); co-author, Neutron Transport in Moving Medium (ibid, vol. 21, No. 2, 1966); co-author, On Velocity Spectrum of Neutron in Heavy Nuclei Medium (ibid, vol. 24, No. 1, 1968).
Address: Academy of Sciences of the U.S.S.R., 14 Leninsky Prospekt, Moscow V-71, U.S.S.R.

KOSTUIK, John, B.Sc. Educ.: Queen's Univ., Kingston. Vice Pres. and Gen. Manager, Denison Mines Ltd. Vice Pres. and Director, Mining Assoc. of Canada.
Nuclear interests: Management; Metallurgy.
Address: Denison Mines Limited, 4 King Street West, Toronto 1, Ontario, Canada.

KOSTYRKO, Andrzej, Chem. Eng., Master of Tech. Sci. Born 1928. Educ.: Polytech. School Gdańsk. Asst. in Chair of Chem. Technol. of Wood and Peat, Polytech. School Gdańsk, 1950-51; Tech. Inspector in Export-Import Head Office, Papexport, Warsaw, 1952-53; Constructor in Factory of Electronic Tubes Telam, Warsaw, 1954-58; Asst. in Sect. of Dosimetry, Inst. of Nucl. Res., Swierk, 1959-67 (in Reactor Exploitation Dept., 1959-60, Adjunct in Health Phys. Div., 1961-67); Adjunct, Lab. of Plasma Phys. and Technol., Inst. of Nucl. Res., Swierk, 1967-.
Nuclear interests: MHD generator electrode materials; Physicochemistry of high-temperature refractories.
Address: 159 m. 65 W. J. Marchlewskiego, Warsaw, Poland.

KOSTYU, Ya. E. Paper: Co-author, Method for Determination of Oil-Water Interface Based on Delayed Neutrons Detection (Atomnaya Energiya, vol. 21, No. 1, 1966).
Address: Academy of Sciences of the U.S.S.R., 14 Leninsky Prospekt, Moscow V-71, U.S.S.R.

KOTELNIKOV, G. A. Papers: Co-author, Fuel Elements Burn-up Determination with Semiconductor Germanium Gamma-Spectra Meter (letter to the Editor, Atomnaya Energiya, vol. 21, No. 5, 1966); co-author, Determination of Oxygen in Germanium and Silicon Using Activation by Helium - 3 (ibid, vol. 23, No. 2, 1967).
Address: Academy of Sciences of the U.S.S.R., 14 Leninsky Prospekt, Moscow V-71, U.S.S.R.

KOTEL'NIKOV, G. N. Papers: Co-author, Some Aspects of Aerial γ-surveys in Wooded Areas (letter to the Editor, Atomnaya Energiya, vol. 8, No. 4, 1960); Features of Shift of Equilibrium in Uranium - Radium Series of Uranium Deposits Containing Hard Bitumens (letter to the Editor, ibid, vol. 19, No. 5, 1965); co-author, On Possibility of Formation of Placer Uranium Deposits (letter to the Editor, ibid, vol. 23, No. 1, 1967); co-author, Vein Solid Bitumens in Uranium Deposits (ibid, vol. 24, No. 6, 1968).
Address: Academy of Sciences of the U.S.S.R., 14 Leninsky Prospekt, Moscow V-71, U.S.S.R.

KOTEL'NIKOV, V. P. Paper: Co-author, New Data about Atmosphere Radioactivity and Deposition Density on the Black Sea (letter to the Editor, Atomnaya Energiya, vol. 19, No. 5, 1965).
Address: Academy of Sciences of the U.S.S.R., 14 Leninsky Prospekt, Moscow V-71, U.S.S.R.

KOTHARI, Laxman Singh, M.Sc. (Delhi), Ph.D. (Bombay). Born 1926. Educ.: Delhi Univ. Jun. Res. Officer (1955-58), Res. Officer (1958-61), Atomic Energy Establishment, Trombay; Reader in Phys., Panjab Univ., Chandigarh, 1961-62; Phys. Prof., Delhi. Univ., 1963-.
Nuclear interests: Interaction of neutrons with solids, in particular their slowing down near thermal equilibrium; Equilibrium neutron energy spectra in finite crystalline moderator assemblies; Effect of lattice structure on neutron diffusion and neutron wave propagation. Neutron diffusion across discontinuities in media; Neutron thermalisation in ice.
Address: Department of Physics and Astrophysics, Delhi University, Delhi 7, India.

KOTHBAUER, A., Dir.Dipl.Ing. Managing Director, Dampfkraftwerk Korneuburg G.m.b.H.
Address: Dampfkraftwerk Korneuburg G.m.b.H., 6A Am Hof, Vienna 1, Austria.

KOTIK, Jack, Dr. Dept. Head and Sen. Sci., Rad. Penetration Group, TRG, Inc.
Address: TRG, Incorporated, Radiation Penetration Group, Route 110, Melville, New York 11749, U.S.A.

KOTLICKI, Henryk. Member, Polish State Council for the Peaceful Use of Nucl. Energy.
Address: State Council for the Peaceful Use of Nuclear Energy, Room 1819, Palace of Culture and Science, Warsaw, Poland.

KOTSCHI, W., Dipl. Phys. Member, Fachausschuss für Strahlenschutz (joint group of the Deutsches Atomforum and Kernenergieverwertung Gesellschaft).
Address: Technischer Uberwachungsverein Bayern e.V. 14-16 Kaiserstrasse, 8000 München 23, Germany.

KOTT, Josef, Ing. (Mech. Eng.). Born 1932. Educ.: Tech. Univ. Editorial Board, Jaderná Energie; Working Group for Reactor Hazards, Czechoslovakian A.E.C. Society: Czechoslovak Sci. and Tech. Soc.
Nuclear interests: Reactor analysis; Reactor engineering science; Experimental reactor physics; Reactor shielding; In-pile dosimetry.
Address: Nuclear Power Division, Skoda Concern, Pilsen, Czechoslovakia.

KOTTWITZ, David Alexander, B.S., Ph.D. Born 1926. Educ.: Texas and Pennsylvania Univs. With G.E.C., Hanford Labs., 1956-65; Pacific Northwest Lab., Battelle Memorial Inst., 1965-. Societies: A.P.S.; A.N.S.
Nuclear interests: Slow neutron scattering; Neutron thermalisation; Neutron diffraction.
Address: Battelle Memorial Institute, Pacific Northwest Laboratory, Richland, Washington, U.S.A.

KOTZER, Peter. Res. Faculty, Dept. of Phys., Washington Univ.
Nuclear interests: Cosmic rays; High energy physics.
Address: Washington University, Department of Physics, Seattle, Washington 98105, U.S.A.

KOULUMIES. Marja, Dr. Born 1912. Educ.: Helsinki Univ.. Asst. to Director, Dept. of Radiotherapy, Univ. Central Hospital, Helsinki. Society: Finnish Radiol. Soc.
Address: Department of Radiotherapy, University Central Hospital, 4 Haartmaninkatu, Helsinki 29, Finland.

KOUMOUTSOS, N., Dr. Member, Sci. Com., "Democritos" Nucl. Res. Centre, Greek A.E.C.
Address: "Democritos" Nuclear Research Centre, Aghia Paraskevi-Attiki, Athens, Greece.

KOURIM, Vaclav, Ing. chem., C.Sc. Born 1927. Educ.: Prague Tech. Univ. At. Nucl. Res. Inst., Czechoslovak Acad. Sci., 1954-. Society: Czechoslovak Chem. Soc.
Nuclear interests: Radiochemistry; Inorganic ion exchanges; Fission products; Waste treatment.
Address: Ustav jaderného výzkumu CSAV Rez u Prahy, Czechoslovakia.

KOUTRAS, Demetrios, M.D. Born 1930. Educ.: Athens Univ. Lecturer in Medicine, Glasgow Univ. (Western Infirmary), 1961; Assoc. Physician, Dept. of Clinical Therapeutics, Athens Univ., 1962-63; Visiting Sci., Nat. Inst. Health, U.S.A., 1963-64; Director, Thyroid Clinic, "Alexandra" Hospital, Athens, Greece. Book: Co-author, Clinical Aspects of Iodine Metabolism (Oxford, Blackwell Sci. Publications, 1964). Society: New York Acad. Sci.
Nuclear interests: The use of radioactive isotopes of iodine in the diagnosis and treatment of thyroid disease, and in research on the thyroid.
Address: Department of Clinical Therapeutics, "Alexandra" Hospital, Vas. Sofias and Lourou Str., Athens, Greece.

KOUTS, Herbert John Cecil, Ph.D. E.O. Lawrence Award, 1963. Born 1919. Educ.: Louisiana State and Princeton Univs. Group Leader, Exptl. Reactor Phys., Brookhaven Nat. Lab., 1950-. Chairman (1964), Member, (1962-63, 65-66), Advisory Com. on Reactor safeguards, U.S.A.E.C. Societies: A.N.S.; A.P.S.
Nuclear interests: Reactor physics; Nuclear physics.
Address: Brookhaven National Laboratory, Upton, New York, U.S.A.

KOUYOUMZELIS, Theodore George, D.Sc. (Athens). Born 1906. Educ.: Athens and Munich Univs. Full Prof. Phys. (1958-61), Dean (1961-), Nat. Tech. Univ. Sec. Gen., Greek A.E.C., 1954-60; Permanent Rep., C.E.R.N., 1955-. Books: Nucl. Phys. for Advanced Univ. Students (Athens, Tzakas, 1947); Elements Nucl. Phys. (Athens, Peristerakis, 1961). Societies: Greek Chem. Union; Greek Physicists Union; A.P.S.; A.N.S.
Nuclear interests: Nuclear physics; Nuclear instrumentation. High energy physics; Elementary particles.
Address: 23 Pindou Street, Filothei, Athens, Greece.

KOVACEVIC JOVANOVIC, Olga, Chem. Eng. Born 1923. Educ.: Zagreb Univ. Sen. Res. Eng., Inst. of Nucl. Sci. Boris Kidrich, Belgrade, 1952-; Lecturer, Isotope School A.E.C., Belgrade, 1957-. Publications: Radioisotopes (Belgrade, Rad, 1957); co-author, Practical Radiochem. (Belgrade, Naucna knjiga, 1959); Radioactive Isotopes (Belgrade, Express Press, 1960); co-author, Tech. Encyclopedia. Societies: Chem. Soc., Belgrade; Soc. Yugoslav Eng. and Tech., Belgrade.
Nuclear interest: Nuclear chemistry.
Address: Institute of Nuclear Sciences Boris Kidric, Belgrade, Yugoslavia.

KOVACS, Istvan, Dr. Born 1913. Educ.: Budapest Sci. Univ. Prof., Atomic Phys. Dept. Budapest Polytech. Univ., 1949-. Pres., Spectroscopical Com., Hungarian Acad. Sci. Societies: Presidency Member, Roland Eötvös Phys. Soc.; Hungarian Acad. Sci.
Nuclear interests: Reactor design; Nuclear physics.
Address: Atomic Physics Department, Polytechnical University, Budafoki ut 8, Budapest 112, Hungary.

KOVACS, Julius Stephen, B.S., M.S., Ph.D. Born 1928. Educ.: Lehigh and Indiana Univs. Prof., Michigan State Univ. Society: A.P.S. Nuclear interests: Theoretical elementary particle interactions.
Address: Physics Department, Michigan State University, East Lansing, Michigan, U.S.A.

KOVACS GABORNE, Mrs., Dr. Sen. Officer, Res. Inst. for Irrigation and Rice Cultivation.
Address: Research Institute for Irrigation and Rice Cultivation, 2 Szabadsag ut., Szarvas, Hungary.

KOVAL'CHENKO, M. S. Papers: Co-author, Investigation of Effect of Neutron Irradiation on Structure and Properties of LaB_6 (ibid, vol. 21, No. 6, 1966); co-author, Calculation of Stored Energy in Irradiated Graphite from X-Ray Data (letter to the Editor, ibid, vol. 22, No. 2, 1967); co-author, Some Features of Rad. Damage of Titanium Diboride (ibid, vol. 23, No. 4, 1967); co-author, Study of Rad. Stabilities of Borides (letter to the Editor, ibid, vol. 24, No. 2, 1968).
Address: Academy of Sciences of the U.S.S.R., 14 Leninsky Prospekt, Moscow V-71, U.S.S.R.

KOVALENKO, S. S. Papers: Co-author, Yields of Delayed Neutrons from Fission in ^{239}Pu and ^{232}Th as Induced by Neutrons of Energy 14.5 MeV (letter to the Editor, Atomnaya Energiya, vol. 11, No. 6, 1961); co-author, The Dependence of Total Kinetic Energy for Fission Fragments of Incident Neutron Energy (ibid, vol. 13, No. 5, 1962); co-author, Total Kinetic Energy of U^{233} and U^{232} Fission Fragments (letter to the Editor, ibid, vol. 15, No. 4, 1963); co-author, U^{235} - U^{238} Ternary Fission Probability Ratio for Neutrons of Different Energies (letter to the Editor, ibid, vol. 16, No. 2, 1964).
Address: Academy of Sciences of the U.S.S.R., 14 Leninsky Prospekt, Moscow V-71, U.S.S.R.

KOVALENKO, V. A. Paper: Co-author, Cryogenic Magnetic Mirror Machine WGL-2 (letter to the Editor, Atomnaya Energiya, vol. 21, No. 2, 1966).
Address: Academy of Sciences of the U.S.S.R., 14 Leninsky Prospekt, Moscow V-71, U.S.S.R.

KOVALEV, A. V. Paper: Co-author, Fabrication of Thin Plates from Refractory Carbides (Atomnaya Energiya, vol. 20, No. 6, 1966).
Address: Academy of Sciences of the U.S.S.R., 14 Leninsky Prospekt, Moscow V-71, U.S.S.R.

KOVALEV, E. E. Papers: Co-author, A Technique for Dubna Synchrocyclotron Neutron Beam Study of Material Shielding Properties (Atomnaya Energiya, vol. 16, No. 5, 1964); co-author, Rad. Shield Calculation for Cylindrical Source (letter to the Editor, ibid, vol. 20, No. 1, 1966); co-author, Secondary Protons Emitted from Thick Aluminium Targets During Bombarding by 340 Mev Protons (ibid, vol. 22, No. 6, 1967); co-author, High Energy Nucleons Emitted from Thick Targets of Aluminium During Bombarding by 660 Mev Protons (ibid, vol. 23, No. 3, 1967).
Address: Academy of Sciences of the U.S.S.R., 14 Leninsky Prospekt, Moscow V-71, U.S.S.R.

KOVAL'SKII, Aleksandr Alekseevich. Born 1906. Educ.: Polytechnic Inst., Leningrad. Prof. Inst. of Chem. Phys., Acad. of Sciences of the U.S.S.R., 1947; Former Director, Inst. of Chem. Kinetics and Combustion, Siberian Branch of the Acad. of Sci. of the U.S.S.R. Paper: Co-author, Cross-sections for Inelastic Interaction between 120 and 380 MeV Neutrons and Nuclei (Dokl. Akad. Nauk SSSR, 1956, 106, No. 2). Society: Corr. Member, Acad. of Sci. of the U.S.S.R.
Nuclear interest: Nuclear physics.
Address: Academy of Sciences of the U.S.S.R., 14 Leninsky Prospekt, Moscow V-71, U.S.S.R.

KOVAL'SKII, V. V. VASKhNIL Corresponding Academician. Director, Lab. of All-Union Inst. for Sci. Res. in Animal Breeding.
Nuclear interest: Biochemistry of animals.
Address: Laboratory of the All-Union Institute for Scientific Research in Animal Breeding, ul. 8 Marta 3, Moscow 83, U.S.S.R.

KOVANIC, Pavel, Ph.D. Born 1928. Educ.: Ural Inst. Technol., Sverdlovsk, U.S.S.R. Head, Res. Group, Nucl. Res. Inst., Czechoslovak Acad. Sci. Society: Czechoslovak Sci. and Tech. Soc.
Nuclear interests: Reactor kinetics and control; Nuclear instrumentation; Automation; Applications of digital techniques; Optimum treatment of experimental data.
Address: Nuclear Research Institute, Czechoslovak Academy of Sciences, Rez, Czechoslovakia.

KOVANITS, P. Paper: Co-author, Digital Follow-Up System for Nucl. Eng. (Atomnaya Energiya, vol. 21, No. 2, 1966).
Address: Academy of Sciences of the U.S.S.R., 14 Leninsky Prospekt, Moscow V-71, U.S.S.R.

KOVAR, Frantisek, Eng. Born 1919. Educ.: Prague High Tech. Univ. Czechoslovak A.E.C. (1954-), Formerly Chairman, now Sec. Gen. Society: Czechoslovak Sci. and Tech. Soc.
Nuclear interest: Nuclear energy and reactors.
Address: 7 Slezská, Prague 2, Czechoslovakia.

KOVAR, Zdenek, Ing. (Diploma of Tech. Univ.), C.Sc. Born 1926. Educ.: Brno Tech. Univ. and Inst. of Tech. Phys. U.S.S.R. Acad. of Sci., Leningrad. Sci. Worker, Radiol. Dosimetry Div., Nucl. Res. Inst., Czechoslovak Acad. of Sci., 1960-. Member, Editorial Board, Jaderna Energie. Societies: Member of Nat. Commissions.
Nuclear interests: Dosimetry, special line: Absolute methods of dosimetric measurements, namely calorimetry.
Address: 100 Na Truhlárce, Prague 8, Czechoslovakia.

KOVYRSHIN, L. A. Paper: Co-author, Equipment for Temperature Control During In-Reactor Irradiation of Construction Materials Samples (Atomnaya Energiya, vol. 22, No. 6, 1967).
Address: Academy of Sciences of the U.S.S.R., 14 Leninsky Prospekt, Moscow V-71, U.S.S.R.

KOWALSKA, Krystyna, M.Sc. Born 1922. Educ.: Lodz Univ. At Inst. of Nucl. Res., 1954-. Society: Polish Phys. Soc.
Nuclear interests: Reactor theory; Neutron transport theory; Neutron thermalisation; Reactor codes.
Address: 14a/117 Al. Wyzwolenia, Warsaw 57, Poland.

KOWARSKI, Lew, D. ès. Sc. physiques (Paris). Born 1907. Educ.: Lyons and Paris Univs. Temporary Civil Servant (Dept. of Sci. and Ind. Res.), stationed in Cambridge and Montreal, 1941-46; Sci. Director, C.E.A., Paris, 1946-54; Div. Director, C.E.R.N., Geneva, 1954-63; Sci. Adviser (1956-63), now Member, Group of Experts on Production of Energy from Radioisotopes, and Member, Study Group on Long Term Role of Nucl. Energy in Western Europe, O.E.C.D., E.N.E.A. Visiting Prof., Nucl. Eng. Dept., Purdue Univ., 1963-. Societies: Sté. de Physique; Phys. Soc.; A.P.S.; A.N.S.
Nuclear interests: Neutron physics; Reactor design; Nuclear applications of information theory; Organisation and international collaboration in nuclear research.
Address: Purdue University, Lafayette, Indiana, U.S.A.

KOYAMA, Seitaro. Born 1912. Educ.: Tokyo Univ. Prof. Chem., Faculty of Sci., Niigata Univ., 1951-. Member, Special Com. on Atomic Energy, Sci. Council of Japan. Societies: Chem. Soc. of Japan; Japan Soc. for Analytical Chem.; Japan Rad. Res. Soc.
Nuclear interests: Radioactive contamination of human environment. Radioactive fallout.
Address: Faculty of Science, Niigata University, 5214 Nishi-Ohata-Machi, Niigata, Japan.

KOZHEVNIKOV, A. V. Paper: Co-author, Exptl. Investigation of Variation of Electron Beam Cross Dimensions in the 1.5 GeV Synchrotron (Atomnaya Energiya, vol. 24, No. 1, 1968).
Address: Academy of Sciences of the U.S.S.R., 14 Leninsky Prospekt, Moscow V-71, U.S.S.R.

KOZHEVNIKOV, D. A. Papers: Age and Area of Neutron Migration from Polyenergetic Sources in Organic and Metal-Containing Moderators (Atomnaya Energiya, vol. 19, No. 4, 1965); Approximate Similarity of Neutron Fields, Formed by Neutron Sources with Different Spectra (letter to the Editor, ibid, vol. 20, No. 2, 1966); A Neutron Age of Hydrocarbons (letter to the Editor, ibid, vol. 24, No. 2, 1968); On Slowing Down of Neutrons in Capturing Media (letter to the Editor, ibid, vol. 24, No. 2, 1968).
Address: Academy of Sciences of the U.S.S.R., 14 Leninsky Prospekt, Moscow V-71, U.S.S.R.

KOZHUHOV, I. V. Paper: Co-author, Collective Linear Acceleration of Ions (Atomnaya Energiya, vol. 24, No. 4, 1968).
Address: Academy of Sciences of the U.S.S.R., 14 Leninsky Prospekt, Moscow V-71, U.S.S.R.

KOZIEL, Jerzy, Eng., M.S. Born 1934. Educ.: Warsaw Tech. Univ.
Nuclear interests: Reactor physics and technology; Physical characteristics of reactor cores; Neutron density distribution measurements on space and spectrum; Zero power and critical experiments.
Address: Apt. 5, 7 Pazinskiego Street, Warsaw 89, Poland.

KOZIK, B. Papers: Correlation of Neutrons in Reflected Reactor (Abstract, Atomnaya Energiya, vol. 20, No. 4, 1966); On Statistical Theory of Reactors (Kernenergie, vol. 9, No. 3, 1966); Correlation of Neutrons in Nucl. Reactors with Respect to Space and Energy Distribution of Neutrons (Atomnaya Energiya, vol. 20, No. 6, 1966); Correlation Theory of Neutron Multiplication in Nuclear Reactors (Kernenergie, vol. 9, No. 12, 1966).
Address: Joint Institute for Nuclear Research, Dubna, Near Moscow, U.S.S.R.

KOZIN, V. P. Papers: Co-author, 100 MeV Neutron Collector (Jaderna Energie, vol. 13, No. 11, 1967); co-author, Storage Ring for 100 MeV Electrons (Atomnaya Energiya, vol. 23, No. 6, 1967).
Address: Physico-Technical Institute, 2 Polytechnicheskaya Ulitsa, Leningrad, U.S.S.R.

KOZLOV, F. A. Papers: Gasometric and Gravimetric Methods of Assaying Sodium for its Oxygen Content. Application to the Analysis of the Contents of Oxide Traps (letter to the Editor, Atomnaya Energiya, vol. 12, No. 4, 1962); co-author, Natrium Decontamination from Interaction Products with Water in Circulation Loop (letter to the Editor, ibid, vol. 19, No. 3, 1965); co-author, Dependence of Heat Conductibility Sodium from Concentration of Oxides (ibid, No. 4); co-author, Leakage Discovery in Sodium-Water Steam Generator (ibid, vol. 20, No. 6, 1966).
Address: Academy of Sciences of the U.S.S.R., 14 Leninsky Prospekt, Moscow V-71, U.S.S.R.

KOZLOV, V. I. Paper: Co-author, Measurements of Slow Neutron Spectra on Reactor Physical Stand at Beloyarskaya I.V. Kurchatov GRES (Atomnaya Energiya, vol. 15, No. 6, 1963).
Address: Academy of Sciences of the U.S.S.R., 14 Leninsky Prospekt, Moscow V-71, U.S.S.R.

KOZLOV, V. Ya. Papers: Co-author, Start-up Conditions in a Uranium-Graphite Power

Reactor Providing Superheated Steam (Atomnaya Energiya, vol. 9, No. 1, 1960); co-author, Chem. Stability of Deposits and Transport of Radioactive Material in Water and Steam in a Water-Steam Loop in the First Soviet Power Station (ibid, No. 2); co-author, Experience with the Operation of The First Nucl. Power Station (ibid, vol. 11, No. 1, 1961); co-author, Operational Experience of the First Nucl. Power Station (ibid, vol. 16, No. 6, 1964).
Address' Academy of Sciences of the U.S.S.R., 14 Leninsky Prospekt, Moscow V-71, U.S.S.R.

KOZLOV, Yu. D. Paper: Co-author, Reliability of Rad. Chem. Installations (letter to the Editor, Atomnaya Energiya, vol. 24, No. 4, 1968).
Address: Academy of Sciences of the U.S.S.R., 14 Leninsky Prospekt, Moscow V-71, U.S.S.R.

KOZLOVA, Mrs. Anna Vasilievna, Prof., Dr. of Medicine. Born 1906. Educ.: Voronezh Univ. Sen. Sci. Worker (1945-48) and Chief of Radiol. Dept. (1948-), Roentgen-Radiol. Inst., Moscow. Chief, Central Postgraduate Medical Inst., Moscow. Books: The Principles of Radium-Therapy (1956); and 3 other books. Papers: 65 scientific articles in the field of medical radiology. Societies: All-Union Soc. of Roentgenologists and Radiologists (member of Directorate); All-Union Com. of Medical Radiol.; Soc. of Soviet-Japan friendship (member of Directorate); Soviet National Com. 'Living Conditions and Health'.
Address: Institute of Roentgen-Radiology, 7 Solyanka Street, Moscow, U.S.S.R.

KOZLOVA, P. S. Paper: Co-author, On Two Genetic Types of Postmagmatic Thorium-Rare Earths Deposits (Atomnaya Energiya, vol. 19, No. 3, 1965).
Address: Academy of Sciences of the U.S.S.R., 14 Leninsky Prospekt, Moscow V-71, U.S.S.R.

KOZLOVSKI, S. A. Papers: Co-author, Streaming of Neutrons through Cylindrical and Flat Neutron Guides (Kernenergie, vol. 11, No. 2, 1968). co-author, Determination of Fast Neutron Flux with Aid of ZnS (Ag) + Plastics Detector and Basson's Detector (Atomnaya Energiya, vol. 24, No. 6, 1968).
Address: Academy of Sciences of the U.S.S.R., 14 Leninsky Prospekt, Moscow V-71, U.S.S.R.

KOZUSZNIK, Boguslaw, B.A. in Medicine (specialised in pediatry, ind. med. and public health). Born 1910. Educ.: Karol's Univ., Prague. Under-Sec. of State, Ministry of Health, 1946-59; Gen. Supervisor of Hygiene and Sanitation in Poland, 1954-59; Lecturer and Head of Chair of Health Organisation, Post-Graduate Medical School, Warsaw; Leader of Polish Deleg., W.H.O., Geneva, 1948, 1956, 1957, 1958, 1959. Societies: State Com. of Peaceful Exploitation of Atomic Energy (Exec. Board, 1959, 1960, 1961);

Vice-Chairman, U.N.I.C.E.F. Nuclear interests: Hygiene; Radiation; Organisation of health-protection in the field of nuclear research.
Address: Aleja I-szej Armii Wojska Polskiego 16/2, Warsaw, Poland.

KOZYREV, A. P. Paper: Co-author, Alkali Metals Boiling Heat Transfer (letter to the Editor, Atomnaya Energiya, vol. 19, No. 2, 1965).
Address: Academy of Sciences of the U.S.S.R., 14 Leninsky Prospekt, Moscow V-71, U.S.S.R.

KOZYREVA-ALEKSANDROVA, L. S. Book: Co-author, Radioaktivnyi izotop ioda J^{131} (Radioactive Isotope of Iodine J^{131}) (M. Atomizdat, 1960).
Address: Academy of Sciences of the U.S.S.R., 14 Leninsky Prospekt, Moscow V-71, U.S.S.R.

KRAAK, W., Drs. Born 1929. Educ.: Amsterdam Municipal Univ. Joint Establishment for Nucl. Energy Res., Kjeller, Norway, 1957-60; Nucl. Res. Centre, Fontenay-aux-Roses, France, 1960-61; Reactor Centrum Nederland, Petten, 1961-64; Inst. for Atomenergie, Kjeller, 1964-. Member, Editorial Com. Atoomenergie en haar toepassingen. Societies: Roy. Netherlands Chem. Soc.; Netherland Phys. Soc.
Nuclear interests: Chemistry of heavy elements; Fuel reprocessing.
Address: Institutt for Atomenergi, P.O. Box 40, Kjeller, Norway.

KRAAN, W. VAN DER. See VAN DER KRAAN, W.

KRACIK, Jiří, Prof., Ing. Dr. Sc., Ph.D. Born 1929. Educ.: Czechoslovak Tech. Univ., Prague. Lecturer (1957), Head, Phys. Dept.; New Method of Conversion of Energy Group, Czechoslovak Atomic Commission. Books: The Elec. Discharges (Prague, S.N.T.L., 1964); Plasma Phys. (Prague, Academia, 1966). Society: Sci. Council, Czechoslovak Tech. Univ.
Nuclear interest: Hot plasma.
Address: 1902 Technická ul., Prague 6-Dejvice, Czechoslovakia.

KRAEVSKY, N. A. Adviser to Russian rep. at 4th. session, U.N. Sci. Com. on Effects of Atomic Rad.; Corresponding Member, Acad. of Medical Sci. of the U.S.S.R.
Address: Academy of Medical Sciences of the U.S.S.R., Moscow, U.S.S.R.

KRAFFT, Pierre, M.Sc., Ing. diplômé. Born 1929. Educ.: Ecole Polytechnique de l'Univ. de Lausanne and Georgia Inst. Technol With Detroit Edison Co., Detroit, 1957-58; Atomelectra, Ltd., Zurich, and Supt., Swiss Exptl. Nucl. Power Plant, Lucens. Society: Sté. Suisse des spécialistes du génie nucléaire. Nuclear interests: Engineering; Economics. Administration; Operation.

Address: Centrale Atomique, Lucens, VD, Switzerland.

KRAIG, Howard Irwin, B.S., M.S. (Mech. Eng.). Born 1926. Educ.: Columbia Univ. Head, Nucl. Eng. Dept., New York Shipbuilding Corp. Societies: A.N.S.; Soc. of Naval Architects and Marine Eng.
Address: 213 Madison Avenue, Merchantville, New Jersey, U.S.A.

KRAINER, Helmut, Prof. Dr. mont. Born 1909. Educ.: Director of Res., Fried. Krupp, Essen; Apl. Prof., Braunschweig T.H. Nuclear interests: Metallurgy and reactor design.
Address: 103 Altendorfestrasse, Essen, Germany.

KRAINII, A. G. Papers: Co-author, Investigation of Effect of Neutron Irradiation on Structure and Properties of LaB_6 (Atomnaya Energiya, vol. 21, No. 6, 1966); co-author, Some Features of Rad. Damage of Titanium Diboride (ibid, vol. 23, No. 4, 1967); co-author, Study of Rad. Stabilities of Borides (letter to the Editor, ibid, vol. 24, No. 2, 1968).
Address: Academy of Sciences of the U.S.S.R., 14 Leninsky Prospekt, Moscow V-71, U.S.S.R.

KRAMER, Andrew William, B.S. (Elec. Eng.), Licensed Professional Eng., Illinois. Born 1893. Educ.: Armour Inst. Technol., Chicago. Formerly Editor Power Eng.; now Consulting Editor, Power Eng. Chairman, Nucl. Div., American Power Conference. Books: Boiling Water Reactors (Addison-Wesley Publishing Co., 1958); Nucl. Propulsion for Merchant Vessels (U.S.A.E.C., 1960). Societies: A.N.S.; I.E.E.E.; A.S.M.E. Nuclear interests: Nuclear physics; Reactor design and operation; Nuclear power development; Ionising radiation.
Address: 667 Rockland Avenue, Lake Bluff, Illinois, U.S.A.

KRAMER, Ernst Lodewijk, Dr. Born 1902. Educ.: Rotterdam Economic Univ. Chairman of the Board, Reactor Centrum Nederland; Member, Nucl. Energy Commission, Ministry of Economic Affairs; Member, Ind. Council for Nucl. Energy; Member. Central Council for Nucl. Energy.
Address: 17 Bloemcamplaan, Wassenaar, Netherlands.

KRAMER, Gordon. At Columbus Labs., Battelle Memorial Inst., 1965-.
Address: Battelle Memorial Institute, 505 King Avenue, Columbus 1, Ohio, U.S.A.

KRAMER, Gustav, Dr.Educ.: Heidelberg Univ. Prof, Phys., Hamburg Univ.
Address: Physikalisches Staatsinstitut, Univ. Hamburg, 149 Luruper Chausee, Hamburg-Bahrenfeld, Germany.

KRAMER, Hans O. R., Dr. Ehemaliges Mitglied, Vorstand, Farbenfabriken Bayer A.G. Member, Specialised Nucl. Sect. for Economic Problems, and Member, Specialised Nucl. Sect. for Social Health and Development Problems, Economic and Social Com., Euratom, 1966-.
Address: Farbenfabriken Bayer A.G., Bayerwerk, 509 Leverkusen, Germany.

KRAMER, L. B. Reactor Design Manager, Bettis Atomic Power Div., SIW, S2Wa, S3W, S4W Projects. Member, Special Administrative Com. on Nucl. Problems, A.S.T.M.
Address: Westinghouse Electric Corp., Bettis Atomic Power Division, Clairton Site, P.O. Box 1526, Pittsburgh 30, Pennsylvania, U.S.A.

KRAMER-AGEEV, E. A. Papers: Co-author, Nomograms for Neutron Water Shielding Calculation (letter to the Editor, Atomnaya Energiya, vol. 15, No. 2, 1963); co-author, Neutron Flux Distribution in Straight Cylindrical Channel (letter to the Editor. ibid, vol. 19, No. 1, 1965); co-author, Angular Distribution of Scattering Neutrons Doses by Screens (letter to the Editor, ibid, vol. 20, No. 2, 1966).
Address: Academy of Sciences of the U.S.S.R., 14 Leninsky Prospekt, Moscow V-71, U.S.S.R.

KRAMISH, Arnold, B.S., M.A. Born 1923. Educ.: Denver, Harvard and Stanford Univs. Consultant and Staff Member, U.S.A.E.C., 1947-51; Res. Manager, Atomic Ind. Forum, New York, 1955; Sen. Staff Member, Rand Corp., 1951-; Res. Fellow, Council on Foreign Relations, New York, 1958-59; Consultant, N.S.F., 1958-60; Consultant, Internat. Bank, 1958; Consultant, European Inst. Univ. Studies, Brussels, 1960-62; Consultant, O.E.E.C., Paris 1961-62;Guggenheim Fellow, 1966-67. Books: Atomic Energy in the Soviet Union (Stanford Univ. Press, Oxford Univ. Press, 1959); co-author, Atomic Energy for your Business(New York, David McKay, 1956); co-author, The Effects of Atomic Weapons (U.S.A.E.C. and McGraw-Hill, 1950); The Peaceful Atom in Foreign Policy (Harper and Row, 1963). Society: Atomic Ind. Forum.
Nuclear interests: Power; Isotope applications; Management; Nuclear physics; Political and economic implications; Arms control: Research and development studies.
Address: 1000 Connecticut Avenue N.W., Washington D.C., U.S.A.

KRANICH, W. L. Member, Nucl. Eng. Education Programme Com., Worcester Polytech. Inst.
Address: Worcester Polytechnic Institute, Worcester, Massachusetts, U.S.A.

KRANTZ, Reinhold J. Chem. Prof., Chem. Dept., Redlands Univ.

Address: Redlands University, Redlands, California, U.S.A.

KRANZ, Jakob, Prof. Sec., Kommission zur friedlichen Nutzung der Atomkräfte, Bavaria. Address: 1 Physikalisches Institut, Munich University, Munich, Germany.

KRASIK, Sidney, B.S.(E.E.), Ph.D. Born 1911. Educ.: Carnegie Inst. Technol. and Cornell Univ. Manager, Phys. Dept., Bettis Plant (1950-54), Manager, P.W.R. Phys., Bettis Plant (1954-56), Manager, Central Physics and Maths., Bettis (1956-), Project Manager, NERVA, Vice-Pres. and Gen. Manager (1962-64), Sen. Consultant, Atomic Defence and Space Group (1964-), Westinghouse Elec. Corp. Societies: Fellow, A.P.S.; A.A.A.S.; Fellow, A.N.S.; Fellow, I.E.E.E. Address: 2081 Beechwood Blvd., Pittsburgh 17, Pennsylvania, U.S.A.

KRASIN, A. K. Member, Editorial Board, Atomnaya Energiya. Nuclear interests: Reactor physics. Address: Nuclear Energy Institute, Academy of Sciences of the U.S.S.R., 14 Leninsky Prospekt, Moscow V-71, U.S.S.R.

KRASNICKI, Szczesny Bogumil Jan, Dr. of Phys. Born 1936. Educ.: Jagellonian Univ. Asst. (1956-65), Lecturer (1958-), Sen. Asst. (1965-66), Jagellonian Univ., Cracow; Sen. Res. Asst., Nucl. Phys. Inst., Cracow, 1965-. Nuclear interest: Scattering of neutrons by solids and liquids. Address: 20 m. 7 ul. Podzamcze, Cracow, Poland.

KRASNOKUTSKII, R. N. Paper: Co-author, Fast Neutron Capture Cross Section for Rhenium (letter to the Editor, Atomnaya Energiya, vol. 19, No. 1, 1965). Address: Academy of Sciences of the U.S.S.R., 14 Leninsky Prospekt, Moscow V-71, U.S.S.R.

KRASNONOSEN'KIH, P. P. Paper: Co-author, 1,5 GeV Electron Synchrotron at the Polytech. Inst. of Tomsk (Atomnaya Energiya, vol. 21, No. 6, 1966). Address: Academy of Sciences of the U.S.S.R., 14 Leninsky Prospekt, Moscow V-71, U.S.S.R.

KRASNOPEROV, V. M. Paper: Co-author, Possibility of Rd^{83} Use for Mössbauer's Effect Investigations with Kr^{83} (letter to the Editor, Atomnaya Energiya, vol. 23, No. 1, 1967). Address: Academy of Sciences of the U.S.S.R., 14 Leninsky Prospekt, Moscow V-71, U.S.S.R.

KRASNORUTSKII, V. S. Paper: Co-author, Impurity State Effect on Deformation Texture of Low-Alloyed Alpha-Uranium (letter to the Editor, Atomnaya Energiya, vol. 24, No. 1, 1968). Address: Academy of Sciences of the U.S.S.R., 14 Leninsky Prospekt, Moscow V-71, U.S.S.R.

KRASNOV, N. N. Papers: Co-author, Yields of Isotope Au^{195} in Nucl. Reactions on Cyclotron (letter to the Editor, Atomnaya Energiya, vol. 23, No. 1, 1967); co-author, Reaction $Li + He_2^4$, $Li + H^2$, and $Li + H_1^1$ as Cyclotron Sources of Fast Neutrons (letter to the Editor, ibid, vol. 24, No. 1, 1968); co-author, Yields of Ce^{139} in Nucl. Reactions La^{139} (p,n) and La^{139} (d,2n) (letter to the Editor, ibid, vol. 24, No. 3, 1968); co-author, Excitation Function of Cu^{65} (p,n) Zn^{65} Reaction (letter to the Editor, ibid, vol. 24, No. 3, 1968). Address: Academy of Sciences of the U.S.S.R., 14 Leninsky Prospekt, Moscow V-71, U.S.S.R.

KRASNOW, Arthur L., B.S., S.B. Born 1924. Educ.: U.S. Naval Acad. and M.I.T. Nucl. Sales Manager, Baird-Atomic, Inc., and Atomic Instrument Co., 1952-57; Eastern Regional Manager, Nucl. Systems Div., Budd Co., 1957-59; Pres., Atomic Personnel, Inc., 1959-. Societies: A.N.S.: Delaware Valley Soc. for Rad. Safety; Pennsylvania Assoc. of Personnel Services; Nat. Employment Assoc. Nuclear interests: Professional personnel and employment consultant in the nuclear and allied scientific fields. Address: 1518 Walnut Street, Philadelphia Pennsylvania 19102, U.S.A.

KRASNOYAROV, N. V. Phys., Inst. of Phys. and Power, Acad. of Sci. of the U.S.S.R. Papers: Co-author, The Behaviour of a Reactor with Temperature Self-regulation (letter to the Editor, Atomnaya Energiya, vol. 7, No. 4, 1959); co-author, A Fast Neutron Burst Reactor (ibid, vol. 10, No. 5, 1961); co-author, Use of Concrete for Shielding of High Temperature Reactors (ibid, vol. 19, No. 6, 1965); co-author, Physical Problems in Developing Fast Power Reactors (Kernenergie, vol. 9, No. 9, 1966). Address: Institute of Physics and Power, Academy of Sciences of the U.S.S.R., Obninsk, near Maloyaroslavets, Moscow, U.S.S.R.

KRASOVEC, Franc, Dr. of Chem. Sci. Born 1924. Educ.: Ljubljana Univ. Inst. Josef Stefan, 1953-. Nuclear interest: Solvent extraction chemistry. Address: Institute Josef Stefan, P.O. Box 199, Ljubljana, Yugoslavia.

KRASOVITSKII, V. B. Papers: Acceleration of Charged Particle in Field of Plane Wave with Changing Phase Velocity (letter to the Editor, Atomnaya Energiya, vol. 20, No. 4, 1966); Energy Gain of Oscillator in Field of Standing Wave Near Cyclotron Resonance (ibid, vol. 22, No. 6, 1967); co-author, Theory of Non-linear Interaction of a Modulated Beam with Plasma (ibid, vol. 24, No. 6, 1968). Address: Academy of Sciences of the U.S.S.R., 14 Leninsky Prospekt, Moscow V-71, U.S.S.R.

KRATSCHMANN, Helmut, Dipl. Ing. Born 1932. Educ.: Vienna Inst. of Technol. Asst. Instructor, Technol. High School.
Nuclear interests: Reactor operation and maintenance.
Address: Österreichische Studiengesellschaft für Atomenergie G.m.b.H., 10 Lenaug., Vienna 8, Austria.

KRATZER, Myron B., B.Chem. Eng. Born 1925. Educ.: Ohio State Univ. Deputy Director Internat. Affairs Div. (1959-64), Director Internat. Affairs Div. (1964-67), Asst. Gen. Manager for Internat. Activities (1967-), U.S.A.E.C.
Address: International Activities Division, United States Atomic Energy Commission, Washington D.C. 20545, U.S.A.

KRAUCH, Carl Heinrich, Dr. rer. nat., Dipl. Chem. Born 1931. Educ.: Heidelberg and Göttingen Univs. Abtl. Strahlenchemie, Max Planck Inst. für Kohlenforschung, 1958-67. Societies: Faraday Soc.; Gesellschaft Deutscher; Cheimker; Gesellschaft für Lichtforschung.
Nuclear interests: Radiation chemistry; Photochemistry.
Address: Basf, Ludwigshafen/Rhein, Germany.

KRAUCH, Helmut, Dr. Chem. Born 1927. Educ.: Heidelberg Univ. At Max-Planck-Inst., Heidelberg, 1953-; Res. Asst., Yale Univ., 1956; Chem. Dept., Brookhaven Nat. Lab., 1957; Sci. Asst. to Board of Directors, German Atomic Centre, Karlsruhe, 1958; Lecturer Nucl. Technol., School of Economics, Mannheim, 1959; Head, Studiengruppe für Angewandte Radio-und Strahlenchemie, 1960; Director, Studiengruppe für Systemforschung, 1964; Visiting Prof. Res. Management,California Univ., Berkeley, 1966-67.
Book: Name Reactions of Organic Chem. (New York,Wiley, 1961; Heidelberg, Hüthig Verlag, 1961).
Nuclear interests: Radiation technology; Research management, planning and administration.
Address: 94 Handschuhsheimer Landstrasse, Heidelberg, Germany.

KRAUS, J. Eng., Inst. for Nucl. Phys. Res., Amsterdam.
Address: Institute for Nuclear Physics Research, 18 Oosterringdijk, Amsterdam O, Netherlands.

KRAUS, Kurt A., B.S. (Harvard), Ph.D. (Johns Hopkins). Born 1914. Educ.: Harvard and Johns Hopkins Univs. Now at Chem. Div., O.R.N.L. Society: A.C.S.
Nuclear interests: Actinide chemistry, radiochemistry, seperations chemistry.
Address: Oak Ridge National Laboratory, Chemistry Division, P.O. Box X, Oak Ridge, Tennessee, U.S.A.

KRAUS, Vladimír, Metal. Eng., Candidate of sci. Born 1931. Educ.: Mining Univ., Ostrava. Head, Phys. Metal. Lab., Inst. of Nucl. Res., Rez.
Nuclear interests: Reactor metallurgy; Physical metallurgy of uranium metal and its alloys.
Address: Institute of Nuclear Research, Rez by Prague, Czechoslovakia.

KRAUSE, A. V. Electron Tubes, 20th Century Electronics Ltd.
Address: 20th Century Electronics Ltd., Centronics Works, King Henry's Drive, New Addington, Croydon CR9 OBG, Surrey, England.

KRAUSE, Robert F. Director, Yankee Atomic Elec. Co.
Address: Yankee Atomic Electric Co., 441 Stuart Street, Boston 16, Massachusetts, U.S.A.

KRAUSHAAR, Jack Jourdan, B.S. (Lafayette Coll.), M.S. (Syracuse), Ph.D. (Syracuse). Born 1923. Educ.: Swarthmore and Lafayette Colls. and Syracuse Univ. Res. Fellowship (1949-51), Res. Assoc. (1951-53), Brookhaven Nat. Lab.; Instructor, Stanford Univ., 1953-56; Asst. Prof. (1956-58), Assoc. Prof. (1958-63), Prof. (1963-), Colorado Univ. Scoieties: Fellow, A.P.S.; Federation of American Sci.
Nuclear interests: Nuclear spectroscopy; Gamma ray angular and polarisation correlation studies; Decay schemes; Nucleon transfer reactions; Elastic and inelastic charged particle studie
Address: Physics and Astrophysics Department, Colorado University, Boulder, Colorado, U.S.A.

KRAVCHENKO, V. S. Paper: Empirical Equation of Temperature Dependence of Heavy Water Density (letter to the Editor, Atomnaya Energiya, vol. 20, No. 2, 1966).
Address: Academy of Sciences of the U.S.S.R., 14 Leninsky Prospekt, Moscow V-71, U.S.S.R.

KRAVCHIN, B. V. Paper: Co-author, Divertor of the Sirius Stellarator (Atomnaya Energiya, vol. 23, No. 2, 1967).
Address: Academy of Sciences of the U.S.S.R., 14 Leninsky Prospekt, Moscow V-71, U.S.S.R.

KRAVTCHENKO, Julien, D. ès Sc. Born 1911. Educ.: Paris Univ. At Faculté des Sci., Grenoble Univ.; At Inst. Polytech., Grenoble. Societies: Soc. Mathématique de France; American Mathematical Soc.
Nuclear interest: Transferts thermiques dans les réacteurs nucléaires.
Address: Laboratoires de Mécanique des Fluides, 44-46 Avenue Félix-Viallet, 38 Grenoble, France.

KRAWCZYK-OBOJSKA, Irena, Dr.Dipl. Eng. Born 1929. Educ.: Warsaw Inst. Technol. Sci. worker, Warsaw Inst. Technol., 1950-51; Sci. worker, Inst. Gen. Chem., Warsaw, 1952-55; Sci. worker, Analytical Dept.

(1955-61), Sci. Worker, Dept. Application of Isotopes in Chem. and Chem. Technol. (1961-), Inst. Nucl. Res., Polish Acad. Sci., Warsaw.
Nuclear interest: Application of ion-exchange resins in nuclear chemistry (analysis and technology).
Address: 16 Ul. Dorodna, Warsaw 9, Poland.

KRAWCZYNSKI, Stefan, Dr. rer.nat., Dipl. Phys. Born 1929. Educ.: Munich Univ. At Munich Univ. 1948-55; Max-Planck-Inst., Göttingen (Nucl. Eng. Technol.), 1955-56; Head, Reprocessing and Waste Disposal Dept., Kernreaktor-Bau-u. Betriebsgesellschaft m.b.H. Karlsruhe, 1958-60; Head Service Reprocessing, C.E.E.A. C.C.R. Ispra, 1961-. Panel, Safety and Rad. Protection in Nucl. Plants, German A.E.C.; Panel, Low-level Waste Disposal, I.A.E.A., Vienna. Books: Co-author, Lexikons Kerntechnik; Radioactive Wastes (C.Thiemig Verlag).
Nuclear interest: Nuclear chemical engineering.
Address: Institut für Reaktor-Werkstoffe, Kernforschungsanlage Jülich, Jülich 517, Germany.

KRAYBILL, Henry Lawrence, S.B., Ph.D. Born 1918. Educ.: Chicago Univ. Instructor (1948-51), Asst. Prof. (1951-57), Assoc. Prof. (1957-), Director of Graduate Studies, Phys. Dept. (1964-65, 1966-67), Yale Univ. Society: A.P.S.
Nuclear interest: High energy nuclear physics.
Address: 960 Benham Street, Hamden, Connecticut, U.S.A.

KREBES, D. T. Van de Graaff Supervisor, IIT Res. Inst.
Address: IIT Research Institute, Technology Centre, 10 West 35th Street, Chicago, Illinois 60616, U.S.A.

KREBS, A., Dr. med. Director, Rad. Hygiene Sect., Federal Ministry of Health.
Address: Federal Ministry of Health, 87 Deutschherrenstr., 532 Bad Godesberg, Germany.

KREBS, Adolph T., Dr. Prof. of Biol., Coll. of Arts and Sci., and Lecturer in Physiology and Assoc. in Radiol., School of Medicine, Louisville Univ.; Head, Radiobiol. Dept., and Chief Res. Physicist, U.S. Army Medical Res. Lab., Fort Knox.
Address: United States Army Medical Research Laboratory, Fort Knox, Kentucky, U.S.A.

KREBS, Carl, M.D. Born 1892. Educ.: Copenhagen Univ. Prof., Radiol., Aarhus Univ.; Director, Radiumstationen for Jutland, Aarhus. Societies: Hon. Member, Nordisk Forening for Medicinsk Radiol.; Norsk Medicinsk Selskab.
Nuclear interests: Significance of general X-ray radiation in lowering cancer resistance of normal organism and for spontaneous occurance of leukemia and leukemia-like diseases.
Address: 41 Kystvej, Aarhus, Denmark.

KREBS, William A. W., Jr., A.B., LL.B. Born 1916. Educ.:Yale Univ. and Yale Law School. Vice-Pres., Arthur D. Little, Inc. 1956-; Sen. Lecturer, Sloan School of Management, M.I.T. Formerly: Counsel, U.S.A.E.C.; Commissioner, Massachusetts Commission on Atomic Energy.
Nuclear interests: Atomic energy management; Government policy; Economic development.
Address: Arthur D. Little, Incorporated, Acorn Park, Cambridge 02140, Massachusetts, U.S.A.

KREGER, William E., Dr., Ph.D. (Phys.). Head, Nucl. Rad. Phys. Branch (1959-61), Head, Nucleonics Div. (1961-65), Head, Cyclotron Project (1965-66), Head, Phys. Sci. Div. (1966-), U.S. Naval Radiol. Defence Lab., San Francisco. Exec. Com., Shielding Div., A.N.S., 1962-. Society: Fellow, A.P.S.
Address: Nucleonics Division, United States Naval Radiological Defence Laboratory, San Francisco, California 94135, U.S.A.

KREH, Edward Joseph, Jr., B.S. (Mech. Eng.). Born 1915. Educ.: Carnegie Mellon and Pittsburgh Univs. Manager, Mech. Equipment Sect. (1951-54), Manager, Equipment Development Subdiv. (1954-57), Div. Manager's Staff (1957-58), Div. Apparatus Eng. Manager (1958-59), Manager, Nucl Core Dept. (1959-61), Manager, Core Materials Dept. (1961-64), Manager, Materials Dept. (1964-65), Manager, Central Labs. (1965-), Westinghouse Bettis Plant. Societies: A.S.M.E.; A.N.S.; N.A.C.E.; American Management Assoc.
Nuclear interests: Reactor plant and equipment design; Nuclear core fabrication; Metallurgy.
Address: 176 Longue Vue Drive, Pittsburgh, Pennsylvania, 15228, U.S.A.

KREIDL, Norbert Joachim, Ph.D. Born 1904. Educ.: Vienna Univ. Director, Materials R. and D., Bausch and Lomb, Inc., 1943-65; Prof., Ceramics, Rutgers, The State Univ., 1964-66; Prof., Ceramic Eng., Missouri Univ. at Rolla, 1966-. Vice Pres., Internat. Commission on Glass (Union of glass socs.); Chairman, Materials Advisory Board, Com. on Irradiated Materials, N.R.C. Book: Chapter 5 in Modern Materials (Academic Press, 1958). Societies: Optical Soc. America; American Ceramic Soc.
Nuclear interests: Radiation effects in ceramics; Solid state dosimetry.
Address: Department of Ceramic Engineering, Missouri University at Rolla, Rolla, Missouri 65401, U.S.A.

KREITZ, Karl, Dr.-Ing. Born 1894. Educ.: Aachen T.H. Beratender Ing., früher Leiter der Versuchsanstalt der Stahl- und

Rohrenwerk Reisholz G.m.b.H. 1925-59; Personal Member, Study Assoc. for the use of Nucl. Energy in Navigation and Industry. Society: Verein Deutscher Eisenhuttenleute. Nuclear interest: Werkstoff Fragen Stahl.
Address: 16 Böcklinstrasse, Düsseldorf, Germany.

KREJCI, Miloslav, Elec. Eng., C.Sc. Born 1926. Educ.: Prague Tech. Univ. Nucl. Res. Inst., Czechoslovak Acad. Sci. 1956-.
Nuclear interests: Reactor control and automation; Applications of digital and analogue computers in reactor engineering.
Address: Nuclear Research Institute C.S.A.V., Rez, Czechoslovakia.

KREKELER, Heinz L., Dr. phil. (Berlin), Hon. Dr. of Laws(South Carolina and Xavier (Cincinnati) Univs.). Born 1906. Educ.: Freiburg im Breisgau, Munich, Göttingen and Berlin Univs. Consul Gen. to U.S.A., 1950-51; Chargé d'Affaires to U.S.A., 1951-53; Ambassador to U.S.A., 1953; Ambassador E. and P. to U.S.A., 1955-58; Commission of C.E.E.A., 1958-64. Society: Deutsche Chemische Gesellschaft.
Nuclear interests: General; Industrial applications.
Address: 4912 Gut Lindemannshof, Post Sylbach in Lippe, Germany.

KREMENEK, Jaroslav, Ph.D. Born 1925. Educ.: Charles' Univ., Prague. Skoda Works, Plzen, 1950-56; A. F. Ioffe Physico-Techn. Inst., U.S.S.R. Acad. Sci., 1956-61; Nucl. Res. Inst., Czechoslovak Acad. Sci., 1961-.
Nuclear interests: Nuclear reactions; Nuclear theory; Instrumentation.
Address: Nuclear Research Institute, Rez, Czechoslovakia.

KREMERS, Howard E., A.B., M.S., Ph.D. Born 1917. Educ.: Western Reserve, Syracuse, and Illinois Univs. Manager, Market Development and Tech. Service, American Potash and Chem. Corp. Societies: A.I.M.E.; A.C.S.; American Ceramic Soc.; Electrochem. Soc.; A.S.M.; A.I.A.A.; American Inst. Chem.; A.A.A.A.
Nuclear interest: Market development and technical service aspects of primary rare earth and thorium materials.
Address: American Potash and Chemical Corporation, 99 Park Avenue, New York, New York 10016, U.S.A.

KREPELKA, Jiří, Ing. Born 1933. Educ.: Prague High School of Chem. Ustav jaderného výzkumu, Czechoslovak Acad. of Sci.
Nuclear interests: Low-level radioactive wastes; Measurement of low level radioactivity; Determination of radionuclides.
Address: 143/198 Belohorská tr., Prague 6, Czechoslovakia.

KRESS, Herwig, Dr.-Ing. Born 1910. Educ.: Munich T.H. and Berlin T.H. Escher-Wyss G.m.b.H., Ravensburg, 1947-58; Chief, Development and Eng., Kernreaktor Bau- und Betriebsgesellschaft m.b.H., Karlsruhe, 1959-60; Chief, Development, Fried. Krupp, Essen, 1961-67; Chairman, Tech. Advisory Board, Brown Boveri/Krupp Reaktorbau G.m.b.H., 1961-65.
Nuclear interests: Reactor design; Management.
Address: 7981 Oberzell, 5 Kloecken, Germany.

KRESS, R. P. Paper: Co-author, Use of Associated Particle Method for Absolute Determination of N for Neutrons Emitted by a Source (letter to the Editor, Atomnaya Energiya, vol. 16, No. 3, 1964).
Address: Academy of Sciences of the U.S.S.R., 14 Leninsky Prospekt, Moscow V-71, U.S.S.R.

KRETCHMAR, Arthur Lockwood, B.S., M.S., Ph.D., M.D. Born 1921. Educ.: Harvard, Wayne State and Michigan Univs. Clinician (1954-59), Chief Sci. (1959-66) Oak Ridge Assoc. Univs.; Assoc. Prof. Res., Tennessee Univ. Memorial Res. Centre, 1966-. Societies: A.A.A.S.; Soc. Exptl. Biol. and Medicine; N.Y.Acad. Sci.; A.C.S.
Nuclear interests: Medical uses of radioisotopes - diagnosis and treatment; Metabolic effects of total-body irradiation.
Address: Tennessee University, Memorial Research Centre, 1924 Alcoa Highway, Knoxville, Tennessee 37920, U.S.A.

KRETZMANN, Reinhard, Dr. rer. nat. Born 1915. Educ.: Cologne Univ. Director, Member of Management, Valvo G.m.b.H., 1962-.
Nuclear interest: Development, production and sales of components of nuclear measuring equipment.
Address: Valvo G.m.b.H., 19 Burchardstrasse, Hamburg 1, Germany.

KREY, Philip William Patrick, B.S., M.S. Born 1927. Educ.: St. Francis Coll. (Brooklyn) and Duquesne Univ. Sect. Chief, Radiochem. Sect., Radiol. Div., Chem. Warfare Lab., 1956-57; Manager Analytical Div. (1957-64), Radiol. Safety Officer (1958-64), Board of Directors (1960-62), Isotopes, Inc.; Environmental Studies Div., Health and Safety Lab., U.S.A.E.C., 1965-. Society: A.C.S.
Nuclear interests: Radiochemistry; Carbon-14 dating; Neutron activation analysis; Nuclear weapons effects testing; World-wide fallout and attendant hazard evaluation; Application of radiotracers to industrial problems.
Address: 7 Bluefield Court, Hillsdale, New Jersey, U.S.A.

KREYBIG, Thomas von. See VON KREYBIG, Thomas.

KRIEGER, Herman L., B.A., M.A., M.S. Born 1918. Educ.: New York Univ. Chem., i/c Nucl. Eng. Lab., Environmental Surveillance and Control Programme, Dept. of

Health, Educ. and Welfare. Com. on Radionuclide Applications, U.S.A.S.I.; Chem. Technol. Advisory Com., Ohio Coll. of Appl. Sci.; Com. on Radiochem. Methodology, A.P.H.A. Societies: A.C.S.; H.P.S.; RESA. Nuclear interests: In fields of assessment of environmental radiation hazard; Development of methodology for identification of radioactive contamination; Measuring the transfer of nuclides to the environment; Treatment of radioactive wastes and waste management.
Address: 4676 Columbia Parkway, Cincinnati, Ohio 45226, U.S.A.

KRIELE, Rudolf Hermann Hurter, Dr. of Law. Born 1900. Educ.: Tübingen, Munich, Berlin and Münster Univs. German Red Cross, Frankfurt-on-Main, 1949-52; Federal Audit Office, Frankfurt-on-Main, 1953-56; Federal Chancellor's Office, Bonn, 1956-59; Formerly Head of Division: Legal and economic aspects of nuclear energy; international co-operation in the fields of nuclear energy and water economy; administrative matters; Federal Ministry of Sci. Res., Bad Godesberg, 1959-. Board of Management, Gesellschaft für Kernforschung m.b.H., Karlsruhe; Geschäftsführer, Gesellschaft für Strahlenforschung m.b.H.; Formerly on Supervisory Board, Gesellschaft für Kernenergieverwertung in Schiffbau und Schiffahrt m.b.H., Hamburg; Board and Main Com., Deutsche Forschungsgemeinschaft, Bad Godesberg.
Address: 38 Germanenstrasse, Bad Godesberg, Germany.

KRIENKE, O. Karl., Jr. Formerly Asst. Prof., now Assoc. Prof., Inst. for Res., Seattle Pacific Coll.
Address: Seattle Pacific College, Institute for Research, Seattle 19, Washington, U.S.A.

KRILL, Arthur M., B.S. (Mech. Eng.), M.S. (Mech. Eng.). Born 1921. Educ.: Colorado Univ. Pres., Falcon R. and D. Co.; Ex-Head, Mech. Div., Denver Res. Inst. Director, American Astronautical Soc. (Rocky Mountain Sect.). Book: Editor, Advances in Hypervelocity Techniques (Plenum Press, 1962). Society: American Astronautical Soc. Nuclear interest: Reactor instrumentation.
Address: 1441 Ogden Street, Denver, Colorado 80218, U.S.A.

KRILL, Walter R. Dean, Coll. of Veterinary Medicine, Ohio State Univ.
Address: College of Veterinary Medicine, Ohio State University, Columbus 10, Ohio, U.S.A.

KRISEMENT, Karl Otto Oswald, Dr. phil. Born 1920. Educ.: Cologne Univ. Wissenschaftlicher Mitarbeiter im Max-Planck-Inst. für Eisenforschung, Düsseldorf, 1951-; Lecturer (1958-), Privatdoz. (1960-) Aachen T.H. Societies: Verein Deutscher Eisenhüttenleute; Deutsche Gesellschaft für Metallkunde; Deutsche Physikalische Gesellschaft. Nuclear interests: Metallurgy; Neutron diffraction.
Address: 10 Schiller Strasse, Roxel 4401, Germany.

KRISHNAMOORTHY, Cunteepuram, B.Sc. (Agriculture, Madras), M.Sc. (Agricultural Entomology, Andhra). Born 1913. Educ.: Madras and Andhra Univs. Asst. Entomologist, Rice Stem Borer, Bapatla, 1955-58; Asst. Entomologist, Storage of Food Grains in Madras, and Andhra States, 1950-54; Asst. Entomologist, Agricultural Research Inst., Rajendranagar, 1958-59; Lecturer in Entomology, Agricultural Coll., Bapatla, Andhra Pradesh, 1949-50, 1954-55; now State Entomologist, Govt. of Andhra Pradesh. Society: Fellow, Royal Entomological Soc., London. Nuclear interests: Biology and control of insect pests with the aid of radioisotopes.
Address: Agricultural Research Institute, Rajendranagar (P.O.), Hyderabad, Andhra Pradesh, India.

KRISHNAMOORTHY, Pallavoor Neelakantan, M.Sc. Born 1927. Educ.: Bombay Univ. Deputy Director, Directorate of Rad. Protection, Bhabha Atomic Res. Centre, Trombay. Staff Consultant, Indian A.E.C. Alternate to the Governor from India to the I.A.E.A., Vienna. Nuclear interests: Health physics and nuclear physics.
Address: Directorate of Radiation Protection, Bhabha Atomic Research Centre, Trombay, Bombay-74 (AS), India.

KRISHNAMURTHI, Sundaram, M.B.B.S., M.S. Born 1919. Educ.: Madras Univ. Director and Sci. Director, Cancer Inst., Madras. Sec., Indian Cancer Soc. (S.E.Div.). Book: Symposium on Cobalt 60 Beam Therapy in Malignant Disease. Nuclear interest: Radiation beam research of malignant disease, especially the role of chemical sensitizers.
Address: Cancer Institute, Madras 20, India.

KRISHNASWAMY, R., M.Sc., Ph.D. Born 1924. Educ.: Banaras Hindu and Ohio State Univs. Lecturer, Banaras Hindu Univ., 1951-53; Sen. Sci. Officer, Atomic Energy Establishment, Trombay, India, 1953-58; Sen. Mineral Technologist, Atomic Minerals Div., Dept. of Atomic Energy, 1958-. Societies: A.I.M.M.E.; Indian Inst. of Metals.
Address: Department of Atomic Energy, Atomic Minerals Division, 9-B/6, N.E.Area, Rajiner Nagar, New Delhi, India.

KRISTAN, Janez, B.E., Ph.D. Born 1921. Educ.: Ljubljana Univ. Asst., Phys. Chem. (1953-), Head, Rad. Protection Group (1959-), J.Stefan Nucl. Inst., Ljubljana. Society: Yugoslav Soc. for Radiol. Protection. Nuclear interests: Radiation protection; General dosimetry; Aerosol research; Waste treatment.

Address: J.Stefan Nuclear Institute, P.O. Box 199/4, Ljubljana, Yugoslavia.

KRISTANOV, Liubomir. Member, Com. for the Peaceful uses of Atomic Energy, Council of Ministers of the Bulgarian People's Republic.
Address: Council of Ministers, Sofia, Bulgaria.

KRISTENSEN, P. V., Prof. Phys. Inst., Aarhus Univ.
Address: The Physical Institute, Aarhus University, Aarhus, Denmark.

KRISTIANOVICH, S. A. Director, Inst. of Theoretical and Appl. Mechanics, Novosibirsk Sci. Centre, Acad. Sci. of the U.S.S.R.
Address: Novosibirsk Science Centre, 20 Sovietskaya Ulitsa, Novosibirsk, Siberia, U.S.S.R.

KRISYUK, I. T. Paper: Evaluation of I and Br Isotope Contribution in Process of Delayed Neutron Emission (letter to the Editor, Atomnaya Energiya, vol. 16, No. 2, 1964).
Address: Academy of Sciences of the U.S.S.R., 14 Leninsky Prospekt, Moscow V-71, U.S.S.R.

KRIVANEK, Miloslav, Ing. Chem., C.Sc. Born 1926. Educ.: Brno Tech. Univ. Head Radiochem. Lab., VTA, Brno; Sen. Radiochem., Joint Inst. of Nucl. Res., Dubna, U.S.S.R., 1960-62; Head Radiochem. Div., Nucl. Res. Inst., Czechoslovak Acad. Sci., Rez. Chairman, Nucl. Chem. Group, Czechoslovak Chem.Soc. Nuclear interests: Activation analysis; Radiochemical methods in analytical chemistry; Radiochemistry of fission products.
Address: Nuclear Research Institute, Rez, Czechoslovakia.

KRIVELEV, G. P. Papers: Co-author, Integral Method of Measurement of $\beta_{eff}/1$ Value (letter to the Editor, Atomnaya Energiya, vol. 20, No. 2, 1966); co-author, Semi- Statistical Method of Effective Delayed Neutrons Fraction Measurement (letter to the Editor, ibid, vol. 22, No. 1, 1967).
Address: Academy of Sciences of the U.S.S.R., 14 Leninsky Prospekt, Moscow V-71, U.S.S.R.

KRIVOHATSKII, L. S. Paper: Co-author, Determination of Probabilities of Spontaneous Fission of U^{233}, U^{235} and Am^{243} (Atomnaya Energiya, vol. 20, No. 4, 1966).
Address: Academy of Sciences of the U.S.S.R., 14 Leninsky Prospekt, Moscow V-71, U.S.S.R.

KRIVOKHATSKII, A. S. Papers: Co-author, Cross-Section and Resonance Integrals of Capture and Fission of Long-Lived Americium (Atomnaya Energiya, vol. 23, No. 4, 1967); co-author, Production of Pu^{238} by U^{235} and Np^{237} Neutron Irradiation (letter to the Editor, ibid, vol. 23, No. 6, 1967); co-author, Evolution of Energy upon Radioactive Decay of some Isotopes of Transuranium Elements (letter to the Editor, ibid, vol. 24, No. 2, 1968); co-author, Formation of Cm^{242} and Cm^{244} by Reactors Neutron Irradiation of Am^{241} (ibid, vol. 24, No. 3, 1968).
Address: Academy of Sciences of the U.S.S.R., 14 Leninsky Prospekt, Moscow V-71, U.S.S.R.

KRIZEK, Vladimír, Dipl. Ing., C.Sc. Born 1924. Educ.: Prague Tech. Univ. Manager, Nucl. Energy Dept., Výzkumný ústav energetických zarízeni, První brnenská strojírna n.p., Brno. Society: Czechoslovak Sci. and Tech. Soc. (Cs. Vedeckotechnika spolecnost). Nuclear interests: Nuclear power plants; Heat exchangers; Components of primary circuits; Heat transfer problems.
Address: 59 C Uvoz, Brno 2, Czechoslovakia.

KROEBEL, Werner Adolf Johannes. Univ. Prof., Dr. phil. habil. Born 1904. Educ.: Berlin and Göttingen Univs. Direktor des Instituts für angewandte Physik and Ordinarius an der Kiel Univ. Societies: Gesellschaft für Kernenergieverwertung in Schiffbau und Schiffahrt m.b.H.; Technisch-Wissenschaftlichen Beirats dieser Gesellschaft; Nordwestdeutsche Physikalische Gesellschaft e.V.; Fernsehtechnische Gesellschaft. Nuclear interests: Reaktor, Regeltechnik und kernphysikalische Messmethoden.
Address: 25 Wehrbergallee, Schellhorn b/ Preetz i.Holst., Germany.

KROEGER, Henry R., B.S. (Mech. Eng.). Born 1913. Educ.: Ohio State Univ. Sen. Eng., Nucl. Energy for the Propulsion of Aircraft (NEPA) Project, Fairchild Engine and Airplane Corp., 1947-51; Project Eng. and Head, Nucl. Technol. Group, Bendix Aviation Co., 1951-55; Assoc. (1955-57), Vice-Pres. (1957-65), Astra Inc.; Member, Tech. Staff, Texas Instr. Inc., 1965-. Society: A.N.S. Nuclear interests: Power plant economics; Heat transfer; Fluid flow; Thermal stress; Space power.
Address: Texas Instruments Incorporated, P.O. Box 5474, Dallas, Texas, U.S.A.

KRÖGER, M., Prof Dr. phil. Haus der Technik, Ständiges Seminar für Kerntechnik, Gruppe 2 Chemie der Kernbrennstoffe.
Address: 173 Laurentiursweg, Essen-Steele, Germany.

KROH, Jerzy, Prof., Ph.D. Born 1924. Educ.: Lodz Tech. Univ. Director, Inst. of Rad. Chem., Lódź. Book: Free Radicals in Rad. Chem. (PWN, 1967). Societies: Polish Chem. Soc.; The Miller Trust for Rad. Chem. Nuclear interest: Radiation chemistry.
Address: Institute of Radiation Chemistry, 15 ul.Wróblewskiego, Lodz, Poland.

KROHMER, Jack Stewart, B.S., M.A., Ph.D. Certified Radiol. Phys., Certified Health Phys. Born 1921. Educ.: Western Reserve and Texas Univs. Sen. Instructor in Radiol., Western Reserve Univ., 1947-57; Asst. Prof. Radiol. (Rad. Phys.) Southwestern Medical School, Texas Univ. 1957-61; Assoc. Prof. (1961-62).

Prof. (1962-63), Res. Prof. Biophys. (1963-66), New York State Univ., Buffalo; Lecturer in Phys., Bucknell Univ., 1966-; Rad. Phys., The Geisinger Medical Centre, Danville, Pennsylvania, 1966-; Consultant Phys., Brooke Army Medical Centre, Fort Sam Houston, Texas, 1960-; Consultant Phys., Lackland Air Force Base, Texas, and Keesler Air Force Base, Mississippi, 1960-; Director, Placement Service, American Assoc. of Phys. in Medicine. Com., N.C.R.P., 1961-. Books: Contributor to Clinical Use of Radioactive Isotopes (Fields and Seeds, 1960); contributor to Handbook of Rad. Hygiene (McGraw-Hill, 1959); contributor to The Handbook of Biochem. and Biophys. (Handbook House, World Publishing Co., 1966). Societies: American Roentgen Ray Soc.; American Coll. of Radiol.; Radiol. Soc. North America; Rad. Res. Soc.; Biophys. Soc.; Soc. Nucl. Medicine; H.P.S.; Soc. Cancer Res.; A.A.A.S.; American Assoc. Phys. in Medicine; Pennsylvania Radiol. Soc.
Nuclear interests: Interactions of radiation with matter; Biological effects of radiation; Radiation dosimetry; Clinical uses of radioisotopes; Radiation protection; Health physics.
Address: Radiology Department, The Geisinger Medical Centre, Danville, Pennsylvania 17821, U.S.A.

KROHNE, Theodore F. Born 1915. Educ.: Elmhurst Coll. and Northwestern Univ. At present Staff Asst., Public Information, Argonne Nat. Lab. Societies: A.N.S.; Nucl. Energy Writers' Assoc.
Nuclear interests: Assisting the lay public in understanding science in general and nuclear science in particular.
Address: Public Information Office, Argonne National Laboratory, Argonne, Illinois, U.S.A.

KROKOWSKI, Ernst, Dr.rer.nat., Dr.med. Educ.: Humboldt (Berlin) and Berlin Freie Univs. Asst. Prof., Dept. of Radiol., Berlin Free Univ. Society: Deutsche Röntgen-Gesellschaft.
Nuclear interests: Metabolism of the thyroid gland with radioactive iodine. Extinctions measurements. Studies of calcium metabolism.
Address: Strahleninstitut Klinik der Freien Universität Berlin im Städt Krankenhaus Westend, 130 Spandauer Damm, 1 Berlin 19, Germany.

KROL, F. P. Chief Eng., Nucl. Div., Steel and Alloy Tank Co,
Address: Steel and Alloy Tank Co., Foot of Bessemer Street, Newark 5, New Jersey, U.S.A.

KROLIKOWSKI, Wojciech, M.Sc., Ph.D., D.Sc. Born 1926. Educ.: Warsaw Univ. Swiss Federal Inst. of Technol., 1956-57; Prof., and Head, Chair of Theory of Elementary Particles, Warsaw Univ., 1958-; Prof., and Head, Group of Theory of Elementary Particles, Inst. for Nucl. Res., Warsaw, 1958-; Inst. for Advanced Study, Princeton, 1961; Internat. Centre for Theoretical Phys., Trieste, 1965-66; Middle East Tech. Univ., Ankara, 1965-66.
Nuclear interest: Theory of interactions of elementary particles, especially relations between strong and weak interactions and partial conservation relations.
Address: Warsaw University, Institute of Theoretical Physics, 69 Hoza Street, Warsaw, Poland.

KROLL, Norman, Prof. Theoretical Phys. Res. (mainly Fundamental Particle Res.) (U.S.A.E.C. contract), Columbia Univ.
Address: Nevis Cyclotron Laboratories, Irvington, New York, U.S.A.

KROLZIG, Alfred, Dipl. Ing. Born 1921. Educ.: Hannover T.H. Nord West Deutscher Rundfunk. Technische Elektronik, DESY, Hamburg, 1957-.
Nuclear interest: Development of special apparatus used in high energy physics.
Address: Deutsches Elektronen - Snychrotron, 56 Flottbeker Drift, Hamburg - Gr. Flottbek 1, Germany.

KROMER, Carl-Theodor, Dr.Ing., Prof. Born 1901. Educ.: Karlsruhe and Stuttgart Univs. Vorsitzer, Vorstand der Badenwerk A.G. Member, Comite d'Etudes de l'Energie Nucléaire, Union Internationale des Producteurs et Distributeurs d'Energie Electrique; Verwaltungsrat, Deutsches Atomforum; Fachkommission V, Deutsche Atomkommission.
Nuclear interest: Nuclear power plants.
Address: 19 Günterstalstrasse, Freiburg/Br, Germany.

KROMHOUT, Robert Andrew, B.S. (Phys. Kansas State Coll.), M.S. (Phys., Illinois), Ph.D. (Phys.,Illinois). Born 1923. Educ.: Kansas State Coll. and Illinois Univ. Asst. Prof., Illinois Univ., 1954-56; Asst. Prof. (1956-60), Head, Phys. Dept. (1960-62), Assoc. Prof. (1960-63), Prof. (1963-), Florida State Univ. Society: A.P.S.
Nuclear interest: Nuclear magnetic resonance.
Address: Physics Department, Florida State University, Tallahassee, Florida, U.S.A.

KRONAUER, Emilio. Directeur-gén. des Ateliers de Secheron S.A., Geneva. Member, Conseil d'Administration, Energie Nucléaire, S.A.; Member, Federal Commission for Atomic Energy, Bern; Member, Board of Directors, Therm-Atom A.G. Zurich.
Address: 14 avenue de Secheron, Geneva, Switzerland.

KRONBERGER, Hans, B.Sc., Ph.D. Born 1920. Educ.: Durham and Birmingham Univs. Head of Lab., Capenhurst (1953), Chief Phys., Risley, Director of R. and D., Ind. Group, Risley (1957), Deputy Managing Director, Reactor Group, Risley (1961-64), Sci. - in - Chief, Reactor Group, Risley, U.K.A.E.A. Societies: F. Inst. P.; F.R.S.
Nuclear interest: Development of prototype reactors.
Address: United Kingdon Atomic Energy

Authority, Reactor Group, Risley, nr. Warrington, Lancs., England.

KRONE, Ralph W., B.S., M.S., Ph.D. Born 1919. Educ.: Antioch Coll., Illinois and Johns Hopkins Univs. Prof. Phys., Kansas Univ. Society: A.P.S.
Nuclear interests: Nuclear structure physics; Gamma ray spectroscopy; Charged particle reactions.
Address: Physics Department, Kansas University, Lawrence, Kansas, U.S.A.

KRONIG, Ralph, B.A., M.A., Ph.D. Born 1904. Educ.: Columbia Univ., New York. Prof. Theoretical Phys., Delft Technol. Univ., 1939-.
Books: Band Spectra and Molecular Structure (1930); The Optical Basis of the Theory of Valency (1935); Leerboek der Natuurkunde (1946); Textbook of Physics (1954). Societies: Royal Netherlands Acad. Sci.; Royal Norwegian Soc. of Sci.; Roy. Inst. Eng. of Netherlands; Netherlands Phys.Soc.
Nuclear interest: Nuclear theory.
Address: 204 Oostsingel, Delft, Netherlands.

KROSHKIN, N. I. Paper: Co-author, Medium Spectra of Neutrons in Double and Triple Fission of U^{235} by Thermal Neutrons (Abstract, Atomnaya Energiya, vol. 20, No. 4, 1966).
Address: Academy of Sciences of the U.S.S.R., 14 Leninsky Prospekt, Moscow V-71, U.S.S.R.

KROT, N. N. Papers: Co-author, The Used Fuel Rods of the First Soviet Nucl. Power Station (letter to the Editor, Atomnaya Energiya, vol. 8, No. 5, 1960); co-author, Study of a Used Fuel Rod from the First Nucl. Power Station (ibid, vol. 11, No. 2, 1961); co-author, Distributions of Capture to Fission Ratio of ^{239}Pu by Height of Reactor BR-5 (ibid, vol. 16, No. 6, 1964).
Address: Academy of Sciences of the U.S.S.R., 14 Leninsky Prospekt, Moscow V-71, U.S.S.R.

KROTENKO, V. T. Paper: Co-author, Study of Inelastic Scattering of Slow Neutrons by Polyethylene (Atomnaya Energiya, vol. 20, No. 1, 1966).
Address: Academy of Sciences of the U.S.S.R., 14 Leninsky Prospekt, Moscow V-71, U.S.S.R.

KROTKOV, F. G. Prof.
Address: Academy of Medical Sciences of the U.S.S.R., Committee of Medical Radiology, 14 Solianka Street, Moscow, U.S.S.R.

KROUSE, L. M. Manager, Nucl. Dept., Joy Manufacturing Co.
Address: Nuclear Department, Joy Manufacturing Co., 3101 Broadway, Buffalo 25, New York, U.S.A.

KRTIL, Josef, Dipl.-Ing., Cand. Chem. Sci. Born 1932. Educ.: Chem. Technol. Univ., Prague. Vice-Head, Fission Products Dept., Nucl. Res. Inst., Czechoslovak Acad. Sci. Nuclear interests: Nuclear chemistry; Separation of fission products.
Address: Nuclear Research Institute, Czechoslovak Academy of Sciences, Rez, Czechoslovakia.

KRUEGER, Gordon, Prof. Chairman, Chem. Dept., Taylor Univ.
Address: Taylor University, Upland, Indiana, U.S.A.

KRUESI, Frank E., Jr. Director, Savannah River Lab., 1967-.
Address: U.S. Atomic Energy Commission, Savannah River Laboratory, Aiken, South Carolina, U.S.A.

KRUESSMANN, Adolf, Dipl.-Ing. Born 1911. Educ.: Hanover T.H. Director, Mess-u. Regeltechnik Abteilung Vertrieb Anlagen (Reaktoren), Hartmann und Braun A.G. Society: Sen. Member, I.S.A.
Nuclear interest: Control.
Address: Hartmann und Braun A.G., 97 Graefstrasse, Postfach 1367, 6 Frankfurt-am-Main W.13, Germany.

KRUG, R. C. Virginia Polytechnic Inst. representative on Council. Oak Ridge Assoc. Universities.
Address: Virginia Polytechnic Institute, Blacksburg, Virginia, U.S.A.

KRÜGER, Friedrich Wolfgang, Dipl.-Phys. Born 1935. Educ.: Greifswald Univ. Books: Chapter on Materialien zur Abschirmung radioaktiver Strahlung (Shielding materials) in Werkstoffe der Kerntechnik, Teil IV (Berlin, VEB Deutscher Verlag der Wissenschaften, 1963).
Nuclear interest: Radiation shielding.
Address: 45-46 Gorschstrasse, 110 Berlin, German Democratic Republic.

KRUGER, George E. Vice Pres. and Tech. Director, Mining Div., Tech. Director, Atomic Energy Div. (1963-), Chase Manhattan Bank.
Address: 1 Chase Manhattan Piaza, New York, New York 10015, U.S.A.

KRÜGER, Hubert, Prof. Dr. Director, Physikalisches Inst., Tübingen Univ.
Address: Physikalisches Institut, Universitat Tübingen, 6 Gmelinstr., Tübingen, Germany.

KRUGER, Paul, Ph.D., B.S. Born 1925. Educ.: Chicago Univ. and M.I.T. At Chicago Univ., 1950-53; Res. Staff, Gen. Motors,1953-54; Manager, Phys. Sci. Dept., Nucl Sci. and Eng. Corp., 1954-61; Manager, Nucl. Projects, Hazleton Nucl. Sci. Corp., 1961-62; Assoc. Prof. Nucl. Chem., Stanford Univ., 1962-. Societies: A.C.S.; A.N.S.
Nuclear interests: Low-level radioactivity technology; Activation analysis; Radiochemistry; Reactor chemistry; Use of radioisotopes in meteorology, hydrology, field tracing, and nuclear explosives engineering.
Address: Civil Engineering Department, Stanford University, Stanford, California 94305, U.S.A.

KRUGLIKOV, A. N. Paper: Co-author, Low-Temperature Canal of WWR-M Reactor of Phys.-Tech. Inst. AS U.S.S.R. (letter to the Editor, Atomnaya Energiya, vol. 20, No. 3, 1966). Address: Physical-Technical Institute, Academy of Sciences of the U.S.S.R., 2 Politechnicheskaya Ulitsa, Leningrad, U.S.S.R.

KRUGLOV, A. A. Paper: Co-author, Phase Diagrams of Plutonium with IIIA, IVA, VIII and IB Group-Metals (Atomnaya Energiya, vol. 23, No. 6, 1967). Address: Academy of Sciences of the U.S.S.R., 14 Leninsky Prospekt, Moscow V-71, U.S.S.R.

KRUGLOV, S. P. Papers:Co-author, Three Methods to Measure Bremsstrahlung Beam Energy in $E\gamma_{Max}$ -15-80 MeV Range (letter to the Editor, Atomnaya Energiya, vol. 16, No. 3, 1964); co-author, Energy Spectra of Electron Produced by High-Energy X-Rays in Light Materials (letter to the Editor, ibid, vol. 23, No. 2, 1967). Address: Academy of Sciences of the U.S.S.R., 14 Leninsky Prospekt, Moscow V-71, U.S.S.R.

KRUGLYI, M. S. Paper: Co-author, Determination of Indium by (γ,γ)- Reaction with 5 MeV Linear Electron Accelerator (letter to the Editor, Atomnaya Energiya, vol. 24, No. 3, 1968). Address: Academy of Sciences of the U.S.S.R., 14 Leninsky Prospekt, Moscow V-71, U.S.S.R.

KRUGTEN, Hans van. See VAN KRUGTEN, Hans.

KRUMBEIN, Aaron Davis, A.B., Ph.D. Born 1921. Educ.: Brooklyn Coll. and New York Univ. Naval Ordnance Lab., 1955; Maryland Univ., 1950-56; Nucl. Development Corp. of America (now part of United Nucl. Corp.), 1956-. Societies: A.P.S.; A.N.S.; A.A.A.S. Nuclear interests: Reactor physics; Shielding; Reactor safety; Nuclear physics; Nuclear detectors.
Address: 1 Tower Lane, Monsey, New York, U.S.A.

KRUMHOLZ, Louis A., B.S., M.S., Ph.D. Born 1909. Educ.: Coll. of St. Thomas (at St. Paul, Minnesota), and Illinois and Michigan Univs. Resident Biol., Lerner Marine Lab. (American Museum of Natural History), Bimini, Bahamas, 1954-56; Aquatic Biol.(i/c Ecological Survey of White Oak Lake), Tennessee Valley Authority, Oak Ridge, Tennessee, 1950-54; Prof. Biol., Biol. Dept., Louisville Univ., 1957-.
Nuclear interests: Accumulation of radioactive wastes by aquatic organisms; Effects of radioactive wastes on aquatic organisms; Radioactive fallout from nuclear weapons tests.
Address: Water Resources Laboratory, Louisville University, Louisville, Kentucky 40208, U.S.A.

KRUMMEICH, Wilhelm, Dr. phil. nat., Dipl.-Phys. Born 1928. Educ.: Justus Liebig (Giessen) and Johannes Gutenberg (Mainz) Univs. Nucleargruppe Degussa, 1956-. Betriebsleiter der Produktion Brennelemente II, Nukem, Nuklear-Chemie und Metallurgie G.m.b.H., 1960-.
Nuclear interests: UO_2: Fertigungsmethoden e.g. Pressen, Sintern u.a. physikal. Eigenschaften von UO_2 - Pellets, insbes. ekektr. Leitfähigkeit, Wärmeleitfähigkeit, therm. Ausdehnung - in Abhängigkeit von Fremdoxydbeigabe, Temperatur, Dichte u. entspr. Messmethoden. UC: Darstellung u. physikalische Eigenschaften. Graphit: Nukleargraphite u. physikal. u. chem. Eigenschaften.
Address: 21 Josefstr., (16) Hanau/Main, Germany.

KRUSE, Hans. Dr. jur. Born 1921. Educ.: Göttingen Univ. Internat. Law Inst., Göttingen Univ. until 1957; Gesellschaft zur Förderung der Kernphysikalischen Forschung e.V., Düsseldorf, 1957-59; GFKF Study Group on Atomic Energy Law and Economics, Münster Univ., 1959-60; Arbeitsgemeinschaft Wissenschaft u. Politik e.V., München, 1961-. Books: Die Atomenergie im Völkerrecht und in den wichtigsten Landesrechten (1955); co-author, Deutsches Atomenergierecht (1955-); Die internationale Zusammenarbeit auf dem Gebiet der friedlichen Verwendung der Atomenergie (1956); Atomenergierecht (1961); Legal Aspects of the Peaceful Utilisation of Atomic Energy (1961). Societies: Gesellschaft für Rechtsvergleichung; Interparlamentarische Arbeitsgemeinschaft Fachbeirat.
Nuclear interests: Atomic energy law and economics; Administration of research projects.
Address: 17 Marienplatz, Munich 2, Germany.

KRUSE, Olan E., B.S., M.A., Ph.D. Born 1921. Educ.: Texas Univ. Prof. and Chairman, Dept. of Phys., Texas A and I Univ. Societies: A.I.P.; A.A.P.T.; A.A.A.S.
Nuclear interest: Nuclear physics, teach introductory nuclear physics and nuclear instrumentation.
Address: Texas A and I University, Department of Physics, Kingsville, Texas, U.S.A.

KRUSEMAN, Jacob Paul. Born 1897. Societies: Chairman, Voorlichtingsinstituut Het Atoom, Amsterdam; Chairman, Stichting Kernvoortstuwing Koopvaardijschepen, The Hague; Member of Board, Stichting Fundamenteel Onderzoek der Materie, Utrecht; Member of Board, Reactor Centrum Nederland, The Hague.
Nuclear interests: General; Ships' propulsion.
Address: 15 Arubalaan Hilversum, Netherlands.

KRUSKAL, Martin David, B.S., M.S., Ph.D. (New York). Born 1925. Educ.: Chicago and New York Univs. Assoc. Head, Theoretical Div., Plasma Phys. Lab., Princeton Univ. Prof. Astrophys. Sci., Princeton Univ. Holds various consulting appointments. Exchange visit to Lebedev Phys. Inst., 1965-66 (3 months). Societies: American Mathematical Soc.; Mathematical Assoc. of America; A.P.S.; Soc. for Ind. and of Appl. Mathematics.

KRU

Nuclear interest: Controlled thermonuclear reactors.
Address: Plasma Physics Laboratory, Forrestal Research Centre, Princeton University, P.O. Box 451, Princeton, New Jersey, U.S.A.

KRUTTSCHNITT, Julius, Ph.D., M.I.M.M., M.A.I.M.M., M.Am.I.M.M.E., Member, Uranium Mining Com., Australian A.E.C.
Address: Mount Isa Mines Ltd., Box 1433T, G.P.O., Brisbane, Queensland, Australia.

KRUYS, Pierre, Ph.D. (Chem.), Graduate in Nucl. Eng. Born 1926. Educ.: Brussels Univ. and Argonne Nat. Lab. Attached to Res. Div. of Euratom. Gen. Sec., Sté. Chimique de Belgique; Member, Euratom/U.S. Agreement Joint R. and D. Board.
Nuclear interest: Radiation chemistry.
Address: 70 avenue des Grenadiers, Brussels 5, Belgium.

KRUZHILIN, G. N. Director, G. M. Krzhizhanovsky Power Inst., Acad. of Sci. of the U.S.S.R. Corresponding Member, Acad. of Sci. of the U.S.S.R.; Deleg. to 2nd I.C.P.U.A.E., Geneva, Sept. 1958.
Address: G. M. Krizhizhanovsky Power Institute, Academy of Sciences of the U.S.S.R., 19 Leninsky Prospekt, Moscow, U.S.S.R.

KRYLOV, B. E. Paper: Co-author, Liner State Control of Rotating Cement Furnaces with Radioactive Isotopes (letter to the Editor, Atomnaya Energiya, vol. 19, No. 2, 1965).
Address: Academy of Sciences of the U.S.S.R., 14 Leninsky Prospekt, Moscow V-71, U.S.S.R.

KRYMM, R. Member, for I.A.E.A., Study Group on Long Term Role of Nucl. Energy in Western Europe, O.E.C.D., E.N.E.A.
Address: O.E.C.D. European Nuclear Energy Agency, 38 boulevard Suchet, Paris 16, France.

KSYCKI, Sister Mary Joecile, S.S.N.D., B.S., M.S., Ph.D. Born 1913. Educ.: St. Louis Univ. Res. Assoc., Rad. Project, Notre Dame Univ., 1958; Prof. Chem. and Director, Chem. Dept. (1954-), Director, Nucl. Sci. Dept. (1958-), Notre Dame Coll. Book: Co-author, Radioactivity: Fundamentals and Experiments (Holt, Rinehart and Winston, Inc., 1963). Societies: A.C.S.; Nucl. Medicine Soc.
Nuclear interests: Use of radioisotopes as tracers in biochemical reactions and in medical diagnosis; Effects of beta and gamma rays on biochemical systems and animal tissues.
Address: Notre Dame College, 320 East Ripa Avenue, St. Louis, Missouri 63125, U.S.A.

KU, P. M., B.S., D.I.C. Born 1915. Educ.: Chiao-Tung Univ., Shanghai, and London Univ. Asst. Prof., M.I.T., 1949-51; Mech. Eng. (1951-55), Chief, Engines and Lubrication Sect. (1955-56), N.B.S., Washington D.C. Manager, Res. Sect. (1956-59), Director, Aerospace Propulsion Res. Dept. (1959-), Southwest Res. Inst. Assoc. Editor, American Soc. Lubricating Eng. Transactions and Lubrication Eng., 1960-; Chairman, A.S.M.E. Res. Com. on Lubrication 1966-; Director (1964-) and Councillor, A.S.L.E. Societies: American Soc. Lubricating Eng.; Fellow, A.S.M.E.; A.S.T.M.; Coordinating Res. Council.
Nuclear interests: Powerplant, liquid-metal technology.
Address: Aerospace Propulsion Research Department, Southwest Research Institute, 8500 Culebra Road, San Antonio, Texas 78206, U.S.A.

KU, Wen Kwei, B.S. (Civil Eng.), M.S. (Civil Eng.). Born 1909. Educ.: Nat. Wu-Han (China) and Illinois Univs. Vice Chairman, Shihmen Reservoir Commission (1963-66), Vice Chairman and Exec. Sec., Tsengwen Reservoir Development Commission (1967-), Taiwan Provincial Govt.; Vice Pres., and Chief Eng., Taiwan Power Co., Ministry of Economic Affairs, 1964-. Pres., Chinese Inst. of Hydraulic Eng. Societies: Chinese Inst. of Eng.; Chinese Inst. of Civil Eng.; A.S.C.E.
Nuclear interest: Nuclear power generation.
Address: 14 Lane 9, Chungking South Road, Sec. 2, Taipei, Formosa.

KUANG, Hao-hwai. Paper: Co-author, Nucl. Interactions of Cosmic-Ray High-Energy Particles with Liquid Scintillator (Scientia Sinica, vol. 13, No. 5, 1964).
Address: Academia Sinica, 3 Wen Tsin Chien, Peking, China.

KUBA, Jaromír, R.N.Dr. Born 1928. Educ.: Masaryk's Univ., Brno. Hutní projekt, Brno (Project, Inst. for Metal.), -1957; Výzkumný ústav hutnictví zeleza, Brno (Res. Inst. for Metal.), -1964. Book: Measurement of ness with Radioisotopes. (Prague, V.T.S., 1959). Society: Commission for Nucl. Technique.
Nuclear interest: Industrial uses and applications of radioactive isotopes.
Address: Troubsko, Rudé armády 42,p. Bosonohy, Czechoslovakia.

KUBA, Josef, D.Sc. Born 1915. Educ.: Charles Univ., Prague. Deputy Prof., Tech. Univ., 1958; Director, Nat. Tech. Museum, Prague, 1959; Director, Exchange and Training Div. (1964-), Acting Director, Div. of Publications (1968-), I.A.E.A. Book: Coincidence tables for atomic spectroscopy (Elsevir, 1964, 1136 p.).
Address: International Atomic Energy Agency, 11 Kaerntnerring, Vienna 1, Austria.

KUBAL, Josef, R.N.Dr. Born 1905. Educ.: Charles Univ., Prague. Sci. Worker, Nucl. Res. Inst., Czechoslovak Acad. Sci.
Nuclear interests: Synthesis of nuclear emulsions; Radiologic dosimetry; Radiation chemistry.
Address: 111 Na Pankráci, Prague 4, Czechoslovakia.

KUBOTA, Denjiro. Managing Director, Mitsubishi Oil Co. Ltd.
Address: Mitsubishi Oil Co. Ltd., Sanyu Building, 1 Shiba-Kotohiracho, Minato-ku, Tokyo, Japan.

KUBOTA, Hirosuke. Res. Eng., Rad. Shielding and Decontamination, Nucl. Sect., Eng. Div., Takenaka Komuten Co. Ltd.
Address: Nuclear Section, Engineering Division, Takenaka Komuten Co. Ltd., 2-30 Dojimanakamachi, Kitaku, Osaka, Japan.

KUBUSHIRO, Kaneyoshi. Born 1903. Educ.: Elec. Eng. Dept., Kyushu Univ. Japan Elec. Generating and Transmission Co., 1939-51; Director, Material Div., Tokyo Elec. Power Co. Ltd., 1951-56; Councillor and Formerly Director of Construction, Japan Atomic Energy Res. Inst., 1956-. Specialist Member, A.E.C. of Japan; Member of Subcom. for Earthquake Countermeasure for Reactors.
Address: 57 1-chome, Shin-machi, Setagaya-ku, Tokyo, Japan.

KUCERA, Zdenek, Ing. Born 1930. Educ.: Tech. Univ., Prague. Head, Designing Dept., Nucl. Power Div., Skoda Concern, Plzen. Society: Czechoslovak Sci. and Tech. Soc. Nuclear interests: Reactor design; Nuclear power economics.
Address: 30 tr. Budovatelu, Pilsen, Czechoslovakia.

KUCHCINSKI, Adam, M.Sc. Born 1925. Educ.: Warsaw Tech. Univ. Geophys. Exploration Establishment, Warsaw, 1953-57; Central Lab. for Radiol. Protection 1957-60; Inst. of Nucl. Res., 1956-60, 1963-; Antarctic Res. Exp., 1966.
Nuclear interests: Radiometry and solid states detectors in technical and industrial radioisotope applications.
Address: 59 m 12 ul. Szcześliwicka, Warsaw 22, Poland.

KUCHEROV, R. Ya. Papers: Co-author, Isotope Separation by Diffusion in a Steam Current (2nd I.C.P.U.A.E., Geneva, Sept. 1958); co-author, A Separate Cascade Consisting of the Diffusion Columns (letter to the Editor, Atomnaya Energiya, vol. 6, No. 2, 1959); co-author, Isotope Separators and the Transport Equation (letter to the Editor, ibid, vol. 7, No. 3, 1959); co-author, On Theory of Multi-component Isotope Separation Cascades (ibid, vol. 19, No. 4, 1965).
Address: Nuclear Energy Institute, Academy of Sciences of the U.S.S.R., 14 Leninsky Prospekt, Moscow V-71, U.S.S.R.

KUCHERYAEV, V. A. Paper: Co-author, Shielding Properties of Iron-Serpentine Concrete (letter to the Editor, Atomnaya Energiya, vol. 20, No. 1, 1966).
Address: Academy of Sciences of the U.S.S.R., 14 Leninsky Prospekt, Moscow V-71, U.S.S.R.

KUCHERYAEV, Yu. A. Paper: Co-author, Dissociation of Fast H_2^+ Ions and Charge Exchange of Fast Protons in Lithium Arc (Abstract, Atomnaya Energiya, vol. 19, No. 4, 1965).
Address: Academy of Sciences of the U.S.S.R., 14 Leninsky Prospekt, Moscow V-71, U.S.S.R.

KÜCHLE, M. Member, European American Com. on Reactor Phys., O.E.C.D. E.N.E.A. Paper: Measurements of the Temperature Dependence of Thermal Neutron Diffusion Parameters in Water and Dowtherm A (Nucl. Sci. and Eng., vol. 8, No. 1, 1960).
Address: Kernforschungszentrum Karlsruhe, Institut für Neutronenphysik, Karlsruhe, Germany.

KÜCHLER, Leopold, Dr. phil. habil., Prof. Born 1910. Educ.: Vienna Univ. Direktor, Farbwerke Hoechst A.G. Vorsitzender, Arbeitskreis III/5, Aufarbeitung bestrahlter Brennstoffe, Deutsche Atomkommision.
Nuclear interests: Radiochemistry; Reprocessing; Heavy water; Graphite; Radiation chemistry.
Address: Farbwerke Hoechst A.G., Frankfurt-am-Main 80, Germany.

KUCHOWICZ, Bronislaw Stanislaw Jerzy, M.Sc., Dr. of Phys. Born 1932. Educ.: Poznań and Warsaw Univs. Inst. of Nucl. Res., Warsaw, -1963; Dept. of Radiochem., Warsaw Univ., 1964-. CINDA-system reader for Poland since 1965. Books: Nuclear Astrophysics. A bibliographical survey. 4 parts. (Warsaw, Nucl. Energy Information Centre, 1965); Bibliography of the Neutrino. 2 volumes. (Warsaw, Nucl. Energy Information Centre, 1966). Society: Polish Phys. Soc.
Nuclear interests: Nuclear physics and elementary particles in astronomy (abundance of chemical elements and nuclear reactions in stars), relativistic astrophysics, neutrino.
Address: Department of Radiochemistry, University, 101 ul. Zwirki i Wigury, Warsaw, Poland.

KUCHTA, Josef, M.M.E. Born 1930. Educ.: Pilzen Tech. Univ. Res. Group, Nucl. Power Plant Div., Skoda Concern, Pilzen.
Nuclear interest: Strength calculations of reactor parts.
Address: 13 Budovatelu, Pilzen, Czechoslovakia.

KUCHTEVIC, V. I.
Nuclear interest: Health physics.
Address: Fyzikalne energeticky institut, Obninsk, near Maloyaroslavets, Moscow, U.S.S.R.

KUCKUCK, Hermann, Dr. agr. habil., Prof. Born 1903. Educ.: Berlin Univ. Direktor, Inst. für Angewandte Genetik Hanover T.H., 1954-.
Nuclear interest: Induction of mutations by X-rays and other waves in relation to genetics and plant-breeding.
Address: 2 Herrenhäuser Strasse, Hanover-Herrenhausen, Germany.

KUCKUK, Theo MAYER-. See MAYER-KUCKUK, Theo.

KUCUK, M., Dr. Member, Faculty of Sci., Nucl. Phys. Dept., Istanbul Univ.
Address: Istanbul University, Faculty of Science, Nuclear Physics Department, Istanbul, Turkey.

KUDA, Radovan, Ing. (Structural Eng.). Born 1928. Educ.: Tech. Univ. Brno, Czechoslovakia. Senior Asst., Dept. of Concrete and Bridges, Tech. Univ., Brno. Section for Application of Nucl. Energy, Sect. for Concrete and Masonry Structures, Czechoslovak Assoc. of Technicians. Nuclear interests: Atomic energy in building research, reactor and laboratory design, nuclear shielding.
Address: 6 Vsetickova, Brno, Czechoslovakia.

KUDELAINEN, V. I. Paper: Co-author, Indication of Electron Beam by Residual Gas glow (Atomnaya Energiya, vol. 24, No. 1, 1968).
Address: Academy of Sciences of the U.S.S.R., 14 Leninsky Prospekt, Moscow V-71, U.S.S.R.

KUDRIN, L. P. Papers: Angular and Energy Distribution for Fission Neutrons (Atomnaya Energiya, vol. 3, No. 7, 1957); co-author, Uranium Atomic Potential and Calculation of Ionisation Energy (ibid, vol. 22, No. 2, 1967); co-author, Calculation of Calcium and Uranium Monoftorids Dissociation Energy (ibid, vol. 22, No. 2, 1967); co-author, Evaluation of Conductivity of U - F Plasma at High Temperatures (ibid, vol. 22, No. 4, 1967).
Address: Academy of Sciences of the U.S.S.R., 14 Leninsky Prospekt, Moscow V-71, U.S.S.R.

KUDRJAVCEVA, A. V. Paper: Co-author, Some Methods of Lowering Fluxes of Penetrating Secondary Gamma Rad. (Jaderna Energie, vol. 13, No. 11, 1967).
Address: Fyzikalne energeticky institut, Obninsk, U.S.S.R.

KUDRYAVTSEV, A. A., Prof. Director, Lab. of All-Union Inst. for Sci. Res. in Exptl. Veterinary Medicine.
Nuclear interest: Animal physiology.
Address: Laboratory of the All-Union Institute for Scientific Research in Experimental Veterinary Medicine, p/o Kuz'minki, Ukhtomsk raion, Moscow province, U.S.S.R.

KUDRYAVTSEVA, A. V. Papers: Co-author, Some Methods of Decrease of Secondary Penetrating Gamma-Rad. Fluxes (Abstract, Atomnaya Energiya, vol. 19, No. 5, 1965); co-author, Application of Multigroup Methods for the Calculation of Neutron Flux Distributions in Inhomogeneous Shields (Kernenergie, vol. 11, No. 3, 1968).
Address: Academy of Sciences of the U.S.S.R., 14 Leninsky Prospekt, Moscow V-71, U.S.S.R.

KUEHN, Matthew N. Born 1914. Educ.: Washington Univ. Uranium Div., Mallinckrodt Chem. Works, 1942-. Formerly Exec. Officer, Inst. Nucl. Materials Management. Book: Chapter V, Management of Nucl. Materials (D. van Nostrand Co., Inc., 1960).
Nuclear interest: Management of nuclear materials.
Address: 4442 Mohegan, Affton 23, Missouri, U.S.A.

KUGLER, Gaston, Dipl. in Economics. Born 1910. Educ.: Graz Univ. Insurance companies, Chamber of Commerce, Building Trade, 1945-55; At (1955-), Managing Director (1960-), Osterreichische Draukraftwerke A.G. Austrian Board of Elec. Suppliers; Carinthian Assoc. for Political and Ind. Economy; Assoc. for the Foundation of a Univ. in Klagenfurt; Hon. Lecturer, Power Economy, Graz Tech. Coll. Societies: Co-operation in Foratom; Arbeitsgemeinschaft Kernkraftwerk der Österreichische Elektrizitätswirtschaft; Österreichische Studiengesellschaft für Atomenergie.
Nuclear interest: Planning and construction of a nuclear power plant in Austria.
Address: 22 Pamperlallee, A-9201 Krumpendorf, Austria.

KUGLER, Josef, Ober-Ing. Born 1904. Deputy Manager, Power Plant Dept., Siemens-Schuckertwerke G.m.b.H., Vienna, -1967; Manager, Power Plant Dept., Siemens G.m.b.H., 1967-.
Nuclear interest: Nuclear power plant projects.
Address: Siemens G.m.b.H., 15 Nibelungengasse, 1010 Vienna, Austria.

KUHARENKO, N. K. Papers: Co-author, New Models of Porons Stratum for Neutron Carrotage (letter to the Editor, Atomnaya Energiya, vol. 15, No. 4, 1963); co-author, On the Use of Ac-Be Neutron Sources in Trade Geophysics (letter to the Editor, ibid, vol. 16, No. 3, 1964).
Address: Academy of Sciences of the U.S.S.R., 14 Leninsky Prospekt, Moscow V-71, U.S.S.R.

KUHL, David Edmund, A.B., M.D. Born 1929. Educ.: Temple and Pennsylvania Univs. Internship, (1955-56), Residency, Dept. of Radiol., (1958-61), Hospital, Pennsylvania Univ.; Assoc. Prof. of Radiol., School of Medicine, Pennsylvania Univ.; Diplomate, American Board of Radiol. Member: Board of Trustees, Soc. of Nucl. Medicine; Task Group on Scanning, I.C.R.U.; Com. on Radiol., Nat. Acad. of Sci., Nat. Res. Council; Advisory Com. on Medical uses of Radioisotopes, U.S.A.E.C.; Com. on Nucl. Medicine, American Coll. of Radiol.; Rad. Study Sect., Nat. Insts. of Health. Societies: American Coll. of Radiol.; Radiol. Soc. of North America; Soc. of Nucl. Medicine; Internat. Soc. of Stereology.
Nuclear interests: Clinical nuclear medicine; Research in fundamental problems of emission imaging of radionuclides in medicine.
Address: Department of Radiology, Hospital of the University of Pennsylvania, Philadelphia, Pennsylvania, U.S.A.

KUHLMAN, Carl William, B.S., Ph.D. Born 1925. Educ.: Harvard and Washington Univs.

Various management positions with Mallinckrodt, 1950-60; Tech. Director, Mallinckrodt Chemical Works, 1960-64; Asst. to Sen. Vice-Pres., United Nucl. Corp., 1964-65; Asst. to Pres., Douglas United Nucl., 1965-67; Vice-Pres. and Asst. Gen. Manager, Tech. Div. Douglas United Nucl., Inc., 1967-. Book: Co-author, Uranium Production Technology (edited by C. D. Harrington and A. E. Ruehle) (Van Nostrand, 1959). Societies: A.C.S.; A.N.S.; A.A.A.S.
Nuclear interests: Management, fuel cycle systems and production.
Address: 2339 Harris, Richland, Washington, U.S.A.

KUHLMANN, Albert, Dr.-Ing., Dipl.-Ing. Born 1925. Educ.: Karlsruhe T.H. Chairman, Board of Directors, Technischer Überwachungs-Verein Rheinland e.V. Asst. Prof., Rad. Protection, Aachen T.H. Book: Einführung in die Probleme der Kernreaktorsicherheit (Introduction into the problems of nuclear reactor safety) (Düsseldorf, V.D.I.-Verlag G.m.b.H., 1967).
Nuclear interests: Nuclear safety, especially safe operation of nuclear power plants.
Address: Technischer Überwachungs-Verein Rheinland e.V., Cologne, 90 Lukasstr., Germany.

KUHLTHAU, Alden Robert, B.S., M.S., Ph.D. Born 1921. Educ.: Wake Forest Coll. and Virginia Univ. Asst. Prof. Phys., New Hampshire Univ., 1948-51; Asst. Director, Ordnance Res. Labs. (1951-54), Director, Ordnance Res. Labs. (1954-59), Director, Eng. Sci. Res. Labs. (1959-67), Prof. Aerospace Eng. (1959-), Assoc. Dean, School of Eng. and Appl. Sci. (1961-67), Assoc. Provost for Res. (1967-), Virginia Univ. Societies: A.P.S.; A.A.P.T.; A.S.E.E.; A.I.A.A.; A.A.A.S.
Nuclear interest: Isotope separation.
Address: The Rotunda, Virginia University, Charlottesville, Virginia, U.S.A.

KUHN, Philip A. Member, Editorial Board, Bulletin of the Atomic Sci.
Address: Bulletin of the Atomic Scientists, 935 East 60th Street, Chicago 37, Illinois, U.S.A.

KUHN, Richard, Prof. Biochem., Dr.phil., Dr. rer.nat.h.c., Dr.phil.h.c. Nobel Prize for Chem., 1938. Born 1900. Educ.: Vienna and Munich Univs. Director, Max-Planck-Inst. for Medical Res., Inst. for Chem.
Address: Max-Planck Institute for Medical Research, Institute for Chemistry, 29 Jahnstrasse, Heidelberg, Germany.

KÜHN, Wilhelm K. G., Dr.rer.nat., Dipl.-Phys. Born 1922. Educ.: Freiburg i. Br. Univ. Kernforschungszentrum Karlsruhe 1957-65; Inst. für Strahlenbiologie, Hanover T.H., 1965-.
Nuclear interest: Application of radioisotopes in industry and agriculture.
Address: 99 Vinnhorstsveg, Hanover, Germany.

KÜHNEL, Roland, Eng. Born 1930. Development, design and construction of chem. plants, Lurgi, Frankfurt-am-Main, 1955-56; Nucl. Dept., German Babcock and Wilcox, Oberhausen/Rhld., 1956-57; Nucl. Dept., A.E.G., Frankfurt-am-Main, 1957-.
Nuclear interests: Nuclear power station overall design, with special view to the nuclear steams supply system, technical project management work, thermodynamic, hydraulic and physics design of reactor cores.
Address: 31 Kurt-Schumacher-Ring, Sprendlingen, Germany.

KUHNS, William G. Vice Pres. (-1967), Pres. and Chief Exec. Officer (1967-), Gen. Public Utilities Corp.
Address: General Public Utilities Corporation, 80 Pine Street, New York, New York 10005, U.S.A.

KUKARIN, A. I. Papers: Co-author, Ionisation Chamber with Silver Electrodes for Measuring of Thermal Neutron Fluxes under High Level Gamma-Rad. (letter to the Editor, Atomnaya Energiya, vol. 21, No. 1, 1966); co-author, Propagation of Air of Biological Neutron Doses from Point Sources (letter to the Editor, ibid, vol. 23, No. 1, 1967).
Address: Academy of Sciences of the U.S.S.R., 14 Leninsky Prospekt, Moscow V-71, U.S.S.R.

KUKAVADZE, G. M.
Nuclear interest: Fission product yields.
Address: Academy of Sciences of the U.S.S.R., 14 Leninsky Prospekt, Moscow V-71, U.S.S.R.

KUKITA, M. Member for Japan, Study Group on Food Irradiation, O.E.C.D., E.N.E.A.
Address: O.E.C.D., European Nuclear Energy Agency, 38 boulevard Suchet, Paris 16, France.

KULAKOV, A. Soviet deleg., Convention on Thermonucl. Processes, Inst. of Elec. Eng., London, April 29-30, 1959.
Address: Academy of Sciences of the U.S.S.R., 14 Leninsky Prospekt, Moscow V-71, U.S.S.R.

KULAKOVSKII, M. Ya. Papers: Co-author, Shielding Properties of Borated Fire-Proof Chromite Concretes (Atomnaya Energiya, vol. 22, No. 2, 1967); co-author, Heat Emission in Borated Concrete Shields (ibid, vol. 22, No. 2, 1967); co-author, Shielding Properties of Borated Concretes (ibid, vol. 23, No. 1, 1967); co-author, On Radiational Stability of Plain Concrete (ibid, vol. 23, No. 4, 1967).
Address: Academy of Sciences of the U.S.S.R., 14 Leninsky Prospekt, Moscow V-71, U.S.S.R.

KUL'BASHNAYA, M. V. Paper: Co-author, Concerning Influence of Surface Active Agents on Bituminisation Process of Radioactive Waste (Atomnaya Energiya, vol. 22, No. 5, 1967).
Address: Academy of Sciences of the U.S.S.R., 14 Leninsky Prospekt, Moscow V-71, U.S.S.R.

KULESH, I. Paper: Co-author, Study of Radiochem. Stability of High Boiling Hydrocarbons in Reactor (Atomnaya Energiya, vol. 20, No. 1, 1966).
Address: Academy of Sciences of the U.S.S.R., 14 Leninsky Prospekt, Moscow V-71, U.S.S.R.

KULESHOVA, Yu. F. Paper: Co-author, Investigation of Electrophoresis Filters with Graphite Anodes (letter to the Editor, Atomnaya Energiya, vol. 23, No. 4, 1967).
Address: Academy of Sciences of the U.S.S.R., 14 Leninsky Prospekt, Moscow V-71, U.S.S.R.

KULICHENKO, V. V. Papers: Co-author, Some Aspects of Means of Locating Radioisotopes and the Problem of Safe Storage (Atomnaya Energiya, vol. 10, No. 1, 1961); co-author, Radiations from Fission-product Preparations (ibid, vol. 10, No. 4, 1961); co-author, Possibility and Conditions of Utilisation of Chemical Reaction Heat for Thermal Working out of Liquid Radioactive Wastes (ibid, vol. 20, No. 3, 1966); co-author, Use of Chem. Reaction Heat for Thermal Treatment of Liquid Radioactive Waste (letter to the Editor, ibid, vol. 24, No. 5, 1968).
Address: Academy of Sciences of the U.S.S.R., 14 Leninsky Prospekt, Moscow V-71, U.S.S.R.

KULIKOV, I. A. Paper: Co-author, New Chem. Method for Determination of Rad. Dose in Reactor (Atomnaya Energiya, vol. 20, No. 6, 1966).
Address: Academy of Sciences of the U.S.S.R., 14 Leninsky Prospekt, Moscow V-71, U.S.S.R.

KULIKOV, O. F. Paper: Co-author, Method of Measuring of Radial Betatron Oscillation Frequency in Synchrotron (letter to the Editor, Atomnaya Energiya, vol. 23, No. 4, 1967).
Address: Academy of Sciences of the U.S.S.R., 14 Leninsky Prospekt, Moscow V-71, U.S.S.R.

KULIPANOV, G. N. Papers: Co-author, System for Control and Measuring of Electron Beams Parameters in VEP-1 Electron-Electron Storage Ring (Atomnaya Energiya, vol. 20, No. 3, 1966); co-author, Non Linear Resonances of Betatron Oscillations in Storage Rings (ibid, vol. 22, No. 3, 1967); co-author, Lifetime and Intensity Measurements for Electron Beams in Storage Rings (ibid, vol. 22, No. 3, 1967); co-author, Lifetime and Beam Dimensions in Storage Rings as Function of Number of Particles (ibid, vol. 22, No. 3, 1967).
Address: Academy of Sciences of the U.S.S.R., 14 Leninsky Prospekt, Moscow V-71, U.S.S.R.

KULISH, E. E. Papers: Co-author, Some Eng. and Technol. Aspects of Radioisotope and Labelled Compound Production in the U.S.S.R. (2nd I.C.P.U.A.E., Geneva, Sept. 1958); co-author, Rad. Characteristics of ^{145}Sm and Enriched ^{75}Se Gamma-Sources (letter to the Editor, Atomnaya Energiya, vol. 15, No. 6, 1963).
Address: Nuclear Energy Institute, Academy of Sciences of the U.S.S.R., 14 Leninsky Prospekt, Moscow V-71, U.S.S.R.

KULJIAN, Harry A. B.S. (Elec. Eng.), B.S. (Mech. Eng.). Born 1894. Educ.: M.I.T. Chairman of the Board of Kuljian Corp. Books: Nucl. Power Plant Design, (A. S. Barnes and Company, Inc., 1968); Man and the World of Sci. (A. S. Barnes and Co. Inc., 1964). Societies: Fellow, A.S.M.E.; Fellow, I.E.E.E.; A.A.A.S. A.N.S.
Nuclear interests: Reactor concepts; Management; Systems.
Address: 131 Raynham Road, Merion, Pa. 19066, U.S.A.

KULL, Donald C., B.A., M.A. (Public Administration). Born 1921. Educ.: Gustavus Adolphus Coll., Minnesota and Minnesota Univ. Exec. Asst. to Gen. Manager, U.S.A.E.C., 1966-. Society: American Soc. for Public Administration.
Nuclear interest: Management.
Address: Executive Assistant to the General Manager, United States Atomic Energy Commission, Washington, D. C. 20545, U.S.A.

KULP, J. Laurence, B.S., M.S., M.A., Ph.D. Born 1921. Educ.: Drew, Wheaton, Ohio State, Princeton and Columbia Univs. Prof. Geochem., Columbia Univ.(on faculty of Columbia 1948-65; full Prof., 1958-65; Adjunct Prof., 1965-67); Pres., Isotopes, Inc., 1964-. Societies: A.C.S.; A.P.S.; American Min. Soc.; A.G.U.; A.A.P.G.; American Geological Soc.
Nuclear interests: Isotope geology and fallout problem.
Address: Isotopes, Inc., 50 Van Buren Avenue, Westwood, New Jersey 07675, U.S.A.

KULP, Stanley W., B.S., M.E. Born 1926. Educ.: California Univ., Berkeley. Project Manager, Nucl., Kaiser Eng. Society: A.N.S.
Nuclear interests: Project management of engineering design and construction of nuclear facilities; Experimental, testing and power reactor.
Address: 300 Lakeside Drive, Oakland, California 94604, U.S.A.

KULVARSKAYA, B. S. Paper: Investigation of Evaporation Velocities of Uranium and Zirconium Carbides Cathodes and Their Solid Solutions (Atomnaya Energiya, vol. 21, No. 5, 1966).
Address: Academy of Sciences of the U.S.S.R., 14 Leninsky Prospekt, Moscow V-71, U.S.S.R.

KULYUKINA, L. A. Paper: Co-author, Cascade Interactions of Particles with Nuclei in High Energy Region (letter to the Editor, Atomnaya Energiya, vol. 16, No. 6, 1964).
Address: Academy of Sciences of the U.S.S.R., 14 Leninsky Prospekt, Moscow V-71, U.S.S.R.

KULYUPINA, N. V. Papers: Co-author, High Power Gamma Irradiation Unit UK-30000 (letter to the Editor, Atomnaya Energiya, vol.

19, No. 1, 1965); co-author, High Power Gamma Irradiation Unit UK-70000 (ibid).
Address: Academy of Sciences of the U.S.S.R., 14 Leninsky Prospekt, Moscow V-71, U.S.S.R.

KUMAGAYA, Tomiyoshi, M.D., Chief Div. of the Hospital, Nat. Inst. of Radiol. Sci.
Address: National Institute of Radiological Sciences, 9-1, 4-chome, Anagawa, Chiba-shi, Chiba-ken, Japan.

KUMANO, Shuichi. Director in Charge, Technological Study Unit, Mitsui Steamship Co. Ltd.
Address: Mitsui Steamship Co. Ltd. (Mitsui Line), Tokyo, Japan.

KUMATORI, Toshiyuki, M.D. Born 1921. Educ.: Tokyo Univ. Medical Staff, Tokyo Univ. Hospital, 1950; Medical Staff, 1st Nat. Hospital of Tokyo, 1952; Sci. Staff (1959), Head, Div. of Rad. Health (1965), Nat. Inst. of Radiol. Sci. Book: Hoshasen no Bogo (Maruzen Co. Ltd., 1965). Societies: Internat. Nucl. Hematology Sect., Inst. Nucl. Eng., London; Japan Rad. Res. Soc.; Japan Hematological Soc; Nippon Societas Radiologica.
Nuclear interest: Radiation effects on human being
Address: 4-9-1 Anagawa, Chiba-shi, Japan.

KUME, Sanshiro. Lecturer, Lab. of Nucl. Studies, Faculty of Sci., Osaka Univ.
Address: Osaka University, Nakanoshima, Kita-ku, Osaka, Japan.

KUMLEBEN, Francis GERARD-. See GERARD-KUMLEBEN, Francis.

KUMPF, Friedrich - Bernhart, C.E. Born 1917. Educ.: Lausanne Ecole Polytech. Manager, Entreprise Génie Civil au Congo, Civil Eng. Co., 1950-; Deputy Manager, (1961-), Manager (1965-), Managing-Director (1967-), Eng. and Design Dept., Electrobel. Societies: Assoc. des Anciens Elèves de l'Ecole Polytech., Lausanne; Soc. des Ing. Civils de France.
Nuclear interest: Generation and transmission of electricity, (namely nuclear energy).
Address: Electrobel S.A., 1 place du Trône, Brussels 1, Belgium.

KUNCHENKO, V. V. Papers: Co-author, Dependence of Hardened Uranium Texture from Heating Nature and Other Parameters of Heat Treatment (Atomnaya Energiya, vol. 16, No. 4, 1964); co-author, Interrelation between Thermal Expansion Coefficient α and Growth index GI (letter to the Editor, ibid, vol. 20, No. 2, 1966); co-author, Texture Distribution along Cross-Section of Alpha- and Gamma-Deformed and Quenched Uranium Rods (ibid, vol. 21, No. 3, 1966); co-author, Impurity State Effect on Deformation Texture of Low-Alloyed Alpha Uranium (letter to the Editor, ibid, vol. 24, No. 1, 1968).
Address: Academy of Sciences of the U.S.S.R., 14 Leninsky Prospekt, Moscow V-71, U.S.S.R.

KUNDU, Dhirendra Nath, B.Sc., M.Sc., Ph.D. Born 1916. Educ.: Calcutta and Ohio State Univs. Asst. Prof. Phys., Ohio State Univ., 1954-55; Professional Assoc., N.A.S., Nat. Res. Council, Washington D.C., 1955-57; Prof., Inst. Nucl. Phys., Calcutta, 1958-. Societies: Fellow, A.P.S.; Fellow, Nat. Inst. Sci., India; Fellow, Indian Phys. Soc.
Address: Saha Institute of Nuclear Physics, 92 Upper Circular Road, Calcutta 9, India.

KUNEGIN, Eugenii Petrovich. Sci. Asst., Acad. of Sci. of the U.S.S.R.; Sen. Sci., Kurchatov Inst. of Atomic Energy; Deleg. to 2nd I.C.P.U.A.E., Geneva,Sept. 1958.
Address: Kurchatov Institute of Atomic Energy, 46 Ulitsa Kurchatova, Post Box 3402, Moscow, U.S.S.R.

KUNICHIKA, Sango, Dr. Prof. i/c, Div. 1, Lab. of Nucl. Study, and Director, Inst. for Chem. Res., Kyoto Univ.
Address: Instytut Fizyki Uniwersytetu Jagiel-Chemical Research, Yoshida-Honmachi, Sakyo-ku, Kyoto, Japan.

KUNISZ, Miss Maria D., Sc.D. Sen. Lecturer, Inst. Phys., Jagellonian Univ., Cracow.
Address: Instytut Fizki Uniwersytetu Jagiellonskiego, 13 ul. Golebia, Cracow 1, Poland.

KÜNKEL, Hans Adam, Prof., Dr.rer.nat. Born 1910. Educ.: Königsberg and Berlin Tech. Univs. Director, Radiobiol. Inst., Hamburg Univ.
Book: Atomschutzfibel (Göttingen, Plesse-Verlag, 1950). Societies: Deutsche Röntgengesellschaft; Deutsche Physikalische Gesellschaft; Deutsche Gesellschaft für Biophysik; Sté. Européenne de Radiobiol.
Nuclear interests: Radiobiology; Biophysics; Radiation protection; Radioactive isotopes.
Address: Strahlenbiologisches Institut, Universitäts-Krankenhaus Eppendorf, Hamburg 20, Germany.

KUNKLER, Peter Bernard, M.A., M.D. (Cantab.), M.R.C.P. (London), F.F.R., D.M.R.T. Born 1920. Educ.: Cambridge Univ. and St. Bartholomews Hospital, London. Physician i/c of Radiotherapy, Charing Cross Hospital; Consultant Radiotherapist, United Birmingham Hospitals. Clinical Lecturer, Radiotherapy, Birmingham Univ.; Director, Radiotherapy, Cardiff Roy. Infirmary and South Wales and Monmouthshire Radiotherapy Service; Lecturer, Radiotherapy, Welsh Nat. School of Medicine. Book: Co-author, Treatment of Cancer in Clinical Practice. Societies: British Inst. of Radiol.; Roy. Soc. of Medicine.
Nuclear interest: Applications of radiation in treatment and diagnosis of malignant disease.
Address: South Wales and Monmouthshire Radiotherapy Service, Velindre Hospital, Whitchurch, Cardiff, Wales.

KUNSTADT, Eugen, Univ. Prof., Dr. Scientiarum. Born 1913. Educ.: Komenský Univ. Bratislava. Chief, Dept. Radiol. and Nucl.

Medicine, Medical School Safarik, Kosice. Society: Sect. of Nucl. Medicine, Czechoslovak Medical Soc. J. E. Purkyne.
Nuclear interest: Nuclear medicine, especially use of radioisotopes in diagnosis and therapy of malignant diseases.
Address: Radiology Department, Safárik University School of Medicine, 53 Rastislavova, Kosice, Czechoslovakia.

KUNTOHADJI, Ir. Deputy Director and Head, Exec. Div., Inst. for Atomic Energy. Sec., Indonesian A.E.C. Deleg. to 2nd I.C.P.U.A.E., Geneva, Sept. 1958.
Address: Institute for Atomic Energy, Djl. Palatehan I No. 26, Blok K5, Kebajoran-Baru, Djakarta, Indonesia.

KUNZ, Emil, M.D. Born 1930. Educ.: Charles Univ., Prague, and Faculty of Medical Hygiene, Leningrad. Hygienist, Municipal Station of Hygiene and Epidemiology, Prague, 1956-57; Head, Div. of Work and Rad. Hygiene, Ministry of Health, Prague, 1957-64; Head, Dept. of Environmental Radioactivity, Inst. of Rad. Hygiene, Prague, 1964-; Lecturer, Faculty of Medical Hygiene, Charles Univ., Prague, 1960-. Books: Rad. Hygiene - Chapter in Environmental Hygiene (1963); Hygiene of Work (Prague, State Pedagogical Publishing House, 1964). Society: Czechoslovak Soc. of Nucl. Medicine and Rad. Hygiene.
Nuclear interests: Radiation hygiene and radiation protection.
Address: Institute of Radiation Hygiene, 48 Srobárova, Prague 10, Czechoslovakia.

KUNZE, Jay F. Ph.D., M.S., B.S. (Nucl. Phys.). Born 1933. Educ.: Carnegie Inst. Technol. Project Phys., Carnegie Inst. Technol. 1955-59; Nucl. Phys., Nucl. Materials and Propulsion Operation, Idaho Test Station (1959), now Manager, Operations and Analysis, Gen. Elec. Co. Societies: A.P.S.; A.N.S.
Nuclear interests: Critical experiments; Power reactor experiments; Reactor physics.
Address: c/o General Electric Co., Box 2147, Idaho Falls, Idaho, U.S.A.

KUNZE, Paul, Prof. Dr. phil. Born 1897. Educ.: Leipzig and Munich Univs. Director, Inst. for Exptl. Nucl. Phys. Consultant, Kernenergie. Book: Grimsehl (chapter nucl. phys., Teubner, Leipzig, 1959). Societies: German Phys. Soc.; Sci. Council of Peaceful Application of Atomic Energy.
Nuclear interest: Nuclear physics.
Address: Institut für Experimental Kernphysik, 19 Zellescher Weg, Dresden A27, German Democratic Republic.

KUNZEL, Herbert. Director, Atomic Development Mutual Fund Inc.
Address: Atomic Development Mutual Fund Inc., 1033 30th Street Northwest, Washington 7, D.C., U.S.A.

KÜNZLI, Heinz. Born 1926. Director, Internat. Potash Inst.; Co-editor, Neue Technik (New Techniques); Chairman, Commission for Information, Swiss Assoc. for Nucl. Energy.
Address: 23 Normannenstrasse, CH-3000 Bern, Switzerland.

KUO, Chi Sheng, B.S. (M.E.), M.S. (M.E.), M.E. Born 1923. Educ.: Chinese Nat. Chiao-Tung, Texas and Columbia Univs. Eng., Chem. Construction Corp., U.S.A., 1952-56; Sen. Eng., Hydrocarbon Res., Inc., U.S.A., 1956-60; Sen. Eng., United Nucl. Corp., U.S.A., 1960-62; Sen. Eng., Bechtel Corp., U.S.A., 1962-. Society: A.S.M.E.
Nuclear interests: Reactor design and analysis related to heat transfer and fluid mechanics.
Address: Bechtel Corp., 101 California Street, San Francisco, California, U.S.A.

KUO, Mo-jo. Pres., Acad. of Sci., Peking.
Address: Academia Sinica, 3 Wen Tsin Chien, Peking, China.

KUPER, J. B. Horner, B.A., Ph.D. Born 1909. Educ.: Williams Coll.and Princeton Univ. Dept. Chairman, Brookhaven Nat. Lab., 1947-; Editor, Review of Scientific Instruments, 1954-. Societies: Fellow, A.P.S.; Fellow, I.E.E.E.; Rad. Res. Soc.; H.P.S.
Nuclear interests: Instrumentation; Health physics.
Address: Brookhaven National Laboratory, Upton, New York 11973, U.S.A.

KÜPFMÜLLER, Karl, Prof., Dr.-Ing. E.h. Born 1897; Books: Einführung in die theoretische Elektrotechnik (7. Auflage. Berlin, Springer-Verlag, 1962); co-author, Dosis und Wirkung (Freiburg, Edition Cantor, 1949); Die Systemtheorie der elektrischen Nachrichtenübertragung 2. Aufl. (Stuttgart, S. Hirzel, 1952).
Nuclear interest: Electronics.
Address: 40 Ohlystrasse, 61 Darmstadt, Germany.

KUPKA, Ivan, B.Sc. (Mech. Eng.). Born 1935. Educ.: Pilsen Tech. Univ. Res. Eng., Skoda Works, Pilsen, 1959-. Society: Czechoslovak Sci. and Tech. Soc.
Nuclear interests: Experimental research and testing of large reactor pressure vessels; Residual stresses in the manufacture of nuclear reactor pressure vessels.
Address: 2 Orechová, Pilsen, Czechoslovakia.

KUPPERMANN, Aron, Ph.D. (Phys. Chem.), Chem. Eng., Civil, Eng. Born 1926. Educ.: Sao Paulo (Brazil) and Notre Dame Univs. Asst. Prof. Chem., Inst. Tecnológico de Aeronáutica, Brazil, 1950-51; Head, Anodising Sect., Industria e Comercio Ajax, Sao Paulo, Brazil, 1952; Res. Assoc., Rad. Lab., Notre Dame Univ., 1953-55; Resident Res. Assoc., Argonne Nat. Lab., 1957; Res. Assoc., Inst. de Energia Atomica, Brazil, 1959-60; Instructor, Phys. Chem. (1955-57), Asst. Prof. Phys. Chem. (1957-61), Assoc. Prof. Phys. Chem. (1961-63),

Illinois Univ.; Prof. Chem. Phys., California Inst. Technol., 1963-. Editorial Board, Phys. Chem.; Councillor for Chem., Rad. Res. Soc.; Councillor for Chem. Internat. Assoc. Rad. Res. Book: Co-author, Actions Chimiques et Biologiques des Radiations, 5th serie (Paris, 1961). Societies: A.C.S.; A.P.S.; A.A.A.S.; Faraday Soc.; Rad. Res. Soc.
Nuclear interests: Radiation chemistry; Diffusion kinetics; Low energy electron scattering.
Address: Chemistry and Chemical Engineering Division, California Institute of Technology, Pasadena, California, U.S.A.

KUPRIANOFF, Johann M., Dr. Eng. Born 1904. Educ.: Karlsruhe Tech. Univ. Prof. Food Technol., Karlsruhe Tech Univ.; Director, Federal Res. Inst. for Food Preservation, Karlsruhe, 1948-. Pres., Gen. Conference, Internat. Inst. Refrigeration, Paris; Chairman, Working Group on Food Irradiation, German Com. for Food Additives; Com. Rad. Protection, German A.E.C. Book: Co-author, Food Sterilisation by Irradiation and Food Contamination (Darmstadt Steinkopff). Societies: Inst. Food Technologists, Chicago; Heidelberg Acad. Sci.
Nuclear interests: Irradiation of biological material and foodstuffs; Food contamination.
Address: 12 Kaiserstrasse, Karlsruhe, Germany.

KURANZ, John L., M.S. (Eng.). Born 1921. Educ.: Marquette and Oklahoma Univs. Vice-Pres. and Sec., Nucl.-Chicago Corp.; Member of A.S.E.E.-A.N.S. Com. on Objective Criteria in Nucl. Eng. Education; Advisory Group Com. E-10, A.S.T.M; Chairman, Advisory Com. on Isotopes and Rad. Development, U.S.A.E.C., 1965-68; Director, The Amersham/Searle Corp., 1968-. Societies: A.P.S.; A.N.S.; A.S.T.M.; Illinois Soc. Professional Eng.
Nuclear interests: Active in research development and design of systems for new applications of isotopes and radiation in fields of industrial, academic and medical research. Active in professional society affairs promoting nuclear education and peaceful uses of isotopes in United States and abroad.
Address: 333 E. Howard Avenue, Des Plaines, Ill., U.S.A.

KURASAWA, Fumio, Prof. Agriculture Dept. Member, Radioisotope Res. Com., Niigata Univ. School of Medicine.
Address: Department of Agriculture, Niigata University School of Medicine, 757, 1-Bancho Asahimachi-Street, Niigata, Japan.

KURASHOV, A. A. Papers: Co-author, A Multichannel Time-of-flight Fast Neutron Spectrometer (Atomnaya Energiya, vol. 5, No. 2, 1958); co-author, Light Pencil (letter to the Editor, ibid, vol. 19, No. 4, 1965).
Address: Academy of Sciences of the U.S.S.R., 14 Leninsky Prospekt, Moscow V-71, U.S.S.R.

KURATA, Chikara, B.A. (Eng.). Born 1889. Educ.: Sendai Coll. Technol. Chairman of Board, Hitachi, Ltd.; Pres., Japan Machinery Federation; Pres., Japan Sci. Foundation; Pres., Tokyo Atomic Ind. Consortium; Pres. Tokyo Atomic Ind. Res. Lab.; Councillor, Japan Atomic Ind. Forum; Councillor and Advisor, Japan Atomic Energy Res. Inst.; Director, Japan Atomic Power Co.; Councillor, Sci. and Technol. Agency; Governing Director, Japan Federation of Employers' Assocs. Society: Councillor, Japan Elec. Machine Ind. Assoc. Nuclear interest: Management.
Address: Hitachi, Ltd., 1-4 Marunouchi, Chiyoda-ku, Tokyo, Japan.

KURATA, Okito. Born 1901. Educ.: Kyushu Imperial Univ. Pres., Mitsui Mining Co., Ltd.; Director, Mitsui Cement Co., Ltd.; Sen. Adviser, Mitsui Construction Co., Ltd. Councillor, Japan Coal Producer's Assoc.; Director, Federation of Economic Organisation.
Address: 2-2 4-Chome, Kita Asagaya, Tokyo, Japan.

KURBATOV, I. M. Paper: Co-author, On Optimal Direction of Nucl. Reactor Thermal Processes (letter to the Editor, Atomnaya Energiya, vol. 19, No. 6, 1965).
Address: Academy of Sciences of the U.S.S.R., 14 Leninsky Prospekt, Moscow V-71, U.S.S.R.

KURIHARA, Yoshiro. Chief, Atomic Power Sect., Gen. Eng. Dept., Lab., Meidensha Elec. Manufacturing Co. Ltd.
Address: Meidensha Electric Manufacturing Co. Ltd., 2-chome, Higashi-Osaki, Shinaga-wa, Tokyo, Japan.

KURIMURA, Ryuzo. Director, Mitsui Mining and Smelting Co., Ltd.; Chief, Metals and Materials Div., Nippon Atomic Energy Industry Group. Vice-Pres., Japan Assoc. of Metals for Atomic Energy.
Address: c/o Daiichi Bussan Kaisha, Ltd.,1-chome, Shibatamura-cho, Minatoku, Tokyo, Japan.

KURIR, Anton, Dipl.-Eng. Agric. Dr., Dr.rer. nat. Born 1909. Educ.: Zagreb, Vienna Agricultural and Vienna Univs. Prof. Vienna Agric. Univ., 1955; Head and Chair, Inst. Forest Entomology and Forest Protection. Societies: Council Court of Patents, Vienna; Internat. Assoc. Univ. Profs. and Lecturers, London; Internat. Union Forest Res. Organisations, London; Deutsche Gesellschaft für angewandte Entomologie, Munich; Zoologisch-Botanische Gesellschaft, Vienna.
Nuclear interests: Applied radiation in biology and entomology; Use of radioisotopes in forest entomology (cobalt and strontium); Insects control with atoms.
Address: Institut für Forstentomologie und Forstschutz der Hochschule für Bodenkultur, 33 Gregor Mendel-Strasse, Vienna 18, Austria.

KÜRKCUOGLU, Nusret, Prof. Sen. Officer, Inst. for Nucl. Energy, Istanbul Tech. Univ.
Address: Istanbul Technical University, Gümüssuyu, Istanbul, Turkey.

KURLAND, George Stanley, A.B., M.D. Born 1919. Educ.: Harvard Univ. Asst. Clinical Prof. Medicine (Harvard); Chief, Cardiac Clinic, Beth Israel Hospital, Boston. Papers: Numerous papers on thyroid function, biologic effects of radioiodine, metabolism of radio-thyroxine and radio therapy of cardiovascular disease. Societies: American Thyroid Assoc.; Amer- n ican Heart Assoc.
Address: Beth Israel Hospital, 330 Brookline Avenue, Boston, Massachusetts 02215, U.S.A.

KUROCHKINA, L. M. Paper: Co-author, Ultrasonic Effect on Plasticity of High Boron Stainless Steel (letter to the Editor, Atomnaya Energiya, vol. 20, No. 5, 1966).
Address: Academy of Sciences of the U.S.S.R., 14 Leninsky Prospekt, Moscow V-71, U.S.S.R.

KURODA, K. Acting Manager, Japan Assoc. of Metals for Atomic Energy.
Address: Japan Association of Metals for Atomic Energy, 1-1 Shiba Tamuracho, Minato-ku, Tokyo, Japan.

KURODA, M., Prof. Reactor Materials Div., Res. Lab. of Nucl. Reactors, Tokyo Inst. Technol.
Address: Reactor Materials Division, Research Laboratory of Nuclear Reactors, Tokyo Institute of Technology, 1 Oh-Okayama, Meguro-ku, Tokyo, Japan.

KURODA, Paul Kazuo, B.S. (Chem.), Sc.D. (Chem.). Born 1917. Educ.: Tokyo Univ. Asst. Prof. (1952), Assoc. Prof. (1955), Prof. (1961), Arkansas Univ. Member, Subcom. on Low Level Contamination of Materials and Reagents, N.A.S.-N.R.C. Societies: A.P.S.; A.C.S.; A.G.U. Nuclear interests: Nuclear radiochemistry, geochemistry and cosmochemistry.
Address: 908 Eva Avenue, Fayetteville, Arkansas, U.S.A.

KURODA, Yoshitesu, Prof. Director of Nucl. Eng. Course, Tokai Univ.
Address: Nuclear Engineering Course, Engineering Department, Tokai University, 1431 Yoyogi-Tomigaya-cho, Shibuya-ku, Tokyo, Japan.

KURODE, Kaichiro. Member, Atomic Energy and Isotope Com., Tokushima Univ.
Address: c/o Medical Faculty of Tokushima University, Kurmoto-cho, Tokushima-shi, Japan.

KUROSAWA, Hiroshi, Dr. Sen. Officer, School of Medicine, Toho Univ.
Address: Toho University, 4-77 Omori, Ota-ku, Tokyo, Japan.

KUROSAWA, M. Manager, Accident Dept., Kyoei Mutual Fire and Marine Insurance Co.
Address: Kyoei Mutual Fire and Marine Insurance Co., 3, 1-chome Shiba-Tamura-cho, Minato-ku, Tokyo, Japan.

KURRLE, Geoffrey Rosevear, M.D., B.S., F.R.A.C.P., F.C.R.A. Born 1906. Educ.: Melbourne Univ. Medical Director, Cancer Inst. Board, Victoria.
Nuclear interests: Clinical medicine and radiotherapy.
Address: Cancer Institute Board, 278 William Street, Melbourne, C.1, Australia.

KURSUNOGLU, Behram, B.Sc., Ph.D. Born 1922. Educ.: Edinburgh and Cambridge Univs. Res. Assoc., Cornell Univ., 1952-54; Visiting Prof. Phys., Miami, Fla., 1954-55; Sen. Res. Fellow, Yale Univ., summer 1955. Director, Faculty of Nucl. Sci. and Technol., Middle East Tech. Univ., Ankara, Turkey. Adviser to Turkish General Staff on Atomic Energy; Formerly Member,Turkish A.E.C.. Book: Advanced Non-Relativistic Quantum Mechanics (1955). Societies: Fellow, Cambridge Philosophical Soc.; A.P.S.
Address: Faculty of Nuclear Sciences and Technology, Middle East Technical University, Ankara, Turkey.

KÜRTEN, Helmut, Ing. Sen. Officer, Nucl. Dept., Deutsche Picker G.m.b.H.
Address: Deutsche Picker G.m.b.H., 7 Peter Bauer-Strasse, Cologne, Germany.

KURTEPOV, M. M. Papers: Co-author, Corrosion Aggressiveness of Fuel Element Solvents in Relation to Structural Materials (Atomnaya Energiya, vol. 15, No. 1, 1963); Corrosion of Austenitic Stainless Steel during Evaporation of Radioactive Solution (ibid, vol. 19, No. 2, 1965).
Address: Academy of Sciences of the U.S.S.R., 14 Leninsky Prospekt, Moscow V-71, U.S.S.R.

KURTZ, Robert. Editor, Martin News; Public Relations Director, Nucl. Div., Martin Co., 1964-.
Address: Martin Co., Mail No. 202, Baltimore 3, Maryland, U.S.A.

KUSA, Mamoru, D.Sc. Born 1919. Educ.: Hokkaido Univ. Prof. Embryology, Yamagata Univ., 1959-. Society: Zoological Soc. Japan. Nuclear interest: Use of radioisotopes in analysis of oogenesis in various teleosts.
Address: Research Institute for Radioisotopes, Yamagata University, Koshirakawa-cho, Yamagata-shi, Japan.

KUSAKABE, Shigeru. Vice Pres., Japan Atomic Energy Res. Inst.
Address: Japan Atomic Energy Research Institute, 1-1-13 Shinbashi, Minato-ku, Tokyo, Japan.

KUSANO, Kazuhito, Dr. Born 1922. Educ.: Meiji Inst. Technol. Prof., Miyazaki Univ. Societies: A.C.S.; Chem. Soc. of Japan. Nuclear interests: Radiation and tracer tech+ niques.

Address: c/o Faculty of Engineering, Miyazaki University, 118 Nishimaruyama-cho, Miyazaki-shi, Japan.

KUSANO, Mitso, Dr. Chief, Meidensha Lab., Meidensha Elec. Manufacturing Co. Ltd.
Address: Meidensha Electric Manufacturing Co. Ltd., 2-chome, Higashi-Osaki, Shinaga-wa, Tokyo, Japan.

KUSCER, Ivan, Ph.D. Born 1918. Educ.: Ljubljana Univ. Lecturer (1950-59), Prof. Phys. (1959-), Ljubljana Univ. Book: Phys. (Ljubljana, 1958).
Nuclear interest: Neutron transport theory.
Address: Institute of Physics, Ljubljana University, 26 Jadranska, Ljubljana, Yugoslavia.

KUSCH, Polykarp, Prof. Chairman, Phys. Dept., Columbia Univ. Member, Board of Directors, EON Corp., 1964-.
Address: Physics Department, Columbia University, New York 27, New York, U.S.A.

KUSHNIRENKO, E. A. Papers: Co-author, Status Report of Electron-Electron Storage Ring VEP-1 (Atomnaya Energiya, vol. 19, No. 6, 1965); co-author, Electron-Electron Scattering at Energy of 2 x 135 MeV (ibid, vol. 22, No. 3, 1967).
Address: Novosibirsk Science Centre, 20 Sovietskaya Ulitsa, Novosibirsk, Siberia, U.S.S.R.

KUSSOVNIKOV, A. S. Paper: Co-author, Experience in Operating the SM-2 Exptl. Reactor (Kernenergie, vol. 9, No. 10, 1966).
Address: Academy of Sciences of the U.S.S.R., 14 Leninsky Prospekt, Moscow V-71, U.S.S.R.

KUSTER, W. J. SCHMIDT-. See SCHMIDT-KUSTER, W. J.

KUSUMEGI, Asao, M.Sc., Ph.D. Born 1924. Educ.: Kyoto Univ. Res. Asst., Lab. of Nucl. Phys., Inst. for Chem. Res., Kyoto Univ., 1956; Res. Fellow, S.C. Div., C.E.R.N., 1957-; Asst. Prof., Inst. for Nucl. Study, Tokyo Univ., 1963-. Societies: Phys. Soc.of Japan; A.P.S. Nuclear interests: High-energy nuclear physics and instrumentation.
Address: Institute for Nuclear Study, Tokyo University, Tanashi, Kitatama, Tokyo, Japan.

KUSUNOKI, Nobuo, M.D. Prof. and Director, Fukushima Radioisotope Res. Com., Fukushima Medical Coll.
Address: Fukushima Medical College, 14 Sugizuma-cho, Fukushima City, Japan.

KUSUNOSE, Kumahiko. Formerly Supt., now Director, Japan Atomic Energy Res. Inst.
Address: Japan Atomic Energy Research Institute, 1-1-13 Shinbashi, Minato-ku, Tokyo, Japan.

KUTACEK, Milan, R.N.Dr., Ph.Mr. (Pharmacy), C.Sc., Docent. Born 1925. Educ.: Charles Univ. Prague. Dept. of Biochem., Univ. Hospital, 1951-52; Dept. of Chem., Coll. of Agronomy, 1952-58; Dept. of Plant Physiology, Inst. of Plant Production, Acad. of Agricultural Sci., 1958-62; Dept. of Radiobiol., Inst. of Exptl. Botany, Acad. of Sci., 1962-; Stages at Justus Liebig Univ., Giessen (1966), and Yale Univ., New Haven (1967). Sec., Plant Physiology Sect., Czechoslovak Botanical Soc. Book: Chapters in Handbuch der Papierchromatographic (Editors, Hais and Macek) (Jena, Fischer Verlag, 1958). Society: Czechoslovak Biochem. Soc.
Nuclear interests:Metabolism of plants, especially their hormonal metabolism followed by tracer techniques; Influence of radiation on plant metabolism.
Address: Institute of Experimental Botany, Czechoslovak Academy of Sciences, Department of Radiobiology, 15 Ke dvoru, Praha-Vokovice, Czechoslovakia.

KUTAITSEV, V. I. Papers: Co-author, Interaction between Plutonium and Other Metals in Connection with their Arrangement in Mendeleev's Periodic Table (2nd I.C.P.U.A.E., Geneva, Sept. 1958); co-author, Metal Studies on Pu, U and their Alloys (Atomnaya Energiya, vol. 5, No. 1, 1958); co-author, Preparation and Studies of Some Plutonium Monocarbide Properties (ibid, vol. 22, No. 6, 1967); co-author, Phase Diagrams of Plutonium with IIIA, IVA, VIII and IB Group-Metals (ibid, vol. 23, No. 6, 1967).
Address: Nuclear Energy Institute, Academy of Sciences of the U.S.S.R., 14 Leninsky Prospekt, Moscow V-71, U.S.S.R.

KUTKA, Jan, Dr.-Ing. Born 1927. Educ.: Dresden Tech. Univ. Ustav jaderného výzkumu, Prague-Rez, 1956; Zentralinst. für Kernforschung, Rossendorf, 1963.
Nuclear interest: Metallurgy.
Address: Deutsche Akademie der Wissenschaften zu Berlin, Zentralinstitut für Kernforschung, Rossendorf, Postfach 19, 8051 Dresden, German Democratic Republic.

KUTOVOI, V. I. Paper: Co-author, On Co^{60} and Cs^{137} Gamma Rays Linear Attenuation Coefficient of Alloys (letter to the Editor, Atomnaya Energiya, vol. 19, No. 2, 1965).
Address: Academy of Sciences of the U.S.S.R., 14 Leninsky Prospekt, Moscow V-71, U.S.S.R.

KUTSIN, Yu. P. MEL'NIK-. See MEL'NIK-KUTSIN, Yu. P.

KUTTIG, Helmut, Privatdoz.,Dr. med. Born 1921. Educ.: Heidelberg Univ. Wissenschaftlicher Rat. Chief of Telecobalt Dept., Strahlenklinik Univ., Czerny-Krankenhaus, Heidelberg, 1951-. Society: Deutsche Röntgen-Gesellschaft.
Nuclear interest: Telecobalt-therapy; Clinical dosimetry.
Address: Universitäts-Strahlenklinik, Czerny-Krankenhaus, Heidelberg, Germany.

KUTUZOV, A. A. Papers: Co-author, Passage of γ-rays through Heterogeneous Media (Atomnaya Energiya, vol. 12, No. 1, 1962); co-author, Use of Multigroup Methods to Compute a Biological Shield (ibid, vol. 12, No. 2, 1962); co-author, Calculations of γ-Rays Build-up Factors for Heterogeneous Media (letter to the Editor, ibid, vol. 13, No. 6, 1962); co-author, Analysis and Generalisation of Cross-Correlation Method for Measurement of Particles Life-Time Distribution in Physical System (ibid, vol. 20, No. 5, 1966).
Address: Academy of Sciences of the U.S.S.R., 14 Leninsky Prospekt, Moscow V-71, U.S.S.R.

KUUSI, Eino Juhani, M.Sc. Born 1938. Educ.: Helsinki Tech. Univ. Res. Worker, Finnish Atomic Energy Commission, 1964-. Chairman, Phys. Club, Eng. Soc., Finland, 1967. Societies: Finnish Nucl. Soc.; Finnish Phys. Soc. Nuclear interests: Application of radioisotopes; Neutron activation analysis.
Address: Helsinki Technical University, Reactor Laboratory, Otaniemi, Helsinki, Finland.

KUVSHINOV, M. I. Paper: Co-author, Exptl. Study of Interaction in Array of Fissile Spheres (letter to the Editor, Atomnaya Energiya, vol. 22, No. 4, 1967).
Address: Academy of Sciences of the U.S.S.R., 14 Leninsky Prospekt, Moscow V-71, U.S.S.R.

KUYPER, Adrian C., Ph.D. Born 1907. Educ.: Iowa State Univ. Assoc. Prof. Physiological Chem., Wayne State Univ. Coll. Medicine, Detroit. Societies: Soc. for Exptl. Biol. and Medicine; American Soc. Biol. Chem.; A.C.S. Nuclear interest: Use of isotopes in biochemical research.
Address: Wayne State University College of Medicine, 1401 Rivard, Detroit 7, Michigan, U.S.A.

KUZAEV, B. I. Paper: Co-author, Use of Associated Particle Method for Absolute Determination of N for Neutrons Emitted by a Source (letter to the Editor, Atomnaya Energiya, vol. 16, No. 3, 1964).
Address: Academy of Sciences of the U.S.S.R., 14 Leninsky Prospekt, Moscow V-71, U.S.S.R.

KUZEMA, A. S. Paper: Co-author, Doubled Mass-Spectrometer with Nonuniform Magnetic Field (letter to the Editor, Atomnaya Energiya, vol. 22, No. 6, 1967).
Address: Academy of Sciences of the U.S.S.R., 14 Leninsky Prospekt, Moscow V-71, U.S.S.R.

KUZICHEVA, V. S. Paper: Co-author, On Prospects of Decontaminating Liquid Radioactive Wastes by Method of "Oil" Flotation (Abstract, Atomnaya Energiya, vol. 20, No. 1, 1966).
Address: Academy of Sciences of the U.S.S.R., 14 Leninsky Prospekt, Moscow V-71, U.S.S.R.

KUZIN, Alexander M. D.Sc., Head, Radiobiol. Lab., Biophys. Inst., Acad. of Sci. of the U.S.S.R.; Member. U.N. Sci. Com. on the Effects of Atomic Rad.; Deleg. to 2nd I.C.P.U.A.E., Geneva, Sept. 1958. Book: Chem. ugrozhayut chelovechestvu yadernye vzryvy (The Threat of Nucl. Explosions to Man) (M. AN SSSR, 1959). Paper: The Biochem. Mechanism of the Disturbance of Cell Division by Rad. (Symposium, The Initial Effects of Ionizing Rad. on Cells, Moscow, October, 1960).
Address: Institute of Biophysics, Academy of Sciences of the U.S.S.R., 14 Leninsky Prospekt, Moscow V-71, U.S.S.R.

KUZIN, N. I. Paper: Co-author, Preparation and Studies of Some Plutonium Monocarbide Properties (Atomnaya Energiya, vol. 22, No. 6, 1967).
Address: Academy of Sciences of the U.S.S.R., 14 Leninsky Prospekt, Moscow V-71, U.S.S.R.

KUZMIAK, Mikuláš, C.Sc. Born 1930. Educ.: Elec. Eng. Faculty, Prague. Sci. worker, Nucl. Res. Inst., 1965-.
Nuclear interests: Sector - focused cyclotrons.
Address: Nuclear Research Institute, Rez u Prahy, 124, Czechoslovakia.

KUZ'MICHEV, Yu. S. Paper: Co-author, Ultrasonic Effect on Plasticity of High Boron Stainless Steel (letter to the Editor, Atomnaya Energiya, vol. 20, No. 5, 1966).
Address: Academy of Sciences of the U.S.S.R., 14 Leninsky Prospekt, Moscow V-71, U.S.S.R.

KUZMIN, A. M. Paper: Co-author, Multigroup Effective Method for Reactor Calculation (letter to the Editor, Atomnaya Energiya, vol. 21, No. 5, 1966).
Address: Academy of Sciences of the U.S.S.R., 14 Leninsky Prospekt, Moscow V-71, U.S.S.R.

KUZ'MIN, V. A. Paper: Co-author, Secondary Protons Emitted from Thick Aluminium Targets During Bombarding by 340 Mev Protons (Atomnaya Energiya, vol. 22, No. 6, 1967).
Address: Academy of Sciences of the U.S.S.R., 14 Leninsky Prospekt, Moscow V-71, U.S.S.R.

KUZ'MIN, V. I. Paper: Co-author, Use of Perturbation Theory Functionals for Different Neutron Spectra Reactor Fuel Loading Minimisation (Atomnaya Energiya, vol. 24, No. 3, 1968).
Address: Academy of Sciences of the U.S.S.R., 14 Leninsky Prospekt, Moscow V- 71, U.S.S.R.

KUZ'MIN, V. M. Paper: Co-author, Investigation of PuO_2 Fuel Element Assembly of BR-5 Reactor (Atomnaya Energiya, vol. 24, No. 2, 1968).
Address: Academy of Sciences of the U.S.S.R., 14 Leninsky Prospekt, Moscow V-71, U.S.S.R.

KUZMIN, V. N. Papers: Co-author, 1,5 GeV Electron Synchrotron at the Polytech. Inst. of Tomsk (Atomnaya Energiya, vol. 21, No. 6, 1966); co-author, 300 MeV Electron

Synchrotron at the Polytech. Inst. of Tomsk (letter to the Editor, ibid, vol. 21, No. 6, 1966).
Address: Academy of Sciences of the U.S.S.R., 14 Leninsky Prospekt, Moscow V-71, U.S.S.R.

KUZ'MIN, V. V. Paper: Co-author, Thermoluminescence Fast Neutron Dosimeter (Atomnaya Energiya, vol. 22, No. 6, 1967).
Address: Academy of Sciences of the U.S.S.R., 14 Leninsky Prospekt, Moscow V-71, U.S.S.R.

KUZNESOV, V. G. Paper: Co-author, Optimum Ratio of Neutron and Gamma Dose behind Shielding of Reactor (letter to the Editor, Atomnaya Energiya, vol. 20, No. 1, 1966).
Address: Academy of Sciences of the U.S.S.R., 14 Leninsky Prospekt, Moscow V-71, U.S.S.R.

KUZNETS, E. D. Paper: Co-author, Tritium Content in Atmospheric Fall over Moscow during 1962-63 (letter to the Editor, Atomnaya Energiya, vol. 20, No. 1, 1966).
Address: Academy of Sciences of the U.S.S.R., 14 Leninsky Prospekt, Moscow V-71, U.S.S.R.

KUZNETSON, V. G. Paper: Neutron Detector with Alternating Thickness of Moderator (Atomnaya Energiya, vol. 22, No. 2, 1967).
Address: Academy of Sciences of the U.S.S.R., 14 Leninsky Prospekt, Moscow V-71, U.S.S.R.

KUZNETSOV, A. B. Papers: Co-author, Characteristics of the Proton Beam in a 10 GeV Synchrophasotron (Atomnaya Energiya, vol. 12, No. 5, 1962); co-author, Investigation of Accelerated Particle Beam Formation at Proton Synchrotron with Induction Electrodes (ibid, vol. 14, No. 2, 1963); co-author, Collective Linear Acceleration of Ions (ibid, vol. 24, No. 4, 1968).
Address: Academy of Sciences of the U.S.S.R., 14 Leninsky Prospekt, Moscow V-71, U.S.S.R.

KUZNETSOV, A. N. Papers: Co-author, Design of 300 kev Sector-Focused Cyclotron with External Ion Injection (Abstract, Atomnaya Energiya, vol. 19, No. 5, 1965); co-author, Magnetic Field of 300 kev Sector-Focused Cyclotron with External Ion Injection (Abstract, ibid); co-author, On Decrease of Neutron Yield from Tritium Targets (letter to the Editor, ibid, vol. 21, No. 5, 1966).
Address: Academy of Sciences of the U.S.S.R., 14 Leninsky Prospekt, Moscow V-71, U.S.S.R.

KUZNETSOV, A. V. Papers: Co-author, Gamma-Field of Dipped in Water Isotropic Co^{60} Source near Interface of Water and Air (letter to the Editor, Atomnaya Energiya, vol. 23, No. 1, 1967); co-author, Back-Scattering of Gamma-Rays Isotropic Sources from Cylindrical Barriers (ibid, vol. 23, No. 3, 1967).
Address: Academy of Sciences of the U.S.S.R., 14 Leninsky Prospekt, Moscow V-71, U.S.S.R.

KUZNETSOV, F. M. Papers: Co-author, Use of a Subcritical Insert to examine the Phys. Parameters of a Uranium-graphite Lattice (Atomnaya Energiya, vol. 11, No. 1, 1961);Interactions in a Subcritical Reactor (ibid, vol. 16, No. 1, 1964); co-author, Critical Parameters for Aqueous Solutions of $UO_2(NO_3)_2$ (ibid); co-author, Heterogeneous Absorbers Efficiency in Homogeneous Uranium-Water Reactors (ibid, vol. 19, No. 1, 1965).
Address: Academy of Sciences of the U.S.S.R., 14 Leninsky Prospekt, Moscow V-71, U.S.S.R.

KUZNETSOV, I. A. Paper: Co-author, BN-350 and BOR Fast Reactors (Atomnaya Energiya, vol. 21, No. 6, 1966).
Address: Academy of Sciences of the U.S.S.R., 14 Leninsky Prospekt, Moscow V-71, U.S.S.R.

KUZNETSOV, M. I. Paper: Co-author, Use of Gamma-Gamma Coincidence Spectrometer with Summation of Pulse Amplitudes for Analysis of Mixture Radioactive Isotopes (Atomnaya Energiya, vol. 19, No. 4, 1965).
Address: Academy of Sciences of the U.S.S.R., 14 Leninsky Prospekt, Moscow V-71, U.S.S.R.

KUZNETSOV, R. A. Paper: Co-author, Photoactivation Determination of Zirconium in Alloys with Beryllium, Magnesium, Aluminium and Niobium (Atomnaya Energiya, vol. 24, No. 2, 1968).
Address: Academy of Sciences of the U.S.S.R., 14 Leninsky Prospekt, Moscow V-71, U.S.S.R.

KUZNETSOV, S. A. Paper: Co-author, 1,5 GeV Electron Synchrotron at the Polytech.Inst. of Tomsk (Atomnaya Energiya, vol. 21, No. 6, 1966).
Address: Academy of Sciences of the U.S.S.R., 14 Leninsky Prospekt, Moscow V-71, U.S.S.R.

KUZNETSOV, V. F. Papers: Co-author, Prompt Neutron Number and the Kinetic Energy of Fragments in ^{235}U Low Energy Fission (letter to the Editor, Atomnaya Energiya, vol. 15, No. 1, 1963); co-author, Mean Number of Neutrons Emitted in Fission of U^{233} and U^{235} by Neutrons with Energies from 0.08 MeV to 1,0 MeV: Part 1. Relative Measurements Part II. Absolutisation of Relative Measurements and Balance of Energy Realised in Fission (ibid, vol. 22, No. 5, 1967).
Address: Academy of Sciences of the U.S.S.R., 14 Leninsky Prospekt, Moscow V-71, U.S.S.R.

KUZNETSOV, V. G. Paper: Co-author, Gamma-Ray Transmission through Oblique Barriers (Atomnaya Energiya, vol. 23, No. 1, 1967).
Address: Academy of Sciences of the U.S.S.R., 14 Leninsky Prospekt, Moscow V-71, U.S.S.R.

KUZNETSOV, V. I. Papers: Co-author, Pu (IV) Coprecipitation with Organic Coprecipitants (letter to the Editor, Atomnaya Energiya, vol. 8, No. 2, 1960); co-author, On Spontaneous Fission of Isotope 102^{254} (letter to the Editor, ibid, vol. 22, No. 6, 1967).
Address: Academy of Sciences of the U.S.S.R., 14 Leninsky Prospekt, Moscow V-71, U.S.S.R.

KUZNETSOV, V. M. Paper: Co-author, 1,5 GeV Electron Synchrotron at the Polytech. Inst. of Tomsk (Atomnaya Energiya, vol. 21, No. 6, 1966).
Address: Academy of Sciences of the U.S.S.R., 14 Leninsky Prospekt, Moscow V-71, U.S.S.R.

KUZNETSOV, Vladimir. Phys., Nucl. Phys. Inst., Novosibirsk Sci. Centre, Acad. Sci. of the U.S.S.R.
Address: Novosibirsk Science Centre, 20 Sovietskaya Ulitsa, Novosibirsk, Siberia, U.S.S.R.

KUZNETSOVA, A. P. Paper: Co-author, Mechanical Properties Change of Irradiated Precipitation-Hardening Al-Alloy (letter to the Editor, Atomnaya Energiya, vol. 21, No. 1, 1966).
Address: Academy of Sciences of the U.S.S.R., 14 Leninsky Prospekt, Moscow V-71, U.S.S.R.

KUZNETSOVA, V. G. Paper: Co-author, Preparation and Studies of Some Plutonium Monocarbide Properties (Atomnaya Energiya, vol. 22, No. 6, 1967).
Address: Academy of Sciences of the U.S.S.R., 14 Leninsky Prospekt, Moscow V-71, U.S.S.R.

KVALE, Eivind, Cand. real. Div. of Nucl. Chem., Oslo Univ.
Address: Oslo University, 1 Forskningsv., Blindern, Norway.

KVARTSKHAVA, Il'ya Filippovich, Prof. Dr. Sci. Asst., Acad. of Sci. of the Georgian S.S.R.; Director, Phys.-Tech. Inst., Acad. of Sci. of the Georgian S.S.R.; Member, Board of Editors, Nucl. Fusion; Deleg. to 2nd I.C.P.U.A.E., Geneva, Sept. 1958; Deleg. to Conference sur la Physique des Plasmas et la Recherche concernant la Fusion Nucléaire Contrôlée; Salzburg, Sept. 1961. Paper: A Spark Source of Multiply-charged Ions (letter to the Editor, Atomnaya Energiya, vol. 3, No. 8, 1957).
Address: Institute of Physics and Engineering, Academy of Sciences of the Georgian S.S.R., Sukhumi, U.S.S.R.

KWIECINSKI, Stanislaw, Dr. Tech. Sc. Born 1933. Educ.: Acad. of Mining and Metal. Cracow. Gen. Phys. Dept. (1955), Inst. of Nucl. Techniques (1963), Acad. of Mining and Metal., Cracow.
Nuclear interests: Reactor physics and engineering; Activation analysis; Application of radioactive isotopes in science and industry.
Address: Institute of Nuclear Techniques, 30 al. Mickiewicza, Cracow, Poland.

KWON, Oh Suk. Sen. Chem., Isotope Lab., Physico-chem. Div., Nat. Chem. Labs., Seoul.
Address: Isotope Laboratory, Physico-chemistry Div., National Chemistry Laboratories, Ministry of Health and Social Affairs, 79 Saichong-No, Chongno-ku, Seoul, Korea.

KYKER, Granvil Charles, B.S., Ph.D. Born 1912. Educ.: Carson Newman Coll. and North Carolina Univ. Assoc. Prof., North Carolina Univ., 1943-50; Prof. and Dept. Head, Puerto Rico Univ., 1950-52; Chief, Preclinical Res., Medical Div. (1952-65), Asst. to Director, Fellowships (1965-), Oak Ridge Assoc. Univs. (formerly O.R.I.N.S.). Trustee, Programme Chiarman, Soc. Nucl. Medicine. Societies:A.C.S.; American Soc. Biol. Chem.; A.A.A.S.; Soc. Exptl. Biol. and Medicine; American Assoc. Clinical Chem.; Elisha Mitchell Sci. Soc.; Soc. Nucl. Medicine; A.A.U.P.
Nuclear interests: Medical radioisotopes; Potentially useful internal emitters; Nuclear minerals: Their radioisotopic distribution, excretion, toxicity, and metabolic effects; Microbiologic uptake of certain lanthanons and of mixed fission products; Lipid metabolism; Nature and mechanism of rare-earth fatty liver.
Address: Fellowship Office, Oak Ridge Associated Universities, Post Office Box 117, Oak Ridge, Tennessee, U.S.A.

KYLES, James, M.A. Born 1914. Educ.: Edinburgh Univ. Sen. Lecturer, Edinburgh Univ., 1946-. Society: Fellow, Roy. Soc. of Edinburgh.
Nuclear interest: β- and γ-ray spectroscopy.
Address: Edinburgh University, Natural Philosophy Department, Drummond Street, Edinburgh, Scotland.

KYNER, Roy E., D.V.M. (Iowa State), M.P.H. (California). Born 1918. Educ.: Iowa State and California Univs. and Reed Coll., Oregan. Lt.-Col., U.S. Air Force, Veterinary Corps, 1943-; Rad. Processed Food Programme Project Officer, Div. of Biol. and Medicine, U.S.A.E.C., Washington, D.C., 1960-. Societies: American Veterinary Medicine Assoc.; H.P.S.; A.P.H.A.; Assoc. Military Surgeons.
Nuclear interests: Whole-body irradiation studies large animals; Fission products inhalation studies; Food preservation by irradiation.
Address: Division of Biology and Medicine, U.S. Atomic Energy Commission, Washington 25, D.C., U.S.A.

KYRS, Miroslav, Dipl.Ing., Cand. Chem. Sci. Born 1930. Educ.: Brno Tech. Univ. and Mendeleef Chem. Technol. Inst., Moscow. Head, Fission Products Dept., Nucl. Res. Inst., Czechoslovak Acad. Sci., Rez. Nucl. Technol. Com., Czechoslovak Sci. and Tech. Soc. Society: Czechoslovak Sci. and Tech. Soc.
Nuclear interests: Radiochemistry; Nuclear chemistry; Fission product extraction.
Address: Nuclear Research Institute, Rez, Czechoslovakia.

KYSELA, Frantisek, R.N.Dr., Ph.D. (Agriculture). Born 1906. Educ.: Charles Univ., Prague. Lecturer Phys., Charles Univ. Eng. School, 1946-51; Chief Eng., Radiobiol. Dept., Inst. Exptl. Botany, Czechoslovak Acad. Sci. Member, A.E.C., Prague.

Nuclear interests: Nuclear physics; Dosimetry; Nuclear metrology.
Address: 16/15 Ke dvoru, Prague 6- Vokovice, Czechoslovakia.

KYZ'YUROV, V. S. Papers: Co-author, Streaming of Neutrons through Cylindrical and Flat Neutron Guides (Kernenergie, vol. 11, No. 2, 1968); co-author, Determination of Fast Neutron Flux with Aid of ZnS (Ag) + Plastics Detector and Basson's Detector (Atomnaya Energiya, vol. 24, No. 6, 1968).
Address: Academy of Sciences of the U.S.S.R., 14 Leninsky Prospekt, Moscow V-71, U.S.S.R.